Tips for Customizing Your RSMeans Cost Data By Location

The cost data found throughout the book reflects national averages.

Customize your cost data for over 930 locations throughout the U.S. and Canada.

- ☑ To adjust costs to your region, use the Location Factors found in the back of the book.
- ☑ To compare costs from region-to-region, use the City Cost Indexes found in the back of the book.
- ☑ For additional guidance, refer to the How RSMeans Data Works section towards the front of the book.

Metric Conversion Formulas

Length

cm = 0.3937 in.	in. = 2.5400 cm.
meter = 3.2808 ft.	ft. = 0.3048 m.
meter = 1.0936 yd.	yd. = 0.9144 m.
Km. = 0.6214 mile	mile = 1.6093 km.

Area

sq. cm. = 0.1550 sq. in.	sq.in. = 6.4516 sq. cm.
sq. m. = 10.7639 sq. ft.	sq. ft. = 0.0929 sq. m.
sq. m. = 1.1960 sq. yd.	sq. yd. = 0.8361 sq. m.
hectare = 2.4710 acres	acre = 0.4047 hectare
sq. km. = 0.3861 sq. mile	sq. mile = 2.5900 sq. km.

Weight

gram = 15.4324 grains	grain = 0.0648 g.
gram = 0.0353 oz.	oz. = 28.3495 g.
kg. = 2.2046 lbs.	lb. = 0.4536 kg.

Weight, cont.

kg. = 0.0011 ton (short)	ton (short) = 907.1848 kg.
ton (met.) = 1.1023 ton (short)	ton (short) = 0.9072 ton (met.)
ton (met.) = 0.9842 ton (large)	ton (large) = 1.0160 ton (met.)

Common Units of Measure

Abbreviation	Measurement	Formula
SF	Square feet	Length (in feet) x Width (in feet)
SY	Square yards	Square feet / 9
CF	Cubic feet	Length (in feet) x Width (in feet) x Depth (in feet)
CY	Cubic yards	Cubic feet / 27
BF	Board foot	Length (in inches) x width (in inches) x thickness (in inches)/144
MBF	Thousand board foot	Board foot / 1,000
LF	Linear feet	12 inches of the item
SFCA	Square foot of contact area	A square foot of concrete formwork
SQ	Square	100 square feet
Ton	Ton	2,000 pounds

$188.95 per copy (in United States)
Price is subject to change without prior notice.

0213

RSMeans® Plumbing Cost Data

36TH ANNUAL EDITION

2013

RSMeans®
A division of Reed Construction Data, LLC
Construction Publishers & Consultants
700 Longwater Drive
Norwell, MA 02061
USA
1-800-334-3509
www.rsmeans.com

Printed in the United States of America
ISSN 1537-8411
ISBN 978-1-936335-70-1

This book is printed using soy-based printing ink.
This book is recyclable.

Related RSMeans Products and Services

The engineers at RSMeans suggest the following products and services as companion information resources to *RSMeans Plumbing Cost Data:*

Construction Cost Data Books

Assemblies Cost Data 2013
Square Foot Costs 2013

Reference Books

Estimating Building Costs
RSMeans Estimating Handbook
Green Building: Project Planning & Estimating
How to Estimate with RSMeans Data
Plan Reading & Material Takeoff
Project Scheduling & Management for Construction
Universal Design Ideas for Style, Comfort & Safety

Seminars and In-House Training

Unit Price Estimating
Means CostWorks® Training
RSMeans Data for Job Order Contracting (JOC)
Practical Project Management for Construction Professionals
Scheduling with MSProject for Construction Professionals
Mechanical & Electrical Estimating

RSMeans on the Internet

Visit RSMeans at **www.rsmeans.com** for a one-stop portal for the most reliable and current resources available on the market. Here you can learn about our more than 20 annual cost data books, also available on CD and online versions, as well as our libary of reference books and RSMeans seminars.

RSMeans Electronic Data

Get the information found in RSMeans cost books electronically at **MeansCostWorks.com**.

RSMeans Business Solutions

Cost engineers and analysts conduct facility life cycle and benchmark research studies and predictive cost modeling, as well as offer consultation services for conceptual estimating, real property management, and job order contracting. Research studies are designed with the application of proprietary cost and project data from Reed Construction Data and RSMeans' extensive North American databases. Analysts offer building product manufacturers qualitative and quantitative market research, as well as time/motion studies for new products, and Web-based dashboards for market opportunity analysis. Clients are from both the public and private sectors, including federal, state, and municipal agencies; corporations; institutions; construction management firms; hospitals; and associations.

RSMeans for Job Order Contracting (JOC)

Best practice job order contracting (JOC) services support the owner in the development of their program. RSMeans develops all required contracting documents, works with stakeholders, and develops qualified and experienced contractor lists. RSMeans JOCWorks® software is the best of class in the functional tools that meet JOC specific requirements. RSMeans data is organized to meet JOC cost estimating requirements.

Construction Costs for Software Applications

More than 25 unit price and assemblies cost databases are available through a number of leading estimating and facilities management software providers (listed below). For more information see the "Other RSMeans Products" pages at the back of this publication.

RSMeansData™ is also available to federal, state, and local government agencies as multi-year, multi-seat licenses.

- 4Clicks-Solutions, LLC
- ArenaSoft Estimating
- Assetworks
- Beck Technology
- BSD – Building Systems Design, Inc.
- CMS – Construction Management Software
- Corecon Technologies, Inc.
- CorVet Systems
- Hard Dollar Corporation
- Maxwell Systems
- MC^2
- Parsons Corporation
- ProEst
- Sage Timberline Office
- UDA Technologies, Inc.
- US Cost, Inc.
- VFA – Vanderweil Facility Advisers
- WinEstimator, Inc.
- R & K Solutions
- Tririga

Table of Contents

Foreword

Our Mission

Since 1942, RSMeans has been actively engaged in construction cost publishing and consulting throughout North America.

Today, more than 70 years after RSMeans began, our primary objective remains the same: to provide you, the construction and facilities professional, with the most current and comprehensive construction cost data possible.

Whether you are a contractor, owner, architect, engineer, facilities manager, or anyone else who needs a reliable construction cost estimate, you'll find this publication to be a highly useful and necessary tool.

With the constant flow of new construction methods and materials today, it's difficult to find the time to look at and evaluate all the different construction cost possibilities. In addition, because labor and material costs keep changing, last year's cost information is not a reliable basis for today's estimate or budget.

That's why so many construction professionals turn to RSMeans. We keep track of the costs for you, along with a wide range of other key information, from city cost indexes . . . to productivity rates . . . to crew composition . . . to contractor's overhead and profit rates.

RSMeans performs these functions by collecting data from all facets of the industry and organizing it in a format that is instantly accessible to you. From the preliminary budget to the detailed unit price estimate, you'll find the data in this book useful for all phases of construction cost determination.

The Staff, the Organization, and Our Services

When you purchase one of RSMeans' publications, you are, in effect, hiring the services of a full-time staff of construction and engineering professionals.

Our thoroughly experienced and highly qualified staff works daily at collecting, analyzing, and disseminating comprehensive cost information for your needs. These staff members have years of practical construction experience and engineering training prior to joining the firm. As a result, you can count on them not only for accurate cost figures, but also for additional background reference information that will help you create a realistic estimate.

The RSMeans organization is always prepared to help you solve construction problems through its variety of data solutions, including online, CD, and print book formats, as well as cost estimating expertise available via our business solutions, training, and seminars.

Besides a full array of construction cost estimating books, RSMeans also publishes a number of other reference works for the construction industry. Subjects include construction estimating and project and business management, special topics such as green building and job order contracting, and a library of facility management references.

In addition, you can access all of our construction cost data electronically in convenient CD format or on the web. Visit **RSMeansonline.com** for more information on our 24/7 online cost data.

What's more, you can increase your knowledge and improve your construction estimating and management performance with an RSMeans construction seminar or in-house training program. These two-day seminar programs offer unparalleled opportunities for everyone in your organization to become updated on a wide variety of construction-related issues.

RSMeans is also a worldwide provider of construction cost management and analysis services for commercial and government owners.

In short, RSMeans can provide you with the tools and expertise for constructing accurate and dependable construction estimates and budgets in a variety of ways.

Robert Snow Means Established a Tradition of Quality That Continues Today

Robert Snow Means spent years building RSMeans, making certain he always delivered a quality product.

Today, at RSMeans, we do more than talk about the quality of our data and the usefulness of our books. We stand behind all of our data, from historical cost indexes to construction materials and techniques to current costs.

If you have any questions about our products or services, please call us toll-free at 1-800-334-3509. Our customer service representatives will be happy to assist you. You can also visit our website at **www.rsmeans.com**

How the Book Is Built: An Overview

The Construction Specifications Institute (CSI) and Construction Specifications Canada (CSC) have produced the 2010 edition of MasterFormat™, a system of titles and numbers used extensively to organize construction information.

All unit price data in the RSMeans cost data books is now arranged in the 50-division MasterFormat™ 2010 system.

A Powerful Construction Tool

You have in your hands one of the most powerful construction tools available today. A successful project is built on the foundation of an accurate and dependable estimate. This book will enable you to construct just such an estimate.

For the casual user the book is designed to be:

- quickly and easily understood so you can get right to your estimate.
- filled with valuable information so you can understand the necessary factors that go into the cost estimate.

For the regular user, the book is designed to be:

- a handy desk reference that can be quickly referred to for key costs.
- a comprehensive, fully reliable source of current construction costs and productivity rates so you'll be prepared to estimate any project.
- a source book for preliminary project cost, product selections, and alternate materials and methods.

To meet all of these requirements, we have organized the book into the following clearly defined sections.

Quick Start

See our "Quick Start" instructions on the following page to get started right away.

Estimating with RSMeans Unit Price Cost Data

Please refer to these steps for guidance on completing an estimate using RSMeans unit price cost data.

How to Use the Book: The Details

This section contains an in-depth explanation of how the book is arranged . . . and how you can use it to determine a reliable construction cost estimate. It includes information about how we develop our cost figures and how to completely prepare your estimate.

Unit Price Section

All cost data has been divided into the 50 divisions according to the MasterFormat system of classification and numbering. For a listing of these divisions and an outline of their subdivisions, see the Unit Price Section Table of Contents.

Estimating tips are included at the beginning of each division.

Assemblies Section

The cost data in this section has been organized in an "Assemblies" format. These assemblies are the functional elements of a building and are arranged according to the 7 elements of the UNIFORMAT II classification system. For a complete explanation of a typical "Assemblies" page, see "How to Use the Assemblies Cost Tables."

Reference Section

This section includes information on Equipment Rental Costs, Crew Listings, Historical Cost Indexes, City Cost Indexes, Location Factors, Reference Tables, Change Orders, Square Foot Costs, and a listing of Abbreviations.

Equipment Rental Costs: This section contains the average costs to rent and operate hundreds of pieces of construction equipment.

Crew Listings: This section lists all of the crews referenced in the book. For the purposes of this book, a crew is composed of more than one trade classification and/or the addition of power equipment to any trade classification. Power equipment is included in the cost of the crew. Costs are shown both with bare labor rates and with the installing contractor's overhead and profit added. For each, the total crew cost per eight-hour day and the composite cost per labor-hour are listed.

Historical Cost Indexes: These indexes provide you with data to adjust construction costs over time.

City Cost Indexes: All costs in this book are U.S. national averages. Costs vary because of the regional economy. You can adjust costs by CSI Division to over 700 locations throughout the U.S. and Canada by using the data in this section.

Location Factors: You can adjust total project costs to over 900 locations throughout the U.S. and Canada by using the data in this section.

Reference Tables: At the beginning of selected major classifications in the Unit Price section are reference numbers shown in a shaded box. These numbers refer you to related information in the Reference Section. In this section, you'll find reference tables, explanations, estimating information that support how we develop the unit price data, technical data, and estimating procedures.

Change Orders: This section includes information on the factors that influence the pricing of change orders.

Square Foot Costs: This section contains costs for 59 different building types that allow you to make a rough estimate for the overall cost of a project or its major components.

Abbreviations: A listing of abbreviations used throughout this book, along with the terms they represent, is included in this section.

Index

A comprehensive listing of all terms and subjects in this book will help you quickly find what you need when you are not sure where it occurs in MasterFormat.

The Scope of This Book

This book is designed to be as comprehensive and as easy to use as possible. To that end we have made certain assumptions and limited its scope in two key ways:

1. We have established material prices based on a national average.
2. We have computed labor costs based on a 30-city national average of union wage rates.

For a more detailed explanation of how the cost data is developed, see "How to Use the Book: The Details."

Project Size/Type

The material prices in Means data cost books are "contractor's prices." They are the prices that contractors can expect to pay at the lumberyards, suppliers/distributers warehouses, etc. Small orders of speciality items would be higher than the costs shown, while very large orders, such as truckload lots, would be less. The variation would depend on the size, timing, and negotiating power of the contractor. The labor costs are primarily for new construction or major renovation rather than repairs or minor alterations.

With reasonable exercise of judgment, the figures can be used for any building work.

Absolute Essentials for a Quick Start

If you feel you are ready to use this book and don't think you will need the detailed instructions that begin on the following page, this Absolute Essentials for a Quick Start page is for you. These steps will allow you to get started estimating in a matter of minutes.

1 Scope

Think through the project that you will be estimating, and identify the many individual work tasks that will need to be covered in your estimate.

2 Quantify

Determine the number of units that will be required for each work task that you identified.

3 Pricing

Locate individual Unit Price line items that match the work tasks you identified. The Unit Price Section Table of Contents that begins on page 1 and the Index in the back of the book will help you find these line items.

4 Multiply

Multiply the Total Incl O&P cost for a Unit Price line item in the book by your quantity for that item. The price you calculate will be an estimate for a completed item of work performed by a subcontractor. Keep adding line items in this manner to build your estimate.

5 Project Overhead

Include project overhead items in your estimate. These items are needed to make the job run and are typically, but not always, provided by the General Contractor. They can be found in Division 1. An alternate method of estimating project overhead costs is to apply a percentage of the total project cost.

Include rented tools not included in crews, waste, rubbish handling, and cleanup.

6 Estimate Summary

Include General Contractor's markup on subcontractors, General Contractor's office overhead and profit, and sales tax on materials and equipment.

Adjust your estimate to the project's location by using the City Cost Indexes or Location Factors found in the Reference Section.

Editors' Note: We urge you to spend time reading and understanding the supporting material in the front of this book. An accurate estimate requires experience, knowledge, and careful calculation. The more you know about how we at RSMeans developed the data, the more accurate your estimate will be. In addition, it is important to take into consideration the reference material in the back of the book such as Equipment Listings, Crew Listings, City Cost Indexes, Location Factors, and Reference Tables.

Estimating with RSMeans Unit Price Cost Data

Following these steps will allow you to complete an accurate estimate using RSMeans Unit Price cost data.

1 Scope Out the Project

- Identify the individual work tasks that will need to be covered in your estimate.
- The Unit Price data in this book has been divided into 50 Divisions according to the CSI MasterFormat 2010—their titles are listed on the back cover of your book.
- Think through the project and identify those CSI Divisions needed in your estimate.
- The Unit Price Section Table of Contents on page 1 may also be helpful when scoping out your project.
- Experienced estimators find it helpful to begin with Division 2 and continue through completion. Division 1 can be estimated after the full project scope is known.

2 Quantify

- Determine the number of units required for each work task that you identified.
- Experienced estimators include an allowance for waste in their quantities. (Waste is not included in RSMeans Unit Price line items unless so stated.)

3 Price the Quantities

- Use the Unit Price Table of Contents, and the Index, to locate individual Unit Price line items for your estimate.
- Reference Numbers indicated within a Unit Price section refer to additional information that you may find useful.
- The crew indicates who is performing the work for that task. Crew codes are expanded in the Crew Listings in the Reference Section to include all trades and equipment that comprise the crew.
- The Daily Output is the amount of work the crew is expected to do in one day.
- The Labor-Hours value is the amount of time it will take for the crew to install one unit of work.
- The abbreviated Unit designation indicates the unit of measure upon which the crew, productivity, and prices are based.
- Bare Costs are shown for materials, labor, and equipment needed to complete the Unit Price line item. Bare costs do not include waste, project overhead, payroll insurance, payroll taxes, main office overhead, or profit.
- The Total Incl O&P cost is the billing rate or invoice amount of the installing contractor or subcontractor who performs the work for the Unit Price line item.

4 Multiply

- Multiply the total number of units needed for your project by the Total Incl O&P cost for each Unit Price line item.
- Be careful that your take off unit of measure matches the unit of measure in the Unit column.
- The price you calculate is an estimate for a completed item of work.
- Keep scoping individual tasks, determining the number of units required for those tasks, matching each task with individual Unit Price line items in the book, and multiply quantities by Total Incl O&P costs.
- An estimate completed in this manner is priced as if a subcontractor, or set of subcontractors, is performing the work. The estimate does not yet include Project Overhead or Estimate Summary components such as General Contractor markups on subcontracted work, General Contractor office overhead and profit, contingency, and location factor.

5 Project Overhead

- Include project overhead items from Division 1 – General Requirements.
- These items are needed to make the job run. They are typically, but not always, provided by the General Contractor. Items include, but are not limited to, field personnel, insurance, performance bond, permits, testing, temporary utilities, field office and storage facilities, temporary scaffolding and platforms, equipment mobilization and demobilization, temporary roads and sidewalks, winter protection, temporary barricades and fencing, temporary security, temporary signs, field engineering and layout, final cleaning and commissioning.
- Each item should be quantified, and matched to individual Unit Price line items in Division 1, and then priced and added to your estimate.
- An alternate method of estimating project overhead costs is to apply a percentage of the total project cost, usually 5% to 15% with an average of 10% (see General Conditions, page ix).
- Include other project related expenses in your estimate such as:
    - Rented equipment not itemized in the Crew Listings
    - Rubbish handling throughout the project (see 02 41 19.19)

6 Estimate Summary

- Include sales tax as required by laws of your state or county.
- Include the General Contractor's markup on self-performed work, usually 5% to 15% with an average of 10%.
- Include the General Contractor's markup on subcontracted work, usually 5% to 15% with an average of 10%.
- Include General Contractor's main office overhead and profit:
  - RSMeans gives general guidelines on the General Contractor's main office overhead (see section 01 31 13.50 and Reference Number R013113-50).
  - RSMeans gives no guidance on the General Contractor's profit.
  - Markups will depend on the size of the General Contractor's operations, his projected annual revenue, the level of risk he is taking on, and on the level of competition in the local area and for this project in particular.
- Include a contingency, usually 3% to 5%, if appropriate.
- Adjust your estimate to the project's location by using the City Cost Indexes or the Location Factors in the Reference Section:
  - Look at the rules on the pages for How to Use the City Cost Indexes to see how to apply the Indexes for your location.
  - When the proper Index or Factor has been identified for the project's location, convert it to a multiplier by dividing it by 100, and then multiply that multiplier by your estimated total cost. The original estimated total cost will now be adjusted up or down from the national average to a total that is appropriate for your location.

Editors' Notes:

1) *We urge you to spend time reading and understanding the supporting material in the front of this book. An accurate estimate requires experience, knowledge, and careful calculation. The more you know about how we at RSMeans developed the data, the more accurate your estimate will be. In addition, it is important to take into consideration the reference material in the back of the book such as Equipment Listings, Crew Listings, City Cost Indexes, Location Factors, and Reference Tables.*
2) *Contractors who are bidding or are involved in JOC, DOC, SABER, or IDIQ type contracts are cautioned that workers' compensation insurance, federal and state payroll taxes, waste, project supervision, project overhead, main office overhead, and profit are not included in bare costs. The coefficient or multiplier must cover these costs.*

How to Use the Book: The Details

What's Behind the Numbers? The Development of Cost Data

The staff at RSMeans continually monitors developments in the construction industry in order to ensure reliable, thorough, and up-to-date cost information. While overall construction costs may vary relative to general economic conditions, price fluctuations within the industry are dependent upon many factors. Individual price variations may, in fact, be opposite to overall economic trends. Therefore, costs are constantly tracked and complete updates are published yearly. Also, new items are frequently added in response to changes in materials and methods.

Costs—$ (U.S.)

All costs represent U.S. national averages and are given in U.S. dollars. The RSMeans City Cost Indexes can be used to adjust costs to a particular location. The City Cost Indexes for Canada can be used to adjust U.S. national averages to local costs in Canadian dollars. No exchange rate conversion is necessary.

G The processes or products identified by the green symbol in our publications have been determined to be environmentally responsible and/or resource-efficient solely by the RSMeans engineering staff. The inclusion of the green symbol does not represent compliance with any specific industry association or standard.

Material Costs

The RSMeans staff contacts manufacturers, dealers, distributors, and contractors all across the U.S. and Canada to determine national average material costs. If you have access to current material costs for your specific location, you may wish to make adjustments to reflect differences from the national average. Included within material costs are fasteners for a normal installation. RSMeans engineers use manufacturers' recommendations, written specifications, and/or standard construction practice for size and spacing of fasteners. Adjustments to material costs may be required for your specific application or location. Material costs do not include sales tax.

Labor Costs

Labor costs are based on the average of wage rates from 30 major U.S. cities. Rates are determined from labor union agreements or prevailing wages for construction trades for the current year. Rates, along with overhead and profit markups, are listed on the inside back cover of this book.

- If wage rates in your area vary from those used in this book, or if rate increases are expected within a given year, labor costs should be adjusted accordingly.

Labor costs reflect productivity based on actual working conditions. In addition to actual installation, these figures include time spent during a normal weekday on tasks such as, material receiving and handling, mobilization at site, site movement, breaks, and cleanup. Productivity data is developed over an extended period so as not to be influenced by abnormal variations and reflects a typical average.

Equipment Costs

Equipment costs include not only rental, but also operating costs for equipment under normal use. The operating costs include parts and labor for routine servicing such as repair and replacement of pumps, filters, and worn lines. Normal operating expendables, such as fuel, lubricants, tires, and electricity (where applicable), are also included. Extraordinary operating expendables with highly variable wear patterns, such as diamond bits and blades, are excluded. These costs are included under materials. Equipment rental rates are obtained from industry sources throughout North America—contractors, suppliers, dealers, manufacturers, and distributors.

Rental rates can also be treated as reimbursement costs for contractor-owned equipment. Owned equipment costs include depreciation, loan payments, interest, taxes, insurance, storage and major repairs.

Equipment costs do not include operators' wages; nor do they include the cost to move equipment to a job site (mobilization) or from a job site (demobilization).

Equipment Cost/Day——The cost of power equipment required for each crew is included in the Crew Listings in the Reference Section (small tools that are considered as essential everyday tools are not listed out separately). The Crew Listings itemize specialized tools and heavy equipment along with labor trades. The daily cost of itemized equipment included in a crew is based on dividing the weekly bare rental rate by 5 (number of working days per week) and then adding the hourly operating cost times 8 (the number of hours per day). This Equipment Cost/Day is shown in the last column of the Equipment Rental Cost pages in the Reference Section.

Mobilization/Demobilization——The cost to move construction equipment from an equipment yard or rental company to the job site and back again is not included in equipment costs. Mobilization (to the site) and demobilization (from the site) costs can be found in the Unit Price Section. If a piece of equipment is already at the job site, it is not appropriate to utilize mob./demob. costs again in an estimate.

Overhead and Profit

Total Cost including O&P for the *Installing Contractor* is shown in the last column on the Unit Price and/or the Assemblies pages of this book. This figure is the sum of the bare material cost plus 10% for profit, the bare labor cost plus total overhead and profit, and the bare equipment cost plus 10% for profit. Details for the calculation of Overhead and Profit on labor are shown on the inside back cover and in the Reference Section of this book. (See "How RSMeans Data Works" for an example of this calculation.)

General Conditions

Cost data in this book is presented in two ways: Bare Costs and Total Cost including O&P (Overhead and Profit). General Conditions, or General Requirements, of the contract should also be added to the Total Cost including O&P when applicable. Costs for General Conditions are listed in Division 1 of the Unit Price Section and the Reference

Section of this book. General Conditions for the *Installing Contractor* may range from 0% to 10% of the Total Cost including O&P. For the *General* or *Prime Contractor*, costs for General Conditions may range from 5% to 15% of the Total Cost including O&P, with a figure of 10% as the most typical allowance. If applicable, the Assemblies and Models sections of this book use costs that include the installing contractor's overhead and profit (O&P).

Factors Affecting Costs

Costs can vary depending upon a number of variables. Here's how we have handled the main factors affecting costs.

Quality——The prices for materials and the workmanship upon which productivity is based represent sound construction work. They are also in line with U.S. government specifications.

Overtime——We have made no allowance for overtime. If you anticipate premium time or work beyond normal working hours, be sure to make an appropriate adjustment to your labor costs.

Productivity——The productivity, daily output, and labor-hour figures for each line item are based on working an eight-hour day in daylight hours in moderate temperatures. For work that extends beyond normal work hours or is performed under adverse conditions, productivity may decrease. (See "How RSMeans Data Works" for more on productivity.)

Size of Project——The size, scope of work, and type of construction project will have a significant impact on cost. Economies of scale can reduce costs for large projects. Unit costs can often run higher for small projects.

Location——Material prices in this book are for metropolitan areas. However, in dense urban areas, traffic and site storage limitations may increase costs. Beyond a 20-mile radius of large cities, extra trucking or transportation charges may also increase the material costs slightly. On the other hand, lower wage rates may be in effect. Be sure to consider both of these factors when preparing an estimate, particularly if the job site is located in a central city or remote rural location. In addition, highly specialized subcontract items may require travel and per-diem expenses for mechanics.

Other Factors—

- season of year
- contractor management
- weather conditions
- local union restrictions
- building code requirements
- availability of:
  - adequate energy
  - skilled labor
  - building materials
- owner's special requirements/restrictions
- safety requirements
- environmental considerations

Unpredictable Factors——General business conditions influence "in-place" costs of all items. Substitute materials and construction methods may have to be employed. These may affect the installed cost and/or life cycle costs. Such factors may be difficult to evaluate and cannot necessarily be predicted on the basis of the job's location in a particular section of the country. Thus, where these factors apply, you may find significant but unavoidable cost variations for which you will have to apply a measure of judgment to your estimate.

Rounding of Costs

In general, all unit prices in excess of $5.00 have been rounded to make them easier to use and still maintain adequate precision of the results. The rounding rules we have chosen are in the following table.

Prices from . . .	Rounded to the nearest . . .
$.01 to $5.00	$.01
$5.01 to $20.00	$.05
$20.01 to $100.00	$.50
$100.01 to $300.00	$1.00
$300.01 to $1,000.00	$5.00
$1,000.01 to $10,000.00	$25.00
$10,000.01 to $50,000.00	$100.00
$50,000.01 and above	$500.00

How Subcontracted Items Affect Costs

A considerable portion of all large construction jobs is usually subcontracted. In fact, the percentage done by subcontractors is constantly increasing and may run over 90%. Since the workers employed by these companies do nothing else but install their particular product, they soon become expert in that line. The result is, installation by these firms is accomplished so efficiently that the total in-place cost, even adding the general contractor's overhead and profit, is no more, and often less, than if the principal contractor had handled the installation himself/herself. Companies that deal with construction specialties are anxious to have their product perform well and, consequently, the installation will be the best possible.

Contingencies

The allowance for contingencies generally provides for unforeseen construction difficulties. On alterations or repair jobs, 20% is not too much. If drawings are final and only field contingencies are being considered, 2% or 3% is probably sufficient, and often nothing need be added. Contractually, changes in plans will be covered by extras. The contractor should consider inflationary price trends and possible material shortages during the course of the job. These escalation factors are dependent upon both economic conditions and the anticipated time between the estimate and actual construction. If drawings are not complete or approved, or a budget cost is wanted, it is wise to add 5% to 10%. Contingencies, then, are a matter of judgment. Additional allowances for contingencies are shown in Division 1.

Important Estimating Considerations

The "productivity," or daily output, of each craftsman includes mobilization and cleanup time, break time and plan layout time, as well as an allowance to carry stock from the storage trailer or location on the job site up to 200′ into the building and to the first or second floor. If material has to be transported over greater distances or to higher floors, an additional allowance should be considered by the estimator. An allowance has also been included in the piping and fittings installation time for leak checking and minor tightening. Equipment installation time includes the following applicable items: positioning, leveling and securing the unit in place, connecting all associated piping, ducts, vents, etc., which shall have been estimated separately, connecting to an adjacent power source, filling/bleeding, startup, adjusting the controls up and down to ensure proper response, setting the integral controls/valves/regulators/thermostats for proper operation (does not include external building type control systems, DDC systems, etc.), explaining/training owner's operator, and warranty. A reasonable breakdown of the labor costs is as follows:

1.	Movement into building, installation/setting of equipment	35%
2.	Connection to piping/duct/power, etc.	25%
3.	Filling/flushing/cleaning/touchup, etc.	15%
4.	Startup/running adjustments	5%
5.	Training owner's representative	5%
6.	Warranty/call back/service	15%

Note that cranes or other lifting equipment are not included on any lines in the Mechanical divisions. For example, if a crane is required to lift a heavy piece of pipe into place high above a gym floor, or to put a rooftop unit on the roof of a four-story building, etc., it must be added. Due to the potential for extreme variation—from nothing additional required to a major crane or helicopter—we feel that including a nominal amount for "lifting contingency" would be useless and detract from the accuracy of the estimate. When using equipment rental from RSMeans, remember to include the cost of the operator(s).

Estimating Labor-Hours

The labor-hours expressed in this publication are based on Average Installation time, using an efficiency level of approximately 60%-65% (see item 7), which has been found reasonable and acceptable by many contractors. The book uses this national efficiency average to establish a consistent benchmark. For bid situations, adjustments to this efficiency level should be the responsibility of the contractor bidding the project. The unit labor-hour is divided in the following manner. A typical day for a journeyman might be:

1.	Study Plans	3%	14.4 min.
2.	Material Procurement	3%	14.4 min.
3.	Receiving and Storing	3%	14.4 min
4.	Mobilization	5%	24.0 min.
5.	Site Movement	5%	24.0 min.
6.	Layout and Marking	8%	38.4 min.
7.	Actual Installation	64%	307.2 min.
8.	Cleanup	3%	14.4 min.
9.	Breaks, Nonproductive	6%	28.8 min.
		100%	480.0 min.

If any of the percentages expressed in this breakdown do not apply to the particular work or project situation, then that percentage or a portion of it may be deducted from or added to labor-hours.

Final Checklist

Estimating can be a straightforward process provided you remember the basics. Here's a checklist of some of the steps you should remember to complete before finalizing your estimate. Did you remember to . . .

- factor in the City Cost Index for your locale?
- take into consideration which items have been marked up and by how much?
- mark up the entire estimate sufficiently for your purposes?
- read the background information on techniques and technical matters that could impact your project time span and cost?
- include all components of your project in the final estimate?
- double check your figures for accuracy?
- call RSMeans if you have any questions about your estimate or the data you've found in our publications? Remember, RSMeans stands behind its publications. If you have any questions about your estimate . . . about the costs you've used from our books . . . or even about the technical aspects of the job that may affect your estimate, feel free to call the RSMeans editors at 1-800-334-3509.

Unit Price Section

Table of Contents

Table of Contents (cont.)

How RSMeans Data Works

All RSMeans unit price data is organized in the same way.

It is important to understand the structure, so that you can find information easily and use it correctly.

RSMeans **Line Numbers** consist of 12 characters, which identify a unique location in the database for each task. The first 6 or 8 digits conform to the Construction Specifications Institute MasterFormat 2010. The remainder of the digits are a further breakdown by RSMeans in order to arrange items in understandable groups of similar tasks. Line numbers are consistent across all RSMeans publications, so a line number in any RSMeans product will always refer to the same unit of work.

RSMeans engineers have created **reference** information to assist you in your estimate. If there is information that applies to a section, it will be indicated at the start of the section. In this case, R033105-10 provides information on the proportionate quantities of formwork, reinforcing, and concrete used in cast-in-place concrete items such as footings, slabs, beams, and columns. The Reference Section is located in the back of the book on the pages with a gray edge.

RSMeans **Descriptions** are shown in a hierarchical structure to make them readable. In order to read a complete description, read up through the indents to the top of the section. Include everything that is above and to the left that is not contradicted by information below. For instance, the complete description for line 03 30 53.40 3550 is "Concrete in place, including forms (4 uses), reinforcing steel, concrete, placement, and finishing unless otherwise indicated; Equipment pad (3000 psi), 4' x 4' x 6" thick."

When using **RSMeans data**, it is important to read through an entire section to ensure that you use the data that most closely matches your work. Note that sometimes there is additional information shown in the section that may improve your price. There are frequently lines that further describe, add to, or adjust data for specific situations.

03 30 Cast-In-Place Concrete

03 30 53 – Miscellaneous Cast-In-Place Concrete

03 30 53.40 Concrete In Place

0010	**CONCRETE IN PLACE**	R033105-10
0020	Including forms (4 uses), Grade 60 rebar, concrete (Portland cement	R033105-20
0050	Type I), placement and finishing unless otherwise indicated	R033105-50
0300	Beams (3500 psi), 5 kip per L.F., 10′ span	R033105-65
0350	25′ span	R033105-70
0500	Chimney foundations (5000 psi), over 5 C.Y.	R033105-85
0510	(3500 psi), under 5 C.Y.	
0700	Columns, square (4000 psi), 12″ x 12″, less than 2% reinforcing	
3450	Over 10,000 S.F.	
3500	Add per floor for 3 to 6 stories high	
3520	For 7 to 20 stories high	
3540	Equipment pad (3000 psi), 3′ x 3′ x 6″ thick	
3550	4′ x 4′ x 6″ thick	
3560	5′ x 5′ x 8″ thick	
3570	6′ x 6′ x 8″ thick	

The data published in RSMeans print books represents a "national average" cost. This data should be modified to the project location using the **City Cost Indexes** or **Location Factors** tables found in the reference section (see pages 596–644). Use the location factors to adjust estimate totals if the project covers multiple trades. Use the city cost indexes (CCI) for single trade projects or projects where a more detailed analysis is required. All figures in the two tables are derived from the same research. The last row of data in the CCI, the weighted average, is the same as the numbers reported for each location in the location factor table.

Crews

Crew No.	Bare Costs		Incl. Subs O&P		Cost Per Labor-Hour	
Crew C-14H	Hr.	Daily	Hr.	Daily	Bare Costs	Incl. O&P
1 Carpenter Foreman (outside)	$46.90	$375.20	$72.25	$578.00	$44.17	$67.88
2 Carpenters	44.90	718.40	69.15	1106.40		
1 Rodman (reinf.)	49.80	398.40	78.70	629.60		
1 Laborer	35.45	283.60	54.60	436.80		
1 Cement Finisher	43.05	344.40	63.40	507.20		
1 Gas Engine Vibrator		33.00		36.30	.69	.76
48 L.H., Daily Totals		$2153.00		$3294.30	$44.85	$68.63

Crews include labor or labor and equipment necessary to accomplish each task. In this case, Crew C-14H is used. RSMeans selects a crew to represent the workers and equipment that are typically used for that task. In this case, Crew C-14H consists of one carpenter foreman (outside), two carpenters, one rodman, one laborer, one cement finisher, and one gas engine vibrator. Details of all crews can be found in the reference section.

The **Daily Output** is the amount of work that the crew can do in a normal 8-hour workday, including mobilization, layout, movement of materials, and cleanup. In this case, crew C-14H can install thirty 4' x 4' x 6" thick concrete pads in a day. Daily output is variable, based on many factors, including the size of the job, location, and environmental conditions. RSMeans data represents work done in daylight (or adequate lighting) and temperate conditions.

Bare Costs are the costs of materials, labor, and equipment that the installing contractor pays. They represent the cost, in U.S. dollars, for one unit of work. They do not include any markups for profit or labor burden.

Crew	Daily Output	Labor-Hours	Unit	2013 Bare Costs Material	Labor	Equipment	Total	Total Incl O&P
C-14A	15.62	12.804	C.Y.	31[illegible]	575	48	938	1,300
"	18.55	10.782		330	485	40.50	855.50	1,150
C-14C	32.22	3.476		144	149	1.01	294.01	390
"	23.71	4.724		170	[illegible]03	1.37	374.37	505
C-14A	11.96	16.722		365	755	62.50	1,182.50	1,625
	2200	.025		.75	1.01	.33	2.09	2.71
	31800	.002			.07	.02	.09	.13
	21200	.003			.10	.[illegible]3	.13	.20
C-14H	45	1.067	Ea.	44.50	47	.74	92.24	122
	30	1.600		66	70.50	1.10	137.60	183
	18	[illegible].667		117	118	1.84	2[illegible]6.84	310
	14	3.429		157	151	2.37	310.[illegible]7	410

The **Total Incl O&P column** is the total cost, including overhead and profit, that the installing contractor will charge the customer. This represents the cost of materials plus 10% profit, the cost of labor plus labor burden and 10% profit, and the cost of equipment plus 10% profit. It does not include the general contractor's overhead and profit. Note: See the inside back cover for details of how RSMeans calculates labor burden.

The **Total column** represents the total bare cost for the installing contractor, in U.S. dollars. In this case, the sum of $66 for material + $70.50 for labor + $1.10 for equipment is $137.60.

The figure in the **Labor Hours** column is the amount of labor required to perform one unit of work—in this case the amount of labor required to construct one 4' x 4' equipment pad. This figure is calculated by dividing the number of hours of labor in the crew by the daily output (48 labor hours divided by 30 pads = 1.6 hours of labor per pad). Multiply 1.600 times 60 to see the value in minutes: 60 x 1.6 = 96 minutes. Note: the labor hour figure is not dependent on the crew size. A change in crew size will result in a corresponding change in daily output, but the labor hours per unit of work will not change.

All RSMeans unit cost data includes the typical **Unit of Measure** used for estimating that item. For concrete-in-place the typical unit is cubic yards (C.Y.) or each (Ea.). For installing broadloom carpet it is square yard, and for gypsum board it is square foot. The estimator needs to take special care that the unit in the data matches the unit in the take-off. Unit conversions may be found in the Reference Section.

How RSMeans Data Works

Sample Estimate

This sample demonstrates the elements of an estimate, including a tally of the RSMeans data lines, and a summary of the markups on a contractor's work to arrive at a total cost to the owner. The RSMeans Location Factor is added at the bottom of the estimate to adjust the cost of the work to a specific location.

Work Performed: The body of the estimate shows the RSMeans data selected, including line number, a brief description of each item, its take-off unit and quantity, and the bare costs of materials, labor, and equipment. This estimate also includes a column titled "SubContract." This data is taken from the RSMeans column "Total Incl O&P," and represents the total that a subcontractor would charge a general contractor for the work, including the sub's markup for overhead and profit.

Division 1, General Requirements: This is the first division numerically, but the last division estimated. Division 1 includes project-wide needs provided by the general contractor. These requirements vary by project, but may include temporary facilities and utilities, security, testing, project cleanup, etc. For small projects a percentage can be used, typically between 5% and 15% of project cost. For large projects the costs may be itemized and priced individually.

Bonds: Bond costs should be added to the estimate. The figures here represent a typical performance bond, ensuring the owner that if the general contractor does not complete the obligations in the construction contract the bonding company will pay the cost for completion of the work.

Location Adjustment: RSMeans published data is based on national average costs. If necessary, adjust the total cost of the project using a location factor from the "Location Factor" table or the "City Cost Index" table. Use location factors if the work is general, covering multiple trades. If the work is by a single trade (e.g., masonry) use the more specific data found in the "City Cost Indexes."

Project Name: Pre-Engineered Steel Building		Architect: As Shown		
Location:	Anywhere, USA			
Line Number	Description	Qty	Unit	Material
03 30 53.40 3940	Strip footing, 12" x 24", reinforced	34	C.Y.	$4,488.00
03 30 53.40 3950	Strip footing, 12" x 36", reinforced	15	C.Y.	$1,905.00
03 11 13.65 3000	Concrete slab edge forms	500	L.F.	$155.00
03 22 05.50 0200	Welded wire fabric reinforcing	150	C.S.F.	$2,602.50
03 31 05.35 0300	Ready mix concrete, 4000 psi for slab on grade	186	C.Y.	$18,972.00
03 31 05.70 4300	Place, strike off & consolidate concrete slab	186	C.Y.	$0.00
03 35 29.30 0250	Machine float & trowel concrete slab	15,000	S.F.	$0.00
03 35 29.35 0160	Cut control joints in concrete slab	950	L.F.	$66.50
03 39 23.13 0300	Sprayed concrete curing membrane	150	C.S.F.	$1,207.50
Division 03	**Subtotal**			**$29,396.50**
08 36 13.10 2650	Manual 10' x 10' steel sectional overhead door	8	Ea.	$8,800.00
08 36 13.10 2860	Insulation and steel back panel for OH door	800	S.F.	$3,680.00
Division 08	**Subtotal**			**$12,480.00**
13 34 19.50 1100	Pre-Engineered Steel Building, 100' x 150' x 24'	15,000	SF Flr.	$0.00
13 34 19.50 6050	Framing for PESB door opening, 3' x 7'	4	Opng.	$0.00
13 34 19.50 6100	Framing for PESB door opening, 10' x 10'	8	Opng.	$0.00
13 34 19.50 6200	Framing for PESB window opening, 4' x 3'	6	Opng.	$0.00
13 34 19.50 5750	PESB door, 3' x 7', single leaf	4	Opng.	$2,320.00
13 34 19.50 7750	PESB sliding window, 4' x 3' with screen	6	Opng.	$2,220.00
13 34 19.50 [illegible]550	PESB gutter, eave type, 26 ga., painted	300	L.F.	$1,950.00
13 34 19.50 86[illegible]0	PESB roof vent, 12" wide x 10' long	15	Ea.	$517.50
13 34 19.50 6900	PESB insulation, vinyl faced, 4" thick	27,400	S.F.	$10,686.00
Division 13	**Subtotal**			**$17,693.50**
		Subtotal		**$59,570.00**
Division 01	**General Requirements @ 7%**			**4,169.90**
		Estimate Subtotal		**$63,739.90**
		Sales Tax @ 5%		**3,187.00**
		Subtotal		**66,926.90**
		GC O & P		**6,692.69**
		Subtotal		**73,619.58**
		Contingency @ 5%		
		Subtotal		
		Bond @ $12/1000 +10% O&P		
		Subtotal		
		Location Adjustment Factor		
		Grand Total		

This estimate is based on an interactive spreadsheet.
A copy of this spreadsheet is located on the RSMeans website at **http://www.reedconstructiondata.com/rsmeans/extras/546011**.
You are free to download it and adjust it to your methodology.

This example shows the cost to construct a pre-engineered steel building. The foundation, doors, windows, and insulation will be installed by the general contractor. A subcontractor will install the structural steel, roofing, and siding.

		01/01/13	STD	
Labor	**Equipment**	**SubContract**	**Estimate Total**	
$3,400.00	$23.12	$0.00		
$1,200.00	$8.10	$0.00		
$1,135.00	$0.00	$0.00		
$3,825.00	$0.00	$0.00		
$0.00	$0.00	$0.00		
$3,003.90	$111.60	$0.00		
$8,550.00	$450.00	$0.00		
$408.50	$104.50	$0.00		
$892.50	$0.00	$0.00		
$22,414.90	**$697.32**	**$0.00**	**$52,508.72**	**Division 03**
$3,200.00	$0.00	$0.00		
$0.00	$0.00	$0.00		
$3,200.00	**$0.00**	**$0.00**	**$15,680.00**	**Division 08**
$0.00	$0.00	$337,500.00		
$0.00	$0.00	$2,220.00		
$0.00	$0.00	$9,000.00		
$0.00	$0.00	$3,270.00		
$640.00	$0.00	$0.00		
$546.00	$66.60	$0.00		
$750.00	$0.00	$0.00		
$3,000.00	$0.00	$0.00		
$8,494.00	$0.00	$0.00		
$13,430.00	**$66.60**	**$351,990.00**	**$383,180.10**	**Division 13**
$39,044.90	**$763.92**	**$351,990.00**	**$451,368.82**	**Subtotal**
2,733.14	**53.47**	**24,639.30**		**Gen. Requirements**
$41,778.04	**$817.39**	**$376,629.30**	**$451,368.82**	**Estimate Subtotal**
	40.87	**9,415.73**		**Sales tax**
41,778.04	**858.26**	**386,045.03**		**Subtotal**
22,810.81	**85.83**	**38,604.50**		**GC O&P**
64,588.85	**944.09**	**424,649.54**	**$563,802.07**	**Subtotal**
			28,190.10	**Contingency**
			$591,992.17	**Subtotal**
			7,814.30	**Bond**
			$599,806.47	**Subtotal**
102.30			**13,795.55**	**Location Adjustment**
			$613,602.01	**Grand Total**

Sales Tax: If the work is subject to state or local sales taxes, the amount must be added to the estimate. Sales tax may be added to material costs, equipment costs, and subcontracted work. In this case, sales tax was added in all three categories. It was assumed that approximately half the subcontracted work would be material cost, so the tax was applied to 50% of the subcontract total.

GC O&P: This entry represents the general contractor's markup on material, labor, equipment, and subcontractor costs. RSMeans' standard markup on materials, equipment, and subcontracted work is 10%. In this estimate, the markup on the labor performed by the GC's workers uses "Skilled Workers Average" shown in Column F on the table "Installing Contractor's Overhead & Profit," which can be found on the inside-back cover of the book.

Contingency: A factor for contingency may be added to any estimate to represent the cost of unknowns that may occur between the time that the estimate is performed and the time the project is constructed. The amount of the allowance will depend on the stage of design at which the estimate is done, and the contractor's assessment of the risk involved. Refer to section 01 21 16.50 for contigency allowances.

Estimating Tips

01 20 00 Price and Payment Procedures

- Allowances that should be added to estimates to cover contingencies and job conditions that are not included in the national average material and labor costs are shown in section 01 21.
- When estimating historic preservation projects (depending on the condition of the existing structure and the owner's requirements), a 15%–20% contingency or allowance is recommended, regardless of the stage of the drawings.

01 30 00 Administrative Requirements

- Before determining a final cost estimate, it is a good practice to review all the items listed in Subdivisions 01 31 and 01 32 to make final adjustments for items that may need customizing to specific job conditions.
- Requirements for initial and periodic submittals can represent a significant cost to the General Requirements of a job. Thoroughly check the submittal specifications when estimating a project to determine any costs that should be included.

01 40 00 Quality Requirements

- All projects will require some degree of Quality Control. This cost is not included in the unit cost of construction listed in each division. Depending upon the terms of the contract, the various costs of inspection and testing can be the responsibility of either the owner or the contractor. Be sure to include the required costs in your estimate.

01 50 00 Temporary Facilities and Controls

- Barricades, access roads, safety nets, scaffolding, security, and many more requirements for the execution of a safe project are elements of direct cost. These costs can easily be overlooked when preparing an estimate. When looking through the major classifications of this subdivision, determine which items apply to each division in your estimate.
- Construction Equipment Rental Costs can be found in the Reference Section in section 01 54 33. Operators' wages are not included in equipment rental costs.
- Equipment mobilization and demobilization costs are not included in equipment rental costs and must be considered separately in section 01 54 36.50.
- The cost of small tools provided by the installing contractor for his workers is covered in the "Overhead" column on the "Installing Contractor's Overhead and Profit" table that lists labor trades, base rates and markups and, therefore, is included in the "Total Incl. O&P" cost of any Unit Price line item. For those users who are constrained by contract terms to use only bare costs, there are two line items in section 01 54 39.70 for small tools as a percentage of bare labor cost. If some of those users are further constrained by contract terms to refrain from using line items from Division 1, you are advised to cover the cost of small tools within your coefficient or multiplier.

01 70 00 Execution and Closeout Requirements

- When preparing an estimate, thoroughly read the specifications to determine the requirements for Contract Closeout. Final cleaning, record documentation, operation and maintenance data, warranties and bonds, and spare parts and maintenance materials can all be elements of cost for the completion of a contract. Do not overlook these in your estimate.

Reference Numbers

Reference numbers are shown in shaded boxes at the beginning of some major classifications. These numbers refer to related items in the Reference Section. The reference information may be an estimating procedure, an alternate pricing method, or technical information.

Note: Not all subdivisions listed here necessarily appear in this publication.

01 11 Summary of Work

01 11 31 – Professional Consultants

01 11 31.10 Architectural Fees		Crew	Daily Output	Labor-Hours	Unit	Material	2013 Bare Costs Labor	Equipment	Total	Total Incl O&P
0010	**ARCHITECTURAL FEES**									
0020	For new construction									
0060	Minimum				Project				4.90%	4.90%
0090	Maximum								16%	16%
0100	For alteration work, to $500,000, add to new construction fee								50%	50%
0150	Over $500,000, add to new construction fee								25%	25%

01 11 31.20 Construction Management Fees

		Crew	Daily Output	Labor-Hours	Unit	Material	Labor	Equipment	Total	Total Incl O&P
0010	**CONSTRUCTION MANAGEMENT FEES**									
0020	$1,000,000 job, minimum				Project				4.50%	4.50%
0050	Maximum								7.50%	7.50%
0300	$50,000,000 job, minimum								2.50%	2.50%
0350	Maximum								4%	4%

01 11 31.30 Engineering Fees

		Crew	Daily Output	Labor-Hours	Unit	Material	Labor	Equipment	Total	Total Incl O&P
0010	**ENGINEERING FEES** R011110-30									
0020	Educational planning consultant, minimum				Project				.50%	.50%
0100	Maximum				"				2.50%	2.50%
0200	Electrical, minimum				Contrct				4.10%	4.10%
0300	Maximum								10.10%	10.10%
0400	Elevator & conveying systems, minimum								2.50%	2.50%
0500	Maximum								5%	5%
0600	Food service & kitchen equipment, minimum								8%	8%
0700	Maximum								12%	12%
0800	Landscaping & site development, minimum								2.50%	2.50%
0900	Maximum								6%	6%
1000	Mechanical (plumbing & HVAC), minimum								4.10%	4.10%
1100	Maximum								10.10%	10.10%
1200	Structural, minimum				Project				1%	1%
1300	Maximum				"				2.50%	2.50%

01 21 Allowances

01 21 16 – Contingency Allowances

01 21 16.50 Contingencies

		Crew	Daily Output	Labor-Hours	Unit	Material	Labor	Equipment	Total	Total Incl O&P
0010	**CONTINGENCIES**, Add to estimate									
0020	Conceptual stage				Project				20%	20%
0050	Schematic stage								15%	15%
0100	Preliminary working drawing stage (Design Dev.)								10%	10%
0150	Final working drawing stage								3%	3%

01 21 53 – Factors Allowance

01 21 53.50 Factors

		Crew	Daily Output	Labor-Hours	Unit	Material	Labor	Equipment	Total	Total Incl O&P
0010	**FACTORS** Cost adjustments R012153-10									
0100	Add to construction costs for particular job requirements									
0500	Cut & patch to match existing construction, add, minimum				Costs	2%	3%			
0550	Maximum					5%	9%			
0800	Dust protection, add, minimum					1%	2%			
0850	Maximum					4%	11%			
1100	Equipment usage curtailment, add, minimum					1%	1%			
1150	Maximum					3%	10%			
1400	Material handling & storage limitation, add, minimum					1%	1%			
1450	Maximum					6%	7%			
1700	Protection of existing work, add, minimum					2%	2%			

01 21 Allowances

01 21 53 – Factors Allowance

01 21 53.50 Factors		Crew	Daily Output	Labor-Hours	Unit	Material	2013 Bare Costs Labor	Equipment	Total	Total Incl O&P
1750	Maximum				Costs	5%	7%			
2000	Shift work requirements, add, minimum						5%			
2050	Maximum						30%			
2300	Temporary shoring and bracing, add, minimum					2%	5%			
2350	Maximum					5%	12%			
2400	Work inside prisons and high security areas, add, minimum						30%			
2450	Maximum						50%			

01 21 55 – Job Conditions Allowance

01 21 55.50 Job Conditions		Crew	Daily Output	Labor-Hours	Unit	Material	Labor	Equipment	Total	Total Incl O&P
0010	**JOB CONDITIONS** Modifications to applicable									
0020	cost summaries									
0100	Economic conditions, favorable, deduct				Project				2%	2%
0200	Unfavorable, add								5%	5%
0300	Hoisting conditions, favorable, deduct								2%	2%
0400	Unfavorable, add								5%	5%
0700	Labor availability, surplus, deduct								1%	1%
0800	Shortage, add								10%	10%
0900	Material storage area, available, deduct								1%	1%
1000	Not available, add								2%	2%
1100	Subcontractor availability, surplus, deduct								5%	5%
1200	Shortage, add								12%	12%
1300	Work space, available, deduct								2%	2%
1400	Not available, add								5%	5%

01 21 57 – Overtime Allowance

01 21 57.50 Overtime		Crew	Daily Output	Labor-Hours	Unit	Material	Labor	Equipment	Total	Total Incl O&P
0010	**OVERTIME** for early completion of projects or where R012909-90									
0020	labor shortages exist, add to usual labor, up to				Costs		100%			

01 21 61 – Cost Indexes

01 21 61.10 Construction Cost Index		Crew	Daily Output	Labor-Hours	Unit	Material	Labor	Equipment	Total	Total Incl O&P
0010	**CONSTRUCTION COST INDEX** (Reference) over 930 zip code locations in									
0020	the U.S. and Canada, total bldg. cost, min. (Guymon, OK)				%					67.40%
0050	Average									100%
0100	Maximum (New York, NY)									131.90%
01 21 61.20 Historical Cost Indexes										
0010	**HISTORICAL COST INDEXES** (See Reference Section)									
01 21 61.30 Labor Index										
0010	**LABOR INDEX** (Reference) For over 930 zip code locations in									
0020	the U.S. and Canada, minimum (Guymon, OK)				%		30.70%			
0050	Average						100%			
0100	Maximum (New York, NY)						167%			
01 21 61.50 Material Index										
0011	**MATERIAL INDEX** For over 930 zip code locations in									
0020	the U.S. and Canada, minimum (Vale, OR)				%	91.80%				
0040	Average					100%				
0060	Maximum (Ketchikan, AK)					132.80%				

01 21 63 – Taxes

01 21 63.10 Taxes		Crew	Daily Output	Labor-Hours	Unit	Material	Labor	Equipment	Total	Total Incl O&P
0010	**TAXES** R012909-80									
0020	Sales tax, State, average				%	5.03%				
0050	Maximum R012909-85					7.25%				

01 21 Allowances

01 21 63 – Taxes

01 21 63.10 Taxes		Crew	Daily Output	Labor-Hours	Unit	Material	2013 Bare Costs Labor	Equipment	Total	Total Incl O&P
0200	Social Security, on first $114,000 of wages				%		7.65%			
0300	Unemployment, combined Federal and State, minimum						.80%			
0350	Average						7.80%			
0400	Maximum						13.50%			

01 31 Project Management and Coordination

01 31 13 – Project Coordination

01 31 13.20 Field Personnel

		Crew	Daily Output	Labor-Hours	Unit	Material	Labor	Equipment	Total	Total Incl O&P
0010	**FIELD PERSONNEL**									
0020	Clerk, average				Week		500		500	775
0100	Field engineer, minimum						1,025		1,025	1,575
0120	Average						1,325		1,325	2,050
0140	Maximum						1,500		1,500	2,325
0160	General purpose laborer, average						1,425		1,425	2,175
0180	Project manager, minimum						1,875		1,875	2,900
0200	Average						2,150		2,150	3,350
0220	Maximum						2,475		2,475	3,825
0240	Superintendent, minimum						1,825		1,825	2,825
0260	Average						2,000		2,000	3,100
0280	Maximum						2,300		2,300	3,550
0290	Timekeeper, average						1,175		1,175	1,800

01 31 13.30 Insurance

		Crew	Daily Output	Labor-Hours	Unit	Material	Labor	Equipment	Total	Total Incl O&P
0010	**INSURANCE** R013113-40									
0020	Builders risk, standard, minimum				Job				.24%	.24%
0050	Maximum R013113-60								.64%	.64%
0200	All-risk type, minimum								.25%	.25%
0250	Maximum								.62%	.62%
0400	Contractor's equipment floater, minimum				Value				.50%	.50%
0450	Maximum				″				1.50%	1.50%
0600	Public liability, average				Job				2.02%	2.02%
0800	Workers' compensation & employer's liability, average									
0850	by trade, carpentry, general				Payroll		15.13%			
1000	Electrical						5.68%			
1150	Insulation						11.44%			
1450	Plumbing						6.66%			
1550	Sheet metal work (HVAC)						8.56%			

01 31 13.40 Main Office Expense

		Crew	Daily Output	Labor-Hours	Unit	Material	Labor	Equipment	Total	Total Incl O&P
0010	**MAIN OFFICE EXPENSE** Average for General Contractors									
0020	As a percentage of their annual volume									
0125	Annual volume under 1 million dollars				% Vol.				17.50%	
0145	Up to 2.5 million dollars								8%	
0150	Up to 4.0 million dollars								6.80%	
0200	Up to 7.0 million dollars								5.60%	
0250	Up to 10 million dollars								5.10%	
0300	Over 10 million dollars								3.90%	

01 31 13.50 General Contractor's Mark-Up

		Crew	Daily Output	Labor-Hours	Unit	Material	Labor	Equipment	Total	Total Incl O&P
0010	**GENERAL CONTRACTOR'S MARK-UP** on Change Orders									
0200	Extra work, by subcontractors, add				%				10%	10%
0250	By General Contractor, add								15%	15%
0400	Omitted work, by subcontractors, deduct all but								5%	5%

01 31 Project Management and Coordination

01 31 13 – Project Coordination

01 31 13.50 General Contractor's Mark-Up		Crew	Daily Output	Labor-Hours	Unit	2013 Bare Costs Material	Labor	Equipment	Total	Total Incl O&P
0450	By General Contractor, deduct all but				%				7.50%	7.50%
0600	Overtime work, by subcontractors, add								15%	15%
0650	By General Contractor, add								10%	10%

01 31 13.60 Installing Contractor's Main Office Overhead

		Crew	Daily Output	Labor-Hours	Unit	Material	Labor	Equipment	Total	Total Incl O&P
0010	**INSTALLING CONTRACTOR'S MAIN OFFICE OVERHEAD**									
0020	As percent of direct costs, minimum				%				5%	
0050	Average								13%	
0100	Maximum								30%	

01 31 13.80 Overhead and Profit

		Crew	Daily Output	Labor-Hours	Unit	Material	Labor	Equipment	Total	Total Incl O&P
0010	**OVERHEAD & PROFIT** Allowance to add to items in this									
0020	book that do not include Subs O&P, average				%				25%	
0100	Allowance to add to items in this book that									
0110	do include Subs O&P, minimum				%				5%	5%
0150	Average								10%	10%
0200	Maximum								15%	15%
0300	Typical, by size of project, under $100,000								30%	
0350	$500,000 project								25%	
0400	$2,000,000 project								20%	
0450	Over $10,000,000 project								15%	

01 31 13.90 Performance Bond

		Crew	Daily Output	Labor-Hours	Unit	Material	Labor	Equipment	Total	Total Incl O&P
0010	**PERFORMANCE BOND** R013113-80									
0020	For buildings, minimum				Job				.60%	.60%
0100	Maximum				"				2.50%	2.50%

01 32 Construction Progress Documentation

01 32 33 – Photographic Documentation

01 32 33.50 Photographs

		Crew	Daily Output	Labor-Hours	Unit	Material	Labor	Equipment	Total	Total Incl O&P
0010	**PHOTOGRAPHS**									
0020	8" x 10", 4 shots, 2 prints ea., std. mounting				Set	475			475	520
0100	Hinged linen mounts					530			530	580
0200	8" x 10", 4 shots, 2 prints each, in color					415			415	460
0300	For I.D. slugs, add to all above					5.30			5.30	5.85
1500	Time lapse equipment, camera and projector, buy				Ea.	2,650			2,650	2,925
1550	Rent per month				"	1,575			1,575	1,725
1700	Cameraman and film, including processing, B.&W.				Day	1,375			1,375	1,525
1720	Color				"	1,375			1,375	1,525

01 41 Regulatory Requirements

01 41 26 – Permit Requirements

01 41 26.50 Permits

		Crew	Daily Output	Labor-Hours	Unit	Material	Labor	Equipment	Total	Total Incl O&P
0010	**PERMITS**									
0020	Rule of thumb, most cities, minimum				Job				.50%	.50%
0100	Maximum				"				2%	2%

01 51 Temporary Utilities

01 51 13 – Temporary Electricity

01 51 13.80 Temporary Utilities		Crew	Daily Output	Labor-Hours	Unit	Material	2013 Bare Costs Labor	Equipment	Total	Total Incl O&P
0010	**TEMPORARY UTILITIES**									
0100	Heat, incl. fuel and operation, per week, 12 hrs. per day	1 Skwk	100	.080	CSF Flr	27.50	3.70		31.20	36
0200	24 hrs. per day	″	60	.133		53	6.15		59.15	67.50
0350	Lighting, incl. service lamps, wiring & outlets, minimum	1 Elec	34	.235		2.80	12.35		15.15	21.50
0360	Maximum	″	17	.471		5.85	24.50		30.35	43.50
0400	Power for temp lighting only, 6.6 KWH, per month								.92	1.01
0430	11.8 KWH, per month								1.65	1.82
0450	23.6 KWH, per month								3.30	3.63
0600	Power for job duration incl. elevator, etc., minimum								47	51.50
0650	Maximum								110	121
1000	Toilet, portable, see Equip. Rental 01 54 33 in Reference Section									

01 52 Construction Facilities

01 52 13 – Field Offices and Sheds

01 52 13.20 Office and Storage Space		Crew	Daily Output	Labor-Hours	Unit	Material	Labor	Equipment	Total	Total Incl O&P
0010	**OFFICE AND STORAGE SPACE**									
0020	Office trailer, furnished, no hookups, 20′ x 8′, buy	2 Skwk	1	16	Ea.	8,550	740		9,290	10,600
0250	Rent per month					153			153	169
0300	32′ x 8′, buy	2 Skwk	.70	22.857		13,600	1,050		14,650	16,600
0350	Rent per month					190			190	209
0400	50′ x 10′, buy	2 Skwk	.60	26.667		22,300	1,225		23,525	26,500
0450	Rent per month					291			291	320
0500	50′ x 12′, buy	2 Skwk	.50	32		27,000	1,475		28,475	32,000
0550	Rent per month					340			340	375
0700	For air conditioning, rent per month, add					46			46	50.50
0800	For delivery, add per mile				Mile	10.60			10.60	11.65
0890	Delivery each way				Ea.	200			200	220
0900	Bunk house trailer, 8′ x 40′ duplex dorm with kitchen, no hookups, buy	2 Carp	1	16		37,300	720		38,020	42,100
0910	9 man with kitchen and bath, no hookups, buy		1	16		38,000	720		38,720	42,900
0920	18 man sleeper with bath, no hookups, buy		1	16		49,000	720		49,720	55,000
1000	Portable buildings, prefab, on skids, economy, 8′ x 8′		265	.060	S.F.	29.50	2.71		32.21	36.50
1100	Deluxe, 8′ x 12′		150	.107	″	24	4.79		28.79	34
1200	Storage boxes, 20′ x 8′, buy	2 Skwk	1.80	8.889	Ea.	3,025	410		3,435	3,950
1250	Rent per month					74			74	81.50
1300	40′ x 8′, buy	2 Skwk	1.40	11.429		3,975	530		4,505	5,200
1350	Rent per month					102			102	112

01 54 Construction Aids

01 54 16 – Temporary Hoists

01 54 16.50 Weekly Forklift Crew		Crew	Daily Output	Labor-Hours	Unit	Material	Labor	Equipment	Total	Total Incl O&P
0010	**WEEKLY FORKLIFT CREW**									
0100	All-terrain forklift, 45′ lift, 35′ reach, 9000 lb. capacity	A-3P	.20	40	Week		1,825	2,550	4,375	5,600

01 54 19 – Temporary Cranes

01 54 19.50 Daily Crane Crews		Crew	Daily Output	Labor-Hours	Unit	Material	Labor	Equipment	Total	Total Incl O&P
0010	**DAILY CRANE CREWS** for small jobs, portal to portal									
0100	12-ton truck-mounted hydraulic crane	A-3H	1	8	Day		390	855	1,245	1,525
0200	25-ton	A-3I	1	8			390	980	1,370	1,675
0300	40-ton	A-3J	1	8			390	1,225	1,615	1,925
0400	55-ton	A-3K	1	16			730	1,650	2,380	2,900

01 54 Construction Aids

01 54 19 – Temporary Cranes

01 54 19.50 Daily Crane Crews		Crew	Daily Output	Labor-Hours	Unit	Material	2013 Bare Costs Labor	Equipment	Total	Total Incl O&P
0500	80-ton	A-3L	1	16	Day		730	2,350	3,080	3,675
0600	100-ton	A-3M	1	16			730	2,350	3,080	3,675
0900	If crane is needed on a Saturday, Sunday or Holiday									
0910	At time-and-a-half, add				Day		50%			
0920	At double time, add				″		100%			

01 54 19.60 Monthly Tower Crane Crew

		Crew	Daily Output	Labor-Hours	Unit	Material	Labor	Equipment	Total	Total Incl O&P
0010	**MONTHLY TOWER CRANE CREW**, excludes concrete footing									
0100	Static tower crane, 130′ high, 106′ jib, 6200 lb. capacity	A-3N	.05	176	Month		8,600	23,600	32,200	38,900

01 54 23 – Temporary Scaffolding and Platforms

01 54 23.70 Scaffolding

		Crew	Daily Output	Labor-Hours	Unit	Material	Labor	Equipment	Total	Total Incl O&P
0010	**SCAFFOLDING** R015423-10									
0015	Steel tube, regular, no plank, labor only to erect & dismantle									
0090	Building exterior, wall face, 1 to 5 stories, 6′-4″ x 5′ frames	3 Carp	8	3	C.S.F.		135		135	207
0200	6 to 12 stories	4 Carp	8	4			180		180	277
0301	13 to 20 stories	5 Clab	8	5			177		177	273
0460	Building interior, wall face area, up to 16′ high	3 Carp	12	2			90		90	138
0560	16′ to 40′ high		10	2.400			108		108	166
0800	Building interior floor area, up to 30′ high		150	.160	C.C.F.		7.20		7.20	11.05
0900	Over 30′ high	4 Carp	160	.200	″		9		9	13.85
0906	Complete system for face of walls, no plank, material only rent/mo				C.S.F.	36			36	39.50
0908	Interior spaces, no plank, material only rent/mo				C.C.F.	3.42			3.42	3.76
0910	Steel tubular, heavy duty shoring, buy									
0920	Frames 5′ high 2′ wide				Ea.	81.50			81.50	90
0925	5′ high 4′ wide					93			93	102
0930	6′ high 2′ wide					93.50			93.50	103
0935	6′ high 4′ wide					109			109	120
0940	Accessories									
0945	Cross braces				Ea.	15.85			15.85	17.40
0950	U-head, 8″ x 8″					19.10			19.10	21
0955	J-head, 4″ x 8″					13.90			13.90	15.30
0960	Base plate, 8″ x 8″					15.50			15.50	17.05
0965	Leveling jack					33.50			33.50	36.50
1000	Steel tubular, regular, buy									
1100	Frames 3′ high 5′ wide				Ea.	58.50			58.50	64.50
1150	5′ high 5′ wide					66.50			66.50	73
1200	6′-4″ high 5′ wide					76			76	83.50
1350	7′-6″ high 6′ wide					158			158	173
1500	Accessories cross braces					15.50			15.50	17.05
1550	Guardrail post					16.40			16.40	18.05
1600	Guardrail 7′ section					6.30			6.30	6.95
1650	Screw jacks & plates					21.50			21.50	24
1700	Sidearm brackets					25.50			25.50	28.50
1750	8″ casters					31			31	34
1800	Plank 2″ x 10″ x 16′-0″					51.50			51.50	56.50
1900	Stairway section					286			286	315
1910	Stairway starter bar					32			32	35
1920	Stairway inside handrail					53			53	58
1930	Stairway outside handrail					81.50			81.50	90
1940	Walk-thru frame guardrail					41.50			41.50	46
2000	Steel tubular, regular, rent/mo.									
2100	Frames 3′ high 5′ wide				Ea.	5			5	5.50
2150	5′ high 5′ wide					5			5	5.50

01 54 Construction Aids

01 54 23 – Temporary Scaffolding and Platforms

01 54 23.70 Scaffolding		Crew	Daily Output	Labor-Hours	Unit	2013 Bare Costs Material	Labor	Equipment	Total	Total Incl O&P
2200	6'-4" high 5' wide				Ea.	5.15			5.15	5.65
2250	7'-6" high 6' wide					7			7	7.70
2500	Accessories, cross braces					1			1	1.10
2550	Guardrail post					1			1	1.10
2600	Guardrail 7' section					1			1	1.10
2650	Screw jacks & plates					2			2	2.20
2700	Sidearm brackets					2			2	2.20
2750	8" casters					8			8	8.80
2800	Outrigger for rolling tower					3			3	3.30
2850	Plank 2" x 10" x 16'-0"					6			6	6.60
2900	Stairway section					40			40	44
2940	Walk-thru frame guardrail					2.50			2.50	2.75
3000	Steel tubular, heavy duty shoring, rent/mo.									
3250	5' high 2' & 4' wide				Ea.	5			5	5.50
3300	6' high 2' & 4' wide					5			5	5.50
3500	Accessories, cross braces					1			1	1.10
3600	U - head, 8" x 8"					1			1	1.10
3650	J - head, 4" x 8"					1			1	1.10
3700	Base plate, 8" x 8"					1			1	1.10
3750	Leveling jack					2			2	2.20
5700	Planks, 2" x 10" x 16'-0", labor only to erect & remove to 50' H	3 Carp	72	.333			14.95		14.95	23
5800	Over 50' high	4 Carp	80	.400			17.95		17.95	27.50
6000	Heavy duty shoring for elevated slab forms to 8'-2" high, floor area									
6100	Labor only to erect & dismantle	4 Carp	16	2	C.S.F.		90		90	138
6110	Materials only, rent.mo				"	29.50			29.50	32.50
6500	To 14'-8" high									
6600	Labor only to erect & dismantle	4 Carp	10	3.200	C.S.F.		144		144	221
6610	Materials only, rent/mo				"	43			43	47.50

01 54 23.75 Scaffolding Specialties

		Crew	Daily Output	Labor-Hours	Unit	Material	Labor	Equipment	Total	Total Incl O&P
0010	**SCAFFOLDING SPECIALTIES**									
1200	Sidewalk bridge, heavy duty steel posts & beams, including									
1210	parapet protection & waterproofing (material cost is rent/month)									
1220	8' to 10' wide, 2 posts	3 Carp	15	1.600	L.F.	34.50	72		106.50	149
1230	3 posts	"	10	2.400	"	53	108		161	224
1500	Sidewalk bridge using tubular steel scaffold frames including									
1510	planking (material cost is rent/month)	3 Carp	45	.533	L.F.	5.60	24		29.60	43
1600	For 2 uses per month, deduct from all above					50%				
1700	For 1 use every 2 months, add to all above					100%				
1900	Catwalks, 20" wide, no guardrails, 7' span, buy				Ea.	133			133	146
2000	10' span, buy					186			186	205
3720	Putlog, standard, 8' span, with hangers, buy					66.50			66.50	73
3730	Rent per month					10			10	11
3750	12' span, buy					99			99	109
3755	Rent per month					15			15	16.50
3760	Trussed type, 16' span, buy					228			228	251
3770	Rent per month					20			20	22
3790	22' span, buy					274			274	300
3795	Rent per month					30			30	33
3800	Rolling ladders with handrails, 30" wide, buy, 2 step					204			204	225
4000	7 step					635			635	700
4050	10 step					880			880	965
4100	Rolling towers, buy, 5' wide, 7' long, 10' high					1,075			1,075	1,200

01 54 Construction Aids

01 54 23 – Temporary Scaffolding and Platforms

01 54 23.75 Scaffolding Specialties		Crew	Daily Output	Labor-Hours	Unit	Material	2013 Bare Costs Labor	Equipment	Total	Total Incl O&P
4200	For 5′ high added sections, to buy, add				Ea.	164			164	180
4300	Complete incl. wheels, railings, outriggers,									
4350	21′ high, to buy				Ea.	1,825			1,825	2,000
4400	Rent/month = 5% of purchase cost				″	113			113	124

01 54 36 – Equipment Mobilization

01 54 36.50 Mobilization

		Crew	Daily Output	Labor-Hours	Unit	Material	Labor	Equipment	Total	Total Incl O&P
0010	**MOBILIZATION** (Use line item again for demobilization) R015433-10									
0015	Up to 25 mi. haul dist (50 mi. RT for mob/demob crew)									
0020	Dozer, loader, backhoe, excav., grader, paver, roller, 70 to 150 H.P.	B-34N	4	2	Ea.		73	142	215	269
0100	Above 150 HP	B-34K	3	2.667			97.50	320	417.50	505
0900	Shovel or dragline, 3/4 C.Y.	″	3.60	2.222			81.50	267	348.50	420
1100	Small equipment, placed in rear of, or towed by pickup truck	A-3A	8	1			35.50	20.50	56	77.50
1150	Equip up to 70 HP, on flatbed trailer behind pickup truck	A-3D	4	2			71.50	69.50	141	186
2000	Crane, truck-mounted, up to 75 ton, (driver only, one-way)	1 Eqhv	7.20	1.111			54		54	82
2100	Crane, truck-mounted, over 75 ton	A-3E	2.50	6.400			273	66	339	490
2200	Crawler-mounted, up to 75 ton	A-3F	2	8			340	500	840	1,075
2300	Over 75 ton	A-3G	1.50	10.667			455	750	1,205	1,525
2500	For each additional 5 miles haul distance, add						10%	10%		
3000	For large pieces of equipment, allow for assembly/knockdown									
3100	For mob/demob of micro-tunneling equip, see Section 33 05 23.19									

01 54 39 – Construction Equipment

01 54 39.70 Small Tools

		Crew	Daily Output	Labor-Hours	Unit	Material	Labor	Equipment	Total	Total Incl O&P
0010	**SMALL TOOLS** R013113-50									
0020	As % of contractor's bare labor cost for project, minimum				Total		.50%			
0100	Maximum				″		2%			

01 55 Vehicular Access and Parking

01 55 23 – Temporary Roads

01 55 23.50 Roads and Sidewalks

		Crew	Daily Output	Labor-Hours	Unit	Material	Labor	Equipment	Total	Total Incl O&P
0010	**ROADS AND SIDEWALKS** Temporary									
0050	Roads, gravel fill, no surfacing, 4″ gravel depth	B-14	715	.067	S.Y.	4.35	2.52	.51	7.38	9.20
0100	8″ gravel depth	″	615	.078	″	8.70	2.93	.60	12.23	14.75
1000	Ramp, 3/4″ plywood on 2″ x 6″ joists, 16″ O.C.	2 Carp	300	.053	S.F.	1.33	2.39		3.72	5.15
1100	On 2″ x 10″ joists, 16″ O.C.	″	275	.058	″	1.85	2.61		4.46	6.05

01 56 Temporary Barriers and Enclosures

01 56 13 – Temporary Air Barriers

01 56 13.60 Tarpaulins

		Crew	Daily Output	Labor-Hours	Unit	Material	Labor	Equipment	Total	Total Incl O&P
0010	**TARPAULINS**									
0020	Cotton duck, 10 oz. to 13.13 oz. per S.Y., 6′x8′				S.F.	.79			.79	.87
0050	30′x30′					1.50			1.50	1.65
0100	Polyvinyl coated nylon, 14 oz. to 18 oz., minimum					1.19			1.19	1.31
0150	Maximum					1.19			1.19	1.31
0200	Reinforced polyethylene 3 mils thick, white					.03			.03	.03
0300	4 mils thick, white, clear or black					.09			.09	.10
0400	5.5 mils thick, clear					.16			.16	.18
0500	White, fire retardant					.41			.41	.45
0600	12 mils, oil resistant, fire retardant					.27			.27	.30

01 56 Temporary Barriers and Enclosures

01 56 13 – Temporary Air Barriers

01 56 13.60 Tarpaulins		Crew	Daily Output	Labor-Hours	Unit	Material	Labor (2013 Bare Costs)	Equipment	Total	Total Incl O&P
0700	8.5 mils, black				S.F.	.57			.57	.63
0710	Woven polyethylene, 6 mils thick					.16			.16	.18
0730	Polyester reinforced w/integral fastening system 11 mils thick					.24			.24	.26
0740	Mylar polyester, non-reinforced, 7 mils thick					1.17			1.17	1.29

01 56 13.90 Winter Protection		Crew	Daily Output	Labor-Hours	Unit	Material	Labor	Equipment	Total	Total Incl O&P
0010	**WINTER PROTECTION**									
0100	Framing to close openings	2 Clab	500	.032	S.F.	.40	1.13		1.53	2.19
0200	Tarpaulins hung over scaffolding, 8 uses, not incl. scaffolding		1500	.011		.25	.38		.63	.86
0250	Tarpaulin polyester reinf. w/integral fastening system 11 mils thick		1600	.010		.24	.35		.59	.81
0300	Prefab fiberglass panels, steel frame, 8 uses		1200	.013		2.20	.47		2.67	3.15

01 56 16 – Temporary Dust Barriers

01 56 16.10 Dust Barriers, Temporary		Crew	Daily Output	Labor-Hours	Unit	Material	Labor	Equipment	Total	Total Incl O&P
0010	**DUST BARRIERS, TEMPORARY**									
0020	Spring loaded telescoping pole & head, to 12', erect and dismantle	1 Clab	240	.033	Ea.		1.18		1.18	1.82
0025	Cost per day (based upon 250 days)				Day	.24			.24	.26
0030	To 21', erect and dismantle	1 Clab	240	.033	Ea.		1.18		1.18	1.82
0035	Cost per day (based upon 250 days)				Day	.39			.39	.42
0040	Accessories, caution tape reel, erect and dismantle	1 Clab	480	.017	Ea.		.59		.59	.91
0045	Cost per day (based upon 250 days)				Day	.05			.05	.05
0060	Foam rail and connector, erect and dismantle	1 Clab	240	.033	Ea.		1.18		1.18	1.82
0065	Cost per day (based upon 250 days)				Day	.10			.10	.11
0070	Caution tape	1 Clab	384	.021	C.L.F.	2.70	.74		3.44	4.11
0080	Zipper, standard duty		60	.133	Ea.	8	4.73		12.73	16.10
0090	Heavy duty		48	.167	"	9.50	5.90		15.40	19.55
0100	Polyethylene sheet, 4 mil		37	.216	Sq.	2.92	7.65		10.57	15
0110	6 mil		37	.216	"	4.03	7.65		11.68	16.25
1000	Dust partition, 6 mil polyethylene, 1" x 3" frame	2 Carp	2000	.008	S.F.	.27	.36		.63	.84
1080	2" x 4" frame	"	2000	.008	"	.31	.36		.67	.89

01 56 23 – Temporary Barricades

01 56 23.10 Barricades		Crew	Daily Output	Labor-Hours	Unit	Material	Labor	Equipment	Total	Total Incl O&P
0010	**BARRICADES**									
0020	5' high, 3 rail @ 2" x 8", fixed	2 Carp	20	.800	L.F.	5.35	36		41.35	61.50
0150	Movable		30	.533		4.47	24		28.47	42
1000	Guardrail, wooden, 3' high, 1" x 6", on 2" x 4" posts		200	.080		1.30	3.59		4.89	7
1100	2" x 6", on 4" x 4" posts		165	.097		2.24	4.35		6.59	9.15
1200	Portable metal with base pads, buy					15.55			15.55	17.10
1250	Typical installation, assume 10 reuses	2 Carp	600	.027		2.55	1.20		3.75	4.65
1300	Barricade tape, polyethylene, 7 mil, 3" wide x 500' long roll				Ea.	25			25	27.50
3000	Detour signs, set up and remove									
3010	Reflective aluminum, MUTCD, 24" x 24", post mounted	1 Clab	20	.400	Ea.	2.24	14.20		16.44	24.50

01 56 26 – Temporary Fencing

01 56 26.50 Temporary Fencing		Crew	Daily Output	Labor-Hours	Unit	Material	Labor	Equipment	Total	Total Incl O&P
0010	**TEMPORARY FENCING**									
0020	Chain link, 11 ga., 4' high	2 Clab	400	.040	L.F.	2.59	1.42		4.01	5.05
0100	6' high		300	.053		2.59	1.89		4.48	5.75
0200	Rented chain link, 6' high, to 1000' (up to 12 mo.)		400	.040		4.39	1.42		5.81	7
0250	Over 1000' (up to 12 mo.)		300	.053		4.29	1.89		6.18	7.65
0350	Plywood, painted, 2" x 4" frame, 4' high	A-4	135	.178		6.20	7.60		13.80	18.50
0400	4" x 4" frame, 8' high	"	110	.218		10.95	9.35		20.30	26.50
0500	Wire mesh on 4" x 4" posts, 4' high	2 Carp	100	.160		9.30	7.20		16.50	21.50
0550	8' high	"	80	.200		14.10	9		23.10	29.50

01 56 Temporary Barriers and Enclosures

01 56 29 – Temporary Protective Walkways

01 56 29.50 Protection		Crew	Daily Output	Labor-Hours	Unit	2013 Bare Costs Material	Labor	Equipment	Total	Total Incl O&P
0010	**PROTECTION**									
0020	Stair tread, 2" x 12" planks, 1 use	1 Carp	75	.107	Tread	3.97	4.79		8.76	11.75
0100	Exterior plywood, 1/2" thick, 1 use		65	.123		1.74	5.55		7.29	10.40
0200	3/4" thick, 1 use		60	.133		2.32	6		8.32	11.75
2200	Sidewalks, 2" x 12" planks, 2 uses		350	.023	S.F.	.66	1.03		1.69	2.31
2300	Exterior plywood, 2 uses, 1/2" thick		750	.011		.29	.48		.77	1.06
2400	5/8" thick		650	.012		.36	.55		.91	1.25
2500	3/4" thick		600	.013		.39	.60		.99	1.35

01 56 32 – Temporary Security

01 56 32.50 Watchman		Crew	Daily Output	Labor-Hours	Unit	Material	Labor	Equipment	Total	Total Incl O&P
0010	**WATCHMAN**									
0020	Service, monthly basis, uniformed person, minimum				Hr.				25	27.50
0100	Maximum								45.50	50
0200	Person and command dog, minimum								31	34
0300	Maximum								54.50	60
0500	Sentry dog, leased, with job patrol (yard dog), 1 dog				Week				290	320
0600	2 dogs				"				390	430
0800	Purchase, trained sentry dog, minimum				Ea.				1,375	1,500
0900	Maximum				"				2,725	3,000

01 58 Project Identification

01 58 13 – Temporary Project Signage

01 58 13.50 Signs		Crew	Daily Output	Labor-Hours	Unit	Material	Labor	Equipment	Total	Total Incl O&P
0010	**SIGNS**									
0020	High intensity reflectorized, no posts, buy				S.F.	34			34	37.50

01 74 Cleaning and Waste Management

01 74 13 – Progress Cleaning

01 74 13.20 Cleaning Up		Crew	Daily Output	Labor-Hours	Unit	Material	Labor	Equipment	Total	Total Incl O&P
0010	**CLEANING UP**									
0020	After job completion, allow, minimum				Job				.30%	.30%
0040	Maximum				"				1%	1%
0050	Cleanup of floor area, continuous, per day, during const.	A-5	24	.750	M.S.F.	1.77	26.50	2.77	31.04	46
0100	Final by GC at end of job	"	11.50	1.565	"	1.87	55.50	5.80	63.17	94
1000	Mechanical demolition, see Section 02 41 19									

01 91 Commissioning

01 91 13 – General Commissioning Requirements

01 91 13.50 Building Commissioning		Crew	Daily Output	Labor-Hours	Unit	Material	Labor	Equipment	Total	Total Incl O&P
0010	**BUILDING COMMISSIONING**									
0100	Basic building commissioning, minimum				%				.25%	.25%
0150	Maximum								.50%	.50%
0200	Enhanced building commissioning, minimum								.50%	.50%
0250	Maximum								1%	1%

01 93 Facility Maintenance

01 93 13 – Facility Maintenance Procedures

01 93 13.15 Mechanical Facilities Maintenance			Crew	Daily Output	Labor-Hours	Unit	Material	Labor	Equipment	Total	Total Incl O&P
							2013 Bare Costs				
0010	**MECHANICAL FACILITIES MAINTENANCE**										
0100	Air conditioning system maintenance										
0130	Belt, replace		1 Stpi	15	.533	Ea.		30		30	45.50
0170	Fan, clean			16	.500			28.50		28.50	42.50
0180	Filter, remove, clean, replace			12	.667			38		38	57
0190	Flexible coupling alignment, inspect			40	.200			11.35		11.35	17.05
0200	Gas leak locate and repair			4	2			113		113	171
0250	Pump packing gland, remove and replace			11	.727			41		41	62
0270	Tighten			32	.250			14.15		14.15	21.50
0290	Pump disassemble and assemble			4	2			113		113	171
0300	Air pressure regulator, disassemble, clean, assemble		1 Skwk	4	2			92.50		92.50	143
0310	Repair or replace part		"	6	1.333			61.50		61.50	95.50
0320	Purging system		1 Stpi	16	.500			28.50		28.50	42.50
0400	Compressor, air, remove or install fan wheel		1 Skwk	20	.400			18.50		18.50	28.50
0410	Disassemble or assemble 2 cylinder, 2 stage			4	2			92.50		92.50	143
0420	4 cylinder, 4 stage			1	8			370		370	570
0430	Repair or replace part			2	4			185		185	286
0700	Demolition, for mech. demolition see Section 23 05 05.10 or 22 05 05.10										
0800	Ductwork, clean										
0810	Rectangular										
0820	6"	G	1 Shee	187.50	.043	L.F.		2.27		2.27	3.47
0830	8"	G		140.63	.057			3.03		3.03	4.63
0840	10"	G		112.50	.071			3.79		3.79	5.80
0850	12"	G		93.75	.085			4.55		4.55	6.95
0860	14"	G		80.36	.100			5.30		5.30	8.10
0870	16"	G		70.31	.114			6.05		6.05	9.25
0900	Round										
0910	4"	G	1 Shee	358.10	.022	L.F.		1.19		1.19	1.82
0920	6"	G		238.73	.034			1.79		1.79	2.72
0930	8"	G		179.05	.045			2.38		2.38	3.63
0940	10"	G		143.24	.056			2.98		2.98	4.54
0950	12"	G		119.37	.067			3.57		3.57	5.45
0960	16"	G		89.52	.089			4.76		4.76	7.25
1000	Expansion joint, not screwed, install or remove		1 Stpi	3	2.667	Ea.		151		151	227
1010	Repack		"	6	1.333	"		75.50		75.50	114
1200	Fire protection equipment										
1220	Fire hydrant, replace		Q-1	3	5.333	Ea.		268		268	405
1230	Service, lubricate, inspect, flush, clean		1 Plum	7	1.143			64		64	96
1240	Test		"	11	.727			40.50		40.50	61
1310	Inspect valves, pressure, nozzle,		1 Spri	4	2	System		109		109	165
1500	Instrumentation preventive maintenance, rule of thumb										
1510	Annual cost per instrument for calibration & trouble shooting										
1520	Includes flow meters, analytical transmitters, flow control										
1530	chemical feed systems, recorders and distrib. control sys.					Ea.					325
1800	Plumbing fixtures, for installation see Section 22 41 00										
1801	Plumbing fixtures										
1840	Open drain with snake		1 Plum	13	.615	Ea.		34.50		34.50	51.50
1850	Open drain with toilet auger			16	.500			28		28	42
1860	Grease trap, clean			2	4			223		223	335
1870	Plaster trap, clean			6	1.333			74.50		74.50	112
1881	Clean commode		1 Clab	20	.400		.30	14.20		14.50	22.50
1882	Clean commode seat			48	.167		.15	5.90		6.05	9.25
1883	Clean single sink			28	.286		.19	10.15		10.34	15.80

01 93 Facility Maintenance

01 93 13 – Facility Maintenance Procedures

01 93 13.15 Mechanical Facilities Maintenance		Crew	Daily Output	Labor-Hours	Unit	Material	2013 Bare Costs Labor	Equipment	Total	Total Incl O&P
1884	Clean double sink	1 Clab	20	.400	Ea.	.30	14.20		14.50	22.50
1885	Clean lavatory		28	.286		.19	10.15		10.34	15.80
1886	Clean bathtub		12	.667		.59	23.50		24.09	37
1887	Clean fiberglass shower		12	.667		.59	23.50		24.09	37
1888	Clean fiberglass tub/shower		8	1		.74	35.50		36.24	55.50
1889	Clean faucet set		32	.250		.12	8.85		8.97	13.80
1891	Clean shower head		80	.100		.01	3.55		3.56	5.45
1893	Clean water heater		16	.500		1.04	17.75		18.79	28.50
1900	Relief valve, test and adjust	1 Stpi	20	.400			22.50		22.50	34
1910	Clean pump, heater, or motor for whirlpool	1 Clab	16	.500		.15	17.75		17.90	27.50
1920	Clean thermal cover for whirlpool	"	16	.500		1.48	17.75		19.23	29
2000	Repair or replace-steam trap	1 Stpi	8	1			56.50		56.50	85.50
2020	Y-type or bell strainer		6	1.333			75.50		75.50	114
2040	Water trap or vacuum breaker, screwed joints		13	.615			35		35	52.50
2100	Steam specialties, clean									
2120	Air separator with automatic trap, 1" fittings	1 Stpi	12	.667	Ea.		38		38	57
2130	Bucket trap, 2" pipe		7	1.143			64.50		64.50	97.50
2150	Drip leg, 2" fitting		45	.178			10.05		10.05	15.15
2200	Thermodynamic trap, 1" fittings		50	.160			9.05		9.05	13.65
2210	Thermostatic		65	.123			6.95		6.95	10.50
2240	Screen and seat in y-type strainer, plug type.		25	.320			18.15		18.15	27.50
2242	Screen and seat in y-type strainer, flange type.		12	.667			38		38	57
2500	Valve, replace broken handwheel		24	.333			18.90		18.90	28.50
3000	Valve, overhaul, regulator, relief, flushometer, mixing									
3040	Cold water, gas	1 Stpi	5	1.600	Ea.		90.50		90.50	136
3050	Hot water, steam		3	2.667			151		151	227
3080	Globe, gate, check up to 4" cold water, gas		10	.800			45.50		45.50	68
3090	Hot water, steam		5	1.600			90.50		90.50	136
3100	Over 4" ID hot or cold line		1.40	5.714			325		325	485
3120	Remove and replace, gate, globe or check up to 4"	Q-5	6	2.667			136		136	205
3130	Over 4"	"	2	8			410		410	615
3150	Repack up to 4"	1 Stpi	13	.615			35		35	52.50
3160	Over 4"	"	4	2			113		113	171

Division Notes

	CREW	DAILY OUTPUT	LABOR-HOURS	UNIT	BARE COSTS MAT.	LABOR	EQUIP.	TOTAL	TOTAL INCL O&P

Estimating Tips

02 30 00 Subsurface Investigation

In preparing estimates on structures involving earthwork or foundations, all information concerning soil characteristics should be obtained. Look particularly for hazardous waste, evidence of prior dumping of debris, and previous stream beds.

02 40 00 Demolition and Structure Moving

The costs shown for selective demolition do not include rubbish handling or disposal. These items should be estimated separately using RSMeans data or other sources.

- Historic preservation often requires that the contractor remove materials from the existing structure, rehab them, and replace them. The estimator must be aware of any related measures and precautions that must be taken when doing selective demolition and cutting and patching. Requirements may include special handling and storage, as well as security.
- In addition to Subdivision 02 41 00, you can find selective demolition items in each division. Example: Roofing demolition is in Division 7.

02 40 00 Building Deconstruction

This section provides costs for the careful dismantling and recycling of most of low-rise building materials.

02 50 00 Containment of Hazardous Waste

This section addresses on-site hazardous waste disposal costs.

02 80 00 Hazardous Material Disposal/ Remediation

This subdivision includes information on hazardous waste handling, asbestos remediation, lead remediation, and mold remediation. See reference R028213-20 and R028319-60 for further guidance in using these unit price lines.

02 90 00 Monitoring Chemical Sampling, Testing Analysis

This section provides costs for on-site sampling and testing hazardous waste.

Reference Numbers

Reference numbers are shown in shaded boxes at the beginning of some major classifications. These numbers refer to related items in the Reference Section. The reference information may be an estimating procedure, an alternate pricing method, or technical information.

Note: Not all subdivisions listed here necessarily appear in this publication.

02 21 Surveys

02 21 13 – Site Surveys

02 21 13.09 Topographical Surveys		Crew	Daily Output	Labor-Hours	Unit	2013 Bare Costs Material	Labor	Equipment	Total	Total Incl O&P
0010	**TOPOGRAPHICAL SURVEYS**									
0020	Topographical surveying, conventional, minimum	A-7	3.30	7.273	Acre	19	355	23.50	397.50	590
0100	Maximum	A-8	.60	53.333	″	58	2,550	129	2,737	4,075

02 21 13.13 Boundary and Survey Markers

Line	Description	Crew	Daily Output	Labor-Hours	Unit	Material	Labor	Equipment	Total	Total Incl O&P
0010	**BOUNDARY AND SURVEY MARKERS**									
0300	Lot location and lines, large quantities, minimum	A-7	2	12	Acre	33.50	590	39	662.50	980
0320	Average	″	1.25	19.200		53.50	940	62	1,055.50	1,575
0400	Small quantities, maximum	A-8	1	32		71	1,525	77.50	1,673.50	2,500
0600	Monuments, 3′ long	A-7	10	2.400	Ea.	35	118	7.75	160.75	227
0800	Property lines, perimeter, cleared land	″	1000	.024	L.F.	.03	1.18	.08	1.29	1.92
0900	Wooded land	A-8	875	.037	″	.05	1.74	.09	1.88	2.82

02 21 13.16 Aerial Surveys

Line	Description	Crew	Daily Output	Labor-Hours	Unit	Material	Labor	Equipment	Total	Total Incl O&P
0010	**AERIAL SURVEYS**									
1500	Aerial surveying, including ground control, minimum fee, 10 acres				Total					4,600
1510	100 acres									9,200
1550	From existing photography, deduct									1,600
1600	2′ contours, 10 acres				Acre					170
1650	20 acres									85
1800	50 acres									37
1850	100 acres									30
2000	1000 acres									30
2050	10,000 acres									30
2150	For 1′ contours and									
2160	dense urban areas, add to above				Acre					20%

02 41 Demolition

02 41 13 – Selective Site Demolition

02 41 13.17 Demolish, Remove Pavement and Curb

Line	Description	Crew	Daily Output	Labor-Hours	Unit	Material	Labor	Equipment	Total	Total Incl O&P
0010	**DEMOLISH, REMOVE PAVEMENT AND CURB** R024119-10									
5010	Pavement removal, bituminous roads, 3″ thick	B-38	690	.058	S.Y.		2.34	1.86	4.20	5.60
5050	4″ to 6″ thick	″	420	.095	″		3.84	3.06	6.90	9.20

02 41 13.33 Minor Site Demolition

Line	Description	Crew	Daily Output	Labor-Hours	Unit	Material	Labor	Equipment	Total	Total Incl O&P
0010	**MINOR SITE DEMOLITION** R024119-10									
0015	No hauling, abandon catch basin or manhole	B-6	7	3.429	Ea.		133	52.50	185.50	262
0020	Remove existing catch basin or manhole, masonry		4	6			233	92	325	455
0030	Catch basin or manhole frames and covers, stored		13	1.846			72	28.50	100.50	141
0040	Remove and reset		7	3.429			133	52.50	185.50	262
0900	Hydrants, fire, remove only	B-21A	5	8			355	94	449	645
0950	Remove and reset	″	2	20			890	236	1,126	1,600
2900	Pipe removal, sewer/water, no excavation, 12″ diameter	B-6	175	.137	L.F.		5.35	2.10	7.45	10.45
2930	15″-18″ diameter	B-12Z	150	.160			6.40	12	18.40	23
2960	21″-24″ diameter		120	.200			8	15	23	28.50
3000	27″-36″ diameter		90	.267			10.65	20	30.65	38.50
3200	Steel, welded connections, 4″ diameter	B-6	160	.150			5.85	2.30	8.15	11.45
3300	10″ diameter		80	.300			11.65	4.59	16.24	23
4000	Sidewalk removal, bituminous, 2″ thick		350	.069	S.Y.		2.67	1.05	3.72	5.25
4010	2-1/2″ thick		325	.074			2.87	1.13	4	5.65
4100	Concrete, plain, 4″		160	.150			5.85	2.30	8.15	11.45
4110	Plain, 5″		140	.171			6.65	2.62	9.27	13.10
4120	Plain, 6″		120	.200			7.80	3.06	10.86	15.25

02 41 19 – Selective Demolition

02 41 19.19 Selective Facility Services Demolition

	02 41 19.19 Selective Facility Services Demolition	Crew	Daily Output	Labor-Hours	Unit	Material	2013 Bare Costs Labor	Equipment	Total	Total Incl O&P
0010	**SELECTIVE FACILITY SERVICES DEMOLITION**, Rubbish Handling R024119-10									
0020	The following are to be added to the demolition prices									
0600	Dumpster, weekly rental, 1 dump/week, 6 C.Y. capacity (2 Tons)				Week	460			460	505
0700	10 C.Y. capacity (3 Tons)					535			535	590
0725	20 C.Y. capacity (5 Tons) R024119-20					630			630	695
0800	30 C.Y. capacity (7 Tons)					810			810	890
0840	40 C.Y. capacity (10 Tons)					860			860	945
2000	Load, haul, dump and return, 50' haul, hand carried	2 Clab	24	.667	C.Y.		23.50		23.50	36.50
2005	Wheeled		37	.432			15.35		15.35	23.50
2040	51' to 100' haul, hand carried		16.50	.970			34.50		34.50	53
2045	Wheeled		25	.640			22.50		22.50	35
2080	Over 100' haul, add per 100 L.F., hand carried		35.50	.451			16		16	24.50
2085	Wheeled		54	.296			10.50		10.50	16.20
2120	In elevators, per 10 floors, add		140	.114			4.05		4.05	6.25
2130	Load, haul, dump and return, up to 50' haul, incl. up to 5 rsr stair, hand		23	.696			24.50		24.50	38
2135	Wheeled		35	.457			16.20		16.20	25
2140	6 – 10 riser stairs, hand carried		22	.727			26		26	39.50
2145	Wheeled		34	.471			16.70		16.70	25.50
2150	11 – 20 riser stairs, hand carried		20	.800			28.50		28.50	43.50
2155	Wheeled		31	.516			18.30		18.30	28
2160	21 – 40 riser stairs, hand carried		16	1			35.50		35.50	54.50
2165	Wheeled		24	.667			23.50		23.50	36.50
2170	100' haul, incl. 5 riser stair, hand carried		15	1.067			38		38	58
2175	Wheeled		23	.696			24.50		24.50	38
2180	6 – 10 riser stair, hand carried		14	1.143			40.50		40.50	62.50
2185	Wheeled		21	.762			27		27	41.50
2190	11 – 20 riser stair, hand carried		12	1.333			47.50		47.50	73
2195	Wheeled		18	.889			31.50		31.50	48.50
2200	21 – 40 riser stair, hand carried		8	2			71		71	109
2205	Wheeled		12	1.333			47.50		47.50	73
2210	Over 100' haul, add per 100 L.F., hand carried		35.50	.451			16		16	24.50
2215	Wheeled		54	.296			10.50		10.50	16.20
2220	For each additional flight of stairs, up to 5 risers, add		550	.029	Flight		1.03		1.03	1.59
2225	6 – 10 risers, add		275	.058			2.06		2.06	3.18
2230	11 – 20 risers, add		138	.116			4.11		4.11	6.35
2235	21 – 40 risers, add		69	.232			8.20		8.20	12.65
3000	Loading & trucking, including 2 mile haul, chute loaded	B-16	45	.711	C.Y.		26	15.40	41.40	56.50
3040	Hand loading truck, 50' haul	"	48	.667			24	14.45	38.45	53
3080	Machine loading truck	B-17	120	.267			10.20	6.55	16.75	23
5000	Haul, per mile, up to 8 C.Y. truck	B-34B	1165	.007			.25	.59	.84	1.04
5100	Over 8 C.Y. truck	"	1550	.005			.19	.45	.64	.78

02 41 19.20 Selective Demolition, Dump Charges

	02 41 19.20 Selective Demolition, Dump Charges	Crew	Daily Output	Labor-Hours	Unit	Material	Labor	Equipment	Total	Total Incl O&P
0010	**SELECTIVE DEMOLITION, DUMP CHARGES** R024119-10									
0100	Building construction materials				Ton	82			82	90
0300	Rubbish only					70			70	77
0500	Reclamation station, usual charge					82			82	90

02 41 19.27 Selective Demolition, Torch Cutting

	02 41 19.27 Selective Demolition, Torch Cutting	Crew	Daily Output	Labor-Hours	Unit	Material	Labor	Equipment	Total	Total Incl O&P
0010	**SELECTIVE DEMOLITION, TORCH CUTTING** R024119-10									
0020	Steel, 1" thick plate	1 Clab	360	.022	L.F.	.38	.79		1.17	1.63
0040	1" diameter bar	"	210	.038	Ea.		1.35		1.35	2.08
1000	Oxygen lance cutting, reinforced concrete walls									
1040	12" to 16" thick walls	1 Clab	10	.800	L.F.		28.50		28.50	43.50

02 41 Demolition

02 41 19 – Selective Demolition

02 41 19.27 Selective Demolition, Torch Cutting		Crew	Daily Output	Labor-Hours	Unit	2013 Bare Costs Material	Labor	Equipment	Total	Total Incl O&P
1080	24" thick walls	1 Clab	6	1.333	L.F.		47.50		47.50	73

02 65 Underground Storage Tank Removal

02 65 10 – Underground Tank and Contaminated Soil Removal

02 65 10.30 Removal of Underground Storage Tanks

		Crew	Daily Output	Labor-Hours	Unit	Material	Labor	Equipment	Total	Total Incl O&P
0010	**REMOVAL OF UNDERGROUND STORAGE TANKS** R026510-20									
0011	Petroleum storage tanks, non-leaking									
0100	Excavate & load onto trailer									
0110	3000 gal. to 5000 gal. tank G	B-14	4	12	Ea.		450	92	542	790
0120	6000 gal. to 8000 gal. tank G	B-3A	3	13.333			505	345	850	1,150
0130	9000 gal. to 12000 gal. tank G	"	2	20			755	520	1,275	1,725
0190	Known leaking tank, add				%				100%	100%
0200	Remove sludge, water and remaining product from tank bottom									
0201	of tank with vacuum truck									
0300	3000 gal. to 5000 gal. tank G	A-13	5	1.600	Ea.		73.50	155	228.50	282
0310	6000 gal. to 8000 gal. tank G		4	2			91.50	194	285.50	350
0320	9000 gal. to 12000 gal. tank G		3	2.667			122	259	381	470
0390	Dispose of sludge off-site, average				Gal.				6	6.60
0400	Insert inert solid CO_2 "dry ice" into tank									
0401	For cleaning/transporting tanks (1.5 lb./100 gal. cap) G	1 Clab	500	.016	Lb.	1.19	.57		1.76	2.18
1020	Haul tank to certified salvage dump, 100 miles round trip									
1023	3000 gal. to 5000 gal. tank				Ea.				745	820
1026	6000 gal. to 8000 gal. tank								850	935
1029	9,000 gal. to 12,000 gal. tank								1,000	1,100
1100	Disposal of contaminated soil to landfill									
1110	Minimum				C.Y.				125	140
1111	Maximum				"				400	440
1120	Disposal of contaminated soil to									
1121	bituminous concrete batch plant									
1130	Minimum				C.Y.				70	77
1131	Maximum				"				110	120
2010	Decontamination of soil on site incl poly tarp on top/bottom									
2011	Soil containment berm, and chemical treatment									
2020	Minimum G	B-11C	100	.160	C.Y.	7.50	6.65	3.67	17.82	22.50
2021	Maximum G	"	100	.160		9.75	6.65	3.67	20.07	25
2050	Disposal of decontaminated soil, minimum								125	140
2055	Maximum								400	440

02 81 Transportation and Disposal of Hazardous Materials

02 81 20 – Hazardous Waste Handling

02 81 20.10 Hazardous Waste Cleanup/Pickup/Disposal

		Crew	Daily Output	Labor-Hours	Unit	Material	Labor	Equipment	Total	Total Incl O&P
0010	**HAZARDOUS WASTE CLEANUP/PICKUP/DISPOSAL**									
0100	For contractor rental equipment, i.e., Dozer,									
0110	Front end loader, Dump truck, etc., see 01 54 33 Reference Section									
1000	Solid pickup									
1100	55 gal. drums				Ea.				225	250
1120	Bulk material, minimum				Ton				170	190
1130	Maximum				"				575	635
1200	Transportation to disposal site									

02 81 Transportation and Disposal of Hazardous Materials

02 81 20 – Hazardous Waste Handling

02 81 20.10 Hazardous Waste Cleanup/Pickup/Disposal		Crew	Daily Output	Labor-Hours	Unit	2013 Bare Costs Material	Labor	Equipment	Total	Total Incl O&P
1220	Truckload = 80 drums or 25 C.Y. or 18 tons									
1260	Minimum				Mile				3.85	4.25
1270	Maximum				"				7	7.10
3000	Liquid pickup, vacuum truck, stainless steel tank									
3100	Minimum charge, 4 hours									
3110	1 compartment, 2200 gallon				Hr.				125	140
3120	2 compartment, 5000 gallon				"				180	200
3400	Transportation in 6900 gallon bulk truck				Mile				7.70	8.50
3410	In teflon lined truck				"				9.90	10.90
5000	Heavy sludge or dry vacuumable material				Hr.				125	140
6000	Dumpsite disposal charge, minimum				Ton				125	140
6020	Maximum				"				400	440

02 82 Asbestos Remediation

02 82 13 – Asbestos Abatement

02 82 13.39 Asbestos Remediation Plans and Methods

Line	Description	Crew	Daily Output	Labor-Hours	Unit	Material	Labor	Equipment	Total	Total Incl O&P
0010	**ASBESTOS REMEDIATION PLANS AND METHODS**									
0100	Building Survey-Commercial Building				Ea.				2,000	2,200
0200	Asbestos Abatement Remediation Plan				"				1,200	1,325

02 82 13.41 Asbestos Abatement Equip.

Line	Description	Crew	Daily Output	Labor-Hours	Unit	Material	Labor	Equipment	Total	Total Incl O&P
0010	**ASBESTOS ABATEMENT EQUIP.** R028213-20									
0011	Equipment and supplies, buy									
0200	Air filtration device, 2000 CFM				Ea.	875			875	965
0250	Large volume air sampling pump, minimum					315			315	345
0260	Maximum					320			320	350
0300	Airless sprayer unit, 2 gun					4,650			4,650	5,125
0350	Light stand, 500 watt					44.50			44.50	49
0400	Personal respirators									
0410	Negative pressure, 1/2 face, dual operation, min.				Ea.	24.50			24.50	27
0420	Maximum					30.50			30.50	34
0450	P.A.P.R., full face, minimum					104			104	114
0460	Maximum					189			189	208
0470	Supplied air, full face, incl. air line, minimum					181			181	199
0480	Maximum					375			375	410
0500	Personnel sampling pump					215			215	236
1500	Power panel, 20 unit, incl. GFI					590			590	650
1600	Shower unit, including pump and filters					1,100			1,100	1,200
1700	Supplied air system (type C)					5,175			5,175	5,675
1750	Vacuum cleaner, HEPA, 16 gal., stainless steel, wet/dry					1,175			1,175	1,300
1760	55 gallon					1,200			1,200	1,325
1800	Vacuum loader, 9 – 18 ton/hr.					91,000			91,000	100,000
1900	Water atomizer unit, including 55 gal. drum					225			225	248
2000	Worker protection, whole body, foot, head cover & gloves, plastic					12.50			12.50	13.75
2500	Respirator, single use					22.50			22.50	24.50
2550	Cartridge for respirator					6.95			6.95	7.65
2570	Glove bag, 7 mil, 50" x 64"					17.50			17.50	19.25
2580	10 mil, 44" x 60"					5.50			5.50	6.05
2590	6 mil, 44" x 60"					5.50			5.50	6.05
3000	HEPA vacuum for work area, minimum					1,425			1,425	1,550
3050	Maximum					1,450			1,450	1,600
6000	Disposable polyethylene bags, 6 mil, 3 C.F.					1.60			1.60	1.76

02 82 13 – Asbestos Abatement

02 82 13.41 Asbestos Abatement Equip.		Crew	Daily Output	Labor-Hours	Unit	Material	2013 Bare Costs Labor	Equipment	Total	Total Incl O&P
6300	Disposable fiber drums, 3 C.F.				Ea.	17.80			17.80	19.60
6400	Pressure sensitive caution labels, 3" x 5"					3.28			3.28	3.61
6450	11" x 17"					7.30			7.30	8.05
6500	Negative air machine, 1800 CFM					840			840	925

02 82 13.42 Preparation of Asbestos Containment Area

Line	Description	Crew	Daily Output	Labor-Hours	Unit	Material	Labor	Equipment	Total	Total Incl O&P
0010	**PREPARATION OF ASBESTOS CONTAINMENT AREA** R028213-20									
0100	Pre-cleaning, HEPA vacuum and wet wipe, flat surfaces	A-9	12000	.005	S.F.	.02	.27		.29	.43
0200	Protect carpeted area, 2 layers 6 mil poly on 3/4" plywood	"	1000	.064		2.10	3.18		5.28	7.25
0300	Separation barrier, 2" x 4" @ 16", 1/2" plywood ea. side, 8' high	2 Carp	400	.040		3.10	1.80		4.90	6.20
0310	12' high		320	.050		3.10	2.25		5.35	6.85
0320	16' high		200	.080		2.10	3.59		5.69	7.85
0400	Personnel decontam. chamber, 2" x 4" @ 16", 3/4" ply ea. side		280	.057		4.10	2.57		6.67	8.45
0450	Waste decontam. chamber, 2" x 4" studs @ 16", 3/4" ply ea. side		360	.044		4.10	2		6.10	7.60
0500	Cover surfaces with polyethylene sheeting									
0501	Including glue and tape									
0550	Floors, each layer, 6 mil	A-9	8000	.008	S.F.	.04	.40		.44	.67
0551	4 mil		9000	.007		.03	.35		.38	.58
0560	Walls, each layer, 6 mil		6000	.011		.04	.53		.57	.87
0561	4 mil		7000	.009		.03	.45		.48	.74
0570	For heights above 12', add						20%			
0575	For heights above 20', add						30%			
0580	For fire retardant poly, add					100%				
0590	For large open areas, deduct					10%	20%			
0600	Seal floor penetrations with foam firestop to 36 sq. in.	2 Carp	200	.080	Ea.	8.10	3.59		11.69	14.45
0610	36 sq. in. to 72 sq. in.		125	.128		16.20	5.75		21.95	26.50
0615	72 sq. in. to 144 sq. in.		80	.200		32.50	9		41.50	49.50
0620	Wall penetrations, to 36 square inches		180	.089		8.10	3.99		12.09	15.05
0630	36 sq. in. to 72 sq. in.		100	.160		16.20	7.20		23.40	29
0640	72 sq. in. to 144 sq. in.		60	.267		32.50	11.95		44.45	54
0800	Caulk seams with latex	1 Carp	230	.035	L.F.	.16	1.56		1.72	2.59
0900	Set up neg. air machine, 1-2k CFM/25 M.C.F. volume	1 Asbe	4.30	1.860	Ea.		92.50		92.50	144
0950	Set up and remove portable shower unit	2 Asbe	4	4	"		199		199	310

02 82 13.43 Bulk Asbestos Removal

Line	Description	Crew	Daily Output	Labor-Hours	Unit	Material	Labor	Equipment	Total	Total Incl O&P
0010	**BULK ASBESTOS REMOVAL**									
0020	Includes disposable tools and 2 suits and 1 respirator filter/day/worker									
0200	Boiler insulation	A-9	480	.133	S.F.	.60	6.65		7.25	10.95
0210	With metal lath, add				%				50%	50%
0300	Boiler breeching or flue insulation	A-9	520	.123	S.F.	.49	6.10		6.59	10.05
0310	For active boiler, add				%				100%	100%
0400	Duct or AHU insulation	A-10B	440	.073	S.F.	.29	3.62		3.91	5.95
0500	Duct vibration isolation joints, up to 24 sq. in. duct	A-9	56	1.143	Ea.	4.56	57		61.56	93.50
0520	25 sq. in. to 48 sq. in. duct		48	1.333		5.35	66.50		71.85	109
0530	49 sq. in. to 76 sq. in. duct		40	1.600		6.40	79.50		85.90	131
0600	Pipe insulation, air cell type, up to 4" diameter pipe		900	.071	L.F.	.28	3.54		3.82	5.80
0610	4" to 8" diameter pipe		800	.080		.32	3.98		4.30	6.55
0620	10" to 12" diameter pipe		700	.091		.37	4.55		4.92	7.45
0630	14" to 16" diameter pipe		550	.116		.46	5.80		6.26	9.50
0650	Over 16" diameter pipe		650	.098	S.F.	.39	4.90		5.29	8.05
0700	With glove bag up to 3" diameter pipe		200	.320	L.F.	8.50	15.90		24.40	34
1000	Pipe fitting insulation up to 4" diameter pipe		320	.200	Ea.	.80	9.95		10.75	16.35
1100	6" to 8" diameter pipe		304	.211		.84	10.50		11.34	17.15
1110	10" to 12" diameter pipe		192	.333		1.33	16.60		17.93	27.50

02 82 Asbestos Remediation

02 82 13 – Asbestos Abatement

02 82 13.43 Bulk Asbestos Removal		Crew	Daily Output	Labor-Hours	Unit	Material	2013 Bare Costs Labor	Equipment	Total	Total Incl O&P
1120	14" to 16" diameter pipe	A-9	128	.500	Ea.	2	25		27	40.50
1130	Over 16" diameter pipe		176	.364	S.F.	1.45	18.10		19.55	29.50
1200	With glove bag, up to 8" diameter pipe		75	.853	L.F.	6.65	42.50		49.15	73.50
2000	Scrape foam fireproofing from flat surface		2400	.027	S.F.	.11	1.33		1.44	2.18
2100	Irregular surfaces		1200	.053		.21	2.65		2.86	4.35
3000	Remove cementitious material from flat surface		1800	.036		.14	1.77		1.91	2.91
3100	Irregular surface		1000	.064		.18	3.18		3.36	5.15
6000	Remove contaminated soil from crawl space by hand		400	.160	C.F.	.64	7.95		8.59	13.05
6100	With large production vacuum loader	A-12	700	.091	"	.37	4.55	1.11	6.03	8.65
7000	Radiator backing, not including radiator removal	A-9	1200	.053	S.F.	.21	2.65		2.86	4.35
9000	For type B (supplied air) respirator equipment, add				%				10%	10%

02 82 13.44 Demolition In Asbestos Contaminated Area		Crew	Daily Output	Labor-Hours	Unit	Material	Labor	Equipment	Total	Total Incl O&P
0010	**DEMOLITION IN ASBESTOS CONTAMINATED AREA**									
0200	Ceiling, including suspension system, plaster and lath	A-9	2100	.030	S.F.	.12	1.52		1.64	2.49
0210	Finished plaster, leaving wire lath		585	.109		.44	5.45		5.89	8.95
0220	Suspended acoustical tile		3500	.018		.07	.91		.98	1.49
0230	Concealed tile grid system		3000	.021		.09	1.06		1.15	1.74
0240	Metal pan grid system		1500	.043		.17	2.12		2.29	3.49
0250	Gypsum board		2500	.026		.10	1.27		1.37	2.09
0260	Lighting fixtures up to 2' x 4'		72	.889	Ea.	3.55	44		47.55	72.50
0400	Partitions, non load bearing									
0410	Plaster, lath, and studs	A-9	690	.093	S.F.	.96	4.62		5.58	8.20
0450	Gypsum board and studs	"	1390	.046	"	.18	2.29		2.47	3.76
9000	For type B (supplied air) respirator equipment, add				%				10%	10%

02 82 13.45 OSHA Testing		Crew	Daily Output	Labor-Hours	Unit	Material	Labor	Equipment	Total	Total Incl O&P
0010	**OSHA TESTING**									
0100	Certified technician, minimum				Day				200	220
0110	Maximum								300	330
0120	Industrial hygienist, minimum								300	330
0130	Maximum								400	440
0200	Asbestos sampling and PCM analysis, NIOSH 7400, minimum	1 Asbe	8	1	Ea.	2.80	49.50		52.30	80
0210	Maximum		4	2		3.10	99.50		102.60	157
1000	Cleaned area samples		8	1		2.68	49.50		52.18	80
1100	PCM air sample analysis, NIOSH 7400, minimum		8	1		31	49.50		80.50	112
1110	Maximum		4	2		3.26	99.50		102.76	158
1200	TEM air sample analysis, NIOSH 7402, minimum								90	120
1210	Maximum								360	450

02 82 13.46 Decontamination of Asbestos Containment Area		Crew	Daily Output	Labor-Hours	Unit	Material	Labor	Equipment	Total	Total Incl O&P
0010	**DECONTAMINATION OF ASBESTOS CONTAINMENT AREA**									
0100	Spray exposed substrate with surfactant (bridging)									
0200	Flat surfaces	A-9	6000	.011	S.F.	.37	.53		.90	1.23
0250	Irregular surfaces		4000	.016	"	.32	.80		1.12	1.59
0300	Pipes, beams, and columns		2000	.032	L.F.	.57	1.59		2.16	3.10
1000	Spray encapsulate polyethylene sheeting		8000	.008	S.F.	.32	.40		.72	.97
1100	Roll down polyethylene sheeting		8000	.008	"		.40		.40	.62
1500	Bag polyethylene sheeting		400	.160	Ea.	.78	7.95		8.73	13.20
2000	Fine clean exposed substrate, with nylon brush		2400	.027	S.F.		1.33		1.33	2.06
2500	Wet wipe substrate		4800	.013			.66		.66	1.03
2600	Vacuum surfaces, fine brush		6400	.010			.50		.50	.77
3000	Structural demolition									
3100	Wood stud walls	A-9	2800	.023	S.F.		1.14		1.14	1.77
3500	Window manifolds, not incl. window replacement		4200	.015			.76		.76	1.18

02 82 Asbestos Remediation

02 82 13 – Asbestos Abatement

02 82 13.46 Decontamination of Asbestos Containment Area		Crew	Daily Output	Labor-Hours	Unit	2013 Bare Costs Material	Labor	Equipment	Total	Total Incl O&P
3600	Plywood carpet protection	A-9	2000	.032	S.F.		1.59		1.59	2.47
4000	Remove custom decontamination facility	A-10A	8	3	Ea.	15.30	150		165.30	249
4100	Remove portable decontamination facility	3 Asbe	12	2	"	13	99.50		112.50	168
5000	HEPA vacuum, shampoo carpeting	A-9	4800	.013	S.F.	.06	.66		.72	1.10
9000	Final cleaning of protected surfaces	A-10A	8000	.003	"		.15		.15	.23

02 82 13.47 Asbestos Waste Pkg., Handling, and Disp.		Crew	Daily Output	Labor-Hours	Unit	Material	Labor	Equipment	Total	Total Incl O&P
0010	ASBESTOS WASTE PACKAGING, HANDLING, AND DISPOSAL									
0100	Collect and bag bulk material, 3 C.F. bags, by hand	A-9	400	.160	Ea.	1.60	7.95		9.55	14.10
0200	Large production vacuum loader	A-12	880	.073		.81	3.62	.88	5.31	7.45
1000	Double bag and decontaminate	A-9	960	.067		1.60	3.32		4.92	6.90
2000	Containerize bagged material in drums, per 3 C.F. drum	"	800	.080		17.80	3.98		21.78	26
3000	Cart bags 50′ to dumpster	2 Asbe	400	.040	↓		1.99		1.99	3.09
5000	Disposal charges, not including haul, minimum				C.Y.				60	66
5020	Maximum				"				350	385
9000	For type B (supplied air) respirator equipment, add				%				10%	10%

02 82 13.48 Asbestos Encapsulation With Sealants		Crew	Daily Output	Labor-Hours	Unit	Material	Labor	Equipment	Total	Total Incl O&P
0010	ASBESTOS ENCAPSULATION WITH SEALANTS									
0100	Ceilings and walls, minimum	A-9	21000	.003	S.F.	.28	.15		.43	.55
0110	Maximum		10600	.006	"	.43	.30		.73	.94
0300	Pipes to 12″ diameter including minor repairs, minimum		800	.080	L.F.	.38	3.98		4.36	6.60
0310	Maximum	↓	400	.160	"	1.06	7.95		9.01	13.50

02 85 Mold Remediation

02 85 16 – Mold Remediation Preparation and Containment

02 85 16.50 Preparation of Mold Containment Area		Crew	Daily Output	Labor-Hours	Unit	Material	Labor	Equipment	Total	Total Incl O&P
0010	PREPARATION OF MOLD CONTAINMENT AREA									
0100	Pre-cleaning, HEPA vacuum and wet wipe, flat surfaces	A-9	12000	.005	S.F.	.02	.27		.29	.43
0300	Separation barrier, 2″ x 4″ @ 16″, 1/2″ plywood ea. side, 8′ high	2 Carp	400	.040		3.10	1.80		4.90	6.20
0310	12′ high		320	.050		3.10	2.25		5.35	6.85
0320	16′ high		200	.080		2.10	3.59		5.69	7.85
0400	Personnel decontam. chamber, 2″ x 4″ @ 16″, 3/4″ ply ea. side		280	.057		4.10	2.57		6.67	8.45
0450	Waste decontam. chamber, 2″ x 4″ studs @ 16″, 3/4″ ply each side	↓	360	.044	↓	4.10	2		6.10	7.60
0500	Cover surfaces with polyethylene sheeting									
0501	Including glue and tape									
0550	Floors, each layer, 6 mil	A-9	8000	.008	S.F.	.04	.40		.44	.67
0551	4 mil		9000	.007		.03	.35		.38	.58
0560	Walls, each layer, 6 mil		6000	.011		.04	.53		.57	.87
0561	4 mil	↓	7000	.009	↓	.03	.45		.48	.74
0570	For heights above 12′, add						20%			
0575	For heights above 20′, add						30%			
0580	For fire retardant poly, add					100%				
0590	For large open areas, deduct					10%	20%			
0600	Seal floor penetrations with foam firestop to 36 sq. in.	2 Carp	200	.080	Ea.	8.10	3.59		11.69	14.45
0610	36 sq. in. to 72 sq. in.		125	.128		16.20	5.75		21.95	26.50
0615	72 sq. in. to 144 sq. in.		80	.200		32.50	9		41.50	49.50
0620	Wall penetrations, to 36 square inches		180	.089		8.10	3.99		12.09	15.05
0630	36 sq. in. to 72 sq. in.		100	.160		16.20	7.20		23.40	29
0640	72 sq. in. to 144 sq. in.	↓	60	.267	↓	32.50	11.95		44.45	54
0800	Caulk seams with latex caulk	1 Carp	230	.035	L.F.	.16	1.56		1.72	2.59
0900	Set up neg. air machine, 1-2k CFM/25 M.C.F. volume	1 Asbe	4.30	1.860	Ea.		92.50		92.50	144

02 85 Mold Remediation

02 85 33 – Removal and Disposal of Materials with Mold

02 85 33.50 Demolition in Mold Contaminated Area		Crew	Daily Output	Labor-Hours	Unit	2013 Bare Costs Material	Labor	Equipment	Total	Total Incl O&P
0010	**DEMOLITION IN MOLD CONTAMINATED AREA**									
0200	Ceiling, including suspension system, plaster and lath	A-9	2100	.030	S.F.	.12	1.52		1.64	2.49
0210	Finished plaster, leaving wire lath		585	.109		.44	5.45		5.89	8.95
0220	Suspended acoustical tile		3500	.018		.07	.91		.98	1.49
0230	Concealed tile grid system		3000	.021		.09	1.06		1.15	1.74
0240	Metal pan grid system		1500	.043		.17	2.12		2.29	3.49
0250	Gypsum board		2500	.026		.10	1.27		1.37	2.09
0255	Plywood		2500	.026		.10	1.27		1.37	2.09
0260	Lighting fixtures up to 2′ x 4′		72	.889	Ea.	3.55	44		47.55	72.50
0400	Partitions, non load bearing									
0410	Plaster, lath, and studs	A-9	690	.093	S.F.	.96	4.62		5.58	8.20
0450	Gypsum board and studs		1390	.046		.18	2.29		2.47	3.76
0465	Carpet & pad		1390	.046		.18	2.29		2.47	3.76
0600	Pipe insulation, air cell type, up to 4″ diameter pipe		900	.071	L.F.	.28	3.54		3.82	5.80
0610	4″ to 8″ diameter pipe		800	.080		.32	3.98		4.30	6.55
0620	10″ to 12″ diameter pipe		700	.091		.37	4.55		4.92	7.45
0630	14″ to 16″ diameter pipe		550	.116		.46	5.80		6.26	9.50
0650	Over 16″ diameter pipe		650	.098	S.F.	.39	4.90		5.29	8.05
9000	For type B (supplied air) respirator equipment, add				%				10%	10%

Division Notes

		CREW	DAILY OUTPUT	LABOR-HOURS	UNIT	BARE COSTS MAT.	LABOR	EQUIP.	TOTAL	TOTAL INCL O&P

Estimating Tips

General

- Carefully check all the plans and specifications. Concrete often appears on drawings other than structural drawings, including mechanical and electrical drawings for equipment pads. The cost of cutting and patching is often difficult to estimate. See Subdivision 03 81 for Concrete Cutting, Subdivision 02 41 19.16 for Cutout Demolition, Subdivision 03 05 05.10 for Concrete Demolition, and Subdivision 02 41 19.19 for Rubbish Handling (handling, loading and hauling of debris).
- Always obtain concrete prices from suppliers near the job site. A volume discount can often be negotiated, depending upon competition in the area. Remember to add for waste, particularly for slabs and footings on grade.

03 10 00 Concrete Forming and Accessories

- A primary cost for concrete construction is forming. Most jobs today are constructed with prefabricated forms. The selection of the forms best suited for the job and the total square feet of forms required for efficient concrete forming and placing are key elements in estimating concrete construction. Enough forms must be available for erection to make efficient use of the concrete placing equipment and crew.
- Concrete accessories for forming and placing depend upon the systems used. Study the plans and specifications to ensure that all special accessory requirements have been included in the cost estimate, such as anchor bolts, inserts, and hangers.
- Included within costs for forms-in-place are all necessary bracing and shoring.

03 20 00 Concrete Reinforcing

- Ascertain that the reinforcing steel supplier has included all accessories, cutting, bending, and an allowance for lapping, splicing, and waste. A good rule of thumb is 10% for lapping, splicing, and waste. Also, 10% waste should be allowed for welded wire fabric.
- The unit price items in the subdivisions for Reinforcing In Place, Glass Fiber Reinforcing, and Welded Wire Fabric include the labor to install accessories such as beam and slab bolsters, high chairs, and bar ties and tie wire. The material cost for these accessories is not included; they may be obtained from the Accessories Division.

03 30 00 Cast-In-Place Concrete

- When estimating structural concrete, pay particular attention to requirements for concrete additives, curing methods, and surface treatments. Special consideration for climate, hot or cold, must be included in your estimate. Be sure to include requirements for concrete placing equipment, and concrete finishing.
- For accurate concrete estimating, the estimator must consider each of the following major components individually: forms, reinforcing steel, ready-mix concrete, placement of the concrete, and finishing of the top surface. For faster estimating, Subdivision 03 30 53.40 for Concrete-In-Place can be used; here, various items of concrete work are presented that include the costs of all five major components (unless specifically stated otherwise).

03 40 00 Precast Concrete
03 50 00 Cast Decks and Underlayment

- The cost of hauling precast concrete structural members is often an important factor. For this reason, it is important to get a quote from the nearest supplier. It may become economically feasible to set up precasting beds on the site if the hauling costs are prohibitive.

Reference Numbers

Reference numbers are shown in shaded boxes at the beginning of some major classifications. These numbers refer to related items in the Reference Section. The reference information may be an estimating procedure, an alternate pricing method, or technical information.

Note: Not all subdivisions listed here necessarily appear in this publication.

03 11 Concrete Forming

03 11 13 – Structural Cast-In-Place Concrete Forming

03 11 13.40 Forms In Place, Equipment Foundations		Crew	Daily Output	Labor-Hours	Unit	Material	Labor	Equipment	Total	Total Incl O&P
						2013 Bare Costs				
0010	FORMS IN PLACE, EQUIPMENT FOUNDATIONS									
0020	1 use	C-2	160	.300	SFCA	3.08	13.10		16.18	23.50
0050	2 use		190	.253		1.70	11.05		12.75	18.85
0100	3 use		200	.240		1.23	10.50		11.73	17.50
0150	4 use		205	.234		1.01	10.20		11.21	16.85

03 11 13.45 Forms In Place, Footings

Line	Description	Crew	Daily Output	Labor-Hours	Unit	Material	Labor	Equipment	Total	Total Incl O&P
0010	FORMS IN PLACE, FOOTINGS									
0020	Continuous wall, plywood, 1 use	C-1	375	.085	SFCA	6.20	3.63		9.83	12.40
0050	2 use		440	.073		3.40	3.09		6.49	8.50
0100	3 use		470	.068		2.47	2.90		5.37	7.20
0150	4 use		485	.066		2.01	2.81		4.82	6.55
5000	Spread footings, job-built lumber, 1 use		305	.105		2	4.46		6.46	9.05
5050	2 use		371	.086		1.11	3.67		4.78	6.85
5100	3 use		401	.080		.80	3.39		4.19	6.15
5150	4 use		414	.077		.65	3.29		3.94	5.75

03 11 13.65 Forms In Place, Slab On Grade

Line	Description	Crew	Daily Output	Labor-Hours	Unit	Material	Labor	Equipment	Total	Total Incl O&P
0010	FORMS IN PLACE, SLAB ON GRADE									
3000	Edge forms, wood, 4 use, on grade, to 6" high	C-1	600	.053	L.F.	.31	2.27		2.58	3.83
6000	Trench forms in floor, wood, 1 use		160	.200	SFCA	1.66	8.50		10.16	14.95
6050	2 use		175	.183		.92	7.80		8.72	13
6100	3 use		180	.178		.67	7.55		8.22	12.40
6150	4 use		185	.173		.54	7.35		7.89	11.95

03 15 Concrete Accessories

03 15 05 – Concrete Forming Accessories

03 15 05.75 Sleeves and Chases

Line	Description	Crew	Daily Output	Labor-Hours	Unit	Material	Labor	Equipment	Total	Total Incl O&P
0010	SLEEVES AND CHASES									
0100	Plastic, 1 use, 12" long, 2" diameter	1 Carp	100	.080	Ea.	2.09	3.59		5.68	7.85
0150	4" diameter		90	.089		6	3.99		9.99	12.75
0200	6" diameter		75	.107		12.20	4.79		16.99	21
0250	12" diameter		60	.133		40	6		46	53
5000	Sheet metal, 2" diameter G		100	.080		1.57	3.59		5.16	7.25
5100	4" diameter G		90	.089		1.96	3.99		5.95	8.30
5150	6" diameter G		75	.107		1.96	4.79		6.75	9.55
5200	12" diameter G		60	.133		3.90	6		9.90	13.50
6000	Steel pipe, 2" diameter G		100	.080		5.10	3.59		8.69	11.15
6100	4" diameter G		90	.089		18	3.99		21.99	26
6150	6" diameter G		75	.107		30.50	4.79		35.29	41
6200	12" diameter G		60	.133		60.50	6		66.50	75.50

03 15 19 – Cast-In Concrete Anchors

03 15 19.10 Anchor Bolts

Line	Description	Crew	Daily Output	Labor-Hours	Unit	Material	Labor	Equipment	Total	Total Incl O&P
0010	ANCHOR BOLTS									
0015	Made from recycled materials G									
0025	Single bolts installed in fresh concrete, no templates									
0030	Hooked w/nut and washer, 1/2" diameter, 8" long G	1 Carp	132	.061	Ea.	1.43	2.72		4.15	5.75
0040	12" long G		131	.061		1.59	2.74		4.33	5.95
0050	5/8" diameter, 8" long G		129	.062		3.11	2.78		5.89	7.70
0060	12" long G		127	.063		3.82	2.83		6.65	8.55
0070	3/4" diameter, 8" long G		127	.063		3.82	2.83		6.65	8.55
0080	12" long G		125	.064		4.78	2.87		7.65	9.70

03 15 19 – Cast-In Concrete Anchors

03 15 19.10 Anchor Bolts			Crew	Daily Output	Labor-Hours	Unit	2013 Bare Costs Material	Labor	Equipment	Total	Total Incl O&P
0090	2-bolt pattern, including job-built 2-hole template, per set										
0100	J-type, incl. hex nut & washer, 1/2" diameter x 6" long	G	1 Carp	21	.381	Set	5.75	17.10		22.85	33
0110	12" long	G		21	.381		6.35	17.10		23.45	33.50
0120	18" long	G		21	.381		7.35	17.10		24.45	34.50
0130	3/4" diameter x 8" long	G		20	.400		10.85	17.95		28.80	39.50
0140	12" long	G		20	.400		12.75	17.95		30.70	41.50
0150	18" long	G		20	.400		15.60	17.95		33.55	44.50
0160	1" diameter x 12" long	G		19	.421		21.50	18.90		40.40	52.50
0170	18" long	G		19	.421		25.50	18.90		44.40	57
0180	24" long	G		19	.421		30	18.90		48.90	62
0190	36" long	G		18	.444		40	19.95		59.95	74.50
0200	1-1/2" diameter x 18" long	G		17	.471		63.50	21		84.50	103
0210	24" long	G		16	.500		75	22.50		97.50	117
0300	L-type, incl. hex nut & washer, 3/4" diameter x 12" long	G		20	.400		12.05	17.95		30	41
0310	18" long	G		20	.400		14.55	17.95		32.50	43.50
0320	24" long	G		20	.400		17.05	17.95		35	46.50
0330	30" long	G		20	.400		21	17.95		38.95	50.50
0340	36" long	G		20	.400		23.50	17.95		41.45	53
0350	1" diameter x 12" long	G		19	.421		18.50	18.90		37.40	49.50
0360	18" long	G		19	.421		22.50	18.90		41.40	53.50
0370	24" long	G		19	.421		26.50	18.90		45.40	58.50
0380	30" long	G		19	.421		31	18.90		49.90	63
0390	36" long	G		18	.444		35	19.95		54.95	69
0400	42" long	G		18	.444		42	19.95		61.95	76.50
0410	48" long	G		18	.444		46.50	19.95		66.45	81.50
0420	1-1/4" diameter x 18" long	G		18	.444		32.50	19.95		52.45	66
0430	24" long	G		18	.444		37.50	19.95		57.45	72
0440	30" long	G		17	.471		43	21		64	80
0450	36" long	G		17	.471		48.50	21		69.50	86
0460	42" long	G	2 Carp	32	.500		54.50	22.50		77	94.50
0470	48" long	G		32	.500		62	22.50		84.50	103
0480	54" long	G		31	.516		72	23		95	115
0490	60" long	G		31	.516		79	23		102	123
0500	1-1/2" diameter x 18" long	G		33	.485		46.50	22		68.50	84.50
0510	24" long	G		32	.500		53.50	22.50		76	93.50
0520	30" long	G		31	.516		60	23		83	102
0530	36" long	G		30	.533		68.50	24		92.50	113
0540	42" long	G		30	.533		78	24		102	123
0550	48" long	G		29	.552		87	25		112	134
0560	54" long	G		28	.571		105	25.50		130.50	156
0570	60" long	G		28	.571		115	25.50		140.50	167
0580	1-3/4" diameter x 18" long	G		31	.516		65	23		88	107
0590	24" long	G		30	.533		76	24		100	121
0600	30" long	G		29	.552		87.50	25		112.50	135
0610	36" long	G		28	.571		99.50	25.50		125	150
0620	42" long	G		27	.593		111	26.50		137.50	164
0630	48" long	G		26	.615		122	27.50		149.50	177
0640	54" long	G		26	.615		151	27.50		178.50	209
0650	60" long	G		25	.640		163	28.50		191.50	224
0660	2" diameter x 24" long	G		27	.593		101	26.50		127.50	152
0670	30" long	G		27	.593		113	26.50		139.50	166
0680	36" long	G		26	.615		124	27.50		151.50	180
0690	42" long	G		25	.640		138	28.50		166.50	197

03 15 19 – Cast-In Concrete Anchors

03 15 19.10 Anchor Bolts			Crew	Daily Output	Labor-Hours	Unit	Material	Labor	Equipment	Total	Total Incl O&P
0700	48" long	G	2 Carp	24	.667	Set	158	30		188	220
0710	54" long	G		23	.696		188	31		219	254
0720	60" long	G		23	.696		202	31		233	270
0730	66" long	G		22	.727		216	32.50		248.50	288
0740	72" long	G		21	.762		236	34		270	310
1000	4-bolt pattern, including job-built 4-hole template, per set										
1100	J-type, incl. hex nut & washer, 1/2" diameter x 6" long	G	1 Carp	19	.421	Set	8.30	18.90		27.20	38
1110	12" long	G		19	.421		9.55	18.90		28.45	39.50
1120	18" long	G		18	.444		11.45	19.95		31.40	43
1130	3/4" diameter x 8" long	G		17	.471		18.50	21		39.50	53
1140	12" long	G		17	.471		22.50	21		43.50	57
1150	18" long	G		17	.471		28	21		49	63.50
1160	1" diameter x 12" long	G		16	.500		40	22.50		62.50	78.50
1170	18" long	G		15	.533		47.50	24		71.50	89
1180	24" long	G		15	.533		57	24		81	100
1190	36" long	G		15	.533		77	24		101	122
1200	1-1/2" diameter x 18" long	G		13	.615		124	27.50		151.50	179
1210	24" long	G		12	.667		147	30		177	208
1300	L-type, incl. hex nut & washer, 3/4" diameter x 12" long	G		17	.471		21	21		42	55.50
1310	18" long	G		17	.471		26	21		47	61
1320	24" long	G		17	.471		31	21		52	66.50
1330	30" long	G		16	.500		38.50	22.50		61	76.50
1340	36" long	G		16	.500		43.50	22.50		66	82
1350	1" diameter x 12" long	G		16	.500		34	22.50		56.50	71.50
1360	18" long	G		15	.533		41.50	24		65.50	82.50
1370	24" long	G		15	.533		50.50	24		74.50	92.50
1380	30" long	G		15	.533		59	24		83	102
1390	36" long	G		15	.533		67	24		91	111
1400	42" long	G		14	.571		80.50	25.50		106	128
1410	48" long	G		14	.571		90	25.50		115.50	139
1420	1-1/4" diameter x 18" long	G		14	.571		61.50	25.50		87	107
1430	24" long	G		14	.571		72	25.50		97.50	119
1440	30" long	G		13	.615		83	27.50		110.50	134
1450	36" long	G		13	.615		94	27.50		121.50	146
1460	42" long	G	2 Carp	25	.640		106	28.50		134.50	161
1470	48" long	G		24	.667		120	30		150	178
1480	54" long	G		23	.696		141	31		172	203
1490	60" long	G		23	.696		155	31		186	218
1500	1-1/2" diameter x 18" long	G		25	.640		89.50	28.50		118	143
1510	24" long	G		24	.667		104	30		134	160
1520	30" long	G		23	.696		117	31		148	177
1530	36" long	G		22	.727		134	32.50		166.50	198
1540	42" long	G		22	.727		152	32.50		184.50	219
1550	48" long	G		21	.762		171	34		205	241
1560	54" long	G		20	.800		207	36		243	284
1570	60" long	G		20	.800		227	36		263	305
1580	1-3/4" diameter x 18" long	G		22	.727		127	32.50		159.50	191
1590	24" long	G		21	.762		148	34		182	216
1600	30" long	G		21	.762		172	34		206	242
1610	36" long	G		20	.800		196	36		232	272
1620	42" long	G		19	.842		220	38		258	300
1630	48" long	G		18	.889		241	40		281	325
1640	54" long	G		18	.889		298	40		338	390

Material, Labor and Equipment are printed under the heading "2013 Bare Costs".

03 15 Concrete Accessories

03 15 19 – Cast-In Concrete Anchors

03 15 19.10 Anchor Bolts			Crew	Daily Output	Labor-Hours	Unit	2013 Bare Costs Material	Labor	Equipment	Total	Total Incl O&P
1650	60" long	G	2 Carp	17	.941	Set	320	42.50		362.50	420
1660	2" diameter x 24" long	G		19	.842		198	38		236	276
1670	30" long	G		18	.889		223	40		263	310
1680	36" long	G		18	.889		245	40		285	330
1690	42" long	G		17	.941		273	42.50		315.50	365
1700	48" long	G		16	1		315	45		360	415
1710	54" long	G		15	1.067		370	48		418	485
1720	60" long	G		15	1.067		400	48		448	515
1730	66" long	G		14	1.143		430	51.50		481.50	550
1740	72" long	G		14	1.143		470	51.50		521.50	595
1990	For galvanized, add					Ea.	75%				

03 15 19.45 Machinery Anchors			Crew	Daily Output	Labor-Hours	Unit	Material	Labor	Equipment	Total	Total Incl O&P
0010	**MACHINERY ANCHORS**, heavy duty, incl. sleeve, floating base nut,										
0020	lower stud & coupling nut, fiber plug, connecting stud, washer & nut.										
0030	For flush mounted embedment in poured concrete heavy equip. pads.										
0200	Stud & bolt, 1/2" diameter	G	E-16	40	.400	Ea.	71	20.50	3.60	95.10	119
0300	5/8" diameter	G		35	.457		79	23.50	4.11	106.61	133
0500	3/4" diameter	G		30	.533		91	27	4.80	122.80	154
0600	7/8" diameter	G		25	.640		99.50	32.50	5.75	137.75	174
0800	1" diameter	G		20	.800		104	41	7.20	152.20	196
0900	1-1/4" diameter	G		15	1.067		139	54.50	9.60	203.10	261

03 21 Reinforcement Bars

03 21 10 – Uncoated Reinforcing Steel

03 21 10.50 Reinforcing Steel, Mill Base Plus Extras			Crew	Daily Output	Labor-Hours	Unit	Material	Labor	Equipment	Total	Total Incl O&P
0010	**REINFORCING STEEL, MILL BASE PLUS EXTRAS**										

03 21 10.60 Reinforcing In Place			Crew	Daily Output	Labor-Hours	Unit	Material	Labor	Equipment	Total	Total Incl O&P
0015	**REINFORCING IN PLACE**, 50-60 ton lots, A615 Grade 60										
0020	Includes labor, but not material cost, to install accessories										
0030	Made from recycled materials	G									
0502	Footings, #4 to #7	G	4 Rodm	4200	.008	Lb.	.50	.38		.88	1.15
0552	#8 to #18	G		7200	.004		.50	.22		.72	.90
0602	Slab on grade, #3 to #7	G		4200	.008		.50	.38		.88	1.15
0900	For other than 50 – 60 ton lots										
1000	Under 10 ton job, #3 to #7, add						25%	10%			
1010	#8 to #18, add						20%	10%			
1050	10 – 50 ton job, #3 to #7, add						10%				
1060	#8 to #18, add						5%				
1100	60 – 100 ton job, #3 to #7, deduct						5%				
1110	#8 to #18, deduct						10%				
1150	Over 100 ton job, #3 to #7, deduct						10%				
1160	#8 to #18, deduct						15%				

03 22 Fabric and Grid Reinforcing

03 22 05 – Uncoated Welded Wire Fabric

03 22 05.50 Welded Wire Fabric		Crew	Daily Output	Labor-Hours	Unit	Material	Labor	Equipment	Total	Total Incl O&P
						2013 Bare Costs				
0010	**WELDED WIRE FABRIC** ASTM A185									
0020	Includes labor, but not material cost, to install accessories									
0030	Made from recycled materials G									
0050	Sheets									
0100	6 x 6 - W1.4 x W1.4 (10 x 10) 21 lb. per C.S.F. G	2 Rodm	35	.457	C.S.F.	14.60	23		37.60	52

03 30 Cast-In-Place Concrete

03 30 53 – Miscellaneous Cast-In-Place Concrete

03 30 53.40 Concrete In Place

		Crew	Daily Output	Labor-Hours	Unit	Material	Labor	Equipment	Total	Total Incl O&P
0010	**CONCRETE IN PLACE**									
0020	Including forms (4 uses), Grade 60 rebar, concrete (Portland cement									
0050	Type I), placement and finishing unless otherwise indicated									
3540	Equipment pad (3000 psi), 3′ x 3′ x 6″ thick	C-14H	45	1.067	Ea.	44.50	47	.74	92.24	122
3550	4′ x 4′ x 6″ thick		30	1.600		66	70.50	1.10	137.60	183
3560	5′ x 5′ x 8″ thick		18	2.667		117	118	1.84	236.84	310
3570	6′ x 6′ x 8″ thick		14	3.429		157	151	2.37	310.37	410
3580	8′ x 8′ x 10″ thick		8	6		330	265	4.14	599.14	775
3590	10′ x 10′ x 12″ thick		5	9.600		560	425	6.60	991.60	1,275
3800	Footings (3000 psi), spread under 1 C.Y.	C-14C	28	4	C.Y.	162	172	1.16	335.16	445
3825	1 C.Y. to 5 C.Y.		43	2.605		197	112	.76	309.76	390
3850	Over 5 C.Y. R033053-60		75	1.493		179	64	.43	243.43	296
3900	Footings, strip (3000 psi), 18″ x 9″, unreinforced	C-14L	40	2.400		117	100	.82	217.82	284
3920	18″ x 9″, reinforced	C-14C	35	3.200		141	137	.93	278.93	370
3925	20″ x 10″, unreinforced	C-14L	45	2.133		114	89	.73	203.73	263
3930	20″ x 10″, reinforced	C-14C	40	2.800		134	120	.81	254.81	335
3935	24″ x 12″, unreinforced	C-14L	55	1.745		112	73	.59	185.59	236
3940	24″ x 12″, reinforced	C-14C	48	2.333		132	100	.68	232.68	300
3945	36″ x 12″, unreinforced	C-14L	70	1.371		109	57.50	.47	166.97	209
3950	36″ x 12″, reinforced	C-14C	60	1.867		127	80	.54	207.54	264
4000	Foundation mat (3000 psi), under 10 C.Y.		38.67	2.896		202	124	.84	326.84	415
4050	Over 20 C.Y.		56.40	1.986		176	85	.58	261.58	325
4650	Slab on grade (3500 psi), not including finish, 4″ thick	C-14E	60.75	1.449		119	64	.55	183.55	231
4700	6″ thick	″	92	.957		114	42	.36	156.36	191
4701	Thickened slab edge (3500 psi), for slab on grade poured									
4702	monolithically with slab; depth is in addition to slab thickness;									
4703	formed vertical outside edge, earthen bottom and inside slope									
4705	8″ deep x 8″ wide bottom, unreinforced	C-14L	2190	.044	L.F.	3.27	1.83	.01	5.11	6.45
4710	8″ x 8″, reinforced	C-14C	1670	.067		5.60	2.88	.02	8.50	10.65
4715	12″ deep x 12″ wide bottom, unreinforced	C-14L	1800	.053		6.65	2.23	.02	8.90	10.75
4720	12″ x 12″, reinforced	C-14C	1310	.086		10.95	3.67	.02	14.64	17.75
4725	16″ deep x 16″ wide bottom, unreinforced	C-14L	1440	.067		11.20	2.78	.02	14	16.65
4730	16″ x 16″, reinforced	C-14C	1120	.100		16.35	4.29	.03	20.67	24.50
4735	20″ deep x 20″ wide bottom, unreinforced	C-14L	1150	.083		17	3.49	.03	20.52	24
4740	20″ x 20″, reinforced	C-14C	920	.122		23.50	5.20	.04	28.74	34
4745	24″ deep x 24″ wide bottom, unreinforced	C-14L	930	.103		24	4.31	.04	28.35	33
4750	24″ x 24″, reinforced	C-14C	740	.151		32.50	6.50	.04	39.04	45.50

03 31 Structural Concrete

03 31 05 – Normal Weight Structural Concrete

03 31 05.35 Normal Weight Concrete, Ready Mix		Crew	Daily Output	Labor-Hours	Unit	2013 Bare Costs Material	Labor	Equipment	Total	Total Incl O&P
0010	**NORMAL WEIGHT CONCRETE, READY MIX**, delivered									
0012	Includes local aggregate, sand, Portland cement (Type I) and water									
0015	Excludes all additives and treatments									
0020	2000 psi				C.Y.	91			91	100
0100	2500 psi					93.50			93.50	103
0150	3000 psi					97			97	107
0200	3500 psi					99			99	109
0300	4000 psi					102			102	112
1000	For high early strength (Portland cement Type III), add					10%				
1300	For winter concrete (hot water), add					4.50			4.50	4.95
1400	For hot weather concrete (ice), add					9.35			9.35	10.25
1410	For mid-range water reducer, add					3.79			3.79	4.17
1420	For high-range water reducer/superplasticizer, add					6.10			6.10	6.70
1430	For retarder, add					2.67			2.67	2.94
1440	For non-Chloride accelerator, add					4.90			4.90	5.40
1450	For Chloride accelerator, per 1%, add					3.31			3.31	3.64
1460	For fiber reinforcing, synthetic (1 lb./C.Y.), add					6.70			6.70	7.40
1500	For Saturday delivery, add					8.85			8.85	9.70
1510	For truck holding/waiting time past 1st hour per load, add				Hr.	82			82	90.50
1520	For short load (less than 4 C.Y.), add per load				Ea.	112			112	123
2000	For all lightweight aggregate, add				C.Y.	45%				

03 31 05.70 Placing Concrete		Crew	Daily Output	Labor-Hours	Unit	Material	Labor	Equipment	Total	Total Incl O&P
0010	**PLACING CONCRETE**									
0020	Includes labor and equipment to place, level (strike off) and consolidate									
1900	Footings, continuous, shallow, direct chute	C-6	120	.400	C.Y.		14.80	.55	15.35	23
1950	Pumped	C-20	150	.427			16.30	5.20	21.50	31
2000	With crane and bucket	C-7	90	.800			31	13.40	44.40	61.50
2100	Footings, continuous, deep, direct chute	C-6	140	.343			12.70	.47	13.17	19.90
2150	Pumped	C-20	160	.400			15.25	4.89	20.14	29
2200	With crane and bucket	C-7	110	.655			25.50	10.95	36.45	50.50
2400	Footings, spread, under 1 C.Y., direct chute	C-6	55	.873			32.50	1.20	33.70	51
2450	Pumped	C-20	65	.985			37.50	12.05	49.55	71
2500	With crane and bucket	C-7	45	1.600			62	27	89	124
2600	Over 5 C.Y., direct chute	C-6	120	.400			14.80	.55	15.35	23
2650	Pumped	C-20	150	.427			16.30	5.20	21.50	31
2700	With crane and bucket	C-7	100	.720			28	12.05	40.05	56
2900	Foundation mats, over 20 C.Y., direct chute	C-6	350	.137			5.10	.19	5.29	7.95
2950	Pumped	C-20	400	.160			6.10	1.96	8.06	11.45
3000	With crane and bucket	C-7	300	.240			9.25	4.02	13.27	18.50

03 35 Concrete Finishing

03 35 29 – Tooled Concrete Finishing

03 35 29.30 Finishing Floors		Crew	Daily Output	Labor-Hours	Unit	Material	Labor	Equipment	Total	Total Incl O&P
0010	**FINISHING FLOORS**									
0012	Finishing requires that concrete first be placed, struck off & consolidated									
0015	Basic finishing for various unspecified flatwork									
0100	Bull float only	C-10	4000	.006	S.F.		.24		.24	.36
0125	Bull float & manual float		2000	.012			.49		.49	.73
0150	Bull float, manual float, & broom finish, w/edging & joints		1850	.013			.53		.53	.78
0200	Bull float, manual float & manual steel trowel		1265	.019			.77		.77	1.15
0210	For specified Random Access Floors in ACI Classes 1, 2, 3 and 4 to achieve									

03 35 Concrete Finishing

03 35 29 – Tooled Concrete Finishing

03 35 29.30 Finishing Floors		Crew	Daily Output	Labor-Hours	Unit	Material	2013 Bare Costs Labor	Equipment	Total	Total Incl O&P
0215	Composite Overall Floor Flatness and Levelness values up to F35/F25									
0250	Bull float, machine float & machine trowel (walk-behind)	C-10C	1715	.014	S.F.		.57	.03	.60	.88
0300	Power screed, bull float, machine float & trowel (walk-behind)	C-10D	2400	.010			.41	.05	.46	.65
0350	Power screed, bull float, machine float & trowel (ride-on)	C-10E	4000	.006	↓		.24	.07	.31	.43
0352	For specified Random Access Floors in ACI Classes 5, 6, 7 and 8 to achieve									
0354	Composite Overall Floor Flatness and Levelness values up to F50/F50									
0356	Add for two-dimensional restraightening after power float	C-10	6000	.004	S.F.		.16		.16	.24
0358	For specified Random or Defined Access Floors in ACI Class 9 to achieve									
0360	Composite Overall Floor Flatness and Levelness values up to F100/F100									
0362	Add for two-dimensional restraightening after bull float & power float	C-10	3000	.008	S.F.		.32		.32	.48
0364	For specified Superflat Defined Access Floors in ACI Class 9 to achieve									
0366	Minimum Floor Flatness and Levelness values of F100/F100									
0368	Add for 2-dim'l restraightening after bull float, power float, power trowel	C-10	2000	.012	S.F.		.49		.49	.73

03 35 29.35 Control Joints, Saw Cut		Crew	Daily Output	Labor-Hours	Unit	Material	Labor	Equipment	Total	Total Incl O&P
0010	**CONTROL JOINTS, SAW CUT**									
0100	Sawcut control joints in green concrete									
0120	1" depth	C-27	2000	.008	L.F.	.03	.34	.09	.46	.65
0140	1-1/2" depth		1800	.009		.05	.38	.10	.53	.72
0160	2" depth	↓	1600	.010	↓	.07	.43	.11	.61	.82
0180	Sawcut joint reservoir in cured concrete									
0182	3/8" wide x 3/4" deep, with single saw blade	C-27	1000	.016	L.F.	.05	.69	.18	.92	1.25
0184	1/2" wide x 1" deep, with double saw blades		900	.018		.10	.77	.19	1.06	1.45
0186	3/4" wide x 1-1/2" deep, with double saw blades	↓	800	.020		.20	.86	.22	1.28	1.73
0190	Water blast joint to wash away laitance, 2 passes	C-29	2500	.003			.11	.03	.14	.20
0200	Air blast joint to blow out debris and air dry, 2 passes	C-28	2000	.004	↓		.17	.01	.18	.26
0300	For backer rod, see Section 07 91 23.10									
0342	For joint sealant, see Section 07 92 13.20									

03 63 Epoxy Grouting

03 63 05 – Grouting of Dowels and Fasteners

03 63 05.10 Epoxy Only		Crew	Daily Output	Labor-Hours	Unit	Material	Labor	Equipment	Total	Total Incl O&P
0010	**EPOXY ONLY**									
1500	Chemical anchoring, epoxy cartridge, excludes layout, drilling, fastener									
1530	For fastener 3/4" diam. x 6" embedment	2 Skwk	72	.222	Ea.	4.61	10.25		14.86	21
1535	1" diam. x 8" embedment		66	.242		6.90	11.20		18.10	25
1540	1-1/4" diam. x 10" embedment		60	.267		13.85	12.30		26.15	34.50
1545	1-3/4" diam. x 12" embedment		54	.296		23	13.70		36.70	46.50
1550	14" embedment		48	.333		27.50	15.40		42.90	54.50
1555	2" diam. x 12" embedment		42	.381		37	17.60		54.60	67.50
1560	18" embedment	↓	32	.500	↓	46	23		69	86

03 82 Concrete Boring

03 82 13 – Concrete Core Drilling

03 82 13.10 Core Drilling		Crew	Daily Output	Labor-Hours	Unit	Material	2013 Bare Costs Labor	Equipment	Total	Total Incl O&P
0010	**CORE DRILLING**									
0020	Reinf. conc slab, up to 6" thick, incl. bit, layout & set up									
0100	1" diameter core	B-89A	17	.941	Ea.	.18	38.50	6.60	45.28	67
0150	Each added inch thick in same hole, add		1440	.011		.03	.45	.08	.56	.82
0200	2" diameter core		16.50	.970		.29	39.50	6.80	46.59	69
0250	Each added inch thick in same hole, add		1080	.015		.05	.60	.10	.75	1.09
0300	3" diameter core		16	1		.41	41	7.05	48.46	71
0350	Each added inch thick in same hole, add		720	.022		.07	.91	.16	1.14	1.65
0500	4" diameter core		15	1.067		.50	43.50	7.50	51.50	76
0550	Each added inch thick in same hole, add		480	.033		.08	1.36	.23	1.67	2.45
0700	6" diameter core		14	1.143		.83	46.50	8.05	55.38	82
0750	Each added inch thick in same hole, add		360	.044		.14	1.81	.31	2.26	3.29
0900	8" diameter core		13	1.231		1.08	50.50	8.65	60.23	88
0950	Each added inch thick in same hole, add		288	.056		.18	2.27	.39	2.84	4.13
1100	10" diameter core		12	1.333		1.58	54.50	9.35	65.43	96
1150	Each added inch thick in same hole, add		240	.067		.26	2.72	.47	3.45	5
1300	12" diameter core		11	1.455		1.73	59.50	10.25	71.48	105
1350	Each added inch thick in same hole, add		206	.078		.29	3.17	.55	4.01	5.80
1500	14" diameter core		10	1.600		2.19	65.50	11.25	78.94	116
1550	Each added inch thick in same hole, add		180	.089		.37	3.63	.62	4.62	6.70
1700	18" diameter core		9	1.778		2.93	72.50	12.50	87.93	129
1750	Each added inch thick in same hole, add		144	.111		.49	4.54	.78	5.81	8.40
1754	24" diameter core		8	2		4.13	81.50	14.05	99.68	146
1756	Each added inch thick in same hole, add		120	.133		.69	5.45	.94	7.08	10.20
1760	For horizontal holes, add to above						20%	20%		
1770	Prestressed hollow core plank, 8" thick									
1780	1" diameter core	B-89A	17.50	.914	Ea.	.25	37.50	6.45	44.20	65
1790	Each added inch thick in same hole, add		3840	.004		.03	.17	.03	.23	.32
1794	2" diameter core		17.25	.928		.39	38	6.50	44.89	66
1796	Each added inch thick in same hole, add		2880	.006		.05	.23	.04	.32	.44
1800	3" diameter core		17	.941		.55	38.50	6.60	45.65	67.50
1810	Each added inch thick in same hole, add		1920	.008		.07	.34	.06	.47	.67
1820	4" diameter core		16.50	.970		.66	39.50	6.80	46.96	69
1830	Each added inch thick in same hole, add		1280	.013		.08	.51	.09	.68	.98
1840	6" diameter core		15.50	1.032		1.10	42	7.25	50.35	74
1850	Each added inch thick in same hole, add		960	.017		.14	.68	.12	.94	1.33
1860	8" diameter core		15	1.067		1.43	43.50	7.50	52.43	77
1870	Each added inch thick in same hole, add		768	.021		.18	.85	.15	1.18	1.67
1880	10" diameter core		14	1.143		2.11	46.50	8.05	56.66	83
1890	Each added inch thick in same hole, add		640	.025		.26	1.02	.18	1.46	2.06
1900	12" diameter core		13.50	1.185		2.30	48.50	8.35	59.15	86
1910	Each added inch thick in same hole, add		548	.029		.29	1.19	.21	1.69	2.39

03 82 16 – Concrete Drilling

03 82 16.10 Concrete Impact Drilling		Crew	Daily Output	Labor-Hours	Unit	Material	Labor	Equipment	Total	Total Incl O&P
0010	**CONCRETE IMPACT DRILLING**									
0050	Up to 4" deep in conc/brick floor/wall, incl. bit & layout, no anchor									
0100	Holes, 1/4" diameter	1 Carp	75	.107	Ea.	.06	4.79		4.85	7.45
0150	For each additional inch of depth, add		430	.019		.01	.84		.85	1.31
0200	3/8" diameter		63	.127		.05	5.70		5.75	8.85
0250	For each additional inch of depth, add		340	.024		.01	1.06		1.07	1.64
0300	1/2" diameter		50	.160		.05	7.20		7.25	11.10
0350	For each additional inch of depth, add		250	.032		.01	1.44		1.45	2.22

03 82 16 – Concrete Drilling

03 82 16.10 Concrete Impact Drilling		Crew	Daily Output	Labor-Hours	Unit	Material	2013 Bare Costs Labor	Equipment	Total	Total Incl O&P
0400	5/8" diameter	1 Carp	48	.167	Ea.	.08	7.50		7.58	11.65
0450	For each additional inch of depth, add		240	.033		.02	1.50		1.52	2.32
0500	3/4" diameter		45	.178		.11	8		8.11	12.40
0550	For each additional inch of depth, add		220	.036		.03	1.63		1.66	2.54
0600	7/8" diameter		43	.186		.14	8.35		8.49	13
0650	For each additional inch of depth, add		210	.038		.04	1.71		1.75	2.67
0700	1" diameter		40	.200		.15	9		9.15	14
0750	For each additional inch of depth, add		190	.042		.04	1.89		1.93	2.95
0800	1-1/4" diameter		38	.211		.23	9.45		9.68	14.80
0850	For each additional inch of depth, add		180	.044		.06	2		2.06	3.13
0900	1-1/2" diameter		35	.229		.35	10.25		10.60	16.20
0950	For each additional inch of depth, add	↓	165	.048	↓	.09	2.18		2.27	3.45
1000	For ceiling installations, add						40%			

Estimating Tips

05 05 00 Common Work Results for Metals

- Nuts, bolts, washers, connection angles, and plates can add a significant amount to both the tonnage of a structural steel job and the estimated cost. As a rule of thumb, add 10% to the total weight to account for these accessories.
- Type 2 steel construction, commonly referred to as "simple construction," consists generally of field-bolted connections with lateral bracing supplied by other elements of the building, such as masonry walls or x-bracing. The estimator should be aware, however, that shop connections may be accomplished by welding or bolting. The method may be particular to the fabrication shop and may have an impact on the estimated cost.

05 10 00 Structural Steel

- Steel items can be obtained from two sources: a fabrication shop or a metals service center. Fabrication shops can fabricate items under more controlled conditions than can crews in the field. They are also more efficient and can produce items more economically. Metal service centers serve as a source of long mill shapes to both fabrication shops and contractors.
- Most line items in this structural steel subdivision, and most items in 05 50 00 Metal Fabrications, are indicated as being shop fabricated. The bare material cost for these shop fabricated items is the "Invoice Cost" from the shop and includes the mill base price of steel plus mill extras, transportation to the shop, shop drawings and detailing where warranted, shop fabrication and handling, sandblasting and a shop coat of primer paint, all necessary structural bolts, and delivery to the job site. The bare labor cost and bare equipment cost for these shop fabricated items is for field installation or erection.
- Line items in Subdivision 05 12 23.40 Lightweight Framing, and other items scattered in Division 5, are indicated as being field fabricated. The bare material cost for these field fabricated items is the "Invoice Cost" from the metals service center and includes the mill base price of steel plus mill extras, transportation to the metals service center, material handling, and delivery of long lengths of mill shapes to the job site. Material costs for structural bolts and welding rods should be added to the estimate. The bare labor cost and bare equipment cost for these items is for both field fabrication and field installation or erection, and include time for cutting, welding and drilling in the fabricated metal items. Drilling into concrete and fasteners to fasten field fabricated items to other work are not included and should be added to the estimate.

05 20 00 Steel Joist Framing

- In any given project the total weight of open web steel joists is determined by the loads to be supported and the design. However, economies can be realized in minimizing the amount of labor used to place the joists. This is done by maximizing the joist spacing, and therefore minimizing the number of joists required to be installed on the job. Certain spacings and locations may be required by the design, but in other cases maximizing the spacing and keeping it as uniform as possible will keep the costs down.

05 30 00 Steel Decking

- The takeoff and estimating of metal deck involves more than simply the area of the floor or roof and the type of deck specified or shown on the drawings. Many different sizes and types of openings may exist. Small openings for individual pipes or conduits may be drilled after the floor/roof is installed, but larger openings may require special deck lengths as well as reinforcing or structural support. The estimator should determine who will be supplying this reinforcing. Additionally, some deck terminations are part of the deck package, such as screed angles and pour stops, and others will be part of the steel contract, such as angles attached to structural members and cast-in-place angles and plates. The estimator must ensure that all pieces are accounted for in the complete estimate.

05 50 00 Metal Fabrications

- The most economical steel stairs are those that use common materials, standard details, and most importantly, a uniform and relatively simple method of field assembly. Commonly available A36 channels and plates are very good choices for the main stringers of the stairs, as are angles and tees for the carrier members. Risers and treads are usually made by specialty shops, and it is most economical to use a typical detail in as many places as possible. The stairs should be pre-assembled and shipped directly to the site. The field connections should be simple and straightforward to be accomplished efficiently, and with minimum equipment and labor.

Reference Numbers

Reference numbers are shown in shaded boxes at the beginning of some major classifications. These numbers refer to related items in the Reference Section. The reference information may be an estimating procedure, an alternate pricing method, or technical information.
Note: Not all subdivisions listed here necessarily appear in this publication.

05 05 19 – Post-Installed Concrete Anchors

05 05 19.10 Chemical Anchors			Crew	Daily Output	Labor-Hours	Unit	Material	2013 Bare Costs Labor	Equipment	Total	Total Incl O&P
0010	CHEMICAL ANCHORS										
0020	Includes layout & drilling										
1430	Chemical anchor, w/rod & epoxy cartridge, 3/4" diam. x 9-1/2" long		B-89A	27	.593	Ea.	8.90	24	4.17	37.07	52
1435	1" diameter x 11-3/4" long			24	.667		16.75	27	4.69	48.44	65.50
1440	1-1/4" diameter x 14" long			21	.762		35.50	31	5.35	71.85	93
1445	1-3/4" diameter x 15" long			20	.800		58	32.50	5.60	96.10	120
1450	18" long			17	.941		69.50	38.50	6.60	114.60	143
1455	2" diameter x 18" long			16	1		87.50	41	7.05	135.55	167
1460	24" long			15	1.067		113	43.50	7.50	164	200

05 05 19.20 Expansion Anchors			Crew	Daily Output	Labor-Hours	Unit	Material	Labor	Equipment	Total	Total Incl O&P
0010	EXPANSION ANCHORS										
0100	Anchors for concrete, brick or stone, no layout and drilling										
0200	Expansion shields, zinc, 1/4" diameter, 1-5/16" long, single	G	1 Carp	90	.089	Ea.	.45	3.99		4.44	6.65
0300	1-3/8" long, double	G		85	.094		.55	4.23		4.78	7.10
0400	3/8" diameter, 1-1/2" long, single	G		85	.094		.77	4.23		5	7.35
0500	2" long, double	G		80	.100		1.24	4.49		5.73	8.25
0600	1/2" diameter, 2-1/16" long, single	G		80	.100		1.47	4.49		5.96	8.50
0700	2-1/2" long, double	G		75	.107		1.98	4.79		6.77	9.60
0800	5/8" diameter, 2-5/8" long, single	G		75	.107		2.67	4.79		7.46	10.35
0900	2-3/4" long, double	G		70	.114		2.82	5.15		7.97	11
1000	3/4" diameter, 2-3/4" long, single	G		70	.114		3.07	5.15		8.22	11.30
1100	3-15/16" long, double	G		65	.123		5.40	5.55		10.95	14.45
2100	Hollow wall anchors for gypsum wall board, plaster or tile										
2300	1/8" diameter, short	G	1 Carp	160	.050	Ea.	.22	2.25		2.47	3.70
2400	Long	G		150	.053		.21	2.39		2.60	3.92
2500	3/16" diameter, short	G		150	.053		.37	2.39		2.76	4.10
2600	Long	G		140	.057		.55	2.57		3.12	4.56
2700	1/4" diameter, short	G		140	.057		.58	2.57		3.15	4.59
2800	Long	G		130	.062		.57	2.76		3.33	4.89
3000	Toggle bolts, bright steel, 1/8" diameter, 2" long	G		85	.094		.20	4.23		4.43	6.70
3100	4" long	G		80	.100		.25	4.49		4.74	7.20
3200	3/16" diameter, 3" long	G		80	.100		.25	4.49		4.74	7.20
3300	6" long	G		75	.107		.34	4.79		5.13	7.75
3400	1/4" diameter, 3" long	G		75	.107		.32	4.79		5.11	7.75
3500	6" long	G		70	.114		.48	5.15		5.63	8.45
3600	3/8" diameter, 3" long	G		70	.114		.78	5.15		5.93	8.75
3700	6" long	G		60	.133		1.13	6		7.13	10.45
3800	1/2" diameter, 4" long	G		60	.133		1.66	6		7.66	11.05
3900	6" long	G		50	.160		2.16	7.20		9.36	13.45
4000	Nailing anchors										
4100	Nylon nailing anchor, 1/4" diameter, 1" long		1 Carp	3.20	2.500	C	13.05	112		125.05	187
4200	1-1/2" long			2.80	2.857		17.60	128		145.60	217
4300	2" long			2.40	3.333		22.50	150		172.50	256
4400	Metal nailing anchor, 1/4" diameter, 1" long	G		3.20	2.500		16.40	112		128.40	191
4500	1-1/2" long	G		2.80	2.857		21	128		149	221
4600	2" long	G		2.40	3.333		26	150		176	260
5000	Screw anchors for concrete, masonry,										
5100	stone & tile, no layout or drilling included										
5700	Lag screw shields, 1/4" diameter, short	G	1 Carp	90	.089	Ea.	.38	3.99		4.37	6.55
5800	Long	G		85	.094		.43	4.23		4.66	6.95
5900	3/8" diameter, short	G		85	.094		.69	4.23		4.92	7.25
6000	Long	G		80	.100		.85	4.49		5.34	7.85

05 05 Common Work Results for Metals

05 05 19 – Post-Installed Concrete Anchors

05 05 19.20 Expansion Anchors			Crew	Daily Output	Labor-Hours	Unit	2013 Bare Costs Material	Labor	Equipment	Total	Total Incl O&P
6100	1/2" diameter, short	G	1 Carp	80	.100	Ea.	1.12	4.49		5.61	8.15
6200	Long	G		75	.107		1.31	4.79		6.10	8.85
6300	5/8" diameter, short	G		70	.114		1.84	5.15		6.99	9.90
6400	Long	G		65	.123		2.18	5.55		7.73	10.90
6600	Lead, #6 & #8, 3/4" long	G		260	.031		.17	1.38		1.55	2.32
6700	#10 - #14, 1-1/2" long	G		200	.040		.27	1.80		2.07	3.07
6800	#16 & #18, 1-1/2" long	G		160	.050		.37	2.25		2.62	3.87
6900	Plastic, #6 & #8, 3/4" long			260	.031		.04	1.38		1.42	2.17
7000	#8 & #10, 7/8" long			240	.033		.04	1.50		1.54	2.34
7100	#10 & #12, 1" long			220	.036		.04	1.63		1.67	2.55
7200	#14 & #16, 1-1/2" long			160	.050		.07	2.25		2.32	3.54
8000	Wedge anchors, not including layout or drilling										
8050	Carbon steel, 1/4" diameter, 1-3/4" long	G	1 Carp	150	.053	Ea.	.39	2.39		2.78	4.12
8100	3-1/4" long	G		140	.057		.52	2.57		3.09	4.52
8150	3/8" diameter, 2-1/4" long	G		145	.055		.51	2.48		2.99	4.38
8200	5" long	G		140	.057		.89	2.57		3.46	4.93
8250	1/2" diameter, 2-3/4" long	G		140	.057		1.01	2.57		3.58	5.05
8300	7" long	G		125	.064		1.72	2.87		4.59	6.30
8350	5/8" diameter, 3-1/2" long	G		130	.062		1.54	2.76		4.30	5.95
8400	8-1/2" long	G		115	.070		3.28	3.12		6.40	8.40
8450	3/4" diameter, 4-1/4" long	G		115	.070		2.64	3.12		5.76	7.70
8500	10" long	G		95	.084		6	3.78		9.78	12.40
8550	1" diameter, 6" long	G		100	.080		9.05	3.59		12.64	15.50
8575	9" long	G		85	.094		11.75	4.23		15.98	19.45
8600	12" long	G		75	.107		12.70	4.79		17.49	21.50
8650	1-1/4" diameter, 9" long	G		70	.114		26.50	5.15		31.65	37
8700	12" long	G		60	.133		34	6		40	46.50
8750	For type 303 stainless steel, add						350%				
8800	For type 316 stainless steel, add						450%				
8950	Self-drilling concrete screw, hex washer head, 3/16" diam. x 1-3/4" long	G	1 Carp	300	.027	Ea.	.20	1.20		1.40	2.06
8960	2-1/4" long	G		250	.032		.21	1.44		1.65	2.44
8970	Phillips flat head, 3/16" diam. x 1-3/4" long	G		300	.027		.19	1.20		1.39	2.05
8980	2-1/4" long	G		250	.032		.21	1.44		1.65	2.44

05 05 21 – Fastening Methods for Metal

05 05 21.15 Drilling Steel		Crew	Daily Output	Labor-Hours	Unit	Material	Labor	Equipment	Total	Total Incl O&P
0010	**DRILLING STEEL**									
1910	Drilling & layout for steel, up to 1/4" deep, no anchor									
1920	Holes, 1/4" diameter	1 Sswk	112	.071	Ea.	.08	3.58		3.66	6.50
1925	For each additional 1/4" depth, add		336	.024		.08	1.19		1.27	2.22
1930	3/8" diameter		104	.077		.08	3.85		3.93	6.95
1935	For each additional 1/4" depth, add		312	.026		.08	1.28		1.36	2.38
1940	1/2" diameter		96	.083		.09	4.17		4.26	7.55
1945	For each additional 1/4" depth, add		288	.028		.09	1.39		1.48	2.58
1950	5/8" diameter		88	.091		.13	4.55		4.68	8.25
1955	For each additional 1/4" depth, add		264	.030		.13	1.52		1.65	2.85
1960	3/4" diameter		80	.100		.16	5		5.16	9.15
1965	For each additional 1/4" depth, add		240	.033		.16	1.67		1.83	3.16
1970	7/8" diameter		72	.111		.21	5.55		5.76	10.15
1975	For each additional 1/4" depth, add		216	.037		.21	1.85		2.06	3.54
1980	1" diameter		64	.125		.21	6.25		6.46	11.40
1985	For each additional 1/4" depth, add		192	.042		.21	2.09		2.30	3.96
1990	For drilling up, add						40%			

05 05 Common Work Results for Metals

05 05 21 – Fastening Methods for Metal

05 05 21.90 Welding Steel

05 05 21.90 Welding Steel		Crew	Daily Output	Labor-Hours	Unit	Material	2013 Bare Costs Labor	Equipment	Total	Total Incl O&P
0010	**WELDING STEEL**, Structural R050521-20									
0020	Field welding, 1/8" E6011, cost per welder, no operating engineer	E-14	8	1	Hr.	4.35	52	18	74.35	118
0200	With 1/2 operating engineer	E-13	8	1.500		4.35	75	18	97.35	152
0300	With 1 operating engineer	E-12	8	2		4.35	98	18	120.35	187
0500	With no operating engineer, 2# weld rod per ton	E-14	8	1	Ton	4.35	52	18	74.35	118
0600	8# E6011 per ton	"	2	4		17.40	208	72	297.40	470
0800	With one operating engineer per welder, 2# E6011 per ton	E-12	8	2		4.35	98	18	120.35	187
0900	8# E6011 per ton	"	2	8		17.40	390	72	479.40	750
1200	Continuous fillet, stick welding, incl. equipment									
1300	Single pass, 1/8" thick, 0.1#/L.F.	E-14	150	.053	L.F.	.22	2.78	.96	3.96	6.25
1400	3/16" thick, 0.2#/L.F.		75	.107		.44	5.55	1.92	7.91	12.50
1500	1/4" thick, 0.3#/L.F.		50	.160		.65	8.35	2.88	11.88	18.75
1610	5/16" thick, 0.4#/L.F.		38	.211		.87	10.95	3.79	15.61	24.50
1800	3 passes, 3/8" thick, 0.5#/L.F.		30	.267		1.09	13.90	4.80	19.79	31.50
2010	4 passes, 1/2" thick, 0.7#/L.F.		22	.364		1.52	18.95	6.55	27.02	43
2200	5 to 6 passes, 3/4" thick, 1.3#/L.F.		12	.667		2.83	34.50	12	49.33	78.50
2400	8 to 11 passes, 1" thick, 2.4#/L.F.		6	1.333		5.20	69.50	24	98.70	156
2600	For all position welding, add, minimum						20%			
2700	Maximum						300%			
2900	For semi-automatic welding, deduct, minimum						5%			
3000	Maximum						15%			
4000	Cleaning and welding plates, bars, or rods									
4010	to existing beams, columns, or trusses	E-14	12	.667	L.F.	1.09	34.50	12	47.59	76.50

05 05 23 – Metal Fastenings

05 05 23.30 Lag Screws

05 05 23.30 Lag Screws		Crew	Daily Output	Labor-Hours	Unit	Material	Labor	Equipment	Total	Total Incl O&P
0010	**LAG SCREWS**									
0020	Steel, 1/4" diameter, 2" long G	1 Carp	200	.040	Ea.	.08	1.80		1.88	2.86
0100	3/8" diameter, 3" long G		150	.053		.28	2.39		2.67	4
0200	1/2" diameter, 3" long G		130	.062		.58	2.76		3.34	4.90
0300	5/8" diameter, 3" long G		120	.067		1.06	2.99		4.05	5.80

05 05 23.35 Machine Screws

05 05 23.35 Machine Screws		Crew	Daily Output	Labor-Hours	Unit	Material	Labor	Equipment	Total	Total Incl O&P
0010	**MACHINE SCREWS**									
0020	Steel, round head, #8 x 1" long G	1 Carp	4.80	1.667	C	4.02	75		79.02	119
0110	#8 x 2" long G		2.40	3.333		6.30	150		156.30	238
0200	#10 x 1" long G		4	2		6.20	90		96.20	145
0300	#10 x 2" long G		2	4		8.65	180		188.65	287

05 05 23.50 Powder Actuated Tools and Fasteners

05 05 23.50 Powder Actuated Tools and Fasteners		Crew	Daily Output	Labor-Hours	Unit	Material	Labor	Equipment	Total	Total Incl O&P
0010	**POWDER ACTUATED TOOLS & FASTENERS**									
0020	Stud driver, .22 caliber, single shot				Ea.	145			145	160
0100	.27 caliber, semi automatic, strip				"	425			425	470
0300	Powder load, single shot, .22 cal, power level 2, brown				C	5.05			5.05	5.55
0400	Strip, .27 cal, power level 4, red					7.25			7.25	8
0600	Drive pin, .300 x 3/4" long G	1 Carp	4.80	1.667		4.63	75		79.63	120
0700	.300 x 3" long with washer G	"	4	2		11.95	90		101.95	151

05 05 23.55 Rivets

05 05 23.55 Rivets		Crew	Daily Output	Labor-Hours	Unit	Material	Labor	Equipment	Total	Total Incl O&P
0010	**RIVETS**									
0100	Aluminum rivet & mandrel, 1/2" grip length x 1/8" diameter G	1 Carp	4.80	1.667	C	6.60	75		81.60	122
0200	3/16" diameter G		4	2		12.35	90		102.35	152
0300	Aluminum rivet, steel mandrel, 1/8" diameter G		4.80	1.667		14.80	75		89.80	131
0400	3/16" diameter G		4	2		21	90		111	162
0500	Copper rivet, steel mandrel, 1/8" diameter G		4.80	1.667		11.90	75		86.90	128

05 05 23 – Metal Fastenings

05 05 23.55 Rivets			Crew	Daily Output	Labor-Hours	Unit	Material	2013 Bare Costs Labor	Equipment	Total	Total Incl O&P
0800	Stainless rivet & mandrel, 1/8" diameter	G	1 Carp	4.80	1.667	C	29.50	75		104.50	148
0900	3/16" diameter	G		4	2		43.50	90		133.50	186
1000	Stainless rivet, steel mandrel, 1/8" diameter	G		4.80	1.667		19.35	75		94.35	137
1100	3/16" diameter	G		4	2		29	90		119	170
1200	Steel rivet and mandrel, 1/8" diameter	G		4.80	1.667		8.05	75		83.05	124
1300	3/16" diameter	G		4	2		13.30	90		103.30	153
1400	Hand riveting tool, standard					Ea.	70			70	77.50
1500	Deluxe						287			287	315
1600	Power riveting tool, standard						545			545	595
1700	Deluxe						2,950			2,950	3,250

05 05 23.70 Structural Blind Bolts

			Crew	Daily Output	Labor-Hours	Unit	Material	Labor	Equipment	Total	Total Incl O&P
0010	**STRUCTURAL BLIND BOLTS**										
0100	1/4" diameter x 1/4" grip	G	1 Sswk	240	.033	Ea.	1.24	1.67		2.91	4.34
0150	1/2" grip	G		216	.037		1.33	1.85		3.18	4.77
0200	3/8" diameter x 1/2" grip	G		232	.034		1.75	1.73		3.48	5
0250	3/4" grip	G		208	.038		1.84	1.92		3.76	5.45
0300	1/2" diameter x 1/2" grip	G		224	.036		3.99	1.79		5.78	7.60
0350	3/4" grip	G		200	.040		5.60	2		7.60	9.70
0400	5/8" diameter x 3/4" grip	G		216	.037		8.25	1.85		10.10	12.40
0450	1" grip	G		192	.042		9.50	2.09		11.59	14.15

05 05 23.90 Welding Rod

			Crew	Daily Output	Labor-Hours	Unit	Material	Labor	Equipment	Total	Total Incl O&P
0010	**WELDING ROD**										
0020	Steel, type 6011, 1/8" diam., less than 500#					Lb.	2.18			2.18	2.39
0100	500# to 2,000#						1.96			1.96	2.16
0200	2,000# to 5,000#						1.84			1.84	2.03
0300	5/32" diameter, less than 500#						2.20			2.20	2.42
0310	500# to 2,000#						1.98			1.98	2.18
0320	2,000# to 5,000#						1.86			1.86	2.05
0400	3/16" diam., less than 500#						2.43			2.43	2.67
0500	500# to 2,000#						2.19			2.19	2.41
0600	2,000# to 5,000#						2.06			2.06	2.26
0620	Steel, type 6010, 1/8" diam., less than 500#						2.21			2.21	2.43
0630	500# to 2,000#						1.99			1.99	2.19
0640	2,000# to 5,000#						1.87			1.87	2.06
0650	Steel, type 7018 Low Hydrogen, 1/8" diam., less than 500#						2.72			2.72	2.99
0660	500# to 2,000#						2.45			2.45	2.70
0670	2,000# to 5,000#						2.30			2.30	2.53
0700	Steel, type 7024 Jet Weld, 1/8" diam., less than 500#						2.50			2.50	2.75
0710	500# to 2,000#						2.25			2.25	2.48
0720	2,000# to 5,000#						2.12			2.12	2.33
1550	Aluminum, type 4043 TIG, 1/8" diam., less than 10#						4.98			4.98	5.50
1560	10# to 60#						4.49			4.49	4.94
1570	Over 60#						4.22			4.22	4.64
1600	Aluminum, type 5356 TIG, 1/8" diam., less than 10#						5.40			5.40	5.90
1610	10# to 60#						4.85			4.85	5.35
1620	Over 60#						4.56			4.56	5
1900	Cast iron, type 8 Nickel, 1/8" diam., less than 500#						27.50			27.50	30
1910	500# to 1,000#						24.50			24.50	27
1920	Over 1,000#						23			23	25.50
2000	Stainless steel, type 316/316L, 1/8" diam., less than 500#						7.55			7.55	8.30
2100	500# to 1000#						6.80			6.80	7.50
2220	Over 1000#						6.40			6.40	7.05

05 12 Structural Steel Framing

05 12 23 – Structural Steel for Buildings

05 12 23.40 Lightweight Framing			Crew	Daily Output	Labor-Hours	Unit	Material	Labor	Equipment	Total	Total Incl O&P
							2013 Bare Costs				
0010	**LIGHTWEIGHT FRAMING**										
0015	Made from recycled materials	G									
0400	Angle framing, field fabricated, 4" and larger	G	E-3	440	.055	Lb.	.75	2.77	.33	3.85	6.15
0450	Less than 4" angles	G		265	.091		.78	4.59	.54	5.91	9.65
0600	Channel framing, field fabricated, 8" and larger	G		500	.048		.78	2.43	.29	3.50	5.50
0650	Less than 8" channels	G		335	.072		.78	3.63	.43	4.84	7.85
1000	Continuous slotted channel framing system, shop fab, minimum	G	2 Sswk	2400	.007		4.03	.33		4.36	5.05
1200	Maximum	G	"	1600	.010		4.55	.50		5.05	5.90
1250	Plate & bar stock for reinforcing beams and trusses	G					1.43			1.43	1.57
1300	Cross bracing, rods, shop fabricated, 3/4" diameter	G	E-3	700	.034		1.56	1.74	.21	3.51	5.05
1310	7/8" diameter	G		850	.028		1.56	1.43	.17	3.16	4.47
1320	1" diameter	G		1000	.024		1.56	1.22	.14	2.92	4.05
1330	Angle, 5" x 5" x 3/8"	G		2800	.009		1.56	.43	.05	2.04	2.56
1350	Hanging lintels, shop fabricated, average	G		850	.028		1.56	1.43	.17	3.16	4.47
1380	Roof frames, shop fabricated, 3'-0" square, 5' span	G	E-2	4200	.013		1.56	.65	.36	2.57	3.24
1400	Tie rod, not upset, 1-1/2" to 4" diameter, with turnbuckle	G	2 Sswk	800	.020		1.69	1		2.69	3.65
1420	No turnbuckle	G		700	.023		1.63	1.14		2.77	3.83
1500	Upset, 1-3/4" to 4" diameter, with turnbuckle	G		800	.020		1.69	1		2.69	3.65
1520	No turnbuckle	G		700	.023		1.63	1.14		2.77	3.83

05 12 23.60 Pipe Support Framing			Crew	Daily Output	Labor-Hours	Unit	Material	Labor	Equipment	Total	Total Incl O&P
0010	**PIPE SUPPORT FRAMING**										
0020	Under 10#/L.F., shop fabricated	G	E-4	3900	.008	Lb.	1.74	.42	.04	2.20	2.70
0200	10.1 to 15#/L.F.	G		4300	.007		1.72	.38	.03	2.13	2.60
0400	15.1 to 20#/L.F.	G		4800	.007		1.69	.34	.03	2.06	2.49
0600	Over 20#/L.F.	G		5400	.006		1.66	.30	.03	1.99	2.39

05 54 Metal Floor Plates

05 54 13 – Floor Plates

05 54 13.20 Checkered Plates			Crew	Daily Output	Labor-Hours	Unit	Material	Labor	Equipment	Total	Total Incl O&P
0010	**CHECKERED PLATES**, steel, field fabricated										
0015	Made from recycled materials	G									
0020	1/4" & 3/8", 2000 to 5000 S.F., bolted	G	E-4	2900	.011	Lb.	.93	.56	.05	1.54	2.06
0100	Welded	G		4400	.007	"	.89	.37	.03	1.29	1.68
0300	Pit or trench cover and frame, 1/4" plate, 2' to 3' wide	G		100	.320	S.F.	12.40	16.20	1.44	30.04	44
0400	For galvanizing, add	G				Lb.	.29			.29	.32
0500	Platforms, 1/4" plate, no handrails included, rectangular	G	E-4	4200	.008		3.20	.39	.03	3.62	4.25
0600	Circular	G	"	2500	.013		4	.65	.06	4.71	5.60

Estimating Tips

06 05 00 Common Work Results for Wood, Plastics, and Composites

- Common to any wood-framed structure are the accessory connector items such as screws, nails, adhesives, hangers, connector plates, straps, angles, and hold-downs. For typical wood-framed buildings, such as residential projects, the aggregate total for these items can be significant, especially in areas where seismic loading is a concern. For floor and wall framing, the material cost is based on 10 to 25 lbs. per MBF. Hold-downs, hangers, and other connectors should be taken off by the piece.

  Included with material costs are fasteners for a normal installation. RSMeans engineers use manufacturer's recommendations, written specifications, and/or standard construction practice for size and spacing of fasteners. Prices for various fasteners are shown for informational purposes only. Adjustments should be made if unusual fastening conditions exist.

06 10 00 Carpentry

- Lumber is a traded commodity and therefore sensitive to supply and demand in the marketplace. Even in "budgetary" estimating of wood-framed projects, it is advisable to call local suppliers for the latest market pricing.
- Common quantity units for wood-framed projects are "thousand board feet" (MBF). A board foot is a volume of wood, 1" x 1' x 1', or 144 cubic inches. Board-foot quantities are generally calculated using nominal material dimensions—dressed sizes are ignored. Board foot per lineal foot of any stick of lumber can be calculated by dividing the nominal cross-sectional area by 12. As an example, 2,000 lineal feet of 2 x 12 equates to 4 MBF by dividing the nominal area, 2 x 12, by 12, which equals 2, and multiplying by 2,000 to give 4,000 board feet. This simple rule applies to all nominal dimensioned lumber.
- Waste is an issue of concern at the quantity takeoff for any area of construction. Framing lumber is sold in even foot lengths, i.e., 10', 12', 14', 16', and depending on spans, wall heights, and the grade of lumber, waste is inevitable. A rule of thumb for lumber waste is 5%–10% depending on material quality and the complexity of the framing.
- Wood in various forms and shapes is used in many projects, even where the main structural framing is steel, concrete, or masonry. Plywood as a back-up partition material and 2x boards used as blocking and cant strips around roof edges are two common examples. The estimator should ensure that the costs of all wood materials are included in the final estimate.

06 20 00 Finish Carpentry

- It is necessary to consider the grade of workmanship when estimating labor costs for erecting millwork and interior finish. In practice, there are three grades: premium, custom, and economy. The RSMeans daily output for base and case moldings is in the range of 200 to 250 L.F. per carpenter per day. This is appropriate for most average custom-grade projects. For premium projects, an adjustment to productivity of 25%–50% should be made, depending on the complexity of the job.

Reference Numbers

Reference numbers are shown in shaded boxes at the beginning of some major classifications. These numbers refer to related items in the Reference Section. The reference information may be an estimating procedure, an alternate pricing method, or technical information.

Note: Not all subdivisions listed here necessarily appear in this publication.

06 11 Wood Framing

06 11 10 – Framing with Dimensional, Engineered or Composite Lumber

06 11 10.24 Miscellaneous Framing		Crew	Daily Output	Labor-Hours	Unit	2013 Bare Costs Material	Labor	Equipment	Total	Total Incl O&P
0010	**MISCELLANEOUS FRAMING**									
8500	Firestops, 2″ x 4″	2 Carp	.51	31.373	M.B.F.	540	1,400		1,940	2,775
8505	Pneumatic nailed		.62	25.806		540	1,150		1,690	2,375
8520	2″ x 6″		.60	26.667		560	1,200		1,760	2,475
8525	Pneumatic nailed		.73	21.858		560	980		1,540	2,125
8540	2″ x 8″		.60	26.667		595	1,200		1,795	2,500
8560	2″ x 12″		.70	22.857		660	1,025		1,685	2,300
8600	Nailers, treated, wood construction, 2″ x 4″		.53	30.189		625	1,350		1,975	2,800
8605	Pneumatic nailed		.64	25.157		625	1,125		1,750	2,450
8620	2″ x 6″		.75	21.333		635	960		1,595	2,175
8625	Pneumatic nailed	↓	.90	17.778	↓	635	800		1,435	1,925

06 16 Sheathing

06 16 36 – Wood Panel Product Sheathing

06 16 36.10 Sheathing		Crew	Daily Output	Labor-Hours	Unit	Material	Labor	Equipment	Total	Total Incl O&P
0010	**SHEATHING**									
0012	Plywood on roofs, CDX									
0050	3/8″ thick	2 Carp	1525	.010	S.F.	.51	.47		.98	1.29
0055	Pneumatic nailed		1860	.009		.51	.39		.90	1.15
0200	5/8″ thick		1300	.012		.73	.55		1.28	1.65
0205	Pneumatic nailed		1586	.010		.73	.45		1.18	1.50
0300	3/4″ thick		1200	.013		.77	.60		1.37	1.77
0305	Pneumatic nailed		1464	.011		.77	.49		1.26	1.61
0500	Plywood on walls with exterior CDX, 3/8″ thick		1200	.013		.51	.60		1.11	1.48
0505	Pneumatic nailed		1488	.011		.51	.48		.99	1.30
0700	5/8″ thick		1050	.015		.73	.68		1.41	1.85
0705	Pneumatic nailed		1302	.012		.73	.55		1.28	1.65
0800	3/4″ thick		975	.016		.77	.74		1.51	1.98
0805	Pneumatic nailed	↓	1209	.013	↓	.77	.59		1.36	1.76

Estimating Tips

07 10 00 Dampproofing and Waterproofing

- Be sure of the job specifications before pricing this subdivision. The difference in cost between waterproofing and dampproofing can be great. Waterproofing will hold back standing water. Dampproofing prevents the transmission of water vapor. Also included in this section are vapor retarding membranes.

07 20 00 Thermal Protection

- Insulation and fireproofing products are measured by area, thickness, volume or R-value. Specifications may give only what the specific R-value should be in a certain situation. The estimator may need to choose the type of insulation to meet that R-value.

07 30 00 Steep Slope Roofing
07 40 00 Roofing and Siding Panels

- Many roofing and siding products are bought and sold by the square. One square is equal to an area that measures 100 square feet.

This simple change in unit of measure could create a large error if the estimator is not observant. Accessories necessary for a complete installation must be figured into any calculations for both material and labor.

07 50 00 Membrane Roofing
07 60 00 Flashing and Sheet Metal
07 70 00 Roofing and Wall Specialties and Accessories

- The items in these subdivisions compose a roofing system. No one component completes the installation, and all must be estimated. Built-up or single-ply membrane roofing systems are made up of many products and installation trades. Wood blocking at roof perimeters or penetrations, parapet coverings, reglets, roof drains, gutters, downspouts, sheet metal flashing, skylights, smoke vents, and roof hatches all need to be considered along with the roofing material. Several different installation trades will need to work together on the roofing system. Inherent difficulties in the scheduling and coordination of various trades must be accounted for when estimating labor costs.

07 90 00 Joint Protection

- To complete the weather-tight shell, the sealants and caulkings must be estimated. Where different materials meet—at expansion joints, at flashing penetrations, and at hundreds of other locations throughout a construction project—they provide another line of defense against water penetration. Often, an entire system is based on the proper location and placement of caulking or sealants. The detailed drawings that are included as part of a set of architectural plans show typical locations for these materials. When caulking or sealants are shown at typical locations, this means the estimator must include them for all the locations where this detail is applicable. Be careful to keep different types of sealants separate, and remember to consider backer rods and primers if necessary.

Reference Numbers

Reference numbers are shown in shaded boxes at the beginning of some major classifications. These numbers refer to related items in the Reference Section. The reference information may be an estimating procedure, an alternate pricing method, or technical information.

Note: Not all subdivisions listed here necessarily appear in this publication.

07 65 Flexible Flashing

07 65 10 – Sheet Metal Flashing

07 65 10.10 Sheet Metal Flashing and Counter Flashing		Crew	Daily Output	Labor-Hours	Unit	2013 Bare Costs Material	Labor	Equipment	Total	Total Incl O&P
0010	**SHEET METAL FLASHING AND COUNTER FLASHING**									
0011	Including up to 4 bends									
0020	Aluminum, mill finish, .013" thick	1 Rofc	145	.055	S.F.	.73	2.11		2.84	4.35
0030	.016" thick		145	.055		.84	2.11		2.95	4.47
0060	.019" thick		145	.055		1.07	2.11		3.18	4.73
0100	.032" thick		145	.055		1.25	2.11		3.36	4.93
0200	.040" thick		145	.055		2.14	2.11		4.25	5.90
0300	.050" thick		145	.055		2.25	2.11		4.36	6.05
0325	Mill finish 5" x 7" step flashing, .016" thick		1920	.004	Ea.	.14	.16		.30	.42
0350	Mill finish 12" x 12" step flashing, .016" thick		1600	.005	"	.50	.19		.69	.87
0400	Painted finish, add				S.F.	.29			.29	.32
1000	Mastic-coated 2 sides, .005" thick	1 Rofc	330	.024		1.67	.93		2.60	3.40
1100	.016" thick		330	.024		1.84	.93		2.77	3.58
1600	Copper, 16 oz., sheets, under 1000 lb.		115	.070		7.55	2.66		10.21	12.85
1700	Over 4000 lb.		155	.052		7.55	1.97		9.52	11.65
1900	20 oz. sheets, under 1000 lb.		110	.073		9.85	2.78		12.63	15.55
2000	Over 4000 lb.		145	.055		9.35	2.11		11.46	13.85
2200	24 oz. sheets, under 1000 lb.		105	.076		12.10	2.91		15.01	18.25
2300	Over 4000 lb.		135	.059		11.50	2.26		13.76	16.45
2500	32 oz. sheets, under 1000 lb.		100	.080		16.15	3.06		19.21	23
2600	Over 4000 lb.		130	.062		15.35	2.35		17.70	21
5800	Lead, 2.5 lb. per S.F., up to 12" wide		135	.059		4.99	2.26		7.25	9.30
5900	Over 12" wide		135	.059		3.99	2.26		6.25	8.20
8900	Stainless steel sheets, 32 ga., .010" thick		155	.052		4	1.97		5.97	7.70
9000	28 ga., .015" thick		155	.052		5.65	1.97		7.62	9.50
9100	26 ga., .018" thick		155	.052		7.25	1.97		9.22	11.30
9200	24 ga., .025" thick		155	.052		8.90	1.97		10.87	13.10
9400	Terne coated stainless steel, .015" thick, 28 ga		155	.052		7.35	1.97		9.32	11.40
9500	.018" thick, 26 ga		155	.052		8.30	1.97		10.27	12.40
9600	Zinc and copper alloy (brass), .020" thick		155	.052		8.55	1.97		10.52	12.70
9700	.027" thick		155	.052		10.70	1.97		12.67	15.05
9800	.032" thick		155	.052		13.65	1.97		15.62	18.35
9900	.040" thick		155	.052		17.10	1.97		19.07	22

07 65 13 – Laminated Sheet Flashing

07 65 13.10 Laminated Sheet Flashing		Crew	Daily Output	Labor-Hours	Unit	Material	Labor	Equipment	Total	Total Incl O&P
0010	**LAMINATED SHEET FLASHING**, Including up to 4 bends									
0500	Aluminum, fabric-backed 2 sides, .004" thick	1 Rofc	330	.024	S.F.	1.41	.93		2.34	3.11
0700	.005" thick		330	.024		1.67	.93		2.60	3.40
0750	Mastic-backed, self adhesive		460	.017		3.46	.66		4.12	4.93
0800	Mastic-coated 2 sides, .004" thick		330	.024		1.41	.93		2.34	3.11
2800	Copper, paperbacked 1 side, 2 oz.		330	.024		1.67	.93		2.60	3.40
2900	3 oz.		330	.024		1.97	.93		2.90	3.73
3100	Paperbacked 2 sides, 2 oz.		330	.024		1.72	.93		2.65	3.45
3150	3 oz.		330	.024		2.07	.93		3	3.84
3200	5 oz.		330	.024		2.69	.93		3.62	4.52
3250	7 oz.		330	.024		4.81	.93		5.74	6.85
3400	Mastic-backed 2 sides, copper, 2 oz.		330	.024		1.65	.93		2.58	3.38
3500	3 oz.		330	.024		2	.93		2.93	3.76
3700	5 oz.		330	.024		3	.93		3.93	4.86
3800	Fabric-backed 2 sides, copper, 2 oz.		330	.024		1.93	.93		2.86	3.68
4000	3 oz.		330	.024		2.19	.93		3.12	3.97
4100	5 oz.		330	.024		3.13	.93		4.06	5

07 65 Flexible Flashing

07 65 13 – Laminated Sheet Flashing

07 65 13.10 Laminated Sheet Flashing		Crew	Daily Output	Labor-Hours	Unit	2013 Bare Costs Material	Labor	Equipment	Total	Total Incl O&P
4300	Copper-clad stainless steel, .015" thick, under 500 lb.	1 Rofc	115	.070	S.F.	5.80	2.66		8.46	10.90
4400	Over 2000 lb.		155	.052		5.55	1.97		7.52	9.40
4600	.018" thick, under 500 lb.		100	.080		6.55	3.06		9.61	12.35
4700	Over 2000 lb.		145	.055		6.20	2.11		8.31	10.40
8550	Shower pan, 3 ply copper and fabric, 3 oz.		155	.052		3.13	1.97		5.10	6.75
8600	7 oz.		155	.052		3.99	1.97		5.96	7.70

07 65 19 – Plastic Sheet Flashing

07 65 19.10 Plastic Sheet Flashing and Counter Flashing

		Crew	Daily Output	Labor-Hours	Unit	Material	Labor	Equipment	Total	Total Incl O&P
0010	**PLASTIC SHEET FLASHING AND COUNTER FLASHING**									
7300	Polyvinyl chloride, black, .010" thick	1 Rofc	285	.028	S.F.	.21	1.07		1.28	2.04
7400	.020" thick		285	.028		.28	1.07		1.35	2.12
7600	.030" thick		285	.028		.35	1.07		1.42	2.20
7700	.056" thick		285	.028		.85	1.07		1.92	2.75
7900	Black or white for exposed roofs, .060" thick		285	.028		1.80	1.07		2.87	3.79
8060	PVC tape, 5" x 45 mils, for joint covers, 100 L.F./roll				Ea.	170			170	187
8850	Polyvinyl chloride, .030" thick	1 Rofc	160	.050	S.F.	.94	1.91		2.85	4.25

07 72 Roof Accessories

07 72 53 – Snow Guards

07 72 53.10 Snow Guard Options

		Crew	Daily Output	Labor-Hours	Unit	Material	Labor	Equipment	Total	Total Incl O&P
0010	**SNOW GUARD OPTIONS**									
0100	Slate & asphalt shingle roofs, fastened with nails	1 Rofc	160	.050	Ea.	12.55	1.91		14.46	17
0200	Standing seam metal roofs, fastened with set screws		48	.167		16.95	6.35		23.30	29.50
0300	Surface mount for metal roofs, fastened with solder		48	.167		6.05	6.35		12.40	17.40
0400	Double rail pipe type, including pipe		130	.062	L.F.	32	2.35		34.35	39

07 72 73 – Pitch Pockets

07 72 73.10 Pitch Pockets, Variable Sizes

		Crew	Daily Output	Labor-Hours	Unit	Material	Labor	Equipment	Total	Total Incl O&P
0010	**PITCH POCKETS, VARIABLE SIZES**									
0100	Adjustable, 4" to 7", welded corners, 4" deep	1 Rofc	48	.167	Ea.	15	6.35		21.35	27.50
0200	Side extenders, 6"	"	240	.033	"	2	1.27		3.27	4.34

07 76 Roof Pavers

07 76 16 – Roof Decking Pavers

07 76 16.10 Roof Pavers and Supports

		Crew	Daily Output	Labor-Hours	Unit	Material	Labor	Equipment	Total	Total Incl O&P
0010	**ROOF PAVERS AND SUPPORTS**									
1000	Roof decking pavers, concrete blocks, 2" thick, natural	1 Clab	115	.070	S.F.	3.73	2.47		6.20	7.90
1100	Colors		115	.070	"	3.77	2.47		6.24	7.95
1200	Support pedestal, bottom cap		960	.008	Ea.	2.99	.30		3.29	3.74
1300	Top cap		960	.008		4.79	.30		5.09	5.70
1400	Leveling shims, 1/16"		1920	.004		1.19	.15		1.34	1.54
1500	1/8"		1920	.004		1.19	.15		1.34	1.54
1600	Buffer pad		960	.008		2.49	.30		2.79	3.19
1700	PVC legs (4" SDR 35)		2880	.003	Inch	.12	.10		.22	.28
2000	Alternate pricing method, system in place		101	.079	S.F.	7.35	2.81		10.16	12.40

07 84 Firestopping

07 84 13 – Penetration Firestopping

07 84 13.10 Firestopping		Crew	Daily Output	Labor-Hours	Unit	2013 Bare Costs Material	Labor	Equipment	Total	Total Incl O&P
0010	**FIRESTOPPING**									
0100	Metallic piping, non insulated									
0110	Through walls, 2″ diameter	1 Carp	16	.500	Ea.	15.60	22.50		38.10	51.50
0120	4″ diameter		14	.571		24	25.50		49.50	65.50
0130	6″ diameter		12	.667		32	30		62	81.50
0140	12″ diameter		10	.800		56.50	36		92.50	118
0150	Through floors, 2″ diameter		32	.250		9.45	11.25		20.70	27.50
0160	4″ diameter		28	.286		13.60	12.85		26.45	34.50
0170	6″ diameter		24	.333		17.85	14.95		32.80	42.50
0180	12″ diameter		20	.400		30	17.95		47.95	60.50
0190	Metallic piping, insulated									
0200	Through walls, 2″ diameter	1 Carp	16	.500	Ea.	22	22.50		44.50	59
0210	4″ diameter		14	.571		30	25.50		55.50	72.50
0220	6″ diameter		12	.667		38.50	30		68.50	88
0230	12″ diameter		10	.800		63	36		99	125
0240	Through floors, 2″ diameter		32	.250		16	11.25		27.25	35
0250	4″ diameter		28	.286		20	12.85		32.85	42
0260	6″ diameter		24	.333		24.50	14.95		39.45	50
0270	12″ diameter		20	.400		30	17.95		47.95	60.50
0280	Non metallic piping, non insulated									
0290	Through walls, 2″ diameter	1 Carp	12	.667	Ea.	64	30		94	117
0300	4″ diameter		10	.800		80.50	36		116.50	144
0310	6″ diameter		8	1		112	45		157	192
0330	Through floors, 2″ diameter		16	.500		50.50	22.50		73	90
0340	4″ diameter		6	1.333		62.50	60		122.50	161
0350	6″ diameter		6	1.333		75	60		135	175
0370	Ductwork, insulated & non insulated, round									
0380	Through walls, 6″ diameter	1 Carp	12	.667	Ea.	32.50	30		62.50	82
0390	12″ diameter		10	.800		65	36		101	127
0400	18″ diameter		8	1		106	45		151	185
0410	Through floors, 6″ diameter		16	.500		17.85	22.50		40.35	54
0420	12″ diameter		14	.571		32.50	25.50		58	75.50
0430	18″ diameter		12	.667		56.50	30		86.50	109
0440	Ductwork, insulated & non insulated, rectangular									
0450	With stiffener/closure angle, through walls, 6″ x 12″	1 Carp	8	1	Ea.	27	45		72	98.50
0460	12″ x 24″		6	1.333		36	60		96	132
0470	24″ x 48″		4	2		103	90		193	251
0480	With stiffener/closure angle, through floors, 6″ x 12″		10	.800		14.65	36		50.65	71.50
0490	12″ x 24″		8	1		26.50	45		71.50	98
0500	24″ x 48″		6	1.333		51.50	60		111.50	149
0510	Multi trade openings									
0520	Through walls, 6″ x 12″	1 Carp	2	4	Ea.	56.50	180		236.50	340
0530	12″ x 24″	″	1	8		229	360		589	805
0540	24″ x 48″	2 Carp	1	16		920	720		1,640	2,100
0550	48″ x 96″	″	.75	21.333		3,675	960		4,635	5,525
0560	Through floors, 6″ x 12″	1 Carp	2	4		56.50	180		236.50	340
0570	12″ x 24″	″	1	8		229	360		589	805
0580	24″ x 48″	2 Carp	.75	21.333		920	960		1,880	2,475
0590	48″ x 96″	″	.50	32		3,675	1,425		5,100	6,275
0600	Structural penetrations, through walls									
0610	Steel beams, W8 x 10	1 Carp	8	1	Ea.	36	45		81	109
0620	W12 x 14		6	1.333		56.50	60		116.50	155
0630	W21 x 44		5	1.600		114	72		186	236

07 84 Firestopping

07 84 13 – Penetration Firestopping

07 84 13.10 Firestopping		Crew	Daily Output	Labor-Hours	Unit	Material	Labor	Equipment	Total	Total Incl O&P
						2013 Bare Costs				
0640	W36 x 135	1 Carp	3	2.667	Ea.	277	120		397	490
0650	Bar joists, 18" deep		6	1.333		52	60		112	150
0660	24" deep		6	1.333		65	60		125	164
0670	36" deep		5	1.600		97.50	72		169.50	218
0680	48" deep	↓	4	2	↓	114	90		204	263
0690	Construction joints, floor slab at exterior wall									
0700	Precast, brick, block or drywall exterior									
0710	2" wide joint	1 Carp	125	.064	L.F.	8.10	2.87		10.97	13.40
0720	4" wide joint	"	75	.107	"	16.25	4.79		21.04	25.50
0730	Metal panel, glass or curtain wall exterior									
0740	2" wide joint	1 Carp	40	.200	L.F.	19.25	9		28.25	35
0750	4" wide joint	"	25	.320	"	26.50	14.35		40.85	51
0760	Floor slab to drywall partition									
0770	Flat joint	1 Carp	100	.080	L.F.	8	3.59		11.59	14.35
0780	Fluted joint		50	.160		16.25	7.20		23.45	29
0790	Etched fluted joint	↓	75	.107	↓	10.55	4.79		15.34	19
0800	Floor slab to concrete/masonry partition									
0810	Flat joint	1 Carp	75	.107	L.F.	17.85	4.79		22.64	27
0820	Fluted joint	"	50	.160	"	21	7.20		28.20	34
0830	Concrete/CMU wall joints									
0840	1" wide	1 Carp	100	.080	L.F.	9.75	3.59		13.34	16.25
0850	2" wide		75	.107		17.85	4.79		22.64	27
0860	4" wide	↓	50	.160	↓	34	7.20		41.20	48.50
0870	Concrete/CMU floor joints									
0880	1" wide	1 Carp	200	.040	L.F.	4.89	1.80		6.69	8.15
0890	2" wide		150	.053		8.90	2.39		11.29	13.50
0900	4" wide	↓	100	.080	↓	17.05	3.59		20.64	24.50

07 91 Preformed Joint Seals

07 91 13 – Compression Seals

07 91 13.10 Compression Seals

Line	Description	Crew	Daily Output	Labor-Hours	Unit	Material	Labor	Equipment	Total	Total Incl O&P
0010	**COMPRESSION SEALS**									
4900	O-ring type cord, 1/4"	1 Bric	472	.017	L.F.	.48	.76		1.24	1.68
4910	1/2"		440	.018		1.79	.81		2.60	3.20
4920	3/4"		424	.019		3.62	.85		4.47	5.25
4930	1"		408	.020		6.95	.88		7.83	9
4940	1-1/4"		384	.021		13.75	.93		14.68	16.50
4960	1-3/4"		352	.023		36.50	1.02		37.52	41.50
4970	2"	↓	344	.023	↓	64.50	1.04		65.54	72

07 91 16 – Joint Gaskets

07 91 16.10 Joint Gaskets

Line	Description	Crew	Daily Output	Labor-Hours	Unit	Material	Labor	Equipment	Total	Total Incl O&P
0010	**JOINT GASKETS**									
4400	Joint gaskets, neoprene, closed cell w/adh, 1/8" x 3/8"	1 Bric	240	.033	L.F.	.30	1.49		1.79	2.59
4500	1/4" x 3/4"		215	.037		.65	1.67		2.32	3.24
4700	1/2" x 1"		200	.040		1.44	1.79		3.23	4.29
4800	3/4" x 1-1/2"	↓	165	.048	↓	1.60	2.17		3.77	5.05

07 91 Preformed Joint Seals

07 91 23 – Backer Rods

07 91 23.10 Backer Rods		Crew	Daily Output	Labor-Hours	Unit	Material	Labor	Equipment	Total	Total Incl O&P
						2013 Bare Costs				
0010	**BACKER RODS**									
0030	Backer rod, polyethylene, 1/4" diameter	1 Bric	4.60	1.739	C.L.F.	2.75	78		80.75	121
0050	1/2" diameter		4.60	1.739		5.35	78		83.35	124
0070	3/4" diameter		4.60	1.739		5.20	78		83.20	124
0090	1" diameter		4.60	1.739		10.75	78		88.75	130

07 91 26 – Joint Fillers

07 91 26.10 Joint Fillers

		Crew	Daily Output	Labor-Hours	Unit	Material	Labor	Equipment	Total	Total Incl O&P
0010	**JOINT FILLERS**									
4360	Butyl rubber filler, 1/4" x 1/4"	1 Bric	290	.028	L.F.	.15	1.24		1.39	2.03
4365	1/2" x 1/2"		250	.032		.59	1.43		2.02	2.82
4370	1/2" x 3/4"		210	.038		.89	1.71		2.60	3.57
4375	3/4" x 3/4"		230	.035		1.33	1.56		2.89	3.83
4380	1" x 1"		180	.044		1.78	1.99		3.77	4.98
4390	For coloring, add					12%				
4980	Polyethylene joint backing, 1/4" x 2"	1 Bric	2.08	3.846	C.L.F.	12	172		184	274
4990	1/4" x 6"		1.28	6.250	"	28	280		308	455
5600	Silicone, room temp vulcanizing foam seal, 1/4" x 1/2"		1312	.006	L.F.	.33	.27		.60	.78
5610	1/2" x 1/2"		656	.012		.67	.55		1.22	1.57
5620	1/2" x 3/4"		442	.018		1	.81		1.81	2.33
5630	3/4" x 3/4"		328	.024		1.50	1.09		2.59	3.30
5640	1/8" x 1"		1312	.006		.33	.27		.60	.78
5650	1/8" x 3"		442	.018		1	.81		1.81	2.33
5670	1/4" x 3"		295	.027		2.01	1.22		3.23	4.05
5680	1/4" x 6"		148	.054		4.01	2.42		6.43	8.10
5690	1/2" x 6"		82	.098		8	4.37		12.37	15.40
5700	1/2" x 9"		52.50	.152		12.05	6.85		18.90	23.50
5710	1/2" x 12"		33	.242		16.05	10.85		26.90	34

07 92 Joint Sealants

07 92 13 – Elastomeric Joint Sealants

07 92 13.20 Caulking and Sealant Options

		Crew	Daily Output	Labor-Hours	Unit	Material	Labor	Equipment	Total	Total Incl O&P
0010	**CAULKING AND SEALANT OPTIONS**									
0050	Latex acrylic based, bulk				Gal.	29.50			29.50	32.50
0055	Bulk in place 1/4" x 1/4" bead	1 Bric	300	.027	L.F.	.09	1.19		1.28	1.91
0065	1/4" x 1/2"		288	.028		.21	1.24		1.45	2.11
0075	3/8" x 3/8"		284	.028		.23	1.26		1.49	2.16
0080	3/8" x 1/2"		280	.029		.31	1.28		1.59	2.28
0085	3/8" x 5/8"		276	.029		.39	1.30		1.69	2.40
0095	3/8" x 3/4"		272	.029		.46	1.32		1.78	2.51
0100	1/2" x 1/2"		275	.029		.41	1.30		1.71	2.43
0105	1/2" x 5/8"		269	.030		.52	1.33		1.85	2.59
0110	1/2" x 3/4"		263	.030		.62	1.36		1.98	2.74
0115	1/2" x 7/8"		256	.031		.72	1.40		2.12	2.92
0120	1/2" x 1"		250	.032		.83	1.43		2.26	3.08
0125	3/4" x 3/4"		244	.033		.93	1.47		2.40	3.25
0130	3/4" x 1"		225	.036		1.24	1.59		2.83	3.78
0135	1" x 1"		200	.040		1.66	1.79		3.45	4.53
0190	Cartridges				Gal.	25			25	27.50
0200	11 fl. oz. cartridge				Ea.	2.14			2.14	2.35
0500	1/4" x 1/2"	1 Bric	288	.028	L.F.	.17	1.24		1.41	2.07

07 92 Joint Sealants

07 92 13 – Elastomeric Joint Sealants

07 92 13.20 Caulking and Sealant Options		Crew	Daily Output	Labor-Hours	Unit	2013 Bare Costs Material	Labor	Equipment	Total	Total Incl O&P
0600	1/2" x 1/2"	1 Bric	275	.029	L.F.	.35	1.30		1.65	2.35
0800	3/4" x 3/4"		244	.033		.79	1.47		2.26	3.09
0900	3/4" x 1"		225	.036		1.05	1.59		2.64	3.56
1000	1" x 1"	↓	200	.040	↓	1.31	1.79		3.10	4.15
1400	Butyl based, bulk				Gal.	31			31	34
1500	Cartridges				"	35			35	38.50
1700	1/4" x 1/2", 154 L.F./gal.	1 Bric	288	.028	L.F.	.20	1.24		1.44	2.10
1800	1/2" x 1/2", 77 L.F./gal.	"	275	.029	"	.40	1.30		1.70	2.41
2300	Polysulfide compounds, 1 component, bulk				Gal.	56			56	61.50
2600	1 or 2 component, in place, 1/4" x 1/4", 308 L.F./gal.	1 Bric	300	.027	L.F.	.18	1.19		1.37	2.01
2700	1/2" x 1/4", 154 L.F./gal.		288	.028		.36	1.24		1.60	2.28
2900	3/4" x 3/8", 68 L.F./gal.		272	.029		.82	1.32		2.14	2.91
3000	1" x 1/2", 38 L.F./gal.	↓	250	.032	↓	1.47	1.43		2.90	3.79
3200	Polyurethane, 1 or 2 component				Gal.	51			51	56
3300	Cartridges				"	70			70	77
3500	Bulk, in place, 1/4" x 1/4"	1 Bric	300	.027	L.F.	.17	1.19		1.36	1.99
3655	1/2" x 1/4"		288	.028		.33	1.24		1.57	2.24
3800	3/4" x 3/8"		272	.029		.75	1.32		2.07	2.83
3900	1" x 1/2"	↓	250	.032	↓	1.33	1.43		2.76	3.63
4100	Silicone rubber, bulk				Gal.	42			42	46
4200	Cartridges				"	41			41	45

07 92 16 – Rigid Joint Sealants

07 92 16.10 Rigid Joint Sealants		Crew	Daily Output	Labor-Hours	Unit	Material	Labor	Equipment	Total	Total Incl O&P
0010	**RIGID JOINT SEALANTS**									
5800	Tapes, sealant, PVC foam adhesive, 1/16" x 1/4"				C.L.F.	6			6	6.60
5900	1/16" x 1/2"					8			8	8.80
5950	1/16" x 1"					15			15	16.50
6000	1/8" x 1/2"				↓	9			9	9.90

07 92 19 – Acoustical Joint Sealants

07 92 19.10 Acoustical Sealant		Crew	Daily Output	Labor-Hours	Unit	Material	Labor	Equipment	Total	Total Incl O&P
0010	**ACOUSTICAL SEALANT**									
0020	Acoustical sealant, elastomeric, cartridges				Ea.	6.25			6.25	6.90
0025	In place, 1/4" x 1/4"	1 Bric	300	.027	L.F.	.26	1.19		1.45	2.09
0030	1/4" x 1/2"		288	.028		.51	1.24		1.75	2.44
0035	1/2" x 1/2"		275	.029		1.02	1.30		2.32	3.09
0040	1/2" x 3/4"		263	.030		1.53	1.36		2.89	3.75
0045	3/4" x 3/4"		244	.033		2.30	1.47		3.77	4.75
0050	1" x 1"	↓	200	.040	↓	4.09	1.79		5.88	7.20

Division Notes

		CREW	DAILY OUTPUT	LABOR-HOURS	UNIT	BARE COSTS MAT.	LABOR	EQUIP.	TOTAL	TOTAL INCL O&P

Estimating Tips

General

- Room Finish Schedule: A complete set of plans should contain a room finish schedule. If one is not available, it would be well worth the time and effort to obtain one.

09 20 00 Plaster and Gypsum Board

- Lath is estimated by the square yard plus a 5% allowance for waste. Furring, channels, and accessories are measured by the linear foot. An extra foot should be allowed for each accessory miter or stop.
- Plaster is also estimated by the square yard. Deductions for openings vary by preference, from zero deduction to 50% of all openings over 2 feet in width. The estimator should allow one extra square foot for each linear foot of horizontal interior or exterior angle located below the ceiling level. Also, double the areas of small radius work.
- Drywall accessories, studs, track, and acoustical caulking are all measured by the linear foot. Drywall taping is figured by the square foot. Gypsum wallboard is estimated by the square foot. No material deductions should be made for door or window openings under 32 S.F.

09 60 00 Flooring

- Tile and terrazzo areas are taken off on a square foot basis. Trim and base materials are measured by the linear foot. Accent tiles are listed per each. Two basic methods of installation are used. Mud set is approximately 30% more expensive than thin set. In terrazzo work, be sure to include the linear footage of embedded decorative strips, grounds, machine rubbing, and power cleanup.
- Wood flooring is available in strip, parquet, or block configuration. The latter two types are set in adhesives with quantities estimated by the square foot. The laying pattern will influence labor costs and material waste. In addition to the material and labor for laying wood floors, the estimator must make allowances for sanding and finishing these areas unless the flooring is prefinished.
- Sheet flooring is measured by the square yard. Roll widths vary, so consideration should be given to use the most economical width, as waste must be figured into the total quantity. Consider also the installation methods available, direct glue down or stretched.

09 70 00 Wall Finishes

- Wall coverings are estimated by the square foot. The area to be covered is measured, length by height of wall above baseboards, to calculate the square footage of each wall. This figure is divided by the number of square feet in the single roll which is being used. Deduct, in full, the areas of openings such as doors and windows. Where a pattern match is required allow 25%–30% waste.

09 80 00 Acoustic Treatment

- Acoustical systems fall into several categories. The takeoff of these materials should be by the square foot of area with a 5% allowance for waste. Do not forget about scaffolding, if applicable, when estimating these systems.

09 90 00 Painting and Coating

- A major portion of the work in painting involves surface preparation. Be sure to include cleaning, sanding, filling, and masking costs in the estimate.
- Protection of adjacent surfaces is not included in painting costs. When considering the method of paint application, an important factor is the amount of protection and masking required. These must be estimated separately and may be the determining factor in choosing the method of application.

Reference Numbers

Reference numbers are shown in shaded boxes at the beginning of some major classifications. These numbers refer to related items in the Reference Section. The reference information may be an estimating procedure, an alternate pricing method, or technical information.

Note: Not all subdivisions listed here necessarily appear in this publication.

09 22 Supports for Plaster and Gypsum Board

09 22 03 – Fastening Methods for Finishes

09 22 03.20 Drilling Plaster/Drywall		Crew	Daily Output	Labor-Hours	Unit	2013 Bare Costs Material	Labor	Equipment	Total	Total Incl O&P
0010	**DRILLING PLASTER/DRYWALL**									
1100	Drilling & layout for drywall/plaster walls, up to 1" deep, no anchor									
1200	Holes, 1/4" diameter	1 Carp	150	.053	Ea.	.01	2.39		2.40	3.70
1300	3/8" diameter		140	.057		.01	2.57		2.58	3.96
1400	1/2" diameter		130	.062		.01	2.76		2.77	4.27
1500	3/4" diameter		120	.067		.01	2.99		3	4.62
1600	1" diameter		110	.073		.02	3.27		3.29	5.05
1700	1-1/4" diameter		100	.080		.03	3.59		3.62	5.60
1800	1-1/2" diameter		90	.089		.04	3.99		4.03	6.20
1900	For ceiling installations, add						40%			

09 91 Painting

09 91 23 – Interior Painting

09 91 23.52 Miscellaneous, Interior		Crew	Daily Output	Labor-Hours	Unit	Material	Labor	Equipment	Total	Total Incl O&P
0010	**MISCELLANEOUS, INTERIOR**									
3800	Grilles, per side, oil base, primer coat, brushwork	1 Pord	520	.015	S.F.	.12	.60		.72	1.02
3850	Spray		1140	.007		.13	.27		.40	.55
3880	Paint 1 coat, brushwork		520	.015		.22	.60		.82	1.13
3900	Spray		1140	.007		.24	.27		.51	.67
3920	Paint 2 coats, brushwork		325	.025		.42	.95		1.37	1.89
3940	Spray		650	.012		.48	.48		.96	1.24
3950	Prime & paint 1 coat		325	.025		.34	.95		1.29	1.80
3960	Prime & paint 2 coats		270	.030		.33	1.15		1.48	2.08
4500	Louvers, one side, primer, brushwork		524	.015		.09	.59		.68	.99
4520	Paint one coat, brushwork		520	.015		.10	.60		.70	.99
4530	Spray		1140	.007		.11	.27		.38	.53
4540	Paint two coats, brushwork		325	.025		.19	.95		1.14	1.63
4550	Spray		650	.012		.21	.48		.69	.94
4560	Paint three coats, brushwork		270	.030		.27	1.15		1.42	2.02
4570	Spray		500	.016		.31	.62		.93	1.27
5000	Pipe, 1" - 4" diameter, primer or sealer coat, oil base, brushwork	2 Pord	1250	.013	L.F.	.09	.50		.59	.84
5100	Spray		2165	.007		.09	.29		.38	.53
5200	Paint 1 coat, brushwork		1250	.013		.11	.50		.61	.86
5300	Spray		2165	.007		.10	.29		.39	.54
5350	Paint 2 coats, brushwork		775	.021		.19	.80		.99	1.41
5400	Spray		1240	.013		.22	.50		.72	.99
5420	Paint 3 coats, brushwork		775	.021		.29	.80		1.09	1.52
5450	5" - 8" diameter, primer or sealer coat, brushwork		620	.026		.19	1		1.19	1.71
5500	Spray		1085	.015		.31	.57		.88	1.20
5550	Paint 1 coat, brushwork		620	.026		.30	1		1.30	1.82
5600	Spray		1085	.015		.33	.57		.90	1.22
5650	Paint 2 coats, brushwork		385	.042		.39	1.61		2	2.84
5700	Spray		620	.026		.43	1		1.43	1.97
5720	Paint 3 coats, brushwork		385	.042		.58	1.61		2.19	3.04
5750	9" - 12" diameter, primer or sealer coat, brushwork		415	.039		.28	1.49		1.77	2.55
5800	Spray		725	.022		.35	.85		1.20	1.67
5850	Paint 1 coat, brushwork		415	.039		.30	1.49		1.79	2.57
6000	Spray		725	.022		.33	.85		1.18	1.65
6200	Paint 2 coats, brushwork		260	.062		.58	2.38		2.96	4.21
6250	Spray		415	.039		.65	1.49		2.14	2.95
6270	Paint 3 coats, brushwork		260	.062		.86	2.38		3.24	4.52

09 91 Painting

09 91 23 – Interior Painting

09 91 23.52 Miscellaneous, Interior		Crew	Daily Output	Labor-Hours	Unit	2013 Bare Costs Material	Labor	Equipment	Total	Total Incl O&P
6300	13" - 16" diameter, primer or sealer coat, brushwork	2 Pord	310	.052	L.F.	.38	2		2.38	3.41
6350	Spray		540	.030		.42	1.15		1.57	2.18
6400	Paint 1 coat, brushwork		310	.052		.40	2		2.40	3.44
6450	Spray		540	.030		.44	1.15		1.59	2.21
6500	Paint 2 coats, brushwork		195	.082		.78	3.18		3.96	5.60
6550	Spray		310	.052		.86	2		2.86	3.95
6600	Radiators, per side, primer, brushwork	1 Pord	520	.015	S.F.	.09	.60		.69	.99
6620	Paint, one coat		520	.015		.08	.60		.68	.98
6640	Two coats		340	.024		.19	.91		1.10	1.57
6660	Three coats		283	.028		.27	1.09		1.36	1.94

Division Notes

		CREW	DAILY OUTPUT	LABOR-HOURS	UNIT	BARE COSTS MAT.	LABOR	EQUIP.	TOTAL	TOTAL INCL O&P

Estimating Tips

General

- The items in this division are usually priced per square foot or each.
- Many items in Division 10 require some type of support system or special anchors that are not usually furnished with the item. The required anchors must be added to the estimate in the appropriate division.
- Some items in Division 10, such as lockers, may require assembly before installation. Verify the amount of assembly required. Assembly can often exceed installation time.

10 20 00 Interior Specialties

- Support angles and blocking are not included in the installation of toilet compartments, shower/dressing compartments, or cubicles. Appropriate line items from Divisions 5 or 6 may need to be added to support the installations.
- Toilet partitions are priced by the stall. A stall consists of a side wall, pilaster, and door with hardware. Toilet tissue holders and grab bars are extra.
- The required acoustical rating of a folding partition can have a significant impact on costs. Verify the sound transmission coefficient rating of the panel priced to the specification requirements.
- Grab bar installation does not include supplemental blocking or backing to support the required load. When grab bars are installed at an existing facility, provisions must be made to attach the grab bars to solid structure.

Reference Numbers

Reference numbers are shown in shaded boxes at the beginning of some major classifications. These numbers refer to related items in the Reference Section. The reference information may be an estimating procedure, an alternate pricing method, or technical information.

Note: Not all subdivisions listed here necessarily appear in this publication.

10 21 Compartments and Cubicles

10 21 13 – Toilet Compartments

10 21 13.13 Metal Toilet Compartments		Crew	Daily Output	Labor-Hours	Unit	2013 Bare Costs Material	Labor	Equipment	Total	Total Incl O&P
0010	**METAL TOILET COMPARTMENTS**									
0110	Cubicles, ceiling hung									
0200	Powder coated steel	2 Carp	4	4	Ea.	510	180		690	840
0500	Stainless steel	"	4	4		995	180		1,175	1,375
0600	For handicap units, incl. 52" grab bars, add					430			430	475
0900	Floor and ceiling anchored									
1000	Powder coated steel	2 Carp	5	3.200	Ea.	590	144		734	865
1300	Stainless steel	"	5	3.200		1,100	144		1,244	1,450
1400	For handicap units, incl. 52" grab bars, add					315			315	345
1610	Floor anchored									
1700	Powder coated steel	2 Carp	7	2.286	Ea.	580	103		683	800
2000	Stainless steel	"	7	2.286		1,400	103		1,503	1,675
2100	For handicap units, incl. 52" grab bars, add					310			310	345
2200	For juvenile units, deduct					41.50			41.50	45.50
2450	Floor anchored, headrail braced									
2500	Powder coated steel	2 Carp	6	2.667	Ea.	385	120		505	610
2804	Stainless steel	"	4.60	3.478		1,050	156		1,206	1,400
2900	For handicap units, incl. 52" grab bars, add					370			370	410
3000	Wall hung partitions, powder coated steel	2 Carp	7	2.286		635	103		738	860
3300	Stainless steel	"	7	2.286		1,650	103		1,753	1,975
3400	For handicap units, incl. 52" grab bars, add					370			370	410
4000	Screens, entrance, floor mounted, 58" high, 48" wide									
4200	Powder coated steel	2 Carp	15	1.067	Ea.	242	48		290	340
4500	Stainless steel	"	15	1.067	"	910	48		958	1,075
4650	Urinal screen, 18" wide									
4704	Powder coated steel	2 Carp	6.15	2.602	Ea.	206	117		323	405
5004	Stainless steel	"	6.15	2.602	"	580	117		697	820
5100	Floor mounted, head rail braced									
5300	Powder coated steel	2 Carp	8	2	Ea.	230	90		320	390
5600	Stainless steel	"	8	2	"	595	90		685	795
5750	Pilaster, flush									
5800	Powder coated steel	2 Carp	10	1.600	Ea.	278	72		350	415
6100	Stainless steel		10	1.600		670	72		742	845
6300	Post braced, powder coated steel		10	1.600		153	72		225	279
6600	Stainless steel		10	1.600		450	72		522	605
6700	Wall hung, bracket supported									
6800	Powder coated steel	2 Carp	10	1.600	Ea.	153	72		225	279
7100	Stainless steel		10	1.600		223	72		295	355
7400	Flange supported, powder coated steel		10	1.600		106	72		178	227
7700	Stainless steel		10	1.600		310	72		382	450
7800	Wedge type, powder coated steel		10	1.600		134	72		206	259
8100	Stainless steel		10	1.600		575	72		647	745

10 21 13.16 Plastic-Laminate-Clad Toilet Compartments		Crew	Daily Output	Labor-Hours	Unit	Material	Labor	Equipment	Total	Total Incl O&P
0010	**PLASTIC-LAMINATE-CLAD TOILET COMPARTMENTS**									
0110	Cubicles, ceiling hung									
0300	Plastic laminate on particle board	2 Carp	4	4	Ea.	550	180		730	880
0600	For handicap units, incl. 52" grab bars, add				"	430			430	475
0900	Floor and ceiling anchored									
1100	Plastic laminate on particle board	2 Carp	5	3.200	Ea.	780	144		924	1,075
1400	For handicap units, incl. 52" grab bars, add				"	315			315	345
1610	Floor mounted									
1800	Plastic laminate on particle board	2 Carp	7	2.286	Ea.	485	103		588	695

10 21 Compartments and Cubicles

10 21 13 – Toilet Compartments

10 21 13.16 Plastic-Laminate-Clad Toilet Compartments		Crew	Daily Output	Labor-Hours	Unit	2013 Bare Costs Material	Labor	Equipment	Total	Total Incl O&P
2450	Floor mounted, headrail braced									
2600	Plastic laminate on particle board	2 Carp	6	2.667	Ea.	750	120		870	1,000
3400	For handicap units, incl. 52" grab bars, add					370			370	410
4300	Entrance screen, floor mtd., plas. lam., 58" high, 48" wide	2 Carp	15	1.067		610	48		658	745
4800	Urinal screen, 18" wide, ceiling braced, plastic laminate		8	2		174	90		264	330
5400	Floor mounted, headrail braced		8	2		180	90		270	335
5900	Pilaster, flush, plastic laminate		10	1.600		505	72		577	670
6400	Post braced, plastic laminate		10	1.600		295	72		367	435
6700	Wall hung, bracket supported									
6900	Plastic laminate on particle board	2 Carp	10	1.600	Ea.	90.50	72		162.50	211
7450	Flange supported									
7500	Plastic laminate on particle board	2 Carp	10	1.600	Ea.	230	72		302	365

10 21 13.19 Plastic Toilet Compartments		Crew	Daily Output	Labor-Hours	Unit	Material	Labor	Equipment	Total	Total Incl O&P
0010	**PLASTIC TOILET COMPARTMENTS**									
0110	Cubicles, ceiling hung									
0250	Phenolic	2 Carp	4	4	Ea.	860	180		1,040	1,225
0600	For handicap units, incl. 52" grab bars, add				"	430			430	475
0900	Floor and ceiling anchored									
1050	Phenolic	2 Carp	5	3.200	Ea.	845	144		989	1,150
1400	For handicap units, incl. 52" grab bars, add				"	315			315	345
1610	Floor mounted									
1750	Phenolic	2 Carp	7	2.286	Ea.	765	103		868	1,000
2100	For handicap units, incl. 52" grab bars, add					310			310	345
2200	For juvenile units, deduct					41.50			41.50	45.50
2450	Floor mounted, headrail braced									
2550	Phenolic	2 Carp	6	2.667	Ea.	750	120		870	1,000

10 21 13.40 Stone Toilet Compartments		Crew	Daily Output	Labor-Hours	Unit	Material	Labor	Equipment	Total	Total Incl O&P
0010	**STONE TOILET COMPARTMENTS**									
0100	Cubicles, ceiling hung, marble	2 Marb	2	8	Ea.	1,625	330		1,955	2,275
0600	For handicap units, incl. 52" grab bars, add					430			430	475
0800	Floor & ceiling anchored, marble	2 Marb	2.50	6.400		2,025	265		2,290	2,650
1400	For handicap units, incl. 52" grab bars, add					315			315	345
1600	Floor mounted, marble	2 Marb	3	5.333		1,200	221		1,421	1,650
2400	Floor mounted, headrail braced, marble	"	3	5.333		1,150	221		1,371	1,575
2900	For handicap units, incl. 52" grab bars, add					370			370	410
4100	Entrance screen, floor mounted marble, 58" high, 48" wide	2 Marb	9	1.778		755	73.50		828.50	940
4600	Urinal screen, 18" wide, ceiling braced, marble	D-1	6	2.667		750	108		858	990
5100	Floor mounted, head rail braced									
5200	Marble	D-1	6	2.667	Ea.	640	108		748	865
5700	Pilaster, flush, marble		9	1.778		835	72		907	1,025
6200	Post braced, marble		9	1.778		825	72		897	1,025

10 21 16 – Shower and Dressing Compartments

10 21 16.10 Partitions, Shower		Crew	Daily Output	Labor-Hours	Unit	Material	Labor	Equipment	Total	Total Incl O&P
0010	**PARTITIONS, SHOWER** floor mounted, no plumbing									
0400	Cabinet, one piece, fiberglass, 32" x 32"	2 Carp	5	3.200	Ea.	525	144		669	800
0420	36" x 36"		5	3.200		645	144		789	930
0440	36" x 48"		5	3.200		1,450	144		1,594	1,825
0460	Acrylic, 32" x 32"		5	3.200		325	144		469	575
0480	36" x 36"		5	3.200		1,050	144		1,194	1,375
0500	36" x 48"		5	3.200		1,475	144		1,619	1,850
0520	Shower door for above, clear plastic, 24" wide	1 Carp	8	1		177	45		222	264
0540	28" wide		8	1		201	45		246	291

10 21 Compartments and Cubicles

10 21 16 – Shower and Dressing Compartments

10 21 16.10 Partitions, Shower		Crew	Daily Output	Labor-Hours	Unit	Material	2013 Bare Costs Labor	Equipment	Total	Total Incl O&P
0560	Tempered glass, 24" wide	1 Carp	8	1	Ea.	191	45		236	279
0580	28" wide	↓	8	1		216	45		261	305
2400	Glass stalls, with doors, no receptors, chrome on brass	2 Shee	3	5.333		1,650	284		1,934	2,250
2700	Anodized aluminum	"	4	4		1,150	213		1,363	1,600
2900	Marble shower stall, stock design, with shower door	2 Marb	1.20	13.333		2,475	555		3,030	3,550
3000	With curtain		1.30	12.308		2,200	510		2,710	3,175
3200	Receptors, precast terrazzo, 32" x 32"		14	1.143		273	47.50		320.50	370
3300	48" x 34"		9.50	1.684		475	70		545	630
3500	Plastic, simulated terrazzo receptor, 32" x 32"		14	1.143		148	47.50		195.50	235
3600	32" x 48"		12	1.333		215	55.50		270.50	320
3800	Precast concrete, colors, 32" x 32"		14	1.143		186	47.50		233.50	277
3900	48" x 48"	↓	8	2		254	83		337	405
4100	Shower doors, economy plastic, 24" wide	1 Shee	9	.889		144	47.50		191.50	231
4200	Tempered glass door, economy		8	1		258	53.50		311.50	365
4400	Folding, tempered glass, aluminum frame		6	1.333		390	71		461	540
4500	Sliding, tempered glass, 48" opening		6	1.333		540	71		611	700
4700	Deluxe, tempered glass, chrome on brass frame, minimum		8	1		385	53.50		438.50	500
4800	Maximum		1	8		625	425		1,050	1,325
4850	On anodized aluminum frame, minimum		2	4		540	213		753	915
4900	Maximum	↓	1	8	↓	625	425		1,050	1,325
5100	Shower enclosure, tempered glass, anodized alum. frame									
5120	2 panel & door, corner unit, 32" x 32"	1 Shee	2	4	Ea.	1,025	213		1,238	1,450
5140	Neo-angle corner unit, 16" x 24" x 16"	"	2	4		1,025	213		1,238	1,450
5200	Shower surround, 3 wall, polypropylene, 32" x 32"	1 Carp	4	2		445	90		535	625
5220	PVC, 32" x 32"		4	2		380	90		470	560
5240	Fiberglass		4	2		420	90		510	600
5250	2 wall, polypropylene, 32" x 32"		4	2		315	90		405	490
5270	PVC		4	2		395	90		485	575
5290	Fiberglass	↓	4	2		400	90		490	580
5300	Tub doors, tempered glass & frame, minimum	1 Shee	8	1		229	53.50		282.50	335
5400	Maximum		6	1.333		535	71		606	700
5600	Chrome plated, brass frame, minimum		8	1		305	53.50		358.50	415
5700	Maximum		6	1.333		745	71		816	925
5900	Tub/shower enclosure, temp. glass, alum. frame, minimum		2	4		410	213		623	775
6200	Maximum		1.50	5.333		855	284		1,139	1,375
6500	On chrome-plated brass frame, minimum		2	4		565	213		778	950
6600	Maximum	↓	1.50	5.333		1,200	284		1,484	1,750
6800	Tub surround, 3 wall, polypropylene	1 Carp	4	2		256	90		346	420
6900	PVC		4	2		390	90		480	570
7000	Fiberglass, minimum		4	2		400	90		490	580
7100	Maximum	↓	3	2.667	↓	680	120		800	935

10 28 Toilet, Bath, and Laundry Accessories

10 28 13 – Toilet Accessories

10 28 13.13 Commercial Toilet Accessories		Crew	Daily Output	Labor-Hours	Unit	Material	Labor	Equipment	Total	Total Incl O&P
						2013 Bare Costs				
0010	**COMMERCIAL TOILET ACCESSORIES**									
0200	Curtain rod, stainless steel, 5′ long, 1″ diameter	1 Carp	13	.615	Ea.	34	27.50		61.50	79.50
0300	1-1/4″ diameter	″	13	.615	″	27.50	27.50		55	72.50
0500	Dispenser units, combined soap & towel dispensers,									
0510	mirror and shelf, flush mounted	1 Carp	10	.800	Ea.	310	36		346	395
0600	Towel dispenser and waste receptacle,									
0610	18 gallon capacity	1 Carp	10	.800	Ea.	286	36		322	370
0800	Grab bar, straight, 1-1/4″ diameter, stainless steel, 18″ long		24	.333		29	14.95		43.95	55
0900	24″ long		23	.348		29.50	15.60		45.10	56.50
1000	30″ long		22	.364		31.50	16.35		47.85	59.50
1100	36″ long		20	.400		38.50	17.95		56.45	70
1105	42″ long		20	.400		38	17.95		55.95	69.50
1200	1-1/2″ diameter, 24″ long		23	.348		31	15.60		46.60	58
1300	36″ long		20	.400		36	17.95		53.95	67
1310	42″ long		18	.444		43	19.95		62.95	78
1500	Tub bar, 1-1/4″ diameter, 24″ x 36″		14	.571		92.50	25.50		118	142
1600	Plus vertical arm		12	.667		97.50	30		127.50	153
1900	End tub bar, 1″ diameter, 90° angle, 16″ x 32″		12	.667		109	30		139	166
2300	Hand dryer, surface mounted, electric, 115 volt, 20 amp		4	2		430	90		520	615
2400	230 volt, 10 amp		4	2		745	90		835	960
2600	Hat and coat strip, stainless steel, 4 hook, 36″ long		24	.333		75	14.95		89.95	106
2700	6 hook, 60″ long		20	.400		117	17.95		134.95	157
3000	Mirror, with stainless steel 3/4″ square frame, 18″ x 24″		20	.400		43.50	17.95		61.45	75.50
3100	36″ x 24″		15	.533		97.50	24		121.50	144
3200	48″ x 24″		10	.800		146	36		182	217
3300	72″ x 24″		6	1.333		230	60		290	345
3500	With 5″ stainless steel shelf, 18″ x 24″		20	.400		186	17.95		203.95	233
3600	36″ x 24″		15	.533		217	24		241	276
3700	48″ x 24″		10	.800		245	36		281	325
3800	72″ x 24″		6	1.333		300	60		360	420
4100	Mop holder strip, stainless steel, 5 holders, 48″ long		20	.400		89	17.95		106.95	126
4200	Napkin/tampon dispenser, recessed		15	.533		525	24		549	610
4250	Napkin receptacle, recessed		6.50	1.231		165	55.50		220.50	267
4300	Robe hook, single, regular		36	.222		18.10	10		28.10	35.50
4400	Heavy duty, concealed mounting		36	.222		18.10	10		28.10	35.50
4600	Soap dispenser, chrome, surface mounted, liquid		20	.400		46.50	17.95		64.45	78.50
4700	Powder		20	.400		55	17.95		72.95	88
5000	Recessed stainless steel, liquid		10	.800		137	36		173	207
5600	Shelf, stainless steel, 5″ wide, 18 ga., 24″ long		24	.333		67.50	14.95		82.45	97
5700	48″ long		16	.500		106	22.50		128.50	152
5800	8″ wide shelf, 18 ga., 24″ long		22	.364		69	16.35		85.35	101
5900	48″ long		14	.571		121	25.50		146.50	173
6000	Toilet seat cover dispenser, stainless steel, recessed		20	.400		167	17.95		184.95	211
6050	Surface mounted		15	.533		34	24		58	74.50
6100	Toilet tissue dispenser, surface mounted, SS, single roll		30	.267		17.80	11.95		29.75	38
6200	Double roll		24	.333		23.50	14.95		38.45	49
6240	Plastic, twin/jumbo dbl. roll		24	.333		26	14.95		40.95	51.50
6400	Towel bar, stainless steel, 18″ long		23	.348		42	15.60		57.60	70
6500	30″ long		21	.381		111	17.10		128.10	150
6700	Towel dispenser, stainless steel, surface mounted		16	.500		44.50	22.50		67	83.50
6800	Flush mounted, recessed		10	.800		227	36		263	305
6900	Plastic, touchless, battery operated		16	.500		85.50	22.50		108	129
7000	Towel holder, hotel type, 2 guest size		20	.400		48.50	17.95		66.45	81

10 28 Toilet, Bath, and Laundry Accessories

10 28 13 – Toilet Accessories

10 28 13.13 Commercial Toilet Accessories		Crew	Daily Output	Labor-Hours	Unit	Material	Labor	Equipment	Total	Total Incl O&P
						2013 Bare Costs				
7200	Towel shelf, stainless steel, 24" long, 8" wide	1 Carp	20	.400	Ea.	58	17.95		75.95	91.50
7400	Tumbler holder, tumbler only		30	.267		43.50	11.95		55.45	66.50
7500	Soap, tumbler & toothbrush		30	.267		19.60	11.95		31.55	40
7700	Wall urn ash receiver, surface mount, 11" long		12	.667		95	30		125	150
7800	7-1/2", long		18	.444		120	19.95		139.95	163
8000	Waste receptacles, stainless steel, with top, 13 gallon		10	.800		295	36		331	380
8100	36 gallon		8	1		375	45		420	480

10 28 16 – Bath Accessories

10 28 16.20 Medicine Cabinets		Crew	Daily Output	Labor-Hours	Unit	Material	Labor	Equipment	Total	Total Incl O&P
0010	**MEDICINE CABINETS**									
0020	With mirror, sst frame, 16" x 22", unlighted	1 Carp	14	.571	Ea.	81.50	25.50		107	130
0100	Wood frame		14	.571		136	25.50		161.50	190
0300	Sliding mirror doors, 20" x 16" x 4-3/4", unlighted		7	1.143		125	51.50		176.50	216
0400	24" x 19" x 8-1/2", lighted		5	1.600		179	72		251	310
0600	Triple door, 30" x 32", unlighted, plywood body		7	1.143		320	51.50		371.50	430
0700	Steel body		7	1.143		350	51.50		401.50	465
0900	Oak door, wood body, beveled mirror, single door		7	1.143		195	51.50		246.50	293
1000	Double door		6	1.333		390	60		450	520
1200	Hotel cabinets, stainless, with lower shelf, unlighted		10	.800		200	36		236	275
1300	Lighted		5	1.600		305	72		377	445

10 44 Fire Protection Specialties

10 44 13 – Fire Protection Cabinets

10 44 13.53 Fire Equipment Cabinets		Crew	Daily Output	Labor-Hours	Unit	Material	Labor	Equipment	Total	Total Incl O&P
0010	**FIRE EQUIPMENT CABINETS**, not equipped, 20 ga. steel box D4020-310									
0040	recessed, D.S. glass in door, box size given									
1000	Portable extinguisher, single, 8" x 12" x 27", alum. door & frame	Q-12	8	2	Ea.	144	98.50		242.50	305
1100	Steel door and frame		8	2		99.50	98.50		198	258
1200	Stainless steel door and frame		8	2		186	98.50		284.50	355
2000	Portable extinguisher, large, 8" x 12" x 36", alum. door & frame		8	2		234	98.50		332.50	405
2100	Steel door and frame D4020-330		8	2		148	98.50		246.50	310
2200	Stainless steel door and frame		8	2		245	98.50		343.50	420
2500	8" x 16" x 38", aluminum door & frame		8	2		255	98.50		353.50	430
2600	Steel door and frame		8	2		223	98.50		321.50	395
2700	Fire blanket & extinguisher cab, inc blanket, rec stl., 14" x 40" x 8"		7	2.286		171	112		283	355
2800	Fire blanket cab, inc blanket, surf mtd, stl, 15"x10"x5", w/pwdr coat fin		8	2		86	98.50		184.50	243
3000	Hose rack assy., 1-1/2" valve & 100' hose, 24" x 40" x 5-1/2"									
3100	Aluminum door and frame	Q-12	6	2.667	Ea.	335	131		466	570
3200	Steel door and frame D4020-410		6	2.667		223	131		354	445
3300	Stainless steel door and frame		6	2.667		420	131		551	660
4000	Hose rack assy., 2-1/2" x 1-1/2" valve, 100' hose, 24" x 40" x 8"									
4100	Aluminum door and frame	Q-12	6	2.667	Ea.	340	131		471	575
4200	Steel door and frame R211226-10		6	2.667		229	131		360	450
4300	Stainless steel door and frame		6	2.667		450	131		581	695
5000	Hose rack assy., 2-1/2" x 1-1/2" valve, 100' hose									
5010	and extinguisher, 30" x 40" x 8"									
5100	Aluminum door and frame	Q-12	5	3.200	Ea.	430	157		587	710
5200	Steel door and frame		5	3.200		286	157		443	550
5300	Stainless steel door and frame		5	3.200		525	157		682	810
6000	Hose rack assy., 1-1/2" valve, 100' hose									

10 44 13 – Fire Protection Cabinets

10 44 13.53 Fire Equipment Cabinets		Crew	Daily Output	Labor-Hours	Unit	Material	2013 Bare Costs Labor	Equipment	Total	Total Incl O&P
6010	and 2-1/2" FD valve, 24" x 44" x 8"									
6100	Aluminum door and frame	Q-12	5	3.200	Ea.	365	157		522	635
6200	Steel door and frame		5	3.200		234	157		391	495
6300	Stainless steel door and frame		5	3.200		510	157		667	795
7000	Hose rack assy., 1-1/2" valve & 100' hose, 2-1/2" FD valve R211226-20									
7010	and extinguisher, 30" x 44" x 8"									
7100	Aluminum door and frame	Q-12	5	3.200	Ea.	455	157		612	735
7200	Steel door and frame		5	3.200		300	157		457	565
7300	Stainless steel door and frame		5	3.200		605	157		762	900
8000	Valve cabinet for 2-1/2" FD angle valve, 18" x 18" x 8"									
8100	Aluminum door and frame	Q-12	12	1.333	Ea.	149	65.50		214.50	263
8200	Steel door and frame		12	1.333		123	65.50		188.50	234
8300	Stainless steel door and frame		12	1.333		200	65.50		265.50	320

10 44 16 – Fire Extinguishers

10 44 16.13 Portable Fire Extinguishers

Code	Description	Crew	Daily Output	Labor-Hours	Unit	Material	Labor	Equipment	Total	Total Incl O&P
0010	**PORTABLE FIRE EXTINGUISHERS**									
0140	CO_2, with hose and "H" horn, 10 lb.				Ea.	261			261	287
0160	15 lb.					335			335	365
0180	20 lb.					375			375	415
1000	Dry chemical, pressurized									
1040	Standard type, portable, painted, 2-1/2 lb.				Ea.	36			36	39.50
1060	5 lb.					49			49	54
1080	10 lb.					77.50			77.50	85.50
1100	20 lb.					131			131	144
1120	30 lb.					395			395	435
1300	Standard type, wheeled, 150 lb.					2,250			2,250	2,450
2000	ABC all purpose type, portable, 2-1/2 lb.					21			21	23.50
2060	5 lb.					26.50			26.50	29
2080	9-1/2 lb.					44			44	48
2100	20 lb.					79.50			79.50	87.50
5000	Pressurized water, 2-1/2 gallon, stainless steel					102			102	112
5060	With anti-freeze					106			106	116
9400	Installation of extinguishers, 12 or more, on nailable surface	1 Carp	30	.267			11.95		11.95	18.45
9420	On masonry or concrete	"	15	.533			24		24	37

10 44 16.16 Wheeled Fire Extinguisher Units

Code	Description	Crew	Daily Output	Labor-Hours	Unit	Material	Labor	Equipment	Total	Total Incl O&P
0010	**WHEELED FIRE EXTINGUISHER UNITS**									
0350	CO_2, portable, with swivel horn									
0360	Wheeled type, cart mounted, 50 lb.				Ea.	1,075			1,075	1,175
0400	100 lb.				"	4,025			4,025	4,425
2200	ABC all purpose type									
2300	Wheeled, 45 lb.				Ea.	735			735	810
2360	150 lb.				"	1,850			1,850	2,025

Division Notes

		CREW	DAILY OUTPUT	LABOR-HOURS	UNIT	BARE COSTS MAT.	LABOR	EQUIP.	TOTAL	TOTAL INCL O&P

Estimating Tips

General

- The items in this division are usually priced per square foot or each. Many of these items are purchased by the owner for installation by the contractor. Check the specifications for responsibilities and include time for receiving, storage, installation, and mechanical and electrical hookups in the appropriate divisions.
- Many items in Division 11 require some type of support system that is not usually furnished with the item. Examples of these systems include blocking for the attachment of casework and support angles for ceiling-hung projection screens. The required blocking or supports must be added to the estimate in the appropriate division.
- Some items in Division 11 may require assembly or electrical hookups. Verify the amount of assembly required or the need for a hard electrical connection and add the appropriate costs.

Reference Numbers

Reference numbers are shown in shaded boxes at the beginning of some major classifications. These numbers refer to related items in the Reference Section. The reference information may be an estimating procedure, an alternate pricing method, or technical information.

Note: Not all subdivisions listed here necessarily appear in this publication.

11 11 Vehicle Service Equipment

11 11 13 – Compressed-Air Vehicle Service Equipment

11 11 13.10 Compressed Air Equipment		Crew	Daily Output	Labor-Hours	Unit	2013 Bare Costs Material	Labor	Equipment	Total	Total Incl O&P
0010	**COMPRESSED AIR EQUIPMENT**									
0030	Compressors, electric, 1-1/2 H.P., standard controls	L-4	1.50	16	Ea.	440	675		1,115	1,525
0550	Dual controls		1.50	16		795	675		1,470	1,925
0600	5 H.P., 115/230 volt, standard controls		1	24		1,900	1,000		2,900	3,625
0650	Dual controls	↓	1	24	↓	2,700	1,000		3,700	4,500

11 11 19 – Vehicle Lubrication Equipment

11 11 19.10 Lubrication Equipment		Crew	Daily Output	Labor-Hours	Unit	Material	Labor	Equipment	Total	Total Incl O&P
0010	**LUBRICATION EQUIPMENT**									
3000	Lube equipment, 3 reel type, with pumps, not including piping	L-4	.50	48	Set	8,700	2,025		10,725	12,700
3700	Pump lubrication, pneumatic, not incl. air compressor									
3710	Oil/gear lube	Q-1	9.60	1.667	Ea.	820	83.50		903.50	1,025
3720	Grease	"	9.60	1.667	"	1,050	83.50		1,133.50	1,275

11 11 33 – Vehicle Spray Painting Equipment

11 11 33.10 Spray Painting Equipment		Crew	Daily Output	Labor-Hours	Unit	Material	Labor	Equipment	Total	Total Incl O&P
0010	**SPRAY PAINTING EQUIPMENT**									
4000	Spray painting booth, 26′ long, complete	L-4	.40	60	Ea.	16,200	2,525		18,725	21,700

11 19 Detention Equipment

11 19 30 – Detention Cell Equipment

11 19 30.10 Cell Equipment		Crew	Daily Output	Labor-Hours	Unit	Material	Labor	Equipment	Total	Total Incl O&P
0010	**CELL EQUIPMENT**									
3000	Toilet apparatus including wash basin, average	L-8	1.50	13.333	Ea.	3,350	630		3,980	4,650

11 21 Mercantile and Service Equipment

11 21 53 – Barber and Beauty Shop Equipment

11 21 53.10 Barber Equipment		Crew	Daily Output	Labor-Hours	Unit	Material	Labor	Equipment	Total	Total Incl O&P
0010	**BARBER EQUIPMENT**									
0020	Chair, hydraulic, movable, minimum	1 Carp	24	.333	Ea.	490	14.95		504.95	560
0050	Maximum	"	16	.500		3,125	22.50		3,147.50	3,450
0500	Sink, hair washing basin, rough plumbing not incl.	1 Plum	8	1		445	56		501	575
1000	Sterilizer, liquid solution for tools				↓	145			145	159

11 23 Commercial Laundry and Dry Cleaning Equipment

11 23 13 – Dry Cleaning Equipment

11 23 13.13 Dry Cleaners		Crew	Daily Output	Labor-Hours	Unit	Material	Labor	Equipment	Total	Total Incl O&P
0010	**DRY CLEANERS** Not incl. rough-in									

11 23 16 – Drying and Conditioning Equipment

11 23 16.13 Dryers		Crew	Daily Output	Labor-Hours	Unit	Material	Labor	Equipment	Total	Total Incl O&P
0010	**DRYERS**, Not including rough-in									
1500	Industrial, 30 lb. capacity	1 Plum	2	4	Ea.	3,075	223		3,298	3,725
1600	50 lb. capacity	"	1.70	4.706	"	3,300	263		3,563	4,025

11 23 26 – Commercial Washers and Extractors

11 23 26.13 Washers and Extractors		Crew	Daily Output	Labor-Hours	Unit	Material	Labor	Equipment	Total	Total Incl O&P
0010	**WASHERS AND EXTRACTORS**, not including rough-in									
6000	Combination washer/extractor, 20 lb. capacity	L-6	1.50	8	Ea.	5,775	435		6,210	7,025

11 23 Commercial Laundry and Dry Cleaning Equipment

11 23 26 – Commercial Washers and Extractors

11 23 26.13 Washers and Extractors		Crew	Daily Output	Labor-Hours	Unit	Material	Labor (2013 Bare Costs)	Equipment	Total	Total Incl O&P
6100	30 lb. capacity	L-6	.80	15	Ea.	9,050	820		9,870	11,200
6200	50 lb. capacity	↓	.68	17.647	↓	10,600	965		11,565	13,200
6300	75 lb. capacity	↓	.30	40	↓	20,000	2,175		22,175	25,300
6350	125 lb. capacity	↓	.16	75	↓	27,300	4,100		31,400	36,300

11 23 33 – Coin-Operated Laundry Equipment

11 23 33.13 Coin Operated Washers and Dryers

		Crew	Daily Output	Labor-Hours	Unit	Material	Labor	Equipment	Total	Total Incl O&P
0010	**COIN OPERATED WASHERS AND DRYERS**									
0990	Dryer, gas fired									
1000	Commercial, 30 lb. capacity, coin operated, single	1 Plum	3	2.667	Ea.	3,200	149		3,349	3,750
1100	Double stacked	"	2	4	"	7,100	223		7,323	8,125
5290	Clothes washer									
5300	Commercial, coin operated, average	1 Plum	3	2.667	Ea.	1,250	149		1,399	1,600

11 24 Maintenance Equipment

11 24 19 – Vacuum Cleaning Systems

11 24 19.10 Vacuum Cleaning

		Crew	Daily Output	Labor-Hours	Unit	Material	Labor	Equipment	Total	Total Incl O&P
0010	**VACUUM CLEANING**									
0020	Central, 3 inlet, residential	1 Skwk	.90	8.889	Total	1,050	410		1,460	1,800
0200	Commercial	↓	.70	11.429	↓	1,250	530		1,780	2,200
0400	5 inlet system, residential	↓	.50	16	↓	1,475	740		2,215	2,775
0600	7 inlet system, commercial	↓	.40	20	↓	1,675	925		2,600	3,250
0800	9 inlet system, residential	↓	.30	26.667	↓	3,700	1,225		4,925	5,975
4010	Rule of thumb: First 1200 S.F., installed				↓				1,425	1,575
4020	For each additional S.F., add				S.F.					.26

11 26 Unit Kitchens

11 26 13 – Metal Unit Kitchens

11 26 13.10 Commercial Unit Kitchens

		Crew	Daily Output	Labor-Hours	Unit	Material	Labor	Equipment	Total	Total Incl O&P
0010	**COMMERCIAL UNIT KITCHENS**									
1500	Combination range, refrigerator and sink, 30" wide, minimum	L-1	2	8	Ea.	1,125	435		1,560	1,875
1550	Maximum	↓	1	16	↓	3,875	865		4,740	5,550
1570	60" wide, average	↓	1.40	11.429	↓	2,800	620		3,420	4,000
1590	72" wide, average	↓	1.20	13.333	↓	4,400	720		5,120	5,925
1600	Office model, 48" wide	↓	2	8	↓	2,725	435		3,160	3,650
1620	Refrigerator and sink only	↓	2.40	6.667	↓	2,850	360		3,210	3,675
1640	Combination range, refrigerator, sink, microwave									
1660	Oven and ice maker	L-1	.80	20	Ea.	4,675	1,075		5,750	6,750

11 27 Photographic Processing Equipment

11 27 13 – Darkroom Processing Equipment

11 27 13.10 Darkroom Equipment		Crew	Daily Output	Labor-Hours	Unit	2013 Bare Costs Material	Labor	Equipment	Total	Total Incl O&P
0010	DARKROOM EQUIPMENT									
0020	Developing sink, 5" deep, 24" x 48"	Q-1	2	8	Ea.	5,625	400		6,025	6,775
0050	48" x 52"		1.70	9.412		5,675	475		6,150	6,925
0200	10" deep, 24" x 48"		1.70	9.412		7,775	475		8,250	9,250
0250	24" x 108"		1.50	10.667		10,200	535		10,735	12,100
3500	Washers, round, minimum sheet 11" x 14"		2	8		3,400	400		3,800	4,350
3550	Maximum sheet 20" x 24"		1	16		3,475	805		4,280	5,025
3800	Square, minimum sheet 20" x 24"		1	16		3,125	805		3,930	4,625
3900	Maximum sheet 50" x 56"		.80	20		4,850	1,000		5,850	6,850
4500	Combination tank sink, tray sink, washers, with									
4510	Dry side tables, average	Q-1	.45	35.556	Ea.	10,600	1,775		12,375	14,400

11 31 Residential Appliances

11 31 13 – Residential Kitchen Appliances

11 31 13.13 Cooking Equipment

		Crew	Daily Output	Labor-Hours	Unit	Material	Labor	Equipment	Total	Total Incl O&P
0010	COOKING EQUIPMENT									
0020	Cooking range, 30" free standing, 1 oven, minimum	2 Clab	10	1.600	Ea.	435	56.50		491.50	570
0050	Maximum		4	4		1,700	142		1,842	2,100
0150	2 oven, minimum		10	1.600		1,175	56.50		1,231.50	1,375
0200	Maximum		10	1.600		2,525	56.50		2,581.50	2,875

11 31 13.23 Refrigeration Equipment

		Crew	Daily Output	Labor-Hours	Unit	Material	Labor	Equipment	Total	Total Incl O&P
0010	REFRIGERATION EQUIPMENT									
5200	Icemaker, automatic, 20 lb. per day	1 Plum	7	1.143	Ea.	935	64		999	1,125
5350	51 lb. per day	"	2	4	"	1,450	223		1,673	1,925

11 31 13.33 Kitchen Cleaning Equipment

		Crew	Daily Output	Labor-Hours	Unit	Material	Labor	Equipment	Total	Total Incl O&P
0010	KITCHEN CLEANING EQUIPMENT									
2750	Dishwasher, built-in, 2 cycles, minimum	L-1	4	4	Ea.	227	216		443	575
2800	Maximum		2	8		430	435		865	1,125
2950	4 or more cycles, minimum		4	4		340	216		556	700
2960	Average		4	4		455	216		671	825
3000	Maximum		2	8		1,075	435		1,510	1,825

11 31 13.43 Waste Disposal Equipment

		Crew	Daily Output	Labor-Hours	Unit	Material	Labor	Equipment	Total	Total Incl O&P
0010	WASTE DISPOSAL EQUIPMENT									
3300	Garbage disposal, sink type, minimum	L-1	10	1.600	Ea.	86	86.50		172.50	225
3350	Maximum	"	10	1.600	"	217	86.50		303.50	370

11 31 13.53 Kitchen Ventilation Equipment

		Crew	Daily Output	Labor-Hours	Unit	Material	Labor	Equipment	Total	Total Incl O&P
0010	KITCHEN VENTILATION EQUIPMENT									
4150	Hood for range, 2 speed, vented, 30" wide, minimum	L-3	5	3.200	Ea.	63	156		219	310
4200	Maximum		3	5.333		760	261		1,021	1,225
4300	42" wide, minimum		5	3.200		163	156		319	420
4330	Custom		5	3.200		1,725	156		1,881	2,150
4350	Maximum		3	5.333		2,100	261		2,361	2,725
4500	For ventless hood, 2 speed, add					18.45			18.45	20.50
4650	For vented 1 speed, deduct from maximum					48.50			48.50	53.50

11 31 23 – Residential Laundry Appliances

11 31 23.13 Washers

		Crew	Daily Output	Labor-Hours	Unit	Material	Labor	Equipment	Total	Total Incl O&P
0010	WASHERS									
5000	Residential, 4 cycle, average	1 Plum	3	2.667	Ea.	855	149		1,004	1,175
6650	Washing machine, automatic, minimum		3	2.667		480	149		629	750

11 31 Residential Appliances

11 31 23 – Residential Laundry Appliances

11 31 23.13 Washers		Crew	Daily Output	Labor-Hours	Unit	Material	2013 Bare Costs Labor	Equipment	Total	Total Incl O&P
6700	Maximum	1 Plum	1	8	Ea.	1,450	445		1,895	2,275

11 31 23.23 Dryers

		Crew	Daily Output	Labor-Hours	Unit	Material	Labor	Equipment	Total	Total Incl O&P
0010	**DRYERS**									
0500	Gas fired residential, 16 lb. capacity, average	1 Plum	3	2.667	Ea.	655	149		804	945
7450	Vent kits for dryers	1 Carp	10	.800	"	39.50	36		75.50	99

11 31 33 – Miscellaneous Residential Appliances

11 31 33.13 Sump Pumps

		Crew	Daily Output	Labor-Hours	Unit	Material	Labor	Equipment	Total	Total Incl O&P
0010	**SUMP PUMPS**									
6400	Cellar drainer, pedestal, 1/3 H.P., molded PVC base	1 Plum	3	2.667	Ea.	135	149		284	375
6450	Solid brass	"	2	4	"	289	223		512	655
6460	Sump pump, see also Section 22 14 29.16									

11 31 33.23 Water Heaters

		Crew	Daily Output	Labor-Hours	Unit	Material	Labor	Equipment	Total	Total Incl O&P
0010	**WATER HEATERS**									
6900	Electric, glass lined, 30 gallon, minimum	L-1	5	3.200	Ea.	625	173		798	950
6950	Maximum		3	5.333		870	289		1,159	1,400
7100	80 gallon, minimum		2	8		1,125	435		1,560	1,900
7150	Maximum	↓	1	16		1,575	865		2,440	3,025
7180	Gas, glass lined, 30 gallon, minimum	2 Plum	5	3.200		805	179		984	1,150
7220	Maximum		3	5.333		1,125	298		1,423	1,675
7260	50 gallon, minimum		2.50	6.400		845	355		1,200	1,475
7300	Maximum	↓	1.50	10.667	↓	1,175	595		1,770	2,175
7310	Water heater, see also Section 22 33 30.13									

11 31 33.43 Air Quality

		Crew	Daily Output	Labor-Hours	Unit	Material	Labor	Equipment	Total	Total Incl O&P
0010	**AIR QUALITY**									
2450	Dehumidifier, portable, automatic, 15 pint	1 Elec	4	2	Ea.	150	105		255	320
2550	40 pint					229			229	252
3550	Heater, electric, built-in, 1250 watt, ceiling type, minimum	1 Elec	4	2		106	105		211	273
3600	Maximum		3	2.667		172	140		312	400
3700	Wall type, minimum		4	2		170	105		275	345
3750	Maximum		3	2.667		183	140		323	410
3900	1500 watt wall type, with blower		4	2		170	105		275	345
3950	3000 watt	↓	3	2.667		350	140		490	595
4850	Humidifier, portable, 8 gallons per day					173			173	191
5000	15 gallons per day				↓	208			208	229

11 41 Foodservice Storage Equipment

11 41 13 – Refrigerated Food Storage Cases

11 41 13.20 Refrigerated Food Storage Equipment

		Crew	Daily Output	Labor-Hours	Unit	Material	Labor	Equipment	Total	Total Incl O&P
0010	**REFRIGERATED FOOD STORAGE EQUIPMENT**									
2350	Cooler, reach-in, beverage, 6' long	Q-1	6	2.667	Ea.	3,475	134		3,609	4,000
4300	Freezers, reach-in, 44 C.F.		4	4		3,500	201		3,701	4,150
4500	68 C.F.	↓	3	5.333	↓	4,600	268		4,868	5,475

11 44 Food Cooking Equipment

11 44 13 – Commercial Ranges

11 44 13.10 Cooking Equipment		Crew	Daily Output	Labor-Hours	Unit	Material	Labor	Equipment	Total	Total Incl O&P
						2013 Bare Costs				
0010	COOKING EQUIPMENT									
0020	Bake oven, gas, one section	Q-1	8	2	Ea.	5,050	100		5,150	5,700
0300	Two sections		7	2.286		8,675	115		8,790	9,725
0600	Three sections		6	2.667		14,300	134		14,434	15,900
0900	Electric convection, single deck	L-7	4	7		6,950	305		7,255	8,125
6350	Kettle, w/steam jacket, tilting, w/positive lock, SS, 20 gallons		7	4		7,550	173		7,723	8,575
6600	60 gallons		6	4.667		10,900	202		11,102	12,300

11 47 Ice Machines

11 47 10 – Commercial Ice Machines

11 47 10.10 Commercial Ice Equipment		Crew	Daily Output	Labor-Hours	Unit	Material	Labor	Equipment	Total	Total Incl O&P
0010	COMMERCIAL ICE EQUIPMENT									
5800	Ice cube maker, 50 pounds per day	Q-1	6	2.667	Ea.	1,575	134		1,709	1,925
6050	500 pounds per day	"	4	4	"	2,800	201		3,001	3,400

11 48 Cleaning and Disposal Equipment

11 48 13 – Commercial Dishwashers

11 48 13.10 Dishwashers		Crew	Daily Output	Labor-Hours	Unit	Material	Labor	Equipment	Total	Total Incl O&P
0010	DISHWASHERS									
2700	Dishwasher, commercial, rack type									
2720	10 to 12 racks per hour	Q-1	3.20	5	Ea.	4,475	251		4,726	5,300
2730	Energy star rated, 35 to 40 racks/hour G		1.30	12.308		4,900	620		5,520	6,325
2740	50 to 60 racks/hour G		1.30	12.308		10,700	620		11,320	12,600
2800	Automatic, 190 to 230 racks per hour	L-6	.35	34.286		14,400	1,875		16,275	18,700
2820	235 to 275 racks per hour		.25	48		34,400	2,625		37,025	41,900
2840	8,750 to 12,500 dishes per hour		.10	120		53,500	6,550		60,050	68,500

11 53 Laboratory Equipment

11 53 13 – Laboratory Fume Hoods

11 53 13.13 Recirculating Laboratory Fume Hoods		Crew	Daily Output	Labor-Hours	Unit	Material	Labor	Equipment	Total	Total Incl O&P
0010	RECIRCULATING LABORATORY FUME HOODS									
0670	Service fixtures, average				Ea.	233			233	256
0680	For sink assembly with hot and cold water, add	1 Plum	1.40	5.714	"	725	320		1,045	1,275

11 53 19 – Laboratory Sterilizers

11 53 19.13 Sterilizers		Crew	Daily Output	Labor-Hours	Unit	Material	Labor	Equipment	Total	Total Incl O&P
0010	STERILIZERS									
0700	Glassware washer, undercounter, minimum	L-1	1.80	8.889	Ea.	6,175	480		6,655	7,525
0710	Maximum	"	1	16		12,700	865		13,565	15,300
1850	Utensil washer-sanitizer	1 Plum	2	4		11,100	223		11,323	12,500

11 53 23 – Laboratory Refrigerators

11 53 23.13 Refrigerators		Crew	Daily Output	Labor-Hours	Unit	Material	Labor	Equipment	Total	Total Incl O&P
0010	REFRIGERATORS									
1200	Blood bank, 28.6 C.F. emergency signal				Ea.	9,325			9,325	10,300
1210	Reach-in, 16.9 C.F.				"	8,325			8,325	9,175

11 53 Laboratory Equipment

11 53 33 – Emergency Safety Appliances

11 53 33.13 Emergency Equipment		Crew	Daily Output	Labor-Hours	Unit	Material	2013 Bare Costs Labor	Equipment	Total	Total Incl O&P
0010	EMERGENCY EQUIPMENT									
1400	Safety equipment, eye wash, hand held				Ea.	405			405	445
1450	Deluge shower				"	760			760	840

11 53 43 – Service Fittings and Accessories

11 53 43.13 Fittings		Crew	Daily Output	Labor-Hours	Unit	Material	Labor	Equipment	Total	Total Incl O&P
0010	FITTINGS									
1600	Sink, one piece plastic, flask wash, hose, free standing	1 Plum	1.60	5	Ea.	1,925	279		2,204	2,550
1610	Epoxy resin sink, 25" x 16" x 10"	"	2	4	"	218	223		441	575
8000	Alternate pricing method: as percent of lab furniture									
8050	Installation, not incl. plumbing & duct work				% Furn.				22%	22%
8100	Plumbing, final connections, simple system								10%	10%
8110	Moderately complex system								15%	15%
8120	Complex system								20%	20%
8150	Electrical, simple system								10%	10%
8160	Moderately complex system								20%	20%
8170	Complex system				↓				35%	35%

11 71 Medical Sterilizing Equipment

11 71 10 – Medical Sterilizers & Distillers

11 71 10.10 Sterilizers and Distillers		Crew	Daily Output	Labor-Hours	Unit	Material	Labor	Equipment	Total	Total Incl O&P
0010	STERILIZERS AND DISTILLERS									
0700	Distiller, water, steam heated, 50 gal. capacity	1 Plum	1.40	5.714	Ea.	19,000	320		19,320	21,400
5600	Sterilizers, floor loading, 26" x 62" x 42", single door, steam					121,000			121,000	133,000
5650	Double door, steam					204,000			204,000	224,000
5800	General purpose, 20" x 20" x 38", single door					12,200			12,200	13,400
6000	Portable, counter top, steam, minimum					3,300			3,300	3,625
6020	Maximum					5,850			5,850	6,425
6050	Portable, counter top, gas, 17" x 15" x 32-1/2"					39,300			39,300	43,200
6150	Manual washer/sterilizer, 16" x 16" x 26"	1 Plum	2	4	↓	54,000	223		54,223	59,500
6200	Steam generators, electric 10 kW to 180 kW, freestanding									
6250	Minimum	1 Elec	3	2.667	Ea.	8,675	140		8,815	9,750
6300	Maximum	"	.70	11.429		28,800	600		29,400	32,600
8200	Bed pan washer-sanitizer	1 Plum	2	4	↓	7,425	223		7,648	8,475

11 73 Patient Care Equipment

11 73 10 – Patient Treatment Equipment

11 73 10.10 Treatment Equipment		Crew	Daily Output	Labor-Hours	Unit	Material	Labor	Equipment	Total	Total Incl O&P
0010	TREATMENT EQUIPMENT									
1800	Heat therapy unit, humidified, 26" x 78" x 28"				Ea.	3,500			3,500	3,850
2100	Hubbard tank with accessories, stainless steel,									
2110	125 GPM at 45 psi water pressure				Ea.	26,800			26,800	29,500
2150	For electric overhead hoist, add					2,925			2,925	3,225
3600	Paraffin bath, 126°F, auto controlled					1,100			1,100	1,200
4600	Station, dietary, medium, with ice					16,500			16,500	18,200
8400	Whirlpool bath, mobile, sst, 18" x 24" x 60"					4,850			4,850	5,325
8450	Fixed, incl. mixing valves	1 Plum	2	4	↓	9,575	223		9,798	10,800

11 74 Dental Equipment

11 74 10 – Dental Office Equipment

11 74 10.10 Diagnostic and Treatment Equipment		Crew	Daily Output	Labor-Hours	Unit	Material	Labor	Equipment	Total	Total Incl O&P
						2013 Bare Costs				
0010	**DIAGNOSTIC AND TREATMENT EQUIPMENT**									
0020	Central suction system, minimum	1 Plum	1.20	6.667	Ea.	1,550	370		1,920	2,250
0100	Maximum	"	.90	8.889		4,250	495		4,745	5,425
0300	Air compressor, minimum	1 Skwk	.80	10		2,950	460		3,410	3,975
0400	Maximum		.50	16		7,750	740		8,490	9,675
0600	Chair, electric or hydraulic, minimum		.50	16		2,200	740		2,940	3,550
0700	Maximum		.25	32		8,225	1,475		9,700	11,300
2000	Light, ceiling mounted, minimum		8	1		1,125	46		1,171	1,325
2100	Maximum		8	1		1,900	46		1,946	2,175
2200	Unit light, minimum	2 Skwk	5.33	3.002		690	139		829	970
2210	Maximum		5.33	3.002		1,425	139		1,564	1,775
2220	Track light, minimum		3.20	5		1,600	231		1,831	2,100
2230	Maximum		3.20	5		2,775	231		3,006	3,400
2300	Sterilizers, steam portable, minimum					1,150			1,150	1,250
2350	Maximum					10,400			10,400	11,400
2600	Steam, institutional					3,250			3,250	3,550
2650	Dry heat, electric, portable, 3 trays					1,225			1,225	1,325

11 76 Operating Room Equipment

11 76 10 – Operating Room Equipment

11 76 10.10 Surgical Equipment

		Crew	Daily Output	Labor-Hours	Unit	Material	Labor	Equipment	Total	Total Incl O&P
0010	**SURGICAL EQUIPMENT**									
5000	Scrub, surgical, stainless steel, single station, minimum	1 Plum	3	2.667	Ea.	4,375	149		4,524	5,025
5100	Maximum				"	6,375			6,375	7,000

11 78 Mortuary Equipment

11 78 13 – Mortuary Refrigerators

11 78 13.10 Mortuary and Autopsy Equipment

		Crew	Daily Output	Labor-Hours	Unit	Material	Labor	Equipment	Total	Total Incl O&P
0010	**MORTUARY AND AUTOPSY EQUIPMENT**									
0015	Autopsy table, standard	1 Plum	1	8	Ea.	9,175	445		9,620	10,800
0020	Deluxe	"	.60	13.333	"	15,600	745		16,345	18,200

11 91 Religious Equipment

11 91 13 – Baptisteries

11 91 13.10 Baptistry

		Crew	Daily Output	Labor-Hours	Unit	Material	Labor	Equipment	Total	Total Incl O&P
0010	**BAPTISTRY**									
0150	Fiberglass, 3′-6″ deep, x 13′-7″ long,									
0160	steps at both ends, incl. plumbing, minimum	L-8	1	20	Ea.	5,150	940		6,090	7,100
0200	Maximum	"	.70	28.571		8,375	1,350		9,725	11,300
0250	Add for filter, heater and lights					1,825			1,825	2,025

11 91 23 – Sanctuary Equipment

11 91 23.10 Sanctuary Furnishings

		Crew	Daily Output	Labor-Hours	Unit	Material	Labor	Equipment	Total	Total Incl O&P
0010	**SANCTUARY FURNISHINGS**									
0020	Altar, wood, custom design, plain	1 Carp	1.40	5.714	Ea.	2,300	257		2,557	2,925
0050	Deluxe	"	.20	40	"	11,100	1,800		12,900	15,000

Estimating Tips

General

- The items in this division are usually priced per square foot or each. Most of these items are purchased by the owner and installed by the contractor. Do not assume the items in Division 12 will be purchased and installed by the contractor. Check the specifications for responsibilities and include receiving, storage, installation, and mechanical and electrical hookups in the appropriate divisions.
- Some items in this division require some type of support system that is not usually furnished with the item. Examples of these systems include blocking for the attachment of casework and heavy drapery rods. The required blocking must be added to the estimate in the appropriate division.

Reference Numbers

Reference numbers are shown in shaded boxes at the beginning of some major classifications. These numbers refer to related items in the Reference Section. The reference information may be an estimating procedure, an alternate pricing method, or technical information.

Note: Not all subdivisions listed here necessarily appear in this publication.

12 32 Manufactured Wood Casework

12 32 23 – Hardwood Casework

12 32 23.10 Manufactured Wood Casework, Stock Units		Crew	Daily Output	Labor-Hours	Unit	2013 Bare Costs Material	Labor	Equipment	Total	Total Incl O&P
0010	**MANUFACTURED WOOD CASEWORK, STOCK UNITS**									
0700	Kitchen base cabinets, hardwood, not incl. counter tops,									
0710	24" deep, 35" high, prefinished									
0800	One top drawer, one door below, 12" wide	2 Carp	24.80	.645	Ea.	240	29		269	310
0840	18" wide		23.30	.687		272	31		303	345
0880	24" wide		22.30	.717		330	32		362	410
1000	Four drawers, 12" wide		24.80	.645		252	29		281	325
1040	18" wide		23.30	.687		282	31		313	360
1060	24" wide		22.30	.717		300	32		332	380
1200	Two top drawers, two doors below, 27" wide		22	.727		355	32.50		387.50	440
1260	36" wide		20.30	.788		425	35.50		460.50	520
1300	48" wide		18.90	.847		480	38		518	590
1500	Range or sink base, two doors below, 30" wide		21.40	.748		325	33.50		358.50	410
1540	36" wide		20.30	.788		370	35.50		405.50	460
1580	48" wide		18.90	.847		405	38		443	505
1800	For sink front units, deduct					150			150	165
9000	For deluxe models of all cabinets, add					40%				
9500	For custom built in place, add					25%	10%			
9558	Rule of thumb, kitchen cabinets not including									
9560	appliances & counter top, minimum	2 Carp	30	.533	L.F.	160	24		184	213
9600	Maximum	"	25	.640	"	345	28.50		373.50	425

12 32 23.30 Manufactured Wood Casework Vanities

		Crew	Daily Output	Labor-Hours	Unit	Material	Labor	Equipment	Total	Total Incl O&P
0010	**MANUFACTURED WOOD CASEWORK VANITIES**									
8000	Vanity bases, 2 doors, 30" high, 21" deep, 24" wide	2 Carp	20	.800	Ea.	288	36		324	370
8050	30" wide		16	1		345	45		390	450
8100	36" wide		13.33	1.200		335	54		389	450
8150	48" wide		11.43	1.400		440	63		503	575
9000	For deluxe models of all vanities, add to above					40%				
9500	For custom built in place, add to above					25%	10%			

Estimating Tips

General

- The items and systems in this division are usually estimated, purchased, supplied, and installed as a unit by one or more subcontractors. The estimator must ensure that all parties are operating from the same set of specifications and assumptions, and that all necessary items are estimated and will be provided. Many times the complex items and systems are covered, but the more common ones, such as excavation or a crane, are overlooked for the very reason that everyone assumes nobody could miss them. The estimator should be the central focus and be able to ensure that all systems are complete.
- Another area where problems can develop in this division is at the interface between systems. The estimator must ensure, for instance, that anchor bolts, nuts, and washers are estimated and included for the air-supported structures and pre-engineered buildings to be bolted to their foundations. Utility supply is a common area where essential items or pieces of equipment can be missed or overlooked due to the fact that each subcontractor may feel it is another's responsibility. The estimator should also be aware of certain items which may be supplied as part of a package but installed by others, and ensure that the installing contractor's estimate includes the cost of installation. Conversely, the estimator must also ensure that items are not costed by two different subcontractors, resulting in an inflated overall estimate.

13 30 00 Special Structures

- The foundations and floor slab, as well as rough mechanical and electrical, should be estimated, as this work is required for the assembly and erection of the structure. Generally, as noted in the book, the pre-engineered building comes as a shell. Pricing is based on the size and structural design parameters stated in the reference section. Additional features, such as windows and doors with their related structural framing, must also be included by the estimator. Here again, the estimator must have a clear understanding of the scope of each portion of the work and all the necessary interfaces.

Reference Numbers

Reference numbers are shown in shaded boxes at the beginning of some major classifications. These numbers refer to related items in the Reference Section. The reference information may be an estimating procedure, an alternate pricing method, or technical information.

Note: Not all subdivisions listed here necessarily appear in this publication.

13 11 Swimming Pools

13 11 13 – Below-Grade Swimming Pools

13 11 13.50 Swimming Pools		Crew	Daily Output	Labor-Hours	Unit	Material	Labor	Equipment	Total	Total Incl O&P
						2013 Bare Costs				
0010	**SWIMMING POOLS** Residential in-ground, vinyl lined, concrete									
0020	Sides including equipment, sand bottom	B-52	300	.187	SF Surf	19.30	7.75	1.98	29.03	35.50
0100	Metal or polystyrene sides R131113-20	B-14	410	.117		16.15	4.39	.90	21.44	25.50
0200	Add for vermiculite bottom					1.23			1.23	1.35
0500	Gunite bottom and sides, white plaster finish									
0600	12′ x 30′ pool	B-52	145	.386	SF Surf	36	16	4.09	56.09	68.50
0720	16′ x 32′ pool		155	.361		32.50	14.95	3.83	51.28	62.50
0750	20′ x 40′ pool		250	.224		29	9.30	2.37	40.67	49
0810	Concrete bottom and sides, tile finish									
0820	12′ x 30′ pool	B-52	80	.700	SF Surf	36.50	29	7.40	72.90	92.50
0830	16′ x 32′ pool		95	.589		30	24.50	6.25	60.75	77.50
0840	20′ x 40′ pool		130	.431		24	17.85	4.57	46.42	59
1100	Motel, gunite with plaster finish, incl. medium									
1150	capacity filtration & chlorination	B-52	115	.487	SF Surf	44.50	20	5.15	69.65	85
1200	Municipal, gunite with plaster finish, incl. high									
1250	capacity filtration & chlorination	B-52	100	.560	SF Surf	57	23	5.95	85.95	105
1350	Add for formed gutters				L.F.	84			84	92.50
1360	Add for stainless steel gutters				"	249			249	274
1700	Filtration and deck equipment only, as % of total				Total				20%	20%
1800	Deck equipment, rule of thumb, 20′ x 40′ pool				SF Pool				1.18	1.30
1900	5000 S.F. pool				"				1.73	1.90
3000	Painting pools, preparation + 3 coats, 20′ x 40′ pool, epoxy	2 Pord	.33	48.485	Total	1,725	1,875		3,600	4,725
3100	Rubber base paint, 18 gallons	"	.33	48.485		1,275	1,875		3,150	4,225
3500	42′ x 82′ pool, 75 gallons, epoxy paint	3 Pord	.14	171		7,300	6,625		13,925	18,000
3600	Rubber base paint	"	.14	171		5,275	6,625		11,900	15,800

13 11 46 – Swimming Pool Accessories

13 11 46.50 Swimming Pool Equipment		Crew	Daily Output	Labor-Hours	Unit	Material	Labor	Equipment	Total	Total Incl O&P
0010	**SWIMMING POOL EQUIPMENT**									
0020	Diving stand, stainless steel, 3 meter	2 Carp	.40	40	Ea.	11,700	1,800		13,500	15,600
0300	1 meter	"	2.70	5.926		6,800	266		7,066	7,875
2100	Lights, underwater, 12 volt, with transformer, 300 watt	1 Elec	1	8		290	420		710	945
2200	110 volt, 500 watt, standard		1	8		242	420		662	890
2400	Low water cutoff type		1	8		281	420		701	935
2800	Heaters, see Section 23 52 28.10									

13 17 Tubs and Pools

13 17 13 – Hot Tubs

13 17 13.10 Redwood Hot Tub System		Crew	Daily Output	Labor-Hours	Unit	Material	Labor	Equipment	Total	Total Incl O&P
0010	**REDWOOD HOT TUB SYSTEM**									
7050	4′ diameter x 4′ deep	Q-1	1	16	Ea.	2,925	805		3,730	4,425
7100	5′ diameter x 4′ deep		1	16		3,700	805		4,505	5,275
7150	6′ diameter x 4′ deep		.80	20		4,500	1,000		5,500	6,475
7200	8′ diameter x 4′ deep		.80	20		6,600	1,000		7,600	8,775

13 17 33 – Whirlpool Tubs

13 17 33.10 Whirlpool Bath		Crew	Daily Output	Labor-Hours	Unit	Material	Labor	Equipment	Total	Total Incl O&P
0010	**WHIRLPOOL BATH**									
6000	Whirlpool, bath with vented overflow, molded fiberglass									
6100	66″ x 36″ x 24″	Q-1	1	16	Ea.	3,325	805		4,130	4,850
6400	72″ x 36″ x 21″		1	16		1,775	805		2,580	3,150
6500	60″ x 34″ x 21″		1	16		1,700	805		2,505	3,075

13 17 Tubs and Pools

13 17 33 – Whirlpool Tubs

13 17 33.10 Whirlpool Bath		Crew	Daily Output	Labor-Hours	Unit	Material	2013 Bare Costs Labor	Equipment	Total	Total Incl O&P
6600	72" x 42" x 23"	Q-1	1	16	Ea.	2,050	805		2,855	3,450
6710	For color add					10%				
6711	For designer colors and trim add				↓	25%				

13 24 Special Activity Rooms

13 24 16 – Saunas

13 24 16.50 Saunas and Heaters

		Crew	Daily Output	Labor-Hours	Unit	Material	Labor	Equipment	Total	Total Incl O&P
0010	**SAUNAS AND HEATERS**									
0020	Prefabricated, incl. heater & controls, 7' high, 6' x 4', C/C	L-7	2.20	12.727	Ea.	4,900	550		5,450	6,225
0050	6' x 4', C/P		2	14		4,450	605		5,055	5,825
0400	6' x 5', C/C		2	14		5,400	605		6,005	6,850
0450	6' x 5', C/P		2	14		4,925	605		5,530	6,350
0600	6' x 6', C/C		1.80	15.556		5,725	675		6,400	7,325
0650	6' x 6', C/P		1.80	15.556		5,225	675		5,900	6,775
0800	6' x 9', C/C		1.60	17.500		7,125	755		7,880	8,975
0850	6' x 9', C/P		1.60	17.500		6,450	755		7,205	8,250
1000	8' x 12', C/C		1.10	25.455		11,400	1,100		12,500	14,200
1050	8' x 12', C/P		1.10	25.455		10,300	1,100		11,400	13,100
1200	8' x 8', C/C		1.40	20		8,425	865		9,290	10,600
1250	8' x 8', C/P		1.40	20		7,800	865		8,665	9,900
1400	8' x 10', C/C		1.20	23.333		9,350	1,000		10,350	11,900
1450	8' x 10', C/P		1.20	23.333		8,550	1,000		9,550	11,000
1600	10' x 12', C/C		1	28		12,000	1,200		13,200	15,100
1650	10' x 12', C/P	↓	1	28		10,800	1,200		12,000	13,700
2500	Heaters only (incl. above), wall mounted, to 200 C.F.					630			630	690
2750	To 300 C.F.					830			830	915
3000	Floor standing, to 720 C.F., 10,000 watts, w/controls	1 Elec	3	2.667		2,100	140		2,240	2,500
3250	To 1,000 C.F., 16,000 watts	"	3	2.667	↓	3,275	140		3,415	3,825

13 24 26 – Steam Baths

13 24 26.50 Steam Baths and Components

		Crew	Daily Output	Labor-Hours	Unit	Material	Labor	Equipment	Total	Total Incl O&P
0010	**STEAM BATHS AND COMPONENTS**									
0020	Heater, timer & head, single, to 140 C.F.	1 Plum	1.20	6.667	Ea.	2,200	370		2,570	2,975
0500	To 300 C.F.		1.10	7.273		2,500	405		2,905	3,350
1000	Commercial size, with blow-down assembly, to 800 C.F.		.90	8.889		9,000	495		9,495	10,600
1500	To 2500 C.F.	↓	.80	10		11,400	560		11,960	13,300
2000	Multiple, motels, apts., 2 baths, w/blow-down assm., 500 C.F.	Q-1	1.30	12.308		7,750	620		8,370	9,450
2500	4 baths	"	.70	22.857		9,050	1,150		10,200	11,700
2700	Conversion unit for residential tub, including door				↓	4,300			4,300	4,725

13 34 Fabricated Engineered Structures

13 34 23 – Fabricated Structures

13 34 23.10 Comfort Stations		Crew	Daily Output	Labor-Hours	Unit	Material	2013 Bare Costs Labor	Equipment	Total	Total Incl O&P
0010	**COMFORT STATIONS** Prefab., stock, w/doors, windows & fixt.									
0100	Not incl. interior finish or electrical									
0300	Mobile, on steel frame, 2 unit				S.F.	188			188	207
0350	7 unit					249			249	274
0400	Permanent, including concrete slab, 2 unit	B-12J	50	.320		249	13.50	17.70	280.20	315
0500	6 unit	"	43	.372	▼	191	15.70	20.50	227.20	258
0600	Alternate pricing method, mobile, 2 fixture				Fixture	5,575			5,575	6,150
0650	7 fixture					8,975			8,975	9,875
0700	Permanent, 2 unit	B-12J	.70	22.857		20,500	965	1,275	22,740	25,500
0750	6 unit	"	.50	32	▼	17,400	1,350	1,775	20,525	23,200

13 42 Building Modules

13 42 63 – Detention Cell Modules

13 42 63.16 Steel Detention Cell Modules

		Crew	Daily Output	Labor-Hours	Unit	Material	Labor	Equipment	Total	Total Incl O&P
0010	**STEEL DETENTION CELL MODULES**									
2000	Cells, prefab., 5′ to 6′ wide, 7′ to 8′ high, 7′ to 8′ deep,									
2010	bar front, cot, not incl. plumbing	E-4	1.50	21.333	Ea.	9,650	1,075	96	10,821	12,600

Estimating Tips

General

- Many products in Division 14 will require some type of support or blocking for installation not included with the item itself. Examples are supports for conveyors or tube systems, attachment points for lifts, and footings for hoists or cranes. Add these supports in the appropriate division.

14 10 00 Dumbwaiters
14 20 00 Elevators

- Dumbwaiters and elevators are estimated and purchased in a method similar to buying a car. The manufacturer has a base unit with standard features. Added to this base unit price will be whatever options the owner or specifications require. Increased load capacity, additional vertical travel, additional stops, higher speed, and cab finish options are items to be considered. When developing an estimate for dumbwaiters and elevators, remember that some items needed by the installers may have to be included as part of the general contract.

  Examples are:
  - — shaftway
  - — rail support brackets
  - — machine room
  - — electrical supply
  - — sill angles
  - — electrical connections
  - — pits
  - — roof penthouses
  - — pit ladders

Check the job specifications and drawings before pricing.

- Installation of elevators and handicapped lifts in historic structures can require significant additional costs. The associated structural requirements may involve cutting into and repairing finishes, mouldings, flooring, etc. The estimator must account for these special conditions.

14 30 00 Escalators and Moving Walks

- Escalators and moving walks are specialty items installed by specialty contractors. There are numerous options associated with these items. For specific options, contact a manufacturer or contractor. In a method similar to estimating dumbwaiters and elevators, you should verify the extent of general contract work and add items as necessary.

14 40 00 Lifts
14 90 00 Other Conveying Equipment

- Products such as correspondence lifts, chutes, and pneumatic tube systems, as well as other items specified in this subdivision, may require trained installers. The general contractor might not have any choice as to who will perform the installation or when it will be performed. Long lead times are often required for these products, making early decisions in scheduling necessary.

Reference Numbers

Reference numbers are shown in shaded boxes at the beginning of some major classifications. These numbers refer to related items in the Reference Section. The reference information may be an estimating procedure, an alternate pricing method, or technical information.

Note: Not all subdivisions listed here necessarily appear in this publication.

14 45 Vehicle Lifts

14 45 10 – Hydraulic Vehicle Lifts

14 45 10.10 Hydraulic Lifts		Crew	Daily Output	Labor-Hours	Unit	Material	Labor	Equipment	Total	Total Incl O&P
						2013 Bare Costs				
0010	**HYDRAULIC LIFTS**									
2200	Single post, 8000 lb. capacity	L-4	.40	60	Ea.	5,500	2,525		8,025	9,950
2810	Double post, 6000 lb. capacity		2.67	8.989		7,750	380		8,130	9,100
2815	9000 lb. capacity		2.29	10.480		18,400	440		18,840	20,900
2820	15,000 lb. capacity		2	12		20,800	505		21,305	23,600
2822	26,000 lb. capacity		1.80	13.333		29,200	560		29,760	33,100
2825	30,000 lb. capacity		1.60	15		45,900	630		46,530	51,500
2830	Ramp style, 4 post, 25,000 lb. capacity		2	12		18,000	505		18,505	20,600
2835	35,000 lb. capacity		1	24		83,500	1,000		84,500	93,500
2840	50,000 lb. capacity		1	24		93,500	1,000		94,500	104,500
2845	75,000 lb. capacity	↓	1	24		108,500	1,000		109,500	121,000
2850	For drive thru tracks, add, minimum					1,150			1,150	1,275
2855	Maximum					1,975			1,975	2,175
2860	Ramp extensions, 3′ (set of 2)					945			945	1,050
2865	Rolling jack platform					3,300			3,300	3,625
2870	Electric/hydraulic jacking beam					8,825			8,825	9,700
2880	Scissor lift, portable, 6000 lb. capacity				↓	8,650			8,650	9,525

Estimating Tips

Pipe for fire protection and all uses is located in Subdivisions 21 11 13 and 22 11 13.

The labor adjustment factors listed in Subdivision 22 01 02.20 also apply to Division 21.

Many, but not all, areas in the U.S. require backflow protection in the fire system. It is advisable to check local building codes for specific requirements.

For your reference, the following is a list of the most applicable Fire Codes and Standards which may be purchased from the NFPA, 1 Batterymarch Park, Quincy, MA 02169-7471.

NFPA 1: Uniform Fire Code

NFPA 10: Portable Fire Extinguishers

NFPA 11: Low-, Medium-, and High-Expansion Foam

NFPA 12: Carbon Dioxide Extinguishing Systems (Also companion 12A)

NFPA 13: Installation of Sprinkler Systems (Also companion 13D, 13E, and 13R)

NFPA 14: Installation of Standpipe and Hose Systems

NFPA 15: Water Spray Fixed Systems for Fire Protection

NFPA 16: Installation of Foam-Water Sprinkler and Foam-Water Spray Systems

NFPA 17: Dry Chemical Extinguishing Systems (Also companion 17A)

NFPA 18: Wetting Agents

NFPA 20: Installation of Stationary Pumps for Fire Protection

NFPA 22: Water Tanks for Private Fire Protection

NFPA 24: Installation of Private Fire Service Mains and their Appurtenances

NFPA 25: Inspection, Testing and Maintenance of Water-Based Fire Protection

Reference Numbers

Reference numbers are shown in shaded boxes at the beginning of some major classifications. These numbers refer to related items in the Reference Section. The reference information may be an estimating procedure, an alternate pricing method, or technical information.

Note: Not all subdivisions listed here necessarily appear in this publication.

Note: **Trade Service,** *in part, has been used as a reference source for some of the material prices used in Division 21.*

21 05 Common Work Results for Fire Suppression

21 05 23 – General-Duty Valves for Water-Based Fire-Suppression Piping

	21 05 23.50 General-Duty Valves	Crew	Daily Output	Labor-Hours	Unit	2013 Bare Costs Material	Labor	Equipment	Total	Total Incl O&P
0010	**GENERAL-DUTY VALVES**, for water-based fire suppression									
6200	Valves and components									
6210	Alarm, includes									
6220	retard chamber, trim, gauges, alarm line strainer									
6260	3″ size	Q-12	3	5.333	Ea.	1,475	262		1,737	2,025
6280	4″ size	″	2	8		1,525	395		1,920	2,275
6300	6″ size	Q-13	4	8		1,700	415		2,115	2,500
6320	8″ size	″	3	10.667		2,050	555		2,605	3,075
6500	Check, swing, C.I. body, brass fittings, auto. ball drip									
6520	4″ size	Q-12	3	5.333	Ea.	296	262		558	720
6540	6″ size	Q-13	4	8		550	415		965	1,225
6580	8″ size	″	3	10.667		1,050	555		1,605	1,975
6800	Check, wafer, butterfly type, C.I. body, bronze fittings									
6820	4″ size	Q-12	4	4	Ea.	435	197		632	775
6840	6″ size	Q-13	5.50	5.818		750	305		1,055	1,275
6860	8″ size		5	6.400		1,350	335		1,685	1,975
6880	10″ size		4.50	7.111		2,050	370		2,420	2,800
8800	Flow control valve, includes trim and gauges, 2″ size	Q-12	2	8		4,225	395		4,620	5,250
8820	3″ size	″	1.50	10.667		4,650	525		5,175	5,900
8840	4″ size	Q-13	2.80	11.429		5,200	595		5,795	6,625
8860	6″ size	″	2	16		5,850	835		6,685	7,700
9200	Pressure operated relief valve, brass body	1 Spri	18	.444		520	24.50		544.50	605
9600	Waterflow indicator, with recycling retard and									
9610	two single pole retard switches, 2″ thru 6″ pipe size	1 Spri	8	1	Ea.	121	54.50		175.50	216

21 11 Facility Fire-Suppression Water-Service Piping

21 11 13 – Facility Water Distribution Piping

21 11 13.16 Pipe, Plastic

	21 11 13.16 Pipe, Plastic	Crew	Daily Output	Labor-Hours	Unit	Material	Labor	Equipment	Total	Total Incl O&P
0010	**PIPE, PLASTIC**									
0020	CPVC, fire suppression, (C-UL-S. FM, NFPA 13, 13D & 13R)									
0030	Socket joint, no couplings or hangers									
0100	SDR 13.5, (ASTM F442)									
0120	3/4″ diameter	Q-12	420	.038	L.F.	1.42	1.87		3.29	4.38
0130	1″ diameter		340	.047		2.18	2.31		4.49	5.90
0140	1-1/4″ diameter		260	.062		3.42	3.03		6.45	8.30
0150	1-1/2″ diameter		190	.084		4.72	4.14		8.86	11.45
0160	2″ diameter		140	.114		7.10	5.60		12.70	16.25
0170	2-1/2″ diameter		130	.123		12.90	6.05		18.95	23.50
0180	3″ diameter		120	.133		19.60	6.55		26.15	31.50

21 11 13.18 Pipe Fittings, Plastic

	21 11 13.18 Pipe Fittings, Plastic	Crew	Daily Output	Labor-Hours	Unit	Material	Labor	Equipment	Total	Total Incl O&P
0010	**PIPE FITTINGS, PLASTIC**									
0020	CPVC, fire suppression, (C-UL-S. FM, NFPA 13, 13D & 13R)									
0030	Socket joint									
0100	90° Elbow									
0120	3/4″	1 Plum	26	.308	Ea.	2.10	17.15		19.25	28.50
0130	1″		22.70	.352		4.58	19.65		24.23	34.50
0140	1-1/4″		20.20	.396		5.80	22		27.80	40
0150	1-1/2″		18.20	.440		8.20	24.50		32.70	46
0160	2″	Q-1	33.10	.483		10.20	24.50		34.70	47.50
0170	2-1/2″		24.20	.661		19.60	33		52.60	71.50
0180	3″		20.80	.769		26.50	38.50		65	87.50

21 11 13.18 Pipe Fittings, Plastic		Crew	Daily Output	Labor-Hours	Unit	2013 Bare Costs Material	Labor	Equipment	Total	Total Incl O&P
0200	45° Elbow									
0210	3/4"	1 Plum	26	.308	Ea.	2.85	17.15		20	29
0220	1"		22.70	.352		3.38	19.65		23.03	33
0230	1-1/4"		20.20	.396		4.88	22		26.88	39
0240	1-1/2"		18.20	.440		6.75	24.50		31.25	44.50
0250	2"	Q-1	33.10	.483		8.40	24.50		32.90	46
0260	2-1/2"		24.20	.661		15.10	33		48.10	66.50
0270	3"		20.80	.769		21.50	38.50		60	82
0300	Tee									
0310	3/4"	1 Plum	17.30	.462	Ea.	2.85	26		28.85	42
0320	1"		15.20	.526		5.65	29.50		35.15	50
0330	1-1/4"		13.50	.593		8.50	33		41.50	59.50
0340	1-1/2"		12.10	.661		12.45	37		49.45	69
0350	2"	Q-1	20	.800		18.45	40		58.45	81
0360	2-1/2"		16.20	.988		30	49.50		79.50	108
0370	3"		13.90	1.151		35.50	58		93.50	126
0400	Tee, reducing x any size									
0420	1"	1 Plum	15.20	.526	Ea.	4.80	29.50		34.30	49.50
0430	1-1/4"		13.50	.593		8.80	33		41.80	59.50
0440	1-1/2"		12.10	.661		10.65	37		47.65	67
0450	2"	Q-1	20	.800		16.95	40		56.95	79
0460	2-1/2"		16.20	.988		23.50	49.50		73	100
0470	3"		13.90	1.151		27	58		85	117
0500	Coupling									
0510	3/4"	1 Plum	26	.308	Ea.	2.03	17.15		19.18	28
0520	1"		22.70	.352		2.63	19.65		22.28	32.50
0530	1-1/4"		20.20	.396		3.83	22		25.83	37.50
0540	1-1/2"		18.20	.440		5.50	24.50		30	43
0550	2"	Q-1	33.10	.483		7.45	24.50		31.95	44.50
0560	2-1/2"		24.20	.661		11.35	33		44.35	62.50
0570	3"		20.80	.769		14.80	38.50		53.30	74.50
0600	Coupling, reducing									
0610	1" x 3/4"	1 Plum	22.70	.352	Ea.	2.63	19.65		22.28	32.50
0620	1-1/4" x 1"		20.20	.396		3.98	22		25.98	38
0630	1-1/2" x 3/4"		18.20	.440		6	24.50		30.50	43.50
0640	1-1/2" x 1"		18.20	.440		5.80	24.50		30.30	43.50
0650	1-1/2" x 1-1/4"		18.20	.440		5.50	24.50		30	43
0660	2" x 1"	Q-1	33.10	.483		7.75	24.50		32.25	45
0670	2" x 1-1/2"	"	33.10	.483		7.45	24.50		31.95	44.50
0700	Cross									
0720	3/4"	1 Plum	13	.615	Ea.	4.50	34.50		39	56.50
0730	1"		11.30	.708		5.65	39.50		45.15	65.50
0740	1-1/4"		10.10	.792		7.80	44		51.80	75
0750	1-1/2"		9.10	.879		10.65	49		59.65	85.50
0760	2"	Q-1	16.60	.964		17.55	48.50		66.05	92.50
0770	2-1/2"	"	12.10	1.322		38.50	66.50		105	142
0800	Cap									
0820	3/4"	1 Plum	52	.154	Ea.	1.20	8.60		9.80	14.25
0830	1"		45	.178		1.73	9.90		11.63	16.85
0840	1-1/4"		40	.200		2.78	11.15		13.93	19.85
0850	1-1/2"		36.40	.220		3.83	12.25		16.08	22.50
0860	2"	Q-1	66	.242		5.80	12.20		18	24.50
0870	2-1/2"		48.40	.331		8.35	16.60		24.95	34

21 11 Facility Fire-Suppression Water-Service Piping

21 11 13 – Facility Water Distribution Piping

21 11 13.18 Pipe Fittings, Plastic		Crew	Daily Output	Labor-Hours	Unit	Material	2013 Bare Costs Labor	Equipment	Total	Total Incl O&P
0880	3″	Q-1	41.60	.385	Ea.	13.50	19.30		32.80	44
0900	Adapter, sprinkler head, female w/metal thd. insert, (s x FNPT)									
0920	3/4″ x 1/2″	1 Plum	52	.154	Ea.	5.25	8.60		13.85	18.75
0930	1″ x 1/2″		45	.178		5.55	9.90		15.45	21
0940	1″ x 3/4″		45	.178		8.80	9.90		18.70	24.50

21 11 16 – Facility Fire Hydrants

21 11 16.50 Fire Hydrants for Buildings		Crew	Daily Output	Labor-Hours	Unit	Material	Labor	Equipment	Total	Total Incl O&P
0010	**FIRE HYDRANTS FOR BUILDINGS**									
3750	Hydrants, wall, w/caps, single, flush, polished brass									
3800	2-1/2″ x 2-1/2″	Q-12	5	3.200	Ea.	193	157		350	450
3840	2-1/2″ x 3″		5	3.200		390	157		547	665
3860	3″ x 3″		4.80	3.333		315	164		479	595
3900	For polished chrome, add					20%				
3950	Double, flush, polished brass									
4000	2-1/2″ x 2-1/2″ x 4″	Q-12	5	3.200	Ea.	520	157		677	805
4040	2-1/2″ x 2-1/2″ x 6″		4.60	3.478		745	171		916	1,075
4080	3″ x 3″ x 4″		4.90	3.265		1,100	161		1,261	1,475
4120	3″ x 3″ x 6″		4.50	3.556		1,150	175		1,325	1,525
4200	For polished chrome, add					10%				
4350	Double, projecting, polished brass									
4400	2-1/2″ x 2-1/2″ x 4″	Q-12	5	3.200	Ea.	230	157		387	490
4450	2-1/2″ x 2-1/2″ x 6″	″	4.60	3.478	″	470	171		641	775
4460	Valve control, dbl. flush/projecting hydrant, cap &									
4470	chain, extension rod & cplg., escutcheon, polished brass	Q-12	8	2	Ea.	274	98.50		372.50	450
4480	Four-way square, flush, polished brass									
4540	2-1/2″(4) x 6″	Q-12	3.60	4.444	Ea.	2,900	219		3,119	3,525

21 11 19 – Fire-Department Connections

21 11 19.50 Connections for the Fire-Department		Crew	Daily Output	Labor-Hours	Unit	Material	Labor	Equipment	Total	Total Incl O&P
0010	**CONNECTIONS FOR THE FIRE-DEPARTMENT**									
6000	Roof manifold, horiz., brass, without valves & caps									
6040	2-1/2″ x 2-1/2″ x 4″	Q-12	4.80	3.333	Ea.	160	164		324	425
6060	2-1/2″ x 2-1/2″ x 6″		4.60	3.478		173	171		344	450
6080	2-1/2″ x 2-1/2″ x 2-1/2″ x 4″		4.60	3.478		256	171		427	540
6090	2-1/2″ x 2-1/2″ x 2-1/2″ x 6″		4.60	3.478		263	171		434	550
7000	Sprinkler line tester, cast brass					23.50			23.50	25.50
7140	Standpipe connections, wall, w/plugs & chains									
7160	Single, flush, brass, 2-1/2″ x 2-1/2″	Q-12	5	3.200	Ea.	144	157		301	395
7180	2-1/2″ x 3″	″	5	3.200	″	149	157		306	400
7240	For polished chrome, add					15%				
7280	Double, flush, polished brass									
7300	2-1/2″ x 2-1/2″ x 4″	Q-12	5	3.200	Ea.	470	157		627	750
7330	2-1/2″ x 2-1/2″ x 6″		4.60	3.478		655	171		826	980
7340	3″ x 3″ x 4″		4.90	3.265		855	161		1,016	1,175
7370	3″ x 3″ x 6″		4.50	3.556		1,250	175		1,425	1,650
7400	For polished chrome, add					15%				
7440	For sill cock combination, add				Ea.	82			82	90
7580	Double projecting, polished brass									
7600	2-1/2″ x 2-1/2″ x 4″	Q-12	5	3.200	Ea.	430	157		587	710
7630	2-1/2″ x 2-1/2″ x 6″	″	4.60	3.478	″	725	171		896	1,050
7680	For polished chrome, add					15%				
7900	Three way, flush, polished brass									
7920	2-1/2″ (3) x 4″	Q-12	4.80	3.333	Ea.	1,500	164		1,664	1,900

21 11 Facility Fire-Suppression Water-Service Piping

21 11 19 – Fire-Department Connections

21 11 19.50 Connections for the Fire-Department		Crew	Daily Output	Labor-Hours	Unit	Material	Labor	Equipment	Total	Total Incl O&P
						2013 Bare Costs				
7930	2-1/2" (3) x 6"	Q-12	4.80	3.333	Ea.	1,500	164		1,664	1,900
8000	For polished chrome, add				↓	9%				
8020	Three way, projecting, polished brass									
8040	2-1/2"(3) x 4"	Q-12	4.80	3.333	Ea.	1,000	164		1,164	1,350
8070	2-1/2" (3) x 6"	"	4.60	3.478		1,000	171		1,171	1,350
8100	For polished chrome, add				↓	12%				
8200	Four way, square, flush, polished brass,									
8240	2-1/2"(4) x 6"	Q-12	3.60	4.444	Ea.	1,250	219		1,469	1,700
8300	For polished chrome, add				"	10%				
8550	Wall, vertical, flush, cast brass									
8600	Two way, 2-1/2" x 2-1/2" x 4"	Q-12	5	3.200	Ea.	395	157		552	670
8660	Four way, 2-1/2"(4) x 6"		3.80	4.211		1,300	207		1,507	1,725
8680	Six way, 2-1/2"(6) x 6"	↓	3.40	4.706		1,550	231		1,781	2,075
8700	For polished chrome, add				↓	10%				
8800	Sidewalk siamese unit, polished brass, two way									
8820	2-1/2" x 2-1/2" x 4"	Q-12	2.50	6.400	Ea.	575	315		890	1,100
8850	2-1/2" x 2-1/2" x 6"		2	8		740	395		1,135	1,400
8860	3" x 3" x 4"		2.50	6.400		820	315		1,135	1,375
8890	3" x 3" x 6"	↓	2	8	↓	1,100	395		1,495	1,825
8940	For polished chrome, add					12%				
9100	Sidewalk siamese unit, polished brass, three way									
9120	2-1/2" x 2-1/2" x 2-1/2" x 6"	Q-12	2	8	Ea.	910	395		1,305	1,600
9160	For polished chrome, add				"	15%				

21 12 Fire-Suppression Standpipes

21 12 13 – Fire-Suppression Hoses and Nozzles

21 12 13.50 Fire Hoses and Nozzles

		Crew	Daily Output	Labor-Hours	Unit	Material	Labor	Equipment	Total	Total Incl O&P
0010	**FIRE HOSES AND NOZZLES**									
0200	Adapters, rough brass, straight hose threads R211226-10									
0220	One piece, female to male, rocker lugs									
0240	1" x 1" R211226-20				Ea.	43.50			43.50	48
0260	1-1/2" x 1"					43.50			43.50	47.50
0280	1-1/2" x 1-1/2"					12.90			12.90	14.20
0300	2" x 1-1/2"					59.50			59.50	65.50
0320	2" x 2"					42			42	46
0340	2-1/2" x 1-1/2"					13.25			13.25	14.55
0360	2-1/2" x 2"					34			34	37.50
0380	2-1/2" x 2-1/2"					19			19	21
0400	3" x 2-1/2"					113			113	125
0420	3" x 3"					83.50			83.50	92
0500	For polished brass, add					50%				
0520	For polished chrome, add				↓	75%				
0700	One piece, female to male, hexagon									
0740	1-1/2" x 3/4"				Ea.	37			37	40.50
0760	2" x 1-1/2"					76.50			76.50	84
0780	2-1/2" x 1"					146			146	161
0800	2-1/2" x 1-1/2"					50.50			50.50	55.50
0820	2-1/2" x 2"					34			34	37.50
0840	3" x 2-1/2"					63.50			63.50	70
0900	For polished chrome, add				↓	75%				
1100	Swivel, female to female, pin lugs									

21 12 Fire-Suppression Standpipes

21 12 13 – Fire-Suppression Hoses and Nozzles

21 12 13.50 Fire Hoses and Nozzles		Crew	Daily Output	Labor-Hours	Unit	Material	2013 Bare Costs Labor	Equipment	Total	Total Incl O&P
1120	1-1/2" x 1-1/2"				Ea.	55.50			55.50	61
1200	2-1/2" x 2-1/2"					108			108	119
1260	For polished brass, add					50%				
1280	For polished chrome, add					75%				
1400	Couplings, sngl. & dbl. jacket, pin lug or rocker lug, cast brass									
1410	1-1/2"				Ea.	48			48	53
1420	2-1/2"				"	63			63	69.50
1500	For polished brass, add					20%				
1520	For polished chrome, add					40%				
1580	Reducing, F x M, interior installation, cast brass									
1590	2" x 1-1/2"				Ea.	59.50			59.50	65.50
1600	2-1/2" x 1-1/2"					13.25			13.25	14.55
1680	For polished brass, add					50%				
1720	For polished chrome, add					75%				
2200	Hose, less couplings									
2260	Synthetic jacket, lined, 300 lb. test, 1-1/2" diameter	Q-12	2600	.006	L.F.	2.94	.30		3.24	3.69
2280	2-1/2" diameter		2200	.007		5.05	.36		5.41	6.15
2360	High strength, 500 lb. test, 1-1/2" diameter		2600	.006		3.03	.30		3.33	3.79
2380	2-1/2" diameter		2200	.007		5.35	.36		5.71	6.45
5000	Nipples, straight hose to tapered iron pipe, brass									
5060	Female to female, 1-1/2" x 1-1/2"				Ea.	17.65			17.65	19.40
5100	2-1/2" x 2-1/2"					33			33	36.50
5190	For polished chrome, add					75%				
5200	Double male or male to female, 1" x 1"					38			38	42
5220	1-1/2" x 1"					56			56	61.50
5230	1-1/2" x 1-1/2"					12.85			12.85	14.15
5260	2" x 1-1/2"					61			61	67
5270	2" x 2"					76.50			76.50	84
5280	2-1/2" x 1-1/2"					50.50			50.50	55.50
5300	2-1/2" x 2"					41			41	45.50
5310	2-1/2" x 2-1/2"					23.50			23.50	26
5340	For polished chrome, add					75%				
5600	Nozzles, brass									
5620	Adjustable fog, 3/4" booster line				Ea.	96.50			96.50	106
5630	1" booster line					125			125	137
5640	1-1/2" leader line					75.50			75.50	83
5660	2-1/2" direct connection					136			136	149
5680	2-1/2" playpipe nozzle					209			209	230
5780	For chrome plated, add					8%				
5850	Electrical fire, adjustable fog, no shock									
5900	1-1/2"				Ea.	390			390	430
5920	2-1/2"					525			525	575
5980	For polished chrome, add					6%				
6200	Heavy duty, comb. adj. fog and str. stream, with handle									
6210	1" booster line				Ea.	375			375	410
6240	1-1/2"					420			420	465
6260	2-1/2", for playpipe					750			750	825
6280	2-1/2" direct connection					525			525	575
6300	2-1/2" playpipe combination					525			525	575
6480	For polished chrome, add					7%				
6500	Plain fog, polished brass, 1-1/2"					95			95	105
6540	Chrome plated, 1-1/2"					91.50			91.50	101
6700	Plain stream, polished brass, 1-1/2" x 10"					43.50			43.50	47.50

21 12 Fire-Suppression Standpipes

21 12 13 – Fire-Suppression Hoses and Nozzles

21 12 13.50 Fire Hoses and Nozzles		Crew	Daily Output	Labor-Hours	Unit	Material	2013 Bare Costs Labor	Equipment	Total	Total Incl O&P
6760	2-1/2" x 15" x 7/8" or 1-1/2"				Ea.	78.50			78.50	86.50
6860	For polished chrome, add					20%				
7000	Underwriters playpipe, 2-1/2" x 30" with 1-1/8" tip				Ea.	233			233	256
7040	Less tip					170			170	187
9200	Storage house, hose only, primed steel					775			775	850
9220	Aluminum					1,700			1,700	1,875
9280	Hose and hydrant house, primed steel					1,050			1,050	1,150
9300	Aluminum					1,525			1,525	1,675
9340	Tools, crowbar and brackets	1 Carp	12	.667		63	30		93	116
9360	Combination hydrant wrench and spanner					28.50			28.50	31.50
9380	Fire axe and brackets									
9400	6 lb.	1 Carp	12	.667	Ea.	110	30		140	167

21 12 16 – Fire-Suppression Hose Reels

21 12 16.50 Fire-Suppression Hose Reels		Crew	Daily Output	Labor-Hours	Unit	Material	Labor	Equipment	Total	Total Incl O&P
0010	**FIRE-SUPPRESSION HOSE REELS**									
2990	Hose reel, swinging, for 1-1/2" polyester neoprene lined hose									
3000	50' long	Q-12	14	1.143	Ea.	119	56		175	216
3020	100' long		14	1.143		167	56		223	268
3060	For 2-1/2" cotton rubber hose, 75' long		14	1.143		193	56		249	297
3100	150' long		14	1.143		225	56		281	330

21 12 19 – Fire-Suppression Hose Racks

21 12 19.50 Fire Hose Racks		Crew	Daily Output	Labor-Hours	Unit	Material	Labor	Equipment	Total	Total Incl O&P
0010	**FIRE HOSE RACKS**									
2600	Hose rack, swinging, for 1-1/2" diameter hose,									
2620	Enameled steel, 50' & 75' lengths of hose	Q-12	20	.800	Ea.	55	39.50		94.50	120
2640	100' and 125' lengths of hose		20	.800		55	39.50		94.50	120
2680	Chrome plated, 50' and 75' lengths of hose		20	.800		89.50	39.50		129	158
2700	100' and 125' lengths of hose		20	.800		94.50	39.50		134	164
2780	For hose rack nipple, 1-1/2" polished brass, add					26.50			26.50	29
2820	2-1/2" polished brass, add					49			49	54
2840	1-1/2" polished chrome, add					37			37	41
2860	2-1/2" polished chrome, add					52			52	57.50

21 12 23 – Fire-Suppression Hose Valves

21 12 23.70 Fire Hose Valves		Crew	Daily Output	Labor-Hours	Unit	Material	Labor	Equipment	Total	Total Incl O&P
0010	**FIRE HOSE VALVES**									
0020	Angle, combination pressure adjust/restricting, rough brass R211226-10									
0030	1-1/2"	1 Spri	12	.667	Ea.	77	36.50		113.50	140
0040	2-1/2" R211226-20	"	7	1.143	"	158	62.50		220.50	268
0042	Nonpressure adjustable/restricting, rough brass									
0044	1-1/2"	1 Spri	12	.667	Ea.	48	36.50		84.50	108
0046	2-1/2"	"	7	1.143	"	81.50	62.50		144	184
0050	For polished brass, add					30%				
0060	For polished chrome, add					40%				
0080	Wheel handle, 300 lb., 1-1/2"	1 Spri	12	.667	Ea.	86	36.50		122.50	150
0090	2-1/2"	"	7	1.143	"	158	62.50		220.50	268
0100	For polished brass, add					35%				
0110	For polished chrome, add					50%				
1000	Ball drip, automatic, rough brass, 1/2"	1 Spri	20	.400	Ea.	14.70	22		36.70	49
1010	3/4"	"	20	.400	"	16.55	22		38.55	51
1100	Ball, 175 lb., sprinkler system, FM/UL, threaded, bronze									
1120	Slow close									

21 12 Fire-Suppression Standpipes

21 12 23 – Fire-Suppression Hose Valves

21 12 23.70 Fire Hose Valves		Crew	Daily Output	Labor-Hours	Unit	Material	2013 Bare Costs Labor	Equipment	Total	Total Incl O&P
1150	1" size	1 Spri	19	.421	Ea.	222	23		245	280
1160	1-1/4" size		15	.533		241	29		270	310
1170	1-1/2" size		13	.615		315	33.50		348.50	400
1180	2" size		11	.727		385	40		425	485
1190	2-1/2" size	Q-12	15	1.067		520	52.50		572.50	650
1230	For supervisory switch kit, all sizes									
1240	One circuit, add	1 Spri	48	.167	Ea.	168	9.10		177.10	199
1280	Quarter turn for trim									
1300	1/2" size	1 Spri	22	.364	Ea.	29.50	19.85		49.35	62.50
1310	3/4" size		20	.400		31.50	22		53.50	68
1320	1" size		19	.421		35	23		58	73
1330	1-1/4" size		15	.533		57.50	29		86.50	108
1340	1-1/2" size		13	.615		72.50	33.50		106	130
1350	2" size		11	.727		86	40		126	155
1400	Caps, polished brass with chain, 3/4"					38			38	42
1420	1"					48.50			48.50	53.50
1440	1-1/2"					12.85			12.85	14.15
1460	2-1/2"					19.50			19.50	21.50
1480	3"					30.50			30.50	33.50
1900	Escutcheon plate, for angle valves, polished brass, 1-1/2"					15.20			15.20	16.70
1920	2-1/2"					24.50			24.50	27
1940	3"					31			31	34
1980	For polished chrome, add					15%				
2000	Foam, control valve, 3"	1 Spri	6	1.333		1,775	73		1,848	2,050
2020	Supply valve, 2-1/2"		7	1.143		140	62.50		202.50	248
2040	Proportioner, 8"		2	4		2,500	219		2,719	3,050
2060	Oscillating foam monitor with electric remote control	Q-12	5.33	3.002		10,300	148		10,448	11,500
3000	Gate, hose, wheel handle, N.R.S., rough brass, 1-1/2"	1 Spri	12	.667		122	36.50		158.50	189
3040	2-1/2", 300 lb.	"	7	1.143		172	62.50		234.50	283
3080	For polished brass, add					40%				
3090	For polished chrome, add					50%				
3800	Hydrant, screw type, crank handle, brass									
3840	2-1/2" size	Q-12	11	1.455	Ea.	335	71.50		406.50	480
3880	For chrome, same price									
4200	Hydrolator, vent and draining, rough brass, 1-1/2"	1 Spri	12	.667	Ea.	68	36.50		104.50	130
4280	For polished brass, add					50%				
4290	For polished chrome, add					90%				
5000	Pressure restricting, adjustable rough brass, 1-1/2"	1 Spri	12	.667		114	36.50		150.50	180
5020	2-1/2"	"	7	1.143		160	62.50		222.50	270
5080	For polished brass, add					30%				
5090	For polished chrome, add					45%				
8000	Wye, leader line, ball type, swivel female x male x male									
8040	2-1/2" x 1-1/2" x 1-1/2" polished brass				Ea.	245			245	270
8060	2-1/2" x 1-1/2" x 1-1/2" polished chrome				"	245			245	270

21 13 13.50 Wet-Pipe Sprinkler System Components		Crew	Daily Output	Labor-Hours	Unit	2013 Bare Costs Material	Labor	Equipment	Total	Total Incl O&P
0010	**WET-PIPE SPRINKLER SYSTEM COMPONENTS**									
1100	Alarm, electric pressure switch (circuit closer) R211313-10	1 Spri	26	.308	Ea.	78	16.80		94.80	112
1140	For explosion proof, max 20 PSI, contacts close or open		26	.308		510	16.80		526.80	585
1220	Water motor, complete with gong R211313-30		4	2		370	109		479	570
1800	Firecycle system, controls, includes panel,									
1820	batteries, solenoid valves and pressure switches	Q-13	1	32	Ea.	18,500	1,675		20,175	22,800
1860	Detector	1 Spri	16	.500	"	645	27.50		672.50	745
1900	Flexible sprinkler head connectors									
1910	Braided stainless steel hose with mounting bracket									
1920	1/2" and 1" outlet size									
1930	24" length	1 Spri	32	.250	Ea.	26.50	13.65		40.15	49.50
1940	36" length		30	.267		30.50	14.55		45.05	55.50
1950	48" length		26	.308		34.50	16.80		51.30	63.50
1960	60" length		22	.364		38.50	19.85		58.35	72.50
1970	72" length		18	.444		42.50	24.50		67	83.50
1982	May replace hard-pipe armovers									
1984	For wet, pre-action, deluge or dry pipe systems									
2000	Release, emergency, manual, for hydraulic or pneumatic system	1 Spri	12	.667	Ea.	194	36.50		230.50	268
2060	Release, thermostatic, for hydraulic or pneumatic release line		20	.400		600	22		622	695
2200	Sprinkler cabinets, 6 head capacity		16	.500		71	27.50		98.50	119
2260	12 head capacity		16	.500		75	27.50		102.50	124
2340	Sprinkler head escutcheons, standard, brass tone, 1" size		40	.200		2.56	10.95		13.51	19.25
2360	Chrome, 1" size		40	.200		2.74	10.95		13.69	19.45
2400	Recessed type, brass tone		40	.200		3.16	10.95		14.11	19.95
2440	Chrome or white enamel		40	.200		3.38	10.95		14.33	20
2600	Sprinkler heads, not including supply piping									
3700	Standard spray, pendent or upright, brass, 135°F to 286°F									
3720	1/2" NPT, 3/8" orifice	1 Spri	16	.500	Ea.	13.85	27.50		41.35	56
3730	1/2" NPT, 7/16" orifice		16	.500		13.65	27.50		41.15	56
3740	1/2" NPT, 1/2" orifice		16	.500		8.90	27.50		36.40	51
3760	1/2" NPT, 17/32" orifice		16	.500		11.60	27.50		39.10	54
3780	3/4" NPT, 17/32" orifice		16	.500		10.70	27.50		38.20	53
3800	For open sprinklers, deduct					15%				
3840	For chrome, add				Ea.	3.11			3.11	3.42
3860	For wax and lead coating, add					31.50			31.50	35
3880	For wax coating, add					18.90			18.90	21
3900	For lead coating, add					20.50			20.50	22.50
3920	For 360°F, same cost									
3930	For 400°F, add				Ea.	59.50			59.50	65.50
3940	For 500°F, add				"	59.50			59.50	65.50
4200	Sidewall, vertical brass, 135°F to 286°F									
4240	1/2" NPT, 1/2" orifice	1 Spri	16	.500	Ea.	22.50	27.50		50	66
4280	3/4" NPT, 17/32" orifice	"	16	.500		64.50	27.50		92	112
4360	For satin chrome, add					2.57			2.57	2.83
4400	For 360°F, same cost									
4500	Sidewall, horizontal, brass, 135°F to 286°F									
4520	1/2" NPT, 1/2" orifice	1 Spri	16	.500	Ea.	22.50	27.50		50	66
4540	For 360°F, same cost									
4800	Recessed pendent, brass, 135°F to 286°F									
4820	1/2" NPT, 3/8" orifice	1 Spri	10	.800	Ea.	43.50	43.50		87	114
4830	1/2" NPT, 7/16" orifice		10	.800		23	43.50		66.50	91
4840	1/2" NPT, 1/2" orifice		10	.800		17.55	43.50		61.05	85.50
4860	1/2" NPT, 17/32" orifice		10	.800		43.50	43.50		87	114

21 13 13 – Wet-Pipe Sprinkler Systems

21 13 13.50 Wet-Pipe Sprinkler System Components		Crew	Daily Output	Labor-Hours	Unit	Material	Labor	Equipment	Total	Total Incl O&P
						2013 Bare Costs				
4900	For satin chrome, add				Ea.	3.27			3.27	3.60
5000	Recessed-vertical sidewall, brass, 135°F to 286°F									
5020	1/2" NPT, 3/8" orifice	1 Spri	10	.800	Ea.	29	43.50		72.50	98
5030	1/2" NPT, 7/16" orifice		10	.800		29	43.50		72.50	98
5040	1/2" NPT, 1/2" orifice		10	.800		29	43.50		72.50	98
5100	For bright nickel, same cost									
5600	Concealed, complete with cover plate									
5620	1/2" NPT, 1/2" orifice, 135°F to 212°F	1 Spri	9	.889	Ea.	33	48.50		81.50	109
5800	Window, brass, 1/2" NPT, 1/4" orifice		16	.500		30	27.50		57.50	74
5810	1/2" NPT, 5/16" orifice		16	.500		30	27.50		57.50	74
5820	1/2" NPT, 3/8" orifice		16	.500		30	27.50		57.50	74
5830	1/2" NPT, 7/16" orifice		16	.500		30	27.50		57.50	74
5840	1/2" NPT, 1/2" orifice		16	.500		32	27.50		59.50	76
5860	For polished chrome, add					4.44			4.44	4.88
5880	3/4" NPT, 5/8" orifice	1 Spri	16	.500		33.50	27.50		61	77.50
5890	3/4 NPT, 3/4" orifice	"	16	.500		33.50	27.50		61	77.50
6000	Sprinkler head guards, bright zinc, 1/2" NPT					5.35			5.35	5.90
6020	Bright zinc, 3/4" NPT					5.35			5.35	5.90
6025	Residential sprinkler components, (one and two family)									
6026	Water motor alarm, with strainer	1 Spri	4	2	Ea.	365	109		474	565
6027	Fast response, glass bulb, 135°F to 155°F									
6028	1/2" NPT, pendent, brass	1 Spri	16	.500	Ea.	24	27.50		51.50	67.50
6029	1/2" NPT, sidewall, brass		16	.500		24	27.50		51.50	67.50
6030	1/2" NPT, pendent, brass, extended coverage		16	.500		20	27.50		47.50	63
6031	1/2" NPT, sidewall, brass, extended coverage		16	.500		26	27.50		53.50	69.50
6032	3/4" NPT sidewall, brass, extended coverage		16	.500		21.50	27.50		49	64.50
6033	For chrome, add					15%				
6034	For polyester/teflon coating add					20%				
6100	Sprinkler head wrenches, standard head				Ea.	22.50			22.50	24.50
6120	Recessed head					34.50			34.50	38
6160	Tamper switch, (valve supervisory switch)	1 Spri	16	.500		87.50	27.50		115	137

21 13 16 – Dry-Pipe Sprinkler Systems

21 13 16.50 Dry-Pipe Sprinkler System Components		Crew	Daily Output	Labor-Hours	Unit	Material	Labor	Equipment	Total	Total Incl O&P
0010	**DRY-PIPE SPRINKLER SYSTEM COMPONENTS**									
0600	Accelerator	1 Spri	8	1	Ea.	675	54.50		729.50	825
0800	Air compressor for dry pipe system, automatic, complete									
0820	30 gal. system capacity, 3/4 HP	1 Spri	1.30	6.154	Ea.	735	335		1,070	1,325
0860	30 gal. system capacity, 1 HP		1.30	6.154		760	335		1,095	1,350
0910	30 gal. system capacity, 1-1/2 HP		1.30	6.154		795	335		1,130	1,375
0920	30 gal. system capacity, 2 HP		1.30	6.154		840	335		1,175	1,425
0960	Air pressure maintenance control		24	.333		305	18.20		323.20	370
1600	Dehydrator package, incl. valves and nipples R211313-20		12	.667		635	36.50		671.50	750
2600	Sprinkler heads, not including supply piping									
2640	Dry, pendent, 1/2" orifice, 3/4" or 1" NPT									
2660	1/2" to 6" length	1 Spri	14	.571	Ea.	113	31		144	171
2670	6-1/4" to 8" length		14	.571		118	31		149	176
2680	8-1/4" to 12" length		14	.571		123	31		154	182
2690	12-1/4" to 15" length		14	.571		128	31		159	188
2700	15-1/4" to 18" length		14	.571		133	31		164	193
2710	18-1/4" to 21" length		13	.615		138	33.50		171.50	202
2720	21-1/4" to 24" length		13	.615		143	33.50		176.50	208
2730	24-1/4" to 27" length		13	.615		148	33.50		181.50	214

21 13 Fire-Suppression Sprinkler Systems

21 13 16 – Dry-Pipe Sprinkler Systems

21 13 16.50 Dry-Pipe Sprinkler System Components		Crew	Daily Output	Labor-Hours	Unit	2013 Bare Costs Material	Labor	Equipment	Total	Total Incl O&P
2740	27-1/4" to 30" length	1 Spri	13	.615	Ea.	153	33.50		186.50	219
2750	30-1/4" to 33" length		13	.615		158	33.50		191.50	225
2760	33-1/4" to 36" length		13	.615		164	33.50		197.50	231
2780	36-1/4" to 39" length		12	.667		169	36.50		205.50	241
2790	39-1/4" to 42" length		12	.667		173	36.50		209.50	246
2800	For each inch or fraction, add					1.71			1.71	1.88
6330	Valves and components									
6340	Alarm test/shut off valve, 1/2"	1 Spri	20	.400	Ea.	19.25	22		41.25	54
8000	Dry pipe air check valve, 3" size	Q-12	2	8		1,600	395		1,995	2,350
8200	Dry pipe valve, incl. trim and gauges, 3" size		2	8		2,350	395		2,745	3,200
8220	4" size		1	16		2,550	785		3,335	4,000
8240	6" size	Q-13	2	16		3,000	835		3,835	4,550
8280	For accelerator trim with gauges, add	1 Spri	8	1		215	54.50		269.50	320

21 13 20 – Fire-Cycle Sprinkler Systems

21 13 20.50 Firecycle Fire-Suppression Sprinkler Systems		Crew	Daily Output	Labor-Hours	Unit	Material	Labor	Equipment	Total	Total Incl O&P
0010	**FIRECYCLE FIRE-SUPPRESSION SPRINKLER SYSTEMS**									
8400	Firecycle package, includes swing check									
8420	and flow control valves with required trim									
8440	2" size	Q-12	2	8	Ea.	4,150	395		4,545	5,175
8460	3" size		1.50	10.667		4,550	525		5,075	5,825
8480	4" size		1	16		5,100	785		5,885	6,775
8500	6" size	Q-13	1.40	22.857		5,875	1,200		7,075	8,250

21 13 26 – Deluge Fire-Suppression Sprinkler Systems

21 13 26.50 Deluge Fire-Suppression Sprinkler Sys. Comp.		Crew	Daily Output	Labor-Hours	Unit	Material	Labor	Equipment	Total	Total Incl O&P
0010	**DELUGE FIRE-SUPPRESSION SPRINKLER SYSTEM COMPONENTS**									
1400	Deluge system, monitoring panel w/deluge valve & trim	1 Spri	18	.444	Ea.	10,200	24.50		10,224.50	11,200
6200	Valves and components									
7000	Deluge, assembly, incl. trim, pressure									
7020	operated relief, emergency release, gauges									
7040	2" size	Q-12	2	8	Ea.	3,325	395		3,720	4,250
7060	3" size		1.50	10.667		3,725	525		4,250	4,900
7080	4" size		1	16		4,250	785		5,035	5,850
7100	6" size	Q-13	1.80	17.778		5,025	925		5,950	6,925
7800	Pneumatic actuator, bronze, required on all									
7820	pneumatic release systems, any size deluge	1 Spri	18	.444	Ea.	370	24.50		394.50	440

21 13 39 – Foam-Water Systems

21 13 39.50 Foam-Water System Components		Crew	Daily Output	Labor-Hours	Unit	Material	Labor	Equipment	Total	Total Incl O&P
0010	**FOAM-WATER SYSTEM COMPONENTS**									
2600	Sprinkler heads, not including supply piping									
3600	Foam-water, pendent or upright, 1/2" NPT	1 Spri	12	.667	Ea.	169	36.50		205.50	241

21 21 Carbon-Dioxide Fire-Extinguishing Systems

21 21 16 – Carbon-Dioxide Fire-Extinguishing Equipment

21 21 16.50 CO2 Fire Extinguishing System		Crew	Daily Output	Labor-Hours	Unit	Material	2013 Bare Costs Labor	Equipment	Total	Total Incl O&P
0010	**CO_2 FIRE EXTINGUISHING SYSTEM**									
0042	For detectors and control stations, see Section 28 31 23.50									
0100	Control panel, single zone with batteries (2 zones det., 1 suppr.)	1 Elec	1	8	Ea.	1,600	420		2,020	2,375
0150	Multizone (4) with batteries (8 zones det., 4 suppr.)	"	.50	16		3,025	840		3,865	4,600
1000	Dispersion nozzle, CO_2, 3" x 5"	1 Plum	18	.444		62	25		87	106
2000	Extinguisher, CO_2 system, high pressure, 75 lb. cylinder	Q-1	6	2.667		1,175	134		1,309	1,500
2100	100 lb. cylinder	"	5	3.200		1,200	161		1,361	1,575
3000	Electro/mechanical release	L-1	4	4		155	216		371	495
3400	Manual pull station	1 Plum	6	1.333		55.50	74.50		130	174
4000	Pneumatic damper release	"	8	1	↓	206	56		262	310

21 22 Clean-Agent Fire-Extinguishing Systems

21 22 16 – Clean-Agent Fire-Extinguishing Equipment

21 22 16.50 FM200 Fire Extinguishing System		Crew	Daily Output	Labor-Hours	Unit	Material	Labor	Equipment	Total	Total Incl O&P
0010	**FM200 FIRE EXTINGUISHING SYSTEM**									
1100	Dispersion nozzle FM200, 1-1/2"	1 Plum	14	.571	Ea.	62	32		94	116
2400	Extinguisher, FM200 system, filled, with mounting bracket									
2460	26 lb. container	Q-1	8	2	Ea.	2,125	100		2,225	2,475
2480	44 lb. container		7	2.286		2,825	115		2,940	3,300
2500	63 lb. container		6	2.667		3,300	134		3,434	3,825
2520	101 lb. container		5	3.200		4,425	161		4,586	5,125
2540	196 lb. container	↓	4	4	↓	7,175	201		7,376	8,200
6000	Average FM200 system, minimum				C.F.	1.63			1.63	1.79
6020	Maximum				"	3.24			3.24	3.56

21 31 Centrifugal Fire Pumps

21 31 13 – Electric-Drive, Centrifugal Fire Pumps

21 31 13.50 Electric-Drive Fire Pumps		Crew	Daily Output	Labor-Hours	Unit	Material	Labor	Equipment	Total	Total Incl O&P
0010	**ELECTRIC-DRIVE FIRE PUMPS** Including controller, fittings and relief valve									
3100	250 GPM, 55 psi, 15 HP, 3550 RPM, 2" pump	Q-13	.70	45.714	Ea.	14,800	2,375		17,175	19,800
3200	500 GPM, 50 psi, 27 HP, 1770 RPM, 4" pump		.68	47.059		15,100	2,450		17,550	20,300
3250	500 GPM, 100 psi, 47 HP, 3550 RPM, 3" pump		.66	48.485		16,100	2,525		18,625	21,600
3300	500 GPM, 125 psi, 64 HP, 3550 RPM, 3" pump		.62	51.613		17,900	2,675		20,575	23,800
3350	750 GPM, 50 psi, 44 HP, 1770 RPM, 5" pump		.64	50		15,700	2,600		18,300	21,100
3400	750 GPM, 100 psi, 66 HP, 3550 RPM, 4" pump		.58	55.172		18,100	2,875		20,975	24,200
3450	750 GPM, 165 psi, 120 HP, 3550 RPM, 4" pump		.56	57.143		25,000	2,975		27,975	32,000
3500	1000 GPM, 50 psi, 48 HP 1770 RPM, 5" pump		.60	53.333		16,500	2,775		19,275	22,400
3550	1000 GPM, 100 psi, 86 HP, 3550 RPM, 5" pump		.54	59.259		22,800	3,075		25,875	29,800
3600	1000 GPM, 150 psi, 142 HP, 3550 RPM, 5" pump		.50	64		26,900	3,325		30,225	34,600
3650	1000 GPM, 200 psi, 245 HP, 1770 RPM, 6" pump		.36	88.889		39,100	4,625		43,725	50,000
3660	1250 GPM, 75 psi, 75 HP, 1770 RPM, 5" pump		.55	58.182		21,300	3,025		24,325	28,000
3700	1500 GPM, 50 psi, 66 HP, 1770 RPM, 6" pump		.50	64		20,200	3,325		23,525	27,300
3750	1500 GPM, 100 psi, 139 HP, 1770 RPM, 6" pump		.46	69.565		26,200	3,625		29,825	34,300
3800	1500 GPM, 150 psi, 200 HP, 1770 RPM, 6" pump		.36	88.889		42,400	4,625		47,025	53,500
3850	1500 GPM, 200 psi, 279 HP, 1770 RPM, 6" pump		.32	100		45,100	5,200		50,300	57,500
3900	2000 GPM, 100 psi, 167 HP, 1770 RPM, 6" pump		.34	94.118		28,600	4,900		33,500	38,800
3950	2000 GPM, 150 psi, 292 HP, 1770 RPM, 6" pump		.28	114		42,900	5,950		48,850	56,000
4000	2500 GPM, 100 psi, 213 HP, 1770 RPM, 8" pump		.30	106		35,600	5,550		41,150	47,600
4040	2500 GPM, 135 psi, 339 HP, 1770 RPM, 8" pump		.26	123		47,100	6,400		53,500	61,500

21 31 Centrifugal Fire Pumps

21 31 13 – Electric-Drive, Centrifugal Fire Pumps

21 31 13.50 Electric-Drive Fire Pumps		Crew	Daily Output	Labor-Hours	Unit	2013 Bare Costs Material	Labor	Equipment	Total	Total Incl O&P
4100	3000 GPM, 100 psi, 250 HP, 1770 RPM, 8" pump	Q-13	.28	114	Ea.	51,000	5,950		56,950	65,000
4150	3000 GPM, 140 psi, 428 HP, 1770 RPM, 10" pump		.24	133		55,000	6,950		61,950	71,000
4200	3500 GPM, 100 psi, 300 HP, 1770 RPM, 10" pump		.26	123		51,000	6,400		57,400	65,500
4250	3500 GPM, 140 psi, 450 HP, 1770 RPM, 10" pump	↓	.24	133		62,000	6,950		68,950	79,000
5000	For jockey pump 1", 3 HP, with control, add	Q-12	2	8	↓	2,450	395		2,845	3,275

21 31 16 – Diesel-Drive, Centrifugal Fire Pumps

21 31 16.50 Diesel-Drive Fire Pumps		Crew	Daily Output	Labor-Hours	Unit	2013 Bare Costs Material	Labor	Equipment	Total	Total Incl O&P
0010	**DIESEL-DRIVE FIRE PUMPS** Including controller, fittings and relief valve									
0050	500 GPM, 50 psi, 27 HP, 4" pump	Q-13	.64	50	Ea.	34,200	2,600		36,800	41,500
0100	500 GPM, 100 psi, 62 HP, 4" pump		.60	53.333		38,600	2,775		41,375	46,700
0150	500 GPM, 125 psi, 78 HP, 4" pump		.56	57.143		38,600	2,975		41,575	47,000
0200	750 GPM, 50 psi, 44 HP, 5" pump		.60	53.333		35,100	2,775		37,875	42,800
0250	750 GPM, 100 psi, 80 HP, 4" pump		.56	57.143		38,200	2,975		41,175	46,600
0300	750 GPM, 165 psi, 203 HP, 5" pump		.52	61.538		51,500	3,200		54,700	61,500
0350	1000 GPM, 50 psi, 48 HP, 5" pump		.58	55.172		38,500	2,875		41,375	46,600
0400	1,000 GPM, 100 psi, 89 HP, 4" pump		.56	57.143		38,500	2,975		41,475	46,800
0450	1000 GPM, 150 psi, 148 HP, 4" pump		.48	66.667		49,600	3,475		53,075	59,500
0470	1000 GPM, 200 psi, 280 HP, 5" pump		.40	80		60,500	4,175		64,675	73,500
0480	1250 GPM, 75 psi, 75 HP, 5" pump		.54	59.259		39,000	3,075		42,075	47,600
0500	1500 GPM, 50 psi, 66 HP, 6" pump		.50	64		37,400	3,325		40,725	46,200
0550	1500 GPM, 100 psi, 140 HP, 6" pump		.46	69.565		49,000	3,625		52,625	59,500
0600	1500 GPM, 150 psi, 228 HP, 6" pump		.42	76.190		55,000	3,975		58,975	66,500
0650	1500 GPM, 200 psi, 279 HP, 6" pump		.38	84.211		68,500	4,375		72,875	81,500
0700	2,000 GPM, 100 psi, 167 HP, 6" pump		.34	94.118		48,900	4,900		53,800	61,500
0750	2000 GPM, 150 psi, 284 HP, 6" pump		.30	106		62,500	5,550		68,050	77,000
0800	2500 GPM, 100 psi, 213 HP, 8" pump		.32	100		52,500	5,200		57,700	65,500
0820	2500 GPM, 150 psi, 365 HP, 8" pump		.26	123		67,500	6,400		73,900	83,500
0850	3000 GPM, 100 psi, 250 HP, 8" pump		.28	114		78,500	5,950		84,450	95,500
0900	3000 GPM, 150 psi, 384 HP, 10" pump		.20	160		83,000	8,325		91,325	104,000
0950	3,500 GPM, 100 psi, 300 HP, 10" pump		.24	133		68,000	6,950		74,950	85,500
1000	3500 GPM, 150 psi, 518 HP, 10" pump	↓	.20	160	↓	96,000	8,325		104,325	118,000

Division Notes

		CREW	DAILY OUTPUT	LABOR-HOURS	UNIT	BARE COSTS MAT.	BARE COSTS LABOR	BARE COSTS EQUIP.	BARE COSTS TOTAL	TOTAL INCL O&P

Estimating Tips

22 10 00 Plumbing Piping and Pumps

This subdivision is primarily basic pipe and related materials. The pipe may be used by any of the mechanical disciplines, i.e., plumbing, fire protection, heating, and air conditioning.

Note: CPVC plastic piping approved for fire protection is located in 21 11 13.

- The labor adjustment factors listed in Subdivision 22 01 02.20 apply throughout Divisions 21, 22, and 23. CAUTION: the correct percentage may vary for the same items. For example, the percentage add for the basic pipe installation should be based on the maximum height that the craftsman must install for that particular section. If the pipe is to be located 14′ above the floor but it is suspended on threaded rod from beams, the bottom flange of which is 18′ high (4′ rods), then the height is actually 18′ and the add is 20%. The pipe coverer, however, does not have to go above the 14′, and so his or her add should be 10%.
- Most pipe is priced first as straight pipe with a joint (coupling, weld, etc.) every 10′ and a hanger usually every 10′. There are exceptions with hanger spacing such as for cast iron pipe (5′) and plastic pipe (3 per 10′). Following each type of pipe there are several lines listing sizes and the amount to be subtracted to delete couplings and hangers. This is for pipe that is to be buried or supported together on trapeze hangers. The reason that the couplings are deleted is that these runs are usually long, and frequently longer lengths of pipe are used. By deleting the couplings, the estimator is expected to look up and add back the correct reduced number of couplings.
- When preparing an estimate, it may be necessary to approximate the fittings. Fittings usually run between 25% and 50% of the cost of the pipe. The lower percentage is for simpler runs, and the higher number is for complex areas, such as mechanical rooms.
- For historic restoration projects, the systems must be as invisible as possible, and pathways must be sought for pipes, conduit, and ductwork. While installations in accessible spaces (such as basements and attics) are relatively straightforward to estimate, labor costs may be more difficult to determine when delivery systems must be concealed.

22 40 00 Plumbing Fixtures

- Plumbing fixture costs usually require two lines: the fixture itself and its "rough-in, supply, and waste."
- In the Assemblies Section (Plumbing D2010) for the desired fixture, the System Components Group at the center of the page shows the fixture on the first line. The rest of the list (fittings, pipe, tubing, etc.) will total up to what we refer to in the Unit Price section as "Rough-in, supply, waste, and vent." Note that for most fixtures we allow a nominal 5′ of tubing to reach from the fixture to a main or riser.
- Remember that gas- and oil-fired units need venting.

Reference Numbers

Reference numbers are shown in shaded boxes at the beginning of some major classifications. These numbers refer to related items in the Reference Section. The reference information may be an estimating procedure, an alternate pricing method, or technical information.

Note: Not all subdivisions listed here necessarily appear in this publication.

22 01 Operation and Maintenance of Plumbing

22 01 02 – Labor Adjustments

22 01 02.10 Boilers, General		Crew	Daily Output	Labor-Hours	Unit	Material	2013 Bare Costs Labor	Equipment	Total	Total Incl O&P
0010	**BOILERS, GENERAL**, Prices do not include flue piping, elec. wiring,									
0020	gas or oil piping, boiler base, pad, or tankless unless noted									
0100	Boiler H.P.: 10 KW = 34 lb./steam/hr. = 33,475 BTU/hr.									
0150	To convert SFR to BTU rating: Hot water, 150 x SFR;									
0160	Forced hot water, 180 x SFR; steam, 240 x SFR									

22 01 02.20 Labor Adjustment Factors

		Crew	Daily Output	Labor-Hours	Unit	Material	Labor	Equipment	Total	Total Incl O&P
0010	**LABOR ADJUSTMENT FACTORS**, (For Div. 21, 22 and 23) R220102-20									
0100	Labor factors, The below are reasonable suggestions, however									
0110	each project must be evaluated for its own peculiarities, and									
0120	the adjustments be increased or decreased depending on the									
0130	severity of the special conditions.									
1000	Add to labor for elevated installation (Above floor level)									
1080	10′ to 14.5′ high R221113-70						10%			
1100	15′ to 19.5′ high						20%			
1120	20′ to 24.5′ high						25%			
1140	25′ to 29.5′ high						35%			
1160	30′ to 34.5′ high						40%			
1180	35′ to 39.5′ high						50%			
1200	40′ and higher						55%			
2000	Add to labor for crawl space									
2100	3′ high						40%			
2140	4′ high						30%			
3000	Add to labor for multi-story building									
3100	Add per floor for floors 3 thru 19						2%			
3140	Add per floor for floors 20 and up						4%			
4000	Add to labor for working in existing occupied buildings									
4100	Hospital						35%			
4140	Office building						25%			
4180	School						20%			
4220	Factory or warehouse						15%			
4260	Multi dwelling						15%			
5000	Add to labor, miscellaneous									
5100	Cramped shaft						35%			
5140	Congested area						15%			
5180	Excessive heat or cold						30%			
9010	each project should be evaluated for its own peculiarities.									
9100	Other factors to be considered are:									
9140	Movement of material and equipment through finished areas									
9180	Equipment room									
9220	Attic space									
9260	No service road									
9300	Poor unloading/storage area									
9340	Congested site area/heavy traffic									

22 05 05 – Selective Plumbing Demolition

22 05 05.10 Plumbing Demolition		Crew	Daily Output	Labor-Hours	Unit	Material	2013 Bare Costs Labor	Equipment	Total	Total Incl O&P
0010	**PLUMBING DEMOLITION** R220105-10									
0400	Air compressor, up thru 2 H.P.	Q-1	10	1.600	Ea.		80.50		80.50	121
0410	3 H.P. thru 7-1/2 H.P.		5.60	2.857			144		144	216
0420	10 H.P. thru 15 H.P.		1.40	11.429			575		575	865
0430	20 H.P. thru 30 H.P.	Q-2	1.30	18.462			960		960	1,450
0500	Backflow preventer, up thru 2″ diameter	1 Plum	17	.471			26.50		26.50	39.50
0510	2-1/2″ thru 3″ diameter	Q-1	10	1.600			80.50		80.50	121
0520	4″ thru 6″ diameter	″	5	3.200			161		161	242
0530	8″ thru 10″ diameter	Q-2	3	8			415		415	630
0700	Carriers and supports									
0710	Fountains, sinks, lavatories and urinals	1 Plum	14	.571	Ea.		32		32	48
0720	Water closets	″	12	.667	″		37		37	56
0730	Grinder pump or sewage ejector system									
0732	Simplex	Q-1	7	2.286	Ea.		115		115	173
0734	Duplex	″	2.80	5.714			287		287	430
0738	Hot water dispenser	1 Plum	36	.222			12.40		12.40	18.70
0740	Hydrant, wall		26	.308			17.15		17.15	26
0744	Ground		12	.667			37		37	56
0760	Cleanouts and drains, up thru 4″ pipe diameter		10	.800			44.50		44.50	67
0764	5″ thru 8″ pipe diameter	Q-1	10	1.600			80.50		80.50	121
0780	Industrial safety fixtures	1 Plum	8	1			56		56	84
1020	Fixtures, including 10′ piping									
1100	Bathtubs, cast iron	1 Plum	4	2	Ea.		112		112	168
1120	Fiberglass		6	1.333			74.50		74.50	112
1140	Steel		5	1.600			89.50		89.50	134
1150	Bidet	Q-1	7	2.286			115		115	173
1200	Lavatory, wall hung	1 Plum	10	.800			44.50		44.50	67
1220	Counter top		8	1			56		56	84
1300	Sink, single compartment		8	1			56		56	84
1320	Double compartment		7	1.143			64		64	96
1340	Shower, stall and receptor	Q-1	6	2.667			134		134	202
1350	Group	″	7	2.286			115		115	173
1400	Water closet, floor mounted	1 Plum	8	1			56		56	84
1420	Wall mounted	″	7	1.143			64		64	96
1440	Wash fountain, 36″ diameter	Q-2	8	3			156		156	235
1442	54″ diameter	″	7	3.429			179		179	269
1500	Urinal, floor mounted	1 Plum	4	2			112		112	168
1520	Wall mounted	″	7	1.143			64		64	96
1590	Whirl pool or hot tub	Q-1	2.60	6.154			310		310	465
1600	Water fountains, free standing	1 Plum	8	1			56		56	84
1620	Wall or deck mounted		6	1.333			74.50		74.50	112
1800	Medical gas specialties		8	1			56		56	84
1810	Plumbing demo, Floor drain, Remove		12	.667			37		37	56
1900	Piping fittings, single connection, up thru 1-1/2″ diameter		30	.267			14.90		14.90	22.50
1910	2″ thru 4″ diameter		14	.571			32		32	48
1980	Pipe hanger/support removal		80	.100			5.60		5.60	8.40
1990	Glass pipe with fittings, 1″ thru 3″ diameter		200	.040	L.F.		2.23		2.23	3.36
1992	4″ thru 6″ diameter		150	.053			2.98		2.98	4.48
2000	Piping, metal, up thru 1-1/2″ diameter		200	.040			2.23		2.23	3.36
2050	2″ thru 3-1/2″ diameter		150	.053			2.98		2.98	4.48
2100	4″ thru 6″ diameter	2 Plum	100	.160			8.95		8.95	13.45
2150	8″ thru 14″ diameter	″	60	.267			14.90		14.90	22.50
2153	16″ thru 20″ diameter	Q-18	70	.343			18.15	.75	18.90	28.50

22 05 05 – Selective Plumbing Demolition

22 05 05.10 Plumbing Demolition		Crew	Daily Output	Labor-Hours	Unit	Material	2013 Bare Costs Labor	Equipment	Total	Total Incl O&P
2155	24" thru 26" diameter	Q-18	55	.436	L.F.		23	.96	23.96	35.50
2156	30" thru 36" diameter	↓	40	.600			31.50	1.32	32.82	49.50
2160	Plastic pipe with fittings, up thru 1-1/2" diameter	1 Plum	250	.032			1.79		1.79	2.69
2162	2" thru 3" diameter	"	200	.040			2.23		2.23	3.36
2164	4" thru 6" diameter	Q-1	200	.080			4.02		4.02	6.05
2166	8" thru 14" diameter		150	.107			5.35		5.35	8.05
2168	16" diameter		100	.160	↓		8.05		8.05	12.10
2170	Prison fixtures, lavatory or sink		18	.889	Ea.		44.50		44.50	67
2172	Shower		5.60	2.857			144		144	216
2174	Urinal or water closet		13	1.231			62		62	93
2180	Pumps, all fractional horse-power		12	1.333			67		67	101
2184	1 H.P. thru 5 H.P.		6	2.667			134		134	202
2186	7-1/2 H.P. thru 15 H.P.	↓	2.50	6.400			320		320	485
2188	20 H.P. thru 25 H.P.	Q-2	4	6			310		310	470
2190	30 H.P. thru 60 H.P.		.80	30			1,550		1,550	2,350
2192	75 H.P. thru 100 H.P.		.60	40			2,075		2,075	3,150
2194	150 H.P.	↓	.50	48			2,500		2,500	3,775
2198	Pump, sump or submersible	1 Plum	12	.667			37		37	56
2200	Receptors and interceptors, up thru 20 GPM	"	8	1			56		56	84
2204	25 thru 100 GPM	Q-1	6	2.667			134		134	202
2208	125 thru 300 GPM	"	2.40	6.667			335		335	505
2211	325 thru 500 GPM	Q-2	2.60	9.231	↓		480		480	725
2212	Deduct for salvage, aluminum scrap				Ton				700	770
2214	Brass scrap								2,450	2,675
2216	Copper scrap								3,200	3,525
2218	Lead scrap								520	570
2220	Steel scrap				↓				180	200
2230	Temperature maintenance cable	1 Plum	1200	.007	L.F.		.37		.37	.56
2250	Water heater, 40 gal.	"	6	1.333	Ea.		74.50		74.50	112
3100	Tanks, water heaters and liquid containers									
3110	Up thru 45 gallons	Q-1	22	.727	Ea.		36.50		36.50	55
3120	50 thru 120 gallons		14	1.143			57.50		57.50	86.50
3130	130 thru 240 gallons		7.60	2.105			106		106	159
3140	250 thru 500 gallons	↓	5.40	2.963			149		149	224
3150	600 thru 1000 gallons	Q-2	1.60	15			780		780	1,175
3160	1100 thru 2000 gallons		.70	34.286			1,775		1,775	2,700
3170	2100 thru 4000 gallons	↓	.50	48			2,500		2,500	3,775
6000	Remove and reset fixtures, minimum	1 Plum	6	1.333			74.50		74.50	112
6100	Maximum	"	4	2			112		112	168
9100	Valve, metal valves, strainers and similar, up thru 1-1/2" diameter	1 Stpi	28	.286			16.20		16.20	24.50
9110	2" thru 3" diameter	Q-1	11	1.455			73		73	110
9120	4" thru 6" diameter	"	8	2			100		100	151
9130	8" thru 14" diameter	Q-2	8	3			156		156	235
9140	16" thru 20" diameter		2	12			625		625	940
9150	24" diameter	↓	1.20	20			1,050		1,050	1,575
9200	Valve, plastic, up thru 1-1/2" diameter	1 Plum	42	.190			10.65		10.65	16
9210	2" thru 3" diameter		15	.533			30		30	45
9220	4" thru 6" diameter		12	.667			37		37	56
9300	Vent flashing and caps	↓	55	.145			8.10		8.10	12.25
9350	Water filter, commercial, 1" thru 1-1/2"	Q-1	2	8			400		400	605
9360	2" thru 2-1/2"	"	1.60	10	↓		500		500	755
9400	Water heaters									
9410	Up thru 245 GPH	Q-1	2.40	6.667	Ea.		335		335	505

22 05 Common Work Results for Plumbing

22 05 05 – Selective Plumbing Demolition

22 05 05.10 Plumbing Demolition		Crew	Daily Output	Labor-Hours	Unit	Material	2013 Bare Costs Labor	Equipment	Total	Total Incl O&P
9420	250 thru 756 GPH	Q-1	1.60	10	Ea.		500		500	755
9430	775 thru 1640 GPH		.80	20			1,000		1,000	1,525
9440	1650 thru 4000 GPH	Q-2	.50	48			2,500		2,500	3,775
9470	Water softener	Q-1	2	8			400		400	605

22 05 23 – General-Duty Valves for Plumbing Piping

22 05 23.10 Valves, Brass

		Crew	Daily Output	Labor-Hours	Unit	Material	Labor	Equipment	Total	Total Incl O&P
0010	**VALVES, BRASS**									
0032	For motorized valves, see Section 23 09 53.10									
0500	Gas cocks, threaded									
0510	1/4"	1 Plum	26	.308	Ea.	14.70	17.15		31.85	42
0520	3/8"		24	.333		14.70	18.60		33.30	44
0530	1/2"		24	.333		13.15	18.60		31.75	42.50
0540	3/4"		22	.364		15.50	20.50		36	47.50
0550	1"		19	.421		31	23.50		54.50	69.50
0560	1-1/4"		15	.533		44.50	30		74.50	94
0570	1-1/2"		13	.615		61	34.50		95.50	119
0580	2"		11	.727		103	40.50		143.50	175
0672	For larger sizes use lubricated plug valve, Section 23 05 23.70									

22 05 23.20 Valves, Bronze

		Crew	Daily Output	Labor-Hours	Unit	Material	Labor	Equipment	Total	Total Incl O&P
0010	**VALVES, BRONZE** R220523-80									
1020	Angle, 150 lb., rising stem, threaded									
1030	1/8" R220523-90	1 Plum	24	.333	Ea.	127	18.60		145.60	167
1040	1/4"		24	.333		127	18.60		145.60	167
1050	3/8"		24	.333		127	18.60		145.60	167
1060	1/2"		22	.364		127	20.50		147.50	170
1070	3/4"		20	.400		173	22.50		195.50	224
1080	1"		19	.421		249	23.50		272.50	310
1090	1-1/4"		15	.533		320	30		350	395
1100	1-1/2"		13	.615		415	34.50		449.50	510
1110	2"		11	.727		670	40.50		710.50	800
1300	Ball									
1304	Soldered									
1312	3/8"	1 Plum	21	.381	Ea.	13.45	21.50		34.95	47
1316	1/2"		18	.444		13.45	25		38.45	52.50
1320	3/4"		17	.471		22	26.50		48.50	64
1324	1"		15	.533		28	30		58	76
1328	1-1/4"		13	.615		47	34.50		81.50	103
1332	1-1/2"		11	.727		60	40.50		100.50	127
1336	2"		9	.889		75	49.50		124.50	157
1340	2-1/2"		7	1.143		380	64		444	515
1344	3"		5	1.600		505	89.50		594.50	690
1350	Single union end									
1358	3/8"	1 Plum	21	.381	Ea.	19.15	21.50		40.65	53
1362	1/2"		18	.444		20.50	25		45.50	60
1366	3/4"		17	.471		34.50	26.50		61	77.50
1370	1"		15	.533		46.50	30		76.50	96.50
1374	1-1/4"		13	.615		79.50	34.50		114	139
1378	1-1/2"		11	.727		99.50	40.50		140	170
1382	2"		9	.889		155	49.50		204.50	246
1398	Threaded, 150 psi									
1400	1/4"	1 Plum	24	.333	Ea.	12.70	18.60		31.30	42
1430	3/8"		24	.333		12.70	18.60		31.30	42

22 05 23 – General-Duty Valves for Plumbing Piping

22 05 23.20 Valves, Bronze		Crew	Daily Output	Labor-Hours	Unit	Material	Labor	Equipment	Total	Total Incl O&P
						2013 Bare Costs				
1450	1/2"	1 Plum	22	.364	Ea.	12.70	20.50		33.20	44.50
1460	3/4"		20	.400		21	22.50		43.50	56.50
1470	1"		19	.421		30.50	23.50		54	69
1480	1-1/4"		15	.533		53	30		83	104
1490	1-1/2"		13	.615		69	34.50		103.50	128
1500	2"		11	.727		83.50	40.50		124	153
1510	2-1/2"		9	.889		280	49.50		329.50	385
1520	3"		8	1		425	56		481	555
1600	Butterfly, 175 psi, full port, solder or threaded ends									
1610	Stainless steel disc and stem									
1620	1/4"	1 Plum	24	.333	Ea.	15.65	18.60		34.25	45.50
1630	3/8"		24	.333		15.65	18.60		34.25	45.50
1640	1/2"		22	.364		16.50	20.50		37	48.50
1650	3/4"		20	.400		26.50	22.50		49	63
1660	1"		19	.421		32.50	23.50		56	71.50
1670	1-1/4"		15	.533		52.50	30		82.50	103
1680	1-1/2"		13	.615		68	34.50		102.50	126
1690	2"		11	.727		85	40.50		125.50	155
1750	Check, swing, class 150, regrinding disc, threaded									
1800	1/8"	1 Plum	24	.333	Ea.	59.50	18.60		78.10	93
1830	1/4"		24	.333		59.50	18.60		78.10	93
1840	3/8"		24	.333		63	18.60		81.60	97.50
1850	1/2"		24	.333		67.50	18.60		86.10	103
1860	3/4"		20	.400		89.50	22.50		112	132
1870	1"		19	.421		128	23.50		151.50	177
1880	1-1/4"		15	.533		186	30		216	250
1890	1-1/2"		13	.615		216	34.50		250.50	290
1900	2"		11	.727		320	40.50		360.50	410
1910	2-1/2"	Q-1	15	1.067		715	53.50		768.50	865
1920	3"	"	13	1.231		950	62		1,012	1,150
2000	For 200 lb., add					5%	10%			
2040	For 300 lb., add					15%	15%			
2060	Check swing, 300#, sweat, 3/8" size	1 Plum	24	.333	Ea.	40	18.60		58.60	72
2070	1/2"		24	.333		40	18.60		58.60	72
2080	3/4"		20	.400		51	22.50		73.50	89.50
2090	1"		19	.421		71	23.50		94.50	114
2100	1-1/4"		15	.533		92	30		122	146
2110	1-1/2"		13	.615		122	34.50		156.50	186
2120	2"		11	.727		184	40.50		224.50	263
2130	2-1/2"	Q-1	15	1.067		435	53.50		488.50	560
2140	3"	"	13	1.231		585	62		647	735
2850	Gate, N.R.S., soldered, 125 psi									
2900	3/8"	1 Plum	24	.333	Ea.	51	18.60		69.60	84
2920	1/2"		24	.333		44	18.60		62.60	76.50
2940	3/4"		20	.400		49.50	22.50		72	88
2950	1"		19	.421		70.50	23.50		94	113
2960	1-1/4"		15	.533		108	30		138	164
2970	1-1/2"		13	.615		121	34.50		155.50	185
2980	2"		11	.727		171	40.50		211.50	250
2990	2-1/2"	Q-1	15	1.067		430	53.50		483.50	550
3000	3"	"	13	1.231		560	62		622	710
3350	Threaded, class 150									
3410	1/4"	1 Plum	24	.333	Ea.	73.50	18.60		92.10	109

22 05 23 – General-Duty Valves for Plumbing Piping

22 05 23.20 Valves, Bronze		Crew	Daily Output	Labor-Hours	Unit	2013 Bare Costs Material	Labor	Equipment	Total	Total Incl O&P
3420	3/8"	1 Plum	24	.333	Ea.	73.50	18.60		92.10	109
3430	1/2"		24	.333		69	18.60		87.60	104
3440	3/4"		20	.400		79	22.50		101.50	121
3450	1"		19	.421		99.50	23.50		123	145
3460	1-1/4" size		15	.533		135	30		165	193
3470	1-1/2"		13	.615		193	34.50		227.50	264
3480	2"		11	.727		230	40.50		270.50	315
3490	2-1/2"	Q-1	15	1.067		540	53.50		593.50	670
3500	3" size	"	13	1.231		765	62		827	935
3600	Gate, flanged, 150 lb.									
3610	1"	1 Plum	7	1.143	Ea.	1,300	64		1,364	1,525
3620	1-1/2"		6	1.333		1,625	74.50		1,699.50	1,875
3630	2"		5	1.600		2,400	89.50		2,489.50	2,750
3634	2-1/2"	Q-1	5	3.200		3,700	161		3,861	4,325
3640	3"	"	4.50	3.556		4,300	179		4,479	5,025
3850	Rising stem, soldered, 300 psi									
3900	3/8"	1 Plum	24	.333	Ea.	141	18.60		159.60	183
3920	1/2"		24	.333		116	18.60		134.60	156
3940	3/4"		20	.400		132	22.50		154.50	180
3950	1"		19	.421		179	23.50		202.50	233
3960	1-1/4"		15	.533		248	30		278	320
3970	1-1/2"		13	.615		300	34.50		334.50	380
3980	2"		11	.727		480	40.50		520.50	590
3990	2-1/2"	Q-1	15	1.067		1,025	53.50		1,078.50	1,200
4000	3"	"	13	1.231		1,600	62		1,662	1,850
4250	Threaded, class 150									
4310	1/4"	1 Plum	24	.333	Ea.	67	18.60		85.60	102
4320	3/8"		24	.333		67	18.60		85.60	102
4330	1/2"		24	.333		61	18.60		79.60	95.50
4340	3/4"		20	.400		71.50	22.50		94	112
4350	1"		19	.421		96	23.50		119.50	141
4360	1-1/4"		15	.533		130	30		160	188
4370	1-1/2"		13	.615		164	34.50		198.50	233
4380	2"		11	.727		221	40.50		261.50	305
4390	2-1/2"	Q-1	15	1.067		515	53.50		568.50	645
4400	3"	"	13	1.231		720	62		782	885
4500	For 300 psi, threaded, add					100%	15%			
4540	For chain operated type, add					15%				
4850	Globe, class 150, rising stem, threaded									
4920	1/4"	1 Plum	24	.333	Ea.	96	18.60		114.60	134
4940	3/8" size		24	.333		94.50	18.60		113.10	132
4950	1/2"		24	.333		94.50	18.60		113.10	132
4960	3/4"		20	.400		127	22.50		149.50	173
4970	1"		19	.421		199	23.50		222.50	255
4980	1-1/4"		15	.533		315	30		345	395
4990	1-1/2"		13	.615		385	34.50		419.50	475
5000	2"		11	.727		575	40.50		615.50	695
5010	2-1/2"	Q-1	15	1.067		1,150	53.50		1,203.50	1,350
5020	3"	"	13	1.231		1,650	62		1,712	1,925
5120	For 300 lb. threaded, add					50%	15%			
5600	Relief, pressure & temperature, self-closing, ASME, threaded									
5640	3/4"	1 Plum	28	.286	Ea.	155	15.95		170.95	195
5650	1"		24	.333		236	18.60		254.60	288

22 05 23.20 Valves, Bronze		Crew	Daily Output	Labor-Hours	Unit	2013 Bare Costs Material	Labor	Equipment	Total	Total Incl O&P
5660	1-1/4"	1 Plum	20	.400	Ea.	455	22.50		477.50	535
5670	1-1/2"		18	.444		870	25		895	1,000
5680	2"		16	.500		950	28		978	1,100
5950	Pressure, poppet type, threaded									
6000	1/2"	1 Plum	30	.267	Ea.	41	14.90		55.90	67.50
6040	3/4"	"	28	.286	"	44	15.95		59.95	72
6400	Pressure, water, ASME, threaded									
6440	3/4"	1 Plum	28	.286	Ea.	87	15.95		102.95	120
6450	1"		24	.333		186	18.60		204.60	232
6460	1-1/4"		20	.400		294	22.50		316.50	360
6470	1-1/2"		18	.444		405	25		430	485
6480	2"		16	.500		590	28		618	685
6490	2-1/2"		15	.533		2,675	30		2,705	2,975
6900	Reducing, water pressure									
6920	300 psi to 25-75 psi, threaded or sweat									
6940	1/2"	1 Plum	24	.333	Ea.	278	18.60		296.60	335
6950	3/4"		20	.400		278	22.50		300.50	340
6960	1"		19	.421		430	23.50		453.50	510
6970	1-1/4"		15	.533		770	30		800	890
6980	1-1/2"		13	.615		1,150	34.50		1,184.50	1,325
6990	2"		11	.727		1,550	40.50		1,590.50	1,750
7100	For built-in by-pass or 10-35 psi, add					28			28	31
7700	High capacity, 250 psi to 25-75 psi, threaded									
7740	1/2"	1 Plum	24	.333	Ea.	380	18.60		398.60	445
7780	3/4"		20	.400		380	22.50		402.50	450
7790	1"		19	.421		545	23.50		568.50	630
7800	1-1/4"		15	.533		940	30		970	1,075
7810	1-1/2"		13	.615		1,375	34.50		1,409.50	1,575
7820	2"		11	.727		2,000	40.50		2,040.50	2,250
7830	2-1/2"		9	.889		2,875	49.50		2,924.50	3,225
7840	3"		8	1		3,450	56		3,506	3,875
7850	3" flanged (iron body)	Q-1	10	1.600		3,400	80.50		3,480.50	3,850
7860	4" flanged (iron body)	"	8	2		4,400	100		4,500	4,975
7920	For higher pressure, add					25%				
8000	Silent check, bronze trim									
8010	Compact wafer type, for 125 or 150 lb. flanges									
8020	1-1/2"	1 Plum	11	.727	Ea.	320	40.50		360.50	410
8021	2"	"	9	.889		335	49.50		384.50	445
8022	2-1/2"	Q-1	9	1.778		355	89.50		444.50	525
8023	3"		8	2		395	100		495	585
8024	4"		5	3.200		675	161		836	985
8025	5"	Q-2	6	4		895	208		1,103	1,300
8026	6"	"	5	4.800		1,050	250		1,300	1,550
8050	For 250 or 300 lb. flanges, thru 6" no change									
8060	Full flange wafer type, 150 lb.									
8063	1-1/2"	1 Plum	11	.727	Ea.	420	40.50		460.50	525
8064	2"	"	9	.889		565	49.50		614.50	695
8065	2-1/2"	Q-1	9	1.778		635	89.50		724.50	835
8066	3"		8	2		705	100		805	925
8067	4"		5	3.200		1,050	161		1,211	1,400
8068	5"	Q-2	6	4		1,350	208		1,558	1,800
8069	6"	"	5	4.800		1,775	250		2,025	2,325
8080	For 300 lb., add					40%	10%			

22 05 23.20 Valves, Bronze		Crew	Daily Output	Labor-Hours	Unit	Material	2013 Bare Costs Labor	Equipment	Total	Total Incl O&P
8100	Globe type, 150 lb.									
8110	2″	1 Plum	9	.889	Ea.	660	49.50		709.50	800
8111	2-1/2″	Q-1	9	1.778		815	89.50		904.50	1,025
8112	3″		8	2		975	100		1,075	1,225
8113	4″		5	3.200		1,350	161		1,511	1,725
8114	5″	Q-2	6	4		1,650	208		1,858	2,150
8115	6″	″	5	4.800		2,300	250		2,550	2,900
8130	For 300 lb., add					20%	10%			
8140	Screwed end type, 250 lb.									
8141	1/2″	1 Plum	24	.333	Ea.	59.50	18.60		78.10	93.50
8142	3/4″		20	.400		59.50	22.50		82	99
8143	1″		19	.421		68	23.50		91.50	110
8144	1-1/4″		15	.533		94	30		124	148
8145	1-1/2″		13	.615		105	34.50		139.50	168
8146	2″		11	.727		142	40.50		182.50	217
8350	Tempering, water, sweat connections									
8400	1/2″	1 Plum	24	.333	Ea.	93.50	18.60		112.10	131
8440	3/4″	″	20	.400	″	115	22.50		137.50	160
8650	Threaded connections									
8700	1/2″	1 Plum	24	.333	Ea.	115	18.60		133.60	154
8740	3/4″		20	.400		525	22.50		547.50	610
8750	1″		19	.421		590	23.50		613.50	685
8760	1-1/4″		15	.533		910	30		940	1,050
8770	1-1/2″		13	.615		995	34.50		1,029.50	1,150
8780	2″		11	.727		1,500	40.50		1,540.50	1,700

22 05 23.40 Valves, Lined, Corrosion Resistant/High Purity		Crew	Daily Output	Labor-Hours	Unit	Material	Labor	Equipment	Total	Total Incl O&P
0010	**VALVES, LINED, CORROSION RESISTANT/HIGH PURITY**									
2000	Butterfly 150 lb. ductile iron,									
2010	Wafer type									
2030	TFE lined 3″, lever handle	Q-1	8	2	Ea.	1,500	100		1,600	1,800
2050	4″, lever handle	″	5	3.200		1,950	161		2,111	2,375
2070	6″, gear operated	Q-2	4.50	5.333		2,725	278		3,003	3,425
2080	8″, gear operated		4	6		2,825	310		3,135	3,575
2090	10″, gear operated		3.50	6.857		5,000	355		5,355	6,050
2100	12″, gear operated		2.50	9.600		5,975	500		6,475	7,325
3500	Check lift, 125 lb., cast iron flanged									
3510	Horizontal PPL or SL lined									
3530	1″	1 Plum	14	.571	Ea.	670	32		702	790
3540	1-1/2″		11	.727		815	40.50		855.50	955
3550	2″		8	1		945	56		1,001	1,125
3560	2-1/2″	Q-1	5	3.200		1,250	161		1,411	1,600
3570	3″		4.50	3.556		1,550	179		1,729	1,975
3590	4″		3	5.333		2,050	268		2,318	2,650
3610	6″	Q-2	3	8		3,475	415		3,890	4,450
3620	8″	″	2.50	9.600		7,625	500		8,125	9,150
4250	Vertical PPL or SL lined									
4270	1″	1 Plum	14	.571	Ea.	750	32		782	875
4290	1-1/2″		11	.727		875	40.50		915.50	1,025
4300	2″		8	1		1,175	56		1,231	1,375
4310	2-1/2″	Q-1	5	3.200		1,325	161		1,486	1,700
4320	3″		4.50	3.556		1,725	179		1,904	2,150
4340	4″		3	5.333		2,900	268		3,168	3,600

22 05 23.40 Valves, Lined, Corrosion Resistant/High Purity		Crew	Daily Output	Labor-Hours	Unit	2013 Bare Costs Material	Labor	Equipment	Total	Total Incl O&P
4360	6"	Q-2	3	8	Ea.	3,000	415		3,415	3,925
4370	8"	"	2.50	9.600		3,450	500		3,950	4,550
5000	Clamp type, ductile iron, 150 lb. flanged									
5010	TFE lined									
5030	1" size, lever handle	1 Plum	9	.889	Ea.	1,350	49.50		1,399.50	1,550
5050	1-1/2" size, lever handle		6	1.333		1,775	74.50		1,849.50	2,050
5060	2" size, lever handle		5	1.600		2,150	89.50		2,239.50	2,500
5080	3" size, lever handle	Q-1	4.50	3.556		2,725	179		2,904	3,275
5100	4" size, gear operated	"	3	5.333		3,900	268		4,168	4,700
5120	6" size, gear operated	Q-2	3	8		13,500	415		13,915	15,500
5130	8" size, gear operated	"	2.50	9.600		14,600	500		15,100	16,800
6000	Diaphragm type, cast iron, 125 lb. flanged									
6010	PTFE or VITON, lined									
6030	1" size, handwheel operated	1 Plum	9	.889	Ea.	315	49.50		364.50	425
6050	1-1/2" size, handwheel operated		6	1.333		385	74.50		459.50	530
6060	2" size, handwheel operated		5	1.600		460	89.50		549.50	640
6080	3" size, handwheel operated	Q-1	4.50	3.556		760	179		939	1,100
6100	4" size, handwheel operated	"	3	5.333		1,375	268		1,643	1,925
6120	6" size, handwheel operated	Q-2	3	8		2,350	415		2,765	3,225
6130	8" size, handwheel operated	"	2.50	9.600		4,750	500		5,250	5,975

22 05 23.60 Valves, Plastic		Crew	Daily Output	Labor-Hours	Unit	Material	Labor	Equipment	Total	Total Incl O&P
0010	**VALVES, PLASTIC** R220523-90									
1100	Angle, PVC, threaded									
1110	1/4"	1 Plum	26	.308	Ea.	74	17.15		91.15	107
1120	1/2"		26	.308		74	17.15		91.15	107
1130	3/4"		25	.320		87.50	17.85		105.35	124
1140	1"		23	.348		106	19.40		125.40	146
1150	Ball, PVC, socket or threaded, single union									
1230	1/2"	1 Plum	26	.308	Ea.	26.50	17.15		43.65	55.50
1240	3/4"		25	.320		30.50	17.85		48.35	60.50
1250	1"		23	.348		38.50	19.40		57.90	71
1260	1-1/4"		21	.381		51.50	21.50		73	89
1270	1-1/2"		20	.400		61.50	22.50		84	101
1280	2"		17	.471		89	26.50		115.50	138
1290	2-1/2"	Q-1	26	.615		375	31		406	460
1300	3"		24	.667		455	33.50		488.50	550
1310	4"		20	.800		770	40		810	910
1360	For PVC, flanged, add					100%	15%			
1450	Double union 1/2"	1 Plum	26	.308		45	17.15		62.15	75.50
1460	3/4"		25	.320		56.50	17.85		74.35	89
1470	1"		23	.348		67.50	19.40		86.90	103
1480	1-1/4"		21	.381		107	21.50		128.50	150
1490	1-1/2"		20	.400		113	22.50		135.50	158
1500	2"		17	.471		151	26.50		177.50	206
1650	CPVC, socket or threaded, single union									
1700	1/2"	1 Plum	26	.308	Ea.	53	17.15		70.15	84.50
1720	3/4"		25	.320		66.50	17.85		84.35	101
1730	1"		23	.348		80	19.40		99.40	117
1750	1-1/4"		21	.381		127	21.50		148.50	172
1760	1-1/2"		20	.400		127	22.50		149.50	174
1770	2"		17	.471		176	26.50		202.50	233
1780	3"	Q-1	24	.667		725	33.50		758.50	845

22 05 23 – General-Duty Valves for Plumbing Piping

22 05 23.60 Valves, Plastic		Crew	Daily Output	Labor-Hours	Unit	2013 Bare Costs Material	Labor	Equipment	Total	Total Incl O&P
1840	For CPVC, flanged, add				Ea.	65%	15%			
1880	For true union, socket or threaded, add				↓	50%	5%			
2050	Polypropylene, threaded									
2100	1/4"	1 Plum	26	.308	Ea.	42.50	17.15		59.65	72.50
2120	3/8"		26	.308		42.50	17.15		59.65	72.50
2130	1/2"		26	.308		42.50	17.15		59.65	72.50
2140	3/4"		25	.320		52	17.85		69.85	84
2150	1"		23	.348		59	19.40		78.40	94
2160	1-1/4"		21	.381		80	21.50		101.50	120
2170	1-1/2"		20	.400		98.50	22.50		121	142
2180	2"	↓	17	.471		132	26.50		158.50	185
2190	3"	Q-1	24	.667		299	33.50		332.50	380
2200	4"	"	20	.800	↓	530	40		570	640
2550	PVC, three way, socket or threaded									
2600	1/2"	1 Plum	26	.308	Ea.	62.50	17.15		79.65	94.50
2640	3/4"		25	.320		71	17.85		88.85	105
2650	1"		23	.348		76.50	19.40		95.90	113
2660	1-1/2"		20	.400		181	22.50		203.50	233
2670	2"	↓	17	.471		206	26.50		232.50	267
2680	3"	Q-1	24	.667		430	33.50		463.50	525
2740	For flanged, add				↓	60%	15%			
3150	Ball check, PVC, socket or threaded									
3200	1/4"	1 Plum	26	.308	Ea.	41.50	17.15		58.65	72
3220	3/8"		26	.308		41.50	17.15		58.65	72
3240	1/2"		26	.308		41.50	17.15		58.65	72
3250	3/4"		25	.320		46.50	17.85		64.35	78.50
3260	1"		23	.348		58.50	19.40		77.90	93
3270	1-1/4"		21	.381		98	21.50		119.50	140
3280	1-1/2"		20	.400		98	22.50		120.50	142
3290	2"	↓	17	.471		133	26.50		159.50	187
3310	3"	Q-1	24	.667		370	33.50		403.50	455
3320	4"	"	20	.800		525	40		565	635
3360	For PVC, flanged, add				↓	50%	15%			
3750	CPVC, socket or threaded									
3800	1/2"	1 Plum	26	.308	Ea.	64	17.15		81.15	96.50
3840	3/4"		25	.320		75.50	17.85		93.35	110
3850	1"		23	.348		90.50	19.40		109.90	129
3860	1-1/2"		20	.400		156	22.50		178.50	205
3870	2"	↓	17	.471		156	26.50		182.50	211
3880	3"	Q-1	24	.667		565	33.50		598.50	670
3920	4"	"	20	.800		765	40		805	900
3930	For CPVC, flanged, add				↓	40%	15%			
4340	Polypropylene, threaded									
4360	1/2"	1 Plum	26	.308	Ea.	41.50	17.15		58.65	72
4400	3/4"		25	.320		57.50	17.85		75.35	90.50
4440	1"		23	.348		61.50	19.40		80.90	97
4450	1-1/2"		20	.400		119	22.50		141.50	165
4460	2"	↓	17	.471		151	26.50		177.50	206
4500	For polypropylene flanged, add				↓	200%	15%			
4850	Foot valve, PVC, socket or threaded									
4900	1/2"	1 Plum	34	.235	Ea.	69.50	13.15		82.65	96.50
4930	3/4"		32	.250		79	13.95		92.95	108
4940	1"		28	.286		102	15.95		117.95	137

22 05 Common Work Results for Plumbing

22 05 23 – General-Duty Valves for Plumbing Piping

22 05 23.60 Valves, Plastic		Crew	Daily Output	Labor-Hours	Unit	Material	2013 Bare Costs Labor	Equipment	Total	Total Incl O&P
4950	1-1/4"	1 Plum	27	.296	Ea.	197	16.55		213.55	241
4960	1-1/2"		26	.308		197	17.15		214.15	242
4970	2"		24	.333		229	18.60		247.60	280
4980	3"		20	.400		545	22.50		567.50	635
4990	4"		18	.444		960	25		985	1,100
5000	For flanged, add					25%	10%			
5050	CPVC, socket or threaded									
5060	1/2"	1 Plum	34	.235	Ea.	72	13.15		85.15	99.50
5070	3/4"		32	.250		83	13.95		96.95	113
5080	1"		28	.286		102	15.95		117.95	137
5090	1-1/4"		27	.296		164	16.55		180.55	205
5100	1-1/2"		26	.308		164	17.15		181.15	206
5110	2"		24	.333		211	18.60		229.60	260
5120	3"		20	.400		430	22.50		452.50	505
5130	4"		18	.444		780	25		805	895
5140	For flanged, add					25%	10%			
5280	Needle valve, PVC, threaded									
5300	1/4"	1 Plum	26	.308	Ea.	51	17.15		68.15	82
5340	3/8"		26	.308		59	17.15		76.15	91
5360	1/2"		26	.308		59	17.15		76.15	91
5380	For polypropylene, add					10%				
5800	Y check, PVC, socket or threaded									
5820	1/2"	1 Plum	26	.308	Ea.	55	17.15		72.15	86.50
5840	3/4"		25	.320		89	17.85		106.85	125
5850	1"		23	.348		96.50	19.40		115.90	135
5860	1-1/4"		21	.381		151	21.50		172.50	199
5870	1-1/2"		20	.400		165	22.50		187.50	216
5880	2"		17	.471		205	26.50		231.50	266
5890	2-1/2"		15	.533		435	30		465	520
5900	3"	Q-1	24	.667		435	33.50		468.50	525
5910	4"	"	20	.800		710	40		750	840
5960	For PVC flanged, add					45%	15%			
6350	Y sediment strainer, PVC, socket or threaded									
6400	1/2"	1 Plum	26	.308	Ea.	53	17.15		70.15	84
6440	3/4"		24	.333		56	18.60		74.60	89.50
6450	1"		23	.348		67	19.40		86.40	103
6460	1-1/4"		21	.381		112	21.50		133.50	155
6470	1-1/2"		20	.400		112	22.50		134.50	157
6480	2"		17	.471		138	26.50		164.50	192
6490	2-1/2"		15	.533		335	30		365	415
6500	3"	Q-1	24	.667		335	33.50		368.50	420
6510	4"	"	20	.800		560	40		600	675
6560	For PVC, flanged, add					55%	15%			

22 05 29 – Hangers and Supports for Plumbing Piping and Equipment

22 05 29.10 Hangers & Supp. for Plumb'g/HVAC Pipe/Equip.		Crew	Daily Output	Labor-Hours	Unit	Material	Labor	Equipment	Total	Total Incl O&P
0010	**HANGERS AND SUPPORTS FOR PLUMB'G/HVAC PIPE/EQUIP.**									
0011	TYPE numbers per MSS-SP58									
0050	Brackets									
0060	Beam side or wall, malleable iron, TYPE 34									
0070	3/8" threaded rod size	1 Plum	48	.167	Ea.	3.62	9.30		12.92	18
0080	1/2" threaded rod size		48	.167		6.40	9.30		15.70	21
0090	5/8" threaded rod size		48	.167		9.80	9.30		19.10	25

22 05 29.10 Hangers & Supp. for Plumb'g/HVAC Pipe/Equip.		Crew	Daily Output	Labor-Hours	Unit	2013 Bare Costs Material	Labor	Equipment	Total	Total Incl O&P
0100	3/4" threaded rod size	1 Plum	48	.167	Ea.	11.85	9.30		21.15	27
0110	7/8" threaded rod size		48	.167		13.95	9.30		23.25	29.50
0120	For concrete installation, add						30%			
0150	Wall, welded steel, medium, TYPE 32									
0160	0 size, 12" wide, 18" deep	1 Plum	34	.235	Ea.	241	13.15		254.15	285
0170	1 size, 18" wide, 24" deep		34	.235		286	13.15		299.15	335
0180	2 size, 24" wide, 30" deep		34	.235		380	13.15		393.15	435
0200	Beam attachment, welded, TYPE 22									
0202	3/8"	Q-15	80	.200	Ea.	8.85	10.05	.66	19.56	25.50
0203	1/2"		76	.211		8.85	10.55	.69	20.09	26.50
0204	5/8"		72	.222		9.30	11.15	.73	21.18	28
0205	3/4"		68	.235		10.15	11.80	.78	22.73	30
0206	7/8"		64	.250		14.70	12.55	.83	28.08	36
0207	1"		56	.286		20	14.35	.94	35.29	44.50
0300	Clamps									
0310	C-clamp, for mounting on steel beam flange, w/locknut, TYPE 23									
0320	3/8" threaded rod size	1 Plum	160	.050	Ea.	3.51	2.79		6.30	8.05
0330	1/2" threaded rod size		160	.050		4.37	2.79		7.16	9
0340	5/8" threaded rod size		160	.050		6.70	2.79		9.49	11.55
0350	3/4" threaded rod size		160	.050		9	2.79		11.79	14.10
0352	7/8" threaded rod size		140	.057		28.50	3.19		31.69	36.50
0400	High temperature to 1050°F, alloy steel									
0410	4" pipe size	Q-1	106	.151	Ea.	28.50	7.60		36.10	43
0420	6" pipe size		106	.151		48.50	7.60		56.10	64.50
0430	8" pipe size		97	.165		54.50	8.30		62.80	72.50
0440	10" pipe size		84	.190		86.50	9.55		96.05	109
0450	12" pipe size		72	.222		100	11.15		111.15	127
0460	14" pipe size		64	.250		263	12.55		275.55	310
0470	16" pipe size		56	.286		274	14.35		288.35	320
0480	Beam clamp, flange type, TYPE 25									
0482	For 3/8" bolt	1 Plum	48	.167	Ea.	13.25	9.30		22.55	28.50
0483	For 1/2" bolt		44	.182		18.35	10.15		28.50	35.50
0484	For 5/8" bolt		40	.200		18.95	11.15		30.10	38
0485	For 3/4" bolt		36	.222		18.95	12.40		31.35	39.50
0486	For 1" bolt		32	.250		18.95	13.95		32.90	42
0500	I-beam, for mounting on bottom flange, strap iron, TYPE 21									
0510	2" flange size	1 Plum	96	.083	Ea.	15.85	4.65		20.50	24.50
0520	3" flange size		95	.084		17.95	4.70		22.65	27
0530	4" flange size		93	.086		8.30	4.80		13.10	16.40
0540	5" flange size		92	.087		9.15	4.85		14	17.35
0550	6" flange size		90	.089		10.50	4.96		15.46	19
0560	7" flange size		88	.091		11.50	5.05		16.55	20.50
0570	8" flange size		86	.093		12.25	5.20		17.45	21.50
0600	One hole, vertical mounting, malleable iron									
0610	1/2" pipe size	1 Plum	160	.050	Ea.	1.24	2.79		4.03	5.55
0620	3/4" pipe size		145	.055		1.50	3.08		4.58	6.30
0630	1" pipe size		136	.059		2.09	3.28		5.37	7.25
0640	1-1/4" pipe size		128	.063		2.49	3.49		5.98	8
0650	1-1/2" pipe size		120	.067		2.77	3.72		6.49	8.65
0660	2" pipe size		112	.071		4.04	3.99		8.03	10.45
0670	2-1/2" pipe size		104	.077		7.90	4.29		12.19	15.15
0680	3" pipe size		96	.083		11.25	4.65		15.90	19.35
0690	3-1/2" pipe size		90	.089		13.70	4.96		18.66	22.50

22 05 29 – Hangers and Supports for Plumbing Piping and Equipment

22 05 29.10 Hangers & Supp. for Plumb'g/HVAC Pipe/Equip.		Crew	Daily Output	Labor-Hours	Unit	2013 Bare Costs Material	Labor	Equipment	Total	Total Incl O&P
0700	4" pipe size	1 Plum	84	.095	Ea.	15.15	5.30		20.45	24.50
0750	Riser or extension pipe, carbon steel, TYPE 8									
0756	1/2" pipe size	1 Plum	52	.154	Ea.	5.05	8.60		13.65	18.50
0760	3/4" pipe size		48	.167		5.05	9.30		14.35	19.55
0770	1" pipe size		47	.170		5.25	9.50		14.75	20
0780	1-1/4" pipe size		46	.174		6.45	9.70		16.15	21.50
0790	1-1/2" pipe size		45	.178		6.95	9.90		16.85	22.50
0800	2" pipe size		43	.186		7.20	10.40		17.60	23.50
0810	2-1/2" pipe size		41	.195		7.85	10.90		18.75	25
0820	3" pipe size		40	.200		8.70	11.15		19.85	26.50
0830	3-1/2" pipe size		39	.205		10.60	11.45		22.05	29
0840	4" pipe size		38	.211		10.85	11.75		22.60	29.50
0850	5" pipe size		37	.216		14.70	12.05		26.75	34.50
0860	6" pipe size		36	.222		18.20	12.40		30.60	38.50
0870	8" pipe size		34	.235		31	13.15		44.15	54
0880	10" pipe size		32	.250		42.50	13.95		56.45	67.50
0890	12" pipe size		28	.286		60	15.95		75.95	90
0900	For plastic coating 3/4" to 4", add					190%				
0910	For copper plating 3/4" to 4", add					58%				
0950	Two piece, complete, carbon steel, medium weight, TYPE 4									
0960	1/2" pipe size	Q-1	137	.117	Ea.	3.67	5.85		9.52	12.90
0970	3/4" pipe size		134	.119		3.67	6		9.67	13.10
0980	1" pipe size		132	.121		3.70	6.10		9.80	13.20
0990	1-1/4" pipe size		130	.123		4.86	6.20		11.06	14.65
1000	1-1/2" pipe size		126	.127		4.86	6.40		11.26	14.95
1010	2" pipe size		124	.129		5.65	6.50		12.15	15.95
1020	2-1/2" pipe size		120	.133		6	6.70		12.70	16.70
1030	3" pipe size		117	.137		6.70	6.85		13.55	17.70
1040	3-1/2" pipe size		114	.140		9.50	7.05		16.55	21
1050	4" pipe size		110	.145		9.50	7.30		16.80	21.50
1060	5" pipe size		106	.151		18.20	7.60		25.80	31.50
1070	6" pipe size		104	.154		23	7.75		30.75	37
1080	8" pipe size		100	.160		27.50	8.05		35.55	42.50
1090	10" pipe size		96	.167		49.50	8.35		57.85	67
1100	12" pipe size		89	.180		64	9.05		73.05	83.50
1110	14" pipe size		82	.195		123	9.80		132.80	150
1120	16" pipe size		68	.235		133	11.80		144.80	164
1130	For galvanized, add					45%				
1150	Insert, concrete									
1160	Wedge type, carbon steel body, malleable iron nut, galvanized									
1170	1/4" threaded rod size	1 Plum	96	.083	Ea.	2.25	4.65		6.90	9.50
1180	3/8" threaded rod size		96	.083		2.26	4.65		6.91	9.50
1190	1/2" threaded rod size		96	.083		2.54	4.65		7.19	9.80
1200	5/8" threaded rod size		96	.083		3.02	4.65		7.67	10.30
1210	3/4" threaded rod size		96	.083		6.05	4.65		10.70	13.65
1220	7/8" threaded rod size		96	.083		7.85	4.65		12.50	15.65
1250	Pipe guide sized for insulation									
1260	No. 1, 1" pipe size, 1" thick insulation	1 Stpi	26	.308	Ea.	157	17.45		174.45	199
1270	No. 2, 1-1/4"-2" pipe size, 1" thick insulation		23	.348		183	19.70		202.70	232
1280	No. 3, 1-1/4"-2" pipe size, 1-1/2" thick insulation		21	.381		183	21.50		204.50	235
1290	No. 4, 2-1/2"-3-1/2" pipe size, 1-1/2" thick insulation		18	.444		183	25		208	240
1300	No. 5, 4"-5" pipe size, 1-1/2" thick insulation		16	.500		205	28.50		233.50	269
1310	No. 6, 5"-6" pipe size, 2" thick insulation	Q-5	21	.762		228	39		267	310

22 05 29 – Hangers and Supports for Plumbing Piping and Equipment

22 05 29.10	Hangers & Supp. for Plumb'g/HVAC Pipe/Equip.	Crew	Daily Output	Labor-Hours	Unit	2013 Bare Costs Material	Labor	Equipment	Total	Total Incl O&P
1320	No. 7, 8" pipe size, 2" thick insulation	Q-5	16	1	Ea.	315	51		366	420
1330	No. 8, 10" pipe size, 2" thick insulation		12	1.333		520	68		588	670
1340	No. 9, 12" pipe size, 2" thick insulation	Q-6	17	1.412		520	74.50		594.50	680
1350	No. 10, 12"-14" pipe size, 2-1/2" thick insulation		16	1.500		580	79.50		659.50	755
1360	No. 11, 16" pipe size, 2-1/2" thick insulation		10.50	2.286		580	121		701	815
1370	No. 12, 16"-18" pipe size, 3" thick insulation		9	2.667		815	141		956	1,100
1380	No. 13, 20" pipe size, 3" thick insulation		7.50	3.200		815	169		984	1,150
1390	No. 14, 24" pipe size, 3" thick insulation		7	3.429		1,100	181		1,281	1,500
1400	Bands									
1410	Adjustable band, carbon steel, for non-insulated pipe, TYPE 7									
1420	1/2" pipe size	Q-1	142	.113	Ea.	.79	5.65		6.44	9.35
1430	3/4" pipe size		140	.114		.79	5.75		6.54	9.50
1440	1" pipe size		137	.117		.79	5.85		6.64	9.70
1450	1-1/4" pipe size		134	.119		.85	6		6.85	10
1460	1-1/2" pipe size		131	.122		.85	6.15		7	10.20
1470	2" pipe size		129	.124		.85	6.25		7.10	10.35
1480	2-1/2" pipe size		125	.128		1.53	6.45		7.98	11.40
1490	3" pipe size		122	.131		1.66	6.60		8.26	11.75
1500	3-1/2" pipe size		119	.134		2.68	6.75		9.43	13.10
1510	4" pipe size		114	.140		2.68	7.05		9.73	13.55
1520	5" pipe size		110	.145		4.50	7.30		11.80	15.95
1530	6" pipe size		108	.148		5	7.45		12.45	16.70
1540	8" pipe size		104	.154		6.75	7.75		14.50	19.10
1550	For copper plated, add					50%				
1560	For galvanized, add					30%				
1570	For plastic coating, add					30%				
1600	Adjusting nut malleable iron, steel band, TYPE 9									
1610	1/2" pipe size, galvanized band	Q-1	137	.117	Ea.	1.40	5.85		7.25	10.40
1620	3/4" pipe size, galvanized band		135	.119		1.53	5.95		7.48	10.65
1630	1" pipe size, galvanized band		132	.121		1.53	6.10		7.63	10.85
1640	1-1/4" pipe size, galvanized band		129	.124		1.71	6.25		7.96	11.30
1650	1-1/2" pipe size, galvanized band		126	.127		1.71	6.40		8.11	11.50
1660	2" pipe size, galvanized band		124	.129		1.80	6.50		8.30	11.75
1670	2-1/2" pipe size, galvanized band		120	.133		2.93	6.70		9.63	13.30
1680	3" pipe size, galvanized band		117	.137		3.06	6.85		9.91	13.70
1690	3-1/2" pipe size, galvanized band		114	.140		4.10	7.05		11.15	15.10
1700	4" pipe size, cadmium plated band		110	.145		4.10	7.30		11.40	15.50
1740	For plastic coated band, add					35%				
1750	For completely copper coated, add					45%				
1800	Clevis, adjustable, carbon steel, for non-insulated pipe, TYPE 1									
1810	1/2" pipe size	Q-1	137	.117	Ea.	1.87	5.85		7.72	10.90
1820	3/4" pipe size		135	.119		1.87	5.95		7.82	11
1830	1" pipe size		132	.121		1.90	6.10		8	11.25
1840	1-1/4" pipe size		129	.124		2.06	6.25		8.31	11.65
1850	1-1/2" pipe size		126	.127		2.12	6.40		8.52	11.95
1860	2" pipe size		124	.129		2.60	6.50		9.10	12.60
1870	2-1/2" pipe size		120	.133		4.09	6.70		10.79	14.60
1880	3" pipe size		117	.137		4.89	6.85		11.74	15.75
1890	3-1/2" pipe size		114	.140		5.35	7.05		12.40	16.50
1900	4" pipe size		110	.145		6.10	7.30		13.40	17.70
1910	5" pipe size		106	.151		8.10	7.60		15.70	20.50
1920	6" pipe size		104	.154		10.85	7.75		18.60	23.50
1930	8" pipe size		100	.160		16.80	8.05		24.85	30.50

22 05 29 – Hangers and Supports for Plumbing Piping and Equipment

	22 05 29.10 Hangers & Supp. for Plumb'g/HVAC Pipe/Equip.	Crew	Daily Output	Labor-Hours	Unit	Material	2013 Bare Costs Labor	Equipment	Total	Total Incl O&P
1940	10" pipe size	Q-1	96	.167	Ea.	30.50	8.35		38.85	46.50
1950	12" pipe size		89	.180		40.50	9.05		49.55	58
1960	14" pipe size		82	.195		52.50	9.80		62.30	72.50
1970	16" pipe size		68	.235		79	11.80		90.80	105
1971	18" pipe size		54	.296		95.50	14.90		110.40	128
1972	20" pipe size		38	.421		181	21		202	231
1980	For galvanized, add					66%				
1990	For copper plated 1/2" to 4", add					77%				
2000	For light weight 1/2" to 4", deduct					13%				
2010	Insulated pipe type, 3/4" to 12" pipe, add					180%				
2020	Insulated pipe type, chrome-moly U-strap, add					530%				
2250	Split ring, malleable iron, for non-insulated pipe, TYPE 11									
2260	1/2" pipe size	Q-1	137	.117	Ea.	6.40	5.85		12.25	15.90
2270	3/4" pipe size		135	.119		6.50	5.95		12.45	16.10
2280	1" pipe size		132	.121		6.85	6.10		12.95	16.70
2290	1-1/4" pipe size		129	.124		8.50	6.25		14.75	18.75
2300	1-1/2" pipe size		126	.127		10.35	6.40		16.75	21
2310	2" pipe size		124	.129		11.70	6.50		18.20	22.50
2320	2-1/2" pipe size		120	.133		16.90	6.70		23.60	28.50
2330	3" pipe size		117	.137		21	6.85		27.85	33.50
2340	3-1/2" pipe size		114	.140		22	7.05		29.05	34.50
2350	4" pipe size		110	.145		22	7.30		29.30	35
2360	5" pipe size		106	.151		29.50	7.60		37.10	43.50
2370	6" pipe size		104	.154		65.50	7.75		73.25	83.50
2380	8" pipe size		100	.160		108	8.05		116.05	130
2390	For copper plated, add					8%				
2500	Washer, flat steel									
2502	3/8"	1 Plum	240	.033	Ea.	.05	1.86		1.91	2.86
2503	1/2"		220	.036		.14	2.03		2.17	3.21
2504	5/8"		200	.040		.26	2.23		2.49	3.65
2505	3/4"		180	.044		.38	2.48		2.86	4.16
2506	7/8"		160	.050		.53	2.79		3.32	4.78
2507	1"		140	.057		.64	3.19		3.83	5.50
2508	1-1/4"		120	.067		.92	3.72		4.64	6.60
2520	Nut, steel, hex									
2522	3/8"	1 Plum	200	.040	Ea.	.96	2.23		3.19	4.42
2523	1/2"		180	.044		1.86	2.48		4.34	5.80
2524	5/8"		160	.050		3.15	2.79		5.94	7.65
2525	3/4"		140	.057		4.75	3.19		7.94	10.05
2526	7/8"		120	.067		7.55	3.72		11.27	13.95
2527	1"		100	.080		10.95	4.46		15.41	18.70
2528	1-1/4"		80	.100		21	5.60		26.60	31.50
2532	Turnbuckle, TYPE 13									
2534	3/8"	1 Plum	80	.100	Ea.	3.46	5.60		9.06	12.20
2535	1/2"		72	.111		5.45	6.20		11.65	15.35
2536	5/8"		64	.125		9.65	7		16.65	21
2537	3/4"		56	.143		12.55	7.95		20.50	26
2538	7/8"		48	.167		29.50	9.30		38.80	46.50
2539	1"		40	.200		38.50	11.15		49.65	59.50
2540	1-1/4"		32	.250		69.50	13.95		83.45	97
2650	Rods, carbon steel									
2660	Continuous thread									
2670	1/4" thread size	1 Plum	144	.056	L.F.	1.53	3.10		4.63	6.35

22 05 29 – Hangers and Supports for Plumbing Piping and Equipment

22 05 29.10 Hangers & Supp. for Plumb'g/HVAC Pipe/Equip.		Crew	Daily Output	Labor-Hours	Unit	2013 Bare Costs Material	Labor	Equipment	Total	Total Incl O&P
2680	3/8" thread size	1 Plum	144	.056	L.F.	1.63	3.10		4.73	6.45
2690	1/2" thread size		144	.056		2.57	3.10		5.67	7.50
2700	5/8" thread size		144	.056		3.65	3.10		6.75	8.70
2710	3/4" thread size		144	.056		6.40	3.10		9.50	11.70
2720	7/8" thread size		144	.056		8.05	3.10		11.15	13.50
2721	1" thread size	Q-1	160	.100		13.70	5		18.70	22.50
2722	1-1/8" thread size	"	120	.133		21.50	6.70		28.20	34
2725	1/4" thread size, bright finish	1 Plum	144	.056		1.90	3.10		5	6.75
2726	1/2" thread size, bright finish	"	144	.056		3.42	3.10		6.52	8.45
2730	For galvanized, add					40%				
2750	Both ends machine threaded 18" length									
2760	3/8" thread size	1 Plum	240	.033	Ea.	6.25	1.86		8.11	9.70
2770	1/2" thread size		240	.033		10.05	1.86		11.91	13.85
2780	5/8" thread size		240	.033		14.40	1.86		16.26	18.65
2790	3/4" thread size		240	.033		23	1.86		24.86	28
2800	7/8" thread size		240	.033		31	1.86		32.86	37.50
2810	1" thread size		240	.033		37	1.86		38.86	43.50
2820	Rod couplings									
2821	3/8"	1 Plum	60	.133	Ea.	1.98	7.45		9.43	13.40
2822	1/2"		54	.148		2.63	8.25		10.88	15.35
2823	5/8"		48	.167		2.96	9.30		12.26	17.25
2824	3/4"		44	.182		4.16	10.15		14.31	19.90
2825	7/8"		40	.200		6.95	11.15		18.10	24.50
2826	1"		34	.235		9.35	13.15		22.50	30
2827	1-1/8"		30	.267		40.50	14.90		55.40	67
2860	Pipe hanger assy, adj. clevis, saddle, rod, clamp, insul. allowance									
2864	1/2" pipe size	Q-5	35	.457	Ea.	43.50	23.50		67	83
2866	3/4" pipe size		34.80	.460		45	23.50		68.50	85
2868	1" pipe size		34.60	.462		48.50	23.50		72	88.50
2869	1-1/4" pipe size		34.30	.466		48.50	24		72.50	89.50
2870	1-1/2" pipe size		33.90	.472		50	24		74	91
2872	2" pipe size		33.30	.480		55.50	24.50		80	98
2874	2-1/2" pipe size		32.30	.495		61.50	25.50		87	106
2876	3" pipe size		31.20	.513		69.50	26		95.50	116
2880	4" pipe size		30.70	.521		76	26.50		102.50	124
2884	6" pipe size		29.80	.537		94	27.50		121.50	144
2888	8" pipe size		28	.571		129	29		158	186
2892	10" pipe size		25.20	.635		145	32.50		177.50	209
2896	12" pipe size		23.20	.690		209	35		244	282
2900	Rolls									
2910	Adjustable yoke, carbon steel with CI roll, TYPE 43									
2918	2" pipe size	Q-1	140	.114	Ea.	17.60	5.75		23.35	28
2920	2-1/2" pipe size		137	.117		17.60	5.85		23.45	28
2930	3" pipe size		131	.122		19.60	6.15		25.75	31
2940	3-1/2" pipe size		124	.129		26	6.50		32.50	38.50
2950	4" pipe size		117	.137		26	6.85		32.85	39
2960	5" pipe size		110	.145		30.50	7.30		37.80	44.50
2970	6" pipe size		104	.154		40	7.75		47.75	55.50
2980	8" pipe size		96	.167		56.50	8.35		64.85	74.50
2990	10" pipe size		80	.200		71.50	10.05		81.55	94
3010	14" pipe size		56	.286		263	14.35		277.35	310
3020	16" pipe size		48	.333		325	16.75		341.75	380
3050	Chair, carbon steel with CI roll									

22 05 29.10 Hangers & Supp. for Plumb'g/HVAC Pipe/Equip.		Crew	Daily Output	Labor-Hours	Unit	Material	2013 Bare Costs Labor	Equipment	Total	Total Incl O&P
3060	2" pipe size	1 Plum	68	.118	Ea.	23	6.55		29.55	35.50
3070	2-1/2" pipe size		65	.123		25	6.85		31.85	38
3080	3" pipe size		62	.129		25.50	7.20		32.70	39
3090	3-1/2" pipe size		60	.133		32.50	7.45		39.95	47
3100	4" pipe size		58	.138		35	7.70		42.70	50
3110	5" pipe size		56	.143		38	7.95		45.95	54
3120	6" pipe size		53	.151		53	8.40		61.40	71
3130	8" pipe size		50	.160		71.50	8.95		80.45	92.50
3140	10" pipe size		48	.167		90	9.30		99.30	113
3150	12" pipe size		46	.174		131	9.70		140.70	159
3170	Single pipe roll, (see line 2650 for rods), TYPE 41, 1" pipe size	Q-1	137	.117		15.85	5.85		21.70	26.50
3180	1-1/4" pipe size		131	.122		16.40	6.15		22.55	27.50
3190	1-1/2" pipe size		129	.124		16.95	6.25		23.20	28
3200	2" pipe size		124	.129		17.55	6.50		24.05	29
3210	2-1/2" pipe size		118	.136		18.85	6.80		25.65	31
3220	3" pipe size		115	.139		19.90	7		26.90	32.50
3230	3-1/2" pipe size		113	.142		22	7.10		29.10	35
3240	4" pipe size		112	.143		23.50	7.20		30.70	37
3250	5" pipe size		110	.145		28	7.30		35.30	42
3260	6" pipe size		101	.158		40.50	7.95		48.45	56.50
3270	8" pipe size		90	.178		56.50	8.95		65.45	75.50
3280	10" pipe size		80	.200		66.50	10.05		76.55	88.50
3290	12" pipe size		68	.235		103	11.80		114.80	131
3300	Saddles (add vertical pipe riser, usually 3" diameter)									
3310	Pipe support, complete, adjust., CI saddle, TYPE 36									
3320	2-1/2" pipe size	1 Plum	96	.083	Ea.	129	4.65		133.65	149
3330	3" pipe size		88	.091		131	5.05		136.05	152
3340	3-1/2" pipe size		79	.101		134	5.65		139.65	157
3350	4" pipe size		68	.118		189	6.55		195.55	218
3360	5" pipe size		64	.125		192	7		199	222
3370	6" pipe size		59	.136		201	7.55		208.55	232
3380	8" pipe size		53	.151		220	8.40		228.40	254
3390	10" pipe size		50	.160		242	8.95		250.95	280
3400	12" pipe size		48	.167		251	9.30		260.30	291
3450	For standard pipe support, one piece, CI deduct					34%				
3460	For stanchion support, CI with steel yoke, deduct					60%				
3550	Insulation shield 1" thick, 1/2" pipe size, TYPE 40	1 Asbe	100	.080	Ea.	6.60	3.98		10.58	13.45
3560	3/4" pipe size		100	.080		7.60	3.98		11.58	14.55
3570	1" pipe size		98	.082		8.15	4.06		12.21	15.30
3580	1-1/4" pipe size		98	.082		8.75	4.06		12.81	15.90
3590	1-1/2" pipe size		96	.083		8.75	4.14		12.89	16.05
3600	2" pipe size		96	.083		9.05	4.14		13.19	16.40
3610	2-1/2" pipe size		94	.085		9.35	4.23		13.58	16.85
3620	3" pipe size		94	.085		10	4.23		14.23	17.55
3630	2" thick, 3-1/2" pipe size		92	.087		11.20	4.32		15.52	19
3640	4" pipe size		92	.087		15.95	4.32		20.27	24.50
3650	5" pipe size		90	.089		19.80	4.42		24.22	29
3660	6" pipe size		90	.089		23.50	4.42		27.92	33
3670	8" pipe size		88	.091		35.50	4.52		40.02	46
3680	10" pipe size		88	.091		48	4.52		52.52	60
3690	12" pipe size		86	.093		55	4.62		59.62	67.50
3700	14" pipe size		86	.093		106	4.62		110.62	124
3710	16" pipe size		84	.095		113	4.73		117.73	131

22 05 29 – Hangers and Supports for Plumbing Piping and Equipment

22 05 29.10 Hangers & Supp. for Plumb'g/HVAC Pipe/Equip.		Crew	Daily Output	Labor-Hours	Unit	2013 Bare Costs Material	Labor	Equipment	Total	Total Incl O&P
3720	18" pipe size	1 Asbe	84	.095	Ea.	125	4.73		129.73	145
3730	20" pipe size		82	.098		138	4.85		142.85	160
3732	24" pipe size		80	.100		167	4.97		171.97	192
3733	30" pipe size		78	.103		505	5.10		510.10	565
3735	36" pipe size		75	.107		770	5.30		775.30	860
3750	Covering protection saddle, TYPE 39									
3760	1" covering size									
3770	3/4" pipe size	1 Plum	68	.118	Ea.	11.95	6.55		18.50	23
3780	1" pipe size		68	.118		11.95	6.55		18.50	23
3790	1-1/4" pipe size		68	.118		11.95	6.55		18.50	23
3800	1-1/2" pipe size		66	.121		12.95	6.75		19.70	24.50
3810	2" pipe size		66	.121		12.95	6.75		19.70	24.50
3820	2-1/2" pipe size		64	.125		12.95	7		19.95	25
3830	3" pipe size		64	.125		18.10	7		25.10	30.50
3840	3-1/2" pipe size		62	.129		19.25	7.20		26.45	32
3850	4" pipe size		62	.129		19.25	7.20		26.45	32
3860	5" pipe size		60	.133		19.25	7.45		26.70	32
3870	6" pipe size		60	.133		23	7.45		30.45	36.50
3900	1-1/2" covering size									
3910	3/4" pipe size	1 Plum	68	.118	Ea.	14.50	6.55		21.05	26
3920	1" pipe size		68	.118		14.20	6.55		20.75	25.50
3930	1-1/4" pipe size		68	.118		14.50	6.55		21.05	26
3940	1-1/2" pipe size		66	.121		14.50	6.75		21.25	26
3950	2" pipe size		66	.121		15.70	6.75		22.45	27.50
3960	2-1/2" pipe size		64	.125		18.60	7		25.60	31
3970	3" pipe size		64	.125		18.60	7		25.60	31
3980	3-1/2" pipe size		62	.129		18.85	7.20		26.05	31.50
3990	4" pipe size		62	.129		18.85	7.20		26.05	31.50
4000	5" pipe size		60	.133		18.85	7.45		26.30	31.50
4010	6" pipe size		60	.133		27	7.45		34.45	40.50
4020	8" pipe size		58	.138		34	7.70		41.70	49
4022	10" pipe size		56	.143		34	7.95		41.95	49.50
4024	12" pipe size		54	.148		60.50	8.25		68.75	79
4028	2" covering size									
4029	2-1/2" pipe size	1 Plum	62	.129	Ea.	21	7.20		28.20	34
4032	3" pipe size		60	.133		21	7.45		28.45	34
4033	4" pipe size		58	.138		21	7.70		28.70	34.50
4034	6" pipe size		56	.143		30	7.95		37.95	45
4035	8" pipe size		54	.148		35.50	8.25		43.75	51.50
4080	10" pipe size		58	.138		37.50	7.70		45.20	53
4090	12" pipe size		56	.143		68	7.95		75.95	87
4100	14" pipe size		56	.143		68	7.95		75.95	87
4110	16" pipe size		54	.148		98	8.25		106.25	120
4120	18" pipe size		54	.148		98	8.25		106.25	120
4130	20" pipe size		52	.154		109	8.60		117.60	133
4150	24" pipe size		50	.160		127	8.95		135.95	153
4160	30" pipe size		48	.167		140	9.30		149.30	168
4180	36" pipe size		45	.178		154	9.90		163.90	185
4186	2-1/2" covering size									
4187	3" pipe size	1 Plum	58	.138	Ea.	20	7.70		27.70	33.50
4188	4" pipe size		56	.143		22	7.95		29.95	36
4189	6" pipe size		52	.154		33.50	8.60		42.10	50
4190	8" pipe size		48	.167		41	9.30		50.30	59

22 05 Common Work Results for Plumbing

22 05 29 – Hangers and Supports for Plumbing Piping and Equipment

22 05 29.10 Hangers & Supp. for Plumb'g/HVAC Pipe/Equip.		Crew	Daily Output	Labor-Hours	Unit	2013 Bare Costs Material	Labor	Equipment	Total	Total Incl O&P
4191	10″ pipe size	1 Plum	44	.182	Ea.	45	10.15		55.15	65
4192	12″ pipe size		40	.200		74	11.15		85.15	98.50
4193	14″ pipe size		36	.222		74	12.40		86.40	100
4194	16″ pipe size		32	.250		103	13.95		116.95	134
4195	18″ pipe size		28	.286		115	15.95		130.95	150
4200	Sockets									
4210	Rod end, malleable iron, TYPE 16									
4220	1/4″ thread size	1 Plum	240	.033	Ea.	2.20	1.86		4.06	5.20
4230	3/8″ thread size		240	.033		2.20	1.86		4.06	5.20
4240	1/2″ thread size		230	.035		2.60	1.94		4.54	5.80
4250	5/8″ thread size		225	.036		5.10	1.98		7.08	8.65
4260	3/4″ thread size		220	.036		8.35	2.03		10.38	12.25
4270	7/8″ thread size		210	.038		15.25	2.13		17.38	19.95
4290	Strap, 1/2″ pipe size, TYPE 26	Q-1	142	.113		4.58	5.65		10.23	13.55
4300	3/4″ pipe size		140	.114		4.69	5.75		10.44	13.80
4310	1″ pipe size		137	.117		4.89	5.85		10.74	14.25
4320	1-1/4″ pipe size		134	.119		5.40	6		11.40	15
4330	1-1/2″ pipe size		131	.122		6.80	6.15		12.95	16.70
4340	2″ pipe size		129	.124		7	6.25		13.25	17.10
4350	2-1/2″ pipe size		125	.128		12.60	6.45		19.05	23.50
4360	3″ pipe size		122	.131		13.25	6.60		19.85	24.50
4370	3-1/2″ pipe size		119	.134		15.75	6.75		22.50	27.50
4380	4″ pipe size		114	.140		16.75	7.05		23.80	29
4400	U-bolt, carbon steel									
4410	Standard, with nuts, TYPE 42									
4420	1/2″ pipe size	1 Plum	160	.050	Ea.	2.09	2.79		4.88	6.50
4430	3/4″ pipe size		158	.051		2.09	2.83		4.92	6.55
4450	1″ pipe size		152	.053		2.15	2.94		5.09	6.80
4460	1-1/4″ pipe size		148	.054		2.63	3.02		5.65	7.45
4470	1-1/2″ pipe size		143	.056		2.75	3.12		5.87	7.75
4480	2″ pipe size		139	.058		3	3.21		6.21	8.15
4490	2-1/2″ pipe size		134	.060		4.97	3.33		8.30	10.45
4500	3″ pipe size		128	.063		5.20	3.49		8.69	10.95
4510	3-1/2″ pipe size		122	.066		5.50	3.66		9.16	11.55
4520	4″ pipe size		117	.068		5.60	3.82		9.42	11.90
4530	5″ pipe size		114	.070		5.65	3.92		9.57	12.10
4540	6″ pipe size		111	.072		11.55	4.02		15.57	18.75
4550	8″ pipe size		109	.073		13.70	4.10		17.80	21.50
4560	10″ pipe size		107	.075		24.50	4.17		28.67	33.50
4570	12″ pipe size		104	.077		33.50	4.29		37.79	43.50
4580	For plastic coating on 1/2″ thru 6″ size, add					150%				
4700	U-hook, carbon steel, requires mounting screws or bolts									
4710	3/4″ thru 2″ pipe size									
4720	6″ long	1 Plum	96	.083	Ea.	.77	4.65		5.42	7.85
4730	8″ long		96	.083		.99	4.65		5.64	8.10
4740	10″ long		96	.083		1.12	4.65		5.77	8.25
4750	12″ long		96	.083		1.22	4.65		5.87	8.35
4760	For copper plated, add					50%				
7000	Roof supports									
7006	Duct									
7010	Rectangular, open, 12″ off roof									
7020	To 18″ wide	Q-9	26	.615	Ea.	150	29.50		179.50	210
7030	To 24″ wide	″	22	.727		175	35		210	246

22 05 29 – Hangers and Supports for Plumbing Piping and Equipment

22 05 29.10 Hangers & Supp. for Plumb'g/HVAC Pipe/Equip.		Crew	Daily Output	Labor-Hours	Unit	2013 Bare Costs Material	Labor	Equipment	Total	Total Incl O&P
7040	To 36" wide	Q-10	30	.800	Ea.	200	40		240	281
7050	To 48" wide		28	.857		225	42.50		267.50	315
7060	To 60" wide	↓	24	1	↓	250	50		300	350
7100	Equipment									
7120	Equipment support	Q-5	20	.800	Ea.	59	41		100	127
7300	Pipe									
7310	Roller type									
7320	Up to 2-1/2" diam. pipe									
7324	3-1/2" off roof	Q-5	24	.667	Ea.	14	34		48	66.50
7326	Up to 10" off roof	"	20	.800	"	19	41		60	82.50
7340	2-1/2" to 3-1/2" diam. pipe									
7342	Up to 16" off roof	Q-5	18	.889	Ea.	45.50	45.50		91	118
7360	4" to 5" diam. pipe									
7362	Up to 12" off roof	Q-5	16	1	Ea.	65	51		116	149
7400	Strut/channel type									
7410	Up to 2-1/2" diam. pipe									
7424	3-1/2" off roof	Q-5	24	.667	Ea.	12	34		46	64
7426	Up to 10" off roof	"	20	.800	"	17	41		58	80
7440	Strut and roller type									
7452	2-1/2" to 3-1/2" diam. pipe									
7454	Up to 16" off roof	Q-5	18	.889	Ea.	65.50	45.50		111	140
7460	Strut and hanger type									
7470	Up to 3" diam. pipe									
7474	Up to 8" off roof	Q-5	19	.842	Ea.	41.50	43		84.50	111
8000	Pipe clamp, plastic, 1/2" CTS	1 Plum	80	.100		.22	5.60		5.82	8.65
8010	3/4" CTS		73	.110		.23	6.10		6.33	9.45
8020	1" CTS		68	.118		.52	6.55		7.07	10.45
8080	Economy clamp, 1/4" CTS		175	.046		.05	2.55		2.60	3.90
8090	3/8" CTS		168	.048		.07	2.66		2.73	4.08
8100	1/2" CTS		160	.050		.07	2.79		2.86	4.28
8110	3/4" CTS		145	.055		.14	3.08		3.22	4.79
8200	Half clamp, 1/2" CTS		80	.100		.08	5.60		5.68	8.50
8210	3/4" CTS		73	.110		.12	6.10		6.22	9.35
8300	Suspension clamp, 1/2" CTS		80	.100		.22	5.60		5.82	8.65
8310	3/4" CTS		73	.110		.23	6.10		6.33	9.45
8320	1" CTS		68	.118		.52	6.55		7.07	10.45
8400	Insulator, 1/2" CTS		80	.100		.36	5.60		5.96	8.80
8410	3/4" CTS		73	.110		.37	6.10		6.47	9.60
8420	1" CTS		68	.118		.38	6.55		6.93	10.30
8500	J hook clamp with nail 1/2" CTS		240	.033		.13	1.86		1.99	2.94
8501	3/4" CTS	↓	240	.033	↓	.13	1.86		1.99	2.94
8800	Wire cable support system									
8810	Cable with hook terminal and locking device									
8830	2 mm, (.079") dia cable, (100 lb. cap.)									
8840	1 m, (3.3') length, with hook	1 Shee	96	.083	Ea.	3.03	4.44		7.47	10.10
8850	2 m, (6.6') length, with hook		84	.095		3.50	5.10		8.60	11.60
8860	3 m, (9.9') length, with hook	↓	72	.111		3.97	5.90		9.87	13.40
8870	5 m, (16.4') length, with hook	Q-9	60	.267		4.97	12.80		17.77	25
8880	10 m, (32.8') length, with hook	"	30	.533	↓	7.20	25.50		32.70	47
8900	3mm, (.118") dia cable, (200 lb. cap.)									
8910	1 m, (3.3') length, with hook	1 Shee	96	.083	Ea.	3.88	4.44		8.32	11
8920	2 m, (6.6') length, with hook		84	.095		4.36	5.10		9.46	12.55
8930	3 m, (9.9') length, with hook	↓	72	.111		4.81	5.90		10.71	14.35

22 05 29 – Hangers and Supports for Plumbing Piping and Equipment

22 05 29.10 Hangers & Supp. for Plumb'g/HVAC Pipe/Equip.		Crew	Daily Output	Labor-Hours	Unit	2013 Bare Costs Material	Labor	Equipment	Total	Total Incl O&P
8940	5 m, (16.4') length, with hook	Q-9	60	.267	Ea.	5.95	12.80		18.75	26
8950	10 m, (32.8') length, with hook	"	30	.533	↓	8.55	25.50		34.05	48.50
9000	Cable system accessories									
9010	Anchor bolt, 3/8", with nut	1 Shee	140	.057	Ea.	1.16	3.05		4.21	5.95
9020	Air duct corner protector		160	.050		.52	2.67		3.19	4.64
9030	Air duct support attachment	↓	140	.057	↓	.99	3.05		4.04	5.75
9040	Flange clip, hammer-on style									
9044	For flange thickness 3/32" - 9/64", 160 lb. cap.	1 Shee	180	.044	Ea.	.29	2.37		2.66	3.93
9048	For flange thickness 1/8" - 1/4", 200 lb. cap.		160	.050		.29	2.67		2.96	4.39
9052	For flange thickness 5/16" - 1/2", 200 lb. cap.		150	.053		.56	2.84		3.40	4.96
9056	For flange thickness 9/16" - 3/4", 200 lb. cap.		140	.057	↓	.75	3.05		3.80	5.50
9060	Wire insulation protection tube	↓	180	.044	L.F.	.34	2.37		2.71	3.98
9070	Wire cutter				Ea.	41.50			41.50	45.50

22 05 33 – Heat Tracing for Plumbing Piping

22 05 33.20 Temperature Maintenance Cable		Crew	Daily Output	Labor-Hours	Unit	Material	Labor	Equipment	Total	Total Incl O&P
0010	**TEMPERATURE MAINTENANCE CABLE**									
0040	Components									
0080	Heating cable									
0100	208 V									
0140	125°F	Q-1	1060.80	.015	L.F.	9.85	.76		10.61	12
0150	140°F	"	1060.80	.015	"	9.85	.76		10.61	12
0200	120 V									
0220	125°F	Q-1	1060.80	.015	L.F.	8.65	.76		9.41	10.70
0300	Power kit w/1 end seal	1 Elec	48.80	.164	Ea.	75	8.60		83.60	95.50
0310	Splice kit		35.50	.225		89	11.80		100.80	115
0320	End seal		160	.050		9.05	2.62		11.67	13.85
0330	Tee kit w/1 end seal		26.80	.299		94	15.65		109.65	127
0340	Powered splice w/2 end seals		20	.400		105	21		126	147
0350	Powered tee kit w/3 end seals		18	.444		135	23.50		158.50	183
0360	Cross kit w/2 end seals	↓	18.60	.430	↓	139	22.50		161.50	187
0500	Recommended thickness of fiberglass insulation									
0510	Pipe size									
0520	1/2" - 1" use 1" insulation									
0530	1-1/4" - 2" use 1-1/2" insulation									
0540	2-1/2" - 6" use 2" insulation									
0560	NOTE: For pipe sizes 1-1/4" and smaller use 1/4" larger diameter									
0570	insulation to allow room for installation over cable.									

22 05 48 – Vibration and Seismic Controls for Plumbing Piping and Equipment

22 05 48.40 Vibration Absorbers		Crew	Daily Output	Labor-Hours	Unit	Material	Labor	Equipment	Total	Total Incl O&P
0010	**VIBRATION ABSORBERS**									
0100	Hangers, neoprene flex									
0200	10 – 120 lb. capacity				Ea.	24.50			24.50	26.50
0220	75 – 550 lb. capacity					34			34	37.50
0240	250 – 1100 lb. capacity					63			63	69.50
0260	1000 – 4000 lb. capacity					114			114	125
0500	Spring flex, 60 lb. capacity					34			34	37.50
0520	450 lb. capacity					45			45	49.50
0540	900 lb. capacity					63			63	69.50
0560	1100 – 1300 lb. capacity				↓	63			63	69.50
0600	Rubber in shear									
0610	45 – 340 lb., up to 1/2" rod size	1 Stpi	22	.364	Ea.	14.30	20.50		34.80	47
0620	130 – 700 lb., up to 3/4" rod size		20	.400		31	22.50		53.50	68

22 05 Common Work Results for Plumbing

22 05 48 – Vibration and Seismic Controls for Plumbing Piping and Equipment

22 05 48.40 Vibration Absorbers		Crew	Daily Output	Labor-Hours	Unit	Material	Labor	Equipment	Total	Total Incl O&P
						2013 Bare Costs				
0630	50 – 1000 lb., up to 3/4" rod size	1 Stpi	18	.444	Ea.	36.50	25		61.50	78
1000	Mounts, neoprene, 45 – 380 lb. capacity					16.65			16.65	18.35
1020	250 – 1100 lb. capacity					45.50			45.50	50
1040	1000 – 4000 lb. capacity					101			101	112
1100	Spring flex, 60 lb. capacity					68			68	75
1120	165 lb. capacity					69			69	76
1140	260 lb. capacity					71			71	78
1160	450 lb. capacity					90.50			90.50	99.50
1180	600 lb. capacity					93.50			93.50	103
1200	750 lb. capacity					98			98	108
1220	900 lb. capacity					100			100	110
1240	1100 lb. capacity					103			103	113
1260	1300 lb. capacity					103			103	113
1280	1500 lb. capacity					157			157	173
1300	1800 lb. capacity					169			169	186
1320	2200 lb. capacity					169			169	186
1340	2600 lb. capacity				↓	169			169	186
1399	Spring type									
1400	2 Piece									
1410	50 – 1000 lb.	1 Stpi	12	.667	Ea.	130	38		168	200
1420	1100 – 1600 lb.	"	12	.667	"	141	38		179	212
1500	Double spring open									
1510	150 – 450 lb.	1 Stpi	24	.333	Ea.	89	18.90		107.90	127
1520	500 – 1000 lb.		24	.333		89	18.90		107.90	127
1530	1100 – 1600 lb.		24	.333		89	18.90		107.90	127
1540	1700 – 2400 lb.		24	.333		97	18.90		115.90	135
1550	2500 – 3400 lb.	↓	24	.333		165	18.90		183.90	211
2000	Pads, cork rib, 18" x 18" x 1", 10-50 psi					163			163	179
2020	18" x 36" x 1", 10-50 psi					325			325	360
2100	Shear flexible pads, 18" x 18" x 3/8", 20-70 psi					74.50			74.50	81.50
2120	18" x 36" x 3/8", 20-70 psi				↓	163			163	179
2150	Laminated neoprene and cork									
2160	1" thick	1 Stpi	16	.500	S.F.	81	28.50		109.50	132
2200	Neoprene elastomer isolation bearing pad									
2230	464 PSI, 5/8" x 19-11/16" x 39-3/8"	1 Stpi	16	.500	Ea.	260	28.50		288.50	330
3000	Note overlap in capacities due to deflections									

22 05 53 – Identification for Plumbing Piping and Equipment

22 05 53.10 Piping System Identification Labels

		Crew	Daily Output	Labor-Hours	Unit	Material	Labor	Equipment	Total	Total Incl O&P
0010	**PIPING SYSTEM IDENTIFICATION LABELS**									
0100	Indicate contents and flow direction									
0106	Pipe markers									
0110	Plastic snap around									
0114	1/2" pipe	1 Plum	80	.100	Ea.	6.50	5.60		12.10	15.55
0116	3/4" pipe		80	.100		6.50	5.60		12.10	15.55
0118	1" pipe		80	.100		6.50	5.60		12.10	15.55
0120	2" pipe		75	.107		7.70	5.95		13.65	17.40
0122	3" pipe		70	.114		12.60	6.40		19	23.50
0124	4" pipe		60	.133		12.60	7.45		20.05	25
0126	6" pipe		60	.133		12.60	7.45		20.05	25
0128	8" pipe		56	.143		16.70	7.95		24.65	30.50
0130	10" pipe		56	.143		16.70	7.95		24.65	30.50
0200	Over 10" pipe size	↓	50	.160	↓	20.50	8.95		29.45	36

22 05 Common Work Results for Plumbing

22 05 53 – Identification for Plumbing Piping and Equipment

22 05 53.10 Piping System Identification Labels		Crew	Daily Output	Labor-Hours	Unit	Material	2013 Bare Costs Labor	Equipment	Total	Total Incl O&P
1110	Self adhesive									
1114	1″ pipe	1 Plum	80	.100	Ea.	3.20	5.60		8.80	11.90
1116	2″ pipe		75	.107		3.20	5.95		9.15	12.45
1118	3″ pipe		70	.114		4.90	6.40		11.30	15
1120	4″ pipe		60	.133		4.90	7.45		12.35	16.60
1122	6″ pipe		60	.133		4.90	7.45		12.35	16.60
1124	8″ pipe		56	.143		4.90	7.95		12.85	17.40
1126	10″ pipe		56	.143		9.90	7.95		17.85	23
1200	Over 10″ pipe size		50	.160		9.90	8.95		18.85	24.50
2000	Valve tags									
2010	Numbered plus identifying legend									
2100	Brass, 2″ diameter	1 Plum	40	.200	Ea.	2.95	11.15		14.10	20
2200	Plastic, 1-1/2″ diam.	″	40	.200	″	2.70	11.15		13.85	19.75

22 05 76 – Facility Drainage Piping Cleanouts

22 05 76.10 Cleanouts		Crew	Daily Output	Labor-Hours	Unit	Material	Labor	Equipment	Total	Total Incl O&P
0010	**CLEANOUTS**									
0060	Floor type									
0080	Round or square, scoriated nickel bronze top									
0100	2″ pipe size	1 Plum	10	.800	Ea.	183	44.50		227.50	268
0120	3″ pipe size		8	1		274	56		330	385
0140	4″ pipe size		6	1.333		274	74.50		348.50	410
0160	5″ pipe size		4	2		345	112		457	550
0180	6″ pipe size	Q-1	6	2.667		345	134		479	580
0200	8″ pipe size	″	4	4		615	201		816	980
0340	Recessed for tile, same price									
0980	Round top, recessed for terrazzo									
1000	2″ pipe size	1 Plum	9	.889	Ea.	183	49.50		232.50	276
1080	3″ pipe size		6	1.333		274	74.50		348.50	410
1100	4″ pipe size		4	2		274	112		386	470
1120	5″ pipe size	Q-1	6	2.667		345	134		479	580
1140	6″ pipe size		5	3.200		345	161		506	620
1160	8″ pipe size		4	4		615	201		816	980
2000	Round scoriated nickel bronze top, extra heavy duty									
2060	2″ pipe size	1 Plum	9	.889	Ea.	249	49.50		298.50	350
2080	3″ pipe size		6	1.333		340	74.50		414.50	485
2100	4″ pipe size		4	2		340	112		452	545
2120	5″ pipe size	Q-1	6	2.667		415	134		549	655
2140	6″ pipe size		5	3.200		415	161		576	695
2160	8″ pipe size		4	4		680	201		881	1,050
4000	Wall type, square smooth cover, over wall frame									
4060	2″ pipe size	1 Plum	14	.571	Ea.	298	32		330	380
4080	3″ pipe size		12	.667		320	37		357	410
4100	4″ pipe size		10	.800		345	44.50		389.50	445
4120	5″ pipe size		9	.889		500	49.50		549.50	625
4140	6″ pipe size		8	1		555	56		611	695
4160	8″ pipe size	Q-1	11	1.455		740	73		813	920
5000	Extension, C.I.; bronze countersunk plug, 8″ long									
5040	2″ pipe size	1 Plum	16	.500	Ea.	157	28		185	214
5060	3″ pipe size		14	.571		182	32		214	248
5080	4″ pipe size		13	.615		185	34.50		219.50	255
5100	5″ pipe size		12	.667		289	37		326	375
5120	6″ pipe size		11	.727		350	40.50		390.50	445

22 05 Common Work Results for Plumbing

22 05 76 – Facility Drainage Piping Cleanouts

22 05 76.20 Cleanout Tees		Crew	Daily Output	Labor-Hours	Unit	Material	Labor	Equipment	Total	Total Incl O&P
						2013 Bare Costs				
0010	**CLEANOUT TEES**									
0100	Cast iron, B&S, with countersunk plug									
0200	2" pipe size	1 Plum	4	2	Ea.	248	112		360	440
0220	3" pipe size		3.60	2.222		270	124		394	485
0240	4" pipe size		3.30	2.424		335	135		470	575
0260	5" pipe size	Q-1	5.50	2.909		730	146		876	1,025
0280	6" pipe size	"	5	3.200		910	161		1,071	1,250
0300	8" pipe size	Q-3	5	6.400		1,050	340		1,390	1,650
0500	For round smooth access cover, same price									
0600	For round scoriated access cover, same price									
0700	For square smooth access cover, add				Ea.	60%				
2000	Cast iron, no hub									
2010	Cleanout tee, with 2 couplings									
2012	2"	Q-1	22	.727	Ea.	38	36.50		74.50	97
2014	3"		19	.842		50.50	42.50		93	120
2016	4"		16.50	.970		71	48.50		119.50	152
2018	6"	Q-2	20	1.200		169	62.50		231.50	280
2040	Cleanout plug, no hub, with 1 coupling									
2042	2"	Q-1	44	.364	Ea.	11.10	18.25		29.35	39.50
2046	3"		38	.421		17.25	21		38.25	51
2048	4"		33	.485		28.50	24.50		53	68
2050	6"	Q-2	40	.600		102	31.50		133.50	160
2052	8"	"	33	.727		143	38		181	214
4000	Plastic, tees and adapters. Add plugs									
4010	ABS, DWV									
4020	Cleanout tee, 1-1/2" pipe size	1 Plum	15	.533	Ea.	14.10	30		44.10	60.50
4030	2" pipe size	Q-1	27	.593		15.45	30		45.45	62
4040	3" pipe size		21	.762		29	38.50		67.50	89.50
4050	4" pipe size		16	1		62	50		112	144
4100	Cleanout plug, 1-1/2" pipe size	1 Plum	32	.250		2.43	13.95		16.38	23.50
4110	2" pipe size	Q-1	56	.286		3.20	14.35		17.55	25
4120	3" pipe size		36	.444		5.15	22.50		27.65	39
4130	4" pipe size		30	.533		9.05	27		36.05	50.50
4180	Cleanout adapter fitting, 1-1/2" pipe size	1 Plum	32	.250		3.85	13.95		17.80	25
4190	2" pipe size	Q-1	56	.286		5.95	14.35		20.30	28
4200	3" pipe size		36	.444		15.15	22.50		37.65	50
4210	4" pipe size		30	.533		28	27		55	71
5000	PVC, DWV									
5010	Cleanout tee, 1-1/2" pipe size	1 Plum	15	.533	Ea.	10.25	30		40.25	56.50
5020	2" pipe size	Q-1	27	.593		11.95	30		41.95	58
5030	3" pipe size		21	.762		23	38.50		61.50	83
5040	4" pipe size		16	1		41.50	50		91.50	121
5090	Cleanout plug, 1-1/2" pipe size	1 Plum	32	.250		2.44	13.95		16.39	23.50
5100	2" pipe size	Q-1	56	.286		2.71	14.35		17.06	24.50
5110	3" pipe size		36	.444		4.85	22.50		27.35	39
5120	4" pipe size		30	.533		7.15	27		34.15	48.50
5130	6" pipe size		24	.667		24	33.50		57.50	76.50
5170	Cleanout adapter fitting, 1-1/2" pipe size	1 Plum	32	.250		3.25	13.95		17.20	24.50
5180	2" pipe size	Q-1	56	.286		4.20	14.35		18.55	26
5190	3" pipe size		36	.444		11.85	22.50		34.35	46.50
5200	4" pipe size		30	.533		19.40	27		46.40	62
5210	6" pipe size		24	.667		56.50	33.50		90	113

22 05 Common Work Results for Plumbing

22 05 76 – Facility Drainage Piping Cleanouts

22 05 76.20 Cleanout Tees		Crew	Daily Output	Labor-Hours	Unit	Material	Labor	Equipment	Total	Total Incl O&P
						2013 Bare Costs				
5300	Cleanout tee, with plug									
5310	4"	Q-1	14	1.143	Ea.	54	57.50		111.50	146
5320	6"	"	8	2		63	100		163	220
5330	8"	Q-2	11	2.182		132	114		246	315
5340	12"	"	10	2.400	↓	320	125		445	545
5400	Polypropylene, Schedule 40									
5410	Cleanout tee with plug									
5420	1-1/2"	1 Plum	10	.800	Ea.	34	44.50		78.50	105
5430	2"	Q-1	17	.941		36.50	47.50		84	111
5440	3"		11	1.455		77.50	73		150.50	196
5450	4"	↓	9	1.778	↓	105	89.50		194.50	250

22 07 Plumbing Insulation

22 07 16 – Plumbing Equipment Insulation

22 07 16.10 Insulation for Plumbing Equipment		Crew	Daily Output	Labor-Hours	Unit	Material	Labor	Equipment	Total	Total Incl O&P
0010	**INSULATION FOR PLUMBING EQUIPMENT**									
2900	Domestic water heater wrap kit									
2920	1-1/2" with vinyl jacket, 20-60 gal. G	1 Plum	8	1	Ea.	18.20	56		74.20	104
2925	50 to 80 gallons G	"	8	1	"	21.50	56		77.50	108

22 07 19 – Plumbing Piping Insulation

22 07 19.10 Piping Insulation		Crew	Daily Output	Labor-Hours	Unit	Material	Labor	Equipment	Total	Total Incl O&P
0010	**PIPING INSULATION**									
0100	Rule of thumb, as a percentage of total mechanical costs				Job				10%	10%
0110	Insulation req'd. is based on the surface size/area to be covered									
0230	Insulated protectors, (ADA)									
0235	For exposed piping under sinks or lavatories									
0240	Vinyl coated foam, velcro tabs									
0245	P Trap, 1-1/4" or 1-1/2"	1 Plum	32	.250	Ea.	22.50	13.95		36.45	45.50
0260	Valve and supply cover									
0265	1/2", 3/8", and 7/16" pipe size	1 Plum	32	.250	Ea.	21.50	13.95		35.45	44.50
0280	Tailpiece offset (wheelchair)									
0285	1-1/4" pipe size	1 Plum	32	.250	Ea.	18.20	13.95		32.15	41
0600	Pipe covering (price copper tube one size less than IPS)									
1000	Mineral wool									
1010	Preformed, 1200°F, plain									
1014	1" wall									
1016	1/2" iron pipe size G	Q-14	230	.070	L.F.	1.65	3.11		4.76	6.65
1018	3/4" iron pipe size G		220	.073		1.72	3.25		4.97	6.95
1022	1" iron pipe size G		210	.076		1.77	3.41		5.18	7.25
1024	1-1/4" iron pipe size G		205	.078		1.82	3.49		5.31	7.40
1026	1-1/2" iron pipe size G		205	.078		1.90	3.49		5.39	7.50
1028	2" iron pipe size G		200	.080		2.32	3.58		5.90	8.10
1030	2-1/2" iron pipe size G		190	.084		2.46	3.77		6.23	8.55
1032	3" iron pipe size G		180	.089		2.55	3.98		6.53	9
1034	4" iron pipe size G		150	.107		3.14	4.77		7.91	10.85
1036	5" iron pipe size G		140	.114		3.55	5.10		8.65	11.85
1038	6" iron pipe size G		120	.133		3.79	5.95		9.74	13.40
1040	7" iron pipe size G		110	.145		4.43	6.50		10.93	14.95
1042	8" iron pipe size G		100	.160		6.10	7.15		13.25	17.80
1044	9" iron pipe size G	↓	90	.178	↓	6.65	7.95		14.60	19.65

22 07 19.10 Piping Insulation			Crew	Daily Output	Labor-Hours	Unit	Material	Labor	Equipment	Total	Total Incl O&P
							2013 Bare Costs				
1046	10" iron pipe size	G	Q-14	90	.178	L.F.	7.50	7.95		15.45	20.50
1050	1-1/2" wall										
1052	1/2" iron pipe size	G	Q-14	225	.071	L.F.	2.82	3.18		6	8.05
1054	3/4" iron pipe size	G		215	.074		2.89	3.33		6.22	8.35
1056	1" iron pipe size	G		205	.078		2.93	3.49		6.42	8.60
1058	1-1/4" iron pipe size	G		200	.080		2.99	3.58		6.57	8.85
1060	1-1/2" iron pipe size	G		200	.080		3.14	3.58		6.72	9
1062	2" iron pipe size	G		190	.084		3.32	3.77		7.09	9.50
1064	2-1/2" iron pipe size	G		180	.089		3.62	3.98		7.60	10.20
1066	3" iron pipe size	G		165	.097		3.77	4.34		8.11	10.90
1068	4" iron pipe size	G		140	.114		4.53	5.10		9.63	12.95
1070	5" iron pipe size	G		130	.123		4.84	5.50		10.34	13.85
1072	6" iron pipe size	G		110	.145		4.99	6.50		11.49	15.60
1074	7" iron pipe size	G		100	.160		5.90	7.15		13.05	17.60
1076	8" iron pipe size	G		90	.178		6.55	7.95		14.50	19.55
1078	9" iron pipe size	G		85	.188		7.55	8.40		15.95	21.50
1080	10" iron pipe size	G		80	.200		8.30	8.95		17.25	23
1082	12" iron pipe size	G		75	.213		9.45	9.55		19	25
1084	14" iron pipe size	G		70	.229		10.80	10.20		21	28
1086	16" iron pipe size	G		65	.246		12.30	11		23.30	30.50
1088	18" iron pipe size	G		60	.267		13.85	11.95		25.80	34
1090	20" iron pipe size	G		55	.291		15.25	13		28.25	37
1092	22" iron pipe size	G		50	.320		17.45	14.30		31.75	41
1094	24" iron pipe size	G	↓	45	.356	↓	18.65	15.90		34.55	45
1100	2" wall										
1102	1/2" iron pipe size	G	Q-14	220	.073	L.F.	3.65	3.25		6.90	9.05
1104	3/4" iron pipe size	G		210	.076		3.78	3.41		7.19	9.45
1106	1" iron pipe size	G		200	.080		4.01	3.58		7.59	9.95
1108	1-1/4" iron pipe size	G		190	.084		4.30	3.77		8.07	10.60
1110	1-1/2" iron pipe size	G		190	.084		4.48	3.77		8.25	10.80
1112	2" iron pipe size	G		180	.089		4.70	3.98		8.68	11.35
1114	2-1/2" iron pipe size	G		170	.094		5.30	4.21		9.51	12.40
1116	3" iron pipe size	G		160	.100		5.70	4.47		10.17	13.20
1118	4" iron pipe size	G		130	.123		6.60	5.50		12.10	15.85
1120	5" iron pipe size	G		120	.133		7.35	5.95		13.30	17.35
1122	6" iron pipe size	G		100	.160		7.65	7.15		14.80	19.50
1124	8" iron pipe size	G		80	.200		9.25	8.95		18.20	24
1126	10" iron pipe size	G		70	.229		11.35	10.20		21.55	28.50
1128	12" iron pipe size	G		65	.246		12.55	11		23.55	31
1130	14" iron pipe size	G		60	.267		14.05	11.95		26	34
1132	16" iron pipe size	G		55	.291		15.90	13		28.90	37.50
1134	18" iron pipe size	G		50	.320		17.20	14.30		31.50	41
1136	20" iron pipe size	G		45	.356		19.80	15.90		35.70	46.50
1138	22" iron pipe size	G		45	.356		21.50	15.90		37.40	48
1140	24" iron pipe size	G	↓	40	.400	↓	22.50	17.90		40.40	53
1150	4" wall										
1152	1-1/4" iron pipe size	G	Q-14	170	.094	L.F.	11.65	4.21		15.86	19.35
1154	1-1/2" iron pipe size	G		165	.097		12.10	4.34		16.44	20
1156	2" iron pipe size	G		155	.103		12.35	4.62		16.97	20.50
1158	4" iron pipe size	G		105	.152		15.60	6.80		22.40	28
1160	6" iron pipe size	G		75	.213		18.40	9.55		27.95	35
1162	8" iron pipe size	G		60	.267		21	11.95		32.95	41.50
1164	10" iron pipe size	G		50	.320		25	14.30		39.30	49.50

22 07 19.10 Piping Insulation		Crew	Daily Output	Labor-Hours	Unit	Material	Labor	Equipment	Total	Total Incl O&P
						2013 Bare Costs				
1166	12″ iron pipe size G	Q-14	45	.356	L.F.	27.50	15.90		43.40	55
1168	14″ iron pipe size G		40	.400		29.50	17.90		47.40	60.50
1170	16″ iron pipe size G		35	.457		33.50	20.50		54	68.50
1172	18″ iron pipe size G		32	.500		36	22.50		58.50	74
1174	20″ iron pipe size G		30	.533		38.50	24		62.50	79.50
1176	22″ iron pipe size G		28	.571		43.50	25.50		69	87
1178	24″ iron pipe size G	↓	26	.615	↓	46	27.50		73.50	93.50
4280	Cellular glass, closed cell foam, all service jacket, sealant,									
4281	working temp. (-450°F to +900°F), 0 water vapor transmission									
4284	1″ wall									
4286	1/2″ iron pipe size G	Q-14	120	.133	L.F.	5	5.95		10.95	14.75
4300	1-1/2″ wall,									
4301	1″ iron pipe size G	Q-14	105	.152	L.F.	8.70	6.80		15.50	20
4304	2-1/2″ iron pipe size G		90	.178		15.40	7.95		23.35	29.50
4306	3″ iron pipe size G		85	.188		15.40	8.40		23.80	30
4308	4″ iron pipe size G		70	.229		16.30	10.20		26.50	34
4310	5″ iron pipe size G	↓	65	.246	↓	19	11		30	38
4320	2″ wall,									
4322	1″ iron pipe size G	Q-14	100	.160	L.F.	11.60	7.15		18.75	24
4324	2-1/2″ iron pipe size G		85	.188		18.70	8.40		27.10	33.50
4326	3″ iron pipe size G		80	.200		18.70	8.95		27.65	34.50
4328	4″ iron pipe size G		65	.246		20.50	11		31.50	39.50
4330	5″ iron pipe size G		60	.267		23.50	11.95		35.45	44
4332	6″ iron pipe size G		50	.320		27.50	14.30		41.80	52.50
4336	8″ iron pipe size G		40	.400		32	17.90		49.90	63
4338	10″ iron pipe size G	↓	35	.457	↓	39	20.50		59.50	75
4350	2-1/2″ wall,									
4360	12″ iron pipe size G	Q-14	32	.500	L.F.	53	22.50		75.50	93
4362	14″ iron pipe size G	″	28	.571	″	57	25.50		82.50	102
4370	3″ wall,									
4378	6″ iron pipe size G	Q-14	48	.333	L.F.	36.50	14.90		51.40	63
4380	8″ iron pipe size G		38	.421		45.50	18.85		64.35	79.50
4382	10″ iron pipe size G		33	.485		54	21.50		75.50	92.50
4384	16″ iron pipe size G		25	.640		72.50	28.50		101	125
4386	18″ iron pipe size G		22	.727		82.50	32.50		115	141
4388	20″ iron pipe size G	↓	20	.800	↓	89.50	36		125.50	154
4400	3-1/2″ wall,									
4412	12″ iron pipe size G	Q-14	27	.593	L.F.	69	26.50		95.50	117
4414	14″ iron pipe size G	″	25	.640	″	71	28.50		99.50	123
4430	4″ wall,									
4446	16″ iron pipe size G	Q-14	22	.727	L.F.	92.50	32.50		125	153
4448	18″ iron pipe size G		20	.800		101	36		137	167
4450	20″ iron pipe size G	↓	18	.889	↓	114	40		154	187
4480	Fittings, average with fabric and mastic									
4484	1″ wall,									
4486	1/2″ iron pipe size G	1 Asbe	40	.200	Ea.	5.65	9.95		15.60	21.50
4500	1-1/2″ wall,									
4502	1″ iron pipe size G	1 Asbe	38	.211	Ea.	7.50	10.45		17.95	24.50
4504	2-1/2″ iron pipe size G		32	.250		13.60	12.45		26.05	34.50
4506	3″ iron pipe size G		30	.267		13.60	13.25		26.85	35.50
4508	4″ iron pipe size G		28	.286		17.15	14.20		31.35	41
4510	5″ iron pipe size G	↓	24	.333	↓	21	16.55		37.55	48.50
4520	2″ wall,									

22 07 19 – Plumbing Piping Insulation

22 07 19.10 Piping Insulation		Crew	Daily Output	Labor-Hours	Unit	Material	Labor	Equipment	Total	Total Incl O&P
						2013 Bare Costs				
4522	1" iron pipe size	G	1 Asbe	36	.222	Ea.	10.90	11.05		21.95	29
4524	2-1/2" iron pipe size	G		30	.267		17.15	13.25		30.40	39.50
4526	3" iron pipe size	G		28	.286		17.15	14.20		31.35	41
4528	4" iron pipe size	G		24	.333		21	16.55		37.55	48.50
4530	5" iron pipe size	G		22	.364		31	18.05		49.05	62
4532	6" iron pipe size	G		20	.400		44	19.90		63.90	79
4536	8" iron pipe size	G		12	.667		73.50	33		106.50	132
4538	10" iron pipe size	G		8	1		85	49.50		134.50	171
4550	2-1/2" wall,										
4560	12" iron pipe size	G	1 Asbe	6	1.333	Ea.	156	66.50		222.50	275
4562	14" iron pipe size	G	"	4	2	"	206	99.50		305.50	380
4570	3" wall,										
4578	6" iron pipe size	G	1 Asbe	16	.500	Ea.	63	25		88	108
4580	8" iron pipe size	G		10	.800		94.50	40		134.50	166
4582	10" iron pipe size	G		6	1.333		128	66.50		194.50	243
4900	Calcium silicate, with 8 oz. canvas cover										
5100	1" wall, 1/2" iron pipe size	G	Q-14	170	.094	L.F.	3.30	4.21		7.51	10.20
5130	3/4" iron pipe size	G		170	.094		3.33	4.21		7.54	10.20
5140	1" iron pipe size	G		170	.094		3.23	4.21		7.44	10.10
5150	1-1/4" iron pipe size	G		165	.097		3.30	4.34		7.64	10.40
5160	1-1/2" iron pipe size	G		165	.097		3.35	4.34		7.69	10.45
5170	2" iron pipe size	G		160	.100		3.84	4.47		8.31	11.15
5180	2-1/2" iron pipe size	G		160	.100		4.09	4.47		8.56	11.45
5190	3" iron pipe size	G		150	.107		4.42	4.77		9.19	12.25
5200	4" iron pipe size	G		140	.114		5.50	5.10		10.60	14
5210	5" iron pipe size	G		135	.119		5.80	5.30		11.10	14.60
5220	6" iron pipe size	G		130	.123		6.15	5.50		11.65	15.30
5280	1-1/2" wall, 1/2" iron pipe size	G		150	.107		3.57	4.77		8.34	11.35
5310	3/4" iron pipe size	G		150	.107		3.64	4.77		8.41	11.40
5320	1" iron pipe size	G		150	.107		3.96	4.77		8.73	11.75
5330	1-1/4" iron pipe size	G		145	.110		4.26	4.94		9.20	12.35
5340	1-1/2" iron pipe size	G		145	.110		4.57	4.94		9.51	12.70
5350	2" iron pipe size	G		140	.114		5.05	5.10		10.15	13.50
5360	2-1/2" iron pipe size	G		140	.114		5.50	5.10		10.60	14
5370	3" iron pipe size	G		135	.119		5.75	5.30		11.05	14.60
5380	4" iron pipe size	G		125	.128		6.65	5.75		12.40	16.25
5390	5" iron pipe size	G		120	.133		7.50	5.95		13.45	17.50
5400	6" iron pipe size	G		110	.145		7.75	6.50		14.25	18.65
5460	2" wall, 1/2" iron pipe size	G		135	.119		5.50	5.30		10.80	14.30
5490	3/4" iron pipe size	G		135	.119		5.80	5.30		11.10	14.60
5500	1" iron pipe size	G		135	.119		6.10	5.30		11.40	14.95
5510	1-1/4" iron pipe size	G		130	.123		6.55	5.50		12.05	15.75
5520	1-1/2" iron pipe size	G		130	.123		6.80	5.50		12.30	16.05
5530	2" iron pipe size	G		125	.128		7.20	5.75		12.95	16.85
5540	2-1/2" iron pipe size	G		125	.128		8.55	5.75		14.30	18.30
5550	3" iron pipe size	G		120	.133		8.65	5.95		14.60	18.75
5560	4" iron pipe size	G		115	.139		9.95	6.20		16.15	20.50
5570	5" iron pipe size	G		110	.145		11.35	6.50		17.85	22.50
5580	6" iron pipe size	G		105	.152		12.40	6.80		19.20	24.50
5600	Calcium silicate, no cover										
5720	1" wall, 1/2" iron pipe size	G	Q-14	180	.089	L.F.	3.01	3.98		6.99	9.50
5740	3/4" iron pipe size	G		180	.089		3.01	3.98		6.99	9.50
5750	1" iron pipe size	G		180	.089		2.88	3.98		6.86	9.35

22 07 19 – Plumbing Piping Insulation

22 07 19.10 Piping Insulation		Crew	Daily Output	Labor-Hours	Unit	Material	Labor	Equipment	Total	Total Incl O&P
						2013 Bare Costs				
5760	1-1/4" iron pipe size G	Q-14	175	.091	L.F.	2.93	4.09		7.02	9.55
5770	1-1/2" iron pipe size G		175	.091		2.95	4.09		7.04	9.60
5780	2" iron pipe size G		170	.094		3.41	4.21		7.62	10.30
5790	2-1/2" iron pipe size G		170	.094		3.60	4.21		7.81	10.50
5800	3" iron pipe size G		160	.100		3.87	4.47		8.34	11.20
5810	4" iron pipe size G		150	.107		4.84	4.77		9.61	12.70
5820	5" iron pipe size G		145	.110		5.05	4.94		9.99	13.20
5830	6" iron pipe size G		140	.114		5.30	5.10		10.40	13.75
5900	1-1/2" wall, 1/2" iron pipe size G		160	.100		3.17	4.47		7.64	10.45
5920	3/4" iron pipe size G		160	.100		3.23	4.47		7.70	10.50
5930	1" iron pipe size G		160	.100		3.52	4.47		7.99	10.80
5940	1-1/4" iron pipe size G		155	.103		3.79	4.62		8.41	11.30
5950	1-1/2" iron pipe size G		155	.103		4.08	4.62		8.70	11.65
5960	2" iron pipe size G		150	.107		4.49	4.77		9.26	12.35
5970	2-1/2" iron pipe size G		150	.107		4.90	4.77		9.67	12.80
5980	3" iron pipe size G		145	.110		5.10	4.94		10.04	13.30
5990	4" iron pipe size G		135	.119		5.95	5.30		11.25	14.75
6000	5" iron pipe size G		130	.123		6.70	5.50		12.20	15.90
6010	6" iron pipe size G		120	.133		6.80	5.95		12.75	16.75
6020	7" iron pipe size G		115	.139		8	6.20		14.20	18.45
6030	8" iron pipe size G		105	.152		9.05	6.80		15.85	20.50
6040	9" iron pipe size G		100	.160		10.85	7.15		18	23
6050	10" iron pipe size G		95	.168		11.95	7.55		19.50	25
6060	12" iron pipe size G		90	.178		14.20	7.95		22.15	28
6070	14" iron pipe size G		85	.188		16.05	8.40		24.45	31
6080	16" iron pipe size G		80	.200		17.95	8.95		26.90	33.50
6090	18" iron pipe size G		75	.213		19.85	9.55		29.40	37
6120	2" wall, 1/2" iron pipe size G		145	.110		5	4.94		9.94	13.15
6140	3/4" iron pipe size G		145	.110		5.30	4.94		10.24	13.45
6150	1" iron pipe size G		145	.110		5.55	4.94		10.49	13.80
6160	1-1/4" iron pipe size G		140	.114		5.95	5.10		11.05	14.50
6170	1-1/2" iron pipe size G		140	.114		6.25	5.10		11.35	14.80
6180	2" iron pipe size G		135	.119		6.55	5.30		11.85	15.45
6190	2-1/2" iron pipe size G		135	.119		7.85	5.30		13.15	16.90
6200	3" iron pipe size G		130	.123		7.90	5.50		13.40	17.25
6210	4" iron pipe size G		125	.128		9.15	5.75		14.90	18.95
6220	5" iron pipe size G		120	.133		10.40	5.95		16.35	20.50
6230	6" iron pipe size G		115	.139		11.40	6.20		17.60	22
6240	7" iron pipe size G		110	.145		12.35	6.50		18.85	23.50
6250	8" iron pipe size G		105	.152		13.55	6.80		20.35	25.50
6260	9" iron pipe size G		100	.160		15.35	7.15		22.50	28
6270	10" iron pipe size G		95	.168		17.05	7.55		24.60	30.50
6280	12" iron pipe size G		90	.178		18.95	7.95		26.90	33.50
6290	14" iron pipe size G		85	.188		21	8.40		29.40	36
6300	16" iron pipe size G		80	.200		23	8.95		31.95	39.50
6310	18" iron pipe size G		75	.213		25	9.55		34.55	42.50
6320	20" iron pipe size G		65	.246		31	11		42	51
6330	22" iron pipe size G		60	.267		34	11.95		45.95	56
6340	24" iron pipe size G		55	.291		35	13		48	58.50
6360	3" wall, 1/2" iron pipe size G		115	.139		9.10	6.20		15.30	19.65
6380	3/4" iron pipe size G		115	.139		9.15	6.20		15.35	19.75
6390	1" iron pipe size G		115	.139		9.20	6.20		15.40	19.80
6400	1-1/4" iron pipe size G		110	.145		9.35	6.50		15.85	20.50

22 07 19 – Plumbing Piping Insulation

22 07 19.10 Piping Insulation		Crew	Daily Output	Labor-Hours	Unit	2013 Bare Costs Material	Labor	Equipment	Total	Total Incl O&P
6410	1-1/2" iron pipe size	G	Q-14	110	.145	L.F.	9.45	6.50		15.95	20.50
6420	2" iron pipe size	G		105	.152		9.75	6.80		16.55	21.50
6430	2-1/2" iron pipe size	G		105	.152		11.45	6.80		18.25	23
6440	3" iron pipe size	G		100	.160		11.55	7.15		18.70	24
6450	4" iron pipe size	G		95	.168		14.75	7.55		22.30	28
6460	5" iron pipe size	G		90	.178		15.95	7.95		23.90	30
6470	6" iron pipe size	G		90	.178		17.95	7.95		25.90	32
6480	7" iron pipe size	G		85	.188		20	8.40		28.40	35
6490	8" iron pipe size	G		85	.188		21.50	8.40		29.90	36.50
6500	9" iron pipe size	G		80	.200		24.50	8.95		33.45	40.50
6510	10" iron pipe size	G		75	.213		26	9.55		35.55	43.50
6520	12" iron pipe size	G		70	.229		28.50	10.20		38.70	47.50
6530	14" iron pipe size	G		65	.246		32	11		43	52.50
6540	16" iron pipe size	G		60	.267		35.50	11.95		47.45	57.50
6550	18" iron pipe size	G		55	.291		39	13		52	63
6560	20" iron pipe size	G		50	.320		46.50	14.30		60.80	73
6570	22" iron pipe size	G		45	.356		50	15.90		65.90	79.50
6580	24" iron pipe size	G		40	.400		54	17.90		71.90	87.50
6600	Fiberglass, with all service jacket										
6640	1/2" wall, 1/2" iron pipe size	G	Q-14	250	.064	L.F.	.74	2.86		3.60	5.25
6660	3/4" iron pipe size	G		240	.067		.83	2.98		3.81	5.55
6670	1" iron pipe size	G		230	.070		.86	3.11		3.97	5.80
6680	1-1/4" iron pipe size	G		220	.073		.92	3.25		4.17	6.05
6690	1-1/2" iron pipe size	G		220	.073		1.06	3.25		4.31	6.20
6700	2" iron pipe size	G		210	.076		1.13	3.41		4.54	6.55
6710	2-1/2" iron pipe size	G		200	.080		1.18	3.58		4.76	6.85
6840	1" wall, 1/2" iron pipe size	G		240	.067		.89	2.98		3.87	5.60
6860	3/4" iron pipe size	G		230	.070		.97	3.11		4.08	5.90
6870	1" iron pipe size	G		220	.073		1.04	3.25		4.29	6.20
6880	1-1/4" iron pipe size	G		210	.076		1.13	3.41		4.54	6.55
6890	1-1/2" iron pipe size	G		210	.076		1.22	3.41		4.63	6.65
6900	2" iron pipe size	G		200	.080		1.32	3.58		4.90	7
6910	2-1/2" iron pipe size	G		190	.084		1.50	3.77		5.27	7.50
6920	3" iron pipe size	G		180	.089		1.60	3.98		5.58	7.95
6930	3-1/2" iron pipe size	G		170	.094		1.74	4.21		5.95	8.45
6940	4" iron pipe size	G		150	.107		2.12	4.77		6.89	9.75
6950	5" iron pipe size	G		140	.114		2.39	5.10		7.49	10.60
6960	6" iron pipe size	G		120	.133		2.54	5.95		8.49	12.05
6970	7" iron pipe size	G		110	.145		3.02	6.50		9.52	13.40
6980	8" iron pipe size	G		100	.160		4.15	7.15		11.30	15.65
6990	9" iron pipe size	G		90	.178		4.38	7.95		12.33	17.15
7000	10" iron pipe size	G		90	.178		4.40	7.95		12.35	17.20
7010	12" iron pipe size	G		80	.200		4.80	8.95		13.75	19.20
7020	14" iron pipe size	G		80	.200		5.75	8.95		14.70	20.50
7030	16" iron pipe size	G		70	.229		7.30	10.20		17.50	24
7040	18" iron pipe size	G		70	.229		8.10	10.20		18.30	25
7050	20" iron pipe size	G		60	.267		9.05	11.95		21	28.50
7060	24" iron pipe size	G		60	.267		11.05	11.95		23	30.50
7080	1-1/2" wall, 1/2" iron pipe size	G		230	.070		1.69	3.11		4.80	6.70
7100	3/4" iron pipe size	G		220	.073		1.69	3.25		4.94	6.90
7110	1" iron pipe size	G		210	.076		1.81	3.41		5.22	7.30
7120	1-1/4" iron pipe size	G		200	.080		1.97	3.58		5.55	7.70
7130	1-1/2" iron pipe size	G		200	.080		2.07	3.58		5.65	7.85

22 07 19.10 Piping Insulation		Crew	Daily Output	Labor-Hours	Unit	Material	Labor	Equipment	Total	Total Incl O&P
						2013 Bare Costs				
7140	2" iron pipe size	G	Q-14	190	.084	L.F.	2.28	3.77		6.05	8.35
7150	2-1/2" iron pipe size	G		180	.089		2.44	3.98		6.42	8.90
7160	3" iron pipe size	G		170	.094		2.56	4.21		6.77	9.35
7170	3-1/2" iron pipe size	G		160	.100		2.80	4.47		7.27	10.05
7180	4" iron pipe size	G		140	.114		2.91	5.10		8.01	11.15
7190	5" iron pipe size	G		130	.123		3.26	5.50		8.76	12.15
7200	6" iron pipe size	G		110	.145		3.44	6.50		9.94	13.90
7210	7" iron pipe size	G		100	.160		3.82	7.15		10.97	15.30
7220	8" iron pipe size	G		90	.178		4.82	7.95		12.77	17.65
7230	9" iron pipe size	G		85	.188		5	8.40		13.40	18.60
7240	10" iron pipe size	G		80	.200		5.20	8.95		14.15	19.60
7250	12" iron pipe size	G		75	.213		5.90	9.55		15.45	21.50
7260	14" iron pipe size	G		70	.229		7.05	10.20		17.25	23.50
7270	16" iron pipe size	G		65	.246		9.15	11		20.15	27
7280	18" iron pipe size	G		60	.267		10.30	11.95		22.25	30
7290	20" iron pipe size	G		55	.291		10.50	13		23.50	31.50
7300	24" iron pipe size	G		50	.320		12.85	14.30		27.15	36
7320	2" wall, 1/2" iron pipe size	G		220	.073		2.62	3.25		5.87	7.95
7340	3/4" iron pipe size	G		210	.076		2.70	3.41		6.11	8.25
7350	1" iron pipe size	G		200	.080		2.88	3.58		6.46	8.70
7360	1-1/4" iron pipe size	G		190	.084		3.04	3.77		6.81	9.20
7370	1-1/2" iron pipe size	G		190	.084		3.17	3.77		6.94	9.35
7380	2" iron pipe size	G		180	.089		3.34	3.98		7.32	9.85
7390	2-1/2" iron pipe size	G		170	.094		3.60	4.21		7.81	10.50
7400	3" iron pipe size	G		160	.100		3.82	4.47		8.29	11.15
7410	3-1/2" iron pipe size	G		150	.107		4.14	4.77		8.91	11.95
7420	4" iron pipe size	G		130	.123		4.45	5.50		9.95	13.45
7430	5" iron pipe size	G		120	.133		5.10	5.95		11.05	14.85
7440	6" iron pipe size	G		100	.160		5.25	7.15		12.40	16.90
7450	7" iron pipe size	G		90	.178		6	7.95		13.95	18.95
7460	8" iron pipe size	G		80	.200		6.40	8.95		15.35	21
7470	9" iron pipe size	G		75	.213		7	9.55		16.55	22.50
7480	10" iron pipe size	G		70	.229		7.65	10.20		17.85	24.50
7490	12" iron pipe size	G		65	.246		8.55	11		19.55	26.50
7500	14" iron pipe size	G		60	.267		11.05	11.95		23	30.50
7510	16" iron pipe size	G		55	.291		12.10	13		25.10	33.50
7520	18" iron pipe size	G		50	.320		13.50	14.30		27.80	37
7530	20" iron pipe size	G		45	.356		15.35	15.90		31.25	41.50
7540	24" iron pipe size	G		40	.400		16.45	17.90		34.35	46
7560	2-1/2" wall, 1/2" iron pipe size	G		210	.076		3.10	3.41		6.51	8.70
7562	3/4" iron pipe size	G		200	.080		3.24	3.58		6.82	9.10
7564	1" iron pipe size	G		190	.084		3.38	3.77		7.15	9.55
7566	1-1/4" iron pipe size	G		185	.086		3.50	3.87		7.37	9.85
7568	1-1/2" iron pipe size	G		180	.089		3.68	3.98		7.66	10.25
7570	2" iron pipe size	G		170	.094		3.86	4.21		8.07	10.80
7572	2-1/2" iron pipe size	G		160	.100		4.45	4.47		8.92	11.85
7574	3" iron pipe size	G		150	.107		4.68	4.77		9.45	12.55
7576	3-1/2" iron pipe size	G		140	.114		5.10	5.10		10.20	13.55
7578	4" iron pipe size	G		120	.133		5.35	5.95		11.30	15.15
7580	5" iron pipe size	G		110	.145		6.40	6.50		12.90	17.15
7582	6" iron pipe size	G		90	.178		7.70	7.95		15.65	21
7584	7" iron pipe size	G		80	.200		7.75	8.95		16.70	22.50
7586	8" iron pipe size	G		70	.229		8.25	10.20		18.45	25

22 07 19 – Plumbing Piping Insulation

22 07 19.10 Piping Insulation		Crew	Daily Output	Labor-Hours	Unit	Material	Labor	Equipment	Total	Total Incl O&P
7588	9" iron pipe size	G	Q-14	65	.246	L.F.	9	11		20	27
7590	10" iron pipe size	G		60	.267		9.80	11.95		21.75	29.50
7592	12" iron pipe size	G		55	.291		11.90	13		24.90	33
7594	14" iron pipe size	G		50	.320		13.90	14.30		28.20	37.50
7596	16" iron pipe size	G		45	.356		15.95	15.90		31.85	42
7598	18" iron pipe size	G		40	.400		17.30	17.90		35.20	47
7602	24" iron pipe size	G		30	.533		23	24		47	62
7620	3" wall, 1/2" iron pipe size	G		200	.080		3.97	3.58		7.55	9.90
7622	3/4" iron pipe size	G		190	.084		4.20	3.77		7.97	10.45
7624	1" iron pipe size	G		180	.089		4.45	3.98		8.43	11.10
7626	1-1/4" iron pipe size	G		175	.091		4.54	4.09		8.63	11.35
7628	1-1/2" iron pipe size	G		170	.094		4.74	4.21		8.95	11.75
7630	2" iron pipe size	G		160	.100		5.10	4.47		9.57	12.60
7632	2-1/2" iron pipe size	G		150	.107		5.35	4.77		10.12	13.25
7634	3" iron pipe size	G		140	.114		5.65	5.10		10.75	14.20
7636	3-1/2" iron pipe size	G		130	.123		6.25	5.50		11.75	15.45
7638	4" iron pipe size	G		110	.145		6.75	6.50		13.25	17.50
7640	5" iron pipe size	G		100	.160		7.60	7.15		14.75	19.50
7642	6" iron pipe size	G		80	.200		8.15	8.95		17.10	23
7644	7" iron pipe size	G		70	.229		9.30	10.20		19.50	26
7646	8" iron pipe size	G		60	.267		10.15	11.95		22.10	29.50
7648	9" iron pipe size	G		55	.291		10.95	13		23.95	32
7650	10" iron pipe size	G		50	.320		11.75	14.30		26.05	35
7652	12" iron pipe size	G		45	.356		14.60	15.90		30.50	40.50
7654	14" iron pipe size	G		40	.400		16.85	17.90		34.75	46.50
7656	16" iron pipe size	G		35	.457		19.10	20.50		39.60	53
7658	18" iron pipe size	G		32	.500		20.50	22.50		43	57
7660	20" iron pipe size	G		30	.533		22	24		46	61.50
7662	24" iron pipe size	G		28	.571		28	25.50		53.50	70.50
7664	26" iron pipe size	G		26	.615		31	27.50		58.50	77
7666	30" iron pipe size	G		24	.667		37	30		67	87
7800	For fiberglass with standard canvas jacket, deduct						5%				
7802	For fittings, add 3 L.F. for each fitting										
7804	plus 4 L.F. for each flange of the fitting										
7810	Finishes										
7814	For single layer of felt, add						10%	10%			
7816	For roofing paper, 45 lb. to 55 lb., add						25%	10%			
7820	Polyethylene tubing flexible closed cell foam, UV resistant										
7828	Standard temperature (-90°F to +212°F)										
7830	3/8" wall, 1/8" iron pipe size	G	1 Asbe	130	.062	L.F.	.22	3.06		3.28	4.99
7831	1/4" iron pipe size	G		130	.062		.23	3.06		3.29	5
7832	3/8" iron pipe size	G		130	.062		.26	3.06		3.32	5.05
7833	1/2" iron pipe size	G		126	.063		.29	3.16		3.45	5.20
7834	3/4" iron pipe size	G		122	.066		.34	3.26		3.60	5.40
7835	1" iron pipe size	G		120	.067		.39	3.31		3.70	5.60
7836	1-1/4" iron pipe size	G		118	.068		.47	3.37		3.84	5.75
7837	1-1/2" iron pipe size	G		118	.068		.60	3.37		3.97	5.90
7838	2" iron pipe size	G		116	.069		.79	3.43		4.22	6.15
7839	2-1/2" iron pipe size	G		114	.070		1.01	3.49		4.50	6.50
7840	3" iron pipe size	G		112	.071		1.56	3.55		5.11	7.20
7842	1/2" wall, 1/8" iron pipe size	G		120	.067		.33	3.31		3.64	5.50
7843	1/4" iron pipe size	G		120	.067		.35	3.31		3.66	5.55
7844	3/8" iron pipe size	G		120	.067		.39	3.31		3.70	5.60

22 07 19.10 Piping Insulation			Crew	Daily Output	Labor-Hours	Unit	Material	Labor	Equipment	Total	Total Incl O&P
							2013 Bare Costs				
7845	1/2" iron pipe size	G	1 Asbe	118	.068	L.F.	.43	3.37		3.80	5.70
7846	3/4" iron pipe size	G		116	.069		.49	3.43		3.92	5.85
7847	1" iron pipe size	G		114	.070		.55	3.49		4.04	6
7848	1-1/4" iron pipe size	G		112	.071		.65	3.55		4.20	6.20
7849	1-1/2" iron pipe size	G		110	.073		.77	3.61		4.38	6.45
7850	2" iron pipe size	G		108	.074		1.11	3.68		4.79	6.90
7851	2-1/2" iron pipe size	G		106	.075		1.37	3.75		5.12	7.35
7852	3" iron pipe size	G		104	.077		1.89	3.82		5.71	8.05
7853	3-1/2" iron pipe size	G		102	.078		2.14	3.90		6.04	8.40
7854	4" iron pipe size	G		100	.080		2.44	3.98		6.42	8.90
7855	3/4" wall, 1/8" iron pipe size	G		110	.073		.50	3.61		4.11	6.15
7856	1/4" iron pipe size	G		110	.073		.53	3.61		4.14	6.20
7857	3/8" iron pipe size	G		108	.074		.61	3.68		4.29	6.35
7858	1/2" iron pipe size	G		106	.075		.73	3.75		4.48	6.65
7859	3/4" iron pipe size	G		104	.077		.88	3.82		4.70	6.90
7860	1" iron pipe size	G		102	.078		1.01	3.90		4.91	7.15
7861	1-1/4" iron pipe size	G		100	.080		1.38	3.98		5.36	7.70
7862	1-1/2" iron pipe size	G		100	.080		1.56	3.98		5.54	7.90
7863	2" iron pipe size	G		98	.082		1.98	4.06		6.04	8.50
7864	2-1/2" iron pipe size	G		96	.083		2.39	4.14		6.53	9.10
7865	3" iron pipe size	G		94	.085		3.15	4.23		7.38	10
7866	3-1/2" iron pipe size	G		92	.087		3.51	4.32		7.83	10.55
7867	4" iron pipe size	G		90	.089		3.92	4.42		8.34	11.15
7868	1" wall, 1/4" iron pipe size	G		100	.080		1.06	3.98		5.04	7.35
7869	3/8" iron pipe size	G		98	.082		1.08	4.06		5.14	7.50
7870	1/2" iron pipe size	G		96	.083		1.35	4.14		5.49	7.95
7871	3/4" iron pipe size	G		94	.085		1.57	4.23		5.80	8.30
7872	1" iron pipe size	G		92	.087		1.91	4.32		6.23	8.80
7873	1-1/4" iron pipe size	G		90	.089		2.17	4.42		6.59	9.25
7874	1-1/2" iron pipe size	G		90	.089		2.48	4.42		6.90	9.60
7875	2" iron pipe size	G		88	.091		3.23	4.52		7.75	10.55
7876	2-1/2" iron pipe size	G		86	.093		4.22	4.62		8.84	11.85
7877	3" iron pipe size	G		84	.095		4.74	4.73		9.47	12.55
7878	Contact cement, quart can	G				Ea.	10.50			10.50	11.55
7879	Rubber tubing, flexible closed cell foam										
7880	3/8" wall, 1/4" iron pipe size	G	1 Asbe	120	.067	L.F.	.38	3.31		3.69	5.55
7900	3/8" iron pipe size	G		120	.067		.43	3.31		3.74	5.60
7910	1/2" iron pipe size	G		115	.070		.48	3.46		3.94	5.90
7920	3/4" iron pipe size	G		115	.070		.55	3.46		4.01	5.95
7930	1" iron pipe size	G		110	.073		.63	3.61		4.24	6.30
7940	1-1/4" iron pipe size	G		110	.073		.77	3.61		4.38	6.45
7950	1-1/2" iron pipe size	G		110	.073		.85	3.61		4.46	6.55
8100	1/2" wall, 1/4" iron pipe size	G		90	.089		.54	4.42		4.96	7.45
8120	3/8" iron pipe size	G		90	.089		.60	4.42		5.02	7.50
8130	1/2" iron pipe size	G		89	.090		.67	4.47		5.14	7.70
8140	3/4" iron pipe size	G		89	.090		.75	4.47		5.22	7.80
8150	1" iron pipe size	G		88	.091		.82	4.52		5.34	7.90
8160	1-1/4" iron pipe size	G		87	.092		.95	4.57		5.52	8.15
8170	1-1/2" iron pipe size	G		87	.092		1.15	4.57		5.72	8.35
8180	2" iron pipe size	G		86	.093		1.47	4.62		6.09	8.80
8190	2-1/2" iron pipe size	G		86	.093		1.85	4.62		6.47	9.25
8200	3" iron pipe size	G		85	.094		2.06	4.68		6.74	9.50
8210	3-1/2" iron pipe size	G		85	.094		2.86	4.68		7.54	10.40

22 07 19 – Plumbing Piping Insulation

22 07 19.10 Piping Insulation		Crew	Daily Output	Labor-Hours	Unit	Material	Labor	Equipment	Total	Total Incl O&P
8220	4" iron pipe size	G	1 Asbe	80	.100	L.F.	3.06	4.97		8.03	11.05
8230	5" iron pipe size	G		80	.100		4.17	4.97		9.14	12.30
8240	6" iron pipe size	G		75	.107		4.17	5.30		9.47	12.85
8300	3/4" wall, 1/4" iron pipe size	G		90	.089		.85	4.42		5.27	7.80
8320	3/8" iron pipe size	G		90	.089		.92	4.42		5.34	7.85
8330	1/2" iron pipe size	G		89	.090		1.10	4.47		5.57	8.15
8340	3/4" iron pipe size	G		89	.090		1.35	4.47		5.82	8.45
8350	1" iron pipe size	G		88	.091		1.54	4.52		6.06	8.70
8360	1-1/4" iron pipe size	G		87	.092		2.05	4.57		6.62	9.35
8370	1-1/2" iron pipe size	G		87	.092		2.32	4.57		6.89	9.65
8380	2" iron pipe size	G		86	.093		2.68	4.62		7.30	10.15
8390	2-1/2" iron pipe size	G		86	.093		3.58	4.62		8.20	11.15
8400	3" iron pipe size	G		85	.094		4.08	4.68		8.76	11.75
8410	3-1/2" iron pipe size	G		85	.094		4.81	4.68		9.49	12.55
8420	4" iron pipe size	G		80	.100		5.10	4.97		10.07	13.35
8430	5" iron pipe size	G		80	.100		6.15	4.97		11.12	14.50
8440	6" iron pipe size	G		80	.100		7.45	4.97		12.42	15.90
8444	1" wall, 1/2" iron pipe size	G		86	.093		2.05	4.62		6.67	9.45
8445	3/4" iron pipe size	G		84	.095		2.48	4.73		7.21	10.10
8446	1" iron pipe size	G		84	.095		2.89	4.73		7.62	10.55
8447	1-1/4" iron pipe size	G		82	.098		3.23	4.85		8.08	11.10
8448	1-1/2" iron pipe size	G		82	.098		3.76	4.85		8.61	11.70
8449	2" iron pipe size	G		80	.100		4.95	4.97		9.92	13.15
8450	2-1/2" iron pipe size	G	▼	80	.100	▼	6.45	4.97		11.42	14.80
8456	Rubber insulation tape, 1/8" x 2" x 30'	G				Ea.	12.05			12.05	13.25
8460	Polyolefin tubing, flexible closed cell foam, UV stabilized, work										
8462	temp.-165°F to +210°F, 0 water vapor transmission										
8464	3/8" wall, 1/8" iron pipe size	G	1 Asbe	140	.057	L.F.	.47	2.84		3.31	4.93
8466	1/4" iron pipe size	G		140	.057		.49	2.84		3.33	4.95
8468	3/8" iron pipe size	G		140	.057		.54	2.84		3.38	5
8470	1/2" iron pipe size	G		136	.059		.60	2.92		3.52	5.20
8472	3/4" iron pipe size	G		132	.061		.71	3.01		3.72	5.45
8474	1" iron pipe size	G		130	.062		.81	3.06		3.87	5.65
8476	1-1/4" iron pipe size	G		128	.063		.99	3.11		4.10	5.90
8478	1-1/2" iron pipe size	G		128	.063		1.21	3.11		4.32	6.15
8480	2" iron pipe size	G		126	.063		1.35	3.16		4.51	6.40
8482	2-1/2" iron pipe size	G		123	.065		1.96	3.23		5.19	7.15
8484	3" iron pipe size	G		121	.066		2.17	3.29		5.46	7.50
8486	4" iron pipe size	G		118	.068		3.91	3.37		7.28	9.55
8500	1/2" wall, 1/8" iron pipe size	G		130	.062		.68	3.06		3.74	5.50
8502	1/4" iron pipe size	G		130	.062		.73	3.06		3.79	5.55
8504	3/8" iron pipe size	G		130	.062		.79	3.06		3.85	5.60
8506	1/2" iron pipe size	G		128	.063		.86	3.11		3.97	5.80
8508	3/4" iron pipe size	G		126	.063		.99	3.16		4.15	6
8510	1" iron pipe size	G		123	.065		1.14	3.23		4.37	6.25
8512	1-1/4" iron pipe size	G		121	.066		1.35	3.29		4.64	6.60
8514	1-1/2" iron pipe size	G		119	.067		1.62	3.34		4.96	7
8516	2" iron pipe size	G		117	.068		1.90	3.40		5.30	7.40
8518	2-1/2" iron pipe size	G		114	.070		2.56	3.49		6.05	8.20
8520	3" iron pipe size	G		112	.071		3.33	3.55		6.88	9.15
8522	4" iron pipe size	G		110	.073		4.62	3.61		8.23	10.70
8534	3/4" wall, 1/8" iron pipe size	G		120	.067		1.04	3.31		4.35	6.30
8536	1/4" iron pipe size	G		120	.067		1.10	3.31		4.41	6.35

Column group: 2013 Bare Costs (Material, Labor, Equipment, Total)

22 07 19 – Plumbing Piping Insulation

22 07 19.10 Piping Insulation			Crew	Daily Output	Labor-Hours	Unit	Material	Labor	Equipment	Total	Total Incl O&P
8538	3/8" iron pipe size	G	1 Asbe	117	.068	L.F.	1.27	3.40		4.67	6.70
8540	1/2" iron pipe size	G		114	.070		1.40	3.49		4.89	6.95
8542	3/4" iron pipe size	G		112	.071		1.52	3.55		5.07	7.15
8544	1" iron pipe size	G		110	.073		2.13	3.61		5.74	7.95
8546	1-1/4" iron pipe size	G		108	.074		2.85	3.68		6.53	8.85
8548	1-1/2" iron pipe size	G		108	.074		3.33	3.68		7.01	9.35
8550	2" iron pipe size	G		106	.075		3.90	3.75		7.65	10.15
8552	2-1/2" iron pipe size	G		104	.077		4.92	3.82		8.74	11.35
8554	3" iron pipe size	G		102	.078		6.15	3.90		10.05	12.80
8556	4" iron pipe size	G		100	.080		8.10	3.98		12.08	15.10
8570	1" wall, 1/8" iron pipe size	G		110	.073		1.89	3.61		5.50	7.70
8572	1/4" iron pipe size	G		108	.074		1.94	3.68		5.62	7.85
8574	3/8" iron pipe size	G		106	.075		1.98	3.75		5.73	8.05
8576	1/2" iron pipe size	G		104	.077		2.10	3.82		5.92	8.25
8578	3/4" iron pipe size	G		102	.078		2.74	3.90		6.64	9.05
8580	1" iron pipe size	G		100	.080		3.18	3.98		7.16	9.70
8582	1-1/4" iron pipe size	G		97	.082		3.61	4.10		7.71	10.30
8584	1-1/2" iron pipe size	G		97	.082		4.13	4.10		8.23	10.90
8586	2" iron pipe size	G		95	.084		5.40	4.19		9.59	12.45
8588	2-1/2" iron pipe size	G		93	.086		7.05	4.28		11.33	14.40
8590	3" iron pipe size	G		91	.088		8.60	4.37		12.97	16.25
8606	Contact adhesive (R-320)	G				Qt.	26.50			26.50	29.50
8608	Contact adhesive (R-320)	G				Gal.	109			109	120
8610	NOTE: Preslit/preglued vs unslit, same price										

22 07 19.30 Piping Insulation Protective Jacketing, PVC		Crew	Daily Output	Labor-Hours	Unit	Material	Labor	Equipment	Total	Total Incl O&P
0010	**PIPING INSULATION PROTECTIVE JACKETING, PVC**									
0100	PVC, white, 48" lengths cut from roll goods									
0120	20 mil thick									
0140	Size based on OD of insulation									
0150	1-1/2" ID	Q-14	270	.059	L.F.	.20	2.65		2.85	4.34
0152	2" ID		260	.062		.27	2.75		3.02	4.58
0154	2-1/2" ID		250	.064		.32	2.86		3.18	4.80
0156	3" ID		240	.067		.39	2.98		3.37	5.05
0158	3-1/2" ID		230	.070		.44	3.11		3.55	5.30
0160	4" ID		220	.073		.50	3.25		3.75	5.60
0162	4-1/2" ID		210	.076		.56	3.41		3.97	5.90
0164	5" ID		200	.080		.61	3.58		4.19	6.20
0166	5-1/2" ID		190	.084		.67	3.77		4.44	6.60
0168	6" ID		180	.089		.73	3.98		4.71	7
0170	6-1/2" ID		175	.091		.79	4.09		4.88	7.20
0172	7" ID		170	.094		.84	4.21		5.05	7.45
0174	7-1/2" ID		164	.098		.91	4.36		5.27	7.80
0176	8" ID		161	.099		.96	4.45		5.41	7.95
0178	8-1/2" ID		158	.101		1.02	4.53		5.55	8.15
0180	9" ID		155	.103		1.08	4.62		5.70	8.35
0182	9-1/2" ID		152	.105		1.14	4.71		5.85	8.55
0184	10" ID		149	.107		1.19	4.80		5.99	8.75
0186	10-1/2" ID		146	.110		1.25	4.90		6.15	9
0188	11" ID		143	.112		1.31	5		6.31	9.20
0190	11-1/2" ID		140	.114		1.37	5.10		6.47	9.45
0192	12" ID		137	.117		1.42	5.20		6.62	9.65
0194	12-1/2" ID		134	.119		1.49	5.35		6.84	9.95

22 07 19 – Plumbing Piping Insulation

22 07 19.30 Piping Insulation Protective Jacketing, PVC		Crew	Daily Output	Labor-Hours	Unit	2013 Bare Costs Material	Labor	Equipment	Total	Total Incl O&P
0195	13" ID	Q-14	132	.121	L.F.	1.54	5.40		6.94	10.10
0196	13-1/2" ID		132	.121		1.60	5.40		7	10.15
0198	14" ID		130	.123		1.66	5.50		7.16	10.40
0200	15" ID		128	.125		1.78	5.60		7.38	10.65
0202	16" ID		126	.127		1.89	5.70		7.59	10.90
0204	17" ID		124	.129		2.01	5.75		7.76	11.15
0206	18" ID		122	.131		2.13	5.85		7.98	11.45
0208	19" ID		120	.133		2.24	5.95		8.19	11.70
0210	20" ID		118	.136		2.35	6.05		8.40	12
0212	21" ID		116	.138		2.47	6.15		8.62	12.30
0214	22" ID		114	.140		2.58	6.30		8.88	12.60
0216	23" ID		112	.143		2.69	6.40		9.09	12.90
0218	24" ID		110	.145		2.81	6.50		9.31	13.20
0220	25" ID		108	.148		2.93	6.65		9.58	13.50
0222	26" ID		106	.151		3.04	6.75		9.79	13.85
0224	27" ID		104	.154		3.16	6.90		10.06	14.20
0226	28" ID		102	.157		3.27	7		10.27	14.50
0228	29" ID		100	.160		3.39	7.15		10.54	14.85
0230	30" ID	↓	98	.163	↓	3.51	7.30		10.81	15.20
0300	For colors, add				Ea.	10%				
1000	30 mil thick									
1010	Size based on OD of insulation									
1020	2" ID	Q-14	260	.062	L.F.	.40	2.75		3.15	4.72
1022	2-1/2" ID		250	.064		.48	2.86		3.34	4.98
1024	3" ID		240	.067		.58	2.98		3.56	5.25
1026	3-1/2" ID		230	.070		.65	3.11		3.76	5.55
1028	4" ID		220	.073		.74	3.25		3.99	5.85
1030	4-1/2" ID		210	.076		.82	3.41		4.23	6.20
1032	5" ID		200	.080		.91	3.58		4.49	6.55
1034	5-1/2" ID		190	.084		.99	3.77		4.76	6.95
1036	6" ID		180	.089		1.08	3.98		5.06	7.40
1038	6-1/2" ID		175	.091		1.16	4.09		5.25	7.65
1040	7" ID		170	.094		1.25	4.21		5.46	7.95
1042	7-1/2" ID		164	.098		1.34	4.36		5.70	8.25
1044	8" ID		161	.099		1.43	4.45		5.88	8.45
1046	8-1/2" ID		158	.101		1.51	4.53		6.04	8.70
1048	9" ID		155	.103		1.60	4.62		6.22	8.90
1050	9-1/2" ID		152	.105		1.68	4.71		6.39	9.15
1052	10" ID		149	.107		1.77	4.80		6.57	9.40
1054	10-1/2" ID		146	.110		1.85	4.90		6.75	9.65
1056	11" ID		143	.112		1.94	5		6.94	9.90
1058	11-1/2" ID		140	.114		2.02	5.10		7.12	10.15
1060	12" ID		137	.117		2.11	5.20		7.31	10.40
1062	12-1/2" ID		134	.119		2.20	5.35		7.55	10.70
1063	13" ID		132	.121		2.29	5.40		7.69	10.90
1064	13-1/2" ID		132	.121		2.37	5.40		7.77	11
1066	14" ID		130	.123		2.46	5.50		7.96	11.25
1068	15" ID		128	.125		2.63	5.60		8.23	11.60
1070	16" ID		126	.127		2.80	5.70		8.50	11.90
1072	17" ID		124	.129		2.97	5.75		8.72	12.20
1074	18" ID		122	.131		3.15	5.85		9	12.55
1076	19" ID		120	.133		3.32	5.95		9.27	12.90
1078	20" ID		118	.136		3.49	6.05		9.54	13.25

22 07 19.30 Piping Insulation Protective Jacketing, PVC		Crew	Daily Output	Labor-Hours	Unit	2013 Bare Costs Material	Labor	Equipment	Total	Total Incl O&P
1080	21″ ID	Q-14	116	.138	L.F.	3.66	6.15		9.81	13.65
1082	22″ ID		114	.140		3.83	6.30		10.13	13.95
1084	23″ ID		112	.143		4	6.40		10.40	14.35
1086	24″ ID		110	.145		4.17	6.50		10.67	14.70
1088	25″ ID		108	.148		4.34	6.65		10.99	15.05
1090	26″ ID		106	.151		4.51	6.75		11.26	15.45
1092	27″ ID		104	.154		4.68	6.90		11.58	15.85
1094	28″ ID		102	.157		4.85	7		11.85	16.25
1096	29″ ID		100	.160		5	7.15		12.15	16.60
1098	30″ ID		98	.163		5.20	7.30		12.50	17.05
1300	For colors, add				Ea.	10%				
2000	PVC, white, fitting covers									
2020	Fiberglass insulation inserts included with sizes 1-3/4″ thru 9-3/4″									
2030	Size is based on OD of insulation									
2040	90° Elbow fitting									
2060	1-3/4″	Q-14	135	.119	Ea.	.47	5.30		5.77	8.75
2062	2″		130	.123		.58	5.50		6.08	9.20
2064	2-1/4″		128	.125		.65	5.60		6.25	9.40
2068	2-1/2″		126	.127		.71	5.70		6.41	9.60
2070	2-3/4″		123	.130		.77	5.80		6.57	9.90
2072	3″		120	.133		.82	5.95		6.77	10.15
2074	3-3/8″		116	.138		.93	6.15		7.08	10.60
2076	3-3/4″		113	.142		1.04	6.35		7.39	11
2078	4-1/8″		110	.145		1.31	6.50		7.81	11.55
2080	4-3/4″		105	.152		1.58	6.80		8.38	12.35
2082	5-1/4″		100	.160		1.84	7.15		8.99	13.10
2084	5-3/4″		95	.168		2.43	7.55		9.98	14.35
2086	6-1/4″		90	.178		4.05	7.95		12	16.80
2088	6-3/4″		87	.184		4.27	8.25		12.52	17.50
2090	7-1/4″		85	.188		5.35	8.40		13.75	19
2092	7-3/4″		83	.193		5.60	8.60		14.20	19.55
2094	8-3/4″		80	.200		7.20	8.95		16.15	22
2096	9-3/4″		77	.208		9.65	9.30		18.95	25
2098	10-7/8″		74	.216		10.75	9.65		20.40	27
2100	11-7/8″		71	.225		12.35	10.10		22.45	29
2102	12-7/8″		68	.235		15.55	10.50		26.05	33.50
2104	14-1/8″		66	.242		17.80	10.85		28.65	36.50
2106	15-1/8″		64	.250		19.15	11.20		30.35	38.50
2108	16-1/8″		63	.254		21	11.35		32.35	40.50
2110	17-1/8″		62	.258		23	11.55		34.55	43.50
2112	18-1/8″		61	.262		30	11.75		41.75	51
2114	19-1/8″		60	.267		40.50	11.95		52.45	63
2116	20-1/8″		59	.271		52	12.15		64.15	76.50
2200	45° Elbow fitting									
2220	1-3/4″ thru 9-3/4″ same price as 90° Elbow fitting									
2320	10-7/8″	Q-14	74	.216	Ea.	10.75	9.65		20.40	27
2322	11-7/8″		71	.225		11.85	10.10		21.95	28.50
2324	12-7/8″		68	.235		13.25	10.50		23.75	31
2326	14-1/8″		66	.242		15.10	10.85		25.95	33.50
2328	15-1/8″		64	.250		16.25	11.20		27.45	35
2330	16-1/8″		63	.254		18.55	11.35		29.90	38
2332	17-1/8″		62	.258		21	11.55		32.55	41
2334	18-1/8″		61	.262		25.50	11.75		37.25	46

22 07 19.30 Piping Insulation Protective Jacketing, PVC		Crew	Daily Output	Labor-Hours	Unit	Material	2013 Bare Costs Labor	Equipment	Total	Total Incl O&P
2336	19-1/8"	Q-14	60	.267	Ea.	35	11.95		46.95	57
2338	20-1/8"	↓	59	.271	↓	39.50	12.15		51.65	62.50
2400	Tee fitting									
2410	1-3/4"	Q-14	96	.167	Ea.	.88	7.45		8.33	12.55
2412	2"		94	.170		1	7.60		8.60	12.95
2414	2-1/4"		91	.176		1.09	7.85		8.94	13.40
2416	2-1/2"		88	.182		1.19	8.15		9.34	13.95
2418	2-3/4"		85	.188		1.31	8.40		9.71	14.55
2420	3"		82	.195		1.42	8.75		10.17	15.10
2422	3-3/8"		79	.203		1.63	9.05		10.68	15.85
2424	3-3/4"		76	.211		1.85	9.40		11.25	16.70
2426	4-1/8"		73	.219		2.19	9.80		11.99	17.65
2428	4-3/4"		70	.229		2.73	10.20		12.93	18.90
2430	5-1/4"		67	.239		3.28	10.70		13.98	20
2432	5-3/4"		63	.254		4.36	11.35		15.71	22.50
2434	6-1/4"		60	.267		5.75	11.95		17.70	25
2436	6-3/4"		59	.271		7.10	12.15		19.25	26.50
2438	7-1/4"		57	.281		11.45	12.55		24	32
2440	7-3/4"		54	.296		12.55	13.25		25.80	34.50
2442	8-3/4"		52	.308		15.30	13.75		29.05	38.50
2444	9-3/4"		50	.320		18	14.30		32.30	42
2446	10-7/8"		48	.333		18.25	14.90		33.15	43
2448	11-7/8"		47	.340		20.50	15.25		35.75	46
2450	12-7/8"		46	.348		22.50	15.55		38.05	48.50
2452	14-1/8"		45	.356		24.50	15.90		40.40	51.50
2454	15-1/8"		44	.364		26.50	16.25		42.75	54.50
2456	16-1/8"		43	.372		28.50	16.65		45.15	57
2458	17-1/8"		42	.381		30.50	17.05		47.55	60
2460	18-1/8"		41	.390		33.50	17.45		50.95	64
2462	19-1/8"		40	.400		36.50	17.90		54.40	68
2464	20-1/8"	↓	39	.410	↓	40	18.35		58.35	72.50
4000	Mechanical grooved fitting cover, including insert									
4020	90° Elbow fitting									
4030	3/4" & 1"	Q-14	140	.114	Ea.	3.70	5.10		8.80	12
4040	1-1/4" & 1-1/2"		135	.119		4.68	5.30		9.98	13.40
4042	2"		130	.123		7.40	5.50		12.90	16.70
4044	2-1/2"		125	.128		8.20	5.75		13.95	17.90
4046	3"		120	.133		9.20	5.95		15.15	19.35
4048	3-1/2"		115	.139		10.65	6.20		16.85	21.50
4050	4"		110	.145		11.85	6.50		18.35	23
4052	5"		100	.160		14.75	7.15		21.90	27.50
4054	6"		90	.178		22	7.95		29.95	36.50
4056	8"		80	.200		23.50	8.95		32.45	40
4058	10"		75	.213		30.50	9.55		40.05	48.50
4060	12"		68	.235		43.50	10.50		54	64.50
4062	14"		65	.246		43.50	11		54.50	65
4064	16"		63	.254		59.50	11.35		70.85	83
4066	18"	↓	61	.262	↓	81.50	11.75		93.25	108
4100	45° Elbow fitting									
4120	3/4" & 1"	Q-14	140	.114	Ea.	3.29	5.10		8.39	11.55
4130	1-1/4" & 1-1/2"		135	.119		4.19	5.30		9.49	12.85
4140	2"		130	.123		6.60	5.50		12.10	15.80
4142	2-1/2"		125	.128		7.30	5.75		13.05	16.95

22 07 19 – Plumbing Piping Insulation

22 07 19.30 Piping Insulation Protective Jacketing, PVC		Crew	Daily Output	Labor-Hours	Unit	2013 Bare Costs Material	Labor	Equipment	Total	Total Incl O&P
4144	3"	Q-14	120	.133	Ea.	8.20	5.95		14.15	18.25
4146	3-1/2"		115	.139		8.65	6.20		14.85	19.15
4148	4"		110	.145		10.55	6.50		17.05	21.50
4150	5"		100	.160		13.10	7.15		20.25	25.50
4152	6"		90	.178		19.35	7.95		27.30	34
4154	8"		80	.200		21	8.95		29.95	37
4156	10"		75	.213		27	9.55		36.55	44.50
4158	12"		68	.235		39	10.50		49.50	59
4160	14"		65	.246		50.50	11		61.50	72.50
4162	16"		63	.254		61	11.35		72.35	85
4164	18"		61	.262		72	11.75		83.75	97.50
4200	Tee fitting									
4220	3/4" & 1"	Q-14	93	.172	Ea.	4.82	7.70		12.52	17.25
4230	1-1/4" & 1-1/2"		90	.178		6.10	7.95		14.05	19.05
4240	2"		87	.184		9.60	8.25		17.85	23.50
4242	2-1/2"		84	.190		10.70	8.50		19.20	25
4244	3"		80	.200		11.95	8.95		20.90	27
4246	3-1/2"		77	.208		12.65	9.30		21.95	28.50
4248	4"		73	.219		15.45	9.80		25.25	32.50
4250	5"		67	.239		17.05	10.70		27.75	35.50
4252	6"		60	.267		25	11.95		36.95	46
4254	8"		54	.296		27.50	13.25		40.75	50.50
4256	10"		50	.320		39.50	14.30		53.80	65.50
4258	12"		46	.348		56.50	15.55		72.05	86.50
4260	14"		43	.372		68.50	16.65		85.15	102
4262	16"		42	.381		82	17.05		99.05	117
4264	18"		41	.390		90.50	17.45		107.95	127

22 07 19.40 Pipe Insulation Protective Jacketing, Aluminum		Crew	Daily Output	Labor-Hours	Unit	Material	Labor	Equipment	Total	Total Incl O&P
0010	**PIPE INSULATION PROTECTIVE JACKETING, ALUMINUM**									
0100	Metal roll jacketing									
0120	Aluminum with polykraft moisture barrier									
0140	Smooth, based on OD of insulation, .016" thick									
0180	1/2" ID	Q-14	220	.073	L.F.	.20	3.25		3.45	5.25
0190	3/4" ID		215	.074		.26	3.33		3.59	5.45
0200	1" ID		210	.076		.32	3.41		3.73	5.65
0210	1-1/4" ID		205	.078		.38	3.49		3.87	5.80
0220	1-1/2" ID		202	.079		.44	3.54		3.98	6
0230	1-3/4" ID		199	.080		.50	3.60		4.10	6.15
0240	2" ID		195	.082		.56	3.67		4.23	6.30
0250	2-1/4" ID		191	.084		.62	3.75		4.37	6.50
0260	2-1/2" ID		187	.086		.69	3.83		4.52	6.70
0270	2-3/4" ID		184	.087		.75	3.89		4.64	6.90
0280	3" ID		180	.089		.82	3.98		4.80	7.10
0290	3-1/4" ID		176	.091		.88	4.07		4.95	7.25
0300	3-1/2" ID		172	.093		.94	4.16		5.10	7.50
0310	3-3/4" ID		169	.095		1	4.23		5.23	7.70
0320	4" ID		165	.097		1.06	4.34		5.40	7.90
0330	4-1/4" ID		161	.099		1.12	4.45		5.57	8.15
0340	4-1/2" ID		157	.102		1.19	4.56		5.75	8.40
0350	4-3/4" ID		154	.104		1.25	4.65		5.90	8.60
0360	5" ID		150	.107		1.31	4.77		6.08	8.85
0370	5-1/4" ID		146	.110		1.37	4.90		6.27	9.10

22 07 19 – Plumbing Piping Insulation

22 07 19.40 Pipe Insulation Protective Jacketing, Aluminum		Crew	Daily Output	Labor-Hours	Unit	Material	2013 Bare Costs Labor	Equipment	Total	Total Incl O&P
0380	5-1/2" ID	Q-14	143	.112	L.F.	1.43	5		6.43	9.30
0390	5-3/4" ID		139	.115		1.49	5.15		6.64	9.65
0400	6" ID		135	.119		1.55	5.30		6.85	9.95
0410	6-1/4" ID		133	.120		1.61	5.40		7.01	10.10
0420	6-1/2" ID		131	.122		1.67	5.45		7.12	10.35
0430	7" ID		128	.125		1.79	5.60		7.39	10.65
0440	7-1/4" ID		125	.128		1.85	5.75		7.60	10.95
0450	7-1/2" ID		123	.130		1.91	5.80		7.71	11.15
0460	8" ID		121	.132		2.03	5.90		7.93	11.45
0470	8-1/2" ID		119	.134		2.15	6		8.15	11.70
0480	9" ID		116	.138		2.27	6.15		8.42	12.10
0490	9-1/2" ID		114	.140		2.39	6.30		8.69	12.40
0500	10" ID		112	.143		2.51	6.40		8.91	12.70
0510	10-1/2" ID		110	.145		2.63	6.50		9.13	13
0520	11" ID		107	.150		2.75	6.70		9.45	13.45
0530	11-1/2" ID		105	.152		2.88	6.80		9.68	13.75
0540	12" ID		103	.155		3	6.95		9.95	14.10
0550	12-1/2" ID		100	.160		3.12	7.15		10.27	14.55
0560	13" ID		99	.162		3.24	7.25		10.49	14.80
0570	14" ID		98	.163		3.49	7.30		10.79	15.20
0580	15" ID		96	.167		3.73	7.45		11.18	15.70
0590	16" ID		95	.168		3.98	7.55		11.53	16.10
0600	17" ID		93	.172		4.22	7.70		11.92	16.60
0610	18" ID		92	.174		4.38	7.80		12.18	16.90
0620	19" ID		90	.178		4.70	7.95		12.65	17.50
0630	20" ID		89	.180		4.95	8.05		13	17.95
0640	21" ID		87	.184		5.20	8.25		13.45	18.50
0650	22" ID		86	.186		5.45	8.30		13.75	18.90
0660	23" ID		84	.190		5.70	8.50		14.20	19.50
0670	24" ID		83	.193		5.90	8.60		14.50	19.90
0710	For smooth .020" thick, add					27%	10%			
0720	For smooth .024" thick, add					52%	20%			
0730	For smooth .032" thick, add					104%	33%			
0800	For stucco embossed, add					1%				
0820	For corrugated, add					2.50%				
0900	White aluminum with polysurlyn moisture barrier									
0910	Smooth, % is an add to polykraft lines of same thickness									
0940	For smooth .016" thick, add				L.F.	35%				
0960	For smooth .024" thick, add				"	22%				
1000	Aluminum fitting covers									
1010	Size is based on OD of insulation									
1020	90° LR elbow, 2 piece									
1100	1-1/2"	Q-14	140	.114	Ea.	3.45	5.10		8.55	11.75
1110	1-3/4"		135	.119		3.45	5.30		8.75	12.05
1120	2"		130	.123		4.34	5.50		9.84	13.30
1130	2-1/4"		128	.125		4.34	5.60		9.94	13.45
1140	2-1/2"		126	.127		4.34	5.70		10.04	13.55
1150	2-3/4"		123	.130		4.34	5.80		10.14	13.80
1160	3"		120	.133		4.71	5.95		10.66	14.45
1170	3-1/4"		117	.137		4.71	6.10		10.81	14.70
1180	3-1/2"		115	.139		5.55	6.20		11.75	15.80
1190	3-3/4"		113	.142		5.70	6.35		12.05	16.15
1200	4"		110	.145		6	6.50		12.50	16.70

22 07 19.40 Pipe Insulation Protective Jacketing, Aluminum		Crew	Daily Output	Labor-Hours	Unit	2013 Bare Costs Material	Labor	Equipment	Total	Total Incl O&P
1210	4-1/4"	Q-14	108	.148	Ea.	6.15	6.65		12.80	17.05
1220	4-1/2"		106	.151		6.15	6.75		12.90	17.25
1230	4-3/4"		104	.154		6.15	6.90		13.05	17.45
1240	5"		102	.157		6.95	7		13.95	18.50
1250	5-1/4"		100	.160		8	7.15		15.15	19.90
1260	5-1/2"		97	.165		9.95	7.40		17.35	22.50
1270	5-3/4"		95	.168		9.95	7.55		17.50	22.50
1280	6"		92	.174		8.50	7.80		16.30	21.50
1290	6-1/4"		90	.178		8.50	7.95		16.45	21.50
1300	6-1/2"		87	.184		10.25	8.25		18.50	24
1310	7"		85	.188		12.40	8.40		20.80	27
1320	7-1/4"		84	.190		12.40	8.50		20.90	27
1330	7-1/2"		83	.193		18.25	8.60		26.85	33.50
1340	8"		82	.195		13.60	8.75		22.35	28.50
1350	8-1/2"		80	.200		25	8.95		33.95	41.50
1360	9"		78	.205		25	9.20		34.20	42
1370	9-1/2"		77	.208		18.50	9.30		27.80	35
1380	10"		76	.211		18.50	9.40		27.90	35
1390	10-1/2"		75	.213		18.90	9.55		28.45	36
1400	11"		74	.216		18.90	9.65		28.55	36
1410	11-1/2"		72	.222		21.50	9.95		31.45	39
1420	12"		71	.225		21.50	10.10		31.60	39
1430	12-1/2"		69	.232		38	10.35		48.35	58
1440	13"		68	.235		38	10.50		48.50	58.50
1450	14"		66	.242		51.50	10.85		62.35	73.50
1460	15"		64	.250		53.50	11.20		64.70	76.50
1470	16"		63	.254		58.50	11.35		69.85	82
2000	45° Elbow, 2 piece									
2010	2-1/2"	Q-14	126	.127	Ea.	3.59	5.70		9.29	12.75
2020	2-3/4"		123	.130		3.59	5.80		9.39	13
2030	3"		120	.133		4.05	5.95		10	13.70
2040	3-1/4"		117	.137		4.05	6.10		10.15	13.95
2050	3-1/2"		115	.139		4.62	6.20		10.82	14.75
2060	3-3/4"		113	.142		4.62	6.35		10.97	14.95
2070	4"		110	.145		4.70	6.50		11.20	15.25
2080	4-1/4"		108	.148		5.35	6.65		12	16.20
2090	4-1/2"		106	.151		5.35	6.75		12.10	16.40
2100	4-3/4"		104	.154		5.35	6.90		12.25	16.60
2110	5"		102	.157		6.30	7		13.30	17.85
2120	5-1/4"		100	.160		6.30	7.15		13.45	18.05
2130	5-1/2"		97	.165		6.60	7.40		14	18.70
2140	6"		92	.174		6.60	7.80		14.40	19.35
2150	6-1/2"		87	.184		9.10	8.25		17.35	23
2160	7"		85	.188		9.10	8.40		17.50	23
2170	7-1/2"		83	.193		9.20	8.60		17.80	23.50
2180	8"		82	.195		9.20	8.75		17.95	23.50
2190	8-1/2"		80	.200		11.45	8.95		20.40	26.50
2200	9"		78	.205		11.45	9.20		20.65	27
2210	9-1/2"		77	.208		15.90	9.30		25.20	32
2220	10"		76	.211		15.90	9.40		25.30	32
2230	10-1/2"		75	.213		14.85	9.55		24.40	31
2240	11"		74	.216		14.85	9.65		24.50	31.50
2250	11-1/2"		72	.222		17.85	9.95		27.80	35

22 07 19 – Plumbing Piping Insulation

22 07 19.40 Pipe Insulation Protective Jacketing, Aluminum		Crew	Daily Output	Labor-Hours	Unit	2013 Bare Costs Material	Labor	Equipment	Total	Total Incl O&P
2260	12"	Q-14	71	.225	Ea.	17.85	10.10		27.95	35.50
2270	13"		68	.235		21	10.50		31.50	39.50
2280	14"		66	.242		25.50	10.85		36.35	45.50
2290	15"		64	.250		39.50	11.20		50.70	61
2300	16"		63	.254		43	11.35		54.35	65
2310	17"		62	.258		42	11.55		53.55	64
2320	18"		61	.262		48	11.75		59.75	70.50
2330	19"		60	.267		58.50	11.95		70.45	83
2340	20"		59	.271		56.50	12.15		68.65	81
2350	21"		58	.276		61	12.35		73.35	86
3000	Tee, 4 piece									
3010	2-1/2"	Q-14	88	.182	Ea.	29	8.15		37.15	44.50
3020	2-3/4"		86	.186		29	8.30		37.30	45
3030	3"		84	.190		32.50	8.50		41	49.50
3040	3-1/4"		82	.195		32.50	8.75		41.25	49.50
3050	3-1/2"		80	.200		34.50	8.95		43.45	51.50
3060	4"		78	.205		34.50	9.20		43.70	52.50
3070	4-1/4"		76	.211		36.50	9.40		45.90	54.50
3080	4-1/2"		74	.216		36.50	9.65		46.15	55
3090	4-3/4"		72	.222		36.50	9.95		46.45	55.50
3100	5"		70	.229		37.50	10.20		47.70	57.50
3110	5-1/4"		68	.235		37.50	10.50		48	58
3120	5-1/2"		66	.242		39.50	10.85		50.35	60.50
3130	6"		64	.250		39.50	11.20		50.70	61
3140	6-1/2"		60	.267		43.50	11.95		55.45	66
3150	7"		58	.276		43.50	12.35		55.85	66.50
3160	7-1/2"		56	.286		48	12.80		60.80	73
3170	8"		54	.296		48	13.25		61.25	73.50
3180	8-1/2"		52	.308		49.50	13.75		63.25	76
3190	9"		50	.320		49.50	14.30		63.80	76.50
3200	9-1/2"		49	.327		36.50	14.60		51.10	63
3210	10"		48	.333		36.50	14.90		51.40	63.50
3220	10-1/2"		47	.340		39	15.25		54.25	66.50
3230	11"		46	.348		39	15.55		54.55	67
3240	11-1/2"		45	.356		41	15.90		56.90	70
3250	12"		44	.364		41	16.25		57.25	71
3260	13"		43	.372		43	16.65		59.65	73.50
3270	14"		42	.381		45.50	17.05		62.55	76.50
3280	15"		41	.390		49	17.45		66.45	81
3290	16"		40	.400		50.50	17.90		68.40	83.50
3300	17"		39	.410		58	18.35		76.35	92
3310	18"		38	.421		60	18.85		78.85	95.50
3320	19"		37	.432		66.50	19.35		85.85	103
3330	20"		36	.444		68	19.90		87.90	106
3340	22"		35	.457		86.50	20.50		107	127
3350	23"		34	.471		89	21		110	130
3360	24"		31	.516		91	23		114	136

22 07 19.50 Pipe Insulation Protective Jacketing, St. Stl.		Crew	Daily Output	Labor-Hours	Unit	Material	Labor	Equipment	Total	Total Incl O&P
0010	PIPE INSULATION PROTECTIVE JACKETING, STAINLESS STEEL									
0100	Metal roll jacketing									
0120	Type 304 with moisture barrier									
0140	Smooth, based on OD of insulation, .010" thick									

22 07 19 – Plumbing Piping Insulation

22 07 19.50 Pipe Insulation Protective Jacketing, St. Stl.		Crew	Daily Output	Labor-Hours	Unit	2013 Bare Costs Material	Labor	Equipment	Total	Total Incl O&P
0260	2-1/2" ID	Q-14	250	.064	L.F.	2.12	2.86		4.98	6.80
0270	2-3/4" ID		245	.065		2.31	2.92		5.23	7.10
0280	3" ID		240	.067		2.50	2.98		5.48	7.40
0290	3-1/4" ID		235	.068		2.69	3.05		5.74	7.70
0300	3-1/2" ID		230	.070		2.88	3.11		5.99	8
0310	3-3/4" ID		225	.071		3.07	3.18		6.25	8.30
0320	4" ID		220	.073		3.26	3.25		6.51	8.65
0330	4-1/4" ID		215	.074		3.44	3.33		6.77	8.95
0340	4-1/2" ID		210	.076		3.63	3.41		7.04	9.30
0350	5" ID		200	.080		4.01	3.58		7.59	9.95
0360	5-1/2" ID		190	.084		4.39	3.77		8.16	10.70
0370	6" ID		180	.089		4.76	3.98		8.74	11.45
0380	6-1/2" ID		175	.091		5.15	4.09		9.24	12
0390	7" ID		170	.094		5.50	4.21		9.71	12.60
0400	7-1/2" ID		164	.098		5.90	4.36		10.26	13.30
0410	8" ID		161	.099		6.25	4.45		10.70	13.80
0420	8-1/2" ID		158	.101		6.65	4.53		11.18	14.35
0430	9" ID		155	.103		7.05	4.62		11.67	14.90
0440	9-1/2" ID		152	.105		7.40	4.71		12.11	15.45
0450	10" ID		149	.107		7.80	4.80		12.60	16
0460	10-1/2" ID		146	.110		8.15	4.90		13.05	16.60
0470	11" ID		143	.112		8.55	5		13.55	17.15
0480	12" ID		137	.117		9.30	5.20		14.50	18.30
0490	13" ID		132	.121		10.05	5.40		15.45	19.45
0500	14" ID		130	.123		10.80	5.50		16.30	20.50
0700	For smooth .016" thick, add					45%	33%			
1000	Stainless steel, Type 316, fitting covers									
1010	Size is based on OD of insulation									
1020	90° LR Elbow, 2 piece									
1100	1-1/2"	Q-14	126	.127	Ea.	12.50	5.70		18.20	22.50
1110	2-3/4"		123	.130		12.50	5.80		18.30	23
1120	3"		120	.133		13.05	5.95		19	23.50
1130	3-1/4"		117	.137		13.05	6.10		19.15	24
1140	3-1/2"		115	.139		13.75	6.20		19.95	25
1150	3-3/4"		113	.142		14.50	6.35		20.85	26
1160	4"		110	.145		15.95	6.50		22.45	27.50
1170	4-1/4"		108	.148		21	6.65		27.65	34
1180	4-1/2"		106	.151		21	6.75		27.75	34
1190	5"		102	.157		21.50	7		28.50	35
1200	5-1/2"		97	.165		32.50	7.40		39.90	47
1210	6"		92	.174		35.50	7.80		43.30	51
1220	6-1/2"		87	.184		49	8.25		57.25	66.50
1230	7"		85	.188		49	8.40		57.40	66.50
1240	7-1/2"		83	.193		57	8.60		65.60	76
1250	8"		80	.200		57	8.95		65.95	76.50
1260	8-1/2"		80	.200		59	8.95		67.95	79
1270	9"		78	.205		89.50	9.20		98.70	113
1280	9-1/2"		77	.208		88	9.30		97.30	111
1290	10"		76	.211		88	9.40		97.40	112
1300	10-1/2"		75	.213		105	9.55		114.55	131
1310	11"		74	.216		101	9.65		110.65	126
1320	12"		71	.225		114	10.10		124.10	142
1330	13"		68	.235		158	10.50		168.50	190

22 07 Plumbing Insulation

22 07 19 – Plumbing Piping Insulation

22 07 19.50 Pipe Insulation Protective Jacketing, St. Stl.		Crew	Daily Output	Labor-Hours	Unit	Material	Labor	Equipment	Total	Total Incl O&P
						2013 Bare Costs				
1340	14"	Q-14	66	.242	Ea.	159	10.85		169.85	192
2000	45° Elbow, 2 piece									
2010	2-1/2"	Q-14	126	.127	Ea.	10.40	5.70		16.10	20.50
2020	2-3/4"		123	.130		10.40	5.80		16.20	20.50
2030	3"		120	.133		11.20	5.95		17.15	21.50
2040	3-1/4"		117	.137		11.20	6.10		17.30	22
2050	3-1/2"		115	.139		11.35	6.20		17.55	22
2060	3-3/4"		113	.142		11.35	6.35		17.70	22.50
2070	4"		110	.145		14.60	6.50		21.10	26
2080	4-1/4"		108	.148		20.50	6.65		27.15	33
2090	4-1/2"		106	.151		20.50	6.75		27.25	33
2100	4-3/4"		104	.154		20.50	6.90		27.40	33
2110	5"		102	.157		21	7		28	34
2120	5-1/2"		97	.165		21	7.40		28.40	34.50
2130	6"		92	.174		24.50	7.80		32.30	39
2140	6-1/2"		87	.184		24.50	8.25		32.75	40
2150	7"		85	.188		42	8.40		50.40	59
2160	7-1/2"		83	.193		42.50	8.60		51.10	60.50
2170	8"		82	.195		42.50	8.75		51.25	60.50
2180	8-1/2"		80	.200		50	8.95		58.95	69
2190	9"		78	.205		50	9.20		59.20	69.50
2200	9-1/2"		77	.208		58.50	9.30		67.80	79
2210	10"		76	.211		58.50	9.40		67.90	79
2220	10-1/2"		75	.213		70.50	9.55		80.05	92.50
2230	11"		74	.216		70.50	9.65		80.15	92.50
2240	12"		71	.225		76.50	10.10		86.60	99.50
2250	13"	▼	68	.235	▼	90	10.50		100.50	115

22 11 Facility Water Distribution

22 11 13 – Facility Water Distribution Piping

22 11 13.14 Pipe, Brass

		Crew	Daily Output	Labor-Hours	Unit	Material	Labor	Equipment	Total	Total Incl O&P
0010	**PIPE, BRASS**, Plain end									
0900	Field threaded, coupling & clevis hanger assembly 10' O.C.									
0920	Regular weight									
1120	1/2" diameter	1 Plum	48	.167	L.F.	6.50	9.30		15.80	21
1140	3/4" diameter		46	.174		8.65	9.70		18.35	24
1160	1" diameter	▼	43	.186		12.60	10.40		23	29.50
1180	1-1/4" diameter	Q-1	72	.222		19.15	11.15		30.30	38
1200	1-1/2" diameter		65	.246		23	12.35		35.35	44
1220	2" diameter		53	.302		32.50	15.15		47.65	58.50
1240	2-1/2" diameter		41	.390		50.50	19.60		70.10	85
1260	3" diameter	▼	31	.516		71	26		97	117
1300	4" diameter	Q-2	37	.649	▼	125	34		159	188
1930	To delete coupling & hanger, subtract									
1940	1/2" diam.					14%	46%			
1950	3/4" diam. to 1-1/2" diam.					8%	47%			
1960	2" diam. to 4" diam.					10%	37%			

22 11 13.16 Pipe Fittings, Brass

		Crew	Daily Output	Labor-Hours	Unit	Material	Labor	Equipment	Total	Total Incl O&P
0010	**PIPE FITTINGS, BRASS**, Rough bronze, threaded.									
1000	Standard wt., 90° Elbow									
1040	1/8"	1 Plum	13	.615	Ea.	18	34.50		52.50	71.50

22 11 13.16 Pipe Fittings, Brass		Crew	Daily Output	Labor-Hours	Unit	Material	Labor	Equipment	Total	Total Incl O&P
						2013 Bare Costs				
1060	1/4"	1 Plum	13	.615	Ea.	18	34.50		52.50	71.50
1080	3/8"		13	.615		18	34.50		52.50	71.50
1100	1/2"		12	.667		18	37		55	76
1120	3/4"		11	.727		24	40.50		64.50	87.50
1140	1"		10	.800		39	44.50		83.50	110
1160	1-1/4"	Q-1	17	.941		63.50	47.50		111	141
1180	1-1/2"		16	1		78.50	50		128.50	162
1200	2"		14	1.143		126	57.50		183.50	226
1220	2-1/2"		11	1.455		305	73		378	445
1240	3"		8	2		465	100		565	665
1260	4"	Q-2	11	2.182		945	114		1,059	1,225
1280	5"		8	3		2,625	156		2,781	3,100
1300	6"		7	3.429		3,850	179		4,029	4,525
1500	45° Elbow, 1/8"	1 Plum	13	.615		22	34.50		56.50	76
1540	1/4"		13	.615		22	34.50		56.50	76
1560	3/8"		13	.615		22	34.50		56.50	76
1580	1/2"		12	.667		22	37		59	80.50
1600	3/4"		11	.727		31.50	40.50		72	95.50
1620	1"		10	.800		53.50	44.50		98	126
1640	1-1/4"	Q-1	17	.941		85	47.50		132.50	165
1660	1-1/2"		16	1		107	50		157	194
1680	2"		14	1.143		173	57.50		230.50	278
1700	2-1/2"		11	1.455		330	73		403	475
1720	3"		8	2		510	100		610	710
1740	4"	Q-2	11	2.182		1,125	114		1,239	1,425
1760	5"		8	3		2,125	156		2,281	2,575
1780	6"		7	3.429		2,950	179		3,129	3,525
2000	Tee, 1/8"	1 Plum	9	.889		21	49.50		70.50	98
2040	1/4"		9	.889		21	49.50		70.50	98
2060	3/8"		9	.889		21	49.50		70.50	98
2080	1/2"		8	1		21	56		77	108
2100	3/4"		7	1.143		30	64		94	129
2120	1"		6	1.333		54.50	74.50		129	172
2140	1-1/4"	Q-1	10	1.600		93.50	80.50		174	224
2160	1-1/2"		9	1.778		105	89.50		194.50	250
2180	2"		8	2		175	100		275	345
2200	2-1/2"		7	2.286		415	115		530	635
2220	3"		5	3.200		635	161		796	940
2240	4"	Q-2	7	3.429		1,575	179		1,754	2,000
2260	5"		5	4.800		3,275	250		3,525	3,975
2280	6"		4	6		5,150	310		5,460	6,125
2500	Coupling, 1/8"	1 Plum	26	.308		15.05	17.15		32.20	42.50
2540	1/4"		22	.364		15.05	20.50		35.55	47
2560	3/8"		18	.444		15.05	25		40.05	54
2580	1/2"		15	.533		15.05	30		45.05	61.50
2600	3/4"		14	.571		21	32		53	71.50
2620	1"		13	.615		36	34.50		70.50	91
2640	1-1/4"	Q-1	22	.727		60	36.50		96.50	122
2660	1-1/2"		20	.800		78.50	40		118.50	147
2680	2"		18	.889		130	44.50		174.50	209
2700	2-1/2"		14	1.143		221	57.50		278.50	330
2720	3"		10	1.600		305	80.50		385.50	455
2740	4"	Q-2	12	2		630	104		734	850

22 11 13.16 Pipe Fittings, Brass		Crew	Daily Output	Labor-Hours	Unit	2013 Bare Costs Material	Labor	Equipment	Total	Total Incl O&P
2760	5″	Q-2	10	2.400	Ea.	1,175	125		1,300	1,500
2780	6″		9	2.667		1,675	139		1,814	2,050
3000	Union, 125 lb.									
3020	1/8″	1 Plum	12	.667	Ea.	51.50	37		88.50	113
3040	1/4″		12	.667		51.50	37		88.50	113
3060	3/8″		12	.667		51.50	37		88.50	113
3080	1/2″		11	.727		51.50	40.50		92	118
3100	3/4″		10	.800		70.50	44.50		115	145
3120	1″		9	.889		106	49.50		155.50	192
3140	1-1/4″	Q-1	16	1		154	50		204	245
3160	1-1/2″		15	1.067		183	53.50		236.50	282
3180	2″		13	1.231		247	62		309	365
3200	2-1/2″		10	1.600		675	80.50		755.50	860
3220	3″		7	2.286		1,050	115		1,165	1,325
3240	4″	Q-2	10	2.400		3,250	125		3,375	3,775

22 11 13.23 Pipe/Tube, Copper		Crew	Daily Output	Labor-Hours	Unit	Material	Labor	Equipment	Total	Total Incl O&P
0010	**PIPE/TUBE, COPPER**, Solder joints R221113-50									
0100	Solder									
0120	Solder, lead free, roll				Lb.	9.35			9.35	10.30
1000	Type K tubing, couplings & clevis hanger assemblies 10′ O.C.									
1100	1/4″ diameter R221113-70	1 Plum	84	.095	L.F.	5	5.30		10.30	13.50
1120	3/8″ diameter		82	.098		4.83	5.45		10.28	13.50
1140	1/2″ diameter		78	.103		5.55	5.70		11.25	14.70
1160	5/8″ diameter		77	.104		7.10	5.80		12.90	16.60
1180	3/4″ diameter		74	.108		9.75	6.05		15.80	19.85
1200	1″ diameter		66	.121		13.10	6.75		19.85	24.50
1220	1-1/4″ diameter		56	.143		16.30	7.95		24.25	30
1240	1-1/2″ diameter		50	.160		21	8.95		29.95	37
1260	2″ diameter		40	.200		32.50	11.15		43.65	53
1280	2-1/2″ diameter	Q-1	60	.267		50	13.40		63.40	75
1300	3″ diameter		54	.296		70	14.90		84.90	99.50
1320	3-1/2″ diameter		42	.381		97	19.15		116.15	136
1330	4″ diameter		38	.421		120	21		141	164
1340	5″ diameter		32	.500		265	25		290	330
1360	6″ diameter	Q-2	38	.632		395	33		428	480
1380	8″ diameter	″	34	.706		655	37		692	775
1390	For other than full hard temper, add					13%				
1440	For silver solder, add						15%			
1800	For medical clean, (oxygen class), add					12%				
1950	To delete cplgs. & hngrs., 1/4″-1″ pipe, subtract					27%	60%			
1960	1-1/4″ -3″ pipe, subtract					14%	52%			
1970	3-1/2″-5″ pipe, subtract					10%	60%			
1980	6″-8″ pipe, subtract					19%	53%			
2000	Type L tubing, couplings & clevis hanger assemblies 10′ O.C.									
2100	1/4″ diameter	1 Plum	88	.091	L.F.	2.75	5.05		7.80	10.65
2120	3/8″ diameter		84	.095		3.81	5.30		9.11	12.20
2140	1/2″ diameter		81	.099		4.24	5.50		9.74	12.95
2160	5/8″ diameter		79	.101		6.40	5.65		12.05	15.50
2180	3/4″ diameter		76	.105		6.50	5.85		12.35	16.05
2200	1″ diameter		68	.118		9.60	6.55		16.15	20.50
2220	1-1/4″ diameter		58	.138		13.70	7.70		21.40	26.50
2240	1-1/2″ diameter		52	.154		17.55	8.60		26.15	32.50

22 11 13 – Facility Water Distribution Piping

22 11 13.23 Pipe/Tube, Copper		Crew	Daily Output	Labor-Hours	Unit	Material	Labor	Equipment	Total	Total Incl O&P
						2013 Bare Costs				
2260	2" diameter	1 Plum	42	.190	L.F.	27.50	10.65		38.15	46.50
2280	2-1/2" diameter	Q-1	62	.258		42.50	12.95		55.45	66.50
2300	3" diameter		56	.286		58	14.35		72.35	85.50
2320	3-1/2" diameter		43	.372		82	18.70		100.70	119
2340	4" diameter		39	.410		100	20.50		120.50	142
2360	5" diameter		34	.471		214	23.50		237.50	271
2380	6" diameter	Q-2	40	.600		305	31.50		336.50	380
2400	8" diameter	"	36	.667		495	34.50		529.50	595
2410	For other than full hard temper, add					21%				
2590	For silver solder, add						15%			
2900	For medical clean, (oxygen class), add					12%				
2940	To delete cplgs. & hngrs., 1/4"-1" pipe, subtract					37%	63%			
2960	1-1/4"-3" pipe, subtract					12%	53%			
2970	3-1/2"-5" pipe, subtract					12%	63%			
2980	6"-8" pipe, subtract					24%	55%			
3000	Type M tubing, couplings & clevis hanger assemblies 10' O.C.									
3100	1/4" diameter	1 Plum	90	.089	L.F.	3.07	4.96		8.03	10.80
3120	3/8" diameter		87	.092		3.14	5.15		8.29	11.20
3140	1/2" diameter		84	.095		3.27	5.30		8.57	11.60
3160	5/8" diameter		81	.099		4.96	5.50		10.46	13.75
3180	3/4" diameter		78	.103		4.97	5.70		10.67	14.05
3200	1" diameter		70	.114		7.80	6.40		14.20	18.20
3220	1-1/4" diameter		60	.133		11.45	7.45		18.90	24
3240	1-1/2" diameter		54	.148		15.45	8.25		23.70	29.50
3260	2" diameter		44	.182		25	10.15		35.15	43
3280	2-1/2" diameter	Q-1	64	.250		37.50	12.55		50.05	60.50
3300	3" diameter		58	.276		50.50	13.85		64.35	77
3320	3-1/2" diameter		45	.356		74.50	17.85		92.35	109
3340	4" diameter		40	.400		92.50	20		112.50	133
3360	5" diameter		36	.444		214	22.50		236.50	269
3370	6" diameter	Q-2	42	.571		305	30		335	380
3380	8" diameter	"	38	.632		495	33		528	590
3440	For silver solder, add						15%			
3960	To delete cplgs. & hngrs., 1/4"-1" pipe, subtract					35%	65%			
3970	1-1/4"-3" pipe, subtract					19%	56%			
3980	3-1/2"-5" pipe, subtract					13%	65%			
3990	6"-8" pipe, subtract					28%	58%			
4000	Type DWV tubing, couplings & clevis hanger assemblies 10' O.C.									
4100	1-1/4" diameter	1 Plum	60	.133	L.F.	12.15	7.45		19.60	24.50
4120	1-1/2" diameter		54	.148		15.20	8.25		23.45	29
4140	2" diameter		44	.182		20.50	10.15		30.65	38.50
4160	3" diameter	Q-1	58	.276		39.50	13.85		53.35	64.50
4180	4" diameter		40	.400		71	20		91	109
4200	5" diameter		36	.444		194	22.50		216.50	247
4220	6" diameter	Q-2	42	.571		284	30		314	355
4240	8" diameter	"	38	.632		615	33		648	725
4730	To delete cplgs. & hngrs., 1-1/4"-2" pipe, subtract					16%	53%			
4740	3"-4" pipe, subtract					13%	60%			
4750	5"-8" pipe, subtract					23%	58%			
5200	ACR tubing, type L, hard temper, cleaned and									
5220	capped, no couplings or hangers									
5240	3/8" OD				L.F.	2.09			2.09	2.30
5250	1/2" OD					3.18			3.18	3.50

22 11 13.23 Pipe/Tube, Copper

	22 11 13.23 Pipe/Tube, Copper	Crew	Daily Output	Labor-Hours	Unit	2013 Bare Costs Material	Labor	Equipment	Total	Total Incl O&P
5260	5/8" OD				L.F.	3.88			3.88	4.27
5270	3/4" OD					5.45			5.45	6
5280	7/8" OD					6.05			6.05	6.70
5290	1-1/8" OD					8.80			8.80	9.70
5300	1-3/8" OD					11.85			11.85	13
5310	1-5/8" OD					15.15			15.15	16.70
5320	2-1/8" OD					24			24	26
5330	2-5/8" OD					35.50			35.50	39
5340	3-1/8" OD					47.50			47.50	52.50
5350	3-5/8" OD					62			62	68.50
5360	4-1/8" OD					79			79	87
5380	ACR tubing, type L, hard, cleaned and capped									
5381	No couplings or hangers									
5384	3/8"	1 Stpi	160	.050	L.F.	2.09	2.83		4.92	6.55
5385	1/2"		160	.050		3.18	2.83		6.01	7.75
5386	5/8"		160	.050		3.88	2.83		6.71	8.55
5387	3/4"		130	.062		5.45	3.49		8.94	11.25
5388	7/8"		130	.062		6.05	3.49		9.54	11.95
5389	1-1/8"		115	.070		8.80	3.94		12.74	15.65
5390	1-3/8"		100	.080		11.85	4.53		16.38	19.80
5391	1-5/8"		90	.089		15.15	5.05		20.20	24.50
5392	2-1/8"		80	.100		24	5.65		29.65	34.50
5393	2-5/8"	Q-5	125	.128		35.50	6.55		42.05	49
5394	3-1/8"		105	.152		47.50	7.75		55.25	64
5395	4-1/8"		95	.168		79	8.60		87.60	100
5800	Refrigeration tubing, dryseal, 50' coils									
5840	1/8" OD				Coil	44.50			44.50	49
5850	3/16" OD					54			54	59.50
5860	1/4" OD					64			64	70.50
5870	5/16" OD					83.50			83.50	91.50
5880	3/8" OD					92.50			92.50	102
5890	1/2" OD					125			125	137
5900	5/8" OD					169			169	186
5910	3/4" OD					204			204	224
5920	7/8" OD					305			305	335
5930	1-1/8" OD					440			440	485
5940	1-3/8" OD					660			660	725
5950	1-5/8" OD					835			835	920

22 11 13.25 Pipe/Tube Fittings, Copper

	22 11 13.25 Pipe/Tube Fittings, Copper	Crew	Daily Output	Labor-Hours	Unit	Material	Labor	Equipment	Total	Total Incl O&P
0010	**PIPE/TUBE FITTINGS, COPPER**, Wrought unless otherwise noted									
0040	Solder joints, copper x copper									
0070	90° elbow, 1/4"	1 Plum	22	.364	Ea.	8.95	20.50		29.45	40.50
0090	3/8"		22	.364		8.50	20.50		29	40
0100	1/2"		20	.400		2.83	22.50		25.33	36.50
0110	5/8"		19	.421		9.55	23.50		33.05	46
0120	3/4"		19	.421		6.35	23.50		29.85	42.50
0130	1"		16	.500		15.65	28		43.65	59
0140	1-1/4"		15	.533		23.50	30		53.50	71
0150	1-1/2"		13	.615		37	34.50		71.50	92
0160	2"		11	.727		67	40.50		107.50	135
0170	2-1/2"	Q-1	13	1.231		134	62		196	241
0180	3"		11	1.455		179	73		252	305

22 11 13 – Facility Water Distribution Piping

22 11 13.25	Pipe/Tube Fittings, Copper	Crew	Daily Output	Labor-Hours	Unit	Material	Labor	Equipment	Total	Total Incl O&P
0190	3-1/2"	Q-1	10	1.600	Ea.	625	80.50		705.50	810
0200	4"		9	1.778		460	89.50		549.50	640
0210	5"		6	2.667		1,925	134		2,059	2,325
0220	6"	Q-2	9	2.667		2,575	139		2,714	3,025
0230	8"	"	8	3		9,500	156		9,656	10,600
0250	45° elbow, 1/4"	1 Plum	22	.364		15.95	20.50		36.45	48
0270	3/8"		22	.364		13.65	20.50		34.15	45.50
0280	1/2"		20	.400		5.20	22.50		27.70	39.50
0290	5/8"		19	.421		26	23.50		49.50	64
0300	3/4"		19	.421		9.10	23.50		32.60	45.50
0310	1"		16	.500		23	28		51	67
0320	1-1/4"		15	.533		31	30		61	79
0330	1-1/2"		13	.615		37.50	34.50		72	92.50
0340	2"		11	.727		62.50	40.50		103	130
0350	2-1/2"	Q-1	13	1.231		133	62		195	239
0360	3"		13	1.231		197	62		259	310
0370	3-1/2"		10	1.600		350	80.50		430.50	505
0380	4"		9	1.778		420	89.50		509.50	595
0390	5"		6	2.667		1,625	134		1,759	2,000
0400	6"	Q-2	9	2.667		2,550	139		2,689	3,025
0410	8"	"	8	3		8,700	156		8,856	9,800
0450	Tee, 1/4"	1 Plum	14	.571		17.85	32		49.85	67.50
0470	3/8"		14	.571		14.35	32		46.35	64
0480	1/2"		13	.615		4.84	34.50		39.34	57
0490	5/8"		12	.667		31	37		68	90.50
0500	3/4"		12	.667		11.70	37		48.70	69
0510	1"		10	.800		36	44.50		80.50	107
0520	1-1/4"		9	.889		49.50	49.50		99	129
0530	1-1/2"		8	1		76	56		132	168
0540	2"		7	1.143		119	64		183	226
0550	2-1/2"	Q-1	8	2		239	100		339	415
0560	3"		7	2.286		365	115		480	575
0570	3-1/2"		6	2.667		1,050	134		1,184	1,375
0580	4"		5	3.200		885	161		1,046	1,200
0590	5"		4	4		2,900	201		3,101	3,475
0600	6"	Q-2	6	4		3,950	208		4,158	4,675
0610	8"	"	5	4.800		15,300	250		15,550	17,200
0612	Tee, reducing on the outlet, 1/4"	1 Plum	15	.533		30	30		60	78
0613	3/8"		15	.533		28.50	30		58.50	76
0614	1/2"		14	.571		25	32		57	75.50
0615	5/8"		13	.615		50.50	34.50		85	107
0616	3/4"		12	.667		11.25	37		48.25	68.50
0617	1"		11	.727		36	40.50		76.50	101
0618	1-1/4"		10	.800		53.50	44.50		98	126
0619	1-1/2"		9	.889		56.50	49.50		106	137
0620	2"		8	1		92.50	56		148.50	186
0621	2-1/2"	Q-1	9	1.778		279	89.50		368.50	440
0622	3"		8	2		335	100		435	520
0623	4"		6	2.667		630	134		764	890
0624	5"		5	3.200		2,900	161		3,061	3,425
0625	6"	Q-2	7	3.429		3,950	179		4,129	4,625
0626	8"	"	6	4		15,300	208		15,508	17,100
0630	Tee, reducing on the run, 1/4"	1 Plum	15	.533		38	30		68	86.50

Material, Labor, Equipment and Total columns are under "2013 Bare Costs".

22 11 13.25 Pipe/Tube Fittings, Copper		Crew	Daily Output	Labor-Hours	Unit	Material	Labor	Equipment	Total	Total Incl O&P
						2013 Bare Costs				
0631	3/8"	1 Plum	15	.533	Ea.	43	30		73	92.50
0632	1/2"		14	.571		34.50	32		66.50	86
0633	5/8"		13	.615		50.50	34.50		85	108
0634	3/4"		12	.667		13.80	37		50.80	71
0635	1"		11	.727		43	40.50		83.50	109
0636	1-1/4"		10	.800		70	44.50		114.50	144
0637	1-1/2"		9	.889		121	49.50		170.50	208
0638	2"		8	1		154	56		210	254
0639	2-1/2"	Q-1	9	1.778		360	89.50		449.50	530
0640	3"		8	2		530	100		630	735
0641	4"		6	2.667		1,175	134		1,309	1,500
0642	5"		5	3.200		2,750	161		2,911	3,275
0643	6"	Q-2	7	3.429		4,175	179		4,354	4,875
0644	8"	"	6	4		15,300	208		15,508	17,100
0650	Coupling, 1/4"	1 Plum	24	.333		2.06	18.60		20.66	30.50
0670	3/8"		24	.333		2.74	18.60		21.34	31
0680	1/2"		22	.364		2.14	20.50		22.64	33
0690	5/8"		21	.381		6.50	21.50		28	39
0700	3/4"		21	.381		4.32	21.50		25.82	37
0710	1"		18	.444		8.60	25		33.60	47
0715	1-1/4"		17	.471		15.30	26.50		41.80	56.50
0716	1-1/2"		15	.533		20	30		50	67
0718	2"		13	.615		33.50	34.50		68	88.50
0721	2-1/2"	Q-1	15	1.067		71.50	53.50		125	160
0722	3"		13	1.231		108	62		170	211
0724	3-1/2"		8	2		206	100		306	380
0726	4"		7	2.286		226	115		341	420
0728	5"		6	2.667		560	134		694	815
0731	6"	Q-2	8	3		920	156		1,076	1,250
0732	8"	"	7	3.429		2,975	179		3,154	3,550
0741	Coupling, reducing, concentric									
0743	1/2"	1 Plum	23	.348	Ea.	5.60	19.40		25	35
0745	3/4"		21.50	.372		10.05	21		31.05	42.50
0747	1"		19.50	.410		15.35	23		38.35	51.50
0748	1-1/4"		18	.444		21.50	25		46.50	61.50
0749	1-1/2"		16	.500		28.50	28		56.50	73
0751	2"		14	.571		59.50	32		91.50	114
0752	2-1/2"		13	.615		123	34.50		157.50	188
0753	3"	Q-1	14	1.143		149	57.50		206.50	251
0755	4"	"	8	2		315	100		415	495
0757	5"	Q-2	7.50	3.200		1,675	167		1,842	2,075
0759	6"		7	3.429		2,775	179		2,954	3,350
0761	8"		6.50	3.692		5,500	192		5,692	6,325
0771	Cap, sweat									
0773	1/2"	1 Plum	40	.200	Ea.	2.06	11.15		13.21	19.05
0775	3/4"		38	.211		3.84	11.75		15.59	22
0777	1"		32	.250		9.10	13.95		23.05	31
0778	1-1/4"		29	.276		12.20	15.40		27.60	36.50
0779	1-1/2"		26	.308		17.85	17.15		35	45.50
0781	2"		22	.364		33	20.50		53.50	66.50
0791	Flange, sweat									
0793	3"	Q-1	22	.727	Ea.	440	36.50		476.50	540
0795	4"		18	.889		615	44.50		659.50	740

22 11 13.25 Pipe/Tube Fittings, Copper		Crew	Daily Output	Labor-Hours	Unit	Material	Labor	Equipment	Total	Total Incl O&P
						2013 Bare Costs				
0797	5"	Q-1	12	1.333	Ea.	1,150	67		1,217	1,350
0799	6"	Q-2	18	1.333		1,200	69.50		1,269.50	1,425
0801	8"	"	16	1.500		2,975	78		3,053	3,400
0850	Unions, 1/4"	1 Plum	21	.381		63	21.50		84.50	101
0870	3/8"		21	.381		63.50	21.50		85	102
0880	1/2"		19	.421		34	23.50		57.50	72.50
0890	5/8"		18	.444		144	25		169	196
0900	3/4"		18	.444		42.50	25		67.50	84
0910	1"		15	.533		72.50	30		102.50	125
0920	1-1/4"		14	.571		126	32		158	187
0930	1-1/2"		12	.667		166	37		203	239
0940	2"		10	.800		283	44.50		327.50	375
0950	2-1/2"	Q-1	12	1.333		620	67		687	780
0960	3"	"	10	1.600		1,600	80.50		1,680.50	1,875
0980	Adapter, copper x male IPS, 1/4"	1 Plum	20	.400		29.50	22.50		52	66
0990	3/8"		20	.400		14.65	22.50		37.15	49.50
1000	1/2"		18	.444		5.90	25		30.90	44
1010	3/4"		17	.471		9.95	26.50		36.45	50.50
1020	1"		15	.533		25.50	30		55.50	73.50
1030	1-1/4"		13	.615		38	34.50		72.50	93.50
1040	1-1/2"		12	.667		44	37		81	104
1050	2"		11	.727		74	40.50		114.50	143
1060	2-1/2"	Q-1	10.50	1.524		284	76.50		360.50	425
1070	3"		10	1.600		375	80.50		455.50	535
1080	3-1/2"		9	1.778		465	89.50		554.50	645
1090	4"		8	2		510	100		610	710
1200	5"		6	2.667		3,025	134		3,159	3,525
1210	6"	Q-2	8.50	2.824		3,375	147		3,522	3,925
1214	Adapter, copper x female IPS									
1216	1/2"	1 Plum	18	.444	Ea.	9.40	25		34.40	48
1218	3/4"		17	.471		12.95	26.50		39.45	54
1220	1"		15	.533		30	30		60	78
1221	1-1/4"		13	.615		44	34.50		78.50	99.50
1222	1-1/2"		12	.667		68.50	37		105.50	132
1224	2"		11	.727		93	40.50		133.50	163
1250	Cross, 1/2"		10	.800		45.50	44.50		90	117
1260	3/4"		9.50	.842		88.50	47		135.50	169
1270	1"		8	1		151	56		207	250
1280	1-1/4"		7.50	1.067		216	59.50		275.50	330
1290	1-1/2"		6.50	1.231		310	68.50		378.50	445
1300	2"		5.50	1.455		585	81		666	760
1310	2-1/2"	Q-1	6.50	2.462		1,350	124		1,474	1,650
1320	3"	"	5.50	2.909		1,675	146		1,821	2,075
1500	Tee fitting, mechanically formed,(Type 1, 'branch sizes up to 2 in.')									
1520	1/2" run size, 3/8" to 1/2" branch size	1 Plum	80	.100	Ea.		5.60		5.60	8.40
1530	3/4" run size, 3/8" to 3/4" branch size		60	.133			7.45		7.45	11.20
1540	1" run size, 3/8" to 1" branch size		54	.148			8.25		8.25	12.45
1550	1-1/4" run size, 3/8" to 1-1/4" branch size		48	.167			9.30		9.30	14
1560	1-1/2" run size, 3/8" to 1-1/2" branch size		40	.200			11.15		11.15	16.80
1570	2" run size, 3/8" to 2" branch size		35	.229			12.75		12.75	19.20
1580	2-1/2" run size, 1/2" to 2" branch size		32	.250			13.95		13.95	21
1590	3" run size, 1" to 2" branch size		26	.308			17.15		17.15	26
1600	4" run size, 1" to 2" branch size		24	.333			18.60		18.60	28

22 11 13.25 Pipe/Tube Fittings, Copper		Crew	Daily Output	Labor-Hours	Unit	Material	2013 Bare Costs Labor	Equipment	Total	Total Incl O&P
1640	Tee fitting, mechanically formed, (Type 2, branches 2-1/2" thru 4")									
1650	2-1/2" run size, 2-1/2" branch size	1 Plum	12.50	.640	Ea.		35.50		35.50	54
1660	3" run size, 2-1/2" to 3" branch size		12	.667			37		37	56
1670	3-1/2" run size, 2-1/2" to 3-1/2" branch size		11	.727			40.50		40.50	61
1680	4" run size, 2-1/2" to 4" branch size		10.50	.762			42.50		42.50	64
1698	5" run size, 2" to 4" branch size		9.50	.842			47		47	71
1700	6" run size, 2" to 4" branch size		8.50	.941			52.50		52.50	79
1710	8" run size, 2" to 4" branch size	↓	7	1.143	↓		64		64	96
1800	ACR fittings, OD size									
1802	Tee, straight									
1808	5/8"	1 Stpi	12	.667	Ea.	4.84	38		42.84	62.50
1810	3/4"		12	.667		31	38		69	91.50
1812	7/8"		10	.800		11.70	45.50		57.20	81
1813	1"		10	.800		84.50	45.50		130	161
1814	1-1/8"		10	.800		36	45.50		81.50	108
1816	1-3/8"		9	.889		49.50	50.50		100	131
1818	1-5/8"		8	1		76	56.50		132.50	169
1820	2-1/8"	↓	7	1.143		119	64.50		183.50	228
1822	2-5/8"	Q-5	8	2		239	102		341	415
1824	3-1/8"		7	2.286		365	117		482	575
1826	4-1/8"	↓	5	3.200	↓	885	163		1,048	1,225
1830	90° elbow									
1836	5/8"	1 Stpi	19	.421	Ea.	9.50	24		33.50	46.50
1838	3/4"		19	.421		17.25	24		41.25	55
1840	7/8"		16	.500		17.15	28.50		45.65	61.50
1842	1-1/8"		16	.500		23	28.50		51.50	67.50
1844	1-3/8"		15	.533		23.50	30		53.50	71.50
1846	1-5/8"		13	.615		37	35		72	93
1848	2-1/8"	↓	11	.727		67	41		108	136
1850	2-5/8"	Q-5	13	1.231		134	62.50		196.50	243
1852	3-1/8"		11	1.455		179	74		253	310
1854	4-1/8"	↓	9	1.778	↓	460	90.50		550.50	640
1860	Coupling									
1866	5/8"	1 Stpi	21	.381	Ea.	2.14	21.50		23.64	35
1868	3/4"		21	.381		6.15	21.50		27.65	39.50
1870	7/8"		18	.444		4.32	25		29.32	43
1871	1"		18	.444		14.65	25		39.65	54
1872	1-1/8"		18	.444		8.70	25		33.70	47.50
1874	1-3/8"		17	.471		15.30	26.50		41.80	57
1876	1-5/8"		15	.533		20	30		50	67.50
1878	2-1/8"	↓	13	.615		33.50	35		68.50	89.50
1880	2-5/8"	Q-5	15	1.067		76	54.50		130.50	166
1882	3-1/8"		13	1.231		110	62.50		172.50	216
1884	4-1/8"	↓	7	2.286	↓	250	117		367	450
2000	DWV, solder joints, copper x copper									
2030	90° Elbow, 1-1/4"	1 Plum	13	.615	Ea.	27.50	34.50		62	81.50
2050	1-1/2"		12	.667		41	37		78	101
2070	2"	↓	10	.800		53.50	44.50		98	126
2090	3"	Q-1	10	1.600		133	80.50		213.50	267
2100	4"	"	9	1.778		930	89.50		1,019.50	1,150
2150	45° Elbow, 1-1/4"	1 Plum	13	.615		25.50	34.50		60	79.50
2170	1-1/2"		12	.667		21	37		58	79
2180	2"	↓	10	.800		48	44.50		92.50	120

22 11 13.25 Pipe/Tube Fittings, Copper		Crew	Daily Output	Labor-Hours	Unit	Material	Labor	Equipment	Total	Total Incl O&P
						2013 Bare Costs				
2190	3"	Q-1	10	1.600	Ea.	102	80.50		182.50	233
2200	4"	"	9	1.778		490	89.50		579.50	675
2250	Tee, Sanitary, 1-1/4"	1 Plum	9	.889		54	49.50		103.50	134
2270	1-1/2"		8	1		67.50	56		123.50	158
2290	2"		7	1.143		78.50	64		142.50	183
2310	3"	Q-1	7	2.286		320	115		435	530
2330	4"	"	6	2.667		815	134		949	1,100
2400	Coupling, 1-1/4"	1 Plum	14	.571		12.85	32		44.85	62
2420	1-1/2"		13	.615		15.95	34.50		50.45	69
2440	2"		11	.727		22	40.50		62.50	85.50
2460	3"	Q-1	11	1.455		43	73		116	157
2480	4"	"	10	1.600		137	80.50		217.50	271
2602	Traps, see Section 22 13 16.60									
3500	Compression joint fittings									
3510	As used for plumbing and oil burner work									
3520	Fitting price includes nuts and sleeves									
3540	Sleeve, 1/8"				Ea.	.22			.22	.24
3550	3/16"					.22			.22	.24
3560	1/4"					.08			.08	.09
3570	5/16"					.23			.23	.25
3580	3/8"					.46			.46	.51
3600	1/2"					.61			.61	.67
3620	Nut, 1/8"					.32			.32	.35
3630	3/16"					.34			.34	.37
3640	1/4"					.33			.33	.36
3650	5/16"					.49			.49	.54
3660	3/8"					.54			.54	.59
3670	1/2"					.95			.95	1.05
3710	Union, 1/8"	1 Plum	26	.308		2	17.15		19.15	28
3720	3/16"		24	.333		1.92	18.60		20.52	30
3730	1/4"		24	.333		2.01	18.60		20.61	30
3740	5/16"		23	.348		2.58	19.40		21.98	32
3750	3/8"		22	.364		2.60	20.50		23.10	33.50
3760	1/2"		22	.364		4	20.50		24.50	35
3780	5/8"		21	.381		4.43	21.50		25.93	37
3820	Union tee, 1/8"		17	.471		7.60	26.50		34.10	48
3830	3/16"		16	.500		4.42	28		32.42	47
3840	1/4"		15	.533		4	30		34	49.50
3850	5/16"		15	.533		5.65	30		35.65	51
3860	3/8"		15	.533		5.70	30		35.70	51.50
3870	1/2"		15	.533		9.65	30		39.65	55.50
3910	Union elbow, 1/4"		24	.333		2.47	18.60		21.07	30.50
3920	5/16"		23	.348		4.34	19.40		23.74	34
3930	3/8"		22	.364		4.19	20.50		24.69	35
3940	1/2"		22	.364		6.85	20.50		27.35	38
3980	Female connector, 1/8"		26	.308		1.55	17.15		18.70	27.50
4000	3/16" x 1/8"		24	.333		1.98	18.60		20.58	30
4010	1/4" x 1/8"		24	.333		1.75	18.60		20.35	30
4020	1/4"		24	.333		2.35	18.60		20.95	30.50
4030	3/8" x 1/4"		22	.364		2.35	20.50		22.85	33
4040	1/2" x 3/8"		22	.364		3.75	20.50		24.25	34.50
4050	5/8" x 1/2"		21	.381		5.30	21.50		26.80	38
4090	Male connector, 1/8"		26	.308		1.25	17.15		18.40	27.50

22 11 13 – Facility Water Distribution Piping

22 11 13.25 Pipe/Tube Fittings, Copper		Crew	Daily Output	Labor-Hours	Unit	Material	2013 Bare Costs Labor	Equipment	Total	Total Incl O&P
4100	3/16" x 1/8"	1 Plum	24	.333	Ea.	1.29	18.60		19.89	29.50
4110	1/4" x 1/8"		24	.333		1.34	18.60		19.94	29.50
4120	1/4"		24	.333		1.34	18.60		19.94	29.50
4130	5/16" x 1/8"		23	.348		1.58	19.40		20.98	30.50
4140	5/16" x 1/4"		23	.348		1.79	19.40		21.19	31
4150	3/8" x 1/8"		22	.364		1.87	20.50		22.37	32.50
4160	3/8" x 1/4"		22	.364		2.15	20.50		22.65	33
4170	3/8"		22	.364		2.15	20.50		22.65	33
4180	3/8" x 1/2"		22	.364		2.95	20.50		23.45	34
4190	1/2" x 3/8"		22	.364		2.86	20.50		23.36	33.50
4200	1/2"		22	.364		3.21	20.50		23.71	34
4210	5/8" x 1/2"		21	.381		3.71	21.50		25.21	36
4240	Male elbow, 1/8"		26	.308		2.58	17.15		19.73	29
4250	3/16" x 1/8"		24	.333		2.85	18.60		21.45	31
4260	1/4" x 1/8"		24	.333		2.15	18.60		20.75	30.50
4270	1/4"		24	.333		2.45	18.60		21.05	30.50
4280	3/8" x 1/4"		22	.364		3.11	20.50		23.61	34
4290	3/8"		22	.364		5.35	20.50		25.85	36.50
4300	1/2" x 1/4"		22	.364		5.15	20.50		25.65	36
4310	1/2" x 3/8"		22	.364		8.10	20.50		28.60	39.50
4340	Female elbow, 1/8"		26	.308		5.55	17.15		22.70	32
4350	1/4" x 1/8"		24	.333		3.15	18.60		21.75	31.50
4360	1/4"		24	.333		4.13	18.60		22.73	32.50
4370	3/8" x 1/4"		22	.364		4.07	20.50		24.57	35
4380	1/2" x 3/8"		22	.364		6.95	20.50		27.45	38
4390	1/2"		22	.364		9.35	20.50		29.85	41
4420	Male run tee, 1/4" x 1/8"		15	.533		3.86	30		33.86	49.50
4430	5/16" x 1/8"		15	.533		6.65	30		36.65	52.50
4440	3/8" x 1/4"		15	.533		5.95	30		35.95	51.50
4480	Male branch tee, 1/4" x 1/8"		15	.533		4.47	30		34.47	50
4490	1/4"		15	.533		4.81	30		34.81	50.50
4500	3/8" x 1/4"		15	.533		5.60	30		35.60	51
4510	1/2" x 3/8"		15	.533		8.70	30		38.70	54.50
4520	1/2"		15	.533		13.95	30		43.95	60.50
4800	Flare joint fittings									
4810	Refrigeration fittings									
4820	Flare joint nuts and labor not incl. in price. Add 1 nut per jnt.									
4830	90° Elbow, 1/4"				Ea.	2.32			2.32	2.55
4840	3/8"					3.32			3.32	3.65
4850	1/2"					6.20			6.20	6.85
4860	5/8"					9.10			9.10	10
4870	3/4"					13.80			13.80	15.15
5030	Tee, 1/4"					2.99			2.99	3.29
5040	5/16"					3.58			3.58	3.94
5050	3/8"					3.89			3.89	4.28
5060	1/2"					5.85			5.85	6.45
5070	5/8"					7.45			7.45	8.20
5080	3/4"					8.50			8.50	9.35
5140	Union, 3/16"					1.37			1.37	1.51
5150	1/4"					.94			.94	1.03
5160	5/16"					1.70			1.70	1.87
5170	3/8"					1.58			1.58	1.74
5180	1/2"					2.35			2.35	2.59

22 11 13 – Facility Water Distribution Piping

22 11 13.25 Pipe/Tube Fittings, Copper		Crew	Daily Output	Labor-Hours	Unit	2013 Bare Costs Material	Labor	Equipment	Total	Total Incl O&P
5190	5/8"				Ea.	3.78			3.78	4.16
5200	3/4"					10.55			10.55	11.60
5260	Long flare nut, 3/16"	1 Stpi	42	.190		1.74	10.80		12.54	18.15
5270	1/4"		41	.195		.96	11.05		12.01	17.70
5280	5/16"		40	.200		1.74	11.35		13.09	18.95
5290	3/8"		39	.205		2.09	11.60		13.69	19.80
5300	1/2"		38	.211		3.34	11.95		15.29	21.50
5310	5/8"		37	.216		5.85	12.25		18.10	25
5320	3/4"		34	.235		11.15	13.35		24.50	32.50
5380	Short flare nut, 3/16"		42	.190		2.92	10.80		13.72	19.45
5390	1/4"		41	.195		.68	11.05		11.73	17.40
5400	5/16"		40	.200		.81	11.35		12.16	17.95
5410	3/8"		39	.205		1.06	11.60		12.66	18.65
5420	1/2"		38	.211		1.58	11.95		13.53	19.70
5430	5/8"		36	.222		2.20	12.60		14.80	21.50
5440	3/4"		34	.235		7.35	13.35		20.70	28
5500	90° Elbow flare by MIPS, 1/4"					2.59			2.59	2.85
5510	3/8"					2.82			2.82	3.10
5520	1/2"					4.97			4.97	5.45
5530	5/8"					8.20			8.20	9
5540	3/4"					13.35			13.35	14.70
5600	Flare by FIPS, 1/4"					4.73			4.73	5.20
5610	3/8"					5.15			5.15	5.65
5620	1/2"					7.35			7.35	8.10
5670	Flare by sweat, 1/4"					2.64			2.64	2.90
5680	3/8"					3.43			3.43	3.77
5690	1/2"					7.05			7.05	7.75
5700	5/8"					22.50			22.50	25
5760	Tee flare by IPS, 1/4"					4.73			4.73	5.20
5770	3/8"					6.65			6.65	7.30
5780	1/2"					11.10			11.10	12.25
5790	5/8"					7.55			7.55	8.30
5850	Connector, 1/4"					1.02			1.02	1.12
5860	3/8"					1.53			1.53	1.68
5870	1/2"					2.49			2.49	2.74
5880	5/8"					2.15			2.15	2.37
5890	3/4"					7.05			7.05	7.75
5950	Seal cap, 1/4"					.84			.84	.92
5960	3/8"					1.01			1.01	1.11
5970	1/2"					1.26			1.26	1.39
5980	5/8"					2.62			2.62	2.88
5990	3/4"					7.65			7.65	8.40
6000	Water service fittings									
6010	Flare joints nut and labor are included in the fitting price.									
6020	90° Elbow, C x C, 3/8"	1 Plum	19	.421	Ea.	113	23.50		136.50	160
6030	1/2"		18	.444		113	25		138	162
6040	3/4"		16	.500		144	28		172	200
6050	1"		15	.533		273	30		303	345
6080	2"		10	.800		940	44.50		984.50	1,100
6090	90° Elbow, C x MPT, 3/8"		19	.421		81.50	23.50		105	125
6100	1/2"		18	.444		81.50	25		106.50	127
6110	3/4"		16	.500		94.50	28		122.50	146
6120	1"		15	.533		257	30		287	330

22 11 13.25 Pipe/Tube Fittings, Copper		Crew	Daily Output	Labor-Hours	Unit	Material	Labor	Equipment	Total	Total Incl O&P
						2013 Bare Costs				
6130	1-1/4"	1 Plum	13	.615	Ea.	530	34.50		564.50	630
6140	1-1/2"		12	.667		530	37		567	635
6150	2"		10	.800		690	44.50		734.50	820
6160	90° Elbow, C x FPT, 3/8"		19	.421		87	23.50		110.50	131
6170	1/2"		18	.444		87.50	25		112.50	134
6180	3/4"		16	.500		108	28		136	161
6190	1"		15	.533		249	30		279	320
6200	1-1/4"		13	.615		278	34.50		312.50	355
6210	1-1/2"		12	.667		1,075	37		1,112	1,225
6220	2"		10	.800		1,550	44.50		1,594.50	1,775
6230	Tee, C x C x C, 3/8"		13	.615		165	34.50		199.50	233
6240	1/2"		12	.667		165	37		202	237
6250	3/4"		11	.727		266	40.50		306.50	355
6260	1"		10	.800		345	44.50		389.50	440
6330	Tube nut, C x nut seat, 3/8"		40	.200		33.50	11.15		44.65	54
6340	1/2"		38	.211		33.50	11.75		45.25	54.50
6350	3/4"		34	.235		33.50	13.15		46.65	57
6360	1"		32	.250		60.50	13.95		74.45	88
6380	Coupling, C x C, 3/8"		19	.421		101	23.50		124.50	147
6390	1/2"		18	.444		101	25		126	149
6400	3/4"		16	.500		128	28		156	183
6410	1"		15	.533		237	30		267	305
6420	1-1/4"		12	.667		385	37		422	475
6430	1-1/2"		12	.667		580	37		617	695
6440	2"		10	.800		870	44.50		914.50	1,025
6450	Adapter, C x FPT, 3/8"		19	.421		76.50	23.50		100	120
6460	1/2"		18	.444		76.50	25		101.50	122
6470	3/4"		16	.500		92.50	28		120.50	144
6480	1"		15	.533		197	30		227	262
6490	1-1/4"		13	.615		420	34.50		454.50	510
6500	1-1/2"		12	.667		420	37		457	515
6510	2"		10	.800		545	44.50		589.50	665
6520	Adapter, C x MPT, 3/8"		19	.421		68.50	23.50		92	111
6530	1/2"		18	.444		68.50	25		93.50	113
6540	3/4"		16	.500		94.50	28		122.50	146
6550	1"		15	.533		167	30		197	228
6560	1-1/4"		13	.615		415	34.50		449.50	510
6570	1-1/2"		12	.667		415	37		452	515
6580	2"		10	.800		565	44.50		609.50	685
6992	Tube connector fittings, See Section 22 11 13.76 for plastic ftng.									
7000	Insert type Brass/copper, 100 psi @ 180°F, CTS									
7010	Adapter MPT 3/8" x 3/8" CTS	1 Plum	29	.276	Ea.	2.93	15.40		18.33	26
7020	1/2" x 1/2"		26	.308		2.95	17.15		20.10	29.50
7030	3/4" x 1/2"		26	.308		3.77	17.15		20.92	30
7040	3/4" x 3/4"		25	.320		4.35	17.85		22.20	32
7050	Adapter CTS 1/2" x 1/2" sweat		24	.333		3.96	18.60		22.56	32.50
7060	3/4" x 3/4" sweat		22	.364		1.39	20.50		21.89	32
7070	Coupler center set 3/8" CTS		25	.320		1.44	17.85		19.29	28.50
7080	1/2" CTS		23	.348		3.96	19.40		23.36	33.50
7090	3/4" CTS		22	.364		1.39	20.50		21.89	32
7100	Elbow 90°, copper 3/8"		25	.320		3.16	17.85		21.01	30.50
7110	1/2" CTS		23	.348		2.02	19.40		21.42	31
7120	3/4" CTS		22	.364		2.46	20.50		22.96	33

22 11 13.25 Pipe/Tube Fittings, Copper		Crew	Daily Output	Labor-Hours	Unit	2013 Bare Costs Material	Labor	Equipment	Total	Total Incl O&P
7130	Tee copper 3/8" CTS	1 Plum	17	.471	Ea.	3.75	26.50		30.25	43.50
7140	1/2" CTS		15	.533		2.58	30		32.58	48
7150	3/4" CTS		14	.571		3.96	32		35.96	52.50
7160	3/8" x 3/8" x 1/2"		16	.500		2.98	28		30.98	45.50
7170	1/2" x 3/8" x 1/2"		15	.533		1.93	30		31.93	47
7180	3/4" x 1/2" x 3/4"		14	.571		2.98	32		34.98	51.50

22 11 13.27 Pipe/Tube, Grooved Joint for Copper

		Crew	Daily Output	Labor-Hours	Unit	Material	Labor	Equipment	Total	Total Incl O&P
0010	**PIPE/TUBE, GROOVED JOINT FOR COPPER**									
4000	Fittings: coupling material required at joints not incl. in fitting price.									
4001	Add 1 selected coupling, material only, per joint for installed price.									
4010	Coupling, rigid style									
4018	2" diameter	1 Plum	50	.160	Ea.	17.20	8.95		26.15	32.50
4020	2-1/2" diameter	Q-1	80	.200		19.60	10.05		29.65	36.50
4022	3" diameter		67	.239		22	12		34	42
4024	4" diameter		50	.320		33	16.05		49.05	60
4026	5" diameter		40	.400		55.50	20		75.50	91.50
4028	6" diameter	Q-2	50	.480		73.50	25		98.50	118
4100	Elbow, 90° or 45°									
4108	2" diameter	1 Plum	25	.320	Ea.	30.50	17.85		48.35	60.50
4110	2-1/2" diameter	Q-1	40	.400		33	20		53	67
4112	3" diameter		33	.485		46.50	24.50		71	88
4114	4" diameter		25	.640		107	32		139	167
4116	5" diameter		20	.800		305	40		345	395
4118	6" diameter	Q-2	25	.960		485	50		535	610
4200	Tee									
4208	2" diameter	1 Plum	17	.471	Ea.	50.50	26.50		77	95
4210	2-1/2" diameter	Q-1	27	.593		53.50	30		83.50	104
4212	3" diameter		22	.727		80	36.50		116.50	143
4214	4" diameter		17	.941		175	47.50		222.50	264
4216	5" diameter		13	1.231		490	62		552	635
4218	6" diameter	Q-2	17	1.412		605	73.50		678.50	780
4300	Reducer, concentric									
4310	3" x 2-1/2" diameter	Q-1	35	.457	Ea.	44	23		67	83
4312	4" x 2-1/2" diameter		32	.500		90.50	25		115.50	138
4314	4" x 3" diameter		29	.552		90.50	27.50		118	141
4316	5" x 3" diameter		25	.640		255	32		287	330
4318	5" x 4" diameter		22	.727		255	36.50		291.50	335
4320	6" x 3" diameter	Q-2	28	.857		276	44.50		320.50	370
4322	6" x 4" diameter		26	.923		276	48		324	380
4324	6" x 5" diameter		24	1		276	52		328	385
4350	Flange, w/groove gasket									
4351	ANSI class 125 and 150									
4355	2" diameter	1 Plum	23	.348	Ea.	133	19.40		152.40	176
4356	2-1/2" diameter	Q-1	37	.432		139	21.50		160.50	185
4358	3" diameter		31	.516		145	26		171	199
4360	4" diameter		23	.696		159	35		194	228
4362	5" diameter		19	.842		214	42.50		256.50	300
4364	6" diameter	Q-2	23	1.043		233	54.50		287.50	340

22 11 13.44 Pipe, Steel

		Crew	Daily Output	Labor-Hours	Unit	Material	Labor	Equipment	Total	Total Incl O&P
0010	**PIPE, STEEL** R221113-50									
0020	All pipe sizes are to Spec. A-53 unless noted otherwise R221113-70									
0032	Schedule 10, see Line 22 11 13.48 0500									

22 11 Facility Water Distribution

22 11 13 – Facility Water Distribution Piping

22 11 13.44 Pipe, Steel		Crew	Daily Output	Labor-Hours	Unit	Material	2013 Bare Costs Labor	Equipment	Total	Total Incl O&P
0050	Schedule 40, threaded, with couplings, and clevis hanger									
0060	assemblies sized for covering, 10′ O.C.									
0540	Black, 1/4″ diameter	1 Plum	66	.121	L.F.	4.43	6.75		11.18	15.05
0550	3/8″ diameter		65	.123		4.96	6.85		11.81	15.80
0560	1/2″ diameter		63	.127		2.95	7.10		10.05	13.90
0570	3/4″ diameter		61	.131		3.46	7.30		10.76	14.80
0580	1″ diameter		53	.151		5.10	8.40		13.50	18.30
0590	1-1/4″ diameter	Q-1	89	.180		6.30	9.05		15.35	20.50
0600	1-1/2″ diameter		80	.200		9.50	10.05		19.55	25.50
0610	2″ diameter		64	.250		9.80	12.55		22.35	29.50
0620	2-1/2″ diameter		50	.320		15.30	16.05		31.35	41
0630	3″ diameter		43	.372		19.50	18.70		38.20	49.50
0640	3-1/2″ diameter		40	.400		27	20		47	60
0650	4″ diameter		36	.444		29.50	22.50		52	66
0809	A-106, gr. A/B, seamless w/cplgs. & clevis hanger assemblies									
0811	1/4″ diameter	1 Plum	66	.121	L.F.	7.85	6.75		14.60	18.80
0812	3/8″ diameter		65	.123		7.80	6.85		14.65	18.90
0813	1/2″ diameter		63	.127		9	7.10		16.10	20.50
0814	3/4″ diameter		61	.131		11	7.30		18.30	23
0815	1″ diameter		53	.151		12.05	8.40		20.45	26
0816	1-1/4″ diameter	Q-1	89	.180		14.60	9.05		23.65	29.50
0817	1-1/2″ diameter		80	.200		22	10.05		32.05	39.50
0819	2″ diameter		64	.250		24.50	12.55		37.05	45.50
0821	2-1/2″ diameter		50	.320		29	16.05		45.05	56
0822	3″ diameter		43	.372		32	18.70		50.70	63
0823	4″ diameter		36	.444		49	22.50		71.50	87
1220	To delete coupling & hanger, subtract									
1230	1/4″ diam. to 3/4″ diam.					31%	56%			
1240	1″ diam. to 1-1/2″ diam.					23%	51%			
1250	2″ diam. to 4″ diam.					23%	41%			
1280	All pipe sizes are to Spec. A-53 unless noted otherwise									
1281	Schedule 40, threaded, with couplings and clevis hanger									
1282	assemblies sized for covering, 10′ O. C.									
1290	Galvanized, 1/4″ diameter	1 Plum	66	.121	L.F.	6.25	6.75		13	17.05
1300	3/8″ diameter		65	.123		6.85	6.85		13.70	17.90
1310	1/2″ diameter		63	.127		4.10	7.10		11.20	15.15
1320	3/4″ diameter		61	.131		4.87	7.30		12.17	16.35
1330	1″ diameter		53	.151		6.75	8.40		15.15	20
1340	1-1/4″ diameter	Q-1	89	.180		8.50	9.05		17.55	23
1350	1-1/2″ diameter		80	.200		9.95	10.05		20	26
1360	2″ diameter		64	.250		13.30	12.55		25.85	33.50
1370	2-1/2″ diameter		50	.320		22	16.05		38.05	48.50
1380	3″ diameter		43	.372		28	18.70		46.70	59
1390	3-1/2″ diameter		40	.400		36	20		56	70
1400	4″ diameter		36	.444		41	22.50		63.50	79
1750	To delete coupling & hanger, subtract									
1760	1/4″ diam. to 3/4″ diam.					31%	56%			
1770	1″ diam. to 1-1/2″ diam.					23%	51%			
1780	2″ diam. to 4″ diam.					23%	41%			
1900	Pipe nipple std black 2″lg 1/2″	1 Plum	19	.421	Ea.	.97	23.50		24.47	36.50
1910	Pipe nipple std black 2″lg 1″ diameter	″	15	.533		1.78	30		31.78	47
1920	3″lg 2-1/2″	Q-1	16	1		11.10	50		61.10	88
2000	Welded, sch. 40, on yoke & roll hanger assy's, sized for covering, 10′ O.C.									

22 11 13.44 Pipe, Steel		Crew	Daily Output	Labor-Hours	Unit	2013 Bare Costs Material	Labor	Equipment	Total	Total Incl O&P
2040	Black, 1″ diameter	Q-15	93	.172	L.F.	4.96	8.65	.57	14.18	19.05
2050	1-1/4″ diameter		84	.190		6.45	9.55	.63	16.63	22
2060	1-1/2″ diameter		76	.211		7.05	10.55	.69	18.29	24.50
2070	2″ diameter		61	.262		8.55	13.20	.87	22.62	30
2080	2-1/2″ diameter		47	.340		14.20	17.10	1.12	32.42	43
2090	3″ diameter		43	.372		17.20	18.70	1.23	37.13	48.50
2100	3-1/2″ diameter		39	.410		22	20.50	1.35	43.85	56.50
2110	4″ diameter		37	.432		23.50	21.50	1.43	46.43	60
2120	5″ diameter		32	.500		36	25	1.65	62.65	79.50
2130	6″ diameter	Q-16	36	.667		37.50	34.50	1.47	73.47	95.50
2140	8″ diameter		29	.828		55.50	43	1.82	100.32	128
2150	10″ diameter		24	1		65	52	2.20	119.20	152
2160	12″ diameter		19	1.263		89	66	2.78	157.78	200
2170	14″ diameter, (two rod roll type hanger for 14″ diam. and up)		15	1.600		108	83.50	3.52	195.02	249
2180	16″ diameter, (two rod roll type hanger)		13	1.846		125	96	4.06	225.06	287
2190	18″ diameter, (two rod roll type hanger)		11	2.182		147	114	4.80	265.80	335
2200	20″ diameter, (two rod roll type hanger)		9	2.667		173	139	5.85	317.85	405
2220	24″ diameter, (two rod roll type hanger)		8	3		241	156	6.60	403.60	505
2560	To delete hanger, subtract									
2570	1″ diam. to 1-1/2″ diam.					15%	34%			
2580	2″ diam. to 3-1/2″ diam.					9%	21%			
2590	4″ diam. to 12″ diam.					5%	12%			
2596	14″ diam. to 36″ diam.					3%	10%			
3250	Flanged, 150 lb. weld neck, on yoke & roll hangers									
3260	sized for covering, 10′ O.C.									
3290	Black, 1″ diameter	Q-15	70	.229	L.F.	10.15	11.50	.75	22.40	29.50
3300	1-1/4″ diameter		64	.250		11.60	12.55	.83	24.98	32.50
3310	1-1/2″ diameter		58	.276		12.25	13.85	.91	27.01	35.50
3320	2″ diameter		45	.356		14.55	17.85	1.17	33.57	44.50
3330	2-1/2″ diameter		36	.444		21	22.50	1.47	44.97	58
3340	3″ diameter		32	.500		24.50	25	1.65	51.15	67
3350	3-1/2″ diameter		29	.552		31	27.50	1.82	60.32	77.50
3360	4″ diameter		26	.615		32.50	31	2.03	65.53	84
3370	5″ diameter		21	.762		50	38.50	2.51	91.01	115
3380	6″ diameter	Q-16	25	.960		53.50	50	2.11	105.61	137
3390	8″ diameter		19	1.263		83	66	2.78	151.78	193
3400	10″ diameter		16	1.500		120	78	3.30	201.30	254
3410	12″ diameter		14	1.714		151	89.50	3.77	244.27	305
3470	For 300 lb. flanges, add					63%				
3480	For 600 lb. flanges, add					310%				
3960	To delete flanges & hanger, subtract									
3970	1″ diam. to 2″ diam.					76%	65%			
3980	2-1/2″ diam. to 4″ diam.					62%	59%			
3990	5″ diam. to 12″ diam.					60%	46%			
4750	Schedule 80, threaded, with couplings, and clevis hanger assemblies									
4760	sized for covering, 10′ O.C.									
4790	Black, 1/4″ diameter	1 Plum	54	.148	L.F.	11.05	8.25		19.30	24.50
4800	3/8″ diameter		53	.151		14.25	8.40		22.65	28.50
4810	1/2″ diameter		52	.154		4.42	8.60		13.02	17.80
4820	3/4″ diameter		50	.160		5.30	8.95		14.25	19.30
4830	1″ diameter		45	.178		7.25	9.90		17.15	23
4840	1-1/4″ diameter	Q-1	75	.213		9.55	10.70		20.25	26.50
4850	1-1/2″ diameter		69	.232		11.05	11.65		22.70	29.50

22 11 13.44 Pipe, Steel		Crew	Daily Output	Labor-Hours	Unit	Material	Labor (2013 Bare Costs)	Equipment	Total	Total Incl O&P
4860	2" diameter	Q-1	56	.286	L.F.	14.90	14.35		29.25	38
4870	2-1/2" diameter		44	.364		23	18.25		41.25	52.50
4880	3" diameter		38	.421		30	21		51	65
4890	3-1/2" diameter		35	.457		37	23		60	75
4900	4" diameter		32	.500		35	25		60	76.50
5430	To delete coupling & hanger, subtract									
5440	1/4" diam. to 1/2" diam.					31%	54%			
5450	3/4" diam. to 1-1/2" diam.					28%	49%			
5460	2" diam. to 4" diam.					21%	40%			
5510	Galvanized, 1/4" diameter	1 Plum	54	.148	L.F.	8	8.25		16.25	21.50
5520	3/8" diameter		53	.151		8.85	8.40		17.25	22.50
5530	1/2" diameter		52	.154		5.70	8.60		14.30	19.20
5540	3/4" diameter		50	.160		6.85	8.95		15.80	21
5550	1" diameter		45	.178		9.45	9.90		19.35	25.50
5560	1-1/4" diameter	Q-1	75	.213		12.60	10.70		23.30	30
5570	1-1/2" diameter		69	.232		14.75	11.65		26.40	34
5580	2" diameter		56	.286		20	14.35		34.35	44
5590	2-1/2" diameter		44	.364		31	18.25		49.25	62
5600	3" diameter		38	.421		41	21		62	77
5610	3-1/2" diameter		35	.457		49	23		72	88
5620	4" diameter		32	.500		58	25		83	102
5930	To delete coupling & hanger, subtract									
5940	1/4" diam. to 1/2" diam.					31%	54%			
5950	3/4" diam. to 1-1/2" diam.					28%	49%			
5960	2" diam. to 4" diam.					21%	40%			
6000	Welded, on yoke & roller hangers									
6010	sized for covering, 10' O.C.									
6040	Black, 1" diameter	Q-15	85	.188	L.F.	7.95	9.45	.62	18.02	23.50
6050	1-1/4" diameter		79	.203		10.40	10.15	.67	21.22	27.50
6060	1-1/2" diameter		72	.222		11.90	11.15	.73	23.78	30.50
6070	2" diameter		57	.281		15.05	14.10	.93	30.08	38.50
6080	2-1/2" diameter		44	.364		22.50	18.25	1.20	41.95	53.50
6090	3" diameter		40	.400		29	20	1.32	50.32	64
6100	3-1/2" diameter		34	.471		35.50	23.50	1.55	60.55	76
6110	4" diameter		33	.485		33	24.50	1.60	59.10	74.50
6120	5" diameter, A-106B		26	.615		52	31	2.03	85.03	106
6130	6" diameter, A-106B	Q-16	30	.800		73.50	41.50	1.76	116.76	146
6140	8" diameter, A-106B		25	.960		109	50	2.11	161.11	198
6150	10" diameter, A-106B		20	1.200		204	62.50	2.64	269.14	320
6160	12" diameter, A-106B		15	1.600		400	83.50	3.52	487.02	570
6540	To delete hanger, subtract									
6550	1" diam. to 1-1/2" diam.					30%	14%			
6560	2" diam. to 3" diam.					23%	9%			
6570	3-1/2" diam. to 5" diam.					12%	6%			
6580	6" diam. to 12" diam.					10%	4%			
7250	Flanged, 300 lb. weld neck, on yoke & roll hangers									
7260	sized for covering, 10' O.C.									
7290	Black, 1" diameter	Q-15	66	.242	L.F.	14.35	12.20	.80	27.35	35
7300	1-1/4" diameter		61	.262		16.80	13.20	.87	30.87	39.50
7310	1-1/2" diameter		54	.296		18.30	14.90	.98	34.18	43.50
7320	2" diameter		42	.381		23	19.15	1.26	43.41	56
7330	2-1/2" diameter		33	.485		31.50	24.50	1.60	57.60	73.50
7340	3" diameter		29	.552		38.50	27.50	1.82	67.82	85.50

22 11 13 – Facility Water Distribution Piping

22 11 13.44 Pipe, Steel		Crew	Daily Output	Labor-Hours	Unit	Material	2013 Bare Costs Labor	Equipment	Total	Total Incl O&P
7350	3-1/2" diameter	Q-15	24	.667	L.F.	51	33.50	2.20	86.70	109
7360	4" diameter		23	.696		48.50	35	2.30	85.80	109
7370	5" diameter		19	.842		76	42.50	2.78	121.28	150
7380	6" diameter	Q-16	23	1.043		95.50	54.50	2.30	152.30	190
7390	8" diameter		17	1.412		148	73.50	3.11	224.61	277
7400	10" diameter		14	1.714		278	89.50	3.77	371.27	445
7410	12" diameter		12	2		495	104	4.40	603.40	705
7470	For 600 lb. flanges, add					100%				
7940	To delete flanges & hanger, subtract									
7950	1" diam. to 1-1/2" diam.					75%	66%			
7960	2" diam. to 3" diam.					62%	60%			
7970	3-1/2" diam. to 5" diam.					54%	66%			
7980	6" diam. to 12" diam.					55%	62%			
9000	Threading pipe labor, one end, all schedules through 80									
9010	1/4" through 3/4" pipe size	1 Plum	80	.100	Ea.		5.60		5.60	8.40
9020	1" through 2" pipe size		73	.110			6.10		6.10	9.20
9030	2-1/2" pipe size		53	.151			8.40		8.40	12.70
9040	3" pipe size		50	.160			8.95		8.95	13.45
9050	3-1/2" pipe size	Q-1	89	.180			9.05		9.05	13.60
9060	4" pipe size		73	.219			11		11	16.60
9070	5" pipe size		53	.302			15.15		15.15	23
9080	6" pipe size		46	.348			17.45		17.45	26.50
9090	8" pipe size		29	.552			27.50		27.50	41.50
9100	10" pipe size		21	.762			38.50		38.50	57.50
9110	12" pipe size		13	1.231			62		62	93
9200	Welding labor per joint									
9210	Schedule 40,									
9230	1/2" pipe size	Q-15	32	.500	Ea.		25	1.65	26.65	40
9240	3/4" pipe size		27	.593			30	1.96	31.96	47
9250	1" pipe size		23	.696			35	2.30	37.30	55
9260	1-1/4" pipe size		20	.800			40	2.64	42.64	63.50
9270	1-1/2" pipe size		19	.842			42.50	2.78	45.28	66.50
9280	2" pipe size		16	1			50	3.30	53.30	79
9290	2-1/2" pipe size		13	1.231			62	4.06	66.06	97.50
9300	3" pipe size		12	1.333			67	4.40	71.40	106
9310	4" pipe size		10	1.600			80.50	5.30	85.80	127
9320	5" pipe size		9	1.778			89.50	5.85	95.35	140
9330	6" pipe size		8	2			100	6.60	106.60	158
9340	8" pipe size		5	3.200			161	10.55	171.55	254
9350	10" pipe size		4	4			201	13.20	214.20	320
9360	12" pipe size		3	5.333			268	17.60	285.60	425
9370	14" pipe size		2.60	6.154			310	20.50	330.50	490
9380	16" pipe size		2.20	7.273			365	24	389	575
9390	18" pipe size		2	8			400	26.50	426.50	635
9400	20" pipe size		1.80	8.889			445	29.50	474.50	705
9410	22" pipe size		1.70	9.412			475	31	506	745
9420	24" pipe size		1.50	10.667			535	35	570	845
9450	Schedule 80,									
9460	1/2" pipe size	Q-15	27	.593	Ea.		30	1.96	31.96	47
9470	3/4" pipe size		23	.696			35	2.30	37.30	55
9480	1" pipe size		20	.800			40	2.64	42.64	63.50
9490	1-1/4" pipe size		19	.842			42.50	2.78	45.28	66.50
9500	1-1/2" pipe size		18	.889			44.50	2.93	47.43	70

22 11 13.44 Pipe, Steel		Crew	Daily Output	Labor-Hours	Unit	Material	2013 Bare Costs Labor	Equipment	Total	Total Incl O&P
9510	2″ pipe size	Q-15	15	1.067	Ea.		53.50	3.52	57.02	84.50
9520	2-1/2″ pipe size		12	1.333			67	4.40	71.40	106
9530	3″ pipe size		11	1.455			73	4.80	77.80	115
9540	4″ pipe size		8	2			100	6.60	106.60	158
9550	5″ pipe size		6	2.667			134	8.80	142.80	212
9560	6″ pipe size		5	3.200			161	10.55	171.55	254
9570	8″ pipe size		4	4			201	13.20	214.20	320
9580	10″ pipe size		3	5.333			268	17.60	285.60	425
9590	12″ pipe size	↓	2	8			400	26.50	426.50	635
9600	14″ pipe size	Q-16	2.60	9.231			480	20.50	500.50	750
9610	16″ pipe size		2.30	10.435			545	23	568	845
9620	18″ pipe size		2	12			625	26.50	651.50	970
9630	20″ pipe size		1.80	13.333			695	29.50	724.50	1,075
9640	22″ pipe size		1.60	15			780	33	813	1,200
9650	24″ pipe size	↓	1.50	16	↓		835	35	870	1,300

22 11 13.45 Pipe Fittings, Steel, Threaded		Crew	Daily Output	Labor-Hours	Unit	Material	Labor	Equipment	Total	Total Incl O&P
0010	**PIPE FITTINGS, STEEL, THREADED**									
0020	Cast Iron									
0040	Standard weight, black									
0060	90° Elbow, straight									
0070	1/4″	1 Plum	16	.500	Ea.	8.50	28		36.50	51.50
0080	3/8″		16	.500		12.25	28		40.25	55.50
0090	1/2″		15	.533		5.40	30		35.40	51
0100	3/4″		14	.571		5.60	32		37.60	54
0110	1″	↓	13	.615		6.65	34.50		41.15	59
0120	1-1/4″	Q-1	22	.727		9.40	36.50		45.90	65.50
0130	1-1/2″		20	.800		13	40		53	75
0140	2″		18	.889		20.50	44.50		65	89.50
0150	2-1/2″		14	1.143		48.50	57.50		106	140
0160	3″		10	1.600		80	80.50		160.50	209
0170	3-1/2″		8	2		217	100		317	390
0180	4″	↓	6	2.667	↓	148	134		282	365
0250	45° Elbow, straight									
0260	1/4″	1 Plum	16	.500	Ea.	10.45	28		38.45	53.50
0270	3/8″		16	.500		11.25	28		39.25	54.50
0280	1/2″		15	.533		8.20	30		38.20	54
0300	3/4″		14	.571		8.25	32		40.25	57
0320	1″	↓	13	.615		9.70	34.50		44.20	62
0330	1-1/4″	Q-1	22	.727		13.05	36.50		49.55	69.50
0340	1-1/2″		20	.800		21.50	40		61.50	84.50
0350	2″		18	.889		25	44.50		69.50	94.50
0360	2-1/2″		14	1.143		65	57.50		122.50	158
0370	3″		10	1.600		103	80.50		183.50	234
0380	3-1/2″		8	2		234	100		334	410
0400	4″	↓	6	2.667	↓	214	134		348	435
0500	Tee, straight									
0510	1/4″	1 Plum	10	.800	Ea.	13.25	44.50		57.75	81.50
0520	3/8″		10	.800		12.90	44.50		57.40	81
0530	1/2″		9	.889		8.35	49.50		57.85	83.50
0540	3/4″		9	.889		9.80	49.50		59.30	85.50
0550	1″	↓	8	1		8.70	56		64.70	93.50
0560	1-1/4″	Q-1	14	1.143		15.85	57.50		73.35	104

22 11 13.45 Pipe Fittings, Steel, Threaded		Crew	Daily Output	Labor-Hours	Unit	2013 Bare Costs Material	Labor	Equipment	Total	Total Incl O&P
0570	1-1/2"	Q-1	13	1.231	Ea.	20.50	62		82.50	116
0580	2"		11	1.455		28.50	73		101.50	142
0590	2-1/2"		9	1.778		74.50	89.50		164	216
0600	3"		6	2.667		114	134		248	330
0610	3-1/2"		5	3.200		231	161		392	495
0620	4"		4	4		223	201		424	550
0660	Tee, reducing, run or outlet									
0661	1/2"	1 Plum	9	.889	Ea.	20.50	49.50		70	97
0662	3/4"		9	.889		17.95	49.50		67.45	94.50
0663	1"		8	1		20	56		76	106
0664	1-1/4"	Q-1	14	1.143		28.50	57.50		86	118
0665	1-1/2"		13	1.231		33.50	62		95.50	130
0666	2"		11	1.455		52.50	73		125.50	168
0667	2-1/2"		9	1.778		98	89.50		187.50	242
0668	3"		6	2.667		200	134		334	420
0669	3-1/2"		5	3.200		360	161		521	640
0670	4"		4	4		390	201		591	730
0674	Reducer, concentric									
0675	3/4"	1 Plum	18	.444	Ea.	15	25		40	54
0676	1"	"	15	.533		10.15	30		40.15	56
0677	1-1/4"	Q-1	26	.615		30.50	31		61.50	80
0678	1-1/2"		24	.667		47.50	33.50		81	103
0679	2"		21	.762		54	38.50		92.50	117
0680	2-1/2"		18	.889		77	44.50		121.50	152
0681	3"		14	1.143		120	57.50		177.50	219
0682	3-1/2"		12	1.333		207	67		274	330
0683	4"		10	1.600		226	80.50		306.50	370
0687	Reducer, eccentric									
0688	3/4"	1 Plum	16	.500	Ea.	31.50	28		59.50	77
0689	1"	"	14	.571		34	32		66	85.50
0690	1-1/4"	Q-1	25	.640		54.50	32		86.50	109
0691	1-1/2"		22	.727		72	36.50		108.50	135
0692	2"		20	.800		104	40		144	175
0693	2-1/2"		16	1		143	50		193	233
0694	3"		12	1.333		226	67		293	350
0695	3-1/2"		10	1.600		315	80.50		395.50	465
0696	4"		9	1.778		405	89.50		494.50	580
0700	Standard weight, galvanized cast iron									
0720	90° Elbow, straight									
0730	1/4"	1 Plum	16	.500	Ea.	13.15	28		41.15	56.50
0740	3/8"		16	.500		13.15	28		41.15	56.50
0750	1/2"		15	.533		15	30		45	61.50
0760	3/4"		14	.571		14.70	32		46.70	64
0770	1"		13	.615		17	34.50		51.50	70
0780	1-1/4"	Q-1	22	.727		26.50	36.50		63	84
0790	1-1/2"		20	.800		36.50	40		76.50	101
0800	2"		18	.889		53.50	44.50		98	126
0810	2-1/2"		14	1.143		110	57.50		167.50	208
0820	3"		10	1.600		167	80.50		247.50	305
0830	3-1/2"		8	2		267	100		367	445
0840	4"		6	2.667		305	134		439	540
0900	45° Elbow, straight									
0910	1/4"	1 Plum	16	.500	Ea.	15	28		43	58.50

22 11 13.45 Pipe Fittings, Steel, Threaded		Crew	Daily Output	Labor-Hours	Unit	Material (2013 Bare Costs)	Labor (2013 Bare Costs)	Equipment (2013 Bare Costs)	Total (2013 Bare Costs)	Total Incl O&P
0920	3/8"	1 Plum	16	.500	Ea.	15	28		43	58.50
0930	1/2"		15	.533		15	30		45	61.50
0940	3/4"		14	.571		17.15	32		49.15	67
0950	1"		13	.615		21	34.50		55.50	74.50
0960	1-1/4"	Q-1	22	.727		31.50	36.50		68	89.50
0970	1-1/2"		20	.800		44.50	40		84.50	110
0980	2"		18	.889		62.50	44.50		107	136
0990	2-1/2"		14	1.143		123	57.50		180.50	223
1000	3"		10	1.600		197	80.50		277.50	335
1010	3-1/2"		8	2		315	100		415	495
1020	4"		6	2.667		360	134		494	595
1100	Tee, straight									
1110	1/4"	1 Plum	10	.800	Ea.	15.85	44.50		60.35	84.50
1120	3/8"		10	.800		15.85	44.50		60.35	84.50
1130	1/2"		9	.889		15.85	49.50		65.35	92
1140	3/4"		9	.889		21	49.50		70.50	98
1150	1"		8	1		23	56		79	110
1160	1-1/4"	Q-1	14	1.143		40	57.50		97.50	131
1170	1-1/2"		13	1.231		53	62		115	152
1180	2"		11	1.455		66.50	73		139.50	183
1190	2-1/2"		9	1.778		135	89.50		224.50	282
1200	3"		6	2.667		350	134		484	585
1210	3-1/2"		5	3.200		380	161		541	655
1220	4"		4	4		425	201		626	775
1300	Extra heavy weight, black									
1310	Couplings, steel straight									
1320	1/4"	1 Plum	19	.421	Ea.	3.77	23.50		27.27	39.50
1330	3/8"		19	.421		4.10	23.50		27.60	40
1340	1/2"		19	.421		5.55	23.50		29.05	41.50
1350	3/4"		18	.444		5.90	25		30.90	44
1360	1"		15	.533		7.55	30		37.55	53.50
1370	1-1/4"	Q-1	26	.615		12.10	31		43.10	60
1380	1-1/2"		24	.667		12.10	33.50		45.60	64
1390	2"		21	.762		18.45	38.50		56.95	78
1400	2-1/2"		18	.889		27.50	44.50		72	97
1410	3"		14	1.143		32.50	57.50		90	123
1420	3-1/2"		12	1.333		43.50	67		110.50	149
1430	4"		10	1.600		51.50	80.50		132	178
1510	90° Elbow, straight									
1520	1/2"	1 Plum	15	.533	Ea.	27.50	30		57.50	75
1530	3/4"		14	.571		28	32		60	79
1540	1"		13	.615		34	34.50		68.50	89
1550	1-1/4"	Q-1	22	.727		50.50	36.50		87	111
1560	1-1/2"		20	.800		63	40		103	130
1580	2"		18	.889		77.50	44.50		122	153
1590	2-1/2"		14	1.143		187	57.50		244.50	292
1600	3"		10	1.600		247	80.50		327.50	390
1610	4"		6	2.667		515	134		649	765
1650	45° Elbow, straight									
1660	1/2"	1 Plum	15	.533	Ea.	39	30		69	88
1670	3/4"		14	.571		37.50	32		69.50	89.50
1680	1"		13	.615		45.50	34.50		80	102
1690	1-1/4"	Q-1	22	.727		74.50	36.50		111	137

22 11 13.45 Pipe Fittings, Steel, Threaded		Crew	Daily Output	Labor-Hours	Unit	Material	2013 Bare Costs Labor	Equipment	Total	Total Incl O&P
1700	1-1/2"	Q-1	20	.800	Ea.	82	40		122	151
1710	2"		18	.889		117	44.50		161.50	195
1720	2-1/2"		14	1.143		202	57.50		259.50	310
1800	Tee, straight									
1810	1/2"	1 Plum	9	.889	Ea.	43	49.50		92.50	122
1820	3/4"		9	.889		43	49.50		92.50	122
1830	1"		8	1		52	56		108	142
1840	1-1/4"	Q-1	14	1.143		78	57.50		135.50	172
1850	1-1/2"		13	1.231		100	62		162	203
1860	2"		11	1.455		124	73		197	247
1870	2-1/2"		9	1.778		265	89.50		354.50	425
1880	3"		6	2.667		360	134		494	595
1890	4"		4	4		705	201		906	1,075
4000	Standard weight, black									
4010	Couplings, steel straight, merchants									
4030	1/4"	1 Plum	19	.421	Ea.	1.01	23.50		24.51	36.50
4040	3/8"		19	.421		1.22	23.50		24.72	37
4050	1/2"		19	.421		1.30	23.50		24.80	37
4060	3/4"		18	.444		1.64	25		26.64	39.50
4070	1"		15	.533		2.31	30		32.31	47.50
4080	1-1/4"	Q-1	26	.615		2.95	31		33.95	50
4090	1-1/2"		24	.667		3.73	33.50		37.23	54.50
4100	2"		21	.762		5.35	38.50		43.85	63.50
4110	2-1/2"		18	.889		16.75	44.50		61.25	85.50
4120	3"		14	1.143		23.50	57.50		81	113
4130	3-1/2"		12	1.333		41.50	67		108.50	147
4140	4"		10	1.600		41.50	80.50		122	167
4166	Plug, 1/4"	1 Plum	38	.211		2.39	11.75		14.14	20.50
4167	3/8"		38	.211		2.39	11.75		14.14	20.50
4168	1/2"		38	.211		2.39	11.75		14.14	20.50
4169	3/4"		32	.250		7.20	13.95		21.15	29
4170	1"		30	.267		7.70	14.90		22.60	31
4171	1-1/4"	Q-1	52	.308		8.85	15.45		24.30	33
4172	1-1/2"		48	.333		12.55	16.75		29.30	39
4173	2"		42	.381		16.20	19.15		35.35	47
4176	2-1/2"		36	.444		24	22.50		46.50	60
4180	4"		20	.800		41.50	40		81.50	106
4200	Standard weight, galvanized									
4210	Couplings, steel straight, merchants									
4230	1/4"	1 Plum	19	.421	Ea.	1.16	23.50		24.66	37
4240	3/8"		19	.421		1.49	23.50		24.99	37
4250	1/2"		19	.421		1.58	23.50		25.08	37
4260	3/4"		18	.444		1.99	25		26.99	39.50
4270	1"		15	.533		2.78	30		32.78	48
4280	1-1/4"	Q-1	26	.615		3.55	31		34.55	50.50
4290	1-1/2"		24	.667		4.41	33.50		37.91	55.50
4300	2"		21	.762		6.60	38.50		45.10	65
4310	2-1/2"		18	.889		20.50	44.50		65	89.50
4320	3"		14	1.143		27.50	57.50		85	117
4330	3-1/2"		12	1.333		48	67		115	154
4340	4"		10	1.600		48	80.50		128.50	174
4370	Plug, galvanized, square head									
4374	1/2"	1 Plum	38	.211	Ea.	5.75	11.75		17.50	24

22 11 13.45 Pipe Fittings, Steel, Threaded		Crew	Daily Output	Labor-Hours	Unit	Material	2013 Bare Costs Labor	Equipment	Total	Total Incl O&P
4375	3/4"	1 Plum	32	.250	Ea.	5.75	13.95		19.70	27.50
4376	1"		30	.267		5.75	14.90		20.65	29
4377	1-1/4"	Q-1	52	.308		9.65	15.45		25.10	34
4378	1-1/2"		48	.333		13	16.75		29.75	39.50
4379	2"		42	.381		16.15	19.15		35.30	47
4380	2-1/2"		36	.444		34	22.50		56.50	71
4381	3"		28	.571		43.50	28.50		72	91
4382	4"		20	.800		120	40		160	193
4700	Nipple, black									
4710	1/2" x 4" long	1 Plum	19	.421	Ea.	2.21	23.50		25.71	38
4712	3/4" x 4" long		18	.444		2.67	25		27.67	40.50
4714	1" x 4" long		15	.533		3.70	30		33.70	49
4716	1-1/4" x 4" long	Q-1	26	.615		4.63	31		35.63	51.50
4718	1-1/2" x 4" long		24	.667		5.45	33.50		38.95	56.50
4720	2" x 4" long		21	.762		7.60	38.50		46.10	66
4722	2-1/2" x 4" long		18	.889		20.50	44.50		65	90
4724	3" x 4" long		14	1.143		26	57.50		83.50	116
4726	4" x 4" long		10	1.600		35.50	80.50		116	160
4800	Nipple, galvanized									
4810	1/2" x 4" long	1 Plum	19	.421	Ea.	2.71	23.50		26.21	38.50
4812	3/4" x 4" long		18	.444		3.35	25		28.35	41
4814	1" x 4" long		15	.533		4.49	30		34.49	50
4816	1-1/4" x 4" long	Q-1	26	.615		5.50	31		36.50	52.50
4818	1-1/2" x 4" long		24	.667		7	33.50		40.50	58
4820	2" x 4" long		21	.762		8.85	38.50		47.35	67.50
4822	2-1/2" x 4" long		18	.889		23.50	44.50		68	92.50
4824	3" x 4" long		14	1.143		31	57.50		88.50	121
4826	4" x 4" long		10	1.600		41.50	80.50		122	167
5000	Malleable iron, 150 lb.									
5020	Black									
5040	90° elbow, straight									
5060	1/4"	1 Plum	16	.500	Ea.	4.17	28		32.17	46.50
5070	3/8"		16	.500		4.17	28		32.17	46.50
5080	1/2"		15	.533		2.89	30		32.89	48
5090	3/4"		14	.571		3.49	32		35.49	52
5100	1"		13	.615		6.05	34.50		40.55	58
5110	1-1/4"	Q-1	22	.727		10	36.50		46.50	66
5120	1-1/2"		20	.800		13.15	40		53.15	75
5130	2"		18	.889		22.50	44.50		67	92
5140	2-1/2"		14	1.143		50.50	57.50		108	142
5150	3"		10	1.600		74	80.50		154.50	202
5160	3-1/2"		8	2		203	100		303	375
5170	4"		6	2.667		159	134		293	375
5250	45° elbow, straight									
5270	1/4"	1 Plum	16	.500	Ea.	6.30	28		34.30	49
5280	3/8"		16	.500		6.30	28		34.30	49
5290	1/2"		15	.533		4.77	30		34.77	50.50
5300	3/4"		14	.571		5.90	32		37.90	54.50
5310	1"		13	.615		7.45	34.50		41.95	59.50
5320	1-1/4"	Q-1	22	.727		13.35	36.50		49.85	69.50
5330	1-1/2"		20	.800		16.25	40		56.25	78.50
5340	2"		18	.889		24.50	44.50		69	94
5350	2-1/2"		14	1.143		71	57.50		128.50	165

22 11 13 – Facility Water Distribution Piping

22 11 13.45 Pipe Fittings, Steel, Threaded		Crew	Daily Output	Labor-Hours	Unit	Material	Labor	Equipment	Total	Total Incl O&P
5360	3"	Q-1	10	1.600	Ea.	92.50	80.50		173	223
5370	3-1/2"		8	2		181	100		281	350
5380	4"		6	2.667		181	134		315	400
5450	Tee, straight									
5470	1/4"	1 Plum	10	.800	Ea.	6.05	44.50		50.55	73.50
5480	3/8"		10	.800		6.05	44.50		50.55	73.50
5490	1/2"		9	.889		3.86	49.50		53.36	79
5500	3/4"		9	.889		5.55	49.50		55.05	80.50
5510	1"		8	1		9.50	56		65.50	94.50
5520	1-1/4"	Q-1	14	1.143		15.40	57.50		72.90	103
5530	1-1/2"		13	1.231		19.10	62		81.10	114
5540	2"		11	1.455		32.50	73		105.50	146
5550	2-1/2"		9	1.778		70.50	89.50		160	212
5560	3"		6	2.667		104	134		238	315
5570	3-1/2"		5	3.200		240	161		401	505
5580	4"		4	4		250	201		451	580
5601	Tee, reducing, on outlet									
5602	1/2"	1 Plum	9	.889	Ea.	9.50	49.50		59	85
5603	3/4"		9	.889		9.15	49.50		58.65	84.50
5604	1"		8	1		15.20	56		71.20	101
5605	1-1/4"	Q-1	14	1.143		27	57.50		84.50	116
5606	1-1/2"		13	1.231		27	62		89	123
5607	2"		11	1.455		37	73		110	151
5608	2-1/2"		9	1.778		107	89.50		196.50	251
5609	3"		6	2.667		145	134		279	360
5610	3-1/2"		5	3.200		345	161		506	620
5611	4"		4	4		296	201		497	630
5650	Coupling									
5670	1/4"	1 Plum	19	.421	Ea.	5.20	23.50		28.70	41
5680	3/8"		19	.421		5.20	23.50		28.70	41
5690	1/2"		19	.421		3.99	23.50		27.49	40
5700	3/4"		18	.444		4.68	25		29.68	42.50
5710	1"		15	.533		7	30		37	52.50
5720	1-1/4"	Q-1	26	.615		9.10	31		40.10	56.50
5730	1-1/2"		24	.667		12.25	33.50		45.75	64
5740	2"		21	.762		18.15	38.50		56.65	77.50
5750	2-1/2"		18	.889		50	44.50		94.50	122
5760	3"		14	1.143		68	57.50		125.50	161
5770	3-1/2"		12	1.333		137	67		204	251
5780	4"		10	1.600		137	80.50		217.50	271
5840	Reducer, concentric, 1/4"	1 Plum	19	.421		5.45	23.50		28.95	41.50
5850	3/8"		19	.421		7	23.50		30.50	43
5860	1/2"		19	.421		5.50	23.50		29	41.50
5870	3/4"		16	.500		6.55	28		34.55	49
5880	1"		15	.533		11	30		41	57
5890	1-1/4"	Q-1	26	.615		12.35	31		43.35	60
5900	1-1/2"		24	.667		17.70	33.50		51.20	70
5910	2"		21	.762		25.50	38.50		64	85.50
5911	2-1/2"		18	.889		57	44.50		101.50	130
5912	3"		14	1.143		70.50	57.50		128	164
5913	3-1/2"		12	1.333		219	67		286	340
5914	4"		10	1.600		166	80.50		246.50	305
5981	Bushing, 1/4"	1 Plum	19	.421		1.16	23.50		24.66	37

22 11 13 – Facility Water Distribution Piping

22 11 13.45 Pipe Fittings, Steel, Threaded		Crew	Daily Output	Labor-Hours	Unit	Material	2013 Bare Costs Labor	Equipment	Total	Total Incl O&P
5982	3/8"	1 Plum	19	.421	Ea.	5	23.50		28.50	41
5983	1/2"		19	.421		5	23.50		28.50	41
5984	3/4"		16	.500		5.20	28		33.20	48
5985	1"		15	.533		7.60	30		37.60	53.50
5986	1-1/4"	Q-1	26	.615		9.45	31		40.45	57
5987	1-1/2"		24	.667		8.15	33.50		41.65	59.50
5988	2"		21	.762		10.30	38.50		48.80	69
5989	Cap, 1/4"	1 Plum	38	.211		3.84	11.75		15.59	22
5991	3/8"		38	.211		3.23	11.75		14.98	21.50
5992	1/2"		38	.211		2.92	11.75		14.67	21
5993	3/4"		32	.250		3.96	13.95		17.91	25.50
5994	1"		30	.267		4.78	14.90		19.68	28
5995	1-1/4"	Q-1	52	.308		6.20	15.45		21.65	30.50
5996	1-1/2"		48	.333		8.65	16.75		25.40	34.50
5997	2"		42	.381		12.65	19.15		31.80	43
6000	For galvanized elbows, tees, and couplings add					20%				
6058	For galvanized reducers, caps and bushings add					20%				
6100	90° Elbow, galvanized, 150 lb., reducing									
6110	3/4" x 1/2"	1 Plum	15.40	.519	Ea.	8.30	29		37.30	52.50
6112	1" x 3/4"		14	.571		10.65	32		42.65	60
6114	1" x 1/2"		14.50	.552		11.30	31		42.30	59
6116	1-1/4" x 1"	Q-1	24.20	.661		17.85	33		50.85	69.50
6118	1-1/4" x 3/4"		25.40	.630		21.50	31.50		53	71
6120	1-1/4" x 1/2"		26.20	.611		23	30.50		53.50	71
6122	1-1/2" x 1-1/4"		21.60	.741		28.50	37		65.50	87.50
6124	1-1/2" x 1"		23.50	.681		28.50	34		62.50	83
6126	1-1/2" x 3/4"		24.60	.650		28.50	32.50		61	80.50
6128	2" x 1-1/2"		20.50	.780		33.50	39		72.50	95.50
6130	2" x 1-1/4"		21	.762		38.50	38.50		77	99.50
6132	2" x 1"		22.80	.702		39.50	35.50		75	96.50
6134	2" x 3/4"		23.90	.669		40.50	33.50		74	95
6136	2-1/2" x 2"		12.30	1.301		113	65.50		178.50	223
6138	2-1/2" x 1-1/2"		12.50	1.280		126	64.50		190.50	236
6140	3" x 2-1/2"		8.60	1.860		206	93.50		299.50	365
6142	3" x 2"		11.80	1.356		179	68		247	299
6144	4" x 3"		8.20	1.951		465	98		563	660
6160	90° Elbow, black, 150 lb., reducing									
6170	1" x 3/4"	1 Plum	14	.571	Ea.	7.30	32		39.30	56
6174	1-1/2" x 1"	Q-1	23.50	.681		17.25	34		51.25	70.50
6178	1-1/2" x 3/4"		24.60	.650		19.75	32.50		52.25	70.50
6182	2" x 1-1/2"		20.50	.780		25	39		64	86.50
6186	2" x 1"		22.80	.702		28.50	35.50		64	84.50
6190	2" x 3/4"		23.90	.669		30	33.50		63.50	83.50
6194	2-1/2" x 2"		12.30	1.301		68.50	65.50		134	174
7000	Union, with brass seat									
7010	1/4"	1 Plum	15	.533	Ea.	20.50	30		50.50	67.50
7020	3/8"		15	.533		14.10	30		44.10	60.50
7030	1/2"		14	.571		12.75	32		44.75	62
7040	3/4"		13	.615		14.70	34.50		49.20	67.50
7050	1"		12	.667		19.15	37		56.15	77
7060	1-1/4"	Q-1	21	.762		27.50	38.50		66	87.50
7070	1-1/2"		19	.842		34	42.50		76.50	101
7080	2"		17	.941		40	47.50		87.50	115

22 11 13 – Facility Water Distribution Piping

22 11 13.45 Pipe Fittings, Steel, Threaded		Crew	Daily Output	Labor-Hours	Unit	Material	Labor	Equipment	Total	Total Incl O&P
7090	2-1/2"	Q-1	13	1.231	Ea.	119	62		181	223
7100	3"	↓	9	1.778	↓	143	89.50		232.50	291
7120	Union, galvanized									
7124	1/2"	1 Plum	14	.571	Ea.	16.95	32		48.95	66.50
7125	3/4"		13	.615		19.50	34.50		54	73
7126	1"	↓	12	.667		25.50	37		62.50	84
7127	1-1/4"	Q-1	21	.762		37	38.50		75.50	98
7128	1-1/2"		19	.842		44.50	42.50		87	113
7129	2"		17	.941		51.50	47.50		99	128
7130	2-1/2"		13	1.231		178	62		240	289
7131	3"	↓	9	1.778	↓	250	89.50		339.50	410
7500	Malleable iron, 300 lb.									
7520	Black									
7540	90° Elbow, straight, 1/4"	1 Plum	16	.500	Ea.	14.70	28		42.70	58
7560	3/8"		16	.500		13	28		41	56.50
7570	1/2"		15	.533		16.85	30		46.85	63.50
7580	3/4"		14	.571		19.05	32		51.05	69
7590	1"	↓	13	.615		24.50	34.50		59	78.50
7600	1-1/4"	Q-1	22	.727		35.50	36.50		72	94
7610	1-1/2"		20	.800		42	40		82	107
7620	2"		18	.889		60	44.50		104.50	134
7630	2-1/2"		14	1.143		156	57.50		213.50	259
7640	3"		10	1.600		179	80.50		259.50	320
7650	4"	↓	6	2.667		420	134		554	665
7700	45° Elbow, straight, 1/4"	1 Plum	16	.500		22	28		50	66
7720	3/8"		16	.500		22	28		50	66
7730	1/2"		15	.533		24	30		54	71.50
7740	3/4"		14	.571		27	32		59	77.50
7750	1"	↓	13	.615		30	34.50		64.50	84.50
7760	1-1/4"	Q-1	22	.727		48	36.50		84.50	108
7770	1-1/2"		20	.800		62.50	40		102.50	129
7780	2"		18	.889		94	44.50		138.50	171
7790	2-1/2"		14	1.143		210	57.50		267.50	320
7800	3"		10	1.600		277	80.50		357.50	425
7810	4"	↓	6	2.667		630	134		764	890
7850	Tee, straight, 1/4"	1 Plum	10	.800		18.45	44.50		62.95	87.50
7870	3/8"		10	.800		19.50	44.50		64	88.50
7880	1/2"		9	.889		25	49.50		74.50	102
7890	3/4"		9	.889		27	49.50		76.50	104
7900	1"	↓	8	1		32.50	56		88.50	120
7910	1-1/4"	Q-1	14	1.143		35.50	57.50		93	126
7920	1-1/2"		13	1.231		54	62		116	153
7930	2"		11	1.455		79.50	73		152.50	198
7940	2-1/2"		9	1.778		194	89.50		283.50	345
7950	3"		6	2.667		276	134		410	505
7960	4"	↓	4	4		785	201		986	1,175
8050	Couplings, straight, 1/4"	1 Plum	19	.421		13.20	23.50		36.70	50
8070	3/8"		19	.421		13.20	23.50		36.70	50
8080	1/2"		19	.421		14.75	23.50		38.25	51.50
8090	3/4"		18	.444		16.95	25		41.95	56
8100	1"	↓	15	.533		19.40	30		49.40	66.50
8110	1-1/4"	Q-1	26	.615		23	31		54	72
8120	1-1/2"		24	.667		34.50	33.50		68	88.50

(Material, Labor, Equipment and Total columns: 2013 Bare Costs)

22 11 13 – Facility Water Distribution Piping

22 11 13.45 Pipe Fittings, Steel, Threaded		Crew	Daily Output	Labor-Hours	Unit	2013 Bare Costs Material	Labor	Equipment	Total	Total Incl O&P
8130	2"	Q-1	21	.762	Ea.	49.50	38.50		88	112
8140	2-1/2"		18	.889		91.50	44.50		136	168
8150	3"		14	1.143		132	57.50		189.50	232
8160	4"		10	1.600		240	80.50		320.50	385
8200	Galvanized									
8220	90° Elbow, straight, 1/4"	1 Plum	16	.500	Ea.	26	28		54	71
8222	3/8"		16	.500		13	28		41	56.50
8224	1/2"		15	.533		16.85	30		46.85	63.50
8226	3/4"		14	.571		19.05	32		51.05	69
8228	1"		13	.615		24.50	34.50		59	78.50
8230	1-1/4"	Q-1	22	.727		35.50	36.50		72	94
8232	1-1/2"		20	.800		42	40		82	107
8234	2"		18	.889		60	44.50		104.50	134
8236	2-1/2"		14	1.143		289	57.50		346.50	405
8238	3"		10	1.600		345	80.50		425.50	500
8240	4"		6	2.667		1,025	134		1,159	1,325
8280	45° Elbow, straight									
8282	1/2"	1 Plum	15	.533	Ea.	47.50	30		77.50	97
8284	3/4"		14	.571		54.50	32		86.50	108
8286	1"		13	.615		59.50	34.50		94	117
8288	1-1/4"	Q-1	22	.727		92.50	36.50		129	157
8290	1-1/2"		20	.800		120	40		160	193
8292	2"		18	.889		166	44.50		210.50	250
8310	Tee, straight, 1/4"	1 Plum	10	.800		41	44.50		85.50	112
8312	3/8"		10	.800		38	44.50		82.50	109
8314	1/2"		9	.889		47.50	49.50		97	127
8316	3/4"		9	.889		52.50	49.50		102	132
8318	1"		8	1		64.50	56		120.50	155
8320	1-1/4"	Q-1	14	1.143		92	57.50		149.50	188
8322	1-1/2"		13	1.231		96	62		158	198
8324	2"		11	1.455		153	73		226	279
8326	2-1/2"		9	1.778		485	89.50		574.50	670
8328	3"		6	2.667		550	134		684	805
8330	4"		4	4		1,400	201		1,601	1,850
8380	Couplings, straight, 1/4"	1 Plum	19	.421		17.20	23.50		40.70	54.50
8382	3/8"		19	.421		24.50	23.50		48	62.50
8384	1/2"		19	.421		25.50	23.50		49	63.50
8386	3/4"		18	.444		28.50	25		53.50	69
8388	1"		15	.533		38	30		68	86.50
8390	1-1/4"	Q-1	26	.615		49.50	31		80.50	101
8392	1-1/2"		24	.667		70	33.50		103.50	127
8394	2"		21	.762		84.50	38.50		123	150
8396	2-1/2"		18	.889		222	44.50		266.50	310
8398	3"		14	1.143		264	57.50		321.50	380
8399	4"		10	1.600		425	80.50		505.50	585
8529	Black									
8530	Reducer, concentric, 1/4"	1 Plum	19	.421	Ea.	17.45	23.50		40.95	54.50
8531	3/8"		19	.421		17.45	23.50		40.95	54.50
8532	1/2"		17	.471		21	26.50		47.50	62.50
8533	3/4"		16	.500		30	28		58	75
8534	1"		15	.533		31.50	30		61.50	79.50
8535	1-1/4"	Q-1	26	.615		53.50	31		84.50	106
8536	1-1/2"		24	.667		55.50	33.50		89	112

22 11 13 – Facility Water Distribution Piping

22 11 13.45 Pipe Fittings, Steel, Threaded		Crew	Daily Output	Labor-Hours	Unit	Material	2013 Bare Costs Labor	Equipment	Total	Total Incl O&P
8537	2″	Q-1	21	.762	Ea.	80.50	38.50		119	146
8550	Cap, 1/4″	1 Plum	38	.211		13.55	11.75		25.30	32.50
8551	3/8″		38	.211		13.55	11.75		25.30	32.50
8552	1/2″		34	.235		13.45	13.15		26.60	34.50
8553	3/4″		32	.250		16.95	13.95		30.90	39.50
8554	1″		30	.267		22	14.90		36.90	47
8555	1-1/4″	Q-1	52	.308		24.50	15.45		39.95	50.50
8556	1-1/2″		48	.333		36	16.75		52.75	64.50
8557	2″		42	.381		50	19.15		69.15	84
8570	Plug, 1/4″	1 Plum	38	.211		3.28	11.75		15.03	21.50
8571	3/8″		38	.211		3.44	11.75		15.19	21.50
8572	1/2″		34	.235		3.60	13.15		16.75	24
8573	3/4″		32	.250		4.49	13.95		18.44	26
8574	1″		30	.267		6.85	14.90		21.75	30
8575	1-1/4″	Q-1	52	.308		14.15	15.45		29.60	39
8576	1-1/2″		48	.333		16.20	16.75		32.95	43
8577	2″		42	.381		25.50	19.15		44.65	57
9500	Union with brass seat, 1/4″	1 Plum	15	.533		30	30		60	78
9530	3/8″		15	.533		30.50	30		60.50	78.50
9540	1/2″		14	.571		24.50	32		56.50	75
9550	3/4″		13	.615		27	34.50		61.50	81.50
9560	1″		12	.667		35.50	37		72.50	95
9570	1-1/4″	Q-1	21	.762		57.50	38.50		96	121
9580	1-1/2″		19	.842		60.50	42.50		103	130
9590	2″		17	.941		75	47.50		122.50	154
9600	2-1/2″		13	1.231		241	62		303	360
9610	3″		9	1.778		310	89.50		399.50	480
9620	4″		5	3.200		990	161		1,151	1,350
9630	Union, all iron, 1/4″	1 Plum	15	.533		46.50	30		76.50	96
9650	3/8″		15	.533		46.50	30		76.50	96
9660	1/2″		14	.571		49	32		81	102
9670	3/4″		13	.615		52.50	34.50		87	109
9680	1″		12	.667		64	37		101	127
9690	1-1/4″	Q-1	21	.762		98.50	38.50		137	166
9700	1-1/2″		19	.842		120	42.50		162.50	196
9710	2″		17	.941		152	47.50		199.50	239
9720	2-1/2″		13	1.231		270	62		332	390
9730	3″		9	1.778		515	89.50		604.50	700
9750	For galvanized unions, add					15%				
9757	Forged steel, 3000 lb.									
9758	Black									
9760	90° Elbow, 1/4″	1 Plum	16	.500	Ea.	19.35	28		47.35	63.50
9761	3/8″		16	.500		19.35	28		47.35	63.50
9762	1/2″		15	.533		14.80	30		44.80	61.50
9763	3/4″		14	.571		18.60	32		50.60	68.50
9764	1″		13	.615		27.50	34.50		62	82
9765	1-1/4″	Q-1	22	.727		52.50	36.50		89	113
9766	1-1/2″		20	.800		67.50	40		107.50	135
9767	2″		18	.889		82	44.50		126.50	158
9780	45° Elbow 1/4″	1 Plum	16	.500		24.50	28		52.50	69
9781	3/8″		16	.500		24.50	28		52.50	69
9782	1/2″		15	.533		24	30		54	71.50
9783	3/4″		14	.571		28	32		60	79

22 11 13 – Facility Water Distribution Piping

22 11 13.45 Pipe Fittings, Steel, Threaded		Crew	Daily Output	Labor-Hours	Unit	Material	2013 Bare Costs Labor	Equipment	Total	Total Incl O&P
9784	1"	1 Plum	13	.615	Ea.	38	34.50		72.50	93.50
9785	1-1/4"	Q-1	22	.727		52.50	36.50		89	113
9786	1-1/2"		20	.800		75.50	40		115.50	144
9787	2"		18	.889		104	44.50		148.50	182
9800	Tee, 1/4"	1 Plum	10	.800		23.50	44.50		68	93
9801	3/8"		10	.800		23.50	44.50		68	93
9802	1/2"		9	.889		21	49.50		70.50	98
9803	3/4"		9	.889		28.50	49.50		78	106
9804	1"		8	1		37.50	56		93.50	126
9805	1-1/4"	Q-1	14	1.143		73	57.50		130.50	167
9806	1-1/2"		13	1.231		85	62		147	187
9807	2"		11	1.455		110	73		183	230
9820	Reducer, concentric, 1/4"	1 Plum	19	.421		11.80	23.50		35.30	48.50
9821	3/8"		19	.421		12.30	23.50		35.80	49
9822	1/2"		17	.471		12.30	26.50		38.80	53
9823	3/4"		16	.500		14.60	28		42.60	58
9824	1"		15	.533		19.05	30		49.05	66
9825	1-1/4"	Q-1	26	.615		33.50	31		64.50	83
9826	1-1/2"		24	.667		36	33.50		69.50	90.50
9827	2"		21	.762		52.50	38.50		91	116
9840	Cap, 1/4"	1 Plum	38	.211		6.90	11.75		18.65	25.50
9841	3/8"		38	.211		7.25	11.75		19	25.50
9842	1/2"		34	.235		7.05	13.15		20.20	27.50
9843	3/4"		32	.250		9.85	13.95		23.80	32
9844	1"		30	.267		15.10	14.90		30	39
9845	1-1/4"	Q-1	52	.308		24.50	15.45		39.95	50.50
9846	1-1/2"		48	.333		29	16.75		45.75	57
9847	2"		42	.381		42	19.15		61.15	75
9860	Plug, 1/4"	1 Plum	38	.211		3.28	11.75		15.03	21.50
9861	3/8"		38	.211		3.44	11.75		15.19	21.50
9862	1/2"		34	.235		3.60	13.15		16.75	24
9863	3/4"		32	.250		4.49	13.95		18.44	26
9864	1"		30	.267		6.85	14.90		21.75	30
9865	1-1/4"	Q-1	52	.308		14.15	15.45		29.60	39
9866	1-1/2"		48	.333		16.20	16.75		32.95	43
9867	2"		42	.381		25.50	19.15		44.65	57
9880	Union, bronze seat, 1/4"	1 Plum	15	.533		53.50	30		83.50	104
9881	3/8"		15	.533		53.50	30		83.50	104
9882	1/2"		14	.571		51.50	32		83.50	105
9883	3/4"		13	.615		69	34.50		103.50	127
9884	1"		12	.667		77.50	37		114.50	142
9885	1-1/4"	Q-1	21	.762		151	38.50		189.50	224
9886	1-1/2"		19	.842		155	42.50		197.50	235
9887	2"		17	.941		184	47.50		231.50	274
9900	Coupling, 1/4"	1 Plum	19	.421		7.55	23.50		31.05	44
9901	3/8"		19	.421		7.55	23.50		31.05	44
9902	1/2"		17	.471		6.20	26.50		32.70	46.50
9903	3/4"		16	.500		8.05	28		36.05	51
9904	1"		15	.533		14	30		44	60.50
9905	1-1/4"	Q-1	26	.615		23.50	31		54.50	72
9906	1-1/2"		24	.667		30	33.50		63.50	83.50
9907	2"		21	.762		37.50	38.50		76	99

22 11 13.47 Pipe Fittings, Steel		Crew	Daily Output	Labor-Hours	Unit	Material	2013 Bare Costs Labor	Equipment	Total	Total Incl O&P
0010	**PIPE FITTINGS, STEEL**, Flanged, Welded & Special									
0020	Flanged joints, C.I., standard weight, black. One gasket & bolt									
0040	set, mat'l only, required at each joint, not included (see line 0620)									
0060	90° Elbow, straight, 1-1/2" pipe size	Q-1	14	1.143	Ea.	475	57.50		532.50	610
0080	2" pipe size		13	1.231		283	62		345	405
0090	2-1/2" pipe size		12	1.333		305	67		372	435
0100	3" pipe size		11	1.455		255	73		328	390
0110	4" pipe size		8	2		315	100		415	495
0120	5" pipe size		7	2.286		745	115		860	995
0130	6" pipe size	Q-2	9	2.667		495	139		634	755
0140	8" pipe size		8	3		855	156		1,011	1,175
0150	10" pipe size		7	3.429		1,875	179		2,054	2,325
0160	12" pipe size		6	4		3,775	208		3,983	4,475
0171	90° Elbow, reducing									
0172	2-1/2" by 2" pipe size	Q-1	12	1.333	Ea.	925	67		992	1,125
0173	3" by 2-1/2" pipe size		11	1.455		945	73		1,018	1,150
0174	4" by 3" pipe size		8	2		700	100		800	920
0175	5" by 3" pipe size		7	2.286		1,500	115		1,615	1,800
0176	6" by 4" pipe size	Q-2	9	2.667		870	139		1,009	1,175
0177	8" by 6" pipe size		8	3		1,275	156		1,431	1,625
0178	10" by 8" pipe size		7	3.429		2,400	179		2,579	2,925
0179	12" by 10" pipe size		6	4		4,625	208		4,833	5,400
0200	45° Elbow, straight, 1-1/2" pipe size	Q-1	14	1.143		575	57.50		632.50	720
0220	2" pipe size		13	1.231		410	62		472	545
0230	2-1/2" pipe size		12	1.333		435	67		502	580
0240	3" pipe size		11	1.455		425	73		498	575
0250	4" pipe size		8	2		480	100		580	675
0260	5" pipe size		7	2.286		1,150	115		1,265	1,425
0270	6" pipe size	Q-2	9	2.667		780	139		919	1,075
0280	8" pipe size		8	3		1,150	156		1,306	1,475
0290	10" pipe size		7	3.429		2,425	179		2,604	2,925
0300	12" pipe size		6	4		3,700	208		3,908	4,375
0310	Cross, straight									
0311	2-1/2" pipe size	Q-1	6	2.667	Ea.	900	134		1,034	1,200
0312	3" pipe size		5	3.200		945	161		1,106	1,275
0313	4" pipe size		4	4		1,225	201		1,426	1,650
0314	5" pipe size		3	5.333		2,575	268		2,843	3,225
0315	6" pipe size	Q-2	5	4.800		2,575	250		2,825	3,200
0316	8" pipe size		4	6		4,150	310		4,460	5,050
0317	10" pipe size		3	8		4,925	415		5,340	6,050
0318	12" pipe size		2	12		7,525	625		8,150	9,225
0350	Tee, straight, 1-1/2" pipe size	Q-1	10	1.600		550	80.50		630.50	725
0370	2" pipe size		9	1.778		310	89.50		399.50	475
0380	2-1/2" pipe size		8	2		450	100		550	645
0390	3" pipe size		7	2.286		315	115		430	520
0400	4" pipe size		5	3.200		480	161		641	770
0410	5" pipe size		4	4		1,275	201		1,476	1,725
0420	6" pipe size	Q-2	6	4		695	208		903	1,075
0430	8" pipe size		5	4.800		1,200	250		1,450	1,675
0440	10" pipe size		4	6		3,200	310		3,510	4,000
0450	12" pipe size		3	8		5,025	415		5,440	6,175
0459	Tee, reducing on outlet									

22 11 13.47 Pipe Fittings, Steel		Crew	Daily Output	Labor-Hours	Unit	2013 Bare Costs Material	Labor	Equipment	Total	Total Incl O&P
0460	2-1/2″ by 2″ pipe size	Q-1	8	2	Ea.	940	100		1,040	1,175
0461	3″ by 2-1/2″ pipe size		7	2.286		940	115		1,055	1,200
0462	4″ by 3″ pipe size		5	3.200		990	161		1,151	1,350
0463	5″ by 4″ pipe size		4	4		2,150	201		2,351	2,650
0464	6″ by 4″ pipe size	Q-2	6	4		950	208		1,158	1,375
0465	8″ by 6″ pipe size		5	4.800		1,475	250		1,725	2,000
0466	10″ by 8″ pipe size		4	6		3,400	310		3,710	4,225
0467	12″ by 10″ pipe size		3	8		5,875	415		6,290	7,100
0476	Reducer, concentric									
0477	3″ by 2-1/2″	Q-1	12	1.333	Ea.	565	67		632	725
0478	4″ by 3″		9	1.778		635	89.50		724.50	835
0479	5″ by 4″		8	2		985	100		1,085	1,225
0480	6″ by 4″	Q-2	10	2.400		850	125		975	1,125
0481	8″ by 6″		9	2.667		1,075	139		1,214	1,375
0482	10″ by 8″		8	3		2,200	156		2,356	2,650
0483	12″ by 10″		7	3.429		3,825	179		4,004	4,475
0492	Reducer, eccentric									
0493	4″ by 3″	Q-1	8	2	Ea.	1,025	100		1,125	1,275
0494	5″ by 4″	″	7	2.286		1,575	115		1,690	1,900
0495	6″ by 4″	Q-2	9	2.667		975	139		1,114	1,275
0496	8″ by 6″		8	3		1,225	156		1,381	1,575
0497	10″ by 8″		7	3.429		3,150	179		3,329	3,725
0498	12″ by 10″		6	4		4,275	208		4,483	5,025
0500	For galvanized elbows and tees, add					100%				
0520	For extra heavy weight elbows and tees, add					140%				
0620	Gasket and bolt set, 150#, 1/2″ pipe size	1 Plum	20	.400		2.50	22.50		25	36.50
0622	3/4″ pipe size		19	.421		2.64	23.50		26.14	38.50
0624	1″ pipe size		18	.444		2.67	25		27.67	40.50
0626	1-1/4″ pipe size		17	.471		2.80	26.50		29.30	42.50
0628	1-1/2″ pipe size		15	.533		2.98	30		32.98	48.50
0630	2″ pipe size		13	.615		4.85	34.50		39.35	57
0640	2-1/2″ pipe size		12	.667		5.60	37		42.60	62
0650	3″ pipe size		11	.727		5.80	40.50		46.30	67.50
0660	3-1/2″ pipe size		9	.889		9.40	49.50		58.90	85
0670	4″ pipe size		8	1		10.40	56		66.40	95.50
0680	5″ pipe size		7	1.143		17.10	64		81.10	115
0690	6″ pipe size		6	1.333		18.10	74.50		92.60	132
0700	8″ pipe size		5	1.600		19.55	89.50		109.05	156
0710	10″ pipe size		4.50	1.778		35.50	99		134.50	188
0720	12″ pipe size		4.20	1.905		37.50	106		143.50	202
0730	14″ pipe size		4	2		39.50	112		151.50	212
0740	16″ pipe size		3	2.667		41.50	149		190.50	270
0750	18″ pipe size		2.70	2.963		80.50	165		245.50	340
0760	20″ pipe size		2.30	3.478		131	194		325	435
0780	24″ pipe size		1.90	4.211		164	235		399	535
0790	26″ pipe size		1.60	5		222	279		501	665
0810	30″ pipe size		1.40	5.714		430	320		750	955
0830	36″ pipe size		1.10	7.273		810	405		1,215	1,500
0850	For 300 lb. gasket set, add					40%				
2000	Flanged unions, 125 lb., black, 1/2″ pipe size	1 Plum	17	.471	Ea.	72	26.50		98.50	119
2040	3/4″ pipe size		17	.471		95	26.50		121.50	145
2050	1″ pipe size		16	.500		92	28		120	143
2060	1-1/4″ pipe size	Q-1	28	.571		109	28.50		137.50	163

22 11 13 – Facility Water Distribution Piping

22 11 13.47 Pipe Fittings, Steel		Crew	Daily Output	Labor-Hours	Unit	Material	2013 Bare Costs Labor	Equipment	Total	Total Incl O&P
2070	1-1/2" pipe size	Q-1	27	.593	Ea.	101	30		131	156
2080	2" pipe size		26	.615		118	31		149	177
2090	2-1/2" pipe size		24	.667		160	33.50		193.50	227
2100	3" pipe size		22	.727		181	36.50		217.50	254
2110	3-1/2" pipe size		18	.889		305	44.50		349.50	405
2120	4" pipe size		16	1		246	50		296	345
2130	5" pipe size		14	1.143		575	57.50		632.50	715
2140	6" pipe size	Q-2	19	1.263		540	66		606	695
2150	8" pipe size	"	16	1.500		1,225	78		1,303	1,475
2200	For galvanized unions, add					150%				
2290	Threaded flange									
2300	Cast iron									
2310	Black, 125 lb., per flange									
2320	1" pipe size	1 Plum	27	.296	Ea.	41	16.55		57.55	70.50
2330	1-1/4" pipe size	Q-1	44	.364		49.50	18.25		67.75	82
2340	1-1/2" pipe size		40	.400		45.50	20		65.50	80.50
2350	2" pipe size		36	.444		45.50	22.50		68	83.50
2360	2-1/2" pipe size		28	.571		53	28.50		81.50	102
2370	3" pipe size		20	.800		68.50	40		108.50	136
2380	3-1/2" pipe size		16	1		96	50		146	182
2390	4" pipe size		12	1.333		92.50	67		159.50	203
2400	5" pipe size		10	1.600		131	80.50		211.50	265
2410	6" pipe size	Q-2	14	1.714		148	89.50		237.50	296
2420	8" pipe size		12	2		233	104		337	415
2430	10" pipe size		10	2.400		415	125		540	645
2440	12" pipe size		8	3		910	156		1,066	1,225
2460	For galvanized flanges, add					95%				
2490	Blind flange									
2492	Cast iron									
2494	Black, 125 lb., per flange									
2496	1" pipe size	1 Plum	27	.296	Ea.	65.50	16.55		82.05	97
2500	1-1/2" pipe size	Q-1	40	.400		73.50	20		93.50	112
2502	2" pipe size		36	.444		83	22.50		105.50	125
2504	2-1/2" pipe size		28	.571		91.50	28.50		120	144
2506	3" pipe size		20	.800		110	40		150	181
2508	4" pipe size		12	1.333		140	67		207	255
2510	5" pipe size		10	1.600		228	80.50		308.50	370
2512	6" pipe size	Q-2	14	1.714		248	89.50		337.50	405
2514	8" pipe size		12	2		390	104		494	585
2516	10" pipe size		10	2.400		575	125		700	825
2518	12" pipe size		8	3		1,075	156		1,231	1,425
2520	For galvanized flanges, add					80%				
2570	Threaded flange									
2580	Forged steel,									
2590	Black 150 lb., per flange									
2600	1/2" pipe size	1 Plum	30	.267	Ea.	26.50	14.90		41.40	51.50
2610	3/4" pipe size		28	.286		26.50	15.95		42.45	53
2620	1" pipe size		27	.296		26.50	16.55		43.05	54
2630	1-1/4" pipe size	Q-1	44	.364		26.50	18.25		44.75	56.50
2640	1-1/2" pipe size		40	.400		26.50	20		46.50	59.50
2650	2" pipe size		36	.444		29.50	22.50		52	66
2660	2-1/2" pipe size		28	.571		36	28.50		64.50	83
2670	3" pipe size		20	.800		36.50	40		76.50	101

22 11 13.47 Pipe Fittings, Steel		Crew	Daily Output	Labor-Hours	Unit	Material	2013 Bare Costs Labor	Equipment	Total	Total Incl O&P
2690	4" pipe size	Q-1	12	1.333	Ea.	42.50	67		109.50	148
2700	5" pipe size		10	1.600		66.50	80.50		147	195
2710	6" pipe size	Q-2	14	1.714		73	89.50		162.50	215
2720	8" pipe size		12	2		126	104		230	295
2730	10" pipe size		10	2.400		228	125		353	440
2860	Black 300 lb., per flange									
2870	1/2" pipe size	1 Plum	30	.267	Ea.	29	14.90		43.90	54.50
2880	3/4" pipe size		28	.286		29	15.95		44.95	56
2890	1" pipe size		27	.296		29	16.55		45.55	57
2900	1-1/4" pipe size	Q-1	44	.364		29	18.25		47.25	59.50
2910	1-1/2" pipe size		40	.400		29	20		49	62.50
2920	2" pipe size		36	.444		34	22.50		56.50	71
2930	2-1/2" pipe size		28	.571		47.50	28.50		76	95
2940	3" pipe size		20	.800		50.50	40		90.50	116
2960	4" pipe size		12	1.333		73	67		140	181
2970	6" pipe size	Q-2	14	1.714		136	89.50		225.50	284
3000	Weld joint, butt, carbon steel, standard weight									
3040	90° elbow, long radius									
3050	1/2" pipe size	Q-15	16	1	Ea.	45	50	3.30	98.30	129
3060	3/4" pipe size		16	1		45	50	3.30	98.30	129
3070	1" pipe size		16	1		21	50	3.30	74.30	102
3080	1-1/4" pipe size		14	1.143		21	57.50	3.77	82.27	114
3090	1-1/2" pipe size		13	1.231		21	62	4.06	87.06	120
3100	2" pipe size		10	1.600		22.50	80.50	5.30	108.30	152
3110	2-1/2" pipe size		8	2		27.50	100	6.60	134.10	189
3120	3" pipe size		7	2.286		29	115	7.55	151.55	213
3130	4" pipe size		5	3.200		48	161	10.55	219.55	305
3136	5" pipe size		4	4		103	201	13.20	317.20	435
3140	6" pipe size	Q-16	5	4.800		106	250	10.55	366.55	505
3150	8" pipe size		3.75	6.400		200	335	14.10	549.10	735
3160	10" pipe size		3	8		400	415	17.60	832.60	1,075
3170	12" pipe size		2.50	9.600		590	500	21	1,111	1,425
3180	14" pipe size		2	12		970	625	26.50	1,621.50	2,050
3190	16" pipe size		1.50	16		1,325	835	35	2,195	2,775
3191	18" pipe size		1.25	19.200		1,525	1,000	42	2,567	3,225
3192	20" pipe size		1.15	20.870		2,450	1,075	46	3,571	4,375
3194	24" pipe size		1.02	23.529		3,450	1,225	52	4,727	5,700
3200	45° Elbow, long									
3210	1/2" pipe size	Q-15	16	1	Ea.	62.50	50	3.30	115.80	148
3220	3/4" pipe size		16	1		62.50	50	3.30	115.80	148
3230	1" pipe size		16	1		22	50	3.30	75.30	104
3240	1-1/4" pipe size		14	1.143		22	57.50	3.77	83.27	115
3250	1-1/2" pipe size		13	1.231		22	62	4.06	88.06	122
3260	2" pipe size		10	1.600		22	80.50	5.30	107.80	151
3270	2-1/2" pipe size		8	2		26.50	100	6.60	133.10	187
3280	3" pipe size		7	2.286		27.50	115	7.55	150.05	212
3290	4" pipe size		5	3.200		49.50	161	10.55	221.05	310
3296	5" pipe size		4	4		75	201	13.20	289.20	400
3300	6" pipe size	Q-16	5	4.800		97.50	250	10.55	358.05	495
3310	8" pipe size		3.75	6.400		163	335	14.10	512.10	695
3320	10" pipe size		3	8		325	415	17.60	757.60	1,000
3330	12" pipe size		2.50	9.600		460	500	21	981	1,275
3340	14" pipe size		2	12		615	625	26.50	1,266.50	1,650

22 11 13 – Facility Water Distribution Piping

22 11 13.47 Pipe Fittings, Steel		Crew	Daily Output	Labor-Hours	Unit	2013 Bare Costs Material	Labor	Equipment	Total	Total Incl O&P
3341	16" pipe size	Q-16	1.50	16	Ea.	1,100	835	35	1,970	2,525
3342	18" pipe size		1.25	19.200		1,575	1,000	42	2,617	3,275
3343	20" pipe size		1.15	20.870		1,625	1,075	46	2,746	3,450
3345	24" pipe size		1.05	22.857		2,450	1,200	50.50	3,700.50	4,550
3346	26" pipe size		.85	28.235		2,700	1,475	62	4,237	5,275
3347	30" pipe size		.45	53.333		2,975	2,775	117	5,867	7,575
3349	36" pipe size		.38	63.158		3,275	3,300	139	6,714	8,725
3350	Tee, straight									
3352	For reducing tees and concentrics see starting line 4600									
3360	1/2" pipe size	Q-15	10	1.600	Ea.	110	80.50	5.30	195.80	248
3370	3/4" pipe size		10	1.600		110	80.50	5.30	195.80	248
3380	1" pipe size		10	1.600		54.50	80.50	5.30	140.30	187
3390	1-1/4" pipe size		9	1.778		68.50	89.50	5.85	163.85	216
3400	1-1/2" pipe size		8	2		68.50	100	6.60	175.10	234
3410	2" pipe size		6	2.667		54.50	134	8.80	197.30	272
3420	2-1/2" pipe size		5	3.200		75.50	161	10.55	247.05	335
3430	3" pipe size		4	4		84	201	13.20	298.20	410
3440	4" pipe size		3	5.333		118	268	17.60	403.60	555
3446	5" pipe size		2.50	6.400		195	320	21	536	720
3450	6" pipe size	Q-16	3	8		203	415	17.60	635.60	870
3460	8" pipe size		2.50	9.600		355	500	21	876	1,175
3470	10" pipe size		2	12		695	625	26.50	1,346.50	1,725
3480	12" pipe size		1.60	15		975	780	33	1,788	2,275
3481	14" pipe size		1.30	18.462		1,700	960	40.50	2,700.50	3,375
3482	16" pipe size		1	24		1,900	1,250	53	3,203	4,025
3483	18" pipe size		.80	30		3,025	1,550	66	4,641	5,750
3484	20" pipe size		.75	32		4,750	1,675	70.50	6,495.50	7,800
3486	24" pipe size		.70	34.286		6,150	1,775	75.50	8,000.50	9,525
3487	26" pipe size		.55	43.636		6,775	2,275	96	9,146	11,000
3488	30" pipe size		.30	80		7,425	4,175	176	11,776	14,600
3490	36" pipe size		.25	96		8,200	5,000	211	13,411	16,800
3491	Eccentric reducer, 1-1/2" pipe size	Q-15	14	1.143		63.50	57.50	3.77	124.77	161
3492	2" pipe size		11	1.455		43	73	4.80	120.80	163
3493	2-1/2" pipe size		9	1.778		50	89.50	5.85	145.35	195
3494	3" pipe size		8	2		59	100	6.60	165.60	223
3495	4" pipe size		6	2.667		75.50	134	8.80	218.30	295
3496	6" pipe size	Q-16	5	4.800		232	250	10.55	492.55	640
3497	8" pipe size		4	6		345	310	13.20	668.20	865
3498	10" pipe size		3	8		435	415	17.60	867.60	1,125
3499	12" pipe size		2.50	9.600		385	500	21	906	1,200
3501	Cap, 1-1/2" pipe size	Q-15	28	.571		24.50	28.50	1.89	54.89	72
3502	2" pipe size		22	.727		22	36.50	2.40	60.90	81.50
3503	2-1/2" pipe size		18	.889		30	44.50	2.93	77.43	103
3504	3" pipe size		16	1		33.50	50	3.30	86.80	116
3505	4" pipe size		12	1.333		45	67	4.40	116.40	155
3506	6" pipe size	Q-16	10	2.400		92	125	5.30	222.30	295
3507	8" pipe size		8	3		138	156	6.60	300.60	395
3508	10" pipe size		6	4		212	208	8.80	428.80	560
3509	12" pipe size		5	4.800		283	250	10.55	543.55	695
3511	14" pipe size		4	6		293	310	13.20	616.20	810
3512	16" pipe size		4	6		455	310	13.20	778.20	985
3513	18" pipe size		3	8		665	415	17.60	1,097.60	1,375
3517	Weld joint, butt, carbon steel, extra strong									

22 11 13.47 Pipe Fittings, Steel		Crew	Daily Output	Labor-Hours	Unit	2013 Bare Costs Material	Labor	Equipment	Total	Total Incl O&P
3519	90° elbow, long									
3520	1/2" pipe size	Q-15	13	1.231	Ea.	55.50	62	4.06	121.56	159
3530	3/4" pipe size		12	1.333		55.50	67	4.40	126.90	167
3540	1" pipe size		11	1.455		27	73	4.80	104.80	145
3550	1-1/4" pipe size		10	1.600		27	80.50	5.30	112.80	157
3560	1-1/2" pipe size		9	1.778		27	89.50	5.85	122.35	170
3570	2" pipe size		8	2		27.50	100	6.60	134.10	189
3580	2-1/2" pipe size		7	2.286		38.50	115	7.55	161.05	224
3590	3" pipe size		6	2.667		49.50	134	8.80	192.30	266
3600	4" pipe size		4	4		81.50	201	13.20	295.70	410
3606	5" pipe size		3.50	4.571		195	230	15.10	440.10	575
3610	6" pipe size	Q-16	4.50	5.333		206	278	11.75	495.75	660
3620	8" pipe size		3.50	6.857		390	355	15.10	760.10	985
3630	10" pipe size		2.50	9.600		830	500	21	1,351	1,700
3640	12" pipe size		2.25	10.667		1,025	555	23.50	1,603.50	1,975
3650	45° Elbow, long									
3660	1/2" pipe size	Q-15	13	1.231	Ea.	62	62	4.06	128.06	165
3670	3/4" pipe size		12	1.333		62	67	4.40	133.40	174
3680	1" pipe size		11	1.455		29	73	4.80	106.80	147
3690	1-1/4" pipe size		10	1.600		29	80.50	5.30	114.80	158
3700	1-1/2" pipe size		9	1.778		29	89.50	5.85	124.35	172
3710	2" pipe size		8	2		29	100	6.60	135.60	190
3720	2-1/2" pipe size		7	2.286		64	115	7.55	186.55	252
3730	3" pipe size		6	2.667		37	134	8.80	179.80	253
3740	4" pipe size		4	4		58.50	201	13.20	272.70	385
3746	5" pipe size		3.50	4.571		139	230	15.10	384.10	515
3750	6" pipe size	Q-16	4.50	5.333		160	278	11.75	449.75	610
3760	8" pipe size		3.50	6.857		277	355	15.10	647.10	860
3770	10" pipe size		2.50	9.600		535	500	21	1,056	1,375
3780	12" pipe size		2.25	10.667		790	555	23.50	1,368.50	1,725
3800	Tee, straight									
3810	1/2" pipe size	Q-15	9	1.778	Ea.	136	89.50	5.85	231.35	289
3820	3/4" pipe size		8.50	1.882		131	94.50	6.20	231.70	293
3830	1" pipe size		8	2		52	100	6.60	158.60	216
3840	1-1/4" pipe size		7	2.286		52	115	7.55	174.55	239
3850	1-1/2" pipe size		6	2.667		52	134	8.80	194.80	269
3860	2" pipe size		5	3.200		59.50	161	10.55	231.05	320
3870	2-1/2" pipe size		4	4		99	201	13.20	313.20	430
3880	3" pipe size		3.50	4.571		124	230	15.10	369.10	500
3890	4" pipe size		2.50	6.400		149	320	21	490	670
3896	5" pipe size		2.25	7.111		375	355	23.50	753.50	975
3900	6" pipe size	Q-16	2.25	10.667		286	555	23.50	864.50	1,175
3910	8" pipe size		2	12		545	625	26.50	1,196.50	1,575
3920	10" pipe size		1.75	13.714		845	715	30	1,590	2,050
3930	12" pipe size		1.50	16		1,225	835	35	2,095	2,650
4000	Eccentric reducer, 1-1/2" pipe size	Q-15	10	1.600		17.30	80.50	5.30	103.10	146
4010	2" pipe size		9	1.778		47.50	89.50	5.85	142.85	193
4020	2-1/2" pipe size		8	2		71.50	100	6.60	178.10	237
4030	3" pipe size		7	2.286		59	115	7.55	181.55	246
4040	4" pipe size		5	3.200		95.50	161	10.55	267.05	360
4046	5" pipe size		4.70	3.404		266	171	11.25	448.25	560
4050	6" pipe size	Q-16	4.50	5.333		259	278	11.75	548.75	720
4060	8" pipe size		3.50	6.857		390	355	15.10	760.10	985

22 11 13 – Facility Water Distribution Piping

22 11 13.47 Pipe Fittings, Steel		Crew	Daily Output	Labor-Hours	Unit	Material	Labor	Equipment	Total	Total Incl O&P
						2013 Bare Costs				
4070	10" pipe size	Q-16	2.50	9.600	Ea.	665	500	21	1,186	1,500
4080	12" pipe size		2.25	10.667		845	555	23.50	1,423.50	1,775
4090	14" pipe size		2.10	11.429		1,575	595	25	2,195	2,650
4100	16" pipe size		1.90	12.632		2,025	660	28	2,713	3,250
4151	Cap, 1-1/2" pipe size	Q-15	24	.667		24.50	33.50	2.20	60.20	80
4152	2" pipe size		18	.889		22	44.50	2.93	69.43	94
4153	2-1/2" pipe size		16	1		30	50	3.30	83.30	112
4154	3" pipe size		14	1.143		33.50	57.50	3.77	94.77	127
4155	4" pipe size		10	1.600		45	80.50	5.30	130.80	176
4156	6" pipe size	Q-16	9	2.667		92	139	5.85	236.85	315
4157	8" pipe size		7	3.429		138	179	7.55	324.55	430
4158	10" pipe size		5	4.800		212	250	10.55	472.55	620
4159	12" pipe size		4	6		283	310	13.20	606.20	795
4190	Weld fittings, reducing, standard weight									
4200	Welding ring w/spacer pins, 2" pipe size				Ea.	3.67			3.67	4.04
4210	2-1/2" pipe size					4.44			4.44	4.88
4220	3" pipe size					4.60			4.60	5.05
4230	4" pipe size					4.78			4.78	5.25
4236	5" pipe size					5.60			5.60	6.20
4240	6" pipe size					5.60			5.60	6.20
4250	8" pipe size					6.65			6.65	7.30
4260	10" pipe size					8.15			8.15	8.95
4270	12" pipe size					9.35			9.35	10.25
4280	14" pipe size					10.85			10.85	11.95
4290	16" pipe size					12.25			12.25	13.45
4300	18" pipe size					14.10			14.10	15.55
4310	20" pipe size					15			15	16.50
4330	24" pipe size					17.45			17.45	19.15
4340	26" pipe size					23.50			23.50	25.50
4350	30" pipe size					23.50			23.50	25.50
4370	36" pipe size					26			26	28.50
4600	Tee, reducing on outlet									
4601	2-1/2" x 2" pipe size	Q-15	5	3.200	Ea.	100	161	10.55	271.55	365
4602	3" x 2-1/2" pipe size		4	4		118	201	13.20	332.20	450
4604	4" x 3" pipe size		3	5.333		124	268	17.60	409.60	560
4605	5" x 4" pipe size		2.50	6.400		268	320	21	609	805
4606	6" x 5" pipe size	Q-16	3	8		370	415	17.60	802.60	1,050
4607	8" x 6" pipe size		2.50	9.600		460	500	21	981	1,275
4608	10" x 8" pipe size		2	12		850	625	26.50	1,501.50	1,900
4609	12" x 10" pipe size		1.60	15		1,250	780	33	2,063	2,575
4610	16" x 12" pipe size		1.50	16		2,000	835	35	2,870	3,500
4611	14" x 12" pipe size		1.52	15.789		1,900	820	34.50	2,754.50	3,375
4618	Reducer, concentric									
4619	2-1/2" by 2" pipe size	Q-15	10	1.600	Ea.	42.50	80.50	5.30	128.30	174
4620	3" by 2-1/2" pipe size		9	1.778		35	89.50	5.85	130.35	179
4621	3-1/2" by 3" pipe size		8	2		89.50	100	6.60	196.10	256
4622	4" by 2-1/2" pipe size		7	2.286		54.50	115	7.55	177.05	241
4623	5" by 3" pipe size		7	2.286		114	115	7.55	236.55	305
4624	6" by 4" pipe size	Q-16	6	4		94	208	8.80	310.80	430
4625	8" by 6" pipe size		5	4.800		124	250	10.55	384.55	525
4626	10" by 8" pipe size		4	6		240	310	13.20	563.20	750
4627	12" by 10" pipe size		3	8		296	415	17.60	728.60	975
4660	Reducer, eccentric									

22 11 13.47 Pipe Fittings, Steel		Crew	Daily Output	Labor-Hours	Unit	Material	2013 Bare Costs Labor	Equipment	Total	Total Incl O&P
4662	3" x 2" pipe size	Q-15	8	2	Ea.	59	100	6.60	165.60	223
4664	4" x 3" pipe size		6	2.667		71.50	134	8.80	214.30	291
4666	4" x 2" pipe size		6	2.667		87.50	134	8.80	230.30	310
4670	6" x 4" pipe size	Q-16	5	4.800		148	250	10.55	408.55	550
4672	6" x 3" pipe size		5	4.800		232	250	10.55	492.55	640
4676	8" x 6" pipe size		4	6		188	310	13.20	511.20	690
4678	8" x 4" pipe size		4	6		345	310	13.20	668.20	865
4682	10" x 8" pipe size		3	8		247	415	17.60	679.60	920
4684	10" x 6" pipe size		3	8		435	415	17.60	867.60	1,125
4688	12" x 10" pipe size		2.50	9.600		385	500	21	906	1,200
4690	12" x 8" pipe size		2.50	9.600		605	500	21	1,126	1,450
4691	14" x 12" pipe size		2.20	10.909		675	570	24	1,269	1,625
4693	16" x 14" pipe size		1.80	13.333		1,025	695	29.50	1,749.50	2,200
4694	16" x 12" pipe size		2	12		1,375	625	26.50	2,026.50	2,475
4696	18" x 16" pipe size		1.60	15		1,100	780	33	1,913	2,425
5000	Weld joint, socket, forged steel, 3000 lb., schedule 40 pipe									
5010	90° elbow, straight									
5020	1/4" pipe size	Q-15	22	.727	Ea.	23.50	36.50	2.40	62.40	83.50
5030	3/8" pipe size		22	.727		23.50	36.50	2.40	62.40	83.50
5040	1/2" pipe size		20	.800		13.20	40	2.64	55.84	78
5050	3/4" pipe size		20	.800		13.70	40	2.64	56.34	78.50
5060	1" pipe size		20	.800		17.60	40	2.64	60.24	83
5070	1-1/4" pipe size		18	.889		34	44.50	2.93	81.43	108
5080	1-1/2" pipe size		16	1		39.50	50	3.30	92.80	123
5090	2" pipe size		12	1.333		58	67	4.40	129.40	169
5100	2-1/2" pipe size		10	1.600		165	80.50	5.30	250.80	310
5110	3" pipe size		8	2		284	100	6.60	390.60	475
5120	4" pipe size		6	2.667		750	134	8.80	892.80	1,025
5130	45° Elbow, straight									
5134	1/4" pipe size	Q-15	22	.727	Ea.	24	36.50	2.40	62.90	83.50
5135	3/8" pipe size		22	.727		24	36.50	2.40	62.90	83.50
5136	1/2" pipe size		20	.800		17.55	40	2.64	60.19	83
5137	3/4" pipe size		20	.800		20.50	40	2.64	63.14	86
5140	1" pipe size		20	.800		26.50	40	2.64	69.14	93
5150	1-1/4" pipe size		18	.889		36.50	44.50	2.93	83.93	111
5160	1-1/2" pipe size		16	1		45	50	3.30	98.30	129
5170	2" pipe size		12	1.333		72	67	4.40	143.40	185
5180	2-1/2" pipe size		10	1.600		190	80.50	5.30	275.80	335
5190	3" pipe size		8	2		315	100	6.60	421.60	505
5200	4" pipe size		6	2.667		620	134	8.80	762.80	895
5250	Tee, straight									
5254	1/4" pipe size	Q-15	15	1.067	Ea.	26	53.50	3.52	83.02	113
5255	3/8" pipe size		15	1.067		26	53.50	3.52	83.02	113
5256	1/2" pipe size		13	1.231		16.35	62	4.06	82.41	115
5257	3/4" pipe size		13	1.231		19.95	62	4.06	86.01	119
5260	1" pipe size		13	1.231		27	62	4.06	93.06	127
5270	1-1/4" pipe size		12	1.333		41.50	67	4.40	112.90	151
5280	1-1/2" pipe size		11	1.455		55	73	4.80	132.80	176
5290	2" pipe size		8	2		80	100	6.60	186.60	246
5300	2-1/2" pipe size		6	2.667		229	134	8.80	371.80	465
5310	3" pipe size		5	3.200		550	161	10.55	721.55	855
5320	4" pipe size		4	4		880	201	13.20	1,094.20	1,300
5350	For reducing sizes, add					60%				

22 11 13 – Facility Water Distribution Piping

22 11 13.47 Pipe Fittings, Steel		Crew	Daily Output	Labor-Hours	Unit	Material	Labor	Equipment	Total	Total Incl O&P
						2013 Bare Costs				
5450	Couplings									
5451	1/4" pipe size	Q-15	23	.696	Ea.	15	35	2.30	52.30	71.50
5452	3/8" pipe size		23	.696		15	35	2.30	52.30	71.50
5453	1/2" pipe size		21	.762		6.95	38.50	2.51	47.96	68
5454	3/4" pipe size		21	.762		8.95	38.50	2.51	49.96	70
5460	1" pipe size		20	.800		9.90	40	2.64	52.54	74.50
5470	1-1/4" pipe size		20	.800		17.55	40	2.64	60.19	82.50
5480	1-1/2" pipe size		18	.889		19.40	44.50	2.93	66.83	91.50
5490	2" pipe size		14	1.143		31	57.50	3.77	92.27	125
5500	2-1/2" pipe size		12	1.333		69.50	67	4.40	140.90	182
5510	3" pipe size		9	1.778		156	89.50	5.85	251.35	310
5520	4" pipe size		7	2.286		234	115	7.55	356.55	440
5570	Union, 1/4" pipe size		21	.762		32	38.50	2.51	73.01	96
5571	3/8" pipe size		21	.762		32	38.50	2.51	73.01	96
5572	1/2" pipe size		19	.842		27	42.50	2.78	72.28	96
5573	3/4" pipe size		19	.842		31.50	42.50	2.78	76.78	101
5574	1" pipe size		19	.842		40.50	42.50	2.78	85.78	111
5575	1-1/4" pipe size		17	.941		65.50	47.50	3.11	116.11	146
5576	1-1/2" pipe size		15	1.067		71	53.50	3.52	128.02	162
5577	2" pipe size		11	1.455		102	73	4.80	179.80	227
5600	Reducer, 1/4" pipe size		23	.696		38.50	35	2.30	75.80	97
5601	3/8" pipe size		23	.696		40.50	35	2.30	77.80	99.50
5602	1/2" pipe size		21	.762		24.50	38.50	2.51	65.51	87.50
5603	3/4" pipe size		21	.762		24.50	38.50	2.51	65.51	87.50
5604	1" pipe size		21	.762		34.50	38.50	2.51	75.51	98.50
5605	1-1/4" pipe size		19	.842		44	42.50	2.78	89.28	115
5607	1-1/2" pipe size		17	.941		47	47.50	3.11	97.61	126
5608	2" pipe size		13	1.231		52	62	4.06	118.06	155
5612	Cap, 1/4" pipe size		46	.348		14.30	17.45	1.15	32.90	43.50
5613	3/8" pipe size		46	.348		14.30	17.45	1.15	32.90	43.50
5614	1/2" pipe size		42	.381		8.75	19.15	1.26	29.16	40
5615	3/4" pipe size		42	.381		10.20	19.15	1.26	30.61	41.50
5616	1" pipe size		42	.381		15.45	19.15	1.26	35.86	47.50
5617	1-1/4" pipe size		38	.421		18.20	21	1.39	40.59	53.50
5618	1-1/2" pipe size		34	.471		25.50	23.50	1.55	50.55	65
5619	2" pipe size		26	.615		39	31	2.03	72.03	91.50
5630	T-O-L, 1/4" pipe size, nozzle		23	.696		7.10	35	2.30	44.40	63
5631	3/8" pipe size, nozzle		23	.696		7.20	35	2.30	44.50	63
5632	1/2" pipe size, nozzle		22	.727		7.10	36.50	2.40	46	65.50
5633	3/4" pipe size, nozzle		21	.762		8.20	38.50	2.51	49.21	69.50
5634	1" pipe size, nozzle		20	.800		9.55	40	2.64	52.19	74
5635	1-1/4" pipe size, nozzle		18	.889		15.10	44.50	2.93	62.53	87
5636	1-1/2" pipe size, nozzle		16	1		15.10	50	3.30	68.40	95.50
5637	2" pipe size, nozzle		12	1.333		17.20	67	4.40	88.60	125
5638	2-1/2" pipe size, nozzle		10	1.600		57.50	80.50	5.30	143.30	190
5639	4" pipe size, nozzle		6	2.667		131	134	8.80	273.80	355
5640	W-O-L, 1/4" pipe size, nozzle		23	.696		16.80	35	2.30	54.10	73.50
5641	3/8" pipe size, nozzle		23	.696		15.85	35	2.30	53.15	72.50
5642	1/2" pipe size, nozzle		22	.727		14.70	36.50	2.40	53.60	74
5643	3/4" pipe size, nozzle		21	.762		15.50	38.50	2.51	56.51	77.50
5644	1" pipe size, nozzle		20	.800		16.20	40	2.64	58.84	81.50
5645	1-1/4" pipe size, nozzle		18	.889		19.25	44.50	2.93	66.68	91
5646	1-1/2" pipe size, nozzle	Q-15	16	1	Ea.	19.25	50	3.30	72.55	100

22 11 Facility Water Distribution

22 11 13 – Facility Water Distribution Piping

22 11 13.47 Pipe Fittings, Steel		Crew	Daily Output	Labor-Hours	Unit	2013 Bare Costs Material	Labor	Equipment	Total	Total Incl O&P
5647	2" pipe size, nozzle	Q-15	12	1.333	Ea.	19.45	67	4.40	90.85	127
5648	2-1/2" pipe size, nozzle		10	1.600		44.50	80.50	5.30	130.30	176
5649	3" pipe size, nozzle		8	2		48.50	100	6.60	155.10	212
5650	4" pipe size, nozzle		6	2.667		61.50	134	8.80	204.30	279
5651	5" pipe size, nozzle		5	3.200		152	161	10.55	323.55	420
5652	6" pipe size, nozzle		4	4		177	201	13.20	391.20	515
5653	8" pipe size, nozzle		3	5.333		340	268	17.60	625.60	800
5654	10" pipe size, nozzle		2.60	6.154		490	310	20.50	820.50	1,025
5655	12" pipe size, nozzle		2.20	7.273		930	365	24	1,319	1,600
5674	S-O-L, 1/4" pipe size, outlet		23	.696		9.45	35	2.30	46.75	65.50
5675	3/8" pipe size, outlet		23	.696		9.45	35	2.30	46.75	65.50
5676	1/2" pipe size, outlet		22	.727		8.80	36.50	2.40	47.70	67.50
5677	3/4" pipe size, outlet		21	.762		8.95	38.50	2.51	49.96	70
5678	1" pipe size, outlet		20	.800		9.90	40	2.64	52.54	74.50
5679	1-1/4" pipe size, outlet		18	.889		16.55	44.50	2.93	63.98	88.50
5680	1-1/2" pipe size, outlet		16	1		16.55	50	3.30	69.85	97.50
5681	2" pipe size, outlet	↓	12	1.333	↓	18.80	67	4.40	90.20	126
6000	Weld-on flange, forged steel									
6020	Slip-on, 150 lb. flange (welded front and back)									
6050	1/2" pipe size	Q-15	18	.889	Ea.	17.50	44.50	2.93	64.93	89.50
6060	3/4" pipe size		18	.889		17.50	44.50	2.93	64.93	89.50
6070	1" pipe size		17	.941		17.50	47.50	3.11	68.11	93.50
6080	1-1/4" pipe size		16	1		17.50	50	3.30	70.80	98.50
6090	1-1/2" pipe size		15	1.067		17.50	53.50	3.52	74.52	104
6100	2" pipe size		12	1.333		19.45	67	4.40	90.85	127
6110	2-1/2" pipe size		10	1.600		25	80.50	5.30	110.80	154
6120	3" pipe size		9	1.778		26.50	89.50	5.85	121.85	169
6130	3-1/2" pipe size		7	2.286		33	115	7.55	155.55	218
6140	4" pipe size		6	2.667		33	134	8.80	175.80	248
6150	5" pipe size	↓	5	3.200		57.50	161	10.55	229.05	315
6160	6" pipe size	Q-16	6	4		54.50	208	8.80	271.30	385
6170	8" pipe size		5	4.800		82.50	250	10.55	343.05	480
6180	10" pipe size		4	6		143	310	13.20	466.20	640
6190	12" pipe size		3	8		212	415	17.60	644.60	885
6191	14" pipe size		2.50	9.600		281	500	21	802	1,100
6192	16" pipe size	↓	1.80	13.333	↓	440	695	29.50	1,164.50	1,575
6200	300 lb. flange									
6210	1/2" pipe size	Q-15	17	.941	Ea.	23.50	47.50	3.11	74.11	100
6220	3/4" pipe size		17	.941		23.50	47.50	3.11	74.11	100
6230	1" pipe size		16	1		23.50	50	3.30	76.80	105
6240	1-1/4" pipe size		13	1.231		23.50	62	4.06	89.56	123
6250	1-1/2" pipe size		12	1.333		23.50	67	4.40	94.90	131
6260	2" pipe size		11	1.455		31	73	4.80	108.80	149
6270	2-1/2" pipe size		9	1.778		35	89.50	5.85	130.35	179
6280	3" pipe size		7	2.286		38	115	7.55	160.55	223
6290	4" pipe size		6	2.667		55.50	134	8.80	198.30	273
6300	5" pipe size	↓	4	4		95	201	13.20	309.20	425
6310	6" pipe size	Q-16	5	4.800		95.50	250	10.55	356.05	490
6320	8" pipe size		4	6		163	310	13.20	486.20	665
6330	10" pipe size		3.40	7.059		278	370	15.55	663.55	875
6340	12" pipe size	↓	2.80	8.571	↓	340	445	18.85	803.85	1,075
6400	Welding neck, 150 lb. flange									
6410	1/2" pipe size	Q-15	40	.400	Ea.	25.50	20	1.32	46.82	60

22 11 13.47 Pipe Fittings, Steel		Crew	Daily Output	Labor-Hours	Unit	Material	Labor	Equipment	Total	Total Incl O&P
						2013 Bare Costs				
6420	3/4" pipe size	Q-15	36	.444	Ea.	25.50	22.50	1.47	49.47	63
6430	1" pipe size		32	.500		25.50	25	1.65	52.15	68
6440	1-1/4" pipe size		29	.552		25.50	27.50	1.82	54.82	71.50
6450	1-1/2" pipe size		26	.615		25.50	31	2.03	58.53	76.50
6460	2" pipe size		20	.800		29.50	40	2.64	72.14	95.50
6470	2-1/2" pipe size		16	1		32	50	3.30	85.30	114
6480	3" pipe size		14	1.143		35	57.50	3.77	96.27	129
6500	4" pipe size		10	1.600		42.50	80.50	5.30	128.30	173
6510	5" pipe size		8	2		66.50	100	6.60	173.10	232
6520	6" pipe size	Q-16	10	2.400		64	125	5.30	194.30	264
6530	8" pipe size		7	3.429		113	179	7.55	299.55	400
6540	10" pipe size		6	4		180	208	8.80	396.80	525
6550	12" pipe size		5	4.800		262	250	10.55	522.55	675
6551	14" pipe size		4.50	5.333		380	278	11.75	669.75	850
6552	16" pipe size		3	8		570	415	17.60	1,002.60	1,275
6553	18" pipe size		2.50	9.600		800	500	21	1,321	1,650
6554	20" pipe size		2.30	10.435		975	545	23	1,543	1,925
6556	24" pipe size		2	12		1,300	625	26.50	1,951.50	2,425
6557	26" pipe size		1.70	14.118		1,425	735	31	2,191	2,700
6558	30" pipe size		.90	26.667		1,650	1,400	58.50	3,108.50	3,975
6559	36" pipe size		.75	32		1,900	1,675	70.50	3,645.50	4,650
6560	300 lb. flange									
6570	1/2" pipe size	Q-15	36	.444	Ea.	31	22.50	1.47	54.97	69
6580	3/4" pipe size		34	.471		31	23.50	1.55	56.05	71
6590	1" pipe size		30	.533		31	27	1.76	59.76	76.50
6600	1-1/4" pipe size		28	.571		31	28.50	1.89	61.39	79
6610	1-1/2" pipe size		24	.667		31	33.50	2.20	66.70	87
6620	2" pipe size		18	.889		38.50	44.50	2.93	85.93	112
6630	2-1/2" pipe size		14	1.143		44	57.50	3.77	105.27	139
6640	3" pipe size		12	1.333		44.50	67	4.40	115.90	155
6650	4" pipe size		8	2		73	100	6.60	179.60	238
6660	5" pipe size		7	2.286		111	115	7.55	233.55	305
6670	6" pipe size	Q-16	9	2.667		113	139	5.85	257.85	340
6680	8" pipe size		6	4		195	208	8.80	411.80	540
6690	10" pipe size		5	4.800		355	250	10.55	615.55	775
6700	12" pipe size		4	6		455	310	13.20	778.20	985
7740	Plain ends for plain end pipe, mechanically coupled									
7750	Cplg. & labor required at joints not included, add 1 per									
7760	joint for installed price, see line 9180									
7770	Malleable iron, painted, unless noted otherwise									
7800	90° Elbow 1"				Ea.	92.50			92.50	102
7810	1-1/2"					109			109	120
7820	2"					163			163	179
7830	2-1/2"					191			191	210
7840	3"					198			198	218
7860	4"					224			224	247
7870	5" welded steel					258			258	283
7880	6"					325			325	360
7890	8" welded steel					610			610	675
7900	10" welded steel					810			810	890
7910	12" welded steel					900			900	990
7970	45° Elbow 1"					55			55	60.50
7980	1-1/2"					82			82	90.50

22 11 13 – Facility Water Distribution Piping

22 11 13.47 Pipe Fittings, Steel		Crew	Daily Output	Labor-Hours	Unit	Material	Labor	Equipment	Total	Total Incl O&P
						2013 Bare Costs				
7990	2"				Ea.	173			173	190
8000	2-1/2"					173			173	190
8010	3"					198			198	218
8030	4"					206			206	227
8040	5" welded steel					258			258	283
8050	6"					295			295	325
8060	8"					345			345	380
8070	10" welded steel					370			370	410
8080	12" welded steel					670			670	740
8140	Tee, straight 1"					103			103	114
8150	1-1/2"					133			133	146
8160	2"					133			133	146
8170	2-1/2"					173			173	191
8180	3"					266			266	293
8200	4"					380			380	420
8210	5" welded steel					520			520	575
8220	6"					455			455	500
8230	8" welded steel					655			655	720
8240	10" welded steel					1,025			1,025	1,125
8250	12" welded steel				▼	1,250			1,250	1,375
8340	Segmentally welded steel, painted									
8390	Wye 2"				Ea.	166			166	182
8400	2-1/2"					166			166	182
8410	3"					185			185	204
8430	4"					277			277	305
8440	5"					335			335	370
8450	6"					465			465	510
8460	8"					600			600	660
8470	10"					885			885	975
8480	12"					1,375			1,375	1,500
8540	Wye, lateral 2"					180			180	198
8550	2-1/2"					207			207	227
8560	3"					245			245	270
8580	4"					340			340	370
8590	5"					575			575	630
8600	6"					590			590	650
8610	8"					995			995	1,100
8620	10"					1,025			1,025	1,125
8630	12"					1,900			1,900	2,075
8690	Cross, 2"					175			175	193
8700	2-1/2"					175			175	193
8710	3"					209			209	230
8730	4"					286			286	315
8740	5"					405			405	445
8750	6"					535			535	590
8760	8"					685			685	755
8770	10"					1,000			1,000	1,100
8780	12"					1,450			1,450	1,600
8800	Tees, reducing 2" x 1"					113			113	125
8810	2" x 1-1/2"					113			113	125
8820	3" x 1"					113			113	125
8830	3" x 1-1/2"					114			114	125
8840	3" x 2"					115			115	126

22 11 13 – Facility Water Distribution Piping

22 11 13.47 Pipe Fittings, Steel		Crew	Daily Output	Labor-Hours	Unit	Material	2013 Bare Costs Labor	Equipment	Total	Total Incl O&P
8850	4" x 1"				Ea.	167			167	184
8860	4" x 1-1/2"					167			167	184
8870	4" x 2"					165			165	181
8880	4" x 2-1/2"					170			170	187
8890	4" x 3"					170			170	187
8900	6" x 2"					247			247	272
8910	6" x 3"					266			266	293
8920	6" x 4"					325			325	355
8930	8" x 2"					335			335	370
8940	8" x 3"					335			335	365
8950	8" x 4"					335			335	365
8960	8" x 5"					355			355	390
8970	8" x 6"					390			390	430
8980	10" x 4"					510			510	560
8990	10" x 6"					510			510	560
9000	10" x 8"					540			540	595
9010	12" x 6"					785			785	865
9020	12" x 8"					790			790	865
9030	12" x 10"				↓	815			815	900
9080	Adapter nipples 3" long									
9090	1"				Ea.	15.80			15.80	17.40
9100	1-1/2"					15.80			15.80	17.40
9110	2"					15.80			15.80	17.40
9120	2-1/2"					18.20			18.20	20
9130	3"					22			22	24.50
9140	4"					36.50			36.50	40
9150	6"				↓	96.50			96.50	106
9180	Coupling, mechanical, plain end pipe to plain end pipe or fitting									
9190	1"	Q-1	29	.552	Ea.	55.50	27.50		83	103
9200	1-1/2"		28	.571		55.50	28.50		84	104
9210	2"		27	.593		55.50	30		85.50	106
9220	2-1/2"		26	.615		55.50	31		86.50	108
9230	3"		25	.640		81.50	32		113.50	138
9240	3-1/2"		24	.667		94	33.50		127.50	155
9250	4"	↓	22	.727		94	36.50		130.50	159
9260	5"	Q-2	28	.857		134	44.50		178.50	215
9270	6"		24	1		164	52		216	259
9280	8"		19	1.263		289	66		355	420
9290	10"		16	1.500		375	78		453	535
9300	12"	↓	12	2	↓	470	104		574	675
9310	Outlets for precut holes through pipe wall									
9331	Strapless type, with gasket									
9332	4" to 8" pipe x 1/2"	1 Plum	13	.615	Ea.	60.50	34.50		95	118
9333	4" to 8" pipe x 3/4"		13	.615		65	34.50		99.50	123
9334	10" pipe and larger x 1/2"		11	.727		60.50	40.50		101	128
9335	10" pipe and larger x 3/4"	↓	11	.727	↓	65	40.50		105.50	133
9341	Thermometer wells with gasket									
9342	4" to 8" pipe, 6" stem	1 Plum	14	.571	Ea.	91.50	32		123.50	149
9343	8" pipe and larger, 6" stem	"	13	.615	"	91.50	34.50		126	153
9940	For galvanized fittings for plain end pipe, add					20%				

22 11 13.48 Pipe, Fittings and Valves, Steel, Grooved-Joint		Crew	Daily Output	Labor-Hours	Unit	2013 Bare Costs Material	Labor	Equipment	Total	Total Incl O&P
0010	**PIPE, FITTINGS AND VALVES, STEEL, GROOVED-JOINT** R221113-70									
0012	Fittings are ductile iron. Steel fittings noted.									
0020	Pipe includes coupling & clevis type hanger assemblies, 10′ O.C.									
0500	Schedule 10, black									
0550	2″ diameter	1 Plum	43	.186	L.F.	5.90	10.40		16.30	22
0560	2-1/2″ diameter	Q-1	61	.262		7.60	13.20		20.80	28
0570	3″ diameter		55	.291		8.85	14.60		23.45	32
0580	3-1/2″ diameter		53	.302		11.25	15.15		26.40	35.50
0590	4″ diameter		49	.327		11.90	16.40		28.30	37.50
0600	5″ diameter		40	.400		14.25	20		34.25	46
0610	6″ diameter	Q-2	46	.522		18.80	27		45.80	61.50
0620	8″ diameter	″	41	.585		26	30.50		56.50	74.50
0700	To delete couplings & hangers, subtract									
0710	2″ diam. to 5″ diam.					25%	20%			
0720	6″ diam. to 8″ diam.					27%	15%			
1000	Schedule 40, black									
1040	3/4″ diameter	1 Plum	71	.113	L.F.	4.67	6.30		10.97	14.60
1050	1″ diameter		63	.127		4.46	7.10		11.56	15.55
1060	1-1/4″ diameter		58	.138		5.75	7.70		13.45	17.95
1070	1-1/2″ diameter		51	.157		6.55	8.75		15.30	20.50
1080	2″ diameter		40	.200		8.05	11.15		19.20	25.50
1090	2-1/2″ diameter	Q-1	57	.281		13.95	14.10		28.05	36.50
1100	3″ diameter		50	.320		17.30	16.05		33.35	43
1110	4″ diameter		45	.356		23.50	17.85		41.35	53
1120	5″ diameter		37	.432		36.50	21.50		58	73
1130	6″ diameter	Q-2	42	.571		40.50	30		70.50	89.50
1140	8″ diameter		37	.649		61.50	34		95.50	119
1150	10″ diameter		31	.774		77	40.50		117.50	146
1160	12″ diameter		27	.889		89	46.50		135.50	168
1170	14″ diameter		20	1.200		114	62.50		176.50	219
1180	16″ diameter		17	1.412		138	73.50		211.50	263
1190	18″ diameter		14	1.714		160	89.50		249.50	310
1200	20″ diameter		12	2		201	104		305	380
1210	24″ diameter		10	2.400		226	125		351	435
1740	To delete coupling & hanger, subtract									
1750	3/4″ diam. to 2″ diam.					65%	27%			
1760	2-1/2″ diam. to 5″ diam.					41%	18%			
1770	6″ diam. to 12″ diam.					31%	13%			
1780	14″ diam. to 24″ diam.					35%	10%			
1800	Galvanized									
1840	3/4″ diameter	1 Plum	71	.113	L.F.	6.05	6.30		12.35	16.10
1850	1″ diameter		63	.127		6.05	7.10		13.15	17.35
1860	1-1/4″ diameter		58	.138		7.95	7.70		15.65	20.50
1870	1-1/2″ diameter		51	.157		9.15	8.75		17.90	23.50
1880	2″ diameter		40	.200		11.60	11.15		22.75	29.50
1890	2-1/2″ diameter	Q-1	57	.281		19.25	14.10		33.35	42
1900	3″ diameter		50	.320		24.50	16.05		40.55	51
1910	4″ diameter		45	.356		33.50	17.85		51.35	64
1920	5″ diameter		37	.432		58.50	21.50		80	97
1930	6″ diameter	Q-2	42	.571		66.50	30		96.50	118
1940	8″ diameter		37	.649		78	34		112	137
1950	10″ diameter		31	.774		125	40.50		165.50	198

22 11 13 – Facility Water Distribution Piping

22 11 13.48 Pipe, Fittings and Valves, Steel, Grooved-Joint		Crew	Daily Output	Labor-Hours	Unit	2013 Bare Costs Material	Labor	Equipment	Total	Total Incl O&P
1960	12″ diameter	Q-2	27	.889	L.F.	153	46.50		199.50	238
2540	To delete coupling & hanger, subtract									
2550	3/4″ diam. to 2″ diam.					36%	27%			
2560	2-1/2″ diam. to 5″ diam.					19%	18%			
2570	6″ diam. to 12″ diam.					14%	13%			
2600	Schedule 80, black									
2610	3/4″ diameter	1 Plum	65	.123	L.F.	6.10	6.85		12.95	17.05
2650	1″ diameter		61	.131		7.60	7.30		14.90	19.40
2660	1-1/4″ diameter		55	.145		9.90	8.10		18	23
2670	1-1/2″ diameter		49	.163		11.55	9.10		20.65	26.50
2680	2″ diameter		38	.211		14.95	11.75		26.70	34
2690	2-1/2″ diameter	Q-1	54	.296		22.50	14.90		37.40	47
2700	3″ diameter		48	.333		29.50	16.75		46.25	57
2710	4″ diameter		44	.364		33.50	18.25		51.75	64.50
2720	5″ diameter		35	.457		53.50	23		76.50	93
2730	6″ diameter	Q-2	40	.600		72.50	31.50		104	127
2740	8″ diameter		35	.686		110	35.50		145.50	175
2750	10″ diameter		29	.828		200	43		243	285
2760	12″ diameter		24	1		395	52		447	510
3240	To delete coupling & hanger, subtract									
3250	3/4″ diam. to 2″ diam.					30%	25%			
3260	2-1/2″ diam. to 5″ diam.					14%	17%			
3270	6″ diam. to 12″ diam.					12%	12%			
3300	Galvanized									
3310	3/4″ diameter	1 Plum	65	.123	L.F.	7.60	6.85		14.45	18.70
3350	1″ diameter		61	.131		9.85	7.30		17.15	22
3360	1-1/4″ diameter		55	.145		13	8.10		21.10	26.50
3370	1-1/2″ diameter		46	.174		15.25	9.70		24.95	31.50
3380	2″ diameter		38	.211		20	11.75		31.75	40
3390	2-1/2″ diameter	Q-1	54	.296		30.50	14.90		45.40	56
3400	3″ diameter		48	.333		40.50	16.75		57.25	69.50
3410	4″ diameter		44	.364		56.50	18.25		74.75	90
3420	5″ diameter		35	.457		68	23		91	110
3430	6″ diameter	Q-2	40	.600		94	31.50		125.50	150
3440	8″ diameter		35	.686		142	35.50		177.50	210
3450	10″ diameter		29	.828		260	43		303	350
3460	12″ diameter		24	1		520	52		572	650
3920	To delete coupling & hanger, subtract									
3930	3/4″ diam. to 2″ diam.					30%	25%			
3940	2-1/2″ diam. to 5″ diam.					15%	17%			
3950	6″ diam. to 12″ diam.					11%	12%			
3990	Fittings: coupling material required at joints not incl. in fitting price.									
3994	Add 1 selected coupling, material only, per joint for installed price.									
4000	Elbow, 90° or 45°, painted									
4030	3/4″ diameter	1 Plum	50	.160	Ea.	49	8.95		57.95	67.50
4040	1″ diameter		50	.160		26	8.95		34.95	42.50
4050	1-1/4″ diameter		40	.200		26	11.15		37.15	46
4060	1-1/2″ diameter		33	.242		26	13.55		39.55	49.50
4070	2″ diameter		25	.320		26	17.85		43.85	56
4080	2-1/2″ diameter	Q-1	40	.400		26	20		46	59.50
4090	3″ diameter		33	.485		46.50	24.50		71	87.50
4100	4″ diameter		25	.640		50.50	32		82.50	104
4110	5″ diameter		20	.800		120	40		160	193

22 11 13.48 Pipe, Fittings and Valves, Steel, Grooved-Joint		Crew	Daily Output	Labor-Hours	Unit	Material	Labor	Equipment	Total	Total Incl O&P
						2013 Bare Costs				
4120	6″ diameter	Q-2	25	.960	Ea.	141	50		191	232
4130	8″ diameter		21	1.143		295	59.50		354.50	415
4140	10″ diameter		18	1.333		535	69.50		604.50	695
4150	12″ diameter		15	1.600		855	83.50		938.50	1,075
4170	14″ diameter		12	2		860	104		964	1,100
4180	16″ diameter		11	2.182		1,125	114		1,239	1,400
4190	18″ diameter	Q-3	15	2.133		1,425	113		1,538	1,725
4200	20″ diameter		13	2.462		1,875	131		2,006	2,250
4210	24″ diameter		11	2.909		2,700	155		2,855	3,200
4250	For galvanized elbows, add					26%				
4690	Tee, painted									
4700	3/4″ diameter	1 Plum	38	.211	Ea.	53	11.75		64.75	75.50
4740	1″ diameter		33	.242		40.50	13.55		54.05	65
4750	1-1/4″ diameter		27	.296		40.50	16.55		57.05	69.50
4760	1-1/2″ diameter		22	.364		40.50	20.50		61	75
4770	2″ diameter		17	.471		40.50	26.50		67	84
4780	2-1/2″ diameter	Q-1	27	.593		40.50	30		70.50	89.50
4790	3″ diameter		22	.727		56	36.50		92.50	117
4800	4″ diameter		17	.941		85	47.50		132.50	165
4810	5″ diameter		13	1.231		198	62		260	310
4820	6″ diameter	Q-2	17	1.412		229	73.50		302.50	365
4830	8″ diameter		14	1.714		500	89.50		589.50	685
4840	10″ diameter		12	2		1,050	104		1,154	1,300
4850	12″ diameter		10	2.400		1,450	125		1,575	1,800
4851	14″ diameter		9	2.667		1,500	139		1,639	1,875
4852	16″ diameter		8	3		1,550	156		1,706	1,925
4853	18″ diameter	Q-3	11	2.909		1,625	155		1,780	2,025
4854	20″ diameter		10	3.200		2,325	170		2,495	2,825
4855	24″ diameter		8	4		3,550	213		3,763	4,225
4900	For galvanized tees, add					24%				
4906	Couplings, rigid style, painted									
4908	1″ diameter	1 Plum	100	.080	Ea.	19.80	4.46		24.26	28.50
4909	1-1/4″ diameter		100	.080		19.80	4.46		24.26	28.50
4910	1-1/2″ diameter		67	.119		19.80	6.65		26.45	32
4912	2″ diameter		50	.160		22	8.95		30.95	37.50
4914	2-1/2″ diameter	Q-1	80	.200		25	10.05		35.05	42.50
4916	3″ diameter		67	.239		29	12		41	50
4918	4″ diameter		50	.320		41	16.05		57.05	69
4920	5″ diameter		40	.400		53	20		73	88.50
4922	6″ diameter	Q-2	50	.480		69.50	25		94.50	114
4924	8″ diameter		42	.571		110	30		140	166
4926	10″ diameter		35	.686		181	35.50		216.50	253
4928	12″ diameter		32	.750		202	39		241	282
4930	14″ diameter		24	1		225	52		277	325
4931	16″ diameter		20	1.200		293	62.50		355.50	420
4932	18″ diameter		18	1.333		340	69.50		409.50	480
4933	20″ diameter		16	1.500		465	78		543	630
4934	24″ diameter		13	1.846		595	96		691	800
4940	Flexible, standard, painted									
4950	3/4″ diameter	1 Plum	100	.080	Ea.	14.40	4.46		18.86	22.50
4960	1″ diameter		100	.080		14.40	4.46		18.86	22.50
4970	1-1/4″ diameter		80	.100		19	5.60		24.60	29.50
4980	1-1/2″ diameter		67	.119		20.50	6.65		27.15	32.50

22 11 Facility Water Distribution

22 11 13 – Facility Water Distribution Piping

22 11 13.48 Pipe, Fittings and Valves, Steel, Grooved-Joint		Crew	Daily Output	Labor-Hours	Unit	Material	2013 Bare Costs Labor	Equipment	Total	Total Incl O&P
4990	2" diameter	1 Plum	50	.160	Ea.	22	8.95		30.95	38
5000	2-1/2" diameter	Q-1	80	.200		26	10.05		36.05	43.50
5010	3" diameter		67	.239		29	12		41	49.50
5020	3-1/2" diameter		57	.281		41	14.10		55.10	66
5030	4" diameter		50	.320		41.50	16.05		57.55	69.50
5040	5" diameter		40	.400		63	20		83	99.50
5050	6" diameter	Q-2	50	.480		74.50	25		99.50	120
5070	8" diameter		42	.571		121	30		151	178
5090	10" diameter		35	.686		197	35.50		232.50	271
5110	12" diameter		32	.750		224	39		263	305
5120	14" diameter		24	1		315	52		367	425
5130	16" diameter		20	1.200		410	62.50		472.50	545
5140	18" diameter		18	1.333		480	69.50		549.50	635
5150	20" diameter		16	1.500		755	78		833	950
5160	24" diameter		13	1.846		830	96		926	1,050
5176	Lightweight style, painted									
5178	1-1/2" diameter	1 Plum	67	.119	Ea.	18.20	6.65		24.85	30
5180	2" diameter	"	50	.160		22	8.95		30.95	38
5182	2-1/2" diameter	Q-1	80	.200		26	10.05		36.05	43.50
5184	3" diameter		67	.239		29	12		41	49.50
5186	3-1/2" diameter		57	.281		35	14.10		49.10	59.50
5188	4" diameter		50	.320		41.50	16.05		57.55	69.50
5190	5" diameter		40	.400		63	20		83	99.50
5192	6" diameter	Q-2	50	.480		74.50	25		99.50	120
5194	8" diameter		42	.571		121	30		151	178
5196	10" diameter		35	.686		245	35.50		280.50	325
5198	12" diameter		32	.750		274	39		313	360
5200	For galvanized couplings, add					33%				
5220	Tee, reducing, painted									
5225	2" x 1-1/2" diameter	Q-1	38	.421	Ea.	86	21		107	127
5226	2-1/2" x 2" diameter		28	.571		86	28.50		114.50	138
5227	3" x 2-1/2" diameter		23	.696		76	35		111	136
5228	4" x 3" diameter		18	.889		102	44.50		146.50	180
5229	5" x 4" diameter		15	1.067		219	53.50		272.50	320
5230	6" x 4" diameter	Q-2	18	1.333		241	69.50		310.50	370
5231	8" x 6" diameter		15	1.600		500	83.50		583.50	675
5232	10" x 8" diameter		13	1.846		660	96		756	870
5233	12" x 10" diameter		11	2.182		1,000	114		1,114	1,275
5234	14" x 12" diameter		10	2.400		840	125		965	1,125
5235	16" x 12" diameter		9	2.667		1,050	139		1,189	1,350
5236	18" x 12" diameter	Q-3	12	2.667		1,250	142		1,392	1,600
5237	18" x 16" diameter		11	2.909		1,575	155		1,730	1,950
5238	20" x 16" diameter		10	3.200		2,050	170		2,220	2,525
5239	24" x 20" diameter		9	3.556		3,250	189		3,439	3,850
5240	Reducer, concentric, painted									
5241	2-1/2" x 2" diameter	Q-1	43	.372	Ea.	30.50	18.70		49.20	61.50
5242	3" x 2-1/2" diameter		35	.457		37	23		60	75
5243	4" x 3" diameter		29	.552		44.50	27.50		72	90.50
5244	5" x 4" diameter		22	.727		61.50	36.50		98	123
5245	6" x 4" diameter	Q-2	26	.923		71.50	48		119.50	151
5246	8" x 6" diameter		23	1.043		185	54.50		239.50	285
5247	10" x 8" diameter		20	1.200		375	62.50		437.50	510
5248	12" x 10" diameter		16	1.500		675	78		753	860

22 11 13.48	Pipe, Fittings and Valves, Steel, Grooved-Joint	Crew	Daily Output	Labor-Hours	Unit	2013 Bare Costs Material	Labor	Equipment	Total	Total Incl O&P
5255	Eccentric, painted									
5256	2-1/2″ x 2″ diameter	Q-1	42	.381	Ea.	64	19.15		83.15	99.50
5257	3″ x 2-1/2″ diameter		34	.471		73	23.50		96.50	116
5258	4″ x 3″ diameter		28	.571		89	28.50		117.50	141
5259	5″ x 4″ diameter		21	.762		121	38.50		159.50	191
5260	6″ x 4″ diameter	Q-2	25	.960		141	50		191	231
5261	8″ x 6″ diameter		22	1.091		285	57		342	400
5262	10″ x 8″ diameter		19	1.263		780	66		846	960
5263	12″ x 10″ diameter		15	1.600		1,075	83.50		1,158.50	1,300
5270	Coupling, reducing, painted									
5272	2″ x 1-1/2″ diameter	1 Plum	52	.154	Ea.	29.50	8.60		38.10	45.50
5274	2-1/2″ x 2″ diameter	Q-1	82	.195		38	9.80		47.80	57
5276	3″ x 2″ diameter		69	.232		43.50	11.65		55.15	65.50
5278	4″ x 2″ diameter		52	.308		69	15.45		84.45	99
5280	5″ x 4″ diameter		42	.381		77.50	19.15		96.65	115
5282	6″ x 4″ diameter	Q-2	52	.462		117	24		141	165
5284	8″ x 6″ diameter	″	44	.545		175	28.50		203.50	236
5290	Outlet coupling, painted									
5294	1-1/2″ x 1″ pipe size	1 Plum	65	.123	Ea.	39	6.85		45.85	53.50
5296	2″ x 1″ pipe size	″	48	.167		40	9.30		49.30	58
5298	2-1/2″ x 1″ pipe size	Q-1	78	.205		61.50	10.30		71.80	83
5300	2-1/2″ x 1-1/4″ pipe size	1 Plum	70	.114		69	6.40		75.40	85.50
5302	3″ x 1″ pipe size	Q-1	65	.246		78	12.35		90.35	105
5304	4″ x 3/4″ pipe size		48	.333		87	16.75		103.75	121
5306	4″ x 1-1/2″ pipe size		46	.348		123	17.45		140.45	163
5308	6″ x 1-1/2″ pipe size	Q-2	44	.545		174	28.50		202.50	234
5750	Flange, w/groove gasket, black steel									
5754	See Line 22 11 13.47 0620 for gasket & bolt set									
5760	ANSI class 125 and 150, painted									
5780	2″ pipe size	1 Plum	23	.348	Ea.	87	19.40		106.40	125
5790	2-1/2″ pipe size	Q-1	37	.432		108	21.50		129.50	152
5800	3″ pipe size		31	.516		117	26		143	167
5820	4″ pipe size		23	.696		156	35		191	224
5830	5″ pipe size		19	.842		181	42.50		223.50	263
5840	6″ pipe size	Q-2	23	1.043		197	54.50		251.50	299
5850	8″ pipe size		17	1.412		223	73.50		296.50	355
5860	10″ pipe size		14	1.714		350	89.50		439.50	525
5870	12″ pipe size		12	2		460	104		564	660
5880	14″ pipe size		10	2.400		900	125		1,025	1,175
5890	16″ pipe size		9	2.667		1,050	139		1,189	1,350
5900	18″ pipe size		6	4		1,300	208		1,508	1,750
5910	20″ pipe size		5	4.800		1,550	250		1,800	2,075
5920	24″ pipe size		4.50	5.333		1,975	278		2,253	2,600
5940	ANSI class 350, painted									
5946	2″ pipe size	1 Plum	23	.348	Ea.	109	19.40		128.40	149
5948	2-1/2″ pipe size	Q-1	37	.432		126	21.50		147.50	172
5950	3″ pipe size		31	.516		172	26		198	229
5952	4″ pipe size		23	.696		229	35		264	305
5954	5″ pipe size		19	.842		261	42.50		303.50	350
5956	6″ pipe size	Q-2	23	1.043		305	54.50		359.50	415
5958	8″ pipe size		17	1.412		350	73.50		423.50	495
5960	10″ pipe size		14	1.714		555	89.50		644.50	745
5962	12″ pipe size	1 Plum	12	.667		590	37		627	705

22 11 13 – Facility Water Distribution Piping

22 11 13.48 Pipe, Fittings and Valves, Steel, Grooved-Joint		Crew	Daily Output	Labor-Hours	Unit	2013 Bare Costs Material	Labor	Equipment	Total	Total Incl O&P
6100	Cross, painted									
6110	2" diameter	1 Plum	12.50	.640	Ea.	72.50	35.50		108	134
6112	2-1/2" diameter	Q-1	20	.800		72.50	40		112.50	141
6114	3" diameter		15.50	1.032		129	52		181	220
6116	4" diameter		12.50	1.280		215	64.50		279.50	335
6118	6" diameter	Q-2	12.50	1.920		560	100		660	770
6120	8" diameter		10.50	2.286		740	119		859	995
6122	10" diameter		9	2.667		1,250	139		1,389	1,600
6124	12" diameter		7.50	3.200		1,825	167		1,992	2,250
7400	Suction diffuser									
7402	Grooved end inlet x flanged outlet									
7410	3" x 3"	Q-1	50	.320	Ea.	705	16.05		721.05	805
7412	4" x 4"		38	.421		960	21		981	1,075
7414	5" x 5"		30	.533		1,125	27		1,152	1,275
7416	6" x 6"	Q-2	38	.632		1,425	33		1,458	1,600
7418	8" x 8"		27	.889		2,625	46.50		2,671.50	2,975
7420	10" x 10"		20	1.200		3,575	62.50		3,637.50	4,025
7422	12" x 12"		16	1.500		5,900	78		5,978	6,600
7424	14" x 14"		15	1.600		5,475	83.50		5,558.50	6,150
7426	16" x 14"		14	1.714		6,850	89.50		6,939.50	7,650
7500	Strainer, tee type, painted									
7506	2" pipe size	1 Plum	38	.211	Ea.	470	11.75		481.75	540
7508	2-1/2" pipe size	Q-1	62	.258		495	12.95		507.95	565
7510	3" pipe size		50	.320		555	16.05		571.05	635
7512	4" pipe size		38	.421		630	21		651	725
7514	5" pipe size		30	.533		910	27		937	1,050
7516	6" pipe size	Q-2	38	.632		990	33		1,023	1,150
7518	8" pipe size		27	.889		1,525	46.50		1,571.50	1,750
7520	10" pipe size		20	1.200		2,225	62.50		2,287.50	2,550
7522	12" pipe size		16	1.500		2,875	78		2,953	3,275
7524	14" pipe size		15	1.600		8,975	83.50		9,058.50	10,000
7526	16" pipe size		14	1.714		11,200	89.50		11,289.50	12,400
7570	Expansion joint, max. 3" travel									
7572	2" diameter	1 Plum	38	.211	Ea.	595	11.75		606.75	675
7574	3" diameter	Q-1	50	.320		685	16.05		701.05	775
7576	4" diameter	"	38	.421		900	21		921	1,025
7578	6" diameter	Q-2	38	.632		1,400	33		1,433	1,600
7800	Ball valve w/handle, carbon steel trim									
7810	1-1/2" pipe size	1 Plum	50	.160	Ea.	135	8.95		143.95	161
7812	2" pipe size	"	38	.211		149	11.75		160.75	182
7814	2-1/2" pipe size	Q-1	62	.258		325	12.95		337.95	380
7816	3" pipe size		50	.320		525	16.05		541.05	605
7818	4" pipe size		38	.421		815	21		836	925
7820	6" pipe size	Q-2	30	.800		2,500	41.50		2,541.50	2,825
7830	With gear operator									
7834	2-1/2" pipe size	Q-1	62	.258	Ea.	660	12.95		672.95	745
7836	3" pipe size		50	.320		935	16.05		951.05	1,050
7838	4" pipe size		38	.421		1,200	21		1,221	1,325
7840	6" pipe size	Q-2	30	.800		2,800	41.50		2,841.50	3,150
7870	Check valve									
7874	2-1/2" pipe size	Q-1	62	.258	Ea.	225	12.95		237.95	268
7876	3" pipe size		50	.320		266	16.05		282.05	315
7878	4" pipe size		38	.421		281	21		302	340

22 11 13 – Facility Water Distribution Piping

22 11 13.48 Pipe, Fittings and Valves, Steel, Grooved-Joint		Crew	Daily Output	Labor-Hours	Unit	Material	2013 Bare Costs Labor	Equipment	Total	Total Incl O&P
7880	5″ pipe size	Q-1	30	.533	Ea.	470	27		497	555
7882	6″ pipe size	Q-2	38	.632		555	33		588	660
7884	8″ pipe size		27	.889		755	46.50		801.50	905
7886	10″ pipe size		20	1.200		2,200	62.50		2,262.50	2,500
7888	12″ pipe size		16	1.500		2,600	78		2,678	2,975
7900	Plug valve, balancing, w/lever operator									
7906	3″ pipe size	Q-1	50	.320	Ea.	560	16.05		576.05	640
7908	4″ pipe size	″	38	.421		685	21		706	785
7909	6″ pipe size	Q-2	30	.800		1,050	41.50		1,091.50	1,225
7916	With gear operator									
7920	3″ pipe size	Q-1	50	.320	Ea.	1,050	16.05		1,066.05	1,200
7922	4″ pipe size	″	38	.421		1,100	21		1,121	1,225
7926	8″ pipe size	Q-2	27	.889		2,275	46.50		2,321.50	2,575
7928	10″ pipe size		20	1.200		3,100	62.50		3,162.50	3,525
7930	12″ pipe size		16	1.500		4,725	78		4,803	5,325
8000	Butterfly valve, 2 position handle, with standard trim									
8010	1-1/2″ pipe size	1 Plum	50	.160	Ea.	212	8.95		220.95	246
8020	2″ pipe size	″	38	.211		212	11.75		223.75	251
8030	3″ pipe size	Q-1	50	.320		305	16.05		321.05	360
8050	4″ pipe size	″	38	.421		335	21		356	400
8070	6″ pipe size	Q-2	38	.632		675	33		708	795
8080	8″ pipe size		27	.889		910	46.50		956.50	1,075
8090	10″ pipe size		20	1.200		1,600	62.50		1,662.50	1,875
8200	With stainless steel trim									
8240	1-1/2″ pipe size	1 Plum	50	.160	Ea.	269	8.95		277.95	310
8250	2″ pipe size	″	38	.211		269	11.75		280.75	315
8270	3″ pipe size	Q-1	50	.320		360	16.05		376.05	420
8280	4″ pipe size	″	38	.421		390	21		411	460
8300	6″ pipe size	Q-2	38	.632		730	33		763	855
8310	8″ pipe size		27	.889		1,800	46.50		1,846.50	2,075
8320	10″ pipe size		20	1.200		3,275	62.50		3,337.50	3,700
8322	12″ pipe size		16	1.500		3,750	78		3,828	4,250
8324	14″ pipe size		15	1.600		4,450	83.50		4,533.50	5,025
8326	16″ pipe size		14	1.714		6,425	89.50		6,514.50	7,200
8328	18″ pipe size	Q-3	12	2.667		7,900	142		8,042	8,925
8330	20″ pipe size		11	2.909		9,725	155		9,880	10,900
8332	24″ pipe size		10	3.200		13,000	170		13,170	14,600
8336	Note: sizes 8″ up w/manual gear operator									
9000	Cut one groove, labor									
9010	3/4″ pipe size	Q-1	152	.105	Ea.		5.30		5.30	7.95
9020	1″ pipe size		140	.114			5.75		5.75	8.65
9030	1-1/4″ pipe size		124	.129			6.50		6.50	9.75
9040	1-1/2″ pipe size		114	.140			7.05		7.05	10.60
9050	2″ pipe size		104	.154			7.75		7.75	11.65
9060	2-1/2″ pipe size		96	.167			8.35		8.35	12.60
9070	3″ pipe size		88	.182			9.15		9.15	13.75
9080	3-1/2″ pipe size		83	.193			9.70		9.70	14.60
9090	4″ pipe size		78	.205			10.30		10.30	15.50
9100	5″ pipe size		72	.222			11.15		11.15	16.80
9110	6″ pipe size		70	.229			11.50		11.50	17.30
9120	8″ pipe size		54	.296			14.90		14.90	22.50
9130	10″ pipe size		38	.421			21		21	32
9140	12″ pipe size		30	.533			27		27	40.50

22 11 13.48 Pipe, Fittings and Valves, Steel, Grooved-Joint		Crew	Daily Output	Labor-Hours	Unit	Material	2013 Bare Costs Labor	Equipment	Total	Total Incl O&P
9150	14" pipe size	Q-1	20	.800	Ea.		40		40	60.50
9160	16" pipe size		19	.842			42.50		42.50	63.50
9170	18" pipe size		18	.889			44.50		44.50	67
9180	20" pipe size		17	.941			47.50		47.50	71
9190	24" pipe size		15	1.067			53.50		53.50	80.50
9210	Roll one groove									
9220	3/4" pipe size	Q-1	266	.060	Ea.		3.02		3.02	4.55
9230	1" pipe size		228	.070			3.53		3.53	5.30
9240	1-1/4" pipe size		200	.080			4.02		4.02	6.05
9250	1-1/2" pipe size		178	.090			4.52		4.52	6.80
9260	2" pipe size		116	.138			6.95		6.95	10.45
9270	2-1/2" pipe size		110	.145			7.30		7.30	11
9280	3" pipe size		100	.160			8.05		8.05	12.10
9290	3-1/2" pipe size		94	.170			8.55		8.55	12.90
9300	4" pipe size		86	.186			9.35		9.35	14.05
9310	5" pipe size		84	.190			9.55		9.55	14.40
9320	6" pipe size		80	.200			10.05		10.05	15.15
9330	8" pipe size		66	.242			12.20		12.20	18.35
9340	10" pipe size		58	.276			13.85		13.85	21
9350	12" pipe size		46	.348			17.45		17.45	26.50
9360	14" pipe size		30	.533			27		27	40.50
9370	16" pipe size		28	.571			28.50		28.50	43
9380	18" pipe size		27	.593			30		30	45
9390	20" pipe size		25	.640			32		32	48.50
9400	24" pipe size		23	.696			35		35	52.50

22 11 13.60 Tubing, Stainless Steel		Crew	Daily Output	Labor-Hours	Unit	Material	Labor	Equipment	Total	Total Incl O&P
0010	**TUBING, STAINLESS STEEL**									
5010	Type 304, no joints, no hangers									
5020	.035 wall									
5021	1/4"	1 Plum	160	.050	L.F.	2.46	2.79		5.25	6.90
5022	3/8"		160	.050		3.10	2.79		5.89	7.60
5023	1/2"		160	.050		3.99	2.79		6.78	8.60
5024	5/8"		160	.050		4.89	2.79		7.68	9.60
5025	3/4"		133	.060		5.90	3.36		9.26	11.55
5026	7/8"		133	.060		7	3.36		10.36	12.75
5027	1"		114	.070		6.85	3.92		10.77	13.40
5040	.049 wall									
5041	1/4"	1 Plum	160	.050	L.F.	2.88	2.79		5.67	7.35
5042	3/8"		160	.050		3.99	2.79		6.78	8.60
5043	1/2"		160	.050		4.42	2.79		7.21	9.05
5044	5/8"		160	.050		6	2.79		8.79	10.80
5045	3/4"		133	.060		6.35	3.36		9.71	12.05
5046	7/8"		133	.060		7.45	3.36		10.81	13.25
5047	1"		114	.070		7.95	3.92		11.87	14.60
5060	.065 wall									
5061	1/4"	1 Plum	160	.050	L.F.	3.23	2.79		6.02	7.75
5062	3/8"		160	.050		4.46	2.79		7.25	9.10
5063	1/2"		160	.050		5.40	2.79		8.19	10.15
5064	5/8"		160	.050		6.55	2.79		9.34	11.40
5065	3/4"		133	.060		6.95	3.36		10.31	12.70
5066	7/8"		133	.060		9	3.36		12.36	14.95
5067	1"		114	.070		8.80	3.92		12.72	15.55

22 11 Facility Water Distribution

22 11 13 – Facility Water Distribution Piping

22 11 13.60 Tubing, Stainless Steel		Crew	Daily Output	Labor-Hours	Unit	2013 Bare Costs Material	Labor	Equipment	Total	Total Incl O&P
5210	Type 316									
5220	.035 wall									
5221	1/4"	1 Plum	160	.050	L.F.	3.71	2.79		6.50	8.30
5222	3/8"		160	.050		4.74	2.79		7.53	9.40
5223	1/2"		160	.050		6.45	2.79		9.24	11.30
5224	5/8"		160	.050		10.20	2.79		12.99	15.40
5225	3/4"		133	.060		8.95	3.36		12.31	14.90
5226	7/8"		133	.060		14.30	3.36		17.66	21
5227	1"		114	.070		9.35	3.92		13.27	16.20
5240	.049 wall									
5241	1/4"	1 Plum	160	.050	L.F.	5.55	2.79		8.34	10.35
5242	3/8"		160	.050		5.95	2.79		8.74	10.75
5243	1/2"		160	.050		6.95	2.79		9.74	11.85
5244	5/8"		160	.050		9.95	2.79		12.74	15.10
5245	3/4"		133	.060		7.70	3.36		11.06	13.55
5246	7/8"		133	.060		11.95	3.36		15.31	18.20
5247	1"		114	.070		12	3.92		15.92	19.10
5260	.065 wall									
5261	1/4"	1 Plum	160	.050	L.F.	4.59	2.79		7.38	9.25
5262	3/8"		160	.050		9.70	2.79		12.49	14.90
5263	1/2"		160	.050		9.75	2.79		12.54	14.90
5264	5/8"		160	.050		12.80	2.79		15.59	18.30
5265	3/4"		133	.060		12.90	3.36		16.26	19.25
5266	7/8"		133	.060		14	3.36		17.36	20.50
5267	1"		114	.070		13.30	3.92		17.22	20.50

22 11 13.61 Tubing Fittings, Stainless Steel

Line	Description	Crew	Daily Output	Labor-Hours	Unit	Material	Labor	Equipment	Total	Total Incl O&P
0010	**TUBING FITTINGS, STAINLESS STEEL**									
8200	Tube fittings, compression type									
8202	Type 316									
8204	90° elbow									
8206	1/4"	1 Plum	24	.333	Ea.	17.60	18.60		36.20	47.50
8207	3/8"		22	.364		21.50	20.50		42	54.50
8208	1/2"		22	.364		35.50	20.50		56	69.50
8209	5/8"		21	.381		40	21.50		61.50	76
8210	3/4"		21	.381		63	21.50		84.50	101
8211	7/8"		20	.400		96.50	22.50		119	140
8212	1"		20	.400		120	22.50		142.50	166
8220	Union tee									
8222	1/4"	1 Plum	15	.533	Ea.	25	30		55	72.50
8224	3/8"		15	.533		32	30		62	80.50
8225	1/2"		15	.533		49.50	30		79.50	99.50
8226	5/8"		14	.571		55.50	32		87.50	109
8227	3/4"		14	.571		74	32		106	130
8228	7/8"		13	.615		139	34.50		173.50	205
8229	1"		13	.615		159	34.50		193.50	227
8234	Union									
8236	1/4"	1 Plum	24	.333	Ea.	12.20	18.60		30.80	41.50
8237	3/8"		22	.364		17.40	20.50		37.90	49.50
8238	1/2"		22	.364		26	20.50		46.50	59
8239	5/8"		21	.381		34	21.50		55.50	69
8240	3/4"		21	.381		42	21.50		63.50	78
8241	7/8"		20	.400		68.50	22.50		91	109

22 11 13 – Facility Water Distribution Piping

22 11 13.61 Tubing Fittings, Stainless Steel		Crew	Daily Output	Labor-Hours	Unit	Material	Labor	Equipment	Total	Total Incl O&P
						2013 Bare Costs				
8242	1"	1 Plum	20	.400	Ea.	72.50	22.50		95	114
8250	Male connector									
8252	1/4" x 1/4"	1 Plum	24	.333	Ea.	7.80	18.60		26.40	36.50
8253	3/8" x 3/8"		22	.364		12.20	20.50		32.70	44
8254	1/2" x 1/2"		22	.364		18.05	20.50		38.55	50.50
8256	3/4" x 3/4"		21	.381		27.50	21.50		49	62.50
8258	1" x 1"	↓	20	.400	↓	48.50	22.50		71	86.50

22 11 13.64 Pipe, Stainless Steel		Crew	Daily Output	Labor-Hours	Unit	Material	Labor	Equipment	Total	Total Incl O&P
0010	**PIPE, STAINLESS STEEL** R221113-70									
0020	Welded, with clevis type hanger assemblies, 10' O.C.									
0500	Schedule 5, type 304									
0540	1/2" diameter	Q-15	128	.125	L.F.	6.05	6.30	.41	12.76	16.55
0550	3/4" diameter		116	.138		7.80	6.95	.46	15.21	19.50
0560	1" diameter		103	.155		9.45	7.80	.51	17.76	22.50
0570	1-1/4" diameter		93	.172		13.10	8.65	.57	22.32	28
0580	1-1/2" diameter		85	.188		17.55	9.45	.62	27.62	34
0590	2" diameter		69	.232		29.50	11.65	.77	41.92	51
0600	2-1/2" diameter		53	.302		32	15.15	1	48.15	59.50
0610	3" diameter		48	.333		39	16.75	1.10	56.85	68.50
0620	4" diameter		44	.364		49.50	18.25	1.20	68.95	83.50
0630	5" diameter	↓	36	.444		101	22.50	1.47	124.97	146
0640	6" diameter	Q-16	42	.571		93.50	30	1.26	124.76	149
0650	8" diameter		34	.706		144	37	1.55	182.55	215
0660	10" diameter		26	.923		199	48	2.03	249.03	294
0670	12" diameter	↓	21	1.143	↓	262	59.50	2.51	324.01	380
0700	To delete hangers, subtract									
0710	1/2" diam. to 1-1/2" diam.					8%	19%			
0720	2" diam. to 5" diam.					4%	9%			
0730	6" diam. to 12" diam.					3%	4%			
0750	For small quantities, add				L.F.	10%				
1250	Schedule 5, type 316									
1290	1/2" diameter	Q-15	128	.125	L.F.	10.40	6.30	.41	17.11	21.50
1300	3/4" diameter		116	.138		13.55	6.95	.46	20.96	26
1310	1" diameter		103	.155		13.95	7.80	.51	22.26	27.50
1320	1-1/4" diameter		93	.172		16.80	8.65	.57	26.02	32
1330	1-1/2" diameter		85	.188		24	9.45	.62	34.07	41
1340	2" diameter		69	.232		41	11.65	.77	53.42	63.50
1350	2-1/2" diameter		53	.302		53	15.15	1	69.15	82
1360	3" diameter		48	.333		67	16.75	1.10	84.85	100
1370	4" diameter		44	.364		83.50	18.25	1.20	102.95	120
1380	5" diameter	↓	36	.444		159	22.50	1.47	182.97	210
1390	6" diameter	Q-16	42	.571		156	30	1.26	187.26	217
1400	8" diameter		34	.706		234	37	1.55	272.55	315
1410	10" diameter		26	.923		335	48	2.03	385.03	445
1420	12" diameter	↓	21	1.143		430	59.50	2.51	492.01	565
1490	For small quantities, add				↓	10%				
1940	To delete hanger, subtract									
1950	1/2" diam. to 1-1/2" diam.					5%	19%			
1960	2" diam. to 5" diam.					3%	9%			
1970	6" diam. to 12" diam.					2%	4%			
2000	Schedule 10, type 304									
2040	1/4" diameter	Q-15	131	.122	L.F.	3.87	6.15	.40	10.42	13.95

22 11 13.64 Pipe, Stainless Steel		Crew	Daily Output	Labor-Hours	Unit	Material	2013 Bare Costs Labor	Equipment	Total	Total Incl O&P
2050	3/8" diameter	Q-15	128	.125	L.F.	5.10	6.30	.41	11.81	15.50
2060	1/2" diameter		125	.128		10.05	6.45	.42	16.92	21
2070	3/4" diameter		113	.142		10.20	7.10	.47	17.77	22.50
2080	1" diameter		100	.160		10.25	8.05	.53	18.83	24
2090	1-1/4" diameter		91	.176		11.75	8.85	.58	21.18	27
2100	1-1/2" diameter		83	.193		13.15	9.70	.64	23.49	30
2110	2" diameter		67	.239		16.10	12	.79	28.89	36.50
2120	2-1/2" diameter		51	.314		21	15.75	1.04	37.79	47.50
2130	3" diameter		46	.348		25.50	17.45	1.15	44.10	56
2140	4" diameter		42	.381		31.50	19.15	1.26	51.91	65
2150	5" diameter		35	.457		40.50	23	1.51	65.01	81
2160	6" diameter	Q-16	40	.600		50	31.50	1.32	82.82	103
2170	8" diameter		33	.727		87	38	1.60	126.60	155
2180	10" diameter		25	.960		121	50	2.11	173.11	211
2190	12" diameter		21	1.143		151	59.50	2.51	213.01	258
2250	For small quantities, add					10%				
2650	To delete hanger, subtract									
2660	1/4" diam. to 3/4" diam.					9%	22%			
2670	1" diam. to 2" diam.					4%	15%			
2680	2-1/2" diam. to 5" diam.					3%	8%			
2690	6" diam. to 12" diam.					3%	4%			
2750	Schedule 10, type 316									
2790	1/4" diameter	Q-15	131	.122	L.F.	5.25	6.15	.40	11.80	15.45
2800	3/8" diameter		128	.125		7	6.30	.41	13.71	17.60
2810	1/2" diameter		125	.128		7.60	6.45	.42	14.47	18.50
2820	3/4" diameter		113	.142		9.10	7.10	.47	16.67	21
2830	1" diameter		100	.160		13.15	8.05	.53	21.73	27
2840	1-1/4" diameter		91	.176		16.85	8.85	.58	26.28	32.50
2850	1-1/2" diameter		83	.193		18.70	9.70	.64	29.04	36
2860	2" diameter		67	.239		23	12	.79	35.79	44
2870	2-1/2" diameter		51	.314		30	15.75	1.04	46.79	57.50
2880	3" diameter		46	.348		35.50	17.45	1.15	54.10	67
2890	4" diameter		42	.381		45	19.15	1.26	65.41	80
2900	5" diameter		35	.457		55	23	1.51	79.51	96.50
2910	6" diameter	Q-16	40	.600		67.50	31.50	1.32	100.32	122
2920	8" diameter		33	.727		128	38	1.60	167.60	200
2930	10" diameter		25	.960		169	50	2.11	221.11	264
2940	12" diameter		21	1.143		212	59.50	2.51	274.01	325
2990	For small quantities, add					10%				
3430	To delete hanger, subtract									
3440	1/4" diam. to 3/4" diam.					6%	22%			
3450	1" diam. to 2" diam.					3%	15%			
3460	2-1/2" diam. to 5" diam.					2%	8%			
3470	6" diam. to 12" diam.					2%	4%			
3500	Threaded, couplings and clevis hanger assemblies, 10' O.C.									
3520	Schedule 40, type 304									
3540	1/4" diameter	1 Plum	54	.148	L.F.	8.70	8.25		16.95	22
3550	3/8" diameter		53	.151		9.35	8.40		17.75	23
3560	1/2" diameter		52	.154		10.95	8.60		19.55	25
3570	3/4" diameter		51	.157		12.45	8.75		21.20	27
3580	1" diameter		45	.178		17.55	9.90		27.45	34.50
3590	1-1/4" diameter	Q-1	76	.211		24	10.55		34.55	42.50
3600	1-1/2" diameter		69	.232		27	11.65		38.65	47.50

22 11 13 – Facility Water Distribution Piping

22 11 13.64 Pipe, Stainless Steel		Crew	Daily Output	Labor-Hours	Unit	Material	Labor	Equipment	Total	Total Incl O&P
						2013 Bare Costs				
3610	2" diameter	Q-1	57	.281	L.F.	39.50	14.10		53.60	64
3620	2-1/2" diameter		44	.364		70	18.25		88.25	105
3630	3" diameter		38	.421		88.50	21		109.50	129
3640	4" diameter	Q-2	51	.471		118	24.50		142.50	166
3740	For small quantities, add					10%				
4200	To delete couplings & hangers, subtract									
4210	1/4" diam. to 3/4" diam.					15%	56%			
4220	1" diam. to 2" diam.					18%	49%			
4230	2-1/2" diam. to 4" diam.					34%	40%			
4250	Schedule 40, type 316									
4290	1/4" diameter	1 Plum	54	.148	L.F.	10.70	8.25		18.95	24.50
4300	3/8" diameter		53	.151		11.75	8.40		20.15	25.50
4310	1/2" diameter		52	.154		14.40	8.60		23	29
4320	3/4" diameter		51	.157		16.80	8.75		25.55	31.50
4330	1" diameter		45	.178		24	9.90		33.90	41.50
4340	1-1/4" diameter	Q-1	76	.211		32.50	10.55		43.05	51.50
4350	1-1/2" diameter		69	.232		37	11.65		48.65	58
4360	2" diameter		57	.281		51.50	14.10		65.60	77.50
4370	2-1/2" diameter		44	.364		89	18.25		107.25	125
4380	3" diameter		38	.421		113	21		134	156
4390	4" diameter	Q-2	51	.471		150	24.50		174.50	202
4490	For small quantities, add					10%				
4900	To delete couplings & hangers, subtract									
4910	1/4" diam. to 3/4" diam.					12%	56%			
4920	1" diam. to 2" diam.					14%	49%			
4930	2-1/2" diam. to 4" diam.					27%	40%			
5000	Schedule 80, type 304									
5040	1/4" diameter	1 Plum	53	.151	L.F.	17.20	8.40		25.60	31.50
5050	3/8" diameter		52	.154		19.70	8.60		28.30	34.50
5060	1/2" diameter		51	.157		24	8.75		32.75	39
5070	3/4" diameter		48	.167		25.50	9.30		34.80	42
5080	1" diameter		43	.186		32	10.40		42.40	50.50
5090	1-1/4" diameter	Q-1	73	.219		41.50	11		52.50	62.50
5100	1-1/2" diameter		67	.239		54.50	12		66.50	78
5110	2" diameter		54	.296		68	14.90		82.90	97.50
5190	For small quantities, add					10%				
5700	To delete couplings & hangers, subtract									
5710	1/4" diam. to 3/4" diam.					10%	53%			
5720	1" diam. to 2" diam.					14%	47%			
5750	Schedule 80, type 316									
5790	1/4" diameter	1 Plum	53	.151	L.F.	20.50	8.40		28.90	35
5800	3/8" diameter		52	.154		25	8.60		33.60	41
5810	1/2" diameter		51	.157		30	8.75		38.75	46
5820	3/4" diameter		48	.167		33	9.30		42.30	50.50
5830	1" diameter		43	.186		43.50	10.40		53.90	63
5840	1-1/4" diameter	Q-1	73	.219		62	11		73	85
5850	1-1/2" diameter		67	.239		67	12		79	91.50
5860	2" diameter		54	.296		84.50	14.90		99.40	116
5950	For small quantities, add					10%				
7000	To delete couplings & hangers, subtract									
7010	1/4" diam. to 3/4" diam.					9%	53%			
7020	1" diam. to 2" diam.					14%	47%			
8000	Weld joints with clevis type hanger assemblies, 10' O.C.									

22 11 Facility Water Distribution

22 11 13 – Facility Water Distribution Piping

22 11 13.64 Pipe, Stainless Steel		Crew	Daily Output	Labor-Hours	Unit	2013 Bare Costs Material	Labor	Equipment	Total	Total Incl O&P
8010	Schedule 40, type 304									
8050	1/8″ pipe size	Q-15	126	.127	L.F.	7.25	6.40	.42	14.07	18.05
8060	1/4″ pipe size		125	.128		7.50	6.45	.42	14.37	18.40
8070	3/8″ pipe size		122	.131		7.90	6.60	.43	14.93	19.10
8080	1/2″ pipe size		118	.136		9.05	6.80	.45	16.30	20.50
8090	3/4″ pipe size		109	.147		9.85	7.35	.48	17.68	22.50
8100	1″ pipe size		95	.168		13.40	8.45	.56	22.41	28
8110	1-1/4″ pipe size		86	.186		17.45	9.35	.61	27.41	34
8120	1-1/2″ pipe size		78	.205		19.75	10.30	.68	30.73	37.50
8130	2″ pipe size		62	.258		27	12.95	.85	40.80	50
8140	2-1/2″ pipe size		49	.327		41	16.40	1.08	58.48	71
8150	3″ pipe size		44	.364		49	18.25	1.20	68.45	83
8160	3-1/2″ pipe size		44	.364		62.50	18.25	1.20	81.95	97.50
8170	4″ pipe size		39	.410		63	20.50	1.35	84.85	101
8180	5″ pipe size		32	.500		89.50	25	1.65	116.15	138
8190	6″ pipe size	Q-16	37	.649		100	34	1.43	135.43	163
8200	8″ pipe size		29	.828		146	43	1.82	190.82	228
8210	10″ pipe size		24	1		247	52	2.20	301.20	355
8220	12″ pipe size		20	1.200		400	62.50	2.64	465.14	535
8300	Schedule 40, type 316									
8310	1/8″ pipe size	Q-15	126	.127	L.F.	8.35	6.40	.42	15.17	19.25
8320	1/4″ pipe size		125	.128		9.25	6.45	.42	16.12	20.50
8330	3/8″ pipe size		122	.131		10	6.60	.43	17.03	21.50
8340	1/2″ pipe size		118	.136		12.10	6.80	.45	19.35	24
8350	3/4″ pipe size		109	.147		13.75	7.35	.48	21.58	26.50
8360	1″ pipe size		95	.168		18.95	8.45	.56	27.96	34.50
8370	1-1/4″ pipe size		86	.186		24.50	9.35	.61	34.46	41
8380	1-1/2″ pipe size		78	.205		28	10.30	.68	38.98	47
8390	2″ pipe size		62	.258		36.50	12.95	.85	50.30	60.50
8400	2-1/2″ pipe size		49	.327		54.50	16.40	1.08	71.98	85.50
8410	3″ pipe size		44	.364		66	18.25	1.20	85.45	101
8420	3-1/2″ pipe size		44	.364		74.50	18.25	1.20	93.95	111
8430	4″ pipe size		39	.410		84.50	20.50	1.35	106.35	125
8440	5″ pipe size		32	.500		117	25	1.65	143.65	168
8450	6″ pipe size	Q-16	37	.649		135	34	1.43	170.43	202
8460	8″ pipe size		29	.828		168	43	1.82	212.82	252
8470	10″ pipe size		24	1		225	52	2.20	279.20	330
8480	12″ pipe size		20	1.200		305	62.50	2.64	370.14	430
8500	Schedule 80, type 304									
8510	1/4″ pipe size	Q-15	110	.145	L.F.	15.80	7.30	.48	23.58	29
8520	3/8″ pipe size		109	.147		18.10	7.35	.48	25.93	31.50
8530	1/2″ pipe size		106	.151		22	7.60	.50	30.10	36
8540	3/4″ pipe size		96	.167		23	8.35	.55	31.90	38
8550	1″ pipe size		87	.184		27.50	9.25	.61	37.36	44.50
8560	1-1/4″ pipe size		81	.198		31	9.90	.65	41.55	49.50
8570	1-1/2″ pipe size		74	.216		42	10.85	.71	53.56	63
8580	2″ pipe size		58	.276		51.50	13.85	.91	66.26	79
8590	2-1/2″ pipe size		46	.348		67.50	17.45	1.15	86.10	102
8600	3″ pipe size		41	.390		76.50	19.60	1.29	97.39	115
8610	4″ pipe size		33	.485		94	24.50	1.60	120.10	142
8630	6″ pipe size	Q-16	30	.800		168	41.50	1.76	211.26	249
8640	Schedule 80, type 316									
8650	1/4″ pipe size	Q-15	110	.145	L.F.	18.95	7.30	.48	26.73	32.50

22 11 13 – Facility Water Distribution Piping

22 11 13.64 Pipe, Stainless Steel		Crew	Daily Output	Labor-Hours	Unit	Material	2013 Bare Costs Labor	Equipment	Total	Total Incl O&P
8660	3/8" pipe size	Q-15	109	.147	L.F.	23.50	7.35	.48	31.33	37.50
8670	1/2" pipe size		106	.151		27.50	7.60	.50	35.60	42.50
8680	3/4" pipe size		96	.167		29.50	8.35	.55	38.40	45.50
8690	1" pipe size		87	.184		37.50	9.25	.61	47.36	55.50
8700	1-1/4" pipe size		81	.198		48.50	9.90	.65	59.05	69
8710	1-1/2" pipe size		74	.216		51	10.85	.71	62.56	73.50
8720	2" pipe size		58	.276		62.50	13.85	.91	77.26	90.50
8730	2-1/2" pipe size		46	.348		78	17.45	1.15	96.60	113
8740	3" pipe size		41	.390		90	19.60	1.29	110.89	130
8760	4" pipe size		33	.485		119	24.50	1.60	145.10	169
8770	6" pipe size	Q-16	30	.800		215	41.50	1.76	258.26	300
9100	Threading pipe labor, sst, one end, schedules 40 & 80									
9110	1/4" through 3/4" pipe size	1 Plum	61.50	.130	Ea.		7.25		7.25	10.95
9120	1" through 2" pipe size		55.90	.143			8		8	12.05
9130	2-1/2" pipe size		41.50	.193			10.75		10.75	16.20
9140	3" pipe size		38.50	.208			11.60		11.60	17.45
9150	3-1/2" pipe size	Q-1	68.40	.234			11.75		11.75	17.70
9160	4" pipe size		73	.219			11		11	16.60
9170	5" pipe size		40.70	.393			19.75		19.75	29.50
9180	6" pipe size		35.40	.452			22.50		22.50	34
9190	8" pipe size		22.30	.717			36		36	54.50
9200	10" pipe size		16.10	.994			50		50	75
9210	12" pipe size		12.30	1.301			65.50		65.50	98.50
9250	Welding labor per joint for stainless steel									
9260	Schedule 5 and 10									
9270	1/4" pipe size	Q-15	36	.444	Ea.		22.50	1.47	23.97	35
9280	3/8" pipe size		35	.457			23	1.51	24.51	36
9290	1/2" pipe size		35	.457			23	1.51	24.51	36
9300	3/4" pipe size		28	.571			28.50	1.89	30.39	45
9310	1" pipe size		25	.640			32	2.11	34.11	51
9320	1-1/4" pipe size		22	.727			36.50	2.40	38.90	57.50
9330	1-1/2" pipe size		21	.762			38.50	2.51	41.01	60.50
9340	2" pipe size		18	.889			44.50	2.93	47.43	70
9350	2-1/2" pipe size		12	1.333			67	4.40	71.40	106
9360	3" pipe size		9.73	1.644			82.50	5.45	87.95	130
9370	4" pipe size		7.37	2.171			109	7.15	116.15	172
9380	5" pipe size		6.15	2.602			131	8.60	139.60	206
9390	6" pipe size		5.71	2.802			141	9.25	150.25	222
9400	8" pipe size		3.69	4.336			218	14.30	232.30	345
9410	10" pipe size		2.91	5.498			276	18.15	294.15	435
9420	12" pipe size		2.31	6.926			350	23	373	550
9500	Schedule 40									
9510	1/4" pipe size	Q-15	28	.571	Ea.		28.50	1.89	30.39	45
9520	3/8" pipe size		27	.593			30	1.96	31.96	47
9530	1/2" pipe size		25.40	.630			31.50	2.08	33.58	50
9540	3/4" pipe size		22.22	.720			36	2.38	38.38	57
9550	1" pipe size		20.25	.790			39.50	2.61	42.11	63
9560	1-1/4" pipe size		18.82	.850			42.50	2.81	45.31	67.50
9570	1-1/2" pipe size		17.78	.900			45	2.97	47.97	71.50
9580	2" pipe size		15.09	1.060			53.50	3.50	57	84
9590	2-1/2" pipe size		7.96	2.010			101	6.65	107.65	159
9600	3" pipe size		6.43	2.488			125	8.20	133.20	197
9610	4" pipe size		4.88	3.279			165	10.80	175.80	260

22 11 13 – Facility Water Distribution Piping

22 11 13.64 Pipe, Stainless Steel		Crew	Daily Output	Labor-Hours	Unit	Material	2013 Bare Costs Labor	Equipment	Total	Total Incl O&P
9620	5" pipe size	Q-15	4.26	3.756	Ea.		189	12.40	201.40	298
9630	6" pipe size		3.77	4.244			213	14	227	335
9640	8" pipe size		2.44	6.557			330	21.50	351.50	520
9650	10" pipe size		1.92	8.333			420	27.50	447.50	660
9660	12" pipe size		1.52	10.526			530	34.50	564.50	835
9750	Schedule 80									
9760	1/4" pipe size	Q-15	21.55	.742	Ea.		37.50	2.45	39.95	58.50
9770	3/8" pipe size		20.75	.771			38.50	2.54	41.04	61.50
9780	1/2" pipe size		19.54	.819			41	2.70	43.70	65
9790	3/4" pipe size		17.09	.936			47	3.09	50.09	74.50
9800	1" pipe size		15.58	1.027			51.50	3.39	54.89	81
9810	1-1/4" pipe size		14.48	1.105			55.50	3.65	59.15	87.50
9820	1-1/2" pipe size		13.68	1.170			59	3.86	62.86	93
9830	2" pipe size		11.61	1.378			69	4.55	73.55	109
9840	2-1/2" pipe size		6.12	2.614			131	8.65	139.65	208
9850	3" pipe size		4.94	3.239			163	10.70	173.70	257
9860	4" pipe size		3.75	4.267			214	14.10	228.10	340
9870	5" pipe size		3.27	4.893			246	16.15	262.15	390
9880	6" pipe size		2.90	5.517			277	18.20	295.20	435
9890	8" pipe size		1.87	8.556			430	28	458	675
9900	10" pipe size		1.48	10.811			545	35.50	580.50	860
9910	12" pipe size		1.17	13.675			685	45	730	1,075
9920	Schedule 160, 1/2" pipe size		17	.941			47.50	3.11	50.61	74.50
9930	3/4" pipe size		14.81	1.080			54.50	3.57	58.07	85.50
9940	1" pipe size		13.50	1.185			59.50	3.91	63.41	94
9950	1-1/4" pipe size		12.55	1.275			64	4.21	68.21	101
9960	1-1/2" pipe size		11.85	1.350			68	4.46	72.46	107
9970	2" pipe size		10	1.600			80.50	5.30	85.80	127
9980	3" pipe size		4.28	3.738			188	12.35	200.35	297
9990	4" pipe size		3.25	4.923			247	16.25	263.25	390

22 11 13.66 Pipe Fittings, Stainless Steel		Crew	Daily Output	Labor-Hours	Unit	Material	Labor	Equipment	Total	Total Incl O&P
0010	**PIPE FITTINGS, STAINLESS STEEL**									
0100	Butt weld joint, schedule 5, type 304									
0120	90° Elbow, long									
0140	1/2"	Q-15	17.50	.914	Ea.	19.80	46	3.02	68.82	94.50
0150	3/4"		14	1.143		19.80	57.50	3.77	81.07	113
0160	1"		12.50	1.280		21	64.50	4.22	89.72	125
0170	1-1/4"		11	1.455		28.50	73	4.80	106.30	147
0180	1-1/2"		10.50	1.524		24	76.50	5.05	105.55	147
0190	2"		9	1.778		28.50	89.50	5.85	123.85	172
0200	2-1/2"		6	2.667		66	134	8.80	208.80	284
0210	3"		4.86	3.292		72.50	165	10.85	248.35	340
0220	3-1/2"		4.27	3.747		182	188	12.35	382.35	495
0230	4"		3.69	4.336		99	218	14.30	331.30	455
0240	5"		3.08	5.195		385	261	17.15	663.15	840
0250	6"	Q-16	4.29	5.594		297	291	12.30	600.30	780
0260	8"		2.76	8.696		625	455	19.15	1,099.15	1,400
0270	10"		2.18	11.009		970	575	24	1,569	1,975
0280	12"		1.73	13.873		1,375	725	30.50	2,130.50	2,650
0320	For schedule 5, type 316, add					30%				
0600	45° Elbow, long									
0620	1/2"	Q-15	17.50	.914	Ea.	19.80	46	3.02	68.82	94.50

22 11 13.66 Pipe Fittings, Stainless Steel		Crew	Daily Output	Labor-Hours	Unit	Material	2013 Bare Costs Labor	Equipment	Total	Total Incl O&P
0630	3/4"	Q-15	14	1.143	Ea.	19.80	57.50	3.77	81.07	113
0640	1"		12.50	1.280		21	64.50	4.22	89.72	125
0650	1-1/4"		11	1.455		28.50	73	4.80	106.30	147
0660	1-1/2"		10.50	1.524		24	76.50	5.05	105.55	147
0670	2"		9	1.778		28.50	89.50	5.85	123.85	172
0680	2-1/2"		6	2.667		66	134	8.80	208.80	284
0690	3"		4.86	3.292		72.50	165	10.85	248.35	340
0700	3-1/2"		4.27	3.747		182	188	12.35	382.35	495
0710	4"		3.69	4.336		83.50	218	14.30	315.80	440
0720	5"		3.08	5.195		310	261	17.15	588.15	755
0730	6"	Q-16	4.29	5.594		209	291	12.30	512.30	685
0740	8"		2.76	8.696		440	455	19.15	914.15	1,175
0750	10"		2.18	11.009		770	575	24	1,369	1,725
0760	12"		1.73	13.873		965	725	30.50	1,720.50	2,175
0800	For schedule 5, type 316, add					25%				
1100	Tee, straight									
1130	1/2"	Q-15	11.66	1.372	Ea.	59.50	69	4.53	133.03	174
1140	3/4"		9.33	1.715		59.50	86	5.65	151.15	202
1150	1"		8.33	1.921		62.50	96.50	6.35	165.35	221
1160	1-1/4"		7.33	2.183		50.50	110	7.20	167.70	228
1170	1-1/2"		7	2.286		49.50	115	7.55	172.05	236
1180	2"		6	2.667		51.50	134	8.80	194.30	269
1190	2-1/2"		4	4		124	201	13.20	338.20	455
1200	3"		3.24	4.938		198	248	16.30	462.30	610
1210	3-1/2"		2.85	5.614		260	282	18.55	560.55	730
1220	4"		2.46	6.504		141	325	21.50	487.50	670
1230	5"		2	8		445	400	26.50	871.50	1,125
1240	6"	Q-16	2.85	8.421		350	440	18.55	808.55	1,075
1250	8"		1.84	13.043		750	680	28.50	1,458.50	1,875
1260	10"		1.45	16.552		1,200	860	36.50	2,096.50	2,675
1270	12"		1.15	20.870		1,675	1,075	46	2,796	3,525
1320	For schedule 5, type 316, add					25%				
2000	Butt weld joint, schedule 10, type 304									
2020	90° elbow, long									
2040	1/2"	Q-15	17	.941	Ea.	13.50	47.50	3.11	64.11	89.50
2050	3/4"		14	1.143		13.50	57.50	3.77	74.77	106
2060	1"		12.50	1.280		14.25	64.50	4.22	82.97	117
2070	1-1/4"		11	1.455		19.50	73	4.80	97.30	137
2080	1-1/2"		10.50	1.524		16.50	76.50	5.05	98.05	139
2090	2"		9	1.778		19.50	89.50	5.85	114.85	162
2100	2-1/2"		6	2.667		45	134	8.80	187.80	261
2110	3"		4.86	3.292		42	165	10.85	217.85	305
2120	3-1/2"		4.27	3.747		124	188	12.35	324.35	435
2130	4"		3.69	4.336		70.50	218	14.30	302.80	425
2140	5"		3.08	5.195		263	261	17.15	541.15	705
2150	6"	Q-16	4.29	5.594		203	291	12.30	506.30	675
2160	8"		2.76	8.696		430	455	19.15	904.15	1,175
2170	10"		2.18	11.009		660	575	24	1,259	1,625
2180	12"		1.73	13.873		940	725	30.50	1,695.50	2,150
2500	45° elbow, long									
2520	1/2"	Q-15	17.50	.914	Ea.	13.50	46	3.02	62.52	87
2530	3/4"		14	1.143		13.50	57.50	3.77	74.77	106
2540	1"		12.50	1.280		14.25	64.50	4.22	82.97	117

22 11 13.66 Pipe Fittings, Stainless Steel		Crew	Daily Output	Labor-Hours	Unit	2013 Bare Costs Material	Labor	Equipment	Total	Total Incl O&P
2550	1-1/4"	Q-15	11	1.455	Ea.	19.50	73	4.80	97.30	137
2560	1-1/2"		10.50	1.524		16.50	76.50	5.05	98.05	139
2570	2"		9	1.778		19.50	89.50	5.85	114.85	162
2580	2-1/2"		6	2.667		45	134	8.80	187.80	261
2590	3"		4.86	3.292		34	165	10.85	209.85	298
2600	3-1/2"		4.27	3.747		124	188	12.35	324.35	435
2610	4"		3.69	4.336		57	218	14.30	289.30	410
2620	5"		3.08	5.195		210	261	17.15	488.15	645
2630	6"	Q-16	4.29	5.594		143	291	12.30	446.30	610
2640	8"		2.76	8.696		300	455	19.15	774.15	1,025
2650	10"		2.18	11.009		525	575	24	1,124	1,475
2660	12"		1.73	13.873		655	725	30.50	1,410.50	1,850
2670	Reducer, concentric									
2674	1" x 3/4"	Q-15	13.25	1.208	Ea.	29.50	60.50	3.98	93.98	128
2676	2" x 1-1/2"	"	9.75	1.641		24	82.50	5.40	111.90	156
2678	6" x 4"	Q-16	4.91	4.888		72	255	10.75	337.75	475
2680	Caps									
2682	1"	Q-15	25	.640	Ea.	30	32	2.11	64.11	84
2684	1-1/2"		21	.762		36	38.50	2.51	77.01	100
2685	2"		18	.889		34	44.50	2.93	81.43	107
2686	4"		7.38	2.168		54	109	7.15	170.15	231
2687	6"	Q-16	8.58	2.797		86.50	146	6.15	238.65	320
3000	Tee, straight									
3030	1/2"	Q-15	11.66	1.372	Ea.	40.50	69	4.53	114.03	153
3040	3/4"		9.33	1.715		40.50	86	5.65	132.15	181
3050	1"		8.33	1.921		43	96.50	6.35	145.85	199
3060	1-1/4"		7.33	2.183		49	110	7.20	166.20	226
3070	1-1/2"		7	2.286		49	115	7.55	171.55	235
3080	2"		6	2.667		51	134	8.80	193.80	268
3090	2-1/2"		4	4		35.50	201	13.20	249.70	360
3100	3"		3.24	4.938		67	248	16.30	331.30	465
3110	3-1/2"		2.85	5.614		98.50	282	18.55	399.05	555
3120	4"		2.46	6.504		96	325	21.50	442.50	620
3130	5"		2	8		305	400	26.50	731.50	970
3140	6"	Q-16	2.85	8.421		240	440	18.55	698.55	945
3150	8"		1.84	13.043		510	680	28.50	1,218.50	1,625
3151	10"		1.45	16.552		825	860	36.50	1,721.50	2,250
3152	12"		1.15	20.870		1,150	1,075	46	2,271	2,925
3154	For schedule 10, type 316, add					25%				
3281	Butt weld joint, schedule 40, type 304									
3284	90° Elbow, long, 1/2"	Q-15	12.70	1.260	Ea.	15.75	63.50	4.16	83.41	117
3288	3/4"		11.10	1.441		15.75	72.50	4.76	93.01	132
3289	1"		10.13	1.579		16.50	79.50	5.20	101.20	143
3290	1-1/4"		9.40	1.702		22	85.50	5.60	113.10	159
3300	1-1/2"		8.89	1.800		17.25	90.50	5.95	113.70	162
3310	2"		7.55	2.119		25	106	7	138	195
3320	2-1/2"		3.98	4.020		47.50	202	13.25	262.75	370
3330	3"		3.21	4.984		61	250	16.45	327.45	460
3340	3-1/2"		2.83	5.654		225	284	18.65	527.65	700
3350	4"		2.44	6.557		105	330	21.50	456.50	635
3360	5"		2.13	7.512		355	375	25	755	990
3370	6"	Q-16	2.83	8.481		310	440	18.65	768.65	1,025
3380	8"		1.83	13.115		615	685	29	1,329	1,725

22 11 13 – Facility Water Distribution Piping

22 11 13.66 Pipe Fittings, Stainless Steel		Crew	Daily Output	Labor-Hours	Unit	Material	Labor	Equipment	Total	Total Incl O&P
						2013 Bare Costs				
3390	10"	Q-16	1.44	16.667	Ea.	1,300	870	36.50	2,206.50	2,775
3400	12"		1.14	21.053		1,650	1,100	46.50	2,796.50	3,525
3410	For schedule 40, type 316, add					25%				
3460	45° Elbow, long, 1/2"	Q-15	12.70	1.260	Ea.	15.75	63.50	4.16	83.41	117
3470	3/4"		11.10	1.441		15.75	72.50	4.76	93.01	132
3480	1"		10.13	1.579		16.50	79.50	5.20	101.20	143
3490	1-1/4"		9.40	1.702		22	85.50	5.60	113.10	159
3500	1-1/2"		8.89	1.800		17.25	90.50	5.95	113.70	162
3510	2"		7.55	2.119		25	106	7	138	195
3520	2-1/2"		3.98	4.020		47.50	202	13.25	262.75	370
3530	3"		3.21	4.984		48	250	16.45	314.45	445
3540	3-1/2"		2.83	5.654		225	284	18.65	527.65	700
3550	4"		2.44	6.557		75	330	21.50	426.50	600
3560	5"		2.13	7.512		248	375	25	648	870
3570	6"	Q-16	2.83	8.481		218	440	18.65	676.65	925
3580	8"		1.83	13.115		430	685	29	1,144	1,525
3590	10"		1.44	16.667		910	870	36.50	1,816.50	2,350
3600	12"		1.14	21.053		1,150	1,100	46.50	2,296.50	2,975
3610	For schedule 40, type 316, add					25%				
3660	Tee, straight 1/2"	Q-15	8.46	1.891	Ea.	40.50	95	6.25	141.75	194
3670	3/4"		7.40	2.162		40.50	109	7.15	156.65	216
3680	1"		6.74	2.374		43	119	7.85	169.85	236
3690	1-1/4"		6.27	2.552		100	128	8.40	236.40	310
3700	1-1/2"		5.92	2.703		46	136	8.90	190.90	264
3710	2"		5.03	3.181		68.50	160	10.50	239	330
3720	2-1/2"		2.65	6.038		86.50	305	19.90	411.40	570
3730	3"		2.14	7.477		87	375	24.50	486.50	690
3740	3-1/2"		1.88	8.511		164	425	28	617	855
3750	4"		1.62	9.877		164	495	32.50	691.50	960
3760	5"		1.42	11.268		380	565	37	982	1,300
3770	6"	Q-16	1.88	12.766		365	665	28	1,058	1,425
3780	8"		1.22	19.672		735	1,025	43.50	1,803.50	2,400
3790	10"		.96	25		1,425	1,300	55	2,780	3,575
3800	12"		.76	31.579		1,900	1,650	69.50	3,619.50	4,650
3810	For schedule 40, type 316, add					25%				
3820	Tee, reducing on outlet, 3/4" x 1/2"	Q-15	7.73	2.070	Ea.	59.50	104	6.85	170.35	230
3822	1" x 1/2"		7.24	2.210		58.50	111	7.30	176.80	240
3824	1" x 3/4"		6.96	2.299		54	115	7.60	176.60	242
3826	1-1/4" x 1"		6.43	2.488		132	125	8.20	265.20	340
3828	1-1/2" x 1/2"		6.58	2.432		79.50	122	8	209.50	280
3830	1-1/2" x 3/4"		6.35	2.520		74.50	127	8.30	209.80	282
3832	1-1/2" x 1"		6.18	2.589		57	130	8.55	195.55	268
3834	2" x 1"		5.50	2.909		102	146	9.60	257.60	345
3836	2" x 1-1/2"		5.30	3.019		85.50	152	9.95	247.45	335
3838	2-1/2" x 2"		3.15	5.079		129	255	16.75	400.75	545
3840	3" x 1-1/2"		2.72	5.882		131	295	19.40	445.40	610
3842	3" x 2"		2.65	6.038		109	305	19.90	433.90	595
3844	4" x 2"		2.10	7.619		272	385	25	682	900
3846	4" x 3"		1.77	9.040		197	455	30	682	935
3848	5" x 4"		1.48	10.811		455	545	35.50	1,035.50	1,350
3850	6" x 3"	Q-16	2.19	10.959		505	570	24	1,099	1,450
3852	6" x 4"		2.04	11.765		435	615	26	1,076	1,425
3854	8" x 4"		1.46	16.438		1,025	855	36	1,916	2,475

22 11 13 – Facility Water Distribution Piping

22 11 13.66 Pipe Fittings, Stainless Steel		Crew	Daily Output	Labor-Hours	Unit	2013 Bare Costs Material	Labor	Equipment	Total	Total Incl O&P
3856	10" x 8"	Q-16	.69	34.783	Ea.	1,725	1,800	76.50	3,601.50	4,675
3858	12" x 10"		.55	43.636		2,275	2,275	96	4,646	6,050
3950	Reducer, concentric, 3/4" x 1/2"	Q-15	11.85	1.350		30	68	4.46	102.46	140
3952	1" x 3/4"		10.60	1.509		31.50	76	4.98	112.48	154
3954	1-1/4" x 3/4"		10.19	1.570		77.50	79	5.20	161.70	210
3956	1-1/4" x 1"		9.76	1.639		39	82.50	5.40	126.90	173
3958	1-1/2" x 3/4"		9.88	1.619		65.50	81.50	5.35	152.35	201
3960	1-1/2" x 1"		9.47	1.690		52.50	85	5.60	143.10	192
3962	2" x 1"		8.65	1.850		30	93	6.10	129.10	180
3964	2" x 1-1/2"		8.16	1.961		26.50	98.50	6.45	131.45	184
3966	2-1/2" x 1"		5.71	2.802		113	141	9.25	263.25	345
3968	2-1/2" x 2"		5.21	3.071		58	154	10.15	222.15	305
3970	3" x 1"		4.88	3.279		78	165	10.80	253.80	345
3972	3" x 1-1/2"		4.72	3.390		40	170	11.20	221.20	310
3974	3" x 2"		4.51	3.548		37.50	178	11.70	227.20	320
3976	4" x 2"		3.69	4.336		53.50	218	14.30	285.80	405
3978	4" x 3"		2.77	5.776		40	290	19.05	349.05	500
3980	5" x 3"		2.56	6.250		237	315	20.50	572.50	760
3982	5" x 4"		2.27	7.048		191	355	23.50	569.50	770
3984	6" x 3"	Q-16	3.57	6.723		122	350	14.80	486.80	675
3986	6" x 4"		3.19	7.524		101	390	16.55	507.55	720
3988	8" x 4"		2.44	9.836		315	510	21.50	846.50	1,150
3990	8" x 6"		2.22	10.811		236	565	24	825	1,125
3992	10" x 6"		1.91	12.565		440	655	27.50	1,122.50	1,500
3994	10" x 8"		1.61	14.907		365	775	33	1,173	1,600
3995	12" x 6"		1.63	14.724		675	765	32.50	1,472.50	1,925
3996	12" x 8"		1.41	17.021		620	885	37.50	1,542.50	2,050
3997	12" x 10"		1.27	18.898		435	985	41.50	1,461.50	2,000
4000	Socket weld joint, 3000 lb., type 304									
4100	90° Elbow									
4140	1/4"	Q-15	13.47	1.188	Ea.	44	59.50	3.92	107.42	143
4150	3/8"		12.97	1.234		57	62	4.07	123.07	161
4160	1/2"		12.21	1.310		62.50	66	4.32	132.82	173
4170	3/4"		10.68	1.498		72	75.50	4.94	152.44	197
4180	1"		9.74	1.643		108	82.50	5.40	195.90	249
4190	1-1/4"		9.05	1.768		189	89	5.85	283.85	350
4200	1-1/2"		8.55	1.871		229	94	6.20	329.20	400
4210	2"		7.26	2.204		370	111	7.25	488.25	580
4300	45° Elbow									
4340	1/4"	Q-15	13.47	1.188	Ea.	83	59.50	3.92	146.42	185
4350	3/8"		12.97	1.234		83	62	4.07	149.07	189
4360	1/2"		12.21	1.310		83	66	4.32	153.32	195
4370	3/4"		10.68	1.498		94	75.50	4.94	174.44	221
4380	1"		9.74	1.643		136	82.50	5.40	223.90	280
4390	1-1/4"		9.05	1.768		224	89	5.85	318.85	385
4400	1-1/2"		8.55	1.871		225	94	6.20	325.20	395
4410	2"		7.26	2.204		410	111	7.25	528.25	625
4500	Tee									
4540	1/4"	Q-15	8.97	1.784	Ea.	58	89.50	5.90	153.40	205
4550	3/8"		8.64	1.852		69	93	6.10	168.10	223
4560	1/2"		8.13	1.968		85	99	6.50	190.50	250
4570	3/4"		7.12	2.247		98	113	7.40	218.40	286
4580	1"		6.48	2.469		132	124	8.15	264.15	340

	22 11 13.66 Pipe Fittings, Stainless Steel	Crew	Daily Output	Labor-Hours	Unit	2013 Bare Costs Material	Labor	Equipment	Total	Total Incl O&P
4590	1-1/4"	Q-15	6.03	2.653	Ea.	234	133	8.75	375.75	470
4600	1-1/2"		5.69	2.812		335	141	9.30	485.30	595
4610	2"		4.83	3.313		510	166	10.95	686.95	825
5000	Socket weld joint, 3000 lb., type 316									
5100	90° Elbow									
5140	1/4"	Q-15	13.47	1.188	Ea.	54	59.50	3.92	117.42	153
5150	3/8"		12.97	1.234		63	62	4.07	129.07	167
5160	1/2"		12.21	1.310		76	66	4.32	146.32	187
5170	3/4"		10.68	1.498		100	75.50	4.94	180.44	228
5180	1"		9.74	1.643		142	82.50	5.40	229.90	286
5190	1-1/4"		9.05	1.768		253	89	5.85	347.85	420
5200	1-1/2"		8.55	1.871		286	94	6.20	386.20	465
5210	2"		7.26	2.204		485	111	7.25	603.25	710
5300	45° Elbow									
5340	1/4"	Q-15	13.47	1.188	Ea.	104	59.50	3.92	167.42	208
5350	3/8"		12.97	1.234		104	62	4.07	170.07	212
5360	1/2"		12.21	1.310		104	66	4.32	174.32	218
5370	3/4"		10.68	1.498		121	75.50	4.94	201.44	251
5380	1"		9.74	1.643		182	82.50	5.40	269.90	330
5390	1-1/4"		9.05	1.768		258	89	5.85	352.85	425
5400	1-1/2"		8.55	1.871		292	94	6.20	392.20	470
5410	2"		7.26	2.204		435	111	7.25	553.25	655
5500	Tee									
5540	1/4"	Q-15	8.97	1.784	Ea.	73	89.50	5.90	168.40	222
5550	3/8"		8.64	1.852		89.50	93	6.10	188.60	245
5560	1/2"		8.13	1.968		99.50	99	6.50	205	265
5570	3/4"		7.12	2.247		123	113	7.40	243.40	315
5580	1"		6.48	2.469		188	124	8.15	320.15	405
5590	1-1/4"		6.03	2.653		299	133	8.75	440.75	540
5600	1-1/2"		5.69	2.812		420	141	9.30	570.30	690
5610	2"		4.83	3.313		670	166	10.95	846.95	1,000
5700	For socket weld joint, 6000 lb., type 304 and 316, add					100%				
6000	Threaded companion flange									
6010	Stainless steel, 150 lb., type 304									
6020	1/2" diam.	1 Plum	30	.267	Ea.	45	14.90		59.90	72
6030	3/4" diam.		28	.286		50	15.95		65.95	79
6040	1" diam.		27	.296		55	16.55		71.55	85.50
6050	1-1/4" diam.	Q-1	44	.364		71	18.25		89.25	106
6060	1-1/2" diam.		40	.400		71	20		91	109
6070	2" diam.		36	.444		93	22.50		115.50	136
6080	2-1/2" diam.		28	.571		130	28.50		158.50	186
6090	3" diam.		20	.800		136	40		176	211
6110	4" diam.		12	1.333		186	67		253	305
6130	6" diam.	Q-2	14	1.714		335	89.50		424.50	500
6140	8" diam.	"	12	2		645	104		749	865
6150	For type 316 add					40%				
6260	Weld flanges, stainless steel, type 304									
6270	Slip on, 150 lb. (welded, front and back)									
6280	1/2" diam.	Q-15	12.70	1.260	Ea.	40	63.50	4.16	107.66	144
6290	3/4" diam.		11.11	1.440		41	72.50	4.75	118.25	159
6300	1" diam.		10.13	1.579		45	79.50	5.20	129.70	174
6310	1-1/4" diam.		9.41	1.700		61	85.50	5.60	152.10	202
6320	1-1/2" diam.		8.89	1.800		61	90.50	5.95	157.45	210

22 11 13.66 Pipe Fittings, Stainless Steel		Crew	Daily Output	Labor-Hours	Unit	Material	Labor (2013 Bare Costs)	Equipment	Total	Total Incl O&P
6330	2" diam.	Q-15	7.55	2.119	Ea.	79	106	7	192	255
6340	2-1/2" diam.		3.98	4.020		111	202	13.25	326.25	440
6350	3" diam.		3.21	4.984		119	250	16.45	385.45	525
6370	4" diam.		2.44	6.557		162	330	21.50	513.50	695
6390	6" diam.	Q-16	1.89	12.698		247	660	28	935	1,300
6400	8" diam.	"	1.22	19.672		465	1,025	43.50	1,533.50	2,125
6410	For type 316, add					40%				
6530	Weld neck 150 lb.									
6540	1/2" diam.	Q-15	25.40	.630	Ea.	20.50	31.50	2.08	54.08	72.50
6550	3/4" diam.		22.22	.720		24	36	2.38	62.38	83.50
6560	1" diam.		20.25	.790		26.50	39.50	2.61	68.61	92.50
6570	1-1/4" diam.		18.82	.850		43.50	42.50	2.81	88.81	115
6580	1-1/2" diam.		17.78	.900		34.50	45	2.97	82.47	109
6590	2" diam.		15.09	1.060		41.50	53.50	3.50	98.50	129
6600	2-1/2" diam.		7.96	2.010		60	101	6.65	167.65	225
6610	3" diam.		6.43	2.488		68.50	125	8.20	201.70	273
6630	4" diam.		4.88	3.279		92.50	165	10.80	268.30	360
6640	5" diam.		4.26	3.756		134	189	12.40	335.40	445
6650	6" diam.	Q-16	5.66	4.240		150	221	9.35	380.35	510
6652	8" diam.		3.65	6.575		266	340	14.45	620.45	825
6654	10" diam.		2.88	8.333		375	435	18.35	828.35	1,075
6656	12" diam.		2.28	10.526		580	550	23	1,153	1,500
6670	For type 316 add					23%				
7000	Threaded joint, 150 lb., type 304									
7030	90° elbow									
7040	1/8"	1 Plum	13	.615	Ea.	25.50	34.50		60	79.50
7050	1/4"		13	.615		25.50	34.50		60	79.50
7070	3/8"		13	.615		29.50	34.50		64	84
7080	1/2"		12	.667		26.50	37		63.50	85
7090	3/4"		11	.727		31.50	40.50		72	95.50
7100	1"		10	.800		43	44.50		87.50	115
7110	1-1/4"	Q-1	17	.941		66.50	47.50		114	145
7120	1-1/2"		16	1		77	50		127	161
7130	2"		14	1.143		111	57.50		168.50	210
7140	2-1/2"		11	1.455		271	73		344	410
7150	3"		8	2		395	100		495	585
7160	4"	Q-2	11	2.182		670	114		784	905
7180	45° elbow									
7190	1/8"	1 Plum	13	.615	Ea.	37	34.50		71.50	92.50
7200	1/4"		13	.615		37	34.50		71.50	92.50
7210	3/8"		13	.615		37.50	34.50		72	92.50
7220	1/2"		12	.667		38	37		75	97.50
7230	3/4"		11	.727		41	40.50		81.50	106
7240	1"		10	.800		47.50	44.50		92	119
7250	1-1/4"	Q-1	17	.941		66	47.50		113.50	144
7260	1-1/2"		16	1		85.50	50		135.50	170
7270	2"		14	1.143		121	57.50		178.50	220
7280	2-1/2"		11	1.455		380	73		453	530
7290	3"		8	2		560	100		660	765
7300	4"	Q-2	11	2.182		1,000	114		1,114	1,275
7320	Tee, straight									
7330	1/8"	1 Plum	9	.889	Ea.	39	49.50		88.50	118
7340	1/4"		9	.889		39	49.50		88.50	118

22 11 13.66 Pipe Fittings, Stainless Steel		Crew	Daily Output	Labor-Hours	Unit	Material	Labor	Equipment	Total	Total Incl O&P
						2013 Bare Costs				
7350	3/8"	1 Plum	9	.889	Ea.	41.50	49.50		91	121
7360	1/2"		8	1		39.50	56		95.50	128
7370	3/4"		7	1.143		42	64		106	143
7380	1"		6.50	1.231		54	68.50		122.50	163
7390	1-1/4"	Q-1	11	1.455		91	73		164	210
7400	1-1/2"		10	1.600		116	80.50		196.50	249
7410	2"		9	1.778		145	89.50		234.50	293
7420	2-1/2"		7	2.286		390	115		505	605
7430	3"		5	3.200		590	161		751	890
7440	4"	Q-2	7	3.429		1,475	179		1,654	1,875
7460	Coupling, straight									
7470	1/8"	1 Plum	19	.421	Ea.	10.25	23.50		33.75	47
7480	1/4"		19	.421		12.10	23.50		35.60	49
7490	3/8"		19	.421		14.50	23.50		38	51.50
7500	1/2"		19	.421		19.30	23.50		42.80	56.50
7510	3/4"		18	.444		25.50	25		50.50	66
7520	1"		15	.533		41.50	30		71.50	90.50
7530	1-1/4"	Q-1	26	.615		66.50	31		97.50	120
7540	1-1/2"		24	.667		74.50	33.50		108	133
7550	2"		21	.762		123	38.50		161.50	194
7560	2-1/2"		18	.889		287	44.50		331.50	380
7570	3"		14	1.143		390	57.50		447.50	515
7580	4"	Q-2	16	1.500		550	78		628	725
7600	Reducer, concentric, 1/2"	1 Plum	12	.667		20.50	37		57.50	79
7610	3/4"		11	.727		26.50	40.50		67	90.50
7612	1"		10	.800		44.50	44.50		89	116
7614	1-1/4"	Q-1	17	.941		94.50	47.50		142	175
7616	1-1/2"		16	1		103	50		153	189
7618	2"		14	1.143		161	57.50		218.50	264
7620	2-1/2"		11	1.455		400	73		473	550
7622	3"		8	2		445	100		545	640
7624	4"	Q-2	11	2.182		735	114		849	980
7710	Union									
7720	1/8"	1 Plum	12	.667	Ea.	47.50	37		84.50	109
7730	1/4"		12	.667		47.50	37		84.50	109
7740	3/8"		12	.667		55.50	37		92.50	117
7750	1/2"		11	.727		69.50	40.50		110	138
7760	3/4"		10	.800		96	44.50		140.50	172
7770	1"		9	.889		138	49.50		187.50	227
7780	1-1/4"	Q-1	16	1		340	50		390	450
7790	1-1/2"		15	1.067		370	53.50		423.50	485
7800	2"		13	1.231		465	62		527	610
7810	2-1/2"		10	1.600		880	80.50		960.50	1,100
7820	3"		7	2.286		1,175	115		1,290	1,475
7830	4"	Q-2	10	2.400		1,625	125		1,750	1,975
7838	Caps									
7840	1/2"	1 Plum	24	.333	Ea.	12.90	18.60		31.50	42
7841	3/4"		22	.364		18.80	20.50		39.30	51
7842	1"		20	.400		30	22.50		52.50	66.50
7843	1-1/2"	Q-1	32	.500		76	25		101	122
7844	2"	"	28	.571		96	28.50		124.50	149
7845	4"	Q-2	22	1.091		380	57		437	500
7850	For 150 lb., type 316, add					25%				

22 11 Facility Water Distribution

22 11 13 – Facility Water Distribution Piping

	22 11 13.74 Pipe, Plastic	Crew	Daily Output	Labor-Hours	Unit	Material	2013 Bare Costs Labor	Equipment	Total	Total Incl O&P
0010	**PIPE, PLASTIC** R221113-70									
0020	Fiberglass reinforced, couplings 10′ O.C., clevis hanger assy's, 3 per 10′									
0080	General service									
0120	2″ diameter	Q-1	59	.271	L.F.	8.70	13.60		22.30	30
0140	3″ diameter		52	.308		12.60	15.45		28.05	37.50
0150	4″ diameter		48	.333		17.10	16.75		33.85	44
0160	6″ diameter		39	.410		31.50	20.50		52	66
0170	8″ diameter	Q-2	49	.490		50.50	25.50		76	94
0180	10″ diameter		41	.585		74.50	30.50		105	128
0190	12″ diameter		36	.667		96.50	34.50		131	159
0600	PVC, high impact/pressure, cplgs. 10′ O.C., clevis hanger assy's, 3 per 10′									
1020	Schedule 80									
1070	1/2″ diameter	1 Plum	50	.160	L.F.	4.77	8.95		13.72	18.70
1080	3/4″ diameter		47	.170		5.80	9.50		15.30	20.50
1090	1″ diameter		43	.186		7.40	10.40		17.80	24
1100	1-1/4″ diameter		39	.205		9.60	11.45		21.05	28
1110	1-1/2″ diameter		34	.235		11.10	13.15		24.25	32
1120	2″ diameter	Q-1	55	.291		14.45	14.60		29.05	38
1140	3″ diameter		50	.320		29	16.05		45.05	56
1150	4″ diameter		46	.348		41.50	17.45		58.95	72
1170	6″ diameter		38	.421		93.50	21		114.50	135
1730	To delete coupling & hangers, subtract									
1740	1/4″ diam. to 1/2″ diam.					62%	80%			
1750	3/4″ diam. to 1-1/4″ diam.					58%	73%			
1760	1-1/2″ diam. to 6″ diam.					40%	57%			
1770	8″ diam. to 12″ diam.					34%	50%			
1800	PVC, couplings 10′ O.C., clevis hanger assemblies, 3 per 10′									
1820	Schedule 40									
1860	1/2″ diameter	1 Plum	54	.148	L.F.	3.09	8.25		11.34	15.85
1870	3/4″ diameter		51	.157		3.33	8.75		12.08	16.85
1880	1″ diameter		46	.174		3.87	9.70		13.57	18.85
1890	1-1/4″ diameter		42	.190		4.77	10.65		15.42	21.50
1900	1-1/2″ diameter		36	.222		4.61	12.40		17.01	24
1910	2″ diameter	Q-1	59	.271		5.75	13.60		19.35	27
1920	2-1/2″ diameter		56	.286		8.75	14.35		23.10	31
1930	3″ diameter		53	.302		10.70	15.15		25.85	35
1940	4″ diameter		48	.333		14.05	16.75		30.80	40.50
1950	5″ diameter		43	.372		26	18.70		44.70	56.50
1960	6″ diameter		39	.410		28	20.50		48.50	62
1970	8″ diameter	Q-2	48	.500		41	26		67	84
1980	10″ diameter		43	.558		80	29		109	132
1990	12″ diameter		42	.571		101	30		131	156
2000	14″ diameter		31	.774		158	40.50		198.50	234
2010	16″ diameter		23	1.043		224	54.50		278.50	330
2340	To delete coupling & hangers, subtract									
2360	1/2″ diam. to 1-1/4″ diam.					65%	74%			
2370	1-1/2″ diam. to 6″ diam.					44%	57%			
2380	8″ diam. to 12″ diam.					41%	53%			
2390	14″ diam. to 16″ diam.					48%	45%			
2420	Schedule 80									
2440	1/4″ diameter	1 Plum	58	.138	L.F.	2.35	7.70		10.05	14.20
2450	3/8″ diameter		55	.145		2.35	8.10		10.45	14.85

22 11 13.74 Pipe, Plastic		Crew	Daily Output	Labor-Hours	Unit	Material	Labor	Equipment	Total	Total Incl O&P
						2013 Bare Costs				
2460	1/2″ diameter	1 Plum	50	.160	L.F.	3.13	8.95		12.08	16.90
2470	3/4″ diameter		47	.170		3.53	9.50		13.03	18.20
2480	1″ diameter		43	.186		4.12	10.40		14.52	20
2490	1-1/4″ diameter		39	.205		5.10	11.45		16.55	23
2500	1-1/2″ diameter		34	.235		5.45	13.15		18.60	26
2510	2″ diameter	Q-1	55	.291		6.60	14.60		21.20	29.50
2520	2-1/2″ diameter		52	.308		9.15	15.45		24.60	33.50
2530	3″ diameter		50	.320		13.25	16.05		29.30	38.50
2540	4″ diameter		46	.348		18.05	17.45		35.50	46.50
2550	5″ diameter		42	.381		26	19.15		45.15	58
2560	6″ diameter		38	.421		41	21		62	77
2570	8″ diameter	Q-2	47	.511		50.50	26.50		77	96
2580	10″ diameter		42	.571		95	30		125	150
2590	12″ diameter		38	.632		120	33		153	182
2830	To delete coupling & hangers, subtract									
2840	1/4″ diam. to 1/2″ diam.					66%	80%			
2850	3/4″ diam. to 1-1/4″ diam.					61%	73%			
2860	1-1/2″ diam. to 6″ diam.					41%	57%			
2870	8″ diam. to 12″ diam.					31%	50%			
2900	Schedule 120									
2910	1/2″ diameter	1 Plum	50	.160	L.F.	3.32	8.95		12.27	17.10
2950	3/4″ diameter		47	.170		3.77	9.50		13.27	18.45
2960	1″ diameter		43	.186		4.43	10.40		14.83	20.50
2970	1-1/4″ diameter		39	.205		5.50	11.45		16.95	23.50
2980	1-1/2″ diameter		33	.242		6.15	13.55		19.70	27.50
2990	2″ diameter	Q-1	54	.296		7.65	14.90		22.55	31
3000	2-1/2″ diameter		52	.308		12.40	15.45		27.85	37
3010	3″ diameter		49	.327		15.35	16.40		31.75	41.50
3020	4″ diameter		45	.356		22.50	17.85		40.35	52
3030	6″ diameter		37	.432		43.50	21.50		65	80
3240	To delete coupling & hangers, subtract									
3250	1/2″ diam. to 1-1/4″ diam.					52%	74%			
3260	1-1/2″ diam. to 4″ diam.					30%	57%			
3270	6″ diam.					17%	50%			
3300	PVC, pressure, couplings 10′ O.C., clevis hanger assy's, 3 per 10′									
3310	SDR 26, 160 psi									
3350	1-1/4″ diameter	1 Plum	42	.190	L.F.	3.78	10.65		14.43	20
3360	1-1/2″ diameter	″	36	.222		4.04	12.40		16.44	23
3370	2″ diameter	Q-1	59	.271		4.79	13.60		18.39	26
3380	2-1/2″ diameter		56	.286		7.20	14.35		21.55	29.50
3390	3″ diameter		53	.302		9.65	15.15		24.80	33.50
3400	4″ diameter		48	.333		13.45	16.75		30.20	40
3420	6″ diameter		39	.410		27.50	20.50		48	61.50
3430	8″ diameter	Q-2	48	.500		41.50	26		67.50	84.50
3660	To delete coupling & clevis hanger assy's, subtract									
3670	1-1/4″ diam.					63%	68%			
3680	1-1/2″ diam. to 4″ diam.					48%	57%			
3690	6″ diam. to 8″ diam.					60%	54%			
3720	SDR 21, 200 psi, 1/2″ diameter	1 Plum	54	.148	L.F.	2.70	8.25		10.95	15.40
3740	3/4″ diameter		51	.157		3	8.75		11.75	16.50
3750	1″ diameter		46	.174		3.23	9.70		12.93	18.15
3760	1-1/4″ diameter		42	.190		3.94	10.65		14.59	20.50
3770	1-1/2″ diameter		36	.222		4.24	12.40		16.64	23.50

22 11 13.74 Pipe, Plastic		Crew	Daily Output	Labor-Hours	Unit	Material	Labor (2013 Bare Costs)	Equipment	Total	Total Incl O&P
3780	2" diameter	Q-1	59	.271	L.F.	5.10	13.60		18.70	26
3790	2-1/2" diameter		56	.286		9.10	14.35		23.45	31.50
3800	3" diameter		53	.302		10.30	15.15		25.45	34.50
3810	4" diameter		48	.333		16.60	16.75		33.35	43.50
3830	6" diameter		39	.410		30	20.50		50.50	64
3840	8" diameter	Q-2	48	.500		46	26		72	89.50
4000	To delete coupling & hangers, subtract									
4010	1/2" diam. to 3/4" diam.					71%	77%			
4020	1" diam. to 1-1/4" diam.					63%	70%			
4030	1-1/2" diam. to 6" diam.					44%	57%			
4040	8" diam.					46%	54%			
4100	DWV type, schedule 40, couplings 10' O.C., clevis hanger assy's, 3 per 10'									
4210	ABS, schedule 40, foam core type									
4212	Plain end black									
4214	1-1/2" diameter	1 Plum	39	.205	L.F.	3.09	11.45		14.54	20.50
4216	2" diameter	Q-1	62	.258		3.59	12.95		16.54	23.50
4218	3" diameter		56	.286		6.50	14.35		20.85	28.50
4220	4" diameter		51	.314		9.10	15.75		24.85	33.50
4222	6" diameter		42	.381		20.50	19.15		39.65	51.50
4240	To delete coupling & hangers, subtract									
4244	1-1/2" diam. to 6" diam.					43%	48%			
4400	PVC									
4410	1-1/4" diameter	1 Plum	42	.190	L.F.	3.44	10.65		14.09	19.80
4420	1-1/2" diameter	"	36	.222		3.12	12.40		15.52	22
4460	2" diameter	Q-1	59	.271		3.59	13.60		17.19	24.50
4470	3" diameter		53	.302		6.45	15.15		21.60	30
4480	4" diameter		48	.333		8.80	16.75		25.55	34.50
4490	6" diameter		39	.410		17.65	20.50		38.15	50.50
4500	8" diameter	Q-2	48	.500		28	26		54	69.50
4510	To delete coupling & hangers, subtract									
4520	1-1/4" diam. to 1-1/2" diam.					48%	60%			
4530	2" diam. to 8" diam.					42%	54%			
4550	PVC, schedule 40, foam core type									
4552	Plain end, white									
4554	1-1/2" diameter	1 Plum	39	.205	L.F.	2.86	11.45		14.31	20.50
4556	2" diameter	Q-1	62	.258		3.27	12.95		16.22	23
4558	3" diameter		56	.286		5.85	14.35		20.20	28
4560	4" diameter		51	.314		7.85	15.75		23.60	32
4562	6" diameter		42	.381		15.80	19.15		34.95	46.50
4564	8" diameter	Q-2	51	.471		25	24.50		49.50	64.50
4568	10" diameter		48	.500		29.50	26		55.50	71.50
4570	12" diameter		46	.522		34	27		61	78.50
4580	To delete coupling & hangers, subtract									
4582	1-1/2" dia to 2" dia					58%	54%			
4584	3" dia to 12" dia					46%	42%			
4800	PVC, clear pipe, cplgs. 10' O.C., clevis hanger assy's 3 per 10', Sched. 40									
4840	1/4" diameter	1 Plum	59	.136	L.F.	2.92	7.55		10.47	14.60
4850	3/8" diameter		56	.143		3.26	7.95		11.21	15.60
4860	1/2" diameter		54	.148		3.91	8.25		12.16	16.75
4870	3/4" diameter		51	.157		4.60	8.75		13.35	18.25
4880	1" diameter		46	.174		5.95	9.70		15.65	21
4890	1-1/4" diameter		42	.190		7.40	10.65		18.05	24
4900	1-1/2" diameter		36	.222		8.45	12.40		20.85	28

22 11 13.74 Pipe, Plastic		Crew	Daily Output	Labor-Hours	Unit	Material	Labor	Equipment	Total	Total Incl O&P
						2013 Bare Costs				
4910	2" diameter	Q-1	59	.271	L.F.	10.85	13.60		24.45	32.50
4920	2-1/2" diameter		56	.286		16.95	14.35		31.30	40
4930	3" diameter		53	.302		21.50	15.15		36.65	46.50
4940	3-1/2" diameter		50	.320		30	16.05		46.05	57.50
4950	4" diameter		48	.333		31.50	16.75		48.25	60
5250	To delete coupling & hangers, subtract									
5260	1/4" diam. to 3/8" diam.					60%	81%			
5270	1/2" diam. to 3/4" diam.					41%	77%			
5280	1" diam. to 1-1/2" diam.					26%	67%			
5290	2" diam. to 4" diam.					16%	58%			
5300	CPVC, socket joint, couplings 10' O.C., clevis hanger assemblies, 3 per 10'									
5302	Schedule 40									
5304	1/2" diameter	1 Plum	54	.148	L.F.	3.65	8.25		11.90	16.45
5305	3/4" diameter		51	.157		4.24	8.75		12.99	17.85
5306	1" diameter		46	.174		5.30	9.70		15	20.50
5307	1-1/4" diameter		42	.190		6.65	10.65		17.30	23.50
5308	1-1/2" diameter		36	.222		7.65	12.40		20.05	27
5309	2" diameter	Q-1	59	.271		9.45	13.60		23.05	31
5310	2-1/2" diameter		56	.286		15.75	14.35		30.10	39
5311	3" diameter		53	.302		19	15.15		34.15	44
5312	4" diameter		48	.333		26	16.75		42.75	53.50
5314	6" diameter		43	.372		50	18.70		68.70	83
5318	To delete coupling & hangers, subtract									
5319	1/2" diam. to 3/4" diam.					37%	77%			
5320	1" diam. to 1-1/4" diam.					27%	70%			
5321	1-1/2" diam. to 3" diam.					21%	57%			
5322	4" diam. to 6" diam.					16%	57%			
5324	Schedule 80									
5325	1/2" diameter	1 Plum	50	.160	L.F.	3.86	8.95		12.81	17.70
5326	3/4" diameter		47	.170		4.57	9.50		14.07	19.35
5327	1" diameter		43	.186		5.80	10.40		16.20	22
5328	1-1/4" diameter		39	.205		7.40	11.45		18.85	25.50
5329	1-1/2" diameter		34	.235		8.65	13.15		21.80	29.50
5330	2" diameter	Q-1	55	.291		10.95	14.60		25.55	34
5331	2-1/2" diameter		52	.308		17.80	15.45		33.25	43
5332	3" diameter		50	.320		20.50	16.05		36.55	46.50
5333	4" diameter		46	.348		29	17.45		46.45	58
5334	6" diameter		38	.421		57.50	21		78.50	95.50
5335	8" diameter	Q-2	47	.511		111	26.50		137.50	162
5339	To delete couplings & hangers, subtract									
5340	1/2" diam. to 3/4" diam.					44%	77%			
5341	1" diam. to 1-1/4" diam.					32%	71%			
5342	1-1/2" diam. to 4" diam.					25%	58%			
5343	6" diam. to 8" diam.					20%	53%			
5360	CPVC, threaded, couplings 10' O.C., clevis hanger assemblies, 3 per 10'									
5380	Schedule 40									
5460	1/2" diameter	1 Plum	54	.148	L.F.	4.40	8.25		12.65	17.30
5470	3/4" diameter		51	.157		5.55	8.75		14.30	19.30
5480	1" diameter		46	.174		6.70	9.70		16.40	22
5490	1-1/4" diameter		42	.190		7.70	10.65		18.35	24.50
5500	1-1/2" diameter		36	.222		8.55	12.40		20.95	28
5510	2" diameter	Q-1	59	.271		10.55	13.60		24.15	32
5520	2-1/2" diameter		56	.286		16.95	14.35		31.30	40

22 11 13 – Facility Water Distribution Piping

22 11 13.74 Pipe, Plastic		Crew	Daily Output	Labor-Hours	Unit	Material	Labor	Equipment	Total	Total Incl O&P
						2013 Bare Costs				
5530	3" diameter	Q-1	53	.302	L.F.	20.50	15.15		35.65	45.50
5540	4" diameter		48	.333		33	16.75		49.75	61
5550	6" diameter		43	.372		53.50	18.70		72.20	86.50
5730	To delete coupling & hangers, subtract									
5740	1/2" diam. to 3/4" diam.					37%	77%			
5750	1" diam. to 1-1/4" diam.					27%	70%			
5760	1-1/2" diam. to 3" diam.					21%	57%			
5770	4" diam. to 6" diam.					16%	57%			
5800	Schedule 80									
5860	1/2" diameter	1 Plum	50	.160	L.F.	4.61	8.95		13.56	18.50
5870	3/4" diameter		47	.170		5.90	9.50		15.40	21
5880	1" diameter		43	.186		7.15	10.40		17.55	23.50
5890	1-1/4" diameter		39	.205		8.50	11.45		19.95	26.50
5900	1-1/2" diameter		34	.235		9.55	13.15		22.70	30.50
5910	2" diameter	Q-1	55	.291		12.05	14.60		26.65	35.50
5920	2-1/2" diameter		52	.308		18.95	15.45		34.40	44.50
5930	3" diameter		50	.320		22.50	16.05		38.55	48.50
5940	4" diameter		46	.348		35.50	17.45		52.95	65.50
5950	6" diameter		38	.421		61	21		82	99
5960	8" diameter	Q-2	47	.511		110	26.50		136.50	161
6060	To delete couplings & hangers, subtract									
6070	1/2" diam. to 3/4" diam.					44%	77%			
6080	1" diam. to 1-1/4" diam.					32%	71%			
6090	1-1/2" diam. to 4" diam.					25%	58%			
6100	6" diam. to 8" diam.					20%	53%			
6240	CTS, 1/2" diameter	1 Plum	54	.148	L.F.	2.38	8.25		10.63	15.05
6250	3/4" diameter		51	.157		2.97	8.75		11.72	16.45
6260	1" diameter		46	.174		4.38	9.70		14.08	19.40
6270	1 1/4"		42	.190		5.65	10.65		16.30	22.50
6280	1 1/2" diameter		36	.222		7	12.40		19.40	26.50
6290	2" diameter	Q-1	59	.271		10.80	13.60		24.40	32.50
6370	To delete coupling & hangers, subtract									
6380	1/2" diam.					51%	79%			
6390	3/4" diam.					40%	76%			
6392	1" thru 2" diam.					72%	68%			
6500	Residential installation, plastic pipe									
6510	Couplings 10' O.C., strap hangers 3 per 10'									
6520	PVC, Schedule 40									
6530	1/2" diameter	1 Plum	138	.058	L.F.	2.14	3.23		5.37	7.25
6540	3/4" diameter		128	.063		2.28	3.49		5.77	7.75
6550	1" diameter		119	.067		2.84	3.75		6.59	8.75
6560	1-1/4" diameter		111	.072		3.56	4.02		7.58	9.95
6570	1-1/2" diameter		104	.077		3.73	4.29		8.02	10.55
6580	2" diameter	Q-1	197	.081		4.36	4.08		8.44	10.95
6590	2-1/2" diameter		162	.099		8.20	4.96		13.16	16.45
6600	4" diameter		123	.130		11.85	6.55		18.40	23
6700	PVC, DWV, Schedule 40									
6720	1-1/4" diameter	1 Plum	100	.080	L.F.	3.03	4.46		7.49	10.05
6730	1-1/2" diameter	"	94	.085		3.11	4.75		7.86	10.55
6740	2" diameter	Q-1	178	.090		3.50	4.52		8.02	10.65
6760	4" diameter	"	110	.145		9.55	7.30		16.85	21.50
7280	PEX, flexible, no couplings or hangers									
7282	Note: For labor costs add 25% to the couplings and fittings labor total.									

	22 11 13.74 Pipe, Plastic	Crew	Daily Output	Labor-Hours	Unit	2013 Bare Costs Material	Labor	Equipment	Total	Total Incl O&P
7285	For fittings see section 23 83 16.10 7000									
7300	Non-barrier type, hot/cold tubing rolls									
7310	1/4" diameter x 100'				L.F.	.46			.46	.51
7350	3/8" diameter x 100'					.51			.51	.56
7360	1/2" diameter x 100'					.57			.57	.63
7370	1/2" diameter x 500'					.57			.57	.63
7380	1/2" diameter x 1000'					.57			.57	.63
7400	3/4" diameter x 100'					1.05			1.05	1.16
7410	3/4" diameter x 500'					1.05			1.05	1.16
7420	3/4" diameter x 1000'					1.05			1.05	1.16
7460	1" diameter x 100'					1.80			1.80	1.98
7470	1" diameter x 300'					1.80			1.80	1.98
7480	1" diameter x 500'					1.78			1.78	1.96
7500	1-1/4" diameter x 100'					3.04			3.04	3.34
7510	1-1/4" diameter x 300'					3.04			3.04	3.34
7540	1-1/2" diameter x 100'					4.15			4.15	4.57
7550	1-1/2" diameter x 300'				↓	4.15			4.15	4.57
7596	Most sizes available in red or blue									
7700	Non-barrier type, hot/cold tubing straight lengths									
7710	1/2" diameter x 20'				L.F.	.57			.57	.63
7750	3/4" diameter x 20'					1.05			1.05	1.16
7760	1" diameter x 20'					1.80			1.80	1.98
7770	1-1/4" diameter x 20'					3.04			3.04	3.34
7780	1-1/2" diameter x 20'				↓	4.15			4.15	4.57
7796	Most sizes available in red or blue									
9000	Polypropylene pipe									
9002	For fusion weld fittings and accessories see line 22 11 13.76 9400									
9004	Note: sizes 1/2" thru 4" use socket fusion									
9005	Sizes 6" thru 10" use butt fusion									
9010	SDR 7.4, (domestic hot water piping)									
9011	Enhanced to minimize thermal expansion and high temperature life									
9016	13' lengths, size is ID, includes joints 13' O.C. and hangers 3 per 10'									
9020	3/8" diameter	1 Plum	53	.151	L.F.	2.98	8.40		11.38	16
9022	1/2" diameter		52	.154		3.37	8.60		11.97	16.65
9024	3/4" diameter		50	.160		3.87	8.95		12.82	17.70
9026	1" diameter		45	.178		4.83	9.90		14.73	20.50
9028	1-1/4" diameter		40	.200		6.80	11.15		17.95	24.50
9030	1-1/2" diameter	↓	35	.229		9.20	12.75		21.95	29.50
9032	2" diameter	Q-1	58	.276		12.70	13.85		26.55	35
9034	2-1/2" diameter		55	.291		17.25	14.60		31.85	41
9036	3" diameter		52	.308		23.50	15.45		38.95	49.50
9038	3-1/2" diameter		49	.327		33	16.40		49.40	61
9040	4" diameter		46	.348		37.50	17.45		54.95	67.50
9042	6" diameter	↓	39	.410		44.50	20.50		65	80
9044	8" diameter	Q-2	48	.500		68	26		94	114
9046	10" diameter	"	43	.558	↓	109	29		138	164
9050	To delete joint & hangers, subtract									
9052	3/8" diam. to 1" diam.					45%	65%			
9054	1-1/4" diam. to 4" diam.					15%	45%			
9056	6" diam. to 10" diam.					5%	24%			
9060	SDR 11, (domestic cold water piping)									
9062	13' lengths, size is ID, includes joints 13' O.C. and hangers 3 per 10'									
9064	1/2" diameter	1 Plum	57	.140	L.F.	3.07	7.85		10.92	15.20

22 11 13.74 Pipe, Plastic		Crew	Daily Output	Labor-Hours	Unit	Material	Labor	Equipment	Total	Total Incl O&P
						2013 Bare Costs				
9066	3/4" diameter	1 Plum	54	.148	L.F.	3.38	8.25		11.63	16.15
9068	1" diameter		49	.163		3.95	9.10		13.05	18.05
9070	1-1/4" diameter		45	.178		5.50	9.90		15.40	21
9072	1-1/2" diameter		40	.200		6.95	11.15		18.10	24.50
9074	2" diameter	Q-1	62	.258		9.50	12.95		22.45	30
9076	2-1/2" diameter		59	.271		12.35	13.60		25.95	34
9078	3" diameter		56	.286		16.95	14.35		31.30	40
9080	3-1/2" diameter		53	.302		24	15.15		39.15	49.50
9082	4" diameter		50	.320		27	16.05		43.05	54
9084	6" diameter		47	.340		31	17.10		48.10	60
9086	8" diameter	Q-2	51	.471		50.50	24.50		75	92.50
9088	10" diameter	"	46	.522		77	27		104	126
9090	To delete joint & hangers, subtract									
9092	3/8" diam. to 1" diam.					45%	65%			
9094	1-1/4" diam. to 4" diam.					15%	45%			
9096	6" diam. to 10" diam.					5%	24%			

22 11 13.76 Pipe Fittings, Plastic		Crew	Daily Output	Labor-Hours	Unit	Material	Labor	Equipment	Total	Total Incl O&P
0010	**PIPE FITTINGS, PLASTIC**									
0030	Epoxy resin, fiberglass reinforced, general service									
0090	Elbow, 90°, 2"	Q-1	33.10	.483	Ea.	102	24.50		126.50	149
0100	3"		20.80	.769		113	38.50		151.50	182
0110	4"		16.50	.970		115	48.50		163.50	200
0120	6"		10.10	1.584		225	79.50		304.50	370
0130	8"	Q-2	9.30	2.581		415	134		549	655
0140	10"		8.50	2.824		520	147		667	795
0150	12"		7.60	3.158		750	164		914	1,075
0160	45° Elbow, same as 90°									
0170	Elbow, 90°, flanged									
0172	2"	Q-1	23	.696	Ea.	164	35		199	234
0173	3"		16	1		188	50		238	283
0174	4"		13	1.231		245	62		307	365
0176	6"		8	2		445	100		545	640
0177	8"	Q-2	9	2.667		805	139		944	1,100
0178	10"		7	3.429		1,100	179		1,279	1,475
0179	12"		5	4.800		1,475	250		1,725	2,000
0186	Elbow, 45°, flanged									
0188	2"	Q-1	23	.696	Ea.	164	35		199	234
0189	3"		16	1		188	50		238	283
0190	4"		13	1.231		245	62		307	365
0192	6"		8	2		445	100		545	640
0193	8"	Q-2	9	2.667		760	139		899	1,050
0194	10"		7	3.429		970	179		1,149	1,350
0195	12"		5	4.800		1,250	250		1,500	1,750
0290	Tee, 2"	Q-1	20	.800		54	40		94	120
0300	3"		13.90	1.151		64.50	58		122.50	158
0310	4"		11	1.455		89.50	73		162.50	208
0320	6"		6.70	2.388		225	120		345	430
0330	8"	Q-2	6.20	3.871		1,600	202		1,802	2,050
0340	10"		5.70	4.211		1,800	219		2,019	2,325
0350	12"		5.10	4.706		2,200	245		2,445	2,800
0352	Tee, flanged									
0354	2"	Q-1	17	.941	Ea.	222	47.50		269.50	315

22 11 Facility Water Distribution

22 11 13 – Facility Water Distribution Piping

22 11 13.76 Pipe Fittings, Plastic		Crew	Daily Output	Labor-Hours	Unit	Material	2013 Bare Costs Labor	Equipment	Total	Total Incl O&P
0355	3″	Q-1	10	1.600	Ea.	294	80.50		374.50	445
0356	4″		8	2		330	100		430	515
0358	6″		5	3.200		565	161		726	865
0359	8″	Q-2	6	4		1,075	208		1,283	1,500
0360	10″		5	4.800		1,550	250		1,800	2,100
0361	12″		4	6		2,150	310		2,460	2,850
0365	Wye, flanged									
0367	2″	Q-1	17	.941	Ea.	440	47.50		487.50	555
0368	3″		10	1.600		605	80.50		685.50	785
0369	4″		8	2		810	100		910	1,050
0371	6″		5	3.200		1,125	161		1,286	1,500
0372	8″	Q-2	6	4		1,875	208		2,083	2,375
0373	10″		5	4.800		2,900	250		3,150	3,575
0374	12″		4	6		4,225	310		4,535	5,125
0380	Couplings									
0410	2″	Q-1	33.10	.483	Ea.	19.05	24.50		43.55	57.50
0420	3″		20.80	.769		20.50	38.50		59	80.50
0430	4″		16.50	.970		27	48.50		75.50	104
0440	6″		10.10	1.584		65.50	79.50		145	192
0450	8″	Q-2	9.30	2.581		111	134		245	325
0460	10″		8.50	2.824		163	147		310	400
0470	12″		7.60	3.158		218	164		382	490
0473	High corrosion resistant couplings, add					30%				
0474	Reducer, concentric, flanged									
0475	2″ x 1-1/2″	Q-1	30	.533	Ea.	425	27		452	510
0476	3″ x 2″		24	.667		465	33.50		498.50	560
0477	4″ x 3″		19	.842		520	42.50		562.50	640
0479	6″ x 4″		15	1.067		520	53.50		573.50	655
0480	8″ x 6″	Q-2	16	1.500		860	78		938	1,075
0481	10″ x 8″		13	1.846		795	96		891	1,025
0482	12″ x 10″		11	2.182		1,225	114		1,339	1,525
0486	Adapter, bell x male or female									
0488	2″	Q-1	28	.571	Ea.	21.50	28.50		50	66.50
0489	3″		20	.800		31.50	40		71.50	95
0491	4″		17	.941		42	47.50		89.50	118
0492	6″		12	1.333		88.50	67		155.50	199
0493	8″	Q-2	15	1.600		128	83.50		211.50	267
0494	10″	″	11	2.182		185	114		299	375
0528	Flange									
0532	2″	Q-1	46	.348	Ea.	31.50	17.45		48.95	61
0533	3″		32	.500		38	25		63	80
0534	4″		26	.615		50.50	31		81.50	102
0536	6″		16	1		95.50	50		145.50	181
0537	8″	Q-2	18	1.333		146	69.50		215.50	265
0538	10″		14	1.714		207	89.50		296.50	360
0539	12″		10	2.400		260	125		385	475
2100	PVC schedule 80, socket joint									
2110	90° elbow, 1/2″	1 Plum	30.30	.264	Ea.	2.36	14.75		17.11	24.50
2130	3/4″		26	.308		3.01	17.15		20.16	29.50
2140	1″		22.70	.352		4.84	19.65		24.49	35
2150	1-1/4″		20.20	.396		6.50	22		28.50	40.50
2160	1-1/2″		18.20	.440		6.95	24.50		31.45	44.50
2170	2″	Q-1	33.10	.483		8.40	24.50		32.90	46

22 11 13.76 Pipe Fittings, Plastic		Crew	Daily Output	Labor-Hours	Unit	2013 Bare Costs Material	Labor	Equipment	Total	Total Incl O&P
2180	3"	Q-1	20.80	.769	Ea.	22	38.50		60.50	82.50
2190	4"		16.50	.970		33.50	48.50		82	111
2200	6"		10.10	1.584		95.50	79.50		175	225
2210	8"	Q-2	9.30	2.581		264	134		398	490
2250	45° elbow, 1/2"	1 Plum	30.30	.264		4.43	14.75		19.18	27
2270	3/4"		26	.308		6.75	17.15		23.90	33.50
2280	1"		22.70	.352		10.15	19.65		29.80	40.50
2290	1-1/4"		20.20	.396		12.95	22		34.95	47.50
2300	1-1/2"		18.20	.440		15.30	24.50		39.80	54
2310	2"	Q-1	33.10	.483		19.85	24.50		44.35	58.50
2320	3"		20.80	.769		50.50	38.50		89	114
2330	4"		16.50	.970		91	48.50		139.50	174
2340	6"		10.10	1.584		115	79.50		194.50	246
2350	8"	Q-2	9.30	2.581		249	134		383	475
2400	Tee, 1/2"	1 Plum	20.20	.396		6.65	22		28.65	41
2420	3/4"		17.30	.462		6.95	26		32.95	46.50
2430	1"		15.20	.526		8.70	29.50		38.20	53.50
2440	1-1/4"		13.50	.593		24	33		57	76.50
2450	1-1/2"		12.10	.661		24	37		61	82
2460	2"	Q-1	20	.800		30	40		70	93.50
2470	3"		13.90	1.151		40.50	58		98.50	132
2480	4"		11	1.455		47	73		120	162
2490	6"		6.70	2.388		161	120		281	360
2500	8"	Q-2	6.20	3.871		375	202		577	715
2510	Flange, socket, 150 lb., 1/2"	1 Plum	55.60	.144		12.90	8.05		20.95	26.50
2514	3/4"		47.60	.168		13.80	9.40		23.20	29.50
2518	1"		41.70	.192		15.35	10.70		26.05	33
2522	1-1/2"		33.30	.240		16.15	13.40		29.55	38
2526	2"	Q-1	60.60	.264		21.50	13.25		34.75	43.50
2530	4"		30.30	.528		46.50	26.50		73	91
2534	6"		18.50	.865		73	43.50		116.50	146
2538	8"	Q-2	17.10	1.404		131	73		204	254
2550	Coupling, 1/2"	1 Plum	30.30	.264		4.26	14.75		19.01	26.50
2570	3/4"		26	.308		5.75	17.15		22.90	32.50
2580	1"		22.70	.352		5.95	19.65		25.60	36
2590	1-1/4"		20.20	.396		9.05	22		31.05	43.50
2600	1-1/2"		18.20	.440		9.75	24.50		34.25	47.50
2610	2"	Q-1	33.10	.483		10.45	24.50		34.95	48
2620	3"		20.80	.769		29.50	38.50		68	90.50
2630	4"		16.50	.970		37	48.50		85.50	114
2640	6"		10.10	1.584		79.50	79.50		159	208
2650	8"	Q-2	9.30	2.581		108	134		242	320
2660	10"		8.50	2.824		370	147		517	630
2670	12"		7.60	3.158		430	164		594	720
2700	PVC (white), schedule 40, socket joints									
2760	90° elbow, 1/2"	1 Plum	33.30	.240	Ea.	.46	13.40		13.86	20.50
2770	3/4"		28.60	.280		.51	15.60		16.11	24
2780	1"		25	.320		.92	17.85		18.77	28
2790	1-1/4"		22.20	.360		1.63	20		21.63	32.50
2800	1-1/2"		20	.400		1.76	22.50		24.26	35.50
2810	2"	Q-1	36.40	.440		2.75	22		24.75	36.50
2820	2-1/2"		26.70	.599		8.60	30		38.60	55
2830	3"		22.90	.699		10	35		45	64

22 11 13.76 Pipe Fittings, Plastic		Crew	Daily Output	Labor-Hours	Unit	Material	Labor	Equipment	Total	Total Incl O&P
						2013 Bare Costs				
2840	4"	Q-1	18.20	.879	Ea.	17.90	44		61.90	86
2850	5"		12.10	1.322		46	66.50		112.50	151
2860	6"		11.10	1.441		57	72.50		129.50	172
2870	8"	Q-2	10.30	2.330		146	121		267	345
2980	45° elbow, 1/2"	1 Plum	33.30	.240		.76	13.40		14.16	21
2990	3/4"		28.60	.280		1.17	15.60		16.77	25
3000	1"		25	.320		1.41	17.85		19.26	28.50
3010	1-1/4"		22.20	.360		1.97	20		21.97	32.50
3020	1-1/2"		20	.400		2.47	22.50		24.97	36
3030	2"	Q-1	36.40	.440		3.21	22		25.21	37
3040	2-1/2"		26.70	.599		8.35	30		38.35	54.50
3050	3"		22.90	.699		13	35		48	67.50
3060	4"		18.20	.879		23.50	44		67.50	92
3070	5"		12.10	1.322		46	66.50		112.50	151
3080	6"		11.10	1.441		57.50	72.50		130	173
3090	8"	Q-2	10.30	2.330		138	121		259	335
3180	Tee, 1/2"	1 Plum	22.20	.360		.57	20		20.57	31
3190	3/4"		19	.421		.65	23.50		24.15	36
3200	1"		16.70	.479		1.22	26.50		27.72	42
3210	1-1/4"		14.80	.541		1.91	30		31.91	47.50
3220	1-1/2"		13.30	.602		2.33	33.50		35.83	53
3230	2"	Q-1	24.20	.661		3.38	33		36.38	53.50
3240	2-1/2"		17.80	.899		11.15	45		56.15	80.50
3250	3"		15.20	1.053		14.65	53		67.65	95.50
3260	4"		12.10	1.322		26.50	66.50		93	129
3270	5"		8.10	1.975		64	99		163	220
3280	6"		7.40	2.162		89.50	109		198.50	263
3290	8"	Q-2	6.80	3.529		207	184		391	505
3380	Coupling, 1/2"	1 Plum	33.30	.240		.30	13.40		13.70	20.50
3390	3/4"		28.60	.280		.41	15.60		16.01	24
3400	1"		25	.320		.73	17.85		18.58	28
3410	1-1/4"		22.20	.360		1.01	20		21.01	31.50
3420	1-1/2"		20	.400		1.07	22.50		23.57	34.50
3430	2"	Q-1	36.40	.440		1.64	22		23.64	35.50
3440	2-1/2"		26.70	.599		3.62	30		33.62	49.50
3450	3"		22.90	.699		5.65	35		40.65	59.50
3460	4"		18.20	.879		8.20	44		52.20	75.50
3470	5"		12.10	1.322		15	66.50		81.50	117
3480	6"		11.10	1.441		26	72.50		98.50	138
3490	8"	Q-2	10.30	2.330		48.50	121		169.50	237
3600	Cap, schedule 40, PVC socket 1/2"	1 Plum	60.60	.132		.41	7.35		7.76	11.55
3610	3/4"		51.90	.154		.48	8.60		9.08	13.50
3620	1"		45.50	.176		.76	9.80		10.56	15.65
3630	1-1/4"		40.40	.198		1.07	11.05		12.12	17.85
3640	1-1/2"		36.40	.220		1.17	12.25		13.42	19.75
3650	2"	Q-1	66.10	.242		1.41	12.15		13.56	19.85
3660	2-1/2"		48.50	.330		4.49	16.55		21.04	30
3670	3"		41.60	.385		4.91	19.30		24.21	34.50
3680	4"		33.10	.483		11.15	24.50		35.65	49
3690	6"		20.20	.792		27	40		67	89.50
3700	8"	Q-2	18.60	1.290		67.50	67		134.50	175
3710	Reducing insert, schedule 40, socket weld									
3712	3/4"	1 Plum	31.50	.254	Ea.	.48	14.15		14.63	22

22 11 13 – Facility Water Distribution Piping

22 11 13.76 Pipe Fittings, Plastic		Crew	Daily Output	Labor-Hours	Unit	Material	2013 Bare Costs Labor	Equipment	Total	Total Incl O&P
3713	1"	1 Plum	27.50	.291	Ea.	.87	16.25		17.12	25.50
3715	1-1/2"		22	.364		1.24	20.50		21.74	32
3716	2"	Q-1	40	.400		2.05	20		22.05	33
3717	4"		20	.800		10.85	40		50.85	72.50
3718	6"		12.20	1.311		27	66		93	129
3719	8"	Q-2	11.30	2.124		106	111		217	284
3730	Reducing insert, socket weld x female/male thread									
3732	1/2"	1 Plum	38.30	.209	Ea.	2.17	11.65		13.82	19.95
3733	3/4"		32.90	.243		1.34	13.55		14.89	22
3734	1"		28.80	.278		1.88	15.50		17.38	25.50
3736	1-1/2"		23	.348		3.38	19.40		22.78	32.50
3737	2"	Q-1	41.90	.382		3.62	19.20		22.82	33
3738	4"	"	20.90	.766		34.50	38.50		73	96
3742	Male adapter, socket weld x male thread									
3744	1/2"	1 Plum	38.30	.209	Ea.	.41	11.65		12.06	18
3745	3/4"		32.90	.243		.46	13.55		14.01	21
3746	1"		28.80	.278		.83	15.50		16.33	24.50
3748	1-1/2"		23	.348		1.34	19.40		20.74	30.50
3749	2"	Q-1	41.90	.382		1.76	19.20		20.96	31
3750	4"	"	20.90	.766		9.70	38.50		48.20	68.50
3754	Female adapter, socket weld x female thread									
3756	1/2"	1 Plum	38.30	.209	Ea.	.51	11.65		12.16	18.10
3757	3/4"		32.90	.243		.65	13.55		14.20	21
3758	1"		28.80	.278		.76	15.50		16.26	24.50
3760	1-1/2"		23	.348		1.34	19.40		20.74	30.50
3761	2"	Q-1	41.90	.382		1.80	19.20		21	31
3762	4"	"	20.90	.766		10.15	38.50		48.65	69
3800	PVC, schedule 80, socket joints									
3810	Reducing insert									
3812	3/4"	1 Plum	28.60	.280	Ea.	1.37	15.60		16.97	25
3813	1"		25	.320		3.93	17.85		21.78	31.50
3815	1-1/2"		20	.400		8.40	22.50		30.90	43
3816	2"	Q-1	36.40	.440		12	22		34	46.50
3817	4"		18.20	.879		45.50	44		89.50	117
3818	6"		11.10	1.441		63.50	72.50		136	179
3819	8"	Q-2	10.20	2.353		365	123		488	585
3830	Reducing insert, socket weld x female/male thread									
3832	1/2"	1 Plum	34.80	.230	Ea.	8.60	12.85		21.45	29
3833	3/4"		29.90	.268		5.25	14.95		20.20	28.50
3834	1"		26.10	.307		8.25	17.10		25.35	35
3836	1-1/2"		20.90	.383		10.40	21.50		31.90	43.50
3837	2"	Q-1	38	.421		15.20	21		36.20	49
3838	4"	"	19	.842		73.50	42.50		116	145
3844	Adapter, male socket x male thread									
3846	1/2"	1 Plum	34.80	.230	Ea.	3.32	12.85		16.17	23
3847	3/4"		29.90	.268		3.66	14.95		18.61	26.50
3848	1"		26.10	.307		6.35	17.10		23.45	33
3850	1-1/2"		20.90	.383		10.65	21.50		32.15	43.50
3851	2"	Q-1	38	.421		15.40	21		36.40	49
3852	4"	"	19	.842		34.50	42.50		77	102
3860	Adapter, female socket x female thread									
3862	1/2"	1 Plum	34.80	.230	Ea.	3.99	12.85		16.84	23.50
3863	3/4"		29.90	.268		5.95	14.95		20.90	29

22 11 13 – Facility Water Distribution Piping

22 11 13.76 Pipe Fittings, Plastic		Crew	Daily Output	Labor-Hours	Unit	Material	Labor	Equipment	Total	Total Incl O&P
						2013 Bare Costs				
3864	1"	1 Plum	26.10	.307	Ea.	8.75	17.10		25.85	35.50
3866	1-1/2"		20.90	.383		17.35	21.50		38.85	51
3867	2"	Q-1	38	.421		30.50	21		51.50	65.50
3868	4"	"	19	.842		92.50	42.50		135	166
3872	Union, socket joints									
3874	1/2"	1 Plum	25.80	.310	Ea.	8.75	17.30		26.05	35.50
3875	3/4"		22.10	.362		11.10	20		31.10	42.50
3876	1"		19.30	.415		12.65	23		35.65	49
3878	1-1/2"		15.50	.516		28.50	29		57.50	75
3879	2"	Q-1	28.10	.569		38.50	28.50		67	85.50
3888	Cap									
3890	1/2"	1 Plum	54.50	.147	Ea.	4.17	8.20		12.37	16.95
3891	3/4"		46.70	.171		4.40	9.55		13.95	19.25
3892	1"		41	.195		7.80	10.90		18.70	25
3894	1-1/2"		32.80	.244		9.40	13.60		23	31
3895	2"	Q-1	59.50	.269		25	13.50		38.50	48
3896	4"		30	.533		75	27		102	123
3897	6"		18.20	.879		186	44		230	272
3898	8"	Q-2	16.70	1.437		239	75		314	375
4500	DWV, ABS, non pressure, socket joints									
4540	1/4 Bend, 1-1/4"	1 Plum	20.20	.396	Ea.	5.05	22		27.05	39
4560	1-1/2"	"	18.20	.440		3.85	24.50		28.35	41
4570	2"	Q-1	33.10	.483		6.10	24.50		30.60	43.50
4580	3"		20.80	.769		15.40	38.50		53.90	75
4590	4"		16.50	.970		31	48.50		79.50	108
4600	6"		10.10	1.584		135	79.50		214.50	268
4650	1/8 Bend, same as 1/4 Bend									
4800	Tee, sanitary									
4820	1-1/4"	1 Plum	13.50	.593	Ea.	6.70	33		39.70	57.50
4830	1-1/2"	"	12.10	.661		5.85	37		42.85	62
4840	2"	Q-1	20	.800		9.05	40		49.05	70.50
4850	3"		13.90	1.151		25	58		83	115
4860	4"		11	1.455		44	73		117	159
4862	Tee, sanitary, reducing, 2" x 1-1/2"		22	.727		9.15	36.50		45.65	65
4864	3" x 2"		15.30	1.046		18.10	52.50		70.60	99
4868	4" x 3"		12.10	1.322		43	66.50		109.50	148
4870	Combination Y and 1/8 bend									
4872	1-1/2"	1 Plum	12.10	.661	Ea.	14	37		51	71
4874	2"	Q-1	20	.800		15.70	40		55.70	78
4876	3"		13.90	1.151		36	58		94	127
4878	4"		11	1.455		71.50	73		144.50	189
4880	3" x 1-1/2"		15.50	1.032		36	52		88	118
4882	4" x 3"		12.10	1.322		56	66.50		122.50	162
4900	Wye, 1-1/4"	1 Plum	13.50	.593		7.70	33		40.70	58.50
4902	1-1/2"	"	12.10	.661		9.55	37		46.55	66
4904	2"	Q-1	20	.800		11.70	40		51.70	73.50
4906	3"		13.90	1.151		27.50	58		85.50	118
4908	4"		11	1.455		56	73		129	172
4910	6"		6.70	2.388		167	120		287	365
4918	3" x 1-1/2"		15.50	1.032		22	52		74	103
4920	4" x 3"		12.10	1.322		44.50	66.50		111	149
4922	6" x 4"		6.90	2.319		138	116		254	325
4930	Double Wye, 1-1/2"	1 Plum	9.10	.879		27	49		76	104

22 11 Facility Water Distribution

22 11 13 – Facility Water Distribution Piping

22 11 13.76 Pipe Fittings, Plastic		Crew	Daily Output	Labor-Hours	Unit	2013 Bare Costs Material	Labor	Equipment	Total	Total Incl O&P
4932	2"	Q-1	16.60	.964	Ea.	32	48.50		80.50	109
4934	3"		10.40	1.538		73.50	77.50		151	197
4936	4"		8.25	1.939		144	97.50		241.50	305
4940	2" x 1-1/2"		16.80	.952		28	48		76	103
4942	3" x 2"		10.60	1.509		54	76		130	174
4944	4" x 3"		8.45	1.893		114	95		209	268
4946	6" x 4"		7.25	2.207		176	111		287	360
4950	Reducer bushing, 2" x 1-1/2"		36.40	.440		3.15	22		25.15	37
4952	3" x 1-1/2"		27.30	.586		13.30	29.50		42.80	59
4954	4" x 2"		18.20	.879		25.50	44		69.50	94.50
4956	6" x 4"		11.10	1.441		71.50	72.50		144	188
4960	Couplings, 1-1/2"	1 Plum	18.20	.440		1.91	24.50		26.41	39
4962	2"	Q-1	33.10	.483		2.57	24.50		27.07	39.50
4963	3"		20.80	.769		7.45	38.50		45.95	66
4964	4"		16.50	.970		13.40	48.50		61.90	88.50
4966	6"		10.10	1.584		56	79.50		135.50	182
4970	2" x 1-1/2"		33.30	.480		5.50	24		29.50	42.50
4972	3" x 1-1/2"		21	.762		15.30	38.50		53.80	74.50
4974	4" x 3"		16.70	.958		24.50	48		72.50	99.50
4978	Closet flange, 4"	1 Plum	32	.250		13.30	13.95		27.25	35.50
4980	4" x 3"	"	34	.235		15.20	13.15		28.35	36.50
5000	DWV, PVC, schedule 40, socket joints									
5040	1/4 bend, 1-1/4"	1 Plum	20.20	.396	Ea.	10.10	22		32.10	44.50
5060	1-1/2"	"	18.20	.440		2.77	24.50		27.27	40
5070	2"	Q-1	33.10	.483		4.37	24.50		28.87	41.50
5080	3"		20.80	.769		12.90	38.50		51.40	72
5090	4"		16.50	.970		25.50	48.50		74	102
5100	6"		10.10	1.584		89	79.50		168.50	218
5105	8"	Q-2	9.30	2.581		180	134		314	400
5106	10"	"	8.50	2.824		234	147		381	480
5110	1/4 bend, long sweep, 1-1/2"	1 Plum	18.20	.440		6.65	24.50		31.15	44.50
5112	2"	Q-1	33.10	.483		7.20	24.50		31.70	44.50
5114	3"		20.80	.769		16.95	38.50		55.45	76.50
5116	4"		16.50	.970		32.50	48.50		81	109
5150	1/8 bend, 1-1/4"	1 Plum	20.20	.396		6.60	22		28.60	41
5170	1-1/2"	"	18.20	.440		2.71	24.50		27.21	40
5180	2"	Q-1	33.10	.483		3.91	24.50		28.41	41
5190	3"		20.80	.769		11.50	38.50		50	70.50
5200	4"		16.50	.970		20	48.50		68.50	95.50
5210	6"		10.10	1.584		81.50	79.50		161	210
5215	8"	Q-2	9.30	2.581		153	134		287	370
5216	10"		8.50	2.824		196	147		343	440
5217	12"		7.60	3.158		270	164		434	545
5250	Tee, sanitary 1-1/4"	1 Plum	13.50	.593		10.15	33		43.15	61
5254	1-1/2"	"	12.10	.661		5	37		42	61
5255	2"	Q-1	20	.800		7.40	40		47.40	68.50
5256	3"		13.90	1.151		19.40	58		77.40	109
5257	4"		11	1.455		34.50	73		107.50	148
5259	6"		6.70	2.388		144	120		264	340
5261	8"	Q-2	6.20	3.871		430	202		632	780
5276	Tee, sanitary, reducing									
5281	2" x 1-1/2" x 1-1/2"	Q-1	23	.696	Ea.	6.50	35		41.50	59.50
5282	2" x 1-1/2" x 2"		22	.727		7.85	36.50		44.35	63.50

22 11 13.76	Pipe Fittings, Plastic	Crew	Daily Output	Labor-Hours	Unit	2013 Bare Costs Material	Labor	Equipment	Total	Total Incl O&P
5283	2" x 2" x 1-1/2"	Q-1	22	.727	Ea.	6.50	36.50		43	62
5284	3" x 3" x 1-1/2"		15.50	1.032		13.70	52		65.70	93
5285	3" x 3" x 2"		15.30	1.046		14.50	52.50		67	95
5286	4" x 4" x 1-1/2"		12.30	1.301		37	65.50		102.50	140
5287	4" x 4" x 2"		12.20	1.311		31	66		97	133
5288	4" x 4" x 3"		12.10	1.322		42	66.50		108.50	146
5291	6" x 6" x 4"		6.90	2.319		139	116		255	330
5294	Tee, double sanitary									
5295	1-1/2"	1 Plum	9.10	.879	Ea.	11.15	49		60.15	86.50
5296	2"	Q-1	16.60	.964		15.05	48.50		63.55	89.50
5297	3"		10.40	1.538		42	77.50		119.50	163
5298	4"		8.25	1.939		67.50	97.50		165	222
5303	Wye, reducing									
5304	2" x 1-1/2" x 1-1/2"	Q-1	23	.696	Ea.	12.70	35		47.70	66.50
5305	2" x 2" x 1-1/2"		22	.727		11.05	36.50		47.55	67
5306	3" x 3" x 2"		15.30	1.046		36	52.50		88.50	119
5307	4" x 4" x 2"		12.20	1.311		26.50	66		92.50	128
5309	4" x 4" x 3"		12.10	1.322		35.50	66.50		102	140
5314	Combination Y & 1/8 bend, 1-1/2"	1 Plum	12.10	.661		12.10	37		49.10	69
5315	2"	Q-1	20	.800		12.95	40		52.95	75
5317	3"		13.90	1.151		32.50	58		90.50	123
5318	4"		11	1.455		64	73		137	181
5319	6"		6.70	2.388		256	120		376	460
5320	8"	Q-2	6.20	3.871		505	202		707	860
5321	10"		5.70	4.211		705	219		924	1,100
5322	12"		5.10	4.706		820	245		1,065	1,275
5324	Combination Y & 1/8 bend, reducing									
5325	2" x 2" x 1-1/2"	Q-1	22	.727	Ea.	17.05	36.50		53.55	74
5327	3" x 3" x 1-1/2"		15.50	1.032		30.50	52		82.50	112
5328	3" x 3" x 2"		15.30	1.046		21.50	52.50		74	103
5329	4" x 4" x 2"		12.20	1.311		34.50	66		100.50	137
5331	Wye, 1-1/4"	1 Plum	13.50	.593		12.95	33		45.95	64.50
5332	1-1/2"	"	12.10	.661		9.45	37		46.45	66
5333	2"	Q-1	20	.800		9	40		49	70.50
5334	3"		13.90	1.151		24.50	58		82.50	114
5335	4"		11	1.455		44	73		117	159
5336	6"		6.70	2.388		133	120		253	325
5337	8"	Q-2	6.20	3.871		310	202		512	645
5338	10"		5.70	4.211		670	219		889	1,075
5339	12"		5.10	4.706		1,100	245		1,345	1,575
5341	2" x 1-1/2"	Q-1	22	.727		11.05	36.50		47.55	67
5342	3" x 1-1/2"		15.50	1.032		15.90	52		67.90	95.50
5343	4" x 3"		12.10	1.322		35.50	66.50		102	140
5344	6" x 4"		6.90	2.319		97.50	116		213.50	282
5345	8" x 6"	Q-2	6.40	3.750		234	195		429	550
5347	Double wye, 1-1/2"	1 Plum	9.10	.879		20.50	49		69.50	96.50
5348	2"	Q-1	16.60	.964		23.50	48.50		72	99
5349	3"		10.40	1.538		48	77.50		125.50	169
5350	4"		8.25	1.939		97.50	97.50		195	254
5353	Double wye, reducing									
5354	2" x 2" x 1-1/2" x 1-1/2"	Q-1	16.80	.952	Ea.	21	48		69	95
5355	3" x 3" x 2" x 2"		10.60	1.509		36	76		112	154
5356	4" x 4" x 3" x 3"		8.45	1.893		77.50	95		172.50	228

22 11 13 – Facility Water Distribution Piping

22 11 13.76 Pipe Fittings, Plastic		Crew	Daily Output	Labor-Hours	Unit	2013 Bare Costs Material	Labor	Equipment	Total	Total Incl O&P
5357	6" x 6" x 4" x 4"	Q-1	7.25	2.207	Ea.	259	111		370	450
5374	Coupling, 1-1/4"	1 Plum	20.20	.396		6.10	22		28.10	40
5376	1-1/2"	"	18.20	.440		1.38	24.50		25.88	38.50
5378	2"	Q-1	33.10	.483		1.80	24.50		26.30	38.50
5380	3"		20.80	.769		6.20	38.50		44.70	65
5390	4"		16.50	.970		10.95	48.50		59.45	85.50
5400	6"		10.10	1.584		35.50	79.50		115	160
5402	8"	Q-2	9.30	2.581		87.50	134		221.50	298
5404	2" x 1-1/2"	Q-1	33.30	.480		4	24		28	41
5406	3" x 1-1/2"		21	.762		12.20	38.50		50.70	71
5408	4" x 3"		16.70	.958		19.75	48		67.75	94.50
5410	Reducer bushing, 2" x 1-1/4"		36.50	.438		5.65	22		27.65	39.50
5411	2" x 1-1/2"		36.40	.440		2.37	22		24.37	36
5412	3" x 1-1/2"		27.30	.586		11.40	29.50		40.90	57
5413	3" x 2"		27.10	.590		5.95	29.50		35.45	51
5414	4" x 2"		18.20	.879		19.80	44		63.80	88.50
5415	4" x 3"		16.70	.958		10.10	48		58.10	83.50
5416	6" x 4"		11.10	1.441		52.50	72.50		125	167
5418	8" x 6"	Q-2	10.20	2.353		114	123		237	310
5425	Closet flange 4"	Q-1	32	.500		13.95	25		38.95	53.50
5426	4" x 3"	"	34	.471		13.95	23.50		37.45	51
5450	Solvent cement for PVC, industrial grade, per quart				Qt.	24.50			24.50	26.50
5500	CPVC, Schedule 80, threaded joints									
5540	90° Elbow, 1/4"	1 Plum	32	.250	Ea.	11.55	13.95		25.50	33.50
5560	1/2"		30.30	.264		6.70	14.75		21.45	29.50
5570	3/4"		26	.308		10	17.15		27.15	37
5580	1"		22.70	.352		14.05	19.65		33.70	45
5590	1-1/4"		20.20	.396		27	22		49	63.50
5600	1-1/2"		18.20	.440		29	24.50		53.50	69
5610	2"	Q-1	33.10	.483		39	24.50		63.50	79.50
5620	2-1/2"		24.20	.661		121	33		154	183
5630	3"		20.80	.769		130	38.50		168.50	201
5640	4"		16.50	.970		203	48.50		251.50	298
5650	6"		10.10	1.584		240	79.50		319.50	385
5660	45° Elbow same as 90° Elbow									
5700	Tee, 1/4"	1 Plum	22	.364	Ea.	22.50	20.50		43	55
5702	1/2"		20.20	.396		22.50	22		44.50	58
5704	3/4"		17.30	.462		32.50	26		58.50	74.50
5706	1"		15.20	.526		35	29.50		64.50	82.50
5708	1-1/4"		13.50	.593		35	33		68	88.50
5710	1-1/2"		12.10	.661		36.50	37		73.50	96
5712	2"	Q-1	20	.800		41	40		81	106
5714	2-1/2"		16.20	.988		156	49.50		205.50	246
5716	3"		13.90	1.151		233	58		291	345
5718	4"		11	1.455		550	73		623	715
5720	6"		6.70	2.388		635	120		755	880
5730	Coupling, 1/4"	1 Plum	32	.250		14.70	13.95		28.65	37
5732	1/2"		30.30	.264		12.10	14.75		26.85	35.50
5734	3/4"		26	.308		19.55	17.15		36.70	47.50
5736	1"		22.70	.352		22	19.65		41.65	54
5738	1-1/4"		20.20	.396		23.50	22		45.50	59.50
5740	1-1/2"		18.20	.440		25.50	24.50		50	65
5742	2"	Q-1	33.10	.483		30	24.50		54.50	69.50

22 11 13 – Facility Water Distribution Piping

22 11 13.76 Pipe Fittings, Plastic		Crew	Daily Output	Labor-Hours	Unit	Material	2013 Bare Costs Labor	Equipment	Total	Total Incl O&P
5744	2-1/2"	Q-1	24.20	.661	Ea.	53.50	33		86.50	109
5746	3"		20.80	.769		62	38.50		100.50	127
5748	4"		16.50	.970		127	48.50		175.50	214
5750	6"		10.10	1.584		173	79.50		252.50	310
5752	8"	Q-2	9.30	2.581		370	134		504	605
5900	CPVC, Schedule 80, socket joints									
5904	90° Elbow, 1/4"	1 Plum	32	.250	Ea.	11	13.95		24.95	33
5906	1/2"		30.30	.264		4.31	14.75		19.06	26.50
5908	3/4"		26	.308		5.50	17.15		22.65	32
5910	1"		22.70	.352		8.70	19.65		28.35	39
5912	1-1/4"		20.20	.396		18.85	22		40.85	54.50
5914	1-1/2"		18.20	.440		21	24.50		45.50	60
5916	2"	Q-1	33.10	.483		25.50	24.50		50	64.50
5918	2-1/2"		24.20	.661		58.50	33		91.50	115
5920	3"		20.80	.769		66	38.50		104.50	131
5922	4"		16.50	.970		119	48.50		167.50	205
5924	6"		10.10	1.584		240	79.50		319.50	385
5926	8"		9.30	1.720		585	86.50		671.50	775
5930	45° Elbow, 1/4"	1 Plum	32	.250		16.35	13.95		30.30	39
5932	1/2"		30.30	.264		5.25	14.75		20	28
5934	3/4"		26	.308		7.60	17.15		24.75	34.50
5936	1"		22.70	.352		12.10	19.65		31.75	43
5938	1-1/4"		20.20	.396		24	22		46	59.50
5940	1-1/2"		18.20	.440		24.50	24.50		49	64
5942	2"	Q-1	33.10	.483		27.50	24.50		52	66.50
5944	2-1/2"		24.20	.661		56	33		89	112
5946	3"		20.80	.769		72	38.50		110.50	137
5948	4"		16.50	.970		98.50	48.50		147	182
5950	6"		10.10	1.584		305	79.50		384.50	455
5952	8"		9.30	1.720		630	86.50		716.50	825
5960	Tee, 1/4"	1 Plum	22	.364		10.10	20.50		30.60	41.50
5962	1/2"		20.20	.396		10.10	22		32.10	44.50
5964	3/4"		17.30	.462		10.25	26		36.25	50.50
5966	1"		15.20	.526		12.60	29.50		42.10	58
5968	1-1/4"		13.50	.593		26.50	33		59.50	79
5970	1-1/2"		12.10	.661		30.50	37		67.50	89
5972	2"	Q-1	20	.800		34	40		74	97.50
5974	2-1/2"		16.20	.988		86	49.50		135.50	170
5976	3"		13.90	1.151		86	58		144	182
5978	4"		11	1.455		115	73		188	236
5980	6"		6.70	2.388		298	120		418	510
5982	8"	Q-2	6.20	3.871		835	202		1,037	1,225
5990	Coupling, 1/4"	1 Plum	32	.250		11.70	13.95		25.65	34
5992	1/2"		30.30	.264		4.55	14.75		19.30	27
5994	3/4"		26	.308		6.35	17.15		23.50	33
5996	1"		22.70	.352		8.55	19.65		28.20	39
5998	1-1/4"		20.20	.396		12.80	22		34.80	47.50
6000	1-1/2"		18.20	.440		16.15	24.50		40.65	55
6002	2"	Q-1	33.10	.483		18.75	24.50		43.25	57
6004	2-1/2"		24.20	.661		41.50	33		74.50	96
6006	3"		20.80	.769		45.50	38.50		84	108
6008	4"		16.50	.970		59.50	48.50		108	139
6010	6"		10.10	1.584		140	79.50		219.50	274

22 11 13 – Facility Water Distribution Piping

22 11 13.76 Pipe Fittings, Plastic		Crew	Daily Output	Labor-Hours	Unit	Material	Labor	Equipment	Total	Total Incl O&P
						2013 Bare Costs				
6012	8"	Q-2	9.30	2.581	Ea.	380	134		514	615
6200	CTS, 100 psi at 180°F, hot and cold water									
6230	90° Elbow, 1/2"	1 Plum	20	.400	Ea.	.46	22.50		22.96	34
6250	3/4"		19	.421		.76	23.50		24.26	36.50
6251	1"		16	.500		2.22	28		30.22	44.50
6252	1-1/4"		15	.533		4.42	30		34.42	50
6253	1-1/2"		14	.571		8	32		40	57
6254	2"	Q-1	23	.696		15.30	35		50.30	69.50
6260	45° Elbow, 1/2"	1 Plum	20	.400		.58	22.50		23.08	34
6280	3/4"		19	.421		1.01	23.50		24.51	36.50
6281	1"		16	.500		3.77	28		31.77	46
6282	1-1/4"		15	.533		5.45	30		35.45	51
6283	1-1/2"		14	.571		7.90	32		39.90	56.50
6284	2"	Q-1	23	.696		16.55	35		51.55	70.50
6290	Tee, 1/2"	1 Plum	13	.615		.59	34.50		35.09	52
6310	3/4"		12	.667		1.11	37		38.11	57
6311	1"		11	.727		5.80	40.50		46.30	67.50
6312	1-1/4"		10	.800		8.90	44.50		53.40	77
6313	1-1/2"		10	.800		11.60	44.50		56.10	80
6314	2"	Q-1	17	.941		18.80	47.50		66.30	91.50
6320	Coupling, 1/2"	1 Plum	22	.364		.38	20.50		20.88	31
6340	3/4"		21	.381		.50	21.50		22	32.50
6341	1"		18	.444		2.26	25		27.26	40
6342	1-1/4"		17	.471		2.92	26.50		29.42	42.50
6343	1-1/2"		16	.500		4.12	28		32.12	46.50
6344	2"	Q-1	28	.571		7.85	28.50		36.35	51.50
6360	Solvent cement for CPVC, commercial grade, per quart				Qt.	40			40	44
7340	PVC flange, slip-on, Sch 80 std., 1/2"	1 Plum	22	.364	Ea.	12.90	20.50		33.40	44.50
7350	3/4"		21	.381		13.75	21.50		35.25	47
7360	1"		18	.444		15.35	25		40.35	54.50
7370	1-1/4"		17	.471		15.80	26.50		42.30	57
7380	1-1/2"		16	.500		16.15	28		44.15	60
7390	2"	Q-1	26	.615		21.50	31		52.50	70
7400	2-1/2"		24	.667		33	33.50		66.50	87
7410	3"		18	.889		36.50	44.50		81	108
7420	4"		15	1.067		46.50	53.50		100	132
7430	6"		10	1.600		73	80.50		153.50	202
7440	8"	Q-2	11	2.182		131	114		245	315
7550	Union, schedule 40, socket joints, 1/2"	1 Plum	19	.421		4.49	23.50		27.99	40.50
7560	3/4"		18	.444		5.10	25		30.10	43
7570	1"		15	.533		5.25	30		35.25	51
7580	1-1/4"		14	.571		15.85	32		47.85	65.50
7590	1-1/2"		13	.615		17.65	34.50		52.15	71
7600	2"	Q-1	20	.800		24	40		64	86.50
7992	Polybutyl/polyethyl pipe, for copper fittings see Line 22 11 13.25 7000									
8000	Compression type, PVC, 160 psi cold water									
8010	Coupling, 3/4" CTS	1 Plum	21	.381	Ea.	3.48	21.50		24.98	36
8020	1" CTS		18	.444		4.32	25		29.32	42.50
8030	1-1/4" CTS		17	.471		6.10	26.50		32.60	46
8040	1-1/2" CTS		16	.500		8.30	28		36.30	51
8050	2" CTS		15	.533		11.65	30		41.65	58
8060	Female adapter, 3/4" FPT x 3/4" CTS		23	.348		5.70	19.40		25.10	35.50
8070	3/4" FPT x 1" CTS		21	.381		6.60	21.50		28.10	39.50

22 11 13.76 Pipe Fittings, Plastic		Crew	Daily Output	Labor-Hours	Unit	Material	2013 Bare Costs Labor	Equipment	Total	Total Incl O&P
8080	1" FPT x 1" CTS	1 Plum	20	.400	Ea.	6.60	22.50		29.10	41
8090	1-1/4" FPT x 1-1/4" CTS		18	.444		8.65	25		33.65	47
8100	1-1/2" FPT x 1-1/2" CTS		16	.500		9.90	28		37.90	53
8110	2" FPT x 2" CTS		13	.615		14.35	34.50		48.85	67.50
8130	Male adapter, 3/4" MPT x 3/4" CTS		23	.348		4.74	19.40		24.14	34
8140	3/4" MPT x 1" CTS		21	.381		5.65	21.50		27.15	38
8150	1" MPT x 1" CTS		20	.400		5.65	22.50		28.15	39.50
8160	1-1/4" MPT x 1-1/4" CTS		18	.444		7.60	25		32.60	46
8170	1-1/2" MPT x 1-1/2" CTS		16	.500		9.15	28		37.15	52
8180	2" MPT x 2" CTS		13	.615		11.85	34.50		46.35	64.50
8200	Spigot adapter, 3/4" IPS x 3/4" CTS		23	.348		2.70	19.40		22.10	32
8210	3/4" IPS x 1" CTS		21	.381		3.30	21.50		24.80	35.50
8220	1" IPS x 1" CTS		20	.400		3.30	22.50		25.80	37
8230	1-1/4" IPS x 1-1/4" CTS		18	.444		4.99	25		29.99	43
8240	1-1/2" IPS x 1-1/2" CTS		16	.500		5.25	28		33.25	48
8250	2" IPS x 2" CTS	↓	13	.615	↓	6.45	34.50		40.95	58.50
8270	Price includes insert stiffeners									
8280	250 psi is same price as 160 psi									
8300	Insert type, nylon, 160 & 250 psi, cold water									
8310	Clamp ring stainless steel, 3/4" IPS	1 Plum	115	.070	Ea.	2.38	3.88		6.26	8.45
8320	1" IPS		107	.075		2.43	4.17		6.60	8.95
8330	1-1/4" IPS		101	.079		2.45	4.42		6.87	9.35
8340	1-1/2" IPS		95	.084		3.28	4.70		7.98	10.70
8350	2" IPS		85	.094		3.77	5.25		9.02	12.05
8370	Coupling, 3/4" IPS		22	.364		.84	20.50		21.34	31.50
8380	1" IPS		19	.421		.88	23.50		24.38	36.50
8390	1-1/4" IPS		18	.444		1.30	25		26.30	39
8400	1-1/2" IPS		17	.471		1.54	26.50		28.04	41
8410	2" IPS		16	.500		2.91	28		30.91	45
8430	Elbow, 90°, 3/4" IPS		22	.364		1.68	20.50		22.18	32.50
8440	1" IPS		19	.421		1.86	23.50		25.36	37.50
8450	1-1/4" IPS		18	.444		2.08	25		27.08	40
8460	1-1/2" IPS		17	.471		2.45	26.50		28.95	42
8470	2" IPS		16	.500		3.42	28		31.42	46
8490	Male adapter, 3/4" IPS x 3/4" MPT		25	.320		.84	17.85		18.69	28
8500	1" IPS x 1" MPT		21	.381		.87	21.50		22.37	33
8510	1-1/4" IPS x 1-1/4" MPT		20	.400		1.37	22.50		23.87	35
8520	1-1/2" IPS x 1-1/2" MPT		18	.444		1.54	25		26.54	39
8530	2" IPS x 2" MPT		15	.533		2.95	30		32.95	48.50
8550	Tee, 3/4" IPS		14	.571		1.63	32		33.63	50
8560	1" IPS		13	.615		2.12	34.50		36.62	54
8570	1-1/4" IPS		12	.667		3.31	37		40.31	59.50
8580	1-1/2" IPS		11	.727		3.75	40.50		44.25	65
8590	2" IPS	↓	10	.800	↓	7.40	44.50		51.90	75
8610	Insert type, PVC, 100 psi @ 180°F, hot & cold water									
8620	Coupler, male, 3/8" CTS x 3/8" MPT	1 Plum	29	.276	Ea.	.68	15.40		16.08	24
8630	3/8" CTS x 1/2" MPT		28	.286		.68	15.95		16.63	25
8640	1/2" CTS x 1/2" MPT		27	.296		.68	16.55		17.23	26
8650	1/2" CTS x 3/4" MPT		26	.308		2.18	17.15		19.33	28.50
8660	3/4" CTS x 1/2" MPT		25	.320		1.98	17.85		19.83	29
8670	3/4" CTS x 3/4" MPT		25	.320		2.34	17.85		20.19	29.50
8700	Coupling, 3/8" CTS x 1/2" CTS		25	.320		4.58	17.85		22.43	32
8710	1/2" CTS		23	.348		5.50	19.40		24.90	35

22 11 13.76 Pipe Fittings, Plastic		Crew	Daily Output	Labor-Hours	Unit	Material (2013 Bare Costs)	Labor (2013 Bare Costs)	Equipment (2013 Bare Costs)	Total (2013 Bare Costs)	Total Incl O&P
8730	3/4″ CTS	1 Plum	22	.364	Ea.	11.60	20.50		32.10	43.50
8750	Elbow 90°, 3/8″ CTS		25	.320		4	17.85		21.85	31.50
8760	1/2″ CTS		23	.348		4.73	19.40		24.13	34
8770	3/4″ CTS		22	.364		7.25	20.50		27.75	38.50
8800	Rings, crimp, copper, 3/8″ CTS		120	.067		.22	3.72		3.94	5.85
8810	1/2″ CTS		117	.068		.22	3.82		4.04	6
8820	3/4″ CTS		115	.070		.29	3.88		4.17	6.15
8850	Reducer tee, bronze, 3/8″ x 3/8″ x 1/2″ CTS		17	.471		4.35	26.50		30.85	44.50
8860	1/2″ x 1/2″ x 3/4″ CTS		15	.533		5.55	30		35.55	51
8870	3/4″ x 1/2″ x 1/2″ CTS		14	.571		5.55	32		37.55	54
8890	3/4″ x 3/4″ x 1/2″ CTS		14	.571		5.65	32		37.65	54.50
8900	1″ x 1/2″ x 1/2″ CTS		14	.571		9.35	32		41.35	58.50
8930	Tee, 3/8″ CTS		17	.471		3.18	26.50		29.68	43
8940	1/2″ CTS		15	.533		2.86	30		32.86	48
8950	3/4″ CTS		14	.571		3.05	32		35.05	51.50
8960	Copper rings included in fitting price									
9000	Flare type, assembled, acetal, hot & cold water									
9010	Coupling, 1/4″ & 3/8″ CTS	1 Plum	24	.333	Ea.	3.36	18.60		21.96	31.50
9020	1/2″ CTS		22	.364		3.84	20.50		24.34	34.50
9030	3/4″ CTS		21	.381		5.70	21.50		27.20	38.50
9040	1″ CTS		18	.444		7.25	25		32.25	45.50
9050	Elbow 90°, 1/4″ CTS		26	.308		3.73	17.15		20.88	30
9060	3/8″ CTS		24	.333		4	18.60		22.60	32.50
9070	1/2″ CTS		22	.364		4.73	20.50		25.23	35.50
9080	3/4″ CTS		21	.381		7.25	21.50		28.75	40
9090	1″ CTS		18	.444		9.15	25		34.15	47.50
9110	Tee, 1/4″ CTS		16	.500		4.09	28		32.09	46.50
9114	3/8″ CTS		15	.533		4.13	30		34.13	49.50
9120	1/2″ CTS		14	.571		5.25	32		37.25	54
9130	3/4″ CTS		13	.615		8.25	34.50		42.75	60.50
9140	1″ CTS		12	.667		11.05	37		48.05	68
9400	Polypropylene, fittings and accessories									
9404	Fittings fusion welded, sizes are I.D.									
9408	Note: sizes 1/2″ thru 4″ use socket fusion									
9410	Sizes 6″ thru 10″ use butt fusion									
9416	Coupling									
9420	3/8″	1 Plum	39	.205	Ea.	.84	11.45		12.29	18.15
9422	1/2″		37.40	.214		1.09	11.95		13.04	19.20
9424	3/4″		35.40	.226		1.22	12.60		13.82	20.50
9426	1″		29.70	.269		1.61	15.05		16.66	24.50
9428	1-1/4″		27.60	.290		1.93	16.15		18.08	26.50
9430	1-1/2″		24.80	.323		4.05	18		22.05	31.50
9432	2″	Q-1	43	.372		8.10	18.70		26.80	37
9434	2-1/2″		35.60	.449		9.05	22.50		31.55	44
9436	3″		30.90	.518		19.90	26		45.90	61
9438	3-1/2″		27.80	.576		32.50	29		61.50	79
9440	4″		25	.640		42.50	32		74.50	95.50
9442	Reducing coupling, female to female									
9446	2″ to 1-1/2″	Q-1	49	.327	Ea.	11.65	16.40		28.05	37.50
9448	2-1/2″ to 2″		41.20	.388		12.80	19.50		32.30	43.50
9450	3″ to 2-1/2″		33.10	.483		17.25	24.50		41.75	55.50
9470	Reducing bushing, female to female									
9472	1/2″ to 3/8″	1 Plum	38.20	.209	Ea.	1.09	11.70		12.79	18.80

22 11 13 – Facility Water Distribution Piping

22 11 13.76 Pipe Fittings, Plastic		Crew	Daily Output	Labor-Hours	Unit	2013 Bare Costs Material	Labor	Equipment	Total	Total Incl O&P
9474	3/4" to 3/8" or 1/2"	1 Plum	36.50	.219	Ea.	1.22	12.25		13.47	19.75
9476	1" to 3/4" or 1/2"		33.10	.242		1.63	13.50		15.13	22.50
9478	1-1/4" to 3/4" or 1"		28.70	.279		2.51	15.55		18.06	26.50
9480	1-1/2" to 1/2" thru 1-1/4"	↓	26.20	.305		4.15	17.05		21.20	30
9482	2" to 1/2" thru 1-1/2"	Q-1	43	.372		8.30	18.70		27	37
9484	2-1/2" to 1/2" thru 2"		41.20	.388		9.30	19.50		28.80	39.50
9486	3" to 1-1/2" thru 2-1/2"		33.10	.483		20.50	24.50		45	59
9488	3-1/2" to 2" thru 3"		29.20	.548		33	27.50		60.50	78
9490	4" to 2-1/2" thru 3-1/2"		26.20	.611		40.50	30.50		71	90.50
9491	6" to 4" SDR 7.4		16.50	.970		62	48.50		110.50	142
9492	6" to 4" SDR 11		16.50	.970		62	48.50		110.50	142
9493	8" to 6" SDR 7.4		10.10	1.584		91	79.50		170.50	220
9494	8" to 6" SDR 11		10.10	1.584		83.50	79.50		163	212
9495	10" to 8" SDR 7.4		7.80	2.051		125	103		228	292
9496	10" to 8" SDR 11	↓	7.80	2.051	↓	109	103		212	275
9500	90° Elbow									
9504	3/8"	1 Plum	39	.205	Ea.	.92	11.45		12.37	18.25
9506	1/2"		37.40	.214		1.17	11.95		13.12	19.30
9508	3/4"		35.40	.226		1.50	12.60		14.10	20.50
9510	1"		29.70	.269		2.17	15.05		17.22	25
9512	1-1/4"		27.60	.290		3.34	16.15		19.49	28
9514	1-1/2"	↓	24.80	.323		7.20	18		25.20	35
9516	2"	Q-1	43	.372		11.05	18.70		29.75	40
9518	2-1/2"		35.60	.449		24.50	22.50		47	61
9520	3"		30.90	.518		45.50	26		71.50	89
9522	3-1/2"		27.80	.576		64.50	29		93.50	115
9524	4"		25	.640		99.50	32		131.50	158
9526	6" SDR 7.4		5.55	2.883		114	145		259	345
9528	6" SDR 11	↓	5.55	2.883		94.50	145		239.50	320
9530	8" SDR 7.4	Q-2	8.10	2.963		395	154		549	665
9532	8" SDR 11		8.10	2.963		305	154		459	565
9534	10" SDR 7.4		7.50	3.200		555	167		722	860
9536	10" SDR 11	↓	7.50	3.200	↓	510	167		677	815
9551	45° Elbow									
9554	3/8"	1 Plum	39	.205	Ea.	.90	11.45		12.35	18.25
9556	1/2"		37.40	.214		1.17	11.95		13.12	19.30
9558	3/4"		35.40	.226		1.50	12.60		14.10	20.50
9564	1"		29.70	.269		2.17	15.05		17.22	25
9566	1-1/4"		27.60	.290		3.34	16.15		19.49	28
9568	1-1/2"	↓	24.80	.323		7.20	18		25.20	35
9570	2"	Q-1	43	.372		10.95	18.70		29.65	40
9572	2-1/2"		35.60	.449		24	22.50		46.50	60.50
9574	3"		30.90	.518		44.50	26		70.50	88
9576	3-1/2"		27.80	.576		63.50	29		92.50	114
9578	4"		25	.640		98	32		130	157
9580	6" SDR 7.4		5.55	2.883		113	145		258	340
9582	6" SDR 11	↓	5.55	2.883		93.50	145		238.50	320
9584	8" SDR 7.4	Q-2	8.10	2.963		310	154		464	570
9586	8" SDR 11		8.10	2.963		271	154		425	530
9588	10" SDR 7.4		7.50	3.200		500	167		667	800
9590	10" SDR 11	↓	7.50	3.200	↓	410	167		577	700
9600	Tee									
9604	3/8"	1 Plum	26	.308	Ea.	1.16	17.15		18.31	27.50

22 11 13 – Facility Water Distribution Piping

22 11 13.76 Pipe Fittings, Plastic		Crew	Daily Output	Labor-Hours	Unit	Material	Labor	Equipment	Total	Total Incl O&P
						2013 Bare Costs				
9606	1/2"	1 Plum	24.90	.321	Ea.	1.58	17.95		19.53	28.50
9608	3/4"		23.70	.338		2.17	18.85		21.02	31
9610	1"		19.90	.402		2.76	22.50		25.26	37
9612	1-1/4"		18.50	.432		4.21	24		28.21	41
9614	1-1/2"		16.60	.482		12.05	27		39.05	54
9616	2"	Q-1	26.80	.597		17.25	30		47.25	64
9618	2-1/2"		23.80	.672		29	34		63	83
9620	3"		20.60	.777		53	39		92	117
9622	3-1/2"		18.40	.870		82.50	43.50		126	157
9624	4"		16.70	.958		105	48		153	188
9626	6" SDR 7.4		3.70	4.324		123	217		340	460
9628	6" SDR 11		3.70	4.324		93.50	217		310.50	430
9630	8" SDR 7.4	Q-2	5.40	4.444		335	231		566	720
9632	8" SDR 11		5.40	4.444		291	231		522	670
9634	10" SDR 7.4		5	4.800		580	250		830	1,025
9636	10" SDR 11		5	4.800		470	250		720	890
9638	For reducing tee use same tee price									
9660	End cap									
9662	3/8"	1 Plum	78	.103	Ea.	1.16	5.70		6.86	9.90
9664	1/2"		74.60	.107		1.72	6		7.72	10.90
9666	3/4"		71.40	.112		2.17	6.25		8.42	11.80
9668	1"		59.50	.134		2.63	7.50		10.13	14.20
9670	1-1/4"		55.60	.144		4.15	8.05		12.20	16.65
9672	1-1/2"		49.50	.162		5.70	9		14.70	19.90
9674	2"	Q-1	80	.200		9.55	10.05		19.60	25.50
9676	2-1/2"		71.40	.224		13.90	11.25		25.15	32
9678	3"		61.70	.259		31.50	13.05		44.55	54
9680	3-1/2"		55.20	.290		37.50	14.55		52.05	63.50
9682	4"		50	.320		57.50	16.05		73.55	87
9684	6" SDR 7.4		26.50	.604		79	30.50		109.50	132
9686	6" SDR 11		26.50	.604		79	30.50		109.50	132
9688	8" SDR 7.4	Q-2	16.30	1.472		79	76.50		155.50	203
9690	8" SDR 11		16.30	1.472		68.50	76.50		145	192
9692	10" SDR 7.4		14.90	1.611		118	84		202	256
9694	10" SDR 11		14.90	1.611		101	84		185	237
9800	Accessories and tools									
9802	Pipe clamps for suspension, not including rod or beam clamp									
9804	3/8"	1 Plum	74	.108	Ea.	2.03	6.05		8.08	11.35
9805	1/2"		70	.114		2.54	6.40		8.94	12.40
9806	3/4"		68	.118		2.93	6.55		9.48	13.10
9807	1"		66	.121		3.19	6.75		9.94	13.70
9808	1-1/4"		64	.125		3.28	7		10.28	14.10
9809	1-1/2"		62	.129		3.58	7.20		10.78	14.80
9810	2"	Q-1	110	.145		4.44	7.30		11.74	15.90
9811	2-1/2"		104	.154		5.70	7.75		13.45	17.90
9812	3"		98	.163		6.10	8.20		14.30	19.10
9813	3-1/2"		92	.174		6.70	8.75		15.45	20.50
9814	4"		86	.186		7.55	9.35		16.90	22.50
9815	6"		70	.229		9.20	11.50		20.70	27.50
9816	8"	Q-2	100	.240		33	12.50		45.50	55
9817	10"	"	94	.255		37.50	13.30		50.80	61.50
9820	Pipe cutter									
9822	For 3/8" thru 1-1/4"				Ea.	112			112	123

22 11 13 – Facility Water Distribution Piping

22 11 13.76 Pipe Fittings, Plastic		Crew	Daily Output	Labor-Hours	Unit	Material	2013 Bare Costs Labor	Equipment	Total	Total Incl O&P
9824	For 1-1/2" thru 4"				Ea.	296			296	325
9826	Note: Pipes may be cut with standard									
9827	iron saw with blades for plastic.									
9830	Manual welding device, 800 W, 110 V									
9831	For 3/8" thru 2"				Ea.	355			355	390
9834	Manual welding device, 1400 W, 110 V									
9835	For 1-1/2" thru 4"				Ea.	695			695	765
9838	Butt welding machine									
9839	For 6" thru 10"				Ea.	24,200			24,200	26,600
9982	For plastic hangers see Line 22 05 29.10 8000									
9986	For copper/brass fittings see Line 22 11 13.25 7000									

22 11 13.78 Pipe, High Density Polyethylene Plastic (HDPE)		Crew	Daily Output	Labor-Hours	Unit	Material	Labor	Equipment	Total	Total Incl O&P
0010	**PIPE, HIGH DENSITY POLYETHYLENE PLASTIC (HDPE)**									
0020	Not incl. hangers, trenching, backfill, hoisting or digging equipment.									
0030	Standard length is 40', add a weld for each joint									
0040	Single wall									
0050	Straight									
0054	1" diameter DR 11				L.F.	.79			.79	.87
0058	1-1/2" diameter DR 11					1			1	1.10
0062	2" diameter DR 11					1.67			1.67	1.84
0066	3" diameter DR 11					2.01			2.01	2.21
0070	3" diameter DR 17					1.61			1.61	1.77
0074	4" diameter DR 11					3.36			3.36	3.70
0078	4" diameter DR 17					3.36			3.36	3.70
0082	6" diameter DR 11					8.35			8.35	9.20
0086	6" diameter DR 17					5.55			5.55	6.10
0090	8" diameter DR 11					13.90			13.90	15.30
0094	8" diameter DR 26					6.50			6.50	7.15
0098	10" diameter DR 11					22			22	24
0102	10" diameter DR 26					10.05			10.05	11.05
0106	12" diameter DR 11					32			32	35
0110	12" diameter DR 26					15.05			15.05	16.60
0114	16" diameter DR 11					48.50			48.50	53.50
0118	16" diameter DR 26					22			22	24
0122	18" diameter DR 11					62			62	68
0126	18" diameter DR 26					28.50			28.50	31.50
0130	20" diameter DR 11					75.50			75.50	83
0134	20" diameter DR 26					33.50			33.50	37
0138	22" diameter DR 11					92			92	101
0142	22" diameter DR 26					42			42	46
0146	24" diameter DR 11					109			109	120
0150	24" diameter DR 26					48.50			48.50	53.50
0154	28" diameter DR 17					100			100	111
0158	28" diameter DR 26					67			67	73.50
0162	30" diameter DR 21					94			94	103
0166	30" diameter DR 26					77			77	84.50
0170	36" diameter DR 26					77			77	84.50
0174	42" diameter DR 26					149			149	164
0178	48" diameter DR 26					196			196	216
0182	54" diameter DR 26				▼	246			246	271
0300	90° Elbow									
0304	1" diameter DR 11				Ea.	5.60			5.60	6.15

22 11 13 – Facility Water Distribution Piping

22 11 13.78 Pipe, High Density Polyethylene Plastic (HDPE)		Crew	Daily Output	Labor-Hours	Unit	Material	Labor	Equipment	Total	Total Incl O&P
						2013 Bare Costs				
0308	1-1/2" diameter DR 11				Ea.	7			7	7.70
0312	2" diameter DR 11					7			7	7.70
0316	3" diameter DR 11					14			14	15.40
0320	3" diameter DR 17					14			14	15.40
0324	4" diameter DR 11					19.60			19.60	21.50
0328	4" diameter DR 17					19.60			19.60	21.50
0332	6" diameter DR 11					45			45	49.50
0336	6" diameter DR 17					45			45	49.50
0340	8" diameter DR 11					111			111	122
0344	8" diameter DR 26					98			98	108
0348	10" diameter DR 11					415			415	455
0352	10" diameter DR 26					380			380	420
0356	12" diameter DR 11					435			435	480
0360	12" diameter DR 26					395			395	435
0364	16" diameter DR 11					515			515	570
0368	16" diameter DR 26					510			510	560
0372	18" diameter DR 11					650			650	715
0376	18" diameter DR 26					610			610	670
0380	20" diameter DR 11					765			765	840
0384	20" diameter DR 26					740			740	810
0388	22" diameter DR 11					790			790	870
0392	22" diameter DR 26					765			765	840
0396	24" diameter DR 11					890			890	980
0400	24" diameter DR 26					865			865	950
0404	28" diameter DR 17					1,075			1,075	1,200
0408	28" diameter DR 26					1,025			1,025	1,125
0412	30" diameter DR 17					1,525			1,525	1,675
0416	30" diameter DR 26					1,400			1,400	1,550
0420	36" diameter DR 26					1,775			1,775	1,950
0424	42" diameter DR 26					2,300			2,300	2,525
0428	48" diameter DR 26					2,675			2,675	2,950
0432	54" diameter DR 26				▼	6,375			6,375	7,000
0500	45° Elbow									
0512	2" diameter DR 11				Ea.	5.60			5.60	6.15
0516	3" diameter DR 11					14			14	15.40
0520	3" diameter DR 17					14			14	15.40
0524	4" diameter DR 11					19.60			19.60	21.50
0528	4" diameter DR 17					19.60			19.60	21.50
0532	6" diameter DR 11					45			45	49.50
0536	6" diameter DR 17					45			45	49.50
0540	8" diameter DR 11					111			111	122
0544	8" diameter DR 26					64.50			64.50	71
0548	10" diameter DR 11					415			415	455
0552	10" diameter DR 26					380			380	420
0556	12" diameter DR 11					435			435	480
0560	12" diameter DR 26					395			395	435
0564	16" diameter DR 11					229			229	252
0568	16" diameter DR 26					216			216	238
0572	18" diameter DR 11					243			243	268
0576	18" diameter DR 26					223			223	245
0580	20" diameter DR 11					370			370	405
0584	20" diameter DR 26					350			350	385
0588	22" diameter DR 11					510			510	560

22 11 13 – Facility Water Distribution Piping

22 11 13.78 Pipe, High Density Polyethylene Plastic (HDPE)		Crew	Daily Output	Labor-Hours	Unit	2013 Bare Costs Material	Labor	Equipment	Total	Total Incl O&P
0592	22" diameter DR 26				Ea.	485			485	530
0596	24" diameter DR 11					625			625	685
0600	24" diameter DR 26					600			600	660
0604	28" diameter DR 17					710			710	780
0608	28" diameter DR 26					690			690	755
0612	30" diameter DR 17					880			880	965
0616	30" diameter DR 26					855			855	940
0620	36" diameter DR 26					1,075			1,075	1,200
0624	42" diameter DR 26					1,400			1,400	1,525
0628	48" diameter DR 26					1,525			1,525	1,675
0632	54" diameter DR 26					2,025			2,025	2,225
0700	Tee									
0704	1" diameter DR 11				Ea.	7.30			7.30	8.05
0708	1-1/2" diameter DR 11					10.25			10.25	11.30
0712	2" diameter DR 11					8.80			8.80	9.65
0716	3" diameter DR 11					16.10			16.10	17.70
0720	3" diameter DR 17					16.10			16.10	17.70
0724	4" diameter DR 11					23.50			23.50	26
0728	4" diameter DR 17					23.50			23.50	26
0732	6" diameter DR 11					58.50			58.50	64.50
0736	6" diameter DR 17					58.50			58.50	64.50
0740	8" diameter DR 11					145			145	159
0744	8" diameter DR 17					145			145	159
0748	10" diameter DR 11					430			430	475
0752	10" diameter DR 17					430			430	475
0756	12" diameter DR 11					575			575	635
0760	12" diameter DR 17					575			575	635
0764	16" diameter DR 11					305			305	335
0768	16" diameter DR 17					251			251	276
0772	18" diameter DR 11					430			430	470
0776	18" diameter DR 17					355			355	390
0780	20" diameter DR 11					525			525	575
0784	20" diameter DR 17					430			430	470
0788	22" diameter DR 11					670			670	740
0792	22" diameter DR 17					530			530	585
0796	24" diameter DR 11					835			835	915
0800	24" diameter DR 17					705			705	775
0804	28" diameter DR 17					1,350			1,350	1,500
0812	30" diameter DR 17					1,550			1,550	1,725
0820	36" diameter DR 17					2,575			2,575	2,825
0824	42" diameter DR 26					2,875			2,875	3,150
0828	48" diameter DR 26					3,075			3,075	3,375
1000	Flange adptr, w/back-up ring and 1/2 cost of plated bolt set									
1004	1" diameter DR 11				Ea.	28			28	30.50
1008	1-1/2" diameter DR 11					28			28	30.50
1012	2" diameter DR 11					17.55			17.55	19.35
1016	3" diameter DR 11					20.50			20.50	22.50
1020	3" diameter DR 17					20.50			20.50	22.50
1024	4" diameter DR 11					28			28	30.50
1028	4" diameter DR 17					28			28	30.50
1032	6" diameter DR 11					39.50			39.50	43.50
1036	6" diameter DR 17					39.50			39.50	43.50
1040	8" diameter DR 11					57			57	63

22 11 Facility Water Distribution

22 11 13 – Facility Water Distribution Piping

22 11 13.78 Pipe, High Density Polyethylene Plastic (HDPE)		Crew	Daily Output	Labor-Hours	Unit	Material	Labor	Equipment	Total	Total Incl O&P
						2013 Bare Costs				
1044	8″ diameter DR 26				Ea.	57			57	63
1048	10″ diameter DR 11					91			91	100
1052	10″ diameter DR 26					91			91	100
1056	12″ diameter DR 11					133			133	147
1060	12″ diameter DR 26					133			133	147
1064	16″ diameter DR 11					286			286	315
1068	16″ diameter DR 26					286			286	315
1072	18″ diameter DR 11					370			370	410
1076	18″ diameter DR 26					370			370	410
1080	20″ diameter DR 11					515			515	565
1084	20″ diameter DR 26					515			515	565
1088	22″ diameter DR 11					560			560	615
1092	22″ diameter DR 26					555			555	610
1096	24″ diameter DR 17					605			605	665
1100	24″ diameter DR 32.5					605			605	665
1104	28″ diameter DR 15.5					820			820	900
1108	28″ diameter DR 32.5					820			820	900
1112	30″ diameter DR 11					950			950	1,050
1116	30″ diameter DR 21					950			950	1,050
1120	36″ diameter DR 26					1,050			1,050	1,150
1124	42″ diameter DR 26					1,175			1,175	1,300
1128	48″ diameter DR 26					1,450			1,450	1,600
1132	54″ diameter DR 26					1,750			1,750	1,925
1200	Reducer									
1208	2″ x 1-1/2″ diameter DR 11				Ea.	8.40			8.40	9.25
1212	3″ x 2″ diameter DR 11					8.40			8.40	9.25
1216	4″ x 2″ diameter DR 11					9.80			9.80	10.80
1220	4″ x 3″ diameter DR 11					12.60			12.60	13.85
1224	6″ x 4″ diameter DR 11					29.50			29.50	32.50
1228	8″ x 6″ diameter DR 11					45			45	49.50
1232	10″ x 8″ diameter DR 11					77			77	85
1236	12″ x 8″ diameter DR 11					126			126	139
1240	12″ x 10″ diameter DR 11					101			101	111
1244	14″ x 12″ diameter DR 11					112			112	123
1248	16″ x 14″ diameter DR 11					143			143	157
1252	18″ x 16″ diameter DR 11					174			174	191
1256	20″ x 18″ diameter DR 11					345			345	380
1260	22″ x 20″ diameter DR 11					425			425	465
1264	24″ x 22″ diameter DR 11					475			475	525
1268	26″ x 24″ diameter DR 11					560			560	615
1272	28″ x 24″ diameter DR 11					715			715	785
1276	32″ x 28″ diameter DR 17					925			925	1,025
1280	36″ x 32″ diameter DR 17					1,250			1,250	1,375
4000	Welding labor per joint, not including welding machine									
4010	Pipe joint size (cost based on thickest wall for each diam.)									
4030	1″ pipe size	4 Skwk	273	.117	Ea.		5.40		5.40	8.40
4040	1-1/2″ pipe size		175	.183			8.45		8.45	13.05
4050	2″ pipe size		128	.250			11.55		11.55	17.85
4060	3″ pipe size		100	.320			14.80		14.80	23
4070	4″ pipe size		77	.416			19.20		19.20	29.50
4080	6″ pipe size	5 Skwk	63	.635			29.50		29.50	45.50
4090	8″ pipe size		48	.833			38.50		38.50	59.50
4100	10″ pipe size		40	1			46		46	71.50

22 11 13.78 Pipe, High Density Polyethylene Plastic (HDPE)		Crew	Daily Output	Labor-Hours	Unit	Material	2013 Bare Costs Labor	Equipment	Total	Total Incl O&P
4110	12" pipe size	6 Skwk	41	1.171	Ea.		54		54	83.50
4120	16" pipe size		34	1.412			65		65	101
4130	18" pipe size		32	1.500			69.50		69.50	107
4140	20" pipe size	8 Skwk	37	1.730			80		80	124
4150	22" pipe size		35	1.829			84.50		84.50	131
4160	24" pipe size		34	1.882			87		87	134
4170	28" pipe size		33	1.939			89.50		89.50	139
4180	30" pipe size		32	2			92.50		92.50	143
4190	36" pipe size		31	2.065			95.50		95.50	148
4200	42" pipe size		30	2.133			98.50		98.50	152
4210	48" pipe size	9 Skwk	33	2.182			101		101	156
4220	54" pipe size	"	31	2.323			107		107	166
4300	Note: Cost for set up each time welder is moved.									
4301	Add 50% of a weld cost									
4310	Welder usually remains stationary with pipe moved through it.									
4340	Weld machine, rental per day based on diam. capacity									
4350	1" thru 2" diameter				Ea.			40.50	40.50	44.50
4360	3" thru 4" diameter							46	46	50.50
4370	6" thru 8" diameter							103	103	113
4380	10" thru 12" diameter							178	178	196
4390	16" thru 18" diameter							259	259	285
4400	20" thru 24" diameter							500	500	550
4410	28" thru 32" diameter							545	545	600
4420	36" diameter							570	570	625
4430	42" thru 54" diameter							890	890	980
5000	Dual wall contained pipe									
5040	Straight									
5054	1" DR 11 x 3" DR 11				L.F.	6.75			6.75	7.40
5058	1" DR 11 x 4" DR 11					7.20			7.20	7.90
5062	1-1/2" DR 11 x 4" DR 17					7.40			7.40	8.15
5066	2" DR 11 x 4" DR 17					7.95			7.95	8.75
5070	2" DR 11 x 6" DR 17					11.90			11.90	13.10
5074	3" DR 11 x 6" DR 17					13.30			13.30	14.60
5078	3" DR 11 x 6" DR 26					11.40			11.40	12.55
5086	3" DR 17 x 8" DR 17					16.85			16.85	18.55
5090	4" DR 11 x 8" DR 17					19.75			19.75	21.50
5094	4" DR 17 x 8" DR 26					15.60			15.60	17.15
5098	6" DR 11 x 10" DR 17					30.50			30.50	33.50
5102	6" DR 17 x 10" DR 26					23			23	25
5106	6" DR 26 x 10" DR 26					21.50			21.50	24
5110	8" DR 17 x 12" DR 26					32.50			32.50	36
5114	8" DR 26 x 12" DR 32.5					27.50			27.50	30.50
5118	10" DR 17 x 14" DR 26					45			45	50
5122	10" DR 17 x 16" DR 26					49.50			49.50	54.50
5126	10" DR 26 x 16" DR 26					45			45	49.50
5130	12" DR 26 x 16" DR 26					51			51	56.50
5134	12" DR 17 x 18" DR 26					65.50			65.50	72
5138	12" DR 26 x 18" DR 26					58.50			58.50	64.50
5142	14" DR 26 x 20" DR 32.5					64			64	70.50
5146	16" DR 26 x 22" DR 32.5					97			97	107
5150	18" DR 26 x 24" DR 32.5					89			89	98
5154	20" DR 32.5 x 28" DR 32.5					115			115	127
5158	22" DR 32.5 x 30" DR 32.5					115			115	127

22 11 13.78 Pipe, High Density Polyethylene Plastic (HDPE)		Crew	Daily Output	Labor-Hours	Unit	2013 Bare Costs Material	Labor	Equipment	Total	Total Incl O&P
5162	24" DR 32.5 x 32" DR 32.5				L.F.	147			147	162
5166	36" DR 32.5 x 42" DR 32.5				↓	249			249	274
5300	Force transfer coupling									
5354	1" DR 11 x 3" DR 11				Ea.	243			243	267
5358	1" DR 11 x 4" DR 17					255			255	280
5362	1-1/2" DR 11 x 4" DR 17					267			267	293
5366	2" DR 11 x 4" DR 17					279			279	305
5370	2" DR 11 x 6" DR 17					405			405	445
5374	3" DR 11 x 6" DR 17					405			405	445
5378	3" DR 11 x 6" DR 26					405			405	445
5382	3" DR 11 x 8" DR 11					430			430	470
5386	3" DR 11 x 8" DR 17					430			430	470
5390	4" DR 11 x 8" DR 17					430			430	470
5394	4" DR 17 x 8" DR 26					375			375	415
5398	6" DR 11 x 10" DR 17					545			545	595
5402	6" DR 17 x 10" DR 26					445			445	490
5406	6" DR 26 x 10" DR 26					445			445	490
5410	8" DR 17 x 12" DR 26					595			595	655
5414	8" DR 26 x 12" DR 32.5					480			480	530
5418	10" DR 17 x 14" DR 26					780			780	855
5422	10" DR 17 x 16" DR 26					780			780	855
5426	10" DR 26 x 16" DR 26					780			780	855
5430	12" DR 26 x 16" DR 26					975			975	1,075
5434	12" DR 17 x 18" DR 26					1,000			1,000	1,100
5438	12" DR 26 x 18" DR 26					1,000			1,000	1,100
5442	14" DR 26 x 20" DR 32.5					1,050			1,050	1,150
5446	16" DR 26 x 22" DR 32.5					1,100			1,100	1,200
5450	18" DR 26 x 24" DR 32.5					1,175			1,175	1,300
5454	20" DR 32.5 x 28" DR 32.5					1,425			1,425	1,575
5458	22" DR 32.5 x 30" DR 32.5					1,575			1,575	1,750
5462	24" DR 32.5 x 32" DR 32.5					1,725			1,725	1,900
5466	36" DR 32.5 x 42" DR 32.5				↓	2,850			2,850	3,125
5600	90° Elbow									
5654	1" DR 11 x 3" DR 11				Ea.	200			200	220
5658	1" DR 11 x 4" DR 17					192			192	211
5662	1-1/2" DR 11 x 4" DR 17					210			210	231
5666	2" DR 11 x 4" DR 17					222			222	244
5670	2" DR 11 x 6" DR 17					283			283	310
5674	3" DR 11 x 6" DR 17					325			325	355
5678	3" DR 17 x 6" DR 26					258			258	284
5682	3" DR 17 x 8" DR 11					520			520	575
5686	3" DR 17 x 8" DR 17					405			405	445
5690	4" DR 11 x 8" DR 17					480			480	525
5694	4" DR 17 x 8" DR 26					380			380	415
5698	6" DR 11 x 10" DR 17					635			635	700
5702	6" DR 17 x 10" DR 26					485			485	535
5706	6" DR 26 x 10" DR 26					455			455	500
5710	8" DR 17 x 12" DR 26					725			725	795
5714	8" DR 26 x 12" DR 32.5					670			670	735
5718	10" DR 17 x 14" DR 26					1,150			1,150	1,250
5722	10" DR 17 x 16" DR 26					1,100			1,100	1,200
5726	10" DR 26 x 16" DR 26					1,000			1,000	1,100
5730	12" DR 26 x 16" DR 26					1,125			1,125	1,225

22 11 13 – Facility Water Distribution Piping

22 11 13.78 Pipe, High Density Polyethylene Plastic (HDPE)		Crew	Daily Output	Labor-Hours	Unit	2013 Bare Costs Material	Labor	Equipment	Total	Total Incl O&P
5734	12" DR 17 x 18" DR 26				Ea.	1,325			1,325	1,475
5738	12" DR 26 x 18" DR 26					1,225			1,225	1,350
5742	14" DR 26 x 20" DR 32.5					1,625			1,625	1,775
5746	16" DR 26 x 22" DR 32.5					1,625			1,625	1,800
5750	18" DR 26 x 24" DR 32.5					2,000			2,000	2,200
5754	20" DR 32.5 x 28" DR 32.5					2,300			2,300	2,525
5758	22" DR 32.5 x 30" DR 32.5					3,150			3,150	3,450
5762	24" DR 32.5 x 32" DR 32.5					3,050			3,050	3,350
5766	36" DR 32.5 x 42" DR 32.5				↓	4,950			4,950	5,450
5800	45° Elbow									
5804	1" DR 11 x 3" DR 11				Ea.	120			120	132
5808	1" DR 11 x 4" DR 17					119			119	131
5812	1-1/2" DR 11 x 4" DR 17					128			128	141
5816	2" DR 11 x 4" DR 17					140			140	154
5820	2" DR 11 x 6" DR 17					175			175	193
5824	3" DR 11 x 6" DR 17					195			195	214
5828	3" DR 17 x 6" DR 26					161			161	177
5832	3" DR 17 x 8" DR 11					298			298	330
5836	3" DR 17 x 8" DR 17					238			238	262
5840	4" DR 11 x 8" DR 17					278			278	305
5844	4" DR 17 x 8" DR 26					227			227	250
5848	6" DR 11 x 10" DR 17					365			365	405
5852	6" DR 17 x 10" DR 26					288			288	315
5856	6" DR 26 x 10" DR 26					271			271	299
5860	8" DR 17 x 12" DR 26					440			440	480
5864	8" DR 26 x 12" DR 32.5					410			410	450
5868	10" DR 17 x 14" DR 26					685			685	755
5872	10" DR 17 x 16" DR 26					670			670	740
5876	10" DR 26 x 16" DR 26					630			630	690
5880	12" DR 26 x 16" DR 26					710			710	780
5884	12" DR 17 x 18" DR 26					835			835	920
5888	12" DR 26 x 18" DR 26					780			780	855
5892	14" DR 26 x 20" DR 32.5					1,000			1,000	1,100
5896	16" DR 26 x 22" DR 32.5					1,050			1,050	1,150
5900	18" DR 26 x 24" DR 32.5					1,250			1,250	1,375
5904	20" DR 32.5 x 28" DR 32.5					1,525			1,525	1,675
5908	22" DR 32.5 x 30" DR 32.5					1,925			1,925	2,125
5912	24" DR 32.5 x 32" DR 32.5					1,925			1,925	2,125
5916	36" DR 32.5 x 42" DR 32.5				↓	3,000			3,000	3,275
6000	Access port with 4" riser									
6050	1" DR 11 x 4" DR 17				Ea.	262			262	288
6054	1-1/2" DR 11 x 4" DR 17					267			267	294
6058	2" DR 11 x 6" DR 17					325			325	360
6062	3" DR 11 x 6" DR 17					330			330	360
6066	3" DR 17 x 6" DR 26					320			320	355
6070	3" DR 17 x 8" DR 11					370			370	405
6074	3" DR 17 x 8" DR 17					350			350	385
6078	4" DR 11 x 8" DR 17					365			365	400
6082	4" DR 17 x 8" DR 26					350			350	385
6086	6" DR 11 x 10" DR 17					455			455	500
6090	6" DR 17 x 10" DR 26					420			420	460
6094	6" DR 26 x 10" DR 26					410			410	450
6098	8" DR 17 x 12" DR 26					440			440	485

22 11 13.78 Pipe, High Density Polyethylene Plastic (HDPE)		Crew	Daily Output	Labor-Hours	Unit	2013 Bare Costs Material	Labor	Equipment	Total	Total Incl O&P
6102	8″ DR 26 x 12″ DR 32.5				Ea.	420			420	460
6200	End termination with vent plug									
6204	1″ DR 11 x 3″ DR 11				Ea.	217			217	238
6208	1″ DR 11 x 4″ DR 17					232			232	255
6212	1-1/2″ DR 11 x 4″ DR 17					232			232	255
6216	2″ DR 11 x 4″ DR 17					252			252	277
6220	2″ DR 11 x 6″ DR 17					287			287	315
6224	3″ DR 11 x 6″ DR 17					287			287	315
6228	3″ DR 17 x 6″ DR 26					287			287	315
6232	3″ DR 17 x 8″ DR 11					400			400	435
6236	3″ DR 17 x 8″ DR 17					400			400	435
6240	4″ DR 11 x 8″ DR 17					400			400	435
6244	4″ DR 17 x 8″ DR 26					345			345	380
6248	6″ DR 11 x 10″ DR 17					470			470	520
6252	6″ DR 17 x 10″ DR 26					390			390	425
6256	6″ DR 26 x 10″ DR 26					390			390	425
6260	8″ DR 17 x 12″ DR 26					575			575	630
6264	8″ DR 26 x 12″ DR 32.5					460			460	505
6268	10″ DR 17 x 14″ DR 26					595			595	655
6272	10″ DR 17 x 16″ DR 26					595			595	655
6276	10″ DR 26 x 16″ DR 26					595			595	655
6280	12″ DR 26 x 16″ DR 26					595			595	655
6284	12″ DR 17 x 18″ DR 26					785			785	865
6288	12″ DR 26 x 18″ DR 26					785			785	865
6292	14″ DR 26 x 20″ DR 32.5					800			800	880
6296	16″ DR 26 x 22″ DR 32.5					930			930	1,025
6300	18″ DR 26 x 24″ DR 32.5					960			960	1,050
6304	20″ DR 32.5 x 28″ DR 32.5					1,225			1,225	1,350
6308	22″ DR 32.5 x 30″ DR 32.5					1,350			1,350	1,475
6312	24″ DR 32.5 x 32″ DR 32.5					1,500			1,500	1,650
6316	36″ DR 32.5 x 42″ DR 32.5				▼	2,175			2,175	2,400
6600	Tee									
6604	1″ DR 11 x 3″ DR 11				Ea.	221			221	243
6608	1″ DR 11 x 4″ DR 17					274			274	300
6612	1-1/2″ DR 11 x 4″ DR 17					305			305	335
6616	2″ DR 11 x 4″ DR 17					335			335	365
6620	2″ DR 11 x 6″ DR 17					405			405	445
6624	3″ DR 11 x 6″ DR 17					465			465	510
6628	3″ DR 17 x 6″ DR 26					455			455	500
6632	3″ DR 17 x 8″ DR 11					550			550	605
6636	3″ DR 17 x 8″ DR 17					520			520	575
6640	4″ DR 11 x 8″ DR 17					570			570	625
6644	4″ DR 17 x 8″ DR 26					545			545	600
6648	6″ DR 11 x 10″ DR 17					710			710	780
6652	6″ DR 17 x 10″ DR 26					675			675	740
6656	6″ DR 26 x 10″ DR 26					665			665	735
6660	8″ DR 17 x 12″ DR 26					665			665	735
6664	8″ DR 26 x 12″ DR 32.5					870			870	955
6668	10″ DR 17 x 14″ DR 26					1,000			1,000	1,100
6672	10″ DR 17 x 16″ DR 26					1,050			1,050	1,150
6676	10″ DR 26 x 16″ DR 26					1,050			1,050	1,150
6680	12″ DR 26 x 16″ DR 26					1,150			1,150	1,250
6684	12″ DR 17 x 18″ DR 26					1,150			1,150	1,250

22 11 13.78 Pipe, High Density Polyethylene Plastic (HDPE)		Crew	Daily Output	Labor-Hours	Unit	Material (2013 Bare Costs)	Labor (2013 Bare Costs)	Equipment (2013 Bare Costs)	Total (2013 Bare Costs)	Total Incl O&P
6688	12" DR 26 x 18" DR 26				Ea.	1,250			1,250	1,400
6692	14" DR 26 x 20" DR 32.5					1,675			1,675	1,825
6696	16" DR 26 x 22" DR 32.5					2,050			2,050	2,250
6700	18" DR 26 x 24" DR 32.5					2,225			2,225	2,450
6704	20" DR 32.5 x 28" DR 32.5					2,800			2,800	3,100
6708	22" DR 32.5 x 30" DR 32.5					3,325			3,325	3,675
6712	24" DR 32.5 x 32" DR 32.5					4,000			4,000	4,375
6716	36" DR 32.5 x 42" DR 32.5				↓	8,125			8,125	8,950
6800	Wye									
6816	2" DR 11 x 4" DR 17				Ea.	420			420	460
6820	2" DR 11 x 6" DR 17					475			475	525
6824	3" DR 11 x 6" DR 17					480			480	530
6828	3" DR 17 x 6" DR 26					465			465	515
6832	3" DR 17 x 8" DR 11					585			585	645
6836	3" DR 17 x 8" DR 17					545			545	600
6840	4" DR 11 x 8" DR 17					595			595	655
6844	4" DR 17 x 8" DR 26					565			565	625
6848	6" DR 11 x 10" DR 17					740			740	815
6852	6" DR 17 x 10" DR 26					685			685	755
6856	6" DR 26 x 10" DR 26					685			685	750
6860	8" DR 17 x 12" DR 26					980			980	1,075
6864	8" DR 26 x 12" DR 32.5					945			945	1,050
6868	10" DR 17 x 14" DR 26					1,175			1,175	1,300
6872	10" DR 17 x 16" DR 26					1,250			1,250	1,375
6876	10" DR 26 x 16" DR 26					1,225			1,225	1,350
6880	12" DR 26 x 16" DR 26					1,375			1,375	1,525
6884	12" DR 17 x 18" DR 26					1,550			1,550	1,725
6888	12" DR 26 x 18" DR 26					1,575			1,575	1,750
6892	14" DR 26 x 20" DR 32.5					1,925			1,925	2,125
6896	16" DR 26 x 22" DR 32.5					2,775			2,775	3,050
6900	18" DR 26 x 24" DR 32.5					3,250			3,250	3,575
6904	20" DR 32.5 x 28" DR 32.5					4,075			4,075	4,500
6908	22" DR 32.5 x 30" DR 32.5					4,650			4,650	5,125
6912	24" DR 32.5 x 32" DR 32.5				↓	5,300			5,300	5,825
9000	Welding labor per joint, not including welding machine									
9010	Pipe joint size, outer pipe (cost based on the thickest walls)									
9020	Straight pipe									
9050	3" pipe size	4 Skwk	96	.333	Ea.		15.40		15.40	24
9060	4" pipe size	"	77	.416			19.20		19.20	29.50
9070	6" pipe size	5 Skwk	60	.667			31		31	47.50
9080	8" pipe size	"	40	1			46		46	71.50
9090	10" pipe size	6 Skwk	41	1.171			54		54	83.50
9100	12" pipe size		39	1.231			57		57	88
9110	14" pipe size		38	1.263			58.50		58.50	90.50
9120	16" pipe size	↓	35	1.371			63.50		63.50	98
9130	18" pipe size	8 Skwk	45	1.422			65.50		65.50	102
9140	20" pipe size		42	1.524			70.50		70.50	109
9150	22" pipe size		40	1.600			74		74	114
9160	24" pipe size		38	1.684			78		78	120
9170	28" pipe size		37	1.730			80		80	124
9180	30" pipe size		36	1.778			82		82	127
9190	32" pipe size		35	1.829			84.50		84.50	131
9200	42" pipe size	↓	32	2	↓		92.50		92.50	143

22 11 Facility Water Distribution

22 11 13 – Facility Water Distribution Piping

22 11 13.78 Pipe, High Density Polyethylene Plastic (HDPE)		Crew	Daily Output	Labor-Hours	Unit	Material	2013 Bare Costs Labor	Equipment	Total	Total Incl O&P
9300	Note: Cost for set up each time welder is moved.									
9301	Add 100% of weld labor cost									
9310	For handling between fitting welds add 50% of weld labor cost									
9320	Welder usually remains stationary with pipe moved through it									
9360	Weld machine, rental per day based on diam. capacity									
9380	3″ thru 4″ diameter				Ea.			63.50	63.50	70
9390	6″ thru 8″ diameter							207	207	228
9400	10″ thru 12″ diameter							310	310	340
9410	14″ thru 18″ diameter							420	420	460
9420	20″ thru 24″ diameter							750	750	825
9430	28″ thru 32″ diameter							880	880	970
9440	42″ thru 58″ diameter							910	910	1,000

22 11 19 – Domestic Water Piping Specialties

22 11 19.10 Flexible Connectors

Line	Description	Crew	Daily Output	Labor-Hours	Unit	Material	Labor	Equipment	Total	Total Incl O&P
0010	**FLEXIBLE CONNECTORS**, Corrugated, 7/8″ O.D., 1/2″ I.D.									
0050	Gas, seamless brass, steel fittings									
0200	12″ long	1 Plum	36	.222	Ea.	17.40	12.40		29.80	38
0220	18″ long		36	.222		21.50	12.40		33.90	42.50
0240	24″ long		34	.235		25.50	13.15		38.65	48
0260	30″ long		34	.235		27.50	13.15		40.65	50.50
0280	36″ long		32	.250		30.50	13.95		44.45	54.50
0320	48″ long		30	.267		38.50	14.90		53.40	65
0340	60″ long		30	.267		46	14.90		60.90	73
0360	72″ long		30	.267		53	14.90		67.90	81
2000	Water, copper tubing, dielectric separators									
2100	12″ long	1 Plum	36	.222	Ea.	17.05	12.40		29.45	37.50
2220	15″ long		36	.222		18.95	12.40		31.35	39.50
2240	18″ long		36	.222		20.50	12.40		32.90	41
2260	24″ long		34	.235		25.50	13.15		38.65	48

22 11 19.14 Flexible Metal Hose

Line	Description	Crew	Daily Output	Labor-Hours	Unit	Material	Labor	Equipment	Total	Total Incl O&P
0010	**FLEXIBLE METAL HOSE**, Connectors, standard lengths									
0100	Bronze braided, bronze ends									
0120	3/8″ diameter x 12″	1 Stpi	26	.308	Ea.	18.85	17.45		36.30	47.50
0140	1/2″ diameter x 12″		24	.333		17.95	18.90		36.85	48.50
0160	3/4″ diameter x 12″		20	.400		26.50	22.50		49	63
0180	1″ diameter x 18″		19	.421		34	24		58	73.50
0200	1-1/2″ diameter x 18″		13	.615		53.50	35		88.50	112
0220	2″ diameter x 18″		11	.727		64.50	41		105.50	133
1000	Carbon steel ends									
1020	1/4″ diameter x 12″	1 Stpi	28	.286	Ea.	15.50	16.20		31.70	41.50
1040	3/8″ diameter x 12″		26	.308		15.90	17.45		33.35	44
1060	1/2″ diameter x 12″		24	.333		17.55	18.90		36.45	48
1080	1/2″ diameter x 24″		24	.333		42	18.90		60.90	74.50
1120	3/4″ diameter x 12″		20	.400		25	22.50		47.50	61.50
1140	3/4″ diameter x 24″		20	.400		52	22.50		74.50	91
1160	3/4″ diameter x 36″		20	.400		61.50	22.50		84	102
1180	1″ diameter x 18″		19	.421		31.50	24		55.50	70.50
1200	1″ diameter x 30″		19	.421		60	24		84	102
1220	1″ diameter x 36″		19	.421		98.50	24		122.50	144
1240	1-1/4″ diameter x 18″		15	.533		43	30		73	93
1260	1-1/4″ diameter x 36″		15	.533		97	30		127	152
1280	1-1/2″ diameter x 18″		13	.615		69.50	35		104.50	129

22 11 19 – Domestic Water Piping Specialties

22 11 19.14 Flexible Metal Hose		Crew	Daily Output	Labor-Hours	Unit	Material	Labor	Equipment	Total	Total Incl O&P
						2013 Bare Costs				
1300	1-1/2" diameter x 36"	1 Stpi	13	.615	Ea.	105	35		140	169
1320	2" diameter x 24"		11	.727		96.50	41		137.50	168
1340	2" diameter x 36"		11	.727		130	41		171	205
1360	2-1/2" diameter x 24"		9	.889		221	50.50		271.50	320
1380	2-1/2" diameter x 36"		9	.889		253	50.50		303.50	355
1400	3" diameter x 24"		7	1.143		305	64.50		369.50	435
1420	3" diameter x 36"		7	1.143		96	64.50		160.50	204
2000	Carbon steel braid, carbon steel solid ends									
2100	1/2" diameter x 12"	1 Stpi	24	.333	Ea.	38.50	18.90		57.40	70.50
2120	3/4" diameter x 12"		20	.400		58	22.50		80.50	98
2140	1" diameter x 12"		19	.421		82	24		106	126
2160	1-1/4" diameter x 12"		15	.533		27	30		57	75
2180	1-1/2" diameter x 12"		13	.615		29	35		64	84.50
3000	Stainless steel braid, welded on carbon steel ends									
3100	1/2" diameter x 12"	1 Stpi	24	.333	Ea.	50.50	18.90		69.40	84
3120	3/4" diameter x 12"		20	.400		62.50	22.50		85	103
3140	3/4" diameter x 24"		20	.400		72	22.50		94.50	114
3160	3/4" diameter x 36"		20	.400		81.50	22.50		104	124
3180	1" diameter x 12"		19	.421		59.50	24		83.50	102
3200	1" diameter x 24"		19	.421		85	24		109	130
3220	1" diameter x 36"		19	.421		102	24		126	148
3240	1-1/4" diameter x 12"		15	.533		111	30		141	168
3260	1-1/4" diameter x 24"		15	.533		121	30		151	179
3280	1-1/4" diameter x 36"		15	.533		137	30		167	196
3300	1-1/2" diameter x 12"		13	.615		43	35		78	100
3320	1-1/2" diameter x 24"		13	.615		132	35		167	198
3340	1-1/2" diameter x 36"		13	.615		154	35		189	223
3400	Metal stainless steel braid, over corrugated stainless steel, flanged ends									
3410	150 PSI									
3420	1/2" diameter x 12"	1 Stpi	24	.333	Ea.	144	18.90		162.90	187
3430	1" diameter x 12"		20	.400		180	22.50		202.50	232
3440	1-1/2" diameter x 12"		15	.533		204	30		234	270
3450	2-1/2" diameter x 9"		12	.667		108	38		146	176
3460	3" diameter x 9"		9	.889		93	50.50		143.50	178
3470	4" diameter x 9"		7	1.143		113	64.50		177.50	222
3480	4" diameter x 30"		5	1.600		560	90.50		650.50	750
3490	4" diameter x 36"		4.80	1.667		580	94.50		674.50	780
3500	6" diameter x 11"		5	1.600		188	90.50		278.50	345
3510	6" diameter x 36"		3.80	2.105		685	119		804	930
3520	8" diameter x 12"		4	2		410	113		523	620
3530	10" diameter x 13"		3	2.667		625	151		776	915
3540	12" diameter x 14"	Q-5	4	4		950	204		1,154	1,350
6000	Molded rubber with helical wire reinforcement									
6010	150 PSI									
6020	1-1/2" diameter x 12"	1 Stpi	15	.533	Ea.	88	30		118	142
6030	2" diameter x 12"		12	.667		240	38		278	320
6040	3" diameter x 12"		8	1		251	56.50		307.50	360
6050	4" diameter x 12"		6	1.333		320	75.50		395.50	465
6060	6" diameter x 18"		4	2		475	113		588	695
6070	8" diameter x 24"		3	2.667		665	151		816	960
6080	10" diameter x 24"		2	4		810	227		1,037	1,225
6090	12" diameter x 24"	Q-5	3	5.333		925	272		1,197	1,425
7000	Molded teflon with stainless steel flanges									

22 11 Facility Water Distribution

22 11 19 – Domestic Water Piping Specialties

22 11 19.14 Flexible Metal Hose		Crew	Daily Output	Labor-Hours	Unit	Material	2013 Bare Costs Labor	Equipment	Total	Total Incl O&P
7010	150 PSI									
7020	2-1/2" diameter x 3-3/16"	Q-1	7.80	2.051	Ea.	2,325	103		2,428	2,700
7030	3" diameter x 3-5/8"		6.50	2.462		1,625	124		1,749	1,950
7040	4" diameter x 3-5/8"		5	3.200		2,100	161		2,261	2,575
7050	6" diameter x 4"		4.30	3.721		2,950	187		3,137	3,525
7060	8" diameter x 6"		3.80	4.211		4,700	211		4,911	5,500

22 11 19.18 Mixing Valve

		Crew	Daily Output	Labor-Hours	Unit	Material	Labor	Equipment	Total	Total Incl O&P
0010	**MIXING VALVE**, Automatic, water tempering.									
0040	1/2" size	1 Stpi	19	.421	Ea.	535	24		559	625
0050	3/4" size		18	.444		535	25		560	630
0100	1" size		16	.500		790	28.50		818.50	915
0120	1-1/4" size		13	.615		1,100	35		1,135	1,250
0140	1-1/2" size		10	.800		1,300	45.50		1,345.50	1,500
0160	2" size		8	1		1,625	56.50		1,681.50	1,850
0170	2-1/2" size		6	1.333		1,625	75.50		1,700.50	1,900
0180	3" size		4	2		3,825	113		3,938	4,375
0190	4" size		3	2.667		3,825	151		3,976	4,425

22 11 19.22 Pressure Reducing Valve

		Crew	Daily Output	Labor-Hours	Unit	Material	Labor	Equipment	Total	Total Incl O&P
0010	**PRESSURE REDUCING VALVE**, Steam, pilot operated.									
0100	Threaded, iron body									
0200	1-1/2" size	1 Stpi	8	1	Ea.	1,775	56.50		1,831.50	2,025
0220	2" size	"	5	1.600	"	2,025	90.50		2,115.50	2,350
1000	Flanged, iron body, 125 lb. flanges									
1020	2" size	1 Stpi	8	1	Ea.	2,050	56.50		2,106.50	2,325
1040	2-1/2" size	"	4	2		2,475	113		2,588	2,875
1060	3" size	Q-5	4.50	3.556		3,000	181		3,181	3,550
1080	4" size	"	3	5.333		4,300	272		4,572	5,125
1500	For 250 lb. flanges, add					5%				

22 11 19.26 Pressure Regulators

		Crew	Daily Output	Labor-Hours	Unit	Material	Labor	Equipment	Total	Total Incl O&P
0010	**PRESSURE REGULATORS**									
0100	Gas appliance regulators									
0106	Main burner and pilot applications									
0108	Rubber seat poppet type									
0109	1/8" pipe size	1 Stpi	24	.333	Ea.	15	18.90		33.90	45
0110	1/4" pipe size		24	.333		17.30	18.90		36.20	47.50
0112	3/8" pipe size		24	.333		18.35	18.90		37.25	48.50
0113	1/2" pipe size		24	.333		20.50	18.90		39.40	51
0114	3/4" pipe size		20	.400		24	22.50		46.50	60.50
0122	Lever action type									
0123	3/8" pipe size	1 Stpi	24	.333	Ea.	22.50	18.90		41.40	53
0124	1/2" pipe size		24	.333		22.50	18.90		41.40	53
0125	3/4" pipe size		20	.400		44	22.50		66.50	82.50
0126	1" pipe size		19	.421		44	24		68	84.50
0132	Double diaphragm type									
0133	3/8" pipe size	1 Stpi	24	.333	Ea.	33.50	18.90		52.40	65.50
0134	1/2" pipe size		24	.333		46.50	18.90		65.40	79.50
0135	3/4" pipe size		20	.400		76.50	22.50		99	119
0136	1" pipe size		19	.421		94.50	24		118.50	140
0137	1-1/4" pipe size		15	.533		330	30		360	405
0138	1-1/2" pipe size		13	.615		605	35		640	720
0139	2" pipe size		11	.727		605	41		646	725
0140	2-1/2" pipe size	Q-5	15	1.067		1,175	54.50		1,229.50	1,375

22 11 19 – Domestic Water Piping Specialties

22 11 19.26 Pressure Regulators		Crew	Daily Output	Labor-Hours	Unit	2013 Bare Costs Material	Labor	Equipment	Total	Total Incl O&P
0141	3" pipe size	Q-5	13	1.231	Ea.	1,175	62.50		1,237.50	1,400
0142	4" pipe size (flanged)	↓	8	2	↓	2,075	102		2,177	2,450
0160	Main burner only									
0162	Straight-thru-flow design									
0163	1/2" pipe size	1 Stpi	24	.333	Ea.	37.50	18.90		56.40	70
0164	3/4" pipe size		20	.400		51.50	22.50		74	90.50
0165	1" pipe size		19	.421		73	24		97	117
0166	1-1/4" pipe size	↓	15	.533	↓	73	30		103	126
0200	Oil, light, hot water, ordinary steam, threaded									
0220	Bronze body, 1/4" size	1 Stpi	24	.333	Ea.	149	18.90		167.90	193
0230	3/8" size		24	.333		161	18.90		179.90	206
0240	1/2" size		24	.333		200	18.90		218.90	249
0250	3/4" size		20	.400		239	22.50		261.50	296
0260	1" size		19	.421		365	24		389	435
0270	1-1/4" size		15	.533		490	30		520	585
0320	Iron body, 1/4" size		24	.333		117	18.90		135.90	157
0330	3/8" size		24	.333		137	18.90		155.90	180
0340	1/2" size		24	.333		145	18.90		163.90	189
0350	3/4" size		20	.400		177	22.50		199.50	228
0360	1" size		19	.421		230	24		254	289
0370	1-1/4" size	↓	15	.533	↓	320	30		350	395
9002	For water pressure regulators, see Section 22 05 23.20									

22 11 19.34 Sleeves and Escutcheons		Crew	Daily Output	Labor-Hours	Unit	Material	Labor	Equipment	Total	Total Incl O&P
0010	**SLEEVES & ESCUTCHEONS**									
0100	Pipe sleeve									
0110	Steel, w/water stop, 12" long, with link seal									
0120	2" diam. for 1/2" carrier pipe	1 Plum	8.40	.952	Ea.	50	53		103	135
0130	2-1/2" diam. for 3/4" carrier pipe		8	1		57	56		113	147
0140	2-1/2" diam. for 1" carrier pipe		8	1		53.50	56		109.50	143
0150	3" diam. for 1-1/4" carrier pipe		7.20	1.111		69.50	62		131.50	170
0160	3-1/2" diam. for 1-1/2" carrier pipe		6.80	1.176		69	65.50		134.50	175
0170	4" diam. for 2" carrier pipe		6	1.333		75	74.50		149.50	195
0180	4" diam. for 2-1/2" carrier pipe		6	1.333		76.50	74.50		151	196
0190	5" diam. for 3" carrier pipe		5.40	1.481		89.50	82.50		172	224
0200	6" diam. for 4" carrier pipe	↓	4.80	1.667		101	93		194	251
0210	10" diam. for 6" carrier pipe	Q-1	8	2		174	100		274	345
0220	12" diam. for 8" carrier pipe		7.20	2.222		244	112		356	435
0230	14" diam. for 10" carrier pipe		6.40	2.500		281	126		407	500
0240	16" diam. for 12" carrier pipe		5.80	2.759		320	139		459	565
0250	18" diam. for 14" carrier pipe		5.20	3.077		470	155		625	755
0260	24" diam. for 18" carrier pipe		4	4		735	201		936	1,125
0270	24" diam. for 20" carrier pipe		4	4		640	201		841	1,000
0280	30" diam. for 24" carrier pipe	↓	3.20	5	↓	975	251		1,226	1,450
0500	Wall sleeve									
0510	Ductile iron with rubber gasket seal									
0520	3"	1 Plum	8.40	.952	Ea.	700	53		753	850
0530	4"		7.20	1.111		750	62		812	920
0540	6"		6	1.333		920	74.50		994.50	1,125
0550	8"		4	2		1,125	112		1,237	1,425
0560	10"		3	2.667		1,375	149		1,524	1,750
0570	12"	↓	2.40	3.333	↓	1,625	186		1,811	2,075
5000	Escutcheon									

22 11 19 – Domestic Water Piping Specialties

22 11 19.34 Sleeves and Escutcheons		Crew	Daily Output	Labor-Hours	Unit	2013 Bare Costs Material	Labor	Equipment	Total	Total Incl O&P
5100	Split ring, pipe									
5110	Chrome plated									
5120	1/2"	1 Plum	160	.050	Ea.	1.05	2.79		3.84	5.35
5130	3/4"		160	.050		1.16	2.79		3.95	5.50
5140	1"		135	.059		1.25	3.31		4.56	6.35
5150	1-1/2"		115	.070		1.76	3.88		5.64	7.80
5160	2"		100	.080		1.96	4.46		6.42	8.85
5170	4"		80	.100		4.35	5.60		9.95	13.20
5180	6"	↓	68	.118	↓	7.25	6.55		13.80	17.90
5400	Shallow flange type									
5410	Chrome plated steel									
5420	1/2" CTS	1 Plum	180	.044	Ea.	.32	2.48		2.80	4.09
5430	3/4" CTS		180	.044		.25	2.48		2.73	4.02
5440	1/2" IPS		180	.044		.36	2.48		2.84	4.14
5450	3/4" IPS		180	.044		.40	2.48		2.88	4.18
5460	1" IPS		175	.046		.46	2.55		3.01	4.35
5470	1-1/2" IPS		170	.047		.76	2.63		3.39	4.80
5480	2" IPS	↓	160	.050	↓	.91	2.79		3.70	5.20

22 11 19.38 Water Supply Meters		Crew	Daily Output	Labor-Hours	Unit	Material	Labor	Equipment	Total	Total Incl O&P
0010	**WATER SUPPLY METERS**									
1000	Detector, serves dual systems such as fire and domestic or									
1020	process water, wide range cap., UL and FM approved									
1100	3" mainline x 2" by-pass, 400 GPM	Q-1	3.60	4.444	Ea.	7,025	223		7,248	8,050
1140	4" mainline x 2" by-pass, 700 GPM	"	2.50	6.400		7,025	320		7,345	8,200
1180	6" mainline x 3" by-pass, 1600 GPM	Q-2	2.60	9.231		10,700	480		11,180	12,500
1220	8" mainline x 4" by-pass, 2800 GPM		2.10	11.429		15,900	595		16,495	18,400
1260	10" mainline x 6" by-pass, 4400 GPM		2	12		22,700	625		23,325	25,900
1300	10" x 12" mainlines x 6" by-pass, 5400 GPM	↓	1.70	14.118	↓	30,900	735		31,635	35,000
2000	Domestic/commercial, bronze									
2020	Threaded									
2060	5/8" diameter, to 20 GPM	1 Plum	16	.500	Ea.	47	28		75	94
2080	3/4" diameter, to 30 GPM		14	.571		86	32		118	143
2100	1" diameter, to 50 GPM	↓	12	.667	↓	131	37		168	200
2300	Threaded/flanged									
2340	1-1/2" diameter, to 100 GPM	1 Plum	8	1	Ea.	320	56		376	435
2360	2" diameter, to 160 GPM	"	6	1.333	"	430	74.50		504.50	585
2600	Flanged, compound									
2640	3" diameter, 320 GPM	Q-1	3	5.333	Ea.	2,925	268		3,193	3,625
2660	4" diameter, to 500 GPM		1.50	10.667		4,700	535		5,235	5,975
2680	6" diameter, to 1,000 GPM		1	16		7,500	805		8,305	9,450
2700	8" diameter, to 1,800 GPM	↓	.80	20	↓	11,800	1,000		12,800	14,400
7000	Turbine									
7260	Flanged									
7300	2" diameter, to 160 GPM	1 Plum	7	1.143	Ea.	575	64		639	725
7320	3" diameter, to 450 GPM	Q-1	3.60	4.444		1,200	223		1,423	1,650
7340	4" diameter, to 650 GPM	"	2.50	6.400		1,975	320		2,295	2,650
7360	6" diameter, to 1800 GPM	Q-2	2.60	9.231		3,550	480		4,030	4,625
7380	8" diameter, to 2500 GPM		2.10	11.429		5,625	595		6,220	7,100
7400	10" diameter, to 5500 GPM	↓	1.70	14.118	↓	7,600	735		8,335	9,450

22 11 19 – Domestic Water Piping Specialties

22 11 19.42 Backflow Preventers		Crew	Daily Output	Labor-Hours	Unit	Material	2013 Bare Costs Labor	Equipment	Total	Total Incl O&P
0010	**BACKFLOW PREVENTERS**, Includes valves									
0020	and four test cocks, corrosion resistant, automatic operation									
1000	Double check principle									
1010	Threaded, with ball valves									
1020	3/4" pipe size	1 Plum	16	.500	Ea.	220	28		248	284
1030	1" pipe size		14	.571		252	32		284	325
1040	1-1/2" pipe size		10	.800		535	44.50		579.50	655
1050	2" pipe size		7	1.143		620	64		684	780
1080	Threaded, with gate valves									
1100	3/4" pipe size	1 Plum	16	.500	Ea.	1,025	28		1,053	1,175
1120	1" pipe size		14	.571		1,050	32		1,082	1,200
1140	1-1/2" pipe size		10	.800		1,350	44.50		1,394.50	1,550
1160	2" pipe size		7	1.143		1,650	64		1,714	1,900
1200	Flanged, valves are gate									
1210	3" pipe size	Q-1	4.50	3.556	Ea.	2,650	179		2,829	3,200
1220	4" pipe size	"	3	5.333		2,750	268		3,018	3,450
1230	6" pipe size	Q-2	3	8		4,075	415		4,490	5,100
1240	8" pipe size		2	12		7,400	625		8,025	9,100
1250	10" pipe size		1	24		11,000	1,250		12,250	14,000
1300	Flanged, valves are OS&Y									
1370	1" pipe size	1 Plum	5	1.600	Ea.	1,125	89.50		1,214.50	1,375
1374	1-1/2" pipe size		5	1.600		1,450	89.50		1,539.50	1,700
1378	2" pipe size		4.80	1.667		1,750	93		1,843	2,075
1380	3" pipe size	Q-1	4.50	3.556		3,775	179		3,954	4,425
1400	4" pipe size	"	3	5.333		5,725	268		5,993	6,700
1420	6" pipe size	Q-2	3	8		8,950	415		9,365	10,500
1430	8" pipe size	"	2	12		17,700	625		18,325	20,300
4000	Reduced pressure principle									
4100	Threaded, bronze, valves are ball									
4120	3/4" pipe size	1 Plum	16	.500	Ea.	420	28		448	505
4140	1" pipe size		14	.571		455	32		487	550
4150	1-1/4" pipe size		12	.667		835	37		872	975
4160	1-1/2" pipe size		10	.800		915	44.50		959.50	1,075
4180	2" pipe size		7	1.143		1,025	64		1,089	1,225
5000	Flanged, bronze, valves are OS&Y									
5060	2-1/2" pipe size	Q-1	5	3.200	Ea.	6,650	161		6,811	7,575
5080	3" pipe size		4.50	3.556		7,750	179		7,929	8,800
5100	4" pipe size		3	5.333		9,075	268		9,343	10,400
5120	6" pipe size	Q-2	3	8		14,500	415		14,915	16,500
5200	Flanged, iron, valves are gate									
5210	2-1/2" pipe size	Q-1	5	3.200	Ea.	2,525	161		2,686	3,025
5220	3" pipe size		4.50	3.556		2,650	179		2,829	3,175
5230	4" pipe size		3	5.333		3,575	268		3,843	4,325
5240	6" pipe size	Q-2	3	8		5,025	415		5,440	6,150
5250	8" pipe size		2	12		9,025	625		9,650	10,900
5260	10" pipe size		1	24		12,700	1,250		13,950	15,900
5600	Flanged, iron, valves are OS&Y									
5660	2-1/2" pipe size	Q-1	5	3.200	Ea.	3,150	161		3,311	3,725
5680	3" pipe size		4.50	3.556		3,425	179		3,604	4,050
5700	4" pipe size		3	5.333		3,800	268		4,068	4,575
5720	6" pipe size	Q-2	3	8		5,500	415		5,915	6,675
5740	8" pipe size		2	12		9,675	625		10,300	11,500

22 11 19 – Domestic Water Piping Specialties

22 11 19.42 Backflow Preventers		Crew	Daily Output	Labor-Hours	Unit	Material	2013 Bare Costs Labor	Equipment	Total	Total Incl O&P
5760	10" pipe size	Q-2	1	24	Ea.	12,900	1,250		14,150	16,100

22 11 19.50 Vacuum Breakers

		Crew	Daily Output	Labor-Hours	Unit	Material	Labor	Equipment	Total	Total Incl O&P
0010	**VACUUM BREAKERS** R221113-40									
0013	See also backflow preventers Section 22 11 19.42									
1000	Anti-siphon continuous pressure type									
1010	Max. 150 PSI - 210°F									
1020	Bronze body									
1030	1/2" size	1 Stpi	24	.333	Ea.	183	18.90		201.90	230
1040	3/4" size		20	.400		183	22.50		205.50	235
1050	1" size		19	.421		189	24		213	244
1060	1-1/4" size		15	.533		375	30		405	455
1070	1-1/2" size		13	.615		460	35		495	560
1080	2" size		11	.727		475	41		516	580
1200	Max. 125 PSI with atmospheric vent									
1210	Brass, in-line construction									
1220	1/4" size	1 Stpi	24	.333	Ea.	72.50	18.90		91.40	109
1230	3/8" size	"	24	.333		72.50	18.90		91.40	109
1260	For polished chrome finish, add					13%				
2000	Anti-siphon, non-continuous pressure type									
2010	Hot or cold water 125 PSI - 210°F									
2020	Bronze body									
2030	1/4" size	1 Stpi	24	.333	Ea.	46	18.90		64.90	79
2040	3/8" size		24	.333		46	18.90		64.90	79
2050	1/2" size		24	.333		52	18.90		70.90	86
2060	3/4" size		20	.400		61.50	22.50		84	102
2070	1" size		19	.421		96.50	24		120.50	142
2080	1-1/4" size		15	.533		169	30		199	232
2090	1-1/2" size		13	.615		198	35		233	271
2100	2" size		11	.727		310	41		351	400
2110	2-1/2" size		8	1		890	56.50		946.50	1,050
2120	3" size		6	1.333		1,175	75.50		1,250.50	1,425
2150	For polished chrome finish, add					50%				

22 11 19.54 Water Hammer Arresters/Shock Absorbers

		Crew	Daily Output	Labor-Hours	Unit	Material	Labor	Equipment	Total	Total Incl O&P
0010	**WATER HAMMER ARRESTERS/SHOCK ABSORBERS**									
0490	Copper									
0500	3/4" male I.P.S. For 1 to 11 fixtures	1 Plum	12	.667	Ea.	28	37		65	87
0600	1" male I.P.S. For 12 to 32 fixtures		8	1		45.50	56		101.50	134
0700	1-1/4" male I.P.S. For 33 to 60 fixtures		8	1		47	56		103	136
0800	1-1/2" male I.P.S. For 61 to 113 fixtures		8	1		67.50	56		123.50	158
0900	2" male I.P.S. For 114 to 154 fixtures		8	1		98.50	56		154.50	192
1000	2-1/2" male I.P.S. For 155 to 330 fixtures		4	2		305	112		417	505
4000	Bellows type									
4010	3/4" FNPT, to 11 fixture units	1 Plum	10	.800	Ea.	241	44.50		285.50	330
4020	1" FNPT, to 32 fixture units		8.80	.909		485	50.50		535.50	610
4030	1" FNPT, to 60 fixture units		8.80	.909		730	50.50		780.50	880
4040	1" FNPT, to 113 fixture units		8.80	.909		1,825	50.50		1,875.50	2,100
4050	1" FNPT, to 154 fixture units		6.60	1.212		2,075	67.50		2,142.50	2,375
4060	1-1/2" FNPT, to 300 fixture units		5	1.600		2,550	89.50		2,639.50	2,925

22 11 19.64 Hydrants

		Crew	Daily Output	Labor-Hours	Unit	Material	Labor	Equipment	Total	Total Incl O&P
0010	**HYDRANTS**									
0050	Wall type, moderate climate, bronze, encased									
0200	3/4" IPS connection	1 Plum	16	.500	Ea.	705	28		733	820

22 11 19 – Domestic Water Piping Specialties

22 11 19.64 Hydrants		Crew	Daily Output	Labor-Hours	Unit	Material	2013 Bare Costs Labor	Equipment	Total	Total Incl O&P
0300	1" IPS connection	1 Plum	14	.571	Ea.	810	32		842	940
0500	Anti-siphon type, 3/4" connection	↓	16	.500	↓	610	28		638	710
1000	Non-freeze, bronze, exposed									
1100	3/4" IPS connection, 4" to 9" thick wall	1 Plum	14	.571	Ea.	480	32		512	575
1120	10" to 14" thick wall		12	.667		515	37		552	625
1140	15" to 19" thick wall		12	.667		580	37		617	690
1160	20" to 24" thick wall	↓	10	.800	↓	625	44.50		669.50	755
1200	For 1" IPS connection, add					15%	10%			
1240	For 3/4" adapter type vacuum breaker, add				Ea.	60			60	66
1280	For anti-siphon type, add				"	126			126	139
2000	Non-freeze bronze, encased, anti-siphon type									
2100	3/4" IPS connection, 5" to 9" thick wall	1 Plum	14	.571	Ea.	1,225	32		1,257	1,375
2120	10" to 14" thick wall		12	.667		1,250	37		1,287	1,425
2140	15" to 19" thick wall		12	.667		1,300	37		1,337	1,500
2160	20" to 24" thick wall	↓	10	.800	↓	1,350	44.50		1,394.50	1,575
2200	For 1" IPS connection, add					10%	10%			
3000	Ground box type, bronze frame, 3/4" IPS connection									
3080	Non-freeze, all bronze, polished face, set flush									
3100	2 feet depth of bury	1 Plum	8	1	Ea.	905	56		961	1,075
3120	3 feet depth of bury		8	1		970	56		1,026	1,150
3140	4 feet depth of bury		8	1		1,050	56		1,106	1,225
3160	5 feet depth of bury		7	1.143		1,125	64		1,189	1,325
3180	6 feet depth of bury		7	1.143		1,175	64		1,239	1,400
3200	7 feet depth of bury		6	1.333		1,250	74.50		1,324.50	1,475
3220	8 feet depth of bury		5	1.600		1,325	89.50		1,414.50	1,575
3240	9 feet depth of bury		4	2		1,400	112		1,512	1,700
3260	10 feet depth of bury	↓	4	2		1,450	112		1,562	1,775
3400	For 1" IPS connection, add					15%	10%			
3450	For 1-1/4" IPS connection, add					325%	14%			
3500	For 1-1/2" connection, add					370%	18%			
3550	For 2" connection, add					445%	24%			
3600	For tapped drain port in box, add				↓	82			82	90
4000	Non-freeze, CI body, bronze frame & scoriated cover									
4010	with hose storage									
4100	2 feet depth of bury	1 Plum	7	1.143	Ea.	1,675	64		1,739	1,925
4120	3 feet depth of bury		7	1.143		1,750	64		1,814	2,000
4140	4 feet depth of bury		7	1.143		1,800	64		1,864	2,075
4160	5 feet depth of bury		6.50	1.231		1,825	68.50		1,893.50	2,100
4180	6 feet depth of bury		6	1.333		1,850	74.50		1,924.50	2,150
4200	7 feet depth of bury		5.50	1.455		1,925	81		2,006	2,250
4220	8 feet depth of bury		5	1.600		2,000	89.50		2,089.50	2,325
4240	9 feet depth of bury		4.50	1.778		2,050	99		2,149	2,400
4260	10 feet depth of bury	↓	4	2		2,125	112		2,237	2,500
4280	For 1" IPS connection, add					365			365	405
4300	For tapped drain port in box, add				↓	82			82	90
5000	Moderate climate, all bronze, polished face									
5020	and scoriated cover, set flush									
5100	3/4" IPS connection	1 Plum	16	.500	Ea.	625	28		653	730
5120	1" IPS connection	"	14	.571		770	32		802	895
5200	For tapped drain port in box, add				↓	82			82	90
6000	Ground post type, all non-freeze, all bronze, aluminum casing									
6010	guard, exposed head, 3/4" IPS connection									
6100	2 feet depth of bury	1 Plum	8	1	Ea.	885	56		941	1,050

22 11 19 – Domestic Water Piping Specialties

22 11 19.64 Hydrants		Crew	Daily Output	Labor-Hours	Unit	Material	2013 Bare Costs Labor	Equipment	Total	Total Incl O&P
6120	3 feet depth of bury	1 Plum	8	1	Ea.	955	56		1,011	1,125
6140	4 feet depth of bury		8	1		1,025	56		1,081	1,200
6160	5 feet depth of bury		7	1.143		1,100	64		1,164	1,300
6180	6 feet depth of bury		7	1.143		1,175	64		1,239	1,400
6200	7 feet depth of bury		6	1.333		1,250	74.50		1,324.50	1,475
6220	8 feet depth of bury		5	1.600		1,350	89.50		1,439.50	1,600
6240	9 feet depth of bury		4	2		1,425	112		1,537	1,725
6260	10 feet depth of bury	↓	4	2	↓	1,500	112		1,612	1,825
6300	For 1" IPS connection, add					40%	10%			
6350	For 1-1/4" IPS connection, add					140%	14%			
6400	For 1-1/2" IPS connection, add					225%	18%			
6450	For 2" IPS connection, add					315%	24%			

22 11 23 – Domestic Water Pumps

22 11 23.10 General Utility Pumps		Crew	Daily Output	Labor-Hours	Unit	Material	Labor	Equipment	Total	Total Incl O&P
0010	**GENERAL UTILITY PUMPS**									
2000	Single stage									
3000	Double suction,									
3190	75 HP, to 2500 GPM	Q-3	.28	114	Ea.	20,100	6,075		26,175	31,300
3220	100 HP, to 3000 GPM		.26	123		25,500	6,550		32,050	38,000
3240	150 HP, to 4000 GPM	↓	.24	133	↓	36,000	7,075		43,075	50,500
4000	Centrifugal, end suction, mounted on base									
4010	Horizontal mounted, with drip proof motor, rated @ 100' head									
4020	Vertical split case, single stage									
4040	100 GPM, 5 HP, 1-1/2" discharge	Q-1	1.70	9.412	Ea.	3,675	475		4,150	4,725
4050	200 GPM, 10 HP, 2" discharge		1.30	12.308		4,000	620		4,620	5,325
4060	250 GPM, 10 HP, 3" discharge	↓	1.28	12.500		4,450	630		5,080	5,850
4070	300 GPM, 15 HP, 2" discharge	Q-2	1.56	15.385		5,225	800		6,025	6,950
4080	500 GPM, 20 HP, 4" discharge		1.44	16.667		6,025	870		6,895	7,925
4090	750 GPM, 30 HP, 4" discharge		1.20	20		6,350	1,050		7,400	8,550
4100	1050 GPM, 40 HP, 5" discharge		1	24		9,175	1,250		10,425	12,000
4110	1500 GPM, 60 HP, 6" discharge		.60	40		13,400	2,075		15,475	17,900
4120	2000 GPM, 75 HP, 6" discharge		.50	48		14,600	2,500		17,100	19,800
4130	3000 GPM, 100 HP, 8" discharge	↓	.40	60	↓	18,900	3,125		22,025	25,500
4200	Horizontal split case, single stage									
4210	100 GPM, 7.5 HP, 1-1/2" discharge	Q-1	1.70	9.412	Ea.	3,725	475		4,200	4,800
4220	250 GPM, 15 HP, 2-1/2" discharge	"	1.30	12.308		7,900	620		8,520	9,600
4230	500 GPM, 20 HP, 4" discharge	Q-2	1.60	15		9,625	780		10,405	11,800
4240	750 GPM, 25 HP, 5" discharge		1.54	15.584		10,100	810		10,910	12,300
4250	1000 GPM, 40 HP, 5" discharge	↓	1.20	20		14,000	1,050		15,050	16,900
4260	1500 GPM, 50 HP, 6" discharge	Q-3	1.42	22.535		16,400	1,200		17,600	19,900
4270	2000 GPM, 75 HP, 8" discharge		1.14	28.070		22,300	1,500		23,800	26,900
4280	3000 GPM, 100 HP, 10" discharge		.96	33.333		25,000	1,775		26,775	30,200
4290	3500 GPM, 150 HP, 10" discharge		.86	37.209		32,900	1,975		34,875	39,200
4300	4000 GPM, 200 HP, 10" discharge	↓	.66	48.485	↓	39,300	2,575		41,875	47,100
4330	Horizontal split case, two stage, 500' head									
4340	100 GPM, 40 HP, 1-1/2" discharge	Q-2	1.70	14.118	Ea.	16,100	735		16,835	18,900
4350	200 GPM, 50 HP, 1-1/2" discharge	"	1.44	16.667		16,900	870		17,770	19,900
4360	300 GPM, 75 HP, 2" discharge	Q-3	1.57	20.382		23,200	1,075		24,275	27,200
4370	400 GPM, 100 HP, 3" discharge		1.14	28.070		27,500	1,500		29,000	32,500
4380	800 GPM, 200 HP, 4" discharge	↓	.86	37.209	↓	36,400	1,975		38,375	43,000
5000	Centrifugal, in-line									
5006	Vertical mount, iron body, 125 lb. flgd, 1800 RPM TEFC mtr									

22 11 Facility Water Distribution

22 11 23 – Domestic Water Pumps

22 11 23.10 General Utility Pumps		Crew	Daily Output	Labor-Hours	Unit	2013 Bare Costs Material	Labor	Equipment	Total	Total Incl O&P
5010	Single stage									
5012	.5 HP, 1-1/2" suction & discharge	Q-1	3.20	5	Ea.	1,325	251		1,576	1,850
5014	.75 HP, 2" suction & discharge		2.80	5.714		2,450	287		2,737	3,125
5015	1 HP, 3" suction & discharge		2.50	6.400		2,850	320		3,170	3,625
5016	1.5 HP, 3" suction & discharge		2.30	6.957		2,900	350		3,250	3,725
5018	2 HP, 4" suction & discharge		2.10	7.619		3,625	385		4,010	4,550
5020	3 HP, 5" suction & discharge		1.90	8.421		4,250	425		4,675	5,300
5030	5 HP, 6" suction & discharge		1.60	10		4,825	500		5,325	6,050
5040	7.5 HP, 6" suction & discharge		1.30	12.308		5,550	620		6,170	7,025
5050	10 HP, 8" suction & discharge	Q-2	1.80	13.333		7,400	695		8,095	9,175
5060	15 HP, 8" suction & discharge		1.70	14.118		8,650	735		9,385	10,600
5064	20 HP, 8" suction & discharge		1.60	15		10,000	780		10,780	12,200
5070	30 HP, 8" suction & discharge		1.50	16		12,800	835		13,635	15,300
5080	40 HP, 8" suction & discharge		1.40	17.143		15,300	895		16,195	18,200
5090	50 HP, 8" suction & discharge		1.30	18.462		15,500	960		16,460	18,600
5094	60 HP, 8" suction & discharge		.90	26.667		16,300	1,400		17,700	20,100
5100	75 HP, 8" suction & discharge	Q-3	.60	53.333		18,700	2,825		21,525	24,900
5110	100 HP, 8" suction & discharge	"	.50	64		21,300	3,400		24,700	28,500

22 11 23.11 Miscellaneous Pumps		Crew	Daily Output	Labor-Hours	Unit	Material	Labor	Equipment	Total	Total Incl O&P
0010	**MISCELLANEOUS PUMPS**									
0020	Water pump, portable, gasoline powered									
0100	170 GPH, 2" discharge	Q-1	11	1.455	Ea.	600	73		673	770
0110	343 GPH, 3" discharge		10.50	1.524		735	76.50		811.50	920
0120	608 GPH, 4" discharge		10	1.600		1,800	80.50		1,880.50	2,100
0500	Pump, propylene body, housing and impeller									
0510	22 GPM, 1/3 HP, 40' HD	1 Plum	5	1.600	Ea.	720	89.50		809.50	930
0520	33 GPM, 1/2 HP, 40' HD	Q-1	5	3.200		660	161		821	965
0530	53 GPM, 3/4 HP, 40' HD	"	4	4		1,050	201		1,251	1,450
0600	Rotary pump, CI									
0614	282 GPH, 3/4 HP, 1" discharge	Q-1	4.50	3.556	Ea.	1,200	179		1,379	1,600
0618	277 GPH, 1 HP, 1" discharge		4	4		1,200	201		1,401	1,625
0624	1100 GPH, 1.5 HP, 1-1/4" discharge		3.60	4.444		3,025	223		3,248	3,650
0628	1900 GPH, 2 HP, 1-1/4" discharge		3.20	5		4,175	251		4,426	4,975
1000	Turbine pump, CI									
1010	50 GPM, 2 HP, 3" discharge	Q-1	.80	20	Ea.	3,500	1,000		4,500	5,375
1020	100 GPM, 3 HP, 4" discharge	Q-2	.96	25		5,750	1,300		7,050	8,275
1030	250 GPM, 15 HP, 6" discharge	"	.94	25.532		7,125	1,325		8,450	9,850
1040	500 GPM, 25 HP, 6" discharge	Q-3	1.22	26.230		7,200	1,400		8,600	10,000
1050	1000 GPM, 50 HP, 8" discharge		1.14	28.070		9,925	1,500		11,425	13,200
1060	2000 GPM, 100 HP, 10" discharge		1	32		11,800	1,700		13,500	15,500
1070	3000 GPM, 150 HP, 10" discharge		.80	40		18,900	2,125		21,025	24,000
1080	4000 GPM, 200 HP, 12" discharge		.70	45.714		20,700	2,425		23,125	26,500
1090	6000 GPM, 300 HP, 14" discharge		.60	53.333		31,600	2,825		34,425	39,000
1100	10,000 GPM, 300 HP, 18" discharge		.58	55.172		36,500	2,925		39,425	44,500
2000	Centrifugal stainless steel pumps									
2100	100 GPM, 100' TDH, 3 HP	Q-1	1.80	8.889	Ea.	10,400	445		10,845	12,200
2130	250 GPM, 100' TDH, 10 HP	"	1.28	12.500		14,600	630		15,230	16,900
2160	500 GPM, 100' TDH, 20 HP	Q-2	1.44	16.667		16,700	870		17,570	19,600
2200	Vertical turbine stainless steel pumps									
2220	100 GPM, 100' TDH, 7.5 HP	Q-2	.95	25.263	Ea.	58,500	1,325		59,825	66,000
2240	250 GPM, 100' TDH, 15 HP		.94	25.532		64,500	1,325		65,825	73,000
2260	500 GPM, 100' TDH, 20 HP		.93	25.806		71,000	1,350		72,350	80,000

22 11 Facility Water Distribution

22 11 23 – Domestic Water Pumps

22 11 23.11 Miscellaneous Pumps		Crew	Daily Output	Labor-Hours	Unit	Material	Labor	Equipment	Total	Total Incl O&P
						2013 Bare Costs				
2280	750 GPM, 100′ TDH, 30 HP	Q-3	1.19	26.891	Ea.	77,000	1,425		78,425	87,000
2300	1000 GPM, 100′ TDH, 40 HP	″	1.16	27.586		83,500	1,475		84,975	93,500
2320	100 GPM, 200′ TDH, 15 HP	Q-2	.94	25.532		89,500	1,325		90,825	100,500
2340	250 GPM, 200′ TDH, 25 HP	Q-3	1.22	26.230		96,000	1,400		97,400	107,500
2360	500 GPM, 200′ TDH, 40 HP		1.16	27.586		102,000	1,475		103,475	114,500
2380	750 GPM, 200′ TDH, 60 HP		1.13	28.319		103,000	1,500		104,500	115,500
2400	1000 GPM, 200′ TDH, 75 HP		1.12	28.571		106,000	1,525		107,525	119,000

22 11 23.13 Domestic-Water Packaged Booster Pumps

		Crew	Daily Output	Labor-Hours	Unit	Material	Labor	Equipment	Total	Total Incl O&P
0010	DOMESTIC-WATER PACKAGED BOOSTER PUMPS									
0200	Pump system, with diaphragm tank, control, press. switch									
0300	1 HP pump	Q-1	1.30	12.308	Ea.	6,275	620		6,895	7,825
0400	1-1/2 HP pump		1.25	12.800		6,350	645		6,995	7,950
0420	2 HP pump		1.20	13.333		6,500	670		7,170	8,150
0440	3 HP pump		1.10	14.545		6,600	730		7,330	8,350
0460	5 HP pump	Q-2	1.50	16		7,300	835		8,135	9,275
0480	7-1/2 HP pump		1.42	16.901		8,125	880		9,005	10,300
0500	10 HP pump		1.34	17.910		8,500	935		9,435	10,800
2000	Pump system, variable speed, base, controls, starter									
2010	Duplex, 100′ head									
2020	400 GPM, 7-1/2HP, 4″ discharge	Q-2	.70	34.286	Ea.	37,000	1,775		38,775	43,400
2025	Triplex, 100′ head									
2030	1000 GPM, 15HP, 6″ discharge	Q-2	.50	48	Ea.	54,000	2,500		56,500	63,500
2040	1700 GPM, 30HP, 6″ discharge	″	.30	80	″	65,000	4,175		69,175	78,000

22 12 Facility Potable-Water Storage Tanks

22 12 21 – Facility Underground Potable-Water Storage Tanks

22 12 21.13 Fiberglass, Underground Potable-Water Storage Tanks

		Crew	Daily Output	Labor-Hours	Unit	Material	Labor	Equipment	Total	Total Incl O&P
0010	FIBERGLASS, UNDERGROUND POTABLE-WATER STORAGE TANKS									
0020	Excludes excavation, backfill, & piping									
2000	600 gallon capacity	B-21B	3.75	10.667	Ea.	3,000	410	172	3,582	4,125
2010	1,000 gallon capacity		3.50	11.429		3,850	440	185	4,475	5,125
2020	2,000 gallon capacity		3.25	12.308		5,625	475	199	6,299	7,125
2030	4,000 gallon capacity		3	13.333		7,775	515	215	8,505	9,575
2040	6,000 gallon capacity		2.65	15.094		8,750	580	244	9,574	10,800
2050	8,000 gallon capacity		2.30	17.391		10,500	670	281	11,451	12,800
2060	10,000 gallon capacity		2	20		12,000	770	325	13,095	14,700
2070	12,000 gallon capacity		1.50	26.667		16,600	1,025	430	18,055	20,300
2080	15,000 gallon capacity		1	40		19,200	1,550	645	21,395	24,300
2090	20,000 gallon capacity		.75	53.333		24,700	2,050	860	27,610	31,300
2100	25,000 gallon capacity		.50	80		37,000	3,075	1,300	41,375	46,900
2110	30,000 gallon capacity		.35	114		72,500	4,400	1,850	78,750	89,000
2120	40,000 gallon capacity		.30	133		80,500	5,125	2,150	87,775	99,000

22 12 23 – Facility Indoor Potable-Water Storage Tanks

22 12 23.13 Facility Steel, Indoor Pot.-Water Storage Tanks

		Crew	Daily Output	Labor-Hours	Unit	Material	Labor	Equipment	Total	Total Incl O&P
0010	FACILITY STEEL, INDOOR POT.-WATER STORAGE TANKS									
2000	Galvanized steel, 15 gal., 14″ diam., 26″ LOA	1 Plum	12	.667	Ea.	1,325	37		1,362	1,525
2060	30 gal., 14″ diam. x 49″ LOA		11	.727		1,575	40.50		1,615.50	1,775
2080	80 gal., 20″ diam. x 64″ LOA		9	.889		2,450	49.50		2,499.50	2,775
2100	135 gal., 24″ diam. x 75″ LOA		6	1.333		3,575	74.50		3,649.50	4,025
2120	240 gal., 30″ diam. x 86″ LOA		4	2		7,000	112		7,112	7,875

22 12 Facility Potable-Water Storage Tanks

22 12 23 – Facility Indoor Potable-Water Storage Tanks

22 12 23.13	Facility Steel, Indoor Pot.-Water Storage Tanks	Crew	Daily Output	Labor-Hours	Unit	2013 Bare Costs Material	Labor	Equipment	Total	Total Incl O&P
2140	300 gal., 36" diam. x 76" LOA	1 Plum	3	2.667	Ea.	9,950	149		10,099	11,100
2160	400 gal., 36" diam. x 100" LOA	Q-1	4	4		12,200	201		12,401	13,800
2180	500 gal., 36" diam., x 126" LOA	"	3	5.333		15,100	268		15,368	17,000
3000	Glass lined, P.E., 80 gal., 20" diam. x 60" LOA	1 Plum	9	.889		4,725	49.50		4,774.50	5,275
3060	140 gal., 24" diam. x 80" LOA		6	1.333		6,550	74.50		6,624.50	7,325
3080	225 gal., 30" diam. x 78" LOA		4	2		5,200	112		5,312	5,900
3100	325 gal., 36" diam. x 81" LOA		3	2.667		6,650	149		6,799	7,525
3120	460 gal., 42" diam. x 84" LOA	Q-1	4	4		7,000	201		7,201	8,000
3140	605 gal., 48" diam. x 87" LOA		3	5.333		13,400	268		13,668	15,100
3160	740 gal., 54" diam. x 91" LOA		3	5.333		15,100	268		15,368	17,000
3180	940 gal., 60" diam. x 93" LOA		2.50	6.400		17,400	320		17,720	19,600
3200	1330 gal., 66" diam. x 107" LOA		2	8		23,800	400		24,200	26,800
3220	1615 gal., 72" diam. x 110" LOA		1.50	10.667		27,100	535		27,635	30,600
3240	2285 gal., 84" diam. x 128" LOA		1	16		33,500	805		34,305	38,000
3260	3440 gal., 96" diam. x 157" LOA	Q-2	1.50	16		49,600	835		50,435	56,000

22 13 Facility Sanitary Sewerage

22 13 16 – Sanitary Waste and Vent Piping

22 13 16.20 Pipe, Cast Iron

Code	Description	Crew	Daily Output	Labor-Hours	Unit	Material	Labor	Equipment	Total	Total Incl O&P
0010	**PIPE, CAST IRON**, Soil, on clevis hanger assemblies, 5' O.C. R221113-70									
0020	Single hub, service wt., lead & oakum joints 10' O.C.									
2120	2" diameter R221316-10	Q-1	63	.254	L.F.	9.10	12.75		21.85	29
2140	3" diameter		60	.267		12.80	13.40		26.20	34
2160	4" diameter R221316-20		55	.291		16.90	14.60		31.50	40.50
2180	5" diameter	Q-2	76	.316		22.50	16.45		38.95	50
2200	6" diameter	"	73	.329		29	17.10		46.10	57.50
2220	8" diameter	Q-3	59	.542		44	29		73	92
2240	10" diameter		54	.593		71	31.50		102.50	126
2260	12" diameter		48	.667		101	35.50		136.50	165
2261	15" diameter		40	.800		149	42.50		191.50	228
2320	For service weight, double hub, add					10%				
2340	For extra heavy, single hub, add					48%	4%			
2360	For extra heavy, double hub, add					71%	4%			
2400	Lead for caulking, (1#/diam. in.)	Q-1	160	.100	Lb.	1.04	5		6.04	8.70
2420	Oakum for caulking, (1/8#/diam. in.)	"	40	.400	"	4.20	20		24.20	35
2960	To delete hangers, subtract									
2970	2" diam. to 4" diam.					16%	19%			
2980	5" diam. to 8" diam.					14%	14%			
2990	10" diam. to 15" diam.					13%	19%			
3000	Single hub, service wt., push-on gasket joints 10' O.C.									
3010	2" diameter	Q-1	66	.242	L.F.	10.05	12.20		22.25	29.50
3020	3" diameter		63	.254		14	12.75		26.75	34.50
3030	4" diameter		57	.281		18.45	14.10		32.55	41.50
3040	5" diameter	Q-2	79	.304		25	15.80		40.80	51.50
3050	6" diameter	"	75	.320		31	16.65		47.65	59.50
3060	8" diameter	Q-3	62	.516		49.50	27.50		77	96
3070	10" diameter		56	.571		80.50	30.50		111	134
3080	12" diameter		49	.653		113	34.50		147.50	177
3082	15" diameter		40	.800		163	42.50		205.50	244
3100	For service weight, double hub, add					65%				
3110	For extra heavy, single hub, add					48%	4%			

22 13 16 – Sanitary Waste and Vent Piping

22 13 16.20 Pipe, Cast Iron

	22 13 16.20 Pipe, Cast Iron	Crew	Daily Output	Labor-Hours	Unit	Material	2013 Bare Costs Labor	Equipment	Total	Total Incl O&P
3120	For extra heavy, double hub, add					29%	4%			
3130	To delete hangers, subtract									
3140	2″ diam. to 4″ diam.					12%	21%			
3150	5″ diam. to 8″ diam.					10%	16%			
3160	10″ diam. to 15″ diam.					9%	21%			
4000	No hub, couplings 10′ O.C.									
4100	1-1/2″ diameter	Q-1	71	.225	L.F.	9	11.30		20.30	27
4120	2″ diameter		67	.239		9.20	12		21.20	28
4140	3″ diameter		64	.250		12.80	12.55		25.35	33
4160	4″ diameter		58	.276		16.70	13.85		30.55	39.50
4180	5″ diameter	Q-2	83	.289		22.50	15.05		37.55	47
4200	6″ diameter	″	79	.304		29	15.80		44.80	56
4220	8″ diameter	Q-3	69	.464		50	24.50		74.50	92
4240	10″ diameter		61	.525		83	28		111	134
4244	12″ diameter		58	.552		108	29.50		137.50	162
4248	15″ diameter		52	.615		159	32.50		191.50	225
4280	To delete hangers, subtract									
4290	1-1/2″ diam. to 6″ diam.					22%	47%			
4300	8″ diam. to 10″ diam.					21%	44%			
4310	12″ diam. to 15″ diam.					19%	40%			

22 13 16.30 Pipe Fittings, Cast Iron

	22 13 16.30 Pipe Fittings, Cast Iron	Crew	Daily Output	Labor-Hours	Unit	Material	Labor	Equipment	Total	Total Incl O&P
0010	**PIPE FITTINGS, CAST IRON**, Soil									
0040	Hub and spigot, service weight, lead & oakum joints									
0080	1/4 bend, 2″	Q-1	16	1	Ea.	18.20	50		68.20	95.50
0120	3″		14	1.143		24.50	57.50		82	113
0140	4″		13	1.231		38	62		100	135
0160	5″	Q-2	18	1.333		53	69.50		122.50	164
0180	6″	″	17	1.412		66	73.50		139.50	184
0200	8″	Q-3	11	2.909		199	155		354	450
0220	10″		10	3.200		291	170		461	575
0224	12″		9	3.556		395	189		584	720
0226	15″		7	4.571		1,350	243		1,593	1,875
0242	Short sweep, CI, 90°, 2″	Q-1	16	1		17.65	50		67.65	95
0243	3″		14	1.143		32	57.50		89.50	122
0244	4″		13	1.231		48.50	62		110.50	147
0245	6″	Q-2	17	1.412		98.50	73.50		172	220
0246	8″	Q-3	11	2.909		214	155		369	470
0247	10″		10	3.200		440	170		610	740
0248	12″		9	3.556		1,025	189		1,214	1,400
0251	Long sweep elbow									
0252	2″	Q-1	16	1	Ea.	27.50	50		77.50	106
0253	3″		14	1.143		39.50	57.50		97	130
0254	4″		13	1.231		58	62		120	157
0255	6″	Q-2	17	1.412		116	73.50		189.50	239
0256	8″	Q-3	11	2.909		263	155		418	520
0257	10″		10	3.200		445	170		615	745
0258	12″		9	3.556		720	189		909	1,075
0259	15″		7	4.571		1,875	243		2,118	2,450
0266	Closet bend, 3″ diameter with flange 10″ x 16″	Q-1	14	1.143		105	57.50		162.50	202
0268	16″ x 16″		12	1.333		126	67		193	240
0270	Closet bend, 4″ diameter, 1″ x 4″ ring, 6″ x 16″		13	1.231		101	62		163	205
0280	8″ x 16″		13	1.231		93.50	62		155.50	196

22 13 16 – Sanitary Waste and Vent Piping

22 13 16.30	Pipe Fittings, Cast Iron	Crew	Daily Output	Labor-Hours	Unit	2013 Bare Costs Material	Labor	Equipment	Total	Total Incl O&P
0290	10" x 12"	Q-1	12	1.333	Ea.	87.50	67		154.50	197
0300	10" x 18"		11	1.455		124	73		197	246
0310	12" x 16"		11	1.455		107	73		180	228
0330	16" x 16"		10	1.600		134	80.50		214.50	268
0340	1/8 bend, 2"		16	1		12.95	50		62.95	90
0350	3"		14	1.143		20.50	57.50		78	109
0360	4"		13	1.231		30	62		92	126
0380	5"	Q-2	18	1.333		41	69.50		110.50	150
0400	6"	"	17	1.412		50.50	73.50		124	167
0420	8"	Q-3	11	2.909		150	155		305	400
0440	10"		10	3.200		216	170		386	495
0460	12"		9	3.556		410	189		599	735
0461	15"		7	4.571		955	243		1,198	1,425
0500	Sanitary tee, 2"	Q-1	10	1.600		25.50	80.50		106	149
0540	3"		9	1.778		41	89.50		130.50	179
0620	4"		8	2		50.50	100		150.50	207
0700	5"	Q-2	12	2		100	104		204	267
0800	6"	"	11	2.182		114	114		228	296
0880	8"	Q-3	7	4.571		300	243		543	695
0881	10"		7	4.571		550	243		793	970
0882	12"		6	5.333		975	283		1,258	1,500
0883	15"		4	8		1,950	425		2,375	2,800
0900	Sanitary tee, tapped									
0901	2" x 2"	Q-1	9	1.778	Ea.	34	89.50		123.50	172
0902	3" x 2"		8	2		38	100		138	193
0903	4" x 2"		7	2.286		62	115		177	242
0910	Sanitary cross, tapped [double tapped sanitary tee]									
0911	2" x 2"	Q-1	7	2.286	Ea.	54.50	115		169.50	233
0912	3" x 2"		6	2.667		67	134		201	276
0913	4" x 2"		5	3.200		65.50	161		226.50	315
0940	Sanitary Tee, reducing									
0942	3" x 2"	Q-1	10	1.600	Ea.	35.50	80.50		116	160
0943	4" x 3"		9	1.778		46	89.50		135.50	185
0944	4" x 2"		9	1.778		43	89.50		132.50	182
0945	5" x 3"	Q-2	12.50	1.920		119	100		219	282
0946	6" x 4"		10	2.400		128	125		253	330
0947	6" x 3"		11	2.182		92	114		206	272
0948	6" x 2"		12.50	1.920		89	100		189	249
0949	8" x 6"	Q-3	9	3.556		254	189		443	565
0950	8" x 5"		9	3.556		254	189		443	565
0951	8" x 4"		10	3.200		193	170		363	470
0954	10" x 6"		8	4		405	213		618	765
0958	12" x 8"		7	4.571		730	243		973	1,175
0962	15" x 10"		5	6.400		1,550	340		1,890	2,200
1000	Tee, 2"	Q-1	10	1.600		36.50	80.50		117	162
1060	3"		9	1.778		54.50	89.50		144	194
1120	4"		8	2		70.50	100		170.50	229
1200	5"	Q-2	12	2		149	104		253	320
1300	6"	"	11	2.182		147	114		261	330
1380	8"	Q-3	7	4.571		265	243		508	655
1400	Combination Y and 1/8 bend									
1420	2"	Q-1	10	1.600	Ea.	32	80.50		112.50	156
1460	3"		9	1.778		48.50	89.50		138	187

22 13 16.30 Pipe Fittings, Cast Iron		Crew	Daily Output	Labor-Hours	Unit	2013 Bare Costs Material	Labor	Equipment	Total	Total Incl O&P
1520	4"	Q-1	8	2	Ea.	66.50	100		166.50	225
1540	5"	Q-2	12	2		126	104		230	296
1560	6"		11	2.182		160	114		274	345
1580	8"		7	3.429		395	179		574	705
1582	12"	Q-3	6	5.333		810	283		1,093	1,325
1584	Combination Y & 1/8 bend, reducing									
1586	3" x 2"	Q-1	10	1.600	Ea.	36.50	80.50		117	161
1587	4" x 2"		9.50	1.684		49.50	84.50		134	181
1588	4" x 3"		9	1.778		57.50	89.50		147	197
1589	6" x 2"	Q-2	12.50	1.920		103	100		203	265
1590	6" x 3"		12	2		110	104		214	278
1591	6" x 4"		11	2.182		115	114		229	298
1592	8" x 2"	Q-3	11	2.909		218	155		373	470
1593	8" x 4"		10	3.200		192	170		362	465
1594	8" x 6"		9	3.556		263	189		452	575
1600	Double Y, 2"	Q-1	8	2		55.50	100		155.50	212
1610	3"		7	2.286		70.50	115		185.50	251
1620	4"		6.50	2.462		92	124		216	287
1630	5"	Q-2	9	2.667		165	139		304	390
1640	6"	"	8	3		237	156		393	495
1650	8"	Q-3	5.50	5.818		580	310		890	1,100
1660	10"		5	6.400		1,225	340		1,565	1,825
1670	12"		4.50	7.111		1,475	380		1,855	2,200
1676	Combination double Y & 1/8 bend									
1678	2"	Q-1	8	2	Ea.	65	100		165	223
1680	3"		7	2.286		83.50	115		198.50	265
1682	4"		6.50	2.462		137	124		261	335
1684	6"	Q-2	8	3		475	156		631	755
1690	Combination double Y & 1/8 bend, CI, reducing									
1692	3" x 2"	Q-1	8	2	Ea.	71.50	100		171.50	230
1694	4" x 2"		7	2.286		84.50	115		199.50	266
1696	4" x 3"		7	2.286		96	115		211	279
1698	6" x 4"	Q-2	9	2.667		310	139		449	550
1700	Double Y, CI, reducing									
1702	3" x 2"	Q-1	8	2	Ea.	60	100		160	217
1703	4" x 2"		7	2.286		72.50	115		187.50	253
1704	4" x 3"		7	2.286		76	115		191	257
1706	6" x 3"	Q-2	9.50	2.526		184	132		316	400
1707	6" x 4"	"	9	2.667		171	139		310	395
1708	8" x 4"	Q-3	8.50	3.765		400	200		600	740
1709	8" x 6"		8	4		415	213		628	775
1711	10" x 6"		8	4		865	213		1,078	1,275
1712	10" x 8"		6.50	4.923		1,225	262		1,487	1,750
1713	12" x 6"		7.50	4.267		1,475	227		1,702	1,975
1714	12" x 8"		6	5.333		1,550	283		1,833	2,125
1740	Reducer, 3" x 2"	Q-1	15	1.067		17.85	53.50		71.35	100
1750	4" x 2"		14.50	1.103		20.50	55.50		76	106
1760	4" x 3"		14	1.143		23	57.50		80.50	112
1770	5" x 2"		14	1.143		48	57.50		105.50	140
1780	5" x 3"		13.50	1.185		51	59.50		110.50	146
1790	5" x 4"		13	1.231		30	62		92	126
1800	6" x 2"		13.50	1.185		46	59.50		105.50	140
1810	6" x 3"		13	1.231		47.50	62		109.50	145

22 13 Facility Sanitary Sewerage

22 13 16 – Sanitary Waste and Vent Piping

22 13 16.30 Pipe Fittings, Cast Iron		Crew	Daily Output	Labor-Hours	Unit	2013 Bare Costs Material	Labor	Equipment	Total	Total Incl O&P
1830	6″ x 4″	Q-1	12.50	1.280	Ea.	47	64.50		111.50	149
1840	6″ x 5″		11	1.455		50.50	73		123.50	166
1880	8″ x 3″	Q-2	13.50	1.778		89.50	92.50		182	238
1900	8″ x 4″		13	1.846		78	96		174	231
1920	8″ x 5″		12	2		81	104		185	246
1940	8″ x 6″		12	2		80	104		184	245
1942	10″ x 4″		12.50	1.920		117	100		217	280
1943	10″ x 6″		11.50	2.087		131	109		240	310
1944	10″ x 8″		9.50	2.526		132	132		264	345
1945	12″ x 4″		11.50	2.087		191	109		300	375
1946	12″ x 6″		11	2.182		205	114		319	395
1947	12″ x 8″		9	2.667		211	139		350	440
1948	12″ x 10″		8.50	2.824		214	147		361	455
1949	15″ x 6″	Q-3	12	2.667		400	142		542	655
1950	15″ x 8″		10	3.200		435	170		605	735
1951	15″ x 10″		9.50	3.368		450	179		629	765
1952	15″ x 12″		9	3.556		455	189		644	785
1960	Increaser, 2″ x 3″	Q-1	15	1.067		42.50	53.50		96	128
1980	2″ x 4″		14	1.143		42.50	57.50		100	134
2000	2″ x 5″		13	1.231		51.50	62		113.50	150
2020	3″ x 4″		13	1.231		47	62		109	145
2040	3″ x 5″		13	1.231		51.50	62		113.50	150
2060	3″ x 6″		12	1.333		63	67		130	171
2070	4″ x 5″		13	1.231		55.50	62		117.50	154
2080	4″ x 6″		12	1.333		63.50	67		130.50	171
2090	4″ x 8″	Q-2	13	1.846		129	96		225	287
2100	5″ x 6″	Q-1	11	1.455		100	73		173	220
2110	5″ x 8″	Q-2	12	2		152	104		256	325
2120	6″ x 8″		12	2		152	104		256	325
2130	6″ x 10″		8	3		274	156		430	535
2140	8″ x 10″		6.50	3.692		275	192		467	595
2150	10″ x 12″		5.50	4.364		490	227		717	880
2500	Y, 2″	Q-1	10	1.600		23	80.50		103.50	147
2510	3″		9	1.778		43	89.50		132.50	182
2520	4″		8	2		57.50	100		157.50	215
2530	5″	Q-2	12	2		102	104		206	269
2540	6″	″	11	2.182		133	114		247	315
2550	8″	Q-3	7	4.571		325	243		568	720
2560	10″		6	5.333		520	283		803	1,000
2570	12″		5	6.400		1,125	340		1,465	1,725
2580	15″		4	8		2,600	425		3,025	3,500
2581	Y, reducing									
2582	3″ x 2″	Q-1	10	1.600	Ea.	33.50	80.50		114	158
2584	4″ x 2″		9.50	1.684		44.50	84.50		129	176
2586	4″ x 3″		9	1.778		49.50	89.50		139	188
2588	6″ x 2″	Q-2	12.50	1.920		88	100		188	248
2590	6″ x 3″		12.25	1.959		90	102		192	253
2592	6″ x 4″		12	2		89	104		193	255
2594	8″ x 2″	Q-3	11	2.909		198	155		353	450
2596	8″ x 3″		10.50	3.048		197	162		359	460
2598	8″ x 4″		10	3.200		170	170		340	445
2600	8″ x 6″		9	3.556		211	189		400	515
2602	10″ x 3″		10	3.200		296	170		466	580

22 13 16 – Sanitary Waste and Vent Piping

22 13 16.30 Pipe Fittings, Cast Iron		Crew	Daily Output	Labor-Hours	Unit	Material	2013 Bare Costs Labor	Equipment	Total	Total Incl O&P
2604	10″ x 4″	Q-3	9.50	3.368	Ea.	291	179		470	590
2606	10″ x 6″		8	4		320	213		533	670
2608	10″ x 8″		8	4		435	213		648	800
2610	12″ x 4″		9	3.556		490	189		679	825
2612	12″ x 6″		8.50	3.765		505	200		705	860
2614	12″ x 8″		8	4		610	213		823	990
2616	12″ x 10″		7.50	4.267		970	227		1,197	1,425
2618	15″ x 4″		7	4.571		1,575	243		1,818	2,125
2620	15″ x 6″		6.50	4.923		1,650	262		1,912	2,225
2622	15″ x 8″		6.50	4.923		1,675	262		1,937	2,250
2624	15″ x 10″		5	6.400		1,725	340		2,065	2,400
2626	15″ x 12″		4.50	7.111		1,850	380		2,230	2,600
3000	For extra heavy, add					44%	4%			
3600	Hub and spigot, service weight gasket joint									
3605	Note: gaskets and joint labor have									
3606	been included with all listed fittings.									
3610	1/4 bend, 2″	Q-1	20	.800	Ea.	27.50	40		67.50	91
3620	3″		17	.941		36.50	47.50		84	111
3630	4″		15	1.067		53.50	53.50		107	139
3640	5″	Q-2	21	1.143		77	59.50		136.50	175
3650	6″	″	19	1.263		91	66		157	199
3660	8″	Q-3	12	2.667		254	142		396	495
3670	10″		11	2.909		385	155		540	660
3680	12″		10	3.200		515	170		685	825
3690	15″		8	4		1,500	213		1,713	1,975
3692	Short sweep, CI, 90°, 2″	Q-1	20	.800		27	40		67	90
3693	3″		17	.941		44	47.50		91.50	120
3694	4″		15	1.067		64	53.50		117.50	151
3695	6″	Q-2	19	1.263		124	66		190	235
3696	8″	Q-3	12	2.667		269	142		411	510
3697	10″		11	2.909		535	155		690	825
3698	12″		10	3.200		1,150	170		1,320	1,500
3700	Closet bend, 3″ diameter with ring 10″ x 16″	Q-1	17	.941		117	47.50		164.50	200
3710	16″ x 16″		15	1.067		138	53.50		191.50	233
3730	Closet bend, 4″ diameter, 1″ x 4″ ring, 6″ x 16″		15	1.067		117	53.50		170.50	209
3740	8″ x 16″		15	1.067		109	53.50		162.50	201
3750	10″ x 12″		14	1.143		103	57.50		160.50	200
3760	10″ x 18″		13	1.231		139	62		201	246
3770	12″ x 16″		13	1.231		122	62		184	227
3780	16″ x 16″		12	1.333		149	67		216	265
3786	Long sweep elbow									
3787	2″	Q-1	20	.800	Ea.	37	40		77	101
3788	3″		17	.941		51.50	47.50		99	128
3789	4″		15	1.067		73	53.50		126.50	161
3790	6″	Q-2	19	1.263		141	66		207	254
3791	8″	Q-3	12	2.667		320	142		462	565
3792	10″		11	2.909		540	155		695	830
3793	12″		10	3.200		845	170		1,015	1,175
3794	15″		8	4		2,025	213		2,238	2,550
3800	1/8 bend, 2″	Q-1	20	.800		22.50	40		62.50	85
3810	3″		17	.941		32.50	47.50		80	107
3820	4″		15	1.067		45	53.50		98.50	130
3830	5″	Q-2	21	1.143		65	59.50		124.50	161

22 13 16.30 Pipe Fittings, Cast Iron		Crew	Daily Output	Labor-Hours	Unit	Material	Labor	Equipment	Total	Total Incl O&P
						2013 Bare Costs				
3840	6"	Q-2	19	1.263	Ea.	75	66		141	182
3850	8"	Q-3	12	2.667		205	142		347	440
3860	10"		11	2.909		310	155		465	580
3870	12"		10	3.200		530	170		700	840
3880	15"		8	4		1,100	213		1,313	1,525
3882	Sanitary tee, tapped									
3884	2" x 2"	Q-1	11	1.455	Ea.	53	73		126	168
3886	3" x 2"		10	1.600		60	80.50		140.50	187
3888	4" x 2"		9	1.778		87	89.50		176.50	230
3890	Sanitary cross, tapped [double tapped sanitary tee]									
3892	2" x 2"	Q-1	9	1.778	Ea.	73.50	89.50		163	215
3894	3" x 2"		8	2		88.50	100		188.50	249
3896	4" x 2"		7	2.286		90	115		205	272
3900	Sanitary Tee, 2"		12	1.333		44	67		111	150
3910	3"		10	1.600		65.50	80.50		146	193
3920	4"		9	1.778		81	89.50		170.50	223
3930	5"	Q-2	13	1.846		148	96		244	310
3940	6"	"	11	2.182		163	114		277	350
3950	8"	Q-3	8.50	3.765		410	200		610	750
3952	10"		8	4		745	213		958	1,125
3954	12"		7	4.571		1,225	243		1,468	1,725
3956	15"		6	5.333		2,250	283		2,533	2,900
3960	Sanitary Tee, reducing									
3961	3" x 2"	Q-1	10.50	1.524	Ea.	57	76.50		133.50	178
3962	4" x 2"		10	1.600		67.50	80.50		148	196
3963	4" x 3"		9.50	1.684		73.50	84.50		158	208
3964	5" x 3"	Q-2	13.50	1.778		156	92.50		248.50	310
3965	6" x 2"		13	1.846		123	96		219	280
3966	6" x 3"		12.50	1.920		129	100		229	293
3967	6" x 4"		12	2		168	104		272	340
3968	8" x 4"	Q-3	10.50	3.048		263	162		425	535
3969	8" x 5"		10	3.200		335	170		505	620
3970	8" x 6"		9.50	3.368		335	179		514	635
3971	10" x 6"		9	3.556		525	189		714	860
3972	12" x 8"		8.50	3.765		910	200		1,110	1,300
3973	15" x 10"		6	5.333		1,775	283		2,058	2,400
3980	Tee, 2"	Q-1	12	1.333		55.50	67		122.50	162
3990	3"		10	1.600		79	80.50		159.50	208
4000	4"		9	1.778		101	89.50		190.50	245
4010	5"	Q-2	13	1.846		197	96		293	360
4020	6"	"	11	2.182		196	114		310	385
4030	8"	Q-3	8	4		375	213		588	730
4060	Combination Y and 1/8 bend									
4070	2"	Q-1	12	1.333	Ea.	50.50	67		117.50	157
4080	3"		10	1.600		73	80.50		153.50	201
4090	4"		9	1.778		97	89.50		186.50	241
4100	5"	Q-2	13	1.846		174	96		270	335
4110	6"	"	11	2.182		210	114		324	400
4120	8"	Q-3	8	4		505	213		718	875
4121	12"	"	7	4.571		1,050	243		1,293	1,525
4130	Combination Y & 1/8 bend, reducing									
4132	3" x 2"	Q-1	10.50	1.524	Ea.	58	76.50		134.50	179
4134	4" x 2"		10	1.600		74	80.50		154.50	203

22 13 16 – Sanitary Waste and Vent Piping

22 13 16.30 Pipe Fittings, Cast Iron		Crew	Daily Output	Labor-Hours	Unit	Material	2013 Bare Costs Labor	Equipment	Total	Total Incl O&P
4136	4" x 3"	Q-1	9.50	1.684	Ea.	85	84.50		169.50	221
4138	6" x 2"	Q-2	13	1.846		137	96		233	296
4140	6" x 3"		12.50	1.920		147	100		247	315
4142	6" x 4"		12	2		155	104		259	330
4144	8" x 2"	Q-3	11	2.909		297	155		452	560
4146	8" x 4"		10.50	3.048		262	162		424	530
4148	8" x 6"		9.50	3.368		345	179		524	645
4160	Double Y, 2"	Q-1	10	1.600		84	80.50		164.50	213
4170	3"		8	2		107	100		207	269
4180	4"		7	2.286		138	115		253	325
4190	5"	Q-2	10	2.400		237	125		362	450
4200	6"	"	9	2.667		310	139		449	550
4210	8"	Q-3	6	5.333		745	283		1,028	1,250
4220	10"		5	6.400		1,500	340		1,840	2,150
4230	12"		4.50	7.111		1,825	380		2,205	2,600
4234	Combination double Y & 1/8 bend									
4235	2"	Q-1	10	1.600	Ea.	93.50	80.50		174	224
4236	3"		8	2		120	100		220	283
4237	4"		7	2.286		183	115		298	375
4238	6"	Q-2	9	2.667		550	139		689	810
4242	Combination double Y & 1/8 bend, CI, reducing									
4243	3" x 2"	Q-1	9	1.778	Ea.	102	89.50		191.50	246
4244	4" x 2"		8	2		119	100		219	282
4245	4" x 3"		7.50	2.133		136	107		243	310
4246	6" x 4"	Q-2	10	2.400		365	125		490	590
4248	Double Y, CI, reducing									
4249	3" x 2"	Q-1	9	1.778	Ea.	91	89.50		180.50	234
4250	4" x 2"		8	2		106	100		206	268
4251	4" x 3"		7.50	2.133		104	107		211	275
4252	6" x 3"	Q-2	10.50	2.286		233	119		352	435
4253	6" x 4"	"	10	2.400		226	125		351	435
4254	8" x 4"	Q-3	9.50	3.368		485	179		664	805
4255	8" x 6"		9	3.556		520	189		709	855
4256	10" x 6"		8.50	3.765		1,000	200		1,200	1,425
4257	10" x 8"		7.50	4.267		1,450	227		1,677	1,925
4258	12" x 6"		8	4		1,650	213		1,863	2,125
4259	12" x 8"		7	4.571		1,775	243		2,018	2,325
4260	Reducer, 3" x 2"	Q-1	17	.941		39.50	47.50		87	115
4270	4" x 2"		16.50	.970		45	48.50		93.50	123
4280	4" x 3"		16	1		50.50	50		100.50	131
4290	5" x 2"		16	1		81.50	50		131.50	166
4300	5" x 3"		15.50	1.032		87.50	52		139.50	174
4310	5" x 4"		15	1.067		69	53.50		122.50	157
4320	6" x 2"		15.50	1.032		80	52		132	166
4330	6" x 3"		15	1.067		84.50	53.50		138	174
4336	6" x 4"		14	1.143		87	57.50		144.50	182
4340	6" x 5"		13	1.231		99	62		161	202
4360	8" x 3"	Q-2	15	1.600		157	83.50		240.50	298
4370	8" x 4"		15	1.600		148	83.50		231.50	289
4380	8" x 5"		14	1.714		160	89.50		249.50	310
4390	8" x 6"		14	1.714		160	89.50		249.50	310
4394	10" x 4"		13.50	1.778		228	92.50		320.50	390
4395	10" x 6"		13	1.846		252	96		348	420

22 13 16.30 Pipe Fittings, Cast Iron		Crew	Daily Output	Labor-Hours	Unit	Material	2013 Bare Costs Labor	Equipment	Total	Total Incl O&P
4396	10″ x 8″	Q-2	12.50	1.920	Ea.	282	100		382	460
4397	12″ x 4″		12	2		330	104		434	515
4398	12″ x 6″		11.50	2.087		350	109		459	550
4399	12″ x 8″		11	2.182		390	114		504	595
4400	12″ x 10″		10.50	2.286		430	119		549	655
4401	15″ x 6″	Q-3	13	2.462		570	131		701	825
4402	15″ x 8″		12	2.667		635	142		777	915
4403	15″ x 10″		11	2.909		690	155		845	995
4404	15″ x 12″		10	3.200		725	170		895	1,050
4430	Increaser, 2″ x 3″	Q-1	17	.941		55	47.50		102.50	132
4440	2″ x 4″		16	1		58	50		108	139
4450	2″ x 5″		15	1.067		75.50	53.50		129	164
4460	3″ x 4″		15	1.067		62	53.50		115.50	149
4470	3″ x 5″		15	1.067		75.50	53.50		129	164
4480	3″ x 6″		14	1.143		88	57.50		145.50	183
4490	4″ x 5″		15	1.067		79.50	53.50		133	168
4500	4″ x 6″		14	1.143		88	57.50		145.50	184
4510	4″ x 8″	Q-2	15	1.600		184	83.50		267.50	330
4520	5″ x 6″	Q-1	13	1.231		125	62		187	231
4530	5″ x 8″	Q-2	14	1.714		207	89.50		296.50	360
4540	6″ x 8″		14	1.714		207	89.50		296.50	360
4550	6″ x 10″		10	2.400		370	125		495	595
4560	8″ x 10″		8.50	2.824		370	147		517	630
4570	10″ x 12″		7.50	3.200		610	167		777	925
4600	Y, 2″	Q-1	12	1.333		42	67		109	147
4610	3″		10	1.600		67.50	80.50		148	195
4620	4″		9	1.778		88	89.50		177.50	231
4630	5″	Q-2	13	1.846		150	96		246	310
4640	6″	″	11	2.182		183	114		297	370
4650	8″	Q-3	8	4		435	213		648	795
4660	10″		7	4.571		715	243		958	1,150
4670	12″		6	5.333		1,375	283		1,658	1,925
4672	15″		5	6.400		2,875	340		3,215	3,675
4680	Y, reducing									
4681	3″ x 2″	Q-1	10.50	1.524	Ea.	55	76.50		131.50	176
4682	4″ x 2″		10	1.600		69.50	80.50		150	198
4683	4″ x 3″		9.50	1.684		77	84.50		161.50	212
4684	6″ x 2″	Q-2	13	1.846		122	96		218	280
4685	6″ x 3″		12.50	1.920		127	100		227	291
4686	6″ x 4″		12	2		129	104		233	299
4687	8″ x 2″	Q-3	11	2.909		263	155		418	520
4688	8″ x 3″		10.75	2.977		264	158		422	530
4689	8″ x 4″		10.50	3.048		240	162		402	510
4690	8″ x 6″		9.50	3.368		291	179		470	590
4691	10″ x 3″		10	3.200		405	170		575	700
4692	10″ x 4″		9.50	3.368		400	179		579	710
4693	10″ x 6″		9	3.556		440	189		629	765
4694	10″ x 8″		8.80	3.636		585	193		778	935
4695	12″ x 4″		9.50	3.368		630	179		809	965
4696	12″ x 6″		9	3.556		655	189		844	1,000
4697	12″ x 8″		8.50	3.765		785	200		985	1,175
4698	12″ x 10″		8	4		1,200	213		1,413	1,625
4699	15″ x 4″		8	4		1,750	213		1,963	2,250

	22 13 16.30 Pipe Fittings, Cast Iron	Crew	Daily Output	Labor-Hours	Unit	2013 Bare Costs Material	Labor	Equipment	Total	Total Incl O&P
4700	15" x 6"	Q-3	7.50	4.267	Ea.	1,825	227		2,052	2,350
4701	15" x 8"		7	4.571		1,875	243		2,118	2,450
4702	15" x 10"		6	5.333		1,975	283		2,258	2,575
4703	15" x 12"	↓	5	6.400	↓	2,125	340		2,465	2,825
4900	For extra heavy, add					44%	4%			
4940	Gasket and making push-on joint									
4950	2"	Q-1	40	.400	Ea.	9.40	20		29.40	41
4960	3"		35	.457		12.20	23		35.20	48
4970	4"	↓	32	.500		15.30	25		40.30	55
4980	5"	Q-2	43	.558		24	29		53	70.50
4990	6"	"	40	.600		25	31.50		56.50	74.50
5000	8"	Q-3	32	1		55	53		108	141
5010	10"		29	1.103		95.50	58.50		154	194
5020	12"		25	1.280		122	68		190	236
5022	15"	↓	21	1.524	↓	146	81		227	283
5030	Note: gaskets and joint labor have									
5040	Been included with all listed fittings.									
5990	No hub									
6000	Cplg. & labor required at joints not incl. in fitting									
6010	price. Add 1 coupling per joint for installed price									
6020	1/4 Bend, 1-1/2"				Ea.	9.40			9.40	10.30
6060	2"					10.15			10.15	11.15
6080	3"					14.20			14.20	15.65
6120	4"					21			21	23
6140	5"					50.50			50.50	56
6160	6"					51			51	56
6180	8"					143			143	157
6181	10"					252			252	278
6182	12"					895			895	985
6183	15"					1,175			1,175	1,275
6184	1/4 Bend, long sweep, 1-1/2"					22.50			22.50	24.50
6186	2"					22.50			22.50	24.50
6188	3"					27			27	29.50
6189	4"					43			43	47.50
6190	5"					83			83	91.50
6191	6"					94.50			94.50	104
6192	8"					257			257	282
6193	10"					440			440	480
6200	1/8 Bend, 1-1/2"					7.90			7.90	8.70
6210	2"					8.75			8.75	9.65
6212	3"					11.75			11.75	12.95
6214	4"					15.40			15.40	16.95
6216	5"					32			32	35.50
6218	6"					34			34	37.50
6220	8"					98.50			98.50	108
6222	10"				↓	187			187	206
6364	Closet flange									
6366	4"				Ea.	14.20			14.20	15.65
6370	Closet bend, no hub									
6376	4" x 16"				Ea.	75.50			75.50	83
6380	Sanitary Tee, tapped, 1-1/2"					18.60			18.60	20.50
6382	2" x 1-1/2"					16.45			16.45	18.05
6384	2"					17.65			17.65	19.40

22 13 16.30 Pipe Fittings, Cast Iron		Crew	Daily Output	Labor-Hours	Unit	Material	Labor	Equipment	Total	Total Incl O&P
						2013 Bare Costs				
6386	3" x 2"				Ea.	26.50			26.50	29
6388	3"					45			45	49.50
6390	4" x 1-1/2"					23.50			23.50	25.50
6392	4" x 2"					26.50			26.50	29
6393	4"					26.50			26.50	29
6394	6" x 1-1/2"					60.50			60.50	67
6396	6" x 2"					62			62	68
6459	Sanitary Tee, 1-1/2"					13.15			13.15	14.50
6460	2"					14			14	15.40
6470	3"					17.35			17.35	19.10
6472	4"					33			33	36
6474	5"					76.50			76.50	84.50
6476	6"					78.50			78.50	86
6478	8"					315			315	350
6480	10"				↓	390			390	430
6724	Sanitary Tee, reducing									
6725	3" x 2"				Ea.	15.35			15.35	16.85
6726	4" x 3"					25.50			25.50	28
6727	5" x 4"					59.50			59.50	65
6728	6" x 3"					57.50			57.50	63.50
6729	8" x 4"					167			167	184
6730	Y, 1-1/2"					13.30			13.30	14.65
6740	2"					13			13	14.30
6750	3"					18.95			18.95	21
6760	4"					30			30	33.50
6762	5"					72			72	79
6764	6"					80.50			80.50	88.50
6768	8"					190			190	209
6769	10"					420			420	465
6770	12"					830			830	915
6771	15"					1,850			1,850	2,050
6791	Y, reducing, 3" x 2"					14			14	15.40
6792	4" x 2"					20			20	22.50
6793	5" x 2"					44.50			44.50	49
6794	6" x 2"					49.50			49.50	54.50
6795	6" x 4"					64.50			64.50	71
6796	8" x 4"					111			111	122
6797	8" x 6"					136			136	150
6798	10" x 6"					305			305	335
6799	10" x 8"					365			365	405
6800	Double Y, 2"					20.50			20.50	22.50
6920	3"					38			38	42
7000	4"					77.50			77.50	85
7100	6"					137			137	150
7120	8"				↓	390			390	430
7200	Combination Y and 1/8 Bend									
7220	1-1/2"				Ea.	14.15			14.15	15.55
7260	2"					14.85			14.85	16.30
7320	3"					23.50			23.50	25.50
7400	4"					45			45	49.50
7480	5"					92			92	101
7500	6"					124			124	136
7520	8"					289			289	320

22 13 16 – Sanitary Waste and Vent Piping

22 13 16.30 Pipe Fittings, Cast Iron		Crew	Daily Output	Labor-Hours	Unit	2013 Bare Costs Material	Labor	Equipment	Total	Total Incl O&P
7800	Reducer, 3" x 2"				Ea.	7.20			7.20	7.95
7820	4" x 2"					11.15			11.15	12.25
7840	4" x 3"					11.15			11.15	12.25
7842	6" x 3"					30			30	33
7844	6" x 4"					30			30	33
7846	6" x 5"					30.50			30.50	33.50
7848	8" x 2"					47.50			47.50	52
7850	8" x 3"					44			44	48
7852	8" x 4"					46			46	50.50
7854	8" x 5"					52			52	57.50
7856	8" x 6"					51			51	56.50
7858	10" x 4"					90.50			90.50	99.50
7860	10" x 6"					95.50			95.50	105
7862	10" x 8"					112			112	124
7864	12" x 4"					187			187	206
7866	12" x 6"					201			201	221
7868	12" x 8"					206			206	227
7870	12" x 10"					210			210	231
7872	15" x 4"					390			390	430
7874	15" x 6"					370			370	405
7876	15" x 8"					425			425	470
7878	15" x 10"					440			440	485
7880	15" x 12"					445			445	490
8000	Coupling, standard (by CISPI Mfrs.)									
8020	1-1/2"	Q-1	48	.333	Ea.	13	16.75		29.75	39.50
8040	2"		44	.364		13	18.25		31.25	42
8080	3"		38	.421		15.55	21		36.55	49
8120	4"		33	.485		18.35	24.50		42.85	56.50
8160	5"	Q-2	44	.545		44.50	28.50		73	92
8180	6"	"	40	.600		46.50	31.50		78	98
8200	8"	Q-3	33	.970		88	51.50		139.50	174
8220	10"	"	26	1.231		116	65.50		181.50	227
8300	Coupling, cast iron clamp & neoprene gasket (by MG)									
8310	1-1/2"	Q-1	48	.333	Ea.	7.40	16.75		24.15	33
8320	2"		44	.364		8.35	18.25		26.60	36.50
8330	3"		38	.421		9.95	21		30.95	43
8340	4"		33	.485		13.75	24.50		38.25	51.50
8350	5"	Q-2	44	.545		24	28.50		52.50	69.50
8360	6"	"	40	.600		25	31.50		56.50	74.50
8380	8"	Q-3	33	.970		95.50	51.50		147	183
8400	10"	"	26	1.231		166	65.50		231.50	282
8410	Reducing, no hub									
8416	2" x 1-1/2"	Q-1	44	.364	Ea.	8.85	18.25		27.10	37.50
8600	Coupling, Stainless steel, heavy duty									
8620	1-1/2"	Q-1	48	.333	Ea.	9.55	16.75		26.30	35.50
8630	2"		44	.364		9.90	18.25		28.15	38.50
8640	2" x 1-1/2"		44	.364		12.05	18.25		30.30	41
8650	3"		38	.421		10.75	21		31.75	44
8660	4"		33	.485		12.15	24.50		36.65	50
8670	4" x 3"		33	.485		17.50	24.50		42	56
8680	5"	Q-2	44	.545		26	28.50		54.50	71.50
8690	6"	"	40	.600		29.50	31.50		61	79
8700	8"	Q-3	33	.970		49.50	51.50		101	132

22 13 16 – Sanitary Waste and Vent Piping

22 13 16.30 Pipe Fittings, Cast Iron		Crew	Daily Output	Labor-Hours	Unit	Material	2013 Bare Costs Labor	Equipment	Total	Total Incl O&P
8710	10"	Q-3	26	1.231	Ea.	62.50	65.50		128	167
8712	12"		22	1.455		97	77.50		174.50	222
8715	15"		18	1.778		114	94.50		208.50	268

22 13 16.40 Pipe Fittings, Cast Iron for Drainage

		Crew	Daily Output	Labor-Hours	Unit	Material	Labor	Equipment	Total	Total Incl O&P
0010	**PIPE FITTINGS, CAST IRON FOR DRAINAGE**, Special									
0020	Cast iron, drainage, threaded, black									
0030	90° Elbow, straight									
0031	1-1/2" pipe size	Q-1	20	.800	Ea.	32.50	40		72.50	96
0032	2" pipe size		18	.889		48.50	44.50		93	121
0033	3" pipe size		10	1.600		183	80.50		263.50	320
0034	4" pipe size		6	2.667		289	134		423	520
0040	90° Long turn elbow, straight									
0041	1-1/2" pipe size	Q-1	20	.800	Ea.	45	40		85	110
0042	2" pipe size		18	.889		67	44.50		111.50	141
0043	3" pipe size		10	1.600		245	80.50		325.50	390
0044	4" pipe size		6	2.667		420	134		554	660
0050	90° Street elbow, straight									
0051	1-1/2" pipe size	Q-1	20	.800	Ea.	46	40		86	111
0052	2" pipe size	"	18	.889	"	61	44.50		105.50	134
0060	45° Elbow									
0061	1-1/2" pipe size	Q-1	20	.800	Ea.	31.50	40		71.50	95
0062	2" pipe size		18	.889		45.50	44.50		90	117
0063	3" pipe size		10	1.600		178	80.50		258.50	315
0064	4" pipe size		6	2.667		279	134		413	505
0070	45° Street elbow									
0071	1-1/2" pipe size	Q-1	20	.800	Ea.	43	40		83	108
0072	2" pipe size	"	18	.889	"	69.50	44.50		114	144
0092	Tees, straight									
0093	1-1/2" pipe size	Q-1	13	1.231	Ea.	53	62		115	152
0094	2" pipe size	"	11	1.455	"	87.50	73		160.50	207
0100	TY's, straight									
0101	1-1/2" pipe size	Q-1	13	1.231	Ea.	52	62		114	151
0102	2" pipe size		11	1.455		86	73		159	205
0103	3" pipe size		6	2.667		315	134		449	545
0104	4" pipe size		4	4		470	201		671	820
0120	45° Y branch, straight									
0121	1-1/2" pipe size	Q-1	13	1.231	Ea.	64	62		126	163
0122	2" pipe size		11	1.455		119	73		192	241
0123	3" pipe size		6	2.667		410	134		544	650
0124	4" pipe size		4	4		610	201		811	980
0147	Double Y branch, straight									
0148	1-1/2" pipe size	Q-1	10	1.600	Ea.	158	80.50		238.50	295
0149	2" pipe size	"	7	2.286	"	147	115		262	335
0160	P trap									
0161	1-1/2" pipe size	Q-1	15	1.067	Ea.	95	53.50		148.50	185
0162	2" pipe size		13	1.231		152	62		214	260
0163	3" pipe size		7	2.286		490	115		605	715
0164	4" pipe size		5	3.200		1,100	161		1,261	1,450
0180	Tucker connection									
0181	1-1/2" pipe size	Q-1	24	.667	Ea.	91.50	33.50		125	152
0182	2" pipe size	"	21	.762	"	113	38.50		151.50	182
0205	Cast iron, drainage, threaded, galvanized									

22 13 16.40 Pipe Fittings, Cast Iron for Drainage		Crew	Daily Output	Labor-Hours	Unit	2013 Bare Costs Material	Labor	Equipment	Total	Total Incl O&P
0206	90° Elbow, straight									
0207	1-1/2" pipe size	Q-1	20	.800	Ea.	41.50	40		81.50	107
0208	2" pipe size		18	.889		69	44.50		113.50	143
0209	3" pipe size		10	1.600		273	80.50		353.50	420
0210	4" pipe size		6	2.667		530	134		664	785
0216	90° Long turn elbow, straight									
0217	1-1/2" pipe size	Q-1	20	.800	Ea.	54.50	40		94.50	121
0218	2" pipe size		18	.889		93.50	44.50		138	170
0219	3" pipe size		10	1.600		305	80.50		385.50	455
0220	4" pipe size		6	2.667		530	134		664	780
0226	90° Street elbow, straight									
0227	1-1/2" pipe size	Q-1	20	.800	Ea.	60.50	40		100.50	127
0228	2" pipe size	"	18	.889	"	95	44.50		139.50	171
0236	45° Elbow									
0237	1-1/2" pipe size	Q-1	20	.800	Ea.	47.50	40		87.50	113
0238	2" pipe size		18	.889		67.50	44.50		112	142
0239	3" pipe size		10	1.600		235	80.50		315.50	380
0240	4" pipe size		6	2.667		385	134		519	625
0246	45° Street elbow									
0247	1-1/2" pipe size	Q-1	20	.800	Ea.	62.50	40		102.50	130
0248	2" pipe size	"	18	.889	"	96.50	44.50		141	173
0268	Tees, straight									
0269	1-1/2" pipe size	Q-1	13	1.231	Ea.	72	62		134	172
0270	2" pipe size	"	11	1.455	"	120	73		193	242
0276	TY's, straight									
0277	1-1/2" pipe size	Q-1	13	1.231	Ea.	73	62		135	173
0278	2" pipe size		11	1.455		119	73		192	241
0279	3" pipe size		6	2.667		475	134		609	720
0280	4" pipe size		4	4		575	201		776	935
0296	45° Y branch, straight									
0297	1-1/2" pipe size	Q-1	13	1.231	Ea.	80.50	62		142.50	182
0298	2" pipe size		11	1.455		163	73		236	290
0299	3" pipe size		6	2.667		575	134		709	830
0300	4" pipe size		4	4		880	201		1,081	1,275
0323	Double Y branch, straight									
0324	1-1/2" pipe size	Q-1	10	1.600	Ea.	161	80.50		241.50	298
0325	2" pipe size	"	7	2.286	"	275	115		390	475
0336	P Trap									
0337	1-1/2" pipe size	Q-1	15	1.067	Ea.	131	53.50		184.50	225
0338	2" pipe size		13	1.231		229	62		291	345
0339	3" pipe size		7	2.286		720	115		835	970
0340	4" pipe size		5	3.200		1,725	161		1,886	2,150
0356	Tucker connection									
0357	1-1/2" pipe size	Q-1	24	.667	Ea.	209	33.50		242.50	281
0358	2" pipe size	"	21	.762	"	225	38.50		263.50	305
1000	Drip pan elbow, (safety valve discharge elbow)									
1010	Cast iron, threaded inlet									
1014	2-1/2"	Q-1	8	2	Ea.	1,125	100		1,225	1,400
1015	3"		6.40	2.500		1,200	126		1,326	1,525
1017	4"		4.80	3.333		1,625	167		1,792	2,050
1018	Cast iron, flanged inlet									
1019	6"	Q-2	3.60	6.667	Ea.	1,700	345		2,045	2,400
1020	8"	"	2.60	9.231	"	1,800	480		2,280	2,700

22 13 16.50 Shower Drains

	22 13 16.50 Shower Drains	Crew	Daily Output	Labor-Hours	Unit	2013 Bare Costs Material	Labor	Equipment	Total	Total Incl O&P
0010	**SHOWER DRAINS**									
2780	Shower, with strainer, uniform diam. trap, bronze top									
2800	2" and 3" pipe size	Q-1	8	2	Ea.	395	100		495	585
2820	4" pipe size	"	7	2.286		435	115		550	650
2840	For galvanized body, add					175			175	193
2860	With strainer, backwater valve, drum trap									
2880	1-1/2", 2" & 3" pipe size	Q-1	8	2	Ea.	350	100		450	535
2890	4" pipe size	"	7	2.286		490	115		605	715
2900	For galvanized body, add					175			175	193

22 13 16.60 Traps

	22 13 16.60 Traps	Crew	Daily Output	Labor-Hours	Unit	Material	Labor	Equipment	Total	Total Incl O&P
0010	**TRAPS**									
0030	Cast iron, service weight									
0050	Running P trap, without vent									
1100	2"	Q-1	16	1	Ea.	131	50		181	220
1140	3"		14	1.143		131	57.50		188.50	231
1150	4"		13	1.231		131	62		193	237
1160	6"	Q-2	17	1.412		570	73.50		643.50	740
1180	Running trap, single hub, with vent									
2080	3" pipe size, 3" vent	Q-1	14	1.143	Ea.	104	57.50		161.50	201
2120	4" pipe size, 4" vent	"	13	1.231		141	62		203	248
2140	5" pipe size, 4" vent	Q-2	11	2.182		225	114		339	420
2160	6" pipe size, 4" vent		10	2.400		605	125		730	855
2180	6" pipe size, 6" vent		8	3		655	156		811	960
2200	8" pipe size, 4" vent	Q-3	10	3.200		2,750	170		2,920	3,275
2220	8" pipe size, 6" vent	"	8	4		2,175	213		2,388	2,700
2300	For double hub, vent, add					10%	20%			
2800	S trap,									
2850	4" pipe size	Q-1	13	1.231	Ea.	63.50	62		125.50	163
3000	P trap, B&S, 2" pipe size		16	1		31	50		81	110
3040	3" pipe size		14	1.143		46.50	57.50		104	138
3060	4" pipe size		13	1.231		67	62		129	167
3080	5" pipe size	Q-2	18	1.333		148	69.50		217.50	268
3100	6" pipe size	"	17	1.412		206	73.50		279.50	340
3120	8" pipe size	Q-3	11	2.909		620	155		775	920
3130	10" pipe size	"	10	3.200		1,175	170		1,345	1,550
3150	P trap, no hub, 1-1/2" pipe size	Q-1	17	.941		16.80	47.50		64.30	89.50
3160	2" pipe size		16	1		15.90	50		65.90	93
3170	3" pipe size		14	1.143		35	57.50		92.50	125
3180	4" pipe size		13	1.231		62	62		124	161
3190	6" pipe size	Q-2	17	1.412		150	73.50		223.50	276
3350	Deep seal trap, B&S									
3400	1-1/4" pipe size	Q-1	14	1.143	Ea.	49	57.50		106.50	140
3410	1-1/2" pipe size		14	1.143		49	57.50		106.50	140
3420	2" pipe size		14	1.143		44	57.50		101.50	135
3440	3" pipe size		12	1.333		57	67		124	164
3460	4" pipe size		11	1.455		90.50	73		163.50	210
3500	For trap primer connection, add					125			125	137
3540	For trap with floor cleanout, add					70%	5%			
3580	For trap with adjustable cleanout, add	Q-1	10	1.600	Ea.	284	80.50		364.50	430
4700	Copper, drainage, drum trap									
4800	3" x 5" solid, 1-1/2" pipe size	1 Plum	16	.500	Ea.	255	28		283	320
4840	3" x 6" swivel, 1-1/2" pipe size	"	16	.500	"	405	28		433	485

22 13 Facility Sanitary Sewerage

22 13 16 – Sanitary Waste and Vent Piping

22 13 16.60 Traps		Crew	Daily Output	Labor-Hours	Unit	2013 Bare Costs Material	Labor	Equipment	Total	Total Incl O&P
5100	P trap, standard pattern									
5200	1-1/4" pipe size	1 Plum	18	.444	Ea.	188	25		213	245
5240	1-1/2" pipe size		17	.471		182	26.50		208.50	240
5260	2" pipe size		15	.533		281	30		311	355
5280	3" pipe size		11	.727		675	40.50		715.50	800
5340	With cleanout, swivel joint and slip joint									
5360	1-1/4" pipe size	1 Plum	18	.444	Ea.	132	25		157	184
5400	1-1/2" pipe size		17	.471		277	26.50		303.50	345
5420	2" pipe size		15	.533		445	30		475	535
5750	Chromed brass, tubular, P trap, without cleanout, 20 Ga.									
5800	1-1/4" pipe size	1 Plum	18	.444	Ea.	18.70	25		43.70	58
5840	1-1/2" pipe size	"	17	.471	"	21.50	26.50		48	63
5900	With cleanout, 20 Ga.									
5940	1-1/4" pipe size	1 Plum	18	.444	Ea.	30	25		55	70.50
6000	1-1/2" pipe size	"	17	.471	"	32.50	26.50		59	75
6350	S trap, without cleanout, 20 Ga.									
6400	1-1/4" pipe size	1 Plum	18	.444	Ea.	48.50	25		73.50	91
6440	1-1/2" pipe size	"	17	.471	"	55.50	26.50		82	101
6550	With cleanout, 20 Ga.									
6600	1-1/4" pipe size	1 Plum	18	.444	Ea.	57	25		82	100
6640	1-1/2" pipe size	"	17	.471		62	26.50		88.50	108
6660	Corrosion resistant, glass, P trap, 1-1/2" pipe size	Q-1	17	.941		69.50	47.50		117	148
6670	2" pipe size		16	1		91.50	50		141.50	176
6680	3" pipe size		14	1.143		187	57.50		244.50	293
6690	4" pipe size		13	1.231		280	62		342	405
6700	6" pipe size	Q-2	17	1.412		1,075	73.50		1,148.50	1,275
6710	ABS DWV P trap, solvent weld joint									
6720	1-1/2" pipe size	1 Plum	18	.444	Ea.	12.65	25		37.65	51.50
6722	2" pipe size		17	.471		16.65	26.50		43.15	58
6724	3" pipe size		15	.533		65.50	30		95.50	117
6726	4" pipe size		14	.571		131	32		163	192
6732	PVC DWV P trap, solvent weld joint									
6733	1-1/2" pipe size	1 Plum	18	.444	Ea.	9.60	25		34.60	48
6734	2" pipe size		17	.471		12.95	26.50		39.45	54
6735	3" pipe size		15	.533		44	30		74	93.50
6736	4" pipe size		14	.571		100	32		132	158
6760	PP DWV, dilution trap, 1-1/2" pipe size		16	.500		211	28		239	274
6770	P trap, 1-1/2" pipe size		17	.471		52.50	26.50		79	97.50
6780	2" pipe size		16	.500		77.50	28		105.50	128
6790	3" pipe size		14	.571		143	32		175	205
6800	4" pipe size		13	.615		238	34.50		272.50	315
6830	S trap, 1-1/2" pipe size		16	.500		43.50	28		71.50	89.50
6840	2" pipe size		15	.533		65	30		95	117
6850	Universal trap, 1-1/2" pipe size		14	.571		85.50	32		117.50	142
6860	PVC DWV hub x hub, basin trap, 1-1/4" pipe size		18	.444		11.65	25		36.65	50.50
6870	Sink P trap, 1-1/2" pipe size		18	.444		11.65	25		36.65	50.50
6880	Tubular S trap, 1-1/2" pipe size		17	.471		21.50	26.50		48	63.50
6890	PVC sch. 40 DWV, drum trap									
6900	1-1/2" pipe size	1 Plum	16	.500	Ea.	48.50	28		76.50	95
6910	P trap, 1-1/2" pipe size		18	.444		10.70	25		35.70	49.50
6920	2" pipe size		17	.471		14.40	26.50		40.90	55.50
6930	3" pipe size		15	.533		49	30		79	99
6940	4" pipe size		14	.571		111	32		143	171

22 13 16 – Sanitary Waste and Vent Piping

22 13 16.60 Traps		Crew	Daily Output	Labor-Hours	Unit	Material	Labor	Equipment	Total	Total Incl O&P
						2013 Bare Costs				
6950	P trap w/clean out, 1-1/2" pipe size	1 Plum	18	.444	Ea.	17.65	25		42.65	57
6960	2" pipe size		17	.471		30	26.50		56.50	72.50
6970	P trap adjustable, 1-1/2" pipe size		17	.471		18.90	26.50		45.40	60.50
6980	P trap adj. w/union & cleanout, 1-1/2" pipe size		16	.500		52	28		80	99.50
7000	Trap primer, flow through type, 1/2" diameter		24	.333		45.50	18.60		64.10	78
7100	With sediment strainer	↓	22	.364	↓	49.50	20.50		70	85
7450	Trap primer distribution unit									
7500	2 openings	1 Plum	18	.444	Ea.	23.50	25		48.50	63.50
7540	3 openings		17	.471		30	26.50		56.50	72.50
7560	4 openings	↓	16	.500	↓	35.50	28		63.50	81
7850	Trap primer manifold									
7900	2 outlet	1 Plum	18	.444	Ea.	58.50	25		83.50	102
7940	4 outlet		16	.500		94.50	28		122.50	146
7960	6 outlet		15	.533		131	30		161	189
7980	8 outlet	↓	13	.615	↓	166	34.50		200.50	235

22 13 16.80 Vent Flashing and Caps

		Crew	Daily Output	Labor-Hours	Unit	Material	Labor	Equipment	Total	Total Incl O&P
0010	**VENT FLASHING AND CAPS**									
0120	Vent caps									
0140	Cast iron									
0160	1-1/4" - 1-1/2" pipe	1 Plum	23	.348	Ea.	32.50	19.40		51.90	64.50
0170	2" - 2-1/8" pipe		22	.364		37.50	20.50		58	71.50
0180	2-1/2" - 3-5/8" pipe		21	.381		42.50	21.50		64	79
0190	4" - 4-1/8" pipe		19	.421		52	23.50		75.50	92.50
0200	5" - 6" pipe	↓	17	.471	↓	77.50	26.50		104	125
0300	PVC									
0320	1-1/4" - 1-1/2" pipe	1 Plum	24	.333	Ea.	8.20	18.60		26.80	37
0330	2" - 2-1/8" pipe	"	23	.348	"	8.85	19.40		28.25	39
0900	Vent flashing									
1000	Aluminum with lead ring									
1020	1-1/4" pipe	1 Plum	20	.400	Ea.	11.20	22.50		33.70	46
1030	1-1/2" pipe		20	.400		13.40	22.50		35.90	48
1040	2" pipe		18	.444		12.05	25		37.05	51
1050	3" pipe		17	.471		13.30	26.50		39.80	54
1060	4" pipe	↓	16	.500	↓	16.05	28		44.05	59.50
1350	Copper with neoprene ring									
1400	1-1/4" pipe	1 Plum	20	.400	Ea.	19.95	22.50		42.45	55.50
1430	1-1/2" pipe		20	.400		19.95	22.50		42.45	55.50
1440	2" pipe		18	.444		21	25		46	60.50
1450	3" pipe		17	.471		24.50	26.50		51	66.50
1460	4" pipe	↓	16	.500	↓	27	28		55	72
2000	Galvanized with neoprene ring									
2020	1-1/4" pipe	1 Plum	20	.400	Ea.	16.95	22.50		39.45	52
2030	1-1/2" pipe		20	.400		16.95	22.50		39.45	52
2040	2" pipe		18	.444		17.55	25		42.55	57
2050	3" pipe		17	.471		18.55	26.50		45.05	60
2060	4" pipe	↓	16	.500	↓	21.50	28		49.50	65.50
2980	Neoprene, one piece									
3000	1-1/4" pipe	1 Plum	24	.333	Ea.	7.25	18.60		25.85	36
3030	1-1/2" pipe		24	.333		7.95	18.60		26.55	37
3040	2" pipe		23	.348		7.30	19.40		26.70	37
3050	3" pipe		21	.381		8.40	21.50		29.90	41
3060	4" pipe	↓	20	.400	↓	12.45	22.50		34.95	47

22 13 Facility Sanitary Sewerage

22 13 16 – Sanitary Waste and Vent Piping

22 13 16.80 Vent Flashing and Caps		Crew	Daily Output	Labor-Hours	Unit	Material	2013 Bare Costs Labor	Equipment	Total	Total Incl O&P
4000	Lead, 4#, 8" skirt, vent through roof									
4100	2" pipe	1 Plum	18	.444	Ea.	39.50	25		64.50	81
4110	3" pipe		17	.471		47.50	26.50		74	91.50
4120	4" pipe		16	.500		54	28		82	102
4130	6" pipe		14	.571		77	32		109	133

22 13 19 – Sanitary Waste Piping Specialties

22 13 19.13 Sanitary Drains

		Crew	Daily Output	Labor-Hours	Unit	Material	Labor	Equipment	Total	Total Incl O&P
0010	**SANITARY DRAINS**									
0400	Deck, auto park, C.I., 13" top									
0440	3", 4", 5", and 6" pipe size	Q-1	8	2	Ea.	1,325	100		1,425	1,625
0480	For galvanized body, add				"	720			720	790
0800	Promenade, heelproof grate, C.I., 14" top									
0840	2", 3", and 4" pipe size	Q-1	10	1.600	Ea.	510	80.50		590.50	680
0860	5" and 6" pipe size		9	1.778		635	89.50		724.50	835
0880	8" pipe size		8	2		750	100		850	975
0940	For galvanized body, add					380			380	415
0960	With polished bronze top, 2"-3"-4" diam.					860			860	945
1200	Promenade, heelproof grate, C.I., lateral, 14" top									
1240	2", 3" and 4" pipe size	Q-1	10	1.600	Ea.	665	80.50		745.50	855
1260	5" and 6" pipe size		9	1.778		790	89.50		879.50	1,000
1280	8" pipe size		8	2		910	100		1,010	1,150
1340	For galvanized body, add					380			380	415
1360	For polished bronze top, add					350			350	385
1500	Promenade, slotted grate, C.I., 11" top									
1540	2", 3", 4", 5", and 6" pipe size	Q-1	12	1.333	Ea.	460	67		527	605
1600	For galvanized body, add					213			213	234
1640	With polished bronze top					730			730	805
2000	Floor, medium duty, C.I., deep flange, 7" diam. top									
2040	2" and 3" pipe size	Q-1	12	1.333	Ea.	186	67		253	305
2080	For galvanized body, add					89.50			89.50	98.50
2120	With polished bronze top					291			291	320
2160	Heavy duty, C.I., 12" dia anti-tilt grate									
2180	2", 3", 4", 5" and 6" pipe size	Q-1	10	1.600	Ea.	540	80.50		620.50	715
2220	For galvanized body, add					274			274	300
2240	With polished bronze top					810			810	890
2300	X-Heavy duty, C.I., 15" antitilt grate									
2320	4", 5", 6", and 8" pipe size	Q-1	8	2	Ea.	1,150	100		1,250	1,425
2360	For galvanized body, add					445			445	490
2380	With polished bronze top					1,625			1,625	1,775
2400	Heavy duty, with sediment bucket, C.I., 12" diam. loose grate									
2420	2", 3", 4", 5", and 6" pipe size	Q-1	9	1.778	Ea.	640	89.50		729.50	840
2440	For galvanized body, add					405			405	450
2460	With polished bronze top					905			905	995
2500	Heavy duty, cleanout & trap w/bucket, C.I., 15" top									
2540	2", 3", and 4" pipe size	Q-1	6	2.667	Ea.	6,075	134		6,209	6,900
2560	For galvanized body, add					1,550			1,550	1,725
2580	With polished bronze top					6,575			6,575	7,250
2600	Medium duty, with perforated SS basket, C.I., body,									
2610	18" top for refuse container washing area									
2620	2" thru 6" pipe size	Q-1	4	4	Ea.	4,025	201		4,226	4,725
2630	Acid resistant									
2638	PVC									

22 13 19 – Sanitary Waste Piping Specialties

22 13 19.13 Sanitary Drains		Crew	Daily Output	Labor-Hours	Unit	Material	Labor	Equipment	Total	Total Incl O&P
						2013 Bare Costs				
2640	2", 3" and 4" pipe size	Q-1	16	1	Ea.	297	50		347	400
2644	Cast iron, epoxy coated									
2646	2", 3" and 4" pipe size	Q-1	14	1.143	Ea.	435	57.50		492.50	565
2650	PVC or ABS thermoplastic									
2660	3" and 4" pipe size	Q-1	16	1	Ea.	320	50		370	425
2680	Extra heavy duty, oil intercepting, gas seal cone,									
2690	with cleanout, loose grate, C.I., body 16" top									
2700	3" and 4" diameter outlet, 4" slab depth	Q-1	4	4	Ea.	6,575	201		6,776	7,550
2720	4" diameter outlet, 8" slab depth		3	5.333		6,575	268		6,843	7,650
2740	4" diam. outlet, 10"-12" slab depth, 16"top		2	8		7,400	400		7,800	8,750
2910	Prison cell, vandal-proof, 1-1/2", and 2" diam. pipe		12	1.333		410	67		477	555
2920	3" pipe size	↓	10	1.600	↓	460	80.50		540.50	625
2930	Trap drain, light duty, backwater valve C.I. top									
2950	8" diameter top, 2" pipe size	Q-1	12	1.333	Ea.	360	67		427	495
2960	10" diameter top, 3" pipe size		10	1.600		495	80.50		575.50	665
2970	12" diameter top, 4" pipe size	↓	8	2	↓	690	100		790	910

22 13 19.14 Floor Receptors		Crew	Daily Output	Labor-Hours	Unit	Material	Labor	Equipment	Total	Total Incl O&P
0010	**FLOOR RECEPTORS**, For connection to 2", 3" & 4" diameter pipe									
0200	12-1/2" square top, 25 sq. in. open area	Q-1	10	1.600	Ea.	900	80.50		980.50	1,100
0300	For grate with 4" diam. x 3-3/4" high funnel, add					161			161	177
0400	For grate with 6" diameter x 6" high funnel, add					207			207	227
0500	For full hinged grate with open center, add					75			75	82.50
0600	For aluminum bucket, add					135			135	149
0700	For acid-resisting bucket, add					253			253	279
0900	For stainless steel mesh bucket liner, add					200			200	220
1000	For bronze antisplash dome strainer, add					96.50			96.50	106
1100	For partial solid cover, add					48.50			48.50	53
1200	For trap primer connection, add					82			82	90
2000	12-5/8" diameter top, 40 sq. in. open area	Q-1	10	1.600	↓	715	80.50		795.50	905
2100	For options, add same prices as square top									
3000	8" x 4" rectangular top, 7.5 sq. in. open area	Q-1	14	1.143	Ea.	690	57.50		747.50	845
3100	For trap primer connections, add					82			82	90
4000	24" x 16" rectangular top, 70 sq. in. open area	Q-1	4	4		4,025	201		4,226	4,750
4100	For trap primer connection, add				↓	161			161	177

22 13 19.15 Sink Waste Treatment		Crew	Daily Output	Labor-Hours	Unit	Material	Labor	Equipment	Total	Total Incl O&P
0010	**SINK WASTE TREATMENT**, System for commercial kitchens									
0100	includes clock timer, & fittings									
0200	System less chemical, wall mounted cabinet	1 Plum	16	.500	Ea.	525	28		553	620
2000	Chemical, 1 gallon, add					68.50			68.50	75.50
2100	6 gallons, add					310			310	340
2200	15 gallons, add					840			840	925
2300	30 gallons, add					1,575			1,575	1,725
2400	55 gallons, add				↓	2,675			2,675	2,950

22 13 23 – Sanitary Waste Interceptors

22 13 23.10 Interceptors		Crew	Daily Output	Labor-Hours	Unit	Material	Labor	Equipment	Total	Total Incl O&P
0010	**INTERCEPTORS**									
0150	Grease, fabricated steel, 4 GPM, 8 lb. fat capacity	1 Plum	4	2	Ea.	1,100	112		1,212	1,375
0200	7 GPM, 14 lb. fat capacity		4	2		1,525	112		1,637	1,850
1000	10 GPM, 20 lb. fat capacity		4	2		1,800	112		1,912	2,150
1040	15 GPM, 30 lb. fat capacity		4	2		2,650	112		2,762	3,100
1060	20 GPM, 40 lb. fat capacity	↓	3	2.667		3,250	149		3,399	3,800
1080	25 GPM, 50 lb. fat capacity	Q-1	3.50	4.571		3,650	230		3,880	4,375

22 13 23 – Sanitary Waste Interceptors

22 13 23.10 Interceptors		Crew	Daily Output	Labor-Hours	Unit	Material	2013 Bare Costs Labor	Equipment	Total	Total Incl O&P
1100	35 GPM, 70 lb. fat capacity	Q-1	3	5.333	Ea.	4,525	268		4,793	5,375
1120	50 GPM, 100 lb. fat capacity		2	8		9,575	400		9,975	11,100
1140	75 GPM, 150 lb. fat capacity		2	8		12,100	400		12,500	13,900
1160	100 GPM, 200 lb. fat capacity		2	8		13,600	400		14,000	15,500
1180	150 GPM, 300 lb. fat capacity		2	8		15,100	400		15,500	17,200
1200	200 GPM, 400 lb. fat capacity		1.50	10.667		21,900	535		22,435	24,900
1220	250 GPM, 500 lb. fat capacity		1.30	12.308		25,500	620		26,120	28,900
1240	300 GPM, 600 lb. fat capacity	↓	1	16		30,000	805		30,805	34,200
1260	400 GPM, 800 lb. fat capacity	Q-2	1.20	20		36,700	1,050		37,750	42,000
1280	500 GPM, 1000 lb. fat capacity	"	1	24	↓	43,500	1,250		44,750	49,700
1580	For seepage pan, add					7%				
3000	Hair, cast iron, 1-1/4" and 1-1/2" pipe connection	1 Plum	8	1	Ea.	415	56		471	540
3100	For chrome-plated cast iron, add					253			253	279
3200	For polished bronze, add				↓	253			253	279
3400	Lint interceptor, fabricated steel									
3410	Size based on 10 GPM per machine									
3420	30 GPM, 2" pipe size	Q-1	3	5.333	Ea.	4,850	268		5,118	5,725
3430	70 GPM, 3" pipe size		2.50	6.400		5,625	320		5,945	6,675
3440	100 GPM, 4" pipe size		2	8		6,625	400		7,025	7,900
3450	200 GPM, 4" pipe size		1.50	10.667		8,025	535		8,560	9,625
3460	300 GPM, 6" pipe size	↓	1	16		9,325	805		10,130	11,500
3470	400 GPM, 6" pipe size	Q-2	1.20	20		10,600	1,050		11,650	13,300
3480	500 GPM, 6" pipe size	"	1	24		11,900	1,250		13,150	15,000
4000	Oil, fabricated steel, 10 GPM, 2" pipe size	1 Plum	4	2		2,450	112		2,562	2,875
4100	15 GPM, 2" or 3" pipe size		4	2		3,375	112		3,487	3,875
4120	20 GPM, 2" or 3" pipe size	↓	3	2.667		4,075	149		4,224	4,700
4140	25 GPM, 2" or 3" pipe size	Q-1	3.50	4.571		4,425	230		4,655	5,225
4160	35 GPM, 2", 3", or 4" pipe size		3	5.333		5,375	268		5,643	6,325
4180	50 GPM, 2", 3", or 4" pipe size		2	8		7,225	400		7,625	8,550
4200	75 GPM, 3" pipe size		2	8		12,500	400		12,900	14,400
4220	100 GPM, 3" pipe size		2	8		13,600	400		14,000	15,500
4240	150 GPM, 4" pipe size		2	8		16,800	400		17,200	19,100
4260	200 GPM, 4" pipe size		1.50	10.667		23,800	535		24,335	26,900
4280	250 GPM, 5" pipe size		1.30	12.308		27,500	620		28,120	31,200
4300	300 GPM, 5" pipe size	↓	1	16		31,200	805		32,005	35,500
4320	400 GPM, 6" pipe size	Q-2	1.20	20		40,100	1,050		41,150	45,700
4340	500 GPM, 6" pipe size	"	1	24	↓	50,500	1,250		51,750	57,500
5000	Sand interceptor, fabricated steel									
5020	20 GPM, 4" pipe size	Q-1	3	5.333	Ea.	6,050	268		6,318	7,075
5030	50 GPM, 4" pipe size		2.50	6.400		10,300	320		10,620	11,800
5040	150 GPM, 4" pipe size		2	8		31,000	400		31,400	34,800
5050	250 GPM, 6" pipe size	↓	1.30	12.308		32,300	620		32,920	36,400
5060	500 GPM, 6" pipe size	Q-2	1	24		41,600	1,250		42,850	47,600
6000	Solids, precious metals recovery, C.I., 1-1/4" to 2" pipe	1 Plum	4	2		620	112		732	855
6100	Dental Lab., large, C.I., 1-1/2" to 2" pipe	"	3	2.667	↓	2,175	149		2,324	2,625

22 13 26 – Sanitary Waste Separators

22 13 26.10 Separators		Crew	Daily Output	Labor-Hours	Unit	Material	Labor	Equipment	Total	Total Incl O&P
0010	**SEPARATORS**, Entrainment eliminator, steel body, 150 PSIG									
0100	1/4" size	1 Stpi	24	.333	Ea.	259	18.90		277.90	315
0120	1/2" size		24	.333		268	18.90		286.90	325
0140	3/4" size		20	.400		280	22.50		302.50	345
0160	1" size		19	.421		289	24		313	355

22 13 Facility Sanitary Sewerage

22 13 26 – Sanitary Waste Separators

22 13 26.10 Separators		Crew	Daily Output	Labor-Hours	Unit	Material	Labor	Equipment	Total	Total Incl O&P
						2013 Bare Costs				
0180	1-1/4" size	1 Stpi	15	.533	Ea.	305	30		335	385
0200	1-1/2" size		13	.615		340	35		375	425
0220	2" size		11	.727		375	41		416	475

22 13 29 – Sanitary Sewerage Pumps

22 13 29.13 Wet-Pit-Mounted, Vertical Sewerage Pumps

Line	Description	Crew	Daily Output	Labor-Hours	Unit	Material	Labor	Equipment	Total	Total Incl O&P
0010	**WET-PIT-MOUNTED, VERTICAL SEWERAGE PUMPS**									
0020	Controls incl. alarm/disconnect panel w/wire. Excavation not included									
0260	Simplex, 9 GPM at 60 PSIG, 91 gal. tank				Ea.	3,225			3,225	3,550
0300	Unit with manway, 26" I.D., 18" high					3,575			3,575	3,950
0340	26" I.D., 36" high					3,625			3,625	4,000
0380	43" I.D., 4' high					3,850			3,850	4,250
0600	Simplex, 9 GPM at 60 PSIG, 150 gal. tank, indoor					3,425			3,425	3,775
0700	Unit with manway, 26" I.D., 36" high					4,100			4,100	4,500
0740	26" I.D., 4' high					4,325			4,325	4,750
2000	Duplex, 18 GPM at 60 PSIG, 150 gal. tank, indoor					6,825			6,825	7,525
2060	Unit with manway, 43" I.D., 4' high					7,850			7,850	8,650
2400	For core only					1,825			1,825	2,000
3000	Indoor residential type installation									
3020	Simplex, 9 GPM at 60 PSIG, 91 gal. HDPE tank				Ea.	3,225			3,225	3,550

22 13 29.14 Sewage Ejector Pumps

Line	Description	Crew	Daily Output	Labor-Hours	Unit	Material	Labor	Equipment	Total	Total Incl O&P
0010	**SEWAGE EJECTOR PUMPS**, With operating and level controls									
0100	Simplex system incl. tank, cover, pump 15' head									
0500	37 gal. PE tank, 12 GPM, 1/2 HP, 2" discharge	Q-1	3.20	5	Ea.	480	251		731	905
0510	3" discharge		3.10	5.161		520	259		779	960
0530	87 GPM, .7 HP, 2" discharge		3.20	5		735	251		986	1,175
0540	3" discharge		3.10	5.161		795	259		1,054	1,275
0600	45 gal. coated stl. tank, 12 GPM, 1/2 HP, 2" discharge		3	5.333		855	268		1,123	1,350
0610	3" discharge		2.90	5.517		890	277		1,167	1,400
0630	87 GPM, .7 HP, 2" discharge		3	5.333		1,100	268		1,368	1,600
0640	3" discharge		2.90	5.517		1,150	277		1,427	1,700
0660	134 GPM, 1 HP, 2" discharge		2.80	5.714		1,175	287		1,462	1,725
0680	3" discharge		2.70	5.926		1,250	298		1,548	1,825
0700	70 gal. PE tank, 12 GPM, 1/2 HP, 2" discharge		2.60	6.154		920	310		1,230	1,500
0710	3" discharge		2.40	6.667		980	335		1,315	1,575
0730	87 GPM, 0.7 HP, 2" discharge		2.50	6.400		1,200	320		1,520	1,775
0740	3" discharge		2.30	6.957		1,275	350		1,625	1,925
0760	134 GPM, 1 HP, 2" discharge		2.20	7.273		1,300	365		1,665	1,975
0770	3" discharge		2	8		1,375	400		1,775	2,125
0800	75 gal. coated stl. tank, 12 GPM, 1/2 HP, 2" discharge		2.40	6.667		1,025	335		1,360	1,625
0810	3" discharge		2.20	7.273		1,075	365		1,440	1,725
0830	87 GPM, .7 HP, 2" discharge		2.30	6.957		1,300	350		1,650	1,950
0840	3" discharge		2.10	7.619		1,350	385		1,735	2,075
0860	134 GPM, 1 HP, 2" discharge		2	8		1,375	400		1,775	2,125
0880	3" discharge		1.80	8.889		1,450	445		1,895	2,275
1040	Duplex system incl. tank, covers, pumps									
1060	110 gal. fiberglass tank, 24 GPM, 1/2 HP, 2" discharge	Q-1	1.60	10	Ea.	1,850	500		2,350	2,775
1080	3" discharge		1.40	11.429		1,950	575		2,525	3,000
1100	174 GPM, .7 HP, 2" discharge		1.50	10.667		2,400	535		2,935	3,425
1120	3" discharge		1.30	12.308		2,475	620		3,095	3,650
1140	268 GPM, 1 HP, 2" discharge		1.20	13.333		2,600	670		3,270	3,850
1160	3" discharge		1	16		2,700	805		3,505	4,150
1260	135 gal. coated stl. tank, 24 GPM, 1/2 HP, 2" discharge	Q-2	1.70	14.118		1,900	735		2,635	3,200

22 13 Facility Sanitary Sewerage

22 13 29 – Sanitary Sewerage Pumps

22 13 29.14 Sewage Ejector Pumps		Crew	Daily Output	Labor-Hours	Unit	Material	Labor	Equipment	Total	Total Incl O&P
						2013 Bare Costs				
2000	3" discharge	Q-2	1.60	15	Ea.	2,025	780		2,805	3,400
2640	174 GPM, .7 HP, 2" discharge		1.60	15		2,500	780		3,280	3,925
2660	3" discharge		1.50	16		2,625	835		3,460	4,150
2700	268 GPM, 1 HP, 2" discharge		1.30	18.462		2,700	960		3,660	4,425
3040	3" discharge		1.10	21.818		2,875	1,125		4,000	4,875
3060	275 gal. coated stl. tank, 24 GPM, 1/2 HP, 2" discharge		1.50	16		2,375	835		3,210	3,875
3080	3" discharge		1.40	17.143		2,425	895		3,320	4,000
3100	174 GPM, .7 HP, 2" discharge		1.40	17.143		3,075	895		3,970	4,725
3120	3" discharge		1.30	18.462		3,250	960		4,210	5,025
3140	268 GPM, 1 HP, 2" discharge		1.10	21.818		3,375	1,125		4,500	5,400
3160	3" discharge	↓	.90	26.667	↓	3,550	1,400		4,950	6,000
3260	Pump system accessories, add									
3300	Alarm horn and lights, 115 V mercury switch	Q-1	8	2	Ea.	97	100		197	257
3340	Switch, mag. contactor, alarm bell, light, 3 level control		5	3.200		475	161		636	765
3380	Alternator, mercury switch activated	↓	4	4	↓	855	201		1,056	1,250

22 14 Facility Storm Drainage

22 14 23 – Storm Drainage Piping Specialties

22 14 23.33 Backwater Valves

		Crew	Daily Output	Labor-Hours	Unit	Material	Labor	Equipment	Total	Total Incl O&P
0010	**BACKWATER VALVES**, C.I. Body									
6980	Bronze gate and automatic flapper valves									
7000	3" and 4" pipe size	Q-1	13	1.231	Ea.	1,900	62		1,962	2,175
7100	5" and 6" pipe size	"	13	1.231	"	2,900	62		2,962	3,300
7240	Bronze flapper valve, bolted cover									
7260	2" pipe size	Q-1	16	1	Ea.	555	50		605	685
7280	3" pipe size		14.50	1.103		865	55.50		920.50	1,025
7300	4" pipe size	↓	13	1.231		1,075	62		1,137	1,275
7320	5" pipe size	Q-2	18	1.333		1,275	69.50		1,344.50	1,500
7340	6" pipe size	"	17	1.412		1,525	73.50		1,598.50	1,775
7360	8" pipe size	Q-3	10	3.200		2,150	170		2,320	2,625
7380	10" pipe size	"	9	3.556	↓	3,525	189		3,714	4,150
7500	For threaded cover, same cost									
7540	Revolving disk type, same cost as flapper type									

22 14 26 – Facility Storm Drains

22 14 26.13 Roof Drains

		Crew	Daily Output	Labor-Hours	Unit	Material	Labor	Equipment	Total	Total Incl O&P
0010	**ROOF DRAINS**									
0140	Cornice, C.I., 45° or 90° outlet									
0200	3" and 4" pipe size	Q-1	12	1.333	Ea.	360	67		427	495
0260	For galvanized body, add					70			70	77
0280	For polished bronze dome, add				↓	87.50			87.50	96.50
3860	Roof, flat metal deck, C.I. body, 12" C.I. dome									
3880	2" pipe size	Q-1	15	1.067	Ea.	266	53.50		319.50	375
3890	3" pipe size		14	1.143		365	57.50		422.50	485
3900	4" pipe size		13	1.231		365	62		427	495
3910	5" pipe size		12	1.333		495	67		562	645
3920	6" pipe size	↓	10	1.600	↓	635	80.50		715.50	820
4280	Integral expansion joint, C.I. body, 12" C.I. dome									
4300	2" pipe size	Q-1	8	2	Ea.	545	100		645	750
4320	3" pipe size		7	2.286		565	115		680	795
4340	4" pipe size		6	2.667		605	134		739	865

22 14 26 – Facility Storm Drains

22 14 26.13 Roof Drains		Crew	Daily Output	Labor-Hours	Unit	Material	Labor	Equipment	Total	Total Incl O&P
						2013 Bare Costs				
4360	5" pipe size	Q-1	4	4	Ea.	760	201		961	1,150
4380	6" pipe size		3	5.333		830	268		1,098	1,325
4400	8" pipe size		3	5.333		1,275	268		1,543	1,800
4440	For galvanized body, add					340			340	375
4620	Main, all aluminum, 12" low profile dome									
4640	2", 3" and 4" pipe size	Q-1	14	1.143	Ea.	405	57.50		462.50	530
4660	5" and 6" pipe size		13	1.231		535	62		597	680
4680	8" pipe size		10	1.600		665	80.50		745.50	850
4690	Main, CI body, 12" poly. dome, 2", 3", & 4" pipe		8	2		300	100		400	480
4710	5" and 6" pipe size		6	2.667		430	134		564	675
4720	8" pipe size		4	4		555	201		756	920
4730	For underdeck clamp, add		22	.727		207	36.50		243.50	283
4740	For vandalproof dome, add					52.50			52.50	58
4750	For galvanized body, add					405			405	450
4760	Main, PVC body and dome, 2" pipe size	Q-1	14	1.143		131	57.50		188.50	231
4780	3" pipe size		14	1.143		131	57.50		188.50	231
4800	4" pipe size		14	1.143		131	57.50		188.50	231
4820	For underdeck clamp, add		24	.667		27	33.50		60.50	80.50
4900	Terrace planting area, with perforated overflow, C.I.									
4920	2", 3" and 4" pipe size	Q-1	8	2	Ea.	570	100		670	775
22 14 26.16 Facility Area Drains										
0010	**FACILITY AREA DRAINS**									
4980	Scupper floor, oblique strainer, C.I.									
5000	6" x 7" top, 2", 3" and 4" pipe size	Q-1	16	1	Ea.	262	50		312	365
5100	8" x 12" top, 5" and 6" pipe size	"	14	1.143		510	57.50		567.50	645
5160	For galvanized body, add					40%				
5200	For polished bronze strainer, add					85%				
22 14 26.19 Facility Trench Drains										
0010	**FACILITY TRENCH DRAINS**									
5980	Trench, floor, heavy duty, modular, C.I., 12" x 12" top									
6000	2", 3", 4", 5", & 6" pipe size	Q-1	8	2	Ea.	830	100		930	1,075
6100	For unit with polished bronze top		8	2		1,250	100		1,350	1,500
6200	For 12" extension section, C.I. top		8	2		830	100		930	1,075
6240	For 12" extension section, polished bronze top		8	2		1,300	100		1,400	1,575
6600	Trench, floor, for cement concrete encasement									
6610	Not including trenching or concrete									
6640	Polyester polymer concrete									
6650	4" internal width, with grate									
6660	Light duty steel grate	Q-1	120	.133	L.F.	32	6.70		38.70	45.50
6670	Medium duty steel grate		115	.139		42.50	7		49.50	57
6680	Heavy duty iron grate		110	.145		56.50	7.30		63.80	73.50
6700	12" internal width, with grate									
6770	Heavy duty galvanized grate	Q-1	80	.200	L.F.	154	10.05		164.05	185
6800	Fiberglass									
6810	8" internal width, with grate									
6820	Medium duty galvanized grate	Q-1	115	.139	L.F.	96.50	7		103.50	117
6830	Heavy duty iron grate	"	110	.145	"	120	7.30		127.30	143

22 14 29 – Sump Pumps

22 14 29.13 Wet-Pit-Mounted, Vertical Sump Pumps		Crew	Daily Output	Labor-Hours	Unit	Material	Labor	Equipment	Total	Total Incl O&P
0010	**WET-PIT-MOUNTED, VERTICAL SUMP PUMPS**									
0400	Molded PVC base, 21 GPM at 15' head, 1/3 HP	1 Plum	5	1.600	Ea.	135	89.50		224.50	283
0800	Iron base, 21 GPM at 15' head, 1/3 HP		5	1.600		164	89.50		253.50	315

22 14 Facility Storm Drainage

22 14 29 – Sump Pumps

22 14 29.13 Wet-Pit-Mounted, Vertical Sump Pumps		Crew	Daily Output	Labor-Hours	Unit	Material	2013 Bare Costs Labor	Equipment	Total	Total Incl O&P
1200	Solid brass, 21 GPM at 15' head, 1/3 HP	1 Plum	5	1.600	Ea.	289	89.50		378.50	455

22 14 29.16 Submersible Sump Pumps

		Crew	Daily Output	Labor-Hours	Unit	Material	Labor	Equipment	Total	Total Incl O&P
0010	**SUBMERSIBLE SUMP PUMPS**									
7000	Sump pump, automatic									
7100	Plastic, 1-1/4" discharge, 1/4 HP	1 Plum	6	1.333	Ea.	130	74.50		204.50	255
7140	1/3 HP		5	1.600		195	89.50		284.50	350
7160	1/2 HP		5	1.600		231	89.50		320.50	390
7180	1-1/2" discharge, 1/2 HP		4	2		265	112		377	460
7500	Cast iron, 1-1/4" discharge, 1/4 HP		6	1.333		184	74.50		258.50	315
7540	1/3 HP		6	1.333		216	74.50		290.50	350
7560	1/2 HP		5	1.600		261	89.50		350.50	420

22 14 53 – Rainwater Storage Tanks

22 14 53.13 Fiberglass, Rainwater Storage Tank

		Crew	Daily Output	Labor-Hours	Unit	Material	Labor	Equipment	Total	Total Incl O&P
0010	**FIBERGLASS, RAINWATER STORAGE TANK**									
2000	600 gallon	B-21B	3.75	10.667	Ea.	3,000	410	172	3,582	4,125
2010	1,000 gallon		3.50	11.429		3,850	440	185	4,475	5,125
2020	2,000 gallon		3.25	12.308		5,625	475	199	6,299	7,125
2030	4,000 gallon		3	13.333		7,775	515	215	8,505	9,575
2040	6,000 gallon		2.65	15.094		8,750	580	244	9,574	10,800
2050	8,000 gallon		2.30	17.391		10,500	670	281	11,451	12,800
2060	10,000 gallon		2	20		12,000	770	325	13,095	14,700
2070	12,000 gallon		1.50	26.667		16,600	1,025	430	18,055	20,300
2080	15,000 gallon		1	40		19,200	1,550	645	21,395	24,300
2090	20,000 gallon		.75	53.333		24,700	2,050	860	27,610	31,300
2100	25,000 gallon		.50	80		37,000	3,075	1,300	41,375	46,900
2110	30,000 gallon		.35	114		72,500	4,400	1,850	78,750	89,000
2120	40,000 gallon		.30	133		80,500	5,125	2,150	87,775	99,000

22 15 General Service Compressed-Air Systems

22 15 13 – General Service Compressed-Air Piping

22 15 13.10 Compressor Accessories

		Crew	Daily Output	Labor-Hours	Unit	Material	Labor	Equipment	Total	Total Incl O&P
0010	**COMPRESSOR ACCESSORIES**									
4000	Couplers, air line, sleeve type									
4010	Female, connection size NPT									
4020	1/4"	1 Stpi	38	.211	Ea.	8.65	11.95		20.60	27.50
4030	3/8"		36	.222		14.45	12.60		27.05	35
4040	1/2"		35	.229		21	12.95		33.95	42.50
4050	3/4"		34	.235		21	13.35		34.35	43
4100	Male									
4110	1/4"	1 Stpi	38	.211	Ea.	9	11.95		20.95	28
4120	3/8"		36	.222		9.85	12.60		22.45	30
4130	1/2"		35	.229		21	12.95		33.95	42.50
4140	3/4"		34	.235		22	13.35		35.35	44.50
4150	Coupler, combined male and female halves									
4160	1/2"	1 Stpi	17	.471	Ea.	42	26.50		68.50	86
4170	3/4"	"	15	.533	"	43	30		73	93

22 15 General Service Compressed-Air Systems

22 15 19 – General Service Packaged Air Compressors and Receivers

22 15 19.10 Air Compressors		Crew	Daily Output	Labor-Hours	Unit	2013 Bare Costs Material	Labor	Equipment	Total	Total Incl O&P
0010	**AIR COMPRESSORS**									
5250	Air, reciprocating air cooled, splash lubricated, tank mounted									
5300	Single stage, 1 phase, 140 psi									
5303	1/2 HP, 30 gal. tank	1 Stpi	3	2.667	Ea.	1,800	151		1,951	2,200
5305	3/4 HP, 30 gal. tank		2.60	3.077		1,825	174		1,999	2,275
5307	1 HP, 30 gal. tank		2.20	3.636		2,250	206		2,456	2,775
5309	2 HP, 30 gal. tank	Q-5	4	4		2,725	204		2,929	3,300
5310	3 HP, 30 gal. tank		3.60	4.444		3,075	227		3,302	3,750
5314	3 HP, 60 gal. tank		3.50	4.571		3,475	233		3,708	4,175
5320	5 HP, 60 gal. tank		3.20	5		3,650	255		3,905	4,400
5330	5 HP, 80 gal. tank		3	5.333		4,025	272		4,297	4,825
5340	7.5 HP, 80 gal. tank		2.60	6.154		5,200	315		5,515	6,200
5600	2 stage pkg., 3 phase									
5650	6 CFM at 125 psi 1-1/2 HP, 60 gal. tank	Q-5	3	5.333	Ea.	3,300	272		3,572	4,025
5670	10.9 CFM at 125 psi, 3 HP, 80 gal. tank		1.50	10.667		3,725	545		4,270	4,925
5680	38.7 CFM at 125 psi, 10 HP, 120 gal. tank		.60	26.667		6,700	1,350		8,050	9,425
5690	105 CFM at 125 psi, 25 HP, 250 gal. tank	Q-6	.60	40		13,900	2,125		16,025	18,400
5800	With single stage pump									
5850	8.3 CFM at 125 psi, 2 HP, 80 gal. tank	Q-6	3.50	6.857	Ea.	3,700	365		4,065	4,600
5860	38.7 CFM at 125 psi, 10 HP, 120 gal. tank	"	.90	26.667	"	7,375	1,400		8,775	10,200
6000	Reciprocating, 2 stage, tank mtd, 3 Ph., Cap. rated @175 PSIG									
6050	Pressure lubricated, hvy. duty, 9.7 CFM, 3 HP, 120 gal. tank	Q-5	1.30	12.308	Ea.	5,675	625		6,300	7,175
6054	5 CFM, 1-1/2 HP, 80 gal. tank		2.80	5.714		3,525	291		3,816	4,325
6056	6.4 CFM, 2 HP, 80 gal. tank		2	8		3,700	410		4,110	4,675
6058	8.1 CFM, 3 HP, 80 gal. tank		1.70	9.412		3,725	480		4,205	4,825
6059	14.8 CFM, 5 HP, 80 gal. tank		1	16		4,025	815		4,840	5,650
6060	16.5 CFM, 5 HP, 120 gal. tank		1	16		4,500	815		5,315	6,175
6063	13 CFM, 6 HP, 80 gal. tank		.90	17.778		5,775	905		6,680	7,725
6066	19.8 CFM, 7.5 HP, 80 gal. tank		.80	20		5,775	1,025		6,800	7,875
6070	25.8 CFM, 7-1/2 HP, 120 gal. tank		.80	20		8,700	1,025		9,725	11,100
6078	34.8 CFM, 10 HP, 80 gal. tank		.70	22.857		7,875	1,175		9,050	10,400
6080	34.8 CFM, 10 HP, 120 gal. tank		.60	26.667		8,350	1,350		9,700	11,200
6090	53.7 CFM, 15 HP, 120 gal. tank	Q-6	.80	30		10,100	1,575		11,675	13,500
6100	76.7 CFM, 20 HP, 120 gal. tank		.70	34.286		13,200	1,825		15,025	17,200
6104	76.7 CFM, 20 HP, 240 gal. tank		.68	35.294		14,500	1,875		16,375	18,700
6110	90.1 CFM, 25 HP, 120 gal. tank		.63	38.095		13,400	2,025		15,425	17,800
6120	101 CFM, 30 HP, 120 gal. tank		.57	42.105		14,900	2,225		17,125	19,700
6130	101 CFM, 30 HP, 250 gal. tank		.52	46.154		16,100	2,450		18,550	21,400
6200	Oil-less, 13.6 CFM, 5 HP, 120 gal. tank	Q-5	.88	18.182		16,600	925		17,525	19,600
6210	13.6 CFM, 5 HP, 250 gal. tank		.80	20		17,800	1,025		18,825	21,100
6220	18.2 CFM, 7.5 HP, 120 gal. tank		.73	21.918		16,600	1,125		17,725	19,900
6230	18.2 CFM, 7.5 HP, 250 gal. tank		.67	23.881		17,800	1,225		19,025	21,400
6250	30.5 CFM, 10 HP, 120 gal. tank		.57	28.070		19,600	1,425		21,025	23,700
6260	30.5 CFM, 10 HP, 250 gal. tank		.53	30.189		20,800	1,550		22,350	25,200
6270	41.3 CFM, 15 HP, 120 gal. tank	Q-6	.70	34.286		21,300	1,825		23,125	26,100
6280	41.3 CFM, 15 HP, 250 gal. tank	"	.67	35.821		22,500	1,900		24,400	27,600

22 31 Domestic Water Softeners

22 31 13 – Residential Domestic Water Softeners

22 31 13.10 Residential Water Softeners		Crew	Daily Output	Labor-Hours	Unit	Material	2013 Bare Costs Labor	Equipment	Total	Total Incl O&P
0010	RESIDENTIAL WATER SOFTENERS									
7350	Water softener, automatic, to 30 grains per gallon	2 Plum	5	3.200	Ea.	380	179		559	690
7400	To 100 grains per gallon	"	4	4	"	650	223		873	1,050

22 31 16 – Commercial Domestic Water Softeners

22 31 16.10 Water Softeners		Crew	Daily Output	Labor-Hours	Unit	Material	Labor	Equipment	Total	Total Incl O&P
0010	WATER SOFTENERS									
5800	Softener systems, automatic, intermediate sizes									
5820	available, may be used in multiples.									
6000	Hardness capacity between regenerations and flow									
6060	40,000 grains, 14 GPM	Q-1	4	4	Ea.	1,500	201		1,701	1,950
6070	50,000 grains, 17 GPM		3.60	4.444		2,325	223		2,548	2,875
6080	90,000 grains, 25 GPM		2.80	5.714		4,875	287		5,162	5,800
6100	150,000 grains, 37 GPM cont., 51 GPM peak		1.20	13.333		4,875	670		5,545	6,375
6200	300,000 grains, 81 GPM cont., 113 GPM peak		1	16		9,200	805		10,005	11,300
6300	750,000 grains, 160 GPM cont., 230 GPM peak		.80	20		12,000	1,000		13,000	14,700
6400	900,000 grains, 185 GPM cont., 270 GPM peak	↓	.70	22.857	↓	19,300	1,150		20,450	22,900
8000	Water treatment, salts, 50 lb. bag									
8020	Salt, water softener, bag, pelletized				Lb.	.31			.31	.34
8030	Salt, water softener, bag, crystal rock salt				"	.28			.28	.31

22 32 Domestic Water Filtration Equipment

22 32 19 – Domestic-Water Off-Floor Cartridge Filters

22 32 19.10 Water Filters		Crew	Daily Output	Labor-Hours	Unit	Material	Labor	Equipment	Total	Total Incl O&P
0010	**WATER FILTERS**, Purification and treatment.									
1000	Cartridge style, dirt and rust type	1 Plum	12	.667	Ea.	179	37		216	253
1200	Replacement cartridge		32	.250		21.50	13.95		35.45	44.50
1600	Taste and odor type		12	.667		210	37		247	287
1700	Replacement cartridge		32	.250		40.50	13.95		54.45	65.50
3000	Central unit, dirt/rust/odor/taste/scale		4	2		1,275	112		1,387	1,575
3100	Replacement cartridge, standard		20	.400		250	22.50		272.50	310
3600	Replacement cartridge, heavy duty	↓	20	.400	↓	280	22.50		302.50	345
8000	Commercial, fully automatic or push button automatic									
8200	Iron removal, 660 GPH, 1" pipe size	Q-1	1.50	10.667	Ea.	2,775	535		3,310	3,875
8240	1500 GPH, 1-1/4" pipe size		1	16		4,700	805		5,505	6,350
8280	2340 GPH, 1-1/2" pipe size		.80	20		5,150	1,000		6,150	7,200
8320	3420 GPH, 2" pipe size		.60	26.667		9,500	1,350		10,850	12,500
8360	4620 GPH, 2-1/2" pipe size		.50	32		15,100	1,600		16,700	19,000
8500	Neutralizer for acid water, 780 GPH, 1" pipe size		1.50	10.667		2,700	535		3,235	3,750
8540	1140 GPH, 1-1/4" pipe size		1	16		3,025	805		3,830	4,525
8580	1740 GPH, 1-1/2" pipe size		.80	20		4,400	1,000		5,400	6,350
8620	2520 GPH, 2" pipe size		.60	26.667		5,700	1,350		7,050	8,300
8660	3480 GPH, 2-1/2" pipe size		.50	32		9,500	1,600		11,100	12,800
8800	Sediment removal, 780 GPH, 1" pipe size		1.50	10.667		2,575	535		3,110	3,625
8840	1140 GPH, 1-1/4" pipe size		1	16		3,075	805		3,880	4,575
8880	1740 GPH, 1-1/2" pipe size		.80	20		4,075	1,000		5,075	6,000
8920	2520 GPH, 2" pipe size		.60	26.667		5,900	1,350		7,250	8,500
8960	3480 GPH, 2-1/2" pipe size		.50	32		9,250	1,600		10,850	12,600
9200	Taste and odor removal, 660 GPH, 1" pipe size		1.50	10.667		3,600	535		4,135	4,750
9240	1500 GPH, 1-1/4" pipe size		1	16		6,150	805		6,955	7,975
9280	2340 GPH, 1-1/2" pipe size		.80	20		7,025	1,000		8,025	9,250

22 32 Domestic Water Filtration Equipment

22 32 19 – Domestic-Water Off-Floor Cartridge Filters

22 32 19.10 Water Filters		Crew	Daily Output	Labor-Hours	Unit	Material	2013 Bare Costs Labor	Equipment	Total	Total Incl O&P
9320	3420 GPH, 2" pipe size	Q-1	.60	26.667	Ea.	10,800	1,350		12,150	13,900
9360	4620 GPH, 2-1/2" pipe size	↓	.50	32	↓	16,600	1,600		18,200	20,700

22 33 Electric Domestic Water Heaters

22 33 13 – Instantaneous Electric Domestic Water Heaters

22 33 13.10 Hot Water Dispensers

		Crew	Daily Output	Labor-Hours	Unit	Material	Labor	Equipment	Total	Total Incl O&P
0010	HOT WATER DISPENSERS									
0160	Commercial, 100 cup, 11.3 amp	1 Plum	14	.571	Ea.	510	32		542	615
3180	Household, 60 cup	"	14	.571	"	269	32		301	345

22 33 13.20 Instantaneous Electric Point-Of-use Water Heaters

		Crew	Daily Output	Labor-Hours	Unit	Material	Labor	Equipment	Total	Total Incl O&P
0010	INSTANTANEOUS ELECTRIC POINT-OF-USE WATER HEATERS									
8965	Point of use, electric, glass lined									
8969	Energy saver									
8970	2.5 gal. single element	1 Plum	2.80	2.857	Ea.	227	159		386	490
8971	4 gal. single element		2.80	2.857		233	159		392	495
8974	6 gal. single element		2.50	3.200		242	179		421	535
8975	10 gal. single element		2.50	3.200		305	179		484	605
8976	15 gal. single element		2.40	3.333		340	186		526	655
8977	20 gal. single element		2.40	3.333		380	186		566	695
8978	30 gal. single element		2.30	3.478		440	194		634	775
8979	40 gal. single element	↓	2.20	3.636	↓	740	203		943	1,125
8988	Commercial (ASHRAE energy std. 90)									
8989	6 gallon	1 Plum	2.50	3.200	Ea.	630	179		809	965
8990	10 gallon		2.50	3.200		675	179		854	1,000
8991	15 gallon		2.40	3.333		715	186		901	1,075
8992	20 gallon		2.40	3.333		765	186		951	1,125
8993	30 gallon	↓	2.30	3.478	↓	1,625	194		1,819	2,075
8995	Under the sink, copper, w/bracket									
8996	2.5 gallon	1 Plum	4	2	Ea.	435	112		547	645

22 33 30 – Residential, Electric Domestic Water Heaters

22 33 30.13 Residential, Small-Capacity Elec. Water Heaters

			Crew	Daily Output	Labor-Hours	Unit	Material	Labor	Equipment	Total	Total Incl O&P
0010	RESIDENTIAL, SMALL-CAPACITY ELECTRIC DOMESTIC WATER HEATERS										
1000	Residential, electric, glass lined tank, 5 yr., 10 gal., single element	D2020–210	1 Plum	2.30	3.478	Ea.	305	194		499	625
1040	20 gallon, single element			2.20	3.636		380	203		583	720
1060	30 gallon, double element	D2020–220		2.20	3.636		695	203		898	1,075
1080	40 gallon, double element			2	4		740	223		963	1,150
1100	52 gallon, double element	D2020–230		2	4		830	223		1,053	1,250
1120	66 gallon, double element			1.80	4.444		1,125	248		1,373	1,600
1140	80 gallon, double element			1.60	5		1,250	279		1,529	1,800
1180	120 gallon, double element		↓	1.40	5.714	↓	1,750	320		2,070	2,400

22 33 33 – Light-Commercial Electric Domestic Water Heaters

22 33 33.10 Commercial Electric Water Heaters

		Crew	Daily Output	Labor-Hours	Unit	Material	Labor	Equipment	Total	Total Incl O&P
0010	COMMERCIAL ELECTRIC WATER HEATERS									
4000	Commercial, 100° rise. NOTE: for each size tank, a range of									
4010	heaters between the ones shown are available									
4020	Electric									
4100	5 gal., 3 kW, 12 GPH, 208 volt	1 Plum	2	4	Ea.	2,625	223		2,848	3,200
4120	10 gal., 6 kW, 25 GPH, 208 volt		2	4		2,900	223		3,123	3,525
4140	50 gal., 9 kW, 37 GPH, 208 volt		1.80	4.444		3,975	248		4,223	4,750
4160	50 gal., 36 kW, 148 GPH, 208 volt		1.80	4.444		6,075	248		6,323	7,075

22 33 Electric Domestic Water Heaters

22 33 33 – Light-Commercial Electric Domestic Water Heaters

22 33 33.10 Commercial Electric Water Heaters		Crew	Daily Output	Labor-Hours	Unit	Material	2013 Bare Costs Labor	Equipment	Total	Total Incl O&P
4180	80 gal., 12 kW, 49 GPH, 208 volt	1 Plum	1.50	5.333	Ea.	4,900	298		5,198	5,850
4200	80 gal., 36 kW, 148 GPH, 208 volt		1.50	5.333		6,775	298		7,073	7,925
4220	100 gal., 36 kW, 148 GPH, 208 volt		1.20	6.667		7,075	370		7,445	8,325
4240	120 gal., 36 kW, 148 GPH, 208 volt		1.20	6.667		7,375	370		7,745	8,650
4260	150 gal., 15 kW , 61 GPH, 480 volt		1	8		17,300	445		17,745	19,700
4280	150 gal., 120 kW, 490 GPH, 480 volt		1	8		24,300	445		24,745	27,500
4300	200 gal., 15 kW, 61 GPH, 480 volt	Q-1	1.70	9.412		18,800	475		19,275	21,300
4320	200 gal., 120 kW , 490 GPH, 480 volt		1.70	9.412		25,600	475		26,075	28,900
4340	250 gal., 15 kW, 61 GPH, 480 volt		1.50	10.667		19,300	535		19,835	22,100
4360	250 gal., 150 kW, 615 GPH, 480 volt		1.50	10.667		28,200	535		28,735	31,800
4380	300 gal., 30 kW, 123 GPH, 480 volt		1.30	12.308		21,300	620		21,920	24,300
4400	300 gal., 180 kW, 738 GPH, 480 volt		1.30	12.308		36,800	620		37,420	41,300
4420	350 gal., 30 kW, 123 GPH, 480 volt		1.10	14.545		22,500	730		23,230	25,900
4440	350 gal., 180 kW, 738 GPH, 480 volt		1.10	14.545		31,700	730		32,430	36,000
4460	400 gal., 30 kW, 123 GPH, 480 volt		1	16		25,500	805		26,305	29,300
4480	400 gal., 210 kW, 860 GPH, 480 volt		1	16		36,600	805		37,405	41,500
4500	500 gal., 30 kW, 123 GPH, 480 volt		.80	20		29,800	1,000		30,800	34,200
4520	500 gal., 240 kW, 984 GPH, 480 volt		.80	20		52,500	1,000		53,500	59,000
4540	600 gal., 30 kW, 123 GPH, 480 volt	Q-2	1.20	20		26,000	1,050		27,050	30,200
4560	600 gal., 300 kW, 1230 GPH, 480 volt		1.20	20		34,400	1,050		35,450	39,400
4580	700 gal., 30 kW, 123 GPH, 480 volt		1	24		23,800	1,250		25,050	28,100
4600	700 gal., 300 kW, 1230 GPH, 480 volt		1	24		35,800	1,250		37,050	41,300
4620	800 gal., 60 kW, 245 GPH, 480 volt		.90	26.667		25,800	1,400		27,200	30,500
4640	800 gal., 300 kW, 1230 GPH, 480 volt		.90	26.667		36,600	1,400		38,000	42,400
4660	1000 gal., 60 kW, 245 GPH, 480 volt		.70	34.286		28,200	1,775		29,975	33,700
4680	1000 gal., 480 kW, 1970 GPH, 480 volt		.70	34.286		46,800	1,775		48,575	54,000
4700	1200 gal., 60 kW, 245 GPH, 480 volt		.60	40		48,000	2,075		50,075	56,000
4720	1200 gal., 480 kW, 1970 GPH, 480 volt		.60	40		73,000	2,075		75,075	83,000
4740	1500 gal., 60 kW, 245 GPH, 480 volt		.50	48		63,000	2,500		65,500	73,500
4760	1500 gal., 480 kW, 1970 GPH, 480 volt		.50	48		87,500	2,500		90,000	100,500
4770	2000 gal., 480 kW, 1970 GPH, 480 volt		.48	50		98,500	2,600		101,100	112,500
5400	Modulating step control, 2-5 steps	1 Elec	5.30	1.509		900	79		979	1,100
5440	6-10 steps		3.20	2.500		1,150	131		1,281	1,475
5460	11-15 steps		2.70	2.963		1,625	155		1,780	2,025
5480	16-20 steps		1.60	5		2,400	262		2,662	3,050

22 34 Fuel-Fired Domestic Water Heaters

22 34 13 – Instantaneous, Tankless, Gas Domestic Water Heaters

22 34 13.10 Instantaneous, Tankless, Gas Water Heaters

		Crew	Daily Output	Labor-Hours	Unit	Material	Labor	Equipment	Total	Total Incl O&P
0010	**INSTANTANEOUS, TANKLESS, GAS WATER HEATERS**									
9410	Natural gas/propane, 3.2 GPM G	1 Plum	2	4	Ea.	345	223		568	715
9420	6.4 GPM		1.90	4.211		615	235		850	1,025
9430	8.4 GPM		1.80	4.444		745	248		993	1,200
9440	9.5 GPM		1.60	5		890	279		1,169	1,400

22 34 30 – Residential Gas Domestic Water Heaters

22 34 30.13 Residential, Atmos, Gas Domestic Wtr Heaters

		Crew	Daily Output	Labor-Hours	Unit	Material	Labor	Equipment	Total	Total Incl O&P
0010	**RESIDENTIAL, ATMOSPHERIC, GAS DOMESTIC WATER HEATERS**									
2000	Gas fired, foam lined tank, 10 yr., vent not incl.									
2040	30 gallon	1 Plum	2	4	Ea.	895	223		1,118	1,325
2060	40 gallon		1.90	4.211		895	235		1,130	1,350

22 34 Fuel-Fired Domestic Water Heaters

22 34 30 – Residential Gas Domestic Water Heaters

22 34 30.13 Residential, Atmos, Gas Domestic Wtr Heaters		Crew	Daily Output	Labor-Hours	Unit	Material	2013 Bare Costs Labor	Equipment	Total	Total Incl O&P
2080	50 gallon	1 Plum	1.80	4.444	Ea.	935	248		1,183	1,400
2090	60 gallon		1.70	4.706		1,275	263		1,538	1,800
2100	75 gallon		1.50	5.333		1,350	298		1,648	1,950
2120	100 gallon		1.30	6.154		1,525	345		1,870	2,200
2900	Water heater, safety-drain pan, 26" round		20	.400		37	22.50		59.50	74.50

22 34 36 – Commercial Gas Domestic Water Heaters

22 34 36.13 Commercial, Atmos., Gas Domestic Water Htrs.		Crew	Daily Output	Labor-Hours	Unit	Material	Labor	Equipment	Total	Total Incl O&P
0010	**COMMERCIAL, ATMOSPHERIC, GAS DOMESTIC WATER HEATERS**									
6000	Gas fired, flush jacket, std. controls, vent not incl.									
6040	75 MBH input, 73 GPH	1 Plum	1.40	5.714	Ea.	3,150	320		3,470	3,950
6060	98 MBH input, 95 GPH		1.40	5.714		5,050	320		5,370	6,025
6080	120 MBH input, 110 GPH		1.20	6.667		5,225	370		5,595	6,300
6100	120 MBH input, 115 GPH		1.10	7.273		6,325	405		6,730	7,550
6120	140 MBH input, 130 GPH		1	8		8,025	445		8,470	9,500
6140	155 MBH input, 150 GPH		.80	10		7,800	560		8,360	9,425
6160	180 MBH input, 170 GPH		.70	11.429		8,275	640		8,915	10,100
6180	200 MBH input, 192 GPH		.60	13.333		8,325	745		9,070	10,300
6200	250 MBH input, 245 GPH		.50	16		8,875	895		9,770	11,100
6220	260 MBH input, 250 GPH	Q-1	.80	20		9,425	1,000		10,425	11,900
6240	360 MBH input, 360 GPH		.80	20		11,100	1,000		12,100	13,700
6260	500 MBH input, 480 GPH		.70	22.857		15,700	1,150		16,850	18,900
6280	725 MBH input, 690 GPH		.60	26.667		19,100	1,350		20,450	23,000
6900	For low water cutoff, add	1 Plum	8	1		325	56		381	445
6960	For bronze body hot water circulator, add	"	4	2		1,875	112		1,987	2,250

22 34 36.45 Commercial Packaged Water Heater Systems		Crew	Daily Output	Labor-Hours	Unit	Material	Labor	Equipment	Total	Total Incl O&P
0010	**COMMERCIAL PACKAGED WATER HEATER SYSTEMS**									
1000	Car wash package, continuous duty, high recovery, gas fired									
1040	100° rise, 180 MBH input, 174 GPH	1 Plum	3	2.667	Ea.	5,625	149		5,774	6,400
1060	280 MBH input, 270 GPH		2.50	3.200		6,000	179		6,179	6,875
1080	400 MBH input, 386 GPH		2.50	3.200		6,925	179		7,104	7,875
1100	480 MBH input, 464 GPH		2	4		7,300	223		7,523	8,350
1120	605 MBH input, 584 GPH		1.50	5.333		8,325	298		8,623	9,600
1140	700 MBH input, 676 GPH		1	8		8,650	445		9,095	10,200
1160	1000 MBH input, 966 GPH	Q-1	1.60	10		9,625	500		10,125	11,400
1180	1200 MBH input, 1159 GPH		1.40	11.429		13,100	575		13,675	15,300
1200	1400 MBH input, 1353 GPH		1.20	13.333		14,300	670		14,970	16,700
3000	Combination dishwasher & general purpose, 2 temp. gas fired									
3040	124 GPH @ 140° rise; 434 GPH @ 40° rise	1 Plum	2	4	Ea.	5,725	223		5,948	6,625
3060	193 GPH @ 140° rise; 677 GPH @ 40° rise		1.60	5		6,550	279		6,829	7,625
3080	276 GPH @ 140° rise; 969 GPH @ 40° rise		1.40	5.714		7,450	320		7,770	8,675
3100	330 GPH @ 140° rise; 1160 GPH @ 40° rise		1.20	6.667		7,900	370		8,270	9,250
3120	417 GPH @ 140° rise; 1460 GPH @ 40° rise		1	8		8,925	445		9,370	10,500
3140	481 GPH @ 140° rise; 1685 GPH @ 40° rise	Q-1	1.40	11.429		10,300	575		10,875	12,200
3160	688 GPH @ 140° rise; 2410 GPH @ 40° rise		1.20	13.333		11,300	670		11,970	13,400
3180	828 GPH @ 140° rise; 2790 GPH @ 40° rise		1	16		14,800	805		15,605	17,500
3200	965 GPH @ 140° rise; 3370 GPH @ 40° rise		.80	20		16,000	1,000		17,000	19,100
3960	For unit base, add					440			440	485
5000	Coin laundry units, gas fired, 100° rise									
5020	Single heater,									
5040	280 MBH input, 270 GPH	1 Plum	1.60	5	Ea.	6,550	279		6,829	7,625
5060	400 MBH input, 386 GPH		1.30	6.154		8,650	345		8,995	10,000
5080	480 MBH input, 464 GPH		1	8		9,100	445		9,545	10,700

22 34 Fuel-Fired Domestic Water Heaters

22 34 36 – Commercial Gas Domestic Water Heaters

22 34 36.45 Commercial Packaged Water Heater Systems		Crew	Daily Output	Labor-Hours	Unit	2013 Bare Costs Material	Labor	Equipment	Total	Total Incl O&P
5100	605 MBH input, 584 GPH	Q-1	1.40	11.429	Ea.	10,700	575		11,275	12,700
5120	700 MBH input, 676 GPH		1.20	13.333		11,300	670		11,970	13,400
5140	1000 MBH input, 966 GPH		1	16		12,300	805		13,105	14,700
5160	1200 MBH input, 1159 GPH		.90	17.778		15,900	895		16,795	18,900
5180	1400 MBH input, 1353 GPH		.70	22.857		17,700	1,150		18,850	21,200
6000	Multiple heater									
6040	560 MBH input, 540 GPH	1 Plum	1	8	Ea.	10,400	445		10,845	12,100
6060	800 MBH input, 772 GPH	Q-1	1.30	12.308		13,500	620		14,120	15,700
6080	960 MBH input, 928 GPH		1.20	13.333		15,200	670		15,870	17,700
6100	1210 MBH input, 1168 GPH		1	16		17,100	805		17,905	20,000
6120	1400 MBH input, 1352 GPH		1	16		17,900	805		18,705	20,800
6140	1700 MBH input, 1642 GPH		.80	20		20,400	1,000		21,400	23,900
6160	2000 MBH input, 1932 GPH		.70	22.857		24,600	1,150		25,750	28,800
6180	2400 MBH input, 2318 GPH	Q-2	.90	26.667		27,700	1,400		29,100	32,500
6200	2600 MBH input, 2512 GPH		.80	30		30,900	1,550		32,450	36,300
6220	2800 MBH input, 2706 GPH		.70	34.286		34,800	1,775		36,575	41,000
6800	Standard system sizing is based on									
6820	80% of the washers operating at one time.									

22 34 46 – Oil-Fired Domestic Water Heaters

22 34 46.10 Residential Oil-Fired Water Heaters

		Crew	Daily Output	Labor-Hours	Unit	Material	Labor	Equipment	Total	Total Incl O&P
0010	**RESIDENTIAL OIL-FIRED WATER HEATERS**									
3000	Oil fired, glass lined tank, 5 yr., vent not included, 30 gallon D2020–260	1 Plum	2	4	Ea.	1,050	223		1,273	1,475
3040	50 gallon		1.80	4.444		1,250	248		1,498	1,750
3060	70 gallon		1.50	5.333		1,775	298		2,073	2,425

22 34 46.20 Commercial Oil-Fired Water Heaters

		Crew	Daily Output	Labor-Hours	Unit	Material	Labor	Equipment	Total	Total Incl O&P
0010	**COMMERCIAL OIL-FIRED WATER HEATERS**									
8000	Oil fired, glass lined, UL listed, std. controls, vent not incl.									
8060	140 gal., 140 MBH input, 134 GPH	Q-1	2.13	7.512	Ea.	19,000	375		19,375	21,500
8080	140 gal., 199 MBH input, 191 GPH		2	8		19,600	400		20,000	22,200
8100	140 gal., 255 MBH input, 247 GPH		1.60	10		20,200	500		20,700	23,000
8120	140 gal., 270 MBH input, 259 GPH		1.20	13.333		24,900	670		25,570	28,400
8140	140 gal., 400 MBH input, 384 GPH		1	16		25,600	805		26,405	29,300
8160	140 gal., 540 MBH input, 519 GPH		.96	16.667		26,800	835		27,635	30,700
8180	140 gal., 720 MBH input, 691 GPH		.92	17.391		27,200	875		28,075	31,300
8200	221 gal., 300 MBH input, 288 GPH		.88	18.182		36,000	915		36,915	41,000
8220	221 gal., 600 MBH input, 576 GPH		.86	18.605		40,200	935		41,135	45,600
8240	221 gal., 800 MBH input, 768 GPH		.82	19.512		40,500	980		41,480	46,000
8260	201 gal., 1000 MBH input, 960 GPH	Q-2	1.26	19.048		41,400	990		42,390	47,100
8280	201 gal., 1250 MBH input, 1200 GPH		1.22	19.672		41,800	1,025		42,825	47,600
8300	201 gal., 1500 MBH input, 1441 GPH		1.16	20.690		45,500	1,075		46,575	51,500
8320	411 gal., 600 MBH input, 576 GPH		1.12	21.429		45,900	1,125		47,025	52,000
8340	411 gal., 800 MBH input, 768 GPH		1.08	22.222		46,000	1,150		47,150	52,500
8360	411 gal., 1000 MBH input, 960 GPH		1.04	23.077		48,600	1,200		49,800	55,500
8380	411 gal., 1250 MBH input, 1200 GPH		.98	24.490		49,700	1,275		50,975	56,500
8400	397 gal., 1500 MBH input, 1441 GPH		.92	26.087		50,000	1,350		51,350	57,500
8420	397 gal., 1750 MBH input, 1681 GPH		.86	27.907		54,500	1,450		55,950	62,000
8430	397 gal., 2000 MBH input, 1921 GPH		.82	29.268		59,000	1,525		60,525	67,000
8440	375 gal., 2250 MBH input, 2161 GPH		.76	31.579		60,500	1,650		62,150	69,000
8450	375 gal., 2500 MBH input, 2401 GPH		.82	29.268		62,500	1,525		64,025	71,500
8500	Oil fired, polymer lined									
8510	400 MBH, 125 gallon	Q-2	1	24	Ea.	24,400	1,250		25,650	28,800
8520	400 MBH, 600 gallon		.80	30		45,800	1,550		47,350	53,000

22 34 Fuel-Fired Domestic Water Heaters

22 34 46 – Oil-Fired Domestic Water Heaters

22 34 46.20 Commercial Oil-Fired Water Heaters		Crew	Daily Output	Labor-Hours	Unit	Material	Labor	Equipment	Total	Total Incl O&P
						2013 Bare Costs				
8530	800 MBH, 400 gallon	Q-2	.67	35.821	Ea.	41,700	1,875		43,575	48,700
8540	800 MBH, 600 gallon		.60	40		54,000	2,075		56,075	62,500
8550	1000 MBH, 600 gallon		.50	48		56,000	2,500		58,500	65,500
8560	1200 MBH, 900 gallon		.40	60		67,500	3,125		70,625	79,000
8900	For low water cutoff, add	1 Plum	8	1		325	56		381	445
8960	For bronze body hot water circulator, add	"	4	2		710	112		822	950

22 41 Residential Plumbing Fixtures

22 41 06 – Plumbing Fixtures General

22 41 06.10 Plumbing Fixture Notes		Crew	Daily Output	Labor-Hours	Unit	Material	Labor	Equipment	Total	Total Incl O&P
0010	**PLUMBING FIXTURE NOTES**, Incl. trim fittings unless otherwise noted R224000-30									
0080	For rough-in, supply, waste, and vent, see add for each type									
0122	For electric water coolers, see Section 22 47 16.10									
0160	For color, unless otherwise noted, add				Ea.	20%				

22 41 13 – Residential Water Closets, Urinals, and Bidets

22 41 13.10 Bidets		Crew	Daily Output	Labor-Hours	Unit	Material	Labor	Equipment	Total	Total Incl O&P
0010	**BIDETS**									
0180	Vitreous china, with trim on fixture	Q-1	5	3.200	Ea.	540	161		701	835
0200	With trim for wall mounting	"	5	3.200		645	161		806	950
9590	For color add					40%				
9591	For designer colors and trim add					50%				
9600	For rough-in, supply, waste and vent, add	Q-1	1.78	8.989		375	450		825	1,100

22 41 13.40 Water Closets		Crew	Daily Output	Labor-Hours	Unit	Material	Labor	Equipment	Total	Total Incl O&P
0010	**WATER CLOSETS** D2010–110									
0022	For seats, see Section 22 41 13.44									
0032	For automatic flush, see Line 22 42 39.10 0972									
0150	Tank type, vitreous china, incl. seat, supply pipe w/stop, 1.6 gpf or noted									
0200	Wall hung R224000-30									
0400	Two piece, close coupled	Q-1	5.30	3.019	Ea.	630	152		782	920
0960	For rough-in, supply, waste, vent and carrier	"	2.73	5.861	"	815	294		1,109	1,350
0999	Floor mounted									
1020	One piece, low profile	Q-1	5.30	3.019	Ea.	480	152		632	755
1050	One piece		5.30	3.019		1,025	152		1,177	1,350
1100	Two piece, close coupled		5.30	3.019		230	152		382	480
1102	Economy		5.30	3.019		129	152		281	370
1110	Two piece, close coupled, dual flush		5.30	3.019		294	152		446	555
1140	Two piece, close coupled, 1.28 gpf, ADA G		5.30	3.019		305	152		457	565
1960	For color, add					30%				
1961	For designer colors and trim, add					55%				
1980	For rough-in, supply, waste and vent	Q-1	3.05	5.246	Ea.	320	264		584	750

22 41 13.44 Toilet Seats		Crew	Daily Output	Labor-Hours	Unit	Material	Labor	Equipment	Total	Total Incl O&P
0010	**TOILET SEATS**									
0100	Molded composition, white									
0150	Industrial, w/o cover, open front, regular bowl	1 Plum	24	.333	Ea.	21	18.60		39.60	51
0200	With self-sustaining hinge		24	.333		22.50	18.60		41.10	52.50
0220	With self-sustaining check hinge		24	.333		22.50	18.60		41.10	52.50
0240	Extra heavy, with check hinge		24	.333		27.50	18.60		46.10	58.50
0260	Elongated bowl, same price									
0300	Junior size, w/o cover, open front	1 Plum	24	.333	Ea.	40.50	18.60		59.10	72.50
0320	Regular primary bowl, open front		24	.333		40.50	18.60		59.10	72.50

22 41 Residential Plumbing Fixtures

22 41 13 – Residential Water Closets, Urinals, and Bidets

22 41 13.44 Toilet Seats		Crew	Daily Output	Labor-Hours	Unit	2013 Bare Costs Material	Labor	Equipment	Total	Total Incl O&P
0340	Regular baby bowl, open front, check hinge	1 Plum	24	.333	Ea.	37	18.60		55.60	68.50
0380	Open back & front, w/o cover, reg. or elongated bowl	↓	24	.333	↓	29	18.60		47.60	60
0400	Residential									
0420	Regular bowl, w/cover, closed front	1 Plum	24	.333	Ea.	29	18.60		47.60	59.50
0440	Open front	"	24	.333	"	26	18.60		44.60	57
0460	Elongated bowl, add					25%				
0500	Self-raising hinge, w/o cover, open front									
0520	Regular bowl	1 Plum	24	.333	Ea.	101	18.60		119.60	139
0540	Elongated bowl	"	24	.333	"	23	18.60		41.60	53.50
0700	Molded wood, white, with cover									
0720	Closed front, regular bowl, square back	1 Plum	24	.333	Ea.	10.90	18.60		29.50	40
0740	Extended back		24	.333		13.80	18.60		32.40	43
0780	Elongated bowl, square back		24	.333		14.10	18.60		32.70	43.50
0800	Open front	↓	24	.333	↓	15.05	18.60		33.65	44.50
0850	Decorator styles									
0890	Vinyl top, patterned	1 Plum	24	.333	Ea.	22	18.60		40.60	52
0900	Vinyl padded, plain colors, regular bowl		24	.333		22	18.60		40.60	52
0930	Elongated bowl	↓	24	.333	↓	24	18.60		42.60	54.50
1000	Solid plastic, white									
1030	Industrial, w/o cover, open front, regular bowl	1 Plum	24	.333	Ea.	26	18.60		44.60	57
1080	Extra heavy, concealed check hinge		24	.333		18.75	18.60		37.35	48.50
1100	Self-sustaining hinge		24	.333		23	18.60		41.60	53.50
1150	Elongated bowl		24	.333		29	18.60		47.60	60
1170	Concealed check		24	.333		17.25	18.60		35.85	47
1190	Self-sustaining hinge, concealed check		24	.333		51.50	18.60		70.10	84.50
1220	Residential, with cover, closed front, regular bowl		24	.333		45	18.60		63.60	77.50
1240	Elongated bowl		24	.333		55	18.60		73.60	88
1260	Open front, regular bowl		24	.333		40	18.60		58.60	72
1280	Elongated bowl	↓	24	.333	↓	48.50	18.60		67.10	81.50

22 41 16 – Residential Lavatories and Sinks

22 41 16.10 Lavatories		Crew	Daily Output	Labor-Hours	Unit	Material	Labor	Equipment	Total	Total Incl O&P
0010	**LAVATORIES**, With trim, white unless noted otherwise D2010-310									
0500	Vanity top, porcelain enamel on cast iron									
0600	20" x 18" R224000-30	Q-1	6.40	2.500	Ea.	325	126		451	545
0640	33" x 19" oval		6.40	2.500		680	126		806	935
0680	20" x 17" oval		6.40	2.500		169	126		295	375
0720	19" round		6.40	2.500		238	126		364	450
0760	20" x 12" triangular bowl	↓	6.40	2.500	↓	257	126		383	470
0860	For color, add					25%				
0861	For designer colors and trim, add					70%				
1000	Cultured marble, 19" x 17", single bowl	Q-1	6.40	2.500	Ea.	185	126		311	390
1040	25" x 19", single bowl		6.40	2.500		212	126		338	420
1080	31" x 19", single bowl		6.40	2.500		231	126		357	445
1120	25" x 22", single bowl		6.40	2.500		229	126		355	440
1160	37" x 22", single bowl		6.40	2.500		264	126		390	480
1200	49" x 22", single bowl	↓	6.40	2.500	↓	320	126		446	540
1580	For color, same price									
1900	Stainless steel, self-rimming, 25" x 22", single bowl, ledge	Q-1	6.40	2.500	Ea.	360	126		486	585
1960	17" x 22", single bowl		6.40	2.500		350	126		476	575
2040	18-3/4" round		6.40	2.500		810	126		936	1,075
2600	Steel, enameled, 20" x 17", single bowl		5.80	2.759		195	139		334	425
2660	19" round		5.80	2.759		169	139		308	395

22 41 16 – Residential Lavatories and Sinks

22 41 16.10 Lavatories		Crew	Daily Output	Labor-Hours	Unit	Material	Labor	Equipment	Total	Total Incl O&P
						2013 Bare Costs				
2720	18″ round	Q-1	5.80	2.759	Ea.	138	139		277	360
2860	For color, add					10%				
2861	For designer colors and trim, add					20%				
2900	Vitreous china, 20″ x 16″, single bowl	Q-1	5.40	2.963	Ea.	260	149		409	510
2960	20″ x 17″, single bowl		5.40	2.963		176	149		325	420
3020	19″ round, single bowl		5.40	2.963		174	149		323	415
3080	19″ x 16″, single bowl		5.40	2.963		267	149		416	520
3140	17″ x 14″, single bowl		5.40	2.963		214	149		363	460
3200	22″ x 13″, single bowl		5.40	2.963		267	149		416	520
3560	For color, add					50%				
3561	For designer colors and trim, add					100%				
3580	Rough-in, supply, waste and vent for all above lavatories	Q-1	2.30	6.957	Ea.	400	350		750	965
4000	Wall hung									
4040	Porcelain enamel on cast iron, 16″ x 14″, single bowl	Q-1	8	2	Ea.	500	100		600	700
4060	18″ x 15″ single bowl		8	2		395	100		495	585
4120	19″ x 17″, single bowl		8	2		415	100		515	605
4180	20″ x 18″, single bowl		8	2		273	100		373	450
4240	22″ x 19″, single bowl		8	2		675	100		775	890
4580	For color, add					30%				
4581	For designer colors and trim, add					75%				
6000	Vitreous china, 18″ x 15″, single bowl with backsplash	Q-1	7	2.286	Ea.	231	115		346	425
6060	19″ x 17″, single bowl		7	2.286		190	115		305	380
6120	20″ x 18″, single bowl		7	2.286		273	115		388	475
6210	27″ x 20″, wheelchair type		7	2.286		460	115		575	680
6500	For color, add					30%				
6501	For designer colors and trim, add					50%				
6960	Rough-in, supply, waste and vent for above lavatories	Q-1	1.66	9.639	Ea.	480	485		965	1,250
7000	Pedestal type									
7600	Vitreous china, 27″ x 21″, white	Q-1	6.60	2.424	Ea.	595	122		717	840
7610	27″ x 21″, colored		6.60	2.424		715	122		837	970
7620	27″ x 21″, premium color		6.60	2.424		835	122		957	1,100
7660	26″ x 20″, white		6.60	2.424		535	122		657	775
7670	26″ x 20″, colored		6.60	2.424		640	122		762	890
7680	26″ x 20″, premium color		6.60	2.424		675	122		797	930
7700	24″ x 20″, white		6.60	2.424		520	122		642	755
7710	24″ x 20″, colored		6.60	2.424		620	122		742	870
7720	24″ x 20″, premium color		6.60	2.424		725	122		847	980
7760	21″ x 18″, white		6.60	2.424		246	122		368	455
7770	21″ x 18″, colored		6.60	2.424		280	122		402	495
7990	Rough-in, supply, waste and vent for pedestal lavatories		1.66	9.639		480	485		965	1,250

22 41 16.30 Sinks		Crew	Daily Output	Labor-Hours	Unit	Material	Labor	Equipment	Total	Total Incl O&P
0010	**SINKS**, With faucets and drain D2010–410									
2000	Kitchen, counter top style, P.E. on C.I., 24″ x 21″ single bowl	Q-1	5.60	2.857	Ea.	276	144		420	520
2100	31″ x 22″ single bowl		5.60	2.857		560	144		704	830
2200	32″ x 21″ double bowl		4.80	3.333		310	167		477	590
2300	42″ x 21″ double bowl		4.80	3.333		1,150	167		1,317	1,500
2310	For color, add					20%				
2311	For designer colors and trim, add					50%				
3000	Stainless steel, self rimming, 19″ x 18″ single bowl	Q-1	5.60	2.857	Ea.	580	144		724	855
3100	25″ x 22″ single bowl		5.60	2.857		645	144		789	925
3200	33″ x 22″ double bowl		4.80	3.333		945	167		1,112	1,300
3300	43″ x 22″ double bowl		4.80	3.333		1,100	167		1,267	1,450

22 41 Residential Plumbing Fixtures

22 41 16 – Residential Lavatories and Sinks

22 41 16.30 Sinks		Crew	Daily Output	Labor-Hours	Unit	Material	Labor	Equipment	Total	Total Incl O&P
						2013 Bare Costs				
3400	22" x 43" triple bowl	Q-1	4.40	3.636	Ea.	1,325	183		1,508	1,725
3500	Corner double bowl each 14" x 16"		4.80	3.333		815	167		982	1,150
4000	Steel, enameled, with ledge, 24" x 21" single bowl		5.60	2.857		470	144		614	730
4100	32" x 21" double bowl		4.80	3.333		485	167		652	780
4960	For color sinks except stainless steel, add					10%				
4961	For designer colors and trim add					20%				
4980	For rough-in, supply, waste and vent, counter top sinks	Q-1	2.14	7.477		450	375		825	1,050
5000	Kitchen, raised deck, P.E. on C.I.									
5100	32" x 21", dual level, double bowl	Q-1	2.60	6.154	Ea.	385	310		695	890
5200	42" x 21", double bowl & disposer well	"	2.20	7.273		1,150	365		1,515	1,825
5700	For color, add					20%				
5701	For designer colors and trim add					50%				
5790	For rough-in, supply, waste & vent, sinks	Q-1	1.85	8.649		450	435		885	1,150

22 41 19 – Residential Bathtubs

22 41 19.10 Baths		Crew	Daily Output	Labor-Hours	Unit	Material	Labor	Equipment	Total	Total Incl O&P
0010	**BATHS** D2010–510									
0100	Tubs, recessed porcelain enamel on cast iron, with trim									
0180	48" x 42"	Q-1	4	4	Ea.	2,575	201		2,776	3,125
0220	72" x 36"	"	3	5.333	"	2,675	268		2,943	3,325
0300	Mat bottom									
0340	4'-6" long	Q-1	5	3.200	Ea.	1,300	161		1,461	1,675
0380	5' long		4.40	3.636		1,100	183		1,283	1,475
0420	5'-6" long		4	4		1,750	201		1,951	2,225
0480	Above floor drain, 5' long		4	4		855	201		1,056	1,250
0560	Corner 48" x 44"		4.40	3.636		2,575	183		2,758	3,100
0750	For color, add					30%				
0760	For designer colors & trim, add					60%				
2000	Enameled formed steel, 4'-6" long	Q-1	5.80	2.759	Ea.	495	139		634	755
2300	Above floor drain, 5' long	"	5.50	2.909	"	480	146		626	750
2350	For color, add					10%				
4000	Soaking, acrylic, w/pop-up drain 66" x 36" x 20" deep	Q-1	5.50	2.909	Ea.	1,775	146		1,921	2,175
4100	60" x 48" x 18-1/2" deep		5	3.200		1,075	161		1,236	1,425
4200	72" x 42" x 23" deep		4.80	3.333		1,750	167		1,917	2,175
4310	For color, add					5%				
4311	For designer colors & trim, add					20%				
4600	Module tub & showerwall surround, molded fiberglass									
4610	5' long x 34" wide x 76" high	Q-1	4	4	Ea.	815	201		1,016	1,200
4620	For color add					10%				
4621	For designer colors and trim add					25%				
4750	Handicap with 1-1/2" OD grab bar, antiskid bottom									
4760	60" x 32-3/4" x 72" high	Q-1	4	4	Ea.	810	201		1,011	1,200
4770	60" x 30" x 71" high with molded seat		3.50	4.571		795	230		1,025	1,225
9600	Rough-in, supply, waste and vent, for all above tubs, add		2.07	7.729		475	390		865	1,100

22 41 23 – Residential Showers

22 41 23.20 Showers		Crew	Daily Output	Labor-Hours	Unit	Material	Labor	Equipment	Total	Total Incl O&P
0010	**SHOWERS** D2010–710									
1500	Stall, with drain only. Add for valve and door/curtain									
1510	Baked enamel, molded stone receptor, 30" square	Q-1	5.20	3.077	Ea.	1,150	155		1,305	1,500
1520	32" square		5	3.200		1,175	161		1,336	1,525
1530	36" square		4.80	3.333		2,925	167		3,092	3,450
1540	Terrazzo receptor, 32" square		5	3.200		1,350	161		1,511	1,725
1560	36" square		4.80	3.333		1,475	167		1,642	1,875

22 41 Residential Plumbing Fixtures

22 41 23 – Residential Showers

22 41 23.20 Showers		Crew	Daily Output	Labor-Hours	Unit	2013 Bare Costs Material	Labor	Equipment	Total	Total Incl O&P
1580	36″ corner angle	Q-1	4.80	3.333	Ea.	1,725	167		1,892	2,150
1600	For color, add					10%				
1601	For designer colors and trim, add					15%				
1604	For thermostatic valve add				Ea.	585			585	645
3000	Fiberglass, one piece, with 3 walls, 32″ x 32″ square	Q-1	5.50	2.909		550	146		696	825
3100	36″ x 36″ square	″	5.50	2.909	↓	565	146		711	840
3200	Handicap, 1-1/2″ O.D. grab bars, nonskid floor									
3210	48″ x 34-1/2″ x 72″ corner seat	Q-1	5	3.200	Ea.	750	161		911	1,050
3220	60″ x 34-1/2″ x 72″ corner seat		4	4		760	201		961	1,150
3230	48″ x 34-1/2″ x 72″ fold up seat		5	3.200		2,175	161		2,336	2,650
3250	64″ x 65-3/4″ x 81-1/2″ fold. seat, whlchr.	↓	3.80	4.211		2,325	211		2,536	2,875
3260	For thermostatic valve add					585			585	645
4000	Polypropylene, stall only, w/molded-stone floor, 30″ x 30″	Q-1	2	8		635	400		1,035	1,300
4100	32″ x 32″	″	2	8		650	400		1,050	1,325
4110	For thermostatic valve add					585			585	645
4200	Rough-in, supply, waste and vent for above showers	Q-1	2.05	7.805	↓	595	390		985	1,250

22 41 23.40 Shower System Components

		Crew	Daily Output	Labor-Hours	Unit	Material	Labor	Equipment	Total	Total Incl O&P
0010	**SHOWER SYSTEM COMPONENTS**									
4500	Receptor only									
4510	For tile, 36″ x 36″	1 Plum	4	2	Ea.	365	112		477	575
4520	Fiberglass receptor only, 32″ x 32″		8	1		107	56		163	202
4530	34″ x 34″		7.80	1.026		126	57		183	225
4540	36″ x 36″	↓	7.60	1.053	↓	132	58.50		190.50	234
4600	Rectangular									
4620	32″ x 48″	1 Plum	7.40	1.081	Ea.	149	60.50		209.50	254
4630	32″ x 54″		7.20	1.111		196	62		258	310
4640	32″ x 60″		7	1.143		208	64		272	325
5000	Built-in, head, arm, 2.5 GPM valve		4	2		80	112		192	256
5200	Head, arm, by-pass, integral stops, handles		3.60	2.222		230	124		354	440
5500	Head, water economizer, 1.6 GPM G		24	.333		52	18.60		70.60	85
5800	Mixing valve, built-in		6	1.333		153	74.50		227.50	281
5900	Exposed	↓	6	1.333	↓	630	74.50		704.50	805

22 41 36 – Residential Laundry Trays

22 41 36.10 Laundry Sinks

		Crew	Daily Output	Labor-Hours	Unit	Material	Labor	Equipment	Total	Total Incl O&P
0010	**LAUNDRY SINKS**, With trim D2010–420									
0020	Porcelain enamel on cast iron, black iron frame									
0050	24″ x 21″, single compartment	Q-1	6	2.667	Ea.	425	134		559	670
0100	26″ x 21″, single compartment		6	2.667		450	134		584	695
0200	48″ x 20″, double compartment	↓	5	3.200	↓	830	161		991	1,150
2000	Molded stone, on wall hanger or legs									
2020	22″ x 23″, single compartment	Q-1	6	2.667	Ea.	191	134		325	410
2100	45″ x 21″, double compartment	″	5	3.200	″	330	161		491	605
3000	Plastic, on wall hanger or legs									
3020	18″ x 23″, single compartment	Q-1	6.50	2.462	Ea.	152	124		276	355
3100	20″ x 24″, single compartment		6.50	2.462		152	124		276	355
3200	36″ x 23″, double compartment		5.50	2.909		182	146		328	420
3300	40″ x 24″, double compartment		5.50	2.909		272	146		418	520
5000	Stainless steel, counter top, 22″ x 17″ single compartment		6	2.667		64	134		198	273
5200	33″ x 22″, double compartment		5	3.200		79	161		240	330
9600	Rough-in, supply, waste and vent, for all laundry sinks	↓	2.14	7.477	↓	450	375		825	1,050

22 41 39.10 Faucets and Fittings		Crew	Daily Output	Labor-Hours	Unit	2013 Bare Costs Material	Labor	Equipment	Total	Total Incl O&P
0010	**FAUCETS AND FITTINGS**									
0150	Bath, faucets, diverter spout combination, sweat	1 Plum	8	1	Ea.	86.50	56		142.50	179
0200	For integral stops, IPS unions, add					109			109	120
0300	Three valve combinations, spout, head, arm, flange, sweat	1 Plum	6	1.333		67.50	74.50		142	186
0400	For integral stops, IPS unions, add				Pr.	64.50			64.50	71
0420	Bath, press-bal mix valve w/diverter, spout, shower head, arm/flange	1 Plum	8	1	Ea.	168	56		224	269
0500	Drain, central lift, 1-1/2" IPS male		20	.400		71	22.50		93.50	112
0600	Trip lever, 1-1/2" IPS male		20	.400		45	22.50		67.50	83
0700	Pop up, 1-1/2" IPS male		18	.444		53	25		78	96
0800	Chain and stopper, 1-1/2" IPS male		24	.333		32.50	18.60		51.10	63.50
0810	Bidet									
0812	Fitting, over the rim, swivel spray/pop-up drain	1 Plum	8	1	Ea.	190	56		246	293
1000	Kitchen sink faucets, top mount, cast spout		10	.800		61.50	44.50		106	135
1100	For spray, add		24	.333		16.15	18.60		34.75	46
1110	For basket strainer w/tail piece, add		24	.333		14.85	18.60		33.45	44.50
1200	Wall type, swing tube spout		10	.800		74.50	44.50		119	149
1240	For soap dish, add					3.60			3.60	3.96
1250	For basket strainer w/tail piece, add					43			43	47
1300	Single control lever handle									
1310	With pull out spray									
1320	Polished chrome	1 Plum	10	.800	Ea.	178	44.50		222.50	262
2000	Laundry faucets, shelf type, IPS or copper unions		12	.667		49.50	37		86.50	111
2100	Lavatory faucet, centerset, without drain		10	.800		44.50	44.50		89	116
2120	With pop-up drain		6.66	1.201		62.50	67		129.50	170
2130	For acrylic handles, add					5.15			5.15	5.65
2150	Concealed, 12" centers	1 Plum	10	.800		101	44.50		145.50	178
2160	With pop-up drain	"	6.66	1.201		118	67		185	231
2210	Porcelain cross handles and pop-up drain									
2220	Polished chrome	1 Plum	6.66	1.201	Ea.	163	67		230	280
2230	Polished brass	"	6.66	1.201	"	244	67		311	370
2260	Single lever handle and pop-up drain									
2280	Satin nickel	1 Plum	6.66	1.201	Ea.	263	67		330	390
2290	Polished chrome		6.66	1.201		188	67		255	305
2600	Shelfback, 4" to 6" centers, 17 Ga. tailpiece		10	.800		78	44.50		122.50	153
2650	With pop-up drain		6.66	1.201		95	67		162	205
2700	Shampoo faucet with supply tube		24	.333		48.50	18.60		67.10	81.50
2800	Self-closing, center set		10	.800		131	44.50		175.50	211
2810	Automatic sensor and operator, with faucet head G		6.15	1.301		370	72.50		442.50	520
4000	Shower by-pass valve with union		18	.444		68.50	25		93.50	113
4100	Shower arm with flange and head		22	.364		79	20.50		99.50	117
4140	Shower, hand held, pin mount, massage action, chrome		22	.364		70	20.50		90.50	107
4142	Polished brass		22	.364		134	20.50		154.50	178
4144	Shower, hand held, wall mtd, adj. spray, 2 wall mounts, chrome		20	.400		94	22.50		116.50	137
4146	Polished brass		20	.400		182	22.50		204.50	234
4148	Shower, hand held head, bar mounted 24", adj. spray, chrome		20	.400		151	22.50		173.50	200
4150	Polished brass		20	.400		320	22.50		342.50	385
4200	Shower thermostatic mixing valve, concealed, with shower head trim kit		8	1		330	56		386	445
4220	Shower pressure balancing mixing valve,									
4230	With shower head, arm, flange and diverter tub spout									
4240	Chrome	1 Plum	6.14	1.303	Ea.	350	72.50		422.50	495
4250	Satin nickel		6.14	1.303		475	72.50		547.50	630
4260	Polished graphite		6.14	1.303		475	72.50		547.50	630

22 41 Residential Plumbing Fixtures

22 41 39 – Residential Faucets, Supplies and Trim

22 41 39.10 Faucets and Fittings		Crew	Daily Output	Labor-Hours	Unit	2013 Bare Costs Material	Labor	Equipment	Total	Total Incl O&P
5000	Sillcock, compact, brass, IPS or copper to hose	1 Plum	24	.333	Ea.	9.15	18.60		27.75	38
6000	Stop and waste valves, bronze									
6100	Angle, solder end 1/2″	1 Plum	24	.333	Ea.	11.35	18.60		29.95	40.50
6110	3/4″		20	.400		13.30	22.50		35.80	48
6300	Straightway, solder end 3/8″		24	.333		11.35	18.60		29.95	40.50
6310	1/2″		24	.333		11.60	18.60		30.20	41
6320	3/4″		20	.400		17.20	22.50		39.70	52.50
6410	Straightway, threaded 1/2″		24	.333		15	18.60		33.60	44.50
6420	3/4″		20	.400		21	22.50		43.50	56.50
6430	1″		19	.421		12.60	23.50		36.10	49.50
7800	Water closet, wax gasket		96	.083		1.83	4.65		6.48	9
7820	Gasket toilet tank to bowl		32	.250		2.38	13.95		16.33	23.50
7830	Replacement diaphragm washer assy for ballcock valve		12	.667		2.38	37		39.38	58.50
7850	Dual flush valve		12	.667		30	37		67	89
8000	Water supply stops, polished chrome plate									
8200	Angle, 3/8″	1 Plum	24	.333	Ea.	9.90	18.60		28.50	39
8300	1/2″		22	.364		9.90	20.50		30.40	41.50
8400	Straight, 3/8″		26	.308		10.40	17.15		27.55	37.50
8500	1/2″		24	.333		10.40	18.60		29	39.50
8600	Water closet, angle, w/flex riser, 3/8″		24	.333		38	18.60		56.60	70
9100	Miscellaneous									
9720	Teflon tape, 1/2″ x 520″ roll				Ea.	.97			.97	1.07

22 41 39.70 Washer/Dryer Accessories		Crew	Daily Output	Labor-Hours	Unit	Material	Labor	Equipment	Total	Total Incl O&P
0010	**WASHER/DRYER ACCESSORIES**									
1020	Valves ball type single lever									
1030	1/2″ diam., IPS	1 Plum	21	.381	Ea.	54	21.50		75.50	91.50
1040	1/2″ diam., solder	″	21	.381	″	54	21.50		75.50	91.50
1050	Recessed box, 16 ga., two hose valves and drain									
1060	1/2″ size, 1-1/2″ drain	1 Plum	18	.444	Ea.	106	25		131	154
1070	1/2″ size, 2″ drain	″	17	.471	″	104	26.50		130.50	154
1080	With grounding electric receptacle									
1090	1/2″ size, 1-1/2″ drain	1 Plum	18	.444	Ea.	118	25		143	168
1100	1/2″ size, 2″ drain	″	17	.471	″	125	26.50		151.50	178
1110	With grounding and dryer receptacle									
1120	1/2″ size, 1-1/2″ drain	1 Plum	18	.444	Ea.	143	25		168	195
1130	1/2″ size, 2″ drain	″	17	.471	″	144	26.50		170.50	199
1140	Recessed box 16 ga., ball valves with single lever and drain									
1150	1/2″ size, 1-1/2″ drain	1 Plum	19	.421	Ea.	199	23.50		222.50	255
1160	1/2″ size, 2″ drain	″	18	.444	″	190	25		215	246
1170	With grounding electric receptacle									
1180	1/2″ size, 1-1/2″ drain	1 Plum	19	.421	Ea.	182	23.50		205.50	237
1190	1/2″ size, 2″ drain	″	18	.444	″	192	25		217	249
1200	With grounding and dryer receptacles									
1210	1/2″ size, 1-1/2″ drain	1 Plum	19	.421	Ea.	199	23.50		222.50	255
1220	1/2″ size, 2″ drain	″	18	.444	″	210	25		235	269
1300	Recessed box, 20 ga., two hose valves and drain (economy type)									
1310	1/2″ size, 1-1/2″ drain	1 Plum	19	.421	Ea.	82.50	23.50		106	126
1320	1/2″ size, 2″ drain		18	.444		80.50	25		105.50	126
1330	Box with drain only		24	.333		47.50	18.60		66.10	80.50
1340	1/2″ size, 1-1/2″ ABS/PVC drain		19	.421		95.50	23.50		119	141
1350	1/2″ size, 2″ ABS/PVC drain		18	.444		125	25		150	176
1352	Box with drain and 15 A receptacle		24	.333		90.50	18.60		109.10	128

22 41 Residential Plumbing Fixtures

22 41 39 – Residential Faucets, Supplies and Trim

22 41 39.70	Washer/Dryer Accessories	Crew	Daily Output	Labor-Hours	Unit	2013 Bare Costs Material	Labor	Equipment	Total	Total Incl O&P
1360	1/2" size, 2" drain ABS/PVC, 15 A receptacle	1 Plum	24	.333	Ea.	119	18.60		137.60	159
1400	Wall mounted									
1410	1/2" size, 1-1/2" plastic drain	1 Plum	19	.421	Ea.	24.50	23.50		48	62.50
1420	1/2" size, 2" plastic drain	"	18	.444	"	21	25		46	60.50
1500	Dryer vent kit									
1510	8' flex duct, clamps and outside hood	1 Plum	20	.400	Ea.	15.30	22.50		37.80	50.50
1980	Rough-in, supply, waste, and vent for washer boxes	"	3.46	2.310	"	510	129		639	755

22 42 Commercial Plumbing Fixtures

22 42 13 – Commercial Water Closets, Urinals, and Bidets

22 42 13.30 Urinals

Line	Description	Crew	Daily Output	Labor-Hours	Unit	Material	Labor	Equipment	Total	Total Incl O&P
0010	**URINALS** D2010–210									
0102	For automatic flush see Line 22 42 39.10 0972									
3000	Wall hung, vitreous china, with hanger & self-closing valve R224000-30									
3100	Siphon jet type	Q-1	3	5.333	Ea.	305	268		573	740
3120	Blowout type		3	5.333		430	268		698	875
3140	Water saving .5 gpf G		3	5.333		690	268		958	1,150
3300	Rough-in, supply, waste & vent		2.83	5.654		585	284		869	1,075
5000	Stall type, vitreous china, includes valve		2.50	6.400		680	320		1,000	1,225
6980	Rough-in, supply, waste and vent		1.99	8.040		560	405		965	1,225
8000	Waterless (no flush) urinal									
8010	Wall hung									
8014	Fiberglass reinforced polyester									
8020	Standard unit G	Q-1	21.30	.751	Ea.	385	37.50		422.50	475
8030	ADA compliant unit G G	"	21.30	.751		400	37.50		437.50	495
8070	For solid color, add G					60			60	66
8080	For 2" brass flange, (new const.), add G	Q-1	96	.167		19.20	8.35		27.55	33.50
8200	Vitreous china									
8220	ADA compliant unit, 14" G	Q-1	21.30	.751	Ea.	198	37.50		235.50	275
8240	ADA compliant unit, 18" G		21.30	.751		320	37.50		357.50	405
8250	ADA compliant unit, 15.5"		21.30	.751		272	37.50		309.50	355
8270	For solid color, add G					60			60	66
8290	Rough-in, supply, waste & vent G	Q-1	2.92	5.479		530	275		805	995
8400	Trap liquid									
8410	1 quart G				Ea.	15.15			15.15	16.70
8420	1 gallon G				"	56			56	61.50

22 42 13.40 Water Closets

Line	Description	Crew	Daily Output	Labor-Hours	Unit	Material	Labor	Equipment	Total	Total Incl O&P
0010	**WATER CLOSETS**									
3000	Bowl only, with flush valve, seat, 1.6 gpf unless noted									
3100	Wall hung	Q-1	5.80	2.759	Ea.	705	139		844	990
3200	For rough-in, supply, waste and vent, single WC		2.56	6.250		895	315		1,210	1,450
3300	Floor mounted		5.80	2.759		325	139		464	570
3350	With wall outlet		5.80	2.759		500	139		639	760
3360	With floor outlet, 1.28 gpf G		5.80	2.759		585	139		724	855
3362	With floor outlet, 1.28 gpf, ADA G		5.80	2.759		620	139		759	895
3400	For rough-in, supply, waste and vent, single WC		2.84	5.634		400	283		683	865
3500	Gang side by side carrier system, rough-in, supply, waste & vent									
3510	For single hook-up	Q-1	1.97	8.122	Ea.	1,100	410		1,510	1,825
3520	For each additional hook-up, add	"	2.14	7.477	"	1,050	375		1,425	1,725
3550	Gang back to back carrier system, rough-in, supply, waste & vent									

22 42 Commercial Plumbing Fixtures

22 42 13 – Commercial Water Closets, Urinals, and Bidets

22 42 13.40 Water Closets		Crew	Daily Output	Labor-Hours	Unit	2013 Bare Costs Material	Labor	Equipment	Total	Total Incl O&P
3560	For pair hook-up	Q-1	1.76	9.091	Pr.	1,700	455		2,155	2,575
3570	For each additional pair hook-up, add	"	1.81	8.840	"	1,675	445		2,120	2,500

22 42 16 – Commercial Lavatories and Sinks

22 42 16.10 Handwasher-Dryer Module

		Crew	Daily Output	Labor-Hours	Unit	Material	Labor	Equipment	Total	Total Incl O&P
0010	**HANDWASHER-DRYER MODULE** D2010–610									
0030	Wall mounted									
0040	With electric dryer									
0050	Sensor operated	Q-1	8	2	Ea.	3,650	100		3,750	4,150
0110	Sensor operated (ADA)		8	2		3,650	100		3,750	4,175
0140	Sensor operated (ADA), surface mounted	↓	8	2	↓	4,400	100		4,500	5,000
0150	With paper towels									
0180	Sensor operated	Q-1	8	2	Ea.	3,550	100		3,650	4,050

22 42 16.14 Lavatories

		Crew	Daily Output	Labor-Hours	Unit	Material	Labor	Equipment	Total	Total Incl O&P
0010	**LAVATORIES**, With trim, white unless noted otherwise									
0020	Commercial lavatories same as residential. See Section 22 41 16.10									

22 42 16.20 Commercial Sinks

		Crew	Daily Output	Labor-Hours	Unit	Material	Labor	Equipment	Total	Total Incl O&P
0010	**COMMERCIAL SINKS**									
5900	Scullery sink, stainless steel									
5910	1 bowl and drain board, 43" x 22" O.D.	Q-1	5.40	2.963	Ea.	3,800	149		3,949	4,400
5920	2 bowls and drain board, 49" x 22" O.D.		4.60	3.478		6,225	175		6,400	7,125
5930	3 bowls and drain board, 43" x 22" O.D.		4.20	3.810		6,375	191		6,566	7,325
5940	1 bowl and drain board, 50" x 28" O.D., with legs	↓	5.40	2.963	↓	2,825	149		2,974	3,350

22 42 16.30 Classroom Sinks

		Crew	Daily Output	Labor-Hours	Unit	Material	Labor	Equipment	Total	Total Incl O&P
0010	**CLASSROOM SINKS**									
6020	Countertop, stainless steel									
6024	with faucet, bubbler and strainer, ADA compliant									
6036	25" x 17" single bowl	Q-1	5.20	3.077	Ea.	1,200	155		1,355	1,550
6040	28" x 22" single bowl		5.20	3.077		1,225	155		1,380	1,575
6044	31" x 19" single bowl		5.20	3.077		1,325	155		1,480	1,700
6070	37" x 17" double bowl		4.40	3.636		1,725	183		1,908	2,175
6100	For rough-in, supply, waste and vent, counter top classroom sinks	↓	2.14	7.477	↓	450	375		825	1,050

22 42 16.34 Laboratory Countertops and Sinks

		Crew	Daily Output	Labor-Hours	Unit	Material	Labor	Equipment	Total	Total Incl O&P
0010	**LABORATORY COUNTERTOPS AND SINKS**									
0050	Laboratory sinks, corrosion resistant									
1000	Stainless steel sink, bench mounted, with									
1020	plug & waste fitting with 1-1/2" straight threads									
1030	Single bowl, 2 drainboards, backnut & strainer									
1050	18-1/2" x 15-1/2" x 12-1/2" sink, 54" x 24" O.D.	Q-1	3	5.333	Ea.	1,650	268		1,918	2,225
1100	Single bowl, single drainboard, backnut & strainer									
1130	18-1/2" x 15-1/2" x 12-1/2" sink, 47" x 24" O.D.	Q-1	3	5.333	Ea.	1,175	268		1,443	1,675
1146	Double bowl, single drainboard, backnut & strainer									
1150	18-1/2" x 15-1/2" x 12-1/2" sink, 70" x 24" O.D.	Q-1	3	5.333	Ea.	1,825	268		2,093	2,425
1280	Polypropylene									
1290	Flanged 1-1/4" wide, rectangular with strainer									
1300	plug & waste fitting, 1-1/2" straight threads									
1320	12" x 12" x 8" sink, 14-1/2" x 14-1/2" O.D.	Q-1	4	4	Ea.	228	201		429	555
1340	16" x 16" x 8" sink, 18-1/2" x 18-1/2" O.D.		4	4		320	201		521	655
1360	21" x 18" x 10" sink, 23-1/2" x 20-1/2" O.D.		4	4		385	201		586	730
1490	For rough-in, supply, waste & vent, add	↓	2.02	7.921	↓	214	400		614	835
1600	Polypropylene									
1620	Cup sink, oval, integral strainers									

22 42 Commercial Plumbing Fixtures

22 42 16 – Commercial Lavatories and Sinks

22 42 16.34 Laboratory Countertops and Sinks		Crew	Daily Output	Labor-Hours	Unit	2013 Bare Costs Material	Labor	Equipment	Total	Total Incl O&P
1640	6" x 3" I.D., 7" x 4" O.D.	Q-1	6	2.667	Ea.	113	134		247	325
1660	9" x 3" I.D., 10" x 4-1/2" O.D.	"	6	2.667		133	134		267	350
1740	1-1/2" diam. x 11" long					50			50	55
1980	For rough-in, supply, waste & vent, add	Q-1	1.70	9.412	↓	224	475		699	955

22 42 16.40 Service Sinks

		Crew	Daily Output	Labor-Hours	Unit	Material	Labor	Equipment	Total	Total Incl O&P
0010	**SERVICE SINKS**									
6650	Service, floor, corner, P.E. on C.I., 28" x 28"	Q-1	4.40	3.636	Ea.	965	183		1,148	1,325
6750	Vinyl coated rim guard, add					67.50			67.50	74
6760	Mop sink, molded stone, 24" x 36"	1 Plum	3.33	2.402		273	134		407	500
6770	Mop sink, molded stone, 24" x 36", w/rim 3 sides	"	3.33	2.402		278	134		412	505
6790	For rough-in, supply, waste & vent, floor service sinks	Q-1	1.64	9.756		1,400	490		1,890	2,275
7000	Service, wall, P.E. on C.I., roll rim, 22" x 18"		4	4		735	201		936	1,125
7100	24" x 20"	↓	4	4		810	201		1,011	1,200
7600	For stainless steel rim guard, two sides only, add					80			80	88
7800	For stainless steel rim guard, front only, add					55			55	60.50
8600	Vitreous china, 22" x 20"	Q-1	4	4		565	201		766	925
8960	For stainless steel rim guard, front or one side, add					43.50			43.50	48
8980	For rough-in, supply, waste & vent, wall service sinks	Q-1	1.30	12.308	↓	1,700	620		2,320	2,800

22 42 23 – Commercial Showers

22 42 23.30 Group Showers

		Crew	Daily Output	Labor-Hours	Unit	Material	Labor	Equipment	Total	Total Incl O&P
0010	**GROUP SHOWERS**									
6000	Group, w/pressure balancing valve, rough-in and rigging not included									
6800	Column, 6 heads, no receptors, less partitions	Q-1	3	5.333	Ea.	3,000	268		3,268	3,700
6900	With stainless steel partitions		1	16		9,250	805		10,055	11,400
7600	5 heads, no receptors, less partitions		3	5.333		2,475	268		2,743	3,125
7620	4 heads (1 handicap) no receptors, less partitions ♿		3	5.333		3,975	268		4,243	4,775
7700	With stainless steel partitions		1	16		7,600	805		8,405	9,550
8000	Wall, 2 heads, no receptors, less partitions		4	4		1,375	201		1,576	1,825
8100	With stainless steel partitions	↓	2	8	↓	3,225	400		3,625	4,150

22 42 33 – Wash Fountains

22 42 33.20 Commercial Wash Fountains

		Crew	Daily Output	Labor-Hours	Unit	Material	Labor	Equipment	Total	Total Incl O&P
0010	**COMMERCIAL WASH FOUNTAINS** D2010–610									
1900	Group, foot control									
2000	Precast terrazzo, circular, 36" diam., 5 or 6 persons	Q-2	3	8	Ea.	5,450	415		5,865	6,625
2100	54" diameter for 8 or 10 persons		2.50	9.600		6,800	500		7,300	8,225
2400	Semi-circular, 36" diam. for 3 persons		3	8		4,775	415		5,190	5,875
2500	54" diam. for 4 or 5 persons		2.50	9.600		6,050	500		6,550	7,400
2700	Quarter circle (corner), 54" for 3 persons		3.50	6.857		6,000	355		6,355	7,150
3000	Stainless steel, circular, 36" diameter		3.50	6.857		5,450	355		5,805	6,550
3100	54" diameter		2.80	8.571		7,050	445		7,495	8,425
3400	Semi-circular, 36" diameter		3.50	6.857		4,750	355		5,105	5,775
3500	54" diameter		2.80	8.571		6,000	445		6,445	7,275
5000	Thermoplastic, pre-assembled, circular, 36" diameter		6	4		3,925	208		4,133	4,650
5100	54" diameter		4	6		4,575	310		4,885	5,500
5400	Semi-circular, 36" diameter		6	4		3,000	208		3,208	3,650
5600	54" diameter	↓	4	6	↓	4,400	310		4,710	5,325
5610	Group, infrared control, barrier free ♿									
5614	Precast terrazzo									
5620	Semi-circular 36" diam. for 3 persons	Q-2	3	8	Ea.	7,150	415		7,565	8,500
5630	46" diam. for 4 persons ♿		2.80	8.571		7,725	445		8,170	9,175
5640	Circular, 54" diam. for 8 persons, button control	↓	2.50	9.600		9,125	500		9,625	10,800

22 42 Commercial Plumbing Fixtures

22 42 33 – Wash Fountains

22 42 33.20 Commercial Wash Fountains		Crew	Daily Output	Labor-Hours	Unit	2013 Bare Costs Material	Labor	Equipment	Total	Total Incl O&P
5700	Rough-in, supply, waste and vent for above wash fountains	Q-1	1.82	8.791	Ea.	670	440		1,110	1,400
6200	Duo for small washrooms, stainless steel		2	8		2,725	400		3,125	3,575
6400	Bowl with backsplash		2	8		2,250	400		2,650	3,075
6500	Rough-in, supply, waste & vent for duo fountains		2.02	7.921		400	400		800	1,050

22 42 39 – Commercial Faucets, Supplies, and Trim

22 42 39.10 Faucets and Fittings		Crew	Daily Output	Labor-Hours	Unit	Material	Labor	Equipment	Total	Total Incl O&P
0010	**FAUCETS AND FITTINGS**									
0840	Flush valves, with vacuum breaker									
0850	Water closet									
0860	Exposed, rear spud	1 Plum	8	1	Ea.	137	56		193	235
0870	Top spud		8	1		126	56		182	223
0880	Concealed, rear spud		8	1		180	56		236	282
0890	Top spud		8	1		149	56		205	248
0900	Wall hung		8	1		159	56		215	259
0910	Dual flush flushometer		12	.667		197	37		234	272
0912	Flushometer retrofit kit		18	.444		14.75	25		39.75	54
0920	Urinal									
0930	Exposed, stall	1 Plum	8	1	Ea.	126	56		182	223
0940	Wall, (washout)		8	1		126	56		182	223
0950	Pedestal, top spud		8	1		129	56		185	226
0960	Concealed, stall		8	1		145	56		201	243
0970	Wall (washout)		8	1		156	56		212	256
0971	Automatic flush sensor and operator for G									
0972	urinals or water closets, standard G	1 Plum	8	1	Ea.	415	56		471	545
0980	High efficiency water saving									
0984	Water closets, 1.28 gpf G	1 Plum	8	1	Ea.	415	56		471	545
0988	Urinals, .5 gpf G	"	8	1	"	415	56		471	545
2790	Faucets for lavatories									
2800	Self-closing, center set	1 Plum	10	.800	Ea.	131	44.50		175.50	211
2810	Automatic sensor and operator, with faucet head		6.15	1.301		370	72.50		442.50	520
3000	Service sink faucet, cast spout, pail hook, hose end		14	.571		80	32		112	136

22 42 39.30 Carriers and Supports		Crew	Daily Output	Labor-Hours	Unit	Material	Labor	Equipment	Total	Total Incl O&P
0010	**CARRIERS AND SUPPORTS**, For plumbing fixtures									
0500	Drinking fountain, wall mounted									
0600	Plate type with studs, top back plate	1 Plum	7	1.143	Ea.	83.50	64		147.50	188
0700	Top front and back plate		7	1.143		102	64		166	208
0800	Top & bottom, front & back plates, w/bearing jacks		7	1.143		148	64		212	258
3000	Lavatory, concealed arm									
3050	Floor mounted, single									
3100	High back fixture	1 Plum	6	1.333	Ea.	465	74.50		539.50	620
3200	Flat slab fixture		6	1.333		400	74.50		474.50	550
3220	Paraplegic		6	1.333		515	74.50		589.50	680
3250	Floor mounted, back to back									
3300	High back fixtures	1 Plum	5	1.600	Ea.	660	89.50		749.50	860
3400	Flat slab fixtures		5	1.600		810	89.50		899.50	1,025
3430	Paraplegic		5	1.600		755	89.50		844.50	965
3500	Wall mounted, in stud or masonry									
3600	High back fixture	1 Plum	6	1.333	Ea.	274	74.50		348.50	410
3700	Flat slab fixture	"	6	1.333	"	237	74.50		311.50	370
4000	Exposed arm type, floor mounted									
4100	Single high back or flat slab fixture	1 Plum	6	1.333	Ea.	660	74.50		734.50	835
4200	Back to back, high back or flat slab fixtures		5	1.600		1,025	89.50		1,114.50	1,275

22 42 39 – Commercial Faucets, Supplies, and Trim

22 42 39.30 Carriers and Supports		Crew	Daily Output	Labor-Hours	Unit	Material	Labor	Equipment	Total	Total Incl O&P
						2013 Bare Costs				
4300	Wall mounted, High back or flat slab lavatory	1 Plum	6	1.333	Ea.	455	74.50		529.50	615
4600	Sink, floor mounted									
4650	Exposed arm system									
4700	Single heavy fixture	1 Plum	5	1.600	Ea.	770	89.50		859.50	985
4750	Single heavy sink with slab		5	1.600		990	89.50		1,079.50	1,225
4800	Back to back, standard fixtures		5	1.600		570	89.50		659.50	760
4850	Back to back, heavy fixtures		5	1.600		875	89.50		964.50	1,100
4900	Back to back, heavy sink with slab		5	1.600		875	89.50		964.50	1,100
4950	Exposed offset arm system									
5000	Single heavy deep fixture	1 Plum	5	1.600	Ea.	735	89.50		824.50	945
5100	Plate type system									
5200	With bearing jacks, single fixture	1 Plum	5	1.600	Ea.	845	89.50		934.50	1,075
5300	With exposed arms, single heavy fixture		5	1.600		1,075	89.50		1,164.50	1,325
5400	Wall mounted, exposed arms, single heavy fixture		5	1.600		395	89.50		484.50	570
6000	Urinal, floor mounted, 2″ or 3″ coupling, blowout type		6	1.333		490	74.50		564.50	650
6100	With fixture or hanger bolts, blowout or washout		6	1.333		350	74.50		424.50	495
6200	With bearing plate		6	1.333		390	74.50		464.50	540
6300	Wall mounted, plate type system		6	1.333		300	74.50		374.50	445
6980	Water closet, siphon jet									
7000	Horizontal, adjustable, caulk									
7040	Single, 4″ pipe size	1 Plum	5.33	1.501	Ea.	655	84		739	845
7050	4″ pipe size, paraplegic		5.33	1.501		655	84		739	845
7060	5″ pipe size		5.33	1.501		845	84		929	1,050
7100	Double, 4″ pipe size		5	1.600		1,225	89.50		1,314.50	1,475
7110	4″ pipe size, paraplegic		5	1.600		1,225	89.50		1,314.50	1,475
7120	5″ pipe size		5	1.600		1,475	89.50		1,564.50	1,750
7160	Horizontal, adjustable, extended, caulk									
7180	Single, 4″ pipe size	1 Plum	5.33	1.501	Ea.	885	84		969	1,100
7200	5″ pipe size		5.33	1.501		1,125	84		1,209	1,350
7240	Double, 4″ pipe size		5	1.600		1,550	89.50		1,639.50	1,825
7260	5″ pipe size		5	1.600		1,875	89.50		1,964.50	2,175
7400	Vertical, adjustable, caulk or thread									
7440	Single, 4″ pipe size	1 Plum	5.33	1.501	Ea.	905	84		989	1,125
7460	5″ pipe size		5.33	1.501		1,125	84		1,209	1,350
7480	6″ pipe size		5	1.600		1,275	89.50		1,364.50	1,525
7520	Double, 4″ pipe size		5	1.600		1,350	89.50		1,439.50	1,600
7540	5″ pipe size		5	1.600		1,550	89.50		1,639.50	1,825
7560	6″ pipe size		4	2		1,725	112		1,837	2,075
7600	Vertical, adjustable, extended, caulk									
7620	Single, 4″ pipe size	1 Plum	5.33	1.501	Ea.	905	84		989	1,125
7640	5″ pipe size		5.33	1.501		1,125	84		1,209	1,350
7680	6″ pipe size		5	1.600		1,275	89.50		1,364.50	1,525
7720	Double, 4″ pipe size		5	1.600		1,350	89.50		1,439.50	1,600
7740	5″ pipe size		5	1.600		1,550	89.50		1,639.50	1,825
7760	6″ pipe size		4	2		1,725	112		1,837	2,075
7780	Water closet, blow out									
7800	Vertical offset, caulk or thread									
7820	Single, 4″ pipe size	1 Plum	5.33	1.501	Ea.	665	84		749	860
7840	Double, 4″ pipe size	″	5	1.600	″	1,150	89.50		1,239.50	1,375
7880	Vertical offset, extended, caulk									
7900	Single, 4″ pipe size	1 Plum	5.33	1.501	Ea.	835	84		919	1,050
7920	Double, 4″ pipe size	″	5	1.600	″	1,300	89.50		1,389.50	1,575
7960	Vertical, for floor mounted back-outlet									

22 42 Commercial Plumbing Fixtures

22 42 39 – Commercial Faucets, Supplies, and Trim

	22 42 39.30 Carriers and Supports	Crew	Daily Output	Labor-Hours	Unit	2013 Bare Costs Material	Labor	Equipment	Total	Total Incl O&P
7980	Single, 4″ thread, 2″ vent	1 Plum	5.33	1.501	Ea.	585	84		669	770
8000	Double, 4″ thread, 2″ vent	″	6	1.333	″	1,725	74.50		1,799.50	2,000
8040	Vertical, for floor mounted back-outlet, extended									
8060	Single, 4″ caulk, 2″ vent	1 Plum	6	1.333	Ea.	585	74.50		659.50	755
8080	Double, 4″ caulk, 2″ vent	″	6	1.333	″	1,725	74.50		1,799.50	2,000
8200	Water closet, residential									
8220	Vertical centerline, floor mount									
8240	Single, 3″ caulk, 2″ or 3″ vent	1 Plum	6	1.333	Ea.	535	74.50		609.50	695
8260	4″ caulk, 2″ or 4″ vent		6	1.333		690	74.50		764.50	865
8280	3″ copper sweat, 3″ vent		6	1.333		480	74.50		554.50	635
8300	4″ copper sweat, 4″ vent	↓	6	1.333	↓	580	74.50		654.50	750
8400	Vertical offset, floor mount									
8420	Single, 3″ or 4″ caulk, vent	1 Plum	4	2	Ea.	665	112		777	905
8440	3″ or 4″ copper sweat, vent		5	1.600		665	89.50		754.50	870
8460	Double, 3″ or 4″ caulk, vent		4	2		1,150	112		1,262	1,425
8480	3″ or 4″ copper sweat, vent	↓	5	1.600	↓	1,150	89.50		1,239.50	1,375
9000	Water cooler (electric), floor mounted									
9100	Plate type with bearing plate, single	1 Plum	6	1.333	Ea.	335	74.50		409.50	480
9140	Plate type with bearing plate, back to back	″	4	2	″	470	112		582	690

22 43 Healthcare Plumbing Fixtures

22 43 13 – Healthcare Water Closets

	22 43 13.40 Water Closets	Crew	Daily Output	Labor-Hours	Unit	Material	Labor	Equipment	Total	Total Incl O&P
0010	**WATER CLOSETS**									
1000	Bowl only, 1 piece, w/seat and flush valve, ADA compliant 18″ high									
1030	Floor mounted									
1150	With wall outlet	Q-1	5.30	3.019	Ea.	385	152		537	655
1180	For rough-in, supply, waste and vent		2.84	5.634		400	283		683	865
1200	With floor outlet		5.30	3.019		325	152		477	590
1800	For rough-in, supply, waste and vent		3.05	5.246		320	264		584	750
3100	Wall hung	↓	5.80	2.759	↓	705	139		844	990
3150	Hospital type, slotted rim for bed pan									
3156	Elongated bowl, top spud	Q-1	5.80	2.759	Ea.	485	139		624	745
3160	Elongated bowl, rear spud		5.80	2.759		340	139		479	580
3200	For rough-in, supply, waste and vent, single WC	↓	2.56	6.250	↓	895	315		1,210	1,450
3300	Floor mounted									
3320	Bariatric, (1,200 lb. capacity), elongated bowl, ADA	Q-1	4.60	3.478	Ea.	2,775	175		2,950	3,325
3360	Hospital type, slotted rim for bed pan									
3370	Elongated bowl, top spud	Q-1	5	3.200	Ea.	360	161		521	640
3380	Elongated bowl, rear spud		5	3.200		385	161		546	665
3500	For rough-in, supply, waste and vent	↓	3.05	5.246	↓	320	264		584	750

22 43 16 – Healthcare Sinks

	22 43 16.10 Sinks	Crew	Daily Output	Labor-Hours	Unit	Material	Labor	Equipment	Total	Total Incl O&P
0010	**SINKS**									
0020	Vitreous china									
6702	Hospital type, without trim (see Section 22 41 39.10)									
6710	20″ x 18″, contoured splash shield	Q-1	8	2	Ea.	89.50	100		189.50	249
6730	28″ x 20″, surgeon, side decks		8	2		555	100		655	765
6740	28″ x 22″, surgeon scrub-up, deep bowl		8	2		860	100		960	1,100
6750	20″ x 27″, patient, wheelchair		7	2.286		475	115		590	695

22 43 Healthcare Plumbing Fixtures

22 43 16 – Healthcare Sinks

22 43 16.10 Sinks		Crew	Daily Output	Labor-Hours	Unit	Material	2013 Bare Costs Labor	Equipment	Total	Total Incl O&P
6760	30" x 22", all purpose	Q-1	7	2.286	Ea.	710	115		825	955
6770	30" x 22", plaster work		7	2.286		710	115		825	955
6820	20" x 24" clinic service, liquid/solid waste		6	2.667		945	134		1,079	1,250

22 43 19 – Healthcare Bathtubs

22 43 19.10 Bathtubs

		Crew	Daily Output	Labor-Hours	Unit	Material	Labor	Equipment	Total	Total Incl O&P
0010	**BATHTUBS**									
5002	Hospital type, with trim (see Section 22 41 39.10)									
5050	Bathing pool, porcelain enamel on cast iron, grab bars									
5060	pop-up drain, 72" x 36"	Q-1	3	5.333	Ea.	3,275	268		3,543	4,000
5100	Perineal (sitz), vitreous china		3	5.333		1,250	268		1,518	1,775
5120	For pedestal, vitreous china, add		8	2		236	100		336	410
5180	Pier tub, porcelain enamel on cast iron, 66-3/4" x 30"		3	5.333		2,825	268		3,093	3,500
5200	Base, porcelain enamel on cast iron		8	2		1,325	100		1,425	1,600
5300	Whirlpool, porcelain enamel on cast iron, 72" x 36"		1	16		3,100	805		3,905	4,625
5310	For color add					5%				
5311	For designer colors and trim add					15%				

22 43 23 – Healthcare Showers

22 43 23.10 Showers

		Crew	Daily Output	Labor-Hours	Unit	Material	Labor	Equipment	Total	Total Incl O&P
0010	**SHOWERS**									
5950	Module, handicap, SS panel, fixed & hand held head, control									
5960	valves, grab bar, curtain & rod, folding seat	1 Plum	4	2	Ea.	1,900	112		2,012	2,275

22 43 39 – Healthcare Faucets

22 43 39.10 Faucets and Fittings

		Crew	Daily Output	Labor-Hours	Unit	Material	Labor	Equipment	Total	Total Incl O&P
0010	**FAUCETS AND FITTINGS**									
2850	Medical, bedpan cleanser, with pedal valve,	1 Plum	12	.667	Ea.	785	37		822	920
2860	With screwdriver stop valve		12	.667		405	37		442	500
2870	With self-closing spray valve		12	.667		250	37		287	330
2900	Faucet, gooseneck spout, wrist handles, grid drain		10	.800		183	44.50		227.50	268
2940	Mixing valve, knee action, screwdriver stops		4	2		440	112		552	650

22 45 Emergency Plumbing Fixtures

22 45 13 – Emergency Showers

22 45 13.10 Emergency Showers

		Crew	Daily Output	Labor-Hours	Unit	Material	Labor	Equipment	Total	Total Incl O&P
0010	**EMERGENCY SHOWERS**, Rough-in not included									
5000	Shower, single head, drench, ball valve, pull, freestanding	Q-1	4	4	Ea.	360	201		561	700
5200	Horizontal or vertical supply		4	4		520	201		721	875
6000	Multi-nozzle, eye/face wash combination		4	4		625	201		826	995
6400	Multi-nozzle, 12 spray, shower only		4	4		1,950	201		2,151	2,450
6600	For freeze-proof, add		6	2.667		440	134		574	685
8000	Walk-thru decontamination with eye-face wash		2	8		3,275	400		3,675	4,225
8200	For freeze proof, add		4	4		540	201		741	895

22 45 16 – Eyewash Equipment

22 45 16.10 Eyewash Safety Equipment

		Crew	Daily Output	Labor-Hours	Unit	Material	Labor	Equipment	Total	Total Incl O&P
0010	**EYEWASH SAFETY EQUIPMENT**, Rough-in not included									
1000	Eye wash fountain									
1400	Plastic bowl, pedestal mounted	Q-1	4	4	Ea.	269	201		470	600
1600	Unmounted		4	4		234	201		435	560
1800	Wall mounted		4	4		435	201		636	785
2000	Stainless steel, pedestal mounted		4	4		415	201		616	760

22 45 Emergency Plumbing Fixtures

22 45 16 – Eyewash Equipment

22 45 16.10 Eyewash Safety Equipment		Crew	Daily Output	Labor-Hours	Unit	Material	Labor	Equipment	Total	Total Incl O&P
						2013 Bare Costs				
2200	Unmounted	Q-1	4	4	Ea.	259	201		460	590
2400	Wall mounted	↓	4	4	↓	300	201		501	635

22 45 19 – Self-Contained Eyewash Equipment

22 45 19.10 Self-Contained Eyewash Safety Equipment		Crew	Daily Output	Labor-Hours	Unit	Material	Labor	Equipment	Total	Total Incl O&P
0010	**SELF-CONTAINED EYEWASH SAFETY EQUIPMENT**									
3000	Eye wash, portable, self-contained				Ea.	990			990	1,100

22 45 26 – Eye/Face Wash Equipment

22 45 26.10 Eye/Face Wash Safety Equipment		Crew	Daily Output	Labor-Hours	Unit	Material	Labor	Equipment	Total	Total Incl O&P
0010	**EYE/FACE WASH SAFETY EQUIPMENT**, Rough-in not included									
4000	Eye and face wash, combination fountain									
4200	Stainless steel, pedestal mounted	Q-1	4	4	Ea.	1,025	201		1,226	1,425
4400	Unmounted		4	4		259	201		460	590
4600	Wall mounted	↓	4	4	↓	275	201		476	610

22 46 Security Plumbing Fixtures

22 46 13 – Security Water Closets and Urinals

22 46 13.10 Security Water Closets and Urinals		Crew	Daily Output	Labor-Hours	Unit	Material	Labor	Equipment	Total	Total Incl O&P
0010	**SECURITY WATER CLOSETS AND URINALS**, Stainless steel									
2000	Urinal, back supply and flush									
2200	Wall hung	Q-1	4	4	Ea.	1,225	201		1,426	1,650
2240	Stall		2.50	6.400		2,100	320		2,420	2,775
2300	For urinal rough-in, supply, waste and vent	↓	1.49	10.738	↓	380	540		920	1,225
3000	Water closet, integral seat, back supply and flush									
3300	Wall hung, wall outlet	Q-1	5.80	2.759	Ea.	985	139		1,124	1,275
3400	Floor mount, wall outlet		5.80	2.759		1,275	139		1,414	1,600
3440	Floor mount, floor outlet	↓	5.80	2.759		1,300	139		1,439	1,625
3480	For recessed tissue holder, add					88			88	96.50
3500	For water closet rough-in, supply, waste and vent	Q-1	1.19	13.445	↓	350	675		1,025	1,400
5000	Water closet and lavatory units, push button filler valves,									
5010	soap & paper holders, seat									
5300	Wall hung	Q-1	5	3.200	Ea.	2,100	161		2,261	2,575
5400	Floor mount		5	3.200		2,000	161		2,161	2,450
6300	For unit rough-in, supply, waste and vent	↓	1	16	↓	410	805		1,215	1,650

22 46 16 – Security Lavatories and Sinks

22 46 16.10 Security Lavatories		Crew	Daily Output	Labor-Hours	Unit	Material	Labor	Equipment	Total	Total Incl O&P
0010	**SECURITY LAVATORIES**, Stainless steel									
1000	Lavatory, wall hung, push button filler valve									
1100	Rectangular bowl	Q-1	8	2	Ea.	1,075	100		1,175	1,350
1200	Oval bowl		8	2		1,125	100		1,225	1,375
1240	Oval bowl, corner mount		8	2		1,225	100		1,325	1,500
1300	For lavatory rough-in, supply, waste and vent	↓	1.50	10.667	↓	485	535		1,020	1,350

22 46 63 – Security Service Sink

22 46 63.10 Security Service Sink		Crew	Daily Output	Labor-Hours	Unit	Material	Labor	Equipment	Total	Total Incl O&P
0010	**SECURITY SERVICE SINK**, Stainless steel									
1700	Service sink, with soap dish									
1740	24" x 19" size	Q-1	3	5.333	Ea.	2,175	268		2,443	2,800
1790	For sink rough-in, supply, waste and vent	"	.89	17.978	"	1,175	905		2,080	2,625

22 46 Security Plumbing Fixtures

22 46 73 – Security Shower

22 46 73.10 Security Shower		Crew	Daily Output	Labor-Hours	Unit	Material	2013 Bare Costs Labor	Equipment	Total	Total Incl O&P
0010	**SECURITY SHOWER**, Stainless steel									
1800	Shower cabinet, unitized									
1840	36" x 36" x 88"	Q-1	2.20	7.273	Ea.	4,950	365		5,315	6,000
1900	Shower package for built-in									
1940	Hot & cold valves, recessed soap dish	Q-1	6	2.667	Ea.	540	134		674	790

22 47 Drinking Fountains and Water Coolers

22 47 13 – Drinking Fountains

22 47 13.10 Drinking Water Fountains

Line	Description	Crew	Daily Output	Labor-Hours	Unit	Material	Labor	Equipment	Total	Total Incl O&P
0010	**DRINKING WATER FOUNTAINS**, For connection to cold water supply									
0802	For remote water chiller, see Section 22 47 23.10									
1000	Wall mounted, non-recessed R224000-30									
1200	Aluminum,									
1280	Dual bubbler type R224000-50	1 Plum	3.20	2.500	Ea.	2,300	140		2,440	2,725
1400	Bronze, with no back		4	2		1,025	112		1,137	1,325
1600	Cast iron, enameled, low back, single bubbler D2010–810		4	2		900	112		1,012	1,150
1640	Dual bubbler type		3.20	2.500		1,450	140		1,590	1,800
1680	Triple bubbler type		3.20	2.500		1,875	140		2,015	2,275
1800	Cast aluminum, enameled, for correctional institutions		4	2		1,375	112		1,487	1,675
2000	Fiberglass, 12" back, single bubbler unit		4	2		1,550	112		1,662	1,875
2040	Dual bubbler		3.20	2.500		2,100	140		2,240	2,525
2080	Triple bubbler		3.20	2.500		2,325	140		2,465	2,750
2200	Polymarble, no back, single bubbler		4	2		705	112		817	945
2240	Dual bubbler		3.20	2.500		1,700	140		1,840	2,075
2280	Triple bubbler		3.20	2.500		2,025	140		2,165	2,425
2400	Precast stone, no back		4	2		915	112		1,027	1,175
2700	Stainless steel, single bubbler, no back		4	2		1,050	112		1,162	1,325
2740	With back		4	2		540	112		652	760
2780	Dual handle & wheelchair projection type		4	2		730	112		842	975
2820	Dual level for handicapped type		3.20	2.500		1,500	140		1,640	1,825
2840	Vandal resistant type		4	2		650	112		762	885
3300	Vitreous china									
3340	7" back	1 Plum	4	2	Ea.	610	112		722	840
3940	For vandal-resistant bottom plate, add					73.50			73.50	81
3960	For freeze-proof valve system, add	1 Plum	2	4		650	223		873	1,050
3980	For rough-in, supply and waste, add	"	2.21	3.620		345	202		547	685
4000	Wall mounted, semi-recessed									
4200	Poly-marble, single bubbler	1 Plum	4	2	Ea.	925	112		1,037	1,200
4600	Stainless steel, satin finish, single bubbler		4	2		1,125	112		1,237	1,425
4900	Vitreous china, single bubbler		4	2		885	112		997	1,150
5980	For rough-in, supply and waste, add		1.83	4.372		345	244		589	745
6000	Wall mounted, fully recessed									
6400	Poly-marble, single bubbler	1 Plum	4	2	Ea.	1,600	112		1,712	1,925
6440	For water glass filler, add					80			80	88
6800	Stainless steel, single bubbler	1 Plum	4	2		1,525	112		1,637	1,850
6900	Fountain and cuspidor combination		2	4		2,950	223		3,173	3,575
7560	For freeze-proof valve system, add		2	4		780	223		1,003	1,200
7580	For rough-in, supply and waste, add		1.83	4.372		345	244		589	745
7600	Floor mounted, pedestal type									
7700	Aluminum, architectural style, C.I. base	1 Plum	2	4	Ea.	2,250	223		2,473	2,800

22 47 Drinking Fountains and Water Coolers

22 47 13 – Drinking Fountains

22 47 13.10 Drinking Water Fountains		Crew	Daily Output	Labor-Hours	Unit	2013 Bare Costs Material	Labor	Equipment	Total	Total Incl O&P
7780	Wheelchair handicap unit	1 Plum	2	4	Ea.	1,575	223		1,798	2,075
8000	Bronze, architectural style		2	4		1,925	223		2,148	2,450
8040	Enameled steel cylindrical column style		2	4		2,075	223		2,298	2,600
8200	Precast stone/concrete, cylindrical column		1	8		1,325	445		1,770	2,125
8240	Wheelchair handicap unit		1	8		2,725	445		3,170	3,675
8400	Stainless steel, architectural style		2	4		1,750	223		1,973	2,250
8600	Enameled iron, heavy duty service, 2 bubblers		2	4		2,575	223		2,798	3,175
8660	4 bubblers		2	4		3,950	223		4,173	4,650
8880	For freeze-proof valve system, add		2	4		650	223		873	1,050
8900	For rough-in, supply and waste, add		1.83	4.372		345	244		589	745
9100	Deck mounted									
9500	Stainless steel, circular receptor	1 Plum	4	2	Ea.	415	112		527	625
9540	14″ x 9″ receptor		4	2		345	112		457	545
9580	25″ x 17″ deep receptor, with water glass filler		3	2.667		294	149		443	550
9760	White enameled steel, 14″ x 9″ receptor		4	2		360	112		472	565
9860	White enameled cast iron, 24″ x 16″ receptor		3	2.667		430	149		579	695
9980	For rough-in, supply and waste, add		1.83	4.372		345	244		589	745

22 47 16 – Pressure Water Coolers

22 47 16.10 Electric Water Coolers		Crew	Daily Output	Labor-Hours	Unit	Material	Labor	Equipment	Total	Total Incl O&P
0010	**ELECTRIC WATER COOLERS** D2010–820									
0100	Wall mounted, non-recessed									
0140	4 GPH	Q-1	4	4	Ea.	680	201		881	1,050
0160	8 GPH, barrier free, sensor operated		4	4		1,025	201		1,226	1,450
0180	8.2 GPH		4	4		725	201		926	1,100
0220	14.3 GPH		4	4		830	201		1,031	1,225
0600	8 GPH hot and cold water		4	4		1,175	201		1,376	1,600
0640	For stainless steel cabinet, add R224000-50					84			84	92
1000	Dual height, 8.2 GPH	Q-1	3.80	4.211		1,975	211		2,186	2,500
1040	14.3 GPH	″	3.80	4.211		975	211		1,186	1,400
1240	For stainless steel cabinet, add					171			171	188
2600	Wheelchair type, 8 GPH	Q-1	4	4		915	201		1,116	1,300
3000	Simulated recessed, 8 GPH		4	4		630	201		831	995
3040	11.5 GPH		4	4		1,050	201		1,251	1,450
3200	For glass filler, add					99			99	109
3240	For stainless steel cabinet, add					73.50			73.50	81
3300	Semi-recessed, 8.1 GPH	Q-1	4	4		715	201		916	1,100
3320	12 GPH	″	4	4		820	201		1,021	1,200
3340	For glass filler, add					129			129	142
3360	For stainless steel cabinet, add					145			145	160
3400	Full recessed, stainless steel, 8 GPH	Q-1	3.50	4.571		1,925	230		2,155	2,475
3420	11.5 GPH	″	3.50	4.571		2,200	230		2,430	2,775
3460	For glass filler, add					139			139	153
3600	For mounting can only					243			243	268
4600	Floor mounted, flush-to-wall									
4640	4 GPH	1 Plum	3	2.667	Ea.	705	149		854	1,000
4680	8.2 GPH		3	2.667		740	149		889	1,050
4720	14.3 GPH		3	2.667		850	149		999	1,150
4960	14 GPH hot and cold water		3	2.667		1,050	149		1,199	1,375
4980	For stainless steel cabinet, add					128			128	141
5000	Dual height, 8.2 GPH	1 Plum	2	4		1,100	223		1,323	1,550
5040	14.3 GPH	″	2	4		1,150	223		1,373	1,600
5120	For stainless steel cabinet, add					186			186	205

22 47 Drinking Fountains and Water Coolers

22 47 16 – Pressure Water Coolers

22 47 16.10 Electric Water Coolers		Crew	Daily Output	Labor-Hours	Unit	2013 Bare Costs Material	Labor	Equipment	Total	Total Incl O&P
5600	Explosion Proof, 16 GPH	1 Plum	3	2.667	Ea.	1,975	149		2,124	2,400
6000	Refrigerator Compartment Type, 4.5 GPH		3	2.667		1,425	149		1,574	1,775
6600	Bottle Supply Type, 1.0 GPH		4	2		375	112		487	585
6640	Hot and cold, 1.0 GPH		4	2		510	112		622	730
9800	For supply, waste & vent, all coolers		2.21	3.620		345	202		547	685

22 47 23 – Remote Water Coolers

22 47 23.10 Remote Water Coolers		Crew	Daily Output	Labor-Hours	Unit	Material	Labor	Equipment	Total	Total Incl O&P
0010	**REMOTE WATER COOLERS**, 80°F inlet									
0100	Air cooled, 50°F outlet, 115 V, 4.1 GPH	1 Plum	6	1.333	Ea.	525	74.50		599.50	690
0200	5.7 GPH		5.50	1.455		835	81		916	1,050
0300	8.0 GPH		5	1.600		625	89.50		714.50	825
0400	10.0 GPH		4.50	1.778		920	99		1,019	1,150
0500	13.4 GPH		4	2		1,425	112		1,537	1,750
0700	29 GPH	Q-1	5	3.200		1,675	161		1,836	2,100
1000	230V, 32 GPH	"	5	3.200		1,750	161		1,911	2,175

22 51 Swimming Pool Plumbing Systems

22 51 19 – Swimming Pool Water Treatment Equipment

22 51 19.50 Swimming Pool Filtration Equipment		Crew	Daily Output	Labor-Hours	Unit	Material	Labor	Equipment	Total	Total Incl O&P
0010	**SWIMMING POOL FILTRATION EQUIPMENT**									
0900	Filter system, sand or diatomite type, incl. pump, 6,000 gal./hr.	2 Plum	1.80	8.889	Total	1,875	495		2,370	2,825
1020	Add for chlorination system, 800 S.F. pool		3	5.333	Ea.	105	298		403	565
1040	5,000 S.F. pool		3	5.333	"	1,875	298		2,173	2,525

22 52 Fountain Plumbing Systems

22 52 16 – Fountain Pumps

22 52 16.10 Fountain Water Pumps		Crew	Daily Output	Labor-Hours	Unit	Material	Labor	Equipment	Total	Total Incl O&P
0010	**FOUNTAIN WATER PUMPS**									
0100	Pump w/controls									
0200	Single phase, 100′ cord, 1/2 H.P. pump	2 Skwk	4.40	3.636	Ea.	1,275	168		1,443	1,650
0300	3/4 H.P. pump		4.30	3.721		2,175	172		2,347	2,675
0400	1 H.P. pump		4.20	3.810		2,350	176		2,526	2,875
0500	1-1/2 H.P. pump		4.10	3.902		2,800	180		2,980	3,350
0600	2 H.P. pump		4	4		3,775	185		3,960	4,450
0700	Three phase, 200′ cord, 5 H.P. pump		3.90	4.103		5,175	190		5,365	6,000
0800	7-1/2 H.P. pump		3.80	4.211		9,000	195		9,195	10,200
0900	10 H.P. pump		3.70	4.324		9,825	200		10,025	11,100
1000	15 H.P. pump		3.60	4.444		11,900	205		12,105	13,300
2000	DESIGN NOTE: Use two horsepower per surface acre.									

22 52 33 – Fountain Ancillary

22 52 33.10 Fountain Miscellaneous		Crew	Daily Output	Labor-Hours	Unit	Material	Labor	Equipment	Total	Total Incl O&P
0010	**FOUNTAIN MISCELLANEOUS**									
1100	Nozzles, minimum	2 Skwk	8	2	Ea.	149	92.50		241.50	305
1200	Maximum		8	2		320	92.50		412.50	495
1300	Lights w/mounting kits, 200 watt		18	.889		1,050	41		1,091	1,225
1400	300 watt		18	.889		1,275	41		1,316	1,475
1500	500 watt		18	.889		1,425	41		1,466	1,650
1600	Color blender		12	1.333		555	61.50		616.50	705

22 62 Vacuum Systems for Laboratory and Healthcare Facilities

22 62 19 – Vacuum Equipment for Laboratory and Healthcare Facilities

22 62 19.70 Healthcare Vacuum Equipment		Crew	Daily Output	Labor-Hours	Unit	2013 Bare Costs Material	Labor	Equipment	Total	Total Incl O&P
0010	**HEALTHCARE VACUUM EQUIPMENT**									
0100	Medical, with receiver									
0110	Duplex									
0120	20 SCFM	Q-1	1.14	14.035	Ea.	17,600	705		18,305	20,500
0130	60 SCFM, 10 HP	Q-2	1.20	20	"	32,000	1,050		33,050	36,800
0200	Triplex									
0220	180 SCFM	Q-2	.86	27.907	Ea.	49,600	1,450		51,050	56,500
0300	Dental oral									
0310	Duplex									
0330	165 SCFM with 77 gal. separator	Q-2	1.30	18.462	Ea.	52,000	960		52,960	58,500
1100	Vacuum system									
1110	Vacuum outlet alarm panel	1 Plum	3.20	2.500	Ea.	1,100	140		1,240	1,425

22 63 Gas Systems for Laboratory and Healthcare Facilities

22 63 13 – Gas Piping for Laboratory and Healthcare Facilities

22 63 13.70 Healthcare Gas Piping		Crew	Daily Output	Labor-Hours	Unit	Material	Labor	Equipment	Total	Total Incl O&P
0010	**HEALTHCARE GAS PIPING**									
1000	Nitrogen or oxygen system									
1010	Cylinder manifold									
1020	5 cylinder	1 Plum	.80	10	Ea.	5,950	560		6,510	7,400
1026	10 cylinder	"	.40	20	"	7,175	1,125		8,300	9,550
3000	Outlets and valves									
3010	Recessed, wall mounted									
3012	Single outlet	1 Plum	3.20	2.500	Ea.	71.50	140		211.50	289
3100	Ceiling outlet									
3190	Zone valve with box									
3210	2" valve size	1 Plum	3.20	2.500	Ea.	420	140		560	670
4000	Alarm panel, medical gases and vacuum									
4010	Alarm panel	1 Plum	3.20	2.500	Ea.	1,100	140		1,240	1,425

22 66 Chemical-Waste Systems for Lab. and Healthcare Facilities

22 66 53 – Laboratory Chemical-Waste and Vent Piping

22 66 53.30 Glass Pipe		Crew	Daily Output	Labor-Hours	Unit	Material	Labor	Equipment	Total	Total Incl O&P
0010	**GLASS PIPE**, Borosilicate, couplings & clevis hanger assemblies, 10' O.C. R221113-70									
0020	Drainage									
1100	1-1/2" diameter	Q-1	52	.308	L.F.	10.30	15.45		25.75	35
1120	2" diameter		44	.364		13.45	18.25		31.70	42.50
1140	3" diameter		39	.410		18.20	20.50		38.70	51
1160	4" diameter		30	.533		32.50	27		59.50	76.50
1180	6" diameter	↓	26	.615	↓	60.50	31		91.50	113
1870	To delete coupling & hanger, subtract									
1880	1-1/2" diam. to 2" diam.					19%	22%			
1890	3" diam. to 6" diam.					20%	17%			
2000	Process supply (pressure), beaded joints									
2040	1/2" diameter	1 Plum	36	.222	L.F.	5.25	12.40		17.65	24.50
2060	3/4" diameter		31	.258		5.90	14.40		20.30	28
2080	1" diameter	↓	27	.296		15.80	16.55		32.35	42.50
2100	1-1/2" diameter	Q-1	47	.340		13.90	17.10		31	41.50
2120	2" diameter		39	.410		18.80	20.50		39.30	51.50

	22 66 53.30 Glass Pipe	Crew	Daily Output	Labor-Hours	Unit	2013 Bare Costs Material	Labor	Equipment	Total	Total Incl O&P
2140	3" diameter	Q-1	34	.471	L.F.	25	23.50		48.50	63
2160	4" diameter		25	.640		38	32		70	90.50
2180	6" diameter		21	.762		106	38.50		144.50	175
2860	To delete coupling & hanger, subtract									
2870	1/2" diam. to 1" diam.					25%	33%			
2880	1-1/2" diam. to 3" diam.					22%	21%			
2890	4" diam. to 6" diam.					23%	15%			
3800	Conical joint, transparent									
3980	6" diameter	Q-1	21	.762	L.F.	143	38.50		181.50	216
4500	To delete couplings & hangers, subtract									
4530	6" diam.					22%	26%			

22 66 53.40 Pipe Fittings, Glass

	22 66 53.40 Pipe Fittings, Glass	Crew	Daily Output	Labor-Hours	Unit	Material	Labor	Equipment	Total	Total Incl O&P
0010	**PIPE FITTINGS, GLASS**									
0020	Drainage, beaded ends									
0040	Coupling & labor required at joints not incl. in fitting									
0050	price. Add 1 per joint for installed price									
0070	90° Bend or sweep, 1-1/2"				Ea.	30.50			30.50	34
0090	2"					39			39	42.50
0100	3"					64			64	70.50
0110	4"					102			102	113
0120	6" (sweep only)					310			310	345
0200	45° Bend or sweep same as 90°									
0350	Tee, single sanitary, 1-1/2"				Ea.	49.50			49.50	54.50
0370	2"					49.50			49.50	54.50
0380	3"					65.50			65.50	72
0390	4"					133			133	146
0400	6"					355			355	390
0410	Tee, straight, 1-1/2"					61.50			61.50	67.50
0430	2"					61.50			61.50	67.50
0440	3"					88.50			88.50	97.50
0450	4"					156			156	171
0460	6"					385			385	420
0500	Coupling, stainless steel, TFE seal ring									
0520	1-1/2"	Q-1	32	.500	Ea.	22	25		47	62.50
0530	2"		30	.533		28	27		55	71
0540	3"		25	.640		37.50	32		69.50	90
0550	4"		23	.696		65	35		100	124
0560	6"		20	.800		145	40		185	221
0600	Coupling, stainless steel, bead to plain end									
0610	1-1/2"	Q-1	36	.444	Ea.	28.50	22.50		51	65
0620	2"		34	.471		35.50	23.50		59	74.50
0630	3"		29	.552		60.50	27.50		88	108
0640	4"		27	.593		90	30		120	144
0650	6"		24	.667		305	33.50		338.50	385
2350	Coupling, Viton liner, for temperatures to 400°F									
2370	1/2"	Q-1	40	.400	Ea.	46.50	20		66.50	82
2380	3/4"		37	.432		53	21.50		74.50	91
2390	1"		35	.457		63	23		86	104
2400	1-1/2"		32	.500		22.50	25		47.50	63
2410	2"		30	.533		28	27		55	71
2420	3"		25	.640		37.50	32		69.50	90
2430	4"		23	.696		65	35		100	124

22 66 53 – Laboratory Chemical-Waste and Vent Piping

22 66 53.40 Pipe Fittings, Glass		Crew	Daily Output	Labor-Hours	Unit	Material	2013 Bare Costs Labor	Equipment	Total	Total Incl O&P
2440	6"	Q-1	20	.800	Ea.	145	40		185	221
2550	For beaded joint armored fittings, add					200%				
2600	Conical ends. Flange set, gasket & labor not incl. in fitting									
2620	price. Add 1 per joint for installed price.									
2650	90° Sweep elbow, 1"				Ea.	105			105	116
2670	1-1/2"					211			211	232
2680	2"					219			219	241
2690	3"					340			340	375
2700	4"					605			605	665
2710	6"					1,050			1,050	1,175
2750	Cross (straight), add					55%				
2850	Tee, add					20%				

22 66 53.60 Corrosion Resistant Pipe		Crew	Daily Output	Labor-Hours	Unit	Material	Labor	Equipment	Total	Total Incl O&P
0010	**CORROSION RESISTANT PIPE**, No couplings or hangers R221113-70									
0020	Iron alloy, drain, mechanical joint									
1000	1-1/2" diameter	Q-1	70	.229	L.F.	46	11.50		57.50	68
1100	2" diameter		66	.242		47	12.20		59.20	70
1120	3" diameter		60	.267		60.50	13.40		73.90	86.50
1140	4" diameter		52	.308		77.50	15.45		92.95	109
1980	Iron alloy, drain, B&S joint									
2000	2" diameter	Q-1	54	.296	L.F.	55.50	14.90		70.40	84
2100	3" diameter		52	.308		71	15.45		86.45	102
2120	4" diameter		48	.333		98	16.75		114.75	133
2140	6" diameter	Q-2	59	.407		158	21		179	205
2160	8" diameter	"	54	.444		297	23		320	360
2980	Plastic, epoxy, fiberglass filament wound, B&S joint									
3000	2" diameter	Q-1	62	.258	L.F.	12	12.95		24.95	32.50
3100	3" diameter		51	.314		14	15.75		29.75	39
3120	4" diameter		45	.356		20	17.85		37.85	49
3140	6" diameter		32	.500		28	25		53	69
3160	8" diameter	Q-2	38	.632		44	33		77	98
3180	10" diameter		32	.750		60	39		99	125
3200	12" diameter		28	.857		72	44.50		116.50	147
3980	Polyester, fiberglass filament wound, B&S joint									
4000	2" diameter	Q-1	62	.258	L.F.	13.05	12.95		26	34
4100	3" diameter		51	.314		17	15.75		32.75	42
4120	4" diameter		45	.356		25	17.85		42.85	54.50
4140	6" diameter		32	.500		36	25		61	77.50
4160	8" diameter	Q-2	38	.632		85.50	33		118.50	144
4180	10" diameter		32	.750		105	39		144	174
4200	12" diameter		28	.857		125	44.50		169.50	205
4980	Polypropylene, acid resistant, fire retardant, schedule 40									
5000	1-1/2" diameter	Q-1	68	.235	L.F.	7.60	11.80		19.40	26
5100	2" diameter		62	.258		10.40	12.95		23.35	31
5120	3" diameter		51	.314		21	15.75		36.75	47
5140	4" diameter		45	.356		27	17.85		44.85	56.50
5160	6" diameter		32	.500		54	25		79	97.50
5980	Proxylene, fire retardant, Schedule 40									
6000	1-1/2" diameter	Q-1	68	.235	L.F.	11.90	11.80		23.70	31
6100	2" diameter		62	.258		16.30	12.95		29.25	37.50
6120	3" diameter		51	.314		29.50	15.75		45.25	56
6140	4" diameter		45	.356		41.50	17.85		59.35	73

22 66 53 – Laboratory Chemical-Waste and Vent Piping

22 66 53.60 Corrosion Resistant Pipe		Crew	Daily Output	Labor-Hours	Unit	Material	2013 Bare Costs Labor	Equipment	Total	Total Incl O&P
6160	6" diameter	Q-1	32	.500	L.F.	70.50	25		95.50	116
6820	For Schedule 80, add				↓	35%	2%			

22 66 53.70 Pipe Fittings, Corrosion Resistant		Crew	Daily Output	Labor-Hours	Unit	Material	Labor	Equipment	Total	Total Incl O&P
0010	PIPE FITTINGS, CORROSION RESISTANT									
0030	Iron alloy									
0050	Mechanical joint									
0060	1/4 Bend, 1-1/2"	Q-1	12	1.333	Ea.	77	67		144	186
0080	2"		10	1.600		126	80.50		206.50	260
0090	3"		9	1.778		151	89.50		240.50	300
0100	4"		8	2		174	100		274	340
0110	1/8 Bend, 1-1/2"		12	1.333		49	67		116	155
0130	2"		10	1.600		84	80.50		164.50	214
0140	3"		9	1.778		112	89.50		201.50	257
0150	4"	↓	8	2	↓	150	100		250	315
0160	Tee and Y, sanitary, straight									
0170	1-1/2"	Q-1	8	2	Ea.	84	100		184	244
0180	2"		7	2.286		112	115		227	296
0190	3"		6	2.667		174	134		308	395
0200	4"		5	3.200		320	161		481	590
0360	Coupling, 1-1/2"		14	1.143		47.50	57.50		105	139
0380	2"		12	1.333		54	67		121	161
0390	3"		11	1.455		56.50	73		129.50	173
0400	4"	↓	10	1.600	↓	64.50	80.50		145	192
0500	Bell & Spigot									
0510	1/4 and 1/16 bend, 2"	Q-1	16	1	Ea.	99.50	50		149.50	185
0520	3"		14	1.143		232	57.50		289.50	345
0530	4"	↓	13	1.231		238	62		300	355
0540	6"	Q-2	17	1.412		470	73.50		543.50	630
0550	8"	"	12	2		1,950	104		2,054	2,275
0620	1/8 bend, 2"	Q-1	16	1		112	50		162	199
0640	3"		14	1.143		207	57.50		264.50	315
0650	4"	↓	13	1.231		238	62		300	355
0660	6"	Q-2	17	1.412		395	73.50		468.50	545
0680	8"	"	12	2		1,550	104		1,654	1,850
0700	Tee, sanitary, 2"	Q-1	10	1.600		213	80.50		293.50	355
0710	3"		9	1.778		690	89.50		779.50	895
0720	4"	↓	8	2		575	100		675	780
0730	6"	Q-2	11	2.182		715	114		829	960
0740	8"	"	8	3		1,950	156		2,106	2,375
1800	Y, sanitary, 2"	Q-1	10	1.600		224	80.50		304.50	365
1820	3"		9	1.778		400	89.50		489.50	575
1830	4"	↓	8	2		360	100		460	545
1840	6"	Q-2	11	2.182		1,200	114		1,314	1,500
1850	8"	"	8	3	↓	3,300	156		3,456	3,850
3000	Epoxy, filament wound									
3030	Quick-lock joint									
3040	90° Elbow, 2"	Q-1	28	.571	Ea.	97.50	28.50		126	150
3060	3"		16	1		112	50		162	199
3070	4"		13	1.231		153	62		215	261
3080	6"	↓	8	2		223	100		323	395
3090	8"	Q-2	9	2.667		410	139		549	660
3100	10"		7	3.429		520	179		699	840

22 66 53.70 Pipe Fittings, Corrosion Resistant		Crew	Daily Output	Labor-Hours	Unit	Material	Labor	Equipment	Total	Total Incl O&P
						2013 Bare Costs				
3110	12"	Q-2	6	4	Ea.	740	208		948	1,125
3120	45° Elbow, 2"	Q-1	28	.571		75	28.50		103.50	126
3130	3"		16	1		106	50		156	193
3140	4"		13	1.231		109	62		171	213
3150	6"		8	2		223	100		323	395
3160	8"	Q-2	9	2.667		410	139		549	660
3170	10"		7	3.429		515	179		694	840
3180	12"		6	4		740	208		948	1,125
3190	Tee, 2"	Q-1	19	.842		233	42.50		275.50	320
3200	3"		11	1.455		280	73		353	420
3210	4"		9	1.778		335	89.50		424.50	505
3220	6"		5	3.200		575	161		736	870
3230	8"	Q-2	6	4		645	208		853	1,025
3240	10"		5	4.800		900	250		1,150	1,375
3250	12"		4	6		1,275	310		1,585	1,875
4000	Polypropylene, acid resistant									
4020	Non-pressure, electrofusion joints									
4050	1/4 bend, 1-1/2"	1 Plum	16	.500	Ea.	24	28		52	68.50
4060	2"	Q-1	28	.571		25.50	28.50		54	71
4080	3"		17	.941		27.50	47.50		75	101
4090	4"		14	1.143		44.50	57.50		102	136
4110	6"		8	2		106	100		206	268
4150	1/4 Bend, long sweep									
4170	1-1/2"	1 Plum	16	.500	Ea.	14.40	28		42.40	58
4180	2"	Q-1	28	.571		25.50	28.50		54	71
4200	3"		17	.941		31.50	47.50		79	106
4210	4"		14	1.143		46	57.50		103.50	137
4250	1/8 bend, 1-1/2"	1 Plum	16	.500		12.30	28		40.30	55.50
4260	2"	Q-1	28	.571		15.75	28.50		44.25	60.50
4280	3"		17	.941		29	47.50		76.50	103
4290	4"		14	1.143		32	57.50		89.50	122
4310	6"		8	2		89	100		189	249
4400	Tee, sanitary									
4420	1-1/2"	1 Plum	10	.800	Ea.	16.45	44.50		60.95	85
4430	2"	Q-1	17	.941		19.20	47.50		66.70	92
4450	3"		11	1.455		38.50	73		111.50	152
4460	4"		9	1.778		57.50	89.50		147	198
4480	6"		5	3.200		385	161		546	660
4490	Tee, sanitary reducing, 2" x 2" x 1-1/2"		17	.941		19.20	47.50		66.70	92
4491	3" x 3" x 1-1/2"		11	1.455		34	73		107	148
4492	3" x 3" x 2"		11	1.455		38.50	73		111.50	152
4493	4" x 4" x 2"		10	1.600		51.50	80.50		132	178
4494	4" x 4" x 3"		9	1.778		56	89.50		145.50	196
4496	6" x 6" x 4"		5	3.200		188	161		349	450
4500	Tee/wye, long turn									
4520	1-1/2"	1 Plum	10	.800	Ea.	21.50	44.50		66	90.50
4530	2"	Q-1	17	.941		27.50	47.50		75	101
4550	3"		11	1.455		46.50	73		119.50	161
4570	4"		9	1.778		63.50	89.50		153	204
4650	Wye 45°, 1-1/2"	1 Plum	10	.800		17.80	44.50		62.30	86.50
4652	2"	Q-1	17	.941		24.50	47.50		72	98
4653	3"		11	1.455		41	73		114	155
4654	4"		9	1.778		59.50	89.50		149	200

22 66 53 – Laboratory Chemical-Waste and Vent Piping

22 66 53.70 Pipe Fittings, Corrosion Resistant		Crew	Daily Output	Labor-Hours	Unit	Material	Labor	Equipment	Total	Total Incl O&P
						2013 Bare Costs				
4656	6"	Q-1	5	3.200	Ea.	154	161		315	410
4660	Wye, reducing									
4662	2" x 2" x 1-1/2"	Q-1	17	.941	Ea.	23.50	47.50		71	96.50
4666	3" x 3" x 2"		11	1.455		46.50	73		119.50	161
4668	4" x 4" x 2"		10	1.600		55	80.50		135.50	182
4669	4" x 4" x 3"		9	1.778		58	89.50		147.50	198
4671	6" x 6" x 2"		6	2.667		93	134		227	305
4673	6" x 6" x 3"		5.50	2.909		103	146		249	335
4675	6" x 6" x 4"	↓	5	3.200	↓	107	161		268	360
4678	Combination Y & 1/8 bend									
4681	1-1/2"	1 Plum	10	.800	Ea.	21.50	44.50		66	90.50
4683	2"	Q-1	17	.941		27.50	47.50		75	101
4684	3"		11	1.455		46.50	73		119.50	161
4685	4"	↓	9	1.778	↓	63.50	89.50		153	204
4689	Combination Y & 1/8 bend, reducing									
4692	2" x 2" x 1-1/2"	Q-1	17	.941	Ea.	39	47.50		86.50	114
4694	3" x 3" x 1-1/2"		12	1.333		37	67		104	142
4695	3" x 3" x 2"		11	1.455		41	73		114	155
4697	4" x 4" x 2"		10	1.600		58	80.50		138.50	185
4699	4" x 4" x 3"	↓	9	1.778	↓	60.50	89.50		150	201
4710	Hub adapter									
4712	1-1/2"	1 Plum	16	.500	Ea.	30	28		58	75
4713	2"	Q-1	28	.571		34	28.50		62.50	80.50
4714	3"		17	.941		43	47.50		90.50	119
4715	4"	↓	14	1.143	↓	56	57.50		113.50	149
4719	Mechanical joint adapter									
4721	1-1/2"	1 Plum	16	.500	Ea.	18.50	28		46.50	62.50
4722	2"	Q-1	28	.571		19.20	28.50		47.70	64
4723	3"		17	.941		27.50	47.50		75	101
4724	4"	↓	14	1.143	↓	41	57.50		98.50	132
4728	Couplings									
4731	1-1/2"	1 Plum	16	.500	Ea.	10.25	28		38.25	53.50
4732	2"	Q-1	28	.571		13	28.50		41.50	57.50
4733	3"		17	.941		16.45	47.50		63.95	89
4734	4"		14	1.143		23.50	57.50		81	112
4736	6"	↓	8	2	↓	37	100		137	192

22 66 83 – Chemical-Waste Tanks

22 66 83.13 Chemical-Waste Dilution Tanks		Crew	Daily Output	Labor-Hours	Unit	Material	Labor	Equipment	Total	Total Incl O&P
0010	**CHEMICAL-WASTE DILUTION TANKS**									
7000	Tanks, covers included									
7800	Polypropylene									
7810	Continuous service to 200°F									
7830	2 gallon, 8" x 8" x 8"	Q-1	20	.800	Ea.	131	40		171	205
7850	7 gallon, 12" x 12" x 12"		20	.800		199	40		239	280
7870	16 gallon, 18" x 12" x 18"		17	.941		261	47.50		308.50	360
8010	33 gallon, 24" x 18" x 18"		12	1.333		340	67		407	475
8070	44 gallon, 24" x 18" x 24"		10	1.600		410	80.50		490.50	575
8080	89 gallon, 36" x 24" x 24"	↓	8	2	↓	580	100		680	790
8150	Polyethylene, heavy duty walls									
8160	Continuous service to 180°F									
8180	5 gallon, 12" X 6" X 18"	Q-1	20	.800	Ea.	49	40		89	114
8210	15 gallon, 14" I.D. x 27" deep		17	.941		75	47.50		122.50	154

22 66 83 – Chemical-Waste Tanks

22 66 83.13 Chemical-Waste Dilution Tanks		Crew	Daily Output	Labor-Hours	Unit	2013 Bare Costs Material	Labor	Equipment	Total	Total Incl O&P
8230	55 gallon, 22" I.D. x 36" deep	Q-1	10	1.600	Ea.	142	80.50		222.50	277
8250	100 gallon 28" I.D. x 42" deep		8	2		355	100		455	540
8270	200 gallon 36" I.D. x 48" deep		6	2.667		555	134		689	810
8290	360 gallon 48" I.D. x 48" deep		5	3.200		645	161		806	950

Estimating Tips

The labor adjustment factors listed in Subdivision 22 01 02.20 also apply to Division 23.

23 10 00 Facility Fuel Systems

- The prices in this subdivision for above- and below-ground storage tanks do not include foundations or hold-down slabs, unless noted. The estimator should refer to Divisions 3 and 31 for foundation system pricing. In addition to the foundations, required tank accessories, such as tank gauges, leak detection devices, and additional manholes and piping, must be added to the tank prices.

23 50 00 Central Heating Equipment

- When estimating the cost of an HVAC system, check to see who is responsible for providing and installing the temperature control system. It is possible to overlook controls, assuming that they would be included in the electrical estimate.
- When looking up a boiler, be careful on specified capacity. Some manufacturers rate their products on output while others use input.
- Include HVAC insulation for pipe, boiler, and duct (wrap and liner).
- Be careful when looking up mechanical items to get the correct pressure rating and connection type (thread, weld, flange).

23 70 00 Central HVAC Equipment

- Combination heating and cooling units are sized by the air conditioning requirements. (See Reference No. R236000-20 for preliminary sizing guide.)
- A ton of air conditioning is nominally 400 CFM.
- Rectangular duct is taken off by the linear foot for each size, but its cost is usually estimated by the pound. Remember that SMACNA standards now base duct on internal pressure.
- Prefabricated duct is estimated and purchased like pipe: straight sections and fittings.
- Note that cranes or other lifting equipment are not included on any lines in Division 23. For example, if a crane is required to lift a heavy piece of pipe into place high above a gym floor, or to put a rooftop unit on the roof of a four-story building, etc., it must be added. Due to the potential for extreme variation—from nothing additional required to a major crane or helicopter—we feel that including a nominal amount for "lifting contingency" would be useless and detract from the accuracy of the estimate. When using equipment rental cost data from RSMeans, do not forget to include the cost of the operator(s).

Reference Numbers

Reference numbers are shown in shaded boxes at the beginning of some major classifications. These numbers refer to related items in the Reference Section. The reference information may be an estimating procedure, an alternate pricing method, or technical information.

Note: Not all subdivisions listed here necessarily appear in this publication.

Note: **Trade Service,** *in part, has been used as a reference source for some of the material prices used in Division 23.*

23 05 02 – HVAC General

23 05 02.10 Air Conditioning, General		Crew	Daily Output	Labor-Hours	Unit	Material	2013 Bare Costs Labor	Equipment	Total	Total Incl O&P
0010	**AIR CONDITIONING, GENERAL** Prices are for standard efficiencies (SEER 13)									
0020	for upgrade to SEER 14 add					10%				

23 05 05 – Selective HVAC Demolition

23 05 05.10 HVAC Demolition

		Crew	Daily Output	Labor-Hours	Unit	Material	Labor	Equipment	Total	Total Incl O&P
0010	**HVAC DEMOLITION** R220105-10									
0100	Air conditioner, split unit, 3 ton	Q-5	2	8	Ea.		410		410	615
0150	Package unit, 3 ton	Q-6	3	8			425		425	635
0190	Rooftop, self contained, up to 5 ton	1 Plum	1.20	6.667			370		370	560
0250	Air curtain	Q-9	20	.800	L.F.		38.50		38.50	58.50
0254	Air filters, up thru 16,000 CFM		20	.800	Ea.		38.50		38.50	58.50
0256	20,000 thru 60,000 CFM		16	1			48		48	73
0297	Boiler blowdown	Q-5	8	2			102		102	154
0298	Boilers									
0300	Electric, up thru 148 kW	Q-19	2	12	Ea.		615		615	930
0310	150 thru 518 kW	"	1	24			1,225		1,225	1,850
0320	550 thru 2000 kW	Q-21	.40	80			4,225		4,225	6,350
0330	2070 kW and up	"	.30	106			5,625		5,625	8,450
0340	Gas and/or oil, up thru 150 MBH	Q-7	2.20	14.545			785		785	1,175
0350	160 thru 2000 MBH		.80	40			2,150		2,150	3,250
0360	2100 thru 4500 MBH		.50	64			3,450		3,450	5,200
0370	4600 thru 7000 MBH		.30	106			5,750		5,750	8,650
0380	7100 thru 12,000 MBH		.16	200			10,800		10,800	16,200
0390	12,200 thru 25,000 MBH		.12	266			14,400		14,400	21,700
0400	Central station air handler unit, up thru 15 ton	Q-5	1.60	10			510		510	770
0410	17.5 thru 30 ton	"	.80	20			1,025		1,025	1,525
0430	Computer room unit									
0434	Air cooled split, up thru 10 ton	Q-5	.67	23.881	Ea.		1,225		1,225	1,825
0436	12 thru 23 ton		.53	30.189			1,550		1,550	2,325
0440	Chilled water, up thru 10 ton		1.30	12.308			625		625	945
0444	12 thru 23 ton		1	16			815		815	1,225
0450	Glycol system, up thru 10 ton		.53	30.189			1,550		1,550	2,325
0454	12 thru 23 ton		.40	40			2,050		2,050	3,075
0460	Water cooled, not including condenser, up thru 10 ton		.80	20			1,025		1,025	1,525
0464	12 thru 23 ton		.60	26.667			1,350		1,350	2,050
0600	Condenser, up thru 50 ton		1	16			815		815	1,225
0610	51 thru 100 ton	Q-6	.80	30			1,575		1,575	2,400
0620	101 thru 1000 ton	"	.16	150			7,925		7,925	11,900
0660	Condensing unit, up thru 10 ton	Q-5	1.25	12.800			655		655	980
0670	11 thru 50 ton	"	.40	40			2,050		2,050	3,075
0680	60 thru 100 ton	Q-6	.30	80			4,225		4,225	6,375
0700	Cooling tower, up thru 400 ton		.80	30			1,575		1,575	2,400
0710	450 thru 600 ton		.53	45.283			2,400		2,400	3,600
0720	700 thru 1300 ton		.40	60			3,175		3,175	4,775
0780	Dehumidifier, up thru 155 lb./hr.	Q-1	8	2			100		100	151
0790	240 lb./hr. and up	"	2	8			400		400	605
1560	Ductwork									
1570	Metal, steel, sst, fabricated	Q-9	1000	.016	Lb.		.77		.77	1.17
1580	Aluminum, fabricated		485	.033	"		1.58		1.58	2.41
1590	Spiral, prefabricated		400	.040	L.F.		1.92		1.92	2.93
1600	Fiberglass, prefabricated		400	.040			1.92		1.92	2.93
1610	Flex, prefabricated		500	.032			1.54		1.54	2.34
1620	Glass fiber reinforced plastic, prefabricated		280	.057			2.74		2.74	4.18

23 05 Common Work Results for HVAC

23 05 05 – Selective HVAC Demolition

23 05 05.10 HVAC Demolition		Crew	Daily Output	Labor-Hours	Unit	Material	2013 Bare Costs Labor	Equipment	Total	Total Incl O&P
1630	Diffusers, registers or grills, up thru 20" max dimension	1 Shee	50	.160	Ea.		8.55		8.55	13
1640	21 thru 36" max dimension		36	.222			11.85		11.85	18.05
1650	Above 36" max dimension		30	.267			14.20		14.20	21.50
1700	Evaporator, up thru 12,000 BTUH	Q-5	5.30	3.019			154		154	232
1710	12,500 thru 30,000 BTUH	"	2.70	5.926			300		300	455
1720	31,000 BTUH and up	Q-6	1.50	16			845		845	1,275
1730	Evaporative cooler, up thru 5 H.P.	Q-9	2.70	5.926			284		284	435
1740	10 thru 30 H.P.	"	.67	23.881			1,150		1,150	1,750
1750	Exhaust systems									
1760	Exhaust components	1 Shee	8	1	System		53.50		53.50	81.50
1770	Weld fume hoods	"	20	.400	Ea.		21.50		21.50	32.50
2120	Fans, up thru 1 H.P. or 2000 CFM	Q-9	8	2			96		96	146
2124	1-1/2 thru 10 H.P. or 20,000 CFM		5.30	3.019			145		145	221
2128	15 thru 30 H.P. or above 20,000 CFM		4	4			192		192	293
2150	Fan coil air conditioner, chilled water, up thru 7.5 ton	Q-5	14	1.143			58.50		58.50	87.50
2154	Direct expansion, up thru 10 ton		8	2			102		102	154
2158	11 thru 30 ton		2	8			410		410	615
2170	Flue shutter damper	Q-9	8	2			96		96	146
2200	Furnace, electric	Q-20	2	10			490		490	740
2300	Gas or oil, under 120 MBH	Q-9	4	4			192		192	293
2340	Over 120 MBH	"	3	5.333			256		256	390
2730	Heating and ventilating unit	Q-5	2.70	5.926			300		300	455
2740	Heater, electric, wall, baseboard and quartz	1 Elec	10	.800			42		42	62.50
2750	Heater, electric, unit, cabinet, fan and convector	"	8	1			52.50		52.50	78.50
2760	Heat exchanger, shell and tube type	Q-5	1.60	10			510		510	770
2770	Plate type	Q-6	.60	40			2,125		2,125	3,175
2810	Heat pump									
2820	Air source, split, up thru 10 ton	Q-5	.90	17.778	Ea.		905		905	1,375
2830	15 thru 25 ton	Q-6	.80	30			1,575		1,575	2,400
2850	Single package, up thru 12 ton	Q-5	1	16			815		815	1,225
2860	Water source, up thru 15 ton	"	.90	17.778			905		905	1,375
2870	20 thru 50 ton	Q-6	.80	30			1,575		1,575	2,400
2910	Heat recovery package, up thru 20,000 CFM	Q-5	2	8			410		410	615
2920	25,000 CFM and up		1.20	13.333			680		680	1,025
2930	Heat transfer package, up thru 130 GPM		.80	20			1,025		1,025	1,525
2934	255 thru 800 GPM		.42	38.095			1,950		1,950	2,925
2940	Humidifier		10.60	1.509			77		77	116
2961	Hydronic unit heaters, up thru 200 MBH		14	1.143			58.50		58.50	87.50
2962	Above 200 MBH		8	2			102		102	154
2964	Valance units		32	.500			25.50		25.50	38.50
2966	Radiant floor heating									
2967	System valves, controls, manifolds	Q-5	16	1	Ea.		51		51	77
2968	Per room distribution		8	2			102		102	154
2970	Hydronic heating, baseboard radiation		16	1			51		51	77
2976	Convectors and free standing radiators		18	.889			45.50		45.50	68
2980	Induced draft fan, up thru 1 H.P.	Q-9	4.60	3.478			167		167	255
2984	1-1/2 H.P. thru 7-1/2 H.P.	"	2.20	7.273			350		350	530
2988	Infrared unit	Q-5	16	1			51		51	77
2992	Louvers	1 Shee	46	.174	S.F.		9.25		9.25	14.15
3000	Mechanical equipment, light items. Unit is weight, not cooling.	Q-5	.90	17.778	Ton		905		905	1,375
3600	Heavy items	"	1.10	14.545	"		740		740	1,125
3700	Deduct for salvage (when applicable), minimum				Job				73	80
3710	Maximum				"				455	500

23 05 Common Work Results for HVAC

23 05 05 – Selective HVAC Demolition

23 05 05.10 HVAC Demolition		Crew	Daily Output	Labor-Hours	Unit	Material	2013 Bare Costs Labor	Equipment	Total	Total Incl O&P
3720	Make up air unit, up thru 6000 CFM	Q-5	3	5.333	Ea.		272		272	410
3730	6500 thru 30,000 CFM	″	1.60	10			510		510	770
3740	35,000 thru 75,000 CFM	Q-6	1	24			1,275		1,275	1,900
3800	Mixing boxes, constant and VAV	Q-9	18	.889			42.50		42.50	65
4000	Packaged terminal air conditioner, up thru 18,000 BTUH	Q-5	8	2			102		102	154
4010	24,000 thru 48,000 BTUH	″	2.80	5.714			291		291	440
5000	Refrigerant compressor, reciprocating or scroll									
5010	Up thru 5 ton	1 Stpi	6	1.333	Ea.		75.50		75.50	114
5020	5.08 thru 10 ton	Q-5	6	2.667			136		136	205
5030	15 thru 50 ton	″	.40	40			2,050		2,050	3,075
5040	60 thru 130 ton	Q-6	.48	50			2,650		2,650	3,975
5090	Remove refrigerant from system	1 Stpi	40	.200	Lb.		11.35		11.35	17.05
5100	Roof top air conditioner, up thru 10 ton	Q-5	1.40	11.429	Ea.		585		585	875
5110	12 thru 40 ton	Q-6	1	24			1,275		1,275	1,900
5120	50 thru 140 ton		.50	48			2,550		2,550	3,825
5130	150 thru 300 ton		.30	80			4,225		4,225	6,375
6000	Self contained single package air conditioner, up thru 10 ton	Q-5	1.60	10			510		510	770
6010	15 thru 60 ton	Q-6	1.20	20			1,050		1,050	1,600
6100	Space heaters, up thru 200 MBH	Q-5	10	1.600			81.50		81.50	123
6110	Over 200 MBH		5	3.200			163		163	246
6200	Split ductless, both sections		8	2			102		102	154
6300	Steam condensate meter	1 Stpi	11	.727			41		41	62
6600	Thru-the-wall air conditioner	L-2	8	2			78.50		78.50	121
7000	Vent chimney, prefabricated, up thru 12″ diameter	Q-9	94	.170	V.L.F.		8.15		8.15	12.45
7010	14″ thru 36″ diameter		40	.400			19.20		19.20	29.50
7020	38″ thru 48″ diameter		32	.500			24		24	36.50
7030	54″ thru 60″ diameter	Q-10	14	1.714			85.50		85.50	130
7400	Ventilators, up thru 14″ neck diameter	Q-9	58	.276	Ea.		13.25		13.25	20
7410	16″ thru 50″ neck diameter		40	.400			19.20		19.20	29.50
7450	Relief vent, up thru 24″ x 96″		22	.727			35		35	53
7460	48″ x 60″ thru 96″ x 144″		10	1.600			77		77	117
8000	Water chiller up thru 10 ton	Q-5	2.50	6.400			325		325	490
8010	15 thru 100 ton	Q-6	.48	50			2,650		2,650	3,975
8020	110 thru 500 ton	Q-7	.29	110			5,950		5,950	8,950
8030	600 thru 1000 ton		.23	139			7,500		7,500	11,300
8040	1100 ton and up		.20	160			8,625		8,625	13,000
8400	Window air conditioner	1 Carp	16	.500			22.50		22.50	34.50

23 05 23 – General-Duty Valves for HVAC Piping

23 05 23.30 Valves, Iron Body		Crew	Daily Output	Labor-Hours	Unit	Material	Labor	Equipment	Total	Total Incl O&P
0010	**VALVES, IRON BODY** R220523-90									
0022	For grooved joint, see Section 22 11 13.48									
0100	Angle, 125 lb.									
0110	Flanged									
0116	2″	1 Plum	5	1.600	Ea.	1,200	89.50		1,289.50	1,425
0118	4″	Q-1	3	5.333		1,975	268		2,243	2,575
0120	6″	Q-2	3	8		3,850	415		4,265	4,875
0122	8″	″	2.50	9.600		6,900	500		7,400	8,325
1020	Butterfly, wafer type, gear actuator, 200 lb.									
1030	2″	1 Plum	14	.571	Ea.	91.50	32		123.50	148
1040	2-1/2″	Q-1	9	1.778		93	89.50		182.50	236
1050	3″		8	2		96.50	100		196.50	257
1060	4″		5	3.200		107	161		268	360

23 05 23 – General-Duty Valves for HVAC Piping

23 05 23.30 Valves, Iron Body		Crew	Daily Output	Labor-Hours	Unit	Material	Labor	Equipment	Total	Total Incl O&P
						2013 Bare Costs				
1070	5″	Q-2	5	4.800	Ea.	120	250		370	510
1080	6″		5	4.800		136	250		386	525
1090	8″		4.50	5.333		169	278		447	605
1100	10″		4	6		228	310		538	720
1110	12″		3	8		345	415		760	1,000
1200	Wafer type, lever actuator, 200 lb.									
1220	2″	1 Plum	14	.571	Ea.	155	32		187	219
1230	2-1/2″	Q-1	9	1.778		160	89.50		249.50	310
1240	3″		8	2		170	100		270	335
1250	4″		5	3.200		206	161		367	470
1260	5″	Q-2	5	4.800		287	250		537	690
1270	6″		5	4.800		345	250		595	755
1280	8″		4.50	5.333		535	278		813	1,000
1290	10″		4	6		750	310		1,060	1,300
1300	12″		3	8		1,375	415		1,790	2,125
1600	Gate, threaded, 125 lb.									
1603	1-1/2″	1 Plum	13	.615	Ea.	247	34.50		281.50	325
1604	2″		11	.727		875	40.50		915.50	1,025
1605	2-1/2″		10	.800		950	44.50		994.50	1,125
1606	3″		8	1		1,125	56		1,181	1,325
1607	4″		5	1.600		1,600	89.50		1,689.50	1,875
1650	Gate, 125 lb., N.R.S.									
2150	Flanged									
2200	2″	1 Plum	5	1.600	Ea.	645	89.50		734.50	845
2240	2-1/2″	Q-1	5	3.200		665	161		826	970
2260	3″		4.50	3.556		745	179		924	1,100
2280	4″		3	5.333		1,075	268		1,343	1,575
2290	5″	Q-2	3.40	7.059		1,800	370		2,170	2,550
2300	6″		3	8		1,800	415		2,215	2,625
2320	8″		2.50	9.600		3,100	500		3,600	4,175
2340	10″		2.20	10.909		5,475	570		6,045	6,875
2360	12″		1.70	14.118		7,525	735		8,260	9,375
2420	For 250 lb. flanged, add					200%	10%			
3500	OS&Y, 125 lb., (225# noted), threaded									
3504	3/4″ (225 lb.)	1 Plum	18	.444	Ea.	440	25		465	525
3505	1″ (225 lb.)		16	.500		500	28		528	590
3506	1-1/2″ (225 lb.)		13	.615		730	34.50		764.50	850
3507	2″		11	.727		1,275	40.50		1,315.50	1,450
3550	OS&Y, 125 lb., flanged									
3600	2″	1 Plum	5	1.600	Ea.	425	89.50		514.50	605
3640	2-1/2″	Q-1	5	3.200		430	161		591	710
3660	3″		4.50	3.556		470	179		649	790
3670	3-1/2″		3	5.333		570	268		838	1,025
3680	4″		3	5.333		690	268		958	1,150
3690	5″	Q-2	3.40	7.059		1,100	370		1,470	1,775
3700	6″		3	8		1,100	415		1,515	1,850
3720	8″		2.50	9.600		1,975	500		2,475	2,925
3740	10″		2.20	10.909		3,600	570		4,170	4,825
3760	12″		1.70	14.118		4,925	735		5,660	6,525
3900	For 175 lb., flanged, add					200%	10%			
4350	Globe, OS&Y									
4540	Class 125, flanged									
4550	2″	1 Plum	5	1.600	Ea.	835	89.50		924.50	1,050

23 05 23 – General-Duty Valves for HVAC Piping

23 05 23.30	Valves, Iron Body	Crew	Daily Output	Labor-Hours	Unit	Material	Labor	Equipment	Total	Total Incl O&P
						2013 Bare Costs				
4560	2-1/2"	Q-1	5	3.200	Ea.	840	161		1,001	1,175
4570	3"		4.50	3.556		1,025	179		1,204	1,400
4580	4"		3	5.333		1,450	268		1,718	2,000
4590	5"	Q-2	3.40	7.059		2,675	370		3,045	3,475
4600	6"		3	8		2,675	415		3,090	3,550
4610	8"		2.50	9.600		5,200	500		5,700	6,475
4612	10"		2.20	10.909		8,150	570		8,720	9,825
4614	12"		1.70	14.118		7,950	735		8,685	9,850
5040	Class 250, flanged									
5050	2"	1 Plum	4.50	1.778	Ea.	1,325	99		1,424	1,625
5060	2-1/2"	Q-1	4.50	3.556		1,725	179		1,904	2,175
5070	3"		4	4		1,800	201		2,001	2,275
5080	4"		2.70	5.926		2,625	298		2,923	3,350
5090	5"	Q-2	3	8		4,875	415		5,290	6,000
5100	6"		2.70	8.889		4,725	465		5,190	5,900
5110	8"		2.20	10.909		8,000	570		8,570	9,650
5120	10"		2	12		13,100	625		13,725	15,300
5130	12"		1.60	15		20,400	780		21,180	23,600
5240	Valve sprocket rim w/chain, for 2" valve	1 Stpi	30	.267		121	15.10		136.10	157
5250	2-1/2" valve		27	.296		121	16.80		137.80	160
5260	3-1/2" valve		25	.320		121	18.15		139.15	162
5270	6" valve		20	.400		121	22.50		143.50	168
5280	8" valve		18	.444		148	25		173	201
5290	12" valve		16	.500		148	28.50		176.50	206
5300	16" valve		12	.667		205	38		243	282
5310	20" valve		10	.800		205	45.50		250.50	293
5320	36" valve		8	1		365	56.50		421.50	485
5450	Swing check, 125 lb., threaded									
5470	1"	1 Plum	13	.615	Ea.	635	34.50		669.50	750
5500	2"	"	11	.727		390	40.50		430.50	490
5540	2-1/2"	Q-1	15	1.067		500	53.50		553.50	630
5550	3"		13	1.231		540	62		602	690
5560	4"		10	1.600		860	80.50		940.50	1,075
5950	Flanged									
5994	1"	1 Plum	7	1.143	Ea.	182	64		246	296
5998	1-1/2"		6	1.333		255	74.50		329.50	390
6000	2"		5	1.600		375	89.50		464.50	545
6040	2-1/2"	Q-1	5	3.200		340	161		501	615
6050	3"		4.50	3.556		365	179		544	670
6060	4"		3	5.333		570	268		838	1,025
6070	6"	Q-2	3	8		980	415		1,395	1,700
6080	8"		2.50	9.600		1,850	500		2,350	2,775
6090	10"		2.20	10.909		3,150	570		3,720	4,300
6100	12"		1.70	14.118		5,100	735		5,835	6,700
6110	18"		1.30	18.462		12,400	960		13,360	15,100
6114	24"		.75	32		23,400	1,675		25,075	28,200
6160	For 250 lb. flanged, add					200%	20%			
6600	Silent check, bronze trim									
6610	Compact wafer type, for 125 or 150 lb. flanges									
6630	1-1/2"	1 Plum	11	.727	Ea.	124	40.50		164.50	198
6640	2"	"	9	.889		149	49.50		198.50	239
6650	2-1/2"	Q-1	9	1.778		163	89.50		252.50	315
6660	3"		8	2		175	100		275	345

23 05 23 – General-Duty Valves for HVAC Piping

23 05 23.30 Valves, Iron Body		Crew	Daily Output	Labor-Hours	Unit	Material	Labor	Equipment	Total	Total Incl O&P
						2013 Bare Costs				
6670	4"	Q-1	5	3.200	Ea.	223	161		384	485
6680	5"	Q-2	6	4		291	208		499	635
6690	6"		6	4		390	208		598	745
6700	8"		4.50	5.333		695	278		973	1,175
6710	10"		4	6		1,200	310		1,510	1,800
6720	12"		3	8		2,300	415		2,715	3,150
6740	For 250 or 300 lb. flanges, thru 6" no change									
6741	For 8" and 10", add				Ea.	11%	10%			
6750	Twin disc									
6752	2"	1 Plum	9	.889	Ea.	290	49.50		339.50	395
6754	4"	Q-1	5	3.200		475	161		636	760
6756	6"	Q-2	5	4.800		700	250		950	1,150
6758	8"		4.50	5.333		1,075	278		1,353	1,600
6760	10"		4	6		1,725	310		2,035	2,375
6762	12"		3	8		2,225	415		2,640	3,075
6764	18"		1.50	16		9,650	835		10,485	11,900
6766	24"		.75	32		13,600	1,675		15,275	17,500
6800	Full flange type, 150 lb.									
6900	Globe type, 125 lb.									
6911	2-1/2"	Q-1	9	1.778	Ea.	370	89.50		459.50	540
6912	3"		8	2		395	100		495	585
6913	4"		5	3.200		535	161		696	825
6914	5"	Q-2	6	4		685	208		893	1,075
6915	6"		5	4.800		860	250		1,110	1,325
6916	8"		4.50	5.333		1,575	278		1,853	2,150
6917	10"		4	6		1,975	310		2,285	2,650
6918	12"		3	8		3,300	415		3,715	4,250
6940	For 250 lb., add					40%	10%			
6980	Screwed end type, 125 lb.									
6981	1"	1 Plum	19	.421	Ea.	51.50	23.50		75	92
6982	1-1/4"		15	.533		66.50	30		96.50	119
6983	1-1/2"		13	.615		81.50	34.50		116	141
6984	2"		11	.727		112	40.50		152.50	184

23 05 23.70 Valves, Semi-Steel		Crew	Daily Output	Labor-Hours	Unit	Material	Labor	Equipment	Total	Total Incl O&P
0010	**VALVES, SEMI-STEEL** R220523-90									
1020	Lubricated plug valve, threaded, 200 psi									
1030	1/2"	1 Plum	18	.444	Ea.	90	25		115	137
1040	3/4"		16	.500		90	28		118	141
1050	1"		14	.571		115	32		147	174
1060	1-1/4"		12	.667		137	37		174	206
1070	1-1/2"		11	.727		147	40.50		187.50	223
1080	2"		8	1		175	56		231	276
1090	2-1/2"	Q-1	5	3.200		269	161		430	535
1100	3"	"	4.50	3.556		330	179		509	635
6990	Flanged, 200 psi									
7000	2"	1 Plum	8	1	Ea.	217	56		273	320
7010	2-1/2"	Q-1	5	3.200		310	161		471	580
7020	3"		4.50	3.556		375	179		554	685
7030	4"		3	5.333		470	268		738	925
7036	5"		2.50	6.400		935	320		1,255	1,500
7040	6"	Q-2	3	8		780	415		1,195	1,475
7050	8"		2.50	9.600		1,575	500		2,075	2,475

23 05 23 – General-Duty Valves for HVAC Piping

23 05 23.70 Valves, Semi-Steel		Crew	Daily Output	Labor-Hours	Unit	2013 Bare Costs Material	Labor	Equipment	Total	Total Incl O&P
7060	10″	Q-2	2.20	10.909	Ea.	2,450	570		3,020	3,525
7070	12″		1.70	14.118		4,700	735		5,435	6,275

23 05 23.90 Valves, Stainless Steel		Crew	Daily Output	Labor-Hours	Unit	Material	Labor	Equipment	Total	Total Incl O&P
0010	**VALVES, STAINLESS STEEL** R220523-90									
1700	Check, 200 lb., threaded									
1710	1/4″	1 Plum	24	.333	Ea.	147	18.60		165.60	190
1720	1/2″		22	.364		147	20.50		167.50	193
1730	3/4″		20	.400		161	22.50		183.50	211
1750	1″		19	.421		205	23.50		228.50	261
1760	1-1/2″		13	.615		395	34.50		429.50	485
1770	2″		11	.727		660	40.50		700.50	790
1800	150 lb., flanged									
1810	2-1/2″	Q-1	5	3.200	Ea.	1,675	161		1,836	2,100
1820	3″		4.50	3.556		1,675	179		1,854	2,125
1830	4″		3	5.333		2,500	268		2,768	3,150
1840	6″	Q-2	3	8		4,450	415		4,865	5,525
1850	8″	″	2.50	9.600		7,450	500		7,950	8,925
2100	Gate, OS&Y, 150 lb., flanged									
2120	1/2″	1 Plum	18	.444	Ea.	470	25		495	560
2140	3/4″		16	.500		500	28		528	590
2150	1″		14	.571		620	32		652	730
2160	1-1/2″		11	.727		860	40.50		900.50	1,000
2170	2″		8	1		1,000	56		1,056	1,175
2180	2-1/2″	Q-1	5	3.200		1,600	161		1,761	2,025
2190	3″		4.50	3.556		1,600	179		1,779	2,050
2200	4″		3	5.333		2,450	268		2,718	3,100
2205	5″		2.80	5.714		4,250	287		4,537	5,100
2210	6″	Q-2	3	8		4,250	415		4,665	5,300
2220	8″		2.50	9.600		6,800	500		7,300	8,250
2230	10″		2.30	10.435		10,800	545		11,345	12,700
2240	12″		1.90	12.632		13,800	660		14,460	16,200
2260	For 300 lb., flanged, add					120%	15%			
2600	600 lb., flanged									
2620	1/2″	1 Plum	16	.500	Ea.	213	28		241	276
2640	3/4″		14	.571		230	32		262	300
2650	1″		12	.667		277	37		314	360
2660	1-1/2″		10	.800		445	44.50		489.50	550
2670	2″		7	1.143		610	64		674	765
2680	2-1/2″	Q-1	4	4		5,575	201		5,776	6,450
2690	3″	″	3.60	4.444		5,575	223		5,798	6,475
3100	Globe, OS&Y, 150 lb., flanged									
3120	1/2″	1 Plum	18	.444	Ea.	470	25		495	560
3140	3/4″		16	.500		525	28		553	615
3150	1″		14	.571		650	32		682	765
3160	1-1/2″		11	.727		970	40.50		1,010.50	1,125
3170	2″		8	1		1,175	56		1,231	1,375
3180	2-1/2″	Q-1	5	3.200		2,550	161		2,711	3,050
3190	3″		4.50	3.556		2,550	179		2,729	3,075
3200	4″		3	5.333		3,125	268		3,393	3,825
3210	6″	Q-2	3	8		5,475	415		5,890	6,650

23 05 Common Work Results for HVAC

23 05 93 – Testing, Adjusting, and Balancing for HVAC

23 05 93.50 Piping, Testing		Crew	Daily Output	Labor-Hours	Unit	2013 Bare Costs Material	Labor	Equipment	Total	Total Incl O&P
0010	**PIPING, TESTING**									
0100	Nondestructive testing									
0110	Nondestructive hydraulic pressure test, isolate & 1 hr. hold									
0120	1″ - 4″ pipe									
0140	0 – 250 L.F.	1 Stpi	1.33	6.015	Ea.		340		340	515
0160	250 – 500 L.F.	″	.80	10			565		565	855
0180	500 – 1000 L.F.	Q-5	1.14	14.035			715		715	1,075
0200	1000 – 2000 L.F.	″	.80	20			1,025		1,025	1,525
0300	6″ - 10″ pipe									
0320	0 – 250 L.F.	Q-5	1	16	Ea.		815		815	1,225
0340	250 – 500 L.F.		.73	21.918			1,125		1,125	1,675
0360	500 – 1000 L.F.		.53	30.189			1,550		1,550	2,325
0380	1000 – 2000 L.F.		.38	42.105			2,150		2,150	3,225
1000	Pneumatic pressure test, includes soaping joints									
1120	1″ - 4″ pipe									
1140	0 – 250 L.F.	Q-5	2.67	5.993	Ea.	12.20	305		317.20	475
1160	250 – 500 L.F.		1.33	12.030		24.50	615		639.50	950
1180	500 – 1000 L.F.		.80	20		36.50	1,025		1,061.50	1,575
1200	1000 – 2000 L.F.		.50	32		49	1,625		1,674	2,500
1300	6″ - 10″ pipe									
1320	0 – 250 L.F.	Q-5	1.33	12.030	Ea.	12.20	615		627.20	940
1340	250 – 500 L.F.		.67	23.881		24.50	1,225		1,249.50	1,850
1360	500 – 1000 L.F.		.40	40		49	2,050		2,099	3,125
1380	1000 – 2000 L.F.		.25	64		61	3,275		3,336	4,975
2000	X-Ray of welds									
2110	2″ diam.	1 Stpi	8	1	Ea.	12	56.50		68.50	98.50
2120	3″ diam.		8	1		12	56.50		68.50	98.50
2130	4″ diam.		8	1		18	56.50		74.50	105
2140	6″ diam.		8	1		18	56.50		74.50	105
2150	8″ diam.		6.60	1.212		18	68.50		86.50	123
2160	10″ diam.		6	1.333		24	75.50		99.50	141
3000	Liquid penetration of welds									
3110	2″ diam.	1 Stpi	14	.571	Ea.	3.70	32.50		36.20	52.50
3120	3″ diam.		13.60	.588		3.70	33.50		37.20	54
3130	4″ diam.		13.40	.597		3.70	34		37.70	55
3140	6″ diam.		13.20	.606		3.70	34.50		38.20	55.50
3150	8″ diam.		13	.615		5.55	35		40.55	58.50
3160	10″ diam.		12.80	.625		5.55	35.50		41.05	59.50

23 07 HVAC Insulation

23 07 13 – Duct Insulation

23 07 13.10 Duct Thermal Insulation

		Crew	Daily Output	Labor-Hours	Unit	Material	Labor	Equipment	Total	Total Incl O&P
0010	**DUCT THERMAL INSULATION**									
0100	Rule of thumb, as a percentage of total mechanical costs				Job				10%	10%
0110	Insulation req'd. is based on the surface size/area to be covered									
3730	Sheet insulation									
3760	Polyethylene foam, closed cell, UV resistant									
3770	Standard temperature (-90°F to +212°F)									
3771	1/4″ thick G	Q-14	450	.036	S.F.	1.74	1.59		3.33	4.38
3772	3/8″ thick G		440	.036		2.48	1.63		4.11	5.25

23 07 HVAC Insulation

23 07 13 – Duct Insulation

23 07 13.10 Duct Thermal Insulation			Crew	Daily Output	Labor-Hours	Unit	Material	2013 Bare Costs Labor	Equipment	Total	Total Incl O&P
3773	1/2" thick	G	Q-14	420	.038	S.F.	3.05	1.70		4.75	6
3774	3/4" thick	G		400	.040		4.36	1.79		6.15	7.60
3775	1" thick	G		380	.042		5.90	1.88		7.78	9.40
3776	1-1/2" thick	G		360	.044		9.30	1.99		11.29	13.35
3777	2" thick	G		340	.047		12.30	2.11		14.41	16.80
3778	2-1/2" thick	G	↓	320	.050	↓	15.80	2.24		18.04	21
3779	Adhesive (see line 7878)										
3780	Foam, rubber										
3782	1" thick	G	1 Stpi	50	.160	S.F.	2.89	9.05		11.94	16.85
7000	Board insulation										
7020	Mineral wool, 1200° F										
7022	6 lb. density, plain										
7024	1" thick	G	Q-14	370	.043	S.F.	.33	1.93		2.26	3.36
7026	1-1/2" thick	G		350	.046		.50	2.04		2.54	3.73
7028	2" thick	G		330	.048		.67	2.17		2.84	4.11
7030	3" thick	G		300	.053		.83	2.39		3.22	4.62
7032	4" thick	G	↓	280	.057	↓	1.33	2.56		3.89	5.45
7038	8 lb. density, plain										
7040	1" thick	G	Q-14	360	.044	S.F.	.40	1.99		2.39	3.53
7042	1-1/2" thick	G		340	.047		.61	2.11		2.72	3.94
7044	2" thick	G		320	.050		.81	2.24		3.05	4.36
7046	3" thick	G		290	.055		1.21	2.47		3.68	5.15
7048	4" thick	G	↓	270	.059	↓	1.61	2.65		4.26	5.90
7060	10 lb. density, plain										
7062	1" thick	G	Q-14	350	.046	S.F.	.55	2.04		2.59	3.79
7064	1-1/2" thick	G		330	.048		.83	2.17		3	4.28
7066	2" thick	G		310	.052		1.11	2.31		3.42	4.81
7068	3" thick	G		280	.057		1.66	2.56		4.22	5.80
7070	4" thick	G	↓	260	.062	↓	2.21	2.75		4.96	6.70
7878	Contact cement, quart can					Ea.	10.50			10.50	11.55

23 07 16 – HVAC Equipment Insulation

23 07 16.10 HVAC Equipment Thermal Insulation			Crew	Daily Output	Labor-Hours	Unit	Material	Labor	Equipment	Total	Total Incl O&P
0010	**HVAC EQUIPMENT THERMAL INSULATION**										
0100	Rule of thumb, as a percentage of total mechanical costs					Job				10%	10%
0110	Insulation req'd. is based on the surface size/area to be covered										
1000	Boiler, 1-1/2" calcium silicate only	G	Q-14	110	.145	S.F.	3.77	6.50		10.27	14.25
1020	Plus 2" fiberglass	G	"	80	.200	"	4.74	8.95		13.69	19.10
2000	Breeching, 2" calcium silicate										
2020	Rectangular	G	Q-14	42	.381	S.F.	7.40	17.05		24.45	34.50
2040	Round	G	"	38.70	.413	"	7.70	18.50		26.20	37
2300	Calcium silicate block, + 200°F to + 1200°F										
2310	On irregular surfaces, valves and fittings										
2340	1" thick	G	Q-14	30	.533	S.F.	3.34	24		27.34	40.50
2360	1-1/2" thick	G		25	.640		3.77	28.50		32.27	48.50
2380	2" thick	G		22	.727		4.94	32.50		37.44	56
2400	3" thick	G	↓	18	.889	↓	7.60	40		47.60	70.50
2410	On plane surfaces										
2420	1" thick	G	Q-14	126	.127	S.F.	3.34	5.70		9.04	12.45
2430	1-1/2" thick	G		120	.133		3.77	5.95		9.72	13.40
2440	2" thick	G		100	.160		4.94	7.15		12.09	16.55
2450	3" thick	G	↓	70	.229	↓	7.60	10.20		17.80	24.50

23 09 13 – Instrumentation and Control Devices for HVAC

23 09 13.60 Water Level Controls		Crew	Daily Output	Labor-Hours	Unit	Material	2013 Bare Costs Labor	Equipment	Total	Total Incl O&P
0010	**WATER LEVEL CONTROLS**									
3000	Low water cut-off for hot water boiler, 50 psi maximum									
3100	1" top & bottom equalizing pipes, manual reset	1 Stpi	14	.571	Ea.	325	32.50		357.50	410
3200	1" top & bottom equalizing pipes		14	.571		325	32.50		357.50	410
3300	2-1/2" side connection for nipple-to-boiler		14	.571		296	32.50		328.50	375

23 09 53 – Pneumatic and Electric Control System for HVAC

23 09 53.10 Control Components		Crew	Daily Output	Labor-Hours	Unit	Material	2013 Bare Costs Labor	Equipment	Total	Total Incl O&P
0010	**CONTROL COMPONENTS** R230500-10									
2000	Gauges, pressure or vacuum									
2100	2" diameter dial	1 Stpi	32	.250	Ea.	16.35	14.15		30.50	39.50
2200	2-1/2" diameter dial		32	.250		17.95	14.15		32.10	41
2300	3-1/2" diameter dial		32	.250		35.50	14.15		49.65	61
2400	4-1/2" diameter dial		32	.250		52.50	14.15		66.65	79.50
2700	Flanged iron case, black ring									
2800	3-1/2" diameter dial	1 Stpi	32	.250	Ea.	102	14.15		116.15	134
2900	4-1/2" diameter dial		32	.250		105	14.15		119.15	138
3000	6" diameter dial		32	.250		167	14.15		181.15	205
3010	Steel case, 0 – 300 psi									
3012	2" diam. dial	1 Stpi	16	.500	Ea.	8.50	28.50		37	52
3014	4" diam. dial	"	16	.500	"	39.50	28.50		68	86
3020	Aluminum case, 0 – 300 psi									
3022	3-1/2" diam. dial	1 Stpi	16	.500	Ea.	33.50	28.50		62	79.50
3024	4-1/2" diam. dial		16	.500		106	28.50		134.50	159
3026	6" diam. dial		16	.500		173	28.50		201.50	234
3028	8-1/2" diam. dial		16	.500		295	28.50		323.50	370
3030	Brass case, 0 – 300 psi									
3032	2" diam. dial	1 Stpi	16	.500	Ea.	42	28.50		70.50	88.50
3034	4-1/2" diam. dial	"	16	.500	"	107	28.50		135.50	161
3040	Steel case, high pressure, 0 -10,000 psi									
3042	4-1/2" diam. dial	1 Stpi	16	.500	Ea.	274	28.50		302.50	345
3044	6-1/2" diam. dial		16	.500		298	28.50		326.50	375
3046	8-1/2" diam. dial		16	.500		370	28.50		398.50	450
3080	Pressure gauge, differential, magnehelic									
3084	0 – 2" W.C., with air filter kit	1 Stpi	6	1.333	Ea.	98.50	75.50		174	222
3300	For compound pressure-vacuum, add					18%				
4000	Thermometers									
4100	Dial type, 3-1/2" diameter, vapor type, union connection	1 Stpi	32	.250	Ea.	238	14.15		252.15	284
4120	Liquid type, union connection		32	.250		460	14.15		474.15	525
4500	Stem type, 6-1/2" case, 2" stem, 1/2" NPT		32	.250		64.50	14.15		78.65	92.50
4520	4" stem, 1/2" NPT		32	.250		79.50	14.15		93.65	109
4600	9" case, 3-1/2" stem, 3/4" NPT		28	.286		115	16.20		131.20	151
4620	6" stem, 3/4" NPT		28	.286		133	16.20		149.20	171
4640	8" stem, 3/4" NPT		28	.286		220	16.20		236.20	267
4660	12" stem, 1" NPT		26	.308		184	17.45		201.45	229
4670	Bi-metal, dial type, steel case brass stem									
4672	2" dial, 4" - 9" stem	1 Stpi	16	.500	Ea.	34	28.50		62.50	79.50
4673	2-1/2" dial, 4" - 9" stem		16	.500		34	28.50		62.50	79.50
4674	3-1/2" dial, 4" - 9" stem		16	.500		41.50	28.50		70	88
4680	Mercury filled, industrial, union connection type									
4682	Angle stem, 7" scale	1 Stpi	16	.500	Ea.	67.50	28.50		96	117
4683	9" scale		16	.500		181	28.50		209.50	242
4684	12" scale		16	.500		234	28.50		262.50	300

23 09 Instrumentation and Control for HVAC

23 09 53 – Pneumatic and Electric Control System for HVAC

23 09 53.10 Control Components		Crew	Daily Output	Labor-Hours	Unit	2013 Bare Costs Material	Labor	Equipment	Total	Total Incl O&P
4686	Straight stem, 7" scale	1 Stpi	16	.500	Ea.	179	28.50		207.50	240
4687	9" scale		16	.500		167	28.50		195.50	227
4688	12" scale		16	.500		206	28.50		234.50	270
4690	Mercury filled, industrial, separable socket type, with well									
4692	Angle stem, with socket, 7" scale	1 Stpi	16	.500	Ea.	69	28.50		97.50	119
4693	9" scale		16	.500		72	28.50		100.50	122
4694	12" scale		16	.500		224	28.50		252.50	290
4696	Straight stem, with socket, 7" scale		16	.500		37.50	28.50		66	84
4697	9" scale		16	.500		90.50	28.50		119	142
4698	12" scale		16	.500		98	28.50		126.50	151
6000	Valves, motorized zone									
6100	Sweat connections, 1/2" C x C	1 Stpi	20	.400	Ea.	166	22.50		188.50	216
6110	3/4" C x C		20	.400		166	22.50		188.50	217
6120	1" C x C		19	.421		217	24		241	274
6140	1/2" C x C, with end switch, 2 wire		20	.400		164	22.50		186.50	214
6150	3/4" C x C, with end switch, 2 wire		20	.400		175	22.50		197.50	227
6160	1" C x C, with end switch, 2 wire		19	.421		200	24		224	256

23 11 Facility Fuel Piping

23 11 13 – Facility Fuel-Oil Piping

23 11 13.10 Fuel Oil Specialties

Code	Description	Crew	Daily Output	Labor-Hours	Unit	Material	Labor	Equipment	Total	Total Incl O&P
0010	**FUEL OIL SPECIALTIES**									
0020	Foot valve, single poppet, metal to metal construction									
0040	Bevel seat, 1/2" diameter	1 Stpi	20	.400	Ea.	61.50	22.50		84	102
0060	3/4" diameter		18	.444		71	25		96	117
1000	Oil filters, 3/8" IPT, 20 gal. per hour		20	.400		19.25	22.50		41.75	55
2000	Remote tank gauging system, self contained									
3000	Valve, ball check, globe type, 3/8" diameter	1 Stpi	24	.333	Ea.	12.80	18.90		31.70	42.50
3500	Fusible, 3/8" diameter		24	.333		13.10	18.90		32	43
3600	1/2" diameter		24	.333		32	18.90		50.90	63.50
3610	3/4" diameter		20	.400		70	22.50		92.50	111
3620	1" diameter		19	.421		205	24		229	262
4000	Nonfusible, 3/8" diameter		24	.333		22.50	18.90		41.40	53
4500	Shutoff, gate type, lever handle, spring-fusible kit									
4520	1/4" diameter	1 Stpi	14	.571	Ea.	34.50	32.50		67	86.50
4540	3/8" diameter		12	.667		33.50	38		71.50	94
4560	1/2" diameter		10	.800		49.50	45.50		95	123
4570	3/4" diameter		8	1		86	56.50		142.50	180
5000	Vent alarm, whistling signal					26			26	28.50
5500	Vent protector/breather, 1-1/4" diameter	1 Stpi	32	.250		12.80	14.15		26.95	35.50

23 11 23 – Facility Natural-Gas Piping

23 11 23.10 Gas Meters

Code	Description	Crew	Daily Output	Labor-Hours	Unit	Material	Labor	Equipment	Total	Total Incl O&P
0010	**GAS METERS**									
4000	Residential									
4010	Gas meter, residential, 3/4" pipe size	1 Plum	14	.571	Ea.	161	32		193	225
4020	Gas meter, residential, 1" pipe size		12	.667		165	37		202	237
4030	Gas meter, residential, 1-1/4" pipe size		10	.800		166	44.50		210.50	249

23 12 Facility Fuel Pumps

23 12 13 – Facility Fuel-Oil Pumps

23 12 13.10 Pump and Motor Sets		Crew	Daily Output	Labor-Hours	Unit	2013 Bare Costs Material	Labor	Equipment	Total	Total Incl O&P
0010	**PUMP AND MOTOR SETS**									
1810	Light fuel and diesel oils									
1820	20 GPH 1/3 HP	Q-5	6	2.667	Ea.	1,025	136		1,161	1,325
1850	145 GPH, 1/2 HP	″	4	4	″	1,075	204		1,279	1,500

23 13 Facility Fuel-Storage Tanks

23 13 13 – Facility Underground Fuel-Oil, Storage Tanks

23 13 13.09 Single-Wall Steel Fuel-Oil Tanks

		Crew	Daily Output	Labor-Hours	Unit	Material	Labor	Equipment	Total	Total Incl O&P
0010	**SINGLE-WALL STEEL FUEL-OIL TANKS**									
5000	Tanks, steel ugnd., sti-p3, not incl. hold-down bars									
5500	Excavation, pad, pumps and piping not included									
5510	Single wall, 500 gallon capacity, 7 ga. shell	Q-5	2.70	5.926	Ea.	2,050	300		2,350	2,700
5520	1,000 gallon capacity, 7 ga. shell	″	2.50	6.400		3,825	325		4,150	4,700
5530	2,000 gallon capacity, 1/4″ thick shell	Q-7	4.60	6.957		6,200	375		6,575	7,400
5535	2,500 gallon capacity, 7 ga. shell	Q-5	3	5.333		6,850	272		7,122	7,925
5540	5,000 gallon capacity, 1/4″ thick shell	Q-7	3.20	10		11,500	540		12,040	13,500
5560	10,000 gallon capacity, 1/4″ thick shell		2	16		12,300	865		13,165	14,900
5580	15,000 gallon capacity, 5/16″ thick shell		1.70	18.824		19,000	1,025		20,025	22,400
5600	20,000 gallon capacity, 5/16″ thick shell		1.50	21.333		36,300	1,150		37,450	41,700
5610	25,000 gallon capacity, 3/8″ thick shell		1.30	24.615		33,900	1,325		35,225	39,300
5620	30,000 gallon capacity, 3/8″ thick shell		1.10	29.091		38,400	1,575		39,975	44,600
5630	40,000 gallon capacity, 3/8″ thick shell		.90	35.556		35,600	1,925		37,525	42,100
5640	50,000 gallon capacity, 3/8″ thick shell		.80	40		39,500	2,150		41,650	46,800

23 13 13.13 Dbl-Wall Steel, Undrgrnd Fuel-Oil, Stor. Tanks

		Crew	Daily Output	Labor-Hours	Unit	Material	Labor	Equipment	Total	Total Incl O&P
0010	**DOUBLE-WALL STEEL, UNDERGROUND FUEL-OIL, STORAGE TANKS**									
6200	Steel, underground, 360°, double wall, U.L. listed,									
6210	with sti-P3 corrosion protection,									
6220	(dielectric coating, cathodic protection, electrical									
6230	isolation) 30 year warranty,									
6240	not incl. manholes or hold-downs.									
6250	500 gallon capacity	Q-5	2.40	6.667	Ea.	4,725	340		5,065	5,700
6260	1,000 gallon capactiy	″	2.25	7.111		6,900	365		7,265	8,125
6270	2,000 gallon capacity	Q-7	4.16	7.692		9,050	415		9,465	10,600
6280	3,000 gallon capacity		3.90	8.205		10,900	445		11,345	12,700
6290	4,000 gallon capacity		3.64	8.791		13,900	475		14,375	16,000
6300	5,000 gallon capacity		2.91	10.997		19,900	595		20,495	22,800
6310	6,000 gallon capacity		2.42	13.223		23,600	715		24,315	27,000
6320	8,000 gallon capacity		2.08	15.385		24,600	830		25,430	28,300
6330	10,000 gallon capacity		1.82	17.582		27,700	950		28,650	31,900
6340	12,000 gallon capacity		1.70	18.824		30,600	1,025		31,625	35,200
6350	15,000 gallon capacity		1.33	24.060		31,900	1,300		33,200	37,100
6360	20,000 gallon capacity		1.33	24.060		32,700	1,300		34,000	37,900
6370	25,000 gallon capacity		1.16	27.586		72,500	1,500		74,000	82,500
6380	30,000 gallon capacity		1.03	31.068		87,500	1,675		89,175	98,500
6390	40,000 gallon capacity		.80	40		112,500	2,150		114,650	127,000
6395	50,000 gallon capacity		.73	43.836		136,500	2,375		138,875	154,000
6400	For hold-downs 500-2000 gal., add		16	2	Set	183	108		291	365
6410	For hold-downs 3000-6000 gal., add		12	2.667		370	144		514	625
6420	For hold-downs 8000-12,000 gal., add		11	2.909		445	157		602	725
6430	For hold-downs 15,000 gal., add		9	3.556		640	192		832	995
6440	For hold-downs 20,000 gal., add		8	4		735	216		951	1,125

23 13 Facility Fuel-Storage Tanks

23 13 13 – Facility Underground Fuel-Oil, Storage Tanks

23 13 13.13 Dbl-Wall Steel, Undrgrnd Fuel-Oil, Stor. Tanks		Crew	Daily Output	Labor-Hours	Unit	2013 Bare Costs Material	Labor	Equipment	Total	Total Incl O&P
6450	For hold-downs 20,000 gal. plus, add	Q-7	6	5.333	Set	960	288		1,248	1,475
6500	For manways, add				Ea.	1,800			1,800	2,000
6600	In place with hold-downs									
6652	550 gallon capacity	Q-5	1.84	8.696	Ea.	4,925	445		5,370	6,075
23 13 13.23 Glass-Fiber-Reinfcd-Plastic, Fuel-Oil, Storage										
0010	**GLASS-FIBER-REINFCD-PLASTIC, UNDERGRND FUEL-OIL, STORAGE**									
0210	Fiberglass, underground, single wall, U.L. listed, not including									
0220	manway or hold-down strap									
0225	550 gallon capacity	Q-5	2.67	5.993	Ea.	3,950	305		4,255	4,775
0230	1,000 gallon capacity	"	2.46	6.504		4,800	330		5,130	5,775
0240	2,000 gallon capacity	Q-7	4.57	7.002		6,700	380		7,080	7,950
0250	4,000 gallon capacity		3.55	9.014		9,525	485		10,010	11,200
0260	6,000 gallon capacity		2.67	11.985		10,600	645		11,245	12,600
0270	8,000 gallon capacity		2.29	13.974		12,500	755		13,255	14,900
0280	10,000 gallon capacity		2	16		14,200	865		15,065	16,900
0282	12,000 gallon capacity		1.88	17.021		23,900	920		24,820	27,700
0284	15,000 gallon capacity		1.68	19.048		26,300	1,025		27,325	30,600
0290	20,000 gallon capacity		1.45	22.069		28,900	1,200		30,100	33,600
0300	25,000 gallon capacity		1.28	25		42,800	1,350		44,150	49,100
0320	30,000 gallon capacity		1.14	28.070		51,500	1,525		53,025	59,500
0340	40,000 gallon capacity		.89	35.955		73,500	1,950		75,450	84,000
0360	48,000 gallon capacity	↓	.81	39.506		86,500	2,125		88,625	98,500
0500	For manway, fittings and hold-downs, add					20%	15%			
0600	For manways, add					2,125			2,125	2,350
1000	For helical heating coil, add	Q-5	2.50	6.400	↓	4,325	325		4,650	5,275
1020	Fiberglass, underground, double wall, U.L. listed									
1030	includes manways, not incl. hold-down straps									
1040	600 gallon capacity	Q-5	2.42	6.612	Ea.	7,875	335		8,210	9,150
1050	1,000 gallon capacity	"	2.25	7.111		10,600	365		10,965	12,100
1060	2,500 gallon capacity	Q-7	4.16	7.692		15,100	415		15,515	17,200
1070	3,000 gallon capacity		3.90	8.205		16,400	445		16,845	18,700
1080	4,000 gallon capacity		3.64	8.791		16,500	475		16,975	18,900
1090	6,000 gallon capacity		2.42	13.223		21,200	715		21,915	24,500
1100	8,000 gallon capacity		2.08	15.385		23,500	830		24,330	27,200
1110	10,000 gallon capacity		1.82	17.582		27,000	950		27,950	31,100
1120	12,000 gallon capacity		1.70	18.824		33,100	1,025		34,125	37,900
1122	15,000 gallon capacity		1.52	21.053		43,300	1,125		44,425	49,300
1124	20,000 gallon capacity		1.33	24.060		53,000	1,300		54,300	60,000
1126	25,000 gallon capacity		1.16	27.586		69,500	1,500		71,000	79,000
1128	30,000 gallon capacity	↓	1.03	31.068		83,500	1,675		85,175	94,500
1140	For hold-down straps, add				↓	2%	10%			
1150	For hold-downs 500-4000 gal., add	Q-7	16	2	Set	405	108		513	605
1160	For hold-downs 5000-15000 gal., add		8	4		810	216		1,026	1,225
1170	For hold-downs 20,000 gal., add		5.33	6.004		1,225	325		1,550	1,825
1180	For hold-downs 25,000 gal., add		4	8		1,625	430		2,055	2,425
1190	For hold-downs 30,000 gal., add	↓	2.60	12.308	↓	2,425	665		3,090	3,675
2210	Fiberglass, underground, single wall, U.L. listed, including									
2220	hold-down straps, no manways									
2225	550 gallon capacity	Q-5	2	8	Ea.	4,350	410		4,760	5,400
2230	1,000 gallon capacity	"	1.88	8.511		5,200	435		5,635	6,375
2240	2,000 gallon capacity	Q-7	3.55	9.014		7,100	485		7,585	8,525
2250	4,000 gallon capacity		2.90	11.034		9,925	595		10,520	11,800

23 13 Facility Fuel-Storage Tanks

23 13 13 – Facility Underground Fuel-Oil, Storage Tanks

23 13 13.23	Glass-Fiber-Reinfcd-Plastic, Fuel-Oil, Storage	Crew	Daily Output	Labor-Hours	Unit	2013 Bare Costs Material	Labor	Equipment	Total	Total Incl O&P
2260	6,000 gallon capacity	Q-7	2	16	Ea.	11,400	865		12,265	13,800
2270	8,000 gallon capacity		1.78	17.978		13,400	970		14,370	16,200
2280	10,000 gallon capacity		1.60	20		15,000	1,075		16,075	18,100
2282	12,000 gallon capacity		1.52	21.053		24,800	1,125		25,925	28,900
2284	15,000 gallon capacity		1.39	23.022		27,100	1,250		28,350	31,800
2290	20,000 gallon capacity		1.14	28.070		30,100	1,525		31,625	35,400
2300	25,000 gallon capacity		.96	33.333		44,400	1,800		46,200	51,500
2320	30,000 gallon capacity		.80	40		54,000	2,150		56,150	63,000
3020	Fiberglass, underground, double wall, U.L. listed									
3030	includes manways and hold-down straps									
3040	600 gallon capacity	Q-5	1.86	8.602	Ea.	8,275	440		8,715	9,750
3050	1,000 gallon capacity	"	1.70	9.412		11,000	480		11,480	12,800
3060	2,500 gallon capacity	Q-7	3.29	9.726		15,500	525		16,025	17,900
3070	3,000 gallon capacity		3.13	10.224		16,800	550		17,350	19,300
3080	4,000 gallon capacity		2.93	10.922		16,900	590		17,490	19,500
3090	6,000 gallon capacity		1.86	17.204		22,000	930		22,930	25,700
3100	8,000 gallon capacity		1.65	19.394		24,300	1,050		25,350	28,400
3110	10,000 gallon capacity		1.48	21.622		27,800	1,175		28,975	32,400
3120	12,000 gallon capacity		1.40	22.857		33,900	1,225		35,125	39,200
3122	15,000 gallon capacity		1.28	25		44,100	1,350		45,450	50,500
3124	20,000 gallon capacity		1.06	30.189		54,000	1,625		55,625	62,000
3126	25,000 gallon capacity		.90	35.556		71,000	1,925		72,925	81,000
3128	30,000 gallon capacity		.74	43.243		86,000	2,325		88,325	98,000

23 13 23 – Facility Aboveground Fuel-Oil, Storage Tanks

23 13 23.16	Horizontal, Stl, Abvgrd Fuel-Oil, Storage Tanks	Crew	Daily Output	Labor-Hours	Unit	Material	Labor	Equipment	Total	Total Incl O&P
0010	**HORIZONTAL, STEEL, ABOVEGROUND FUEL-OIL, STORAGE TANKS**									
3000	Steel, storage, above ground, including cradles, coating,									
3020	fittings, not including foundation, pumps or piping									
3040	Single wall, 275 gallon	Q-5	5	3.200	Ea.	490	163		653	785
3060	550 gallon	"	2.70	5.926		3,100	300		3,400	3,850
3080	1,000 gallon	Q-7	5	6.400		5,075	345		5,420	6,125
3100	1,500 gallon		4.75	6.737		7,250	365		7,615	8,525
3120	2,000 gallon		4.60	6.957		8,800	375		9,175	10,200
3320	Double wall, 500 gallon capacity	Q-5	2.40	6.667		2,625	340		2,965	3,375
3330	2000 gallon capacity	Q-7	4.15	7.711		9,975	415		10,390	11,600
3340	4000 gallon capacity		3.60	8.889		17,800	480		18,280	20,300
3350	6000 gallon capacity		2.40	13.333		21,000	720		21,720	24,200
3360	8000 gallon capacity		2	16		27,000	865		27,865	31,000
3370	10000 gallon capacity		1.80	17.778		30,200	960		31,160	34,700
3380	15000 gallon capacity		1.50	21.333		45,900	1,150		47,050	52,000
3390	20000 gallon capacity		1.30	24.615		52,500	1,325		53,825	59,500
3400	25000 gallon capacity		1.15	27.826		63,500	1,500		65,000	72,500
3410	30000 gallon capacity		1	32		69,500	1,725		71,225	79,000

23 13 23.26	Horizontal, Conc., Abvgrd Fuel-Oil, Stor. Tanks	Crew	Daily Output	Labor-Hours	Unit	Material	Labor	Equipment	Total	Total Incl O&P
0010	**HORIZONTAL, CONCRETE, ABOVEGROUND FUEL-OIL, STORAGE TANKS**									
0050	Concrete, storage, above ground, including pad & pump									
0100	500 gallon	F-3	2	20	Ea.	10,000	915	325	11,240	12,800
0200	1,000 gallon	"	2	20		14,000	915	325	15,240	17,200
0300	2,000 gallon	F-4	2	24		18,000	1,075	560	19,635	22,100

23 21 20.10 Air Control

	23 21 20.10 Air Control	Crew	Daily Output	Labor-Hours	Unit	2013 Bare Costs Material	Labor	Equipment	Total	Total Incl O&P
0010	**AIR CONTROL**									
0030	Air separator, with strainer									
0040	2" diameter	Q-5	6	2.667	Ea.	1,150	136		1,286	1,450
0080	2-1/2" diameter		5	3.200		1,275	163		1,438	1,650
0100	3" diameter		4	4		1,975	204		2,179	2,475
1000	Micro-bubble separator for total air removal, closed loop system									
1010	Requires bladder type tank in system.									
1020	Water (hot or chilled) or glycol system									
1030	Threaded									
1040	3/4" diameter	1 Stpi	20	.400	Ea.	88.50	22.50		111	131
1050	1" diameter		19	.421		99.50	24		123.50	145
1060	1-1/4" diameter		16	.500		138	28.50		166.50	194
1070	1-1/2" diameter		13	.615		179	35		214	250
1080	2" diameter		11	.727		865	41		906	1,025
1090	2-1/2" diameter	Q-5	15	1.067		965	54.50		1,019.50	1,125
1100	3" diameter		13	1.231		1,375	62.50		1,437.50	1,625
1110	4" diameter		10	1.600		1,500	81.50		1,581.50	1,775

23 21 20.14 Air Purging Scoop

	23 21 20.14 Air Purging Scoop	Crew	Daily Output	Labor-Hours	Unit	Material	Labor	Equipment	Total	Total Incl O&P
0010	**AIR PURGING SCOOP**, With tappings.									
0020	For air vent and expansion tank connection									
0100	1" pipe size, threaded	1 Stpi	19	.421	Ea.	23	24		47	61
0110	1-1/4" pipe size, threaded		15	.533		23	30		53	70.50
0120	1-1/2" pipe size, threaded		13	.615		50.50	35		85.50	108
0130	2" pipe size, threaded		11	.727		57.50	41		98.50	126

23 21 20.18 Automatic Air Vent

	23 21 20.18 Automatic Air Vent	Crew	Daily Output	Labor-Hours	Unit	Material	Labor	Equipment	Total	Total Incl O&P
0010	**AUTOMATIC AIR VENT**									
0020	Cast iron body, stainless steel internals, float type									
0180	1/2" NPT inlet, 250 psi	1 Stpi	10	.800	Ea.	310	45.50		355.50	410
0220	3/4" NPT inlet, 250 psi	"	10	.800	"	310	45.50		355.50	410
0600	Forged steel body, stainless steel internals, float type									
0640	1/2" NPT inlet, 750 psi	1 Stpi	12	.667	Ea.	1,025	38		1,063	1,175
0680	3/4" NPT inlet, 750 psi	"	12	.667	"	1,025	38		1,063	1,175
1100	Formed steel body, noncorrosive									
1110	1/8" NPT inlet 150 psi	1 Stpi	32	.250	Ea.	11.75	14.15		25.90	34.50
1120	1/4" NPT inlet 150 psi		32	.250		39.50	14.15		53.65	65
1130	3/4" NPT inlet 150 psi		32	.250		39.50	14.15		53.65	65
1300	Chrome plated brass, automatic/manual, for radiators									
1310	1/8" NPT inlet, nickel plated brass	1 Stpi	32	.250	Ea.	6.85	14.15		21	29

23 21 20.26 Circuit Setter

	23 21 20.26 Circuit Setter	Crew	Daily Output	Labor-Hours	Unit	Material	Labor	Equipment	Total	Total Incl O&P
0010	**CIRCUIT SETTER**, Balance valve									
0018	Threaded									
0019	1/2" pipe size	1 Stpi	22	.364	Ea.	70.50	20.50		91	109
0020	3/4" pipe size	"	20	.400	"	75	22.50		97.50	117

23 21 20.30 Cocks, Drains and Specialties

	23 21 20.30 Cocks, Drains and Specialties	Crew	Daily Output	Labor-Hours	Unit	Material	Labor	Equipment	Total	Total Incl O&P
0010	**COCKS, DRAINS AND SPECIALTIES**									
1000	Boiler drain									
1010	Pipe thread to hose									
1020	Bronze									
1030	1/2" size	1 Stpi	36	.222	Ea.	11.15	12.60		23.75	31.50
1040	3/4" size	"	34	.235	"	12.15	13.35		25.50	33.50
1100	Solder to hose									
1110	Bronze									

23 21 20 – Hydronic HVAC Piping Specialties

23 21 20.30 Cocks, Drains and Specialties		Crew	Daily Output	Labor-Hours	Unit	Material	Labor	Equipment	Total	Total Incl O&P
						2013 Bare Costs				
1120	1/2" size	1 Stpi	46	.174	Ea.	11.15	9.85		21	27
1130	3/4" size	"	44	.182	"	12.15	10.30		22.45	29
1600	With built-in vacuum breaker									
1610	1/2" I.P. or solder	1 Stpi	36	.222	Ea.	34	12.60		46.60	56.50
1630	With tamper proof vacuum breaker									
1640	1/2" I.P. or solder	1 Stpi	36	.222	Ea.	39	12.60		51.60	61.50
1650	3/4" I.P. or solder	"	34	.235	"	39.50	13.35		52.85	63.50
3000	Cocks									
3010	Air, lever or tee handle									
3020	Bronze, single thread									
3030	1/8" size	1 Stpi	52	.154	Ea.	9.75	8.70		18.45	24
3040	1/4" size		46	.174		10.30	9.85		20.15	26
3050	3/8" size		40	.200		10.40	11.35		21.75	28.50
3060	1/2" size		36	.222		12.30	12.60		24.90	32.50
3100	Bronze, double thread									
3110	1/8" size	1 Stpi	26	.308	Ea.	12.20	17.45		29.65	40
3120	1/4" size		22	.364		12.80	20.50		33.30	45
3130	3/8" size		18	.444		13.30	25		38.30	52.50
3140	1/2" size		15	.533		16.30	30		46.30	63.50
4500	Gauge cock, brass									
4510	1/4" FPT	1 Stpi	24	.333	Ea.	10.30	18.90		29.20	40
4512	1/4" MPT	"	24	.333	"	12.95	18.90		31.85	43
4600	Pigtail, steam syphon									
4604	1/4"	1 Stpi	24	.333	Ea.	18.80	18.90		37.70	49
4650	Snubber valve									
4654	1/4"	1 Stpi	22	.364	Ea.	12.15	20.50		32.65	44.50
4660	Nipple, black steel									
4664	1/4" x 3"	1 Stpi	37	.216	Ea.	2.14	12.25		14.39	21

23 21 20.34 Dielectric Unions		Crew	Daily Output	Labor-Hours	Unit	Material	Labor	Equipment	Total	Total Incl O&P
0010	**DIELECTRIC UNIONS**, Standard gaskets for water and air									
0020	250 psi maximum pressure									
0280	Female IPT to sweat, straight									
0300	1/2" pipe size	1 Plum	24	.333	Ea.	5.45	18.60		24.05	34
0340	3/4" pipe size		20	.400		6.10	22.50		28.60	40
0360	1" pipe size		19	.421		8.30	23.50		31.80	44.50
0380	1-1/4" pipe size		15	.533		12.85	30		42.85	59
0400	1-1/2" pipe size		13	.615		19.30	34.50		53.80	72.50
0420	2" pipe size		11	.727		26	40.50		66.50	89.50
0580	Female IPT to brass pipe thread, straight									
0600	1/2" pipe size	1 Plum	24	.333	Ea.	11.80	18.60		30.40	41
0640	3/4" pipe size		20	.400		13.05	22.50		35.55	48
0660	1" pipe size		19	.421		24	23.50		47.50	62
0680	1-1/4" pipe size		15	.533		29.50	30		59.50	77.50
0700	1-1/2" pipe size		13	.615		43	34.50		77.50	98.50
0720	2" pipe size		11	.727		84	40.50		124.50	154
0780	Female IPT to female IPT, straight									
0800	1/2" pipe size	1 Plum	24	.333	Ea.	11.05	18.60		29.65	40
0840	3/4" pipe size		20	.400		12.50	22.50		35	47.50
0860	1" pipe size		19	.421		16.70	23.50		40.20	54
0880	1-1/4" pipe size		15	.533		22.50	30		52.50	70
0900	1-1/2" pipe size		13	.615		34.50	34.50		69	89.50
0920	2" pipe size		11	.727		50.50	40.50		91	117

23 21 Hydronic Piping and Pumps

23 21 20 – Hydronic HVAC Piping Specialties

23 21 20.34 Dielectric Unions		Crew	Daily Output	Labor-Hours	Unit	Material	Labor	Equipment	Total	Total Incl O&P
						2013 Bare Costs				
2000	175 psi maximum pressure									
2180	Female IPT to sweat									
2240	2″ pipe size	1 Plum	9	.889	Ea.	150	49.50		199.50	239
2260	2-1/2″ pipe size	Q-1	15	1.067		162	53.50		215.50	260
2280	3″ pipe size		14	1.143		223	57.50		280.50	330
2300	4″ pipe size		11	1.455		590	73		663	760
2480	Female IPT to brass pipe									
2500	1-1/2″ pipe size	1 Plum	11	.727	Ea.	176	40.50		216.50	255
2540	2″ pipe size	″	9	.889		207	49.50		256.50	305
2560	2-1/2″ pipe size	Q-1	15	1.067		305	53.50		358.50	415
2580	3″ pipe size		14	1.143		355	57.50		412.50	475
2600	4″ pipe size		11	1.455		635	73		708	810
23 21 20.42 Expansion Joints										
0010	**EXPANSION JOINTS**									
0100	Bellows type, neoprene cover, flanged spool									
0140	6″ face to face, 1-1/4″ diameter	1 Stpi	11	.727	Ea.	255	41		296	345
0160	1-1/2″ diameter	″	10.60	.755		255	43		298	345
0180	2″ diameter	Q-5	13.30	1.203		258	61.50		319.50	375
0190	2-1/2″ diameter		12.40	1.290		267	66		333	395
0200	3″ diameter		11.40	1.404		299	71.50		370.50	440
0210	4″ diameter		8.40	1.905		325	97		422	500
0220	5″ diameter		7.60	2.105		395	107		502	595
0230	6″ diameter		6.80	2.353		405	120		525	630
0240	8″ diameter		5.40	2.963		475	151		626	745
0250	10″ diameter		5	3.200		650	163		813	960
0260	12″ diameter		4.60	3.478		745	177		922	1,075
0480	10″ face to face, 2″ diameter		13	1.231		370	62.50		432.50	505
0500	2-1/2″ diameter		12	1.333		390	68		458	530
0520	3″ diameter		11	1.455		400	74		474	550
0540	4″ diameter		8	2		455	102		557	655
0560	5″ diameter		7	2.286		540	117		657	770
0580	6″ diameter		6	2.667		560	136		696	820
0600	8″ diameter		5	3.200		670	163		833	980
0620	10″ diameter		4.60	3.478		735	177		912	1,075
0640	12″ diameter		4	4		910	204		1,114	1,300
0660	14″ diameter		3.80	4.211		1,125	215		1,340	1,575
0680	16″ diameter		2.90	5.517		1,300	281		1,581	1,850
0700	18″ diameter		2.50	6.400		1,475	325		1,800	2,125
0720	20″ diameter		2.10	7.619		1,550	390		1,940	2,275
0740	24″ diameter		1.80	8.889		1,800	455		2,255	2,650
0760	26″ diameter		1.40	11.429		2,025	585		2,610	3,100
0780	30″ diameter		1.20	13.333		2,250	680		2,930	3,500
0800	36″ diameter		1	16		2,800	815		3,615	4,300
23 21 20.46 Expansion Tanks										
0010	**EXPANSION TANKS**									
1502	Plastic, corrosion resistant, see Section 22 66 83.13									
1505	Aboveground fuel-oil, storage tanks, see Section 23 13 23									
1507	Underground fuel-oil storage tanks, see Section 23 13 13									
1512	Tank leak detection systems, see Section 28 33 33.50									
2000	Steel, liquid expansion, ASME, painted, 15 gallon capacity	Q-5	17	.941	Ea.	595	48		643	725
2020	24 gallon capacity		14	1.143		635	58.50		693.50	790
2040	30 gallon capacity		12	1.333		665	68		733	835

23 21 20 – Hydronic HVAC Piping Specialties

23 21 20.46 Expansion Tanks		Crew	Daily Output	Labor-Hours	Unit	Material	Labor	Equipment	Total	Total Incl O&P
						2013 Bare Costs				
2060	40 gallon capacity	Q-5	10	1.600	Ea.	780	81.50		861.50	985
2360	Galvanized									
2370	15 gallon capacity	Q-5	17	.941	Ea.	1,050	48		1,098	1,225
2380	24 gallon capacity		14	1.143		1,075	58.50		1,133.50	1,275
2390	30 gallon capacity		12	1.333		1,250	68		1,318	1,475
3000	Steel ASME expansion, rubber diaphragm, 19 gal. cap. accept.		12	1.333		2,800	68		2,868	3,175
3020	31 gallon capacity		8	2		3,050	102		3,152	3,500
3040	61 gallon capacity		6	2.667		3,650	136		3,786	4,225

23 21 20.50 Float Valves		Crew	Daily Output	Labor-Hours	Unit	Material	Labor	Equipment	Total	Total Incl O&P
0010	**FLOAT VALVES**									
0020	With ball and bracket									
0030	Single seat, threaded									
0040	Brass body									
0050	1/2"	1 Stpi	11	.727	Ea.	81	41		122	151
0060	3/4"		9	.889		92	50.50		142.50	177
0070	1"		7	1.143		122	64.50		186.50	232
0080	1-1/2"		4.50	1.778		185	101		286	355
0090	2"		3.60	2.222		189	126		315	395
0300	For condensate receivers, CI, in-line mount									
0320	1" inlet	1 Stpi	7	1.143	Ea.	122	64.50		186.50	232
0360	For condensate receiver, CI, external float, flanged tank mount									
0370	3/4" inlet	1 Stpi	5	1.600	Ea.	92	90.50		182.50	237

23 21 20.54 Flow Check Control		Crew	Daily Output	Labor-Hours	Unit	Material	Labor	Equipment	Total	Total Incl O&P
0010	**FLOW CHECK CONTROL**									
0100	Bronze body, soldered									
0110	3/4" size	1 Stpi	20	.400	Ea.	60	22.50		82.50	101
0120	1" size	"	19	.421	"	72.50	24		96.50	116
0200	Cast iron body, threaded									
0210	3/4" size	1 Stpi	20	.400	Ea.	46.50	22.50		69	85
0220	1" size		19	.421		52.50	24		76.50	93.50
0230	1-1/4" size		15	.533		63.50	30		93.50	116
0240	1-1/2" size		13	.615		97	35		132	160
0250	2" size		11	.727		140	41		181	216

23 21 20.58 Hydronic Heating Control Valves		Crew	Daily Output	Labor-Hours	Unit	Material	Labor	Equipment	Total	Total Incl O&P
0010	**HYDRONIC HEATING CONTROL VALVES**									
0050	Hot water, nonelectric, thermostatic									
0100	Radiator supply, 1/2" diameter	1 Stpi	24	.333	Ea.	65.50	18.90		84.40	101
0120	3/4" diameter	"	20	.400	"	68.50	22.50		91	109
1000	Manual, radiator supply									
1010	1/2" pipe size, angle union	1 Stpi	24	.333	Ea.	44.50	18.90		63.40	77.50
1020	3/4" pipe size, angle union	"	20	.400	"	56	22.50		78.50	96
1100	Radiator, balancing, straight, sweat connections									
1110	1/2" pipe size	1 Stpi	24	.333	Ea.	17	18.90		35.90	47
1120	3/4" pipe size	"	20	.400	"	24	22.50		46.50	60
1200	Steam, radiator, supply									
1210	1/2" pipe size, angle union	1 Stpi	24	.333	Ea.	43.50	18.90		62.40	76.50
1220	3/4" pipe size, angle union	"	20	.400	"	47.50	22.50		70	86
8000	System balancing and shut-off									
8020	Butterfly, quarter turn, calibrated, threaded or solder									
8040	Bronze, -30°F to +350°F, pressure to 175 psi									
8060	1/2" size	1 Stpi	22	.364	Ea.	16.50	20.50		37	49
8070	3/4" size	"	20	.400	"	26.50	22.50		49	63.50

23 21 20 – Hydronic HVAC Piping Specialties

23 21 20.66 Monoflow Tee Fitting		Crew	Daily Output	Labor-Hours	Unit	Material	Labor	Equipment	Total	Total Incl O&P
0010	**MONOFLOW TEE FITTING**									
1100	For one pipe hydronic, supply and return									
1110	Copper, soldered									
1120	3/4" x 1/2" size	1 Stpi	13	.615	Ea.	19.50	35		54.50	74

23 21 20.74 Strainers, Basket Type		Crew	Daily Output	Labor-Hours	Unit	Material	Labor	Equipment	Total	Total Incl O&P
0010	**STRAINERS, BASKET TYPE**, Perforated stainless steel basket									
0100	Brass or monel available									
2000	Simplex style									
2300	Bronze body									
2320	Screwed, 3/8" pipe size	1 Stpi	22	.364	Ea.	164	20.50		184.50	211
2340	1/2" pipe size		20	.400		266	22.50		288.50	325
2360	3/4" pipe size		17	.471		355	26.50		381.50	430
2380	1" pipe size		15	.533		355	30		385	435
2400	1-1/4" pipe size		13	.615		510	35		545	615
2420	1-1/2" pipe size		12	.667		510	38		548	620
2440	2" pipe size		10	.800		755	45.50		800.50	900
2460	2-1/2" pipe size	Q-5	15	1.067		735	54.50		789.50	890
2480	3" pipe size	"	14	1.143		1,075	58.50		1,133.50	1,275
2600	Flanged, 2" pipe size	1 Stpi	6	1.333		780	75.50		855.50	975
2620	2-1/2" pipe size	Q-5	4.50	3.556		1,225	181		1,406	1,625
2640	3" pipe size		3.50	4.571		1,375	233		1,608	1,875
2660	4" pipe size		3	5.333		2,325	272		2,597	2,975
2680	5" pipe size	Q-6	3.40	7.059		3,575	375		3,950	4,475
2700	6" pipe size		3	8		4,550	425		4,975	5,625
2710	8" pipe size		2.50	9.600		7,400	510		7,910	8,900
3600	Iron body									
3700	Screwed, 3/8" pipe size	1 Stpi	22	.364	Ea.	98.50	20.50		119	140
3720	1/2" pipe size		20	.400		102	22.50		124.50	146
3740	3/4" pipe size		17	.471		130	26.50		156.50	183
3760	1" pipe size		15	.533		133	30		163	192
3780	1-1/4" pipe size		13	.615		173	35		208	243
3800	1-1/2" pipe size		12	.667		191	38		229	267
3820	2" pipe size		10	.800		228	45.50		273.50	320
3840	2-1/2" pipe size	Q-5	15	1.067		305	54.50		359.50	415
3860	3" pipe size	"	14	1.143		370	58.50		428.50	495
4000	Flanged, 2" pipe size	1 Stpi	6	1.333		355	75.50		430.50	505
4020	2-1/2" pipe size	Q-5	4.50	3.556		475	181		656	800
4040	3" pipe size		3.50	4.571		500	233		733	900
4060	4" pipe size		3	5.333		760	272		1,032	1,250
4080	5" pipe size	Q-6	3.40	7.059		1,150	375		1,525	1,825
4100	6" pipe size		3	8		1,475	425		1,900	2,250
4120	8" pipe size		2.50	9.600		2,675	510		3,185	3,725
4140	10" pipe size		2.20	10.909		5,875	575		6,450	7,350
7000	Stainless steel body									
7200	Screwed, 1" pipe size	1 Stpi	15	.533	Ea.	283	30		313	355
7210	1-1/4" pipe size		13	.615		440	35		475	540
7220	1-1/2" pipe size		12	.667		440	38		478	540
7240	2" pipe size		10	.800		655	45.50		700.50	790
7260	2-1/2" pipe size	Q-5	15	1.067		925	54.50		979.50	1,100
7280	3" pipe size	"	14	1.143		1,275	58.50		1,333.50	1,500
7400	Flanged, 2" pipe size	1 Stpi	6	1.333		1,075	75.50		1,150.50	1,300
7420	2-1/2" pipe size	Q-5	4.50	3.556		1,950	181		2,131	2,400

23 21 20.74 Strainers, Basket Type		Crew	Daily Output	Labor-Hours	Unit	Material	2013 Bare Costs Labor	Equipment	Total	Total Incl O&P
7440	3" pipe size	Q-5	3.50	4.571	Ea.	1,975	233		2,208	2,525
7460	4" pipe size		3	5.333		3,125	272		3,397	3,825
7480	6" pipe size	Q-6	3	8		5,425	425		5,850	6,600
7500	8" pipe size	"	2.50	9.600		15,000	510		15,510	17,300
8100	Duplex style									
8200	Bronze body									
8240	Screwed, 3/4" pipe size	1 Stpi	16	.500	Ea.	1,100	28.50		1,128.50	1,275
8260	1" pipe size		14	.571		1,100	32.50		1,132.50	1,275
8280	1-1/4" pipe size		12	.667		2,175	38		2,213	2,450
8300	1-1/2" pipe size		11	.727		2,175	41		2,216	2,450
8320	2" pipe size		9	.889		3,525	50.50		3,575.50	3,950
8340	2-1/2" pipe size	Q-5	14	1.143		4,550	58.50		4,608.50	5,100
8420	Flanged, 2" pipe size	1 Stpi	6	1.333		3,775	75.50		3,850.50	4,275
8440	2-1/2" pipe size	Q-5	4.50	3.556		5,225	181		5,406	6,025
8460	3" pipe size		3.50	4.571		5,700	233		5,933	6,600
8480	4" pipe size		3	5.333		8,175	272		8,447	9,375
8500	5" pipe size	Q-6	3.40	7.059		18,200	375		18,575	20,600
8520	6" pipe size	"	3	8		18,200	425		18,625	20,600
8700	Iron body									
8740	Screwed, 3/4" pipe size	1 Stpi	16	.500	Ea.	665	28.50		693.50	775
8760	1" pipe size		14	.571		665	32.50		697.50	780
8780	1-1/4" pipe size		12	.667		1,175	38		1,213	1,325
8800	1-1/2" pipe size		11	.727		1,175	41		1,216	1,325
8820	2" pipe size		9	.889		1,975	50.50		2,025.50	2,250
8840	2-1/2" pipe size	Q-5	14	1.143		2,175	58.50		2,233.50	2,500
9000	Flanged, 2" pipe size	1 Stpi	6	1.333		2,100	75.50		2,175.50	2,450
9020	2-1/2" pipe size	Q-5	4.50	3.556		2,250	181		2,431	2,750
9040	3" pipe size		3.50	4.571		2,450	233		2,683	3,050
9060	4" pipe size		3	5.333		4,150	272		4,422	4,950
9080	5" pipe size	Q-6	3.40	7.059		8,200	375		8,575	9,575
9100	6" pipe size		3	8		8,200	425		8,625	9,650
9120	8" pipe size		2.50	9.600		15,000	510		15,510	17,300
9140	10" pipe size		2.20	10.909		20,800	575		21,375	23,800
9160	12" pipe size		1.70	14.118		23,100	745		23,845	26,500
9170	14" pipe size		1.40	17.143		28,000	905		28,905	32,200
9180	16" pipe size		1	24		35,100	1,275		36,375	40,500
9700	Stainless steel body									
9740	Screwed, 1" pipe size	1 Stpi	14	.571	Ea.	1,900	32.50		1,932.50	2,125
9760	1-1/2" pipe size		11	.727		2,850	41		2,891	3,200
9780	2" pipe size		9	.889		4,000	50.50		4,050.50	4,475
9860	Flanged, 2" pipe size		6	1.333		4,350	75.50		4,425.50	4,925
9880	2-1/2" pipe size	Q-5	4.50	3.556		7,300	181		7,481	8,300
9900	3" pipe size		3.50	4.571		7,950	233		8,183	9,100
9920	4" pipe size		3	5.333		10,400	272		10,672	11,800
9940	6" pipe size	Q-6	3	8		15,800	425		16,225	18,000
9960	8" pipe size	"	2.50	9.600		44,200	510		44,710	49,400

23 21 20.76 Strainers, Y Type, Bronze Body		Crew	Daily Output	Labor-Hours	Unit	Material	Labor	Equipment	Total	Total Incl O&P
0010	**STRAINERS, Y TYPE, BRONZE BODY**									
0050	Screwed, 125 lb., 1/4" pipe size	1 Stpi	24	.333	Ea.	22	18.90		40.90	52.50
0070	3/8" pipe size		24	.333		29	18.90		47.90	60
0100	1/2" pipe size		20	.400		29	22.50		51.50	65.50
0120	3/4" pipe size		19	.421		32.50	24		56.50	72

23 21 20.76 Strainers, Y Type, Bronze Body		Crew	Daily Output	Labor-Hours	Unit	Material	2013 Bare Costs Labor	Equipment	Total	Total Incl O&P
0140	1" pipe size	1 Stpi	17	.471	Ea.	38.50	26.50		65	82
0150	1-1/4" pipe size		15	.533		77.50	30		107.50	131
0160	1-1/2" pipe size		14	.571		83	32.50		115.50	140
0180	2" pipe size		13	.615		110	35		145	174
0182	3" pipe size		12	.667		835	38		873	975
0200	300 lb., 2-1/2" pipe size	Q-5	17	.941		485	48		533	605
0220	3" pipe size		16	1		960	51		1,011	1,125
0240	4" pipe size		15	1.067		2,200	54.50		2,254.50	2,475
0500	For 300 lb. rating 1/4" thru 2", add					15%				
1000	Flanged, 150 lb., 1-1/2" pipe size	1 Stpi	11	.727	Ea.	450	41		491	555
1020	2" pipe size	"	8	1		610	56.50		666.50	755
1030	2-1/2" pipe size	Q-5	5	3.200		905	163		1,068	1,250
1040	3" pipe size		4.50	3.556		1,125	181		1,306	1,500
1060	4" pipe size		3	5.333		1,700	272		1,972	2,250
1080	5" pipe size	Q-6	3.40	7.059		1,700	375		2,075	2,400
1100	6" pipe size		3	8		3,225	425		3,650	4,175
1106	8" pipe size		2.60	9.231		3,550	490		4,040	4,625
1500	For 300 lb. rating, add					40%				

23 21 20.78 Strainers, Y Type, Iron Body		Crew	Daily Output	Labor-Hours	Unit	Material	Labor	Equipment	Total	Total Incl O&P
0010	**STRAINERS, Y TYPE, IRON BODY**									
0050	Screwed, 250 lb., 1/4" pipe size	1 Stpi	20	.400	Ea.	10.55	22.50		33.05	45.50
0070	3/8" pipe size		20	.400		10.55	22.50		33.05	45.50
0100	1/2" pipe size		20	.400		10.55	22.50		33.05	45.50
0120	3/4" pipe size		18	.444		12.50	25		37.50	52
0140	1" pipe size		16	.500		17.30	28.50		45.80	61.50
0150	1-1/4" pipe size		15	.533		23	30		53	71
0160	1-1/2" pipe size		12	.667		28.50	38		66.50	88
0180	2" pipe size		8	1		42.50	56.50		99	133
0200	2-1/2" pipe size	Q-5	12	1.333		255	68		323	380
0220	3" pipe size		11	1.455		275	74		349	415
0240	4" pipe size		5	3.200		465	163		628	760
0500	For galvanized body, add					50%				
1000	Flanged, 125 lb., 1-1/2" pipe size	1 Stpi	11	.727	Ea.	147	41		188	224
1020	2" pipe size	"	8	1		108	56.50		164.50	205
1030	2-1/2" pipe size	Q-5	5	3.200		110	163		273	365
1040	3" pipe size		4.50	3.556		143	181		324	430
1060	4" pipe size		3	5.333		260	272		532	695
1080	5" pipe size	Q-6	3.40	7.059		405	375		780	1,000
1100	6" pipe size		3	8		495	425		920	1,175
1120	8" pipe size		2.50	9.600		830	510		1,340	1,675
1140	10" pipe size		2	12		1,600	635		2,235	2,700
1160	12" pipe size		1.70	14.118		2,375	745		3,120	3,750
1170	14" pipe size		1.30	18.462		4,400	975		5,375	6,325
1180	16" pipe size		1	24		6,225	1,275		7,500	8,750
1500	For 250 lb. rating, add					20%				
2000	For galvanized body, add					50%				
2500	For steel body, add					40%				

23 21 20.80 Suction Diffusers		Crew	Daily Output	Labor-Hours	Unit	Material	Labor	Equipment	Total	Total Incl O&P
0010	**SUCTION DIFFUSERS**									
0100	Cast iron body with integral straightening vanes, strainer									
1000	Flanged									
1010	2" inlet, 1-1/2" pump side	1 Stpi	6	1.333	Ea.	305	75.50		380.50	450

23 21 20 – Hydronic HVAC Piping Specialties

23 21 20.80 Suction Diffusers		Crew	Daily Output	Labor-Hours	Unit	Material	Labor	Equipment	Total	Total Incl O&P
						2013 Bare Costs				
1020	2" pump side	1 Stpi	5	1.600	Ea.	315	90.50		405.50	485
1030	3" inlet, 2" pump side	Q-5	6.50	2.462	↓	380	125		505	610

23 21 20.84 Thermoflo Indicator

		Crew	Daily Output	Labor-Hours	Unit	Material	Labor	Equipment	Total	Total Incl O&P
0010	**THERMOFLO INDICATOR**, For balancing									
1000	Sweat connections, 1-1/4" pipe size	1 Stpi	12	.667	Ea.	610	38		648	725
1020	1-1/2" pipe size	"	10	.800	"	620	45.50		665.50	750

23 21 20.88 Venturi Flow

		Crew	Daily Output	Labor-Hours	Unit	Material	Labor	Equipment	Total	Total Incl O&P
0010	**VENTURI FLOW**, Measuring device									
0050	1/2" diameter	1 Stpi	24	.333	Ea.	281	18.90		299.90	340
0100	3/4" diameter		20	.400		257	22.50		279.50	315
0120	1" diameter		19	.421		276	24		300	340
0140	1-1/4" diameter		15	.533		340	30		370	420
0160	1-1/2" diameter		13	.615		355	35		390	445
0180	2" diameter	↓	11	.727		365	41		406	465
0200	2-1/2" diameter	Q-5	16	1		500	51		551	625
0220	3" diameter		14	1.143		515	58.50		573.50	660
0240	4" diameter	↓	11	1.455		775	74		849	960
0260	5" diameter	Q-6	4	6		1,025	315		1,340	1,600
0280	6" diameter		3.50	6.857		1,125	365		1,490	1,800
0300	8" diameter		3	8		1,450	425		1,875	2,225
0320	10" diameter		2	12		3,425	635		4,060	4,725
0330	12" diameter		1.80	13.333		4,950	705		5,655	6,500
0340	14" diameter		1.60	15		5,600	795		6,395	7,375
0350	16" diameter	↓	1.40	17.143		6,250	905		7,155	8,250
0500	For meter, add				↓	2,125			2,125	2,350

23 21 23 – Hydronic Pumps

23 21 23.13 In-Line Centrifugal Hydronic Pumps

		Crew	Daily Output	Labor-Hours	Unit	Material	Labor	Equipment	Total	Total Incl O&P
0010	**IN-LINE CENTRIFUGAL HYDRONIC PUMPS**									
0600	Bronze, sweat connections, 1/40 HP, in line									
0640	3/4" size	Q-1	16	1	Ea.	206	50		256	300
1000	Flange connection, 3/4" to 1-1/2" size									
1040	1/12 HP	Q-1	6	2.667	Ea.	530	134		664	785
1060	1/8 HP		6	2.667		915	134		1,049	1,200
1100	1/3 HP		6	2.667		1,025	134		1,159	1,325
1140	2" size, 1/6 HP		5	3.200		1,325	161		1,486	1,700
1180	2-1/2" size, 1/4 HP		5	3.200		1,700	161		1,861	2,125
1220	3" size, 1/4 HP		4	4		1,800	201		2,001	2,275
1260	1/3 HP		4	4		2,200	201		2,401	2,700
1300	1/2 HP		4	4		2,225	201		2,426	2,750
1340	3/4 HP		4	4		2,450	201		2,651	2,975
1380	1 HP	↓	4	4	↓	3,950	201		4,151	4,625
2000	Cast iron, flange connection									
2040	3/4" to 1-1/2" size, in line, 1/12 HP	Q-1	6	2.667	Ea.	345	134		479	580
2060	1/8 HP		6	2.667		575	134		709	830
2100	1/3 HP		6	2.667		640	134		774	905
2140	2" size, 1/6 HP		5	3.200		700	161		861	1,000
2180	2-1/2" size, 1/4 HP		5	3.200		925	161		1,086	1,275
2220	3" size, 1/4 HP		4	4		935	201		1,136	1,325
2260	1/3 HP		4	4		1,275	201		1,476	1,700
2300	1/2 HP		4	4		1,325	201		1,526	1,750
2340	3/4 HP		4	4		1,525	201		1,726	1,975
2380	1 HP	↓	4	4	↓	2,200	201		2,401	2,700

23 21 23 – Hydronic Pumps

23 21 23.13 In-Line Centrifugal Hydronic Pumps		Crew	Daily Output	Labor-Hours	Unit	2013 Bare Costs Material	Labor	Equipment	Total	Total Incl O&P
2600	For non-ferrous impeller, add					3%				
3000	High head, bronze impeller									
3030	1-1/2" size 1/2 HP	Q-1	5	3.200	Ea.	1,100	161		1,261	1,475
3040	1-1/2" size 3/4 HP		5	3.200		1,200	161		1,361	1,575
3050	2" size 1 HP		4	4		1,475	201		1,676	1,925
3090	2" size 1-1/2 HP		4	4		1,825	201		2,026	2,300
4000	Close coupled, end suction, bronze impeller									
4040	1-1/2" size, 1-1/2 HP, to 40 GPM	Q-1	3	5.333	Ea.	2,125	268		2,393	2,750
4090	2" size, 2 HP, to 50 GPM		3	5.333		2,525	268		2,793	3,175
4100	2" size, 3 HP, to 90 GPM		2.30	6.957		2,625	350		2,975	3,425
4190	2-1/2" size, 3 HP, to 150 GPM		2	8		2,850	400		3,250	3,725
4300	3" size, 5 HP, to 225 GPM		1.80	8.889		3,250	445		3,695	4,250
4410	3" size, 10 HP, to 350 GPM		1.60	10		4,775	500		5,275	6,000
4420	4" size, 7-1/2 HP, to 350 GPM		1.60	10		4,250	500		4,750	5,400
4520	4" size, 10 HP, to 600 GPM	Q-2	1.70	14.118		4,850	735		5,585	6,425
4530	5" size, 15 HP, to 1000 GPM		1.70	14.118		4,850	735		5,585	6,450
4610	5" size, 20 HP, to 1350 GPM		1.50	16		5,175	835		6,010	6,925
4620	5" size, 25 HP, to 1550 GPM		1.50	16		7,050	835		7,885	9,025
5000	Base mounted, bronze impeller, coupling guard									
5040	1-1/2" size, 1-1/2 HP, to 40 GPM	Q-1	2.30	6.957	Ea.	5,625	350		5,975	6,725
5090	2" size, 2 HP, to 50 GPM		2.30	6.957		6,375	350		6,725	7,550
5100	2" size, 3 HP, to 90 GPM		2	8		6,550	400		6,950	7,800
5190	2-1/2" size, 3 HP, to 150 GPM		1.80	8.889		6,975	445		7,420	8,350
5300	3" size, 5 HP, to 225 GPM		1.60	10		7,450	500		7,950	8,950
5410	4" size, 5 HP, to 350 GPM		1.50	10.667		7,650	535		8,185	9,200
5420	4" size, 7-1/2 HP, to 350 GPM		1.50	10.667		8,400	535		8,935	10,100
5520	5" size, 10 HP, to 600 GPM	Q-2	1.60	15		10,200	780		10,980	12,400
5530	5" size, 15 HP, to 1000 GPM		1.60	15		11,100	780		11,880	13,400
5610	6" size, 20 HP, to 1350 GPM		1.40	17.143		12,500	895		13,395	15,100
5620	6" size, 25 HP, to 1550 GPM		1.40	17.143		13,800	895		14,695	16,500
5800	The above pump capacities are based on 1800 RPM,									
5810	at a 60 foot head. Increasing the RPM									
5820	or decreasing the head will increase the GPM.									

23 21 29 – Automatic Condensate Pump Units

23 21 29.10 Condensate Removal Pump System		Crew	Daily Output	Labor-Hours	Unit	Material	Labor	Equipment	Total	Total Incl O&P
0010	**CONDENSATE REMOVAL PUMP SYSTEM**									
0020	Pump with 1 gal. ABS tank G									
0100	115 V G									
0120	1/50 HP, 200 GPH G	1 Stpi	12	.667	Ea.	191	38		229	267
0140	1/18 HP, 270 GPH G		10	.800		204	45.50		249.50	292
0160	1/5 HP, 450 GPH G		8	1		455	56.50		511.50	585
0200	230 V G									
0240	1/18 HP, 270 GPH G	1 Stpi	10	.800	Ea.	211	45.50		256.50	300
0260	1/5 HP, 450 GPH G	″	8	1	″	505	56.50		561.50	640

23 22 Steam and Condensate Piping and Pumps

23 22 23 – Steam Condensate Pumps

23 22 23.23 Pumps, Pneumatic Ejector		Crew	Daily Output	Labor-Hours	Unit	Material	2013 Bare Costs Labor	Equipment	Total	Total Incl O&P
0010	**PUMPS, PNEUMATIC EJECTOR**									
0020	With steel receiver, level controls, inlet/outlet gate/check valves									
0030	Cross connect. not incl. compressor, fittings or piping									
0040	Duplex									
0050	30 GPM	Q-2	1.70	14.118	Ea.	30,800	735		31,535	35,000
0060	50 GPM		1.56	15.385		31,900	800		32,700	36,300
0070	100 GPM		1.30	18.462		62,500	960		63,460	70,500
0080	150 GPM		1.10	21.818		73,000	1,125		74,125	82,000
0090	200 GPM		.85	28.235		78,500	1,475		79,975	88,000
0100	250 GPM	Q-3	.91	35.165		83,500	1,875		85,375	95,000
0110	300 GPM	"	.57	56.140		136,000	2,975		138,975	154,000

23 23 Refrigerant Piping

23 23 23 – Refrigerants

23 23 23.10 Anti-Freeze		Crew	Daily Output	Labor-Hours	Unit	Material	Labor	Equipment	Total	Total Incl O&P
0010	**ANTI-FREEZE**, Inhibited									
0900	Ethylene glycol concentrated									
1000	55 gallon drums, small quantities				Gal.	10.60			10.60	11.65
1200	Large quantities					8.80			8.80	9.70
2000	Propylene glycol, for solar heat, small quantities					14.50			14.50	15.95
2100	Large quantities					12.70			12.70	13.95

23 34 HVAC Fans

23 34 14 – Blower HVAC Fans

23 34 14.10 Blower Type HVAC Fans		Crew	Daily Output	Labor-Hours	Unit	Material	Labor	Equipment	Total	Total Incl O&P
0010	**BLOWER TYPE HVAC FANS**									
2500	Ceiling fan, right angle, extra quiet, 0.10" S.P.									
2520	95 CFM	Q-20	20	1	Ea.	273	49		322	375
2540	210 CFM		19	1.053		325	51.50		376.50	435
2560	385 CFM		18	1.111		410	54.50		464.50	535
2640	For wall or roof cap, add	1 Shee	16	.500		273	26.50		299.50	340
2660	For straight thru fan, add					10%				
2680	For speed control switch, add	1 Elec	16	.500		149	26		175	203

23 34 23 – HVAC Power Ventilators

23 34 23.10 HVAC Power Circulators and Ventilators		Crew	Daily Output	Labor-Hours	Unit	Material	Labor	Equipment	Total	Total Incl O&P
0010	**HVAC POWER CIRCULATORS AND VENTILATORS**									
6650	Residential, bath exhaust, grille, back draft damper									
6660	50 CFM	Q-20	24	.833	Ea.	57	40.50		97.50	125
6670	110 CFM		22	.909		90.50	44.50		135	168
6680	Light combination, squirrel cage, 100 watt, 70 CFM		24	.833		100	40.50		140.50	172
6700	Light/heater combination, ceiling mounted									
6710	70 CFM, 1450 watt	Q-20	24	.833	Ea.	149	40.50		189.50	226
6800	Heater combination, recessed, 70 CFM		24	.833		60.50	40.50		101	129
6820	With 2 infrared bulbs		23	.870		94.50	42.50		137	169
6940	Residential roof jacks and wall caps									
6944	Wall cap with back draft damper									
6946	3" & 4" diam. round duct	1 Shee	11	.727	Ea.	24	39		63	85
6948	6" diam. round duct	"	11	.727	"	59	39		98	124
6958	Roof jack with bird screen and back draft damper									

23 34 HVAC Fans

23 34 23 – HVAC Power Ventilators

23 34 23.10 HVAC Power Circulators and Ventilators		Crew	Daily Output	Labor-Hours	Unit	Material	2013 Bare Costs Labor	Equipment	Total	Total Incl O&P
6960	3" & 4" diam. round duct	1 Shee	11	.727	Ea.	24	39		63	85.50
6962	3-1/4" x 10" rectangular duct	"	10	.800	"	44.50	42.50		87	114
6980	Transition									
6982	3-1/4" x 10" to 6" diam. round	1 Shee	20	.400	Ea.	29	21.50		50.50	64

23 35 Special Exhaust Systems

23 35 16 – Engine Exhaust Systems

23 35 16.10 Engine Exhaust Removal Systems

		Crew	Daily Output	Labor-Hours	Unit	Material	Labor	Equipment	Total	Total Incl O&P
0010	**ENGINE EXHAUST REMOVAL SYSTEMS** D3090–320									
0500	Engine exhaust, garage, in-floor system									
0510	Single tube outlet assemblies									
0520	For transite pipe ducting, self-storing tube									
0530	3" tubing adapter plate	1 Shee	16	.500	Ea.	239	26.50		265.50	305
0540	4" tubing adapter plate		16	.500		243	26.50		269.50	310
0550	5" tubing adapter plate		16	.500		243	26.50		269.50	310
0600	For vitrified tile ducting									
0610	3" tubing adapter plate, self-storing tube	1 Shee	16	.500	Ea.	239	26.50		265.50	305
0620	4" tubing adapter plate, self-storing tube		16	.500		243	26.50		269.50	310
0660	5" tubing adapter plate, self-storing tube		16	.500		243	26.50		269.50	310
0800	Two tube outlet assemblies									
0810	For transite pipe ducting, self-storing tube									
0820	3" tubing, dual exhaust adapter plate	1 Shee	16	.500	Ea.	239	26.50		265.50	305
0850	For vitrified tile ducting									
0860	3" tubing, dual exhaust, self-storing tube	1 Shee	16	.500	Ea.	248	26.50		274.50	315
0870	3" tubing, double outlet, non-storing tubes	"	16	.500	"	248	26.50		274.50	315
0900	Accessories for metal tubing, (overhead systems also)									
0910	Adapters, for metal tubing end									
0920	3" tail pipe type				Ea.	49.50			49.50	54.50
0930	4" tail pipe type					53			53	58.50
0940	5" tail pipe type					53			53	58.50
0990	5" diesel stack type					315			315	345
1000	6" diesel stack type					325			325	360
1100	Bullnose (guide) required for in-floor assemblies									
1110	3" tubing size				Ea.	27			27	29.50
1120	4" tubing size					29			29	31.50
1130	5" tubing size					31.50			31.50	34.50
1150	Plain rings, for tubing end									
1160	3" tubing size				Ea.	23.50			23.50	25.50
1170	4" tubing size				"	39.50			39.50	43.50
1200	Tubing, galvanized, flexible, (for overhead systems also)									
1210	3" ID				L.F.	11.50			11.50	12.65
1220	4" ID					14.15			14.15	15.55
1230	5" ID					16.65			16.65	18.30
1240	6" ID					19.25			19.25	21
1250	Stainless steel, flexible, (for overhead system, also)									
1260	3" ID				L.F.	25.50			25.50	28
1270	4" ID					35			35	38.50
1280	5" ID					39.50			39.50	43.50
1290	6" ID					46			46	50.50
1500	Engine exhaust, garage, overhead components, for neoprene tubing									
1510	Alternate metal tubing & accessories see above									

23 35 Special Exhaust Systems

23 35 16 – Engine Exhaust Systems

23 35 16.10 Engine Exhaust Removal Systems		Crew	Daily Output	Labor-Hours	Unit	Material	2013 Bare Costs Labor	Equipment	Total	Total Incl O&P
1550	Adapters, for neoprene tubing end									
1560	3" tail pipe, adjustable, neoprene				Ea.	53			53	58.50
1570	3" tail pipe, heavy wall neoprene					58.50			58.50	64.50
1580	4" tail pipe, heavy wall neoprene					100			100	110
1590	5" tail pipe, heavy wall neoprene				↓	105			105	116
1650	Connectors, tubing									
1660	3" interior, aluminum				Ea.	23.50			23.50	25.50
1670	4" interior, aluminum					39.50			39.50	43.50
1710	5" interior, neoprene					58.50			58.50	64.50
1750	3" spiralock, neoprene					23.50			23.50	25.50
1760	4" spiralock, neoprene					39.50			39.50	43.50
1780	Y for 3" ID tubing, neoprene, dual exhaust					195			195	215
1790	Y for 4" ID tubing, aluminum, dual exhaust				↓	195			195	215
1850	Elbows, aluminum, splice into tubing for strap									
1860	3" neoprene tubing size				Ea.	48.50			48.50	53.50
1870	4" neoprene tubing size					40.50			40.50	44.50
1900	Flange assemblies, connect tubing to overhead duct				↓	47.50			47.50	52.50
2000	Hardware and accessories									
2020	Cable, galvanized, 1/8" diameter				L.F.	.50			.50	.55
2040	Cleat, tie down cable or rope				Ea.	5.30			5.30	5.85
2060	Pulley					7.30			7.30	8.05
2080	Pulley hook, universal				↓	5.30			5.30	5.85
2100	Rope, nylon, 1/4" diameter				L.F.	.40			.40	.44
2120	Winch, 1" diameter				Ea.	113			113	124
2150	Lifting strap, mounts on neoprene									
2160	3" tubing size				Ea.	26.50			26.50	29
2170	4" tubing size					26.50			26.50	29
2180	5" tubing size					26.50			26.50	29
2190	6" tubing size				↓	26.50			26.50	29
2200	Tubing, neoprene, 11' lengths									
2210	3" ID				L.F.	10.30			10.30	11.35
2220	4" ID					17.20			17.20	18.90
2230	5" ID				↓	25.50			25.50	28
2500	Engine exhaust, thru-door outlet									
2510	3" tube size	1 Carp	16	.500	Ea.	55	22.50		77.50	95
2530	4" tube size	"	16	.500	"	70	22.50		92.50	112
3000	Tubing, exhaust, flex hose, with									
3010	coupler, damper and tail pipe adapter									
3020	Neoprene									
3040	3" x 20'	1 Shee	6	1.333	Ea.	315	71		386	455
3050	4" x 15'		5.40	1.481		425	79		504	585
3060	4" x 20'		5	1.600		510	85.50		595.50	690
3070	5" x 15'	↓	4.40	1.818	↓	575	97		672	780
3100	Galvanized									
3110	3" x 20'	1 Shee	6	1.333	Ea.	330	71		401	470
3120	4" x 17'		5.60	1.429		360	76		436	510
3130	4" x 20'		5	1.600		400	85.50		485.50	570
3140	5" x 17'	↓	4.60	1.739	↓	425	92.50		517.50	610

23 35 Special Exhaust Systems

23 35 43 – Welding Fume Elimination Systems

23 35 43.10 Welding Fume Elimination System Components		Crew	Daily Output	Labor-Hours	Unit	Material	2013 Bare Costs Labor	Equipment	Total	Total Incl O&P
0010	**WELDING FUME ELIMINATION SYSTEM COMPONENTS**									
7500	Welding fume elimination accessories for garage exhaust systems									
7600	Cut off (blast gate)									
7610	3" tubing size, 3" x 6" opening	1 Shee	24	.333	Ea.	25	17.75		42.75	54.50
7620	4" tubing size, 4" x 8" opening		24	.333		26	17.75		43.75	55.50
7630	5" tubing size, 5" x 10" opening		24	.333		30.50	17.75		48.25	60.50
7640	6" tubing size		24	.333		32.50	17.75		50.25	62.50
7650	8" tubing size		24	.333		44	17.75		61.75	75.50
7700	Hoods, magnetic, with handle & screen									
7710	3" tubing size, 3" x 6" opening	1 Shee	24	.333	Ea.	106	17.75		123.75	144
7720	4" tubing size, 4" x 8" opening		24	.333		97	17.75		114.75	134
7730	5" tubing size, 5" x 10" opening		24	.333		96.50	17.75		114.25	133

23 38 Ventilation Hoods

23 38 13 – Commercial-Kitchen Hoods

23 38 13.10 Hood and Ventilation Equipment

		Crew	Daily Output	Labor-Hours	Unit	Material	Labor	Equipment	Total	Total Incl O&P
0010	**HOOD AND VENTILATION EQUIPMENT**									
2970	Exhaust hood, sst, gutter on all sides, 4' x 4' x 2'	1 Carp	1.80	4.444	Ea.	4,675	200		4,875	5,425
2980	4' x 4' x 7'	"	1.60	5	"	7,425	225		7,650	8,525
7800	Vent hood, wall canopy with fire protection	L-3A	9	1.333	L.F.	445	65.50		510.50	590
7810	Without fire protection		10	1.200		335	59		394	455
7820	Island canopy with fire protection		7	1.714		445	84		529	620
7830	Without fire protection		8	1.500		335	73.50		408.50	480
7840	Back shelf with fire protection		11	1.091		445	53.50		498.50	570
7850	Without fire protection		12	1		335	49		384	440
7860	Range hood & CO_2 system, minimum	1 Carp	2.50	3.200	Ea.	3,325	144		3,469	3,900
7870	Maximum	"	1	8		40,000	360		40,360	44,600
7950	Hood fire protection system, minimum	Q-1	3	5.333		5,225	268		5,493	6,150
8050	Maximum	"	1	16		38,200	805		39,005	43,200

23 51 Breechings, Chimneys, and Stacks

23 51 13 – Draft Control Devices

23 51 13.16 Vent Dampers

		Crew	Daily Output	Labor-Hours	Unit	Material	Labor	Equipment	Total	Total Incl O&P
0010	**VENT DAMPERS**									
5000	Vent damper, bi-metal, gas, 3" diameter	Q-9	24	.667	Ea.	44	32		76	97.50
5010	4" diameter	"	24	.667	"	44	32		76	97.50

23 51 13.19 Barometric Dampers

		Crew	Daily Output	Labor-Hours	Unit	Material	Labor	Equipment	Total	Total Incl O&P
0010	**BAROMETRIC DAMPERS**									
1000	Barometric, gas fired system only, 6" size for 5" and 6" pipes	1 Shee	20	.400	Ea.	65	21.50		86.50	104
1020	7" size, for 6" and 7" pipes		19	.421		69.50	22.50		92	111
1040	8" size, for 7" and 8" pipes		18	.444		90	23.50		113.50	135
1060	9" size, for 8" and 9" pipes		16	.500		101	26.50		127.50	152
2000	All fuel, oil, oil/gas, coal									
2020	10" for 9" and 10" pipes	1 Shee	15	.533	Ea.	150	28.50		178.50	209
2040	12" for 11" and 12" pipes		15	.533		196	28.50		224.50	260
2060	14" for 13" and 14" pipes		14	.571		255	30.50		285.50	325
2080	16" for 15" and 16" pipes		13	.615		355	33		388	445
2100	18" for 17" and 18" pipes		12	.667		475	35.50		510.50	575

23 51 Breechings, Chimneys, and Stacks

23 51 13 – Draft Control Devices

23 51 13.19 Barometric Dampers		Crew	Daily Output	Labor-Hours	Unit	Material	2013 Bare Costs Labor	Equipment	Total	Total Incl O&P
2120	20″ for 19″ and 21″ pipes	1 Shee	10	.800	Ea.	565	42.50		607.50	690
2140	24″ for 22″ and 25″ pipes	Q-9	12	1.333		695	64		759	865
2160	28″ for 26″ and 30″ pipes		10	1.600		860	77		937	1,075
2180	32″ for 31″ and 34″ pipes	↓	8	2		1,100	96		1,196	1,375
3260	For thermal switch for above, add	1 Shee	24	.333	↓	62	17.75		79.75	95

23 51 23 – Gas Vents

23 51 23.10 Gas Chimney Vents		Crew	Daily Output	Labor-Hours	Unit	Material	Labor	Equipment	Total	Total Incl O&P
0010	**GAS CHIMNEY VENTS**, Prefab metal, U.L. listed									
0020	Gas, double wall, galvanized steel									
0080	3″ diameter	Q-9	72	.222	V.L.F.	4.84	10.65		15.49	21.50
0100	4″ diameter		68	.235		6.60	11.30		17.90	24.50
0120	5″ diameter		64	.250		7.50	12		19.50	26.50
0140	6″ diameter		60	.267		9.40	12.80		22.20	30
0160	7″ diameter		56	.286		10.80	13.70		24.50	33
0180	8″ diameter		52	.308		16.45	14.75		31.20	40.50
0200	10″ diameter		48	.333		31.50	16		47.50	59
0220	12″ diameter		44	.364		37.50	17.45		54.95	67.50
0240	14″ diameter		42	.381		62	18.30		80.30	96.50
0260	16″ diameter		40	.400		89.50	19.20		108.70	128
0280	18″ diameter	↓	38	.421		111	20		131	153
0300	20″ diameter	Q-10	36	.667		130	33		163	194
0320	22″ diameter		34	.706		165	35		200	236
0340	24″ diameter	↓	32	.750	↓	204	37.50		241.50	282
0600	For 4″, 5″ and 6″ oval, add					50%				
0650	Gas, double wall, galvanized steel, fittings									
0660	Elbow 45°, 3″ diameter	Q-9	36	.444	Ea.	12.40	21.50		33.90	46
0670	4″ diameter		34	.471		14.50	22.50		37	50.50
0680	5″ diameter		32	.500		17.50	24		41.50	56
0690	6″ diameter		30	.533		21.50	25.50		47	62.50
0700	7″ diameter		28	.571		32.50	27.50		60	78
0710	8″ diameter		26	.615		43.50	29.50		73	93
0720	10″ diameter		24	.667		92.50	32		124.50	151
0730	12″ diameter		22	.727		99	35		134	162
0740	14″ diameter		21	.762		155	36.50		191.50	227
0750	16″ diameter		20	.800		201	38.50		239.50	280
0760	18″ diameter	↓	19	.842		264	40.50		304.50	355
0770	20″ diameter	Q-10	18	1.333		295	66.50		361.50	425
0780	22″ diameter		17	1.412		470	70		540	625
0790	24″ diameter	↓	16	1.500	↓	600	74.50		674.50	775
0916	Adjustable length									
0918	3″ diameter, to 12″	Q-9	36	.444	Ea.	11.40	21.50		32.90	45
0920	4″ diameter, to 12″		34	.471		13.25	22.50		35.75	49
0924	6″ diameter, to 12″		30	.533		17.10	25.50		42.60	58
0928	8″ diameter, to 12″		26	.615		30.50	29.50		60	78.50
0930	10″ diameter, to 18″		24	.667		93.50	32		125.50	152
0932	12″ diameter, to 18″		22	.727		95	35		130	158
0936	16″ diameter, to 18″		20	.800		192	38.50		230.50	270
0938	18″ diameter, to 18″	↓	19	.842		298	40.50		338.50	390
0944	24″ diameter, to 18″	Q-10	16	1.500		420	74.50		494.50	575
0950	Elbow 90°, adjustable, 3″ diameter	Q-9	36	.444		21	21.50		42.50	55.50
0960	4″ diameter		34	.471		24.50	22.50		47	61.50
0970	5″ diameter		32	.500		31	24		55	70.50

23 51 Breechings, Chimneys, and Stacks

23 51 23 – Gas Vents

23 51 23.10 Gas Chimney Vents		Crew	Daily Output	Labor-Hours	Unit	Material	Labor	Equipment	Total	Total Incl O&P
						2013 Bare Costs				
0980	6" diameter	Q-9	30	.533	Ea.	35.50	25.50		61	78.50
0990	7" diameter		28	.571		51.50	27.50		79	99
1010	8" diameter		26	.615		59	29.50		88.50	110
1020	Wall thimble, 4 to 7" adjustable, 3" diameter		36	.444		13.20	21.50		34.70	47
1022	4" diameter		34	.471		14.85	22.50		37.35	51
1024	5" diameter		32	.500		22	24		46	60.50
1026	6" diameter		30	.533		25	25.50		50.50	66.50
1028	7" diameter		28	.571		29	27.50		56.50	74
1030	8" diameter		26	.615		45	29.50		74.50	94.50
1040	Roof flashing, 3" diameter		36	.444		7.80	21.50		29.30	41
1050	4" diameter		34	.471		9.05	22.50		31.55	44.50
1060	5" diameter		32	.500		16.70	24		40.70	55
1070	6" diameter		30	.533		21.50	25.50		47	63
1080	7" diameter		28	.571		27	27.50		54.50	72
1090	8" diameter		26	.615		33	29.50		62.50	81.50
1100	10" diameter		24	.667		39	32		71	92
1110	12" diameter		22	.727		54.50	35		89.50	113
1120	14" diameter		20	.800		155	38.50		193.50	230
1130	16" diameter		18	.889		192	42.50		234.50	276
1140	18" diameter		16	1		204	48		252	298
1150	20" diameter	Q-10	18	1.333		335	66.50		401.50	470
1160	22" diameter		14	1.714		425	85.50		510.50	600
1170	24" diameter		12	2		470	99.50		569.50	665
1200	Tee, 3" diameter	Q-9	27	.593		32.50	28.50		61	79
1210	4" diameter		26	.615		35	29.50		64.50	83.50
1220	5" diameter		25	.640		38.50	30.50		69	89.50
1230	6" diameter		24	.667		42	32		74	95
1240	7" diameter		23	.696		55	33.50		88.50	111
1250	8" diameter		22	.727		62	35		97	121
1260	10" diameter		21	.762		165	36.50		201.50	238
1270	12" diameter		20	.800		170	38.50		208.50	246
1280	14" diameter		18	.889		430	42.50		472.50	540
1290	16" diameter		16	1		450	48		498	570
1300	18" diameter		14	1.143		540	55		595	680
1310	20" diameter	Q-10	17	1.412		1,050	70		1,120	1,250
1320	22" diameter		13	1.846		1,075	92		1,167	1,325
1330	24" diameter		12	2		1,100	99.50		1,199.50	1,350
1460	Tee cap, 3" diameter	Q-9	45	.356		3.24	17.05		20.29	29.50
1470	4" diameter		42	.381		3.65	18.30		21.95	32
1480	5" diameter		40	.400		4.82	19.20		24.02	35
1490	6" diameter		37	.432		5.65	21		26.65	37.50
1500	7" diameter		35	.457		9.10	22		31.10	43.50
1510	8" diameter		34	.471		10.70	22.50		33.20	46.50
1520	10" diameter		32	.500		22	24		46	61
1530	12" diameter		30	.533		29.50	25.50		55	71.50
1540	14" diameter		28	.571		60.50	27.50		88	109
1550	16" diameter		25	.640		70.50	30.50		101	125
1560	18" diameter		24	.667		79.50	32		111.50	137
1570	20" diameter	Q-10	27	.889		85	44		129	161
1580	22" diameter		22	1.091		97.50	54.50		152	191
1590	24" diameter		21	1.143		110	57		167	208
1750	Top, 3" diameter	Q-9	46	.348		17	16.70		33.70	44
1760	4" diameter		44	.364		22.50	17.45		39.95	51

23 51 Breechings, Chimneys, and Stacks

23 51 23 – Gas Vents

23 51 23.10 Gas Chimney Vents		Crew	Daily Output	Labor-Hours	Unit	Material	2013 Bare Costs Labor	Equipment	Total	Total Incl O&P
1770	5" diameter	Q-9	42	.381	Ea.	25	18.30		43.30	55.50
1780	6" diameter		40	.400		31	19.20		50.20	64
1790	7" diameter		38	.421		46	20		66	81.50
1800	8" diameter		36	.444		60	21.50		81.50	98.50
1810	10" diameter		34	.471		108	22.50		130.50	154
1820	12" diameter		32	.500		123	24		147	172
1830	14" diameter		30	.533		157	25.50		182.50	212
1840	16" diameter		28	.571		228	27.50		255.50	293
1850	18" diameter		26	.615		315	29.50		344.50	390
1860	20" diameter	Q-10	28	.857		830	42.50		872.50	980
1870	22" diameter		22	1.091		895	54.50		949.50	1,075
1880	24" diameter		20	1.200		910	59.50		969.50	1,100
1900	Gas, double wall, galvanized steel, oval									
1904	4" x 1'	Q-9	68	.235	V.L.F.	16.20	11.30		27.50	35
1906	5"		64	.250		38.50	12		50.50	61
1908	5"/6" x 1'		60	.267		35.50	12.80		48.30	58.50
1910	Oval fittings									
1912	Adjustable length									
1914	4" diameter to 12" lg.	Q-9	34	.471	Ea.	15.65	22.50		38.15	52
1916	5" diameter to 12" lg.		32	.500		43.50	24		67.50	84.50
1918	5"/6" diameter to 12"		30	.533		39.50	25.50		65	82.50
1920	Elbow 45°									
1922	4"	Q-9	34	.471	Ea.	28.50	22.50		51	66
1924	5"		32	.500		54	24		78	96
1926	5"/6"		30	.533		51.50	25.50		77	95.50
1930	Elbow 45°, flat									
1932	4"	Q-9	34	.471	Ea.	28.50	22.50		51	66
1934	5"		32	.500		54	24		78	96
1936	5"/6"		30	.533		51.50	25.50		77	95.50
1940	Top									
1942	4"	Q-9	44	.364	Ea.	34	17.45		51.45	64
1944	5"		42	.381		24	18.30		42.30	54
1946	5"/6"		40	.400		57.50	19.20		76.70	93
1950	Adjustable flashing									
1952	4"	Q-9	34	.471	Ea.	8.90	22.50		31.40	44.50
1954	5"		32	.500		28	24		52	67.50
1956	5"/6"		30	.533		29.50	25.50		55	71.50
1960	Tee									
1962	4"	Q-9	26	.615	Ea.	42.50	29.50		72	91.50
1964	5"		25	.640		89	30.50		119.50	145
1966	5"/6"		24	.667		82.50	32		114.50	140
1970	Tee with short snout									
1972	4"	Q-9	26	.615	Ea.	42.50	29.50		72	91.50

23 51 26 – All-Fuel Vent Chimneys

23 51 26.10 All-Fuel Vent Chimneys, Press. Tight, Dbl. Wall		Crew	Daily Output	Labor-Hours	Unit	Material	Labor	Equipment	Total	Total Incl O&P
0010	**ALL-FUEL VENT CHIMNEYS, PRESSURE TIGHT, DOUBLE WALL**									
3200	All fuel, pressure tight, double wall, 1" insulation, U.L. listed, 1400°F.									
3210	304 stainless steel liner, aluminized steel outer jacket									
3220	6" diameter	Q-9	60	.267	L.F.	56	12.80		68.80	81
3221	8" diameter		52	.308		59	14.75		73.75	87.50
3222	10" diameter		48	.333		66	16		82	97
3223	12" diameter		44	.364		75.50	17.45		92.95	110

23 51 26 – All-Fuel Vent Chimneys

23 51 26.10 All-Fuel Vent Chimneys, Press. Tight, Dbl. Wall		Crew	Daily Output	Labor-Hours	Unit	Material	2013 Bare Costs Labor	Equipment	Total	Total Incl O&P
3224	14" diameter	Q-9	42	.381	L.F.	85	18.30		103.30	122
3225	16" diameter		40	.400		95.50	19.20		114.70	135
3226	18" diameter		38	.421		109	20		129	151
3227	20" diameter	Q-10	36	.667		123	33		156	186
3228	24" diameter	"	32	.750		158	37.50		195.50	230
3260	For 316 stainless steel liner add					30%				
3280	All fuel, pressure tight, double wall fittings									
3284	304 stainless steel inner, aluminized steel jacket									
3288	Adjustable 20"/29" section									
3292	6" diameter	Q-9	30	.533	Ea.	187	25.50		212.50	245
3293	8" diameter		26	.615		195	29.50		224.50	259
3294	10" diameter		24	.667		220	32		252	291
3295	12" diameter		22	.727		248	35		283	325
3296	14" diameter		21	.762		280	36.50		316.50	365
3297	16" diameter		20	.800		315	38.50		353.50	405
3298	18" diameter		19	.842		355	40.50		395.50	450
3299	20" diameter	Q-10	18	1.333		405	66.50		471.50	545
3300	24" diameter	"	16	1.500		520	74.50		594.50	685
3350	Elbow 90° fixed									
3354	6" diameter	Q-9	30	.533	Ea.	390	25.50		415.50	470
3355	8" diameter		26	.615		440	29.50		469.50	530
3356	10" diameter		24	.667		500	32		532	600
3357	12" diameter		22	.727		565	35		600	680
3358	14" diameter		21	.762		640	36.50		676.50	760
3359	16" diameter		20	.800		720	38.50		758.50	855
3360	18" diameter		19	.842		815	40.50		855.50	955
3361	20" diameter	Q-10	18	1.333		925	66.50		991.50	1,125
3362	24" diameter	"	16	1.500		1,175	74.50		1,249.50	1,425
3380	For 316 stainless steel liner, add					30%				
3400	Elbow 45°									
3404	6" diameter	Q-9	30	.533	Ea.	195	25.50		220.50	253
3405	8" diameter		26	.615		219	29.50		248.50	286
3406	10" diameter		24	.667		249	32		281	325
3407	12" diameter		22	.727		285	35		320	370
3408	14" diameter		21	.762		320	36.50		356.50	405
3409	16" diameter		20	.800		360	38.50		398.50	455
3410	18" diameter		19	.842		410	40.50		450.50	510
3411	20" diameter	Q-10	18	1.333		460	66.50		526.50	610
3412	24" diameter	"	16	1.500		590	74.50		664.50	765
3430	For 316 stainless steel liner, add					30%				
3450	Tee 90°									
3454	6" diameter	Q-9	24	.667	Ea.	229	32		261	300
3455	8" diameter		22	.727		251	35		286	330
3456	10" diameter		21	.762		283	36.50		319.50	365
3457	12" diameter		20	.800		330	38.50		368.50	420
3458	14" diameter		18	.889		375	42.50		417.50	475
3459	16" diameter		16	1		410	48		458	525
3460	18" diameter		14	1.143		485	55		540	615
3461	20" diameter	Q-10	17	1.412		560	70		630	720
3462	24" diameter	"	12	2		700	99.50		799.50	920
3480	For Tee Cap, add					35%	20%			
3500	For 316 stainless steel liner, add					30%				
3520	Plate support, galvanized									

23 51 26 – All-Fuel Vent Chimneys

23 51 26.10 All-Fuel Vent Chimneys, Press. Tight, Dbl. Wall		Crew	Daily Output	Labor-Hours	Unit	2013 Bare Costs Material	Labor	Equipment	Total	Total Incl O&P
3524	6″ diameter	Q-9	26	.615	Ea.	115	29.50		144.50	171
3525	8″ diameter		22	.727		134	35		169	200
3526	10″ diameter		20	.800		146	38.50		184.50	219
3527	12″ diameter		18	.889		153	42.50		195.50	233
3528	14″ diameter		17	.941		181	45		226	268
3529	16″ diameter		16	1		191	48		239	283
3530	18″ diameter		15	1.067		201	51		252	299
3531	20″ diameter	Q-10	16	1.500		211	74.50		285.50	345
3532	24″ diameter	″	14	1.714		219	85.50		304.50	370
3570	Bellows, lined									
3574	6″ diameter	Q-9	30	.533	Ea.	1,250	25.50		1,275.50	1,425
3575	8″ diameter		26	.615		1,300	29.50		1,329.50	1,500
3576	10″ diameter		24	.667		1,325	32		1,357	1,525
3577	12″ diameter		22	.727		1,375	35		1,410	1,550
3578	14″ diameter		21	.762		1,425	36.50		1,461.50	1,600
3579	16″ diameter		20	.800		1,475	38.50		1,513.50	1,675
3580	18″ diameter		19	.842		1,525	40.50		1,565.50	1,725
3581	20″ diameter	Q-10	18	1.333		1,575	66.50		1,641.50	1,825
3590	For all 316 stainless steel construction, add					55%				
3600	Ventilated roof thimble, 304 stainless steel									
3620	6″ diameter	Q-9	26	.615	Ea.	239	29.50		268.50	310
3624	8″ diameter		22	.727		248	35		283	325
3625	10″ diameter		20	.800		256	38.50		294.50	340
3626	12″ diameter		18	.889		263	42.50		305.50	355
3627	14″ diameter		17	.941		273	45		318	370
3628	16″ diameter		16	1		294	48		342	400
3629	18″ diameter		15	1.067		320	51		371	430
3630	20″ diameter	Q-10	16	1.500		335	74.50		409.50	480
3631	24″ diameter	″	14	1.714		365	85.50		450.50	530
3650	For 316 stainless steel, add					30%				
3670	Exit cone, 316 stainless steel only									
3674	6″ diameter	Q-9	46	.348	Ea.	181	16.70		197.70	225
3675	8″ diameter		42	.381		187	18.30		205.30	233
3676	10″ diameter		40	.400		197	19.20		216.20	247
3677	12″ diameter		38	.421		210	20		230	262
3678	14″ diameter		37	.432		225	21		246	279
3679	16″ diameter		36	.444		282	21.50		303.50	345
3680	18″ diameter		35	.457		305	22		327	370
3681	20″ diameter	Q-10	28	.857		365	42.50		407.50	465
3682	24″ diameter	″	26	.923		480	46		526	595
3720	Roof guide, 304 stainless steel									
3724	6″ diameter	Q-9	25	.640	Ea.	83	30.50		113.50	138
3725	8″ diameter		21	.762		96.50	36.50		133	162
3726	10″ diameter		19	.842		106	40.50		146.50	178
3727	12″ diameter		17	.941		109	45		154	189
3728	14″ diameter		16	1		127	48		175	213
3729	16″ diameter		15	1.067		135	51		186	226
3730	18″ diameter		14	1.143		143	55		198	241
3731	20″ diameter	Q-10	15	1.600		150	79.50		229.50	286
3732	24″ diameter	″	13	1.846		157	92		249	310
3750	For 316 stainless steel, add					30%				
3770	Rain cap with bird screen									
3774	6″ diameter	Q-9	46	.348	Ea.	280	16.70		296.70	335

23 51 26 – All-Fuel Vent Chimneys

23 51 26.10 All-Fuel Vent Chimneys, Press. Tight, Dbl. Wall		Crew	Daily Output	Labor-Hours	Unit	Material	Labor	Equipment	Total	Total Incl O&P
						2013 Bare Costs				
3775	8" diameter	Q-9	42	.381	Ea.	325	18.30		343.30	385
3776	10" diameter		40	.400		380	19.20		399.20	445
3777	12" diameter		38	.421		440	20		460	510
3778	14" diameter		37	.432		495	21		516	575
3779	16" diameter		36	.444		565	21.50		586.50	655
3780	18" diameter		35	.457		655	22		677	755
3781	20" diameter	Q-10	28	.857		740	42.50		782.50	880
3782	24" diameter	"	26	.923		890	46		936	1,050

23 51 26.30 All-Fuel Vent Chimneys, Double Wall, St. Stl.		Crew	Daily Output	Labor-Hours	Unit	Material	Labor	Equipment	Total	Total Incl O&P
0010	**ALL-FUEL VENT CHIMNEYS, DOUBLE WALL, STAINLESS STEEL**									
7780	All fuel, pressure tight, double wall, 4" insulation, U.L. listed, 1400°F.									
7790	304 stainless steel liner, aluminized steel outer jacket									
7800	6" diameter	Q-9	60	.267	V.L.F.	62	12.80		74.80	87.50
7804	8" diameter		52	.308		71	14.75		85.75	101
7806	10" diameter		48	.333		79	16		95	112
7808	12" diameter		44	.364		90.50	17.45		107.95	126
7810	14" diameter		42	.381		102	18.30		120.30	140
7880	For 316 stainless steel liner add				L.F.	30%				
8000	All fuel, double wall, stainless steel fittings									
8010	Roof support 6" diameter	Q-9	30	.533	Ea.	109	25.50		134.50	159
8030	8" diameter		26	.615		127	29.50		156.50	185
8040	10" diameter		24	.667		135	32		167	197
8050	12" diameter		22	.727		143	35		178	210
8060	14" diameter		21	.762		150	36.50		186.50	221
8100	Elbow 45°, 6" diameter		30	.533		233	25.50		258.50	295
8140	8" diameter		26	.615		262	29.50		291.50	335
8160	10" diameter		24	.667		298	32		330	380
8180	12" diameter		22	.727		340	35		375	430
8200	14" diameter		21	.762		380	36.50		416.50	475
8300	Insulated tee, 6" diameter		30	.533		274	25.50		299.50	340
8360	8" diameter		26	.615		300	29.50		329.50	375
8380	10" diameter		24	.667		335	32		367	420
8400	12" diameter		22	.727		380	35		415	475
8420	14" diameter		20	.800		445	38.50		483.50	550
8500	Boot tee, 6" diameter		28	.571		525	27.50		552.50	615
8520	8" diameter		24	.667		570	32		602	675
8530	10" diameter		22	.727		665	35		700	785
8540	12" diameter		20	.800		785	38.50		823.50	925
8550	14" diameter		18	.889		895	42.50		937.50	1,050
8600	Rain cap with bird screen, 6" diameter		30	.533		280	25.50		305.50	350
8640	8" diameter		26	.615		325	29.50		354.50	400
8660	10" diameter		24	.667		380	32		412	465
8680	12" diameter		22	.727		440	35		475	535
8700	14" diameter		21	.762		495	36.50		531.50	600
8800	Flat roof flashing, 6" diameter		30	.533		103	25.50		128.50	152
8840	8" diameter		26	.615		112	29.50		141.50	168
8860	10" diameter		24	.667		120	32		152	181
8880	12" diameter		22	.727		132	35		167	198
8900	14" diameter		21	.762		135	36.50		171.50	204

23 51 33 – Insulated Sectional Chimneys

	23 51 33.10 Prefabricated Insulated Sectional Chimneys	Crew	Daily Output	Labor-Hours	Unit	Material	2013 Bare Costs Labor	Equipment	Total	Total Incl O&P
0010	**PREFABRICATED INSULATED SECTIONAL CHIMNEYS**									
9000	High temp. (2000°F), steel jacket, acid resistant refractory lining									
9010	11 ga. galvanized jacket, U.L. listed									
9020	Straight section, 48" long, 10" diameter	Q-10	13.30	1.805	Ea.	400	90		490	575
9030	12" diameter		11.20	2.143		440	107		547	645
9040	18" diameter		7.40	3.243		615	161		776	920
9050	24" diameter		4.60	5.217		840	260		1,100	1,325
9120	Tee section, 10" diameter		4.40	5.455		855	271		1,126	1,350
9130	12" diameter		3.70	6.486		900	325		1,225	1,475
9140	18" diameter		2.40	10		1,225	500		1,725	2,100
9150	24" diameter		1.50	16		1,675	795		2,470	3,075
9220	Cleanout pier section, 10" diameter		3.50	6.857		655	340		995	1,250
9230	12" diameter		2.50	9.600		690	480		1,170	1,500
9240	18" diameter		1.90	12.632		885	630		1,515	1,925
9250	24" diameter		1.30	18.462		1,125	920		2,045	2,625
9320	For drain, add					53%				
9330	Elbow, 30° and 45°, 10" diameter	Q-10	6.60	3.636		555	181		736	885
9340	12" diameter		5.60	4.286		600	213		813	980
9350	18" diameter		3.70	6.486		900	325		1,225	1,475
9360	24" diameter		2.30	10.435		1,350	520		1,870	2,275
9430	For 60° and 90° elbow, add					112%				
9440	End cap, 10" diameter	Q-10	26	.923		955	46		1,001	1,125
9450	12" diameter		22	1.091		1,100	54.50		1,154.50	1,275
9460	18" diameter		15	1.600		1,350	79.50		1,429.50	1,600
9470	24" diameter		9	2.667		1,650	133		1,783	2,025
9540	Increaser (1 diameter), 10" diameter		6.60	3.636		475	181		656	795
9550	12" diameter		5.60	4.286		505	213		718	880
9560	18" diameter		3.70	6.486		740	325		1,065	1,300
9570	24" diameter		2.30	10.435		1,050	520		1,570	1,975
9600	For expansion joints, add to straight section					7%				
9610	For 1/4" hot rolled steel jacket, add					157%				
9620	For 2950°F very high temperature, add					89%				
9630	26 ga. aluminized jacket, straight section, 48" long									
9640	10" diameter	Q-10	15.30	1.569	V.L.F.	237	78		315	380
9650	12" diameter		12.90	1.860		252	92.50		344.50	420
9660	18" diameter		8.50	2.824		365	140		505	620
9670	24" diameter		5.30	4.528		520	225		745	915
9700	Accessories (all models)									
9710	Guy band, 10" diameter	Q-10	32	.750	Ea.	74	37.50		111.50	138
9720	12" diameter		30	.800		78	40		118	146
9730	18" diameter		26	.923		102	46		148	182
9740	24" diameter		24	1		149	50		199	239
9810	Draw band, galv. stl., 11 ga., 10" diameter		32	.750		65	37.50		102.50	129
9820	12" diameter		30	.800		69	40		109	136
9830	18" diameter		26	.923		86	46		132	165
9840	24" diameter		24	1		110	50		160	197
9910	Draw band, aluminized stl., 26 ga., 10" diameter		32	.750		18.95	37.50		56.45	78
9920	12" diameter		30	.800		18.95	40		58.95	81.50
9930	18" diameter		26	.923		24	46		70	96.50
9940	24" diameter		24	1		28	50		78	107

23 52 Heating Boilers

23 52 13 – Electric Boilers

23 52 13.10 Electric Boilers, ASME		Crew	Daily Output	Labor-Hours	Unit	Material	2013 Bare Costs Labor	Equipment	Total	Total Incl O&P
0010	**ELECTRIC BOILERS, ASME**, Standard controls and trim D3020–102									
1000	Steam, 6 KW, 20.5 MBH	Q-19	1.20	20	Ea.	3,800	1,025		4,825	5,725
1040	9 KW, 30.7 MBH		1.20	20		3,900	1,025		4,925	5,850
1060	18 KW, 61.4 MBH		1.20	20		4,025	1,025		5,050	5,975
1080	24 KW, 81.8 MBH		1.10	21.818		4,675	1,125		5,800	6,800
1120	36 KW, 123 MBH		1.10	21.818		5,150	1,125		6,275	7,350
2000	Hot water, 7.5 KW, 25.6 MBH		1.30	18.462		4,800	950		5,750	6,700
2020	15 KW, 51.2 MBH		1.30	18.462		4,825	950		5,775	6,750
2040	30 KW, 102 MBH		1.20	20		5,150	1,025		6,175	7,225
2060	45 KW, 164 MBH		1.20	20		5,250	1,025		6,275	7,325
2070	60 KW, 205 MBH		1.20	20		5,375	1,025		6,400	7,450
2080	75 KW, 256 MBH		1.10	21.818		5,725	1,125		6,850	7,975

23 52 16 – Condensing Boilers

23 52 16.24 Condensing Boilers		Crew	Daily Output	Labor-Hours	Unit	Material	Labor	Equipment	Total	Total Incl O&P
0010	**CONDENSING BOILERS**, Cast iron									
0020	Packaged with standard controls, circulator and trim									
0030	Intermittent (spark) pilot, natural or LP gas									
0040	Hot water, DOE MBH output, (AFUE)									
0100	42 MBH, (84.0%) G	Q-5	1.80	8.889	Ea.	1,525	455		1,980	2,350
0120	57 MBH, (84.3%) G		1.60	10		1,700	510		2,210	2,625
0140	85 MBH, (84.0%) G		1.40	11.429		1,850	585		2,435	2,925
0160	112 MBH, (83.7%) G		1.20	13.333		2,075	680		2,755	3,325
0180	140 MBH, (83.3%) G	Q-6	1.60	15		2,350	795		3,145	3,775
0200	167 MBH, (83.0%) G		1.40	17.143		2,650	905		3,555	4,275
0220	194 MBH, (82.7%) G		1.20	20		2,900	1,050		3,950	4,775

23 52 19 – Pulse Combustion Boilers

23 52 19.20 Pulse Type Combustion Boilers		Crew	Daily Output	Labor-Hours	Unit	Material	Labor	Equipment	Total	Total Incl O&P
0010	**PULSE TYPE COMBUSTION BOILERS**									
7990	Special feature gas fired boilers									
8000	Pulse combustion, standard controls/trim									
8050	88,000 BTU G	Q-5	1.40	11.429	Ea.	5,050	585		5,635	6,425
8080	134,000 BTU G	″	1.20	13.333	″	5,675	680		6,355	7,275

23 52 23 – Cast-Iron Boilers

23 52 23.20 Gas-Fired Boilers		Crew	Daily Output	Labor-Hours	Unit	Material	Labor	Equipment	Total	Total Incl O&P
0010	**GAS-FIRED BOILERS**, Natural or propane, standard controls, packaged									
1000	Cast iron, with insulated jacket									
2000	Steam, gross output, 81 MBH	Q-7	1.40	22.857	Ea.	1,800	1,225		3,025	3,825
2020	102 MBH		1.30	24.615		2,075	1,325		3,400	4,275
2040	122 MBH		1	32		2,250	1,725		3,975	5,075
2060	163 MBH		.90	35.556		2,750	1,925		4,675	5,900
2080	203 MBH		.90	35.556		3,050	1,925		4,975	6,225
2100	240 MBH		.85	37.647		3,075	2,025		5,100	6,425
2120	280 MBH		.80	40		3,950	2,150		6,100	7,575
2140	320 MBH		.70	45.714		4,300	2,475		6,775	8,425
3000	Hot water, gross output, 80 MBH		1.46	21.918		1,600	1,175		2,775	3,550
3020	100 MBH		1.35	23.704		1,900	1,275		3,175	4,000
3040	122 MBH		1.10	29.091		2,075	1,575		3,650	4,625
3060	163 MBH		1	32		2,575	1,725		4,300	5,425
3080	203 MBH		1	32		2,875	1,725		4,600	5,750
3100	240 MBH		.95	33.684		2,875	1,825		4,700	5,900
3120	280 MBH		.90	35.556		3,650	1,925		5,575	6,900

23 52 Heating Boilers

23 52 23 – Cast-Iron Boilers

23 52 23.20 Gas-Fired Boilers		Crew	Daily Output	Labor-Hours	Unit	Material	2013 Bare Costs Labor	Equipment	Total	Total Incl O&P
3140	320 MBH	Q-7	.80	40	Ea.	4,000	2,150		6,150	7,650
7000	For tankless water heater, add					10%				
7050	For additional zone valves up to 312 MBH add					162			162	178

23 52 23.40 Oil-Fired Boilers		Crew	Daily Output	Labor-Hours	Unit	Material	Labor	Equipment	Total	Total Incl O&P
0010	**OIL-FIRED BOILERS**, Standard controls, flame retention burner, packaged									
1000	Cast iron, with insulated flush jacket									
2000	Steam, gross output, 109 MBH	Q-7	1.20	26.667	Ea.	2,175	1,450		3,625	4,575
2020	144 MBH		1.10	29.091		2,475	1,575		4,050	5,075
2040	173 MBH		1	32		2,750	1,725		4,475	5,650
2060	207 MBH		.90	35.556		3,000	1,925		4,925	6,175
3000	Hot water, same price as steam									
4000	For tankless coil in smaller sizes, add				Ea.	15%				

23 52 26 – Steel Boilers

23 52 26.40 Oil-Fired Boilers		Crew	Daily Output	Labor-Hours	Unit	Material	Labor	Equipment	Total	Total Incl O&P
0010	**OIL-FIRED BOILERS**, Standard controls, flame retention burner									
5000	Steel, with insulated flush jacket									
7000	Hot water, gross output, 103 MBH	Q-6	1.60	15	Ea.	1,925	795		2,720	3,325
7020	122 MBH		1.45	16.506		2,050	875		2,925	3,575
7040	137 MBH		1.36	17.595		2,175	930		3,105	3,800
7060	168 MBH		1.30	18.405		2,275	975		3,250	3,975
7080	225 MBH		1.22	19.704		2,875	1,050		3,925	4,750
7340	For tankless coil in steam or hot water, add					7%				

23 52 28 – Swimming Pool Boilers

23 52 28.10 Swimming Pool Heaters		Crew	Daily Output	Labor-Hours	Unit	Material	Labor	Equipment	Total	Total Incl O&P
0010	**SWIMMING POOL HEATERS**, Not including wiring, external									
0020	piping, base or pad,									
0160	Gas fired, input, 155 MBH	Q-6	1.50	16	Ea.	1,850	845		2,695	3,300
0200	199 MBH		1	24		1,975	1,275		3,250	4,075
0220	250 MBH		.70	34.286		2,150	1,825		3,975	5,075
0240	300 MBH		.60	40		2,250	2,125		4,375	5,650
0260	399 MBH		.50	48		2,525	2,550		5,075	6,600
0280	500 MBH		.40	60		8,050	3,175		11,225	13,600
0300	650 MBH		.35	68.571		8,550	3,625		12,175	14,900
0320	750 MBH		.33	72.727		9,350	3,850		13,200	16,100
0360	990 MBH		.22	109		12,500	5,775		18,275	22,500
0370	1,260 MBH		.21	114		14,700	6,050		20,750	25,300
0380	1,440 MBH		.19	126		14,700	6,675		21,375	26,300
0400	1,800 MBH		.14	171		17,500	9,075		26,575	32,800
0410	2,070 MBH		.13	184		20,600	9,750		30,350	37,400
2000	Electric, 12 KW, 4,800 gallon pool	Q-19	3	8		2,075	410		2,485	2,900
2020	15 KW, 7,200 gallon pool		2.80	8.571		2,100	440		2,540	3,000
2040	24 KW, 9,600 gallon pool		2.40	10		2,425	515		2,940	3,450
2060	30 KW, 12,000 gallon pool		2	12		2,475	615		3,090	3,650
2080	36 KW, 14,400 gallon pool		1.60	15		2,850	770		3,620	4,275
2100	57 KW, 24,000 gallon pool		1.20	20		3,575	1,025		4,600	5,475
9000	To select pool heater: 12 BTUH x S.F. pool area									
9010	X temperature differential = required output									
9050	For electric, KW = gallons x 2.5 divided by 1000									
9100	For family home type pool, double the									
9110	Rated gallon capacity = 1/2°F rise per hour									

23 52 Heating Boilers

23 52 88 – Burners

23 52 88.10 Replacement Type Burners		Crew	Daily Output	Labor-Hours	Unit	Material	Labor	Equipment	Total	Total Incl O&P
						2013 Bare Costs				
0010	**REPLACEMENT TYPE BURNERS**									
0990	Residential, conversion, gas fired, LP or natural									
1000	Gun type, atmospheric input 50 to 225 MBH	Q-1	2.50	6.400	Ea.	720	320		1,040	1,275
1020	100 to 400 MBH	″	2	8	″	1,200	400		1,600	1,925
3000	Flame retention oil fired assembly, input									
3020	.50 to 2.25 GPH	Q-1	2.40	6.667	Ea.	289	335		624	825
3040	2.0 to 5.0 GPH	″	2	8		345	400		745	985
4600	Gas safety, shut off valve, 3/4″ threaded	1 Stpi	20	.400		178	22.50		200.50	230
4610	1″ threaded		19	.421		172	24		196	226
4620	1-1/4″ threaded		15	.533		194	30		224	259
4630	1-1/2″ threaded		13	.615		210	35		245	284
4640	2″ threaded		11	.727		235	41		276	320
4650	2-1/2″ threaded	Q-1	15	1.067		269	53.50		322.50	375
4660	3″ threaded		13	1.231		370	62		432	500
4670	4″ flanged		3	5.333		2,625	268		2,893	3,300
4680	6″ flanged	Q-2	3	8		5,875	415		6,290	7,100

23 56 Solar Energy Heating Equipment

23 56 16 – Packaged Solar Heating Equipment

23 56 16.40 Solar Heating Systems

		Crew	Daily Output	Labor-Hours	Unit	Material	Labor	Equipment	Total	Total Incl O&P
0010	**SOLAR HEATING SYSTEMS** D2020–265									
0020	System/Package prices, not including connecting									
0030	pipe, insulation, or special heating/plumbing fixtures D2020–270									
0152	For solar ultraviolet pipe insulation see Section 22 07 19.10									
0500	Hot water, standard package, low temperature D2020–275									
0540	1 collector, circulator, fittings, 65 gal. tank G	Q-1	.50	32	Ea.	1,675	1,600		3,275	4,250
0580	2 collectors, circulator, fittings, 120 gal. tank D2020–280 G		.40	40		2,425	2,000		4,425	5,700
0620	3 collectors, circulator, fittings, 120 gal. tank G		.34	47.059		3,325	2,375		5,700	7,200
0700	Medium temperature package D2020–285									
0720	1 collector, circulator, fittings, 80 gal. tank G	Q-1	.50	32	Ea.	2,425	1,600		4,025	5,100
0740	2 collectors, circulator, fittings, 120 gal. tank D2020–290 G		.40	40		3,550	2,000		5,550	6,925
0780	3 collectors, circulator, fittings, 120 gal. tank G		.30	53.333		4,600	2,675		7,275	9,075
0980	For each additional 120 gal. tank, add G D2020–295					1,600			1,600	1,750

23 56 19 – Solar Heating Components

23 56 19.50 Solar Heating Ancillary

		Crew	Daily Output	Labor-Hours	Unit	Material	Labor	Equipment	Total	Total Incl O&P
0010	**SOLAR HEATING ANCILLARY**									
2300	Circulators, air D3010–650 G									
2310	Blowers									
2330	100-300 S.F. system, 1/10 HP D3010–660 G	Q-9	16	1	Ea.	249	48		297	345
2340	300-500 S.F. system, 1/5 HP G		15	1.067		330	51		381	440
2350	Two speed, 100-300 S.F., 1/10 HP D3010–675 G		14	1.143		142	55		197	241
2400	Reversible fan, 20″ diameter, 2 speed G		18	.889		113	42.50		155.50	189
2550	Booster fan 6″ diameter, 120 CFM G		16	1		36.50	48		84.50	114
2570	6″ diameter, 225 CFM G		16	1		45	48		93	123
2580	8″ diameter, 150 CFM G		16	1		41	48		89	119
2590	8″ diameter, 310 CFM G		14	1.143		63.50	55		118.50	153
2600	8″ diameter, 425 CFM G		14	1.143		71	55		126	162
2650	Rheostat G		32	.500		15.45	24		39.45	53.50
2660	Shutter/damper G		12	1.333		52	64		116	155
2670	Shutter motor G		16	1		129	48		177	215

23 56 19 – Solar Heating Components

23 56 19.50 Solar Heating Ancillary		Crew	Daily Output	Labor-Hours	Unit	2013 Bare Costs Material	Labor	Equipment	Total	Total Incl O&P
2800	Circulators, liquid, 1/25 HP, 5.3 GPM G	Q-1	14	1.143	Ea.	268	57.50		325.50	380
2820	1/20 HP, 17 GPM G		12	1.333		161	67		228	278
2850	1/20 HP, 17 GPM, stainless steel G		12	1.333		254	67		321	380
2870	1/12 HP, 30 GPM G		10	1.600		345	80.50		425.50	500
3000	Collector panels, air with aluminum absorber plate									
3010	Wall or roof mount									
3040	Flat black, plastic glazing									
3080	4′ x 8′ G	Q-9	6	2.667	Ea.	660	128		788	925
3100	4′ x 10′ G		5	3.200	"	815	154		969	1,125
3200	Flush roof mount, 10′ to 16′ x 22″ wide G		96	.167	L.F.	485	8		493	545
3210	Manifold, by L.F. width of collectors G		160	.100	"	131	4.80		135.80	151
3300	Collector panels, liquid with copper absorber plate									
3320	Black chrome, tempered glass glazing									
3330	Alum. frame, 4′ x 8′, 5/32″ single glazing G	Q-1	9.50	1.684	Ea.	1,025	84.50		1,109.50	1,250
3390	Alum. frame, 4′ x 10′, 5/32″ single glazing G		6	2.667		1,175	134		1,309	1,500
3450	Flat black, alum. frame, 3.5′ x 7.5′ G		9	1.778		710	89.50		799.50	915
3500	4′ x 8′ G		5.50	2.909		890	146		1,036	1,200
3520	4′ x 10′ G		10	1.600		1,000	80.50		1,080.50	1,225
3540	4′ x 12.5′ G		5	3.200		1,200	161		1,361	1,575
3550	Liquid with fin tube absorber plate									
3560	Alum. frame 4′ x 8′ tempered glass G	Q-1	10	1.600	Ea.	530	80.50		610.50	705
3580	Liquid with vacuum tubes, 4′ x 6′-10″ G		9	1.778		875	89.50		964.50	1,100
3600	Liquid, full wetted, plastic, alum. frame, 4′ x 10′ G		5	3.200		241	161		402	505
3650	Collector panel mounting, flat roof or ground rack G		7	2.286		235	115		350	430
3670	Roof clamps G		70	.229	Set	2.54	11.50		14.04	20
3700	Roof strap, teflon G	1 Plum	205	.039	L.F.	20.50	2.18		22.68	26
3900	Differential controller with two sensors									
3930	Thermostat, hard wired G	1 Plum	8	1	Ea.	86	56		142	179
3950	Line cord and receptacle G		12	.667		570	37		607	680
4050	Pool valve system G		2.50	3.200		287	179		466	585
4070	With 12 VAC actuator G		2	4		300	223		523	665
4080	Pool pump system, 2″ pipe size G		6	1.333		187	74.50		261.50	320
4100	Five station with digital read-out G		3	2.667		236	149		385	485
4150	Sensors									
4200	Brass plug, 1/2″ MPT G	1 Plum	32	.250	Ea.	19.45	13.95		33.40	42.50
4210	Brass plug, reversed G		32	.250		26.50	13.95		40.45	50
4220	Freeze prevention G		32	.250		23.50	13.95		37.45	47
4240	Screw attached G		32	.250		9	13.95		22.95	31
4250	Brass, immersion G		32	.250		28.50	13.95		42.45	52.50
4300	Heat exchanger									
4330	Fluid to air coil, up flow, 45 MBH G	Q-1	4	4	Ea.	310	201		511	645
4380	70 MBH G		3.50	4.571		350	230		580	725
4400	80 MBH G		3	5.333		465	268		733	915
4580	Fluid to fluid package includes two circulating pumps									
4590	expansion tank, check valve, relief valve									
4600	controller, high temperature cutoff and sensors G	Q-1	2.50	6.400	Ea.	745	320		1,065	1,300
4650	Heat transfer fluid									
4700	Propylene glycol, inhibited anti-freeze G	1 Plum	28	.286	Gal.	14.50	15.95		30.45	40
4800	Solar storage tanks, knocked down									
4810	Air, galvanized steel clad, double wall, 4″ fiberglass insulation									
5120	45 Mil reinforced polypropylene lining,									
5140	4′ high, 4′ x 4′ = 64 C.F./450 gallons G	Q-9	2	8	Ea.	3,650	385		4,035	4,575
5150	4′ x 8′ = 128 C.F./900 gallons G		1.50	10.667		5,450	510		5,960	6,775

23 56 Solar Energy Heating Equipment

23 56 19 – Solar Heating Components

23 56 19.50 Solar Heating Ancillary			Crew	Daily Output	Labor-Hours	Unit	2013 Bare Costs Material	Labor	Equipment	Total	Total Incl O&P
5160	4′ x 12′ = 190 C.F./1300 gallons	G	Q-9	1.30	12.308	Ea.	7,275	590		7,865	8,900
5170	8′ x 8′ = 250 C.F./1700 gallons	G		1	16		7,275	770		8,045	9,175
5190	6′-3″ high, 7′ x 7′ = 306 C.F./2000 gallons	G	Q-10	1.20	20		14,200	995		15,195	17,200
5200	7′ x 10′-6″ = 459 C.F./3000 gallons	G		.80	30		17,800	1,500		19,300	21,900
5210	7′ x 14′ = 613 C.F./4000 gallons	G		.60	40		21,400	2,000		23,400	26,500
5220	10′-6″ x 10′-6″ = 689 C.F./4500 gallons	G		.50	48		21,400	2,400		23,800	27,200
5230	10′-6″ x 14′ = 919 C.F./6000 gallons	G		.40	60		24,900	2,975		27,875	32,000
5240	14′ x 14′ = 1225 C.F./8000 gallons	G	Q-11	.40	80		28,500	4,050		32,550	37,500
5250	14′ x 17′-6″ = 1531 C.F./10,000 gallons	G		.30	106		32,100	5,425		37,525	43,600
5260	17′-6″ x 17′-6″ = 1914 C.F./12,500 gallons	G		.25	128		35,600	6,500		42,100	49,100
5270	17′-6″ x 21′ = 2297 C.F./15,000 gallons	G		.20	160		39,100	8,125		47,225	55,500
5280	21′ x 21′ = 2756 C.F./18,000 gallons	G		.18	177		42,700	9,025		51,725	61,000
5290	30 Mil reinforced Hypalon lining, add						.02%				
7000	Solar control valves and vents										
7050	Air purger, 1″ pipe size	G	1 Plum	12	.667	Ea.	46.50	37		83.50	108
7070	Air eliminator, automatic 3/4″ size	G		32	.250		30	13.95		43.95	54
7090	Air vent, automatic, 1/8″ fitting	G		32	.250		22.50	13.95		36.45	46
7100	Manual, 1/8″ NPT	G		32	.250		14.45	13.95		28.40	37
7120	Backflow preventer, 1/2″ pipe size	G		16	.500		126	28		154	180
7130	3/4″ pipe size	G		16	.500		131	28		159	186
7150	Balancing valve, 3/4″ pipe size	G		20	.400		53	22.50		75.50	92
7180	Draindown valve, 1/2″ copper tube	G		9	.889		212	49.50		261.50	310
7200	Flow control valve, 1/2″ pipe size	G		22	.364		134	20.50		154.50	179
7220	Expansion tank, up to 5 gal.	G		32	.250		66	13.95		79.95	93.50
7250	Hydronic controller (aquastat)	G		8	1		149	56		205	248
7400	Pressure gauge, 2″ dial	G		32	.250		23	13.95		36.95	46.50
7450	Relief valve, temp. and pressure 3/4″ pipe size	G		30	.267		20	14.90		34.90	44.50
7500	Solenoid valve, normally closed										
7520	Brass, 3/4″ NPT, 24V	G	1 Plum	9	.889	Ea.	115	49.50		164.50	202
7530	1″ NPT, 24V	G		9	.889		1,075	49.50		1,124.50	1,250
7750	Vacuum relief valve, 3/4″ pipe size	G		32	.250		28.50	13.95		42.45	52.50
7800	Thermometers										
7820	Digital temperature monitoring, 4 locations	G	1 Plum	2.50	3.200	Ea.	137	179		316	420
7900	Upright, 1/2″ NPT	G		8	1		28	56		84	115
7970	Remote probe, 2″ dial	G		8	1		32	56		88	119
7990	Stem, 2″ dial, 9″ stem	G		16	.500		21	28		49	65
8250	Water storage tank with heat exchanger and electric element										
8270	66 gal. with 2″ x 2 lb. density insulation	G	1 Plum	1.60	5	Ea.	1,550	279		1,829	2,125
8300	80 gal. with 2″ x 2 lb. density insulation	G		1.60	5		1,550	279		1,829	2,125
8380	120 gal. with 2″ x 2 lb. density insulation	G		1.40	5.714		1,775	320		2,095	2,425
8400	120 gal. with 2″ x 2 lb. density insul., 40 S.F. heat coil	G		1.40	5.714		2,250	320		2,570	2,950
8500	Water storage module, plastic										
8600	Tubular, 12″ diameter, 4′ high	G	1 Carp	48	.167	Ea.	103	7.50		110.50	125
8610	12″ diameter, 8′ high	G		40	.200		159	9		168	189
8620	18″ diameter, 5′ high	G		38	.211		174	9.45		183.45	206
8630	18″ diameter, 10′ high	G		32	.250		229	11.25		240.25	269
8640	58″ diameter, 5′ high	G	2 Carp	32	.500		490	22.50		512.50	570
8650	Cap, 12″ diameter	G					19			19	21
8660	18″ diameter	G					24			24	26.50

23 57 Heat Exchangers for HVAC

23 57 16 – Steam-to-Water Heat Exchangers

23 57 16.10 Shell/Tube Type Steam-to-Water Heat Exch.		Crew	Daily Output	Labor-Hours	Unit	Material	2013 Bare Costs Labor	Equipment	Total	Total Incl O&P
0010	SHELL AND TUBE TYPE STEAM-TO-WATER HEAT EXCHANGERS									
0016	Shell & tube type, 2 or 4 pass, 3/4" O.D. copper tubes,									
0020	C.I. heads, C.I. tube sheet, steel shell									
0100	Hot water 40°F to 180°F, by steam at 10 PSI									
0120	8 GPM	Q-5	6	2.667	Ea.	2,075	136		2,211	2,475
0140	10 GPM		5	3.200		3,125	163		3,288	3,675
0160	40 GPM	↓	4	4	↓	4,825	204		5,029	5,625
0500	For bronze head and tube sheet, add					50%				

23 57 19 – Liquid-to-Liquid Heat Exchangers

23 57 19.13 Plate-Type, Liquid-to-Liquid Heat Exchangers		Crew	Daily Output	Labor-Hours	Unit	Material	Labor	Equipment	Total	Total Incl O&P
0010	PLATE-TYPE, LIQUID-TO-LIQUID HEAT EXCHANGERS									
3000	Plate type,									
3100	400 GPM	Q-6	.80	30	Ea.	34,500	1,575		36,075	40,300
3120	800 GPM	"	.50	48		59,500	2,550		62,050	69,500
3140	1200 GPM	Q-7	.34	94.118		88,500	5,075		93,575	104,500
3160	1800 GPM	"	.24	133	↓	117,000	7,200		124,200	140,000

23 57 19.16 Shell-Type, Liquid-to-Liquid Heat Exchangers		Crew	Daily Output	Labor-Hours	Unit	Material	Labor	Equipment	Total	Total Incl O&P
0010	SHELL-TYPE, LIQUID-TO-LIQUID HEAT EXCHANGERS									
1000	Hot water 40°F to 140°F, by water at 200°F									
1020	7 GPM	Q-5	6	2.667	Ea.	2,550	136		2,686	3,000
1040	16 GPM		5	3.200		3,600	163		3,763	4,200
1060	34 GPM		4	4		5,450	204		5,654	6,300
1080	55 GPM		3	5.333		7,900	272		8,172	9,100
1100	74 GPM		1.50	10.667		9,875	545		10,420	11,700
1120	86 GPM	↓	1.40	11.429		13,200	585		13,785	15,400
1140	112 GPM	Q-6	2	12		16,500	635		17,135	19,100
1160	126 GPM		1.80	13.333		20,500	705		21,205	23,700
1180	152 GPM	↓	1	24	↓	26,200	1,275		27,475	30,700

23 72 Air-to-Air Energy Recovery Equipment

23 72 16 – Heat-Pipe Air-To-Air Energy-Recovery Equipment

23 72 16.10 Heat Pipes		Crew	Daily Output	Labor-Hours	Unit	Material	Labor	Equipment	Total	Total Incl O&P
0010	HEAT PIPES									
8000	Heat pipe type, glycol, 50% efficient									
8010	100 MBH, 1700 CFM	1 Stpi	.80	10	Ea.	4,100	565		4,665	5,375
8020	160 MBH, 2700 CFM		.60	13.333		5,525	755		6,280	7,200
8030	620 MBH, 4000 CFM	↓	.40	20	↓	8,325	1,125		9,450	10,900

23 81 Decentralized Unitary HVAC Equipment

23 81 13 – Packaged Terminal Air-Conditioners

23 81 13.10 Packaged Cabinet Type Air-Conditioners		Crew	Daily Output	Labor-Hours	Unit	Material	Labor	Equipment	Total	Total Incl O&P
0010	**PACKAGED CABINET TYPE AIR-CONDITIONERS**, Cabinet, wall sleeve,									
0100	louver, electric heat, thermostat, manual changeover, 208 V									
0200	6,000 BTUH cooling, 8800 BTU heat	Q-5	6	2.667	Ea.	720	136		856	995
0220	9,000 BTUH cooling, 13,900 BTU heat		5	3.200		785	163		948	1,100
0240	12,000 BTUH cooling, 13,900 BTU heat		4	4		850	204		1,054	1,250
0260	15,000 BTUH cooling, 13,900 BTU heat	↓	3	5.333	↓	1,050	272		1,322	1,550

23 81 Decentralized Unitary HVAC Equipment

23 81 19 – Self-Contained Air-Conditioners

23 81 19.10 Window Unit Air Conditioners		Crew	Daily Output	Labor-Hours	Unit	Material	2013 Bare Costs Labor	Equipment	Total	Total Incl O&P
0010	**WINDOW UNIT AIR CONDITIONERS**									
4000	Portable/window, 15 amp 125 V grounded receptacle required									
4060	5000 BTUH	1 Carp	8	1	Ea.	284	45		329	380
4340	6000 BTUH		8	1		325	45		370	430
4480	8000 BTUH		6	1.333		445	60		505	580
4500	10,000 BTUH		6	1.333		570	60		630	715
4520	12,000 BTUH	L-2	8	2		695	78.50		773.50	885
4600	Window/thru-the-wall, 15 amp 230 V grounded receptacle required									
4780	18,000 BTUH	L-2	6	2.667	Ea.	930	105		1,035	1,175
4940	25,000 BTUH		4	4		1,175	157		1,332	1,550
4960	29,000 BTUH		4	4		1,300	157		1,457	1,675

23 81 43 – Air-Source Unitary Heat Pumps

23 81 43.10 Air-Source Heat Pumps

		Crew	Daily Output	Labor-Hours	Unit	Material	Labor	Equipment	Total	Total Incl O&P
0010	**AIR-SOURCE HEAT PUMPS**, Not including interconnecting tubing									
1000	Air to air, split system, not including curbs, pads, fan coil and ductwork									
1010	For curbs/pads see Section 23 91 00									
1012	Outside condensing unit only, for fan coil see Section 23 82 19.10									
1015	1.5 ton cooling, 7 MBH heat @ 0°F	Q-5	2.40	6.667	Ea.	2,100	340		2,440	2,825
1020	2 ton cooling, 8.5 MBH heat @ 0°F		2	8		2,225	410		2,635	3,075
1030	2.5 ton cooling, 10 MBH heat @ 0°F		1.60	10		2,450	510		2,960	3,475
1040	3 ton cooling, 13 MBH heat @ 0°F		1.20	13.333		2,700	680		3,380	4,000
1050	3.5 ton cooling, 18 MBH heat @ 0°F		1	16		2,900	815		3,715	4,425
1054	4 ton cooling, 24 MBH heat @ 0°F		.80	20		3,200	1,025		4,225	5,050
1060	5 ton cooling, 27 MBH heat @ 0°F		.50	32		3,500	1,625		5,125	6,300
1500	Single package, not including curbs, pads, or plenums									
1502	1/2 ton cooling, supplementary heat included	Q-5	8	2	Ea.	2,275	102		2,377	2,650
1504	3/4 ton cooling, supplementary heat supplementary heat included		6	2.667		2,400	136		2,536	2,825
1506	1 ton cooling, supplementary heat included		4	4		2,700	204		2,904	3,275
1510	1.5 ton cooling, 5 MBH heat @ 0°F		1.55	10.323		2,650	525		3,175	3,725
1520	2 ton cooling, 6.5 MBH heat @ 0°F		1.50	10.667		2,850	545		3,395	3,950
1540	2.5 ton cooling, 8 MBH heat @ 0°F		1.40	11.429		3,075	585		3,660	4,250
1560	3 ton cooling, 10 MBH heat @ 0°F		1.20	13.333		3,350	680		4,030	4,700
1570	3.5 ton cooling, 11 MBH heat @ 0°F		1	16		3,675	815		4,490	5,250
1580	4 ton cooling, 13 MBH heat @ 0°F		.96	16.667		4,000	850		4,850	5,675
1620	5 ton cooling, 27 MBH heat @ 0°F		.65	24.615		4,225	1,250		5,475	6,550
1640	7.5 ton cooling, 35 MBH heat @ 0°F		.40	40		7,000	2,050		9,050	10,800

23 81 46 – Water-Source Unitary Heat Pumps

23 81 46.10 Water Source Heat Pumps

		Crew	Daily Output	Labor-Hours	Unit	Material	Labor	Equipment	Total	Total Incl O&P
0010	**WATER SOURCE HEAT PUMPS**, Not incl. connecting tubing or water source									
2000	Water source to air, single package									
2100	1 ton cooling, 13 MBH heat @ 75°F	Q-5	2	8	Ea.	1,750	410		2,160	2,550
2120	1.5 ton cooling, 17 MBH heat @ 75°F		1.80	8.889		1,850	455		2,305	2,700
2140	2 ton cooling, 19 MBH heat @ 75°F		1.70	9.412		1,975	480		2,455	2,900
2160	2.5 ton cooling, 25 MBH heat @ 75°F		1.60	10		2,075	510		2,585	3,075
2180	3 ton cooling, 27 MBH heat @ 75°F		1.40	11.429		2,150	585		2,735	3,250
2190	3.5 ton cooling, 29 MBH heat @ 75°F		1.30	12.308		2,450	625		3,075	3,625
2200	4 ton cooling, 31 MBH heat @ 75°F		1.20	13.333		2,600	680		3,280	3,900
2220	5 ton cooling, 29 MBH heat @ 75°F		.90	17.778		2,925	905		3,830	4,575
3960	For supplementary heat coil, add					10%				

23 82 Convection Heating and Cooling Units

23 82 19 – Fan Coil Units

23 82 19.10 Fan Coil Air Conditioning		Crew	Daily Output	Labor-Hours	Unit	2013 Bare Costs Material	Labor	Equipment	Total	Total Incl O&P
0010	**FAN COIL AIR CONDITIONING** Cabinet mounted, filters, controls									
0030	Fan coil AC, cabinet mounted, filters and controls									
0940	Direct expansion, for use w/air cooled condensing unit, 1.5 ton cooling	Q-5	5	3.200	Ea.	730	163		893	1,050
0950	2 ton cooling		4.80	3.333		785	170		955	1,125
0960	2.5 ton cooling		4.40	3.636		805	185		990	1,175
0970	3 ton cooling		3.80	4.211		1,025	215		1,240	1,450
0980	3.5 ton cooling		3.60	4.444		1,050	227		1,277	1,500
0990	4 ton cooling		3.40	4.706		1,175	240		1,415	1,650
1000	5 ton cooling		3	5.333		1,400	272		1,672	1,950

23 82 27 – Infrared Units

23 82 27.10 Infrared Type Heating Units		Crew	Daily Output	Labor-Hours	Unit	Material	Labor	Equipment	Total	Total Incl O&P
0010	**INFRARED TYPE HEATING UNITS**									
0020	Gas fired, unvented, electric ignition, 100% shutoff.									
0030	Piping and wiring not included									
0060	Input, 15 MBH	Q-5	7	2.286	Ea.	535	117		652	765
0100	30 MBH		6	2.667		825	136		961	1,100
0120	45 MBH		5	3.200		1,100	163		1,263	1,475
0140	50 MBH		4.50	3.556		1,125	181		1,306	1,500
0160	60 MBH		4	4		1,175	204		1,379	1,575
0180	75 MBH		3	5.333		1,175	272		1,447	1,700
0200	90 MBH		2.50	6.400		1,300	325		1,625	1,925
0220	105 MBH		2	8		1,400	410		1,810	2,175
0240	120 MBH		2	8		1,450	410		1,860	2,200
2000	Electric, single or three phase									
2050	6 kW, 20,478 BTU	1 Elec	2.30	3.478	Ea.	575	182		757	910
2100	13.5 KW, 40,956 BTU		2.20	3.636		830	191		1,021	1,200
2150	24 KW, 81,912 BTU		2	4		1,675	210		1,885	2,175
3000	Oil fired, pump, controls, fusible valve, oil supply tank									
3050	91,000 BTU	Q-5	2.50	6.400	Ea.	6,425	325		6,750	7,575
3080	105,000 BTU		2.25	7.111		6,800	365		7,165	8,050
3110	119,000 BTU		2	8		7,450	410		7,860	8,825

23 82 29 – Radiators

23 82 29.10 Hydronic Heating		Crew	Daily Output	Labor-Hours	Unit	Material	Labor	Equipment	Total	Total Incl O&P
0010	**HYDRONIC HEATING**, Terminal units, not incl. main supply pipe									
1000	Radiation									
1100	Panel, baseboard, C.I., including supports, no covers	Q-5	46	.348	L.F.	36.50	17.75		54.25	66.50
3000	Radiators, cast iron									
3100	Free standing or wall hung, 6 tube, 25" high	Q-5	96	.167	Section	39.50	8.50		48	56.50
3150	4 tube 25" high		96	.167		29.50	8.50		38	45.50
3200	4 tube, 19" high		96	.167		27.50	8.50		36	43.50
3250	Adj. brackets, 2 per wall radiator up to 30 sections	1 Stpi	32	.250	Ea.	30.50	14.15		44.65	55
3500	Recessed, 20" high x 5" deep, without grille	Q-5	60	.267	Section	33.50	13.60		47.10	57
3600	For inlet grille, add				"	2.33			2.33	2.56
9500	To convert SFR to BTU rating: Hot water, 150 x SFR									
9510	Forced hot water, 180 x SFR; steam, 240 x SFR									

23 82 33 – Convectors

23 82 33.10 Convector Units		Crew	Daily Output	Labor-Hours	Unit	Material	Labor	Equipment	Total	Total Incl O&P
0010	**CONVECTOR UNITS**, Terminal units, not incl. main supply pipe									
2204	Convector, multifin, 2 pipe w/cabinet									
2210	17" H x 24" L	Q-5	10	1.600	Ea.	99.50	81.50		181	233
2214	17" H x 36" L		8.60	1.860		149	95		244	305

23 82 Convection Heating and Cooling Units

23 82 33 – Convectors

23 82 33.10 Convector Units		Crew	Daily Output	Labor-Hours	Unit	2013 Bare Costs Material	Labor	Equipment	Total	Total Incl O&P
2218	17" H x 48" L	Q-5	7.40	2.162	Ea.	199	110		309	385
2222	21" H x 24" L		9	1.778		99.50	90.50		190	246
2228	21" H x 48" L	↓	6.80	2.353	↓	199	120		319	400
2240	For knob operated damper, add					140%				
2241	For metal trim strips, add	Q-5	64	.250	Ea.	19.25	12.75		32	40
2243	For snap-on inlet grille, add					10%	10%			
2245	For hinged access door, add	Q-5	64	.250	Ea.	35.50	12.75		48.25	58
2246	For air chamber, auto-venting, add	"	58	.276	"	6.85	14.05		20.90	28.50

23 82 36 – Finned-Tube Radiation Heaters

23 82 36.10 Finned Tube Radiation		Crew	Daily Output	Labor-Hours	Unit	Material	Labor	Equipment	Total	Total Incl O&P
0010	**FINNED TUBE RADIATION**, Terminal units, not incl. main supply pipe									
1310	Baseboard, pkgd, 1/2" copper tube, alum. fin, 7" high	Q-5	60	.267	L.F.	7.65	13.60		21.25	29
1320	3/4" copper tube, alum. fin, 7" high	"	58	.276	"	8.40	14.05		22.45	30.50
1381	Rough in baseboard panel & fin tube, supply & balance valves	1 Stpi	1.06	7.547	Ea.	315	430		745	990
1500	Note: fin tube may also require corners, caps, etc.									

23 83 Radiant Heating Units

23 83 16 – Radiant-Heating Hydronic Piping

23 83 16.10 Radiant Floor Heating		Crew	Daily Output	Labor-Hours	Unit	Material	Labor	Equipment	Total	Total Incl O&P
0010	**RADIANT FLOOR HEATING**									
0100	Tubing, PEX (cross-linked polyethylene)									
0110	Oxygen barrier type for systems with ferrous materials									
0120	1/2"	Q-5	800	.020	L.F.	1.01	1.02		2.03	2.65
0130	3/4"		535	.030		1.43	1.52		2.95	3.87
0140	1"	↓	400	.040	↓	2.23	2.04		4.27	5.50
0200	Non barrier type for ferrous free systems									
0210	1/2"	Q-5	800	.020	L.F.	.57	1.02		1.59	2.17
0220	3/4"		535	.030		1.05	1.52		2.57	3.46
0230	1"	↓	400	.040	↓	1.80	2.04		3.84	5.05
1000	Manifolds									
1110	Brass									
1120	With supply and return valves, flow meter, thermometer,									
1122	auto air vent and drain/fill valve.									
1130	1", 2 circuit	Q-5	14	1.143	Ea.	245	58.50		303.50	355
1140	1", 3 circuit		13.50	1.185		280	60.50		340.50	400
1150	1", 4 circuit		13	1.231		305	62.50		367.50	430
1154	1", 5 circuit		12.50	1.280		365	65.50		430.50	500
1158	1", 6 circuit		12	1.333		395	68		463	535
1162	1", 7 circuit		11.50	1.391		430	71		501	580
1166	1", 8 circuit		11	1.455		480	74		554	635
1172	1", 9 circuit		10.50	1.524		515	77.50		592.50	680
1174	1", 10 circuit		10	1.600		555	81.50		636.50	735
1178	1", 11 circuit		9.50	1.684		575	86		661	765
1182	1", 12 circuit	↓	9	1.778	↓	635	90.50		725.50	835
1610	Copper manifold header, (cut to size)									
1620	1" header, 12 – 1/2" sweat outlets	Q-5	3.33	4.805	Ea.	91	245		336	470
1630	1-1/4" header, 12 – 1/2" sweat outlets		3.20	5		106	255		361	500
1640	1-1/4" header, 12 – 3/4" sweat outlets		3	5.333		114	272		386	535
1650	1-1/2" header, 12 – 3/4" sweat outlets		3.10	5.161		139	263		402	550
1660	2" header, 12 – 3/4" sweat outlets	↓	2.90	5.517	↓	201	281		482	645

23 83 16 – Radiant-Heating Hydronic Piping

23 83 16.10 Radiant Floor Heating		Crew	Daily Output	Labor-Hours	Unit	Material	2013 Bare Costs Labor	Equipment	Total	Total Incl O&P
3000	Valves									
3110	Thermostatic zone valve actuator with end switch	Q-5	40	.400	Ea.	38.50	20.50		59	73
3114	Thermostatic zone valve actuator	"	36	.444	"	81	22.50		103.50	123
3120	Motorized straight zone valve with operator complete									
3130	3/4"	Q-5	35	.457	Ea.	130	23.50		153.50	178
3140	1"		32	.500		141	25.50		166.50	194
3150	1-1/4"	↓	29.60	.541	↓	179	27.50		206.50	239
3500	4 Way mixing valve, manual, brass									
3530	1"	Q-5	13.30	1.203	Ea.	180	61.50		241.50	291
3540	1-1/4"		11.40	1.404		195	71.50		266.50	320
3550	1-1/2"		11	1.455		249	74		323	385
3560	2"		10.60	1.509		350	77		427	505
3800	Mixing valve motor, 4 way for valves, 1" and 1-1/4"		34	.471		310	24		334	375
3810	Mixing valve motor, 4 way for valves, 1-1/2" and 2"	↓	30	.533	↓	355	27		382	430
5000	Radiant floor heating, zone control panel									
5120	4 Zone actuator valve control, expandable	Q-5	20	.800	Ea.	163	41		204	241
5130	6 Zone actuator valve control, expandable		18	.889		226	45.50		271.50	315
6070	Thermal track, straight panel for long continuous runs, 5.333 S.F.		40	.400		26.50	20.50		47	60
6080	Thermal track, utility panel, for direction reverse at run end, 5.333 S.F.		40	.400		26.50	20.50		47	60
6090	Combination panel, for direction reverse plus straight run, 5.333 S.F.	↓	40	.400	↓	26.50	20.50		47	60
7000	PEX tubing fittings									
7100	Compression type									
7116	Coupling									
7120	1/2" x 1/2"	1 Stpi	27	.296	Ea.	6.70	16.80		23.50	33
7124	3/4" x 3/4"	"	23	.348	"	10.60	19.70		30.30	41
7130	Adapter									
7132	1/2" x female sweat 1/2"	1 Stpi	27	.296	Ea.	4.36	16.80		21.16	30.50
7134	1/2" x female sweat 3/4"		26	.308		4.88	17.45		22.33	32
7136	5/8" x female sweat 3/4"	↓	24	.333	↓	7	18.90		25.90	36
7140	Elbow									
7142	1/2" x female sweat 1/2"	1 Stpi	27	.296	Ea.	6.65	16.80		23.45	33
7144	1/2" x female sweat 3/4"		26	.308		7.80	17.45		25.25	35
7146	5/8" x female sweat 3/4"	↓	24	.333	↓	8.75	18.90		27.65	38
7200	Insert type									
7206	PEX x male NPT									
7210	1/2" x 1/2"	1 Stpi	29	.276	Ea.	2.48	15.65		18.13	26
7220	3/4" x 3/4"		27	.296		3.67	16.80		20.47	29.50
7230	1" x 1"	↓	26	.308	↓	6.20	17.45		23.65	33.50
7300	PEX coupling									
7310	1/2" x 1/2"	1 Stpi	30	.267	Ea.	.94	15.10		16.04	24
7320	3/4" x 3/4"		29	.276		1.17	15.65		16.82	25
7330	1" x 1"	↓	28	.286	↓	2.20	16.20		18.40	27
7400	PEX stainless crimp ring									
7410	1/2" x 1/2"	1 Stpi	86	.093	Ea.	.34	5.25		5.59	8.30
7420	3/4" x 3/4"		84	.095		.47	5.40		5.87	8.60
7430	1" x 1"	↓	82	.098	↓	.68	5.55		6.23	9.05

23 83 33 – Electric Radiant Heaters

23 83 33.10 Electric Heating		Crew	Daily Output	Labor-Hours	Unit	Material	Labor	Equipment	Total	Total Incl O&P
0010	**ELECTRIC HEATING**, not incl. conduit or feed wiring									
1100	Rule of thumb: Baseboard units, including control	1 Elec	4.40	1.818	kW	101	95.50		196.50	254
1300	Baseboard heaters, 2' long, 350 watt		8	1	Ea.	30	52.50		82.50	112
1400	3' long, 750 watt		8	1		46	52.50		98.50	129

23 83 33 – Electric Radiant Heaters

23 83 33.10 Electric Heating		Crew	Daily Output	Labor-Hours	Unit	Material	Labor	Equipment	Total	Total Incl O&P
						2013 Bare Costs				
1600	4′ long, 1000 watt	1 Elec	6.70	1.194	Ea.	41	62.50		103.50	139
1800	5′ long, 935 watt		5.70	1.404		51	73.50		124.50	166
2000	6′ long, 1500 watt		5	1.600		56	84		140	187
2200	7′ long, 1310 watt		4.40	1.818		84	95.50		179.50	236
2400	8′ long, 2000 watt		4	2		69.50	105		174.50	233
2600	9′ long, 1680 watt		3.60	2.222		104	116		220	288
2800	10′ long, 1875 watt		3.30	2.424		193	127		320	400
2950	Wall heaters with fan, 120 to 277 volt									
3160	Recessed, residential, 750 watt	1 Elec	6	1.333	Ea.	134	70		204	252
3170	1000 watt		6	1.333		134	70		204	252
3180	1250 watt		5	1.600		134	84		218	272
3190	1500 watt		4	2		134	105		239	305
3600	Thermostats, integral		16	.500		30	26		56	72
3800	Line voltage, 1 pole		8	1		28.50	52.50		81	110
3810	2 pole		8	1		29.50	52.50		82	111
4000	Heat trace system, 400 degree									
4020	115 V, 2.5 watts per L.F.	1 Elec	530	.015	L.F.	7.70	.79		8.49	9.70
4030	5 watts per L.F.		530	.015		7.70	.79		8.49	9.70
4050	10 watts per L.F.		530	.015		7.70	.79		8.49	9.70
4060	208 V, 5 watts per L.F.		530	.015		7.70	.79		8.49	9.70
4080	480 V, 8 watts per L.F.		530	.015		7.70	.79		8.49	9.70
4200	Heater raceway									
4260	Heat transfer cement									
4280	1 gallon				Ea.	57			57	62.50
4300	5 gallon				"	220			220	242
4320	Snap band, clamp									
4340	3/4″ pipe size	1 Elec	470	.017	Ea.		.89		.89	1.33
4360	1″ pipe size		444	.018			.94		.94	1.41
4380	1-1/4″ pipe size		400	.020			1.05		1.05	1.57
4400	1-1/2″ pipe size		355	.023			1.18		1.18	1.77
4420	2″ pipe size		320	.025			1.31		1.31	1.96
4440	3″ pipe size		160	.050			2.62		2.62	3.92
4460	4″ pipe size		100	.080			4.19		4.19	6.25
4480	Thermostat NEMA 3R, 22 amp, 0-150 Deg, 10′ cap.		8	1		241	52.50		293.50	345
4500	Thermostat NEMA 4X, 25 amp, 40 Deg, 5-1/2′ cap.		7	1.143		241	60		301	355
4520	Thermostat NEMA 4X, 22 amp, 25-325 Deg, 10′ cap.		7	1.143		650	60		710	805
4540	Thermostat NEMA 4X, 22 amp, 15-140 Deg,		6	1.333		545	70		615	705
4580	Thermostat NEMA 4,7,9, 22 amp, 25-325 Deg, 10′ cap.		3.60	2.222		800	116		916	1,050
4600	Thermostat NEMA 4,7,9, 22 amp, 15-140 Deg,		3	2.667		795	140		935	1,075
4720	Fiberglass application tape, 36 yard roll		11	.727		77.50	38		115.50	143
5000	Radiant heating ceiling panels, 2′ x 4′, 500 watt		16	.500		279	26		305	345
5050	750 watt		16	.500		300	26		326	370
5200	For recessed plaster frame, add		32	.250		106	13.10		119.10	136
5300	Infrared quartz heaters, 120 volts, 1000 watts		6.70	1.194		230	62.50		292.50	345
5350	1500 watt		5	1.600		230	84		314	380
5400	240 volts, 1500 watt		5	1.600		230	84		314	380
5450	2000 watt		4	2		230	105		335	410
5500	3000 watt		3	2.667		255	140		395	490
5550	4000 watt		2.60	3.077		255	161		416	520
5570	Modulating control		.80	10		275	525		800	1,100
5600	Unit heaters, heavy duty, with fan & mounting bracket									
5650	Single phase, 208-240-277 volt, 3 kW	1 Elec	3.20	2.500	Ea.	420	131		551	655
5750	5 kW		2.40	3.333		440	175		615	745

23 83 33 – Electric Radiant Heaters

	23 83 33.10 Electric Heating	Crew	Daily Output	Labor-Hours	Unit	Material	Labor	Equipment	Total	Total Incl O&P
5800	7 kW	1 Elec	1.90	4.211	Ea.	680	221		901	1,075
5850	10 kW		1.30	6.154		775	320		1,095	1,325
5950	15 kW		.90	8.889		1,250	465		1,715	2,075
6000	480 volt, 3 kW		3.30	2.424		465	127		592	700
6020	4 kW		3	2.667		470	140		610	730
6040	5 kW		2.60	3.077		490	161		651	775
6060	7 kW		2	4		730	210		940	1,125
6080	10 kW		1.40	5.714		815	299		1,114	1,350
6100	13 kW		1.10	7.273		1,250	380		1,630	1,950
6120	15 kW		1	8		1,250	420		1,670	2,000
6140	20 kW		.90	8.889		1,625	465		2,090	2,500
6300	3 phase, 208-240 volt, 5 kW		2.40	3.333		415	175		590	715
6320	7 kW		1.90	4.211		645	221		866	1,025
6340	10 kW		1.30	6.154		720	320		1,040	1,275
6360	15 kW		.90	8.889		1,325	465		1,790	2,150
6380	20 kW		.70	11.429		1,650	600		2,250	2,725
6400	25 kW		.50	16		1,950	840		2,790	3,400
6500	480 volt, 5 kW		2.60	3.077		585	161		746	880
6520	7 kW		2	4		735	210		945	1,125
6540	10 kW		1.40	5.714		775	299		1,074	1,300
6560	13 kW		1.10	7.273		1,425	380		1,805	2,150
6580	15 kW		1	8		1,425	420		1,845	2,200
6600	20 kW		.90	8.889		1,850	465		2,315	2,725
6800	Vertical discharge heaters, with fan									
6820	Single phase, 208-240-277 volt, 10 kW	1 Elec	1.30	6.154	Ea.	720	320		1,040	1,275
6840	15 kW		.90	8.889		1,200	465		1,665	2,025
6900	3 phase, 208-240 volt, 10 kW		1.30	6.154		720	320		1,040	1,275
6920	15 kW		.90	8.889		1,200	465		1,665	2,025
6940	20 kW		.70	11.429		1,725	600		2,325	2,800
7100	480 volt, 10 kW		1.40	5.714		775	299		1,074	1,300
7120	15 kW		1	8		1,250	420		1,670	2,000
7140	20 kW		.90	8.889		1,625	465		2,090	2,500
7160	25 kW		.60	13.333		1,950	700		2,650	3,200
7900	Cabinet convector heaters, 240 volt									
7920	3′ long, 2000 watt	1 Elec	5.30	1.509	Ea.	1,900	79		1,979	2,225
7940	3000 watt		5.30	1.509		2,000	79		2,079	2,300
7960	4000 watt		5.30	1.509		2,050	79		2,129	2,375
7980	6000 watt		4.60	1.739		2,125	91		2,216	2,450
8000	8000 watt		4.60	1.739		2,200	91		2,291	2,550
8020	4′ long, 4000 watt		4.60	1.739		2,075	91		2,166	2,400
8040	6000 watt		4	2		2,150	105		2,255	2,525
8060	8000 watt		4	2		2,225	105		2,330	2,600
8080	10,000 watt		4	2		2,250	105		2,355	2,625
8100	Available also in 208 or 277 volt									
8200	Cabinet unit heaters, 120 to 277 volt, 1 pole,									
8220	wall mounted, 2 kW	1 Elec	4.60	1.739	Ea.	1,850	91		1,941	2,150
8230	3 kW		4.60	1.739		1,875	91		1,966	2,175
8240	4 kW		4.40	1.818		1,925	95.50		2,020.50	2,275
8250	5 kW		4.40	1.818		2,000	95.50		2,095.50	2,350
8260	6 kW		4.20	1.905		2,000	100		2,100	2,350
8270	8 kW		4	2		2,075	105		2,180	2,450
8280	10 kW		3.80	2.105		2,125	110		2,235	2,525
8290	12 kW		3.50	2.286		2,300	120		2,420	2,725

(Material, Labor, Equipment and Total columns: 2013 Bare Costs)

23 83 33 – Electric Radiant Heaters

23 83 33.10 Electric Heating		Crew	Daily Output	Labor-Hours	Unit	Material	Labor	Equipment	Total	Total Incl O&P
						2013 Bare Costs				
8300	13.5 kW	1 Elec	2.90	2.759	Ea.	2,350	145		2,495	2,800
8310	16 kW		2.70	2.963		2,400	155		2,555	2,875
8320	20 kW		2.30	3.478		3,475	182		3,657	4,100
8330	24 kW		1.90	4.211		3,525	221		3,746	4,200
8350	Recessed, 2 kW		4.40	1.818		1,800	95.50		1,895.50	2,125
8370	3 kW		4.40	1.818		1,875	95.50		1,970.50	2,200
8380	4 kW		4.20	1.905		1,875	100		1,975	2,200
8390	5 kW		4.20	1.905		2,100	100		2,200	2,450
8400	6 kW		4	2		2,200	105		2,305	2,575
8410	8 kW		3.80	2.105		2,200	110		2,310	2,600
8420	10 kW		3.50	2.286		2,675	120		2,795	3,125
8430	12 kW		2.90	2.759		2,700	145		2,845	3,200
8440	13.5 kW		2.70	2.963		2,850	155		3,005	3,375
8450	16 kW		2.30	3.478		2,875	182		3,057	3,450
8460	20 kW		1.90	4.211		2,900	221		3,121	3,525
8470	24 kW		1.60	5		3,450	262		3,712	4,200
8490	Ceiling mounted, 2 kW		3.20	2.500		1,875	131		2,006	2,250
8510	3 kW		3.20	2.500		1,950	131		2,081	2,350
8520	4 kW		3	2.667		1,975	140		2,115	2,375
8530	5 kW		3	2.667		2,075	140		2,215	2,475
8540	6 kW		2.80	2.857		2,075	150		2,225	2,525
8550	8 kW		2.40	3.333		2,150	175		2,325	2,625
8560	10 kW		2.20	3.636		2,125	191		2,316	2,625
8570	12 kW		2	4		2,250	210		2,460	2,800
8580	13.5 kW		1.50	5.333		2,275	279		2,554	2,925
8590	16 kW		1.30	6.154		2,400	320		2,720	3,125
8600	20 kW		.90	8.889		3,375	465		3,840	4,400
8610	24 kW		.60	13.333		3,450	700		4,150	4,850
8630	208 to 480 V, 3 pole									
8650	Wall mounted, 2 kW	1 Elec	4.60	1.739	Ea.	2,000	91		2,091	2,325
8670	3 kW		4.60	1.739		2,075	91		2,166	2,425
8680	4 kW		4.40	1.818		2,125	95.50		2,220.50	2,500
8690	5 kW		4.40	1.818		2,200	95.50		2,295.50	2,575
8700	6 kW		4.20	1.905		2,225	100		2,325	2,600
8710	8 kW		4	2		2,300	105		2,405	2,675
8720	10 kW		3.80	2.105		2,325	110		2,435	2,750
8730	12 kW		3.50	2.286		2,375	120		2,495	2,800
8740	13.5 kW		2.90	2.759		2,425	145		2,570	2,875
8750	16 kW		2.70	2.963		2,425	155		2,580	2,875
8760	20 kW		2.30	3.478		3,075	182		3,257	3,650
8770	24 kW		1.90	4.211		3,600	221		3,821	4,300
8790	Recessed, 2 kW		4.40	1.818		2,000	95.50		2,095.50	2,350
8810	3 kW		4.40	1.818		2,075	95.50		2,170.50	2,450
8820	4 kW		4.20	1.905		1,975	100		2,075	2,325
8830	5 kW		4.20	1.905		2,200	100		2,300	2,575
8840	6 kW		4	2		2,225	105		2,330	2,600
8850	8 kW		3.80	2.105		2,300	110		2,410	2,700
8860	10 kW		3.50	2.286		2,325	120		2,445	2,750
8870	12 kW		2.90	2.759		2,375	145		2,520	2,850
8880	13.5 kW		2.70	2.963		2,425	155		2,580	2,875
8890	16 kW		2.30	3.478		3,500	182		3,682	4,125
8900	20 kW		1.90	4.211		3,600	221		3,821	4,300
8920	24 kW		1.60	5		3,650	262		3,912	4,400

23 83 Radiant Heating Units

23 83 33 – Electric Radiant Heaters

23 83 33.10 Electric Heating		Crew	Daily Output	Labor-Hours	Unit	Material	2013 Bare Costs Labor	Equipment	Total	Total Incl O&P
8940	Ceiling mount, 2 kW	1 Elec	3.20	2.500	Ea.	2,000	131		2,131	2,400
8950	3 kW		3.20	2.500		2,075	131		2,206	2,500
8960	4 kW		3	2.667		2,125	140		2,265	2,550
8970	5 kW		3	2.667		2,200	140		2,340	2,625
8980	6 kW		2.80	2.857		2,225	150		2,375	2,675
8990	8 kW		2.40	3.333		2,300	175		2,475	2,775
9000	10 kW		2.20	3.636		2,325	191		2,516	2,850
9020	13.5 kW		1.50	5.333		2,425	279		2,704	3,075
9030	16 kW		1.30	6.154		3,550	320		3,870	4,400
9040	20 kW		.90	8.889		3,550	465		4,015	4,600
9060	24 kW	↓	.60	13.333	↓	3,425	700		4,125	4,825

23 91 Prefabricated Equipment Supports

23 91 10 – Prefabricated Curbs, Pads and Stands

23 91 10.10 Prefabricated Pads and Stands		Crew	Daily Output	Labor-Hours	Unit	Material	Labor	Equipment	Total	Total Incl O&P
0010	**PREFABRICATED PADS AND STANDS**									
6000	Pad, fiberglass reinforced concrete with polystyrene foam core									
6050	Condenser, 2" thick, 20" x 38"	1 Shee	8	1	Ea.	26.50	53.50		80	111
6220	30" x 36"	"	8	1		41.50	53.50		95	127
6340	36" x 54"	Q-9	6	2.667	↓	77.50	128		205.50	281

Division Notes

	CREW	DAILY OUTPUT	LABOR-HOURS	UNIT	BARE COSTS MAT.	BARE COSTS LABOR	BARE COSTS EQUIP.	BARE COSTS TOTAL	TOTAL INCL O&P

Estimating Tips

26 05 00 Common Work Results for Electrical

- Conduit should be taken off in three main categories—power distribution, branch power, and branch lighting—so the estimator can concentrate on systems and components, therefore making it easier to ensure all items have been accounted for.
- For cost modifications for elevated conduit installation, add the percentages to labor according to the height of installation, and only to the quantities exceeding the different height levels, not to the total conduit quantities.
- Remember that aluminum wiring of equal ampacity is larger in diameter than copper and may require larger conduit.
- If more than three wires at a time are being pulled, deduct percentages from the labor hours of that grouping of wires.
- When taking off grounding systems, identify separately the type and size of wire, and list each unique type of ground connection.
- The estimator should take the weights of materials into consideration when completing a takeoff. Topics to consider include: How will the materials be supported? What methods of support are available? How high will the support structure have to reach? Will the final support structure be able to withstand the total burden? Is the support material included or separate from the fixture, equipment, and material specified?
- Do not overlook the costs for equipment used in the installation. If scaffolding or highlifts are available in the field, contractors may use them in lieu of the proposed ladders and rolling staging.

26 20 00 Low-Voltage Electrical Transmission

- Supports and concrete pads may be shown on drawings for the larger equipment, or the support system may be only a piece of plywood for the back of a panelboard. In either case, it must be included in the costs.

26 40 00 Electrical and Cathodic Protection

- When taking off cathodic protections systems, identify the type and size of cable, and list each unique type of anode connection.

26 50 00 Lighting

- Fixtures should be taken off room by room, using the fixture schedule, specifications, and the ceiling plan. For large concentrations of lighting fixtures in the same area, deduct the percentages from labor hours.

Reference Numbers

Reference numbers are shown in shaded boxes at the beginning of some major classifications. These numbers refer to related items in the Reference Section. The reference information may be an estimating procedure, an alternate pricing method, or technical information.

Note: Not all subdivisions listed here necessarily appear in this publication.

Note: **Trade Service,** *in part, has been used as a reference source for some of the material prices used in Division 26.*

26 05 33 – Raceway and Boxes for Electrical Systems

26 05 33.95 Cutting and Drilling		Crew	Daily Output	Labor-Hours	Unit	2013 Bare Costs Material	Labor	Equipment	Total	Total Incl O&P
0010	**CUTTING AND DRILLING**									
0100	Hole drilling to 10′ high, concrete wall									
0110	8″ thick, 1/2″ pipe size	R-31	12	.667	Ea.	.25	35	4.01	39.26	57
0120	3/4″ pipe size		12	.667		.25	35	4.01	39.26	57
0130	1″ pipe size		9.50	.842		.39	44	5.05	49.44	72
0140	1-1/4″ pipe size		9.50	.842		.39	44	5.05	49.44	72
0150	1-1/2″ pipe size		9.50	.842		.39	44	5.05	49.44	72
0160	2″ pipe size		4.40	1.818		.55	95.50	10.95	107	156
0170	2-1/2″ pipe size		4.40	1.818		.55	95.50	10.95	107	156
0180	3″ pipe size		4.40	1.818		.55	95.50	10.95	107	156
0190	3-1/2″ pipe size		3.30	2.424		.66	127	14.55	142.21	207
0200	4″ pipe size		3.30	2.424		.66	127	14.55	142.21	207
0500	12″ thick, 1/2″ pipe size		9.40	.851		.37	44.50	5.10	49.97	72.50
0520	3/4″ pipe size		9.40	.851		.37	44.50	5.10	49.97	72.50
0540	1″ pipe size		7.30	1.096		.59	57.50	6.60	64.69	94
0560	1-1/4″ pipe size		7.30	1.096		.59	57.50	6.60	64.69	94
0570	1-1/2″ pipe size		7.30	1.096		.59	57.50	6.60	64.69	94
0580	2″ pipe size		3.60	2.222		.82	116	13.35	130.17	190
0590	2-1/2″ pipe size		3.60	2.222		.82	116	13.35	130.17	190
0600	3″ pipe size		3.60	2.222		.82	116	13.35	130.17	190
0610	3-1/2″ pipe size		2.80	2.857		.99	150	17.15	168.14	244
0630	4″ pipe size		2.50	3.200		.99	168	19.25	188.24	273
0650	16″ thick, 1/2″ pipe size		7.60	1.053		.49	55	6.35	61.84	90
0670	3/4″ pipe size		7	1.143		.49	60	6.85	67.34	97.50
0690	1″ pipe size		6	1.333		.78	70	8	78.78	115
0710	1-1/4″ pipe size		5.50	1.455		.78	76	8.75	85.53	124
0730	1-1/2″ pipe size		5.50	1.455		.78	76	8.75	85.53	124
0750	2″ pipe size		3	2.667		1.09	140	16.05	157.14	228
0770	2-1/2″ pipe size		2.70	2.963		1.09	155	17.80	173.89	253
0790	3″ pipe size		2.50	3.200		1.09	168	19.25	188.34	273
0810	3-1/2″ pipe size		2.30	3.478		1.32	182	21	204.32	297
0830	4″ pipe size		2	4		1.32	210	24	235.32	345
0850	20″ thick, 1/2″ pipe size		6.40	1.250		.62	65.50	7.50	73.62	107
0870	3/4″ pipe size		6	1.333		.62	70	8	78.62	114
0890	1″ pipe size		5	1.600		.98	84	9.60	94.58	137
0910	1-1/4″ pipe size		4.80	1.667		.98	87.50	10	98.48	143
0930	1-1/2″ pipe size		4.60	1.739		.98	91	10.45	102.43	149
0950	2″ pipe size		2.70	2.963		1.37	155	17.80	174.17	253
0970	2-1/2″ pipe size		2.40	3.333		1.37	175	20	196.37	285
0990	3″ pipe size		2.20	3.636		1.37	191	22	214.37	310
1010	3-1/2″ pipe size		2	4		1.65	210	24	235.65	345
1030	4″ pipe size		1.70	4.706		1.65	247	28.50	277.15	405
1050	24″ thick, 1/2″ pipe size		5.50	1.455		.74	76	8.75	85.49	124
1070	3/4″ pipe size		5.10	1.569		.74	82	9.45	92.19	134
1090	1″ pipe size		4.30	1.860		1.17	97.50	11.20	109.87	160
1110	1-1/4″ pipe size		4	2		1.17	105	12	118.17	171
1130	1-1/2″ pipe size		4	2		1.17	105	12	118.17	171
1150	2″ pipe size		2.40	3.333		1.64	175	20	196.64	285
1170	2-1/2″ pipe size		2.20	3.636		1.64	191	22	214.64	310
1190	3″ pipe size		2	4		1.64	210	24	235.64	345
1210	3-1/2″ pipe size		1.80	4.444		1.99	233	26.50	261.49	380
1230	4″ pipe size		1.50	5.333		1.99	279	32	312.99	460
1500	Brick wall, 8″ thick, 1/2″ pipe size		18	.444		.25	23.50	2.67	26.42	38

26 05 33.95 Cutting and Drilling		Crew	Daily Output	Labor-Hours	Unit	Material	2013 Bare Costs Labor	Equipment	Total	Total Incl O&P
1520	3/4" pipe size	R-31	18	.444	Ea.	.25	23.50	2.67	26.42	38
1540	1" pipe size		13.30	.602		.39	31.50	3.62	35.51	51.50
1560	1-1/4" pipe size		13.30	.602		.39	31.50	3.62	35.51	51.50
1580	1-1/2" pipe size		13.30	.602		.39	31.50	3.62	35.51	51.50
1600	2" pipe size		5.70	1.404		.55	73.50	8.45	82.50	120
1620	2-1/2" pipe size		5.70	1.404		.55	73.50	8.45	82.50	120
1640	3" pipe size		5.70	1.404		.55	73.50	8.45	82.50	120
1660	3-1/2" pipe size		4.40	1.818		.66	95.50	10.95	107.11	156
1680	4" pipe size		4	2		.66	105	12	117.66	171
1700	12" thick, 1/2" pipe size		14.50	.552		.37	29	3.32	32.69	47.50
1720	3/4" pipe size		14.50	.552		.37	29	3.32	32.69	47.50
1740	1" pipe size		11	.727		.59	38	4.37	42.96	62.50
1760	1-1/4" pipe size		11	.727		.59	38	4.37	42.96	62.50
1780	1-1/2" pipe size		11	.727		.59	38	4.37	42.96	62.50
1800	2" pipe size		5	1.600		.82	84	9.60	94.42	137
1820	2-1/2" pipe size		5	1.600		.82	84	9.60	94.42	137
1840	3" pipe size		5	1.600		.82	84	9.60	94.42	137
1860	3-1/2" pipe size		3.80	2.105		.99	110	12.65	123.64	180
1880	4" pipe size		3.30	2.424		.99	127	14.55	142.54	207
1900	16" thick, 1/2" pipe size		12.30	.650		.49	34	3.91	38.40	56
1920	3/4" pipe size		12.30	.650		.49	34	3.91	38.40	56
1940	1" pipe size		9.30	.860		.78	45	5.15	50.93	74
1960	1-1/4" pipe size		9.30	.860		.78	45	5.15	50.93	74
1980	1-1/2" pipe size		9.30	.860		.78	45	5.15	50.93	74
2000	2" pipe size		4.40	1.818		1.09	95.50	10.95	107.54	156
2010	2-1/2" pipe size		4.40	1.818		1.09	95.50	10.95	107.54	156
2030	3" pipe size		4.40	1.818		1.09	95.50	10.95	107.54	156
2050	3-1/2" pipe size		3.30	2.424		1.32	127	14.55	142.87	207
2070	4" pipe size		3	2.667		1.32	140	16.05	157.37	228
2090	20" thick, 1/2" pipe size		10.70	.748		.62	39	4.49	44.11	64
2110	3/4" pipe size		10.70	.748		.62	39	4.49	44.11	64
2130	1" pipe size		8	1		.98	52.50	6	59.48	86
2150	1-1/4" pipe size		8	1		.98	52.50	6	59.48	86
2170	1-1/2" pipe size		8	1		.98	52.50	6	59.48	86
2190	2" pipe size		4	2		1.37	105	12	118.37	172
2210	2-1/2" pipe size		4	2		1.37	105	12	118.37	172
2230	3" pipe size		4	2		1.37	105	12	118.37	172
2250	3-1/2" pipe size		3	2.667		1.65	140	16.05	157.70	228
2270	4" pipe size		2.70	2.963		1.65	155	17.80	174.45	253
2290	24" thick, 1/2" pipe size		9.40	.851		.74	44.50	5.10	50.34	73
2310	3/4" pipe size		9.40	.851		.74	44.50	5.10	50.34	73
2330	1" pipe size		7.10	1.127		1.17	59	6.75	66.92	97
2350	1-1/4" pipe size		7.10	1.127		1.17	59	6.75	66.92	97
2370	1-1/2" pipe size		7.10	1.127		1.17	59	6.75	66.92	97
2390	2" pipe size		3.60	2.222		1.64	116	13.35	130.99	191
2410	2-1/2" pipe size		3.60	2.222		1.64	116	13.35	130.99	191
2430	3" pipe size		3.60	2.222		1.64	116	13.35	130.99	191
2450	3-1/2" pipe size		2.80	2.857		1.99	150	17.15	169.14	245
2470	4" pipe size	▼	2.50	3.200	▼	1.99	168	19.25	189.24	274
3000	Knockouts to 8' high, metal boxes & enclosures									
3020	With hole saw, 1/2" pipe size	1 Elec	53	.151	Ea.		7.90		7.90	11.85
3040	3/4" pipe size		47	.170			8.90		8.90	13.35
3050	1" pipe size		40	.200			10.50		10.50	15.70

26 05 33 – Raceway and Boxes for Electrical Systems

	26 05 33.95 Cutting and Drilling	Crew	Daily Output	Labor-Hours	Unit	Material	2013 Bare Costs Labor	Equipment	Total	Total Incl O&P
3060	1-1/4" pipe size	1 Elec	36	.222	Ea.		11.65		11.65	17.40
3070	1-1/2" pipe size		32	.250			13.10		13.10	19.60
3080	2" pipe size		27	.296			15.55		15.55	23
3090	2-1/2" pipe size		20	.400			21		21	31.50
4010	3" pipe size		16	.500			26		26	39
4030	3-1/2" pipe size		13	.615			32.50		32.50	48.50
4050	4" pipe size		11	.727			38		38	57
4070	With hand punch set, 1/2" pipe size		40	.200			10.50		10.50	15.70
4090	3/4" pipe size		32	.250			13.10		13.10	19.60
4110	1" pipe size		30	.267			13.95		13.95	21
4130	1-1/4" pipe size		28	.286			14.95		14.95	22.50
4150	1-1/2" pipe size		26	.308			16.10		16.10	24
4170	2" pipe size		20	.400			21		21	31.50
4190	2-1/2" pipe size		17	.471			24.50		24.50	37
4200	3" pipe size		15	.533			28		28	42
4220	3-1/2" pipe size		12	.667			35		35	52.50
4240	4" pipe size		10	.800			42		42	62.50
4260	With hydraulic punch, 1/2" pipe size		44	.182			9.55		9.55	14.25
4280	3/4" pipe size		38	.211			11.05		11.05	16.50
4300	1" pipe size		38	.211			11.05		11.05	16.50
4320	1-1/4" pipe size		38	.211			11.05		11.05	16.50
4340	1-1/2" pipe size		38	.211			11.05		11.05	16.50
4360	2" pipe size		32	.250			13.10		13.10	19.60
4380	2-1/2" pipe size		27	.296			15.55		15.55	23
4400	3" pipe size		23	.348			18.25		18.25	27.50
4420	3-1/2" pipe size		20	.400			21		21	31.50
4440	4" pipe size		18	.444			23.50		23.50	35

26 05 80 – Wiring Connections

26 05 80.10 Motor Connections

		Crew	Daily Output	Labor-Hours	Unit	Material	Labor	Equipment	Total	Total Incl O&P
0010	**MOTOR CONNECTIONS**									
0020	Flexible conduit and fittings, 115 volt, 1 phase, up to 1 HP motor	1 Elec	8	1	Ea.	8.45	52.50		60.95	88
0050	2 HP motor		6.50	1.231		15.95	64.50		80.45	114
0100	3 HP motor		5.50	1.455		14.20	76		90.20	130
0110	230 volt, 3 phase, 3 HP motor		6.78	1.180		10	62		72	104
0112	5 HP motor		5.47	1.463		8.60	76.50		85.10	124
0114	7-1/2 HP motor		4.61	1.735		12.15	91		103.15	149
0120	10 HP motor		4.20	1.905		27.50	100		127.50	179
0150	15 HP motor		3.30	2.424		27.50	127		154.50	220
0200	25 HP motor		2.70	2.963		38.50	155		193.50	275
0400	50 HP motor		2.20	3.636		77.50	191		268.50	370
0600	100 HP motor		1.50	5.333		182	279		461	620

26 05 90 – Residential Applications

26 05 90.10 Residential Wiring

		Crew	Daily Output	Labor-Hours	Unit	Material	Labor	Equipment	Total	Total Incl O&P
0010	**RESIDENTIAL WIRING**									
0020	20' avg. runs and #14/2 wiring incl. unless otherwise noted									
1000	Service & panel, includes 24' SE-AL cable, service eye, meter,									
1010	Socket, panel board, main bkr., ground rod, 15 or 20 amp									
1020	1-pole circuit breakers, and misc. hardware									
1100	100 amp, with 10 branch breakers	1 Elec	1.19	6.723	Ea.	555	350		905	1,150
1110	With PVC conduit and wire		.92	8.696		625	455		1,080	1,375
1120	With RGS conduit and wire		.73	10.959		780	575		1,355	1,725
1150	150 amp, with 14 branch breakers		1.03	7.767		860	405		1,265	1,550

26 05 90 – Residential Applications

26 05 90.10 Residential Wiring		Crew	Daily Output	Labor-Hours	Unit	2013 Bare Costs Material	Labor	Equipment	Total	Total Incl O&P
1170	With PVC conduit and wire	1 Elec	.82	9.756	Ea.	995	510		1,505	1,875
1180	With RGS conduit and wire		.67	11.940		1,300	625		1,925	2,350
1200	200 amp, with 18 branch breakers	2 Elec	1.80	8.889		1,125	465		1,590	1,925
1220	With PVC conduit and wire		1.46	10.959		1,250	575		1,825	2,225
1230	With RGS conduit and wire		1.24	12.903		1,675	675		2,350	2,825
1800	Lightning surge suppressor for above services, add	1 Elec	32	.250		50	13.10		63.10	74.50
2000	Switch devices									
2100	Single pole, 15 amp, Ivory, with a 1-gang box, cover plate,									
2110	Type NM (Romex) cable	1 Elec	17.10	.468	Ea.	12.20	24.50		36.70	50
2120	Type MC (BX) cable		14.30	.559		26	29.50		55.50	73
2130	EMT & wire		5.71	1.401		33.50	73.50		107	147
2150	3-way, #14/3, type NM cable		14.55	.550		16.20	29		45.20	61
2170	Type MC cable		12.31	.650		35	34		69	89
2180	EMT & wire		5	1.600		37	84		121	166
2200	4-way, #14/3, type NM cable		14.55	.550		24.50	29		53.50	69.50
2220	Type MC cable		12.31	.650		43	34		77	98
2230	EMT & wire		5	1.600		45	84		129	175
2250	S.P., 20 amp, #12/2, type NM cable		13.33	.600		21	31.50		52.50	70
2270	Type MC cable		11.43	.700		32.50	36.50		69	90.50
2280	EMT & wire		4.85	1.649		43.50	86.50		130	177
2290	S.P. rotary dimmer, 600W, no wiring		17	.471		25	24.50		49.50	64
2300	S.P. rotary dimmer, 600W, type NM cable		14.55	.550		30	29		59	76
2320	Type MC cable		12.31	.650		43.50	34		77.50	99
2330	EMT & wire		5	1.600		52.50	84		136.50	183
2350	3-way rotary dimmer, type NM cable		13.33	.600		34.50	31.50		66	85
2370	Type MC cable		11.43	.700		48.50	36.50		85	109
2380	EMT & wire		4.85	1.649		57.50	86.50		144	192
2400	Interval timer wall switch, 20 amp, 1-30 min., #12/2									
2410	Type NM cable	1 Elec	14.55	.550	Ea.	54	29		83	103
2420	Type MC cable		12.31	.650		60.50	34		94.50	118
2430	EMT & wire		5	1.600		76.50	84		160.50	210
2500	Decorator style									
2510	S.P., 15 amp, type NM cable	1 Elec	17.10	.468	Ea.	16.30	24.50		40.80	54.50
2520	Type MC cable		14.30	.559		30	29.50		59.50	77
2530	EMT & wire		5.71	1.401		37.50	73.50		111	151
2550	3-way, #14/3, type NM cable		14.55	.550		20.50	29		49.50	65.50
2570	Type MC cable		12.31	.650		39	34		73	93.50
2580	EMT & wire		5	1.600		41	84		125	170
2600	4-way, #14/3, type NM cable		14.55	.550		28.50	29		57.50	74
2620	Type MC cable		12.31	.650		47	34		81	103
2630	EMT & wire		5	1.600		49	84		133	179
2650	S.P., 20 amp, #12/2, type NM cable		13.33	.600		25	31.50		56.50	74.50
2670	Type MC cable		11.43	.700		36.50	36.50		73	95
2680	EMT & wire		4.85	1.649		47.50	86.50		134	182
2700	S.P., slide dimmer, type NM cable		17.10	.468		32.50	24.50		57	72
2720	Type MC cable		14.30	.559		46.50	29.50		76	95
2730	EMT & wire		5.71	1.401		55	73.50		128.50	171
2750	S.P., touch dimmer, type NM cable		17.10	.468		29	24.50		53.50	68.50
2770	Type MC cable		14.30	.559		43	29.50		72.50	91.50
2780	EMT & wire		5.71	1.401		52	73.50		125.50	167
2800	3-way touch dimmer, type NM cable		13.33	.600		51	31.50		82.50	103
2820	Type MC cable		11.43	.700		65	36.50		101.50	127
2830	EMT & wire		4.85	1.649		74	86.50		160.50	210

26 05 90.10 Residential Wiring		Crew	Daily Output	Labor-Hours	Unit	2013 Bare Costs Material	Labor	Equipment	Total	Total Incl O&P
3000	Combination devices									
3100	S.P. switch/15 amp recpt., Ivory, 1-gang box, plate									
3110	Type NM cable	1 Elec	11.43	.700	Ea.	24	36.50		60.50	81.50
3120	Type MC cable		10	.800		38	42		80	105
3130	EMT & wire		4.40	1.818		46.50	95.50		142	195
3150	S.P. switch/pilot light, type NM cable		11.43	.700		24	36.50		60.50	81.50
3170	Type MC cable		10	.800		38	42		80	105
3180	EMT & wire		4.43	1.806		47	94.50		141.50	194
3190	2-S.P. switches, 2-#14/2, no wiring		14	.571		7.60	30		37.60	53.50
3200	2-S.P. switches, 2-#14/2, type NM cables		10	.800		27.50	42		69.50	92.50
3220	Type MC cable		8.89	.900		48.50	47		95.50	124
3230	EMT & wire		4.10	1.951		50	102		152	208
3250	3-way switch/15 amp recpt., #14/3, type NM cable		10	.800		31.50	42		73.50	97
3270	Type MC cable		8.89	.900		50	47		97	126
3280	EMT & wire		4.10	1.951		52	102		154	210
3300	2-3 way switches, 2-#14/3, type NM cables		8.89	.900		41	47		88	116
3320	Type MC cable		8	1		71.50	52.50		124	157
3330	EMT & wire		4	2		59.50	105		164.50	223
3350	S.P. switch/20 amp recpt., #12/2, type NM cable		10	.800		34	42		76	100
3370	Type MC cable		8.89	.900		40	47		87	115
3380	EMT & wire		4.10	1.951		56.50	102		158.50	216
3400	Decorator style									
3410	S.P. switch/15 amp recpt., type NM cable	1 Elec	11.43	.700	Ea.	28	36.50		64.50	86
3420	Type MC cable		10	.800		42	42		84	109
3430	EMT & wire		4.40	1.818		51	95.50		146.50	199
3450	S.P. switch/pilot light, type NM cable		11.43	.700		28	36.50		64.50	86
3470	Type MC cable		10	.800		42	42		84	109
3480	EMT & wire		4.40	1.818		51	95.50		146.50	199
3500	2-S.P. switches, 2-#14/2, type NM cables		10	.800		31.50	42		73.50	97
3520	Type MC cable		8.89	.900		52.50	47		99.50	129
3530	EMT & wire		4.10	1.951		54	102		156	213
3550	3-way/15 amp recpt., #14/3, type NM cable		10	.800		35.50	42		77.50	102
3570	Type MC cable		8.89	.900		54	47		101	130
3580	EMT & wire		4.10	1.951		56	102		158	215
3650	2-3 way switches, 2-#14/3, type NM cables		8.89	.900		45	47		92	120
3670	Type MC cable		8	1		75.50	52.50		128	162
3680	EMT & wire		4	2		63.50	105		168.50	227
3700	S.P. switch/20 amp recpt., #12/2, type NM cable		10	.800		38	42		80	105
3720	Type MC cable		8.89	.900		44.50	47		91.50	119
3730	EMT & wire		4.10	1.951		61	102		163	220
4000	Receptacle devices									
4010	Duplex outlet, 15 amp recpt., Ivory, 1-gang box, plate									
4015	Type NM cable	1 Elec	14.55	.550	Ea.	10.95	29		39.95	55
4020	Type MC cable		12.31	.650		25	34		59	78.50
4030	EMT & wire		5.33	1.501		32	78.50		110.50	153
4050	With #12/2, type NM cable		12.31	.650		13.55	34		47.55	66
4070	Type MC cable		10.67	.750		25	39.50		64.50	86.50
4080	EMT & wire		4.71	1.699		36	89		125	173
4100	20 amp recpt., #12/2, type NM cable		12.31	.650		22	34		56	75
4120	Type MC cable		10.67	.750		33.50	39.50		73	95.50
4130	EMT & wire		4.71	1.699		44.50	89		133.50	182
4140	For GFI see Section 26 05 90.10 line 4300 below									
4150	Decorator style, 15 amp recpt., type NM cable	1 Elec	14.55	.550	Ea.	15	29		44	59.50

26 05 90 – Residential Applications

26 05 90.10 Residential Wiring		Crew	Daily Output	Labor-Hours	Unit	Material	Labor	Equipment	Total	Total Incl O&P
						2013 Bare Costs				
4170	Type MC cable	1 Elec	12.31	.650	Ea.	29	34		63	83
4180	EMT & wire		5.33	1.501		36	78.50		114.50	158
4200	With #12/2, type NM cable		12.31	.650		17.60	34		51.60	70.50
4220	Type MC cable		10.67	.750		29	39.50		68.50	91
4230	EMT & wire		4.71	1.699		40	89		129	177
4250	20 amp recpt. #12/2, type NM cable		12.31	.650		26	34		60	79.50
4270	Type MC cable		10.67	.750		37.50	39.50		77	100
4280	EMT & wire		4.71	1.699		48.50	89		137.50	187
4300	GFI, 15 amp recpt., type NM cable		12.31	.650		47	34		81	103
4320	Type MC cable		10.67	.750		61	39.50		100.50	126
4330	EMT & wire		4.71	1.699		68	89		157	208
4350	GFI with #12/2, type NM cable		10.67	.750		49.50	39.50		89	114
4370	Type MC cable		9.20	.870		61	45.50		106.50	135
4380	EMT & wire		4.21	1.900		72	99.50		171.50	229
4400	20 amp recpt., #12/2 type NM cable		10.67	.750		51.50	39.50		91	116
4420	Type MC cable		9.20	.870		63	45.50		108.50	138
4430	EMT & wire		4.21	1.900		74.50	99.50		174	231
4500	Weather-proof cover for above receptacles, add		32	.250		4.79	13.10		17.89	25
4550	Air conditioner outlet, 20 amp-240 volt recpt.									
4560	30′ of #12/2, 2 pole circuit breaker									
4570	Type NM cable	1 Elec	10	.800	Ea.	62	42		104	131
4580	Type MC cable		9	.889		75.50	46.50		122	153
4590	EMT & wire		4	2		84.50	105		189.50	250
4600	Decorator style, type NM cable		10	.800		66.50	42		108.50	136
4620	Type MC cable		9	.889		80.50	46.50		127	158
4630	EMT & wire		4	2		89	105		194	255
4650	Dryer outlet, 30 amp-240 volt recpt., 20′ of #10/3									
4660	2 pole circuit breaker									
4670	Type NM cable	1 Elec	6.41	1.248	Ea.	65	65.50		130.50	170
4680	Type MC cable		5.71	1.401		73	73.50		146.50	190
4690	EMT & wire		3.48	2.299		79.50	120		199.50	267
4700	Range outlet, 50 amp-240 volt recpt., 30′ of #8/3									
4710	Type NM cable	1 Elec	4.21	1.900	Ea.	93	99.50		192.50	251
4720	Type MC cable		4	2		147	105		252	320
4730	EMT & wire		2.96	2.703		110	142		252	335
4750	Central vacuum outlet, Type NM cable		6.40	1.250		61.50	65.50		127	166
4770	Type MC cable		5.71	1.401		82	73.50		155.50	200
4780	EMT & wire		3.48	2.299		87.50	120		207.50	277
4800	30 amp-110 volt locking recpt., #10/2 circ. bkr.									
4810	Type NM cable	1 Elec	6.20	1.290	Ea.	71.50	67.50		139	180
4830	EMT & wire	"	3.20	2.500	"	98	131		229	305
4900	Low voltage outlets									
4910	Telephone recpt., 20′ of 4/C phone wire	1 Elec	26	.308	Ea.	11.15	16.10		27.25	36.50
4920	TV recpt., 20′ of RG59U coax wire, F type connector	"	16	.500	"	19.75	26		45.75	60.50
4950	Door bell chime, transformer, 2 buttons, 60′ of bellwire									
4970	Economy model	1 Elec	11.50	.696	Ea.	64.50	36.50		101	126
4980	Custom model		11.50	.696		118	36.50		154.50	185
4990	Luxury model, 3 buttons		9.50	.842		310	44		354	405
6000	Lighting outlets									
6050	Wire only (for fixture), type NM cable	1 Elec	32	.250	Ea.	7.10	13.10		20.20	27.50
6070	Type MC cable		24	.333		15.10	17.45		32.55	42.50
6080	EMT & wire		10	.800		21	42		63	85.50
6100	Box (4″), and wire (for fixture), type NM cable		25	.320		17.60	16.75		34.35	44.50

26 05 90.10 Residential Wiring		Crew	Daily Output	Labor-Hours	Unit	Material	2013 Bare Costs Labor	Equipment	Total	Total Incl O&P
6120	Type MC cable	1 Elec	20	.400	Ea.	25.50	21		46.50	59.50
6130	EMT & wire		11	.727		31.50	38		69.50	91.50
6200	Fixtures (use with lines 6050 or 6100 above)									
6210	Canopy style, economy grade	1 Elec	40	.200	Ea.	37	10.50		47.50	56.50
6220	Custom grade		40	.200		57	10.50		67.50	78
6250	Dining room chandelier, economy grade		19	.421		92.50	22		114.50	135
6260	Custom grade		19	.421		272	22		294	330
6270	Luxury grade		15	.533		610	28		638	715
6310	Kitchen fixture (fluorescent), economy grade		30	.267		68.50	13.95		82.45	96.50
6320	Custom grade		25	.320		188	16.75		204.75	232
6350	Outdoor, wall mounted, economy grade		30	.267		35	13.95		48.95	59.50
6360	Custom grade		30	.267		120	13.95		133.95	153
6370	Luxury grade		25	.320		272	16.75		288.75	325
6410	Outdoor PAR floodlights, 1 lamp, 150 watt		20	.400		37.50	21		58.50	73
6420	2 lamp, 150 watt each		20	.400		62.50	21		83.50	100
6430	For infrared security sensor, add		32	.250		126	13.10		139.10	158
6450	Outdoor, quartz-halogen, 300 watt flood		20	.400		45.50	21		66.50	81.50
6600	Recessed downlight, round, pre-wired, 50 or 75 watt trim		30	.267		46.50	13.95		60.45	72
6610	With shower light trim		30	.267		57.50	13.95		71.45	84
6620	With wall washer trim		28	.286		69.50	14.95		84.45	98.50
6630	With eye-ball trim		28	.286		69.50	14.95		84.45	98.50
6640	For direct contact with insulation, add					2.47			2.47	2.72
6700	Porcelain lamp holder	1 Elec	40	.200		4.12	10.50		14.62	20
6710	With pull switch		40	.200		5.65	10.50		16.15	22
6750	Fluorescent strip, 1-20 watt tube, wrap around diffuser, 24"		24	.333		58	17.45		75.45	90
6770	2-34 watt tubes, 48"		20	.400		88.50	21		109.50	129
6780	With residential ballast		20	.400		99.50	21		120.50	141
6800	Bathroom heat lamp, 1-250 watt		28	.286		51	14.95		65.95	78.50
6810	2-250 watt lamps		28	.286		81.50	14.95		96.45	112
6820	For timer switch, see Section 26 05 90.10 line 2400									
6900	Outdoor post lamp, incl. post, fixture, 35' of #14/2									
6910	Type NMC cable	1 Elec	3.50	2.286	Ea.	203	120		323	400
6920	Photo-eye, add		27	.296		37.50	15.55		53.05	64.50
6950	Clock dial time switch, 24 hr., w/enclosure, type NM cable		11.43	.700		70.50	36.50		107	133
6970	Type MC cable		11	.727		84.50	38		122.50	150
6980	EMT & wire		4.85	1.649		91.50	86.50		178	230
7000	Alarm systems									
7050	Smoke detectors, box, #14/3, type NM cable	1 Elec	14.55	.550	Ea.	35.50	29		64.50	82
7070	Type MC cable		12.31	.650		49	34		83	105
7080	EMT & wire		5	1.600		51	84		135	181
7090	For relay output to security system, add					14			14	15.40
8000	Residential equipment									
8050	Disposal hook-up, incl. switch, outlet box, 3' of flex									
8060	20 amp-1 pole circ. bkr., and 25' of #12/2									
8070	Type NM cable	1 Elec	10	.800	Ea.	32.50	42		74.50	98.50
8080	Type MC cable		8	1		45	52.50		97.50	128
8090	EMT & wire		5	1.600		57.50	84		141.50	189
8100	Trash compactor or dishwasher hook-up, incl. outlet box,									
8110	3' of flex, 15 amp-1 pole circ. bkr., and 25' of #14/2									
8120	Type NM cable	1 Elec	10	.800	Ea.	24	42		66	89
8130	Type MC cable		8	1		40	52.50		92.50	123
8140	EMT & wire		5	1.600		49.50	84		133.50	179
8150	Hot water sink dispensor hook-up, use line 8100									

26 05 90 – Residential Applications

26 05 90.10 Residential Wiring		Crew	Daily Output	Labor-Hours	Unit	Material	2013 Bare Costs Labor	Equipment	Total	Total Incl O&P
8200	Vent/exhaust fan hook-up, type NM cable	1 Elec	32	.250	Ea.	7.10	13.10		20.20	27.50
8220	Type MC cable		24	.333		15.10	17.45		32.55	42.50
8230	EMT & wire		10	.800		21	42		63	85.50
8250	Bathroom vent fan, 50 CFM (use with above hook-up)									
8260	Economy model	1 Elec	15	.533	Ea.	27	28		55	71.50
8270	Low noise model		15	.533		36.50	28		64.50	82
8280	Custom model		12	.667		126	35		161	191
8300	Bathroom or kitchen vent fan, 110 CFM									
8310	Economy model	1 Elec	15	.533	Ea.	74.50	28		102.50	124
8320	Low noise model	"	15	.533	"	92.50	28		120.50	144
8350	Paddle fan, variable speed (w/o lights)									
8360	Economy model (AC motor)	1 Elec	10	.800	Ea.	124	42		166	199
8362	With light kit		10	.800		164	42		206	244
8370	Custom model (AC motor)		10	.800		195	42		237	278
8372	With light kit		10	.800		236	42		278	325
8380	Luxury model (DC motor)		8	1		385	52.50		437.50	505
8382	With light kit		8	1		425	52.50		477.50	550
8390	Remote speed switch for above, add		12	.667		32	35		67	88
8500	Whole house exhaust fan, ceiling mount, 36", variable speed									
8510	Remote switch, incl. shutters, 20 amp-1 pole circ. bkr.									
8520	30' of #12/2, type NM cable	1 Elec	4	2	Ea.	1,225	105		1,330	1,500
8530	Type MC cable		3.50	2.286		1,225	120		1,345	1,525
8540	EMT & wire		3	2.667		1,250	140		1,390	1,575
8600	Whirlpool tub hook-up, incl. timer switch, outlet box									
8610	3' of flex, 20 amp-1 pole GFI circ. bkr.									
8620	30' of #12/2, type NM cable	1 Elec	5	1.600	Ea.	124	84		208	262
8630	Type MC cable		4.20	1.905		132	100		232	294
8640	EMT & wire		3.40	2.353		143	123		266	340
8650	Hot water heater hook-up, incl. 1-2 pole circ. bkr., box;									
8660	3' of flex, 20' of #10/2, type NM cable	1 Elec	5	1.600	Ea.	34	84		118	163
8670	Type MC cable		4.20	1.905		51	100		151	205
8680	EMT & wire		3.40	2.353		52	123		175	241
9000	Heating/air conditioning									
9050	Furnace/boiler hook-up, incl. firestat, local on-off switch									
9060	Emergency switch, and 40' of type NM cable	1 Elec	4	2	Ea.	56	105		161	219
9070	Type MC cable		3.50	2.286		77	120		197	264
9080	EMT & wire		1.50	5.333		90	279		369	520
9100	Air conditioner hook-up, incl. local 60 amp disc. switch									
9110	3' sealtite, 40 amp, 2 pole circuit breaker									
9130	40' of #8/2, type NM cable	1 Elec	3.50	2.286	Ea.	202	120		322	400
9140	Type MC cable		3	2.667		283	140		423	520
9150	EMT & wire		1.30	6.154		242	320		562	745
9200	Heat pump hook-up, 1-40 & 1-100 amp 2 pole circ. bkr.									
9210	Local disconnect switch, 3' sealtite									
9220	40' of #8/2 & 30' of #3/2									
9230	Type NM cable	1 Elec	1.30	6.154	Ea.	540	320		860	1,075
9240	Type MC cable		1.08	7.407		660	390		1,050	1,300
9250	EMT & wire		.94	8.511		620	445		1,065	1,350
9500	Thermostat hook-up, using low voltage wire									
9520	Heating only, 25' of #18-3	1 Elec	24	.333	Ea.	10.20	17.45		27.65	37.50
9530	Heating/cooling, 25' of #18-4	"	20	.400	"	12.35	21		33.35	45

26 24 19.40 Motor Starters and Controls		Crew	Daily Output	Labor-Hours	Unit	Material	Labor	Equipment	Total	Total Incl O&P
						2013 Bare Costs				
0010	**MOTOR STARTERS AND CONTROLS**									
0050	Magnetic, FVNR, with enclosure and heaters, 480 volt									
0080	2 HP, size 00	1 Elec	3.50	2.286	Ea.	193	120		313	390
0100	5 HP, size 0		2.30	3.478		243	182		425	540
0200	10 HP, size 1		1.60	5		273	262		535	690
0300	25 HP, size 2	2 Elec	2.20	7.273		515	380		895	1,125
0400	50 HP, size 3		1.80	8.889		835	465		1,300	1,625
0500	100 HP, size 4		1.20	13.333		1,850	700		2,550	3,100
0600	200 HP, size 5		.90	17.778		4,350	930		5,280	6,175
0610	400 HP, size 6		.80	20		19,500	1,050		20,550	23,000
0620	NEMA 7, 5 HP, size 0	1 Elec	1.60	5		1,450	262		1,712	2,000
0630	10 HP, size 1	"	1.10	7.273		1,500	380		1,880	2,225
0640	25 HP, size 2	2 Elec	1.80	8.889		2,450	465		2,915	3,375
0650	50 HP, size 3		1.20	13.333		3,675	700		4,375	5,075
0660	100 HP, size 4		.90	17.778		5,925	930		6,855	7,900
0670	200 HP, size 5		.50	32		14,100	1,675		15,775	18,000
0700	Combination, with motor circuit protectors, 5 HP, size 0	1 Elec	1.80	4.444		755	233		988	1,175
0800	10 HP, size 1	"	1.30	6.154		785	320		1,105	1,350
0900	25 HP, size 2	2 Elec	2	8		1,100	420		1,520	1,850
1000	50 HP, size 3		1.32	12.121		1,600	635		2,235	2,700
1200	100 HP, size 4		.80	20		3,475	1,050		4,525	5,375
1220	NEMA 7, 5 HP, size 0	1 Elec	1.30	6.154		2,625	320		2,945	3,350
1230	10 HP, size 1	"	1	8		2,675	420		3,095	3,575
1240	25 HP, size 2	2 Elec	1.32	12.121		3,575	635		4,210	4,875
1250	50 HP, size 3		.80	20		5,900	1,050		6,950	8,075
1260	100 HP, size 4		.60	26.667		9,200	1,400		10,600	12,200
1270	200 HP, size 5		.40	40		20,000	2,100		22,100	25,100
1400	Combination, with fused switch, 5 HP, size 0	1 Elec	1.80	4.444		580	233		813	985
1600	10 HP, size 1	"	1.30	6.154		620	320		940	1,150
1800	25 HP, size 2	2 Elec	2	8		1,000	420		1,420	1,725
2000	50 HP, size 3		1.32	12.121		1,700	635		2,335	2,825
2200	100 HP, size 4		.80	20		2,975	1,050		4,025	4,850
3500	Magnetic FVNR with NEMA 12, enclosure & heaters, 480 volt									
3600	5 HP, size 0	1 Elec	2.20	3.636	Ea.	220	191		411	525
3700	10 HP, size 1	"	1.50	5.333		330	279		609	785
3800	25 HP, size 2	2 Elec	2	8		620	420		1,040	1,300
3900	50 HP, size 3		1.60	10		955	525		1,480	1,825
4000	100 HP, size 4		1	16		2,275	840		3,115	3,750
4100	200 HP, size 5		.80	20		5,475	1,050		6,525	7,600
4200	Combination, with motor circuit protectors, 5 HP, size 0	1 Elec	1.70	4.706		730	247		977	1,175
4300	10 HP, size 1	"	1.20	6.667		755	350		1,105	1,350
4400	25 HP, size 2	2 Elec	1.80	8.889		1,125	465		1,590	1,950
4500	50 HP, size 3		1.20	13.333		1,850	700		2,550	3,075
4600	100 HP, size 4		.74	21.622		4,175	1,125		5,300	6,275
4700	Combination, with fused switch, 5 HP, size 0	1 Elec	1.70	4.706		705	247		952	1,150
4800	10 HP, size 1	"	1.20	6.667		735	350		1,085	1,325
4900	25 HP, size 2	2 Elec	1.80	8.889		1,125	465		1,590	1,925
5000	50 HP, size 3		1.20	13.333		1,800	700		2,500	3,025
5100	100 HP, size 4		.74	21.622		3,625	1,125		4,750	5,700
5200	Factory installed controls, adders to size 0 thru 5									
5300	Start-stop push button	1 Elec	32	.250	Ea.	46	13.10		59.10	70
5400	Hand-off-auto-selector switch		32	.250		46	13.10		59.10	70
5500	Pilot light		32	.250		86.50	13.10		99.60	115

26 24 Switchboards and Panelboards

26 24 19 – Motor-Control Centers

26 24 19.40 Motor Starters and Controls		Crew	Daily Output	Labor-Hours	Unit	Material	2013 Bare Costs Labor	Equipment	Total	Total Incl O&P
5600	Start-stop-pilot	1 Elec	32	.250	Ea.	132	13.10		145.10	166
5700	Auxiliary contact, NO or NC		32	.250		63.50	13.10		76.60	89
5800	NO-NC	↓	32	.250	↓	127	13.10		140.10	159

26 29 Low-Voltage Controllers

26 29 13 – Enclosed Controllers

26 29 13.20 Control Stations

		Crew	Daily Output	Labor-Hours	Unit	Material	Labor	Equipment	Total	Total Incl O&P
0010	**CONTROL STATIONS**									
0050	NEMA 1, heavy duty, stop/start	1 Elec	8	1	Ea.	132	52.50		184.50	225
0100	Stop/start, pilot light		6.20	1.290		180	67.50		247.50	300
0200	Hand/off/automatic		6.20	1.290		98	67.50		165.50	209
0400	Stop/start/reverse		5.30	1.509		179	79		258	315
0500	NEMA 7, heavy duty, stop/start		6	1.333		390	70		460	535
0600	Stop/start, pilot light	↓	4	2	↓	480	105		585	680

26 42 Cathodic Protection

26 42 16 – Passive Cathodic Protection for Underground Storage Tank

26 42 16.50 Cathodic Protection Wiring Methods

		Crew	Daily Output	Labor-Hours	Unit	Material	Labor	Equipment	Total	Total Incl O&P
0010	**CATHODIC PROTECTION WIRING METHODS**									
1000	Anodes, magnesium type, 9 #	R-15	18.50	2.595	Ea.	38.50	133	17.95	189.45	262
1010	17 #		13	3.692		70	190	25.50	285.50	390
1020	32 #		10	4.800		122	247	33	402	540
1030	48 #	↓	7.20	6.667		160	345	46	551	740
1100	Graphite type w/epoxy cap, 3" x 60" (32 #)	R-22	8.40	4.438		125	197		322	435
1110	4" x 80" (68 #)		6	6.213		237	276		513	675
1120	6" x 72" (80 #)		5.20	7.169		1,475	320		1,795	2,100
1130	6" x 36" (45 #)	↓	9.60	3.883		745	173		918	1,075
2000	Rectifiers, silicon type, air cooled, 28 V/10 A	R-19	3.50	5.714		2,275	300		2,575	2,950
2010	20 V/20 A		3.50	5.714		2,325	300		2,625	3,000
2100	Oil immersed, 28 V/10 A		3	6.667		3,100	350		3,450	3,925
2110	20 V/20 A	↓	3	6.667	↓	3,125	350		3,475	3,950
3000	Anode backfill, coke breeze	R-22	3850	.010	Lb.	.22	.43		.65	.89
4000	Cable, HMWPE, No. 8		2.40	15.533	M.L.F.	490	690		1,180	1,600
4010	No. 6		2.40	15.533		730	690		1,420	1,850
4020	No. 4		2.40	15.533		1,100	690		1,790	2,275
4030	No. 2		2.40	15.533		1,725	690		2,415	2,950
4040	No. 1		2.20	16.945		2,350	755		3,105	3,750
4050	No. 1/0		2.20	16.945		2,925	755		3,680	4,350
4060	No. 2/0		2.20	16.945		4,525	755		5,280	6,150
4070	No. 4/0	↓	2	18.640	↓	5,875	830		6,705	7,725
5000	Test station, 7 terminal box, flush curb type w/lockable cover	R-19	12	1.667	Ea.	70.50	87.50		158	209
5010	Reference cell, 2" dia PVC conduit, cplg., plug, set flush	"	4.80	4.167	"	148	219		367	490

Division Notes

		CREW	DAILY OUTPUT	LABOR-HOURS	UNIT	BARE COSTS MAT.	LABOR	EQUIP.	TOTAL	TOTAL INCL O&P

Division 28 - Electronic Safety and Security

Estimating Tips

- When estimating material costs for electronic safety and security systems, it is always prudent to obtain manufacturers' quotations for equipment prices and special installation requirements that affect the total cost.
- Fire alarm systems consist of control panels, annunciator panels, battery with rack, charger, and fire alarm actuating and indicating devices. Some fire alarm systems include speakers, telephone lines, door closer controls, and other components. Be careful not to overlook the costs related to installation for these items. Also be aware of costs for integrated automation instrumentation and terminal devices, control equipment, control wiring, and programming.
- Security equipment includes items such as CCTV, access control, and other detection and identification systems to perform alert and alarm functions. Be sure to consider the costs related to installation for this security equipment, such as for integrated automation instrumentation and terminal devices, control equipment, control wiring, and programming.

Reference Numbers

Reference numbers are shown in shaded boxes at the beginning of some major classifications. These numbers refer to related items in the Reference Section. The reference information may be an estimating procedure, an alternate pricing method, or technical information.

Note: Not all subdivisions listed here necessarily appear in this publication.

28 16 Intrusion Detection

28 16 16 – Intrusion Detection Systems Infrastructure

28 16 16.50 Intrusion Detection		Crew	Daily Output	Labor-Hours	Unit	Material	2013 Bare Costs Labor	Equipment	Total	Total Incl O&P
0010	**INTRUSION DETECTION**, not including wires & conduits									
0100	Burglar alarm, battery operated, mechanical trigger	1 Elec	4	2	Ea.	272	105		377	455
0200	Electrical trigger		4	2		325	105		430	510
0400	For outside key control, add		8	1		82	52.50		134.50	169
0600	For remote signaling circuitry, add		8	1		122	52.50		174.50	214
0800	Card reader, flush type, standard		2.70	2.963		910	155		1,065	1,225
1000	Multi-code	↓	2.70	2.963	↓	1,175	155		1,330	1,525

28 31 Fire Detection and Alarm

28 31 23 – Fire Detection and Alarm Annunciation Panels and Fire Stations

28 31 23.50 Alarm Panels and Devices

		Crew	Daily Output	Labor-Hours	Unit	Material	Labor	Equipment	Total	Total Incl O&P
0010	**ALARM PANELS AND DEVICES**, not including wires & conduits									
3594	Fire, alarm control panel									
3600	4 zone	2 Elec	2	8	Ea.	390	420		810	1,050
3800	8 zone		1	16		765	840		1,605	2,100
4000	12 zone	↓	.67	23.988		2,350	1,250		3,600	4,450
4020	Alarm device	1 Elec	8	1		233	52.50		285.50	335
4050	Actuating device		8	1		330	52.50		382.50	440
4200	Battery and rack		4	2		405	105		510	600
4400	Automatic charger		8	1		570	52.50		622.50	710
4600	Signal bell		8	1		77	52.50		129.50	164
4800	Trouble buzzer or manual station		8	1		81.50	52.50		134	169
5600	Strobe and horn		5.30	1.509		149	79		228	282
5800	Fire alarm horn		6.70	1.194		59.50	62.50		122	159
6000	Door holder, electro-magnetic		4	2		101	105		206	268
6200	Combination holder and closer		3.20	2.500		121	131		252	330
6600	Drill switch		8	1		365	52.50		417.50	480
6800	Master box		2.70	2.963		6,275	155		6,430	7,125
7000	Break glass station		8	1		54.50	52.50		107	139
7800	Remote annunciator, 8 zone lamp	↓	1.80	4.444		205	233		438	575
8000	12 zone lamp	2 Elec	2.60	6.154		330	320		650	840
8200	16 zone lamp	"	2.20	7.273	↓	410	380		790	1,025

28 31 43 – Fire Detection Sensors

28 31 43.50 Fire and Heat Detectors

		Crew	Daily Output	Labor-Hours	Unit	Material	Labor	Equipment	Total	Total Incl O&P
0010	**FIRE & HEAT DETECTORS**									
5000	Detector, rate of rise	1 Elec	8	1	Ea.	50	52.50		102.50	134
5100	Fixed temperature	"	8	1	"	50	52.50		102.50	134

28 31 46 – Smoke Detection Sensors

28 31 46.50 Smoke Detectors

		Crew	Daily Output	Labor-Hours	Unit	Material	Labor	Equipment	Total	Total Incl O&P
0010	**SMOKE DETECTORS**									
5200	Smoke detector, ceiling type	1 Elec	6.20	1.290	Ea.	108	67.50		175.50	220
5400	Duct type	"	3.20	2.500	"	320	131		451	550

28 33 Gas Detection and Alarm

28 33 33 – Gas Detection Sensors

28 33 33.50 Tank Leak Detection Systems		Crew	Daily Output	Labor-Hours	Unit	2013 Bare Costs Material	Labor	Equipment	Total	Total Incl O&P
0010	**TANK LEAK DETECTION SYSTEMS** Liquid and vapor									
0100	For hydrocarbons and hazardous liquids/vapors									
0120	Controller, data acquisition, incl. printer, modem, RS232 port									
0140	24 channel, for use with all probes				Ea.	4,075			4,075	4,500
0160	9 channel, for external monitoring				"	900			900	990
0200	Probes									
0210	Well monitoring									
0220	Liquid phase detection				Ea.	465			465	515
0230	Hydrocarbon vapor, fixed position					465			465	515
0240	Hydrocarbon vapor, float mounted					465			465	515
0250	Both liquid and vapor hydrocarbon				↓	465			465	515
0300	Secondary containment, liquid phase									
0310	Pipe trench/manway sump				Ea.	805			805	885
0320	Double wall pipe and manual sump					805			805	890
0330	Double wall fiberglass annular space					315			315	350
0340	Double wall steel tank annular space				↓	330			330	360
0500	Accessories									
0510	Modem, non-dedicated phone line				Ea.	297			297	325
0600	Monitoring, internal									
0610	Automatic tank gauge, incl. overfill				Ea.	1,125			1,125	1,250
0620	Product line				"	1,125			1,125	1,250
0700	Monitoring, special									
0710	Cathodic protection				Ea.	710			710	780
0720	Annular space chemical monitor				"	965			965	1,050

Division Notes

	CREW	DAILY OUTPUT	LABOR-HOURS	UNIT	BARE COSTS MAT.	BARE COSTS LABOR	BARE COSTS EQUIP.	BARE COSTS TOTAL	TOTAL INCL O&P

Estimating Tips

31 05 00 Common Work Results for Earthwork

- Estimating the actual cost of performing earthwork requires careful consideration of the variables involved. This includes items such as type of soil, whether water will be encountered, dewatering, whether banks need bracing, disposal of excavated earth, and length of haul to fill or spoil sites, etc. If the project has large quantities of cut or fill, consider raising or lowering the site to reduce costs, while paying close attention to the effect on site drainage and utilities.
- If the project has large quantities of fill, creating a borrow pit on the site can significantly lower the costs.
- It is very important to consider what time of year the project is scheduled for completion. Bad weather can create large cost overruns from dewatering, site repair, and lost productivity from cold weather.
- New lines have been added for marine piling and rough grading.

Reference Numbers

Reference numbers are shown in shaded boxes at the beginning of some major classifications. These numbers refer to related items in the Reference Section. The reference information may be an estimating procedure, an alternate pricing method, or technical information.

Note: Not all subdivisions listed here necessarily appear in this publication.

31 23 16.13 Excavating, Trench		Crew	Daily Output	Labor-Hours	Unit	Material	2013 Bare Costs Labor	Equipment	Total	Total Incl O&P
0010	**EXCAVATING, TRENCH** G1030–805									
0011	Or continuous footing									
0020	Common earth with no sheeting or dewatering included									
0050	1′ to 4′ deep, 3/8 C.Y. excavator	B-11C	150	.107	B.C.Y.		4.42	2.45	6.87	9.45
0060	1/2 C.Y. excavator	B-11M	200	.080			3.32	2	5.32	7.25
0090	4′ to 6′ deep, 1/2 C.Y. excavator	"	200	.080			3.32	2	5.32	7.25
0100	5/8 C.Y. excavator	B-12Q	250	.064			2.70	2.38	5.08	6.75
0300	1/2 C.Y. excavator, truck mounted	B-12J	200	.080			3.37	4.42	7.79	10
0500	6′ to 10′ deep, 3/4 C.Y. excavator	B-12F	225	.071			3	3.09	6.09	7.95
0600	1 C.Y. excavator, truck mounted	B-12K	400	.040			1.69	2.53	4.22	5.35
0900	10′ to 14′ deep, 3/4 C.Y. excavator	B-12F	200	.080			3.37	3.47	6.84	8.95
1000	1-1/2 C.Y. excavator	B-12B	540	.030			1.25	1.92	3.17	4.01
1300	14′ to 20′ deep, 1 C.Y. excavator	B-12A	320	.050			2.11	2.55	4.66	6
1340	20′ to 24′ deep, 1 C.Y. excavator	"	288	.056			2.34	2.83	5.17	6.70
1352	4′ to 6′ deep, 1/2 C.Y. excavator w/trench box	B-13H	188	.085			3.59	5.30	8.89	11.30
1354	5/8 C.Y. excavator	"	235	.068			2.87	4.24	7.11	9.05
1362	6′ to 10′ deep, 3/4 C.Y. excavator w/trench box	B-13G	212	.075			3.18	3.81	6.99	9.05
1374	10′ to 14′ deep, 3/4 C.Y. excavator w/trench box	"	188	.085			3.59	4.30	7.89	10.20
1376	1-1/2 C.Y. excavator	B-13E	508	.032			1.33	2.26	3.59	4.51
1381	14′ to 20′ deep, 1 C.Y. excavator w/trench box	B-13D	301	.053			2.24	3.09	5.33	6.80
1386	20′ to 24′ deep, 1 C.Y. excavator w/trench box	"	271	.059			2.49	3.43	5.92	7.55
1400	By hand with pick and shovel 2′ to 6′ deep, light soil	1 Clab	8	1			35.50		35.50	54.50
1500	Heavy soil	"	4	2			71		71	109
1700	For tamping backfilled trenches, air tamp, add	A-1G	100	.080	E.C.Y.		2.84	.54	3.38	4.96
1900	Vibrating plate, add	B-18	180	.133	"		4.82	.25	5.07	7.65
2100	Trim sides and bottom for concrete pours, common earth		1500	.016	S.F.		.58	.03	.61	.92
2300	Hardpan		600	.040	"		1.44	.07	1.51	2.30
5020	Loam & Sandy clay with no sheeting or dewatering included									
5050	1′ to 4′ deep, 3/8 C.Y. tractor loader/backhoe	B-11C	162	.099	B.C.Y.		4.10	2.27	6.37	8.75
5060	1/2 C.Y. excavator	B-11M	216	.074			3.07	1.85	4.92	6.70
5080	4′ to 6′ deep, 1/2 C.Y. excavator	"	216	.074			3.07	1.85	4.92	6.70
5090	5/8 C.Y. excavator	B-12Q	276	.058			2.44	2.16	4.60	6.10
5130	1/2 C.Y. excavator, truck mounted	B-12J	216	.074			3.12	4.09	7.21	9.25
5140	6′ to 10′ deep, 3/4 C.Y. excavator	B-12F	243	.066			2.77	2.86	5.63	7.40
5160	1 C.Y. excavator, truck mounted	B-12K	432	.037			1.56	2.34	3.90	4.95
5190	10′ to 14′ deep, 3/4 C.Y. excavator	B-12F	216	.074			3.12	3.22	6.34	8.30
5210	1-1/2 C.Y. excavator	B-12B	583	.027			1.16	1.78	2.94	3.71
5250	14′ to 20′ deep, 1 C.Y. excavator	B-12A	346	.046			1.95	2.36	4.31	5.55
5300	20′ to 24′ deep, 1 C.Y. excavator	"	311	.051			2.17	2.62	4.79	6.20
5352	4′ to 6′ deep, 1/2 C.Y. excavator w/trench box	B-13H	205	.078			3.29	4.86	8.15	10.35
5354	5/8 C.Y. excavator	"	257	.062			2.62	3.88	6.50	8.25
5362	6′ to 10′ deep, 3/4 C.Y. excavator w/trench box	B-13G	231	.069			2.92	3.50	6.42	8.30
5370	10′ to 14′ deep, 3/4 C.Y. excavator w/trench box	"	205	.078			3.29	3.94	7.23	9.35
5374	1-1/2 C.Y. excavator	B-13E	554	.029			1.22	2.07	3.29	4.13
5382	14′ to 20′ deep, 1 C.Y. excavator w/trench box	B-13D	329	.049			2.05	2.82	4.87	6.25
5392	20′ to 24′ deep, 1 C.Y. excavator w/trench box	"	295	.054			2.29	3.15	5.44	6.95
6020	Sand & gravel with no sheeting or dewatering included									
6050	1′ to 4′ deep, 3/8 C.Y. excavator	B-11C	165	.097	B.C.Y.		4.02	2.23	6.25	8.60
6060	1/2 C.Y. excavator	B-11M	220	.073			3.02	1.82	4.84	6.60
6080	4′ to 6′ deep, 1/2 C.Y. excavator	"	220	.073			3.02	1.82	4.84	6.60
6090	5/8 C.Y. excavator	B-12Q	275	.058			2.45	2.17	4.62	6.10
6130	1/2 C.Y. excavator, truck mounted	B-12J	220	.073			3.06	4.02	7.08	9.10
6140	6′ to 10′ deep, 3/4 C.Y. excavator	B-12F	248	.065			2.72	2.80	5.52	7.20

31 23 16.13 Excavating, Trench		Crew	Daily Output	Labor-Hours	Unit	Material	2013 Bare Costs Labor	Equipment	Total	Total Incl O&P
6160	1 C.Y. excavator, truck mounted	B-12K	440	.036	B.C.Y.		1.53	2.30	3.83	4.85
6190	10' to 14' deep, 3/4 C.Y. excavator	B-12F	220	.073			3.06	3.16	6.22	8.15
6210	1-1/2 C.Y. excavator	B-12B	594	.027			1.14	1.74	2.88	3.65
6250	14' to 20' deep, 1 C.Y. excavator	B-12A	352	.045			1.91	2.32	4.23	5.45
6300	20' to 24' deep, 1 C.Y. excavator	"	317	.050			2.13	2.57	4.70	6.05
6352	4' to 6' deep, 1/2 C.Y. excavator w/trench box	B-13H	209	.077			3.23	4.77	8	10.15
6354	5/8 C.Y. excavator	"	261	.061			2.58	3.82	6.40	8.15
6362	6' to 10' deep, 3/4 C.Y. excavator w/trench box	B-13G	236	.068			2.86	3.42	6.28	8.10
6370	10' to 14' deep, 3/4 C.Y. excavator w/trench box	"	209	.077			3.23	3.87	7.10	9.15
6374	1-1/2 C.Y. excavator	B-13E	564	.028			1.20	2.04	3.24	4.06
6382	14' to 20' deep, 1 C.Y. excavator w/trench box	B-13D	334	.048			2.02	2.78	4.80	6.15
6392	20' to 24' deep, 1 C.Y. excavator w/trench box	"	301	.053	↓		2.24	3.09	5.33	6.80
7020	Dense hard clay with no sheeting or dewatering included									
7050	1' to 4' deep, 3/8 C.Y. excavator	B-11C	132	.121	B.C.Y.		5.05	2.78	7.83	10.70
7060	1/2 C.Y. excavator	B-11M	176	.091			3.77	2.28	6.05	8.25
7080	4' to 6' deep, 1/2 C.Y. excavator	"	176	.091			3.77	2.28	6.05	8.25
7090	5/8 C.Y. excavator	B-12Q	220	.073			3.06	2.71	5.77	7.65
7130	1/2 C.Y. excavator, truck mounted	B-12J	176	.091			3.83	5	8.83	11.35
7140	6' to 10' deep, 3/4 C.Y. excavator	B-12F	198	.081			3.40	3.51	6.91	9.05
7160	1 C.Y. excavator, truck mounted	B-12K	352	.045			1.91	2.87	4.78	6.10
7190	10' to 14' deep, 3/4 C.Y. excavator	B-12F	176	.091			3.83	3.95	7.78	10.20
7210	1-1/2 C.Y. excavator	B-12B	475	.034			1.42	2.18	3.60	4.56
7250	14' to 20' deep, 1 C.Y. excavator	B-12A	282	.057			2.39	2.89	5.28	6.80
7300	20' to 24' deep, 1 C.Y. excavator	"	254	.063	↓		2.65	3.21	5.86	7.55

31 23 16.14 Excavating, Utility Trench		Crew	Daily Output	Labor-Hours	Unit	Material	Labor	Equipment	Total	Total Incl O&P
0010	**EXCAVATING, UTILITY TRENCH** G1030-805									
0011	Common earth									
0050	Trenching with chain trencher, 12 H.P., operator walking									
0100	4" wide trench, 12" deep	B-53	800	.010	L.F.		.46	.09	.55	.79
0150	18" deep		750	.011			.49	.09	.58	.84
0200	24" deep		700	.011			.52	.10	.62	.90
0300	6" wide trench, 12" deep		650	.012			.56	.11	.67	.97
0350	18" deep		600	.013			.61	.12	.73	1.05
0400	24" deep		550	.015			.67	.13	.80	1.15
0450	36" deep		450	.018			.81	.16	.97	1.40
0600	8" wide trench, 12" deep		475	.017			.77	.15	.92	1.33
0650	18" deep		400	.020			.92	.18	1.10	1.57
0700	24" deep		350	.023			1.05	.20	1.25	1.80
0750	36" deep	↓	300	.027	↓		1.22	.23	1.45	2.11
1000	Backfill by hand including compaction, add									
1050	4" wide trench, 12" deep	A-1G	800	.010	L.F.		.35	.07	.42	.62
1100	18" deep		530	.015			.53	.10	.63	.93
1150	24" deep		400	.020			.71	.13	.84	1.24
1300	6" wide trench, 12" deep		540	.015			.53	.10	.63	.92
1350	18" deep		405	.020			.70	.13	.83	1.23
1400	24" deep		270	.030			1.05	.20	1.25	1.84
1450	36" deep		180	.044			1.58	.30	1.88	2.76
1600	8" wide trench, 12" deep		400	.020			.71	.13	.84	1.24
1650	18" deep		265	.030			1.07	.20	1.27	1.87
1700	24" deep		200	.040			1.42	.27	1.69	2.47
1750	36" deep	↓	135	.059	↓		2.10	.40	2.50	3.68
2000	Chain trencher, 40 H.P. operator riding									

31 23 Excavation and Fill

31 23 16 – Excavation

31 23 16.14 Excavating, Utility Trench		Crew	Daily Output	Labor-Hours	Unit	Material	2013 Bare Costs Labor	Equipment	Total	Total Incl O&P
2050	6″ wide trench and backfill, 12″ deep	B-54	1200	.007	L.F.		.31	.28	.59	.77
2100	18″ deep		1000	.008			.37	.34	.71	.92
2150	24″ deep		975	.008			.38	.35	.73	.95
2200	36″ deep		900	.009			.41	.38	.79	1.03
2250	48″ deep		750	.011			.49	.45	.94	1.24
2300	60″ deep		650	.012			.56	.52	1.08	1.42
2400	8″ wide trench and backfill, 12″ deep		1000	.008			.37	.34	.71	.92
2450	18″ deep		950	.008			.39	.36	.75	.97
2500	24″ deep		900	.009			.41	.38	.79	1.03
2550	36″ deep		800	.010			.46	.42	.88	1.16
2600	48″ deep		650	.012			.56	.52	1.08	1.42
2700	12″ wide trench and backfill, 12″ deep		975	.008			.38	.35	.73	.95
2750	18″ deep		860	.009			.43	.39	.82	1.07
2800	24″ deep		800	.010			.46	.42	.88	1.16
2850	36″ deep		725	.011			.51	.47	.98	1.27
3000	16″ wide trench and backfill, 12″ deep		835	.010			.44	.41	.85	1.11
3050	18″ deep		750	.011			.49	.45	.94	1.24
3100	24″ deep	↓	700	.011	↓		.52	.48	1	1.32
3200	Compaction with vibratory plate, add								35%	35%
5100	Hand excavate and trim for pipe bells after trench excavation									
5200	8″ pipe	1 Clab	155	.052	L.F.		1.83		1.83	2.82
5300	18″ pipe	″	130	.062	″		2.18		2.18	3.36

31 23 19 – Dewatering

31 23 19.20 Dewatering Systems

Line	Description	Crew	Daily Output	Labor-Hours	Unit	Material	Labor	Equipment	Total	Total Incl O&P
0010	**DEWATERING SYSTEMS**									
0020	Excavate drainage trench, 2′ wide, 2′ deep	B-11C	90	.178	C.Y.		7.35	4.08	11.43	15.75
0100	2′ wide, 3′ deep, with backhoe loader	″	135	.119			4.92	2.72	7.64	10.50
0200	Excavate sump pits by hand, light soil	1 Clab	7.10	1.127			40		40	61.50
0300	Heavy soil	″	3.50	2.286	↓		81		81	125
0500	Pumping 8 hr., attended 2 hrs. per day, including 20 L.F.									
0550	of suction hose & 100 L.F. discharge hose									
0600	2″ diaphragm pump used for 8 hours	B-10H	4	3	Day		130	19.25	149.25	219
0650	4″ diaphragm pump used for 8 hours	B-10I	4	3			130	32	162	233
0800	8 hrs. attended, 2″ diaphragm pump	B-10H	1	12			520	77	597	875
0900	3″ centrifugal pump	B-10J	1	12			520	87.50	607.50	885
1000	4″ diaphragm pump	B-10I	1	12			520	127	647	930
1100	6″ centrifugal pump	B-10K	1	12	↓		520	375	895	1,200
1300	CMP, incl. excavation 3′ deep, 12″ diameter	B-6	115	.209	L.F.	8.25	8.10	3.20	19.55	25
1400	18″ diameter		100	.240	″	12.55	9.35	3.67	25.57	32
1600	Sump hole construction, incl. excavation and gravel, pit		1250	.019	C.F.	1.06	.75	.29	2.10	2.62
1700	With 12″ gravel collar, 12″ pipe, corrugated, 16 ga.		70	.343	L.F.	25.50	13.35	5.25	44.10	54.50
1800	15″ pipe, corrugated, 16 ga.		55	.436		32.50	16.95	6.70	56.15	69.50
1900	18″ pipe, corrugated, 16 ga.		50	.480		38.50	18.65	7.35	64.50	78.50
2000	24″ pipe, corrugated, 14 ga.		40	.600	↓	46	23.50	9.20	78.70	96
2200	Wood lining, up to 4′ x 4′, add	↓	300	.080	SFCA	17.65	3.11	1.22	21.98	25.50
9950	See Section 31 23 19.40 for wellpoints									
9960	See Section 31 23 19.30 for deep well systems									

31 23 19.30 Wells

Line	Description	Crew	Daily Output	Labor-Hours	Unit	Material	Labor	Equipment	Total	Total Incl O&P
0010	**WELLS**									
0011	For dewatering 10′ to 20′ deep, 2′ diameter									
0020	with steel casing, minimum	B-6	165	.145	V.L.F.	39	5.65	2.23	46.88	54
0050	Average		98	.245		44.50	9.55	3.75	57.80	67.50

31 23 19 – Dewatering

31 23 19.30 Wells		Crew	Daily Output	Labor-Hours	Unit	Material	Labor (2013 Bare Costs)	Equipment (2013 Bare Costs)	Total (2013 Bare Costs)	Total Incl O&P
0100	Maximum	B-6	49	.490	V.L.F.	50	19.05	7.50	76.55	92.50
0300	For dewatering pumps see 01 54 33 in Reference Section									
0500	For domestic water wells, see Section 33 21 13.10									

31 23 19.40 Wellpoints

		Crew	Daily Output	Labor-Hours	Unit	Material	Labor	Equipment	Total	Total Incl O&P
0010	**WELLPOINTS** R312319-90									
0011	For equipment rental, see 01 54 33 in Reference Section									
0100	Installation and removal of single stage system									
0110	Labor only, .75 labor-hours per L.F.	1 Clab	10.70	.748	LF Hdr		26.50		26.50	41
0200	2.0 labor-hours per L.F.	″	4	2	″		71		71	109
0400	Pump operation, 4 @ 6 hr. shifts									
0410	Per 24 hour day	4 Eqlt	1.27	25.197	Day		1,150		1,150	1,750
0500	Per 168 hour week, 160 hr. straight, 8 hr. double time		.18	177	Week		8,150		8,150	12,300
0550	Per 4.3 week month		.04	800	Month		36,600		36,600	55,500
0600	Complete installation, operation, equipment rental, fuel &									
0610	removal of system with 2″ wellpoints 5′ O.C.									
0700	100′ long header, 6″ diameter, first month	4 Eqlt	3.23	9.907	LF Hdr	164	455		619	865
0800	Thereafter, per month		4.13	7.748		131	355		486	680
1000	200′ long header, 8″ diameter, first month		6	5.333		148	244		392	535
1100	Thereafter, per month		8.39	3.814		74	175		249	345
1300	500′ long header, 8″ diameter, first month		10.63	3.010		57.50	138		195.50	271
1400	Thereafter, per month		20.91	1.530		41	70		111	151
1600	1,000′ long header, 10″ diameter, first month		11.62	2.754		49	126		175	245
1700	Thereafter, per month		41.81	.765		24.50	35		59.50	80
1900	Note: above figures include pumping 168 hrs. per week									
1910	and include the pump operator and one stand-by pump.									

31 23 23 – Fill

31 23 23.15 Borrow, Loading And/Or Spreading

		Crew	Daily Output	Labor-Hours	Unit	Material	Labor	Equipment	Total	Total Incl O&P
0010	**BORROW, LOADING AND/OR SPREADING**									
0020	Material only, bank run gravel				Ton	17.10			17.10	18.85
0500	Haul 2 mi. spread, 200 HP dozer, bank run gravel	B-15	1100	.025			1.01	2.47	3.48	4.26
1000	Hand spread, bank run gravel	A-5	33	.545			19.35	2.01	21.36	32
1800	Delivery charge, minimum 20 tons, 1 hr. round trip, add	B-34B	130	.062			2.25	5.35	7.60	9.30
1820	1-1/2 hr. round trip, add		93	.086			3.15	7.45	10.60	13
1840	2 hr. round trip, add		65	.123			4.50	10.65	15.15	18.60

31 23 23.16 Fill By Borrow and Utility Bedding

		Crew	Daily Output	Labor-Hours	Unit	Material	Labor	Equipment	Total	Total Incl O&P
0010	**FILL BY BORROW AND UTILITY BEDDING**									
0049	Utility bedding, for pipe & conduit, not incl. compaction G1030–805									
0050	Crushed or screened bank run gravel	B-6	150	.160	L.C.Y.	27	6.20	2.45	35.65	42
0100	Crushed stone 3/4″ to 1/2″		150	.160		30.50	6.20	2.45	39.15	46
0200	Sand, dead or bank		150	.160		18	6.20	2.45	26.65	32
0500	Compacting bedding in trench	A-1D	90	.089	E.C.Y.		3.15	.38	3.53	5.25
0600	If material source exceeds 2 miles, add for extra mileage.									
0610	See Section 31 23 23.20 for hauling mileage add.									

31 23 23.17 General Fill

		Crew	Daily Output	Labor-Hours	Unit	Material	Labor	Equipment	Total	Total Incl O&P
0010	**GENERAL FILL**									
0011	Spread dumped material, no compaction									
0020	By dozer, no compaction	B-10B	1000	.012	L.C.Y.		.52	1.33	1.85	2.26
0100	By hand	1 Clab	12	.667	″		23.50		23.50	36.50
0500	Gravel fill, compacted, under floor slabs, 4″ deep	B-37	10000	.005	S.F.	.42	.18	.02	.62	.76
0600	6″ deep		8600	.006		.63	.21	.02	.86	1.03
0700	9″ deep		7200	.007		1.05	.25	.02	1.32	1.56

31 23 23.17 General Fill		Crew	Daily Output	Labor-Hours	Unit	Material	2013 Bare Costs Labor	Equipment	Total	Total Incl O&P
0800	12″ deep	B-37	6000	.008	S.F.	1.47	.30	.03	1.80	2.11
1000	Alternate pricing method, 4″ deep		120	.400	E.C.Y.	31.50	15	1.29	47.79	59
1100	6″ deep		160	.300		31.50	11.25	.97	43.72	53
1200	9″ deep		200	.240		31.50	9	.77	41.27	49
1300	12″ deep		220	.218		31.50	8.20	.70	40.40	48
1400	Granular fill				L.C.Y.	21			21	23

31 23 23.20 Hauling		Crew	Daily Output	Labor-Hours	Unit	Material	Labor	Equipment	Total	Total Incl O&P
0010	**HAULING**									
0011	Excavated or borrow, loose cubic yards									
0012	no loading equipment, including hauling,waiting, loading/dumping									
0013	time per cycle (wait, load, travel, unload or dump & return)									
0014	8 C.Y. truck, 15 MPH ave, cycle 0.5 miles, 10 min. wait/Ld./Uld.	B-34A	320	.025	L.C.Y.		.92	1.30	2.22	2.83
0016	cycle 1 mile		272	.029			1.08	1.53	2.61	3.33
0018	cycle 2 miles		208	.038			1.41	2	3.41	4.36
0020	cycle 4 miles		144	.056			2.03	2.89	4.92	6.30
0022	cycle 6 miles		112	.071			2.61	3.72	6.33	8.10
0024	cycle 8 miles		88	.091			3.33	4.73	8.06	10.30
0026	20 MPH ave,cycle 0.5 mile		336	.024			.87	1.24	2.11	2.69
0028	cycle 1 mile		296	.027			.99	1.41	2.40	3.07
0030	cycle 2 miles		240	.033			1.22	1.74	2.96	3.78
0032	cycle 4 miles		176	.045			1.66	2.37	4.03	5.15
0034	cycle 6 miles		136	.059			2.15	3.06	5.21	6.65
0036	cycle 8 miles		112	.071			2.61	3.72	6.33	8.10
0044	25 MPH ave, cycle 4 miles		192	.042			1.53	2.17	3.70	4.73
0046	cycle 6 miles		160	.050			1.83	2.60	4.43	5.65
0048	cycle 8 miles		128	.063			2.29	3.26	5.55	7.10
0050	30 MPH ave, cycle 4 miles		216	.037			1.36	1.93	3.29	4.20
0052	cycle 6 miles		176	.045			1.66	2.37	4.03	5.15
0054	cycle 8 miles		144	.056			2.03	2.89	4.92	6.30
0114	15 MPH ave, cycle 0.5 mile, 15 min. wait/Ld./Uld.		224	.036			1.31	1.86	3.17	4.05
0116	cycle 1 mile		200	.040			1.46	2.08	3.54	4.53
0118	cycle 2 miles		168	.048			1.74	2.48	4.22	5.40
0120	cycle 4 miles		120	.067			2.44	3.47	5.91	7.55
0122	cycle 6 miles		96	.083			3.05	4.34	7.39	9.45
0124	cycle 8 miles		80	.100			3.66	5.20	8.86	11.35
0126	20 MPH ave, cycle 0.5 mile		232	.034			1.26	1.80	3.06	3.91
0128	cycle 1 mile		208	.038			1.41	2	3.41	4.36
0130	cycle 2 miles		184	.043			1.59	2.26	3.85	4.93
0132	cycle 4 miles		144	.056			2.03	2.89	4.92	6.30
0134	cycle 6 miles		112	.071			2.61	3.72	6.33	8.10
0136	cycle 8 miles		96	.083			3.05	4.34	7.39	9.45
0144	25 MPH ave, cycle 4 miles		152	.053			1.93	2.74	4.67	5.95
0146	cycle 6 miles		128	.063			2.29	3.26	5.55	7.10
0148	cycle 8 miles		112	.071			2.61	3.72	6.33	8.10
0150	30 MPH ave, cycle 4 miles		168	.048			1.74	2.48	4.22	5.40
0152	cycle 6 miles		144	.056			2.03	2.89	4.92	6.30
0154	cycle 8 miles		120	.067			2.44	3.47	5.91	7.55
0214	15 MPH ave, cycle 0.5 mile, 20 min wait/Ld./Uld.		176	.045			1.66	2.37	4.03	5.15
0216	cycle 1 mile		160	.050			1.83	2.60	4.43	5.65
0218	cycle 2 miles		136	.059			2.15	3.06	5.21	6.65
0220	cycle 4 miles		104	.077			2.82	4.01	6.83	8.70
0222	cycle 6 miles		88	.091			3.33	4.73	8.06	10.30

31 23 23 – Fill

31 23 23.20 Hauling		Crew	Daily Output	Labor-Hours	Unit	Material	2013 Bare Costs Labor	Equipment	Total	Total Incl O&P
0224	cycle 8 miles	B-34A	72	.111	L.C.Y.		4.07	5.80	9.87	12.60
0226	20 MPH ave, cycle 0.5 mile		176	.045			1.66	2.37	4.03	5.15
0228	cycle 1 mile		168	.048			1.74	2.48	4.22	5.40
0230	cycle 2 miles		144	.056			2.03	2.89	4.92	6.30
0232	cycle 4 miles		120	.067			2.44	3.47	5.91	7.55
0234	cycle 6 miles		96	.083			3.05	4.34	7.39	9.45
0236	cycle 8 miles		88	.091			3.33	4.73	8.06	10.30
0244	25 MPH ave, cycle 4 miles		128	.063			2.29	3.26	5.55	7.10
0246	cycle 6 miles		112	.071			2.61	3.72	6.33	8.10
0248	cycle 8 miles		96	.083			3.05	4.34	7.39	9.45
0250	30 MPH ave, cycle 4 miles		136	.059			2.15	3.06	5.21	6.65
0252	cycle 6 miles		120	.067			2.44	3.47	5.91	7.55
0254	cycle 8 miles		104	.077			2.82	4.01	6.83	8.70
0314	15 MPH ave, cycle 0.5 mile, 25 min wait/Ld./Uld.		144	.056			2.03	2.89	4.92	6.30
0316	cycle 1 mile		128	.063			2.29	3.26	5.55	7.10
0318	cycle 2 miles		112	.071			2.61	3.72	6.33	8.10
0320	cycle 4 miles		96	.083			3.05	4.34	7.39	9.45
0322	cycle 6 miles		80	.100			3.66	5.20	8.86	11.35
0324	cycle 8 miles		64	.125			4.58	6.50	11.08	14.15
0326	20 MPH ave, cycle 0.5 mile		144	.056			2.03	2.89	4.92	6.30
0328	cycle 1 mile		136	.059			2.15	3.06	5.21	6.65
0330	cycle 2 miles		120	.067			2.44	3.47	5.91	7.55
0332	cycle 4 miles		104	.077			2.82	4.01	6.83	8.70
0334	cycle 6 miles		88	.091			3.33	4.73	8.06	10.30
0336	cycle 8 miles		80	.100			3.66	5.20	8.86	11.35
0344	25 MPH ave, cycle 4 miles		112	.071			2.61	3.72	6.33	8.10
0346	cycle 6 miles		96	.083			3.05	4.34	7.39	9.45
0348	cycle 8 miles		88	.091			3.33	4.73	8.06	10.30
0350	30 MPH ave, cycle 4 miles		112	.071			2.61	3.72	6.33	8.10
0352	cycle 6 miles		104	.077			2.82	4.01	6.83	8.70
0354	cycle 8 miles		96	.083			3.05	4.34	7.39	9.45
0414	15 MPH ave, cycle 0.5 mile, 30 min wait/Ld./Uld.		120	.067			2.44	3.47	5.91	7.55
0416	cycle 1 mile		112	.071			2.61	3.72	6.33	8.10
0418	cycle 2 miles		96	.083			3.05	4.34	7.39	9.45
0420	cycle 4 miles		80	.100			3.66	5.20	8.86	11.35
0422	cycle 6 miles		72	.111			4.07	5.80	9.87	12.60
0424	cycle 8 miles		64	.125			4.58	6.50	11.08	14.15
0426	20 MPH ave, cycle 0.5 mile		120	.067			2.44	3.47	5.91	7.55
0428	cycle 1 mile		112	.071			2.61	3.72	6.33	8.10
0430	cycle 2 miles		104	.077			2.82	4.01	6.83	8.70
0432	cycle 4 miles		88	.091			3.33	4.73	8.06	10.30
0434	cycle 6 miles		80	.100			3.66	5.20	8.86	11.35
0436	cycle 8 miles		72	.111			4.07	5.80	9.87	12.60
0444	25 MPH ave, cycle 4 miles		96	.083			3.05	4.34	7.39	9.45
0446	cycle 6 miles		88	.091			3.33	4.73	8.06	10.30
0448	cycle 8 miles		80	.100			3.66	5.20	8.86	11.35
0450	30 MPH ave, cycle 4 miles		96	.083			3.05	4.34	7.39	9.45
0452	cycle 6 miles		88	.091			3.33	4.73	8.06	10.30
0454	cycle 8 miles		80	.100			3.66	5.20	8.86	11.35
0514	15 MPH ave, cycle 0.5 mile, 35 min wait/Ld./Uld.		104	.077			2.82	4.01	6.83	8.70
0516	cycle 1 mile		96	.083			3.05	4.34	7.39	9.45
0518	cycle 2 miles		88	.091			3.33	4.73	8.06	10.30
0520	cycle 4 miles		72	.111			4.07	5.80	9.87	12.60

31 23 23.20 Hauling		Crew	Daily Output	Labor-Hours	Unit	Material	2013 Bare Costs Labor	Equipment	Total	Total Incl O&P
0522	cycle 6 miles	B-34A	64	.125	L.C.Y.		4.58	6.50	11.08	14.15
0524	cycle 8 miles		56	.143			5.25	7.45	12.70	16.20
0526	20 MPH ave, cycle 0.5 mile		104	.077			2.82	4.01	6.83	8.70
0528	cycle 1 mile		96	.083			3.05	4.34	7.39	9.45
0530	cycle 2 miles		96	.083			3.05	4.34	7.39	9.45
0532	cycle 4 miles		80	.100			3.66	5.20	8.86	11.35
0534	cycle 6 miles		72	.111			4.07	5.80	9.87	12.60
0536	cycle 8 miles		64	.125			4.58	6.50	11.08	14.15
0544	25 MPH ave, cycle 4 miles		88	.091			3.33	4.73	8.06	10.30
0546	cycle 6 miles		80	.100			3.66	5.20	8.86	11.35
0548	cycle 8 miles		72	.111			4.07	5.80	9.87	12.60
0550	30 MPH ave, cycle 4 miles		88	.091			3.33	4.73	8.06	10.30
0552	cycle 6 miles		80	.100			3.66	5.20	8.86	11.35
0554	cycle 8 miles		72	.111			4.07	5.80	9.87	12.60
1014	12 C.Y. truck, cycle 0.5 mile, 15 MPH ave, 15 min. wait/Ld./Uld.	B-34B	336	.024			.87	2.06	2.93	3.60
1016	cycle 1 mile		300	.027			.98	2.31	3.29	4.03
1018	cycle 2 miles		252	.032			1.16	2.75	3.91	4.80
1020	cycle 4 miles		180	.044			1.63	3.85	5.48	6.70
1022	cycle 6 miles		144	.056			2.03	4.81	6.84	8.40
1024	cycle 8 miles		120	.067			2.44	5.75	8.19	10.10
1025	cycle 10 miles		96	.083			3.05	7.20	10.25	12.60
1026	20 MPH ave, cycle 0.5 mile		348	.023			.84	1.99	2.83	3.48
1028	cycle 1 mile		312	.026			.94	2.22	3.16	3.88
1030	cycle 2 miles		276	.029			1.06	2.51	3.57	4.38
1032	cycle 4 miles		216	.037			1.36	3.21	4.57	5.60
1034	cycle 6 miles		168	.048			1.74	4.12	5.86	7.20
1036	cycle 8 miles		144	.056			2.03	4.81	6.84	8.40
1038	cycle 10 miles		120	.067			2.44	5.75	8.19	10.10
1040	25 MPH ave, cycle 4 miles		228	.035			1.28	3.04	4.32	5.30
1042	cycle 6 miles		192	.042			1.53	3.61	5.14	6.30
1044	cycle 8 miles		168	.048			1.74	4.12	5.86	7.20
1046	cycle 10 miles		144	.056			2.03	4.81	6.84	8.40
1050	30 MPH ave, cycle 4 miles		252	.032			1.16	2.75	3.91	4.80
1052	cycle 6 miles		216	.037			1.36	3.21	4.57	5.60
1054	cycle 8 miles		180	.044			1.63	3.85	5.48	6.70
1056	cycle 10 miles		156	.051			1.88	4.44	6.32	7.75
1060	35 MPH ave, cycle 4 miles		264	.030			1.11	2.62	3.73	4.58
1062	cycle 6 miles		228	.035			1.28	3.04	4.32	5.30
1064	cycle 8 miles		204	.039			1.44	3.39	4.83	5.95
1066	cycle 10 miles		180	.044			1.63	3.85	5.48	6.70
1068	cycle 20 miles		120	.067			2.44	5.75	8.19	10.10
1069	cycle 30 miles		84	.095			3.49	8.25	11.74	14.40
1070	cycle 40 miles		72	.111			4.07	9.60	13.67	16.85
1072	40 MPH ave, cycle 6 miles		240	.033			1.22	2.88	4.10	5.05
1074	cycle 8 miles		216	.037			1.36	3.21	4.57	5.60
1076	cycle 10 miles		192	.042			1.53	3.61	5.14	6.30
1078	cycle 20 miles		120	.067			2.44	5.75	8.19	10.10
1080	cycle 30 miles		96	.083			3.05	7.20	10.25	12.60
1082	cycle 40 miles		72	.111			4.07	9.60	13.67	16.85
1084	cycle 50 miles		60	.133			4.88	11.55	16.43	20
1094	45 MPH ave, cycle 8 miles		216	.037			1.36	3.21	4.57	5.60
1096	cycle 10 miles		204	.039			1.44	3.39	4.83	5.95
1098	cycle 20 miles		132	.061			2.22	5.25	7.47	9.15

31 23 23.20 Hauling		Crew	Daily Output	Labor-Hours	Unit	Material	2013 Bare Costs Labor	Equipment	Total	Total Incl O&P
1100	cycle 30 miles	B-34B	108	.074	L.C.Y.		2.71	6.40	9.11	11.20
1102	cycle 40 miles		84	.095			3.49	8.25	11.74	14.40
1104	cycle 50 miles		72	.111			4.07	9.60	13.67	16.85
1106	50 MPH ave, cycle 10 miles		216	.037			1.36	3.21	4.57	5.60
1108	cycle 20 miles		144	.056			2.03	4.81	6.84	8.40
1110	cycle 30 miles		108	.074			2.71	6.40	9.11	11.20
1112	cycle 40 miles		84	.095			3.49	8.25	11.74	14.40
1114	cycle 50 miles		72	.111			4.07	9.60	13.67	16.85
1214	15 MPH ave, cycle 0.5 mile, 20 min. wait/Ld./Uld.		264	.030			1.11	2.62	3.73	4.58
1216	cycle 1 mile		240	.033			1.22	2.88	4.10	5.05
1218	cycle 2 miles		204	.039			1.44	3.39	4.83	5.95
1220	cycle 4 miles		156	.051			1.88	4.44	6.32	7.75
1222	cycle 6 miles		132	.061			2.22	5.25	7.47	9.15
1224	cycle 8 miles		108	.074			2.71	6.40	9.11	11.20
1225	cycle 10 miles		96	.083			3.05	7.20	10.25	12.60
1226	20 MPH ave, cycle 0.5 mile		264	.030			1.11	2.62	3.73	4.58
1228	cycle 1 mile		252	.032			1.16	2.75	3.91	4.80
1230	cycle 2 miles		216	.037			1.36	3.21	4.57	5.60
1232	cycle 4 miles		180	.044			1.63	3.85	5.48	6.70
1234	cycle 6 miles		144	.056			2.03	4.81	6.84	8.40
1236	cycle 8 miles		132	.061			2.22	5.25	7.47	9.15
1238	cycle 10 miles		108	.074			2.71	6.40	9.11	11.20
1240	25 MPH ave, cycle 4 miles		192	.042			1.53	3.61	5.14	6.30
1242	cycle 6 miles		168	.048			1.74	4.12	5.86	7.20
1244	cycle 8 miles		144	.056			2.03	4.81	6.84	8.40
1246	cycle 10 miles		132	.061			2.22	5.25	7.47	9.15
1250	30 MPH ave, cycle 4 miles		204	.039			1.44	3.39	4.83	5.95
1252	cycle 6 miles		180	.044			1.63	3.85	5.48	6.70
1254	cycle 8 miles		156	.051			1.88	4.44	6.32	7.75
1256	cycle 10 miles		144	.056			2.03	4.81	6.84	8.40
1260	35 MPH ave, cycle 4 miles		216	.037			1.36	3.21	4.57	5.60
1262	cycle 6 miles		192	.042			1.53	3.61	5.14	6.30
1264	cycle 8 miles		168	.048			1.74	4.12	5.86	7.20
1266	cycle 10 miles		156	.051			1.88	4.44	6.32	7.75
1268	cycle 20 miles		108	.074			2.71	6.40	9.11	11.20
1269	cycle 30 miles		72	.111			4.07	9.60	13.67	16.85
1270	cycle 40 miles		60	.133			4.88	11.55	16.43	20
1272	40 MPH ave, cycle 6 miles		192	.042			1.53	3.61	5.14	6.30
1274	cycle 8 miles		180	.044			1.63	3.85	5.48	6.70
1276	cycle 10 miles		156	.051			1.88	4.44	6.32	7.75
1278	cycle 20 miles		108	.074			2.71	6.40	9.11	11.20
1280	cycle 30 miles		84	.095			3.49	8.25	11.74	14.40
1282	cycle 40 miles		72	.111			4.07	9.60	13.67	16.85
1284	cycle 50 miles		60	.133			4.88	11.55	16.43	20
1294	45 MPH ave, cycle 8 miles		180	.044			1.63	3.85	5.48	6.70
1296	cycle 10 miles		168	.048			1.74	4.12	5.86	7.20
1298	cycle 20 miles		120	.067			2.44	5.75	8.19	10.10
1300	cycle 30 miles		96	.083			3.05	7.20	10.25	12.60
1302	cycle 40 miles		72	.111			4.07	9.60	13.67	16.85
1304	cycle 50 miles		60	.133			4.88	11.55	16.43	20
1306	50 MPH ave, cycle 10 miles		180	.044			1.63	3.85	5.48	6.70
1308	cycle 20 miles		132	.061			2.22	5.25	7.47	9.15
1310	cycle 30 miles		96	.083			3.05	7.20	10.25	12.60

31 23 23.20 Hauling		Crew	Daily Output	Labor-Hours	Unit	Material	2013 Bare Costs Labor	Equipment	Total	Total Incl O&P
1312	cycle 40 miles	B-34B	84	.095	L.C.Y.		3.49	8.25	11.74	14.40
1314	cycle 50 miles		72	.111			4.07	9.60	13.67	16.85
1414	15 MPH ave, cycle 0.5 mile, 25 min. wait/Ld./Uld.		204	.039			1.44	3.39	4.83	5.95
1416	cycle 1 mile		192	.042			1.53	3.61	5.14	6.30
1418	cycle 2 miles		168	.048			1.74	4.12	5.86	7.20
1420	cycle 4 miles		132	.061			2.22	5.25	7.47	9.15
1422	cycle 6 miles		120	.067			2.44	5.75	8.19	10.10
1424	cycle 8 miles		96	.083			3.05	7.20	10.25	12.60
1425	cycle 10 miles		84	.095			3.49	8.25	11.74	14.40
1426	20 MPH ave, cycle 0.5 mile		216	.037			1.36	3.21	4.57	5.60
1428	cycle 1 mile		204	.039			1.44	3.39	4.83	5.95
1430	cycle 2 miles		180	.044			1.63	3.85	5.48	6.70
1432	cycle 4 miles		156	.051			1.88	4.44	6.32	7.75
1434	cycle 6 miles		132	.061			2.22	5.25	7.47	9.15
1436	cycle 8 miles		120	.067			2.44	5.75	8.19	10.10
1438	cycle 10 miles		96	.083			3.05	7.20	10.25	12.60
1440	25 MPH ave, cycle 4 miles		168	.048			1.74	4.12	5.86	7.20
1442	cycle 6 miles		144	.056			2.03	4.81	6.84	8.40
1444	cycle 8 miles		132	.061			2.22	5.25	7.47	9.15
1446	cycle 10 miles		108	.074			2.71	6.40	9.11	11.20
1450	30 MPH ave, cycle 4 miles		168	.048			1.74	4.12	5.86	7.20
1452	cycle 6 miles		156	.051			1.88	4.44	6.32	7.75
1454	cycle 8 miles		132	.061			2.22	5.25	7.47	9.15
1456	cycle 10 miles		120	.067			2.44	5.75	8.19	10.10
1460	35 MPH ave, cycle 4 miles		180	.044			1.63	3.85	5.48	6.70
1462	cycle 6 miles		156	.051			1.88	4.44	6.32	7.75
1464	cycle 8 miles		144	.056			2.03	4.81	6.84	8.40
1466	cycle 10 miles		132	.061			2.22	5.25	7.47	9.15
1468	cycle 20 miles		96	.083			3.05	7.20	10.25	12.60
1469	cycle 30 miles		72	.111			4.07	9.60	13.67	16.85
1470	cycle 40 miles		60	.133			4.88	11.55	16.43	20
1472	40 MPH ave, cycle 6 miles		168	.048			1.74	4.12	5.86	7.20
1474	cycle 8 miles		156	.051			1.88	4.44	6.32	7.75
1476	cycle 10 miles		144	.056			2.03	4.81	6.84	8.40
1478	cycle 20 miles		96	.083			3.05	7.20	10.25	12.60
1480	cycle 30 miles		84	.095			3.49	8.25	11.74	14.40
1482	cycle 40 miles		60	.133			4.88	11.55	16.43	20
1484	cycle 50 miles		60	.133			4.88	11.55	16.43	20
1494	45 MPH ave, cycle 8 miles		156	.051			1.88	4.44	6.32	7.75
1496	cycle 10 miles		144	.056			2.03	4.81	6.84	8.40
1498	cycle 20 miles		108	.074			2.71	6.40	9.11	11.20
1500	cycle 30 miles		84	.095			3.49	8.25	11.74	14.40
1502	cycle 40 miles		72	.111			4.07	9.60	13.67	16.85
1504	cycle 50 miles		60	.133			4.88	11.55	16.43	20
1506	50 MPH ave, cycle 10 miles		156	.051			1.88	4.44	6.32	7.75
1508	cycle 20 miles		120	.067			2.44	5.75	8.19	10.10
1510	cycle 30 miles		96	.083			3.05	7.20	10.25	12.60
1512	cycle 40 miles		72	.111			4.07	9.60	13.67	16.85
1514	cycle 50 miles		60	.133			4.88	11.55	16.43	20
1614	15 MPH, cycle 0.5 mile, 30 min. wait/Ld./Uld.		180	.044			1.63	3.85	5.48	6.70
1616	cycle 1 mile		168	.048			1.74	4.12	5.86	7.20
1618	cycle 2 miles		144	.056			2.03	4.81	6.84	8.40
1620	cycle 4 miles		120	.067			2.44	5.75	8.19	10.10

31 23 23.20 Hauling		Crew	Daily Output	Labor-Hours	Unit	Material	2013 Bare Costs Labor	Equipment	Total	Total Incl O&P
1622	cycle 6 miles	B-34B	108	.074	L.C.Y.		2.71	6.40	9.11	11.20
1624	cycle 8 miles		84	.095			3.49	8.25	11.74	14.40
1625	cycle 10 miles		84	.095			3.49	8.25	11.74	14.40
1626	20 MPH ave, cycle 0.5 mile		180	.044			1.63	3.85	5.48	6.70
1628	cycle 1 mile		168	.048			1.74	4.12	5.86	7.20
1630	cycle 2 miles		156	.051			1.88	4.44	6.32	7.75
1632	cycle 4 miles		132	.061			2.22	5.25	7.47	9.15
1634	cycle 6 miles		120	.067			2.44	5.75	8.19	10.10
1636	cycle 8 miles		108	.074			2.71	6.40	9.11	11.20
1638	cycle 10 miles		96	.083			3.05	7.20	10.25	12.60
1640	25 MPH ave, cycle 4 miles		144	.056			2.03	4.81	6.84	8.40
1642	cycle 6 miles		132	.061			2.22	5.25	7.47	9.15
1644	cycle 8 miles		108	.074			2.71	6.40	9.11	11.20
1646	cycle 10 miles		108	.074			2.71	6.40	9.11	11.20
1650	30 MPH ave, cycle 4 miles		144	.056			2.03	4.81	6.84	8.40
1652	cycle 6 miles		132	.061			2.22	5.25	7.47	9.15
1654	cycle 8 miles		120	.067			2.44	5.75	8.19	10.10
1656	cycle 10 miles		108	.074			2.71	6.40	9.11	11.20
1660	35 MPH ave, cycle 4 miles		156	.051			1.88	4.44	6.32	7.75
1662	cycle 6 miles		144	.056			2.03	4.81	6.84	8.40
1664	cycle 8 miles		132	.061			2.22	5.25	7.47	9.15
1666	cycle 10 miles		120	.067			2.44	5.75	8.19	10.10
1668	cycle 20 miles		84	.095			3.49	8.25	11.74	14.40
1669	cycle 30 miles		72	.111			4.07	9.60	13.67	16.85
1670	cycle 40 miles		60	.133			4.88	11.55	16.43	20
1672	40 MPH, cycle 6 miles		144	.056			2.03	4.81	6.84	8.40
1674	cycle 8 miles		132	.061			2.22	5.25	7.47	9.15
1676	cycle 10 miles		120	.067			2.44	5.75	8.19	10.10
1678	cycle 20 miles		96	.083			3.05	7.20	10.25	12.60
1680	cycle 30 miles		72	.111			4.07	9.60	13.67	16.85
1682	cycle 40 miles		60	.133			4.88	11.55	16.43	20
1684	cycle 50 miles		48	.167			6.10	14.45	20.55	25
1694	45 MPH ave, cycle 8 miles		144	.056			2.03	4.81	6.84	8.40
1696	cycle 10 miles		132	.061			2.22	5.25	7.47	9.15
1698	cycle 20 miles		96	.083			3.05	7.20	10.25	12.60
1700	cycle 30 miles		84	.095			3.49	8.25	11.74	14.40
1702	cycle 40 miles		60	.133			4.88	11.55	16.43	20
1704	cycle 50 miles		60	.133			4.88	11.55	16.43	20
1706	50 MPH ave, cycle 10 miles		132	.061			2.22	5.25	7.47	9.15
1708	cycle 20 miles		108	.074			2.71	6.40	9.11	11.20
1710	cycle 30 miles		84	.095			3.49	8.25	11.74	14.40
1712	cycle 40 miles		72	.111			4.07	9.60	13.67	16.85
1714	cycle 50 miles		60	.133			4.88	11.55	16.43	20
2000	Hauling, 8 C.Y. truck, small project cost per hour	B-34A	8	1	Hr.		36.50	52	88.50	114
2100	12 C.Y. Truck	B-34B	8	1			36.50	86.50	123	151
2150	16.5 C.Y. Truck	B-34C	8	1			36.50	92.50	129	158
2175	18 C.Y. 8 wheel Truck	B-34I	8	1			36.50	108	144.50	175
2200	20 C.Y. Truck	B-34D	8	1			36.50	94	130.50	160
2300	Grading at dump, or embankment if required, by dozer	B-10B	1000	.012	L.C.Y.		.52	1.33	1.85	2.26
2310	Spotter at fill or cut, if required	1 Clab	8	1	Hr.		35.50		35.50	54.50
9014	18 C.Y. truck, 8 wheels,15 min. wait/Ld./Uld.,15 MPH, cycle 0.5 mi.	B-34I	504	.016	L.C.Y.		.58	1.72	2.30	2.78
9016	cycle 1 mile		450	.018			.65	1.93	2.58	3.12
9018	cycle 2 miles		378	.021			.77	2.29	3.06	3.71

31 23 23.20 Hauling		Crew	Daily Output	Labor-Hours	Unit	Material	2013 Bare Costs Labor	Equipment	Total	Total Incl O&P
9020	cycle 4 miles	B-34I	270	.030	L.C.Y.		1.08	3.21	4.29	5.20
9022	cycle 6 miles		216	.037			1.36	4.01	5.37	6.50
9024	cycle 8 miles		180	.044			1.63	4.82	6.45	7.80
9025	cycle 10 miles		144	.056			2.03	6	8.03	9.70
9026	20 MPH ave, cycle 0.5 mile		522	.015			.56	1.66	2.22	2.69
9028	cycle 1 mile		468	.017			.63	1.85	2.48	3
9030	cycle 2 miles		414	.019			.71	2.09	2.80	3.38
9032	cycle 4 miles		324	.025			.90	2.68	3.58	4.32
9034	cycle 6 miles		252	.032			1.16	3.44	4.60	5.55
9036	cycle 8 miles		216	.037			1.36	4.01	5.37	6.50
9038	cycle 10 miles		180	.044			1.63	4.82	6.45	7.80
9040	25 MPH ave, cycle 4 miles		342	.023			.86	2.54	3.40	4.10
9042	cycle 6 miles		288	.028			1.02	3.01	4.03	4.87
9044	cycle 8 miles		252	.032			1.16	3.44	4.60	5.55
9046	cycle 10 miles		216	.037			1.36	4.01	5.37	6.50
9050	30 MPH ave, cycle 4 miles		378	.021			.77	2.29	3.06	3.71
9052	cycle 6 miles		324	.025			.90	2.68	3.58	4.32
9054	cycle 8 miles		270	.030			1.08	3.21	4.29	5.20
9056	cycle 10 miles		234	.034			1.25	3.71	4.96	6
9060	35 MPH ave, cycle 4 miles		396	.020			.74	2.19	2.93	3.54
9062	cycle 6 miles		342	.023			.86	2.54	3.40	4.10
9064	cycle 8 miles		288	.028			1.02	3.01	4.03	4.87
9066	cycle 10 miles		270	.030			1.08	3.21	4.29	5.20
9068	cycle 20 miles		162	.049			1.81	5.35	7.16	8.65
9070	cycle 30 miles		126	.063			2.32	6.90	9.22	11.10
9072	cycle 40 miles		90	.089			3.25	9.65	12.90	15.60
9074	40 MPH ave, cycle 6 miles		360	.022			.81	2.41	3.22	3.90
9076	cycle 8 miles		324	.025			.90	2.68	3.58	4.32
9078	cycle 10 miles		288	.028			1.02	3.01	4.03	4.87
9080	cycle 20 miles		180	.044			1.63	4.82	6.45	7.80
9082	cycle 30 miles		144	.056			2.03	6	8.03	9.70
9084	cycle 40 miles		108	.074			2.71	8.05	10.76	13
9086	cycle 50 miles		90	.089			3.25	9.65	12.90	15.60
9094	45 MPH ave, cycle 8 miles		324	.025			.90	2.68	3.58	4.32
9096	cycle 10 miles		306	.026			.96	2.83	3.79	4.59
9098	cycle 20 miles		198	.040			1.48	4.38	5.86	7.10
9100	cycle 30 miles		144	.056			2.03	6	8.03	9.70
9102	cycle 40 miles		126	.063			2.32	6.90	9.22	11.10
9104	cycle 50 miles		108	.074			2.71	8.05	10.76	13
9106	50 MPH ave, cycle 10 miles		324	.025			.90	2.68	3.58	4.32
9108	cycle 20 miles		216	.037			1.36	4.01	5.37	6.50
9110	cycle 30 miles		162	.049			1.81	5.35	7.16	8.65
9112	cycle 40 miles		126	.063			2.32	6.90	9.22	11.10
9114	cycle 50 miles		108	.074			2.71	8.05	10.76	13
9214	20 min. wait/Ld./Uld.,15 MPH, cycle 0.5 mi.		396	.020			.74	2.19	2.93	3.54
9216	cycle 1 mile		360	.022			.81	2.41	3.22	3.90
9218	cycle 2 miles		306	.026			.96	2.83	3.79	4.59
9220	cycle 4 miles		234	.034			1.25	3.71	4.96	6
9222	cycle 6 miles		198	.040			1.48	4.38	5.86	7.10
9224	cycle 8 miles		162	.049			1.81	5.35	7.16	8.65
9225	cycle 10 miles		144	.056			2.03	6	8.03	9.70
9226	20 MPH ave, cycle 0.5 mile		396	.020			.74	2.19	2.93	3.54
9228	cycle 1 mile		378	.021			.77	2.29	3.06	3.71

31 23 23.20 Hauling		Crew	Daily Output	Labor-Hours	Unit	Material	2013 Bare Costs Labor	Equipment	Total	Total Incl O&P
9230	cycle 2 miles	B-34I	324	.025	L.C.Y.		.90	2.68	3.58	4.32
9232	cycle 4 miles		270	.030			1.08	3.21	4.29	5.20
9234	cycle 6 miles		216	.037			1.36	4.01	5.37	6.50
9236	cycle 8 miles		198	.040			1.48	4.38	5.86	7.10
9238	cycle 10 miles		162	.049			1.81	5.35	7.16	8.65
9240	25 MPH ave, cycle 4 miles		288	.028			1.02	3.01	4.03	4.87
9242	cycle 6 miles		252	.032			1.16	3.44	4.60	5.55
9244	cycle 8 miles		216	.037			1.36	4.01	5.37	6.50
9246	cycle 10 miles		198	.040			1.48	4.38	5.86	7.10
9250	30 MPH ave, cycle 4 miles		306	.026			.96	2.83	3.79	4.59
9252	cycle 6 miles		270	.030			1.08	3.21	4.29	5.20
9254	cycle 8 miles		234	.034			1.25	3.71	4.96	6
9256	cycle 10 miles		216	.037			1.36	4.01	5.37	6.50
9260	35 MPH ave, cycle 4 miles		324	.025			.90	2.68	3.58	4.32
9262	cycle 6 miles		288	.028			1.02	3.01	4.03	4.87
9264	cycle 8 miles		252	.032			1.16	3.44	4.60	5.55
9266	cycle 10 miles		234	.034			1.25	3.71	4.96	6
9268	cycle 20 miles		162	.049			1.81	5.35	7.16	8.65
9270	cycle 30 miles		108	.074			2.71	8.05	10.76	13
9272	cycle 40 miles		90	.089			3.25	9.65	12.90	15.60
9274	40 MPH ave, cycle 6 miles		288	.028			1.02	3.01	4.03	4.87
9276	cycle 8 miles		270	.030			1.08	3.21	4.29	5.20
9278	cycle 10 miles		234	.034			1.25	3.71	4.96	6
9280	cycle 20 miles		162	.049			1.81	5.35	7.16	8.65
9282	cycle 30 miles		126	.063			2.32	6.90	9.22	11.10
9284	cycle 40 miles		108	.074			2.71	8.05	10.76	13
9286	cycle 50 miles		90	.089			3.25	9.65	12.90	15.60
9294	45 MPH ave, cycle 8 miles		270	.030			1.08	3.21	4.29	5.20
9296	cycle 10 miles		252	.032			1.16	3.44	4.60	5.55
9298	cycle 20 miles		180	.044			1.63	4.82	6.45	7.80
9300	cycle 30 miles		144	.056			2.03	6	8.03	9.70
9302	cycle 40 miles		108	.074			2.71	8.05	10.76	13
9304	cycle 50 miles		90	.089			3.25	9.65	12.90	15.60
9306	50 MPH ave, cycle 10 miles		270	.030			1.08	3.21	4.29	5.20
9308	cycle 20 miles		198	.040			1.48	4.38	5.86	7.10
9310	cycle 30 miles		144	.056			2.03	6	8.03	9.70
9312	cycle 40 miles		126	.063			2.32	6.90	9.22	11.10
9314	cycle 50 miles		108	.074			2.71	8.05	10.76	13
9414	25 min. wait/Ld./Uld.,15 MPH, cycle 0.5 mi.		306	.026			.96	2.83	3.79	4.59
9416	cycle 1 mile		288	.028			1.02	3.01	4.03	4.87
9418	cycle 2 miles		252	.032			1.16	3.44	4.60	5.55
9420	cycle 4 miles		198	.040			1.48	4.38	5.86	7.10
9422	cycle 6 miles		180	.044			1.63	4.82	6.45	7.80
9424	cycle 8 miles		144	.056			2.03	6	8.03	9.70
9425	cycle 10 miles		126	.063			2.32	6.90	9.22	11.10
9426	20 MPH ave, cycle 0.5 mile		324	.025			.90	2.68	3.58	4.32
9428	cycle 1 mile		306	.026			.96	2.83	3.79	4.59
9430	cycle 2 miles		270	.030			1.08	3.21	4.29	5.20
9432	cycle 4 miles		234	.034			1.25	3.71	4.96	6
9434	cycle 6 miles		198	.040			1.48	4.38	5.86	7.10
9436	cycle 8 miles		180	.044			1.63	4.82	6.45	7.80
9438	cycle 10 miles		144	.056			2.03	6	8.03	9.70
9440	25 MPH ave, cycle 4 miles		252	.032			1.16	3.44	4.60	5.55

31 23 23.20 Hauling		Crew	Daily Output	Labor-Hours	Unit	Material	2013 Bare Costs Labor	Equipment	Total	Total Incl O&P
9442	cycle 6 miles	B-34I	216	.037	L.C.Y.		1.36	4.01	5.37	6.50
9444	cycle 8 miles		198	.040			1.48	4.38	5.86	7.10
9446	cycle 10 miles		180	.044			1.63	4.82	6.45	7.80
9450	30 MPH ave, cycle 4 miles		252	.032			1.16	3.44	4.60	5.55
9452	cycle 6 miles		234	.034			1.25	3.71	4.96	6
9454	cycle 8 miles		198	.040			1.48	4.38	5.86	7.10
9456	cycle 10 miles		180	.044			1.63	4.82	6.45	7.80
9460	35 MPH ave, cycle 4 miles		270	.030			1.08	3.21	4.29	5.20
9462	cycle 6 miles		234	.034			1.25	3.71	4.96	6
9464	cycle 8 miles		216	.037			1.36	4.01	5.37	6.50
9466	cycle 10 miles		198	.040			1.48	4.38	5.86	7.10
9468	cycle 20 miles		144	.056			2.03	6	8.03	9.70
9470	cycle 30 miles		108	.074			2.71	8.05	10.76	13
9472	cycle 40 miles		90	.089			3.25	9.65	12.90	15.60
9474	40 MPH ave, cycle 6 miles		252	.032			1.16	3.44	4.60	5.55
9476	cycle 8 miles		234	.034			1.25	3.71	4.96	6
9478	cycle 10 miles		216	.037			1.36	4.01	5.37	6.50
9480	cycle 20 miles		144	.056			2.03	6	8.03	9.70
9482	cycle 30 miles		126	.063			2.32	6.90	9.22	11.10
9484	cycle 40 miles		90	.089			3.25	9.65	12.90	15.60
9486	cycle 50 miles		90	.089			3.25	9.65	12.90	15.60
9494	45 MPH ave, cycle 8 miles		234	.034			1.25	3.71	4.96	6
9496	cycle 10 miles		216	.037			1.36	4.01	5.37	6.50
9498	cycle 20 miles		162	.049			1.81	5.35	7.16	8.65
9500	cycle 30 miles		126	.063			2.32	6.90	9.22	11.10
9502	cycle 40 miles		108	.074			2.71	8.05	10.76	13
9504	cycle 50 miles		90	.089			3.25	9.65	12.90	15.60
9506	50 MPH ave, cycle 10 miles		234	.034			1.25	3.71	4.96	6
9508	cycle 20 miles		180	.044			1.63	4.82	6.45	7.80
9510	cycle 30 miles		144	.056			2.03	6	8.03	9.70
9512	cycle 40 miles		108	.074			2.71	8.05	10.76	13
9514	cycle 50 miles		90	.089			3.25	9.65	12.90	15.60
9614	30 min. wait/Ld./Uld.,15 MPH, cycle 0.5 mi.		270	.030			1.08	3.21	4.29	5.20
9616	cycle 1 mile		252	.032			1.16	3.44	4.60	5.55
9618	cycle 2 miles		216	.037			1.36	4.01	5.37	6.50
9620	cycle 4 miles		180	.044			1.63	4.82	6.45	7.80
9622	cycle 6 miles		162	.049			1.81	5.35	7.16	8.65
9624	cycle 8 miles		126	.063			2.32	6.90	9.22	11.10
9625	cycle 10 miles		126	.063			2.32	6.90	9.22	11.10
9626	20 MPH ave, cycle 0.5 mile		270	.030			1.08	3.21	4.29	5.20
9628	cycle 1 mile		252	.032			1.16	3.44	4.60	5.55
9630	cycle 2 miles		234	.034			1.25	3.71	4.96	6
9632	cycle 4 miles		198	.040			1.48	4.38	5.86	7.10
9634	cycle 6 miles		180	.044			1.63	4.82	6.45	7.80
9636	cycle 8 miles		162	.049			1.81	5.35	7.16	8.65
9638	cycle 10 miles		144	.056			2.03	6	8.03	9.70
9640	25 MPH ave, cycle 4 miles		216	.037			1.36	4.01	5.37	6.50
9642	cycle 6 miles		198	.040			1.48	4.38	5.86	7.10
9644	cycle 8 miles		180	.044			1.63	4.82	6.45	7.80
9646	cycle 10 miles		162	.049			1.81	5.35	7.16	8.65
9650	30 MPH ave, cycle 4 miles		216	.037			1.36	4.01	5.37	6.50
9652	cycle 6 miles		198	.040			1.48	4.38	5.86	7.10
9654	cycle 8 miles		180	.044			1.63	4.82	6.45	7.80

31 23 23 – Fill

31 23 23.20 Hauling		Crew	Daily Output	Labor-Hours	Unit	Material	2013 Bare Costs Labor	Equipment	Total	Total Incl O&P
9656	cycle 10 miles	B-34I	162	.049	L.C.Y.		1.81	5.35	7.16	8.65
9660	35 MPH ave, cycle 4 miles		234	.034			1.25	3.71	4.96	6
9662	cycle 6 miles		216	.037			1.36	4.01	5.37	6.50
9664	cycle 8 miles		198	.040			1.48	4.38	5.86	7.10
9666	cycle 10 miles		180	.044			1.63	4.82	6.45	7.80
9668	cycle 20 miles		126	.063			2.32	6.90	9.22	11.10
9670	cycle 30 miles		108	.074			2.71	8.05	10.76	13
9672	cycle 40 miles		90	.089			3.25	9.65	12.90	15.60
9674	40 MPH ave, cycle 6 miles		216	.037			1.36	4.01	5.37	6.50
9676	cycle 8 miles		198	.040			1.48	4.38	5.86	7.10
9678	cycle 10 miles		180	.044			1.63	4.82	6.45	7.80
9680	cycle 20 miles		144	.056			2.03	6	8.03	9.70
9682	cycle 30 miles		108	.074			2.71	8.05	10.76	13
9684	cycle 40 miles		90	.089			3.25	9.65	12.90	15.60
9686	cycle 50 miles		72	.111			4.07	12.05	16.12	19.50
9694	45 MPH ave, cycle 8 miles		216	.037			1.36	4.01	5.37	6.50
9696	cycle 10 miles		198	.040			1.48	4.38	5.86	7.10
9698	cycle 20 miles		144	.056			2.03	6	8.03	9.70
9700	cycle 30 miles		126	.063			2.32	6.90	9.22	11.10
9702	cycle 40 miles		108	.074			2.71	8.05	10.76	13
9704	cycle 50 miles		90	.089			3.25	9.65	12.90	15.60
9706	50 MPH ave, cycle 10 miles		198	.040			1.48	4.38	5.86	7.10
9708	cycle 20 miles		162	.049			1.81	5.35	7.16	8.65
9710	cycle 30 miles		126	.063			2.32	6.90	9.22	11.10
9712	cycle 40 miles		108	.074			2.71	8.05	10.76	13
9714	cycle 50 miles		90	.089			3.25	9.65	12.90	15.60

Division Notes

		CREW	DAILY OUTPUT	LABOR-HOURS	UNIT	BARE COSTS MAT.	LABOR	EQUIP.	TOTAL	TOTAL INCL O&P

Estimating Tips

32 01 00 Operations and Maintenance of Exterior Improvements

- Recycling of asphalt pavement is becoming very popular and is an alternative to removal and replacement. It can be a good value engineering proposal if removed pavement can be recycled, either at the project site or at another site that is reasonably close to the project site. Two new sections on repair of flexible and rigid pavement have been added.

32 10 00 Bases, Ballasts, and Paving

- When estimating paving, keep in mind the project schedule. If an asphaltic paving project is in a colder climate and runs through to the spring, consider placing the base course in the autumn and then topping it in the spring, just prior to completion. This could save considerable costs in spring repair. Keep in mind that prices for asphalt and concrete are generally higher in the cold seasons. New lines have been added for pavement markings including tactile warning systems and new fence lines.

32 90 00 Planting

- The timing of planting and guarantee specifications often dictate the costs for establishing tree and shrub growth and a stand of grass or ground cover. Establish the work performance schedule to coincide with the local planting season. Maintenance and growth guarantees can add from 20%–100% to the total landscaping cost. The cost to replace trees and shrubs can be as high as 5% of the total cost, depending on the planting zone, soil conditions, and time of year.

Reference Numbers

Reference numbers are shown in shaded boxes at the beginning of some major classifications. These numbers refer to related items in the Reference Section. The reference information may be an estimating procedure, an alternate pricing method, or technical information.

Note: Not all subdivisions listed here necessarily appear in this publication.

32 12 16 – Asphalt Paving

32 12 16.13 Plant-Mix Asphalt Paving		Crew	Daily Output	Labor-Hours	Unit	2013 Bare Costs Material	Labor	Equipment	Total	Total Incl O&P
0010	**PLANT-MIX ASPHALT PAVING**									
0020	And large paved areas with no hauling included									
0025	See Section 31 23 23.20 for hauling costs									
0080	Binder course, 1-1/2" thick	B-25	7725	.011	S.Y.	5.40	.44	.35	6.19	7
0120	2" thick		6345	.014		7.20	.54	.42	8.16	9.20
0130	2-1/2" thick		5620	.016		8.95	.61	.48	10.04	11.30
0160	3" thick		4905	.018		10.75	.70	.55	12	13.50
0170	3-1/2" thick		4520	.019		12.55	.76	.59	13.90	15.60
0200	4" thick		4140	.021		14.35	.83	.65	15.83	17.80
0300	Wearing course, 1" thick	B-25B	10575	.009		3.56	.36	.28	4.20	4.77
0340	1-1/2" thick		7725	.012		6	.49	.38	6.87	7.75
0380	2" thick		6345	.015		8.05	.60	.46	9.11	10.30
0420	2-1/2" thick		5480	.018		9.90	.69	.53	11.12	12.55
0460	3" thick		4900	.020		11.80	.78	.59	13.17	14.85
0470	3-1/2" thick		4520	.021		13.85	.84	.64	15.33	17.25
0480	4" thick		4140	.023		15.85	.92	.70	17.47	19.60
0500	Open graded friction course	B-25C	5000	.010		2.22	.38	.47	3.07	3.54
0800	Alternate method of figuring paving costs									
0810	Binder course, 1-1/2" thick	B-25	630	.140	Ton	66	5.45	4.26	75.71	85.50
0811	2" thick		690	.128		66	4.96	3.89	74.85	84.50
0812	3" thick		800	.110		66	4.28	3.35	73.63	82.50
0813	4" thick		900	.098		66	3.81	2.98	72.79	81.50
0850	Wearing course, 1" thick	B-25B	575	.167		66	6.60	5.05	77.65	88
0851	1-1/2" thick		630	.152		66	6.05	4.62	76.67	87
0852	2" thick		690	.139		66	5.50	4.22	75.72	85.50
0853	2-1/2" thick		765	.125		66	4.97	3.81	74.78	84.50
0854	3" thick		800	.120		66	4.76	3.64	74.40	84
1000	Pavement replacement over trench, 2" thick	B-37	90	.533	S.Y.	7.40	20	1.72	29.12	40.50
1050	4" thick		70	.686		14.65	25.50	2.21	42.36	58
1080	6" thick		55	.873		23.50	32.50	2.81	58.81	78.50

32 12 16.14 Asphaltic Concrete Paving		Crew	Daily Output	Labor-Hours	Unit	Material	Labor	Equipment	Total	Total Incl O&P
0011	**ASPHALTIC CONCRETE PAVING**, parking lots & driveways									
0015	No asphalt hauling included									
0018	Use 6.05 C.Y. per inch per M.S.F. for hauling									
0020	6" stone base, 2" binder course, 1" topping	B-25C	9000	.005	S.F.	1.75	.21	.26	2.22	2.54
0025	2" binder course, 2" topping		9000	.005		2.15	.21	.26	2.62	2.97
0030	3" binder course, 2" topping		9000	.005		2.55	.21	.26	3.02	3.42
0035	4" binder course, 2" topping		9000	.005		2.95	.21	.26	3.42	3.85
0040	1.5" binder course, 1" topping		9000	.005		1.55	.21	.26	2.02	2.32
0042	3" binder course, 1" topping		9000	.005		2.15	.21	.26	2.62	2.98
0045	3" binder course, 3" topping		9000	.005		2.96	.21	.26	3.43	3.86
0050	4" binder course, 3" topping		9000	.005		3.35	.21	.26	3.82	4.30
0055	4" binder course, 4" topping		9000	.005		3.75	.21	.26	4.22	4.73
0300	Binder course, 1-1/2" thick		35000	.001		.60	.05	.07	.72	.81
0400	2" thick		25000	.002		.78	.08	.09	.95	1.07
0500	3" thick		15000	.003		1.20	.13	.16	1.49	1.68
0600	4" thick		10800	.004		1.57	.18	.22	1.97	2.24
0800	Sand finish course, 3/4" thick		41000	.001		.31	.05	.06	.42	.48
0900	1" thick		34000	.001		.39	.06	.07	.52	.60
1000	Fill pot holes, hot mix, 2" thick	B-16	4200	.008		.81	.28	.16	1.25	1.50
1100	4" thick		3500	.009		1.19	.33	.20	1.72	2.04
1120	6" thick		3100	.010		1.60	.37	.22	2.19	2.59

32 12 Flexible Paving

32 12 16 – Asphalt Paving

32 12 16.14 Asphaltic Concrete Paving		Crew	Daily Output	Labor-Hours	Unit	Material	Labor	Equipment	Total	Total Incl O&P
						2013 Bare Costs				
1140	Cold patch, 2" thick	B-51	3000	.016	S.F.	.83	.57	.09	1.49	1.89
1160	4" thick		2700	.018		1.57	.64	.10	2.31	2.82
1180	6" thick		1900	.025		2.44	.90	.14	3.48	4.23

32 84 Planting Irrigation

32 84 13 – Drip Irrigation

32 84 13.10 Subsurface Drip Irrigation

			Crew	Daily Output	Labor-Hours	Unit	Material	Labor	Equipment	Total	Total Incl O&P
0010	**SUBSURFACE DRIP IRRIGATION**										
0011	Looped grid, pressure compensating										
0100	Preinserted PE emitter, line, hand bury, irregular area, small	G	3 Skwk	1200	.020	L.F.	.28	.92		1.20	1.74
0150	Medium	G		1800	.013		.28	.62		.90	1.26
0200	Large	G		2520	.010		.28	.44		.72	.99
0250	Rectangular area, small	G		2040	.012		.28	.54		.82	1.15
0300	Medium	G		2640	.009		.28	.42		.70	.96
0350	Large	G		3600	.007		.28	.31		.59	.79
0400	Install in trench, irregular area, small	G		4050	.006		.28	.27		.55	.73
0450	Medium	G		7488	.003		.28	.15		.43	.54
0500	Large	G		16560	.001		.28	.07		.35	.41
0550	Rectangular area, small	G		8100	.003		.28	.14		.42	.52
0600	Medium			21960	.001		.28	.05		.33	.39
0650	Large	G		33264	.001		.28	.03		.31	.36
0700	Trenching and backfill	G	B-53	500	.016			.73	.14	.87	1.26
0750	For vinyl tubing, 1/4", add to above						25%	10%			
0800	Vinyl tubing, 1/4", material only	G					.04			.04	.04
0850	Supply tubing, 1/2", material only, 100' coil	G					.10			.10	.11
0900	500' coil	G					.09			.09	.10
0950	Compression fittings	G	1 Skwk	90	.089	Ea.	.69	4.11		4.80	7.10
1000	Barbed fittings, 1/4"	G		360	.022		.15	1.03		1.18	1.76
1100	Flush risers	G		60	.133		3.29	6.15		9.44	13.15
1150	Flush ends, figure eight	G		180	.044		.50	2.05		2.55	3.73
1200	Ball valve, 4-1/2"	G		20	.400		9	18.50		27.50	38.50
1250	4-3/4"	G		20	.400		7	18.50		25.50	36
1300	Auto flush, spring loaded	G		90	.089		2.10	4.11		6.21	8.65
1350	Volumetric	G		90	.089		7.25	4.11		11.36	14.35
1400	Air relief valve, inline with compensation tee, 1/2"	G		45	.178		13.10	8.20		21.30	27
1450	1"	G		30	.267		16.45	12.30		28.75	37
1500	Round box for flush ends, 6"	G		30	.267		7.40	12.30		19.70	27
1550	Fertilizer injector, non-proportional	G		4	2		11.15	92.50		103.65	155
1600	Screen filter, 3/4" screen	G		12	.667		10.30	31		41.30	59
1650	1" disk	G		8	1		13.10	46		59.10	86
1700	1-1/2" disk	G		4	2		71	92.50		163.50	221
1750	2" disk	G		3	2.667		93	123		216	294
1800	Typical installation 18" O.C., small, minimum					S.F.				1.50	2.30
1850	Maximum									1.75	2.70
1900	Large, minimum									1.45	2.23
2000	Maximum									1.68	2.59
2100	For non-pressure compensating systems, deduct									10%	10%

32 84 Planting Irrigation

32 84 23 – Underground Sprinklers

	32 84 23.10 Sprinkler Irrigation System	Crew	Daily Output	Labor-Hours	Unit	2013 Bare Costs Material	Labor	Equipment	Total	Total Incl O&P
0010	**SPRINKLER IRRIGATION SYSTEM** G2050–710									
0011	For lawns									
0100	Golf course with fully automatic system	C-17	.05	1600	9 holes	97,000	74,500		171,500	222,000
0200	24′ diam. head at 15′ O.C. incl. piping, auto oper., minimum	B-20	70	.343	Head	23	13.60		36.60	46.50
0300	Maximum		40	.600		55	24		79	97
0500	60′ diam. head at 40′ O.C. incl. piping, auto oper., minimum		28	.857		69	34		103	129
0600	Maximum		23	1.043		190	41.50		231.50	273
0800	Residential system, custom, 1″ supply		2000	.012	S.F.	.25	.48		.73	1.01
0900	1-1/2″ supply		1800	.013	″	.48	.53		1.01	1.35
0990	For renovation work, add to above						50%			
1020	Pop up spray head w/risers, hi-pop, full circle pattern, 4″	2 Skwk	76	.211	Ea.	4.41	9.75		14.16	19.90
1030	1/2 circle pattern, 4″		76	.211		4.41	9.75		14.16	19.90
1040	6″, full circle pattern		76	.211		9.70	9.75		19.45	25.50
1050	1/2 circle pattern, 6″		76	.211		8.35	9.75		18.10	24.50
1060	12″, full circle pattern		76	.211		11.35	9.75		21.10	27.50
1070	1/2 circle pattern, 12″		76	.211		11.25	9.75		21	27.50
1080	Pop up bubbler head w/risers, hi-pop bubbler head, 4″		76	.211		4.25	9.75		14	19.75
1090	6″		76	.211		8.95	9.75		18.70	25
1100	12″		76	.211		10.70	9.75		20.45	27
1110	Impact full/part circle sprinklers, 28′-54′ 25-60 PSI		37	.432		19.30	20		39.30	52
1120	Spaced 37′-49′ @ 25-50 PSI		37	.432		26.50	20		46.50	60.50
1130	Spaced 43′-61′ @ 30-60 PSI		37	.432		75	20		95	114
1140	Spaced 54′-78′ @ 40-80 PSI		37	.432		131	20		151	175
1145	Impact rotor pop-up full/part commercial circle sprinklers									
1150	Spaced 42′-65′ 35-80 PSI	2 Skwk	25	.640	Ea.	16.65	29.50		46.15	64
1160	Spaced 48′-76′ 45-85 PSI	″	25	.640	″	18.30	29.50		47.80	65.50
1165	Impact rotor pop-up part. circle comm., 53′-75′, 55-100 PSI, w/accessories									
1170	Plastic case, metal cover	2 Skwk	25	.640	Ea.	81	29.50		110.50	135
1180	Rubber cover		25	.640		62	29.50		91.50	114
1190	Iron case, metal cover		22	.727		134	33.50		167.50	199
1200	Rubber cover		22	.727		142	33.50		175.50	208
1250	Plastic case, 2 nozzle, metal cover		25	.640		129	29.50		158.50	188
1260	Rubber cover		25	.640		129	29.50		158.50	188
1270	Iron case, 2 nozzle, metal cover		22	.727		184	33.50		217.50	254
1280	Rubber cover		22	.727		185	33.50		218.50	255
1282	Impact rotor pop-up full circle commercial, 39′-99′, 30-100 PSI									
1284	Plastic case, metal cover	2 Skwk	25	.640	Ea.	139	29.50		168.50	199
1286	Rubber cover		25	.640		156	29.50		185.50	218
1288	Iron case, metal cover		22	.727		202	33.50		235.50	274
1290	Rubber cover		22	.727		209	33.50		242.50	282
1292	Plastic case, 2 nozzle, metal cover		22	.727		149	33.50		182.50	216
1294	Rubber cover		22	.727		149	33.50		182.50	216
1296	Iron case, 2 nozzle, metal cover		20	.800		197	37		234	273
1298	Rubber cover		20	.800		204	37		241	281
1305	Electric remote control valve, plastic, 3/4″		18	.889		16.95	41		57.95	82
1310	1″		18	.889		32.50	41		73.50	99.50
1320	1-1/2″		18	.889		59.50	41		100.50	129
1330	2″		18	.889		84.50	41		125.50	157
1335	Quick coupling valves, brass, locking cover									
1340	Inlet coupling valve, 3/4″	2 Skwk	18.75	.853	Ea.	21.50	39.50		61	84.50
1350	1″		18.75	.853		30	39.50		69.50	94
1360	Controller valve boxes, 6″ round boxes		18.75	.853		6.85	39.50		46.35	68.50

32 84 23.10 Sprinkler Irrigation System		Crew	Daily Output	Labor-Hours	Unit	Material	Labor	Equipment	Total	Total Incl O&P
						2013 Bare Costs				
1370	10" round boxes	2 Skwk	14.25	1.123	Ea.	11.05	52		63.05	92
1380	12" square box	↓	9.75	1.641	↓	15.55	76		91.55	134
1388	Electromech. control, 14 day 3-60 min., auto start to 23/day									
1390	4 station	2 Skwk	1.04	15.385	Ea.	183	710		893	1,300
1400	7 station		.64	25		197	1,150		1,347	2,000
1410	12 station		.40	40		217	1,850		2,067	3,100
1420	Dual programs, 18 station		.24	66.667		610	3,075		3,685	5,450
1430	23 station	↓	.16	100	↓	675	4,625		5,300	7,900
1435	Backflow preventer, bronze, 0-175 PSI, w/valves, test cocks									
1440	3/4"	2 Skwk	2	8	Ea.	99	370		469	680
1450	1"		2	8		114	370		484	695
1460	1-1/2"		2	8		277	370		647	875
1470	2"	↓	2	8	↓	340	370		710	940
1475	Pressure vacuum breaker, brass, 15-150 PSI									
1480	3/4"	2 Skwk	2	8	Ea.	59.50	370		429.50	635
1490	1"		2	8		65.50	370		435.50	640
1500	1-1/2"		2	8		266	370		636	860
1510	2"	↓	2	8		320	370		690	925
2000	Pop-up spray, head & nozzle, low/medium volume, plastic	1 Skwk	30	.267		3.18	12.30		15.48	22.50
2200	Brass, economy		30	.267		11.30	12.30		23.60	31.50
2400	Heavy duty		30	.267		19.30	12.30		31.60	40
3000	Riser mounted spray, head, low/medium volume, plastic		30	.267		3.18	12.30		15.48	22.50
3200	Brass		30	.267		10.45	12.30		22.75	30.50
4000	Pop-up impact sprinkler, body & case, plastic low/medium volume		30	.267		22	12.30		34.30	43.50
4200	Brass, high/medium volume		30	.267		48.50	12.30		60.80	72
4400	High volume		30	.267		131	12.30		143.30	163
5000	Quick coupling valve and key, brass, thread type		25	.320		57	14.80		71.80	85.50
5200	Lug type		25	.320		60	14.80		74.80	89
5400	Pop-up gear drive sprinkler, nozzle & case plastic, medium volume	↓	30	.267	↓	25.50	12.30		37.80	47
6000	Riser mounted gear drive sprinkler, nozzle and case									
6200	Plastic, medium volume	1 Skwk	30	.267	Ea.	14.50	12.30		26.80	35
7000	Riser mounted impact sprinkler, body, part or full circle									
7200	Full circle plastic, low/medium volume	1 Skwk	25	.320	Ea.	14.50	14.80		29.30	39
7400	Brass, low/medium volume		25	.320		38	14.80		52.80	65
7600	Medium volume		25	.320		49	14.80		63.80	76.50
7800	Female thread	↓	25	.320	↓	90	14.80		104.80	122
8000	Riser mounted impact sprinkler, body, full circle only									
8200	Plastic, low/medium volume	1 Skwk	25	.320	Ea.	8.50	14.80		23.30	32.50
8400	Brass, low/medium volume		25	.320		37	14.80		51.80	63.50
8600	High volume		25	.320		75	14.80		89.80	106
8800	Very high volume	↓	25	.320	↓	146	14.80		160.80	184

Division Notes

		CREW	DAILY OUTPUT	LABOR-HOURS	UNIT	BARE COSTS MAT.	LABOR	EQUIP.	TOTAL	TOTAL INCL O&P

Estimating Tips

33 10 00 Water Utilities
33 30 00 Sanitary Sewerage Utilities
33 40 00 Storm Drainage Utilities

- Never assume that the water, sewer, and drainage lines will go in at the early stages of the project. Consider the site access needs before dividing the site in half with open trenches, loose pipe, and machinery obstructions. Always inspect the site to establish that the site drawings are complete. Check off all existing utilities on your drawings as you locate them. Be especially careful with underground utilities because appurtenances are sometimes buried during regrading or repaving operations. If you find any discrepancies, mark up the site plan for further research. Differing site conditions can be very costly if discovered later in the project.
- See also Section 33 01 00 for restoration of pipe where removal/replacement may be undesirable. Use of new types of piping materials can reduce the overall project cost. Owners/design engineers should consider the installing contractor as a valuable source of current information on utility products and local conditions that could lead to significant cost savings.

Reference Numbers

Reference numbers are shown in shaded boxes at the beginning of some major classifications. These numbers refer to related items in the Reference Section. The reference information may be an estimating procedure, an alternate pricing method, or technical information.

Note: Not all subdivisions listed here necessarily appear in this publication.

Note: **Trade Service,** *in part, has been used as a reference source for some of the material prices used in Division 33.*

33 01 Operation and Maintenance of Utilities

33 01 10 – Operation and Maintenance of Water Utilities

33 01 10.10 Corrosion Resistance		Crew	Daily Output	Labor-Hours	Unit	2013 Bare Costs Material	Labor	Equipment	Total	Total Incl O&P
0010	**CORROSION RESISTANCE**									
0012	Wrap & coat, add to pipe, 4" diameter				L.F.	2.30			2.30	2.53
0020	5" diameter					2.89			2.89	3.18
0040	6" diameter					3.46			3.46	3.81
0060	8" diameter					4.49			4.49	4.94
0080	10" diameter					5.65			5.65	6.20
0100	12" diameter					6.80			6.80	7.45
0120	14" diameter					7.80			7.80	8.60
0140	16" diameter					8.85			8.85	9.75
0160	18" diameter					9.75			9.75	10.70
0180	20" diameter					10.75			10.75	11.85
0200	24" diameter					13.10			13.10	14.40
0220	Small diameter pipe, 1" diameter, add					1.50			1.50	1.65
0240	2" diameter					1.65			1.65	1.82
0260	2-1/2" diameter					1.81			1.81	1.99
0280	3" diameter				▼	2.17			2.17	2.39
0300	Fittings, field covered, add				S.F.	9.50			9.50	10.45
0500	Coating, bituminous, per diameter inch, 1 coat, add				L.F.	.58			.58	.64
0540	3 coat					1.75			1.75	1.93
0560	Coal tar epoxy, per diameter inch, 1 coat, add					.21			.21	.23
0600	3 coat					.63			.63	.69
1000	Polyethylene H.D. extruded, .025" thk., 1/2" diameter add					.32			.32	.35
1020	3/4" diameter					.36			.36	.40
1040	1" diameter					.41			.41	.45
1060	1-1/4" diameter					.51			.51	.56
1080	1-1/2" diameter					.53			.53	.58
1100	.030" thk., 2" diameter					.57			.57	.63
1120	2-1/2" diameter					.66			.66	.73
1140	.035" thk., 3" diameter					.83			.83	.91
1160	3-1/2" diameter					.94			.94	1.03
1180	4" diameter					1.05			1.05	1.16
1200	.040" thk, 5" diameter					1.29			1.29	1.42
1220	6" diameter					1.37			1.37	1.51
1240	8" diameter					1.80			1.80	1.98
1260	10" diameter					2.25			2.25	2.48
1280	12" diameter					2.70			2.70	2.97
1300	.060" thk., 14" diameter					3.74			3.74	4.11
1320	16" diameter					4.45			4.45	4.90
1340	18" diameter					5.15			5.15	5.65
1360	20" diameter				▼	5.85			5.85	6.45
1380	Fittings, field wrapped, add				S.F.	6			6	6.60

33 01 10.20 Pipe Repair

		Crew	Daily Output	Labor-Hours	Unit	Material	Labor	Equipment	Total	Total Incl O&P
0010	**PIPE REPAIR**									
0020	Not including excavation or backfill									
0100	Clamp, stainless steel, lightweight, for steel pipe									
0110	3" long, 1/2" diameter pipe	1 Plum	34	.235	Ea.	11.85	13.15		25	33
0120	3/4" diameter pipe		32	.250		12.40	13.95		26.35	34.50
0130	1" diameter pipe		30	.267		13.10	14.90		28	37
0140	1-1/4" diameter pipe		28	.286		13.70	15.95		29.65	39
0150	1-1/2" diameter pipe		26	.308		14.60	17.15		31.75	42
0160	2" diameter pipe		24	.333		16	18.60		34.60	45.50
0170	2-1/2" diameter pipe		23	.348		21.50	19.40		40.90	52.50

33 01 Operation and Maintenance of Utilities

33 01 10 – Operation and Maintenance of Water Utilities

33 01 10.20 Pipe Repair		Crew	Daily Output	Labor-Hours	Unit	2013 Bare Costs Material	Labor	Equipment	Total	Total Incl O&P
0180	3″ diameter pipe	1 Plum	22	.364	Ea.	19.25	20.50		39.75	51.50
0190	3-1/2″ diameter pipe		21	.381		25	21.50		46.50	59.50
0200	4″ diameter pipe	B-20	44	.545		25	21.50		46.50	61
0210	5″ diameter pipe		42	.571		29.50	22.50		52	67.50
0220	6″ diameter pipe		38	.632		34	25		59	76
0230	8″ diameter pipe		30	.800		40.50	32		72.50	93.50
0240	10″ diameter pipe		28	.857		125	34		159	191
0250	12″ diameter pipe		24	1		175	39.50		214.50	254
0260	14″ diameter pipe		22	1.091		250	43.50		293.50	340
0270	16″ diameter pipe		20	1.200		270	47.50		317.50	370
0280	18″ diameter pipe		18	1.333		297	53		350	405
0290	20″ diameter pipe		16	1.500		300	59.50		359.50	420
0300	24″ diameter pipe		14	1.714		335	68		403	470
0360	For 6″ long, add					100%	40%			
0370	For 9″ long, add					200%	100%			
0380	For 12″ long, add					300%	150%			
0390	For 18″ long, add					500%	200%			
0400	Pipe Freezing for live repairs of systems 3/8 inch to 6 inch									
0410	Note: Pipe Freezing can also be used to install a valve into a live system									
0420	Pipe Freezing each side 3/8 inch	2 Skwk	8	2	Ea.	515	92.50		607.50	710
0425	Pipe Freezing each side 3/8 inch, second location same kit		8	2		22	92.50		114.50	167
0430	Pipe Freezing each side 3/4 inch		8	2		500	92.50		592.50	695
0435	Pipe Freezing each side 3/4 inch, second location same kit		8	2		22	92.50		114.50	167
0440	Pipe Freezing each side 1-1/2 inch		6	2.667		500	123		623	740
0445	Pipe Freezing each side 1-1/2 inch , second location same kit		6	2.667		22	123		145	215
0450	Pipe Freezing each side 2 inch		6	2.667		890	123		1,013	1,175
0455	Pipe Freezing each side 2 inch, second location same kit		6	2.667		22	123		145	215
0460	Pipe Freezing each side 2-1/2 inch-3 inch		6	2.667		900	123		1,023	1,175
0465	Pipe Freeze each side 2-1/2 -3 inch, second location same kit		6	2.667		44	123		167	240
0470	Pipe Freezing each side 4 inch		4	4		1,500	185		1,685	1,925
0475	Pipe Freezing each side 4 inch, second location same kit		4	4		70	185		255	365
0480	Pipe Freezing each side 5-6 inch		4	4		3,975	185		4,160	4,650
0485	Pipe Freezing each side 5-6 inch, second location same kit		4	4		210	185		395	515
0490	Pipe Frz extra 20 lb. CO_2 cylinders (3/8″ to 2″ - 1 ea, 3″ -2 ea)					210			210	231
0500	Pipe Freezing extra 50 lb. CO_2 cylinders (4″ - 2 ea, 5″-6″ -6 ea)					470			470	520
1000	Clamp, stainless steel, with threaded service tap									
1040	Full seal for iron, steel, PVC pipe									
1100	6″ long, 2″ diameter pipe	1 Plum	17	.471	Ea.	83.50	26.50		110	132
1110	2-1/2″ diameter pipe		16	.500		85.50	28		113.50	136
1120	3″ diameter pipe		15.60	.513		190	28.50		218.50	251
1130	3-1/2″ diameter pipe		15	.533		192	30		222	256
1140	4″ diameter pipe	B-20	32	.750		194	30		224	260
1150	6″ diameter pipe		28	.857		202	34		236	275
1160	8″ diameter pipe		21	1.143		211	45.50		256.50	300
1170	10″ diameter pipe		20	1.200		222	47.50		269.50	320
1180	12″ diameter pipe		17	1.412		234	56		290	345
1205	For 9″ long, add					20%	45%			
1210	For 12″ long, add					40%	80%			
1220	For 18″ long, add					70%	110%			
1600	Clamp, stainless steel, single section									
1640	Full seal for iron, steel, PVC pipe									
1700	6″ long, 2″ diameter pipe	1 Plum	17	.471	Ea.	87	26.50		113.50	135
1710	2-1/2″ diameter pipe		16	.500		90	28		118	141

33 01 Operation and Maintenance of Utilities

33 01 10 – Operation and Maintenance of Water Utilities

33 01 10.20 Pipe Repair		Crew	Daily Output	Labor-Hours	Unit	Material	2013 Bare Costs Labor	Equipment	Total	Total Incl O&P
1720	3″ diameter pipe	1 Plum	15.60	.513	Ea.	104	28.50		132.50	158
1730	3-1/2″ diameter pipe		15	.533		110	30		140	166
1740	4″ diameter pipe	B-20	32	.750		117	30		147	174
1750	6″ diameter pipe		27	.889		145	35.50		180.50	215
1760	8″ diameter pipe		21	1.143		171	45.50		216.50	258
1770	10″ diameter pipe		20	1.200		224	47.50		271.50	320
1780	12″ diameter pipe		17	1.412		261	56		317	375
1800	For 9″ long, add					40%	45%			
1810	For 12″ long, add					60%	80%			
1820	For 18″ long, add					120%	110%			
2000	Clamp, stainless steel, two section									
2040	Full seal, for iron, steel, PVC pipe									
2100	6″ long, 4″ diameter pipe	B-20	24	1	Ea.	230	39.50		269.50	315
2110	6″ diameter pipe		20	1.200		264	47.50		311.50	365
2120	8″ diameter pipe		13	1.846		300	73.50		373.50	445
2130	10″ diameter pipe		12	2		305	79.50		384.50	455
2140	12″ diameter pipe		10	2.400		390	95.50		485.50	575
2200	9″ long, 4″ diameter pipe		16	1.500		300	59.50		359.50	420
2210	6″ diameter pipe		13	1.846		340	73.50		413.50	490
2220	8″ diameter pipe		9	2.667		375	106		481	580
2230	10″ diameter pipe		8	3		495	119		614	730
2240	12″ diameter pipe		7	3.429		565	136		701	835
2250	14″ diameter pipe		6.40	3.750		760	149		909	1,075
2260	16″ diameter pipe		6	4		705	159		864	1,025
2270	18″ diameter pipe		5	4.800		800	191		991	1,175
2280	20″ diameter pipe		4.60	5.217		885	207		1,092	1,300
2290	24″ diameter pipe		4	6		1,175	238		1,413	1,675
2320	For 12″ long, add to 9″					15%	25%			
2330	For 18″ long, add to 9″					70%	55%			
8000	For internal cleaning and inspection, see Section 33 01 30.16									
8100	For pipe testing, see Section 23 05 93.50									

33 01 30 – Operation and Maintenance of Sewer Utilities

33 01 30.16 TV Inspection of Sewer Pipelines		Crew	Daily Output	Labor-Hours	Unit	Material	Labor	Equipment	Total	Total Incl O&P
0010	**TV INSPECTION OF SEWER PIPELINES**									
0100	Pipe internal cleaning & inspection, cleaning, pressure pipe systems									
0120	Pig method, lengths 1000′ to 10,000′									
0140	4″ diameter thru 24″ diameter, minimum				L.F.				4	4.60
0160	Maximum				″				18	21
6000	Sewage/sanitary systems									
6100	Power rodder with header & cutters									
6110	Mobilization charge, minimum				Total				695	800
6120	Mobilization charge, maximum				″				9,125	10,600
6140	Cleaning 4″-12″ diameter				L.F.				3.77	4.33
6190	14″-24″ diameter								4.42	5.10
6240	30″ diameter								6.45	7.40
6250	36″ diameter								7.50	8.60
6260	48″ diameter								8.55	9.85
6270	60″ diameter								9.65	11.05
6280	72″ diameter								10.70	12.35
9000	Inspection, television camera with video									
9060	up to 500 linear feet				Total				795	910

33 01 Operation and Maintenance of Utilities

33 01 30 – Operation and Maintenance of Sewer Utilities

33 01 30.71 Pipebursting		Crew	Daily Output	Labor-Hours	Unit	Material	2013 Bare Costs Labor	Equipment	Total	Total Incl O&P
0010	**PIPEBURSTING**									
0011	300′ runs, replace with HDPE pipe									
0020	Not including excavation, backfill, shoring, or dewatering									
0100	6″ to 15″ diameter, minimum				L.F.				100	110
0200	Maximum								206	227
0300	18″ to 36″ diameter, minimum								200	220
0400	Maximum								420	465
0500	Mobilize and demobilize, minimum				Job				3,000	3,300
0600	Maximum				"				30,900	34,000

33 01 30.72 Relining Sewers		Crew	Daily Output	Labor-Hours	Unit	Material	Labor	Equipment	Total	Total Incl O&P
0010	**RELINING SEWERS**									
0011	With cement incl. bypass & cleaning									
0020	Less than 10,000 L.F., urban, 6″ to 10″	C-17E	130	.615	L.F.	9.35	28.50	.71	38.56	55.50
0200	24″ to 36″		90	.889		14.95	41.50	1.03	57.48	81.50
0300	48″ to 72″		80	1		24	46.50	1.16	71.66	99.50

33 01 30.74 HDPE Pipe Lining		Crew	Daily Output	Labor-Hours	Unit	Material	Labor	Equipment	Total	Total Incl O&P
0010	**HDPE PIPE LINING**, excludes cleaning and video inspection									
0020	Pipe relined with one pipe size smaller than original (4″ for 6″)									
0100	6″ diameter, original size	B-6B	600	.080	L.F.	3.75	2.89	1.54	8.18	10.25
0150	8″ diameter, original size		600	.080		8.45	2.89	1.54	12.88	15.40
0200	10″ diameter, original size		600	.080		10.90	2.89	1.54	15.33	18.10
0250	12″ diameter, original size		400	.120		19.40	4.33	2.31	26.04	30.50
0300	14″ diameter, original size		400	.120		20	4.33	2.31	26.64	31

33 05 Common Work Results for Utilities

33 05 16 – Utility Structures

33 05 16.13 Precast Concrete Utility Boxes		Crew	Daily Output	Labor-Hours	Unit	Material	Labor	Equipment	Total	Total Incl O&P
0010	**PRECAST CONCRETE UTILITY BOXES**, 6″ thick									
0050	5′ x 10′ x 6′ high, I.D.	B-13	2	28	Ea.	3,600	1,075	365	5,040	6,000
0100	6′ x 10′ x 6′ high, I.D.		2	28		3,725	1,075	365	5,165	6,150
0150	5′ x 12′ x 6′ high, I.D.		2	28		3,950	1,075	365	5,390	6,400
0200	6′ x 12′ x 6′ high, I.D.		1.80	31.111		4,425	1,200	405	6,030	7,175
0250	6′ x 13′ x 6′ high, I.D.		1.50	37.333		5,800	1,450	485	7,735	9,100
0300	8′ x 14′ x 7′ high, I.D.		1	56		6,275	2,150	730	9,155	11,000
0350	Hand hole, precast concrete, 1-1/2″ thick									
0400	1′-0″ x 2′-0″ x 1′-9″, I.D., light duty	B-1	4	6	Ea.	405	217		622	780
0450	4′-6″ x 3′-2″ x 2′-0″, O.D., heavy duty	B-6	3	8	"	1,450	310	122	1,882	2,200

33 05 23 – Trenchless Utility Installation

33 05 23.19 Microtunneling		Crew	Daily Output	Labor-Hours	Unit	Material	Labor	Equipment	Total	Total Incl O&P
0010	**MICROTUNNELING**									
0011	Not including excavation, backfill, shoring,									
0020	or dewatering, average 50′/day, slurry method									
0100	24″ to 48″ outside diameter, minimum				L.F.				850	935
0110	Adverse conditions, add				%				50%	50%
1000	Rent microtunneling machine, average monthly lease				Month				94,500	104,000
1010	Operating technician				Day				620	690
1100	Mobilization and demobilization, minimum				Job				40,000	45,000
1110	Maximum				"				432,500	476,000

33 05 Common Work Results for Utilities

33 05 23 – Trenchless Utility Installation

33 05 23.20 Horizontal Boring		Crew	Daily Output	Labor-Hours	Unit	Material	2013 Bare Costs Labor	Equipment	Total	Total Incl O&P
0010	**HORIZONTAL BORING**									
0011	Casing only, 100′ minimum,									
0020	not incl. jacking pits or dewatering									
0100	Roadwork, 1/2″ thick wall, 24″ diameter casing	B-42	20	3.200	L.F.	132	128	67	327	420
0200	36″ diameter		16	4		242	160	83.50	485.50	610
0300	48″ diameter		15	4.267		340	171	89	600	735
0500	Railroad work, 24″ diameter		15	4.267		132	171	89	392	510
0600	36″ diameter		14	4.571		242	183	95.50	520.50	660
0700	48″ diameter	↓	12	5.333		340	214	111	665	830
0900	For ledge, add								20%	20%
1000	Small diameter boring, 3″, sandy soil	B-82	900	.018		24	.72	.06	24.78	27.50
1040	Rocky soil	″	500	.032	↓	24	1.30	.12	25.42	28.50
1100	Prepare jacking pits, incl. mobilization & demobilization, minimum				Ea.				3,150	3,625
1101	Maximum				″				21,600	25,000

33 05 26 – Utility Identification

33 05 26.10 Utility Accessories

		Crew	Daily Output	Labor-Hours	Unit	Material	Labor	Equipment	Total	Total Incl O&P
0010	**UTILITY ACCESSORIES**									
0400	Underground tape, detectable, reinforced, alum. foil core, 2″	1 Clab	150	.053	C.L.F.	2.37	1.89		4.26	5.50
0500	6″	″	140	.057	″	12	2.03		14.03	16.30

33 11 Water Utility Distribution Piping

33 11 13 – Public Water Utility Distribution Piping

33 11 13.15 Water Supply, Ductile Iron Pipe

		Crew	Daily Output	Labor-Hours	Unit	Material	Labor	Equipment	Total	Total Incl O&P
0010	**WATER SUPPLY, DUCTILE IRON PIPE** R331113-80									
0020	Not including excavation or backfill									
2000	Pipe, class 50 water piping, 18′ lengths									
2020	Mechanical joint, 4″ diameter	B-21A	200	.200	L.F.	14.55	8.90	2.36	25.81	32
2040	6″ diameter		160	.250		16.55	11.10	2.95	30.60	38.50
2060	8″ diameter		133.33	.300		22	13.35	3.53	38.88	48
2080	10″ diameter		114.29	.350		28.50	15.55	4.12	48.17	59.50
2100	12″ diameter		105.26	.380		36	16.90	4.48	57.38	70
2120	14″ diameter		100	.400		59.50	17.75	4.71	81.96	97.50
2140	16″ diameter		72.73	.550		61.50	24.50	6.50	92.50	112
2160	18″ diameter		68.97	.580		60	26	6.85	92.85	113
2170	20″ diameter		57.14	.700		78	31	8.25	117.25	142
2180	24″ diameter		47.06	.850		95	38	10	143	174
3000	Tyton, push-on joint, 4″ diameter		400	.100		16.35	4.44	1.18	21.97	26
3020	6″ diameter		333.33	.120		16.85	5.35	1.41	23.61	28
3040	8″ diameter		200	.200		22.50	8.90	2.36	33.76	41
3060	10″ diameter		181.82	.220		33.50	9.75	2.59	45.84	54
3080	12″ diameter		160	.250		35	11.10	2.95	49.05	58.50
3100	14″ diameter		133.33	.300		38.50	13.35	3.53	55.38	66.50
3120	16″ diameter		114.29	.350		52.50	15.55	4.12	72.17	85.50
3140	18″ diameter		100	.400		58	17.75	4.71	80.46	96
3160	20″ diameter		88.89	.450		60.50	20	5.30	85.80	103
3180	24″ diameter	↓	76.92	.520	↓	73.50	23	6.15	102.65	123
8000	Fittings, mechanical joint									
8006	90° bend, 4″ diameter	B-20A	16	2	Ea.	258	86.50		344.50	415
8020	6″ diameter		12.80	2.500		385	108		493	590
8040	8″ diameter	↓	10.67	2.999		685	130		815	950

33 11 13.15 Water Supply, Ductile Iron Pipe		Crew	Daily Output	Labor-Hours	Unit	2013 Bare Costs Material	Labor	Equipment	Total	Total Incl O&P
8060	10" diameter	B-21A	11.43	3.500	Ea.	920	155	41	1,116	1,275
8080	12" diameter		10.53	3.799		1,275	169	45	1,489	1,700
8100	14" diameter		10	4		1,725	178	47	1,950	2,225
8120	16" diameter		7.27	5.502		2,250	244	65	2,559	2,925
8140	18" diameter		6.90	5.797		3,000	258	68.50	3,326.50	3,800
8160	20" diameter		5.71	7.005		3,800	310	82.50	4,192.50	4,750
8180	24" diameter		4.70	8.511		5,850	380	100	6,330	7,125
8200	Wye or tee, 4" diameter	B-20A	10.67	2.999		415	130		545	660
8220	6" diameter		8.53	3.751		630	163		793	940
8240	8" diameter		7.11	4.501		1,000	195		1,195	1,400
8260	10" diameter	B-21A	7.62	5.249		1,450	233	62	1,745	2,000
8280	12" diameter		7.02	5.698		1,900	253	67	2,220	2,525
8300	14" diameter		6.67	5.997		2,825	266	70.50	3,161.50	3,600
8320	16" diameter		4.85	8.247		3,650	365	97	4,112	4,675
8340	18" diameter		4.60	8.696		4,525	385	102	5,012	5,675
8360	20" diameter		3.81	10.499		6,375	465	124	6,964	7,875
8380	24" diameter		3.14	12.739		10,800	565	150	11,515	12,900
8398	45° bends, 4" diameter	B-20A	16	2		241	86.50		327.50	395
8400	6" diameter	"	12.80	2.500		350	108		458	550
8410	12" diameter	B-21A	10.53	3.799		1,100	169	45	1,314	1,500
8420	16" diameter		7.27	5.502		1,975	244	65	2,284	2,625
8430	20" diameter		5.71	7.005		3,100	310	82.50	3,492.50	3,975
8440	24" diameter		4.70	8.511		4,300	380	100	4,780	5,400
8450	Decreaser, 6" x 4" diameter	B-20A	14.22	2.250		340	97.50		437.50	525
8460	8" x 6" diameter	"	11.64	2.749		500	119		619	730
8470	10" x 6" diameter	B-21A	13.33	3.001		610	133	35.50	778.50	910
8480	12" x 6" diameter		12.70	3.150		750	140	37	927	1,075
8490	16" x 6" diameter		10	4		1,350	178	47	1,575	1,800
8500	20" x 6" diameter		8.42	4.751		2,850	211	56	3,117	3,500
8552	For water utility valves see Section 33 12 16									
8700	Joint restraint, ductile iron mechanical joints									
8710	4" diameter	B-20A	32	1	Ea.	35.50	43.50		79	105
8720	6" diameter		25.60	1.250		42.50	54		96.50	129
8730	8" diameter		21.33	1.500		62	65		127	167
8740	10" diameter		18.28	1.751		89.50	76		165.50	213
8750	12" diameter		16.84	1.900		128	82.50		210.50	265
8760	14" diameter		16	2		165	86.50		251.50	315
8770	16" diameter		11.64	2.749		220	119		339	425
8780	18" diameter		11.03	2.901		305	126		431	525
8785	20" diameter		9.14	3.501		375	152		527	645
8790	24" diameter		7.53	4.250		490	184		674	820
9600	Steel sleeve with tap, 4" diameter	B-20	3	8		445	320		765	980
9620	6" diameter		2	12		485	475		960	1,275
9630	8" diameter		2	12		535	475		1,010	1,325

33 11 13.20 Water Supply, Polyethylene Pipe, C901		Crew	Daily Output	Labor-Hours	Unit	Material	Labor	Equipment	Total	Total Incl O&P
0010	**WATER SUPPLY, POLYETHYLENE PIPE, C901**									
0020	Not including excavation or backfill									
1000	Piping, 160 PSI, 3/4" diameter	Q-1A	525	.019	L.F.	.44	1.07		1.51	2.09
1120	1" diameter		485	.021		.60	1.16		1.76	2.41
1140	1-1/2" diameter		450	.022		1.24	1.25		2.49	3.24
1160	2" diameter		365	.027		2.06	1.54		3.60	4.59
2000	Fittings, insert type, nylon, 160 & 250 psi, cold water									

33 11 13.20 Water Supply, Polyethylene Pipe, C901		Crew	Daily Output	Labor-Hours	Unit	Material	2013 Bare Costs Labor	Equipment	Total	Total Incl O&P
2220	Clamp ring, stainless steel, 3/4" diameter	Q-1A	345	.029	Ea.	1.49	1.63		3.12	4.09
2240	1" diameter		321	.031		1.68	1.75		3.43	4.49
2260	1-1/2" diameter		285	.035		2.39	1.97		4.36	5.60
2280	2" diameter		255	.039		2.57	2.20		4.77	6.15
2300	Coupling, 3/4" diameter		66	.152		1.12	8.50		9.62	14.10
2320	1" diameter		57	.175		1.18	9.85		11.03	16.15
2340	1-1/2" diameter		51	.196		2.05	11		13.05	18.85
2360	2" diameter		48	.208		3.91	11.70		15.61	22
2400	Elbow, 90°, 3/4" diameter		66	.152		2.24	8.50		10.74	15.30
2420	1" diameter		57	.175		2.47	9.85		12.32	17.55
2440	1-1/2" diameter		51	.196		3.27	11		14.27	20
2460	2" diameter	↓	48	.208	↓	4.57	11.70		16.27	22.50

33 11 13.25 Water Supply, Polyvinyl Chloride Pipe		Crew	Daily Output	Labor-Hours	Unit	Material	Labor	Equipment	Total	Total Incl O&P
0010	**WATER SUPPLY, POLYVINYL CHLORIDE PIPE** R331113-80									
0020	Not including excavation or backfill, unless specified									
2100	PVC pipe, Class 150, 1-1/2" diameter	Q-1A	750	.013	L.F.	.42	.75		1.17	1.59
2120	2" diameter		686	.015		.68	.82		1.50	1.98
2140	2-1/2" diameter	↓	500	.020		1.26	1.12		2.38	3.08
2160	3" diameter	B-20	430	.056	↓	1.40	2.22		3.62	4.96
3010	AWWA C905, PR 100, DR 25									
3030	14" diameter	B-20A	213	.150	L.F.	14.15	6.50		20.65	25.50
3040	16" diameter		200	.160		14.15	6.95		21.10	26
3050	18" diameter		160	.200		24.50	8.65		33.15	40
3060	20" diameter		133	.241		30	10.45		40.45	49
3070	24" diameter		107	.299		43.50	12.95		56.45	67
3080	30" diameter		80	.400		81	17.35		98.35	116
3090	36" diameter		80	.400		125	17.35		142.35	165
3100	42" diameter		60	.533		168	23		191	219
3200	48" diameter	↓	60	.533		219	23		242	276
3960	Pressure pipe, class 200, ASTM 2241, SDR 21, 3/4" diameter	Q-1A	1000	.010		.17	.56		.73	1.04
3980	1" diameter		900	.011		.25	.62		.87	1.22
4000	1-1/2" diameter		750	.013		.67	.75		1.42	1.87
4010	2" diameter		686	.015		.72	.82		1.54	2.02
4020	2-1/2" diameter	↓	500	.020		1.22	1.12		2.34	3.03
4030	3" diameter	B-20A	430	.074		1.68	3.23		4.91	6.75
4040	4" diameter		375	.085		2.66	3.70		6.36	8.55
4050	6" diameter		316	.101		5.95	4.39		10.34	13.15
4060	8" diameter	↓	260	.123		9.50	5.35		14.85	18.55
4090	Including trenching to 3' deep, 3/4" diameter	Q-1C	300	.080		.17	3.95	11.20	15.32	18.45
4100	1" diameter		280	.086		.25	4.23	12	16.48	19.85
4110	1-1/2" diameter		260	.092		.67	4.55	12.90	18.12	22
4120	2" diameter		220	.109		.72	5.40	15.25	21.37	25.50
4130	2-1/2" diameter		200	.120		1.22	5.90	16.80	23.92	28.50
4140	3" diameter		175	.137		1.68	6.75	19.20	27.63	33
4150	4" diameter		150	.160		2.66	7.90	22.50	33.06	39.50
4160	6" diameter	↓	125	.192	↓	5.95	9.45	27	42.40	50.50
4165	Fittings									
4170	Elbow, 90°, 3/4"	Q-1A	114	.088	Ea.	1.84	4.93		6.77	9.45
4180	1"		100	.100		5.40	5.60		11	14.35
4190	1-1/2"		80	.125		19	7.05		26.05	31.50
4200	2"	↓	72	.139		30	7.80		37.80	45
4210	3"	B-20A	46	.696		36.50	30		66.50	86

33 11 13.25 Water Supply, Polyvinyl Chloride Pipe		Crew	Daily Output	Labor-Hours	Unit	Material	Labor	Equipment	Total	Total Incl O&P
						2013 Bare Costs				
4220	4"	B-20A	36	.889	Ea.	45	38.50		83.50	108
4230	6"		24	1.333		78	58		136	174
4240	8"		14	2.286		144	99		243	310
4250	Elbow, 45°, 3/4"	Q-1A	114	.088		3.11	4.93		8.04	10.85
4260	1"		100	.100		7.45	5.60		13.05	16.65
4270	1-1/2"		80	.125		22	7.05		29.05	34.50
4280	2"		72	.139		23.50	7.80		31.30	37.50
4290	2-1/2"		54	.185		24.50	10.40		34.90	42.50
4300	3"	B-20A	46	.696		29.50	30		59.50	78.50
4310	4"		36	.889		43.50	38.50		82	107
4320	6"		24	1.333		75.50	58		133.50	171
4330	8"		14	2.286		144	99		243	310
4340	Tee, 3/4"	Q-1A	76	.132		3.22	7.40		10.62	14.70
4350	1"		66	.152		16.90	8.50		25.40	31.50
4360	1-1/2"		54	.185		28	10.40		38.40	46
4370	2"		48	.208		31	11.70		42.70	51.50
4380	2-1/2"		36	.278		32	15.60		47.60	59
4390	3"	B-20A	30	1.067		33.50	46		79.50	108
4400	4"		24	1.333		56.50	58		114.50	150
4410	6"		14.80	2.162		130	93.50		223.50	285
4420	8"		9	3.556		240	154		394	500
4430	Coupling, 3/4"	Q-1A	114	.088		1.46	4.93		6.39	9.05
4440	1"		100	.100		6.55	5.60		12.15	15.65
4450	1-1/2"		80	.125		10.50	7.05		17.55	22
4460	2"		72	.139		11.50	7.80		19.30	24.50
4470	2-1/2"		54	.185		12.65	10.40		23.05	29.50
4480	3"	B-20A	46	.696		15.80	30		45.80	63.50
4490	4"		36	.889		24.50	38.50		63	85.50
4500	6"		24	1.333		37.50	58		95.50	129
4510	8"		14	2.286		86	99		185	246
4520	Pressure pipe Class 150, SDR 18, AWWA C900, 4" diameter		380	.084	L.F.	2.63	3.65		6.28	8.45
4530	6" diameter		316	.101		5.55	4.39		9.94	12.75
4540	8" diameter		264	.121		8.20	5.25		13.45	17.05
4550	10" diameter		220	.145		12.85	6.30		19.15	24
4560	12" diameter		186	.172		20	7.45		27.45	33.50
8000	Fittings with rubber gasket									
8003	Class 150, DR 18									
8006	90° Bend , 4" diameter	B-20	100	.240	Ea.	51.50	9.55		61.05	71
8020	6" diameter		90	.267		83	10.60		93.60	107
8040	8" diameter		80	.300		149	11.90		160.90	181
8060	10" diameter		50	.480		335	19.05		354.05	395
8080	12" diameter		30	.800		425	32		457	520
8100	Tee, 4" diameter		90	.267		58.50	10.60		69.10	81
8120	6" diameter		80	.300		146	11.90		157.90	179
8140	8" diameter		70	.343		189	13.60		202.60	229
8160	10" diameter		40	.600		590	24		614	685
8180	12" diameter		20	1.200		765	47.50		812.50	915
8200	45° Bend, 4" diameter		100	.240		43	9.55		52.55	62
8220	6" diameter		90	.267		83	10.60		93.60	107
8240	8" diameter		50	.480		149	19.05		168.05	193
8260	10" diameter		50	.480		335	19.05		354.05	395
8280	12" diameter		30	.800		425	32		457	520
8300	Reducing tee 6" x 4"		100	.240		130	9.55		139.55	158

33 11 Water Utility Distribution Piping

33 11 13 – Public Water Utility Distribution Piping

33 11 13.25 Water Supply, Polyvinyl Chloride Pipe		Crew	Daily Output	Labor-Hours	Unit	2013 Bare Costs Material	Labor	Equipment	Total	Total Incl O&P
8320	8″ x 6″	B-20	90	.267	Ea.	234	10.60		244.60	273
8330	10″ x 6″		90	.267		380	10.60		390.60	430
8340	10″ x 8″		90	.267		390	10.60		400.60	445
8350	12″ x 6″		90	.267		445	10.60		455.60	505
8360	12″ x 8″		90	.267		475	10.60		485.60	535
8400	Tapped service tee (threaded type) 6″ x 6″ x 3/4″		100	.240		96.50	9.55		106.05	121
8430	6″ x 6″ x 1″		90	.267		96.50	10.60		107.10	122
8440	6″ x 6″ x 1-1/2″		90	.267		96.50	10.60		107.10	122
8450	6″ x 6″ x 2″		90	.267		96.50	10.60		107.10	122
8460	8″ x 8″ x 3/4″		90	.267		142	10.60		152.60	172
8470	8″ x 8″ x 1″		90	.267		142	10.60		152.60	172
8480	8″ x 8″ x 1-1/2″		90	.267		142	10.60		152.60	172
8490	8″ x 8″ x 2″		90	.267		142	10.60		152.60	172
8500	Repair coupling 4″		100	.240		26.50	9.55		36.05	43.50
8520	6″ diameter		90	.267		40.50	10.60		51.10	61
8540	8″ diameter		50	.480		97	19.05		116.05	137
8560	10″ diameter		50	.480		204	19.05		223.05	255
8580	12″ diameter		50	.480		298	19.05		317.05	360
8600	Plug end 4″		100	.240		23	9.55		32.55	39.50
8620	6″ diameter		90	.267		41	10.60		51.60	61.50
8640	8″ diameter		50	.480		69.50	19.05		88.55	106
8660	10″ diameter		50	.480		97.50	19.05		116.55	137
8680	12″ diameter	↓	50	.480	↓	119	19.05		138.05	161
8700	PVC pipe, joint restraint									
8710	4″ diameter	B-20A	32	1	Ea.	43.50	43.50		87	114
8720	6″ diameter		25.60	1.250		53.50	54		107.50	142
8730	8″ diameter		21.33	1.500		76.50	65		141.50	183
8740	10″ diameter		18.28	1.751		136	76		212	265
8750	12″ diameter		16.84	1.900		143	82.50		225.50	282
8760	14″ diameter		16	2		220	86.50		306.50	375
8770	16″ diameter		11.64	2.749		274	119		393	480
8780	18″ diameter		11.03	2.901		330	126		456	555
8785	20″ diameter		9.14	3.501		400	152		552	670
8790	24″ diameter	↓	7.53	4.250	↓	465	184		649	790

33 11 13.35 Water Supply, HDPE		Crew	Daily Output	Labor-Hours	Unit	Material	Labor	Equipment	Total	Total Incl O&P
0010	**WATER SUPPLY, HDPE**									
0011	Butt fusion joints, SDR 21 40′ lengths not including excavation or backfill									
0100	4″ diameter	B-22A	400	.100	L.F.	3.75	4.07	1.64	9.46	12.20
0200	6″ diameter		380	.105		8.45	4.28	1.73	14.46	17.70
0300	8″ diameter		320	.125		10.90	5.10	2.05	18.05	22
0400	10″ diameter		300	.133		19.40	5.40	2.19	26.99	32
0500	12″ diameter	↓	260	.154		20	6.25	2.53	28.78	34.50
0600	14″ diameter	B-22B	220	.182		36	7.40	5.05	48.45	56.50
0700	16″ diameter		180	.222		41	9.05	6.15	56.20	65.50
0800	18″ diameter		140	.286		46	11.60	7.90	65.50	77
0900	24″ diameter	↓	100	.400	↓	66	16.25	11.05	93.30	110
1000	Fittings									
1100	Elbows, 90 degrees									
1200	4″ diameter	B-22A	32	1.250	Ea.	25	51	20.50	96.50	128
1300	6″ diameter		28	1.429		58	58	23.50	139.50	179
1400	8″ diameter		24	1.667		165	68	27.50	260.50	315
1500	10″ diameter		18	2.222		545	90.50	36.50	672	775

33 11 Water Utility Distribution Piping

33 11 13 – Public Water Utility Distribution Piping

33 11 13.35 Water Supply, HDPE		Crew	Daily Output	Labor-Hours	Unit	Material	2013 Bare Costs Labor	Equipment	Total	Total Incl O&P
1600	12″ diameter	B-22A	12	3.333	Ea.	545	136	54.50	735.50	870
1700	14″ diameter	B-22B	9	4.444		750	181	123	1,054	1,225
1800	16″ diameter		6	6.667		860	271	184	1,315	1,575
1900	18″ diameter		4	10		975	405	277	1,657	2,000
2000	24″ diameter		3	13.333		1,325	540	370	2,235	2,675
2100	Tees									
2200	4″ diameter	B-22A	30	1.333	Ea.	28	54	22	104	138
2300	6″ diameter		26	1.538		67.50	62.50	25.50	155.50	199
2400	8″ diameter		22	1.818		215	74	30	319	385
2500	10″ diameter		15	2.667		284	108	44	436	525
2600	12″ diameter		10	4		895	163	65.50	1,123.50	1,300
2700	14″ diameter	B-22B	8	5		1,050	203	138	1,391	1,600
2800	16″ diameter		6	6.667		1,250	271	184	1,705	2,000
2900	18″ diameter		4	10		1,400	405	277	2,082	2,450
3000	24″ diameter		2	20		1,850	815	555	3,220	3,900
4100	Caps									
4110	4″ diameter	B-22A	34	1.176	Ea.	17.45	48	19.30	84.75	114
4120	6″ diameter		30	1.333		40.50	54	22	116.50	152
4130	8″ diameter		26	1.538		70	62.50	25.50	158	201
4150	10″ diameter		20	2		310	81.50	33	424.50	500
4160	12″ diameter		14	2.857		315	116	47	478	575

33 11 13.45 Water Supply, Copper Pipe		Crew	Daily Output	Labor-Hours	Unit	Material	Labor	Equipment	Total	Total Incl O&P
0010	**WATER SUPPLY, COPPER PIPE**									
0020	Not including excavation or backfill									
2000	Tubing, type K, 20′ joints, 3/4″ diameter	Q-1	400	.040	L.F.	8.70	2.01		10.71	12.60
2200	1″ diameter		320	.050		11.55	2.51		14.06	16.55
3000	1-1/2″ diameter		265	.060		18.45	3.03		21.48	25
3020	2″ diameter		230	.070		28.50	3.49		31.99	36.50
3040	2-1/2″ diameter		146	.110		42	5.50		47.50	54.50
3060	3″ diameter		134	.119		58	6		64	73
4012	4″ diameter		95	.168		96	8.45		104.45	119
4016	6″ diameter	Q-2	80	.300		239	15.60		254.60	286
5000	Tubing, type L									
5108	2″ diameter	Q-1	230	.070	L.F.	23.50	3.49		26.99	31.50
6010	3″ diameter		134	.119		46	6		52	60
6012	4″ diameter		95	.168		76.50	8.45		84.95	97
6016	6″ diameter	Q-2	80	.300		209	15.60		224.60	254
7020	Fittings, brass, corporation stops, 3/4″ diameter	1 Plum	19	.421	Ea.	38	23.50		61.50	77.50
7040	1″ diameter		16	.500		45	28		73	91.50
7060	1-1/2″ diameter		13	.615		140	34.50		174.50	206
7080	2″ diameter		11	.727		210	40.50		250.50	292
7100	Curb stops, 3/4″ diameter		19	.421		44	23.50		67.50	84
7120	1″ diameter		16	.500		72	28		100	122
7140	1-1/2″ diameter		13	.615		189	34.50		223.50	259
7160	2″ diameter		11	.727		278	40.50		318.50	365
7165	Fittings, brass, corporation stops, no lead, 3/4″ diameter		19	.421		74	23.50		97.50	117
7166	1″ diameter		16	.500		98	28		126	150
7167	1-1/2″ diameter		13	.615		197	34.50		231.50	269
7168	2″ diameter		11	.727		330	40.50		370.50	425
7170	Curb stops, no lead, 3/4″ diameter		19	.421		100	23.50		123.50	146
7171	1″ diameter		16	.500		151	28		179	208
7172	1-1/2″ diameter		13	.615		225	34.50		259.50	300

33 11 Water Utility Distribution Piping

33 11 13 – Public Water Utility Distribution Piping

33 11 13.45 Water Supply, Copper Pipe		Crew	Daily Output	Labor-Hours	Unit	Material	2013 Bare Costs Labor	Equipment	Total	Total Incl O&P
7173	2″ diameter	1 Plum	11	.727	Ea.	230	40.50		270.50	315

33 12 Water Utility Distribution Equipment

33 12 13 – Water Service Connections

33 12 13.15 Tapping, Crosses and Sleeves

		Crew	Daily Output	Labor-Hours	Unit	Material	Labor	Equipment	Total	Total Incl O&P
0010	**TAPPING, CROSSES AND SLEEVES**									
4000	Drill and tap pressurized main (labor only)									
4100	6″ main, 1″ to 2″ service	Q-1	3	5.333	Ea.		268		268	405
4150	8″ main, 1″ to 2″ service	″	2.75	5.818	″		292		292	440
4500	Tap and insert gate valve									
4600	8″ main, 4″ branch	B-21	3.20	8.750	Ea.		360	43	403	595
4650	6″ branch		2.70	10.370			425	51	476	710
4700	10″ Main, 4″ branch		2.70	10.370			425	51	476	710
4750	6″ branch		2.35	11.915			490	58.50	548.50	815
4800	12″ main, 6″ branch		2.35	11.915			490	58.50	548.50	815
4850	8″ branch		2.35	11.915			490	58.50	548.50	815
7020	Crosses, 4″ x 4″		37	.757		1,400	31	3.71	1,434.71	1,600
7030	6″ x 4″		25	1.120		1,200	46	5.50	1,251.50	1,400
7040	6″ x 6″		25	1.120		1,200	46	5.50	1,251.50	1,400
7060	8″ x 6″		21	1.333		1,450	54.50	6.55	1,511.05	1,700
7080	8″ x 8″		21	1.333		1,525	54.50	6.55	1,586.05	1,775
7100	10″ x 6″		21	1.333		2,400	54.50	6.55	2,461.05	2,725
7120	10″ x 10″		21	1.333		2,550	54.50	6.55	2,611.05	2,925
7140	12″ x 6″		18	1.556		2,400	64	7.60	2,471.60	2,725
7160	12″ x 12″		18	1.556		3,325	64	7.60	3,396.60	3,775
7180	14″ x 6″		16	1.750		6,500	72	8.60	6,580.60	7,275
7200	14″ x 14″		16	1.750		6,875	72	8.60	6,955.60	7,675
7220	16″ x 6″		14	2		6,875	82	9.80	6,966.80	7,675
7240	16″ x 10″		14	2		7,050	82	9.80	7,141.80	7,875
7260	16″ x 16″		14	2		7,425	82	9.80	7,516.80	8,300
7280	18″ x 6″		10	2.800		10,400	115	13.70	10,528.70	11,700
7300	18″ x 12″		10	2.800		10,500	115	13.70	10,628.70	11,700
7320	18″ x 18″		10	2.800		9,575	115	13.70	9,703.70	10,700
7340	20″ x 6″		8	3.500		8,425	144	17.15	8,586.15	9,500
7360	20″ x 12″		8	3.500		9,025	144	17.15	9,186.15	10,200
7380	20″ x 20″		8	3.500		13,200	144	17.15	13,361.15	14,700
7400	24″ x 6″		6	4.667		10,900	191	23	11,114	12,300
7420	24″ x 12″		6	4.667		11,000	191	23	11,214	12,400
7440	24″ x 18″		6	4.667		16,500	191	23	16,714	18,400
7460	24″ x 24″		6	4.667		17,100	191	23	17,314	19,100
7600	Cut-in sleeves with rubber gaskets, 4″		18	1.556		258	64	7.60	329.60	390
7620	6″		12	2.333		310	95.50	11.45	416.95	500
7640	8″		10	2.800		405	115	13.70	533.70	635
7660	10″		10	2.800		515	115	13.70	643.70	760
7680	12″		9	3.111		690	128	15.25	833.25	970
7800	Cut-in valves with rubber gaskets, 4″		18	1.556		370	64	7.60	441.60	515
7820	6″		12	2.333		500	95.50	11.45	606.95	710
7840	8″		10	2.800		775	115	13.70	903.70	1,050
7860	10″		10	2.800		810	115	13.70	938.70	1,075
7880	12″		9	3.111		965	128	15.25	1,108.25	1,275
7900	Tapping Valve 4 inch, MJ, ductile iron		18	1.556		645	64	7.60	716.60	815

33 12 13 – Water Service Connections

33 12 13.15 Tapping, Crosses and Sleeves		Crew	Daily Output	Labor-Hours	Unit	Material	Labor	Equipment	Total	Total Incl O&P
						2013 Bare Costs				
7920	6 inch, MJ, ductile iron	B-21	12	2.333	Ea.	875	95.50	11.45	981.95	1,125
8000	Sleeves with rubber gaskets, 4″ x 4″		37	.757		820	31	3.71	854.71	950
8010	6″ x 4″		25	1.120		800	46	5.50	851.50	955
8020	6″ x 6″		25	1.120		895	46	5.50	946.50	1,050
8030	8″ x 4″		21	1.333		810	54.50	6.55	871.05	980
8040	8″ x 6″		21	1.333		975	54.50	6.55	1,036.05	1,175
8060	8″ x 8″		21	1.333		1,275	54.50	6.55	1,336.05	1,500
8070	10″ x 4″		21	1.333		885	54.50	6.55	946.05	1,075
8080	10″ x 6″		21	1.333		1,050	54.50	6.55	1,111.05	1,250
8090	10″ x 8″		21	1.333		1,250	54.50	6.55	1,311.05	1,475
8100	10″ x 10″		21	1.333		1,950	54.50	6.55	2,011.05	2,225
8110	12″ x 4″		18	1.556		905	64	7.60	976.60	1,100
8120	12″ x 6″		18	1.556		1,150	64	7.60	1,221.60	1,350
8130	12″ x 8″		18	1.556		1,400	64	7.60	1,471.60	1,650
8135	12″ x 10″		18	1.556		1,975	64	7.60	2,046.60	2,275
8140	12″ x 12″		18	1.556		2,775	64	7.60	2,846.60	3,150
8160	14″ x 6″		16	1.750		1,150	72	8.60	1,230.60	1,400
8180	14″ x 14″		16	1.750		3,150	72	8.60	3,230.60	3,575
8200	16″ x 6″		14	2		1,100	82	9.80	1,191.80	1,325
8220	16″ x 10″		14	2		1,900	82	9.80	1,991.80	2,225
8240	16″ x 16″		14	2		6,775	82	9.80	6,866.80	7,575
8260	18″ x 6″		10	2.800		1,950	115	13.70	2,078.70	2,350
8280	18″ x 12″		10	2.800		3,350	115	13.70	3,478.70	3,875
8300	18″ x 18″		10	2.800		9,200	115	13.70	9,328.70	10,300
8320	20″ x 6″		8	3.500		1,075	144	17.15	1,236.15	1,425
8340	20″ x 12″		8	3.500		2,825	144	17.15	2,986.15	3,350
8360	20″ x 20″		8	3.500		11,000	144	17.15	11,161.15	12,300
8380	24″ x 6″		6	4.667		1,175	191	23	1,389	1,625
8400	24″ x 12″		6	4.667		2,900	191	23	3,114	3,525
8420	24″ x 18″		6	4.667		9,875	191	23	10,089	11,200
8440	24″ x 24″		6	4.667		14,400	191	23	14,614	16,200
8800	Hydrant valve box, 6′ long	B-20	20	1.200		160	47.50		207.50	250
8820	8′ long		18	1.333		225	53		278	330
8830	Valve box w/lid 4′ deep		14	1.714		103	68		171	218
8840	Valve box and large base w/lid		14	1.714		305	68		373	440

33 12 16 – Water Utility Distribution Valves

33 12 16.10 Valves		Crew	Daily Output	Labor-Hours	Unit	Material	Labor	Equipment	Total	Total Incl O&P
0010	**VALVES**, water distribution									
0011	See Sections 22 05 23.20 and 22 05 23.60									
3000	Butterfly valves with boxes, cast iron, mech. jt.									
3100	4″ diameter	B-6	6	4	Ea.	785	156	61	1,002	1,175
3140	6″ diameter		6	4		910	156	61	1,127	1,300
3180	8″ diameter		6	4		1,075	156	61	1,292	1,475
3300	10″ diameter		6	4		1,525	156	61	1,742	2,000
3340	12″ diameter		6	4		1,975	156	61	2,192	2,450
3400	14″ diameter		4	6		4,000	233	92	4,325	4,850
3440	16″ diameter		4	6		4,600	233	92	4,925	5,525
3460	18″ diameter		4	6		6,125	233	92	6,450	7,175
3480	20″ diameter		4	6		7,275	233	92	7,600	8,450
3500	24″ diameter		4	6		9,425	233	92	9,750	10,900
3600	With lever operator									
3610	4″ diameter	B-6	6	4	Ea.	685	156	61	902	1,050

33 12 16 – Water Utility Distribution Valves

33 12 16.10 Valves		Crew	Daily Output	Labor-Hours	Unit	Material	Labor	Equipment	Total	Total Incl O&P
						2013 Bare Costs				
3614	6" diameter	B-6	6	4	Ea.	805	156	61	1,022	1,200
3616	8" diameter		6	4		975	156	61	1,192	1,375
3618	10" diameter		6	4		1,425	156	61	1,642	1,875
3620	12" diameter		6	4		1,850	156	61	2,067	2,350
3622	14" diameter		4	6		3,700	233	92	4,025	4,525
3624	16" diameter		4	6		4,300	233	92	4,625	5,175
3626	18" diameter		4	6		5,825	233	92	6,150	6,850
3628	20" diameter		4	6		6,975	233	92	7,300	8,125
3630	24" diameter		4	6		9,125	233	92	9,450	10,500
3700	Check valves, flanged									
3710	4" diameter	B-6	6	4	Ea.	900	156	61	1,117	1,300
3714	6" diameter		6	4		1,225	156	61	1,442	1,650
3716	8" diameter		6	4		1,725	156	61	1,942	2,200
3718	10" diameter		6	4		3,150	156	61	3,367	3,775
3720	12" diameter		6	4		4,650	156	61	4,867	5,400
3722	14" diameter		4	6		6,500	233	92	6,825	7,600
3724	16" diameter		4	6		8,000	233	92	8,325	9,250
3726	18" diameter		4	6		11,900	233	92	12,225	13,500
3728	20" diameter		4	6		13,800	233	92	14,125	15,700
3730	24" diameter		4	6		19,700	233	92	20,025	22,200
3800	Gate valves, C.I., 250 PSI, mechanical joint, w/boxes									
3810	4" diameter	B-6	6	4	Ea.	420	156	61	637	770
3814	6" diameter		6	4		635	156	61	852	1,000
3816	8" diameter		6	4		825	156	61	1,042	1,200
3818	10" diameter		6	4		1,100	156	61	1,317	1,525
3820	12" diameter		6	4		1,400	156	61	1,617	1,850
3822	14" diameter		4	6		4,350	233	92	4,675	5,225
3824	16" diameter		4	6		4,725	233	92	5,050	5,625
3826	18" diameter		4	6		7,650	233	92	7,975	8,875
3828	20" diameter		4	6		10,800	233	92	11,125	12,400
3830	24" diameter		4	6		13,500	233	92	13,825	15,400
3831	30" diameter		4	6		35,600	233	92	35,925	39,600
3832	36" diameter		4	6		52,000	233	92	52,325	57,500
3880	Sleeve, for tapping mains, 8" x 4", add					810			810	890
3884	10" x 6", add					1,050			1,050	1,150
3888	12" x 6", add					1,150			1,150	1,250
3892	12" x 8", add					1,400			1,400	1,550

33 12 16.20 Valves		Crew	Daily Output	Labor-Hours	Unit	Material	Labor	Equipment	Total	Total Incl O&P
0010	**VALVES**									
0011	Special trim or use									
9000	Valves, gate valve, N.R.S. PIV with post, 4" diameter	B-6	6	4	Ea.	1,475	156	61	1,692	1,900
9020	6" diameter		6	4		1,675	156	61	1,892	2,150
9040	8" diameter		6	4		1,875	156	61	2,092	2,350
9060	10" diameter		6	4		2,150	156	61	2,367	2,650
9080	12" diameter		6	4		2,450	156	61	2,667	3,000
9100	14" diameter		6	4		5,200	156	61	5,417	6,000
9120	OS&Y, 4" diameter		6	4		620	156	61	837	985
9140	6" diameter		6	4		885	156	61	1,102	1,275
9160	8" diameter		6	4		1,450	156	61	1,667	1,900
9180	10" diameter		6	4		2,375	156	61	2,592	2,900
9200	12" diameter		6	4		3,500	156	61	3,717	4,150
9220	14" diameter		4	6		4,700	233	92	5,025	5,625

33 12 16 – Water Utility Distribution Valves

33 12 16.20 Valves		Crew	Daily Output	Labor-Hours	Unit	Material	2013 Bare Costs Labor	Equipment	Total	Total Incl O&P
9400	Check valves, rubber disc, 2-1/2″ diameter	B-6	6	4	Ea.	740	156	61	957	1,125
9420	3″ diameter		6	4		750	156	61	967	1,125
9440	4″ diameter		6	4		900	156	61	1,117	1,300
9480	6″ diameter		6	4		1,225	156	61	1,442	1,650
9500	8″ diameter		6	4		1,725	156	61	1,942	2,200
9520	10″ diameter		6	4		3,150	156	61	3,367	3,775
9540	12″ diameter		6	4		4,650	156	61	4,867	5,400
9542	14″ diameter		4	6		6,500	233	92	6,825	7,600
9700	Detector check valves, reducing, 4″ diameter		6	4		1,350	156	61	1,567	1,775
9720	6″ diameter		6	4		1,800	156	61	2,017	2,275
9740	8″ diameter		6	4		2,625	156	61	2,842	3,175
9760	10″ diameter		6	4		4,650	156	61	4,867	5,425
9800	Galvanized, 4″ diameter		6	4		1,900	156	61	2,117	2,400
9820	6″ diameter		6	4		2,600	156	61	2,817	3,150
9840	8″ diameter		6	4		4,050	156	61	4,267	4,775
9860	10″ diameter	↓	6	4	↓	4,425	156	61	4,642	5,150

33 12 19 – Water Utility Distribution Fire Hydrants

33 12 19.10 Fire Hydrants		Crew	Daily Output	Labor-Hours	Unit	Material	Labor	Equipment	Total	Total Incl O&P
0010	**FIRE HYDRANTS** G3010–410									
0020	Mechanical joints unless otherwise noted									
1000	Fire hydrants, two way; excavation and backfill not incl.									
1100	4-1/2″ valve size, depth 2′-0″	B-21	10	2.800	Ea.	1,525	115	13.70	1,653.70	1,875
1200	4′-6″		9	3.111		1,825	128	15.25	1,968.25	2,250
1260	6′-0″		7	4		1,925	164	19.60	2,108.60	2,400
1340	8′-0″		6	4.667		2,075	191	23	2,289	2,600
1420	10′-0″		5	5.600		2,200	230	27.50	2,457.50	2,800
2000	5-1/4″ valve size, depth 2′-0″		10	2.800		1,750	115	13.70	1,878.70	2,125
2080	4′-0″		9	3.111		1,850	128	15.25	1,993.25	2,275
2160	6′-0″		7	4		2,025	164	19.60	2,208.60	2,500
2240	8′-0″		6	4.667		2,125	191	23	2,339	2,650
2320	10′-0″	↓	5	5.600		2,450	230	27.50	2,707.50	3,050
2350	For threeway valves, add					7%				
2400	Lower barrel extensions with stems, 1′-0″	B-20	14	1.714		335	68		403	470
2440	2′-0″		13	1.846		405	73.50		478.50	560
2480	3′-0″		12	2		765	79.50		844.50	960
2520	4′-0″	↓	10	2.400	↓	770	95.50		865.50	990
5000	Indicator post									
5020	Adjustable, valve size 4″ to 14″, 4′ bury	B-21	10	2.800	Ea.	820	115	13.70	948.70	1,100
5060	8′ bury		7	4		1,150	164	19.60	1,333.60	1,525
5080	10′ bury		6	4.667		1,200	191	23	1,414	1,650
5100	12′ bury		5	5.600		1,450	230	27.50	1,707.50	1,975
5120	14′ bury		4	7		1,575	287	34.50	1,896.50	2,200
5500	Non-adjustable, valve size 4″ to 14″, 3′ bury		10	2.800		820	115	13.70	948.70	1,100
5520	3′-6″ bury		10	2.800		820	115	13.70	948.70	1,100
5540	4′ bury	↓	9	3.111	↓	820	128	15.25	963.25	1,125

33 16 Water Utility Storage Tanks

33 16 13 – Aboveground Water Utility Storage Tanks

33 16 13.16 Prestressed Conc. Water Storage Tanks		Crew	Daily Output	Labor-Hours	Unit	2013 Bare Costs Material	Labor	Equipment	Total	Total Incl O&P
0010	**PRESTRESSED CONC. WATER STORAGE TANKS**									
0020	Not including fdn., pipe or pumps, 250,000 gallons				Ea.				294,500	324,500

33 21 Water Supply Wells

33 21 13 – Public Water Supply Wells

33 21 13.10 Wells and Accessories

33 21 13.10 Wells and Accessories		Crew	Daily Output	Labor-Hours	Unit	2013 Bare Costs Material	Labor	Equipment	Total	Total Incl O&P
0010	**WELLS & ACCESSORIES**									
0011	Domestic									
0100	Drilled, 4″ to 6″ diameter	B-23	120	.333	L.F.		11.95	23.50	35.45	44
0200	8″ diameter	″	95.20	.420	″		15.05	29.50	44.55	55.50
0400	Gravel pack well, 40′ deep, incl. gravel & casing, complete									
0500	24″ diameter casing x 18″ diameter screen	B-23	.13	307	Total	38,100	11,000	21,500	70,600	82,500
0600	36″ diameter casing x 18″ diameter screen		.12	333	″	39,300	12,000	23,300	74,600	87,000
0800	Observation wells, 1-1/4″ riser pipe		163	.245	V.L.F.	20	8.80	17.15	45.95	54.50
0900	For flush Buffalo roadway box, add	1 Skwk	16.60	.482	Ea.	50	22.50		72.50	89.50
1200	Test well, 2-1/2″ diameter, up to 50′ deep (15 to 50 GPM)	B-23	1.51	26.490	″	785	950	1,850	3,585	4,375
1300	Over 50′ deep, add	″	121.80	.328	L.F.	21	11.75	23	55.75	66.50
1500	Pumps, installed in wells to 100′ deep, 4″ submersible									
1510	1/2 H.P.	Q-1	3.22	4.969	Ea.	465	250		715	885
1520	3/4 H.P.		2.66	6.015		600	300		900	1,125
1600	1 H.P.		2.29	6.987		650	350		1,000	1,250
1700	1-1/2 H.P.	Q-22	1.60	10		820	500	405	1,725	2,100
1800	2 H.P.		1.33	12.030		890	605	485	1,980	2,425
1900	3 H.P.		1.14	14.035		1,400	705	565	2,670	3,225
2000	5 H.P.		1.14	14.035		2,350	705	565	3,620	4,250
3000	Pump, 6″ submersible, 25′ to 150′ deep, 25 H.P., 249 to 297 GPM		.89	17.978		7,650	905	725	9,280	10,600
3100	25′ to 500′ deep, 30 H.P., 100 to 300 GPM		.73	21.918		10,400	1,100	885	12,385	14,100
8000	Steel well casing	B-23A	3020	.008	Lb.	.92	.32	.87	2.11	2.46
8110	Well screen assembly, stainless steel, 2″ diameter		273	.088	L.F.	80	3.53	9.60	93.13	104
8120	3″ diameter		253	.095		140	3.81	10.35	154.16	171
8130	4″ diameter		200	.120		175	4.82	13.10	192.92	215
8140	5″ diameter		168	.143		185	5.75	15.60	206.35	230
8150	6″ diameter		126	.190		210	7.65	21	238.65	266
8160	8″ diameter		98.50	.244		285	9.80	26.50	321.30	360
8170	10″ diameter		73	.329		355	13.20	36	404.20	450
8180	12″ diameter		62.50	.384		405	15.40	42	462.40	515
8190	14″ diameter		54.30	.442		455	17.75	48.50	521.25	580
8200	16″ diameter		48.30	.497		505	19.95	54.50	579.45	645
8210	18″ diameter		39.20	.612		630	24.50	67	721.50	805
8220	20″ diameter		31.20	.769		705	31	84	820	915
8230	24″ diameter		23.80	1.008		845	40.50	110	995.50	1,125
8240	26″ diameter		21	1.143		920	46	125	1,091	1,200
8300	Slotted PVC, 1-1/4″ diameter		521	.046		2.34	1.85	5.05	9.24	10.95
8310	1-1/2″ diameter		488	.049		2.85	1.97	5.40	10.22	12.05
8320	2″ diameter		273	.088		3.90	3.53	9.60	17.03	20
8330	3″ diameter		253	.095		5.70	3.81	10.35	19.86	23.50
8340	4″ diameter		200	.120		6.40	4.82	13.10	24.32	29
8350	5″ diameter		168	.143		13.40	5.75	15.60	34.75	40.50
8360	6″ diameter		126	.190		15.90	7.65	21	44.55	52
8370	8″ diameter		98.50	.244		25	9.80	26.50	61.30	72
8400	Artificial gravel pack, 2″ screen, 6″ casing	B-23B	174	.138		3.50	5.55	17.05	26.10	31

33 21 Water Supply Wells

33 21 13 – Public Water Supply Wells

33 21 13.10 Wells and Accessories		Crew	Daily Output	Labor-Hours	Unit	Material	2013 Bare Costs Labor	Equipment	Total	Total Incl O&P
8405	8″ casing	B-23B	111	.216	L.F.	4.50	8.70	26.50	39.70	47.50
8410	10″ casing		74.50	.322		6.40	12.95	40	59.35	71
8415	12″ casing		60	.400		9	16.05	49.50	74.55	89
8420	14″ casing		50.20	.478		10.45	19.20	59	88.65	106
8425	16″ casing		40.70	.590		13.25	23.50	73	109.75	131
8430	18″ casing		36	.667		15.90	27	82.50	125.40	149
8435	20″ casing		29.50	.814		18.70	32.50	101	152.20	182
8440	24″ casing		25.70	.934		21.50	37.50	115	174	208
8445	26″ casing		24.60	.976		23.50	39	121	183.50	219
8450	30″ casing		20	1.200		27	48	148	223	267
8455	36″ casing		16.40	1.463		30.50	58.50	181	270	325
8500	Develop well		8	3	Hr.	320	120	370	810	950
8550	Pump test well		8	3		84.50	120	370	574.50	685
8560	Standby well	B-23A	8	3		83	120	330	533	635
8570	Standby, drill rig		8	3			120	330	450	545
8580	Surface seal well, concrete filled		1	24	Ea.	830	965	2,625	4,420	5,250
8590	Well test pump, install & remove	B-23	1	40			1,425	2,800	4,225	5,275
8600	Well sterilization, chlorine	2 Clab	1	16		125	565		690	1,025
8610	Well water pressure switch	1 Clab	12	.667		48	23.50		71.50	89.50
8630	Well, water pressure switch with manual reset	″	12	.667		34	23.50		57.50	73.50
9950	See Section 31 23 19.40 for wellpoints									
9960	See Section 31 23 19.30 for drainage wells									

33 21 13.20 Water Supply Wells, Pumps

		Crew	Daily Output	Labor-Hours	Unit	Material	Labor	Equipment	Total	Total Incl O&P
0010	**WATER SUPPLY WELLS, PUMPS**									
0011	With pressure control									
1000	Deep well, jet, 42 gal. galvanized tank									
1040	3/4 HP	1 Plum	.80	10	Ea.	1,400	560		1,960	2,400
3000	Shallow well, jet, 30 gal. galvanized tank									
3040	1/2 HP	1 Plum	2	4	Ea.	1,100	223		1,323	1,550

33 31 Sanitary Utility Sewerage Piping

33 31 13 – Public Sanitary Utility Sewerage Piping

33 31 13.10 Sewage Collection, Valves

		Crew	Daily Output	Labor-Hours	Unit	Material	Labor	Equipment	Total	Total Incl O&P
0010	**SEWAGE COLLECTION, VALVES**									
1000	Backwater sewer line valve									
1010	Offset type, bronze swing check assy and bronze cover									
1020	B&S connections									
1040	2″ size	2 Skwk	14	1.143	Ea.	530	53		583	665
1050	3″ size		12.50	1.280		825	59		884	1,000
1060	4″ size		11	1.455		1,075	67		1,142	1,275
1070	6″ size	3 Skwk	15	1.600		1,525	74		1,599	1,800
1080	8″ size	″	7.50	3.200		2,150	148		2,298	2,600
1110	Backwater drainage control									
1120	Offset type, w/bronze manually operated shear gate									
1130	B&S connections									
1160	4″ size	2 Skwk	11	1.455	Ea.	1,775	67		1,842	2,075
1170	6″ size	3 Skwk	15	1.600	″	2,400	74		2,474	2,750

33 31 Sanitary Utility Sewerage Piping

33 31 13 – Public Sanitary Utility Sewerage Piping

33 31 13.15 Sewage Collection, Concrete Pipe		Crew	Daily Output	Labor-Hours	Unit	Material	2013 Bare Costs Labor	Equipment	Total	Total Incl O&P
0010	SEWAGE COLLECTION, CONCRETE PIPE									
0020	See Section 33 41 13.60 for sewage/drainage collection, concrete pipe									

33 31 13.25 Sewage Collection, Polyvinyl Chloride Pipe		Crew	Daily Output	Labor-Hours	Unit	Material	Labor	Equipment	Total	Total Incl O&P
0010	SEWAGE COLLECTION, POLYVINYL CHLORIDE PIPE									
0020	Not including excavation or backfill									
2000	20′ lengths, SDR 35, B&S, 4″ diameter	B-20	375	.064	L.F.	1.45	2.54		3.99	5.50
2040	6″ diameter	"	350	.069		3.26	2.72		5.98	7.80
2120	10″ diameter	B-21	330	.085		11.40	3.48	.42	15.30	18.35
2160	12″ diameter		320	.088		12.75	3.59	.43	16.77	20
2200	15″ diameter		240	.117		14.30	4.78	.57	19.65	23.50
4000	Piping, DWV PVC, no exc./bkfill., 10′ L, Sch 40, 4″ diameter	B-20	375	.064		4.37	2.54		6.91	8.75
4010	6″ diameter		350	.069		9.40	2.72		12.12	14.50
4020	8″ diameter		335	.072		17.60	2.84		20.44	23.50

33 36 Utility Septic Tanks

33 36 13 – Utility Septic Tank and Effluent Wet Wells

33 36 13.13 Concrete Utility Septic Tank		Crew	Daily Output	Labor-Hours	Unit	Material	Labor	Equipment	Total	Total Incl O&P
0010	CONCRETE UTILITY SEPTIC TANK G3020-300									
0011	Not including excavation or piping									
0015	Septic tanks, precast, 1,000 gallon	B-21	8	3.500	Ea.	995	144	17.15	1,156.15	1,350
0060	1,500 gallon		7	4		1,600	164	19.60	1,783.60	2,050
0100	2,000 gallon		5	5.600		2,050	230	27.50	2,307.50	2,650
0200	5,000 gallon	B-13	3.50	16		9,375	620	208	10,203	11,500
0300	15,000 gallon, 4 piece	B-13B	1.70	32.941		21,500	1,275	655	23,430	26,400
0400	25,000 gallon, 4 piece		1.10	50.909		41,800	1,975	1,025	44,800	50,000
0500	40,000 gallon, 4 piece		.80	70		53,500	2,700	1,400	57,600	64,500
0520	50,000 gallon, 5 piece	B-13C	.60	93.333		61,500	3,600	2,775	67,875	76,000
0640	75,000 gallon, cast in place	C-14C	.25	448		75,000	19,200	130	94,330	112,000
0660	100,000 gallon	"	.15	746		92,500	32,000	217	124,717	151,500
1150	Leaching field chambers, 13′ x 3′-7″ x 1′-4″, standard	B-13	16	3.500		485	135	45.50	665.50	785
1200	Heavy duty, 8′ x 4′ x 1′-6″		14	4		286	154	52	492	610
1300	13′ x 3′-9″ x 1′-6″		12	4.667		1,075	180	60.50	1,315.50	1,550
1350	20′ x 4′ x 1′-6″		5	11.200		1,150	430	146	1,726	2,075
1400	Leaching pit, precast concrete, 3′ diameter, 3′ deep	B-21	8	3.500		680	144	17.15	841.15	985
1500	6′ diameter, 3′ section		4.70	5.957		835	244	29	1,108	1,325
2000	Velocity reducing pit, precast conc., 6′ diameter, 3′ deep		4.70	5.957		1,575	244	29	1,848	2,150

33 36 13.19 Polyethylene Utility Septic Tank		Crew	Daily Output	Labor-Hours	Unit	Material	Labor	Equipment	Total	Total Incl O&P
0010	POLYETHYLENE UTILITY SEPTIC TANK									
0015	High density polyethylene, 1,000 gallon	B-21	8	3.500	Ea.	1,150	144	17.15	1,311.15	1,525
0020	1,250 gallon		8	3.500		1,375	144	17.15	1,536.15	1,775
0025	1,500 gallon		7	4		1,650	164	19.60	1,833.60	2,100

33 36 19 – Utility Septic Tank Effluent Filter

33 36 19.13 Utility Septic Tank Effluent Tube Filter		Crew	Daily Output	Labor-Hours	Unit	Material	Labor	Equipment	Total	Total Incl O&P
0010	UTILITY SEPTIC TANK EFFLUENT TUBE FILTER									
3000	Effluent filter, 4″ diameter	1 Skwk	8	1	Ea.	51	46		97	128
3020	6″ diameter	"	7	1.143	"	206	53		259	310

33 36 Utility Septic Tanks

33 36 33 – Utility Septic Tank Drainage Field

33 36 33.13 Utility Septic Tank Tile Drainage Field

	33 36 33.13 Utility Septic Tank Tile Drainage Field	Crew	Daily Output	Labor-Hours	Unit	2013 Bare Costs Material	Labor	Equipment	Total	Total Incl O&P
0010	**UTILITY SEPTIC TANK TILE DRAINAGE FIELD**									
0015	Distribution box, concrete, 5 outlets	B-21	20	1.400	Ea.	76.50	57.50	6.85	140.85	180
0020	7 outlets	2 Clab	16	1		76.50	35.50		112	139
0025	9 outlets		8	2		470	71		541	630
0115	Distribution boxes, HDPE, 5 outlets		20	.800		46.50	28.50		75	94.50
0120	8 outlets		10	1.600		58	56.50		114.50	152
0240	Distribution boxes, Outlet Flow Leveler	1 Clab	50	.160		2.20	5.65		7.85	11.15
0300	Precast concrete, galley, 4′ x 4′ x 4′	B-21	16	1.750		240	72	8.60	320.60	385

33 36 50 – Drainage Field Systems

33 36 50.10 Drainage Field Excavation and Fill

	33 36 50.10 Drainage Field Excavation and Fill	Crew	Daily Output	Labor-Hours	Unit	Material	Labor	Equipment	Total	Total Incl O&P
0010	**DRAINAGE FIELD EXCAVATION AND FILL**									
2200	Excavation for septic tank, 3/4 C.Y. backhoe	B-12F	145	.110	C.Y.		4.65	4.79	9.44	12.35
2400	4′ trench for disposal field, 3/4 C.Y. backhoe	″	335	.048	L.F.		2.01	2.07	4.08	5.35
2600	Gravel fill, run of bank	B-6	150	.160	C.Y.	22	6.20	2.45	30.65	36
2800	Crushed stone, 3/4″	″	150	.160	″	41	6.20	2.45	49.65	57

33 41 Storm Utility Drainage Piping

33 41 13 – Public Storm Utility Drainage Piping

33 41 13.40 Piping, Storm Drainage, Corrugated Metal

	33 41 13.40 Piping, Storm Drainage, Corrugated Metal	Crew	Daily Output	Labor-Hours	Unit	Material	Labor	Equipment	Total	Total Incl O&P
0010	**PIPING, STORM DRAINAGE, CORRUGATED METAL**									
0020	Not including excavation or backfill									
2000	Corrugated metal pipe, galvanized									
2020	Bituminous coated with paved invert, 20′ lengths									
2040	8″ diameter, 16 ga.	B-14	330	.145	L.F.	8.60	5.45	1.11	15.16	19
2060	10″ diameter, 16 ga.		260	.185		8.95	6.95	1.41	17.31	22
2080	12″ diameter, 16 ga.		210	.229		11	8.55	1.75	21.30	27
2100	15″ diameter, 16 ga.		200	.240		15.05	9	1.84	25.89	32.50
2120	18″ diameter, 16 ga.		190	.253		16.70	9.50	1.93	28.13	35
2140	24″ diameter, 14 ga.		160	.300		20	11.25	2.30	33.55	42
2160	30″ diameter, 14 ga.	B-13	120	.467		26.50	18	6.05	50.55	63.50
2180	36″ diameter, 12 ga.		120	.467		33.50	18	6.05	57.55	70.50
2200	48″ diameter, 12 ga.		100	.560		50.50	21.50	7.30	79.30	96.50
2220	60″ diameter, 10 ga.	B-13B	75	.747		76	29	14.85	119.85	144
2240	72″ diameter, 8 ga.	″	45	1.244		91	48	25	164	201
2250	End sections, 8″ diameter, 16 ga.	B-14	20	2.400	Ea.	43.50	90	18.35	151.85	206
2255	10″ diameter, 16 ga.		20	2.400		60	90	18.35	168.35	224
2260	12″ diameter, 16 ga.		18	2.667		123	100	20.50	243.50	310
2265	15″ diameter, 16 ga.		18	2.667		205	100	20.50	325.50	400
2270	18″ diameter, 16 ga.		16	3		242	113	23	378	465
2275	24″ diameter, 16 ga.	B-13	16	3.500		284	135	45.50	464.50	570
2280	30″ diameter, 16 ga.		14	4		490	154	52	696	835
2285	36″ diameter, 14 ga.		14	4		730	154	52	936	1,100
2290	48″ diameter, 14 ga.		10	5.600		1,850	216	73	2,139	2,425
2292	60″ diameter, 14 ga.		6	9.333		1,950	360	121	2,431	2,825
2294	72″ diameter, 14 ga.	B-13B	5	11.200		3,125	430	223	3,778	4,350
2300	Bends or elbows, 8″ diameter	B-14	28	1.714		116	64.50	13.10	193.60	241
2320	10″ diameter		25	1.920		144	72	14.70	230.70	285
2340	12″ diameter, 16 ga.		23	2.087		167	78.50	15.95	261.45	320
2342	18″ diameter, 16 ga.		20	2.400		241	90	18.35	349.35	425
2344	24″ diameter, 14 ga.		16	3		330	113	23	466	565

33 41 13.40 Piping, Storm Drainage, Corrugated Metal		Crew	Daily Output	Labor-Hours	Unit	Material	2013 Bare Costs Labor	Equipment	Total	Total Incl O&P
2346	30″ diameter, 14 ga.	B-14	15	3.200	Ea.	410	120	24.50	554.50	665
2348	36″ diameter, 14 ga.	B-13	15	3.733		555	144	48.50	747.50	885
2350	48″ diameter, 12 ga.	"	12	4.667		770	180	60.50	1,010.50	1,200
2352	60″ diameter, 10 ga.	B-13B	10	5.600		1,025	216	112	1,353	1,575
2354	72″ diameter, 10 ga.	"	6	9.333		1,275	360	186	1,821	2,150
2360	Wyes or tees, 8″ diameter	B-14	25	1.920		163	72	14.70	249.70	305
2380	10″ diameter		21	2.286		203	85.50	17.50	306	375
2400	12″ diameter, 16 ga.		19	2.526		241	95	19.35	355.35	430
2410	18″ diameter, 16 ga.		16	3		320	113	23	456	555
2412	24″ diameter, 14 ga.		16	3		480	113	23	616	725
2414	30″ diameter, 14 ga.	B-13	12	4.667		620	180	60.50	860.50	1,025
2416	36″ diameter, 14 ga.		11	5.091		765	197	66	1,028	1,225
2418	48″ diameter, 12 ga.		10	5.600		1,125	216	73	1,414	1,625
2420	60″ diameter, 10 ga.	B-13B	8	7		1,650	270	139	2,059	2,375
2422	72″ diameter, 10 ga.	"	5	11.200		1,975	430	223	2,628	3,075
2500	Galvanized, uncoated, 20′ lengths									
2520	8″ diameter, 16 ga.	B-14	355	.135	L.F.	7.75	5.05	1.03	13.83	17.50
2540	10″ diameter, 16 ga.		280	.171		8.90	6.45	1.31	16.66	21
2560	12″ diameter, 16 ga.		220	.218		9.90	8.20	1.67	19.77	25.50
2580	15″ diameter, 16 ga.		220	.218		11.90	8.20	1.67	21.77	27.50
2600	18″ diameter, 16 ga.		205	.234		15.05	8.80	1.79	25.64	32
2620	24″ diameter, 14 ga.		175	.274		18	10.30	2.10	30.40	38
2640	30″ diameter, 14 ga.	B-13	130	.431		24	16.65	5.60	46.25	58
2660	36″ diameter, 12 ga.		130	.431		30	16.65	5.60	52.25	64.50
2680	48″ diameter, 12 ga.		110	.509		45.50	19.65	6.60	71.75	87.50
2690	60″ diameter, 10 ga.	B-13B	78	.718		68.50	27.50	14.30	110.30	134
2695	72″ diameter, 10 ga.	"	60	.933		82	36	18.60	136.60	166
2711	Bends or elbows, 12″ diameter, 16 ga.	B-14	30	1.600	Ea.	144	60	12.25	216.25	264
2712	15″ diameter, 16 ga.		25.04	1.917		178	72	14.65	264.65	320
2714	18″ diameter, 16 ga.		20	2.400		200	90	18.35	308.35	380
2716	24″ diameter, 14 ga.		16	3		287	113	23	423	515
2718	30″ diameter, 14 ga.		15	3.200		360	120	24.50	504.50	605
2720	36″ diameter, 14 ga.	B-13	15	3.733		495	144	48.50	687.50	820
2722	48″ diameter, 12 ga.		12	4.667		655	180	60.50	895.50	1,075
2724	60″ diameter, 10 ga.		10	5.600		1,000	216	73	1,289	1,500
2726	72″ diameter, 10 ga.		6	9.333		1,300	360	121	1,781	2,100
2728	Wyes or tees, 12″ diameter, 16 ga.	B-14	22.48	2.135		191	80	16.35	287.35	350
2730	18″ diameter, 16 ga.		15	3.200		279	120	24.50	423.50	515
2732	24″ diameter, 14 ga.		15	3.200		430	120	24.50	574.50	685
2734	30″ diameter, 14 ga.		14	3.429		560	129	26	715	840
2736	36″ diameter, 14 ga.	B-13	14	4		725	154	52	931	1,100
2738	48″ diameter, 12 ga.		12	4.667		1,050	180	60.50	1,290.50	1,500
2740	60″ diameter, 10 ga.		10	5.600		1,500	216	73	1,789	2,050
2742	72″ diameter, 10 ga.		6	9.333		1,800	360	121	2,281	2,650
2780	End sections, 8″ diameter	B-14	35	1.371		66	51.50	10.50	128	164
2785	10″ diameter		35	1.371		70	51.50	10.50	132	168
2790	12″ diameter		35	1.371		103	51.50	10.50	165	205
2800	18″ diameter		30	1.600		126	60	12.25	198.25	244
2810	24″ diameter	B-13	25	2.240		190	86.50	29	305.50	375
2820	30″ diameter		25	2.240		299	86.50	29	414.50	495
2825	36″ diameter		20	2.800		460	108	36.50	604.50	710
2830	48″ diameter		10	5.600		905	216	73	1,194	1,400
2835	60″ diameter	B-13B	5	11.200		1,575	430	223	2,228	2,625

33 41 13.40 Piping, Storm Drainage, Corrugated Metal		Crew	Daily Output	Labor-Hours	Unit	2013 Bare Costs Material	Labor	Equipment	Total	Total Incl O&P
2840	72″ diameter	B-13B	4	14	Ea.	1,875	540	279	2,694	3,175
2850	Couplings, 12″ diameter					10.50			10.50	11.55
2855	18″ diameter					15			15	16.50
2860	24″ diameter					19.50			19.50	21.50
2865	30″ diameter					24.50			24.50	27
2870	36″ diameter					29			29	32
2875	48″ diameter					45			45	49.50
2880	60″ diameter					56			56	61.50
2885	72″ diameter					69			69	76

33 41 13.60 Sewage/Drainage Collection, Concrete Pipe		Crew	Daily Output	Labor-Hours	Unit	Material	Labor	Equipment	Total	Total Incl O&P
0010	**SEWAGE/DRAINAGE COLLECTION, CONCRETE PIPE**									
0020	Not including excavation or backfill									
1000	Non-reinforced pipe, extra strength, B&S or T&G joints									
1010	6″ diameter	B-14	265.04	.181	L.F.	6.60	6.80	1.39	14.79	19.15
1020	8″ diameter		224	.214		7.25	8.05	1.64	16.94	22
1030	10″ diameter		216	.222		8.05	8.35	1.70	18.10	23.50
1040	12″ diameter		200	.240		9.60	9	1.84	20.44	26.50
1050	15″ diameter		180	.267		13.20	10	2.04	25.24	32
1060	18″ diameter		144	.333		15.55	12.50	2.55	30.60	39
1070	21″ diameter		112	.429		19.15	16.10	3.28	38.53	49
1080	24″ diameter		100	.480		25.50	18	3.67	47.17	60
2000	Reinforced culvert, class 3, no gaskets									
2010	12″ diameter	B-14	150	.320	L.F.	11	12	2.45	25.45	33
2020	15″ diameter		150	.320		13.95	12	2.45	28.40	36.50
2030	18″ diameter		132	.364		18.45	13.65	2.78	34.88	44.50
2035	21″ diameter		120	.400		23	15	3.06	41.06	51.50
2040	24″ diameter		100	.480		26	18	3.67	47.67	60.50
2045	27″ diameter	B-13	92	.609		40	23.50	7.90	71.40	88.50
2050	30″ diameter		88	.636		47.50	24.50	8.30	80.30	99
2060	36″ diameter		72	.778		62	30	10.10	102.10	125
2070	42″ diameter	B-13B	72	.778		79.50	30	15.50	125	151
2080	48″ diameter		64	.875		97	34	17.40	148.40	178
2090	60″ diameter		48	1.167		133	45	23	201	241
2100	72″ diameter		40	1.400		204	54	28	286	340
2120	84″ diameter		32	1.750		275	67.50	35	377.50	445
2140	96″ diameter		24	2.333		330	90	46.50	466.50	555
2200	With gaskets, class 3, 12″ diameter	B-21	168	.167		12.10	6.85	.82	19.77	24.50
2220	15″ diameter		160	.175		15.35	7.20	.86	23.41	29
2230	18″ diameter		152	.184		20.50	7.55	.90	28.95	35
2240	24″ diameter		136	.206		32.50	8.45	1.01	41.96	50
2260	30″ diameter	B-13	88	.636		55	24.50	8.30	87.80	107
2270	36″ diameter	″	72	.778		71	30	10.10	111.10	136
2290	48″ diameter	B-13B	64	.875		109	34	17.40	160.40	191
2310	72″ diameter	″	40	1.400		223	54	28	305	360
2330	Flared ends, 6′-1″ long, 12″ diameter	B-21	190	.147		38.50	6.05	.72	45.27	52.50
2340	15″ diameter		155	.181		41.50	7.40	.89	49.79	58
2400	6′-2″ long, 18″ diameter		122	.230		53.50	9.40	1.12	64.02	74.50
2420	24″ diameter		88	.318		56.50	13.05	1.56	71.11	84
2440	36″ diameter	B-13	60	.933		118	36	12.15	166.15	198
3080	Radius pipe, add to pipe prices, 12″ to 60″ diameter					50%				
3090	Over 60″ diameter, add					20%				
3500	Reinforced elliptical, 8′ lengths, C507 class 3									

33 41 Storm Utility Drainage Piping

33 41 13 – Public Storm Utility Drainage Piping

33 41 13.60 Sewage/Drainage Collection, Concrete Pipe		Crew	Daily Output	Labor-Hours	Unit	Material	2013 Bare Costs Labor	Equipment	Total	Total Incl O&P
3520	14″ x 23″ inside, round equivalent 18″ diameter	B-21	82	.341	L.F.	52	14	1.67	67.67	80.50
3530	24″ x 38″ inside, round equivalent 30″ diameter	B-13	58	.966		78	37.50	12.55	128.05	157
3540	29″ x 45″ inside, round equivalent 36″ diameter		52	1.077		93	41.50	14	148.50	181
3550	38″ x 60″ inside, round equivalent 48″ diameter		38	1.474		154	57	19.15	230.15	277
3560	48″ x 76″ inside, round equivalent 60″ diameter		26	2.154		195	83	28	306	375
3570	58″ x 91″ inside, round equivalent 72″ diameter		22	2.545		310	98.50	33	441.50	525
3780	Concrete slotted pipe, class 4 mortar joint									
3800	12″ diameter	B-21	168	.167	L.F.	30	6.85	.82	37.67	44.50
3840	18″ diameter	″	152	.184	″	35	7.55	.90	43.45	51
3900	Concrete slotted pipe, Class 4 O-ring joint									
3940	12″ diameter	B-21	168	.167	L.F.	30	6.85	.82	37.67	44.50
3960	18″ diameter	″	152	.184	″	35	7.55	.90	43.45	51

33 42 Culverts

33 42 16 – Concrete Culverts

33 42 16.15 Oval Arch Culverts

Line	Description	Crew	Daily Output	Labor-Hours	Unit	Material	Labor	Equipment	Total	Total Incl O&P
0010	**OVAL ARCH CULVERTS**									
3000	Corrugated galvanized or aluminum, coated & paved									
3020	17″ x 13″, 16 ga., 15″ equivalent	B-14	200	.240	L.F.	12.20	9	1.84	23.04	29
3040	21″ x 15″, 16 ga., 18″ equivalent		150	.320		15.50	12	2.45	29.95	38
3060	28″ x 20″, 14 ga., 24″ equivalent		125	.384		21	14.40	2.94	38.34	48
3080	35″ x 24″, 14 ga., 30″ equivalent		100	.480		29	18	3.67	50.67	63.50
3100	42″ x 29″, 12 ga., 36″ equivalent	B-13	100	.560		38	21.50	7.30	66.80	83
3120	49″ x 33″, 12 ga., 42″ equivalent		90	.622		44.50	24	8.10	76.60	95
3140	57″ x 38″, 12 ga., 48″ equivalent		75	.747		56	29	9.70	94.70	116
3160	Steel, plain oval arch culverts, plain									
3180	17″ x 13″, 16 ga., 15″ equivalent	B-14	225	.213	L.F.	11	8	1.63	20.63	26
3200	21″ x 15″, 16 ga., 18″ equivalent		175	.274		14	10.30	2.10	26.40	33.50
3220	28″ x 20″, 14 ga., 24″ equivalent		150	.320		19	12	2.45	33.45	42
3240	35″ x 24″, 14 ga., 30″ equivalent	B-13	108	.519		26	20	6.75	52.75	66.50
3260	42″ x 29″, 12 ga., 36″ equivalent		108	.519		34	20	6.75	60.75	75.50
3280	49″ x 33″, 12 ga., 42″ equivalent		92	.609		40	23.50	7.90	71.40	88.50
3300	57″ x 38″, 12 ga., 48″ equivalent		75	.747		50	29	9.70	88.70	110
3320	End sections, 17″ x 13″		22	2.545	Ea.	131	98.50	33	262.50	330
3340	42″ x 29″		17	3.294	″	360	127	43	530	635
3360	Multi-plate arch, steel	B-20	1690	.014	Lb.	1.27	.56		1.83	2.27

33 44 Storm Utility Water Drains

33 44 13 – Utility Area Drains

33 44 13.13 Catchbasins

Line	Description	Crew	Daily Output	Labor-Hours	Unit	Material	Labor	Equipment	Total	Total Incl O&P
0010	**CATCHBASINS**									
0011	Not including footing & excavation									
1600	Frames & covers, C.I., 24″ square, 500 lb.	B-6	7.80	3.077	Ea.	320	120	47	487	590
1700	26″ D shape, 600 lb.		7	3.429		475	133	52.50	660.50	785
1800	Light traffic, 18″ diameter, 100 lb.		10	2.400		115	93.50	36.50	245	310
1900	24″ diameter, 300 lb.		8.70	2.759		185	107	42	334	415
2000	36″ diameter, 900 lb.		5.80	4.138		535	161	63.50	759.50	905
2100	Heavy traffic, 24″ diameter, 400 lb.		7.80	3.077		230	120	47	397	490
2200	36″ diameter, 1150 lb.		3	8		750	310	122	1,182	1,425

33 44 Storm Utility Water Drains

33 44 13 – Utility Area Drains

33 44 13.13 Catchbasins		Crew	Daily Output	Labor-Hours	Unit	Material	2013 Bare Costs Labor	Equipment	Total	Total Incl O&P
2300	Mass. State standard, 26" diameter, 475 lb.	B-6	7	3.429	Ea.	250	133	52.50	435.50	535
2400	30" diameter, 620 lb.		7	3.429		325	133	52.50	510.50	620
2500	Watertight, 24" diameter, 350 lb.		7.80	3.077		300	120	47	467	565
2600	26" diameter, 500 lb.		7	3.429		400	133	52.50	585.50	700
2700	32" diameter, 575 lb.	↓	6	4	↓	800	156	61	1,017	1,175
2800	3 piece cover & frame, 10" deep,									
2900	1200 lb., for heavy equipment	B-6	3	8	Ea.	995	310	122	1,427	1,700
3000	Raised for paving 1-1/4" to 2" high									
3100	4 piece expansion ring									
3200	20" to 26" diameter	1 Clab	3	2.667	Ea.	148	94.50		242.50	310
3300	30" to 36" diameter	"	3	2.667	"	204	94.50		298.50	370
3320	Frames and covers, existing, raised for paving, 2", including									
3340	row of brick, concrete collar, up to 12" wide frame	B-6	18	1.333	Ea.	45	52	20.50	117.50	152
3360	20" to 26" wide frame		11	2.182		67	85	33.50	185.50	241
3380	30" to 36" wide frame	↓	9	2.667		83	104	41	228	296
3400	Inverts, single channel brick	D-1	3	5.333		97	215		312	430
3500	Concrete		5	3.200		102	129		231	310
3600	Triple channel, brick		2	8		148	325		473	655
3700	Concrete	↓	3	5.333	↓	137	215		352	475

33 46 Subdrainage

33 46 16 – Subdrainage Piping

33 46 16.25 Piping, Subdrainage, Corrugated Metal

		Crew	Daily Output	Labor-Hours	Unit	Material	Labor	Equipment	Total	Total Incl O&P
0010	**PIPING, SUBDRAINAGE, CORRUGATED METAL** G1030–805									
0021	Not including excavation and backfill									
2010	Aluminum, perforated									
2020	6" diameter, 18 ga.	B-20	380	.063	L.F.	6.50	2.51		9.01	11
2200	8" diameter, 16 ga.	"	370	.065		8.50	2.57		11.07	13.30
2220	10" diameter, 16 ga.	B-21	360	.078		10.65	3.19	.38	14.22	17
2240	12" diameter, 16 ga.		285	.098		11.90	4.03	.48	16.41	19.85
2260	18" diameter, 16 ga.	↓	205	.137	↓	17.85	5.60	.67	24.12	29
3000	Uncoated galvanized, perforated									
3020	6" diameter, 18 ga.	B-20	380	.063	L.F.	6	2.51		8.51	10.45
3200	8" diameter, 16 ga.	"	370	.065		8.25	2.57		10.82	13.05
3220	10" diameter, 16 ga.	B-21	360	.078		8.75	3.19	.38	12.32	14.90
3240	12" diameter, 16 ga.		285	.098		9.75	4.03	.48	14.26	17.45
3260	18" diameter, 16 ga.	↓	205	.137	↓	14.90	5.60	.67	21.17	25.50
4000	Steel, perforated, asphalt coated									
4020	6" diameter 18 ga.	B-20	380	.063	L.F.	6.50	2.51		9.01	11
4030	8" diameter 18 ga.	"	370	.065		8.50	2.57		11.07	13.30
4040	10" diameter 16 ga.	B-21	360	.078		10	3.19	.38	13.57	16.30
4050	12" diameter 16 ga.		285	.098		11	4.03	.48	15.51	18.85
4060	18" diameter 16 ga.	↓	205	.137	↓	17	5.60	.67	23.27	28

33 46 16.30 Piping, Subdrainage, Plastic

		Crew	Daily Output	Labor-Hours	Unit	Material	Labor	Equipment	Total	Total Incl O&P
0010	**PIPING, SUBDRAINAGE, PLASTIC**									
0020	Not including excavation and backfill									
2100	Perforated PVC, 4" diameter	B-14	314	.153	L.F.	1.45	5.75	1.17	8.37	11.70
2110	6" diameter		300	.160		3.26	6	1.22	10.48	14.15
2120	8" diameter		290	.166		6.35	6.20	1.27	13.82	17.90
2130	10" diameter		280	.171		9.15	6.45	1.31	16.91	21.50
2140	12" diameter	↓	270	.178	↓	12.75	6.65	1.36	20.76	26

33 49 Storm Drainage Structures

33 49 13 – Storm Drainage Manholes, Frames, and Covers

33 49 13.10 Storm Drainage Manholes, Frames and Covers		Crew	Daily Output	Labor-Hours	Unit	Material	Labor	Equipment	Total	Total Incl O&P
						2013 Bare Costs				
0010	**STORM DRAINAGE MANHOLES, FRAMES & COVERS**									
0020	Excludes footing, excavation, backfill (See line items for frame & cover)									
0050	Brick, 4′ inside diameter, 4′ deep	D-1	1	16	Ea.	395	645		1,040	1,400
0100	6′ deep		.70	22.857		550	925		1,475	2,000
0150	8′ deep		.50	32		705	1,300		2,005	2,725
0200	For depths over 8′, add		4	4	V.L.F.	84	162		246	340
0400	Concrete blocks (radial), 4′ I.D., 4′ deep		1.50	10.667	Ea.	360	430		790	1,050
0500	6′ deep		1	16		485	645		1,130	1,500
0600	8′ deep		.70	22.857		605	925		1,530	2,075
0700	For depths over 8′, add		5.50	2.909	V.L.F.	63.50	117		180.50	248
0800	Concrete, cast in place, 4′ x 4′, 8″ thick, 4′ deep	C-14H	2	24	Ea.	500	1,050	16.55	1,566.55	2,200
0900	6′ deep		1.50	32		720	1,425	22	2,167	3,000
1000	8′ deep		1	48		1,025	2,125	33	3,183	4,400
1100	For depths over 8′, add		8	6	V.L.F.	117	265	4.14	386.14	540
1110	Precast, 4′ I.D., 4′ deep	B-22	4.10	7.317	Ea.	730	305	50	1,085	1,325
1120	6′ deep		3	10		930	415	68.50	1,413.50	1,725
1130	8′ deep		2	15		1,075	625	103	1,803	2,250
1140	For depths over 8′, add		16	1.875	V.L.F.	123	78	12.85	213.85	269
1150	5′ I.D., 4′ deep	B-6	3	8	Ea.	1,525	310	122	1,957	2,275
1160	6′ deep		2	12		1,850	465	184	2,499	2,950
1170	8′ deep		1.50	16		2,275	620	245	3,140	3,725
1180	For depths over 8′, add		12	2	V.L.F.	272	78	30.50	380.50	450
1190	6′ I.D., 4′ deep		2	12	Ea.	2,150	465	184	2,799	3,300
1200	6′ deep		1.50	16		2,575	620	245	3,440	4,075
1210	8′ deep		1	24		3,175	935	365	4,475	5,300
1220	For depths over 8′, add		8	3	V.L.F.	365	117	46	528	635
1250	Slab tops, precast, 8″ thick									
1300	4′ diameter manhole	B-6	8	3	Ea.	246	117	46	409	500
1400	5′ diameter manhole		7.50	3.200		405	124	49	578	690
1500	6′ diameter manhole		7	3.429		610	133	52.50	795.50	930
3800	Steps, heavyweight cast iron, 7″ x 9″	1 Bric	40	.200		18	8.95		26.95	33.50
3900	8″ x 9″		40	.200		21.50	8.95		30.45	37.50
3928	12″ x 10-1/2″		40	.200		25	8.95		33.95	41
4000	Standard sizes, galvanized steel		40	.200		24	8.95		32.95	39.50
4100	Aluminum		40	.200		24	8.95		32.95	39.50
4150	Polyethylene		40	.200		22	8.95		30.95	37.50

33 51 Natural-Gas Distribution

33 51 13 – Natural-Gas Piping

33 51 13.10 Piping, Gas Service and Distribution, P.E.		Crew	Daily Output	Labor-Hours	Unit	Material	Labor	Equipment	Total	Total Incl O&P
0010	**PIPING, GAS SERVICE AND DISTRIBUTION, POLYETHYLENE**									
0020	Not including excavation or backfill									
1000	60 psi coils, compression coupling @ 100′, 1/2″ diameter, SDR 11	B-20A	608	.053	L.F.	.60	2.28		2.88	4.13
1010	1″ diameter, SDR 11		544	.059		1.40	2.55		3.95	5.40
1040	1-1/4″ diameter, SDR 11		544	.059		1.74	2.55		4.29	5.80
1100	2″ diameter, SDR 11		488	.066		2.84	2.84		5.68	7.45
1160	3″ diameter, SDR 11		408	.078		5.55	3.40		8.95	11.25
1500	60 PSI 40′ joints with coupling, 3″ diameter, SDR 11	B-21A	408	.098		5.80	4.36	1.15	11.31	14.20
1540	4″ diameter, SDR 11		352	.114		12.10	5.05	1.34	18.49	22.50
1600	6″ diameter, SDR 11		328	.122		33	5.40	1.44	39.84	46.50
1640	8″ diameter, SDR 11		272	.147		51	6.55	1.73	59.28	68

33 52 Liquid Fuel Distribution

33 52 16 – Gasoline Distribution

33 52 16.13 Gasoline Piping		Crew	Daily Output	Labor-Hours	Unit	2013 Bare Costs Material	Labor	Equipment	Total	Total Incl O&P
0010	**GASOLINE PIPING**									
0020	Primary containment pipe, fiberglass-reinforced									
0030	Plastic pipe 15′ & 30′ lengths									
0040	2″ diameter	Q-6	425	.056	L.F.	5.85	2.99		8.84	10.90
0050	3″ diameter		400	.060		10.15	3.17		13.32	16
0060	4″ diameter	↓	375	.064	↓	13.40	3.38		16.78	19.85
0100	Fittings									
0110	Elbows, 90° & 45°, bell-ends, 2″	Q-6	24	1	Ea.	42.50	53		95.50	126
0120	3″ diameter		22	1.091		53.50	57.50		111	146
0130	4″ diameter		20	1.200		68.50	63.50		132	171
0200	Tees, bell ends, 2″		21	1.143		59.50	60.50		120	156
0210	3″ diameter		18	1.333		63	70.50		133.50	175
0230	Flanges bell ends, 2″		24	1		33	53		86	116
0240	3″ diameter		22	1.091		38.50	57.50		96	129
0250	4″ diameter		20	1.200		44	63.50		107.50	144
0260	Sleeve couplings, 2″		21	1.143		12.15	60.50		72.65	104
0270	3″ diameter		18	1.333		17.45	70.50		87.95	125
0280	4″ diameter		15	1.600		22.50	84.50		107	152
0290	Threaded adapters 2″		21	1.143		17.65	60.50		78.15	110
0300	3″ diameter		18	1.333		33.50	70.50		104	143
0310	4″ diameter		15	1.600		37	84.50		121.50	168
0320	Reducers, 2″		27	.889		27	47		74	101
0330	3″ diameter		22	1.091		27	57.50		84.50	117
0340	4″ diameter	↓	20	1.200	↓	36	63.50		99.50	136
1010	Gas station product line for secondary containment (double wall)									
1100	Fiberglass reinforced plastic pipe 25′ lengths									
1120	Pipe, plain end, 3″ diameter	Q-6	375	.064	L.F.	26	3.38		29.38	33.50
1130	4″ diameter		350	.069		31.50	3.63		35.13	40
1140	5″ diameter		325	.074		39	3.90		42.90	48.50
1150	6″ diameter	↓	300	.080	↓	38.50	4.23		42.73	49
1200	Fittings									
1230	Elbows, 90° & 45°, 3″ diameter	Q-6	18	1.333	Ea.	133	70.50		203.50	253
1240	4″ diameter		16	1.500		164	79.50		243.50	299
1250	5″ diameter		14	1.714		180	90.50		270.50	335
1260	6″ diameter		12	2		200	106		306	380
1270	Tees, 3″ diameter		15	1.600		162	84.50		246.50	305
1280	4″ diameter		12	2		198	106		304	375
1290	5″ diameter		9	2.667		310	141		451	550
1300	6″ diameter		6	4		370	211		581	725
1310	Couplings, 3″ diameter		18	1.333		54	70.50		124.50	166
1320	4″ diameter		16	1.500		117	79.50		196.50	248
1330	5″ diameter		14	1.714		210	90.50		300.50	365
1340	6″ diameter		12	2		310	106		416	500
1350	Cross-over nipples, 3″ diameter		18	1.333		10.40	70.50		80.90	117
1360	4″ diameter		16	1.500		12.50	79.50		92	133
1370	5″ diameter		14	1.714		15.60	90.50		106.10	153
1380	6″ diameter		12	2		18.70	106		124.70	180
1400	Telescoping, reducers, concentric 4″ x 3″		18	1.333		47	70.50		117.50	158
1410	5″ x 4″		17	1.412		93.50	74.50		168	215
1420	6″ x 5″	↓	16	1.500	↓	229	79.50		308.50	370

33 61 Hydronic Energy Distribution

33 61 13 – Underground Hydronic Energy Distribution

33 61 13.20 Pipe Conduit, Prefabricated/Preinsulated		Crew	Daily Output	Labor-Hours	Unit	2013 Bare Costs Material	Labor	Equipment	Total	Total Incl O&P
0010	**PIPE CONDUIT, PREFABRICATED/PREINSULATED** R221113-70									
0020	Does not include trenching, fittings or crane.									
0300	For cathodic protection, add 12 to 14%									
0310	of total built-up price (casing plus service pipe)									
0580	Polyurethane insulated system, 250°F. max. temp.									
0620	Black steel service pipe, standard wt., 1/2" insulation									
0660	3/4" diam. pipe size	Q-17	54	.296	L.F.	62	15.10	.98	78.08	92
0670	1" diam. pipe size		50	.320		68	16.30	1.06	85.36	101
0680	1-1/4" diam. pipe size		47	.340		76	17.35	1.12	94.47	111
0690	1-1/2" diam. pipe size		45	.356		82.50	18.15	1.17	101.82	120
0700	2" diam. pipe size		42	.381		85.50	19.40	1.26	106.16	125
0710	2-1/2" diam. pipe size		34	.471		87	24	1.55	112.55	134
0720	3" diam. pipe size		28	.571		101	29	1.89	131.89	157
0730	4" diam. pipe size		22	.727		127	37	2.40	166.40	199
0740	5" diam. pipe size		18	.889		162	45.50	2.93	210.43	249
0750	6" diam. pipe size	Q-18	23	1.043		189	55	2.30	246.30	293
0760	8" diam. pipe size		19	1.263		276	67	2.78	345.78	410
0770	10" diam. pipe size		16	1.500		350	79.50	3.30	432.80	510
0780	12" diam. pipe size		13	1.846		435	97.50	4.06	536.56	630
0790	14" diam. pipe size		11	2.182		485	115	4.80	604.80	715
0800	16" diam. pipe size		10	2.400		555	127	5.30	687.30	805
0810	18" diam. pipe size		8	3		640	159	6.60	805.60	950
0820	20" diam. pipe size		7	3.429		715	181	7.55	903.55	1,075
0830	24" diam. pipe size		6	4		875	211	8.80	1,094.80	1,300
0900	For 1" thick insulation, add					10%				
0940	For 1-1/2" thick insulation, add					13%				
0980	For 2" thick insulation, add					20%				
1500	Gland seal for system, 3/4" diam. pipe size	Q-17	32	.500	Ea.	840	25.50	1.65	867.15	965
1510	1" diam. pipe size		32	.500		840	25.50	1.65	867.15	965
1540	1-1/4" diam. pipe size		30	.533		900	27	1.76	928.76	1,025
1550	1-1/2" diam. pipe size		30	.533		900	27	1.76	928.76	1,025
1560	2" diam. pipe size		28	.571		1,075	29	1.89	1,105.89	1,225
1570	2-1/2" diam. pipe size		26	.615		1,150	31.50	2.03	1,183.53	1,325
1580	3" diam. pipe size		24	.667		1,225	34	2.20	1,261.20	1,400
1590	4" diam. pipe size		22	.727		1,450	37	2.40	1,489.40	1,650
1600	5" diam. pipe size		19	.842		1,800	43	2.78	1,845.78	2,050
1610	6" diam. pipe size	Q-18	26	.923		1,900	49	2.03	1,951.03	2,175
1620	8" diam. pipe size		25	.960		2,225	51	2.11	2,278.11	2,525
1630	10" diam. pipe size		23	1.043		2,600	55	2.30	2,657.30	2,925
1640	12" diam. pipe size		21	1.143		2,875	60.50	2.51	2,938.01	3,250
1650	14" diam. pipe size		19	1.263		3,225	67	2.78	3,294.78	3,625
1660	16" diam. pipe size		18	1.333		3,800	70.50	2.93	3,873.43	4,275
1670	18" diam. pipe size		16	1.500		4,025	79.50	3.30	4,107.80	4,550
1680	20" diam. pipe size		14	1.714		4,575	90.50	3.77	4,669.27	5,175
1690	24" diam. pipe size		12	2		5,075	106	4.40	5,185.40	5,750
2000	Elbow, 45° for system									
2020	3/4" diam. pipe size	Q-17	14	1.143	Ea.	530	58.50	3.77	592.27	675
2040	1" diam. pipe size		13	1.231		545	62.50	4.06	611.56	700
2050	1-1/4" diam. pipe size		11	1.455		620	74	4.80	698.80	795
2060	1-1/2" diam. pipe size		9	1.778		645	90.50	5.85	741.35	850
2070	2" diam. pipe size		6	2.667		675	136	8.80	819.80	960
2080	2-1/2" diam. pipe size		4	4		730	204	13.20	947.20	1,125
2090	3" diam. pipe size		3.50	4.571		845	233	15.10	1,093.10	1,300

33 61 Hydronic Energy Distribution

33 61 13 – Underground Hydronic Energy Distribution

33 61 13.20 Pipe Conduit, Prefabricated/Preinsulated		Crew	Daily Output	Labor-Hours	Unit	2013 Bare Costs Material	Labor	Equipment	Total	Total Incl O&P
2100	4" diam. pipe size	Q-17	3	5.333	Ea.	985	272	17.60	1,274.60	1,500
2110	5" diam. pipe size	↓	2.80	5.714		1,275	291	18.85	1,584.85	1,850
2120	6" diam. pipe size	Q-18	4	6		1,425	315	13.20	1,753.20	2,075
2130	8" diam. pipe size		3	8		2,075	425	17.60	2,517.60	2,925
2140	10" diam. pipe size		2.40	10		2,550	530	22	3,102	3,625
2150	12" diam. pipe size		2	12		3,350	635	26.50	4,011.50	4,675
2160	14" diam. pipe size		1.80	13.333		4,175	705	29.50	4,909.50	5,650
2170	16" diam. pipe size		1.60	15		4,950	795	33	5,778	6,675
2180	18" diam. pipe size		1.30	18.462		6,225	975	40.50	7,240.50	8,375
2190	20" diam. pipe size		1	24		7,825	1,275	53	9,153	10,600
2200	24" diam. pipe size	↓	.70	34.286		9,850	1,825	75.50	11,750.50	13,600
2260	For elbow, 90°, add					25%				
2300	For tee, straight, add					85%	30%			
2340	For tee, reducing, add					170%	30%			
2380	For weldolet, straight, add				↓	50%				
2400	Polyurethane insulation, 1"									
2410	FRP carrier and casing									
2420	4"	Q-5	18	.889	L.F.	53.50	45.50		99	127
2422	6"	Q-6	23	1.043		92.50	55		147.50	185
2424	8"		19	1.263		151	67		218	267
2426	10"		16	1.500		203	79.50		282.50	345
2428	12"	↓	13	1.846	↓	265	97.50		362.50	440
2430	FRP carrier and PVC casing									
2440	4"	Q-5	18	.889	L.F.	21	45.50		66.50	91
2444	8"	Q-6	19	1.263		44	67		111	150
2446	10"		16	1.500		63	79.50		142.50	189
2448	12"	↓	13	1.846	↓	80.50	97.50		178	236
2450	PVC carrier and casing									
2460	4"	Q-1	36	.444	L.F.	10.40	22.50		32.90	45
2462	6"	"	29	.552		14.55	27.50		42.05	57.50
2464	8"	Q-2	36	.667		20	34.50		54.50	75
2466	10"		32	.750		27.50	39		66.50	89.50
2468	12"	↓	31	.774	↓	32	40.50		72.50	96

33 63 Steam Energy Distribution

33 63 13 – Underground Steam and Condensate Distribution Piping

33 63 13.10 Calcium Silicate Insulated System		Crew	Daily Output	Labor-Hours	Unit	Material	Labor	Equipment	Total	Total Incl O&P
0010	**CALCIUM SILICATE INSULATED SYSTEM**									
0011	High temp. (1200 degrees F)									
2840	Steel casing with protective exterior coating									
2850	6-5/8" diameter	Q-18	52	.462	L.F.	115	24.50	1.02	140.52	164
2860	8-5/8" diameter		50	.480		125	25.50	1.06	151.56	177
2870	10-3/4" diameter		47	.511		147	27	1.12	175.12	203
2880	12-3/4" diameter		44	.545		160	29	1.20	190.20	221
2890	14" diameter		41	.585		180	31	1.29	212.29	246
2900	16" diameter		39	.615		193	32.50	1.35	226.85	263
2910	18" diameter		36	.667		213	35.50	1.47	249.97	290
2920	20" diameter		34	.706		240	37.50	1.55	279.05	320
2930	22" diameter		32	.750		335	39.50	1.65	376.15	425
2940	24" diameter		29	.828		380	44	1.82	425.82	490
2950	26" diameter		26	.923		435	49	2.03	486.03	550

33 63 Steam Energy Distribution

33 63 13 – Underground Steam and Condensate Distribution Piping

33 63 13.10 Calcium Silicate Insulated System		Crew	Daily Output	Labor-Hours	Unit	Material	2013 Bare Costs Labor	Equipment	Total	Total Incl O&P
2960	28″ diameter	Q-18	23	1.043	L.F.	540	55	2.30	597.30	680
2970	30″ diameter		21	1.143		575	60.50	2.51	638.01	725
2980	32″ diameter		19	1.263		645	67	2.78	714.78	815
2990	34″ diameter		18	1.333		655	70.50	2.93	728.43	830
3000	36″ diameter		16	1.500		700	79.50	3.30	782.80	895
3040	For multi-pipe casings, add					10%				
3060	For oversize casings, add					2%				
3400	Steel casing gland seal, single pipe									
3420	6-5/8″ diameter	Q-18	25	.960	Ea.	1,600	51	2.11	1,653.11	1,825
3440	8-5/8″ diameter		23	1.043		1,875	55	2.30	1,932.30	2,125
3450	10-3/4″ diameter		21	1.143		2,100	60.50	2.51	2,163.01	2,400
3460	12-3/4″ diameter		19	1.263		2,450	67	2.78	2,519.78	2,800
3470	14″ diameter		17	1.412		2,725	74.50	3.11	2,802.61	3,125
3480	16″ diameter		16	1.500		3,200	79.50	3.30	3,282.80	3,650
3490	18″ diameter		15	1.600		3,575	84.50	3.52	3,663.02	4,050
3500	20″ diameter		13	1.846		3,925	97.50	4.06	4,026.56	4,475
3510	22″ diameter		12	2		4,425	106	4.40	4,535.40	5,050
3520	24″ diameter		11	2.182		4,975	115	4.80	5,094.80	5,650
3530	26″ diameter		10	2.400		5,675	127	5.30	5,807.30	6,425
3540	28″ diameter		9.50	2.526		6,400	134	5.55	6,539.55	7,250
3550	30″ diameter		9	2.667		6,525	141	5.85	6,671.85	7,400
3560	32″ diameter		8.50	2.824		7,325	149	6.20	7,480.20	8,300
3570	34″ diameter		8	3		8,000	159	6.60	8,165.60	9,050
3580	36″ diameter		7	3.429		8,500	181	7.55	8,688.55	9,625
3620	For multi-pipe casings, add					5%				
4000	Steel casing anchors, single pipe									
4020	6-5/8″ diameter	Q-18	8	3	Ea.	1,425	159	6.60	1,590.60	1,825
4040	8-5/8″ diameter		7.50	3.200		1,500	169	7.05	1,676.05	1,925
4050	10-3/4″ diameter		7	3.429		1,975	181	7.55	2,163.55	2,450
4060	12-3/4″ diameter		6.50	3.692		2,100	195	8.10	2,303.10	2,600
4070	14″ diameter		6	4		2,475	211	8.80	2,694.80	3,050
4080	16″ diameter		5.50	4.364		2,875	231	9.60	3,115.60	3,500
4090	18″ diameter		5	4.800		3,225	254	10.55	3,489.55	3,950
4100	20″ diameter		4.50	5.333		3,575	282	11.75	3,868.75	4,375
4110	22″ diameter		4	6		3,975	315	13.20	4,303.20	4,875
4120	24″ diameter		3.50	6.857		4,325	365	15.10	4,705.10	5,325
4130	26″ diameter		3	8		4,900	425	17.60	5,342.60	6,050
4140	28″ diameter		2.50	9.600		5,375	510	21	5,906	6,700
4150	30″ diameter		2	12		5,700	635	26.50	6,361.50	7,250
4160	32″ diameter		1.50	16		6,825	845	35	7,705	8,850
4170	34″ diameter		1	24		7,675	1,275	53	9,003	10,400
4180	36″ diameter		1	24		8,325	1,275	53	9,653	11,100
4220	For multi-pipe, add					5%	20%			
4800	Steel casing elbow									
4820	6-5/8″ diameter	Q-18	15	1.600	Ea.	1,975	84.50	3.52	2,063.02	2,300
4830	8-5/8″ diameter		15	1.600		2,100	84.50	3.52	2,188.02	2,425
4850	10-3/4″ diameter		14	1.714		2,525	90.50	3.77	2,619.27	2,925
4860	12-3/4″ diameter		13	1.846		3,000	97.50	4.06	3,101.56	3,450
4870	14″ diameter		12	2		3,200	106	4.40	3,310.40	3,700
4880	16″ diameter		11	2.182		3,500	115	4.80	3,619.80	4,025
4890	18″ diameter		10	2.400		4,000	127	5.30	4,132.30	4,600
4900	20″ diameter		9	2.667		4,275	141	5.85	4,421.85	4,925
4910	22″ diameter		8	3		4,600	159	6.60	4,765.60	5,300

33 63 13 – Underground Steam and Condensate Distribution Piping

33 63 13.10 Calcium Silicate Insulated System		Crew	Daily Output	Labor-Hours	Unit	2013 Bare Costs Material	Labor	Equipment	Total	Total Incl O&P
4920	24″ diameter	Q-18	7	3.429	Ea.	5,075	181	7.55	5,263.55	5,850
4930	26″ diameter		6	4		5,525	211	8.80	5,744.80	6,400
4940	28″ diameter		5	4.800		5,975	254	10.55	6,239.55	6,975
4950	30″ diameter		4	6		6,000	315	13.20	6,328.20	7,100
4960	32″ diameter		3	8		6,825	425	17.60	7,267.60	8,175
4970	34″ diameter		2	12		7,500	635	26.50	8,161.50	9,225
4980	36″ diameter		2	12		8,000	635	26.50	8,661.50	9,775
5500	Black steel service pipe, std. wt., 1″ thick insulation									
5510	3/4″ diameter pipe size	Q-17	54	.296	L.F.	52	15.10	.98	68.08	80.50
5540	1″ diameter pipe size		50	.320		54	16.30	1.06	71.36	85
5550	1-1/4″ diameter pipe size		47	.340		60.50	17.35	1.12	78.97	94
5560	1-1/2″ diameter pipe size		45	.356		66	18.15	1.17	85.32	102
5570	2″ diameter pipe size		42	.381		73.50	19.40	1.26	94.16	111
5580	2-1/2″ diameter pipe size		34	.471		77.50	24	1.55	103.05	123
5590	3″ diameter pipe size		28	.571		88	29	1.89	118.89	143
5600	4″ diameter pipe size		22	.727		113	37	2.40	152.40	183
5610	5″ diameter pipe size		18	.889		153	45.50	2.93	201.43	239
5620	6″ diameter pipe size	Q-18	23	1.043		166	55	2.30	223.30	269
6000	Black steel service pipe, std. wt., 1-1/2″ thick insul.									
6010	3/4″ diameter pipe size	Q-17	54	.296	L.F.	53	15.10	.98	69.08	81.50
6040	1″ diameter pipe size		50	.320		59	16.30	1.06	76.36	90.50
6050	1-1/4″ diameter pipe size		47	.340		66	17.35	1.12	84.47	100
6060	1-1/2″ diameter pipe size		45	.356		72	18.15	1.17	91.32	108
6070	2″ diameter pipe size		42	.381		78.50	19.40	1.26	99.16	116
6080	2-1/2″ diameter pipe size		34	.471		83.50	24	1.55	109.05	130
6090	3″ diameter pipe size		28	.571		93.50	29	1.89	124.39	149
6100	4″ diameter pipe size		22	.727		120	37	2.40	159.40	191
6110	5″ diameter pipe size		18	.889		153	45.50	2.93	201.43	239
6120	6″ diameter pipe size	Q-18	23	1.043		173	55	2.30	230.30	277
6130	8″ diameter pipe size		19	1.263		250	67	2.78	319.78	380
6140	10″ diameter pipe size		16	1.500		315	79.50	3.30	397.80	470
6150	12″ diameter pipe size		13	1.846		375	97.50	4.06	476.56	560
6190	For 2″ thick insulation, add					15%				
6220	For 2-1/2″ thick insulation, add					25%				
6260	For 3″ thick insulation, add					30%				
6800	Black steel service pipe, ex. hvy. wt., 1″ thick insul.									
6820	3/4″ diameter pipe size	Q-17	50	.320	L.F.	54.50	16.30	1.06	71.86	85.50
6840	1″ diameter pipe size		47	.340		58	17.35	1.12	76.47	91
6850	1-1/4″ diameter pipe size		44	.364		67.50	18.55	1.20	87.25	103
6860	1-1/2″ diameter pipe size		42	.381		71.50	19.40	1.26	92.16	109
6870	2″ diameter pipe size		40	.400		75.50	20.50	1.32	97.32	115
6880	2-1/2″ diameter pipe size		31	.516		93.50	26.50	1.70	121.70	144
6890	3″ diameter pipe size		27	.593		105	30	1.96	136.96	164
6900	4″ diameter pipe size		21	.762		139	39	2.51	180.51	214
6910	5″ diameter pipe size		17	.941		194	48	3.11	245.11	289
6920	6″ diameter pipe size	Q-18	22	1.091		197	57.50	2.40	256.90	305
7400	Black steel service pipe, ex. hvy. wt., 1-1/2″ thick insul.									
7420	3/4″ diameter pipe size	Q-17	50	.320	L.F.	45.50	16.30	1.06	62.86	75.50
7440	1″ diameter pipe size		47	.340		52	17.35	1.12	70.47	84
7450	1-1/4″ diameter pipe size		44	.364		60.50	18.55	1.20	80.25	96
7460	1-1/2″ diameter pipe size		42	.381		66	19.40	1.26	86.66	103
7470	2″ diameter pipe size		40	.400		73.50	20.50	1.32	95.32	112
7480	2-1/2″ diameter pipe size		31	.516		79	26.50	1.70	107.20	128

33 63 Steam Energy Distribution

33 63 13 – Underground Steam and Condensate Distribution Piping

33 63 13.10 Calcium Silicate Insulated System		Crew	Daily Output	Labor-Hours	Unit	Material	2013 Bare Costs Labor	Equipment	Total	Total Incl O&P
7490	3" diameter pipe size	Q-17	27	.593	L.F.	93	30	1.96	124.96	150
7500	4" diameter pipe size		21	.762		121	39	2.51	162.51	195
7510	5" diameter pipe size		17	.941		168	48	3.11	219.11	259
7520	6" diameter pipe size	Q-18	22	1.091		185	57.50	2.40	244.90	293
7530	8" diameter pipe size		18	1.333		277	70.50	2.93	350.43	415
7540	10" diameter pipe size		15	1.600		330	84.50	3.52	418.02	490
7550	12" diameter pipe size		13	1.846		405	97.50	4.06	506.56	595
7590	For 2" thick insulation, add					13%				
7640	For 2-1/2" thick insulation, add					18%				
7680	For 3" thick insulation, add					24%				

33 63 13.20 Combined Steam Pipe With Condensate Return		Crew	Daily Output	Labor-Hours	Unit	Material	Labor	Equipment	Total	Total Incl O&P
0010	**COMBINED STEAM PIPE WITH CONDENSATE RETURN**									
9010	8" and 4" in 24" case	Q-18	100	.240	L.F.	585	12.70	.53	598.23	665
9020	6" and 3" in 20" case		104	.231		400	12.20	.51	412.71	460
9030	3" and 1-1/2" in 16" case		110	.218		299	11.55	.48	311.03	350
9040	2" and 1-1/4" in 12-3/4" case		114	.211		260	11.15	.46	271.61	305
9050	1-1/2" and 1-1/4" in 10-3/4" case		116	.207		258	10.95	.46	269.41	300
9100	Steam pipe only (no return)									
9110	6" in 18" case	Q-18	108	.222	L.F.	330	11.75	.49	342.24	380
9120	4" in 14" case		112	.214		260	11.35	.47	271.82	305
9130	3" in 14" case		114	.211		240	11.15	.46	251.61	281
9140	2-1/2" in 12-3/4" case		118	.203		224	10.75	.45	235.20	263
9150	2" in 12-3/4" case		122	.197		214	10.40	.43	224.83	252

Estimating Tips

Products such as conveyors, material handling cranes and hoists, as well as other items specified in this division, require trained installers. The general contractor may not have any choice as to who will perform the installation or when it will be performed. Long lead times are often required for these products, making early decisions in purchasing and scheduling necessary.

The installation of this type of equipment may require the embedment of mounting hardware during construction of floors, structural walls, or interior walls/partitions. Electrical connections will require coordination with the electrical contractor.

Reference Numbers

Reference numbers are shown in shaded boxes at the beginning of some major classifications. These numbers refer to related items in the Reference Section. The reference information may be an estimating procedure, an alternate pricing method, or technical information.

Note: Not all subdivisions listed here necessarily appear in this publication.

41 22 Cranes and Hoists

41 22 23 – Hoists

41 22 23.10 Material Handling		Crew	Daily Output	Labor-Hours	Unit	Material	2013 Bare Costs Labor	Equipment	Total	Total Incl O&P
0010	**MATERIAL HANDLING**, cranes, hoists and lifts									
1500	Cranes, portable hydraulic, floor type, 2,000 lb. capacity				Ea.	3,325			3,325	3,675
1600	4,000 lb. capacity					3,825			3,825	4,200
1800	Movable gantry type, 12′ to 15′ range, 2,000 lb. capacity					1,625			1,625	1,775
1900	6,000 lb. capacity					4,000			4,000	4,400
2100	Hoists, electric overhead, chain, hook hung, 15′ lift, 1 ton cap.					2,775			2,775	3,050
2200	3 ton capacity					3,250			3,250	3,575
2500	5 ton capacity				▼	7,050			7,050	7,750
2600	For hand-pushed trolley, add					15%				
2700	For geared trolley, add					30%				
2800	For motor trolley, add					75%				
3000	For lifts over 15′, 1 ton, add				L.F.	25			25	27.50
3100	5 ton, add				"	52			52	57
3300	Lifts, scissor type, portable, electric, 36″ high, 2,000 lb.				Ea.	3,225			3,225	3,550
3400	48″ high, 4,000 lb.				"	4,000			4,000	4,400

Estimating Tips

This section involves equipment and construction costs for air noise and odor pollution control systems.

These systems may be interrelated and care must be taken that the complete systems are estimated. For example, air pollution equipment may include dust and air-entrained particles that have to be collected. The vacuum systems could be noisy, requiring silencers to reduce noise pollution, and the collected solids have to be disposed of to prevent solid pollution.

Reference Numbers

Reference numbers are shown in shaded boxes at the beginning of some major classifications. These numbers refer to related items in the Reference Section. The reference information may be an estimating procedure, an alternate pricing method, or technical information.

Note: Not all subdivisions listed here necessarily appear in this publication.

44 11 16.10 Dust Collection Systems		Crew	Daily Output	Labor-Hours	Unit	Material	2013 Bare Costs Labor	Equipment	Total	Total Incl O&P
0010	**DUST COLLECTION SYSTEMS** Commercial/Industrial									
0120	Central vacuum units									
0130	Includes stand, filters and motorized shaker									
0200	500 CFM, 10" inlet, 2 HP	Q-20	2.40	8.333	Ea.	4,175	405		4,580	5,225
0220	1000 CFM, 10" inlet, 3 HP		2.20	9.091		4,400	445		4,845	5,500
0240	1500 CFM, 10" inlet, 5 HP		2	10		4,625	490		5,115	5,850
0260	3000 CFM, 13" inlet, 10 HP		1.50	13.333		12,900	650		13,550	15,200
0280	5000 CFM, 16" inlet, 2 @ 10 HP	↓	1	20	↓	13,700	975		14,675	16,600
1000	Vacuum tubing, galvanized									
1100	2-1/8" OD, 16 ga.	Q-9	440	.036	L.F.	2.95	1.74		4.69	5.90
1110	2-1/2" OD, 16 ga.		420	.038		3.24	1.83		5.07	6.35
1120	3" OD, 16 ga.		400	.040		4.13	1.92		6.05	7.45
1130	3-1/2" OD, 16 ga.		380	.042		5.85	2.02		7.87	9.55
1140	4" OD, 16 ga.		360	.044		6	2.13		8.13	9.85
1150	5" OD, 14 ga.		320	.050		10.80	2.40		13.20	15.50
1160	6" OD, 14 ga.		280	.057		12.10	2.74		14.84	17.50
1170	8" OD, 14 ga.		200	.080		18.30	3.84		22.14	26
1180	10" OD, 12 ga.		160	.100		35.50	4.80		40.30	46.50
1190	12" OD, 12 ga.		120	.133		47	6.40		53.40	61.50
1200	14" OD, 12 ga.	↓	80	.200	↓	53.50	9.60		63.10	73
1940	Hose, flexible wire reinforced rubber									
1956	3" diam.	Q-9	400	.040	L.F.	9.50	1.92		11.42	13.40
1960	4" diam.		360	.044		11.50	2.13		13.63	15.90
1970	5" diam.		320	.050		14.75	2.40		17.15	19.85
1980	6" diam.	↓	280	.057	↓	15.65	2.74		18.39	21.50
2000	90° Elbow, slip fit									
2110	2-1/8" diam.	Q-9	70	.229	Ea.	10.85	10.95		21.80	28.50
2120	2-1/2" diam.		65	.246		15.20	11.80		27	34.50
2130	3" diam.		60	.267		20.50	12.80		33.30	42
2140	3-1/2" diam.		55	.291		25.50	13.95		39.45	49.50
2150	4" diam.		50	.320		31.50	15.35		46.85	58
2160	5" diam.		45	.356		60	17.05		77.05	92
2170	6" diam.		40	.400		82.50	19.20		101.70	121
2180	8" diam.	↓	30	.533	↓	157	25.50		182.50	211
2400	45° Elbow, slip fit									
2410	2-1/8" diam.	Q-9	70	.229	Ea.	9.50	10.95		20.45	27
2420	2-1/2" diam.		65	.246		14	11.80		25.80	33.50
2430	3" diam.		60	.267		16.90	12.80		29.70	38
2440	3-1/2" diam.		55	.291		21.50	13.95		35.45	45
2450	4" diam.		50	.320		27.50	15.35		42.85	54
2460	5" diam.		45	.356		46.50	17.05		63.55	77
2470	6" diam.		40	.400		63	19.20		82.20	98.50
2480	8" diam.	↓	35	.457	↓	123	22		145	170
2800	90° TY, slip fit thru 6" diam.									
2810	2-1/8" diam.	Q-9	42	.381	Ea.	17.75	18.30		36.05	47.50
2820	2-1/2" diam.		39	.410		26.50	19.70		46.20	59
2830	3" diam.		36	.444		37	21.50		58.50	73
2840	3-1/2" diam.		33	.485		39	23.50		62.50	78
2850	4" diam.		30	.533		41.50	25.50		67	85
2860	5" diam.		27	.593		70.50	28.50		99	121
2870	6" diam.	↓	24	.667		101	32		133	160
2880	8" diam., butt end					155			155	170
2890	10" diam., butt end					475			475	525

44 11 16 – Fugitive Dust Barrier Systems

44 11 16.10 Dust Collection Systems		Crew	Daily Output	Labor-Hours	Unit	2013 Bare Costs Material	Labor	Equipment	Total	Total Incl O&P
2900	12" diam., butt end				Ea.	475			475	525
2910	14" diam., butt end					855			855	940
2920	6" x 4" diam., butt end					106			106	116
2930	8" x 4" diam., butt end					200			200	220
2940	10" x 4" diam., butt end					263			263	289
2950	12" x 4" diam., butt end				↓	300			300	330
3100	90° Elbow, butt end segmented									
3110	8" diam., butt end, segmented				Ea.	157			157	172
3120	10" diam., butt end, segmented					390			390	430
3130	12" diam., butt end, segmented					495			495	545
3140	14" diam., butt end, segmented				↓	620			620	680
3200	45° Elbow, butt end segmented									
3210	8" diam., butt end, segmented				Ea.	123			123	136
3220	10" diam., butt end, segmented					282			282	310
3230	12" diam., butt end, segmented					288			288	315
3240	14" diam., butt end, segmented				↓	585			585	645
3400	All butt end fittings require one coupling per joint.									
3410	Labor for fitting included with couplings.									
3460	Compression coupling, galvanized, neoprene gasket									
3470	2-1/8" diam.	Q-9	44	.364	Ea.	10.85	17.45		28.30	38.50
3480	2-1/2" diam.		44	.364		10.85	17.45		28.30	38.50
3490	3" diam.		38	.421		15.60	20		35.60	48
3500	3-1/2" diam.		35	.457		17.45	22		39.45	52.50
3510	4" diam.		33	.485		18.95	23.50		42.45	56.50
3520	5" diam.		29	.552		21.50	26.50		48	64.50
3530	6" diam.		26	.615		25.50	29.50		55	73
3540	8" diam.		22	.727		46	35		81	104
3550	10" diam.		20	.800		67.50	38.50		106	133
3560	12" diam.		18	.889		84	42.50		126.50	158
3570	14" diam.	↓	16	1	↓	119	48		167	204
3800	Air gate valves, galvanized									
3810	2-1/8" diam.	Q-9	30	.533	Ea.	111	25.50		136.50	161
3820	2-1/2" diam.		28	.571		119	27.50		146.50	173
3830	3" diam.		26	.615		127	29.50		156.50	185
3840	4" diam.		23	.696		147	33.50		180.50	213
3850	6" diam.	↓	18	.889	↓	205	42.50		247.50	290

Division Notes

		CREW	DAILY OUTPUT	LABOR-HOURS	UNIT	BARE COSTS MAT.	BARE COSTS LABOR	BARE COSTS EQUIP.	BARE COSTS TOTAL	TOTAL INCL O&P

Estimating Tips

This new division contains information about water and wastewater equipment and systems, which was formerly located in Division 44.

The main areas of focus are total wastewater treatment plants and components of wastewater treatment plants. In addition, there are assemblies such as sewage treatment lagoons that can be found in some publications under G30 Site Mechanical Utilities.

Also included in this section are oil/water separators for wastewater treatment.

New sections added for this year are for activated sludge cells and air diffuser systems for wastewater treatment.

Reference Numbers

Reference numbers are shown in shaded boxes at the beginning of some major classifications. These numbers refer to related items in the Reference Section. The reference information may be an estimating procedure, an alternate pricing method, or technical information.

Note: Not all subdivisions listed here necessarily appear in this publication.

46 07 Packaged Water and Wastewater Treatment Equipment

46 07 53 – Packaged Wastewater Treatment Equipment

46 07 53.10 Biological Packaged Water Treatment Plants		Crew	Daily Output	Labor-Hours	Unit	Material	Labor	Equipment	Total	Total Incl O&P
						2013 Bare Costs				
0010	**BIOLOGICAL PACKAGED WATER TREATMENT PLANTS**									
0011	Not including fencing or external piping									
0020	Steel packaged, blown air aeration plants									
0100	1,000 GPD				Gal.				50	55
0200	5,000 GPD								20	22
0300	15,000 GPD								20	22
0400	30,000 GPD								14	15.40
0500	50,000 GPD								10	11
0600	100,000 GPD								9	9.90
0700	200,000 GPD								8	8.80
0800	500,000 GPD				↓				7	7.70
1000	Concrete, extended aeration, primary and secondary treatment									
1010	10,000 GPD				Gal.				20	22
1100	30,000 GPD								14	15.40
1200	50,000 GPD								10	11
1400	100,000 GPD								9	9.90
1500	500,000 GPD				↓				7	7.70
1700	Municipal wastewater treatment facility									
1720	1.0 MGD				Gal.				9.70	11
1740	1.5 MGD								9.20	10.60
1760	2.0 MGD								8.75	10
1780	3.0 MGD								6.80	7.80
1800	5.0 MGD				↓				5.80	6.70
2000	Holding tank system, not incl. excavation or backfill									
2010	Recirculating chemical water closet	2 Plum	4	4	Ea.	500	223		723	885
2100	For voltage converter, add	"	16	1		283	56		339	395
2200	For high level alarm, add	1 Plum	7.80	1.026	↓	110	57		167	207

46 07 53.20 Wastewater Treatment System		Crew	Daily Output	Labor-Hours	Unit	Material	Labor	Equipment	Total	Total Incl O&P
0010	**WASTEWATER TREATMENT SYSTEM**									
0020	Fiberglass, 1,000 gallon	B-21	1.29	21.705	Ea.	4,000	890	106	4,996	5,900
0100	1,500 gallon	"	1.03	27.184	"	8,000	1,125	133	9,258	10,700

46 23 Grit Removal And Handling Equipment

46 23 23 – Vortex Grit Removal Equipment

46 23 23.10 Rainwater Filters		Crew	Daily Output	Labor-Hours	Unit	Material	Labor	Equipment	Total	Total Incl O&P
0010	**RAINWATER FILTERS**									
0100	42 gal./min	B-21	3.50	8	Ea.	40,000	330	39	40,369	44,500
0200	65 gal./min		3.50	8		40,000	330	39	40,369	44,500
0300	208 gal./min	↓	3.50	8	↓	40,000	330	39	40,369	44,500

46 25 Oil and Grease Separation and Removal Equipment

46 25 13 – Coalescing Oil-Water Separators

46 25 13.20 Oil/Water Separators		Crew	Daily Output	Labor-Hours	Unit	2013 Bare Costs Material	Labor	Equipment	Total	Total Incl O&P
0010	**OIL/WATER SEPARATORS**									
0020	Underground, tank only									
0030	Excludes excavation, backfill, & piping									
0100	200 GPM	B-21	3.50	8	Ea.	39,800	330	39	40,169	44,300
0110	400 GPM		3.25	8.615		54,500	355	42	54,897	60,500
0120	600 GPM		2.75	10.182		54,500	415	50	54,965	60,500
0130	800 GPM		2.50	11.200		68,500	460	55	69,015	76,000
0140	1000 GPM		2	14		81,500	575	68.50	82,143.50	91,000
0150	1200 GPM		1.50	18.667		87,500	765	91.50	88,356.50	97,500
0160	1500 GPM	↓	1	28	↓	107,500	1,150	137	108,787	120,000

Division Notes

		CREW	DAILY OUTPUT	LABOR-HOURS	UNIT	BARE COSTS MAT.	BARE COSTS LABOR	BARE COSTS EQUIP.	BARE COSTS TOTAL	TOTAL INCL O&P

Assemblies Section

Table of Contents

How RSMeans Assemblies Data Works

Assemblies estimating provides a fast and reasonably accurate way to develop construction costs. An assembly is the grouping of individual work items, with appropriate quantities, to provide a cost for a major construction component in a convenient unit of measure.

An assemblies estimate is often used during early stages of design development to compare the cost impact of various design alternatives on total building cost.

Assemblies estimates are also used as an efficient tool to verify construction estimates.

Assemblies estimates do not require a completed design or detailed drawings. Instead, they are based on the general size of the structure and other known parameters of the project. The degree of accuracy of an assemblies estimate is generally within +/- 15%.

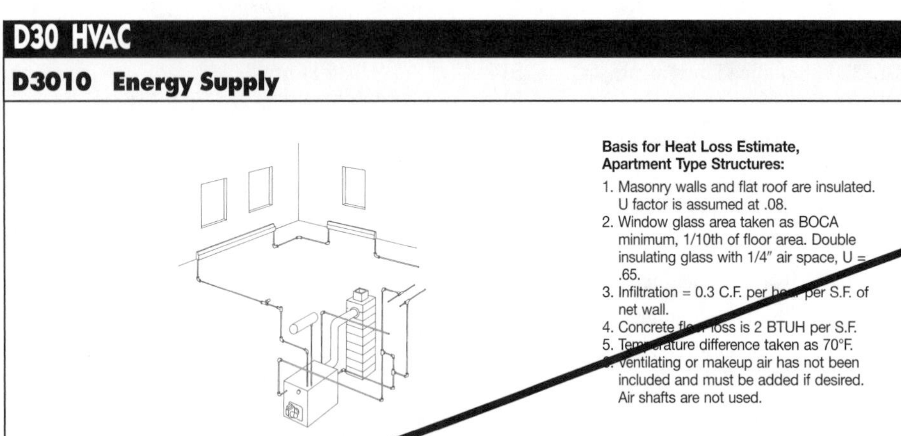

D30 HVAC
D3010 Energy Supply
Basis for Heat Loss Estimate, Apartment Type Structures:
1. Masonry walls and flat roof are insulated. U factor is assumed at .08.
2. Window glass area taken as BOCA minimum, 1/10th of floor area. Double insulating glass with 1/4" air space, U = .65.
3. Infiltration = 0.3 C.F. per hour per S.F. of net wall.
4. Concrete floor loss is 2 BTUH per S.F.
5. Temperature difference taken as 70°F.
6. Ventilating or makeup air has not been included and must be added if desired. Air shafts are not used.

System Components	QUANTITY	UNIT	COST EACH MAT.	COST EACH INST.	COST EACH TOTAL
SYSTEM D3010 510 1760					
HEATING SYSTEM, FIN TUBE RADIATION, FORCED HOT WATER					
1,000 S.F. AREA, 10,000 C.F. VOLUME					
Boiler, oil fired, CI, burner, ctrls/insul/breech/pipe/ftng/valves, 109 MBH	1.000	Ea.	4,200	3,806.25	8,006.25
Circulating pump, CI flange connection, 1/12 HP	1.000	Ea.	380	202	582
Expansion tank, painted steel, ASME 18 Gal capacity	1.000	Ea.	700	87.50	787.50
Storage tank, steel, above ground, 275 Gal capacity w/supports	1.000	Ea.	540	246	786
Copper tubing type L, solder joint, hanger 10' OC, 3/4" diam	100.000	L.F.	720	885	1,605
Radiation, 3/4" copper tube w/alum fin baseboard pkg, 7" high	30.000	L.F.	277.50	630	907.50
Pipe covering, calcium silicate w/cover, 1' wall, 3/4' diam	100.000	L.F.	363	655	1,018
TOTAL			7,180.50	6,511.75	13,692.25
COST PER S.F.			7.18	6.51	13.69

D3010 510	Apartment Building Heating - Fin Tube Radiation		COST PER S.F. MAT.	COST PER S.F. INST.	COST PER S.F. TOTAL
1740	Heating systems, fin tube radiation, forced hot water				
1760	1,000 S.F. area, 10,000 C.F. volume		7.18	6.50	13.68
1800	10,000 S.F. area, 100,000 C.F. volume	R235000 -10	3.48	3.93	7.41
1840	20,000 S.F. area, 200,000 C.F. volume		3.99	4.39	8.38
1880	30,000 S.F. area, 300,000 C.F. volume	R235000 -20	3.82	4.25	8.07
1890					

RSMeans assemblies are identified by a **unique 12-character identifier**. The assemblies are numbered using UNIFORMAT II, ASTM Standard E1557. The first 6 characters represent this system to Level 4. The last 6 characters represent further breakdown by RSMeans in order to arrange items in understandable groups of similar tasks. Line numbers are consistent across all RSMeans publications, so a line number in any RSMeans assemblies data set will always refer to the same work.

Information is available in the Reference Section to assist the estimator with estimating procedures, alternate pricing methods, and additional technical information.

The **Reference Box** indicates the exact location of this information in the Reference Section at the back of the book. The "R" stands for "reference," and the remaining characters are the RSMeans line numbers.

RSMeans assemblies **Narrative descriptions** are shown in a hierarchical structure to make them readable. In order to read a complete description, read up through the indents to the top of the section. Include everything that is above and to the left that is not contradicted by information below.

Most assemblies consist of three major elements: a graphic, the system components, and the cost data itself. The **Graphic** is a visual representation showing the typical appearance of the assembly in question, frequently accompanied by additional explanatory technical information describing the class of items. The **System Components** is a listing of the individual tasks that make up the assembly, including the quantity and unit of measure for each item, along with the cost of material and installation. The **Assemblies Data** below lists prices for other similar systems with dimensional and/or size variations.

All RSMeans assemblies costs represent the cost for the installing contractor. An allowance for profit has been added to all material, labor, and equipment rental costs. A markup for labor burdens, including workers' compensation, fixed overhead, and business overhead, is included with installation costs.

The data published in RSMeans print books represents a "national average" cost. This data should be modified to the project location using the **City Cost Indexes** or **Location Factors** tables found in the Reference Section in the back of the book.

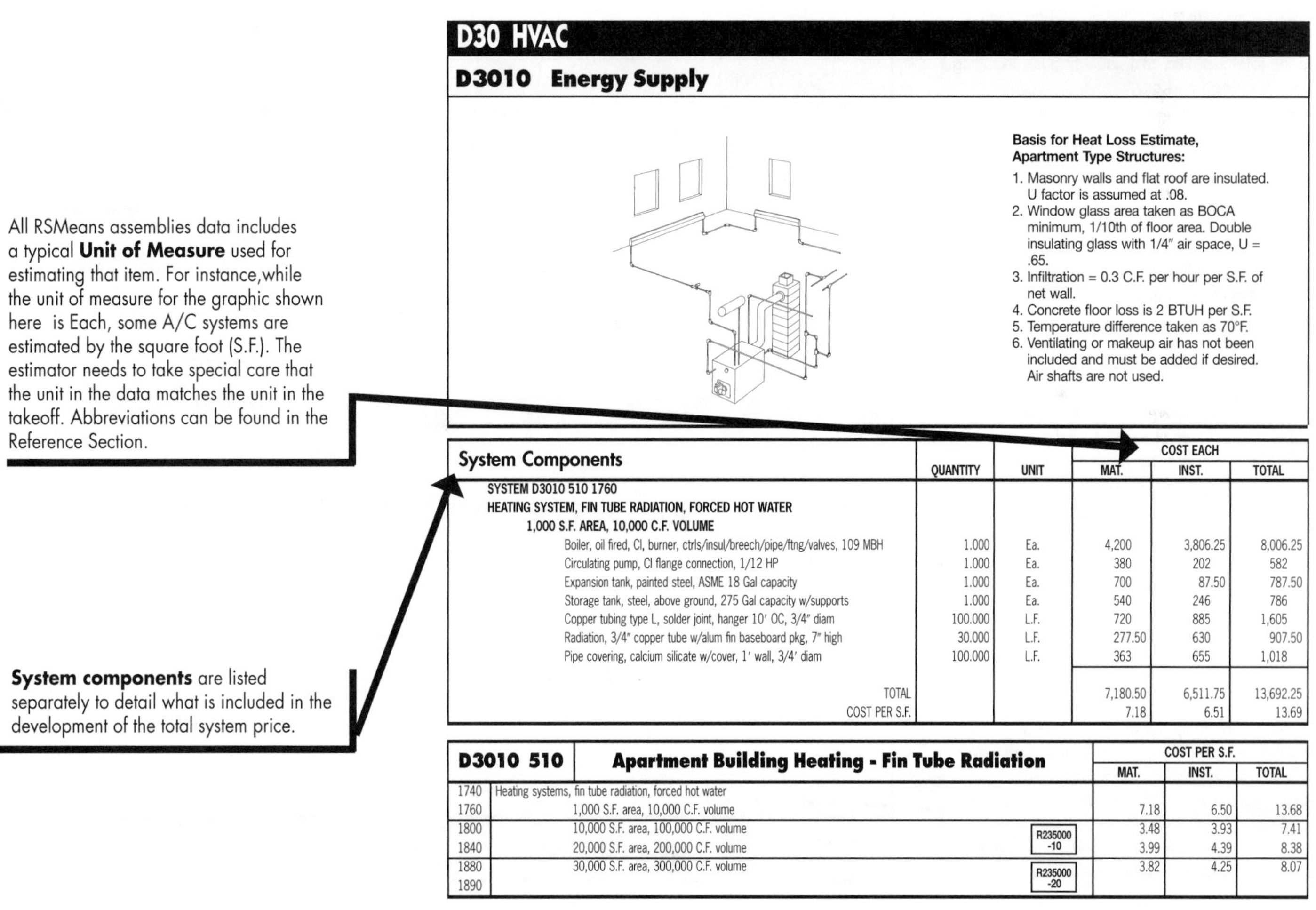

D30 HVAC

D3010 Energy Supply

Basis for Heat Loss Estimate, Apartment Type Structures:

1. Masonry walls and flat roof are insulated. U factor is assumed at .08.
2. Window glass area taken as BOCA minimum, 1/10th of floor area. Double insulating glass with 1/4" air space, U = .65.
3. Infiltration = 0.3 C.F. per hour per S.F. of net wall.
4. Concrete floor loss is 2 BTUH per S.F.
5. Temperature difference taken as 70°F.
6. Ventilating or makeup air has not been included and must be added if desired. Air shafts are not used.

System Components	QUANTITY	UNIT	COST EACH MAT.	COST EACH INST.	COST EACH TOTAL
SYSTEM D3010 510 1760					
HEATING SYSTEM, FIN TUBE RADIATION, FORCED HOT WATER					
1,000 S.F. AREA, 10,000 C.F. VOLUME					
Boiler, oil fired, CI, burner, ctrls/insul/breech/pipe/ftng/valves, 109 MBH	1.000	Ea.	4,200	3,806.25	8,006.25
Circulating pump, CI flange connection, 1/12 HP	1.000	Ea.	380	202	582
Expansion tank, painted steel, ASME 18 Gal capacity	1.000	Ea.	700	87.50	787.50
Storage tank, steel, above ground, 275 Gal capacity w/supports	1.000	Ea.	540	246	786
Copper tubing type L, solder joint, hanger 10′ OC, 3/4″ diam	100.000	L.F.	720	885	1,605
Radiation, 3/4″ copper tube w/alum fin baseboard pkg, 7″ high	30.000	L.F.	277.50	630	907.50
Pipe covering, calcium silicate w/cover, 1′ wall, 3/4′ diam	100.000	L.F.	363	655	1,018
TOTAL			7,180.50	6,511.75	13,692.25
COST PER S.F.			7.18	6.51	13.69

D3010 510	Apartment Building Heating - Fin Tube Radiation		COST PER S.F. MAT.	COST PER S.F. INST.	COST PER S.F. TOTAL
1740	Heating systems, fin tube radiation, forced hot water				
1760	1,000 S.F. area, 10,000 C.F. volume		7.18	6.50	13.68
1800	10,000 S.F. area, 100,000 C.F. volume	R235000 -10	3.48	3.93	7.41
1840	20,000 S.F. area, 200,000 C.F. volume		3.99	4.39	8.38
1880	30,000 S.F. area, 300,000 C.F. volume	R235000 -20	3.82	4.25	8.07
1890					

All RSMeans assemblies data includes a typical **Unit of Measure** used for estimating that item. For instance, while the unit of measure for the graphic shown here is Each, some A/C systems are estimated by the square foot (S.F.). The estimator needs to take special care that the unit in the data matches the unit in the takeoff. Abbreviations can be found in the Reference Section.

System components are listed separately to detail what is included in the development of the total system price.

How RSMeans Assemblies Data Works (continued)

Sample Estimate

This sample demonstrates the elements of an estimate, including a tally of the RSMeans data lines. Published assemblies costs include all markups for labor burden and profit for the installing contractor. This estimate adds a summary of the markups applied by a general contractor on the installing contractors' work. These figures represent the total cost to the owner. The RSMeans location factor is added at the bottom of the estimate to adjust the cost of the work to a specific location

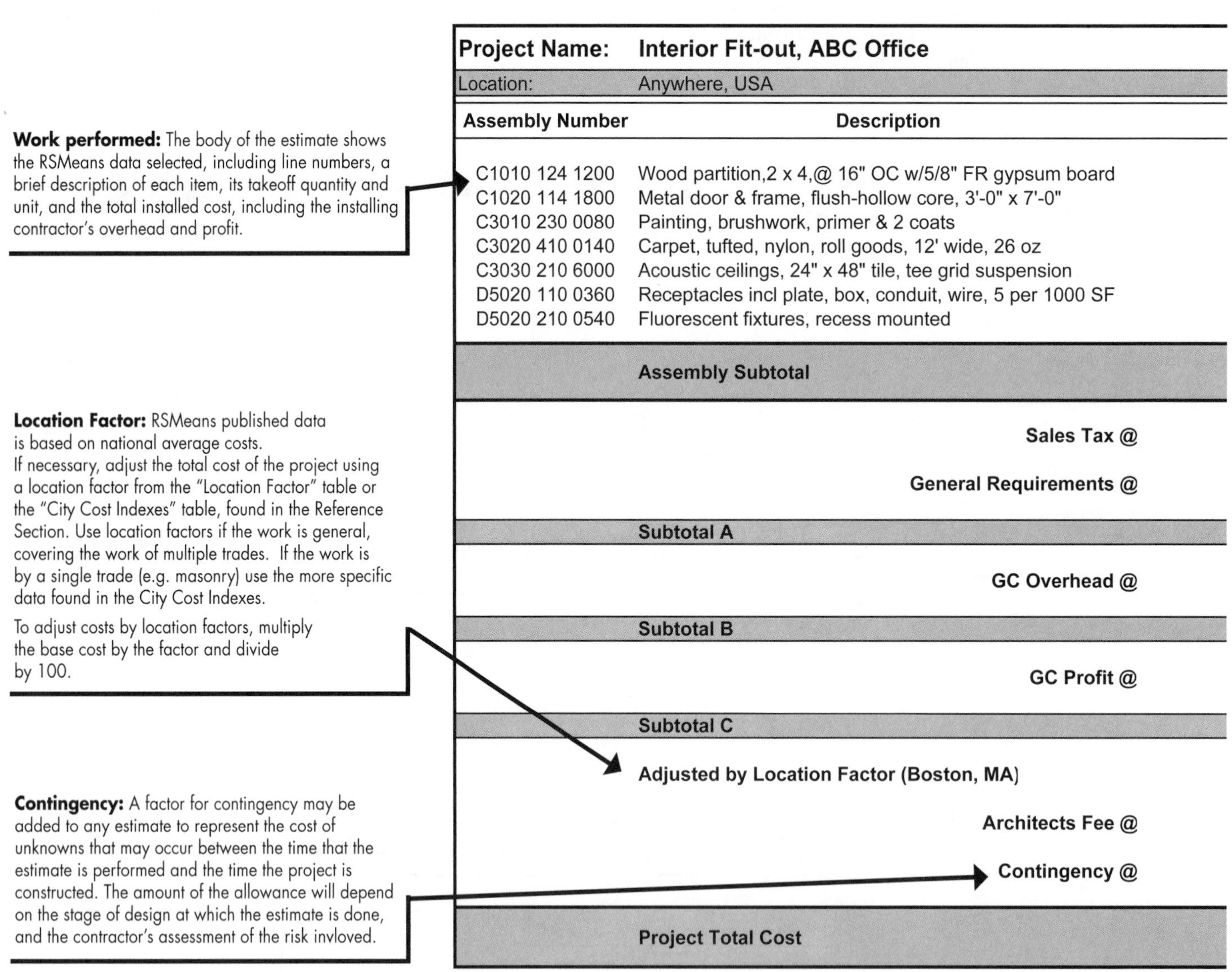

Project Name:	Interior Fit-out, ABC Office
Location:	Anywhere, USA
Assembly Number	**Description**
C1010 124 1200	Wood partition,2 x 4,@ 16" OC w/5/8" FR gypsum board
C1020 114 1800	Metal door & frame, flush-hollow core, 3'-0" x 7'-0"
C3010 230 0080	Painting, brushwork, primer & 2 coats
C3020 410 0140	Carpet, tufted, nylon, roll goods, 12' wide, 26 oz
C3030 210 6000	Acoustic ceilings, 24" x 48" tile, tee grid suspension
D5020 110 0360	Receptacles incl plate, box, conduit, wire, 5 per 1000 SF
D5020 210 0540	Fluorescent fixtures, recess mounted
	Assembly Subtotal
	Sales Tax @
	General Requirements @
	Subtotal A
	GC Overhead @
	Subtotal B
	GC Profit @
	Subtotal C
	Adjusted by Location Factor (Boston, MA)
	Architects Fee @
	Contingency @
	Project Total Cost

Work performed: The body of the estimate shows the RSMeans data selected, including line numbers, a brief description of each item, its takeoff quantity and unit, and the total installed cost, including the installing contractor's overhead and profit.

Location Factor: RSMeans published data is based on national average costs. If necessary, adjust the total cost of the project using a location factor from the "Location Factor" table or the "City Cost Indexes" table, found in the Reference Section. Use location factors if the work is general, covering the work of multiple trades. If the work is by a single trade (e.g. masonry) use the more specific data found in the City Cost Indexes.

To adjust costs by location factors, multiply the base cost by the factor and divide by 100.

Contingency: A factor for contingency may be added to any estimate to represent the cost of unknowns that may occur between the time that the estimate is performed and the time the project is constructed. The amount of the allowance will depend on the stage of design at which the estimate is done, and the contractor's assessment of the risk invloved.

This estimate is based on an interactive spreadsheet.
A copy of this spreadsheet is located on the RSMeans website at **http://www.reedconstructiondata.com/rsmeans/extras/546011**.
You are free to download it and adjust it to your methodology.

Date:		
Qty.	**Unit**	**Subtotal**
600.000	S.F.	$2,682.00
2.000	Ea.	$2,550.00
600.000	S.F.	$696.00
240.000	S.F.	$1,708.80
200.000	S.F.	$870.00
6.000	S.F.	$14.76
200.000	S.F.	$1,580.00
		$10,101.56
5 %		**$ 252.54**
7 %		**$ 707.11**
		$11,061.21
5 %		**$ 553.06**
		$11,614.27
5 %		**$ 580.71**
		$12,194.98
118.1		**$ 14,402.27**
8 %		**$ 1,152.18**
15 %		**$ 2,160.34**
		$ 17,714.80

Sales Tax: If the work is subject to state or local sales taxes, the amount must be added to the estimate. In a conceptual estimate it can be assumed that one half of the total represents material costs. Therefore, apply the sales tax rate to 50% of the assembly subtotal.

General Requirements: This item covers project-wide needs provided by the General Contractor. These items vary by project, but may include temporary facilities and utilities, security, testing, project cleanup, etc. In assemblies estimates a percentage is used, typically between 5% and 15% of project cost.

General Contractor Overhead: This entry represents the General Contractor's markup on all work to cover project administration costs.

General Contractor Profit: This entry represents the GC's profit on all work performed. The value included here can vary widely by project, and is influenced by the GC's perception of the project's financial risk and market conditions.

Architects Fee: If appropriate, add the design cost to the project estimate. These fees vary based on project complexity and size. Typical design and engineering fees can be found in the reference section.

D Services

D2010 Plumbing Fixtures

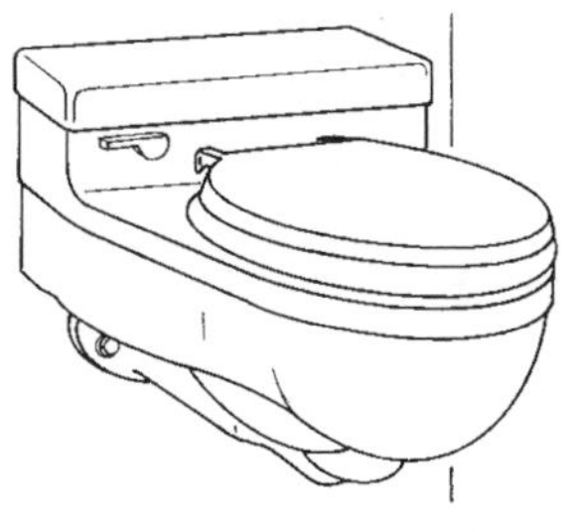
One Piece Wall Hung

Systems are complete with trim seat and rough-in (supply, waste and vent) for connection to supply branches and waste mains.

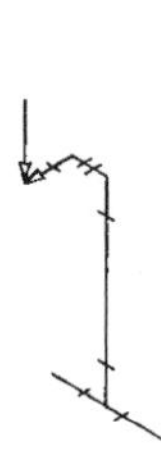
Supply

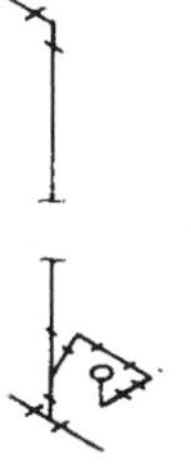
Waste/Vent

Floor Mount

System Components	QUANTITY	UNIT	COST EACH MAT.	INST.	TOTAL
SYSTEM D2010 110 1880					
WATER CLOSET, VITREOUS CHINA					
TANK TYPE, WALL HUNG, TWO PIECE					
Water closet, tank type vit china wall hung 2 pc. w/seat supply & stop	1.000	Ea.	690	228	918
Pipe Steel galvanized, schedule 40, threaded, 2″ diam.	4.000	L.F.	58.60	75.60	134.20
Pipe, CI soil, no hub, cplg 10′ OC, hanger 5′ OC, 4″ diam.	2.000	L.F.	36.70	42	78.70
Pipe, coupling, standard coupling, CI soil, no hub, 4″ diam.	2.000	Ea.	40	73	113
Copper tubing type L solder joint, hangar 10′ O.C., 1/2″ diam.	6.000	L.F.	27.96	49.80	77.76
Wrought copper 90° elbow for solder joints 1/2″ diam.	2.000	Ea.	6.22	67	73.22
Wrought copper Tee for solder joints 1/2″ diam.	1.000	Ea.	5.30	51.50	56.80
Supports/carrier, water closet, siphon jet, horiz, single, 4″ waste	1.000	Ea.	720	126	846
TOTAL			1,584.78	712.90	2,297.68

D2010 110	Water Closet Systems		COST EACH MAT.	INST.	TOTAL
1800	Water closet, vitreous china				
1840	Tank type, wall hung				
1880	Close coupled two piece	R224000 -30	1,575	715	2,290
1920	Floor mount, one piece		1,525	760	2,285
1960	One piece low profile	R224000 -40	920	760	1,680
2000	Two piece close coupled		645	760	1,405
2040	Bowl only with flush valve				
2080	Wall hung		1,900	810	2,710
2120	Floor mount		840	770	1,610
2160	Floor mount, ADA compliant with 18″ high bowl		840	790	1,630

D2010 Plumbing Fixtures

Systems are complete with trim, seat, flush valve and rough-in (supply, waste and vent) for connection to supply branches and waste mains.

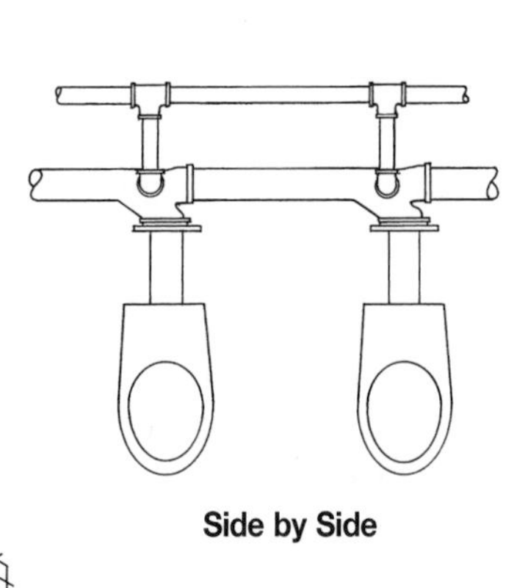

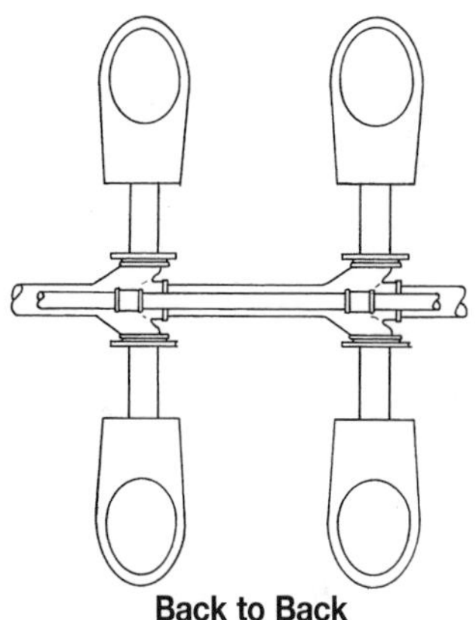

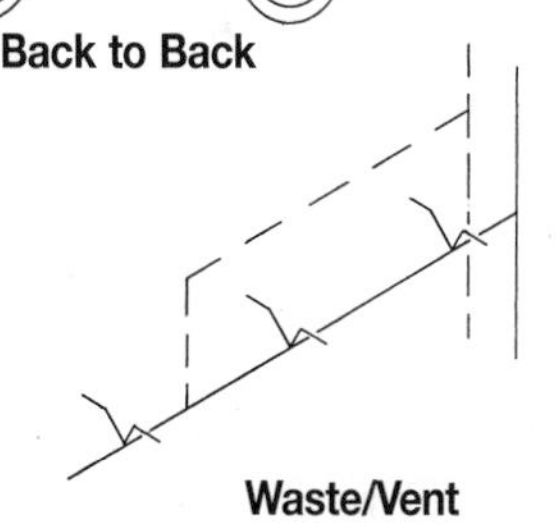

Supply

Waste/Vent

Supply

System Components	QUANTITY	UNIT	COST EACH MAT.	INST.	TOTAL
SYSTEM D2010 120 1760					
WATER CLOSETS, BATTERY MOUNT, WALL HUNG, SIDE BY SIDE, FIRST CLOSET					
Water closet, bowl only w/flush valve, seat, wall hung	1.000	Ea.	780	209	989
Pipe, CI soil, no hub, cplg 10′ OC, hanger 5′ OC, 4″ diam.	3.000	L.F.	55.05	63	118.05
Coupling, standard, CI, soil, no hub, 4″ diam.	2.000	Ea.	40	73	113
Copper tubing, type L, solder joints, hangers 10′ OC, 1″ diam.	6.000	L.F.	63.30	59.40	122.70
Copper tubing, type DWV, solder joints, hangers 10′OC, 2″ diam.	6.000	L.F.	138	91.80	229.80
Wrought copper 90° elbow for solder joints 1″ diam.	1.000	Ea.	17.20	42	59.20
Wrought copper Tee for solder joints 1″ diam.	1.000	Ea.	39.50	67	106.50
Support/carrier, siphon jet, horiz, adjustable single, 4″ pipe	1.000	Ea.	720	126	846
Valve, gate, bronze, 125 lb, NRS, soldered 1″ diam.	1.000	Ea.	77.50	35.50	113
Wrought copper, DWV, 90° elbow, 2″ diam.	1.000	Ea.	59	67	126
TOTAL			1,989.55	833.70	2,823.25

D2010 120	Water Closets, Group		COST EACH MAT.	INST.	TOTAL
1760	Water closets, battery mount, wall hung, side by side, first closet	R224000-30	2,000	835	2,835
1800	Each additional water closet, add		1,900	790	2,690
3000	Back to back, first pair of closets	R224000-40	3,450	1,100	4,550
3100	Each additional pair of closets, back to back		3,400	1,100	4,500

D2010 Plumbing Fixtures

Systems are complete with trim, flush valve and rough-in (supply, waste and vent) for connection to supply branches and waste mains.

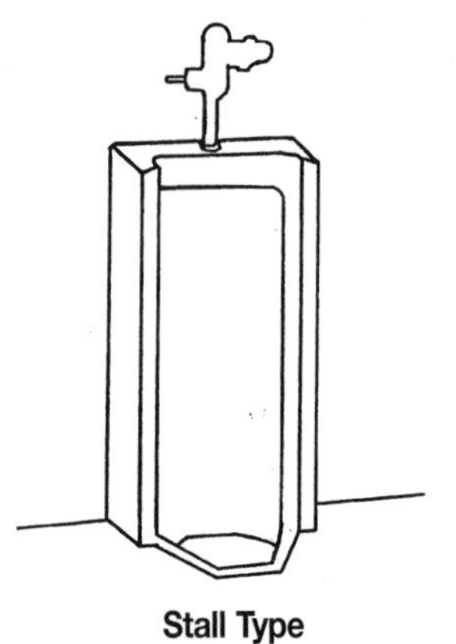
Stall Type

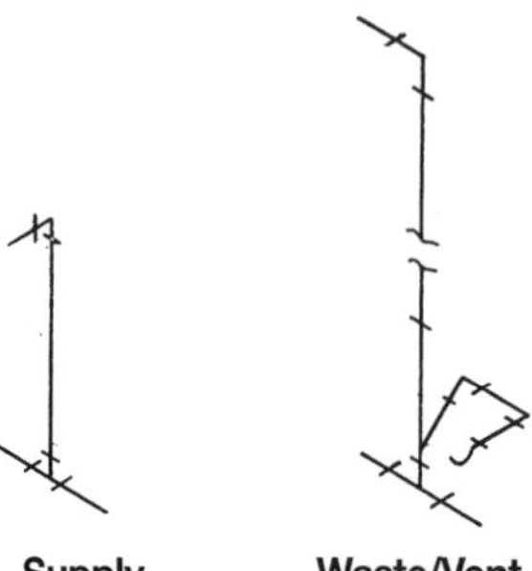
Supply Waste/Vent

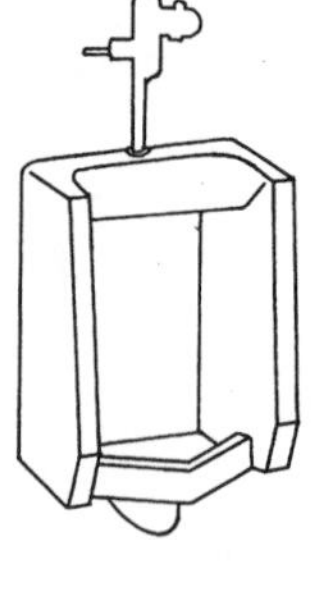
Wall Hung

System Components	QUANTITY	UNIT	COST EACH MAT.	COST EACH INST.	COST EACH TOTAL
SYSTEM D2010 210 2000					
URINAL, VITREOUS CHINA, WALL HUNG					
Urinal, wall hung, vitreous china, incl. hanger	1.000	Ea.	335	405	740
Pipe, steel, galvanized, schedule 40, threaded, 1-1/2" diam.	5.000	L.F.	54.75	75.75	130.50
Copper tubing type DWV, solder joint, hangers 10' OC, 2" diam.	3.000	L.F.	69	45.90	114.90
Combination Y & 1/8 bend for CI soil pipe, no hub, 3" diam.	1.000	Ea.	19.10		19.10
Pipe, CI, no hub, cplg. 10' OC, hanger 5' OC, 3" diam.	4.000	L.F.	56.20	75.60	131.80
Pipe coupling standard, CI soil, no hub, 3" diam.	3.000	Ea.	34.20	64	98.20
Copper tubing type L, solder joint, hanger 10' OC 3/4" diam.	5.000	L.F.	36	44.25	80.25
Wrought copper 90° elbow for solder joints 3/4" diam.	1.000	Ea.	7	35.50	42.50
Wrought copper Tee for solder joints, 3/4" diam.	1.000	Ea.	12.85	56	68.85
TOTAL			624.10	802	1,426.10

D2010 210	Urinal Systems		COST EACH MAT.	COST EACH INST.	COST EACH TOTAL
2000	Urinal, vitreous china, wall hung	R224000-30	625	800	1,425
2040	Stall type		1,375	960	2,335

D2010 Plumbing Fixtures

Systems are complete with trim, flush valve and rough-in (supply, waste and vent) for connection to supply branches and waste mains.

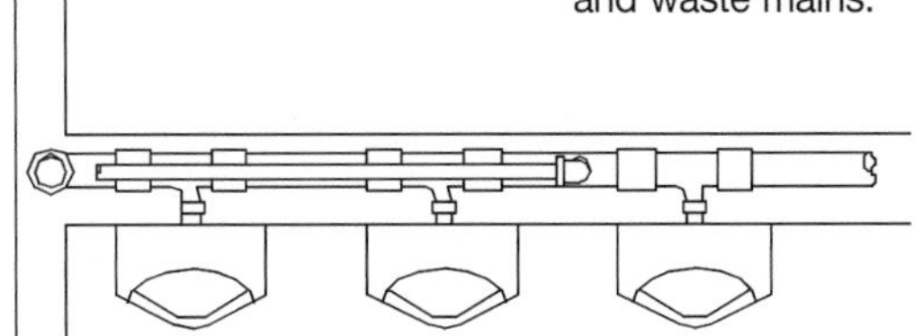

Side by Side

Back to Back

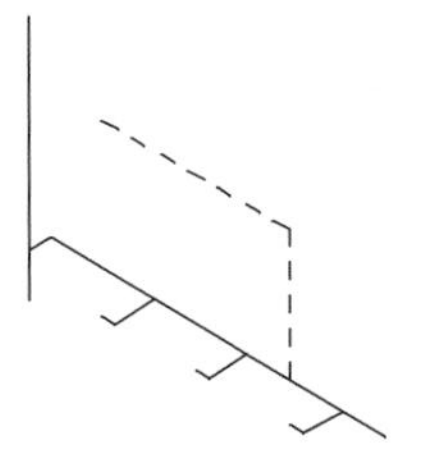

Waste/Vent

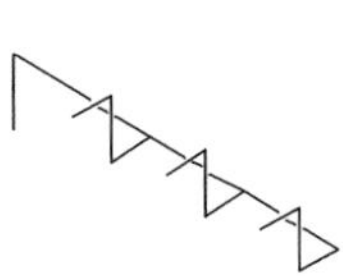

Supply

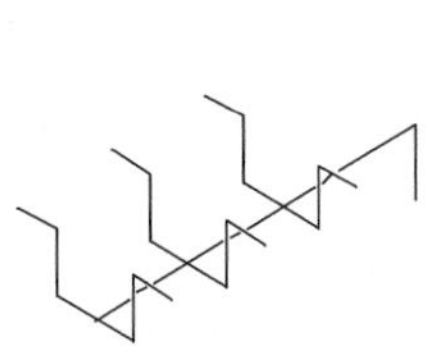

Supply

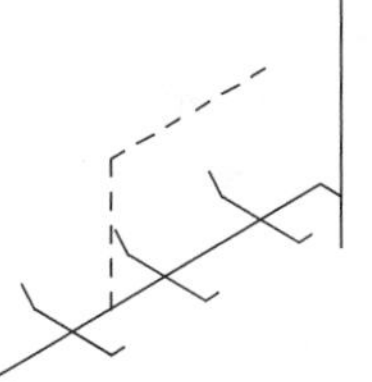

Waste/Vent

System Components	QUANTITY	UNIT	COST EACH MAT.	COST EACH INST.	COST EACH TOTAL
SYSTEM D2010 220 1760					
URINALS, BATTERY MOUNT, WALL HUNG, SIDE BY SIDE, FIRST URINAL					
Urinal, wall hung, vitreous china, with hanger & trim	1.000	Ea.	335	405	740
No hub cast iron soil pipe, 3″ diameter	4.000	L.F.	56.20	75.60	131.80
No hub cast iron sanitary tee, 3″ diameter	1.000	Ea.	19.10		19.10
No hub coupling, 3″ diameter	2.000	Ea.	34.20	64	98.20
Copper tubing, type L, 3/4″ diameter	5.000	L.F.	36	44.25	80.25
Copper tubing, type DWV, 2″ diameter	2.000	L.F.	46	30.60	76.60
Copper 90° elbow, 3/4″ diameter	2.000	Ea.	14	71	85
Copper 90° elbow, type DWV, 2″ diameter	2.000	Ea.	59	67	126
Galvanized steel pipe, 1-1/2″ diameter	5.000	L.F.	54.75	75.75	130.50
Cast iron drainage elbow, 90°, 1-1/2″ diameter	1.000	Ea.	12.85	56	68.85
TOTAL			667.10	889.20	1,556.30

D2010 220	Urinal Systems, Battery Mount		COST EACH MAT.	COST EACH INST.	COST EACH TOTAL
1760	Urinals, battery mount, side by side, first urinal		665	890	1,555
1800	Each additional urinal, add		685	865	1,550
2000	Back to back, first pair of urinals	R224000 -30	1,250	1,450	2,700
2100	Each additional pair of urinals, back to back		950	1,150	2,100

D2010 Plumbing Fixtures

Systems are complete with trim and rough-in (supply, waste and vent) to connect to supply branches and waste mains.

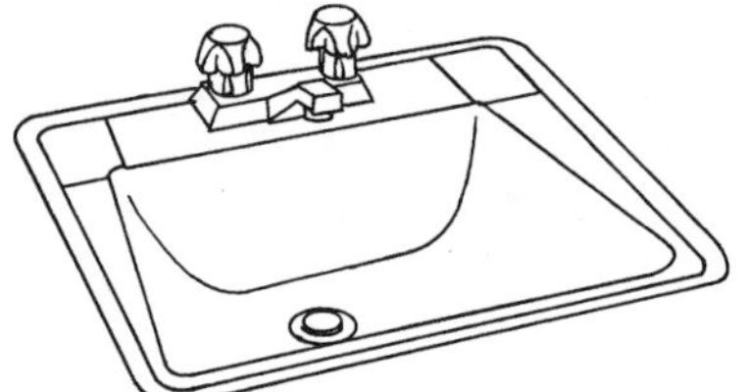

Vanity Top

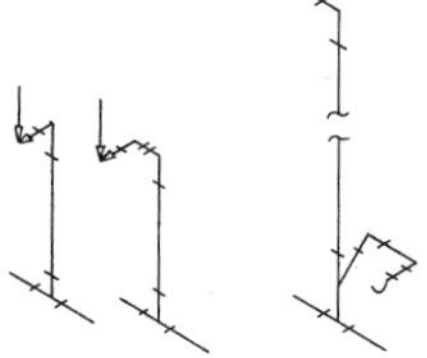

Supply Waste/Vent

Wall Hung

System Components	QUANTITY	UNIT	COST EACH MAT.	COST EACH INST.	COST EACH TOTAL
SYSTEM D2010 310 1560					
LAVATORY W/TRIM, VANITY TOP, P.E. ON C.I., 20″ X 18″					
Lavatory w/trim, PE on CI, white, vanity top, 20″ x 18″ oval	1.000	Ea.	355	189	544
Pipe, steel, galvanized, schedule 40, threaded, 1-1/4″ diam.	4.000	L.F.	37.40	54.40	91.80
Copper tubing type DWV, solder joint, hanger 10′ OC 1-1/4″ diam.	4.000	L.F.	53.40	44.80	98.20
Wrought copper DWV, Tee, sanitary, 1-1/4″ diam.	1.000	Ea.	59.50	74.50	134
P trap w/cleanout, 20 ga., 1-1/4″ diam.	1.000	Ea.	207	37.50	244.50
Copper tubing type L, solder joint, hanger 10′ OC 1/2″ diam.	10.000	L.F.	46.60	83	129.60
Wrought copper 90° elbow for solder joints 1/2″ diam.	2.000	Ea.	6.22	67	73.22
Wrought copper Tee for solder joints, 1/2″ diam.	2.000	Ea.	10.60	103	113.60
Stop, chrome, angle supply, 1/2″ diam.	2.000	Ea.	21.80	61	82.80
TOTAL			797.52	714.20	1,511.72

D2010 310	Lavatory Systems	COST EACH MAT.	COST EACH INST.	COST EACH TOTAL
1560	Lavatory w/trim, vanity top, PE on CI, 20″ x 18″, Vanity top by others.	800	715	1,515
1600	19″ x 16″ oval	630	715	1,345
1640	18″ round	705	715	1,420
1680	Cultured marble, 19″ x 17″	645	715	1,360
1720	25″ x 19″	675	715	1,390
1760	Stainless, self-rimming, 25″ x 22″	840	715	1,555
1800	17″ x 22″ R224000-30	830	715	1,545
1840	Steel enameled, 20″ x 17″	660	735	1,395
1880	19″ round	630	735	1,365
1920	Vitreous china, 20″ x 16″	730	750	1,480
1960	19″ x 16″	735	750	1,485
2000	22″ x 13″	735	750	1,485
2040	Wall hung, PE on CI, 18″ x 15″	965	790	1,755
2080	19″ x 17″	985	790	1,775
2120	20″ x 18″	830	790	1,620
2160	Vitreous china, 18″ x 15″	785	810	1,595
2200	19″ x 17″	740	810	1,550
2240	24″ x 20″	830	810	1,640
2300	20″ x 27″, handicap	1,100	875	1,975

D2010 Plumbing Fixtures

Systems are complete with trim, flush valve and rough-in (supply, waste and vent) for connection to supply branches and waste mains.

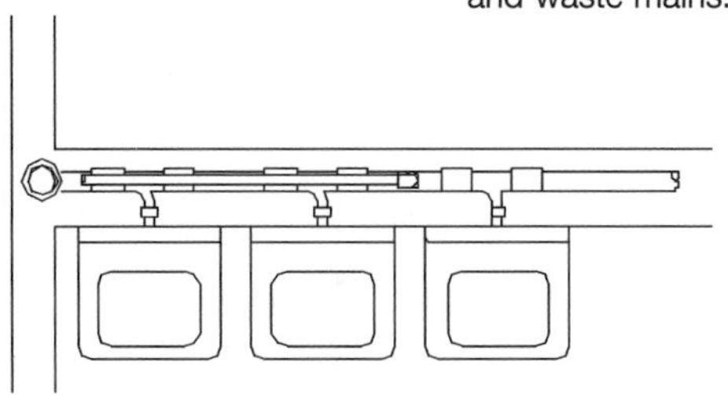

Side by Side

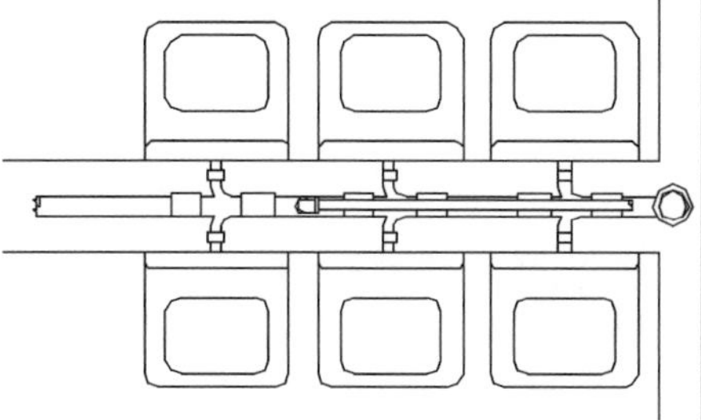

Back to Back

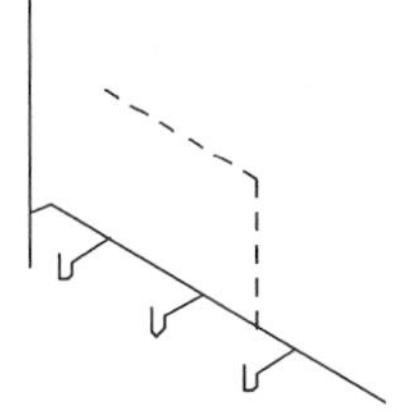

Waste/Vent

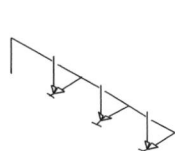

Supply
(Two supply systems required)

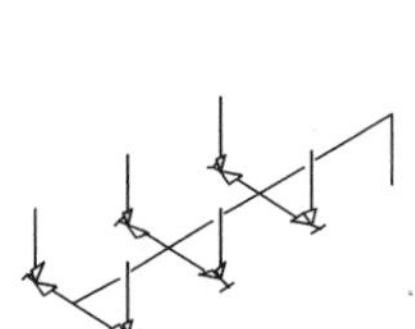

Supply
(Two supply systems required)

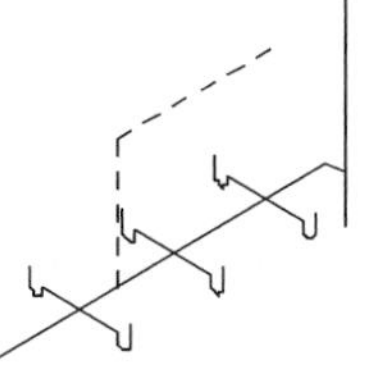

Waste/Vent

System Components	QUANTITY	UNIT	COST EACH		
			MAT.	INST.	TOTAL
SYSTEM D2010 320 1760					
LAVATORIES, BATTERY MOUNT, WALL HUNG, SIDE BY SIDE, FIRST LAVATORY					
Lavatory w/trim wall hung PE on CI 20″ x 18″	1.000	Ea.	300	151	451
Stop, chrome, angle supply, 3/8″ diameter	2.000	Ea.	21.80	56	77.80
Concealed arm support	1.000	Ea.	510	112	622
P trap w/cleanout, 20 ga. C.P., 1-1/4″ diameter	1.000	Ea.	33	37.50	70.50
Copper tubing, type L, 1/2″ diameter	10.000	L.F.	46.60	83	129.60
Copper tubing, type DWV, 1-1/4″ diameter	4.000	L.F.	53.40	44.80	98.20
Copper 90° elbow, 1/2″ diameter	2.000	Ea.	6.22	67	73.22
Copper tee, 1/2″ diameter	2.000	Ea.	10.60	103	113.60
DWV copper sanitary tee, 1-1/4″ diameter	2.000	Ea.	119	149	268
Galvanized steel pipe, 1-1/4″ diameter	4.000	L.F.	37.40	54.40	91.80
Black cast iron 90° elbow, 1-1/4″ diameter	1.000	Ea.	10.35	55	65.35
TOTAL			1,148.37	912.70	2,061.07

D2010 320	Lavatory Systems, Battery Mount	COST EACH		
		MAT.	INST.	TOTAL
1760	Lavatories, battery mount, side by side, first lavatory	1,150	915	2,065
1800	Each additional lavatory, add	1,000	635	1,635
2000	Back to back, first pair of lavatories	1,825	1,450	3,275
2100	Each additional pair of lavatories, back to back	1,725	1,200	2,925

D2010 Plumbing Fixtures

Systems are complete with trim and rough-in (supply, waste and vent) to connect to supply branches and waste mains.

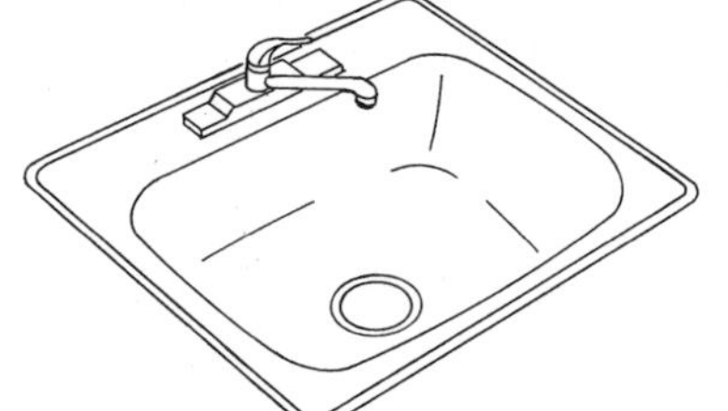

Countertop Single Bowl

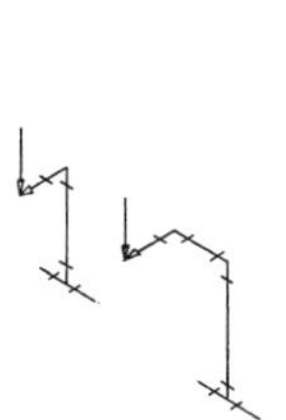

Supply

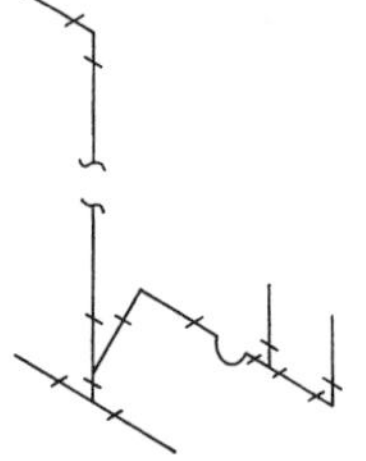

Waste/Vent

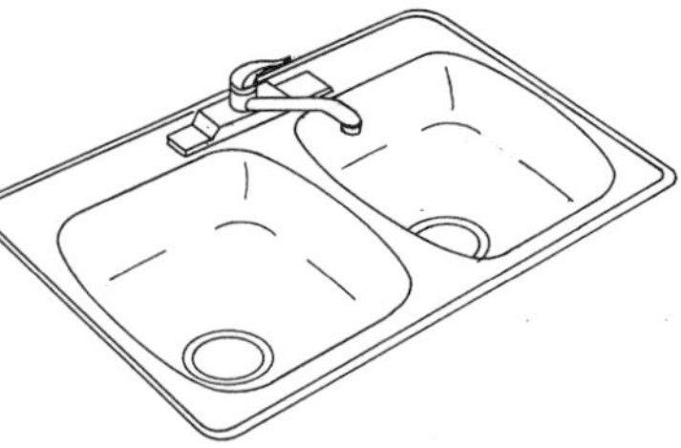

Countertop Double Bowl

System Components	QUANTITY	UNIT	COST EACH MAT.	COST EACH INST.	COST EACH TOTAL
SYSTEM D2010 410 1720					
KITCHEN SINK W/TRIM, COUNTERTOP, P.E. ON C.I., 24″ X 21″, SINGLE BOWL					
Kitchen sink, counter top, PE on CI, 1 bowl, 24″ x 21″ OD	1.000	Ea.	305	216	521
Pipe, steel, galvanized, schedule 40, threaded, 1-1/4″ diam.	4.000	L.F.	37.40	54.40	91.80
Copper tubing, type DWV, solder, hangers 10′ OC 1-1/2″ diam.	6.000	L.F.	100.50	74.70	175.20
Wrought copper, DWV, Tee, sanitary, 1-1/2″ diam.	1.000	Ea.	74	84	158
P trap, standard, copper, 1-1/2″ diam.	1.000	Ea.	200	39.50	239.50
Copper tubing, type L, solder joints, hangers 10′ OC 1/2″ diam.	10.000	L.F.	46.60	83	129.60
Wrought copper 90° elbow for solder joints 1/2″ diam.	2.000	Ea.	6.22	67	73.22
Wrought copper Tee for solder joints, 1/2″ diam.	2.000	Ea.	10.60	103	113.60
Stop, angle supply, chrome, 1/2″ CTS	2.000	Ea.	21.80	61	82.80
TOTAL			802.12	782.60	1,584.72

D2010 410	Kitchen Sink Systems		COST EACH MAT.	COST EACH INST.	COST EACH TOTAL
1720	Kitchen sink w/trim, countertop, PE on CI, 24″x21″, single bowl		800	785	1,585
1760	30″ x 21″ single bowl		1,100	785	1,885
1800	32″ x 21″ double bowl	R221316 -10	870	845	1,715
1840	42″ x 21″ double bowl		1,800	855	2,655
1880	Stainless steel, 19″ x 18″ single bowl	R221316 -20	1,125	785	1,910
1920	25″ x 22″ single bowl		1,200	785	1,985
1960	33″ x 22″ double bowl	R224000 -20	1,575	845	2,420
2000	43″ x 22″ double bowl		1,750	855	2,605
2040	44″ x 22″ triple bowl		2,025	890	2,915
2080	44″ x 24″ corner double bowl		1,450	855	2,305
2120	Steel, enameled, 24″ x 21″ single bowl		1,000	785	1,785
2160	32″ x 21″ double bowl		1,050	845	1,895
2240	Raised deck, PE on CI, 32″ x 21″, dual level, double bowl		970	1,075	2,045
2280	42″ x 21″ dual level, triple bowl		1,850	1,175	3,025

D2010 Plumbing Fixtures

Systems are complete with trim and rough-in (supply, waste and vent) to connect to supply branches and waste mains.

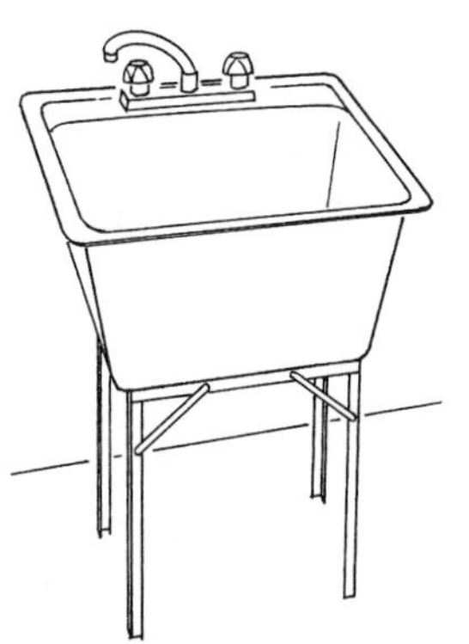

Single Compartment Sink

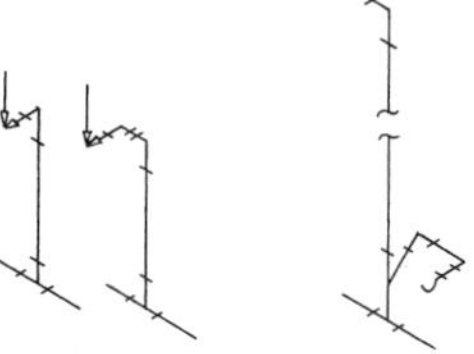

Supply Waste/Vent

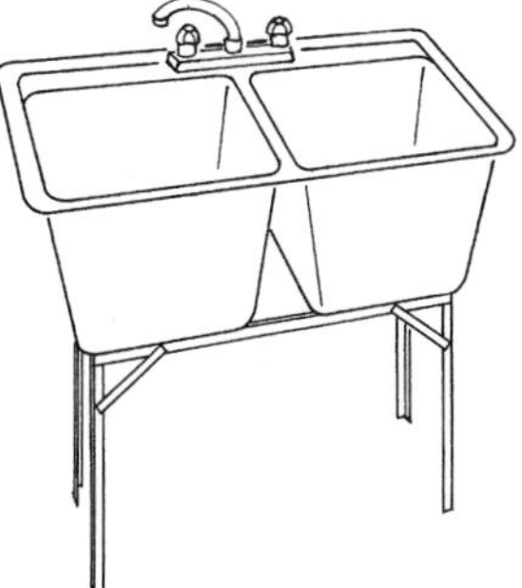

Double Compartment Sink

System Components	QUANTITY	UNIT	COST EACH MAT.	COST EACH INST.	COST EACH TOTAL
SYSTEM D2010 420 1760					
LAUNDRY SINK W/TRIM, PE ON CI, BLACK IRON FRAME					
24″ X 20″ OD, SINGLE COMPARTMENT					
Laundry sink PE on CI w/trim & frame, 24″ x 21″ OD, 1 compartment	1.000	Ea.	470	202	672
Pipe, steel, galvanized, schedule 40, threaded, 1-1/4″ diam	4.000	L.F.	37.40	54.40	91.80
Copper tubing, type DWV, solder joint, hanger 10′ OC 1-1/2″diam	6.000	L.F.	100.50	74.70	175.20
Wrought copper, DWV, Tee, sanitary, 1-1/2″ diam	1.000	Ea.	74	84	158
P trap, standard, copper, 1-1/2″ diam	1.000	Ea.	200	39.50	239.50
Copper tubing type L, solder joints, hangers 10′ OC, 1/2″ diam	10.000	L.F.	46.60	83	129.60
Wrought copper 90° elbow for solder joints 1/2″ diam	2.000	Ea.	6.22	67	73.22
Wrought copper Tee for solder joints, 1/2″ diam	2.000	Ea.	10.60	103	113.60
Stop, angle supply, 1/2″ diam	2.000	Ea.	21.80	61	82.80
TOTAL			967.12	768.60	1,735.72

D2010 420	Laundry Sink Systems		COST EACH MAT.	COST EACH INST.	COST EACH TOTAL
1740	Laundry sink w/trim, PE on CI, black iron frame				
1760	24″ x 20″, single compartment		965	770	1,735
1800	24″ x 23″ single compartment	R224000-30	990	770	1,760
1840	48″ x 21″ double compartment		1,450	835	2,285
1920	Molded stone, on wall, 22″ x 21″ single compartment		705	770	1,475
1960	45″x 21″ double compartment		895	835	1,730
2040	Plastic, on wall or legs, 18″ x 23″ single compartment		665	755	1,420
2080	20″ x 24″ single compartment		665	755	1,420
2120	36″ x 23″ double compartment		730	810	1,540
2160	40″ x 24″ double compartment		830	810	1,640

D2010 Plumbing Fixtures

Corrosion resistant laboratory sink systems are complete with trim and rough-in (supply, waste and vent) to connect to supply branches and waste mains.

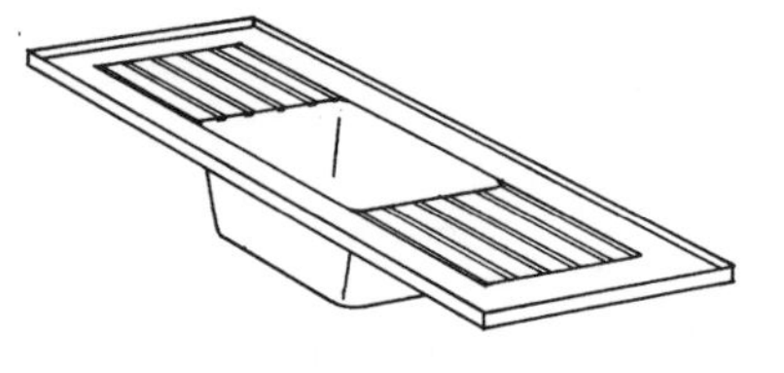

Laboratory Sink

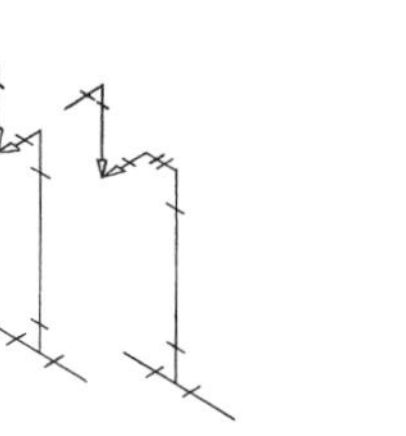

Supply Waste/Vent

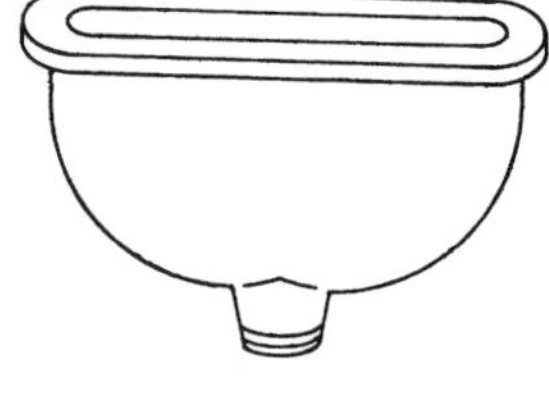

Polypropylene Cup Sink

System Components	QUANTITY	UNIT	COST EACH		
			MAT.	INST.	TOTAL
SYSTEM D2010 430 1600					
LABORATORY SINK W/TRIM, STAINLESS STEEL, SINGLE BOWL					
DOUBLE DRAINBOARD, 54" X 24" O.D.					
Sink w/trim, stainless steel, 1 bowl, 2 drainboards 54" x 24" OD	1.000	Ea.	1,825	405	2,230
Pipe, polypropylene, schedule 40, acid resistant 1-1/2" diam.	10.000	L.F.	83.50	178	261.50
Tee, sanitary, polypropylene, acid resistant, 1-1/2" diam.	1.000	Ea.	18.05	67	85.05
P trap, polypropylene, acid resistant, 1-1/2" diam.	1.000	Ea.	58	39.50	97.50
Copper tubing type L, solder joint, hanger 10' O.C. 1/2" diam.	10.000	L.F.	46.60	83	129.60
Wrought copper 90° elbow for solder joints 1/2" diam.	2.000	Ea.	6.22	67	73.22
Wrought copper Tee for solder joints, 1/2" diam.	2.000	Ea.	10.60	103	113.60
Stop, angle supply, chrome, 1/2" diam.	2.000	Ea.	21.80	61	82.80
TOTAL			2,069.77	1,003.50	3,073.27

D2010 430	Laboratory Sink Systems	COST EACH		
		MAT.	INST.	TOTAL
1580	Laboratory sink w/trim, stainless steel, single bowl,			
1590	Stainless steel, single bowl,			
1600	Double drainboard, 54" x 24" O.D.	2,075	1,000	3,075
1640	Single drainboard, 47" x 24"O.D.	1,525	1,000	2,525
1670	Stainless steel, double bowl,			
1680	70" x 24" O.D.	2,275	1,000	3,275
1750	Polyethylene, single bowl,			
1760	Flanged, 14-1/2" x 14-1/2" O.D.	495	905	1,400
1800	18-1/2" x 18-1/2" O.D.	595	905	1,500
1840	23-1/2" x 20-1/2" O.D.	670	905	1,575
1920	Polypropylene, cup sink, oval, 7" x 4" O.D.	370	795	1,165
1960	10" x 4-1/2" O.D.	395	795	1,190

D2010 Plumbing Fixtures

Corrosion resistant laboratory sink systems are complete with trim and rough–in (supply, waste and vent) to connect to supply branches and waste mains.

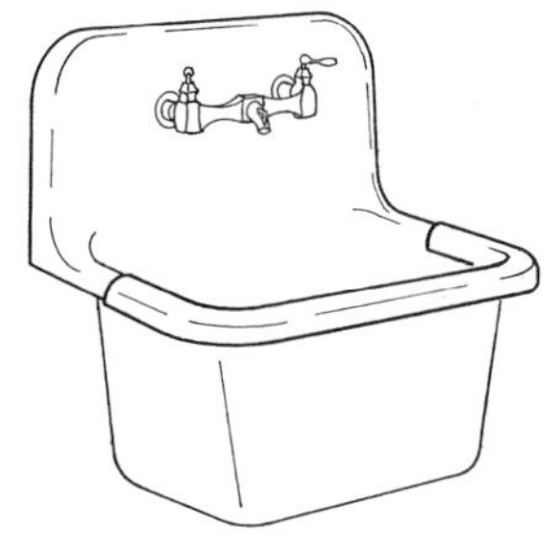
Wall Hung

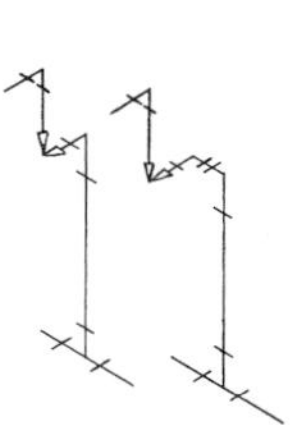
Supply

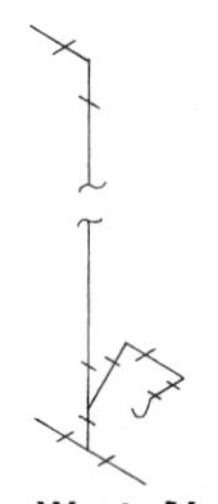
Waste/Vent

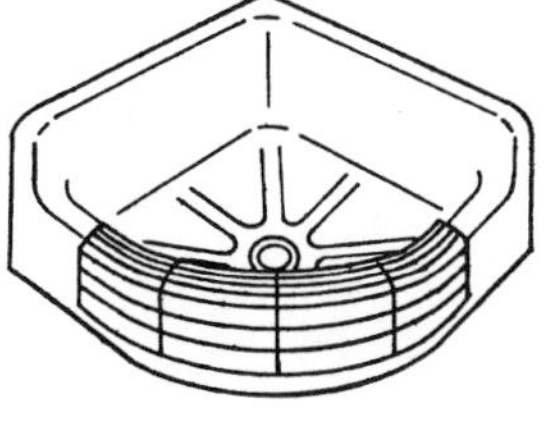
Corner, Floor

System Components	QUANTITY	UNIT	COST EACH MAT.	COST EACH INST.	COST EACH TOTAL
SYSTEM D2010 440 4260					
SERVICE SINK, PE ON CI, CORNER FLOOR, 28″X28″, W/RIM GUARD & TRIM					
Service sink, corner floor, PE on CI, 28″ x 28″, w/rim guard & trim	1.000	Ea.	1,050	275	1,325
Copper tubing type DWV, solder joint, hanger 10′OC 3″ diam.	6.000	L.F.	261	126	387
Copper tubing type DWV, solder joint, hanger 10′OC 2″ diam	4.000	L.F.	92	61.20	153.20
Wrought copper DWV, Tee, sanitary, 3″ diam.	1.000	Ea.	355	173	528
P trap with cleanout & slip joint, copper 3″ diam	1.000	Ea.	740	61	801
Copper tubing, type L, solder joints, hangers 10′ OC, 1/2″ diam	10.000	L.F.	46.60	83	129.60
Wrought copper 90° elbow for solder joints 1/2″ diam	2.000	Ea.	6.22	67	73.22
Wrought copper Tee for solder joints, 1/2″ diam	2.000	Ea.	10.60	103	113.60
Stop, angle supply, chrome, 1/2″ diam	2.000	Ea.	21.80	61	82.80
TOTAL			2,583.22	1,010.20	3,593.42

D2010 440	Service Sink Systems		COST EACH MAT.	COST EACH INST.	COST EACH TOTAL
4260	Service sink w/trim, PE on CI, corner floor, 28″ x 28″, w/rim guard	R224000 -30	2,575	1,000	3,575
4300	Wall hung w/rim guard, 22″ x 18″		2,775	1,175	3,950
4340	24″ x 20″		2,875	1,175	4,050
4380	Vitreous china, wall hung 22″ x 20″		2,600	1,175	3,775

D2010 Plumbing Fixtures

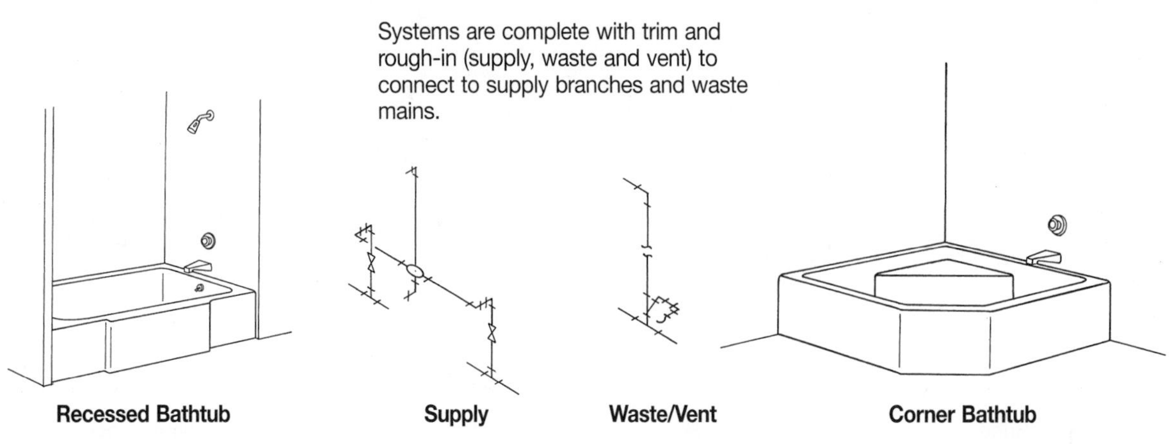
Systems are complete with trim and rough-in (supply, waste and vent) to connect to supply branches and waste mains.
Recessed Bathtub
Supply
Waste/Vent
Corner Bathtub

System Components	QUANTITY	UNIT	COST EACH MAT.	COST EACH INST.	COST EACH TOTAL
SYSTEM D2010 510 2000					
BATHTUB, RECESSED, PORCELAIN ENAMEL ON CAST IRON,, 48″ x 42″					
Bath tub, porcelain enamel on cast iron, w/fittings, 48″ x 42″	1.000	Ea.	2,825	305	3,130
Pipe, steel, galvanized, schedule 40, threaded, 1-1/4″ diam.	4.000	L.F.	37.40	54.40	91.80
Pipe, CI no hub soil w/couplings 10′ OC, hangers 5′ OC, 4″ diam.	3.000	L.F.	55.05	63	118.05
Combination Y and 1/8 bend for C.I. soil pipe, no hub, 4″ pipe size	1.000	Ea.	49.50		49.50
Drum trap, 3″ x 5″, copper, 1-1/2″ diam.	1.000	Ea.	280	42	322
Copper tubing type L, solder joints, hangers 10′ OC 1/2″ diam.	10.000	L.F.	46.60	83	129.60
Wrought copper 90° elbow, solder joints, 1/2″ diam.	2.000	Ea.	6.22	67	73.22
Wrought copper Tee, solder joints, 1/2″ diam.	2.000	Ea.	10.60	103	113.60
Stop, angle supply, 1/2″ diameter	2.000	Ea.	21.80	61	82.80
Copper tubing type DWV, solder joints, hanger 10′ OC 1-1/2″ diam.	3.000	L.F.	50.25	37.35	87.60
Pipe coupling, standard, C.I. soil no hub, 4″ pipe size	2.000	Ea.	40	73	113
TOTAL			3,422.42	888.75	4,311.17

D2010 510	Bathtub Systems	COST EACH MAT.	COST EACH INST.	COST EACH TOTAL
2000	Bathtub, recessed, P.E. on CI., 48″ x 42″ R224000-30	3,425	890	4,315
2040	72″ x 36″	3,525	990	4,515
2080	Mat bottom, 5′ long	1,800	860	2,660
2120	5′-6″ long	2,525	890	3,415
2160	Corner, 48″ x 42″	3,425	860	4,285
2200	Formed steel, enameled, 4′-6″ long	1,150	795	1,945

D2010 Plumbing Fixtures

Systems are complete with trim, flush valve and rough-in (supply, waste and vent) for connection to supply branches and waste mains.

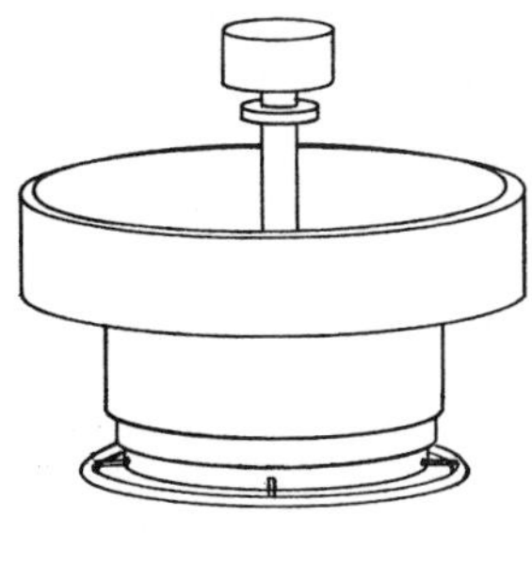

Circular Fountain

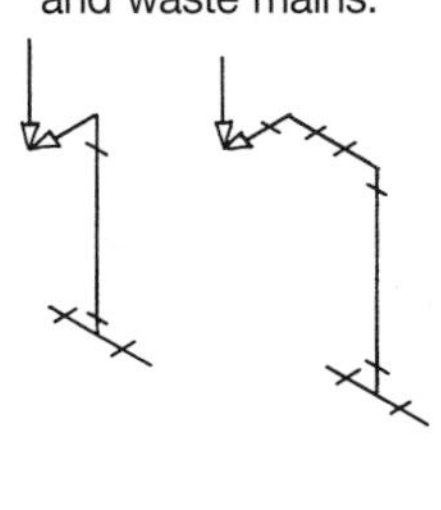

Supply

Waste/Vent

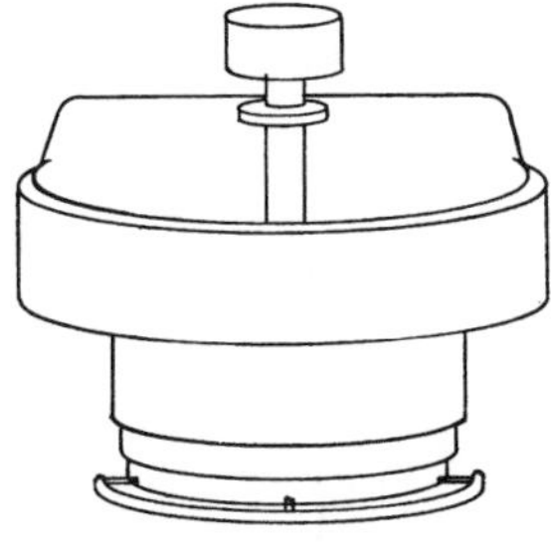

Semi-Circular Fountain

System Components	QUANTITY	UNIT	COST EACH MAT.	COST EACH INST.	COST EACH TOTAL
SYSTEM D2010 610 1760					
GROUP WASH FOUNTAIN, PRECAST TERRAZZO					
CIRCULAR, 36″ DIAMETER					
Wash fountain, group, precast terrazzo, foot control 36″ diam.	1.000	Ea.	6,000	630	6,630
Copper tubing type DWV, solder joint, hanger 10′ OC, 2″ diam.	10.000	L.F.	230	153	383
P trap, standard, copper, 2″ diam.	1.000	Ea.	310	45	355
Wrought copper, Tee, sanitary, 2″ diam.	1.000	Ea.	86.50	96	182.50
Copper tubing type L, solder joint, hanger 10′ OC 1/2″ diam.	20.000	L.F.	93.20	166	259.20
Wrought copper 90° elbow for solder joints 1/2″ diam.	3.000	Ea.	9.33	100.50	109.83
Wrought copper Tee for solder joints, 1/2″ diam.	2.000	Ea.	10.60	103	113.60
TOTAL			6,739.63	1,293.50	8,033.13

D2010 610	Group Wash Fountain Systems	COST EACH MAT.	COST EACH INST.	COST EACH TOTAL
1740	Group wash fountain, precast terrazzo			
1760	Circular, 36″ diameter	6,750	1,300	8,050
1800	54″ diameter R224000 -30	8,225	1,425	9,650
1840	Semi-circular, 36″ diameter	6,000	1,300	7,300
1880	54″ diameter	7,400	1,425	8,825
1960	Stainless steel, circular, 36″ diameter	6,750	1,200	7,950
2000	54″ diameter	8,500	1,325	9,825
2040	Semi-circular, 36″ diameter	5,975	1,200	7,175
2080	54″ diameter	7,350	1,325	8,675
2160	Thermoplastic, circular, 36″ diameter	5,075	980	6,055
2200	54″ diameter	5,775	1,125	6,900
2240	Semi-circular, 36″ diameter	4,075	980	5,055
2280	54″ diameter	5,600	1,125	6,725

Systems are complete with trim and rough-in (supply, waste and vent) for connection to supply branches and waste mains.

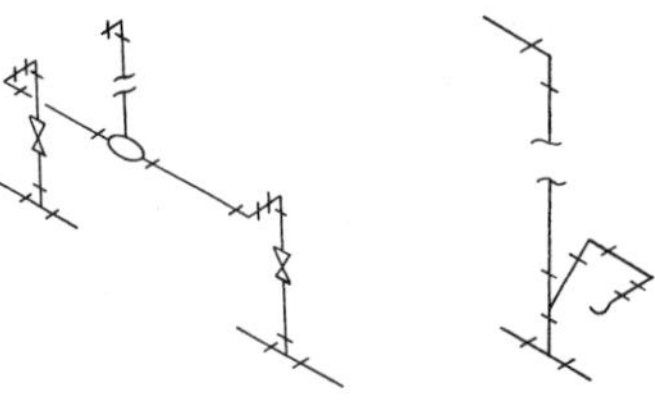

Three Wall

Supply

Waste/Vent

Corner Angle

System Components	QUANTITY	UNIT	COST EACH MAT.	COST EACH INST.	COST EACH TOTAL
SYSTEM D2010 710 1560					
SHOWER, STALL, BAKED ENAMEL, MOLDED STONE RECEPTOR, 30″ SQUARE					
Shower stall, enameled steel, molded stone receptor, 30″ square	1.000	Ea.	1,275	233	1,508
Copper tubing type DWV, solder joints, hangers 10′ OC, 2″ diam.	6.000	L.F.	100.50	74.70	175.20
Wrought copper DWV, Tee, sanitary, 2″ diam.	1.000	Ea.	74	84	158
Trap, standard, copper, 2″ diam.	1.000	Ea.	200	39.50	239.50
Copper tubing type L, solder joint, hanger 10′ OC 1/2″ diam.	16.000	L.F.	74.56	132.80	207.36
Wrought copper 90° elbow for solder joints 1/2″ diam.	3.000	Ea.	9.33	100.50	109.83
Wrought copper Tee for solder joints, 1/2″ diam.	2.000	Ea.	10.60	103	113.60
Stop and waste, straightway, bronze, solder joint 1/2″ diam.	2.000	Ea.	25.50	56	81.50
TOTAL			1,769.49	823.50	2,592.99

D2010 710	Shower Systems		COST EACH MAT.	COST EACH INST.	COST EACH TOTAL
1560	Shower, stall, baked enamel, molded stone receptor, 30″ square		1,775	825	2,600
1600	32″ square		1,775	835	2,610
1640	Terrazzo receptor, 32″ square	R224000 -30	1,975	835	2,810
1680	36″ square		2,125	845	2,970
1720	36″ corner angle		2,400	845	3,245
1800	Fiberglass one piece, three walls, 32″ square		1,100	810	1,910
1840	36″ square		1,125	810	1,935
1880	Polypropylene, molded stone receptor, 30″ square		1,200	1,200	2,400
1920	32″ square		1,200	1,200	2,400
1960	Built-in head, arm, bypass, stops and handles		121	310	431
2050	Shower, stainless steel panels, handicap				
2100	w/fixed and handheld head, control valves, grab bar, and seat		6,025	3,700	9,725
2500	Shower, group with six heads, thermostatic mix valves & balancing valve		5,450	910	6,360
2520	Five heads		4,525	825	5,350

D2010 Plumbing Fixtures

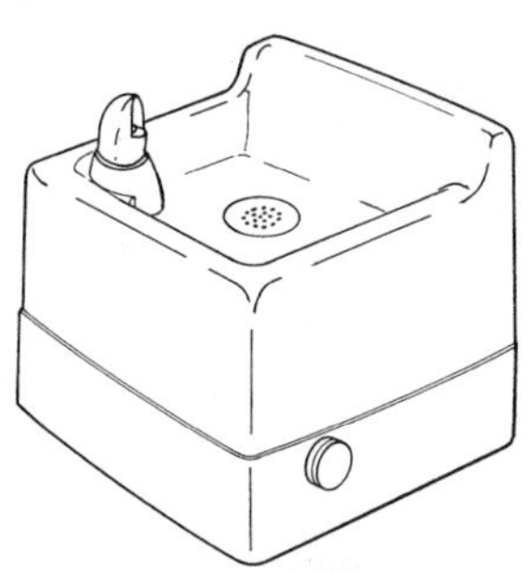

Wall Mounted, No Back

Systems are complete with trim and rough-in (supply, waste and vent) to connect to supply branches and waste mains.

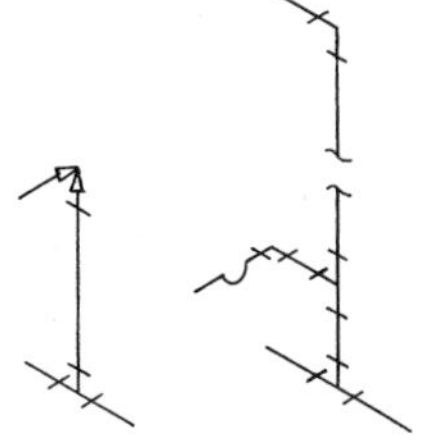

Supply Waste/Vent

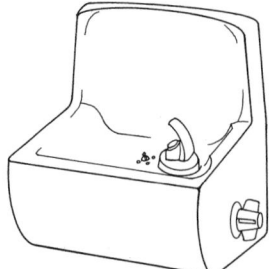

Wall Mounted, Low Back

System Components	QUANTITY	UNIT	COST EACH MAT.	COST EACH INST.	COST EACH TOTAL
SYSTEM D2010 810 1800					
DRINKING FOUNTAIN, ONE BUBBLER, WALL MOUNTED					
NON RECESSED, BRONZE, NO BACK					
Drinking fountain, wall mount, bronze, 1 bubbler	1.000	Ea.	1,150	168	1,318
Copper tubing, type L, solder joint, hanger 10′ OC 3/8″ diam.	5.000	L.F.	20.95	40	60.95
Stop, supply, straight, chrome, 3/8″ diam.	1.000	Ea.	12.45	28	40.45
Wrought copper 90° elbow for solder joints 3/8″ diam.	1.000	Ea.	9.35	30.50	39.85
Wrought copper Tee for solder joints, 3/8″ diam.	1.000	Ea.	15.75	48	63.75
Copper tubing, type DWV, solder joint, hanger 10′ OC 1-1/4″ diam.	4.000	L.F.	53.40	44.80	98.20
P trap, standard, copper drainage, 1-1/4″ diam.	1.000	Ea.	207	37.50	244.50
Wrought copper, DWV, Tee, sanitary, 1-1/4″ diam.	1.000	Ea.	59.50	74.50	134
TOTAL			1,528.40	471.30	1,999.70

D2010 810	Drinking Fountain Systems	COST EACH MAT.	COST EACH INST.	COST EACH TOTAL
1740	Drinking fountain, one bubbler, wall mounted			
1760	Non recessed			
1800	Bronze, no back R224000-30	1,525	470	1,995
1840	Cast iron, enameled, low back	1,375	470	1,845
1880	Fiberglass, 12″ back	2,075	470	2,545
1920	Stainless steel, no back	1,525	470	1,995
1960	Semi-recessed, poly marble	1,400	470	1,870
2040	Stainless steel	1,625	470	2,095
2080	Vitreous china	1,350	470	1,820
2120	Full recessed, poly marble	2,125	470	2,595
2200	Stainless steel	2,050	470	2,520
2240	Floor mounted, pedestal type, aluminum	2,850	640	3,490
2320	Bronze	2,500	640	3,140
2360	Stainless steel	2,300	640	2,940

D2010 Plumbing Fixtures

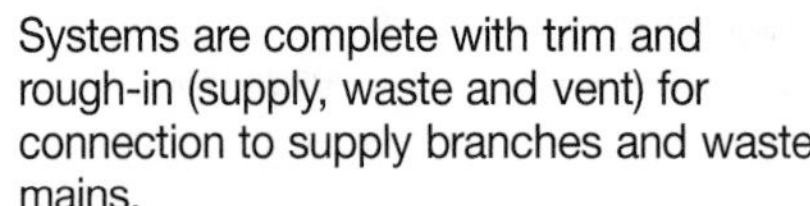

Systems are complete with trim and rough-in (supply, waste and vent) for connection to supply branches and waste mains.

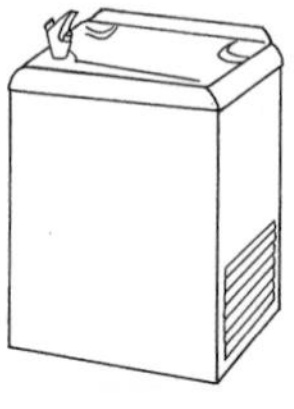
Wall Hung

Supply

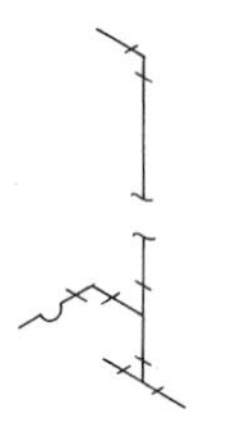
Waste/Vent

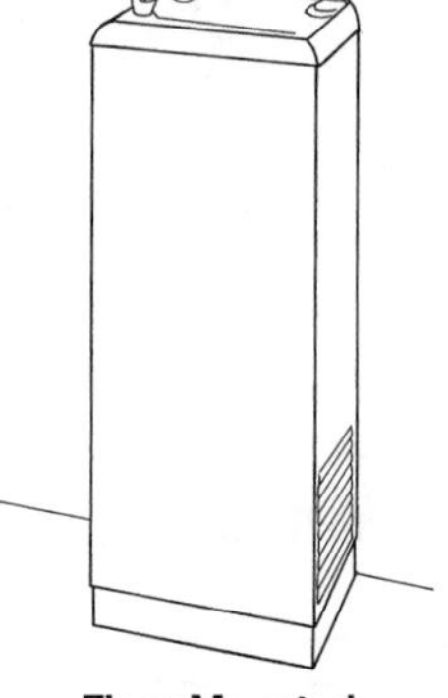
Floor Mounted

System Components	QUANTITY	UNIT	COST EACH MAT.	INST.	TOTAL
SYSTEM D2010 820 1840					
WATER COOLER, ELECTRIC, SELF CONTAINED, WALL HUNG, 8.2 G.P.H.					
Water cooler, wall mounted, 8.2 GPH	1.000	Ea.	795	305	1,100
Copper tubing type DWV, solder joint, hanger 10′ OC 1-1/4″ diam.	4.000	L.F.	53.40	44.80	98.20
Wrought copper DWV, Tee, sanitary 1-1/4″ diam.	1.000	Ea.	59.50	74.50	134
P trap, copper drainage, 1-1/4″ diam.	1.000	Ea.	207	37.50	244.50
Copper tubing type L, solder joint, hanger 10′ OC 3/8″ diam.	5.000	L.F.	20.95	40	60.95
Wrought copper 90° elbow for solder joints 3/8″ diam.	1.000	Ea.	9.35	30.50	39.85
Wrought copper Tee for solder joints, 3/8″ diam.	1.000	Ea.	15.75	48	63.75
Stop and waste, straightway, bronze, solder, 3/8″ diam.	1.000	Ea.	12.45	28	40.45
TOTAL			1,173.40	608.30	1,781.70

D2010 820	Water Cooler Systems		COST EACH MAT.	INST.	TOTAL
1840	Water cooler, electric, wall hung, 8.2 G.P.H.	R224000 -30	1,175	610	1,785
1880	Dual height, 14.3 G.P.H.		1,450	625	2,075
1920	Wheelchair type, 7.5 G.P.H.		1,375	610	1,985
1960	Semi recessed, 8.1 G.P.H.		1,175	610	1,785
2000	Full recessed, 8 G.P.H.	R224000 -50	2,500	650	3,150
2040	Floor mounted, 14.3 G.P.H.		1,325	525	1,850
2080	Dual height, 14.3 G.P.H.		1,650	640	2,290
2120	Refrigerated compartment type, 1.5 G.P.H.		1,925	525	2,450

D2010 Plumbing Fixtures

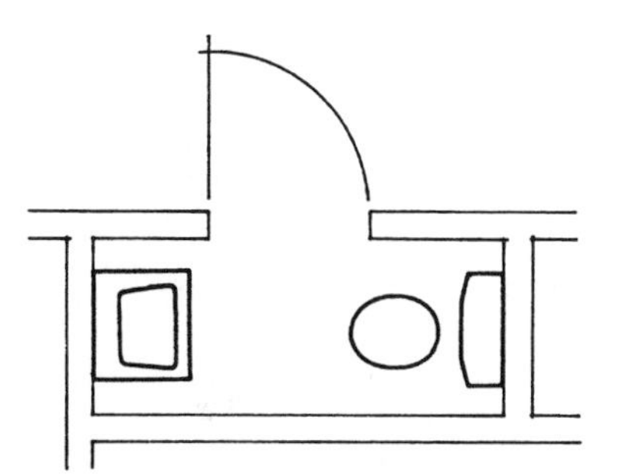

Two Fixture Bathroom Systems consisting of a lavatory, water closet, and rough-in service piping.

- Prices for plumbing and fixtures only.

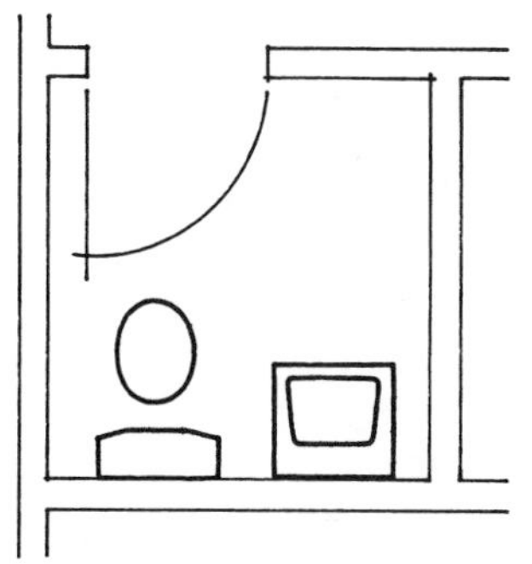

*Common wall is with an adjacent bathroom.

System Components	QUANTITY	UNIT	COST EACH MAT.	COST EACH INST.	COST EACH TOTAL
SYSTEM D2010 920 1180					
BATHROOM, LAVATORY & WATER CLOSET, 2 WALL PLUMBING, STAND ALONE					
Water closet, 2 Pc. close cpld. vit .china flr. mntd. w/seat, supply & stop	1.000	Ea.	253	228	481
Water closet, rough-in waste & vent	1.000	Set	355	395	750
Lavatory w/ftngs., wall hung, white, PE on CI, 20″ x 18″	1.000	Ea.	300	151	451
Lavatory, rough-in waste & vent	1.000	Set	530	730	1,260
Copper tubing type L, solder joint, hanger 10′ OC 1/2″ diam.	10.000	L.F.	46.60	83	129.60
Pipe, steel, galvanized, schedule 40, threaded, 2″ diam.	12.000	L.F.	175.80	226.80	402.60
Pipe, CI soil, no hub, coupling 10′ OC, hanger 5′ OC, 4″ diam.	7.000	L.F.	130.20	154	284.20
TOTAL			1,790.60	1,967.80	3,758.40

D2010 920	Two Fixture Bathroom, Two Wall Plumbing	COST EACH MAT.	COST EACH INST.	COST EACH TOTAL
1180	Bathroom, lavatory & water closet, 2 wall plumbing, stand alone	1,800	1,975	3,775
1200	Share common plumbing wall*	1,600	1,700	3,300

D2010 922	Two Fixture Bathroom, One Wall Plumbing		COST EACH MAT.	COST EACH INST.	COST EACH TOTAL
2220	Bathroom, lavatory & water closet, one wall plumbing, stand alone	R224000-30	1,650	1,775	3,425
2240	Share common plumbing wall*		1,450	1,500	2,950
2260		R224000-40			
2280					

D2010 Plumbing Fixtures

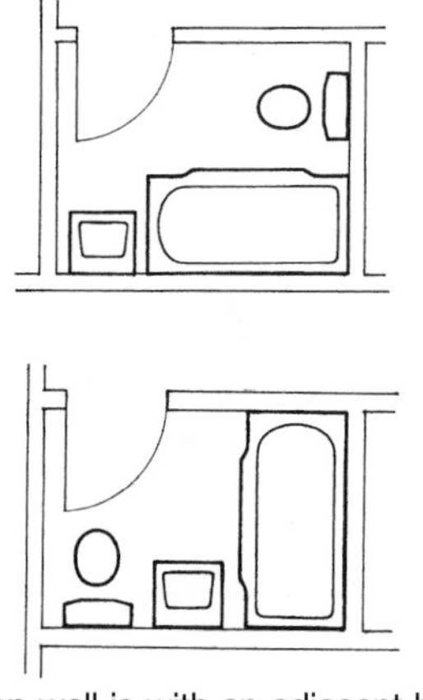

Three Fixture Bathroom Systems consisting of a lavatory, water closet, bathtub or shower and rough-in service piping.

- Prices for plumbing and fixtures only.

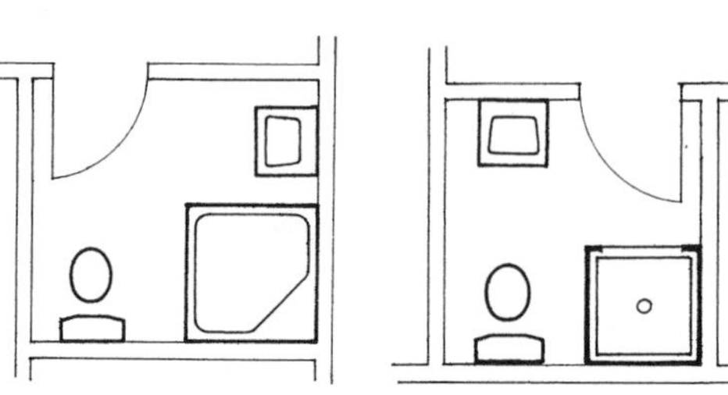

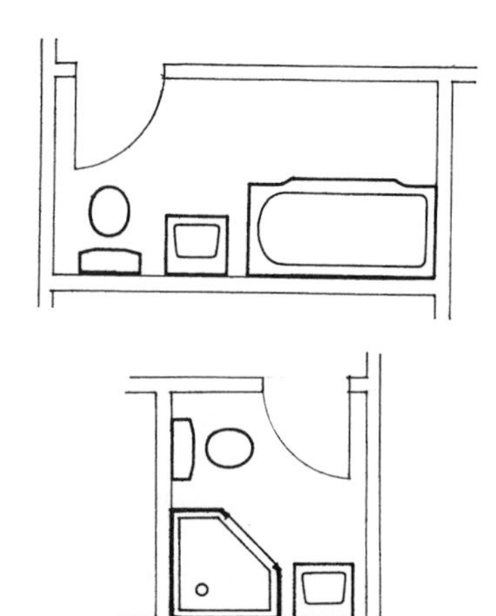

*Common wall is with an adjacent bathroom.

System Components	QUANTITY	UNIT	COST EACH MAT.	COST EACH INST.	COST EACH TOTAL
SYSTEM D2010 924 1170					
BATHROOM, LAVATORY, WATER CLOSET & BATHTUB					
ONE WALL PLUMBING, STAND ALONE					
Wtr closet, 2 pc close cpld vit china flr mntd w/seat supply & stop	1.000	Ea.	253	228	481
Water closet, rough-in waste & vent	1.000	Set	355	395	750
Lavatory w/ftngs, wall hung, white, PE on CI, 20" x 18"	1.000	Ea.	300	151	451
Lavatory, rough-in waste & vent	1.000	Set	530	730	1,260
Bathtub, white PE on CI, w/ftgs, mat bottom, recessed, 5' long	1.000	Ea.	1,200	275	1,475
Baths, rough-in waste and vent	1.000	Set	468	526.50	994.50
TOTAL			3,106	2,305.50	5,411.50

D2010 924	Three Fixture Bathroom, One Wall Plumbing		COST EACH MAT.	COST EACH INST.	COST EACH TOTAL
1150	Bathroom, three fixture, one wall plumbing	R224000-30			
1160	Lavatory, water closet & bathtub				
1170	Stand alone	R224000-40	3,100	2,300	5,400
1180	Share common plumbing wall *		2,650	1,675	4,325

D2010 926	Three Fixture Bathroom, Two Wall Plumbing	COST EACH MAT.	COST EACH INST.	COST EACH TOTAL
2130	Bathroom, three fixture, two wall plumbing			
2140	Lavatory, water closet & bathtub			
2160	Stand alone	3,125	2,325	5,450
2180	Long plumbing wall common *	2,800	1,850	4,650
3610	Lavatory, bathtub & water closet			
3620	Stand alone	3,425	2,650	6,075
3640	Long plumbing wall common *	3,175	2,400	5,575
4660	Water closet, corner bathtub & lavatory			
4680	Stand alone	4,775	2,375	7,150
4700	Long plumbing wall common *	4,275	1,775	6,050
6100	Water closet, stall shower & lavatory			
6120	Stand alone	3,650	2,650	6,300
6140	Long plumbing wall common *	3,450	2,450	5,900
7060	Lavatory, corner stall shower & water closet			
7080	Stand alone	4,000	2,350	6,350
7100	Short plumbing wall common *	3,200	1,575	4,775

D2010 Plumbing Fixtures

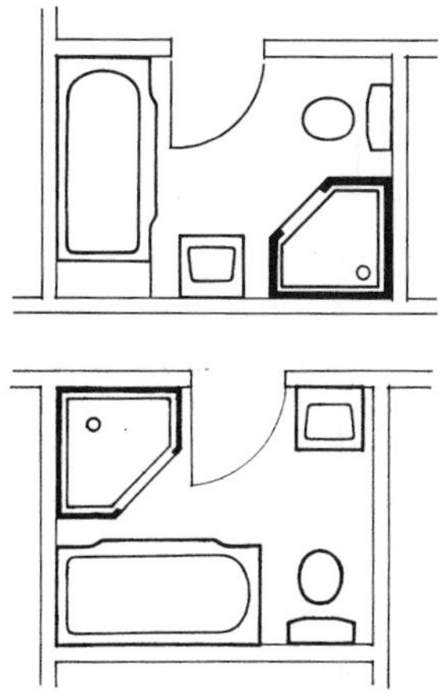

Four Fixture Bathroom Systems consisting of a lavatory, water closet, bathtub, shower and rough-in service piping.

- Prices for plumbing and fixtures only.

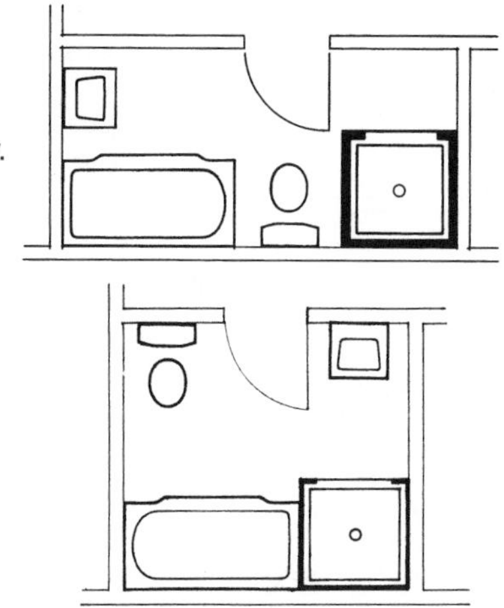

*Common wall is with an adjacent bathroom.

System Components	QUANTITY	UNIT	COST EACH MAT.	COST EACH INST.	COST EACH TOTAL
SYSTEM D2010 928 1160					
BATHROOM, BATHTUB, WATER CLOSET, STALL SHOWER & LAVATORY					
TWO WALL PLUMBING, STAND ALONE					
Wtr closet, 2 pc close cpld vit china flr mntd w/seat supply & stop	1.000	Ea.	253	228	481
Water closet, rough-in waste & vent	1.000	Set	355	395	750
Lavatory w/ftngs, wall hung, white PE on CI, 20″ x 18″	1.000	Ea.	300	151	451
Lavatory, rough-in waste & vent	1.000	Set	53	73	126
Bathtub, white PE on CI, w/ftgs, mat bottom, recessed, 5′ long	1.000	Ea.	1,200	275	1,475
Baths, rough-in waste and vent	1.000	Set	520	585	1,105
Shower stall, bkd enam, molded stone receptor, door & trim 32″ sq.	1.000	Ea.	1,275	242	1,517
Shower stall, rough-in supply, waste & vent	1.000	Set	655	590	1,245
TOTAL			4,611	2,539	7,150

D2010 928	Four Fixture Bathroom, Two Wall Plumbing		COST EACH MAT.	COST EACH INST.	COST EACH TOTAL
1140	Bathroom, four fixture, two wall plumbing	R224000-30			
1150	Bathtub, water closet, stall shower & lavatory				
1160	Stand alone	R224000-40	4,600	2,550	7,150
1180	Long plumbing wall common *		4,100	1,950	6,050
2260	Bathtub, lavatory, corner stall shower & water closet				
2280	Stand alone		5,225	2,550	7,775
2320	Long plumbing wall common *		4,725	1,975	6,700
3620	Bathtub, stall shower, lavatory & water closet				
3640	Stand alone		5,100	3,200	8,300
3660	Long plumbing wall (opp. door) common *		4,575	2,600	7,175

D2010 930	Four Fixture Bathroom, Three Wall Plumbing	COST EACH MAT.	COST EACH INST.	COST EACH TOTAL
4680	Bathroom, four fixture, three wall plumbing			
4700	Bathtub, stall shower, lavatory & water closet			
4720	Stand alone	5,925	3,525	9,450
4760	Long plumbing wall (opposite door) common *	5,750	3,250	9,000

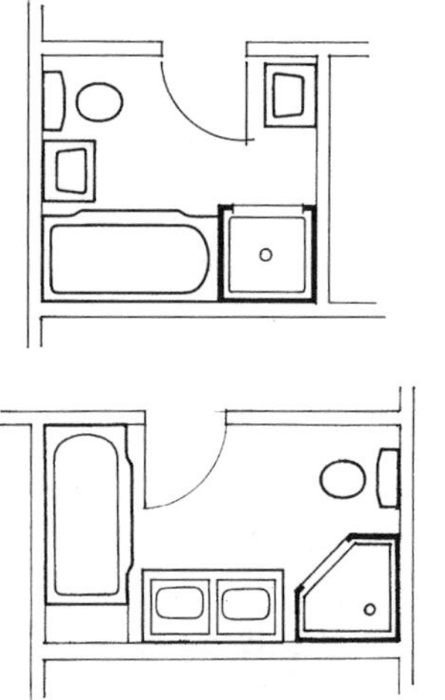

Five Fixture Bathroom Systems consisting of two lavatories, a water closet, bathtub, shower and rough-in service piping.

- Prices for plumbing and fixtures only.

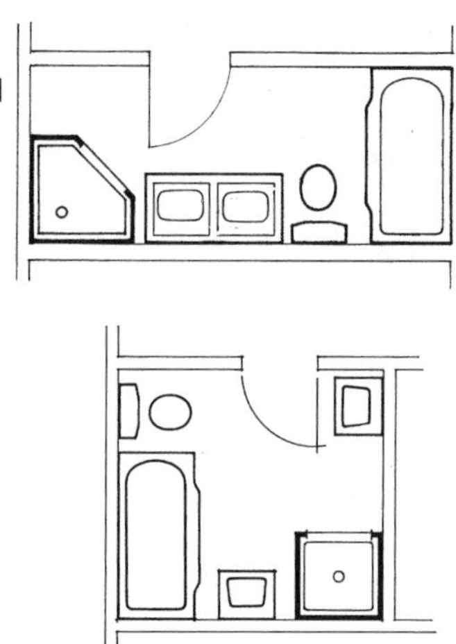

*Common wall is with an adjacent bathroom.

System Components	QUANTITY	UNIT	COST EACH MAT.	COST EACH INST.	COST EACH TOTAL
SYSTEM D2010 932 1360					
BATHROOM, BATHTUB, WATER CLOSET, STALL SHOWER & TWO LAVATORIES					
TWO WALL PLUMBING, STAND ALONE					
Wtr closet, 2 pc close cpld vit china flr mntd incl seat,supply & stop	1.000	Ea.	253	228	481
Water closet, rough-in waste & vent	1.000	Set	355	395	750
Lavatory w/ftngs, wall hung, white PE on CI, 20″ x 18″	2.000	Ea.	600	302	902
Lavatory, rough-in waste & vent	2.000	Set	1,060	1,460	2,520
Bathtub, white PE on CI, w/ftgs, mat bottom, recessed, 5′ long	1.000	Ea.	1,200	275	1,475
Baths, rough-in waste and vent	1.000	Set	520	585	1,105
Shower stall, bkd enam molded stone receptor, door & ftng, 32″ sq.	1.000	Ea.	1,275	242	1,517
Shower stall, rough-in supply, waste & vent	1.000	Set	655	590	1,245
TOTAL			5,918	4,077	9,995

D2010 932	Five Fixture Bathroom, Two Wall Plumbing	COST EACH MAT.	COST EACH INST.	COST EACH TOTAL
1320	Bathroom, five fixture, two wall plumbing R224000-30			
1340	Bathtub, water closet, stall shower & two lavatories			
1360	Stand alone	5,925	4,075	10,000
1400	One short plumbing wall common *	5,400	3,500	8,900
1500	Bathtub, two lavatories, corner stall shower & water closet			
1520	Stand alone	6,550	4,075	10,625
1540	Long plumbing wall common*	5,875	3,175	9,050

D2010 934	Five Fixture Bathroom, Three Wall Plumbing	COST EACH MAT.	COST EACH INST.	COST EACH TOTAL
2360	Bathroom, five fixture, three wall plumbing			
2380	Water closet, bathtub, two lavatories & stall shower			
2400	Stand alone	6,550	4,075	10,625
2440	One short plumbing wall common *	6,025	3,500	9,525

D2010 936	Five Fixture Bathroom, One Wall Plumbing	COST EACH MAT.	COST EACH INST.	COST EACH TOTAL
4080	Bathroom, five fixture, one wall plumbing			
4100	Bathtub, two lavatories, corner stall shower & water closet			
4120	Stand alone	6,225	3,650	9,875
4160	Share common wall *	5,250	2,350	7,600

D2010 Plumbing Fixtures

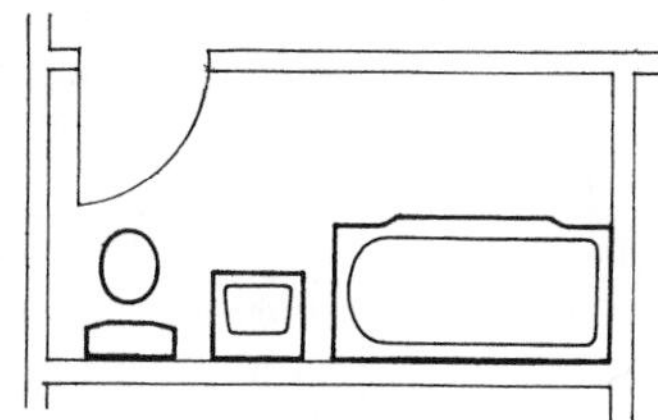

Example of Plumbing Cost Calculations: The bathroom system includes the individual fixtures such as bathtub, lavatory, shower and water closet. These fixtures are listed below as separate items merely as a checklist.

D2010 951 Plumbing Systems 20 Unit, 2 Story Apartment Building

	FIXTURE	SYSTEM	LINE	QUANTITY	UNIT	COST EACH		
						MAT.	INST.	TOTAL
0440	Bathroom	D2010 926	3640	20	Ea.	63,500	48,000	111,500
0480	Bathtub							
0520	Booster pump[1]	not req'd.						
0560	Drinking fountain							
0600	Garbage disposal[1]	not incl.						
0660								
0680	Grease interceptor							
0720	Water heater	D2020 250	2140	1	Ea.	13,500	3,275	16,775
0760	Kitchen sink	D2010 410	1960	20	Ea.	31,600	16,900	48,500
0800	Laundry sink	D2010 420	1840	4	Ea.	5,775	3,325	9,100
0840	Lavatory							
0900								
0920	Roof drain, 1 floor	D2040 210	4200	2	Ea.	1,850	1,750	3,600
0960	Roof drain, add'l floor	D2040 210	4240	20	L.F.	370	440	810
1000	Service sink	D2010 440	4300	1	Ea.	2,775	1,175	3,950
1040	Sewage ejector[1]	not req'd.						
1080	Shower							
1100								
1160	Sump pump							
1200	Urinal							
1240	Water closet							
1320								
1360	**SUB TOTAL**					119,500	75,000	194,500
1481	Water controls	R22113-40		10%[2]		12,000	7,475	19,475
1521	Pipe & fittings[3]	R221113-40		30%[2]		35,900	22,500	58,400
1560	Other							
1601	Quality/complexity	R221113-40		15%[2]		17,900	11,200	29,100
1680								
1720	**TOTAL**					185,500	116,000	301,500
1741								

[1]**Note:** Cost for items such as booster pumps, backflow preventers, sewage ejectors, water meters, etc., may be obtained from the unit price section in the front of this book.

Water controls, pipe and fittings, and the Quality/Complexity factors come from Table R221113-40.

[2]Percentage of subtotal.

[3]Long, easily discernable runs of pipe would be more accurately priced from Unit Price Section 22 11 13. If this is done, reduce the miscellaneous percentage in proportion.

D2020 Domestic Water Distribution

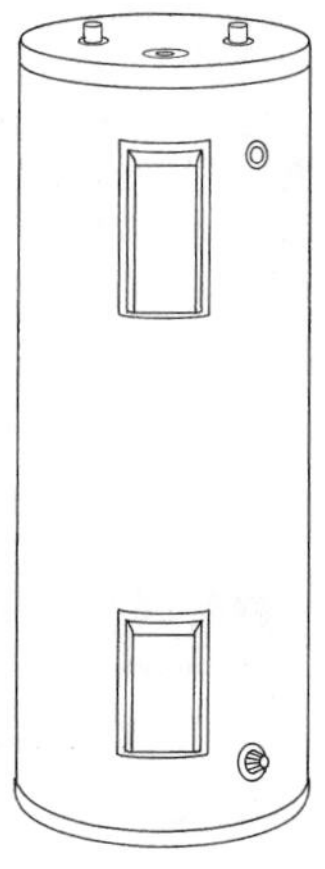

Installation includes piping and fittings within 10′ of heater. Electric water heaters do not require venting.

1 Kilowatt hour will raise:			
Gallons of Water	Degrees F	Gallons of Water	Degrees F
4.1	100°	6.8	60°
4.5	90°	8.2	50°
5.1	80°	10.0	40°
5.9	70°		

System Components	QUANTITY	UNIT	COST EACH MAT.	COST EACH INST.	COST EACH TOTAL
SYSTEM D2020 210 1780					
ELECTRIC WATER HEATER, RESIDENTIAL, 100°F RISE					
10 GALLON TANK, 7 GPH					
Water heater, residential electric, glass lined tank, 10 gal.	1.000	Ea.	335	292	627
Copper tubing, type L, solder joint, hanger 10′ OC 1/2″ diam.	30.000	L.F.	139.80	249	388.80
Wrought copper 90° elbow for solder joints 1/2″ diam.	4.000	Ea.	12.44	134	146.44
Wrought copper Tee for solder joints, 1/2″ diam.	2.000	Ea.	10.60	103	113.60
Union, wrought copper, 1/2″ diam.	2.000	Ea.	74	71	145
Valve, gate, bronze, 125 lb, NRS, soldered 1/2″ diam.	2.000	Ea.	97	56	153
Relief valve, bronze, press & temp, self-close, 3/4″ IPS	1.000	Ea.	171	24	195
Wrought copper adapter, CTS to MPT 3/4″ IPS	1.000	Ea.	10.95	39.50	50.45
Copper tubing, type L, solder joints, 3/4″ diam.	1.000	L.F.	7.20	8.85	16.05
Wrought copper 90° elbow for solder joints 3/4″ diam.	1.000	Ea.	7	35.50	42.50
TOTAL			864.99	1,012.85	1,877.84

D2020 210	Electric Water Heaters - Residential Systems		COST EACH MAT.	COST EACH INST.	COST EACH TOTAL
1760	Electric water heater, residential, 100°F rise	R224000-10			
1780	10 gallon tank, 7 GPH		865	1,025	1,890
1820	20 gallon tank, 7 GPH	R224000-20	1,100	1,075	2,175
1860	30 gallon tank, 7 GPH		1,600	1,125	2,725
1900	40 gallon tank, 8 GPH	R224000-40	1,950	1,250	3,200
1940	52 gallon tank, 10 GPH		2,050	1,250	3,300
1980	66 gallon tank, 13 GPH		3,075	1,450	4,525
2020	80 gallon tank, 16 GPH		3,250	1,500	4,750
2060	120 gallon tank, 23 GPH		4,750	1,750	6,500

D2020 Domestic Water Distribution

Installation includes piping and fittings within 10′ of heater. Gas heaters require vent piping (not included with these units).

System Components	QUANTITY	UNIT	COST EACH MAT.	COST EACH INST.	COST EACH TOTAL
SYSTEM D2020 220 2260					
GAS FIRED WATER HEATER, RESIDENTIAL, 100°F RISE					
30 GALLON TANK, 32 GPH					
Water heater, residential, gas, glass lined tank, 30 gallon	1.000	Ea.	985	335	1,320
Copper tubing, type L, solder joint, hanger 10′ OC, 3/4″ diam	33.000	L.F.	237.60	292.05	529.65
Wrought copper 90° elbow for solder joints, 3/4″ diam	5.000	Ea.	35	177.50	212.50
Wrought copper Tee for solder joints, 3/4″ diam	2.000	Ea.	25.70	112	137.70
Wrought copper union for soldered joints, 3/4″ diam.	2.000	Ea.	93	75	168
Valve bronze, 125 lb., NRS, soldered 3/4″ diam	2.000	Ea.	109	67	176
Relief valve, press & temp, bronze, self-close, 3/4″ diam	1.000	Ea.	171	24	195
Wrought copper, adapter, CTS to MPT 3/4″ IPS	1.000	Ea.	10.95	39.50	50.45
Pipe steel black, schedule 40, threaded, 1/2″ diam	10.000	L.F.	32.50	106.50	139
Pipe, 90° elbow, malleable iron black, 150 lb., threaded, 1/2″ diam	2.000	Ea.	6.36	90	96.36
Pipe, union with brass seat, malleable iron black, 1/2″ diam	1.000	Ea.	14.05	48	62.05
Valve, gas stop w/o check, brass, 1/2″ IPS	1.000	Ea.	14.50	28	42.50
TOTAL			1,734.66	1,394.55	3,129.21

D2020 220	Gas Fired Water Heaters - Residential Systems		COST EACH MAT.	COST EACH INST.	COST EACH TOTAL
2200	Gas fired water heater, residential, 100°F rise	R224000 -10			
2260	30 gallon tank, 32 GPH		1,725	1,400	3,125
2300	40 gallon tank, 32 GPH	R224000 -20	2,075	1,575	3,650
2340	50 gallon tank, 63 GPH		2,100	1,575	3,675
2380	75 gallon tank, 63 GPH	R224000 -40	3,000	1,750	4,750
2420	100 gallon tank, 63 GPH		3,175	1,825	5,000

D2020 Domestic Water Distribution

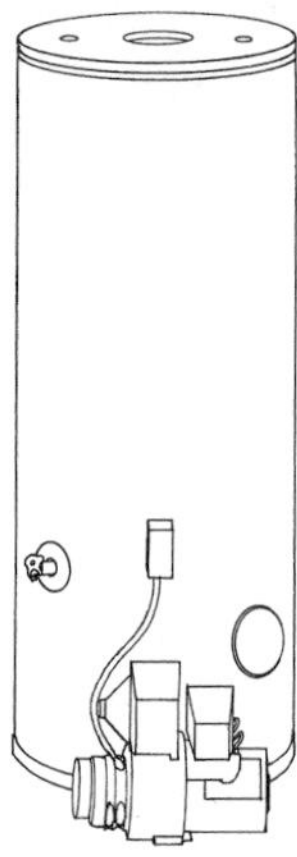

Installation includes piping and fittings within 10′ of heater. Oil fired heaters require vent piping (not included in these prices).

System Components	QUANTITY	UNIT	COST EACH MAT.	COST EACH INST.	COST EACH TOTAL
SYSTEM D2020 230 2220					
OIL FIRED WATER HEATER, RESIDENTIAL, 100°F RISE					
30 GALLON TANK, 103 GPH					
Water heater, residential, oil glass lined tank, 30 Gal	1.000	Ea.	1,150	335	1,485
Copper tubing, type L, solder joint, hanger 10′ O.C. 3/4″ diam.	33.000	L.F.	237.60	292.05	529.65
Wrought copper 90° elbow for solder joints 3/4″ diam.	5.000	Ea.	35	177.50	212.50
Wrought copper Tee for solder joints, 3/4″ diam.	2.000	Ea.	25.70	112	137.70
Wrought copper union for soldered joints, 3/4″ diam.	2.000	Ea.	93	75	168
Valve, gate, bronze, 125 lb, NRS, soldered 3/4″ diam.	2.000	Ea.	109	67	176
Relief valve, bronze, press & temp, self-close, 3/4″ IPS	1.000	Ea.	171	24	195
Wrought copper adapter, CTS to MPT, 3/4″ IPS	1.000	Ea.	10.95	39.50	50.45
Copper tubing, type L, solder joint, hanger 10′ OC 3/8″ diam.	10.000	L.F.	41.90	80	121.90
Wrought copper 90° elbow for solder joints 3/8″ diam.	2.000	Ea.	18.70	61	79.70
Valve, globe, fusible, 3/8″ diam.	1.000	Ea.	14.40	28.50	42.90
TOTAL			1,907.25	1,291.55	3,198.80

D2020 230	Oil Fired Water Heaters - Residential Systems		COST EACH MAT.	COST EACH INST.	COST EACH TOTAL
2200	Oil fired water heater, residential, 100°F rise				
2220	30 gallon tank, 103 GPH		1,900	1,300	3,200
2260	50 gallon tank, 145 GPH	R224000 -20	2,450	1,450	3,900
2300	70 gallon tank, 164 GPH		3,475	1,625	5,100
2340	85 gallon tank, 181 GPH		10,700	1,675	12,375

D2020 Domestic Water Distribution

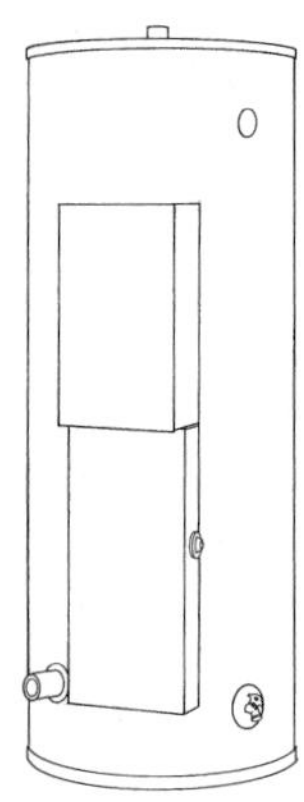

Systems below include piping and fittings within 10′ of heater. Electric water heaters do not require venting.

System Components	QUANTITY	UNIT	COST EACH MAT.	COST EACH INST.	COST EACH TOTAL
SYSTEM D2020 240 1820					
ELECTRIC WATER HEATER, COMMERCIAL, 100°F RISE					
50 GALLON TANK, 9 KW, 37 GPH					
Water heater, commercial, electric, 50 Gal, 9 KW, 37 GPH	1.000	Ea.	4,375	375	4,750
Copper tubing, type L, solder joint, hanger 10′ OC, 3/4″ diam	34.000	L.F.	244.80	300.90	545.70
Wrought copper 90° elbow for solder joints 3/4″ diam	5.000	Ea.	35	177.50	212.50
Wrought copper Tee for solder joints, 3/4″ diam	2.000	Ea.	25.70	112	137.70
Wrought copper union for soldered joints, 3/4″ diam.	2.000	Ea.	93	75	168
Valve, gate, bronze, 125 lb, NRS, soldered 3/4″ diam	2.000	Ea.	109	67	176
Relief valve, bronze, press & temp, self-close, 3/4″ IPS	1.000	Ea.	171	24	195
Wrought copper adapter, copper tubing to male, 3/4″ IPS	1.000	Ea.	10.95	39.50	50.45
TOTAL			5,064.45	1,170.90	6,235.35

D2020 240	Electric Water Heaters - Commercial Systems		COST EACH MAT.	COST EACH INST.	COST EACH TOTAL
1800	Electric water heater, commercial, 100°F rise				
1820	50 gallon tank, 9 KW 37 GPH		5,075	1,175	6,250
1860	80 gal, 12 KW 49 GPH	R224000 -10	6,800	1,450	8,250
1900	36 KW 147 GPH		9,225	1,550	10,775
1940	120 gal, 36 KW 147 GPH	R224000 -20	9,875	1,675	11,550
1980	150 gal, 120 KW 490 GPH		28,600	1,800	30,400
2020	200 gal, 120 KW 490 GPH	R224000 -40	30,000	1,850	31,850
2060	250 gal, 150 KW 615 GPH		33,800	2,150	35,950
2100	300 gal, 180 KW 738 GPH		43,200	2,275	45,475
2140	350 gal, 30 KW 123 GPH		27,600	2,450	30,050
2180	180 KW 738 GPH		37,700	2,450	40,150
2220	500 gal, 30 KW 123 GPH		35,600	2,900	38,500
2260	240 KW 984 GPH		60,500	2,900	63,400
2300	700 gal, 30 KW 123 GPH		29,400	3,300	32,700
2340	300 KW 1230 GPH		42,600	3,300	45,900
2380	1000 gal, 60 KW 245 GPH		37,100	4,600	41,700
2420	480 KW 1970 GPH		57,500	4,625	62,125
2460	1500 gal, 60 KW 245 GPH		75,500	5,700	81,200
2500	480 KW 1970 GPH		102,500	5,700	108,200

D2020 Domestic Water Distribution

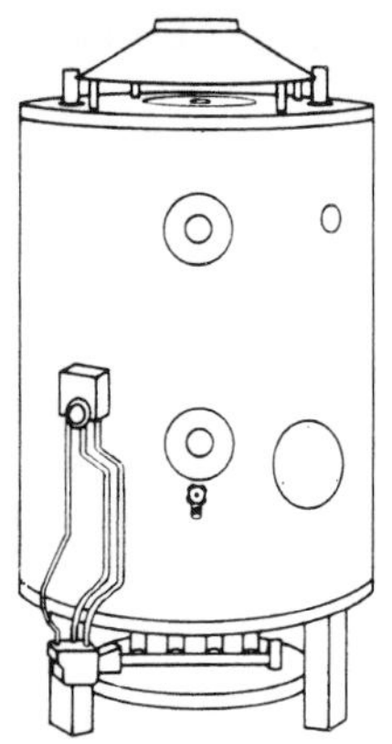

Units may be installed in multiples for increased capacity.

Included below is the heater with self-energizing gas controls, safety pilots, insulated jacket, hi-limit aquastat and pressure relief valve.

Installation includes piping and fittings within 10′ of heater. Gas heaters require vent piping (not included in these prices).

System Components	QUANTITY	UNIT	COST EACH MAT.	COST EACH INST.	COST EACH TOTAL
SYSTEM D2020 250 1780					
GAS FIRED WATER HEATER, COMMERCIAL, 100°F RISE					
75.5 MBH INPUT, 63 GPH					
Water heater, commercial, gas, 75.5 MBH, 63 GPH	1.000	Ea.	3,475	480	3,955
Copper tubing, type L, solder joint, hanger 10′ OC, 1-1/4″ diam	30.000	L.F.	451.50	348	799.50
Wrought copper 90° elbow for solder joints 1-1/4″ diam	4.000	Ea.	104	180	284
Wrought copper tee for solder joints, 1-1/4″ diam	2.000	Ea.	109	149	258
Wrought copper union for soldered joints, 1-1/4″ diam	2.000	Ea.	278	96	374
Valve, gate, bronze, 125 lb, NRS, soldered 1-1/4″ diam	2.000	Ea.	238	90	328
Relief valve, bronze, press & temp, self-close, 3/4″ IPS	1.000	Ea.	171	24	195
Copper tubing, type L, solder joints, 3/4″ diam	8.000	L.F.	57.60	70.80	128.40
Wrought copper 90° elbow for solder joints 3/4″ diam	1.000	Ea.	7	35.50	42.50
Wrought copper, adapter, CTS to MPT, 3/4″ IPS	1.000	Ea.	10.95	39.50	50.45
Pipe steel black, schedule 40, threaded, 3/4″ diam	10.000	L.F.	38.10	110	148.10
Pipe, 90° elbow, malleable iron black, 150 lb threaded, 3/4″ diam	2.000	Ea.	7.68	96	103.68
Pipe, union with brass seat, malleable iron black, 3/4″ diam	1.000	Ea.	16.15	51.50	67.65
Valve, gas stop w/o check, brass, 3/4″ IPS	1.000	Ea.	17.05	30.50	47.55
TOTAL			4,981.03	1,800.80	6,781.83

D2020 250	Gas Fired Water Heaters - Commercial Systems		COST EACH MAT.	COST EACH INST.	COST EACH TOTAL
1760	Gas fired water heater, commercial, 100°F rise				
1780	75.5 MBH input, 63 GPH		4,975	1,800	6,775
1860	100 MBH input, 91 GPH	R224000 -10	7,250	1,875	9,125
1980	155 MBH input, 150 GPH		10,100	2,175	12,275
2060	200 MBH input, 192 GPH	R224000 -20	11,200	2,650	13,850
2140	300 MBH input, 278 GPH		13,500	3,275	16,775
2180	390 MBH input, 374 GPH	R224000 -40	15,600	3,325	18,925
2220	500 MBH input, 480 GPH		20,700	3,550	24,250
2260	600 MBH input, 576 GPH		24,500	3,850	28,350

D2020 Domestic Water Distribution

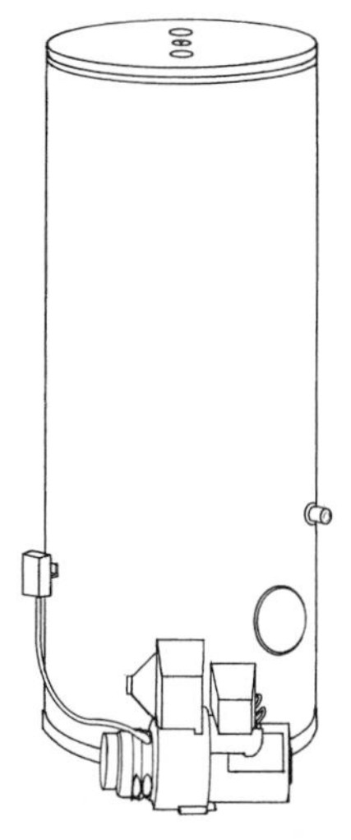

Units may be installed in multiples for increased capacity.

Included below is the heater, wired-in flame retention burners, cadmium cell primary controls, hi-limit controls, ASME pressure relief valves, draft controls, and insulated jacket.

Oil fired water heater systems include piping and fittings within 10′ of heater. Oil fired heaters require vent piping (not included in these systems).

System Components	QUANTITY	UNIT	COST EACH MAT.	COST EACH INST.	COST EACH TOTAL
SYSTEM D2020 260 1820					
OIL FIRED WATER HEATER, COMMERCIAL, 100°F RISE					
140 GAL., 140 MBH INPUT, 134 GPH					
Water heater, commercial, oil, 140 gal., 140 MBH input, 134 GPH	1.000	Ea.	20,900	570	21,470
Copper tubing, type L, solder joint, hanger 10′ OC, 3/4″ diam.	34.000	L.F.	244.80	300.90	545.70
Wrought copper 90° elbow for solder joints 3/4″ diam.	5.000	Ea.	35	177.50	212.50
Wrought copper Tee for solder joints, 3/4″ diam.	2.000	Ea.	25.70	112	137.70
Wrought copper union for soldered joints, 3/4″ diam.	2.000	Ea.	93	75	168
Valve, bronze, 125 lb, NRS, soldered 3/4″ diam.	2.000	Ea.	109	67	176
Relief valve, bronze, press & temp, self-close, 3/4″ IPS	1.000	Ea.	171	24	195
Wrought copper adapter, copper tubing to male, 3/4″ IPS	1.000	Ea.	10.95	39.50	50.45
Copper tubing, type L, solder joint, hanger 10′ OC, 3/8″ diam.	10.000	L.F.	41.90	80	121.90
Wrought copper 90° elbow for solder joints 3/8″ diam.	2.000	Ea.	18.70	61	79.70
Valve, globe, fusible, 3/8″ IPS	1.000	Ea.	14.40	28.50	42.90
TOTAL			21,664.45	1,535.40	23,199.85

D2020 260	Oil Fired Water Heaters - Commercial Systems		COST EACH MAT.	COST EACH INST.	COST EACH TOTAL
1800	Oil fired water heater, commercial, 100°F rise				
1820	140 gal., 140 MBH input, 134 GPH		21,700	1,525	23,225
1900	140 gal., 255 MBH input, 247 GPH	R224000 -10	23,700	1,925	25,625
1940	140 gal., 270 MBH input, 259 GPH		28,900	2,175	31,075
1980	140 gal., 400 MBH input, 384 GPH	R224000 -20	30,100	2,525	32,625
2060	140 gal., 720 MBH input, 691 GPH		32,000	2,650	34,650
2100	221 gal., 300 MBH input, 288 GPH	R224000 -40	42,500	2,900	45,400
2140	221 gal., 600 MBH input, 576 GPH		47,100	2,925	50,025
2180	221 gal., 800 MBH input, 768 GPH		47,800	3,050	50,850
2220	201 gal., 1000 MBH input, 960 GPH		48,900	3,075	51,975
2260	201 gal., 1250 MBH input, 1200 GPH		49,800	3,150	52,950
2300	201 gal., 1500 MBH input, 1441 GPH		54,000	3,225	57,225
2340	411 gal., 600 MBH input, 576 GPH		54,500	3,300	57,800
2380	411 gal., 800 MBH input, 768 GPH		54,500	3,400	57,900
2420	411 gal., 1000 MBH input, 960 GPH		60,000	3,900	63,900
2460	411 gal., 1250 MBH input, 1200 GPH		61,000	4,025	65,025
2500	397 gal., 1500 MBH input, 1441 GPH		62,000	4,150	66,150

D2020 Domestic Water Distribution

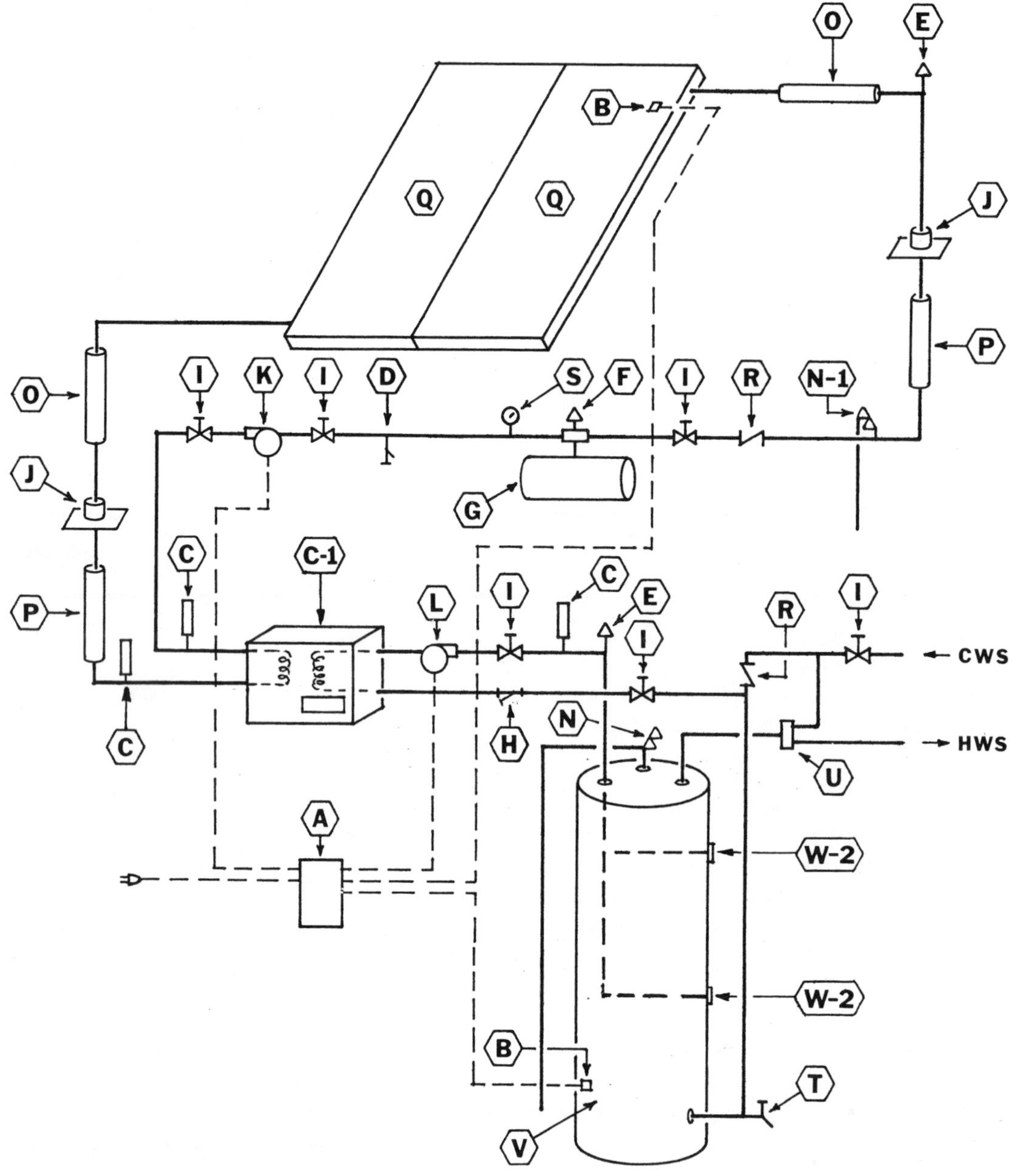

In this closed-loop indirect collection system, fluid with a low freezing temperature, such as propylene glycol, transports heat from the collectors to water storage. The transfer fluid is contained in a closed-loop consisting of collectors, supply and return piping, and a remote heat exchanger. The heat exchanger transfers heat energy from the fluid in the collector loop to potable water circulated in a storage loop. A typical two-or-three panel system contains 5 to 6 gallons of heat transfer fluid.

When the collectors become approximately 20°F warmer than the storage temperature, a controller activates the circulator on the collector and storage loops. The circulators will move the fluid and potable water through the heat exchanger until heat collection no longer occurs. At that point, the system shuts down. Since the heat transfer medium is a fluid with a very low freezing temperature, there is no need for it to be drained from the system between periods of collection.

D2020 Domestic Water Distribution

System Components		QUANTITY	UNIT	COST EACH MAT.	INST.	TOTAL
	SYSTEM D2020 265 2760					
	SOLAR, CLOSED LOOP, ADD-ON HOT WATER SYS., EXTERNAL HEAT EXCHANGER					
	3/4″ TUBING, TWO 3′X7′ BLACK CHROME COLLECTORS					
A,B,G,L,K,M	Heat exchanger fluid-fluid pkg incl 2 circulators, expansion tank, Check valve, relief valve, controller, hi temp cutoff, & 2 sensors	1.000	Ea.	820	485	1,305
C	Thermometer, 2″ dial	3.000	Ea.	69	126	195
D, T	Fill & drain valve, brass, 3/4″ connection	1.000	Ea.	10.05	28	38.05
E	Air vent, manual, 1/8″ fitting	2.000	Ea.	31.80	42	73.80
F	Air purger	1.000	Ea.	51.50	56	107.50
H	Strainer, Y type, bronze body, 3/4″ IPS	1.000	Ea.	36	36	72
I	Valve, gate, bronze, NRS, soldered 3/4″ diam	6.000	Ea.	327	201	528
J	Neoprene vent flashing	2.000	Ea.	27.40	67	94.40
N-1, N	Relief valve temp & press, 150 psi 210°F self-closing 3/4″ IPS	1.000	Ea.	22	22.50	44.50
O	Pipe covering, urethane, ultraviolet cover, 1″ wall 3/4″ diam	20.000	L.F.	60.20	121	181.20
P	Pipe covering, fiberglass, all service jacket, 1″ wall, 3/4″ diam	50.000	L.F.	53.50	241.50	295
Q	Collector panel solar energy blk chrome on copper, 1/8″ temp glass 3′x7′	2.000	Ea.	2,250	254	2,504
	Roof clamps for solar energy collector panels	2.000	Set	5.58	34.60	40.18
R	Valve, swing check, bronze, regrinding disc, 3/4″ diam	2.000	Ea.	196	67	263
S	Pressure gauge, 60 psi, 2″ dial	1.000	Ea.	25.50	21	46.50
U	Valve, water tempering, bronze, sweat connections, 3/4″ diam	1.000	Ea.	126	33.50	159.50
W-2, V	Tank water storage w/heating element, drain, relief valve, existing	1.000	Ea.			
	Copper tubing type L, solder joint, hanger 10′ OC 3/4″ diam	20.000	L.F.	144	177	321
	Copper tubing, type M, solder joint, hanger 10′ OC 3/4″ diam	70.000	L.F.	381.50	602	983.50
	Sensor wire, #22-2 conductor multistranded	.500	C.L.F.	7.83	31.25	39.08
	Solar energy heat transfer fluid, propylene glycol anti-freeze	6.000	Gal.	95.70	144	239.70
	Wrought copper fittings & solder, 3/4″ diam	76.000	Ea.	532	2,698	3,230
	TOTAL			5,272.56	5,488.35	10,760.91

D2020 265	Solar, Closed Loop, Add-On Hot Water Systems	COST EACH MAT.	INST.	TOTAL
2550	Solar, closed loop, add-on hot water system, external heat exchanger			
2570	3/8″ tubing, 3 ea. 4′ x 4′-4″ vacuum tube collectors	5,800	5,125	10,925
2580	1/2″ tubing, 4 ea. 4 x 4′-4″ vacuum tube collectors R235616 -60	6,275	5,525	11,800
2600	2 ea. 3′x7′ black chrome collectors	4,675	5,225	9,900
2620	3 ea. 3′x7′ black chrome collectors	5,800	5,375	11,175
2640	2 ea. 3′x7′ flat black collectors	3,975	5,225	9,200
2660	3 ea. 3′x7′ flat black collectors	4,750	5,375	10,125
2700	3/4″ tubing, 3 ea. 3′x7′ black chrome collectors	6,400	5,625	12,025
2720	3 ea. 3′x7′ flat black absorber plate collectors	5,375	5,650	11,025
2740	2 ea. 4′x9′ flat black w/plastic glazing collectors	4,975	5,675	10,650
2760	2 ea. 3′x7′ black chrome collectors	5,275	5,500	10,775
2780	1″ tubing, 4 ea 2′x9′ plastic absorber & glazing collectors	6,675	6,375	13,050
2800	4 ea. 3′x7′ black chrome absorber collectors	8,825	6,400	15,225
2820	4 ea. 3′x7′ flat black absorber collectors	7,450	6,425	13,875

D2020 Domestic Water Distribution

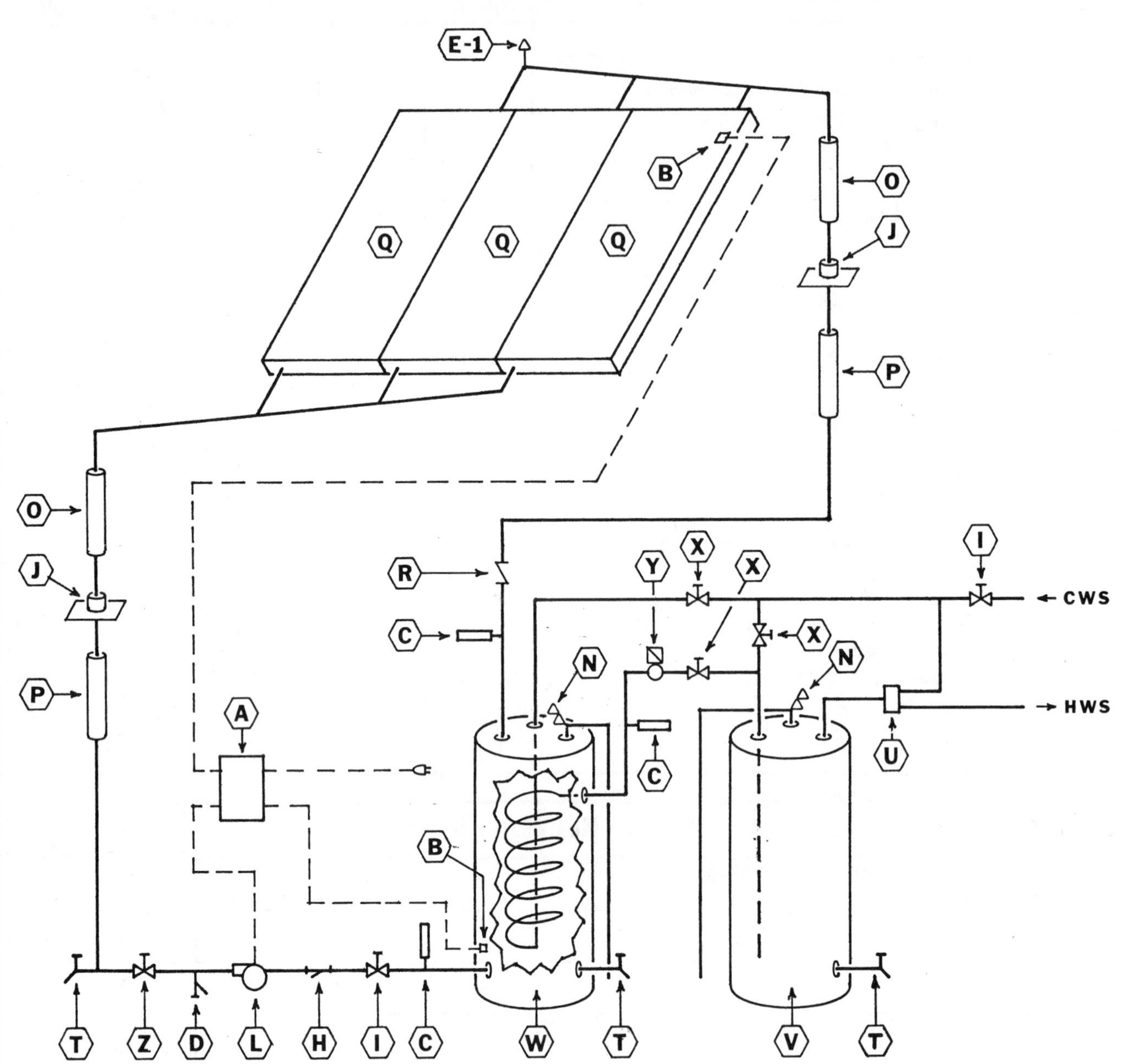

In the drainback indirect-collection system, the heat transfer fluid is distilled water contained in a loop consisting of collectors, supply and return piping, and an unpressurized holding tank. A large heat exchanger containing incoming potable water is immersed in the holding tank. When a controller activates solar collection, the distilled water is pumped through the collectors and heated and pumped back down to the holding tank. When the temperature differential between the water in the collectors and water in storage is such that collection no longer occurs, the pump turns off and gravity causes the distilled water in the collector loop to drain back to the holding tank. All the loop piping is pitched so that the water can drain out of the collectors and piping and not freeze there. As hot water is needed in the home, incoming water first flows through the holding tank with the immersed heat exchanger and is warmed and then flows through a conventional heater for any supplemental heating that is necessary.

D2020 Domestic Water Distribution

System Components		QUANTITY	UNIT	COST EACH MAT.	INST.	TOTAL
SYSTEM D2020 270 2760						
SOLAR, DRAINBACK, ADD ON, HOT WATER, IMMERSED HEAT EXCHANGER						
3/4″ TUBING, THREE EA 3′X7′ BLACK CHROME COLLECTOR						
A, B	Differential controller 2 sensors, thermostat, solar energy system	1.000	Ea.	625	56	681
C	Thermometer 2″ dial	3.000	Ea.	69	126	195
D, T	Fill & drain valve, brass, 3/4″ connection	1.000	Ea.	10.05	28	38.05
E-1	Automatic air vent 1/8″ fitting	1.000	Ea.	25	21	46
H	Strainer, Y type, bronze body, 3/4″ IPS	1.000	Ea.	36	36	72
I	Valve, gate, bronze, NRS, soldered 3/4″ diam	2.000	Ea.	109	67	176
J	Neoprene vent flashing	2.000	Ea.	27.40	67	94.40
L	Circulator, solar heated liquid, 1/20 HP	1.000	Ea.	177	101	278
N	Relief valve temp. & press. 150 psi 210°F self-closing 3/4″ IPS	1.000	Ea.	22	22.50	44.50
O	Pipe covering, urethane, ultraviolet cover, 1″ wall, 3/4″ diam	20.000	L.F.	60.20	121	181.20
P	Pipe covering, fiberglass, all service jacket, 1″ wall, 3/4″ diam	50.000	L.F.	53.50	241.50	295
Q	Collector panel solar energy blk chrome on copper, 1/8″ temp glas 3′x7′	3.000	Ea.	3,375	381	3,756
	Roof clamps for solar energy collector panels	3.000	Set	8.37	51.90	60.27
R	Valve, swing check, bronze, regrinding disc, 3/4″ diam	1.000	Ea.	98	33.50	131.50
U	Valve, water tempering, bronze sweat connections, 3/4″ diam	1.000	Ea.	126	33.50	159.50
V	Tank, water storage w/heating element, drain, relief valve, existing	1.000	Ea.			
W	Tank, water storage immersed heat exchr elec 2″x1/2# insul 120 gal	1.000	Ea.	1,950	480	2,430
X	Valve, globe, bronze, rising stem, 3/4″ diam, soldered	3.000	Ea.	417	100.50	517.50
Y	Flow control valve	1.000	Ea.	148	30.50	178.50
Z	Valve, ball, bronze, solder 3/4″ diam, solar loop flow control	1.000	Ea.	23	33.50	56.50
	Copper tubing, type L, solder joint, hanger 10′ OC 3/4″ diam	20.000	L.F.	144	177	321
	Copper tubing, type M, solder joint, hanger 10′ OC 3/4″ diam	70.000	L.F.	381.50	602	983.50
	Sensor wire, #22-2 conductor, multistranded	.500	C.L.F.	7.83	31.25	39.08
	Wrought copper fittings & solder, 3/4″ diam	76.000	Ea.	532	2,698	3,230
	TOTAL			8,424.85	5,539.65	13,964.50

D2020 270	Solar, Drainback, Hot Water Systems	COST EACH MAT.	INST.	TOTAL
2550	Solar, drainback, hot water, immersed heat exchanger			
2560	3/8″ tubing, 3 ea. 4′ x 4′-4″ vacuum tube collectors	7,500	4,975	12,475
2580	1/2″ tubing, 4 ea 4′ x 4′-4″ vacuum tube collectors, 80 gal tank R235616 -60	8,000	5,400	13,400
2600	120 gal tank	8,250	5,450	13,700
2640	2 ea. 3′x7′ blk chrome collectors, 80 gal tank	6,375	5,075	11,450
2660	3 ea. 3′x7′ blk chrome collectors, 120 gal tank	7,750	5,275	13,025
2700	2 ea. 3′x7′ flat blk collectors, 120 gal tank	5,925	5,150	11,075
2720	3 ea. 3′x7′ flat blk collectors, 120 gal tank	6,725	5,300	12,025
2760	3/4″ tubing, 3 ea 3′x7′ black chrome collectors, 120 gal tank	8,425	5,550	13,975
2780	3 ea. 3′x7′ flat black absorber collectors, 120 gal tank	7,400	5,550	12,950
2800	2 ea. 4′x9′ flat blk w/plastic glazing collectors 120 gal tank	7,000	5,575	12,575
2840	1″ tubing, 4 ea. 2′x9′ plastic absorber & glazing collectors, 120 gal tank	8,825	6,275	15,100
2860	4 ea. 3′x7′ black chrome absorber collectors, 120 gal tank	11,000	6,300	17,300
2880	4 ea. 3′x7′ flat black absorber collectors, 120 gal tank	9,600	6,325	15,925

D2020 Domestic Water Distribution

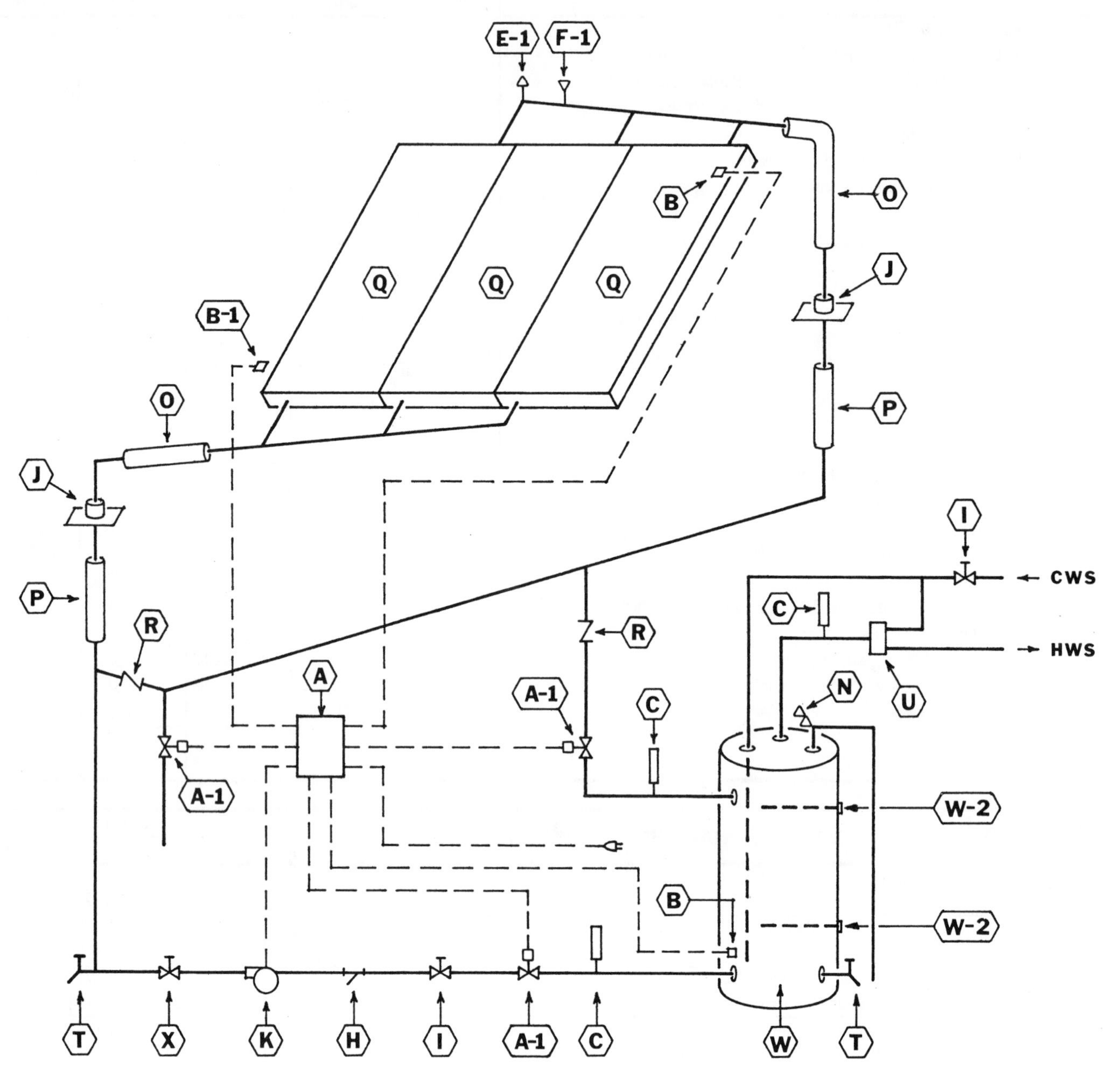

In the draindown direct-collection system, incoming domestic water is heated in the collectors. When the controller activates solar collection, domestic water is first heated as it flows through the collectors and is then pumped to storage. When conditions are no longer suitable for heat collection, the pump shuts off and the water in the loop drains down and out of the system by means of solenoid valves and properly pitched piping.

D2020 Domestic Water Distribution

System Components

System Components		QUANTITY	UNIT	COST EACH MAT.	INST.	TOTAL
	SYSTEM D2020 275 2760					
	SOLAR, DRAINDOWN, HOT WATER, DIRECT COLLECTION					
	3/4" TUBING, THREE 3'X7' BLACK CHROME COLLECTORS					
A, B	Differential controller, 2 sensors, thermostat, solar energy system	1.000	Ea.	625	56	681
A-1	Solenoid valve, solar heating loop, brass, 3/4" diam, 24 volts	3.000	Ea.	381	223.50	604.50
B-1	Solar energy sensor, freeze prevention	1.000	Ea.	26	21	47
C	Thermometer, 2" dial	3.000	Ea.	69	126	195
E-1	Vacuum relief valve, 3/4" diam	1.000	Ea.	31.50	21	52.50
F-1	Air vent, automatic, 1/8" fitting	1.000	Ea.	25	21	46
H	Strainer, Y type, bronze body, 3/4" IPS	1.000	Ea.	36	36	72
I	Valve, gate, bronze, NRS, soldered, 3/4" diam	2.000	Ea.	109	67	176
J	Vent flashing neoprene	2.000	Ea.	27.40	67	94.40
K	Circulator, solar heated liquid, 1/25 HP	1.000	Ea.	295	86.50	381.50
N	Relief valve temp & press 150 psi 210°F self-closing 3/4" IPS	1.000	Ea.	22	22.50	44.50
O	Pipe covering, urethane, ultraviolet cover, 1" wall, 3/4" diam	20.000	L.F.	60.20	121	181.20
P	Pipe covering, fiberglass, all service jacket, 1" wall, 3/4" diam	50.000	L.F.	53.50	241.50	295
	Roof clamps for solar energy collector panels	3.000	Set	8.37	51.90	60.27
Q	Collector panel solar energy blk chrome on copper, 1/8" temp glass 3'x7'	3.000	Ea.	3,375	381	3,756
R	Valve, swing check, bronze, regrinding disc, 3/4" diam, soldered	2.000	Ea.	196	67	263
T	Drain valve, brass, 3/4" connection	2.000	Ea.	20.10	56	76.10
U	Valve, water tempering, bronze, sweat connections, 3/4" diam	1.000	Ea.	126	33.50	159.50
W-2, W	Tank, water storage elec elem 2"x1/2# insul 120 gal	1.000	Ea.	1,950	480	2,430
X	Valve, globe, bronze, rising stem, 3/4" diam, soldered	1.000	Ea.	139	33.50	172.50
	Copper tubing, type L, solder joints, hangers 10' OC 3/4" diam	20.000	L.F.	144	177	321
	Copper tubing, type M, solder joints, hangers 10' OC 3/4" diam	70.000	L.F.	381.50	602	983.50
	Sensor wire, #22-2 conductor, multistranded	.500	C.L.F.	7.83	31.25	39.08
	Wrought copper fittings & solder, 3/4" diam	76.000	Ea.	532	2,698	3,230
	TOTAL			8,640.40	5,721.15	14,361.55

D2020 275	Solar, Draindown, Hot Water Systems	COST EACH MAT.	INST.	TOTAL
2550	Solar, draindown, hot water			
2560	3/8" tubing, 3 ea. 4' x 4'-4" vacuum tube collectors, 80 gal tank	7,300	5,425	12,725
2580	1/2" tubing, 4 ea. 4' x 4'-4" vacuum tube collectors, 80 gal tank R235616-60	8,250	5,575	13,825
2600	120 gal tank	8,500	5,650	14,150
2640	2 ea. 3'x7' black chrome collectors, 80 gal tank	6,650	5,275	11,925
2660	3 ea. 3'x7' black chrome collectors, 120 gal tank	8,025	5,475	13,500
2700	2 ea. 3'x7' flat black collectors, 120 gal tank	6,200	5,350	11,550
2720	3 ea. 3'x7' flat black collectors, 120 gal tank	6,925	5,450	12,375
2760	3/4" tubing, 3 ea. 3'x7' black chrome collectors, 120 gal tank	8,650	5,725	14,375
2780	3 ea. 3'x7' flat collectors, 120 gal tank	7,600	5,750	13,350
2800	2 ea. 4'x9' flat black & plastic glazing collectors, 120 gal tank	7,225	5,775	13,000
2840	1" tubing, 4 ea. 2'x9' plastic absorber & glazing collectors, 120 gal tank	12,000	6,450	18,450
2860	4 ea. 3'x7' black chrome absorber collectors, 120 gal tank	14,200	6,475	20,675
2880	4 ea. 3'x7' flat black absorber collectors, 120 gal tank	12,800	6,500	19,300

D2020 Domestic Water Distribution

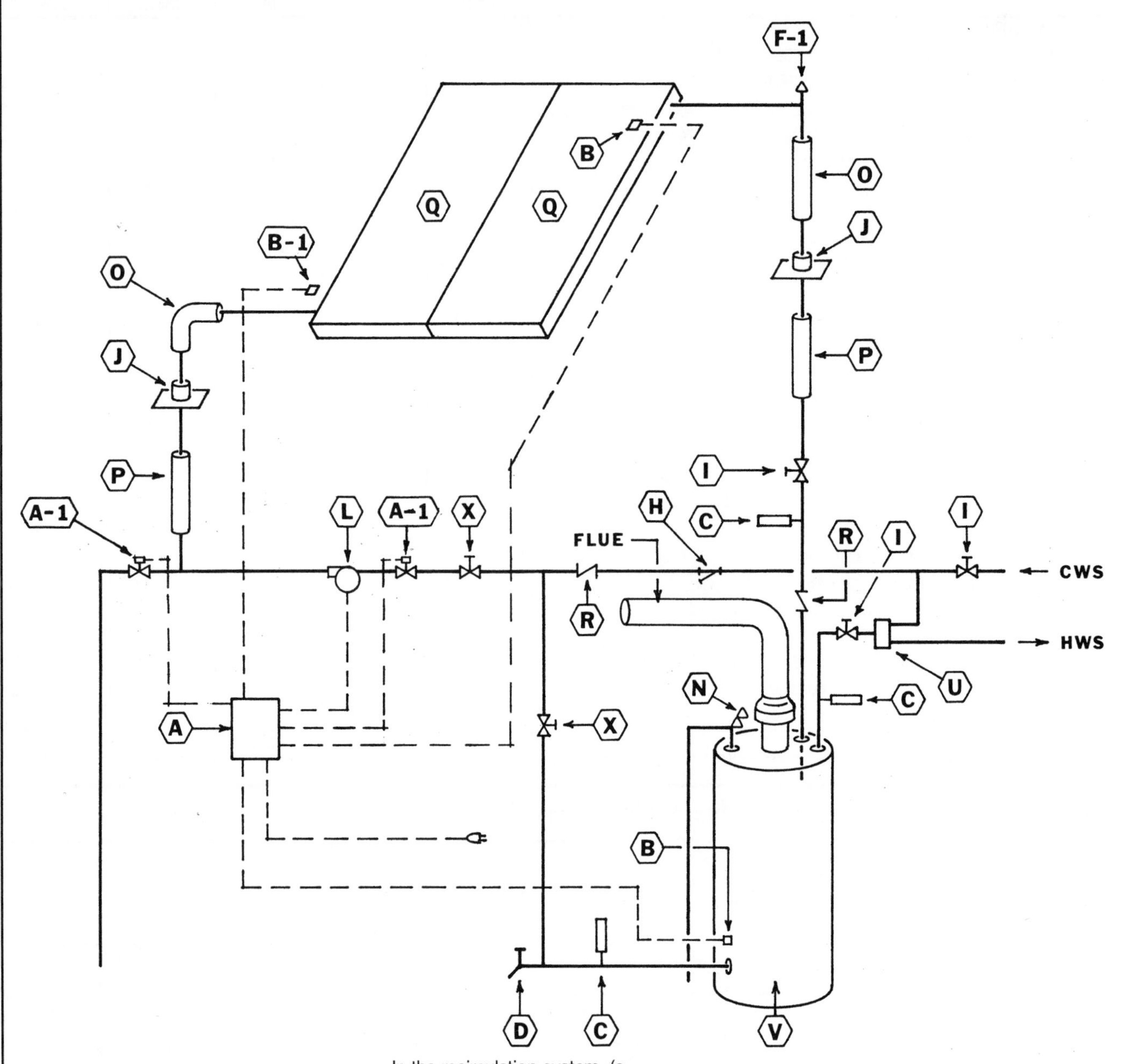

In the recirculation system, (a direct-collection system), incoming domestic water is heated in the collectors. When the controller activates solar collection, domestic water is heated as it flows through the collectors and then it flows back to storage. When conditions are not suitable for heat collection, the pump shuts off and the flow of the water stops. In this type of system, water remains in the collector loop at all times. A "frost sensor" at the collector activates the circulation of warm water from storage through the collectors when protection from freezing is required.

D2020 Domestic Water Distribution

System Components		QUANTITY	UNIT	COST EACH MAT.	COST EACH INST.	COST EACH TOTAL
	SYSTEM D2020 280 2820					
	SOLAR, RECIRCULATION, HOT WATER					
	3/4″ TUBING, TWO 3′X7′ BLACK CHROME COLLECTORS					
A, B	Differential controller 2 sensors, thermostat, for solar energy system	1.000	Ea.	625	56	681
A-1	Solenoid valve, solar heating loop, brass, 3/4″ IPS, 24 volts	2.000	Ea.	254	149	403
B-1	Solar energy sensor freeze prevention	1.000	Ea.	26	21	47
C	Thermometer, 2″ dial	3.000	Ea.	69	126	195
D	Drain valve, brass, 3/4″ connection	1.000	Ea.	10.05	28	38.05
F-1	Air vent, automatic, 1/8″ fitting	1.000	Ea.	25	21	46
H	Strainer, Y type, bronze body, 3/4″ IPS	1.000	Ea.	36	36	72
I	Valve, gate, bronze, 125 lb, soldered 3/4″ diam	3.000	Ea.	163.50	100.50	264
J	Vent flashing, neoprene	2.000	Ea.	27.40	67	94.40
L	Circulator, solar heated liquid, 1/20 HP	1.000	Ea.	177	101	278
N	Relief valve, temp & press 150 psi 210° F self-closing 3/4″ IPS	2.000	Ea.	44	45	89
O	Pipe covering, urethane, ultraviolet cover, 1″ wall, 3/4″ diam	20.000	L.F.	60.20	121	181.20
P	Pipe covering, fiberglass, all service jacket, 1″ wall 3/4″ diam	50.000	L.F.	53.50	241.50	295
Q	Collector panel solar energy blk chrome on copper, 1/8″ temp glass 3′x7′	2.000	Ea.	2,250	254	2,504
	Roof clamps for solar energy collector panels	2.000	Set	5.58	34.60	40.18
R	Valve, swing check, bronze, 125 lb, regrinding disc, soldered 3/4″ diam	2.000	Ea.	196	67	263
U	Valve, water tempering, bronze, sweat connections, 3/4″ diam	1.000	Ea.	126	33.50	159.50
V	Tank, water storage, w/heating element, drain, relief valve, existing	1.000	Ea.			
X	Valve, globe, bronze, 125 lb, 3/4″ diam	2.000	Ea.	278	67	345
	Copper tubing, type L, solder joints, hangers 10′ OC 3/4″ diam	20.000	L.F.	144	177	321
	Copper tubing, type M, solder joints, hangers 10′ OC 3/4″ diam	70.000	L.F.	381.50	602	983.50
	Wrought copper fittings & solder, 3/4″ diam	76.000	Ea.	532	2,698	3,230
	Sensor wire, #22-2 conductor, multistranded	.500	C.L.F.	7.83	31.25	39.08
	TOTAL			5,491.56	5,077.35	10,568.91

D2020 280	Solar Recirculation, Domestic Hot Water Systems	COST EACH MAT.	COST EACH INST.	COST EACH TOTAL
2550	Solar recirculation, hot water			
2560	3/8″ tubing, 3 ea. 4′ x 4′-4″ vacuum tube collectors	5,950	4,725	10,675
2580	1/2″ tubing, 4 ea. 4′ x 4′-4″ vacuum tube collectors R235616-60	6,450	5,125	11,575
2640	2 ea. 3′x7′ black chrome collectors	4,825	4,825	9,650
2660	3 ea. 3′x7′ black chrome collectors	5,950	4,950	10,900
2700	2 ea. 3′x7′ flat black collectors	4,150	4,825	8,975
2720	3 ea. 3′x7′ flat black collectors	4,925	4,975	9,900
2760	3/4″ tubing, 3 ea. 3′x7′ black chrome collectors	6,625	5,225	11,850
2780	3 ea. 3′x7′ flat black absorber plate collectors	5,575	5,250	10,825
2800	2 ea. 4′x9′ flat black w/plastic glazing collectors	5,200	5,275	10,475
2820	2 ea. 3′x7′ black chrome collectors	5,500	5,075	10,575
2840	1″ tubing, 4 ea. 2′x9′ black plastic absorber & glazing collectors	9,075	5,950	15,025
2860	4 ea. 3′x7′ black chrome absorber collectors	11,200	5,975	17,175
2880	4 ea. 3′x7′ flat black absorber collectors	9,850	6,000	15,850

D2020 Domestic Water Distribution

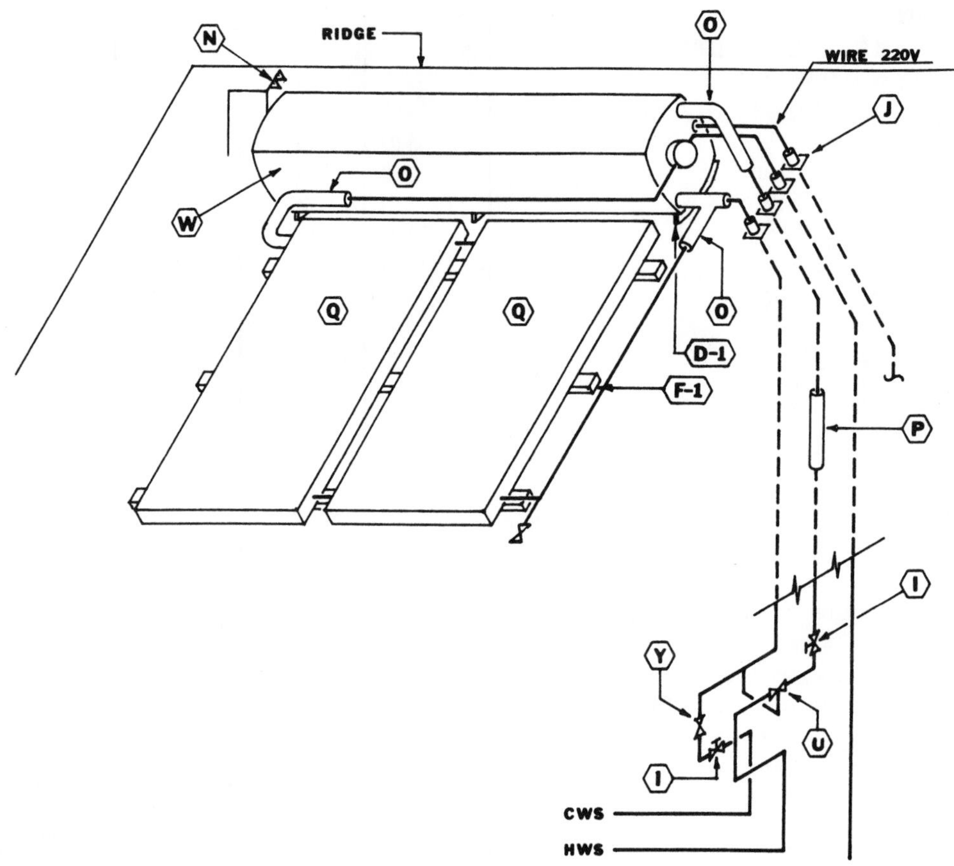

The thermosyphon domestic hot water system, a direct collection system, operates under city water pressure and does not require pumps for system operation. An insulated water storage tank is located above the collectors. As the sun heats the collectors, warm water in them rises by means of natural convection; the colder water in the storage tank flows into the collectors by means of gravity. As long as the sun is shining the water continues to flow through the collectors and to become warmer.

To prevent freezing, the system must be drained or the collectors covered with an insulated lid when the temperature drops below 32°F.

D2020 Domestic Water Distribution

System Components		QUANTITY	UNIT	COST EACH MAT.	COST EACH INST.	COST EACH TOTAL
SYSTEM D2020 285 0960						
SOLAR, THERMOSYPHON, WATER HEATER						
3/4" TUBING, TWO 3'X7' BLACK CHROME COLLECTORS						
D-1	Framing lumber, fir, 2" x 6" x 8', tank cradle	.008	M.B.F.	4.92	7.08	12
F-1	Framing lumber, fir, 2" x 4" x 24', sleepers	.016	M.B.F.	9.52	21.20	30.72
I	Valve, gate, bronze, 125 lb, soldered 1/2" diam	2.000	Ea.	97	56	153
J	Vent flashing, neoprene	4.000	Ea.	54.80	134	188.80
O	Pipe covering, urethane, ultraviolet cover, 1" wall, 1/2"diam	40.000	L.F.	92.40	238	330.40
P	Pipe covering fiberglass all service jacket 1" wall 1/2" diam	160.000	L.F.	156.80	740.80	897.60
Q	Collector panel solar, blk chrome on copper, 3/16" temp glass 3'-6" x 7.5'	2.000	Ea.	1,560	268	1,828
Y	Flow control valve, globe, bronze, 125#, soldered, 1/2" diam	1.000	Ea.	104	28	132
U	Valve, water tempering, bronze,sweat connections, 1/2" diam	1.000	Ea.	103	28	131
W	Tank, water storage, solar energy system, 80 Gal, 2" x 1/2 lb insul	1.000	Ea.	1,700	420	2,120
	Copper tubing type L, solder joints, hangers 10' OC 1/2" diam	150.000	L.F.	699	1,245	1,944
	Copper tubing type M, solder joints, hangers 10' OC 1/2" diam	50.000	L.F.	179.50	400	579.50
	Sensor wire, #22-2 gauge multistranded	.500	C.L.F.	7.83	31.25	39.08
	Wrought copper fittings & solder, 1/2" diam	75.000	Ea.	233.25	2,512.50	2,745.75
	TOTAL			5,002.02	6,129.83	11,131.85

D2020 285	Thermosyphon, Hot Water		COST EACH MAT.	COST EACH INST.	COST EACH TOTAL
0960	Solar, thermosyphon, hot water, two collector system	R235616 -60	5,000	6,125	11,125
0970					

D2020 Domestic Water Distribution

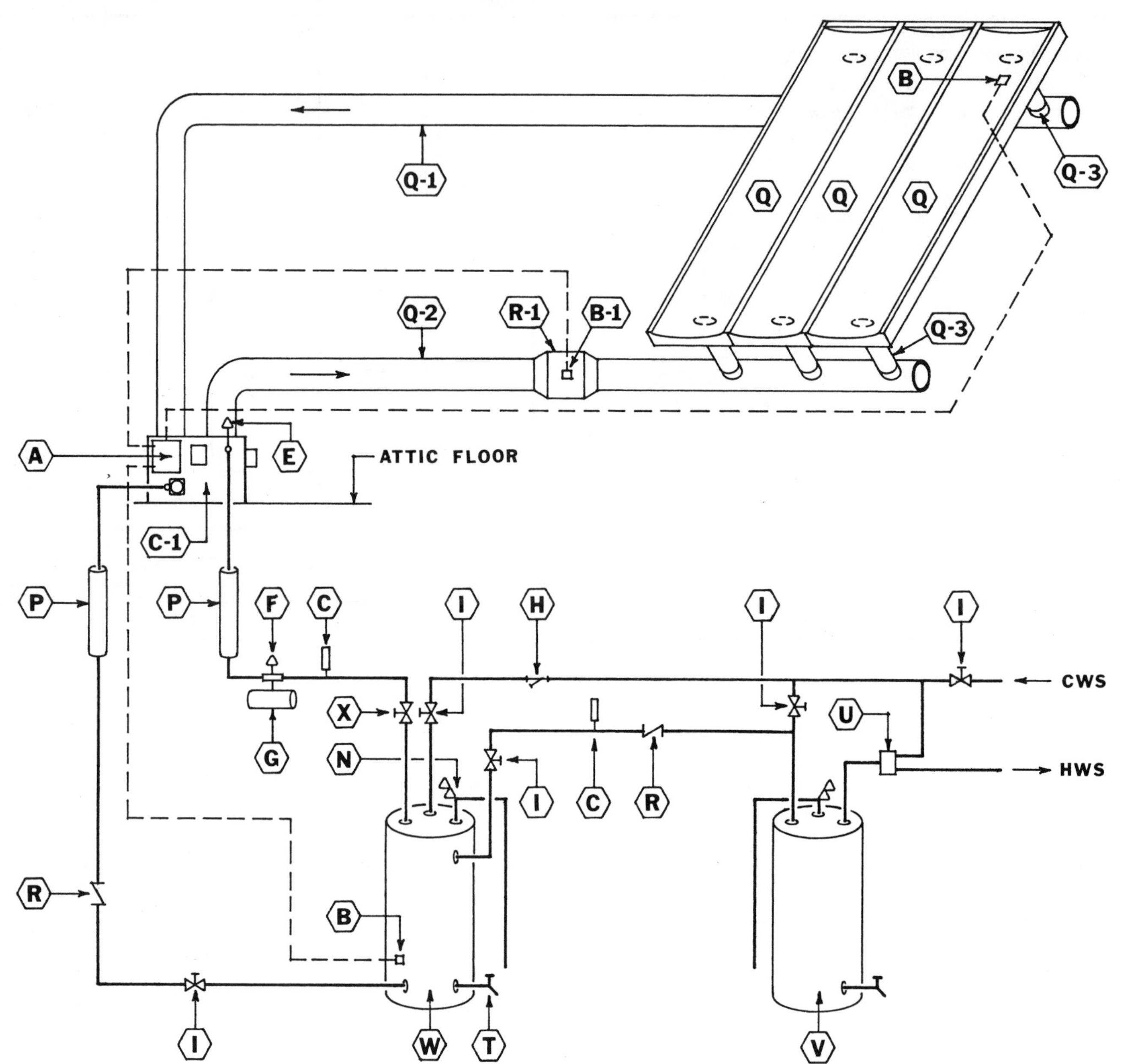

This domestic hot water pre-heat system includes heat exchanger with a circulating pump, blower, air-to-water, coil and controls, mounted in the upper collector manifold. Heat from the hot air coming out of the collectors is transferred through the heat exchanger. For each degree of DHW preheating gained, one degree less heating is needed from the fuel fired water heater. The system is simple, inexpensive to operate and can provide a substantial portion of DHW requirements for modest additional cost.

D2020 Domestic Water Distribution

System Components		QUANTITY	UNIT	COST EACH MAT.	COST EACH INST.	COST EACH TOTAL
	SYSTEM D2020 290 2560					
	SOLAR, HOT WATER, AIR TO WATER HEAT EXCHANGE					
	THREE COLLECTORS, OPTICAL BLACK ON ALUMINUM, 10'X2',80 GAL TANK					
A, B	Differential controller, 2 sensors, thermostat, solar energy system	1.000	Ea.	625	56	681
C	Thermometer, 2" dial	2.000	Ea.	46	84	130
C-1	Heat exchanger, air to fluid, up flow 70 MBH	1.000	Ea.	380	345	725
E	Air vent, manual, for solar energy system 1/8" fitting	1.000	Ea.	15.90	21	36.90
F	Air purger	1.000	Ea.	51.50	56	107.50
G	Expansion tank	1.000	Ea.	72.50	21	93.50
H	Strainer, Y type, bronze body, 1/2" IPS	1.000	Ea.	31.50	34	65.50
I	Valve, gate, bronze, 125 lb, NRS, soldered 1/2" diam	5.000	Ea.	242.50	140	382.50
N	Relief valve, temp & pressure solar 150 psi 210°F self-closing	1.000	Ea.	22	22.50	44.50
P	Pipe covering, fiberglass, all service jacket, 1" wall, 1/2" diam	60.000	L.F.	58.80	277.80	336.60
Q	Collector panel solar energy, air, black on alum plate, flush mount, 10'x2'	30.000	L.F.	16,050	366	16,416
B-1, R-1	Shutter damper for solar heater circulator	1.000	Ea.	57	97.50	154.50
B-1, R-1	Shutter motor for solar heater circulator	1.000	Ea.	142	73	215
R	Backflow preventer, 1/2" pipe size	2.000	Ea.	276	84	360
T	Drain valve, brass, 3/4" connection	1.000	Ea.	10.05	28	38.05
U	Valve, water tempering, bronze, sweat connections, 1/2 " diam	1.000	Ea.	103	28	131
W	Tank, water storage, solar energy system, 80 Gal, 2" x 2 lb insul	1.000	Ea.	1,700	420	2,120
V	Tank, water storage, w/heating element, drain, relief valve, existing	1.000	System			
X	Valve, globe, bronze, 125 lb, rising stem, 1/2" diam	1.000	Ea.	104	28	132
	Copper tubing type M, solder joints, hangers 10' OC, 1/2" diam	50.000	L.F.	179.50	400	579.50
	Copper tubing type L, solder joints, hangers 10' OC 1/2" diam	10.000	L.F.	46.60	83	129.60
	Wrought copper fittings & solder, 1/2" diam	10.000	Ea.	31.10	335	366.10
	Sensor wire, #22-2 conductor multistranded	.500	C.L.F.	7.83	31.25	39.08
Q-1, Q-2	Ductwork, fiberglass, aluminized jacket, 1-1/2" thick, 8" diam	32.000	S.F.	159.36	208	367.36
Q-3	Manifold for flush mount solar energy collector panels, air	6.000	L.F.	864	43.80	907.80
	TOTAL			21,276.14	3,282.85	24,558.99

D2020 290	Solar Hot Water, Air To Water Heat Exchange		COST EACH MAT.	COST EACH INST.	COST EACH TOTAL
2550	Solar hot water, air to water heat exchange				
2560	Three collectors, optical black on aluminum, 10' x 2', 80 Gal tank		21,300	3,275	24,575
2580	Four collectors, optical black on aluminum, 10' x 2', 80 Gal tank	R235616-60	27,000	3,475	30,475
2600	Four collectors, optical black on aluminum, 10' x 2', 120 Gal tank		27,200	3,525	30,725

D2020 Domestic Water Distribution

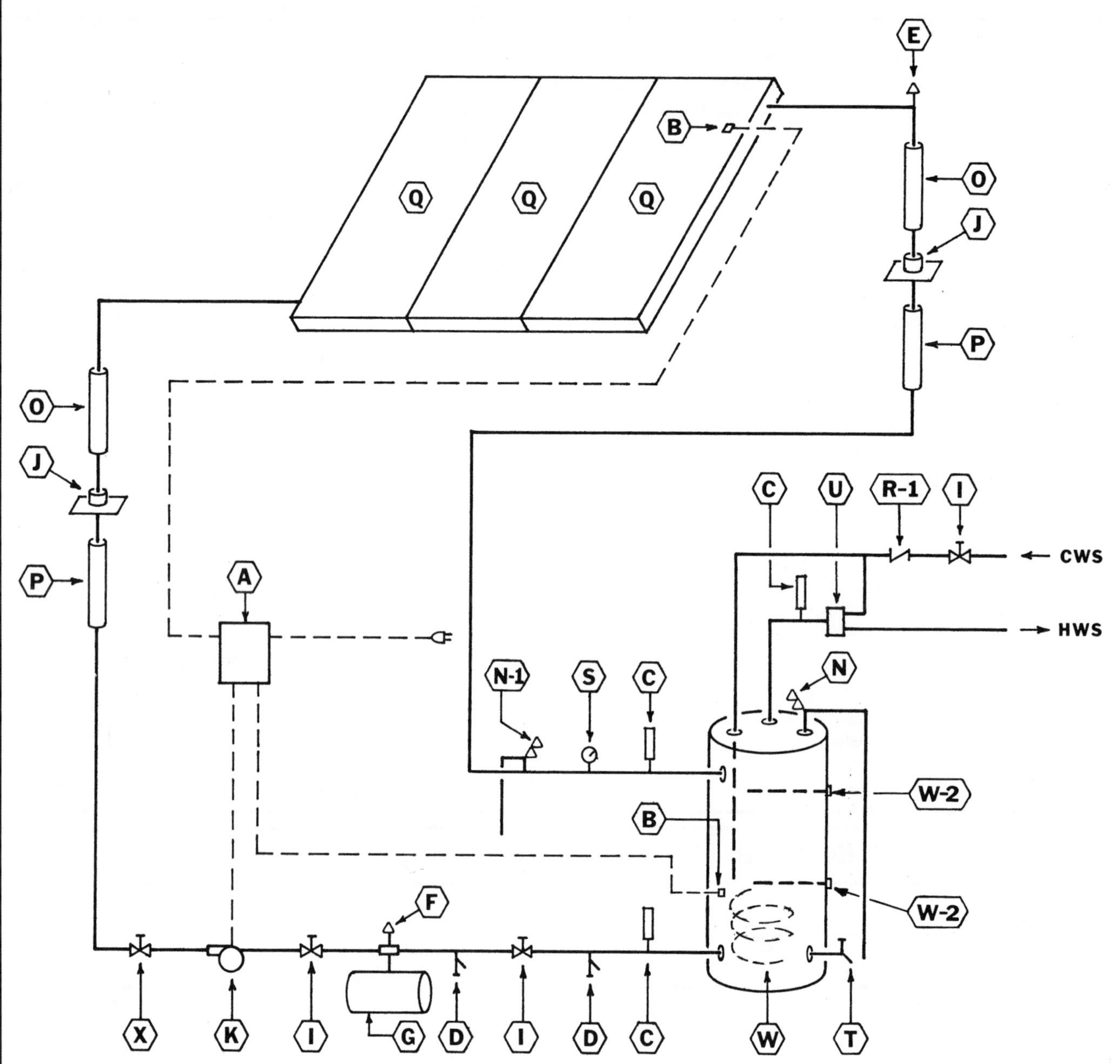

In this closed-loop indirect collection system, fluid with a low freezing temperature, such as propylene glycol, transports heat from the collectors to water storage. The transfer fluid is contained in a closed-loop consisting of collectors, supply and return piping, and a heat exchanger immersed in the storage tank. A typical two-or-three panel system contains 5 to 6 gallons of heat transfer fluid.

When the collectors become approximately 20°F warmer than the storage temperature, a controller activates the circulator. The circulator moves the fluid continuously through the collectors until the temperature difference between the collectors and storage is such that heat collection no longer occurs; at that point, the circulator shuts off. Since the heat transfer fluid has a very low freezing temperature, there is no need for it to be drained from the collectors between periods of collection.

D2020 Domestic Water Distribution

System Components		QUANTITY	UNIT	COST EACH MAT.	INST.	TOTAL
	SYSTEM D2020 295 2760					
	SOLAR, CLOSED LOOP, HOT WATER SYSTEM, IMMERSED HEAT EXCHANGER					
	3/4″ TUBING, THREE 3′ X 7′ BLACK CHROME COLLECTORS					
A, B	Differential controller, 2 sensors, thermostat, solar energy system	1.000	Ea.	625	56	681
C	Thermometer 2″ dial	3.000	Ea.	69	126	195
D, T	Fill & drain valves, brass, 3/4″ connection	3.000	Ea.	30.15	84	114.15
E	Air vent, manual, 1/8″ fitting	1.000	Ea.	15.90	21	36.90
F	Air purger	1.000	Ea.	51.50	56	107.50
G	Expansion tank	1.000	Ea.	72.50	21	93.50
I	Valve, gate, bronze, NRS, soldered 3/4″ diam	3.000	Ea.	163.50	100.50	264
J	Neoprene vent flashing	2.000	Ea.	27.40	67	94.40
K	Circulator, solar heated liquid, 1/25 HP	1.000	Ea.	295	86.50	381.50
N-1, N	Relief valve, temp & press 150 psi 210°F self-closing 3/4″ IPS	2.000	Ea.	44	45	89
O	Pipe covering, urethane ultraviolet cover, 1″ wall, 3/4″ diam	20.000	L.F.	60.20	121	181.20
P	Pipe covering, fiberglass, all service jacket, 1″ wall, 3/4″ diam	50.000	L.F.	53.50	241.50	295
	Roof clamps for solar energy collector panel	3.000	Set	8.37	51.90	60.27
Q	Collector panel solar blk chrome on copper, 1/8″ temp glass, 3′x7′	3.000	Ea.	3,375	381	3,756
R-1	Valve, swing check, bronze, regrinding disc, 3/4″ diam, soldered	1.000	Ea.	98	33.50	131.50
S	Pressure gauge, 60 psi, 2-1/2″ dial	1.000	Ea.	25.50	21	46.50
U	Valve, water tempering, bronze, sweat connections, 3/4″ diam	1.000	Ea.	126	33.50	159.50
W-2, W	Tank, water storage immersed heat exchr elec elem 2″x2# insul 120 Gal	1.000	Ea.	1,950	480	2,430
X	Valve, globe, bronze, rising stem, 3/4″ diam, soldered	1.000	Ea.	139	33.50	172.50
	Copper tubing type L, solder joint, hanger 10′ OC 3/4″ diam	20.000	L.F.	144	177	321
	Copper tubing, type M, solder joint, hanger 10′ OC 3/4″ diam	70.000	L.F.	381.50	602	983.50
	Sensor wire, #22-2 conductor multistranded	.500	C.L.F.	7.83	31.25	39.08
	Solar energy heat transfer fluid, propylene glycol, anti-freeze	6.000	Gal.	95.70	144	239.70
	Wrought copper fittings & solder, 3/4″ diam	76.000	Ea.	532	2,698	3,230
	TOTAL			8,390.55	5,712.15	14,102.70

D2020 295	Solar, Closed Loop, Hot Water Systems		COST EACH MAT.	INST.	TOTAL
2550	Solar, closed loop, hot water system, immersed heat exchanger				
2560	3/8″ tubing, 3 ea. 4′ x 4′-4″ vacuum tube collectors, 80 gal. tank		7,550	5,175	12,725
2580	1/2″ tubing, 4 ea. 4′ x 4′-4″ vacuum tube collectors, 80 gal. tank		8,025	5,575	13,600
2600	120 gal. tank		8,275	5,625	13,900
2640	2 ea. 3′x7′ black chrome collectors, 80 gal. tank	R235616 -60	6,475	5,325	11,800
2660	120 gal. tank		6,675	5,325	12,000
2700	2 ea. 3′x7′ flat black collectors, 120 gal. tank		5,975	5,325	11,300
2720	3 ea. 3′x7′ flat black collectors, 120 gal. tank		6,750	5,475	12,225
2760	3/4″ tubing, 3 ea. 3′x7′ black chrome collectors, 120 gal. tank		8,400	5,700	14,100
2780	3 ea. 3′x7′ flat black collectors, 120 gal. tank		7,350	5,725	13,075
2800	2 ea. 4′x9′ flat black w/plastic glazing collectors 120 gal. tank		6,975	5,750	12,725
2840	1″ tubing, 4 ea. 2′x9′ plastic absorber & glazing collectors 120 gal. tank		8,550	6,425	14,975
2860	4 ea. 3′x7′ black chrome collectors, 120 gal. tank		10,700	6,450	17,150
2880	4 ea. 3′x7′ flat black absorber collectors, 120 gal. tank		9,350	6,475	15,825

D2040 Rain Water Drainage

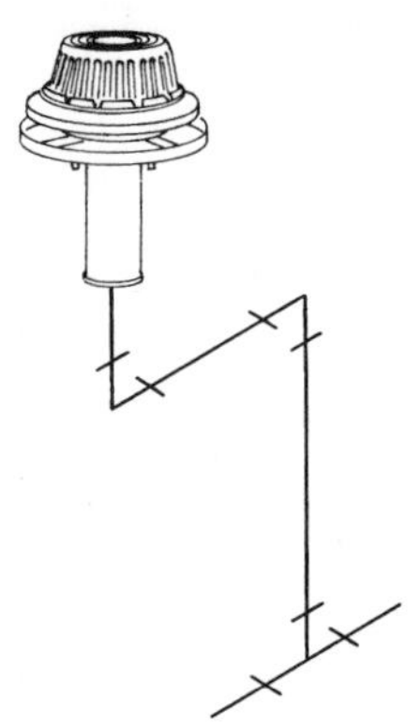

Design Assumptions: Vertical conductor size is based on a maximum rate of rainfall of 4″ per hour. To convert roof area to other rates multiply "Max. S.F. Roof Area" shown by four and divide the result by desired local rate. The answer is the local roof area that may be handled by the indicated pipe diameter.

Basic cost is for roof drain, 10′ of vertical leader and 10′ of horizontal, plus connection to the main.

Pipe Dia.	Max. S.F. Roof Area	Gallons per Min.
2″	544	23
3″	1610	67
4″	3460	144
5″	6280	261
6″	10,200	424
8″	22,000	913

System Components	QUANTITY	UNIT	COST EACH		
			MAT.	INST.	TOTAL
SYSTEM D2040 210 1880					
ROOF DRAIN, DWV PVC PIPE, 2″ DIAM., 10′ HIGH					
Drain, roof, main, PVC, dome type 2″ pipe size	1.000	Ea.	144	86.50	230.50
Clamp, roof drain, underdeck	1.000	Ea.	30	50.50	80.50
Pipe, Tee, PVC DWV, schedule 40, 2″ pipe size	1.000	Ea.	8.10	60.50	68.60
Pipe, PVC, DWV, schedule 40, 2″ diam.	20.000	L.F.	79	410	489
Pipe, elbow, PVC schedule 40, 2″ diam.	2.000	Ea.	6.06	67	73.06
TOTAL			267.16	674.50	941.66

D2040 210	Roof Drain Systems	COST EACH		
		MAT.	INST.	TOTAL
1880	Roof drain, DWV PVC, 2″ diam., piping, 10′ high	267	675	942
1920	For each additional foot add	3.95	20.50	24.45
1960	3″ diam., 10′ high	360	790	1,150
2000	For each additional foot add	7.10	23	30.10
2040	4″ diam., 10′ high	445	880	1,325
2080	For each additional foot add	9.65	25	34.65
2120	5″ diam., 10′ high	1,325	1,025	2,350
2160	For each additional foot add	28.50	28	56.50
2200	6″ diam., 10′ high	1,350	1,125	2,475
2240	For each additional foot add	19.40	31	50.40
2280	8″ diam., 10′ high	2,950	1,925	4,875
2320	For each additional foot add	45	39	84
3940	C.I., soil, single hub, service wt., 2″ diam. piping, 10′ high	560	735	1,295
3980	For each additional foot add	10	19.20	29.20
4120	3″ diam., 10′ high	780	795	1,575
4160	For each additional foot add	14.05	20	34.05
4200	4″ diam., 10′ high	920	870	1,790
4240	For each additional foot add	18.60	22	40.60
4280	5″ diam., 10′ high	1,275	970	2,245
4320	For each additional foot add	25	25	50
4360	6″ diam., 10′ high	1,625	1,025	2,650
4400	For each additional foot add	31.50	26	57.50
4440	8″ diam., 10′ high	3,175	2,100	5,275
4480	For each additional foot add	48.50	43.50	92
6040	Steel galv. sch 40 threaded, 2″ diam. piping, 10′ high	730	715	1,445
6080	For each additional foot add	14.65	18.90	33.55
6120	3″ diam., 10′ high	1,425	1,025	2,450
6160	For each additional foot add	31	28	59

D2040 Rain Water Drainage

D2040 210	Roof Drain Systems	COST EACH		
		MAT.	INST.	TOTAL
6200	4″ diam., 10′ high	2,050	1,325	3,375
6240	For each additional foot add	45.50	33.50	79
6280	5″ diam., 10′ high	2,325	1,100	3,425
6320	For each additional foot add	64.50	32.50	97
6360	6″ diam, 10′ high	2,775	1,425	4,200
6400	For each additional foot add	73	45	118
6440	8″ diam., 10′ high	4,375	2,050	6,425
6480	For each additional foot add	86	51	137

D3010 Energy Supply

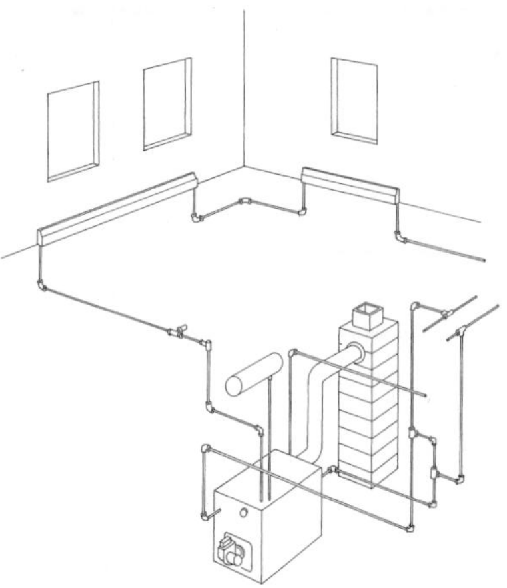

Basis for Heat Loss Estimate, Apartment Type Structures:

1. Masonry walls and flat roof are insulated. U factor is assumed at .08.
2. Window glass area taken as BOCA minimum, 1/10th of floor area. Double insulating glass with 1/4″ air space, U = .65.
3. Infiltration = 0.3 C.F. per hour per S.F. of net wall.
4. Concrete floor loss is 2 BTUH per S.F.
5. Temperature difference taken as 70°F.
6. Ventilating or makeup air has not been included and must be added if desired. Air shafts are not used.

System Components	QUANTITY	UNIT	COST EACH MAT.	INST.	TOTAL
SYSTEM D3010 510 1760					
HEATING SYSTEM, FIN TUBE RADIATION, FORCED HOT WATER					
1,000 S.F. AREA, 10,000 C.F. VOLUME					
Boiler, oil fired, CI, burner, ctrls/insul/breech/pipe/ftng/valves, 109 MBH	1.000	Ea.	4,200	3,806.25	8,006.25
Circulating pump, CI flange connection, 1/12 HP	1.000	Ea.	380	202	582
Expansion tank, painted steel, ASME 18 Gal capacity	1.000	Ea.	700	87.50	787.50
Storage tank, steel, above ground, 275 Gal capacity w/supports	1.000	Ea.	540	246	786
Copper tubing type L, solder joint, hanger 10′ OC, 3/4″ diam	100.000	L.F.	720	885	1,605
Radiation, 3/4″ copper tube w/alum fin baseboard pkg, 7″ high	30.000	L.F.	277.50	630	907.50
Pipe covering, calcium silicate w/cover, 1′ wall, 3/4′ diam	100.000	L.F.	363	655	1,018
TOTAL			7,180.50	6,511.75	13,692.25
COST PER S.F.			7.18	6.51	13.69

D3010 510	Apartment Building Heating - Fin Tube Radiation	COST PER S.F. MAT.	INST.	TOTAL
1740	Heating systems, fin tube radiation, forced hot water			
1760	1,000 S.F. area, 10,000 C.F. volume	7.18	6.50	13.68
1800	10,000 S.F. area, 100,000 C.F. volume	3.48	3.93	7.41
1840	20,000 S.F. area, 200,000 C.F. volume	3.99	4.39	8.38
1880	30,000 S.F. area, 300,000 C.F. volume	3.82	4.25	8.07

D3010 Energy Supply

Many styles of active solar energy systems exist. Those shown on the following page represent the majority of systems now being installed in different regions of the country.

The five active domestic hot water (DHW) systems which follow are typically specified as two or three panel systems with additional variations being the type of glazing and size of the storage tanks. Various combinations have been costed for the user's evaluation and comparison. The basic specifications from which the following systems were developed satisfy the construction detail requirements specified in the HUD Intermediate Minimum Property Standards (IMPS). If these standards are not complied with, the renewable energy system's costs could be significantly lower than shown.

To develop the system's specifications and costs it was necessary to make a number of assumptions about the systems. Certain systems are more appropriate to one climatic region than another or the systems may require modifications to be usable in particular locations. Specific instances in which the systems are not appropriate throughout the country as specified, or in which modification will be needed include the following:

- The freeze protection mechanisms provided in the DHW systems vary greatly. In harsh climates, where freeze is a major concern, a closed-loop indirect collection system may be more appropriate than a direct collection system.
- The thermosyphon water heater system described cannot be used when temperatures drop below 32°F.
- In warm climates it may be necessary to modify the systems installed to prevent overheating.

For each renewable resource (solar) system a schematic diagram and descriptive summary of the system is provided along with a list of all the components priced as part of the system. The costs were developed based on these specifications.

Considerations affecting costs which may increase or decrease beyond the estimates presented here include the following:

- Special structural qualities (allowance for earthquake, future expansion, high winds, and unusual spans or shapes);
- Isolated building site or rough terrain that would affect the transportation of personnel, material, or equipment;
- Unusual climatic conditions during the construction process;
- Substitution of other materials or system components for those used in the system specifications.

D3010 Energy Supply

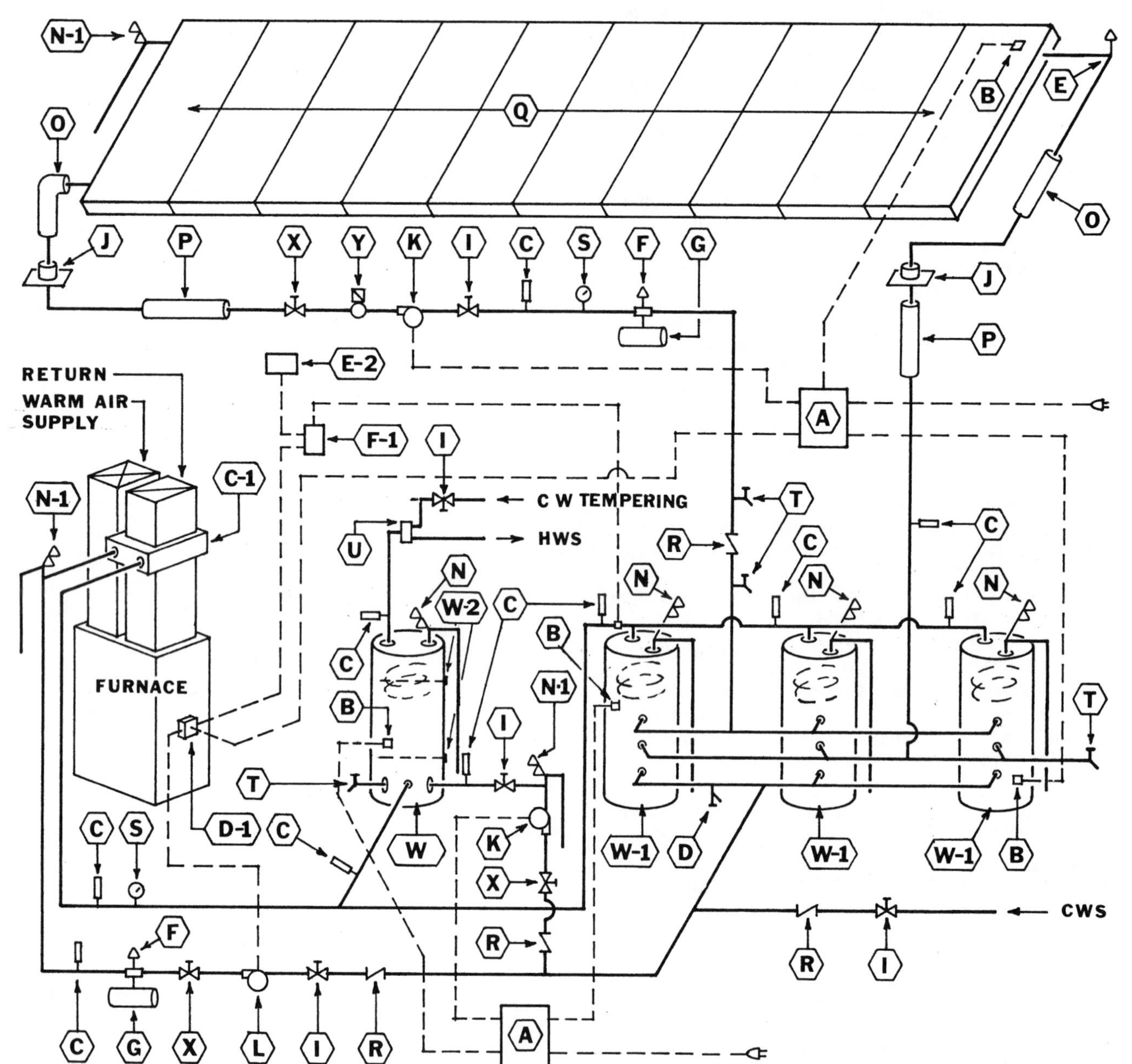

In this closed-loop indirect collection system, fluid with a low freezing temperature, propylene glycol, transports heat from the collectors to water storage. The transfer fluid is contained in a closed loop consisting of collectors, supply and return piping, and a heat exchanger immersed in the storage tank.

When the collectors become approximately 20°F warmer than the storage temperature, the controller activates the circulator. The circulator moves the fluid continuously until the temperature difference between fluid in the collectors and storage is such that the collection will no longer occur and then the circulator turns off. Since the heat transfer fluid has a very low freezing temperature, there is no need for it to be drained from the collectors between periods of collection.

D3010 Energy Supply

System Components	QUANTITY	UNIT	COST EACH MAT.	COST EACH INST.	COST EACH TOTAL
SYSTEM D3010 650 2750					
SOLAR, CLOSED LOOP, SPACE/HOT WATER					
1″ TUBING, TEN 3′X7′ BLK CHROME ON COPPER ABSORBER COLLECTORS					
A, B Differential controller 2 sensors, thermostat, solar energy system	2.000	Ea.	1,250	112	1,362
C Thermometer, 2″ dial	10.000	Ea.	230	420	650
C-1 Heat exchanger, solar energy system, fluid to air, up flow, 80 MBH	1.000	Ea.	510	405	915
D, T Fill & drain valves, brass, 3/4″ connection	5.000	Ea.	50.25	140	190.25
D-1 Fan center	1.000	Ea.	127	139	266
E Air vent, manual, 1/8″ fitting	1.000	Ea.	15.90	21	36.90
E-2 Thermostat, 2 stage for sensing room temperature	1.000	Ea.	177	85.50	262.50
F Air purger	2.000	Ea.	103	112	215
F-1 Controller, liquid temperature, solar energy system	1.000	Ea.	114	134	248
G Expansion tank	2.000	Ea.	145	42	187
I Valve, gate, bronze, 125 lb, soldered, 1″ diam	5.000	Ea.	387.50	177.50	565
J Vent flashing, neoprene	2.000	Ea.	27.40	67	94.40
K Circulator, solar heated liquid, 1/25 HP	2.000	Ea.	590	173	763
L Circulator, solar heated liquid, 1/20 HP	1.000	Ea.	177	101	278
N Relief valve, temp & pressure 150 psi 210°F self-closing	4.000	Ea.	88	90	178
N-1 Relief valve, pressure poppet, bronze, 30 psi, 3/4″ IPS	3.000	Ea.	144	72	216
O Pipe covering, urethane, ultraviolet cover, 1″ wall, 1″ diam	50.000	L.F.	150.50	302.50	453
P Pipe covering, fiberglass, all service jacket, 1″ wall, 1″ diam	60.000	L.F.	64.20	289.80	354
Q Collector panel solar energy blk chrome on copper 1/8″ temp glass 3′x7′	10.000	Ea.	11,250	1,270	12,520
Roof clamps for solar energy collector panel	10.000	Set	27.90	173	200.90
R Valve, swing check, bronze, 125 lb, regrinding disc, 3/4″ & 1″ diam	4.000	Ea.	564	142	706
S Pressure gage, 0-60 psi, for solar energy system	2.000	Ea.	51	42	93
U Valve, water tempering, bronze, sweat connections, 3/4″ diam	1.000	Ea.	126	33.50	159.50
W-2, W-1, W Tank, water storage immersed heat xchr elec elem 2″x1/2# ins 120 gal	4.000	Ea.	7,800	1,920	9,720
X Valve, globe, bronze, 125 lb, rising stem, 1″ diam	3.000	Ea.	657	106.50	763.50
Y Valve, flow control	1.000	Ea.	148	30.50	178.50
Copper tubing, type M, solder joint, hanger 10′ OC 1″ diam	110.000	L.F.	946	1,056	2,002
Copper tubing, type L, solder joint, hanger 10′ OC 3/4″ diam	20.000	L.F.	144	177	321
Wrought copper fittings & solder, 3/4″ & 1″ diam	121.000	Ea.	2,081.20	5,082	7,163.20
Sensor, wire, #22-2 conductor, multistranded	.700	C.L.F.	10.96	43.75	54.71
Ductwork, galvanized steel, for heat exchanger	8.000	Lb.	6.16	62	68.16
Solar energy heat transfer fluid propylene glycol, anti-freeze	25.000	Gal.	398.75	600	998.75
TOTAL			28,561.72	13,621.55	42,183.27

D3010 650	Solar, Closed Loop, Space/Hot Water Systems	COST EACH MAT.	COST EACH INST.	COST EACH TOTAL
2540	Solar, closed loop, space/hot water			
2550	1/2″ tubing, 12 ea. 4′x4′4″ vacuum tube collectors	25,800	12,700	38,500
2600	3/4″ tubing, 12 ea. 4′x4′4″ vacuum tube collectors (R235616-60)	27,200	13,100	40,300
2650	10 ea. 3′ x 7′ black chrome absorber collectors	26,900	12,700	39,600
2700	10 ea. 3′ x 7′ flat black absorber collectors	23,400	12,800	36,200
2750	1″ tubing, 10 ea. 3′ x 7′ black chrome absorber collectors	28,600	13,600	42,200
2800	10 ea. 3′ x 7′ flat black absorber collectors	25,200	13,700	38,900
2850	6 ea. 4′ x 9′ flat black w/plastic glazing collectors	23,300	13,600	36,900
2900	12 ea. 2′ x 9′ plastic absorber and glazing collectors	24,400	13,900	38,300

D3010 Energy Supply

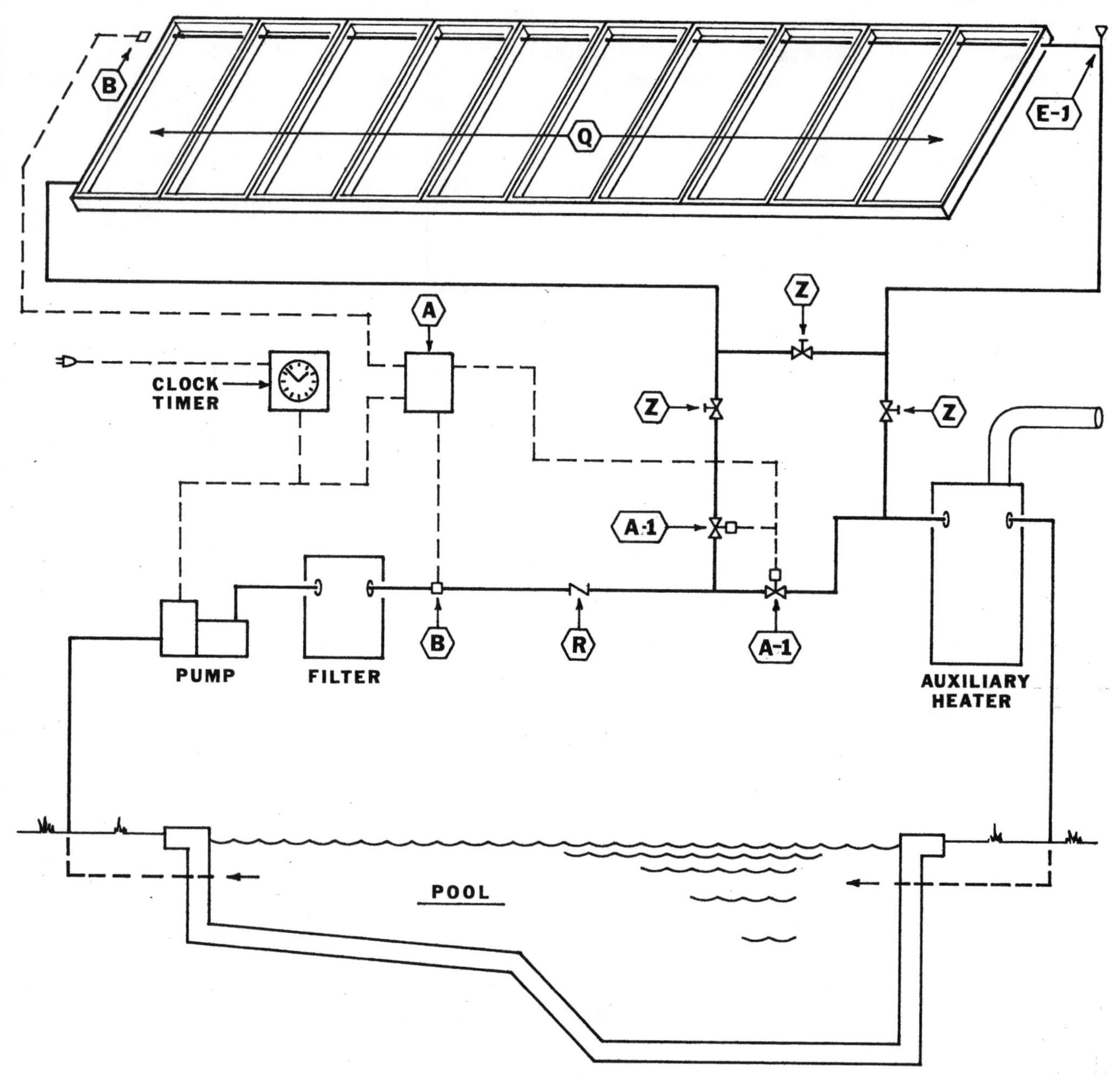

This draindown pool system uses a differential thermostat similar to those used in solar domestic hot water and space heating applications. To heat the pool, the pool water passes through the conventional pump-filter loop and then flows through the collectors. When collection is not possible, or when the pool temperature is reached, all water drains from the solar loop back to the pool through the existing piping. The modes are controlled by solenoid valves or other automatic valves in conjunction with a vacuum breaker relief valve, which facilitates draindown.

D3010 Energy Supply

System Components	QUANTITY	UNIT	COST EACH MAT.	COST EACH INST.	COST EACH TOTAL
SYSTEM D3010 660 2640					
SOLAR SWIMMING POOL HEATER, ROOF MOUNTED COLLECTORS					
TEN 4′ X 10′ FULLY WETTED UNGLAZED PLASTIC ABSORBERS					
A Differential thermostat/controller, 110V, adj pool pump system	1.000	Ea.	330	335	665
A-1 Solenoid valve, PVC, normally 1 open 1 closed (included)	2.000	Ea.			
B Sensor, thermistor type (included)	2.000	Ea.			
E-1 Valve, vacuum relief	1.000	Ea.	31.50	21	52.50
Q Collector panel, solar energy, plastic, liquid full wetted, 4′ x 10′	10.000	Ea.	2,650	2,420	5,070
R Valve, ball check, PVC, socket, 1-1/2″ diam	1.000	Ea.	108	33.50	141.50
Z Valve, ball, PVC, socket, 1-1/2″ diam	3.000	Ea.	202.50	100.50	303
Pipe, PVC, sch 40, 1-1/2″ diam	80.000	L.F.	404	1,496	1,900
Pipe fittings, PVC sch 40, socket joint, 1-1/2″ diam	10.000	Ea.	19.40	335	354.40
Sensor wire, #22-2 conductor, multistranded	.500	C.L.F.	7.83	31.25	39.08
Roof clamps for solar energy collector panels	10.000	Set	27.90	173	200.90
Roof strap, teflon for solar energy collector panels	26.000	L.F.	585	85.28	670.28
TOTAL			4,366.13	5,030.53	9,396.66

D3010 660	Solar Swimming Pool Heater Systems	COST EACH MAT.	COST EACH INST.	COST EACH TOTAL
2530	Solar swimming pool heater systems, roof mounted collectors			
2540	10 ea. 3′x7′ black chrome absorber, 1/8″ temp. glass	13,000	3,875	16,875
2560	10 ea. 4′x8′ black chrome absorber, 3/16″ temp. glass R235616-60	14,700	4,625	19,325
2580	10 ea. 3′8″x6′ flat black absorber, 3/16″ temp. glass	9,525	3,950	13,475
2600	10 ea. 4′x9′ flat black absorber, plastic glazing	11,500	4,800	16,300
2620	10 ea. 2′x9′ rubber absorber, plastic glazing	13,400	5,200	18,600
2640	10 ea. 4′x10′ fully wetted unglazed plastic absorber	4,375	5,025	9,400
2660	Ground mounted collectors			
2680	10 ea. 3′x7′ black chrome absorber, 1/8″ temp. glass	13,100	4,325	17,425
2700	10 ea. 4′x8′ black chrome absorber, 3/16″ temp. glass	14,800	5,075	19,875
2720	10 ea. 3′8″x6′ flat blk absorber, 3/16″ temp. glass	9,625	4,400	14,025
2740	10 ea. 4′x9′ flat blk absorber, plastic glazing	11,600	5,275	16,875
2760	10 ea. 2′x9′ rubber absorber, plastic glazing	13,500	5,475	18,975
2780	10 ea. 4′x10′ fully wetted unglazed plastic absorber	4,475	5,475	9,950

D3010 Energy Supply

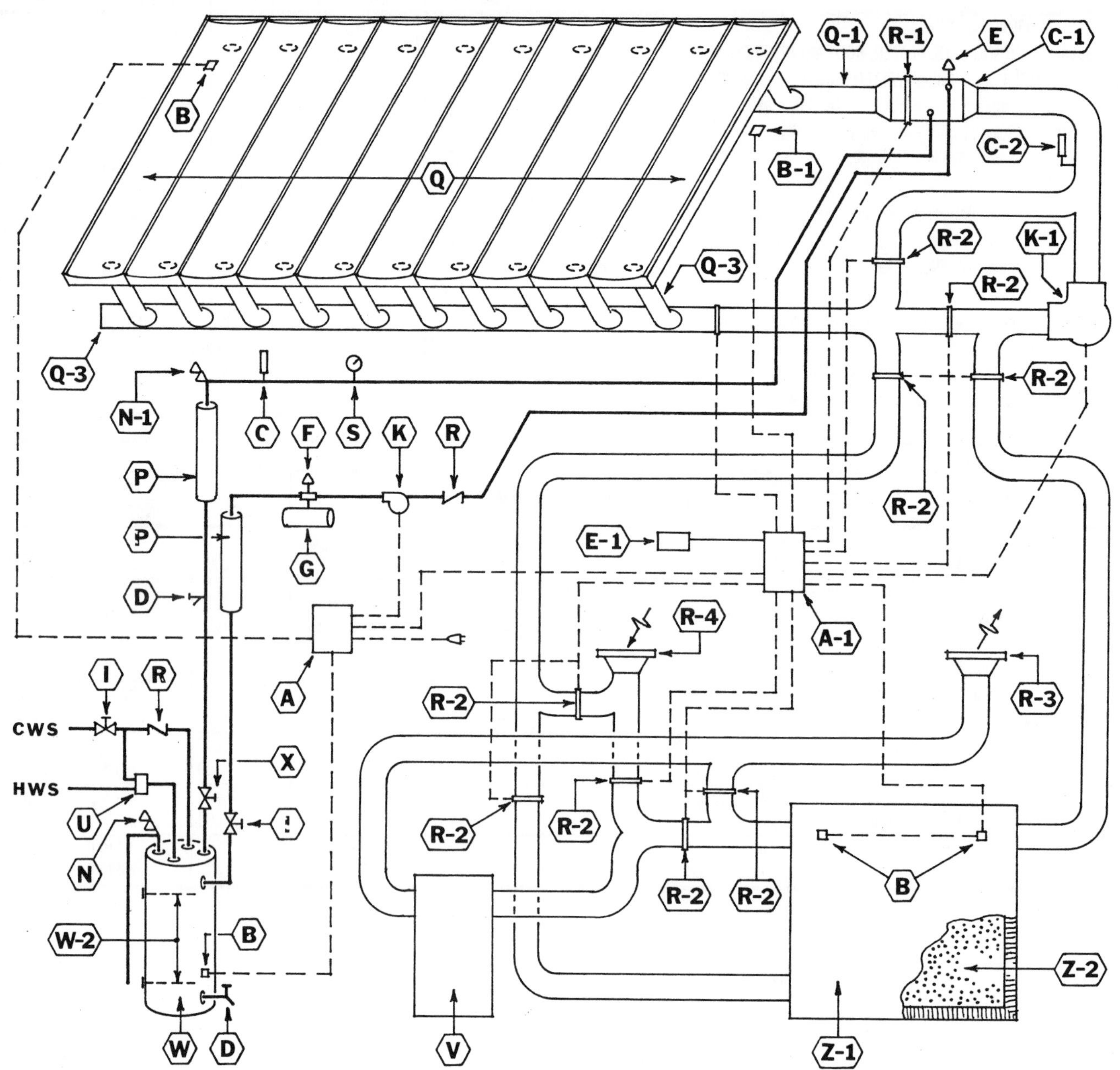

The complete Solar Air Heating System provides maximum savings of conventional fuel with both space heating and year-round domestic hot water heating. It allows for the home air conditioning to operate simultaneously and independently from the solar domestic water heating in summer. The system's modes of operation are:

Mode 1: The building is heated directly from the collectors with air circulated by the Solar Air Mover.

Mode 2: When heat is not needed in the building, the dampers change within the air mover to circulate the air from the collectors to the rock storage bin.

Mode 3: When heat is not available from the collector array and is available in rock storage, the air mover draws heated air from rock storage and directs it into the building. When heat is not available from the collectors or the rock storage bin, the auxiliary heating unit will provide heat for the building. The size of the collector array is typically 25% the size of the main floor area.

D3010 Energy Supply

System Components	QUANTITY	UNIT	COST EACH MAT.	COST EACH INST.	COST EACH TOTAL
SYSTEM D3010 675 1210					
SOLAR, SPACE/HOT WATER, AIR TO WATER HEAT EXCHANGE					
A, B Differential controller 2 sensors thermos., solar energy sys liquid loop	1.000	Ea.	625	56	681
A-1, B Differential controller 2 sensors 6 station solar energy sys air loop	1.000	Ea.	259	224	483
B-1 Solar energy sensor, freeze prevention	1.000	Ea.	26	21	47
C Thermometer for solar energy system, 2″ dial	2.000	Ea.	46	84	130
C-1 Heat exchanger, solar energy system, air to fluid, up flow, 70 MBH	1.000	Ea.	380	345	725
D Drain valve, brass, 3/4″ connection	2.000	Ea.	20.10	56	76.10
E Air vent, manual, for solar energy system 1/8″ fitting	1.000	Ea.	15.90	21	36.90
E-1 Thermostat, 2 stage for sensing room temperature	1.000	Ea.	177	85.50	262.50
F Air purger	1.000	Ea.	51.50	56	107.50
G Expansion tank, for solar energy system	1.000	Ea.	72.50	21	93.50
I Valve, gate, bronze, 125 lb, NRS, soldered 3/4″ diam	2.000	Ea.	109	67	176
K Circulator, solar heated liquid, 1/25 HP	1.000	Ea.	295	86.50	381.50
N Relief valve temp & press 150 psi 210°F self-closing, 3/4″ IPS	1.000	Ea.	22	22.50	44.50
N-1 Relief valve, pressure, poppet, bronze, 30 psi, 3/4″ IPS	1.000	Ea.	48	24	72
P Pipe covering, fiberglass, all service jacket, 1″ wall, 3/4″ diam	60.000	L.F.	64.20	289.80	354
Q-3 Manifold for flush mount solar energy collector panels	20.000	L.F.	2,880	146	3,026
Q Collector panel solar energy, air, black on alum. plate, flush mount 10′x2′	100.000	L.F.	53,500	1,220	54,720
R Valve, swing check, bronze, 125 lb, regrinding disc, 3/4″ diam	2.000	Ea.	196	67	263
S Pressure gage, 2″ dial, for solar energy system	1.000	Ea.	25.50	21	46.50
U Valve, water tempering, bronze, sweat connections, 3/4″ diam	1.000	Ea.	126	33.50	159.50
W-2, W Tank, water storage, solar, elec element 2″x1/2# insul, 80 Gal	1.000	Ea.	1,700	420	2,120
X Valve, globe, bronze, 125 lb, soldered, 3/4″ diam	1.000	Ea.	139	33.50	172.50
Copper tubing type L, solder joints, hangers 10′ OC 3/4″ diam	10.000	L.F.	72	88.50	160.50
Copper tubing type M, solder joints, hangers 10′ OC 3/4″ diam	60.000	L.F.	327	516	843
Wrought copper fittings & solder, 3/4″ diam	26.000	Ea.	182	923	1,105
Sensor wire, #22-2 conductor multistranded	1.200	C.L.F.	18.78	75	93.78
Q-1 Duct work, rigid fiberglass, rectangular	400.000	S.F.	340	2,080	2,420
Duct work, spiral preformed, steel, PVC coated both sides, 12″ x 10″	8.000	Ea.	208	392	600
R-2 Shutter/damper for solar heater circulator	9.000	Ea.	513	877.50	1,390.50
K-1 Shutter motor for solar heater circulator blower	9.000	Ea.	1,278	657	1,935
K-1 Fan, solar energy heated air circulator, space & DHW system	1.000	Ea.	1,700	2,350	4,050
Z-1 Tank, solar energy air storage, 6′-3″H 7′x7′ = 306 CF/2000 Gal	1.000	Ea.	15,700	1,525	17,225
Z-2 Crushed stone 1-1/2″	11.000	C.Y.	374	85.80	459.80
C-2 Themometer, remote probe, 2″ dial	1.000	Ea.	35	84	119
R-1 Solenoid valve	1.000	Ea.	127	74.50	201.50
V, R-4, R-3 Furnace, supply diffusers, return grilles, existing	1.000	Ea.	15,700	1,525	17,225
TOTAL			97,352.48	14,653.60	112,006.08

D3010 675	Air To Water Heat Exchange		COST EACH MAT.	COST EACH INST.	COST EACH TOTAL
1210	Solar, air to water heat exchange, for space/hot water heating	R235616 -60	97,500	14,700	112,200
1220					

D3020 Heat Generating Systems

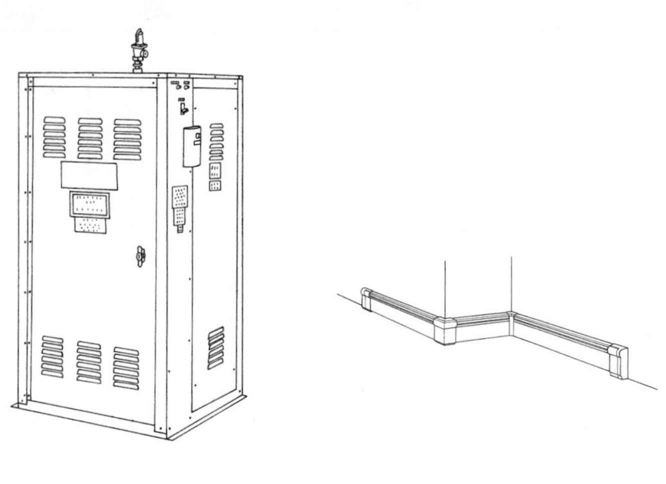

Boiler   Baseboard Radiation

Small Electric Boiler System Considerations:

1. Terminal units are fin tube baseboard radiation rated at 720 BTU/hr with 200° water temperature or 820 BTU/hr steam.
2. Primary use being for residential or smaller supplementary areas, the floor levels are based on 7-1/2′ ceiling heights.
3. All distribution piping is copper for boilers through 205 MBH. All piping for larger systems is steel pipe.

System Components	QUANTITY	UNIT	COST EACH		
			MAT.	INST.	TOTAL
SYSTEM D3020 102 1120					
SMALL HEATING SYSTEM, HYDRONIC, ELECTRIC BOILER					
1,480 S.F., 61 MBH, STEAM, 1 FLOOR					
Boiler, electric steam, std cntrls, trim, ftngs and valves, 18 KW, 61.4 MBH	1.000	Ea.	4,867.50	1,705	6,572.50
Copper tubing type L, solder joint, hanger 10′OC, 1-1/4″ diam	160.000	L.F.	2,408	1,856	4,264
Radiation, 3/4″ copper tube w/alum fin baseboard pkg 7″ high	60.000	L.F.	555	1,260	1,815
Rough in baseboard panel or fin tube with valves & traps	10.000	Set	3,450	6,450	9,900
Pipe covering, calcium silicate w/cover, 1″ wall 1-1/4″ diam	160.000	L.F.	580.80	1,080	1,660.80
Low water cut-off, quick hookup, in gage glass tappings	1.000	Ea.	260	42.50	302.50
TOTAL			12,121.30	12,393.50	24,514.80
COST PER S.F.			8.19	8.37	16.56

D3020 102	Small Heating Systems, Hydronic, Electric Boilers	COST PER S.F.		
		MAT.	INST.	TOTAL
1100	Small heating systems, hydronic, electric boilers			
1120	Steam, 1 floor, 1480 S.F., 61 M.B.H.	8.18	8.38	16.56
1160	3,000 S.F., 123 M.B.H.	6.45	7.35	13.80
1200	5,000 S.F., 205 M.B.H.	5.75	6.80	12.55
1240	2 floors, 12,400 S.F., 512 M.B.H.	4.58	6.75	11.33
1280	3 floors, 24,800 S.F., 1023 M.B.H.	5	6.65	11.65
1360	Hot water, 1 floor, 1,000 S.F., 41 M.B.H.	12.90	4.64	17.54
1400	2,500 S.F., 103 M.B.H.	9.85	8.35	18.20
1440	2 floors, 4,850 S.F., 205 M.B.H.	9.90	10.05	19.95
1480	3 floors, 9,700 S.F., 410 M.B.H.	10.95	10.35	21.30

D3090 Other HVAC Systems/Equip

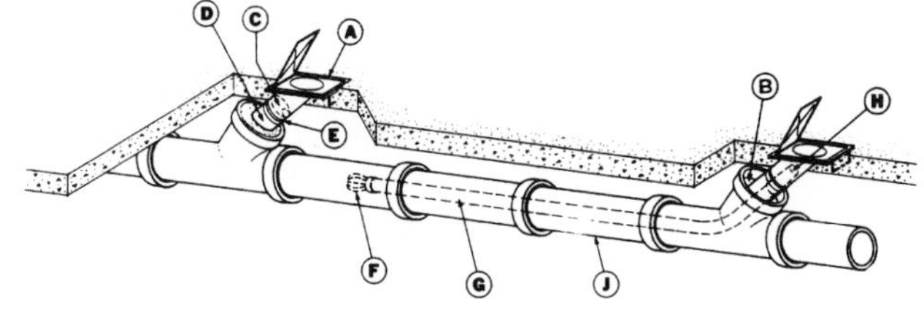

Cast Iron Garage Exhaust System

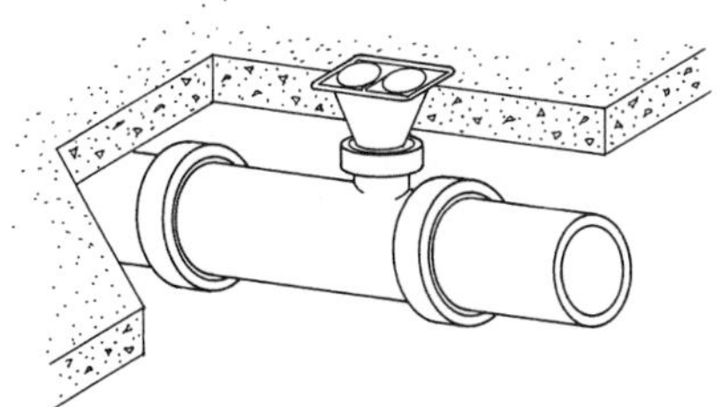
Dual Exhaust System

System Components	QUANTITY	UNIT	COST EACH MAT.	COST EACH INST.	COST EACH TOTAL
SYSTEM D3090 320 1040					
GARAGE, EXHAUST, SINGLE 3″ EXHAUST OUTLET, CARS & LIGHT TRUCKS					
A Outlet top assy, for engine exhaust system with adapters and ftngs, 3″ diam	1.000	Ea.	263	40.50	303.50
F Bullnose (guide) for engine exhaust system, 3″ diam	1.000	Ea.	29.50		29.50
G Galvanized flexible tubing for engine exhaust system, 3″ diam	8.000	L.F.	101.20		101.20
H Adapter for metal tubing end of engine exhaust system, 3″ tail pipe	1.000	Ea.	54.50		54.50
J Pipe, sewer, cast iron, push-on joint, 8″ diam.	18.000	L.F.	981	747	1,728
Excavating utility trench, chain trencher, 8″ wide, 24″ deep	18.000	L.F.		32.40	32.40
Backfill utility trench by hand, incl. compaction, 8″ wide 24″ deep	18.000	L.F.		44.46	44.46
Stand for blower, concrete over polystyrene core, 6″ high	1.000	Ea.	5.70	20.50	26.20
AC&V duct spiral reducer 10″x8″	1.000	Ea.	16.15	36.50	52.65
AC&V duct spiral reducer 12″x10″	1.000	Ea.	18.50	49	67.50
AC&V duct spiral preformed 45° elbow, 8″ diam	1.000	Ea.	8.90	42	50.90
AC&V utility fan, belt drive, 3 phase, 2000 CFM, 1 HP	2.000	Ea.	2,550	650	3,200
Safety switch, heavy duty fused, 240V, 3 pole, 30 amp	1.000	Ea.	180	196	376
TOTAL			4,208.45	1,858.36	6,066.81

D3090 320	Garage Exhaust Systems	COST PER BAY MAT.	COST PER BAY INST.	COST PER BAY TOTAL
1040	Garage, single 3″ exhaust outlet, cars & light trucks, one bay	4,208.45	1,858.36	6,067.06
1060	Additional bays up to seven bays	995	500	1,495
1500	4″ outlet, trucks, one bay	4,250	1,850	6,100
1520	Additional bays up to six bays	1,025	500	1,525
1600	5″ outlet, diesel trucks, one bay	4,550	1,850	6,400
1650	Additional single bays up to six	1,450	590	2,040
1700	Two adjoining bays	4,550	1,850	6,400
2000	Dual exhaust, 3″ outlets, pair of adjoining bays	5,225	2,325	7,550
2100	Additional pairs of adjoining bays	1,575	590	2,165

D4010 Sprinklers

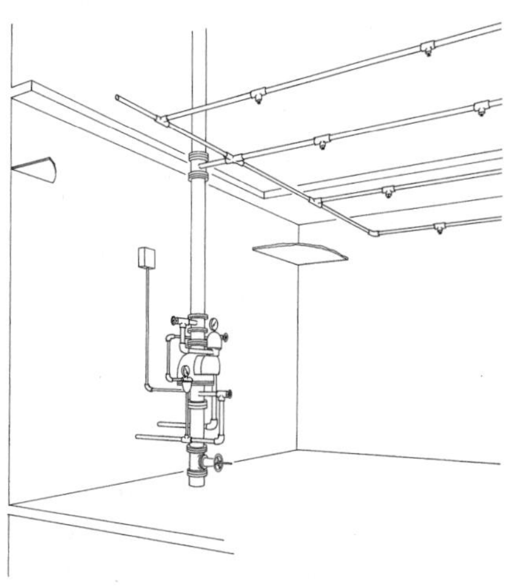

Dry Pipe System: A system employing automatic sprinklers attached to a piping system containing air under pressure, the release of which from the opening of sprinklers permits the water pressure to open a valve known as a "dry pipe valve". The water then flows into the piping system and out the opened sprinklers.

All areas are assumed to be open.

System Components	QUANTITY	UNIT	COST EACH		
			MAT.	INST.	TOTAL
SYSTEM D4010 310 0580					
DRY PIPE SPRINKLER, STEEL, BLACK, SCH. 40 PIPE					
LIGHT HAZARD, ONE FLOOR, 2000 S.F.					
Valve, gate, iron body 125 lb., OS&Y, flanged, 4" pipe size	1.000	Ea.	566.25	303.75	870
Valve, swing check, bronze, 125 lb, regrinding disc, 2-1/2" pipe size	1.000	Ea.	588.75	60.38	649.13
Valve, angle, bronze, 150 lb., rising stem, threaded, 2" pipe size	1.000	Ea.	555	45.75	600.75
*Alarm valve, 2-1/2" pipe size	1.000	Ea.	1,218.75	296.25	1,515
Alarm, water motor, complete with gong	1.000	Ea.	303.75	123.75	427.50
Fire alarm horn, electric	1.000	Ea.	49.13	70.13	119.26
Valve swing check w/balldrip CI with brass trim, 4" pipe size	1.000	Ea.	243.75	296.25	540
Pipe, steel, black, schedule 40, 4" diam.	10.000	L.F.	195	255.53	450.53
Dry pipe valve, trim & gauges, 4" pipe size	1.000	Ea.	2,118.75	881.25	3,000
Pipe, steel, black, schedule 40, threaded, cplg & hngr 10'OC 2-1/2" diam.	20.000	L.F.	252	360	612
Pipe, steel, black, schedule 40, threaded, cplg & hngr 10'OC 2" diam.	12.500	L.F.	100.78	177.19	277.97
Pipe, steel, black, schedule 40, threaded, cplg & hngr 10'OC 1-1/4" diam.	37.500	L.F.	195.47	382.50	577.97
Pipe, steel, black, schedule 40, threaded, cplg & hngr 10'OC 1" diam.	112.000	L.F.	470.40	1,066.80	1,537.20
Pipe Tee, malleable iron black, 150 lb. threaded, 4" pipe size	2.000	Ea.	412.50	457.50	870
Pipe Tee, malleable iron black, 150 lb. threaded, 2-1/2" pipe size	2.000	Ea.	116.25	201	317.25
Pipe Tee, malleable iron black, 150 lb. threaded, 2" pipe size	1.000	Ea.	27	82.50	109.50
Pipe Tee, malleable iron black, 150 lb. threaded, 1-1/4" pipe size	5.000	Ea.	63.38	324.38	387.76
Pipe Tee, malleable iron black, 150 lb. threaded, 1" pipe size	4.000	Ea.	31.35	252	283.35
Pipe 90° elbow malleable iron black, 150 lb. threaded, 1" pipe size	6.000	Ea.	29.93	231.75	261.68
Sprinkler head dry 1/2" orifice 1" NPT, 3" to 4-3/4" length	12.000	Ea.	1,488	564	2,052
Air compressor, 200 Gal sprinkler system capacity, 1/3 HP	1.000	Ea.	607.50	378.75	986.25
*Standpipe connection, wall, flush, brs. w/plug & chain 2-1/2"x2-1/2"	1.000	Ea.	119.25	177.75	297
Valve gate bronze, 300 psi, NRS, class 150, threaded, 1" pipe size	1.000	Ea.	81.75	26.63	108.38
TOTAL			9,834.69	7,015.79	16,850.48
COST PER S.F.			4.92	3.51	8.43
*Not included in systems under 2000 S.F.					

D4010 310	Dry Pipe Sprinkler Systems		COST PER S.F.		
			MAT.	INST.	TOTAL
0520	Dry pipe sprinkler systems, steel, black, sch. 40 pipe				
0530	Light hazard, one floor, 500 S.F.		9	5.95	14.95
0560	1000 S.F.	R211313-10	5.20	3.50	8.70
0580	2000 S.F.		4.91	3.50	8.41
0600	5000 S.F.	R211313-20	2.64	2.39	5.03
0620	10,000 S.F.		1.92	1.98	3.90

D4010 Sprinklers

D4010 310	Dry Pipe Sprinkler Systems	COST PER S.F.		
		MAT.	INST.	TOTAL
0640	50,000 S.F. R211313-30	1.53	1.75	3.28
0660	Each additional floor, 500 S.F.	2.08	2.90	4.98
0680	1000 S.F.	1.80	2.40	4.20
0700	2000 S.F.	1.74	2.21	3.95
0720	5000 S.F.	1.45	1.90	3.35
0740	10,000 S.F.	1.36	1.75	3.11
0760	50,000 S.F.	1.22	1.55	2.77
1000	Ordinary hazard, one floor, 500 S.F.	9.20	6	15.20
1020	1000 S.F.	5.30	3.52	8.82
1040	2000 S.F.	5.05	3.65	8.70
1060	5000 S.F.	3.04	2.54	5.58
1080	10,000 S.F.	2.51	2.60	5.11
1100	50,000 S.F.	2.28	2.45	4.73
1140	Each additional floor, 500 S.F.	2.26	2.97	5.23
1160	1000 S.F.	2.07	2.67	4.74
1180	2000 S.F.	2.06	2.42	4.48
1200	5000 S.F.	1.92	2.08	4
1220	10,000 S.F.	1.78	2.05	3.83
1240	50,000 S.F.	1.66	1.78	3.44
1500	Extra hazard, one floor, 500 S.F.	12.65	7.45	20.10
1520	1000 S.F.	7.80	5.40	13.20
1540	2000 S.F.	5.75	4.70	10.45
1560	5000 S.F.	3.63	3.57	7.20
1580	10,000 S.F.	3.69	3.40	7.09
1600	50,000 S.F.	4	3.28	7.28
1660	Each additional floor, 500 S.F.	3.12	3.68	6.80
1680	1000 S.F.	3.06	3.50	6.56
1700	2000 S.F.	2.93	3.50	6.43
1720	5000 S.F.	2.51	3.07	5.58
1740	10,000 S.F.	2.82	2.80	5.62
1760	50,000 S.F.	2.90	2.69	5.59
2020	Grooved steel, black, sch. 40 pipe, light hazard, one floor, 2000 S.F.	4.64	3	7.64
2060	10,000 S.F.	1.76	1.73	3.49
2100	Each additional floor, 2000 S.F.	1.65	1.77	3.42
2150	10,000 S.F.	1.20	1.50	2.70
2200	Ordinary hazard, one floor, 2000 S.F.	4.88	3.20	8.08
2250	10,000 S.F.	2.24	2.19	4.43
2300	Each additional floor, 2000 S.F.	1.89	1.97	3.86
2350	10,000 S.F.	1.69	1.96	3.65
2400	Extra hazard, one floor, 2000 S.F.	5.50	4	9.50
2450	10,000 S.F.	3.14	2.84	5.98
2500	Each additional floor, 2000 S.F.	2.69	2.86	5.55
2550	10,000 S.F.	2.44	2.46	4.90
3050	Grooved steel, black, sch. 10 pipe, light hazard, one floor, 2000 S.F.	4.52	2.97	7.49
3100	10,000 S.F.	1.69	1.70	3.39
3150	Each additional floor, 2000 S.F.	1.53	1.74	3.27
3200	10,000 S.F.	1.13	1.47	2.60
3250	Ordinary hazard, one floor, 2000 S.F.	4.78	3.18	7.96
3300	10,000 S.F.	2.12	2.15	4.27
3350	Each additional floor, 2000 S.F.	1.79	1.95	3.74
3400	10,000 S.F.	1.57	1.92	3.49
3450	Extra hazard, one floor, 2000 S.F.	5.45	3.98	9.43
3500	10,000 S.F.	2.93	2.79	5.72
3550	Each additional floor, 2000 S.F.	2.61	2.84	5.45
3600	10,000 S.F.	2.32	2.43	4.75
4050	Copper tubing, type M, light hazard, one floor, 2000 S.F.	6.05	2.97	9.02
4100	10,000 S.F.	2.99	1.72	4.71
4150	Each additional floor, 2000 S.F.	3.09	1.78	4.87

D4010 Sprinklers

D4010 310	Dry Pipe Sprinkler Systems	COST PER S.F.		
		MAT.	INST.	TOTAL
4200	10,000 S.F.	2.43	1.50	3.93
4250	Ordinary hazard, one floor, 2000 S.F.	6.50	3.33	9.83
4300	10,000 S.F.	3.69	2.04	5.73
4350	Each additional floor, 2000 S.F.	4.04	2.07	6.11
4400	10,000 S.F.	3	1.78	4.78
4450	Extra hazard, one floor, 2000 S.F.	7.75	4.05	11.80
4500	10,000 S.F.	8.10	3.08	11.18
4550	Each additional floor, 2000 S.F.	4.92	2.91	7.83
4600	10,000 S.F.	6.15	2.68	8.83
5050	Copper tubing, type M, T-drill system, light hazard, one floor			
5060	2000 S.F.	5.95	2.78	8.73
5100	10,000 S.F.	2.62	1.43	4.05
5150	Each additional floor, 2000 S.F.	2.97	1.59	4.56
5200	10,000 S.F.	2.06	1.21	3.27
5250	Ordinary hazard, one floor, 2000 S.F.	5.90	2.85	8.75
5300	10,000 S.F.	3.37	1.81	5.18
5350	Each additional floor, 2000 S.F.	2.92	1.62	4.54
5400	10,000 S.F.	2.68	1.50	4.18
5450	Extra hazard, one floor, 2000 S.F.	6.70	3.35	10.05
5500	10,000 S.F.	5.85	2.26	8.11
5550	Each additional floor, 2000 S.F.	3.86	2.21	6.07
5600	10,000 S.F.	3.93	1.86	5.79

D4010 Sprinklers

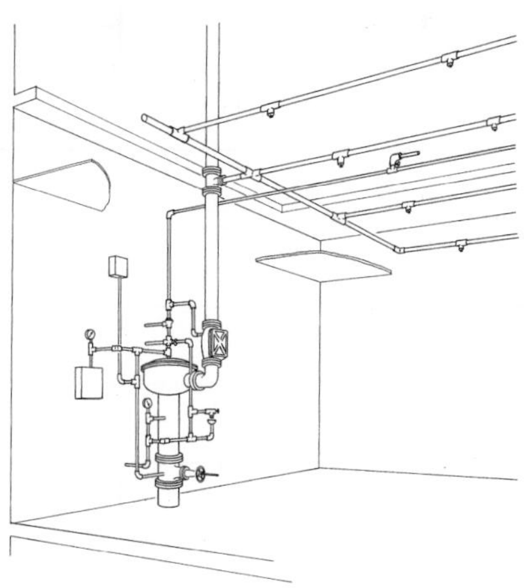

Pre-Action System: A system employing automatic sprinklers attached to a piping system containing air that may or may not be under pressure, with a supplemental heat responsive system of generally more sensitive characteristics than the automatic sprinklers themselves, installed in the same areas as the sprinklers. Actuation of the heat responsive system, as from a fire, opens a valve which permits water to flow into the sprinkler piping system and to be discharged from those sprinklers which were opened by heat from the fire.

All areas are assumed to be open.

System Components	QUANTITY	UNIT	COST EACH MAT.	COST EACH INST.	COST EACH TOTAL
SYSTEM D4010 350 0580					
PREACTION SPRINKLER SYSTEM, STEEL BLACK SCH. 40 PIPE					
LIGHT HAZARD, 1 FLOOR, 2000 S.F.					
Valve, gate, iron body 125 lb., OS&Y, flanged, 4″ pipe size	1.000	Ea.	566.25	303.75	870
*Valve, swing check w/ball drip CI with brass trim 4″ pipe size	1.000	Ea.	243.75	296.25	540
Valve, swing check, bronze, 125 lb, regrinding disc, 2-1/2″ pipe size	1.000	Ea.	588.75	60.38	649.13
Valve, angle, bronze, 150 lb., rising stem, threaded, 2″ pipe size	1.000	Ea.	555	45.75	600.75
*Alarm valve, 2-1/2″ pipe size	1.000	Ea.	1,218.75	296.25	1,515
Alarm, water motor, complete with gong	1.000	Ea.	303.75	123.75	427.50
Fire alarm horn, electric	1.000	Ea.	49.13	70.13	119.26
Thermostatic release for release line	2.000	Ea.	990	49.50	1,039.50
Pipe, steel, black, schedule 40, 4″ diam.	10.000	L.F.	195	255.53	450.53
Dry pipe valve, trim & gauges, 4″ pipe size	1.000	Ea.	2,118.75	881.25	3,000
Pipe, steel, black, schedule 40, threaded, cplg. & hngr. 10′OC 2-1/2″ diam.	20.000	L.F.	252	360	612
Pipe steel black, schedule 40, threaded, cplg. & hngr. 10′OC 2″ diam.	12.500	L.F.	100.78	177.19	277.97
Pipe, steel, black, schedule 40, threaded, cplg. & hngr. 10′OC 1-1/4″ diam.	37.500	L.F.	195.47	382.50	577.97
Pipe, steel, black, schedule 40, threaded, cplg. & hngr. 10′OC 1″ diam.	112.000	L.F.	470.40	1,066.80	1,537.20
Pipe, Tee, malleable iron, black, 150 lb. threaded, 4″ diam.	2.000	Ea.	412.50	457.50	870
Pipe, Tee, malleable iron, black, 150 lb. threaded, 2-1/2″ pipe size	2.000	Ea.	116.25	201	317.25
Pipe, Tee, malleable iron, black, 150 lb. threaded, 2″ pipe size	1.000	Ea.	27	82.50	109.50
Pipe, Tee, malleable iron, black, 150 lb. threaded, 1-1/4″ pipe size	5.000	Ea.	63.38	324.38	387.76
Pipe, Tee, malleable iron, black, 150 lb. threaded, 1″ pipe size	4.000	Ea.	31.35	252	283.35
Pipe, 90° elbow, malleable iron, blk., 150 lb. threaded, 1″ pipe size	6.000	Ea.	29.93	231.75	261.68
Sprinkler head, std. spray, brass 135°-286°F 1/2″ NPT, 3/8″ orifice	12.000	Ea.	182.40	492	674.40
Air compressor auto complete 200 Gal sprinkler sys. cap., 1/3 HP	1.000	Ea.	607.50	378.75	986.25
*Standpipe conn.,wall, flush, brass w/plug & chain 2-1/2″ x 2-1/2″	1.000	Ea.	119.25	177.75	297
Valve, gate, bronze, 300 psi, NRS, class 150, threaded, 1″ pipe size	1.000	Ea.	81.75	26.63	108.38
TOTAL			9,519.09	6,993.29	16,512.38
COST PER S.F.			4.76	3.50	8.26
*Not included in systems under 2000 S.F.					

D4010 350	Preaction Sprinkler Systems		COST PER S.F. MAT.	COST PER S.F. INST.	COST PER S.F. TOTAL
0520	Preaction sprinkler systems, steel, black, sch. 40 pipe				
0530	Light hazard, one floor, 500 S.F.		8.95	4.76	13.71
0560	1000 S.F.	R211313 -10	5.20	3.57	8.77
0580	2000 S.F.		4.76	3.49	8.25
0600	5000 S.F.	R211313 -20	2.51	2.38	4.89
0620	10,000 S.F.		1.81	1.97	3.78

D4010 Sprinklers

D4010 350	Preaction Sprinkler Systems	COST PER S.F.		
		MAT.	INST.	TOTAL
0640	50,000 S.F. R211313-30	1.43	1.75	3.18
0660	Each additional floor, 500 S.F.	2.34	2.59	4.93
0680	1000 S.F.	1.81	2.39	4.20
0700	2000 S.F.	1.74	2.20	3.94
0720	5000 S.F.	1.32	1.89	3.21
0740	10,000 S.F.	1.25	1.74	2.99
0760	50,000 S.F.	1.23	1.61	2.84
1000	Ordinary hazard, one floor, 500 S.F.	9.20	5.15	14.35
1020	1000 S.F.	5.15	3.51	8.66
1040	2000 S.F.	5.15	3.66	8.81
1060	5000 S.F.	2.79	2.53	5.32
1080	10,000 S.F.	2.21	2.59	4.80
1100	50,000 S.F.	1.96	2.43	4.39
1140	Each additional floor, 500 S.F.	2.60	2.99	5.59
1160	1000 S.F.	1.78	2.41	4.19
1180	2000 S.F.	1.66	2.40	4.06
1200	5000 S.F.	1.80	2.25	4.05
1220	10,000 S.F.	1.66	2.36	4.02
1240	50,000 S.F.	1.55	2.10	3.65
1500	Extra hazard, one floor, 500 S.F.	12.35	6.60	18.95
1520	1000 S.F.	7.15	4.94	12.09
1540	2000 S.F.	5.25	4.67	9.92
1560	5000 S.F.	3.44	3.85	7.29
1580	10,000 S.F.	3.33	3.76	7.09
1600	50,000 S.F.	3.54	3.66	7.20
1660	Each additional floor, 500 S.F.	3.13	3.68	6.81
1680	1000 S.F.	2.58	3.47	6.05
1700	2000 S.F.	2.44	3.47	5.91
1720	5000 S.F.	2.07	3.08	5.15
1740	10,000 S.F.	2.22	2.82	5.04
1760	50,000 S.F.	2.19	2.66	4.85
2020	Grooved steel, black, sch. 40 pipe, light hazard, one floor, 2000 S.F.	4.64	2.99	7.63
2060	10,000 S.F.	1.65	1.72	3.37
2100	Each additional floor of 2000 S.F.	1.65	1.76	3.41
2150	10,000 S.F.	1.09	1.49	2.58
2200	Ordinary hazard, one floor, 2000 S.F.	4.72	3.19	7.91
2250	10,000 S.F.	1.94	2.18	4.12
2300	Each additional floor, 2000 S.F.	1.73	1.96	3.69
2350	10,000 S.F.	1.39	1.95	3.34
2400	Extra hazard, one floor, 2000 S.F.	5	3.97	8.97
2450	10,000 S.F.	2.54	2.80	5.34
2500	Each additional floor, 2000 S.F.	2.20	2.83	5.03
2550	10,000 S.F.	1.80	2.42	4.22
3050	Grooved steel, black, sch. 10 pipe light hazard, one floor, 2000 S.F.	4.52	2.96	7.48
3100	10,000 S.F.	1.58	1.69	3.27
3150	Each additional floor, 2000 S.F.	1.53	1.73	3.26
3200	10,000 S.F.	1.02	1.46	2.48
3250	Ordinary hazard, one floor, 2000 S.F.	4.55	2.97	7.52
3300	10,000 S.F.	1.57	2.12	3.69
3350	Each additional floor, 2000 S.F.	1.63	1.94	3.57
3400	10,000 S.F.	1.27	1.91	3.18
3450	Extra hazard, one floor, 2000 S.F.	4.94	3.95	8.89
3500	10,000 S.F.	2.29	2.75	5.04
3550	Each additional floor, 2000 S.F.	2.12	2.81	4.93
3600	10,000 S.F.	1.68	2.39	4.07
4050	Copper tubing, type M, light hazard, one floor, 2000 S.F.	6.05	2.96	9.01
4100	10,000 S.F.	2.88	1.71	4.59
4150	Each additional floor, 2000 S.F.	3.10	1.77	4.87

D4010 Sprinklers

D4010 350	Preaction Sprinkler Systems	COST PER S.F.		
		MAT.	INST.	TOTAL
4200	10,000 S.F.	2.07	1.47	3.54
4250	Ordinary hazard, one floor, 2000 S.F.	6.35	3.32	9.67
4300	10,000 S.F.	3.39	2.03	5.42
4350	Each additional floor, 2000 S.F.	3.11	1.83	4.94
4400	10,000 S.F.	2.45	1.63	4.08
4450	Extra hazard, one floor, 2000 S.F.	7.25	4.02	11.27
4500	10,000 S.F.	7.40	3.04	10.44
4550	Each additional floor, 2000 S.F.	4.43	2.88	7.31
4600	10,000 S.F.	5.50	2.64	8.14
5050	Copper tubing, type M, T-drill system, light hazard, one floor			
5060	2000 S.F.	5.95	2.77	8.72
5100	10,000 S.F.	2.51	1.42	3.93
5150	Each additional floor, 2000 S.F.	2.97	1.58	4.55
5200	10,000 S.F.	1.95	1.20	3.15
5250	Ordinary hazard, one floor, 2000 S.F.	5.75	2.84	8.59
5300	10,000 S.F.	3.07	1.80	4.87
5350	Each additional floor, 2000 S.F.	2.78	1.62	4.40
5400	10,000 S.F.	2.52	1.57	4.09
5450	Extra hazard, one floor, 2000 S.F.	6.20	3.32	9.52
5500	10,000 S.F.	5.15	2.22	7.37
5550	Each additional floor, 2000 S.F.	3.37	2.18	5.55
5600	10,000 S.F.	3.29	1.82	5.11

D4010 Sprinklers

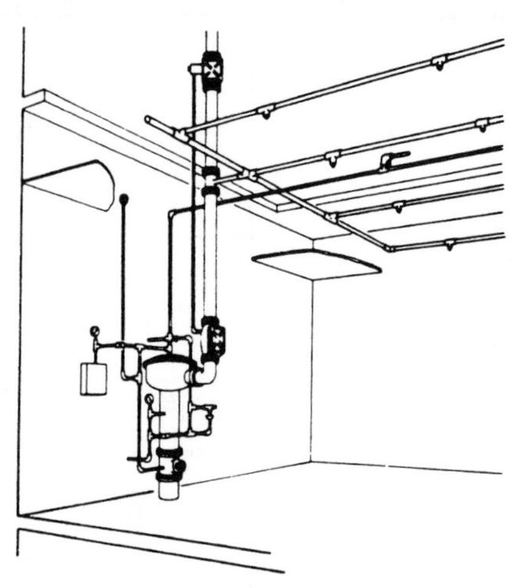

Deluge System: A system employing open sprinklers attached to a piping system connected to a water supply through a valve which is opened by the operation of a heat responsive system installed in the same areas as the sprinklers. When this valve opens, water flows into the piping system and discharges from all sprinklers attached thereto.

All areas are assumed to be open.

System Components	QUANTITY	UNIT	COST EACH MAT.	COST EACH INST.	COST EACH TOTAL
SYSTEM D4010 370 0580					
DELUGE SPRINKLER SYSTEM, STEEL BLACK SCH. 40 PIPE					
LIGHT HAZARD, 1 FLOOR, 2000 S.F.					
Valve, gate, iron body 125 lb., OS&Y, flanged, 4″ pipe size	1.000	Ea.	566.25	303.75	870
Valve, swing check w/ball drip, CI w/brass ftngs., 4″ pipe size	1.000	Ea.	243.75	296.25	540
Valve, swing check, bronze, 125 lb, regrinding disc, 2-1/2″ pipe size	1.000	Ea.	588.75	60.38	649.13
Valve, angle, bronze, 150 lb., rising stem, threaded, 2″ pipe size	1.000	Ea.	555	45.75	600.75
*Alarm valve, 2-1/2″ pipe size	1.000	Ea.	1,218.75	296.25	1,515
Alarm, water motor, complete with gong	1.000	Ea.	303.75	123.75	427.50
Fire alarm horn, electric	1.000	Ea.	49.13	70.13	119.26
Thermostatic release for release line	2.000	Ea.	990	49.50	1,039.50
Pipe, steel, black, schedule 40, 4″ diam.	10.000	L.F.	195	255.53	450.53
Deluge valve trim, pressure relief, emergency release, gauge, 4″ pipe size	1.000	Ea.	3,506.25	881.25	4,387.50
Deluge system, monitoring panel w/deluge valve & trim	1.000	Ea.	8,400	27.38	8,427.38
Pipe, steel, black, schedule 40, threaded, cplg & hngr 10′ OC 2-1/2″ diam.	20.000	L.F.	252	360	612
Pipe, steel, black, schedule 40, threaded, cplg & hngr 10′ OC 2″ diam.	12.500	L.F.	100.78	177.19	277.97
Pipe, steel, black, schedule 40, threaded, cplg & hngr 10′ OC 1-1/4″ diam.	37.500	L.F.	195.47	382.50	577.97
Pipe, steel, black, schedule 40, threaded, cplg & hngr 10′ OC 1″ diam.	112.000	L.F.	470.40	1,066.80	1,537.20
Pipe, Tee, malleable iron, black, 150 lb. threaded, 4″ pipe size	2.000	Ea.	412.50	457.50	870
Pipe, Tee, malleable iron, black, 150 lb. threaded, 2-1/2″ pipe size	2.000	Ea.	116.25	201	317.25
Pipe, Tee, malleable iron, black, 150 lb. threaded, 2″ pipe size	1.000	Ea.	27	82.50	109.50
Pipe, Tee, malleable iron, black, 150 lb. threaded, 1-1/4″ pipe size	5.000	Ea.	63.38	324.38	387.76
Pipe, Tee, malleable iron, black, 150 lb. threaded, 1″ pipe size	4.000	Ea.	31.35	252	283.35
Pipe, 90° elbow, malleable iron, black, 150 lb. threaded 1″ pipe size	6.000	Ea.	29.93	231.75	261.68
Sprinkler head, std spray, brass 135°-286°F 1/2″ NPT, 3/8″ orifice	9.720	Ea.	182.40	492	674.40
Air compressor, auto, complete, 200 Gal sprinkler sys. cap., 1/3 HP	1.000	Ea.	607.50	378.75	986.25
*Standpipe connection, wall, flush w/plug & chain 2-1/2″ x 2-1/2″	1.000	Ea.	119.25	177.75	297
Valve, gate, bronze, 300 psi, NRS, class 150, threaded, 1″ pipe size	1.000	Ea.	81.75	26.63	108.38
TOTAL			19,306.59	7,020.67	26,327.26
COST PER S.F.			9.65	3.51	13.16
*Not included in systems under 2000 S.F.					

D4010 370	Deluge Sprinkler Systems		COST PER S.F. MAT.	COST PER S.F. INST.	COST PER S.F. TOTAL
0520	Deluge sprinkler systems, steel, black, sch. 40 pipe				
0530	Light hazard, one floor, 500 S.F.		27.50	4.81	32.31
0560	1000 S.F.	R211313 -10	14.35	3.43	17.78
0580	2000 S.F.		9.65	3.50	13.15

D4010 Sprinklers

D4010 370	Deluge Sprinkler Systems		MAT.	INST.	TOTAL
			COST PER S.F.		
0600	5000 S.F.	R211313 -20	4.47	2.39	6.86
0620	10,000 S.F.		2.79	1.97	4.76
0640	50,000 S.F.	R211313 -30	1.63	1.75	3.38
0660	Each additional floor, 500 S.F.		2.34	2.59	4.93
0680	1000 S.F.		1.81	2.39	4.20
0700	2000 S.F.		1.74	2.20	3.94
0720	5000 S.F.		1.32	1.89	3.21
0740	10,000 S.F.		1.25	1.74	2.99
0760	50,000 S.F.		1.23	1.61	2.84
1000	Ordinary hazard, one floor, 500 S.F.		28	5.50	33.50
1020	1000 S.F.		14.35	3.54	17.89
1040	2000 S.F.		10.05	3.67	13.72
1060	5000 S.F.		4.75	2.54	7.29
1080	10,000 S.F.		3.19	2.59	5.78
1100	50,000 S.F.		2.21	2.46	4.67
1140	Each additional floor, 500 S.F.		2.60	2.99	5.59
1160	1000 S.F.		1.78	2.41	4.19
1180	2000 S.F.		1.66	2.40	4.06
1200	5000 S.F.		1.67	2.07	3.74
1220	10,000 S.F.		1.60	2.09	3.69
1240	50,000 S.F.		1.46	1.92	3.38
1500	Extra hazard, one floor, 500 S.F.		30.50	6.65	37.15
1520	1000 S.F.		16.75	5.15	21.90
1540	2000 S.F.		10.15	4.68	14.83
1560	5000 S.F.		5.10	3.55	8.65
1580	10,000 S.F.		4.10	3.45	7.55
1600	50,000 S.F.		4.13	3.36	7.49
1660	Each additional floor, 500 S.F.		3.13	3.68	6.81
1680	1000 S.F.		2.58	3.47	6.05
1700	2000 S.F.		2.44	3.47	5.91
1720	5000 S.F.		2.07	3.08	5.15
1740	10,000 S.F.		2.30	2.93	5.23
1760	50,000 S.F.		2.31	2.84	5.15
2000	Grooved steel, black, sch. 40 pipe, light hazard, one floor				
2020	2000 S.F.		9.55	3	12.55
2060	10,000 S.F.		2.65	1.74	4.39
2100	Each additional floor, 2,000 S.F.		1.65	1.76	3.41
2150	10,000 S.F.		1.09	1.49	2.58
2200	Ordinary hazard, one floor, 2000 S.F.		4.72	3.19	7.91
2250	10,000 S.F.		2.92	2.18	5.10
2300	Each additional floor, 2000 S.F.		1.73	1.96	3.69
2350	10,000 S.F.		1.39	1.95	3.34
2400	Extra hazard, one floor, 2000 S.F.		9.90	3.98	13.88
2450	10,000 S.F.		3.54	2.82	6.36
2500	Each additional floor, 2000 S.F.		2.20	2.83	5.03
2550	10,000 S.F.		1.80	2.42	4.22
3000	Grooved steel, black, sch. 10 pipe, light hazard, one floor				
3050	2000 S.F.		8.85	2.83	11.68
3100	10,000 S.F.		2.56	1.69	4.25
3150	Each additional floor, 2000 S.F.		1.53	1.73	3.26
3200	10,000 S.F.		1.02	1.46	2.48
3250	Ordinary hazard, one floor, 2000 S.F.		9.50	3.18	12.68
3300	10,000 S.F.		2.55	2.12	4.67
3350	Each additional floor, 2000 S.F.		1.63	1.94	3.57
3400	10,000 S.F.		1.27	1.91	3.18
3450	Extra hazard, one floor, 2000 S.F.		9.85	3.96	13.81

D4010 Sprinklers

D4010 370	Deluge Sprinkler Systems	COST PER S.F.		
		MAT.	INST.	TOTAL
3500	10,000 S.F.	3.27	2.75	6.02
3550	Each additional floor, 2000 S.F.	2.12	2.81	4.93
3600	10,000 S.F.	1.68	2.39	4.07
4000	Copper tubing, type M, light hazard, one floor			
4050	2000 S.F.	10.95	2.97	13.92
4100	10,000 S.F.	3.86	1.71	5.57
4150	Each additional floor, 2000 S.F.	3.09	1.77	4.86
4200	10,000 S.F.	2.07	1.47	3.54
4250	Ordinary hazard, one floor, 2000 S.F.	11.20	3.33	14.53
4300	10,000 S.F.	4.37	2.03	6.40
4350	Each additional floor, 2000 S.F.	3.11	1.83	4.94
4400	10,000 S.F.	2.45	1.63	4.08
4450	Extra hazard, one floor, 2000 S.F.	12.15	4.03	16.18
4500	10,000 S.F.	8.45	3.06	11.51
4550	Each additional floor, 2000 S.F.	4.43	2.88	7.31
4600	10,000 S.F.	5.50	2.64	8.14
5000	Copper tubing, type M, T-drill system, light hazard, one floor			
5050	2000 S.F.	10.85	2.78	13.63
5100	10,000 S.F.	3.49	1.42	4.91
5150	Each additional floor, 2000 S.F.	2.61	1.60	4.21
5200	10,000 S.F.	1.95	1.20	3.15
5250	Ordinary hazard, one floor, 2000 S.F.	10.65	2.85	13.50
5300	10,000 S.F.	4.05	1.80	5.85
5350	Each additional floor, 2000 S.F.	2.76	1.61	4.37
5400	10,000 S.F.	2.52	1.57	4.09
5450	Extra hazard, one floor, 2000 S.F.	11.10	3.33	14.43
5500	10,000 S.F.	6.15	2.22	8.37
5550	Each additional floor, 2000 S.F.	3.37	2.18	5.55
5600	10,000 S.F.	3.29	1.82	5.11

D4010 Sprinklers

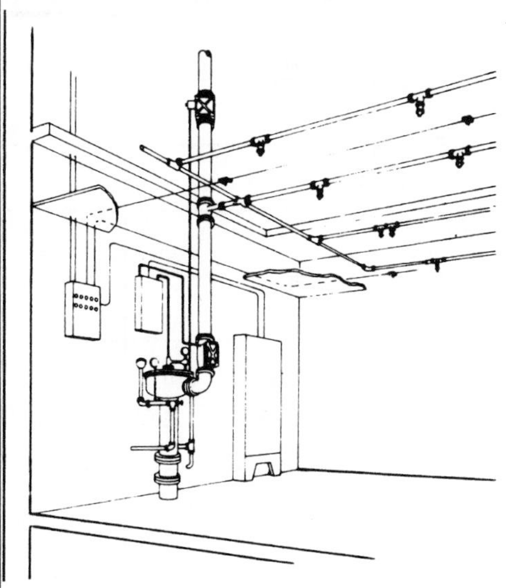

Firecycle is a fixed fire protection sprinkler system utilizing water as its extinguishing agent. It is a time delayed, recycling, preaction type which automatically shuts the water off when heat is reduced below the detector operating temperature and turns the water back on when that temperature is exceeded.

The system senses a fire condition through a closed circuit electrical detector system which controls water flow to the fire automatically. Batteries supply up to 90 hour emergency power supply for system operation. The piping system is dry (until water is required) and is monitored with pressurized air. Should any leak in the system piping occur, an alarm will sound, but water will not enter the system until heat is sensed by a Firecycle detector.

All areas are assumed to be open.

System Components	QUANTITY	UNIT	COST EACH MAT.	COST EACH INST.	COST EACH TOTAL
SYSTEM D4010 390 0580					
FIRECYCLE SPRINKLER SYSTEM, STEEL BLACK SCH. 40 PIPE					
LIGHT HAZARD, ONE FLOOR, 2000 S.F.					
Valve, gate, iron body 125 lb., OS&Y, flanged, 4″ pipe size	1.000	Ea.	566.25	303.75	870
Valve, angle, bronze, 150 lb., rising stem, threaded, 2″ pipe size	1.000	Ea.	555	45.75	600.75
Valve, swing check, bronze, 125 lb, regrinding disc, 2-1/2″ pipe size	1.000	Ea.	588.75	60.38	649.13
*Alarm valve, 2-1/2″ pipe size	1.000	Ea.	1,218.75	296.25	1,515
Alarm, water motor, complete with gong	1.000	Ea.	303.75	123.75	427.50
Pipe, steel, black, schedule 40, 4″ diam.	10.000	L.F.	195	255.53	450.53
Fire alarm, horn, electric	1.000	Ea.	49.13	70.13	119.26
Pipe, steel, black, schedule 40, threaded, cplg & hngr 10′ OC 2-1/2″ diam.	20.000	L.F.	252	360	612
Pipe, steel, black, schedule 40, threaded, cplg & hngr 10′ OC 2″ diam.	12.500	L.F.	100.78	177.19	277.97
Pipe, steel, black, schedule 40, threaded, cplg & hngr 10′ OC 1-1/4″ diam.	37.500	L.F.	195.47	382.50	577.97
Pipe, steel, black, schedule 40, threaded, cplg & hngr 10′ OC 1″ diam.	112.000	L.F.	470.40	1,066.80	1,537.20
Pipe, Tee, malleable iron, black, 150 lb. threaded, 4″ pipe size	2.000	Ea.	412.50	457.50	870
Pipe, Tee, malleable iron, black, 150 lb. threaded, 2-1/2″ pipe size	2.000	Ea.	116.25	201	317.25
Pipe, Tee, malleable iron, black, 150 lb. threaded, 2″ pipe size	1.000	Ea.	27	82.50	109.50
Pipe, Tee, malleable iron, black, 150 lb. threaded, 1-1/4″ pipe size	5.000	Ea.	63.38	324.38	387.76
Pipe, Tee, malleable iron, black, 150 lb. threaded, 1″ pipe size	4.000	Ea.	31.35	252	283.35
Pipe, 90° elbow, malleable iron, black, 150 lb. threaded, 1″ pipe size	6.000	Ea.	29.93	231.75	261.68
Sprinkler head std spray, brass 135°-286°F 1/2″ NPT, 3/8″ orifice	12.000	Ea.	182.40	492	674.40
Firecycle controls, incls panel, battery, solenoid valves, press switches	1.000	Ea.	15,225	1,875	17,100
Detector, firecycle system	2.000	Ea.	1,057.50	61.50	1,119
Firecycle pkg, swing check & flow control valves w/trim 4″ pipe size	1.000	Ea.	4,200	881.25	5,081.25
Air compressor, auto, complete, 200 Gal sprinkler sys. cap., 1/3 HP	1.000	Ea.	607.50	378.75	986.25
*Standpipe connection, wall, flush, brass w/plug & chain 2-1/2″x2-1/2″	1.000	Ea.	119.25	177.75	297
Valve, gate, bronze 300 psi, NRS, class 150, threaded, 1″ diam.	1.000	Ea.	81.75	26.63	108.38
TOTAL			26,649.09	8,584.04	35,233.13
COST PER S.F.			13.32	4.29	17.61

*Not included in systems under 2000 S.F.

D4010 Sprinklers

D4010 390	Firecycle Sprinkler Systems		COST PER S.F.		
			MAT.	INST.	TOTAL
0520	Firecycle sprinkler systems, steel black sch. 40 pipe				
0530	Light hazard, one floor, 500 S.F.		43.50	9.30	52.80
0560	1000 S.F.	R211313 -10	22.50	5.85	28.35
0580	2000 S.F.		13.30	4.29	17.59
0600	5000 S.F.	R211313 -20	5.95	2.69	8.64
0620	10,000 S.F.		3.59	2.13	5.72
0640	50,000 S.F.	R211313 -30	1.82	1.78	3.60
0660	Each additional floor of 500 S.F.		2.41	2.60	5.01
0680	1000 S.F.		1.84	2.40	4.24
0700	2000 S.F.		1.51	2.20	3.71
0720	5000 S.F.		1.34	1.89	3.23
0740	10,000 S.F.		1.32	1.74	3.06
0760	50,000 S.F.		1.28	1.60	2.88
1000	Ordinary hazard, one floor, 500 S.F.		44	9.70	53.70
1020	1000 S.F.		22.50	5.80	28.30
1040	2000 S.F.		13.45	4.44	17.89
1060	5000 S.F.		6.20	2.84	9.04
1080	10,000 S.F.		3.99	2.75	6.74
1100	50,000 S.F.		2.62	2.75	5.37
1140	Each additional floor, 500 S.F.		2.67	3	5.67
1160	1000 S.F.		1.81	2.42	4.23
1180	2000 S.F.		1.84	2.21	4.05
1200	5000 S.F.		1.69	2.07	3.76
1220	10,000 S.F.		1.55	2.04	3.59
1240	50,000 S.F.		1.47	1.82	3.29
1500	Extra hazard, one floor, 500 S.F.		47	11.15	58.15
1520	1000 S.F.		24	7.20	31.20
1540	2000 S.F.		13.85	5.45	19.30
1560	5000 S.F.		6.55	3.85	10.40
1580	10,000 S.F.		5.05	3.90	8.95
1600	50,000 S.F.		4.26	4.34	8.60
1660	Each additional floor, 500 S.F.		3.20	3.69	6.89
1680	1000 S.F.		2.61	3.48	6.09
1700	2000 S.F.		2.48	3.48	5.96
1720	5000 S.F.		2.09	3.08	5.17
1740	10,000 S.F.		2.29	2.82	5.11
1760	50,000 S.F.		2.31	2.74	5.05
2020	Grooved steel, black, sch. 40 pipe, light hazard, one floor				
2030	2000 S.F.		13.20	3.79	16.99
2060	10,000 S.F.		3.71	2.60	6.31
2100	Each additional floor, 2000 S.F.		1.69	1.77	3.46
2150	10,000 S.F.		1.16	1.49	2.65
2200	Ordinary hazard, one floor, 2000 S.F.		13.30	3.99	17.29
2250	10,000 S.F.		3.96	2.47	6.43
2300	Each additional floor, 2000 S.F.		1.77	1.97	3.74
2350	10,000 S.F.		1.46	1.95	3.41
2400	Extra hazard, one floor, 2000 S.F.		13.60	4.77	18.37
2450	10,000 S.F.		4.25	2.94	7.19
2500	Each additional floor, 2000 S.F.		2.24	2.84	5.08
2550	10,000 S.F.		1.87	2.42	4.29
3050	Grooved steel, black, sch. 10 pipe light hazard, one floor,				
3060	2000 S.F.		13.10	3.76	16.86
3100	10,000 S.F.		3.36	1.85	5.21
3150	Each additional floor, 2000 S.F.		1.57	1.74	3.31
3200	10,000 S.F.		1.09	1.46	2.55
3250	Ordinary hazard, one floor, 2000 S.F.		13.20	3.97	17.17
3300	10,000 S.F.		3.60	2.30	5.90
3350	Each additional floor, 2000 S.F.		1.67	1.95	3.62

D4010 Sprinklers

D4010 390	Firecycle Sprinkler Systems	COST PER S.F.		
		MAT.	INST.	TOTAL
3400	10,000 S.F.	1.34	1.91	3.25
3450	Extra hazard, one floor, 2000 S.F.	13.50	4.75	18.25
3500	10,000 S.F.	4.04	2.89	6.93
3550	Each additional floor, 2000 S.F.	2.16	2.82	4.98
3600	10,000 S.F.	1.75	2.39	4.14
4060	Copper tubing, type M, light hazard, one floor, 2000 S.F.	14.65	3.76	18.41
4100	10,000 S.F.	4.66	1.87	6.53
4150	Each additional floor, 2000 S.F.	3.13	1.78	4.91
4200	10,000 S.F.	2.39	1.49	3.88
4250	Ordinary hazard, one floor, 2000 S.F.	14.90	4.12	19.02
4300	10,000 S.F.	5.15	2.19	7.34
4350	Each additional floor, 2000 S.F.	3.15	1.84	4.99
4400	10,000 S.F.	2.48	1.61	4.09
4450	Extra hazard, one floor, 2000 S.F.	15.80	4.82	20.62
4500	10,000 S.F.	9.20	3.23	12.43
4550	Each additional floor, 2000 S.F.	4.47	2.89	7.36
4600	10,000 S.F.	5.60	2.64	8.24
5060	Copper tubing, type M, T-drill system, light hazard, one floor 2000 S.F.	14.50	3.57	18.07
5100	10,000 S.F.	4.29	1.58	5.87
5150	Each additional floor, 2000 S.F.	3.18	1.68	4.86
5200	10,000 S.F.	2.02	1.20	3.22
5250	Ordinary hazard, one floor, 2000 S.F.	14.30	3.64	17.94
5300	10,000 S.F.	4.85	1.96	6.81
5350	Each additional floor, 2000 S.F.	2.80	1.62	4.42
5400	10,000 S.F.	2.59	1.57	4.16
5450	Extra hazard, one floor, 2000 S.F.	14.75	4.12	18.87
5500	10,000 S.F.	6.90	2.36	9.26
5550	Each additional floor, 2000 S.F.	3.41	2.19	5.60
5600	10,000 S.F.	3.36	1.82	5.18

D4010 Sprinklers

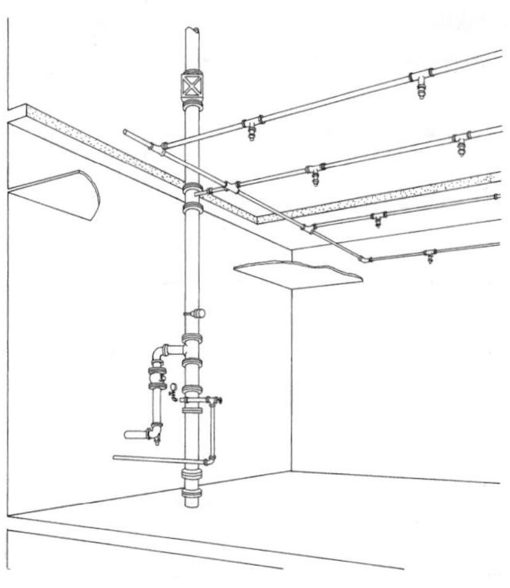

Wet Pipe System. A system employing automatic sprinklers attached to a piping system containing water and connected to a water supply so that water discharges immediately from sprinklers opened by heat from a fire.

All areas are assumed to be open.

System Components	QUANTITY	UNIT	COST EACH MAT.	COST EACH INST.	COST EACH TOTAL
SYSTEM D4010 410 0580					
WET PIPE SPRINKLER, STEEL, BLACK, SCH. 40 PIPE					
LIGHT HAZARD, ONE FLOOR, 2000 S.F.					
Valve, gate, iron body, 125 lb., OS&Y, flanged, 4″ diam.	1.000	Ea.	566.25	303.75	870
Valve, swing check, bronze, 125 lb, regrinding disc, 2-1/2″ pipe size	1.000	Ea.	588.75	60.38	649.13
Valve, angle, bronze, 150 lb., rising stem, threaded, 2″ diam.	1.000	Ea.	555	45.75	600.75
*Alarm valve, 2-1/2″ pipe size	1.000	Ea.	1,218.75	296.25	1,515
Alarm, water motor, complete with gong	1.000	Ea.	303.75	123.75	427.50
Valve, swing check, w/balldrip CI with brass trim 4″ pipe size	1.000	Ea.	243.75	296.25	540
Pipe, steel, black, schedule 40, 4″ diam.	10.000	L.F.	195	255.53	450.53
*Flow control valve, trim & gauges, 4″ pipe size	1.000	Set	4,293.75	671.25	4,965
Fire alarm horn, electric	1.000	Ea.	49.13	70.13	119.26
Pipe, steel, black, schedule 40, threaded, cplg & hngr 10′ OC, 2-1/2″ diam.	20.000	L.F.	252	360	612
Pipe, steel, black, schedule 40, threaded, cplg & hngr 10′ OC, 2″ diam.	12.500	L.F.	100.78	177.19	277.97
Pipe, steel, black, schedule 40, threaded, cplg & hngr 10′ OC, 1-1/4″ diam.	37.500	L.F.	195.47	382.50	577.97
Pipe, steel, black, schedule 40, threaded cplg & hngr 10′ OC, 1″ diam.	112.000	L.F.	470.40	1,066.80	1,537.20
Pipe Tee, malleable iron black, 150 lb. threaded, 4″ pipe size	2.000	Ea.	412.50	457.50	870
Pipe Tee, malleable iron black, 150 lb. threaded, 2-1/2″ pipe size	2.000	Ea.	116.25	201	317.25
Pipe Tee, malleable iron black, 150 lb. threaded, 2″ pipe size	1.000	Ea.	27	82.50	109.50
Pipe Tee, malleable iron black, 150 lb. threaded, 1-1/4″ pipe size	5.000	Ea.	63.38	324.38	387.76
Pipe Tee, malleable iron black, 150 lb. threaded, 1″ pipe size	4.000	Ea.	31.35	252	283.35
Pipe 90° elbow, malleable iron black, 150 lb. threaded, 1″ pipe size	6.000	Ea.	29.93	231.75	261.68
Sprinkler head, standard spray, brass 135°-286°F 1/2″ NPT, 3/8″ orifice	12.000	Ea.	182.40	492	674.40
Valve, gate, bronze, NRS, class 150, threaded, 1″ pipe size	1.000	Ea.	81.75	26.63	108.38
*Standpipe connection, wall, single, flush w/plug & chain 2-1/2″x2-1/2″	1.000	Ea.	119.25	177.75	297
TOTAL			10,096.59	6,355.04	16,451.63
COST PER S.F.			5.05	3.18	8.23
*Not included in systems under 2000 S.F.					

D4010 410	Wet Pipe Sprinkler Systems		COST PER S.F. MAT.	COST PER S.F. INST.	COST PER S.F. TOTAL
0520	Wet pipe sprinkler systems, steel, black, sch. 40 pipe				
0530	Light hazard, one floor, 500 S.F.		2.83	3.06	5.89
0560	1000 S.F.	R211313 -10	5.65	3.17	8.82
0580	2000 S.F.		5.05	3.18	8.23
0600	5000 S.F.	R211313 -20	2.43	2.23	4.66
0620	10,000 S.F.		1.62	1.89	3.51
0640	50,000 S.F.	R211313 -30	1.15	1.71	2.86
0660	Each additional floor, 500 S.F.		1.37	2.61	3.98

D4010 Sprinklers

D4010 410	Wet Pipe Sprinkler Systems	COST PER S.F.		
		MAT.	INST.	TOTAL
0680	1000 S.F. R211313 -40	1.33	2.44	3.77
0700	2000 S.F.	1.25	2.18	3.43
0720	5000 S.F.	.92	1.87	2.79
0740	10,000 S.F.	.90	1.72	2.62
0760	50,000 S.F.	.75	1.35	2.10
1000	Ordinary hazard, one floor, 500 S.F.	3.02	3.28	6.30
1020	1000 S.F.	5.60	3.11	8.71
1040	2000 S.F.	5.20	3.33	8.53
1060	5000 S.F.	2.71	2.38	5.09
1080	10,000 S.F.	2.02	2.51	4.53
1100	50,000 S.F.	1.65	2.36	4.01
1140	Each additional floor, 500 S.F.	1.61	2.94	4.55
1160	1000 S.F.	1.28	2.39	3.67
1180	2000 S.F.	1.41	2.39	3.80
1200	5000 S.F.	1.43	2.26	3.69
1220	10,000 S.F.	1.31	2.34	3.65
1240	50,000 S.F.	1.23	2.08	3.31
1500	Extra hazard, one floor, 500 S.F.	10.75	5.05	15.80
1520	1000 S.F.	6.70	4.38	11.08
1540	2000 S.F.	5.60	4.49	10.09
1560	5000 S.F.	3.57	3.92	7.49
1580	10,000 S.F.	3.07	3.73	6.80
1600	50,000 S.F.	3.38	3.60	6.98
1660	Each additional floor, 500 S.F.	2.14	3.63	5.77
1680	1000 S.F.	2.08	3.45	5.53
1700	2000 S.F.	1.95	3.45	5.40
1720	5000 S.F.	1.67	3.06	4.73
1740	10,000 S.F.	1.87	2.80	4.67
1760	50,000 S.F.	1.88	2.69	4.57
2020	Grooved steel, black sch. 40 pipe, light hazard, one floor, 2000 S.F.	4.94	2.68	7.62
2060	10,000 S.F.	1.89	1.70	3.59
2100	Each additional floor, 2000 S.F.	1.16	1.74	2.90
2150	10,000 S.F.	.74	1.47	2.21
2200	Ordinary hazard, one floor, 2000 S.F.	5	2.88	7.88
2250	10,000 S.F.	1.75	2.10	3.85
2300	Each additional floor, 2000 S.F.	1.24	1.94	3.18
2350	10,000 S.F.	1.04	1.93	2.97
2400	Extra hazard, one floor, 2000 S.F.	5.30	3.66	8.96
2450	10,000 S.F.	2.31	2.72	5.03
2500	Each additional floor, 2000 S.F.	1.71	2.81	4.52
2550	10,000 S.F.	1.45	2.40	3.85
3050	Grooved steel, black sch. 10 pipe, light hazard, one floor, 2000 S.F.	4.82	2.65	7.47
3100	10,000 S.F.	1.39	1.61	3
3150	Each additional floor, 2000 S.F.	1.04	1.71	2.75
3200	10,000 S.F.	.67	1.44	2.11
3250	Ordinary hazard, one floor, 2000 S.F.	4.92	2.86	7.78
3300	10,000 S.F.	1.63	2.06	3.69
3350	Each additional floor, 2000 S.F.	1.14	1.92	3.06
3400	10,000 S.F.	.92	1.89	2.81
3450	Extra hazard, one floor, 2000 S.F.	5.25	3.64	8.89
3500	10,000 S.F.	2.10	2.67	4.77
3550	Each additional floor, 2000 S.F.	1.63	2.79	4.42
3600	10,000 S.F.	1.33	2.37	3.70
4050	Copper tubing, type M, light hazard, one floor, 2000 S.F.	6.35	2.65	9
4100	10,000 S.F.	2.69	1.63	4.32
4150	Each additional floor, 2000 S.F.	2.60	1.75	4.35

D4010 Sprinklers

D4010 410	Wet Pipe Sprinkler Systems	COST PER S.F.		
		MAT.	INST.	TOTAL
4200	10,000 S.F.	1.97	1.47	3.44
4250	Ordinary hazard, one floor, 2000 S.F.	6.65	3.01	9.66
4300	10,000 S.F.	3.20	1.95	5.15
4350	Each additional floor, 2000 S.F.	2.96	1.96	4.92
4400	10,000 S.F.	2.35	1.75	4.10
4450	Extra hazard, one floor, 2000 S.F.	7.55	3.71	11.26
4500	10,000 S.F.	7.20	2.96	10.16
4550	Each additional floor, 2000 S.F.	3.94	2.86	6.80
4600	10,000 S.F.	5.15	2.62	7.77
5050	Copper tubing, type M, T-drill system, light hazard, one floor			
5060	2000 S.F.	6.25	2.46	8.71
5100	10,000 S.F.	2.32	1.34	3.66
5150	Each additional floor, 2000 S.F.	2.48	1.56	4.04
5200	10,000 S.F.	1.60	1.18	2.78
5250	Ordinary hazard, one floor, 2000 S.F.	6.05	2.53	8.58
5300	10,000 S.F.	2.88	1.72	4.60
5350	Each additional floor, 2000 S.F.	2.27	1.59	3.86
5400	10,000 S.F.	2.17	1.55	3.72
5450	Extra hazard, one floor, 2000 S.F.	6.50	3.01	9.51
5500	10,000 S.F.	4.98	2.14	7.12
5550	Each additional floor, 2000 S.F.	2.97	2.21	5.18
5600	10,000 S.F.	2.94	1.80	4.74

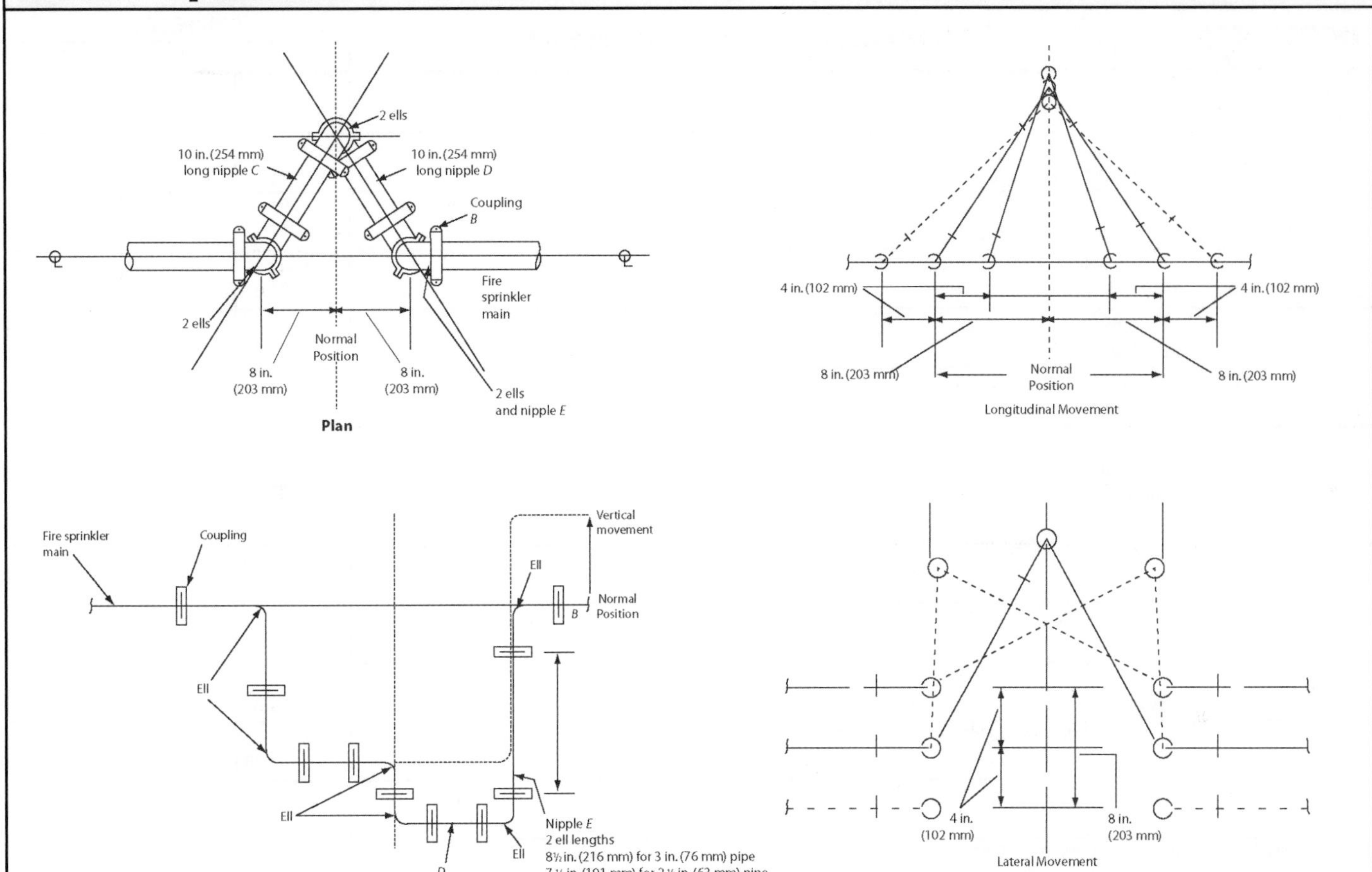

System Components	QUANTITY	UNIT	COST PER EACH MAT.	COST PER EACH INST.	COST PER EACH TOTAL
SYSTEM D4010 412 1000					
WET SPRINKLER SYSTEM, SCHEDULE 10 BLACK STEEL GROOVED PIPE					
SEPARATION ASSEMBLY, 2 INCH DIAMETER					
Hanger support for 2 inch pipe	2.000	Ea.	122	74	196
Pipe roller support yoke	2.000	Ea.	38.70	17.30	56
Spool pieces 2 inch	3.000	Ea.	19.50	46.95	66.45
2 inch grooved elbows	6.000	Ea.	174	162	336
Coupling grooved joint 2 inch pipe	10.000	Ea.	240	134.50	374.50
Roll grooved joint labor only	8.000	Ea.		83.60	83.60
TOTAL			594.20	518.35	1,112.55

D4010 412	Wet Sprinkler Seismic Components	COST PER EACH MAT.	COST PER EACH INST.	COST PER EACH TOTAL
1000	Wet sprinkler sys, Sch. 10, blk steel grooved, separation assembly , 2 inch	595	520	1,115
1100	2-1/2 inch	650	575	1,225
1200	3 inch	850	660	1,510
1300	4 inch	1,050	820	1,870
1400	6 inch	2,050	1,175	3,225
1500	8 inch	3,650	1,375	5,025
2000	Wet sprinkler sys, Sch. 10, blk steel grooved, flexible coupling,2-1/2 inch	425	286	711
2100	3 inch	435	335	770

D4010 Sprinklers

D4010 412	Wet Sprinkler Seismic Components	COST PER EACH		
		MAT.	INST.	TOTAL
2200	4 inch	580	445	1,025
2300	6 inch	835	630	1,465
2400	8 inch	1,225	790	2,015
2990	Allowance for seismic movement by providing annular space in concrete wall			
2995	If additional cores are performed for the same size and at the same time reduce			
2996	price of additional cores by 50 %. If different size cores reduce by 30 %.			
3000	Wet pipe sprinkler systems, 2 inch black steel, annular space 6 inch wall	17.15	600	617.15
3010	8 inch wall	17.40	610	627.40
3020	10 inch wall	17.60	615	632.60
3030	12 inch wall	17.85	620	637.85
3100	Wet pipe sprinkler systems, 2-1/2/3 inch black steel, annular space 6" wall	17.15	620	637.15
3110	8 inch wall	17.40	625	642.40
3120	10 inch wall	17.60	630	647.60
3130	12 inch wall	17.85	635	652.85
3200	Wet pipe sprinkler systems, 4 inch black steel, annular space 6" wall	17.15	685	702.15
3210	8 inch wall	17.40	690	707.40
3220	10 inch wall	17.60	700	717.60
3230	12 inch wall	17.85	705	722.85
3300	Wet pipe sprinkler systems, 6 inch black steel, annular space 6" wall	17.15	715	732.15
3310	8 inch wall	17.40	720	737.40
3320	10 inch wall	17.60	725	742.60
3330	12 inch wall	17.85	730	747.85
3400	Wet pipe sprinkler systems, 8 inch black steel, annular space 6" wall	17.15	745	762.15
3410	8 inch wall	17.40	750	767.40
3420	10 inch wall	17.60	755	772.60
3430	12 inch wall	17.85	760	777.85
3996	Lateral, longitudinal and 4-way seismic strut braces			
4000	Wet pipe sprinkler systems, 2 inch pipe double lateral strut brace	188	282	470
4010	2 -1/2 inch	194	282	476
4020	3 inch	198	283	481
4030	4 inch	223	286	509
4040	6 inch	271	287	558
4050	8 inch	315	287	602
4100	Wet pipe sprinkler systems, 2 inch pipe longitudinal strut brace	188	282	470
4110	2 -1/2 inch	194	282	476
4120	3 inch	198	283	481
4130	4 inch	223	286	509
4140	6 inch	271	287	558
4150	8 inch	315	287	602
4200	Wet pipe sprinkler systems, 2 inch pipe single lateral strut brace	129	186	315
4210	2 -1/2 inch	135	186	321
4220	3 inch	139	186	325
4230	4 inch	162	189	351
4240	6 inch	210	190	400
4250	8 inch	256	191	447
4300	Wet pipe sprinkler systems, 2 inch pipe longitudinal wire brace	98.50	195	293.50
4310	2 -1/2 inch	104	195	299
4320	3 inch	109	196	305
4330	4 inch	135	202	337
4340	6 inch	182	200	382
4350	8 inch	228	200	428
4400	Wet pipe sprinkler systems, 2 inch pipe 4-way wire brace	126	300	426
4410	2 -1/2 inch	132	300	432
4420	3 inch	136	300	436
4430	4 inch	169	310	479
4440	6 inch	215	305	520
4450	8 inch	261	305	566
4500	Wet pipe sprinkler systems, 2 inch pipe 4-way strut brace	305	475	780

D4010 Sprinklers

D4010 412	Wet Sprinkler Seismic Components	COST PER EACH		
		MAT.	INST.	TOTAL
4510	2 -1/2 inch	310	475	785
4520	3 inch	315	475	790
4530	4 inch	345	480	825
4540	6 inch	395	480	875
4550	8 inch	440	480	920

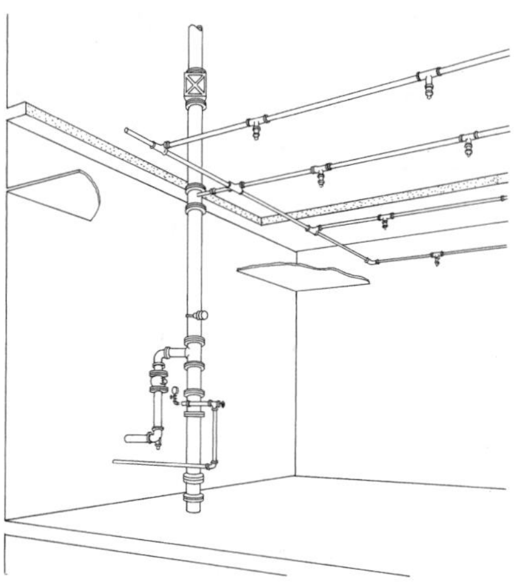

Wet Pipe System. A system employing automatic sprinklers attached to a piping system containing water and connected to a water supply so that water discharges immediately from sprinklers opened by heat from a fire.

All areas are assumed to be open.

System Components	QUANTITY	UNIT	COST PER S.F.		
			MAT.	INST.	TOTAL
SYSTEM D4010 413 0580					
SEISMIC WET PIPE SPRINKLER, STEEL, BLACK, SCH. 40 PIPE					
LIGHT HAZARD, ONE FLOOR, 3′ FLEXIBLE FEED, 2000 S.F.					
Valve, gate, iron body, 125 lb., OS&Y, flanged, 4″ diam.	1.000	Ea.	.21	.11	.32
Valve, swing check, bronze, 125 lb, regrinding disc, 2-1/2″ pipe size	1.000	Ea.	.22	.02	.24
Valve, angle, bronze, 150 lb., rising stem, threaded, 2″ diam.	1.000	Ea.	.21	.02	.23
*Alarm valve, 2-1/2″ pipe size	1.000	Ea.	.46	.11	.57
Alarm, water motor, complete with gong	1.000	Ea.	.11	.05	.16
Valve, swing check, w/balldrip CI with brass trim 4″ pipe size	1.000	Ea.	.09	.11	.20
Pipe, steel, black, schedule 40, 4″ diam.	10.000	L.F.	.07	.09	.16
*Flow control valve, trim & gauges, 4″ pipe size	1.000	Set	1.61	.25	1.86
Fire alarm horn, electric	1.000	Ea.	.02	.03	.05
Pipe, steel, black, schedule 40, threaded, cplg & hngr 10′ OC, 2-1/2″ diam.	20.000	L.F.	.09	.13	.22
Pipe, steel, black, schedule 40, threaded, cplg & hngr 10′ OC, 2″ diam.	12.500	L.F.	.04	.07	.11
Pipe, steel, black, schedule 40, threaded, cplg & hngr 10′ OC, 1-1/4″ diam.	37.500	L.F.	.07	.14	.21
Pipe Tee, malleable iron black, 150 lb. threaded, 4″ pipe size	2.000	Ea.	.15	.17	.32
Pipe Tee, malleable iron black, 150 lb. threaded, 2-1/2″ pipe size	2.000	Ea.	.04	.08	.12
Pipe Tee, malleable iron black, 150 lb. threaded, 2″ pipe size	1.000	Ea.	.01	.03	.04
Pipe Tee, malleable iron black, 150 lb. threaded, 1-1/4″ pipe size	5.000	Ea.	.02	.12	.14
Pipe Tee, malleable iron black, 150 lb. threaded, 1″ pipe size	4.000	Ea.	.01	.09	.10
Pipe 90° elbow, malleable iron black, 150 lb. threaded, 1″ pipe size	6.000	Ea.	.01	.09	.10
Sprinkler head, standard spray, brass 135°-286°F 1/2″ NPT, 3/8″ orifice	12.000	Ea.	.07	.18	.25
Valve, gate, bronze, NRS, class 150, threaded, 1″ pipe size	1.000	Ea.	.03	.01	.04
*Standpipe connection, wall, single, flush w/plug & chain 2-1/2″x2-1/2″	1.000	Ea.	.04	.07	.11
Flexible connector 36 inches long	12.000	Ea.	.15	.10	.25
TOTAL			3.73	2.07	5.80
*Not included in systems under 2000 S.F.					

D4010 413	Seismic Wet Pipe Flexible Feed Sprinkler Systems	COST PER S.F.		
		MAT.	INST.	TOTAL
0520	Seismic wet pipe 3′ flexible feed sprinkler systems, steel, black, sch. 40 pipe			
0525	Note: See D4010-412 for strut braces, flexible joints and separation assemblies			
0530	Light hazard, one floor, 3′ flexible feed, 500 S.F.	2.68	2.47	5.15
0560	1000 S.F.	5.60	2.79	8.39
0580	2000 S.F.	3.73	2.07	5.80
0600	5000 S.F.	2.42	1.96	4.38
0620	10,000 S.F.	1.66	1.76	3.42
0640	50,000 S.F.	1.14	1.49	2.63

D4010 413	Seismic Wet Pipe Flexible Feed Sprinkler Systems	COST PER S.F.		
		MAT.	INST.	TOTAL
0660	Each additional floor, 500 S.F.	1.27	2.05	3.32
0680	1000 S.F.	1.32	2.09	3.41
0700	2000 S.F.	1.16	1.75	2.91
0720	5000 S.F.	.91	1.60	2.51
0740	10,000 S.F.	.94	1.59	2.53
0760	50,000 S.F.	.80	1.27	2.07
1000	Ordinary hazard, one floor, 500 S.F.	2.92	2.72	5.64
1020	1000 S.F.	5.55	2.70	8.25
1040	2000 S.F.	5.20	3.08	8.28
1060	5000 S.F.	2.78	2.21	4.99
1080	10,000 S.F.	2.11	2.39	4.50
1100	50,000 S.F.	1.69	2.13	3.82
1140	Each additional floor, 500 S.F.	1.51	2.38	3.89
1160	1000 S.F.	1.24	1.98	3.22
1180	2000 S.F.	1.44	2.14	3.58
1200	5000 S.F.	1.50	2.09	3.59
1220	10,000 S.F.	1.40	2.22	3.62
1240	50,000 S.F.	1.27	1.85	3.12
1500	Extra hazard, one floor, 500 S.F.	10.55	4.19	14.74
1520	1000 S.F.	6.50	3.54	10.04
1540	2000 S.F.	5.60	4.09	9.69
1560	5000 S.F.	3.65	3.67	7.32
1580	10,000 S.F.	3.19	3.54	6.73
1600	50,000 S.F.	3.49	3.38	6.87
1660	Each additional floor, 500 S.F.	1.94	2.77	4.71
1680	1000 S.F.	1.89	2.61	4.50
1700	2000 S.F.	1.96	3.05	5.01
1720	5000 S.F.	1.75	2.81	4.56
1740	10,000 S.F.	1.99	2.61	4.60
1760	50,000 S.F.	1.99	2.47	4.46
2020	Grooved steel, black sch. 40 pipe, light hazard, one floor, 2000 S.F.	4.88	2.33	7.21
2060	10,000 S.F.	1.85	1.55	3.40
2100	Each additional floor, 2000 S.F.	1.10	1.39	2.49
2150	10,000 S.F.	.70	1.32	2.02
2200	Ordinary hazard, one floor, 2000 S.F.	5.05	2.69	7.74
2250	10,000 S.F.	1.85	2.02	3.87
2300	Each additional floor, 2000 S.F.	1.29	1.75	3.04
2350	10,000 S.F.	1.14	1.85	2.99
2400	Extra hazard, one floor, 2000 S.F.	5.35	3.36	8.71
2450	10,000 S.F.	2.45	2.59	5.04
2500	Each additional floor, 2000 S.F.	1.76	2.51	4.27
2550	10,000 S.F.	1.59	2.27	3.86
3050	Grooved steel black sch. 10 pipe, light hazard, one floor, 2000 S.F.	4.76	2.30	7.06
3100	10,000 S.F.	1.45	1.52	2.97
3150	Each additional floor, 2000 S.F.	.98	1.36	2.34
3200	10,000 S.F.	.63	1.29	1.92
3250	Ordinary hazard, one floor, 2000 S.F.	4.97	2.67	7.64
3300	10,000 S.F.	1.73	1.98	3.71
3350	Each additional floor, 2000 S.F.	1.19	1.73	2.92
3400	10,000 S.F.	1.02	1.81	2.83
3450	Extra hazard, one floor, 2000 S.F.	5.30	3.34	8.64
3500	10,000 S.F.	2.24	2.54	4.78
3550	Each additional floor, 2000 S.F.	1.68	2.49	4.17
3600	10,000 S.F.	1.47	2.24	3.71

D4010 Sprinklers

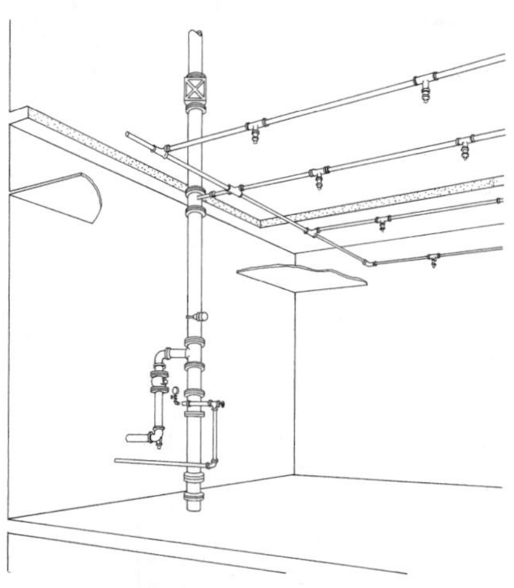

Wet Pipe System. A system employing automatic sprinklers attached to a piping system containing water and connected to a water supply so that water discharges immediately from sprinklers opened by heat from a fire.

All areas are assumed to be open.

System Components	QUANTITY	UNIT	COST PER S.F. MAT.	COST PER S.F. INST.	COST PER S.F. TOTAL
SYSTEM D4010 414 0580					
SEISMIC WET PIPE SPRINKLER, STEEL, BLACK, SCH. 40 PIPE					
LIGHT HAZARD, ONE FLOOR, 4′ FLEXIBLE FEED, 2000 S.F.					
Valve, gate, iron body, 125 lb., OS&Y, flanged, 4″ diam.	1.000	Ea.	.21	.11	.32
Valve, swing check, bronze, 125 lb, regrinding disc, 2-1/2″ pipe size	1.000	Ea.	.22	.02	.24
Valve, angle, bronze, 150 lb., rising stem, threaded, 2″ diam.	1.000	Ea.	.21	.02	.23
*Alarm valve, 2-1/2″ pipe size	1.000	Ea.	.46	.11	.57
Alarm, water motor, complete with gong	1.000	Ea.	.11	.05	.16
Valve, swing check, w/balldrip CI with brass trim 4″ pipe size	1.000	Ea.	.09	.11	.20
Pipe, steel, black, schedule 40, 4″ diam.	10.000	L.F.	.07	.09	.16
*Flow control valve, trim & gauges, 4″ pipe size	1.000	Set	1.61	.25	1.86
Fire alarm horn, electric	1.000	Ea.	.02	.03	.05
Pipe, steel, black, schedule 40, threaded, cplg & hngr 10′ OC, 2-1/2″ diam.	20.000	L.F.	.09	.13	.22
Pipe, steel, black, schedule 40, threaded, cplg & hngr 10′ OC, 2″ diam.	12.500	L.F.	.04	.07	.11
Pipe, steel, black, schedule 40, threaded, cplg & hngr 10′ OC, 1-1/4″ diam.	37.500	L.F.	.07	.14	.21
Pipe Tee, malleable iron black, 150 lb. threaded, 4″ pipe size	2.000	Ea.	.15	.17	.32
Pipe Tee, malleable iron black, 150 lb. threaded, 2-1/2″ pipe size	2.000	Ea.	.04	.08	.12
Pipe Tee, malleable iron black, 150 lb. threaded, 2″ pipe size	1.000	Ea.	.01	.03	.04
Pipe Tee, malleable iron black, 150 lb. threaded, 1-1/4″ pipe size	5.000	Ea.	.02	.12	.14
Pipe Tee, malleable iron black, 150 lb. threaded, 1″ pipe size	4.000	Ea.	.01	.09	.10
Pipe 90° elbow, malleable iron black, 150 lb. threaded, 1″ pipe size	6.000	Ea.	.01	.09	.10
Sprinkler head, standard spray, brass 135°-286°F 1/2″ NPT, 3/8″ orifice	12.000	Ea.	.07	.18	.25
Valve, gate, bronze, NRS, class 150, threaded, 1″ pipe size	1.000	Ea.	.03	.01	.04
*Standpipe connection, wall, single, flush w/plug & chain 2-1/2″x2-1/2″	1.000	Ea.	.04	.07	.11
Flexible connector 48 inches long	12.000	Ea.	.17	.11	.28
TOTAL			3.75	2.08	5.83
*Not included in systems under 2000 S.F.					

D4010 414	Seismic Wet Pipe Flexible Feed Sprinkler Systems	COST PER S.F. MAT.	COST PER S.F. INST.	COST PER S.F. TOTAL
0520	Seismic wet pipe 4′ flexible feed sprinkler systems, steel, black, sch. 40 pipe			
0525	Note: See D4010-412 for strut braces, flexible joints and separation assemblies			
0530	Light hazard, one floor, 4′ flexible feed, 500 S.F.	2.70	2.48	5.18
0560	1000 S.F.	5.60	2.80	8.40
0580	2000 S.F.	3.75	2.08	5.83
0600	5000 S.F.	2.44	1.97	4.41
0620	10,000 S.F.	1.68	1.78	3.46
0640	50,000 S.F.	1.16	1.50	2.66

D4010 Sprinklers

D4010 414	Seismic Wet Pipe Flexible Feed Sprinkler Systems	COST PER S.F.		
		MAT.	INST.	TOTAL
0660	Each additional floor, 500 S.F.	1.30	2.07	3.37
0680	1000 S.F.	1.35	2.11	3.46
0700	2000 S.F.	1.18	1.76	2.94
0720	5000 S.F.	.93	1.61	2.54
0740	10,000 S.F.	.96	1.61	2.57
0760	50,000 S.F.	.82	1.28	2.10
1000	Ordinary hazard, one floor, 500 S.F.	2.95	2.74	5.69
1020	1000 S.F.	5.60	2.72	8.32
1040	2000 S.F.	5.25	3.10	8.35
1060	5000 S.F.	2.81	2.23	5.04
1080	10,000 S.F.	2.14	2.41	4.55
1100	50,000 S.F.	1.71	2.15	3.86
1140	Each additional floor, 500 S.F.	1.54	2.40	3.94
1160	1000 S.F.	1.27	2	3.27
1180	2000 S.F.	1.47	2.16	3.63
1200	5000 S.F.	1.53	2.11	3.64
1220	10,000 S.F.	1.43	2.24	3.67
1240	50,000 S.F.	1.29	1.87	3.16
1500	Extra hazard, one floor, 500 S.F.	10.60	4.22	14.82
1520	1000 S.F.	6.55	3.57	10.12
1540	2000 S.F.	5.65	4.12	9.77
1560	5000 S.F.	3.68	3.70	7.38
1580	10,000 S.F.	3.23	3.57	6.80
1600	50,000 S.F.	3.53	3.41	6.94
1660	Each additional floor, 500 S.F.	1.98	2.80	4.78
1680	1000 S.F.	1.93	2.64	4.57
1700	2000 S.F.	2	3.08	5.08
1720	5000 S.F.	1.78	2.84	4.62
1740	10,000 S.F.	2.03	2.64	4.67
1760	50,000 S.F.	2.03	2.50	4.53
2020	Grooved steel, black sch. 40 pipe, light hazard, one floor, 2000 S.F.	4.90	2.34	7.24
2060	10,000 S.F.	1.86	1.55	3.41
2100	Each additional floor, 2000 S.F.	1.12	1.40	2.52
2150	10,000 S.F.	.71	1.32	2.03
2200	Ordinary hazard, one floor, 2000 S.F.	5.10	2.71	7.81
2250	10,000 S.F.	1.88	2.04	3.92
2300	Each additional floor, 2000 S.F.	1.32	1.77	3.09
2350	10,000 S.F.	1.17	1.87	3.04
2400	Extra hazard, one floor, 2000 S.F.	5.40	3.39	8.79
2450	10,000 S.F.	2.49	2.62	5.11
2500	Each additional floor, 2000 S.F.	1.80	2.54	4.34
2550	10,000 S.F.	1.63	2.30	3.93
3050	Grooved steel black sch. 10 pipe, light hazard, one floor, 2000 S.F.	4.78	2.31	7.09
3100	10,000 S.F.	1.47	1.54	3.01
3150	Each additional floor, 2000 S.F.	1	1.37	2.37
3200	10,000 S.F.	.64	1.29	1.93
3250	Ordinary hazard, one floor, 2000 S.F.	5	2.69	7.69
3300	10,000 S.F.	1.76	2	3.76
3350	Each additional floor, 2000 S.F.	1.22	1.75	2.97
3400	10,000 S.F.	1.05	1.83	2.88
3450	Extra hazard, one floor, 2000 S.F.	5.35	3.37	8.72
3500	10,000 S.F.	2.28	2.57	4.85
3550	Each additional floor, 2000 S.F.	1.72	2.52	4.24
3600	10,000 S.F.	1.51	2.27	3.78

D4020 Standpipes

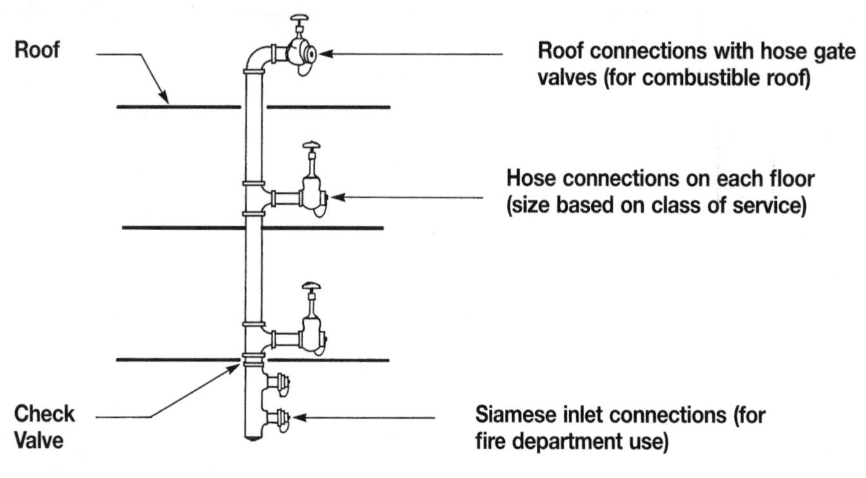

System Components	QUANTITY	UNIT	COST PER FLOOR MAT.	INST.	TOTAL
SYSTEM D4020 310 0560					
WET STANDPIPE RISER, CLASS I, STEEL, BLACK, SCH. 40 PIPE, 10′ HEIGHT					
4″ DIAMETER PIPE, ONE FLOOR					
Pipe, steel, black, schedule 40, threaded, 4″ diam.	20.000	L.F.	650	670	1,320
Pipe, Tee, malleable iron, black, 150 lb. threaded, 4″ pipe size	2.000	Ea.	550	610	1,160
Pipe, 90° elbow, malleable iron, black, 150 lb threaded 4″ pipe size	1.000	Ea.	174	202	376
Pipe, nipple, steel, black, schedule 40, 2-1/2″ pipe size x 3″ long	2.000	Ea.	24.50	151	175.50
Fire valve, gate, 300 lb., brass w/handwheel, 2-1/2″ pipe size	1.000	Ea.	189	94	283
Fire valve, pressure restricting, adj, rgh brs, 2-1/2″ pipe size	1.000	Ea.	352	188	540
Valve, swing check, w/ball drip, CI w/brs. ftngs., 4″ pipe size	1.000	Ea.	325	395	720
Standpipe conn wall dble. flush brs. w/plugs & chains 2-1/2″x2-1/2″x4″	1.000	Ea.	515	237	752
Valve, swing check, bronze, 125 lb, regrinding disc, 2-1/2″ pipe size	1.000	Ea.	785	80.50	865.50
Roof manifold, fire, w/valves & caps, horiz/vert brs 2-1/2″x2-1/2″x4″	1.000	Ea.	176	247	423
Fire, hydrolator, vent & drain, 2-1/2″ pipe size	1.000	Ea.	74.50	55	129.50
Valve, gate, iron body 125 lb., OS&Y, threaded, 4″ pipe size	1.000	Ea.	755	405	1,160
TOTAL			4,570	3,334.50	7,904.50

D4020 310	Wet Standpipe Risers, Class I		COST PER FLOOR MAT.	INST.	TOTAL
0550	Wet standpipe risers, Class I, steel, black sch. 40, 10′ height				
0560	4″ diameter pipe, one floor		4,575	3,325	7,900
0580	Additional floors		1,075	1,025	2,100
0600	6″ diameter pipe, one floor		7,600	5,850	13,450
0620	Additional floors	R211226 -10	1,850	1,650	3,500
0640	8″ diameter pipe, one floor		11,900	7,025	18,925
0660	Additional floors	R211226 -20	2,700	1,975	4,675
0680					

D4020 310	Wet Standpipe Risers, Class II	COST PER FLOOR MAT.	INST.	TOTAL
1030	Wet standpipe risers, Class II, steel, black sch. 40, 10′ height			
1040	2″ diameter pipe, one floor	1,825	1,200	3,025
1060	Additional floors	545	465	1,010
1080	2-1/2″ diameter pipe, one floor	2,675	1,750	4,425
1100	Additional floors	635	540	1,175
1120				

D4020 Standpipes

D4020 310	Wet Standpipe Risers, Class III	COST PER FLOOR		
		MAT.	INST.	TOTAL
1530	Wet standpipe risers, Class III, steel, black sch. 40, 10′ height			
1540	4″ diameter pipe, one floor	4,675	3,325	8,000
1560	Additional floors	920	865	1,785
1580	6″ diameter pipe, one floor	7,725	5,850	13,575
1600	Additional floors	1,900	1,650	3,550
1620	8″ diameter pipe, one floor	12,000	7,025	19,025
1640	Additional floors	2,750	1,975	4,725

D4020 Standpipes

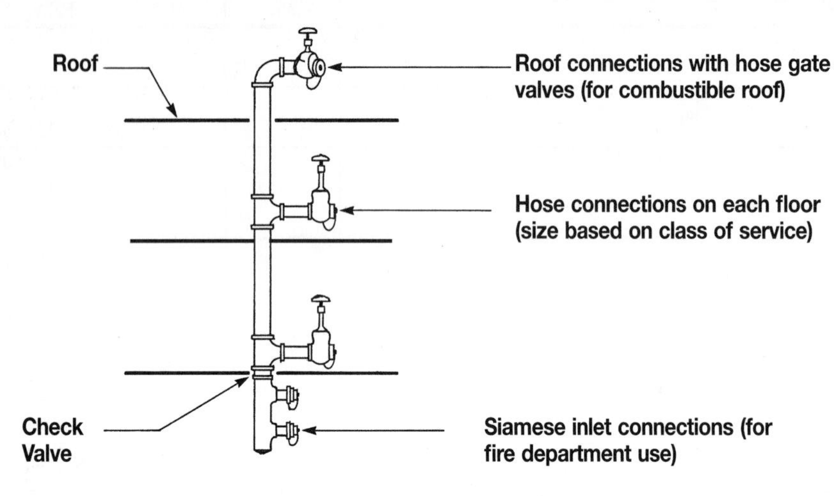

System Components	QUANTITY	UNIT	COST PER FLOOR MAT.	COST PER FLOOR INST.	COST PER FLOOR TOTAL
SYSTEM D4020 330 0540					
DRY STANDPIPE RISER, CLASS I, PIPE, STEEL, BLACK, SCH 40, 10′ HEIGHT					
4″ DIAMETER PIPE, ONE FLOOR					
Pipe, steel, black, schedule 40, threaded, 4″ diam.	20.000	L.F.	650	670	1,320
Pipe, Tee, malleable iron, black, 150 lb. threaded, 4″ pipe size	2.000	Ea.	550	610	1,160
Pipe, 90° elbow, malleable iron, black, 150 lb threaded 4″ pipe size	1.000	Ea.	174	202	376
Pipe, nipple, steel, black, schedule 40, 2-1/2″ pipe size x 3″ long	2.000	Ea.	24.50	151	175.50
Fire valve gate NRS 300 lb., brass w/handwheel, 2-1/2″ pipe size	1.000	Ea.	189	94	283
Fire valve, pressure restricting, adj, rgh brs, 2-1/2″ pipe size	1.000	Ea.	176	94	270
Standpipe conn wall dble. flush brs. w/plugs & chains 2-1/2″x2-1/2″x4″	1.000	Ea.	515	237	752
Valve swing check w/ball drip CI w/brs. ftngs., 4″pipe size	1.000	Ea.	325	395	720
Roof manifold, fire, w/valves & caps, horiz/vert brs 2-1/2″x2-1/2″x4″	1.000	Ea.	176	247	423
TOTAL			2,779.50	2,700	5,479.50

D4020 330	Dry Standpipe Risers, Class I		COST PER FLOOR MAT.	COST PER FLOOR INST.	COST PER FLOOR TOTAL
0530	Dry standpipe riser, Class I, steel, black sch. 40, 10′ height				
0540	4″ diameter pipe, one floor		2,775	2,700	5,475
0560	Additional floors		990	980	1,970
0580	6″ diameter pipe, one floor		5,450	4,625	10,075
0600	Additional floors	R211226 -10	1,775	1,575	3,350
0620	8″ diameter pipe, one floor		8,750	5,600	14,350
0640	Additional floors	R211226 -20	2,625	1,925	4,550
0660					

D4020 330	Dry Standpipe Risers, Class II	COST PER FLOOR MAT.	COST PER FLOOR INST.	COST PER FLOOR TOTAL
1030	Dry standpipe risers, Class II, steel, black sch. 40, 10′ height			
1040	2″ diameter pipe, one floor	1,500	1,250	2,750
1060	Additional floor	470	410	880
1080	2-1/2″ diameter pipe, one floor	2,125	1,450	3,575
1100	Additional floors	560	485	1,045
1120				

D4020 Standpipes

D4020 330	Dry Standpipe Risers, Class III	COST PER FLOOR		
		MAT.	INST.	TOTAL
1530	Dry standpipe risers, Class III, steel, black sch. 40, 10′ height			
1540	4″ diameter pipe, one floor	2,825	2,650	5,475
1560	Additional floors	855	885	1,740
1580	6″ diameter pipe, one floor	5,500	4,625	10,125
1600	Additional floors	1,825	1,575	3,400
1620	8″ diameter pipe, one floor	8,800	5,600	14,400
1640	Additional floor	2,675	1,925	4,600

D4020 Standpipes

D4020 410	Fire Hose Equipment	COST EACH		
		MAT.	INST.	TOTAL
0100	Adapters, reducing, 1 piece, FxM, hexagon, cast brass, 2-1/2" x 1-1/2"	55.50		55.50
0200	Pin lug, 1-1/2" x 1"	47.50		47.50
0250	3" x 2-1/2"	125		125
0300	For polished chrome, add 75% mat.			
0400	Cabinets, D.S. glass in door, recessed, steel box, not equipped			
0500	Single extinguisher, steel door & frame	110	148	258
0550	Stainless steel door & frame	205	148	353
0600	Valve, 2-1/2" angle, steel door & frame	135	99	234
0650	Aluminum door & frame	164	99	263
0700	Stainless steel door & frame	220	99	319
0750	Hose rack assy, 2-1/2" x 1-1/2" valve & 100' hose, steel door & frame	252	198	450
0800	Aluminum door & frame	375	198	573
0850	Stainless steel door & frame	495	198	693
0900	Hose rack assy & extinguisher,2-1/2"x1-1/2" valve & hose,steel door & frame	315	237	552
0950	Aluminum	475	237	712
1000	Stainless steel	575	237	812
1550	Compressor, air, dry pipe system, automatic, 200 gal., 3/4 H.P.	810	505	1,315
1600	520 gal., 1 H.P.	835	505	1,340
1650	Alarm, electric pressure switch (circuit closer)	86	25.50	111.50
2500	Couplings, hose, rocker lug, cast brass, 1-1/2"	53		53
2550	2-1/2"	69.50		69.50
3000	Escutcheon plate, for angle valves, polished brass, 1-1/2"	16.70		16.70
3050	2-1/2"	27		27
3500	Fire pump, electric, w/controller, fittings, relief valve			
3550	4" pump, 30 H.P., 500 G.P.M.	16,600	3,700	20,300
3600	5" pump, 40 H.P., 1000 G.P.M.	18,200	4,175	22,375
3650	5" pump, 100 H.P., 1000 G.P.M.	25,100	4,650	29,750
3700	For jockey pump system, add	2,675	595	3,270
5000	Hose, per linear foot, synthetic jacket, lined,			
5100	300 lb. test, 1-1/2" diameter	3.23	.46	3.69
5150	2-1/2" diameter	5.60	.54	6.14
5200	500 lb. test, 1-1/2" diameter	3.33	.46	3.79
5250	2-1/2" diameter	5.90	.54	6.44
5500	Nozzle, plain stream, polished brass, 1-1/2" x 10"	47.50		47.50
5550	2-1/2" x 15" x 13/16" or 1-1/2"	86.50		86.50
5600	Heavy duty combination adjustable fog and straight stream w/handle 1-1/2"	465		465
5650	2-1/2" direct connection	575		575
6000	Rack, for 1-1/2" diameter hose 100 ft. long, steel	60.50	59.50	120
6050	Brass	104	59.50	163.50
6500	Reel, steel, for 50 ft. long 1-1/2" diameter hose	131	84.50	215.50
6550	For 75 ft. long 2-1/2" diameter hose	212	84.50	296.50
7050	Siamese, w/plugs & chains, polished brass, sidewalk, 4" x 2-1/2" x 2-1/2"	630	475	1,105
7100	6" x 2-1/2" x 2-1/2"	810	595	1,405
7200	Wall type, flush, 4" x 2-1/2" x 2-1/2"	515	237	752
7250	6" x 2-1/2" x 2-1/2"	720	258	978
7300	Projecting, 4" x 2-1/2" x 2-1/2"	475	237	712
7350	6" x 2-1/2" x 2-1/2"	800	258	1,058
7400	For chrome plate, add 15% mat.			
8000	Valves, angle, wheel handle, 300 Lb., rough brass, 1-1/2"	94.50	55	149.50
8050	2-1/2"	174	94	268
8100	Combination pressure restricting, 1-1/2"	85	55	140
8150	2-1/2"	174	94	268
8200	Pressure restricting, adjustable, satin brass, 1-1/2"	125	55	180
8250	2-1/2"	176	94	270
8300	Hydrolator, vent and drain, rough brass, 1-1/2"	74.50	55	129.50
8350	2-1/2"	74.50	55	129.50
8400	Cabinet assy, incls. adapter, rack, hose, and nozzle	780	360	1,140

D4090 Other Fire Protection Systems

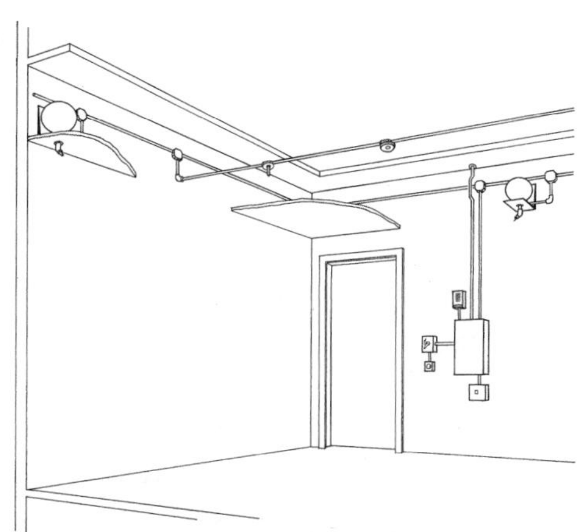

General: Automatic fire protection (suppression) systems other than water sprinklers may be desired for special environments, high risk areas, isolated locations or unusual hazards. Some typical applications would include:

Paint dip tanks
Securities vaults
Electronic data processing
Tape and data storage
Transformer rooms
Spray booths
Petroleum storage
High rack storage

Piping and wiring costs are dependent on the individual application and must be added to the component costs shown below.

All areas are assumed to be open.

D4090 910	**Fire Suppression Unit Components**	COST EACH		
		MAT.	INST.	TOTAL
0020	Detectors with brackets			
0040	Fixed temperature heat detector	55	78.50	133.50
0060	Rate of temperature rise detector	55	78.50	133.50
0080	Ion detector (smoke) detector	119	101	220
0200	Extinguisher agent			
0240	200 lb FM200, container	7,900	305	8,205
0280	75 lb carbon dioxide cylinder	1,300	202	1,502
0320	Dispersion nozzle			
0340	FM200 1-1/2" dispersion nozzle	68	48	116
0380	Carbon dioxide 3" x 5" dispersion nozzle	68	37.50	105.50
0420	Control station			
0440	Single zone control station with batteries	1,750	625	2,375
0470	Multizone (4) control station with batteries	3,350	1,250	4,600
0490				
0500	Electric mechanical release	170	325	495
0520				
0550	Manual pull station	61.50	112	173.50
0570				
0640	Battery standby power 10" x 10" x 17"	445	157	602
0700				
0740	Bell signalling device	85	78.50	163.50

D4090 920	**FM200 Systems**	COST PER C.F.		
		MAT.	INST.	TOTAL
0820	Average FM200 system, minimum			1.79
0840	Maximum			3.56

G Building Sitework

G1030 Site Earthwork

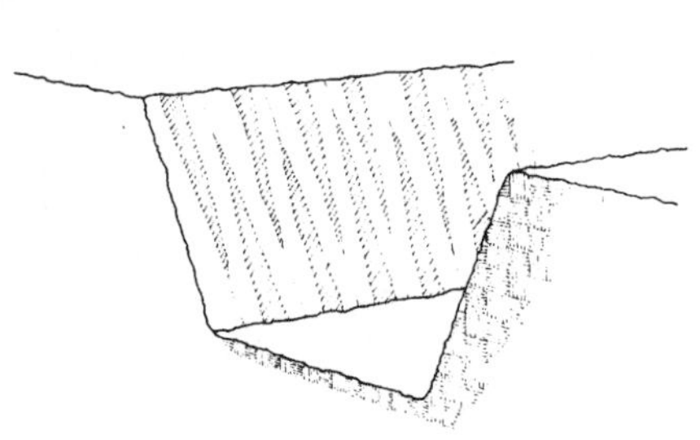

Trenching Systems are shown on a cost per linear foot basis. The systems include: excavation; backfill and removal of spoil; and compaction for various depths and trench bottom widths. The backfill has been reduced to accommodate a pipe of suitable diameter and bedding.

The slope for trench sides varies from none to 1:1.

The Expanded System Listing shows Trenching Systems that range from 2′ to 12′ in width. Depths range from 2′ to 25′.

System Components	QUANTITY	UNIT	COST PER L.F. EQUIP.	LABOR	TOTAL
SYSTEM G1030 805 1310					
TRENCHING COMMON EARTH, NO SLOPE, 2′ WIDE, 2′ DP, 3/8 C.Y. BUCKET					
Excavation, trench, hyd. backhoe, track mtd., 3/8 C.Y. bucket	.148	B.C.Y.	.40	1	1.40
Backfill and load spoil, from stockpile	.153	L.C.Y.	.12	.30	.42
Compaction by vibrating plate, 6″ lifts, 4 passes	.118	E.C.Y.	.03	.37	.40
Remove excess spoil, 8 C.Y. dump truck, 2 mile roundtrip	.040	L.C.Y.	.15	.15	.30
TOTAL			.70	1.82	2.52

G1030 805	Trenching Common Earth	COST PER L.F. EQUIP.	LABOR	TOTAL
1310	Trenching, common earth, no slope, 2′ wide, 2′ deep, 3/8 C.Y. bucket	.70	1.82	2.52
1320	3′ deep, 3/8 C.Y. bucket	1	2.75	3.75
1330	4′ deep, 3/8 C.Y. bucket	1.30	3.67	4.97
1340	6′ deep, 3/8 C.Y. bucket	1.68	4.75	6.43
1350	8′ deep, 1/2 C.Y. bucket	2.21	6.30	8.51
1360	10′ deep, 1 C.Y. bucket	3.59	7.50	11.09
1400	4′ wide, 2′ deep, 3/8 C.Y. bucket	1.62	3.58	5.20
1410	3′ deep, 3/8 C.Y. bucket	2.21	5.40	7.61
1420	4′ deep, 1/2 C.Y. bucket	2.57	6.10	8.67
1430	6′ deep, 1/2 C.Y. bucket	4.12	9.80	13.92
1440	8′ deep, 1/2 C.Y. bucket	6.80	12.65	19.45
1450	10′ deep, 1 C.Y. bucket	8.20	15.65	23.85
1460	12′ deep, 1 C.Y. bucket	10.55	20	30.55
1470	15′ deep, 1-1/2 C.Y. bucket	9.30	17.85	27.15
1480	18′ deep, 2-1/2 C.Y. bucket	12.90	25	37.90
1520	6′ wide, 6′ deep, 5/8 C.Y. bucket w/trench box	8.95	14.50	23.45
1530	8′ deep, 3/4 C.Y. bucket	11.90	19.10	31
1540	10′ deep, 1 C.Y. bucket	11.65	19.80	31.45
1550	12′ deep, 1-1/2 C.Y. bucket	12.50	21.50	34
1560	16′ deep, 2-1/2 C.Y. bucket	18	26.50	44.50
1570	20′ deep, 3-1/2 C.Y. bucket	22	31.50	53.50
1580	24′ deep, 3-1/2 C.Y. bucket	26	38	64
1640	8′ wide, 12′ deep, 1-1/2 C.Y. bucket w/trench box	17.50	27	44.50
1650	15′ deep, 1-1/2 C.Y. bucket	23	35.50	58.50
1660	18′ deep, 2-1/2 C.Y. bucket	26	35.50	61.50
1680	24′ deep, 3-1/2 C.Y. bucket	35.50	49.50	85
1730	10′ wide, 20′ deep, 3-1/2 C.Y. bucket w/trench box	28.50	47	75.50
1740	24′ deep, 3-1/2 C.Y. bucket	42.50	56.50	99
1780	12′ wide, 20′ deep, 3-1/2 C.Y. bucket w/trench box	45	59.50	104.50
1790	25′ deep, bucket	56	75.50	131.50
1800	1/2 to 1 slope, 2′ wide, 2′ deep, 3/8 C.Y. bucket	1	2.75	3.75
1810	3′ deep, 3/8 C.Y. bucket	1.68	4.82	6.50

G1030 Site Earthwork

G1030 805	Trenching Common Earth	COST PER L.F.		
		EQUIP.	LABOR	TOTAL
1820	4' deep, 3/8 C.Y. bucket	2.49	7.35	9.84
1840	6' deep, 3/8 C.Y. bucket	4.02	11.90	15.92
1860	8' deep, 1/2 C.Y. bucket	6.40	19.05	25.45
1880	10' deep, 1 C.Y. bucket	12.35	26.50	38.85
2300	4' wide, 2' deep, 3/8 C.Y. bucket	1.92	4.51	6.43
2310	3' deep, 3/8 C.Y. bucket	2.87	7.50	10.37
2320	4' deep, 1/2 C.Y. bucket	3.63	9.30	12.93
2340	6' deep, 1/2 C.Y. bucket	6.90	17.35	24.25
2360	8' deep, 1/2 C.Y. bucket	13.10	25.50	38.60
2380	10' deep, 1 C.Y. bucket	18.15	36	54.15
2400	12' deep, 1 C.Y. bucket	24	48.50	72.50
2430	15' deep, 1-1/2 C.Y. bucket	26.50	52	78.50
2460	18' deep, 2-1/2 C.Y. bucket	49	81.50	130.50
2840	6' wide, 6' deep, 5/8 C.Y. bucket w/trench box	13.25	21.50	34.75
2860	8' deep, 3/4 C.Y. bucket	19.15	32.50	51.65
2880	10' deep, 1 C.Y. bucket	18.75	33	51.75
2900	12' deep, 1-1/2 C.Y. bucket	23.50	43	66.50
2940	16' deep, 2-1/2 C.Y. bucket	41	63.50	104.50
2980	20' deep, 3-1/2 C.Y. bucket	55	85.50	140.50
3020	24' deep, 3-1/2 C.Y. bucket	77.50	117	194.50
3100	8' wide, 12' deep, 1-1/2 C.Y. bucket w/trench box	29.50	49	78.50
3120	15' deep, 1-1/2 C.Y. bucket	42.50	71.50	114
3140	18' deep, 2-1/2 C.Y. bucket	56.50	85	141.50
3180	24' deep, 3-1/2 C.Y. bucket	87	128	215
3270	10' wide, 20' deep, 3-1/2 C.Y. bucket w/trench box	55.50	98.50	154
3280	24' deep, 3-1/2 C.Y. bucket	96	140	236
3370	12' wide, 20' deep, 3-1/2 C.Y. bucket w/trench box	81	114	195
3380	25' deep, 3-1/2 C.Y. bucket	112	163	275
3500	1 to 1 slope, 2' wide, 2' deep, 3/8 C.Y. bucket	1.30	3.67	4.97
3520	3' deep, 3/8 C.Y. bucket	3.72	8.25	11.97
3540	4' deep, 3/8 C.Y. bucket	3.68	11.05	14.73
3560	6' deep, 3/8 C.Y. bucket	4.02	11.90	15.92
3580	8' deep, 1/2 C.Y. bucket	7.95	24	31.95
3600	10' deep, 1 C.Y. bucket	21	45.50	66.50
3800	4' wide, 2' deep, 3/8 C.Y. bucket	2.21	5.45	7.66
3820	3' deep, 3/8 C.Y. bucket	3.55	9.55	13.10
3840	4' deep, 1/2 C.Y. bucket	4.67	12.45	17.12
3860	6' deep, 1/2 C.Y. bucket	9.65	25	34.65
3880	8' deep, 1/2 C.Y. bucket	19.50	38.50	58
3900	10' deep, 1 C.Y. bucket	28	56	84
3920	12' deep, 1 C.Y. bucket	41.50	81.50	123
3940	15' deep, 1-1/2 C.Y. bucket	43.50	86.50	130
3960	18' deep, 2-1/2 C.Y. bucket	67	112	179
4030	6' wide, 6' deep, 5/8 C.Y. bucket w/trench box	17.35	29	46.35
4040	8' deep, 3/4 C.Y. bucket	25	40	65
4050	10' deep, 1 C.Y. bucket	27	48.50	75.50
4060	12' deep, 1-1/2 C.Y. bucket	36	66	102
4070	16' deep, 2-1/2 C.Y. bucket	64.50	100	164.50
4080	20' deep, 3-1/2 C.Y. bucket	88.50	139	227.50
4090	24' deep, 3-1/2 C.Y. bucket	129	196	325
4500	8' wide, 12' deep, 1-1/2 C.Y. bucket w/trench box	41.50	71.50	113
4550	15' deep, 1-1/2 C.Y. bucket	62.50	107	169.50
4600	18' deep, 2-1/2 C.Y. bucket	86	132	218
4650	24' deep, 3-1/2 C.Y. bucket	138	207	345
4800	10' wide, 20' deep, 3-1/2 C.Y. bucket w/trench box	82.50	150	232.50
4850	24' deep, 3-1/2 C.Y. bucket	147	219	366
4950	12' wide, 20' deep, 3-1/2 C.Y. bucket w/trench box	116	169	285
4980	25' deep, 3-1/2 C.Y. bucket	167	247	414

G1030 Site Earthwork

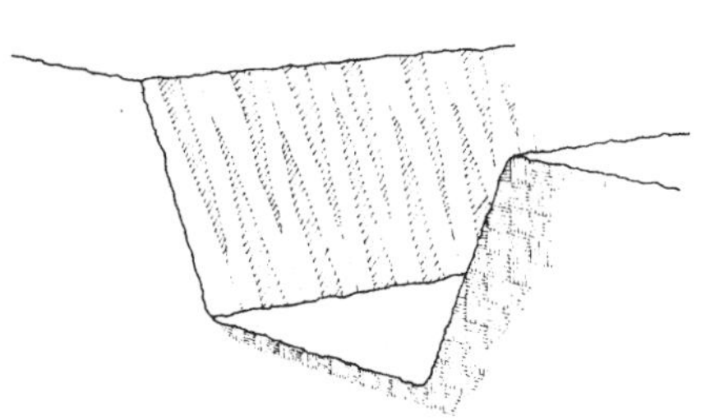

Trenching Systems are shown on a cost per linear foot basis. The systems include: excavation; backfill and removal of spoil; and compaction for various depths and trench bottom widths. The backfill has been reduced to accommodate a pipe of suitable diameter and bedding.

The slope for trench sides varies from none to 1:1.

The Expanded System Listing shows Trenching Systems that range from 2′ to 12′ in width. Depths range from 2′ to 25′.

System Components	QUANTITY	UNIT	COST PER L.F. EQUIP.	COST PER L.F. LABOR	COST PER L.F. TOTAL
SYSTEM G1030 806 1310					
TRENCHING LOAM & SANDY CLAY, NO SLOPE, 2′ WIDE, 2′ DP, 3/8 C.Y. BUCKET					
Excavation, trench, hyd. backhoe, track mtd., 3/8 C.Y. bucket	.148	B.C.Y.	.37	.93	1.30
Backfill and load spoil, from stockpile	.165	L.C.Y.	.13	.33	.46
Compaction by vibrating plate 18″ wide, 6″ lifts, 4 passes	.118	E.C.Y.	.03	.37	.40
Remove excess spoil, 8 C.Y. dump truck, 2 mile roundtrip	.042	L.C.Y.	.16	.16	.32
TOTAL			.70	1.82	2.52

G1030 806	Trenching Loam & Sandy Clay	COST PER L.F. EQUIP.	COST PER L.F. LABOR	COST PER L.F. TOTAL
1310	Trenching, loam & sandy clay, no slope, 2′ wide, 2′ deep, 3/8 C.Y. bucket	.69	1.79	2.48
1320	3′ deep, 3/8 C.Y. bucket	1.08	2.94	4.02
1330	4′ deep, 3/8 C.Y. bucket	1.27	3.58	4.85
1340	6′ deep, 3/8 C.Y. bucket	1.80	4.25	6.05
1350	8′ deep, 1/2 C.Y. bucket	2.38	5.65	8.05
1360	10′ deep, 1 C.Y. bucket	2.68	6.05	8.75
1400	4′ wide, 2′ deep, 3/8 C.Y. bucket	1.61	3.52	5.15
1410	3′ deep, 3/8 C.Y. bucket	2.19	5.30	7.50
1420	4′ deep, 1/2 C.Y. bucket	2.57	6	8.55
1430	6′ deep, 1/2 C.Y. bucket	4.39	8.75	13.15
1440	8′ deep, 1/2 C.Y. bucket	6.65	12.45	19.10
1450	10′ deep, 1 C.Y. bucket	6.40	12.75	19.15
1460	12′ deep, 1 C.Y. bucket	8.05	15.85	24
1470	15′ deep, 1-1/2 C.Y. bucket	9.70	18.45	28
1480	18′ deep, 2-1/2 C.Y. bucket	12.05	20.50	32.50
1520	6′ wide, 6′ deep, 5/8 C.Y. bucket w/trench box	8.65	14.25	23
1530	8′ deep, 3/4 C.Y. bucket	11.45	18.75	30
1540	10′ deep, 1 C.Y. bucket	10.70	19	29.50
1550	12′ deep, 1-1/2 C.Y. bucket	12.20	21.50	33.50
1560	16′ deep, 2-1/2 C.Y. bucket	17.50	27	44.50
1570	20′ deep, 3-1/2 C.Y. bucket	20	32	52
1580	24′ deep, 3-1/2 C.Y. bucket	25.50	39	64.50
1640	8′ wide, 12′ deep, 1-1/4 C.Y. bucket w/trench box	17.25	27	44.50
1650	15′ deep, 1-1/2 C.Y. bucket	21	34.50	55.50
1660	18′ deep, 2-1/2 C.Y. bucket	26.50	38.50	65
1680	24′ deep, 3-1/2 C.Y. bucket	34.50	50	84.50
1730	10′ wide, 20′ deep, 3-1/2 C.Y. bucket w/trench box	35	50	85
1740	24′ deep, 3-1/2 C.Y. bucket	43.50	62	106
1780	12′ wide, 20′ deep, 3-1/2 C.Y. bucket w/trench box	42	59.50	102
1790	25′ deep, bucket	54.50	77	132
1800	1/2:1 slope, 2′ wide, 2′ deep, 3/8 C.Y. bucket	.98	2.68	3.66
1810	3′ deep, 3/8 C.Y. bucket	1.64	4.70	6.35
1820	4′ deep, 3/8 C.Y. bucket	2.43	7.20	9.65
1840	6′ deep, 3/8 C.Y. bucket	4.32	10.65	14.95

G1030 Site Earthwork

G1030 806	Trenching Loam & Sandy Clay	COST PER L.F.		
		EQUIP.	LABOR	TOTAL
1860	8' deep, 1/2 C.Y. bucket	6.85	17	24
1880	10' deep, 1 C.Y. bucket	9.15	21.50	30.50
2300	4' wide, 2' deep, 3/8 C.Y. bucket	1.90	4.43	6.35
2310	3' deep, 3/8 C.Y. bucket	2.83	7.35	10.20
2320	4' deep, 1/2 C.Y. bucket	3.59	9.10	12.70
2340	6' deep, 1/2 C.Y. bucket	7.35	15.60	23
2360	8' deep, 1/2 C.Y. bucket	12.75	25	38
2380	10' deep, 1 C.Y. bucket	14.05	29.50	43.50
2400	12' deep, 1 C.Y. bucket	23.50	48.50	72
2430	15' deep, 1-1/2 C.Y. bucket	27.50	54	81.50
2460	18' deep, 2-1/2 C.Y. bucket	47.50	82.50	130
2840	6' wide, 6' deep, 5/8 C.Y. bucket w/trench box	12.60	21.50	34
2860	8' deep, 3/4 C.Y. bucket	18.45	32	50.50
2880	10' deep, 1 C.Y. bucket	19.10	36	55
2900	12' deep, 1-1/2 C.Y. bucket	23.50	43.50	67
2940	16' deep, 2-1/2 C.Y. bucket	40	64	104
2980	20' deep, 3-1/2 C.Y. bucket	53	86.50	140
3020	24' deep, 3-1/2 C.Y. bucket	75	119	194
3100	8' wide, 12' deep, 1-1/2 C.Y. bucket w/trench box	29	49.50	78.50
3120	15' deep, 1-1/2 C.Y. bucket	39	69.50	109
3140	18' deep, 2-1/2 C.Y. bucket	55	86	141
3180	24' deep, 3-1/2 C.Y. bucket	84.50	130	215
3270	10' wide, 20' deep, 3-1/2 C.Y. bucket w/trench box	67.50	105	173
3280	24' deep, 3-1/2 C.Y. bucket	93	142	235
3320	12' wide, 20' deep, 3-1/2 C.Y. bucket w/trench box	75	114	189
3380	25' deep, 3-1/2 C.Y. bucket w/trench box	102	153	255
3500	1:1 slope, 2' wide, 2' deep, 3/8 C.Y. bucket	1.27	3.58	4.85
3520	3' deep, 3/8 C.Y. bucket	2.28	6.75	9.05
3540	4' deep, 3/8 C.Y. bucket	3.57	10.80	14.35
3560	6' deep, 3/8 C.Y. bucket	4.32	10.65	14.95
3580	8' deep, 1/2 C.Y. bucket	11.35	28.50	40
3600	10' deep, 1 C.Y. bucket	15.60	36.50	52
3800	4' wide, 2' deep, 3/8 C.Y. bucket	2.19	5.30	7.50
3820	3' deep, 3/8 C.Y. bucket	3.49	9.35	12.85
3840	4' deep, 1/2 C.Y. bucket	4.61	12.25	16.85
3860	6' deep, 1/2 C.Y. bucket	10.30	22.50	33
3880	8' deep, 1/2 C.Y. bucket	18.95	38	57
3900	10' deep, 1 C.Y. bucket	21.50	46	67.50
3920	12' deep, 1 C.Y. bucket	31.50	65	96.50
3940	15' deep, 1-1/2 C.Y. bucket	45	89.50	135
3960	18' deep, 2-1/2 C.Y. bucket	65.50	114	180
4030	6' wide, 6' deep, 5/8 C.Y. bucket w/trench box	16.55	29	45.50
4040	8' deep, 3/4 C.Y. bucket	25.50	45	70.50
4050	10' deep, 1 C.Y. bucket	27.50	52.50	80
4060	12' deep, 1-1/2 C.Y. bucket	35	66	101
4070	16' deep, 2-1/2 C.Y. bucket	62.50	101	164
4080	20' deep, 3-1/2 C.Y. bucket	86	142	228
4090	24' deep, 3-1/2 C.Y. bucket	125	199	325
4500	8' wide, 12' deep, 1-1/4 C.Y. bucket w/trench box	40.50	71.50	112
4550	15' deep, 1-1/2 C.Y. bucket	57	104	161
4600	18' deep, 2-1/2 C.Y. bucket	83.50	133	217
4650	24' deep, 3-1/2 C.Y. bucket	134	210	345
4800	10' wide, 20' deep, 3-1/2 C.Y. bucket w/trench box	100	160	260
4850	24' deep, 3-1/2 C.Y. bucket	143	222	365
4950	12' wide, 20' deep, 3-1/2 C.Y. bucket w/trench box	108	169	277
4980	25' deep, 3-1/2 C.Y. bucket	162	250	410

G1030 Site Earthwork

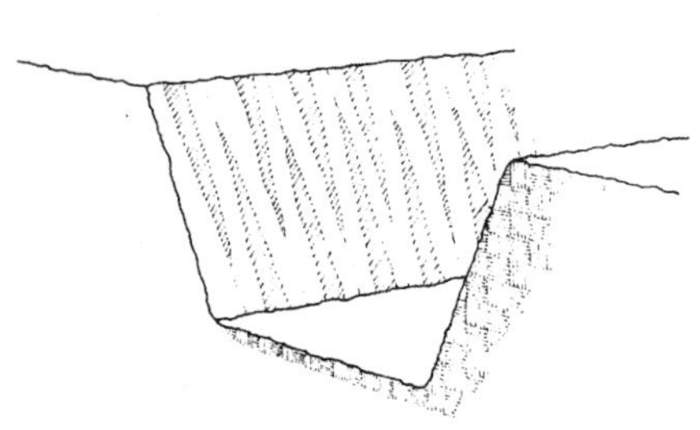

Trenching Systems are shown on a cost per linear foot basis. The systems include: excavation; backfill and removal of spoil; and compaction for various depths and trench bottom widths. The backfill has been reduced to accommodate a pipe of suitable diameter and bedding.

The slope for trench sides varies from none to 1:1.

The Expanded System Listing shows Trenching Systems that range from 2′ to 12′ in width. Depths range from 2′ to 25′.

System Components	QUANTITY	UNIT	COST PER L.F. EQUIP.	COST PER L.F. LABOR	COST PER L.F. TOTAL
SYSTEM G1030 807 1310					
TRENCHING SAND & GRAVEL, NO SLOPE, 2′ WIDE, 2′ DEEP, 3/8 C.Y. BUCKET					
Excavation, trench, hyd. backhoe, track mtd., 3/8 C.Y. bucket	.148	B.C.Y.	.36	.91	1.27
Backfill and load spoil, from stockpile	.140	L.C.Y.	.11	.28	.39
Compaction by vibrating plate 18″ wide, 6″ lifts, 4 passes	.118	E.C.Y.	.03	.37	.40
Remove excess spoil, 8 C.Y. dump truck, 2 mile roundtrip	.035	L.C.Y.	.14	.13	.27
TOTAL			.64	1.69	2.33

G1030 807	Trenching Sand & Gravel	COST PER L.F. EQUIP.	COST PER L.F. LABOR	COST PER L.F. TOTAL
1310	Trenching, sand & gravel, no slope, 2′ wide, 2′ deep, 3/8 C.Y. bucket	.64	1.69	2.33
1320	3′ deep, 3/8 C.Y. bucket	1.03	2.82	3.85
1330	4′ deep, 3/8 C.Y. bucket	1.19	3.40	4.59
1340	6′ deep, 3/8 C.Y. bucket	1.71	4.05	5.75
1350	8′ deep, 1/2 C.Y. bucket	2.25	5.35	7.60
1360	10′ deep, 1 C.Y. bucket	2.50	5.65	8.15
1400	4′ wide, 2′ deep, 3/8 C.Y. bucket	1.47	3.31	4.78
1410	3′ deep, 3/8 C.Y. bucket	2.01	5.05	7.05
1420	4′ deep, 1/2 C.Y. bucket	2.36	5.65	8
1430	6′ deep, 1/2 C.Y. bucket	4.14	8.40	12.55
1440	8′ deep, 1/2 C.Y. bucket	6.25	11.80	18.05
1450	10′ deep, 1 C.Y. bucket	6	12.05	18.05
1460	12′ deep, 1 C.Y. bucket	7.60	15	22.50
1470	15′ deep, 1-1/2 C.Y. bucket	9.15	17.40	26.50
1480	18′ deep, 2-1/2 C.Y. bucket	11.45	19.05	30.50
1520	6′ wide, 6′ deep, 5/8 C.Y. bucket w/trench box	8.20	13.50	21.50
1530	8′ deep, 3/4 C.Y. bucket	10.85	17.75	28.50
1540	10′ deep, 1 C.Y. bucket	10.10	18	28
1550	12′ deep, 1-1/2 C.Y. bucket	11.50	20	31.50
1560	16′ deep, 2 C.Y. bucket	16.75	25.50	42.50
1570	20′ deep, 3-1/2 C.Y. bucket	19.10	30	49
1580	24′ deep, 3-1/2 C.Y. bucket	24	36.50	60.50
1640	8′ wide, 12′ deep, 1-1/2 C.Y. bucket w/trench box	16.20	25.50	41.50
1650	15′ deep, 1-1/2 C.Y. bucket	19.70	32.50	52
1660	18′ deep, 2-1/2 C.Y. bucket	25	36.50	61.50
1680	24′ deep, 3-1/2 C.Y. bucket	32.50	47	79.50
1730	10′ wide, 20′ deep, 3-1/2 C.Y. bucket w/trench box	32.50	47	79.50
1740	24′ deep, 3-1/2 C.Y. bucket	41	58	99
1780	12′ wide, 20′ deep, 3-1/2 C.Y. bucket w/trench box	39.50	55.50	95
1790	25′ deep, 3-1/2 C.Y. bucket	51.50	72	124
1800	1/2:1 slope, 2′ wide, 2′ deep, 3/8 C.Y. bucket	.91	2.55	3.46
1810	3′ deep, 3/8 C.Y. bucket	1.53	4.48	6
1820	4′ deep, 3/8 C.Y. bucket	2.28	6.85	9.15
1840	6′ deep, 3/8 C.Y. bucket	4.10	10.15	14.25

G1030 Site Earthwork

G1030 807	Trenching Sand & Gravel	COST PER L.F.		
		EQUIP.	LABOR	TOTAL
1860	8' deep, 1/2 C.Y. bucket	6.50	16.25	23
1880	10' deep, 1 C.Y. bucket	8.55	20	28.50
2300	4' wide, 2' deep, 3/8 C.Y. bucket	1.74	4.17	5.90
2310	3' deep, 3/8 C.Y. bucket	2.62	6.95	9.55
2320	4' deep, 1/2 C.Y. bucket	3.31	8.65	11.95
2340	6' deep, 1/2 C.Y. bucket	6.95	14.95	22
2360	8' deep, 1/2 C.Y. bucket	12.10	24	36
2380	10' deep, 1 C.Y. bucket	13.25	27.50	41
2400	12' deep, 1 C.Y. bucket	22.50	46	68.50
2430	15' deep, 1-1/2 C.Y. bucket	26	51	77
2460	18' deep, 2-1/2 C.Y. bucket	45	77.50	123
2840	6' wide, 6' deep, 5/8 C.Y. bucket w/trench box	11.95	20.50	32.50
2860	8' deep, 3/4 C.Y. bucket	17.50	30	47.50
2880	10' deep, 1 C.Y. bucket	18.05	34	52
2900	12' deep, 1-1/2 C.Y. bucket	22.50	41	63.50
2940	16' deep, 2 C.Y. bucket	38.50	60.50	99
2980	20' deep, 3-1/2 C.Y. bucket	50.50	81.50	132
3020	24' deep, 3-1/2 C.Y. bucket	71.50	112	184
3100	8' wide, 12' deep, 1-1/4 C.Y. bucket w/trench box	27.50	46.50	74
3120	15' deep, 1-1/2 C.Y. bucket	37	65.50	103
3140	18' deep, 2-1/2 C.Y. bucket	52.50	81	134
3180	24' deep, 3-1/2 C.Y. bucket	80	122	202
3270	10' wide, 20' deep, 3-1/2 C.Y. bucket w/trench box	64	98.50	163
3280	24' deep, 3-1/2 C.Y. bucket	88.50	133	222
3370	12' wide, 20' deep, 3-1/2 C.Y. bucket w/trench box	71.50	109	181
3380	25' deep, 3-1/2 C.Y. bucket	103	154	257
3500	1:1 slope, 2' wide, 2' deep, 3/8 C.Y. bucket	2.04	4.18	6.20
3520	3' deep, 3/8 C.Y. bucket	2.14	6.40	8.55
3540	4' deep, 3/8 C.Y. bucket	3.37	10.30	13.65
3560	6' deep, 3/8 C.Y. bucket	4.10	10.15	14.25
3580	8' deep, 1/2 C.Y. bucket	10.80	27	38
3600	10' deep, 1 C.Y. bucket	14.60	34.50	49
3800	4' wide, 2' deep, 3/8 C.Y. bucket	2.01	5.05	7.05
3820	3' deep, 3/8 C.Y. bucket	3.24	8.90	12.15
3840	4' deep, 1/2 C.Y. bucket	4.26	11.65	15.90
3860	6' deep, 1/2 C.Y. bucket	9.80	21.50	31.50
3880	8' deep, 1/2 C.Y. bucket	17.95	36	54
3900	10' deep, 1 C.Y. bucket	20.50	43.50	64
3920	12' deep, 1 C.Y. bucket	29.50	61.50	91
3940	15' deep, 1-1/2 C.Y. bucket	42.50	84.50	127
3960	18' deep, 2-1/2 C.Y. bucket	62	107	169
4030	6' wide, 6' deep, 5/8 C.Y. bucket w/trench box	15.70	27.50	43
4040	8' deep, 3/4 C.Y. bucket	24	42.50	66.50
4050	10' deep, 1 C.Y. bucket	26	49.50	75.50
4060	12' deep, 1-1/2 C.Y. bucket	33.50	62.50	96
4070	16' deep, 2 C.Y. bucket	60	96	156
4080	20' deep, 3-1/2 C.Y. bucket	81.50	133	215
4090	24' deep, 3-1/2 C.Y. bucket	119	187	305
4500	8' wide, 12' deep, 1-1/2 C.Y. bucket w/trench box	38.50	68	107
4550	15' deep, 1-1/2 C.Y. bucket	54	98.50	153
4600	18' deep, 2-1/2 C.Y. bucket	79.50	125	205
4650	24' deep, 3-1/2 C.Y. bucket	127	197	325
4800	10' wide, 20' deep, 3-1/2 C.Y. bucket w/trench box	95	150	245
4850	24' deep, 3-1/2 C.Y. bucket	136	208	345
4950	12' wide, 20' deep, 3-1/2 C.Y. bucket w/trench box	102	159	261
4980	25' deep, 3-1/2 C.Y. bucket	154	235	390

G1030 Site Earthwork

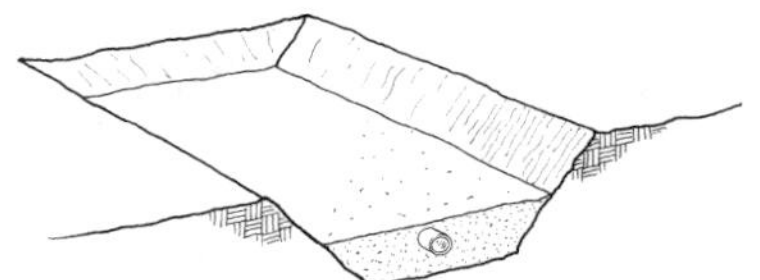

The Pipe Bedding System is shown for various pipe diameters. Compacted bank sand is used for pipe bedding and to fill 12″ over the pipe. No backfill is included. Various side slopes are shown to accommodate different soil conditions. Pipe sizes vary from 6″ to 84″ diameter.

System Components	QUANTITY	UNIT	COST PER L.F. MAT.	INST.	TOTAL
SYSTEM G1030 815 1440					
PIPE BEDDING, SIDE SLOPE 0 TO 1, 1′ WIDE, PIPE SIZE 6″ DIAMETER					
Borrow, bank sand, 2 mile haul, machine spread	.086	C.Y.	1.69	.67	2.36
Compaction, vibrating plate	.086	C.Y.		.21	.21
TOTAL			1.69	.88	2.57

G1030 815	Pipe Bedding	COST PER L.F. MAT.	INST.	TOTAL
1440	Pipe bedding, side slope 0 to 1, 1′ wide, pipe size 6″ diameter	1.69	.88	2.57
1460	2′ wide, pipe size 8″ diameter	3.66	1.88	5.54
1480	Pipe size 10″ diameter	3.74	1.92	5.66
1500	Pipe size 12″ diameter	3.82	1.96	5.78
1520	3′ wide, pipe size 14″ diameter	6.20	3.18	9.38
1540	Pipe size 15″ diameter	6.25	3.21	9.46
1560	Pipe size 16″ diameter	6.30	3.25	9.55
1580	Pipe size 18″ diameter	6.45	3.31	9.76
1600	4′ wide, pipe size 20″ diameter	9.15	4.70	13.85
1620	Pipe size 21″ diameter	9.25	4.74	13.99
1640	Pipe size 24″ diameter	9.45	4.84	14.29
1660	Pipe size 30″ diameter	9.60	4.94	14.54
1680	6′ wide, pipe size 32″ diameter	16.45	8.45	24.90
1700	Pipe size 36″ diameter	16.80	8.65	25.45
1720	7′ wide, pipe size 48″ diameter	26.50	13.75	40.25
1740	8′ wide, pipe size 60″ diameter	32.50	16.70	49.20
1760	10′ wide, pipe size 72″ diameter	45.50	23.50	69
1780	12′ wide, pipe size 84″ diameter	60	30.50	90.50
2140	Side slope 1/2 to 1, 1′ wide, pipe size 6″ diameter	3.16	1.62	4.78
2160	2′ wide, pipe size 8″ diameter	5.35	2.75	8.10
2180	Pipe size 10″ diameter	5.75	2.96	8.71
2200	Pipe size 12″ diameter	6.10	3.13	9.23
2220	3′ wide, pipe size 14″ diameter	8.80	4.52	13.32
2240	Pipe size 15″ diameter	9	4.63	13.63
2260	Pipe size 16″ diameter	9.25	4.76	14.01
2280	Pipe size 18″ diameter	9.70	4.99	14.69
2300	4′ wide, pipe size 20″ diameter	12.85	6.60	19.45
2320	Pipe size 21″ diameter	13.10	6.75	19.85
2340	Pipe size 24″ diameter	13.95	7.15	21.10
2360	Pipe size 30″ diameter	15.45	7.95	23.40
2380	6′ wide, pipe size 32″ diameter	23	11.70	34.70
2400	Pipe size 36″ diameter	24.50	12.45	36.95
2420	7′ wide, pipe size 48″ diameter	38	19.45	57.45
2440	8′ wide, pipe size 60″ diameter	48	24.50	72.50
2460	10′ wide, pipe size 72″ diameter	66	34	100
2480	12′ wide, pipe size 84″ diameter	86.50	44.50	131
2620	Side slope 1 to 1, 1′ wide, pipe size 6″ diameter	4.62	2.37	6.99
2640	2′ wide, pipe size 8″ diameter	7.15	3.66	10.81

G1030 Site Earthwork

G1030 815	Pipe Bedding	COST PER L.F.		
		MAT.	INST.	TOTAL
2660	Pipe size 10″ diameter	7.75	3.97	11.72
2680	Pipe size 12″ diameter	8.45	4.33	12.78
2700	3′ wide, pipe size 14″ diameter	11.40	5.85	17.25
2720	Pipe size 15″ diameter	11.80	6.05	17.85
2740	Pipe size 16″ diameter	12.20	6.25	18.45
2760	Pipe size 18″ diameter	13	6.65	19.65
2780	4′ wide, pipe size 20″ diameter	16.50	8.50	25
2800	Pipe size 21″ diameter	16.95	8.70	25.65
2820	Pipe size 24″ diameter	18.40	9.45	27.85
2840	Pipe size 30″ diameter	21.50	10.90	32.40
2860	6′ wide, pipe size 32″ diameter	29	15.05	44.05
2880	Pipe size 36″ diameter	31.50	16.25	47.75
2900	7′ wide, pipe size 48″ diameter	49	25	74
2920	8′ wide, pipe size 60″ diameter	63.50	32.50	96
2940	10′ wide, pipe size 72″ diameter	86.50	44.50	131
2960	12′ wide, pipe size 84″ diameter	113	58	171

G2040 Site Development

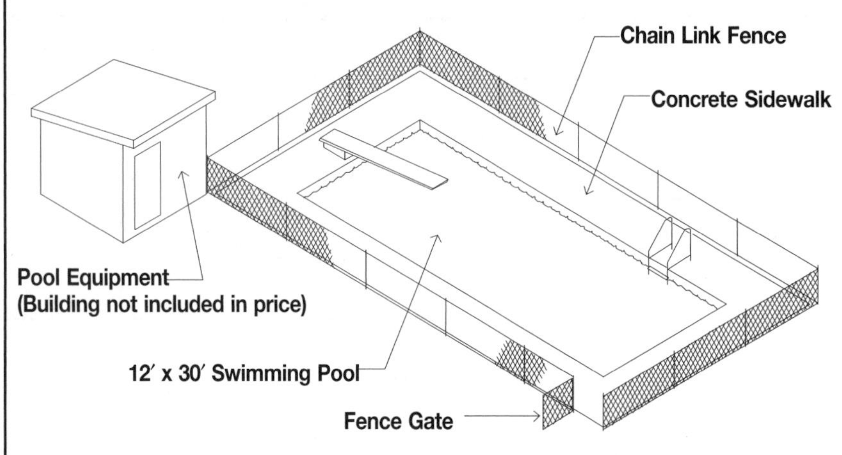

The Swimming Pool System is a complete package. Everything from excavation to deck hardware is included in system costs. Below are three basic types of pool systems: residential, motel, and municipal. Systems elements include: excavation, pool materials, installation, deck hardware, pumps and filters, sidewalk, and fencing.
The Expanded System Listing shows three basic types of pools with a variety of finishes and basic materials. These systems are either vinyl lined with metal sides; gunite shell with a cement plaster finish or tile finish; or concrete sided with vinyl lining. Pool sizes listed here vary from 12′ x 30′ to 60′ x 82.5′. All costs are on a per unit basis.

System Components	QUANTITY	UNIT	COST EACH MAT.	COST EACH INST.	COST EACH TOTAL
SYSTEM G2040 920 1000					
SWIMMING POOL, RESIDENTIAL, CONC. SIDES, VINYL LINED, 12′X 30′					
Swimming pool, residential, in ground including equipment	360.000	S.F.	7,740	5,050.80	12,790.80
4″ thick reinforced concrete sidewalk, broom finish, no base	400.000	S.F.	712	1,000	1,712
Chain link fence, residential, 3′ high	124.000	L.F.	286.44	399.28	685.72
Fence gate, chain link	1.000	Ea.	91.50	134	225.50
TOTAL			8,829.94	6,584.08	15,414.02

G2040 920	Swimming Pools	COST EACH MAT.	COST EACH INST.	COST EACH TOTAL
1000	Swimming pool, residential class, concrete sides, vinyl lined, 12′ x 30′	8,825	6,575	15,400
1100	16′ x 32′	10,000	7,450	17,450
1200	20′ x 40′	12,900	9,525	22,425
1500	Tile finish, 12′ x 30′	19,900	27,700	47,600
1600	16′ x 32′	20,300	29,100	49,400
1700	20′ x 40′	30,500	41,100	71,600
2000	Metal sides, vinyl lined, 12′ x 30′	7,500	4,325	11,825
2100	16′ x 32′	8,500	4,875	13,375
2200	20′ x 40′	11,000	6,150	17,150
3000	Gunite shell, cement plaster finish, 12′ x 30′	15,300	12,000	27,300
3100	16′ x 32′	19,400	15,700	35,100
3200	20′ x 40′	27,000	15,400	42,400
4000	Motel class, concrete sides, vinyl lined, 20′ x 40′	18,600	13,300	31,900
4100	28′ x 60′	26,100	18,500	44,600
4500	Tile finish, 20′ x 40′	35,800	47,600	83,400
4600	28′ x 60′	57,000	77,000	134,000
5000	Metal sides, vinyl lined, 20′ x 40′	19,200	9,725	28,925
5100	28′ x 60′	25,800	13,100	38,900
6000	Gunite shell, cement plaster finish, 20′ x 40′	49,900	38,700	88,600
6100	28′ x 60′	74,500	58,000	132,500
7000	Municipal class, gunite shell, cement plaster finish, 42′ x 75′	201,000	136,000	337,000
7100	60′ x 82.5′	258,000	174,000	432,000
7500	Concrete walls, tile finish, 42′ x 75′	228,000	181,000	409,000
7600	60′ x 82.5′	297,000	239,000	536,000
7700	Tile finish and concrete gutter, 42′ x 75′	249,500	181,000	430,500
7800	60′ x 82.5′	323,500	239,000	562,500
7900	Tile finish and stainless gutter, 42′ x 75′	292,000	181,000	473,000
8000	60′x 82.5′	432,000	277,000	709,000

G2050 Landscaping

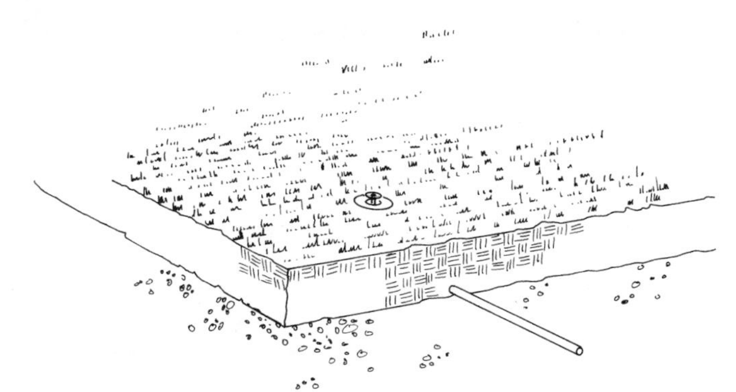

There are three basic types of Site Irrigation Systems: pop-up, riser mounted and quick coupling. Sprinkler heads are spray, impact or gear driven. Each system includes: the hardware for spraying the water; the pipe and fittings needed to deliver the water; and all other accessory equipment such as valves, couplings, nipples, and nozzles. Excavation heads and backfill costs are also included in the system.
The Expanded System Listing shows a wide variety of Site Irrigation Systems.

System Components	QUANTITY	UNIT	COST PER S.F.		
			MAT.	INST.	TOTAL
SYSTEM G2050 710 1000					
SITE IRRIGATION, POP UP SPRAY, PLASTIC, 10′ RADIUS, 1000 S.F., PVC PIPE					
Excavation, chain trencher	54.000	L.F.		42.66	42.66
Pipe, fittings & nipples, PVC Schedule 40, 1″ diameter	5.000	Ea.	4	105	109
Fittings, bends or elbows, 4″ diameter	5.000	Ea.	52.25	237.50	289.75
Couplings PVC plastic, high pressure, 1″ diameter	5.000	Ea.	4	135	139
Valves, bronze, globe, 125 lb. rising stem, threaded, 1″ diameter	1.000	Ea.	219	35.50	254.50
Head & nozzle, pop-up spray, PVC plastic	5.000	Ea.	17.50	95.25	112.75
Backfill by hand with compaction	54.000	L.F.		33.48	33.48
Total cost per 1,000 S.F.			296.75	684.39	981.14
Total cost per S.F.			.30	.68	.98

G2050 710	Site Irrigation	COST PER S.F.		
		MAT.	INST.	TOTAL
1000	Site irrigation, pop up spray, 10′ radius, 1000 S.F., PVC pipe	.30	.69	.99
1100	Polyethylene pipe	.33	.61	.94
1200	14′ radius, 8000 S.F., PVC pipe	.09	.29	.38
1300	Polyethylene pipe	.09	.27	.36
1400	18′ square, 1000 S.F. PVC pipe	.29	.47	.76
1500	Polyethylene pipe	.30	.41	.71
1600	24′ square, 8000 S.F., PVC pipe	.08	.18	.26
1700	Polyethylene pipe	.08	.17	.25
1800	4′ x 30′ strip, 200 S.F., PVC pipe	1.39	1.95	3.34
1900	Polyethylene pipe	1.41	1.81	3.22
2000	6′ x 40′ strip, 800 S.F., PVC pipe	.37	.65	1.02
2200	Economy brass, 11′ radius, 1000 S.F., PVC pipe	.36	.68	1.04
2300	Polyethylene pipe	.37	.61	.98
2400	14′ radius, 8000 S.F., PVC pipe	.11	.29	.40
2500	Polyethylene pipe	.11	.27	.38
2600	3′ x 28′ strip, 200 S.F., PVC pipe	1.53	1.95	3.48
2700	Polyethylene pipe	1.55	1.81	3.36
2800	7′ x 36′ strip, 1000 S.F., PVC pipe	.33	.52	.85
2900	Polyethylene pipe	.33	.48	.81
3000	Hd brass, 11′ radius, 1000 S.F., PVC pipe	.41	.68	1.09
3100	Polyethylene pipe	.42	.61	1.03
3200	14′ radius, 8000 S.F., PVC pipe	.13	.29	.42
3300	Polyethylene pipe	.13	.27	.40
3400	Riser mounted spray, plastic,10′radius, 1000S.F., PVC pipe	.32	.68	1
3500	Polyethylene pipe	.33	.61	.94
3600	12′ radius, 5000 S.F., PVC pipe	.15	.38	.53
3700	Polyethylene pipe	.15	.35	.50
3800	19′ square, 2000 S.F., PVC pipe	.15	.21	.36

G2050 Landscaping

G2050 710	Site Irrigation	COST PER S.F.		
		MAT.	INST.	TOTAL
3900	Polyethylene pipe	.15	.19	.34
4000	24′ square, 8000 S.F., PVC pipe	.08	.18	.26
4100	Polyethylene pipe	.08	.17	.25
4200	5′ x 32′ strip, 300 S.F., PVC pipe	.93	1.31	2.24
4300	Polyethylene pipe	.94	1.22	2.16
4400	6′ x 40′ strip, 800 S.F., PVC pipe	.37	.65	1.02
4500	Polyethylene pipe	.38	.59	.97
4600	Brass, 11′ radius, 1000 S.F., PVC pipe	.36	.68	1.04
4700	Polyethylene pipe	.37	.61	.98
4800	14′ radius, 8000 S.F., PVC pipe	.11	.29	.40
4900	Polythylene pipe	.11	.27	.38
5000	Pop up gear drive stream type,plastic, 30′radius, 10,000 S.F.,PVC pipe	.25	.57	.82
5020	Polyethylene pipe	.13	.17	.30
5040	40,000 S.F., PVC pipe	.18	.58	.76
5060	Polyethylene pipe	.08	.17	.25
5080	Riser mounted gear drive stream type,plas.,30′rad.,10,000S.F.,PVC pipe	.24	.57	.81
5100	Polyethylene pipe	.12	.17	.29
5120	40,000 S.F., PVC pipe	.18	.58	.76
5140	Polyethylene pipe	.08	.17	.25
5200	Q.C. valve thread type w/impact head,brass,75′rad,20,000S.F.,PVC pipe	.10	.21	.31
5250	Polyethylene pipe	.06	.05	.11
5300	100,000 S.F., PVC pipe	.13	.35	.48
5350	Polyethylene pipe	.06	.07	.13
5400	Q.C. valve lug type w/impact head,brass, 75′rad, 20,000 S.F.,PVC pipe	.10	.21	.31
5450	Polyethylene pipe	.06	.05	.11
5500	100,000 S.F. PVC pipe	.13	.35	.48
5550	Polyethylene pipe	.06	.07	.13
6000	Site irrigation,pop up impact type,plastic,40′rad.,10000 S.F.,PVC pipe	.15	.36	.51
6100	Polyethylene pipe	.07	.10	.17
6200	45,000 S.F., PVC pipe	.13	.40	.53
6300	Polyethylene pipe	.05	.10	.15
6400	High medium volume brass, 40′ radius, 10000 S.F., PVC pipe	.16	.36	.52
6500	Polyethylene pipe	.08	.10	.18
6600	45,000 S.F. PVC pipe	.14	.40	.54
6700	Polyethylene pipe	.06	.10	.16
6800	High volume brass, 60′ radius, 25,000 S.F., PVC pipe	.11	.28	.39
6900	Polyethylene pipe	.05	.06	.11
7000	100,000 S.F., PVC pipe	.11	.29	.40
7100	Polyethylene pipe	.06	.07	.13
7200	Riser mounted part/full impact type,plas.,40′rad.,10,000 S.F.,PVC pipe	.14	.36	.50
7300	Polyethylene pipe	.07	.20	.27
7400	45,000 S.F., PVC pipe	.12	.40	.52
7500	Polyethylene	.04	.10	.14
7600	Low medium volume brass, 40′ radius, 10,000 S.F., PVC pipe	.15	.36	.51
7700	Polyethylene pipe	.07	.10	.17
7800	45,000 S.F., PVC pipe	.13	.40	.53
7900	Polyethylene pipe	.05	.10	.15
8000	Medium volume brass, 50′ radius, 30,000 S.F., PVC pipe	.12	.30	.42
8100	Polyethylene pipe	.05	.08	.13
8200	70,000 S.F., PVC pipe	.16	.49	.65
8300	Polyethylene pipe	.07	.11	.18
8400	Riser mounted full only impact type,plas.,40′rad.,10000 S.F., PVC pipe	.14	.36	.50
8500	Polyethylene pipe	.06	.10	.16
8600	45,000 S.F., PVC pipe	.12	.40	.52
8700	Polyethylene pipe	.04	.10	.14
8800	Low medium volume brass, 40′ radius, 10,000 S.F., PVC pipe	.15	.36	.51
8900	Polyethylene pipe	.07	.10	.17
9000	45,000 S.F., PVC pipe	.13	.40	.53

G2050 Landscaping

G2050 710	Site Irrigation	COST PER S.F.		
		MAT.	INST.	TOTAL
9100	Polyethylene pipe	.05	.10	.15
9200	High volume brass, 80′ radius, 75,000 S.F., PVC pipe	.08	.20	.28
9300	Polyethylene pipe	.01	.02	.03
9400	150,000 S.F., PVC pipe	.08	.19	.27
9500	Polyethylene pipe	.01	.02	.03
9600	Very high volume brass, 100′ radius, 100,000 S.F., PVC pipe	.07	.15	.22
9700	Polyethylene pipe	.01	.02	.03
9800	200,000 S.F., PVC pipe	.07	.15	.22
9900	Polyethylene pipe	.01	.02	.03

G2050 Landscaping

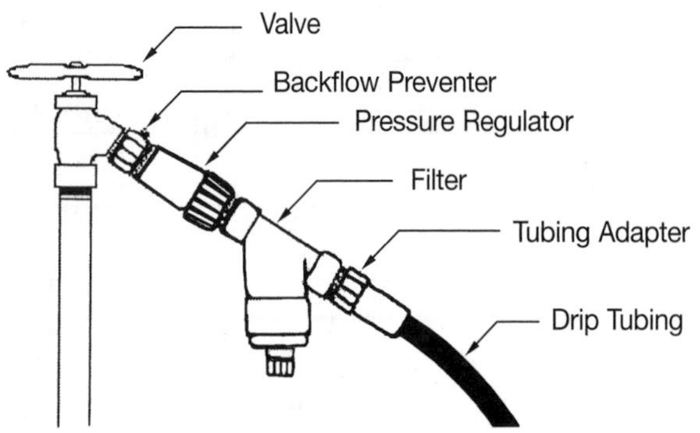

Drip irrigation systems can be for either lawns or individual plants. Shown below are drip irrigation systems for lawns. These drip type systems are most cost effective where there are water restrictions on use or the cost of water is very high. The supply uses PVC piping with the laterals polyethylene drip lines.

System Components	QUANTITY	UNIT	COST PER S.F. MAT.	INST.	TOTAL
SYSTEM G2050 720 1000					
SITE IRRIGATION, DRIP SYSTEM, 800 S.F. (20′ X40′)					
Excavate supply and header lines 4″ x 18″ deep	50.000	L.F.		42	42
Excavate drip lines 4″ x 4″ deep	5.000	L.F.		705.60	705.60
Install supply manifold (stop valve, strainer, PRV, adapters)	1.000	Ea.	18.60	143	161.60
Install 3/4″ PVC supply line and risers to 4 ″ drip level	75.000	L.F.	29.25	143.25	172.50
Install 3/4″ PVC fittings for supply line	33.000	Ea.	61.71	471.90	533.61
Install PVC to drip line adapters	14.000	Ea.	15.12	200.20	215.32
Install 1/4 inch drip lines with hose clamps	560.000	L.F.	106.40	268.80	375.20
Backfill trenches	3.180	C.Y.		125.61	125.61
Cleanup area when completed	1.000	Ea.		109	109
Total cost per 800 S.F.			231.08	2,209.36	2,440.44
Total cost per S.F.			.29	2.76	3.05

G2050 720	Site Irrigation	COST PER S.F. MAT.	INST.	TOTAL
1000	Site irrigation, drip lawn watering system, 800 S.F. (20′ x 40′ area)	.29	2.78	3.07
1100	1000 S.F. (20′ x 50′ area)	.26	2.49	2.75
1200	1600 S.F. (40′ x 40′ area), 2 zone	.30	2.65	2.95
1300	2000 S.F. (40′ x 50′ area), 2 zone	.27	2.39	2.66
1400	2400 S.F. (60′ x 40′ area), 3 zone	.30	2.63	2.93
1500	3000 S.F. (60′ x 50′ area), 3 zone	.28	2.35	2.63
1600	3200 S.F. (2 x 40′ x 40′), 4 zone	.30	2.60	2.90
1700	4000 S.F. (2 x 40′ x 50′), 4 zone	.26	2.36	2.62
1800	4800 S.F. (2 x 60′ x 40′), 6 zone with control	.30	2.60	2.90
1900	6000 S.F. (2 x 60′ x 50′), 6 zone with control	.28	2.32	2.60
2000	6400 S.F. (4 x 40′ x 40′), 8 zone with control	.31	2.61	2.92
2100	8000 S.F. (4 x 40′ x 50′), 8 zone with control	.27	2.38	2.65
2200	3200 S.F. (2 x 40′ x 40′), 4 zone with control	.32	2.65	2.97
2300	4000 S.F. (2 x 40′ x 50′), 4 zone with control	.27	2.41	2.68
2400	2400 S.F. (60′ x 40′), 3 zone with control	.32	2.72	3.04
2500	3000 S.F. (60′ x 50′), 3 zone with control	.30	2.42	2.72
2600	4800 S.F. (2 x 60′ x 40′), 6 zone, manual	.29	2.57	2.86
2700	6000 S.F. (2 x 60′ x 50′), 6 zone, manual	.27	2.30	2.57
2800	1000 S.F. (10′ x 100′ area)	.17	1.53	1.70
2900	2000 S.F. (2 x 10′ x 100′ area)	.18	1.47	1.65

G3010 Water Supply

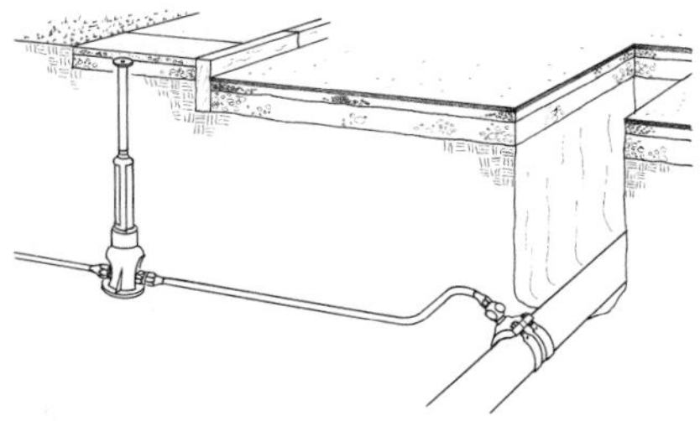

The Water Service Systems are for copper service taps from 1″ to 2″ diameter into pressurized mains from 6″ to 8″ diameter. Costs are given for two offsets with depths varying from 2′ to 10′. Included in system components are excavation and backfill and required curb stops with boxes.

System Components	QUANTITY	UNIT	COST EACH MAT.	COST EACH INST.	COST EACH TOTAL
SYSTEM G3010 120 1000					
WATER SERVICE, 6″ MAIN, 1″ COPPER SERVICE, 10′ OFFSET, 2′ DEEP					
Trench excavation, 1/2 C.Y. backhoe, 1 laborer	2.220	C.Y.		16.09	16.09
Drill & tap pressurized main, 6″, 1″ to 2″ service	1.000	Ea.		405	405
Corporation stop, 1″ diameter	1.000	Ea.	49.50	42	91.50
Saddles, cast iron, 3/4″ & 1″, add	1.000	Ea.	98.50		98.50
Copper tubing, type K, 1″ diameter	11.000	L.F.	140.25	41.58	181.83
Curb stop, brass, 1″ diameter	1.000	Ea.	79.50	42	121.50
Curb box, cast iron, 1″ diameter	1.000	Ea.	47.50	56	103.50
Backfill by hand, no compaction, heavy soil	2.220	C.Y.		5.57	5.57
Compaction, vibrating plate	2.220	C.Y.		17.03	17.03
TOTAL			415.25	625.27	1,040.52

G3010 120	Water Service	COST EACH MAT.	COST EACH INST.	COST EACH TOTAL
1000	Water service, 6″ main, 1″ copper service, 10′ offset, 2′ deep	415	625	1,040
1040	4′ deep	415	690	1,105
1060	6′ deep	415	780	1,195
1080	8′ deep	415	900	1,315
1100	10′ deep	415	955	1,370
1200	20′ offset, 2′ deep	545	705	1,250
1240	4′ deep	545	830	1,375
1260	6′ deep	545	1,000	1,545
1280	8′ deep	545	1,250	1,795
1300	10′ deep	545	1,550	2,095
2000	1-1/2″ copper service, 10′ offset, 2′ deep	775	680	1,455
2040	4′ deep	775	745	1,520
2060	6′ deep	775	835	1,610
2080	8′ deep	775	955	1,730
2100	10′ deep	775	1,100	1,875
2200	20′ offset, 2′ deep	980	765	1,745
2240	4′ deep	980	895	1,875
2260	6′ deep	980	1,075	2,055
2280	8′ deep	980	1,300	2,280
2300	10′ deep	980	1,600	2,580
3000	6″ main, 2″ copper service, 10′ offset, 2′ deep	1,075	710	1,785
3040	4′ deep	1,075	775	1,850
3060	6′ deep	1,075	860	1,935
3080	8′ deep	1,075	980	2,055
3100	10′ deep	1,075	1,125	2,200
3200	20′ offset, 2′ deep	1,375	800	2,175
3240	4′ deep	1,375	930	2,305
3260	6′ deep	1,375	1,125	2,500

G3010 Water Supply

G3010 120	Water Service	COST EACH		
		MAT.	INST.	TOTAL
3280	8′ deep	1,375	1,350	2,725
3300	10′ deep	1,375	1,650	3,025
4000	8″ main, 1″ copper service, 10′ offset, 2′ deep	415	660	1,075
4040	4′ deep	415	725	1,140
4060	6′ deep	415	815	1,230
4080	8′ deep	415	935	1,350
4100	10′ deep	415	1,075	1,490
4200	20′ offset, 2′ deep	545	740	1,285
4240	4′ deep	545	865	1,410
4260	6′ deep	545	1,050	1,595
4280	8′ deep	545	1,275	1,820
4300	10′ deep	545	1,575	2,120
5000	1-1/2″ copper service, 10′ offset, 2′ deep	775	715	1,490
5040	4′ deep	775	780	1,555
5060	6′ deep	775	870	1,645
5080	8′ deep	775	990	1,765
5100	10′ deep	775	1,125	1,900
5200	20′ offset, 2′ deep	980	800	1,780
5240	4′ deep	980	930	1,910
5260	6′ deep	980	1,125	2,105
5280	8′ deep	980	1,350	2,330
5300	10′ deep	980	1,650	2,630
6000	2″ copper service, 10′ offset, 2′ deep	1,075	745	1,820
6040	4′ deep	1,075	810	1,885
6060	6′ deep	1,075	895	1,970
6080	8′ deep	1,075	1,025	2,100
6100	10′ deep	1,075	1,150	2,225
6200	20′ offset, 2′ deep	1,375	835	2,210
6240	4′ deep	1,375	965	2,340
6260	6′ deep	1,375	1,150	2,525
6280	8′ deep	1,375	1,375	2,750
6300	10′ deep	1,375	1,675	3,050

G3010 Water Supply

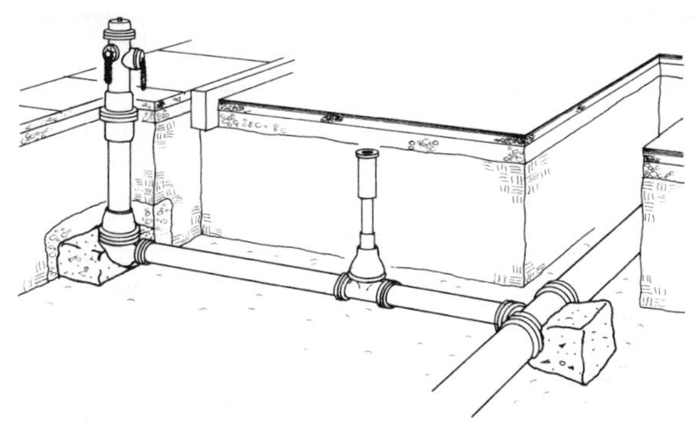

The Fire Hydrant Systems include: four different hydrants with three different lengths of offsets, at several depths of trenching. Excavation and backfill is included with each system as well as thrust blocks as necessary and pipe bedding. Finally spreading of excess material and fine grading complete the components.

System Components	QUANTITY	UNIT	COST PER EACH MAT.	INST.	TOTAL
SYSTEM G3010 410 1500					
HYDRANT, 4-1/2″ VALVE SIZE, TWO WAY, 10′ OFFSET, 2′ DEEP					
Excavation, trench, 1 C.Y. hydraulic backhoe	3.556	C.Y.		28.31	28.31
Sleeve, 12″ x 6″	1.000	Ea.	1,250	106.40	1,356.40
Ductile iron pipe, 6″ diameter	8.000	L.F.	145.60	160.72	306.32
Gate valve, 6″	1.000	Ea.	1,850	305.50	2,155.50
Hydrant valve box, 6′ long	1.000	Ea.	176	73.50	249.50
4-1/2″ valve size, depth 2′-0″	1.000	Ea.	1,675	191.10	1,866.10
Thrust blocks, at valve, shoe and sleeve	1.600	C.Y.	284.80	426.05	710.85
Borrow, crushed stone, 3/8″	4.089	C.Y.	120	31.20	151.20
Backfill and compact, dozer, air tamped	4.089	C.Y.		67.22	67.22
Spread excess excavated material	.204	C.Y.		8.06	8.06
Fine grade, hand	400.000	S.F.		255.50	255.50
TOTAL			5,501.40	1,653.56	7,154.96

G3010 410	Fire Hydrants	COST PER EACH MAT.	INST.	TOTAL
1000	Hydrant, 4-1/2″ valve size, two way, 0′ offset, 2′ deep	5,350	1,150	6,500
1100	4′ deep	5,650	1,175	6,825
1200	6′ deep	5,800	1,225	7,025
1300	8′ deep	6,025	1,275	7,300
1400	10′ deep	6,175	1,325	7,500
1500	10′ offset, 2′ deep	5,500	1,650	7,150
1600	4′ deep	5,800	1,950	7,750
1700	6′ deep	5,950	2,450	8,400
1800	8′ deep	6,175	3,225	9,400
1900	10′ deep	6,325	4,850	11,175
2000	20′ offset, 2′ deep	5,675	2,200	7,875
2100	4′ deep	5,975	2,650	8,625
2200	6′ deep	6,125	3,425	9,550
2300	8′ deep	6,350	4,550	10,900
2400	10′ deep	6,500	6,075	12,575
2500	Hydrant, 4-1/2″ valve size, three way, 0′ offset, 2′ deep	5,475	1,150	6,625
2600	4′ deep	5,800	1,175	6,975
2700	6′ deep	5,950	1,250	7,200
2800	8′ deep	6,175	1,300	7,475
2900	10′ deep	6,350	1,350	7,700
3000	10′ offset, 2′ deep	5,625	1,675	7,300
3100	4′ deep	5,950	1,950	7,900
3200	6′ deep	6,100	2,475	8,575
3300	8′ deep	6,325	3,250	9,575

G3010 Water Supply

G3010 410	Fire Hydrants	COST PER EACH		
		MAT.	INST.	TOTAL
3400	10' deep	6,500	4,900	11,400
3500	20' offset, 2' deep	5,800	2,200	8,000
3600	4' deep	6,125	2,675	8,800
3700	6' deep	6,275	3,450	9,725
3800	8' deep	6,525	4,575	11,100
3900	10' deep	6,675	6,100	12,775
5000	5-1/4" valve size, two way, 0' offset, 2' deep	5,600	1,150	6,750
5100	4' deep	5,725	1,175	6,900
5200	6' deep	5,900	1,225	7,125
5300	8' deep	6,075	1,275	7,350
5400	10' deep	6,425	1,325	7,750
5500	10' offset, 2' deep	5,750	1,650	7,400
5600	4' deep	5,875	1,950	7,825
5700	6' deep	6,050	2,450	8,500
5800	8' deep	6,225	3,225	9,450
5900	10' deep	6,575	4,850	11,425
6000	20' offset, 2' deep	5,925	2,200	8,125
6100	4' deep	6,050	2,650	8,700
6200	6' deep	6,225	3,425	9,650
6300	8' deep	6,400	4,550	10,950
6400	10' deep	6,750	6,075	12,825
6500	Three way, 0' offset, 2' deep	5,750	1,150	6,900
6600	4' deep	5,875	1,175	7,050
6700	6' deep	6,050	1,250	7,300
6800	8' deep	6,250	1,300	7,550
6900	10' deep	6,625	1,350	7,975
7000	10' offset, 2' deep	5,875	1,675	7,550
7100	4' deep	6,025	1,950	7,975
7200	6' deep	6,200	2,475	8,675
7300	8' deep	6,375	3,250	9,625
7400	10' deep	6,750	4,900	11,650
7500	20' offset, 2' deep	6,075	2,200	8,275
7600	4' deep	6,200	2,425	8,625
7700	6' deep	6,400	3,450	9,850

G3020 Sanitary Sewer

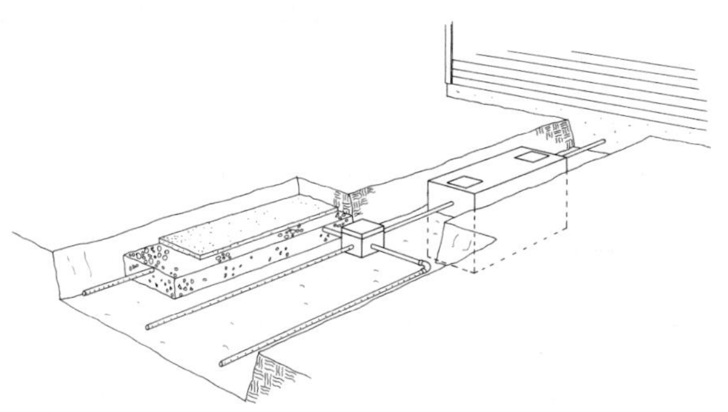

The Septic System includes: a septic tank; leaching field; concrete distribution boxes; plus excavation and gravel backfill.

The Expanded System Listing shows systems with tanks ranging from 1000 gallons to 2000 gallons. Tanks are either concrete or fiberglass. Cost is on a complete unit basis.

System Components	QUANTITY	UNIT	COST EACH MAT.	COST EACH INST.	COST EACH TOTAL
SYSTEM G3020 300 2010					
1000 GALLON TANK, LEACHING FIELD, 300 S.F., 50′ FROM BLDG.					
Septic tank, precast, 1000 gallon	1.000	Ea.	1,100	239.85	1,339.85
Distribution box, precast, 5 outlets	1.000	Ea.	84.50	54.50	139
Sch. 40 sewer pipe, PVC, 4″ diameter	40.000	L.F.	160	392	552
Sch. 40 fitting, wye, & tee, PVC, 4″ diameter	4.000	Ea.	258	65.40	323.40
1-1/2 C.Y. cap. = 125 C.Y./hr.	90.000	B.C.Y.		195.30	195.30
Backfill trench	90.000	L.C.Y.		251.10	251.10
Stone fill, 3/4″ to 1-1/2″	15.000	L.C.Y.	307.50		307.50
Spread fill, 200 HP dozer	15.000	L.C.Y.		117	117
Delivery charge	15.000	L.C.Y.		432.75	432.75
Haul spoil, 12 C.Y. dump truck, 2 mile round trip	25.000	L.C.Y.		168	168
TOTAL			1,910	1,915.90	3,825.90

G3020 300	Septic Systems	COST EACH MAT.	COST EACH INST.	COST EACH TOTAL
2010	1000 gal. septic tank, leaching field, 300 S.F., 50′ from bldg.	1,900	1,925	3,825
2110	200′ from bldg.	2,225	2,800	5,025
2210	6′ deep precast pit, one at 6′-6″ diameter, 50′ from bldg.	2,450	2,175	4,625
2310	200′ from bldg.	2,775	3,025	5,800
2410	4′ deep precast galley, 12′ long, 50′ from bldg.	2,400	2,275	4,675
2510	200′ from bldg.	2,725	3,125	5,850
3010	1250 gal. septic tank, leaching field, 400 S.F., 50′ from bldg.	2,200	2,250	4,450
3110	200′ from bldg.	2,525	3,100	5,625
3210	6′ deep precast pit, one at 8′-0″ diameter, 50′ from bldg.	4,350	2,475	6,825
3310	200′ from bldg.	4,675	3,325	8,000
3410	4′ deep precast galley, 16′ long, 50′ from bldg.	2,925	2,650	5,575
3510	200′ from bldg.	3,250	3,475	6,725
4010	1500 gal. septic tank, leaching field, 500 S.F., 50′ from bldg.	2,750	2,475	5,225
4015	1500 gal. septic tank, leaching field, 1000 S.F., 50′ from bldg.	3,450	4,050	7,500
4110	200′ from bldg., 500 S.F. Field	3,000	3,250	6,250
4115	200′ from bldg., 1000 S.F. Field	3,550	5,000	8,550
4210	6′ deep precast pit, two at 6′-6″ diameter, 50′ from bldg.	5,200	3,675	8,875
4310	200′ from bldg.	5,525	4,500	10,025
4410	4′ deep precast galley, 20′ long, 50′ from bldg.	3,750	3,050	6,800
4510	200′ from bldg.	4,075	3,875	7,950
5010	2000 gal. septic tank, leaching field, 600 S.F., 50′ from bldg.	3,350	2,925	6,275
5015	2000 gal. septic tank, leaching field, 1000 S.F., 50′ from bldg.	3,900	4,500	8,400
5110	200′ from bldg. 500 S.F. Field	3,675	3,775	7,450
5115	200′ from bldg., 1000 S.F. Field	4,050	5,125	9,175
5210	6′ deep precast pit, two at 8′-0″ diameter, 50′ from bldg.	8,575	4,450	13,025
5310	200′ from bldg.	8,575	5,300	13,875

G3020 Sanitary Sewer

G3020 300	Septic Systems	COST EACH		
		MAT.	INST.	TOTAL
5410	4′ deep precast galley, 24 ft. long, 50′ from bldg.	4,600	3,675	8,275
5510	200′ from bldg.	4,925	4,500	9,425
6010	1000 gal. 2 compart. septic tank, leaching field, 300 S.F., 50′ from bldg.	2,100	1,925	4,025
6110	200′ from bldg.	2,425	2,800	5,225
6210	6′ deep precast pit, one at 6′-6″ diameter, 50′ from bldg.	2,650	2,175	4,825
6310	200′ from bldg.	2,975	3,025	6,000
6410	4′ deep precast galley, 12′ long, 50′ from bldg.	2,600	2,275	4,875
6510	200′ from bldg.	2,925	3,125	6,050
8010	1500 gal. septic tank 2 compart. , leaching field, 500 S.F., 50′ from bldg.	3,125	2,475	5,600
8015	1500 gal. 2 compart. septic tank, leaching field, 1000 S.F., 50′ from bldg.	3,825	4,050	7,875
8110	200′ from bldg., 500 S.F. Field	3,375	3,250	6,625
8115	200′ from bldg., 1000 S.F. Field	3,925	5,000	8,925
8210	6′ deep precast pit, two at 6′-6″ diameter, 50′ from bldg.	5,575	3,675	9,250
8310	200′ from bldg.	5,900	4,500	10,400
8410	4′ deep precast galley, 20′ long, 50′ from bldg.	4,125	3,050	7,175
8510	200′ from bldg.	4,450	3,875	8,325
9010	2000 gal. 2 compart. septic tank, leaching field, 600 S.F., 50′ from bldg.	3,675	2,925	6,600
9015	2000 gal. 2 compart. septic tank, leaching field, 1000 S.F., 50′ from bldg.	4,225	4,500	8,725
9110	200′ from bldg. 500 S.F. Field	4,000	3,775	7,775
9115	200′ from bldg., 1000 S.F. Field	4,375	5,125	9,500
9210	6′ deep precast pit, two at 8′-0″ diameter, 50′ from bldg.	8,575	4,450	13,025
9310	200′ from bldg.	8,900	5,300	14,200
9410	4′ deep precast galley, 24 ft. long, 50′ from bldg.	4,925	3,675	8,600
9510	200′ from bldg.	5,250	4,500	9,750

G3020 Sanitary Sewer

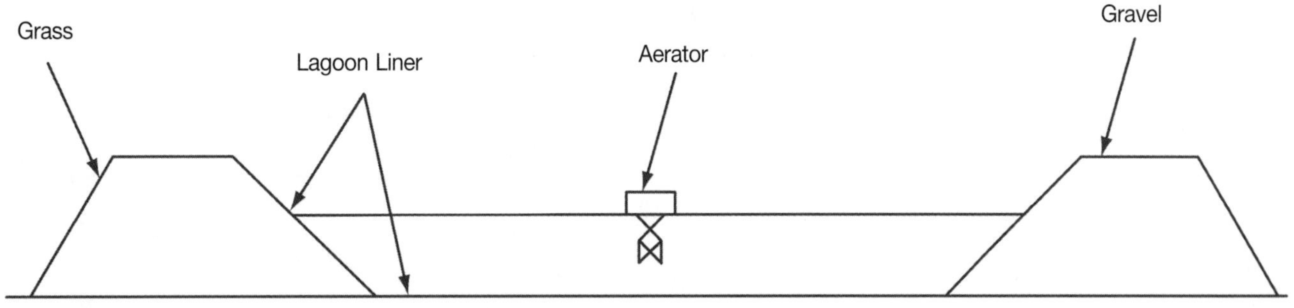

Grass
Lagoon Liner
Aerator
Gravel

System Components	QUANTITY	UNIT	COST EACH MAT.	COST EACH INST.	COST EACH TOTAL
SYSTEM G3020 710 1000					
AERATED DOMESTIC SEWAGE LAGOON, COMMON EARTH, 500,000 GPD					
Supervision of lagoon construction	45.000	Day		25,650	25,650
Quality Control of Earthwork per day	20.000	Day		8,700	8,700
Anchoring aerators per cell, 6 anchors per cell	2.000	Ea.	7,700	7,320	15,020
Excavate lagoon	7778.000	B.C.Y.		18,433.86	18,433.86
Finish grade slopes for seeding	25.000	M.S.F.		701.25	701.25
Spread & grade top of lagoon banks for compaction	454.000	M.S.F.		27,013	27,013
Compact lagoon banks	7127.000	E.C.Y.		5,202.71	5,202.71
Install pond liner	71.000	M.S.F.	33,725	76,325	110,050
Provide water for compaction and dust control	1556.000	E.C.Y.	2,038.36	3,158.68	5,197.04
Dispose of excess material on-site	846.000	L.C.Y.		2,944.08	2,944.08
Seed non-lagoon side of banks	25.000	M.S.F.	650	700	1,350
Place gravel on top of banks	1269.000	S.Y.	8,375.40	1,814.67	10,190.07
Provide 6 ' high fence around lagoon	1076.000	L.F.	23,134	7,219.96	30,353.96
Provide 12' wide fence gate	1.000	Opng.	595	504	1,099
Surface aerator, 30 HP, 50 lbs/HR, 900 RPM	4.000	Ea.	149,600	12,240	161,840
Temporary Fencing	1269.000	L.F.	3,616.65	2,766.42	6,383.07
TOTAL			229,434.41	200,693.63	430,128.04

G3020 710	Aerated Sewage Lagoon	COST EACH MAT.	COST EACH INST.	COST EACH TOTAL
1000	500,000 GPD Domestic Sewage Aerated Lagoon, common earth, excluding power	229,500	201,000	430,500
1100	600,000 GPD	311,500	232,500	544,000
1200	700,000 GPD	316,000	250,500	566,500
1300	800,000 GPD	396,500	279,000	675,500
1400	900,000 GPD	400,500	298,500	699,000
1500	1 MGD	443,500	376,500	820,000
2000	500,000 GPD sandy clay & loam	229,000	198,500	427,500
2100	600,000 GPD	311,500	230,500	542,000
2200	700,000 GPD	316,000	246,000	562,000
2300	800,000 GPD	396,500	276,000	672,500
2400	900,000 GPD	400,500	295,000	695,500
2500	1 MGD	543,500	466,500	1,010,000
3000	500,000 GPD sand & gravel	229,000	196,500	425,500
3100	600,000 GPD	311,500	228,500	540,000
3200	700,000 GPD	316,000	245,000	561,000
3300	800,000 GPD	402,500	274,000	676,500
3400	900,000 GPD	400,500	290,000	690,500
3500	1 MGD	444,000	371,500	815,500
5000	500,000 GPD Lagoon aerators power supply	9,400	11,200	20,600
5100	600,000 GPD	11,300	13,700	25,000

G3020 Sanitary Sewer

G3020 710	Aerated Sewage Lagoon	COST EACH		
		MAT.	INST.	TOTAL
5200	700,000 GPD	11,300	13,700	25,000
5300	800,000 GPD	18,800	22,400	41,200
5400	900,000 GPD	18,800	22,400	41,200
5500	1 MGD	18,800	22,400	41,200
5600	Lagoons, aerated without liner			
6000	500,000 GPD Domestic Sewage Aerated Lagoon, common earth, no power/liner	195,000	118,500	313,500
6100	600,000 GPD	274,500	143,000	417,500
6200	700,000 GPD	276,000	151,500	427,500
6300	800,000 GPD	352,500	173,500	526,000
6400	900,000 GPD	352,500	183,500	536,000
6500	1 MGD	376,000	212,500	588,500
7000	500,000 GPD sandy clay & loam	195,000	115,500	310,500
7100	600,000 GPD	274,500	141,000	415,500
7200	700,000 GPD	276,000	149,000	425,000
7300	800,000 GPD	352,500	170,000	522,500
7400	900,000 GPD	362,000	182,500	544,500
7500	1 MGD	376,000	208,500	584,500

G3030 Storm Sewer

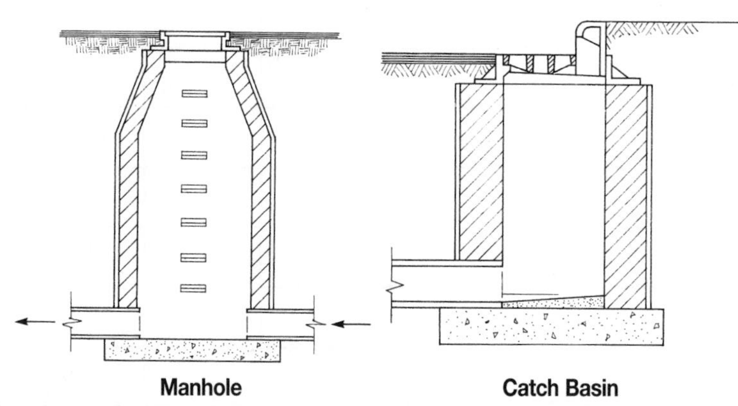

The Manhole and Catch Basin System includes: excavation with a backhoe; a formed concrete footing; frame and cover; cast iron steps and compacted backfill.

The Expanded System Listing shows manholes that have a 4′, 5′ and 6′ inside diameter riser. Depths range from 4′ to 14′. Construction material shown is either concrete, concrete block, precast concrete, or brick.

System Components	QUANTITY	UNIT	COST PER EACH MAT.	INST.	TOTAL
SYSTEM G3030 210 1920					
MANHOLE/CATCH BASIN, BRICK, 4′ I.D. RISER, 4′ DEEP					
Excavation, hydraulic backhoe, 3/8 C.Y. bucket	14.815	B.C.Y.		117.93	117.93
Trim sides and bottom of excavation	64.000	S.F.		58.88	58.88
Forms in place, manhole base, 4 uses	20.000	SFCA	14.20	101	115.20
Reinforcing in place footings, #4 to #7	.019	Ton	20.90	22.80	43.70
Concrete, 3000 psi	.925	C.Y.	98.98		98.98
Place and vibrate concrete, footing, direct chute	.925	C.Y.		47.01	47.01
Catch basin or MH, brick, 4′ ID, 4′ deep	1.000	Ea.	430	980	1,410
Catch basin or MH steps; heavy galvanized cast iron	1.000	Ea.	19.80	13.55	33.35
Catch basin or MH frame and cover	1.000	Ea.	253	235	488
Fill, granular	12.954	L.C.Y.	297.94		297.94
Backfill, spread with wheeled front end loader	12.954	L.C.Y.		32.52	32.52
Backfill compaction, 12″ lifts, air tamp	12.954	E.C.Y.		111.79	111.79
TOTAL			1,134.82	1,720.48	2,855.30

G3030 210	Manholes & Catch Basins	COST PER EACH MAT.	INST.	TOTAL
1920	Manhole/catch basin, brick, 4′ I.D. riser, 4′ deep	1,125	1,725	2,850
1940	6′ deep	1,600	2,375	3,975
1960	8′ deep	2,150	3,250	5,400
1980	10′ deep	2,600	4,050	6,650
3000	12′ deep	3,300	4,375	7,675
3020	14′ deep	4,150	6,150	10,300
3200	Block, 4′ I.D. riser, 4′ deep	1,100	1,400	2,500
3220	6′ deep	1,525	1,950	3,475
3240	8′ deep	2,050	2,700	4,750
3260	10′ deep	2,425	3,350	5,775
3280	12′ deep	3,100	4,275	7,375
3300	14′ deep	3,900	5,200	9,100
4620	Concrete, cast-in-place, 4′ I.D. riser, 4′ deep	1,250	2,375	3,625
4640	6′ deep	1,800	3,175	4,975
4660	8′ deep	2,500	4,575	7,075
4680	10′ deep	3,000	5,700	8,700
4700	12′ deep	3,800	7,075	10,875
4720	14′ deep	4,725	8,475	13,200
5820	Concrete, precast, 4′ I.D. riser, 4′ deep	1,500	1,275	2,775
5840	6′ deep	2,025	1,700	3,725

G3030 Storm Sewer

G3030 210	Manholes & Catch Basins	COST PER EACH		
		MAT.	INST.	TOTAL
5860	8' deep	2,550	2,375	4,925
5880	10' deep	3,075	2,925	6,000
5900	12' deep	3,875	3,600	7,475
5920	14' deep	4,800	4,625	9,425
6000	5' I.D. riser, 4' deep	2,450	1,425	3,875
6020	6' deep	3,100	2,025	5,125
6040	8' deep	3,975	2,675	6,650
6060	10' deep	5,075	3,425	8,500
6080	12' deep	6,300	4,325	10,625
6100	14' deep	7,650	5,300	12,950
6200	6' I.D. riser, 4' deep	3,300	1,900	5,200
6220	6' deep	4,125	2,500	6,625
6240	8' deep	5,225	3,500	8,725
6260	10' deep	6,600	4,450	11,050
6280	12' deep	8,100	5,600	13,700
6300	14' deep	9,725	6,800	16,525

Reference Section

All the reference information is in one section, making it easy to find what you need to know . . . and easy to use the book on a daily basis. This section is visually identified by a vertical gray bar on the page edges.

In this Reference Section, we've included Equipment Rental Costs, a listing of rental and operating costs; Crew Listings, a full listing of all crews and equipment, and their costs; Historical Cost Indexes for cost comparisons over time; City Cost Indexes and Location Factors for adjusting costs to the region you are in; Reference Tables, where you will find explanations, estimating information and procedures, or technical data; Change Orders, information on pricing changes to contract documents; Square Foot Costs that allow you to make a rough estimate for the overall cost of a project; and an explanation of all the Abbreviations in the book.

Table of Contents

Equipment Rental Costs

Estimating Tips

- This section contains the average costs to rent and operate hundreds of pieces of construction equipment. This is useful information when estimating the time and material requirements of any particular operation in order to establish a unit or total cost. Equipment costs include not only rental, but also operating costs for equipment under normal use.

Rental Costs

- Equipment rental rates are obtained from industry sources throughout North America–contractors, suppliers, dealers, manufacturers, and distributors.
- Rental rates vary throughout the country, with larger cities generally having lower rates. Lease plans for new equipment are available for periods in excess of six months, with a percentage of payments applying toward purchase.
- Monthly rental rates vary from 2% to 5% of the purchase price of the equipment depending on the anticipated life of the equipment and its wearing parts.
- Weekly rental rates are about 1/3 the monthly rates, and daily rental rates are about 1/3 the weekly rate.
- Rental rates can also be treated as reimbursement costs for contractor-owned equipment. Owned equipment costs include depreciation, loan payments, interest, taxes, insurance, storage, and major repairs.

Operating Costs

- The operating costs include parts and labor for routine servicing, such as repair and replacement of pumps, filters and worn lines. Normal operating expendables, such as fuel, lubricants, tires and electricity (where applicable), are also included.
- Extraordinary operating expendables with highly variable wear patterns, such as diamond bits and blades, are excluded. These costs can be found as material costs in the Unit Price section.
- The hourly operating costs listed do not include the operator's wages.

Equipment Cost/Day

- Any power equipment required by a crew is shown in the Crew Listings with a daily cost.
- The daily cost of equipment needed by a crew is based on dividing the weekly rental rate by 5 (number of working days in the week), and then adding the hourly operating cost times 8 (the number of hours in a day). This "Equipment Cost/ Day" is shown in the far right column of the Equipment Rental pages.
- If equipment is needed for only one or two days, it is best to develop your own cost by including components for daily rent and hourly operating cost. This is important when the listed Crew for a task does not contain the equipment needed, such as a crane for lifting mechanical heating/cooling equipment up onto a roof.
- If the quantity of work is less than the crew's Daily Output shown for a Unit Price line item that includes a bare unit equipment cost, it is recommended to estimate one day's rental cost and operating cost for equipment shown in the Crew Listing for that line item.

Mobilization/ Demobilization

- The cost to move construction equipment from an equipment yard or rental company to the jobsite and back again is not included in equipment rental costs listed in the Reference section, nor in the bare equipment cost of any Unit Price line item, nor in any equipment costs shown in the Crew listings.
- Mobilization (to the site) and demobilization (from the site) costs can be found in the Unit Price section.
- If a piece of equipment is already at the jobsite, it is not appropriate to utilize mobil./ demob. costs again in an estimate.

01 54 33 | Equipment Rental

		01 54 33 Equipment Rental	UNIT	HOURLY OPER. COST	RENT PER DAY	RENT PER WEEK	RENT PER MONTH	EQUIPMENT COST/DAY	
10	0010	**CONCRETE EQUIPMENT RENTAL** without operators R015433-10							10
	0200	Bucket, concrete lightweight, 1/2 C.Y.	Ea.	.80	23.50	70	210	20.40	
	0300	1 C.Y.		.85	27.50	83	249	23.40	
	0400	1-1/2 C.Y.		1.10	36.50	110	330	30.80	
	0500	2 C.Y.		1.20	45	135	405	36.60	
	0580	8 C.Y.		5.90	253	760	2,275	199.20	
	0600	Cart, concrete, self-propelled, operator walking, 10 C.F.		3.35	56.50	170	510	60.80	
	0700	Operator riding, 18 C.F.		5.55	93.50	280	840	100.40	
	0800	Conveyer for concrete, portable, gas, 16" wide, 26' long		13.30	122	365	1,100	179.40	
	0900	46' long		13.65	147	440	1,325	197.20	
	1000	56' long		13.80	155	465	1,400	203.40	
	1100	Core drill, electric, 2-1/2 H.P., 1" to 8" bit diameter		1.54	59.50	179	535	48.10	
	1150	11 H.P., 8" to 18" cores		5.80	110	330	990	112.40	
	1200	Finisher, concrete floor, gas, riding trowel, 96" wide		13.80	143	430	1,300	196.40	
	1300	Gas, walk-behind, 3 blade, 36" trowel		2.30	20	60	180	30.40	
	1400	4 blade, 48" trowel		4.60	27	81	243	53	
	1500	Float, hand-operated (Bull float) 48" wide		.08	14	42	126	9.05	
	1570	Curb builder, 14 H.P., gas, single screw		14.45	250	750	2,250	265.60	
	1590	Double screw		15.10	287	860	2,575	292.80	
	1600	Floor grinder, concrete and terrazzo, electric, 22" path		2.04	125	374	1,125	91.10	
	1700	Edger, concrete, electric, 7" path		1.01	51.50	155	465	39.10	
	1750	Vacuum pick-up system for floor grinders, wet/dry		1.46	81.50	245	735	60.70	
	1800	Mixer, powered, mortar and concrete, gas, 6 C.F., 18 H.P.		9.05	117	350	1,050	142.40	
	1900	10 C.F., 25 H.P.		11.40	142	425	1,275	176.20	
	2000	16 C.F.		11.75	163	490	1,475	192	
	2100	Concrete, stationary, tilt drum, 2 C.Y.		6.70	228	685	2,050	190.60	
	2120	Pump, concrete, truck mounted 4" line 80' boom		24.35	870	2,605	7,825	715.80	
	2140	5" line, 110' boom		31.50	1,150	3,450	10,400	942	
	2160	Mud jack, 50 C.F. per hr.		7.10	123	370	1,100	130.80	
	2180	225 C.F. per hr.		9.30	142	425	1,275	159.40	
	2190	Shotcrete pump rig, 12 C.Y./hr.		14.70	218	655	1,975	248.60	
	2200	35 C.Y./hr.		17.25	235	705	2,125	279	
	2600	Saw, concrete, manual, gas, 18 H.P.		7.25	43.50	130	390	84	
	2650	Self-propelled, gas, 30 H.P.		14.40	100	300	900	175.20	
	2700	Vibrators, concrete, electric, 60 cycle, 2 H.P.		.41	8.65	26	78	8.50	
	2800	3 H.P.		.59	11.65	35	105	11.70	
	2900	Gas engine, 5 H.P.		2.20	16	48	144	27.20	
	3000	8 H.P.		3.00	15	45	135	33	
	3050	Vibrating screed, gas engine, 8 H.P.		2.91	71.50	215	645	66.30	
	3120	Concrete transit mixer, 6 x 4, 250 H.P., 8 C.Y., rear discharge		62.00	565	1,700	5,100	836	
	3200	Front discharge		73.05	695	2,080	6,250	1,000	
	3300	6 x 6, 285 H.P., 12 C.Y., rear discharge		72.10	655	1,965	5,900	969.80	
	3400	Front discharge	▼	75.40	700	2,105	6,325	1,024	
20	0010	**EARTHWORK EQUIPMENT RENTAL** without operators R015433-10							20
	0040	Aggregate spreader, push type 8' to 12' wide	Ea.	3.20	25.50	76	228	40.80	
	0045	Tailgate type, 8' wide		3.00	32.50	98	294	43.60	
	0055	Earth auger, truck-mounted, for fence & sign posts, utility poles		19.65	455	1,370	4,100	431.20	
	0060	For borings and monitoring wells		45.00	640	1,920	5,750	744	
	0070	Portable, trailer mounted		3.15	31.50	95	285	44.20	
	0075	Truck-mounted, for caissons, water wells		96.25	2,825	8,495	25,500	2,469	
	0080	Horizontal boring machine, 12" to 36" diameter, 45 H.P.		23.80	188	565	1,700	303.40	
	0090	12" to 48" diameter, 65 H.P.		33.50	330	985	2,950	465	
	0095	Auger, for fence posts, gas engine, hand held		.55	6	18	54	8	
	0100	Excavator, diesel hydraulic, crawler mounted, 1/2 C.Y. cap.		21.30	385	1,150	3,450	400.40	
	0120	5/8 C.Y. capacity		32.75	555	1,670	5,000	596	
	0140	3/4 C.Y. capacity		40.10	625	1,870	5,600	694.80	
	0150	1 C.Y. capacity	▼	50.20	690	2,070	6,200	815.60	

01 54 33 | Equipment Rental

			UNIT	HOURLY OPER. COST	RENT PER DAY	RENT PER WEEK	RENT PER MONTH	EQUIPMENT COST/DAY	
20	0200	1-1/2 C.Y. capacity	Ea.	60.35	920	2,765	8,300	1,036	20
	0300	2 C.Y. capacity		81.05	1,200	3,575	10,700	1,363	
	0320	2-1/2 C.Y. capacity		109.55	1,550	4,620	13,900	1,800	
	0325	3-1/2 C.Y. capacity		146.35	2,150	6,415	19,200	2,454	
	0330	4-1/2 C.Y. capacity		175.40	2,750	8,270	24,800	3,057	
	0335	6 C.Y. capacity		223.00	2,925	8,785	26,400	3,541	
	0340	7 C.Y. capacity		225.25	3,025	9,070	27,200	3,616	
	0342	Excavator attachments, bucket thumbs		3.10	240	720	2,150	168.80	
	0345	Grapples		2.65	188	565	1,700	134.20	
	0347	Hydraulic hammer for boom mounting, 5000 ft lb.		14.10	425	1,280	3,850	368.80	
	0349	11,000 ft lb.		22.90	755	2,260	6,775	635.20	
	0350	Gradall type, truck mounted, 3 ton @ 15′ radius, 5/8 C.Y.		43.10	900	2,695	8,075	883.80	
	0370	1 C.Y. capacity		47.25	1,050	3,160	9,475	1,010	
	0400	Backhoe-loader, 40 to 45 H.P., 5/8 C.Y. capacity		14.85	237	710	2,125	260.80	
	0450	45 H.P. to 60 H.P., 3/4 C.Y. capacity		23.80	295	885	2,650	367.40	
	0460	80 H.P., 1-1/4 C.Y. capacity		26.05	320	960	2,875	400.40	
	0470	112 H.P., 1-1/2 C.Y. capacity		41.40	655	1,970	5,900	725.20	
	0482	Backhoe-loader attachment, compactor, 20,000 lb.		5.65	142	425	1,275	130.20	
	0485	Hydraulic hammer, 750 ft lb.		3.20	95	285	855	82.60	
	0486	Hydraulic hammer, 1200 ft lb.		6.20	217	650	1,950	179.60	
	0500	Brush chipper, gas engine, 6″ cutter head, 35 H.P.		11.75	107	320	960	158	
	0550	Diesel engine, 12″ cutter head, 130 H.P.		28.05	287	860	2,575	396.40	
	0600	15″ cutter head, 165 H.P.		33.75	330	990	2,975	468	
	0750	Bucket, clamshell, general purpose, 3/8 C.Y.		1.25	38.50	115	345	33	
	0800	1/2 C.Y.		1.35	45	135	405	37.80	
	0850	3/4 C.Y.		1.50	55	165	495	45	
	0900	1 C.Y.		1.55	58.50	175	525	47.40	
	0950	1-1/2 C.Y.		2.50	80	240	720	68	
	1000	2 C.Y.		2.65	88.50	265	795	74.20	
	1010	Bucket, dragline, medium duty, 1/2 C.Y.		.70	23.50	70	210	19.60	
	1020	3/4 C.Y.		.75	24.50	74	222	20.80	
	1030	1 C.Y.		.75	26	78	234	21.60	
	1040	1-1/2 C.Y.		1.20	40	120	360	33.60	
	1050	2 C.Y.		1.25	45	135	405	37	
	1070	3 C.Y.		1.90	61.50	185	555	52.20	
	1200	Compactor, manually guided 2-drum vibratory smooth roller, 7.5 H.P.		7.00	197	590	1,775	174	
	1250	Rammer/tamper, gas, 8″		2.65	43.50	130	390	47.20	
	1260	15″		2.95	50	150	450	53.60	
	1300	Vibratory plate, gas, 18″ plate, 3000 lb. blow		2.60	23	69	207	34.60	
	1350	21″ plate, 5000 lb. blow		3.25	30.50	92	276	44.40	
	1370	Curb builder/extruder, 14 H.P., gas, single screw		14.45	250	750	2,250	265.60	
	1390	Double screw		15.10	287	860	2,575	292.80	
	1500	Disc harrow attachment, for tractor		.43	71	213	640	46.05	
	1810	Feller buncher, shearing & accumulating trees, 100 H.P.		35.95	550	1,645	4,925	616.60	
	1860	Grader, self-propelled, 25,000 lb.		39.35	605	1,815	5,450	677.80	
	1910	30,000 lb.		43.05	615	1,840	5,525	712.40	
	1920	40,000 lb.		64.35	1,025	3,070	9,200	1,129	
	1930	55,000 lb.		84.30	1,625	4,900	14,700	1,654	
	1950	Hammer, pavement breaker, self-propelled, diesel, 1000 to 1250 lb.		29.95	355	1,060	3,175	451.60	
	2000	1300 to 1500 lb.		44.93	705	2,120	6,350	783.45	
	2050	Pile driving hammer, steam or air, 4150 ft lb. @ 225 bpm		10.25	480	1,435	4,300	369	
	2100	8750 ft lb. @ 145 bpm		12.20	665	2,000	6,000	497.60	
	2150	15,000 ft lb. @ 60 bpm		13.80	805	2,410	7,225	592.40	
	2200	24,450 ft lb. @ 111 bpm		14.80	890	2,675	8,025	653.40	
	2250	Leads, 60′ high for pile driving hammers up to 20,000 ft lb.		3.25	81	243	730	74.60	
	2300	90′ high for hammers over 20,000 ft lb.		4.95	142	426	1,275	124.80	
	2350	Diesel type hammer, 22,400 ft lb.		31.25	615	1,840	5,525	618	
	2400	41,300 ft lb.	↓	40.50	635	1,900	5,700	704	

	01 54 33 | Equipment Rental	UNIT	HOURLY OPER. COST	RENT PER DAY	RENT PER WEEK	RENT PER MONTH	EQUIPMENT COST/DAY	
20	2450 141,000 ft lb.	Ea.	59.15	1,150	3,450	10,400	1,163	20
	2500 Vib. elec. hammer/extractor, 200 kW diesel generator, 34 H.P.		55.90	645	1,935	5,800	834.20	
	2550 80 H.P.		102.85	945	2,830	8,500	1,389	
	2600 150 H.P.		195.65	1,800	5,435	16,300	2,652	
	2800 Log chipper, up to 22" diameter, 600 H.P.		64.65	640	1,915	5,750	900.20	
	2850 Logger, for skidding & stacking logs, 150 H.P.		53.05	825	2,470	7,400	918.40	
	2860 Mulcher, diesel powered, trailer mounted		24.55	215	645	1,925	325.40	
	2900 Rake, spring tooth, with tractor		18.16	335	1,012	3,025	347.70	
	3000 Roller, vibratory, tandem, smooth drum, 20 H.P.		8.55	143	430	1,300	154.40	
	3050 35 H.P.		10.65	243	730	2,200	231.20	
	3100 Towed type vibratory compactor, smooth drum, 50 H.P.		25.65	335	1,010	3,025	407.20	
	3150 Sheepsfoot, 50 H.P.		27.00	375	1,120	3,350	440	
	3170 Landfill compactor, 220 H.P.		86.90	1,425	4,265	12,800	1,548	
	3200 Pneumatic tire roller, 80 H.P.		16.20	330	985	2,950	326.60	
	3250 120 H.P.		24.15	545	1,630	4,900	519.20	
	3300 Sheepsfoot vibratory roller, 240 H.P.		66.00	1,050	3,130	9,400	1,154	
	3320 340 H.P.		89.30	1,550	4,630	13,900	1,640	
	3350 Smooth drum vibratory roller, 75 H.P.		23.80	565	1,695	5,075	529.40	
	3400 125 H.P.		31.05	680	2,040	6,125	656.40	
	3410 Rotary mower, brush, 60", with tractor		22.40	305	920	2,750	363.20	
	3420 Rototiller, walk-behind, gas, 5 H.P.		2.30	67.50	203	610	59	
	3422 8 H.P.		3.57	103	310	930	90.55	
	3440 Scrapers, towed type, 7 C.Y. capacity		5.60	110	330	990	110.80	
	3450 10 C.Y. capacity		6.35	150	450	1,350	140.80	
	3500 15 C.Y. capacity		6.75	173	520	1,550	158	
	3525 Self-propelled, single engine, 14 C.Y. capacity		112.95	1,550	4,680	14,000	1,840	
	3550 Dual engine, 21 C.Y. capacity		173.55	2,025	6,110	18,300	2,610	
	3600 31 C.Y. capacity		230.75	2,900	8,690	26,100	3,584	
	3640 44 C.Y. capacity		281.25	3,550	10,635	31,900	4,377	
	3650 Elevating type, single engine, 11 C.Y. capacity		69.90	1,025	3,050	9,150	1,169	
	3700 22 C.Y. capacity		137.35	2,275	6,800	20,400	2,459	
	3710 Screening plant 110 H.P. w/5′ x 10′ screen		37.95	435	1,300	3,900	563.60	
	3720 5′ x 16′ screen		40.20	540	1,615	4,850	644.60	
	3850 Shovel, crawler-mounted, front-loading, 7 C.Y. capacity		238.20	2,650	7,945	23,800	3,495	
	3855 12 C.Y. capacity		321.20	3,500	10,530	31,600	4,676	
	3860 Shovel/backhoe bucket, 1/2 C.Y.		2.40	65	195	585	58.20	
	3870 3/4 C.Y.		2.45	71.50	215	645	62.60	
	3880 1 C.Y.		2.55	81.50	245	735	69.40	
	3890 1-1/2 C.Y.		2.70	95	285	855	78.60	
	3910 3 C.Y.		3.00	128	385	1,150	101	
	3950 Stump chipper, 18" deep, 30 H.P.		7.47	188	565	1,700	172.75	
	4110 Dozer, crawler, torque converter, diesel 80 H.P.		29.45	410	1,225	3,675	480.60	
	4150 105 H.P.		35.80	510	1,530	4,600	592.40	
	4200 140 H.P.		51.75	790	2,375	7,125	889	
	4260 200 H.P.		76.90	1,200	3,590	10,800	1,333	
	4310 300 H.P.		99.95	1,725	5,145	15,400	1,829	
	4360 410 H.P.		134.95	2,250	6,725	20,200	2,425	
	4370 500 H.P.		173.50	2,925	8,745	26,200	3,137	
	4380 700 H.P.		256.95	4,000	12,015	36,000	4,459	
	4400 Loader, crawler, torque conv., diesel, 1-1/2 C.Y., 80 H.P.		31.20	475	1,420	4,250	533.60	
	4450 1-1/2 to 1-3/4 C.Y., 95 H.P.		34.90	590	1,770	5,300	633.20	
	4510 1-3/4 to 2-1/4 C.Y., 130 H.P.		54.25	870	2,605	7,825	955	
	4530 2-1/2 to 3-1/4 C.Y., 190 H.P.		66.60	1,100	3,270	9,800	1,187	
	4560 3-1/2 to 5 C.Y., 275 H.P.		84.60	1,425	4,265	12,800	1,530	
	4610 Front end loader, 4WD, articulated frame, diesel, 1 to 1-1/4 C.Y., 70 H.P.		19.75	230	690	2,075	296	
	4620 1-1/2 to 1-3/4 C.Y., 95 H.P.		25.15	290	870	2,600	375.20	
	4650 1-3/4 to 2 C.Y., 130 H.P.		20.25	375	1,120	3,350	386	
	4710 2-1/2 to 3-1/2 C.Y., 145 H.P.	↓	33.15	435	1,305	3,925	526.20	

01 54 33 | Equipment Rental

			UNIT	HOURLY OPER. COST	RENT PER DAY	RENT PER WEEK	RENT PER MONTH	EQUIPMENT COST/DAY	
20	4730	3 to 4-1/2 C.Y., 185 H.P.	Ea.	42.60	565	1,690	5,075	678.80	20
	4760	5-1/4 to 5-3/4 C.Y., 270 H.P.		67.35	860	2,585	7,750	1,056	
	4810	7 to 9 C.Y., 475 H.P.		115.70	1,800	5,410	16,200	2,008	
	4870	9 - 11 C.Y., 620 H.P.		155.05	2,500	7,470	22,400	2,734	
	4880	Skid steer loader, wheeled, 10 C.F., 30 H.P. gas		10.25	150	450	1,350	172	
	4890	1 C.Y., 78 H.P., diesel		20.20	245	735	2,200	308.60	
	4892	Skid-steer attachment, auger		.55	92.50	277	830	59.80	
	4893	Backhoe		.67	112	336	1,000	72.55	
	4894	Broom		.73	122	367	1,100	79.25	
	4895	Forks		.24	40	120	360	25.90	
	4896	Grapple		.55	91	273	820	59	
	4897	Concrete hammer		1.06	176	528	1,575	114.10	
	4898	Tree spade		.94	156	468	1,400	101.10	
	4899	Trencher		.78	130	391	1,175	84.45	
	4900	Trencher, chain, boom type, gas, operator walking, 12 H.P.		5.30	46.50	140	420	70.40	
	4910	Operator riding, 40 H.P.		20.40	293	880	2,650	339.20	
	5000	Wheel type, diesel, 4′ deep, 12″ wide		86.65	840	2,525	7,575	1,198	
	5100	6′ deep, 20″ wide		92.15	1,925	5,785	17,400	1,894	
	5150	Chain type, diesel, 5′ deep, 8″ wide		38.35	560	1,680	5,050	642.80	
	5200	Diesel, 8′ deep, 16″ wide		153.95	3,550	10,625	31,900	3,357	
	5202	Rock trencher, wheel type, 6″ wide x 18″ deep		21.10	325	975	2,925	363.80	
	5206	Chain type, 18″ wide x 7′ deep		111.50	2,850	8,520	25,600	2,596	
	5210	Tree spade, self-propelled		14.38	267	800	2,400	275.05	
	5250	Truck, dump, 2-axle, 12 ton, 8 C.Y. payload, 220 H.P.		34.95	228	685	2,050	416.60	
	5300	Three axle dump, 16 ton, 12 C.Y. payload, 400 H.P.		62.05	325	980	2,950	692.40	
	5310	Four axle dump, 25 ton, 18 C.Y. payload, 450 H.P.		72.75	475	1,425	4,275	867	
	5350	Dump trailer only, rear dump, 16-1/2 C.Y.		5.35	140	420	1,250	126.80	
	5400	20 C.Y.		5.80	158	475	1,425	141.40	
	5450	Flatbed, single axle, 1-1/2 ton rating		28.10	68.50	205	615	265.80	
	5500	3 ton rating		33.60	98.50	295	885	327.80	
	5550	Off highway rear dump, 25 ton capacity		74.10	1,225	3,640	10,900	1,321	
	5600	35 ton capacity		82.60	1,350	4,080	12,200	1,477	
	5610	50 ton capacity		102.90	1,625	4,870	14,600	1,797	
	5620	65 ton capacity		105.75	1,600	4,805	14,400	1,807	
	5630	100 ton capacity		153.60	2,825	8,455	25,400	2,920	
	6000	Vibratory plow, 25 H.P., walking	↓	9.05	61.50	185	555	109.40	
40	0010	**GENERAL EQUIPMENT RENTAL** without operators R015433-10							40
	0150	Aerial lift, scissor type, to 15′ high, 1000 lb. cap., electric	Ea.	3.05	51.50	155	465	55.40	
	0160	To 25′ high, 2000 lb. capacity		3.50	68.50	205	615	69	
	0170	Telescoping boom to 40′ high, 500 lb. capacity, gas		18.65	305	915	2,750	332.20	
	0180	To 45′ high, 500 lb. capacity		20.80	355	1,060	3,175	378.40	
	0190	To 60′ high, 600 lb. capacity		23.15	460	1,385	4,150	462.20	
	0195	Air compressor, portable, 6.5 CFM, electric		.42	12.65	38	114	10.95	
	0196	Gasoline		.81	19	57	171	17.90	
	0200	Towed type, gas engine, 60 CFM		13.70	46.50	140	420	137.60	
	0300	160 CFM		16.00	50	150	450	158	
	0400	Diesel engine, rotary screw, 250 CFM		16.35	108	325	975	195.80	
	0500	365 CFM		22.10	132	395	1,175	255.80	
	0550	450 CFM		28.05	163	490	1,475	322.40	
	0600	600 CFM		49.35	227	680	2,050	530.80	
	0700	750 CFM	↓	49.55	235	705	2,125	537.40	
	0800	For silenced models, small sizes, add to rent		3%	5%	5%	5%		
	0900	Large sizes, add to rent		5%	7%	7%	7%		
	0930	Air tools, breaker, pavement, 60 lb.	Ea.	.50	9.65	29	87	9.80	
	0940	80 lb.		.50	10.35	31	93	10.20	
	0950	Drills, hand (jackhammer) 65 lb.		.60	16.65	50	150	14.80	
	0960	Track or wagon, swing boom, 4″ drifter		61.00	870	2,605	7,825	1,009	
	0970	5″ drifter	↓	76.60	990	2,970	8,900	1,207	

01 54 33 | Equipment Rental

			UNIT	HOURLY OPER. COST	RENT PER DAY	RENT PER WEEK	RENT PER MONTH	EQUIPMENT COST/DAY	
40	0975	Track mounted quarry drill, 6" diameter drill	Ea.	117.20	1,425	4,275	12,800	1,793	40
	0980	Dust control per drill		.90	21.50	64	192	20	
	0990	Hammer, chipping, 12 lb.		.55	26	78	234	20	
	1000	Hose, air with couplings, 50' long, 3/4" diameter		.03	5	15	45	3.25	
	1100	1" diameter		.04	6.35	19	57	4.10	
	1200	1-1/2" diameter		.05	9	27	81	5.80	
	1300	2" diameter		.07	12	36	108	7.75	
	1400	2-1/2" diameter		.11	19	57	171	12.30	
	1410	3" diameter		.14	23	69	207	14.90	
	1450	Drill, steel, 7/8" x 2'		.05	8.65	26	78	5.60	
	1460	7/8" x 6'		.05	9	27	81	5.80	
	1520	Moil points		.02	4	12	36	2.55	
	1525	Pneumatic nailer w/accessories		.48	31.50	95	285	22.85	
	1530	Sheeting driver for 60 lb. breaker		.04	6	18	54	3.90	
	1540	For 90 lb. breaker		.12	8	24	72	5.75	
	1550	Spade, 25 lb.		.45	6.65	20	60	7.60	
	1560	Tamper, single, 35 lb.		.55	36.50	109	325	26.20	
	1570	Triple, 140 lb.		.82	54.50	164	490	39.35	
	1580	Wrenches, impact, air powered, up to 3/4" bolt		.40	12.65	38	114	10.80	
	1590	Up to 1-1/4" bolt		.50	23.50	70	210	18	
	1600	Barricades, barrels, reflectorized, 1 to 99 barrels		.03	4.60	13.80	41.50	3	
	1610	100 to 200 barrels		.02	3.53	10.60	32	2.30	
	1620	Barrels with flashers, 1 to 99 barrels		.03	5.25	15.80	47.50	3.40	
	1630	100 to 200 barrels		.03	4.20	12.60	38	2.75	
	1640	Barrels with steady burn type C lights		.04	7	21	63	4.50	
	1650	Illuminated board, trailer mounted, with generator		3.50	132	395	1,175	107	
	1670	Portable barricade, stock, with flashers, 1 to 6 units		.03	5.25	15.80	47.50	3.40	
	1680	25 to 50 units		.03	4.90	14.70	44	3.20	
	1685	Butt fusion machine, wheeled, 1.5 HP electric, 2" - 8" diameter pipe		2.62	167	500	1,500	120.95	
	1690	Tracked, 20 HP diesel, 4"-12" diameter pipe		11.20	490	1,465	4,400	382.60	
	1695	83 HP diesel, 8" - 24" diameter pipe		30.71	975	2,930	8,800	831.70	
	1700	Carts, brick, hand powered, 1000 lb. capacity		.44	72.50	218	655	47.10	
	1800	Gas engine, 1500 lb., 7-1/2' lift		4.17	115	345	1,025	102.35	
	1822	Dehumidifier, medium, 6 lb./hr., 150 CFM		.96	60	180	540	43.70	
	1824	Large, 18 lb./hr., 600 CFM		1.95	122	366	1,100	88.80	
	1830	Distributor, asphalt, trailer mounted, 2000 gal., 38 H.P. diesel		9.90	325	980	2,950	275.20	
	1840	3000 gal., 38 H.P. diesel		11.40	355	1,070	3,200	305.20	
	1850	Drill, rotary hammer, electric		.88	26	78	234	22.65	
	1860	Carbide bit, 1-1/2" diameter, add to electric rotary hammer		.02	3.98	11.95	36	2.55	
	1865	Rotary, crawler, 250 H.P.		144.40	2,050	6,120	18,400	2,379	
	1870	Emulsion sprayer, 65 gal., 5 H.P. gas engine		2.91	97	291	875	81.50	
	1880	200 gal., 5 H.P. engine		7.75	162	485	1,450	159	
	1900	Floor auto-scrubbing machine, walk-behind, 28" path		4.02	260	780	2,350	188.15	
	1930	Floodlight, mercury vapor, or quartz, on tripod, 1000 watt		.40	18.65	56	168	14.40	
	1940	2000 watt		.74	36.50	110	330	27.90	
	1950	Floodlights, trailer mounted with generator, 1 - 300 watt light		3.60	71.50	215	645	71.80	
	1960	2 - 1000 watt lights		4.85	98.50	295	885	97.80	
	2000	4 - 300 watt lights		4.55	93.50	280	840	92.40	
	2005	Foam spray rig, incl. box trailer, compressor, generator, proportioner		33.98	485	1,455	4,375	562.85	
	2020	Forklift, straight mast, 12' lift, 5000 lb., 2 wheel drive, gas		25.50	202	605	1,825	325	
	2040	21' lift, 5000 lb., 4 wheel drive, diesel		20.40	245	735	2,200	310.20	
	2050	For rough terrain, 42' lift, 35' reach, 9000 lb., 110 H.P.		28.05	480	1,440	4,325	512.40	
	2060	For plant, 4 ton capacity, 80 H.P., 2 wheel drive, gas		15.60	93.50	280	840	180.80	
	2080	10 ton capacity, 120 H.P., 2 wheel drive, diesel		23.25	163	490	1,475	284	
	2100	Generator, electric, gas engine, 1.5 kW to 3 kW		3.60	11.35	34	102	35.60	
	2200	5 kW		4.65	15	45	135	46.20	
	2300	10 kW		8.80	36.50	110	330	92.40	
	2400	25 kW	↓	10.25	83.50	250	750	132	

01 54 33 | Equipment Rental

			UNIT	HOURLY OPER. COST	RENT PER DAY	RENT PER WEEK	RENT PER MONTH	EQUIPMENT COST/DAY	
40	2500	Diesel engine, 20 kW	Ea.	11.75	68.50	205	615	135	40
	2600	50 kW		22.65	103	310	930	243.20	
	2700	100 kW		41.85	132	395	1,175	413.80	
	2800	250 kW		82.40	235	705	2,125	800.20	
	2850	Hammer, hydraulic, for mounting on boom, to 500 ft lb.		2.50	75	225	675	65	
	2860	1000 ft lb.		4.25	127	380	1,150	110	
	2900	Heaters, space, oil or electric, 50 MBH		1.98	7.65	23	69	20.45	
	3000	100 MBH		3.57	10.65	32	96	34.95	
	3100	300 MBH		11.46	38.50	115	345	114.70	
	3150	500 MBH		23.09	45	135	405	211.70	
	3200	Hose, water, suction with coupling, 20′ long, 2″ diameter		.02	3	9	27	1.95	
	3210	3″ diameter		.03	4.33	13	39	2.85	
	3220	4″ diameter		.03	5	15	45	3.25	
	3230	6″ diameter		.10	17.35	52	156	11.20	
	3240	8″ diameter		.20	33.50	100	300	21.60	
	3250	Discharge hose with coupling, 50′ long, 2″ diameter		.02	2.67	8	24	1.75	
	3260	3″ diameter		.03	4.67	14	42	3.05	
	3270	4″ diameter		.04	7.35	22	66	4.70	
	3280	6″ diameter		.11	18.65	56	168	12.10	
	3290	8″ diameter		.20	33.50	100	300	21.60	
	3295	Insulation blower		.68	6	18	54	9.05	
	3300	Ladders, extension type, 16′ to 36′ long		.14	24	72	216	15.50	
	3400	40′ to 60′ long		.20	32.50	98	294	21.20	
	3405	Lance for cutting concrete		2.73	104	313	940	84.45	
	3407	Lawn mower, rotary, 22″, 5 H.P.		2.21	63.50	190	570	55.70	
	3408	48″ self propelled		2.98	75	225	675	68.85	
	3410	Level, electronic, automatic, with tripod and leveling rod		1.62	108	323	970	77.55	
	3430	Laser type, for pipe and sewer line and grade		.73	48.50	145	435	34.85	
	3440	Rotating beam for interior control		1.16	77	231	695	55.50	
	3460	Builder's optical transit, with tripod and rod		.09	15.65	47	141	10.10	
	3500	Light towers, towable, with diesel generator, 2000 watt		4.55	93.50	280	840	92.40	
	3600	4000 watt		4.85	98.50	295	885	97.80	
	3700	Mixer, powered, plaster and mortar, 6 C.F., 7 H.P.		2.85	19.65	59	177	34.60	
	3800	10 C.F., 9 H.P.		3.00	30.50	92	276	42.40	
	3850	Nailer, pneumatic		.48	31.50	95	285	22.85	
	3900	Paint sprayers complete, 8 CFM		.83	55.50	166	500	39.85	
	4000	17 CFM		1.47	98	294	880	70.55	
	4020	Pavers, bituminous, rubber tires, 8′ wide, 50 H.P., diesel		31.05	490	1,475	4,425	543.40	
	4030	10′ wide, 150 H.P.		104.70	1,800	5,395	16,200	1,917	
	4050	Crawler, 8′ wide, 100 H.P., diesel		88.45	1,825	5,495	16,500	1,807	
	4060	10′ wide, 150 H.P.		111.70	2,200	6,635	19,900	2,221	
	4070	Concrete paver, 12′ to 24′ wide, 250 H.P.		101.10	1,550	4,640	13,900	1,737	
	4080	Placer-spreader-trimmer, 24′ wide, 300 H.P.		146.00	2,500	7,495	22,500	2,667	
	4100	Pump, centrifugal gas pump, 1-1/2″ diam., 65 GPM		3.90	50	150	450	61.20	
	4200	2″ diameter, 130 GPM		5.30	55	165	495	75.40	
	4300	3″ diameter, 250 GPM		5.60	56.50	170	510	78.80	
	4400	6″ diameter, 1500 GPM		29.55	175	525	1,575	341.40	
	4500	Submersible electric pump, 1-1/4″ diameter, 55 GPM		.37	16.35	49	147	12.75	
	4600	1-1/2″ diameter, 83 GPM		.41	18.65	56	168	14.50	
	4700	2″ diameter, 120 GPM		1.35	23.50	70	210	24.80	
	4800	3″ diameter, 300 GPM		2.30	41.50	125	375	43.40	
	4900	4″ diameter, 560 GPM		9.75	158	475	1,425	173	
	5000	6″ diameter, 1590 GPM		14.35	213	640	1,925	242.80	
	5100	Diaphragm pump, gas, single, 1-1/2″ diameter		1.20	50.50	152	455	40	
	5200	2″ diameter		4.20	63.50	190	570	71.60	
	5300	3″ diameter		4.25	63.50	190	570	72	
	5400	Double, 4″ diameter		6.30	107	320	960	114.40	
	5450	Pressure washer 5 GPM, 3000 psi	↓	4.80	51.50	155	465	69.40	

01 54 33 | Equipment Rental

			UNIT	HOURLY OPER. COST	RENT PER DAY	RENT PER WEEK	RENT PER MONTH	EQUIPMENT COST/DAY	
40	5460	7 GPM, 3000 psi	Ea.	6.30	60	180	540	86.40	40
	5500	Trash pump, self-priming, gas, 2″ diameter		4.50	21.50	64	192	48.80	
	5600	Diesel, 4″ diameter		8.95	90	270	810	125.60	
	5650	Diesel, 6″ diameter		24.45	150	450	1,350	285.60	
	5655	Grout Pump		26.60	262	785	2,350	369.80	
	5700	Salamanders, L.P. gas fired, 100,000 Btu		3.59	13.65	41	123	36.90	
	5705	50,000 Btu		2.69	10.35	31	93	27.70	
	5720	Sandblaster, portable, open top, 3 C.F. capacity		.55	26	78	234	20	
	5730	6 C.F. capacity		.90	38.50	115	345	30.20	
	5740	Accessories for above		.13	21	63	189	13.65	
	5750	Sander, floor		.76	19.65	59	177	17.90	
	5760	Edger		.62	21.50	64	192	17.75	
	5800	Saw, chain, gas engine, 18″ long		2.25	21	63	189	30.60	
	5900	Hydraulic powered, 36″ long		.75	65	195	585	45	
	5950	60″ long		.75	66.50	200	600	46	
	6000	Masonry, table mounted, 14″ diameter, 5 H.P.		1.31	56.50	170	510	44.50	
	6050	Portable cut-off, 8 H.P.		2.45	32.50	97	291	39	
	6100	Circular, hand held, electric, 7-1/4″ diameter		.18	4.33	13	39	4.05	
	6200	12″ diameter		.25	7.65	23	69	6.60	
	6250	Wall saw, w/hydraulic power, 10 H.P.		10.45	60	180	540	119.60	
	6275	Shot blaster, walk-behind, 20″ wide		4.65	288	865	2,600	210.20	
	6280	Sidewalk broom, walk-behind		2.15	60.50	182	545	53.60	
	6300	Steam cleaner, 100 gallons per hour		3.70	76.50	230	690	75.60	
	6310	200 gallons per hour		5.25	95	285	855	99	
	6340	Tar Kettle/Pot, 400 gallons		5.81	76.50	230	690	92.50	
	6350	Torch, cutting, acetylene-oxygen, 150′ hose, excludes gases		.30	15	45	135	11.40	
	6360	Hourly operating cost includes tips and gas		16.65				133.20	
	6410	Toilet, portable chemical		.12	20.50	61	183	13.15	
	6420	Recycle flush type		.15	24.50	73	219	15.80	
	6430	Toilet, fresh water flush, garden hose,		.17	29	87	261	18.75	
	6440	Hoisted, non-flush, for high rise		.14	24	72	216	15.50	
	6465	Tractor, farm with attachment		21.70	282	845	2,525	342.60	
	6500	Trailers, platform, flush deck, 2 axle, 25 ton capacity		5.40	117	350	1,050	113.20	
	6600	40 ton capacity		6.95	162	485	1,450	152.60	
	6700	3 axle, 50 ton capacity		7.50	180	540	1,625	168	
	6800	75 ton capacity		9.30	235	705	2,125	215.40	
	6810	Trailer mounted cable reel for high voltage line work		5.24	250	749	2,250	191.70	
	6820	Trailer mounted cable tensioning rig		10.43	495	1,490	4,475	381.45	
	6830	Cable pulling rig		70.67	2,775	8,340	25,000	2,233	
	6900	Water tank trailer, engine driven discharge, 5000 gallons		6.90	147	440	1,325	143.20	
	6925	10,000 gallons		9.45	202	605	1,825	196.60	
	6950	Water truck, off highway, 6000 gallons		89.15	790	2,375	7,125	1,188	
	7010	Tram car for high voltage line work, powered, 2 conductor		6.04	136	407	1,225	129.70	
	7020	Transit (builder's level) with tripod		.09	15.65	47	141	10.10	
	7030	Trench box, 3000 lb., 6′ x 8′		.51	84.50	254	760	54.90	
	7040	7200 lb., 6′ x 20′		1.05	175	525	1,575	113.40	
	7050	8000 lb., 8′ x 16′		1.07	178	533	1,600	115.15	
	7060	9500 lb., 8′ x 20′		1.19	199	597	1,800	128.90	
	7065	11,000 lb., 8′ x 24′		1.26	209	628	1,875	135.70	
	7070	12,000 lb., 10′ x 20′		1.36	227	680	2,050	146.90	
	7100	Truck, pickup, 3/4 ton, 2 wheel drive		14.95	58.50	175	525	154.60	
	7200	4 wheel drive		15.20	73.50	220	660	165.60	
	7250	Crew carrier, 9 passenger		21.15	86.50	260	780	221.20	
	7290	Flat bed truck, 20,000 lb. GVW		22.20	127	380	1,150	253.60	
	7300	Tractor, 4 x 2, 220 H.P.		30.85	200	600	1,800	366.80	
	7410	330 H.P.		45.80	275	825	2,475	531.40	
	7500	6 x 4, 380 H.P.		52.55	320	960	2,875	612.40	
	7600	450 H.P.		63.65	385	1,160	3,475	741.20	

01 54 33 | Equipment Rental

		01 54 33 | Equipment Rental	UNIT	HOURLY OPER. COST	RENT PER DAY	RENT PER WEEK	RENT PER MONTH	EQUIPMENT COST/DAY	
40	7610	Tractor, with A frame, boom and winch, 225 H.P.	Ea.	33.65	275	825	2,475	434.20	40
	7620	Vacuum truck, hazardous material, 2500 gallons		11.85	295	885	2,650	271.80	
	7625	5,000 gallons		14.49	415	1,240	3,725	363.90	
	7650	Vacuum, HEPA, 16 gallon, wet/dry		.82	18	54	162	17.35	
	7655	55 gallon, wet/dry		.75	27	81	243	22.20	
	7660	Water tank, portable		.17	28.50	85.50	257	18.45	
	7690	Sewer/catch basin vacuum, 14 C.Y., 1500 gallons		18.83	620	1,860	5,575	522.65	
	7700	Welder, electric, 200 amp		3.48	16.35	49	147	37.65	
	7800	300 amp		5.13	19.65	59	177	52.85	
	7900	Gas engine, 200 amp		14.10	23.50	70	210	126.80	
	8000	300 amp		16.15	24.50	74	222	144	
	8100	Wheelbarrow, any size		.08	12.65	38	114	8.25	
	8200	Wrecking ball, 4000 lb.		2.30	70	210	630	60.40	
50	0010	**HIGHWAY EQUIPMENT RENTAL** without operators R015433-10							50
	0050	Asphalt batch plant, portable drum mixer, 100 ton/hr.	Ea.	71.30	1,425	4,285	12,900	1,427	
	0060	200 ton/hr.		79.90	1,525	4,550	13,700	1,549	
	0070	300 ton/hr.		92.55	1,775	5,350	16,100	1,810	
	0100	Backhoe attachment, long stick, up to 185 H.P., 10.5′ long		.34	22.50	68	204	16.30	
	0140	Up to 250 H.P., 12′ long		.37	24.50	73	219	17.55	
	0180	Over 250 H.P., 15′ long		.51	33.50	101	305	24.30	
	0200	Special dipper arm, up to 100 H.P., 32′ long		1.04	69	207	620	49.70	
	0240	Over 100 H.P., 33′ long		1.29	86	258	775	61.90	
	0280	Catch basin/sewer cleaning truck, 3 ton, 9 C.Y., 1000 gal.		45.45	395	1,190	3,575	601.60	
	0300	Concrete batch plant, portable, electric, 200 C.Y./hr.		29.60	510	1,535	4,600	543.80	
	0520	Grader/dozer attachment, ripper/scarifier, rear mounted, up to 135 H.P.		3.45	65	195	585	66.60	
	0540	Up to 180 H.P.		4.00	83.50	250	750	82	
	0580	Up to 250 H.P.		4.40	95	285	855	92.20	
	0700	Pvmt. removal bucket, for hyd. excavator, up to 90 H.P.		1.85	51.50	155	465	45.80	
	0740	Up to 200 H.P.		2.10	73.50	220	660	60.80	
	0780	Over 200 H.P.		2.20	85	255	765	68.60	
	0900	Aggregate spreader, self-propelled, 187 H.P.		53.20	690	2,065	6,200	838.60	
	1000	Chemical spreader, 3 C.Y.		3.35	43.50	130	390	52.80	
	1900	Hammermill, traveling, 250 H.P.		78.38	2,000	6,000	18,000	1,827	
	2000	Horizontal borer, 3″ diameter, 13 H.P. gas driven		3.20	53.50	160	480	57.60	
	2150	Horizontal directional drill, 20,000 lb. thrust, 78 H.P. diesel		30.30	655	1,965	5,900	635.40	
	2160	30,000 lb. thrust, 115 H.P.		37.80	1,000	3,010	9,025	904.40	
	2170	50,000 lb. thrust, 170 H.P.		54.30	1,275	3,845	11,500	1,203	
	2190	Mud trailer for HDD, 1500 gallons, 175 H.P., gas		34.65	155	465	1,400	370.20	
	2200	Hydromulcher, diesel, 3000 gallon, for truck mounting		23.90	250	750	2,250	341.20	
	2300	Gas, 600 gallon		9.00	98.50	295	885	131	
	2400	Joint & crack cleaner, walk behind, 25 H.P.		4.25	50	150	450	64	
	2500	Filler, trailer mounted, 400 gallons, 20 H.P.		9.55	213	640	1,925	204.40	
	3000	Paint striper, self-propelled, 40 gallon, 22 H.P.		7.50	157	470	1,400	154	
	3100	120 gallon, 120 H.P.		24.45	405	1,210	3,625	437.60	
	3200	Post drivers, 6″ I-Beam frame, for truck mounting		19.55	390	1,175	3,525	391.40	
	3400	Road sweeper, self-propelled, 8′ wide, 90 H.P.		39.85	575	1,720	5,150	662.80	
	3450	Road sweeper, vacuum assisted, 4 C.Y., 220 gallons		76.95	625	1,875	5,625	990.60	
	4000	Road mixer, self-propelled, 130 H.P.		46.55	765	2,295	6,875	831.40	
	4100	310 H.P.		81.85	2,100	6,285	18,900	1,912	
	4220	Cold mix paver, incl. pug mill and bitumen tank, 165 H.P.		96.15	2,275	6,850	20,600	2,139	
	4240	Pavement brush, towed		3.15	93.50	280	840	81.20	
	4250	Paver, asphalt, wheel or crawler, 130 H.P., diesel		95.80	2,275	6,790	20,400	2,124	
	4300	Paver, road widener, gas 1′ to 6′, 67 H.P.		48.00	895	2,685	8,050	921	
	4400	Diesel, 2′ to 14′, 88 H.P.		61.75	1,050	3,135	9,400	1,121	
	4600	Slipform pavers, curb and gutter, 2 track, 75 H.P.		46.65	715	2,140	6,425	801.20	
	4700	4 track, 165 H.P.		41.30	735	2,210	6,625	772.40	
	4800	Median barrier, 215 H.P.		47.30	740	2,225	6,675	823.40	

01 54 33 | Equipment Rental

			UNIT	HOURLY OPER. COST	RENT PER DAY	RENT PER WEEK	RENT PER MONTH	EQUIPMENT COST/DAY	
50	4901	Trailer, low bed, 75 ton capacity	Ea.	10.05	233	700	2,100	220.40	50
	5000	Road planer, walk behind, 10″ cutting width, 10 H.P.		3.90	33.50	100	300	51.20	
	5100	Self-propelled, 12″ cutting width, 64 H.P.		11.25	115	345	1,025	159	
	5120	Traffic line remover, metal ball blaster, truck mounted, 115 H.P.		47.40	740	2,220	6,650	823.20	
	5140	Grinder, truck mounted, 115 H.P.		53.80	800	2,405	7,225	911.40	
	5160	Walk-behind, 11 H.P.		4.45	53.50	160	480	67.60	
	5200	Pavement profiler, 4′ to 6′ wide, 450 H.P.		249.60	3,325	9,960	29,900	3,989	
	5300	8′ to 10′ wide, 750 H.P.		391.40	4,350	13,070	39,200	5,745	
	5400	Roadway plate, steel, 1″ x 8′ x 20′		.08	12.65	38	114	8.25	
	5600	Stabilizer, self-propelled, 150 H.P.		48.35	605	1,815	5,450	749.80	
	5700	310 H.P.		96.45	1,700	5,075	15,200	1,787	
	5800	Striper, truck mounted, 120 gallon paint, 460 H.P.		69.75	490	1,475	4,425	853	
	5900	Thermal paint heating kettle, 115 gallons		3.88	25.50	77	231	46.45	
	6000	Tar kettle, 330 gallon, trailer mounted		5.46	58.50	175	525	78.70	
	7000	Tunnel locomotive, diesel, 8 to 12 ton		31.25	565	1,695	5,075	589	
	7005	Electric, 10 ton		25.70	645	1,940	5,825	593.60	
	7010	Muck cars, 1/2 C.Y. capacity		2.00	23.50	71	213	30.20	
	7020	1 C.Y. capacity		2.20	31.50	95	285	36.60	
	7030	2 C.Y. capacity		2.35	36.50	110	330	40.80	
	7040	Side dump, 2 C.Y. capacity		2.55	43.50	130	390	46.40	
	7050	3 C.Y. capacity		3.45	50	150	450	57.60	
	7060	5 C.Y. capacity		4.90	63.50	190	570	77.20	
	7100	Ventilating blower for tunnel, 7-1/2 H.P.		1.93	48.50	145	435	44.45	
	7110	10 H.P.		2.14	50	150	450	47.10	
	7120	20 H.P.		3.23	65	195	585	64.85	
	7140	40 H.P.		5.35	93.50	280	840	98.80	
	7160	60 H.P.		8.18	145	435	1,300	152.45	
	7175	75 H.P.		10.59	193	580	1,750	200.70	
	7180	200 H.P.		21.46	290	870	2,600	345.70	
	7800	Windrow loader, elevating	↓	46.55	1,300	3,925	11,800	1,157	
60	0010	**LIFTING AND HOISTING EQUIPMENT RENTAL** without operators (R015433-10)							60
	0120	Aerial lift truck, 2 person, to 80′	Ea.	29.10	695	2,090	6,275	650.80	
	0140	Boom work platform, 40′ snorkel (R015433-15)		18.10	273	820	2,450	308.80	
	0150	Crane, flatbed mounted, 3 ton capacity		16.00	192	575	1,725	243	
	0200	Crane, climbing, 106′ jib, 6000 lb. capacity, 410 fpm (R312316-45)		36.56	1,575	4,760	14,300	1,244	
	0300	101′ jib, 10,250 lb. capacity, 270 fpm		42.96	2,025	6,040	18,100	1,552	
	0500	Tower, static, 130′ high, 106′ jib, 6200 lb. capacity at 400 fpm		40.31	1,825	5,510	16,500	1,424	
	0600	Crawler mounted, lattice boom, 1/2 C.Y., 15 tons at 12′ radius		36.34	630	1,890	5,675	668.70	
	0700	3/4 C.Y., 20 tons at 12′ radius		48.45	790	2,370	7,100	861.60	
	0800	1 C.Y., 25 tons at 12′ radius		64.60	1,050	3,155	9,475	1,148	
	0900	1-1/2 C.Y., 40 tons at 12′ radius		64.55	1,050	3,180	9,550	1,152	
	1000	2 C.Y., 50 tons at 12′ radius		68.45	1,250	3,720	11,200	1,292	
	1100	3 C.Y., 75 tons at 12′ radius		73.30	1,450	4,355	13,100	1,457	
	1200	100 ton capacity, 60′ boom		82.95	1,675	5,010	15,000	1,666	
	1300	165 ton capacity, 60′ boom		105.60	1,950	5,865	17,600	2,018	
	1400	200 ton capacity, 70′ boom		127.95	2,450	7,335	22,000	2,491	
	1500	350 ton capacity, 80′ boom		178.85	3,675	11,040	33,100	3,639	
	1600	Truck mounted, lattice boom, 6 x 4, 20 tons at 10′ radius		37.54	1,100	3,290	9,875	958.30	
	1700	25 tons at 10′ radius		40.54	1,200	3,580	10,700	1,040	
	1800	8 x 4, 30 tons at 10′ radius		44.09	1,275	3,810	11,400	1,115	
	1900	40 tons at 12′ radius		47.21	1,325	3,980	11,900	1,174	
	2000	60 tons at 15′ radius		53.66	1,400	4,210	12,600	1,271	
	2050	82 tons at 15′ radius		60.53	1,500	4,500	13,500	1,384	
	2100	90 tons at 15′ radius		68.16	1,625	4,900	14,700	1,525	
	2200	115 tons at 15′ radius		77.06	1,825	5,480	16,400	1,712	
	2300	150 tons at 18′ radius		88.40	1,925	5,770	17,300	1,861	
	2350	165 tons at 18′ radius		91.22	2,050	6,120	18,400	1,954	
	2400	Truck mounted, hydraulic, 12 ton capacity	↓	41.40	525	1,575	4,725	646.20	

01 54 33 | Equipment Rental

		01 54 33 | Equipment Rental	UNIT	HOURLY OPER. COST	RENT PER DAY	RENT PER WEEK	RENT PER MONTH	EQUIPMENT COST/DAY	
60	2500	25 ton capacity	Ea.	43.70	630	1,895	5,675	728.60	60
	2550	33 ton capacity		44.25	650	1,945	5,825	743	
	2560	40 ton capacity		57.30	755	2,270	6,800	912.40	
	2600	55 ton capacity		74.65	865	2,590	7,775	1,115	
	2700	80 ton capacity		97.80	1,375	4,155	12,500	1,613	
	2720	100 ton capacity		91.75	1,425	4,290	12,900	1,592	
	2740	120 ton capacity		106.10	1,525	4,610	13,800	1,771	
	2760	150 ton capacity		124.25	2,025	6,075	18,200	2,209	
	2800	Self-propelled, 4 x 4, with telescoping boom, 5 ton		17.15	228	685	2,050	274.20	
	2900	12-1/2 ton capacity		31.50	365	1,095	3,275	471	
	3000	15 ton capacity		32.15	385	1,150	3,450	487.20	
	3050	20 ton capacity		35.00	445	1,340	4,025	548	
	3100	25 ton capacity		36.55	500	1,505	4,525	593.40	
	3150	40 ton capacity		44.85	565	1,690	5,075	696.80	
	3200	Derricks, guy, 20 ton capacity, 60′ boom, 75′ mast		27.52	390	1,167	3,500	453.55	
	3300	100′ boom, 115′ mast		43.10	670	2,010	6,025	746.80	
	3400	Stiffleg, 20 ton capacity, 70′ boom, 37′ mast		29.99	505	1,520	4,550	543.90	
	3500	100′ boom, 47′ mast		46.04	810	2,430	7,300	854.30	
	3550	Helicopter, small, lift to 1250 lb. maximum, w/pilot		101.04	3,150	9,420	28,300	2,692	
	3600	Hoists, chain type, overhead, manual, 3/4 ton		.10	.33	1	3	1	
	3900	10 ton		.70	6	18	54	9.20	
	4000	Hoist and tower, 5000 lb. cap., portable electric, 40′ high		4.64	224	672	2,025	171.50	
	4100	For each added 10′ section, add		.11	17.65	53	159	11.50	
	4200	Hoist and single tubular tower, 5000 lb. electric, 100′ high		6.29	315	938	2,825	237.90	
	4300	For each added 6′-6″ section, add		.18	30.50	91	273	19.65	
	4400	Hoist and double tubular tower, 5000 lb., 100′ high		6.76	345	1,033	3,100	260.70	
	4500	For each added 6′-6″ section, add		.20	33.50	101	305	21.80	
	4550	Hoist and tower, mast type, 6000 lb., 100′ high		7.27	355	1,072	3,225	272.55	
	4570	For each added 10′ section, add		.13	21	63	189	13.65	
	4600	Hoist and tower, personnel, electric, 2000 lb., 100′ @ 125 fpm		15.53	950	2,850	8,550	694.25	
	4700	3000 lb., 100′ @ 200 fpm		17.75	1,075	3,230	9,700	788	
	4800	3000 lb., 150′ @ 300 fpm		19.70	1,200	3,620	10,900	881.60	
	4900	4000 lb., 100′ @ 300 fpm		20.36	1,225	3,690	11,100	900.90	
	5000	6000 lb., 100′ @ 275 fpm		21.90	1,300	3,870	11,600	949.20	
	5100	For added heights up to 500′, add	L.F.	.01	1.67	5	15	1.10	
	5200	Jacks, hydraulic, 20 ton	Ea.	.05	2	6	18	1.60	
	5500	100 ton		.40	11.65	35	105	10.20	
	6100	Jacks, hydraulic, climbing w/50′ jackrods, control console, 30 ton cap.		1.93	129	386	1,150	92.65	
	6150	For each added 10′ jackrod section, add		.05	3.33	10	30	2.40	
	6300	50 ton capacity		3.11	207	621	1,875	149.10	
	6350	For each added 10′ jackrod section, add		.06	4	12	36	2.90	
	6500	125 ton capacity		8.15	545	1,630	4,900	391.20	
	6550	For each added 10′ jackrod section, add		.56	37	111	335	26.70	
	6600	Cable jack, 10 ton capacity with 200′ cable		1.62	108	323	970	77.55	
	6650	For each added 50′ of cable, add		.19	12.65	38	114	9.10	
70	0010	**WELLPOINT EQUIPMENT RENTAL** without operators R015433-10							70
	0020	Based on 2 months rental							
	0100	Combination jetting & wellpoint pump, 60 H.P. diesel	Ea.	18.31	320	957	2,875	337.90	
	0200	High pressure gas jet pump, 200 H.P., 300 psi	"	44.43	273	818	2,450	519.05	
	0300	Discharge pipe, 8″ diameter	L.F.	.01	.52	1.56	4.68	.40	
	0350	12″ diameter		.01	.76	2.28	6.85	.55	
	0400	Header pipe, flows up to 150 GPM, 4″ diameter		.01	.47	1.41	4.23	.35	
	0500	400 GPM, 6″ diameter		.01	.55	1.66	4.98	.40	
	0600	800 GPM, 8″ diameter		.01	.76	2.28	6.85	.55	
	0700	1500 GPM, 10″ diameter		.01	.80	2.41	7.25	.55	
	0800	2500 GPM, 12″ diameter		.02	1.52	4.56	13.70	1.05	
	0900	4500 GPM, 16″ diameter		.03	1.94	5.83	17.50	1.40	

01 54 33 | Equipment Rental

			UNIT	HOURLY OPER. COST	RENT PER DAY	RENT PER WEEK	RENT PER MONTH	EQUIPMENT COST/DAY	
70	0950	For quick coupling aluminum and plastic pipe, add	L.F.	.03	2.01	6.04	18.10	1.45	70
	1100	Wellpoint, 25′ long, with fittings & riser pipe, 1-1/2″ or 2″ diameter	Ea.	.06	4.02	12.05	36	2.90	
	1200	Wellpoint pump, diesel powered, 4″ suction, 20 H.P.		7.73	184	552	1,650	172.25	
	1300	6″ suction, 30 H.P.		10.59	228	684	2,050	221.50	
	1400	8″ suction, 40 H.P.		14.31	315	938	2,825	302.10	
	1500	10″ suction, 75 H.P.		22.19	365	1,097	3,300	396.90	
	1600	12″ suction, 100 H.P.		31.53	580	1,740	5,225	600.25	
	1700	12″ suction, 175 H.P.	▼	47.37	645	1,930	5,800	764.95	
80	0010	**MARINE EQUIPMENT RENTAL** without operators R015433-10							80
	0200	Barge, 400 Ton, 30′ wide x 90′ long	Ea.	17.10	1,075	3,235	9,700	783.80	
	0240	800 Ton, 45′ wide x 90′ long		20.75	1,300	3,930	11,800	952	
	2000	Tugboat, diesel, 100 H.P.		38.60	218	655	1,975	439.80	
	2040	250 H.P.		81.00	400	1,195	3,575	887	
	2080	380 H.P.		158.50	1,200	3,565	10,700	1,981	
	3000	Small work boat, gas, 16-foot, 50 H.P.		18.80	61.50	185	555	187.40	
	4000	Large, diesel, 48-foot, 200 H.P.	▼	89.95	1,250	3,760	11,300	1,472	

Crew No.	Bare Costs		Incl. Subs O&P		Cost Per Labor-Hour	
Crew A-1	**Hr.**	**Daily**	**Hr.**	**Daily**	**Bare Costs**	**Incl. O&P**
1 Building Laborer	$35.45	$283.60	$54.60	$436.80	$35.45	$54.60
1 Concrete Saw, Gas Manual		84.00		92.40	10.50	11.55
8 L.H., Daily Totals		$367.60		$529.20	$45.95	$66.15
Crew A-1A	**Hr.**	**Daily**	**Hr.**	**Daily**	**Bare Costs**	**Incl. O&P**
1 Skilled Worker	$46.20	$369.60	$71.45	$571.60	$46.20	$71.45
1 Shot Blaster, 20"		210.20		231.22	26.27	28.90
8 L.H., Daily Totals		$579.80		$802.82	$72.47	$100.35
Crew A-1B	**Hr.**	**Daily**	**Hr.**	**Daily**	**Bare Costs**	**Incl. O&P**
1 Building Laborer	$35.45	$283.60	$54.60	$436.80	$35.45	$54.60
1 Concrete Saw		175.20		192.72	21.90	24.09
8 L.H., Daily Totals		$458.80		$629.52	$57.35	$78.69
Crew A-1C	**Hr.**	**Daily**	**Hr.**	**Daily**	**Bare Costs**	**Incl. O&P**
1 Building Laborer	$35.45	$283.60	$54.60	$436.80	$35.45	$54.60
1 Chain Saw, Gas, 18"		30.60		33.66	3.83	4.21
8 L.H., Daily Totals		$314.20		$470.46	$39.27	$58.81
Crew A-1D	**Hr.**	**Daily**	**Hr.**	**Daily**	**Bare Costs**	**Incl. O&P**
1 Building Laborer	$35.45	$283.60	$54.60	$436.80	$35.45	$54.60
1 Vibrating Plate, Gas, 18"		34.60		38.06	4.33	4.76
8 L.H., Daily Totals		$318.20		$474.86	$39.77	$59.36
Crew A-1E	**Hr.**	**Daily**	**Hr.**	**Daily**	**Bare Costs**	**Incl. O&P**
1 Building Laborer	$35.45	$283.60	$54.60	$436.80	$35.45	$54.60
1 Vibrating Plate, Gas, 21"		44.40		48.84	5.55	6.11
8 L.H., Daily Totals		$328.00		$485.64	$41.00	$60.70
Crew A-1F	**Hr.**	**Daily**	**Hr.**	**Daily**	**Bare Costs**	**Incl. O&P**
1 Building Laborer	$35.45	$283.60	$54.60	$436.80	$35.45	$54.60
1 Rammer/Tamper, Gas, 8"		47.20		51.92	5.90	6.49
8 L.H., Daily Totals		$330.80		$488.72	$41.35	$61.09
Crew A-1G	**Hr.**	**Daily**	**Hr.**	**Daily**	**Bare Costs**	**Incl. O&P**
1 Building Laborer	$35.45	$283.60	$54.60	$436.80	$35.45	$54.60
1 Rammer/Tamper, Gas, 15"		53.60		58.96	6.70	7.37
8 L.H., Daily Totals		$337.20		$495.76	$42.15	$61.97
Crew A-1H	**Hr.**	**Daily**	**Hr.**	**Daily**	**Bare Costs**	**Incl. O&P**
1 Building Laborer	$35.45	$283.60	$54.60	$436.80	$35.45	$54.60
1 Exterior Steam Cleaner		75.60		83.16	9.45	10.40
8 L.H., Daily Totals		$359.20		$519.96	$44.90	$65.00
Crew A-1J	**Hr.**	**Daily**	**Hr.**	**Daily**	**Bare Costs**	**Incl. O&P**
1 Building Laborer	$35.45	$283.60	$54.60	$436.80	$35.45	$54.60
1 Cultivator, Walk-Behind, 5 H.P.		59.00		64.90	7.38	8.11
8 L.H., Daily Totals		$342.60		$501.70	$42.83	$62.71
Crew A-1K	**Hr.**	**Daily**	**Hr.**	**Daily**	**Bare Costs**	**Incl. O&P**
1 Building Laborer	$35.45	$283.60	$54.60	$436.80	$35.45	$54.60
1 Cultivator, Walk-Behind, 8 H.P.		90.55		99.61	11.32	12.45
8 L.H., Daily Totals		$374.15		$536.40	$46.77	$67.05
Crew A-1M	**Hr.**	**Daily**	**Hr.**	**Daily**	**Bare Costs**	**Incl. O&P**
1 Building Laborer	$35.45	$283.60	$54.60	$436.80	$35.45	$54.60
1 Snow Blower, Walk-Behind		53.60		58.96	6.70	7.37
8 L.H., Daily Totals		$337.20		$495.76	$42.15	$61.97

Crew No.	Bare Costs		Incl. Subs O&P		Cost Per Labor-Hour	
Crew A-2	**Hr.**	**Daily**	**Hr.**	**Daily**	**Bare Costs**	**Incl. O&P**
2 Laborers	$35.45	$567.20	$54.60	$873.60	$35.52	$54.60
1 Truck Driver (light)	35.65	285.20	54.60	436.80		
1 Flatbed Truck, Gas, 1.5 Ton		265.80		292.38	11.07	12.18
24 L.H., Daily Totals		$1118.20		$1602.78	$46.59	$66.78
Crew A-2A	**Hr.**	**Daily**	**Hr.**	**Daily**	**Bare Costs**	**Incl. O&P**
2 Laborers	$35.45	$567.20	$54.60	$873.60	$35.52	$54.60
1 Truck Driver (light)	35.65	285.20	54.60	436.80		
1 Flatbed Truck, Gas, 1.5 Ton		265.80		292.38		
1 Concrete Saw		175.20		192.72	18.38	20.21
24 L.H., Daily Totals		$1293.40		$1795.50	$53.89	$74.81
Crew A-2B	**Hr.**	**Daily**	**Hr.**	**Daily**	**Bare Costs**	**Incl. O&P**
1 Truck Driver (light)	$35.65	$285.20	$54.60	$436.80	$35.65	$54.60
1 Flatbed Truck, Gas, 1.5 Ton		265.80		292.38	33.23	36.55
8 L.H., Daily Totals		$551.00		$729.18	$68.88	$91.15
Crew A-3A	**Hr.**	**Daily**	**Hr.**	**Daily**	**Bare Costs**	**Incl. O&P**
1 Truck Driver (light)	$35.65	$285.20	$54.60	$436.80	$35.65	$54.60
1 Pickup Truck, 4 x 4, 3/4 Ton		165.60		182.16	20.70	22.77
8 L.H., Daily Totals		$450.80		$618.96	$56.35	$77.37
Crew A-3B	**Hr.**	**Daily**	**Hr.**	**Daily**	**Bare Costs**	**Incl. O&P**
1 Equip. Oper. (med.)	$47.50	$380.00	$71.75	$574.00	$42.05	$63.90
1 Truck Driver (heav.)	36.60	292.80	56.05	448.40		
1 Dump Truck, 12 C.Y., 400 H.P.		692.40		761.64		
1 F.E. Loader, W.M.,2.5 C.Y.		526.20		578.82	76.16	83.78
16 L.H., Daily Totals		$1891.40		$2362.86	$118.21	$147.68
Crew A-3C	**Hr.**	**Daily**	**Hr.**	**Daily**	**Bare Costs**	**Incl. O&P**
1 Equip. Oper. (light)	$45.80	$366.40	$69.20	$553.60	$45.80	$69.20
1 Loader, Skid Steer, 78 H.P.		308.60		339.46	38.58	42.43
8 L.H., Daily Totals		$675.00		$893.06	$84.38	$111.63
Crew A-3D	**Hr.**	**Daily**	**Hr.**	**Daily**	**Bare Costs**	**Incl. O&P**
1 Truck Driver, Light	$35.65	$285.20	$54.60	$436.80	$35.65	$54.60
1 Pickup Truck, 4 x 4, 3/4 Ton		165.60		182.16		
1 Flatbed Trailer, 25 Ton		113.20		124.52	34.85	38.34
8 L.H., Daily Totals		$564.00		$743.48	$70.50	$92.94
Crew A-3E	**Hr.**	**Daily**	**Hr.**	**Daily**	**Bare Costs**	**Incl. O&P**
1 Equip. Oper. (crane)	$48.80	$390.40	$73.75	$590.00	$42.70	$64.90
1 Truck Driver (heavy)	36.60	292.80	56.05	448.40		
1 Pickup Truck, 4 x 4, 3/4 Ton		165.60		182.16	10.35	11.39
16 L.H., Daily Totals		$848.80		$1220.56	$53.05	$76.28
Crew A-3F	**Hr.**	**Daily**	**Hr.**	**Daily**	**Bare Costs**	**Incl. O&P**
1 Equip. Oper. (crane)	$48.80	$390.40	$73.75	$590.00	$42.70	$64.90
1 Truck Driver (heavy)	36.60	292.80	56.05	448.40		
1 Pickup Truck, 4 x 4, 3/4 Ton		165.60		182.16		
1 Truck Tractor, 6x4, 380 H.P.		612.40		673.64		
1 Lowbed Trailer, 75 Ton		220.40		242.44	62.40	68.64
16 L.H., Daily Totals		$1681.60		$2136.64	$105.10	$133.54

Crew No.	Bare Costs		Incl. Subs O&P		Cost Per Labor-Hour	

Crew A-3G	Hr.	Daily	Hr.	Daily	Bare Costs	Incl. O&P
1 Equip. Oper. (crane)	$48.80	$390.40	$73.75	$590.00	$42.70	$64.90
1 Truck Driver (heavy)	36.60	292.80	56.05	448.40		
1 Pickup Truck, 4 x 4, 3/4 Ton		165.60		182.16		
1 Truck Tractor, 6x4, 450 H.P.		741.20		815.32		
1 Lowbed Trailer, 75 Ton		220.40		242.44	70.45	77.50
16 L.H., Daily Totals		$1810.40		$2278.32	$113.15	$142.40

Crew A-3H	Hr.	Daily	Hr.	Daily	Bare Costs	Incl. O&P
1 Equip. Oper. (crane)	$48.80	$390.40	$73.75	$590.00	$48.80	$73.75
1 Hyd. Crane, 12 Ton (Daily)		856.20		941.82	107.03	117.73
8 L.H., Daily Totals		$1246.60		$1531.82	$155.82	$191.48

Crew A-3I	Hr.	Daily	Hr.	Daily	Bare Costs	Incl. O&P
1 Equip. Oper. (crane)	$48.80	$390.40	$73.75	$590.00	$48.80	$73.75
1 Hyd. Crane, 25 Ton (Daily)		979.60		1077.56	122.45	134.69
8 L.H., Daily Totals		$1370.00		$1667.56	$171.25	$208.44

Crew A-3J	Hr.	Daily	Hr.	Daily	Bare Costs	Incl. O&P
1 Equip. Oper. (crane)	$48.80	$390.40	$73.75	$590.00	$48.80	$73.75
1 Hyd. Crane, 40 Ton (Daily)		1213.00		1334.30	151.63	166.79
8 L.H., Daily Totals		$1603.40		$1924.30	$200.43	$240.54

Crew A-3K	Hr.	Daily	Hr.	Daily	Bare Costs	Incl. O&P
1 Equip. Oper. (crane)	$48.80	$390.40	$73.75	$590.00	$45.52	$68.80
1 Equip. Oper. Oiler	42.25	338.00	63.85	510.80		
1 Hyd. Crane, 55 Ton (Daily)		1462.00		1608.20		
1 P/U Truck, 3/4 Ton (Daily)		179.60		197.56	102.60	112.86
16 L.H., Daily Totals		$2370.00		$2906.56	$148.13	$181.66

Crew A-3L	Hr.	Daily	Hr.	Daily	Bare Costs	Incl. O&P
1 Equip. Oper. (crane)	$48.80	$390.40	$73.75	$590.00	$45.52	$68.80
1 Equip. Oper. Oiler	42.25	338.00	63.85	510.80		
1 Hyd. Crane, 80 Ton (Daily)		2167.00		2383.70		
1 P/U Truck, 3/4 Ton (Daily)		179.60		197.56	146.66	161.33
16 L.H., Daily Totals		$3075.00		$3682.06	$192.19	$230.13

Crew A-3M	Hr.	Daily	Hr.	Daily	Bare Costs	Incl. O&P
1 Equip. Oper. (crane)	$48.80	$390.40	$73.75	$590.00	$45.52	$68.80
1 Equip. Oper. Oiler	42.25	338.00	63.85	510.80		
1 Hyd. Crane, 100 Ton (Daily)		2164.00		2380.40		
1 P/U Truck, 3/4 Ton (Daily)		179.60		197.56	146.47	161.12
16 L.H., Daily Totals		$3072.00		$3678.76	$192.00	$229.92

Crew A-3N	Hr.	Daily	Hr.	Daily	Bare Costs	Incl. O&P
1 Equip. Oper. (crane)	$48.80	$390.40	$73.75	$590.00	$48.80	$73.75
1 Tower Cane (Monthly)		1072.00		1179.20	134.00	147.40
8 L.H., Daily Totals		$1462.40		$1769.20	$182.80	$221.15

Crew A-3P	Hr.	Daily	Hr.	Daily	Bare Costs	Incl. O&P
1 Equip. Oper., Light	$45.80	$366.40	$69.20	$553.60	$45.80	$69.20
1 A.T. Forklift, 42' lift		512.40		563.64	64.05	70.45
8 L.H., Daily Totals		$878.80		$1117.24	$109.85	$139.66

Crew A-4	Hr.	Daily	Hr.	Daily	Bare Costs	Incl. O&P
2 Carpenters	$44.90	$718.40	$69.15	$1106.40	$42.83	$65.45
1 Painter, Ordinary	38.70	309.60	58.05	464.40		
24 L.H., Daily Totals		$1028.00		$1570.80	$42.83	$65.45

Crew A-5	Hr.	Daily	Hr.	Daily	Bare Costs	Incl. O&P
2 Laborers	$35.45	$567.20	$54.60	$873.60	$35.47	$54.60
.25 Truck Driver (light)	35.65	71.30	54.60	109.20		
.25 Flatbed Truck, Gas, 1.5 Ton		66.45		73.09	3.69	4.06
18 L.H., Daily Totals		$704.95		$1055.90	$39.16	$58.66

Crew A-6	Hr.	Daily	Hr.	Daily	Bare Costs	Incl. O&P
1 Instrument Man	$46.20	$369.60	$71.45	$571.60	$44.75	$68.50
1 Rodman/Chainman	43.30	346.40	65.55	524.40		
1 Level, Electronic		77.55		85.31	4.85	5.33
16 L.H., Daily Totals		$793.55		$1181.31	$49.60	$73.83

Crew A-7	Hr.	Daily	Hr.	Daily	Bare Costs	Incl. O&P
1 Chief of Party	$57.50	$460.00	$88.50	$708.00	$49.00	$75.17
1 Instrument Man	46.20	369.60	71.45	571.60		
1 Rodman/Chainman	43.30	346.40	65.55	524.40		
1 Level, Electronic		77.55		85.31	3.23	3.55
24 L.H., Daily Totals		$1253.55		$1889.31	$52.23	$78.72

Crew A-8	Hr.	Daily	Hr.	Daily	Bare Costs	Incl. O&P
1 Chief of Party	$57.50	$460.00	$88.50	$708.00	$47.58	$72.76
1 Instrument Man	46.20	369.60	71.45	571.60		
2 Rodmen/Chainmen	43.30	692.80	65.55	1048.80		
1 Level, Electronic		77.55		85.31	2.42	2.67
32 L.H., Daily Totals		$1599.95		$2413.70	$50.00	$75.43

Crew A-9	Hr.	Daily	Hr.	Daily	Bare Costs	Incl. O&P
1 Asbestos Foreman	$50.20	$401.60	$77.95	$623.60	$49.76	$77.29
7 Asbestos Workers	49.70	2783.20	77.20	4323.20		
64 L.H., Daily Totals		$3184.80		$4946.80	$49.76	$77.29

Crew A-10A	Hr.	Daily	Hr.	Daily	Bare Costs	Incl. O&P
1 Asbestos Foreman	$50.20	$401.60	$77.95	$623.60	$49.87	$77.45
2 Asbestos Workers	49.70	795.20	77.20	1235.20		
24 L.H., Daily Totals		$1196.80		$1858.80	$49.87	$77.45

Crew A-10B	Hr.	Daily	Hr.	Daily	Bare Costs	Incl. O&P
1 Asbestos Foreman	$50.20	$401.60	$77.95	$623.60	$49.83	$77.39
3 Asbestos Workers	49.70	1192.80	77.20	1852.80		
32 L.H., Daily Totals		$1594.40		$2476.40	$49.83	$77.39

Crew A-10C	Hr.	Daily	Hr.	Daily	Bare Costs	Incl. O&P
3 Asbestos Workers	$49.70	$1192.80	$77.20	$1852.80	$49.70	$77.20
1 Flatbed Truck, Gas, 1.5 Ton		265.80		292.38	11.07	12.18
24 L.H., Daily Totals		$1458.60		$2145.18	$60.77	$89.38

Crew A-10D	Hr.	Daily	Hr.	Daily	Bare Costs	Incl. O&P
2 Asbestos Workers	$49.70	$795.20	$77.20	$1235.20	$47.61	$73.00
1 Equip. Oper. (crane)	48.80	390.40	73.75	590.00		
1 Equip. Oper. Oiler	42.25	338.00	63.85	510.80		
1 Hydraulic Crane, 33 Ton		743.00		817.30	23.22	25.54
32 L.H., Daily Totals		$2266.60		$3153.30	$70.83	$98.54

Crew A-11	Hr.	Daily	Hr.	Daily	Bare Costs	Incl. O&P
1 Asbestos Foreman	$50.20	$401.60	$77.95	$623.60	$49.76	$77.29
7 Asbestos Workers	49.70	2783.20	77.20	4323.20		
2 Chip. Hammers, 12 Lb., Elec.		40.00		44.00	.63	.69
64 L.H., Daily Totals		$3224.80		$4990.80	$50.39	$77.98

Crew A-12	Hr.	Daily	Hr.	Daily	Bare Costs	Incl. O&P
1 Asbestos Foreman	$50.20	$401.60	$77.95	$623.60	$49.76	$77.29
7 Asbestos Workers	49.70	2783.20	77.20	4323.20		
1 Trk-Mtd Vac, 14 CY, 1500 Gal.		522.65		574.91		
1 Flatbed Truck, 20,000 GVW		253.60		278.96	12.13	13.34
64 L.H., Daily Totals		$3961.05		$5800.68	$61.89	$90.64

Crew A-13	Hr.	Daily	Hr.	Daily	Bare Costs	Incl. O&P
1 Equip. Oper. (light)	$45.80	$366.40	$69.20	$553.60	$45.80	$69.20
1 Trk-Mtd Vac, 14 CY, 1500 Gal.		522.65		574.91		
1 Flatbed Truck, 20,000 GVW		253.60		278.96	97.03	106.73
8 L.H., Daily Totals		$1142.65		$1407.47	$142.83	$175.93

Crew B-1	Hr.	Daily	Hr.	Daily	Bare Costs	Incl. O&P
1 Labor Foreman (outside)	$37.45	$299.60	$57.65	$461.20	$36.12	$55.62
2 Laborers	35.45	567.20	54.60	873.60		
24 L.H., Daily Totals		$866.80		$1334.80	$36.12	$55.62

Crew B-1A	Hr.	Daily	Hr.	Daily	Bare Costs	Incl. O&P
1 Labor Foreman (outside)	$37.45	$299.60	$57.65	$461.20	$36.12	$55.62
2 Laborers	35.45	567.20	54.60	873.60		
2 Cutting Torches		22.80		25.08		
2 Sets of Gases		266.40		293.04	12.05	13.26
24 L.H., Daily Totals		$1156.00		$1652.92	$48.17	$68.87

Crew B-1B	Hr.	Daily	Hr.	Daily	Bare Costs	Incl. O&P
1 Labor Foreman (outside)	$37.45	$299.60	$57.65	$461.20	$39.29	$60.15
2 Laborers	35.45	567.20	54.60	873.60		
1 Equip. Oper. (crane)	48.80	390.40	73.75	590.00		
2 Cutting Torches		22.80		25.08		
2 Sets of Gases		266.40		293.04		
1 Hyd. Crane, 12 Ton		646.20		710.82	29.23	32.15
32 L.H., Daily Totals		$2192.60		$2953.74	$68.52	$92.30

Crew B-1C	Hr.	Daily	Hr.	Daily	Bare Costs	Incl. O&P
1 Labor Foreman (outside)	$37.45	$299.60	$57.65	$461.20	$36.12	$55.62
2 Laborers	35.45	567.20	54.60	873.60		
1 Aerial Lift Truck, 60' Boom		462.20		508.42	19.26	21.18
24 L.H., Daily Totals		$1329.00		$1843.22	$55.38	$76.80

Crew B-1D	Hr.	Daily	Hr.	Daily	Bare Costs	Incl. O&P
2 Laborers	$35.45	$567.20	$54.60	$873.60	$35.45	$54.60
1 Small Work Boat, Gas, 50 H.P.		187.40		206.14		
1 Pressure Washer, 7 GPM		86.40		95.04	17.11	18.82
16 L.H., Daily Totals		$841.00		$1174.78	$52.56	$73.42

Crew B-1E	Hr.	Daily	Hr.	Daily	Bare Costs	Incl. O&P
1 Labor Foreman (outside)	$37.45	$299.60	$57.65	$461.20	$35.95	$55.36
3 Laborers	35.45	850.80	54.60	1310.40		
1 Work Boat, Diesel, 200 H.P.		1472.00		1619.20		
2 Pressure Washer, 7 GPM		172.80		190.08	51.40	56.54
32 L.H., Daily Totals		$2795.20		$3580.88	$87.35	$111.90

Crew B-1F	Hr.	Daily	Hr.	Daily	Bare Costs	Incl. O&P
2 Skilled Workers	$46.20	$739.20	$71.45	$1143.20	$42.62	$65.83
1 Laborer	35.45	283.60	54.60	436.80		
1 Small Work Boat, Gas, 50 H.P.		187.40		206.14		
1 Pressure Washer, 7 GPM		86.40		95.04	11.41	12.55
24 L.H., Daily Totals		$1296.60		$1881.18	$54.02	$78.38

Crew B-1G	Hr.	Daily	Hr.	Daily	Bare Costs	Incl. O&P
2 Laborers	$35.45	$567.20	$54.60	$873.60	$35.45	$54.60
1 Small Work Boat, Gas, 50 H.P.		187.40		206.14	11.71	12.88
16 L.H., Daily Totals		$754.60		$1079.74	$47.16	$67.48

Crew B-1H	Hr.	Daily	Hr.	Daily	Bare Costs	Incl. O&P
2 Skilled Workers	$46.20	$739.20	$71.45	$1143.20	$42.62	$65.83
1 Laborer	35.45	283.60	54.60	436.80		
1 Small Work Boat, Gas, 50 H.P.		187.40		206.14	7.81	8.59
24 L.H., Daily Totals		$1210.20		$1786.14	$50.42	$74.42

Crew B-1J	Hr.	Daily	Hr.	Daily	Bare Costs	Incl. O&P
1 Labor Foreman (inside)	$35.95	$287.60	$55.35	$442.80	$35.70	$54.98
1 Laborer	35.45	283.60	54.60	436.80		
16 L.H., Daily Totals		$571.20		$879.60	$35.70	$54.98

Crew B-1K	Hr.	Daily	Hr.	Daily	Bare Costs	Incl. O&P
1 Carpenter Foreman (inside)	$45.40	$363.20	$69.90	$559.20	$45.15	$69.53
1 Carpenter	44.90	359.20	69.15	553.20		
16 L.H., Daily Totals		$722.40		$1112.40	$45.15	$69.53

Crew B-2	Hr.	Daily	Hr.	Daily	Bare Costs	Incl. O&P
1 Labor Foreman (outside)	$37.45	$299.60	$57.65	$461.20	$35.85	$55.21
4 Laborers	35.45	1134.40	54.60	1747.20		
40 L.H., Daily Totals		$1434.00		$2208.40	$35.85	$55.21

Crew B-2A	Hr.	Daily	Hr.	Daily	Bare Costs	Incl. O&P
1 Labor Foreman (outside)	$37.45	$299.60	$57.65	$461.20	$36.12	$55.62
2 Laborers	35.45	567.20	54.60	873.60		
1 Aerial Lift Truck, 60' Boom		462.20		508.42	19.26	21.18
24 L.H., Daily Totals		$1329.00		$1843.22	$55.38	$76.80

Crew B-3	Hr.	Daily	Hr.	Daily	Bare Costs	Incl. O&P
1 Labor Foreman (outside)	$37.45	$299.60	$57.65	$461.20	$38.17	$58.45
2 Laborers	35.45	567.20	54.60	873.60		
1 Equip. Oper. (med.)	47.50	380.00	71.75	574.00		
2 Truck Drivers (heavy)	36.60	585.60	56.05	896.80		
1 Crawler Loader, 3 C.Y.		1187.00		1305.70		
2 Dump Trucks, 12 C.Y., 400 H.P.		1384.80		1523.28	53.58	58.94
48 L.H., Daily Totals		$4404.20		$5634.58	$91.75	$117.39

Crew B-3A	Hr.	Daily	Hr.	Daily	Bare Costs	Incl. O&P
4 Laborers	$35.45	$1134.40	$54.60	$1747.20	$37.86	$58.03
1 Equip. Oper. (med.)	47.50	380.00	71.75	574.00		
1 Hyd. Excavator, 1.5 C.Y.		1036.00		1139.60	25.90	28.49
40 L.H., Daily Totals		$2550.40		$3460.80	$63.76	$86.52

Crew B-3B	Hr.	Daily	Hr.	Daily	Bare Costs	Incl. O&P
2 Laborers	$35.45	$567.20	$54.60	$873.60	$38.75	$59.25
1 Equip. Oper. (med.)	47.50	380.00	71.75	574.00		
1 Truck Driver (heavy)	36.60	292.80	56.05	448.40		
1 Backhoe Loader, 80 H.P.		400.40		440.44		
1 Dump Truck, 12 C.Y., 400 H.P.		692.40		761.64	34.15	37.56
32 L.H., Daily Totals		$2332.80		$3098.08	$72.90	$96.81

Crew B-3C	Hr.	Daily	Hr.	Daily	Bare Costs	Incl. O&P
3 Laborers	$35.45	$850.80	$54.60	$1310.40	$38.46	$58.89
1 Equip. Oper. (med.)	47.50	380.00	71.75	574.00		
1 Crawler Loader, 4 C.Y.		1530.00		1683.00	47.81	52.59
32 L.H., Daily Totals		$2760.80		$3567.40	$86.28	$111.48

Crew B-4	Hr.	Daily	Hr.	Daily	Bare Costs	Incl. O&P
1 Labor Foreman (outside)	$37.45	$299.60	$57.65	$461.20	$35.98	$55.35
4 Laborers	35.45	1134.40	54.60	1747.20		
1 Truck Driver (heavy)	36.60	292.80	56.05	448.40		
1 Truck Tractor, 220 H.P.		366.80		403.48		
1 Flatbed Trailer, 40 Ton		152.60		167.86	10.82	11.90
48 L.H., Daily Totals		$2246.20		$3228.14	$46.80	$67.25

Crew B-5	Hr.	Daily	Hr.	Daily	Bare Costs	Incl. O&P
1 Labor Foreman (outside)	$37.45	$299.60	$57.65	$461.20	$39.18	$59.94
4 Laborers	35.45	1134.40	54.60	1747.20		
2 Equip. Oper. (med.)	47.50	760.00	71.75	1148.00		
1 Air Compressor, 250 cfm		195.80		215.38		
2 Breakers, Pavement, 60 lb.		19.60		21.56		
2 -50' Air Hoses, 1.5"		11.60		12.76		
1 Crawler Loader, 3 C.Y.		1187.00		1305.70	25.25	27.77
56 L.H., Daily Totals		$3608.00		$4911.80	$64.43	$87.71

Crew B-5A	Hr.	Daily	Hr.	Daily	Bare Costs	Incl. O&P
1 Labor Foreman (outside)	$37.45	$299.60	$57.65	$461.20	$38.68	$59.17
6 Laborers	35.45	1701.60	54.60	2620.80		
2 Equip. Oper. (med.)	47.50	760.00	71.75	1148.00		
1 Equip. Oper. (light)	45.80	366.40	69.20	553.60		
2 Truck Drivers (heavy)	36.60	585.60	56.05	896.80		
1 Air Compressor, 365 cfm		255.80		281.38		
2 Breakers, Pavement, 60 lb.		19.60		21.56		
8 -50' Air Hoses, 1"		32.80		36.08		
2 Dump Trucks, 8 C.Y., 220 H.P.		833.20		916.52	11.89	13.08
96 L.H., Daily Totals		$4854.60		$6935.94	$50.57	$72.25

Crew B-5B	Hr.	Daily	Hr.	Daily	Bare Costs	Incl. O&P
1 Powderman	$46.20	$369.60	$71.45	$571.60	$41.83	$63.85
2 Equip. Oper. (med.)	47.50	760.00	71.75	1148.00		
3 Truck Drivers (heavy)	36.60	878.40	56.05	1345.20		
1 F.E. Loader, W.M.,2.5 C.Y.		526.20		578.82		
3 Dump Trucks, 12 C.Y., 400 H.P.		2077.20		2284.92		
1 Air Compressor, 365 CFM		255.80		281.38	59.57	65.52
48 L.H., Daily Totals		$4867.20		$6209.92	$101.40	$129.37

Crew B-5C	Hr.	Daily	Hr.	Daily	Bare Costs	Incl. O&P
3 Laborers	$35.45	$850.80	$54.60	$1310.40	$39.76	$60.66
1 Equip. Oper. (med.)	47.50	380.00	71.75	574.00		
2 Truck Drivers (heav.)	36.60	585.60	56.05	896.80		
1 Equip. Oper. (crane)	48.80	390.40	73.75	590.00		
1 Equip. Oper. Oiler	42.25	338.00	63.85	510.80		
2 Dump Trucks, 12 C.Y., 400 H.P.		1384.80		1523.28		
1 Crawler Loader, 4 C.Y.		1530.00		1683.00		
1 S.P. Crane, 4x4, 25 Ton		593.40		652.74	54.82	60.30
64 L.H., Daily Totals		$6053.00		$7741.02	$94.58	$120.95

Crew B-5D	Hr.	Daily	Hr.	Daily	Bare Costs	Incl. O&P
1 Labor Foreman (outside)	$37.45	$299.60	$57.65	$461.20	$38.86	$59.45
4 Laborers	35.45	1134.40	54.60	1747.20		
2 Equip. Oper. (med.)	47.50	760.00	71.75	1148.00		
1 Truck Driver (heavy)	36.60	292.80	56.05	448.40		
1 Air Compressor, 250 cfm		195.80		215.38		
2 Breakers, Pavement, 60 lb.		19.60		21.56		
2 -50' Air Hoses, 1.5"		11.60		12.76		
1 Crawler Loader, 3 C.Y.		1187.00		1305.70		
1 Dump Truck, 12 C.Y., 400 H.P.		692.40		761.64	32.91	36.20
64 L.H., Daily Totals		$4593.20		$6121.84	$71.77	$95.65

Crew B-6	Hr.	Daily	Hr.	Daily	Bare Costs	Incl. O&P
2 Laborers	$35.45	$567.20	$54.60	$873.60	$38.90	$59.47
1 Equip. Oper. (light)	45.80	366.40	69.20	553.60		
1 Backhoe Loader, 48 H.P.		367.40		404.14	15.31	16.84
24 L.H., Daily Totals		$1301.00		$1831.34	$54.21	$76.31

Crew B-6A	Hr.	Daily	Hr.	Daily	Bare Costs	Incl. O&P
.5 Labor Foreman (outside)	$37.45	$149.80	$57.65	$230.60	$40.67	$62.07
1 Laborer	35.45	283.60	54.60	436.80		
1 Equip. Oper. (med.)	47.50	380.00	71.75	574.00		
1 Vacuum Truck, 5000 Gal.		363.90		400.29	18.20	20.01
20 L.H., Daily Totals		$1177.30		$1641.69	$58.87	$82.08

Crew B-6B	Hr.	Daily	Hr.	Daily	Bare Costs	Incl. O&P
2 Labor Foremen (outside)	$37.45	$599.20	$57.65	$922.40	$36.12	$55.62
4 Laborers	35.45	1134.40	54.60	1747.20		
1 S.P. Crane, 4x4, 5 Ton		274.20		301.62		
1 Flatbed Truck, Gas, 1.5 Ton		265.80		292.38		
1 Butt Fusion Mach., 4"-12" diam.		382.60		420.86	19.22	21.14
48 L.H., Daily Totals		$2656.20		$3684.46	$55.34	$76.76

Crew B-6C	Hr.	Daily	Hr.	Daily	Bare Costs	Incl. O&P
2 Labor Foremen (outside)	$37.45	$599.20	$57.65	$922.40	$36.12	$55.62
4 Laborers	35.45	1134.40	54.60	1747.20		
1 S.P. Crane, 4x4, 12 Ton		471.00		518.10		
1 Flatbed Truck, Gas, 3 Ton		327.80		360.58		
1 Butt Fusion Mach., 8"-24" diam.		831.70		914.87	33.97	37.37
48 L.H., Daily Totals		$3364.10		$4463.15	$70.09	$92.98

Crew B-7	Hr.	Daily	Hr.	Daily	Bare Costs	Incl. O&P
1 Labor Foreman (outside)	$37.45	$299.60	$57.65	$461.20	$37.79	$57.97
4 Laborers	35.45	1134.40	54.60	1747.20		
1 Equip. Oper. (med.)	47.50	380.00	71.75	574.00		
1 Brush Chipper, 12", 130 H.P.		396.40		436.04		
1 Crawler Loader, 3 C.Y.		1187.00		1305.70		
2 Chain Saws, Gas, 36" Long		90.00		99.00	34.86	38.35
48 L.H., Daily Totals		$3487.40		$4623.14	$72.65	$96.32

Crew B-7A	Hr.	Daily	Hr.	Daily	Bare Costs	Incl. O&P
2 Laborers	$35.45	$567.20	$54.60	$873.60	$38.90	$59.47
1 Equip. Oper. (light)	45.80	366.40	69.20	553.60		
1 Rake w/Tractor		347.70		382.47		
2 Chain Saw, Gas, 18"		61.20		67.32	17.04	18.74
24 L.H., Daily Totals		$1342.50		$1876.99	$55.94	$78.21

Crew B-7B	Hr.	Daily	Hr.	Daily	Bare Costs	Incl. O&P
1 Labor Foreman (outside)	$37.45	$299.60	$57.65	$461.20	$37.62	$57.69
4 Laborers	35.45	1134.40	54.60	1747.20		
1 Equip. Oper. (med.)	47.50	380.00	71.75	574.00		
1 Truck Driver (heavy)	36.60	292.80	56.05	448.40		
1 Brush Chipper, 12", 130 H.P.		396.40		436.04		
1 Crawler Loader, 3 C.Y.		1187.00		1305.70		
2 Chain Saws, Gas, 36" Long		90.00		99.00		
1 Dump Truck, 8 C.Y., 220 H.P.		416.60		458.26	37.32	41.05
56 L.H., Daily Totals		$4196.80		$5529.80	$74.94	$98.75

Crews

Crew B-7C	Bare Costs Hr.	Bare Costs Daily	Incl. Subs O&P Hr.	Incl. Subs O&P Daily	Cost Per Labor-Hour Bare Costs	Cost Per Labor-Hour Incl. O&P
1 Labor Foreman (outside)	$37.45	$299.60	$57.65	$461.20	$37.62	$57.69
4 Laborers	35.45	1134.40	54.60	1747.20		
1 Equip. Oper. (med.)	47.50	380.00	71.75	574.00		
1 Truck Driver (heavy)	36.60	292.80	56.05	448.40		
1 Brush Chipper, 12", 130 H.P.		396.40		436.04		
1 Crawler Loader, 3 C.Y.		1187.00		1305.70		
2 Chain Saws, Gas, 36" Long		90.00		99.00		
1 Dump Truck, 12 C.Y., 400 H.P.		692.40		761.64	42.25	46.47
56 L.H., Daily Totals		$4472.60		$5833.18	$79.87	$104.16

Crew B-8	Hr.	Daily	Hr.	Daily	Bare Costs	Incl. O&P
1 Labor Foreman (outside)	$37.45	$299.60	$57.65	$461.20	$39.85	$60.79
2 Laborers	35.45	567.20	54.60	873.60		
2 Equip. Oper. (med.)	47.50	760.00	71.75	1148.00		
1 Equip. Oper. Oiler	42.25	338.00	63.85	510.80		
2 Truck Drivers (heavy)	36.60	585.60	56.05	896.80		
1 Hyd. Crane, 25 Ton		728.60		801.46		
1 Crawler Loader, 3 C.Y.		1187.00		1305.70		
2 Dump Trucks, 12 C.Y., 400 H.P.		1384.80		1523.28	51.57	56.73
64 L.H., Daily Totals		$5850.80		$7520.84	$91.42	$117.51

Crew B-9	Hr.	Daily	Hr.	Daily	Bare Costs	Incl. O&P
1 Labor Foreman (outside)	$37.45	$299.60	$57.65	$461.20	$35.85	$55.21
4 Laborers	35.45	1134.40	54.60	1747.20		
1 Air Compressor, 250 cfm		195.80		215.38		
2 Breakers, Pavement, 60 lb.		19.60		21.56		
2 -50' Air Hoses, 1.5"		11.60		12.76	5.67	6.24
40 L.H., Daily Totals		$1661.00		$2458.10	$41.52	$61.45

Crew B-9A	Hr.	Daily	Hr.	Daily	Bare Costs	Incl. O&P
2 Laborers	$35.45	$567.20	$54.60	$873.60	$35.83	$55.08
1 Truck Driver (heavy)	36.60	292.80	56.05	448.40		
1 Water Tank Trailer, 5000 Gal.		143.20		157.52		
1 Truck Tractor, 220 H.P.		366.80		403.48		
2 -50' Discharge Hoses, 3"		6.10		6.71	21.50	23.65
24 L.H., Daily Totals		$1376.10		$1889.71	$57.34	$78.74

Crew B-9B	Hr.	Daily	Hr.	Daily	Bare Costs	Incl. O&P
2 Laborers	$35.45	$567.20	$54.60	$873.60	$35.83	$55.08
1 Truck Driver (heavy)	36.60	292.80	56.05	448.40		
2 -50' Discharge Hoses, 3"		6.10		6.71		
1 Water Tank Trailer, 5000 Gal.		143.20		157.52		
1 Truck Tractor, 220 H.P.		366.80		403.48		
1 Pressure Washer		69.40		76.34	24.40	26.84
24 L.H., Daily Totals		$1445.50		$1966.05	$60.23	$81.92

Crew B-9D	Hr.	Daily	Hr.	Daily	Bare Costs	Incl. O&P
1 Labor Foreman (Outside)	$37.45	$299.60	$57.65	$461.20	$35.85	$55.21
4 Common Laborers	35.45	1134.40	54.60	1747.20		
1 Air Compressor, 250 cfm		195.80		215.38		
2 -50' Air Hoses, 1.5"		11.60		12.76		
2 Air Powered Tampers		52.40		57.64	6.50	7.14
40 L.H., Daily Totals		$1693.80		$2494.18	$42.34	$62.35

Crew B-10	Hr.	Daily	Hr.	Daily	Bare Costs	Incl. O&P
1 Equip. Oper. (med.)	$47.50	$380.00	$71.75	$574.00	$43.48	$66.03
.5 Laborer	35.45	141.80	54.60	218.40		
12 L.H., Daily Totals		$521.80		$792.40	$43.48	$66.03

Crew B-10A	Hr.	Daily	Hr.	Daily	Bare Costs	Incl. O&P
1 Equip. Oper. (med.)	$47.50	$380.00	$71.75	$574.00	$43.48	$66.03
.5 Laborer	35.45	141.80	54.60	218.40		
1 Roller, 2-Drum, W.B., 7.5 H.P.		174.00		191.40	14.50	15.95
12 L.H., Daily Totals		$695.80		$983.80	$57.98	$81.98

Crew B-10B	Hr.	Daily	Hr.	Daily	Bare Costs	Incl. O&P
1 Equip. Oper. (med.)	$47.50	$380.00	$71.75	$574.00	$43.48	$66.03
.5 Laborer	35.45	141.80	54.60	218.40		
1 Dozer, 200 H.P.		1333.00		1466.30	111.08	122.19
12 L.H., Daily Totals		$1854.80		$2258.70	$154.57	$188.22

Crew B-10C	Hr.	Daily	Hr.	Daily	Bare Costs	Incl. O&P
1 Equip. Oper. (med.)	$47.50	$380.00	$71.75	$574.00	$43.48	$66.03
.5 Laborer	35.45	141.80	54.60	218.40		
1 Dozer, 200 H.P.		1333.00		1466.30		
1 Vibratory Roller, Towed, 23 Ton		407.20		447.92	145.02	159.52
12 L.H., Daily Totals		$2262.00		$2706.62	$188.50	$225.55

Crew B-10D	Hr.	Daily	Hr.	Daily	Bare Costs	Incl. O&P
1 Equip. Oper. (med.)	$47.50	$380.00	$71.75	$574.00	$43.48	$66.03
.5 Laborer	35.45	141.80	54.60	218.40		
1 Dozer, 200 H.P.		1333.00		1466.30		
1 Sheepsft. Roller, Towed		440.00		484.00	147.75	162.53
12 L.H., Daily Totals		$2294.80		$2742.70	$191.23	$228.56

Crew B-10E	Hr.	Daily	Hr.	Daily	Bare Costs	Incl. O&P
1 Equip. Oper. (med.)	$47.50	$380.00	$71.75	$574.00	$43.48	$66.03
.5 Laborer	35.45	141.80	54.60	218.40		
1 Tandem Roller, 5 Ton		154.40		169.84	12.87	14.15
12 L.H., Daily Totals		$676.20		$962.24	$56.35	$80.19

Crew B-10F	Hr.	Daily	Hr.	Daily	Bare Costs	Incl. O&P
1 Equip. Oper. (med.)	$47.50	$380.00	$71.75	$574.00	$43.48	$66.03
.5 Laborer	35.45	141.80	54.60	218.40		
1 Tandem Roller, 10 Ton		231.20		254.32	19.27	21.19
12 L.H., Daily Totals		$753.00		$1046.72	$62.75	$87.23

Crew B-10G	Hr.	Daily	Hr.	Daily	Bare Costs	Incl. O&P
1 Equip. Oper. (med.)	$47.50	$380.00	$71.75	$574.00	$43.48	$66.03
.5 Laborer	35.45	141.80	54.60	218.40		
1 Sheepsfoot Roller, 240 H.P.		1154.00		1269.40	96.17	105.78
12 L.H., Daily Totals		$1675.80		$2061.80	$139.65	$171.82

Crew B-10H	Hr.	Daily	Hr.	Daily	Bare Costs	Incl. O&P
1 Equip. Oper. (med.)	$47.50	$380.00	$71.75	$574.00	$43.48	$66.03
.5 Laborer	35.45	141.80	54.60	218.40		
1 Diaphragm Water Pump, 2"		71.60		78.76		
1 -20' Suction Hose, 2"		1.95		2.15		
2 -50' Discharge Hoses, 2"		3.50		3.85	6.42	7.06
12 L.H., Daily Totals		$598.85		$877.15	$49.90	$73.10

Crew B-10I	Hr.	Daily	Hr.	Daily	Bare Costs	Incl. O&P
1 Equip. Oper. (med.)	$47.50	$380.00	$71.75	$574.00	$43.48	$66.03
.5 Laborer	35.45	141.80	54.60	218.40		
1 Diaphragm Water Pump, 4"		114.40		125.84		
1 -20' Suction Hose, 4"		3.25		3.58		
2 -50' Discharge Hoses, 4"		9.40		10.34	10.59	11.65
12 L.H., Daily Totals		$648.85		$932.15	$54.07	$77.68

Crew No.	Bare Costs		Incl. Subs O&P		Cost Per Labor-Hour	
Crew B-10J	**Hr.**	**Daily**	**Hr.**	**Daily**	**Bare Costs**	**Incl. O&P**
1 Equip. Oper. (med.)	$47.50	$380.00	$71.75	$574.00	$43.48	$66.03
.5 Laborer	35.45	141.80	54.60	218.40		
1 Centrifugal Water Pump, 3"		78.80		86.68		
1 -20' Suction Hose, 3"		2.85		3.13		
2 -50' Discharge Hoses, 3"		6.10		6.71	7.31	8.04
12 L.H., Daily Totals		$609.55		$888.92	$50.80	$74.08
Crew B-10K	**Hr.**	**Daily**	**Hr.**	**Daily**	**Bare Costs**	**Incl. O&P**
1 Equip. Oper. (med.)	$47.50	$380.00	$71.75	$574.00	$43.48	$66.03
.5 Laborer	35.45	141.80	54.60	218.40		
1 Centr. Water Pump, 6"		341.40		375.54		
1 -20' Suction Hose, 6"		11.20		12.32		
2 -50' Discharge Hoses, 6"		24.20		26.62	31.40	34.54
12 L.H., Daily Totals		$898.60		$1206.88	$74.88	$100.57
Crew B-10L	**Hr.**	**Daily**	**Hr.**	**Daily**	**Bare Costs**	**Incl. O&P**
1 Equip. Oper. (med.)	$47.50	$380.00	$71.75	$574.00	$43.48	$66.03
.5 Laborer	35.45	141.80	54.60	218.40		
1 Dozer, 80 H.P.		480.60		528.66	40.05	44.06
12 L.H., Daily Totals		$1002.40		$1321.06	$83.53	$110.09
Crew B-10M	**Hr.**	**Daily**	**Hr.**	**Daily**	**Bare Costs**	**Incl. O&P**
1 Equip. Oper. (med.)	$47.50	$380.00	$71.75	$574.00	$43.48	$66.03
.5 Laborer	35.45	141.80	54.60	218.40		
1 Dozer, 300 H.P.		1829.00		2011.90	152.42	167.66
12 L.H., Daily Totals		$2350.80		$2804.30	$195.90	$233.69
Crew B-10N	**Hr.**	**Daily**	**Hr.**	**Daily**	**Bare Costs**	**Incl. O&P**
1 Equip. Oper. (med.)	$47.50	$380.00	$71.75	$574.00	$43.48	$66.03
.5 Laborer	35.45	141.80	54.60	218.40		
1 F.E. Loader, T.M., 1.5 C.Y		533.60		586.96	44.47	48.91
12 L.H., Daily Totals		$1055.40		$1379.36	$87.95	$114.95
Crew B-10O	**Hr.**	**Daily**	**Hr.**	**Daily**	**Bare Costs**	**Incl. O&P**
1 Equip. Oper. (med.)	$47.50	$380.00	$71.75	$574.00	$43.48	$66.03
.5 Laborer	35.45	141.80	54.60	218.40		
1 F.E. Loader, T.M., 2.25 C.Y.		955.00		1050.50	79.58	87.54
12 L.H., Daily Totals		$1476.80		$1842.90	$123.07	$153.57
Crew B-10P	**Hr.**	**Daily**	**Hr.**	**Daily**	**Bare Costs**	**Incl. O&P**
1 Equip. Oper. (med.)	$47.50	$380.00	$71.75	$574.00	$43.48	$66.03
.5 Laborer	35.45	141.80	54.60	218.40		
1 Crawler Loader, 3 C.Y.		1187.00		1305.70	98.92	108.81
12 L.H., Daily Totals		$1708.80		$2098.10	$142.40	$174.84
Crew B-10Q	**Hr.**	**Daily**	**Hr.**	**Daily**	**Bare Costs**	**Incl. O&P**
1 Equip. Oper. (med.)	$47.50	$380.00	$71.75	$574.00	$43.48	$66.03
.5 Laborer	35.45	141.80	54.60	218.40		
1 Crawler Loader, 4 C.Y.		1530.00		1683.00	127.50	140.25
12 L.H., Daily Totals		$2051.80		$2475.40	$170.98	$206.28
Crew B-10R	**Hr.**	**Daily**	**Hr.**	**Daily**	**Bare Costs**	**Incl. O&P**
1 Equip. Oper. (med.)	$47.50	$380.00	$71.75	$574.00	$43.48	$66.03
.5 Laborer	35.45	141.80	54.60	218.40		
1 F.E. Loader, W.M., 1 C.Y.		296.00		325.60	24.67	27.13
12 L.H., Daily Totals		$817.80		$1118.00	$68.15	$93.17

Crew No.	Bare Costs		Incl. Subs O&P		Cost Per Labor-Hour	
Crew B-10S	**Hr.**	**Daily**	**Hr.**	**Daily**	**Bare Costs**	**Incl. O&P**
1 Equip. Oper. (med.)	$47.50	$380.00	$71.75	$574.00	$43.48	$66.03
.5 Laborer	35.45	141.80	54.60	218.40		
1 F.E. Loader, W.M., 1.5 C.Y.		375.20		412.72	31.27	34.39
12 L.H., Daily Totals		$897.00		$1205.12	$74.75	$100.43
Crew B-10T	**Hr.**	**Daily**	**Hr.**	**Daily**	**Bare Costs**	**Incl. O&P**
1 Equip. Oper. (med.)	$47.50	$380.00	$71.75	$574.00	$43.48	$66.03
.5 Laborer	35.45	141.80	54.60	218.40		
1 F.E. Loader, W.M.,2.5 C.Y.		526.20		578.82	43.85	48.23
12 L.H., Daily Totals		$1048.00		$1371.22	$87.33	$114.27
Crew B-10U	**Hr.**	**Daily**	**Hr.**	**Daily**	**Bare Costs**	**Incl. O&P**
1 Equip. Oper. (med.)	$47.50	$380.00	$71.75	$574.00	$43.48	$66.03
.5 Laborer	35.45	141.80	54.60	218.40		
1 F.E. Loader, W.M., 5.5 C.Y.		1056.00		1161.60	88.00	96.80
12 L.H., Daily Totals		$1577.80		$1954.00	$131.48	$162.83
Crew B-10V	**Hr.**	**Daily**	**Hr.**	**Daily**	**Bare Costs**	**Incl. O&P**
1 Equip. Oper. (med.)	$47.50	$380.00	$71.75	$574.00	$43.48	$66.03
.5 Laborer	35.45	141.80	54.60	218.40		
1 Dozer, 700 H.P.		4459.00		4904.90	371.58	408.74
12 L.H., Daily Totals		$4980.80		$5697.30	$415.07	$474.77
Crew B-10W	**Hr.**	**Daily**	**Hr.**	**Daily**	**Bare Costs**	**Incl. O&P**
1 Equip. Oper. (med.)	$47.50	$380.00	$71.75	$574.00	$43.48	$66.03
.5 Laborer	35.45	141.80	54.60	218.40		
1 Dozer, 105 H.P.		592.40		651.64	49.37	54.30
12 L.H., Daily Totals		$1114.20		$1444.04	$92.85	$120.34
Crew B-10X	**Hr.**	**Daily**	**Hr.**	**Daily**	**Bare Costs**	**Incl. O&P**
1 Equip. Oper. (med.)	$47.50	$380.00	$71.75	$574.00	$43.48	$66.03
.5 Laborer	35.45	141.80	54.60	218.40		
1 Dozer, 410 H.P.		2425.00		2667.50	202.08	222.29
12 L.H., Daily Totals		$2946.80		$3459.90	$245.57	$288.32
Crew B-10Y	**Hr.**	**Daily**	**Hr.**	**Daily**	**Bare Costs**	**Incl. O&P**
1 Equip. Oper. (med.)	$47.50	$380.00	$71.75	$574.00	$43.48	$66.03
.5 Laborer	35.45	141.80	54.60	218.40		
1 Vibr. Roller, Towed, 12 Ton		529.40		582.34	44.12	48.53
12 L.H., Daily Totals		$1051.20		$1374.74	$87.60	$114.56
Crew B-11A	**Hr.**	**Daily**	**Hr.**	**Daily**	**Bare Costs**	**Incl. O&P**
1 Equipment Oper. (med.)	$47.50	$380.00	$71.75	$574.00	$41.48	$63.17
1 Laborer	35.45	283.60	54.60	436.80		
1 Dozer, 200 H.P.		1333.00		1466.30	83.31	91.64
16 L.H., Daily Totals		$1996.60		$2477.10	$124.79	$154.82
Crew B-11B	**Hr.**	**Daily**	**Hr.**	**Daily**	**Bare Costs**	**Incl. O&P**
1 Equipment Oper. (light)	$45.80	$366.40	$69.20	$553.60	$40.63	$61.90
1 Laborer	35.45	283.60	54.60	436.80		
1 Air Powered Tamper		26.20		28.82		
1 Air Compressor, 365 cfm		255.80		281.38		
2 -50' Air Hoses, 1.5"		11.60		12.76	18.35	20.18
16 L.H., Daily Totals		$943.60		$1313.36	$58.98	$82.08
Crew B-11C	**Hr.**	**Daily**	**Hr.**	**Daily**	**Bare Costs**	**Incl. O&P**
1 Equipment Oper. (med.)	$47.50	$380.00	$71.75	$574.00	$41.48	$63.17
1 Laborer	35.45	283.60	54.60	436.80		
1 Backhoe Loader, 48 H.P.		367.40		404.14	22.96	25.26
16 L.H., Daily Totals		$1031.00		$1414.94	$64.44	$88.43

Crew B-11J	Hr.	Daily	Hr.	Daily	Bare Costs	Incl. O&P
1 Equipment Oper. (med.)	$47.50	$380.00	$71.75	$574.00	$41.48	$63.17
1 Laborer	35.45	283.60	54.60	436.80		
1 Grader, 30,000 Lbs.		712.40		783.64		
1 Ripper, Beam & 1 Shank		82.00		90.20	49.65	54.62
16 L.H., Daily Totals		$1458.00		$1884.64	$91.13	$117.79

Crew B-11K	Hr.	Daily	Hr.	Daily	Bare Costs	Incl. O&P
1 Equipment Oper. (med.)	$47.50	$380.00	$71.75	$574.00	$41.48	$63.17
1 Laborer	35.45	283.60	54.60	436.80		
1 Trencher, Chain Type, 8' D		3357.00		3692.70	209.81	230.79
16 L.H., Daily Totals		$4020.60		$4703.50	$251.29	$293.97

Crew B-11L	Hr.	Daily	Hr.	Daily	Bare Costs	Incl. O&P
1 Equipment Oper. (med.)	$47.50	$380.00	$71.75	$574.00	$41.48	$63.17
1 Laborer	35.45	283.60	54.60	436.80		
1 Grader, 30,000 Lbs.		712.40		783.64	44.52	48.98
16 L.H., Daily Totals		$1376.00		$1794.44	$86.00	$112.15

Crew B-11M	Hr.	Daily	Hr.	Daily	Bare Costs	Incl. O&P
1 Equipment Oper. (med.)	$47.50	$380.00	$71.75	$574.00	$41.48	$63.17
1 Laborer	35.45	283.60	54.60	436.80		
1 Backhoe Loader, 80 H.P.		400.40		440.44	25.02	27.53
16 L.H., Daily Totals		$1064.00		$1451.24	$66.50	$90.70

Crew B-11N	Hr.	Daily	Hr.	Daily	Bare Costs	Incl. O&P
1 Labor Foreman (outside)	$37.45	$299.60	$57.65	$461.20	$39.12	$59.72
2 Equipment Operators (med.)	47.50	760.00	71.75	1148.00		
6 Truck Drivers (hvy.)	36.60	1756.80	56.05	2690.40		
1 F.E. Loader, W.M., 5.5 C.Y.		1056.00		1161.60		
1 Dozer, 410 H.P.		2425.00		2667.50		
6 Dump Trucks, Off Hwy., 50 Ton		10782.00		11860.20	198.10	217.91
72 L.H., Daily Totals		$17079.40		$19988.90	$237.21	$277.62

Crew B-11Q	Hr.	Daily	Hr.	Daily	Bare Costs	Incl. O&P
1 Equipment Operator (med.)	$47.50	$380.00	$71.75	$574.00	$43.48	$66.03
.5 Laborer	35.45	141.80	54.60	218.40		
1 Dozer, 140 H.P.		889.00		977.90	74.08	81.49
12 L.H., Daily Totals		$1410.80		$1770.30	$117.57	$147.53

Crew B-11R	Hr.	Daily	Hr.	Daily	Bare Costs	Incl. O&P
1 Equipment Operator (med.)	$47.50	$380.00	$71.75	$574.00	$43.48	$66.03
.5 Laborer	35.45	141.80	54.60	218.40		
1 Dozer, 200 H.P.		1333.00		1466.30	111.08	122.19
12 L.H., Daily Totals		$1854.80		$2258.70	$154.57	$188.22

Crew B-11S	Hr.	Daily	Hr.	Daily	Bare Costs	Incl. O&P
1 Equipment Operator (med.)	$47.50	$380.00	$71.75	$574.00	$43.48	$66.03
.5 Laborer	35.45	141.80	54.60	218.40		
1 Dozer, 300 H.P.		1829.00		2011.90		
1 Ripper, Beam & 1 Shank		82.00		90.20	159.25	175.18
12 L.H., Daily Totals		$2432.80		$2894.50	$202.73	$241.21

Crew B-11T	Hr.	Daily	Hr.	Daily	Bare Costs	Incl. O&P
1 Equipment Operator (med.)	$47.50	$380.00	$71.75	$574.00	$43.48	$66.03
.5 Laborer	35.45	141.80	54.60	218.40		
1 Dozer, 410 H.P.		2425.00		2667.50		
1 Ripper, Beam & 2 Shanks		92.20		101.42	209.77	230.74
12 L.H., Daily Totals		$3039.00		$3561.32	$253.25	$296.78

Crew B-11U	Hr.	Daily	Hr.	Daily	Bare Costs	Incl. O&P
1 Equipment Operator (med.)	$47.50	$380.00	$71.75	$574.00	$43.48	$66.03
.5 Laborer	35.45	141.80	54.60	218.40		
1 Dozer, 520 H.P.		3137.00		3450.70	261.42	287.56
12 L.H., Daily Totals		$3658.80		$4243.10	$304.90	$353.59

Crew B-11V	Hr.	Daily	Hr.	Daily	Bare Costs	Incl. O&P
3 Laborers	$35.45	$850.80	$54.60	$1310.40	$35.45	$54.60
1 Roller, 2-Drum, W.B., 7.5 H.P.		174.00		191.40	7.25	7.97
24 L.H., Daily Totals		$1024.80		$1501.80	$42.70	$62.58

Crew B-11W	Hr.	Daily	Hr.	Daily	Bare Costs	Incl. O&P
1 Equipment Operator (med.)	$47.50	$380.00	$71.75	$574.00	$37.41	$57.24
1 Common Laborer	35.45	283.60	54.60	436.80		
10 Truck Drivers (hvy.)	36.60	2928.00	56.05	4484.00		
1 Dozer, 200 H.P.		1333.00		1466.30		
1 Vibratory Roller, Towed, 23 Ton		407.20		447.92		
10 Dump Trucks, 8 C.Y., 220 H.P.		4166.00		4582.60	61.52	67.68
96 L.H., Daily Totals		$9497.80		$11991.62	$98.94	$124.91

Crew B-11Y	Hr.	Daily	Hr.	Daily	Bare Costs	Incl. O&P
1 Labor Foreman (Outside)	$37.45	$299.60	$57.65	$461.20	$39.69	$60.66
5 Common Laborers	35.45	1418.00	54.60	2184.00		
3 Equipment Operators (med.)	47.50	1140.00	71.75	1722.00		
1 Dozer, 80 H.P.		480.60		528.66		
2 Roller, 2-Drum, W.B., 7.5 H.P.		348.00		382.80		
4 Vibrating Plate, Gas, 21"		177.60		195.36	13.98	15.37
72 L.H., Daily Totals		$3863.80		$5474.02	$53.66	$76.03

Crew B-12A	Hr.	Daily	Hr.	Daily	Bare Costs	Incl. O&P
1 Equip. Oper. (crane)	$48.80	$390.40	$73.75	$590.00	$42.13	$64.17
1 Laborer	35.45	283.60	54.60	436.80		
1 Hyd. Excavator, 1 C.Y.		815.60		897.16	50.98	56.07
16 L.H., Daily Totals		$1489.60		$1923.96	$93.10	$120.25

Crew B-12B	Hr.	Daily	Hr.	Daily	Bare Costs	Incl. O&P
1 Equip. Oper. (crane)	$48.80	$390.40	$73.75	$590.00	$42.13	$64.17
1 Laborer	35.45	283.60	54.60	436.80		
1 Hyd. Excavator, 1.5 C.Y.		1036.00		1139.60	64.75	71.22
16 L.H., Daily Totals		$1710.00		$2166.40	$106.88	$135.40

Crew B-12C	Hr.	Daily	Hr.	Daily	Bare Costs	Incl. O&P
1 Equip. Oper. (crane)	$48.80	$390.40	$73.75	$590.00	$42.13	$64.17
1 Laborer	35.45	283.60	54.60	436.80		
1 Hyd. Excavator, 2 C.Y.		1363.00		1499.30	85.19	93.71
16 L.H., Daily Totals		$2037.00		$2526.10	$127.31	$157.88

Crew B-12D	Hr.	Daily	Hr.	Daily	Bare Costs	Incl. O&P
1 Equip. Oper. (crane)	$48.80	$390.40	$73.75	$590.00	$42.13	$64.17
1 Laborer	35.45	283.60	54.60	436.80		
1 Hyd. Excavator, 3.5 C.Y.		2454.00		2699.40	153.38	168.71
16 L.H., Daily Totals		$3128.00		$3726.20	$195.50	$232.89

Crew B-12E	Hr.	Daily	Hr.	Daily	Bare Costs	Incl. O&P
1 Equip. Oper. (crane)	$48.80	$390.40	$73.75	$590.00	$42.13	$64.17
1 Laborer	35.45	283.60	54.60	436.80		
1 Hyd. Excavator, .5 C.Y.		400.40		440.44	25.02	27.53
16 L.H., Daily Totals		$1074.40		$1467.24	$67.15	$91.70

Crew No.	Bare Costs		Incl. Subs O&P		Cost Per Labor-Hour	

Crew B-12F	Hr.	Daily	Hr.	Daily	Bare Costs	Incl. O&P
1 Equip. Oper. (crane)	$48.80	$390.40	$73.75	$590.00	$42.13	$64.17
1 Laborer	35.45	283.60	54.60	436.80		
1 Hyd. Excavator, .75 C.Y.		694.80		764.28	43.42	47.77
16 L.H., Daily Totals		$1368.80		$1791.08	$85.55	$111.94

Crew B-12G	Hr.	Daily	Hr.	Daily	Bare Costs	Incl. O&P
1 Equip. Oper. (crane)	$48.80	$390.40	$73.75	$590.00	$42.13	$64.17
1 Laborer	35.45	283.60	54.60	436.80		
1 Crawler Crane, 15 Ton		668.70		735.57		
1 Clamshell Bucket, .5 C.Y.		37.80		41.58	44.16	48.57
16 L.H., Daily Totals		$1380.50		$1803.95	$86.28	$112.75

Crew B-12H	Hr.	Daily	Hr.	Daily	Bare Costs	Incl. O&P
1 Equip. Oper. (crane)	$48.80	$390.40	$73.75	$590.00	$42.13	$64.17
1 Laborer	35.45	283.60	54.60	436.80		
1 Crawler Crane, 25 Ton		1148.00		1262.80		
1 Clamshell Bucket, 1 C.Y.		47.40		52.14	74.71	82.18
16 L.H., Daily Totals		$1869.40		$2341.74	$116.84	$146.36

Crew B-12I	Hr.	Daily	Hr.	Daily	Bare Costs	Incl. O&P
1 Equip. Oper. (crane)	$48.80	$390.40	$73.75	$590.00	$42.13	$64.17
1 Laborer	35.45	283.60	54.60	436.80		
1 Crawler Crane, 20 Ton		861.60		947.76		
1 Dragline Bucket, .75 C.Y.		20.80		22.88	55.15	60.66
16 L.H., Daily Totals		$1556.40		$1997.44	$97.28	$124.84

Crew B-12J	Hr.	Daily	Hr.	Daily	Bare Costs	Incl. O&P
1 Equip. Oper. (crane)	$48.80	$390.40	$73.75	$590.00	$42.13	$64.17
1 Laborer	35.45	283.60	54.60	436.80		
1 Gradall, 5/8 C.Y.		883.80		972.18	55.24	60.76
16 L.H., Daily Totals		$1557.80		$1998.98	$97.36	$124.94

Crew B-12K	Hr.	Daily	Hr.	Daily	Bare Costs	Incl. O&P
1 Equip. Oper. (crane)	$48.80	$390.40	$73.75	$590.00	$42.13	$64.17
1 Laborer	35.45	283.60	54.60	436.80		
1 Gradall, 3 Ton, 1 C.Y.		1010.00		1111.00	63.13	69.44
16 L.H., Daily Totals		$1684.00		$2137.80	$105.25	$133.61

Crew B-12L	Hr.	Daily	Hr.	Daily	Bare Costs	Incl. O&P
1 Equip. Oper. (crane)	$48.80	$390.40	$73.75	$590.00	$42.13	$64.17
1 Laborer	35.45	283.60	54.60	436.80		
1 Crawler Crane, 15 Ton		668.70		735.57		
1 F.E. Attachment, .5 C.Y.		58.20		64.02	45.43	49.97
16 L.H., Daily Totals		$1400.90		$1826.39	$87.56	$114.15

Crew B-12M	Hr.	Daily	Hr.	Daily	Bare Costs	Incl. O&P
1 Equip. Oper. (crane)	$48.80	$390.40	$73.75	$590.00	$42.13	$64.17
1 Laborer	35.45	283.60	54.60	436.80		
1 Crawler Crane, 20 Ton		861.60		947.76		
1 F.E. Attachment, .75 C.Y.		62.60		68.86	57.76	63.54
16 L.H., Daily Totals		$1598.20		$2043.42	$99.89	$127.71

Crew B-12N	Hr.	Daily	Hr.	Daily	Bare Costs	Incl. O&P
1 Equip. Oper. (crane)	$48.80	$390.40	$73.75	$590.00	$42.13	$64.17
1 Laborer	35.45	283.60	54.60	436.80		
1 Crawler Crane, 25 Ton		1148.00		1262.80		
1 F.E. Attachment, 1 C.Y.		69.40		76.34	76.09	83.70
16 L.H., Daily Totals		$1891.40		$2365.94	$118.21	$147.87

Crew B-12O	Hr.	Daily	Hr.	Daily	Bare Costs	Incl. O&P
1 Equip. Oper. (crane)	$48.80	$390.40	$73.75	$590.00	$42.13	$64.17
1 Laborer	35.45	283.60	54.60	436.80		
1 Crawler Crane, 40 Ton		1152.00		1267.20		
1 F.E. Attachment, 1.5 C.Y.		78.60		86.46	76.91	84.60
16 L.H., Daily Totals		$1904.60		$2380.46	$119.04	$148.78

Crew B-12P	Hr.	Daily	Hr.	Daily	Bare Costs	Incl. O&P
1 Equip. Oper. (crane)	$48.80	$390.40	$73.75	$590.00	$42.13	$64.17
1 Laborer	35.45	283.60	54.60	436.80		
1 Crawler Crane, 40 Ton		1152.00		1267.20		
1 Dragline Bucket, 1.5 C.Y.		33.60		36.96	74.10	81.51
16 L.H., Daily Totals		$1859.60		$2330.96	$116.22	$145.69

Crew B-12Q	Hr.	Daily	Hr.	Daily	Bare Costs	Incl. O&P
1 Equip. Oper. (crane)	$48.80	$390.40	$73.75	$590.00	$42.13	$64.17
1 Laborer	35.45	283.60	54.60	436.80		
1 Hyd. Excavator, 5/8 C.Y.		596.00		655.60	37.25	40.98
16 L.H., Daily Totals		$1270.00		$1682.40	$79.38	$105.15

Crew B-12S	Hr.	Daily	Hr.	Daily	Bare Costs	Incl. O&P
1 Equip. Oper. (crane)	$48.80	$390.40	$73.75	$590.00	$42.13	$64.17
1 Laborer	35.45	283.60	54.60	436.80		
1 Hyd. Excavator, 2.5 C.Y.		1800.00		1980.00	112.50	123.75
16 L.H., Daily Totals		$2474.00		$3006.80	$154.63	$187.93

Crew B-12T	Hr.	Daily	Hr.	Daily	Bare Costs	Incl. O&P
1 Equip. Oper. (crane)	$48.80	$390.40	$73.75	$590.00	$42.13	$64.17
1 Laborer	35.45	283.60	54.60	436.80		
1 Crawler Crane, 75 Ton		1457.00		1602.70		
1 F.E. Attachment, 3 C.Y.		101.00		111.10	97.38	107.11
16 L.H., Daily Totals		$2232.00		$2740.60	$139.50	$171.29

Crew B-12V	Hr.	Daily	Hr.	Daily	Bare Costs	Incl. O&P
1 Equip. Oper. (crane)	$48.80	$390.40	$73.75	$590.00	$42.13	$64.17
1 Laborer	35.45	283.60	54.60	436.80		
1 Crawler Crane, 75 Ton		1457.00		1602.70		
1 Dragline Bucket, 3 C.Y.		52.20		57.42	94.33	103.76
16 L.H., Daily Totals		$2183.20		$2686.92	$136.45	$167.93

Crew B-12Y	Hr.	Daily	Hr.	Daily	Bare Costs	Incl. O&P
1 Equip. Oper. (crane)	$48.80	$390.40	$73.75	$590.00	$39.90	$60.98
2 Laborers	35.45	567.20	54.60	873.60		
1 Hyd. Excavator, 3.5 C.Y.		2454.00		2699.40	102.25	112.47
24 L.H., Daily Totals		$3411.60		$4163.00	$142.15	$173.46

Crew B-12Z	Hr.	Daily	Hr.	Daily	Bare Costs	Incl. O&P
1 Equip. Oper. (crane)	$48.80	$390.40	$73.75	$590.00	$39.90	$60.98
2 Laborers	35.45	567.20	54.60	873.60		
1 Hyd. Excavator, 2.5 C.Y.		1800.00		1980.00	75.00	82.50
24 L.H., Daily Totals		$2757.60		$3443.60	$114.90	$143.48

Crew B-13	Hr.	Daily	Hr.	Daily	Bare Costs	Incl. O&P
1 Labor Foreman (outside)	$37.45	$299.60	$57.65	$461.20	$38.61	$59.09
4 Laborers	35.45	1134.40	54.60	1747.20		
1 Equip. Oper. (crane)	48.80	390.40	73.75	590.00		
1 Equip. Oper. Oiler	42.25	338.00	63.85	510.80		
1 Hyd. Crane, 25 Ton		728.60		801.46	13.01	14.31
56 L.H., Daily Totals		$2891.00		$4110.66	$51.63	$73.40

Crew No.	Bare Costs		Incl. Subs O&P		Cost Per Labor-Hour	
Crew B-13A	**Hr.**	**Daily**	**Hr.**	**Daily**	**Bare Costs**	**Incl. O&P**
1 Labor Foreman (outside)	$37.45	$299.60	$57.65	$461.20	$39.51	$60.35
2 Laborers	35.45	567.20	54.60	873.60		
2 Equipment Operators (med.)	47.50	760.00	71.75	1148.00		
2 Truck Drivers (heavy)	36.60	585.60	56.05	896.80		
1 Crawler Crane, 75 Ton		1457.00		1602.70		
1 Crawler Loader, 4 C.Y.		1530.00		1683.00		
2 Dump Trucks, 8 C.Y., 220 H.P.		833.20		916.52	68.22	75.04
56 L.H., Daily Totals		$6032.60		$7581.82	$107.72	$135.39
Crew B-13B	**Hr.**	**Daily**	**Hr.**	**Daily**	**Bare Costs**	**Incl. O&P**
1 Labor Foreman (outside)	$37.45	$299.60	$57.65	$461.20	$38.61	$59.09
4 Laborers	35.45	1134.40	54.60	1747.20		
1 Equip. Oper. (crane)	48.80	390.40	73.75	590.00		
1 Equip. Oper. Oiler	42.25	338.00	63.85	510.80		
1 Hyd. Crane, 55 Ton		1115.00		1226.50	19.91	21.90
56 L.H., Daily Totals		$3277.40		$4535.70	$58.52	$80.99
Crew B-13C	**Hr.**	**Daily**	**Hr.**	**Daily**	**Bare Costs**	**Incl. O&P**
1 Labor Foreman (outside)	$37.45	$299.60	$57.65	$461.20	$38.61	$59.09
4 Laborers	35.45	1134.40	54.60	1747.20		
1 Equip. Oper. (crane)	48.80	390.40	73.75	590.00		
1 Equip. Oper. Oiler	42.25	338.00	63.85	510.80		
1 Crawler Crane, 100 Ton		1666.00		1832.60	29.75	32.73
56 L.H., Daily Totals		$3828.40		$5141.80	$68.36	$91.82
Crew B-13D	**Hr.**	**Daily**	**Hr.**	**Daily**	**Bare Costs**	**Incl. O&P**
1 Laborer	$35.45	$283.60	$54.60	$436.80	$42.13	$64.17
1 Equip. Oper. (crane)	48.80	390.40	73.75	590.00		
1 Hyd. Excavator, 1 C.Y.		815.60		897.16		
1 Trench Box		113.40		124.74	58.06	63.87
16 L.H., Daily Totals		$1603.00		$2048.70	$100.19	$128.04
Crew B-13E	**Hr.**	**Daily**	**Hr.**	**Daily**	**Bare Costs**	**Incl. O&P**
1 Laborer	$35.45	$283.60	$54.60	$436.80	$42.13	$64.17
1 Equip. Oper. (crane)	48.80	390.40	73.75	590.00		
1 Hyd. Excavator, 1.5 C.Y.		1036.00		1139.60		
1 Trench Box		113.40		124.74	71.84	79.02
16 L.H., Daily Totals		$1823.40		$2291.14	$113.96	$143.20
Crew B-13F	**Hr.**	**Daily**	**Hr.**	**Daily**	**Bare Costs**	**Incl. O&P**
1 Laborer	$35.45	$283.60	$54.60	$436.80	$42.13	$64.17
1 Equip. Oper. (crane)	48.80	390.40	73.75	590.00		
1 Hyd. Excavator, 3.5 C.Y.		2454.00		2699.40		
1 Trench Box		113.40		124.74	160.46	176.51
16 L.H., Daily Totals		$3241.40		$3850.94	$202.59	$240.68
Crew B-13G	**Hr.**	**Daily**	**Hr.**	**Daily**	**Bare Costs**	**Incl. O&P**
1 Laborer	$35.45	$283.60	$54.60	$436.80	$42.13	$64.17
1 Equip. Oper. (crane)	48.80	390.40	73.75	590.00		
1 Hyd. Excavator, .75 C.Y.		694.80		764.28		
1 Trench Box		113.40		124.74	50.51	55.56
16 L.H., Daily Totals		$1482.20		$1915.82	$92.64	$119.74
Crew B-13H	**Hr.**	**Daily**	**Hr.**	**Daily**	**Bare Costs**	**Incl. O&P**
1 Laborer	$35.45	$283.60	$54.60	$436.80	$42.13	$64.17
1 Equip. Oper. (crane)	48.80	390.40	73.75	590.00		
1 Gradall, 5/8 C.Y.		883.80		972.18		
1 Trench Box		113.40		124.74	62.33	68.56
16 L.H., Daily Totals		$1671.20		$2123.72	$104.45	$132.73

Crew No.	Bare Costs		Incl. Subs O&P		Cost Per Labor-Hour	
Crew B-13I	**Hr.**	**Daily**	**Hr.**	**Daily**	**Bare Costs**	**Incl. O&P**
1 Laborer	$35.45	$283.60	$54.60	$436.80	$42.13	$64.17
1 Equip. Oper. (crane)	48.80	390.40	73.75	590.00		
1 Gradall, 3 Ton, 1 C.Y.		1010.00		1111.00		
1 Trench Box		113.40		124.74	70.21	77.23
16 L.H., Daily Totals		$1797.40		$2262.54	$112.34	$141.41
Crew B-13J	**Hr.**	**Daily**	**Hr.**	**Daily**	**Bare Costs**	**Incl. O&P**
1 Laborer	$35.45	$283.60	$54.60	$436.80	$42.13	$64.17
1 Equip. Oper. (crane)	48.80	390.40	73.75	590.00		
1 Hyd. Excavator, 2.5 C.Y.		1800.00		1980.00		
1 Trench Box		113.40		124.74	119.59	131.55
16 L.H., Daily Totals		$2587.40		$3131.54	$161.71	$195.72
Crew B-14	**Hr.**	**Daily**	**Hr.**	**Daily**	**Bare Costs**	**Incl. O&P**
1 Labor Foreman (outside)	$37.45	$299.60	$57.65	$461.20	$37.51	$57.54
4 Laborers	35.45	1134.40	54.60	1747.20		
1 Equip. Oper. (light)	45.80	366.40	69.20	553.60		
1 Backhoe Loader, 48 H.P.		367.40		404.14	7.65	8.42
48 L.H., Daily Totals		$2167.80		$3166.14	$45.16	$65.96
Crew B-14A	**Hr.**	**Daily**	**Hr.**	**Daily**	**Bare Costs**	**Incl. O&P**
1 Equip. Oper. (crane)	$48.80	$390.40	$73.75	$590.00	$44.35	$67.37
.5 Laborer	35.45	141.80	54.60	218.40		
1 Hyd. Excavator, 4.5 C.Y.		3057.00		3362.70	254.75	280.23
12 L.H., Daily Totals		$3589.20		$4171.10	$299.10	$347.59
Crew B-14B	**Hr.**	**Daily**	**Hr.**	**Daily**	**Bare Costs**	**Incl. O&P**
1 Equip. Oper. (crane)	$48.80	$390.40	$73.75	$590.00	$44.35	$67.37
.5 Laborer	35.45	141.80	54.60	218.40		
1 Hyd. Excavator, 6 C.Y.		3541.00		3895.10	295.08	324.59
12 L.H., Daily Totals		$4073.20		$4703.50	$339.43	$391.96
Crew B-14C	**Hr.**	**Daily**	**Hr.**	**Daily**	**Bare Costs**	**Incl. O&P**
1 Equip. Oper. (crane)	$48.80	$390.40	$73.75	$590.00	$44.35	$67.37
.5 Laborer	35.45	141.80	54.60	218.40		
1 Hyd. Excavator, 7 C.Y.		3616.00		3977.60	301.33	331.47
12 L.H., Daily Totals		$4148.20		$4786.00	$345.68	$398.83
Crew B-14F	**Hr.**	**Daily**	**Hr.**	**Daily**	**Bare Costs**	**Incl. O&P**
1 Equip. Oper. (crane)	$48.80	$390.40	$73.75	$590.00	$44.35	$67.37
.5 Laborer	35.45	141.80	54.60	218.40		
1 Hyd. Shovel, 7 C.Y.		3495.00		3844.50	291.25	320.38
12 L.H., Daily Totals		$4027.20		$4652.90	$335.60	$387.74
Crew B-14G	**Hr.**	**Daily**	**Hr.**	**Daily**	**Bare Costs**	**Incl. O&P**
1 Equip. Oper. (crane)	$48.80	$390.40	$73.75	$590.00	$44.35	$67.37
.5 Laborer	35.45	141.80	54.60	218.40		
1 Hyd. Shovel, 12 C.Y.		4676.00		5143.60	389.67	428.63
12 L.H., Daily Totals		$5208.20		$5952.00	$434.02	$496.00
Crew B-14J	**Hr.**	**Daily**	**Hr.**	**Daily**	**Bare Costs**	**Incl. O&P**
1 Equip. Oper. (med.)	$47.50	$380.00	$71.75	$574.00	$43.48	$66.03
.5 Laborer	35.45	141.80	54.60	218.40		
1 F.E. Loader, 8 C.Y.		2008.00		2208.80	167.33	184.07
12 L.H., Daily Totals		$2529.80		$3001.20	$210.82	$250.10

Crew B-14K	Bare Costs Hr.	Bare Costs Daily	Incl. Subs O&P Hr.	Incl. Subs O&P Daily	Cost Per Labor-Hour Bare Costs	Cost Per Labor-Hour Incl. O&P
1 Equip. Oper. (med.)	$47.50	$380.00	$71.75	$574.00	$43.48	$66.03
.5 Laborer	35.45	141.80	54.60	218.40		
1 F.E. Loader, 10 C.Y.		2734.00		3007.40	227.83	250.62
12 L.H., Daily Totals		$3255.80		$3799.80	$271.32	$316.65

Crew B-15	Hr.	Daily	Hr.	Daily	Bare Costs	Incl. O&P
1 Equipment Oper. (med.)	$47.50	$380.00	$71.75	$574.00	$39.55	$60.33
.5 Laborer	35.45	141.80	54.60	218.40		
2 Truck Drivers (heavy)	36.60	585.60	56.05	896.80		
2 Dump Trucks, 12 C.Y., 400 H.P.		1384.80		1523.28		
1 Dozer, 200 H.P.		1333.00		1466.30	97.06	106.77
28 L.H., Daily Totals		$3825.20		$4678.78	$136.61	$167.10

Crew B-16	Hr.	Daily	Hr.	Daily	Bare Costs	Incl. O&P
1 Labor Foreman (outside)	$37.45	$299.60	$57.65	$461.20	$36.24	$55.73
2 Laborers	35.45	567.20	54.60	873.60		
1 Truck Driver (heavy)	36.60	292.80	56.05	448.40		
1 Dump Truck, 12 C.Y., 400 H.P.		692.40		761.64	21.64	23.80
32 L.H., Daily Totals		$1852.00		$2544.84	$57.88	$79.53

Crew B-17	Hr.	Daily	Hr.	Daily	Bare Costs	Incl. O&P
2 Laborers	$35.45	$567.20	$54.60	$873.60	$38.33	$58.61
1 Equip. Oper. (light)	45.80	366.40	69.20	553.60		
1 Truck Driver (heavy)	36.60	292.80	56.05	448.40		
1 Backhoe Loader, 48 H.P.		367.40		404.14		
1 Dump Truck, 8 C.Y., 220 H.P.		416.60		458.26	24.50	26.95
32 L.H., Daily Totals		$2010.40		$2738.00	$62.83	$85.56

Crew B-17A	Hr.	Daily	Hr.	Daily	Bare Costs	Incl. O&P
2 Labor Foremen (outside)	$37.45	$599.20	$57.65	$922.40	$38.20	$58.88
6 Laborers	35.45	1701.60	54.60	2620.80		
1 Skilled Worker Foreman (out)	48.20	385.60	74.50	596.00		
1 Skilled Worker	46.20	369.60	71.45	571.60		
80 L.H., Daily Totals		$3056.00		$4710.80	$38.20	$58.88

Crew B-17B	Hr.	Daily	Hr.	Daily	Bare Costs	Incl. O&P
2 Laborers	$35.45	$567.20	$54.60	$873.60	$38.33	$58.61
1 Equip. Oper. (light)	45.80	366.40	69.20	553.60		
1 Truck Driver (heavy)	36.60	292.80	56.05	448.40		
1 Backhoe Loader, 48 H.P.		367.40		404.14		
1 Dump Truck, 12 C.Y., 400 H.P.		692.40		761.64	33.12	36.43
32 L.H., Daily Totals		$2286.20		$3041.38	$71.44	$95.04

Crew B-18	Hr.	Daily	Hr.	Daily	Bare Costs	Incl. O&P
1 Labor Foreman (outside)	$37.45	$299.60	$57.65	$461.20	$36.12	$55.62
2 Laborers	35.45	567.20	54.60	873.60		
1 Vibrating Plate, Gas, 21"		44.40		48.84	1.85	2.04
24 L.H., Daily Totals		$911.20		$1383.64	$37.97	$57.65

Crew B-19	Hr.	Daily	Hr.	Daily	Bare Costs	Incl. O&P
1 Pile Driver Foreman (outside)	$45.15	$361.20	$71.90	$575.20	$44.70	$69.76
4 Pile Drivers	43.15	1380.80	68.70	2198.40		
2 Equip. Oper. (crane)	48.80	780.80	73.75	1180.00		
1 Equip. Oper. Oiler	42.25	338.00	63.85	510.80		
1 Crawler Crane, 40 Ton		1152.00		1267.20		
1 Lead, 90' High		124.80		137.28		
1 Hammer, Diesel, 22k ft-lb		618.00		679.80	29.61	32.57
64 L.H., Daily Totals		$4755.60		$6548.68	$74.31	$102.32

Crew B-19A	Hr.	Daily	Hr.	Daily	Bare Costs	Incl. O&P
1 Pile Driver Foreman (outside)	$45.15	$361.20	$71.90	$575.20	$44.70	$69.76
4 Pile Drivers	43.15	1380.80	68.70	2198.40		
2 Equip. Oper. (crane)	48.80	780.80	73.75	1180.00		
1 Equip. Oper. Oiler	42.25	338.00	63.85	510.80		
1 Crawler Crane, 75 Ton		1457.00		1602.70		
1 Lead, 90' high		124.80		137.28		
1 Hammer, Diesel, 41k ft-lb		704.00		774.40	35.72	39.29
64 L.H., Daily Totals		$5146.60		$6978.78	$80.42	$109.04

Crew B-19B	Hr.	Daily	Hr.	Daily	Bare Costs	Incl. O&P
1 Pile Driver Foreman (outside)	$45.15	$361.20	$71.90	$575.20	$44.70	$69.76
4 Pile Drivers	43.15	1380.80	68.70	2198.40		
2 Equip. Oper. (crane)	48.80	780.80	73.75	1180.00		
1 Equip. Oper. Oiler	42.25	338.00	63.85	510.80		
1 Crawler Crane, 40 Ton		1152.00		1267.20		
1 Lead, 90' High		124.80		137.28		
1 Hammer, Diesel, 22k ft-lb		618.00		679.80		
1 Barge, 400 Ton		783.80		862.18	41.85	46.04
64 L.H., Daily Totals		$5539.40		$7410.86	$86.55	$115.79

Crew B-19C	Hr.	Daily	Hr.	Daily	Bare Costs	Incl. O&P
1 Pile Driver Foreman (outside)	$45.15	$361.20	$71.90	$575.20	$44.70	$69.76
4 Pile Drivers	43.15	1380.80	68.70	2198.40		
2 Equip. Oper. (crane)	48.80	780.80	73.75	1180.00		
1 Equip. Oper. Oiler	42.25	338.00	63.85	510.80		
1 Crawler Crane, 75 Ton		1457.00		1602.70		
1 Lead, 90' High		124.80		137.28		
1 Hammer, Diesel, 41k ft-lb		704.00		774.40		
1 Barge, 400 Ton		783.80		862.18	47.96	52.76
64 L.H., Daily Totals		$5930.40		$7840.96	$92.66	$122.52

Crew B-20	Hr.	Daily	Hr.	Daily	Bare Costs	Incl. O&P
1 Labor Foreman (outside)	$37.45	$299.60	$57.65	$461.20	$39.70	$61.23
1 Skilled Worker	46.20	369.60	71.45	571.60		
1 Laborer	35.45	283.60	54.60	436.80		
24 L.H., Daily Totals		$952.80		$1469.60	$39.70	$61.23

Crew B-20A	Hr.	Daily	Hr.	Daily	Bare Costs	Incl. O&P
1 Labor Foreman (outside)	$37.45	$299.60	$57.65	$461.20	$43.34	$65.89
1 Laborer	35.45	283.60	54.60	436.80		
1 Plumber	55.80	446.40	84.05	672.40		
1 Plumber Apprentice	44.65	357.20	67.25	538.00		
32 L.H., Daily Totals		$1386.80		$2108.40	$43.34	$65.89

Crew B-21	Hr.	Daily	Hr.	Daily	Bare Costs	Incl. O&P
1 Labor Foreman (outside)	$37.45	$299.60	$57.65	$461.20	$41.00	$63.02
1 Skilled Worker	46.20	369.60	71.45	571.60		
1 Laborer	35.45	283.60	54.60	436.80		
.5 Equip. Oper. (crane)	48.80	195.20	73.75	295.00		
.5 S.P. Crane, 4x4, 5 Ton		137.10		150.81	4.90	5.39
28 L.H., Daily Totals		$1285.10		$1915.41	$45.90	$68.41

Crew B-21A	Hr.	Daily	Hr.	Daily	Bare Costs	Incl. O&P
1 Labor Foreman (outside)	$37.45	$299.60	$57.65	$461.20	$44.43	$67.46
1 Laborer	35.45	283.60	54.60	436.80		
1 Plumber	55.80	446.40	84.05	672.40		
1 Plumber Apprentice	44.65	357.20	67.25	538.00		
1 Equip. Oper. (crane)	48.80	390.40	73.75	590.00		
1 S.P. Crane, 4x4, 12 Ton		471.00		518.10	11.78	12.95
40 L.H., Daily Totals		$2248.20		$3216.50	$56.20	$80.41

Crew B-21B	Hr.	Daily	Hr.	Daily	Bare Costs	Incl. O&P
1 Labor Foreman (outside)	$37.45	$299.60	$57.65	$461.20	$38.52	$59.04
3 Laborers	35.45	850.80	54.60	1310.40		
1 Equip. Oper. (crane)	48.80	390.40	73.75	590.00		
1 Hyd. Crane, 12 Ton		646.20		710.82	16.16	17.77
40 L.H., Daily Totals		$2187.00		$3072.42	$54.67	$76.81

Crew B-21C	Hr.	Daily	Hr.	Daily	Bare Costs	Incl. O&P
1 Labor Foreman (outside)	$37.45	$299.60	$57.65	$461.20	$38.61	$59.09
4 Laborers	35.45	1134.40	54.60	1747.20		
1 Equip. Oper. (crane)	48.80	390.40	73.75	590.00		
1 Equip. Oper. Oiler	42.25	338.00	63.85	510.80		
2 Cutting Torches		22.80		25.08		
2 Sets of Gases		266.40		293.04		
1 Lattice Boom Crane, 90 Ton		1525.00		1677.50	32.40	35.64
56 L.H., Daily Totals		$3976.60		$5304.82	$71.01	$94.73

Crew B-22	Hr.	Daily	Hr.	Daily	Bare Costs	Incl. O&P
1 Labor Foreman (outside)	$37.45	$299.60	$57.65	$461.20	$41.52	$63.74
1 Skilled Worker	46.20	369.60	71.45	571.60		
1 Laborer	35.45	283.60	54.60	436.80		
.75 Equip. Oper. (crane)	48.80	292.80	73.75	442.50		
.75 S.P. Crane, 4x4, 5 Ton		205.65		226.22	6.86	7.54
30 L.H., Daily Totals		$1451.25		$2138.32	$48.38	$71.28

Crew B-22A	Hr.	Daily	Hr.	Daily	Bare Costs	Incl. O&P
1 Labor Foreman (outside)	$37.45	$299.60	$57.65	$461.20	$40.67	$62.41
1 Skilled Worker	46.20	369.60	71.45	571.60		
2 Laborers	35.45	567.20	54.60	873.60		
1 Equipment Oper. (crane)	48.80	390.40	73.75	590.00		
1 S.P. Crane, 4x4, 5 Ton		274.20		301.62		
1 Butt Fusion Mach., 4"-12" diam.		382.60		420.86	16.42	18.06
40 L.H., Daily Totals		$2283.60		$3218.88	$57.09	$80.47

Crew B-22B	Hr.	Daily	Hr.	Daily	Bare Costs	Incl. O&P
1 Labor Foreman (outside)	$37.45	$299.60	$57.65	$461.20	$40.67	$62.41
1 Skilled Worker	46.20	369.60	71.45	571.60		
2 Laborers	35.45	567.20	54.60	873.60		
1 Equip. Oper. (crane)	48.80	390.40	73.75	590.00		
1 S.P. Crane, 4x4, 5 Ton		274.20		301.62		
1 Butt Fusion Mach., 8"-24" diam.		831.70		914.87	27.65	30.41
40 L.H., Daily Totals		$2732.70		$3712.89	$68.32	$92.82

Crew B-22C	Hr.	Daily	Hr.	Daily	Bare Costs	Incl. O&P
1 Skilled Worker	$46.20	$369.60	$71.45	$571.60	$40.83	$63.02
1 Laborer	35.45	283.60	54.60	436.80		
1 Butt Fusion Mach., 2"-8" diam.		120.95		133.04	7.56	8.32
16 L.H., Daily Totals		$774.15		$1141.44	$48.38	$71.34

Crew B-23	Hr.	Daily	Hr.	Daily	Bare Costs	Incl. O&P
1 Labor Foreman (outside)	$37.45	$299.60	$57.65	$461.20	$35.85	$55.21
4 Laborers	35.45	1134.40	54.60	1747.20		
1 Drill Rig, Truck-Mounted		2469.00		2715.90		
1 Flatbed Truck, Gas, 3 Ton		327.80		360.58	69.92	76.91
40 L.H., Daily Totals		$4230.80		$5284.88	$105.77	$132.12

Crew B-23A	Hr.	Daily	Hr.	Daily	Bare Costs	Incl. O&P
1 Labor Foreman (outside)	$37.45	$299.60	$57.65	$461.20	$40.13	$61.33
1 Laborer	35.45	283.60	54.60	436.80		
1 Equip. Operator (med.)	47.50	380.00	71.75	574.00		
1 Drill Rig, Truck-Mounted		2469.00		2715.90		
1 Pickup Truck, 3/4 Ton		154.60		170.06	109.32	120.25
24 L.H., Daily Totals		$3586.80		$4357.96	$149.45	$181.58

Crew B-23B	Hr.	Daily	Hr.	Daily	Bare Costs	Incl. O&P
1 Labor Foreman (outside)	$37.45	$299.60	$57.65	$461.20	$40.13	$61.33
1 Laborer	35.45	283.60	54.60	436.80		
1 Equip. Operator (med.)	47.50	380.00	71.75	574.00		
1 Drill Rig, Truck-Mounted		2469.00		2715.90		
1 Pickup Truck, 3/4 Ton		154.60		170.06		
1 Centr. Water Pump, 6"		341.40		375.54	123.54	135.90
24 L.H., Daily Totals		$3928.20		$4733.50	$163.68	$197.23

Crew B-24	Hr.	Daily	Hr.	Daily	Bare Costs	Incl. O&P
1 Cement Finisher	$43.05	$344.40	$63.40	$507.20	$41.13	$62.38
1 Laborer	35.45	283.60	54.60	436.80		
1 Carpenter	44.90	359.20	69.15	553.20		
24 L.H., Daily Totals		$987.20		$1497.20	$41.13	$62.38

Crew B-25	Hr.	Daily	Hr.	Daily	Bare Costs	Incl. O&P
1 Labor Foreman (outside)	$37.45	$299.60	$57.65	$461.20	$38.92	$59.55
7 Laborers	35.45	1985.20	54.60	3057.60		
3 Equip. Oper. (med.)	47.50	1140.00	71.75	1722.00		
1 Asphalt Paver, 130 H.P.		2124.00		2336.40		
1 Tandem Roller, 10 Ton		231.20		254.32		
1 Roller, Pneum. Whl., 12 Ton		326.60		359.26	30.48	33.52
88 L.H., Daily Totals		$6106.60		$8190.78	$69.39	$93.08

Crew B-25B	Hr.	Daily	Hr.	Daily	Bare Costs	Incl. O&P
1 Labor Foreman (outside)	$37.45	$299.60	$57.65	$461.20	$39.63	$60.57
7 Laborers	35.45	1985.20	54.60	3057.60		
4 Equip. Oper. (med.)	47.50	1520.00	71.75	2296.00		
1 Asphalt Paver, 130 H.P.		2124.00		2336.40		
2 Tandem Rollers, 10 Ton		462.40		508.64		
1 Roller, Pneum. Whl., 12 Ton		326.60		359.26	30.34	33.38
96 L.H., Daily Totals		$6717.80		$9019.10	$69.98	$93.95

Crew B-25C	Hr.	Daily	Hr.	Daily	Bare Costs	Incl. O&P
1 Labor Foreman (outside)	$37.45	$299.60	$57.65	$461.20	$39.80	$60.83
3 Laborers	35.45	850.80	54.60	1310.40		
2 Equip. Oper. (med.)	47.50	760.00	71.75	1148.00		
1 Asphalt Paver, 130 H.P.		2124.00		2336.40		
1 Tandem Roller, 10 Ton		231.20		254.32	49.07	53.97
48 L.H., Daily Totals		$4265.60		$5510.32	$88.87	$114.80

Crew B-25D	Hr.	Daily	Hr.	Daily	Bare Costs	Incl. O&P
1 Labor Foreman (outside)	$37.45	$299.60	$57.65	$461.20	$39.89	$60.95
3 Laborers	35.45	850.80	54.60	1310.40		
2.125 Equip. Oper. (med.)	47.50	807.50	71.75	1219.75		
.125 Truck Driver (hvy.)	36.60	36.60	56.05	56.05		
.125 Truck Tractor, 6x4, 380 H.P.		76.55		84.20		
.125 Dist. Tanker, 3000 Gallon		38.15		41.97		
1 Asphalt Paver, 130 H.P.		2124.00		2336.40		
1 Tandem Roller, 10 Ton		231.20		254.32	49.40	54.34
50 L.H., Daily Totals		$4464.40		$5764.29	$89.29	$115.29

Crew No.	Bare Costs		Incl. Subs O&P		Cost Per Labor-Hour	
Crew B-25E	**Hr.**	**Daily**	**Hr.**	**Daily**	**Bare Costs**	**Incl. O&P**
1 Labor Foreman (outside)	$37.45	$299.60	$57.65	$461.20	$39.97	$61.06
3 Laborers	35.45	850.80	54.60	1310.40		
2.250 Equip. Oper. (med.)	47.50	855.00	71.75	1291.50		
.25 Truck Driver (hvy.)	36.60	73.20	56.05	112.10		
.25 Truck Tractor, 6x4, 380 H.P.		153.10		168.41		
.25 Dist. Tanker, 3000 Gallon		76.30		83.93		
1 Asphalt Paver, 130 H.P.		2124.00		2336.40		
1 Tandem Roller, 10 Ton		231.20		254.32	49.70	54.67
52 L.H., Daily Totals		$4663.20		$6018.26	$89.68	$115.74

Crew B-26	Hr.	Daily	Hr.	Daily	Bare Costs	Incl. O&P
1 Labor Foreman (outside)	$37.45	$299.60	$57.65	$461.20	$39.82	$60.99
6 Laborers	35.45	1701.60	54.60	2620.80		
2 Equip. Oper. (med.)	47.50	760.00	71.75	1148.00		
1 Rodman (reinf.)	49.80	398.40	78.70	629.60		
1 Cement Finisher	43.05	344.40	63.40	507.20		
1 Grader, 30,000 Lbs.		712.40		783.64		
1 Paving Mach. & Equip.		2667.00		2933.70	38.40	42.24
88 L.H., Daily Totals		$6883.40		$9084.14	$78.22	$103.23

Crew B-26A	Hr.	Daily	Hr.	Daily	Bare Costs	Incl. O&P
1 Labor Foreman (outside)	$37.45	$299.60	$57.65	$461.20	$39.82	$60.99
6 Laborers	35.45	1701.60	54.60	2620.80		
2 Equip. Oper. (med.)	47.50	760.00	71.75	1148.00		
1 Rodman (reinf.)	49.80	398.40	78.70	629.60		
1 Cement Finisher	43.05	344.40	63.40	507.20		
1 Grader, 30,000 Lbs.		712.40		783.64		
1 Paving Mach. & Equip.		2667.00		2933.70		
1 Concrete Saw		175.20		192.72	40.39	44.43
88 L.H., Daily Totals		$7058.60		$9276.86	$80.21	$105.42

Crew B-26B	Hr.	Daily	Hr.	Daily	Bare Costs	Incl. O&P
1 Labor Foreman (outside)	$37.45	$299.60	$57.65	$461.20	$40.46	$61.88
6 Laborers	35.45	1701.60	54.60	2620.80		
3 Equip. Oper. (med.)	47.50	1140.00	71.75	1722.00		
1 Rodman (reinf.)	49.80	398.40	78.70	629.60		
1 Cement Finisher	43.05	344.40	63.40	507.20		
1 Grader, 30,000 Lbs.		712.40		783.64		
1 Paving Mach. & Equip.		2667.00		2933.70		
1 Concrete Pump, 110' Boom		942.00		1036.20	45.01	49.52
96 L.H., Daily Totals		$8205.40		$10694.34	$85.47	$111.40

Crew B-26C	Hr.	Daily	Hr.	Daily	Bare Costs	Incl. O&P
1 Labor Foreman (outside)	$37.45	$299.60	$57.65	$461.20	$39.05	$59.91
6 Laborers	35.45	1701.60	54.60	2620.80		
1 Equip. Oper. (med.)	47.50	380.00	71.75	574.00		
1 Rodman (reinf.)	49.80	398.40	78.70	629.60		
1 Cement Finisher	43.05	344.40	63.40	507.20		
1 Paving Mach. & Equip.		2667.00		2933.70		
1 Concrete Saw		175.20		192.72	35.53	39.08
80 L.H., Daily Totals		$5966.20		$7919.22	$74.58	$98.99

Crew B-27	Hr.	Daily	Hr.	Daily	Bare Costs	Incl. O&P
1 Labor Foreman (outside)	$37.45	$299.60	$57.65	$461.20	$35.95	$55.36
3 Laborers	35.45	850.80	54.60	1310.40		
1 Berm Machine		292.80		322.08	9.15	10.07
32 L.H., Daily Totals		$1443.20		$2093.68	$45.10	$65.43

Crew No.	Bare Costs		Incl. Subs O&P		Cost Per Labor-Hour	
Crew B-28	**Hr.**	**Daily**	**Hr.**	**Daily**	**Bare Costs**	**Incl. O&P**
2 Carpenters	$44.90	$718.40	$69.15	$1106.40	$41.75	$64.30
1 Laborer	35.45	283.60	54.60	436.80		
24 L.H., Daily Totals		$1002.00		$1543.20	$41.75	$64.30

Crew B-29	Hr.	Daily	Hr.	Daily	Bare Costs	Incl. O&P
1 Labor Foreman (outside)	$37.45	$299.60	$57.65	$461.20	$38.61	$59.09
4 Laborers	35.45	1134.40	54.60	1747.20		
1 Equip. Oper. (crane)	48.80	390.40	73.75	590.00		
1 Equip. Oper. Oiler	42.25	338.00	63.85	510.80		
1 Gradall, 5/8 C.Y.		883.80		972.18	15.78	17.36
56 L.H., Daily Totals		$3046.20		$4281.38	$54.40	$76.45

Crew B-30	Hr.	Daily	Hr.	Daily	Bare Costs	Incl. O&P
1 Equip. Oper. (med.)	$47.50	$380.00	$71.75	$574.00	$40.23	$61.28
2 Truck Drivers (heavy)	36.60	585.60	56.05	896.80		
1 Hyd. Excavator, 1.5 C.Y.		1036.00		1139.60		
2 Dump Trucks, 12 C.Y., 400 H.P.		1384.80		1523.28	100.87	110.95
24 L.H., Daily Totals		$3386.40		$4133.68	$141.10	$172.24

Crew B-31	Hr.	Daily	Hr.	Daily	Bare Costs	Incl. O&P
1 Labor Foreman (outside)	$37.45	$299.60	$57.65	$461.20	$37.74	$58.12
3 Laborers	35.45	850.80	54.60	1310.40		
1 Carpenter	44.90	359.20	69.15	553.20		
1 Air Compressor, 250 cfm		195.80		215.38		
1 Sheeting Driver		5.75		6.33		
2 -50' Air Hoses, 1.5"		11.60		12.76	5.33	5.86
40 L.H., Daily Totals		$1722.75		$2559.26	$43.07	$63.98

Crew B-32	Hr.	Daily	Hr.	Daily	Bare Costs	Incl. O&P
1 Laborer	$35.45	$283.60	$54.60	$436.80	$44.49	$67.46
3 Equip. Oper. (med.)	47.50	1140.00	71.75	1722.00		
1 Grader, 30,000 Lbs.		712.40		783.64		
1 Tandem Roller, 10 Ton		231.20		254.32		
1 Dozer, 200 H.P.		1333.00		1466.30	71.14	78.26
32 L.H., Daily Totals		$3700.20		$4663.06	$115.63	$145.72

Crew B-32A	Hr.	Daily	Hr.	Daily	Bare Costs	Incl. O&P
1 Laborer	$35.45	$283.60	$54.60	$436.80	$43.48	$66.03
2 Equip. Oper. (med.)	47.50	760.00	71.75	1148.00		
1 Grader, 30,000 Lbs.		712.40		783.64		
1 Roller, Vibratory, 25 Ton		656.40		722.04	57.03	62.74
24 L.H., Daily Totals		$2412.40		$3090.48	$100.52	$128.77

Crew B-32B	Hr.	Daily	Hr.	Daily	Bare Costs	Incl. O&P
1 Laborer	$35.45	$283.60	$54.60	$436.80	$43.48	$66.03
2 Equip. Oper. (med.)	47.50	760.00	71.75	1148.00		
1 Dozer, 200 H.P.		1333.00		1466.30		
1 Roller, Vibratory, 25 Ton		656.40		722.04	82.89	91.18
24 L.H., Daily Totals		$3033.00		$3773.14	$126.38	$157.21

Crew B-32C	Hr.	Daily	Hr.	Daily	Bare Costs	Incl. O&P
1 Labor Foreman (outside)	$37.45	$299.60	$57.65	$461.20	$41.81	$63.68
2 Laborers	35.45	567.20	54.60	873.60		
3 Equip. Oper. (med.)	47.50	1140.00	71.75	1722.00		
1 Grader, 30,000 Lbs.		712.40		783.64		
1 Tandem Roller, 10 Ton		231.20		254.32		
1 Dozer, 200 H.P.		1333.00		1466.30	47.43	52.17
48 L.H., Daily Totals		$4283.40		$5561.06	$89.24	$115.86

Crew No.	Bare Costs		Incl. Subs O&P		Cost Per Labor-Hour	
Crew B-33A	**Hr.**	**Daily**	**Hr.**	**Daily**	**Bare Costs**	**Incl. O&P**
1 Equip. Oper. (med.)	$47.50	$380.00	$71.75	$574.00	$44.06	$66.85
.5 Laborer	35.45	141.80	54.60	218.40		
.25 Equip. Oper. (med.)	47.50	95.00	71.75	143.50		
1 Scraper, Towed, 7 C.Y.		110.80		121.88		
1.25 Dozers, 300 H.P.		2286.25		2514.88	171.22	188.34
14 L.H., Daily Totals		$3013.85		$3572.66	$215.28	$255.19
Crew B-33B	**Hr.**	**Daily**	**Hr.**	**Daily**	**Bare Costs**	**Incl. O&P**
1 Equip. Oper. (med.)	$47.50	$380.00	$71.75	$574.00	$44.06	$66.85
.5 Laborer	35.45	141.80	54.60	218.40		
.25 Equip. Oper. (med.)	47.50	95.00	71.75	143.50		
1 Scraper, Towed, 10 C.Y.		140.80		154.88		
1.25 Dozers, 300 H.P.		2286.25		2514.88	173.36	190.70
14 L.H., Daily Totals		$3043.85		$3605.66	$217.42	$257.55
Crew B-33C	**Hr.**	**Daily**	**Hr.**	**Daily**	**Bare Costs**	**Incl. O&P**
1 Equip. Oper. (med.)	$47.50	$380.00	$71.75	$574.00	$44.06	$66.85
.5 Laborer	35.45	141.80	54.60	218.40		
.25 Equip. Oper. (med.)	47.50	95.00	71.75	143.50		
1 Scraper, Towed, 15 C.Y.		158.00		173.80		
1.25 Dozers, 300 H.P.		2286.25		2514.88	174.59	192.05
14 L.H., Daily Totals		$3061.05		$3624.57	$218.65	$258.90
Crew B-33D	**Hr.**	**Daily**	**Hr.**	**Daily**	**Bare Costs**	**Incl. O&P**
1 Equip. Oper. (med.)	$47.50	$380.00	$71.75	$574.00	$44.06	$66.85
.5 Laborer	35.45	141.80	54.60	218.40		
.25 Equip. Oper. (med.)	47.50	95.00	71.75	143.50		
1 S.P. Scraper, 14 C.Y.		1840.00		2024.00		
.25 Dozer, 300 H.P.		457.25		502.98	164.09	180.50
14 L.H., Daily Totals		$2914.05		$3462.88	$208.15	$247.35
Crew B-33E	**Hr.**	**Daily**	**Hr.**	**Daily**	**Bare Costs**	**Incl. O&P**
1 Equip. Oper. (med.)	$47.50	$380.00	$71.75	$574.00	$44.06	$66.85
.5 Laborer	35.45	141.80	54.60	218.40		
.25 Equip. Oper. (med.)	47.50	95.00	71.75	143.50		
1 S.P. Scraper, 21 C.Y.		2610.00		2871.00		
.25 Dozer, 300 H.P.		457.25		502.98	219.09	241.00
14 L.H., Daily Totals		$3684.05		$4309.88	$263.15	$307.85
Crew B-33F	**Hr.**	**Daily**	**Hr.**	**Daily**	**Bare Costs**	**Incl. O&P**
1 Equip. Oper. (med.)	$47.50	$380.00	$71.75	$574.00	$44.06	$66.85
.5 Laborer	35.45	141.80	54.60	218.40		
.25 Equip. Oper. (med.)	47.50	95.00	71.75	143.50		
1 Elev. Scraper, 11 C.Y.		1169.00		1285.90		
.25 Dozer, 300 H.P.		457.25		502.98	116.16	127.78
14 L.H., Daily Totals		$2243.05		$2724.78	$160.22	$194.63
Crew B-33G	**Hr.**	**Daily**	**Hr.**	**Daily**	**Bare Costs**	**Incl. O&P**
1 Equip. Oper. (med.)	$47.50	$380.00	$71.75	$574.00	$44.06	$66.85
.5 Laborer	35.45	141.80	54.60	218.40		
.25 Equip. Oper. (med.)	47.50	95.00	71.75	143.50		
1 Elev. Scraper, 22 C.Y.		2459.00		2704.90		
.25 Dozer, 300 H.P.		457.25		502.98	208.30	229.13
14 L.H., Daily Totals		$3533.05		$4143.77	$252.36	$295.98

Crew No.	Bare Costs		Incl. Subs O&P		Cost Per Labor-Hour	
Crew B-33H	**Hr.**	**Daily**	**Hr.**	**Daily**	**Bare Costs**	**Incl. O&P**
.5 Laborer	$35.45	$141.80	$54.60	$218.40	$44.06	$66.85
1 Equipment Operator (med.)	47.50	380.00	71.75	574.00		
.25 Equipment Operator (med.)	47.50	95.00	71.75	143.50		
1 S.P. Scraper, 44 C.Y.		4377.00		4814.70		
.25 Dozer, 410 H.P.		606.25		666.88	355.95	391.54
14 L.H., Daily Totals		$5600.05		$6417.48	$400.00	$458.39
Crew B-33J	**Hr.**	**Daily**	**Hr.**	**Daily**	**Bare Costs**	**Incl. O&P**
1 Equipment Operator (med.)	$47.50	$380.00	$71.75	$574.00	$47.50	$71.75
1 S.P. Scraper, 14 C.Y.		1840.00		2024.00	230.00	253.00
8 L.H., Daily Totals		$2220.00		$2598.00	$277.50	$324.75
Crew B-33K	**Hr.**	**Daily**	**Hr.**	**Daily**	**Bare Costs**	**Incl. O&P**
1 Equipment Operator (med.)	$47.50	$380.00	$71.75	$574.00	$44.06	$66.85
.25 Equipment Operator (med.)	47.50	95.00	71.75	143.50		
.5 Laborer	35.45	141.80	54.60	218.40		
1 S.P. Scraper, 31 C.Y.		3584.00		3942.40		
.25 Dozer, 410 H.P.		606.25		666.88	299.30	329.23
14 L.H., Daily Totals		$4807.05		$5545.18	$343.36	$396.08
Crew B-34A	**Hr.**	**Daily**	**Hr.**	**Daily**	**Bare Costs**	**Incl. O&P**
1 Truck Driver (heavy)	$36.60	$292.80	$56.05	$448.40	$36.60	$56.05
1 Dump Truck, 8 C.Y., 220 H.P.		416.60		458.26	52.08	57.28
8 L.H., Daily Totals		$709.40		$906.66	$88.67	$113.33
Crew B-34B	**Hr.**	**Daily**	**Hr.**	**Daily**	**Bare Costs**	**Incl. O&P**
1 Truck Driver (heavy)	$36.60	$292.80	$56.05	$448.40	$36.60	$56.05
1 Dump Truck, 12 C.Y., 400 H.P.		692.40		761.64	86.55	95.20
8 L.H., Daily Totals		$985.20		$1210.04	$123.15	$151.26
Crew B-34C	**Hr.**	**Daily**	**Hr.**	**Daily**	**Bare Costs**	**Incl. O&P**
1 Truck Driver (heavy)	$36.60	$292.80	$56.05	$448.40	$36.60	$56.05
1 Truck Tractor, 6x4, 380 H.P.		612.40		673.64		
1 Dump Trailer, 16.5 C.Y.		126.80		139.48	92.40	101.64
8 L.H., Daily Totals		$1032.00		$1261.52	$129.00	$157.69
Crew B-34D	**Hr.**	**Daily**	**Hr.**	**Daily**	**Bare Costs**	**Incl. O&P**
1 Truck Driver (heavy)	$36.60	$292.80	$56.05	$448.40	$36.60	$56.05
1 Truck Tractor, 6x4, 380 H.P.		612.40		673.64		
1 Dump Trailer, 20 C.Y.		141.40		155.54	94.22	103.65
8 L.H., Daily Totals		$1046.60		$1277.58	$130.82	$159.70
Crew B-34E	**Hr.**	**Daily**	**Hr.**	**Daily**	**Bare Costs**	**Incl. O&P**
1 Truck Driver (heavy)	$36.60	$292.80	$56.05	$448.40	$36.60	$56.05
1 Dump Truck, Off Hwy., 25 Ton		1321.00		1453.10	165.13	181.64
8 L.H., Daily Totals		$1613.80		$1901.50	$201.72	$237.69
Crew B-34F	**Hr.**	**Daily**	**Hr.**	**Daily**	**Bare Costs**	**Incl. O&P**
1 Truck Driver (heavy)	$36.60	$292.80	$56.05	$448.40	$36.60	$56.05
1 Dump Truck, Off Hwy., 35 Ton		1477.00		1624.70	184.63	203.09
8 L.H., Daily Totals		$1769.80		$2073.10	$221.22	$259.14
Crew B-34G	**Hr.**	**Daily**	**Hr.**	**Daily**	**Bare Costs**	**Incl. O&P**
1 Truck Driver (heavy)	$36.60	$292.80	$56.05	$448.40	$36.60	$56.05
1 Dump Truck, Off Hwy., 50 Ton		1797.00		1976.70	224.63	247.09
8 L.H., Daily Totals		$2089.80		$2425.10	$261.23	$303.14

Crew No.	Bare Costs		Incl. Subs O&P		Cost Per Labor-Hour	
Crew B-34H	**Hr.**	**Daily**	**Hr.**	**Daily**	**Bare Costs**	**Incl. O&P**
1 Truck Driver (heavy)	$36.60	$292.80	$56.05	$448.40	$36.60	$56.05
1 Dump Truck, Off Hwy., 65 Ton		1807.00		1987.70	225.88	248.46
8 L.H., Daily Totals		$2099.80		$2436.10	$262.48	$304.51

Crew B-34I	Hr.	Daily	Hr.	Daily	Bare Costs	Incl. O&P
1 Truck Driver (heavy)	$36.60	$292.80	$56.05	$448.40	$36.60	$56.05
1 Dump Truck, 18 C.Y., 450 H.P.		867.00		953.70	108.38	119.21
8 L.H., Daily Totals		$1159.80		$1402.10	$144.97	$175.26

Crew B-34J	Hr.	Daily	Hr.	Daily	Bare Costs	Incl. O&P
1 Truck Driver (heavy)	$36.60	$292.80	$56.05	$448.40	$36.60	$56.05
1 Dump Truck, Off Hwy., 100 Ton		2920.00		3212.00	365.00	401.50
8 L.H., Daily Totals		$3212.80		$3660.40	$401.60	$457.55

Crew B-34K	Hr.	Daily	Hr.	Daily	Bare Costs	Incl. O&P
1 Truck Driver (heavy)	$36.60	$292.80	$56.05	$448.40	$36.60	$56.05
1 Truck Tractor, 6x4, 450 H.P.		741.20		815.32		
1 Lowbed Trailer, 75 Ton		220.40		242.44	120.20	132.22
8 L.H., Daily Totals		$1254.40		$1506.16	$156.80	$188.27

Crew B-34L	Hr.	Daily	Hr.	Daily	Bare Costs	Incl. O&P
1 Equip. Oper. (light)	$45.80	$366.40	$69.20	$553.60	$45.80	$69.20
1 Flatbed Truck, Gas, 1.5 Ton		265.80		292.38	33.23	36.55
8 L.H., Daily Totals		$632.20		$845.98	$79.03	$105.75

Crew B-34M	Hr.	Daily	Hr.	Daily	Bare Costs	Incl. O&P
1 Equip. Oper. (light)	$45.80	$366.40	$69.20	$553.60	$45.80	$69.20
1 Flatbed Truck, Gas, 3 Ton		327.80		360.58	40.98	45.07
8 L.H., Daily Totals		$694.20		$914.18	$86.78	$114.27

Crew B-34N	Hr.	Daily	Hr.	Daily	Bare Costs	Incl. O&P
1 Truck Driver (heavy)	$36.60	$292.80	$56.05	$448.40	$36.60	$56.05
1 Dump Truck, 8 C.Y., 220 H.P.		416.60		458.26		
1 Flatbed Trailer, 40 Ton		152.60		167.86	71.15	78.27
8 L.H., Daily Totals		$862.00		$1074.52	$107.75	$134.32

Crew B-34P	Hr.	Daily	Hr.	Daily	Bare Costs	Incl. O&P
1 Pipe Fitter	$56.65	$453.20	$85.30	$682.40	$46.60	$70.55
1 Truck Driver (light)	35.65	285.20	54.60	436.80		
1 Equip. Oper. (med.)	47.50	380.00	71.75	574.00		
1 Flatbed Truck, Gas, 3 Ton		327.80		360.58		
1 Backhoe Loader, 48 H.P.		367.40		404.14	28.97	31.86
24 L.H., Daily Totals		$1813.60		$2457.92	$75.57	$102.41

Crew B-34Q	Hr.	Daily	Hr.	Daily	Bare Costs	Incl. O&P
1 Pipe Fitter	$56.65	$453.20	$85.30	$682.40	$47.03	$71.22
1 Truck Driver (light)	35.65	285.20	54.60	436.80		
1 Equip. Oper. (crane)	48.80	390.40	73.75	590.00		
1 Flatbed Trailer, 25 Ton		113.20		124.52		
1 Dump Truck, 8 C.Y., 220 H.P.		416.60		458.26		
1 Hyd. Crane, 25 Ton		728.60		801.46	52.43	57.68
24 L.H., Daily Totals		$2387.20		$3093.44	$99.47	$128.89

Crew B-34R	Hr.	Daily	Hr.	Daily	Bare Costs	Incl. O&P
1 Pipe Fitter	$56.65	$453.20	$85.30	$682.40	$47.03	$71.22
1 Truck Driver (light)	35.65	285.20	54.60	436.80		
1 Equip. Oper. (crane)	48.80	390.40	73.75	590.00		
1 Flatbed Trailer, 25 Ton		113.20		124.52		
1 Dump Truck, 8 C.Y., 220 H.P.		416.60		458.26		
1 Hyd. Crane, 25 Ton		728.60		801.46		
1 Hyd. Excavator, 1 C.Y.		815.60		897.16	86.42	95.06
24 L.H., Daily Totals		$3202.80		$3990.60	$133.45	$166.28

Crew B-34S	Hr.	Daily	Hr.	Daily	Bare Costs	Incl. O&P
2 Pipe Fitters	$56.65	$906.40	$85.30	$1364.80	$49.67	$75.10
1 Truck Driver (heavy)	36.60	292.80	56.05	448.40		
1 Equip. Oper. (crane)	48.80	390.40	73.75	590.00		
1 Flatbed Trailer, 40 Ton		152.60		167.86		
1 Truck Tractor, 6x4, 380 H.P.		612.40		673.64		
1 Hyd. Crane, 80 Ton		1613.00		1774.30		
1 Hyd. Excavator, 2 C.Y.		1363.00		1499.30	116.91	128.60
32 L.H., Daily Totals		$5330.60		$6518.30	$166.58	$203.70

Crew B-34T	Hr.	Daily	Hr.	Daily	Bare Costs	Incl. O&P
2 Pipe Fitters	$56.65	$906.40	$85.30	$1364.80	$49.67	$75.10
1 Truck Driver (heavy)	36.60	292.80	56.05	448.40		
1 Equip. Oper. (crane)	48.80	390.40	73.75	590.00		
1 Flatbed Trailer, 40 Ton		152.60		167.86		
1 Truck Tractor, 6x4, 380 H.P.		612.40		673.64		
1 Hyd. Crane, 80 Ton		1613.00		1774.30	74.31	81.74
32 L.H., Daily Totals		$3967.60		$5019.00	$123.99	$156.84

Crew B-35	Hr.	Daily	Hr.	Daily	Bare Costs	Incl. O&P
1 Labor Foreman (outside)	$37.45	$299.60	$57.65	$461.20	$44.33	$67.56
1 Skilled Worker	46.20	369.60	71.45	571.60		
1 Welder (plumber)	55.80	446.40	84.05	672.40		
1 Laborer	35.45	283.60	54.60	436.80		
1 Equip. Oper. (crane)	48.80	390.40	73.75	590.00		
1 Equip. Oper. Oiler	42.25	338.00	63.85	510.80		
1 Welder, Electric, 300 amp		52.85		58.13		
1 Hyd. Excavator, .75 C.Y.		694.80		764.28	15.58	17.13
48 L.H., Daily Totals		$2875.25		$4065.22	$59.90	$84.69

Crew B-35A	Hr.	Daily	Hr.	Daily	Bare Costs	Incl. O&P
1 Labor Foreman (outside)	$37.45	$299.60	$57.65	$461.20	$43.06	$65.71
2 Laborers	35.45	567.20	54.60	873.60		
1 Skilled Worker	46.20	369.60	71.45	571.60		
1 Welder (plumber)	55.80	446.40	84.05	672.40		
1 Equip. Oper. (crane)	48.80	390.40	73.75	590.00		
1 Equip. Oper. Oiler	42.25	338.00	63.85	510.80		
1 Welder, Gas Engine, 300 amp		144.00		158.40		
1 Crawler Crane, 75 Ton		1457.00		1602.70	28.59	31.45
56 L.H., Daily Totals		$4012.20		$5440.70	$71.65	$97.16

Crew B-36	Hr.	Daily	Hr.	Daily	Bare Costs	Incl. O&P
1 Labor Foreman (outside)	$37.45	$299.60	$57.65	$461.20	$40.67	$62.07
2 Laborers	35.45	567.20	54.60	873.60		
2 Equip. Oper. (med.)	47.50	760.00	71.75	1148.00		
1 Dozer, 200 H.P.		1333.00		1466.30		
1 Aggregate Spreader		40.80		44.88		
1 Tandem Roller, 10 Ton		231.20		254.32	40.13	44.14
40 L.H., Daily Totals		$3231.80		$4248.30	$80.80	$106.21

Crew No.	Bare Costs	Incl. Subs O&P	Cost Per Labor-Hour

Crew B-36A	Hr.	Daily	Hr.	Daily	Bare Costs	Incl. O&P
1 Labor Foreman (outside)	$37.45	$299.60	$57.65	$461.20	$42.62	$64.84
2 Laborers	35.45	567.20	54.60	873.60		
4 Equip. Oper. (med.)	47.50	1520.00	71.75	2296.00		
1 Dozer, 200 H.P.		1333.00		1466.30		
1 Aggregate Spreader		40.80		44.88		
1 Tandem Roller, 10 Ton		231.20		254.32		
1 Roller, Pneum. Whl., 12 Ton		326.60		359.26	34.49	37.94
56 L.H., Daily Totals		$4318.40		$5755.56	$77.11	$102.78

Crew B-36B	Hr.	Daily	Hr.	Daily	Bare Costs	Incl. O&P
1 Labor Foreman (outside)	$37.45	$299.60	$57.65	$461.20	$41.87	$63.74
2 Laborers	35.45	567.20	54.60	873.60		
4 Equip. Oper. (med.)	47.50	1520.00	71.75	2296.00		
1 Truck Driver, Heavy	36.60	292.80	56.05	448.40		
1 Grader, 30,000 Lbs.		712.40		783.64		
1 F.E. Loader, Crl, 1.5 C.Y.		633.20		696.52		
1 Dozer, 300 H.P.		1829.00		2011.90		
1 Roller, Vibratory, 25 Ton		656.40		722.04		
1 Truck Tractor, 6x4, 450 H.P.		741.20		815.32		
1 Water Tank Trailer, 5000 Gal.		143.20		157.52	73.68	81.05
64 L.H., Daily Totals		$7395.00		$9266.14	$115.55	$144.78

Crew B-36C	Hr.	Daily	Hr.	Daily	Bare Costs	Incl. O&P
1 Labor Foreman (outside)	$37.45	$299.60	$57.65	$461.20	$43.31	$65.79
3 Equip. Oper. (med.)	47.50	1140.00	71.75	1722.00		
1 Truck Driver, Heavy	36.60	292.80	56.05	448.40		
1 Grader, 30,000 Lbs.		712.40		783.64		
1 Dozer, 300 H.P.		1829.00		2011.90		
1 Roller, Vibratory, 25 Ton		656.40		722.04		
1 Truck Tractor, 6x4, 450 H.P.		741.20		815.32		
1 Water Tank Trailer, 5000 Gal.		143.20		157.52	102.06	112.26
40 L.H., Daily Totals		$5814.60		$7122.02	$145.37	$178.05

Crew B-36D	Hr.	Daily	Hr.	Daily	Bare Costs	Incl. O&P
1 Labor Foreman (outside)	$37.45	$299.60	$57.65	$461.20	$44.99	$68.22
3 Equip. Oper. (medium)	47.50	1140.00	71.75	1722.00		
1 Grader, 30,000 Lbs.		712.40		783.64		
1 Dozer, 300 H.P.		1829.00		2011.90		
1 Roller, Vibratory, 25 Ton		656.40		722.04	99.93	109.92
32 L.H., Daily Totals		$4637.40		$5700.78	$144.92	$178.15

Crew B-37	Hr.	Daily	Hr.	Daily	Bare Costs	Incl. O&P
1 Labor Foreman (outside)	$37.45	$299.60	$57.65	$461.20	$37.51	$57.54
4 Laborers	35.45	1134.40	54.60	1747.20		
1 Equip. Oper. (light)	45.80	366.40	69.20	553.60		
1 Tandem Roller, 5 Ton		154.40		169.84	3.22	3.54
48 L.H., Daily Totals		$1954.80		$2931.84	$40.73	$61.08

Crew B-37A	Hr.	Daily	Hr.	Daily	Bare Costs	Incl. O&P
2 Laborers	$35.45	$567.20	$54.60	$873.60	$35.52	$54.60
1 Truck Driver (light)	35.65	285.20	54.60	436.80		
1 Flatbed Truck, Gas, 1.5 Ton		265.80		292.38		
1 Tar Kettle, T.M.		78.70		86.57	14.35	15.79
24 L.H., Daily Totals		$1196.90		$1689.35	$49.87	$70.39

Crew B-37B	Hr.	Daily	Hr.	Daily	Bare Costs	Incl. O&P
3 Laborers	$35.45	$850.80	$54.60	$1310.40	$35.50	$54.60
1 Truck Driver (light)	35.65	285.20	54.60	436.80		
1 Flatbed Truck, Gas, 1.5 Ton		265.80		292.38		
1 Tar Kettle, T.M.		78.70		86.57	10.77	11.84
32 L.H., Daily Totals		$1480.50		$2126.15	$46.27	$66.44

Crew No.	Bare Costs	Incl. Subs O&P	Cost Per Labor-Hour

Crew B-37C	Hr.	Daily	Hr.	Daily	Bare Costs	Incl. O&P
2 Laborers	$35.45	$567.20	$54.60	$873.60	$35.55	$54.60
2 Truck Drivers (light)	35.65	570.40	54.60	873.60		
2 Flatbed Trucks, Gas, 1.5 Ton		531.60		584.76		
1 Tar Kettle, T.M.		78.70		86.57	19.07	20.98
32 L.H., Daily Totals		$1747.90		$2418.53	$54.62	$75.58

Crew B-37D	Hr.	Daily	Hr.	Daily	Bare Costs	Incl. O&P
1 Laborer	$35.45	$283.60	$54.60	$436.80	$35.55	$54.60
1 Truck Driver (light)	35.65	285.20	54.60	436.80		
1 Pickup Truck, 3/4 Ton		154.60		170.06	9.66	10.63
16 L.H., Daily Totals		$723.40		$1043.66	$45.21	$65.23

Crew B-37E	Hr.	Daily	Hr.	Daily	Bare Costs	Incl. O&P
3 Laborers	$35.45	$850.80	$54.60	$1310.40	$38.71	$59.14
1 Equip. Oper. (light)	45.80	366.40	69.20	553.60		
1 Equip. Oper. (medium)	47.50	380.00	71.75	574.00		
2 Truck Drivers (light)	35.65	570.40	54.60	873.60		
4 Barrels w/ Flasher		13.60		14.96		
1 Concrete Saw		175.20		192.72		
1 Rotary Hammer Drill		22.65		24.91		
1 Hammer Drill Bit		2.55		2.81		
1 Loader, Skid Steer, 30 H.P.		172.00		189.20		
1 Conc. Hammer Attach.		114.10		125.51		
1 Vibrating Plate, Gas, 18"		34.60		38.06		
2 Flatbed Trucks, Gas, 1.5 Ton		531.60		584.76	19.04	20.95
56 L.H., Daily Totals		$3233.90		$4484.53	$57.75	$80.08

Crew B-37F	Hr.	Daily	Hr.	Daily	Bare Costs	Incl. O&P
3 Laborers	$35.45	$850.80	$54.60	$1310.40	$35.50	$54.60
1 Truck Driver (light)	35.65	285.20	54.60	436.80		
4 Barrels w/ Flasher		13.60		14.96		
1 Concrete Mixer, 10 C.F.		176.20		193.82		
1 Air Compressor, 60 cfm		137.60		151.36		
1 -50' Air Hose, 3/4"		3.25		3.58		
1 Spade (Chipper)		7.60		8.36		
1 Flatbed Truck, Gas, 1.5 Ton		265.80		292.38	18.88	20.76
32 L.H., Daily Totals		$1740.05		$2411.66	$54.38	$75.36

Crew B-37G	Hr.	Daily	Hr.	Daily	Bare Costs	Incl. O&P
1 Labor Foreman (outside)	$37.45	$299.60	$57.65	$461.20	$37.51	$57.54
4 Laborers	35.45	1134.40	54.60	1747.20		
1 Equip. Oper. (light)	45.80	366.40	69.20	553.60		
1 Berm Machine		292.80		322.08		
1 Tandem Roller, 5 Ton		154.40		169.84	9.32	10.25
48 L.H., Daily Totals		$2247.60		$3253.92	$46.83	$67.79

Crew B-37H	Hr.	Daily	Hr.	Daily	Bare Costs	Incl. O&P
1 Labor Foreman (outside)	$37.45	$299.60	$57.65	$461.20	$37.51	$57.54
4 Laborers	35.45	1134.40	54.60	1747.20		
1 Equip. Oper. (light)	45.80	366.40	69.20	553.60		
1 Tandem Roller, 5 Ton		154.40		169.84		
1 Flatbed Trucks, Gas, 1.5 Ton		265.80		292.38		
1 Tar Kettle, T.M.		78.70		86.57	10.39	11.43
48 L.H., Daily Totals		$2299.30		$3310.79	$47.90	$68.97

Crew No.	Bare Costs		Incl. Subs O&P		Cost Per Labor-Hour	
Crew B-37I	**Hr.**	**Daily**	**Hr.**	**Daily**	**Bare Costs**	**Incl. O&P**
3 Laborers	$35.45	$850.80	$54.60	$1310.40	$38.71	$59.14
1 Equip. Oper. (light)	45.80	366.40	69.20	553.60		
1 Equip. Oper. (medium)	47.50	380.00	71.75	574.00		
2 Truck Drivers (light)	35.65	570.40	54.60	873.60		
4 Barrels w/ Flasher		13.60		14.96		
1 Concrete Saw		175.20		192.72		
1 Rotary Hammer Drill		22.65		24.91		
1 Hammer Drill Bit		2.55		2.81		
1 Air Compressor, 60 cfm		137.60		151.36		
1 -50' Air Hose, 3/4"		3.25		3.58		
1 Spade (Chipper)		7.60		8.36		
1 Loader, Skid Steer, 30 H.P.		172.00		189.20		
1 Conc. Hammer Attach.		114.10		125.51		
1 Concrete Mixer, 10 C.F.		176.20		193.82		
1 Vibrating Plate, Gas, 18"		34.60		38.06		
2 Flatbed Trucks, Gas, 1.5 Ton		531.60		584.76	24.84	27.32
56 L.H., Daily Totals		$3558.55		$4841.65	$63.55	$86.46
Crew B-37J	**Hr.**	**Daily**	**Hr.**	**Daily**	**Bare Costs**	**Incl. O&P**
1 Labor Foreman (outside)	$37.45	$299.60	$57.65	$461.20	$37.51	$57.54
4 Laborers	35.45	1134.40	54.60	1747.20		
1 Equip. Oper. (light)	45.80	366.40	69.20	553.60		
1 Air Compressor, 60 cfm		137.60		151.36		
1 -50' Air Hose, 3/4"		3.25		3.58		
2 Concrete Mixer, 10 C.F.		352.40		387.64		
2 Flatbed Trucks, Gas, 1.5 Ton		531.60		584.76		
1 Shot Blaster, 20"		210.20		231.22	25.73	28.30
48 L.H., Daily Totals		$3035.45		$4120.56	$63.24	$85.84
Crew B-37K	**Hr.**	**Daily**	**Hr.**	**Daily**	**Bare Costs**	**Incl. O&P**
1 Labor Foreman (outside)	$37.45	$299.60	$57.65	$461.20	$37.51	$57.54
4 Laborers	35.45	1134.40	54.60	1747.20		
1 Equip. Oper. (light)	45.80	366.40	69.20	553.60		
1 Air Compressor, 60 cfm		137.60		151.36		
1 -50' Air Hose, 3/4"		3.25		3.58		
2 Flatbed Trucks, Gas, 1.5 Ton		531.60		584.76		
1 Shot Blaster, 20"		210.20		231.22	18.39	20.23
48 L.H., Daily Totals		$2683.05		$3732.92	$55.90	$77.77
Crew B-38	**Hr.**	**Daily**	**Hr.**	**Daily**	**Bare Costs**	**Incl. O&P**
1 Labor Foreman (outside)	$37.45	$299.60	$57.65	$461.20	$40.33	$61.56
2 Laborers	35.45	567.20	54.60	873.60		
1 Equip. Oper. (light)	45.80	366.40	69.20	553.60		
1 Equip. Oper. (med.)	47.50	380.00	71.75	574.00		
1 Backhoe Loader, 48 H.P.		367.40		404.14		
1 Hyd. Hammer, (1200 lb.)		179.60		197.56		
1 F.E. Loader, W.M., 4 C.Y.		678.80		746.68		
1 Pvmt. Rem. Bucket		60.80		66.88	32.16	35.38
40 L.H., Daily Totals		$2899.80		$3877.66	$72.50	$96.94
Crew B-39	**Hr.**	**Daily**	**Hr.**	**Daily**	**Bare Costs**	**Incl. O&P**
1 Labor Foreman (outside)	$37.45	$299.60	$57.65	$461.20	$37.51	$57.54
4 Laborers	35.45	1134.40	54.60	1747.20		
1 Equip. Oper. (light)	45.80	366.40	69.20	553.60		
1 Air Compressor, 250 cfm		195.80		215.38		
2 Breakers, Pavement, 60 lb.		19.60		21.56		
2 -50' Air Hoses, 1.5"		11.60		12.76	4.73	5.20
48 L.H., Daily Totals		$2027.40		$3011.70	$42.24	$62.74

Crew No.	Bare Costs		Incl. Subs O&P		Cost Per Labor-Hour	
Crew B-40	**Hr.**	**Daily**	**Hr.**	**Daily**	**Bare Costs**	**Incl. O&P**
1 Pile Driver Foreman (outside)	$45.15	$361.20	$71.90	$575.20	$44.70	$69.76
4 Pile Drivers	43.15	1380.80	68.70	2198.40		
2 Equip. Oper. (crane)	48.80	780.80	73.75	1180.00		
1 Equip. Oper. Oiler	42.25	338.00	63.85	510.80		
1 Crawler Crane, 40 Ton		1152.00		1267.20		
1 Vibratory Hammer & Gen.		2652.00		2917.20	59.44	65.38
64 L.H., Daily Totals		$6664.80		$8648.80	$104.14	$135.14
Crew B-40B	**Hr.**	**Daily**	**Hr.**	**Daily**	**Bare Costs**	**Incl. O&P**
1 Labor Foreman (outside)	$37.45	$299.60	$57.65	$461.20	$39.14	$59.84
3 Laborers	35.45	850.80	54.60	1310.40		
1 Equip. Oper. (crane)	48.80	390.40	73.75	590.00		
1 Equip. Oper. Oiler	42.25	338.00	63.85	510.80		
1 Lattice Boom Crane, 40 Ton		1174.00		1291.40	24.46	26.90
48 L.H., Daily Totals		$3052.80		$4163.80	$63.60	$86.75
Crew B-41	**Hr.**	**Daily**	**Hr.**	**Daily**	**Bare Costs**	**Incl. O&P**
1 Labor Foreman (outside)	$37.45	$299.60	$57.65	$461.20	$36.73	$56.45
4 Laborers	35.45	1134.40	54.60	1747.20		
.25 Equip. Oper. (crane)	48.80	97.60	73.75	147.50		
.25 Equip. Oper. Oiler	42.25	84.50	63.85	127.70		
.25 Crawler Crane, 40 Ton		288.00		316.80	6.55	7.20
44 L.H., Daily Totals		$1904.10		$2800.40	$43.27	$63.65
Crew B-42	**Hr.**	**Daily**	**Hr.**	**Daily**	**Bare Costs**	**Incl. O&P**
1 Labor Foreman (outside)	$37.45	$299.60	$57.65	$461.20	$40.04	$62.87
4 Laborers	35.45	1134.40	54.60	1747.20		
1 Equip. Oper. (crane)	48.80	390.40	73.75	590.00		
1 Equip. Oper. Oiler	42.25	338.00	63.85	510.80		
1 Welder	50.05	400.40	89.30	714.40		
1 Hyd. Crane, 25 Ton		728.60		801.46		
1 Welder, Gas Engine, 300 amp		144.00		158.40		
1 Horz. Boring Csg. Mch.		465.00		511.50	20.90	22.99
64 L.H., Daily Totals		$3900.40		$5494.96	$60.94	$85.86
Crew B-43	**Hr.**	**Daily**	**Hr.**	**Daily**	**Bare Costs**	**Incl. O&P**
1 Labor Foreman (outside)	$37.45	$299.60	$57.65	$461.20	$39.14	$59.84
3 Laborers	35.45	850.80	54.60	1310.40		
1 Equip. Oper. (crane)	48.80	390.40	73.75	590.00		
1 Equip. Oper. Oiler	42.25	338.00	63.85	510.80		
1 Drill Rig, Truck-Mounted		2469.00		2715.90	51.44	56.58
48 L.H., Daily Totals		$4347.80		$5588.30	$90.58	$116.42
Crew B-44	**Hr.**	**Daily**	**Hr.**	**Daily**	**Bare Costs**	**Incl. O&P**
1 Pile Driver Foreman (outside)	$45.15	$361.20	$71.90	$575.20	$43.85	$68.60
4 Pile Drivers	43.15	1380.80	68.70	2198.40		
2 Equip. Oper. (crane)	48.80	780.80	73.75	1180.00		
1 Laborer	35.45	283.60	54.60	436.80		
1 Crawler Crane, 40 Ton		1152.00		1267.20		
1 Lead, 60' High		74.60		82.06		
1 Hammer, Diesel, 15K ft.-lbs.		592.40		651.64	28.42	31.26
64 L.H., Daily Totals		$4625.40		$6391.30	$72.27	$99.86
Crew B-45	**Hr.**	**Daily**	**Hr.**	**Daily**	**Bare Costs**	**Incl. O&P**
1 Equip. Oper. (med.)	$47.50	$380.00	$71.75	$574.00	$42.05	$63.90
1 Truck Driver (heavy)	36.60	292.80	56.05	448.40		
1 Dist. Tanker, 3000 Gallon		305.20		335.72		
1 Truck Tractor, 6x4, 380 H.P.		612.40		673.64	57.35	63.09
16 L.H., Daily Totals		$1590.40		$2031.76	$99.40	$126.99

Crew No.	Bare Costs		Incl. Subs O&P		Cost Per Labor-Hour	

Crew B-46	Hr.	Daily	Hr.	Daily	Bare Costs	Incl. O&P
1 Pile Driver Foreman (outside)	$45.15	$361.20	$71.90	$575.20	$39.63	$62.18
2 Pile Drivers	43.15	690.40	68.70	1099.20		
3 Laborers	35.45	850.80	54.60	1310.40		
1 Chain Saw, Gas, 36″ Long		45.00		49.50	.94	1.03
48 L.H., Daily Totals		$1947.40		$3034.30	$40.57	$63.21

Crew B-47	Hr.	Daily	Hr.	Daily	Bare Costs	Incl. O&P
1 Blast Foreman (outside)	$37.45	$299.60	$57.65	$461.20	$39.57	$60.48
1 Driller	35.45	283.60	54.60	436.80		
1 Equip. Oper. (light)	45.80	366.40	69.20	553.60		
1 Air Track Drill, 4″		1009.00		1109.90		
1 Air Compressor, 600 cfm		530.80		583.88		
2 -50′ Air Hoses, 3″		29.80		32.78	65.40	71.94
24 L.H., Daily Totals		$2519.20		$3178.16	$104.97	$132.42

Crew B-47A	Hr.	Daily	Hr.	Daily	Bare Costs	Incl. O&P
1 Drilling Foreman (outside)	$37.45	$299.60	$57.65	$461.20	$42.83	$65.08
1 Equip. Oper. (heavy)	48.80	390.40	73.75	590.00		
1 Oiler	42.25	338.00	63.85	510.80		
1 Air Track Drill, 5″		1207.00		1327.70	50.29	55.32
24 L.H., Daily Totals		$2235.00		$2889.70	$93.13	$120.40

Crew B-47C	Hr.	Daily	Hr.	Daily	Bare Costs	Incl. O&P
1 Laborer	$35.45	$283.60	$54.60	$436.80	$40.63	$61.90
1 Equip. Oper. (light)	45.80	366.40	69.20	553.60		
1 Air Compressor, 750 cfm		537.40		591.14		
2 -50′ Air Hoses, 3″		29.80		32.78		
1 Air Track Drill, 4″		1009.00		1109.90	98.51	108.36
16 L.H., Daily Totals		$2226.20		$2724.22	$139.14	$170.26

Crew B-47E	Hr.	Daily	Hr.	Daily	Bare Costs	Incl. O&P
1 Labor Foreman (outside)	$37.45	$299.60	$57.65	$461.20	$35.95	$55.36
3 Laborers	35.45	850.80	54.60	1310.40		
1 Flatbed Truck, Gas, 3 Ton		327.80		360.58	10.24	11.27
32 L.H., Daily Totals		$1478.20		$2132.18	$46.19	$66.63

Crew B-47G	Hr.	Daily	Hr.	Daily	Bare Costs	Incl. O&P
1 Labor Foreman (outside)	$37.45	$299.60	$57.65	$461.20	$38.54	$59.01
2 Laborers	35.45	567.20	54.60	873.60		
1 Equip. Oper. (light)	45.80	366.40	69.20	553.60		
1 Air Track Drill, 4″		1009.00		1109.90		
1 Air Compressor, 600 cfm		530.80		583.88		
2 -50′ Air Hoses, 3″		29.80		32.78		
1 Gunite Pump Rig		369.80		406.78	60.61	66.67
32 L.H., Daily Totals		$3172.60		$4021.74	$99.14	$125.68

Crew B-47H	Hr.	Daily	Hr.	Daily	Bare Costs	Incl. O&P
1 Skilled Worker Foreman (out)	$48.20	$385.60	$74.50	$596.00	$46.70	$72.21
3 Skilled Workers	46.20	1108.80	71.45	1714.80		
1 Flatbed Truck, Gas, 3 Ton		327.80		360.58	10.24	11.27
32 L.H., Daily Totals		$1822.20		$2671.38	$56.94	$83.48

Crew B-48	Hr.	Daily	Hr.	Daily	Bare Costs	Incl. O&P
1 Labor Foreman (outside)	$37.45	$299.60	$57.65	$461.20	$40.09	$61.18
3 Laborers	35.45	850.80	54.60	1310.40		
1 Equip. Oper. (crane)	48.80	390.40	73.75	590.00		
1 Equip. Oper. Oiler	42.25	338.00	63.85	510.80		
1 Equip. Oper. (light)	45.80	366.40	69.20	553.60		
1 Centr. Water Pump, 6″		341.40		375.54		
1 -20′ Suction Hose, 6″		11.20		12.32		
1 -50′ Discharge Hose, 6″		12.10		13.31		
1 Drill Rig, Truck-Mounted		2469.00		2715.90	50.60	55.66
56 L.H., Daily Totals		$5078.90		$6543.07	$90.69	$116.84

Crew B-49	Hr.	Daily	Hr.	Daily	Bare Costs	Incl. O&P
1 Labor Foreman (outside)	$37.45	$299.60	$57.65	$461.20	$41.64	$63.93
3 Laborers	35.45	850.80	54.60	1310.40		
2 Equip. Oper. (crane)	48.80	780.80	73.75	1180.00		
2 Equip. Oper. Oilers	42.25	676.00	63.85	1021.60		
1 Equip. Oper. (light)	45.80	366.40	69.20	553.60		
2 Pile Drivers	43.15	690.40	68.70	1099.20		
1 Hyd. Crane, 25 Ton		728.60		801.46		
1 Centr. Water Pump, 6″		341.40		375.54		
1 -20′ Suction Hose, 6″		11.20		12.32		
1 -50′ Discharge Hose, 6″		12.10		13.31		
1 Drill Rig, Truck-Mounted		2469.00		2715.90	40.48	44.53
88 L.H., Daily Totals		$7226.30		$9544.53	$82.12	$108.46

Crew B-50	Hr.	Daily	Hr.	Daily	Bare Costs	Incl. O&P
2 Pile Driver Foremen (outside)	$45.15	$722.40	$71.90	$1150.40	$42.53	$66.51
6 Pile Drivers	43.15	2071.20	68.70	3297.60		
2 Equip. Oper. (crane)	48.80	780.80	73.75	1180.00		
1 Equip. Oper. Oiler	42.25	338.00	63.85	510.80		
3 Laborers	35.45	850.80	54.60	1310.40		
1 Crawler Crane, 40 Ton		1152.00		1267.20		
1 Lead, 60′ High		74.60		82.06		
1 Hammer, Diesel, 15K ft.-lbs.		592.40		651.64		
1 Air Compressor, 600 cfm		530.80		583.88		
2 -50′ Air Hoses, 3″		29.80		32.78		
1 Chain Saw, Gas, 36″ Long		45.00		49.50	21.65	23.81
112 L.H., Daily Totals		$7187.80		$10116.26	$64.18	$90.32

Crew B-51	Hr.	Daily	Hr.	Daily	Bare Costs	Incl. O&P
1 Labor Foreman (outside)	$37.45	$299.60	$57.65	$461.20	$35.82	$55.11
4 Laborers	35.45	1134.40	54.60	1747.20		
1 Truck Driver (light)	35.65	285.20	54.60	436.80		
1 Flatbed Truck, Gas, 1.5 Ton		265.80		292.38	5.54	6.09
48 L.H., Daily Totals		$1985.00		$2937.58	$41.35	$61.20

Crew B-52	Hr.	Daily	Hr.	Daily	Bare Costs	Incl. O&P
1 Carpenter Foreman (outside)	$46.90	$375.20	$72.25	$578.00	$41.41	$63.40
1 Carpenter	44.90	359.20	69.15	553.20		
3 Laborers	35.45	850.80	54.60	1310.40		
1 Cement Finisher	43.05	344.40	63.40	507.20		
.5 Rodman (reinf.)	49.80	199.20	78.70	314.80		
.5 Equip. Oper. (med.)	47.50	190.00	71.75	287.00		
.5 Crawler Loader, 3 C.Y.		593.50		652.85	10.60	11.66
56 L.H., Daily Totals		$2912.30		$4203.45	$52.01	$75.06

Crew B-53	Hr.	Daily	Hr.	Daily	Bare Costs	Incl. O&P
1 Equip. Oper. (light)	$45.80	$366.40	$69.20	$553.60	$45.80	$69.20
1 Trencher, Chain, 12 H.P.		70.40		77.44	8.80	9.68
8 L.H., Daily Totals		$436.80		$631.04	$54.60	$78.88

Crew B-54	Hr.	Daily	Hr.	Daily	Bare Costs	Incl. O&P
1 Equip. Oper. (light)	$45.80	$366.40	$69.20	$553.60	$45.80	$69.20
1 Trencher, Chain, 40 H.P.		339.20		373.12	42.40	46.64
8 L.H., Daily Totals		$705.60		$926.72	$88.20	$115.84

Crew B-54A	Hr.	Daily	Hr.	Daily	Bare Costs	Incl. O&P
.17 Labor Foreman (outside)	$37.45	$50.93	$57.65	$78.40	$46.04	$69.70
1 Equipment Operator (med.)	47.50	380.00	71.75	574.00		
1 Wheel Trencher, 67 H.P.		1198.00		1317.80	127.99	140.79
9.36 L.H., Daily Totals		$1628.93		$1970.20	$174.03	$210.49

Crew B-54B	Hr.	Daily	Hr.	Daily	Bare Costs	Incl. O&P
.25 Labor Foreman (outside)	$37.45	$74.90	$57.65	$115.30	$45.49	$68.93
1 Equipment Operator (med.)	47.50	380.00	71.75	574.00		
1 Wheel Trencher, 150 H.P.		1894.00		2083.40	189.40	208.34
10 L.H., Daily Totals		$2348.90		$2772.70	$234.89	$277.27

Crew B-54C	Hr.	Daily	Hr.	Daily	Bare Costs	Incl. O&P
1 Laborer	$35.45	$283.60	$54.60	$436.80	$41.48	$63.17
1 Equipment Operator (med.)	47.50	380.00	71.75	574.00		
1 Wheel Trencher, 67 H.P.		1198.00		1317.80	74.88	82.36
16 L.H., Daily Totals		$1861.60		$2328.60	$116.35	$145.54

Crew B-54D	Hr.	Daily	Hr.	Daily	Bare Costs	Incl. O&P
1 Laborer	$35.45	$283.60	$54.60	$436.80	$41.48	$63.17
1 Equipment Operator (med.)	47.50	380.00	71.75	574.00		
1 Rock Trencher, 6" Width		363.80		400.18	22.74	25.01
16 L.H., Daily Totals		$1027.40		$1410.98	$64.21	$88.19

Crew B-54E	Hr.	Daily	Hr.	Daily	Bare Costs	Incl. O&P
1 Laborer	$35.45	$283.60	$54.60	$436.80	$41.48	$63.17
1 Equipment Operator (med.)	47.50	380.00	71.75	574.00		
1 Rock Trencher, 18" Width		2596.00		2855.60	162.25	178.47
16 L.H., Daily Totals		$3259.60		$3866.40	$203.72	$241.65

Crew B-55	Hr.	Daily	Hr.	Daily	Bare Costs	Incl. O&P
2 Laborers	$35.45	$567.20	$54.60	$873.60	$35.52	$54.60
1 Truck Driver (light)	35.65	285.20	54.60	436.80		
1 Truck-Mounted Earth Auger		744.00		818.40		
1 Flatbed Truck, Gas, 3 Ton		327.80		360.58	44.66	49.12
24 L.H., Daily Totals		$1924.20		$2489.38	$80.17	$103.72

Crew B-56	Hr.	Daily	Hr.	Daily	Bare Costs	Incl. O&P
1 Laborer	$35.45	$283.60	$54.60	$436.80	$40.63	$61.90
1 Equip. Oper. (light)	45.80	366.40	69.20	553.60		
1 Air Track Drill, 4"		1009.00		1109.90		
1 Air Compressor, 600 cfm		530.80		583.88		
1 -50' Air Hose, 3"		14.90		16.39	97.17	106.89
16 L.H., Daily Totals		$2204.70		$2700.57	$137.79	$168.79

Crew B-57	Hr.	Daily	Hr.	Daily	Bare Costs	Incl. O&P
1 Labor Foreman (outside)	$37.45	$299.60	$57.65	$461.20	$40.87	$62.27
2 Laborers	35.45	567.20	54.60	873.60		
1 Equip. Oper. (crane)	48.80	390.40	73.75	590.00		
1 Equip. Oper. (light)	45.80	366.40	69.20	553.60		
1 Equip. Oper. Oiler	42.25	338.00	63.85	510.80		
1 Crawler Crane, 25 Ton		1148.00		1262.80		
1 Clamshell Bucket, 1 C.Y.		47.40		52.14		
1 Centr. Water Pump, 6"		341.40		375.54		
1 -20' Suction Hose, 6"		11.20		12.32		
20 -50' Discharge Hoses, 6"		242.00		266.20	37.29	41.02
48 L.H., Daily Totals		$3751.60		$4958.20	$78.16	$103.30

Crew B-58	Hr.	Daily	Hr.	Daily	Bare Costs	Incl. O&P
2 Laborers	$35.45	$567.20	$54.60	$873.60	$38.90	$59.47
1 Equip. Oper. (light)	45.80	366.40	69.20	553.60		
1 Backhoe Loader, 48 H.P.		367.40		404.14		
1 Small Helicopter, w/ Pilot		2692.00		2961.20	127.47	140.22
24 L.H., Daily Totals		$3993.00		$4792.54	$166.38	$199.69

Crew B-59	Hr.	Daily	Hr.	Daily	Bare Costs	Incl. O&P
1 Truck Driver (heavy)	$36.60	$292.80	$56.05	$448.40	$36.60	$56.05
1 Truck Tractor, 220 H.P.		366.80		403.48		
1 Water Tank Trailer, 5000 Gal.		143.20		157.52	63.75	70.13
8 L.H., Daily Totals		$802.80		$1009.40	$100.35	$126.18

Crew B-59A	Hr.	Daily	Hr.	Daily	Bare Costs	Incl. O&P
2 Laborers	$35.45	$567.20	$54.60	$873.60	$35.83	$55.08
1 Truck Driver (heavy)	36.60	292.80	56.05	448.40		
1 Water Tank Trailer, 5000 Gal.		143.20		157.52		
1 Truck Tractor, 220 H.P.		366.80		403.48	21.25	23.38
24 L.H., Daily Totals		$1370.00		$1883.00	$57.08	$78.46

Crew B-60	Hr.	Daily	Hr.	Daily	Bare Costs	Incl. O&P
1 Labor Foreman (outside)	$37.45	$299.60	$57.65	$461.20	$41.57	$63.26
2 Laborers	35.45	567.20	54.60	873.60		
1 Equip. Oper. (crane)	48.80	390.40	73.75	590.00		
2 Equip. Oper. (light)	45.80	732.80	69.20	1107.20		
1 Equip. Oper. Oiler	42.25	338.00	63.85	510.80		
1 Crawler Crane, 40 Ton		1152.00		1267.20		
1 Lead, 60' High		74.60		82.06		
1 Hammer, Diesel, 15K ft.-lbs.		592.40		651.64		
1 Backhoe Loader, 48 H.P.		367.40		404.14	39.04	42.95
56 L.H., Daily Totals		$4514.40		$5947.84	$80.61	$106.21

Crew B-61	Hr.	Daily	Hr.	Daily	Bare Costs	Incl. O&P
1 Labor Foreman (outside)	$37.45	$299.60	$57.65	$461.20	$37.92	$58.13
3 Laborers	35.45	850.80	54.60	1310.40		
1 Equip. Oper. (light)	45.80	366.40	69.20	553.60		
1 Cement Mixer, 2 C.Y.		190.60		209.66		
1 Air Compressor, 160 cfm		158.00		173.80	8.71	9.59
40 L.H., Daily Totals		$1865.40		$2708.66	$46.63	$67.72

Crew B-62	Hr.	Daily	Hr.	Daily	Bare Costs	Incl. O&P
2 Laborers	$35.45	$567.20	$54.60	$873.60	$38.90	$59.47
1 Equip. Oper. (light)	45.80	366.40	69.20	553.60		
1 Loader, Skid Steer, 30 H.P.		172.00		189.20	7.17	7.88
24 L.H., Daily Totals		$1105.60		$1616.40	$46.07	$67.35

Crew No.	Bare Costs		Incl. Subs O&P		Cost Per Labor-Hour	
Crew B-63	**Hr.**	**Daily**	**Hr.**	**Daily**	**Bare Costs**	**Incl. O&P**
4 Laborers	$35.45	$1134.40	$54.60	$1747.20	$37.52	$57.52
1 Equip. Oper. (light)	45.80	366.40	69.20	553.60		
1 Loader, Skid Steer, 30 H.P.		172.00		189.20	4.30	4.73
40 L.H., Daily Totals		$1672.80		$2490.00	$41.82	$62.25
Crew B-63B	**Hr.**	**Daily**	**Hr.**	**Daily**	**Bare Costs**	**Incl. O&P**
1 Labor Foreman (inside)	$35.95	$287.60	$55.35	$442.80	$38.16	$58.44
2 Laborers	35.45	567.20	54.60	873.60		
1 Equip. Oper. (light)	45.80	366.40	69.20	553.60		
1 Loader, Skid Steer, 78 H.P.		308.60		339.46	9.64	10.61
32 L.H., Daily Totals		$1529.80		$2209.46	$47.81	$69.05
Crew B-64	**Hr.**	**Daily**	**Hr.**	**Daily**	**Bare Costs**	**Incl. O&P**
1 Laborer	$35.45	$283.60	$54.60	$436.80	$35.55	$54.60
1 Truck Driver (light)	35.65	285.20	54.60	436.80		
1 Power Mulcher (Small)		158.00		173.80		
1 Flatbed Truck, Gas, 1.5 Ton		265.80		292.38	26.49	29.14
16 L.H., Daily Totals		$992.60		$1339.78	$62.04	$83.74
Crew B-65	**Hr.**	**Daily**	**Hr.**	**Daily**	**Bare Costs**	**Incl. O&P**
1 Laborer	$35.45	$283.60	$54.60	$436.80	$35.55	$54.60
1 Truck Driver (light)	35.65	285.20	54.60	436.80		
1 Power Mulcher (Large)		325.40		357.94		
1 Flatbed Truck, Gas, 1.5 Ton		265.80		292.38	36.95	40.65
16 L.H., Daily Totals		$1160.00		$1523.92	$72.50	$95.25
Crew B-66	**Hr.**	**Daily**	**Hr.**	**Daily**	**Bare Costs**	**Incl. O&P**
1 Equip. Oper. (light)	$45.80	$366.40	$69.20	$553.60	$45.80	$69.20
1 Loader-Backhoe, 40 H.P.		260.80		286.88	32.60	35.86
8 L.H., Daily Totals		$627.20		$840.48	$78.40	$105.06
Crew B-67	**Hr.**	**Daily**	**Hr.**	**Daily**	**Bare Costs**	**Incl. O&P**
1 Millwright	$46.50	$372.00	$68.35	$546.80	$46.15	$68.78
1 Equip. Oper. (light)	45.80	366.40	69.20	553.60		
1 Forklift, R/T, 4,000 Lb.		310.20		341.22	19.39	21.33
16 L.H., Daily Totals		$1048.60		$1441.62	$65.54	$90.10
Crew B-67B	**Hr.**	**Daily**	**Hr.**	**Daily**	**Bare Costs**	**Incl. O&P**
1 Millwright Foreman (inside)	$47.00	$376.00	$69.10	$552.80	$46.75	$68.72
1 Millwright	46.50	372.00	68.35	546.80		
16 L.H., Daily Totals		$748.00		$1099.60	$46.75	$68.72
Crew B-68	**Hr.**	**Daily**	**Hr.**	**Daily**	**Bare Costs**	**Incl. O&P**
2 Millwrights	$46.50	$744.00	$68.35	$1093.60	$46.27	$68.63
1 Equip. Oper. (light)	45.80	366.40	69.20	553.60		
1 Forklift, R/T, 4,000 Lb.		310.20		341.22	12.93	14.22
24 L.H., Daily Totals		$1420.60		$1988.42	$59.19	$82.85
Crew B-68A	**Hr.**	**Daily**	**Hr.**	**Daily**	**Bare Costs**	**Incl. O&P**
1 Millwright Foreman (inside)	$47.00	$376.00	$69.10	$552.80	$46.67	$68.60
2 Millwrights	46.50	744.00	68.35	1093.60		
1 Forklift, 8,000 Lb.		180.80		198.88	7.53	8.29
24 L.H., Daily Totals		$1300.80		$1845.28	$54.20	$76.89

Crew No.	Bare Costs		Incl. Subs O&P		Cost Per Labor-Hour	
Crew B-68B	**Hr.**	**Daily**	**Hr.**	**Daily**	**Bare Costs**	**Incl. O&P**
1 Millwright Foreman (inside)	$47.00	$376.00	$69.10	$552.80	$50.91	$75.81
2 Millwrights	46.50	744.00	68.35	1093.60		
2 Electricians	52.40	838.40	78.40	1254.40		
2 Plumbers	55.80	892.80	84.05	1344.80		
1 Forklift, 5,000 Lb.		325.00		357.50	5.80	6.38
56 L.H., Daily Totals		$3176.20		$4603.10	$56.72	$82.20
Crew B-68C	**Hr.**	**Daily**	**Hr.**	**Daily**	**Bare Costs**	**Incl. O&P**
1 Millwright Foreman (inside)	$47.00	$376.00	$69.10	$552.80	$50.42	$74.97
1 Millwright	46.50	372.00	68.35	546.80		
1 Electrician	52.40	419.20	78.40	627.20		
1 Plumber	55.80	446.40	84.05	672.40		
1 Forklift, 5,000 Lb.		325.00		357.50	10.16	11.17
32 L.H., Daily Totals		$1938.60		$2756.70	$60.58	$86.15
Crew B-68D	**Hr.**	**Daily**	**Hr.**	**Daily**	**Bare Costs**	**Incl. O&P**
1 Labor Foreman (inside)	$35.95	$287.60	$55.35	$442.80	$39.07	$59.72
1 Laborer	35.45	283.60	54.60	436.80		
1 Equip. Oper. (light)	45.80	366.40	69.20	553.60		
1 Forklift, 5,000 Lb.		325.00		357.50	13.54	14.90
24 L.H., Daily Totals		$1262.60		$1790.70	$52.61	$74.61
Crew B-68E	**Hr.**	**Daily**	**Hr.**	**Daily**	**Bare Costs**	**Incl. O&P**
1 Struc. Steel Foreman (inside)	$50.55	$404.40	$90.20	$721.60	$50.15	$89.48
3 Struc. Steel Workers	50.05	1201.20	89.30	2143.20		
1 Welder	50.05	400.40	89.30	714.40		
1 Forklift, 8,000 Lb.		180.80		198.88	4.52	4.97
40 L.H., Daily Totals		$2186.80		$3778.08	$54.67	$94.45
Crew B-69	**Hr.**	**Daily**	**Hr.**	**Daily**	**Bare Costs**	**Incl. O&P**
1 Labor Foreman (outside)	$37.45	$299.60	$57.65	$461.20	$39.14	$59.84
3 Laborers	35.45	850.80	54.60	1310.40		
1 Equip Oper. (crane)	48.80	390.40	73.75	590.00		
1 Equip Oper. Oiler	42.25	338.00	63.85	510.80		
1 Hyd. Crane, 80 Ton		1613.00		1774.30	33.60	36.96
48 L.H., Daily Totals		$3491.80		$4646.70	$72.75	$96.81
Crew B-69A	**Hr.**	**Daily**	**Hr.**	**Daily**	**Bare Costs**	**Incl. O&P**
1 Labor Foreman (outside)	$37.45	$299.60	$57.65	$461.20	$39.06	$59.43
3 Laborers	35.45	850.80	54.60	1310.40		
1 Equip. Oper. (med.)	47.50	380.00	71.75	574.00		
1 Concrete Finisher	43.05	344.40	63.40	507.20		
1 Curb/Gutter Paver, 2-Track		801.20		881.32	16.69	18.36
48 L.H., Daily Totals		$2676.00		$3734.12	$55.75	$77.79
Crew B-69B	**Hr.**	**Daily**	**Hr.**	**Daily**	**Bare Costs**	**Incl. O&P**
1 Labor Foreman (outside)	$37.45	$299.60	$57.65	$461.20	$39.06	$59.43
3 Laborers	35.45	850.80	54.60	1310.40		
1 Equip. Oper. (med.)	47.50	380.00	71.75	574.00		
1 Cement Finisher	43.05	344.40	63.40	507.20		
1 Curb/Gutter Paver, 4-Track		772.40		849.64	16.09	17.70
48 L.H., Daily Totals		$2647.20		$3702.44	$55.15	$77.13

Crew No.	Bare Costs		Incl. Subs O&P		Cost Per Labor-Hour	
Crew B-70	**Hr.**	**Daily**	**Hr.**	**Daily**	**Bare Costs**	**Incl. O&P**
1 Labor Foreman (outside)	$37.45	$299.60	$57.65	$461.20	$40.90	$62.39
3 Laborers	35.45	850.80	54.60	1310.40		
3 Equip. Oper. (med.)	47.50	1140.00	71.75	1722.00		
1 Grader, 30,000 Lbs.		712.40		783.64		
1 Ripper, Beam & 1 Shank		82.00		90.20		
1 Road Sweeper, S.P., 8' wide		662.80		729.08		
1 F.E. Loader, W.M., 1.5 C.Y.		375.20		412.72	32.72	35.99
56 L.H., Daily Totals		$4122.80		$5509.24	$73.62	$98.38
Crew B-70A	**Hr.**	**Daily**	**Hr.**	**Daily**	**Bare Costs**	**Incl. O&P**
1 Laborer	$35.45	$283.60	$54.60	$436.80	$45.09	$68.32
4 Equip. Oper. (med.)	47.50	1520.00	71.75	2296.00		
1 Grader, 40,000 Lbs.		1129.00		1241.90		
1 F.E. Loader, W.M.,2.5 C.Y.		526.20		578.82		
1 Dozer, 80 H.P.		480.60		528.66		
1 Roller, Pneum. Whl., 12 Ton		326.60		359.26	61.56	67.72
40 L.H., Daily Totals		$4266.00		$5441.44	$106.65	$136.04
Crew B-71	**Hr.**	**Daily**	**Hr.**	**Daily**	**Bare Costs**	**Incl. O&P**
1 Labor Foreman (outside)	$37.45	$299.60	$57.65	$461.20	$40.90	$62.39
3 Laborers	35.45	850.80	54.60	1310.40		
3 Equip. Oper. (med.)	47.50	1140.00	71.75	1722.00		
1 Pvmt. Profiler, 750 H.P.		5745.00		6319.50		
1 Road Sweeper, S.P., 8' wide		662.80		729.08		
1 F.E. Loader, W.M., 1.5 C.Y.		375.20		412.72	121.13	133.24
56 L.H., Daily Totals		$9073.40		$10954.90	$162.03	$195.62
Crew B-72	**Hr.**	**Daily**	**Hr.**	**Daily**	**Bare Costs**	**Incl. O&P**
1 Labor Foreman (outside)	$37.45	$299.60	$57.65	$461.20	$41.73	$63.56
3 Laborers	35.45	850.80	54.60	1310.40		
4 Equip. Oper. (med.)	47.50	1520.00	71.75	2296.00		
1 Pvmt. Profiler, 750 H.P.		5745.00		6319.50		
1 Hammermill, 250 H.P.		1827.00		2009.70		
1 Windrow Loader		1157.00		1272.70		
1 Mix Paver 165 H.P.		2139.00		2352.90		
1 Roller, Pneum. Whl., 12 Ton		326.60		359.26	174.92	192.41
64 L.H., Daily Totals		$13865.00		$16381.66	$216.64	$255.96
Crew B-73	**Hr.**	**Daily**	**Hr.**	**Daily**	**Bare Costs**	**Incl. O&P**
1 Labor Foreman (outside)	$37.45	$299.60	$57.65	$461.20	$43.23	$65.70
2 Laborers	35.45	567.20	54.60	873.60		
5 Equip. Oper. (med.)	47.50	1900.00	71.75	2870.00		
1 Road Mixer, 310 H.P.		1912.00		2103.20		
1 Tandem Roller, 10 Ton		231.20		254.32		
1 Hammermill, 250 H.P.		1827.00		2009.70		
1 Grader, 30,000 Lbs.		712.40		783.64		
.5 F.E. Loader, W.M., 1.5 C.Y.		187.60		206.36		
.5 Truck Tractor, 220 H.P.		183.40		201.74		
.5 Water Tank Trailer, 5000 Gal.		71.60		78.76	80.08	88.09
64 L.H., Daily Totals		$7892.00		$9842.52	$123.31	$153.79

Crew No.	Bare Costs		Incl. Subs O&P		Cost Per Labor-Hour	
Crew B-74	**Hr.**	**Daily**	**Hr.**	**Daily**	**Bare Costs**	**Incl. O&P**
1 Labor Foreman (outside)	$37.45	$299.60	$57.65	$461.20	$42.01	$63.92
1 Laborer	35.45	283.60	54.60	436.80		
4 Equip. Oper. (med.)	47.50	1520.00	71.75	2296.00		
2 Truck Drivers (heavy)	36.60	585.60	56.05	896.80		
1 Grader, 30,000 Lbs.		712.40		783.64		
1 Ripper, Beam & 1 Shank		82.00		90.20		
2 Stabilizers, 310 H.P.		3574.00		3931.40		
1 Flatbed Truck, Gas, 3 Ton		327.80		360.58		
1 Chem. Spreader, Towed		52.80		58.08		
1 Roller, Vibratory, 25 Ton		656.40		722.04		
1 Water Tank Trailer, 5000 Gal.		143.20		157.52		
1 Truck Tractor, 220 H.P.		366.80		403.48	92.43	101.67
64 L.H., Daily Totals		$8604.20		$10597.74	$134.44	$165.59
Crew B-75	**Hr.**	**Daily**	**Hr.**	**Daily**	**Bare Costs**	**Incl. O&P**
1 Labor Foreman (outside)	$37.45	$299.60	$57.65	$461.20	$42.79	$65.04
1 Laborer	35.45	283.60	54.60	436.80		
4 Equip. Oper. (med.)	47.50	1520.00	71.75	2296.00		
1 Truck Driver (heavy)	36.60	292.80	56.05	448.40		
1 Grader, 30,000 Lbs.		712.40		783.64		
1 Ripper, Beam & 1 Shank		82.00		90.20		
2 Stabilizers, 310 H.P.		3574.00		3931.40		
1 Dist. Tanker, 3000 Gallon		305.20		335.72		
1 Truck Tractor, 6x4, 380 H.P.		612.40		673.64		
1 Roller, Vibratory, 25 Ton		656.40		722.04	106.11	116.73
56 L.H., Daily Totals		$8338.40		$10179.04	$148.90	$181.77
Crew B-76	**Hr.**	**Daily**	**Hr.**	**Daily**	**Bare Costs**	**Incl. O&P**
1 Dock Builder Foreman (outside)	$45.15	$361.20	$71.90	$575.20	$44.53	$69.64
5 Dock Builders	43.15	1726.00	68.70	2748.00		
2 Equip. Oper. (crane)	48.80	780.80	73.75	1180.00		
1 Equip. Oper. Oiler	42.25	338.00	63.85	510.80		
1 Crawler Crane, 50 Ton		1292.00		1421.20		
1 Barge, 400 Ton		783.80		862.18		
1 Hammer, Diesel, 15K ft.-lbs.		592.40		651.64		
1 Lead, 60' High		74.60		82.06		
1 Air Compressor, 600 cfm		530.80		583.88		
2 -50' Air Hoses, 3"		29.80		32.78	45.88	50.47
72 L.H., Daily Totals		$6509.40		$8647.74	$90.41	$120.11
Crew B-76A	**Hr.**	**Daily**	**Hr.**	**Daily**	**Bare Costs**	**Incl. O&P**
1 Labor Foreman (outside)	$37.45	$299.60	$57.65	$461.20	$38.22	$58.53
5 Laborers	35.45	1418.00	54.60	2184.00		
1 Equip. Oper. (crane)	48.80	390.40	73.75	590.00		
1 Equip. Oper. Oiler	42.25	338.00	63.85	510.80		
1 Crawler Crane, 50 Ton		1292.00		1421.20		
1 Barge, 400 Ton		783.80		862.18	32.43	35.68
64 L.H., Daily Totals		$4521.80		$6029.38	$70.65	$94.21
Crew B-77	**Hr.**	**Daily**	**Hr.**	**Daily**	**Bare Costs**	**Incl. O&P**
1 Labor Foreman (outside)	$37.45	$299.60	$57.65	$461.20	$35.89	$55.21
3 Laborers	35.45	850.80	54.60	1310.40		
1 Truck Driver (light)	35.65	285.20	54.60	436.80		
1 Crack Cleaner, 25 H.P.		64.00		70.40		
1 Crack Filler, Trailer Mtd.		204.40		224.84		
1 Flatbed Truck, Gas, 3 Ton		327.80		360.58	14.90	16.40
40 L.H., Daily Totals		$2031.80		$2864.22	$50.80	$71.61

Crew B-78	Hr.	Daily	Hr.	Daily	Bare Costs	Incl. O&P
1 Labor Foreman (outside)	$37.45	$299.60	$57.65	$461.20	$35.82	$55.11
4 Laborers	35.45	1134.40	54.60	1747.20		
1 Truck Driver (light)	35.65	285.20	54.60	436.80		
1 Paint Striper, S.P., 40 Gallon		154.00		169.40		
1 Flatbed Truck, Gas, 3 Ton		327.80		360.58		
1 Pickup Truck, 3/4 Ton		154.60		170.06	13.26	14.58
48 L.H., Daily Totals		$2355.60		$3345.24	$49.08	$69.69

Crew B-78A	Hr.	Daily	Hr.	Daily	Bare Costs	Incl. O&P
1 Equip. Oper. (light)	$45.80	$366.40	$69.20	$553.60	$45.80	$69.20
1 Line Rem. (Metal Balls) 115 H.P.		823.20		905.52	102.90	113.19
8 L.H., Daily Totals		$1189.60		$1459.12	$148.70	$182.39

Crew B-78B	Hr.	Daily	Hr.	Daily	Bare Costs	Incl. O&P
2 Laborers	$35.45	$567.20	$54.60	$873.60	$36.60	$56.22
.25 Equip. Oper. (light)	45.80	91.60	69.20	138.40		
1 Pickup Truck, 3/4 Ton		154.60		170.06		
1 Line Rem.,11 H.P.,Walk Behind		67.60		74.36		
.25 Road Sweeper, S.P., 8' wide		165.70		182.27	21.55	23.70
18 L.H., Daily Totals		$1046.70		$1438.69	$58.15	$79.93

Crew B-78C	Hr.	Daily	Hr.	Daily	Bare Costs	Incl. O&P
1 Labor Foreman (outside)	$37.45	$299.60	$57.65	$461.20	$35.82	$55.11
4 Laborers	35.45	1134.40	54.60	1747.20		
1 Truck Driver (light)	35.65	285.20	54.60	436.80		
1 Paint Striper, T.M., 120 Gal.		853.00		938.30		
1 Flatbed Truck, Gas, 3 Ton		327.80		360.58		
1 Pickup Truck, 3/4 Ton		154.60		170.06	27.82	30.60
48 L.H., Daily Totals		$3054.60		$4114.14	$63.64	$85.71

Crew B-78D	Hr.	Daily	Hr.	Daily	Bare Costs	Incl. O&P
2 Labor Foremen (Outside)	$37.45	$599.20	$57.65	$922.40	$35.87	$55.21
7 Laborers	35.45	1985.20	54.60	3057.60		
1 Truck Driver (light)	35.65	285.20	54.60	436.80		
1 Paint Striper, T.M., 120 Gal.		853.00		938.30		
1 Flatbed Truck, Gas, 3 Ton		327.80		360.58		
3 Pickup Trucks, 3/4 Ton		463.80		510.18		
1 Air Compressor, 60 cfm		137.60		151.36		
1 -50' Air Hose, 3/4"		3.25		3.58		
1 Breakers, Pavement, 60 lb.		9.80		10.78	22.44	24.68
80 L.H., Daily Totals		$4664.85		$6391.57	$58.31	$79.89

Crew B-78E	Hr.	Daily	Hr.	Daily	Bare Costs	Incl. O&P
2 Labor Foremen (Outside)	$37.45	$599.20	$57.65	$922.40	$35.80	$55.11
9 Laborers	35.45	2552.40	54.60	3931.20		
1 Truck Driver (light)	35.65	285.20	54.60	436.80		
1 Paint Striper, T.M., 120 Gal.		853.00		938.30		
1 Flatbed Truck, Gas, 3 Ton		327.80		360.58		
4 Pickup Trucks, 3/4 Ton		618.40		680.24		
2 Air Compressor, 60 cfm		275.20		302.72		
2 -50' Air Hose, 3/4"		6.50		7.15		
2 Breakers, Pavement, 60 lb.		19.60		21.56	21.88	24.07
96 L.H., Daily Totals		$5537.30		$7600.95	$57.68	$79.18

Crew B-78F	Hr.	Daily	Hr.	Daily	Bare Costs	Incl. O&P
2 Labor Foremen (Outside)	$37.45	$599.20	$57.65	$922.40	$35.75	$55.04
11 Laborers	35.45	3119.60	54.60	4804.80		
1 Truck Driver (light)	35.65	285.20	54.60	436.80		
1 Paint Striper, T.M., 120 Gal.		853.00		938.30		
1 Flatbed Truck, Gas, 3 Ton		327.80		360.58		
7 Pickup Trucks, 3/4 Ton		1082.20		1190.42		
3 Air Compressor, 60 cfm		412.80		454.08		
3 -50' Air Hose, 3/4"		9.75		10.73		
3 Breakers, Pavement, 60 lb.		29.40		32.34	24.24	26.66
112 L.H., Daily Totals		$6718.95		$9150.44	$59.99	$81.70

Crew B-79	Hr.	Daily	Hr.	Daily	Bare Costs	Incl. O&P
1 Labor Foreman (outside)	$37.45	$299.60	$57.65	$461.20	$35.89	$55.21
3 Laborers	35.45	850.80	54.60	1310.40		
1 Truck Driver (light)	35.65	285.20	54.60	436.80		
1 Paint Striper, T.M., 120 Gal.		853.00		938.30		
1 Heating Kettle, 115 Gallon		46.45		51.09		
1 Flatbed Truck, Gas, 3 Ton		327.80		360.58		
2 Pickup Trucks, 3/4 Ton		309.20		340.12	38.41	42.25
40 L.H., Daily Totals		$2972.05		$3898.49	$74.30	$97.46

Crew B-79A	Hr.	Daily	Hr.	Daily	Bare Costs	Incl. O&P
1.5 Equip. Oper. (light)	$45.80	$549.60	$69.20	$830.40	$45.80	$69.20
.5 Line Remov. (Grinder) 115 H.P.		455.70		501.27		
1 Line Rem. (Metal Balls) 115 H.P.		823.20		905.52	106.58	117.23
12 L.H., Daily Totals		$1828.50		$2237.19	$152.38	$186.43

Crew B-79B	Hr.	Daily	Hr.	Daily	Bare Costs	Incl. O&P
1 Laborer	$35.45	$283.60	$54.60	$436.80	$35.45	$54.60
1 Set of Gases		133.20		146.52	16.65	18.32
8 L.H., Daily Totals		$416.80		$583.32	$52.10	$72.92

Crew B-79C	Hr.	Daily	Hr.	Daily	Bare Costs	Incl. O&P
1 Labor Foreman (outside)	$37.45	$299.60	$57.65	$461.20	$35.76	$55.04
5 Laborers	35.45	1418.00	54.60	2184.00		
1 Truck Driver (light)	35.65	285.20	54.60	436.80		
1 Paint Striper, T.M., 120 Gal.		853.00		938.30		
1 Heating Kettle, 115 Gallon		46.45		51.09		
1 Flatbed Truck, Gas, 3 Ton		327.80		360.58		
3 Pickup Trucks, 3/4 Ton		463.80		510.18		
1 Air Compressor, 60 cfm		137.60		151.36		
1 -50' Air Hose, 3/4"		3.25		3.58		
1 Breakers, Pavement, 60 lb.		9.80		10.78	32.89	36.18
56 L.H., Daily Totals		$3844.50		$5107.87	$68.65	$91.21

Crew B-79D	Hr.	Daily	Hr.	Daily	Bare Costs	Incl. O&P
2 Labor Foremen (Outside)	$37.45	$599.20	$57.65	$922.40	$35.98	$55.36
5 Laborers	35.45	1418.00	54.60	2184.00		
1 Truck Driver (light)	35.65	285.20	54.60	436.80		
1 Paint Striper, T.M., 120 Gal.		853.00		938.30		
1 Heating Kettle, 115 Gallon		46.45		51.09		
1 Flatbed Truck, Gas, 3 Ton		327.80		360.58		
4 Pickup Trucks, 3/4 Ton		618.40		680.24		
1 Air Compressor, 60 cfm		137.60		151.36		
1 -50' Air Hose, 3/4"		3.25		3.58		
1 Breakers, Pavement, 60 lb.		9.80		10.78	31.19	34.31
64 L.H., Daily Totals		$4298.70		$5739.13	$67.17	$89.67

Crew No.	Bare Costs		Incl. Subs O&P		Cost Per Labor-Hour	
Crew B-79E	**Hr.**	**Daily**	**Hr.**	**Daily**	**Bare Costs**	**Incl. O&P**
2 Labor Foremen (Outside)	$37.45	$599.20	$57.65	$922.40	$35.87	$55.21
7 Laborers	35.45	1985.20	54.60	3057.60		
1 Truck Driver (light)	35.65	285.20	54.60	436.80		
1 Paint Striper, T.M., 120 Gal.		853.00		938.30		
1 Heating Kettle, 115 Gallon		46.45		51.09		
1 Flatbed Truck, Gas, 3 Ton		327.80		360.58		
5 Pickup Trucks, 3/4 Ton		773.00		850.30		
2 Air Compressors, 60 cfm		275.20		302.72		
2 -50' Air Hoses, 3/4"		6.50		7.15		
2 Breakers, Pavement, 60 lb.		19.60		21.56	28.77	31.65
80 L.H., Daily Totals		$5171.15		$6948.51	$64.64	$86.86
Crew B-80	**Hr.**	**Daily**	**Hr.**	**Daily**	**Bare Costs**	**Incl. O&P**
1 Labor Foreman (outside)	$37.45	$299.60	$57.65	$461.20	$38.59	$59.01
1 Laborer	35.45	283.60	54.60	436.80		
1 Truck Driver (light)	35.65	285.20	54.60	436.80		
1 Equip. Oper. (light)	45.80	366.40	69.20	553.60		
1 Flatbed Truck, Gas, 3 Ton		327.80		360.58		
1 Earth Auger, Truck-Mtd.		431.20		474.32	23.72	26.09
32 L.H., Daily Totals		$1993.80		$2723.30	$62.31	$85.10
Crew B-80A	**Hr.**	**Daily**	**Hr.**	**Daily**	**Bare Costs**	**Incl. O&P**
3 Laborers	$35.45	$850.80	$54.60	$1310.40	$35.45	$54.60
1 Flatbed Truck, Gas, 3 Ton		327.80		360.58	13.66	15.02
24 L.H., Daily Totals		$1178.60		$1670.98	$49.11	$69.62
Crew B-80B	**Hr.**	**Daily**	**Hr.**	**Daily**	**Bare Costs**	**Incl. O&P**
3 Laborers	$35.45	$850.80	$54.60	$1310.40	$38.04	$58.25
1 Equip. Oper. (light)	45.80	366.40	69.20	553.60		
1 Crane, Flatbed Mounted, 3 Ton		243.00		267.30	7.59	8.35
32 L.H., Daily Totals		$1460.20		$2131.30	$45.63	$66.60
Crew B-80C	**Hr.**	**Daily**	**Hr.**	**Daily**	**Bare Costs**	**Incl. O&P**
2 Laborers	$35.45	$567.20	$54.60	$873.60	$35.52	$54.60
1 Truck Driver (light)	35.65	285.20	54.60	436.80		
1 Flatbed Truck, Gas, 1.5 Ton		265.80		292.38		
1 Manual Fence Post Auger, Gas		8.00		8.80	11.41	12.55
24 L.H., Daily Totals		$1126.20		$1611.58	$46.92	$67.15
Crew B-81	**Hr.**	**Daily**	**Hr.**	**Daily**	**Bare Costs**	**Incl. O&P**
1 Laborer	$35.45	$283.60	$54.60	$436.80	$39.85	$60.80
1 Equip. Oper. (med.)	47.50	380.00	71.75	574.00		
1 Truck Driver (heavy)	36.60	292.80	56.05	448.40		
1 Hydromulcher, T.M., 3000 Gal.		341.20		375.32		
1 Truck Tractor, 220 H.P.		366.80		403.48	29.50	32.45
24 L.H., Daily Totals		$1664.40		$2238.00	$69.35	$93.25
Crew B-81A	**Hr.**	**Daily**	**Hr.**	**Daily**	**Bare Costs**	**Incl. O&P**
1 Laborer	$35.45	$283.60	$54.60	$436.80	$35.55	$54.60
1 Truck Driver (light)	35.65	285.20	54.60	436.80		
1 Hydromulcher, T.M., 600 Gal.		131.00		144.10		
1 Flatbed Truck, Gas, 3 Ton		327.80		360.58	28.68	31.54
16 L.H., Daily Totals		$1027.60		$1378.28	$64.22	$86.14
Crew B-82	**Hr.**	**Daily**	**Hr.**	**Daily**	**Bare Costs**	**Incl. O&P**
1 Laborer	$35.45	$283.60	$54.60	$436.80	$40.63	$61.90
1 Equip. Oper. (light)	45.80	366.40	69.20	553.60		
1 Horiz. Borer, 6 H.P.		57.60		63.36	3.60	3.96
16 L.H., Daily Totals		$707.60		$1053.76	$44.23	$65.86

Crew No.	Bare Costs		Incl. Subs O&P		Cost Per Labor-Hour	
Crew B-82A	**Hr.**	**Daily**	**Hr.**	**Daily**	**Bare Costs**	**Incl. O&P**
1 Laborer	$35.45	$283.60	$54.60	$436.80	$40.63	$61.90
1 Equip. Oper. (light)	45.80	366.40	69.20	553.60		
1 Flatbed Truck, Gas, 3 Ton		327.80		360.58		
1 Flatbed Trailer, 25 Ton		113.20		124.52		
1 Horiz. Dir. Drill, 20k lb. Thrust		635.40		698.94	67.28	74.00
16 L.H., Daily Totals		$1726.40		$2174.44	$107.90	$135.90
Crew B-82B	**Hr.**	**Daily**	**Hr.**	**Daily**	**Bare Costs**	**Incl. O&P**
2 Laborers	$35.45	$567.20	$54.60	$873.60	$38.90	$59.47
1 Equip. Oper. (light)	45.80	366.40	69.20	553.60		
1 Flatbed Truck, Gas, 3 Ton		327.80		360.58		
1 Flatbed Trailer, 25 Ton		113.20		124.52		
1 Horiz. Dir. Drill, 30k lb. Thrust		904.40		994.84	56.06	61.66
24 L.H., Daily Totals		$2279.00		$2907.14	$94.96	$121.13
Crew B-82C	**Hr.**	**Daily**	**Hr.**	**Daily**	**Bare Costs**	**Incl. O&P**
2 Laborers	$35.45	$567.20	$54.60	$873.60	$38.90	$59.47
1 Equip. Oper. (light)	45.80	366.40	69.20	553.60		
1 Flatbed Truck, Gas, 3 Ton		327.80		360.58		
1 Flatbed Trailer, 25 Ton		113.20		124.52		
1 Horiz. Dir. Drill, 50k lb. Thrust		1203.00		1323.30	68.50	75.35
24 L.H., Daily Totals		$2577.60		$3235.60	$107.40	$134.82
Crew B-82D	**Hr.**	**Daily**	**Hr.**	**Daily**	**Bare Costs**	**Incl. O&P**
1 Equip. Oper. (light)	$45.80	$366.40	$69.20	$553.60	$45.80	$69.20
1 Mud Trailer for HDD, 1500 Gal.		370.20		407.22	46.27	50.90
8 L.H., Daily Totals		$736.60		$960.82	$92.08	$120.10
Crew B-83	**Hr.**	**Daily**	**Hr.**	**Daily**	**Bare Costs**	**Incl. O&P**
1 Tugboat Captain	$47.50	$380.00	$71.75	$574.00	$41.48	$63.17
1 Tugboat Hand	35.45	283.60	54.60	436.80		
1 Tugboat, 250 H.P.		887.00		975.70	55.44	60.98
16 L.H., Daily Totals		$1550.60		$1986.50	$96.91	$124.16
Crew B-84	**Hr.**	**Daily**	**Hr.**	**Daily**	**Bare Costs**	**Incl. O&P**
1 Equip. Oper. (med.)	$47.50	$380.00	$71.75	$574.00	$47.50	$71.75
1 Rotary Mower/Tractor		363.20		399.52	45.40	49.94
8 L.H., Daily Totals		$743.20		$973.52	$92.90	$121.69
Crew B-85	**Hr.**	**Daily**	**Hr.**	**Daily**	**Bare Costs**	**Incl. O&P**
3 Laborers	$35.45	$850.80	$54.60	$1310.40	$38.09	$58.32
1 Equip. Oper. (med.)	47.50	380.00	71.75	574.00		
1 Truck Driver (heavy)	36.60	292.80	56.05	448.40		
1 Aerial Lift Truck, 80'		650.80		715.88		
1 Brush Chipper, 12", 130 H.P.		396.40		436.04		
1 Pruning Saw, Rotary		6.60		7.26	26.34	28.98
40 L.H., Daily Totals		$2577.40		$3491.98	$64.44	$87.30
Crew B-86	**Hr.**	**Daily**	**Hr.**	**Daily**	**Bare Costs**	**Incl. O&P**
1 Equip. Oper. (med.)	$47.50	$380.00	$71.75	$574.00	$47.50	$71.75
1 Stump Chipper, S.P.		172.75		190.03	21.59	23.75
8 L.H., Daily Totals		$552.75		$764.02	$69.09	$95.50
Crew B-86A	**Hr.**	**Daily**	**Hr.**	**Daily**	**Bare Costs**	**Incl. O&P**
1 Equip. Oper. (med.)	$47.50	$380.00	$71.75	$574.00	$47.50	$71.75
1 Grader, 30,000 Lbs.		712.40		783.64	89.05	97.95
8 L.H., Daily Totals		$1092.40		$1357.64	$136.55	$169.71

Crew No.	Bare Costs		Incl. Subs O&P		Cost Per Labor-Hour	

Crew B-86B	Hr.	Daily	Hr.	Daily	Bare Costs	Incl. O&P
1 Equip. Oper. (med.)	$47.50	$380.00	$71.75	$574.00	$47.50	$71.75
1 Dozer, 200 H.P.		1333.00		1466.30	166.63	183.29
8 L.H., Daily Totals		$1713.00		$2040.30	$214.13	$255.04

Crew B-87	Hr.	Daily	Hr.	Daily	Bare Costs	Incl. O&P
1 Laborer	$35.45	$283.60	$54.60	$436.80	$45.09	$68.32
4 Equip. Oper. (med.)	47.50	1520.00	71.75	2296.00		
2 Feller Bunchers, 100 H.P.		1233.20		1356.52		
1 Log Chipper, 22" Tree		900.20		990.22		
1 Dozer, 105 H.P.		592.40		651.64		
1 Chain Saw, Gas, 36" Long		45.00		49.50	69.27	76.20
40 L.H., Daily Totals		$4574.40		$5780.68	$114.36	$144.52

Crew B-88	Hr.	Daily	Hr.	Daily	Bare Costs	Incl. O&P
1 Laborer	$35.45	$283.60	$54.60	$436.80	$45.78	$69.30
6 Equip. Oper. (med.)	47.50	2280.00	71.75	3444.00		
2 Feller Bunchers, 100 H.P.		1233.20		1356.52		
1 Log Chipper, 22" Tree		900.20		990.22		
2 Log Skidders, 50 H.P.		1836.80		2020.48		
1 Dozer, 105 H.P.		592.40		651.64		
1 Chain Saw, Gas, 36" Long		45.00		49.50	82.28	90.51
56 L.H., Daily Totals		$7171.20		$8949.16	$128.06	$159.81

Crew B-89	Hr.	Daily	Hr.	Daily	Bare Costs	Incl. O&P
1 Equip. Oper. (light)	$45.80	$366.40	$69.20	$553.60	$40.73	$61.90
1 Truck Driver (light)	35.65	285.20	54.60	436.80		
1 Flatbed Truck, Gas, 3 Ton		327.80		360.58		
1 Concrete Saw		175.20		192.72		
1 Water Tank, 65 Gal.		18.45		20.30	32.59	35.85
16 L.H., Daily Totals		$1173.05		$1563.99	$73.32	$97.75

Crew B-89A	Hr.	Daily	Hr.	Daily	Bare Costs	Incl. O&P
1 Skilled Worker	$46.20	$369.60	$71.45	$571.60	$40.83	$63.02
1 Laborer	35.45	283.60	54.60	436.80		
1 Core Drill (Large)		112.40		123.64	7.03	7.73
16 L.H., Daily Totals		$765.60		$1132.04	$47.85	$70.75

Crew B-89B	Hr.	Daily	Hr.	Daily	Bare Costs	Incl. O&P
1 Equip. Oper. (light)	$45.80	$366.40	$69.20	$553.60	$40.73	$61.90
1 Truck Driver, Light	35.65	285.20	54.60	436.80		
1 Wall Saw, Hydraulic, 10 H.P.		119.60		131.56		
1 Generator, Diesel, 100 kW		413.80		455.18		
1 Water Tank, 65 Gal.		18.45		20.30		
1 Flatbed Truck, Gas, 3 Ton		327.80		360.58	54.98	60.48
16 L.H., Daily Totals		$1531.25		$1958.02	$95.70	$122.38

Crew B-90	Hr.	Daily	Hr.	Daily	Bare Costs	Incl. O&P
1 Labor Foreman (outside)	$37.45	$299.60	$57.65	$461.20	$38.58	$58.99
3 Laborers	35.45	850.80	54.60	1310.40		
2 Equip. Oper. (light)	45.80	732.80	69.20	1107.20		
2 Truck Drivers (heavy)	36.60	585.60	56.05	896.80		
1 Road Mixer, 310 H.P.		1912.00		2103.20		
1 Dist. Truck, 2000 Gal.		275.20		302.72	34.17	37.59
64 L.H., Daily Totals		$4656.00		$6181.52	$72.75	$96.59

Crew B-90A	Hr.	Daily	Hr.	Daily	Bare Costs	Incl. O&P
1 Labor Foreman (outside)	$37.45	$299.60	$57.65	$461.20	$42.62	$64.84
2 Laborers	35.45	567.20	54.60	873.60		
4 Equip. Oper. (med.)	47.50	1520.00	71.75	2296.00		
2 Graders, 30,000 Lbs.		1424.80		1567.28		
1 Tandem Roller, 10 Ton		231.20		254.32		
1 Roller, Pneum. Whl., 12 Ton		326.60		359.26	35.40	38.94
56 L.H., Daily Totals		$4369.40		$5811.66	$78.03	$103.78

Crew B-90B	Hr.	Daily	Hr.	Daily	Bare Costs	Incl. O&P
1 Labor Foreman (outside)	$37.45	$299.60	$57.65	$461.20	$41.81	$63.68
2 Laborers	35.45	567.20	54.60	873.60		
3 Equip. Oper. (med.)	47.50	1140.00	71.75	1722.00		
1 Roller, Pneum. Whl., 12 Ton		326.60		359.26		
1 Road Mixer, 310 H.P.		1912.00		2103.20	46.64	51.30
48 L.H., Daily Totals		$4245.40		$5519.26	$88.45	$114.98

Crew B-90C	Hr.	Daily	Hr.	Daily	Bare Costs	Incl. O&P
1 Labor Foreman (outside)	$37.45	$299.60	$57.65	$461.20	$39.23	$59.95
4 Laborers	35.45	1134.40	54.60	1747.20		
3 Equip. Oper. (med.)	47.50	1140.00	71.75	1722.00		
3 Truck Drivers (heavy)	36.60	878.40	56.05	1345.20		
3 Road Mixers, 310 H.P.		5736.00		6309.60	65.18	71.70
88 L.H., Daily Totals		$9188.40		$11585.20	$104.41	$131.65

Crew B-90D	Hr.	Daily	Hr.	Daily	Bare Costs	Incl. O&P
1 Labor Foreman (outside)	$37.45	$299.60	$57.65	$461.20	$38.65	$59.13
6 Laborers	35.45	1701.60	54.60	2620.80		
3 Equip. Oper. (med.)	47.50	1140.00	71.75	1722.00		
3 Truck Drivers (heavy)	36.60	878.40	56.05	1345.20		
3 Road Mixers, 310 H.P.		5736.00		6309.60	55.15	60.67
104 L.H., Daily Totals		$9755.60		$12458.80	$93.80	$119.80

Crew B-90E	Hr.	Daily	Hr.	Daily	Bare Costs	Incl. O&P
1 Labor Foreman (outside)	$37.45	$299.60	$57.65	$461.20	$39.82	$60.82
4 Laborers	35.45	1134.40	54.60	1747.20		
3 Equip. Oper. (med.)	47.50	1140.00	71.75	1722.00		
1 Truck Driver (heavy)	36.60	292.80	56.05	448.40		
1 Road Mixers, 310 H.P.		1912.00		2103.20	26.56	29.21
72 L.H., Daily Totals		$4778.80		$6482.00	$66.37	$90.03

Crew B-91	Hr.	Daily	Hr.	Daily	Bare Costs	Incl. O&P
1 Labor Foreman (outside)	$37.45	$299.60	$57.65	$461.20	$41.87	$63.74
2 Laborers	35.45	567.20	54.60	873.60		
4 Equip. Oper. (med.)	47.50	1520.00	71.75	2296.00		
1 Truck Driver (heavy)	36.60	292.80	56.05	448.40		
1 Dist. Tanker, 3000 Gallon		305.20		335.72		
1 Truck Tractor, 6x4, 380 H.P.		612.40		673.64		
1 Aggreg. Spreader, S.P.		838.60		922.46		
1 Roller, Pneum. Whl., 12 Ton		326.60		359.26		
1 Tandem Roller, 10 Ton		231.20		254.32	36.16	39.77
64 L.H., Daily Totals		$4993.60		$6624.60	$78.03	$103.51

Crew B-91B	Hr.	Daily	Hr.	Daily	Bare Costs	Incl. O&P
1 Laborer	$35.45	$283.60	$54.60	$436.80	$41.48	$63.17
1 Equipment Oper. (med.)	47.50	380.00	71.75	574.00		
1 Road Sweeper, Vac. Assist.		990.60		1089.66	61.91	68.10
16 L.H., Daily Totals		$1654.20		$2100.46	$103.39	$131.28

Crew No.	Bare Costs		Incl. Subs O&P		Cost Per Labor-Hour	
Crew B-91C	**Hr.**	**Daily**	**Hr.**	**Daily**	**Bare Costs**	**Incl. O&P**
1 Laborer	$35.45	$283.60	$54.60	$436.80	$35.55	$54.60
1 Truck Driver (light)	35.65	285.20	54.60	436.80		
1 Catch Basin Cleaning Truck		601.60		661.76	37.60	41.36
16 L.H., Daily Totals		$1170.40		$1535.36	$73.15	$95.96

Crew B-91D	**Hr.**	**Daily**	**Hr.**	**Daily**	**Bare Costs**	**Incl. O&P**
1 Labor Foreman (outside)	$37.45	$299.60	$57.65	$461.20	$40.42	$61.65
5 Laborers	35.45	1418.00	54.60	2184.00		
5 Equip. Oper. (med.)	47.50	1900.00	71.75	2870.00		
2 Truck Drivers (heavy)	36.60	585.60	56.05	896.80		
1 Aggreg. Spreader, S.P.		838.60		922.46		
2 Truck Tractor, 6x4, 380 H.P.		1224.80		1347.28		
2 Dist. Tanker, 3000 Gallon		610.40		671.44		
2 Pavement Brush, Towed		162.40		178.64		
2 Roller, Pneum. Whl., 12 Ton		653.20		718.52	33.55	36.91
104 L.H., Daily Totals		$7692.60		$10250.34	$73.97	$98.56

Crew B-92	**Hr.**	**Daily**	**Hr.**	**Daily**	**Bare Costs**	**Incl. O&P**
1 Labor Foreman (outside)	$37.45	$299.60	$57.65	$461.20	$35.95	$55.36
3 Laborers	35.45	850.80	54.60	1310.40		
1 Crack Cleaner, 25 H.P.		64.00		70.40		
1 Air Compressor, 60 cfm		137.60		151.36		
1 Tar Kettle, T.M.		78.70		86.57		
1 Flatbed Truck, Gas, 3 Ton		327.80		360.58	19.00	20.90
32 L.H., Daily Totals		$1758.50		$2440.51	$54.95	$76.27

Crew B-93	**Hr.**	**Daily**	**Hr.**	**Daily**	**Bare Costs**	**Incl. O&P**
1 Equip. Oper. (med.)	$47.50	$380.00	$71.75	$574.00	$47.50	$71.75
1 Feller Buncher, 100 H.P.		616.60		678.26	77.08	84.78
8 L.H., Daily Totals		$996.60		$1252.26	$124.58	$156.53

Crew B-94A	**Hr.**	**Daily**	**Hr.**	**Daily**	**Bare Costs**	**Incl. O&P**
1 Laborer	$35.45	$283.60	$54.60	$436.80	$35.45	$54.60
1 Diaphragm Water Pump, 2″		71.60		78.76		
1 -20′ Suction Hose, 2″		1.95		2.15		
2 -50′ Discharge Hoses, 2″		3.50		3.85	9.63	10.59
8 L.H., Daily Totals		$360.65		$521.55	$45.08	$65.19

Crew B-94B	**Hr.**	**Daily**	**Hr.**	**Daily**	**Bare Costs**	**Incl. O&P**
1 Laborer	$35.45	$283.60	$54.60	$436.80	$35.45	$54.60
1 Diaphragm Water Pump, 4″		114.40		125.84		
1 -20′ Suction Hose, 4″		3.25		3.58		
2 -50′ Discharge Hoses, 4″		9.40		10.34	15.88	17.47
8 L.H., Daily Totals		$410.65		$576.55	$51.33	$72.07

Crew B-94C	**Hr.**	**Daily**	**Hr.**	**Daily**	**Bare Costs**	**Incl. O&P**
1 Laborer	$35.45	$283.60	$54.60	$436.80	$35.45	$54.60
1 Centrifugal Water Pump, 3″		78.80		86.68		
1 -20′ Suction Hose, 3″		2.85		3.13		
2 -50′ Discharge Hoses, 3″		6.10		6.71	10.97	12.07
8 L.H., Daily Totals		$371.35		$533.33	$46.42	$66.67

Crew B-94D	**Hr.**	**Daily**	**Hr.**	**Daily**	**Bare Costs**	**Incl. O&P**
1 Laborer	$35.45	$283.60	$54.60	$436.80	$35.45	$54.60
1 Centr. Water Pump, 6″		341.40		375.54		
1 -20′ Suction Hose, 6″		11.20		12.32		
2 -50′ Discharge Hoses, 6″		24.20		26.62	47.10	51.81
8 L.H., Daily Totals		$660.40		$851.28	$82.55	$106.41

Crew No.	Bare Costs		Incl. Subs O&P		Cost Per Labor-Hour	
Crew C-1	**Hr.**	**Daily**	**Hr.**	**Daily**	**Bare Costs**	**Incl. O&P**
3 Carpenters	$44.90	$1077.60	$69.15	$1659.60	$42.54	$65.51
1 Laborer	35.45	283.60	54.60	436.80		
32 L.H., Daily Totals		$1361.20		$2096.40	$42.54	$65.51

Crew C-2	**Hr.**	**Daily**	**Hr.**	**Daily**	**Bare Costs**	**Incl. O&P**
1 Carpenter Foreman (outside)	$46.90	$375.20	$72.25	$578.00	$43.66	$67.24
4 Carpenters	44.90	1436.80	69.15	2212.80		
1 Laborer	35.45	283.60	54.60	436.80		
48 L.H., Daily Totals		$2095.60		$3227.60	$43.66	$67.24

Crew C-2A	**Hr.**	**Daily**	**Hr.**	**Daily**	**Bare Costs**	**Incl. O&P**
1 Carpenter Foreman (outside)	$46.90	$375.20	$72.25	$578.00	$43.35	$66.28
3 Carpenters	44.90	1077.60	69.15	1659.60		
1 Cement Finisher	43.05	344.40	63.40	507.20		
1 Laborer	35.45	283.60	54.60	436.80		
48 L.H., Daily Totals		$2080.80		$3181.60	$43.35	$66.28

Crew C-3	**Hr.**	**Daily**	**Hr.**	**Daily**	**Bare Costs**	**Incl. O&P**
1 Rodman Foreman (outside)	$51.80	$414.40	$81.85	$654.80	$45.96	$71.88
4 Rodmen (reinf.)	49.80	1593.60	78.70	2518.40		
1 Equip. Oper. (light)	45.80	366.40	69.20	553.60		
2 Laborers	35.45	567.20	54.60	873.60		
3 Stressing Equipment		30.60		33.66		
.5 Grouting Equipment		79.70		87.67	1.72	1.90
64 L.H., Daily Totals		$3051.90		$4721.73	$47.69	$73.78

Crew C-4	**Hr.**	**Daily**	**Hr.**	**Daily**	**Bare Costs**	**Incl. O&P**
1 Rodman Foreman (outside)	$51.80	$414.40	$81.85	$654.80	$50.30	$79.49
3 Rodmen (reinf.)	49.80	1195.20	78.70	1888.80		
3 Stressing Equipment		30.60		33.66	.96	1.05
32 L.H., Daily Totals		$1640.20		$2577.26	$51.26	$80.54

Crew C-4A	**Hr.**	**Daily**	**Hr.**	**Daily**	**Bare Costs**	**Incl. O&P**
2 Rodmen (reinf.)	$49.80	$796.80	$78.70	$1259.20	$49.80	$78.70
4 Stressing Equipment		40.80		44.88	2.55	2.81
16 L.H., Daily Totals		$837.60		$1304.08	$52.35	$81.50

Crew C-5	**Hr.**	**Daily**	**Hr.**	**Daily**	**Bare Costs**	**Incl. O&P**
1 Rodman Foreman (outside)	$51.80	$414.40	$81.85	$654.80	$48.86	$76.32
4 Rodmen (reinf.)	49.80	1593.60	78.70	2518.40		
1 Equip. Oper. (crane)	48.80	390.40	73.75	590.00		
1 Equip. Oper. Oiler	42.25	338.00	63.85	510.80		
1 Hyd. Crane, 25 Ton		728.60		801.46	13.01	14.31
56 L.H., Daily Totals		$3465.00		$5075.46	$61.88	$90.63

Crew C-6	**Hr.**	**Daily**	**Hr.**	**Daily**	**Bare Costs**	**Incl. O&P**
1 Labor Foreman (outside)	$37.45	$299.60	$57.65	$461.20	$37.05	$56.58
4 Laborers	35.45	1134.40	54.60	1747.20		
1 Cement Finisher	43.05	344.40	63.40	507.20		
2 Gas Engine Vibrators		66.00		72.60	1.38	1.51
48 L.H., Daily Totals		$1844.40		$2788.20	$38.42	$58.09

Crew C-7	Hr.	Daily	Hr.	Daily	Bare Costs	Incl. O&P
1 Labor Foreman (outside)	$37.45	$299.60	$57.65	$461.20	$38.61	$58.85
5 Laborers	35.45	1418.00	54.60	2184.00		
1 Cement Finisher	43.05	344.40	63.40	507.20		
1 Equip. Oper. (med.)	47.50	380.00	71.75	574.00		
1 Equip. Oper. (oiler)	42.25	338.00	63.85	510.80		
2 Gas Engine Vibrators		66.00		72.60		
1 Concrete Bucket, 1 C.Y.		23.40		25.74		
1 Hyd. Crane, 55 Ton		1115.00		1226.50	16.73	18.40
72 L.H., Daily Totals		$3984.40		$5562.04	$55.34	$77.25

Crew C-7A	Hr.	Daily	Hr.	Daily	Bare Costs	Incl. O&P
1 Labor Foreman (outside)	$37.45	$299.60	$57.65	$461.20	$35.99	$55.34
5 Laborers	35.45	1418.00	54.60	2184.00		
2 Truck Drivers (Heavy)	36.60	585.60	56.05	896.80		
2 Conc. Transit Mixers		2048.00		2252.80	32.00	35.20
64 L.H., Daily Totals		$4351.20		$5794.80	$67.99	$90.54

Crew C-7B	Hr.	Daily	Hr.	Daily	Bare Costs	Incl. O&P
1 Labor Foreman (outside)	$37.45	$299.60	$57.65	$461.20	$38.22	$58.53
5 Laborers	35.45	1418.00	54.60	2184.00		
1 Equipment Oper. (crane)	48.80	390.40	73.75	590.00		
1 Equipment Oiler	42.25	338.00	63.85	510.80		
1 Conc. Bucket, 2 C.Y.		36.60		40.26		
1 Lattice Boom Crane, 165 Ton		1954.00		2149.40	31.10	34.21
64 L.H., Daily Totals		$4436.60		$5935.66	$69.32	$92.74

Crew C-7C	Hr.	Daily	Hr.	Daily	Bare Costs	Incl. O&P
1 Labor Foreman (outside)	$37.45	$299.60	$57.65	$461.20	$38.71	$59.27
5 Laborers	35.45	1418.00	54.60	2184.00		
2 Equipment Operators (med.)	47.50	760.00	71.75	1148.00		
2 F.E. Loaders, W.M., 4 C.Y.		1357.60		1493.36	21.21	23.33
64 L.H., Daily Totals		$3835.20		$5286.56	$59.92	$82.60

Crew C-7D	Hr.	Daily	Hr.	Daily	Bare Costs	Incl. O&P
1 Labor Foreman (outside)	$37.45	$299.60	$57.65	$461.20	$37.46	$57.49
5 Laborers	35.45	1418.00	54.60	2184.00		
1 Equip. Oper. (med.)	47.50	380.00	71.75	574.00		
1 Concrete Conveyer		203.40		223.74	3.63	4.00
56 L.H., Daily Totals		$2301.00		$3442.94	$41.09	$61.48

Crew C-8	Hr.	Daily	Hr.	Daily	Bare Costs	Incl. O&P
1 Labor Foreman (outside)	$37.45	$299.60	$57.65	$461.20	$39.63	$60.00
3 Laborers	35.45	850.80	54.60	1310.40		
2 Cement Finishers	43.05	688.80	63.40	1014.40		
1 Equip. Oper. (med.)	47.50	380.00	71.75	574.00		
1 Concrete Pump (Small)		715.80		787.38	12.78	14.06
56 L.H., Daily Totals		$2935.00		$4147.38	$52.41	$74.06

Crew C-8A	Hr.	Daily	Hr.	Daily	Bare Costs	Incl. O&P
1 Labor Foreman (outside)	$37.45	$299.60	$57.65	$461.20	$38.32	$58.04
3 Laborers	35.45	850.80	54.60	1310.40		
2 Cement Finishers	43.05	688.80	63.40	1014.40		
48 L.H., Daily Totals		$1839.20		$2786.00	$38.32	$58.04

Crew C-8B	Hr.	Daily	Hr.	Daily	Bare Costs	Incl. O&P
1 Labor Foreman (outside)	$37.45	$299.60	$57.65	$461.20	$38.26	$58.64
3 Laborers	35.45	850.80	54.60	1310.40		
1 Equip. Oper. (med.)	47.50	380.00	71.75	574.00		
1 Vibrating Power Screed		66.30		72.93		
1 Roller, Vibratory, 25 Ton		656.40		722.04		
1 Dozer, 200 H.P.		1333.00		1466.30	51.39	56.53
40 L.H., Daily Totals		$3586.10		$4606.87	$89.65	$115.17

Crew C-8C	Hr.	Daily	Hr.	Daily	Bare Costs	Incl. O&P
1 Labor Foreman (outside)	$37.45	$299.60	$57.65	$461.20	$39.06	$59.43
3 Laborers	35.45	850.80	54.60	1310.40		
1 Cement Finisher	43.05	344.40	63.40	507.20		
1 Equip. Oper. (med.)	47.50	380.00	71.75	574.00		
1 Shotcrete Rig, 12 C.Y./hr		248.60		273.46		
1 Air Compressor, 160 cfm		158.00		173.80		
4 -50' Air Hoses, 1"		16.40		18.04		
4 -50' Air Hoses, 2"		31.00		34.10	9.46	10.40
48 L.H., Daily Totals		$2328.80		$3352.20	$48.52	$69.84

Crew C-8D	Hr.	Daily	Hr.	Daily	Bare Costs	Incl. O&P
1 Labor Foreman (outside)	$37.45	$299.60	$57.65	$461.20	$40.44	$61.21
1 Laborer	35.45	283.60	54.60	436.80		
1 Cement Finisher	43.05	344.40	63.40	507.20		
1 Equipment Oper. (light)	45.80	366.40	69.20	553.60		
1 Air Compressor, 250 cfm		195.80		215.38		
2 -50' Air Hoses, 1"		8.20		9.02	6.38	7.01
32 L.H., Daily Totals		$1498.00		$2183.20	$46.81	$68.22

Crew C-8E	Hr.	Daily	Hr.	Daily	Bare Costs	Incl. O&P
1 Labor Foreman (outside)	$37.45	$299.60	$57.65	$461.20	$38.77	$59.01
3 Laborers	35.45	850.80	54.60	1310.40		
1 Cement Finisher	43.05	344.40	63.40	507.20		
1 Equipment Oper. (light)	45.80	366.40	69.20	553.60		
1 Shotcrete Rig, 35 C.Y./hr		279.00		306.90		
1 Air Compressor, 250 cfm		195.80		215.38		
4 -50' Air Hoses, 1"		16.40		18.04		
4 -50' Air Hoses, 2"		31.00		34.10	10.88	11.97
48 L.H., Daily Totals		$2383.40		$3406.82	$49.65	$70.98

Crew C-10	Hr.	Daily	Hr.	Daily	Bare Costs	Incl. O&P
1 Laborer	$35.45	$283.60	$54.60	$436.80	$40.52	$60.47
2 Cement Finishers	43.05	688.80	63.40	1014.40		
24 L.H., Daily Totals		$972.40		$1451.20	$40.52	$60.47

Crew C-10B	Hr.	Daily	Hr.	Daily	Bare Costs	Incl. O&P
3 Laborers	$35.45	$850.80	$54.60	$1310.40	$38.49	$58.12
2 Cement Finishers	43.05	688.80	63.40	1014.40		
1 Concrete Mixer, 10 C.F.		176.20		193.82		
2 Trowels, 48" Walk-Behind		106.00		116.60	7.05	7.76
40 L.H., Daily Totals		$1821.80		$2635.22	$45.55	$65.88

Crew C-10C	Hr.	Daily	Hr.	Daily	Bare Costs	Incl. O&P
1 Laborer	$35.45	$283.60	$54.60	$436.80	$40.52	$60.47
2 Cement Finishers	43.05	688.80	63.40	1014.40		
1 Trowel, 48" Walk-Behind		53.00		58.30	2.21	2.43
24 L.H., Daily Totals		$1025.40		$1509.50	$42.73	$62.90

Crew C-10D	Hr.	Daily	Hr.	Daily	Bare Costs	Incl. O&P
1 Laborer	$35.45	$283.60	$54.60	$436.80	$40.52	$60.47
2 Cement Finishers	43.05	688.80	63.40	1014.40		
1 Vibrating Power Screed		66.30		72.93		
1 Trowel, 48" Walk-Behind		53.00		58.30	4.97	5.47
24 L.H., Daily Totals		$1091.70		$1582.43	$45.49	$65.93

Crew C-10E	Hr.	Daily	Hr.	Daily	Bare Costs	Incl. O&P
1 Laborer	$35.45	$283.60	$54.60	$436.80	$40.52	$60.47
2 Cement Finishers	43.05	688.80	63.40	1014.40		
1 Vibrating Power Screed		66.30		72.93		
1 Cement Trowel, 96" Ride-On		196.40		216.04	10.95	12.04
24 L.H., Daily Totals		$1235.10		$1740.17	$51.46	$72.51

Crew C-10F	Hr.	Daily	Hr.	Daily	Bare Costs	Incl. O&P
1 Laborer	$35.45	$283.60	$54.60	$436.80	$40.52	$60.47
2 Cement Finishers	43.05	688.80	63.40	1014.40		
1 Aerial Lift Truck, 60' Boom		462.20		508.42	19.26	21.18
24 L.H., Daily Totals		$1434.60		$1959.62	$59.77	$81.65

Crew C-11	Hr.	Daily	Hr.	Daily	Bare Costs	Incl. O&P
1 Struc. Steel Foreman (outside)	$52.05	$416.40	$92.85	$742.80	$49.27	$85.14
6 Struc. Steel Workers	50.05	2402.40	89.30	4286.40		
1 Equip. Oper. (crane)	48.80	390.40	73.75	590.00		
1 Equip. Oper. Oiler	42.25	338.00	63.85	510.80		
1 Lattice Boom Crane, 150 Ton		1861.00		2047.10	25.85	28.43
72 L.H., Daily Totals		$5408.20		$8177.10	$75.11	$113.57

Crew C-12	Hr.	Daily	Hr.	Daily	Bare Costs	Incl. O&P
1 Carpenter Foreman (outside)	$46.90	$375.20	$72.25	$578.00	$44.31	$68.01
3 Carpenters	44.90	1077.60	69.15	1659.60		
1 Laborer	35.45	283.60	54.60	436.80		
1 Equip. Oper. (crane)	48.80	390.40	73.75	590.00		
1 Hyd. Crane, 12 Ton		646.20		710.82	13.46	14.81
48 L.H., Daily Totals		$2773.00		$3975.22	$57.77	$82.82

Crew C-13	Hr.	Daily	Hr.	Daily	Bare Costs	Incl. O&P
1 Struc. Steel Worker	$50.05	$400.40	$89.30	$714.40	$48.33	$82.58
1 Welder	50.05	400.40	89.30	714.40		
1 Carpenter	44.90	359.20	69.15	553.20		
1 Welder, Gas Engine, 300 amp		144.00		158.40	6.00	6.60
24 L.H., Daily Totals		$1304.00		$2140.40	$54.33	$89.18

Crew C-14	Hr.	Daily	Hr.	Daily	Bare Costs	Incl. O&P
1 Carpenter Foreman (outside)	$46.90	$375.20	$72.25	$578.00	$43.86	$67.53
5 Carpenters	44.90	1796.00	69.15	2766.00		
4 Laborers	35.45	1134.40	54.60	1747.20		
4 Rodmen (reinf.)	49.80	1593.60	78.70	2518.40		
2 Cement Finishers	43.05	688.80	63.40	1014.40		
1 Equip. Oper. (crane)	48.80	390.40	73.75	590.00		
1 Equip. Oper. Oiler	42.25	338.00	63.85	510.80		
1 Hyd. Crane, 80 Ton		1613.00		1774.30	11.20	12.32
144 L.H., Daily Totals		$7929.40		$11499.10	$55.07	$79.85

Crew C-14A	Hr.	Daily	Hr.	Daily	Bare Costs	Incl. O&P
1 Carpenter Foreman (outside)	$46.90	$375.20	$72.25	$578.00	$45.04	$69.51
16 Carpenters	44.90	5747.20	69.15	8851.20		
4 Rodmen (reinf.)	49.80	1593.60	78.70	2518.40		
2 Laborers	35.45	567.20	54.60	873.60		
1 Cement Finisher	43.05	344.40	63.40	507.20		
1 Equip. Oper. (med.)	47.50	380.00	71.75	574.00		
1 Gas Engine Vibrator		33.00		36.30		
1 Concrete Pump (Small)		715.80		787.38	3.74	4.12
200 L.H., Daily Totals		$9756.40		$14726.08	$48.78	$73.63

Crew C-14B	Hr.	Daily	Hr.	Daily	Bare Costs	Incl. O&P
1 Carpenter Foreman (outside)	$46.90	$375.20	$72.25	$578.00	$44.96	$69.28
16 Carpenters	44.90	5747.20	69.15	8851.20		
4 Rodmen (reinf.)	49.80	1593.60	78.70	2518.40		
2 Laborers	35.45	567.20	54.60	873.60		
2 Cement Finishers	43.05	688.80	63.40	1014.40		
1 Equip. Oper. (med.)	47.50	380.00	71.75	574.00		
1 Gas Engine Vibrator		33.00		36.30		
1 Concrete Pump (Small)		715.80		787.38	3.60	3.96
208 L.H., Daily Totals		$10100.80		$15233.28	$48.56	$73.24

Crew C-14C	Hr.	Daily	Hr.	Daily	Bare Costs	Incl. O&P
1 Carpenter Foreman (outside)	$46.90	$375.20	$72.25	$578.00	$42.91	$66.17
6 Carpenters	44.90	2155.20	69.15	3319.20		
2 Rodmen (reinf.)	49.80	796.80	78.70	1259.20		
4 Laborers	35.45	1134.40	54.60	1747.20		
1 Cement Finisher	43.05	344.40	63.40	507.20		
1 Gas Engine Vibrator		33.00		36.30	.29	.32
112 L.H., Daily Totals		$4839.00		$7447.10	$43.21	$66.49

Crew C-14D	Hr.	Daily	Hr.	Daily	Bare Costs	Incl. O&P
1 Carpenter Foreman (outside)	$46.90	$375.20	$72.25	$578.00	$44.65	$68.75
18 Carpenters	44.90	6465.60	69.15	9957.60		
2 Rodmen (reinf.)	49.80	796.80	78.70	1259.20		
2 Laborers	35.45	567.20	54.60	873.60		
1 Cement Finisher	43.05	344.40	63.40	507.20		
1 Equip. Oper. (med.)	47.50	380.00	71.75	574.00		
1 Gas Engine Vibrator		33.00		36.30		
1 Concrete Pump (Small)		715.80		787.38	3.74	4.12
200 L.H., Daily Totals		$9678.00		$14573.28	$48.39	$72.87

Crew C-14E	Hr.	Daily	Hr.	Daily	Bare Costs	Incl. O&P
1 Carpenter Foreman (outside)	$46.90	$375.20	$72.25	$578.00	$44.12	$68.41
2 Carpenters	44.90	718.40	69.15	1106.40		
4 Rodmen (reinf.)	49.80	1593.60	78.70	2518.40		
3 Laborers	35.45	850.80	54.60	1310.40		
1 Cement Finisher	43.05	344.40	63.40	507.20		
1 Gas Engine Vibrator		33.00		36.30	.38	.41
88 L.H., Daily Totals		$3915.40		$6056.70	$44.49	$68.83

Crew C-14F	Hr.	Daily	Hr.	Daily	Bare Costs	Incl. O&P
1 Labor Foreman (outside)	$37.45	$299.60	$57.65	$461.20	$40.74	$60.81
2 Laborers	35.45	567.20	54.60	873.60		
6 Cement Finishers	43.05	2066.40	63.40	3043.20		
1 Gas Engine Vibrator		33.00		36.30	.46	.50
72 L.H., Daily Totals		$2966.20		$4414.30	$41.20	$61.31

Crew No.	Bare Costs		Incl. Subs O&P		Cost Per Labor-Hour	
Crew C-14G	**Hr.**	**Daily**	**Hr.**	**Daily**	**Bare Costs**	**Incl. O&P**
1 Labor Foreman (outside)	$37.45	$299.60	$57.65	$461.20	$40.08	$60.06
2 Laborers	35.45	567.20	54.60	873.60		
4 Cement Finishers	43.05	1377.60	63.40	2028.80		
1 Gas Engine Vibrator		33.00		36.30	.59	.65
56 L.H., Daily Totals		$2277.40		$3399.90	$40.67	$60.71
Crew C-14H	**Hr.**	**Daily**	**Hr.**	**Daily**	**Bare Costs**	**Incl. O&P**
1 Carpenter Foreman (outside)	$46.90	$375.20	$72.25	$578.00	$44.17	$67.88
2 Carpenters	44.90	718.40	69.15	1106.40		
1 Rodman (reinf.)	49.80	398.40	78.70	629.60		
1 Laborer	35.45	283.60	54.60	436.80		
1 Cement Finisher	43.05	344.40	63.40	507.20		
1 Gas Engine Vibrator		33.00		36.30	.69	.76
48 L.H., Daily Totals		$2153.00		$3294.30	$44.85	$68.63
Crew C-14L	**Hr.**	**Daily**	**Hr.**	**Daily**	**Bare Costs**	**Incl. O&P**
1 Carpenter Foreman (outside)	$46.90	$375.20	$72.25	$578.00	$41.76	$64.08
6 Carpenters	44.90	2155.20	69.15	3319.20		
4 Laborers	35.45	1134.40	54.60	1747.20		
1 Cement Finisher	43.05	344.40	63.40	507.20		
1 Gas Engine Vibrator		33.00		36.30	.34	.38
96 L.H., Daily Totals		$4042.20		$6187.90	$42.11	$64.46
Crew C-14M	**Hr.**	**Daily**	**Hr.**	**Daily**	**Bare Costs**	**Incl. O&P**
1 Carpenter Foreman (outside)	$46.90	$375.20	$72.25	$578.00	$43.49	$66.70
2 Carpenters	44.90	718.40	69.15	1106.40		
1 Rodman (reinf.)	49.80	398.40	78.70	629.60		
2 Laborers	35.45	567.20	54.60	873.60		
1 Cement Finisher	43.05	344.40	63.40	507.20		
1 Equip. Oper. (med.)	47.50	380.00	71.75	574.00		
1 Gas Engine Vibrator		33.00		36.30		
1 Concrete Pump (Small)		715.80		787.38	11.70	12.87
64 L.H., Daily Totals		$3532.40		$5092.48	$55.19	$79.57
Crew C-15	**Hr.**	**Daily**	**Hr.**	**Daily**	**Bare Costs**	**Incl. O&P**
1 Carpenter Foreman (outside)	$46.90	$375.20	$72.25	$578.00	$42.11	$64.43
2 Carpenters	44.90	718.40	69.15	1106.40		
3 Laborers	35.45	850.80	54.60	1310.40		
2 Cement Finishers	43.05	688.80	63.40	1014.40		
1 Rodman (reinf.)	49.80	398.40	78.70	629.60		
72 L.H., Daily Totals		$3031.60		$4638.80	$42.11	$64.43
Crew C-16	**Hr.**	**Daily**	**Hr.**	**Daily**	**Bare Costs**	**Incl. O&P**
1 Labor Foreman (outside)	$37.45	$299.60	$57.65	$461.20	$39.63	$60.00
3 Laborers	35.45	850.80	54.60	1310.40		
2 Cement Finishers	43.05	688.80	63.40	1014.40		
1 Equip. Oper. (med.)	47.50	380.00	71.75	574.00		
1 Gunite Pump Rig		369.80		406.78		
2 -50' Air Hoses, 3/4"		6.50		7.15		
2 -50' Air Hoses, 2"		15.50		17.05	7.00	7.70
56 L.H., Daily Totals		$2611.00		$3790.98	$46.63	$67.70
Crew C-16A	**Hr.**	**Daily**	**Hr.**	**Daily**	**Bare Costs**	**Incl. O&P**
1 Laborer	$35.45	$283.60	$54.60	$436.80	$42.26	$63.29
2 Cement Finishers	43.05	688.80	63.40	1014.40		
1 Equip. Oper. (med.)	47.50	380.00	71.75	574.00		
1 Gunite Pump Rig		369.80		406.78		
2 -50' Air Hoses, 3/4"		6.50		7.15		
2 -50' Air Hoses, 2"		15.50		17.05		
1 Aerial Lift Truck, 60' Boom		462.20		508.42	26.69	29.36
32 L.H., Daily Totals		$2206.40		$2964.60	$68.95	$92.64

Crew No.	Bare Costs		Incl. Subs O&P		Cost Per Labor-Hour	
Crew C-17	**Hr.**	**Daily**	**Hr.**	**Daily**	**Bare Costs**	**Incl. O&P**
2 Skilled Worker Foremen (out)	$48.20	$771.20	$74.50	$1192.00	$46.60	$72.06
8 Skilled Workers	46.20	2956.80	71.45	4572.80		
80 L.H., Daily Totals		$3728.00		$5764.80	$46.60	$72.06
Crew C-17A	**Hr.**	**Daily**	**Hr.**	**Daily**	**Bare Costs**	**Incl. O&P**
2 Skilled Worker Foremen (out)	$48.20	$771.20	$74.50	$1192.00	$46.63	$72.08
8 Skilled Workers	46.20	2956.80	71.45	4572.80		
.125 Equip. Oper. (crane)	48.80	48.80	73.75	73.75		
.125 Hyd. Crane, 80 Ton		201.63		221.79	2.49	2.74
81 L.H., Daily Totals		$3978.43		$6060.34	$49.12	$74.82
Crew C-17B	**Hr.**	**Daily**	**Hr.**	**Daily**	**Bare Costs**	**Incl. O&P**
2 Skilled Worker Foremen (out)	$48.20	$771.20	$74.50	$1192.00	$46.65	$72.10
8 Skilled Workers	46.20	2956.80	71.45	4572.80		
.25 Equip. Oper. (crane)	48.80	97.60	73.75	147.50		
.25 Hyd. Crane, 80 Ton		403.25		443.57		
.25 Trowel, 48" Walk-Behind		13.25		14.57	5.08	5.59
82 L.H., Daily Totals		$4242.10		$6370.45	$51.73	$77.69
Crew C-17C	**Hr.**	**Daily**	**Hr.**	**Daily**	**Bare Costs**	**Incl. O&P**
2 Skilled Worker Foremen (out)	$48.20	$771.20	$74.50	$1192.00	$46.68	$72.12
8 Skilled Workers	46.20	2956.80	71.45	4572.80		
.375 Equip. Oper. (crane)	48.80	146.40	73.75	221.25		
.375 Hyd. Crane, 80 Ton		604.88		665.36	7.29	8.02
83 L.H., Daily Totals		$4479.27		$6651.41	$53.97	$80.14
Crew C-17D	**Hr.**	**Daily**	**Hr.**	**Daily**	**Bare Costs**	**Incl. O&P**
2 Skilled Worker Foremen (out)	$48.20	$771.20	$74.50	$1192.00	$46.70	$72.14
8 Skilled Workers	46.20	2956.80	71.45	4572.80		
.5 Equip. Oper. (crane)	48.80	195.20	73.75	295.00		
.5 Hyd. Crane, 80 Ton		806.50		887.15	9.60	10.56
84 L.H., Daily Totals		$4729.70		$6946.95	$56.31	$82.70
Crew C-17E	**Hr.**	**Daily**	**Hr.**	**Daily**	**Bare Costs**	**Incl. O&P**
2 Skilled Worker Foremen (out)	$48.20	$771.20	$74.50	$1192.00	$46.60	$72.06
8 Skilled Workers	46.20	2956.80	71.45	4572.80		
1 Hyd. Jack with Rods		92.65		101.92	1.16	1.27
80 L.H., Daily Totals		$3820.65		$5866.72	$47.76	$73.33
Crew C-18	**Hr.**	**Daily**	**Hr.**	**Daily**	**Bare Costs**	**Incl. O&P**
.125 Labor Foreman (outside)	$37.45	$37.45	$57.65	$57.65	$35.67	$54.94
1 Laborer	35.45	283.60	54.60	436.80		
1 Concrete Cart, 10 C.F.		60.80		66.88	6.76	7.43
9 L.H., Daily Totals		$381.85		$561.33	$42.43	$62.37
Crew C-19	**Hr.**	**Daily**	**Hr.**	**Daily**	**Bare Costs**	**Incl. O&P**
.125 Labor Foreman (outside)	$37.45	$37.45	$57.65	$57.65	$35.67	$54.94
1 Laborer	35.45	283.60	54.60	436.80		
1 Concrete Cart, 18 C.F.		100.40		110.44	11.16	12.27
9 L.H., Daily Totals		$421.45		$604.89	$46.83	$67.21
Crew C-20	**Hr.**	**Daily**	**Hr.**	**Daily**	**Bare Costs**	**Incl. O&P**
1 Labor Foreman (outside)	$37.45	$299.60	$57.65	$461.20	$38.16	$58.23
5 Laborers	35.45	1418.00	54.60	2184.00		
1 Cement Finisher	43.05	344.40	63.40	507.20		
1 Equip. Oper. (med.)	47.50	380.00	71.75	574.00		
2 Gas Engine Vibrators		66.00		72.60		
1 Concrete Pump (Small)		715.80		787.38	12.22	13.44
64 L.H., Daily Totals		$3223.80		$4586.38	$50.37	$71.66

Crew C-21	Bare Costs Hr.	Bare Costs Daily	Incl. Subs O&P Hr.	Incl. Subs O&P Daily	Cost Per Labor-Hour Bare Costs	Cost Per Labor-Hour Incl. O&P
1 Labor Foreman (outside)	$37.45	$299.60	$57.65	$461.20	$38.16	$58.23
5 Laborers	35.45	1418.00	54.60	2184.00		
1 Cement Finisher	43.05	344.40	63.40	507.20		
1 Equip. Oper. (med.)	47.50	380.00	71.75	574.00		
2 Gas Engine Vibrators		66.00		72.60		
1 Concrete Conveyer		203.40		223.74	4.21	4.63
64 L.H., Daily Totals		$2711.40		$4022.74	$42.37	$62.86

Crew C-22	Hr.	Daily	Hr.	Daily	Bare Costs	Incl. O&P
1 Rodman Foreman (outside)	$51.80	$414.40	$81.85	$654.80	$49.98	$78.83
4 Rodmen (reinf.)	49.80	1593.60	78.70	2518.40		
.125 Equip. Oper. (crane)	48.80	48.80	73.75	73.75		
.125 Equip. Oper. Oiler	42.25	42.25	63.85	63.85		
.125 Hyd. Crane, 25 Ton		91.08		100.18	2.17	2.39
42 L.H., Daily Totals		$2190.13		$3410.98	$52.15	$81.21

Crew C-23	Hr.	Daily	Hr.	Daily	Bare Costs	Incl. O&P
2 Skilled Worker Foremen (out)	$48.20	$771.20	$74.50	$1192.00	$46.47	$71.53
6 Skilled Workers	46.20	2217.60	71.45	3429.60		
1 Equip. Oper. (crane)	48.80	390.40	73.75	590.00		
1 Equip. Oper. Oiler	42.25	338.00	63.85	510.80		
1 Lattice Boom Crane, 90 Ton		1525.00		1677.50	19.06	20.97
80 L.H., Daily Totals		$5242.20		$7399.90	$65.53	$92.50

Crew C-23A	Hr.	Daily	Hr.	Daily	Bare Costs	Incl. O&P
1 Labor Foreman (outside)	$37.45	$299.60	$57.65	$461.20	$39.88	$60.89
2 Laborers	35.45	567.20	54.60	873.60		
1 Equip. Oper. (crane)	48.80	390.40	73.75	590.00		
1 Equip. Oper. Oiler	42.25	338.00	63.85	510.80		
1 Crawler Crane, 100 Ton		1666.00		1832.60		
3 Conc. Buckets, 8 C.Y.		597.60		657.36	56.59	62.25
40 L.H., Daily Totals		$3858.80		$4925.56	$96.47	$123.14

Crew C-24	Hr.	Daily	Hr.	Daily	Bare Costs	Incl. O&P
2 Skilled Worker Foremen (out)	$48.20	$771.20	$74.50	$1192.00	$46.47	$71.53
6 Skilled Workers	46.20	2217.60	71.45	3429.60		
1 Equip. Oper. (crane)	48.80	390.40	73.75	590.00		
1 Equip. Oper. Oiler	42.25	338.00	63.85	510.80		
1 Lattice Boom Crane, 150 Ton		1861.00		2047.10	23.26	25.59
80 L.H., Daily Totals		$5578.20		$7769.50	$69.73	$97.12

Crew C-25	Hr.	Daily	Hr.	Daily	Bare Costs	Incl. O&P
2 Rodmen (reinf.)	$49.80	$796.80	$78.70	$1259.20	$39.10	$63.27
2 Rodmen Helpers	28.40	454.40	47.85	765.60		
32 L.H., Daily Totals		$1251.20		$2024.80	$39.10	$63.27

Crew C-27	Hr.	Daily	Hr.	Daily	Bare Costs	Incl. O&P
2 Cement Finishers	$43.05	$688.80	$63.40	$1014.40	$43.05	$63.40
1 Concrete Saw		175.20		192.72	10.95	12.05
16 L.H., Daily Totals		$864.00		$1207.12	$54.00	$75.44

Crew C-28	Hr.	Daily	Hr.	Daily	Bare Costs	Incl. O&P
1 Cement Finisher	$43.05	$344.40	$63.40	$507.20	$43.05	$63.40
1 Portable Air Compressor, Gas		17.90		19.69	2.24	2.46
8 L.H., Daily Totals		$362.30		$526.89	$45.29	$65.86

Crew C-29	Hr.	Daily	Hr.	Daily	Bare Costs	Incl. O&P
1 Laborer	$35.45	$283.60	$54.60	$436.80	$35.45	$54.60
1 Pressure Washer		69.40		76.34	8.68	9.54
8 L.H., Daily Totals		$353.00		$513.14	$44.13	$64.14

Crew C-30	Hr.	Daily	Hr.	Daily	Bare Costs	Incl. O&P
1 Laborer	$35.45	$283.60	$54.60	$436.80	$35.45	$54.60
1 Concrete Mixer, 10 C.F.		176.20		193.82	22.02	24.23
8 L.H., Daily Totals		$459.80		$630.62	$57.48	$78.83

Crew C-31	Hr.	Daily	Hr.	Daily	Bare Costs	Incl. O&P
1 Cement Finisher	$43.05	$344.40	$63.40	$507.20	$43.05	$63.40
1 Grout Pump		369.80		406.78	46.23	50.85
8 L.H., Daily Totals		$714.20		$913.98	$89.28	$114.25

Crew D-1	Hr.	Daily	Hr.	Daily	Bare Costs	Incl. O&P
1 Bricklayer	$44.80	$358.40	$67.85	$542.80	$40.38	$61.15
1 Bricklayer Helper	35.95	287.60	54.45	435.60		
16 L.H., Daily Totals		$646.00		$978.40	$40.38	$61.15

Crew D-2	Hr.	Daily	Hr.	Daily	Bare Costs	Incl. O&P
3 Bricklayers	$44.80	$1075.20	$67.85	$1628.40	$41.59	$63.10
2 Bricklayer Helpers	35.95	575.20	54.45	871.20		
.5 Carpenter	44.90	179.60	69.15	276.60		
44 L.H., Daily Totals		$1830.00		$2776.20	$41.59	$63.10

Crew D-3	Hr.	Daily	Hr.	Daily	Bare Costs	Incl. O&P
3 Bricklayers	$44.80	$1075.20	$67.85	$1628.40	$41.43	$62.81
2 Bricklayer Helpers	35.95	575.20	54.45	871.20		
.25 Carpenter	44.90	89.80	69.15	138.30		
42 L.H., Daily Totals		$1740.20		$2637.90	$41.43	$62.81

Crew D-4	Hr.	Daily	Hr.	Daily	Bare Costs	Incl. O&P
1 Bricklayer	$44.80	$358.40	$67.85	$542.80	$40.63	$61.49
2 Bricklayer Helpers	35.95	575.20	54.45	871.20		
1 Equip. Oper. (light)	45.80	366.40	69.20	553.60		
1 Grout Pump, 50 C.F./hr.		130.80		143.88	4.09	4.50
32 L.H., Daily Totals		$1430.80		$2111.48	$44.71	$65.98

Crew D-5	Hr.	Daily	Hr.	Daily	Bare Costs	Incl. O&P
1 Bricklayer	$44.80	$358.40	$67.85	$542.80	$44.80	$67.85
8 L.H., Daily Totals		$358.40		$542.80	$44.80	$67.85

Crew D-6	Hr.	Daily	Hr.	Daily	Bare Costs	Incl. O&P
3 Bricklayers	$44.80	$1075.20	$67.85	$1628.40	$40.56	$61.47
3 Bricklayer Helpers	35.95	862.80	54.45	1306.80		
.25 Carpenter	44.90	89.80	69.15	138.30		
50 L.H., Daily Totals		$2027.80		$3073.50	$40.56	$61.47

Crew D-7	Hr.	Daily	Hr.	Daily	Bare Costs	Incl. O&P
1 Tile Layer	$41.25	$330.00	$60.60	$484.80	$36.98	$54.33
1 Tile Layer Helper	32.70	261.60	48.05	384.40		
16 L.H., Daily Totals		$591.60		$869.20	$36.98	$54.33

Crew D-8	Hr.	Daily	Hr.	Daily	Bare Costs	Incl. O&P
3 Bricklayers	$44.80	$1075.20	$67.85	$1628.40	$41.26	$62.49
2 Bricklayer Helpers	35.95	575.20	54.45	871.20		
40 L.H., Daily Totals		$1650.40		$2499.60	$41.26	$62.49

Crew D-9	Hr.	Daily	Hr.	Daily	Bare Costs	Incl. O&P
3 Bricklayers	$44.80	$1075.20	$67.85	$1628.40	$40.38	$61.15
3 Bricklayer Helpers	35.95	862.80	54.45	1306.80		
48 L.H., Daily Totals		$1938.00		$2935.20	$40.38	$61.15

Crew No.	Bare Costs		Incl. Subs O&P		Cost Per Labor-Hour	
Crew D-10	**Hr.**	**Daily**	**Hr.**	**Daily**	**Bare Costs**	**Incl. O&P**
1 Bricklayer Foreman (outside)	$46.80	$374.40	$70.85	$566.80	$44.09	$66.72
1 Bricklayer	44.80	358.40	67.85	542.80		
1 Bricklayer Helper	35.95	287.60	54.45	435.60		
1 Equip. Oper. (crane)	48.80	390.40	73.75	590.00		
1 S.P. Crane, 4x4, 12 Ton		471.00		518.10	14.72	16.19
32 L.H., Daily Totals		$1881.80		$2653.30	$58.81	$82.92
Crew D-11	**Hr.**	**Daily**	**Hr.**	**Daily**	**Bare Costs**	**Incl. O&P**
1 Bricklayer Foreman (outside)	$46.80	$374.40	$70.85	$566.80	$42.52	$64.38
1 Bricklayer	44.80	358.40	67.85	542.80		
1 Bricklayer Helper	35.95	287.60	54.45	435.60		
24 L.H., Daily Totals		$1020.40		$1545.20	$42.52	$64.38
Crew D-12	**Hr.**	**Daily**	**Hr.**	**Daily**	**Bare Costs**	**Incl. O&P**
1 Bricklayer Foreman (outside)	$46.80	$374.40	$70.85	$566.80	$40.88	$61.90
1 Bricklayer	44.80	358.40	67.85	542.80		
2 Bricklayer Helpers	35.95	575.20	54.45	871.20		
32 L.H., Daily Totals		$1308.00		$1980.80	$40.88	$61.90
Crew D-13	**Hr.**	**Daily**	**Hr.**	**Daily**	**Bare Costs**	**Incl. O&P**
1 Bricklayer Foreman (outside)	$46.80	$374.40	$70.85	$566.80	$42.87	$65.08
1 Bricklayer	44.80	358.40	67.85	542.80		
2 Bricklayer Helpers	35.95	575.20	54.45	871.20		
1 Carpenter	44.90	359.20	69.15	553.20		
1 Equip. Oper. (crane)	48.80	390.40	73.75	590.00		
1 S.P. Crane, 4x4, 12 Ton		471.00		518.10	9.81	10.79
48 L.H., Daily Totals		$2528.60		$3642.10	$52.68	$75.88
Crew E-1	**Hr.**	**Daily**	**Hr.**	**Daily**	**Bare Costs**	**Incl. O&P**
1 Welder Foreman (outside)	$52.05	$416.40	$92.85	$742.80	$49.30	$83.78
1 Welder	50.05	400.40	89.30	714.40		
1 Equip. Oper. (light)	45.80	366.40	69.20	553.60		
1 Welder, Gas Engine, 300 amp		144.00		158.40	6.00	6.60
24 L.H., Daily Totals		$1327.20		$2169.20	$55.30	$90.38
Crew E-2	**Hr.**	**Daily**	**Hr.**	**Daily**	**Bare Costs**	**Incl. O&P**
1 Struc. Steel Foreman (outside)	$52.05	$416.40	$92.85	$742.80	$49.04	$83.95
4 Struc. Steel Workers	50.05	1601.60	89.30	2857.60		
1 Equip. Oper. (crane)	48.80	390.40	73.75	590.00		
1 Equip. Oper. Oiler	42.25	338.00	63.85	510.80		
1 Lattice Boom Crane, 90 Ton		1525.00		1677.50	27.23	29.96
56 L.H., Daily Totals		$4271.40		$6378.70	$76.28	$113.91
Crew E-3	**Hr.**	**Daily**	**Hr.**	**Daily**	**Bare Costs**	**Incl. O&P**
1 Struc. Steel Foreman (outside)	$52.05	$416.40	$92.85	$742.80	$50.72	$90.48
1 Struc. Steel Worker	50.05	400.40	89.30	714.40		
1 Welder	50.05	400.40	89.30	714.40		
1 Welder, Gas Engine, 300 amp		144.00		158.40	6.00	6.60
24 L.H., Daily Totals		$1361.20		$2330.00	$56.72	$97.08
Crew E-3A	**Hr.**	**Daily**	**Hr.**	**Daily**	**Bare Costs**	**Incl. O&P**
1 Struc. Steel Foreman (outside)	$52.05	$416.40	$92.85	$742.80	$50.72	$90.48
1 Struc. Steel Worker	50.05	400.40	89.30	714.40		
1 Welder	50.05	400.40	89.30	714.40		
1 Welder, Gas Engine, 300 amp		144.00		158.40		
1 Aerial Lift Truck, 40' Boom		332.20		365.42	19.84	21.83
24 L.H., Daily Totals		$1693.40		$2695.42	$70.56	$112.31

Crew No.	Bare Costs		Incl. Subs O&P		Cost Per Labor-Hour	
Crew E-4	**Hr.**	**Daily**	**Hr.**	**Daily**	**Bare Costs**	**Incl. O&P**
1 Struc. Steel Foreman (outside)	$52.05	$416.40	$92.85	$742.80	$50.55	$90.19
3 Struc. Steel Workers	50.05	1201.20	89.30	2143.20		
1 Welder, Gas Engine, 300 amp		144.00		158.40	4.50	4.95
32 L.H., Daily Totals		$1761.60		$3044.40	$55.05	$95.14
Crew E-5	**Hr.**	**Daily**	**Hr.**	**Daily**	**Bare Costs**	**Incl. O&P**
2 Struc. Steel Foremen (outside)	$52.05	$832.80	$92.85	$1485.60	$49.55	$85.91
5 Struc. Steel Workers	50.05	2002.00	89.30	3572.00		
1 Equip. Oper. (crane)	48.80	390.40	73.75	590.00		
1 Welder	50.05	400.40	89.30	714.40		
1 Equip. Oper. Oiler	42.25	338.00	63.85	510.80		
1 Lattice Boom Crane, 90 Ton		1525.00		1677.50		
1 Welder, Gas Engine, 300 amp		144.00		158.40	20.86	22.95
80 L.H., Daily Totals		$5632.60		$8708.70	$70.41	$108.86
Crew E-6	**Hr.**	**Daily**	**Hr.**	**Daily**	**Bare Costs**	**Incl. O&P**
3 Struc. Steel Foremen (outside)	$52.05	$1249.20	$92.85	$2228.40	$49.59	$86.15
9 Struc. Steel Workers	50.05	3603.60	89.30	6429.60		
1 Equip. Oper. (crane)	48.80	390.40	73.75	590.00		
1 Welder	50.05	400.40	89.30	714.40		
1 Equip. Oper. Oiler	42.25	338.00	63.85	510.80		
1 Equip. Oper. (light)	45.80	366.40	69.20	553.60		
1 Lattice Boom Crane, 90 Ton		1525.00		1677.50		
1 Welder, Gas Engine, 300 amp		144.00		158.40		
1 Air Compressor, 160 cfm		158.00		173.80		
2 Impact Wrenches		36.00		39.60	14.55	16.01
128 L.H., Daily Totals		$8211.00		$13076.10	$64.15	$102.16
Crew E-7	**Hr.**	**Daily**	**Hr.**	**Daily**	**Bare Costs**	**Incl. O&P**
1 Struc. Steel Foreman (outside)	$52.05	$416.40	$92.85	$742.80	$49.55	$85.91
4 Struc. Steel Workers	50.05	1601.60	89.30	2857.60		
1 Equip. Oper. (crane)	48.80	390.40	73.75	590.00		
1 Equip. Oper. Oiler	42.25	338.00	63.85	510.80		
1 Welder Foreman (outside)	52.05	416.40	92.85	742.80		
2 Welders	50.05	800.80	89.30	1428.80		
1 Lattice Boom Crane, 90 Ton		1525.00		1677.50		
2 Welder, Gas Engine, 300 amp		288.00		316.80	22.66	24.93
80 L.H., Daily Totals		$5776.60		$8867.10	$72.21	$110.84
Crew E-8	**Hr.**	**Daily**	**Hr.**	**Daily**	**Bare Costs**	**Incl. O&P**
1 Struc. Steel Foreman (outside)	$52.05	$416.40	$92.85	$742.80	$49.33	$85.15
4 Struc. Steel Workers	50.05	1601.60	89.30	2857.60		
1 Welder Foreman (outside)	52.05	416.40	92.85	742.80		
4 Welders	50.05	1601.60	89.30	2857.60		
1 Equip. Oper. (crane)	48.80	390.40	73.75	590.00		
1 Equip. Oper. Oiler	42.25	338.00	63.85	510.80		
1 Equip. Oper. (light)	45.80	366.40	69.20	553.60		
1 Lattice Boom Crane, 90 Ton		1525.00		1677.50		
4 Welder, Gas Engine, 300 amp		576.00		633.60	20.20	22.22
104 L.H., Daily Totals		$7231.80		$11166.30	$69.54	$107.37
Crew E-9	**Hr.**	**Daily**	**Hr.**	**Daily**	**Bare Costs**	**Incl. O&P**
2 Struc. Steel Foremen (outside)	$52.05	$832.80	$92.85	$1485.60	$49.59	$86.15
5 Struc. Steel Workers	50.05	2002.00	89.30	3572.00		
1 Welder Foreman (outside)	52.05	416.40	92.85	742.80		
5 Welders	50.05	2002.00	89.30	3572.00		
1 Equip. Oper. (crane)	48.80	390.40	73.75	590.00		
1 Equip. Oper. Oiler	42.25	338.00	63.85	510.80		
1 Equip. Oper. (light)	45.80	366.40	69.20	553.60		
1 Lattice Boom Crane, 90 Ton		1525.00		1677.50		
5 Welder, Gas Engine, 300 amp		720.00		792.00	17.54	19.29
128 L.H., Daily Totals		$8593.00		$13496.30	$67.13	$105.44

Crew E-10	Hr.	Daily	Hr.	Daily	Bare Costs	Incl. O&P
1 Welder Foreman (outside)	$52.05	$416.40	$92.85	$742.80	$51.05	$91.08
1 Welder	50.05	400.40	89.30	714.40		
1 Welder, Gas Engine, 300 amp		144.00		158.40		
1 Flatbed Truck, Gas, 3 Ton		327.80		360.58	29.49	32.44
16 L.H., Daily Totals		$1288.60		$1976.18	$80.54	$123.51

Crew E-11	Hr.	Daily	Hr.	Daily	Bare Costs	Incl. O&P
2 Painters, Struc. Steel	$39.70	$635.20	$71.90	$1150.40	$40.16	$66.90
1 Building Laborer	35.45	283.60	54.60	436.80		
1 Equip. Oper. (light)	45.80	366.40	69.20	553.60		
1 Air Compressor, 250 cfm		195.80		215.38		
1 Sandblaster, Portable, 3 C.F.		20.00		22.00		
1 Set Sand Blasting Accessories		13.65		15.02	7.17	7.89
32 L.H., Daily Totals		$1514.65		$2393.20	$47.33	$74.79

Crew E-11A	Hr.	Daily	Hr.	Daily	Bare Costs	Incl. O&P
2 Painters, Struc. Steel	$39.70	$635.20	$71.90	$1150.40	$40.16	$66.90
1 Building Laborer	35.45	283.60	54.60	436.80		
1 Equip. Oper. (light)	45.80	366.40	69.20	553.60		
1 Air Compressor, 250 cfm		195.80		215.38		
1 Sandblaster, Portable, 3 C.F.		20.00		22.00		
1 Set Sand Blasting Accessories		13.65		15.02		
1 Aerial Lift Truck, 60' Boom		462.20		508.42	21.61	23.78
32 L.H., Daily Totals		$1976.85		$2901.61	$61.78	$90.68

Crew E-11B	Hr.	Daily	Hr.	Daily	Bare Costs	Incl. O&P
2 Painters, Struc. Steel	$39.70	$635.20	$71.90	$1150.40	$38.28	$66.13
1 Building Laborer	35.45	283.60	54.60	436.80		
2 Paint Sprayer, 8 C.F.M.		79.70		87.67		
1 Aerial Lift Truck, 60' Boom		462.20		508.42	22.58	24.84
24 L.H., Daily Totals		$1460.70		$2183.29	$60.86	$90.97

Crew E-12	Hr.	Daily	Hr.	Daily	Bare Costs	Incl. O&P
1 Welder Foreman (outside)	$52.05	$416.40	$92.85	$742.80	$48.92	$81.03
1 Equip. Oper. (light)	45.80	366.40	69.20	553.60		
1 Welder, Gas Engine, 300 amp		144.00		158.40	9.00	9.90
16 L.H., Daily Totals		$926.80		$1454.80	$57.92	$90.92

Crew E-13	Hr.	Daily	Hr.	Daily	Bare Costs	Incl. O&P
1 Welder Foreman (outside)	$52.05	$416.40	$92.85	$742.80	$49.97	$84.97
.5 Equip. Oper. (light)	45.80	183.20	69.20	276.80		
1 Welder, Gas Engine, 300 amp		144.00		158.40	12.00	13.20
12 L.H., Daily Totals		$743.60		$1178.00	$61.97	$98.17

Crew E-14	Hr.	Daily	Hr.	Daily	Bare Costs	Incl. O&P
1 Welder Foreman (outside)	$52.05	$416.40	$92.85	$742.80	$52.05	$92.85
1 Welder, Gas Engine, 300 amp		144.00		158.40	18.00	19.80
8 L.H., Daily Totals		$560.40		$901.20	$70.05	$112.65

Crew E-16	Hr.	Daily	Hr.	Daily	Bare Costs	Incl. O&P
1 Welder Foreman (outside)	$52.05	$416.40	$92.85	$742.80	$51.05	$91.08
1 Welder	50.05	400.40	89.30	714.40		
1 Welder, Gas Engine, 300 amp		144.00		158.40	9.00	9.90
16 L.H., Daily Totals		$960.80		$1615.60	$60.05	$100.97

Crew E-17	Hr.	Daily	Hr.	Daily	Bare Costs	Incl. O&P
1 Struc. Steel Foreman (outside)	$52.05	$416.40	$92.85	$742.80	$51.05	$91.08
1 Structural Steel Worker	50.05	400.40	89.30	714.40		
16 L.H., Daily Totals		$816.80		$1457.20	$51.05	$91.08

Crew E-18	Hr.	Daily	Hr.	Daily	Bare Costs	Incl. O&P
1 Struc. Steel Foreman (outside)	$52.05	$416.40	$92.85	$742.80	$49.94	$86.50
3 Structural Steel Workers	50.05	1201.20	89.30	2143.20		
1 Equipment Operator (med.)	47.50	380.00	71.75	574.00		
1 Lattice Boom Crane, 20 Ton		958.30		1054.13	23.96	26.35
40 L.H., Daily Totals		$2955.90		$4514.13	$73.90	$112.85

Crew E-19	Hr.	Daily	Hr.	Daily	Bare Costs	Incl. O&P
1 Structural Steel Worker	$50.05	$400.40	$89.30	$714.40	$49.30	$83.78
1 Struc. Steel Foreman (outside)	52.05	416.40	92.85	742.80		
1 Equip. Oper. (light)	45.80	366.40	69.20	553.60		
1 Lattice Boom Crane, 20 Ton		958.30		1054.13	39.93	43.92
24 L.H., Daily Totals		$2141.50		$3064.93	$89.23	$127.71

Crew E-20	Hr.	Daily	Hr.	Daily	Bare Costs	Incl. O&P
1 Struc. Steel Foreman (outside)	$52.05	$416.40	$92.85	$742.80	$49.17	$84.62
5 Structural Steel Workers	50.05	2002.00	89.30	3572.00		
1 Equip. Oper. (crane)	48.80	390.40	73.75	590.00		
1 Oiler	42.25	338.00	63.85	510.80		
1 Lattice Boom Crane, 40 Ton		1174.00		1291.40	18.34	20.18
64 L.H., Daily Totals		$4320.80		$6707.00	$67.51	$104.80

Crew E-22	Hr.	Daily	Hr.	Daily	Bare Costs	Incl. O&P
1 Skilled Worker Foreman (out)	$48.20	$385.60	$74.50	$596.00	$46.87	$72.47
2 Skilled Workers	46.20	739.20	71.45	1143.20		
24 L.H., Daily Totals		$1124.80		$1739.20	$46.87	$72.47

Crew E-24	Hr.	Daily	Hr.	Daily	Bare Costs	Incl. O&P
3 Structural Steel Workers	$50.05	$1201.20	$89.30	$2143.20	$49.41	$84.91
1 Equipment Operator (med.)	47.50	380.00	71.75	574.00		
1 Hyd. Crane, 25 Ton		728.60		801.46	22.77	25.05
32 L.H., Daily Totals		$2309.80		$3518.66	$72.18	$109.96

Crew E-25	Hr.	Daily	Hr.	Daily	Bare Costs	Incl. O&P
1 Welder Foreman (outside)	$52.05	$416.40	$92.85	$742.80	$52.05	$92.85
1 Cutting Torch		11.40		12.54		
1 Set of Gases		133.20		146.52	18.07	19.88
8 L.H., Daily Totals		$561.00		$901.86	$70.13	$112.73

Crew F-3	Hr.	Daily	Hr.	Daily	Bare Costs	Incl. O&P
4 Carpenters	$44.90	$1436.80	$69.15	$2212.80	$45.68	$70.07
1 Equip. Oper. (crane)	48.80	390.40	73.75	590.00		
1 Hyd. Crane, 12 Ton		646.20		710.82	16.16	17.77
40 L.H., Daily Totals		$2473.40		$3513.62	$61.84	$87.84

Crew F-4	Hr.	Daily	Hr.	Daily	Bare Costs	Incl. O&P
4 Carpenters	$44.90	$1436.80	$69.15	$2212.80	$45.11	$69.03
1 Equip. Oper. (crane)	48.80	390.40	73.75	590.00		
1 Equip. Oper. Oiler	42.25	338.00	63.85	510.80		
1 Hyd. Crane, 55 Ton		1115.00		1226.50	23.23	25.55
48 L.H., Daily Totals		$3280.20		$4540.10	$68.34	$94.59

Crew F-5	Hr.	Daily	Hr.	Daily	Bare Costs	Incl. O&P
1 Carpenter Foreman (outside)	$46.90	$375.20	$72.25	$578.00	$45.40	$69.92
3 Carpenters	44.90	1077.60	69.15	1659.60		
32 L.H., Daily Totals		$1452.80		$2237.60	$45.40	$69.92

Crew F-6	Hr.	Daily	Hr.	Daily	Bare Costs	Incl. O&P
2 Carpenters	$44.90	$718.40	$69.15	$1106.40	$41.90	$64.25
2 Building Laborers	35.45	567.20	54.60	873.60		
1 Equip. Oper. (crane)	48.80	390.40	73.75	590.00		
1 Hyd. Crane, 12 Ton		646.20		710.82	16.16	17.77
40 L.H., Daily Totals		$2322.20		$3280.82	$58.06	$82.02

Crew F-7	Hr.	Daily	Hr.	Daily	Bare Costs	Incl. O&P
2 Carpenters	$44.90	$718.40	$69.15	$1106.40	$40.17	$61.88
2 Building Laborers	35.45	567.20	54.60	873.60		
32 L.H., Daily Totals		$1285.60		$1980.00	$40.17	$61.88

Crew G-1	Hr.	Daily	Hr.	Daily	Bare Costs	Incl. O&P
1 Roofer Foreman (outside)	$40.20	$321.60	$67.75	$542.00	$35.69	$60.12
4 Roofers, Composition	38.20	1222.40	64.35	2059.20		
2 Roofer Helpers	28.40	454.40	47.85	765.60		
1 Application Equipment		197.20		216.92		
1 Tar Kettle/Pot		92.50		101.75		
1 Crew Truck		221.20		243.32	9.12	10.04
56 L.H., Daily Totals		$2509.30		$3928.79	$44.81	$70.16

Crew G-2	Hr.	Daily	Hr.	Daily	Bare Costs	Incl. O&P
1 Plasterer	$41.45	$331.60	$62.10	$496.80	$37.67	$56.93
1 Plasterer Helper	36.10	288.80	54.10	432.80		
1 Building Laborer	35.45	283.60	54.60	436.80		
1 Grout Pump, 50 C.F./hr.		130.80		143.88	5.45	6.00
24 L.H., Daily Totals		$1034.80		$1510.28	$43.12	$62.93

Crew G-2A	Hr.	Daily	Hr.	Daily	Bare Costs	Incl. O&P
1 Roofer, composition	$38.20	$305.60	$64.35	$514.80	$34.02	$55.60
1 Roofer Helper	28.40	227.20	47.85	382.80		
1 Building Laborer	35.45	283.60	54.60	436.80		
1 Foam Spray Rig, Trailer-Mtd.		562.85		619.13		
1 Pickup Truck, 3/4 Ton		154.60		170.06	29.89	32.88
24 L.H., Daily Totals		$1533.85		$2123.59	$63.91	$88.48

Crew G-3	Hr.	Daily	Hr.	Daily	Bare Costs	Incl. O&P
2 Sheet Metal Workers	$53.30	$852.80	$81.30	$1300.80	$44.38	$67.95
2 Building Laborers	35.45	567.20	54.60	873.60		
32 L.H., Daily Totals		$1420.00		$2174.40	$44.38	$67.95

Crew G-4	Hr.	Daily	Hr.	Daily	Bare Costs	Incl. O&P
1 Labor Foreman (outside)	$37.45	$299.60	$57.65	$461.20	$36.12	$55.62
2 Building Laborers	35.45	567.20	54.60	873.60		
1 Flatbed Truck, Gas, 1.5 Ton		265.80		292.38		
1 Air Compressor, 160 cfm		158.00		173.80	17.66	19.42
24 L.H., Daily Totals		$1290.60		$1800.98	$53.77	$75.04

Crew G-5	Hr.	Daily	Hr.	Daily	Bare Costs	Incl. O&P
1 Roofer Foreman (outside)	$40.20	$321.60	$67.75	$542.00	$34.68	$58.43
2 Roofers, Composition	38.20	611.20	64.35	1029.60		
2 Roofer Helpers	28.40	454.40	47.85	765.60		
1 Application Equipment		197.20		216.92	4.93	5.42
40 L.H., Daily Totals		$1584.40		$2554.12	$39.61	$63.85

Crew G-6A	Hr.	Daily	Hr.	Daily	Bare Costs	Incl. O&P
2 Roofers Composition	$38.20	$611.20	$64.35	$1029.60	$38.20	$64.35
1 Small Compressor, Electric		10.95		12.05		
2 Pneumatic Nailers		45.70		50.27	3.54	3.89
16 L.H., Daily Totals		$667.85		$1091.92	$41.74	$68.24

Crew G-7	Hr.	Daily	Hr.	Daily	Bare Costs	Incl. O&P
1 Carpenter	$44.90	$359.20	$69.15	$553.20	$44.90	$69.15
1 Small Compressor, Electric		10.95		12.05		
1 Pneumatic Nailer		22.85		25.14	4.22	4.65
8 L.H., Daily Totals		$393.00		$590.38	$49.13	$73.80

Crew H-1	Hr.	Daily	Hr.	Daily	Bare Costs	Incl. O&P
2 Glaziers	$43.30	$692.80	$65.55	$1048.80	$46.67	$77.42
2 Struc. Steel Workers	50.05	800.80	89.30	1428.80		
32 L.H., Daily Totals		$1493.60		$2477.60	$46.67	$77.42

Crew H-2	Hr.	Daily	Hr.	Daily	Bare Costs	Incl. O&P
2 Glaziers	$43.30	$692.80	$65.55	$1048.80	$40.68	$61.90
1 Building Laborer	35.45	283.60	54.60	436.80		
24 L.H., Daily Totals		$976.40		$1485.60	$40.68	$61.90

Crew H-3	Hr.	Daily	Hr.	Daily	Bare Costs	Incl. O&P
1 Glazier	$43.30	$346.40	$65.55	$524.40	$38.52	$58.77
1 Helper	33.75	270.00	52.00	416.00		
16 L.H., Daily Totals		$616.40		$940.40	$38.52	$58.77

Crew H-4	Hr.	Daily	Hr.	Daily	Bare Costs	Incl. O&P
1 Carpenter	$44.90	$359.20	$69.15	$553.20	$41.94	$64.14
1 Carpenter Helper	33.75	270.00	52.00	416.00		
.5 Electrician	52.40	209.60	78.40	313.60		
20 L.H., Daily Totals		$838.80		$1282.80	$41.94	$64.14

Crew J-1	Hr.	Daily	Hr.	Daily	Bare Costs	Incl. O&P
3 Plasterers	$41.45	$994.80	$62.10	$1490.40	$39.31	$58.90
2 Plasterer Helpers	36.10	577.60	54.10	865.60		
1 Mixing Machine, 6 C.F.		142.40		156.64	3.56	3.92
40 L.H., Daily Totals		$1714.80		$2512.64	$42.87	$62.82

Crew J-2	Hr.	Daily	Hr.	Daily	Bare Costs	Incl. O&P
3 Plasterers	$41.45	$994.80	$62.10	$1490.40	$39.38	$58.80
2 Plasterer Helpers	36.10	577.60	54.10	865.60		
1 Lather	39.70	317.60	58.30	466.40		
1 Mixing Machine, 6 C.F.		142.40		156.64	2.97	3.26
48 L.H., Daily Totals		$2032.40		$2979.04	$42.34	$62.06

Crew J-3	Hr.	Daily	Hr.	Daily	Bare Costs	Incl. O&P
1 Terrazzo Worker	$41.00	$328.00	$60.25	$482.00	$37.73	$55.42
1 Terrazzo Helper	34.45	275.60	50.60	404.80		
1 Floor Grinder, 22" Path		91.10		100.21		
1 Terrazzo Mixer		192.00		211.20	17.69	19.46
16 L.H., Daily Totals		$886.70		$1198.21	$55.42	$74.89

Crew J-4	Hr.	Daily	Hr.	Daily	Bare Costs	Incl. O&P
2 Cement Finishers	$43.05	$688.80	$63.40	$1014.40	$40.52	$60.47
1 Laborer	35.45	283.60	54.60	436.80		
1 Floor Grinder, 22" Path		91.10		100.21		
1 Floor Edger, 7" Path		39.10		43.01		
1 Vacuum Pick-Up System		60.70		66.77	7.95	8.75
24 L.H., Daily Totals		$1163.30		$1661.19	$48.47	$69.22

Column groups: Crew No. · Bare Costs (Hr., Daily) · Incl. Subs O&P (Hr., Daily) · Cost Per Labor-Hour (Bare Costs, Incl. O&P)

Crew No.	Bare Costs		Incl. Subs O&P		Cost Per Labor-Hour	

Crew J-4A	Hr.	Daily	Hr.	Daily	Bare Costs	Incl. O&P
2 Cement Finishers	$43.05	$688.80	$63.40	$1014.40	$39.25	$59.00
2 Laborers	35.45	567.20	54.60	873.60		
1 Floor Grinder, 22" Path		91.10		100.21		
1 Floor Edger, 7" Path		39.10		43.01		
1 Vacuum Pick-Up System		60.70		66.77		
1 Floor Auto Scrubber		188.15		206.97	11.85	13.03
32 L.H., Daily Totals		$1635.05		$2304.95	$51.10	$72.03

Crew J-4B	Hr.	Daily	Hr.	Daily	Bare Costs	Incl. O&P
1 Laborer	$35.45	$283.60	$54.60	$436.80	$35.45	$54.60
1 Floor Auto Scrubber		188.15		206.97	23.52	25.87
8 L.H., Daily Totals		$471.75		$643.76	$58.97	$80.47

Crew J-6	Hr.	Daily	Hr.	Daily	Bare Costs	Incl. O&P
2 Painters	$38.70	$619.20	$58.05	$928.80	$39.66	$59.98
1 Building Laborer	35.45	283.60	54.60	436.80		
1 Equip. Oper. (light)	45.80	366.40	69.20	553.60		
1 Air Compressor, 250 cfm		195.80		215.38		
1 Sandblaster, Portable, 3 C.F.		20.00		22.00		
1 Set Sand Blasting Accessories		13.65		15.02	7.17	7.89
32 L.H., Daily Totals		$1498.65		$2171.59	$46.83	$67.86

Crew K-1	Hr.	Daily	Hr.	Daily	Bare Costs	Incl. O&P
1 Carpenter	$44.90	$359.20	$69.15	$553.20	$40.27	$61.88
1 Truck Driver (light)	35.65	285.20	54.60	436.80		
1 Flatbed Truck, Gas, 3 Ton		327.80		360.58	20.49	22.54
16 L.H., Daily Totals		$972.20		$1350.58	$60.76	$84.41

Crew K-2	Hr.	Daily	Hr.	Daily	Bare Costs	Incl. O&P
1 Struc. Steel Foreman (outside)	$52.05	$416.40	$92.85	$742.80	$45.92	$78.92
1 Struc. Steel Worker	50.05	400.40	89.30	714.40		
1 Truck Driver (light)	35.65	285.20	54.60	436.80		
1 Flatbed Truck, Gas, 3 Ton		327.80		360.58	13.66	15.02
24 L.H., Daily Totals		$1429.80		$2254.58	$59.58	$93.94

Crew L-1	Hr.	Daily	Hr.	Daily	Bare Costs	Incl. O&P
1 Electrician	$52.40	$419.20	$78.40	$627.20	$54.10	$81.22
1 Plumber	55.80	446.40	84.05	672.40		
16 L.H., Daily Totals		$865.60		$1299.60	$54.10	$81.22

Crew L-2	Hr.	Daily	Hr.	Daily	Bare Costs	Incl. O&P
1 Carpenter	$44.90	$359.20	$69.15	$553.20	$39.33	$60.58
1 Carpenter Helper	33.75	270.00	52.00	416.00		
16 L.H., Daily Totals		$629.20		$969.20	$39.33	$60.58

Crew L-3	Hr.	Daily	Hr.	Daily	Bare Costs	Incl. O&P
1 Carpenter	$44.90	$359.20	$69.15	$553.20	$48.88	$74.50
.5 Electrician	52.40	209.60	78.40	313.60		
.5 Sheet Metal Worker	53.30	213.20	81.30	325.20		
16 L.H., Daily Totals		$782.00		$1192.00	$48.88	$74.50

Crew L-3A	Hr.	Daily	Hr.	Daily	Bare Costs	Incl. O&P
1 Carpenter Foreman (outside)	$46.90	$375.20	$72.25	$578.00	$49.03	$75.27
.5 Sheet Metal Worker	53.30	213.20	81.30	325.20		
12 L.H., Daily Totals		$588.40		$903.20	$49.03	$75.27

Crew L-4	Hr.	Daily	Hr.	Daily	Bare Costs	Incl. O&P
2 Skilled Workers	$46.20	$739.20	$71.45	$1143.20	$42.05	$64.97
1 Helper	33.75	270.00	52.00	416.00		
24 L.H., Daily Totals		$1009.20		$1559.20	$42.05	$64.97

Crew L-5	Hr.	Daily	Hr.	Daily	Bare Costs	Incl. O&P
1 Struc. Steel Foreman (outside)	$52.05	$416.40	$92.85	$742.80	$50.16	$87.59
5 Struc. Steel Workers	50.05	2002.00	89.30	3572.00		
1 Equip. Oper. (crane)	48.80	390.40	73.75	590.00		
1 Hyd. Crane, 25 Ton		728.60		801.46	13.01	14.31
56 L.H., Daily Totals		$3537.40		$5706.26	$63.17	$101.90

Crew L-5A	Hr.	Daily	Hr.	Daily	Bare Costs	Incl. O&P
1 Struc. Steel Foreman (outside)	$52.05	$416.40	$92.85	$742.80	$50.24	$86.30
2 Structural Steel Workers	50.05	800.80	89.30	1428.80		
1 Equip. Oper. (crane)	48.80	390.40	73.75	590.00		
1 S.P. Crane, 4x4, 25 Ton		593.40		652.74	18.54	20.40
32 L.H., Daily Totals		$2201.00		$3414.34	$68.78	$106.70

Crew L-5B	Hr.	Daily	Hr.	Daily	Bare Costs	Incl. O&P
1 Struc. Steel Foreman (outside)	$52.05	$416.40	$92.85	$742.80	$51.26	$81.83
2 Structural Steel Workers	50.05	800.80	89.30	1428.80		
2 Electricians	52.40	838.40	78.40	1254.40		
2 Steamfitters/Pipefitters	56.65	906.40	85.30	1364.80		
1 Equip. Oper. (crane)	48.80	390.40	73.75	590.00		
1 Equip. Oper. Oiler	42.25	338.00	63.85	510.80		
1 Hyd. Crane, 80 Ton		1613.00		1774.30	22.40	24.64
72 L.H., Daily Totals		$5303.40		$7665.90	$73.66	$106.47

Crew L-6	Hr.	Daily	Hr.	Daily	Bare Costs	Incl. O&P
1 Plumber	$55.80	$446.40	$84.05	$672.40	$54.67	$82.17
.5 Electrician	52.40	209.60	78.40	313.60		
12 L.H., Daily Totals		$656.00		$986.00	$54.67	$82.17

Crew L-7	Hr.	Daily	Hr.	Daily	Bare Costs	Incl. O&P
2 Carpenters	$44.90	$718.40	$69.15	$1106.40	$43.27	$66.31
1 Building Laborer	35.45	283.60	54.60	436.80		
.5 Electrician	52.40	209.60	78.40	313.60		
28 L.H., Daily Totals		$1211.60		$1856.80	$43.27	$66.31

Crew L-8	Hr.	Daily	Hr.	Daily	Bare Costs	Incl. O&P
2 Carpenters	$44.90	$718.40	$69.15	$1106.40	$47.08	$72.13
.5 Plumber	55.80	223.20	84.05	336.20		
20 L.H., Daily Totals		$941.60		$1442.60	$47.08	$72.13

Crew L-9	Hr.	Daily	Hr.	Daily	Bare Costs	Incl. O&P
1 Labor Foreman (inside)	$35.95	$287.60	$55.35	$442.80	$40.69	$65.12
2 Building Laborers	35.45	567.20	54.60	873.60		
1 Struc. Steel Worker	50.05	400.40	89.30	714.40		
.5 Electrician	52.40	209.60	78.40	313.60		
36 L.H., Daily Totals		$1464.80		$2344.40	$40.69	$65.12

Crew L-10	Hr.	Daily	Hr.	Daily	Bare Costs	Incl. O&P
1 Struc. Steel Foreman (outside)	$52.05	$416.40	$92.85	$742.80	$50.30	$85.30
1 Structural Steel Worker	50.05	400.40	89.30	714.40		
1 Equip. Oper. (crane)	48.80	390.40	73.75	590.00		
1 Hyd. Crane, 12 Ton		646.20		710.82	26.93	29.62
24 L.H., Daily Totals		$1853.40		$2758.02	$77.22	$114.92

Crew L-11	Hr.	Daily	Hr.	Daily	Bare Costs	Incl. O&P
2 Wreckers	$35.45	$567.20	$60.35	$965.60	$41.38	$65.91
1 Equip. Oper. (crane)	48.80	390.40	73.75	590.00		
1 Equip. Oper. (light)	45.80	366.40	69.20	553.60		
1 Hyd. Excavator, 2.5 C.Y.		1800.00		1980.00		
1 Loader, Skid Steer, 78 H.P.		308.60		339.46	65.89	72.48
32 L.H., Daily Totals		$3432.60		$4428.66	$107.27	$138.40

Crew No.	Bare Costs		Incl. Subs O&P		Cost Per Labor-Hour	
Crew M-1	**Hr.**	**Daily**	**Hr.**	**Daily**	**Bare Costs**	**Incl. O&P**
3 Elevator Constructors	$71.10	$1706.40	$106.30	$2551.20	$67.55	$100.99
1 Elevator Apprentice	56.90	455.20	85.05	680.40		
5 Hand Tools		46.00		50.60	1.44	1.58
32 L.H., Daily Totals		$2207.60		$3282.20	$68.99	$102.57
Crew M-3	**Hr.**	**Daily**	**Hr.**	**Daily**	**Bare Costs**	**Incl. O&P**
1 Electrician Foreman (outside)	$54.40	$435.20	$81.40	$651.20	$54.05	$81.24
1 Common Laborer	35.45	283.60	54.60	436.80		
.25 Equipment Operator, Med.	47.50	95.00	71.75	143.50		
1 Elevator Constructor	71.10	568.80	106.30	850.40		
1 Elevator Apprentice	56.90	455.20	85.05	680.40		
.25 S.P. Crane, 4x4, 20 Ton		137.00		150.70	4.03	4.43
34 L.H., Daily Totals		$1974.80		$2913.00	$58.08	$85.68
Crew M-4	**Hr.**	**Daily**	**Hr.**	**Daily**	**Bare Costs**	**Incl. O&P**
1 Electrician Foreman (outside)	$54.40	$435.20	$81.40	$651.20	$53.47	$80.39
1 Common Laborer	35.45	283.60	54.60	436.80		
.25 Equipment Operator, Crane	48.80	97.60	73.75	147.50		
.25 Equipment Operator, Oiler	42.25	84.50	63.85	127.70		
1 Elevator Constructor	71.10	568.80	106.30	850.40		
1 Elevator Apprentice	56.90	455.20	85.05	680.40		
.25 S.P. Crane, 4x4, 40 Ton		174.20		191.62	4.84	5.32
36 L.H., Daily Totals		$2099.10		$3085.62	$58.31	$85.71
Crew Q-1	**Hr.**	**Daily**	**Hr.**	**Daily**	**Bare Costs**	**Incl. O&P**
1 Plumber	$55.80	$446.40	$84.05	$672.40	$50.23	$75.65
1 Plumber Apprentice	44.65	357.20	67.25	538.00		
16 L.H., Daily Totals		$803.60		$1210.40	$50.23	$75.65
Crew Q-1A	**Hr.**	**Daily**	**Hr.**	**Daily**	**Bare Costs**	**Incl. O&P**
.25 Plumber Foreman (outside)	$57.80	$115.60	$87.05	$174.10	$56.20	$84.65
1 Plumber	55.80	446.40	84.05	672.40		
10 L.H., Daily Totals		$562.00		$846.50	$56.20	$84.65
Crew Q-1C	**Hr.**	**Daily**	**Hr.**	**Daily**	**Bare Costs**	**Incl. O&P**
1 Plumber	$55.80	$446.40	$84.05	$672.40	$49.32	$74.35
1 Plumber Apprentice	44.65	357.20	67.25	538.00		
1 Equip. Oper. (medium)	47.50	380.00	71.75	574.00		
1 Trencher, Chain Type, 8' D		3357.00		3692.70	139.88	153.86
24 L.H., Daily Totals		$4540.60		$5477.10	$189.19	$228.21
Crew Q-2	**Hr.**	**Daily**	**Hr.**	**Daily**	**Bare Costs**	**Incl. O&P**
2 Plumbers	$55.80	$892.80	$84.05	$1344.80	$52.08	$78.45
1 Plumber Apprentice	44.65	357.20	67.25	538.00		
24 L.H., Daily Totals		$1250.00		$1882.80	$52.08	$78.45
Crew Q-3	**Hr.**	**Daily**	**Hr.**	**Daily**	**Bare Costs**	**Incl. O&P**
1 Plumber Foreman (inside)	$56.30	$450.40	$84.80	$678.40	$53.14	$80.04
2 Plumbers	55.80	892.80	84.05	1344.80		
1 Plumber Apprentice	44.65	357.20	67.25	538.00		
32 L.H., Daily Totals		$1700.40		$2561.20	$53.14	$80.04
Crew Q-4	**Hr.**	**Daily**	**Hr.**	**Daily**	**Bare Costs**	**Incl. O&P**
1 Plumber Foreman (inside)	$56.30	$450.40	$84.80	$678.40	$53.14	$80.04
1 Plumber	55.80	446.40	84.05	672.40		
1 Welder (plumber)	55.80	446.40	84.05	672.40		
1 Plumber Apprentice	44.65	357.20	67.25	538.00		
1 Welder, Electric, 300 amp		52.85		58.13	1.65	1.82
32 L.H., Daily Totals		$1753.25		$2619.34	$54.79	$81.85

Crew No.	Bare Costs		Incl. Subs O&P		Cost Per Labor-Hour	
Crew Q-5	**Hr.**	**Daily**	**Hr.**	**Daily**	**Bare Costs**	**Incl. O&P**
1 Steamfitter	$56.65	$453.20	$85.30	$682.40	$50.98	$76.75
1 Steamfitter Apprentice	45.30	362.40	68.20	545.60		
16 L.H., Daily Totals		$815.60		$1228.00	$50.98	$76.75
Crew Q-6	**Hr.**	**Daily**	**Hr.**	**Daily**	**Bare Costs**	**Incl. O&P**
2 Steamfitters	$56.65	$906.40	$85.30	$1364.80	$52.87	$79.60
1 Steamfitter Apprentice	45.30	362.40	68.20	545.60		
24 L.H., Daily Totals		$1268.80		$1910.40	$52.87	$79.60
Crew Q-7	**Hr.**	**Daily**	**Hr.**	**Daily**	**Bare Costs**	**Incl. O&P**
1 Steamfitter Foreman (inside)	$57.15	$457.20	$86.05	$688.40	$53.94	$81.21
2 Steamfitters	56.65	906.40	85.30	1364.80		
1 Steamfitter Apprentice	45.30	362.40	68.20	545.60		
32 L.H., Daily Totals		$1726.00		$2598.80	$53.94	$81.21
Crew Q-8	**Hr.**	**Daily**	**Hr.**	**Daily**	**Bare Costs**	**Incl. O&P**
1 Steamfitter Foreman (inside)	$57.15	$457.20	$86.05	$688.40	$53.94	$81.21
1 Steamfitter	56.65	453.20	85.30	682.40		
1 Welder (steamfitter)	56.65	453.20	85.30	682.40		
1 Steamfitter Apprentice	45.30	362.40	68.20	545.60		
1 Welder, Electric, 300 amp		52.85		58.13	1.65	1.82
32 L.H., Daily Totals		$1778.85		$2656.93	$55.59	$83.03
Crew Q-9	**Hr.**	**Daily**	**Hr.**	**Daily**	**Bare Costs**	**Incl. O&P**
1 Sheet Metal Worker	$53.30	$426.40	$81.30	$650.40	$47.98	$73.17
1 Sheet Metal Apprentice	42.65	341.20	65.05	520.40		
16 L.H., Daily Totals		$767.60		$1170.80	$47.98	$73.17
Crew Q-10	**Hr.**	**Daily**	**Hr.**	**Daily**	**Bare Costs**	**Incl. O&P**
2 Sheet Metal Workers	$53.30	$852.80	$81.30	$1300.80	$49.75	$75.88
1 Sheet Metal Apprentice	42.65	341.20	65.05	520.40		
24 L.H., Daily Totals		$1194.00		$1821.20	$49.75	$75.88
Crew Q-11	**Hr.**	**Daily**	**Hr.**	**Daily**	**Bare Costs**	**Incl. O&P**
1 Sheet Metal Foreman (inside)	$53.80	$430.40	$82.05	$656.40	$50.76	$77.42
2 Sheet Metal Workers	53.30	852.80	81.30	1300.80		
1 Sheet Metal Apprentice	42.65	341.20	65.05	520.40		
32 L.H., Daily Totals		$1624.40		$2477.60	$50.76	$77.42
Crew Q-12	**Hr.**	**Daily**	**Hr.**	**Daily**	**Bare Costs**	**Incl. O&P**
1 Sprinkler Installer	$54.65	$437.20	$82.35	$658.80	$49.17	$74.10
1 Sprinkler Apprentice	43.70	349.60	65.85	526.80		
16 L.H., Daily Totals		$786.80		$1185.60	$49.17	$74.10
Crew Q-13	**Hr.**	**Daily**	**Hr.**	**Daily**	**Bare Costs**	**Incl. O&P**
1 Sprinkler Foreman (inside)	$55.15	$441.20	$83.10	$664.80	$52.04	$78.41
2 Sprinkler Installers	54.65	874.40	82.35	1317.60		
1 Sprinkler Apprentice	43.70	349.60	65.85	526.80		
32 L.H., Daily Totals		$1665.20		$2509.20	$52.04	$78.41
Crew Q-14	**Hr.**	**Daily**	**Hr.**	**Daily**	**Bare Costs**	**Incl. O&P**
1 Asbestos Worker	$49.70	$397.60	$77.20	$617.60	$44.73	$69.47
1 Asbestos Apprentice	39.75	318.00	61.75	494.00		
16 L.H., Daily Totals		$715.60		$1111.60	$44.73	$69.47

Crew No.	Bare Costs		Incl. Subs O&P		Cost Per Labor-Hour	
Crew Q-15	**Hr.**	**Daily**	**Hr.**	**Daily**	**Bare Costs**	**Incl. O&P**
1 Plumber	$55.80	$446.40	$84.05	$672.40	$50.23	$75.65
1 Plumber Apprentice	44.65	357.20	67.25	538.00		
1 Welder, Electric, 300 amp		52.85		58.13	3.30	3.63
16 L.H., Daily Totals		$856.45		$1268.54	$53.53	$79.28
Crew Q-16	**Hr.**	**Daily**	**Hr.**	**Daily**	**Bare Costs**	**Incl. O&P**
2 Plumbers	$55.80	$892.80	$84.05	$1344.80	$52.08	$78.45
1 Plumber Apprentice	44.65	357.20	67.25	538.00		
1 Welder, Electric, 300 amp		52.85		58.13	2.20	2.42
24 L.H., Daily Totals		$1302.85		$1940.93	$54.29	$80.87
Crew Q-17	**Hr.**	**Daily**	**Hr.**	**Daily**	**Bare Costs**	**Incl. O&P**
1 Steamfitter	$56.65	$453.20	$85.30	$682.40	$50.98	$76.75
1 Steamfitter Apprentice	45.30	362.40	68.20	545.60		
1 Welder, Electric, 300 amp		52.85		58.13	3.30	3.63
16 L.H., Daily Totals		$868.45		$1286.14	$54.28	$80.38
Crew Q-17A	**Hr.**	**Daily**	**Hr.**	**Daily**	**Bare Costs**	**Incl. O&P**
1 Steamfitter	$56.65	$453.20	$85.30	$682.40	$50.25	$75.75
1 Steamfitter Apprentice	45.30	362.40	68.20	545.60		
1 Equip. Oper. (crane)	48.80	390.40	73.75	590.00		
1 Hyd. Crane, 12 Ton		646.20		710.82		
1 Welder, Electric, 300 amp		52.85		58.13	29.13	32.04
24 L.H., Daily Totals		$1905.05		$2586.95	$79.38	$107.79
Crew Q-18	**Hr.**	**Daily**	**Hr.**	**Daily**	**Bare Costs**	**Incl. O&P**
2 Steamfitters	$56.65	$906.40	$85.30	$1364.80	$52.87	$79.60
1 Steamfitter Apprentice	45.30	362.40	68.20	545.60		
1 Welder, Electric, 300 amp		52.85		58.13	2.20	2.42
24 L.H., Daily Totals		$1321.65		$1968.54	$55.07	$82.02
Crew Q-19	**Hr.**	**Daily**	**Hr.**	**Daily**	**Bare Costs**	**Incl. O&P**
1 Steamfitter	$56.65	$453.20	$85.30	$682.40	$51.45	$77.30
1 Steamfitter Apprentice	45.30	362.40	68.20	545.60		
1 Electrician	52.40	419.20	78.40	627.20		
24 L.H., Daily Totals		$1234.80		$1855.20	$51.45	$77.30
Crew Q-20	**Hr.**	**Daily**	**Hr.**	**Daily**	**Bare Costs**	**Incl. O&P**
1 Sheet Metal Worker	$53.30	$426.40	$81.30	$650.40	$48.86	$74.22
1 Sheet Metal Apprentice	42.65	341.20	65.05	520.40		
.5 Electrician	52.40	209.60	78.40	313.60		
20 L.H., Daily Totals		$977.20		$1484.40	$48.86	$74.22
Crew Q-21	**Hr.**	**Daily**	**Hr.**	**Daily**	**Bare Costs**	**Incl. O&P**
2 Steamfitters	$56.65	$906.40	$85.30	$1364.80	$52.75	$79.30
1 Steamfitter Apprentice	45.30	362.40	68.20	545.60		
1 Electrician	52.40	419.20	78.40	627.20		
32 L.H., Daily Totals		$1688.00		$2537.60	$52.75	$79.30
Crew Q-22	**Hr.**	**Daily**	**Hr.**	**Daily**	**Bare Costs**	**Incl. O&P**
1 Plumber	$55.80	$446.40	$84.05	$672.40	$50.23	$75.65
1 Plumber Apprentice	44.65	357.20	67.25	538.00		
1 Hyd. Crane, 12 Ton		646.20		710.82	40.39	44.43
16 L.H., Daily Totals		$1449.80		$1921.22	$90.61	$120.08

Crew No.	Bare Costs		Incl. Subs O&P		Cost Per Labor-Hour	
Crew Q-22A	**Hr.**	**Daily**	**Hr.**	**Daily**	**Bare Costs**	**Incl. O&P**
1 Plumber	$55.80	$446.40	$84.05	$672.40	$46.17	$69.91
1 Plumber Apprentice	44.65	357.20	67.25	538.00		
1 Laborer	35.45	283.60	54.60	436.80		
1 Equip. Oper. (crane)	48.80	390.40	73.75	590.00		
1 Hyd. Crane, 12 Ton		646.20		710.82	20.19	22.21
32 L.H., Daily Totals		$2123.80		$2948.02	$66.37	$92.13
Crew Q-23	**Hr.**	**Daily**	**Hr.**	**Daily**	**Bare Costs**	**Incl. O&P**
1 Plumber Foreman (outside)	$57.80	$462.40	$87.05	$696.40	$53.70	$80.95
1 Plumber	55.80	446.40	84.05	672.40		
1 Equip. Oper. (med.)	47.50	380.00	71.75	574.00		
1 Lattice Boom Crane, 20 Ton		958.30		1054.13	39.93	43.92
24 L.H., Daily Totals		$2247.10		$2996.93	$93.63	$124.87
Crew R-1	**Hr.**	**Daily**	**Hr.**	**Daily**	**Bare Costs**	**Incl. O&P**
1 Electrician Foreman	$52.90	$423.20	$79.15	$633.20	$46.27	$69.72
3 Electricians	52.40	1257.60	78.40	1881.60		
2 Helpers	33.75	540.00	52.00	832.00		
48 L.H., Daily Totals		$2220.80		$3346.80	$46.27	$69.72
Crew R-1A	**Hr.**	**Daily**	**Hr.**	**Daily**	**Bare Costs**	**Incl. O&P**
1 Electrician	$52.40	$419.20	$78.40	$627.20	$43.08	$65.20
1 Helper	33.75	270.00	52.00	416.00		
16 L.H., Daily Totals		$689.20		$1043.20	$43.08	$65.20
Crew R-2	**Hr.**	**Daily**	**Hr.**	**Daily**	**Bare Costs**	**Incl. O&P**
1 Electrician Foreman	$52.90	$423.20	$79.15	$633.20	$46.63	$70.30
3 Electricians	52.40	1257.60	78.40	1881.60		
2 Helpers	33.75	540.00	52.00	832.00		
1 Equip. Oper. (crane)	48.80	390.40	73.75	590.00		
1 S.P. Crane, 4x4, 5 Ton		274.20		301.62	4.90	5.39
56 L.H., Daily Totals		$2885.40		$4238.42	$51.52	$75.69
Crew R-3	**Hr.**	**Daily**	**Hr.**	**Daily**	**Bare Costs**	**Incl. O&P**
1 Electrician Foreman	$52.90	$423.20	$79.15	$633.20	$51.88	$77.77
1 Electrician	52.40	419.20	78.40	627.20		
.5 Equip. Oper. (crane)	48.80	195.20	73.75	295.00		
.5 S.P. Crane, 4x4, 5 Ton		137.10		150.81	6.86	7.54
20 L.H., Daily Totals		$1174.70		$1706.21	$58.73	$85.31
Crew R-4	**Hr.**	**Daily**	**Hr.**	**Daily**	**Bare Costs**	**Incl. O&P**
1 Struc. Steel Foreman (outside)	$52.05	$416.40	$92.85	$742.80	$50.92	$87.83
3 Struc. Steel Workers	50.05	1201.20	89.30	2143.20		
1 Electrician	52.40	419.20	78.40	627.20		
1 Welder, Gas Engine, 300 amp		144.00		158.40	3.60	3.96
40 L.H., Daily Totals		$2180.80		$3671.60	$54.52	$91.79
Crew R-5	**Hr.**	**Daily**	**Hr.**	**Daily**	**Bare Costs**	**Incl. O&P**
1 Electrician Foreman	$52.90	$423.20	$79.15	$633.20	$45.66	$68.87
4 Electrician Linemen	52.40	1676.80	78.40	2508.80		
2 Electrician Operators	52.40	838.40	78.40	1254.40		
4 Electrician Groundmen	33.75	1080.00	52.00	1664.00		
1 Crew Truck		221.20		243.32		
1 Flatbed Truck, 20,000 GVW		253.60		278.96		
1 Pickup Truck, 3/4 Ton		154.60		170.06		
.2 Hyd. Crane, 55 Ton		223.00		245.30		
.2 Hyd. Crane, 12 Ton		129.24		142.16		
.2 Earth Auger, Truck-Mtd.		86.24		94.86		
1 Tractor w/Winch		434.20		477.62	17.07	18.78
88 L.H., Daily Totals		$5520.48		$7712.69	$62.73	$87.64

Crew No.	Bare Costs		Incl. Subs O&P		Cost Per Labor-Hour	
Crew R-6	**Hr.**	**Daily**	**Hr.**	**Daily**	**Bare Costs**	**Incl. O&P**
1 Electrician Foreman	$52.90	$423.20	$79.15	$633.20	$45.66	$68.87
4 Electrician Linemen	52.40	1676.80	78.40	2508.80		
2 Electrician Operators	52.40	838.40	78.40	1254.40		
4 Electrician Groundmen	33.75	1080.00	52.00	1664.00		
1 Crew Truck		221.20		243.32		
1 Flatbed Truck, 20,000 GVW		253.60		278.96		
1 Pickup Truck, 3/4 Ton		154.60		170.06		
.2 Hyd. Crane, 55 Ton		223.00		245.30		
.2 Hyd. Crane, 12 Ton		129.24		142.16		
.2 Earth Auger, Truck-Mtd.		86.24		94.86		
1 Tractor w/Winch		434.20		477.62		
3 Cable Trailers		575.10		632.61		
.5 Tensioning Rig		190.72		209.80		
.5 Cable Pulling Rig		1116.50		1228.15	38.46	42.31
88 L.H., Daily Totals		$7402.81		$9783.25	$84.12	$111.17

Crew R-7	Hr.	Daily	Hr.	Daily	Bare Costs	Incl. O&P
1 Electrician Foreman	$52.90	$423.20	$79.15	$633.20	$36.94	$56.52
5 Electrician Groundmen	33.75	1350.00	52.00	2080.00		
1 Crew Truck		221.20		243.32	4.61	5.07
48 L.H., Daily Totals		$1994.40		$2956.52	$41.55	$61.59

Crew R-8	Hr.	Daily	Hr.	Daily	Bare Costs	Incl. O&P
1 Electrician Foreman	$52.90	$423.20	$79.15	$633.20	$46.27	$69.72
3 Electrician Linemen	52.40	1257.60	78.40	1881.60		
2 Electrician Groundmen	33.75	540.00	52.00	832.00		
1 Pickup Truck, 3/4 Ton		154.60		170.06		
1 Crew Truck		221.20		243.32	7.83	8.61
48 L.H., Daily Totals		$2596.60		$3760.18	$54.10	$78.34

Crew R-9	Hr.	Daily	Hr.	Daily	Bare Costs	Incl. O&P
1 Electrician Foreman	$52.90	$423.20	$79.15	$633.20	$43.14	$65.29
1 Electrician Lineman	52.40	419.20	78.40	627.20		
2 Electrician Operators	52.40	838.40	78.40	1254.40		
4 Electrician Groundmen	33.75	1080.00	52.00	1664.00		
1 Pickup Truck, 3/4 Ton		154.60		170.06		
1 Crew Truck		221.20		243.32	5.87	6.46
64 L.H., Daily Totals		$3136.60		$4592.18	$49.01	$71.75

Crew R-10	Hr.	Daily	Hr.	Daily	Bare Costs	Incl. O&P
1 Electrician Foreman	$52.90	$423.20	$79.15	$633.20	$49.38	$74.13
4 Electrician Linemen	52.40	1676.80	78.40	2508.80		
1 Electrician Groundman	33.75	270.00	52.00	416.00		
1 Crew Truck		221.20		243.32		
3 Tram Cars		389.10		428.01	12.71	13.99
48 L.H., Daily Totals		$2980.30		$4229.33	$62.09	$88.11

Crew R-11	Hr.	Daily	Hr.	Daily	Bare Costs	Incl. O&P
1 Electrician Foreman	$52.90	$423.20	$79.15	$633.20	$49.54	$74.44
4 Electricians	52.40	1676.80	78.40	2508.80		
1 Equip. Oper. (crane)	48.80	390.40	73.75	590.00		
1 Common Laborer	35.45	283.60	54.60	436.80		
1 Crew Truck		221.20		243.32		
1 Hyd. Crane, 12 Ton		646.20		710.82	15.49	17.04
56 L.H., Daily Totals		$3641.40		$5122.94	$65.03	$91.48

Crew R-12	Hr.	Daily	Hr.	Daily	Bare Costs	Incl. O&P
1 Carpenter Foreman (inside)	$45.40	$363.20	$69.90	$559.20	$42.21	$66.00
4 Carpenters	44.90	1436.80	69.15	2212.80		
4 Common Laborers	35.45	1134.40	54.60	1747.20		
1 Equip. Oper. (med.)	47.50	380.00	71.75	574.00		
1 Steel Worker	50.05	400.40	89.30	714.40		
1 Dozer, 200 H.P.		1333.00		1466.30		
1 Pickup Truck, 3/4 Ton		154.60		170.06	16.90	18.59
88 L.H., Daily Totals		$5202.40		$7443.96	$59.12	$84.59

Crew R-13	Hr.	Daily	Hr.	Daily	Bare Costs	Incl. O&P
1 Electrician Foreman	$52.90	$423.20	$79.15	$633.20	$50.39	$75.55
3 Electricians	52.40	1257.60	78.40	1881.60		
.25 Equip. Oper. (crane)	48.80	97.60	73.75	147.50		
1 Equipment Oiler	42.25	338.00	63.85	510.80		
.25 Hydraulic Crane, 33 Ton		185.75		204.32	4.42	4.86
42 L.H., Daily Totals		$2302.15		$3377.43	$54.81	$80.41

Crew R-15	Hr.	Daily	Hr.	Daily	Bare Costs	Incl. O&P
1 Electrician Foreman	$52.90	$423.20	$79.15	$633.20	$51.38	$76.99
4 Electricians	52.40	1676.80	78.40	2508.80		
1 Equipment Oper. (light)	45.80	366.40	69.20	553.60		
1 Aerial Lift Truck, 40' Boom		332.20		365.42	6.92	7.61
48 L.H., Daily Totals		$2798.60		$4061.02	$58.30	$84.60

Crew R-18	Hr.	Daily	Hr.	Daily	Bare Costs	Incl. O&P
.25 Electrician Foreman	$52.90	$105.80	$79.15	$158.30	$40.96	$62.21
1 Electrician	52.40	419.20	78.40	627.20		
2 Helpers	33.75	540.00	52.00	832.00		
26 L.H., Daily Totals		$1065.00		$1617.50	$40.96	$62.21

Crew R-19	Hr.	Daily	Hr.	Daily	Bare Costs	Incl. O&P
.5 Electrician Foreman	$52.90	$211.60	$79.15	$316.60	$52.50	$78.55
2 Electricians	52.40	838.40	78.40	1254.40		
20 L.H., Daily Totals		$1050.00		$1571.00	$52.50	$78.55

Crew R-21	Hr.	Daily	Hr.	Daily	Bare Costs	Incl. O&P
1 Electrician Foreman	$52.90	$423.20	$79.15	$633.20	$52.40	$78.42
3 Electricians	52.40	1257.60	78.40	1881.60		
.1 Equip. Oper. (med.)	47.50	38.00	71.75	57.40		
.1 S.P. Crane, 4x4, 25 Ton		59.34		65.27	1.81	1.99
32.8 L.H., Daily Totals		$1778.14		$2637.47	$54.21	$80.41

Crew R-22	Hr.	Daily	Hr.	Daily	Bare Costs	Incl. O&P
.66 Electrician Foreman	$52.90	$279.31	$79.15	$417.91	$44.47	$67.18
2 Helpers	33.75	540.00	52.00	832.00		
2 Electricians	52.40	838.40	78.40	1254.40		
37.28 L.H., Daily Totals		$1657.71		$2504.31	$44.47	$67.18

Crew R-30	Hr.	Daily	Hr.	Daily	Bare Costs	Incl. O&P
.25 Electrician Foreman (outside)	$54.40	$108.80	$81.40	$162.80	$42.12	$63.98
1 Electrician	52.40	419.20	78.40	627.20		
2 Laborers, (Semi-Skilled)	35.45	567.20	54.60	873.60		
26 L.H., Daily Totals		$1095.20		$1663.60	$42.12	$63.98

Crew R-31	Hr.	Daily	Hr.	Daily	Bare Costs	Incl. O&P
1 Electrician	$52.40	$419.20	$78.40	$627.20	$52.40	$78.40
1 Core Drill, Electric, 2.5 H.P.		48.10		52.91	6.01	6.61
8 L.H., Daily Totals		$467.30		$680.11	$58.41	$85.01

Crew No.	Bare Costs		Incl. Subs O & P		Cost Per Labor-Hour	
Crew W-41E	Hr.	Daily	Hr.	Daily	Bare Costs	Incl. O&P
.5 Plumber Foreman (outside)	$57.80	$231.20	$87.05	$348.20	$48.06	$72.87
1 Plumber	55.80	446.40	84.05	672.40		
1 Laborer	35.45	283.60	54.60	436.80		
20 L.H., Daily Totals		$961.20		$1457.40	$48.06	$72.87

Historical Cost Indexes

The table below lists both the RSMeans® historical cost index based on Jan. 1, 1993 = 100 as well as the computed value of an index based on Jan. 1, 2013 costs. Since the Jan. 1, 2013 figure is estimated, space is left to write in the actual index figures as they become available through either the quarterly *RSMeans Construction Cost Indexes* or as printed in the *Engineering News-Record.* To compute the actual index based on Jan. 1, 2013 = 100, divide the historical cost index for a particular year by the actual Jan. 1, 2013 construction cost index. Space has been left to advance the index figures as the year progresses.

Year	Historical Cost Index Jan. 1, 1993 = 100		Current Index Based on Jan. 1, 2013 = 100		Year	Historical Cost Index Jan. 1, 1993 = 100	Current Index Based on Jan. 1, 2013 = 100		Year	Historical Cost Index Jan. 1, 1993 = 100	Current Index Based on Jan. 1, 2013 = 100	
	Est.	Actual	Est.	Actual		Actual	Est.	Actual		Actual	Est.	Actual
Oct 2013*					July 1998	115.1	59.7		July 1980	62.9	31.8	
July 2013*					1997	112.8	57.1		1979	57.8	29.3	
April 2013*					1996	110.2	55.8		1978	53.5	27.1	
Jan 2013*	197.6		100.0	100.0	1995	107.6	54.5		1977	49.5	25.1	
July 2012		194.6	98.5		1994	104.4	52.8		1976	46.9	23.7	
2011		191.2	96.8		1993	101.7	51.5		1975	44.8	22.7	
2010		183.5	92.9		1992	99.4	50.3		1974	41.4	21.0	
2009		180.1	91.1		1991	96.8	49.0		1973	37.7	19.1	
2008		180.4	91.3		1990	94.3	47.7		1972	34.8	17.6	
2007		169.4	85.7		1989	92.1	46.6		1971	32.1	16.2	
2006		162.0	82.0		1988	89.9	45.5		1970	28.7	14.5	
2005		151.6	76.7		1987	87.7	44.4		1969	26.9	13.6	
2004		143.7	72.7		1986	84.2	42.6		1968	24.9	12.6	
2003		132.0	66.8		1985	82.6	41.8		1967	23.5	11.9	
2002		128.7	65.1		1984	82.0	41.5		1966	22.7	11.5	
2001		125.1	63.3		1983	80.2	40.6		1965	21.7	11.0	
2000		120.9	61.2		1982	76.1	38.5		1964	21.2	10.7	
1999		117.6	59.5		1981	70.0	35.4		1963	20.7	10.5	

Adjustments to Costs

The "Historical Cost Index" can be used to convert national average building costs at a particular time to the approximate building costs for some other time.

Example:

Estimate and compare construction costs for different years in the same city.

To estimate the national average construction cost of a building in 1970, knowing that it cost \$900,000 in 2013:

INDEX in 1970 = 28.7

INDEX in 2013 = 197.6

Time Adjustment Using the Historical Cost Indexes:

$$\frac{\text{Index for Year A}}{\text{Index for Year B}} \times \text{Cost in Year B} = \text{Cost in Year A}$$

$$\frac{\text{INDEX 1970}}{\text{INDEX 2013}} \times \text{Cost 2013} = \text{Cost 1970}$$

$$\frac{28.7}{197.6} \times \$900{,}000 = .145 \times \$900{,}000 = \$130{,}500$$

The construction cost of the building in 1970 is \$130,500.

Note: The city cost indexes for Canada can be used to convert U.S. national averages to local costs in Canadian dollars.

*Historical Cost Index updates and other resources are provided on the following website.
http://www.reedconstructiondata.com/rsmeans/chgnotice/878796

How to Use the City Cost Indexes

What you should know before you begin

RSMeans City Cost Indexes (CCI) are an extremely useful tool to use when you want to compare costs from city to city and region to region.

This publication contains average construction cost indexes for 731 U.S. and Canadian cities covering over 930 three-digit zip code locations, as listed directly under each city.

Keep in mind that a City Cost Index number is a percentage ratio of a specific city's cost to the national average cost of the same item at a stated time period.

In other words, these index figures represent relative construction factors (or, if you prefer, multipliers) for Material and Installation costs, as well as the weighted average for Total In Place costs for each CSI MasterFormat division. Installation costs include both labor and equipment rental costs. When estimating equipment rental rates only, for a specific location, use 01 54 33 EQUIPMENT RENTAL COSTS in the Reference Section at the back of the book.

The 30 City Average Index is the average of 30 major U.S. cities and serves as a National Average.

Index figures for both material and installation are based on the 30 major city average of 100 and represent the cost relationship as of July 1, 2012. The index for each division is computed from representative material and labor quantities for that division. The weighted average for each city is a weighted total of the components listed above it, but does not include relative productivity between trades or cities.

As changes occur in local material prices, labor rates, and equipment rental rates, (including fuel costs) the impact of these changes should be accurately measured by the change in the City Cost Index for each particular city (as compared to the 30 City Average).

Therefore, if you know (or have estimated) building costs in one city today, you can easily convert those costs to expected building costs in another city.

In addition, by using the Historical Cost Index, you can easily convert National Average building costs at a particular time to the approximate building costs for some other time. The City Cost Indexes can then be applied to calculate the costs for a particular city.

Quick Calculations

Location Adjustment Using the City Cost Indexes:

$$\frac{\text{Index for City A}}{\text{Index for City B}} \times \text{Cost in City B} = \text{Cost in City A}$$

Time Adjustment for the National Average Using the Historical Cost Index:

$$\frac{\text{Index for Year A}}{\text{Index for Year B}} \times \text{Cost in Year B} = \text{Cost in Year A}$$

Adjustment from the National Average:

$$\frac{\text{Index for City A}}{100} \times \text{National Average Cost} = \text{Cost in City A}$$

Since each of the other RSMeans publications contains many different items, any *one* item multiplied by the particular city index may give incorrect results. However, the larger the number of items compiled, the closer the results should be to actual costs for that particular city.

The City Cost Indexes for Canadian cities are calculated using Canadian material and equipment prices and labor rates, in Canadian dollars. Therefore, indexes for Canadian cities can be used to convert U.S. National Average prices to local costs in Canadian dollars.

How to use this section

1. Compare costs from city to city.

In using the RSMeans Indexes, remember that an index number is not a fixed number but a ratio: It's a percentage ratio of a building component's cost at any stated time to the National Average cost of that same component at the same time period. Put in the form of an equation:

$$\frac{\text{Specific City Cost}}{\text{National Average Cost}} \times 100 = \text{City Index Number}$$

Therefore, when making cost comparisons between cities, do not subtract one city's index number from the index number of another city and read the result as a percentage difference. Instead, divide one city's index number by that of the other city. The resulting number may then be used as a multiplier to calculate cost differences from city to city.

The formula used to find cost differences between cities for the purpose of comparison is as follows:

$$\frac{\text{City A Index}}{\text{City B Index}} \times \text{City B Cost (Known)} = \text{City A Cost (Unknown)}$$

In addition, you can use RSMeans CCI to calculate and compare costs division by division between cities using the same basic formula. (Just be sure that you're comparing similar divisions.)

2. Compare a specific city's construction costs with the National Average.

When you're studying construction location feasibility, it's advisable to compare a prospective project's cost index with an index of the National Average cost.

For example, divide the weighted average index of construction costs of a specific city by that of the 30 City Average, which = 100.

$$\frac{\text{City Index}}{100} = \text{\% of National Average}$$

As a result, you get a ratio that indicates the relative cost of construction in that city in comparison with the National Average.

3. Convert U.S. National Average to actual costs in Canadian City.

$$\frac{\text{Index for Canadian City}}{100} \times \text{National Average Cost} = \text{Cost in Canadian City in \$ CAN}$$

4. Adjust construction cost data based on a National Average.

When you use a source of construction cost data which is based on a National Average (such as RSMeans cost data publications), it is necessary to adjust those costs to a specific location.

$$\frac{\text{City Index}}{100} \times \frac{\text{"Book" Cost Based on}}{\text{National Average Costs}} = \frac{\text{City Cost}}{\text{(Unknown)}}$$

5. When applying the City Cost Indexes to demolition projects, use the appropriate division installation index. For example, for removal of existing doors and windows, use Division 8 (Openings) index.

What you might like to know about how we developed the Indexes

The information presented in the CCI is organized according to the Construction Specifications Institute (CSI) MasterFormat 2010 classification system.

To create a reliable index, RSMeans researched the building type most often constructed in the United States and Canada. Because it was concluded that no one type of building completely represented the building construction industry, nine different types of buildings were combined to create a composite model.

The exact material, labor, and equipment quantities are based on detailed analyses of these nine building types, and then each quantity is weighted in proportion to expected usage. These various material items, labor hours, and equipment rental rates are thus combined to form a composite building representing as closely as possible the actual usage of materials, labor and equipment used in the North American building construction industry.

The following structures were chosen to make up that composite model:

1. Factory, 1 story
2. Office, 2–4 story
3. Store, Retail
4. Town Hall, 2–3 story
5. High School, 2–3 story
6. Hospital, 4–8 story
7. Garage, Parking
8. Apartment, 1–3 story
9. Hotel/Motel, 2–3 story

For the purposes of ensuring the timeliness of the data, the components of the index for the composite model have been streamlined. They currently consist of:

- specific quantities of 66 commonly used construction materials;
- specific labor-hours for 21 building construction trades; and
- specific days of equipment rental for 6 types of construction equipment (normally used to install the 66 material items by the 21 trades.) Fuel costs and routine maintenance costs are included in the equipment cost.

A sophisticated computer program handles the updating of all costs for each city on a quarterly basis. Material and equipment price quotations are gathered quarterly from 731 cities in the United States and Canada. These prices and the latest negotiated labor wage rates for 21 different building trades are used to compile the quarterly update of the City Cost Index.

The 30 major U.S. cities used to calculate the National Average are:

Atlanta, GA	Memphis, TN
Baltimore, MD	Milwaukee, WI
Boston, MA	Minneapolis, MN
Buffalo, NY	Nashville, TN
Chicago, IL	New Orleans, LA
Cincinnati, OH	New York, NY
Cleveland, OH	Philadelphia, PA
Columbus, OH	Phoenix, AZ
Dallas, TX	Pittsburgh, PA
Denver, CO	St. Louis, MO
Detroit, MI	San Antonio, TX
Houston, TX	San Diego, CA
Indianapolis, IN	San Francisco, CA
Kansas City, MO	Seattle, WA
Los Angeles, CA	Washington, DC

What the CCI does not indicate

The weighted average for each city is a total of the divisional components weighted to reflect typical usage, but it does not include the productivity variations between trades or cities.

In addition, the CCI does not take into consideration factors such as the following:

- managerial efficiency
- competitive conditions
- automation
- restrictive union practices
- unique local requirements
- regional variations due to specific building codes

	DIVISION	UNITED STATES			ALABAMA																	
		30 CITY			ANNISTON			BIRMINGHAM			BUTLER			DECATUR			DOTHAN					
		AVERAGE			362			350 - 352			369			356			363					
		MAT.	INST.	TOTAL	MAT.	INST.	TOTAL	MAT.	INST.	TOTAL	MAT.	INST.	TOTAL	MAT.	INST.	TOTAL	MAT.	INST.	TOTAL			
015433	**CONTRACTOR EQUIPMENT**		100.0	100.0		101.7	101.7		101.8	101.8		98.9	98.9		101.7	101.7		98.9	98.9			
0241, 31 - 34	**SITE & INFRASTRUCTURE, DEMOLITION**	100.0	100.0	100.0	88.1	94.4	92.5	86.3	95.0	92.4	102.3	88.4	92.5	81.6	93.1	89.7	99.9	88.4	91.8			
0310	Concrete Forming & Accessories	100.0	100.0	100.0	92.8	49.1	55.0	93.5	71.3	74.3	89.3	44.5	50.5	97.0	47.3	54.0	97.9	43.9	51.2			
0320	Concrete Reinforcing	100.0	100.0	100.0	92.3	82.2	87.3	92.3	82.6	87.5	97.6	48.2	73.0	92.3	77.1	84.8	97.6	47.6	72.7			
0330	Cast-in-Place Concrete	100.0	100.0	100.0	94.9	51.3	77.0	101.3	65.9	86.7	92.6	56.7	77.8	94.4	64.2	82.0	92.6	47.7	74.1			
03	**CONCRETE**	100.0	100.0	100.0	95.3	57.7	76.7	94.2	72.6	83.5	96.1	51.3	74.0	91.1	60.4	75.9	95.6	47.8	72.0			
04	**MASONRY**	100.0	100.0	100.0	91.9	69.5	78.2	94.9	77.9	84.5	96.8	50.5	68.6	93.1	53.4	68.9	98.0	41.3	63.4			
05	**METALS**	100.0	100.0	100.0	93.6	90.8	92.7	94.5	92.6	93.9	92.7	78.6	88.2	95.8	87.4	93.2	92.7	77.3	87.9			
06	**WOOD, PLASTICS & COMPOSITES**	100.0	100.0	100.0	92.5	44.6	65.2	94.4	69.8	80.3	87.4	44.3	62.8	97.6	44.9	67.5	99.1	44.3	67.8			
07	**THERMAL & MOISTURE PROTECTION**	100.0	100.0	100.0	95.2	54.9	79.0	95.5	81.0	89.7	95.2	58.8	80.6	95.0	60.6	81.2	95.2	50.9	77.4			
08	**OPENINGS**	100.0	100.0	100.0	93.6	50.3	83.1	97.1	72.2	91.0	93.7	46.1	82.1	97.4	51.8	86.3	93.7	46.1	82.1			
0920	Plaster & Gypsum Board	100.0	100.0	100.0	93.6	43.3	58.4	96.8	69.3	77.5	91.1	43.0	57.4	98.4	43.6	60.0	101.5	43.0	60.6			
0950, 0980	Ceilings & Acoustic Treatment	100.0	100.0	100.0	88.9	43.3	58.5	96.4	69.3	78.3	88.9	43.0	58.3	92.6	43.6	59.9	88.9	43.0	58.3			
0960	Flooring	100.0	100.0	100.0	102.4	41.7	84.3	104.4	62.5	91.9	107.8	58.0	93.0	104.8	52.4	89.2	114.2	28.5	88.7			
0970, 0990	Wall Finishes & Painting/Coating	100.0	100.0	100.0	93.2	32.9	57.1	93.2	58.1	72.1	93.2	55.7	70.7	93.2	59.3	72.9	93.2	55.7	70.7			
09	**FINISHES**	100.0	100.0	100.0	94.6	44.9	66.7	97.2	67.8	80.7	98.1	47.7	69.9	96.1	48.7	69.5	101.0	41.9	67.9			
COVERS	**DIVS. 10 - 14, 25, 28, 41, 43, 44, 46**	100.0	100.0	100.0	100.0	71.3	94.1	100.0	88.8	97.7	100.0	43.1	88.4	100.0	45.0	88.8	100.0	43.2	88.4			
21, 22, 23	**FIRE SUPPRESSION, PLUMBING & HVAC**	100.0	100.0	100.0	99.3	58.3	82.5	99.9	64.7	85.5	96.7	35.1	71.5	99.9	42.9	76.6	96.7	35.0	71.4			
26, 27, 3370	**ELECTRICAL, COMMUNICATIONS & UTIL.**	100.0	100.0	100.0	95.8	59.7	76.8	103.3	61.3	81.2	97.9	41.9	68.4	97.6	64.4	80.1	96.7	59.8	77.3			
MF2010	**WEIGHTED AVERAGE**	100.0	100.0	100.0	95.9	63.8	81.7	97.4	73.8	87.0	96.3	51.4	76.5	96.5	59.4	80.1	96.4	51.4	76.6			

	DIVISION	ALABAMA																	
		EVERGREEN			GADSDEN			HUNTSVILLE			JASPER			MOBILE			MONTGOMERY		
		364			359			357 - 358			355			365 - 366			360 - 361		
		MAT.	INST.	TOTAL	MAT.	INST.	TOTAL	MAT.	INST.	TOTAL	MAT.	INST.	TOTAL	MAT.	INST.	TOTAL	MAT.	INST.	TOTAL
015433	**CONTRACTOR EQUIPMENT**		98.9	98.9		101.7	101.7		101.7	101.7		101.7	101.7		98.9	98.9		98.9	98.9
0241, 31 - 34	**SITE & INFRASTRUCTURE, DEMOLITION**	102.7	88.7	92.9	86.9	94.0	91.8	81.4	94.1	90.3	86.4	93.7	91.6	94.9	89.4	91.0	93.6	89.3	90.6
0310	Concrete Forming & Accessories	85.8	46.2	51.5	89.4	43.2	49.4	97.0	63.3	67.9	94.5	35.9	43.8	97.1	56.1	61.6	96.6	45.2	52.1
0320	Concrete Reinforcing	97.7	48.2	73.0	97.9	81.4	89.7	92.3	77.6	85.0	92.3	81.3	86.9	95.3	80.7	88.0	95.3	81.3	88.3
0330	Cast-in-Place Concrete	92.6	59.3	78.9	94.4	50.4	76.3	91.9	62.6	79.9	104.6	45.4	80.2	97.1	68.6	85.3	99.1	48.1	78.1
03	**CONCRETE**	96.3	52.9	74.9	95.1	54.7	75.1	90.0	67.0	78.6	98.5	49.6	74.4	92.9	66.5	79.8	93.8	54.7	74.5
04	**MASONRY**	96.8	55.3	71.5	91.4	44.7	62.9	94.5	59.4	73.1	88.4	42.9	60.6	95.2	54.9	70.6	93.4	40.9	61.4
05	**METALS**	92.7	78.5	88.2	93.7	90.0	92.6	95.8	89.0	93.7	93.7	89.3	92.3	94.7	91.1	93.5	94.5	89.7	93.0
06	**WOOD, PLASTICS & COMPOSITES**	83.8	44.3	61.2	88.0	40.4	60.9	97.6	64.1	78.5	94.4	33.0	59.3	97.8	55.2	73.5	94.0	44.3	65.6
07	**THERMAL & MOISTURE PROTECTION**	95.2	59.1	80.7	95.1	65.3	83.1	94.9	73.7	86.4	95.1	47.1	75.8	95.0	72.4	85.9	94.8	64.1	82.5
08	**OPENINGS**	93.7	46.1	82.1	93.6	49.2	82.8	97.4	62.2	88.8	93.6	48.1	82.5	97.4	59.8	88.2	97.4	53.5	86.7
0920	Plaster & Gypsum Board	90.3	43.0	57.2	90.7	39.0	54.6	98.4	63.4	73.9	94.0	31.3	50.2	98.4	54.2	67.5	97.3	43.0	59.3
0950, 0980	Ceilings & Acoustic Treatment	88.9	43.0	58.3	88.9	39.0	55.7	94.6	63.4	73.8	88.9	31.3	50.6	94.6	54.2	67.7	98.2	43.0	61.4
0960	Flooring	105.4	58.0	91.3	100.7	46.6	84.6	104.8	56.0	90.3	103.1	41.7	84.8	113.6	58.0	97.1	111.4	30.7	87.4
0970, 0990	Wall Finishes & Painting/Coating	93.2	55.7	70.7	93.2	63.4	75.4	93.2	61.1	74.0	93.2	40.3	61.5	96.9	56.7	72.8	93.2	55.7	70.7
09	**FINISHES**	97.4	48.9	70.2	93.6	44.3	66.0	96.6	61.3	76.8	94.6	36.0	61.7	101.2	55.4	75.5	100.7	42.2	67.9
COVERS	**DIVS. 10 - 14, 25, 28, 41, 43, 44, 46**	100.0	44.7	88.7	100.0	80.8	96.1	100.0	85.4	97.0	100.0	41.4	88.0	100.0	82.0	96.3	100.0	80.7	96.1
21, 22, 23	**FIRE SUPPRESSION, PLUMBING & HVAC**	96.7	37.5	72.5	101.9	37.9	75.7	99.9	54.5	81.3	101.9	55.4	82.9	99.9	63.5	85.0	100.0	35.8	73.7
26, 27, 3370	**ELECTRICAL, COMMUNICATIONS & UTIL.**	95.1	41.9	67.1	97.6	61.3	78.5	98.8	64.4	80.7	97.1	61.3	78.3	98.8	59.6	78.2	99.4	63.4	80.5
MF2010	**WEIGHTED AVERAGE**	95.9	52.8	77.0	96.5	57.1	79.1	96.6	67.4	83.7	96.8	56.9	79.2	97.5	66.8	84.0	97.5	56.1	79.2

	DIVISION	ALABAMA									ALASKA								
		PHENIX CITY			SELMA			TUSCALOOSA			ANCHORAGE			FAIRBANKS			JUNEAU		
		368			367			354			995 - 996			997			998		
		MAT.	INST.	TOTAL	MAT.	INST.	TOTAL	MAT.	INST.	TOTAL	MAT.	INST.	TOTAL	MAT.	INST.	TOTAL	MAT.	INST.	TOTAL
015433	**CONTRACTOR EQUIPMENT**		98.9	98.9		98.9	98.9		101.7	101.7		114.3	114.3		114.3	114.3		114.3	114.3
0241, 31 - 34	**SITE & INFRASTRUCTURE, DEMOLITION**	106.5	89.3	94.5	99.7	89.2	92.4	81.9	94.1	90.5	123.0	129.2	127.4	115.6	129.2	125.2	122.5	129.2	127.2
0310	Concrete Forming & Accessories	92.8	40.1	47.2	90.5	44.4	50.6	96.9	48.8	55.3	140.2	118.7	121.6	140.3	118.4	121.4	139.4	118.6	121.4
0320	Concrete Reinforcing	97.6	64.3	81.0	97.6	81.0	89.4	92.3	81.4	86.9	147.5	111.2	129.5	142.8	111.3	127.1	110.8	111.2	111.0
0330	Cast-in-Place Concrete	92.6	48.7	74.5	92.6	47.9	74.2	95.7	50.4	77.1	143.0	117.5	132.5	135.2	117.9	128.1	142.9	117.5	132.5
03	**CONCRETE**	98.8	49.5	74.5	95.1	54.2	74.9	91.8	57.1	74.7	131.1	116.1	123.7	118.8	116.2	117.5	121.3	116.1	118.8
04	**MASONRY**	96.8	42.7	63.8	100.1	40.2	63.6	93.3	47.8	65.5	188.3	123.9	149.0	190.1	123.9	149.7	173.8	123.9	143.4
05	**METALS**	92.6	83.4	89.7	92.6	88.6	91.3	95.0	90.0	93.4	121.6	102.8	115.7	118.4	103.0	113.6	118.4	102.8	113.5
06	**WOOD, PLASTICS & COMPOSITES**	92.2	37.0	60.7	89.2	44.3	63.5	97.6	46.4	68.4	116.8	117.4	117.1	121.5	116.8	118.9	116.8	117.4	117.1
07	**THERMAL & MOISTURE PROTECTION**	95.5	64.0	82.8	95.1	54.1	78.6	95.1	67.4	83.9	185.7	116.4	157.8	179.2	117.6	154.4	176.5	116.4	152.3
08	**OPENINGS**	93.6	45.5	81.9	93.6	53.5	83.9	97.4	59.5	88.1	117.2	114.3	116.5	114.1	114.0	114.0	113.3	114.3	113.5
0920	Plaster & Gypsum Board	94.9	35.5	53.3	92.8	43.0	57.9	98.4	45.1	61.1	137.7	117.7	123.7	149.1	117.2	126.8	124.0	117.7	119.6
0950, 0980	Ceilings & Acoustic Treatment	88.9	35.5	53.3	88.9	43.0	58.3	94.6	45.1	61.7	138.3	117.7	124.6	123.7	117.2	119.4	120.1	117.7	118.5
0960	Flooring	110.1	30.7	86.5	108.5	29.5	84.9	104.8	38.9	85.2	159.5	130.6	150.9	150.4	130.6	144.5	150.6	130.6	144.7
0970, 0990	Wall Finishes & Painting/Coating	93.2	55.7	70.7	93.2	55.7	70.7	93.2	48.0	66.1	154.3	115.8	131.2	153.8	124.6	136.3	154.3	115.8	131.2
09	**FINISHES**	99.7	38.2	65.2	98.2	42.0	66.7	96.5	45.7	68.0	149.2	121.0	133.4	142.7	121.7	130.9	139.2	121.0	129.0
COVERS	**DIVS. 10 - 14, 25, 28, 41, 43, 44, 46**	100.0	80.3	96.0	100.0	43.1	88.4	100.0	82.6	96.4	100.0	105.8	101.2	100.0	105.7	101.2	100.0	105.8	101.2
21, 22, 23	**FIRE SUPPRESSION, PLUMBING & HVAC**	96.7	36.2	71.9	96.7	35.3	71.6	99.9	35.6	73.6	100.3	104.3	101.9	100.3	107.6	103.3	100.3	103.2	101.5
26, 27, 3370	**ELECTRICAL, COMMUNICATIONS & UTIL.**	97.3	72.5	84.3	96.4	41.9	67.7	98.3	61.3	78.8	133.9	116.5	124.8	145.4	116.5	130.2	130.4	116.5	123.1
MF2010	**WEIGHTED AVERAGE**	96.8	55.4	78.5	96.1	51.3	76.4	96.5	58.1	79.6	123.8	114.6	119.8	122.0	115.4	119.1	119.7	114.3	117.3

DIVISION		ALASKA			ARIZONA														
		KETCHIKAN			CHAMBERS			FLAGSTAFF			GLOBE			KINGMAN			MESA/TEMPE		
		999			865			860			855			864			852		
		MAT.	INST.	TOTAL	MAT.	INST.	TOTAL	MAT.	INST.	TOTAL	MAT.	INST.	TOTAL	MAT.	INST.	TOTAL	MAT.	INST.	TOTAL
015433	CONTRACTOR EQUIPMENT		114.3	114.3		92.2	92.2		92.2	92.2		93.6	93.6		92.2	92.2		93.6	93.6
0241, 31 - 34	SITE & INFRASTRUCTURE, DEMOLITION	171.1	129.2	141.7	66.2	94.0	85.7	84.1	95.9	92.4	99.6	95.7	96.8	66.2	96.0	87.1	90.1	97.3	95.1
0310	Concrete Forming & Accessories	132.1	118.6	120.5	97.8	44.8	51.9	103.0	67.0	71.9	97.7	44.6	51.8	96.1	66.4	70.4	100.5	66.2	70.8
0320	Concrete Reinforcing	111.4	111.2	111.3	102.1	77.0	89.6	101.9	84.1	93.1	105.5	76.9	91.3	102.2	84.1	93.2	106.2	84.1	95.2
0330	Cast-in-Place Concrete	273.3	117.5	209.2	91.5	53.2	75.8	91.6	74.3	84.5	91.7	53.2	75.9	91.3	70.8	82.9	92.5	70.8	83.6
03	CONCRETE	199.2	116.1	158.1	96.9	54.1	75.8	115.6	72.8	94.5	109.9	54.1	82.3	96.5	71.3	84.0	101.5	71.2	86.5
04	MASONRY	198.9	123.9	153.1	92.6	40.6	60.8	92.8	61.9	73.9	106.0	40.5	66.0	92.6	60.4	73.0	106.2	60.4	78.2
05	METALS	118.6	102.8	113.6	94.7	67.6	86.2	95.2	75.8	89.1	91.7	67.9	84.2	95.3	75.7	89.1	92.0	75.9	87.0
06	WOOD, PLASTICS & COMPOSITES	113.0	117.4	115.5	96.4	42.6	65.7	102.0	67.0	82.0	94.0	42.7	64.7	92.2	67.0	77.8	97.4	67.1	80.1
07	THERMAL & MOISTURE PROTECTION	182.3	116.4	155.8	94.3	50.0	76.5	96.1	67.0	84.4	101.0	47.6	79.5	94.3	63.3	81.8	100.3	63.9	85.6
08	OPENINGS	116.0	114.3	115.6	106.0	50.0	92.4	106.2	72.1	97.9	102.3	50.1	89.6	106.2	70.9	97.6	102.4	70.9	94.8
0920	Plaster & Gypsum Board	137.8	117.7	123.8	86.5	40.9	54.6	89.7	66.1	73.2	92.2	40.9	56.3	80.1	66.1	70.3	94.7	66.1	74.7
0950, 0980	Ceilings & Acoustic Treatment	117.6	117.7	117.7	110.4	40.9	64.1	111.3	66.1	81.2	95.8	40.9	59.2	111.3	66.1	81.2	95.8	66.1	76.0
0960	Flooring	150.4	130.6	144.5	92.5	41.0	77.1	94.7	41.2	78.7	99.1	41.0	81.8	91.5	55.6	80.8	100.4	51.3	85.8
0970, 0990	Wall Finishes & Painting/Coating	153.8	115.8	131.1	93.2	38.6	60.5	93.2	57.2	71.6	101.5	38.6	63.8	93.2	57.2	71.6	101.5	57.2	74.9
09	FINISHES	144.8	121.0	131.5	96.1	42.2	65.9	99.4	60.5	77.6	95.9	42.3	65.8	95.2	62.6	76.9	95.4	61.8	76.6
COVERS	DIVS. 10 - 14, 25, 28, 41, 43, 44, 46	100.0	105.8	101.2	100.0	79.0	95.7	100.0	83.9	96.7	100.0	79.2	95.7	100.0	83.4	96.6	100.0	83.6	96.7
21, 22, 23	FIRE SUPPRESSION, PLUMBING & HVAC	98.4	103.2	100.4	96.3	72.7	86.7	100.2	81.6	92.6	94.2	72.4	85.3	96.3	80.6	89.9	100.2	75.5	90.1
26, 27, 3370	ELECTRICAL, COMMUNICATIONS & UTIL.	145.3	116.5	130.2	101.4	67.2	83.4	100.3	61.1	79.6	96.0	62.5	78.4	101.4	61.1	80.2	92.7	61.1	76.0
MF2010	WEIGHTED AVERAGE	132.8	114.3	124.7	96.8	62.0	81.5	100.7	72.7	88.4	98.2	61.4	82.0	96.8	72.3	86.0	98.1	71.3	86.3

DIVISION		ARIZONA												ARKANSAS					
		PHOENIX			PRESCOTT			SHOW LOW			TUCSON			BATESVILLE			CAMDEN		
		850,853			863			859			856 - 857			725			717		
		MAT.	INST.	TOTAL	MAT.	INST.	TOTAL	MAT.	INST.	TOTAL	MAT.	INST.	TOTAL	MAT.	INST.	TOTAL	MAT.	INST.	TOTAL
015433	CONTRACTOR EQUIPMENT		94.2	94.2		92.2	92.2		93.6	93.6		93.6	93.6		88.7	88.7		88.7	88.7
0241, 31 - 34	SITE & INFRASTRUCTURE, DEMOLITION	90.6	97.8	95.6	72.5	94.0	87.6	101.7	95.7	97.5	85.9	97.5	94.1	71.8	84.9	81.0	72.4	84.4	80.8
0310	Concrete Forming & Accessories	101.4	69.4	73.7	99.2	44.5	51.9	104.3	44.9	52.9	101.0	68.9	73.2	83.9	46.8	51.8	83.7	31.8	38.8
0320	Concrete Reinforcing	104.5	84.4	94.5	101.9	77.2	89.6	106.2	77.1	91.7	87.7	84.1	85.9	91.8	63.7	77.8	93.1	63.5	78.4
0330	Cast-in-Place Concrete	92.5	74.5	85.1	91.6	53.1	75.8	91.8	53.3	75.9	95.2	74.3	86.6	81.1	46.8	67.0	83.2	39.5	65.2
03	CONCRETE	101.1	73.9	87.7	101.9	54.0	78.3	112.1	54.3	83.5	99.6	73.6	86.7	87.0	50.8	69.1	88.7	41.6	65.4
04	MASONRY	94.1	64.9	76.3	92.7	45.7	64.0	106.0	42.0	67.0	92.1	61.9	73.7	105.3	42.5	66.9	113.5	32.7	64.2
05	METALS	93.5	77.3	88.4	95.2	67.7	86.6	91.5	68.2	84.2	92.8	76.0	87.5	97.3	66.2	87.5	97.3	65.7	87.4
06	WOOD, PLASTICS & COMPOSITES	98.4	69.9	82.1	97.8	42.6	66.3	101.9	42.7	68.1	97.7	69.9	81.8	88.3	47.5	65.0	88.0	31.0	55.5
07	THERMAL & MOISTURE PROTECTION	99.9	67.6	86.9	94.8	49.4	76.6	101.2	50.0	80.6	101.3	63.4	86.1	97.3	45.6	76.5	97.1	37.0	72.9
08	OPENINGS	104.0	73.6	96.6	106.2	50.0	92.5	101.4	50.1	88.9	98.4	73.6	92.4	95.1	46.2	83.2	91.5	40.7	79.1
0920	Plaster & Gypsum Board	97.1	69.0	77.4	86.7	40.9	54.7	97.2	40.9	57.8	100.8	69.0	78.5	77.2	46.4	55.6	77.2	29.3	43.7
0950, 0980	Ceilings & Acoustic Treatment	103.1	69.0	80.4	109.5	40.9	63.8	95.8	40.9	59.2	96.8	69.0	78.2	88.2	46.4	60.3	88.2	29.3	49.0
0960	Flooring	100.6	53.6	86.6	93.3	41.0	77.7	101.9	46.7	85.5	91.1	41.2	76.3	102.2	61.2	90.0	102.1	41.0	83.9
0970, 0990	Wall Finishes & Painting/Coating	101.5	63.4	78.6	93.2	38.6	60.5	101.5	38.6	63.8	102.5	57.2	75.4	100.7	43.0	66.1	100.7	49.4	70.0
09	FINISHES	97.4	65.1	79.3	96.8	42.2	66.2	97.6	43.2	67.1	93.6	62.2	76.0	90.2	48.8	67.0	90.3	34.6	59.1
COVERS	DIVS. 10 - 14, 25, 28, 41, 43, 44, 46	100.0	84.5	96.8	100.0	78.9	95.7	100.0	79.2	95.7	100.0	84.4	96.8	100.0	41.1	88.0	100.0	37.0	87.1
21, 22, 23	FIRE SUPPRESSION, PLUMBING & HVAC	100.1	81.6	92.5	100.2	69.9	87.8	94.2	72.7	85.4	100.2	72.2	88.7	94.2	52.7	77.2	94.2	50.0	76.1
26, 27, 3370	ELECTRICAL, COMMUNICATIONS & UTIL.	99.8	69.5	83.9	99.9	61.0	79.4	93.1	62.4	76.9	94.9	58.7	75.8	97.8	62.1	79.0	94.0	59.2	75.7
MF2010	WEIGHTED AVERAGE	98.7	75.4	88.4	98.5	61.1	82.0	98.3	61.8	82.2	96.9	70.9	85.5	94.6	55.3	77.3	94.4	49.5	74.6

DIVISION		ARKANSAS																	
		FAYETTEVILLE			FORT SMITH			HARRISON			HOT SPRINGS			JONESBORO			LITTLE ROCK		
		727			729			726			719			724			720 - 722		
		MAT.	INST.	TOTAL	MAT.	INST.	TOTAL	MAT.	INST.	TOTAL	MAT.	INST.	TOTAL	MAT.	INST.	TOTAL	MAT.	INST.	TOTAL
015433	CONTRACTOR EQUIPMENT		88.7	88.7		88.7	88.7		88.7	88.7		88.7	88.7		107.3	107.3		88.7	88.7
0241, 31 - 34	SITE & INFRASTRUCTURE, DEMOLITION	71.3	86.3	81.8	76.0	86.2	83.2	76.2	84.9	82.3	75.5	85.8	82.7	94.2	100.4	98.5	82.3	86.3	85.1
0310	Concrete Forming & Accessories	79.6	40.2	45.5	98.0	57.3	62.8	88.3	46.8	52.4	81.2	37.1	43.1	87.4	50.2	55.2	93.2	61.6	65.9
0320	Concrete Reinforcing	91.8	68.5	80.2	92.9	69.2	81.1	91.4	63.6	77.5	91.3	63.6	77.5	88.8	65.9	77.4	93.1	64.4	78.8
0330	Cast-in-Place Concrete	81.1	51.1	68.8	92.7	65.7	81.6	90.0	44.8	71.4	85.1	40.0	66.5	88.3	56.8	75.4	93.2	65.8	81.9
03	CONCRETE	86.7	50.2	68.6	94.9	62.9	79.1	94.4	50.0	72.5	92.5	44.1	68.6	91.6	56.9	74.5	94.8	64.0	79.6
04	MASONRY	95.0	43.1	63.3	100.5	51.2	70.4	105.7	40.7	66.0	85.3	34.9	54.5	96.4	43.7	64.2	97.4	51.2	69.2
05	METALS	97.3	68.3	88.2	99.6	70.0	90.3	98.4	65.9	88.2	97.3	66.2	87.5	94.0	79.5	89.4	95.6	68.5	87.1
06	WOOD, PLASTICS & COMPOSITES	84.0	37.1	57.2	104.7	59.6	79.0	93.7	47.5	67.3	85.1	37.3	57.8	91.7	50.4	68.2	99.2	65.3	79.8
07	THERMAL & MOISTURE PROTECTION	98.1	47.1	77.6	98.4	55.1	81.0	97.6	44.3	76.1	97.3	39.3	74.0	102.1	51.0	81.6	96.0	55.7	79.8
08	OPENINGS	95.1	45.7	83.1	96.0	56.1	86.3	96.0	46.8	84.0	91.4	42.5	79.5	98.2	54.6	87.6	96.0	59.3	87.1
0920	Plaster & Gypsum Board	74.7	35.6	47.4	82.6	58.8	66.0	81.4	46.4	56.9	75.5	35.8	47.8	92.6	49.1	62.2	88.7	64.7	71.9
0950, 0980	Ceilings & Acoustic Treatment	88.2	35.6	53.2	90.0	58.8	69.2	90.0	46.4	60.9	88.2	35.8	53.3	92.4	49.1	63.5	95.5	64.7	75.0
0960	Flooring	99.4	61.2	88.0	109.0	63.0	95.3	104.6	61.2	91.7	100.8	61.2	89.0	73.0	55.4	67.7	109.6	63.0	95.7
0970, 0990	Wall Finishes & Painting/Coating	100.7	30.7	58.7	100.7	56.2	74.0	100.7	43.0	66.1	100.7	49.4	70.0	90.2	50.5	66.4	100.7	57.7	74.9
09	FINISHES	89.2	41.8	62.6	93.5	58.4	73.9	92.2	48.8	67.9	90.0	41.9	63.1	87.4	50.7	66.8	95.9	62.0	76.9
COVERS	DIVS. 10 - 14, 25, 28, 41, 43, 44, 46	100.0	50.7	89.9	100.0	77.8	95.5	100.0	51.2	90.0	100.0	38.0	87.3	100.0	47.0	89.2	100.0	78.4	95.6
21, 22, 23	FIRE SUPPRESSION, PLUMBING & HVAC	94.3	51.2	76.7	100.3	50.7	80.0	94.2	50.4	76.3	94.2	49.5	75.9	100.5	53.6	81.3	100.1	58.2	82.9
26, 27, 3370	ELECTRICAL, COMMUNICATIONS & UTIL.	91.2	53.4	71.3	94.9	67.3	80.3	96.2	38.6	65.9	96.3	65.9	80.3	102.3	62.2	81.2	102.6	68.4	84.6
MF2010	WEIGHTED AVERAGE	93.3	53.4	75.7	97.4	61.8	81.7	95.8	51.6	76.4	93.7	52.2	75.4	97.0	60.0	80.7	97.5	64.2	82.8

	DIVISION	ARKANSAS												CALIFORNIA					
		PINE BLUFF			RUSSELLVILLE			TEXARKANA			WEST MEMPHIS			ALHAMBRA			ANAHEIM		
		716			728			718			723			917 - 918			928		
		MAT.	INST.	TOTAL	MAT.	INST.	TOTAL	MAT.	INST.	TOTAL	MAT.	INST.	TOTAL	MAT.	INST.	TOTAL	MAT.	INST.	TOTAL
015433	**CONTRACTOR EQUIPMENT**		88.7	88.7		88.7	88.7		89.4	89.4		107.3	107.3		98.2	98.2		100.5	100.5
0241, 31 - 34	**SITE & INFRASTRUCTURE, DEMOLITION**	77.5	86.2	83.6	72.9	84.8	81.2	89.1	86.9	87.6	100.2	100.4	100.3	96.7	109.2	105.5	96.1	107.9	104.4
0310	Concrete Forming & Accessories	80.9	61.5	64.1	84.6	55.6	59.5	87.2	41.9	48.0	92.6	50.4	56.1	116.6	118.0	117.8	106.5	121.7	119.6
0320	Concrete Reinforcing	93.0	64.4	78.7	92.5	63.5	78.0	92.6	63.7	78.2	88.8	65.9	77.4	109.9	116.7	113.3	95.4	116.7	106.0
0330	Cast-in-Place Concrete	85.1	65.8	77.1	85.0	46.8	69.3	92.7	44.7	73.0	92.6	56.9	77.9	93.8	119.8	104.5	92.0	122.7	104.6
03	**CONCRETE**	93.3	63.9	78.8	90.2	54.6	72.6	91.1	47.8	69.7	99.1	57.0	78.3	104.4	117.4	110.8	100.6	120.1	110.2
04	**MASONRY**	122.0	51.2	78.8	99.8	38.9	62.6	100.5	34.5	60.2	81.8	43.7	58.5	117.4	116.5	116.9	79.3	115.8	101.5
05	**METALS**	98.1	68.4	88.8	97.3	65.8	87.4	91.0	66.3	83.3	93.0	79.7	88.8	83.2	100.3	88.6	101.8	101.6	101.7
06	**WOOD, PLASTICS & COMPOSITES**	84.6	65.3	73.6	89.6	59.6	72.5	93.1	43.8	65.0	97.7	50.4	70.7	97.3	115.5	107.7	97.0	120.1	110.2
07	**THERMAL & MOISTURE PROTECTION**	97.4	55.7	80.6	98.3	45.4	77.0	98.1	45.9	77.1	102.5	51.0	81.8	96.6	116.4	104.6	100.0	121.3	108.6
08	**OPENINGS**	91.4	59.3	83.6	95.1	55.4	85.5	96.5	46.9	84.4	98.2	54.6	87.6	88.2	117.8	95.4	100.9	120.5	105.6
0920	Plaster & Gypsum Board	75.1	64.7	67.8	77.2	58.8	64.3	79.4	42.6	53.7	95.6	49.1	63.0	98.6	116.2	110.9	100.2	120.7	114.5
0950, 0980	Ceilings & Acoustic Treatment	88.2	64.7	72.6	88.2	58.8	68.6	91.9	42.6	59.0	90.4	49.1	62.8	96.4	116.2	109.6	107.6	120.7	116.3
0960	Flooring	100.6	63.0	89.4	101.8	61.2	89.7	102.8	52.8	87.9	75.3	55.4	69.3	97.5	103.9	99.4	108.7	112.6	109.8
0970, 0990	Wall Finishes & Painting/Coating	100.7	57.7	74.9	100.7	36.5	62.3	100.7	31.7	59.3	90.2	50.5	66.4	110.6	108.8	109.5	104.9	108.6	107.1
09	**FINISHES**	90.0	62.0	74.3	90.4	55.2	70.7	92.5	42.1	64.3	88.7	50.7	67.4	97.5	114.3	106.9	103.3	118.5	111.8
COVERS	**DIVS. 10 - 14, 25, 28, 41, 43, 44, 46**	100.0	78.4	95.6	100.0	42.4	88.2	100.0	33.9	86.5	100.0	47.0	89.2	100.0	110.2	102.1	100.0	111.2	102.3
21, 22, 23	**FIRE SUPPRESSION, PLUMBING & HVAC**	100.3	53.0	80.9	94.3	52.5	77.2	100.3	56.3	82.3	94.5	65.1	82.5	94.0	109.1	100.2	100.0	116.6	106.8
26, 27, 3370	**ELECTRICAL, COMMUNICATIONS & UTIL.**	94.2	68.4	80.6	94.9	46.6	69.4	96.2	37.5	65.3	104.1	59.1	80.4	117.5	113.6	115.5	92.4	105.2	99.1
MF2010	**WEIGHTED AVERAGE**	97.1	63.1	82.1	94.5	54.6	76.9	96.0	50.6	76.0	96.0	62.0	81.0	97.3	112.1	103.8	98.8	113.7	105.4

	DIVISION	CALIFORNIA																	
		BAKERSFIELD			BERKELEY			EUREKA			FRESNO			INGLEWOOD			LONG BEACH		
		932 - 933			947			955			936 - 938			903 - 905			906 - 908		
		MAT.	INST.	TOTAL	MAT.	INST.	TOTAL	MAT.	INST.	TOTAL	MAT.	INST.	TOTAL	MAT.	INST.	TOTAL	MAT.	INST.	TOTAL
015433	**CONTRACTOR EQUIPMENT**		98.4	98.4		98.1	98.1		98.1	98.1		98.4	98.4		98.4	98.4		98.4	98.4
0241, 31 - 34	**SITE & INFRASTRUCTURE, DEMOLITION**	97.8	105.5	103.2	116.2	105.0	108.3	107.9	102.0	103.8	100.6	104.9	103.6	91.9	103.9	100.3	99.0	103.9	102.4
0310	Concrete Forming & Accessories	106.4	121.1	119.1	119.8	147.8	144.0	116.1	130.0	128.2	102.4	131.1	127.3	111.4	118.2	117.3	106.6	118.3	116.7
0320	Concrete Reinforcing	107.3	116.6	111.9	92.2	118.1	105.1	104.1	117.3	110.7	91.1	117.3	104.2	108.2	116.7	112.4	107.3	116.7	112.0
0330	Cast-in-Place Concrete	93.8	121.6	105.2	124.0	120.4	122.5	99.8	116.4	106.6	97.0	117.2	105.3	86.6	121.6	101.0	98.5	121.6	108.0
03	**CONCRETE**	101.0	119.4	110.1	110.2	130.8	120.4	111.3	121.6	116.4	102.0	122.4	112.1	97.7	118.3	107.9	107.4	118.3	112.8
04	**MASONRY**	98.2	115.2	108.6	124.7	134.9	130.9	101.0	125.4	115.9	101.0	123.1	114.5	75.3	116.6	100.5	83.6	116.6	103.7
05	**METALS**	97.1	100.7	98.2	105.9	100.7	104.3	101.6	99.2	100.8	101.8	101.1	101.6	91.5	103.0	95.1	91.4	103.0	95.0
06	**WOOD, PLASTICS & COMPOSITES**	92.0	120.2	108.1	113.3	153.1	136.0	110.5	134.8	124.4	103.6	134.8	121.4	103.1	115.8	110.3	96.8	115.8	107.6
07	**THERMAL & MOISTURE PROTECTION**	101.2	110.9	105.1	109.7	135.9	120.3	104.1	113.7	108.0	93.9	114.8	102.3	102.8	117.9	108.9	103.1	118.2	109.2
08	**OPENINGS**	98.6	116.1	102.9	97.8	140.8	108.3	100.1	112.8	103.2	100.3	124.0	106.1	92.6	117.9	98.7	92.6	117.9	98.7
0920	Plaster & Gypsum Board	97.0	120.7	113.5	106.1	154.3	139.8	105.4	135.7	126.6	95.4	135.7	123.6	102.7	116.2	112.1	98.2	116.2	110.8
0950, 0980	Ceilings & Acoustic Treatment	116.4	120.7	119.3	109.6	154.3	139.4	113.1	135.7	128.2	110.6	135.7	127.3	102.1	116.2	111.5	102.1	116.2	111.5
0960	Flooring	109.1	110.2	109.5	113.6	126.8	117.5	113.1	120.8	115.4	115.9	138.3	122.6	99.7	103.9	100.9	97.1	103.9	99.1
0970, 0990	Wall Finishes & Painting/Coating	107.2	97.6	101.5	105.5	140.3	126.3	106.5	48.3	71.6	122.1	110.0	114.8	102.9	108.8	106.4	102.9	108.8	106.4
09	**FINISHES**	106.5	117.0	112.4	107.9	145.3	128.8	109.3	122.3	116.6	108.7	132.1	121.8	100.7	114.4	108.4	99.9	114.4	108.0
COVERS	**DIVS. 10 - 14, 25, 28, 41, 43, 44, 46**	100.0	103.8	100.8	100.0	126.2	105.4	100.0	122.3	104.6	100.0	122.3	104.6	100.0	110.7	102.2	100.0	110.7	102.2
21, 22, 23	**FIRE SUPPRESSION, PLUMBING & HVAC**	100.0	107.7	103.2	94.1	145.6	115.1	94.0	101.7	97.1	100.1	111.3	104.7	93.6	109.2	100.0	93.6	109.2	100.0
26, 27, 3370	**ELECTRICAL, COMMUNICATIONS & UTIL.**	102.5	100.1	101.3	107.0	135.8	122.2	100.2	114.6	107.8	91.8	101.5	96.9	101.6	113.6	108.0	101.3	113.6	107.8
MF2010	**WEIGHTED AVERAGE**	100.1	109.8	104.4	103.8	132.5	116.5	101.3	112.7	106.3	100.3	115.0	106.8	95.0	112.1	102.5	96.5	112.2	103.4

	DIVISION	CALIFORNIA																	
		LOS ANGELES			MARYSVILLE			MODESTO			MOJAVE			OAKLAND			OXNARD		
		900 - 902			959			953			935			946			930		
		MAT.	INST.	TOTAL	MAT.	INST.	TOTAL	MAT.	INST.	TOTAL	MAT.	INST.	TOTAL	MAT.	INST.	TOTAL	MAT.	INST.	TOTAL
015433	**CONTRACTOR EQUIPMENT**		102.1	102.1		98.1	98.1		98.1	98.1		98.4	98.4		98.1	98.1		97.2	97.2
0241, 31 - 34	**SITE & INFRASTRUCTURE, DEMOLITION**	99.4	107.6	105.2	104.2	104.5	104.4	99.0	104.8	103.1	94.3	105.5	102.2	122.4	105.0	110.2	100.9	103.6	102.8
0310	Concrete Forming & Accessories	108.4	121.9	120.1	106.5	131.1	127.8	103.1	131.4	127.6	115.5	114.4	114.5	109.0	147.8	142.6	107.1	121.7	119.7
0320	Concrete Reinforcing	109.0	116.9	113.0	104.1	117.0	110.5	107.9	117.4	112.6	109.3	116.4	112.8	94.3	118.1	106.2	107.3	116.5	111.9
0330	Cast-in-Place Concrete	93.5	122.4	105.4	111.5	117.7	114.0	99.8	117.8	107.2	83.9	121.4	99.4	117.6	122.2	119.5	97.7	122.0	107.6
03	**CONCRETE**	103.2	120.3	111.7	112.7	122.6	117.6	103.8	122.8	113.2	94.7	116.4	105.4	109.9	131.5	120.6	102.9	119.8	111.3
04	**MASONRY**	89.9	119.7	108.1	101.9	117.5	111.4	101.9	124.3	115.5	99.5	113.5	108.0	132.8	134.9	134.1	102.2	112.9	108.7
05	**METALS**	98.3	104.8	100.3	101.1	102.1	101.4	98.3	102.9	99.8	98.6	100.2	99.1	100.4	100.7	100.5	96.9	101.1	98.2
06	**WOOD, PLASTICS & COMPOSITES**	102.9	120.2	112.8	98.3	134.8	119.1	94.3	134.8	117.4	102.9	111.6	107.9	101.4	153.1	130.9	98.0	120.2	110.7
07	**THERMAL & MOISTURE PROTECTION**	102.2	121.2	109.9	103.6	116.1	108.6	103.2	116.9	108.7	100.0	108.7	103.5	107.7	135.9	119.0	103.1	118.0	109.1
08	**OPENINGS**	99.6	120.7	104.7	99.3	124.0	105.3	98.0	124.6	104.5	95.0	111.4	99.0	97.9	140.8	108.3	97.8	120.6	103.3
0920	Plaster & Gypsum Board	102.3	120.7	115.1	98.1	135.7	124.4	100.6	135.7	125.2	107.3	111.8	110.4	100.2	154.3	138.0	100.2	120.7	114.5
0950, 0980	Ceilings & Acoustic Treatment	112.3	120.7	117.9	112.1	135.7	127.8	107.6	135.7	126.3	112.1	111.8	111.9	112.5	154.3	140.4	113.7	120.7	118.3
0960	Flooring	97.4	112.6	101.9	108.7	109.4	108.9	109.1	115.5	111.0	115.2	96.4	109.6	108.7	126.8	114.1	107.6	112.6	109.1
0970, 0990	Wall Finishes & Painting/Coating	101.9	108.8	106.0	106.5	114.2	111.1	106.5	117.3	113.0	106.5	92.9	98.3	105.5	140.3	126.3	106.5	103.5	104.7
09	**FINISHES**	102.3	118.6	111.4	106.4	127.7	118.3	105.3	129.0	118.6	108.6	109.0	108.8	106.9	145.3	128.4	106.0	118.0	112.7
COVERS	**DIVS. 10 - 14, 25, 28, 41, 43, 44, 46**	100.0	111.5	102.3	100.0	122.3	104.6	100.0	122.3	104.6	100.0	102.8	100.6	100.0	126.3	105.4	100.0	111.4	102.3
21, 22, 23	**FIRE SUPPRESSION, PLUMBING & HVAC**	100.1	116.7	106.9	94.0	107.0	99.3	100.0	111.3	104.6	94.0	99.2	96.1	100.1	145.6	118.7	100.0	116.7	106.8
26, 27, 3370	**ELECTRICAL, COMMUNICATIONS & UTIL.**	99.8	120.8	110.8	96.6	101.8	99.3	99.3	108.4	104.1	90.4	96.4	93.6	106.1	140.8	124.4	96.4	105.4	101.2
MF2010	**WEIGHTED AVERAGE**	99.8	116.6	107.2	100.5	113.1	106.1	100.4	116.0	107.3	96.8	105.4	100.6	104.6	133.3	117.3	99.9	112.9	105.6

	DIVISION	CALIFORNIA																	
		PALM SPRINGS			PALO ALTO			PASADENA			REDDING			RICHMOND			RIVERSIDE		
		922			943			910 - 912			960			948			925		
		MAT.	INST.	TOTAL	MAT.	INST.	TOTAL	MAT.	INST.	TOTAL	MAT.	INST.	TOTAL	MAT.	INST.	TOTAL	MAT.	INST.	TOTAL
015433	CONTRACTOR EQUIPMENT		99.3	99.3		98.1	98.1		98.2	98.2		98.1	98.1		98.1	98.1		99.3	99.3
0241, 31 - 34	SITE & INFRASTRUCTURE, DEMOLITION	87.7	105.9	100.5	112.4	104.9	107.1	93.8	109.2	104.6	104.4	104.5	104.5	121.7	104.9	109.9	94.5	106.0	102.5
0310	Concrete Forming & Accessories	103.1	118.0	116.0	107.1	138.3	134.1	106.9	118.1	116.6	106.3	131.1	127.7	122.3	147.4	144.0	106.8	121.7	119.7
0320	Concrete Reinforcing	109.5	116.5	113.0	92.2	117.6	104.9	110.9	116.7	113.8	104.1	117.0	110.5	92.2	117.7	104.9	106.4	116.7	111.5
0330	Cast-in-Place Concrete	87.7	122.5	102.0	104.9	121.9	111.9	89.0	119.8	101.7	110.5	117.6	113.5	120.6	122.0	121.2	95.4	122.7	106.6
03	CONCRETE	95.7	118.4	106.9	99.2	127.1	113.0	99.9	117.4	108.6	111.6	122.6	117.0	111.9	131.1	121.4	101.6	120.1	110.8
04	MASONRY	77.0	113.7	99.4	106.2	128.1	119.6	103.0	116.5	111.3	102.0	117.5	111.4	124.4	128.1	126.7	78.0	115.4	100.8
05	METALS	102.4	101.0	102.0	97.8	99.6	98.4	83.2	100.4	88.6	101.4	102.1	101.6	97.9	99.8	98.5	102.0	101.6	101.9
06	WOOD, PLASTICS & COMPOSITES	91.9	115.8	105.5	98.9	140.9	122.9	84.8	115.5	102.3	98.0	134.8	119.0	117.1	153.1	137.6	97.0	120.1	110.2
07	THERMAL & MOISTURE PROTECTION	99.7	117.4	106.8	107.3	127.2	115.3	96.4	116.9	104.6	103.6	115.7	108.4	107.9	133.1	118.0	100.2	121.2	108.6
08	OPENINGS	96.7	116.4	101.5	97.9	132.5	106.3	88.3	117.8	95.4	100.2	124.0	106.0	97.9	139.2	107.9	99.5	120.5	104.6
0920	Plaster & Gypsum Board	94.9	116.2	109.8	98.1	141.7	128.6	91.1	116.2	108.7	98.4	135.7	124.5	108.5	154.3	140.5	99.5	120.7	114.3
0950, 0980	Ceilings & Acoustic Treatment	104.9	116.2	112.4	110.5	141.7	131.3	96.4	116.2	109.6	119.2	135.7	130.2	110.5	154.3	139.7	111.0	120.7	117.4
0960	Flooring	111.1	103.9	109.0	107.8	119.5	111.3	92.7	103.9	96.1	108.7	109.4	108.9	115.5	119.5	116.7	112.7	112.6	112.7
0970, 0990	Wall Finishes & Painting/Coating	103.2	116.2	111.0	105.5	137.9	124.9	110.6	108.8	109.5	106.5	114.2	111.1	105.5	137.9	124.9	103.2	108.6	106.4
09	FINISHES	101.9	115.3	109.4	105.2	136.6	122.8	95.0	114.3	105.8	107.8	127.7	118.9	109.5	143.8	128.7	104.7	118.5	112.4
COVERS	DIVS. 10 - 14, 25, 28, 41, 43, 44, 46	100.0	110.7	102.2	100.0	124.9	105.1	100.0	110.2	102.1	100.0	122.3	104.6	100.0	126.2	105.4	100.0	111.2	102.3
21, 22, 23	FIRE SUPPRESSION, PLUMBING & HVAC	94.0	105.4	98.7	94.1	139.4	112.6	94.0	109.2	100.2	100.0	107.0	102.9	94.1	134.7	110.7	100.0	116.6	106.8
26, 27, 3370	ELECTRICAL, COMMUNICATIONS & UTIL.	96.2	102.4	99.5	105.9	139.3	123.5	114.4	113.6	114.0	99.7	101.8	100.8	106.5	132.4	120.2	92.5	106.5	99.9
MF2010	WEIGHTED AVERAGE	96.3	109.5	102.2	99.8	128.5	112.5	95.4	112.1	102.8	102.4	113.1	107.1	102.9	128.7	114.3	98.8	113.7	105.4

	DIVISION	CALIFORNIA																	
		SACRAMENTO			SALINAS			SAN BERNARDINO			SAN DIEGO			SAN FRANCISCO			SAN JOSE		
		942,956 - 958			939			923 - 924			919 - 921			940 - 941			951		
		MAT.	INST.	TOTAL	MAT.	INST.	TOTAL	MAT.	INST.	TOTAL	MAT.	INST.	TOTAL	MAT.	INST.	TOTAL	MAT.	INST.	TOTAL
015433	CONTRACTOR EQUIPMENT		97.7	97.7		98.4	98.4		99.3	99.3		98.2	98.2		107.9	107.9		98.8	98.8
0241, 31 - 34	SITE & INFRASTRUCTURE, DEMOLITION	98.7	111.0	107.3	113.9	105.4	108.0	74.7	105.9	96.6	100.3	102.8	102.0	124.7	111.3	115.3	130.9	100.0	109.2
0310	Concrete Forming & Accessories	107.2	134.2	130.6	109.8	134.6	131.2	110.2	117.9	116.9	104.8	109.1	108.5	108.6	148.8	143.4	109.0	148.0	142.7
0320	Concrete Reinforcing	87.3	117.5	102.4	107.9	117.8	112.8	106.4	116.5	111.4	107.4	116.6	112.0	107.4	118.4	112.9	95.0	118.1	106.5
0330	Cast-in-Place Concrete	99.5	118.3	107.2	96.5	118.1	105.4	66.0	122.5	89.2	97.5	105.5	100.8	120.6	123.9	121.9	115.9	122.4	118.6
03	CONCRETE	100.0	123.9	111.8	112.8	124.4	118.6	76.5	118.3	97.1	105.0	108.6	106.8	113.4	133.1	123.2	109.8	132.1	120.8
04	MASONRY	108.2	124.3	118.0	99.0	129.5	117.6	84.8	113.7	102.5	93.4	112.9	105.3	133.3	138.2	136.3	129.7	135.1	133.0
05	METALS	96.0	97.1	96.4	101.2	104.0	102.1	102.0	100.8	101.6	98.5	101.9	99.6	106.6	110.6	107.9	96.5	108.6	100.3
06	WOOD, PLASTICS & COMPOSITES	94.8	138.1	119.5	102.3	137.9	122.6	100.7	115.8	109.3	98.4	105.2	102.3	101.4	153.3	131.0	106.6	153.0	133.0
07	THERMAL & MOISTURE PROTECTION	116.3	122.5	118.8	100.7	127.8	111.6	98.8	117.4	106.3	101.2	104.8	102.7	109.5	138.1	121.0	99.6	138.9	115.4
08	OPENINGS	110.8	126.4	114.6	99.0	132.6	107.2	96.8	116.4	101.6	95.9	110.1	99.3	102.4	140.9	111.7	90.4	140.7	102.6
0920	Plaster & Gypsum Board	94.7	139.0	125.6	101.5	139.0	127.7	101.5	116.2	111.8	98.9	105.3	103.4	102.9	154.3	138.8	95.4	154.3	136.6
0950, 0980	Ceilings & Acoustic Treatment	110.5	139.0	129.5	112.1	139.0	130.0	107.6	116.2	113.3	103.7	105.3	104.8	120.5	154.3	143.0	103.7	154.3	137.4
0960	Flooring	107.7	121.8	111.9	109.9	125.5	114.5	114.7	103.9	111.5	100.5	112.6	104.1	108.7	126.8	114.1	101.8	126.8	109.2
0970, 0990	Wall Finishes & Painting/Coating	103.1	117.3	111.6	107.6	140.3	127.2	103.2	104.0	103.6	107.1	108.8	108.1	105.5	152.7	133.8	106.8	140.3	126.9
09	FINISHES	103.6	132.1	119.5	108.5	135.1	123.4	103.1	113.9	109.2	100.5	109.0	105.3	109.2	146.8	130.2	103.6	145.2	126.9
COVERS	DIVS. 10 - 14, 25, 28, 41, 43, 44, 46	100.0	123.0	104.7	100.0	122.7	104.6	100.0	103.1	100.6	100.0	108.7	101.8	100.0	126.8	105.5	100.0	126.0	105.3
21, 22, 23	FIRE SUPPRESSION, PLUMBING & HVAC	100.0	119.8	108.1	94.0	117.9	103.8	94.0	105.5	98.7	100.0	115.2	106.2	100.1	167.3	127.6	100.0	144.8	118.3
26, 27, 3370	ELECTRICAL, COMMUNICATIONS & UTIL.	101.2	113.3	107.6	91.8	123.6	108.5	96.2	101.6	99.0	99.2	99.1	99.1	106.1	157.1	133.0	102.5	153.2	129.2
MF2010	WEIGHTED AVERAGE	101.7	119.2	109.5	100.3	121.8	109.8	94.3	109.0	100.8	99.6	108.0	103.3	106.8	142.4	122.5	102.5	135.3	116.9

	DIVISION	CALIFORNIA																	
		SAN LUIS OBISPO			SAN MATEO			SAN RAFAEL			SANTA ANA			SANTA BARBARA			SANTA CRUZ		
		934			944			949			926 - 927			931			950		
		MAT.	INST.	TOTAL	MAT.	INST.	TOTAL	MAT.	INST.	TOTAL	MAT.	INST.	TOTAL	MAT.	INST.	TOTAL	MAT.	INST.	TOTAL
015433	CONTRACTOR EQUIPMENT		98.4	98.4		98.1	98.1		98.2	98.2		99.3	99.3		98.4	98.4		98.8	98.8
0241, 31 - 34	SITE & INFRASTRUCTURE, DEMOLITION	106.0	105.5	105.7	119.2	104.9	109.2	111.4	110.9	111.0	86.3	105.9	100.1	100.8	105.6	104.2	130.4	99.7	108.9
0310	Concrete Forming & Accessories	117.1	117.8	117.7	112.9	147.4	142.7	116.2	147.2	143.1	110.4	118.0	117.0	108.0	121.7	119.8	109.0	134.7	131.2
0320	Concrete Reinforcing	109.3	116.4	112.9	92.2	117.8	105.0	92.8	117.8	105.3	110.0	116.5	113.3	107.3	116.7	112.0	117.6	117.9	117.7
0330	Cast-in-Place Concrete	103.6	121.5	111.0	116.8	122.0	118.9	135.9	121.4	129.9	84.2	122.5	99.9	97.3	121.8	107.4	115.1	120.0	117.1
03	CONCRETE	110.8	117.9	114.3	108.3	131.2	119.6	129.7	130.8	130.2	93.4	118.4	105.7	102.8	119.8	111.2	113.1	125.4	119.2
04	MASONRY	100.9	113.5	108.6	124.1	131.5	128.6	101.8	131.6	120.0	74.4	114.0	98.6	99.6	116.1	109.7	133.5	129.6	131.1
05	METALS	99.2	100.4	99.6	97.7	100.0	98.4	99.2	98.3	98.9	102.0	101.0	101.7	97.3	101.4	98.6	103.7	107.4	104.8
06	WOOD, PLASTICS & COMPOSITES	105.1	115.9	111.3	106.5	153.1	133.1	102.3	152.9	131.2	102.3	115.8	110.0	98.0	120.2	110.7	106.6	138.1	124.5
07	THERMAL & MOISTURE PROTECTION	100.9	114.1	106.2	107.7	134.1	118.3	111.7	130.3	119.2	100.0	117.5	107.0	100.2	117.0	107.0	99.5	129.5	111.6
08	OPENINGS	97.1	113.7	101.1	97.8	139.2	107.9	108.2	139.0	115.7	96.1	116.4	101.0	98.6	120.6	104.0	91.7	132.6	101.7
0920	Plaster & Gypsum Board	108.9	116.2	114.0	103.9	154.3	139.2	105.5	154.3	139.6	102.3	116.2	112.0	100.2	120.7	114.5	101.4	139.0	127.7
0950, 0980	Ceilings & Acoustic Treatment	112.1	116.2	114.8	110.5	154.3	139.7	118.7	154.3	142.4	107.6	116.2	113.3	113.7	120.7	118.3	109.4	139.0	129.1
0960	Flooring	116.2	96.4	110.3	110.6	119.5	113.2	120.6	119.5	120.3	115.3	103.9	111.9	109.1	110.2	109.5	106.1	125.5	111.9
0970, 0990	Wall Finishes & Painting/Coating	106.5	98.8	101.9	105.5	137.9	124.9	101.8	137.9	123.4	103.2	104.0	103.6	106.5	103.5	104.7	107.0	140.3	126.9
09	FINISHES	110.2	111.3	110.8	107.3	143.3	127.5	109.9	143.6	128.8	104.7	113.9	109.8	106.7	117.6	112.8	106.9	135.2	122.8
COVERS	DIVS. 10 - 14, 25, 28, 41, 43, 44, 46	100.0	113.2	102.7	100.0	126.3	105.4	100.0	125.6	105.2	100.0	110.7	102.2	100.0	111.4	102.3	100.0	123.0	104.7
21, 22, 23	FIRE SUPPRESSION, PLUMBING & HVAC	94.0	105.5	98.7	94.1	136.7	111.5	94.0	144.2	114.6	94.0	105.5	98.7	100.0	116.7	106.8	100.0	117.9	107.3
26, 27, 3370	ELECTRICAL, COMMUNICATIONS & UTIL.	90.4	96.2	93.5	105.9	138.4	123.0	102.7	114.0	108.7	96.2	103.0	99.8	89.4	109.7	100.1	101.4	123.6	113.1
MF2010	WEIGHTED AVERAGE	99.4	107.9	103.2	102.1	130.3	114.5	104.4	128.7	115.1	96.1	109.5	102.0	99.2	113.9	105.7	104.4	121.9	112.1

DIVISION		CALIFORNIA															COLORADO		
		SANTA ROSA			STOCKTON			SUSANVILLE			VALLEJO			VAN NUYS			ALAMOSA		
		954			952			961			945			913 - 916			811		
		MAT.	INST.	TOTAL	MAT.	INST.	TOTAL	MAT.	INST.	TOTAL	MAT.	INST.	TOTAL	MAT.	INST.	TOTAL	MAT.	INST.	TOTAL
015433	**CONTRACTOR EQUIPMENT**		98.6	98.6		98.1	98.1		98.1	98.1		98.2	98.2		98.2	98.2		93.3	93.3
0241, 31 - 34	**SITE & INFRASTRUCTURE, DEMOLITION**	99.7	104.8	103.3	98.8	104.8	103.0	110.3	104.5	106.2	98.5	111.0	107.2	110.2	109.2	109.5	131.8	88.9	101.7
0310	Concrete Forming & Accessories	104.1	146.5	140.8	106.8	131.5	128.1	107.5	131.1	127.9	106.8	146.1	140.8	112.7	118.0	117.3	105.5	70.4	75.2
0320	Concrete Reinforcing	105.0	118.3	111.6	107.9	117.4	112.6	104.1	117.0	110.5	93.9	118.2	106.0	110.9	116.7	113.8	110.3	79.7	95.1
0330	Cast-in-Place Concrete	109.5	119.4	113.6	97.3	117.8	105.7	100.6	117.6	107.6	108.3	119.5	112.9	93.9	119.8	104.5	102.1	80.8	93.3
03	**CONCRETE**	112.3	130.3	121.2	102.8	122.9	112.7	113.2	122.6	117.8	105.2	129.7	117.3	114.1	117.4	115.8	115.6	76.1	96.1
04	**MASONRY**	100.3	134.7	121.3	101.9	124.3	115.5	100.0	117.5	110.6	78.1	134.6	112.6	117.4	116.5	116.9	126.4	74.4	94.7
05	**METALS**	102.3	105.7	103.4	98.5	103.0	99.9	100.5	102.1	101.0	99.2	99.0	99.1	82.4	100.3	88.0	96.8	81.8	92.1
06	**WOOD, PLASTICS & COMPOSITES**	93.4	152.7	127.3	99.6	134.8	119.7	99.9	134.8	119.8	91.9	152.9	126.7	92.6	115.5	105.7	94.7	70.5	80.9
07	**THERMAL & MOISTURE PROTECTION**	100.4	133.5	113.7	103.3	120.1	110.1	104.1	115.7	108.7	109.6	133.2	119.1	97.3	116.4	105.0	104.6	79.6	94.6
08	**OPENINGS**	97.7	140.6	108.1	98.0	124.6	104.5	100.1	124.0	105.9	110.3	140.7	117.7	88.1	117.8	95.3	96.5	76.8	91.7
0920	Plaster & Gypsum Board	96.9	154.3	137.0	100.6	135.7	125.2	99.4	135.7	124.8	100.1	154.3	138.0	96.2	116.2	110.2	76.8	69.4	71.6
0950, 0980	Ceilings & Acoustic Treatment	107.6	154.3	138.7	113.7	135.7	128.4	112.1	135.7	127.8	120.7	154.3	143.1	93.7	116.2	108.7	106.7	69.4	81.9
0960	Flooring	111.6	119.1	113.9	109.1	115.5	111.0	109.1	109.4	109.2	116.0	126.8	119.2	95.4	103.9	97.9	112.2	56.8	95.7
0970, 0990	Wall Finishes & Painting/Coating	103.2	140.3	125.4	106.5	117.3	113.0	106.5	114.2	111.1	102.7	140.3	125.2	110.6	108.8	109.5	107.6	24.7	57.9
09	**FINISHES**	104.0	142.8	125.7	106.6	129.0	119.2	107.5	127.7	118.8	106.8	144.2	127.8	97.2	114.3	106.8	105.7	62.5	81.5
COVERS	**DIVS. 10 - 14, 25, 28, 41, 43, 44, 46**	100.0	124.2	104.9	100.0	122.3	104.6	100.0	122.3	104.6	100.0	124.5	105.0	100.0	110.2	102.1	100.0	89.8	97.9
21, 22, 23	**FIRE SUPPRESSION, PLUMBING & HVAC**	94.0	165.3	123.2	100.0	111.3	104.6	94.0	107.0	99.3	100.1	132.3	113.3	94.0	109.1	100.2	94.0	73.5	85.6
26, 27, 3370	**ELECTRICAL, COMMUNICATIONS & UTIL.**	96.5	118.7	108.2	99.3	112.1	106.0	100.1	101.8	101.0	98.3	127.4	113.6	114.4	113.6	114.0	95.8	75.1	84.9
MF2010	**WEIGHTED AVERAGE**	100.0	134.2	115.1	100.5	116.6	107.6	101.1	113.1	106.4	101.0	128.5	113.1	98.3	112.1	104.4	101.8	75.7	90.3

DIVISION		COLORADO																	
		BOULDER			COLORADO SPRINGS			DENVER			DURANGO			FORT COLLINS			FORT MORGAN		
		803			808 - 809			800 - 802			813			805			807		
		MAT.	INST.	TOTAL	MAT.	INST.	TOTAL	MAT.	INST.	TOTAL	MAT.	INST.	TOTAL	MAT.	INST.	TOTAL	MAT.	INST.	TOTAL
015433	**CONTRACTOR EQUIPMENT**		94.6	94.6		93.1	93.1		97.5	97.5		93.3	93.3		94.6	94.6		94.6	94.6
0241, 31 - 34	**SITE & INFRASTRUCTURE, DEMOLITION**	92.4	95.1	94.3	94.2	93.1	93.4	93.3	101.3	98.9	125.1	88.9	99.7	104.1	94.7	97.5	95.0	94.6	94.8
0310	Concrete Forming & Accessories	106.6	83.5	86.6	97.1	79.1	81.6	102.8	79.4	82.5	111.6	70.6	76.1	103.8	78.1	81.5	106.9	78.3	82.1
0320	Concrete Reinforcing	104.5	79.8	92.2	103.6	84.1	93.9	103.6	84.1	93.9	110.3	79.7	95.1	104.6	79.9	92.3	104.7	79.7	92.2
0330	Cast-in-Place Concrete	102.3	81.4	93.7	105.1	84.6	96.6	99.4	84.2	93.2	117.3	80.9	102.4	115.5	80.1	100.9	100.4	80.2	92.1
03	**CONCRETE**	102.4	82.2	92.4	105.5	82.2	94.0	100.1	82.2	91.3	118.3	76.2	97.5	112.6	79.3	96.2	101.0	79.3	90.3
04	**MASONRY**	92.5	75.0	81.8	93.3	75.0	82.1	95.3	75.0	82.9	113.5	74.4	89.7	110.1	78.7	90.9	107.0	74.6	87.2
05	**METALS**	99.7	85.3	95.2	102.9	87.5	98.0	105.4	87.6	99.8	96.8	82.0	92.1	101.0	82.1	95.0	99.5	81.9	94.0
06	**WOOD, PLASTICS & COMPOSITES**	95.6	86.7	90.5	86.4	80.8	83.2	93.5	80.7	86.2	102.8	70.5	84.4	93.3	80.6	86.1	95.6	80.6	87.1
07	**THERMAL & MOISTURE PROTECTION**	102.7	81.8	94.3	103.5	81.8	94.8	101.9	75.8	91.4	104.5	79.6	94.5	103.1	73.2	91.1	102.6	81.2	94.0
08	**OPENINGS**	98.9	85.6	95.7	103.4	83.7	98.6	103.8	83.6	98.9	103.8	76.8	97.3	98.9	82.4	94.9	98.9	82.4	94.9
0920	Plaster & Gypsum Board	104.9	86.5	92.0	86.6	80.3	82.2	100.8	80.3	86.5	88.4	69.4	75.1	99.1	80.3	85.9	104.9	80.3	87.7
0950, 0980	Ceilings & Acoustic Treatment	97.3	86.5	90.1	104.6	80.3	88.4	107.8	80.3	89.5	106.7	69.4	81.9	97.3	80.3	86.0	97.3	80.3	86.0
0960	Flooring	108.7	86.9	102.2	100.4	70.7	91.5	105.5	87.9	100.2	116.8	56.8	98.9	105.3	56.8	90.9	109.1	56.8	93.5
0970, 0990	Wall Finishes & Painting/Coating	99.1	69.6	81.4	98.8	42.3	65.0	99.1	79.0	87.1	107.6	24.7	57.9	99.1	42.0	64.9	99.1	55.9	73.2
09	**FINISHES**	104.2	82.9	92.3	101.1	73.6	85.7	104.9	80.7	91.3	107.6	62.5	82.3	103.1	70.4	84.8	104.2	71.9	86.1
COVERS	**DIVS. 10 - 14, 25, 28, 41, 43, 44, 46**	100.0	90.8	98.1	100.0	90.5	98.1	100.0	90.2	98.0	100.0	89.8	97.9	100.0	90.1	98.0	100.0	90.1	98.0
21, 22, 23	**FIRE SUPPRESSION, PLUMBING & HVAC**	94.0	79.4	88.0	100.2	74.2	89.5	99.9	79.5	91.6	94.0	87.5	91.3	100.0	79.3	91.5	94.0	79.3	88.0
26, 27, 3370	**ELECTRICAL, COMMUNICATIONS & UTIL.**	99.3	85.7	92.1	103.2	83.6	92.8	104.9	85.8	94.8	95.1	72.3	83.1	99.3	85.1	91.8	99.6	85.7	92.3
MF2010	**WEIGHTED AVERAGE**	98.4	83.3	91.7	101.4	80.6	92.2	101.7	83.4	93.6	102.1	78.3	91.6	102.3	80.8	92.8	99.0	80.9	91.0

DIVISION		COLORADO																	
		GLENWOOD SPRINGS			GOLDEN			GRAND JUNCTION			GREELEY			MONTROSE			PUEBLO		
		816			804			815			806			814			810		
		MAT.	INST.	TOTAL	MAT.	INST.	TOTAL	MAT.	INST.	TOTAL	MAT.	INST.	TOTAL	MAT.	INST.	TOTAL	MAT.	INST.	TOTAL
015433	**CONTRACTOR EQUIPMENT**		96.1	96.1		94.6	94.6		96.1	96.1		94.6	94.6		94.7	94.7		93.3	93.3
0241, 31 - 34	**SITE & INFRASTRUCTURE, DEMOLITION**	140.7	95.8	109.2	104.9	94.9	97.9	124.7	95.5	104.2	91.4	94.1	93.3	134.3	92.0	104.7	116.7	91.4	98.9
0310	Concrete Forming & Accessories	102.7	78.4	81.7	99.5	78.2	81.1	110.3	77.8	82.2	101.9	82.0	84.6	102.3	78.1	81.4	107.5	79.5	83.2
0320	Concrete Reinforcing	109.2	79.7	94.5	104.7	79.7	92.2	109.5	79.6	94.6	104.5	78.3	91.4	109.0	79.7	94.4	105.5	84.1	94.9
0330	Cast-in-Place Concrete	102.0	80.3	93.1	100.4	80.2	92.1	112.9	79.6	99.2	96.5	59.5	81.3	102.0	80.2	93.1	101.3	85.3	94.7
03	**CONCRETE**	120.5	79.5	100.2	110.3	79.3	95.0	114.7	78.9	97.0	97.8	73.6	85.9	111.8	79.3	95.7	104.7	82.6	93.8
04	**MASONRY**	99.2	74.6	84.2	109.7	74.7	88.3	132.8	74.4	97.2	104.2	50.8	71.6	106.6	74.4	87.0	95.6	75.0	83.0
05	**METALS**	96.5	82.5	92.1	99.7	82.0	94.1	98.1	81.2	92.8	100.9	79.3	94.1	95.7	81.6	91.2	99.6	88.3	96.0
06	**WOOD, PLASTICS & COMPOSITES**	90.0	80.7	84.7	88.5	80.6	84.0	100.7	80.7	89.3	90.8	86.7	88.4	91.2	80.9	85.3	97.1	81.0	87.9
07	**THERMAL & MOISTURE PROTECTION**	104.5	80.7	94.9	103.5	75.3	92.2	103.5	69.8	89.9	102.5	65.9	87.8	104.6	80.7	95.0	103.0	81.3	94.3
08	**OPENINGS**	102.9	82.4	97.9	98.9	82.4	94.9	103.7	82.4	98.5	98.9	85.6	95.7	104.0	82.5	98.8	98.4	83.8	94.9
0920	Plaster & Gypsum Board	108.5	80.3	88.8	96.6	80.3	85.2	120.1	80.3	92.3	97.4	86.5	89.8	76.0	80.3	79.0	81.3	80.3	80.6
0950, 0980	Ceilings & Acoustic Treatment	105.8	80.3	88.8	97.3	80.3	86.0	105.8	80.3	88.8	97.3	86.5	90.1	106.7	80.3	89.1	115.9	80.3	92.2
0960	Flooring	111.5	52.3	93.8	103.2	56.8	89.4	116.2	56.8	98.5	104.3	56.8	90.2	114.4	47.0	94.3	113.2	87.9	105.6
0970, 0990	Wall Finishes & Painting/Coating	107.6	55.9	76.7	99.1	69.6	81.4	107.6	69.6	84.8	99.1	26.2	55.4	107.6	24.7	57.9	107.6	39.2	66.7
09	**FINISHES**	109.9	71.2	88.2	103.0	74.2	86.8	110.6	73.5	89.8	101.7	72.2	85.2	106.3	66.7	84.1	106.5	76.5	89.7
COVERS	**DIVS. 10 - 14, 25, 28, 41, 43, 44, 46**	100.0	90.3	98.0	100.0	90.1	98.0	100.0	90.2	98.0	100.0	90.8	98.1	100.0	90.6	98.1	100.0	91.0	98.2
21, 22, 23	**FIRE SUPPRESSION, PLUMBING & HVAC**	94.0	87.5	91.4	94.0	78.7	87.8	99.9	87.0	94.6	100.0	79.2	91.5	94.0	87.5	91.3	99.9	74.2	89.4
26, 27, 3370	**ELECTRICAL, COMMUNICATIONS & UTIL.**	92.4	72.4	81.9	99.6	85.7	92.3	94.8	56.1	74.4	99.3	85.7	92.1	94.8	58.8	75.8	95.8	75.1	84.9
MF2010	**WEIGHTED AVERAGE**	101.8	80.9	92.6	100.4	80.9	91.8	104.4	78.2	92.9	99.8	77.3	89.9	101.0	78.0	90.8	100.7	79.8	91.5

DIVISION		COLORADO			CONNECTICUT														
		SALIDA			BRIDGEPORT			BRISTOL			HARTFORD			MERIDEN			NEW BRITAIN		
		812			066			060			061			064			060		
		MAT.	INST.	TOTAL	MAT.	INST.	TOTAL	MAT.	INST.	TOTAL	MAT.	INST.	TOTAL	MAT.	INST.	TOTAL	MAT.	INST.	TOTAL
015433	CONTRACTOR EQUIPMENT		94.7	94.7		100.5	100.5		100.5	100.5		100.5	100.5		100.9	100.9		100.5	100.5
0241, 31 - 34	SITE & INFRASTRUCTURE, DEMOLITION	124.8	92.3	102.0	116.1	103.9	107.5	115.4	103.9	107.3	108.0	103.9	105.1	113.4	104.6	107.2	115.5	103.9	107.4
0310	Concrete Forming & Accessories	110.2	78.0	82.4	95.7	123.7	119.9	95.7	123.6	119.8	96.1	123.6	119.9	95.4	123.6	119.8	96.1	123.5	119.8
0320	Concrete Reinforcing	108.8	79.6	94.3	110.7	127.4	119.0	110.7	127.3	119.0	110.7	127.3	119.0	110.7	127.3	119.0	110.7	127.3	119.0
0330	Cast-in-Place Concrete	116.9	80.2	101.8	106.5	126.9	114.9	99.8	126.8	110.9	105.8	126.8	114.4	96.0	126.8	108.7	101.4	126.8	111.9
03	CONCRETE	113.5	79.2	96.5	107.0	125.2	116.0	103.8	125.1	114.3	106.7	125.1	115.8	102.0	125.1	113.4	104.6	125.1	114.7
04	MASONRY	134.9	74.4	98.0	102.6	132.5	120.8	95.3	132.5	118.0	95.7	132.5	118.2	94.9	132.5	117.8	96.3	132.5	118.4
05	METALS	95.4	81.5	91.0	98.8	124.1	106.8	98.8	124.0	106.7	103.6	124.0	110.0	96.4	124.0	105.1	95.6	124.0	104.5
06	WOOD, PLASTICS & COMPOSITES	98.2	80.9	88.3	95.5	122.5	110.9	95.5	122.5	110.9	94.1	122.5	110.3	95.5	122.5	110.9	95.5	122.5	110.9
07	THERMAL & MOISTURE PROTECTION	103.5	80.7	94.4	97.0	129.3	110.0	97.1	125.9	108.7	102.9	125.9	112.2	97.1	125.9	108.7	97.1	125.0	108.3
08	OPENINGS	96.6	82.5	93.2	103.4	131.2	110.2	103.4	131.2	110.2	104.0	131.2	110.7	106.0	131.2	112.1	103.4	131.2	110.2
0920	Plaster & Gypsum Board	76.4	80.3	79.1	98.4	122.6	115.3	98.4	122.6	115.3	92.9	122.6	113.6	100.1	122.6	115.8	98.4	122.6	115.3
0950, 0980	Ceilings & Acoustic Treatment	106.7	80.3	89.1	97.7	122.6	114.3	97.7	122.6	114.3	102.2	122.6	115.8	102.0	122.6	115.7	97.7	122.6	114.3
0960	Flooring	119.2	47.0	97.7	98.3	133.0	108.6	98.3	133.0	108.6	98.1	133.0	108.5	98.3	133.0	108.6	98.3	133.0	108.6
0970, 0990	Wall Finishes & Painting/Coating	107.6	24.7	57.9	96.1	116.4	108.3	96.1	116.4	108.3	96.1	116.4	108.3	96.1	116.4	108.3	96.1	116.4	108.3
09	FINISHES	106.6	66.7	84.2	101.4	124.3	114.2	101.5	124.3	114.3	100.5	124.3	113.9	102.6	124.3	114.8	101.5	124.3	114.3
COVERS	DIVS. 10 - 14, 25, 28, 41, 43, 44, 46	100.0	90.6	98.1	100.0	108.5	101.7	100.0	108.5	101.7	100.0	108.6	101.7	100.0	108.5	101.7	100.0	108.5	101.7
21, 22, 23	FIRE SUPPRESSION, PLUMBING & HVAC	94.0	74.0	85.8	100.0	117.6	107.2	100.0	117.6	107.2	100.0	117.6	107.2	94.0	117.6	103.6	100.0	117.6	107.2
26, 27, 3370	ELECTRICAL, COMMUNICATIONS & UTIL.	95.1	75.1	84.5	101.9	107.8	105.0	101.9	110.4	106.4	101.3	111.4	106.6	101.8	110.4	106.4	102.0	110.4	106.4
MF2010	WEIGHTED AVERAGE	101.5	77.4	90.9	101.7	119.7	109.6	100.9	120.0	109.3	102.0	120.1	109.9	99.2	120.0	108.4	100.6	119.9	109.1

DIVISION		CONNECTICUT																	
		NEW HAVEN			NEW LONDON			NORWALK			STAMFORD			WATERBURY			WILLIMANTIC		
		065			063			068			069			067			062		
		MAT.	INST.	TOTAL	MAT.	INST.	TOTAL	MAT.	INST.	TOTAL	MAT.	INST.	TOTAL	MAT.	INST.	TOTAL	MAT.	INST.	TOTAL
015433	CONTRACTOR EQUIPMENT		100.9	100.9		100.9	100.9		100.5	100.5		100.5	100.5		100.5	100.5		100.5	100.5
0241, 31 - 34	SITE & INFRASTRUCTURE, DEMOLITION	115.3	104.6	107.8	107.0	104.6	105.3	115.9	103.9	107.5	116.7	103.9	107.7	115.7	103.9	107.4	116.2	103.9	107.5
0310	Concrete Forming & Accessories	95.4	123.6	119.8	95.4	123.6	119.8	95.7	124.1	120.3	95.6	124.1	120.3	95.7	123.5	119.8	95.6	123.5	119.8
0320	Concrete Reinforcing	110.7	127.3	119.0	86.8	127.3	107.0	110.7	127.5	119.1	110.7	127.5	119.1	110.7	127.3	119.0	110.7	127.3	119.0
0330	Cast-in-Place Concrete	103.2	126.8	112.9	88.0	126.8	103.9	104.8	128.4	114.5	106.5	128.4	115.5	106.5	126.8	114.9	99.5	126.8	110.7
03	CONCRETE	118.7	125.1	121.9	91.5	125.1	108.1	106.2	125.9	115.9	107.0	125.9	116.3	107.0	125.1	115.9	103.7	125.1	114.2
04	MASONRY	95.5	132.5	118.1	93.8	132.5	117.4	95.0	134.0	118.8	95.8	134.0	119.1	95.8	132.5	118.2	95.1	132.5	117.9
05	METALS	95.8	124.0	104.6	95.5	124.0	104.4	98.8	124.6	106.9	98.8	124.6	106.9	98.8	124.0	106.7	98.6	124.0	106.6
06	WOOD, PLASTICS & COMPOSITES	95.5	122.5	110.9	95.5	122.5	110.9	95.5	122.5	110.9	95.5	122.5	110.9	95.5	122.5	110.9	95.5	122.5	110.9
07	THERMAL & MOISTURE PROTECTION	97.2	125.9	108.7	97.1	125.9	108.7	97.2	129.9	110.4	97.1	129.9	110.3	97.1	125.9	108.7	97.3	125.1	108.5
08	OPENINGS	103.4	131.2	110.2	106.4	131.2	112.5	103.4	131.2	110.2	103.4	131.2	110.2	103.4	131.2	110.2	106.4	131.2	112.5
0920	Plaster & Gypsum Board	98.4	122.6	115.3	98.4	122.6	115.3	98.4	122.6	115.3	98.4	122.6	115.3	98.4	122.6	115.3	98.4	122.6	115.3
0950, 0980	Ceilings & Acoustic Treatment	97.7	122.6	114.3	95.6	122.6	113.6	97.7	122.6	114.3	97.7	122.6	114.3	97.7	122.6	114.3	95.6	122.6	113.6
0960	Flooring	98.3	133.0	108.6	98.3	133.0	108.6	98.3	133.0	108.6	98.3	133.0	108.6	98.3	133.0	108.6	98.3	138.1	110.1
0970, 0990	Wall Finishes & Painting/Coating	96.1	116.4	108.3	96.1	116.4	108.3	96.1	116.4	108.3	96.1	116.4	108.3	96.1	116.4	108.3	96.1	116.4	108.3
09	FINISHES	101.5	124.3	114.3	100.4	124.3	113.8	101.5	124.3	114.3	101.6	124.3	114.3	101.3	124.3	114.2	101.2	125.2	114.6
COVERS	DIVS. 10 - 14, 25, 28, 41, 43, 44, 46	100.0	108.5	101.7	100.0	108.5	101.7	100.0	108.8	101.8	100.0	108.8	101.8	100.0	108.5	101.7	100.0	108.6	101.7
21, 22, 23	FIRE SUPPRESSION, PLUMBING & HVAC	100.0	117.6	107.2	94.0	117.6	103.6	100.0	117.7	107.2	100.0	117.7	107.2	100.0	117.6	107.2	100.0	117.6	107.2
26, 27, 3370	ELECTRICAL, COMMUNICATIONS & UTIL.	101.8	110.4	106.4	98.2	110.4	104.7	101.9	159.2	132.1	101.9	159.2	132.1	101.3	107.8	104.7	101.9	111.4	106.9
MF2010	WEIGHTED AVERAGE	102.1	120.0	110.0	97.2	120.0	107.2	101.2	127.1	112.6	101.4	127.1	112.7	101.3	119.6	109.3	101.2	120.2	109.6

DIVISION		D.C.			DELAWARE									FLORIDA					
		WASHINGTON			DOVER			NEWARK			WILMINGTON			DAYTONA BEACH			FORT LAUDERDALE		
		200 - 205			199			197			198			321			333		
		MAT.	INST.	TOTAL	MAT.	INST.	TOTAL	MAT.	INST.	TOTAL	MAT.	INST.	TOTAL	MAT.	INST.	TOTAL	MAT.	INST.	TOTAL
015433	CONTRACTOR EQUIPMENT		103.7	103.7		118.8	118.8		118.8	118.8		119.0	119.0		98.9	98.9		91.6	91.6
0241, 31 - 34	SITE & INFRASTRUCTURE, DEMOLITION	108.0	93.7	98.0	101.3	113.6	109.9	102.0	113.6	110.1	93.5	113.9	107.8	114.2	90.6	97.6	98.8	78.4	84.5
0310	Concrete Forming & Accessories	99.6	80.2	82.8	95.4	103.1	102.0	96.2	103.1	102.1	97.1	103.1	102.3	97.8	71.0	74.6	96.4	71.6	75.0
0320	Concrete Reinforcing	105.7	89.4	97.6	98.6	100.9	99.7	99.4	100.9	100.2	99.4	100.9	100.2	95.3	76.0	85.7	95.3	75.7	85.5
0330	Cast-in-Place Concrete	125.6	89.1	110.6	91.9	100.9	95.6	82.6	100.9	90.1	89.1	100.9	94.0	94.2	73.3	85.6	98.8	80.1	91.1
03	CONCRETE	112.6	86.3	99.6	94.7	102.8	98.7	90.4	102.8	96.6	93.6	102.8	98.2	91.7	73.9	82.9	93.8	76.5	85.3
04	MASONRY	93.9	80.9	86.0	108.4	96.2	101.0	106.8	96.2	100.3	113.1	96.2	102.8	93.7	68.4	78.3	93.9	71.1	80.0
05	METALS	98.0	106.8	100.8	99.9	116.8	105.2	99.9	116.8	105.2	97.0	116.8	103.2	96.2	92.0	94.9	96.1	92.7	95.0
06	WOOD, PLASTICS & COMPOSITES	96.8	78.4	86.3	95.2	103.3	99.8	97.9	103.3	101.0	100.0	103.3	101.9	98.6	72.3	83.6	94.4	70.2	80.6
07	THERMAL & MOISTURE PROTECTION	100.3	85.8	94.5	99.2	111.4	104.1	100.9	111.4	105.1	99.2	111.4	104.1	97.2	76.5	88.9	97.2	84.1	91.9
08	OPENINGS	98.2	87.7	95.7	94.2	111.2	98.4	94.2	111.2	98.4	93.9	111.2	98.1	99.6	69.4	92.3	97.4	68.5	90.3
0920	Plaster & Gypsum Board	107.2	77.9	86.7	107.3	103.2	104.4	106.9	103.2	104.3	109.8	103.2	105.2	98.4	71.9	79.8	97.5	69.7	78.0
0950, 0980	Ceilings & Acoustic Treatment	106.6	77.9	87.4	107.7	103.2	104.7	101.3	103.2	102.6	105.8	103.2	104.1	94.6	71.9	79.5	94.6	69.7	78.0
0960	Flooring	110.1	97.0	106.2	97.5	108.2	100.7	97.9	108.2	101.0	97.3	108.2	100.5	117.7	76.3	105.4	117.7	71.9	104.1
0970, 0990	Wall Finishes & Painting/Coating	109.8	85.9	95.5	98.4	105.3	102.6	98.4	105.3	102.6	98.4	105.3	102.6	106.8	77.4	89.2	101.6	73.4	84.7
09	FINISHES	104.5	83.0	92.5	102.7	103.9	103.4	101.7	103.9	102.9	102.8	103.9	103.4	105.5	72.4	86.9	102.5	71.1	84.9
COVERS	DIVS. 10 - 14, 25, 28, 41, 43, 44, 46	100.0	99.3	99.9	100.0	90.7	98.1	100.0	90.7	98.1	100.0	90.7	98.1	100.0	75.5	95.0	100.0	84.8	96.9
21, 22, 23	FIRE SUPPRESSION, PLUMBING & HVAC	100.1	94.6	97.9	99.9	116.9	106.9	100.1	116.9	107.0	100.1	116.9	107.0	99.9	75.6	90.0	99.9	70.2	87.8
26, 27, 3370	ELECTRICAL, COMMUNICATIONS & UTIL.	99.5	103.9	101.8	98.5	111.0	105.1	98.8	111.0	105.2	97.3	111.0	104.5	97.4	57.3	76.3	97.4	72.9	84.5
MF2010	WEIGHTED AVERAGE	101.1	92.4	97.3	99.3	108.8	103.5	98.8	108.8	103.2	98.6	108.9	103.1	98.6	74.3	87.9	97.9	75.2	87.9

	DIVISION	FLORIDA																	
		FORT MYERS			GAINESVILLE			JACKSONVILLE			LAKELAND			MELBOURNE			MIAMI		
		339,341			326,344			320,322			338			329			330 - 332,340		
		MAT.	INST.	TOTAL	MAT.	INST.	TOTAL	MAT.	INST.	TOTAL	MAT.	INST.	TOTAL	MAT.	INST.	TOTAL	MAT.	INST.	TOTAL
015433	CONTRACTOR EQUIPMENT		98.9	98.9		98.9	98.9		98.9	98.9		98.9	98.9		98.9	98.9		91.6	91.6
0241, 31 - 34	SITE & INFRASTRUCTURE, DEMOLITION	111.2	89.8	96.2	123.8	89.6	99.9	114.2	90.0	97.2	113.3	90.1	97.0	121.9	89.9	99.5	100.8	78.2	84.9
0310	Concrete Forming & Accessories	91.9	78.3	80.1	92.8	56.6	61.5	97.5	57.0	62.4	88.1	78.9	80.1	94.0	72.7	75.6	102.9	72.1	76.2
0320	Concrete Reinforcing	96.4	97.3	96.9	101.2	66.7	84.0	95.3	66.8	81.1	98.8	98.2	98.5	96.4	76.0	86.2	95.3	75.8	85.6
0330	Cast-in-Place Concrete	103.1	70.8	89.8	108.2	64.8	90.3	95.1	70.8	85.1	105.4	72.3	91.8	113.5	75.7	98.0	100.7	81.1	92.7
03	CONCRETE	94.7	80.3	87.5	102.6	62.9	83.0	92.1	65.1	78.8	96.5	81.1	88.9	103.1	75.5	89.5	95.1	77.0	86.2
04	MASONRY	87.9	64.5	73.6	108.7	63.7	81.3	93.7	63.8	75.4	105.5	79.0	89.3	92.6	72.7	80.5	94.7	74.9	82.6
05	METALS	98.2	99.9	98.8	95.1	86.9	92.5	94.8	87.3	92.4	98.1	100.9	99.0	104.2	92.1	100.4	100.9	91.4	97.9
06	WOOD, PLASTICS & COMPOSITES	91.2	80.9	85.3	92.4	54.1	70.5	98.6	54.1	73.2	86.7	80.9	83.4	94.0	72.3	81.6	100.3	70.2	83.1
07	THERMAL & MOISTURE PROTECTION	97.0	83.1	91.4	97.5	63.7	83.9	97.4	64.4	84.1	96.9	87.9	93.3	97.6	79.0	90.1	102.7	78.2	92.9
08	OPENINGS	98.3	77.8	93.3	97.7	54.8	87.3	99.6	58.1	89.5	98.2	78.5	93.4	98.8	72.8	92.4	97.4	68.4	90.3
0920	Plaster & Gypsum Board	92.8	80.7	84.4	93.6	53.1	65.3	98.4	53.1	66.7	90.3	80.7	83.6	93.6	71.9	78.4	94.5	69.7	77.1
0950, 0980	Ceilings & Acoustic Treatment	88.9	80.7	83.5	88.9	53.1	65.0	94.6	53.1	66.9	88.9	80.7	83.5	91.0	71.9	78.2	98.2	69.7	79.2
0960	Flooring	114.3	57.1	97.3	114.9	44.7	94.0	117.7	66.9	102.6	111.8	58.6	96.0	115.1	76.3	103.5	125.2	75.6	110.5
0970, 0990	Wall Finishes & Painting/Coating	106.8	63.5	80.8	106.8	63.5	80.8	106.8	64.7	81.6	106.8	63.5	80.8	106.8	96.0	100.3	101.6	73.4	84.7
09	FINISHES	102.3	73.0	85.9	103.9	54.0	76.0	105.5	58.6	79.2	101.4	73.3	85.7	104.0	75.5	88.1	105.3	72.3	86.8
COVERS	DIVS. 10 - 14, 25, 28, 41, 43, 44, 46	100.0	75.8	95.1	100.0	73.9	94.7	100.0	72.0	94.3	100.0	75.8	95.1	100.0	77.0	95.3	100.0	85.4	97.0
21, 22, 23	FIRE SUPPRESSION, PLUMBING & HVAC	96.7	65.2	83.8	98.8	67.4	85.9	99.9	67.4	86.6	96.7	85.2	92.0	99.9	77.8	90.9	99.9	70.1	87.7
26, 27, 3370	ELECTRICAL, COMMUNICATIONS & UTIL.	99.9	65.5	81.8	97.8	74.8	85.7	97.1	64.9	80.1	97.8	64.2	80.1	98.8	66.7	81.9	102.0	76.9	88.8
MF2010	WEIGHTED AVERAGE	97.9	75.1	87.8	100.0	68.9	86.3	98.4	68.6	85.2	98.7	81.0	90.9	101.2	77.3	90.7	99.8	76.0	89.3

	DIVISION	FLORIDA																	
		ORLANDO			PANAMA CITY			PENSACOLA			SARASOTA			ST. PETERSBURG			TALLAHASSEE		
		327 - 328,347			324			325			342			337			323		
		MAT.	INST.	TOTAL	MAT.	INST.	TOTAL	MAT.	INST.	TOTAL	MAT.	INST.	TOTAL	MAT.	INST.	TOTAL	MAT.	INST.	TOTAL
015433	CONTRACTOR EQUIPMENT		98.9	98.9		98.9	98.9		98.9	98.9		98.9	98.9		98.9	98.9		98.9	98.9
0241, 31 - 34	SITE & INFRASTRUCTURE, DEMOLITION	114.6	89.8	97.2	128.0	88.9	100.5	127.4	89.3	100.7	115.6	89.9	97.5	114.9	89.5	97.1	110.6	89.1	95.5
0310	Concrete Forming & Accessories	98.6	75.2	78.4	96.8	45.9	52.8	94.5	54.1	59.6	97.8	78.4	81.0	95.7	53.3	59.0	100.1	46.1	53.4
0320	Concrete Reinforcing	95.3	75.3	85.3	99.5	76.1	87.9	102.1	76.6	89.4	95.3	98.1	96.7	98.8	85.1	92.0	95.3	66.6	81.0
0330	Cast-in-Place Concrete	118.0	73.1	99.6	100.0	58.8	83.1	123.5	68.3	100.8	107.7	72.1	93.1	106.6	66.8	90.2	102.9	58.4	84.6
03	CONCRETE	104.4	75.6	90.2	100.2	57.8	79.3	110.9	64.8	88.1	98.2	80.9	89.6	98.1	65.5	82.0	96.0	56.0	76.2
04	MASONRY	99.4	68.4	80.5	98.6	49.2	68.5	116.3	56.7	79.9	94.5	79.0	85.0	147.1	51.0	88.4	96.9	56.1	72.0
05	METALS	101.7	91.4	98.4	95.9	89.4	93.9	97.0	90.6	95.0	98.9	100.5	99.4	99.1	93.7	97.4	89.8	86.7	88.8
06	WOOD, PLASTICS & COMPOSITES	98.2	78.6	87.0	97.4	43.6	66.7	95.4	53.8	71.7	98.6	80.9	88.5	95.8	52.3	71.0	96.0	43.2	65.8
07	THERMAL & MOISTURE PROTECTION	97.1	77.2	89.1	97.6	58.7	82.0	97.6	64.7	84.4	97.2	87.9	93.5	97.2	60.0	82.2	102.8	74.3	91.3
08	OPENINGS	99.6	71.8	92.9	97.3	48.6	85.4	97.2	59.5	88.0	99.6	77.5	94.2	98.2	62.3	89.5	100.1	49.1	87.7
0920	Plaster & Gypsum Board	102.5	78.3	85.6	96.5	42.3	58.6	96.2	52.8	65.9	98.4	80.7	86.0	95.3	51.3	64.5	107.2	41.8	61.5
0950, 0980	Ceilings & Acoustic Treatment	104.6	78.3	87.1	91.0	42.3	58.5	91.0	52.8	65.6	92.6	80.7	84.7	91.0	51.3	64.5	104.6	41.8	62.8
0960	Flooring	117.5	76.3	105.2	117.2	44.6	95.6	112.5	65.2	98.4	117.7	60.0	100.5	116.4	57.2	98.8	117.5	64.3	101.7
0970, 0990	Wall Finishes & Painting/Coating	106.8	73.4	86.7	106.8	67.8	83.4	106.8	67.8	83.4	106.8	63.5	80.8	106.8	63.5	80.8	106.8	67.8	83.4
09	FINISHES	107.9	75.7	89.8	105.9	46.8	72.8	104.0	56.9	77.6	105.0	73.6	87.4	103.9	54.0	76.0	108.1	50.4	75.8
COVERS	DIVS. 10 - 14, 25, 28, 41, 43, 44, 46	100.0	76.2	95.1	100.0	48.1	89.4	100.0	48.2	89.4	100.0	75.8	95.1	100.0	58.2	91.5	100.0	68.7	93.6
21, 22, 23	FIRE SUPPRESSION, PLUMBING & HVAC	99.9	59.6	83.4	99.9	54.8	81.5	99.9	55.2	81.6	99.8	66.1	86.0	99.9	61.5	84.2	100.0	41.1	75.9
26, 27, 3370	ELECTRICAL, COMMUNICATIONS & UTIL.	101.9	62.5	81.2	96.0	61.8	78.0	100.3	56.1	77.0	97.4	64.2	79.9	97.8	64.2	80.1	103.1	61.8	81.4
MF2010	WEIGHTED AVERAGE	101.8	72.3	88.8	99.8	60.2	82.4	102.2	63.4	85.1	99.7	76.9	89.7	102.2	65.6	86.1	99.1	59.1	81.5

	DIVISION	FLORIDA						GEORGIA											
		TAMPA			WEST PALM BEACH			ALBANY			ATHENS			ATLANTA			AUGUSTA		
		335 - 336,346			334,349			317,398			306			300 - 303,399			308 - 309		
		MAT.	INST.	TOTAL	MAT.	INST.	TOTAL	MAT.	INST.	TOTAL	MAT.	INST.	TOTAL	MAT.	INST.	TOTAL	MAT.	INST.	TOTAL
015433	CONTRACTOR EQUIPMENT		98.9	98.9		91.6	91.6		92.5	92.5		93.6	93.6		94.1	94.1		93.6	93.6
0241, 31 - 34	SITE & INFRASTRUCTURE, DEMOLITION	115.4	90.0	97.6	95.6	78.4	83.6	99.0	79.8	85.6	98.6	92.8	94.6	95.2	94.7	94.8	92.0	92.6	92.4
0310	Concrete Forming & Accessories	98.6	78.9	81.6	100.0	71.3	75.1	97.0	45.1	52.1	95.0	42.8	49.8	98.4	74.6	77.8	95.9	67.9	71.6
0320	Concrete Reinforcing	95.3	98.2	96.7	98.0	75.4	86.8	95.0	82.4	88.7	98.6	81.7	90.1	97.9	83.8	90.9	98.9	75.1	87.1
0330	Cast-in-Place Concrete	104.2	72.3	91.1	93.9	77.1	87.0	100.6	47.4	78.7	114.4	52.4	88.9	114.4	71.5	96.7	108.0	49.4	83.9
03	CONCRETE	96.5	81.2	88.9	90.9	75.2	83.2	94.6	54.5	74.8	105.6	54.2	80.2	103.3	75.3	89.5	98.7	63.1	81.1
04	MASONRY	95.0	79.0	85.2	93.4	69.3	78.7	96.1	44.0	64.3	81.0	56.0	65.8	94.9	67.9	78.4	95.2	40.7	61.9
05	METALS	98.1	100.9	99.0	95.2	92.1	94.2	95.4	86.5	92.6	91.1	74.5	85.9	92.0	78.7	87.8	90.8	72.3	85.0
06	WOOD, PLASTICS & COMPOSITES	99.9	80.9	89.1	99.3	70.2	82.7	97.6	39.7	64.5	90.6	36.7	59.8	94.7	76.6	84.4	92.0	74.1	81.8
07	THERMAL & MOISTURE PROTECTION	97.4	87.9	93.6	96.9	72.9	87.3	94.9	60.8	81.2	93.3	53.7	77.4	93.2	73.9	85.4	93.0	58.1	79.0
08	OPENINGS	99.6	80.9	95.1	96.5	68.5	89.7	97.4	45.9	84.8	90.8	44.5	79.5	96.5	73.2	90.8	90.8	65.0	84.5
0920	Plaster & Gypsum Board	98.4	80.7	86.0	101.5	69.7	79.3	98.1	38.2	56.2	96.1	35.1	53.4	98.6	76.3	83.0	97.3	73.6	80.8
0950, 0980	Ceilings & Acoustic Treatment	94.6	80.7	85.4	88.9	69.7	76.1	93.7	38.2	56.7	96.5	35.1	55.6	96.5	76.3	83.0	97.6	73.6	81.6
0960	Flooring	117.7	58.6	100.1	119.9	68.9	104.7	117.7	37.1	93.7	89.7	55.8	79.6	91.0	67.6	84.0	90.0	40.7	75.3
0970, 0990	Wall Finishes & Painting/Coating	106.8	63.5	80.8	101.6	73.4	84.7	101.6	55.2	73.8	91.9	36.6	58.7	91.9	83.8	87.0	91.9	38.2	59.7
09	FINISHES	105.5	73.3	87.5	102.1	70.5	84.4	102.4	42.7	68.9	92.5	42.9	64.7	92.8	74.3	82.4	92.3	60.6	74.6
COVERS	DIVS. 10 - 14, 25, 28, 41, 43, 44, 46	100.0	76.3	95.2	100.0	84.8	96.9	100.0	81.4	96.2	100.0	40.7	87.9	100.0	86.3	97.2	100.0	52.0	90.2
21, 22, 23	FIRE SUPPRESSION, PLUMBING & HVAC	99.9	85.2	93.9	96.7	66.1	84.2	99.9	67.7	86.7	94.0	71.6	84.8	99.9	72.7	88.8	100.0	64.5	85.5
26, 27, 3370	ELECTRICAL, COMMUNICATIONS & UTIL.	97.5	64.2	80.0	98.8	72.9	85.2	92.8	59.5	75.3	101.0	72.5	86.0	100.2	74.1	86.4	101.6	64.4	82.0
MF2010	WEIGHTED AVERAGE	99.5	81.1	91.4	96.6	73.5	86.4	97.4	61.1	81.4	95.0	63.3	81.0	97.5	75.9	88.0	96.3	64.2	82.1

DIVISION		GEORGIA																	
		COLUMBUS			DALTON			GAINESVILLE			MACON			SAVANNAH			STATESBORO		
		318 - 319			307			305			310 - 312			313 - 314			304		
		MAT.	INST.	TOTAL	MAT.	INST.	TOTAL	MAT.	INST.	TOTAL	MAT.	INST.	TOTAL	MAT.	INST.	TOTAL	MAT.	INST.	TOTAL
015433	CONTRACTOR EQUIPMENT		92.5	92.5		105.3	105.3		93.6	93.6		103.0	103.0		93.4	93.4		92.8	92.8
0241, 31 - 34	SITE & INFRASTRUCTURE, DEMOLITION	99.0	80.0	85.6	99.0	97.0	97.6	98.4	92.8	94.5	99.9	95.0	96.5	99.9	80.8	86.5	99.7	77.2	84.0
0310	Concrete Forming & Accessories	96.9	56.3	61.8	88.1	48.2	53.6	98.1	44.6	51.8	96.3	53.5	59.3	98.1	50.4	56.8	82.8	41.8	47.3
0320	Concrete Reinforcing	95.3	83.2	89.2	98.0	77.3	87.7	98.4	81.6	90.0	96.6	82.5	89.5	96.2	75.3	85.8	97.6	43.4	70.6
0330	Cast-in-Place Concrete	100.2	55.7	81.9	111.1	46.5	84.5	120.2	52.3	92.3	98.9	49.7	78.6	110.2	49.4	85.2	114.2	61.0	92.3
03	CONCRETE	94.4	62.4	78.6	104.5	54.6	79.9	107.7	55.0	81.7	93.9	59.0	76.7	99.4	56.2	78.1	104.9	50.5	78.0
04	MASONRY	96.0	57.5	72.5	82.4	33.4	52.5	89.8	56.4	69.4	108.4	38.9	65.9	95.1	52.0	68.8	83.9	41.0	57.7
05	METALS	95.0	88.2	92.8	91.9	84.3	89.5	90.3	74.2	85.3	91.1	87.3	89.9	91.5	84.1	89.1	95.4	71.9	88.0
06	WOOD, PLASTICS & COMPOSITES	97.6	54.4	73.0	74.4	49.5	60.2	94.5	39.5	63.1	106.5	53.5	76.2	112.5	46.7	74.9	67.8	41.1	52.6
07	THERMAL & MOISTURE PROTECTION	94.9	64.7	82.7	95.3	51.9	77.8	93.3	56.6	78.5	93.4	62.5	80.9	96.0	58.2	80.8	94.0	50.2	76.4
08	OPENINGS	97.4	55.8	87.3	92.1	46.4	81.0	90.8	41.7	78.8	95.9	54.4	85.8	98.6	47.9	86.2	92.9	32.9	78.3
0920	Plaster & Gypsum Board	98.1	53.5	66.9	81.7	48.3	58.4	98.6	38.0	56.2	105.4	52.4	68.3	95.5	45.5	60.5	82.3	39.7	52.5
0950, 0980	Ceilings & Acoustic Treatment	93.7	53.5	66.9	109.7	48.3	68.8	96.5	38.0	57.5	88.5	52.4	64.5	101.9	45.5	64.3	104.8	39.7	61.4
0960	Flooring	117.7	52.2	98.2	90.2	37.1	74.4	91.0	37.1	74.9	92.3	37.6	76.0	118.2	48.1	97.3	106.5	46.5	88.6
0970, 0990	Wall Finishes & Painting/Coating	101.6	69.4	82.3	83.1	55.8	66.8	91.9	36.6	58.7	104.0	55.2	74.7	103.0	60.9	77.8	89.8	34.3	56.6
09	FINISHES	102.3	55.9	76.3	100.9	46.3	70.3	93.0	40.3	63.5	90.4	49.6	67.5	104.1	50.1	73.9	103.5	41.1	68.5
COVERS	DIVS. 10 - 14, 25, 28, 41, 43, 44, 46	100.0	83.1	96.5	100.0	23.4	84.3	100.0	41.0	87.9	100.0	81.2	96.2	100.0	51.1	90.0	100.0	37.8	87.3
21, 22, 23	FIRE SUPPRESSION, PLUMBING & HVAC	99.9	49.1	79.1	94.0	58.1	79.3	94.0	71.5	84.8	99.9	65.6	85.9	100.0	63.2	85.0	94.5	55.7	78.6
26, 27, 3370	ELECTRICAL, COMMUNICATIONS & UTIL.	92.8	72.5	82.1	111.7	70.6	90.1	100.9	72.5	86.0	91.9	62.1	76.2	97.7	59.8	77.7	101.1	56.3	77.5
MF2010	WEIGHTED AVERAGE	97.4	63.9	82.6	96.9	59.3	80.3	95.6	63.1	81.3	96.2	63.9	82.0	98.2	61.1	81.9	96.9	53.3	77.7

DIVISION		GEORGIA						HAWAII									IDAHO		
		VALDOSTA			WAYCROSS			HILO			HONOLULU			STATES & POSS., GUAM			BOISE		
		316			315			967			968			969			836 - 837		
		MAT.	INST.	TOTAL	MAT.	INST.	TOTAL	MAT.	INST.	TOTAL	MAT.	INST.	TOTAL	MAT.	INST.	TOTAL	MAT.	INST.	TOTAL
015433	CONTRACTOR EQUIPMENT		92.5	92.5		92.5	92.5		99.0	99.0		99.0	99.0		161.6	161.6		97.5	97.5
0241, 31 - 34	SITE & INFRASTRUCTURE, DEMOLITION	108.3	79.9	88.4	105.2	78.6	86.5	121.6	106.7	111.2	124.4	106.7	112.0	161.6	103.8	121.0	81.1	95.6	91.2
0310	Concrete Forming & Accessories	86.5	44.6	50.3	88.5	67.5	70.3	109.2	136.5	132.8	118.0	136.5	134.0	111.5	66.0	72.1	100.8	76.1	79.4
0320	Concrete Reinforcing	97.3	70.5	84.0	97.3	76.2	86.8	108.4	123.5	115.9	107.3	123.5	115.3	192.9	31.6	112.6	104.1	80.1	92.1
0330	Cast-in-Place Concrete	98.4	51.7	79.2	111.2	49.8	86.0	185.7	127.2	161.7	156.7	127.2	144.6	161.0	108.1	139.3	92.4	73.4	84.6
03	CONCRETE	98.4	53.6	76.2	102.4	63.9	83.4	145.4	129.6	137.6	131.9	129.6	130.8	147.6	74.6	111.5	101.0	75.9	88.6
04	MASONRY	102.1	51.5	71.2	102.8	40.9	65.0	118.3	129.6	125.2	111.8	129.6	122.6	170.1	45.4	94.0	121.5	61.6	84.9
05	METALS	94.6	82.7	90.8	93.6	79.5	89.2	108.7	109.1	108.9	130.4	109.1	123.7	140.3	77.2	120.5	100.7	76.4	93.1
06	WOOD, PLASTICS & COMPOSITES	85.0	38.7	58.5	86.9	75.7	80.5	101.4	138.7	122.7	111.7	138.7	127.1	112.2	69.9	88.0	89.5	79.3	83.7
07	THERMAL & MOISTURE PROTECTION	95.0	64.3	82.7	94.8	52.3	77.7	103.0	126.1	112.3	115.4	126.1	119.7	118.7	68.8	98.6	94.5	66.0	83.0
08	OPENINGS	93.0	42.1	80.6	93.0	59.8	84.9	97.7	131.9	106.0	102.7	131.9	109.8	102.3	56.2	91.0	97.3	73.6	91.5
0920	Plaster & Gypsum Board	90.3	37.2	53.2	90.7	75.3	80.0	100.7	139.7	128.0	131.4	139.7	137.2	197.1	58.1	99.9	81.6	78.6	79.5
0950, 0980	Ceilings & Acoustic Treatment	91.0	37.2	55.2	88.9	75.3	79.9	109.4	139.7	129.6	122.8	139.7	134.1	215.4	58.1	110.6	109.7	78.6	89.0
0960	Flooring	111.0	48.2	92.3	112.2	31.0	88.0	130.8	135.8	132.3	155.3	135.8	149.5	154.3	48.1	122.7	96.2	65.9	87.2
0970, 0990	Wall Finishes & Painting/Coating	101.6	54.5	73.4	101.6	41.7	65.7	100.9	145.4	127.5	105.5	145.4	129.4	106.5	37.5	65.2	98.3	41.2	64.1
09	FINISHES	99.7	44.7	68.9	99.3	59.3	76.9	113.4	138.5	127.4	128.3	138.5	134.0	185.6	61.1	115.8	97.9	71.7	83.2
COVERS	DIVS. 10 - 14, 25, 28, 41, 43, 44, 46	100.0	50.3	89.8	100.0	41.5	88.1	100.0	115.5	103.2	100.0	115.5	103.2	100.0	77.1	95.3	100.0	60.4	91.9
21, 22, 23	FIRE SUPPRESSION, PLUMBING & HVAC	99.9	67.5	86.7	96.0	57.1	80.1	100.0	112.9	105.3	100.1	112.9	105.3	102.0	39.3	76.3	100.0	68.3	87.0
26, 27, 3370	ELECTRICAL, COMMUNICATIONS & UTIL.	90.8	58.6	73.9	95.3	59.3	76.4	107.5	124.7	116.6	108.8	124.7	117.2	156.9	42.9	96.9	95.9	75.9	85.4
MF2010	WEIGHTED AVERAGE	97.4	60.4	81.1	97.1	60.6	81.0	109.5	122.3	115.2	113.6	122.3	117.4	130.4	60.0	99.4	99.6	73.3	88.0

DIVISION		IDAHO															ILLINOIS		
		COEUR D'ALENE			IDAHO FALLS			LEWISTON			POCATELLO			TWIN FALLS			BLOOMINGTON		
		838			834			835			832			833			617		
		MAT.	INST.	TOTAL	MAT.	INST.	TOTAL	MAT.	INST.	TOTAL	MAT.	INST.	TOTAL	MAT.	INST.	TOTAL	MAT.	INST.	TOTAL
015433	CONTRACTOR EQUIPMENT		92.0	92.0		97.5	97.5		92.0	92.0		97.5	97.5		97.5	97.5		100.6	100.6
0241, 31 - 34	SITE & INFRASTRUCTURE, DEMOLITION	78.4	91.6	87.6	79.0	94.7	90.0	84.7	92.1	89.9	81.9	95.6	91.5	88.4	93.9	92.3	97.5	96.8	97.0
0310	Concrete Forming & Accessories	115.3	57.8	65.5	95.2	40.7	48.0	120.2	58.8	67.0	101.0	75.7	79.1	101.8	29.2	39.0	86.8	118.1	113.9
0320	Concrete Reinforcing	111.9	96.9	104.4	106.1	79.1	92.6	111.9	97.1	104.5	104.5	80.2	92.4	106.4	79.3	92.9	97.8	106.2	102.0
0330	Cast-in-Place Concrete	99.9	86.8	94.5	88.0	51.1	72.8	103.8	87.2	97.0	95.0	73.2	86.0	97.5	39.1	73.5	96.7	112.5	103.2
03	CONCRETE	108.0	75.6	92.0	93.4	52.4	73.1	111.6	76.2	94.1	100.3	75.7	88.2	107.7	43.2	75.9	96.7	113.9	105.2
04	MASONRY	123.1	83.2	98.7	116.5	34.0	66.2	123.5	83.8	99.3	119.2	63.4	85.2	121.7	32.1	67.0	116.3	117.5	117.0
05	METALS	94.7	84.8	91.6	108.5	73.7	97.6	94.2	85.9	91.6	108.6	76.2	98.4	108.6	73.2	97.5	95.6	110.1	100.1
06	WOOD, PLASTICS & COMPOSITES	94.4	49.6	68.8	83.8	39.1	58.2	99.3	49.6	70.9	89.5	79.3	83.7	90.3	27.7	54.6	84.2	117.7	103.3
07	THERMAL & MOISTURE PROTECTION	146.7	74.6	117.7	94.1	43.9	73.9	146.9	74.5	117.8	94.6	60.7	81.0	95.3	38.1	72.3	99.2	112.1	104.4
08	OPENINGS	116.0	56.2	101.5	101.2	46.7	87.9	116.0	58.9	102.1	98.0	68.5	90.8	101.2	38.4	85.9	90.0	105.3	93.7
0920	Plaster & Gypsum Board	148.9	48.1	78.4	76.2	37.1	48.9	149.7	48.1	78.6	78.8	78.6	78.7	79.5	25.4	41.7	96.3	118.2	111.6
0950, 0980	Ceilings & Acoustic Treatment	143.3	48.1	79.9	107.7	37.1	60.7	143.3	48.1	79.9	115.9	78.6	91.0	110.4	25.4	53.8	94.6	118.2	110.3
0960	Flooring	136.7	47.5	110.1	96.2	44.7	80.9	139.6	89.7	124.7	99.4	65.9	89.4	100.5	44.7	83.9	98.3	111.1	102.1
0970, 0990	Wall Finishes & Painting/Coating	115.9	67.8	87.1	98.4	33.9	59.7	115.9	67.8	87.1	98.3	42.9	65.1	98.4	25.3	54.6	95.3	123.1	111.9
09	FINISHES	165.3	55.2	103.6	96.2	40.0	64.7	166.4	63.6	108.8	99.8	71.9	84.2	99.8	30.5	60.9	97.8	118.0	109.2
COVERS	DIVS. 10 - 14, 25, 28, 41, 43, 44, 46	100.0	51.1	90.0	100.0	37.9	87.3	100.0	62.9	92.4	100.0	60.3	91.9	100.0	34.1	86.5	100.0	104.2	100.9
21, 22, 23	FIRE SUPPRESSION, PLUMBING & HVAC	99.5	86.4	94.1	100.9	62.3	85.1	100.7	89.4	96.1	99.9	68.3	87.0	99.9	58.4	83.0	93.9	104.8	98.4
26, 27, 3370	ELECTRICAL, COMMUNICATIONS & UTIL.	87.4	80.8	83.9	87.4	72.1	79.3	85.3	80.8	83.0	93.3	72.1	82.1	88.9	62.6	75.0	95.0	94.9	94.9
MF2010	WEIGHTED AVERAGE	107.4	77.1	94.0	99.4	58.3	81.3	108.1	79.5	95.5	100.7	72.5	88.3	101.8	52.7	80.1	96.4	107.8	101.4

DIVISION		ILLINOIS																	
		CARBONDALE			CENTRALIA			CHAMPAIGN			CHICAGO			DECATUR			EAST ST. LOUIS		
		629			628			618 - 619			606 - 608			625			620 - 622		
		MAT.	INST.	TOTAL	MAT.	INST.	TOTAL	MAT.	INST.	TOTAL	MAT.	INST.	TOTAL	MAT.	INST.	TOTAL	MAT.	INST.	TOTAL
015433	**CONTRACTOR EQUIPMENT**		106.9	106.9		106.9	106.9		101.4	101.4		92.7	92.7		101.4	101.4		106.9	106.9
0241, 31 - 34	**SITE & INFRASTRUCTURE, DEMOLITION**	102.5	97.4	98.9	102.8	98.7	99.9	106.7	97.7	100.4	106.6	94.2	97.9	92.7	97.9	96.3	104.9	98.7	100.5
0310	Concrete Forming & Accessories	95.3	102.4	101.4	97.0	110.4	108.6	92.8	116.6	113.4	98.0	157.1	149.2	97.2	116.9	114.3	92.8	110.8	108.4
0320	Concrete Reinforcing	96.6	93.4	95.0	96.6	103.1	99.9	97.8	101.3	99.6	96.2	159.9	127.9	94.7	101.8	98.2	96.6	107.1	101.8
0330	Cast-in-Place Concrete	90.5	97.6	93.4	91.0	114.4	100.6	112.0	108.3	110.5	103.4	147.2	121.4	98.8	112.5	104.4	92.5	114.4	101.5
03	**CONCRETE**	87.5	99.8	93.6	88.0	111.1	99.4	108.9	110.8	109.8	100.0	152.9	126.1	97.9	112.5	105.1	88.9	112.0	100.3
04	**MASONRY**	78.2	101.1	92.2	78.2	114.9	100.6	139.9	117.0	125.9	100.7	151.0	131.4	75.0	118.2	101.4	78.5	114.9	100.7
05	**METALS**	94.8	110.0	99.6	94.8	116.8	101.7	95.6	106.3	98.9	95.4	133.1	107.2	98.2	107.0	101.0	95.9	118.6	103.1
06	**WOOD, PLASTICS & COMPOSITES**	98.1	101.6	100.1	100.7	107.9	104.8	90.9	116.0	105.2	99.5	157.6	132.6	98.3	116.0	108.4	95.4	107.9	102.5
07	**THERMAL & MOISTURE PROTECTION**	94.2	97.3	95.5	94.3	111.0	101.0	99.8	112.1	104.8	103.3	143.1	119.3	99.9	109.0	103.6	94.3	110.6	100.8
08	**OPENINGS**	85.7	106.8	90.8	85.7	112.2	92.1	90.6	111.2	95.6	99.4	158.4	113.7	96.3	111.4	99.9	85.7	113.9	92.6
0920	Plaster & Gypsum Board	100.1	101.6	101.1	101.3	108.1	106.0	98.8	116.5	111.2	97.1	159.3	140.6	103.0	116.5	112.4	98.8	108.1	105.3
0950, 0980	Ceilings & Acoustic Treatment	94.6	101.6	99.2	94.6	108.1	103.6	94.6	116.5	109.2	106.2	159.3	141.6	101.0	116.5	111.3	94.6	108.1	103.6
0960	Flooring	116.7	124.5	119.0	117.7	117.3	117.6	101.1	124.5	108.1	95.4	147.0	110.8	104.5	120.9	109.4	115.6	117.3	116.1
0970, 0990	Wall Finishes & Painting/Coating	103.4	100.1	101.4	103.4	107.3	105.8	95.3	111.3	104.9	91.9	159.6	132.5	95.3	109.6	103.8	103.4	107.3	105.8
09	**FINISHES**	101.2	105.4	103.6	101.7	111.3	107.1	99.8	118.0	110.0	99.7	156.7	131.7	101.9	117.9	110.8	100.9	111.5	106.8
COVERS	**DIVS. 10 - 14, 25, 28, 41, 43, 44, 46**	100.0	96.7	99.3	100.0	100.0	100.0	100.0	103.9	100.8	100.0	124.1	104.9	100.0	104.2	100.9	100.0	100.0	100.0
21, 22, 23	**FIRE SUPPRESSION, PLUMBING & HVAC**	93.8	95.0	94.3	93.8	91.2	92.7	93.9	103.6	97.8	99.9	133.3	113.6	99.9	100.8	100.3	99.9	95.3	98.0
26, 27, 3370	**ELECTRICAL, COMMUNICATIONS & UTIL.**	93.7	99.6	96.8	95.2	103.3	99.5	98.5	94.0	96.1	97.8	130.8	115.2	98.0	88.5	93.0	94.7	103.3	99.2
MF2010	**WEIGHTED AVERAGE**	93.1	100.5	96.4	93.4	105.4	98.7	99.8	106.8	102.9	99.3	138.3	116.5	97.6	105.8	101.2	95.0	106.6	100.1

DIVISION		ILLINOIS																	
		EFFINGHAM			GALESBURG			JOLIET			KANKAKEE			LA SALLE			NORTH SUBURBAN		
		624			614			604			609			613			600 - 603		
		MAT.	INST.	TOTAL	MAT.	INST.	TOTAL	MAT.	INST.	TOTAL	MAT.	INST.	TOTAL	MAT.	INST.	TOTAL	MAT.	INST.	TOTAL
015433	**CONTRACTOR EQUIPMENT**		101.4	101.4		100.6	100.6		90.8	90.8		90.8	90.8		100.6	100.6		90.8	90.8
0241, 31 - 34	**SITE & INFRASTRUCTURE, DEMOLITION**	98.7	97.2	97.7	100.0	96.7	97.7	106.4	93.1	97.1	100.0	92.4	94.7	99.3	97.6	98.1	105.6	92.3	96.3
0310	Concrete Forming & Accessories	101.8	109.6	108.6	92.7	113.7	110.9	100.0	152.1	145.1	93.9	133.1	127.8	105.9	119.1	117.3	99.4	146.3	140.0
0320	Concrete Reinforcing	97.8	98.6	98.2	97.3	105.8	101.5	96.2	144.3	120.2	97.1	138.6	117.8	97.4	131.7	114.5	96.2	153.2	124.6
0330	Cast-in-Place Concrete	98.4	103.7	100.6	99.5	103.6	101.2	103.4	140.2	118.5	96.4	125.1	108.2	99.4	112.8	104.9	103.4	137.7	117.5
03	**CONCRETE**	98.8	105.6	102.1	99.5	108.8	104.1	100.1	145.2	122.4	94.4	130.5	112.2	100.3	119.3	109.7	100.0	143.4	121.5
04	**MASONRY**	82.2	102.9	94.8	116.5	112.0	113.8	103.6	144.1	128.3	100.2	130.6	118.8	116.5	115.3	115.7	100.7	140.3	124.8
05	**METALS**	95.5	104.4	98.3	95.6	108.8	99.7	93.3	124.6	103.1	93.3	119.8	101.6	95.6	124.3	104.6	94.4	127.1	104.7
06	**WOOD, PLASTICS & COMPOSITES**	100.8	111.3	106.8	90.8	113.0	103.4	101.0	153.3	130.8	94.2	132.0	115.8	105.7	118.2	112.8	99.5	147.9	127.1
07	**THERMAL & MOISTURE PROTECTION**	99.5	104.3	101.4	99.4	104.9	101.6	103.1	139.0	117.6	102.4	129.8	113.4	99.5	113.1	105.0	103.6	136.3	116.7
08	**OPENINGS**	90.4	108.0	94.7	90.0	107.2	94.2	97.4	151.9	110.6	89.3	137.2	100.9	90.0	120.8	97.5	97.4	149.7	110.1
0920	Plaster & Gypsum Board	103.0	111.6	109.0	98.8	113.3	109.0	93.8	155.0	136.6	91.7	133.0	120.6	105.9	118.7	114.9	97.1	149.4	133.7
0950, 0980	Ceilings & Acoustic Treatment	94.6	111.6	105.9	94.6	113.3	107.1	106.2	155.0	138.7	106.2	133.0	124.1	94.6	118.7	110.7	106.2	149.4	135.0
0960	Flooring	105.6	124.5	111.2	101.0	111.1	104.0	94.9	136.7	107.4	92.1	136.7	105.4	107.4	115.1	109.7	95.4	136.7	107.7
0970, 0990	Wall Finishes & Painting/Coating	95.3	105.0	101.1	95.3	93.1	93.9	90.1	149.8	125.9	90.1	123.1	109.9	95.3	123.1	111.9	91.9	149.8	126.6
09	**FINISHES**	100.9	113.0	107.7	99.1	111.9	106.3	99.1	150.4	127.8	97.6	131.1	116.3	101.8	118.5	111.2	99.7	146.2	125.8
COVERS	**DIVS. 10 - 14, 25, 28, 41, 43, 44, 46**	100.0	69.7	93.8	100.0	97.1	99.4	100.0	122.2	104.5	100.0	110.4	102.1	100.0	97.2	99.4	100.0	112.6	102.6
21, 22, 23	**FIRE SUPPRESSION, PLUMBING & HVAC**	93.9	99.3	96.1	93.9	100.8	96.7	100.0	130.9	112.6	94.0	125.0	106.6	93.9	122.5	105.6	99.9	121.0	108.5
26, 27, 3370	**ELECTRICAL, COMMUNICATIONS & UTIL.**	95.8	99.6	97.8	96.0	86.0	90.8	97.0	135.3	117.2	91.5	134.4	114.1	93.0	132.2	113.6	96.9	130.8	114.7
MF2010	**WEIGHTED AVERAGE**	95.4	102.3	98.4	97.0	103.2	99.7	98.8	134.5	114.5	94.8	125.4	108.3	97.1	119.0	106.8	98.8	130.3	112.7

DIVISION		ILLINOIS																	
		PEORIA			QUINCY			ROCK ISLAND			ROCKFORD			SOUTH SUBURBAN			SPRINGFIELD		
		615 - 616			623			612			610 - 611			605			626 - 627		
		MAT.	INST.	TOTAL	MAT.	INST.	TOTAL	MAT.	INST.	TOTAL	MAT.	INST.	TOTAL	MAT.	INST.	TOTAL	MAT.	INST.	TOTAL
015433	**CONTRACTOR EQUIPMENT**		100.6	100.6		101.4	101.4		100.6	100.6		100.6	100.6		90.8	90.8		101.4	101.4
0241, 31 - 34	**SITE & INFRASTRUCTURE, DEMOLITION**	100.3	96.8	97.8	97.5	97.3	97.4	98.1	96.2	96.7	99.6	98.0	98.5	105.6	92.3	96.3	98.9	97.9	98.2
0310	Concrete Forming & Accessories	95.6	114.9	112.3	99.6	112.6	110.9	94.3	98.5	97.9	99.8	130.1	126.0	99.4	146.3	140.0	98.0	117.3	114.7
0320	Concrete Reinforcing	94.7	105.9	100.3	97.4	98.5	98.0	97.3	98.5	97.9	89.6	138.2	113.8	96.2	153.2	124.6	94.7	101.7	98.2
0330	Cast-in-Place Concrete	96.5	113.6	103.5	98.6	96.2	97.6	97.4	94.6	96.2	98.8	124.1	109.2	103.4	137.7	117.5	88.8	107.6	96.6
03	**CONCRETE**	96.7	112.8	104.6	98.4	104.3	101.3	97.5	97.6	97.6	97.2	129.2	113.0	100.0	143.4	121.5	93.2	110.9	101.9
04	**MASONRY**	116.9	112.7	114.3	102.1	99.6	100.6	116.3	94.2	102.8	91.4	135.3	118.2	100.7	140.3	124.8	79.6	119.3	103.8
05	**METALS**	98.2	109.8	101.9	95.5	104.1	98.2	95.6	105.1	98.6	98.2	127.3	107.4	94.4	127.1	104.7	100.7	106.9	102.6
06	**WOOD, PLASTICS & COMPOSITES**	98.4	113.0	106.7	98.2	116.0	108.4	92.4	97.5	95.3	98.3	128.7	115.7	99.5	147.9	127.1	99.2	116.0	108.8
07	**THERMAL & MOISTURE PROTECTION**	100.0	110.3	104.2	99.5	101.8	100.4	99.3	95.7	97.9	102.4	127.6	112.5	103.6	136.5	116.8	100.7	111.9	105.2
08	**OPENINGS**	96.4	111.1	100.0	91.3	110.7	96.0	90.0	96.9	91.7	96.4	129.9	104.6	97.4	149.7	110.1	97.3	111.4	100.7
0920	Plaster & Gypsum Board	103.0	113.3	110.2	101.3	116.5	111.9	98.8	97.4	97.8	103.0	129.5	121.6	97.1	149.4	133.7	101.2	116.5	111.9
0950, 0980	Ceilings & Acoustic Treatment	101.0	113.3	109.2	94.6	116.5	109.2	94.6	97.4	96.4	101.0	129.5	120.0	106.2	149.4	135.0	107.4	116.5	113.4
0960	Flooring	104.5	111.1	106.4	104.5	102.9	104.0	102.2	99.8	101.5	104.5	115.1	107.6	95.4	136.7	107.7	104.5	107.9	105.5
0970, 0990	Wall Finishes & Painting/Coating	95.3	123.1	111.9	95.3	107.2	102.4	95.3	96.1	95.8	95.3	130.2	116.2	91.9	149.8	126.6	95.3	107.2	102.4
09	**FINISHES**	102.0	115.4	109.5	100.3	111.9	106.8	99.3	98.4	98.8	102.0	127.8	116.4	99.7	146.2	125.8	102.8	115.3	109.8
COVERS	**DIVS. 10 - 14, 25, 28, 41, 43, 44, 46**	100.0	103.9	100.8	100.0	72.6	94.4	100.0	97.5	99.5	100.0	90.8	98.1	100.0	112.6	102.6	100.0	104.6	100.9
21, 22, 23	**FIRE SUPPRESSION, PLUMBING & HVAC**	99.9	103.8	101.5	93.9	96.6	95.0	93.9	100.6	96.6	99.9	113.6	105.5	99.9	121.0	108.5	99.8	103.9	101.5
26, 27, 3370	**ELECTRICAL, COMMUNICATIONS & UTIL.**	97.1	95.3	96.2	93.0	80.1	86.2	87.7	96.6	92.3	97.5	129.3	114.2	96.9	130.8	114.7	101.1	90.7	95.6
MF2010	**WEIGHTED AVERAGE**	99.7	106.7	102.8	96.1	98.6	97.2	96.0	98.3	97.0	98.6	122.3	109.0	98.8	130.3	112.7	98.3	106.5	101.9

	DIVISION	INDIANA																	
		ANDERSON			BLOOMINGTON			COLUMBUS			EVANSVILLE			FORT WAYNE			GARY		
		460			474			472			476 - 477			467 - 468			463 - 464		
		MAT.	INST.	TOTAL	MAT.	INST.	TOTAL	MAT.	INST.	TOTAL	MAT.	INST.	TOTAL	MAT.	INST.	TOTAL	MAT.	INST.	TOTAL
015433	CONTRACTOR EQUIPMENT		97.2	97.2		87.5	87.5		87.5	87.5		116.7	116.7		97.2	97.2		97.2	97.2
0241, 31 - 34	SITE & INFRASTRUCTURE, DEMOLITION	93.4	95.4	94.8	82.4	94.7	91.0	79.1	94.5	89.9	87.6	124.3	113.3	94.0	95.1	94.8	94.0	98.7	97.3
0310	Concrete Forming & Accessories	94.5	80.2	82.1	100.2	81.0	83.6	94.3	78.0	80.2	93.9	79.1	81.1	92.6	73.5	76.1	94.6	108.4	106.6
0320	Concrete Reinforcing	100.6	81.7	91.2	90.6	81.4	86.1	91.1	81.4	86.2	99.3	75.7	87.5	100.6	76.8	88.8	100.6	101.9	101.3
0330	Cast-in-Place Concrete	106.3	78.7	95.0	101.3	79.6	92.3	100.8	73.4	89.5	96.8	86.3	92.4	113.0	74.4	97.1	111.1	106.6	109.3
03	CONCRETE	98.4	80.5	89.5	101.7	80.4	91.2	101.1	76.9	89.1	102.1	81.2	91.7	101.4	75.1	88.4	100.7	106.5	103.5
04	MASONRY	85.0	80.5	82.2	90.0	78.0	82.6	89.8	76.8	81.8	85.8	81.8	83.3	88.6	78.0	82.1	86.2	111.5	101.7
05	METALS	90.8	88.7	90.1	95.5	77.6	89.8	95.5	76.9	89.7	88.8	83.7	87.2	90.8	86.1	89.3	90.8	103.8	94.9
06	WOOD, PLASTICS & COMPOSITES	111.6	80.3	93.7	120.0	80.7	97.5	114.8	77.4	93.4	98.7	77.6	86.7	111.3	72.7	89.3	109.2	106.4	107.6
07	THERMAL & MOISTURE PROTECTION	99.8	78.9	91.4	94.1	81.4	89.0	93.7	80.6	88.5	98.1	83.0	92.0	99.6	81.2	92.2	98.5	107.8	102.2
08	OPENINGS	100.5	80.8	95.7	107.1	81.0	100.7	102.3	79.2	96.7	99.2	77.7	93.9	100.5	73.1	93.8	100.5	109.9	102.8
0920	Plaster & Gypsum Board	94.8	80.1	84.5	101.0	80.7	86.8	97.2	77.3	83.3	95.8	76.4	82.2	94.0	72.3	78.8	88.2	107.0	101.3
0950, 0980	Ceilings & Acoustic Treatment	89.0	80.1	83.1	83.1	80.7	81.5	83.1	77.3	79.3	89.1	76.4	80.6	89.0	72.3	77.9	89.0	107.0	101.0
0960	Flooring	96.8	80.4	91.9	104.7	77.7	96.6	100.1	77.7	93.4	99.2	83.6	94.5	96.8	77.7	91.1	96.8	120.2	103.8
0970, 0990	Wall Finishes & Painting/Coating	93.1	69.4	78.9	88.8	82.6	85.1	88.8	82.6	85.1	93.9	85.0	88.6	93.1	74.5	81.9	93.1	121.5	110.1
09	FINISHES	93.8	79.2	85.6	97.3	80.6	87.9	95.5	78.4	85.9	96.4	80.4	87.5	93.5	74.4	82.8	92.9	112.0	103.6
COVERS	DIVS. 10 - 14, 25, 28, 41, 43, 44, 46	100.0	92.2	98.4	100.0	84.5	96.8	100.0	83.7	96.7	100.0	88.6	97.7	100.0	84.3	96.8	100.0	99.7	99.9
21, 22, 23	FIRE SUPPRESSION, PLUMBING & HVAC	99.9	76.0	90.2	99.8	79.4	91.4	93.7	77.3	87.0	99.9	79.7	91.6	99.9	73.2	89.0	99.9	103.9	101.6
26, 27, 3370	ELECTRICAL, COMMUNICATIONS & UTIL.	86.6	84.7	85.6	99.4	84.0	91.3	98.5	84.6	91.2	95.4	84.0	89.4	87.4	72.7	79.7	99.1	112.1	106.0
MF2010	WEIGHTED AVERAGE	95.7	82.3	89.8	98.8	81.6	91.2	96.4	80.1	89.3	96.5	85.2	91.5	96.3	77.7	88.1	97.2	107.0	101.5

	DIVISION	INDIANA																	
		INDIANAPOLIS			KOKOMO			LAFAYETTE			LAWRENCEBURG			MUNCIE			NEW ALBANY		
		461 - 462			469			479			470			473			471		
		MAT.	INST.	TOTAL	MAT.	INST.	TOTAL	MAT.	INST.	TOTAL	MAT.	INST.	TOTAL	MAT.	INST.	TOTAL	MAT.	INST.	TOTAL
015433	CONTRACTOR EQUIPMENT		92.9	92.9		97.2	97.2		87.5	87.5		105.5	105.5		95.9	95.9		95.4	95.4
0241, 31 - 34	SITE & INFRASTRUCTURE, DEMOLITION	93.7	98.7	97.2	90.0	95.3	93.8	80.0	94.5	90.2	77.7	111.2	101.2	82.2	94.8	91.1	74.9	97.4	90.7
0310	Concrete Forming & Accessories	95.1	84.1	85.5	98.0	75.9	78.8	92.1	80.7	82.2	90.9	74.6	76.8	92.2	80.0	81.6	88.7	72.5	74.6
0320	Concrete Reinforcing	100.0	81.7	90.9	91.1	81.5	86.3	90.6	81.6	86.2	89.8	77.8	83.8	100.3	81.6	91.0	91.3	75.5	83.5
0330	Cast-in-Place Concrete	102.3	85.6	95.4	105.3	82.9	96.1	101.4	81.9	93.4	94.8	72.4	85.6	106.4	78.7	95.0	97.9	71.4	87.0
03	CONCRETE	100.1	83.9	92.1	95.1	79.9	87.6	101.3	81.0	91.3	94.5	75.1	84.9	100.9	80.4	90.8	99.6	73.1	86.5
04	MASONRY	91.4	82.0	85.7	84.6	78.4	80.8	94.9	80.3	86.0	74.9	76.6	76.0	91.9	80.4	84.9	81.9	69.9	74.6
05	METALS	92.8	79.9	88.8	87.4	88.4	87.7	93.9	77.8	88.8	90.6	85.5	89.0	97.2	88.9	94.6	92.5	80.8	88.9
06	WOOD, PLASTICS & COMPOSITES	108.1	84.0	94.3	114.7	74.2	91.6	112.1	80.4	94.0	97.3	73.8	83.9	114.4	80.1	94.8	99.0	72.9	84.1
07	THERMAL & MOISTURE PROTECTION	97.7	83.4	92.0	99.4	78.6	91.0	93.7	81.9	89.0	98.1	78.3	90.1	96.8	79.2	89.7	85.7	69.2	79.1
08	OPENINGS	109.4	83.0	103.0	94.5	77.5	90.3	100.6	80.8	95.8	101.0	75.3	94.8	101.2	80.7	96.2	98.7	74.9	92.9
0920	Plaster & Gypsum Board	92.9	83.7	86.5	100.2	73.8	81.7	94.3	80.4	84.6	73.3	73.5	73.5	95.8	80.1	84.8	93.7	72.4	78.8
0950, 0980	Ceilings & Acoustic Treatment	93.6	83.7	87.0	89.0	73.8	78.9	78.6	80.4	79.8	93.7	73.5	80.2	84.0	80.1	81.4	89.1	72.4	78.0
0960	Flooring	96.6	86.6	93.6	100.7	87.9	96.9	99.1	88.1	95.8	73.6	86.6	77.5	99.2	80.4	93.6	97.1	65.1	87.6
0970, 0990	Wall Finishes & Painting/Coating	93.1	83.7	87.5	93.1	74.9	82.2	88.8	97.2	93.9	89.7	73.7	80.1	88.8	69.4	77.2	93.9	86.4	89.4
09	FINISHES	94.8	84.6	89.1	95.4	77.7	85.5	93.8	83.8	88.2	86.9	77.1	81.4	94.9	79.0	86.0	95.8	72.9	82.9
COVERS	DIVS. 10 - 14, 25, 28, 41, 43, 44, 46	100.0	93.6	98.7	100.0	84.0	96.7	100.0	91.7	98.3	100.0	44.6	88.7	100.0	91.6	98.3	100.0	44.0	88.6
21, 22, 23	FIRE SUPPRESSION, PLUMBING & HVAC	99.9	82.0	92.6	93.9	79.3	87.9	93.7	78.2	87.4	94.5	74.3	86.2	99.8	76.0	90.0	93.9	75.6	86.4
26, 27, 3370	ELECTRICAL, COMMUNICATIONS & UTIL.	101.0	84.6	92.4	91.4	77.7	84.2	98.0	81.8	89.5	93.2	74.2	83.2	90.9	78.8	84.5	93.8	74.9	83.8
MF2010	WEIGHTED AVERAGE	98.9	84.7	92.6	93.3	81.1	87.9	96.1	82.0	89.9	93.0	78.4	86.6	97.5	81.4	90.4	94.1	75.4	85.9

	DIVISION	INDIANA									IOWA								
		SOUTH BEND			TERRE HAUTE			WASHINGTON			BURLINGTON			CARROLL			CEDAR RAPIDS		
		465 - 466			478			475			526			514			522 - 524		
		MAT.	INST.	TOTAL	MAT.	INST.	TOTAL	MAT.	INST.	TOTAL	MAT.	INST.	TOTAL	MAT.	INST.	TOTAL	MAT.	INST.	TOTAL
015433	CONTRACTOR EQUIPMENT		105.5	105.5		116.7	116.7		116.7	116.7		100.1	100.1		100.1	100.1		96.7	96.7
0241, 31 - 34	SITE & INFRASTRUCTURE, DEMOLITION	96.2	95.6	95.8	89.4	124.6	114.1	89.1	122.4	112.5	100.3	96.6	97.7	88.1	96.7	94.2	102.1	96.1	97.9
0310	Concrete Forming & Accessories	100.7	77.5	80.6	94.7	79.2	81.3	95.5	77.9	80.3	98.7	75.9	79.0	87.3	51.9	56.7	104.3	80.3	83.5
0320	Concrete Reinforcing	100.6	76.9	88.8	99.3	81.8	90.6	91.8	51.1	71.5	98.9	81.3	90.2	99.7	80.7	90.3	99.7	83.7	91.8
0330	Cast-in-Place Concrete	103.5	78.5	93.2	93.7	87.7	91.2	101.9	86.9	95.8	108.0	56.4	86.8	108.0	61.2	88.8	108.3	79.5	96.5
03	CONCRETE	93.4	79.1	86.3	104.6	82.9	93.9	110.0	76.1	93.3	102.5	71.0	87.0	101.4	61.9	81.9	102.7	81.3	92.2
04	MASONRY	84.7	78.8	81.1	93.1	80.9	85.7	85.8	81.9	83.4	98.9	67.3	79.6	101.0	73.8	84.4	105.0	79.1	89.2
05	METALS	90.8	98.6	93.3	89.5	87.1	88.7	84.2	69.9	79.8	88.8	90.9	89.5	88.9	90.5	89.4	91.1	92.8	91.6
06	WOOD, PLASTICS & COMPOSITES	108.9	77.2	90.8	100.6	78.0	87.7	101.0	77.5	87.6	95.6	76.5	84.7	83.3	46.9	62.5	101.7	79.3	88.9
07	THERMAL & MOISTURE PROTECTION	91.9	82.5	88.1	98.2	81.4	91.4	98.3	83.1	92.2	104.5	76.8	93.3	104.8	70.4	90.9	105.3	79.1	94.8
08	OPENINGS	93.0	76.9	89.1	99.8	79.5	94.8	96.1	68.0	89.2	91.2	70.1	86.0	95.7	55.3	85.9	96.7	80.5	92.7
0920	Plaster & Gypsum Board	91.1	76.9	81.1	95.8	76.7	82.4	95.3	76.3	82.0	103.9	75.8	84.3	99.3	45.3	61.5	108.6	78.9	87.8
0950, 0980	Ceilings & Acoustic Treatment	89.9	76.9	81.2	89.1	76.7	80.8	82.7	76.3	78.4	116.3	75.8	89.3	116.3	45.3	69.0	119.9	78.9	92.6
0960	Flooring	94.9	82.5	91.2	99.2	86.6	95.4	100.1	79.2	93.9	114.7	39.2	92.2	108.5	34.4	86.4	132.4	81.1	117.1
0970, 0990	Wall Finishes & Painting/Coating	92.6	86.4	88.9	93.9	84.4	88.2	93.9	85.0	88.6	97.3	82.0	88.1	97.3	77.6	85.5	99.2	71.7	82.7
09	FINISHES	92.9	79.4	85.3	96.4	80.9	87.7	95.5	79.6	86.6	110.8	69.8	87.9	106.5	49.6	74.6	117.3	79.5	96.1
COVERS	DIVS. 10 - 14, 25, 28, 41, 43, 44, 46	100.0	68.5	93.6	100.0	94.1	98.8	100.0	88.5	97.7	100.0	84.2	96.8	100.0	67.4	93.3	100.0	85.2	97.0
21, 22, 23	FIRE SUPPRESSION, PLUMBING & HVAC	99.9	76.0	90.1	99.9	78.5	91.2	93.9	78.9	87.8	94.3	76.6	87.0	94.3	69.8	84.3	100.3	80.5	92.2
26, 27, 3370	ELECTRICAL, COMMUNICATIONS & UTIL.	100.6	82.6	91.1	93.5	84.0	88.5	94.1	83.9	88.7	101.4	70.5	85.1	102.1	80.9	90.9	98.8	79.4	88.6
MF2010	WEIGHTED AVERAGE	95.5	81.8	89.5	97.2	85.7	92.1	94.7	82.4	89.3	97.2	76.2	88.0	97.0	71.4	85.7	100.3	82.8	92.6

DIVISION		IOWA																	
		COUNCIL BLUFFS			CRESTON			DAVENPORT			DECORAH			DES MOINES			DUBUQUE		
		515			508			527 - 528			521			500 - 503,509			520		
		MAT.	INST.	TOTAL	MAT.	INST.	TOTAL	MAT.	INST.	TOTAL	MAT.	INST.	TOTAL	MAT.	INST.	TOTAL	MAT.	INST.	TOTAL
015433	**CONTRACTOR EQUIPMENT**		96.2	96.2		100.1	100.1		100.1	100.1		100.1	100.1		101.7	101.7		95.6	95.6
0241, 31 - 34	**SITE & INFRASTRUCTURE, DEMOLITION**	108.0	92.3	97.0	88.1	96.3	93.9	100.4	99.0	99.5	98.9	96.2	97.0	89.6	99.8	96.8	99.9	93.3	95.3
0310	Concrete Forming & Accessories	86.8	71.0	73.1	87.9	67.6	70.4	103.6	86.7	88.9	96.4	45.7	52.6	106.8	77.1	81.1	88.0	69.7	72.2
0320	Concrete Reinforcing	101.7	79.1	90.4	99.0	80.8	89.9	99.7	94.5	97.1	98.9	80.3	89.6	99.7	81.0	90.4	98.3	83.6	91.0
0330	Cast-in-Place Concrete	112.6	75.9	97.5	107.7	65.2	90.2	104.3	86.0	96.7	105.1	57.9	85.7	90.8	89.5	90.3	106.0	86.7	98.1
03	**CONCRETE**	105.1	75.1	90.2	101.2	70.2	85.9	100.7	88.5	94.7	100.5	57.9	79.5	93.4	82.9	88.2	99.5	79.1	89.4
04	**MASONRY**	106.3	74.7	87.0	101.0	80.8	88.7	101.9	84.5	91.3	120.0	70.1	89.5	95.7	81.9	87.2	105.9	75.5	87.3
05	**METALS**	96.5	90.4	94.6	88.8	90.7	89.4	91.1	100.8	94.2	89.0	88.7	88.9	91.3	93.0	91.8	89.8	92.3	90.6
06	**WOOD, PLASTICS & COMPOSITES**	82.2	70.2	75.4	83.7	64.4	72.7	101.8	85.6	92.5	92.9	38.7	61.9	101.6	74.8	86.3	83.7	66.1	73.6
07	**THERMAL & MOISTURE PROTECTION**	104.8	68.7	90.3	105.5	77.7	94.3	104.9	84.6	96.7	104.7	54.0	84.3	104.9	78.8	94.4	105.1	71.0	91.4
08	**OPENINGS**	95.7	76.1	90.9	106.8	65.2	96.7	96.7	87.7	94.5	94.2	50.8	83.7	96.7	80.9	92.9	95.7	75.0	90.7
0920	Plaster & Gypsum Board	99.3	69.5	78.5	99.3	63.3	74.1	108.6	85.1	92.2	102.6	36.8	56.6	96.4	74.1	80.8	99.3	65.2	75.5
0950, 0980	Ceilings & Acoustic Treatment	116.3	69.5	85.1	116.3	63.3	81.0	119.9	85.1	96.7	116.3	36.8	63.3	119.9	74.1	89.4	116.3	65.2	82.3
0960	Flooring	107.3	83.6	100.3	108.8	34.4	86.6	117.2	91.6	109.6	114.4	48.9	94.9	116.4	83.7	106.6	122.5	78.9	109.5
0970, 0990	Wall Finishes & Painting/Coating	94.1	66.9	77.8	97.3	77.6	85.5	97.3	87.8	91.6	97.3	34.4	59.6	97.3	81.7	88.0	98.2	66.6	79.3
09	**FINISHES**	107.7	72.5	88.0	106.6	61.7	81.5	112.9	87.5	98.7	110.5	43.4	72.9	109.6	78.4	92.1	112.2	70.3	88.7
COVERS	**DIVS. 10 - 14, 25, 28, 41, 43, 44, 46**	100.0	85.1	97.0	100.0	71.6	94.2	100.0	87.1	97.4	100.0	79.0	95.7	100.0	85.1	97.0	100.0	83.3	96.6
21, 22, 23	**FIRE SUPPRESSION, PLUMBING & HVAC**	100.3	76.9	90.7	94.3	79.3	88.1	100.3	88.0	95.3	94.3	72.0	85.2	100.1	79.9	91.8	100.3	75.1	90.0
26, 27, 3370	**ELECTRICAL, COMMUNICATIONS & UTIL.**	104.5	82.1	92.7	96.5	80.9	88.3	95.9	85.0	90.2	98.8	46.2	71.1	109.5	80.9	94.4	102.9	77.6	89.6
MF2010	**WEIGHTED AVERAGE**	101.2	79.1	91.5	97.6	77.7	88.8	99.3	89.2	94.8	98.1	64.8	83.4	98.9	83.5	92.1	99.5	78.6	90.3

DIVISION		IOWA																	
		FORT DODGE			MASON CITY			OTTUMWA			SHENANDOAH			SIBLEY			SIOUX CITY		
		505			504			525			516			512			510 - 511		
		MAT.	INST.	TOTAL	MAT.	INST.	TOTAL	MAT.	INST.	TOTAL	MAT.	INST.	TOTAL	MAT.	INST.	TOTAL	MAT.	INST.	TOTAL
015433	**CONTRACTOR EQUIPMENT**		100.1	100.1		100.1	100.1		95.6	95.6		96.2	96.2		100.1	100.1		100.1	100.1
0241, 31 - 34	**SITE & INFRASTRUCTURE, DEMOLITION**	96.0	95.2	95.4	96.0	96.2	96.2	100.4	91.5	94.1	106.1	91.4	95.8	111.3	95.0	99.9	112.8	95.8	100.9
0310	Concrete Forming & Accessories	88.4	46.8	52.4	92.9	46.5	52.7	94.6	58.4	63.3	88.3	58.4	62.4	88.7	39.1	45.7	104.3	64.7	70.1
0320	Concrete Reinforcing	99.0	63.6	81.4	98.9	80.2	89.6	98.9	81.5	90.3	101.7	67.8	84.8	101.7	63.5	82.7	99.7	63.9	81.9
0330	Cast-in-Place Concrete	101.0	45.6	78.3	101.0	58.5	83.6	108.7	55.7	86.9	108.8	61.5	89.4	106.7	46.1	81.8	107.3	56.7	86.5
03	**CONCRETE**	96.8	51.0	74.2	97.0	58.4	78.0	102.1	63.0	82.8	102.5	62.4	82.7	101.5	47.7	74.9	102.1	62.9	82.7
04	**MASONRY**	99.6	39.4	62.9	113.2	68.9	86.1	102.3	58.5	75.6	105.9	74.7	86.9	124.6	39.7	72.8	98.9	55.5	72.4
05	**METALS**	88.9	80.5	86.2	88.9	89.0	88.9	88.8	90.4	89.3	95.5	84.1	92.0	89.0	79.3	86.0	91.1	82.0	88.3
06	**WOOD, PLASTICS & COMPOSITES**	84.2	47.2	63.0	88.3	38.7	60.0	90.4	59.8	72.9	83.7	55.1	67.4	84.5	37.4	57.6	101.7	64.8	80.6
07	**THERMAL & MOISTURE PROTECTION**	104.9	58.7	86.3	104.4	64.0	88.2	105.2	63.5	88.4	104.2	67.0	89.2	104.5	49.8	82.5	104.9	64.2	88.5
08	**OPENINGS**	100.3	49.7	88.0	92.0	50.7	82.0	95.7	65.3	88.3	86.7	56.1	79.2	92.3	44.4	80.6	96.7	61.1	88.0
0920	Plaster & Gypsum Board	99.3	45.6	61.7	99.3	36.8	55.6	100.5	58.8	71.3	99.3	54.0	67.6	99.3	35.5	54.7	108.6	63.7	77.2
0950, 0980	Ceilings & Acoustic Treatment	116.3	45.6	69.2	116.3	36.8	63.3	116.3	58.8	78.0	116.3	54.0	74.8	116.3	35.5	62.5	119.9	63.7	82.5
0960	Flooring	110.5	48.9	92.2	112.8	48.9	93.8	125.8	52.0	103.8	108.1	35.5	86.5	109.4	35.2	87.3	117.4	55.6	99.0
0970, 0990	Wall Finishes & Painting/Coating	97.3	61.4	75.8	97.3	31.4	57.8	98.2	80.5	87.6	94.1	66.9	77.8	97.3	61.4	75.8	97.8	61.4	76.0
09	**FINISHES**	108.7	47.8	74.6	109.4	43.3	72.3	113.5	58.7	82.8	107.8	54.0	77.7	110.5	39.4	70.7	114.9	62.6	85.6
COVERS	**DIVS. 10 - 14, 25, 28, 41, 43, 44, 46**	100.0	75.1	94.9	100.0	79.4	95.8	100.0	77.3	95.4	100.0	68.5	93.6	100.0	74.1	94.7	100.0	81.3	96.2
21, 22, 23	**FIRE SUPPRESSION, PLUMBING & HVAC**	94.3	65.6	82.5	94.3	73.0	85.6	94.3	71.3	84.9	94.3	70.8	84.7	94.3	61.6	80.9	100.3	68.1	87.1
26, 27, 3370	**ELECTRICAL, COMMUNICATIONS & UTIL.**	104.0	43.6	72.2	102.9	58.0	79.3	101.2	66.6	83.0	98.8	80.4	89.1	98.8	43.2	69.5	98.8	72.9	85.2
MF2010	**WEIGHTED AVERAGE**	97.5	58.9	80.5	97.3	66.9	83.9	98.0	69.7	85.5	97.8	71.3	86.1	98.5	55.8	79.7	100.0	69.7	86.7

DIVISION		IOWA						KANSAS											
		SPENCER			WATERLOO			BELLEVILLE			COLBY			DODGE CITY			EMPORIA		
		513			506 - 507			669			677			678			668		
		MAT.	INST.	TOTAL	MAT.	INST.	TOTAL	MAT.	INST.	TOTAL	MAT.	INST.	TOTAL	MAT.	INST.	TOTAL	MAT.	INST.	TOTAL
015433	**CONTRACTOR EQUIPMENT**		100.1	100.1		100.1	100.1		102.3	102.3		102.3	102.3		102.3	102.3		100.6	100.6
0241, 31 - 34	**SITE & INFRASTRUCTURE, DEMOLITION**	111.4	95.0	99.9	100.9	96.2	97.6	114.4	94.0	100.1	111.8	94.0	99.3	114.3	93.6	99.8	105.7	91.7	95.9
0310	Concrete Forming & Accessories	94.5	39.1	46.5	104.7	48.9	56.4	95.4	55.6	61.0	101.6	60.0	65.6	95.0	60.0	64.7	87.3	67.3	70.0
0320	Concrete Reinforcing	101.7	63.5	82.7	99.7	83.4	91.6	102.5	54.8	78.7	105.1	54.8	80.0	102.5	54.9	78.8	101.1	55.4	78.3
0330	Cast-in-Place Concrete	106.7	46.1	81.8	108.3	55.9	86.7	117.0	57.0	92.3	112.5	56.9	89.7	114.5	56.6	90.7	113.1	47.7	86.2
03	**CONCRETE**	101.9	47.7	75.1	102.7	59.3	81.3	116.7	57.3	87.3	114.0	59.2	86.9	115.2	59.0	87.4	109.2	59.4	84.6
04	**MASONRY**	124.6	39.7	72.8	100.3	74.7	84.7	100.1	57.6	74.2	101.5	57.6	74.7	109.0	60.0	79.1	105.9	67.7	82.6
05	**METALS**	89.0	79.3	86.0	91.1	91.8	91.3	93.8	77.6	88.7	94.1	77.6	89.0	95.6	76.6	89.7	93.5	78.9	88.9
06	**WOOD, PLASTICS & COMPOSITES**	90.3	37.4	60.1	102.3	38.7	66.0	92.5	54.4	70.7	99.8	60.4	77.3	91.8	60.4	73.9	84.4	69.6	75.9
07	**THERMAL & MOISTURE PROTECTION**	105.3	49.8	83.0	104.6	71.0	91.1	100.7	58.3	83.6	100.7	58.9	83.9	100.7	59.9	84.2	99.3	73.1	88.8
08	**OPENINGS**	103.9	44.4	89.4	93.0	53.3	83.3	96.0	49.8	84.8	96.1	53.1	85.6	96.0	53.1	85.6	93.7	58.1	85.1
0920	Plaster & Gypsum Board	100.5	35.5	55.1	108.6	36.8	58.4	100.9	52.9	67.4	105.9	59.1	73.2	100.1	59.1	71.4	97.6	68.6	77.3
0950, 0980	Ceilings & Acoustic Treatment	116.3	35.5	62.5	119.9	36.8	64.6	94.6	52.9	66.8	94.6	59.1	71.0	94.6	59.1	71.0	94.6	68.6	77.3
0960	Flooring	112.3	35.2	89.3	118.9	78.2	106.8	104.0	40.6	85.1	107.4	40.6	87.5	103.8	40.6	85.0	99.4	37.7	81.0
0970, 0990	Wall Finishes & Painting/Coating	97.3	61.4	75.8	97.8	81.2	87.9	95.3	40.7	62.5	95.3	40.7	62.5	95.3	40.7	62.5	95.3	40.7	62.5
09	**FINISHES**	111.5	39.4	71.1	113.4	55.6	81.0	101.6	51.1	73.3	103.0	54.6	75.9	101.4	54.6	75.2	98.8	59.5	76.8
COVERS	**DIVS. 10 - 14, 25, 28, 41, 43, 44, 46**	100.0	74.1	94.7	100.0	80.6	96.0	100.0	41.9	88.1	100.0	42.5	88.3	100.0	42.6	88.3	100.0	41.6	88.1
21, 22, 23	**FIRE SUPPRESSION, PLUMBING & HVAC**	94.3	61.6	80.9	100.3	79.8	91.9	93.9	69.5	83.9	93.9	66.2	82.6	99.9	66.2	86.1	93.9	69.7	84.0
26, 27, 3370	**ELECTRICAL, COMMUNICATIONS & UTIL.**	100.6	43.2	70.4	98.8	58.0	77.3	100.9	66.0	82.5	100.9	66.0	82.5	97.4	72.3	84.2	97.8	72.2	84.3
MF2010	**WEIGHTED AVERAGE**	100.1	55.8	80.5	99.4	71.2	86.9	99.5	64.6	84.1	99.5	64.8	84.2	101.2	65.8	85.6	97.9	68.7	85.0

DIVISION		KANSAS																	
		FORT SCOTT			HAYS			HUTCHINSON			INDEPENDENCE			KANSAS CITY			LIBERAL		
		667			676			675			673			660 - 662			679		
		MAT.	INST.	TOTAL	MAT.	INST.	TOTAL	MAT.	INST.	TOTAL	MAT.	INST.	TOTAL	MAT.	INST.	TOTAL	MAT.	INST.	TOTAL
015433	CONTRACTOR EQUIPMENT		101.6	101.6		102.3	102.3		102.3	102.3		102.3	102.3		99.0	99.0		102.3	102.3
0241, 31 - 34	SITE & INFRASTRUCTURE, DEMOLITION	102.2	92.2	95.2	118.4	94.0	101.3	95.5	94.1	94.5	115.6	94.2	100.6	95.5	91.8	92.9	118.1	94.1	101.3
0310	Concrete Forming & Accessories	102.3	79.5	82.6	99.2	60.0	65.3	89.8	59.9	63.9	109.9	68.3	73.9	99.2	95.5	96.0	95.4	59.9	64.7
0320	Concrete Reinforcing	100.4	87.9	94.2	102.5	54.8	78.7	102.5	54.9	78.8	101.9	59.7	80.9	97.4	93.1	95.3	104.0	54.9	79.5
0330	Cast-in-Place Concrete	104.9	55.8	84.7	88.8	56.9	75.7	82.3	47.1	67.8	115.0	47.3	87.2	90.0	94.6	91.9	88.8	47.1	71.7
03	CONCRETE	104.3	73.5	89.1	105.6	59.2	82.7	88.6	55.7	72.4	116.4	60.4	88.8	96.4	95.0	95.7	107.7	55.8	82.1
04	MASONRY	106.6	58.5	77.2	109.6	57.6	77.9	101.4	59.7	75.9	98.6	62.3	76.4	109.5	96.0	101.3	107.0	59.7	78.1
05	METALS	93.5	90.2	92.5	93.7	77.6	88.7	93.6	76.5	88.2	93.5	79.0	88.9	100.6	97.6	99.7	94.0	76.5	88.5
06	WOOD, PLASTICS & COMPOSITES	101.9	87.2	93.5	96.9	60.4	76.1	86.8	60.4	71.7	109.7	71.0	87.6	97.9	95.9	96.8	92.5	60.4	74.2
07	THERMAL & MOISTURE PROTECTION	100.0	72.4	88.9	100.9	58.9	84.1	99.7	59.8	83.6	100.7	71.8	89.1	99.6	95.7	98.1	101.1	59.8	84.5
08	OPENINGS	93.7	79.3	90.2	96.0	53.1	85.6	96.0	53.1	85.5	93.7	60.0	85.5	95.1	89.4	93.7	96.1	53.1	85.6
0920	Plaster & Gypsum Board	103.8	86.8	91.9	103.8	59.1	72.6	98.8	59.1	71.1	112.1	70.1	82.7	95.9	95.7	95.8	100.9	59.1	71.7
0950, 0980	Ceilings & Acoustic Treatment	94.6	86.8	89.4	94.6	59.1	71.0	94.6	59.1	71.0	94.6	70.1	78.3	94.6	95.7	95.4	94.6	59.1	71.0
0960	Flooring	115.1	40.9	93.0	106.3	40.6	86.7	101.0	40.6	83.0	111.7	40.6	90.5	91.8	97.3	93.4	104.0	40.6	85.1
0970, 0990	Wall Finishes & Painting/Coating	97.1	40.7	63.3	95.3	40.7	62.5	95.3	40.7	62.5	95.3	40.7	62.5	103.4	70.7	83.8	95.3	40.7	62.5
09	FINISHES	104.3	69.7	84.9	103.1	54.6	75.9	98.6	54.6	74.0	105.2	60.9	80.4	98.5	92.4	95.1	102.4	54.6	75.6
COVERS	DIVS. 10 - 14, 25, 28, 41, 43, 44, 46	100.0	47.8	89.3	100.0	42.5	88.3	100.0	42.5	88.3	100.0	43.7	88.5	100.0	63.8	92.6	100.0	42.5	88.3
21, 22, 23	FIRE SUPPRESSION, PLUMBING & HVAC	93.9	67.8	83.2	93.9	66.2	82.6	93.9	66.2	82.6	93.9	68.1	83.3	99.8	97.3	98.8	93.9	66.2	82.6
26, 27, 3370	ELECTRICAL, COMMUNICATIONS & UTIL.	97.0	72.2	83.9	99.7	66.0	81.9	94.3	66.0	79.4	96.6	72.1	83.7	103.3	94.1	98.5	97.4	68.3	82.1
MF2010	WEIGHTED AVERAGE	97.8	72.9	86.8	99.0	64.8	83.9	95.0	64.5	81.5	99.2	68.4	85.6	99.7	93.9	97.2	98.8	64.8	83.8

DIVISION		KANSAS									KENTUCKY								
		SALINA			TOPEKA			WICHITA			ASHLAND			BOWLING GREEN			CAMPTON		
		674			664 - 666			670 - 672			411 - 412			421 - 422			413 - 414		
		MAT.	INST.	TOTAL	MAT.	INST.	TOTAL	MAT.	INST.	TOTAL	MAT.	INST.	TOTAL	MAT.	INST.	TOTAL	MAT.	INST.	TOTAL
015433	CONTRACTOR EQUIPMENT		102.3	102.3		100.6	100.6		102.3	102.3		97.9	97.9		95.4	95.4		101.7	101.7
0241, 31 - 34	SITE & INFRASTRUCTURE, DEMOLITION	104.3	94.0	97.1	97.0	90.8	92.6	96.9	92.8	94.0	109.1	86.9	93.5	75.2	98.1	91.3	83.6	99.3	94.6
0310	Concrete Forming & Accessories	91.7	56.0	60.8	98.7	42.1	49.7	98.1	45.6	52.7	88.0	104.1	102.0	85.0	78.8	79.6	89.1	86.2	86.6
0320	Concrete Reinforcing	101.9	63.5	82.8	94.7	98.7	96.7	94.7	74.8	84.8	92.9	105.3	99.1	90.0	88.3	89.1	90.9	104.2	97.5
0330	Cast-in-Place Concrete	99.4	50.6	79.3	89.5	46.8	72.0	87.5	49.7	72.0	89.1	99.8	93.5	88.4	98.4	92.5	98.6	62.8	83.9
03	CONCRETE	102.3	56.9	79.9	93.6	56.0	75.0	92.5	53.9	73.5	95.5	103.3	99.3	93.7	87.5	90.6	97.8	81.5	89.7
04	MASONRY	124.3	57.6	83.6	103.3	55.9	74.4	97.0	52.4	69.7	91.9	103.3	98.8	93.7	81.5	86.2	91.4	71.9	79.5
05	METALS	95.5	81.2	91.0	100.7	96.3	99.3	100.7	83.1	95.2	91.7	109.2	97.2	93.3	86.2	91.1	92.5	93.1	92.7
06	WOOD, PLASTICS & COMPOSITES	88.3	54.4	68.9	93.8	38.8	62.4	95.3	44.3	66.2	80.2	104.3	93.9	94.0	77.3	84.5	92.6	93.0	92.8
07	THERMAL & MOISTURE PROTECTION	100.2	58.3	83.4	100.4	62.2	85.1	97.0	53.7	79.6	90.2	98.0	93.4	85.6	82.3	84.3	98.4	71.1	87.4
08	OPENINGS	96.0	50.7	85.0	96.4	56.4	86.7	96.4	52.2	85.6	97.9	98.8	98.2	98.7	75.8	93.1	100.3	91.0	98.0
0920	Plaster & Gypsum Board	98.8	52.9	66.7	100.0	36.9	55.9	97.0	42.6	58.9	61.0	104.5	91.4	89.2	76.9	80.6	89.2	92.2	91.3
0950, 0980	Ceilings & Acoustic Treatment	94.6	52.9	66.8	102.8	36.9	58.9	101.9	42.6	62.4	83.0	104.5	97.3	89.1	76.9	81.0	89.1	92.2	91.2
0960	Flooring	102.3	64.2	91.0	105.0	76.4	96.5	104.3	65.3	92.6	78.5	103.9	86.0	95.1	89.2	93.3	96.8	42.7	80.7
0970, 0990	Wall Finishes & Painting/Coating	95.3	40.7	62.5	95.3	54.9	71.1	95.3	50.1	68.2	95.2	94.7	94.9	93.9	69.2	79.1	93.9	68.3	78.6
09	FINISHES	99.9	55.6	75.1	101.9	47.9	71.7	101.3	49.1	72.1	82.9	103.6	94.5	94.4	79.6	86.1	95.2	77.3	85.1
COVERS	DIVS. 10 - 14, 25, 28, 41, 43, 44, 46	100.0	52.0	90.2	100.0	49.0	89.6	100.0	48.3	89.4	100.0	96.4	99.3	100.0	59.1	91.6	100.0	58.1	91.4
21, 22, 23	FIRE SUPPRESSION, PLUMBING & HVAC	99.9	66.3	86.2	99.9	56.4	82.1	99.7	62.4	84.5	93.8	89.7	92.1	99.9	82.4	92.8	94.0	79.4	88.0
26, 27, 3370	ELECTRICAL, COMMUNICATIONS & UTIL.	97.2	72.3	84.1	101.9	72.2	86.3	102.1	72.3	86.4	91.5	95.4	93.6	94.3	82.1	87.9	91.5	68.9	79.6
MF2010	WEIGHTED AVERAGE	100.1	65.9	85.0	99.4	63.8	83.7	98.8	63.2	83.1	93.7	98.0	95.6	95.6	83.3	90.1	94.9	80.0	88.3

DIVISION		KENTUCKY																	
		CORBIN			COVINGTON			ELIZABETHTOWN			FRANKFORT			HAZARD			HENDERSON		
		407 - 409			410			427			406			417 - 418			424		
		MAT.	INST.	TOTAL	MAT.	INST.	TOTAL	MAT.	INST.	TOTAL	MAT.	INST.	TOTAL	MAT.	INST.	TOTAL	MAT.	INST.	TOTAL
015433	CONTRACTOR EQUIPMENT		101.7	101.7		105.5	105.5		95.4	95.4		101.7	101.7		101.7	101.7		116.7	116.7
0241, 31 - 34	SITE & INFRASTRUCTURE, DEMOLITION	81.0	99.1	93.7	79.1	113.2	103.0	70.1	97.9	89.6	86.2	100.0	95.9	81.5	100.8	95.0	78.2	124.9	111.0
0310	Concrete Forming & Accessories	86.1	68.0	70.4	85.0	94.9	93.5	80.5	77.8	78.1	98.3	74.0	77.3	86.1	87.2	87.0	92.1	81.7	83.1
0320	Concrete Reinforcing	90.4	67.9	79.2	89.4	94.3	91.9	90.4	91.3	90.9	91.7	91.3	91.5	91.3	104.2	97.7	90.1	88.8	89.5
0330	Cast-in-Place Concrete	93.3	60.3	79.7	94.3	92.6	93.6	79.9	74.9	77.9	90.2	70.7	82.2	94.8	86.1	91.2	78.1	88.1	82.2
03	CONCRETE	92.9	65.8	79.5	96.0	94.0	95.0	86.1	79.5	82.9	92.5	76.4	84.6	94.6	90.0	92.3	90.8	85.4	88.1
04	MASONRY	91.8	64.5	75.1	105.0	105.8	105.5	78.1	78.5	78.3	88.6	80.6	83.7	89.0	74.6	80.2	97.5	91.1	93.6
05	METALS	92.5	77.8	87.9	90.6	92.6	91.2	92.4	87.5	90.9	99.7	88.4	96.1	92.5	93.5	92.9	84.0	88.6	85.4
06	WOOD, PLASTICS & COMPOSITES	89.8	71.8	79.5	90.9	87.8	89.1	89.5	77.3	82.5	101.6	71.8	84.6	89.8	93.0	91.6	96.5	79.2	86.6
07	THERMAL & MOISTURE PROTECTION	98.2	66.4	85.4	98.3	96.4	97.6	85.3	77.8	82.3	99.4	77.0	90.4	98.3	73.1	88.2	97.8	91.5	95.3
08	OPENINGS	99.6	63.1	90.7	102.1	87.0	98.4	98.7	81.5	94.5	98.8	78.2	93.8	100.7	81.3	96.0	96.6	81.7	93.0
0920	Plaster & Gypsum Board	88.3	70.3	75.7	70.2	88.0	82.6	88.3	76.9	80.4	98.7	70.3	78.8	88.3	92.2	91.0	92.0	77.9	82.2
0950, 0980	Ceilings & Acoustic Treatment	89.1	70.3	76.6	92.8	88.0	89.6	89.1	76.9	81.0	97.3	70.3	79.3	89.1	92.2	91.2	82.7	77.9	79.5
0960	Flooring	95.4	42.7	79.7	71.3	87.2	76.1	92.7	89.2	91.6	99.5	56.4	86.6	95.4	43.5	79.9	98.5	89.2	95.7
0970, 0990	Wall Finishes & Painting/Coating	93.9	51.6	68.6	89.7	84.7	86.7	93.9	77.6	84.1	93.9	83.4	87.6	93.9	68.3	78.6	93.9	92.9	93.3
09	FINISHES	94.4	61.9	76.2	85.9	92.2	89.4	93.2	79.2	85.3	99.0	71.2	83.4	94.4	78.1	85.3	93.8	84.2	88.4
COVERS	DIVS. 10 - 14, 25, 28, 41, 43, 44, 46	100.0	49.5	89.7	100.0	103.9	100.8	100.0	87.0	97.4	100.0	67.5	93.4	100.0	58.9	91.6	100.0	67.0	93.3
21, 22, 23	FIRE SUPPRESSION, PLUMBING & HVAC	93.9	68.3	83.4	94.6	92.4	93.7	94.3	80.9	88.8	100.0	82.1	92.7	94.0	80.4	88.4	94.3	76.8	87.2
26, 27, 3370	ELECTRICAL, COMMUNICATIONS & UTIL.	91.7	82.1	86.6	95.6	79.6	87.2	91.3	82.1	86.5	100.9	90.8	95.6	91.5	60.8	75.3	93.6	82.3	87.6
MF2010	WEIGHTED AVERAGE	94.2	71.4	84.1	95.0	94.1	94.6	91.9	82.6	87.8	98.0	82.3	91.1	94.4	80.4	88.2	92.8	86.6	90.1

	DIVISION	KENTUCKY																	
		LEXINGTON			LOUISVILLE			OWENSBORO			PADUCAH			PIKEVILLE			SOMERSET		
		403 - 405			400 - 402			423			420			415 - 416			425 - 426		
		MAT.	INST.	TOTAL	MAT.	INST.	TOTAL	MAT.	INST.	TOTAL	MAT.	INST.	TOTAL	MAT.	INST.	TOTAL	MAT.	INST.	TOTAL
015433	CONTRACTOR EQUIPMENT		101.7	101.7		95.4	95.4		116.7	116.7		116.7	116.7		97.9	97.9		101.7	101.7
0241, 31 - 34	SITE & INFRASTRUCTURE, DEMOLITION	82.9	101.0	95.6	73.1	97.9	90.5	87.6	124.8	113.7	80.8	124.4	111.4	119.8	86.2	96.2	74.9	99.7	92.3
0310	Concrete Forming & Accessories	97.4	74.8	77.9	96.3	77.8	80.3	90.6	81.4	82.7	88.6	77.9	79.4	96.3	92.7	93.2	86.8	72.8	74.7
0320	Concrete Reinforcing	99.3	91.0	95.2	99.3	91.3	95.3	90.1	88.5	89.3	90.6	85.0	87.8	93.4	105.1	99.2	90.4	91.3	90.9
0330	Cast-in-Place Concrete	95.4	85.5	91.4	97.5	74.8	88.1	91.1	96.6	93.4	83.2	86.3	84.5	97.9	97.2	97.6	78.1	101.4	87.7
03	CONCRETE	96.2	81.9	89.1	97.1	79.5	88.4	102.4	88.1	95.4	95.1	82.3	88.8	108.5	97.3	103.0	81.6	86.4	84.0
04	MASONRY	90.0	73.4	79.9	91.1	78.2	83.3	90.1	82.5	85.4	93.0	90.2	91.3	89.3	99.5	95.5	84.3	78.3	80.6
05	METALS	94.8	88.5	92.8	101.9	87.6	97.4	85.5	88.4	86.4	82.5	86.5	83.8	91.6	107.9	96.7	92.5	88.0	91.0
06	WOOD, PLASTICS & COMPOSITES	104.3	71.8	85.7	101.6	77.3	87.8	94.7	79.2	85.8	92.4	76.3	83.2	89.0	92.3	90.9	90.3	71.8	79.7
07	THERMAL & MOISTURE PROTECTION	98.4	86.1	93.4	95.8	77.7	88.5	98.1	90.0	94.8	97.8	80.5	90.8	90.8	82.6	87.5	97.8	69.7	86.5
08	OPENINGS	99.8	75.8	93.9	99.8	81.5	95.3	96.6	82.5	93.2	95.7	76.2	90.9	98.7	89.5	96.5	99.6	76.1	93.9
0920	Plaster & Gypsum Board	98.3	70.3	78.7	93.2	76.9	81.8	90.8	77.9	81.8	89.5	75.0	79.4	64.3	92.2	83.8	88.3	70.3	75.7
0950, 0980	Ceilings & Acoustic Treatment	93.7	70.3	78.1	97.3	76.9	83.7	82.7	77.9	79.5	82.7	75.0	77.6	83.0	92.2	89.1	89.1	70.3	76.6
0960	Flooring	99.9	72.8	91.8	98.2	89.2	95.5	97.8	89.2	95.2	96.7	61.5	86.2	82.4	103.9	88.8	95.6	42.7	79.9
0970, 0990	Wall Finishes & Painting/Coating	93.9	83.4	87.6	93.9	77.6	84.1	93.9	96.1	95.3	93.9	80.3	85.7	95.2	84.8	89.0	93.9	77.6	84.1
09	FINISHES	98.0	74.6	84.9	97.2	79.7	87.4	94.0	84.0	88.4	93.2	74.9	83.0	85.4	94.5	90.5	93.8	67.3	79.0
COVERS	DIVS. 10 - 14, 25, 28, 41, 43, 44, 46	100.0	89.2	97.8	100.0	86.9	97.3	100.0	95.0	99.0	100.0	63.0	92.4	100.0	58.7	91.6	100.0	59.5	91.7
21, 22, 23	FIRE SUPPRESSION, PLUMBING & HVAC	99.9	75.4	89.9	100.0	80.8	92.1	99.9	75.8	90.1	94.3	81.1	88.9	93.8	87.2	91.1	94.3	73.8	85.9
26, 27, 3370	ELECTRICAL, COMMUNICATIONS & UTIL.	94.8	82.1	88.1	101.7	79.2	89.9	93.7	82.3	87.7	96.2	79.0	87.1	94.7	76.5	85.1	91.9	82.1	86.7
MF2010	WEIGHTED AVERAGE	97.0	81.0	90.0	98.6	82.2	91.3	95.6	86.7	91.7	93.0	84.5	89.2	96.0	90.2	93.4	92.5	79.3	86.7

	DIVISION	LOUISIANA																	
		ALEXANDRIA			BATON ROUGE			HAMMOND			LAFAYETTE			LAKE CHARLES			MONROE		
		713 - 714			707 - 708			704			705			706			712		
		MAT.	INST.	TOTAL	MAT.	INST.	TOTAL	MAT.	INST.	TOTAL	MAT.	INST.	TOTAL	MAT.	INST.	TOTAL	MAT.	INST.	TOTAL
015433	CONTRACTOR EQUIPMENT		89.4	89.4		89.3	89.3		90.0	90.0		90.0	90.0		89.3	89.3		89.4	89.4
0241, 31 - 34	SITE & INFRASTRUCTURE, DEMOLITION	95.2	86.3	88.9	100.3	86.0	90.3	101.2	87.1	91.3	102.1	87.0	91.5	102.8	85.8	90.8	95.2	85.6	88.5
0310	Concrete Forming & Accessories	82.8	35.8	42.1	94.4	61.1	65.6	77.8	47.9	51.9	94.1	52.9	58.4	95.2	56.5	61.7	82.3	36.1	42.3
0320	Concrete Reinforcing	94.3	56.8	75.6	97.9	59.9	78.9	96.5	60.3	78.5	97.9	59.9	78.9	97.9	57.5	77.8	93.3	56.5	75.0
0330	Cast-in-Place Concrete	96.9	39.0	73.1	96.2	57.2	80.2	99.0	44.6	76.7	98.5	45.3	76.6	103.7	65.9	88.2	96.9	55.1	79.7
03	CONCRETE	96.8	42.2	69.8	95.8	60.1	78.2	95.9	50.0	73.2	96.9	52.3	74.9	99.4	60.6	80.3	96.6	47.7	72.5
04	MASONRY	118.4	48.6	75.8	107.9	50.3	72.8	107.3	52.0	73.6	107.3	49.5	72.0	106.3	55.9	75.5	113.4	45.1	71.7
05	METALS	89.7	71.1	83.9	101.8	70.6	92.0	97.2	70.2	88.7	96.4	70.4	88.2	96.4	70.0	88.1	89.7	68.8	83.1
06	WOOD, PLASTICS & COMPOSITES	88.0	34.5	57.5	99.9	66.7	80.9	85.2	49.1	64.6	106.2	56.0	77.6	104.5	60.0	79.1	87.3	34.9	57.3
07	THERMAL & MOISTURE PROTECTION	98.5	50.1	79.0	98.5	57.0	81.8	96.8	57.9	81.1	97.3	55.5	80.5	96.4	58.0	80.9	98.5	49.3	78.7
08	OPENINGS	98.1	40.9	84.2	100.7	59.1	90.6	96.7	53.5	86.2	100.7	53.3	89.2	100.7	56.0	89.8	98.1	44.8	85.1
0920	Plaster & Gypsum Board	75.7	33.0	45.8	103.5	66.0	77.3	95.0	47.8	62.0	106.1	55.0	70.4	106.1	59.1	73.3	75.3	33.3	46.0
0950, 0980	Ceilings & Acoustic Treatment	89.1	33.0	51.7	106.7	66.0	79.6	99.3	47.8	65.0	97.4	55.0	69.2	98.5	59.1	72.3	89.1	33.3	52.0
0960	Flooring	100.8	66.2	90.5	104.0	66.2	92.8	95.7	66.2	86.9	104.0	66.2	92.8	104.0	75.6	95.5	100.4	59.2	88.1
0970, 0990	Wall Finishes & Painting/Coating	100.7	33.3	60.3	105.1	38.1	65.0	105.1	50.5	72.4	105.1	38.1	65.0	105.1	35.1	63.2	100.7	39.7	64.2
09	FINISHES	91.4	40.1	62.7	102.8	60.2	79.0	97.7	51.6	71.9	101.1	53.7	74.6	101.3	57.9	77.0	91.3	39.8	62.4
COVERS	DIVS. 10 - 14, 25, 28, 41, 43, 44, 46	100.0	49.1	89.6	100.0	80.1	95.9	100.0	47.3	89.2	100.0	78.7	95.7	100.0	79.5	95.8	100.0	45.1	88.8
21, 22, 23	FIRE SUPPRESSION, PLUMBING & HVAC	100.3	50.3	79.8	100.0	58.1	82.8	94.0	45.1	74.0	100.1	58.4	83.0	100.1	58.7	83.2	100.3	49.2	79.4
26, 27, 3370	ELECTRICAL, COMMUNICATIONS & UTIL.	93.0	55.4	73.2	110.6	61.1	84.6	101.3	55.5	77.2	102.6	65.7	83.2	102.1	65.7	82.9	95.1	56.6	74.8
MF2010	WEIGHTED AVERAGE	97.2	52.8	77.6	101.5	62.6	84.4	97.5	55.4	78.9	99.9	61.0	82.7	100.1	63.4	83.9	97.1	52.9	77.6

	DIVISION	LOUISIANA									MAINE								
		NEW ORLEANS			SHREVEPORT			THIBODAUX			AUGUSTA			BANGOR			BATH		
		700 - 701			710 - 711			703			043			044			045		
		MAT.	INST.	TOTAL	MAT.	INST.	TOTAL	MAT.	INST.	TOTAL	MAT.	INST.	TOTAL	MAT.	INST.	TOTAL	MAT.	INST.	TOTAL
015433	CONTRACTOR EQUIPMENT		90.2	90.2		89.4	89.4		90.0	90.0		100.5	100.5		100.5	100.5		100.5	100.5
0241, 31 - 34	SITE & INFRASTRUCTURE, DEMOLITION	105.2	88.6	93.5	97.4	85.5	89.1	103.6	87.0	91.9	91.7	100.8	98.1	94.7	101.0	99.1	92.7	100.8	98.4
0310	Concrete Forming & Accessories	96.5	67.9	71.8	99.1	38.1	46.3	89.6	60.7	64.6	97.0	99.1	98.9	90.0	101.0	99.5	86.3	99.2	97.5
0320	Concrete Reinforcing	97.9	61.0	79.5	92.9	56.9	75.0	96.5	60.2	78.4	90.4	112.0	101.2	90.1	113.3	101.6	89.1	113.0	101.0
0330	Cast-in-Place Concrete	102.7	69.2	88.9	96.4	44.4	75.0	106.3	47.5	82.2	94.2	61.2	80.7	82.6	115.3	96.1	82.6	61.3	73.8
03	CONCRETE	99.0	67.6	83.5	96.7	45.0	71.2	101.2	56.6	79.2	102.1	87.8	95.0	95.6	107.4	101.4	95.6	88.0	91.9
04	MASONRY	107.6	61.1	79.2	103.9	46.7	69.0	135.0	47.8	81.7	93.8	66.3	77.0	107.0	104.5	105.5	113.5	86.0	96.7
05	METALS	109.5	73.8	98.3	86.3	68.9	80.8	97.2	69.9	88.7	86.4	88.9	87.2	86.0	91.1	87.6	84.5	90.3	86.3
06	WOOD, PLASTICS & COMPOSITES	100.8	69.3	82.8	104.0	37.8	66.2	93.3	65.9	77.7	94.1	110.0	103.2	89.2	100.8	95.8	84.7	110.0	99.2
07	THERMAL & MOISTURE PROTECTION	96.7	69.9	85.9	100.0	48.3	79.2	96.7	53.6	79.4	102.1	67.8	88.3	97.3	87.4	93.3	97.3	73.8	87.8
08	OPENINGS	101.8	67.8	93.5	96.0	42.9	83.1	101.7	62.7	92.2	100.4	91.9	98.3	100.3	86.8	97.0	100.3	91.8	98.2
0920	Plaster & Gypsum Board	104.1	68.7	79.3	86.8	36.4	51.6	98.7	65.2	75.3	94.5	109.7	105.1	93.5	100.2	98.2	91.0	109.7	104.1
0950, 0980	Ceilings & Acoustic Treatment	97.4	68.7	78.3	95.5	36.4	56.1	99.3	65.2	76.6	94.9	109.7	104.8	90.4	100.2	96.9	88.3	109.7	102.6
0960	Flooring	104.4	66.2	93.0	108.0	63.5	94.7	101.5	45.7	84.9	98.1	57.3	85.9	96.3	111.6	100.9	94.7	57.3	83.6
0970, 0990	Wall Finishes & Painting/Coating	106.5	67.0	82.8	100.7	33.3	60.3	106.5	50.5	72.9	96.1	66.2	78.2	96.1	46.1	66.2	96.1	36.1	60.1
09	FINISHES	101.5	67.6	82.5	96.5	41.5	65.7	100.2	57.5	76.2	98.0	89.6	93.3	96.8	97.7	97.3	95.5	86.3	90.3
COVERS	DIVS. 10 - 14, 25, 28, 41, 43, 44, 46	100.0	84.0	96.7	100.0	62.5	92.3	100.0	80.6	96.0	100.0	102.0	100.4	100.0	111.8	102.4	100.0	102.0	100.4
21, 22, 23	FIRE SUPPRESSION, PLUMBING & HVAC	100.0	67.1	86.5	100.1	54.1	81.3	94.0	58.6	79.5	99.8	66.7	86.3	100.0	77.2	90.7	94.0	66.7	82.8
26, 27, 3370	ELECTRICAL, COMMUNICATIONS & UTIL.	102.7	72.9	87.0	100.8	65.3	82.1	99.7	68.0	83.0	100.9	78.6	89.2	99.7	78.6	88.6	97.5	78.6	87.5
MF2010	WEIGHTED AVERAGE	102.4	70.5	88.4	97.0	55.6	78.8	100.1	62.7	83.6	97.5	81.7	90.5	97.1	92.1	94.9	95.3	83.6	90.1

	DIVISION	MAINE																	
		HOULTON			KITTERY			LEWISTON			MACHIAS			PORTLAND			ROCKLAND		
		047			039			042			046			040 - 041			048		
		MAT.	INST.	TOTAL	MAT.	INST.	TOTAL	MAT.	INST.	TOTAL	MAT.	INST.	TOTAL	MAT.	INST.	TOTAL	MAT.	INST.	TOTAL
015433	CONTRACTOR EQUIPMENT		100.5	100.5		100.5	100.5		100.5	100.5		100.5	100.5		100.5	100.5		100.5	100.5
0241, 31 - 34	SITE & INFRASTRUCTURE, DEMOLITION	94.5	101.8	99.6	88.3	100.9	97.1	91.5	101.0	98.2	93.8	101.8	99.4	88.9	101.0	97.4	89.7	101.8	98.2
0310	Concrete Forming & Accessories	93.1	105.8	104.1	86.9	99.6	97.9	94.9	101.0	100.2	90.7	105.8	103.7	96.2	101.0	100.3	91.6	105.8	103.9
0320	Concrete Reinforcing	90.1	111.7	100.9	88.6	113.0	100.7	110.7	113.3	112.0	90.1	111.7	100.9	110.7	113.3	112.0	90.1	111.7	100.9
0330	Cast-in-Place Concrete	82.6	73.9	79.0	82.6	62.1	74.2	84.3	115.3	97.0	82.6	73.3	78.8	104.5	115.3	108.9	84.3	73.3	79.8
03	CONCRETE	96.5	95.1	95.8	91.4	88.5	89.9	96.3	107.4	101.8	96.0	94.9	95.4	106.1	107.4	106.7	93.4	94.9	94.1
04	MASONRY	92.1	78.1	83.6	103.7	87.0	93.5	92.8	104.5	99.9	92.1	78.1	83.6	93.1	104.5	100.0	87.1	78.1	81.6
05	METALS	84.7	88.8	86.0	84.4	90.3	86.3	88.9	91.1	89.6	84.7	88.7	86.0	90.2	91.1	90.5	84.6	88.7	85.9
06	WOOD, PLASTICS & COMPOSITES	92.9	110.0	102.7	85.2	110.0	99.4	94.8	100.8	98.2	90.1	110.0	101.5	94.0	100.8	97.9	90.9	110.0	101.8
07	THERMAL & MOISTURE PROTECTION	97.4	75.1	88.4	96.9	73.1	87.3	97.1	87.4	93.2	97.3	74.1	88.0	104.5	87.4	97.6	97.1	74.1	87.8
08	OPENINGS	100.4	90.0	97.9	100.3	91.8	98.2	103.4	86.8	99.3	100.4	90.0	97.9	103.4	86.8	99.3	100.3	90.0	97.8
0920	Plaster & Gypsum Board	96.0	109.7	105.6	91.0	109.7	104.1	99.2	100.2	99.9	94.3	109.7	105.1	98.7	100.2	99.7	94.3	109.7	105.1
0950, 0980	Ceilings & Acoustic Treatment	88.3	109.7	102.6	88.3	109.7	102.6	97.7	100.2	99.3	88.3	109.7	102.6	100.4	100.2	100.2	88.3	109.7	102.6
0960	Flooring	97.5	53.6	84.4	94.8	59.7	84.4	98.9	111.6	102.7	96.6	53.6	83.8	98.1	111.6	102.1	97.0	53.6	84.0
0970, 0990	Wall Finishes & Painting/Coating	96.1	140.6	122.8	96.1	40.2	62.6	96.1	46.1	66.2	96.1	140.6	122.8	96.1	46.1	66.2	96.1	140.6	122.8
09	FINISHES	97.0	101.5	99.6	95.1	87.4	90.8	99.4	97.7	98.4	96.5	101.5	99.3	99.1	97.7	98.3	96.1	101.5	99.2
COVERS	DIVS. 10 - 14, 25, 28, 41, 43, 44, 46	100.0	107.7	101.6	100.0	102.3	100.5	100.0	111.8	102.4	100.0	107.7	101.6	100.0	111.8	102.4	100.0	107.7	101.6
21, 22, 23	FIRE SUPPRESSION, PLUMBING & HVAC	94.0	75.5	86.4	94.0	67.3	83.0	100.0	77.2	90.7	94.0	75.5	86.4	99.8	77.2	90.6	94.0	75.5	86.4
26, 27, 3370	ELECTRICAL, COMMUNICATIONS & UTIL.	102.0	78.6	89.7	97.5	78.6	87.5	102.0	78.6	89.7	102.0	78.6	89.7	104.5	78.6	90.9	101.9	78.6	89.6
MF2010	WEIGHTED AVERAGE	95.0	87.6	91.7	94.2	84.0	89.7	97.6	92.1	95.2	94.9	87.5	91.6	99.2	92.1	96.1	94.2	87.5	91.2

	DIVISION	MAINE			MARYLAND														
		WATERVILLE			ANNAPOLIS			BALTIMORE			COLLEGE PARK			CUMBERLAND			EASTON		
		049			214			210 - 212			207 - 208			215			216		
		MAT.	INST.	TOTAL	MAT.	INST.	TOTAL	MAT.	INST.	TOTAL	MAT.	INST.	TOTAL	MAT.	INST.	TOTAL	MAT.	INST.	TOTAL
015433	CONTRACTOR EQUIPMENT		100.5	100.5		99.4	99.4		102.9	102.9		103.6	103.6		99.4	99.4		99.4	99.4
0241, 31 - 34	SITE & INFRASTRUCTURE, DEMOLITION	94.7	100.8	99.0	99.5	91.1	93.6	99.9	95.1	96.6	105.0	93.7	97.1	91.2	90.9	91.0	98.1	88.3	91.2
0310	Concrete Forming & Accessories	85.8	99.1	97.3	96.5	71.2	74.6	97.0	74.2	77.3	87.0	72.0	74.0	88.2	80.5	81.5	86.3	71.3	73.3
0320	Concrete Reinforcing	90.1	112.0	101.0	102.9	83.4	93.2	102.9	83.5	93.2	105.7	79.8	92.8	90.9	75.6	83.3	90.1	79.7	84.9
0330	Cast-in-Place Concrete	82.6	61.2	73.8	134.3	78.7	111.5	119.6	79.5	103.1	126.7	79.3	107.2	106.8	83.7	97.3	118.6	50.1	90.4
03	CONCRETE	97.1	87.8	92.5	116.8	77.1	97.2	109.8	78.8	94.5	112.5	77.2	95.1	98.0	81.6	89.9	106.4	66.6	86.7
04	MASONRY	102.0	66.3	80.2	98.9	74.7	84.1	100.1	74.7	84.6	106.5	69.5	83.9	98.8	82.8	89.0	112.2	44.7	71.0
05	METALS	84.6	88.8	86.0	97.3	95.0	96.6	98.4	95.6	97.5	85.1	96.2	88.6	94.8	92.0	93.9	95.0	88.3	92.9
06	WOOD, PLASTICS & COMPOSITES	84.2	110.0	99.0	98.0	71.3	82.7	100.8	75.3	86.2	81.9	71.5	76.0	90.9	79.5	84.4	88.8	78.1	82.7
07	THERMAL & MOISTURE PROTECTION	97.4	67.8	85.5	99.2	79.7	91.3	97.4	80.4	90.5	100.6	78.2	91.6	96.8	78.5	89.4	96.9	61.4	82.6
08	OPENINGS	100.4	91.9	98.3	92.5	78.7	89.1	94.8	81.5	91.6	92.7	74.7	88.3	93.5	79.2	90.0	91.8	72.3	87.1
0920	Plaster & Gypsum Board	91.0	109.7	104.1	96.7	70.7	78.5	99.7	74.7	82.2	98.2	70.7	79.0	93.9	79.2	83.6	93.4	77.8	82.5
0950, 0980	Ceilings & Acoustic Treatment	88.3	109.7	102.6	103.1	70.7	81.5	99.5	74.7	82.9	96.5	70.7	79.3	99.5	79.2	86.0	99.5	77.8	85.0
0960	Flooring	94.4	57.3	83.3	91.4	80.0	88.0	91.4	80.0	88.0	103.0	81.2	96.5	87.9	89.7	88.4	87.1	53.6	77.1
0970, 0990	Wall Finishes & Painting/Coating	96.1	66.2	78.2	98.0	82.8	88.8	98.0	82.8	88.8	109.8	78.0	90.7	98.0	76.6	85.2	98.0	78.0	86.0
09	FINISHES	95.7	89.6	92.3	96.8	73.1	83.5	96.6	75.5	84.8	99.0	73.6	84.8	94.2	81.7	87.2	94.4	70.3	80.9
COVERS	DIVS. 10 - 14, 25, 28, 41, 43, 44, 46	100.0	102.0	100.4	100.0	86.7	97.3	100.0	87.5	97.4	100.0	82.2	96.4	100.0	90.9	98.1	100.0	78.2	95.5
21, 22, 23	FIRE SUPPRESSION, PLUMBING & HVAC	94.0	66.7	82.8	100.0	82.3	92.8	100.0	82.4	92.8	94.2	87.0	91.2	93.9	72.8	85.3	93.9	69.3	83.8
26, 27, 3370	ELECTRICAL, COMMUNICATIONS & UTIL.	102.0	78.6	89.7	101.5	91.4	96.2	101.1	93.3	97.0	99.0	101.1	100.1	98.7	83.1	90.5	98.1	64.9	80.6
MF2010	WEIGHTED AVERAGE	95.4	81.7	89.4	100.4	82.7	92.6	100.1	84.1	93.0	97.0	84.5	91.5	95.8	81.9	89.7	97.3	69.7	85.1

	DIVISION	MARYLAND															MASSACHUSETTS		
		ELKTON			HAGERSTOWN			SALISBURY			SILVER SPRING			WALDORF			BOSTON		
		219			217			218			209			206			020 - 022, 024		
		MAT.	INST.	TOTAL	MAT.	INST.	TOTAL	MAT.	INST.	TOTAL	MAT.	INST.	TOTAL	MAT.	INST.	TOTAL	MAT.	INST.	TOTAL
015433	CONTRACTOR EQUIPMENT		99.4	99.4		99.4	99.4		99.4	99.4		96.4	96.4		96.4	96.4		106.1	106.1
0241, 31 - 34	SITE & INFRASTRUCTURE, DEMOLITION	85.6	89.0	88.0	89.5	91.8	91.1	98.0	88.3	91.2	92.0	86.7	88.3	98.8	86.7	90.3	99.6	108.2	105.6
0310	Concrete Forming & Accessories	92.0	84.4	85.4	87.3	77.4	78.7	99.5	54.1	60.2	95.7	70.2	73.6	102.9	70.3	74.7	98.8	143.9	137.8
0320	Concrete Reinforcing	90.1	103.3	96.7	90.9	75.7	83.3	90.1	66.7	78.5	104.4	75.1	89.8	105.1	75.1	90.2	109.5	153.9	131.6
0330	Cast-in-Place Concrete	96.1	78.1	88.7	101.8	83.8	94.4	118.6	48.7	89.9	129.7	78.2	108.5	145.4	78.2	117.8	106.5	155.0	126.5
03	CONCRETE	90.2	86.5	88.4	94.1	80.2	87.3	107.3	56.1	82.0	111.1	75.1	93.3	122.3	75.2	99.0	109.7	148.5	128.9
04	MASONRY	98.1	61.4	75.7	104.0	82.8	91.0	112.0	49.3	73.7	105.4	67.6	82.3	90.8	67.6	76.7	109.8	165.9	144.1
05	METALS	95.1	100.6	96.8	94.9	92.1	94.0	95.1	83.5	91.4	89.3	92.6	90.3	89.3	92.7	90.4	100.8	131.2	110.3
06	WOOD, PLASTICS & COMPOSITES	95.3	90.7	92.7	89.9	75.1	81.5	104.8	57.5	77.8	88.2	70.7	78.2	95.7	70.7	81.4	98.1	143.2	123.9
07	THERMAL & MOISTURE PROTECTION	96.5	74.6	87.7	96.5	76.8	88.6	97.0	66.5	84.8	103.2	81.7	94.6	103.7	81.7	94.9	97.4	153.0	119.8
08	OPENINGS	91.8	83.1	89.7	91.8	75.6	87.8	92.0	63.7	85.1	85.7	74.3	82.9	86.4	74.3	83.5	98.9	145.0	110.1
0920	Plaster & Gypsum Board	96.3	90.8	92.5	93.4	74.7	80.3	102.6	56.5	70.3	103.9	70.7	80.7	107.6	70.7	81.8	101.9	144.0	131.3
0950, 0980	Ceilings & Acoustic Treatment	99.5	90.8	93.7	100.5	74.7	83.3	99.5	56.5	70.8	106.6	70.7	82.7	106.6	70.7	82.7	96.6	144.0	128.2
0960	Flooring	89.4	60.6	80.8	87.6	89.7	88.2	92.6	67.1	85.0	109.3	81.2	100.9	113.0	81.2	103.5	98.7	184.0	124.1
0970, 0990	Wall Finishes & Painting/Coating	98.0	78.0	86.0	98.0	78.0	86.0	98.0	78.0	86.0	116.8	78.0	93.5	116.8	78.0	93.5	98.5	152.4	130.8
09	FINISHES	94.5	80.0	86.4	94.1	79.3	85.8	97.1	58.4	75.4	99.4	71.8	83.9	101.3	72.4	85.1	99.2	152.7	129.2
COVERS	DIVS. 10 - 14, 25, 28, 41, 43, 44, 46	100.0	62.4	92.3	100.0	90.4	98.0	100.0	40.3	87.8	100.0	79.8	95.9	100.0	79.8	95.9	100.0	119.6	104.0
21, 22, 23	FIRE SUPPRESSION, PLUMBING & HVAC	93.9	81.9	89.0	99.9	83.4	93.2	93.9	63.2	81.3	94.2	85.7	90.7	94.2	85.6	90.7	100.0	128.2	111.6
26, 27, 3370	ELECTRICAL, COMMUNICATIONS & UTIL.	100.1	91.4	95.5	98.5	83.1	90.4	96.8	70.0	82.7	96.1	101.1	98.7	93.5	101.1	97.5	101.1	139.7	121.4
MF2010	WEIGHTED AVERAGE	94.8	83.2	89.7	96.8	83.5	90.9	97.7	64.6	83.1	96.3	82.6	90.3	97.0	82.7	90.7	101.5	139.2	118.1

		MASSACHUSETTS																	
	DIVISION	BROCKTON			BUZZARDS BAY			FALL RIVER			FITCHBURG			FRAMINGHAM			GREENFIELD		
		023			025			027			014			017			013		
		MAT.	INST.	TOTAL	MAT.	INST.	TOTAL	MAT.	INST.	TOTAL	MAT.	INST.	TOTAL	MAT.	INST.	TOTAL	MAT.	INST.	TOTAL
015433	**CONTRACTOR EQUIPMENT**		102.2	102.2		102.2	102.2		103.2	103.2		100.5	100.5		101.6	101.6		100.5	100.5
0241, 31 - 34	**SITE & INFRASTRUCTURE, DEMOLITION**	96.9	104.7	102.4	86.9	104.6	99.3	96.0	104.8	102.2	88.8	104.6	99.9	85.1	104.3	98.6	92.7	103.1	100.0
0310	Concrete Forming & Accessories	99.1	133.4	128.8	96.9	133.2	128.3	99.1	133.5	128.9	91.1	128.4	123.4	97.9	133.1	128.4	89.5	110.8	108.0
0320	Concrete Reinforcing	110.7	153.8	132.1	88.7	167.9	128.2	110.7	167.9	139.2	88.7	150.2	119.3	88.7	153.1	120.8	92.5	127.6	110.0
0330	Cast-in-Place Concrete	101.4	155.7	123.7	84.3	155.6	113.6	98.1	156.1	122.0	88.4	145.3	111.8	88.4	141.4	110.2	90.9	125.0	104.9
03	**CONCRETE**	107.0	143.9	125.2	90.3	146.4	118.0	105.4	146.7	125.8	88.8	137.3	112.8	91.6	138.8	114.9	92.6	118.2	105.3
04	**MASONRY**	105.5	163.4	140.9	97.5	163.4	137.7	106.4	163.3	141.1	94.1	152.1	129.5	100.0	152.2	131.9	98.2	128.7	116.8
05	**METALS**	97.8	128.9	107.6	92.9	134.3	105.9	97.8	134.7	109.4	92.8	123.8	102.5	92.9	127.6	103.8	95.0	109.7	99.6
06	**WOOD, PLASTICS & COMPOSITES**	97.0	131.1	116.4	94.2	131.1	115.2	97.0	131.3	116.6	90.7	125.3	110.5	96.6	130.8	116.1	88.8	107.5	99.5
07	**THERMAL & MOISTURE PROTECTION**	97.4	149.4	118.3	96.8	144.1	115.8	97.4	143.3	115.8	96.9	138.7	113.7	96.9	144.3	116.0	96.9	120.0	106.2
08	**OPENINGS**	96.8	137.9	106.8	92.5	136.0	103.1	96.8	136.1	106.3	102.1	134.0	109.9	92.9	137.8	103.8	102.2	112.9	104.8
0920	Plaster & Gypsum Board	96.4	131.4	120.9	92.4	131.4	119.7	96.4	131.4	120.9	93.5	125.5	115.9	96.4	131.4	120.9	94.2	107.1	103.2
0950, 0980	Ceilings & Acoustic Treatment	98.7	131.4	120.5	89.3	131.4	117.4	98.7	131.4	120.5	88.3	125.5	113.1	88.3	131.4	117.0	95.6	107.1	103.3
0960	Flooring	99.3	184.0	124.5	98.2	184.0	123.8	99.1	184.0	124.4	96.3	184.0	122.5	97.7	184.0	123.4	95.6	139.2	108.6
0970, 0990	Wall Finishes & Painting/Coating	98.1	137.9	121.9	98.1	137.9	121.9	98.1	137.9	121.9	96.1	137.9	121.1	97.1	137.9	121.5	96.1	108.4	103.5
09	**FINISHES**	99.2	143.3	123.9	95.6	143.3	122.3	99.1	143.5	124.0	94.9	139.9	120.1	95.4	143.1	122.1	96.8	115.7	107.3
COVERS	**DIVS. 10 - 14, 25, 28, 41, 43, 44, 46**	100.0	117.4	103.6	100.0	117.4	103.6	100.0	118.0	103.7	100.0	104.1	100.8	100.0	117.0	103.5	100.0	106.2	101.3
21, 22, 23	**FIRE SUPPRESSION, PLUMBING & HVAC**	100.0	112.3	105.0	94.0	112.4	101.5	100.0	112.5	105.1	94.9	115.9	103.5	94.9	124.2	106.8	94.9	103.7	98.5
26, 27, 3370	**ELECTRICAL, COMMUNICATIONS & UTIL.**	100.2	99.1	99.6	97.1	99.1	98.1	100.2	99.1	99.6	102.5	99.9	101.1	98.6	126.0	113.0	102.5	94.5	98.3
MF2010	**WEIGHTED AVERAGE**	100.1	127.1	112.0	94.3	127.7	109.0	99.9	127.9	112.3	95.7	124.1	108.2	94.9	131.2	110.9	96.9	109.9	102.6

		MASSACHUSETTS																	
	DIVISION	HYANNIS			LAWRENCE			LOWELL			NEW BEDFORD			PITTSFIELD			SPRINGFIELD		
		026			019			018			027			012			010 - 011		
		MAT.	INST.	TOTAL	MAT.	INST.	TOTAL	MAT.	INST.	TOTAL	MAT.	INST.	TOTAL	MAT.	INST.	TOTAL	MAT.	INST.	TOTAL
015433	**CONTRACTOR EQUIPMENT**		102.2	102.2		102.2	102.2		100.5	100.5		103.2	103.2		100.5	100.5		100.5	100.5
0241, 31 - 34	**SITE & INFRASTRUCTURE, DEMOLITION**	93.2	104.6	101.2	97.6	104.7	102.6	96.9	104.7	102.3	94.5	104.8	101.7	97.8	102.8	101.3	97.3	103.2	101.4
0310	Concrete Forming & Accessories	91.4	133.2	127.6	98.9	133.7	129.0	95.9	133.5	128.4	99.1	133.5	128.9	95.9	108.1	106.5	96.1	110.4	108.5
0320	Concrete Reinforcing	88.7	167.9	128.2	109.7	144.0	126.8	110.7	143.4	127.0	110.7	167.9	139.2	91.8	127.6	109.7	110.7	127.6	119.1
0330	Cast-in-Place Concrete	92.6	155.6	118.5	102.3	152.6	123.0	93.0	152.5	117.5	86.5	156.1	115.1	101.4	121.0	109.5	96.8	124.3	108.1
03	**CONCRETE**	96.9	146.4	121.3	107.2	141.1	124.0	98.4	140.8	119.4	99.8	146.7	123.0	99.4	115.6	107.4	100.3	117.8	108.9
04	**MASONRY**	104.5	163.4	140.4	104.7	163.4	140.5	93.4	152.1	129.2	104.5	163.3	140.4	94.1	121.4	110.8	93.7	127.4	114.3
05	**METALS**	94.4	134.3	106.9	95.4	125.2	104.8	95.4	121.9	103.7	97.8	134.7	109.4	95.2	109.7	99.7	97.7	109.7	101.5
06	**WOOD, PLASTICS & COMPOSITES**	88.1	131.1	112.6	97.0	131.1	116.4	96.4	131.1	116.2	97.0	131.3	116.6	96.4	107.5	102.7	96.4	107.5	102.7
07	**THERMAL & MOISTURE PROTECTION**	97.0	144.1	115.9	97.3	149.0	118.1	97.2	145.0	116.4	97.3	143.3	115.8	97.2	116.8	105.1	97.2	119.4	106.1
08	**OPENINGS**	93.1	136.0	103.5	96.8	135.3	106.1	103.4	135.3	111.2	96.8	136.1	106.3	103.4	112.9	105.7	103.4	112.9	105.7
0920	Plaster & Gypsum Board	88.7	131.4	118.6	99.2	131.4	121.7	99.2	131.4	121.7	96.4	131.4	120.9	99.2	107.1	104.7	99.2	107.1	104.7
0950, 0980	Ceilings & Acoustic Treatment	91.4	131.4	118.1	97.7	131.4	120.1	97.7	131.4	120.1	98.7	131.4	120.5	97.7	107.1	104.0	97.7	107.1	104.0
0960	Flooring	96.2	184.0	122.3	98.3	184.0	123.8	98.3	184.0	123.8	99.1	184.0	124.4	98.5	139.2	110.6	98.0	139.2	110.3
0970, 0990	Wall Finishes & Painting/Coating	98.1	137.9	121.9	96.2	137.9	121.2	96.1	137.9	121.1	98.1	137.9	121.9	96.1	108.4	103.5	97.5	108.4	104.0
09	**FINISHES**	95.4	143.3	122.2	98.9	143.3	123.8	98.8	143.3	123.8	99.0	143.5	123.9	98.9	113.8	107.3	98.9	115.3	108.1
COVERS	**DIVS. 10 - 14, 25, 28, 41, 43, 44, 46**	100.0	117.4	103.6	100.0	117.5	103.6	100.0	117.5	103.6	100.0	118.0	103.7	100.0	103.8	100.8	100.0	105.8	101.2
21, 22, 23	**FIRE SUPPRESSION, PLUMBING & HVAC**	100.0	112.4	105.1	100.0	124.2	109.9	100.0	124.1	109.9	100.0	112.5	105.1	100.0	98.0	99.2	100.0	103.1	101.3
26, 27, 3370	**ELECTRICAL, COMMUNICATIONS & UTIL.**	97.6	99.1	98.4	101.4	126.0	114.4	102.0	126.0	114.6	101.1	99.1	100.0	102.0	94.5	98.0	102.0	94.5	98.0
MF2010	**WEIGHTED AVERAGE**	97.3	127.7	110.7	99.8	132.5	114.2	99.0	130.9	113.1	99.3	127.9	111.9	99.1	107.1	102.7	99.6	109.5	104.0

		MASSACHUSETTS			MICHIGAN														
	DIVISION	WORCESTER			ANN ARBOR			BATTLE CREEK			BAY CITY			DEARBORN			DETROIT		
		015 - 016			481			490			487			481			482		
		MAT.	INST.	TOTAL	MAT.	INST.	TOTAL	MAT.	INST.	TOTAL	MAT.	INST.	TOTAL	MAT.	INST.	TOTAL	MAT.	INST.	TOTAL
015433	**CONTRACTOR EQUIPMENT**		100.5	100.5		110.4	110.4		102.4	102.4		110.4	110.4		110.4	110.4		98.5	98.5
0241, 31 - 34	**SITE & INFRASTRUCTURE, DEMOLITION**	97.2	104.6	102.4	79.8	97.2	92.0	87.9	87.6	87.7	71.4	96.0	88.7	79.6	97.4	92.1	94.8	99.0	97.8
0310	Concrete Forming & Accessories	96.4	128.4	124.1	99.4	108.9	107.6	97.7	88.3	89.6	99.5	87.6	89.2	99.2	113.8	111.9	99.9	113.9	112.0
0320	Concrete Reinforcing	110.7	152.4	131.4	101.9	123.3	112.5	99.3	89.9	94.6	101.9	122.4	112.1	101.9	123.4	112.6	101.2	123.4	112.3
0330	Cast-in-Place Concrete	96.2	145.3	116.4	87.0	110.7	96.7	95.3	101.3	97.8	83.2	91.7	86.7	85.0	111.5	95.9	93.3	111.5	100.8
03	**CONCRETE**	100.0	137.7	118.6	92.3	112.5	102.3	96.1	92.5	94.3	90.5	96.3	93.4	91.4	115.0	103.0	95.3	113.8	104.4
04	**MASONRY**	93.2	152.1	129.2	99.1	106.9	103.9	97.1	85.5	90.0	98.6	87.2	91.6	98.9	113.2	107.6	98.1	113.2	107.3
05	**METALS**	97.8	124.6	106.2	93.6	118.6	101.5	93.7	86.1	91.3	94.2	115.2	100.8	93.6	119.1	101.6	97.6	101.6	98.9
06	**WOOD, PLASTICS & COMPOSITES**	96.8	125.3	113.1	99.5	109.1	105.0	100.6	88.1	93.5	99.5	86.8	92.3	99.5	113.6	107.6	99.9	113.6	107.7
07	**THERMAL & MOISTURE PROTECTION**	97.2	138.7	113.9	101.0	108.4	104.0	94.2	86.0	90.9	98.7	94.4	97.0	99.5	114.3	105.5	98.0	114.3	104.6
08	**OPENINGS**	103.4	134.6	111.0	98.4	108.4	100.9	94.1	82.5	91.3	98.4	94.4	97.4	98.4	110.8	101.4	100.2	110.9	102.8
0920	Plaster & Gypsum Board	99.2	125.5	117.6	106.1	108.4	107.7	97.0	83.9	87.8	106.1	85.4	91.6	106.1	113.1	111.0	106.1	113.1	111.0
0950, 0980	Ceilings & Acoustic Treatment	97.7	125.5	116.2	94.7	108.4	103.8	93.7	83.9	87.1	95.7	85.4	88.9	94.7	113.1	106.9	95.7	113.1	107.3
0960	Flooring	98.3	171.6	120.1	95.7	115.9	101.7	98.8	95.2	97.7	95.5	79.7	90.8	95.3	116.6	101.6	95.4	116.6	101.7
0970, 0990	Wall Finishes & Painting/Coating	96.1	137.9	121.1	88.6	101.1	96.1	93.9	81.9	86.7	88.6	82.6	85.0	88.6	101.2	96.1	90.0	101.2	96.7
09	**FINISHES**	98.8	137.4	120.5	96.5	109.5	103.8	97.4	88.9	92.6	96.2	84.9	89.9	96.4	113.3	105.9	98.0	113.3	106.6
COVERS	**DIVS. 10 - 14, 25, 28, 41, 43, 44, 46**	100.0	110.3	102.1	100.0	106.3	101.3	100.0	99.3	99.9	100.0	93.3	98.6	100.0	108.1	101.6	100.0	108.1	101.6
21, 22, 23	**FIRE SUPPRESSION, PLUMBING & HVAC**	100.0	115.9	106.5	100.0	98.4	99.4	99.9	89.2	95.6	100.0	86.4	94.4	100.0	110.5	104.3	100.0	111.7	104.8
26, 27, 3370	**ELECTRICAL, COMMUNICATIONS & UTIL.**	102.0	100.6	101.3	96.3	108.3	102.6	94.3	81.8	87.7	95.1	89.5	92.1	96.3	107.1	102.0	97.3	107.0	102.4
MF2010	**WEIGHTED AVERAGE**	99.6	124.3	110.4	96.8	106.7	101.2	96.5	87.7	92.6	96.2	92.3	94.5	96.6	111.0	102.9	98.4	109.6	103.4

DIVISION		MICHIGAN																	
		FLINT			GAYLORD			GRAND RAPIDS			IRON MOUNTAIN			JACKSON			KALAMAZOO		
		484 - 485			497			493,495			498 - 499			492			491		
		MAT.	INST.	TOTAL	MAT.	INST.	TOTAL	MAT.	INST.	TOTAL	MAT.	INST.	TOTAL	MAT.	INST.	TOTAL	MAT.	INST.	TOTAL
015433	CONTRACTOR EQUIPMENT		110.4	110.4		105.0	105.0		102.4	102.4		94.4	94.4		105.0	105.0		102.4	102.4
0241, 31 - 34	SITE & INFRASTRUCTURE, DEMOLITION	69.0	96.4	88.2	82.4	85.9	84.9	89.4	87.5	88.1	90.6	94.3	93.2	104.3	87.6	92.6	88.3	87.6	87.8
0310	Concrete Forming & Accessories	102.2	89.9	91.6	95.4	55.1	60.5	100.0	79.7	82.4	87.4	84.6	85.0	92.6	86.5	87.3	97.7	88.0	89.3
0320	Concrete Reinforcing	101.9	122.7	112.3	92.5	91.5	92.0	99.3	89.6	94.5	92.4	93.1	92.8	89.9	122.5	106.2	99.3	89.9	94.6
0330	Cast-in-Place Concrete	87.6	93.8	90.1	95.1	70.4	85.0	101.5	94.4	98.6	112.5	88.0	102.5	95.0	93.2	94.2	97.2	101.1	98.8
03	CONCRETE	92.8	98.1	95.4	92.6	68.7	80.8	99.2	86.3	92.8	102.0	87.5	94.8	87.6	96.2	91.9	99.4	92.3	95.9
04	MASONRY	99.1	94.8	96.5	107.4	81.7	91.7	93.1	80.7	85.5	93.6	87.5	89.9	87.9	91.9	90.3	95.8	85.5	89.5
05	METALS	93.6	115.8	100.6	95.0	99.8	96.5	95.0	85.2	91.9	94.4	89.6	92.9	95.2	113.3	100.9	93.7	85.8	91.2
06	WOOD, PLASTICS & COMPOSITES	102.7	88.1	94.3	93.1	48.0	67.3	97.7	77.8	86.3	87.6	84.5	85.8	91.7	84.8	87.7	100.6	88.1	93.5
07	THERMAL & MOISTURE PROTECTION	98.8	97.7	98.3	92.7	69.7	83.4	97.3	75.8	88.6	96.1	82.7	90.7	92.1	95.0	93.3	94.2	86.0	90.9
08	OPENINGS	98.4	95.1	97.6	95.6	53.1	85.2	97.9	81.0	93.8	101.9	72.2	94.7	94.6	86.2	92.6	94.1	82.5	91.3
0920	Plaster & Gypsum Board	107.7	86.8	93.1	96.8	45.1	60.7	98.7	73.2	80.9	59.3	84.5	77.0	94.7	83.1	86.6	97.0	83.9	87.8
0950, 0980	Ceilings & Acoustic Treatment	94.7	86.8	89.4	92.8	45.1	61.0	97.3	73.2	81.3	91.6	84.5	86.9	92.8	83.1	86.3	93.7	83.9	87.1
0960	Flooring	95.5	82.3	91.6	91.1	49.9	78.8	99.0	69.8	90.3	113.2	95.8	108.0	89.8	68.7	83.5	98.8	75.0	91.7
0970, 0990	Wall Finishes & Painting/Coating	88.6	89.2	89.0	89.7	45.2	63.0	93.9	80.6	85.9	106.7	58.3	77.7	89.7	75.0	80.9	93.9	81.9	86.7
09	FINISHES	95.7	87.7	91.2	97.5	51.1	71.5	98.6	77.3	86.7	99.6	84.0	90.9	98.5	81.2	88.8	97.4	85.0	90.5
COVERS	DIVS. 10 - 14, 25, 28, 41, 43, 44, 46	100.0	94.4	98.9	100.0	88.4	97.6	100.0	98.0	99.6	100.0	90.1	98.0	100.0	94.4	98.9	100.0	99.3	99.9
21, 22, 23	FIRE SUPPRESSION, PLUMBING & HVAC	100.0	92.5	96.9	94.3	81.4	89.0	99.9	81.7	92.5	94.2	85.2	90.6	94.3	90.3	92.7	99.9	81.6	92.4
26, 27, 3370	ELECTRICAL, COMMUNICATIONS & UTIL.	96.3	93.2	94.6	92.2	48.0	69.0	99.7	73.6	86.0	98.8	85.9	92.0	96.3	108.2	102.6	94.2	85.2	89.5
MF2010	WEIGHTED AVERAGE	96.4	95.7	96.1	95.2	71.7	84.8	98.0	81.6	90.8	97.2	86.4	92.4	94.7	94.5	94.6	96.8	86.0	92.1

DIVISION		MICHIGAN															MINNESOTA		
		LANSING			MUSKEGON			ROYAL OAK			SAGINAW			TRAVERSE CITY			BEMIDJI		
		488 - 489			494			480,483			486			496			566		
		MAT.	INST.	TOTAL	MAT.	INST.	TOTAL	MAT.	INST.	TOTAL	MAT.	INST.	TOTAL	MAT.	INST.	TOTAL	MAT.	INST.	TOTAL
015433	CONTRACTOR EQUIPMENT		110.4	110.4		102.4	102.4		95.9	95.9		110.4	110.4		94.4	94.4		98.9	98.9
0241, 31 - 34	SITE & INFRASTRUCTURE, DEMOLITION	88.5	96.1	93.8	86.0	87.5	87.1	83.9	96.3	92.6	72.4	96.0	89.0	77.2	94.0	89.0	97.4	98.3	98.0
0310	Concrete Forming & Accessories	102.9	86.8	89.0	98.0	86.6	88.2	95.5	109.8	107.9	99.4	87.4	89.1	87.4	69.8	72.2	90.6	92.6	92.4
0320	Concrete Reinforcing	101.9	122.5	112.2	100.0	89.3	94.6	92.3	119.0	105.6	101.9	122.4	112.1	93.8	89.0	91.4	101.8	108.0	104.9
0330	Cast-in-Place Concrete	92.4	92.1	92.3	95.0	95.3	95.1	76.2	108.6	89.5	85.9	91.6	88.3	87.8	62.7	77.5	102.5	105.4	103.7
03	CONCRETE	95.2	96.1	95.6	94.6	89.6	92.1	79.9	110.1	94.8	91.8	96.2	94.0	84.4	71.6	78.1	97.8	101.0	99.4
04	MASONRY	92.6	92.4	92.5	94.3	75.1	82.6	94.1	110.3	104.0	100.6	87.2	92.4	91.8	81.5	85.5	99.2	105.5	103.0
05	METALS	92.2	115.2	99.4	91.5	84.8	89.4	96.8	98.4	97.3	93.7	115.0	100.4	94.3	87.8	92.3	92.2	121.5	101.4
06	WOOD, PLASTICS & COMPOSITES	102.1	84.9	92.3	96.6	86.9	91.1	94.8	110.1	103.5	95.3	86.8	90.4	87.6	68.2	76.5	71.8	89.4	81.8
07	THERMAL & MOISTURE PROTECTION	98.5	90.5	95.3	93.3	76.9	86.7	97.5	112.2	103.4	99.6	94.4	97.5	95.3	70.2	85.2	106.2	98.2	103.0
08	OPENINGS	98.4	93.3	97.2	93.4	86.4	91.7	98.5	108.4	100.9	96.1	94.4	95.7	101.9	63.9	92.7	96.4	110.2	99.8
0920	Plaster & Gypsum Board	100.8	83.5	88.7	76.1	82.7	80.7	102.7	109.5	107.4	106.1	85.4	91.6	59.3	67.7	65.2	104.5	89.3	93.9
0950, 0980	Ceilings & Acoustic Treatment	100.2	83.5	89.1	94.6	82.7	86.7	93.9	109.5	104.3	94.7	85.4	88.5	91.6	67.7	75.7	144.5	89.3	107.7
0960	Flooring	105.5	82.3	98.6	97.6	64.7	87.8	93.5	115.2	100.0	95.7	79.7	90.9	113.2	49.9	94.4	111.2	125.2	115.4
0970, 0990	Wall Finishes & Painting/Coating	99.3	83.4	89.7	92.3	80.6	85.3	90.0	101.1	96.6	88.6	82.6	85.0	106.7	45.2	69.8	94.1	100.1	97.7
09	FINISHES	101.5	85.0	92.3	93.8	81.8	87.1	95.6	110.2	103.8	96.1	84.9	89.9	98.5	62.8	78.5	115.4	98.8	106.1
COVERS	DIVS. 10 - 14, 25, 28, 41, 43, 44, 46	100.0	93.8	98.7	100.0	99.1	99.8	100.0	98.8	99.8	100.0	93.3	98.6	100.0	86.9	97.3	100.0	92.3	98.4
21, 22, 23	FIRE SUPPRESSION, PLUMBING & HVAC	99.9	90.5	96.0	99.8	83.7	93.2	94.4	107.6	99.8	100.0	86.0	94.2	94.2	80.9	88.8	94.4	86.5	91.2
26, 27, 3370	ELECTRICAL, COMMUNICATIONS & UTIL.	99.9	91.2	95.3	94.7	78.6	86.2	98.5	92.2	95.2	93.5	89.5	91.4	93.9	75.2	84.1	105.3	102.0	103.6
MF2010	WEIGHTED AVERAGE	97.4	93.8	95.8	95.4	83.6	90.2	94.4	104.5	98.9	96.0	92.2	94.3	94.2	77.4	86.8	98.4	99.8	99.0

DIVISION		MINNESOTA																	
		BRAINERD			DETROIT LAKES			DULUTH			MANKATO			MINNEAPOLIS			ROCHESTER		
		564			565			556 - 558			560			553 - 555			559		
		MAT.	INST.	TOTAL	MAT.	INST.	TOTAL	MAT.	INST.	TOTAL	MAT.	INST.	TOTAL	MAT.	INST.	TOTAL	MAT.	INST.	TOTAL
015433	CONTRACTOR EQUIPMENT		101.6	101.6		98.9	98.9		100.5	100.5		101.6	101.6		104.0	104.0		100.5	100.5
0241, 31 - 34	SITE & INFRASTRUCTURE, DEMOLITION	97.3	103.1	101.4	95.5	98.7	97.8	100.1	100.6	100.5	93.8	103.8	100.8	102.4	105.8	104.8	101.6	100.3	100.6
0310	Concrete Forming & Accessories	93.0	95.4	95.1	87.8	95.4	94.3	101.3	114.5	112.8	100.3	103.0	102.6	102.0	129.9	126.1	102.6	108.9	108.0
0320	Concrete Reinforcing	100.6	108.2	104.4	101.8	108.0	104.9	98.6	108.5	103.5	100.4	120.1	110.2	98.8	120.5	109.6	98.6	120.2	109.4
0330	Cast-in-Place Concrete	111.4	110.3	110.9	99.5	109.6	103.7	103.2	108.5	105.4	102.7	109.4	105.4	115.5	119.3	117.1	113.2	101.8	108.5
03	CONCRETE	101.8	103.9	102.9	95.4	103.6	99.5	99.0	111.9	105.4	97.4	109.3	103.3	106.6	124.8	115.6	103.9	109.4	106.6
04	MASONRY	122.2	115.8	118.3	122.3	113.1	116.7	108.1	119.5	115.0	110.7	112.8	111.9	107.4	127.2	119.5	106.7	115.2	111.9
05	METALS	93.4	121.5	102.3	92.2	121.4	101.4	97.6	124.1	105.9	93.3	128.2	104.3	100.3	132.0	110.2	97.5	130.2	107.7
06	WOOD, PLASTICS & COMPOSITES	91.5	89.2	90.2	69.0	89.4	80.6	110.2	114.3	112.5	100.1	100.8	100.5	113.4	128.7	122.1	113.4	107.5	110.0
07	THERMAL & MOISTURE PROTECTION	105.1	110.5	107.2	106.0	110.7	107.9	105.7	118.7	110.9	105.4	101.6	103.9	104.8	130.8	115.3	105.1	102.9	104.2
08	OPENINGS	84.3	110.1	90.6	96.4	110.2	99.8	92.2	119.3	98.8	88.7	120.1	96.4	95.9	136.0	105.6	92.2	124.5	100.0
0920	Plaster & Gypsum Board	86.4	89.3	88.5	103.3	89.3	93.5	92.9	115.1	108.4	91.0	101.3	98.2	101.6	129.8	121.3	101.1	108.1	106.0
0950, 0980	Ceilings & Acoustic Treatment	70.4	89.3	83.0	144.5	89.3	107.7	102.2	115.1	110.8	70.4	101.3	91.0	99.5	129.8	119.7	97.7	108.1	104.6
0960	Flooring	110.1	125.2	114.6	110.0	125.2	114.5	104.3	126.5	110.9	111.8	125.2	115.8	102.3	125.2	109.1	104.7	125.2	110.8
0970, 0990	Wall Finishes & Painting/Coating	88.5	100.1	95.4	94.1	100.1	97.7	87.4	110.6	101.3	99.6	104.2	102.3	96.1	130.9	117.0	90.7	104.2	98.8
09	FINISHES	96.2	100.5	98.6	114.7	100.7	106.8	101.5	116.7	110.1	97.8	107.1	103.0	102.2	129.7	117.6	102.0	111.5	107.3
COVERS	DIVS. 10 - 14, 25, 28, 41, 43, 44, 46	100.0	94.4	98.9	100.0	94.7	98.9	100.0	101.7	100.3	100.0	94.6	98.9	100.0	109.3	101.9	100.0	103.0	100.6
21, 22, 23	FIRE SUPPRESSION, PLUMBING & HVAC	93.6	90.4	92.3	94.4	90.1	92.7	99.9	99.2	99.6	93.6	90.1	92.2	99.9	120.1	108.2	100.0	97.7	99.1
26, 27, 3370	ELECTRICAL, COMMUNICATIONS & UTIL.	102.7	106.8	104.8	105.0	69.2	86.1	102.1	106.8	104.6	109.8	87.3	98.0	101.7	111.0	106.6	100.8	87.3	93.7
MF2010	WEIGHTED AVERAGE	97.1	103.7	100.0	99.1	97.9	98.6	99.7	110.3	104.3	97.3	103.2	99.9	101.4	122.0	110.5	100.1	106.1	102.7

DIVISION		MINNESOTA															MISSISSIPPI		
		SAINT PAUL			ST. CLOUD			THIEF RIVER FALLS			WILLMAR			WINDOM			BILOXI		
		550 - 551			563			567			562			561			395		
		MAT.	INST.	TOTAL	MAT.	INST.	TOTAL	MAT.	INST.	TOTAL	MAT.	INST.	TOTAL	MAT.	INST.	TOTAL	MAT.	INST.	TOTAL
015433	CONTRACTOR EQUIPMENT		100.5	100.5		101.6	101.6		98.9	98.9		101.6	101.6		101.6	101.6		99.4	99.4
0241, 31 - 34	SITE & INFRASTRUCTURE, DEMOLITION	102.1	101.2	101.5	92.3	105.2	101.3	96.3	98.2	97.6	91.8	102.9	99.6	85.5	101.6	96.8	101.0	89.8	93.2
0310	Concrete Forming & Accessories	99.1	127.2	123.5	90.3	122.9	118.5	91.2	92.0	91.9	90.1	94.8	94.2	94.3	86.9	87.9	95.7	55.2	60.7
0320	Concrete Reinforcing	95.1	120.5	107.8	100.6	120.3	110.4	102.1	107.9	105.0	100.2	120.0	110.1	100.2	119.1	109.6	95.3	59.3	77.4
0330	Cast-in-Place Concrete	115.5	118.6	116.8	98.4	118.1	106.5	101.6	90.3	97.0	99.9	88.3	95.1	86.4	66.4	78.2	109.4	54.6	86.9
03	CONCRETE	108.2	123.4	115.7	93.3	121.2	107.0	96.5	95.5	96.0	93.2	98.4	95.8	84.2	87.2	85.7	98.7	57.4	78.3
04	MASONRY	116.5	127.2	123.0	107.5	124.6	117.9	99.1	105.5	103.0	111.8	115.8	114.3	121.9	88.8	101.7	95.6	45.0	64.7
05	METALS	97.0	131.6	107.9	94.1	129.6	105.2	92.4	120.8	101.3	93.2	127.2	103.9	93.1	125.0	103.1	92.5	83.3	89.6
06	WOOD, PLASTICS & COMPOSITES	106.3	125.4	117.2	89.0	120.1	106.8	72.7	89.4	82.2	88.8	89.4	89.1	92.7	86.6	89.2	97.9	58.6	75.5
07	THERMAL & MOISTURE PROTECTION	106.3	130.3	116.0	105.2	120.9	111.5	107.1	99.9	104.2	105.1	110.2	107.1	105.0	90.1	99.0	95.0	56.1	79.3
08	OPENINGS	90.4	134.2	101.1	88.7	131.4	99.1	96.4	110.2	99.8	86.4	95.4	88.6	90.0	93.9	91.0	97.4	59.8	88.2
0920	Plaster & Gypsum Board	89.1	126.6	115.3	86.4	121.2	110.8	104.1	89.3	93.8	86.4	89.6	88.6	86.4	86.6	86.6	102.5	57.8	71.2
0950, 0980	Ceilings & Acoustic Treatment	97.7	126.6	116.9	70.4	121.2	104.2	144.5	89.3	107.7	70.4	89.6	83.2	70.4	86.6	81.2	94.6	57.8	70.1
0960	Flooring	98.9	125.2	106.7	106.4	125.2	112.0	110.9	125.2	115.2	108.3	125.2	113.3	110.7	125.2	115.1	117.7	54.8	99.0
0970, 0990	Wall Finishes & Painting/Coating	96.1	126.5	114.3	99.6	130.9	118.4	94.1	100.1	97.7	94.1	100.1	97.7	94.1	104.2	100.1	101.6	45.4	67.9
09	FINISHES	99.2	127.2	114.9	95.5	124.4	111.7	115.2	98.8	106.0	95.6	98.2	97.1	95.7	93.7	94.5	103.1	54.3	75.7
COVERS	DIVS. 10 - 14, 25, 28, 41, 43, 44, 46	100.0	108.6	101.8	100.0	100.8	100.2	100.0	92.2	98.4	100.0	94.3	98.8	100.0	89.6	97.9	100.0	57.8	91.4
21, 22, 23	FIRE SUPPRESSION, PLUMBING & HVAC	99.9	115.1	106.1	99.6	119.7	107.9	94.4	86.2	91.1	93.6	108.1	99.5	93.6	84.7	90.0	99.9	49.5	79.3
26, 27, 3370	ELECTRICAL, COMMUNICATIONS & UTIL.	99.2	111.0	105.4	102.7	111.0	107.0	102.3	69.2	84.8	102.7	88.8	95.4	109.8	87.3	98.0	98.2	57.9	77.0
MF2010	WEIGHTED AVERAGE	100.4	119.9	109.0	97.2	119.4	107.0	98.0	94.5	96.4	95.6	103.9	99.2	96.0	92.8	94.6	98.1	59.5	81.0

DIVISION		MISSISSIPPI																	
		CLARKSDALE			COLUMBUS			GREENVILLE			GREENWOOD			JACKSON			LAUREL		
		386			397			387			389			390 - 392			394		
		MAT.	INST.	TOTAL	MAT.	INST.	TOTAL	MAT.	INST.	TOTAL	MAT.	INST.	TOTAL	MAT.	INST.	TOTAL	MAT.	INST.	TOTAL
015433	CONTRACTOR EQUIPMENT		99.4	99.4		99.4	99.4		99.4	99.4		99.4	99.4		99.4	99.4		99.4	99.4
0241, 31 - 34	SITE & INFRASTRUCTURE, DEMOLITION	98.6	89.6	92.3	99.5	89.9	92.8	104.4	89.8	94.2	101.4	89.3	92.9	96.6	89.8	91.9	105.4	89.7	94.4
0310	Concrete Forming & Accessories	86.7	51.5	56.3	84.6	53.6	57.8	83.3	66.6	68.9	95.6	51.5	57.4	95.4	60.1	64.8	84.7	55.4	59.3
0320	Concrete Reinforcing	102.7	43.6	73.3	102.1	44.6	73.5	103.3	63.2	83.3	102.7	55.1	79.0	95.3	58.3	76.9	102.7	41.0	72.0
0330	Cast-in-Place Concrete	106.7	55.5	85.6	111.6	60.7	90.7	109.9	57.7	88.4	114.7	55.1	90.2	93.1	58.4	78.9	109.1	57.2	87.8
03	CONCRETE	96.9	53.1	75.3	100.4	56.0	78.5	101.9	64.2	83.3	103.2	55.0	79.4	90.9	60.7	76.0	102.5	55.0	79.0
04	MASONRY	95.4	46.9	65.8	122.0	53.8	80.3	141.2	50.4	85.8	95.9	46.7	65.9	96.6	50.4	68.4	117.8	50.2	76.5
05	METALS	89.6	75.8	85.3	89.5	76.7	85.5	90.7	85.1	88.9	89.6	77.3	85.7	92.4	83.1	89.5	89.6	75.7	85.2
06	WOOD, PLASTICS & COMPOSITES	85.3	53.5	67.2	83.3	54.5	66.9	82.0	71.2	75.8	98.0	53.5	72.6	96.5	62.3	77.0	84.4	57.1	68.8
07	THERMAL & MOISTURE PROTECTION	94.8	49.0	76.4	94.9	53.4	78.2	95.1	62.0	81.8	95.1	48.9	76.5	96.8	61.0	82.4	95.0	55.1	79.0
08	OPENINGS	96.7	51.6	85.7	96.6	51.7	85.7	96.7	64.2	88.8	96.7	53.0	86.0	97.7	58.0	88.1	93.5	52.5	83.5
0920	Plaster & Gypsum Board	90.7	52.5	64.0	90.3	53.6	64.6	90.3	70.7	76.6	101.5	52.5	67.2	95.5	61.5	71.7	90.3	56.2	66.4
0950, 0980	Ceilings & Acoustic Treatment	88.9	52.5	64.7	88.9	53.6	65.4	91.0	70.7	77.5	88.9	52.5	64.7	98.2	61.5	73.8	88.9	56.2	67.1
0960	Flooring	111.9	54.8	94.9	110.8	61.4	96.1	110.0	54.8	93.6	117.7	54.8	99.0	117.9	54.8	99.1	109.2	54.8	93.0
0970, 0990	Wall Finishes & Painting/Coating	101.6	48.2	69.6	101.6	55.3	73.9	101.6	68.8	81.9	101.6	48.2	69.6	101.6	68.8	81.9	101.6	63.6	78.8
09	FINISHES	98.4	52.1	72.5	98.3	55.6	74.4	99.0	65.7	80.3	101.9	52.1	74.0	103.2	60.4	79.2	98.4	56.7	75.1
COVERS	DIVS. 10 - 14, 25, 28, 41, 43, 44, 46	100.0	58.1	91.4	100.0	59.3	91.7	100.0	61.0	92.0	100.0	58.1	91.4	100.0	60.0	91.8	100.0	39.0	87.5
21, 22, 23	FIRE SUPPRESSION, PLUMBING & HVAC	97.4	43.6	75.4	96.7	39.6	73.3	99.9	50.2	79.6	97.4	41.5	74.5	100.0	55.5	81.8	96.7	48.3	76.9
26, 27, 3370	ELECTRICAL, COMMUNICATIONS & UTIL.	97.4	48.3	71.6	95.5	51.2	72.1	97.4	65.3	80.5	97.4	45.3	70.0	100.0	65.3	81.7	97.1	58.4	76.7
MF2010	WEIGHTED AVERAGE	96.1	54.9	77.9	97.4	56.2	79.3	99.9	64.3	84.2	97.3	54.4	78.4	97.4	63.6	82.5	97.5	58.1	80.1

DIVISION		MISSISSIPPI									MISSOURI								
		MCCOMB			MERIDIAN			TUPELO			BOWLING GREEN			CAPE GIRARDEAU			CHILLICOTHE		
		396			393			388			633			637			646		
		MAT.	INST.	TOTAL	MAT.	INST.	TOTAL	MAT.	INST.	TOTAL	MAT.	INST.	TOTAL	MAT.	INST.	TOTAL	MAT.	INST.	TOTAL
015433	CONTRACTOR EQUIPMENT		99.4	99.4		99.4	99.4		99.4	99.4		105.5	105.5		105.5	105.5		102.2	102.2
0241, 31 - 34	SITE & INFRASTRUCTURE, DEMOLITION	92.5	89.5	90.4	96.5	90.1	92.0	96.0	89.5	91.5	93.4	94.4	94.1	94.9	94.2	94.4	102.1	94.5	96.7
0310	Concrete Forming & Accessories	84.7	53.3	57.6	82.1	66.1	68.3	83.8	53.7	57.7	96.0	90.1	90.9	89.4	75.5	77.4	89.9	96.6	95.7
0320	Concrete Reinforcing	103.3	43.0	73.3	102.1	58.4	80.3	100.5	55.1	77.9	100.4	99.8	100.1	101.6	85.4	93.5	98.2	103.8	101.0
0330	Cast-in-Place Concrete	96.9	55.0	79.6	103.6	60.9	86.1	106.7	56.9	86.2	92.5	86.1	89.9	91.6	84.4	88.6	95.8	88.7	92.9
03	CONCRETE	90.6	53.5	72.3	94.7	64.2	79.6	96.4	56.5	76.7	96.5	91.6	94.1	95.7	81.9	88.9	102.3	95.7	99.0
04	MASONRY	123.4	46.2	76.3	95.3	54.5	70.4	130.9	49.9	81.4	124.6	97.6	108.2	121.5	77.7	94.7	105.5	95.6	99.4
05	METALS	89.7	74.0	84.8	90.6	83.3	88.3	89.5	77.5	85.8	94.8	112.9	100.4	95.8	105.2	98.8	94.9	107.0	98.7
06	WOOD, PLASTICS & COMPOSITES	83.3	56.4	68.0	80.8	68.1	73.6	82.6	54.5	66.6	90.4	89.8	90.1	83.3	72.9	77.4	90.0	97.5	94.3
07	THERMAL & MOISTURE PROTECTION	94.5	48.1	75.9	94.7	63.4	82.1	94.8	54.9	78.8	101.7	99.1	100.7	101.6	81.8	93.6	97.3	96.1	96.8
08	OPENINGS	96.7	53.3	86.1	96.6	66.9	89.4	96.6	52.9	86.0	97.8	99.8	98.3	97.8	75.8	92.4	91.2	98.1	92.9
0920	Plaster & Gypsum Board	90.3	55.5	66.0	90.3	67.5	74.4	90.3	53.6	64.6	103.3	89.5	93.7	101.2	72.2	80.9	101.6	97.2	98.5
0950, 0980	Ceilings & Acoustic Treatment	88.9	55.5	66.7	91.0	67.5	75.3	88.9	53.6	65.4	93.4	89.5	90.8	93.4	72.2	79.2	89.0	97.2	94.5
0960	Flooring	110.8	54.8	94.1	109.2	54.8	92.9	110.3	54.8	93.7	103.0	98.2	101.6	99.8	87.1	96.0	94.9	106.2	98.3
0970, 0990	Wall Finishes & Painting/Coating	101.6	44.2	67.2	101.6	79.5	88.4	101.6	61.3	77.5	104.6	105.1	104.9	104.6	76.5	87.8	93.5	91.6	92.4
09	FINISHES	97.7	53.2	72.8	97.9	66.1	80.1	97.9	54.9	73.8	101.3	92.5	96.4	100.0	76.4	86.8	96.3	98.0	97.3
COVERS	DIVS. 10 - 14, 25, 28, 41, 43, 44, 46	100.0	61.6	92.2	100.0	62.1	92.2	100.0	59.3	91.7	100.0	91.2	98.2	100.0	88.5	97.7	100.0	92.0	98.4
21, 22, 23	FIRE SUPPRESSION, PLUMBING & HVAC	96.7	36.4	72.0	99.9	57.7	82.7	97.7	45.6	76.4	93.9	94.9	94.3	99.9	93.2	97.2	94.1	96.5	95.1
26, 27, 3370	ELECTRICAL, COMMUNICATIONS & UTIL.	93.9	51.5	71.6	97.1	58.1	76.6	97.2	51.3	73.0	99.9	82.3	90.6	99.9	105.8	103.0	95.6	81.6	88.2
MF2010	WEIGHTED AVERAGE	96.1	54.0	77.5	96.5	65.3	82.7	97.7	57.3	79.9	98.2	94.5	96.6	99.5	89.5	95.1	96.6	95.2	96.0

DIVISION		MISSOURI																	
		COLUMBIA			FLAT RIVER			HANNIBAL			HARRISONVILLE			JEFFERSON CITY			JOPLIN		
		652			636			634			647			650 - 651			648		
		MAT.	INST.	TOTAL	MAT.	INST.	TOTAL	MAT.	INST.	TOTAL	MAT.	INST.	TOTAL	MAT.	INST.	TOTAL	MAT.	INST.	TOTAL
015433	CONTRACTOR EQUIPMENT		106.9	106.9		105.5	105.5		105.5	105.5		102.2	102.2		106.9	106.9		105.9	105.9
0241, 31 - 34	SITE & INFRASTRUCTURE, DEMOLITION	103.3	96.3	98.4	96.2	94.3	94.9	91.2	94.4	93.4	93.4	95.3	94.8	100.2	96.2	97.4	102.4	99.6	100.4
0310	Concrete Forming & Accessories	88.0	79.7	80.8	101.7	86.6	88.6	94.3	80.4	82.2	87.2	102.4	100.4	97.9	79.6	82.0	101.3	75.0	78.6
0320	Concrete Reinforcing	98.3	118.2	108.2	101.6	108.0	104.8	99.8	95.9	97.9	97.7	110.4	104.1	97.1	110.2	103.6	101.3	103.4	102.4
0330	Cast-in-Place Concrete	88.5	86.1	87.5	95.5	94.0	94.9	87.6	85.9	86.9	98.1	105.4	101.1	89.0	86.1	87.8	103.7	78.9	93.5
03	CONCRETE	86.1	90.5	88.3	99.5	94.3	96.9	92.8	86.5	89.7	98.5	105.3	101.9	87.7	89.0	88.3	101.8	82.6	92.3
04	MASONRY	136.6	89.8	108.1	121.9	79.1	95.8	115.7	97.6	104.7	99.9	106.3	103.8	99.1	89.8	93.4	98.5	88.2	92.2
05	METALS	93.2	120.8	101.9	94.7	115.3	101.1	94.8	111.0	99.8	95.3	111.4	100.4	92.6	117.5	100.4	98.0	103.4	99.7
06	WOOD, PLASTICS & COMPOSITES	90.7	75.8	82.2	97.9	85.8	91.0	88.3	77.0	81.9	86.5	101.9	95.3	101.5	75.8	86.9	101.9	73.3	85.6
07	THERMAL & MOISTURE PROTECTION	94.4	88.7	92.1	101.9	93.6	98.5	101.5	97.8	100.0	96.6	106.0	100.4	97.2	88.7	93.8	96.8	79.8	89.9
08	OPENINGS	90.3	94.1	91.3	97.8	100.1	98.3	97.8	84.2	94.5	91.4	104.4	94.6	87.7	92.0	88.8	92.4	79.8	89.3
0920	Plaster & Gypsum Board	98.8	75.1	82.2	109.1	85.4	92.5	101.6	76.4	84.0	96.1	101.8	100.1	102.8	75.1	83.4	109.3	72.3	83.4
0950, 0980	Ceilings & Acoustic Treatment	94.6	75.1	81.6	93.4	85.4	88.1	93.4	76.4	82.0	89.0	101.8	97.5	101.0	75.1	83.7	89.9	72.3	78.2
0960	Flooring	113.1	99.2	108.9	106.1	87.1	100.4	102.4	98.2	101.1	90.6	106.2	95.2	120.4	99.2	114.1	119.7	77.7	107.2
0970, 0990	Wall Finishes & Painting/Coating	103.4	78.8	88.7	104.6	81.9	91.0	104.6	105.1	104.9	98.2	106.0	102.9	103.4	78.8	88.7	93.1	78.2	84.2
09	FINISHES	99.9	81.6	89.7	103.1	85.0	92.9	100.6	85.0	91.9	93.9	103.4	99.2	104.2	81.6	91.6	103.0	75.9	87.8
COVERS	DIVS. 10 - 14, 25, 28, 41, 43, 44, 46	100.0	95.8	99.1	100.0	90.4	98.0	100.0	89.8	97.9	100.0	94.1	98.8	100.0	95.8	99.1	100.0	90.8	98.1
21, 22, 23	FIRE SUPPRESSION, PLUMBING & HVAC	99.9	95.1	97.9	93.9	94.4	94.1	93.9	94.9	94.3	94.0	101.4	97.1	99.9	90.6	96.1	100.2	73.5	89.3
26, 27, 3370	ELECTRICAL, COMMUNICATIONS & UTIL.	96.4	87.3	91.6	104.6	105.8	105.2	98.6	82.3	90.0	102.8	99.7	101.2	102.3	87.3	94.4	93.2	67.8	79.8
MF2010	WEIGHTED AVERAGE	97.7	93.2	95.7	99.1	95.2	97.4	97.1	91.8	94.7	96.2	102.9	99.1	96.6	91.7	94.4	98.6	81.7	91.2

DIVISION		MISSOURI																	
		KANSAS CITY			KIRKSVILLE			POPLAR BLUFF			ROLLA			SEDALIA			SIKESTON		
		640 - 641			635			639			654 - 655			653			638		
		MAT.	INST.	TOTAL	MAT.	INST.	TOTAL	MAT.	INST.	TOTAL	MAT.	INST.	TOTAL	MAT.	INST.	TOTAL	MAT.	INST.	TOTAL
015433	CONTRACTOR EQUIPMENT		103.5	103.5		97.8	97.8		100.2	100.2		106.9	106.9		99.0	99.0		100.2	100.2
0241, 31 - 34	SITE & INFRASTRUCTURE, DEMOLITION	95.4	97.9	97.2	94.7	90.4	91.7	81.4	94.2	90.4	102.5	96.1	98.0	96.9	91.9	93.4	85.0	94.3	91.5
0310	Concrete Forming & Accessories	100.2	109.3	108.0	87.1	78.9	80.0	87.2	75.5	77.1	95.2	95.4	95.4	92.4	78.8	80.7	88.3	75.5	77.2
0320	Concrete Reinforcing	96.2	119.0	107.5	100.7	103.5	102.1	103.7	99.1	101.4	98.8	99.4	99.1	97.4	103.7	100.5	103.1	103.0	103.1
0330	Cast-in-Place Concrete	96.5	109.4	101.8	95.4	85.8	91.5	73.5	84.9	78.2	90.5	86.7	88.9	94.5	85.1	90.6	78.6	84.9	81.2
03	CONCRETE	97.9	111.3	104.5	111.0	86.8	99.0	86.5	84.1	85.3	87.9	94.1	91.0	100.0	86.5	93.3	90.1	84.8	87.5
04	MASONRY	101.9	109.3	106.4	128.2	89.7	104.7	120.0	77.7	94.2	112.7	90.5	99.1	118.5	87.2	99.4	119.7	77.7	94.1
05	METALS	105.6	115.8	108.8	94.4	105.3	97.8	94.9	102.8	97.4	92.6	112.3	98.8	91.2	106.5	96.0	95.2	104.3	98.1
06	WOOD, PLASTICS & COMPOSITES	101.2	109.1	105.7	76.6	75.8	76.1	75.8	73.0	74.2	98.0	96.9	97.4	91.0	75.9	82.4	77.4	73.0	74.9
07	THERMAL & MOISTURE PROTECTION	96.7	109.3	101.8	108.0	89.7	100.6	106.6	83.6	97.3	94.5	96.8	95.4	100.2	94.4	97.8	106.8	82.1	96.8
08	OPENINGS	98.6	111.3	101.7	102.5	86.2	98.6	103.6	79.4	97.7	90.3	92.3	90.8	94.0	85.9	92.0	103.6	80.5	98.0
0920	Plaster & Gypsum Board	105.6	109.2	108.1	96.2	75.1	81.4	96.7	72.2	79.5	100.1	96.8	97.8	93.4	75.1	80.6	98.7	72.2	80.2
0950, 0980	Ceilings & Acoustic Treatment	96.3	109.2	104.9	91.5	75.1	80.6	93.4	72.2	79.2	94.6	96.8	96.1	94.6	75.1	81.6	93.4	72.2	79.2
0960	Flooring	95.6	109.6	99.7	79.5	94.9	84.1	94.4	83.1	91.0	116.6	98.1	111.1	88.8	97.5	91.4	94.8	83.1	91.4
0970, 0990	Wall Finishes & Painting/Coating	98.2	115.2	108.4	99.7	78.8	87.2	99.2	76.5	85.6	103.4	105.1	104.4	103.4	104.7	104.2	99.2	76.5	85.6
09	FINISHES	98.1	109.9	104.7	99.5	80.8	89.0	99.7	75.8	86.3	101.2	97.1	98.9	97.5	83.6	89.7	100.5	75.8	86.6
COVERS	DIVS. 10 - 14, 25, 28, 41, 43, 44, 46	100.0	101.8	100.4	100.0	89.7	97.9	100.0	88.5	97.7	100.0	92.5	98.5	100.0	90.8	98.1	100.0	88.5	97.7
21, 22, 23	FIRE SUPPRESSION, PLUMBING & HVAC	100.0	109.6	103.9	93.9	94.5	94.2	93.9	92.8	93.5	93.8	97.1	95.2	93.8	95.5	94.5	93.9	92.8	93.5
26, 27, 3370	ELECTRICAL, COMMUNICATIONS & UTIL.	104.5	101.9	103.1	98.8	82.2	90.1	99.1	105.8	102.6	94.7	87.3	90.8	96.3	99.7	98.1	98.1	105.8	102.1
MF2010	WEIGHTED AVERAGE	100.7	108.2	104.0	100.3	89.4	95.5	97.0	89.7	93.8	95.1	95.7	95.3	96.7	92.4	94.8	97.5	89.9	94.2

DIVISION		MISSOURI									MONTANA								
		SPRINGFIELD			ST. JOSEPH			ST. LOUIS			BILLINGS			BUTTE			GREAT FALLS		
		656 - 658			644 - 645			630 - 631			590 - 591			597			594		
		MAT.	INST.	TOTAL	MAT.	INST.	TOTAL	MAT.	INST.	TOTAL	MAT.	INST.	TOTAL	MAT.	INST.	TOTAL	MAT.	INST.	TOTAL
015433	CONTRACTOR EQUIPMENT		101.6	101.6		102.2	102.2		106.5	106.5		99.2	99.2		98.9	98.9		98.9	98.9
0241, 31 - 34	SITE & INFRASTRUCTURE, DEMOLITION	99.2	94.0	95.6	96.9	93.6	94.6	95.1	97.7	96.9	98.7	97.1	97.6	107.2	96.5	99.7	111.0	97.0	101.2
0310	Concrete Forming & Accessories	100.4	75.8	79.1	100.1	91.0	92.2	101.8	104.6	104.2	102.0	70.1	74.4	89.6	74.2	76.3	101.7	70.1	74.4
0320	Concrete Reinforcing	94.7	117.9	106.3	95.1	110.1	102.5	92.9	110.9	101.9	99.6	83.2	91.4	108.1	83.4	95.8	99.7	83.3	91.5
0330	Cast-in-Place Concrete	96.0	78.0	88.6	96.4	105.0	99.9	91.6	105.4	97.3	111.8	69.3	94.3	123.0	70.8	101.5	130.0	65.2	103.3
03	CONCRETE	96.8	85.3	91.1	97.6	100.0	98.8	95.1	106.8	100.9	104.5	73.2	89.0	108.0	75.5	92.0	112.9	71.8	92.6
04	MASONRY	92.3	88.1	89.7	101.7	96.5	98.5	102.0	110.9	107.4	124.0	75.2	94.2	119.2	78.1	94.1	123.4	76.9	95.1
05	METALS	97.1	108.4	100.6	101.7	110.3	104.4	100.5	119.3	106.4	109.7	91.5	104.0	102.5	91.7	99.1	106.4	91.6	101.8
06	WOOD, PLASTICS & COMPOSITES	98.6	74.3	84.7	101.9	89.5	94.8	97.8	102.9	100.7	96.5	68.5	80.5	85.0	74.0	78.7	97.6	68.5	81.0
07	THERMAL & MOISTURE PROTECTION	98.7	81.0	91.6	97.2	96.5	96.9	101.7	107.0	103.8	105.0	71.0	91.3	104.9	72.6	91.9	105.4	69.5	91.0
08	OPENINGS	96.4	82.9	93.1	97.0	98.4	97.3	97.6	110.6	100.8	95.5	68.3	88.8	93.7	72.5	88.5	96.7	67.6	89.6
0920	Plaster & Gypsum Board	101.3	73.5	81.8	110.8	89.0	95.5	110.2	103.0	105.2	97.7	67.8	76.8	99.8	73.5	81.4	108.6	67.8	80.1
0950, 0980	Ceilings & Acoustic Treatment	94.6	73.5	80.5	95.4	89.0	91.1	98.8	103.0	101.6	111.8	67.8	82.5	118.1	73.5	88.4	119.9	67.8	85.2
0960	Flooring	114.0	77.7	103.2	98.7	106.2	100.9	105.8	98.7	103.7	110.4	72.8	99.2	108.1	76.9	98.8	115.7	76.9	104.1
0970, 0990	Wall Finishes & Painting/Coating	97.1	78.2	85.8	93.5	90.6	91.7	104.6	105.7	105.3	95.9	54.0	70.8	94.1	53.5	69.8	94.1	54.0	70.0
09	FINISHES	103.3	76.4	88.2	99.2	93.2	95.8	104.3	103.0	103.6	107.4	68.4	85.5	108.1	72.4	88.1	112.1	69.1	88.0
COVERS	DIVS. 10 - 14, 25, 28, 41, 43, 44, 46	100.0	93.3	98.6	100.0	97.2	99.4	100.0	102.2	100.4	100.0	94.7	98.9	100.0	95.3	99.0	100.0	94.7	98.9
21, 22, 23	FIRE SUPPRESSION, PLUMBING & HVAC	99.9	73.5	89.1	100.2	90.4	96.2	99.9	103.3	101.3	100.2	74.9	89.9	100.3	70.5	88.1	100.3	73.6	89.4
26, 27, 3370	ELECTRICAL, COMMUNICATIONS & UTIL.	100.7	71.1	85.1	102.8	81.6	91.7	102.5	106.2	104.5	97.6	75.2	85.8	105.7	72.2	88.1	97.7	72.2	84.3
MF2010	WEIGHTED AVERAGE	98.6	82.8	91.7	99.9	94.3	97.4	99.9	106.3	102.7	103.3	77.5	91.9	103.1	77.5	91.8	104.5	76.8	92.3

DIVISION		MONTANA																	
		HAVRE			HELENA			KALISPELL			MILES CITY			MISSOULA			WOLF POINT		
		595			596			599			593			598			592		
		MAT.	INST.	TOTAL	MAT.	INST.	TOTAL	MAT.	INST.	TOTAL	MAT.	INST.	TOTAL	MAT.	INST.	TOTAL	MAT.	INST.	TOTAL
015433	**CONTRACTOR EQUIPMENT**		98.9	98.9		98.9	98.9		98.9	98.9		98.9	98.9		98.9	98.9		98.9	98.9
0241, 31 - 34	**SITE & INFRASTRUCTURE, DEMOLITION**	115.7	96.8	102.4	100.5	96.1	97.4	96.7	96.4	96.5	103.7	96.4	98.6	88.5	96.6	94.2	123.6	96.6	104.6
0310	Concrete Forming & Accessories	83.2	68.3	70.3	103.6	67.2	72.1	92.6	71.8	74.6	100.5	66.8	71.3	92.6	72.1	74.8	93.3	67.1	70.6
0320	Concrete Reinforcing	109.0	83.3	96.2	103.1	83.2	93.2	111.0	83.3	97.2	108.5	83.3	96.0	109.9	83.4	96.7	110.1	83.3	96.7
0330	Cast-in-Place Concrete	132.5	62.9	103.8	100.9	64.3	85.8	106.8	64.0	89.2	116.9	62.7	94.6	90.7	66.6	80.8	131.0	63.1	103.1
03	**CONCRETE**	116.2	70.2	93.5	99.7	70.1	85.1	97.9	72.1	85.2	105.0	69.5	87.4	86.4	73.1	79.9	120.1	69.7	95.2
04	**MASONRY**	120.0	72.8	91.2	115.3	76.9	91.8	118.2	78.2	93.8	125.6	68.1	90.5	143.4	78.1	103.6	126.7	68.8	91.3
05	**METALS**	99.0	91.5	96.6	105.6	90.4	100.8	98.9	91.6	96.6	98.2	91.7	96.2	99.4	91.8	97.0	98.3	91.7	96.2
06	**WOOD, PLASTICS & COMPOSITES**	77.5	68.5	72.4	97.7	65.1	79.1	88.0	71.0	78.3	95.0	68.5	79.9	88.0	71.0	78.3	87.3	68.5	76.6
07	**THERMAL & MOISTURE PROTECTION**	105.2	67.3	90.0	104.6	72.2	91.6	104.5	72.4	91.6	104.8	65.5	89.0	104.2	80.8	94.8	105.6	65.3	89.4
08	**OPENINGS**	93.7	67.6	87.4	96.1	64.7	88.5	93.7	69.0	87.7	93.2	67.6	86.9	93.7	68.7	87.6	93.2	67.4	86.9
0920	Plaster & Gypsum Board	95.2	67.8	76.1	96.7	64.3	74.0	99.8	70.4	79.2	107.6	67.8	79.8	99.8	70.4	79.2	102.2	67.8	78.2
0950, 0980	Ceilings & Acoustic Treatment	118.1	67.8	84.6	120.9	64.3	83.2	118.1	70.4	86.3	116.3	67.8	84.0	118.1	70.4	86.3	116.3	67.8	84.0
0960	Flooring	105.2	76.9	96.8	114.9	61.2	98.9	110.1	76.9	100.2	115.5	72.8	102.8	110.1	76.9	100.2	111.6	72.8	100.0
0970, 0990	Wall Finishes & Painting/Coating	94.1	50.8	68.1	94.1	45.2	64.8	94.1	39.1	61.1	94.1	46.9	65.8	94.1	64.7	76.5	94.1	50.8	68.1
09	**FINISHES**	107.4	67.7	85.1	109.9	63.4	83.9	108.0	69.0	86.1	110.6	65.4	85.2	107.3	71.9	87.4	110.3	65.9	85.4
COVERS	**DIVS. 10 - 14, 25, 28, 41, 43, 44, 46**	100.0	73.4	94.6	100.0	94.3	98.8	100.0	95.0	99.0	100.0	91.8	98.3	100.0	95.0	99.0	100.0	92.0	98.4
21, 22, 23	**FIRE SUPPRESSION, PLUMBING & HVAC**	94.3	71.4	84.9	100.2	70.5	88.0	94.3	70.8	84.7	94.3	70.4	84.5	100.3	70.8	88.2	94.3	70.7	84.6
26, 27, 3370	**ELECTRICAL, COMMUNICATIONS & UTIL.**	97.7	73.2	84.8	104.9	72.4	87.8	102.3	70.6	85.6	97.7	78.7	87.7	103.4	70.6	86.1	97.7	78.7	87.7
MF2010	**WEIGHTED AVERAGE**	101.4	75.0	89.7	102.7	75.0	90.5	99.3	76.3	89.1	100.3	75.1	89.2	100.6	77.0	90.2	102.5	75.4	90.6

DIVISION		NEBRASKA																	
		ALLIANCE			COLUMBUS			GRAND ISLAND			HASTINGS			LINCOLN			MCCOOK		
		693			686			688			689			683 - 685			690		
		MAT.	INST.	TOTAL	MAT.	INST.	TOTAL	MAT.	INST.	TOTAL	MAT.	INST.	TOTAL	MAT.	INST.	TOTAL	MAT.	INST.	TOTAL
015433	**CONTRACTOR EQUIPMENT**		96.9	96.9		100.6	100.6		100.6	100.6		100.6	100.6		100.6	100.6		100.6	100.6
0241, 31 - 34	**SITE & INFRASTRUCTURE, DEMOLITION**	100.0	97.7	98.4	98.9	91.1	93.4	103.6	92.2	95.6	102.5	91.4	94.7	92.4	92.2	92.3	103.4	91.2	94.9
0310	Concrete Forming & Accessories	89.3	55.3	59.9	97.7	58.1	63.4	97.3	74.9	77.9	100.4	73.0	76.7	101.7	59.4	65.1	95.0	55.6	60.9
0320	Concrete Reinforcing	113.2	47.0	80.2	103.7	45.8	74.9	103.1	76.6	89.9	103.1	46.5	74.9	94.7	76.1	85.4	105.1	74.1	89.7
0330	Cast-in-Place Concrete	107.2	57.5	86.8	108.8	54.5	86.5	115.2	68.8	96.1	115.2	63.0	93.7	86.1	71.3	80.0	115.2	57.2	91.4
03	**CONCRETE**	119.5	55.1	87.7	104.2	55.7	80.3	108.7	73.8	91.5	108.9	65.4	87.4	92.1	67.7	80.1	108.9	60.8	85.1
04	**MASONRY**	113.1	65.8	84.2	118.7	70.6	89.3	111.4	77.9	91.0	120.4	87.3	100.2	99.9	68.6	80.8	108.5	65.9	82.5
05	**METALS**	99.4	64.2	88.3	91.6	73.8	86.0	93.3	86.3	91.1	94.1	75.3	88.2	98.3	85.4	94.2	94.3	83.2	90.8
06	**WOOD, PLASTICS & COMPOSITES**	84.6	54.5	67.4	95.9	59.2	75.0	95.1	75.4	83.9	98.9	75.4	85.5	98.1	54.8	73.4	93.0	54.5	71.1
07	**THERMAL & MOISTURE PROTECTION**	105.6	66.7	90.0	99.6	63.7	85.2	99.7	75.1	89.8	99.7	85.3	93.9	99.0	69.5	87.1	99.7	63.9	85.3
08	**OPENINGS**	90.0	50.9	80.5	90.2	55.3	81.7	90.2	72.0	85.8	90.2	64.2	83.9	95.5	56.9	86.1	90.3	57.8	82.4
0920	Plaster & Gypsum Board	88.7	53.1	63.8	99.7	57.9	70.5	98.8	74.6	81.9	100.9	74.6	82.5	102.3	53.4	68.1	98.8	53.1	66.8
0950, 0980	Ceilings & Acoustic Treatment	98.9	53.1	68.4	94.6	57.9	70.2	94.6	74.6	81.3	94.6	74.6	81.3	102.8	53.4	69.9	94.6	53.1	66.9
0960	Flooring	104.9	85.1	99.0	103.2	104.0	103.4	103.0	111.1	105.4	104.3	104.0	104.2	104.5	86.6	99.1	102.1	85.1	97.0
0970, 0990	Wall Finishes & Painting/Coating	175.9	57.2	104.8	95.3	66.6	78.1	95.3	70.6	80.5	95.3	66.6	78.1	95.3	81.4	86.9	95.3	49.9	68.1
09	**FINISHES**	104.1	60.2	79.5	100.3	66.4	81.3	100.4	80.2	89.1	101.1	77.4	87.8	102.5	65.7	81.9	100.2	59.3	77.3
COVERS	**DIVS. 10 - 14, 25, 28, 41, 43, 44, 46**	100.0	65.4	92.9	100.0	74.3	94.8	100.0	85.7	97.1	100.0	78.1	95.5	100.0	83.4	96.6	100.0	64.7	92.8
21, 22, 23	**FIRE SUPPRESSION, PLUMBING & HVAC**	94.1	58.4	79.5	93.9	71.0	84.5	99.9	76.9	90.5	93.9	74.1	85.8	99.8	76.9	90.4	93.9	71.5	84.7
26, 27, 3370	**ELECTRICAL, COMMUNICATIONS & UTIL.**	93.1	72.3	82.2	92.8	80.3	86.2	91.4	66.6	78.4	90.8	87.1	88.8	103.1	66.6	83.9	95.4	72.3	83.2
MF2010	**WEIGHTED AVERAGE**	100.0	64.8	84.5	96.8	70.6	85.3	98.7	77.8	89.5	97.8	78.1	89.1	98.6	73.0	87.3	97.6	69.6	85.3

DIVISION		NEBRASKA												NEVADA					
		NORFOLK			NORTH PLATTE			OMAHA			VALENTINE			CARSON CITY			ELKO		
		687			691			680 - 681			692			897			898		
		MAT.	INST.	TOTAL	MAT.	INST.	TOTAL	MAT.	INST.	TOTAL	MAT.	INST.	TOTAL	MAT.	INST.	TOTAL	MAT.	INST.	TOTAL
015433	**CONTRACTOR EQUIPMENT**		92.0	92.0		100.6	100.6		92.0	92.0		95.0	95.0		97.5	97.5		97.5	97.5
0241, 31 - 34	**SITE & INFRASTRUCTURE, DEMOLITION**	81.5	90.6	87.9	104.6	91.4	95.4	82.1	91.3	88.6	87.2	95.2	92.8	72.3	98.5	90.7	58.8	98.3	86.5
0310	Concrete Forming & Accessories	84.2	72.2	73.8	97.2	72.8	76.1	94.2	72.7	75.6	85.5	54.9	59.0	100.9	96.0	96.6	108.9	82.6	86.2
0320	Concrete Reinforcing	103.9	48.4	76.3	104.5	75.1	89.9	99.2	76.8	88.0	105.2	45.5	75.5	109.5	116.0	112.7	111.2	90.9	101.1
0330	Cast-in-Place Concrete	110.3	62.2	90.5	115.2	64.4	94.3	92.8	77.5	86.5	102.2	54.9	82.8	97.0	86.5	92.7	97.8	82.2	91.4
03	**CONCRETE**	103.1	64.6	84.1	108.9	71.0	90.2	94.7	75.4	85.1	106.3	54.0	80.5	102.1	96.3	99.3	100.0	84.3	92.2
04	**MASONRY**	125.1	68.5	90.6	95.6	87.3	90.5	105.5	80.5	90.3	108.4	66.0	82.5	117.1	75.4	91.7	119.4	74.8	92.2
05	**METALS**	94.9	67.0	86.1	93.5	85.2	90.9	97.6	78.8	91.7	105.4	65.3	92.8	94.8	101.9	97.0	93.9	90.0	92.7
06	**WOOD, PLASTICS & COMPOSITES**	79.6	74.9	76.9	95.0	75.4	83.8	89.9	72.2	79.8	79.3	54.0	64.9	89.3	98.6	94.6	98.7	82.9	89.7
07	**THERMAL & MOISTURE PROTECTION**	99.5	72.5	88.7	99.7	79.3	91.5	95.4	80.8	89.5	100.4	66.5	86.8	99.7	85.6	94.0	98.9	76.3	89.8
08	**OPENINGS**	91.8	64.7	85.2	89.5	70.9	85.0	98.3	72.3	92.0	92.0	52.0	82.2	97.9	110.9	101.0	100.5	80.9	95.8
0920	Plaster & Gypsum Board	99.6	74.6	82.1	98.8	74.6	81.9	105.8	71.8	82.0	101.3	53.1	67.6	89.3	98.5	95.7	89.4	82.3	84.4
0950, 0980	Ceilings & Acoustic Treatment	108.6	74.6	86.0	94.6	74.6	81.3	117.7	71.8	87.1	111.9	53.1	72.7	114.0	98.5	103.7	110.4	82.3	91.7
0960	Flooring	131.9	104.0	123.6	102.9	104.0	103.2	135.9	84.9	120.7	134.2	82.7	118.8	101.4	81.1	95.3	106.0	53.0	90.2
0970, 0990	Wall Finishes & Painting/Coating	173.1	66.6	109.3	95.3	66.6	78.1	173.1	68.3	110.3	173.1	69.0	110.7	99.3	82.5	89.2	98.4	82.5	88.9
09	**FINISHES**	122.1	77.4	97.1	100.4	77.4	87.5	126.1	74.2	97.0	124.2	60.5	88.5	101.6	92.0	96.2	100.5	77.1	87.4
COVERS	**DIVS. 10 - 14, 25, 28, 41, 43, 44, 46**	100.0	76.9	95.3	100.0	68.1	93.5	100.0	84.2	96.8	100.0	62.7	92.4	100.0	93.4	98.7	100.0	65.2	92.9
21, 22, 23	**FIRE SUPPRESSION, PLUMBING & HVAC**	93.6	74.0	85.6	99.9	72.3	88.6	99.7	79.0	91.2	93.6	56.5	78.4	99.9	82.7	92.9	97.1	82.3	91.1
26, 27, 3370	**ELECTRICAL, COMMUNICATIONS & UTIL.**	91.6	81.0	86.0	93.5	80.3	86.5	100.5	82.9	91.2	90.3	90.2	90.3	96.2	96.6	96.4	93.7	96.6	95.2
MF2010	**WEIGHTED AVERAGE**	98.6	74.0	87.8	98.1	78.3	89.4	100.4	79.5	91.2	100.0	66.6	85.3	98.9	91.8	95.8	97.6	84.5	91.8

	DIVISION	NEVADA									NEW HAMPSHIRE								
		ELY			LAS VEGAS			RENO			CHARLESTON			CLAREMONT			CONCORD		
		893			889 - 891			894 - 895			036			037			032 - 033		
		MAT.	INST.	TOTAL	MAT.	INST.	TOTAL	MAT.	INST.	TOTAL	MAT.	INST.	TOTAL	MAT.	INST.	TOTAL	MAT.	INST.	TOTAL
015433	CONTRACTOR EQUIPMENT		97.5	97.5		97.5	97.5		97.5	97.5		100.5	100.5		100.5	100.5		100.5	100.5
0241, 31 - 34	SITE & INFRASTRUCTURE, DEMOLITION	63.6	98.4	88.0	66.0	101.0	90.5	63.6	98.5	88.1	89.8	99.3	96.4	83.3	99.3	94.5	96.1	101.4	99.8
0310	Concrete Forming & Accessories	102.3	103.4	103.2	103.5	112.8	111.6	98.8	95.9	96.3	84.7	81.3	81.7	90.0	81.3	82.4	94.0	92.1	92.4
0320	Concrete Reinforcing	110.0	85.3	97.7	101.4	123.7	112.5	103.9	122.5	113.2	88.6	95.4	92.0	88.6	95.4	92.0	88.6	95.9	92.2
0330	Cast-in-Place Concrete	105.0	101.7	103.6	101.7	109.2	104.8	111.0	86.4	100.9	98.8	73.4	88.4	90.8	73.4	83.7	107.5	94.0	102.0
03	CONCRETE	108.1	99.2	103.7	103.2	113.2	108.1	107.8	97.5	102.7	101.0	81.3	91.3	92.9	81.3	87.2	103.8	93.4	98.6
04	MASONRY	123.5	103.2	111.1	111.6	105.2	107.7	118.4	75.4	92.2	89.3	84.2	86.2	89.0	84.2	86.1	91.7	104.6	99.6
05	METALS	93.8	91.2	93.0	100.8	109.1	103.4	95.2	104.5	98.1	92.2	90.9	91.8	92.2	90.9	91.8	92.9	94.3	93.4
06	WOOD, PLASTICS & COMPOSITES	90.1	104.9	98.6	89.3	112.8	102.7	85.6	98.6	93.0	83.6	89.6	87.0	89.3	89.6	89.5	93.0	89.6	91.1
07	THERMAL & MOISTURE PROTECTION	99.5	93.2	97.0	112.5	100.2	107.6	99.0	85.6	93.6	96.6	73.0	87.1	96.5	73.0	87.0	102.3	92.8	98.5
08	OPENINGS	100.5	91.1	98.2	99.5	120.3	104.6	98.3	106.4	100.3	100.2	76.7	94.5	101.4	76.7	95.4	101.4	86.7	97.8
0920	Plaster & Gypsum Board	86.1	105.0	99.3	85.3	113.1	104.8	78.8	98.5	92.6	91.0	88.6	89.3	92.2	88.6	89.7	93.2	88.6	90.0
0950, 0980	Ceilings & Acoustic Treatment	110.4	105.0	106.8	119.5	113.1	115.3	115.9	98.5	104.3	88.3	88.6	88.5	88.3	88.6	88.5	92.0	88.6	89.7
0960	Flooring	103.6	59.3	90.4	94.8	111.3	99.7	100.6	81.1	94.8	93.7	33.3	75.7	95.6	33.3	77.1	93.3	114.2	99.5
0970, 0990	Wall Finishes & Painting/Coating	98.4	117.2	109.6	101.2	121.0	113.1	98.4	82.5	88.9	96.1	47.8	67.1	96.1	48.0	67.3	96.1	99.7	98.2
09	FINISHES	100.0	97.5	98.6	99.4	113.6	107.4	99.2	92.0	95.1	93.7	70.1	80.4	93.8	70.1	80.5	95.6	96.7	96.2
COVERS	DIVS. 10 - 14, 25, 28, 41, 43, 44, 46	100.0	64.4	92.7	100.0	89.7	97.9	100.0	93.4	98.7	100.0	90.9	98.1	100.0	90.9	98.1	100.0	103.0	100.6
21, 22, 23	FIRE SUPPRESSION, PLUMBING & HVAC	97.1	102.9	99.5	100.1	106.6	102.7	99.9	82.7	92.9	94.0	41.7	72.6	94.0	41.8	72.6	99.8	88.8	95.3
26, 27, 3370	ELECTRICAL, COMMUNICATIONS & UTIL.	94.0	109.9	102.4	98.3	114.3	106.7	94.3	96.6	95.5	99.1	55.1	75.9	99.1	55.1	75.9	98.5	79.9	88.7
MF2010	WEIGHTED AVERAGE	98.7	99.3	99.0	100.2	109.0	104.1	99.0	92.0	95.9	95.8	70.6	84.7	94.9	70.7	84.2	98.4	92.8	95.9

	DIVISION	NEW HAMPSHIRE															NEW JERSEY		
		KEENE			LITTLETON			MANCHESTER			NASHUA			PORTSMOUTH			ATLANTIC CITY		
		034			035			031			030			038			082,084		
		MAT.	INST.	TOTAL	MAT.	INST.	TOTAL	MAT.	INST.	TOTAL	MAT.	INST.	TOTAL	MAT.	INST.	TOTAL	MAT.	INST.	TOTAL
015433	CONTRACTOR EQUIPMENT		100.5	100.5		100.5	100.5		100.5	100.5		100.5	100.5		100.5	100.5		98.7	98.7
0241, 31 - 34	SITE & INFRASTRUCTURE, DEMOLITION	98.3	99.5	99.2	83.2	99.5	94.7	94.0	101.4	99.2	99.4	101.4	100.8	92.9	101.5	98.9	103.1	103.4	103.3
0310	Concrete Forming & Accessories	88.8	83.3	84.0	99.2	83.3	85.4	96.5	92.6	93.1	96.1	92.6	93.0	85.9	92.9	91.9	103.8	126.3	123.2
0320	Concrete Reinforcing	88.6	95.5	92.0	89.3	95.5	92.4	110.7	96.0	103.4	110.7	96.0	103.4	88.6	96.1	92.3	84.4	120.4	102.3
0330	Cast-in-Place Concrete	99.3	75.7	89.6	89.1	75.7	83.6	110.6	115.6	112.6	94.0	115.6	102.9	89.2	115.7	100.1	84.3	131.3	103.6
03	CONCRETE	100.7	83.0	92.0	92.4	83.0	87.8	109.0	101.0	105.1	101.0	101.0	101.0	92.3	101.2	96.7	93.7	125.8	109.5
04	MASONRY	93.0	88.0	90.0	99.1	88.0	92.3	94.1	104.6	100.5	93.3	104.6	100.2	89.4	104.6	98.6	96.1	127.6	115.3
05	METALS	92.9	91.6	92.4	92.9	91.6	92.5	97.7	94.8	96.8	97.7	94.8	96.8	94.4	95.5	94.7	92.6	106.6	97.0
06	WOOD, PLASTICS & COMPOSITES	87.8	89.6	88.8	97.7	89.6	93.1	93.2	89.6	91.1	96.4	89.6	92.5	84.7	89.6	87.5	106.4	126.6	117.9
07	THERMAL & MOISTURE PROTECTION	97.0	76.1	88.6	96.6	74.5	87.7	103.4	96.0	100.5	97.3	96.0	96.8	97.0	115.8	104.6	97.0	124.9	108.2
08	OPENINGS	98.8	81.7	94.6	102.4	77.4	96.3	103.4	86.7	99.3	103.4	86.7	99.3	104.0	83.4	99.0	101.0	124.1	106.6
0920	Plaster & Gypsum Board	91.4	88.6	89.5	103.9	88.6	93.2	98.2	88.6	91.5	99.2	88.6	91.8	91.0	88.6	89.3	103.4	126.8	119.8
0950, 0980	Ceilings & Acoustic Treatment	88.3	88.6	88.5	88.3	88.6	88.5	98.6	88.6	91.9	97.7	88.6	91.6	90.4	88.6	89.2	88.3	126.8	114.0
0960	Flooring	95.3	54.5	83.2	103.6	33.3	82.6	99.0	114.2	103.5	98.3	114.2	103.0	93.8	114.2	99.9	101.6	146.6	115.0
0970, 0990	Wall Finishes & Painting/Coating	96.1	47.8	67.1	96.1	62.6	76.0	96.1	99.7	98.2	96.1	99.7	98.2	96.1	99.7	98.2	96.1	120.0	110.4
09	FINISHES	95.7	75.0	84.1	97.6	72.6	83.6	98.8	96.7	97.6	99.5	96.7	97.9	95.0	96.7	96.0	99.6	130.5	116.9
COVERS	DIVS. 10 - 14, 25, 28, 41, 43, 44, 46	100.0	92.2	98.4	100.0	97.4	99.5	100.0	103.0	100.6	100.0	103.0	100.6	100.0	103.0	100.6	100.0	110.9	102.2
21, 22, 23	FIRE SUPPRESSION, PLUMBING & HVAC	94.0	45.6	74.2	94.0	67.3	83.0	99.8	88.8	95.3	100.0	88.8	95.4	100.0	88.9	95.4	99.5	123.4	109.3
26, 27, 3370	ELECTRICAL, COMMUNICATIONS & UTIL.	99.1	65.9	81.6	100.3	58.3	78.2	102.0	79.9	90.4	101.9	79.9	90.3	99.6	79.9	89.3	96.6	136.9	117.8
MF2010	WEIGHTED AVERAGE	96.4	74.6	86.8	96.0	77.7	88.0	100.6	94.0	97.7	99.8	94.0	97.2	97.3	94.5	96.1	97.6	123.3	108.9

	DIVISION	NEW JERSEY																	
		CAMDEN			DOVER			ELIZABETH			HACKENSACK			JERSEY CITY			LONG BRANCH		
		081			078			072			076			073			077		
		MAT.	INST.	TOTAL	MAT.	INST.	TOTAL	MAT.	INST.	TOTAL	MAT.	INST.	TOTAL	MAT.	INST.	TOTAL	MAT.	INST.	TOTAL
015433	CONTRACTOR EQUIPMENT		98.7	98.7		100.5	100.5		100.5	100.5		100.5	100.5		98.7	98.7		98.3	98.3
0241, 31 - 34	SITE & INFRASTRUCTURE, DEMOLITION	103.9	104.0	103.9	115.6	104.8	108.0	120.5	104.8	109.5	115.9	105.0	108.2	103.9	104.9	104.6	109.3	104.8	106.1
0310	Concrete Forming & Accessories	96.1	126.2	122.1	92.4	126.8	122.1	103.1	126.7	123.6	92.4	126.8	122.1	96.1	126.8	122.6	96.6	122.9	119.3
0320	Concrete Reinforcing	110.7	115.1	112.9	85.3	122.1	103.6	85.3	122.1	103.6	85.3	122.1	103.6	110.7	122.1	116.4	85.3	122.1	103.6
0330	Cast-in-Place Concrete	81.8	129.9	101.6	104.8	129.1	114.8	90.0	129.1	106.1	102.4	129.1	113.4	81.8	129.1	101.2	90.9	131.0	107.4
03	CONCRETE	95.2	124.3	109.6	101.8	125.8	113.7	97.0	125.8	111.2	100.0	125.8	112.7	95.2	125.6	110.2	98.9	124.6	111.6
04	MASONRY	87.8	127.6	112.1	91.9	128.1	114.0	106.1	128.1	119.5	95.5	128.1	115.4	84.8	128.1	111.2	99.3	127.6	116.6
05	METALS	97.6	103.9	99.6	92.6	111.0	98.4	94.1	111.0	99.4	92.7	111.0	98.4	97.7	108.7	101.1	92.7	108.5	97.7
06	WOOD, PLASTICS & COMPOSITES	96.4	126.6	113.7	95.8	126.6	113.4	109.4	126.6	119.2	95.8	126.6	113.4	96.4	126.6	113.7	97.4	121.7	111.3
07	THERMAL & MOISTURE PROTECTION	96.9	123.8	107.7	97.4	126.2	109.0	97.5	126.2	109.1	97.1	125.1	108.4	96.9	126.2	108.7	97.0	124.6	108.1
08	OPENINGS	103.4	122.4	108.0	107.3	124.4	111.5	105.4	124.4	110.0	104.6	124.4	109.4	103.4	124.4	108.5	99.6	121.7	105.0
0920	Plaster & Gypsum Board	99.2	126.8	118.5	96.0	126.8	117.6	103.4	126.8	119.8	96.0	126.8	117.6	99.2	126.8	118.5	98.0	121.8	114.6
0950, 0980	Ceilings & Acoustic Treatment	97.7	126.8	117.1	88.3	126.8	114.0	90.4	126.8	114.7	88.3	126.8	114.0	97.7	126.8	117.1	88.3	121.8	110.6
0960	Flooring	98.3	146.6	112.7	97.4	169.0	118.7	102.1	169.0	122.0	97.4	169.0	118.7	98.3	169.0	119.3	98.5	169.0	119.5
0970, 0990	Wall Finishes & Painting/Coating	96.1	120.0	110.4	96.0	122.0	111.6	96.0	122.0	111.6	96.0	122.0	111.6	96.1	122.0	111.6	96.1	120.0	110.4
09	FINISHES	100.0	130.5	117.1	97.6	133.6	117.8	101.0	133.6	119.3	97.4	133.6	117.7	100.0	133.6	118.8	98.6	131.3	116.9
COVERS	DIVS. 10 - 14, 25, 28, 41, 43, 44, 46	100.0	110.9	102.2	100.0	114.7	103.0	100.0	114.7	103.0	100.0	114.7	103.0	100.0	114.7	103.0	100.0	110.2	102.1
21, 22, 23	FIRE SUPPRESSION, PLUMBING & HVAC	100.0	123.4	109.6	99.6	127.2	110.9	100.0	126.6	110.9	99.6	127.2	110.9	100.0	127.2	111.1	99.6	122.5	109.0
26, 27, 3370	ELECTRICAL, COMMUNICATIONS & UTIL.	102.3	136.9	120.5	98.1	134.4	117.2	98.8	134.4	117.6	98.1	134.3	117.1	103.5	134.3	119.7	97.7	130.1	114.8
MF2010	WEIGHTED AVERAGE	99.0	122.8	109.5	99.2	124.9	110.5	100.1	124.8	111.0	98.9	124.9	110.4	99.0	124.7	110.3	98.4	122.3	108.9

DIVISION		NEW JERSEY																	
		NEW BRUNSWICK			NEWARK			PATERSON			POINT PLEASANT			SUMMIT			TRENTON		
		088 - 089			070 - 071			074 - 075			087			079			085 - 086		
		MAT.	INST.	TOTAL	MAT.	INST.	TOTAL	MAT.	INST.	TOTAL	MAT.	INST.	TOTAL	MAT.	INST.	TOTAL	MAT.	INST.	TOTAL
015433	CONTRACTOR EQUIPMENT		98.3	98.3		100.5	100.5		100.5	100.5		98.3	98.3		100.5	100.5		98.3	98.3
0241, 31 - 34	SITE & INFRASTRUCTURE, DEMOLITION	118.1	104.8	108.8	119.8	104.8	109.3	117.8	105.0	108.8	119.7	104.8	109.2	117.7	104.8	108.6	100.5	104.8	103.5
0310	Concrete Forming & Accessories	98.7	126.6	122.9	94.6	126.7	122.4	94.3	126.8	122.4	93.9	122.7	118.8	95.0	126.7	122.5	96.7	122.8	119.3
0320	Concrete Reinforcing	85.3	122.1	103.6	110.7	122.1	116.4	110.7	122.1	116.4	85.3	122.1	103.6	85.3	122.1	103.6	110.7	117.6	114.1
0330	Cast-in-Place Concrete	104.1	131.3	115.3	108.6	129.1	117.0	104.1	129.1	114.4	104.1	131.0	115.2	87.1	129.1	104.3	100.4	131.0	113.0
03	CONCRETE	110.7	126.3	118.4	107.9	125.8	116.8	105.8	125.8	115.7	110.4	124.5	117.3	94.0	125.8	109.7	104.1	123.7	113.8
04	MASONRY	94.0	128.1	114.8	97.0	128.1	116.0	92.5	128.1	114.2	85.0	127.6	111.0	94.0	128.1	114.8	84.2	127.6	110.7
05	METALS	92.7	108.5	97.6	97.6	111.0	101.8	92.7	111.0	98.5	92.7	108.3	97.6	92.6	111.0	98.4	92.7	106.3	97.0
06	WOOD, PLASTICS & COMPOSITES	100.5	126.6	115.4	95.1	126.6	113.1	98.1	126.6	114.4	94.2	121.7	109.9	99.2	126.6	114.9	93.6	121.7	109.6
07	THERMAL & MOISTURE PROTECTION	97.2	125.3	108.5	102.9	126.2	112.3	97.4	125.1	108.6	97.2	124.6	108.2	97.7	126.2	109.2	101.1	124.2	110.4
08	OPENINGS	95.8	124.4	102.8	110.0	124.4	113.5	110.0	124.4	113.5	97.8	122.6	103.8	112.2	124.4	115.2	103.9	120.2	107.9
0920	Plaster & Gypsum Board	100.1	126.8	118.8	95.4	126.8	117.4	99.2	126.8	118.5	96.0	121.8	114.0	98.0	126.8	118.2	95.9	121.8	114.0
0950, 0980	Ceilings & Acoustic Treatment	88.3	126.8	114.0	98.6	126.8	117.4	97.7	126.8	117.1	88.3	121.8	110.6	88.3	126.8	114.0	100.4	121.8	114.6
0960	Flooring	99.5	169.0	120.2	98.3	169.0	119.3	98.3	169.0	119.3	97.4	146.6	112.1	98.6	169.0	119.6	98.5	169.0	119.5
0970, 0990	Wall Finishes & Painting/Coating	96.1	122.0	111.6	96.0	122.0	111.6	96.0	122.0	111.6	96.1	120.0	110.4	96.0	122.0	111.6	96.1	120.0	110.4
09	FINISHES	100.2	133.5	118.9	99.6	133.6	118.6	100.2	133.6	118.9	99.0	127.5	115.0	98.6	133.6	118.2	99.6	131.3	117.4
COVERS	DIVS. 10 - 14, 25, 28, 41, 43, 44, 46	100.0	114.6	103.0	100.0	114.7	103.0	100.0	114.7	103.0	100.0	105.9	101.2	100.0	114.7	103.0	100.0	110.2	102.1
21, 22, 23	FIRE SUPPRESSION, PLUMBING & HVAC	99.5	122.8	109.0	100.0	126.6	110.9	100.0	127.2	111.1	99.5	122.4	108.9	99.6	126.6	110.6	100.0	122.5	109.2
26, 27, 3370	ELECTRICAL, COMMUNICATIONS & UTIL.	97.4	135.4	117.4	107.7	134.3	121.7	103.5	134.4	119.8	96.6	130.1	114.3	98.8	134.4	117.6	106.1	134.9	121.3
MF2010	WEIGHTED AVERAGE	99.4	123.9	110.2	102.7	124.8	112.4	100.9	124.9	111.5	98.9	121.6	108.9	99.2	124.8	110.5	99.5	122.5	109.6

DIVISION		NEW JERSEY			NEW MEXICO														
		VINELAND			ALBUQUERQUE			CARRIZOZO			CLOVIS			FARMINGTON			GALLUP		
		080,083			870 - 872			883			881			874			873		
		MAT.	INST.	TOTAL	MAT.	INST.	TOTAL	MAT.	INST.	TOTAL	MAT.	INST.	TOTAL	MAT.	INST.	TOTAL	MAT.	INST.	TOTAL
015433	CONTRACTOR EQUIPMENT		98.7	98.7		109.2	109.2		109.2	109.2		109.2	109.2		109.2	109.2		109.2	109.2
0241, 31 - 34	SITE & INFRASTRUCTURE, DEMOLITION	108.2	104.0	105.2	81.3	104.1	97.3	102.3	104.1	103.5	90.2	104.1	99.9	87.6	104.1	99.1	95.4	104.1	101.5
0310	Concrete Forming & Accessories	91.6	126.3	121.6	99.3	65.6	70.1	99.3	65.6	70.1	99.3	65.4	70.0	99.3	65.6	70.1	99.3	65.6	70.1
0320	Concrete Reinforcing	84.4	120.4	102.3	104.7	71.5	88.2	112.3	71.5	92.0	113.5	71.5	92.6	114.4	71.5	93.0	109.5	71.5	90.6
0330	Cast-in-Place Concrete	90.9	131.3	107.5	103.9	72.4	90.9	96.7	72.4	86.7	96.6	72.3	86.6	104.8	72.4	91.5	98.7	72.4	87.9
03	CONCRETE	98.5	125.8	112.0	104.5	70.1	87.5	118.9	70.1	94.8	107.9	70.0	89.2	108.4	70.1	89.5	114.1	70.1	92.4
04	MASONRY	85.1	127.6	111.1	107.6	62.0	79.8	103.5	62.0	78.2	103.5	62.0	78.2	114.9	62.0	82.6	103.6	62.0	78.2
05	METALS	92.6	106.6	97.0	101.3	88.0	97.1	98.1	88.0	94.9	97.7	87.9	94.6	99.0	88.0	95.6	98.1	88.0	94.9
06	WOOD, PLASTICS & COMPOSITES	91.4	126.6	111.5	89.4	66.0	76.1	89.5	66.0	76.1	89.5	66.0	76.1	89.5	66.0	76.1	89.5	66.0	76.1
07	THERMAL & MOISTURE PROTECTION	96.8	124.9	108.1	98.7	73.1	88.5	100.2	73.1	89.3	99.0	73.1	88.6	98.9	73.1	88.5	99.9	73.1	89.1
08	OPENINGS	97.3	124.1	103.8	98.1	69.2	91.1	96.8	69.2	90.1	96.9	69.2	90.2	100.7	69.2	93.1	100.8	69.2	93.1
0920	Plaster & Gypsum Board	94.3	126.8	117.1	83.8	64.7	70.4	76.4	64.7	68.2	76.4	64.7	68.2	76.4	64.7	68.2	76.4	64.7	68.2
0950, 0980	Ceilings & Acoustic Treatment	88.3	126.8	114.0	110.0	64.7	79.8	106.7	64.7	78.7	106.7	64.7	78.7	106.7	64.7	78.7	106.7	64.7	78.7
0960	Flooring	96.6	146.6	111.5	98.8	68.4	89.8	100.6	68.4	91.0	100.6	68.4	91.0	100.6	68.4	91.0	100.6	68.4	91.0
0970, 0990	Wall Finishes & Painting/Coating	96.1	120.0	110.4	104.1	51.2	72.4	98.4	51.2	70.1	98.4	51.2	70.1	98.4	51.2	70.1	98.4	51.2	70.1
09	FINISHES	97.6	130.5	116.0	99.2	64.3	79.7	99.9	64.3	80.0	98.4	64.3	79.3	97.9	64.3	79.1	99.2	64.3	79.7
COVERS	DIVS. 10 - 14, 25, 28, 41, 43, 44, 46	100.0	110.9	102.2	100.0	77.3	95.4	100.0	77.3	95.4	100.0	77.3	95.4	100.0	77.3	95.4	100.0	77.3	95.4
21, 22, 23	FIRE SUPPRESSION, PLUMBING & HVAC	99.5	123.5	109.3	100.1	69.1	87.4	96.5	69.1	85.3	96.5	68.7	85.1	99.9	69.1	87.3	96.3	69.1	85.1
26, 27, 3370	ELECTRICAL, COMMUNICATIONS & UTIL.	96.6	136.9	117.8	89.4	72.9	80.7	89.4	72.9	80.7	86.9	72.9	79.5	87.6	72.9	79.8	86.7	72.9	79.4
MF2010	WEIGHTED AVERAGE	97.1	123.4	108.7	99.1	73.5	87.9	99.7	73.5	88.2	97.7	73.4	87.0	99.7	73.5	88.2	99.0	73.5	87.8

DIVISION		NEW MEXICO																	
		LAS CRUCES			LAS VEGAS			ROSWELL			SANTA FE			SOCORRO			TRUTH/CONSEQUENCES		
		880			877			882			875			878			879		
		MAT.	INST.	TOTAL	MAT.	INST.	TOTAL	MAT.	INST.	TOTAL	MAT.	INST.	TOTAL	MAT.	INST.	TOTAL	MAT.	INST.	TOTAL
015433	CONTRACTOR EQUIPMENT		85.5	85.5		109.2	109.2		109.2	109.2		109.2	109.2		109.2	109.2		85.5	85.5
0241, 31 - 34	SITE & INFRASTRUCTURE, DEMOLITION	91.4	83.6	85.9	87.0	104.1	99.0	92.3	104.1	100.6	85.6	104.1	98.5	83.4	104.1	97.9	102.2	83.6	89.1
0310	Concrete Forming & Accessories	96.5	64.4	68.7	99.3	65.6	70.1	99.3	65.6	70.1	97.8	65.6	69.9	99.3	65.6	70.1	96.5	64.4	68.7
0320	Concrete Reinforcing	108.1	71.4	89.8	111.3	71.5	91.5	113.5	71.5	92.6	112.3	71.5	92.0	113.5	71.5	92.6	106.8	71.4	89.2
0330	Cast-in-Place Concrete	91.2	64.8	80.3	102.0	72.4	89.8	96.6	72.4	86.6	106.1	72.4	92.2	99.9	72.4	88.6	109.5	64.8	91.1
03	CONCRETE	86.8	66.6	76.9	105.6	70.1	88.1	108.6	70.1	89.6	106.7	70.1	88.7	104.5	70.1	87.5	98.5	66.7	82.8
04	MASONRY	99.5	61.7	76.4	103.9	62.0	78.3	114.6	62.0	82.5	107.9	62.0	79.9	103.8	62.0	78.3	101.1	61.7	77.0
05	METALS	98.9	81.4	93.4	97.8	88.0	94.8	99.0	88.0	95.5	99.0	88.0	95.5	98.1	88.0	94.9	97.7	81.5	92.6
06	WOOD, PLASTICS & COMPOSITES	80.9	64.9	71.8	89.5	66.0	76.1	89.5	66.0	76.1	89.0	66.0	75.9	89.5	66.0	76.1	80.9	64.9	71.8
07	THERMAL & MOISTURE PROTECTION	85.8	68.5	78.8	98.5	73.1	88.3	99.1	73.1	88.7	98.7	73.1	88.4	98.5	73.1	88.3	86.3	68.5	79.1
08	OPENINGS	90.7	68.6	85.3	96.9	69.2	90.2	96.7	69.2	90.0	96.9	69.2	90.2	96.8	69.2	90.1	90.6	68.6	85.2
0920	Plaster & Gypsum Board	78.4	64.7	68.8	76.4	64.7	68.2	76.4	64.7	68.2	87.5	64.7	71.5	76.4	64.7	68.2	78.4	64.7	68.8
0950, 0980	Ceilings & Acoustic Treatment	94.6	64.7	74.7	106.7	64.7	78.7	106.7	64.7	78.7	111.3	64.7	80.2	106.7	64.7	78.7	94.6	64.7	74.7
0960	Flooring	132.0	68.4	113.0	100.6	68.4	91.0	100.6	68.4	91.0	101.0	68.4	91.3	100.6	68.4	91.0	132.0	68.4	113.0
0970, 0990	Wall Finishes & Painting/Coating	89.3	51.2	66.5	98.4	51.2	70.1	98.4	51.2	70.1	98.4	51.2	70.1	98.4	51.2	70.1	89.3	51.2	66.5
09	FINISHES	110.3	63.5	84.0	97.8	64.3	79.0	98.5	64.3	79.4	100.6	64.3	80.3	97.7	64.3	79.0	111.0	63.5	84.4
COVERS	DIVS. 10 - 14, 25, 28, 41, 43, 44, 46	100.0	74.6	94.8	100.0	77.3	95.4	100.0	77.3	95.4	100.0	77.3	95.4	100.0	77.3	95.4	100.0	74.6	94.8
21, 22, 23	FIRE SUPPRESSION, PLUMBING & HVAC	100.3	68.8	87.4	96.3	69.1	85.1	99.9	69.1	87.3	99.9	69.1	87.3	96.3	69.1	85.1	96.2	68.8	85.0
26, 27, 3370	ELECTRICAL, COMMUNICATIONS & UTIL.	88.1	66.6	76.8	89.4	72.9	80.7	88.3	72.9	80.2	102.3	72.9	86.8	87.3	72.9	79.7	91.1	72.9	81.5
MF2010	WEIGHTED AVERAGE	96.2	69.4	84.4	97.6	73.5	87.0	99.6	73.5	88.1	100.4	73.5	88.5	97.1	73.5	86.7	97.1	70.2	85.2

	DIVISION	NEW MEXICO			NEW YORK																	
		TUCUMCARI			ALBANY			BINGHAMTON			BRONX			BROOKLYN			BUFFALO					
		884			120 - 122			137 - 139			104			112			140 - 142					
		MAT.	INST.	TOTAL	MAT.	INST.	TOTAL	MAT.	INST.	TOTAL	MAT.	INST.	TOTAL	MAT.	INST.	TOTAL	MAT.	INST.	TOTAL			
015433	CONTRACTOR EQUIPMENT		109.2	109.2		114.1	114.1		114.7	114.7		113.0	113.0		115.0	115.0		97.4	97.4			
0241, 31 - 34	SITE & INFRASTRUCTURE, DEMOLITION	89.9	104.1	99.8	70.7	107.2	96.3	92.6	92.9	92.8	110.9	123.3	119.6	117.6	129.1	125.7	97.4	98.3	98.0			
0310	Concrete Forming & Accessories	99.3	65.4	70.0	97.6	95.3	95.6	100.7	83.9	86.1	103.2	175.2	165.5	106.9	174.9	165.7	97.7	114.2	112.0			
0320	Concrete Reinforcing	111.3	71.5	91.5	97.1	108.1	102.6	96.1	96.1	96.1	94.0	177.6	135.6	97.6	185.2	141.2	100.9	100.5	100.7			
0330	Cast-in-Place Concrete	96.6	72.3	86.6	75.6	102.1	86.5	96.9	130.5	110.7	89.4	177.6	125.7	99.0	176.0	130.7	102.6	116.7	108.4			
03	CONCRETE	107.0	70.0	88.7	93.2	100.7	96.9	95.1	103.8	99.4	94.3	174.8	134.0	107.1	175.2	140.7	106.1	111.8	108.9			
04	MASONRY	114.3	62.0	82.4	93.6	101.0	98.1	108.9	129.6	121.5	93.8	175.3	143.6	118.6	175.3	153.2	105.6	115.1	111.4			
05	METALS	97.7	87.9	94.6	98.7	110.8	102.5	95.3	118.2	102.5	103.6	145.2	116.7	103.6	141.2	115.4	100.5	94.8	98.7			
06	WOOD, PLASTICS & COMPOSITES	89.5	66.0	76.1	94.5	92.7	93.5	104.7	79.8	90.5	100.6	175.2	143.2	106.2	174.8	145.4	96.0	114.5	106.5			
07	THERMAL & MOISTURE PROTECTION	99.0	73.1	88.6	92.3	97.0	94.2	104.4	98.3	102.0	103.5	164.4	128.0	106.5	163.4	129.4	100.9	108.6	104.0			
08	OPENINGS	96.7	69.2	90.0	95.2	91.7	94.3	90.2	81.8	88.1	86.9	172.1	107.7	88.7	169.5	108.3	90.1	102.8	93.2			
0920	Plaster & Gypsum Board	76.4	64.7	68.2	102.9	92.2	95.4	112.9	78.7	89.0	104.4	177.1	155.3	110.4	177.1	157.0	105.2	114.6	111.8			
0950, 0980	Ceilings & Acoustic Treatment	106.7	64.7	78.7	98.7	92.2	94.4	97.8	78.7	85.1	83.8	177.1	146.0	93.4	177.1	149.2	103.4	114.6	110.9			
0960	Flooring	100.6	68.4	91.0	92.2	107.0	96.6	105.1	94.7	102.0	98.6	183.9	124.0	112.3	183.9	133.6	104.5	115.5	107.7			
0970, 0990	Wall Finishes & Painting/Coating	98.4	51.2	70.1	90.9	88.6	89.5	92.4	91.8	92.0	108.1	151.3	134.0	123.9	151.3	140.3	104.1	111.7	108.7			
09	FINISHES	98.3	64.3	79.3	96.6	96.4	96.5	98.4	85.4	91.1	98.9	174.7	141.4	114.1	174.4	147.9	102.8	115.0	109.7			
COVERS	DIVS. 10 - 14, 25, 28, 41, 43, 44, 46	100.0	77.3	95.4	100.0	93.8	98.7	100.0	94.9	99.0	100.0	131.7	106.5	100.0	131.0	106.3	100.0	105.8	101.2			
21, 22, 23	FIRE SUPPRESSION, PLUMBING & HVAC	96.5	68.7	85.1	100.0	97.8	99.1	100.4	86.9	94.9	100.2	167.4	127.7	99.8	167.2	127.4	100.0	96.8	98.7			
26, 27, 3370	ELECTRICAL, COMMUNICATIONS & UTIL.	89.4	72.9	80.7	101.5	97.3	99.3	100.7	109.8	105.5	96.4	178.1	139.4	100.6	178.1	141.5	98.6	97.8	98.2			
MF2010	WEIGHTED AVERAGE	98.4	73.4	87.4	97.0	99.8	98.2	98.2	100.1	99.0	98.3	164.8	127.6	103.0	164.8	130.2	100.0	104.1	101.8			

	DIVISION	NEW YORK																	
		ELMIRA			FAR ROCKAWAY			FLUSHING			GLENS FALLS			HICKSVILLE			JAMAICA		
		148 - 149			116			113			128			115,117,118			114		
		MAT.	INST.	TOTAL	MAT.	INST.	TOTAL	MAT.	INST.	TOTAL	MAT.	INST.	TOTAL	MAT.	INST.	TOTAL	MAT.	INST.	TOTAL
015433	CONTRACTOR EQUIPMENT		116.2	116.2		115.0	115.0		115.0	115.0		114.1	114.1		115.0	115.0		115.0	115.0
0241, 31 - 34	SITE & INFRASTRUCTURE, DEMOLITION	92.9	94.3	93.9	120.9	129.1	126.7	120.9	129.1	126.7	62.8	106.5	93.4	110.6	127.9	122.7	115.2	129.1	125.0
0310	Concrete Forming & Accessories	83.0	84.4	84.2	94.0	174.9	164.0	97.6	174.9	164.5	85.3	87.9	87.6	90.7	145.0	137.7	97.6	174.9	164.5
0320	Concrete Reinforcing	102.2	90.2	96.2	97.6	185.2	141.2	99.3	185.2	142.1	96.7	93.7	95.2	97.6	177.3	137.3	97.6	185.2	141.2
0330	Cast-in-Place Concrete	96.0	96.7	96.3	107.1	176.0	135.4	107.1	176.0	135.4	73.8	98.9	84.1	91.0	167.1	122.3	99.0	176.0	130.7
03	CONCRETE	99.9	91.5	95.8	113.0	175.2	143.7	113.5	175.2	144.0	86.0	93.7	89.8	99.0	157.3	127.8	106.4	175.2	140.4
04	MASONRY	102.2	93.1	96.7	122.5	175.3	154.7	116.5	175.3	152.4	95.0	95.4	95.3	113.2	160.5	142.1	120.7	175.3	154.0
05	METALS	95.9	115.3	102.0	103.6	141.2	115.4	103.6	141.2	115.4	94.9	105.6	98.2	105.1	136.4	114.9	103.6	141.2	115.4
06	WOOD, PLASTICS & COMPOSITES	79.9	83.0	81.7	90.5	174.8	138.7	95.0	174.8	140.6	82.2	86.0	84.4	87.2	143.3	119.2	95.0	174.8	140.6
07	THERMAL & MOISTURE PROTECTION	101.7	84.3	94.7	106.4	163.4	129.4	106.5	163.4	129.4	88.1	93.9	90.4	106.1	152.4	124.7	106.3	163.4	129.3
08	OPENINGS	90.6	82.7	88.7	87.2	169.5	107.2	87.2	169.5	107.2	89.1	84.8	88.1	87.2	150.6	102.6	87.2	169.5	107.2
0920	Plaster & Gypsum Board	99.2	82.2	87.3	99.1	177.1	153.7	101.6	177.1	154.4	93.6	85.3	87.8	98.7	144.6	130.8	101.6	177.1	154.4
0950, 0980	Ceilings & Acoustic Treatment	98.8	82.2	87.7	82.4	177.1	145.5	82.4	177.1	145.5	91.5	85.3	87.4	81.4	144.6	123.5	82.4	177.1	145.5
0960	Flooring	94.4	94.7	94.5	107.4	183.9	130.2	108.8	183.9	131.2	84.3	107.0	91.1	106.5	147.8	118.8	108.8	183.9	131.2
0970, 0990	Wall Finishes & Painting/Coating	102.7	88.1	93.9	123.9	151.3	140.3	123.9	151.3	140.3	90.9	88.6	89.5	123.9	151.3	140.3	123.9	151.3	140.3
09	FINISHES	97.8	86.3	91.4	109.4	174.4	145.8	110.1	174.4	146.2	90.9	91.1	91.0	107.8	145.1	128.7	109.6	174.4	145.9
COVERS	DIVS. 10 - 14, 25, 28, 41, 43, 44, 46	100.0	90.7	98.1	100.0	131.0	106.3	100.0	131.0	106.3	100.0	91.2	98.2	100.0	122.4	104.6	100.0	131.0	106.3
21, 22, 23	FIRE SUPPRESSION, PLUMBING & HVAC	94.2	87.8	91.6	93.7	167.2	123.8	93.7	167.2	123.8	94.2	89.9	92.4	99.8	147.8	119.4	93.7	167.2	123.8
26, 27, 3370	ELECTRICAL, COMMUNICATIONS & UTIL.	97.1	97.0	97.0	108.4	178.1	145.1	108.4	178.1	145.1	95.9	97.3	96.7	99.8	138.3	120.1	98.7	178.1	140.5
MF2010	WEIGHTED AVERAGE	96.3	92.7	94.8	102.6	164.8	130.0	102.4	164.8	129.9	92.2	94.9	93.4	101.0	145.4	120.5	100.7	164.8	129.0

	DIVISION	NEW YORK																	
		JAMESTOWN			KINGSTON			LONG ISLAND CITY			MONTICELLO			MOUNT VERNON			NEW ROCHELLE		
		147			124			111			127			105			108		
		MAT.	INST.	TOTAL	MAT.	INST.	TOTAL	MAT.	INST.	TOTAL	MAT.	INST.	TOTAL	MAT.	INST.	TOTAL	MAT.	INST.	TOTAL
015433	CONTRACTOR EQUIPMENT		94.3	94.3		115.0	115.0		115.0	115.0		115.0	115.0		113.0	113.0		113.0	113.0
0241, 31 - 34	SITE & INFRASTRUCTURE, DEMOLITION	94.1	94.0	94.0	116.8	124.8	122.4	118.7	129.1	126.0	112.4	124.8	121.1	118.0	117.1	117.4	117.1	117.1	117.1
0310	Concrete Forming & Accessories	83.0	80.1	80.5	86.9	107.2	104.5	101.7	174.9	165.0	93.6	107.1	105.3	93.7	127.6	123.0	108.4	127.6	125.0
0320	Concrete Reinforcing	102.3	98.4	100.3	97.0	133.1	115.0	97.6	185.2	141.2	96.3	132.9	114.6	92.8	176.3	134.4	92.9	176.3	134.4
0330	Cast-in-Place Concrete	99.5	93.0	96.8	99.8	132.0	113.0	102.2	176.0	132.5	93.5	129.7	108.4	99.9	130.6	112.5	99.9	130.6	112.5
03	CONCRETE	103.1	88.0	95.7	107.2	120.3	113.7	109.4	175.2	141.9	102.2	119.4	110.7	103.4	136.8	119.9	103.0	136.8	119.7
04	MASONRY	110.2	93.1	99.7	110.0	136.7	126.3	115.4	175.3	151.9	103.3	134.1	122.1	99.6	128.0	117.0	99.6	128.0	117.0
05	METALS	93.2	91.0	92.5	103.6	114.7	107.1	103.6	141.2	115.4	103.5	114.4	107.0	103.3	132.6	112.5	103.6	132.6	112.7
06	WOOD, PLASTICS & COMPOSITES	78.5	76.7	77.4	83.4	100.8	93.3	100.6	174.8	143.0	91.0	100.8	96.6	90.7	130.1	113.2	107.6	130.1	120.4
07	THERMAL & MOISTURE PROTECTION	101.2	85.4	94.8	108.0	136.9	119.6	106.4	163.4	129.3	107.6	135.9	119.0	104.4	136.7	117.4	104.4	136.7	117.4
08	OPENINGS	90.4	80.6	88.0	92.1	117.7	98.3	87.2	169.5	107.2	87.2	117.7	94.7	86.9	143.4	100.7	87.0	143.4	100.7
0920	Plaster & Gypsum Board	94.1	75.7	81.2	95.1	100.7	99.1	105.8	177.1	155.7	97.6	100.7	99.8	99.4	130.6	121.2	111.4	130.6	124.9
0950, 0980	Ceilings & Acoustic Treatment	96.1	75.7	82.5	76.9	100.7	92.8	82.4	177.1	145.5	76.9	100.7	92.8	81.9	130.6	114.4	81.9	130.6	114.4
0960	Flooring	97.5	91.3	95.7	99.8	75.1	92.5	110.4	183.9	132.3	102.2	75.1	94.1	91.1	177.2	116.8	97.2	177.2	121.1
0970, 0990	Wall Finishes & Painting/Coating	104.1	77.5	88.2	121.1	110.9	115.0	123.9	151.3	140.3	121.1	110.9	115.0	106.3	151.3	133.3	106.3	151.3	133.3
09	FINISHES	97.7	80.9	88.3	104.9	99.3	101.8	110.8	174.4	146.5	105.5	99.3	102.1	96.5	139.0	120.3	99.5	139.0	121.6
COVERS	DIVS. 10 - 14, 25, 28, 41, 43, 44, 46	100.0	93.0	98.6	100.0	107.9	101.6	100.0	131.0	106.3	100.0	107.9	101.6	100.0	116.0	103.3	100.0	116.0	103.3
21, 22, 23	FIRE SUPPRESSION, PLUMBING & HVAC	94.0	84.7	90.2	94.1	122.6	105.8	99.8	167.2	127.4	94.1	120.3	104.8	94.2	122.3	105.7	94.2	122.3	105.7
26, 27, 3370	ELECTRICAL, COMMUNICATIONS & UTIL.	95.8	92.0	93.8	97.6	113.5	106.0	99.2	178.1	140.8	97.6	113.5	106.0	94.6	154.5	126.2	94.6	154.5	126.2
MF2010	WEIGHTED AVERAGE	96.5	87.9	92.7	100.4	118.5	108.4	102.5	164.8	130.0	99.0	117.6	107.2	98.0	133.0	113.4	98.3	133.0	113.6

DIVISION		NEW YORK																	
		NEW YORK			NIAGARA FALLS			PLATTSBURGH			POUGHKEEPSIE			QUEENS			RIVERHEAD		
		100 - 102			143			129			125 - 126			110			119		
		MAT.	INST.	TOTAL	MAT.	INST.	TOTAL	MAT.	INST.	TOTAL	MAT.	INST.	TOTAL	MAT.	INST.	TOTAL	MAT.	INST.	TOTAL
015433	CONTRACTOR EQUIPMENT		113.4	113.4		94.3	94.3		99.8	99.8		115.0	115.0		115.0	115.0		115.0	115.0
0241, 31 - 34	SITE & INFRASTRUCTURE, DEMOLITION	119.4	123.9	122.5	96.2	95.5	95.7	85.9	103.2	98.0	113.0	124.6	121.1	113.9	129.1	124.6	111.7	127.9	123.0
0310	Concrete Forming & Accessories	107.5	182.4	172.3	83.0	115.6	111.2	90.0	85.4	86.0	86.9	151.3	142.6	90.9	174.9	163.6	95.0	145.0	138.2
0320	Concrete Reinforcing	99.5	184.8	142.0	100.9	101.0	100.9	101.3	93.3	97.3	97.0	133.4	115.1	99.3	185.2	142.1	99.4	174.4	136.8
0330	Cast-in-Place Concrete	100.6	182.2	134.2	103.0	121.2	110.5	90.5	94.6	92.2	96.6	129.2	110.0	94.1	176.0	127.8	92.5	167.1	123.2
03	CONCRETE	104.2	180.7	142.0	105.4	113.9	109.6	99.3	89.8	94.6	104.5	138.9	121.5	102.3	175.2	138.3	99.8	156.8	127.9
04	MASONRY	103.7	175.6	147.6	118.2	121.6	120.3	90.5	89.4	89.9	103.2	130.1	119.6	109.7	175.3	149.7	118.5	160.5	144.1
05	METALS	116.8	145.4	125.8	96.0	92.9	95.0	99.4	85.8	95.1	103.6	116.9	107.8	103.6	141.2	115.4	105.5	135.3	114.9
06	WOOD, PLASTICS & COMPOSITES	104.9	184.5	150.3	78.4	113.3	98.3	89.4	83.6	86.1	83.4	161.7	128.1	87.3	174.8	137.3	92.2	143.3	121.4
07	THERMAL & MOISTURE PROTECTION	103.5	165.7	128.5	101.2	110.4	104.9	102.1	90.9	97.6	107.9	140.7	121.1	106.1	163.4	129.2	106.7	152.4	125.1
08	OPENINGS	92.6	177.2	113.2	90.4	102.3	93.3	97.2	81.4	93.4	92.1	149.5	106.1	87.2	169.5	107.2	87.2	149.9	102.5
0920	Plaster & Gypsum Board	112.6	186.7	164.4	94.1	113.5	107.6	117.3	82.4	92.9	95.1	163.5	143.0	98.7	177.1	153.5	100.2	144.6	131.2
0950, 0980	Ceilings & Acoustic Treatment	104.7	186.7	159.4	96.1	113.5	107.7	105.8	82.4	90.2	76.9	163.5	134.6	82.4	177.1	145.5	82.3	144.6	123.8
0960	Flooring	99.6	183.9	124.7	97.5	115.8	103.0	108.1	107.0	107.8	99.8	148.7	114.4	106.5	183.9	129.5	107.4	147.8	119.5
0970, 0990	Wall Finishes & Painting/Coating	108.1	154.9	136.1	104.1	112.1	108.9	119.1	78.1	94.6	121.1	110.9	115.0	123.9	151.3	140.3	123.9	151.3	140.3
09	FINISHES	105.5	180.7	147.6	97.8	116.1	108.1	102.2	88.3	94.4	104.7	148.5	129.3	108.2	174.4	145.3	108.4	145.1	129.0
COVERS	DIVS. 10 - 14, 25, 28, 41, 43, 44, 46	100.0	136.1	107.4	100.0	103.5	100.7	100.0	49.6	89.7	100.0	112.3	102.5	100.0	131.0	106.3	100.0	122.4	104.6
21, 22, 23	FIRE SUPPRESSION, PLUMBING & HVAC	100.1	167.5	127.7	94.0	100.1	96.5	94.1	89.9	92.4	94.1	120.6	105.0	99.8	167.2	127.4	99.9	147.7	119.5
26, 27, 3370	ELECTRICAL, COMMUNICATIONS & UTIL.	104.8	178.2	143.4	94.3	98.7	96.6	93.8	88.7	91.1	97.6	113.5	106.0	99.8	178.1	141.1	101.6	138.3	120.9
MF2010	WEIGHTED AVERAGE	104.2	167.0	131.9	97.5	105.5	101.0	96.8	88.6	93.2	99.6	128.6	112.4	101.1	164.8	129.1	101.7	145.1	120.9

DIVISION		NEW YORK																	
		ROCHESTER			SCHENECTADY			STATEN ISLAND			SUFFERN			SYRACUSE			UTICA		
		144 - 146			123			103			109			130 - 132			133 - 135		
		MAT.	INST.	TOTAL	MAT.	INST.	TOTAL	MAT.	INST.	TOTAL	MAT.	INST.	TOTAL	MAT.	INST.	TOTAL	MAT.	INST.	TOTAL
015433	CONTRACTOR EQUIPMENT		116.9	116.9		114.1	114.1		113.0	113.0		113.0	113.0		114.1	114.1		114.1	114.1
0241, 31 - 34	SITE & INFRASTRUCTURE, DEMOLITION	74.7	109.2	98.9	72.1	107.1	96.6	122.7	123.3	123.2	113.9	118.2	116.9	91.6	106.7	102.2	70.5	104.9	94.6
0310	Concrete Forming & Accessories	101.5	92.0	93.3	100.5	94.3	95.1	93.2	175.2	164.2	102.0	124.5	121.5	99.6	89.2	90.6	101.0	83.5	85.9
0320	Concrete Reinforcing	101.8	90.3	96.1	95.7	108.1	101.9	94.0	177.6	135.6	92.9	133.4	113.1	97.1	94.5	95.8	97.1	91.5	94.3
0330	Cast-in-Place Concrete	96.4	100.2	97.9	89.1	100.6	93.8	99.9	177.6	131.8	96.9	127.4	109.4	90.1	104.9	96.2	82.4	92.3	86.5
03	CONCRETE	113.2	95.4	104.4	99.6	99.8	99.7	105.1	174.8	139.5	100.0	126.5	113.1	98.5	96.4	97.5	96.7	89.1	92.9
04	MASONRY	102.5	99.8	100.8	93.0	98.4	96.3	106.5	175.3	148.5	98.9	123.7	114.1	100.2	103.9	102.5	91.7	100.0	96.8
05	METALS	97.1	104.9	99.5	98.8	110.8	102.6	101.4	145.2	115.2	101.4	117.6	106.5	98.7	105.0	100.7	96.8	103.5	98.9
06	WOOD, PLASTICS & COMPOSITES	93.1	90.1	91.4	101.0	92.7	96.3	89.2	175.2	138.3	100.3	130.3	117.4	101.0	86.8	92.9	101.0	81.0	89.6
07	THERMAL & MOISTURE PROTECTION	99.9	98.0	99.1	89.3	95.9	91.9	103.8	164.4	128.2	104.3	134.9	116.6	100.2	96.9	98.9	89.0	94.6	91.2
08	OPENINGS	93.3	86.6	91.7	95.2	91.7	94.3	86.9	172.1	107.7	87.0	134.8	98.6	92.8	80.2	89.7	95.2	79.2	91.3
0920	Plaster & Gypsum Board	111.7	89.7	96.4	104.7	92.2	96.0	99.4	177.1	153.8	103.5	130.9	122.6	104.7	86.1	91.7	104.7	80.2	87.5
0950, 0980	Ceilings & Acoustic Treatment	109.8	89.7	96.5	97.8	92.2	94.1	83.8	177.1	146.0	81.9	130.9	114.5	97.8	86.1	90.0	97.8	80.2	86.0
0960	Flooring	90.8	106.8	95.6	91.0	107.0	95.8	95.0	183.9	121.5	94.0	147.8	110.0	93.6	96.6	94.5	91.0	96.7	92.7
0970, 0990	Wall Finishes & Painting/Coating	104.0	96.8	99.7	90.9	88.6	89.5	108.1	151.3	134.0	106.3	120.3	114.7	98.4	97.6	97.9	90.9	97.6	94.9
09	FINISHES	104.9	95.3	99.5	96.3	95.8	96.0	98.5	174.7	141.2	97.4	125.9	113.4	98.4	91.1	94.3	96.3	87.0	91.1
COVERS	DIVS. 10 - 14, 25, 28, 41, 43, 44, 46	100.0	95.1	99.0	100.0	92.9	98.6	100.0	131.7	106.5	100.0	114.4	102.9	100.0	96.0	99.2	100.0	90.5	98.1
21, 22, 23	FIRE SUPPRESSION, PLUMBING & HVAC	99.9	89.1	95.5	100.2	96.4	98.6	100.2	167.4	127.7	94.2	113.7	102.2	100.2	90.1	96.1	100.2	88.5	95.4
26, 27, 3370	ELECTRICAL, COMMUNICATIONS & UTIL.	101.5	94.0	97.5	100.9	97.3	99.0	96.4	178.1	139.4	102.9	121.4	112.7	100.9	97.3	99.0	99.0	97.3	98.1
MF2010	WEIGHTED AVERAGE	100.1	96.0	98.3	97.6	99.0	98.2	100.0	164.8	128.6	98.1	121.5	108.4	98.7	96.2	97.6	96.7	93.3	95.2

DIVISION		NEW YORK									NORTH CAROLINA								
		WATERTOWN			WHITE PLAINS			YONKERS			ASHEVILLE			CHARLOTTE			DURHAM		
		136			106			107			287 - 288			281 - 282			277		
		MAT.	INST.	TOTAL	MAT.	INST.	TOTAL	MAT.	INST.	TOTAL	MAT.	INST.	TOTAL	MAT.	INST.	TOTAL	MAT.	INST.	TOTAL
015433	CONTRACTOR EQUIPMENT		114.1	114.1		113.0	113.0		113.0	113.0		95.0	95.0		95.0	95.0		100.5	100.5
0241, 31 - 34	SITE & INFRASTRUCTURE, DEMOLITION	77.9	106.5	97.9	110.6	117.1	115.2	118.9	117.0	117.6	107.0	76.2	85.4	108.0	76.3	85.8	107.3	84.9	91.6
0310	Concrete Forming & Accessories	86.9	89.0	88.7	106.8	127.6	124.8	107.1	127.6	124.8	93.2	38.6	45.9	98.7	40.8	48.6	95.7	44.8	51.6
0320	Concrete Reinforcing	97.7	94.5	96.1	92.9	176.3	134.4	96.6	176.3	136.3	98.4	62.0	80.3	98.9	57.4	78.2	100.1	56.7	78.5
0330	Cast-in-Place Concrete	95.9	93.6	94.9	88.7	130.6	105.9	99.2	130.6	112.1	113.7	48.8	87.0	119.9	49.9	91.1	115.5	46.4	87.1
03	CONCRETE	109.0	92.4	100.8	93.9	136.8	115.1	103.2	136.8	119.8	105.9	48.4	77.5	108.7	48.9	79.2	106.6	49.4	78.3
04	MASONRY	92.7	102.6	98.7	98.5	128.0	116.5	103.3	128.0	118.4	83.0	44.1	59.3	88.8	53.3	67.1	88.2	36.3	56.5
05	METALS	96.8	104.8	99.4	102.9	132.6	112.3	112.6	132.6	118.9	89.0	82.3	86.9	93.1	80.8	89.2	102.5	79.8	95.4
06	WOOD, PLASTICS & COMPOSITES	83.8	87.8	86.1	105.9	130.1	119.7	105.8	130.1	119.7	95.6	36.8	62.1	101.5	38.9	65.7	98.5	46.6	68.9
07	THERMAL & MOISTURE PROTECTION	89.3	96.2	92.0	104.2	136.7	117.2	104.4	136.7	117.4	100.2	43.4	77.3	100.6	46.9	79.0	100.9	45.0	78.4
08	OPENINGS	95.2	84.4	92.6	87.0	143.4	100.7	90.3	143.4	103.2	90.5	42.1	78.7	94.2	43.0	81.7	94.2	47.3	82.8
0920	Plaster & Gypsum Board	95.6	87.2	89.7	106.8	130.6	123.5	112.1	130.6	125.1	102.5	34.7	55.1	101.6	36.8	56.3	107.8	44.8	63.7
0950, 0980	Ceilings & Acoustic Treatment	97.8	87.2	90.7	81.9	130.6	114.4	102.9	130.6	121.4	93.4	34.7	54.3	98.8	36.8	57.5	97.0	44.8	62.2
0960	Flooring	84.7	96.7	88.2	95.8	177.2	120.1	95.4	177.2	119.8	100.6	44.2	83.8	102.0	44.4	84.9	103.8	44.2	86.1
0970, 0990	Wall Finishes & Painting/Coating	90.9	91.4	91.2	106.3	151.3	133.3	106.3	151.3	133.3	109.6	42.1	69.1	109.6	42.1	69.1	109.6	35.6	65.2
09	FINISHES	94.1	90.3	92.0	97.8	139.0	120.9	103.7	139.0	123.4	99.6	39.0	65.6	100.6	40.7	67.1	102.0	43.5	69.2
COVERS	DIVS. 10 - 14, 25, 28, 41, 43, 44, 46	100.0	95.3	99.0	100.0	116.0	103.3	100.0	119.2	103.9	100.0	78.4	95.6	100.0	79.3	95.8	100.0	73.9	94.7
21, 22, 23	FIRE SUPPRESSION, PLUMBING & HVAC	100.2	83.4	93.3	100.3	122.3	109.3	100.3	122.3	109.3	100.3	53.7	81.2	99.9	54.6	81.4	100.4	52.3	80.7
26, 27, 3370	ELECTRICAL, COMMUNICATIONS & UTIL.	100.9	90.0	95.1	94.6	154.5	126.2	103.0	154.5	130.1	97.9	58.1	76.9	96.8	57.6	76.2	98.7	46.4	71.2
MF2010	WEIGHTED AVERAGE	98.2	93.1	95.9	98.3	133.0	113.6	102.9	133.1	116.2	97.1	55.1	78.6	98.7	56.5	80.1	100.3	53.9	79.9

	DIVISION	NORTH CAROLINA																	
		ELIZABETH CITY			FAYETTEVILLE			GASTONIA			GREENSBORO			HICKORY			KINSTON		
		279			283			280			270,272 - 274			286			285		
		MAT.	INST.	TOTAL	MAT.	INST.	TOTAL	MAT.	INST.	TOTAL	MAT.	INST.	TOTAL	MAT.	INST.	TOTAL	MAT.	INST.	TOTAL
015433	CONTRACTOR EQUIPMENT		104.6	104.6		100.5	100.5		95.0	95.0		100.5	100.5		100.5	100.5		100.5	100.5
0241, 31 - 34	SITE & INFRASTRUCTURE, DEMOLITION	112.3	86.9	94.5	106.0	85.0	91.3	106.7	76.3	85.4	107.2	85.0	91.7	106.0	85.1	91.3	104.9	85.0	90.9
0310	Concrete Forming & Accessories	82.3	40.3	46.0	92.8	61.4	65.6	99.4	40.2	48.2	95.4	45.8	52.5	89.5	38.6	45.5	85.9	43.0	48.8
0320	Concrete Reinforcing	97.9	48.1	73.1	102.3	56.8	79.7	98.9	57.4	78.2	98.9	56.7	77.9	98.4	57.0	77.8	97.9	48.0	73.1
0330	Cast-in-Place Concrete	115.7	47.9	87.8	119.2	48.8	90.2	111.1	54.1	87.7	114.7	48.1	87.3	113.6	49.4	87.2	109.7	45.8	83.4
03	CONCRETE	106.7	46.3	76.8	108.2	57.6	83.2	104.5	50.1	77.7	106.0	50.4	78.5	105.7	47.7	77.1	102.5	46.8	75.0
04	MASONRY	100.0	49.9	69.4	86.3	40.1	58.1	87.3	52.9	66.3	86.2	38.6	57.1	73.5	45.5	56.4	78.8	50.1	61.3
05	METALS	90.9	77.2	86.6	106.7	79.9	98.3	89.8	80.9	87.0	96.3	79.7	91.1	89.0	79.9	86.2	88.0	76.3	84.3
06	WOOD, PLASTICS & COMPOSITES	82.5	40.2	58.4	94.7	68.0	79.5	104.2	38.2	66.5	98.2	46.8	68.9	90.2	36.7	59.7	87.1	43.6	62.3
07	THERMAL & MOISTURE PROTECTION	100.2	44.4	77.8	100.0	45.7	78.2	100.4	46.9	78.9	100.6	43.1	77.5	100.4	43.3	77.5	100.3	44.4	77.8
08	OPENINGS	91.4	37.3	78.3	90.6	59.0	82.9	94.2	42.3	81.6	94.2	47.4	82.8	90.5	37.9	77.7	90.6	44.4	79.4
0920	Plaster & Gypsum Board	99.7	37.6	56.2	107.1	66.8	78.9	109.7	36.1	58.2	109.7	45.0	64.4	102.5	34.6	55.0	102.6	41.7	60.0
0950, 0980	Ceilings & Acoustic Treatment	97.0	37.6	57.4	94.3	66.8	76.0	97.0	36.1	56.4	97.0	45.0	62.4	93.4	34.6	54.2	97.0	41.7	60.2
0960	Flooring	95.5	24.0	74.2	100.7	44.2	83.9	103.8	44.2	86.1	103.8	40.9	85.1	100.5	34.0	80.7	97.9	23.4	75.7
0970, 0990	Wall Finishes & Painting/Coating	109.6	43.9	70.2	109.6	33.9	64.3	109.6	42.1	69.1	109.6	33.0	63.7	109.6	42.1	69.1	109.6	41.0	68.5
09	FINISHES	98.9	37.2	64.3	100.3	56.5	75.8	102.1	40.2	67.4	102.3	43.3	69.2	99.7	36.9	64.5	99.6	38.9	65.6
COVERS	DIVS. 10 - 14, 25, 28, 41, 43, 44, 46	100.0	72.7	94.4	100.0	77.0	95.3	100.0	79.1	95.7	100.0	78.6	95.6	100.0	78.8	95.7	100.0	74.4	94.8
21, 22, 23	FIRE SUPPRESSION, PLUMBING & HVAC	94.3	53.0	77.4	100.2	53.5	81.1	100.3	54.4	81.5	100.3	53.5	81.2	94.3	54.2	77.9	94.3	53.2	77.4
26, 27, 3370	ELECTRICAL, COMMUNICATIONS & UTIL.	98.4	36.0	65.5	96.6	52.8	73.6	97.2	57.6	76.4	97.6	58.1	76.8	95.2	57.6	75.4	95.0	48.0	70.3
MF2010	WEIGHTED AVERAGE	97.0	52.2	77.2	100.2	59.1	82.1	97.8	56.5	79.6	99.0	56.2	80.2	94.8	55.3	77.4	94.5	54.3	76.8

	DIVISION	NORTH CAROLINA															NORTH DAKOTA		
		MURPHY			RALEIGH			ROCKY MOUNT			WILMINGTON			WINSTON-SALEM			BISMARCK		
		289			275 - 276			278			284			271			585		
		MAT.	INST.	TOTAL	MAT.	INST.	TOTAL	MAT.	INST.	TOTAL	MAT.	INST.	TOTAL	MAT.	INST.	TOTAL	MAT.	INST.	TOTAL
015433	CONTRACTOR EQUIPMENT		95.0	95.0		100.5	100.5		100.5	100.5		95.0	95.0		100.5	100.5		98.9	98.9
0241, 31 - 34	SITE & INFRASTRUCTURE, DEMOLITION	108.4	76.2	85.8	107.5	85.1	91.8	110.2	85.2	92.6	108.2	76.4	85.9	107.5	85.1	91.8	99.3	97.4	98.0
0310	Concrete Forming & Accessories	100.1	41.0	48.9	97.0	48.1	54.7	88.3	46.1	51.8	94.5	50.1	56.0	97.1	49.8	56.1	102.3	42.0	50.2
0320	Concrete Reinforcing	98.0	47.6	72.9	98.9	56.8	77.9	97.9	56.7	77.4	99.2	56.7	78.0	98.9	56.7	77.9	108.1	84.6	96.4
0330	Cast-in-Place Concrete	117.5	45.5	87.9	122.9	52.1	93.8	113.2	48.9	86.7	113.2	49.6	87.1	117.5	48.8	89.3	98.2	49.0	77.9
03	CONCRETE	109.0	45.6	77.7	110.0	52.8	81.8	107.3	50.8	79.4	105.8	52.8	79.6	107.4	52.4	80.3	99.1	53.8	76.7
04	MASONRY	75.9	42.9	55.8	86.9	41.0	58.9	79.2	41.1	55.9	73.8	41.3	54.0	86.4	39.9	58.0	103.2	58.8	76.1
05	METALS	87.0	75.6	83.5	93.9	79.9	89.5	90.1	79.3	86.7	88.7	79.8	85.9	93.9	79.8	89.5	96.1	86.0	92.9
06	WOOD, PLASTICS & COMPOSITES	104.9	41.3	68.6	98.5	49.6	70.6	89.5	46.6	65.0	97.5	51.1	71.0	98.2	51.5	71.5	81.7	36.3	55.8
07	THERMAL & MOISTURE PROTECTION	100.3	42.3	77.0	100.8	45.8	78.7	100.6	42.4	77.2	100.2	47.9	79.1	100.6	44.2	77.9	110.0	51.8	86.6
08	OPENINGS	90.5	40.6	78.3	91.4	48.7	81.0	90.6	43.3	79.1	90.6	50.5	80.9	94.2	50.0	83.4	96.8	48.5	85.0
0920	Plaster & Gypsum Board	108.7	39.4	60.2	101.6	47.9	64.0	101.9	44.8	62.0	104.4	49.4	65.9	109.7	49.8	67.8	97.4	34.5	53.4
0950, 0980	Ceilings & Acoustic Treatment	93.4	39.4	57.4	98.8	47.9	64.9	94.3	44.8	61.3	94.3	49.4	64.4	97.0	49.8	65.6	157.3	34.5	75.5
0960	Flooring	104.2	25.0	80.6	102.6	44.2	85.2	99.2	22.6	76.4	101.3	46.1	84.9	103.8	44.2	86.1	115.6	74.7	103.4
0970, 0990	Wall Finishes & Painting/Coating	109.6	41.2	68.6	109.6	37.4	66.3	109.6	40.2	68.0	109.6	40.4	68.1	109.6	34.6	64.7	94.1	30.7	56.1
09	FINISHES	101.6	37.6	65.7	101.0	46.2	70.3	99.8	41.0	66.9	100.2	48.3	71.1	102.3	47.2	71.4	118.7	44.7	77.2
COVERS	DIVS. 10 - 14, 25, 28, 41, 43, 44, 46	100.0	78.7	95.6	100.0	75.3	95.0	100.0	75.5	95.0	100.0	76.1	95.1	100.0	80.4	96.0	100.0	84.2	96.8
21, 22, 23	FIRE SUPPRESSION, PLUMBING & HVAC	94.3	52.9	77.3	100.0	54.0	81.2	94.3	54.7	78.1	100.3	54.9	81.7	100.3	54.2	81.4	100.2	61.2	84.3
26, 27, 3370	ELECTRICAL, COMMUNICATIONS & UTIL.	98.9	30.6	62.9	101.1	41.9	70.0	100.6	41.4	69.4	98.1	52.8	74.2	97.6	58.1	76.8	102.5	72.4	86.7
MF2010	WEIGHTED AVERAGE	95.7	49.9	75.5	99.0	55.1	79.7	96.1	53.9	77.5	96.7	56.5	79.0	98.8	57.5	80.6	101.0	64.4	84.9

	DIVISION	NORTH DAKOTA																	
		DEVILS LAKE			DICKINSON			FARGO			GRAND FORKS			JAMESTOWN			MINOT		
		583			586			580 - 581			582			584			587		
		MAT.	INST.	TOTAL	MAT.	INST.	TOTAL	MAT.	INST.	TOTAL	MAT.	INST.	TOTAL	MAT.	INST.	TOTAL	MAT.	INST.	TOTAL
015433	CONTRACTOR EQUIPMENT		98.9	98.9		98.9	98.9		98.9	98.9		98.9	98.9		98.9	98.9		98.9	98.9
0241, 31 - 34	SITE & INFRASTRUCTURE, DEMOLITION	107.9	95.2	99.0	117.4	93.9	100.9	99.9	97.4	98.2	112.5	93.9	99.4	106.9	93.9	97.7	109.6	97.4	101.1
0310	Concrete Forming & Accessories	104.0	36.5	45.6	93.3	35.6	43.4	103.0	42.8	50.9	97.0	35.5	43.8	94.7	35.3	43.3	93.1	66.9	70.4
0320	Concrete Reinforcing	108.1	84.9	96.6	109.1	41.6	75.5	99.7	84.6	92.2	106.4	84.7	95.6	108.7	49.2	79.1	110.1	85.0	97.6
0330	Cast-in-Place Concrete	119.2	47.5	89.7	107.9	45.9	82.4	97.2	50.8	78.2	107.8	45.9	82.3	117.7	45.8	88.1	107.8	48.1	83.2
03	CONCRETE	112.1	50.8	81.8	111.7	41.7	77.2	107.1	54.8	81.2	108.7	49.6	79.5	110.6	42.9	77.2	107.3	64.6	86.2
04	MASONRY	111.4	65.9	83.6	113.5	60.0	80.8	105.0	58.8	76.8	106.1	65.4	81.2	123.6	34.1	69.0	105.4	65.1	80.8
05	METALS	96.1	85.0	92.6	96.1	64.6	86.2	98.4	86.3	94.6	96.1	80.1	91.1	96.1	65.6	86.5	96.4	86.8	93.4
06	WOOD, PLASTICS & COMPOSITES	85.2	33.4	55.6	74.4	33.4	51.0	81.6	36.8	56.0	78.2	33.4	52.6	76.1	33.4	51.7	74.1	69.5	71.5
07	THERMAL & MOISTURE PROTECTION	106.0	50.4	83.6	106.5	48.1	83.0	106.9	52.9	85.2	106.2	50.4	83.7	105.8	42.3	80.3	106.0	56.9	86.2
08	OPENINGS	96.8	41.7	83.4	96.7	31.5	80.9	96.7	48.8	85.0	96.8	41.7	83.3	96.8	33.6	81.4	96.9	66.6	89.5
0920	Plaster & Gypsum Board	118.6	31.6	57.8	109.1	31.6	54.9	96.9	35.1	53.7	110.7	31.6	55.4	110.3	31.6	55.3	109.1	68.8	80.9
0950, 0980	Ceilings & Acoustic Treatment	153.6	31.6	72.3	153.6	31.6	72.3	155.5	35.1	75.3	153.6	31.6	72.3	153.6	31.6	72.3	153.6	68.8	97.1
0960	Flooring	119.9	37.2	95.3	112.5	37.2	90.1	115.6	74.7	103.4	114.7	37.2	91.6	113.4	37.2	90.7	112.3	86.9	104.7
0970, 0990	Wall Finishes & Painting/Coating	94.1	23.5	51.8	94.1	23.5	51.8	94.1	73.8	81.9	94.1	23.5	51.8	94.1	23.5	51.8	94.1	25.7	53.1
09	FINISHES	122.6	33.6	72.7	120.3	33.6	71.7	118.3	49.8	79.9	120.5	33.6	71.8	119.5	33.6	71.4	119.3	66.2	89.6
COVERS	DIVS. 10 - 14, 25, 28, 41, 43, 44, 46	100.0	33.7	86.5	100.0	33.9	86.5	100.0	84.3	96.8	100.0	33.8	86.5	100.0	82.2	96.4	100.0	87.9	97.5
21, 22, 23	FIRE SUPPRESSION, PLUMBING & HVAC	94.4	63.8	81.9	94.4	69.2	84.1	100.3	66.9	86.6	100.5	35.0	73.7	94.4	36.9	70.9	100.5	61.3	84.4
26, 27, 3370	ELECTRICAL, COMMUNICATIONS & UTIL.	96.7	37.3	65.4	106.2	73.4	88.9	99.7	67.8	82.9	100.5	55.8	77.0	96.7	37.2	65.4	104.0	78.9	90.8
MF2010	WEIGHTED AVERAGE	101.3	56.9	81.7	102.3	58.7	83.1	101.9	65.8	86.0	102.4	52.6	80.5	101.4	46.0	77.0	102.4	71.6	88.9

DIVISION		NORTH DAKOTA			OHIO														
		WILLISTON			AKRON			ATHENS			CANTON			CHILLICOTHE			CINCINNATI		
		588			442 - 443			457			446 - 447			456			451 - 452		
		MAT.	INST.	TOTAL	MAT.	INST.	TOTAL	MAT.	INST.	TOTAL	MAT.	INST.	TOTAL	MAT.	INST.	TOTAL	MAT.	INST.	TOTAL
015433	CONTRACTOR EQUIPMENT		98.9	98.9		96.3	96.3		90.5	90.5		96.3	96.3		100.3	100.3		100.1	100.1
0241, 31 - 34	SITE & INFRASTRUCTURE, DEMOLITION	110.4	93.9	98.8	94.4	104.7	101.6	111.7	92.6	98.3	94.5	104.2	101.3	96.5	104.1	101.8	93.5	103.6	100.6
0310	Concrete Forming & Accessories	98.8	35.7	44.2	99.7	91.1	92.2	92.7	87.6	88.3	99.7	81.2	83.7	94.6	93.8	93.9	96.4	81.8	83.8
0320	Concrete Reinforcing	111.2	41.6	76.6	96.2	93.1	94.6	89.8	88.0	88.9	96.2	76.3	86.3	86.9	80.0	83.5	92.1	81.9	87.0
0330	Cast-in-Place Concrete	107.8	45.9	82.4	91.2	96.7	93.5	111.7	94.0	104.4	92.1	94.3	93.0	101.4	94.6	98.6	93.4	85.7	90.2
03	CONCRETE	108.8	41.7	75.7	91.1	92.7	91.9	104.6	89.3	97.0	91.5	84.4	88.0	98.9	91.3	95.2	93.6	83.3	88.5
04	MASONRY	100.0	60.0	75.6	89.7	94.2	92.4	79.9	91.7	87.1	90.3	85.2	87.2	87.2	99.0	94.4	86.9	86.3	86.5
05	METALS	96.2	64.7	86.3	95.7	81.7	91.3	101.7	79.2	94.6	95.7	74.6	89.1	93.7	84.4	90.8	96.0	84.3	92.3
06	WOOD, PLASTICS & COMPOSITES	79.4	33.4	53.2	97.2	89.7	92.9	87.3	88.7	88.1	97.6	79.4	87.2	100.6	92.5	96.0	103.0	79.7	89.7
07	THERMAL & MOISTURE PROTECTION	106.1	48.1	82.8	107.9	95.9	103.1	100.6	92.7	97.4	109.0	91.8	102.1	102.1	95.1	99.3	100.4	90.5	96.4
08	OPENINGS	96.8	31.5	80.9	112.3	91.6	107.2	100.0	85.3	96.4	105.9	76.8	98.8	92.0	84.8	90.2	100.4	79.3	95.2
0920	Plaster & Gypsum Board	110.7	31.6	55.4	99.3	89.1	92.2	89.5	88.0	88.5	100.5	78.5	85.1	91.2	92.5	92.1	93.1	79.3	83.5
0950, 0980	Ceilings & Acoustic Treatment	153.6	31.6	72.3	96.4	89.1	91.6	103.6	88.0	93.2	96.4	78.5	84.5	97.7	92.5	94.2	98.6	79.3	85.7
0960	Flooring	115.6	37.2	92.2	98.4	91.1	96.3	128.8	102.5	121.0	98.6	81.3	93.5	104.7	102.5	104.1	105.7	91.6	101.5
0970, 0990	Wall Finishes & Painting/Coating	94.1	32.9	57.4	99.3	101.3	100.5	102.7	95.2	98.2	99.3	82.6	89.3	99.9	91.7	95.0	99.9	86.7	92.0
09	FINISHES	120.6	34.7	72.5	100.1	91.9	95.5	105.4	92.0	97.9	100.4	80.8	89.4	102.7	95.7	98.7	103.1	83.7	92.2
COVERS	DIVS. 10 - 14, 25, 28, 41, 43, 44, 46	100.0	33.9	86.5	100.0	90.3	98.0	100.0	54.1	90.6	100.0	87.9	97.5	100.0	88.7	97.7	100.0	92.3	98.4
21, 22, 23	FIRE SUPPRESSION, PLUMBING & HVAC	94.4	69.2	84.1	100.0	96.9	98.7	94.1	54.0	77.7	100.0	82.4	92.8	94.5	91.7	93.4	100.0	82.7	92.9
26, 27, 3370	ELECTRICAL, COMMUNICATIONS & UTIL.	101.0	73.4	86.4	98.9	92.7	95.6	99.1	100.6	99.9	98.1	89.1	93.4	98.6	83.6	90.7	97.4	79.7	88.1
MF2010	WEIGHTED AVERAGE	100.7	58.8	82.3	99.1	93.6	96.6	98.9	82.2	91.6	98.5	85.0	92.5	96.2	91.9	94.3	97.9	85.1	92.3

DIVISION		OHIO																	
		CLEVELAND			COLUMBUS			DAYTON			HAMILTON			LIMA			LORAIN		
		441			430 - 432			453 - 454			450			458			440		
		MAT.	INST.	TOTAL	MAT.	INST.	TOTAL	MAT.	INST.	TOTAL	MAT.	INST.	TOTAL	MAT.	INST.	TOTAL	MAT.	INST.	TOTAL
015433	CONTRACTOR EQUIPMENT		96.6	96.6		95.0	95.0		95.4	95.4		100.3	100.3		93.0	93.0		96.3	96.3
0241, 31 - 34	SITE & INFRASTRUCTURE, DEMOLITION	94.4	105.0	101.9	94.1	99.2	97.7	92.3	103.4	100.1	92.1	103.6	100.2	105.3	92.3	96.2	93.9	103.6	100.7
0310	Concrete Forming & Accessories	99.8	99.4	99.5	98.1	87.6	89.0	96.4	80.3	82.4	96.4	81.2	83.2	92.6	85.9	86.8	99.8	86.8	88.5
0320	Concrete Reinforcing	96.7	93.5	95.1	105.7	82.0	93.9	92.1	80.2	86.2	92.1	81.9	87.0	89.8	80.4	85.1	96.2	93.4	94.8
0330	Cast-in-Place Concrete	89.5	107.0	96.7	91.5	94.4	92.7	86.9	86.4	86.7	93.1	84.9	89.7	102.6	96.1	100.0	86.9	104.1	93.9
03	CONCRETE	90.3	100.1	95.2	95.0	88.6	91.8	90.4	82.2	86.3	93.4	82.8	88.2	97.7	88.2	93.0	89.0	93.4	91.2
04	MASONRY	94.1	105.3	100.9	96.5	95.4	95.8	86.4	82.5	84.0	86.7	84.8	85.5	109.5	83.6	93.7	86.5	99.4	94.4
05	METALS	97.3	84.6	93.3	97.9	80.7	92.5	95.2	77.1	89.5	95.3	84.3	91.9	101.7	80.3	95.0	96.3	82.6	92.0
06	WOOD, PLASTICS & COMPOSITES	96.3	96.7	96.5	95.8	85.5	89.9	104.4	78.9	89.9	103.0	79.7	89.7	87.2	85.8	86.4	97.2	83.7	89.5
07	THERMAL & MOISTURE PROTECTION	106.7	109.1	107.6	100.9	96.2	99.0	105.7	87.6	98.4	102.3	89.8	97.2	100.1	94.5	97.8	109.0	101.8	106.1
08	OPENINGS	101.8	95.5	100.3	104.2	81.5	98.7	100.2	78.7	95.0	97.7	79.3	93.2	100.0	79.9	95.1	105.9	88.4	101.6
0920	Plaster & Gypsum Board	98.4	96.4	97.0	97.0	85.0	88.6	93.1	78.5	82.9	93.1	79.3	83.5	89.5	85.0	86.4	99.3	83.0	87.9
0950, 0980	Ceilings & Acoustic Treatment	94.6	96.4	95.8	96.2	85.0	88.8	99.6	78.5	85.6	98.6	79.3	85.7	102.5	85.0	90.9	96.4	83.0	87.5
0960	Flooring	98.2	105.3	100.3	95.0	89.5	93.3	108.5	82.2	100.7	105.7	91.6	101.5	128.0	89.9	116.6	98.6	105.3	100.6
0970, 0990	Wall Finishes & Painting/Coating	99.3	107.6	104.3	96.1	95.4	95.7	99.9	85.2	91.1	99.9	85.3	91.2	102.7	84.2	91.6	99.3	107.6	104.3
09	FINISHES	99.7	101.2	100.5	96.4	88.5	92.0	104.1	80.8	91.0	103.0	83.2	91.9	104.4	86.4	94.3	100.2	92.0	95.6
COVERS	DIVS. 10 - 14, 25, 28, 41, 43, 44, 46	100.0	95.2	99.0	100.0	94.5	98.9	100.0	91.6	98.3	100.0	91.8	98.3	100.0	89.8	97.9	100.0	91.7	98.3
21, 22, 23	FIRE SUPPRESSION, PLUMBING & HVAC	100.0	100.2	100.1	100.0	92.4	96.9	100.9	83.3	93.7	100.6	81.9	92.9	94.1	84.7	90.3	100.0	83.7	93.3
26, 27, 3370	ELECTRICAL, COMMUNICATIONS & UTIL.	98.4	108.1	103.5	96.6	86.4	91.2	95.9	82.2	88.7	96.3	80.3	87.9	99.5	79.2	88.8	98.3	88.2	93.0
MF2010	WEIGHTED AVERAGE	98.2	100.7	99.3	98.6	90.0	94.8	97.7	83.9	91.6	97.5	84.7	91.9	99.4	85.0	93.1	98.1	90.8	94.9

DIVISION		OHIO																	
		MANSFIELD			MARION			SPRINGFIELD			STEUBENVILLE			TOLEDO			YOUNGSTOWN		
		448 - 449			433			455			439			434 - 436			444 - 445		
		MAT.	INST.	TOTAL	MAT.	INST.	TOTAL	MAT.	INST.	TOTAL	MAT.	INST.	TOTAL	MAT.	INST.	TOTAL	MAT.	INST.	TOTAL
015433	CONTRACTOR EQUIPMENT		96.3	96.3		94.6	94.6		95.4	95.4		99.4	99.4		97.4	97.4		96.3	96.3
0241, 31 - 34	SITE & INFRASTRUCTURE, DEMOLITION	90.4	104.5	100.3	90.3	98.0	95.7	92.6	102.1	99.3	133.8	107.5	115.4	93.4	99.9	97.9	94.3	105.2	101.9
0310	Concrete Forming & Accessories	90.1	85.1	85.8	94.8	83.0	84.6	96.4	84.9	86.4	96.7	85.4	86.9	98.0	94.4	94.9	99.7	84.8	86.8
0320	Concrete Reinforcing	87.5	76.9	82.2	97.4	82.0	89.7	92.1	80.2	86.2	94.7	84.3	89.5	105.7	88.6	97.2	96.2	87.5	91.9
0330	Cast-in-Place Concrete	84.5	92.9	87.9	83.5	91.5	86.8	89.2	86.5	88.1	90.7	91.9	91.2	91.5	102.2	95.9	90.4	98.5	93.7
03	CONCRETE	83.9	85.7	84.8	87.2	85.4	86.3	91.5	84.2	87.9	90.9	86.9	88.9	95.0	95.7	95.3	90.7	89.5	90.1
04	MASONRY	88.8	95.9	93.2	98.1	94.6	96.0	86.6	82.8	84.3	84.0	89.8	87.5	104.9	98.6	101.1	90.0	92.0	91.2
05	METALS	96.5	75.5	89.9	96.9	78.4	91.1	95.2	76.9	89.5	93.3	78.8	88.8	97.7	86.9	94.3	95.8	79.4	90.6
06	WOOD, PLASTICS & COMPOSITES	85.6	83.7	84.5	91.9	83.5	87.1	106.0	85.2	94.1	88.0	83.6	85.5	95.8	93.4	94.4	97.2	82.3	88.7
07	THERMAL & MOISTURE PROTECTION	108.4	95.1	103.0	100.5	83.7	93.8	105.5	88.4	98.6	112.5	92.8	104.6	102.5	102.0	102.3	109.2	93.7	102.9
08	OPENINGS	105.6	78.3	99.0	98.1	79.0	93.5	98.1	80.2	93.7	98.9	81.6	94.7	101.7	90.0	98.9	105.9	83.7	100.5
0920	Plaster & Gypsum Board	92.9	83.0	85.9	94.5	82.9	86.4	93.1	85.0	87.5	95.0	82.6	86.3	97.0	93.2	94.3	99.3	81.5	86.8
0950, 0980	Ceilings & Acoustic Treatment	97.3	83.0	87.8	96.2	82.9	87.4	99.6	85.0	89.9	93.0	82.6	86.1	96.2	93.2	94.2	96.4	81.5	86.5
0960	Flooring	94.2	103.5	96.9	93.7	72.3	87.3	108.5	82.2	100.7	122.8	94.2	114.3	94.2	103.3	96.9	98.6	92.7	96.9
0970, 0990	Wall Finishes & Painting/Coating	99.3	90.3	93.9	96.1	49.6	68.3	99.9	85.2	91.1	108.6	94.6	100.2	96.1	101.9	99.6	99.3	94.7	96.6
09	FINISHES	98.0	89.0	92.9	95.3	77.7	85.4	104.1	84.6	93.2	113.3	87.4	98.8	96.1	96.6	96.4	100.3	86.6	92.6
COVERS	DIVS. 10 - 14, 25, 28, 41, 43, 44, 46	100.0	88.8	97.7	100.0	55.8	91.0	100.0	92.4	98.4	100.0	92.8	98.5	100.0	97.9	99.6	100.0	88.9	97.7
21, 22, 23	FIRE SUPPRESSION, PLUMBING & HVAC	94.0	81.9	89.1	94.0	87.4	91.3	100.9	83.5	93.8	94.3	89.7	92.4	100.1	100.3	100.2	100.0	90.8	96.3
26, 27, 3370	ELECTRICAL, COMMUNICATIONS & UTIL.	95.7	79.1	87.0	90.4	79.1	84.4	95.9	82.4	88.8	85.5	109.4	98.1	96.7	102.1	99.5	98.3	84.4	91.0
MF2010	WEIGHTED AVERAGE	95.6	86.1	91.4	94.8	84.1	90.1	97.6	84.8	91.9	96.6	92.0	94.6	98.8	97.5	98.2	98.4	89.1	94.3

	DIVISION	OHIO			OKLAHOMA															
		ZANESVILLE			ARDMORE			CLINTON			DURANT			ENID			GUYMON		
		437 - 438			734			736			747			737			739		
		MAT.	INST.	TOTAL	MAT.	INST.	TOTAL	MAT.	INST.	TOTAL	MAT.	INST.	TOTAL	MAT.	INST.	TOTAL	MAT.	INST.	TOTAL
015433	CONTRACTOR EQUIPMENT		94.6	94.6		82.3	82.3		81.4	81.4		81.4	81.4		81.4	81.4		81.4	81.4
0241, 31 - 34	SITE & INFRASTRUCTURE, DEMOLITION	93.0	99.3	97.4	97.4	91.1	93.0	99.0	89.6	92.4	92.7	89.4	90.4	100.5	89.6	92.9	103.2	88.3	92.7
0310	Concrete Forming & Accessories	92.0	83.2	84.3	92.3	39.0	46.1	91.1	33.7	41.4	84.2	38.9	45.0	94.5	33.1	41.4	98.0	21.6	31.9
0320	Concrete Reinforcing	96.8	83.8	90.4	92.8	76.7	84.8	93.3	76.7	85.0	93.6	60.8	77.3	92.7	76.7	84.7	93.3	26.9	60.2
0330	Cast-in-Place Concrete	88.0	90.5	89.0	98.9	42.6	75.7	95.6	42.6	73.8	92.5	42.5	71.9	95.6	44.9	74.7	95.6	30.8	69.0
03	CONCRETE	90.6	85.5	88.1	98.6	47.8	73.5	98.2	45.5	72.1	93.2	44.8	69.3	98.7	46.0	72.7	101.7	26.5	64.6
04	MASONRY	95.4	81.0	86.6	97.5	55.0	71.5	125.3	55.0	82.4	96.1	62.7	75.7	105.9	55.0	74.8	100.5	21.9	52.5
05	METALS	98.4	79.6	92.5	91.4	68.6	84.2	91.5	68.6	84.3	91.4	62.2	82.2	92.9	68.6	85.3	91.9	40.0	75.6
06	WOOD, PLASTICS & COMPOSITES	88.3	83.5	85.5	99.2	37.6	64.0	98.2	30.5	59.5	89.7	37.6	60.0	101.9	29.7	60.7	105.7	19.7	56.6
07	THERMAL & MOISTURE PROTECTION	100.7	87.9	95.5	98.5	59.3	82.7	98.6	58.5	82.5	98.2	56.6	81.5	98.6	58.4	82.5	98.9	27.7	70.3
08	OPENINGS	98.1	80.9	93.9	94.3	47.9	83.0	94.4	44.1	82.1	94.4	44.0	82.1	94.3	43.5	82.0	94.5	21.1	76.6
0920	Plaster & Gypsum Board	92.0	82.9	85.7	80.1	36.3	49.4	79.7	29.0	44.2	75.1	36.3	47.9	80.9	28.1	44.0	81.7	17.9	37.0
0950, 0980	Ceilings & Acoustic Treatment	96.2	82.9	87.4	83.7	36.3	52.1	83.7	29.0	47.2	83.7	36.3	52.1	83.7	28.1	46.7	86.4	17.9	40.7
0960	Flooring	92.2	86.5	90.5	106.3	44.7	87.9	105.0	42.6	86.4	101.5	64.2	90.4	106.9	42.6	87.8	108.7	25.2	83.9
0970, 0990	Wall Finishes & Painting/Coating	96.1	77.7	85.1	100.7	53.6	72.5	100.7	53.6	72.5	100.7	53.6	72.5	100.7	53.6	72.5	100.7	19.1	51.8
09	FINISHES	94.8	83.0	88.2	92.9	39.8	63.1	92.7	35.2	60.5	90.6	43.8	64.3	93.5	34.7	60.5	95.2	21.6	53.9
COVERS	DIVS. 10 - 14, 25, 28, 41, 43, 44, 46	100.0	84.8	96.9	100.0	70.0	93.9	100.0	69.2	93.7	100.0	70.1	93.9	100.0	69.1	93.7	100.0	67.2	93.3
21, 22, 23	FIRE SUPPRESSION, PLUMBING & HVAC	94.0	87.4	91.3	94.3	60.2	80.3	94.3	60.2	80.3	94.3	60.3	80.3	100.3	60.2	83.9	94.3	22.7	65.0
26, 27, 3370	ELECTRICAL, COMMUNICATIONS & UTIL.	90.8	79.6	84.9	93.5	67.2	79.7	94.6	67.2	80.2	96.3	50.7	72.3	94.6	67.2	80.2	96.3	16.0	54.1
MF2010	WEIGHTED AVERAGE	95.3	84.8	90.7	95.1	59.3	79.3	96.6	58.0	79.6	94.3	56.9	77.8	97.5	58.0	80.1	96.4	30.7	67.4

	DIVISION	OKLAHOMA																	
		LAWTON			MCALESTER			MIAMI			MUSKOGEE			OKLAHOMA CITY			PONCA CITY		
		735			745			743			744			730 - 731			746		
		MAT.	INST.	TOTAL	MAT.	INST.	TOTAL	MAT.	INST.	TOTAL	MAT.	INST.	TOTAL	MAT.	INST.	TOTAL	MAT.	INST.	TOTAL
015433	CONTRACTOR EQUIPMENT		82.3	82.3		81.4	81.4		89.4	89.4		89.4	89.4		82.5	82.5		81.4	81.4
0241, 31 - 34	SITE & INFRASTRUCTURE, DEMOLITION	96.6	91.1	92.7	86.3	89.7	88.7	87.6	86.9	87.1	87.8	85.6	86.3	95.2	91.6	92.7	93.1	89.6	90.7
0310	Concrete Forming & Accessories	97.9	44.5	51.7	82.3	40.2	45.9	94.7	68.1	71.6	99.0	29.6	39.0	98.1	38.7	46.7	90.4	39.3	46.2
0320	Concrete Reinforcing	92.9	76.7	84.8	93.3	40.4	66.9	91.8	76.8	84.3	92.7	30.7	61.8	92.9	76.7	84.8	92.7	76.7	84.7
0330	Cast-in-Place Concrete	92.5	44.9	72.9	81.1	44.9	66.2	85.0	44.3	68.3	86.0	35.5	65.2	93.6	47.5	74.6	94.9	37.9	71.5
03	CONCRETE	94.8	51.1	73.2	83.7	42.4	63.3	88.4	62.1	75.4	90.2	33.2	62.0	95.3	49.3	72.6	95.2	46.3	71.1
04	MASONRY	100.5	55.0	72.7	114.9	55.6	78.7	98.8	56.1	72.8	117.4	44.6	73.0	100.1	56.7	73.6	91.2	55.5	69.4
05	METALS	96.5	68.7	87.7	91.4	53.0	79.3	91.3	80.8	88.0	92.8	55.7	81.2	98.3	68.6	89.0	91.3	68.3	84.1
06	WOOD, PLASTICS & COMPOSITES	104.7	45.0	70.7	87.3	40.7	60.7	101.9	76.2	87.2	106.1	29.4	62.3	104.0	36.3	65.3	97.5	38.1	63.6
07	THERMAL & MOISTURE PROTECTION	98.4	60.0	83.0	97.9	55.5	80.8	98.2	60.0	82.9	98.3	40.8	75.2	93.5	59.9	80.0	98.4	58.5	82.3
08	OPENINGS	96.0	51.8	85.3	94.3	39.6	81.0	94.3	69.4	88.3	94.3	28.5	78.3	96.0	47.1	84.1	94.3	48.2	83.1
0920	Plaster & Gypsum Board	82.6	44.0	55.6	73.9	39.5	49.8	80.9	75.9	77.4	82.6	27.7	44.2	88.7	34.9	51.1	79.7	36.8	49.7
0950, 0980	Ceilings & Acoustic Treatment	90.0	44.0	59.3	83.7	39.5	54.2	83.7	75.9	78.5	90.0	27.7	48.5	95.5	34.9	55.2	83.7	36.8	52.5
0960	Flooring	109.0	42.6	89.2	100.4	42.6	83.2	107.6	64.2	94.7	110.1	41.2	89.6	108.6	42.6	88.9	104.6	42.6	86.2
0970, 0990	Wall Finishes & Painting/Coating	100.7	53.6	72.5	100.7	39.5	64.0	100.7	80.9	88.8	100.7	28.7	57.5	100.7	53.6	72.5	100.7	53.6	72.5
09	FINISHES	95.2	43.8	66.4	89.5	39.8	61.7	92.4	69.8	79.7	94.8	31.6	59.4	96.8	39.0	64.4	92.2	39.9	62.9
COVERS	DIVS. 10 - 14, 25, 28, 41, 43, 44, 46	100.0	70.8	94.0	100.0	70.5	94.0	100.0	75.0	94.9	100.0	68.8	93.6	100.0	70.4	94.0	100.0	70.2	93.9
21, 22, 23	FIRE SUPPRESSION, PLUMBING & HVAC	100.3	60.2	83.9	94.3	29.6	67.8	94.3	59.2	79.9	100.3	23.0	68.7	100.0	61.0	84.1	94.3	58.8	79.8
26, 27, 3370	ELECTRICAL, COMMUNICATIONS & UTIL.	96.4	67.2	81.0	94.7	65.1	79.1	96.2	65.2	79.9	94.2	27.5	59.0	102.1	67.2	83.8	94.2	61.4	76.9
MF2010	WEIGHTED AVERAGE	97.7	60.6	81.3	93.7	49.9	74.4	94.0	66.9	82.1	96.8	38.8	71.3	98.5	59.8	81.4	94.3	57.9	78.3

	DIVISION	OKLAHOMA												OREGON					
		POTEAU			SHAWNEE			TULSA			WOODWARD			BEND			EUGENE		
		749			748			740 - 741			738			977			974		
		MAT.	INST.	TOTAL	MAT.	INST.	TOTAL	MAT.	INST.	TOTAL	MAT.	INST.	TOTAL	MAT.	INST.	TOTAL	MAT.	INST.	TOTAL
015433	CONTRACTOR EQUIPMENT		88.7	88.7		81.4	81.4		89.4	89.4		81.4	81.4		98.4	98.4		98.4	98.4
0241, 31 - 34	SITE & INFRASTRUCTURE, DEMOLITION	73.9	85.5	82.1	96.2	89.6	91.5	93.9	86.9	89.0	99.3	89.6	92.5	110.4	103.0	105.2	100.0	103.0	102.1
0310	Concrete Forming & Accessories	88.4	41.5	47.8	84.1	38.0	44.2	99.1	38.5	46.6	91.2	33.7	41.4	110.5	101.2	102.5	107.0	101.1	101.9
0320	Concrete Reinforcing	93.7	76.7	85.3	92.7	44.4	68.6	92.9	76.6	84.8	92.7	76.7	84.7	100.2	103.2	101.7	104.6	103.2	103.9
0330	Cast-in-Place Concrete	85.0	43.9	68.1	98.0	42.0	75.0	93.7	43.9	73.2	95.6	42.6	73.8	98.4	104.9	101.1	95.3	104.8	99.2
03	CONCRETE	90.7	50.1	70.7	96.9	41.2	69.4	95.4	48.8	72.4	98.5	45.5	72.3	110.3	102.5	106.4	101.3	102.4	101.9
04	MASONRY	99.8	55.6	72.8	116.1	54.4	78.4	99.7	56.7	73.5	93.9	55.0	70.2	104.9	107.0	106.2	102.0	107.0	105.1
05	METALS	91.4	80.4	87.9	91.3	54.8	79.8	96.0	79.6	90.8	91.5	68.6	84.3	87.3	97.5	90.5	88.0	97.4	91.0
06	WOOD, PLASTICS & COMPOSITES	94.3	40.8	63.8	89.6	37.3	59.8	105.4	37.4	66.5	98.3	30.4	59.5	102.0	100.7	101.3	97.9	100.7	99.5
07	THERMAL & MOISTURE PROTECTION	98.3	55.8	81.2	98.4	58.5	82.3	98.3	54.3	80.6	98.6	58.5	82.5	103.7	95.5	100.4	102.9	92.6	98.8
08	OPENINGS	94.3	50.2	83.6	94.3	36.8	80.3	96.0	47.3	84.1	94.4	44.0	82.1	96.0	104.6	98.1	96.4	104.6	98.4
0920	Plaster & Gypsum Board	78.0	39.5	51.1	75.1	36.0	47.8	82.6	35.9	50.0	80.4	28.9	44.4	99.8	100.6	100.3	97.3	100.6	99.6
0950, 0980	Ceilings & Acoustic Treatment	83.7	39.5	54.2	83.7	36.0	51.9	90.0	35.9	54.0	86.4	28.9	48.1	96.6	100.6	99.3	97.7	100.6	99.6
0960	Flooring	104.0	64.2	92.1	101.5	33.9	81.3	108.8	44.4	89.6	105.0	44.7	87.0	109.2	107.1	108.6	107.5	107.1	107.4
0970, 0990	Wall Finishes & Painting/Coating	100.7	53.6	72.5	100.7	35.6	61.7	100.7	42.3	65.7	100.7	53.6	72.5	110.6	72.8	88.0	110.6	72.8	88.0
09	FINISHES	90.1	45.7	65.2	90.8	35.6	59.9	94.6	38.8	63.3	93.5	35.5	61.0	104.3	99.1	101.4	102.5	99.1	100.6
COVERS	DIVS. 10 - 14, 25, 28, 41, 43, 44, 46	100.0	70.8	94.0	100.0	69.7	93.8	100.0	70.8	94.0	100.0	69.2	93.7	100.0	103.4	100.7	100.0	103.4	100.7
21, 22, 23	FIRE SUPPRESSION, PLUMBING & HVAC	94.3	58.9	79.8	94.3	59.5	80.0	100.3	46.5	78.3	94.3	60.2	80.3	94.0	99.2	96.1	100.0	99.2	99.7
26, 27, 3370	ELECTRICAL, COMMUNICATIONS & UTIL.	94.3	65.2	79.0	96.4	67.2	81.1	96.4	35.8	64.5	96.2	67.2	80.9	100.9	95.9	98.3	99.5	95.9	97.6
MF2010	WEIGHTED AVERAGE	93.5	60.6	79.0	95.8	55.8	78.2	97.5	52.8	77.8	95.3	58.0	78.9	98.3	100.4	99.3	98.2	100.3	99.1

	DIVISION	OREGON																	
		KLAMATH FALLS			MEDFORD			PENDLETON			PORTLAND			SALEM			VALE		
		976			975			978			970 - 972			973			979		
		MAT.	INST.	TOTAL	MAT.	INST.	TOTAL	MAT.	INST.	TOTAL	MAT.	INST.	TOTAL	MAT.	INST.	TOTAL	MAT.	INST.	TOTAL
015433	CONTRACTOR EQUIPMENT		98.4	98.4		98.4	98.4		95.8	95.8		98.4	98.4		98.4	98.4		95.8	95.8
0241, 31 - 34	SITE & INFRASTRUCTURE, DEMOLITION	114.6	103.0	106.5	107.9	103.0	104.5	106.8	96.6	99.6	102.6	103.0	102.9	95.5	103.0	100.8	94.4	96.5	95.9
0310	Concrete Forming & Accessories	103.8	101.0	101.4	103.0	101.0	101.3	103.6	101.5	101.7	108.1	101.3	102.2	107.8	101.2	102.1	109.7	100.4	101.6
0320	Concrete Reinforcing	100.2	103.1	101.7	101.9	103.1	102.5	99.3	103.2	101.3	105.4	103.2	104.3	105.6	103.2	104.4	96.9	103.0	99.9
0330	Cast-in-Place Concrete	98.4	104.8	101.0	98.4	104.8	101.0	99.2	106.2	102.1	97.9	104.9	100.8	87.4	104.9	94.6	78.2	105.8	89.6
03	CONCRETE	113.2	102.4	107.9	107.6	102.4	105.0	93.0	103.1	98.0	102.8	102.5	102.7	97.8	102.5	100.1	78.4	102.4	90.2
04	MASONRY	118.0	107.0	111.3	99.0	107.0	103.9	108.2	107.1	107.5	102.6	107.0	105.3	106.9	107.0	107.0	106.6	107.1	106.9
05	METALS	87.3	97.3	90.5	87.6	97.3	90.7	93.4	98.1	94.9	88.9	97.6	91.6	88.4	97.5	91.2	93.3	97.0	94.4
06	WOOD, PLASTICS & COMPOSITES	93.4	100.7	97.6	92.3	100.7	97.1	94.9	100.9	98.3	99.0	100.7	100.0	95.2	100.7	98.4	103.0	100.9	101.8
07	THERMAL & MOISTURE PROTECTION	103.9	94.0	99.9	103.5	94.0	99.7	96.6	95.9	96.3	102.9	98.1	101.0	102.1	95.5	99.4	96.0	91.6	94.2
08	OPENINGS	96.1	104.6	98.1	98.7	104.6	100.2	91.2	104.7	94.5	94.2	104.6	96.8	96.2	104.6	98.2	91.2	96.8	92.5
0920	Plaster & Gypsum Board	94.9	100.6	98.9	94.1	100.6	98.6	80.1	100.6	94.4	97.9	100.6	99.8	93.9	100.6	98.6	86.3	100.6	96.3
0950, 0980	Ceilings & Acoustic Treatment	104.9	100.6	102.0	111.0	100.6	104.1	67.3	100.6	89.5	99.7	100.6	100.3	107.3	100.6	102.8	67.3	100.6	89.5
0960	Flooring	106.2	107.1	106.4	105.6	107.1	106.1	73.9	107.1	83.8	105.1	107.1	105.7	108.3	107.1	108.0	75.8	107.1	85.1
0970, 0990	Wall Finishes & Painting/Coating	110.6	60.7	80.7	110.6	60.7	80.7	100.7	68.4	81.3	110.4	68.4	85.2	110.6	72.8	88.0	100.7	72.8	84.0
09	FINISHES	105.1	97.8	101.0	105.3	97.8	101.1	72.5	98.7	87.2	102.4	98.6	100.3	103.9	99.1	101.2	72.7	99.2	87.6
COVERS	DIVS. 10 - 14, 25, 28, 41, 43, 44, 46	100.0	103.3	100.7	100.0	103.3	100.7	100.0	93.5	98.7	100.0	103.4	100.7	100.0	103.4	100.7	100.0	103.6	100.7
21, 22, 23	FIRE SUPPRESSION, PLUMBING & HVAC	94.0	99.2	96.1	100.0	99.2	99.7	95.8	114.3	103.4	100.0	99.2	99.7	100.0	99.2	99.7	95.8	73.7	86.8
26, 27, 3370	ELECTRICAL, COMMUNICATIONS & UTIL.	99.6	82.8	90.7	103.2	82.8	92.4	91.4	96.0	93.8	99.7	101.3	100.5	108.2	95.9	101.7	91.4	75.3	82.9
MF2010	WEIGHTED AVERAGE	99.3	98.3	98.9	99.7	98.3	99.1	93.8	102.9	97.8	98.4	101.2	99.6	98.9	100.4	99.5	91.8	91.1	91.5

	DIVISION	PENNSYLVANIA																	
		ALLENTOWN			ALTOONA			BEDFORD			BRADFORD			BUTLER			CHAMBERSBURG		
		181			166			155			167			160			172		
		MAT.	INST.	TOTAL	MAT.	INST.	TOTAL	MAT.	INST.	TOTAL	MAT.	INST.	TOTAL	MAT.	INST.	TOTAL	MAT.	INST.	TOTAL
015433	CONTRACTOR EQUIPMENT		114.1	114.1		114.1	114.1		110.9	110.9		114.1	114.1		114.1	114.1		113.3	113.3
0241, 31 - 34	SITE & INFRASTRUCTURE, DEMOLITION	90.6	106.0	101.4	94.1	105.9	102.4	103.9	102.6	103.0	89.6	105.3	100.6	85.6	107.3	100.8	83.9	103.2	97.5
0310	Concrete Forming & Accessories	99.0	116.5	114.2	85.1	80.5	81.1	87.2	82.9	83.5	87.1	81.1	81.9	86.5	95.1	94.0	85.3	82.1	82.5
0320	Concrete Reinforcing	97.1	110.3	103.7	94.1	107.7	100.9	92.4	82.8	87.6	96.1	108.1	102.1	94.8	108.8	101.8	94.5	105.2	99.8
0330	Cast-in-Place Concrete	81.7	103.7	90.7	91.0	89.3	90.3	105.4	72.6	91.9	86.9	92.0	89.0	80.4	98.5	87.9	88.3	72.5	81.8
03	CONCRETE	93.3	111.6	102.4	88.6	90.2	89.4	100.3	80.6	90.6	95.1	91.4	93.3	80.9	100.1	90.4	100.7	84.6	92.8
04	MASONRY	96.9	104.0	101.2	98.9	67.2	79.5	108.9	88.8	96.6	96.1	86.6	90.3	101.1	98.5	99.5	97.1	90.4	93.0
05	METALS	99.1	123.4	106.8	93.0	119.9	101.4	97.1	105.6	99.7	97.0	119.8	104.2	92.7	122.1	101.9	97.0	116.5	103.1
06	WOOD, PLASTICS & COMPOSITES	100.4	119.8	111.5	79.9	82.5	81.4	84.9	82.4	83.4	86.3	78.9	82.1	81.2	93.5	88.2	89.5	82.5	85.5
07	THERMAL & MOISTURE PROTECTION	100.2	116.2	106.6	99.7	89.8	95.7	105.5	89.3	99.0	100.1	91.8	96.8	99.4	100.6	99.9	99.4	75.2	89.7
08	OPENINGS	92.8	116.4	98.6	87.0	90.9	87.9	96.4	83.6	93.3	93.0	92.6	92.9	87.0	104.5	91.2	89.7	84.6	88.4
0920	Plaster & Gypsum Board	102.3	120.2	114.8	94.3	81.7	85.5	98.0	81.7	86.6	95.3	78.0	83.2	94.3	93.0	93.4	98.6	81.7	86.8
0950, 0980	Ceilings & Acoustic Treatment	88.7	120.2	109.7	93.3	81.7	85.5	95.1	81.7	86.1	91.5	78.0	82.5	94.3	93.0	93.5	91.5	81.7	85.0
0960	Flooring	93.6	99.4	95.3	87.5	101.0	91.5	90.2	91.4	90.5	88.1	104.9	93.1	88.4	83.9	87.0	91.6	47.8	78.6
0970, 0990	Wall Finishes & Painting/Coating	98.4	73.7	83.6	94.7	109.2	103.4	97.3	85.2	90.1	98.4	92.0	94.6	94.7	109.2	103.4	98.4	85.2	90.5
09	FINISHES	96.1	109.4	103.6	94.7	86.4	90.1	96.5	83.3	89.1	94.3	84.4	88.8	94.6	93.4	93.9	94.7	75.9	84.1
COVERS	DIVS. 10 - 14, 25, 28, 41, 43, 44, 46	100.0	106.3	101.3	100.0	95.9	99.2	100.0	99.2	99.8	100.0	99.1	99.8	100.0	102.8	100.6	100.0	96.6	99.3
21, 22, 23	FIRE SUPPRESSION, PLUMBING & HVAC	100.2	110.5	104.4	99.8	85.6	94.0	93.9	90.6	92.5	94.2	91.0	92.9	93.8	95.1	94.3	94.2	91.8	93.2
26, 27, 3370	ELECTRICAL, COMMUNICATIONS & UTIL.	99.9	98.8	99.3	89.4	108.4	99.4	94.0	108.4	101.6	93.0	108.4	101.1	90.0	104.8	97.8	91.6	86.9	89.1
MF2010	WEIGHTED AVERAGE	97.7	109.5	102.9	94.4	93.2	93.9	97.5	92.8	95.5	95.2	96.3	95.7	92.0	101.5	96.2	95.2	90.6	93.2

	DIVISION	PENNSYLVANIA																	
		DOYLESTOWN			DUBOIS			ERIE			GREENSBURG			HARRISBURG			HAZLETON		
		189			158			164 - 165			156			170 - 171			182		
		MAT.	INST.	TOTAL	MAT.	INST.	TOTAL	MAT.	INST.	TOTAL	MAT.	INST.	TOTAL	MAT.	INST.	TOTAL	MAT.	INST.	TOTAL
015433	CONTRACTOR EQUIPMENT		95.0	95.0		110.9	110.9		114.1	114.1		110.9	110.9		113.3	113.3		114.1	114.1
0241, 31 - 34	SITE & INFRASTRUCTURE, DEMOLITION	103.6	91.6	95.2	109.0	102.9	104.7	90.9	106.3	101.7	100.0	104.9	103.4	81.0	105.0	97.9	84.0	106.3	99.6
0310	Concrete Forming & Accessories	84.1	129.3	123.2	86.6	83.5	84.0	98.6	87.4	88.9	93.1	95.1	94.8	94.1	89.8	90.4	81.8	90.5	89.3
0320	Concrete Reinforcing	93.9	137.2	115.5	91.8	108.3	100.1	96.1	108.2	102.1	91.8	108.8	100.3	97.1	105.6	101.3	94.2	110.4	102.3
0330	Cast-in-Place Concrete	77.2	88.1	81.7	101.6	93.9	98.5	89.4	83.3	86.9	97.8	97.9	97.9	86.0	97.8	90.9	77.2	93.0	83.7
03	CONCRETE	89.2	116.4	102.6	101.5	93.1	97.4	87.6	91.2	89.4	95.5	99.7	97.6	93.4	96.9	95.1	86.2	96.4	91.3
04	MASONRY	100.1	128.2	117.2	109.0	89.7	97.2	88.8	90.8	90.0	119.5	98.4	106.6	96.7	92.1	93.9	109.3	97.3	102.0
05	METALS	96.6	127.3	106.3	97.1	119.0	104.0	93.2	118.7	101.2	97.0	120.6	104.4	101.0	120.6	107.2	98.8	122.3	106.2
06	WOOD, PLASTICS & COMPOSITES	81.9	132.2	110.6	83.8	81.3	82.4	96.6	85.7	90.4	91.3	93.4	92.5	94.3	88.9	91.2	80.6	88.9	85.3
07	THERMAL & MOISTURE PROTECTION	98.5	130.7	111.5	105.8	95.7	101.7	100.1	92.9	97.2	105.4	100.3	103.4	103.5	107.3	105.0	99.7	105.9	102.2
08	OPENINGS	94.9	140.1	105.9	96.4	93.9	95.8	87.1	92.6	88.5	96.3	104.4	98.3	92.8	94.4	93.2	93.5	95.0	93.8
0920	Plaster & Gypsum Board	92.6	132.9	120.8	96.7	80.6	85.5	102.3	85.0	90.2	99.4	93.0	94.9	102.6	88.3	92.6	93.1	88.3	89.7
0950, 0980	Ceilings & Acoustic Treatment	87.9	132.9	117.9	95.1	80.6	85.4	88.7	85.0	86.3	94.2	93.0	93.4	97.8	88.3	91.5	89.7	88.3	88.8
0960	Flooring	78.3	139.2	96.5	90.0	104.9	94.4	92.3	95.1	93.1	93.3	70.5	86.5	94.2	95.0	94.4	85.7	98.0	89.4
0970, 0990	Wall Finishes & Painting/Coating	98.0	72.1	82.4	97.3	105.9	102.5	102.6	92.0	96.2	97.3	109.2	104.4	98.4	88.2	92.3	98.4	99.6	99.1
09	FINISHES	87.2	124.4	108.1	96.8	88.5	92.2	96.5	88.7	92.2	97.0	91.7	94.0	96.9	90.2	93.1	92.4	90.9	91.5
COVERS	DIVS. 10 - 14, 25, 28, 41, 43, 44, 46	100.0	72.0	94.3	100.0	99.1	99.8	100.0	100.9	100.2	100.0	102.6	100.5	100.0	97.9	99.6	100.0	97.3	99.4
21, 22, 23	FIRE SUPPRESSION, PLUMBING & HVAC	93.8	126.3	107.1	93.9	91.0	92.7	99.8	93.3	97.1	93.9	94.3	94.1	100.0	93.6	97.4	94.2	101.5	97.2
26, 27, 3370	ELECTRICAL, COMMUNICATIONS & UTIL.	92.4	119.5	106.6	94.6	108.4	101.9	91.4	89.0	90.1	94.6	108.4	101.9	99.4	87.0	92.9	94.0	84.1	88.8
MF2010	WEIGHTED AVERAGE	94.5	120.2	105.8	97.9	97.2	97.6	94.2	95.2	94.6	97.5	101.2	99.2	97.8	96.5	97.2	94.9	98.6	96.5

DIVISION		PENNSYLVANIA																	
		INDIANA			JOHNSTOWN			KITTANNING			LANCASTER			LEHIGH VALLEY			MONTROSE		
		157			159			162			175 - 176			180			188		
		MAT.	INST.	TOTAL	MAT.	INST.	TOTAL	MAT.	INST.	TOTAL	MAT.	INST.	TOTAL	MAT.	INST.	TOTAL	MAT.	INST.	TOTAL
015433	**CONTRACTOR EQUIPMENT**		110.9	110.9		110.9	110.9		114.1	114.1		113.3	113.3		114.1	114.1		114.1	114.1
0241, 31 - 34	**SITE & INFRASTRUCTURE, DEMOLITION**	98.1	103.6	102.0	104.4	104.2	104.2	88.1	107.3	101.6	76.6	105.0	96.5	87.8	105.7	100.3	86.7	104.0	98.8
0310	Concrete Forming & Accessories	87.7	85.8	86.1	86.6	84.4	84.7	86.5	93.6	92.7	87.2	89.1	88.9	93.1	115.7	112.7	82.7	87.7	87.1
0320	Concrete Reinforcing	91.1	108.9	99.9	92.4	108.6	100.5	94.8	108.9	101.8	94.2	105.5	99.8	94.2	110.3	102.2	98.7	110.2	104.4
0330	Cast-in-Place Concrete	95.9	97.7	96.7	106.3	94.8	101.6	83.5	98.5	89.7	75.2	97.3	84.3	83.5	102.7	91.4	81.9	89.8	85.2
03	**CONCRETE**	93.2	95.6	94.3	101.1	93.9	97.5	83.4	99.4	91.3	88.8	96.4	92.5	92.3	110.9	101.5	91.1	93.8	92.5
04	**MASONRY**	104.8	100.2	102.0	105.6	89.2	95.6	104.8	100.3	102.0	102.4	91.6	95.8	96.8	102.7	100.4	96.8	96.8	96.8
05	**METALS**	97.1	120.2	104.4	97.1	119.2	104.0	92.8	122.3	102.0	97.0	120.1	104.2	98.8	122.2	106.1	97.1	117.0	103.3
06	**WOOD, PLASTICS & COMPOSITES**	85.6	81.3	83.2	83.8	82.4	83.0	81.2	91.3	87.0	92.0	88.9	90.2	92.7	119.8	108.2	81.4	87.0	84.6
07	**THERMAL & MOISTURE PROTECTION**	105.3	98.1	102.4	105.5	94.3	101.0	99.5	100.9	100.0	99.0	96.1	97.8	100.0	104.7	101.9	99.7	93.6	97.3
08	**OPENINGS**	96.4	94.0	95.8	96.4	91.0	95.0	87.0	99.5	90.0	89.7	98.2	91.7	93.5	116.4	99.0	89.9	94.0	90.9
0920	Plaster & Gypsum Board	98.4	80.6	86.0	96.5	81.7	86.1	94.3	90.8	91.9	100.7	88.3	92.0	96.4	120.2	113.0	93.6	86.3	88.5
0950, 0980	Ceilings & Acoustic Treatment	95.1	80.6	85.4	94.2	81.7	85.8	94.3	90.8	92.0	91.5	88.3	89.4	89.7	120.2	110.0	91.5	86.3	88.1
0960	Flooring	90.7	104.9	94.9	90.0	101.8	93.5	88.4	104.9	93.3	92.5	95.0	93.2	90.7	99.4	93.3	86.3	60.0	78.5
0970, 0990	Wall Finishes & Painting/Coating	97.3	109.2	104.4	97.3	109.2	104.4	94.7	109.2	103.4	98.4	59.7	75.2	98.4	69.4	81.0	98.4	99.6	99.1
09	**FINISHES**	96.2	89.8	92.6	96.1	89.2	92.2	94.8	95.7	95.3	94.7	86.9	90.3	94.6	107.6	101.9	93.3	83.6	87.9
COVERS	**DIVS. 10 - 14, 25, 28, 41, 43, 44, 46**	100.0	101.3	100.3	100.0	99.4	99.9	100.0	102.6	100.5	100.0	97.5	99.5	100.0	102.5	100.5	100.0	97.1	99.4
21, 22, 23	**FIRE SUPPRESSION, PLUMBING & HVAC**	93.9	91.3	92.8	93.9	91.2	92.8	93.8	98.4	95.7	94.2	93.2	93.8	94.2	109.8	100.6	94.2	100.8	96.9
26, 27, 3370	**ELECTRICAL, COMMUNICATIONS & UTIL.**	94.6	108.4	101.9	94.6	108.4	101.9	89.4	108.4	99.4	93.1	45.9	68.2	94.0	135.9	116.0	93.0	93.2	93.1
MF2010	**WEIGHTED AVERAGE**	96.4	99.1	97.6	97.5	97.4	97.4	92.5	102.9	97.1	94.1	90.0	92.3	95.3	113.4	103.3	94.2	97.3	95.6

DIVISION		PENNSYLVANIA																	
		NEW CASTLE			NORRISTOWN			OIL CITY			PHILADELPHIA			PITTSBURGH			POTTSVILLE		
		161			194			163			190 - 191			150 - 152			179		
		MAT.	INST.	TOTAL	MAT.	INST.	TOTAL	MAT.	INST.	TOTAL	MAT.	INST.	TOTAL	MAT.	INST.	TOTAL	MAT.	INST.	TOTAL
015433	**CONTRACTOR EQUIPMENT**		114.1	114.1		99.6	99.6		114.1	114.1		99.1	99.1		112.1	112.1		113.3	113.3
0241, 31 - 34	**SITE & INFRASTRUCTURE, DEMOLITION**	86.0	107.3	100.9	95.1	100.4	98.8	84.6	105.5	99.3	101.8	99.8	100.4	103.5	107.0	105.9	79.2	105.1	97.4
0310	Concrete Forming & Accessories	86.5	97.2	95.8	83.3	130.0	123.7	86.5	82.2	82.7	98.3	136.3	131.2	100.2	98.2	98.5	79.4	90.4	88.9
0320	Concrete Reinforcing	93.6	94.7	94.1	98.1	144.4	121.1	94.8	94.5	94.6	101.2	144.5	122.8	92.9	109.2	101.0	93.5	100.7	97.1
0330	Cast-in-Place Concrete	81.1	98.9	88.5	78.0	127.2	98.2	78.9	96.6	86.1	95.6	128.7	109.2	101.6	98.1	100.2	80.0	98.6	87.7
03	**CONCRETE**	81.2	98.5	89.7	85.4	131.4	108.2	79.8	90.9	85.3	97.1	134.8	115.7	98.7	101.3	100.0	92.3	96.5	94.4
04	**MASONRY**	100.5	98.5	99.2	114.3	127.1	122.1	100.9	93.1	96.1	100.7	128.5	117.6	98.4	104.5	102.2	96.5	93.8	94.8
05	**METALS**	92.8	115.1	99.8	95.8	132.1	107.2	92.8	113.6	99.3	99.6	133.1	110.1	98.4	121.5	105.7	97.2	118.4	103.9
06	**WOOD, PLASTICS & COMPOSITES**	81.2	97.2	90.4	83.0	130.0	109.8	81.2	78.9	79.9	100.2	137.9	121.7	100.6	97.1	98.6	83.0	88.9	86.3
07	**THERMAL & MOISTURE PROTECTION**	99.4	97.7	98.7	100.6	130.5	112.6	99.3	94.1	97.2	101.4	131.9	113.6	105.6	102.4	104.3	99.1	105.6	101.7
08	**OPENINGS**	87.0	98.2	89.7	88.8	141.0	101.5	87.0	80.7	85.4	96.9	145.3	108.7	99.7	106.4	101.3	89.7	92.1	90.3
0920	Plaster & Gypsum Board	94.3	96.9	96.1	91.0	130.6	118.7	94.3	78.0	82.9	98.9	138.8	126.8	106.3	96.9	99.7	95.7	88.3	90.5
0950, 0980	Ceilings & Acoustic Treatment	94.3	96.9	96.0	96.8	130.6	119.3	94.3	78.0	83.4	96.8	138.8	124.8	95.1	96.9	96.3	91.5	88.3	89.4
0960	Flooring	88.4	59.5	79.8	93.5	137.9	106.7	88.4	104.9	93.3	100.5	141.7	112.8	96.4	106.2	99.4	89.0	95.0	90.8
0970, 0990	Wall Finishes & Painting/Coating	94.7	109.2	103.4	99.8	145.4	127.1	94.7	109.2	103.4	99.8	146.4	127.7	97.3	113.0	106.7	98.4	99.6	99.1
09	**FINISHES**	94.6	91.2	92.7	96.9	133.1	117.2	94.5	87.1	90.4	100.3	138.5	121.7	99.2	100.7	100.0	93.2	91.8	92.4
COVERS	**DIVS. 10 - 14, 25, 28, 41, 43, 44, 46**	100.0	103.2	100.6	100.0	111.0	102.2	100.0	100.2	100.0	100.0	119.9	104.1	100.0	103.0	100.6	100.0	94.9	99.0
21, 22, 23	**FIRE SUPPRESSION, PLUMBING & HVAC**	93.8	94.8	94.2	94.0	126.1	107.1	93.8	92.8	93.4	100.0	129.0	111.9	99.9	99.3	99.7	94.2	100.9	97.0
26, 27, 3370	**ELECTRICAL, COMMUNICATIONS & UTIL.**	90.0	90.3	90.1	92.8	140.1	117.7	92.1	104.8	98.8	97.5	149.1	124.7	97.1	108.4	103.1	91.1	92.8	92.0
MF2010	**WEIGHTED AVERAGE**	92.0	98.0	94.7	94.7	128.4	109.5	92.1	96.1	93.8	99.2	132.2	113.7	99.4	104.7	101.8	94.0	98.8	96.1

DIVISION		PENNSYLVANIA																	
		READING			SCRANTON			STATE COLLEGE			STROUDSBURG			SUNBURY			UNIONTOWN		
		195 - 196			184 - 185			168			183			178			154		
		MAT.	INST.	TOTAL	MAT.	INST.	TOTAL	MAT.	INST.	TOTAL	MAT.	INST.	TOTAL	MAT.	INST.	TOTAL	MAT.	INST.	TOTAL
015433	**CONTRACTOR EQUIPMENT**		118.8	118.8		114.1	114.1		113.3	113.3		114.1	114.1		114.1	114.1		110.9	110.9
0241, 31 - 34	**SITE & INFRASTRUCTURE, DEMOLITION**	99.8	112.8	108.9	91.0	106.4	101.8	81.3	104.7	97.7	85.6	104.1	98.6	89.1	106.2	101.1	98.7	104.9	103.0
0310	Concrete Forming & Accessories	97.8	90.7	91.6	99.2	89.4	90.7	84.4	80.3	80.8	87.9	88.9	88.7	91.0	89.2	89.5	81.3	95.1	93.3
0320	Concrete Reinforcing	99.4	100.8	100.1	97.1	110.5	103.8	95.4	108.4	101.9	97.4	117.1	107.2	96.1	105.5	100.8	91.8	108.8	100.3
0330	Cast-in-Place Concrete	70.0	96.4	80.8	85.3	93.3	88.6	82.3	67.9	76.3	80.4	74.6	78.0	87.5	97.1	91.4	95.9	98.0	96.8
03	**CONCRETE**	84.5	95.9	90.1	95.1	96.1	95.6	95.3	82.9	89.2	90.0	90.4	90.2	95.7	96.4	96.0	92.8	99.8	96.3
04	**MASONRY**	102.6	94.8	97.9	97.2	97.4	97.3	102.5	80.4	89.0	93.7	102.3	98.9	96.1	91.2	93.1	122.1	98.4	107.7
05	**METALS**	96.1	119.1	103.3	101.1	123.0	108.0	96.8	120.7	104.3	98.9	119.6	105.4	96.9	120.4	104.3	96.8	120.6	104.3
06	**WOOD, PLASTICS & COMPOSITES**	100.3	88.9	93.8	100.4	87.0	92.7	88.3	82.5	85.0	87.2	87.0	87.1	90.4	88.9	89.5	77.9	93.4	86.8
07	**THERMAL & MOISTURE PROTECTION**	101.0	108.4	103.9	100.1	97.5	99.0	99.2	90.3	95.6	99.9	86.9	94.6	100.0	104.5	101.8	105.2	100.3	103.2
08	**OPENINGS**	93.9	96.7	94.6	92.8	94.0	93.1	89.6	90.9	90.0	93.5	89.2	92.4	89.8	93.6	90.7	96.3	104.4	98.3
0920	Plaster & Gypsum Board	103.2	88.3	92.8	104.7	86.3	91.8	96.6	81.7	86.2	94.7	86.3	88.8	95.4	88.3	90.4	94.4	93.0	93.5
0950, 0980	Ceilings & Acoustic Treatment	86.6	88.3	87.7	97.8	86.3	90.1	88.8	81.7	84.0	87.9	86.3	86.8	88.8	88.3	88.4	94.2	93.0	93.4
0960	Flooring	97.9	95.0	97.0	93.6	109.6	98.4	91.2	101.0	94.1	88.7	53.2	78.1	89.6	90.9	90.0	87.5	104.9	92.7
0970, 0990	Wall Finishes & Painting/Coating	98.4	100.0	99.4	98.4	99.6	99.1	98.4	109.2	104.9	98.4	67.5	79.9	98.4	99.6	99.1	97.3	109.2	104.4
09	**FINISHES**	97.9	91.5	94.3	98.3	93.5	95.6	93.5	86.4	89.5	93.2	79.9	85.7	94.1	90.5	92.1	94.6	96.7	95.8
COVERS	**DIVS. 10 - 14, 25, 28, 41, 43, 44, 46**	100.0	100.4	100.1	100.0	97.1	99.4	100.0	93.1	98.6	100.0	63.9	92.6	100.0	94.0	98.8	100.0	102.6	100.5
21, 22, 23	**FIRE SUPPRESSION, PLUMBING & HVAC**	100.1	106.6	102.8	100.2	101.5	100.7	94.2	87.3	91.4	94.2	103.3	97.9	94.2	92.8	93.6	93.9	94.3	94.1
26, 27, 3370	**ELECTRICAL, COMMUNICATIONS & UTIL.**	99.8	92.8	96.1	99.9	93.3	96.4	92.2	108.4	100.7	94.0	139.9	118.2	91.5	85.0	88.1	91.4	108.4	100.3
MF2010	**WEIGHTED AVERAGE**	97.1	101.1	98.8	98.4	99.9	99.1	94.7	93.7	94.3	94.7	102.7	98.2	94.7	95.8	95.2	96.7	101.9	99.0

DIVISION		PENNSYLVANIA																	
		WASHINGTON			WELLSBORO			WESTCHESTER			WILKES-BARRE			WILLIAMSPORT			YORK		
		153			169			193			186 - 187			177			173 - 174		
		MAT.	INST.	TOTAL	MAT.	INST.	TOTAL	MAT.	INST.	TOTAL	MAT.	INST.	TOTAL	MAT.	INST.	TOTAL	MAT.	INST.	TOTAL
015433	CONTRACTOR EQUIPMENT		110.9	110.9		114.1	114.1		99.6	99.6		114.1	114.1		114.1	114.1		113.3	113.3
0241, 31 - 34	SITE & INFRASTRUCTURE, DEMOLITION	98.8	104.9	103.0	93.3	103.9	100.7	101.0	98.2	99.0	83.7	106.3	99.6	81.1	104.6	97.6	79.9	105.0	97.5
0310	Concrete Forming & Accessories	87.8	95.3	94.3	86.5	87.5	87.4	89.5	130.4	124.9	90.3	88.2	88.4	87.9	61.0	64.6	82.6	89.6	88.7
0320	Concrete Reinforcing	91.8	108.9	100.4	95.4	110.1	102.7	97.1	111.5	104.3	96.1	110.5	103.3	95.4	67.7	81.6	96.1	105.6	100.8
0330	Cast-in-Place Concrete	95.9	98.0	96.8	86.2	87.8	86.8	86.3	127.9	103.4	77.2	93.2	83.8	74.1	73.3	73.7	80.4	97.5	87.4
03	CONCRETE	93.3	99.9	96.5	97.5	93.0	95.3	92.5	125.6	108.9	87.1	95.5	91.2	84.0	68.5	76.3	93.6	96.7	95.1
04	MASONRY	104.6	100.2	101.9	102.4	91.2	95.6	108.2	128.2	120.4	109.9	97.3	102.2	88.2	91.0	89.9	97.3	91.6	93.8
05	METALS	96.8	120.9	104.4	96.9	117.1	103.3	95.8	116.5	102.3	97.1	122.7	105.1	97.0	101.6	98.4	98.6	120.6	105.5
06	WOOD, PLASTICS & COMPOSITES	85.7	93.4	90.1	85.7	88.9	87.5	89.8	132.1	114.0	89.5	85.2	87.0	87.2	53.3	67.8	86.4	88.9	87.8
07	THERMAL & MOISTURE PROTECTION	105.3	100.8	103.5	100.3	91.6	96.8	100.9	130.6	112.8	99.7	105.5	102.0	99.6	100.2	99.8	99.2	107.0	102.3
08	OPENINGS	96.3	104.4	98.3	93.0	95.0	93.5	88.8	127.1	98.2	89.9	97.6	91.7	89.8	55.9	81.6	89.7	94.4	90.9
0920	Plaster & Gypsum Board	98.1	93.0	94.6	94.5	88.3	90.2	93.1	132.9	120.9	95.7	84.5	87.9	95.7	51.6	64.8	96.9	88.3	90.9
0950, 0980	Ceilings & Acoustic Treatment	94.2	93.0	93.4	88.8	88.3	88.4	96.8	132.9	120.8	91.5	84.5	86.8	91.5	51.6	64.9	90.5	88.3	89.0
0960	Flooring	90.7	104.9	94.9	87.9	52.6	77.4	96.6	141.7	110.0	89.5	98.0	92.1	88.7	51.5	77.6	90.2	95.0	91.6
0970, 0990	Wall Finishes & Painting/Coating	97.3	109.2	104.4	98.4	99.6	99.1	99.8	145.4	127.1	98.4	102.3	100.7	98.4	102.3	100.7	98.4	88.2	92.3
09	FINISHES	96.0	97.0	96.6	93.9	82.3	87.4	98.6	132.9	117.8	94.2	90.4	92.1	93.8	62.6	76.3	93.5	90.0	91.6
COVERS	DIVS. 10 - 14, 25, 28, 41, 43, 44, 46	100.0	102.6	100.5	100.0	94.0	98.8	100.0	114.3	102.9	100.0	96.9	99.4	100.0	93.1	98.6	100.0	97.7	99.5
21, 22, 23	FIRE SUPPRESSION, PLUMBING & HVAC	93.9	98.4	95.8	94.2	92.1	93.3	94.0	126.4	107.2	94.2	101.6	97.2	94.2	91.8	93.2	100.2	93.3	97.3
26, 27, 3370	ELECTRICAL, COMMUNICATIONS & UTIL.	93.9	108.4	101.6	93.0	84.8	88.7	92.7	118.5	106.2	94.0	84.1	88.8	92.0	48.6	69.1	93.1	87.0	89.9
MF2010	WEIGHTED AVERAGE	96.3	103.0	99.2	95.8	93.4	94.7	95.5	122.7	107.5	94.6	98.5	96.3	92.8	79.3	86.9	96.0	96.4	96.2

DIVISION		PUERTO RICO			RHODE ISLAND						SOUTH CAROLINA								
		SAN JUAN			NEWPORT			PROVIDENCE			AIKEN			BEAUFORT			CHARLESTON		
		009			028			029			298			299			294		
		MAT.	INST.	TOTAL	MAT.	INST.	TOTAL	MAT.	INST.	TOTAL	MAT.	INST.	TOTAL	MAT.	INST.	TOTAL	MAT.	INST.	TOTAL
015433	CONTRACTOR EQUIPMENT		90.1	90.1		102.0	102.0		102.0	102.0		100.1	100.1		100.1	100.1		100.1	100.1
0241, 31 - 34	SITE & INFRASTRUCTURE, DEMOLITION	135.3	90.6	104.0	92.0	103.8	100.3	93.0	103.8	100.6	125.1	87.0	98.4	119.7	85.2	95.5	102.1	84.8	89.9
0310	Concrete Forming & Accessories	90.6	17.0	26.9	99.1	122.0	118.9	101.1	122.0	119.2	97.2	70.5	74.1	96.2	38.7	46.4	95.2	40.5	47.8
0320	Concrete Reinforcing	199.0	12.1	105.9	110.7	156.2	133.4	110.7	156.2	133.4	100.0	68.4	84.3	99.1	27.0	63.2	98.9	58.1	78.6
0330	Cast-in-Place Concrete	106.9	30.3	75.4	82.6	118.7	97.4	101.4	118.7	108.5	89.7	72.2	82.5	89.7	48.0	72.6	105.2	48.6	82.0
03	CONCRETE	113.7	21.7	68.3	97.9	126.7	112.1	107.1	126.7	116.8	108.2	71.7	90.2	105.5	41.8	74.0	101.4	48.3	75.2
04	MASONRY	84.4	16.0	42.7	99.2	128.7	117.2	102.6	128.7	118.5	78.0	64.0	69.4	92.7	34.2	57.0	94.0	38.3	60.0
05	METALS	111.2	34.4	87.0	97.7	127.6	107.1	97.7	127.6	107.1	87.7	86.0	87.2	87.8	69.1	81.9	89.6	76.6	85.5
06	WOOD, PLASTICS & COMPOSITES	89.9	16.8	48.2	96.9	121.3	110.8	96.1	121.3	110.5	101.1	71.6	84.3	99.5	39.4	65.2	98.2	39.4	64.6
07	THERMAL & MOISTURE PROTECTION	125.2	21.0	83.3	97.3	119.3	106.1	102.0	119.3	109.0	101.4	69.1	88.4	101.1	37.2	75.4	100.3	41.6	76.7
08	OPENINGS	147.8	14.3	115.3	96.8	129.2	104.7	97.4	129.2	105.1	90.5	67.7	85.0	90.5	38.3	77.8	94.2	43.1	81.8
0920	Plaster & Gypsum Board	114.4	14.2	44.3	96.0	121.4	113.7	92.5	121.4	112.7	106.6	70.6	81.4	110.5	37.4	59.4	112.4	37.4	59.9
0950, 0980	Ceilings & Acoustic Treatment	228.4	14.2	85.7	94.8	121.4	112.5	98.7	121.4	113.8	93.4	70.6	78.2	97.0	37.4	57.3	97.0	37.4	57.3
0960	Flooring	213.7	13.6	154.1	99.1	139.1	111.0	98.5	139.1	110.6	102.8	70.9	93.3	104.3	53.1	89.0	103.8	60.3	90.9
0970, 0990	Wall Finishes & Painting/Coating	205.5	12.5	89.9	98.1	127.0	115.4	98.1	127.0	115.4	109.6	73.5	87.9	109.6	36.2	65.6	109.6	40.9	68.4
09	FINISHES	206.2	16.1	99.7	98.3	126.1	113.9	98.3	126.1	113.9	103.9	70.9	85.4	105.1	41.4	69.4	103.0	43.7	69.8
COVERS	DIVS. 10 - 14, 25, 28, 41, 43, 44, 46	100.0	16.2	82.9	100.0	104.4	100.9	100.0	104.4	100.9	100.0	74.8	94.8	100.0	45.0	88.8	100.0	67.8	93.4
21, 22, 23	FIRE SUPPRESSION, PLUMBING & HVAC	102.8	13.1	66.1	100.0	111.1	104.5	99.8	111.1	104.4	94.3	66.5	82.9	94.3	24.9	65.9	100.3	39.2	75.3
26, 27, 3370	ELECTRICAL, COMMUNICATIONS & UTIL.	131.0	12.4	68.5	101.1	97.1	99.0	101.2	97.1	99.1	95.3	68.7	81.3	99.0	31.6	63.5	97.4	92.5	94.8
MF2010	WEIGHTED AVERAGE	121.0	23.8	78.2	98.7	116.7	106.6	100.1	116.7	107.4	96.1	71.8	85.4	96.8	42.1	72.7	97.7	56.7	79.6

DIVISION		SOUTH CAROLINA															SOUTH DAKOTA		
		COLUMBIA			FLORENCE			GREENVILLE			ROCK HILL			SPARTANBURG			ABERDEEN		
		290 - 292			295			296			297			293			574		
		MAT.	INST.	TOTAL	MAT.	INST.	TOTAL	MAT.	INST.	TOTAL	MAT.	INST.	TOTAL	MAT.	INST.	TOTAL	MAT.	INST.	TOTAL
015433	CONTRACTOR EQUIPMENT		100.1	100.1		100.1	100.1		100.1	100.1		100.1	100.1		100.1	100.1		98.9	98.9
0241, 31 - 34	SITE & INFRASTRUCTURE, DEMOLITION	99.7	84.7	89.2	113.0	84.7	93.2	107.6	84.4	91.3	105.1	84.3	90.5	107.4	84.4	91.3	97.8	94.6	95.6
0310	Concrete Forming & Accessories	98.1	39.5	47.4	83.3	39.7	45.5	94.7	39.5	47.0	93.2	37.2	44.8	97.8	39.5	47.4	100.2	38.3	46.7
0320	Concrete Reinforcing	98.9	57.9	78.5	98.5	58.1	78.4	98.4	46.3	72.5	99.3	46.1	72.8	98.4	46.3	72.5	106.6	39.9	73.4
0330	Cast-in-Place Concrete	108.0	49.2	83.8	89.7	48.4	72.7	89.7	48.4	72.7	89.7	43.5	70.7	89.7	48.4	72.7	102.8	45.3	79.1
03	CONCRETE	103.0	48.0	75.8	100.1	47.9	74.3	98.8	45.6	72.5	96.9	42.9	70.2	99.0	45.6	72.6	101.6	42.5	72.4
04	MASONRY	88.5	36.2	56.6	78.1	38.3	53.8	76.0	38.3	53.0	99.7	33.6	59.3	78.1	38.3	53.8	107.1	57.1	76.6
05	METALS	89.6	75.8	85.3	88.6	76.1	84.6	88.6	71.9	83.3	87.8	71.4	82.6	88.6	71.9	83.3	102.6	65.8	91.0
06	WOOD, PLASTICS & COMPOSITES	100.7	39.0	65.5	83.7	39.0	58.2	97.8	39.0	64.2	96.1	37.1	62.4	102.2	39.0	66.1	95.5	37.6	62.5
07	THERMAL & MOISTURE PROTECTION	98.7	40.3	75.2	100.6	41.5	76.8	100.5	41.5	76.8	100.4	39.2	75.8	100.6	41.5	76.8	104.8	49.8	82.7
08	OPENINGS	94.2	42.9	81.7	90.6	42.9	79.0	90.5	40.1	78.2	90.5	39.2	78.0	90.5	40.1	78.3	92.9	38.4	79.6
0920	Plaster & Gypsum Board	104.5	36.9	57.3	99.0	36.9	55.6	105.0	36.9	57.4	104.1	35.0	55.8	108.3	36.9	58.4	110.8	35.9	58.4
0950, 0980	Ceilings & Acoustic Treatment	99.7	36.9	57.9	94.3	36.9	56.1	93.4	36.9	55.8	93.4	35.0	54.5	93.4	36.9	55.8	118.5	35.9	63.5
0960	Flooring	99.8	45.2	83.5	96.1	45.2	80.9	101.6	59.2	89.0	100.8	46.1	84.5	103.1	59.2	90.0	116.5	52.1	97.3
0970, 0990	Wall Finishes & Painting/Coating	107.2	40.9	67.5	109.6	40.9	68.4	109.6	40.9	68.4	109.6	40.9	68.4	109.6	40.9	68.4	94.1	38.8	60.9
09	FINISHES	100.7	40.5	66.9	99.8	40.9	66.8	101.5	43.3	68.9	100.9	38.9	66.2	102.4	43.3	69.3	111.8	40.5	71.8
COVERS	DIVS. 10 - 14, 25, 28, 41, 43, 44, 46	100.0	67.5	93.4	100.0	67.5	93.4	100.0	67.5	93.4	100.0	66.7	93.2	100.0	67.5	93.4	100.0	42.3	88.2
21, 22, 23	FIRE SUPPRESSION, PLUMBING & HVAC	99.9	34.7	73.2	100.3	34.7	73.5	100.3	33.3	72.9	94.3	32.3	68.9	100.3	33.3	72.9	100.2	41.9	76.4
26, 27, 3370	ELECTRICAL, COMMUNICATIONS & UTIL.	97.8	47.3	71.2	95.2	45.0	68.8	97.4	48.7	71.7	97.4	48.7	71.7	97.4	48.7	71.7	101.1	53.3	75.9
MF2010	WEIGHTED AVERAGE	97.3	48.8	75.9	96.0	48.7	75.2	96.0	48.4	75.0	95.3	46.6	73.8	96.3	48.4	75.2	101.3	51.7	79.4

DIVISION		SOUTH DAKOTA																	
		MITCHELL			MOBRIDGE			PIERRE			RAPID CITY			SIOUX FALLS			WATERTOWN		
		573			576			575			577			570 - 571			572		
		MAT.	INST.	TOTAL	MAT.	INST.	TOTAL	MAT.	INST.	TOTAL	MAT.	INST.	TOTAL	MAT.	INST.	TOTAL	MAT.	INST.	TOTAL
015433	**CONTRACTOR EQUIPMENT**		98.9	98.9		98.9	98.9		98.9	98.9		98.9	98.9		99.9	99.9		98.9	98.9
0241, 31 - 34	**SITE & INFRASTRUCTURE, DEMOLITION**	94.6	94.6	94.6	94.5	94.6	94.6	93.2	94.7	94.2	96.1	94.8	95.1	94.0	96.3	95.6	94.5	94.6	94.6
0310	Concrete Forming & Accessories	99.4	38.7	46.8	90.6	38.5	45.5	100.5	39.5	47.7	107.7	37.6	47.0	101.2	42.0	50.0	87.4	38.3	44.9
0320	Concrete Reinforcing	106.0	48.6	77.4	108.7	39.9	74.5	106.2	58.4	82.4	99.7	58.5	79.2	99.7	58.5	79.2	103.1	40.0	71.6
0330	Cast-in-Place Concrete	99.9	44.0	76.9	99.9	45.3	77.4	86.2	43.7	68.7	99.1	44.3	76.6	91.7	44.6	72.3	99.9	45.3	77.4
03	**CONCRETE**	99.4	43.7	71.9	99.3	42.5	71.2	92.9	45.9	69.7	98.5	45.3	72.3	93.5	47.4	70.7	98.1	42.5	70.6
04	**MASONRY**	95.7	55.5	71.2	104.5	57.1	75.6	102.8	55.8	74.1	104.6	57.1	75.6	100.3	55.8	73.1	129.5	58.1	85.9
05	**METALS**	101.6	66.0	90.4	101.7	65.8	90.4	102.6	74.4	93.7	104.3	74.9	95.1	104.5	75.0	95.2	101.6	66.2	90.5
06	**WOOD, PLASTICS & COMPOSITES**	94.6	38.1	62.3	84.5	37.8	57.9	93.5	38.4	62.0	98.9	34.8	62.3	94.3	41.0	63.9	81.3	37.6	56.3
07	**THERMAL & MOISTURE PROTECTION**	104.6	47.7	81.7	104.7	49.8	82.6	106.9	48.0	83.2	105.1	49.6	82.8	111.4	50.4	86.8	104.5	50.0	82.6
08	**OPENINGS**	91.6	39.0	78.8	94.1	38.0	80.5	95.7	44.3	83.2	96.7	42.3	83.5	96.9	45.7	84.5	91.6	38.1	78.5
0920	Plaster & Gypsum Board	109.0	36.5	58.3	102.5	36.2	56.1	99.6	36.7	55.6	110.3	33.0	56.3	105.3	39.5	59.3	100.3	35.9	55.3
0950, 0980	Ceilings & Acoustic Treatment	114.8	36.5	62.6	118.5	36.2	63.6	116.7	36.7	63.4	121.2	33.0	62.5	117.6	39.5	65.5	114.8	35.9	62.3
0960	Flooring	116.0	52.1	97.0	111.1	52.1	93.5	115.5	37.2	92.2	115.7	78.5	104.6	115.3	77.4	104.0	109.6	52.1	92.5
0970, 0990	Wall Finishes & Painting/Coating	94.1	42.4	63.1	94.1	43.5	63.8	94.1	46.6	65.6	94.1	46.6	65.6	94.1	46.6	65.6	94.1	38.8	60.9
09	**FINISHES**	110.4	41.2	71.7	109.0	41.2	71.0	109.3	39.0	69.9	112.0	45.2	74.6	110.3	48.5	75.6	107.5	40.5	69.9
COVERS	**DIVS. 10 - 14, 25, 28, 41, 43, 44, 46**	100.0	39.2	87.6	100.0	42.3	88.2	100.0	77.3	95.4	100.0	77.2	95.3	100.0	77.7	95.4	100.0	42.2	88.2
21, 22, 23	**FIRE SUPPRESSION, PLUMBING & HVAC**	94.2	41.6	72.7	94.2	42.1	72.8	100.0	41.7	76.2	100.2	42.2	76.5	100.1	40.5	75.7	94.2	42.1	72.9
26, 27, 3370	**ELECTRICAL, COMMUNICATIONS & UTIL.**	99.3	45.8	71.2	101.1	46.7	72.5	104.2	52.1	76.8	97.4	52.1	73.5	102.0	66.3	83.2	98.4	46.7	71.2
MF2010	**WEIGHTED AVERAGE**	98.4	50.5	77.3	99.1	50.9	77.8	100.5	53.7	79.8	101.2	54.6	80.6	100.9	57.1	81.6	99.5	50.9	78.1

DIVISION		TENNESSEE																	
		CHATTANOOGA			COLUMBIA			COOKEVILLE			JACKSON			JOHNSON CITY			KNOXVILLE		
		373 - 374			384			385			383			376			377 - 379		
		MAT.	INST.	TOTAL	MAT.	INST.	TOTAL	MAT.	INST.	TOTAL	MAT.	INST.	TOTAL	MAT.	INST.	TOTAL	MAT.	INST.	TOTAL
015433	**CONTRACTOR EQUIPMENT**		105.3	105.3		98.9	98.9		98.9	98.9		105.2	105.2		99.0	99.0		99.0	99.0
0241, 31 - 34	**SITE & INFRASTRUCTURE, DEMOLITION**	103.0	100.0	100.9	89.0	88.2	88.5	94.5	87.6	89.7	98.0	98.3	98.2	110.0	88.7	95.0	89.0	89.1	89.1
0310	Concrete Forming & Accessories	97.5	56.2	61.8	83.6	60.3	63.4	83.8	36.8	43.1	90.4	46.9	52.8	84.6	40.1	46.1	96.2	43.2	50.3
0320	Concrete Reinforcing	89.0	61.8	75.5	90.0	61.8	75.9	90.0	61.8	75.9	90.0	62.1	76.1	89.6	60.3	75.0	89.0	60.3	74.7
0330	Cast-in-Place Concrete	99.9	61.5	84.1	95.1	52.7	77.7	107.8	44.7	81.9	105.3	51.0	83.0	80.4	62.1	72.9	93.9	49.6	75.7
03	**CONCRETE**	92.2	60.8	76.7	93.9	59.6	76.9	103.5	46.4	75.3	95.6	53.1	74.6	97.0	53.5	75.5	89.5	50.7	70.3
04	**MASONRY**	104.9	50.9	71.9	120.9	55.7	81.1	115.2	45.5	72.7	121.1	49.1	77.2	120.1	46.6	75.2	82.7	54.2	65.3
05	**METALS**	96.2	87.0	93.3	90.6	87.5	89.6	90.7	86.4	89.3	92.9	87.3	91.1	93.5	85.7	91.0	96.8	86.0	93.4
06	**WOOD, PLASTICS & COMPOSITES**	105.5	57.8	78.3	72.6	61.8	66.4	72.8	35.1	51.3	88.7	47.4	65.1	78.1	38.6	55.6	92.0	38.6	61.5
07	**THERMAL & MOISTURE PROTECTION**	99.4	58.6	83.0	92.6	60.1	79.5	93.0	47.4	74.7	94.8	53.2	78.1	94.7	54.6	78.6	92.9	59.4	79.4
08	**OPENINGS**	101.4	56.4	90.5	91.9	56.8	83.4	91.9	43.3	80.1	100.5	53.1	89.0	97.3	46.2	84.9	94.4	46.3	82.7
0920	Plaster & Gypsum Board	80.4	57.0	64.0	87.8	61.0	69.1	87.8	33.6	49.9	89.7	46.2	59.3	101.3	37.1	56.4	109.0	37.1	58.8
0950, 0980	Ceilings & Acoustic Treatment	96.5	57.0	70.2	82.5	61.0	68.2	82.5	33.6	49.9	91.5	46.2	61.3	92.5	37.1	55.6	93.4	37.1	55.9
0960	Flooring	98.5	58.7	86.6	94.2	21.3	72.5	94.3	60.7	84.3	92.3	42.3	77.4	96.8	42.1	80.5	101.0	57.8	88.1
0970, 0990	Wall Finishes & Painting/Coating	102.1	55.2	74.0	91.5	34.9	57.6	91.5	38.1	59.5	93.1	51.8	68.3	99.3	44.0	66.2	99.3	50.7	70.2
09	**FINISHES**	94.7	56.6	73.3	93.8	50.8	69.7	94.4	40.1	64.0	93.8	45.7	66.8	99.1	39.8	65.9	92.3	45.5	66.1
COVERS	**DIVS. 10 - 14, 25, 28, 41, 43, 44, 46**	100.0	42.9	88.3	100.0	48.9	89.6	100.0	42.7	88.3	100.0	44.8	88.7	100.0	45.0	88.8	100.0	47.6	89.3
21, 22, 23	**FIRE SUPPRESSION, PLUMBING & HVAC**	100.2	62.2	84.6	96.9	76.0	88.3	96.9	70.8	86.2	100.0	70.3	87.9	99.9	55.3	81.7	99.9	65.5	85.9
26, 27, 3370	**ELECTRICAL, COMMUNICATIONS & UTIL.**	102.3	67.9	84.2	94.3	59.9	76.1	96.1	60.7	77.5	101.6	61.0	80.2	91.9	48.3	68.9	97.9	56.9	76.3
MF2010	**WEIGHTED AVERAGE**	98.9	65.5	84.2	95.5	66.2	82.6	96.7	59.5	80.3	98.9	63.2	83.2	98.4	56.0	79.8	95.5	60.8	80.2

DIVISION		TENNESSEE									TEXAS								
		MCKENZIE			MEMPHIS			NASHVILLE			ABILENE			AMARILLO			AUSTIN		
		382			375,380 - 381			370 - 372			795 - 796			790 - 791			786 - 787		
		MAT.	INST.	TOTAL	MAT.	INST.	TOTAL	MAT.	INST.	TOTAL	MAT.	INST.	TOTAL	MAT.	INST.	TOTAL	MAT.	INST.	TOTAL
015433	**CONTRACTOR EQUIPMENT**		98.9	98.9		103.0	103.0		106.2	106.2		89.4	89.4		89.4	89.4		88.8	88.8
0241, 31 - 34	**SITE & INFRASTRUCTURE, DEMOLITION**	94.2	87.9	89.8	91.8	94.6	93.8	98.2	101.0	100.2	94.2	86.6	88.9	94.4	87.7	89.7	100.1	86.9	90.8
0310	Concrete Forming & Accessories	91.5	39.8	46.7	95.0	61.8	66.2	99.1	64.0	68.7	96.4	39.6	47.2	97.4	53.2	59.2	95.7	52.6	58.4
0320	Concrete Reinforcing	90.1	62.1	76.2	95.5	65.6	80.6	95.5	64.7	80.1	90.2	51.1	70.7	90.2	49.4	69.9	84.4	48.2	66.4
0330	Cast-in-Place Concrete	105.5	59.4	86.5	91.4	62.4	79.5	90.4	65.9	80.3	97.2	48.1	77.0	96.3	54.7	79.2	98.4	48.6	77.9
03	**CONCRETE**	102.2	52.9	77.8	89.5	64.3	77.0	89.3	66.3	77.9	96.5	45.9	71.5	96.1	53.9	75.3	91.6	51.2	71.6
04	**MASONRY**	119.4	48.1	75.9	95.6	63.7	76.1	89.7	58.3	70.5	103.3	56.7	74.8	103.3	57.4	75.3	101.9	46.9	68.3
05	**METALS**	90.7	87.5	89.7	96.6	90.1	94.6	98.9	90.1	96.1	99.6	68.2	89.7	99.6	67.9	89.6	94.8	64.3	85.2
06	**WOOD, PLASTICS & COMPOSITES**	81.6	37.9	56.7	97.1	62.8	77.5	102.6	65.1	81.2	101.7	38.0	65.3	100.2	55.6	74.7	98.8	55.6	74.1
07	**THERMAL & MOISTURE PROTECTION**	93.0	50.6	75.9	96.0	64.4	83.3	95.6	61.3	81.8	98.4	50.2	79.0	98.0	53.3	80.0	91.8	50.2	75.1
08	**OPENINGS**	91.9	45.2	80.6	99.1	63.1	90.3	101.9	64.5	92.8	92.7	41.5	80.2	92.7	52.9	83.0	97.9	53.4	87.1
0920	Plaster & Gypsum Board	91.1	36.4	52.8	93.4	62.0	71.4	97.5	64.3	74.3	82.6	36.6	50.4	88.3	54.7	64.8	85.4	54.8	64.0
0950, 0980	Ceilings & Acoustic Treatment	82.5	36.4	51.8	98.6	62.0	74.2	89.5	64.3	72.7	90.0	36.6	54.4	95.5	54.7	68.3	96.5	54.8	68.7
0960	Flooring	97.2	40.8	80.4	94.1	37.4	77.2	101.9	66.6	91.4	109.0	76.5	99.3	108.8	76.5	99.2	94.6	41.0	78.6
0970, 0990	Wall Finishes & Painting/Coating	91.5	51.8	67.7	96.4	54.8	71.4	99.9	68.3	81.0	99.1	51.3	70.5	99.1	50.6	70.0	102.3	41.7	66.0
09	**FINISHES**	95.5	39.9	64.4	95.0	56.1	73.2	96.9	64.7	78.9	94.6	46.8	67.8	96.6	57.3	74.6	93.1	49.2	68.5
COVERS	**DIVS. 10 - 14, 25, 28, 41, 43, 44, 46**	100.0	28.5	85.4	100.0	81.8	96.3	100.0	82.2	96.4	100.0	76.5	95.2	100.0	68.7	93.6	100.0	78.3	95.6
21, 22, 23	**FIRE SUPPRESSION, PLUMBING & HVAC**	96.9	70.6	86.1	100.0	69.5	87.5	100.0	78.8	91.3	100.3	44.4	77.4	100.0	50.2	79.7	100.0	54.3	81.3
26, 27, 3370	**ELECTRICAL, COMMUNICATIONS & UTIL.**	95.8	65.5	79.9	102.1	66.0	83.1	96.9	64.3	79.7	96.5	49.5	71.8	102.0	65.8	82.9	97.1	63.2	79.3
MF2010	**WEIGHTED AVERAGE**	96.8	61.1	81.1	97.4	70.0	85.3	97.6	73.1	86.8	98.1	53.6	78.5	98.6	60.1	81.7	97.1	58.1	79.9

DIVISION		TEXAS																	
		BEAUMONT			BROWNWOOD			BRYAN			CHILDRESS			CORPUS CHRISTI			DALLAS		
		776 - 777			768			778			792			783 - 784			752 - 753		
		MAT.	INST.	TOTAL	MAT.	INST.	TOTAL	MAT.	INST.	TOTAL	MAT.	INST.	TOTAL	MAT.	INST.	TOTAL	MAT.	INST.	TOTAL
015433	**CONTRACTOR EQUIPMENT**		90.1	90.1		89.4	89.4		90.1	90.1		89.4	89.4		96.1	96.1		99.0	99.0
0241, 31 - 34	**SITE & INFRASTRUCTURE, DEMOLITION**	86.7	87.1	87.0	99.6	86.0	90.1	78.4	88.6	85.6	104.1	86.5	91.7	142.0	82.7	100.4	105.3	88.3	93.4
0310	Concrete Forming & Accessories	103.5	49.3	56.6	94.9	33.5	41.8	82.5	41.0	46.6	94.9	60.5	65.1	98.6	37.5	45.7	96.8	64.4	68.8
0320	Concrete Reinforcing	87.3	42.1	64.8	90.5	48.4	69.5	89.5	48.3	68.9	90.3	48.7	69.6	83.8	46.7	65.3	96.2	51.7	74.0
0330	Cast-in-Place Concrete	83.8	52.4	70.9	104.8	43.3	79.5	67.2	65.2	66.4	99.7	49.5	79.1	109.2	45.6	83.0	96.5	55.7	79.7
03	**CONCRETE**	94.2	50.0	72.4	104.1	40.9	72.9	79.0	51.9	65.7	105.6	55.1	80.7	99.6	43.9	72.1	99.2	60.4	80.0
04	**MASONRY**	101.9	57.7	74.9	131.9	49.5	81.6	140.6	56.5	89.2	107.9	50.7	73.0	90.9	48.5	65.0	102.3	57.8	75.1
05	**METALS**	96.9	64.5	86.7	97.3	61.9	86.2	96.5	67.4	87.4	97.0	63.8	86.6	94.4	75.1	88.3	102.0	79.6	94.9
06	**WOOD, PLASTICS & COMPOSITES**	115.3	49.4	77.7	100.3	29.7	60.0	80.4	34.3	54.1	101.1	65.4	80.7	118.8	36.7	71.9	100.1	67.7	81.6
07	**THERMAL & MOISTURE PROTECTION**	102.9	55.9	84.0	98.5	44.9	77.0	95.0	56.0	79.3	99.0	49.4	79.0	98.9	45.3	77.3	94.1	63.2	81.7
08	**OPENINGS**	98.7	46.1	85.9	88.8	36.9	76.2	101.2	42.4	86.9	89.8	55.0	81.3	104.5	38.0	88.3	100.3	58.3	90.1
0920	Plaster & Gypsum Board	93.0	48.4	61.8	82.1	28.1	44.3	81.2	32.8	47.4	82.1	64.9	70.1	86.0	35.1	50.4	93.2	67.0	74.9
0950, 0980	Ceilings & Acoustic Treatment	94.9	48.4	63.9	88.2	28.1	48.1	88.2	32.8	51.3	88.2	64.9	72.7	92.9	35.1	54.4	95.3	67.0	76.5
0960	Flooring	105.0	73.9	95.7	108.1	46.3	89.7	79.9	54.6	72.4	107.1	42.4	87.8	107.6	65.4	95.0	98.3	49.5	83.8
0970, 0990	Wall Finishes & Painting/Coating	92.4	49.7	66.8	99.1	29.9	57.6	88.8	61.9	72.7	99.1	39.5	63.4	115.5	59.0	81.6	105.5	52.7	73.9
09	**FINISHES**	89.4	53.5	69.3	94.4	34.7	60.9	78.5	43.9	59.1	94.9	56.0	73.1	100.4	44.3	69.0	97.6	60.9	77.0
COVERS	**DIVS. 10 - 14, 25, 28, 41, 43, 44, 46**	100.0	79.7	95.9	100.0	37.8	87.3	100.0	79.0	95.7	100.0	70.7	94.0	100.0	77.3	95.4	100.0	82.4	96.4
21, 22, 23	**FIRE SUPPRESSION, PLUMBING & HVAC**	100.2	58.2	83.0	94.3	47.5	75.1	94.1	66.2	82.7	94.3	51.8	76.9	100.1	39.4	75.3	99.9	63.3	84.9
26, 27, 3370	**ELECTRICAL, COMMUNICATIONS & UTIL.**	94.3	71.2	82.2	92.3	44.5	67.1	92.0	68.6	79.7	96.5	44.2	69.0	92.6	53.7	72.1	93.5	66.3	79.2
MF2010	**WEIGHTED AVERAGE**	97.3	61.3	81.4	97.8	48.4	76.0	94.5	61.6	80.0	97.5	56.5	79.4	99.7	51.7	78.6	99.5	66.5	85.0

DIVISION		TEXAS																	
		DEL RIO			DENTON			EASTLAND			EL PASO			FORT WORTH			GALVESTON		
		788			762			764			798 - 799,885			760 - 761			775		
		MAT.	INST.	TOTAL	MAT.	INST.	TOTAL	MAT.	INST.	TOTAL	MAT.	INST.	TOTAL	MAT.	INST.	TOTAL	MAT.	INST.	TOTAL
015433	**CONTRACTOR EQUIPMENT**		88.8	88.8		95.2	95.2		89.4	89.4		89.4	89.4		89.4	89.4		99.0	99.0
0241, 31 - 34	**SITE & INFRASTRUCTURE, DEMOLITION**	121.1	85.2	95.9	99.9	79.2	85.4	102.2	86.0	90.8	91.7	86.0	87.7	96.1	87.0	89.7	105.8	86.4	92.2
0310	Concrete Forming & Accessories	93.3	37.1	44.6	100.8	33.0	42.1	95.5	36.5	44.4	96.4	44.3	51.3	93.1	63.8	67.8	91.6	64.4	68.1
0320	Concrete Reinforcing	84.4	43.8	64.2	91.8	49.7	70.8	90.7	48.4	69.6	90.2	46.6	68.5	90.2	51.6	71.0	89.0	63.3	76.2
0330	Cast-in-Place Concrete	117.5	43.1	86.9	81.2	45.1	66.3	110.7	43.9	83.3	90.8	41.3	70.5	100.0	51.5	80.1	89.1	66.3	79.7
03	**CONCRETE**	118.1	41.5	80.3	82.1	42.3	62.4	108.9	42.4	76.1	93.4	44.7	69.4	97.6	57.9	78.0	95.2	66.3	80.9
04	**MASONRY**	105.8	51.1	72.4	140.4	48.2	84.1	101.7	49.3	69.7	101.1	47.2	68.2	97.6	57.8	73.3	97.6	56.5	72.5
05	**METALS**	93.2	59.1	82.5	96.9	75.2	90.1	97.1	62.3	86.2	99.4	62.8	87.9	96.5	68.1	87.6	98.1	86.5	94.5
06	**WOOD, PLASTICS & COMPOSITES**	98.1	35.3	62.2	110.4	32.5	66.0	107.4	33.4	65.1	101.3	46.7	70.1	102.7	67.5	82.6	98.1	65.4	79.5
07	**THERMAL & MOISTURE PROTECTION**	95.5	44.4	75.0	96.7	44.9	75.9	98.9	44.1	76.9	95.5	52.9	78.4	95.4	53.5	78.6	94.1	65.3	82.5
08	**OPENINGS**	98.7	33.6	82.9	103.3	37.7	87.3	75.1	38.9	66.3	92.7	43.0	80.6	84.5	58.3	78.1	105.8	63.3	95.4
0920	Plaster & Gypsum Board	80.7	33.8	47.9	85.4	30.8	47.2	82.1	31.8	47.0	91.1	45.5	59.2	86.7	67.0	73.0	87.8	64.8	71.7
0950, 0980	Ceilings & Acoustic Treatment	88.3	33.8	52.0	91.0	30.8	50.9	88.2	31.8	50.7	95.5	45.5	62.2	98.2	67.0	77.4	91.0	64.8	73.6
0960	Flooring	91.5	35.3	74.8	101.0	44.3	84.1	140.1	46.3	112.2	108.6	64.7	95.5	140.5	42.2	111.2	93.1	54.6	81.7
0970, 0990	Wall Finishes & Painting/Coating	102.3	26.3	56.8	109.4	30.3	62.0	100.7	31.2	59.1	99.1	35.7	61.1	100.7	52.2	71.6	99.3	48.1	68.6
09	**FINISHES**	92.0	35.3	60.3	90.7	34.1	59.0	104.2	37.0	66.5	96.4	47.3	68.9	106.4	59.3	80.0	87.8	60.8	72.7
COVERS	**DIVS. 10 - 14, 25, 28, 41, 43, 44, 46**	100.0	37.6	87.2	100.0	36.5	87.0	100.0	38.9	87.5	100.0	78.2	95.5	100.0	82.1	96.3	100.0	84.2	96.8
21, 22, 23	**FIRE SUPPRESSION, PLUMBING & HVAC**	94.0	45.4	74.1	94.3	39.1	71.7	94.3	47.9	75.3	100.0	34.8	73.3	100.1	58.8	83.2	94.1	65.0	82.2
26, 27, 3370	**ELECTRICAL, COMMUNICATIONS & UTIL.**	94.2	30.6	60.7	94.8	44.6	68.4	92.1	44.5	67.1	93.1	52.3	71.6	93.7	61.2	76.6	93.8	66.0	79.1
MF2010	**WEIGHTED AVERAGE**	98.9	45.9	75.6	97.3	47.3	75.2	96.3	49.1	75.5	97.2	50.6	76.7	97.1	62.9	82.0	96.6	68.3	84.1

DIVISION		TEXAS																	
		GIDDINGS			GREENVILLE			HOUSTON			HUNTSVILLE			LAREDO			LONGVIEW		
		789			754			770 - 772			773			780			756		
		MAT.	INST.	TOTAL	MAT.	INST.	TOTAL	MAT.	INST.	TOTAL	MAT.	INST.	TOTAL	MAT.	INST.	TOTAL	MAT.	INST.	TOTAL
015433	**CONTRACTOR EQUIPMENT**		88.8	88.8		96.6	96.6		98.8	98.8		90.1	90.1		88.8	88.8		91.5	91.5
0241, 31 - 34	**SITE & INFRASTRUCTURE, DEMOLITION**	106.5	85.5	91.8	98.7	83.2	87.8	104.4	86.2	91.6	93.3	86.8	88.7	100.2	86.7	90.7	95.1	90.5	91.8
0310	Concrete Forming & Accessories	91.1	56.6	61.3	88.9	27.7	36.0	93.5	64.6	68.5	89.3	56.3	60.7	93.3	37.8	45.3	85.9	28.6	36.3
0320	Concrete Reinforcing	84.9	45.6	65.3	96.7	25.3	61.1	88.8	63.8	76.4	89.7	48.1	69.0	84.4	47.1	65.9	95.6	23.1	59.5
0330	Cast-in-Place Concrete	99.6	43.5	76.5	87.9	37.8	67.3	86.4	66.4	78.2	92.3	47.9	74.0	84.1	58.4	73.6	102.2	35.5	74.7
03	**CONCRETE**	95.5	50.7	73.4	91.8	32.9	62.7	92.9	66.5	79.9	101.6	52.6	77.4	90.2	47.8	69.2	108.0	31.3	70.1
04	**MASONRY**	113.9	51.2	75.6	157.9	47.0	90.2	97.4	66.4	78.5	138.5	52.6	86.0	99.5	56.3	73.2	153.6	43.0	86.1
05	**METALS**	92.7	61.4	82.9	99.3	63.4	88.0	101.0	87.4	96.7	96.5	63.8	86.2	95.8	64.1	85.8	92.5	50.8	79.4
06	**WOOD, PLASTICS & COMPOSITES**	97.1	60.7	76.3	91.0	26.8	54.4	100.4	65.4	80.5	88.8	59.3	72.0	98.1	36.6	63.0	85.2	29.0	53.1
07	**THERMAL & MOISTURE PROTECTION**	96.2	48.6	77.1	94.2	40.2	72.5	93.7	67.5	83.1	95.9	49.1	77.1	94.6	51.1	77.1	95.6	35.6	71.5
08	**OPENINGS**	97.9	49.4	86.1	98.8	26.5	81.2	108.3	63.3	97.4	101.2	48.6	88.4	98.6	38.6	84.0	89.3	26.5	74.0
0920	Plaster & Gypsum Board	79.4	60.0	65.8	87.4	25.0	43.8	90.5	64.8	72.5	84.9	58.6	66.6	81.9	35.1	49.2	85.8	27.4	44.9
0950, 0980	Ceilings & Acoustic Treatment	88.3	60.0	69.4	90.7	25.0	46.9	96.4	64.8	75.4	88.2	58.6	68.5	92.9	35.1	54.4	88.0	27.4	47.6
0960	Flooring	91.7	35.3	74.9	94.5	48.6	80.8	94.1	54.6	82.4	83.2	35.3	68.9	91.4	65.4	83.6	98.9	35.7	80.0
0970, 0990	Wall Finishes & Painting/Coating	102.3	31.2	59.7	105.5	25.7	57.7	99.3	61.9	76.9	88.8	32.0	54.8	102.3	55.6	74.3	96.1	30.0	56.5
09	**FINISHES**	90.7	51.2	68.6	94.2	31.0	58.8	95.4	62.3	76.9	80.9	50.9	64.1	91.3	43.8	64.7	99.2	30.0	60.4
COVERS	**DIVS. 10 - 14, 25, 28, 41, 43, 44, 46**	100.0	43.5	88.5	100.0	34.9	86.7	100.0	84.2	96.8	100.0	43.8	88.5	100.0	75.3	94.9	100.0	22.1	84.1
21, 22, 23	**FIRE SUPPRESSION, PLUMBING & HVAC**	94.0	59.1	79.7	94.0	44.9	73.9	100.0	66.4	86.3	94.1	59.9	80.1	100.0	38.1	74.7	93.9	28.4	67.1
26, 27, 3370	**ELECTRICAL, COMMUNICATIONS & UTIL.**	90.9	44.2	66.3	90.1	27.6	57.2	96.0	68.7	81.6	92.0	44.2	66.8	94.4	61.0	76.8	90.5	44.7	66.4
MF2010	**WEIGHTED AVERAGE**	95.8	55.4	78.0	98.6	42.9	74.0	99.3	70.3	86.5	97.6	56.3	79.4	96.7	53.2	77.5	98.4	40.0	72.7

	DIVISION	TEXAS																	
		LUBBOCK			LUFKIN			MCALLEN			MCKINNEY			MIDLAND			ODESSA		
		793 - 794			759			785			750			797			797		
		MAT.	INST.	TOTAL	MAT.	INST.	TOTAL	MAT.	INST.	TOTAL	MAT.	INST.	TOTAL	MAT.	INST.	TOTAL	MAT.	INST.	TOTAL
015433	**CONTRACTOR EQUIPMENT**		97.8	97.8		91.5	91.5		96.1	96.1		96.6	96.6		97.8	97.8		89.4	89.4
0241, 31 - 34	**SITE & INFRASTRUCTURE, DEMOLITION**	117.8	84.9	94.7	90.4	91.5	91.2	146.9	82.6	101.8	95.2	83.7	87.2	120.7	84.9	95.6	94.5	86.9	89.1
0310	Concrete Forming & Accessories	94.9	51.8	57.6	88.7	57.8	62.0	97.7	35.2	43.6	88.0	37.0	43.9	98.6	39.1	47.1	96.3	38.9	46.7
0320	Concrete Reinforcing	91.3	51.1	71.3	97.3	60.6	79.0	84.0	46.6	65.4	96.7	49.7	73.3	92.3	50.4	71.5	90.2	50.4	70.4
0330	Cast-in-Place Concrete	97.3	48.4	77.2	91.4	44.8	72.3	118.6	44.0	87.9	82.5	45.9	67.4	103.4	45.8	79.7	97.2	44.7	75.6
03	**CONCRETE**	95.1	52.2	73.9	100.2	54.5	77.6	105.7	42.3	74.4	87.2	44.3	66.0	99.7	45.6	73.0	96.5	44.3	70.7
04	**MASONRY**	102.6	55.0	73.6	118.7	53.4	78.8	105.4	48.5	70.7	169.6	49.5	96.3	120.8	43.0	73.3	103.3	42.9	66.5
05	**METALS**	103.0	80.0	95.8	99.4	67.7	89.4	93.2	74.6	87.4	99.2	74.3	91.4	101.3	78.8	94.2	98.9	66.8	88.8
06	**WOOD, PLASTICS & COMPOSITES**	101.0	55.7	75.1	93.2	60.7	74.6	117.4	34.2	69.9	89.8	37.1	59.7	105.6	38.6	67.4	101.7	38.5	65.6
07	**THERMAL & MOISTURE PROTECTION**	88.3	52.4	73.8	95.4	49.0	76.7	98.8	44.2	76.8	94.0	46.8	75.0	88.5	44.1	70.7	98.4	43.3	76.2
08	**OPENINGS**	102.7	49.9	89.9	69.9	53.8	66.0	102.9	37.2	86.9	98.8	40.0	84.5	101.6	40.3	86.7	92.7	40.2	79.9
0920	Plaster & Gypsum Board	83.1	54.7	63.2	84.7	60.0	67.4	86.0	32.6	48.6	86.6	35.6	50.9	84.9	37.1	51.4	82.6	37.1	50.8
0950, 0980	Ceilings & Acoustic Treatment	91.9	54.7	67.1	80.7	60.0	66.9	92.9	32.6	52.7	90.7	35.6	54.0	89.1	37.1	54.4	90.0	37.1	54.7
0960	Flooring	102.1	40.0	83.6	131.1	38.8	103.6	107.3	65.4	94.8	94.1	36.2	76.9	103.5	49.4	87.4	109.0	49.4	91.2
0970, 0990	Wall Finishes & Painting/Coating	110.3	50.6	74.6	96.1	38.6	61.7	115.5	29.5	64.0	105.5	42.1	67.5	110.3	50.6	74.6	99.1	50.6	70.0
09	**FINISHES**	97.2	49.6	70.5	106.4	53.1	76.5	100.9	39.5	66.5	93.7	37.4	62.2	97.6	41.4	66.1	94.6	41.3	64.7
COVERS	**DIVS. 10 - 14, 25, 28, 41, 43, 44, 46**	100.0	78.2	95.5	100.0	71.9	94.3	100.0	75.2	94.9	100.0	40.3	87.8	100.0	77.9	95.5	100.0	77.6	95.4
21, 22, 23	**FIRE SUPPRESSION, PLUMBING & HVAC**	99.8	43.5	76.8	93.9	60.0	80.0	94.0	29.9	67.8	94.0	42.7	73.0	93.7	45.7	74.1	100.3	45.7	77.9
26, 27, 3370	**ELECTRICAL, COMMUNICATIONS & UTIL.**	95.1	50.5	71.6	91.9	44.2	66.8	91.7	34.8	61.8	90.2	46.0	66.9	95.1	41.1	66.6	96.7	41.0	67.4
MF2010	**WEIGHTED AVERAGE**	99.7	56.2	80.6	95.5	58.8	79.3	99.4	46.1	75.9	98.5	49.7	77.0	99.5	51.3	78.2	98.0	50.1	76.9

	DIVISION	TEXAS																	
		PALESTINE			SAN ANGELO			SAN ANTONIO			TEMPLE			TEXARKANA			TYLER		
		758			769			781 - 782			765			755			757		
		MAT.	INST.	TOTAL	MAT.	INST.	TOTAL	MAT.	INST.	TOTAL	MAT.	INST.	TOTAL	MAT.	INST.	TOTAL	MAT.	INST.	TOTAL
015433	**CONTRACTOR EQUIPMENT**		91.5	91.5		89.4	89.4		91.3	91.3		89.4	89.4		91.5	91.5		91.5	91.5
0241, 31 - 34	**SITE & INFRASTRUCTURE, DEMOLITION**	96.4	91.3	92.8	96.0	87.3	89.9	99.9	91.5	94.0	85.0	85.8	85.5	85.0	91.2	89.4	94.5	91.3	92.2
0310	Concrete Forming & Accessories	80.8	56.9	60.1	95.2	36.8	44.7	93.4	56.0	61.0	98.1	34.6	43.1	95.3	34.6	42.8	90.3	33.6	41.2
0320	Concrete Reinforcing	94.9	48.5	71.8	90.4	50.4	70.5	90.5	48.0	69.3	90.5	48.1	69.4	94.8	51.3	73.1	95.6	51.5	73.7
0330	Cast-in-Place Concrete	83.6	43.3	67.0	98.9	43.0	75.9	82.6	66.7	76.0	81.1	48.3	67.6	84.2	44.1	67.7	100.3	43.0	76.8
03	**CONCRETE**	102.2	51.2	77.0	99.4	42.7	71.4	90.4	59.0	74.9	85.1	43.1	64.3	93.1	42.3	68.0	107.5	41.6	74.9
04	**MASONRY**	113.8	49.6	74.7	128.5	42.3	75.9	99.4	68.1	80.3	139.8	48.1	83.8	173.8	45.7	95.6	163.5	48.5	93.3
05	**METALS**	99.1	61.9	87.4	97.5	64.0	87.0	96.5	68.6	87.7	97.1	62.9	86.4	92.4	66.7	84.3	99.0	67.2	89.0
06	**WOOD, PLASTICS & COMPOSITES**	83.9	60.7	70.7	100.5	36.8	64.2	98.1	53.9	72.9	109.7	33.4	66.2	96.0	32.9	60.0	94.8	31.1	58.4
07	**THERMAL & MOISTURE PROTECTION**	95.9	47.4	76.4	98.3	42.9	76.1	94.3	65.4	82.6	97.9	44.5	76.5	95.2	48.0	76.3	95.7	52.3	78.3
08	**OPENINGS**	69.8	50.9	65.2	88.8	39.2	76.7	100.6	52.2	88.8	73.1	35.7	64.0	89.3	38.2	76.8	69.8	37.1	61.8
0920	Plaster & Gypsum Board	80.6	60.0	66.2	82.1	35.3	49.4	81.9	52.9	61.6	82.1	31.8	47.0	89.6	31.4	48.9	84.7	29.5	46.1
0950, 0980	Ceilings & Acoustic Treatment	80.7	60.0	66.9	88.2	35.3	53.0	92.9	52.9	66.3	88.2	31.8	50.7	88.0	31.4	50.3	80.7	29.5	46.6
0960	Flooring	123.4	60.0	104.5	108.1	35.3	86.4	91.4	65.5	83.7	142.4	42.4	112.6	106.0	60.4	92.4	132.8	37.7	104.5
0970, 0990	Wall Finishes & Painting/Coating	96.1	31.2	57.2	99.1	50.6	70.0	102.3	55.6	74.3	100.7	34.4	61.0	96.1	40.3	62.6	96.1	40.3	62.6
09	**FINISHES**	104.4	55.6	77.1	94.2	37.2	62.2	91.3	56.6	71.8	103.5	35.5	65.4	101.0	38.9	66.2	107.3	33.3	65.8
COVERS	**DIVS. 10 - 14, 25, 28, 41, 43, 44, 46**	100.0	41.3	88.0	100.0	75.2	94.9	100.0	81.1	96.1	100.0	36.0	86.9	100.0	75.3	95.0	100.0	75.0	94.9
21, 22, 23	**FIRE SUPPRESSION, PLUMBING & HVAC**	93.9	56.7	78.7	94.3	42.9	73.2	99.9	66.2	86.1	94.3	34.3	69.7	93.9	33.8	69.3	93.9	58.8	79.5
26, 27, 3370	**ELECTRICAL, COMMUNICATIONS & UTIL.**	87.9	44.2	64.9	96.5	46.6	70.2	94.8	61.1	77.0	93.4	66.1	79.0	91.7	62.2	76.2	90.5	52.9	70.7
MF2010	**WEIGHTED AVERAGE**	95.0	55.9	77.8	97.5	49.1	76.2	97.0	65.6	83.2	95.0	48.8	74.6	97.8	50.5	77.0	98.6	54.1	79.0

	DIVISION	TEXAS															UTAH		
		VICTORIA			WACO			WAXAHACKIE			WHARTON			WICHITA FALLS			LOGAN		
		779			766 - 767			751			774			763			843		
		MAT.	INST.	TOTAL	MAT.	INST.	TOTAL	MAT.	INST.	TOTAL	MAT.	INST.	TOTAL	MAT.	INST.	TOTAL	MAT.	INST.	TOTAL
015433	**CONTRACTOR EQUIPMENT**		97.8	97.8		89.4	89.4		96.6	96.6		99.0	99.0		89.4	89.4		96.8	96.8
0241, 31 - 34	**SITE & INFRASTRUCTURE, DEMOLITION**	110.5	81.7	90.3	93.2	86.9	88.8	96.8	84.2	88.0	116.1	84.6	94.0	93.8	86.6	88.7	92.5	95.9	94.9
0310	Concrete Forming & Accessories	91.5	35.8	43.3	96.8	39.9	47.6	88.1	58.2	62.2	86.4	37.3	43.9	96.8	37.8	45.7	105.1	59.2	65.4
0320	Concrete Reinforcing	85.5	30.8	58.2	90.2	48.2	69.3	96.7	49.7	73.3	88.9	52.4	70.7	90.2	50.4	70.4	104.5	74.9	89.8
0330	Cast-in-Place Concrete	100.0	41.5	76.0	87.7	52.6	73.3	87.0	42.0	68.5	102.8	51.4	81.6	93.7	45.2	73.8	88.1	74.0	82.3
03	**CONCRETE**	102.3	39.0	71.0	92.0	47.1	69.8	90.8	52.5	71.9	106.5	47.0	77.1	94.8	44.0	69.7	107.8	67.8	88.1
04	**MASONRY**	114.1	34.9	65.8	100.1	55.0	72.6	158.5	49.6	92.0	98.6	45.5	66.2	100.6	56.7	73.8	106.0	61.1	78.6
05	**METALS**	96.6	70.4	88.4	99.5	66.1	89.0	99.3	76.3	92.0	98.1	77.3	91.6	99.4	68.2	89.6	100.6	76.3	93.0
06	**WOOD, PLASTICS & COMPOSITES**	102.0	36.6	64.7	108.2	35.8	66.9	89.8	65.5	75.9	91.9	33.2	58.4	108.2	35.7	66.8	81.6	57.4	67.7
07	**THERMAL & MOISTURE PROTECTION**	97.9	40.2	74.7	98.9	47.9	78.4	94.1	46.8	75.1	94.4	50.3	76.6	98.9	51.1	79.7	97.4	62.7	83.4
08	**OPENINGS**	105.6	35.9	88.6	84.5	36.9	72.9	98.8	55.5	88.2	105.8	37.7	89.2	84.5	40.7	73.8	92.6	56.7	83.8
0920	Plaster & Gypsum Board	85.5	35.1	50.2	82.6	34.3	48.8	87.1	64.9	71.5	83.6	31.6	47.2	82.6	34.2	48.7	76.6	56.0	62.2
0950, 0980	Ceilings & Acoustic Treatment	91.9	35.1	54.0	90.0	34.3	52.9	92.6	64.9	74.1	91.0	31.6	51.4	90.0	34.2	52.8	107.7	56.0	73.2
0960	Flooring	92.5	38.8	76.5	141.3	45.9	112.9	94.1	38.8	77.6	90.8	46.3	77.5	142.2	76.5	122.6	100.6	49.5	85.3
0970, 0990	Wall Finishes & Painting/Coating	99.5	38.6	63.0	100.7	34.4	61.0	105.5	38.6	65.4	99.3	32.0	59.0	103.0	50.6	71.6	98.4	63.0	77.2
09	**FINISHES**	85.9	36.7	58.4	104.1	39.2	67.8	94.4	54.0	71.7	87.4	37.4	59.4	104.6	45.3	71.4	99.3	57.2	75.7
COVERS	**DIVS. 10 - 14, 25, 28, 41, 43, 44, 46**	100.0	37.0	87.1	100.0	78.6	95.6	100.0	40.8	87.9	100.0	39.2	87.6	100.0	66.2	93.1	100.0	54.8	90.8
21, 22, 23	**FIRE SUPPRESSION, PLUMBING & HVAC**	94.1	32.2	68.8	100.3	50.4	79.9	94.0	41.7	72.6	94.1	60.0	80.1	100.3	45.8	78.0	99.9	70.7	88.0
26, 27, 3370	**ELECTRICAL, COMMUNICATIONS & UTIL.**	99.0	40.5	68.2	96.7	66.3	80.7	90.1	66.3	77.6	97.8	44.6	69.8	98.4	56.8	76.5	94.3	72.3	82.7
MF2010	**WEIGHTED AVERAGE**	98.7	43.4	74.3	97.3	55.8	79.0	98.5	56.7	80.1	98.5	53.4	78.6	97.9	54.1	78.6	99.4	69.1	86.1

DIVISION		UTAH												VERMONT					
		OGDEN			PRICE			PROVO			SALT LAKE CITY			BELLOWS FALLS			BENNINGTON		
		842,844			845			846 - 847			840 - 841			051			052		
		MAT.	INST.	TOTAL	MAT.	INST.	TOTAL	MAT.	INST.	TOTAL	MAT.	INST.	TOTAL	MAT.	INST.	TOTAL	MAT.	INST.	TOTAL
015433	CONTRACTOR EQUIPMENT		96.8	96.8		95.8	95.8		95.8	95.8		96.7	96.7		100.5	100.5		100.5	100.5
0241, 31 - 34	SITE & INFRASTRUCTURE, DEMOLITION	81.1	95.9	91.4	90.1	94.2	93.0	88.9	94.2	92.6	80.6	95.8	91.3	86.1	98.6	94.9	85.5	98.1	94.4
0310	Concrete Forming & Accessories	105.1	59.2	65.4	107.2	50.4	58.0	106.4	59.1	65.5	107.6	59.1	65.6	95.9	84.6	86.1	93.6	84.6	85.8
0320	Concrete Reinforcing	104.1	74.9	89.6	112.1	74.8	93.5	113.0	74.9	94.0	106.5	74.9	90.7	88.6	85.0	86.8	88.6	86.8	87.7
0330	Cast-in-Place Concrete	89.5	74.0	83.1	88.2	60.9	77.0	88.2	74.0	82.4	97.8	74.0	88.0	95.8	70.6	85.5	95.8	67.3	84.1
03	CONCRETE	97.9	67.8	83.1	109.3	59.4	84.6	107.8	67.7	88.0	116.7	67.7	92.5	97.4	79.7	88.7	97.3	78.8	88.1
04	MASONRY	99.9	61.1	76.2	111.6	61.1	80.8	111.7	61.1	80.8	113.3	61.1	81.5	92.8	59.8	72.6	100.5	94.4	96.8
05	METALS	101.1	76.3	93.3	97.9	75.8	91.0	98.9	76.2	91.8	105.5	76.2	96.3	92.4	78.7	88.1	92.4	80.0	88.5
06	WOOD, PLASTICS & COMPOSITES	81.6	57.4	67.7	84.3	46.2	62.6	82.9	57.4	68.3	83.3	57.4	68.5	96.4	86.3	90.6	93.6	86.3	89.4
07	THERMAL & MOISTURE PROTECTION	96.3	62.7	82.8	99.1	60.6	83.6	99.1	62.7	84.5	103.3	62.7	86.9	96.6	68.6	85.3	96.5	77.7	88.9
08	OPENINGS	92.6	56.7	83.8	96.7	49.5	85.2	96.8	56.7	87.0	94.5	56.7	85.3	100.3	72.4	93.5	100.3	71.4	93.2
0920	Plaster & Gypsum Board	76.6	56.0	62.2	78.7	44.5	54.8	77.1	56.0	62.3	79.4	56.0	63.0	97.2	85.2	88.8	95.1	85.2	88.2
0950, 0980	Ceilings & Acoustic Treatment	107.7	56.0	73.2	107.7	44.5	65.6	107.7	56.0	73.2	101.5	56.0	71.2	88.3	85.2	86.3	88.3	85.2	86.3
0960	Flooring	98.3	49.5	83.7	101.6	36.8	82.3	101.3	49.5	85.8	102.2	49.5	86.5	98.3	31.6	78.4	97.4	31.6	77.8
0970, 0990	Wall Finishes & Painting/Coating	98.4	63.0	77.2	98.4	39.2	62.9	98.4	50.1	69.5	101.4	50.1	70.7	96.1	50.0	68.5	96.1	50.0	68.5
09	FINISHES	97.2	57.2	74.8	100.4	45.4	69.6	99.9	55.7	75.2	98.1	55.7	74.4	95.4	72.6	82.7	94.9	72.6	82.4
COVERS	DIVS. 10 - 14, 25, 28, 41, 43, 44, 46	100.0	54.8	90.8	100.0	47.6	89.3	100.0	54.8	90.8	100.0	54.8	90.8	100.0	49.4	89.7	100.0	49.3	89.6
21, 22, 23	FIRE SUPPRESSION, PLUMBING & HVAC	99.9	70.7	88.0	96.9	70.3	86.0	99.9	70.7	88.0	100.1	70.7	88.1	94.0	94.3	94.1	94.0	88.2	91.6
26, 27, 3370	ELECTRICAL, COMMUNICATIONS & UTIL.	94.7	72.3	82.9	100.2	72.3	85.5	95.2	62.3	77.9	97.7	62.3	79.1	101.7	93.0	97.1	101.7	63.6	81.6
MF2010	WEIGHTED AVERAGE	97.6	69.1	85.1	99.8	65.5	84.7	100.0	67.4	85.6	101.9	67.5	86.7	96.0	81.8	89.7	96.3	80.1	89.1

DIVISION		VERMONT																	
		BRATTLEBORO			BURLINGTON			GUILDHALL			MONTPELIER			RUTLAND			ST. JOHNSBURY		
		053			054			059			056			057			058		
		MAT.	INST.	TOTAL	MAT.	INST.	TOTAL	MAT.	INST.	TOTAL	MAT.	INST.	TOTAL	MAT.	INST.	TOTAL	MAT.	INST.	TOTAL
015433	CONTRACTOR EQUIPMENT		100.5	100.5		100.5	100.5		100.5	100.5		100.5	100.5		100.5	100.5		100.5	100.5
0241, 31 - 34	SITE & INFRASTRUCTURE, DEMOLITION	87.0	98.6	95.2	88.3	98.6	95.5	85.1	97.2	93.6	88.9	98.4	95.5	89.3	98.4	95.7	85.1	97.2	93.6
0310	Concrete Forming & Accessories	96.1	85.1	86.6	95.7	85.0	86.4	94.5	77.3	79.6	101.0	85.0	87.1	96.3	85.0	86.5	93.6	77.3	79.5
0320	Concrete Reinforcing	87.7	86.9	87.3	110.7	85.8	98.3	89.3	85.0	87.2	88.1	85.8	87.0	110.7	85.8	98.3	87.7	85.0	86.3
0330	Cast-in-Place Concrete	98.8	68.0	86.2	104.5	108.3	106.1	92.9	57.7	78.4	110.6	108.3	109.7	93.9	108.3	99.8	92.9	57.7	78.4
03	CONCRETE	99.6	79.3	89.6	106.0	92.9	99.6	95.0	72.0	83.6	105.6	92.9	99.4	101.0	92.9	97.0	94.6	72.0	83.5
04	MASONRY	99.2	94.6	96.4	103.3	71.1	83.7	99.4	43.6	65.3	88.5	71.1	77.9	84.4	71.1	76.3	122.8	43.6	74.5
05	METALS	92.4	81.7	89.0	99.3	80.9	93.5	92.4	78.0	87.9	92.4	80.9	88.8	97.7	80.9	92.4	92.4	78.0	87.9
06	WOOD, PLASTICS & COMPOSITES	96.7	86.3	90.7	91.7	86.3	88.6	93.2	86.3	89.2	95.7	86.3	90.3	96.8	86.3	90.8	89.3	86.3	87.5
07	THERMAL & MOISTURE PROTECTION	96.6	77.7	89.0	103.6	75.7	92.4	96.4	56.9	80.5	103.3	76.2	92.4	96.7	76.2	88.4	96.4	56.9	80.5
08	OPENINGS	100.3	72.4	93.5	103.4	72.5	95.9	100.3	72.4	93.5	100.3	72.5	93.5	103.4	72.5	95.9	100.3	72.4	93.5
0920	Plaster & Gypsum Board	97.2	85.2	88.8	97.2	85.2	88.8	103.0	85.2	90.6	104.4	85.2	91.0	98.0	85.2	89.0	105.5	85.2	91.3
0950, 0980	Ceilings & Acoustic Treatment	88.3	85.2	86.3	94.9	85.2	88.5	88.3	85.2	86.3	92.0	85.2	87.5	93.1	85.2	87.9	88.3	85.2	86.3
0960	Flooring	98.4	31.6	78.5	98.3	63.4	87.9	101.2	31.6	80.5	104.2	63.4	92.0	98.3	63.4	87.9	104.4	31.6	82.7
0970, 0990	Wall Finishes & Painting/Coating	96.1	50.0	68.5	96.1	50.0	68.5	96.1	50.0	68.5	96.1	50.0	68.5	96.1	50.0	68.5	96.1	50.0	68.5
09	FINISHES	95.6	72.6	82.7	97.0	78.1	86.4	96.9	68.2	80.8	99.0	78.1	87.3	96.7	78.1	86.3	98.2	68.2	81.4
COVERS	DIVS. 10 - 14, 25, 28, 41, 43, 44, 46	100.0	49.4	89.7	100.0	78.3	95.6	100.0	43.3	88.4	100.0	78.3	95.6	100.0	78.3	95.6	100.0	43.3	88.4
21, 22, 23	FIRE SUPPRESSION, PLUMBING & HVAC	94.0	96.4	95.0	99.8	71.5	88.2	94.0	62.3	81.0	93.8	71.5	84.6	100.0	71.5	88.3	94.0	62.3	81.0
26, 27, 3370	ELECTRICAL, COMMUNICATIONS & UTIL.	101.7	93.0	97.1	100.2	63.7	81.0	101.7	63.7	81.7	102.0	63.7	81.8	101.7	63.7	81.7	101.7	63.7	81.7
MF2010	WEIGHTED AVERAGE	96.6	86.2	92.0	100.5	77.8	90.5	96.1	67.1	83.3	97.2	77.8	88.7	98.8	77.8	89.6	97.3	67.1	84.0

DIVISION		VERMONT			VIRGINIA														
		WHITE RIVER JCT.			ALEXANDRIA			ARLINGTON			BRISTOL			CHARLOTTESVILLE			CULPEPER		
		050			223			222			242			229			227		
		MAT.	INST.	TOTAL	MAT.	INST.	TOTAL	MAT.	INST.	TOTAL	MAT.	INST.	TOTAL	MAT.	INST.	TOTAL	MAT.	INST.	TOTAL
015433	CONTRACTOR EQUIPMENT		100.5	100.5		101.8	101.8		100.5	100.5		100.5	100.5		104.7	104.7		100.5	100.5
0241, 31 - 34	SITE & INFRASTRUCTURE, DEMOLITION	89.5	97.3	95.0	117.3	89.2	97.6	128.2	87.1	99.4	111.8	85.7	93.5	116.8	87.6	96.3	115.2	87.0	95.4
0310	Concrete Forming & Accessories	91.1	78.0	79.7	90.7	74.6	76.7	89.9	74.2	76.3	86.2	44.3	50.0	84.7	50.6	55.2	81.8	73.5	74.6
0320	Concrete Reinforcing	88.6	85.0	86.8	87.9	84.8	86.3	99.2	84.7	92.0	99.2	68.4	83.8	98.5	74.1	86.4	99.2	84.6	92.0
0330	Cast-in-Place Concrete	98.8	58.7	82.3	117.6	83.3	103.5	114.5	82.8	101.5	114.0	49.9	87.6	118.5	56.9	93.2	117.2	69.9	97.8
03	CONCRETE	101.4	72.6	87.2	108.2	80.5	94.5	112.7	80.2	96.6	109.0	52.6	81.2	110.0	58.9	84.7	107.1	75.4	91.4
04	MASONRY	110.9	44.7	70.5	89.3	72.1	78.8	102.7	71.4	83.6	91.7	53.6	68.4	116.1	55.4	79.1	104.6	71.3	84.3
05	METALS	92.4	78.1	87.9	97.7	96.5	97.4	96.4	96.4	96.4	95.3	86.7	92.6	95.5	91.1	94.2	95.6	94.6	95.3
06	WOOD, PLASTICS & COMPOSITES	90.7	86.3	88.2	97.3	73.1	83.5	94.0	73.1	82.1	86.7	41.7	61.0	85.2	47.5	63.7	83.9	73.1	77.8
07	THERMAL & MOISTURE PROTECTION	96.7	57.5	80.9	99.2	81.7	92.2	101.1	80.4	92.8	100.6	62.3	85.2	100.3	70.1	88.1	100.5	77.8	91.4
08	OPENINGS	100.3	74.6	94.0	94.2	75.8	89.7	92.3	75.7	88.2	95.2	47.7	83.7	93.5	53.7	83.8	93.8	75.7	89.4
0920	Plaster & Gypsum Board	93.5	85.2	87.7	109.7	72.1	83.4	106.0	72.1	82.3	101.6	39.7	58.3	101.6	45.1	62.1	101.9	72.1	81.1
0950, 0980	Ceilings & Acoustic Treatment	88.3	85.2	86.3	97.0	72.1	80.4	94.3	72.1	79.5	93.4	39.7	57.6	93.4	45.1	61.2	94.3	72.1	79.5
0960	Flooring	96.3	31.6	77.0	103.8	86.0	98.5	102.3	83.5	96.7	98.9	68.9	90.0	97.6	68.9	89.1	97.6	83.5	93.4
0970, 0990	Wall Finishes & Painting/Coating	96.1	50.7	68.9	124.6	80.4	98.1	124.6	80.4	98.1	109.6	56.7	77.9	109.6	79.5	91.5	124.6	79.5	97.6
09	FINISHES	94.8	68.7	80.2	103.3	76.6	88.3	103.2	75.8	87.9	99.6	49.2	71.4	99.3	55.8	74.9	100.1	75.6	86.4
COVERS	DIVS. 10 - 14, 25, 28, 41, 43, 44, 46	100.0	43.9	88.5	100.0	89.3	97.8	100.0	84.4	96.8	100.0	70.5	94.0	100.0	76.3	95.2	100.0	84.4	96.8
21, 22, 23	FIRE SUPPRESSION, PLUMBING & HVAC	94.0	63.2	81.4	100.3	89.3	95.8	100.3	88.3	95.4	94.3	53.2	77.5	94.3	70.9	84.7	94.3	75.5	86.6
26, 27, 3370	ELECTRICAL, COMMUNICATIONS & UTIL.	101.7	63.7	81.7	97.5	100.6	99.1	95.0	100.6	97.9	97.2	36.9	65.4	97.1	72.1	83.9	99.8	100.5	100.2
MF2010	WEIGHTED AVERAGE	97.3	67.7	84.3	99.9	86.0	93.8	100.8	85.3	93.9	97.9	56.7	79.8	99.2	68.4	85.6	98.6	81.6	91.1

DIVISION		VIRGINIA																	
		FAIRFAX			FARMVILLE			FREDERICKSBURG			GRUNDY			HARRISONBURG			LYNCHBURG		
		220 - 221			239			224 - 225			246			228			245		
		MAT.	INST.	TOTAL	MAT.	INST.	TOTAL	MAT.	INST.	TOTAL	MAT.	INST.	TOTAL	MAT.	INST.	TOTAL	MAT.	INST.	TOTAL
015433	CONTRACTOR EQUIPMENT		100.5	100.5		104.7	104.7		100.5	100.5		100.5	100.5		100.5	100.5		100.5	100.5
0241, 31 - 34	SITE & INFRASTRUCTURE, DEMOLITION	127.0	87.2	99.1	112.7	87.2	94.8	114.6	87.2	95.4	109.5	84.4	91.9	123.7	85.6	97.0	110.4	85.9	93.2
0310	Concrete Forming & Accessories	84.6	73.9	75.3	96.1	45.6	52.4	84.6	72.1	73.8	89.2	38.0	44.9	80.9	44.2	49.1	86.2	63.0	66.1
0320	Concrete Reinforcing	99.2	84.7	92.0	99.2	52.4	75.9	100.0	84.7	92.4	97.8	52.9	75.4	99.2	68.0	83.7	98.5	68.5	83.6
0330	Cast-in-Place Concrete	114.5	82.7	101.4	118.5	55.0	92.4	116.2	65.8	95.5	114.0	48.8	87.2	114.5	59.5	91.9	114.0	59.4	91.5
03	CONCRETE	112.3	80.0	96.4	109.2	51.8	80.8	106.7	73.4	90.3	107.8	46.1	77.3	110.2	55.6	83.2	107.8	64.2	86.2
04	MASONRY	102.5	71.4	83.5	99.7	46.7	67.4	103.6	71.4	83.9	94.4	49.3	66.9	100.9	52.3	71.2	108.6	55.4	76.1
05	METALS	95.7	96.1	95.8	95.5	78.9	90.3	95.6	95.7	95.6	95.3	71.6	87.8	95.5	84.8	92.2	95.4	87.2	92.9
06	WOOD, PLASTICS & COMPOSITES	86.7	73.1	78.9	99.3	45.9	68.8	86.7	71.1	77.8	89.5	36.5	59.3	83.0	40.6	58.8	86.7	65.3	74.5
07	THERMAL & MOISTURE PROTECTION	101.0	70.9	88.9	100.3	51.2	80.6	100.5	78.3	91.6	100.6	47.0	79.0	100.8	64.6	86.2	100.4	66.2	86.7
08	OPENINGS	92.3	75.7	88.2	93.8	44.9	81.9	93.5	74.6	88.9	95.2	35.9	80.8	93.8	51.6	83.5	93.7	60.6	85.7
0920	Plaster & Gypsum Board	101.9	72.1	81.1	108.7	43.4	63.0	101.9	70.0	79.6	101.6	34.4	54.6	101.6	38.7	57.6	101.6	64.1	75.4
0950, 0980	Ceilings & Acoustic Treatment	94.3	72.1	79.5	93.4	43.4	60.1	94.3	70.0	78.1	93.4	34.4	54.1	93.4	38.7	56.9	93.4	64.1	73.9
0960	Flooring	99.2	83.5	94.5	103.8	62.5	91.5	99.2	83.5	94.5	100.2	34.9	80.8	97.4	76.7	91.2	98.9	68.9	90.0
0970, 0990	Wall Finishes & Painting/Coating	124.6	80.4	98.1	109.6	40.5	68.2	124.6	79.5	97.6	109.6	36.7	65.9	124.6	56.7	83.9	109.6	56.7	77.9
09	FINISHES	101.8	75.8	87.2	101.6	48.0	71.6	100.5	74.4	85.9	99.8	36.6	64.4	100.6	49.9	72.2	99.4	62.8	78.9
COVERS	DIVS. 10 - 14, 25, 28, 41, 43, 44, 46	100.0	84.4	96.8	100.0	47.4	89.2	100.0	80.4	96.0	100.0	45.4	88.8	100.0	70.3	93.9	100.0	73.8	94.7
21, 22, 23	FIRE SUPPRESSION, PLUMBING & HVAC	94.3	88.3	91.8	94.3	45.7	74.4	94.3	88.0	91.7	94.3	65.5	82.5	94.3	69.3	84.0	94.3	70.6	84.6
26, 27, 3370	ELECTRICAL, COMMUNICATIONS & UTIL.	98.4	100.6	99.5	94.2	50.3	71.1	95.2	100.5	98.0	97.2	46.5	70.5	97.4	100.5	99.0	98.3	54.9	75.5
MF2010	WEIGHTED AVERAGE	99.3	84.9	93.0	98.2	54.4	78.9	98.1	83.8	91.8	97.9	54.5	78.8	98.8	69.3	85.8	98.5	67.4	84.8

DIVISION		VIRGINIA																	
		NEWPORT NEWS			NORFOLK			PETERSBURG			PORTSMOUTH			PULASKI			RICHMOND		
		236			233 - 235			238			237			243			230 - 232		
		MAT.	INST.	TOTAL	MAT.	INST.	TOTAL	MAT.	INST.	TOTAL	MAT.	INST.	TOTAL	MAT.	INST.	TOTAL	MAT.	INST.	TOTAL
015433	CONTRACTOR EQUIPMENT		104.7	104.7		105.3	105.3		104.7	104.7		104.6	104.6		100.5	100.5		104.6	104.6
0241, 31 - 34	SITE & INFRASTRUCTURE, DEMOLITION	111.3	88.3	95.2	109.2	89.3	95.2	115.3	88.5	96.5	109.8	88.0	94.5	108.8	84.7	91.9	108.5	88.5	94.4
0310	Concrete Forming & Accessories	95.4	66.7	70.6	99.8	66.9	71.3	89.2	59.5	63.5	85.4	55.2	59.3	89.2	40.0	46.6	98.3	59.5	64.8
0320	Concrete Reinforcing	98.9	73.9	86.4	98.9	73.9	86.5	98.5	74.3	86.5	98.5	73.9	86.3	97.8	67.8	82.9	98.9	74.3	86.7
0330	Cast-in-Place Concrete	115.2	66.6	95.2	125.8	66.7	101.5	122.3	54.9	94.6	114.0	66.6	94.5	114.0	49.1	87.3	111.0	54.9	87.9
03	CONCRETE	106.2	69.3	88.0	111.6	69.4	90.8	111.8	62.2	87.3	104.9	64.1	84.8	107.8	50.1	79.3	104.4	62.2	83.6
04	MASONRY	94.8	56.2	71.2	100.6	56.2	73.5	107.3	54.8	75.2	100.1	56.2	73.3	88.5	50.1	65.1	92.7	54.8	69.6
05	METALS	97.8	90.9	95.6	96.8	91.0	95.0	95.6	91.7	94.4	96.7	90.2	94.7	95.3	83.1	91.5	99.9	91.7	97.3
06	WOOD, PLASTICS & COMPOSITES	98.2	69.1	81.6	100.3	69.1	82.5	89.9	61.1	73.5	86.5	53.6	67.7	89.5	38.3	60.3	97.3	61.1	76.6
07	THERMAL & MOISTURE PROTECTION	100.3	69.7	88.0	101.7	69.7	88.8	100.3	70.7	88.4	100.3	68.1	87.3	100.6	52.0	81.1	103.4	70.7	90.3
08	OPENINGS	94.2	64.5	87.0	94.2	64.5	87.0	93.5	61.1	85.6	94.3	56.1	85.0	95.3	44.0	82.8	94.2	61.1	86.1
0920	Plaster & Gypsum Board	109.7	67.3	80.0	104.5	67.3	78.5	101.9	59.0	71.9	102.2	51.4	66.6	101.6	36.3	55.9	111.0	59.0	74.6
0950, 0980	Ceilings & Acoustic Treatment	97.0	67.3	77.2	99.7	67.3	78.1	94.3	59.0	70.8	97.0	51.4	66.6	93.4	36.3	55.3	103.4	59.0	73.8
0960	Flooring	103.8	68.9	93.4	102.4	68.9	92.4	99.8	73.8	92.0	96.9	68.9	88.5	100.2	68.9	90.9	102.4	73.8	93.9
0970, 0990	Wall Finishes & Painting/Coating	109.6	43.9	70.2	109.6	79.5	91.5	109.6	79.5	91.5	109.6	79.5	91.5	109.6	56.7	77.9	109.6	79.5	91.5
09	FINISHES	102.3	64.9	81.3	101.4	68.8	83.1	100.0	63.8	79.7	99.4	59.2	76.8	99.8	45.6	69.4	102.9	63.8	81.0
COVERS	DIVS. 10 - 14, 25, 28, 41, 43, 44, 46	100.0	76.5	95.2	100.0	76.5	95.2	100.0	76.6	95.2	100.0	74.8	94.9	100.0	45.6	88.9	100.0	76.6	95.2
21, 22, 23	FIRE SUPPRESSION, PLUMBING & HVAC	100.3	66.0	86.3	100.0	65.5	85.9	94.3	69.0	83.9	100.3	65.4	86.0	94.3	56.3	78.7	100.0	69.0	87.3
26, 27, 3370	ELECTRICAL, COMMUNICATIONS & UTIL.	97.2	61.4	78.4	101.9	58.7	79.2	97.4	72.1	84.1	95.4	58.7	76.1	97.2	56.7	75.9	100.2	72.1	85.4
MF2010	WEIGHTED AVERAGE	99.8	69.3	86.3	100.8	69.4	87.0	99.0	70.1	86.3	99.2	66.7	84.9	97.6	57.3	79.8	100.1	70.1	86.9

DIVISION		VIRGINIA									WASHINGTON								
		ROANOKE			STAUNTON			WINCHESTER			CLARKSTON			EVERETT			OLYMPIA		
		240 - 241			244			226			994			982			985		
		MAT.	INST.	TOTAL	MAT.	INST.	TOTAL	MAT.	INST.	TOTAL	MAT.	INST.	TOTAL	MAT.	INST.	TOTAL	MAT.	INST.	TOTAL
015433	CONTRACTOR EQUIPMENT		100.5	100.5		104.7	104.7		100.5	100.5		90.6	90.6		102.5	102.5		102.5	102.5
0241, 31 - 34	SITE & INFRASTRUCTURE, DEMOLITION	108.9	85.9	92.7	113.2	87.3	95.0	122.4	87.0	97.6	95.3	89.9	91.5	90.8	110.6	104.7	92.1	110.6	105.1
0310	Concrete Forming & Accessories	95.2	63.1	67.4	88.8	52.2	57.1	83.1	72.2	73.6	122.5	69.9	77.0	114.1	98.1	100.3	105.9	98.2	99.3
0320	Concrete Reinforcing	98.9	68.5	83.8	98.5	51.3	75.0	98.6	84.1	91.4	106.0	87.6	96.8	116.1	101.8	109.0	116.7	101.8	109.3
0330	Cast-in-Place Concrete	129.4	59.5	100.7	118.5	52.9	91.5	114.5	62.8	93.2	101.5	83.2	94.0	98.8	104.4	101.1	89.9	104.5	95.9
03	CONCRETE	113.0	64.2	88.9	109.3	54.0	82.0	109.6	72.3	91.2	99.6	77.9	88.9	99.4	100.5	100.0	96.0	100.6	98.3
04	MASONRY	95.7	55.4	71.1	104.5	53.5	73.4	98.5	61.4	75.8	101.0	79.9	88.1	118.2	99.2	106.6	109.1	99.2	103.1
05	METALS	97.5	87.3	94.3	95.5	82.5	91.4	95.6	94.0	95.1	87.7	81.7	85.8	103.2	92.3	99.8	98.0	92.4	96.3
06	WOOD, PLASTICS & COMPOSITES	98.2	65.3	79.4	89.5	52.0	68.1	85.2	71.1	77.2	101.0	66.6	81.4	108.0	97.2	101.9	94.8	97.2	96.2
07	THERMAL & MOISTURE PROTECTION	100.2	68.3	87.4	100.2	54.0	81.7	100.9	75.7	90.8	152.0	77.5	122.0	102.5	100.8	101.8	102.4	98.4	100.8
08	OPENINGS	94.2	60.6	86.0	93.8	50.2	83.2	95.3	75.4	90.5	104.4	70.9	96.2	101.0	98.7	100.4	100.9	98.7	100.4
0920	Plaster & Gypsum Board	109.7	64.1	77.8	101.6	49.7	65.3	101.9	70.0	79.6	130.4	65.4	85.0	109.4	97.1	100.8	100.5	97.1	98.1
0950, 0980	Ceilings & Acoustic Treatment	97.0	64.1	75.1	93.4	49.7	64.3	94.3	70.0	78.1	99.4	65.4	76.8	104.9	97.1	99.7	103.8	97.1	99.3
0960	Flooring	103.8	68.9	93.4	99.7	40.9	82.2	98.6	83.5	94.1	103.9	48.8	87.5	117.1	94.9	110.5	110.9	95.1	106.2
0970, 0990	Wall Finishes & Painting/Coating	109.6	56.7	77.9	109.6	36.1	65.5	124.6	91.1	104.5	112.5	61.1	81.7	106.9	79.3	90.4	106.9	83.4	92.8
09	FINISHES	102.1	63.7	80.6	99.6	48.5	71.0	101.0	75.7	86.8	116.0	63.8	86.7	109.0	95.2	101.3	105.9	95.7	100.2
COVERS	DIVS. 10 - 14, 25, 28, 41, 43, 44, 46	100.0	73.8	94.7	100.0	73.7	94.6	100.0	84.2	96.8	100.0	78.3	95.6	100.0	93.0	98.6	100.0	100.1	100.0
21, 22, 23	FIRE SUPPRESSION, PLUMBING & HVAC	100.3	62.7	84.9	94.3	62.0	81.1	94.3	88.3	91.8	94.3	86.9	91.3	100.1	102.6	101.1	99.9	102.7	101.0
26, 27, 3370	ELECTRICAL, COMMUNICATIONS & UTIL.	97.2	56.7	75.9	96.0	75.9	85.4	95.7	100.5	98.2	100.3	93.9	96.9	103.6	99.0	101.2	101.1	101.4	101.3
MF2010	WEIGHTED AVERAGE	100.4	66.2	85.3	98.4	63.9	83.2	98.6	82.7	91.6	100.0	81.4	91.8	102.4	99.7	101.2	100.1	100.3	100.2

DIVISION		WASHINGTON																	
		RICHLAND			SEATTLE			SPOKANE			TACOMA			VANCOUVER			WENATCHEE		
		993			980 - 981,987			990 - 992			983 - 984			986			988		
		MAT.	INST.	TOTAL	MAT.	INST.	TOTAL	MAT.	INST.	TOTAL	MAT.	INST.	TOTAL	MAT.	INST.	TOTAL	MAT.	INST.	TOTAL
015433	CONTRACTOR EQUIPMENT		90.6	90.6		102.4	102.4		90.6	90.6		102.5	102.5		97.3	97.3		102.5	102.5
0241, 31 - 34	SITE & INFRASTRUCTURE, DEMOLITION	97.7	90.7	92.8	94.6	108.8	104.6	96.9	90.7	92.5	93.8	110.6	105.6	103.9	98.2	99.9	103.3	109.3	107.5
0310	Concrete Forming & Accessories	122.6	79.3	85.1	105.2	98.7	99.6	127.0	78.9	85.4	105.2	98.2	99.2	106.2	92.8	94.6	106.9	77.2	81.2
0320	Concrete Reinforcing	101.7	87.7	94.7	114.8	101.9	108.3	102.4	87.6	95.1	114.8	101.8	108.3	115.7	101.6	108.7	115.6	90.6	103.2
0330	Cast-in-Place Concrete	101.8	85.3	95.0	103.6	104.7	104.1	105.7	85.1	97.3	101.5	104.5	102.8	113.3	100.2	107.9	103.7	78.3	93.2
03	CONCRETE	99.1	83.0	91.1	102.2	101.0	101.6	101.4	82.7	92.2	101.2	100.6	100.9	111.2	96.8	104.1	109.0	80.2	94.8
04	MASONRY	103.1	82.4	90.5	113.2	99.2	104.7	103.7	82.4	90.7	113.0	99.2	104.6	113.7	98.4	104.3	115.7	90.3	100.2
05	METALS	88.1	84.1	86.8	104.9	94.2	101.5	90.4	83.6	88.3	104.9	92.4	101.0	102.3	93.0	99.4	102.4	86.8	97.5
06	WOOD, PLASTICS & COMPOSITES	101.1	77.3	87.5	99.3	97.2	98.1	109.6	77.3	91.1	97.9	97.2	97.5	89.7	92.3	91.2	99.1	75.3	85.5
07	THERMAL & MOISTURE PROTECTION	153.2	79.9	123.8	102.3	100.8	101.7	149.5	79.6	121.4	102.2	98.7	100.8	102.5	93.6	98.9	102.8	82.7	94.7
08	OPENINGS	106.8	72.4	98.4	103.5	98.7	102.4	107.4	72.7	98.9	101.7	98.7	101.0	98.4	94.2	97.4	101.2	71.8	94.0
0920	Plaster & Gypsum Board	130.4	76.5	92.7	103.1	97.1	98.9	125.3	76.5	91.1	105.8	97.1	99.7	103.7	92.3	95.7	107.6	74.5	84.5
0950, 0980	Ceilings & Acoustic Treatment	105.5	76.5	86.1	107.4	97.1	100.5	101.4	76.5	84.8	107.7	97.1	100.6	102.3	92.3	95.6	100.1	74.5	83.1
0960	Flooring	104.1	83.0	97.8	110.4	102.8	108.1	103.8	88.4	99.2	110.9	95.1	106.2	116.9	105.9	113.6	113.5	64.9	99.1
0970, 0990	Wall Finishes & Painting/Coating	112.5	67.7	85.6	106.9	83.4	92.8	112.3	61.1	81.6	106.9	83.4	92.8	109.3	68.9	85.1	106.9	67.7	83.4
09	FINISHES	117.6	78.2	95.5	107.1	97.2	101.5	116.0	78.5	95.0	107.5	95.7	100.9	105.2	92.8	98.2	107.9	73.0	88.3
COVERS	DIVS. 10 - 14, 25, 28, 41, 43, 44, 46	100.0	64.8	92.8	100.0	100.1	100.0	100.0	64.7	92.8	100.0	100.1	100.0	100.0	62.8	92.4	100.0	79.2	95.8
21, 22, 23	FIRE SUPPRESSION, PLUMBING & HVAC	100.4	108.1	103.6	100.0	120.0	108.2	100.3	88.1	95.3	100.1	102.7	101.2	100.2	91.4	96.6	94.1	88.9	92.0
26, 27, 3370	ELECTRICAL, COMMUNICATIONS & UTIL.	97.5	93.9	95.6	103.4	107.2	105.4	95.2	79.1	86.7	103.4	101.4	102.4	108.5	99.1	103.5	104.3	99.0	101.5
MF2010	WEIGHTED AVERAGE	101.8	88.7	96.0	102.8	105.1	103.8	102.0	82.4	93.4	102.5	100.3	101.6	103.5	94.1	99.4	102.1	87.5	95.6

DIVISION		WASHINGTON			WEST VIRGINIA														
		YAKIMA			BECKLEY			BLUEFIELD			BUCKHANNON			CHARLESTON			CLARKSBURG		
		989			258 - 259			247 - 248			262			250 - 253			263 - 264		
		MAT.	INST.	TOTAL	MAT.	INST.	TOTAL	MAT.	INST.	TOTAL	MAT.	INST.	TOTAL	MAT.	INST.	TOTAL	MAT.	INST.	TOTAL
015433	CONTRACTOR EQUIPMENT		102.5	102.5		100.5	100.5		100.5	100.5		100.5	100.5		100.5	100.5		100.5	100.5
0241, 31 - 34	SITE & INFRASTRUCTURE, DEMOLITION	96.2	110.1	105.9	103.4	87.8	92.4	103.2	87.7	92.3	109.7	87.5	94.1	101.4	89.2	92.9	110.3	87.5	94.3
0310	Concrete Forming & Accessories	105.6	93.7	95.3	84.8	85.7	85.6	86.1	85.4	85.5	85.5	83.3	83.6	101.7	87.7	89.6	83.2	83.2	83.2
0320	Concrete Reinforcing	115.3	90.2	102.8	97.5	83.2	90.4	97.5	65.6	81.6	98.1	84.6	91.4	98.9	83.4	91.2	98.1	82.8	90.4
0330	Cast-in-Place Concrete	108.5	84.2	98.5	111.5	102.3	107.7	111.5	102.1	107.6	111.1	100.8	106.9	103.1	100.5	102.0	121.8	98.9	112.4
03	CONCRETE	105.9	89.5	97.8	103.5	91.6	97.6	103.6	88.2	96.0	106.5	90.3	98.5	100.9	92.0	96.5	111.0	89.3	100.3
04	MASONRY	106.4	80.7	90.7	95.2	91.5	92.9	90.1	91.5	90.9	100.9	87.4	92.6	91.7	92.7	92.3	103.4	87.4	93.6
05	METALS	103.1	86.3	97.8	95.5	95.0	95.3	95.5	88.5	93.3	95.7	96.4	95.9	97.8	96.2	97.3	95.7	95.7	95.7
06	WOOD, PLASTICS & COMPOSITES	98.4	97.2	97.7	86.9	86.3	86.5	88.6	86.3	87.3	87.9	81.9	84.5	104.5	86.3	94.1	84.5	81.9	83.0
07	THERMAL & MOISTURE PROTECTION	102.3	81.4	93.9	100.3	87.4	95.1	100.4	87.4	95.2	100.6	86.3	94.9	100.0	88.3	95.3	100.6	86.3	94.8
08	OPENINGS	101.1	80.4	96.1	94.8	80.1	91.2	95.6	76.1	90.8	95.6	78.0	91.4	95.4	80.1	91.7	95.6	79.2	91.6
0920	Plaster & Gypsum Board	105.6	97.1	99.6	99.9	85.7	90.0	100.7	85.7	90.2	101.2	81.2	87.2	104.8	85.7	91.5	98.7	81.2	86.4
0950, 0980	Ceilings & Acoustic Treatment	102.2	97.1	98.8	91.5	85.7	87.7	91.5	85.7	87.7	93.4	81.2	85.2	100.7	85.7	90.7	93.4	81.2	85.2
0960	Flooring	111.8	60.1	96.4	96.3	112.9	101.2	96.9	112.9	101.6	96.6	106.4	99.5	102.6	112.9	105.7	95.5	106.4	98.7
0970, 0990	Wall Finishes & Painting/Coating	106.9	61.1	79.5	109.6	36.6	65.8	109.6	36.6	65.8	109.6	88.9	97.2	109.6	89.2	97.3	109.6	88.9	97.2
09	FINISHES	106.7	84.3	94.1	97.5	86.2	91.2	97.8	86.2	91.3	98.8	87.9	92.7	101.4	92.5	96.4	98.0	87.9	92.3
COVERS	DIVS. 10 - 14, 25, 28, 41, 43, 44, 46	100.0	97.4	99.5	100.0	55.6	90.9	100.0	55.5	90.9	100.0	94.4	98.9	100.0	95.8	99.1	100.0	94.4	98.9
21, 22, 23	FIRE SUPPRESSION, PLUMBING & HVAC	100.1	107.4	103.1	94.3	81.9	89.2	94.3	81.7	89.1	94.3	89.9	92.5	100.0	83.9	93.4	94.3	89.7	92.4
26, 27, 3370	ELECTRICAL, COMMUNICATIONS & UTIL.	106.5	93.9	99.9	93.3	92.6	92.9	96.0	72.8	83.8	97.6	96.9	97.2	98.7	92.6	95.5	97.6	96.9	97.2
MF2010	WEIGHTED AVERAGE	102.7	93.6	98.7	96.7	87.2	92.5	96.8	83.2	90.8	98.1	90.3	94.7	98.9	90.1	95.0	98.7	90.1	94.9

DIVISION		WEST VIRGINIA																	
		GASSAWAY			HUNTINGTON			LEWISBURG			MARTINSBURG			MORGANTOWN			PARKERSBURG		
		266			255 - 257			249			254			265			261		
		MAT.	INST.	TOTAL	MAT.	INST.	TOTAL	MAT.	INST.	TOTAL	MAT.	INST.	TOTAL	MAT.	INST.	TOTAL	MAT.	INST.	TOTAL
015433	CONTRACTOR EQUIPMENT		100.5	100.5		100.5	100.5		100.5	100.5		100.5	100.5		100.5	100.5		100.5	100.5
0241, 31 - 34	SITE & INFRASTRUCTURE, DEMOLITION	106.9	87.9	93.6	108.0	90.1	95.4	119.7	87.7	97.3	107.2	85.6	92.0	104.2	87.7	92.6	112.6	89.2	96.2
0310	Concrete Forming & Accessories	85.0	83.5	83.7	96.3	94.9	95.1	83.5	85.4	85.1	84.9	78.9	79.7	83.5	83.0	83.0	87.5	85.3	85.6
0320	Concrete Reinforcing	98.1	83.4	90.8	98.9	89.4	94.2	98.1	65.6	81.9	97.5	74.9	86.2	98.1	82.7	90.4	97.5	84.6	91.1
0330	Cast-in-Place Concrete	116.3	99.0	109.2	121.5	104.7	114.6	111.5	102.1	107.7	116.7	88.7	105.2	111.1	100.6	106.8	113.6	94.1	105.6
03	CONCRETE	107.2	89.6	98.5	109.3	97.7	103.6	112.8	88.2	100.6	107.3	82.1	94.8	103.3	89.8	96.6	108.5	88.9	98.8
04	MASONRY	104.5	90.0	95.7	95.2	91.9	93.1	93.8	91.5	92.4	96.3	80.9	86.9	122.0	87.4	100.9	80.7	88.8	85.7
05	METALS	95.6	96.3	95.8	97.8	98.2	97.9	95.6	88.5	93.4	95.9	85.7	92.7	95.7	95.4	95.6	96.3	96.4	96.4
06	WOOD, PLASTICS & COMPOSITES	87.0	81.9	84.1	98.2	94.5	96.1	84.8	86.3	85.7	86.9	81.0	83.5	84.8	81.9	83.1	88.0	83.4	85.4
07	THERMAL & MOISTURE PROTECTION	100.4	87.5	95.2	100.4	88.8	95.8	101.2	87.4	95.6	100.6	73.3	89.6	100.4	86.3	94.7	100.3	86.7	94.8
08	OPENINGS	93.8	77.7	89.9	94.2	86.0	92.2	95.6	76.1	90.9	96.8	68.5	89.9	96.9	79.2	92.6	94.6	78.9	90.8
0920	Plaster & Gypsum Board	100.4	81.2	86.9	108.7	94.1	98.5	98.7	85.7	89.6	100.4	80.2	86.3	98.7	81.2	86.4	101.6	82.7	88.4
0950, 0980	Ceilings & Acoustic Treatment	93.4	81.2	85.2	93.4	94.1	93.9	93.4	85.7	88.3	93.4	80.2	84.6	93.4	81.2	85.2	93.4	82.7	86.3
0960	Flooring	96.4	112.9	101.3	103.6	104.0	103.7	95.6	112.9	100.8	96.3	55.6	84.2	95.6	106.4	98.8	99.3	103.8	100.7
0970, 0990	Wall Finishes & Painting/Coating	109.6	89.2	97.3	109.6	82.4	93.3	109.6	36.6	65.8	109.6	43.2	69.8	109.6	88.9	97.2	109.6	87.6	96.4
09	FINISHES	98.3	89.2	93.2	101.1	95.6	98.0	99.3	86.2	91.9	98.2	71.6	83.3	97.6	87.9	92.1	99.7	88.6	93.5
COVERS	DIVS. 10 - 14, 25, 28, 41, 43, 44, 46	100.0	94.4	98.9	100.0	97.7	99.5	100.0	55.5	90.9	100.0	53.8	90.6	100.0	64.2	92.7	100.0	95.3	99.0
21, 22, 23	FIRE SUPPRESSION, PLUMBING & HVAC	94.3	90.8	92.8	100.3	86.0	94.4	94.3	81.7	89.1	94.3	84.8	90.4	94.3	89.7	92.4	100.2	88.1	95.2
26, 27, 3370	ELECTRICAL, COMMUNICATIONS & UTIL.	97.6	92.6	94.9	97.2	93.7	95.4	93.3	72.8	82.5	99.4	83.0	90.8	97.8	96.9	97.3	97.7	92.6	95.0
MF2010	WEIGHTED AVERAGE	98.0	90.2	94.6	99.9	92.5	96.6	98.3	83.2	91.7	98.2	80.3	90.3	98.7	89.2	94.5	98.9	89.6	94.8

DIVISION		WEST VIRGINIA									WISCONSIN								
		PETERSBURG			ROMNEY			WHEELING			BELOIT			EAU CLAIRE			GREEN BAY		
		268			267			260			535			547			541 - 543		
		MAT.	INST.	TOTAL	MAT.	INST.	TOTAL	MAT.	INST.	TOTAL	MAT.	INST.	TOTAL	MAT.	INST.	TOTAL	MAT.	INST.	TOTAL
015433	**CONTRACTOR EQUIPMENT**		100.5	100.5		100.5	100.5		100.5	100.5		102.4	102.4		101.1	101.1		98.9	98.9
0241, 31 - 34	**SITE & INFRASTRUCTURE, DEMOLITION**	103.4	88.1	92.6	106.3	88.1	93.5	113.4	88.3	95.8	96.8	107.8	104.5	97.9	103.2	101.6	101.7	99.4	100.1
0310	Concrete Forming & Accessories	86.5	81.7	82.3	82.8	81.8	81.9	89.2	83.4	84.2	103.6	93.1	94.5	101.4	97.2	97.8	108.0	96.9	98.4
0320	Concrete Reinforcing	97.5	75.3	86.4	98.1	72.5	85.3	96.8	82.8	89.8	98.8	116.9	107.8	98.2	100.4	99.3	96.4	93.6	95.0
0330	Cast-in-Place Concrete	111.1	93.0	103.6	116.3	93.0	106.7	113.6	100.8	108.3	101.7	91.8	97.6	100.6	92.5	97.3	104.0	99.5	102.2
03	**CONCRETE**	103.4	85.2	94.4	107.0	84.7	96.0	108.5	90.0	99.4	100.2	97.4	98.8	99.7	96.4	98.1	102.7	97.4	100.1
04	**MASONRY**	96.4	87.4	90.9	94.2	82.5	87.1	103.7	87.4	93.8	102.8	92.5	96.5	91.6	96.5	94.6	123.0	98.4	108.0
05	**METALS**	95.8	92.2	94.7	95.8	91.1	94.3	96.5	95.8	96.3	97.1	107.8	100.5	95.3	102.3	97.5	97.5	99.5	98.2
06	**WOOD, PLASTICS & COMPOSITES**	88.8	81.0	84.3	84.0	81.0	82.3	89.5	81.9	85.1	98.0	95.3	96.5	106.3	98.7	102.0	108.6	98.7	102.9
07	**THERMAL & MOISTURE PROTECTION**	100.5	74.8	90.1	100.5	73.5	89.7	100.7	86.4	94.9	104.8	89.9	98.8	103.9	83.9	95.9	105.6	85.8	97.6
08	**OPENINGS**	96.9	75.4	91.7	96.8	74.8	91.5	95.4	79.2	91.5	101.2	103.7	101.8	99.3	96.2	98.6	96.5	96.7	96.6
0920	Plaster & Gypsum Board	101.2	80.2	86.5	98.3	80.2	85.7	101.6	81.2	87.3	101.0	95.5	97.2	104.6	98.9	100.6	97.6	98.9	98.5
0950, 0980	Ceilings & Acoustic Treatment	93.4	80.2	84.6	93.4	80.2	84.6	93.4	81.2	85.2	91.8	95.5	94.3	107.2	98.9	101.6	98.0	98.9	98.6
0960	Flooring	97.4	106.4	100.0	95.4	96.8	95.8	100.2	106.4	102.1	107.3	116.3	110.0	97.2	112.1	101.6	117.0	117.0	117.0
0970, 0990	Wall Finishes & Painting/Coating	109.6	88.9	97.2	109.6	88.9	97.2	109.6	88.9	97.2	89.7	96.5	93.8	86.6	90.3	88.8	96.4	82.8	88.2
09	**FINISHES**	98.4	87.3	92.2	97.7	85.5	90.9	99.9	87.9	93.2	101.7	97.9	99.6	102.4	99.8	101.0	106.6	99.8	102.8
COVERS	**DIVS. 10 - 14, 25, 28, 41, 43, 44, 46**	100.0	64.1	92.7	100.0	94.3	98.8	100.0	96.1	99.2	100.0	86.2	97.2	100.0	88.9	97.7	100.0	92.3	98.4
21, 22, 23	**FIRE SUPPRESSION, PLUMBING & HVAC**	94.3	90.8	92.8	94.3	90.6	92.8	100.3	89.7	96.0	99.9	87.5	94.8	100.2	83.3	93.3	100.5	84.3	93.8
26, 27, 3370	**ELECTRICAL, COMMUNICATIONS & UTIL.**	101.1	83.1	91.6	100.3	83.1	91.2	94.7	96.9	95.8	102.1	85.9	93.5	104.7	88.5	96.2	98.8	81.2	89.5
MF2010	**WEIGHTED AVERAGE**	97.8	86.1	92.6	98.0	86.0	92.7	100.0	90.3	95.7	100.2	94.8	97.8	99.5	93.5	96.9	101.5	92.6	97.6

DIVISION		WISCONSIN																	
		KENOSHA			LA CROSSE			LANCASTER			MADISON			MILWAUKEE			NEW RICHMOND		
		531			546			538			537			530,532			540		
		MAT.	INST.	TOTAL	MAT.	INST.	TOTAL	MAT.	INST.	TOTAL	MAT.	INST.	TOTAL	MAT.	INST.	TOTAL	MAT.	INST.	TOTAL
015433	**CONTRACTOR EQUIPMENT**		100.3	100.3		101.1	101.1		102.4	102.4		102.4	102.4		90.5	90.5		101.6	101.6
0241, 31 - 34	**SITE & INFRASTRUCTURE, DEMOLITION**	103.2	104.9	104.4	91.4	103.3	99.8	96.1	108.1	104.5	92.5	108.2	103.5	97.6	98.5	98.2	94.6	103.4	100.8
0310	Concrete Forming & Accessories	108.6	101.7	102.6	89.2	97.3	96.2	102.9	93.9	95.1	105.3	95.7	97.0	105.7	122.5	120.2	97.3	101.4	100.9
0320	Concrete Reinforcing	98.6	99.8	99.2	97.9	89.4	93.7	100.1	89.2	94.7	98.8	89.6	94.2	98.8	100.4	99.6	95.4	100.3	97.9
0330	Cast-in-Place Concrete	110.6	96.4	104.8	90.4	90.0	90.3	101.1	95.4	98.8	93.0	96.1	94.3	99.3	112.1	104.6	104.7	80.2	94.6
03	**CONCRETE**	104.7	99.6	102.2	91.0	93.6	92.3	99.9	93.7	96.9	96.1	94.9	95.5	99.3	113.8	106.5	97.5	94.1	95.8
04	**MASONRY**	100.1	105.4	103.4	90.9	98.4	95.5	102.9	98.1	100.0	100.4	100.6	100.5	102.8	121.1	114.0	117.8	93.0	102.6
05	**METALS**	97.9	101.7	99.1	95.2	97.2	95.8	94.7	94.1	94.5	99.6	96.1	98.5	100.3	95.8	98.9	95.4	101.4	97.3
06	**WOOD, PLASTICS & COMPOSITES**	100.5	100.8	100.7	92.6	98.7	96.1	97.4	95.3	96.2	95.2	95.3	95.3	100.0	123.9	113.7	95.7	107.1	102.2
07	**THERMAL & MOISTURE PROTECTION**	104.5	94.8	100.6	103.5	83.5	95.4	104.6	78.5	94.1	107.7	89.4	100.4	102.7	114.3	107.4	105.2	90.6	99.4
08	**OPENINGS**	95.9	104.9	98.1	99.3	84.4	95.7	96.6	83.4	93.4	101.2	96.0	99.9	103.7	117.5	107.0	86.3	100.8	89.9
0920	Plaster & Gypsum Board	90.3	101.2	97.9	99.1	98.9	98.9	99.5	95.5	96.7	104.7	95.5	98.3	102.2	124.8	118.0	87.4	107.8	101.7
0950, 0980	Ceilings & Acoustic Treatment	91.8	101.2	98.0	105.4	98.9	101.0	86.3	95.5	92.4	102.0	95.5	97.7	96.3	124.8	115.3	69.5	107.8	95.0
0960	Flooring	127.1	111.0	122.3	90.8	112.1	97.1	106.8	105.1	106.3	100.9	105.1	102.1	110.5	118.1	112.8	109.8	112.1	110.5
0970, 0990	Wall Finishes & Painting/Coating	98.0	102.7	100.8	86.6	71.8	77.7	89.7	69.4	77.5	89.7	96.5	93.8	91.5	126.9	112.7	99.6	90.3	94.0
09	**FINISHES**	106.9	103.9	105.2	98.9	98.3	98.6	100.1	93.1	96.2	102.0	97.9	99.7	103.9	123.3	114.8	96.7	102.2	99.8
COVERS	**DIVS. 10 - 14, 25, 28, 41, 43, 44, 46**	100.0	98.7	99.7	100.0	89.5	97.9	100.0	49.3	89.6	100.0	89.0	97.7	100.0	104.2	100.8	100.0	86.3	97.2
21, 22, 23	**FIRE SUPPRESSION, PLUMBING & HVAC**	100.1	93.2	97.3	100.2	84.1	93.6	93.9	83.8	89.8	99.9	93.0	97.0	99.9	102.7	101.0	93.6	80.8	88.4
26, 27, 3370	**ELECTRICAL, COMMUNICATIONS & UTIL.**	102.6	97.5	99.9	105.0	88.7	96.4	101.8	88.7	94.9	101.3	92.3	96.6	103.6	100.9	102.2	102.8	88.7	95.4
MF2010	**WEIGHTED AVERAGE**	100.8	99.8	100.4	98.0	92.4	95.5	97.7	90.3	94.4	99.9	96.0	98.2	101.1	108.6	104.4	97.0	93.0	95.2

DIVISION		WISCONSIN																	
		OSHKOSH			PORTAGE			RACINE			RHINELANDER			SUPERIOR			WAUSAU		
		549			539			534			545			548			544		
		MAT.	INST.	TOTAL	MAT.	INST.	TOTAL	MAT.	INST.	TOTAL	MAT.	INST.	TOTAL	MAT.	INST.	TOTAL	MAT.	INST.	TOTAL
015433	**CONTRACTOR EQUIPMENT**		98.9	98.9		102.4	102.4		102.4	102.4		98.9	98.9		101.6	101.6		98.9	98.9
0241, 31 - 34	**SITE & INFRASTRUCTURE, DEMOLITION**	93.1	98.1	96.6	86.5	108.2	101.7	96.7	108.8	105.2	106.2	99.3	101.3	91.2	103.7	100.0	88.8	99.3	96.2
0310	Concrete Forming & Accessories	91.9	83.2	84.4	95.9	94.9	95.1	103.9	101.7	102.0	89.6	94.0	93.4	95.7	91.2	91.8	91.3	96.6	95.9
0320	Concrete Reinforcing	96.5	76.7	86.6	100.2	89.4	94.8	98.8	99.8	99.3	96.6	91.5	94.1	95.4	92.5	94.0	96.6	89.4	93.1
0330	Cast-in-Place Concrete	96.5	87.3	92.7	86.8	97.7	91.3	99.7	96.1	98.2	109.4	97.7	104.6	98.6	95.8	97.5	89.9	97.0	92.8
03	**CONCRETE**	92.5	84.0	88.3	88.0	95.0	91.5	99.3	99.5	99.4	104.5	95.1	99.9	92.1	93.4	92.7	87.6	95.6	91.6
04	**MASONRY**	104.5	86.6	93.6	101.9	98.1	99.6	102.8	105.4	104.4	121.2	97.6	106.8	117.0	102.4	108.1	104.1	95.8	99.0
05	**METALS**	95.6	89.2	93.6	95.4	95.2	95.3	98.7	101.7	99.7	95.5	97.4	96.1	96.3	99.2	97.2	95.4	97.4	96.0
06	**WOOD, PLASTICS & COMPOSITES**	90.2	84.7	87.1	88.8	95.3	92.5	98.3	100.8	99.7	88.0	95.4	92.2	94.1	89.7	91.6	89.8	98.7	94.9
07	**THERMAL & MOISTURE PROTECTION**	104.8	77.1	93.6	104.0	90.6	98.6	104.8	94.4	100.6	105.5	82.2	96.1	104.9	93.6	100.4	104.6	83.2	96.0
08	**OPENINGS**	92.3	82.0	89.8	96.8	90.2	95.2	101.2	104.9	102.1	92.3	85.5	90.7	86.0	90.9	87.2	92.5	95.6	93.3
0920	Plaster & Gypsum Board	85.9	84.5	84.9	93.6	95.5	94.9	101.0	101.2	101.1	85.9	95.5	92.6	87.3	89.8	89.1	85.9	98.9	95.0
0950, 0980	Ceilings & Acoustic Treatment	98.0	84.5	89.0	89.0	95.5	93.3	91.8	101.2	98.0	98.0	95.5	96.3	70.4	89.8	83.3	98.0	98.9	98.6
0960	Flooring	108.3	98.7	105.4	103.2	111.6	105.7	107.3	111.0	108.4	107.5	112.1	108.9	111.3	121.2	114.3	108.1	112.1	109.3
0970, 0990	Wall Finishes & Painting/Coating	94.1	67.7	78.3	89.7	69.4	77.5	89.6	107.7	100.4	94.1	63.0	75.5	88.5	99.5	95.0	94.1	79.5	85.4
09	**FINISHES**	101.8	84.0	91.8	98.2	95.0	96.4	101.7	104.4	103.2	102.8	94.2	97.9	96.1	97.5	96.9	101.4	98.5	99.7
COVERS	**DIVS. 10 - 14, 25, 28, 41, 43, 44, 46**	100.0	61.8	92.2	100.0	63.6	92.6	100.0	98.7	99.7	100.0	54.9	90.8	100.0	86.0	97.1	100.0	92.3	98.4
21, 22, 23	**FIRE SUPPRESSION, PLUMBING & HVAC**	94.4	73.3	85.8	93.9	88.6	91.7	99.9	93.2	97.2	94.4	83.3	89.9	93.6	90.7	92.4	94.4	82.1	89.4
26, 27, 3370	**ELECTRICAL, COMMUNICATIONS & UTIL.**	103.4	74.7	88.3	105.9	91.9	98.5	101.9	95.5	98.5	102.7	79.8	90.6	108.1	95.0	101.2	104.6	79.8	91.6
MF2010	**WEIGHTED AVERAGE**	96.9	81.4	90.1	96.3	93.4	95.0	100.3	99.9	100.1	99.5	89.2	94.9	96.8	95.5	96.2	96.3	91.0	94.0

DIVISION		WYOMING																	
		CASPER			CHEYENNE			NEWCASTLE			RAWLINS			RIVERTON			ROCK SPRINGS		
		826			820			827			823			825			829 - 831		
		MAT.	INST.	TOTAL	MAT.	INST.	TOTAL	MAT.	INST.	TOTAL	MAT.	INST.	TOTAL	MAT.	INST.	TOTAL	MAT.	INST.	TOTAL
015433	CONTRACTOR EQUIPMENT		97.5	97.5		97.5	97.5		97.5	97.5		97.5	97.5		97.5	97.5		97.5	97.5
0241, 31 - 34	SITE & INFRASTRUCTURE, DEMOLITION	97.6	95.1	95.8	90.8	95.1	93.8	81.9	94.7	90.9	96.4	94.8	95.2	89.7	94.7	93.2	85.5	94.6	91.8
0310	Concrete Forming & Accessories	102.7	44.4	52.2	105.0	55.8	62.5	95.9	63.3	67.7	100.0	63.7	68.5	94.9	63.2	67.5	102.0	62.5	67.8
0320	Concrete Reinforcing	111.4	70.3	90.9	105.0	71.1	88.1	112.9	71.5	92.3	112.5	71.6	92.1	113.6	71.2	92.5	113.6	71.2	92.5
0330	Cast-in-Place Concrete	95.2	73.9	86.4	93.6	74.1	85.5	94.5	63.0	81.5	94.6	63.1	81.7	94.6	53.4	77.6	94.5	52.2	77.1
03	CONCRETE	101.7	60.2	81.2	100.5	65.5	83.2	101.1	65.0	83.3	114.0	65.3	89.9	108.9	61.7	85.6	101.7	60.9	81.5
04	MASONRY	101.3	35.3	61.0	98.0	48.7	67.9	94.6	57.0	71.7	94.6	57.0	71.7	94.6	51.0	68.0	151.4	48.3	88.4
05	METALS	98.1	71.0	89.6	99.3	72.4	90.9	94.6	72.1	87.5	94.6	72.5	87.6	94.7	71.6	87.4	95.4	71.9	88.0
06	WOOD, PLASTICS & COMPOSITES	93.5	41.8	64.0	95.3	56.7	73.3	85.6	68.3	75.7	89.4	68.3	77.4	84.5	68.3	75.3	93.0	68.3	78.9
07	THERMAL & MOISTURE PROTECTION	100.5	50.3	80.3	97.4	55.4	80.5	98.9	56.8	82.0	100.5	56.8	82.9	99.9	60.7	84.1	99.0	59.6	83.1
08	OPENINGS	97.2	49.6	85.6	99.4	57.7	89.2	103.2	60.9	92.9	102.9	60.9	92.7	103.1	60.9	92.8	103.6	60.9	93.2
0920	Plaster & Gypsum Board	93.0	39.9	55.9	83.5	55.4	63.8	78.7	67.3	70.7	79.1	67.3	70.8	78.7	67.3	70.7	87.8	67.3	73.5
0950, 0980	Ceilings & Acoustic Treatment	117.7	39.9	65.9	104.7	55.4	71.8	106.7	67.3	80.4	106.7	67.3	80.4	106.7	67.3	80.4	106.7	67.3	80.4
0960	Flooring	99.9	34.2	80.3	103.8	54.9	89.2	97.5	53.6	84.4	100.3	53.6	86.4	97.1	53.6	84.1	102.5	53.6	87.9
0970, 0990	Wall Finishes & Painting/Coating	98.5	45.0	66.4	101.9	45.0	67.8	97.7	57.8	73.8	97.7	57.8	73.8	97.7	57.8	73.8	97.7	57.8	73.8
09	FINISHES	104.9	41.0	69.1	102.6	53.9	75.3	97.2	61.0	76.9	99.8	61.0	78.1	98.2	61.0	77.3	99.8	60.3	77.7
COVERS	DIVS. 10 - 14, 25, 28, 41, 43, 44, 46	100.0	88.3	97.6	100.0	89.9	97.9	100.0	58.6	91.6	100.0	90.7	98.1	100.0	58.6	91.5	100.0	81.5	96.2
21, 22, 23	FIRE SUPPRESSION, PLUMBING & HVAC	100.0	66.7	86.4	100.0	66.7	86.4	96.9	66.0	84.2	96.9	66.0	84.3	96.9	66.0	84.2	99.9	64.6	85.5
26, 27, 3370	ELECTRICAL, COMMUNICATIONS & UTIL.	94.5	66.8	79.9	97.5	71.5	83.8	95.0	71.9	82.9	95.0	71.9	82.9	95.0	65.6	79.5	93.3	65.6	78.7
MF2010	WEIGHTED AVERAGE	99.4	61.5	82.7	99.4	66.7	85.0	97.2	67.5	84.1	99.3	68.6	85.8	98.4	65.6	84.0	101.2	65.6	85.5

DIVISION		WYOMING												CANADA					
		SHERIDAN			WHEATLAND			WORLAND			YELLOWSTONE NAT'L PA			BARRIE, ONTARIO			BATHURST, NEW BRUNSWICK		
		828			822			824			821								
		MAT.	INST.	TOTAL	MAT.	INST.	TOTAL	MAT.	INST.	TOTAL	MAT.	INST.	TOTAL	MAT.	INST.	TOTAL	MAT.	INST.	TOTAL
015433	CONTRACTOR EQUIPMENT		97.5	97.5		97.5	97.5		97.5	97.5		97.5	97.5		101.6	101.6		101.7	101.7
0241, 31 - 34	SITE & INFRASTRUCTURE, DEMOLITION	90.3	95.1	93.7	86.2	94.7	92.2	83.6	94.7	91.4	83.7	94.9	91.5	116.0	101.1	105.6	96.7	97.4	97.2
0310	Concrete Forming & Accessories	102.9	52.7	59.4	97.9	52.1	58.3	97.9	62.9	67.6	98.0	64.2	68.7	124.1	92.0	96.4	102.8	66.6	71.4
0320	Concrete Reinforcing	113.6	71.6	92.7	112.9	71.6	92.3	113.6	70.9	92.3	115.5	70.9	93.3	180.3	86.5	133.6	148.9	59.3	104.3
0330	Cast-in-Place Concrete	97.8	74.0	88.0	98.7	62.8	83.9	94.5	53.0	77.4	94.5	55.0	78.3	167.6	91.0	136.1	138.2	63.9	107.7
03	CONCRETE	109.0	64.2	86.9	105.8	60.0	83.2	101.4	61.3	81.6	101.7	62.6	82.4	152.9	90.8	122.2	135.7	64.9	100.8
04	MASONRY	94.9	58.6	72.7	94.9	56.6	71.5	94.6	46.0	64.9	94.7	49.6	67.2	166.4	100.5	126.2	157.8	67.7	102.8
05	METALS	98.1	73.0	90.2	94.5	72.3	87.5	94.7	71.2	87.3	95.3	71.1	87.7	107.2	91.7	102.4	104.8	73.3	94.9
06	WOOD, PLASTICS & COMPOSITES	96.5	52.6	71.5	87.3	53.1	67.8	87.4	68.3	76.5	87.4	68.3	76.5	117.9	90.6	102.3	94.6	66.8	78.7
07	THERMAL & MOISTURE PROTECTION	100.0	57.6	82.9	99.2	56.0	81.8	98.9	56.9	82.0	98.5	58.5	82.4	111.7	91.6	103.6	104.9	64.9	88.8
08	OPENINGS	103.9	53.2	91.5	101.9	55.5	90.6	103.4	60.9	93.1	96.7	60.9	88.0	92.1	88.9	91.3	87.5	58.4	80.4
0920	Plaster & Gypsum Board	106.6	51.1	67.8	78.7	51.6	59.8	78.7	67.3	70.7	79.0	67.3	70.8	154.1	90.3	109.5	164.9	65.6	95.5
0950, 0980	Ceilings & Acoustic Treatment	109.8	51.1	70.7	106.7	51.6	70.0	106.7	67.3	80.4	107.7	67.3	80.7	95.6	90.3	92.0	96.5	65.6	76.0
0960	Flooring	100.9	53.6	86.8	99.0	52.6	85.2	99.0	53.6	85.5	99.0	53.6	85.5	130.8	100.7	121.9	115.7	48.9	95.8
0970, 0990	Wall Finishes & Painting/Coating	100.1	45.0	67.1	97.7	36.0	60.7	97.7	57.8	73.8	97.7	57.8	73.8	113.0	92.3	100.6	112.2	53.9	77.3
09	FINISHES	105.6	50.7	74.8	98.0	49.3	70.7	97.6	60.8	77.0	97.9	61.7	77.6	118.3	94.0	104.7	113.5	62.2	84.8
COVERS	DIVS. 10 - 14, 25, 28, 41, 43, 44, 46	100.0	89.5	97.8	100.0	81.4	96.2	100.0	58.4	91.5	100.0	59.7	91.8	139.2	76.3	126.4	131.1	67.9	118.2
21, 22, 23	FIRE SUPPRESSION, PLUMBING & HVAC	96.9	66.7	84.5	96.9	65.8	84.2	96.9	65.7	84.1	96.9	67.3	84.8	102.7	103.0	102.8	102.9	73.6	90.9
26, 27, 3370	ELECTRICAL, COMMUNICATIONS & UTIL.	97.8	65.6	80.8	95.0	71.5	82.6	95.0	65.6	79.5	94.0	65.6	79.0	119.5	92.7	105.4	120.1	64.5	90.8
MF2010	WEIGHTED AVERAGE	100.0	66.1	85.1	97.8	65.5	83.6	97.4	64.9	83.0	96.7	65.9	83.2	118.1	95.4	108.1	112.9	70.0	94.0

DIVISION		CANADA																	
		BRANDON, MANITOBA			BRANTFORD, ONTARIO			BRIDGEWATER, NOVA SCOTIA			CALGARY, ALBERTA			CAP-DE-LA-MADELEINE, QUEBEC			CHARLESBOURG, QUEBEC		
		MAT.	INST.	TOTAL	MAT.	INST.	TOTAL	MAT.	INST.	TOTAL	MAT.	INST.	TOTAL	MAT.	INST.	TOTAL	MAT.	INST.	TOTAL
015433	CONTRACTOR EQUIPMENT		103.8	103.8		101.6	101.6		101.4	101.4		106.4	106.4		102.3	102.3		102.3	102.3
0241, 31 - 34	SITE & INFRASTRUCTURE, DEMOLITION	112.4	100.1	103.8	114.0	101.4	105.2	99.1	98.9	98.9	126.5	106.0	112.1	95.1	100.5	98.9	95.1	100.5	98.9
0310	Concrete Forming & Accessories	123.4	75.7	82.2	124.0	99.7	103.0	96.9	73.7	76.8	129.3	99.0	103.1	129.1	90.5	95.7	129.1	90.5	95.7
0320	Concrete Reinforcing	162.0	55.8	109.1	175.2	85.2	130.3	148.9	48.2	98.8	144.1	75.3	109.8	148.9	81.3	115.3	148.9	81.3	115.3
0330	Cast-in-Place Concrete	151.1	79.3	121.6	163.5	112.9	142.7	168.1	73.8	129.3	201.8	108.6	163.4	132.0	100.1	118.9	132.0	100.1	118.9
03	CONCRETE	135.4	73.9	105.0	150.1	101.5	126.1	147.5	69.7	109.0	165.8	97.9	132.3	131.0	92.2	111.9	131.0	92.2	111.9
04	MASONRY	161.3	70.5	105.9	162.8	105.1	127.6	158.4	73.7	106.7	211.8	92.9	139.2	158.7	89.2	116.3	158.7	89.2	116.3
05	METALS	107.4	78.6	98.4	106.6	92.4	102.2	105.8	75.4	96.2	140.7	90.3	124.8	105.0	89.0	100.0	105.0	89.0	100.0
06	WOOD, PLASTICS & COMPOSITES	113.6	76.7	92.5	115.9	98.3	105.9	87.0	73.0	79.0	114.1	98.6	105.2	127.2	90.4	106.2	127.2	90.4	106.2
07	THERMAL & MOISTURE PROTECTION	105.2	74.9	93.0	111.6	97.3	105.8	106.6	71.9	92.6	123.1	96.2	112.3	105.3	92.4	100.1	105.3	92.4	100.1
08	OPENINGS	93.0	67.7	86.9	92.0	95.2	92.8	84.3	66.6	80.0	93.0	87.9	91.8	93.0	83.3	90.7	93.0	83.3	90.7
0920	Plaster & Gypsum Board	131.3	75.5	92.3	154.1	98.3	115.1	158.4	72.0	98.0	179.7	98.1	122.6	189.1	89.9	119.7	189.1	89.9	119.7
0950, 0980	Ceilings & Acoustic Treatment	95.6	75.5	82.2	95.6	98.3	97.4	95.6	72.0	79.9	148.8	98.1	115.0	95.6	89.9	91.8	95.6	89.9	91.8
0960	Flooring	130.8	72.8	113.5	130.8	100.7	121.9	112.3	68.5	99.3	132.8	96.0	121.8	130.8	101.4	122.1	130.8	101.4	122.1
0970, 0990	Wall Finishes & Painting/Coating	112.3	61.3	81.7	112.2	101.3	105.7	112.2	65.9	84.4	112.1	115.0	113.8	112.2	94.2	101.4	112.2	94.2	101.4
09	FINISHES	114.3	74.1	91.8	116.9	100.6	107.7	111.2	72.1	89.3	134.1	100.6	115.3	120.4	93.3	105.2	120.4	93.3	105.2
COVERS	DIVS. 10 - 14, 25, 28, 41, 43, 44, 46	131.1	70.2	118.7	131.1	78.5	120.4	131.1	68.2	118.3	131.1	100.0	124.8	131.1	87.3	122.2	131.1	87.3	122.2
21, 22, 23	FIRE SUPPRESSION, PLUMBING & HVAC	102.9	88.7	97.1	102.9	106.1	104.2	102.9	87.6	96.6	100.5	95.2	98.3	103.4	94.4	99.7	103.4	94.4	99.7
26, 27, 3370	ELECTRICAL, COMMUNICATIONS & UTIL.	122.1	72.6	96.0	117.4	92.2	104.1	123.6	66.0	93.3	120.9	102.7	111.3	117.0	74.4	94.6	117.0	74.4	94.6
MF2010	WEIGHTED AVERAGE	114.9	78.9	99.0	116.4	99.4	108.9	114.3	76.6	97.7	127.9	97.4	114.5	113.7	89.9	103.2	113.7	89.9	103.2

	DIVISION	CANADA																	
		CHARLOTTETOWN, PRINCE EDWARD ISLAND			CHICOUTIMI, QUEBEC			CORNER BROOK, NEWFOUNDLAND			CORNWALL, ONTARIO			DALHOUSIE, NEW BRUNSWICK			DARTMOUTH, NOVA SCOTIA		
		MAT.	INST.	TOTAL	MAT.	INST.	TOTAL	MAT.	INST.	TOTAL	MAT.	INST.	TOTAL	MAT.	INST.	TOTAL	MAT.	INST.	TOTAL
015433	CONTRACTOR EQUIPMENT		101.4	101.4		102.3	102.3		102.9	102.9		101.6	101.6		101.7	101.7		101.4	101.4
0241, 31 - 34	SITE & INFRASTRUCTURE, DEMOLITION	109.9	96.4	100.5	94.6	99.8	98.2	114.7	97.9	102.9	112.1	100.9	104.3	96.5	97.4	97.1	105.9	98.9	101.0
0310	Concrete Forming & Accessories	100.9	59.9	65.4	129.1	94.9	99.5	104.4	63.4	68.9	121.9	92.2	96.2	102.8	66.8	71.6	97.0	73.7	76.8
0320	Concrete Reinforcing	157.5	48.3	103.1	112.3	98.0	105.2	148.9	50.1	99.7	175.2	84.9	130.2	148.9	59.4	104.3	155.4	48.2	102.0
0330	Cast-in-Place Concrete	160.5	62.4	120.2	130.2	101.3	118.3	172.2	72.1	131.1	147.1	103.0	128.9	138.2	64.0	107.7	169.8	73.8	130.3
03	CONCRETE	154.1	59.4	107.3	124.1	97.5	111.0	161.7	64.9	113.9	142.1	94.6	118.6	135.7	65.1	100.8	149.4	69.7	110.0
04	MASONRY	163.9	64.2	103.0	159.2	96.3	120.8	158.7	65.8	102.0	161.6	96.4	121.8	155.4	67.7	101.9	170.0	73.7	111.2
05	METALS	118.2	68.6	102.7	104.6	93.3	101.0	107.8	74.1	97.3	106.6	91.0	101.7	104.8	73.5	94.9	108.0	75.4	97.7
06	WOOD, PLASTICS & COMPOSITES	87.2	59.3	71.3	127.2	95.2	108.9	95.6	62.4	76.7	114.7	91.4	101.4	94.6	66.8	78.7	87.0	73.0	79.0
07	THERMAL & MOISTURE PROTECTION	114.5	61.4	93.2	105.3	98.0	102.4	108.6	63.6	90.5	111.4	91.7	103.5	109.1	64.9	91.4	106.6	71.9	92.6
08	OPENINGS	95.2	51.3	84.5	92.6	82.0	90.0	98.9	58.7	89.1	93.0	88.6	92.0	87.5	58.4	80.4	84.3	66.6	80.0
0920	Plaster & Gypsum Board	161.2	58.0	89.0	189.1	94.9	123.2	164.9	61.1	92.3	226.5	91.1	131.8	164.9	65.6	95.5	160.8	72.0	98.7
0950, 0980	Ceilings & Acoustic Treatment	114.8	58.0	76.9	95.6	94.9	95.1	96.5	61.1	72.9	98.4	91.1	93.6	96.5	65.6	76.0	104.7	72.0	83.0
0960	Flooring	113.0	64.1	98.4	130.8	101.4	122.1	116.4	58.3	99.1	130.8	99.3	121.4	115.7	74.7	103.5	112.3	68.5	99.3
0970, 0990	Wall Finishes & Painting/Coating	112.2	44.1	71.4	112.2	108.1	109.7	112.2	63.4	83.0	112.2	94.4	101.5	112.2	53.9	77.3	112.2	65.9	84.4
09	FINISHES	117.4	59.4	84.9	120.4	98.3	108.0	114.8	62.5	85.5	126.6	94.0	108.4	113.5	67.3	87.6	113.6	72.1	90.3
COVERS	DIVS. 10 - 14, 25, 28, 41, 43, 44, 46	131.1	66.4	117.9	131.1	88.5	122.4	131.1	67.2	118.1	131.1	75.8	119.8	131.1	67.9	118.2	131.1	68.2	118.3
21, 22, 23	FIRE SUPPRESSION, PLUMBING & HVAC	103.1	66.2	88.0	102.9	95.5	99.9	102.9	74.5	91.3	103.4	103.9	103.6	102.9	73.6	90.9	102.9	87.6	96.6
26, 27, 3370	ELECTRICAL, COMMUNICATIONS & UTIL.	118.6	53.6	84.4	116.6	88.1	101.6	119.2	59.7	87.9	118.4	93.2	105.1	120.2	60.9	89.0	124.1	66.0	93.5
MF2010	WEIGHTED AVERAGE	119.0	64.5	95.0	112.6	94.6	104.7	118.2	69.4	96.7	116.5	95.7	107.3	113.0	70.2	94.1	115.9	76.6	98.6

	DIVISION	CANADA																	
		EDMONTON, ALBERTA			FORT MCMURRAY, ALBERTA			FREDERICTON, NEW BRUNSWICK			GATINEAU, QUEBEC			GRANBY, QUEBEC			HALIFAX, NOVA SCOTIA		
		MAT.	INST.	TOTAL	MAT.	INST.	TOTAL	MAT.	INST.	TOTAL	MAT.	INST.	TOTAL	MAT.	INST.	TOTAL	MAT.	INST.	TOTAL
015433	CONTRACTOR EQUIPMENT		106.4	106.4		103.7	103.7		101.7	101.7		102.3	102.3		102.3	102.3		101.4	101.4
0241, 31 - 34	SITE & INFRASTRUCTURE, DEMOLITION	150.4	106.0	119.3	119.8	102.1	107.4	98.2	97.4	97.7	94.9	100.5	98.8	95.4	100.5	99.0	101.7	98.9	99.8
0310	Concrete Forming & Accessories	129.5	99.0	103.1	122.1	96.5	99.9	124.9	67.1	74.9	129.1	90.4	95.6	129.1	90.3	95.6	99.7	80.0	82.7
0320	Concrete Reinforcing	153.6	75.3	114.6	162.0	75.3	118.8	141.8	59.5	100.8	157.5	81.3	119.6	157.5	81.3	119.6	158.4	63.9	111.3
0330	Cast-in-Place Concrete	210.3	108.6	168.5	217.1	107.1	171.9	132.6	64.1	104.4	130.2	100.0	117.8	134.5	100.0	120.3	172.4	82.6	135.5
03	CONCRETE	171.5	97.9	135.1	173.5	96.3	135.3	133.5	65.3	99.8	131.6	92.2	112.1	133.6	92.1	113.1	151.3	78.4	115.3
04	MASONRY	199.8	92.9	134.6	201.4	91.8	134.5	175.2	69.3	110.6	158.6	89.2	116.3	158.9	89.2	116.4	170.6	85.4	118.6
05	METALS	141.8	90.3	125.6	132.2	90.1	119.0	121.3	74.1	106.5	105.0	88.9	100.0	105.0	88.8	99.9	130.8	82.4	115.6
06	WOOD, PLASTICS & COMPOSITES	110.4	98.6	103.7	110.3	95.8	102.0	116.6	66.8	88.2	127.2	90.4	106.2	127.2	90.4	106.2	87.1	79.0	82.5
07	THERMAL & MOISTURE PROTECTION	126.8	96.2	114.5	118.9	94.6	109.1	112.9	66.0	94.0	105.3	92.4	100.1	105.3	90.8	99.4	115.3	79.6	100.9
08	OPENINGS	93.0	87.9	91.8	93.0	86.4	91.4	87.4	57.2	80.0	93.0	78.4	89.5	93.0	78.4	89.5	84.3	73.3	81.6
0920	Plaster & Gypsum Board	163.1	98.1	117.6	153.1	95.3	112.7	181.1	65.6	100.3	148.1	89.9	107.4	150.9	89.9	108.2	159.5	78.3	102.7
0950, 0980	Ceilings & Acoustic Treatment	154.6	98.1	116.9	104.7	95.3	98.5	108.4	65.6	79.9	95.6	89.9	91.8	95.6	89.9	91.8	108.4	78.3	88.3
0960	Flooring	135.6	96.0	123.8	130.8	96.0	120.4	127.3	78.0	112.6	130.8	101.4	122.1	130.8	101.4	122.1	112.9	90.0	106.1
0970, 0990	Wall Finishes & Painting/Coating	112.3	115.0	113.9	112.3	99.7	104.7	112.2	68.7	86.1	112.2	94.2	101.4	112.2	94.2	101.4	112.2	83.8	95.2
09	FINISHES	136.0	100.6	116.1	120.2	97.0	107.2	121.9	69.5	92.6	115.2	93.3	102.9	115.6	93.3	103.1	114.6	82.7	96.7
COVERS	DIVS. 10 - 14, 25, 28, 41, 43, 44, 46	131.1	100.0	124.8	131.1	99.1	124.6	131.1	67.9	118.2	131.1	87.3	122.2	131.1	87.3	122.2	131.1	69.9	118.6
21, 22, 23	FIRE SUPPRESSION, PLUMBING & HVAC	100.6	95.2	98.4	103.3	104.3	103.7	103.0	83.7	95.1	103.4	94.4	99.7	102.9	94.4	99.4	99.8	80.2	91.8
26, 27, 3370	ELECTRICAL, COMMUNICATIONS & UTIL.	115.3	102.7	108.7	111.0	89.1	99.5	123.0	80.3	100.5	117.0	74.4	94.6	117.7	74.4	94.9	126.0	81.9	102.8
MF2010	WEIGHTED AVERAGE	128.5	97.4	114.8	125.2	96.2	112.4	117.6	75.5	99.1	113.3	89.7	102.9	113.5	89.6	103.0	119.5	82.2	103.0

	DIVISION	CANADA																	
		HAMILTON, ONTARIO			HULL, QUEBEC			JOLIETTE, QUEBEC			KAMLOOPS, BRITISH COLUMBIA			KINGSTON, ONTARIO			KITCHENER, ONTARIO		
		MAT.	INST.	TOTAL	MAT.	INST.	TOTAL	MAT.	INST.	TOTAL	MAT.	INST.	TOTAL	MAT.	INST.	TOTAL	MAT.	INST.	TOTAL
015433	CONTRACTOR EQUIPMENT		108.4	108.4		102.3	102.3		102.3	102.3		105.5	105.5		103.9	103.9		103.8	103.8
0241, 31 - 34	SITE & INFRASTRUCTURE, DEMOLITION	117.5	113.0	114.4	94.9	100.5	98.8	95.5	100.5	99.0	118.5	104.1	108.4	112.1	104.7	106.9	101.8	105.3	104.3
0310	Concrete Forming & Accessories	130.7	96.1	100.7	129.1	90.4	95.6	129.1	90.5	95.7	123.2	95.0	98.8	122.1	92.3	96.3	118.3	88.5	92.5
0320	Concrete Reinforcing	161.9	94.7	128.4	157.5	81.3	119.6	148.9	81.3	115.3	116.8	80.6	98.8	175.2	84.9	130.2	110.6	94.6	102.6
0330	Cast-in-Place Concrete	153.0	102.0	132.0	130.2	100.0	117.8	135.6	100.1	121.0	117.4	105.3	112.4	147.1	102.9	128.9	140.8	92.0	120.7
03	CONCRETE	143.3	97.7	120.8	131.6	92.2	112.1	132.7	92.2	112.7	138.6	95.8	117.5	142.1	94.7	118.7	122.0	90.9	106.6
04	MASONRY	178.5	101.5	131.5	158.6	89.2	116.3	159.0	89.2	116.4	165.2	99.0	124.8	168.1	96.4	124.4	159.4	97.6	121.7
05	METALS	124.6	93.7	114.9	105.0	88.9	100.0	105.0	89.0	100.0	107.2	89.7	101.7	108.2	91.0	102.8	115.4	93.4	108.5
06	WOOD, PLASTICS & COMPOSITES	115.2	95.7	104.1	127.2	90.4	106.2	127.2	90.4	106.2	98.4	93.3	95.5	114.7	91.6	101.5	109.1	87.2	96.6
07	THERMAL & MOISTURE PROTECTION	124.1	96.7	113.1	105.3	92.4	100.1	105.3	92.4	100.1	120.1	90.6	108.2	111.4	92.9	104.0	111.0	93.5	104.0
08	OPENINGS	93.0	93.7	93.2	93.0	78.4	89.5	93.0	83.3	90.7	89.1	89.5	89.2	93.0	88.3	91.9	84.8	87.3	85.4
0920	Plaster & Gypsum Board	193.0	95.5	124.8	148.1	89.9	107.4	189.1	89.9	119.7	135.6	92.6	105.5	230.2	91.3	133.0	151.6	86.7	106.2
0950, 0980	Ceilings & Acoustic Treatment	128.2	95.5	106.4	95.6	89.9	91.8	95.6	89.9	91.8	95.6	92.6	93.6	112.0	91.3	98.2	104.7	86.7	92.7
0960	Flooring	132.8	104.2	124.3	130.8	101.4	122.1	130.8	101.4	122.1	129.5	57.6	108.1	130.8	99.3	121.4	127.2	104.2	120.4
0970, 0990	Wall Finishes & Painting/Coating	112.2	104.8	107.8	112.2	94.2	101.4	112.2	94.2	101.4	112.2	87.7	97.5	112.2	87.2	97.2	112.2	94.6	101.6
09	FINISHES	130.6	98.7	112.7	115.2	93.3	102.9	120.4	93.3	105.2	116.4	87.9	100.4	130.1	93.3	109.5	116.6	91.7	102.7
COVERS	DIVS. 10 - 14, 25, 28, 41, 43, 44, 46	131.1	100.7	124.9	131.1	87.3	122.2	131.1	87.3	122.2	131.1	99.5	124.7	131.1	75.9	119.8	131.1	98.7	124.5
21, 22, 23	FIRE SUPPRESSION, PLUMBING & HVAC	100.8	88.0	95.5	102.9	94.4	99.4	102.9	94.4	99.4	102.9	99.6	101.5	103.4	104.0	103.6	99.7	89.2	95.4
26, 27, 3370	ELECTRICAL, COMMUNICATIONS & UTIL.	117.0	97.8	106.9	119.1	74.4	95.6	117.7	74.4	94.9	121.5	85.4	102.5	118.4	91.8	104.4	117.7	95.3	105.9
MF2010	WEIGHTED AVERAGE	120.4	97.0	110.1	113.4	89.7	102.9	113.8	89.9	103.3	115.7	94.3	106.3	117.3	95.8	107.8	112.6	93.5	104.2

		CANADA																	
	DIVISION	LAVAL, QUEBEC			LETHBRIDGE, ALBERTA			LLOYDMINSTER, ALBERTA			LONDON, ONTARIO			MEDICINE HAT, ALBERTA			MONCTON, NEW BRUNSWICK		
		MAT.	INST.	TOTAL	MAT.	INST.	TOTAL	MAT.	INST.	TOTAL	MAT.	INST.	TOTAL	MAT.	INST.	TOTAL	MAT.	INST.	TOTAL
015433	CONTRACTOR EQUIPMENT		102.3	102.3		103.7	103.7		103.7	103.7		104.0	104.0		103.7	103.7		101.7	101.7
0241, 31 - 34	SITE & INFRASTRUCTURE, DEMOLITION	95.4	100.5	99.0	112.4	102.7	105.6	112.4	102.1	105.2	116.3	105.5	108.7	111.1	102.2	104.8	96.1	97.6	97.2
0310	Concrete Forming & Accessories	129.3	90.4	95.6	123.3	96.5	100.1	121.6	86.3	91.0	130.9	89.7	95.3	123.2	86.2	91.2	102.8	67.3	72.1
0320	Concrete Reinforcing	157.5	81.3	119.6	162.0	75.3	118.8	162.0	75.3	118.8	141.2	94.6	118.0	162.0	75.3	118.8	148.9	62.5	105.9
0330	Cast-in-Place Concrete	134.5	100.0	120.3	162.9	107.1	139.9	151.1	103.2	131.4	164.1	99.8	137.7	151.1	103.2	131.4	133.1	81.5	111.9
03	CONCRETE	133.6	92.2	113.2	147.6	96.3	122.3	141.8	90.4	116.4	145.2	94.1	120.0	141.9	90.4	116.5	133.3	72.2	103.1
04	MASONRY	158.8	89.2	116.4	175.8	91.8	124.6	158.3	84.9	113.5	182.5	98.7	131.4	158.3	84.9	113.5	157.4	67.7	102.7
05	METALS	105.1	88.9	100.0	126.1	90.1	114.8	107.4	90.0	101.9	125.2	93.0	115.1	107.4	89.9	101.9	104.8	82.4	97.7
06	WOOD, PLASTICS & COMPOSITES	127.4	90.4	106.2	113.7	95.8	103.5	110.3	85.6	96.2	115.2	88.2	99.8	113.7	85.6	97.6	94.6	66.8	78.7
07	THERMAL & MOISTURE PROTECTION	105.8	92.4	100.4	116.4	94.6	107.6	113.4	90.0	104.0	124.9	94.4	112.7	119.8	90.0	107.8	109.2	67.2	92.3
08	OPENINGS	93.0	78.4	89.5	93.0	86.4	91.4	93.0	80.8	90.1	94.1	88.5	92.7	93.0	80.8	90.1	87.5	63.7	81.7
0920	Plaster & Gypsum Board	151.3	89.9	108.3	139.4	95.3	108.6	134.3	84.8	99.7	194.7	87.9	120.0	137.2	84.8	100.5	164.9	65.6	95.5
0950, 0980	Ceilings & Acoustic Treatment	95.6	89.9	91.8	103.8	95.3	98.2	95.6	84.8	88.4	134.6	87.9	103.5	95.6	84.8	88.4	96.5	65.6	76.0
0960	Flooring	130.8	101.4	122.1	130.8	96.0	120.4	130.8	96.0	120.4	133.0	104.2	124.4	130.8	96.0	120.4	115.7	74.7	103.5
0970, 0990	Wall Finishes & Painting/Coating	112.2	94.2	101.4	112.1	108.9	110.2	112.3	84.8	95.8	112.2	101.5	105.8	112.1	84.8	95.8	112.2	53.9	77.3
09	FINISHES	115.6	93.3	103.1	117.3	98.0	106.5	115.0	87.6	99.6	131.8	93.4	110.3	115.2	87.6	99.7	113.5	67.3	87.6
COVERS	DIVS. 10 - 14, 25, 28, 41, 43, 44, 46	131.1	87.3	122.2	131.1	99.1	124.6	131.1	95.6	123.9	131.1	99.2	124.6	131.1	95.6	123.9	131.1	67.9	118.2
21, 22, 23	FIRE SUPPRESSION, PLUMBING & HVAC	99.7	94.4	97.5	103.1	100.8	102.2	103.4	100.8	102.3	100.8	85.2	94.4	102.9	97.3	100.6	102.9	73.9	91.0
26, 27, 3370	ELECTRICAL, COMMUNICATIONS & UTIL.	118.9	74.4	95.5	112.5	89.1	100.2	109.9	89.1	99.0	113.5	95.3	103.9	109.9	89.1	98.9	125.1	64.5	93.2
MF2010	WEIGHTED AVERAGE	112.9	89.7	102.7	119.7	95.6	109.1	114.7	92.2	104.8	120.8	93.5	108.8	114.8	91.4	104.5	113.3	72.9	95.5

		CANADA																	
	DIVISION	MONTREAL, QUEBEC			MOOSE JAW, SASKATCHEWAN			NEW GLASGOW, NOVA SCOTIA			NEWCASTLE, NEW BRUNSWICK			NORTH BAY, ONTARIO			OSHAWA, ONTARIO		
		MAT.	INST.	TOTAL	MAT.	INST.	TOTAL	MAT.	INST.	TOTAL	MAT.	INST.	TOTAL	MAT.	INST.	TOTAL	MAT.	INST.	TOTAL
015433	CONTRACTOR EQUIPMENT		103.9	103.9		100.1	100.1		101.4	101.4		101.7	101.7		101.6	101.6		103.8	103.8
0241, 31 - 34	SITE & INFRASTRUCTURE, DEMOLITION	104.7	100.0	101.4	112.0	96.6	101.2	99.3	98.9	99.0	96.7	97.3	97.2	114.0	100.5	104.5	113.9	104.6	107.4
0310	Concrete Forming & Accessories	135.7	95.2	100.6	106.9	62.3	68.3	96.9	73.7	76.8	102.8	66.7	71.6	124.0	89.6	94.2	124.1	91.7	96.1
0320	Concrete Reinforcing	151.0	98.0	124.6	114.3	63.8	89.2	148.9	48.2	98.8	148.9	58.4	103.8	175.2	84.4	130.0	175.2	87.1	131.3
0330	Cast-in-Place Concrete	182.2	102.6	149.5	146.6	72.7	116.2	169.8	73.8	130.3	138.2	64.0	107.7	163.5	89.0	132.9	162.8	89.9	132.8
03	CONCRETE	155.8	98.1	127.3	124.6	66.8	96.0	148.3	69.7	109.5	135.7	64.8	100.7	150.1	88.6	119.7	149.7	90.4	120.4
04	MASONRY	165.7	96.4	123.4	156.9	66.0	101.4	158.5	73.7	106.8	157.8	67.7	102.8	162.8	92.2	119.7	162.4	96.7	122.3
05	METALS	131.1	93.7	119.3	103.9	75.1	94.9	105.8	75.4	96.2	104.8	73.0	94.8	106.6	90.7	101.6	106.7	92.9	102.4
06	WOOD, PLASTICS & COMPOSITES	127.4	95.4	109.1	96.1	60.7	75.9	87.0	73.0	79.0	94.6	66.8	78.7	115.9	90.0	101.1	116.2	90.6	101.6
07	THERMAL & MOISTURE PROTECTION	121.5	98.6	112.3	104.7	66.0	89.1	106.6	71.9	92.6	109.2	64.5	91.2	111.6	88.0	102.1	112.0	89.8	103.0
08	OPENINGS	93.0	83.9	90.8	88.2	58.3	80.9	84.3	66.6	80.0	87.5	58.0	80.3	92.0	86.0	90.5	92.0	90.3	91.6
0920	Plaster & Gypsum Board	172.6	94.9	118.2	131.8	59.5	81.2	158.4	72.0	98.0	164.9	65.6	95.5	154.1	89.7	109.0	156.2	90.3	110.1
0950, 0980	Ceilings & Acoustic Treatment	112.8	94.9	100.9	95.6	59.5	71.5	95.6	72.0	79.9	96.5	65.6	76.0	95.6	89.7	91.6	100.2	90.3	93.6
0960	Flooring	133.2	101.4	123.7	120.5	64.5	103.8	112.3	68.5	99.3	115.7	73.5	103.2	130.8	99.3	121.4	130.8	106.9	123.7
0970, 0990	Wall Finishes & Painting/Coating	112.2	108.1	109.7	112.2	68.6	86.1	112.2	65.9	84.4	112.2	53.0	76.7	112.2	93.7	101.1	112.2	109.0	110.3
09	FINISHES	123.4	98.5	109.4	111.5	63.1	84.3	111.2	72.1	89.3	113.5	67.0	87.4	116.9	92.2	103.1	118.1	96.3	105.9
COVERS	DIVS. 10 - 14, 25, 28, 41, 43, 44, 46	131.1	89.0	122.5	131.1	66.1	117.9	131.1	68.2	118.3	131.1	67.1	118.1	131.1	74.5	119.6	131.1	99.7	124.7
21, 22, 23	FIRE SUPPRESSION, PLUMBING & HVAC	100.7	95.6	98.6	103.4	78.9	93.4	102.9	87.6	96.6	102.9	72.6	90.5	102.9	101.8	102.4	99.7	104.5	101.7
26, 27, 3370	ELECTRICAL, COMMUNICATIONS & UTIL.	122.8	88.1	104.5	120.1	64.3	90.8	119.6	66.0	91.3	119.3	63.4	89.9	119.7	93.1	105.7	119.0	92.7	105.2
MF2010	WEIGHTED AVERAGE	121.8	94.9	110.0	112.0	71.3	94.1	114.0	76.6	97.6	113.0	70.2	94.1	116.6	93.5	106.4	115.8	96.8	107.4

		CANADA																	
	DIVISION	OTTAWA, ONTARIO			OWEN SOUND, ONTARIO			PETERBOROUGH, ONTARIO			PORTAGE LA PRAIRIE, MANITOBA			PRINCE ALBERT, SASKATCHEWAN			PRINCE GEORGE, BRITISH COLUMBIA		
		MAT.	INST.	TOTAL	MAT.	INST.	TOTAL	MAT.	INST.	TOTAL	MAT.	INST.	TOTAL	MAT.	INST.	TOTAL	MAT.	INST.	TOTAL
015433	CONTRACTOR EQUIPMENT		103.8	103.8		101.6	101.6		101.6	101.6		103.8	103.8		100.1	100.1		105.5	105.5
0241, 31 - 34	SITE & INFRASTRUCTURE, DEMOLITION	113.6	105.2	107.7	116.0	101.0	105.5	114.0	100.9	104.8	112.9	100.1	103.9	106.8	96.7	99.8	121.9	104.1	109.4
0310	Concrete Forming & Accessories	130.7	92.7	97.8	124.1	88.0	92.9	124.0	90.6	95.1	123.4	75.3	81.7	106.9	62.1	68.1	112.4	89.4	92.5
0320	Concrete Reinforcing	156.9	94.6	125.9	180.3	86.5	133.5	175.2	84.9	130.2	162.0	55.8	109.1	119.5	63.8	91.7	116.8	80.6	98.8
0330	Cast-in-Place Concrete	168.5	100.8	140.7	167.6	85.0	133.6	163.5	90.7	133.6	151.1	78.7	121.3	132.9	72.6	108.1	147.1	105.3	129.9
03	CONCRETE	149.9	95.8	123.2	152.9	86.9	120.3	150.1	89.8	120.3	135.4	73.4	104.8	118.8	66.7	93.1	152.1	93.4	123.1
04	MASONRY	179.8	98.7	130.3	166.4	98.0	124.6	162.8	99.1	123.9	161.3	69.4	105.2	156.0	66.0	101.0	167.4	99.0	125.6
05	METALS	125.7	92.9	115.4	107.2	91.5	102.3	106.6	91.2	101.8	107.4	78.6	98.4	104.0	74.9	94.9	107.2	89.8	101.8
06	WOOD, PLASTICS & COMPOSITES	115.0	92.5	102.1	117.9	86.6	100.0	115.9	88.6	100.3	113.6	76.7	92.5	96.1	60.7	75.9	98.4	85.7	91.1
07	THERMAL & MOISTURE PROTECTION	127.2	94.4	114.0	111.7	88.8	102.5	111.6	93.7	104.4	105.2	74.4	92.8	104.6	64.8	88.6	114.4	89.8	104.5
08	OPENINGS	93.0	90.7	92.5	92.1	85.3	90.5	92.0	87.8	91.0	93.0	67.7	86.9	86.8	58.3	79.9	89.1	85.4	88.2
0920	Plaster & Gypsum Board	228.2	92.2	133.1	154.1	86.2	106.6	154.1	88.2	108.0	131.3	75.5	92.3	131.8	59.5	81.2	135.6	84.8	100.1
0950, 0980	Ceilings & Acoustic Treatment	138.2	92.2	107.6	95.6	86.2	89.3	95.6	88.2	90.7	95.6	75.5	82.2	95.6	59.5	71.5	95.6	84.8	88.4
0960	Flooring	132.8	99.3	122.8	130.8	100.7	121.9	130.8	99.3	121.4	130.8	72.8	113.5	120.5	64.5	103.8	126.2	79.0	112.1
0970, 0990	Wall Finishes & Painting/Coating	112.2	96.2	102.6	113.0	92.3	100.6	112.2	96.0	102.5	112.3	61.3	81.7	112.2	58.5	80.0	112.2	87.7	97.5
09	FINISHES	136.2	94.4	112.8	118.3	91.0	103.0	116.9	92.9	103.5	114.4	73.8	91.6	111.5	61.9	83.7	115.4	87.0	99.5
COVERS	DIVS. 10 - 14, 25, 28, 41, 43, 44, 46	131.1	97.4	124.2	139.2	75.0	126.1	131.1	76.1	119.9	131.1	69.8	118.6	131.1	66.1	117.9	131.1	98.6	124.5
21, 22, 23	FIRE SUPPRESSION, PLUMBING & HVAC	100.8	85.6	94.6	102.7	101.7	102.3	102.9	105.5	104.0	102.9	88.1	96.8	103.4	71.1	90.2	102.9	99.6	101.5
26, 27, 3370	ELECTRICAL, COMMUNICATIONS & UTIL.	109.7	96.1	102.6	120.9	91.8	105.6	117.4	92.7	104.4	119.8	62.6	89.7	120.1	64.3	90.8	117.6	85.4	100.7
MF2010	WEIGHTED AVERAGE	121.1	94.1	109.2	118.2	93.6	107.4	116.4	95.5	107.2	114.7	77.2	98.2	111.1	69.5	92.7	116.8	93.6	106.5

DIVISION		CANADA																	
		QUEBEC, QUEBEC			RED DEER, ALBERTA			REGINA, SASKATCHEWAN			RIMOUSKI, QUEBEC			ROUYN-NORANDA, QUEBEC			SAINT HYACINTHE, QUEBEC		
		MAT.	INST.	TOTAL	MAT.	INST.	TOTAL	MAT.	INST.	TOTAL	MAT.	INST.	TOTAL	MAT.	INST.	TOTAL	MAT.	INST.	TOTAL
015433	**CONTRACTOR EQUIPMENT**		104.2	104.2		103.7	103.7		100.1	100.1		102.3	102.3		102.3	102.3		102.3	102.3
0241, 31 - 34	**SITE & INFRASTRUCTURE, DEMOLITION**	102.9	100.1	100.9	111.1	102.2	104.8	116.9	98.6	104.0	95.4	99.8	98.5	94.9	100.5	98.8	95.4	100.5	99.0
0310	Concrete Forming & Accessories	136.7	95.4	101.0	138.8	86.2	93.3	110.7	92.4	94.9	129.1	94.9	99.5	129.1	90.4	95.6	129.1	90.4	95.6
0320	Concrete Reinforcing	144.4	98.1	121.3	162.0	75.3	118.8	139.4	84.8	112.2	112.3	98.0	105.2	157.5	81.3	119.6	157.5	81.3	119.6
0330	Cast-in-Place Concrete	149.3	103.0	130.3	151.1	103.2	131.4	170.7	96.9	140.4	137.0	101.3	122.3	130.2	100.0	117.8	134.5	100.0	120.3
03	**CONCRETE**	139.1	98.4	119.0	143.0	90.4	117.0	140.5	92.5	116.8	127.4	97.5	112.6	131.6	92.2	112.1	133.6	92.2	113.1
04	**MASONRY**	173.6	96.4	126.5	158.3	84.9	113.5	167.2	96.7	124.2	158.4	96.3	120.5	158.6	89.2	116.3	158.8	89.2	116.4
05	**METALS**	130.0	94.0	118.7	107.4	89.9	101.9	122.3	86.6	111.1	104.6	93.3	101.0	105.0	88.9	100.0	105.0	88.9	100.0
06	**WOOD, PLASTICS & COMPOSITES**	128.0	95.5	109.4	113.7	85.6	97.6	96.4	93.3	94.6	127.2	95.2	108.9	127.2	90.4	106.2	127.2	90.4	106.2
07	**THERMAL & MOISTURE PROTECTION**	120.5	98.8	111.8	129.4	90.0	113.6	114.2	87.3	103.4	105.3	98.0	102.4	105.3	92.4	100.1	105.6	92.4	100.3
08	**OPENINGS**	93.0	91.8	92.7	93.0	80.8	90.1	88.2	81.7	86.6	92.6	82.0	90.0	93.0	78.4	89.5	93.0	78.4	89.5
0920	Plaster & Gypsum Board	193.4	94.9	124.5	137.2	84.8	100.5	170.8	93.0	116.4	188.9	94.9	123.1	147.9	89.9	107.3	150.6	89.9	108.1
0950, 0980	Ceilings & Acoustic Treatment	117.9	94.9	102.6	95.6	84.8	88.4	125.7	93.0	103.9	94.7	94.9	94.8	94.7	89.9	91.5	94.7	89.9	91.5
0960	Flooring	132.8	101.4	123.5	133.2	96.0	122.1	121.5	66.7	105.2	130.8	101.4	122.1	130.8	101.4	122.1	130.8	101.4	122.1
0970, 0990	Wall Finishes & Painting/Coating	112.7	108.1	109.9	112.1	84.8	95.8	112.2	92.9	100.6	112.2	108.1	109.7	112.2	94.2	101.4	112.2	94.2	101.4
09	**FINISHES**	127.4	98.5	111.2	115.8	87.6	100.0	123.9	88.4	104.0	120.2	98.3	107.9	115.0	93.3	102.8	115.3	93.3	103.0
COVERS	**DIVS. 10 - 14, 25, 28, 41, 43, 44, 46**	131.1	89.1	122.6	131.1	95.6	123.9	131.1	73.8	119.4	131.1	88.5	122.4	131.1	87.3	122.2	131.1	87.3	122.2
21, 22, 23	**FIRE SUPPRESSION, PLUMBING & HVAC**	100.5	95.6	98.5	102.9	97.3	100.6	99.9	86.9	94.6	102.9	95.5	99.9	102.9	94.4	99.4	98.6	94.4	96.9
26, 27, 3370	**ELECTRICAL, COMMUNICATIONS & UTIL.**	113.7	88.1	100.2	109.9	89.1	98.9	118.0	93.1	104.9	117.7	88.1	102.1	117.7	74.4	94.9	118.5	74.4	95.3
MF2010	**WEIGHTED AVERAGE**	119.5	95.3	108.9	115.3	91.4	104.8	117.6	90.1	105.5	113.1	94.6	104.9	113.2	89.7	102.9	112.6	89.7	102.5

DIVISION		CANADA																	
		SAINT JOHN, NEW BRUNSWICK			SARNIA, ONTARIO			SASKATOON, SASKATCHEWAN			SAULT STE MARIE, ONTARIO			SHERBROOKE, QUEBEC			SOREL, QUEBEC		
		MAT.	INST.	TOTAL	MAT.	INST.	TOTAL	MAT.	INST.	TOTAL	MAT.	INST.	TOTAL	MAT.	INST.	TOTAL	MAT.	INST.	TOTAL
015433	**CONTRACTOR EQUIPMENT**		101.7	101.7		101.6	101.6		100.1	100.1		101.6	101.6		102.3	102.3		102.3	102.3
0241, 31 - 34	**SITE & INFRASTRUCTURE, DEMOLITION**	96.8	98.6	98.1	112.6	101.0	104.4	107.8	98.6	101.3	102.7	100.5	101.2	95.4	100.5	99.0	95.5	100.5	99.0
0310	Concrete Forming & Accessories	121.3	64.4	72.0	122.8	98.0	101.4	107.0	92.4	94.4	112.3	88.7	91.9	129.1	90.4	95.6	129.1	90.5	95.7
0320	Concrete Reinforcing	148.9	62.9	106.1	124.3	86.3	105.4	119.5	84.8	102.2	112.5	84.7	98.6	157.5	81.3	119.6	148.9	81.3	115.3
0330	Cast-in-Place Concrete	136.2	82.3	114.1	150.9	104.5	131.8	142.0	96.9	123.5	135.6	89.2	116.5	134.5	100.0	120.3	135.6	100.1	121.0
03	**CONCRETE**	136.2	71.3	104.1	135.6	98.0	117.0	123.2	92.5	108.0	119.4	88.3	104.0	133.6	92.2	113.1	132.7	92.2	112.7
04	**MASONRY**	175.9	73.9	113.7	173.5	101.9	129.8	163.8	96.7	122.8	159.3	96.2	120.8	158.9	89.2	116.4	159.0	89.2	116.4
05	**METALS**	104.7	83.5	98.0	106.6	91.6	101.9	105.0	86.6	99.2	105.8	91.7	101.4	105.0	88.9	100.0	105.0	89.0	100.0
06	**WOOD, PLASTICS & COMPOSITES**	116.6	61.3	85.0	115.1	97.2	104.9	94.8	93.3	93.9	102.9	88.3	94.6	127.2	90.4	106.2	127.2	90.4	106.2
07	**THERMAL & MOISTURE PROTECTION**	109.4	70.6	93.8	111.6	97.3	105.9	105.3	87.3	98.1	110.5	90.5	102.4	105.3	92.4	100.1	105.3	92.4	100.1
08	**OPENINGS**	87.4	59.4	80.6	94.1	91.8	93.5	87.3	81.7	85.9	85.1	86.7	85.5	93.0	78.4	89.5	93.0	83.3	90.7
0920	Plaster & Gypsum Board	182.4	60.0	96.8	181.2	97.1	122.4	144.0	93.0	108.4	143.1	87.9	104.5	150.6	89.9	108.1	188.9	89.9	119.6
0950, 0980	Ceilings & Acoustic Treatment	102.9	60.0	74.3	100.2	97.1	98.2	117.5	93.0	101.2	95.6	87.9	90.5	94.7	89.9	91.5	94.7	89.9	91.5
0960	Flooring	126.7	74.7	111.2	130.8	108.6	124.2	120.5	66.7	104.5	123.5	102.7	117.3	130.8	101.4	122.1	130.8	101.4	122.1
0970, 0990	Wall Finishes & Painting/Coating	112.2	80.6	93.3	112.2	108.6	110.0	112.2	92.9	100.6	112.2	100.6	105.2	112.2	94.2	101.4	112.2	94.2	101.4
09	**FINISHES**	120.4	67.3	90.6	121.3	101.6	110.3	117.8	88.4	101.3	112.5	92.6	101.3	115.3	93.3	103.0	120.2	93.3	105.1
COVERS	**DIVS. 10 - 14, 25, 28, 41, 43, 44, 46**	131.1	67.0	118.0	131.1	77.5	120.2	131.1	74.6	119.6	131.1	98.6	124.5	131.1	87.3	122.2	131.1	87.3	122.2
21, 22, 23	**FIRE SUPPRESSION, PLUMBING & HVAC**	102.9	83.6	95.0	102.9	111.8	106.5	99.8	86.9	94.5	102.9	98.3	101.0	103.4	94.4	99.7	102.9	94.4	99.4
26, 27, 3370	**ELECTRICAL, COMMUNICATIONS & UTIL.**	128.6	91.9	109.3	120.5	95.0	107.1	122.0	93.1	106.8	119.2	93.1	105.4	117.7	74.4	94.9	117.7	74.4	94.9
MF2010	**WEIGHTED AVERAGE**	115.5	79.2	99.5	116.2	100.1	109.1	112.0	90.1	102.4	111.4	94.1	103.8	113.6	89.7	103.1	113.8	89.9	103.3

DIVISION		CANADA																	
		ST CATHARINES, ONTARIO			ST JEROME, QUEBEC			ST JOHNS, NEWFOUNDLAND			SUDBURY, ONTARIO			SUMMERSIDE, PRINCE EDWARD ISLAND			SYDNEY, NOVA SCOTIA		
		MAT.	INST.	TOTAL	MAT.	INST.	TOTAL	MAT.	INST.	TOTAL	MAT.	INST.	TOTAL	MAT.	INST.	TOTAL	MAT.	INST.	TOTAL
015433	**CONTRACTOR EQUIPMENT**		101.6	101.6		102.3	102.3		102.9	102.9		101.6	101.6		101.4	101.4		101.4	101.4
0241, 31 - 34	**SITE & INFRASTRUCTURE, DEMOLITION**	102.7	101.9	102.2	94.9	100.5	98.8	115.1	99.9	104.5	102.7	101.5	101.9	106.9	96.4	99.6	94.9	98.9	97.7
0310	Concrete Forming & Accessories	116.2	95.7	98.5	129.1	90.4	95.6	107.9	84.2	87.4	112.3	89.1	92.2	97.2	60.0	65.0	96.9	73.7	76.8
0320	Concrete Reinforcing	111.6	94.6	103.1	157.5	81.3	119.6	167.5	75.6	121.7	112.5	94.5	103.5	146.7	48.3	97.7	148.9	48.2	98.8
0330	Cast-in-Place Concrete	134.5	101.8	121.0	130.2	100.0	117.8	173.5	88.6	138.6	135.6	97.9	120.1	161.2	62.5	120.6	131.0	73.8	107.4
03	**CONCRETE**	118.9	97.5	108.4	131.6	92.2	112.1	165.6	84.4	125.5	119.4	93.2	106.4	152.4	59.4	106.5	129.7	69.7	100.1
04	**MASONRY**	158.9	101.5	123.8	158.6	89.2	116.3	163.9	91.9	120.0	159.0	96.7	120.9	157.6	64.2	100.6	155.7	73.7	105.7
05	**METALS**	105.8	93.4	101.9	105.0	88.9	100.0	124.2	84.1	111.6	105.8	92.9	101.7	105.8	68.7	94.2	105.8	75.4	96.2
06	**WOOD, PLASTICS & COMPOSITES**	106.8	95.6	100.4	127.2	90.4	106.2	95.9	82.6	88.3	102.9	88.3	94.6	87.2	59.3	71.3	87.0	73.0	79.0
07	**THERMAL & MOISTURE PROTECTION**	111.0	97.7	105.7	105.3	92.4	100.1	118.3	86.8	105.7	110.5	91.9	103.0	105.8	62.9	88.6	106.6	71.9	92.6
08	**OPENINGS**	84.2	92.2	86.2	93.0	78.4	89.5	98.9	74.7	93.0	85.1	86.7	85.5	95.2	51.3	84.5	84.3	66.6	80.0
0920	Plaster & Gypsum Board	137.8	95.5	108.2	147.9	89.9	107.3	175.3	81.9	110.0	143.1	87.9	104.5	158.8	58.0	88.3	158.4	72.0	98.0
0950, 0980	Ceilings & Acoustic Treatment	100.2	95.5	97.0	94.7	89.9	91.5	112.0	81.9	91.9	95.6	87.9	90.5	95.6	58.0	70.6	95.6	72.0	79.9
0960	Flooring	125.7	100.7	118.3	130.8	101.4	122.1	116.9	60.4	100.1	123.5	102.7	117.3	112.4	64.1	98.0	112.3	68.5	99.3
0970, 0990	Wall Finishes & Painting/Coating	112.2	104.8	107.8	112.2	94.2	101.4	112.2	77.5	91.4	112.2	95.6	102.3	112.2	44.1	71.4	112.2	65.9	84.4
09	**FINISHES**	113.4	98.0	104.8	115.0	93.3	102.8	119.7	79.5	97.1	112.5	92.1	101.1	112.3	59.4	82.6	111.2	72.1	89.3
COVERS	**DIVS. 10 - 14, 25, 28, 41, 43, 44, 46**	131.1	76.5	120.0	131.1	87.3	122.2	131.1	73.6	119.4	131.1	98.7	124.5	131.1	66.4	117.9	131.1	68.2	118.3
21, 22, 23	**FIRE SUPPRESSION, PLUMBING & HVAC**	99.7	86.8	94.4	102.9	94.4	99.4	100.1	76.2	90.3	102.9	87.8	96.7	102.9	66.2	87.9	102.9	87.6	96.6
26, 27, 3370	**ELECTRICAL, COMMUNICATIONS & UTIL.**	119.4	96.2	107.2	118.4	74.4	95.2	120.8	73.7	96.0	116.4	96.5	105.9	118.4	53.6	84.3	119.6	66.0	91.3
MF2010	**WEIGHTED AVERAGE**	110.6	94.6	103.6	113.3	89.7	102.9	121.7	81.9	104.2	111.1	93.2	103.2	115.7	64.5	93.2	111.7	76.6	96.3

DIVISION		CANADA																	
		THUNDER BAY, ONTARIO			TIMMINS, ONTARIO			TORONTO, ONTARIO			TROIS RIVIERES, QUEBEC			TRURO, NOVA SCOTIA			VANCOUVER, BRITISH COLUMBIA		
		MAT.	INST.	TOTAL	MAT.	INST.	TOTAL	MAT.	INST.	TOTAL	MAT.	INST.	TOTAL	MAT.	INST.	TOTAL	MAT.	INST.	TOTAL
015433	**CONTRACTOR EQUIPMENT**		101.6	101.6		101.6	101.6		104.0	104.0		102.3	102.3		101.4	101.4		112.0	112.0
0241, 31 - 34	**SITE & INFRASTRUCTURE, DEMOLITION**	107.7	101.8	103.5	114.0	100.5	104.5	138.5	106.0	115.7	95.5	100.5	99.0	99.3	98.9	99.0	121.7	108.4	112.4
0310	Concrete Forming & Accessories	124.1	92.6	96.9	124.0	89.6	94.2	131.5	101.4	105.5	129.1	90.5	95.7	96.9	73.7	76.8	126.2	93.7	98.1
0320	Concrete Reinforcing	99.7	93.4	96.6	175.2	84.4	130.0	161.9	94.8	128.5	148.9	81.3	115.3	148.9	48.2	98.8	151.3	83.1	117.3
0330	Cast-in-Place Concrete	148.0	100.2	128.4	163.5	89.0	132.9	154.4	110.1	136.2	135.6	100.1	121.0	169.8	73.8	130.3	164.1	101.3	138.3
03	**CONCRETE**	127.4	95.4	111.6	150.1	88.6	119.7	144.0	102.9	123.7	132.7	92.2	112.7	148.3	69.7	109.5	152.5	94.6	123.9
04	**MASONRY**	159.6	100.7	123.7	162.8	92.2	119.7	192.7	105.9	139.7	159.0	89.2	116.4	158.5	73.7	106.8	177.1	95.2	127.1
05	**METALS**	105.6	92.3	101.4	106.6	90.7	101.6	125.6	94.0	115.7	105.0	89.0	100.0	105.8	75.5	96.3	152.8	92.5	133.9
06	**WOOD, PLASTICS & COMPOSITES**	116.2	91.4	102.0	115.9	90.0	101.1	116.2	100.8	107.4	127.2	90.4	106.2	87.0	73.0	79.0	114.9	93.6	102.8
07	**THERMAL & MOISTURE PROTECTION**	111.2	94.8	104.6	111.6	88.0	102.1	122.8	101.4	114.2	105.3	92.4	100.1	106.6	71.9	92.6	132.4	90.4	115.5
08	**OPENINGS**	83.3	89.3	84.7	92.0	86.0	90.5	92.0	98.5	93.5	93.0	83.3	90.7	84.3	66.6	80.0	94.8	89.7	93.5
0920	Plaster & Gypsum Board	174.1	91.1	116.1	154.1	89.7	109.0	178.7	100.8	124.2	188.9	89.9	119.6	158.4	72.0	98.0	162.9	92.9	113.9
0950, 0980	Ceilings & Acoustic Treatment	95.6	91.1	92.6	95.6	89.7	91.6	130.1	100.8	110.6	94.7	89.9	91.5	95.6	72.0	79.9	156.0	92.9	113.9
0960	Flooring	130.8	107.9	124.0	130.8	99.3	121.4	135.6	110.5	128.1	130.8	101.4	122.1	112.3	68.5	99.3	133.2	99.4	123.1
0970, 0990	Wall Finishes & Painting/Coating	112.2	94.1	101.4	112.2	93.7	101.1	114.3	109.0	111.1	112.2	94.2	101.4	112.2	65.9	84.4	112.1	99.9	104.8
09	**FINISHES**	119.0	95.7	105.9	116.9	92.2	103.1	131.7	104.2	116.3	120.2	93.3	105.1	111.2	72.1	89.3	133.4	95.6	112.2
COVERS	**DIVS. 10 - 14, 25, 28, 41, 43, 44, 46**	131.1	76.2	119.9	131.1	74.5	119.6	131.1	102.3	125.3	131.1	87.3	122.2	131.1	68.2	118.3	131.1	98.5	124.5
21, 22, 23	**FIRE SUPPRESSION, PLUMBING & HVAC**	99.7	86.7	94.4	102.9	101.8	102.4	100.6	96.0	98.8	102.9	94.4	99.4	102.9	87.6	96.6	100.6	84.9	94.2
26, 27, 3370	**ELECTRICAL, COMMUNICATIONS & UTIL.**	117.7	94.6	105.6	119.2	93.1	105.4	113.7	98.3	105.6	119.9	74.4	96.0	117.4	66.0	90.3	119.2	82.8	100.0
MF2010	**WEIGHTED AVERAGE**	111.9	93.3	103.7	116.6	93.5	106.4	121.6	100.5	112.3	114.0	89.9	103.4	113.8	76.6	97.4	126.8	91.9	111.4

DIVISION		CANADA																	
		VICTORIA, BRITISH COLUMBIA			WHITEHORSE, YUKON			WINDSOR, ONTARIO			WINNIPEG, MANITOBA			YARMOUTH, NOVA SCOTIA			YELLOWKNIFE, NWT		
		MAT.	INST.	TOTAL	MAT.	INST.	TOTAL	MAT.	INST.	TOTAL	MAT.	INST.	TOTAL	MAT.	INST.	TOTAL	MAT.	INST.	TOTAL
015433	**CONTRACTOR EQUIPMENT**		108.7	108.7		101.6	101.6		101.6	101.6		105.8	105.8		101.4	101.4		101.5	101.5
0241, 31 - 34	**SITE & INFRASTRUCTURE, DEMOLITION**	121.8	107.9	112.1	110.0	97.4	101.2	97.5	101.8	100.5	122.7	101.5	107.8	99.1	98.9	98.9	128.4	101.8	109.8
0310	Concrete Forming & Accessories	112.4	93.1	95.7	110.5	61.3	67.9	124.1	91.8	96.1	133.2	69.7	78.2	96.9	73.7	76.8	110.4	83.9	87.5
0320	Concrete Reinforcing	116.8	83.0	100.0	133.7	61.9	97.9	109.4	94.5	102.0	146.5	58.4	102.6	148.9	48.2	98.8	132.3	65.8	99.2
0330	Cast-in-Place Concrete	147.1	100.6	128.0	162.2	71.8	125.0	137.7	102.1	123.1	192.3	75.3	144.1	168.1	73.8	129.3	244.8	91.3	181.7
03	**CONCRETE**	152.1	93.8	123.3	135.4	65.8	101.0	120.7	95.8	108.4	161.9	70.1	116.5	147.5	69.7	109.0	174.8	83.3	129.6
04	**MASONRY**	166.5	95.2	123.0	170.0	64.3	105.5	159.1	99.5	122.7	187.4	71.0	116.3	158.4	73.7	106.7	176.8	78.0	116.5
05	**METALS**	107.2	88.4	101.3	121.9	74.8	107.1	105.7	92.8	101.6	152.3	75.3	128.1	105.8	75.4	96.2	117.6	80.7	106.0
06	**WOOD, PLASTICS & COMPOSITES**	98.4	93.5	95.6	96.1	59.8	75.4	116.2	91.0	101.8	114.0	70.4	89.2	87.0	73.0	79.0	96.1	85.5	90.0
07	**THERMAL & MOISTURE PROTECTION**	114.4	90.2	104.7	113.4	63.8	93.5	111.0	93.9	104.2	118.3	72.2	99.7	106.6	71.9	92.6	113.4	81.8	100.7
08	**OPENINGS**	89.1	85.8	88.3	86.8	57.0	79.6	83.0	89.6	84.6	93.0	62.8	85.7	84.3	66.6	80.0	86.8	72.0	83.2
0920	Plaster & Gypsum Board	135.6	92.9	105.7	147.5	58.4	85.2	157.7	90.7	110.9	159.3	69.0	96.2	158.4	72.0	98.0	184.1	84.9	114.7
0950, 0980	Ceilings & Acoustic Treatment	95.6	92.9	93.8	102.9	58.4	73.3	95.6	90.7	92.4	140.2	69.0	92.8	95.6	72.0	79.9	127.5	84.9	99.1
0960	Flooring	126.2	79.0	112.1	121.5	62.5	103.9	130.8	105.0	123.1	132.4	75.3	115.4	112.3	68.5	99.3	121.5	95.9	113.9
0970, 0990	Wall Finishes & Painting/Coating	112.2	99.9	104.8	112.2	57.5	79.4	112.2	97.5	103.4	112.3	56.7	79.0	112.2	65.9	84.4	112.2	83.9	95.2
09	**FINISHES**	115.4	92.1	102.3	115.5	60.8	84.9	116.5	95.0	104.4	128.9	69.7	95.7	111.2	72.1	89.3	125.5	85.5	103.1
COVERS	**DIVS. 10 - 14, 25, 28, 41, 43, 44, 46**	131.1	74.0	119.5	131.1	65.4	117.7	131.1	75.6	119.8	131.1	68.9	118.4	131.1	68.2	118.3	131.1	69.8	118.6
21, 22, 23	**FIRE SUPPRESSION, PLUMBING & HVAC**	102.9	84.8	95.5	103.1	77.6	92.6	99.7	86.3	94.2	100.5	67.8	87.1	102.9	87.6	96.6	103.3	99.1	101.6
26, 27, 3370	**ELECTRICAL, COMMUNICATIONS & UTIL.**	118.6	82.8	99.7	134.1	63.8	97.1	123.0	96.2	108.9	119.7	68.1	92.6	119.6	66.0	91.3	135.2	88.8	110.8
MF2010	**WEIGHTED AVERAGE**	116.8	90.0	105.0	118.4	70.3	97.2	111.2	93.4	103.3	127.3	72.3	103.1	113.9	76.6	97.5	123.9	87.6	107.9

Location Factors

Costs shown in RSMeans cost data publications are based on national averages for materials and installation. To adjust these costs to a specific location, simply multiply the base cost by the factor and divide by 100 for that city. The data is arranged alphabetically by state and postal zip code numbers. For a city not listed, use the factor for a nearby city with similar economic characteristics.

STATE/ZIP	CITY	MAT.	INST.	TOTAL
ALABAMA				
350-352	Birmingham	97.4	73.8	87.0
354	Tuscaloosa	96.5	58.1	79.6
355	Jasper	96.8	56.9	79.2
356	Decatur	96.5	59.4	80.1
357-358	Huntsville	96.6	67.4	83.7
359	Gadsden	96.5	57.1	79.1
360-361	Montgomery	97.5	56.1	79.2
362	Anniston	95.9	63.8	81.7
363	Dothan	96.4	51.4	76.6
364	Evergreen	95.9	52.8	77.0
365-366	Mobile	97.5	66.8	84.0
367	Selma	96.1	51.3	76.4
368	Phenix City	96.8	55.4	78.5
369	Butler	96.3	51.4	76.5
ALASKA				
995-996	Anchorage	123.8	114.6	119.8
997	Fairbanks	122.0	115.4	119.1
998	Juneau	119.7	114.3	117.3
999	Ketchikan	132.8	114.3	124.7
ARIZONA				
850,853	Phoenix	98.7	75.4	88.4
851,852	Mesa/Tempe	98.1	71.3	86.3
855	Globe	98.2	61.4	82.0
856-857	Tucson	96.9	70.9	85.5
859	Show Low	98.3	61.8	82.2
860	Flagstaff	100.7	72.7	88.4
863	Prescott	98.5	61.1	82.0
864	Kingman	96.8	72.3	86.0
865	Chambers	96.8	62.0	81.5
ARKANSAS				
716	Pine Bluff	97.1	63.1	82.1
717	Camden	94.4	49.5	74.6
718	Texarkana	96.0	50.6	76.0
719	Hot Springs	93.7	52.2	75.4
720-722	Little Rock	97.5	64.2	82.8
723	West Memphis	96.0	62.0	81.0
724	Jonesboro	97.0	60.0	80.7
725	Batesville	94.6	55.3	77.3
726	Harrison	95.8	51.6	76.4
727	Fayetteville	93.3	53.4	75.7
728	Russellville	94.5	54.6	76.9
729	Fort Smith	97.4	61.8	81.7
CALIFORNIA				
900-902	Los Angeles	99.8	116.6	107.2
903-905	Inglewood	95.0	112.1	102.5
906-908	Long Beach	96.5	112.2	103.4
910-912	Pasadena	95.4	112.1	102.8
913-916	Van Nuys	98.3	112.1	104.4
917-918	Alhambra	97.3	112.1	103.8
919-921	San Diego	99.6	108.0	103.3
922	Palm Springs	96.3	109.5	102.2
923-924	San Bernardino	94.3	109.0	100.8
925	Riverside	98.8	113.7	105.4
926-927	Santa Ana	96.1	109.5	102.0
928	Anaheim	98.8	113.7	105.4
930	Oxnard	99.9	112.9	105.6
931	Santa Barbara	99.2	113.9	105.7
932-933	Bakersfield	100.1	109.8	104.4
934	San Luis Obispo	99.4	107.9	103.2
935	Mojave	96.8	105.4	100.6
936-938	Fresno	100.3	115.0	106.8
939	Salinas	100.3	121.8	109.8
940-941	San Francisco	106.8	142.4	122.5
942,956-958	Sacramento	101.7	119.2	109.5
943	Palo Alto	99.8	128.5	112.5
944	San Mateo	102.1	130.3	114.5
945	Vallejo	101.0	128.5	113.1
946	Oakland	104.6	133.3	117.3
947	Berkeley	103.8	132.5	116.5
948	Richmond	102.9	128.7	114.3
949	San Rafael	104.4	128.7	115.1
950	Santa Cruz	104.4	121.9	112.1

STATE/ZIP	CITY	MAT.	INST.	TOTAL
CALIFORNIA (CONT'D)				
951	San Jose	102.5	135.3	116.9
952	Stockton	100.5	116.6	107.6
953	Modesto	100.4	116.0	107.3
954	Santa Rosa	100.0	134.2	115.1
955	Eureka	101.3	112.7	106.3
959	Marysville	100.5	113.1	106.1
960	Redding	102.4	113.1	107.1
961	Susanville	101.1	113.1	106.4
COLORADO				
800-802	Denver	101.7	83.4	93.6
803	Boulder	98.4	83.3	91.7
804	Golden	100.4	80.9	91.8
805	Fort Collins	102.3	80.8	92.8
806	Greeley	99.8	77.3	89.9
807	Fort Morgan	99.0	80.9	91.0
808-809	Colorado Springs	101.4	80.6	92.2
810	Pueblo	100.7	79.8	91.5
811	Alamosa	101.8	75.7	90.3
812	Salida	101.5	77.4	90.9
813	Durango	102.1	78.3	91.6
814	Montrose	101.0	78.0	90.8
815	Grand Junction	104.4	78.2	92.9
816	Glenwood Springs	101.8	80.9	92.6
CONNECTICUT				
060	New Britain	100.6	119.9	109.1
061	Hartford	102.0	120.1	109.9
062	Willimantic	101.2	120.2	109.6
063	New London	97.2	120.0	107.2
064	Meriden	99.2	120.0	108.4
065	New Haven	102.1	120.0	110.0
066	Bridgeport	101.7	119.7	109.6
067	Waterbury	101.3	119.6	109.3
068	Norwalk	101.2	127.1	112.6
069	Stamford	101.4	127.1	112.7
D.C.				
200-205	Washington	101.1	92.4	97.3
DELAWARE				
197	Newark	98.8	108.8	103.2
198	Wilmington	98.6	108.9	103.1
199	Dover	99.3	108.8	103.5
FLORIDA				
320,322	Jacksonville	98.4	68.6	85.2
321	Daytona Beach	98.6	74.3	87.9
323	Tallahassee	99.1	59.1	81.5
324	Panama City	99.8	60.2	82.4
325	Pensacola	102.2	63.4	85.1
326,344	Gainesville	100.0	68.9	86.3
327-328,347	Orlando	101.8	72.3	88.8
329	Melbourne	101.2	77.3	90.7
330-332,340	Miami	99.8	76.0	89.3
333	Fort Lauderdale	97.9	75.2	87.9
334,349	West Palm Beach	96.6	73.5	86.4
335-336,346	Tampa	99.5	81.1	91.4
337	St. Petersburg	102.2	65.6	86.1
338	Lakeland	98.7	81.0	90.9
339,341	Fort Myers	97.9	75.1	87.8
342	Sarasota	99.7	76.9	89.7
GEORGIA				
300-303,399	Atlanta	97.5	75.9	88.0
304	Statesboro	96.9	53.3	77.7
305	Gainesville	95.6	63.1	81.3
306	Athens	95.0	63.3	81.0
307	Dalton	96.9	59.3	80.3
308-309	Augusta	96.3	64.2	82.1
310-312	Macon	96.2	63.9	82.0
313-314	Savannah	98.2	61.1	81.9
315	Waycross	97.1	60.6	81.0
316	Valdosta	97.4	60.4	81.1
317,398	Albany	97.4	61.1	81.4
318-319	Columbus	97.4	63.9	82.6

Location Factors

STATE/ZIP	CITY	MAT.	INST.	TOTAL
HAWAII				
967	Hilo	109.5	122.3	115.2
968	Honolulu	113.6	122.3	117.4
STATES & POSS.				
969	Guam	130.4	60.0	99.4
IDAHO				
832	Pocatello	100.7	72.5	88.3
833	Twin Falls	101.8	52.7	80.1
834	Idaho Falls	99.4	58.3	81.3
835	Lewiston	108.1	79.5	95.5
836-837	Boise	99.6	73.3	88.0
838	Coeur d'Alene	107.4	77.1	94.0
ILLINOIS				
600-603	North Suburban	98.8	130.3	112.7
604	Joliet	98.8	134.5	114.5
605	South Suburban	98.8	130.3	112.7
606-608	Chicago	99.3	138.3	116.5
609	Kankakee	94.8	125.4	108.3
610-611	Rockford	98.6	122.3	109.0
612	Rock Island	96.0	98.3	97.0
613	La Salle	97.1	119.0	106.8
614	Galesburg	97.0	103.2	99.7
615-616	Peoria	99.7	106.7	102.8
617	Bloomington	96.4	107.8	101.4
618-619	Champaign	99.8	106.8	102.9
620-622	East St. Louis	95.0	106.6	100.1
623	Quincy	96.1	98.6	97.2
624	Effingham	95.4	102.3	98.4
625	Decatur	97.6	105.8	101.2
626-627	Springfield	98.3	106.5	101.9
628	Centralia	93.4	105.4	98.7
629	Carbondale	93.1	100.5	96.4
INDIANA				
460	Anderson	95.7	82.3	89.8
461-462	Indianapolis	98.9	84.7	92.6
463-464	Gary	97.2	107.0	101.5
465-466	South Bend	95.5	81.8	89.5
467-468	Fort Wayne	96.3	77.7	88.1
469	Kokomo	93.3	81.1	87.9
470	Lawrenceburg	93.0	78.4	86.6
471	New Albany	94.1	75.4	85.9
472	Columbus	96.4	80.1	89.3
473	Muncie	97.5	81.4	90.4
474	Bloomington	98.8	81.6	91.2
475	Washington	94.7	82.4	89.3
476-477	Evansville	96.5	85.2	91.5
478	Terre Haute	97.2	85.7	92.1
479	Lafayette	96.1	82.0	89.9
IOWA				
500-503,509	Des Moines	98.9	83.5	92.1
504	Mason City	97.3	66.9	83.9
505	Fort Dodge	97.5	58.9	80.5
506-507	Waterloo	99.4	71.2	86.9
508	Creston	97.6	77.7	88.8
510-511	Sioux City	100.0	69.7	86.7
512	Sibley	98.5	55.8	79.7
513	Spencer	100.1	55.8	80.5
514	Carroll	97.0	71.4	85.7
515	Council Bluffs	101.2	79.1	91.5
516	Shenandoah	97.8	71.3	86.1
520	Dubuque	99.5	78.6	90.3
521	Decorah	98.1	64.8	83.4
522-524	Cedar Rapids	100.3	82.8	92.6
525	Ottumwa	98.0	69.7	85.5
526	Burlington	97.2	76.2	88.0
527-528	Davenport	99.3	89.2	94.8
KANSAS				
660-662	Kansas City	99.7	93.9	97.2
664-666	Topeka	99.4	63.8	83.7
667	Fort Scott	97.8	72.9	86.8
668	Emporia	97.9	68.7	85.0
669	Belleville	99.5	64.6	84.1
670-672	Wichita	98.8	63.2	83.1
673	Independence	99.2	68.4	85.6
674	Salina	100.1	65.9	85.0
675	Hutchinson	95.0	64.5	81.5
676	Hays	99.0	64.8	83.9
677	Colby	99.5	64.8	84.2

STATE/ZIP	CITY	MAT.	INST.	TOTAL
KANSAS (CONT'D)				
678	Dodge City	101.2	65.8	85.6
679	Liberal	98.8	64.8	83.8
KENTUCKY				
400-402	Louisville	98.6	82.2	91.3
403-405	Lexington	97.0	81.0	90.0
406	Frankfort	98.0	82.3	91.1
407-409	Corbin	94.2	71.4	84.1
410	Covington	95.0	94.1	94.6
411-412	Ashland	93.7	98.0	95.6
413-414	Campton	94.9	80.0	88.3
415-416	Pikeville	96.0	90.2	93.4
417-418	Hazard	94.4	80.4	88.2
420	Paducah	93.0	84.5	89.2
421-422	Bowling Green	95.6	83.3	90.1
423	Owensboro	95.6	86.7	91.7
424	Henderson	92.8	86.6	90.1
425-426	Somerset	92.5	79.3	86.7
427	Elizabethtown	91.9	82.6	87.8
LOUISIANA				
700-701	New Orleans	102.4	70.5	88.4
703	Thibodaux	100.1	62.7	83.6
704	Hammond	97.5	55.4	78.9
705	Lafayette	99.9	61.0	82.7
706	Lake Charles	100.1	63.4	83.9
707-708	Baton Rouge	101.5	62.6	84.4
710-711	Shreveport	97.0	55.6	78.8
712	Monroe	97.1	52.9	77.6
713-714	Alexandria	97.2	52.8	77.6
MAINE				
039	Kittery	94.2	84.0	89.7
040-041	Portland	99.2	92.1	96.1
042	Lewiston	97.6	92.1	95.2
043	Augusta	97.5	81.7	90.5
044	Bangor	97.1	92.1	94.9
045	Bath	95.3	83.6	90.1
046	Machias	94.9	87.5	91.6
047	Houlton	95.0	87.6	91.7
048	Rockland	94.2	87.5	91.2
049	Waterville	95.4	81.7	89.4
MARYLAND				
206	Waldorf	97.0	82.7	90.7
207-208	College Park	97.0	84.5	91.5
209	Silver Spring	96.3	82.6	90.3
210-212	Baltimore	100.1	84.1	93.0
214	Annapolis	100.4	82.7	92.6
215	Cumberland	95.8	81.9	89.7
216	Easton	97.3	69.7	85.1
217	Hagerstown	96.8	83.5	90.9
218	Salisbury	97.7	64.6	83.1
219	Elkton	94.8	83.2	89.7
MASSACHUSETTS				
010-011	Springfield	99.6	109.5	104.0
012	Pittsfield	99.1	107.1	102.7
013	Greenfield	96.9	109.9	102.6
014	Fitchburg	95.7	124.1	108.2
015-016	Worcester	99.6	124.3	110.4
017	Framingham	94.9	131.2	110.9
018	Lowell	99.0	130.9	113.1
019	Lawrence	99.8	132.5	114.2
020-022, 024	Boston	101.5	139.2	118.1
023	Brockton	100.1	127.1	112.0
025	Buzzards Bay	94.3	127.7	109.0
026	Hyannis	97.3	127.7	110.7
027	New Bedford	99.3	127.9	111.9
MICHIGAN				
480,483	Royal Oak	94.4	104.5	98.9
481	Ann Arbor	96.8	106.7	101.2
482	Detroit	98.4	109.6	103.4
484-485	Flint	96.4	95.7	96.1
486	Saginaw	96.0	92.2	94.3
487	Bay City	96.2	92.3	94.5
488-489	Lansing	97.4	93.8	95.8
490	Battle Creek	96.5	87.7	92.6
491	Kalamazoo	96.8	86.0	92.1
492	Jackson	94.7	94.5	94.6
493,495	Grand Rapids	98.0	81.6	90.8
494	Muskegon	95.4	83.6	90.2

STATE/ZIP	CITY	MAT.	INST.	TOTAL
MICHIGAN (CONT'D)				
496	Traverse City	94.2	77.4	86.8
497	Gaylord	95.2	71.7	84.8
498-499	Iron mountain	97.2	86.4	92.4
MINNESOTA				
550-551	Saint Paul	100.4	119.9	109.0
553-555	Minneapolis	101.4	122.0	110.5
556-558	Duluth	99.7	110.3	104.3
559	Rochester	100.1	106.1	102.7
560	Mankato	97.3	103.2	99.9
561	Windom	96.0	92.8	94.6
562	Willmar	95.6	103.9	99.2
563	St. Cloud	97.2	119.4	107.0
564	Brainerd	97.1	103.7	100.0
565	Detroit Lakes	99.1	97.9	98.6
566	Bemidji	98.4	99.8	99.0
567	Thief River Falls	98.0	94.5	96.4
MISSISSIPPI				
386	Clarksdale	96.1	54.9	77.9
387	Greenville	99.9	64.3	84.2
388	Tupelo	97.7	57.3	79.9
389	Greenwood	97.3	54.4	78.4
390-392	Jackson	97.4	63.6	82.5
393	Meridian	96.5	65.3	82.7
394	Laurel	97.5	58.1	80.1
395	Biloxi	98.1	59.5	81.0
396	Mccomb	96.1	54.0	77.5
397	Columbus	97.4	56.2	79.3
MISSOURI				
630-631	St. Louis	99.9	106.3	102.7
633	Bowling Green	98.2	94.5	96.6
634	Hannibal	97.1	91.8	94.7
635	Kirksville	100.3	89.4	95.5
636	Flat River	99.1	95.2	97.4
637	Cape Girardeau	99.5	89.5	95.1
638	Sikeston	97.5	89.9	94.2
639	Poplar Bluff	97.0	89.7	93.8
640-641	Kansas City	100.7	108.2	104.0
644-645	St. Joseph	99.9	94.3	97.4
646	Chillicothe	96.6	95.2	96.0
647	Harrisonville	96.2	102.9	99.1
648	Joplin	98.6	81.7	91.2
650-651	Jefferson City	96.6	91.7	94.4
652	Columbia	97.7	93.2	95.7
653	Sedalia	96.7	92.4	94.8
654-655	Rolla	95.1	95.7	95.3
656-658	Springfield	98.6	82.8	91.7
MONTANA				
590-591	Billings	103.3	77.5	91.9
592	Wolf Point	102.5	75.4	90.6
593	Miles City	100.3	75.1	89.2
594	Great Falls	104.5	76.8	92.3
595	Havre	101.4	75.0	89.7
596	Helena	102.7	75.0	90.5
597	Butte	103.1	77.5	91.8
598	Missoula	100.6	77.0	90.2
599	Kalispell	99.3	76.3	89.1
NEBRASKA				
680-681	Omaha	100.4	79.5	91.2
683-685	Lincoln	98.6	73.0	87.3
686	Columbus	96.8	70.6	85.3
687	Norfolk	98.6	74.0	87.8
688	Grand Island	98.7	77.8	89.5
689	Hastings	97.8	78.1	89.1
690	Mccook	97.6	69.6	85.3
691	North Platte	98.1	78.3	89.4
692	Valentine	100.0	66.6	85.3
693	Alliance	100.0	64.8	84.5
NEVADA				
889-891	Las Vegas	100.2	109.0	104.1
893	Ely	98.7	99.3	99.0
894-895	Reno	99.0	92.0	95.9
897	Carson City	98.9	91.8	95.8
898	Elko	97.6	84.5	91.8
NEW HAMPSHIRE				
030	Nashua	99.8	94.0	97.2
031	Manchester	100.6	94.0	97.7

STATE/ZIP	CITY	MAT.	INST.	TOTAL
NEW HAMPSHIRE (CONT'D)				
032-033	Concord	98.4	92.8	95.9
034	Keene	96.4	74.6	86.8
035	Littleton	96.0	77.7	88.0
036	Charleston	95.8	70.6	84.7
037	Claremont	94.9	70.7	84.2
038	Portsmouth	97.3	94.5	96.1
NEW JERSEY				
070-071	Newark	102.7	124.8	112.4
072	Elizabeth	100.1	124.8	111.0
073	Jersey City	99.0	124.7	110.3
074-075	Paterson	100.9	124.9	111.5
076	Hackensack	98.9	124.9	110.4
077	Long Branch	98.4	122.3	108.9
078	Dover	99.2	124.9	110.5
079	Summit	99.2	124.8	110.5
080,083	Vineland	97.1	123.4	108.7
081	Camden	99.0	122.8	109.5
082,084	Atlantic City	97.6	123.3	108.9
085-086	Trenton	99.5	122.5	109.6
087	Point Pleasant	98.9	121.6	108.9
088-089	New Brunswick	99.4	123.9	110.2
NEW MEXICO				
870-872	Albuquerque	99.1	73.5	87.9
873	Gallup	99.0	73.5	87.8
874	Farmington	99.7	73.5	88.2
875	Santa Fe	100.4	73.5	88.5
877	Las Vegas	97.6	73.5	87.0
878	Socorro	97.1	73.5	86.7
879	Truth/Consequences	97.1	70.2	85.2
880	Las Cruces	96.2	69.4	84.4
881	Clovis	97.7	73.4	87.0
882	Roswell	99.6	73.5	88.1
883	Carrizozo	99.7	73.5	88.2
884	Tucumcari	98.4	73.4	87.4
NEW YORK				
100-102	New York	104.2	167.0	131.9
103	Staten Island	100.0	164.8	128.6
104	Bronx	98.3	164.8	127.6
105	Mount Vernon	98.0	133.0	113.4
106	White Plains	98.3	133.0	113.6
107	Yonkers	102.9	133.1	116.2
108	New Rochelle	98.3	133.0	113.6
109	Suffern	98.1	121.5	108.4
110	Queens	101.1	164.8	129.1
111	Long Island City	102.5	164.8	130.0
112	Brooklyn	103.0	164.8	130.2
113	Flushing	102.4	164.8	129.9
114	Jamaica	100.7	164.8	129.0
115,117,118	Hicksville	101.0	145.4	120.5
116	Far Rockaway	102.6	164.8	130.0
119	Riverhead	101.7	145.1	120.9
120-122	Albany	97.0	99.8	98.2
123	Schenectady	97.6	99.0	98.2
124	Kingston	100.4	118.5	108.4
125-126	Poughkeepsie	99.6	128.6	112.4
127	Monticello	99.0	117.6	107.2
128	Glens Falls	92.2	94.9	93.4
129	Plattsburgh	96.8	88.6	93.2
130-132	Syracuse	98.7	96.2	97.6
133-135	Utica	96.7	93.3	95.2
136	Watertown	98.2	93.1	95.9
137-139	Binghamton	98.2	100.1	99.0
140-142	Buffalo	100.0	104.1	101.8
143	Niagara Falls	97.5	105.5	101.0
144-146	Rochester	100.1	96.0	98.3
147	Jamestown	96.5	87.9	92.7
148-149	Elmira	96.3	92.7	94.8
NORTH CAROLINA				
270,272-274	Greensboro	99.0	56.2	80.2
271	Winston-Salem	98.8	57.5	80.6
275-276	Raleigh	99.0	55.1	79.7
277	Durham	100.3	53.9	79.9
278	Rocky Mount	96.1	53.9	77.5
279	Elizabeth City	97.0	52.2	77.2
280	Gastonia	97.8	56.5	79.6
281-282	Charlotte	98.7	56.5	80.1
283	Fayetteville	100.2	59.1	82.1
284	Wilmington	96.7	56.5	79.0
285	Kinston	94.5	54.3	76.8

Location Factors

STATE/ZIP	CITY	MAT.	INST.	TOTAL
NORTH CAROLINA (CONT'D)				
286	Hickory	94.8	55.3	77.4
287-288	Asheville	97.1	55.1	78.6
289	Murphy	95.7	49.9	75.5
NORTH DAKOTA				
580-581	Fargo	101.9	65.8	86.0
582	Grand Forks	102.4	52.6	80.5
583	Devils Lake	101.3	56.9	81.7
584	Jamestown	101.4	46.0	77.0
585	Bismarck	101.0	64.4	84.9
586	Dickinson	102.3	58.7	83.1
587	Minot	102.4	71.6	88.9
588	Williston	100.7	58.8	82.3
OHIO				
430-432	Columbus	98.6	90.0	94.8
433	Marion	94.8	84.1	90.1
434-436	Toledo	98.8	97.5	98.2
437-438	Zanesville	95.3	84.8	90.7
439	Steubenville	96.6	92.0	94.6
440	Lorain	98.1	90.8	94.9
441	Cleveland	98.2	100.7	99.3
442-443	Akron	99.1	93.6	96.6
444-445	Youngstown	98.4	89.1	94.3
446-447	Canton	98.5	85.0	92.5
448-449	Mansfield	95.6	86.1	91.4
450	Hamilton	97.5	84.7	91.9
451-452	Cincinnati	97.9	85.1	92.3
453-454	Dayton	97.7	83.9	91.6
455	Springfield	97.6	84.8	91.9
456	Chillicothe	96.2	91.9	94.3
457	Athens	98.9	82.2	91.6
458	Lima	99.4	85.0	93.1
OKLAHOMA				
730-731	Oklahoma City	98.5	59.8	81.4
734	Ardmore	95.1	59.3	79.3
735	Lawton	97.7	60.6	81.3
736	Clinton	96.6	58.0	79.6
737	Enid	97.5	58.0	80.1
738	Woodward	95.3	58.0	78.9
739	Guymon	96.4	30.7	67.4
740-741	Tulsa	97.5	52.8	77.8
743	Miami	94.0	66.9	82.1
744	Muskogee	96.8	38.8	71.3
745	Mcalester	93.7	49.9	74.4
746	Ponca City	94.3	57.9	78.3
747	Durant	94.3	56.9	77.8
748	Shawnee	95.8	55.8	78.2
749	Poteau	93.5	60.6	79.0
OREGON				
970-972	Portland	98.4	101.2	99.6
973	Salem	98.9	100.4	99.5
974	Eugene	98.2	100.3	99.1
975	Medford	99.7	98.3	99.1
976	Klamath Falls	99.3	98.3	98.9
977	Bend	98.3	100.4	99.3
978	Pendleton	93.8	102.9	97.8
979	Vale	91.8	91.1	91.5
PENNSYLVANIA				
150-152	Pittsburgh	99.4	104.7	101.8
153	Washington	96.3	103.0	99.2
154	Uniontown	96.7	101.9	99.0
155	Bedford	97.5	92.8	95.5
156	Greensburg	97.5	101.2	99.2
157	Indiana	96.4	99.1	97.6
158	Dubois	97.9	97.2	97.6
159	Johnstown	97.5	97.4	97.4
160	Butler	92.0	101.5	96.2
161	New Castle	92.0	98.0	94.7
162	Kittanning	92.5	102.9	97.1
163	Oil City	92.1	96.1	93.8
164-165	Erie	94.2	95.2	94.6
166	Altoona	94.4	93.2	93.9
167	Bradford	95.2	96.3	95.7
168	State College	94.7	93.7	94.3
169	Wellsboro	95.8	93.4	94.7
170-171	Harrisburg	97.8	96.5	97.2
172	Chambersburg	95.2	90.6	93.2
173-174	York	96.0	96.4	96.2
175-176	Lancaster	94.1	90.0	92.3

STATE/ZIP	CITY	MAT.	INST.	TOTAL
PENNSYLVANIA (CONT'D)				
177	Williamsport	92.8	79.3	86.9
178	Sunbury	94.7	95.8	95.2
179	Pottsville	94.0	98.8	96.1
180	Lehigh Valley	95.3	113.4	103.3
181	Allentown	97.7	109.5	102.9
182	Hazleton	94.9	98.6	96.5
183	Stroudsburg	94.7	102.7	98.2
184-185	Scranton	98.4	99.9	99.1
186-187	Wilkes-Barre	94.6	98.5	96.3
188	Montrose	94.2	97.3	95.6
189	Doylestown	94.5	120.2	105.8
190-191	Philadelphia	99.2	132.2	113.7
193	Westchester	95.5	122.7	107.5
194	Norristown	94.7	128.4	109.5
195-196	Reading	97.1	101.1	98.8
PUERTO RICO				
009	San Juan	121.0	23.8	78.2
RHODE ISLAND				
028	Newport	98.7	116.7	106.6
029	Providence	100.1	116.7	107.4
SOUTH CAROLINA				
290-292	Columbia	97.3	48.8	75.9
293	Spartanburg	96.3	48.4	75.2
294	Charleston	97.7	56.7	79.6
295	Florence	96.0	48.7	75.2
296	Greenville	96.0	48.4	75.0
297	Rock Hill	95.3	46.6	73.8
298	Aiken	96.1	71.8	85.4
299	Beaufort	96.8	42.1	72.7
SOUTH DAKOTA				
570-571	Sioux Falls	100.9	57.1	81.6
572	Watertown	99.5	50.9	78.1
573	Mitchell	98.4	50.5	77.3
574	Aberdeen	101.3	51.7	79.4
575	Pierre	100.5	53.7	79.8
576	Mobridge	99.1	50.9	77.8
577	Rapid City	101.2	54.6	80.6
TENNESSEE				
370-372	Nashville	97.6	73.1	86.8
373-374	Chattanooga	98.9	65.5	84.2
375,380-381	Memphis	97.4	70.0	85.3
376	Johnson City	98.4	56.0	79.8
377-379	Knoxville	95.5	60.8	80.2
382	Mckenzie	96.8	61.1	81.1
383	Jackson	98.9	63.2	83.2
384	Columbia	95.5	66.2	82.6
385	Cookeville	96.7	59.5	80.3
TEXAS				
750	Mckinney	98.5	49.7	77.0
751	Waxahackie	98.5	56.7	80.1
752-753	Dallas	99.5	66.5	85.0
754	Greenville	98.6	42.9	74.0
755	Texarkana	97.8	50.5	77.0
756	Longview	98.4	40.0	72.7
757	Tyler	98.6	54.1	79.0
758	Palestine	95.0	55.9	77.8
759	Lufkin	95.5	58.8	79.3
760-761	Fort Worth	97.1	62.9	82.0
762	Denton	97.3	47.3	75.2
763	Wichita Falls	97.9	54.1	78.6
764	Eastland	96.3	49.1	75.5
765	Temple	95.0	48.8	74.6
766-767	Waco	97.3	55.8	79.0
768	Brownwood	97.8	48.4	76.0
769	San Angelo	97.5	49.1	76.2
770-772	Houston	99.3	70.3	86.5
773	Huntsville	97.6	56.3	79.4
774	Wharton	98.5	53.4	78.6
775	Galveston	96.6	68.3	84.1
776-777	Beaumont	97.3	61.3	81.4
778	Bryan	94.5	61.6	80.0
779	Victoria	98.7	43.4	74.3
780	Laredo	96.7	53.2	77.5
781-782	San Antonio	97.0	65.6	83.2
783-784	Corpus Christi	99.7	51.7	78.6
785	Mc Allen	99.4	46.1	75.9
786-787	Austin	97.1	58.1	79.9

Location Factors

STATE/ZIP	CITY	MAT.	INST.	TOTAL
TEXAS (CONT'D)				
788	Del Rio	98.9	45.9	75.6
789	Giddings	95.8	55.4	78.0
790-791	Amarillo	98.6	60.1	81.7
792	Childress	97.5	56.5	79.4
793-794	Lubbock	99.7	56.2	80.6
795-796	Abilene	98.1	53.6	78.5
797	Midland	99.5	51.3	78.2
798-799,885	El Paso	97.2	50.6	76.7
UTAH				
840-841	Salt Lake City	101.9	67.5	86.7
842,844	Ogden	97.6	69.1	85.1
843	Logan	99.4	69.1	86.1
845	Price	99.8	65.5	84.7
846-847	Provo	100.0	67.4	85.6
VERMONT				
050	White River Jct.	97.3	67.7	84.3
051	Bellows Falls	96.0	81.8	89.7
052	Bennington	96.3	80.1	89.1
053	Brattleboro	96.6	86.2	92.0
054	Burlington	100.5	77.8	90.5
056	Montpelier	97.2	77.8	88.7
057	Rutland	98.8	77.8	89.6
058	St. Johnsbury	97.3	67.1	84.0
059	Guildhall	96.1	67.1	83.3
VIRGINIA				
220-221	Fairfax	99.3	84.9	93.0
222	Arlington	100.8	85.3	93.9
223	Alexandria	99.9	86.0	93.8
224-225	Fredericksburg	98.1	83.8	91.8
226	Winchester	98.6	82.7	91.6
227	Culpeper	98.6	81.6	91.1
228	Harrisonburg	98.8	69.3	85.8
229	Charlottesville	99.2	68.4	85.6
230-232	Richmond	100.1	70.1	86.9
233-235	Norfolk	100.8	69.4	87.0
236	Newport News	99.8	69.3	86.3
237	Portsmouth	99.2	66.7	84.9
238	Petersburg	99.0	70.1	86.3
239	Farmville	98.2	54.4	78.9
240-241	Roanoke	100.4	66.2	85.3
242	Bristol	97.9	56.7	79.8
243	Pulaski	97.6	57.3	79.8
244	Staunton	98.4	63.9	83.2
245	Lynchburg	98.5	67.4	84.8
246	Grundy	97.9	54.5	78.8
WASHINGTON				
980-981,987	Seattle	102.8	105.1	103.8
982	Everett	102.4	99.7	101.2
983-984	Tacoma	102.5	100.3	101.6
985	Olympia	100.1	100.3	100.2
986	Vancouver	103.5	94.1	99.4
988	Wenatchee	102.1	87.5	95.6
989	Yakima	102.7	93.6	98.7
990-992	Spokane	102.0	82.4	93.4
993	Richland	101.8	88.7	96.0
994	Clarkston	100.0	81.4	91.8
WEST VIRGINIA				
247-248	Bluefield	96.8	83.2	90.8
249	Lewisburg	98.3	83.2	91.7
250-253	Charleston	98.9	90.1	95.0
254	Martinsburg	98.2	80.3	90.3
255-257	Huntington	99.9	92.5	96.6
258-259	Beckley	96.7	87.2	92.5
260	Wheeling	100.0	90.3	95.7
261	Parkersburg	98.9	89.6	94.8
262	Buckhannon	98.1	90.3	94.7
263-264	Clarksburg	98.7	90.1	94.9
265	Morgantown	98.7	89.2	94.5
266	Gassaway	98.0	90.2	94.6
267	Romney	98.0	86.0	92.7
268	Petersburg	97.8	86.1	92.6
WISCONSIN				
530,532	Milwaukee	101.1	108.6	104.4
531	Kenosha	100.8	99.8	100.4
534	Racine	100.3	99.9	100.1
535	Beloit	100.2	94.8	97.8
537	Madison	99.9	96.0	98.2

STATE/ZIP	CITY	MAT.	INST.	TOTAL
WISCONSIN (CONT'D)				
538	Lancaster	97.7	90.3	94.4
539	Portage	96.3	93.4	95.0
540	New Richmond	97.0	93.0	95.2
541-543	Green Bay	101.5	92.6	97.6
544	Wausau	96.3	91.0	94.0
545	Rhinelander	99.5	89.2	94.9
546	La Crosse	98.0	92.4	95.5
547	Eau Claire	99.5	93.5	96.9
548	Superior	96.8	95.5	96.2
549	Oshkosh	96.9	81.4	90.1
WYOMING				
820	Cheyenne	99.4	66.7	85.0
821	Yellowstone Nat'l Park	96.7	65.9	83.2
822	Wheatland	97.8	65.5	83.6
823	Rawlins	99.3	68.6	85.8
824	Worland	97.4	64.9	83.0
825	Riverton	98.4	65.6	84.0
826	Casper	99.4	61.5	82.7
827	Newcastle	97.2	67.5	84.1
828	Sheridan	100.0	66.1	85.1
829-831	Rock Springs	101.2	65.6	85.5
CANADIAN FACTORS (reflect Canadian currency)				
ALBERTA				
	Calgary	127.9	97.4	114.5
	Edmonton	128.5	97.4	114.8
	Fort McMurray	125.2	96.2	112.4
	Lethbridge	119.7	95.6	109.1
	Lloydminster	114.7	92.2	104.8
	Medicine Hat	114.8	91.4	104.5
	Red Deer	115.3	91.4	104.8
BRITISH COLUMBIA				
	Kamloops	115.7	94.3	106.3
	Prince George	116.8	93.6	106.5
	Vancouver	126.8	91.9	111.4
	Victoria	116.8	90.0	105.0
MANITOBA				
	Brandon	114.9	78.9	99.0
	Portage la Prairie	114.7	77.2	98.2
	Winnipeg	127.3	72.3	103.1
NEW BRUNSWICK				
	Bathurst	112.9	70.0	94.0
	Dalhousie	113.0	70.2	94.1
	Fredericton	117.6	75.5	99.1
	Moncton	113.3	72.9	95.5
	Newcastle	113.0	70.2	94.1
	St. John	115.5	79.2	99.5
NEWFOUNDLAND				
	Corner Brook	118.2	69.4	96.7
	St Johns	121.7	81.9	104.2
NORTHWEST TERRITORIES				
	Yellowknife	123.9	87.6	107.9
NOVA SCOTIA				
	Bridgewater	114.3	76.6	97.7
	Dartmouth	115.9	76.6	98.6
	Halifax	119.5	82.2	103.0
	New Glasgow	114.0	76.6	97.6
	Sydney	111.7	76.6	96.3
	Truro	113.8	76.6	97.4
	Yarmouth	113.9	76.6	97.5
ONTARIO				
	Barrie	118.1	95.4	108.1
	Brantford	116.4	99.4	108.9
	Cornwall	116.5	95.7	107.3
	Hamilton	120.4	97.0	110.1
	Kingston	117.3	95.8	107.8
	Kitchener	112.6	93.5	104.2
	London	120.8	93.5	108.8
	North Bay	116.6	93.5	106.4
	Oshawa	115.8	96.8	107.4
	Ottawa	121.1	94.1	109.2
	Owen Sound	118.2	93.6	107.4
	Peterborough	116.4	95.5	107.2
	Sarnia	116.2	100.1	109.1

STATE/ZIP	CITY	MAT.	INST.	TOTAL
ONTARIO (CONT'D)				
	Sault Ste Marie	111.4	94.1	103.8
	St. Catharines	110.6	94.6	103.6
	Sudbury	111.1	93.2	103.2
	Thunder Bay	111.9	93.3	103.7
	Timmins	116.6	93.5	106.4
	Toronto	121.6	100.5	112.3
	Windsor	111.2	93.4	103.3
PRINCE EDWARD ISLAND				
	Charlottetown	119.0	64.5	95.0
	Summerside	115.7	64.5	93.2
QUEBEC				
	Cap-de-la-Madeleine	113.7	89.9	103.2
	Charlesbourg	113.7	89.9	103.2
	Chicoutimi	112.6	94.6	104.7
	Gatineau	113.3	89.7	102.9
	Granby	113.5	89.6	103.0
	Hull	113.4	89.7	102.9
	Joliette	113.8	89.9	103.3
	Laval	112.9	89.7	102.7
	Montreal	121.8	94.9	110.0
	Quebec	119.5	95.3	108.9
	Rimouski	113.1	94.6	104.9
	Rouyn-Noranda	113.2	89.7	102.9
	Saint Hyacinthe	112.6	89.7	102.5
	Sherbrooke	113.6	89.7	103.1
	Sorel	113.8	89.9	103.3
	St Jerome	113.3	89.7	102.9
	Trois Rivieres	114.0	89.9	103.4
SASKATCHEWAN				
	Moose Jaw	112.0	71.3	94.1
	Prince Albert	111.1	69.5	92.7
	Regina	117.6	90.1	105.5
	Saskatoon	112.0	90.1	102.4
YUKON				
	Whitehorse	118.4	70.3	97.2

R011105-05 Tips for Accurate Estimating

1. Use pre-printed or columnar forms for orderly sequence of dimensions and locations and for recording telephone quotations.
2. Use only the front side of each paper or form except for certain pre-printed summary forms.
3. Be consistent in listing dimensions: For example, length x width x height. This helps in rechecking to ensure that, the total length of partitions is appropriate for the building area.
4. Use printed (rather than measured) dimensions where given.
5. Add up multiple printed dimensions for a single entry where possible.
6. Measure all other dimensions carefully.
7. Use each set of dimensions to calculate multiple related quantities.
8. Convert foot and inch measurements to decimal feet when listing. Memorize decimal equivalents to .01 parts of a foot (1/8″ equals approximately .01″).
9. Do not "round off" quantities until the final summary.
10. Mark drawings with different colors as items are taken off.
11. Keep similar items together, different items separate.
12. Identify location and drawing numbers to aid in future checking for completeness.
13. Measure or list everything on the drawings or mentioned in the specifications.
14. It may be necessary to list items not called for to make the job complete.
15. Be alert for: Notes on plans such as N.T.S. (not to scale); changes in scale throughout the drawings; reduced size drawings; discrepancies between the specifications and the drawings.
16. Develop a consistent pattern of performing an estimate. For example:
    a. Start the quantity takeoff at the lower floor and move to the next higher floor.
    b. Proceed from the main section of the building to the wings.
    c. Proceed from south to north or vice versa, clockwise or counterclockwise.
    d. Take off floor plan quantities first, elevations next, then detail drawings.
17. List all gross dimensions that can be either used again for different quantities, or used as a rough check of other quantities for verification (exterior perimeter, gross floor area, individual floor areas, etc.).
18. Utilize design symmetry or repetition (repetitive floors, repetitive wings, symmetrical design around a center line, similar room layouts, etc.). Note: Extreme caution is needed here so as not to omit or duplicate an area.
19. Do not convert units until the final total is obtained. For instance, when estimating concrete work, keep all units to the nearest cubic foot, then summarize and convert to cubic yards.
20. When figuring alternatives, it is best to total all items involved in the basic system, then total all items involved in the alternates. Therefore you work with positive numbers in all cases. When adds and deducts are used, it is often confusing whether to add or subtract a portion of an item; especially on a complicated or involved alternate.

R011105-10 Unit Gross Area Requirements

The figures in the table below indicate typical ranges in square feet as a function of the "occupant" unit. This table is best used in the preliminary design stages to help determine the probable size requirement for the total project.

Building Type	Unit	Gross Area in S.F.		
		1/4	Median	3/4
Apartments	Unit	660	860	1,100
Auditorium & Play Theaters	Seat	18	25	38
Bowling Alleys	Lane		940	
Churches & Synagogues	Seat	20	28	39
Dormitories	Bed	200	230	275
Fraternity & Sorority Houses	Bed	220	315	370
Garages, Parking	Car	325	355	385
Hospitals	Bed	685	850	1,075
Hotels	Rental Unit	475	600	710
Housing for the elderly	Unit	515	635	755
Housing, Public	Unit	700	875	1,030
Ice Skating Rinks	Total	27,000	30,000	36,000
Motels	Rental Unit	360	465	620
Nursing Homes	Bed	290	350	450
Restaurants	Seat	23	29	39
Schools, Elementary	Pupil	65	77	90
Junior High & Middle	↓	85	110	129
Senior High	↓	102	130	145
Vocational	↓	110	135	195
Shooting Ranges	Point		450	
Theaters & Movies	Seat		15	

R011105-20 Floor Area Ratios

Table below lists commonly used gross to net area and net to gross area ratios expressed in % for various building types.

Building Type	Gross to Net Ratio	Net to Gross Ratio	Building Type	Gross to Net Ratio	Net to Gross Ratio
Apartment	156	64	School Buildings (campus type)		
Bank	140	72	Administrative	150	67
Church	142	70	Auditorium	142	70
Courthouse	162	61	Biology	161	62
Department Store	123	81	Chemistry	170	59
Garage	118	85	Classroom	152	66
Hospital	183	55	Dining Hall	138	72
Hotel	158	63	Dormitory	154	65
Laboratory	171	58	Engineering	164	61
Library	132	76	Fraternity	160	63
Office	135	75	Gymnasium	142	70
Restaurant	141	70	Science	167	60
Warehouse	108	93	Service	120	83
			Student Union	172	59

The gross area of a building is the total floor area based on outside dimensions.

The net area of a building is the usable floor area for the function intended and excludes such items as stairways, corridors and mechanical rooms. In the case of a commercial building, it might be considered as the "leasable area."

R011105-30 Occupancy Determinations

Description		S.F. Required per Person		
		BOCA	SBC	UBC
Assembly Areas	Fixed Seats	**	6	7
	Movable Seats		15	15
	Concentrated	7		
	Unconcentrated	15		
	Standing Space	3		
Educational	Unclassified			
	Classrooms	20	40	20
	Shop Areas	50	100	50
Institutional	Unclassified		125	
	In-Patient Areas	240		
	Sleeping Areas	120		
Mercantile	Basement	30	30	20
	Ground Floor	30	30	30
	Upper Floors	60	60	50
Office		100	100	100

BOCA=Building Officials & Code Administrators
SBC=Southern Building Code
UBC=Uniform Building Code

** The occupancy load for assembly area with fixed seats shall be determined by the number of fixed seats installed.

R011105-40 Weather Data and Design Conditions

City	Latitude (1)		Winter Temperatures (1)			Winter Degree Days (2)	Summer (Design Dry Bulb) Temperatures and Relative Humidity		
	0	1′	Med. of Annual Extremes	99%	97½%		1%	2½%	5%
UNITED STATES									
Albuquerque, NM	35	0	5.1	12	16	4,400	96/61	94/61	92/61
Atlanta, GA	33	4	11.9	17	22	3,000	94/74	92/74	90/73
Baltimore, MD	39	2	7	14	17	4,600	94/75	91/75	89/74
Birmingham, AL	33	3	13	17	21	2,600	96/74	94/75	92/74
Bismarck, ND	46	5	-32	-23	-19	8,800	95/68	91/68	88/67
Boise, ID	43	3	1	3	10	5,800	96/65	94/64	91/64
Boston, MA	42	2	-1	6	9	5,600	91/73	88/71	85/70
Burlington, VT	44	3	-17	-12	-7	8,200	88/72	85/70	82/69
Charleston, WV	38	2	3	7	11	4,400	92/74	90/73	87/72
Charlotte, NC	35	1	13	18	22	3,200	95/74	93/74	91/74
Casper, WY	42	5	-21	-11	-5	7,400	92/58	90/57	87/57
Chicago, IL	41	5	-8	-3	2	6,600	94/75	91/74	88/73
Cincinnati, OH	39	1	0	1	6	4,400	92/73	90/72	88/72
Cleveland, OH	41	2	-3	1	5	6,400	91/73	88/72	86/71
Columbia, SC	34	0	16	20	24	2,400	97/76	95/75	93/75
Dallas, TX	32	5	14	18	22	2,400	102/75	100/75	97/75
Denver, CO	39	5	-10	-5	1	6,200	93/59	91/59	89/59
Des Moines, IA	41	3	-14	-10	-5	6,600	94/75	91/74	88/73
Detroit, MI	42	2	-3	3	6	6,200	91/73	88/72	86/71
Great Falls, MT	47	3	-25	-21	-15	7,800	91/60	88/60	85/59
Hartford, CT	41	5	-4	3	7	6,200	91/74	88/73	85/72
Houston, TX	29	5	24	28	33	1,400	97/77	95/77	93/77
Indianapolis, IN	39	4	-7	-2	2	5,600	92/74	90/74	87/73
Jackson, MS	32	2	16	21	25	2,200	97/76	95/76	93/76
Kansas City, MO	39	1	-4	2	6	4,800	99/75	96/74	93/74
Las Vegas, NV	36	1	18	25	28	2,800	108/66	106/65	104/65
Lexington, KY	38	0	-1	3	8	4,600	93/73	91/73	88/72
Little Rock, AR	34	4	11	15	20	3,200	99/76	96/77	94/77
Los Angeles, CA	34	0	36	41	43	2,000	93/70	89/70	86/69
Memphis, TN	35	0	10	13	18	3,200	98/77	95/76	93/76
Miami, FL	25	5	39	44	47	200	91/77	90/77	89/77
Milwaukee, WI	43	0	-11	-8	-4	7,600	90/74	87/73	84/71
Minneapolis, MN	44	5	-22	-16	-12	8,400	92/75	89/73	86/71
New Orleans, LA	30	0	28	29	33	1,400	93/78	92/77	90/77
New York, NY	40	5	6	11	15	5,000	92/74	89/73	87/72
Norfolk, VA	36	5	15	20	22	3,400	93/77	91/76	89/76
Oklahoma City, OK	35	2	4	9	13	3,200	100/74	97/74	95/73
Omaha, NE	41	2	-13	-8	-3	6,600	94/76	91/75	88/74
Philadelphia, PA	39	5	6	10	14	4,400	93/75	90/74	87/72
Phoenix, AZ	33	3	27	31	34	1,800	109/71	107/71	105/71
Pittsburgh, PA	40	3	-1	3	7	6,000	91/72	88/71	86/70
Portland, ME	43	4	-10	-6	-1	7,600	87/72	84/71	81/69
Portland, OR	45	4	18	17	23	4,600	89/68	85/67	81/65
Portsmouth, NH	43	1	-8	-2	2	7,200	89/73	85/71	83/70
Providence, RI	41	4	-1	5	9	6,000	89/73	86/72	83/70
Rochester, NY	43	1	-5	1	5	6,800	91/73	88/71	85/70
Salt Lake City, UT	40	5	0	3	8	6,000	97/62	95/62	92/61
San Francisco, CA	37	5	36	38	40	3,000	74/63	71/62	69/61
Seattle, WA	47	4	22	22	27	5,200	85/68	82/66	78/65
Sioux Falls, SD	43	4	-21	-15	-11	7,800	94/73	91/72	88/71
St. Louis, MO	38	4	-3	3	8	5,000	98/75	94/75	91/75
Tampa, FL	28	0	32	36	40	680	92/77	91/77	90/76
Trenton, NJ	40	1	4	11	14	5,000	91/75	88/74	85/73
Washington, DC	38	5	7	14	17	4,200	93/75	91/74	89/74
Wichita, KS	37	4	-3	3	7	4,600	101/72	98/73	96/73
Wilmington, DE	39	4	5	10	14	5,000	92/74	89/74	87/73
ALASKA									
Anchorage	61	1	-29	-23	-18	10,800	71/59	68/58	66/56
Fairbanks	64	5	-59	-51	-47	14,280	82/62	78/60	75/59
CANADA									
Edmonton, Alta.	53	3	-30	-29	-25	11,000	85/66	82/65	79/63
Halifax, N.S.	44	4	-4	1	5	8,000	79/66	76/65	74/64
Montreal, Que.	45	3	-20	-16	-10	9,000	88/73	85/72	83/71
Saskatoon, Sask.	52	1	-35	-35	-31	11,000	89/68	86/66	83/65
St. John's, N.F.	47	4	1	3	7	8,600	77/66	75/65	73/64
Saint John, N.B.	45	2	-15	-12	-8	8,200	80/67	77/65	75/64
Toronto, Ont.	43	4	-10	-5	-1	7,000	90/73	87/72	85/71
Vancouver, B.C.	49	1	13	15	19	6,000	79/67	77/66	74/65
Winnipeg, Man.	49	5	-31	-30	-27	10,800	89/73	86/71	84/70

(1) Handbook of Fundamentals, ASHRAE, Inc., NY 1989
(2) Local Climatological Annual Survey, USDC Env. Science Services Administration, Asheville, NC

R011105-50 Metric Conversion Factors

Description: This table is primarily for converting customary U.S. units in the left hand column to SI metric units in the right hand column. In addition, conversion factors for some commonly encountered Canadian and non-SI metric units are included.

	If You Know		Multiply By		To Find
	Inches	x	25.4[a]	=	Millimeters
	Feet	x	0.3048[a]	=	Meters
Length	Yards	x	0.9144[a]	=	Meters
	Miles (statute)	x	1.609	=	Kilometers
	Square inches	x	645.2	=	Square millimeters
Area	Square feet	x	0.0929	=	Square meters
	Square yards	x	0.8361	=	Square meters
	Cubic inches	x	16,387	=	Cubic millimeters
	Cubic feet	x	0.02832	=	Cubic meters
	Cubic yards	x	0.7646	=	Cubic meters
Volume	Gallons (U.S. liquids)[b]	x	0.003785	=	Cubic meters[c]
(Capacity)	Gallons (Canadian liquid)[b]	x	0.004546	=	Cubic meters[c]
	Ounces (U.S. liquid)[b]	x	29.57	=	Milliliters[c, d]
	Quarts (U.S. liquid)[b]	x	0.9464	=	Liters[c, d]
	Gallons (U.S. liquid)[b]	x	3.785	=	Liters[c, d]
	Kilograms force[d]	x	9.807	=	Newtons
	Pounds force	x	4.448	=	Newtons
Force	Pounds force	x	0.4536	=	Kilograms force[d]
	Kips	x	4448	=	Newtons
	Kips	x	453.6	=	Kilograms force[d]
	Kilograms force per square centimeter[d]	x	0.09807	=	Megapascals
Pressure,	Pounds force per square inch (psi)	x	0.006895	=	Megapascals
Stress,	Kips per square inch	x	6.895	=	Megapascals
Strength	Pounds force per square inch (psi)	x	0.07031	=	Kilograms force per square centimeter[d]
(Force per unit area)	Pounds force per square foot	x	47.88	=	Pascals
	Pounds force per square foot	x	4.882	=	Kilograms force per square meter[d]
	Inch-pounds force	x	0.01152	=	Meter-kilograms force[d]
Bending	Inch-pounds force	x	0.1130	=	Newton-meters
Moment	Foot-pounds force	x	0.1383	=	Meter-kilograms force[d]
Or Torque	Foot-pounds force	x	1.356	=	Newton-meters
	Meter-kilograms force[d]	x	9.807	=	Newton-meters
	Ounces (avoirdupois)	x	28.35	=	Grams
	Pounds (avoirdupois)	x	0.4536	=	Kilograms
Mass	Tons (metric)	x	1000	=	Kilograms
	Tons, short (2000 pounds)	x	907.2	=	Kilograms
	Tons, short (2000 pounds)	x	0.9072	=	Megagrams[e]
	Pounds mass per cubic foot	x	16.02	=	Kilograms per cubic meter
Mass per	Pounds mass per cubic yard	x	0.5933	=	Kilograms per cubic meter
Unit	Pounds mass per gallon (U.S. liquid)[b]	x	119.8	=	Kilograms per cubic meter
Volume	Pounds mass per gallon (Canadian liquid)[b]	x	99.78	=	Kilograms per cubic meter
	Degrees Fahrenheit		(F-32)/1.8	=	Degrees Celsius
Temperature	Degrees Fahrenheit		(F+459.67)/1.8	=	Degrees Kelvin
	Degrees Celsius		C+273.15	=	Degrees Kelvin

[a]The factor given is exact
[b]One U.S. gallon = 0.8327 Canadian gallon
[c]1 liter = 1000 milliliters = 1000 cubic centimeters
1 cubic decimeter = 0.001 cubic meter
[d]Metric but not SI unit
[e]Called "tonne" in England and "metric ton" in other metric countries

R011105-60 Weights and Measures

Measures of Length
1 Mile = 1760 Yards = 5280 Feet
1 Yard = 3 Feet = 36 inches
1 Foot = 12 Inches
1 Mil = 0.001 Inch
1 Fathom = 2 Yards = 6 Feet
1 Rod = 5.5 Yards = 16.5 Feet
1 Hand = 4 Inches
1 Span = 9 Inches
1 Micro-inch = One Millionth Inch or 0.000001 Inch
1 Micron = One Millionth Meter + 0.00003937 Inch

Surveyor's Measure
1 Mile = 8 Furlongs = 80 Chains
1 Furlong = 10 Chains = 220 Yards
1 Chain = 4 Rods = 22 Yards = 66 Feet = 100 Links
1 Link = 7.92 Inches

Square Measure
1 Square Mile = 640 Acres = 6400 Square Chains
1 Acre = 10 Square Chains = 4840 Square Yards = 43,560 Sq. Ft.
1 Square Chain = 16 Square Rods = 484 Square Yards = 4356 Sq. Ft.
1 Square Rod = 30.25 Square Yards = 272.25 Square Feet = 625 Square Lines
1 Square Yard = 9 Square Feet
1 Square Foot = 144 Square Inches
An Acre equals a Square 208.7 Feet per Side

Cubic Measure
1 Cubic Yard = 27 Cubic Feet
1 Cubic Foot = 1728 Cubic Inches
1 Cord of Wood = 4 x 4 x 8 Feet = 128 Cubic Feet
1 Perch of Masonry = 16½ x 1½ x 1 Foot = 24.75 Cubic Feet

Avoirdupois or Commercial Weight
1 Gross or Long Ton = 2240 Pounds
1 Net or Short ton = 2000 Pounds
1 Pound = 16 Ounces = 7000 Grains
1 Ounce = 16 Drachms = 437.5 Grains
1 Stone = 14 Pounds

Shipping Measure
For Measuring Internal Capacity of a Vessel:
1 Register Ton = 100 Cubic Feet

For Measurement of Cargo:
Approximately 40 Cubic Feet of Merchandise is considered a Shipping Ton, unless that bulk would weigh more than 2000 Pounds, in which case Freight Charge may be based upon weight.

40 Cubic Feet = 32.143 U.S. Bushels = 31.16 Imp. Bushels

Liquid Measure
1 Imperial Gallon = 1.2009 U.S. Gallon = 277.42 Cu. In.
1 Cubic Foot = 7.48 U.S. Gallons

General Requirements — R0111 Summary of Work

R011110-30 Engineering Fees

Typical **Structural Engineering Fees** based on type of construction and total project size. These fees are included in Architectural Fees.

Type of Construction	Total Project Size (in thousands of dollars)			
	$500	$500-$1,000	$1,000-$5,000	Over $5000
Industrial buildings, factories & warehouses	Technical payroll times 2.0 to 2.5	1.60%	1.25%	1.00%
Hotels, apartments, offices, dormitories, hospitals, public buildings, food stores	↓	2.00%	1.70%	1.20%
Museums, banks, churches and cathedrals	↓	2.00%	1.75%	1.25%
Thin shells, prestressed concrete, earthquake resistive	↓	2.00%	1.75%	1.50%
Parking ramps, auditoriums, stadiums, convention halls, hangars & boiler houses	↓	2.50%	2.00%	1.75%
Special buildings, major alterations, underpinning & future expansion	↓	Add to above 0.5%	Add to above 0.5%	Add to above 0.5%

For complex reinforced concrete or unusually complicated structures, add 20% to 50%.

Typical **Mechanical and Electrical Engineering Fees** are based on the size of the subcontract. The fee structure for both are shown below. These fees are included in Architectural Fees.

Type of Construction	Subcontract Size							
	$25,000	$50,000	$100,000	$225,000	$350,000	$500,000	$750,000	$1,000,000
Simple structures	6.4%	5.7%	4.8%	4.5%	4.4%	4.3%	4.2%	4.1%
Intermediate structures	8.0	7.3	6.5	5.6	5.1	5.0	4.9	4.8
Complex structures	10.1	9.0	9.0	8.0	7.5	7.5	7.0	7.0

For renovations, add 15% to 25% to applicable fee.

General Requirements — R0121 Allowances

R012153-10 Repair and Remodeling

Cost figures are based on new construction utilizing the most cost-effective combination of labor, equipment and material with the work scheduled in proper sequence to allow the various trades to accomplish their work in an efficient manner.

The costs for repair and remodeling work must be modified due to the following factors that may be present in any given repair and remodeling project.

1. Equipment usage curtailment due to the physical limitations of the project, with only hand-operated equipment being used.
2. Increased requirement for shoring and bracing to hold up the building while structural changes are being made and to allow for temporary storage of construction materials on above-grade floors.
3. Material handling becomes more costly due to having to move within the confines of an enclosed building. For multi-story construction, low capacity elevators and stairwells may be the only access to the upper floors.
4. Large amount of cutting and patching and attempting to match the existing construction is required. It is often more economical to remove entire walls rather than create many new door and window openings. This sort of trade-off has to be carefully analyzed.
5. Cost of protection of completed work is increased since the usual sequence of construction usually cannot be accomplished.
6. Economies of scale usually associated with new construction may not be present. If small quantities of components must be custom fabricated due to job requirements, unit costs will naturally increase. Also, if only small work areas are available at a given time, job scheduling between trades becomes difficult and subcontractor quotations may reflect the excessive start-up and shut-down phases of the job.
7. Work may have to be done on other than normal shifts and may have to be done around an existing production facility which has to stay in production during the course of the repair and remodeling.
8. Dust and noise protection of adjoining non-construction areas can involve substantial special protection and alter usual construction methods.
9. Job may be delayed due to unexpected conditions discovered during demolition or removal. These delays ultimately increase construction costs.
10. Piping and ductwork runs may not be as simple as for new construction. Wiring may have to be snaked through walls and floors.
11. Matching "existing construction" may be impossible because materials may no longer be manufactured. Substitutions may be expensive.
12. Weather protection of existing structure requires additional temporary structures to protect building at openings.
13. On small projects, because of local conditions, it may be necessary to pay a tradesman for a minimum of four hours for a task that is completed in one hour.

All of the above areas can contribute to increased costs for a repair and remodeling project. Each of the above factors should be considered in the planning, bidding and construction stage in order to minimize the increased costs associated with repair and remodeling jobs.

R012909-80 Sales Tax by State

State sales tax on materials is tabulated below (5 states have no sales tax). Many states allow local jurisdictions, such as a county or city, to levy additional sales tax.

Some projects may be sales tax exempt, particularly those constructed with public funds.

State	Tax (%)	State	Tax (%)	State	Tax (%)	State	Tax (%)
Alabama	4	Illinois	6.25	Montana	0	Rhode Island	7
Alaska	0	Indiana	7	Nebraska	5.5	South Carolina	6
Arizona	6.6	Iowa	6	Nevada	6.85	South Dakota	4
Arkansas	6	Kansas	5.3	New Hampshire	0	Tennessee	7
California	7.25	Kentucky	6	New Jersey	7	Texas	6.25
Colorado	2.9	Louisiana	4	New Mexico	5.13	Utah	4.75
Connecticut	6	Maine	5	New York	4	Vermont	6
Delaware	0	Maryland	6	North Carolina	5.75	Virginia	4
District of Columbia	6	Massachusetts	6.25	North Dakota	5	Washington	6.5
Florida	6	Michigan	6	Ohio	5.5	West Virginia	6
Georgia	4	Minnesota	6.88	Oklahoma	4.5	Wisconsin	5
Hawaii	4	Mississippi	7	Oregon	0	Wyoming	4
Idaho	6	Missouri	4.23	Pennsylvania	6	Average	5.03 %

Sales Tax by Province (Canada)

GST - a value-added tax, which the government imposes on most goods and services provided in or imported into Canada. PST - a retail sales tax, which three of the provinces impose on the price of most goods and some services. QST - a value-added tax, similar to the federal GST, which Quebec imposes. HST - Five provinces have combined their retail sales tax with the federal GST into one harmonized tax.

Province	PST (%)	QST (%)	GST(%)	HST(%)
Alberta	0	0	5	0
British Columbia	0	0	0	12
Manitoba	7	0	5	0
New Brunswick	0	0	0	13
Newfoundland	0	0	0	13
Northwest Territories	0	0	5	0
Nova Scotia	0	0	0	15
Ontario	0	0	0	13
Prince Edward Island	10	0	5	0
Quebec	0	9.5	5	0
Saskatchewan	5	0	5	0
Yukon	0	0	5	0

General Requirements — R0129 Payment Procedures

R012909-85 Unemployment Taxes and Social Security Taxes

State unemployment tax rates vary not only from state to state, but also with the experience rating of the contractor. The federal unemployment tax rate is 6.2% of the first $7,000 of wages. This is reduced by a credit of up to 5.4% for timely payment to the state. The minimum federal unemployment tax is 0.8% after all credits.

Social security (FICA) for 2013 is estimated at time of publication to be 7.65% of wages up to $114,000.

R012909-90 Overtime

One way to improve the completion date of a project or eliminate negative float from a schedule is to compress activity duration times. This can be achieved by increasing the crew size or working overtime with the proposed crew.

To determine the costs of working overtime to compress activity duration times, consider the following examples. Below is an overtime efficiency and cost chart based on a five, six, or seven day week with an eight through twelve hour day. Payroll percentage increases for time and one half and double time are shown for the various working days.

Days per Week	Hours per Day	Production Efficiency					Payroll Cost Factors	
		1st Week	2nd Week	3rd Week	4th Week	Average 4 Weeks	@ 1-1/2 Times	@ 2 Times
5	8	100%	100%	100%	100%	100 %	100 %	100 %
	9	100	100	95	90	96.25	105.6	111.1
	10	100	95	90	85	91.25	110.0	120.0
	11	95	90	75	65	81.25	113.6	127.3
	12	90	85	70	60	76.25	116.7	133.3
6	8	100	100	95	90	96.25	108.3	116.7
	9	100	95	90	85	92.50	113.0	125.9
	10	95	90	85	80	87.50	116.7	133.3
	11	95	85	70	65	78.75	119.7	139.4
	12	90	80	65	60	73.75	122.2	144.4
7	8	100	95	85	75	88.75	114.3	128.6
	9	95	90	80	70	83.75	118.3	136.5
	10	90	85	75	65	78.75	121.4	142.9
	11	85	80	65	60	72.50	124.0	148.1
	12	85	75	60	55	68.75	126.2	152.4

General Requirements — R0131 Project Management & Coordination

R013113-40 Builder's Risk Insurance

Builder's Risk Insurance is insurance on a building during construction. Premiums are paid by the owner or the contractor. Blasting, collapse and underground insurance would raise total insurance costs above those listed. Floater policy for materials delivered to the job runs $.75 to $1.25 per $100 value. Contractor equipment insurance runs $.50 to $1.50 per $100 value. Insurance for miscellaneous tools to $1,500 value runs from $3.00 to $7.50 per $100 value.

Tabulated below are New England Builder's Risk insurance rates in dollars per $100 value for $1,000 deductible. For $25,000 deductible, rates can be reduced 13% to 34%. On contracts over $1,000,000, rates may be lower than those tabulated. Policies are written annually for the total completed value in place. For "all risk" insurance (excluding flood, earthquake and certain other perils) add $.025 to total rates below.

Coverage	Frame Construction (Class 1)		Brick Construction (Class 4)		Fire Resistive (Class 6)	
	Range	Average	Range	Average	Range	Average
Fire Insurance	$.350 to $ .850	$.600	$.158 to $.189	$.174	$.052 to $.080	$.070
Extended Coverage	.115 to .200	.158	.080 to .105	.101	.081 to .105	.100
Vandalism	.012 to .016	.014	.008 to .011	.011	.008 to .011	.010
Total Annual Rate	$.477 to $1.066	$.772	$.246 to $.305	$.286	$.141 to $.196	$.180

R013113-50 General Contractor's Overhead

There are two distinct types of overhead on a construction project: Project Overhead and Main Office Overhead. Project Overhead includes those costs at a construction site not directly associated with the installation of construction materials. Examples of Project Overhead costs include the following:

1. Superintendent
2. Construction office and storage trailers
3. Temporary sanitary facilities
4. Temporary utilities
5. Security fencing
6. Photographs
7. Clean up
8. Performance and payment bonds

The above Project Overhead items are also referred to as General Requirements and therefore are estimated in Division 1. Division 1 is the first division listed in the CSI MasterFormat but it is usually the last division estimated. The sum of the costs in Divisions 1 through 49 is referred to as the sum of the direct costs.

All construction projects also include indirect costs. The primary components of indirect costs are the contractor's Main Office Overhead and profit. The amount of the Main Office Overhead expense varies depending on the the following:

1. Owner's compensation
2. Project managers and estimator's wages
3. Clerical support wages
4. Office rent and utilities
5. Corporate legal and accounting costs
6. Advertising
7. Automobile expenses
8. Association dues
9. Travel and entertainment expenses

These costs are usually calculated as a percentage of annual sales volume. This percentage can range from 35% for a small contractor doing less than $500,000 to 5% for a large contractor with sales in excess of $100 million.

R013113-60 Workers' Compensation Insurance Rates by Trade

The table below tabulates the national averages for workers' compensation insurance rates by trade and type of building. The average "Insurance Rate" is multiplied by the "% of Building Cost" for each trade. This produces the "Workers' Compensation" cost by % of total labor cost, to be added for each trade by building type to determine the weighted average workers' compensation rate for the building types analyzed.

Trade	Insurance Rate (% Labor Cost) Range			Average	% of Building Cost Office Bldgs.	Schools & Apts.	Mfg.	Workers' Compensation Office Bldgs.	Schools & Apts.	Mfg.
Excavation, Grading, etc.	3.9 %	to	18.0%	9.2%	4.8%	4.9%	4.5%	0.44%	0.45%	0.41%
Piles & Foundations	5.5	to	50.2	15.3	7.1	5.2	8.7	1.09	0.80	1.33
Concrete	4.7	to	28.1	12.6	5.0	14.8	3.7	0.63	1.86	0.47
Masonry	5.1	to	33.0	12.5	6.9	7.5	1.9	0.86	0.94	0.24
Structural Steel	5.5	to	125.8	36.5	10.7	3.9	17.6	3.91	1.42	6.42
Miscellaneous & Ornamental Metals	4.5	to	22.6	10.6	2.8	4.0	3.6	0.30	0.42	0.38
Carpentry & Millwork	5.5	to	36.6	15.1	3.7	4.0	0.5	0.56	0.60	0.08
Metal or Composition Siding	4.3	to	61.9	16.4	2.3	0.3	4.3	0.38	0.05	0.71
Roofing	5.5	to	90.6	29.6	2.3	2.6	3.1	0.68	0.77	0.92
Doors & Hardware	3.5	to	36.6	10.8	0.9	1.4	0.4	0.10	0.15	0.04
Sash & Glazing	5.4	to	35.7	12.5	3.5	4.0	1.0	0.44	0.50	0.13
Lath & Plaster	3.3	to	28.8	10.9	3.3	6.9	0.8	0.36	0.75	0.09
Tile, Marble & Floors	3.0	to	17.6	8.0	2.6	3.0	0.5	0.21	0.24	0.04
Acoustical Ceilings	3.3	to	36.7	7.9	2.4	0.2	0.3	0.19	0.02	0.02
Painting	4.6	to	31.5	11.1	1.5	1.6	1.6	0.17	0.18	0.18
Interior Partitions	5.5	to	36.6	15.1	3.9	4.3	4.4	0.59	0.65	0.66
Miscellaneous Items	2.5	to	125.0	14.4	5.2	3.7	9.7	0.75	0.53	1.40
Elevators	1.8	to	13.1	5.6	2.1	1.1	2.2	0.12	0.06	0.12
Sprinklers	2.3	to	17.2	6.8	0.5	—	2.0	0.03	—	0.14
Plumbing	2.6	to	12.9	6.7	4.9	7.2	5.2	0.33	0.48	0.35
Heat., Vent., Air Conditioning	3.7	to	20.5	8.6	13.5	11.0	12.9	1.16	0.95	1.11
Electrical	2.8	to	13.1	5.7	10.1	8.4	11.1	0.58	0.48	0.63
Total	1.8 %	to	125.8%	—	100.0%	100.0%	100.0%	13.88%	12.30%	15.87%
Overall Weighted Average 14.02%										

Workers' Compensation Insurance Rates by States

The table below lists the weighted average workers' compensation base rate for each state with a factor comparing this with the national average of 13.7%.

State	Weighted Average	Factor	State	Weighted Average	Factor	State	Weighted Average	Factor
Alabama	19.5%	142	Kentucky	16.0%	117	North Dakota	9.4%	69
Alaska	16.0	117	Louisiana	16.9	123	Ohio	8.8	64
Arizona	9.5	69	Maine	13.2	96	Oklahoma	14.5	106
Arkansas	9.6	70	Maryland	15.6	114	Oregon	12.2	89
California	26.5	193	Massachusetts	12.3	90	Pennsylvania	17.0	124
Colorado	7.4	54	Michigan	18.9	138	Rhode Island	11.9	87
Connecticut	24.0	175	Minnesota	24.7	180	South Carolina	20.1	147
Delaware	10.4	76	Mississippi	13.8	101	South Dakota	16.1	118
District of Columbia	11.6	85	Missouri	16.9	123	Tennessee	14.6	107
Florida	10.7	78	Montana	11.5	84	Texas	10.5	77
Georgia	30.3	221	Nebraska	18.9	138	Utah	8.4	61
Hawaii	8.8	64	Nevada	9.0	66	Vermont	14.3	104
Idaho	10.7	78	New Hampshire	22.9	167	Virginia	9.4	69
Illinois	22.5	164	New Jersey	14.1	103	Washington	11.2	82
Indiana	5.7	42	New Mexico	16.1	118	West Virginia	10.7	78
Iowa	10.7	78	New York	6.2	45	Wisconsin	14.1	103
Kansas	9.1	66	North Carolina	17.5	128	Wyoming	6.1	45
Weighted Average for U.S. is 14.1% of payroll = 100%								

Rates in the following table are the base or manual costs per $100 of payroll for workers' compensation in each state. Rates are usually applied to straight time wages only and not to premium time wages and bonuses.

The weighted average skilled worker rate for 35 trades is 13.7%. For bidding purposes, apply the full value of workers' compensation directly to total labor costs, or if labor is 38%, materials 42% and overhead and profit 20% of total cost, carry 38/80 × 13.7% =6.5% of cost (before overhead and profit) into overhead. Rates vary not only from state to state but also with the experience rating of the contractor.

Rates are the most current available at the time of publication.

R013113-60 Workers' Compensation Insurance Rates by Trade and State (cont.)

State	Carpentry — 3 stories or less 5651	Carpentry — interior cab. work 5437	Carpentry — general 5403	Concrete Work — NOC 5213	Concrete Work — flat (flr., sdwk.) 5221	Electrical Wiring — inside 5190	Excavation — earth NOC 6217	Excavation — rock 6217	Glaziers 5462	Insulation Work 5479	Lathing 5443	Masonry 5022	Painting & Decorating 5474	Pile Driving 6003	Plastering 5480	Plumbing 5183	Roofing 5551	Sheet Metal Work (HVAC) 5538	Steel Erection — door & sash 5102	Steel Erection — inter., ornam. 5102	Steel Erection — structure 5040	Steel Erection — NOC 5057	Tile Work — (interior ceramic) 5348	Waterproofing 9014	Wrecking 5701
AL	33.16	14.38	24.57	12.56	9.62	9.05	13.51	13.51	17.48	16.31	8.34	22.18	17.10	17.88	15.08	7.53	46.99	9.68	11.01	11.01	42.67	23.81	12.67	9.00	42.67
AK	15.60	10.19	12.62	10.76	10.50	6.72	9.35	9.35	31.18	16.18	10.41	14.44	12.20	30.90	11.89	6.88	28.31	7.92	7.95	7.95	29.64	33.41	5.25	5.38	29.64
AZ	13.84	5.84	16.22	7.78	4.67	4.54	5.48	5.48	6.75	10.41	3.94	8.00	7.92	8.19	5.86	4.82	18.65	6.85	11.02	11.02	20.46	9.79	3.05	3.72	20.46
AR	12.64	5.52	10.90	6.45	5.52	4.56	6.98	6.98	7.80	8.42	4.23	8.42	7.44	9.52	13.32	4.58	18.55	5.18	8.04	8.04	19.86	14.63	6.30	2.90	19.86
CA	36.56	36.56	36.56	18.18	18.18	13.06	15.93	15.93	27.92	15.97	17.81	25.50	23.38	22.56	28.82	6.44	67.92	20.51	16.42	16.42	32.53	20.51	12.92	23.38	20.51
CO	9.14	4.76	5.76	6.29	4.46	2.90	5.10	5.10	5.36	7.05	3.69	7.41	5.85	6.81	5.42	3.94	14.82	3.74	4.51	4.51	26.55	9.93	3.98	2.88	19.25
CT	24.20	17.13	32.36	23.82	14.23	7.81	14.76	14.76	20.15	20.20	10.64	29.90	19.14	20.98	18.42	11.02	51.86	13.90	18.99	18.99	65.52	27.25	14.41	6.82	65.52
DE	10.15	10.15	7.92	8.38	7.06	3.56	6.26	6.26	8.82	7.92	8.82	9.44	10.62	12.26	8.82	5.43	21.11	6.82	10.50	10.50	19.57	10.50	4.61	9.44	19.57
DC	9.43	7.29	8.15	13.63	7.29	4.11	9.26	9.26	12.04	6.67	8.28	10.03	6.01	15.97	8.79	7.69	16.04	6.43	19.51	19.51	22.71	9.28	7.60	4.71	22.71
FL	10.88	8.00	11.48	11.88	5.72	5.65	6.59	6.59	9.56	10.12	4.66	9.77	8.86	17.52	10.48	5.31	17.10	7.37	9.25	9.25	22.79	11.00	5.25	4.78	22.79
GA	61.87	20.95	29.49	21.35	16.06	11.54	17.98	17.98	19.86	18.38	12.52	33.03	31.51	24.06	20.88	12.92	90.59	17.73	22.61	22.61	73.34	31.15	13.64	11.47	73.34
HI	8.48	6.62	14.65	6.90	3.88	4.10	5.15	5.15	8.04	5.53	7.12	7.86	6.88	7.68	7.24	4.84	14.83	4.22	6.73	6.73	27.79	7.53	5.28	5.39	27.79
ID	13.87	7.07	11.31	12.31	5.33	3.88	6.92	6.92	9.01	8.49	4.97	9.39	9.72	8.39	10.73	5.22	22.39	7.91	9.62	9.62	28.90	8.36	6.12	4.95	28.90
IL	23.38	15.06	20.46	28.05	11.77	7.55	10.37	10.37	17.20	16.69	10.75	19.99	12.36	17.62	22.12	9.56	33.65	11.09	20.83	20.83	62.17	19.02	17.57	5.48	62.17
IN	7.76	4.17	6.49	4.81	3.67	2.95	3.93	3.93	5.43	6.15	3.51	5.11	4.63	6.30	3.60	2.64	10.53	4.15	5.10	5.10	11.78	4.77	2.98	2.81	11.78
IA	10.34	8.42	12.90	12.72	8.24	4.47	6.88	6.88	9.20	5.82	4.64	8.40	6.76	8.73	10.98	6.40	19.92	4.80	5.70	5.70	36.26	14.65	6.70	4.60	24.10
KS	14.38	6.41	9.59	8.14	6.41	4.94	4.91	4.91	8.62	7.32	4.14	7.47	8.33	8.31	6.21	5.27	16.38	5.64	6.57	6.57	21.39	14.64	5.42	4.41	14.64
KY	18.75	12.00	24.00	11.00	7.25	5.25	12.50	12.50	19.84	11.27	7.23	7.50	10.00	18.59	12.49	5.50	34.00	12.00	10.67	10.67	48.00	13.78	15.43	4.00	48.00
LA	23.72	13.00	21.51	13.00	10.19	7.26	12.45	12.45	15.57	10.97	6.96	14.17	15.60	20.40	13.16	6.05	35.41	13.54	15.88	15.88	43.26	9.29	8.36	6.48	43.26
ME	12.86	9.34	18.66	15.82	8.40	4.84	10.34	10.34	14.07	13.62	5.16	10.73	13.40	9.74	10.18	6.24	23.06	6.89	7.53	7.53	42.08	13.38	5.77	4.87	42.08
MD	12.84	14.69	13.94	19.92	5.91	6.13	7.34	7.34	13.74	13.95	6.39	10.39	7.28	18.75	9.37	6.93	30.19	11.62	12.36	12.36	62.69	20.35	6.49	5.91	34.01
MA	8.68	5.23	9.61	18.85	6.24	2.84	4.35	4.35	9.58	7.78	5.27	10.55	5.09	12.92	4.68	3.50	30.99	5.72	6.89	6.89	54.08	33.00	5.81	2.48	23.75
MI	18.76	12.20	18.76	17.13	11.09	6.05	14.34	14.34	12.16	12.23	12.20	17.36	15.21	50.15	10.05	7.06	41.77	10.97	12.36	12.36	50.15	11.72	12.17	6.73	50.15
MN	19.08	19.63	32.94	10.69	13.05	5.59	10.82	10.82	35.70	23.79	9.93	14.01	15.96	34.92	9.93	7.70	68.07	18.46	9.02	9.02	125.84	6.99	13.08	7.38	6.99
MS	18.84	9.11	13.10	9.88	6.18	5.44	10.83	10.83	13.36	9.02	5.50	11.97	10.74	14.86	10.19	8.67	34.07	7.00	13.91	13.91	38.39	8.30	8.30	5.37	38.39
MO	20.18	12.02	13.09	14.94	10.39	6.92	11.53	11.53	10.35	12.21	8.01	15.54	11.65	16.04	13.35	9.62	38.70	8.65	12.57	12.57	44.61	30.73	11.93	6.78	44.61
MT	10.19	10.19	16.22	8.28	7.10	4.79	10.08	10.08	10.07	12.78	5.36	9.49	8.64	11.25	8.99	5.69	34.41	5.97	8.60	8.60	20.45	7.77	7.08	6.02	7.77
NE	21.40	15.45	20.10	19.40	10.83	8.43	17.60	17.60	11.63	21.53	8.53	19.73	12.58	15.20	14.93	9.58	36.13	12.95	13.88	13.88	52.18	18.15	8.40	7.08	67.55
NV	9.92	6.70	12.24	9.79	5.97	4.28	8.50	8.50	6.28	7.41	3.32	6.54	5.68	8.18	5.81	5.97	17.37	7.36	7.25	7.25	19.00	11.91	4.01	4.36	11.91
NH	22.39	11.73	18.76	25.48	13.80	5.76	12.76	12.76	13.34	13.60	7.44	18.16	25.71	16.95	11.36	10.77	39.32	7.96	11.58	11.58	125.75	36.22	13.21	7.76	125.75
NJ	16.94	12.03	16.94	13.84	11.18	4.86	9.96	9.96	9.40	11.59	14.39	17.16	11.07	13.95	14.39	6.66	36.02	6.81	11.88	11.88	14.78	12.26	9.40	6.16	16.81
NM	24.37	9.11	19.35	12.09	10.28	7.25	7.11	7.11	11.76	10.64	6.21	14.95	11.19	13.88	14.09	6.83	37.02	9.49	11.71	11.71	53.67	22.25	7.23	6.93	53.67
NY	4.33	3.54	6.47	7.73	5.97	2.75	3.94	3.94	6.58	3.77	4.72	6.95	4.90	7.10	3.33	3.81	11.62	4.70	6.32	6.32	11.22	6.07	3.67	3.28	6.06
NC	15.50	11.55	18.63	17.12	8.14	9.96	13.11	13.11	11.87	12.57	9.34	11.05	11.85	16.82	13.63	8.92	33.88	11.87	11.97	11.97	70.40	20.61	8.12	5.41	70.40
ND	9.54	9.54	9.54	5.43	5.43	3.75	4.13	4.13	9.54	9.54	8.78	6.85	5.64	9.33	8.78	5.76	16.48	5.76	9.33	9.33	9.33	9.33	9.54	16.48	9.04
OH	7.73	4.31	5.52	4.67	4.92	3.43	4.79	4.79	8.83	11.38	36.65	6.93	8.15	6.99	7.80	3.52	14.89	4.00	6.58	6.58	11.33	5.84	4.85	4.58	5.84
OK	18.56	9.92	13.23	13.71	8.14	7.02	11.13	11.13	12.71	12.47	5.73	15.58	11.38	12.79	17.44	7.23	24.19	10.52	11.32	11.32	33.69	18.85	7.23	6.79	33.69
OR	22.03	8.09	13.21	9.91	9.68	5.04	9.25	9.25	11.50	11.30	6.46	13.57	12.29	11.17	10.00	6.02	25.38	6.95	7.37	7.37	21.11	12.91	7.22	6.11	21.11
PA	17.84	17.84	13.93	17.58	12.59	6.86	10.29	10.29	13.07	13.93	13.07	14.91	15.32	17.84	13.07	8.68	34.96	8.55	17.39	17.39	27.42	17.39	10.29	14.94	27.42
RI	9.69	10.13	12.42	13.18	10.51	4.12	8.56	8.56	10.26	13.56	4.93	8.95	9.48	20.95	8.02	5.77	19.37	6.98	6.06	6.06	28.84	20.64	4.60	4.61	24.05
SC	24.05	18.13	22.49	16.89	9.59	11.45	11.78	11.78	16.99	18.58	11.06	13.93	15.50	20.61	15.75	12.36	50.08	13.06	13.91	13.91	41.22	32.53	10.50	6.33	41.22
SD	20.93	15.47	19.14	18.15	9.09	5.14	9.75	9.75	12.75	14.72	6.60	13.21	10.64	14.24	10.47	11.11	35.39	8.50	13.06	13.06	39.43	23.87	5.91	5.78	39.43
TN	23.85	11.59	12.99	12.86	6.71	6.45	12.53	12.53	12.51	11.17	7.83	13.19	9.84	18.04	12.47	6.13	34.24	8.48	12.29	12.29	21.99	22.19	8.37	4.68	21.99
TX	9.68	7.76	9.68	8.61	6.25	5.65	7.33	7.33	7.82	10.32	4.91	8.32	8.04	12.94	8.04	5.47	17.96	13.94	7.16	7.16	29.32	11.95	5.11	5.97	8.11
UT	12.88	5.69	8.30	6.49	5.64	3.19	7.02	7.02	6.89	5.87	5.20	7.33	6.66	6.78	5.02	3.49	20.72	5.58	5.07	5.07	22.57	10.74	4.22	4.58	18.77
VT	15.08	11.19	16.10	12.21	10.23	4.29	10.77	10.77	14.25	15.36	5.94	12.50	8.94	11.44	9.33	7.98	27.37	7.17	10.63	10.63	40.22	21.76	6.92	7.60	40.22
VA	10.37	7.72	8.56	9.91	5.11	4.15	7.23	7.23	7.71	6.23	6.03	7.08	8.23	7.98	6.24	4.64	21.77	5.23	6.70	6.70	28.82	11.93	4.63	2.80	28.82
WA	9.36	9.36	9.36	8.89	8.84	3.25	6.31	6.31	13.96	7.06	9.36	12.14	14.60	21.48	11.32	4.26	20.35	4.48	8.15	8.15	8.15	8.15	7.73	20.35	8.15
WV	13.06	8.27	9.67	9.38	5.25	5.27	7.90	7.90	9.46	7.92	5.44	8.80	8.73	12.01	8.28	5.23	24.07	6.86	7.49	7.49	28.94	13.43	6.48	3.54	29.69
WI	11.59	12.19	16.32	11.25	8.29	5.40	7.88	7.88	12.04	12.28	4.23	15.07	12.86	27.64	10.08	6.29	26.71	9.04	11.42	11.42	20.66	22.20	15.04	5.37	20.66
WY	5.50	5.50	5.50	5.50	5.50	5.50	5.50	5.50	5.50	5.50	5.50	5.50	5.50	5.50	5.50	5.50	5.50	5.50	5.50	5.50	5.50	5.50	5.50	5.50	5.50
AVG.	16.40	10.76	15.13	12.62	8.36	5.68	9.20	9.20	12.46	11.44	7.89	12.48	11.12	15.32	10.91	6.66	29.63	8.56	10.55	10.55	36.45	16.12	7.97	6.57	31.38

R013113-60 Workers' Compensation (cont.) (Canada in Canadian dollars)

Province		Alberta	British Columbia	Manitoba	Ontario	New Brunswick	Newfndld. & Labrador	Northwest Territories	Nova Scotia	Prince Edward Island	Quebec	Saskat-chewan	Yukon
Carpentry—3 stories or less	Rate	5.89	3.35	4.04	4.44	3.80	8.38	4.16	5.71	6.55	12.17	3.54	6.63
	Code	42143	721028	40102	723	238130	4226	4-41	4226	403	80110	B1226	202
Carpentry—interior cab. work	Rate	2.21	2.75	4.04	4.44	3.80	4.10	4.16	5.39	3.89	12.17	2.03	6.63
	Code	42133	721021	40102	723	238350	4279	4-41	4274	402	80110	B1127	202
CARPENTRY—general	Rate	5.89	3.35	4.04	4.44	3.80	4.10	4.16	5.71	6.55	12.17	3.54	6.63
	Code	42143	721028	40102	723	238130	4299	4-41	4226	403	80110	B1202	202
CONCRETE WORK—NOC	Rate	3.86	3.24	6.35	17.86	3.80	8.38	4.16	5.04	3.89	12.32	3.38	6.63
	Code	42104	721010	40110	748	238110	4224	4-41	4224	402	80100	B1314	203
CONCRETE WORK—flat (flr. sidewalk)	Rate	3.86	3.24	6.35	17.86	3.80	8.38	4.16	5.04	3.89	12.32	3.38	6.63
	Code	42104	721010	40110	748	238110	4224	4-41	4224	402	80100	B1314	203
ELECTRICAL Wiring—inside	Rate	1.66	1.81	1.70	3.60	1.73	2.42	2.45	2.14	3.89	4.94	2.03	3.89
	Code	42124	721019	40203	704	238210	4261	4-46	4261	402	80170	B1105	206
EXCAVATION—earth NOC	Rate	2.13	3.55	4.09	5.16	3.80	3.22	3.68	3.38	3.89	7.71	2.17	3.89
	Code	40604	721031	40706	711	238190	4214	4-43	4214	402	80030	R1106	207
EXCAVATION—rock	Rate	2.13	3.55	4.09	5.16	3.80	3.22	3.68	3.38	3.89	7.71	2.17	3.89
	Code	40604	721031	40706	711	238190	4214	4-43	4214	402	80030	R1106	207
GLAZIERS	Rate	2.38	3.15	4.04	10.00	3.80	5.14	4.16	5.71	3.89	13.74	3.38	3.89
	Code	42121	715020	40109	751	238150	4233	4-41	4233	402	80150	B1304	212
INSULATION WORK	Rate	1.81	8.22	4.04	10.00	3.80	5.14	4.16	5.71	3.89	12.17	2.54	6.63
	Code	42184	721029	40102	751	238310	4234	4-41	4234	402	80110	B1207	202
LATHING	Rate	4.16	6.65	4.04	4.44	3.80	4.10	4.16	5.39	3.89	12.17	3.38	6.63
	Code	42135	721042	40102	723	238390	4279	4-41	4271	402	80110	B1316	202
MASONRY	Rate	3.86	2.90	4.04	12.39	3.80	5.14	4.16	5.71	3.89	12.32	3.38	6.63
	Code	42102	721037	40102	741	238140	4231	4-41	4231	402	80100	B1318	202
PAINTING & DECORATING	Rate	3.51	3.30	3.39	7.33	3.80	4.10	4.16	5.39	3.89	12.17	3.54	6.63
	Code	42111	721041	40105	719	238320	4275	4-41	4275	402	80110	B1201	202
PILE DRIVING	Rate	3.82	3.47	4.09	6.86	3.80	3.22	3.68	5.04	6.55	7.71	3.38	6.63
	Code	42159	722004	40706	732	238190	4129	4-43	4221	403	80030	B1310	202
PLASTERING	Rate	4.16	6.65	4.64	7.33	3.80	4.10	4.16	5.39	3.89	12.17	3.54	6.63
	Code	42135	721042	40108	719	238390	4271	4-41	4271	402	80110	B1221	202
PLUMBING	Rate	1.66	2.39	2.93	4.06	1.73	2.50	2.45	3.38	2.20	6.38	2.03	3.89
	Code	42122	721043	40204	707	238220	4241	4-46	4241	401	80160	B1101	214
ROOFING	Rate	5.53	5.19	6.85	14.44	3.80	8.38	4.16	7.06	3.89	19.12	3.38	6.63
	Code	42118	721036	40403	728	238160	4236	4-41	4236	402	80130	B1320	202
SHEET METAL WORK (HVAC)	Rate	1.66	2.39	6.85	4.06	1.73	2.50	2.45	2.78	2.20	6.38	2.03	3.89
	Code	42117	721043	40402	707	238220	4244	4-46	4244	401	80160	B1107	208
STEEL ERECTION—door & sash	Rate	1.81	11.51	7.42	17.86	3.80	8.38	4.16	5.71	3.89	18.60	3.38	6.63
	Code	42106	722005	40502	748	238120	4227	4-41	4227	402	80080	B1322	202
STEEL ERECTION—inter., ornam.	Rate	1.81	11.51	7.42	17.86	3.80	8.38	4.16	5.71	3.89	18.68	3.38	6.63
	Code	42106	722005	40502	748	238120	4227	4-41	4227	402	80080	B1322	202
STEEL ERECTION—structure	Rate	1.81	11.51	7.42	17.86	3.80	8.38	4.16	5.71	3.89	18.68	3.38	6.63
	Code	42106	722005	40502	748	238120	4227	4-41	4227	402	80080	B1322	202
STEEL ERECTION—NOC	Rate	1.81	11.51	7.42	17.86	3.80	8.38	4.16	5.71	3.89	18.68	3.38	6.63
	Code	42106	722005	40502	748	238120	4227	4-41	4227	402	80080	B1322	202
TILE WORK—inter. (ceramic)	Rate	2.84	3.44	1.87	7.33	3.80	4.10	4.16	4.83	3.89	12.17	3.38	6.63
	Code	42113	721054	40103	719	238340	4276	4-41	4276	402	80110	B1301	202
WATERPROOFING	Rate	3.51	3.29	4.04	4.44	3.80	4.10	4.16	5.71	3.89	19.12	3.54	6.63
	Code	42139	721016	40102	723	238190	4299	4-41	4239	402	80130	B1217	202
WRECKING	Rate	2.13	5.32	6.86	17.86	3.80	3.22	3.68	3.38	3.89	12.17	2.17	6.63
	Code	40604	721005	40106	748	238190	4211	4-43	4211	402	80110	R1125	202

R013113-80 Performance Bond

This table shows the cost of a Performance Bond for a construction job scheduled to be completed in 12 months. Add 1% of the premium cost per month for jobs requiring more than 12 months to complete. The rates are "standard" rates offered to contractors that the bonding company considers financially sound and capable of doing the work. Preferred rates are offered by some bonding companies based upon financial strength of the contractor. Actual rates vary from contractor to contractor and from bonding company to bonding company. Contractors should prequalify through a bonding agency before submitting a bid on a contract that requires a bond.

Contract Amount	Building Construction Class B Projects	Highways & Bridges	
		Class A New Construction	Class A-1 Highway Resurfacing
First $ 100,000 bid	$25.00 per M	$15.00 per M	$9.40 per M
Next 400,000 bid	$ 2,500 plus $15.00 per M	$ 1,500 plus $10.00 per M	$ 940 plus $7.20 per M
Next 2,000,000 bid	8,500 plus 10.00 per M	5,500 plus 7.00 per M	3,820 plus 5.00 per M
Next 2,500,000 bid	28,500 plus 7.50 per M	19,500 plus 5.50 per M	15,820 plus 4.50 per M
Next 2,500,000 bid	47,250 plus 7.00 per M	33,250 plus 5.00 per M	28,320 plus 4.50 per M
Over 7,500,000 bid	64,750 plus 6.00 per M	45,750 plus 4.50 per M	39,570 plus 4.00 per M

R015423-10 Steel Tubular Scaffolding

On new construction, tubular scaffolding is efficient up to 60′ high or five stories. Above this it is usually better to use a hung scaffolding if construction permits. Swing scaffolding operations may interfere with tenants. In this case, the tubular is more practical at all heights.

In repairing or cleaning the front of an existing building the cost of tubular scaffolding per S.F. of building front increases as the height increases above the first tier. The first tier cost is relatively high due to leveling and alignment.

The minimum efficient crew for erecting and dismantling is three workers. They can set up and remove 18 frame sections per day up to 5 stories high. For 6 to 12 stories high, a crew of four is most efficient. Use two or more on top and two on the bottom for handing up or hoisting. They can also set up and remove 18 frame sections per day. At 7' horizontal spacing, this will run about 800 S.F. per day of erecting and dismantling. Time for placing and removing planks must be added to the above. A crew of three can place and remove 72 planks per day up to 5 stories. For over 5 stories, a crew of four can place and remove 80 planks per day.

The table below shows the number of pieces required to erect tubular steel scaffolding for 1000 S.F. of building frontage. This area is made up of a scaffolding system that is 12 frames (11 bays) long by 2 frames high.

For jobs under twenty-five frames, add 50% to rental cost. Rental rates will be lower for jobs over three months duration. Large quantities for long periods can reduce rental rates by 20%.

Description of Component	Number of Pieces for 1000 S.F. of Building Front	Unit
5′ Wide Standard Frame, 6′-4″ High	24	Ea.
Leveling Jack & Plate	24	↓
Cross Brace	44	
Side Arm Bracket, 21″	12	
Guardrail Post	12	
Guardrail, 7′ section	22	
Stairway Section	2	
Stairway Starter Bar	1	
Stairway Inside Handrail	2	
Stairway Outside Handrail	2	
Walk-Thru Frame Guardrail	2	▼

Scaffolding is often used as falsework over 15′ high during construction of cast-in-place concrete beams and slabs. Two foot wide scaffolding is generally used for heavy beam construction. The span between frames depends upon the load to be carried with a maximum span of 5′.

Heavy duty shoring frames with a capacity of 10,000#/leg can be spaced up to 10′ O.C. depending upon form support design and loading.

Scaffolding used as horizontal shoring requires less than half the material required with conventional shoring.

On new construction, erection is done by carpenters.

Rolling towers supporting horizontal shores can reduce labor and speed the job. For maintenance work, catwalks with spans up to 70′ can be supported by the rolling towers.

R015433-10 Contractor Equipment

Rental Rates shown elsewhere in the book pertain to late model high quality machines in excellent working condition, rented from equipment dealers. Rental rates from contractors may be substantially lower than the rental rates from equipment dealers depending upon economic conditions; for older, less productive machines, reduce rates by a maximum of 15%. Any overtime must be added to the base rates. For shift work, rates are lower. Usual rule of thumb is 150% of one shift rate for two shifts; 200% for three shifts.

For periods of less than one week, operated equipment is usually more economical to rent than renting bare equipment and hiring an operator.

Costs to move equipment to a job site (mobilization) or from a job site (demobilization) are not included in rental rates, nor in any Equipment costs on any Unit Price line items or crew listings. These costs can be found elsewhere. If a piece of equipment is already at a job site, it is not appropriate to utilize mob/demob costs in an estimate again.

Rental rates vary throughout the country with larger cities generally having lower rates. Lease plans for new equipment are available for periods in excess of six months with a percentage of payments applying toward purchase.

Rental rates can also be treated as reimbursement costs for contractor-owned equipment. Owned equipment costs include depreciation, loan payments, interest, taxes, insurance, storage, and major repairs.

Monthly rental rates vary from 2% to 5% of the cost of the equipment depending on the anticipated life of the equipment and its wearing parts. Weekly rates are about 1/3 the monthly rates and daily rental rates about 1/3 the weekly rate.

The hourly operating costs for each piece of equipment include costs to the user such as fuel, oil, lubrication, normal expendables for the equipment, and a percentage of mechanic's wages chargeable to maintenance. The hourly operating costs listed do not include the operator's wages.

The daily cost for equipment used in the standard crews is figured by dividing the weekly rate by five, then adding eight times the hourly operating cost to give the total daily equipment cost, not including the operator. This figure is in the right hand column of the Equipment listings under Equipment Cost/Day.

Pile Driving rates shown for pile hammer and extractor do not include leads, crane, boiler or compressor. Vibratory pile driving requires an added field specialist during set-up and pile driving operation for the electric model. The hydraulic model requires a field specialist for set-up only. Up to 125 reuses of sheet piling are possible using vibratory drivers. For normal conditions, crane capacity for hammer type and size are as follows.

Crane Capacity	Hammer Type and Size		
	Air or Steam	Diesel	Vibratory
25 ton	to 8,750 ft.-lb.		70 H.P.
40 ton	15,000 ft.-lb.	to 32,000 ft.-lb.	170 H.P.
60 ton	25,000 ft.-lb.		300 H.P.
100 ton		112,000 ft.-lb.	

Cranes should be specified for the job by size, building and site characteristics, availability, performance characteristics, and duration of time required.

Backhoes & Shovels rent for about the same as equivalent size cranes but maintenance and operating expense is higher. Crane operators rate must be adjusted for high boom heights. Average adjustments: for 150′ boom add 2% per hour; over 185′, add 4% per hour; over 210′, add 6% per hour; over 250′, add 8% per hour and over 295′, add 12% per hour.

Tower Cranes of the climbing or static type have jibs from 50′ to 200′ and capacities at maximum reach range from 4,000 to 14,000 pounds. Lifting capacities increase up to maximum load as the hook radius decreases.

Typical rental rates, based on purchase price are about 2% to 3% per month.

Erection and dismantling runs between 500 and 2000 labor hours. Climbing operation takes 10 labor hours per 20′ climb. Crane dead time is about 5 hours per 40′ climb. If crane is bolted to side of the building add cost of ties and extra mast sections. Climbing cranes have from 80′ to 180′ of mast while static cranes have 80′ to 800′ of mast.

Truck Cranes can be converted to tower cranes by using tower attachments. Mast heights over 400′ have been used.

A single 100′ high material **Hoist and Tower** can be erected and dismantled in about 400 labor hours; a double 100′ high hoist and tower in about 600 labor hours. Erection times for additional heights are 3 and 4 labor hours per vertical foot respectively up to 150′, and 4 to 5 labor hours per vertical foot over 150′ high. A 40′ high portable Buck hoist takes about 160 labor hours to erect and dismantle. Additional heights take 2 labor hours per vertical foot to 80′ and 3 labor hours per vertical foot for the next 100′. Most material hoists do not meet local code requirements for carrying personnel.

A 150′ high **Personnel Hoist** requires about 500 to 800 labor hours to erect and dismantle. Budget erection time at 5 labor hours per vertical foot for all trades. Local code requirements or labor scarcity requiring overtime can add up to 50% to any of the above erection costs.

Earthmoving Equipment: The selection of earthmoving equipment depends upon the type and quantity of material, moisture content, haul distance, haul road, time available, and equipment available. Short haul cut and fill operations may require dozers only, while another operation may require excavators, a fleet of trucks, and spreading and compaction equipment. Stockpiled material and granular material are easily excavated with front end loaders. Scrapers are most economically used with hauls between 300′ and 1-1/2 miles if adequate haul roads can be maintained. Shovels are often used for blasted rock and any material where a vertical face of 8′ or more can be excavated. Special conditions may dictate the use of draglines, clamshells, or backhoes. Spreading and compaction equipment must be matched to the soil characteristics, the compaction required and the rate the fill is being supplied.

R024119-10 Demolition Defined

Whole Building Demolition - Demolition of the whole building with no concern for any particular building element, component, or material type being demolished. This type of demolition is accomplished with large pieces of construction equipment that break up the structure, load it into trucks and haul it to a disposal site, but disposal or dump fees are not included. Demolition of below-grade foundation elements, such as footings, foundation walls, grade beams, slabs on grade, etc., is not included. Certain mechanical equipment containing flammable liquids or ozone-depleting refrigerants, electric lighting elements, communication equipment components, and other building elements may contain hazardous waste, and must be removed, either selectively or carefully, as hazardous waste before the building can be demolished.

Foundation Demolition - Demolition of below-grade foundation footings, foundation walls, grade beams, and slabs on grade. This type of demolition is accomplished by hand or pneumatic hand tools, and does not include saw cutting, or handling, loading, hauling, or disposal of the debris.

Gutting - Removal of building interior finishes and electrical/mechanical systems down to the load-bearing and sub-floor elements of the rough building frame, with no concern for any particular building element, component, or material type being demolished. This type of demolition is accomplished by hand or pneumatic hand tools, and includes loading into trucks, but not hauling, disposal or dump fees, scaffolding, or shoring. Certain mechanical equipment containing flammable liquids or ozone-depleting refrigerants, electric lighting elements, communication equipment components, and other building elements may contain hazardous waste, and must be removed, either selectively or carefully, as hazardous waste, before the building is gutted.

Selective Demolition - Demolition of a selected building element, component, or finish, with some concern for surrounding or adjacent elements, components, or finishes (see the first Subdivision (s) at the beginning of appropriate Divisions). This type of demolition is accomplished by hand or pneumatic hand tools, and does not include handling, loading, storing, hauling, or disposal of the debris, scaffolding, or shoring. "Gutting" methods may be used in order to save time, but damage that is caused to surrounding or adjacent elements, components, or finishes may have to be repaired at a later time.

Careful Removal - Removal of a piece of service equipment, building element or component, or material type, with great concern for both the removed item and surrounding or adjacent elements, components or finishes. The purpose of careful removal may be to protect the removed item for later re-use, preserve a higher salvage value of the removed item, or replace an item while taking care to protect surrounding or adjacent elements, components, connections, or finishes from cosmetic and/or structural damage. An approximation of the time required to perform this type of removal is 1/3 to 1/2 the time it would take to install a new item of like kind (see Reference Number R220105-10). This type of removal is accomplished by hand or pneumatic hand tools, and does not include loading, hauling, or storing the removed item, scaffolding, shoring, or lifting equipment.

Cutout Demolition - Demolition of a small quantity of floor, wall, roof, or other assembly, with concern for the appearance and structural integrity of the surrounding materials. This type of demolition is accomplished by hand or pneumatic hand tools, and does not include saw cutting, handling, loading, hauling, or disposal of debris, scaffolding, or shoring.

Rubbish Handling - Work activities that involve handling, loading or hauling of debris. Generally, the cost of rubbish handling must be added to the cost of all types of demolition, with the exception of whole building demolition.

Minor Site Demolition - Demolition of site elements outside the footprint of a building. This type of demolition is accomplished by hand or pneumatic hand tools, or with larger pieces of construction equipment, and may include loading a removed item onto a truck (check the Crew for equipment used). It does not include saw cutting, hauling or disposal of debris, and, sometimes, handling or loading.

R024119-20 Dumpsters

Dumpster rental costs on construction sites are presented in two ways.

The cost per week rental includes the delivery of the dumpster; its pulling or emptying once per week, and its final removal. The assumption is made that the dumpster contractor could choose to empty a dumpster by simply bringing in an empty unit and removing the full one. These costs also include the disposal of the materials in the dumpster.

The Alternate Pricing can be used when actual planned conditions are not approximated by the weekly numbers. For example, these lines can be used when a dumpster is needed for 4 weeks and will need to be emptied 2 or 3 times per week. Conversely the Alternate Pricing lines can be used when a dumpster will be rented for several weeks or months but needs to be emptied only a few times over this period.

Existing Conditions R0265 Underground Storage Tank Removal

R026510-20 Underground Storage Tank Removal

Underground Storage Tank Removal can be divided into two categories: Non-Leaking and Leaking. Prior to removing an underground storage tank, tests should be made, with the proper authorities present, to determine whether a tank has been leaking or the surrounding soil has been contaminated.

To safely remove Liquid Underground Storage Tanks:

1. Excavate to the top of the tank.
2. Disconnect all piping.
3. Open all tank vents and access ports.
4. Remove all liquids and/or sludge.
5. Purge the tank with an inert gas.
6. Provide access to the inside of the tank and clean out the interior using proper personal protective equipment (PPE).
7. Excavate soil surrounding the tank using proper PPE for on-site personnel.
8. Pull and properly dispose of the tank.
9. Clean up the site of all contaminated material.
10. Install new tanks or close the excavation.

Existing Conditions — R0282 Asbestos Remediation

R028213-20 Asbestos Removal Process

Asbestos removal is accomplished by a specialty contractor who understands the federal and state regulations regarding the handling and disposal of the material. The process of asbestos removal is divided into many individual steps. An accurate estimate can be calculated only after all the steps have been priced.

The steps are generally as follows:

1. Obtain an asbestos abatement plan from an industrial hygienist.
2. Monitor the air quality in and around the removal area and along the path of travel between the removal area and transport area. This establishes the background contamination.
3. Construct a two part decontamination chamber at entrance to removal area.
4. Install a HEPA filter to create a negative pressure in the removal area.
5. Install wall, floor and ceiling protection as required by the plan, usually 2 layers of fireproof 6 mil polyethylene.
6. Industrial hygienist visually inspects work area to verify compliance with plan.
7. Provide temporary supports for conduit and piping affected by the removal process.
8. Proceed with asbestos removal and bagging process. Monitor air quality as described in Step #2. Discontinue operations when contaminate levels exceed applicable standards.
9. Document the legal disposal of materials in accordance with EPA standards.
10. Thoroughly clean removal area including all ledges, crevices and surfaces.
11. Post abatement inspection by industrial hygienist to verify plan compliance.
12. Provide a certificate from a licensed industrial hygienist attesting that contaminate levels are within acceptable standards before returning area to regular use.

Concrete — R0330 Cast-In-Place Concrete

R033053-60 Maximum Depth of Frost Penetration in Inches

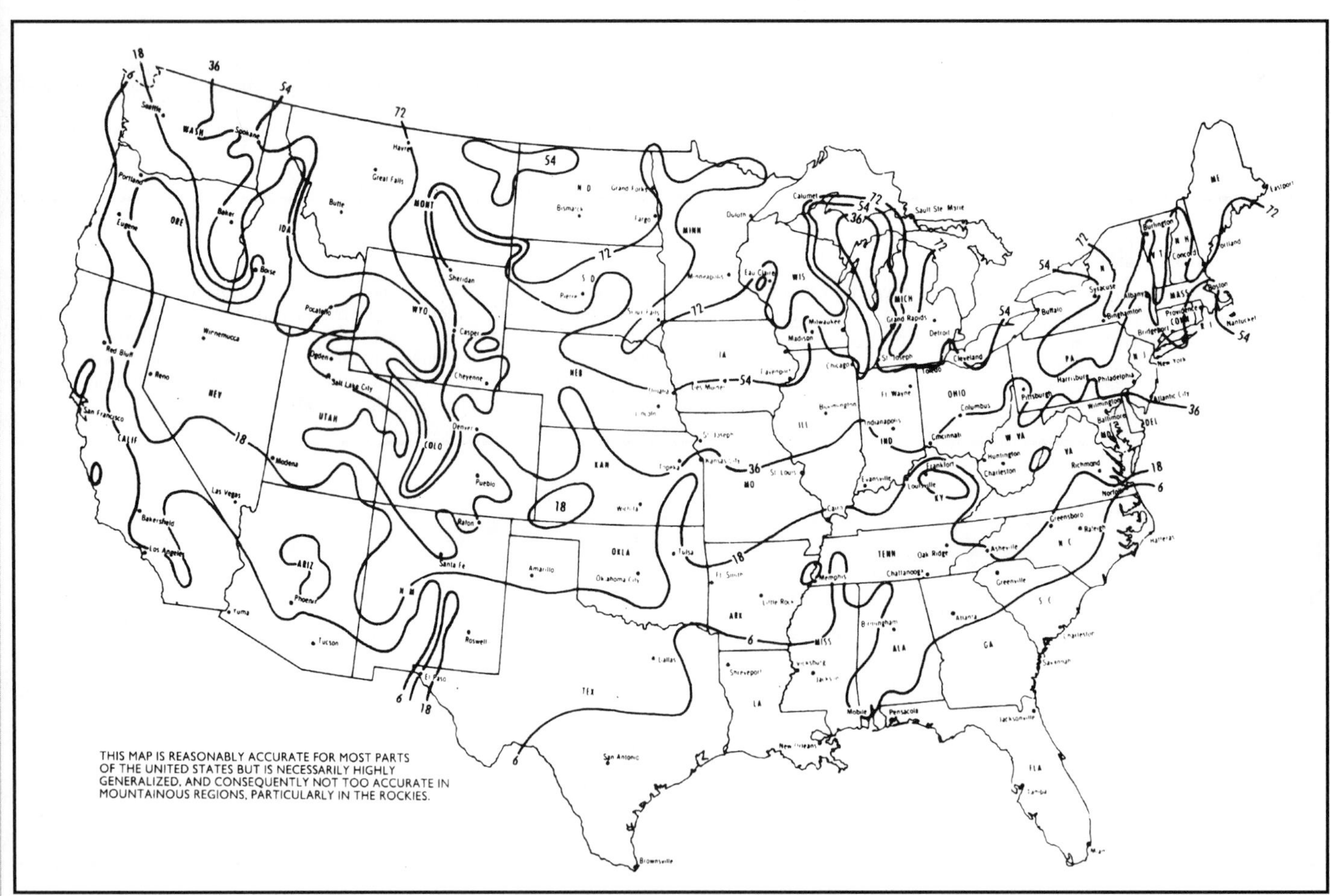

THIS MAP IS REASONABLY ACCURATE FOR MOST PARTS OF THE UNITED STATES BUT IS NECESSARILY HIGHLY GENERALIZED, AND CONSEQUENTLY NOT TOO ACCURATE IN MOUNTAINOUS REGIONS, PARTICULARLY IN THE ROCKIES.

Metals — R0505 Common Work Results for Metals

R050521-20 Welded Structural Steel

Usual weight reductions with welded design run 10% to 20% compared with bolted or riveted connections. This amounts to about the same total cost compared with bolted structures since field welding is more expensive than bolts. For normal spans of 18′ to 24′ figure 6 to 7 connections per ton.

Trusses — For welded trusses add 4% to weight of main members for connections. Up to 15% less steel can be expected in a welded truss compared to one that is shop bolted. Cost of erection is the same whether shop bolted or welded.

General — Typical electrodes for structural steel welding are E6010, E6011, E60T and E70T. Typical buildings vary between 2# to 8# of weld rod per ton of steel. Buildings utilizing continuous design require about three times as much welding as conventional welded structures. In estimating field erection by welding, it is best to use the average linear feet of weld per ton to arrive at the welding cost per ton. The type, size and position of the weld will have a direct bearing on the cost per linear foot. A typical field welder will deposit 1.8# to 2# of weld rod per hour manually. Using semiautomatic methods can increase production by as much as 50% to 75%.

Special Construction — R1311 Swimming Pools

R131113-20 Swimming Pools

Pool prices given per square foot of surface area include pool structure, filter and chlorination equipment, pumps, related piping, ladders/steps, maintenance kit, skimmer and vacuum system. Decks and electrical service to equipment are not included.

Residential in-ground pool construction can be divided into two categories: vinyl lined and gunite. Vinyl lined pool walls are constructed of different materials including wood, concrete, plastic or metal. The bottom is often graded with sand over which the vinyl liner is installed. Vermiculite or soil cement bottoms may be substituted for an added cost.

Gunite pool construction is used both in residential and municipal installations. These structures are steel reinforced for strength and finished with a white cement limestone plaster.

Municipal pools will have a higher cost because plumbing codes require more expensive materials, chlorination equipment and higher filtration rates.

Municipal pools greater than 1,800 S.F. require gutter systems to control waves. This gutter may be formed into the concrete wall. Often a vinyl/stainless steel gutter or gutter/wall system is specified, which will raise the pool cost.

Competition pools usually require tile bottoms and sides with contrasting lane striping, which will also raise the pool cost.

Fire Suppression — R2112 Fire-Suppression Standpipes

R211226-10 Standpipe Systems

The basis for standpipe system design is National Fire Protection Association NFPA 14, however, the authority having jurisdiction should be consulted for special conditions, local requirements and approval.

Standpipe systems, properly designed and maintained, are an effective and valuable time saving aid for extinquishing fires, especially in the upper stories of tall buildings, the interior of large commercial or industrial malls, or other areas where construction features or access make the laying of temporary hose lines time consuming and/or hazardous. Standpipes are frequently installed with automatic sprinkler systems for maximum protection.

There are three general classes of service for standpipe systems:
Class I for use by fire departments and personnel with special training for heavy streams (2-1/2″ hose connections).
Class II for use by building occupants until the arrival of the fire department (1-1/2″ hose connector with hose).
Class III for use by either fire departments and trained personnel or by the building occupants (both 2-1/2″ and 1-1/2″ hose connections or one 2-1/2″ hose valve with an easily removable 2-1/2″ by 1-1/2″ adapter).

Standpipe systems are also classified by the way water is supplied to the system. The four basic types are:
Type 1: Wet standpipe system having supply valve open and water pressure maintained at all times.
Type 2: Standpipe system so arranged through the use of approved devices as to admit water to the system automatically by opening a hose valve.
Type 3: Standpipe system arranged to admit water to the system through manual operation of approved remote control devices located at each hose station.
Type 4: Dry standpipe having no permanent water supply.

R211226-20 NFPA 14 Basic Standpipe Design

Class	Design-Use	Pipe Size Minimums	Water Supply Minimums
Class I	2 1/2″ hose connection on each floor All areas within 150′ of an exit in every exit stairway Fire Department Trained Personnel	Height to 100′, 4″ dia. Heights above 100′, 6″ dia. (275′ max. except with pressure regulators 400′ max.)	For each standpipe riser 500 GPM flow For common supply pipe allow 500 GPM for first standpipe plus 250 GPM for each additional standpipe (2500 GPM max. total) 30 min. duration 65 PSI at 500 GPM
Class II	1 1/2″ hose connection with hose on each floor All areas within 130′ of hose connection measured along path of hose travel Occupant personnel	Height to 50′, 2″ dia. Height above 50′, 2 1/2″ dia.	For each standpipe riser 100 GPM flow For multiple riser common supply pipe 100 GPM 300 min. duration, 65 PSI at 100 GPM
Class III	Both of above. Class I valved connections will meet Class III with additional 2 1/2″ by 1 1/2″ adapter and 1 1/2″ hose.	Same as Class I	Same as Class I

*Note: Where 2 or more standpipes are installed in the same building or section of building they shall be interconnected at the bottom.

Combined Systems

Combined systems are systems where the risers supply both automatic sprinklers and 2-1/2″ hose connection outlets for fire department use. In such a system the sprinkler spacing pattern shall be in accordance with NFPA 13 while the risers and supply piping will be sized in accordance with NFPA 14. When the building is completely sprinklered the risers may be sized by hydraulic calculation.The minimum size riser for buildings not completely sprinklered is 6″.

The minimum water supply of a completely sprinklered, light hazard, high-rise occupancy building will be 500 GPM while the supply required for other types of completely sprinklered high-rise buildings is 1000 GPM.

General System Requirements

1. Approved valves will be provided at the riser for controlling branch lines to hose outlets.
2. A hose valve will be provided at each outlet for attachment of hose.
3. Where pressure at any standpipe outlet exceeds 100 PSI a pressure reducer must be installed to limit the pressure to 100 PSI. Note that the pressure head due to gravity in 100′ of riser is 43.4 PSI. This must be overcome by city pressure, fire pumps, or gravity tanks to provide adequate pressure at the top of the riser.
4. Each hose valve on a wet system having linen hose shall have an automatic drip connection to prevent valve leakage from entering the hose.
5. Each riser will have a valve to isolate it from the rest of the system.
6. One or more fire department connections as an auxiliary supply shall be provided for each Class I or Class III standpipe system. In buildings having two or more zones, a connection will be provided for each zone.
7. There will be no shutoff valve in the fire department connection, but a check valve will be located in the line before it joins the system.
8. All hose connections street side will be identified on a cast plate or fitting as to purpose.

R211313-10 Sprinkler Systems (Automatic)

Sprinkler systems may be classified by type as follows:

1. **Wet Pipe System.** A system employing automatic sprinklers attached to a piping system containing water and connected to a water supply so that water discharges immediately from sprinklers opened by a fire.
2. **Dry Pipe System.** A system employing automatic sprinklers attached to a piping system containing air under pressure, the release of which as from the opening of sprinklers permits the water pressure to open a valve known as a "dry pipe valve". The water then flows into the piping system and out the opened sprinklers.
3. **Pre-Action System.** A system employing automatic sprinklers attached to a piping system containing air that may or may not be under pressure, with a supplemental heat responsive system of generally more sensitive characteristics than the automatic sprinklers themselves, installed in the same areas as the sprinklers; actuation of the heat responsive system, as from a fire, opens a valve which permits water to flow into the sprinkler piping system and to be discharged from any sprinklers which may be open.
4. **Deluge System.** A system employing open sprinklers attached to a piping system connected to a water supply through a valve which is opened by the operation of a heat responsive system installed in the same areas as the sprinklers. When this valve opens, water flows into the piping system and discharges from all sprinklers attached thereto.
5. **Combined Dry Pipe and Pre-Action Sprinkler System.** A system employing automatic sprinklers attached to a piping system containing air under pressure with a supplemental heat responsive system of generally more sensitive characteristics than the automatic sprinklers themselves, installed in the same areas as the sprinklers; operation of the heat responsive system, as from a fire, actuates tripping devices which open dry pipe valves simultaneously and without loss of air pressure in the system. Operation of the heat responsive system also opens approved air exhaust valves at the end of the feed main which facilitates the filling of the system with water which usually precedes the opening of sprinklers. The heat responsive system also serves as an automatic fire alarm system.
6. **Limited Water Supply System.** A system employing automatic sprinklers and conforming to these standards but supplied by a pressure tank of limited capacity.
7. **Chemical Systems.** Systems using halon, carbon dioxide, dry chemical or high expansion foam as selected for special requirements. Agent may extinguish flames by chemically inhibiting flame propagation, suffocate flames by excluding oxygen, interrupting chemical action of oxygen uniting with fuel or sealing and cooling the combustion center.
8. **Firecycle System.** Firecycle is a fixed fire protection sprinkler system utilizing water as its extinguishing agent. It is a time delayed, recycling, preaction type which automatically shuts the water off when heat is reduced below the detector operating temperature and turns the water back on when that temperature is exceeded. The system senses a fire condition through a closed circuit electrical detector system which controls water flow to the fire automatically. Batteries supply up to 90 hour emergency power supply for system operation. The piping system is dry (until water is required) and is monitored with pressurized air. Should any leak in the system piping occur, an alarm will sound, but water will not enter the system until heat is sensed by a firecycle detector.

Area coverage sprinkler systems may be laid out and fed from the supply in any one of several patterns as shown below. It is desirable, if possible, to utilize a central feed and achieve a shorter flow path from the riser to the furthest sprinkler. This permits use of the smallest sizes of pipe possible with resulting savings.

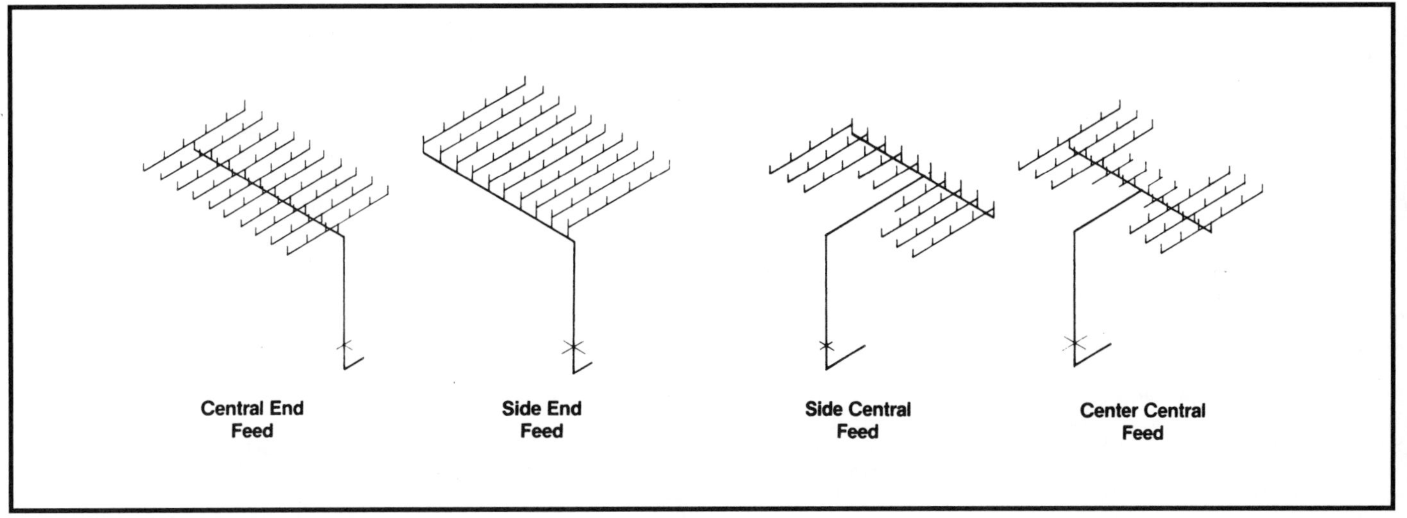

R211313-20 System Classification

System Classification

Rules for installation of sprinkler systems vary depending on the classification of occupancy falling into one of three categories as follows:

Light Hazard Occupancy

The protection area allotted per sprinkler should not exceed 225 S.F., with the maximum distance between lines and sprinklers on lines being 15′. The sprinklers do not need to be staggered. Branch lines should not exceed eight sprinklers on either side of a cross main. Each large area requiring more than 100 sprinklers and without a sub-dividing partition should be supplied by feed mains or risers sized for ordinary hazard occupancy.
Maximum system area = 52,000 S.F.

Included in this group are:

Churches
Clubs
Educational
Hospitals
Institutional
Libraries
(except large stack rooms)
Museums
Nursing Homes
Offices
Residential
Restaurants
Theaters and Auditoriums
(except stages and prosceniums)
Unused Attics

Ordinary Hazard Occupancy

The protection area allotted per sprinkler shall not exceed 130 S.F. of noncombustible ceiling and 130 S.F. of combustible ceiling. The maximum allowable distance between sprinkler lines and sprinklers on line is 15′. Sprinklers shall be staggered if the distance between heads exceeds 12′. Branch lines should not exceed eight sprinklers on either side of a cross main.
Maximum system area = 52,000 S.F.

Included in this group are:

Group 1
Automotive Parking and Showrooms
Bakeries
Beverage manufacturing
Canneries
Dairy Products Manufacturing/Processing
Electronic Plans
Glass and Glass Products Manufacturing
Laundries
Restaurant Service Areas

Group 2
Cereal Mills
Chemical Plants—Ordinary
Confectionery Products
Distilleries
Dry Cleaners
Feed Mills
Horse Stables
Leather Goods Manufacturing
Libraries—Large Stack Room Areas
Machine Shops
Metal Working
Mercantile
Paper and Pulp Mills
Paper Process Plants
Piers and Wharves
Post Offices
Printing and Publishing
Repair Garages
Stages
Textile Manufacturing
Tire Manufacturing
Tobacco Products Manufacturing
Wood Machining
Wood Product Assembly

Extra Hazard Occupancy

The protection area allotted per sprinkler shall not exceed 100 S.F. of noncombustible ceiling and 100 S.F. of combustible ceiling. The maximum allowable distance between lines and between sprinklers on lines is 12′. Sprinklers on alternate lines shall be staggered if the distance between sprinklers on lines exceeds 8′. Branch lines should not exceed six sprinklers on either side of a cross main.
Maximum system area:
  Design by pipe schedule = 25,000 S.F.
  Design by hydraulic calculation = 40,000 S.F.

Included in this group are:

Group 1
Aircraft hangars
Combustible Hydraulic Fluid Use Area
Die Casting
Metal Extruding
Plywood/Particle Board Manufacturing
Printing (inks with flash points < 100 degrees F
Rubber Reclaiming, Compounding, Drying, Milling, Vulcanizing
Saw Mills
Textile Picking, Opening, Blending, Garnetting, Carding, Combing of Cotton, Synthetics, Wood Shoddy, or Burlap
Upholstering with Plastic Foams

Group 2
Asphalt Saturating
Flammable Liquids Spraying
Flow Coating
Manufactured/Modular Home Building Assemblies (where finished enclosure is present and has combustible interiors)
Open Oil Quenching
Plastics Processing
Solvent Cleaning
Varnish and Paint Dipping

R211313-30 Sprinkler Quantities for Various Sizes and Types of Pipe

Sprinkler Quantities: The table below lists the usual maximum number of sprinkler heads for each size of copper and steel pipe for both wet and dry systems. These quantities may be adjusted to meet individual structural needs or local code requirements. Maximum area on any one floor for one system is: light hazard and ordinary hazard 52,000 S.F., extra hazardous 25,000 S.F.

Pipe Size	Light Hazard Occupancy		Ordinary Hazard Occupancy		Extra Hazard Occupancy	
Diameter	Steel Pipe	Copper Pipe	Steel Pipe	Copper Pipe	Steel Pipe	Copper Pipe
1″	2 sprinklers	2 sprinklers	2 sprinklers	2 sprinklers	1 sprinklers	1 sprinklers
1-1/4″	3	3	3	3	2	2
1-1/2″	5	5	5	5	5	5
2″	10	12	10	12	8	8
2-1/2″	30	40	20	25	15	20
3″	60	65	40	45	27	30
3-1/2″	100	115	65	75	40	45
4″			100	115	55	65
5″			160	180	90	100
6″			275	300	150	170

Dry Pipe Systems: A dry pipe system should be installed where a wet pipe system is impractical, as in rooms or buildings which cannot be properly heated.

The use of an approved dry pipe system is more desirable than shutting off the water supply during cold weather.

Not more than 750 gallons of system capacity should be controlled by one dry pipe valve. Where two or more dry pipe valves are used, systems should preferably be divided horizontally.

R211313-40 Adjustment for Sprinkler/Standpipe Installations

Quality/Complexity Multiplier (For all installations)

Economy installation, add .... 0 to 5%
Good quality, medium complexity, add .... 5 to 15%
Above average quality and complexity, add .... 15 to 25%

R220102-20 Labor Adjustment Factors

Labor Adjustment Factors are provided for Divisions 21, 22, and 23 to assist the mechanical estimator account for the various complexities and special conditions of any particular project. While a single percentage has been entered on each line of Division 22 01 02.20, it should be understood that these are just suggested midpoints of ranges of values commonly used by mechanical estimators. They may be increased or decreased depending on the severity of the special conditions.

The group for "existing occupied buildings", has been the subject of requests for explanation. Actually there are two stages to this group: buildings that are existing and "finished" but unoccupied, and those that also are occupied. Buildings that are "finished" may result in higher labor costs due to the workers having to be more careful not to damage finished walls, ceilings, floors, etc. and may necessitate special protective coverings and barriers. Also corridor bends and doorways may not accommodate long pieces of pipe or larger pieces of equipment. Work above an already hung ceiling can be very time consuming. The addition of occupants may force the work to be done on premium time (nights and/or weekends), eliminate the possible use of some preferred tools such as pneumatic drivers, powder charged drivers etc. The estimator should evaluate the access to the work area and just how the work is going to be accomplished to arrive at an increase in labor costs over "normal" new construction productivity.

R220105-10 Demolition (Selective vs. Removal for Replacement)

Demolition can be divided into two basic categories.

One type of demolition involves the removal of material with no concern for its replacement. The labor-hours to estimate this work are found under "Selective Demolition" in the Fire Protection, Plumbing and HVAC Divisions. It is selective in that individual items or all the material installed as a system or trade grouping such as plumbing or heating systems are removed. This may be accomplished by the easiest way possible, such as sawing, torch cutting, or sledge hammer as well as simple unbolting.

The second type of demolition is the removal of some item for repair or replacement. This removal may involve careful draining, opening of unions, disconnecting and tagging of electrical connections, capping of pipes/ducts to prevent entry of debris or leakage of the material contained as well as transport of the item away from its in-place location to a truck/dumpster. An approximation of the time required to accomplish this type of demolition is to use half of the time indicated as necessary to install a new unit. For example; installation of a new pump might be listed as requiring 6 labor-hours so if we had to estimate the removal of the old pump we would allow an additional 3 hours for a total of 9 hours. That is, the complete replacement of a defective pump with a new pump would be estimated to take 9 labor-hours.

R220523-80 Valve Materials

VALVE MATERIALS

Bronze:
Bronze is one of the oldest materials used to make valves. It is most commonly used in hot and cold water systems and other non-corrosive services. It is often used as a seating surface in larger iron body valves to ensure tight closure.

Carbon Steel:
Carbon steel is a high strength material. Therefore, valves made from this metal are used in higher pressure services, such as steam lines up to 600 psi at 850°F. Many steel valves are available with butt-weld ends for economy and are generally used in high pressure steam service as well as other higher pressure non-corrosive services.

Forged Steel:
Valves from tough carbon steel are used in service up to 2000 psi and temperatures up to 1000°F in Gate, Globe and Check valves.

Iron:
Valves are normally used in medium to large pipe lines to control non-corrosive fluid and gases, where pressures do not exceed 250 psi at 450° or 500 psi cold water, oil or gas.

Stainless Steel:
Developed steel alloys can be used in over 90% corrosive services.

Plastic PVC:
This is used in a great variety of valves generally in high corrosive service with lower temperatures and pressures.

VALVE SERVICE PRESSURES

Pressure ratings on valves provide an indication of the safe operating pressure for a valve at some elevated temperature. This temperature is dependent upon the materials used and the fabrication of the valve. When specific data is not available, a good "rule-of-thumb" to follow is the temperature of saturated steam on the primary rating indicated on the valve body. Example: The valve has the number 150S printed on the side indicating 150 psi and hence, a maximum operating temperature of 367°F (temperature of saturated steam and 150 psi).

DEFINITIONS

1. "WOG" – Water, oil, gas (cold working pressures).
2. "SWP" – Steam working pressure.
3. 100% area (full port) – means the area through the valve is equal to or greater than the area of standard pipe.
4. "Standard Opening" – means that the area through the valve is less than the area of standard pipe and therefore these valves should be used only where restriction of flow is unimportant.
5. "Round Port" – means the valve has a full round opening through the plug and body, of the same size and area as standard pipe.
6. "Rectangular Port" – valves have rectangular shaped ports through the plug body. The area of the port is either equal to 100% of the area of standard pipe, or restricted (standard opening). In either case it is clearly marked.
7. "ANSI" – American National Standards Institute.

R220523-90 Valve Selection Considerations

INTRODUCTION: In any piping application, valve performance is critical. Valves should be selected to give the best performance at the lowest cost.

The following is a list of performance characteristics generally expected of valves.

1. Stopping flow or starting it.
2. Throttling flow (Modulation).
3. Flow direction changing.
4. Checking backflow (Permitting flow in only one direction).
5. Relieving or regulating pressure.

In order to properly select the right valve, some facts must be determined.

A. What liquid or gas will flow through the valve?
B. Does the fluid contain suspended particles?
C. Does the fluid remain in liquid form at all times?
D. Which metals does fluid corrode?
E. What are the pressure and temperature limits? (As temperature and pressure rise, so will the price of the valve.)
F. Is there constant line pressure?
G. Is the valve merely an on-off valve?
H. Will checking of backflow be required?
I. Will the valve operate frequently or infrequently?

Valves are classified by design type into such classifications as Gate, Globe, Angle, Check, Ball, Butterfly and Plug. They are also classified by end connection, stem, pressure restrictions and material such as bronze, cast iron, etc. Each valve has a specific use. A quality valve used correctly will provide a lifetime of trouble- free service, but a high quality valve installed in the wrong service may require frequent attention.

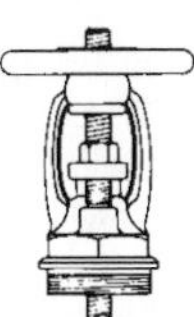

STEM TYPES
(OS & Y)—Rising Stem-Outside Screw and Yoke

Offers a visual indication of whether the valve is open or closed. Recommended where high temperatures, corrosives, and solids in the line might cause damage to inside-valve stem threads. The stem threads are engaged by the yoke bushing so the stem rises through the hand wheel as it is turned.

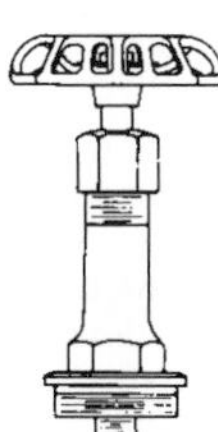

(R.S.)—Rising Stem-Inside Screw

Adequate clearance for operation must be provided because both the hand wheel and the stem rise.
The valve wedge position is indicated by the position of the stem and hand wheel.

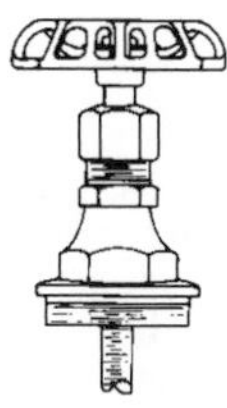

(N.R.S.)—Non-Rising Stem-Inside Screw

A minimum clearance is required for operating this type of valve. Excessive wear or damage to stem threads inside the valve may be caused by heat, corrosion, and solids. Because the hand wheel and stem do not rise, wedge position cannot be visually determined.

VALVE TYPES
Gate Valves

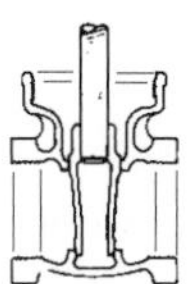

Provide full flow, minute pressure drop, minimum turbulence and minimum fluid trapped in the line.
They are normally used where operation is infrequent.

Globe Valves

Globe valves are designed for throttling and/or frequent operation with positive shut-off. Particular attention must be paid to the several types of seating materials available to avoid unnecessary wear. The seats must be compatible with the fluid in service and may be composition or metal. The configuration of the Globe valve opening causes turbulence which results in increased resistance. Most bronze Globe valves are rising stem-inside screw, but they are also available on O.S. & Y.

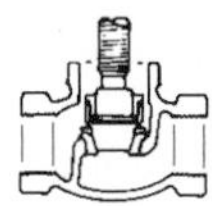

Angle Valves

The fundamental difference between the Angle valve and the Globe valve is the fluid flow through the Angle valve. It makes a 90° turn and offers less resistance to flow than the Globe valve while replacing an elbow. An Angle valve thus reduces the number of joints and installation time.

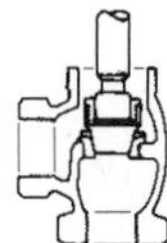

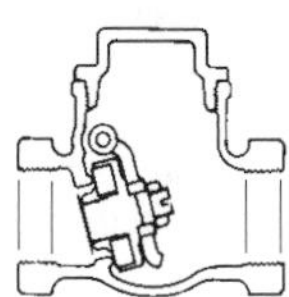

Check Valves

Check valves are designed to prevent backflow by automatically seating when the direction of fluid is reversed.
Swing Check valves are generally installed with Gate-valves, as they provide comparable full flow. Usually recommended for lines where flow velocities are low and should not be used on lines with pulsating flow. Recommended for horizontal installation, or in vertical lines only where flow is upward.

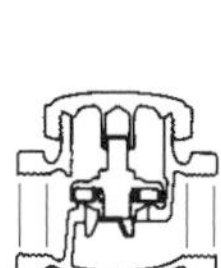

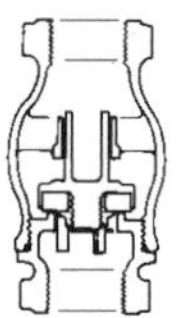

Lift Check Valves

These are commonly used with Globe and Angle valves since they have similar diaphragm seating arrangements and are recommended for preventing backflow of steam, air, gas and water, and on vapor lines with high flow velocities. For horizontal lines, horizontal lift checks should be used and vertical lift checks for vertical lines.

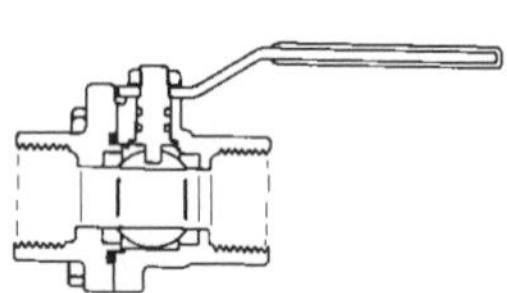

Ball Valves

Ball valves are light and easily installed, yet because of modern elastomeric seats, provide tight closure. Flow is controlled by rotating up to 90° a drilled ball which fits tightly against resilient seals. This ball seats with flow in either direction, and valve handle indicates the degree of opening. Recommended for frequent operation readily adaptable to automation, ideal for installation where space is limited.

Butterfly Valves

Butterfly valves provide bubble-tight closure with excellent throttling characteristics. They can be used for full-open, closed and for throttling applications.

The Butterfly valve consists of a disc within the valve body which is controlled by a shaft. In its closed position, the valve disc seals against a resilient seat. The disc position throughout the full 90° rotation is visually indicated by the position of the operator.

A Butterfly valve is only a fraction of the weight of a Gate valve and requires no gaskets between flanges in most cases. Recommended for frequent operation and adaptable to automation where space is limited.

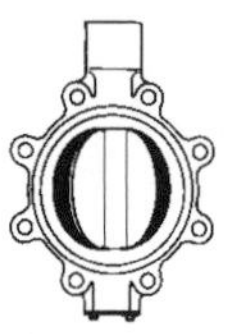

Wafer and Lug type bodies when installed between two pipe flanges, can be easily removed from the line. The pressure of the bolted flanges holds the valve in place.
Locating lugs makes installation easier.

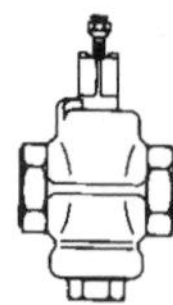

Plug Valves

Lubricated plug valves, because of the wide range of service to which they are adapted, may be classified as all purpose valves. They can be safely used at all pressure and vacuums, and at all temperatures up to the limits of available lubricants. They are the most satisfactory valves for the handling of gritty suspensions and many other destructive, erosive, corrosive and chemical solutions.

R221113-40 Plumbing Approximations for Quick Estimating

Water Control

Water Meter; Backflow Preventer, Shock Absorbers; Vacuum Breakers; Mixer. .... 10 to 15% of Fixtures

Pipe And Fittings .... 30 to 60% of Fixtures

Note: Lower percentage for compact buildings or larger buildings with plumbing in one area.
Larger percentage for large buildings with plumbing spread out.
In extreme cases pipe may be more than 100% of fixtures.
Percentages **do not** include special purpose or process piping.

Plumbing Labor

1 & 2 Story Residential .... Rough-in Labor = 80% of Materials
Apartment Buildings .... Rough-in Labor = 90 to 100% of Materials
Labor for handling and placing fixtures is approximately 25 to 30% of fixtures

Quality/Complexity Multiplier (for all installations)

Economy installation, add.......... 0 to 5%
Good quality, medium complexity, add .... 5 to 15%
Above average quality and complexity, add .... 15 to 25%

R221113-50 Pipe Material Considerations

1. Malleable fittings should be used for gas service.
2. Malleable fittings are used where there are stresses/strains due to expansion and vibration.
3. Cast fittings may be broken as an aid to disassembling of heating lines frozen by long use, temperature and minerals.
4. Cast iron pipe is extensively used for underground and submerged service.
5. Type M (light wall) copper tubing is available in hard temper only and is used for nonpressure and less severe applications than K and L.
6. Type L (medium wall) copper tubing, available hard or soft for interior service.
7. Type K (heavy wall) copper tubing, available in hard or soft temper for use where conditions are severe. For underground and interior service.
8. Hard drawn tubing requires fewer hangers or supports but should not be bent. Silver brazed fittings are recommended, however soft solder is normally used.
9. Type DMV (very light wall) copper tubing designed for drainage, waste and vent plus other non-critical pressure services.

Domestic/Imported Pipe and Fittings Cost

The prices shown in this publication for steel/cast iron pipe and steel, cast iron, malleable iron fittings are based on domestic production sold at the normal trade discounts. The above listed items of foreign manufacture may be available at prices of 1/3 to 1/2 those shown. Some imported items after minor machining or finishing operations are being sold as domestic to further complicate the system.

Caution: Most pipe prices in this book also include a coupling and pipe hangers which for the larger sizes can add significantly to the per foot cost and should be taken into account when comparing "book cost" with quoted supplier's cost.

R221113-70 Piping to 10′ High

When taking off pipe, it is important to identify the different material types and joining procedures, as well as distances between supports and components required for proper support.

During the takeoff, measure through all fittings. Do not subtract the lengths of the fittings, valves, or strainers, etc. This added length plus the final rounding of the totals will compensate for nipples and waste.

When rounding off totals always increase the actual amount to correspond with manufacturer's shipping lengths.

A. Both red brass and yellow brass pipe are normally furnished in 12′ lengths, plain end. The Unit Price section includes in the linear foot costs two field threads and one coupling per 10′ length. A carbon steel clevis type hanger assembly every 10′ is also prorated into the linear foot costs, including both material and labor.

B. Cast iron soil pipe is furnished in either 5′ or 10′ lengths. For pricing purposes, the Unit Price section features 10′ lengths with a joint and a carbon steel clevis hanger assembly every 5′ prorated into the per foot costs of both material and labor.

Three methods of joining are considered: lead and oakum poured joints or push-on gasket type joints for the bell and spigot pipe and a joint clamp for the no-hub soil pipe. The labor and material costs for each of these individual joining procedures are also prorated into the linear costs per foot.

C. Copper tubing covers types K, L, M, and DWV which are furnished in 20′ lengths. Means pricing data is based on a tubing cut each length and a coupling and two soft soldered joints every 10′. A carbon steel, clevis type hanger assembly every 10′ is also prorated into the per foot costs. The prices for refrigeration tubing are for materials only. Labor for full lengths may be based on the type L labor but short cut measures in tight areas can increase the installation labor-hours from 20 to 40%.

D. Corrosion-resistant piping does not lend itself to one particular standard of hanging or support assembly due to its diversity of application and placement. The several varieties of corrosion-resistant piping do not include any material or labor costs for hanger assemblies (See the Unit Price section for appropriate selection).

E. Glass pipe is furnished in standard lengths either 5′ or 10′ long, beaded on one end. Special orders for diverse lengths beaded on both ends are also available. For pricing purposes, R.S. Means features 10′ lengths with a coupling and a carbon steel band hanger assembly every 10′ prorated into the per foot linear costs.

Glass pipe is also available with conical ends and standard lengths ranging from 6″ through 3′ in 6″ increments, then up to 10′ in 12″ increments. Special lengths can be customized for particular installation requirements.

For pricing purposes, Means has based the labor and material pricing on 10′ lengths. Included in these costs per linear foot are the prorated costs for a flanged assembly every 10′ consisting of two flanges, a gasket, two insertable seals, and the required number of bolts and nuts. A carbon steel band hanger assembly based on 10′ center lines has also been prorated into the costs per foot for labor and materials.

F. Plastic pipe of several compositions and joining methods are considered. Fiberglass reinforced pipe (FRP) is priced, based on 10′ lengths (20′ lengths are also available), with coupling and epoxy joints every 10′. FRP is furnished in both "General Service" and "High Strength." A carbon steel clevis hanger assembly, 3 for every 10′, is built into the prorated labor and material costs on a per foot basis.

The PVC and CPVC pipe schedules 40, 80 and 120, plus SDR ratings are all based on 20′ lengths with a coupling installed every 10′, as well as a carbon steel clevis hanger assembly every 3′. The PVC and ABS type DWV piping is based on 10′ lengths with solvent weld couplings every 10′, and with carbon steel clevis hanger assemblies, 3 for every 10′. The rest of the plastic piping in this section is based on flexible 100′ coils and does not include any coupling or supports.

This section ends with PVC drain and sewer piping based on 10′ lengths with bell and spigot ends and 0-ring type, push-on joints.

G. Stainless steel piping includes both weld end and threaded piping, both in the type 304 and 316 specification and in the following schedules, 5, 10, 40, 80, and 160. Although this piping is usually furnished in 20′ lengths, this cost grouping has a joint (either heli-arc butt-welded or threads and coupling) every 10′. A carbon steel clevis type hanger assembly is also included at 10′ intervals and prorated into the linear foot costs.

H. Carbon steel pipe includes both black and galvanized. This section encompasses schedules 40 (standard) and 80 (extra heavy).

Several common methods of joining steel pipe — such as thread and coupled, butt welded, and flanged (150 lb. weld neck flanges) are also included.

For estimating purposes, it is assumed that the piping is purchased in 20′ lengths and that a compatible joint is made up every 10′. These joints are prorated into the labor and material costs per linear foot. The following hanger and support assemblies every 10′ are also included: carbon steel clevis for the T & C pipe, and single rod roll type for both the welded and flanged piping. All of these hangers are oversized to accommodate pipe insulation 3/4″ thick through 5″ pipe size and 1-1/2″ thick from 6″ through 12″ pipe size.

I. Grooved joint steel pipe is priced both black and galvanized, in schedules 10, 40, and 80, furnished in 20′ lengths. This section describes two joining methods: cut groove and roll groove. The schedule 10 piping is roll-grooved, while the heavier schedules are cut-grooved. The labor and material costs are prorated into per linear foot prices, including a coupled joint every 10′, as well as a carbon steel clevis hanger assembly.

Notes:

The pipe hanger assemblies mentioned in the preceding paragraphs include the described hanger; appropriately sized steel, box-type insert and nut; plus 18″ of threaded hanger rod.

C clamps are used when the pipe is to be supported from steel shapes rather than anchored in the slab. C clamps are slightly less costly than inserts. However, to save time in estimating, it is advisable to use the given line number cost, rather than substituting a C clamp for the insert.

Add to piping labor for elevated installation:

10′ to 14.5′ high	10%	30′ to 34.5′ high	40%
15′ to 19.5′ high	20%	35′ to 39.5′ high	50%
20′ to 24.5′ high	25%	40′ and higher	55%
25′ to 29.5′ high	35%		

When using the percentage adds for elevated piping installations as shown above, bear in mind that the given heights are for the pipe supports, even though the insert, anchor, or clamp may be several feet higher than the pipe itself.

An allowance has been included in the piping installation time for testing and minor tightening of leaking joints, fittings, stuffing boxes, packing glands, etc. For extraordinary test requirements such as x-rays, prolonged pressure or demonstration tests, a percentage of the piping labor, based on the estimator's experience, must be added to the labor total. A testing service specializing in weld x-rays should be consulted for pricing if it is an estimate requirement. Equipment installation time includes start-up with associated adjustments.

R221316-10 Drainage Requirements

Drainage lines must have a slope to maintain flow for proper operation. This slope should not be less than 1/4″ per foot for 3″ diameter or smaller pipe and not less than 1/8″ per foot for 4″ diameter or larger pipe. The capacity of building drainage systems is calculated on a basis of "drainage fixture units" (d.f.u.) as per the following chart.

Type of Fixture	d.f.u. Value	Type of Fixture	d.f.u. Value
Automatic clothes washer (2″ standpipe)	3	Service sink (trap standard)	3
Bathroom group (water closet, lavatory and		Service sink (P trap)	2
bathtub or shower) tank type closet	6	Urinal, pedestal, syphon jet blowout	6
Bathtub (with or without overhead shower)	2	Urinal, wall hung	4
Clinic sink	6	Urinal, stall washout	4
Combination sink & tray with food disposal	4	Wash sink (cir. or mult.) per faucet set	2
Dental unit or cuspidor	1	Water closet, tank operated	4
Dental lavatory	1	Water closet, valve operated	6
Drinking fountain	1/2	Fixtures not listed above	
Dishwasher, domestic	2	Trap size 1-1/4″ or smaller	1
Floor drains with 2″ waste	3	Trap size 1-1/2″	2
Kitchen sink, domestic with one 1-1/2″ trap	2	Trap size 2″	3
Kitchen sink, domestic with food disposal	2	Trap size 2-1/2″	4
Lavatory with 1-1/4″ waste	1	Trap size 3″	5
Laundry tray (1 or 2 compartment)	2	Trap size 4″	6
Shower stall, domestic	2		

For continuous or nearly continuous flow into the sytem from a pump, air conditioning equipment or other item, allow 2 fixture units for each gallon per minute of flow.

When the "drainage fixture units" (d.f.u.) for each horizontal branch or vertical stack is computed from the table above, the appropriate pipe size for each branch or stack is determined from the table below.

R221316-20 Allowable Fixture Units (d.f.u.) for Branches and Stacks

Pipe Diam.	Horiz. Branch (not incl. drains)	Stack Size for 3 Stories or 3 Levels	Stack size for Over 3 levels	Maximum for 1 Story building Stack
1-1/2″	3	4	8	2
2″	6	10	24	6
2-1/2″	12	20	42	9
3″	20*	48*	72*	20*
4″	160	240	500	90
5″	360	540	1100	200
6″	620	960	1900	350
8″	1400	2200	3600	600
10″	2500	3800	5600	1000
12″	3900	6000	8400	1500
15″	7000			

*Not more than two water closets or bathroom groups within each branch interval nor more than six water closets or bathroom groups on the stack.

Stacks sized for the total may be reduced as load decreases at each story to a minimum diameter of 1/2 the maximum diameter.

R224000-10 Hot Water Consumption Rates

Type of Building	Size Factor	Maximum Hourly Demand	Average Day Demand
Apartment Dwellings	No. of Apartments:		
	Up to 20	12.0 Gal. per apt.	42.0 Gal. per apt.
	21 to 50	10.0 Gal. per apt.	40.0 Gal. per apt.
	51 to 75	8.5 Gal. per apt.	38.0 Gal. per apt.
	76 to 100	7.0 Gal. per apt.	37.0 Gal. per apt.
	101 to 200	6.0 Gal. per apt.	36.0 Gal. per apt.
	201 up	5.0 Gal. per apt.	35.0 Gal. per apt.
Dormitories	Men	3.8 Gal. per man	13.1 Gal. per man
	Women	5.0 Gal. per woman	12.3 Gal. per woman
Hospitals	Per bed	23.0 Gal. per patient	90.0 Gal. per patient
Hotels	Single room with bath	17.0 Gal. per unit	50.0 Gal. per unit
	Double room with bath	27.0 Gal. per unit	80.0 Gal. per unit
Motels	No. of units:		
	Up to 20	6.0 Gal. per unit	20.0 Gal. per unit
	21 to 100	5.0 Gal. per unit	14.0 Gal. per unit
	101 Up	4.0 Gal. per unit	10.0 Gal. per unit
Nursing Homes		4.5 Gal. per bed	18.4 Gal. per bed
Office buildings		0.4 Gal. per person	1.0 Gal. per person
Restaurants	Full meal type	1.5 Gal./max. meals/hr.	2.4 Gal. per meal
	Drive-in snack type	0.7 Gal./max. meals/hr.	0.7 Gal. per meal
Schools	Elementary	0.6 Gal. per student	0.6 Gal. per student
	Secondary & High	1.0 Gal. per student	1.8 Gal. per student

For evaluation purposes, recovery rate and storage capacity are inversely proportional. Water heaters should be sized so that the maximum hourly demand anticipated can be met in addition to allowance for the heat loss from the pipes and storage tank.

R224000-20 Fixture Demands in Gallons Per Fixture Per Hour

Table below is based on 140° F final temperature except for dishwashers in public places (*) where 180° F water is mandatory.

Fixture	Apartment House	Club	Gym	Hospital	Hotel	Indust. Plant	Office	Private Home	School
Bathtubs	20	20	30	20	20			20	
Dishwashers, automatic	15	50-150*		50-150*	50-200*	20-100*		15	20-100*
Kitchen sink	10	20		20	30	20	20	10	20
Laundry, stationary tubs	20	28		28	28			20	
Laundry, automatic wash	75	75		100	150			75	
Private lavatory	2	2	2	2	2	2	2	2	2
Public lavatory	4	6	8	6	8	12	6		15
Showers	30	150	225	75	75	225	30	30	225
Service sink	20	20		20	30	20	20	15	20
Demand factor	0.30	0.30	0.40	0.25	0.25	0.40	0.30	0.30	0.40
Storage capacity factor	1.25	0.90	1.00	0.60	0.80	1.00	2.00	0.70	1.00

To obtain the probable maximum demand multiply the total demands for the fixtures (gal./fixture/hour) by the demand factor. The heater should have a heating capacity in gallons per hour equal to this maximum. The storage tank should have a capacity in gallons equal to the probable maximum demand multiplied by the storage capacity factor.

R224000-30 Minimum Plumbing Fixture Requirements

Minimum Plumbing Fixture Requirements

Type of Building or Occupancy (2)	Water Closets (14) (Fixtures per Person)		Urinals (5,10) (Fixtures per Person)		Lavatories (Fixtures per Person)		Bathtubs or Showers (Fixtures per Person)	Drinking Fountains (Fixtures per Person) (3, 13)
	Male	Female	Male	Female	Male	Female		
Assembly Places-Theatres, Auditoriums, Convention Halls, etc.-for permanent employee use	1: 1 - 15 2: 16 - 35 3: 36 - 55	1: 1 - 15 2: 16 - 35 3: 36 - 55	0: 1 - 9 1: 10 - 50		1 per 40	1 per 40		
	Over 55, add 1 fixture for each additional 40 persons		Add one fixture for each additional 50 males					
Assembly Places-Theatres, Auditoriums, Convention Halls, etc. - for public use	1: 1 - 100 2: 101 - 200 3: 201 - 400	3: 1 - 50 4: 51 - 100 8: 101 - 200 11: 201 - 400	1: 1 - 100 2: 101 - 200 3: 201 - 400 4: 401 - 600		1: 1 - 200 2: 201 - 400 3: 401 - 750	1: 1 - 200 2: 201 - 400 3: 401 - 750		1: 1 - 150 2: 151 - 400 3: 401 - 750
	Over 400, add 1 fixture for each additional 500 males and 1 for each additional 125 females		Over 600, add 1 fixture for each additional 300 males		Over 750, add 1 fixture for each additional 500 persons			Over 750, add one fixture for each additional 500 persons
Dormitories (9) School or Labor	1 per 10	1 per 8	1 per 25		1 per 12	1 per 12	1 per 8	1 per 150 (12)
	Add 1 fixture for each additional 25 males (over 10) and 1 for each additional 20 females (over 8)		Over 150, add 1 fixture for each additional 50 males		Over 12 add 1 fixture for each additional 20 males and 1 for each 15 additional females		For females add 1 bathtub per 30. Over 150, add 1 per 20	
Dormitories-for Staff Use	1: 1 - 15 2: 16 - 35 3: 36 - 55	1: 1 - 15 3: 16 - 35 4: 36 - 55	1 per 50		1 per 40	1 per 40	1 per 8	
	Over 55, add 1 fixture for each additional 40 persons							
Dwellings: Single Dwelling Multiple Dwelling or Apartment House	1 per dwelling 1 per dwelling or apartment unit				1 per dwelling 1 per dwelling or apartment unit		1 per dwelling 1 per dwelling or apartment unit	
Hospital Waiting rooms	1 per room				1 per room			1 per 150 (12)
Hospitals-for employee use	1: 1 - 15 2: 16 - 35 3: 36 - 55	1: 1 - 15 3: 16 - 35 4: 36 - 55	0: 1 - 9 1: 10 - 50		1 per 40	1 per 40		
	Over 55, add 1 fixture for each additional 40 persons		Add 1 fixture for each additional 50 males					
Hospitals: Individual Room Ward Room	1 per room 1 per 8 patients				1 per room 1 per 10 patients		1 per room 1 per 20 patients	1 per 150 (12)
Industrial (6) Warehouses Workshops, Foundries and similar establishments-for employee use	1: 1 -10 2: 11 - 25 3: 26 - 50 4: 51 - 75 5: 76 - 100	1: 1 -10 2: 11 - 25 3: 26 - 50 4: 51 - 75 5: 76 - 100			Up to 100, per 10 persons Over 100, 1 per 15 persons (7, 8)		1 shower for each 15 persons exposed to excessive heat or to skin contamination with poisonous, infectious or irritating material	1 per 150 (12)
	Over 100, add 1 fixture for each additional 30 persons							
Institutional - Other than Hospitals or Penal Institutions (on each occupied floor)	1 per 25	1 per 20	0: 1 - 9 1: 10 - 50		1 per 10	1 per 10	1 per 8	1 per 150 (12)
			Add 1 fixture for each additional 50 males					
Institutional - Other than Hospitals or Penal Institutions (on each occupied floor)-for employee use	1: 1 - 15 2: 16 - 35 3: 36 - 55	1: 1 - 15 3: 16 - 35 4: 36 - 55	0: 1 - 9 1: 10 - 50		1 per 40	1 per 40	1 per 8	1 per 150 (12)
	Over 55, add 1 fixture for each additional 40 persons		Add 1 fixture for each additional 50 males					
Office or Public Buildings	1: 1 - 100 2: 101 - 200 3: 201 - 400	3: 1 - 50 4: 51 - 100 8: 101 - 200 11: 201 - 400	1: 1 - 100 2: 101 - 200 3: 201 - 400 4: 401 - 600		1: 1 - 200 2: 201 - 400 3: 401 - 750	1: 1 - 200 2: 201 - 400 3: 401 - 750		1 per 150 (12)
	Over 400, add 1 fixture for each additional 500 males and 1 for each additional 150 females		Over 600, add 1 fixture for each additional 300 males		Over 750, add 1 fixture for each additional 500 persons			
Office or Public Buildings - for employee use	1: 1 - 15 2: 16 - 35 3: 36 - 55	1: 1 - 15 3: 16 - 35 4: 36 - 55	0: 1 - 9 1: 10 - 50		1 per 40	1 per 40		
	Over 55, add 1 fixture for each additional 40 persons		Add 1 fixture for each additional 50 males					

R224000-30 Minimum Plumbing Fixture Requirements (cont.)

Minimum Plumbing Fixture Requirements

Type of Building or Occupancy	Water Closets (14) (Fixtures per Person)		Urinals (5, 10) (Fixtures per Person)		Lavatories (Fixtures per Person)		Bathtubs or Showers (Fixtures per Person)	Drinking Fountains (Fixtures per Person) (3, 13)
	Male	Female	Male	Female	Male	Female		
Penal Institutions - for employee use	1: 1 - 15 2: 16 - 35 3: 36 - 55	1: 1 - 15 3: 16 - 35 4: 36 - 55	0: 1 - 9 1: 10 - 50		1 per 40	1 per 40		1 per 150 (12)
	Over 55, add 1 fixture for each additional 40 persons		Add 1 fixture for each additional 50 males					
Penal Institutions - for prison use Cell	1 per cell				1 per cell			1 per cellblock floor
Exercise room	1 per exercise room		1 per exercise room		1 per exercise room			1 per exercise room
Restaurants, Pubs and Lounges (11)	1: 1 - 50 2: 51 - 150 3: 151 - 300	1: 1 - 50 2: 51 - 150 4: 151 - 300	1: 1 - 150		1: 1 - 150 2: 151 - 200 3: 201 - 400	1: 1 - 150 2: 151 - 200 3: 201 - 400		
	Over 300, add 1 fixture for each additional 200 persons		Over 150, add 1 fixture for each additional 150 males		Over 400, add 1 fixture for each additional 400 persons			
Schools - for staff use All Schools	1: 1 - 15 2: 16 - 35 3: 36 - 55	1: 1 - 15 3: 16 - 35 4: 36 - 55	1 per 50		1 per 40	1 per 40		
	Over 55, add 1 fixture for each additional 40 persons							
Schools - for student use: Nursery	1: 1 - 20 2: 21 - 50	1: 1 - 20 2: 21 - 50			1: 1 - 25 2: 26 - 50	1: 1 - 25 2: 26 - 50		1 per 150 (12)
	Over 50, add 1 fixture for each additional 50 persons				Over 50, add 1 fixture for each additional 50 persons			
Elementary	1 per 30	1 per 25	1 per 75		1 per 35	1 per 35		1 per 150 (12)
Secondary	1 per 40	1 per 30	1 per 35		1 per 40	1 per 40		1 per 150 (12)
Others (Colleges, Universities, Adult Centers, etc.	1 per 40	1 per 30	1 per 35		1 per 40	1 per 40		1 per 150 (12)
Worship Places Educational and Activities Unit	1 per 150	1 per 75	1 per 150		1 per 2 water closets			1 per 150 (12)
Worship Places Principal Assembly Place	1 per 150	1 per 75	1 per 150		1 per 2 water closets			1 per 150 (12)

Notes:

1. The figures shown are based upon one (1) fixture being the minimum required for the number of persons indicated or any fraction thereof.
2. Building categories not shown on this table shall be considered separately by the Administrative Authority.
3. Drinking fountains shall not be installed in toilet rooms.
4. Laundry trays. One (1) laundry tray or one (1) automatic washer standpipe for each dwelling unit or one (1) laundry trays or one (1) automatic washer standpipes, or combination thereof, for each twelve (12) apartments. Kitchen sinks, one (1) for each dwelling or apartment unit.
5. For each urinal added in excess of the minimum required, one water closet may be deducted. The number of water closets shall not be reduced to less than two-thirds (2/3) of the minimum requirement.
6. As required by ANSI Z4.1-1968, Sanitation in Places of Employment.
7. Where there is exposure to skin contamination with poisonous, infectious, or irritating materials, provide one (1) lavatory for each five (5) persons.
8. Twenty-four (24) lineal inches of wash sink or eighteen (18) inches of a circular basin, when provided with water outlets for such space shall be considered equivalent to one (1) lavatory.
9. Laundry trays, one (1) for each fifty (50) persons. Service sinks, one (1) for each hundred (100) persons.
10. General. In applying this schedule of facilities, consideration shall be given to the accessibility of the fixtures. Conformity purely on a numerical basis may not result in an installation suited to the need of the individual establishment. For example, schools should be provided with toilet facilities on each floor having classrooms.
    a. Surrounding materials, wall and floor space to a point two (2) feet in front of urinal lip and four (4) feet above the floor, and at least two (2) feet to each side of the urinal shall be lined with non-absorbent materials.
    b. Trough urinals shall be prohibited.
11. A restaurant is defined as a business which sells food to be consumed on the premises.
    a. The number of occupants for a drive-in restaurant shall be considered as equal to the number of parking stalls.
    b. Employee toilet facilities shall not to be included in the above restaurant requirements. Hand washing facilities shall be available in the kitchen for employees.
12. Where food is consumed indoors, water stations may be substituted for drinking fountains. Offices, or public buildings for use by more than six (6) persons shall have one (1) drinking fountain for the first one hundred fifty (150) persons and one additional fountain for each three hundred (300) persons thereafter.
13. There shall be a minimum of one (1) drinking fountain per occupied floor in schools, theaters, auditoriums, dormitories, offices of public building.
14. The total number of water closets for females shall be at least equal to the total number of water closets and urinals required for males.

R224000-40 Plumbing Fixture Installation Time

Item	Rough-In	Set	Total Hours	Item	Rough-In	Set	Total Hours
Bathtub	5	5	10	Shower head only	2	1	3
Bathtub and shower, cast iron	6	6	12	Shower drain	3	1	4
Fire hose reel and cabinet	4	2	6	Shower stall, slate		15	15
Floor drain to 4 inch diameter	3	1	4	Slop sink	5	3	8
Grease trap, single, cast iron	5	3	8	Test 6 fixtures			14
Kitchen gas range		4	4	Urinal, wall	6	2	8
Kitchen sink, single	4	4	8	Urinal, pedestal or floor	6	4	10
Kitchen sink, double	6	6	12	Water closet and tank	4	3	7
Laundry tubs	4	2	6	Water closet and tank, wall hung	5	3	8
Lavatory wall hung	5	3	8	Water heater, 45 gals. gas, automatic	5	2	7
Lavatory pedestal	5	3	8	Water heaters, 65 gals. gas, automatic	5	2	7
Shower and stall	6	4	10	Water heaters, electric, plumbing only	4	2	6

Fixture prices in front of book are based on the cost per fixture set in place. The rough-in cost, which must be added for each fixture, includes carrier, if required, some supply, waste and vent pipe connecting fittings and stops. The lengths of rough-in pipe are nominal runs which would connect to the larger runs and stacks. The supply runs and DWV runs and stacks must be accounted for in separate entries. In the eastern half of the United States it is common for the plumber to carry these to a point 5′ outside the building.

R224000-50 Water Cooler Application

Type of Service	Requirement
Office, School or Hospital	12 persons per gallon per hour
Office, Lobby or Department Store	4 or 5 gallons per hour per fountain
Light manufacturing	7 persons per gallon per hour
Heavy manufacturing	5 persons per gallon per hour
Hot heavy manufacturing	4 persons per gallon per hour
Hotel	.08 gallons per hour per room
Theatre	1 gallon per hour per 100 seats

Heating, Ventilating & A.C. R2305 Common Work Results for HVAC

R230500-10 Subcontractors

On the unit cost pages of the R.S. Means Cost Data books, the last column is entitled "Total Incl. O&P". This is normally the cost of the installing contractor. In the HVAC Division, this is the cost of the mechanical contractor. If the particular work being estimated is to be performed by a sub to the mechanical contractor, the mechanical's profit and handling charge (usually 10%) is added to the total of the last column.

R235616-60 Solar Heating (Space and Hot Water)

Collectors should face as close to due South as possible, however, variations of up to 20 degrees on either side of true South are acceptable. Local climate and collector type may influence the choice between east or west deviations. Obviously they should be located so they are not shaded from the sun's rays. Incline collectors at a slope of latitude minus 5 degrees for domestic hot water and latitude plus 15 degrees for space heating.

Flat plate collectors consist of a number of components as follows: Insulation to reduce heat loss through the bottom and sides of the collector. The enclosure which contains all the components in this assembly is usually weatherproof and prevents dust, wind and water from coming in contact with the absorber plate. The cover plate usually consists of one or more layers of a variety of glass or plastic and reduces the reradiation by creating an air space which traps the heat between the cover and the absorber plates.

The absorber plate must have a good thermal bond with the fluid passages. The absorber plate is usually metallic and treated with a surface coating which improves absorptivity. Black or dark paints or selective coatings are used for this purpose, and the design of this passage and plate combination helps determine a solar system's effectiveness.

Heat transfer fluid passage tubes are attached above and below or integral with an absorber plate for the purpose of transferring thermal energy from the absorber plate to a heat transfer medium. The heat exchanger is a device for transferring thermal energy from one fluid to another.

Piping and storage tanks should be well insulated to minimize heat losses.

Size domestic water heating storage tanks to hold 20 gallons of water per user, minimum, plus 10 gallons per dishwasher or washing machine. For domestic water heating an optimum collector size is approximately 3/4 square foot of area per gallon of water storage. For space heating of residences and small commercial applications the collector is commonly sized between 30% and 50% of the internal floor area. For space heating of large commercial applications, collector areas less than 30% of the internal floor area can still provide significant heat reductions.

A supplementary heat source is recommended for Northern states for December through February.

The solar energy transmission per square foot of collector surface varies greatly with the material used. Initial cost, heat transmittance and useful life are obviously interrelated.

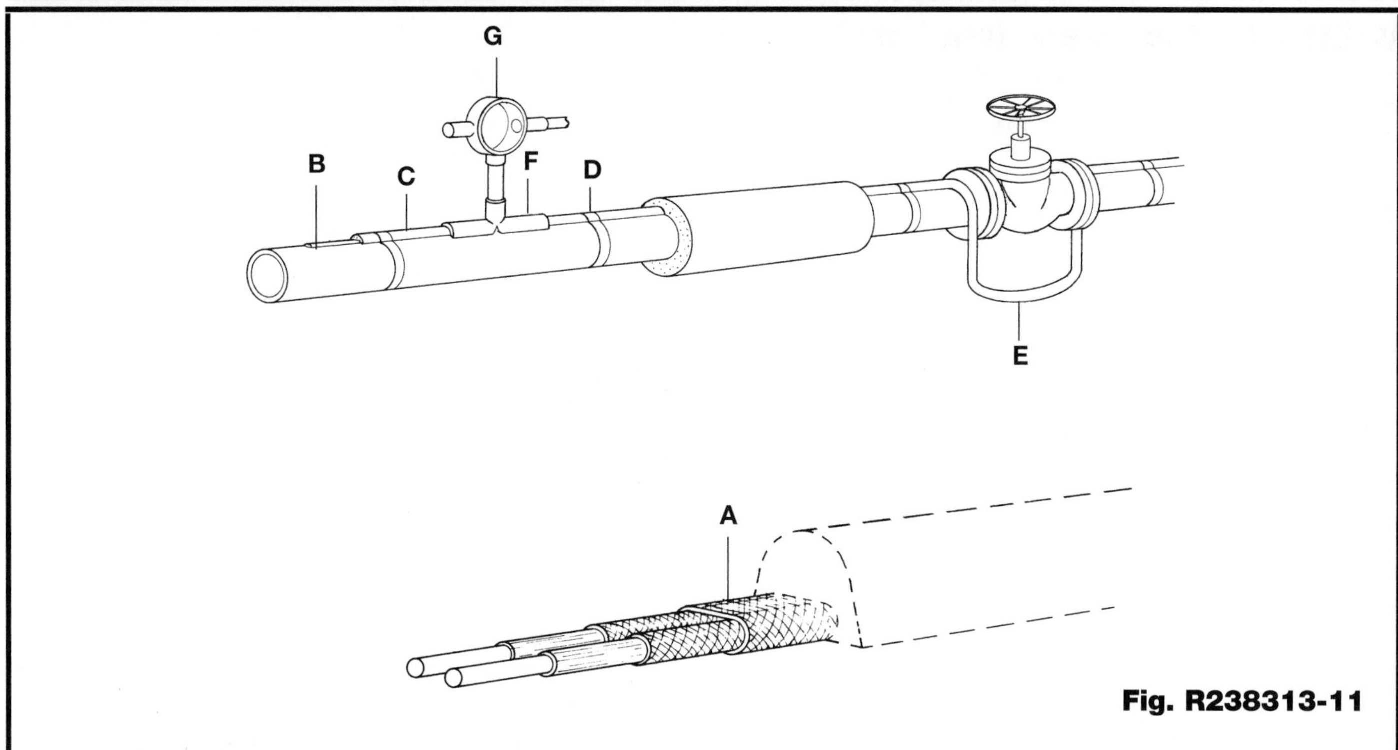

Fig. R238313-11

R238313-10 Heat Trace Systems

Before you can determine the cost of a HEAT TRACE installation, the method of attachment must be established. There are (4) common methods:

1. Cable is simply attached to the pipe with polyester tape every 12′.
2. Cable is attached with a continuous cover of 2″ wide aluminum tape.
3. Cable is attached with factory extruded heat transfer cement and covered with metallic raceway with clips every 10′.
4. Cable is attached between layers of pipe insulation using either clips or polyester tape.

Example: Components for method 3 must include:

A. Heat trace cable by voltage and watts per linear foot.
B. Heat transfer cement, 1 gallon per 60 linear feet of cover.
C. Metallic raceway by size and type.
D. Raceway clips by size of pipe.

When taking off linear foot lengths of cable add the following for each valve in the system. (E)

In all of the above methods each component of the system must be priced individually.

SCREWED OR WELDED VALVE:

Size		Length
1/2″	=	6″
3/4″	=	9″
1″	=	1′ -0″
1-1/2″	=	1′ -6″
2″	=	2′
2-1/2″	=	2′ -6″
3″	=	2′ -6″
4″	=	4′ -0″
6″	=	7′ -0″
8″	=	9′ -6″
10″	=	12′ -6″
12″	=	15′ -0″
14″	=	18′ -0″
16″	=	21′ -6″
18″	=	25′ -6″
20″	=	28′ -6″
24″	=	34′ -0″
30″	=	40′ -0″

FLANGED VALVE:

Size		Length
1/2″	=	1′ -0″
3/4″	=	1′ -6″
1″	=	2′ -0″
1-1/2″	=	2′ -6″
2″	=	2′ -6″
2-1/2″	=	3′ -0″
3″	=	3′ -6″
4″	=	4′ -0″
6″	=	8′ -0″
8″	=	11′ -0″
10″	=	14′ -0″
12″	=	16′ -6″
14″	=	19′ -6″
16″	=	23′ -0″
18″	=	27′ -0″
20″	=	30′ -0″
24″	=	36′ -0″
30″	=	42′ -0″

BUTTERFLY VALVES:

Size		Length
1/2″	=	0′
3/4″	=	0′
1″	=	1′ -0″
1-1/2″	=	1′ -6″
2″	=	2′ -0″
2-1/2″	=	2′ -6″
3″	=	2′ -6″
4″	=	3′ -0″
6″	=	3′ -6″
8″	=	4′ -0″
10″	=	4′ -0″
12″	=	5′ -0″
14″	=	5′ -6″
16″	=	6′ -0″
18″	=	6′ -6″
20″	=	7′ -0″
24″	=	8′ -0″
30″	=	10′ -0″

R238313-10 Heat Trace Systems (cont.)

Add the following quantities of heat transfer cement to linear foot totals for each valve:

Nominal Valve Size	Gallons of Cement per Valve
1/2″	0.14
3/4″	0.21
1″	0.29
1-1/2″	0.36
2″	0.43
2-1/2″	0.70
3″	0.71
4″	1.00
6″	1.43
8″	1.48
10″	1.50
12″	1.60
14″	1.75
16″	2.00
18″	2.25
20″	2.50
24″	3.00
30″	3.75

The following must be added to the list of components to accurately price HEAT TRACE systems:

1. Expediter fitting and clamp fasteners (F)
2. Junction box and nipple connected to expediter fitting (G)
3. Field installed terminal blocks within junction box
4. Ground lugs
5. Piping from power source to expediter fitting
6. Controls
7. Thermostats
8. Branch wiring
9. Cable splices
10. End of cable terminations
11. Branch piping fittings and boxes

Deduct the following percentages from labor if cable lengths in the same area exceed:

150′ to 250′	10%	351′ to 500′	20%
251′ to 350′	15%	Over 500′	25%

Add the following percentages to labor for elevated installations:

15′ to 20′ high	10%	31′ to 35′ high	40%
21′ to 25′ high	20%	36′ to 40′ high	50%
26′ to 30′ high	30%	Over 40′ high	60%

R238313-20 Spiral-Wrapped Heat Trace Cable (Pitch Table)

In order to increase the amount of heat, occasionally heat trace cable is wrapped in a spiral fashion around a pipe; increasing the number of feet of heater cable per linear foot of pipe.

Engineers first determine the heat loss per foot of pipe (based on the insulating material, its thickness, and the temperature differential across it). A ratio is then calculated by the formula:

$$\text{Feet of Heat Trace per Foot of Pipe} = \frac{\text{Watts/Foot of Heat Loss}}{\text{Watts/Foot of the Cable}}$$

The linear distance between wraps (pitch) is then taken from a chart or table. Generally, the pitch is listed on a drawing leaving the estimator to calculate the total length of heat tape required. An approximation may be taken from this table.

Feet of Heat Trace Per Foot of Pipe																
	Nominal Pipe Size in Inches															
Pitch In Inches	1	1¼	1½	2	2½	3	4	6	8	10	12	14	16	18	20	24
3.5	1.80															
4	1.65															
5	1.46	1.60	1.80													
6	1.34	1.45	1.55	1.75												
7	1.25	1.35	1.43	1.57	1.75											
8	1.20	1.28	1.34	1.45	1.60	1.80										
9	1.16	1.23	1.28	1.37	1.51	1.68										
10	1.13	1.19	1.24	1.32	1.44	1.57	1.82									
15	1.06	1.08	1.10	1.15	1.21	1.29	1.42	1.78								
20	1.04	1.05	1.06	1.08	1.13	1.17	1.25	1.49	1.73							
25		1.04	1.04	1.06	1.08	1.11	1.17	1.33	1.51	1.72						
30				1.04	1.05	1.07	1.12	1.24	1.37	1.54	1.70	1.80				
35					1.06	1.06	1.09	1.17	1.28	1.42	1.54	1.64	1.78			
40						1.05	1.07	1.14	1.22	1.33	1.44	1.52	1.64	1.75		
50							1.05	1.09	1.15	1.22	1.29	1.35	1.44	1.53	1.64	1.83
60								1.06	1.11	1.16	1.21	1.25	1.31	1.39	1.46	1.62
70								1.05	1.08	1.12	1.17	1.19	1.24	1.30	1.35	1.47
80									1.06	1.09	1.13	1.15	1.19	1.24	1.28	1.38
90									1.04	1.06	1.10	1.13	1.16	1.19	1.23	1.32
100										1.05	1.08	1.10	1.13	1.15	1.19	1.23

Note: Common practice would normally limit the lower end of the table to 5% of additional heat and above 80% an engineer would likely opt for two (2) parallel cables.

Reference Tables

R312319-90 Wellpoints

A single stage wellpoint system is usually limited to dewatering an average 15′ depth below normal ground water level. Multi-stage systems are employed for greater depth with the pumping equipment installed only at the lowest header level. Ejectors with unlimited lift capacity can be economical when two or more stages of wellpoints can be replaced or when horizontal clearance is restricted, such as in deep trenches or tunneling projects, and where low water flows are expected. Wellpoints are usually spaced on 2-1/2′ to 10′ centers along a header pipe. Wellpoint spacing, header size, and pump size are all determined by the expected flow as dictated by soil conditions.

In almost all soils encountered in wellpoint dewatering, the wellpoints may be jetted into place. Cemented soils and stiff clays may require sand wicks about 12″ in diameter around each wellpoint to increase efficiency and eliminate weeping into the excavation. These sand wicks require 1/2 to 3 C.Y. of washed filter sand and are installed by using a 12″ diameter steel casing and hole puncher jetted into the ground 2′ deeper than the wellpoint. Rock may require predrilled holes.

Labor required for the complete installation and removal of a single stage wellpoint system is in the range of 3/4 to 2 labor-hours per linear foot of header, depending upon jetting conditions, wellpoint spacing, etc.

Continuous pumping is necessary except in some free draining soil where temporary flooding is permissible (as in trenches which are backfilled after each day's work). Good practice requires provision of a stand-by pump during the continuous pumping operation.

Systems for continuous trenching below the water table should be installed three to four times the length of expected daily progress to ensure uninterrupted digging, and header pipe size should not be changed during the job.

For pervious free draining soils, deep wells in place of wellpoints may be economical because of lower installation and maintenance costs. Daily production ranges between two to three wells per day, for 25′ to 40′ depths, to one well per day for depths over 50′.

Detailed analysis and estimating for any dewatering problem is available at no cost from wellpoint manufacturers. Major firms will quote "sufficient equipment" quotes or their affiliates offer lump sum proposals to cover complete dewatering responsibility.

Description for 200′ System with 8″ Header		Quantities
Equipment & Material	Wellpoints 25′ long, 2″ diameter @ 5′ O.C.	40 Each
	Header pipe, 8″ diameter	200 L.F.
	Discharge pipe, 8″ diameter	100 L.F.
	8″ valves	3 Each
	Combination jetting & wellpoint pump (standby)	1 Each
	Wellpoint pump, 8″ diameter	1 Each
	Transportation to and from site	1 Day
	Fuel for 30 days x 60 gal./day	1800 Gallons
	Lubricants for 30 days x 16 lbs./day	480 Lbs.
	Sand for points	40 C.Y.
Labor	Technician to supervise installation	1 Week
	Labor for installation and removal of system	300 Labor-hours
	4 Operators straight time 40 hrs./wk. for 4.33 wks.	693 Hrs.
	4 Operators overtime 2 hrs./wk. for 4.33 wks.	35 Hrs.

R314116-40 Wood Sheet Piling

Wood sheet piling may be used for depths to 20′ where there is no ground water. If moderate ground water is encountered Tongue & Groove sheeting will help to keep it out. When considerable ground water is present, steel sheeting must be used.

For estimating purposes on trench excavation, sizes are as follows:

Depth	Sheeting	Wales	Braces	B.F. per S.F.
To 8′	3 x 12′s	6 x 8′s, 2 line	6 x 8′s, @ 10′	4.0 @ 8′
8′ x 12′	3 x 12′s	10 x 10′s, 2 line	10 x 10′s, @ 9′	5.0 average
12′ to 20′	3 x 12′s	12 x 12′s, 3 line	12 x 12′s, @ 8′	7.0 average

Sheeting to be toed in at least 2′ depending upon soil conditions. A five person crew with an air compressor and sheeting driver can drive and brace 440 SF/day at 8′ deep, 360 SF/day at 12′ deep, and 320 SF/day at 16′ deep. For normal soils, piling can be pulled in 1/3 the time to install. Pulling difficulty increases with the time in the ground. Production can be increased by high pressure jetting.

R314116-45 Steel Sheet Piling

Limiting weights are 22 to 38#/S.F. of wall surface with 27#/S.F. average for usual types and sizes. (Weights of piles themselves are from 30.7#/L.F. to 57#/L.F. but they are 15″ to 21″ wide.) Lightweight sections 12″ to 28″ wide from 3 ga. to 12 ga. thick are also available for shallow excavations. Piles may be driven two at a time with an impact or vibratory hammer (use vibratory to pull) hung from a crane without leads. A reasonable estimate of the life of steel sheet piling is 10 uses with up to 125 uses possible if a vibratory hammer is used. Used piling costs from 50% to 80% of new piling depending on location and market conditions. Sheet piling and H piles can be rented for about 30% of the delivered mill price for the first month and 5% per month thereafter. Allow 1 labor-hour per pile for cleaning and trimming after driving. These costs increase with depth and hydrostatic head. Vibratory drivers are faster in wet granular soils and are excellent for pile extraction. Pulling difficulty increases with the time in the ground and may cost more than driving. It is often economical to abandon the sheet piling, especially if it can be used as the outer wall form. Allow about 1/3 additional length or more for toeing into ground. Add bracing, waler and strut costs. Waler costs can equal the cost per ton of sheeting.

Utilities R3311 Water Utility Distribution Piping

R331113-80 Piping Designations

There are several systems currently in use to describe pipe and fittings. The following paragraphs will help to identify and clarify classifications of piping systems used for water distribution.

Piping may be classified by schedule. Piping schedules include 5S, 10S, 10, 20, 30, Standard, 40, 60, Extra Strong, 80, 100, 120, 140, 160 and Double Extra Strong. These schedules are dependent upon the pipe wall thickness. The wall thickness of a particular schedule may vary with pipe size.

Ductile iron pipe for water distribution is classified by Pressure Classes such as Class 150, 200, 250, 300 and 350. These classes are actually the rated water working pressure of the pipe in pounds per square inch (psi). The pipe in these pressure classes is designed to withstand the rated water working pressure plus a surge allowance of 100 psi.

The American Water Works Association (AWWA) provides standards for various types of **plastic pipe.** C-900 is the specification for polyvinyl chloride (PVC) piping used for water distribution in sizes ranging from 4″ through 12″. C-901 is the specification for polyethylene (PE) pressure pipe, tubing and fittings used for water distribution in sizes ranging from 1/2″ through 3″. C-905 is the specification for PVC piping sizes 14″ and greater.

PVC pressure-rated pipe is identified using the standard dimensional ratio (SDR) method. This method is defined by the American Society for Testing and Materials (ASTM) Standard D 2241. This pipe is available in SDR numbers 64, 41, 32.5, 26, 21, 17, and 13.5. Pipe with an SDR of 64 will have the thinnest wall while pipe with an SDR of 13.5 will have the thickest wall. When the pressure rating (PR) of a pipe is given in psi, it is based on a line supplying water at 73 degrees F.

The National Sanitation Foundation (NSF) seal of approval is applied to products that can be used with potable water. These products have been tested to ANSI/NSF Standard 14.

Valves and strainers are classified by American National Standards Institute (ANSI) Classes. These Classes are 125, 150, 200, 250, 300, 400, 600, 900, 1500 and 2500. Within each class there is an operating pressure range dependent upon temperature. Design parameters should be compared to the appropriate material dependent, pressure-temperature rating chart for accurate valve selection.

Change Orders

Change Order Considerations

A change order is a written document, usually prepared by the design professional, and signed by the owner, the architect/engineer, and the contractor. A change order states the agreement of the parties to: an addition, deletion, or revision in the work; an adjustment in the contract sum, if any; or an adjustment in the contract time, if any. Change orders, or "extras" in the construction process occur after execution of the construction contract and impact architects/engineers, contractors, and owners.

Change orders that are properly recognized and managed can ensure orderly, professional, and profitable progress for all who are involved in the project. There are many causes for change orders and change order requests. In all cases, change orders or change order requests should be addressed promptly and in a precise and prescribed manner. The following paragraphs include information regarding change order pricing and procedures.

The Causes of Change Orders

Reasons for issuing change orders include:

- Unforeseen field conditions that require a change in the work
- Correction of design discrepancies, errors, or omissions in the contract documents
- Owner-requested changes, either by design criteria, scope of work, or project objectives
- Completion date changes for reasons unrelated to the construction process
- Changes in building code interpretations, or other public authority requirements that require a change in the work
- Changes in availability of existing or new materials and products

Procedures

Properly written contract documents must include the correct change order procedures for all parties—owners, design professionals and contractors—to follow in order to avoid costly delays and litigation.

Being "in the right" is not always a sufficient or acceptable defense. The contract provisions requiring notification and documentation must be adhered to within a defined or reasonable time frame.

The appropriate method of handling change orders is by a written proposal and acceptance by all parties involved. Prior to starting work on a project, all parties should identify their authorized agents who may sign and accept change orders, as well as any limits placed on their authority.

Time may be a critical factor when the need for a change arises. For such cases, the contractor might be directed to proceed on a "time and materials" basis, rather than wait for all paperwork to be processed—a delay that could impede progress. In this situation, the contractor must still follow the prescribed change order procedures including, but not limited to, notification and documentation.

All forms used for change orders should be dated and signed by the proper authority. Lack of documentation can be very costly, especially if legal judgments are to be made, and if certain field personnel are no longer available. For time and material change orders, the contractor should keep accurate daily records of all labor and material allocated to the change.

Owners or awarding authorities who do considerable and continual building construction (such as the federal government) realize the inevitability of change orders for numerous reasons, both predictable and unpredictable. As a result, the federal government, the American Institute of Architects (AIA), the Engineers Joint Contract Documents Committee (EJCDC) and other contractor, legal, and technical organizations have developed standards and procedures to be followed by all parties to achieve contract continuance and timely completion, while being financially fair to all concerned.

In addition to the change order standards put forth by industry associations, there are also many books available on the subject.

Pricing Change Orders

When pricing change orders, regardless of their cause, the most significant factor is when the change occurs. The need for a change may be perceived in the field or requested by the architect/engineer *before* any of the actual installation has begun, or may evolve or appear *during* construction when the item of work in question is partially installed. In the latter cases, the original sequence of construction is disrupted, along with all contiguous and supporting systems. Change orders cause the greatest impact when they occur *after* the installation has been completed and must be uncovered, or even replaced. Post-completion changes may be caused by necessary design changes, product failure, or changes in the owner's requirements that are not discovered until the building or the systems begin to function.

Specified procedures of notification and record keeping must be adhered to and enforced regardless of the stage of construction: *before, during,* or *after* installation. Some bidding documents anticipate change orders by requiring that unit prices including overhead and profit percentages—for additional as well as deductible changes—be listed. Generally these unit prices do not fully take into account the ripple effect, or impact on other trades, and should be used for general guidance only.

When pricing change orders, it is important to classify the time frame in which the change occurs. There are two basic time frames for change orders: *pre-installation change orders*, which occur before the start of construction, and *post-installation change orders*, which involve reworking after the original installation. Change orders that occur between these stages may be priced according to the extent of work completed using a combination of techniques developed for pricing *pre-* and *post-installation* changes.

The following factors are the basis for a check list to use when preparing a change order estimate.

Factors To Consider When Pricing Change Orders

As an estimator begins to prepare a change order, the following questions should be reviewed to determine their impact on the final price.

General

- *Is the change order work* pre-installation *or* post-installation?

  Change order work costs vary according to how much of the installation has been completed. Once workers have the project scoped in their minds, even though they have not started, it can be difficult to refocus. Consequently they may spend more than the normal amount of time understanding the change. Also, modifications to work in place, such as trimming or refitting, usually take more time than was initially estimated. The greater the amount of work in place, the more reluctant workers are to change it. Psychologically they may resent the change and as a result the rework takes longer than normal. Post-installation change order estimates must include demolition of existing work as required to accomplish the change. If the work is performed at a later time, additional obstacles, such as building finishes, may be present which must be protected. Regardless of whether the change occurs pre-installation or post-installation, attempt to isolate the identifiable factors and price them separately. For example, add shipping costs that may be required pre-installation or any demolition required post-installation. Then analyze the potential impact on productivity of psychological and/or learning curve factors and adjust the output rates accordingly. One approach is to break down the typical workday into segments and quantify the impact on each segment. The following chart may be useful as a guide:

Activities (Productivity) Expressed as Percentages of a Workday			
Task	Means Mechanical Cost Data (for New Construction)	Pre-Installation Change Orders	Post-Installation Change Orders
1. Study plans	3%	6%	6%
2. Material procurement	3%	3%	3%
3. Receiving and storing	3%	3%	3%
4. Mobilization	5%	5%	5%
5. Site movement	5%	5%	8%
6. Layout and marking	8%	10%	12%
7. Actual installation	64%	59%	54%
8. Clean-up	3%	3%	3%
9. Breaks—non-productive	6%	6%	6%
Total	100%	100%	100%

Change Order Installation Efficiency

The labor-hours expressed (for new construction) are based on average installation time, using an efficiency level of approximately 60%-65%. For change order situations, adjustments to this efficiency level should reflect the daily labor-hour allocation for that particular occurrence.

If any of the specific percentages expressed in the above chart do not apply to a particular project situation, then those percentage points should be reallocated to the appropriate task(s). Example: Using data for new construction, assume there is no new material being utilized. The percentages for Tasks 2 and 3 would therefore be reallocated to other tasks. If the time required for Tasks 2 and 3 can now be applied to installation, we can add the time allocated for *Material Procurement* and *Receiving and Storing* to the *Actual Installation* time for new construction, thereby increasing the Actual Installation percentage.

This chart shows that, due to reduced productivity, labor costs will be higher than those for new construction by 5%-15% for pre-installation change orders and by 15%-25% for post-installation change orders. Each job and change order is unique and must be examined individually. Many factors, covered elsewhere in this section, can each have a significant impact on productivity and change order costs. All such factors should be considered in every case.

- *Will the change substantially delay the original completion date?*

  A significant change in the project may cause the original completion date to be extended. The extended schedule may subject the contractor to new wage rates dictated by relevant labor contracts. Project supervision and other project overhead must also be extended beyond the original completion date. The schedule extension may also put installation into a new weather season. For example, underground piping scheduled for October installation was delayed until January. As a result, frost penetrated the trench area, thereby changing the degree of difficulty of the task. Changes and delays may have a ripple effect throughout the project. This effect must be analyzed and negotiated with the owner.

- *What is the net effect of a deduct change order?*

  In most cases, change orders resulting in a deduction or credit reflect only bare costs. The contractor may retain the overhead and profit based on the original bid.

Materials

- *Will you have to pay more or less for the new material, required by the change order, than you paid for the original purchase?*

  The same material prices or discounts will usually apply to materials purchased for change orders as new construction. In some instances, however, the contractor may forfeit the advantages of competitive pricing for change orders. Consider the following example:

  A contractor purchased over $20,000 worth of fan coil units for an installation, and obtained the maximum discount. Some time later it was determined the project required an additional matching unit. The contractor has to purchase this unit from the original supplier to ensure a match. The supplier at this time may not discount the unit because of the small quantity, and the fact that he is no longer in a competitive situation. The impact of quantity on purchase can add between 0% and 25% to material prices and/or subcontractor quotes.

- *If materials have been ordered or delivered to the job site, will they be subject to a cancellation charge or restocking fee?*

  Check with the supplier to determine if ordered materials are subject to a cancellation charge. Delivered materials not used as result of a change order may be subject to a restocking fee if returned to the supplier. Common restocking charges run between 20% and 40%. Also, delivery charges to return the goods to the supplier must be added.

Labor

- *How efficient is the existing crew at the actual installation?*

  Is the same crew that performed the initial work going to do the change order? Possibly the change consists of the installation of a unit identical to one already installed; therefore, the change should take less time. Be sure to consider this potential productivity increase and modify the productivity rates accordingly.

- *If the crew size is increased, what impact will that have on supervision requirements?*

  Under most bargaining agreements or management practices, there is a point at which a working foreman is replaced by a nonworking foreman. This replacement increases project overhead by adding a nonproductive worker. If additional workers are added to accelerate the project or to perform changes while maintaining the schedule, be sure to add additional supervision time if warranted. Calculate the hours involved and the additional cost directly if possible.

- *What are the other impacts of increased crew size?*

  The larger the crew, the greater the potential for productivity to decrease. Some of the factors that cause this productivity loss are: overcrowding (producing restrictive conditions in the working space), and possibly a shortage of any special tools and equipment required. Such factors affect not only the crew working on the elements directly involved in the change order, but other crews whose movement may also be hampered.

  As the crew increases, check its basic composition for changes by the addition or deletion of apprentices or nonworking foreman, and quantify the potential effects of equipment shortages or other logistical factors.

- *As new crews, unfamiliar with the project, are brought onto the site, how long will it take them to become oriented to the project requirements?*

  The orientation time for a new crew to become 100% effective varies with the site and type of project. Orientation is easiest at a new construction site, and most difficult at existing, very restrictive renovation sites. The type of work also affects orientation time. When all elements of the work are exposed, such as concrete or masonry work, orientation is decreased. When the work is concealed or less visible, such as existing electrical systems, orientation takes longer. Usually orientation can be accomplished in one day or less. Costs for added orientation should be itemized and added to the total estimated cost.

- *How much actual production can be gained by working overtime?*

  Short term overtime can be used effectively to accomplish more work in a day. However, as overtime is scheduled to run beyond several weeks, studies have shown marked decreases in output. The following chart shows the effect of long term overtime on worker efficiency. If the anticipated change requires extended overtime to keep the job on schedule, these factors can be used as a guide to predict the impact on time and cost. Add project overhead, particularly supervision, that may also be incurred.

Days per Week	Hours per Day	Production Efficiency					Payroll Cost Factors	
		1 Week	2 Weeks	3 Weeks	4 Weeks	Average 4 Weeks	@ 1-1/2 Times	@ 2 Times
5	8	100%	100%	100%	100%	100%	100%	100%
	9	100	100	95	90	96.25	105.6	111.1
	10	100	95	90	85	91.25	110.0	120.0
	11	95	90	75	65	81.25	113.6	127.3
	12	90	85	70	60	76.25	116.7	133.3
6	8	100	100	95	90	96.25	108.3	116.7
	9	100	95	90	85	92.50	113.0	125.9
	10	95	90	85	80	87.50	116.7	133.3
	11	95	85	70	65	78.75	119.7	139.4
	12	90	80	65	60	73.75	122.2	144.4
7	8	100	95	85	75	88.75	114.3	128.6
	9	95	90	80	70	83.75	118.3	136.5
	10	90	85	75	65	78.75	121.4	142.9
	11	85	80	65	60	72.50	124.0	148.1
	12	85	75	60	55	68.75	126.2	152.4

Effects of Overtime

Caution: Under many labor agreements, Sundays and holidays are paid at a higher premium than the normal overtime rate.

The use of long-term overtime is counterproductive on almost any construction job; that is, the longer the period of overtime, the lower the actual production rate. Numerous studies have been conducted, and while they have resulted in slightly different numbers, all reach the same conclusion. The figure above tabulates the effects of overtime work on efficiency.

As illustrated, there can be a difference between the *actual* payroll cost per hour and the *effective* cost per hour for overtime work. This is due to the reduced production efficiency with the increase in weekly hours beyond 40. This difference between actual and effective cost results from overtime work over a prolonged period. Short-term overtime work does not result in as great a reduction in efficiency and, in such cases, effective cost may not vary significantly from the actual payroll cost. As the total hours per week are increased on a regular basis, more time is lost due to fatigue, lowered morale, and an increased accident rate.

As an example, assume a project where workers are working 6 days a week, 10 hours per day. From the figure above (based on productivity studies), the average effective productive hours over a 4-week period are:

$$0.875 \times 60 = 52.5$$

Depending upon the locale and day of week, overtime hours may be paid at time and a half or double time. For time and a half, the overall (average) *actual* payroll cost (including regular and overtime hours) is determined as follows:

$$\frac{40 \text{ reg. hrs.} + (20 \text{ overtime hrs.} \times 1.5)}{60 \text{ hrs.}} = 1.167$$

Based on 60 hours, the payroll cost per hour will be 116.7% of the normal rate at 40 hours per week. However, because the effective production (efficiency) for 60 hours is reduced to the equivalent of 52.5 hours, the effective cost of overtime is calculated as follows:

For time and a half:

$$\frac{40 \text{ reg. hrs.} + (20 \text{ overtime hrs.} \times 1.5)}{52.5 \text{ hrs.}} = 1.33$$

Installed cost will be 133% of the normal rate (for labor).

Thus, when figuring overtime, the actual cost per unit of work will be higher than the apparent overtime payroll dollar increase, due to the reduced productivity of the longer workweek. These efficiency calculations are true only for those cost factors determined by hours worked. Costs that are applied weekly or monthly, such as equipment rentals, will not be similarly affected.

Equipment

- *What equipment is required to complete the change order?*

  Change orders may require extending the rental period of equipment already on the job site, or the addition of special equipment brought in to accomplish the change work. In either case, the additional rental charges and operator labor charges must be added.

Summary

The preceding considerations and others you deem appropriate should be analyzed and applied to a change order estimate. The impact of each should be quantified and listed on the estimate to form an audit trail.

Change orders that are properly identified, documented, and managed help to ensure the orderly, professional and profitable progress of the work. They also minimize potential claims or disputes at the end of the project.

Estimating Tips

- The cost figures in this Square Foot Cost section were derived from approximately 11,000 projects contained in the RSMeans database of completed construction projects. They include the contractor's overhead and profit, but do not generally include architectural fees or land costs. The figures have been adjusted to January of the current year. New projects are added to our files each year, and outdated projects are discarded. For this reason, certain costs may not show a uniform annual progression. In no case are all subdivisions of a project listed.
- These projects were located throughout the U.S. and reflect a tremendous variation in square foot (S.F.) and cubic foot (C.F.) costs. This is due to differences, not only in labor and material costs, but also in individual owners' requirements. For instance, a bank in a large city would have different features than one in a rural area. This is true of all the different types of buildings analyzed. Therefore, caution should be exercised when using these square foot costs. For example, for courthouses, costs in the database are local courthouse costs and will not apply to the larger, more elaborate federal courthouses. As a general rule, the projects in the 1/4 column do not include any site work or equipment, while the projects in the 3/4 column may include both equipment and site work. The median figures do not generally include site work.
- None of the figures "go with" any others. All individual cost items were computed and tabulated separately. Thus, the sum of the median figures for plumbing, HVAC and electrical will not normally total up to the total mechanical and electrical costs arrived at by separate analysis and tabulation of the projects.
- Each building was analyzed as to total and component costs and percentages. The figures were arranged in ascending order with the results tabulated as shown. The 1/4 column shows that 25% of the projects had lower costs and 75% had higher. The 3/4 column shows that 75% of the projects had lower costs and 25% had higher. The median column shows that 50% of the projects had lower costs and 50% had higher.
- There are two times when square foot costs are useful. The first is in the conceptual stage when no details are available. Then square foot costs make a useful starting point. The second is after the bids are in and the costs can be worked back into their appropriate categories for information purposes. As soon as details become available in the project design, the square foot approach should be discontinued and the project priced as to its particular components. When more precision is required, or for estimating the replacement cost of specific buildings, the current edition of *RSMeans Square Foot Costs* should be used.
- In using the figures in this section, it is recommended that the median column be used for preliminary figures if no additional information is available. The median figures, when multiplied by the total city construction cost index figures (see city cost indexes) and then multiplied by the project size modifier at the end of this section, should present a fairly accurate base figure, which would then have to be adjusted in view of the estimator's experience, local economic conditions, code requirements, and the owner's particular requirements. There is no need to factor the percentage figures, as these should remain constant from city to city. All tabulations mentioning air conditioning had at least partial air conditioning.
- The editors of this book would greatly appreciate receiving cost figures on one or more of your recent projects, which would then be included in the averages for next year. All cost figures received will be kept confidential, except that they will be averaged with other similar projects to arrive at square foot cost figures for next year's book. See the last page of the book for details and the discount available for submitting one or more of your projects.

	50 17 00 | S.F. Costs		UNIT	UNIT COSTS 1/4	UNIT COSTS MEDIAN	UNIT COSTS 3/4	% OF TOTAL 1/4	% OF TOTAL MEDIAN	% OF TOTAL 3/4	
01	0010	**APARTMENTS Low Rise (1 to 3 story)**	S.F.	71.50	90	119				01
	0020	Total project cost	C.F.	6.40	8.50	10.45				
	0100	Site work	S.F.	6.10	8.35	14.65	6.05%	10.55%	14.05%	
	0500	Masonry		1.40	3.26	5.65	1.54%	3.67%	6.35%	
	1500	Finishes		7.55	10.40	12.85	9.05%	10.75%	12.85%	
	1800	Equipment		2.34	3.54	5.25	2.73%	4.03%	5.95%	
	2720	Plumbing		5.55	7.15	9.10	6.65%	8.95%	10.05%	
	2770	Heating, ventilating, air conditioning		3.54	4.36	6.40	4.20%	5.60%	7.60%	
	2900	Electrical		4.14	5.50	7.45	5.20%	6.65%	8.40%	
	3100	Total: Mechanical & Electrical		14.30	18.20	22.50	15.90%	18.05%	23%	
	9000	Per apartment unit, total cost	Apt.	66,500	101,500	149,500				
	9500	Total: Mechanical & Electrical	"	12,600	19,800	25,800				
02	0010	**APARTMENTS Mid Rise (4 to 7 story)**	S.F.	94.50	114	141				02
	0020	Total project costs	C.F.	7.40	10.20	13.90				
	0100	Site work	S.F.	3.78	7.65	14.80	5.25%	6.70%	9.20%	
	0500	Masonry		6.30	8.65	11.85	5.10%	7.25%	10.50%	
	1500	Finishes		12.35	16.60	19.55	10.70%	13.50%	17.70%	
	1800	Equipment		2.75	4.11	5.40	2.54%	3.47%	4.31%	
	2500	Conveying equipment		2.14	2.64	3.18	2.05%	2.27%	2.69%	
	2720	Plumbing		5.55	8.90	9.40	5.70%	7.20%	8.95%	
	2900	Electrical		6.25	8.45	10.30	6.35%	7.20%	8.95%	
	3100	Total: Mechanical & Electrical		19.95	25	30.50	18.25%	21%	23%	
	9000	Per apartment unit, total cost	Apt.	107,000	126,000	209,000				
	9500	Total: Mechanical & Electrical	"	20,200	23,400	24,400				
03	0010	**APARTMENTS High Rise (8 to 24 story)**	S.F.	107	123	148				03
	0020	Total project costs	C.F.	10.40	12.10	15.50				
	0100	Site work	S.F.	3.89	6.30	8.80	2.58%	4.84%	6.15%	
	0500	Masonry		6.20	11.30	14.05	4.74%	9.65%	11.05%	
	1500	Finishes		11.90	14.85	17.55	9.75%	11.80%	13.70%	
	1800	Equipment		3.45	4.24	5.60	2.78%	3.49%	4.35%	
	2500	Conveying equipment		2.44	3.70	5.05	2.23%	2.78%	3.37%	
	2720	Plumbing		7.90	9.30	11.40	6.80%	7.20%	10.45%	
	2900	Electrical		7.35	9.30	12.60	6.45%	7.65%	8.80%	
	3100	Total: Mechanical & Electrical		22	28	34	17.95%	22.50%	24.50%	
	9000	Per apartment unit, total cost	Apt.	111,500	122,500	170,000				
	9500	Total: Mechanical & Electrical	"	24,100	27,500	29,100				
04	0010	**AUDITORIUMS**	S.F.	111	152	219				04
	0020	Total project costs	C.F.	6.95	9.70	13.90				
	2720	Plumbing	S.F.	6.60	9.75	11.65	5.85%	7.20%	8.70%	
	2900	Electrical		9	12.70	21.50	6.85%	9.45%	11.35%	
	3100	Total: Mechanical & Electrical		59.50	79.50	96.50	24.50%	28%	31%	
05	0010	**AUTOMOTIVE SALES**	S.F.	82.50	112	139				05
	0020	Total project costs	C.F.	5.45	6.50	8.45				
	2720	Plumbing	S.F.	3.76	6.55	7.10	2.89%	6.05%	6.50%	
	2770	Heating, ventilating, air conditioning		5.80	8.85	9.60	4.61%	10%	10.35%	
	2900	Electrical		6.50	10.20	14.15	7.25%	8.80%	12.15%	
	3100	Total: Mechanical & Electrical		21	29.50	35.50	18.90%	20.50%	22%	
06	0010	**BANKS**	S.F.	162	202	256				06
	0020	Total project costs	C.F.	11.55	15.70	20.50				
	0100	Site work	S.F.	18.55	28	41	7.90%	12.95%	16.95%	
	0500	Masonry		8.40	16.10	28.50	3.46%	7.65%	10.10%	
	1500	Finishes		15	22	27.50	5.95%	8.65%	11.70%	
	1800	Equipment		6.10	13.40	27	1.34%	5.55%	10.50%	
	2720	Plumbing		5.05	7.25	10.60	2.82%	3.90%	4.93%	
	2770	Heating, ventilating, air conditioning		9.65	12.85	17.15	4.86%	7.15%	8.50%	
	2900	Electrical		15.30	20.50	26.50	8.20%	10.20%	12.20%	
	3100	Total: Mechanical & Electrical		36.50	49.50	59	16.25%	19.45%	23%	
	3500	See also Divisions 11 16 00 & 11 17 00								

		50 17 00 | S.F. Costs	UNIT	UNIT COSTS 1/4	UNIT COSTS MEDIAN	UNIT COSTS 3/4	% OF TOTAL 1/4	% OF TOTAL MEDIAN	% OF TOTAL 3/4
13	0010	**CHURCHES**	S.F.	109	138	180			
	0020	Total project costs	C.F.	6.75	8.50	11.20			
	1800	Equipment	S.F.	1.30	3.11	6.60	.95%	2.11%	4.31%
	2720	Plumbing		4.24	5.95	8.75	3.51%	4.96%	6.25%
	2770	Heating, ventilating, air conditioning		9.90	12.90	18.30	7.50%	10%	12%
	2900	Electrical		9.15	12.60	17.10	7.30%	8.80%	10.95%
	3100	Total: Mechanical & Electrical	↓	28.50	38	50.50	18.30%	22%	25%
	3500	See also division 11040 (MF2004 11 91 00)							
15	0010	**CLUBS, COUNTRY**	S.F.	117	141	177			
	0020	Total project costs	C.F.	9.40	11.45	15.85			
	2720	Plumbing	S.F.	7.05	10.45	24	5.60%	7.90%	10%
	2900	Electrical		9.20	12.60	16.40	7%	8.95%	11%
	3100	Total: Mechanical & Electrical	↓	49	61	64.50	19%	26.50%	29.50%
17	0010	**CLUBS, SOCIAL Fraternal**	S.F.	98.50	135	179			
	0020	Total project costs	C.F.	5.80	8.85	10.55			
	2720	Plumbing	S.F.	5.85	7.30	11.05	5.60%	6.90%	8.55%
	2770	Heating, ventilating, air conditioning		8.45	10.20	13.15	8.20%	9.25%	14.40%
	2900	Electrical		7	11.50	13.20	5.95%	9.30%	10.55%
	3100	Total: Mechanical & Electrical	↓	36	39	50	21%	23%	23.50%
18	0010	**CLUBS, Y.M.C.A.**	S.F.	122	158	241			
	0020	Total project costs	C.F.	5.40	9	13.45			
	2720	Plumbing	S.F.	7.40	14.70	16.50	5.65%	7.60%	10.85%
	2900	Electrical		9.60	12.10	19.40	6.25%	8%	9.20%
	3100	Total: Mechanical & Electrical	↓	39	44.50	72.50	20.50%	21.50%	28.50%
19	0010	**COLLEGES Classrooms & Administration**	S.F.	117	160	208			
	0020	Total project costs	C.F.	8.30	12.30	19.35			
	0500	Masonry	S.F.	11.65	16.40	20	5.65%	8.05%	10.50%
	2720	Plumbing		6	12.30	23	5.10%	6.60%	8.95%
	2900	Electrical		9.60	14.95	20.50	7.70%	9.75%	12%
	3100	Total: Mechanical & Electrical	↓	40.50	55	74	23.50%	28%	32%
21	0010	**COLLEGES Science, Engineering, Laboratories**	S.F.	220	257	298			
	0020	Total project costs	C.F.	12.60	18.40	21			
	1800	Equipment	S.F.	12.25	27.50	30.50	2%	6.45%	12.65%
	2900	Electrical		18.15	26	39.50	7.10%	9.40%	12.10%
	3100	Total: Mechanical & Electrical	↓	67.50	80	124	28.50%	31.50%	41%
	3500	See also division 11600 (MF2004 11 53 00)							
23	0010	**COLLEGES Student Unions**	S.F.	140	190	230			
	0020	Total project costs	C.F.	7.80	10.25	12.70			
	3100	Total: Mechanical & Electrical	S.F.	52.50	57	67.50	23.50%	26%	29%
25	0010	**COMMUNITY CENTERS**	S.F.	116	144	194			
	0020	Total project costs	C.F.	7.55	10.80	14			
	1800	Equipment	S.F.	2.80	4.59	7.55	1.47%	3.01%	5.30%
	2720	Plumbing		5.50	9.60	13.10	4.85%	7%	8.95%
	2770	Heating, ventilating, air conditioning		8.80	12.80	18.30	6.80%	10.35%	12.90%
	2900	Electrical		9.80	12.95	19	7.20%	8.90%	10.40%
	3100	Total: Mechanical & Electrical	↓	32.50	41	59	19.05%	23.50%	31%
28	0010	**COURT HOUSES**	S.F.	167	197	271			
	0020	Total project costs	C.F.	12.80	15.30	19.30			
	2720	Plumbing	S.F.	7.95	11.10	12.55	5.95%	7.45%	8.20%
	2900	Electrical		17.75	19.70	29	8.90%	10.45%	11.85%
	3100	Total: Mechanical & Electrical	↓	49.50	65	71	22.50%	27.50%	30%
30	0010	**DEPARTMENT STORES**	S.F.	61.50	83.50	105			
	0020	Total project costs	C.F.	3.31	4.29	5.80			
	2720	Plumbing	S.F.	1.92	2.43	3.68	1.82%	4.21%	5.90%
	2770	Heating, ventilating, air conditioning		5.60	8.65	13.05	8.20%	9.10%	14.80%

50 17 00 | S.F. Costs

		50 17 00 | S.F. Costs	UNIT	UNIT COSTS 1/4	UNIT COSTS MEDIAN	UNIT COSTS 3/4	% OF TOTAL 1/4	% OF TOTAL MEDIAN	% OF TOTAL 3/4	
30	2900	Electrical	S.F.	7.10	9.70	11.45	9.05%	12.15%	14.95%	30
	3100	Total: Mechanical & Electrical		12.45	15.95	28	13.20%	21.50%	50%	
31	0010	**DORMITORIES Low Rise (1 to 3 story)**	S.F.	117	162	203				31
	0020	Total project costs	C.F.	6.60	10.70	16				
	2720	Plumbing	S.F.	7.05	9.40	11.90	8.05%	9%	9.65%	
	2770	Heating, ventilating, air conditioning		7.45	8.90	11.90	4.61%	8.05%	10%	
	2900	Electrical		7.75	11.80	16.15	6.40%	8.65%	9.50%	
	3100	Total: Mechanical & Electrical		40.50	43	67.50	22%	25%	27%	
	9000	Per bed, total cost	Bed	49,600	55,000	118,000				
32	0010	**DORMITORIES Mid Rise (4 to 8 story)**	S.F.	144	187	231				32
	0020	Total project costs	C.F.	15.85	17.40	21				
	2900	Electrical	S.F.	15.25	17.35	23.50	8.20%	10.20%	11.95%	
	3100	Total: Mechanical & Electrical	"	42.50	85	86.50	25%	30.50%	35.50%	
	9000	Per bed, total cost	Bed	20,500	46,700	270,000				
34	0010	**FACTORIES**	S.F.	54.50	81	125				34
	0020	Total project costs	C.F.	3.49	5.20	8.65				
	0100	Site work	S.F.	6.20	11.35	17.90	6.95%	11.45%	17.95%	
	2720	Plumbing		2.93	5.45	9.05	3.73%	6.05%	8.10%	
	2770	Heating, ventilating, air conditioning		5.70	8.20	11.05	5.25%	8.45%	11.35%	
	2900	Electrical		6.75	10.70	16.30	8.10%	10.50%	14.20%	
	3100	Total: Mechanical & Electrical		19.30	29.50	39	21%	28.50%	35.50%	
36	0010	**FIRE STATIONS**	S.F.	108	149	199				36
	0020	Total project costs	C.F.	6.30	8.65	11.55				
	0500	Masonry	S.F.	15.85	28.50	37	8.10%	11.30%	15.70%	
	1140	Roofing		3.52	9.55	10.85	1.90%	4.94%	5.05%	
	1580	Painting		2.73	4.08	4.17	1.37%	1.57%	2.07%	
	1800	Equipment		1.34	2.57	4.75	.62%	1.63%	3.42%	
	2720	Plumbing		6.05	9.60	13.60	5.85%	7.35%	9.45%	
	2770	Heating, ventilating, air conditioning		6	9.70	14.95	5.15%	7.40%	9.40%	
	2900	Electrical		7.70	13.45	18.35	6.85%	8.60%	10.60%	
	3100	Total: Mechanical & Electrical		39.50	50.50	57	18.40%	23%	26%	
37	0010	**FRATERNITY HOUSES & Sorority Houses**	S.F.	108	139	190				37
	0020	Total project costs	C.F.	10.75	11.20	13.50				
	2720	Plumbing	S.F.	8.15	9.35	17.10	6.80%	8%	10.85%	
	2900	Electrical		7.10	15.40	18.85	6.60%	9.90%	10.65%	
	3100	Total: Mechanical & Electrical		8.70	27	32.50		15.10%	15.90%	
38	0010	**FUNERAL HOMES**	S.F.	114	155	282				38
	0020	Total project costs	C.F.	11.60	12.95	25				
	2900	Electrical	S.F.	5.05	9.20	10.10	3.58%	4.44%	5.95%	
	3100	Total: Mechanical & Electrical		17.80	26	36.50	12.90%	12.90%	12.90%	
39	0010	**GARAGES, COMMERCIAL (Service)**	S.F.	64.50	99.50	138				39
	0020	Total project costs	C.F.	4.24	6.25	9.10				
	1800	Equipment	S.F.	3.63	8.15	12.70	2.21%	4.62%	6.80%	
	2720	Plumbing		4.46	6.90	12.55	5.45%	7.85%	10.65%	
	2730	Heating & ventilating		5.85	7.60	9.70	5.25%	6.85%	8.20%	
	2900	Electrical		6.10	9.35	13.45	7.15%	9.25%	10.85%	
	3100	Total: Mechanical & Electrical		13.50	26	38	12.35%	17.40%	26%	
40	0010	**GARAGES, MUNICIPAL (Repair)**	S.F.	100	126	178				40
	0020	Total project costs	C.F.	5.90	7.50	12.90				
	0500	Masonry	S.F.	8.90	17.40	27	4.03%	9.15%	12.50%	
	2720	Plumbing		4.25	8.15	15.35	3.59%	6.70%	7.95%	
	2730	Heating & ventilating		7.25	10.55	20.50	6.15%	7.45%	13.50%	
	2900	Electrical		7	11	16	6.65%	8.15%	11.15%	
	3100	Total: Mechanical & Electrical		32.50	45	65	21.50%	23%	28.50%	

	50 17 00 | S.F. Costs		UNIT	UNIT COSTS 1/4	UNIT COSTS MEDIAN	UNIT COSTS 3/4	% OF TOTAL 1/4	% OF TOTAL MEDIAN	% OF TOTAL 3/4	
41	0010	**GARAGES, PARKING**	S.F.	36.50	53.50	92.50				41
	0020	Total project costs	C.F.	3.46	4.70	6.85				
	2720	Plumbing	S.F.	1.04	1.62	2.50	1.72%	2.70%	3.85%	
	2900	Electrical		2.02	2.48	3.90	4.33%	5.20%	6.30%	
	3100	Total: Mechanical & Electrical	↓	4.14	5.75	7.20	7%	8.80%	9.70%	
	3200									
	9000	Per car, total cost	Car	15,500	19,500	24,900				
43	0010	**GYMNASIUMS**	S.F.	103	137	185				43
	0020	Total project costs	C.F.	5.10	6.95	8.50				
	1800	Equipment	S.F.	2.44	4.57	8.75	1.81%	3.30%	6.70%	
	2720	Plumbing		6.50	7.55	9.95	4.65%	6.40%	7.75%	
	2770	Heating, ventilating, air conditioning		7	10.60	21.50	5.15%	9.05%	11.10%	
	2900	Electrical		8.15	10.75	14.30	6.75%	8.50%	10.35%	
	3100	Total: Mechanical & Electrical	↓	29	38	46.50	19.60%	23.50%	29%	
	3500	See also division 11480 (MF2004 11 67 00)								
46	0010	**HOSPITALS**	S.F.	196	246	335				46
	0020	Total project costs	C.F.	14.90	18.55	26.50				
	1800	Equipment	S.F.	4.99	9.60	16.55	.80%	2.53%	4.80%	
	2720	Plumbing		16.95	24	30.50	7.60%	9.10%	10.85%	
	2770	Heating, ventilating, air conditioning		25	32	44.50	7.80%	12.95%	16.65%	
	2900	Electrical		21.50	29	41.50	9.90%	11.75%	14.10%	
	3100	Total: Mechanical & Electrical	↓	63.50	88.50	131	28%	33.50%	37%	
	9000	Per bed or person, total cost	Bed	228,000	314,000	362,000				
	9900	See also division 11700 (MF2004 11 71 00)								
48	0010	**HOUSING For the Elderly**	S.F.	97	123	150				48
	0020	Total project costs	C.F.	6.90	9.60	12.25				
	0100	Site work	S.F.	6.75	10.50	15.35	5.05%	7.90%	12.10%	
	0500	Masonry		2.95	11	16.10	1.30%	6.05%	11%	
	1800	Equipment		2.34	3.22	5.15	1.88%	3.23%	4.43%	
	2510	Conveying systems		2.35	3.16	4.29	1.78%	2.20%	2.81%	
	2720	Plumbing		7.20	9.15	11.55	8.15%	9.55%	10.50%	
	2730	Heating, ventilating, air conditioning		3.69	5.20	7.80	3.30%	5.60%	7.25%	
	2900	Electrical		7.20	9.80	12.55	7.30%	8.50%	10.25%	
	3100	Total: Mechanical & Electrical	↓	26.50	31	39.50	18.10%	22.50%	29%	
	9000	Per rental unit, total cost	Unit	90,000	105,500	117,000				
	9500	Total: Mechanical & Electrical	"	20,000	23,100	26,900				
50	0010	**HOUSING Public (Low Rise)**	S.F.	81.50	113	147				50
	0020	Total project costs	C.F.	7.25	9.05	11.25				
	0100	Site work	S.F.	10.35	14.90	24	8.35%	11.75%	16.50%	
	1800	Equipment		2.21	3.61	5.50	2.26%	3.03%	4.24%	
	2720	Plumbing		5.85	7.75	9.80	7.15%	9.05%	11.60%	
	2730	Heating, ventilating, air conditioning		2.95	5.70	6.25	4.26%	6.05%	6.45%	
	2900	Electrical		4.92	7.35	10.15	5.10%	6.55%	8.25%	
	3100	Total: Mechanical & Electrical	↓	23.50	30	33.50	14.50%	17.55%	26.50%	
	9000	Per apartment, total cost	Apt.	89,500	101,500	127,500				
	9500	Total: Mechanical & Electrical	"	19,100	23,500	26,000				
51	0010	**ICE SKATING RINKS**	S.F.	69.50	163	178				51
	0020	Total project costs	C.F.	5.10	5.25	6.05				
	2720	Plumbing	S.F.	2.60	4.87	4.98	3.12%	3.23%	5.65%	
	2900	Electrical		7.45	11.40	12.10	6.30%	10.15%	15.05%	
	3100	Total: Mechanical & Electrical	↓	31	31	31	18.95%	18.95%	18.95%	
52	0010	**JAILS**	S.F.	211	273	355				52
	0020	Total project costs	C.F.	19.10	26.50	31				
	1800	Equipment	S.F.	8.25	24.50	41.50	2.80%	6.05%	9.80%	
	2720	Plumbing		20.50	27.50	36	7%	8.90%	13.35%	
	2770	Heating, ventilating, air conditioning		19.10	25.50	49.50	7.50%	9.45%	17.75%	
	2900	Electrical		22	30	37.50	8.80%	11.65%	15%	

	50 17 00 | S.F. Costs		UNIT	UNIT COSTS			% OF TOTAL			
				1/4	MEDIAN	3/4	1/4	MEDIAN	3/4	
52	3100	Total: Mechanical & Electrical	S.F.	59.50	105	125	28%	30%	34%	52
53	0010	**LIBRARIES**	S.F.	136	178	232				53
	0020	Total project costs	C.F.	9.15	11.45	14.65				
	0500	Masonry	S.F.	10.70	18.50	31	5.65%	7.45%	11.60%	
	1800	Equipment		1.83	4.92	7.45	.28%	1.39%	4.07%	
	2720	Plumbing		4.94	7	9.75	3.38%	4.60%	5.70%	
	2770	Heating, ventilating, air conditioning		10.95	18.50	22.50	7.80%	10.95%	12.80%	
	2900	Electrical		13.75	18.15	24	8.40%	10.50%	12%	
	3100	Total: Mechanical & Electrical		41.50	52.50	64	20.50%	23%	26.50%	
54	0010	**LIVING, ASSISTED**	S.F.	124	146	172				54
	0020	Total project costs	C.F.	10.40	12.15	13.80				
	0500	Masonry	S.F.	3.59	4.33	5.30	2.36%	3.16%	3.90%	
	1800	Equipment		2.74	3.21	4.19	2.06%	2.45%	2.87%	
	2720	Plumbing		10.35	13.90	14.35	6.05%	8.15%	10.60%	
	2770	Heating, ventilating, air conditioning		12.30	12.85	14.10	7.95%	9.35%	9.70%	
	2900	Electrical		12.10	13.40	15.35	8.95%	9.95%	10.65%	
	3100	Total: Mechanical & Electrical		34	39.50	45.50	25%	28.50%	31.50%	
55	0010	**MEDICAL CLINICS**	S.F.	126	155	197				55
	0020	Total project costs	C.F.	9.25	11.95	15.85				
	1800	Equipment	S.F.	3.40	7.15	11.15	1.05%	2.94%	6.35%	
	2720	Plumbing		8.35	11.80	15.75	6.15%	8.40%	10.10%	
	2770	Heating, ventilating, air conditioning		10	13.10	19.20	6.65%	8.85%	11.35%	
	2900	Electrical		10.80	15.35	20	8.10%	10%	12.25%	
	3100	Total: Mechanical & Electrical		34.50	47	64.50	22.50%	27%	33.50%	
	3500	See also division 11700 (MF2004 11 71 00)								
57	0010	**MEDICAL OFFICES**	S.F.	118	147	180				57
	0020	Total project costs	C.F.	8.85	11.95	16.20				
	1800	Equipment	S.F.	3.91	7.75	11	.68%	4.97%	7.05%	
	2720	Plumbing		6.55	10.10	13.60	5.60%	6.80%	8.50%	
	2770	Heating, ventilating, air conditioning		7.90	11.40	15.10	6.10%	8%	9.70%	
	2900	Electrical		9.50	13.80	19.25	7.60%	9.80%	11.70%	
	3100	Total: Mechanical & Electrical		25.50	37	53.50	19.30%	23%	27.50%	
59	0010	**MOTELS**	S.F.	74.50	108	141				59
	0020	Total project costs	C.F.	6.65	8.90	14.55				
	2720	Plumbing	S.F.	7.60	9.65	11.50	9.45%	10.60%	12.55%	
	2770	Heating, ventilating, air conditioning		4.62	6.90	12.35	5.60%	5.60%	10%	
	2900	Electrical		7.05	8.95	11.15	7.45%	9.05%	10.45%	
	3100	Total: Mechanical & Electrical		22.50	30	51.50	18.50%	24%	25.50%	
	5000									
	9000	Per rental unit, total cost	Unit	38,100	72,500	78,000				
	9500	Total: Mechanical & Electrical	"	7,400	11,200	13,000				
60	0010	**NURSING HOMES**	S.F.	117	151	188				60
	0020	Total project costs	C.F.	9.20	11.50	15.70				
	1800	Equipment	S.F.	3.69	4.89	8.15	2%	3.62%	4.99%	
	2720	Plumbing		10.05	15.20	18.35	8.75%	10.10%	12.70%	
	2770	Heating, ventilating, air conditioning		10.60	16.05	21.50	9.70%	11.45%	11.80%	
	2900	Electrical		11.60	14.55	19.75	9.40%	10.60%	12.50%	
	3100	Total: Mechanical & Electrical		27.50	38.50	65	26%	28%	30%	
	9000	Per bed or person, total cost	Bed	52,000	65,000	84,000				
61	0010	**OFFICES Low Rise (1 to 4 story)**	S.F.	99	129	167				61
	0020	Total project costs	C.F.	7.05	9.70	12.80				
	0100	Site work	S.F.	7.95	13.85	20	6.20%	9.70%	13.55%	
	0500	Masonry		3.93	7.70	14.05	2.62%	5.45%	8.45%	
	1800	Equipment		1.05	2.06	5.60	.60%	1.50%	3.50%	
	2720	Plumbing		3.53	5.45	8	3.66%	4.50%	6.10%	
	2770	Heating, ventilating, air conditioning		7.80	10.90	15.90	7.20%	10.30%	11.70%	
	2900	Electrical		8.05	11.50	16.35	7.45%	9.65%	11.40%	

50 17 00 | S.F. Costs

		50 17 00 | S.F. Costs	UNIT	UNIT COSTS 1/4	UNIT COSTS MEDIAN	UNIT COSTS 3/4	% OF TOTAL 1/4	% OF TOTAL MEDIAN	% OF TOTAL 3/4	
61	3100	Total: Mechanical & Electrical	S.F.	22	30.50	46	18.15%	22%	27%	61
62	0010	**OFFICES Mid Rise (5 to 10 story)**	S.F.	105	127	172				62
	0020	Total project costs	C.F.	7.40	9.45	13.40				
	2720	Plumbing	S.F.	3.16	4.90	7.05	2.83%	3.74%	4.50%	
	2770	Heating, ventilating, air conditioning		7.95	11.35	18.10	7.65%	9.40%	11%	
	2900	Electrical		7.75	9.95	15.05	6.35%	7.80%	10%	
	3100	Total: Mechanical & Electrical		20	26	49	18.95%	21%	27.50%	
63	0010	**OFFICES High Rise (11 to 20 story)**	S.F.	128	162	199				63
	0020	Total project costs	C.F.	9	11.20	16.10				
	2900	Electrical	S.F.	7.80	9.50	14.15	5.80%	7.85%	10.50%	
	3100	Total: Mechanical & Electrical		25	33.50	57	16.90%	23.50%	34%	
64	0010	**POLICE STATIONS**	S.F.	154	202	255				64
	0020	Total project costs	C.F.	12.25	15.05	20.50				
	0500	Masonry	S.F.	14.45	25.50	32	6.70%	9.10%	11.35%	
	1800	Equipment		2.45	10.75	17.05	.98%	3.35%	6.70%	
	2720	Plumbing		8.60	17.15	21.50	5.65%	6.90%	10.75%	
	2770	Heating, ventilating, air conditioning		13.45	17.85	25	5.85%	10.55%	11.70%	
	2900	Electrical		16.85	24	32	9.80%	11.85%	14.80%	
	3100	Total: Mechanical & Electrical		55	66.50	90	25%	31.50%	32.50%	
65	0010	**POST OFFICES**	S.F.	122	150	191				65
	0020	Total project costs	C.F.	7.30	9.25	10.55				
	2720	Plumbing	S.F.	5.45	6.80	8.55	4.24%	5.30%	5.60%	
	2770	Heating, ventilating, air conditioning		8.55	10.55	11.75	6.65%	7.15%	9.35%	
	2900	Electrical		10.05	14.10	16.75	7.25%	9%	11%	
	3100	Total: Mechanical & Electrical		29	38	43	16.25%	18.80%	22%	
66	0010	**POWER PLANTS**	S.F.	845	1,125	2,050				66
	0020	Total project costs	C.F.	23	50.50	108				
	2900	Electrical	S.F.	59.50	126	189	9.30%	12.75%	21.50%	
	8100	Total: Mechanical & Electrical		148	485	1,075	32.50%	32.50%	52.50%	
67	0010	**RELIGIOUS EDUCATION**	S.F.	98	128	157				67
	0020	Total project costs	C.F.	5.45	7.80	9.75				
	2720	Plumbing	S.F.	4.09	5.80	8.15	4.40%	5.30%	7.10%	
	2770	Heating, ventilating, air conditioning		10.35	11.70	16.55	10.05%	11.45%	12.35%	
	2900	Electrical		7.75	10.95	14.50	7.60%	9.05%	10.35%	
	3100	Total: Mechanical & Electrical		31.50	41.50	50.50	22%	23%	27%	
69	0010	**RESEARCH Laboratories & Facilities**	S.F.	150	212	310				69
	0020	Total project costs	C.F.	10.90	21.50	26				
	1800	Equipment	S.F.	6.50	12.80	31	.94%	4.58%	9.10%	
	2720	Plumbing		14.70	18.80	30	6.15%	8.30%	10.80%	
	2770	Heating, ventilating, air conditioning		13.20	44.50	52.50	7.25%	16.50%	17.50%	
	2900	Electrical		17.65	28.50	46.50	9.55%	11.50%	15.40%	
	3100	Total: Mechanical & Electrical		56.50	97.50	140	29.50%	37%	41%	
70	0010	**RESTAURANTS**	S.F.	142	183	238				70
	0020	Total project costs	C.F.	11.95	15.70	20.50				
	1800	Equipment	S.F.	9.20	22.50	34	6.10%	13%	15.65%	
	2720	Plumbing		11.25	13.65	17.95	6.10%	8.15%	9%	
	2770	Heating, ventilating, air conditioning		14.30	19.75	24	9.20%	12%	12.40%	
	2900	Electrical		15	18.55	24	8.35%	10.55%	11.55%	
	3100	Total: Mechanical & Electrical		46.50	50	64.50	21%	25%	29.50%	
	9000	Per seat unit, total cost	Seat	5,225	6,950	8,225				
	9500	Total: Mechanical & Electrical	"	1,300	1,725	2,050				
72	0010	**RETAIL STORES**	S.F.	66.50	89	118				72
	0020	Total project costs	C.F.	4.50	6.40	8.90				
	2720	Plumbing	S.F.	2.41	4.02	6.85	3.26%	4.60%	6.80%	
	2770	Heating, ventilating, air conditioning		5.20	7.10	10.70	6.75%	8.75%	10.15%	
	2900	Electrical		6	8.15	11.80	7.25%	9.90%	11.60%	
	3100	Total: Mechanical & Electrical		15.90	20.50	28	17.05%	21%	23.50%	

	50 17 00 | S.F. Costs		UNIT	UNIT COSTS 1/4	UNIT COSTS MEDIAN	UNIT COSTS 3/4	% OF TOTAL 1/4	% OF TOTAL MEDIAN	% OF TOTAL 3/4	
74	0010	**SCHOOLS Elementary**	S.F.	109	135	165				74
	0020	Total project costs	C.F.	7.10	9.10	11.70				
	0500	Masonry	S.F.	9.40	16.55	24.50	5.25%	10.65%	14.05%	
	1800	Equipment		2.96	4.91	9.25	1.80%	3.13%	4.61%	
	2720	Plumbing		6.25	8.80	11.75	5.70%	7.15%	9.35%	
	2730	Heating, ventilating, air conditioning		9.35	14.85	21	8.15%	10.80%	14.90%	
	2900	Electrical		10.25	13.65	17.40	8.50%	10.05%	11.80%	
	3100	Total: Mechanical & Electrical		37	46	57.50	25%	27.50%	30%	
	9000	Per pupil, total cost	Ea.	12,400	18,500	41,400				
	9500	Total: Mechanical & Electrical	″	3,500	4,450	11,200				
76	0010	**SCHOOLS Junior High & Middle**	S.F.	112	138	169				76
	0020	Total project costs	C.F.	7.10	9.15	10.30				
	0500	Masonry	S.F.	13.90	17.95	22	7.55%	11.10%	14.30%	
	1800	Equipment		3.57	5.75	8.45	1.79%	3.03%	4.80%	
	2720	Plumbing		6.50	8	9.95	5.30%	6.80%	7.25%	
	2770	Heating, ventilating, air conditioning		13	15.80	28	8.90%	11.55%	14.20%	
	2900	Electrical		11.10	13.90	17.75	7.90%	9.40%	10.60%	
	3100	Total: Mechanical & Electrical		35.50	45.50	57	23%	25.50%	29.50%	
	9000	Per pupil, total cost	Ea.	14,200	18,600	25,000				
78	0010	**SCHOOLS Senior High**	S.F.	117	145	181				78
	0020	Total project costs	C.F.	7	10	16.65				
	1800	Equipment	S.F.	3.10	7.20	10	1.88%	2.67%	4.30%	
	2720	Plumbing		6.50	9.80	17.90	5.60%	6.90%	8.30%	
	2770	Heating, ventilating, air conditioning		13.30	15.25	23	8.95%	11.60%	15%	
	2900	Electrical		11.75	15.65	23	8.70%	10.35%	12.50%	
	3100	Total: Mechanical & Electrical		39.50	46	75	24%	26.50%	28.50%	
	9000	Per pupil, total cost	Ea.	11,000	22,300	27,900				
80	0010	**SCHOOLS Vocational**	S.F.	94	137	169				80
	0020	Total project costs	C.F.	5.85	8.40	11.60				
	0500	Masonry	S.F.	5.50	13.65	21	3.20%	4.61%	10.95%	
	1800	Equipment		2.94	7.30	10.10	1.24%	3.10%	4.26%	
	2720	Plumbing		6	8.95	13.20	5.40%	6.90%	8.55%	
	2770	Heating, ventilating, air conditioning		8.40	15.65	26	8.60%	11.90%	14.65%	
	2900	Electrical		9.80	13.40	18.30	8.45%	11%	13.20%	
	3100	Total: Mechanical & Electrical		37	37.50	64.50	27.50%	29.50%	31%	
	9000	Per pupil, total cost	Ea.	13,100	35,100	52,500				
83	0010	**SPORTS ARENAS**	S.F.	82	110	169				83
	0020	Total project costs	C.F.	4.46	8	10.30				
	2720	Plumbing	S.F.	4.77	7.25	15.25	4.35%	6.35%	9.40%	
	2770	Heating, ventilating, air conditioning		10.25	12.15	16.85	8.80%	10.20%	13.55%	
	2900	Electrical		8.55	11.65	15	7.65%	9.75%	11.90%	
	3100	Total: Mechanical & Electrical		21.50	38	51	18.05%	21.50%	25%	
85	0010	**SUPERMARKETS**	S.F.	76	88	103				85
	0020	Total project costs	C.F.	4.23	5.10	7.75				
	2720	Plumbing	S.F.	4.24	5.35	6.20	5.40%	6%	7.45%	
	2770	Heating, ventilating, air conditioning		6.25	8.30	10.10	8.60%	8.65%	9.60%	
	2900	Electrical		9.50	10.85	12.90	10.40%	12.45%	13.60%	
	3100	Total: Mechanical & Electrical		24.50	26	34.50	23.50%	27.50%	28.50%	
86	0010	**SWIMMING POOLS**	S.F.	123	206	440				86
	0020	Total project costs	C.F.	9.85	12.30	13.40				
	2720	Plumbing	S.F.	11.35	13	17.45	4.80%	9.70%	20.50%	
	2900	Electrical		9.25	15	22	5.75%	6.95%	7.60%	
	3100	Total: Mechanical & Electrical		22.50	57.50	77	11.15%	14.10%	23.50%	
87	0010	**TELEPHONE EXCHANGES**	S.F.	163	232	305				87
	0020	Total project costs	C.F.	10.15	16.30	22.50				
	2720	Plumbing	S.F.	6.90	10.65	15.60	4.52%	5.80%	6.90%	
	2770	Heating, ventilating, air conditioning		16	32	40	11.80%	16.05%	18.40%	

		50 17 00 | S.F. Costs	UNIT	UNIT COSTS 1/4	UNIT COSTS MEDIAN	UNIT COSTS 3/4	% OF TOTAL 1/4	% OF TOTAL MEDIAN	% OF TOTAL 3/4	
87	2900	Electrical	S.F.	16.65	26.50	46.50	10.90%	14%	17.85%	87
	3100	Total: Mechanical & Electrical	↓	49	93	132	29.50%	33.50%	44.50%	
91	0010	**THEATERS**	S.F.	103	132	194				91
	0020	Total project costs	C.F.	4.74	7	10.30				
	2720	Plumbing	S.F.	3.42	3.71	15.15	2.92%	4.70%	6.80%	
	2770	Heating, ventilating, air conditioning		9.95	12.10	14.95	8%	12.25%	13.40%	
	2900	Electrical		9	12.10	24.50	8.05%	9.35%	12.25%	
	3100	Total: Mechanical & Electrical	↓	23	31	36.50	21.50%	25.50%	27.50%	
94	0010	**TOWN HALLS City Halls & Municipal Buildings**	S.F.	115	146	190				94
	0020	Total project costs	C.F.	10.45	12.05	17.65				
	2720	Plumbing	S.F.	4.79	8.95	15.70	4.31%	5.95%	7.95%	
	2770	Heating, ventilating, air conditioning		8.65	17.20	25	7.05%	9.05%	13.45%	
	2900	Electrical		10.90	15	20.50	8.05%	9.45%	11.65%	
	3100	Total: Mechanical & Electrical	↓	38	48	73	22%	26.50%	31%	
97	0010	**WAREHOUSES & Storage Buildings**	S.F.	42.50	61.50	91				97
	0020	Total project costs	C.F.	2.24	3.51	5.80				
	0100	Site work	S.F.	4.42	8.75	13.20	6.05%	12.95%	19.55%	
	0500	Masonry		2.53	6.10	13.10	3.60%	7.35%	12%	
	1800	Equipment		.69	1.48	8.30	.72%	1.69%	5.55%	
	2720	Plumbing		1.43	2.56	4.78	2.90%	4.80%	6.55%	
	2730	Heating, ventilating, air conditioning		1.63	4.59	6.15	2.41%	5%	8.90%	
	2900	Electrical		2.53	4.77	7.80	5.15%	7.20%	10.05%	
	3100	Total: Mechanical & Electrical	↓	7.05	10.85	21.50	13.30%	18.90%	26%	
99	0010	**WAREHOUSE & OFFICES Combination**	S.F.	52.50	70	96.50				99
	0020	Total project costs	C.F.	2.69	3.91	5.75				
	1800	Equipment	S.F.	.91	1.76	2.62	.52%	1.19%	2.40%	
	2720	Plumbing		2.03	3.53	5.25	3.74%	4.76%	6.30%	
	2770	Heating, ventilating, air conditioning		3.21	5	7	5%	5.65%	10.05%	
	2900	Electrical		3.59	5.25	8.30	5.85%	8%	10%	
	3100	Total: Mechanical & Electrical	↓	10	15.15	24	14.55%	19.95%	24.50%	

Square Foot Project Size Modifier

One factor that affects the S.F. cost of a particular building is the size. In general, for buildings built to the same specifications in the same locality, the larger building will have the lower S.F. cost. This is due mainly to the decreasing contribution of the exterior walls, plus the economy of scale usually achievable in larger buildings. The area conversion scale shown below will give a factor to convert costs for the typical size building to an adjusted cost for the particular project.

The square foot base size table lists the median costs, most typical project size in our accumulated data, and the range in size of the projects.

The size factor for your project is determined by dividing your project area in S.F. by the typical project size for the particular building type. With this factor, enter the area conversion scale at the appropriate size factor and determine the appropriate cost multiplier for your building size.

Example: Determine the cost per S.F. for a 100,000 S.F. Mid-rise apartment building.

$$\frac{\text{Proposed building area} = 100{,}000\text{ S.F.}}{\text{Typical size from below} = 50{,}000\text{ S.F.}} = 2.00$$

Enter area conversion scale at 2.0, intersect curve, read horizontally the appropriate cost multiplier of .94. Size adjusted cost becomes .94 × $114.00 = $107.00 based on national average costs.

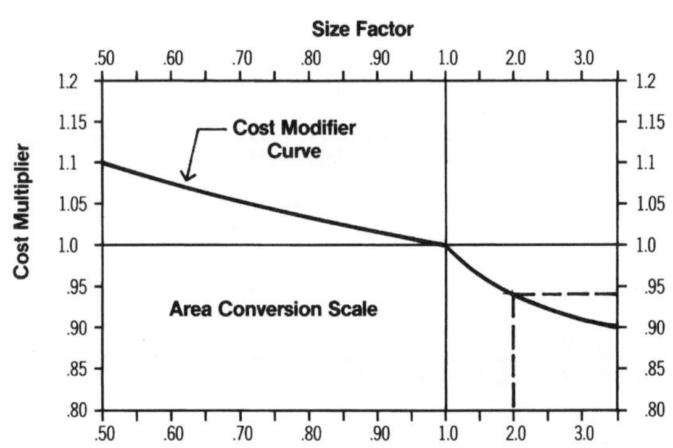

Note: For size factors less than .50, the cost multiplier is 1.1
For size factors greater than 3.5, the cost multiplier is .90

Square Foot Base Size

Building Type	Median Cost per S.F.	Typical Size Gross S.F.	Typical Range Gross S.F.	Building Type	Median Cost per S.F.	Typical Size Gross S.F.	Typical Range Gross S.F.
Apartments, Low Rise	$ 90.00	21,000	9,700 - 37,200	Jails	$ 273.00	40,000	5,500 - 145,000
Apartments, Mid Rise	114.00	50,000	32,000 - 100,000	Libraries	178.00	12,000	7,000 - 31,000
Apartments, High Rise	123.00	145,000	95,000 - 600,000	Living, Assisted	146.00	32,300	23,500 - 50,300
Auditoriums	152.00	25,000	7,600 - 39,000	Medical Clinics	155.00	7,200	4,200 - 15,700
Auto Sales	112.00	20,000	10,800 - 28,600	Medical Offices	147.00	6,000	4,000 - 15,000
Banks	202.00	4,200	2,500 - 7,500	Motels	108.00	40,000	15,800 - 120,000
Churches	138.00	17,000	2,000 - 42,000	Nursing Homes	151.00	23,000	15,000 - 37,000
Clubs, Country	141.00	6,500	4,500 - 15,000	Offices, Low Rise	129.00	20,000	5,000 - 80,000
Clubs, Social	135.00	10,000	6,000 - 13,500	Offices, Mid Rise	127.00	120,000	20,000 - 300,000
Clubs, YMCA	158.00	28,300	12,800 - 39,400	Offices, High Rise	162.00	260,000	120,000 - 800,000
Colleges (Class)	160.00	50,000	15,000 - 150,000	Police Stations	202.00	10,500	4,000 - 19,000
Colleges (Science Lab)	257.00	45,600	16,600 - 80,000	Post Offices	150.00	12,400	6,800 - 30,000
College (Student Union)	190.00	33,400	16,000 - 85,000	Power Plants	1125.00	7,500	1,000 - 20,000
Community Center	144.00	9,400	5,300 - 16,700	Religious Education	128.00	9,000	6,000 - 12,000
Court Houses	197.00	32,400	17,800 - 106,000	Research	212.00	19,000	6,300 - 45,000
Dept. Stores	83.50	90,000	44,000 - 122,000	Restaurants	183.00	4,400	2,800 - 6,000
Dormitories, Low Rise	162.00	25,000	10,000 - 95,000	Retail Stores	89.00	7,200	4,000 - 17,600
Dormitories, Mid Rise	187.00	85,000	20,000 - 200,000	Schools, Elementary	135.00	41,000	24,500 - 55,000
Factories	81.00	26,400	12,900 - 50,000	Schools, Jr. High	138.00	92,000	52,000 - 119,000
Fire Stations	149.00	5,800	4,000 - 8,700	Schools, Sr. High	145.00	101,000	50,500 - 175,000
Fraternity Houses	139.00	12,500	8,200 - 14,800	Schools, Vocational	137.00	37,000	20,500 - 82,000
Funeral Homes	155.00	10,000	4,000 - 20,000	Sports Arenas	110.00	15,000	5,000 - 40,000
Garages, Commercial	99.50	9,300	5,000 - 13,600	Supermarkets	88.00	44,000	12,000 - 60,000
Garages, Municipal	126.00	8,300	4,500 - 12,600	Swimming Pools	206.00	20,000	10,000 - 32,000
Garages, Parking	53.50	163,000	76,400 - 225,300	Telephone Exchange	232.00	4,500	1,200 - 10,600
Gymnasiums	137.00	19,200	11,600 - 41,000	Theaters	132.00	10,500	8,800 - 17,500
Hospitals	246.00	55,000	27,200 - 125,000	Town Halls	146.00	10,800	4,800 - 23,400
House (Elderly)	123.00	37,000	21,000 - 66,000	Warehouses	61.50	25,000	8,000 - 72,000
Housing (Public)	113.00	36,000	14,400 - 74,400	Warehouse & Office	70.00	25,000	8,000 - 72,000
Ice Rinks	163.00	29,000	27,200 - 33,600				

Abbreviations

A	Area Square Feet; Ampere
AAFES	Army and Air Force Exchange Service
ABS	Acrylonitrile Butadiene Stryrene; Asbestos Bonded Steel
A.C., AC	Alternating Current; Air-Conditioning; Asbestos Cement; Plywood Grade A & C
ACI	American Concrete Institute
ACR	Air Conditioning Refrigeration
ADA	Americans with Disabilities Act
AD	Plywood, Grade A & D
Addit.	Additional
Adj.	Adjustable
af	Audio-frequency
AFUE	Annual Fuel Utilization Efficiency
AGA	American Gas Association
Agg.	Aggregate
A.H., Ah	Ampere Hours
A hr.	Ampere-hour
A.H.U., AHU	Air Handling Unit
A.I.A.	American Institute of Architects
AIC	Ampere Interrupting Capacity
Allow.	Allowance
alt., alt	Alternate
Alum.	Aluminum
a.m.	Ante Meridiem
Amp.	Ampere
Anod.	Anodized
ANSI	American National Standards Institute
APA	American Plywood Association
Approx.	Approximate
Apt.	Apartment
Asb.	Asbestos
A.S.B.C.	American Standard Building Code
Asbe.	Asbestos Worker
ASCE.	American Society of Civil Engineers
A.S.H.R.A.E.	American Society of Heating, Refrig. & AC Engineers
ASME	American Society of Mechanical Engineers
ASTM	American Society for Testing and Materials
Attchmt.	Attachment
Avg., Ave.	Average
AWG	American Wire Gauge
AWWA	American Water Works Assoc.
Bbl.	Barrel
B&B, BB	Grade B and Better; Balled & Burlapped
B&S	Bell and Spigot
B.&W.	Black and White
b.c.c.	Body-centered Cubic
B.C.Y.	Bank Cubic Yards
BE	Bevel End
B.F.	Board Feet
Bg. cem.	Bag of Cement
BHP	Boiler Horsepower; Brake Horsepower
B.I.	Black Iron
bidir.	bidirectional
Bit., Bitum.	Bituminous
Bit., Conc.	Bituminous Concrete
Bk.	Backed
Bkrs.	Breakers
Bldg., bldg	Building
Blk.	Block
Bm.	Beam
Boil.	Boilermaker
bpm	Blows per Minute
BR	Bedroom
Brg.	Bearing
Brhe.	Bricklayer Helper
Bric.	Bricklayer
Brk., brk	Brick
brkt	Bracket
Brng.	Bearing
Brs.	Brass
Brz.	Bronze
Bsn.	Basin
Btr.	Better
Btu	British Thermal Unit
BTUH	BTU per Hour
Bu.	bushels
BUR	Built-up Roofing
BX	Interlocked Armored Cable
°C	degree centegrade
c	Conductivity, Copper Sweat
C	Hundred; Centigrade
C/C	Center to Center, Cedar on Cedar
C-C	Center to Center
Cab	Cabinet
Cair.	Air Tool Laborer
Cal.	caliper
Calc	Calculated
Cap.	Capacity
Carp.	Carpenter
C.B.	Circuit Breaker
C.C.A.	Chromate Copper Arsenate
C.C.F.	Hundred Cubic Feet
cd	Candela
cd/sf	Candela per Square Foot
CD	Grade of Plywood Face & Back
CDX	Plywood, Grade C & D, exterior glue
Cefi.	Cement Finisher
Cem.	Cement
CF	Hundred Feet
C.F.	Cubic Feet
CFM	Cubic Feet per Minute
c.g.	Center of Gravity
CHW	Chilled Water; Commercial Hot Water
C.I., CI	Cast Iron
C.I.P., CIP	Cast in Place
Circ.	Circuit
C.L.	Carload Lot
CL	Chain Link
Clab.	Common Laborer
Clam	Common maintenance laborer
C.L.F.	Hundred Linear Feet
CLF	Current Limiting Fuse
CLP	Cross Linked Polyethylene
cm	Centimeter
CMP	Corr. Metal Pipe
CMU	Concrete Masonry Unit
CN	Change Notice
Col.	Column
CO_2	Carbon Dioxide
Comb.	Combination
comm.	Commercial, Communication
Compr.	Compressor
Conc.	Concrete
Cont., cont	Continuous; Continued, Container
Corr.	Corrugated
Cos	Cosine
Cot	Cotangent
Cov.	Cover
C/P	Cedar on Paneling
CPA	Control Point Adjustment
Cplg.	Coupling
CPM	Critical Path Method
CPVC	Chlorinated Polyvinyl Chloride
C.Pr.	Hundred Pair
CRC	Cold Rolled Channel
Creos.	Creosote
Crpt.	Carpet & Linoleum Layer
CRT	Cathode-ray Tube
CS	Carbon Steel, Constant Shear Bar Joist
Csc	Cosecant
C.S.F.	Hundred Square Feet
CSI	Construction Specifications Institute
CT	Current Transformer
CTS	Copper Tube Size
Cu	Copper, Cubic
Cu. Ft.	Cubic Foot
cw	Continuous Wave
C.W.	Cool White; Cold Water
Cwt.	100 Pounds
C.W.X.	Cool White Deluxe
C.Y.	Cubic Yard (27 cubic feet)
C.Y./Hr.	Cubic Yard per Hour
Cyl.	Cylinder
d	Penny (nail size)
D	Deep; Depth; Discharge
Dis., Disch.	Discharge
Db	Decibel
Dbl.	Double
DC	Direct Current
DDC	Direct Digital Control
Demob.	Demobilization
d.f.t.	Dry Film Thickness
d.f.u.	Drainage Fixture Units
D.H.	Double Hung
DHW	Domestic Hot Water
DI	Ductile Iron
Diag.	Diagonal
Diam., Dia	Diameter
Distrib.	Distribution
Div.	Division
Dk.	Deck
D.L.	Dead Load; Diesel
DLH	Deep Long Span Bar Joist
dlx	Deluxe
Do.	Ditto
DOP	Dioctyl Phthalate Penetration Test (Air Filters)
Dp., dp	Depth
D.P.S.T.	Double Pole, Single Throw
Dr.	Drive
DR	Dimension Ratio
Drink.	Drinking
D.S.	Double Strength
D.S.A.	Double Strength A Grade
D.S.B.	Double Strength B Grade
Dty.	Duty
DWV	Drain Waste Vent
DX	Deluxe White, Direct Expansion
dyn	Dyne
e	Eccentricity
E	Equipment Only; East; emissivity
Ea.	Each
EB	Encased Burial
Econ.	Economy
E.C.Y	Embankment Cubic Yards
EDP	Electronic Data Processing
EIFS	Exterior Insulation Finish System
E.D.R.	Equiv. Direct Radiation
Eq.	Equation
EL	elevation
Elec.	Electrician; Electrical
Elev.	Elevator; Elevating
EMT	Electrical Metallic Conduit; Thin Wall Conduit
Eng.	Engine, Engineered
EPDM	Ethylene Propylene Diene Monomer
EPS	Expanded Polystyrene
Eqhv.	Equip. Oper., Heavy
Eqlt.	Equip. Oper., Light
Eqmd.	Equip. Oper., Medium
Eqmm.	Equip. Oper., Master Mechanic
Eqol.	Equip. Oper., Oilers
Equip.	Equipment
ERW	Electric Resistance Welded

Abbreviations

E.S.	Energy Saver
Est.	Estimated
esu	Electrostatic Units
E.W.	Each Way
EWT	Entering Water Temperature
Excav.	Excavation
excl	Excluding
Exp., exp	Expansion, Exposure
Ext., ext	Exterior; Extension
Extru.	Extrusion
f.	Fiber stress
F	Fahrenheit; Female; Fill
Fab., fab	Fabricated; fabric
FBGS	Fiberglass
F.C.	Footcandles
f.c.c.	Face-centered Cubic
f′c.	Compressive Stress in Concrete; Extreme Compressive Stress
F.E.	Front End
FEP	Fluorinated Ethylene Propylene (Teflon)
F.G.	Flat Grain
F.H.A.	Federal Housing Administration
Fig.	Figure
Fin.	Finished
FIPS	Female Iron Pipe Size
Fixt.	Fixture
FJP	Finger jointed and primed
Fl. Oz.	Fluid Ounces
Flr.	Floor
FM	Frequency Modulation; Factory Mutual
Fmg.	Framing
FM/UL	Factory Mutual/Underwriters Labs
Fdn.	Foundation
FNPT	Female National Pipe Thread
Fori.	Foreman, Inside
Foro.	Foreman, Outside
Fount.	Fountain
fpm	Feet per Minute
FPT	Female Pipe Thread
Fr	Frame
F.R.	Fire Rating
FRK	Foil Reinforced Kraft
FSK	Foil/scrim/kraft
FRP	Fiberglass Reinforced Plastic
FS	Forged Steel
FSC	Cast Body; Cast Switch Box
Ft., ft	Foot; Feet
Ftng.	Fitting
Ftg.	Footing
Ft lb.	Foot Pound
Furn.	Furniture
FVNR	Full Voltage Non-Reversing
FVR	Full Voltage Reversing
FXM	Female by Male
Fy.	Minimum Yield Stress of Steel
g	Gram
G	Gauss
Ga.	Gauge
Gal., gal.	Gallon
gpm, GPM	Gallon per Minute
Galv., galv	Galvanized
GC/MS	Gas Chromatograph/Mass Spectrometer
Gen.	General
GFI	Ground Fault Interrupter
GFRC	Glass Fiber Reinforced Concrete
Glaz.	Glazier
GPD	Gallons per Day
gpf	Gallon per flush
GPH	Gallons per Hour
GPM	Gallons per Minute
GR	Grade
Gran.	Granular
Grnd.	Ground
GVW	Gross Vehicle Weight
GWB	Gypsum wall board
H	High Henry
HC	High Capacity
H.D., HD	Heavy Duty; High Density
H.D.O.	High Density Overlaid
HDPE	High density polyethelene plastic
Hdr.	Header
Hdwe.	Hardware
H.I.D., HID	High Intensity Discharge
Help.	Helper Average
HEPA	High Efficiency Particulate Air Filter
Hg	Mercury
HIC	High Interrupting Capacity
HM	Hollow Metal
HMWPE	high molecular weight polyethylene
HO	High Output
Horiz.	Horizontal
H.P., HP	Horsepower; High Pressure
H.P.F.	High Power Factor
Hr.	Hour
Hrs./Day	Hours per Day
HSC	High Short Circuit
Ht.	Height
Htg.	Heating
Htrs.	Heaters
HVAC	Heating, Ventilation & Air-Conditioning
Hvy.	Heavy
HW	Hot Water
Hyd.; Hydr.	Hydraulic
Hz	Hertz (cycles)
I.	Moment of Inertia
IBC	International Building Code
I.C.	Interrupting Capacity
ID	Inside Diameter
I.D.	Inside Dimension; Identification
I.F.	Inside Frosted
I.M.C.	Intermediate Metal Conduit
In.	Inch
Incan.	Incandescent
Incl.	Included; Including
Int.	Interior
Inst.	Installation
Insul., insul	Insulation/Insulated
I.P.	Iron Pipe
I.P.S., IPS	Iron Pipe Size
IPT	Iron Pipe Threaded
I.W.	Indirect Waste
J	Joule
J.I.C.	Joint Industrial Council
K	Thousand; Thousand Pounds; Heavy Wall Copper Tubing, Kelvin
K.A.H.	Thousand Amp. Hours
kcmil	Thousand Circular Mils
KD	Knock Down
K.D.A.T.	Kiln Dried After Treatment
kg	Kilogram
kG	Kilogauss
kgf	Kilogram Force
kHz	Kilohertz
Kip	1000 Pounds
KJ	Kiljoule
K.L.	Effective Length Factor
K.L.F.	Kips per Linear Foot
Km	Kilometer
KO	Knock Out
K.S.F.	Kips per Square Foot
K.S.I.	Kips per Square Inch
kV	Kilovolt
kVA	Kilovolt Ampere
kVAR	Kilovar (Reactance)
KW	Kilowatt
KWh	Kilowatt-hour
L	Labor Only; Length; Long; Medium Wall Copper Tubing
Lab.	Labor
lat	Latitude
Lath.	Lather
Lav.	Lavatory
lb.; #	Pound
L.B., LB	Load Bearing; L Conduit Body
L. & E.	Labor & Equipment
lb./hr.	Pounds per Hour
lb./L.F.	Pounds per Linear Foot
lbf/sq.in.	Pound-force per Square Inch
L.C.L.	Less than Carload Lot
L.C.Y.	Loose Cubic Yard
Ld.	Load
LE	Lead Equivalent
LED	Light Emitting Diode
L.F.	Linear Foot
L.F. Nose	Linear Foot of Stair Nosing
L.F. Rsr	Linear Foot of Stair Riser
Lg.	Long; Length; Large
L & H	Light and Heat
LH	Long Span Bar Joist
L.H.	Labor Hours
L.L., LL	Live Load
L.L.D.	Lamp Lumen Depreciation
lm	Lumen
lm/sf	Lumen per Square Foot
lm/W	Lumen per Watt
LOA	Length Over All
log	Logarithm
L-O-L	Lateralolet
long.	longitude
L.P., LP	Liquefied Petroleum; Low Pressure
L.P.F.	Low Power Factor
LR	Long Radius
L.S.	Lump Sum
Lt.	Light
Lt. Ga.	Light Gauge
L.T.L.	Less than Truckload Lot
Lt. Wt.	Lightweight
L.V.	Low Voltage
M	Thousand; Material; Male; Light Wall Copper Tubing
M^2CA	Meters Squared Contact Area
m/hr.; M.H.	Man-hour
mA	Milliampere
Mach.	Machine
Mag. Str.	Magnetic Starter
Maint.	Maintenance
Marb.	Marble Setter
Mat; Mat'l.	Material
Max.	Maximum
MBF	Thousand Board Feet
MBH	Thousand BTU's per hr.
MC	Metal Clad Cable
MCC	Motor Control Center
M.C.F.	Thousand Cubic Feet
MCFM	Thousand Cubic Feet per Minute
M.C.M.	Thousand Circular Mils
MCP	Motor Circuit Protector
MD	Medium Duty
MDF	Medium-density fibreboard
M.D.O.	Medium Density Overlaid
Med.	Medium
MF	Thousand Feet
M.F.B.M.	Thousand Feet Board Measure
Mfg.	Manufacturing
Mfrs.	Manufacturers
mg	Milligram
MGD	Million Gallons per Day
MGPH	Thousand Gallons per Hour
MH, M.H.	Manhole; Metal Halide; Man-Hour
MHz	Megahertz
Mi.	Mile
MI	Malleable Iron; Mineral Insulated
MIPS	Male Iron Pipe Size
mj	Mechanical Joint
m	Meter
mm	Millimeter
Mill.	Millwright
Min., min.	Minimum, minute

Abbreviations

Misc. Miscellaneous
ml Milliliter, Mainline
M.L.F. Thousand Linear Feet
Mo. Month
Mobil. Mobilization
Mog. Mogul Base
MPH Miles per Hour
MPT Male Pipe Thread
MRGWB Moisture Resistant Gypsum Wallboard
MRT Mile Round Trip
ms Millisecond
M.S.F. Thousand Square Feet
Mstz. Mosaic & Terrazzo Worker
M.S.Y. Thousand Square Yards
Mtd., mtd., mtd Mounted
Mthe. Mosaic & Terrazzo Helper
Mtng. Mounting
Mult. Multi; Multiply
M.V.A. Million Volt Amperes
M.V.A.R. Million Volt Amperes Reactance
MV Megavolt
MW Megawatt
MXM Male by Male
MYD Thousand Yards
N Natural; North
nA Nanoampere
NA Not Available; Not Applicable
N.B.C. National Building Code
NC Normally Closed
NEMA National Electrical Manufacturers Assoc.
NEHB Bolted Circuit Breaker to 600V.
NFPA National Fire Protection Association
NLB Non-Load-Bearing
NM Non-Metallic Cable
nm Nanometer
No. Number
NO Normally Open
N.O.C. Not Otherwise Classified
Nose. Nosing
NPT National Pipe Thread
NQOD Combination Plug-on/Bolt on Circuit Breaker to 240V.
N.R.C., NRC Noise Reduction Coefficient/ Nuclear Regulator Commission
N.R.S. Non Rising Stem
ns Nanosecond
nW Nanowatt
OB Opposing Blade
OC On Center
OD Outside Diameter
O.D. Outside Dimension
ODS Overhead Distribution System
O.G. Ogee
O.H. Overhead
O&P Overhead and Profit
Oper. Operator
Opng. Opening
Orna. Ornamental
OSB Oriented Strand Board
OS&Y Outside Screw and Yoke
OSHA Occupational Safety and Health Act
Ovhd. Overhead
OWG Oil, Water or Gas
Oz. Ounce
P. Pole; Applied Load; Projection
p. Page
Pape. Paperhanger
P.A.P.R. Powered Air Purifying Respirator
PAR Parabolic Reflector
P.B., PB Push Button
Pc., Pcs. Piece, Pieces
P.C. Portland Cement; Power Connector
P.C.F. Pounds per Cubic Foot
PCM Phase Contrast Microscopy

PDCA Painting and Decorating Contractors of America
P.E., PE Professional Engineer; Porcelain Enamel; Polyethylene; Plain End
Perf. Perforated
PEX Cross linked polyethylene
Ph. Phase
P.I. Pressure Injected
Pile. Pile Driver
Pkg. Package
Pl. Plate
Plah. Plasterer Helper
Plas. Plasterer
plf Pounds Per Linear Foot
Pluh. Plumbers Helper
Plum. Plumber
Ply. Plywood
p.m. Post Meridiem
Pntd. Painted
Pord. Painter, Ordinary
pp Pages
PP, PPL Polypropylene
P.P.M. Parts per Million
Pr. Pair
P.E.S.B. Pre-engineered Steel Building
Prefab. Prefabricated
Prefin. Prefinished
Prop. Propelled
PSF, psf Pounds per Square Foot
PSI, psi Pounds per Square Inch
PSIG Pounds per Square Inch Gauge
PSP Plastic Sewer Pipe
Pspr. Painter, Spray
Psst. Painter, Structural Steel
P.T. Potential Transformer
P. & T. Pressure & Temperature
Ptd. Painted
Ptns. Partitions
Pu Ultimate Load
PVC Polyvinyl Chloride
Pvmt. Pavement
PRV Pressure Relief Valve
Pwr. Power
Q Quantity Heat Flow
Qt. Quart
Quan., Qty. Quantity
Q.C. Quick Coupling
r Radius of Gyration
R Resistance
R.C.P. Reinforced Concrete Pipe
Rect. Rectangle
recpt. receptacle
Reg. Regular
Reinf. Reinforced
Req'd. Required
Res. Resistant
Resi. Residential
RF Radio Frequency
RFID Radio-frequency identification
Rgh. Rough
RGS Rigid Galvanized Steel
RHW Rubber, Heat & Water Resistant; Residential Hot Water
rms Root Mean Square
Rnd. Round
Rodm. Rodman
Rofc. Roofer, Composition
Rofp. Roofer, Precast
Rohe. Roofer Helpers (Composition)
Rots. Roofer, Tile & Slate
R.O.W. Right of Way
RPM Revolutions per Minute
R.S. Rapid Start
Rsr Riser
RT Round Trip
S. Suction; Single Entrance; South

SC Screw Cover
SCFM Standard Cubic Feet per Minute
Scaf. Scaffold
Sch., Sched. Schedule
S.C.R. Modular Brick
S.D. Sound Deadening
SDR Standard Dimension Ratio
S.E. Surfaced Edge
Sel. Select
SER, SEU Service Entrance Cable
S.F. Square Foot
S.F.C.A. Square Foot Contact Area
S.F. Flr. Square Foot of Floor
S.F.G. Square Foot of Ground
S.F. Hor. Square Foot Horizontal
SFR Square Feet of Radiation
S.F. Shlf. Square Foot of Shelf
S4S Surface 4 Sides
Shee. Sheet Metal Worker
Sin. Sine
Skwk. Skilled Worker
SL Saran Lined
S.L. Slimline
Sldr. Solder
SLH Super Long Span Bar Joist
S.N. Solid Neutral
SO Stranded with oil resistant inside insulation
S-O-L Socketolet
sp Standpipe
S.P. Static Pressure; Single Pole; Self-Propelled
Spri. Sprinkler Installer
spwg Static Pressure Water Gauge
S.P.D.T. Single Pole, Double Throw
SPF Spruce Pine Fir
S.P.S.T. Single Pole, Single Throw
SPT Standard Pipe Thread
Sq. Square; 100 Square Feet
Sq. Hd. Square Head
Sq. In. Square Inch
S.S. Single Strength; Stainless Steel
S.S.B. Single Strength B Grade
sst, ss Stainless Steel
Sswk. Structural Steel Worker
Sswl. Structural Steel Welder
St.; Stl. Steel
STC Sound Transmission Coefficient
Std. Standard
Stg. Staging
STK Select Tight Knot
STP Standard Temperature & Pressure
Stpi. Steamfitter, Pipefitter
Str. Strength; Starter; Straight
Strd. Stranded
Struct. Structural
Sty. Story
Subj. Subject
Subs. Subcontractors
Surf. Surface
Sw. Switch
Swbd. Switchboard
S.Y. Square Yard
Syn. Synthetic
S.Y.P. Southern Yellow Pine
Sys. System
t. Thickness
T Temperature; Ton
Tan Tangent
T.C. Terra Cotta
T & C Threaded and Coupled
T.D. Temperature Difference
Tdd Telecommunications Device for the Deaf
T.E.M. Transmission Electron Microscopy
temp Temperature, Tempered, Temporary
TFFN Nylon Jacketed Wire

Abbreviations

TFE	Tetrafluoroethylene (Teflon)
T. & G.	Tongue & Groove; Tar & Gravel
Th., Thk.	Thick
Thn.	Thin
Thrded	Threaded
Tilf.	Tile Layer, Floor
Tilh.	Tile Layer, Helper
THHN	Nylon Jacketed Wire
THW.	Insulated Strand Wire
THWN	Nylon Jacketed Wire
T.L., TL	Truckload
T.M.	Track Mounted
Tot.	Total
T-O-L	Threadolet
tmpd	Tempered
T.S.	Trigger Start
Tr.	Trade
Transf.	Transformer
Trhv.	Truck Driver, Heavy
Trlr	Trailer
Trlt.	Truck Driver, Light
TTY	Teletypewriter
TV	Television
T.W.	Thermoplastic Water Resistant Wire
UCI	Uniform Construction Index
UF	Underground Feeder
UGND	Underground Feeder
UHF	Ultra High Frequency
U.I.	United Inch
U.L., UL	Underwriters Laboratory
Uld.	unloading
Unfin.	Unfinished
UPS	Uninterruptible Power Supply
URD	Underground Residential Distribution
US	United States
USGBC	U.S. Green Building Council
USP	United States Primed
UTMCD	Uniform Traffic Manual For Control Devices
UTP	Unshielded Twisted Pair
V	Volt
VA	Volt Amperes
V.C.T.	Vinyl Composition Tile
VAV	Variable Air Volume
VC	Veneer Core
VDC	Volts Direct Current
Vent.	Ventilation
Vert.	Vertical
V.F.	Vinyl Faced
V.G.	Vertical Grain
VHF	Very High Frequency
VHO	Very High Output
Vib.	Vibrating
VLF	Vertical Linear Foot
VOC	Volitile Organic Compound
Vol.	Volume
VRP	Vinyl Reinforced Polyester
W	Wire; Watt; Wide; West
w/	With
W.C., WC	Water Column; Water Closet
W.F.	Wide Flange
W.G.	Water Gauge
Wldg.	Welding
W. Mile	Wire Mile
W-O-L	Weldolet
W.R.	Water Resistant
Wrck.	Wrecker
W.S.P.	Water, Steam, Petroleum
WT., Wt.	Weight
WWF	Welded Wire Fabric
XFER	Transfer
XFMR	Transformer
XHD	Extra Heavy Duty
XHHW, XLPE	Cross-Linked Polyethylene Wire Insulation
XLP	Cross-linked Polyethylene
Xport	Transport
Y	Wye
yd	Yard
yr	Year
Δ	Delta
%	Percent
~	Approximately
Ø	Phase; diameter
@	At
#	Pound; Number
<	Less Than
>	Greater Than
Z	zone

D

G

H

I

Q

R

S

T

U

V

W

X

Y

Notes

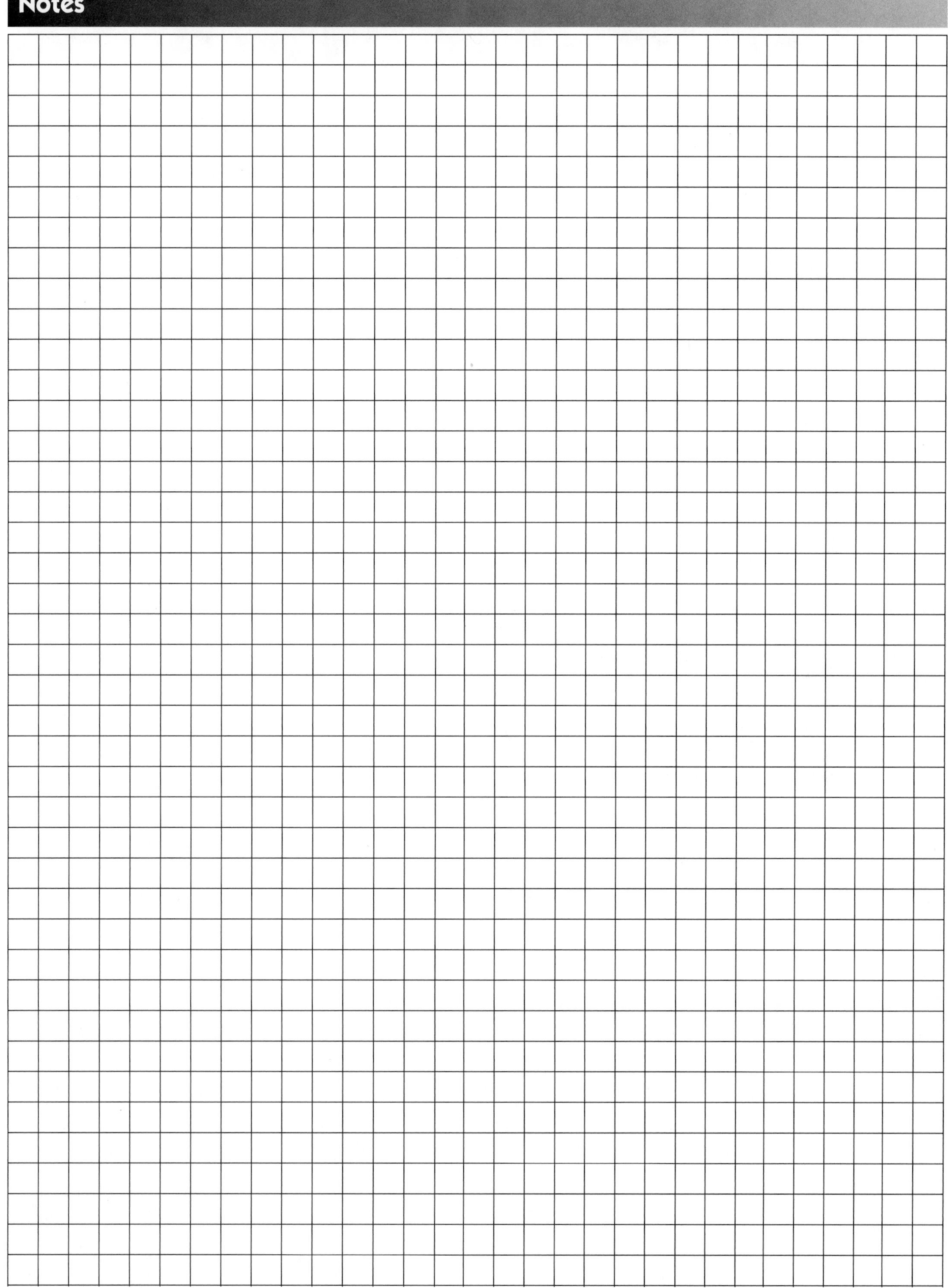

Notes

Notes

Notes

Notes

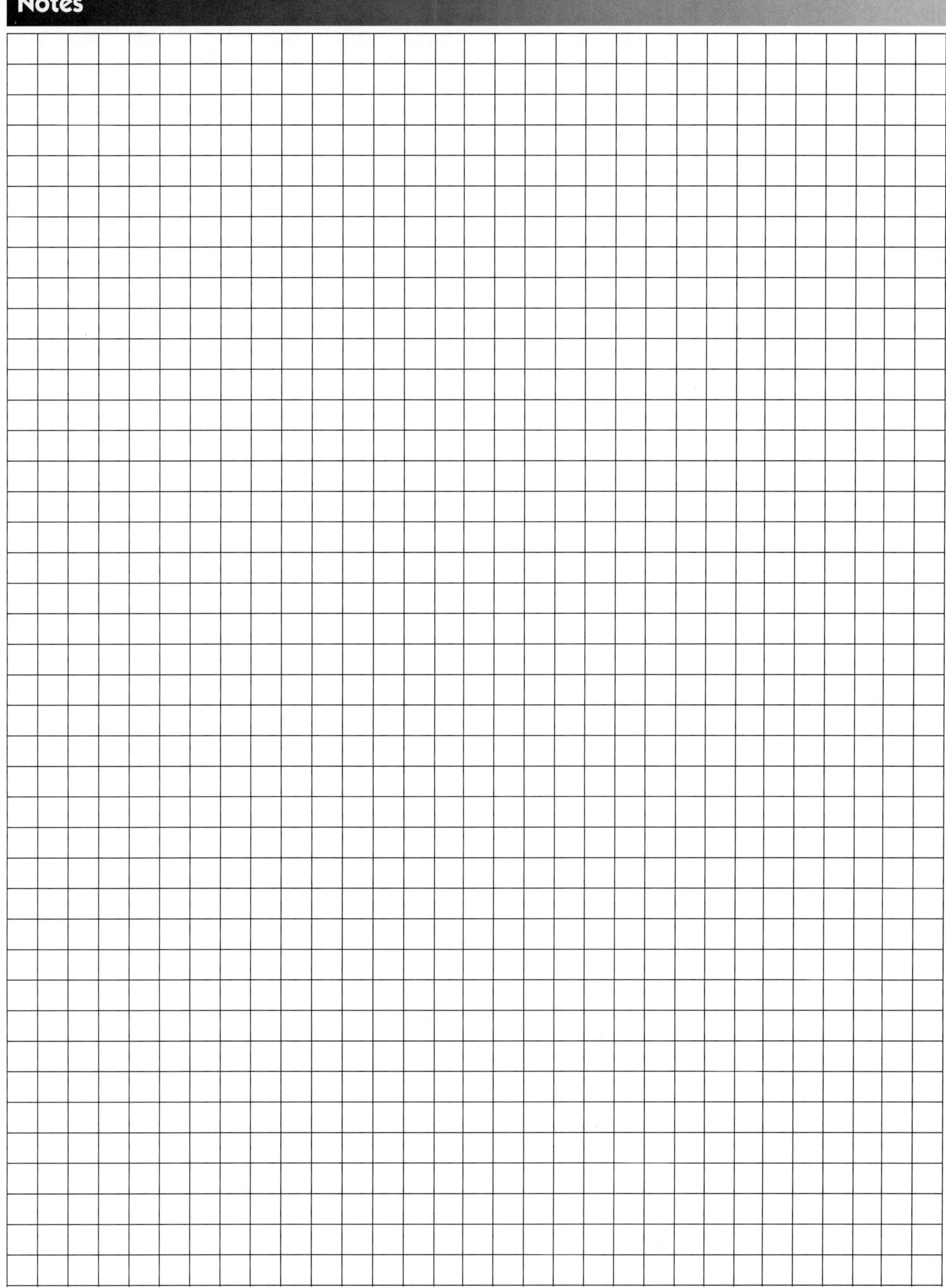

Notes

Division Notes

		CREW	DAILY OUTPUT	LABOR-HOURS	UNIT	BARE COSTS MAT.	LABOR	EQUIP.	TOTAL	TOTAL INCL O&P

Division Notes

		CREW	DAILY OUTPUT	LABOR-HOURS	UNIT	BARE COSTS MAT.	LABOR	EQUIP.	TOTAL	TOTAL INCL O&P

Division Notes

	CREW	DAILY OUTPUT	LABOR-HOURS	UNIT	BARE COSTS MAT.	BARE COSTS LABOR	BARE COSTS EQUIP.	BARE COSTS TOTAL	TOTAL INCL O&P

Division Notes

		CREW	DAILY OUTPUT	LABOR-HOURS	UNIT	BARE COSTS MAT.	LABOR	EQUIP.	TOTAL	TOTAL INCL O&P

Division Notes

		CREW	DAILY OUTPUT	LABOR-HOURS	UNIT	BARE COSTS MAT.	LABOR	EQUIP.	TOTAL	TOTAL INCL O&P

Reed Construction Data/RSMeans—a tradition of excellence in construction cost information and services since 1942

Table of Contents

For more information visit the RSMeans website at **www.rsmeans.com**

Unit prices according to the latest MasterFormat!

Book Selection Guide

The following table provides definitive information on the content of each cost data publication. The number of lines of data provided in each unit price or assemblies division, as well as the number of reference tables and crews, is listed for each book. The presence of other elements such as equipment rental costs, historical cost indexes, city cost indexes, square foot models, or cross-referenced indexes is also indicated. You can use the table to help select the RSMeans book that has the quantity and type of information you most need in your work.

Unit Cost Divisions	Building Construction	Mechanical	Electrical	Commercial Renovation	Square Foot	Site Work Landsc.	Green Building	Interior	Concrete Masonry	Open Shop	Heavy Construction	Light Commercial	Facilities Construction	Plumbing	Residential
1	573	395	410	499		517	124	319	463	572	507	233	1038	411	176
2	755	281	87	706		1005	180	366	223	751	744	447	1197	291	258
3	1678	346	223	1053		1474	1004	323	2037	1678	1680	469	1774	315	376
4	900	22	0	698		701	183	573	1135	876	591	490	1147	0	405
5	1844	159	156	1040		809	1777	1047	712	1835	1031	924	1845	204	720
6	2294	67	69	1881		411	534	1421	300	2273	383	1968	1908	22	2480
7	1467	215	125	1501		563	776	572	497	1464	26	1178	1529	224	942
8	2122	98	48	2351		299	1157	1768	657	2066	0	1790	2581	0	1410
9	1927	72	26	1720		313	449	2012	389	1868	15	1639	2171	54	1416
10	987	17	10	610		215	27	807	156	987	29	497	1073	230	221
11	1022	208	165	486		126	54	878	28	1007	0	231	1042	169	110
12	563	0	2	325		248	150	1698	12	552	19	371	1748	23	324
13	727	117	112	246		346	128	251	66	726	253	82	754	71	80
14	277	36	0	225		28	0	259	0	276	0	12	296	18	6
21	78	0	16	30		0	0	248	0	78	0	60	429	435	224
22	1140	7318	154	1105		1536	1100	605	20	1102	1659	830	7152	9029	659
23	1211	7086	599	846		158	940	695	38	1132	111	752	5176	1860	416
26	1344	499	10087	1019		816	644	1132	55	1329	562	1218	10,011	444	618
27	72	0	266	34		13	0	60	0	72	39	52	268	0	4
28	96	58	124	71		0	21	78	0	99	0	40	139	44	25
31	1482	735	610	803		3207	290	7	1208	1426	3284	600	1545	660	611
32	770	54	0	836		4341	341	362	274	741	1796	381	1647	166	428
33	509	1014	461	207		2027	41	0	231	508	1937	119	1557	1205	117
34	109	0	20	4		192	0	0	31	64	170	0	129	0	0
35	18	0	0	0		327	0	0	0	18	442	0	83	0	0
41	52	0	0	24		7	0	22	0	52	30	0	59	14	0
44	75	83	0	0		0	0	0	0	0	0	0	75	75	0
46	23	16	0	0		274	181	0		23	264	0	33	33	0
48	13	0	28	0		4	28	0		13	4	28	13	0	26
Totals	24128	18896	13798	18320		19957	10,129	15503	8532	23588	15576	14411	48419	15997	12052

Assem Div	Building Construction	Mechanical	Elec-trical	Commercial Renovation	Square Foot	Site Work Landscape	Assemblies	Green Building	Interior	Concrete Masonry	Heavy Construction	Light Commercial	Facilities Construction	Plumbing	Asm Div	Residential
A		15	0	188	150	577	598	0	0	536	571	154	24	0	1	368
B		0	0	848	2483	0	5643	56	329	1966	368	2074	174	0	2	211
C		0	0	647	863	0	1233	0	1558	146	0	763	245	0	3	588
D		1067	886	712	1864	72	2464	265	826	0	0	1348	1084	1083	4	839
E		0	0	86	260	0	298	0	5	0	0	257	5	0	5	392
F		0	0	0	115	0	115	0	0	0	0	115	3	0	6	357
G		527	447	318	202	3456	668	0	0	535	1431	200	293	707	7	296
															8	760
															9	80
															10	0
															11	0
															12	0
Totals		1609	1333	2799	5937	4105	11019	321	2718	3183	2370	4911	1828	1790		3891

Reference Section	Building Construction Costs	Mechanical	Electrical	Commercial Renovation	Square Foot	Site Work Landscape	Assem.	Green Building	Interior	Concrete Masonry	Open Shop	Heavy Construction	Light Commercial	Facilities Construction	Plumbing	Resi.
Reference Tables	yes	yes	yes	yes	no	yes	yes	yes	yes	yes	yes	yes	yes	yes	yes	yes
Models					109								48			28
Crews	558	558	558	536		558		558	558	558	534	558	534	536	558	534
Equipment Rental Costs	yes	yes	yes	yes		yes		yes	yes	yes	yes	yes	yes	yes	yes	yes
Historical Cost Indexes	yes	yes	yes	yes	yes	yes	yes	yes	yes	yes	yes	yes	yes	yes	yes	no
City Cost Indexes	yes	yes	yes	yes	yes	yes	yes	yes	yes	yes	yes	yes	yes	yes	yes	yes

Annual Cost Guides

Cost Guides also available in CD and at rsmeansonline.com

For more information visit the RSMeans website at **www.rsmeans.com**

Unit prices according to the latest MasterFormat!

RSMeans Building Construction Cost Data 2013

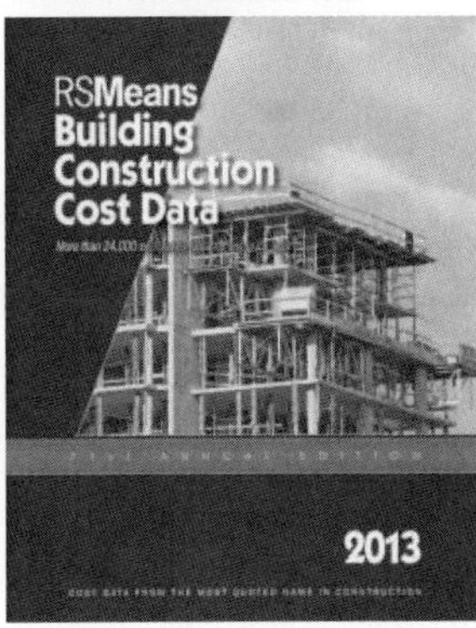

Offers you unchallenged unit price reliability in an easy-to-use format. Whether used for verifying complete, finished estimates or for periodic checks, it supplies more cost facts better and faster than any comparable source. Over 23,000 unit prices have been updated for 2013. The City Cost Indexes and Location Factors cover over 930 areas, for indexing to any project location in North America. Order and get *RSMeans Quarterly Update Service* FREE. You'll have year-long access to the RSMeans Estimating **HOTLINE** FREE with your subscription. Expert assistance when using RSMeans data is just a phone call away.

$186.95 per copy | Available Sept. 2012 | Catalog no. 60013

RSMeans Green Building Cost Data 2013

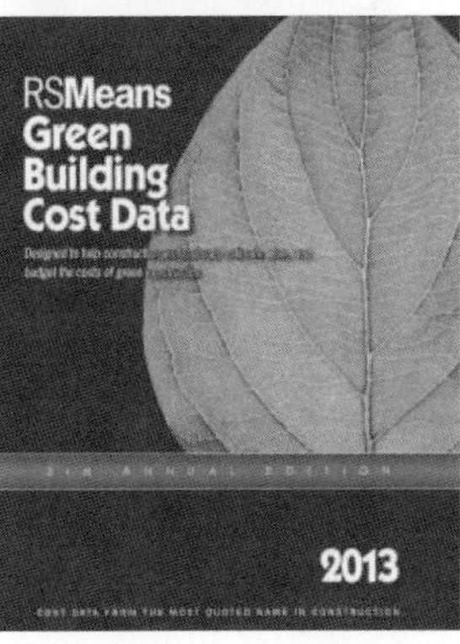

Estimate, plan, and budget the costs of green building for both new commercial construction and renovation work with this first edition of RSMeans Green Building Cost Data. More than 9,000 unit costs for a wide array of green building products plus assemblies costs. Easily identified cross references to LEED and Green Globes building rating systems criteria.

$155.95 per copy | Available Nov. 2012 | Catalog no. 60553

RSMeans Mechanical Cost Data 2013

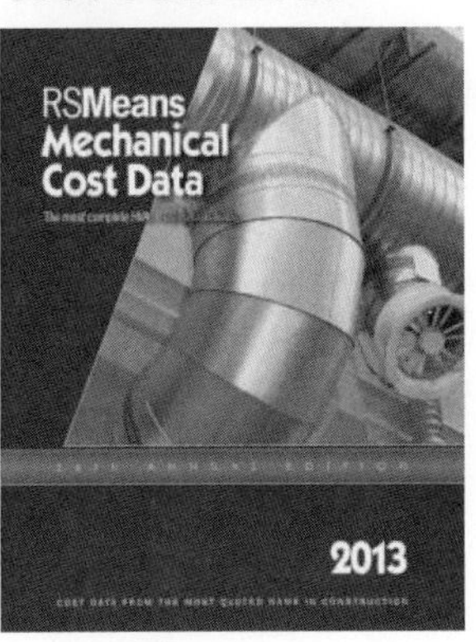

Total unit and systems price guidance for mechanical construction. . . materials, parts, fittings, and complete labor cost information. Includes prices for piping, heating, air conditioning, ventilation, and all related construction.

Plus new 2013 unit costs for:

- Thousands of installed HVAC/ controls assemblies components
- "On-site" Location Factors for over 930 cities and towns in the U.S. and Canada
- Crews, labor, and equipment

$183.95 per copy | Available Oct. 2012 | Catalog no. 60023

RSMeans Facilities Construction Cost Data 2013

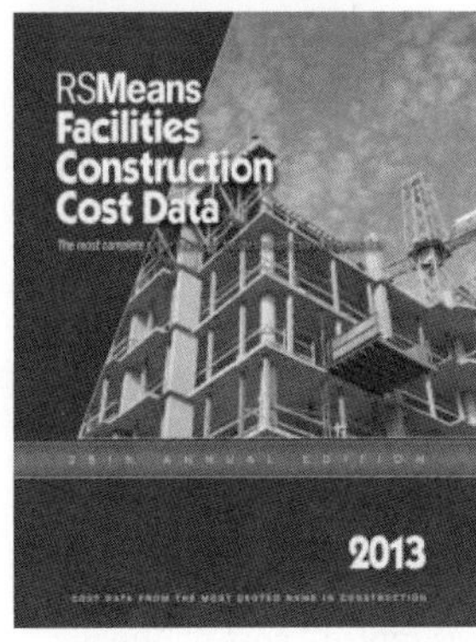

For the maintenance and construction of commercial, industrial, municipal, and institutional properties. Costs are shown for new and remodeling construction and are broken down into materials, labor, equipment, and overhead and profit. Special emphasis is given to sections on mechanical, electrical, furnishings, site work, building maintenance, finish work, and demolition.

More than 47,000 unit costs, plus assemblies costs and a comprehensive Reference Section are included.

$476.95 per copy | Available Nov. 2012 | Catalog no. 60203

RSMeans Square Foot Costs 2013

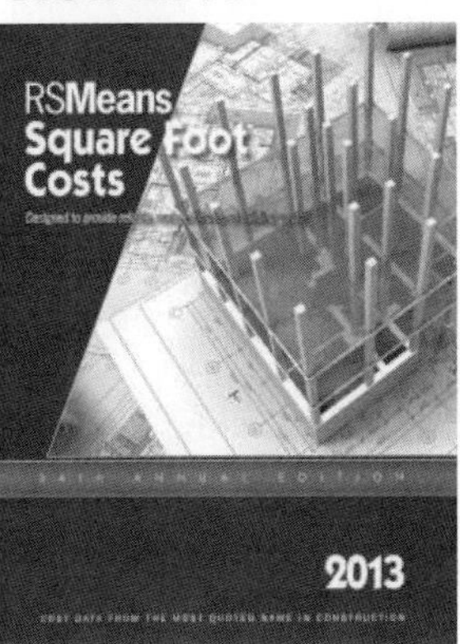

Accurate and Easy To Use

- **Updated price information** based on nationwide figures from suppliers, estimators, labor experts, and contractors
- "How-to-Use" sections, with **clear examples** of commercial, residential, industrial, and institutional structures
- Realistic graphics, offering true-to-life illustrations of building projects
- Extensive information on using square foot cost data, including sample estimates and alternate pricing methods

$197.95 per copy | Available Oct. 2012 | Catalog no. 60053

RSMeans Commercial Renovation Cost Data 2013

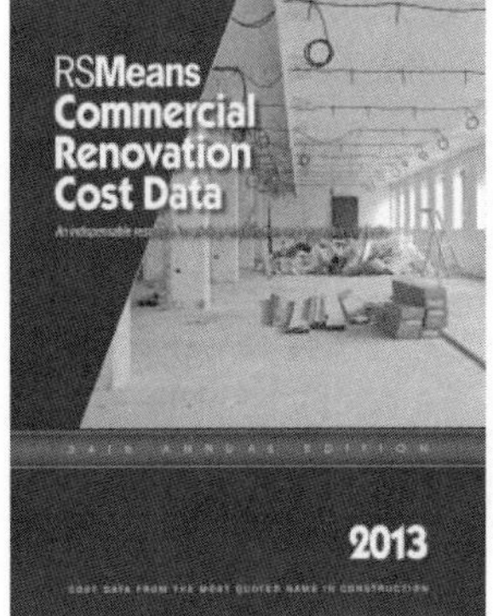

Commercial/Multifamily Residential

Use this valuable tool to estimate commercial and multifamily residential renovation and remodeling.

Includes: New costs for hundreds of unique methods, materials, and conditions that only come up in repair and remodeling, PLUS:

- Unit costs for more than 17,000 construction components
- Installed costs for more than 2,000 assemblies
- More than 930 "on-site" localization factors for the U.S. and Canada

$152.95 per copy | Available Oct. 2012 | Catalog no. 60043

RSMeans Electrical Cost Data 2013

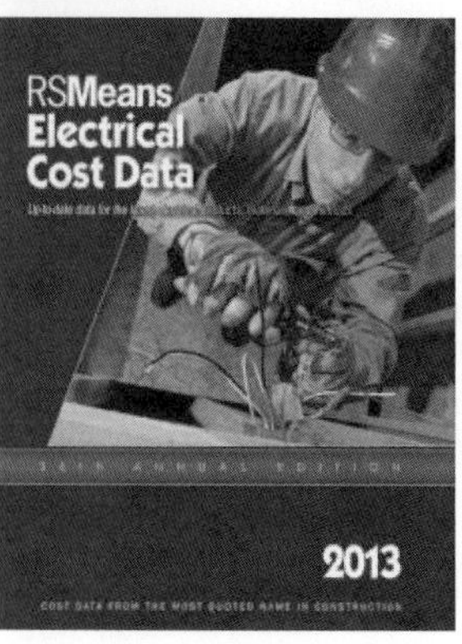

Pricing information for every part of electrical cost planning. More than 13,000 unit and systems costs with design tables; clear specifications and drawings; engineering guides; illustrated estimating procedures; complete labor-hour and materials costs for better scheduling and procurement; and the latest electrical products and construction methods.

- A variety of special electrical systems including cathodic protection
- Costs for maintenance, demolition, HVAC/mechanical, specialties, equipment, and more

$188.95 per copy | Available Oct. 2012 | Catalog no. 60033

RSMeans Electrical Change Order Cost Data 2013

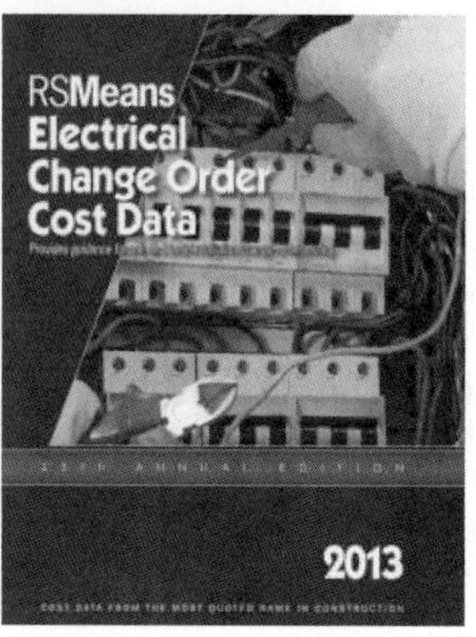

RSMeans Electrical Change Order Cost Data provides you with electrical unit prices exclusively for pricing change orders—based on the recent, direct experience of contractors and suppliers. Analyze and check your own change order estimates against the experience others have had doing the same work. It also covers productivity analysis and change order cost justifications. With useful information for calculating the effects of change orders and dealing with their administration.

$183.95 per copy | Available Dec. 2012 | Catalog no. 60233

RSMeans Assemblies Cost Data 2013

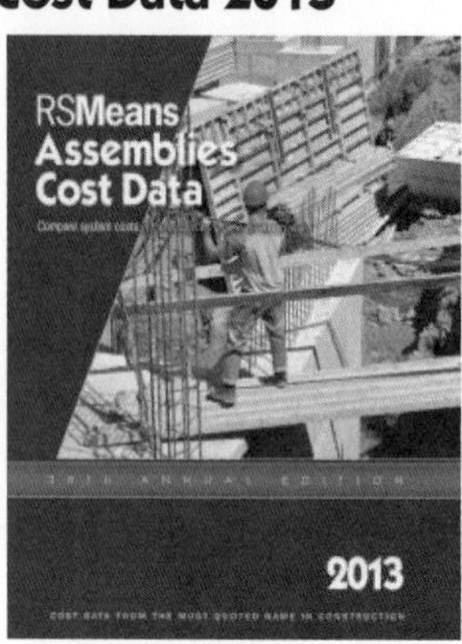

RSMeans Assemblies Cost Data takes the guesswork out of preliminary or conceptual estimates. Now you don't have to try to calculate the assembled cost by working up individual component costs. We've done all the work for you.

Presents detailed illustrations, descriptions, specifications, and costs for every conceivable building assembly—over 350 types in all—arranged in the easy-to-use UNIFORMAT II system. Each illustrated "assembled" cost includes a complete grouping of materials and associated installation costs, including the installing contractor's overhead and profit.

$305.95 per copy | Available Sept. 2012 | Catalog no. 60063

RSMeans Open Shop Building Construction Cost Data 2013

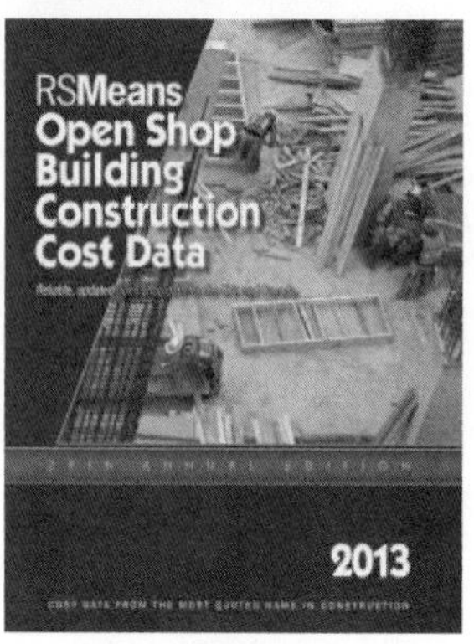

The latest costs for accurate budgeting and estimating of new commercial and residential construction. . . renovation work. . . change orders. . . cost engineering.

RSMeans Open Shop "BCCD" will assist you to:

- Develop benchmark prices for change orders
- Plug gaps in preliminary estimates and budgets
- Estimate complex projects
- Substantiate invoices on contracts
- Price ADA-related renovations

$160.95 per copy | Available Dec. 2012 | Catalog no. 60153

RSMeans Residential Cost Data 2013

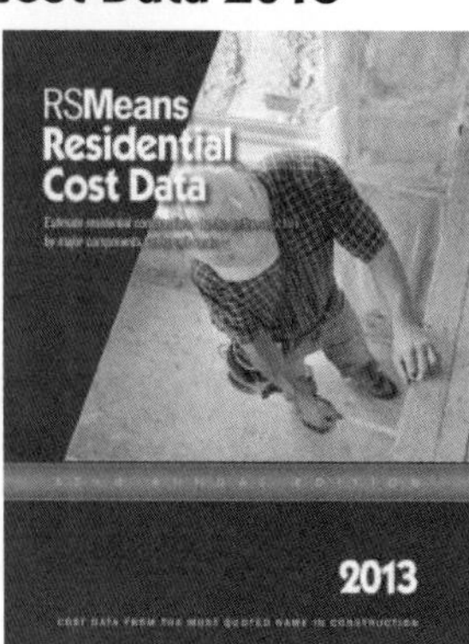

Contains square foot costs for 28 basic home models with the look of today, plus hundreds of custom additions and modifications you can quote right off the page. Includes costs for more than 700 residential systems. Complete with blank estimating forms, sample estimates, and step-by-step instructions.

Now contains line items for cultured stone and brick, PVC trim lumber, and TPO roofing.

$134.95 per copy | Available Oct. 2012 | Catalog no. 60173

RSMeans Site Work & Landscape Cost Data 2013

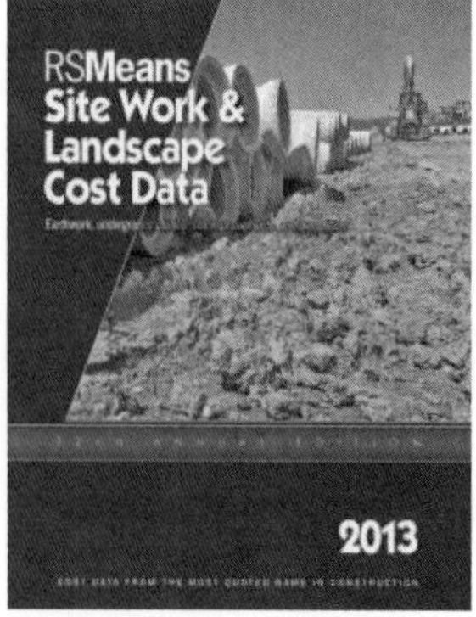

Includes unit and assemblies costs for earthwork, sewerage, piped utilities, site improvements, drainage, paving, trees and shrubs, street openings/repairs, underground tanks, and more. Contains 78 types of assemblies costs for accurate conceptual estimates.

Includes:

- Estimating for infrastructure improvements
- Environmentally-oriented construction
- ADA-mandated handicapped access
- Hazardous waste line items

$181.95 per copy | Available Dec. 2012 | Catalog no. 60283

Annual Cost Guides

Cost Guides also available in CD and at rsmeansonline.com

For more information visit the RSMeans website at **www.rsmeans.com**

Unit prices according to the latest MasterFormat!

RSMeans Facilities Maintenance & Repair Cost Data 2013

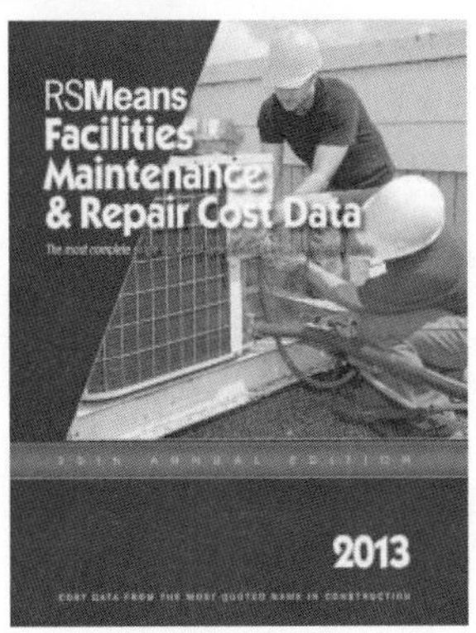

RSMeans Facilities Maintenance & Repair Cost Data gives you a complete system to manage and plan your facility repair and maintenance costs and budget efficiently. Guidelines for auditing a facility and developing an annual maintenance plan. Budgeting is included, along with reference tables on cost and management, and information on frequency and productivity of maintenance operations.

The only nationally recognized source of maintenance and repair costs. Developed in cooperation with the Civil Engineering Research Laboratory (CERL) of the Army Corps of Engineers.

$417.95 per copy | Available Nov. 2012 | Catalog no. 60303

RSMeans Light Commercial Cost Data 2013

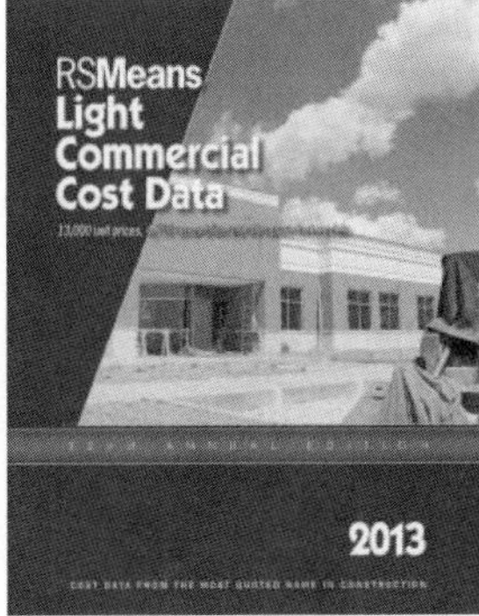

Specifically addresses the light commercial market, which is a specialized niche in the construction industry. Aids you, the owner/designer/contractor, in preparing all types of estimates—from budgets to detailed bids. Includes new advances in methods and materials.

Assemblies Section allows you to evaluate alternatives in the early stages of design/planning.

Over 13,000 unit costs ensure that you have the prices you need. . . when you need them.

$139.95 per copy | Available Nov. 2012 | Catalog no. 60183

RSMeans Concrete & Masonry Cost Data 2013

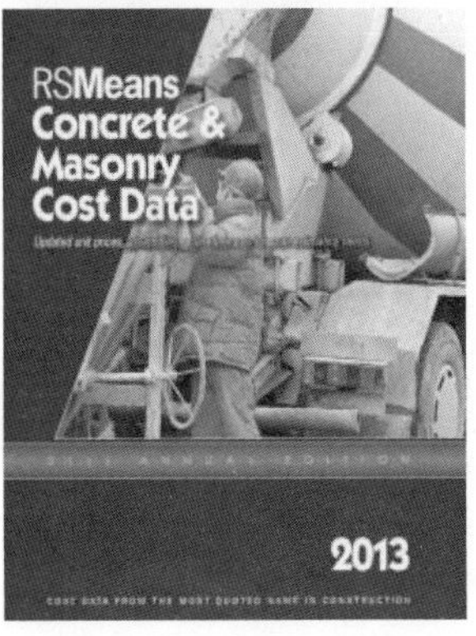

Provides you with cost facts for virtually all concrete/masonry estimating needs, from complicated formwork to various sizes and face finishes of brick and block—all in great detail. The comprehensive Unit Price Section contains more than 8,000 selected entries. Also contains an Assemblies [Cost] Section, and a detailed Reference Section that supplements the cost data.

$171.95 per copy | Available Dec. 2012 | Catalog no. 60113

RSMeans Labor Rates for the Construction Industry 2013

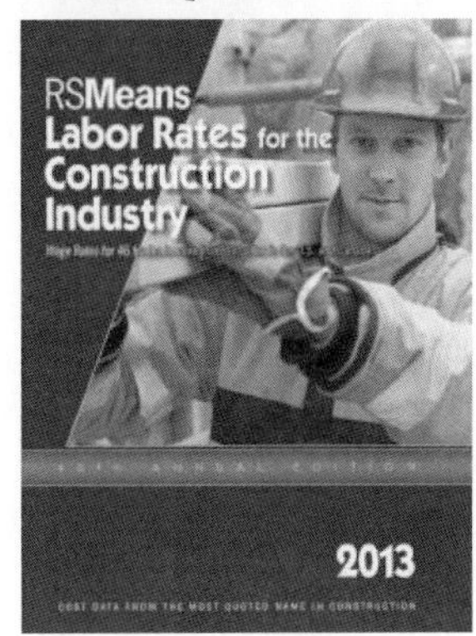

Complete information for estimating labor costs, making comparisons, and negotiating wage rates by trade for more than 300 U.S. and Canadian cities. With 46 construction trades listed by local union number in each city, and historical wage rates included for comparison. Each city chart lists the county and is alphabetically arranged with handy visual flip tabs for quick reference.

$415.95 per copy | Available Dec. 2012 | Catalog no. 60123

RSMeans Construction Cost Indexes 2013

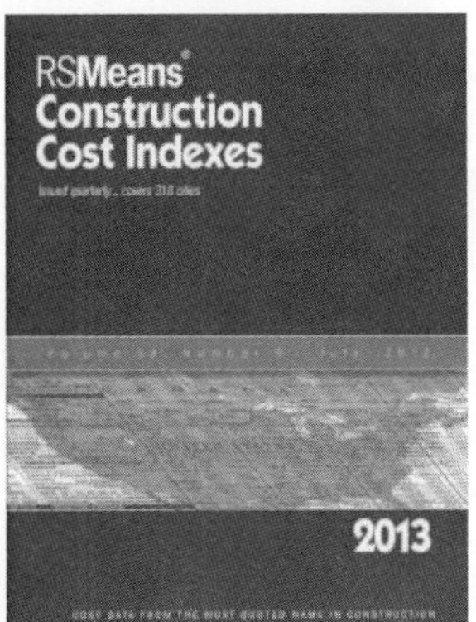

What materials and labor costs will change unexpectedly this year? By how much?

- Breakdowns for 318 major cities
- National averages for 30 key cities
- Expanded five major city indexes
- Historical construction cost indexes

$362.00 per year (subscription) | Catalog no. 50143
$90.50 individual quarters | Catalog no. 60143 A,B,C,D

RSMeans Interior Cost Data 2013

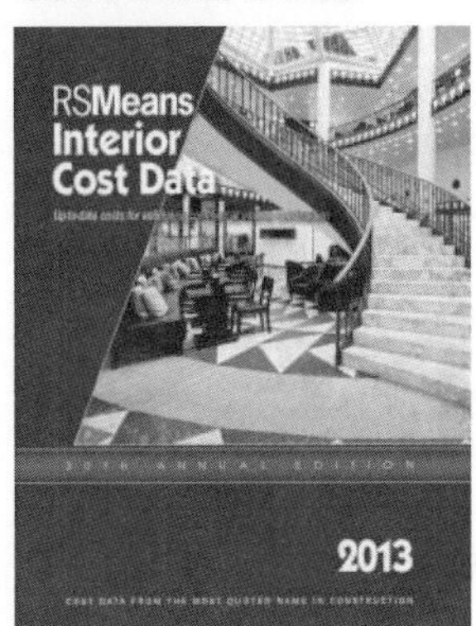

Provides you with prices and guidance needed to make accurate interior work estimates. Contains costs on materials, equipment, hardware, custom installations, furnishings, and labor costs. . . for new and remodel commercial and industrial interior construction, including updated information on office furnishings, and reference information.

$192.95 per copy | Available Nov. 2012 | Catalog no. 60093

Annual Cost Guides

Cost Guides also available in CD and at rsmeansonline.com

For more information visit the RSMeans website at **www.rsmeans.com**

Unit prices according to the latest MasterFormat!

RSMeans Heavy Construction Cost Data 2013

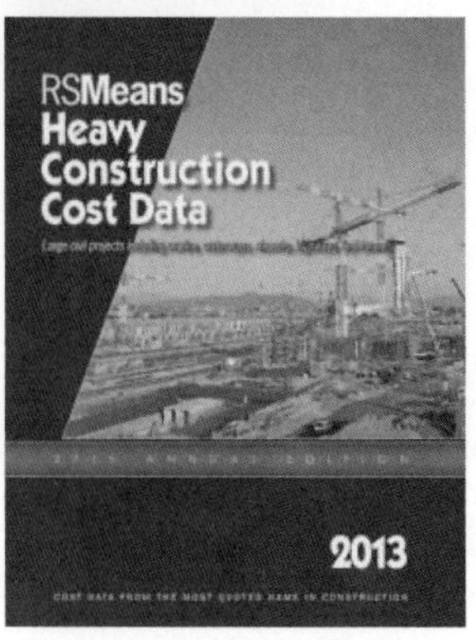

A comprehensive guide to heavy construction costs. Includes costs for highly specialized projects such as tunnels, dams, highways, airports, and waterways. Information on labor rates, equipment, and materials costs is included. Features unit price costs, systems costs, and numerous reference tables for costs and design.

$186.95 per copy | Available Dec. 2012 | Catalog no. 60163

RSMeans Plumbing Cost Data 2013

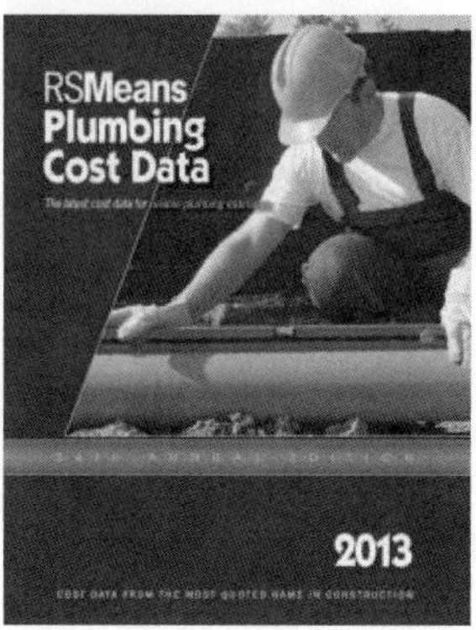

Comprehensive unit prices and assemblies for plumbing, irrigation systems, commercial and residential fire protection, point-of-use water heaters, and the latest approved materials. This publication and its companion, *RSMeans Mechanical Cost Data*, provide full-range cost estimating coverage for all the mechanical trades.

Now contains updated costs for potable water and radiant heat systems and high efficiency fixtures.

$188.95 per copy | Available Oct. 2012 | Catalog no. 60213

2013 RSMeans Seminar Schedule

Note: call for exact dates and details.

Location	Dates
Las Vegas, NV	March
Washington, DC	April
San Francisco, CA	June
Washington, DC	September
Dallas, TX	September
Las Vegas, NV	October
Orlando, FL	November
San Diego, CA	December

1-800-334-3509, Press 1

2013 Regional Seminar Schedule

Note: call for exact dates and details.

Location	Dates
Seattle, WA	January and August
Dallas/Ft. Worth, TX	January
Austin, TX	February
Oklahoma City, OK	February
Anchorage, AK	March and September
Phoenix, AZ	March
San Bernardino, CA	April
Kansas City, MO	April
Toronto	May
Denver, CO	May
Bethesda, MD	June
Columbus, GA	June
El Segundo, CA	August
Jacksonville FL	September
St. Louis, MO	October
Houston, TX	October
Vancouver, BC	November
Baltimore, MD	November
San Antonio, TX	December
Raleigh, NC	December

1-800-334-3509, Press 1

Professional Development

For more information visit the RSMeans website at **www.rsmeans.com**

Registration Information

Register early... Save up to $100! Register 30 days before the start date of a seminar and save $100 off your total fee. *Note: This discount can be applied only once per order. It cannot be applied to team discount registrations or any other special offer.*

How to register Register by phone today! The RSMeans toll-free number for making reservations is **1-800-334-3509, Press 1.**

Individual seminar registration fee - $935. *Means CostWorks*® training registration fee - $375. To register by mail, complete the registration form and return, with your full fee, to: RSMeans Seminars, 700 Longwater Drive, Norwell, MA 02061.

Government pricing All federal government employees save off the regular seminar price. Other promotional discounts cannot be combined with the government discount.

Team discount program for two to four seminar registrations. Call for pricing: 1-800-334-3509, Press 1

Multiple course discounts When signing up for two or more courses, call for pricing.

Refund policy Cancellations will be accepted up to ten business days prior to the seminar start. There are no refunds for cancellations received later than ten working days prior to the first day of the seminar. A $150 processing fee will be applied for all cancellations. Written notice of cancellation is required. Substitutions can be made at any time before the session starts. **No-shows are subject to the full seminar fee.**

AACE approved courses Many seminars described and offered here have been approved for 14 hours (1.4 recertification credits) of credit by the AACE International Certification Board toward meeting the continuing education requirements for recertification as a Certified Cost Engineer/Certified Cost Consultant.

AIA Continuing Education We are registered with the AIA Continuing Education System (AIA/CES) and are committed to developing quality learning activities in accordance with the CES criteria. Many seminars meet the AIA/CES criteria for Quality Level 2. AIA members may receive (14) learning units (LUs) for each two-day RSMeans course.

Daily course schedule The first day of each seminar session begins at 8:30 a.m. and ends at 4:30 p.m. The second day begins at 8:00 a.m. and ends at 4:00 p.m. Participants are urged to bring a hand-held calculator, since many actual problems will be worked out in each session.

Continental breakfast Your registration includes the cost of a continental breakfast, and a morning and afternoon refreshment break. These informal segments allow you to discuss topics of mutual interest with other seminar attendees. (You are free to make your own lunch and dinner arrangements.)

Hotel/transportation arrangements RSMeans arranges to hold a block of rooms at most host hotels. To take advantage of special group rates when making your reservation, be sure to mention that you are attending the RSMeans seminar. You are, of course, free to stay at the lodging place of your choice. **(Hotel reservations and transportation arrangements should be made directly by seminar attendees.)**

Important Class sizes are limited, so please register as soon as possible.

Note: *Pricing subject to change.*

Registration Form

ADDS-1000

Call 1-800-334-3509, Press 1 to register or FAX this form 1-800-632-6732. Visit our website: www.rsmeans.com

Please register the following people for the RSMeans construction seminars as shown here. We understand that we must make our own hotel reservations if overnight stays are necessary.

☐ Full payment of $____________________ enclosed.

☐ Bill me.

Please print name of registrant(s).

(To appear on certificate of completion)

____

____

____

____

P.O. #: __________________________

GOVERNMENT AGENCIES MUST SUPPLY PURCHASE ORDER NUMBER OR TRAINING FORM.

Firm name______________________________

Address______________________________

City/State/Zip______________________________

Telephone no.____________________ Fax no.____________________

E-mail address______________________________

Charge registration(s) to: ☐ MasterCard ☐ VISA ☐ American Express

Account no.____________________ Exp. date____________________

Cardholder's signature______________________________

Seminar name______________________________

Seminar City ______________________________

Please mail check to: RSMeans Seminars, 700 Longwater Drive, Norwell, MA 02061 USA

Professional Development

For more information visit the RSMeans website at **www.rsmeans.com**

RSMeans Online™ Training

Construction estimating is vital to the decision-making process at each state of every project. Means CostWorks.com works the way you do. It's systematic, flexible and intuitive. In this one day class you will see how you can estimate any phase of any project faster and better.

Some of what you'll learn:
- Customizing Means CostWorks.com
- Making the most of RSMeans "Circle Reference" numbers
- How to integrate your cost data
- Generate reports, exporting estimates to MS Excel, sharing, collaborating and more

Also available: RSMeans Online™ training webinar

Maintenance & Repair Estimating for Facilities

This *two-day* course teaches attendees how to plan, budget, and estimate the cost of ongoing and preventive maintenance and repair for existing buildings and grounds.

Some of what you'll learn:
- The most financially favorable maintenance, repair, and replacement scheduling and estimating
- Auditing and value engineering facilities
- Preventive planning and facilities upgrading
- Determining both in-house and contract-out service costs
- Annual, asset-protecting M&R plan

Who should attend: facility managers, maintenance supervisors, buildings and grounds superintendents, plant managers, planners, estimators, and others involved in facilities planning and budgeting.

Facilities Construction Estimating

In this *two-day* course, professionals working in facilities management can get help with their daily challenges to establish budgets for all phases of a project.

Some of what you'll learn:
- Determining the full scope of a project
- Identifying the scope of risks and opportunities
- Creative solutions to estimating issues
- Organizing estimates for presentation and discussion
- Special techniques for repair/remodel and maintenance projects
- Negotiating project change orders

Who should attend: facility managers, engineers, contractors, facility tradespeople, planners, and project managers.

Practical Project Management for Construction Professionals

In this *two-day* course, acquire the essential knowledge and develop the skills to effectively and efficiently execute the day-to-day responsibilities of the construction project manager.

Covers:
- General conditions of the construction contract
- Contract modifications: change orders and construction change directives
- Negotiations with subcontractors and vendors
- Effective writing: notification and communications
- Dispute resolution: claims and liens

Who should attend: architects, engineers, owner's representatives, project managers.

Scheduling with Microsoft Project

Two days of hands-on training gives you a look at the basics to putting it all together with MS Project to create a complete project schedule. Learn to work better and faster, manage changes and updates, enhance tracking, generate actionable reports, and boost control of your time and budget.

Some of what you'll learn:
- Defining task/activity relationships: link tasks, establish predecessor/successor relationships and create logs
- Baseline scheduling: the essential what/when
- Using MS Project to determine the critical path
- Integrate RSMeans data with MS Project
- The dollar-loaded schedule: enhance the reliability of this "must-know" technique for scheduling public projects.

Who should attend: contractors' project-management teams, architects and engineers, project owners and their representatives, and those interested in improving their project planning and scheduling and management skills.

Mechanical & Electrical Estimating

This *two-day* course teaches attendees how to prepare more accurate and complete mechanical/electrical estimates, avoiding the pitfalls of omission and double-counting, while understanding the composition and rationale within the RSMeans mechanical/electrical database.

Some of what you'll learn:
- The unique way mechanical and electrical systems are interrelated
- M&E estimates—conceptual, planning, budgeting, and bidding stages
- Order of magnitude, square foot, assemblies, and unit price estimating
- Comparative cost analysis of equipment and design alternatives

Who should attend: architects, engineers, facilities managers, mechanical and electrical contractors, and others who need a highly reliable method for developing, understanding, and evaluating mechanical and electrical contracts.

Professional Development

For more information visit the RSMeans website at **www.rsmeans.com**

Unit Price Estimating

This interactive *two-day* seminar teaches attendees how to interpret project information and process it into final, detailed estimates with the greatest accuracy level.

The most important credential an estimator can take to the job is the ability to visualize construction and estimate accurately.

Some of what you'll learn:

- Interpreting the design in terms of cost
- The most detailed, time-tested methodology for accurate pricing
- Key cost drivers—material, labor, equipment, staging, and subcontracts
- Understanding direct and indirect costs for accurate job cost accounting and change order management

Who should attend: corporate and government estimators and purchasers, architects, engineers, and others who need to produce accurate project estimates.

Conceptual Estimating Using RSMeans CostWorks® CD

This *two day* class uses the leading industry data and a powerful software package to develop highly accurate conceptual estimates for your construction projects. All attendees must bring a laptop computer loaded with the current year *Square Foot Model Costs* and the *Assemblies Cost Data* CostWorks titles.

Some of what you'll learn:

- Introduction to conceptual estimating
- Types of conceptual estimates
- Helpful hints
- Order of magnitude estimating
- Square foot estimating
- Assemblies estimating

Who should attend: architects, engineers, contractors, construction estimators, owner's representatives, and anyone looking for an electronic method for performing square foot estimating.

RSMeans CostWorks® CD Training

This one-day course helps users become more familiar with the functionality of *RSMeans CostWork*s program. Each menu, icon, screen, and function found in the program is explained in depth. Time is devoted to hands-on estimating exercises.

Some of what you'll learn:

- Searching the database using all navigation methods
- Exporting RSMeans data to your preferred spreadsheet format
- Viewing crews, assembly components, and much more
- Automatically regionalizing the database

This training session requires you to bring a laptop computer to class.

When you register for this course you will receive an outline for your laptop requirements.

Also offering web training for CostWorks CD!

Assessing Scope of Work for Facility Construction Estimating

This *two-day* practical training program addresses the vital importance of understanding the SCOPE of projects in order to produce accurate cost estimates for facility repair and remodeling.

Some of what you'll learn:

- Discussions of site visits, plans/specs, record drawings of facilities, and site-specific lists
- Review of CSI divisions, including means, methods, materials, and the challenges of scoping each topic
- Exercises in SCOPE identification and SCOPE writing for accurate estimating of projects
- Hands-on exercises that require SCOPE, take-off, and pricing

Who should attend: corporate and government estimators, planners, facility managers, and others who need to produce accurate project estimates.

Facilities Est. Using RSMeans CostWorks® CD

Combines hands-on skill building with best estimating practices and real-life problems. Brings you up-to-date with key concepts, and provides tips, pointers, and guidelines to save time and avoid cost oversights and errors.

Some of what you'll learn:

- Estimating process concepts
- Customizing and adapting RSMeans cost data
- Establishing scope of work to account for all known variables
- Budget estimating: when, why, and how
- Site visits: what to look for—what you can't afford to overlook
- How to estimate repair and remodeling variables

This training session requires you to bring a laptop computer to class.

Who should attend: facility managers, architects, engineers, contractors, facility tradespeople, planners, project managers and anyone involved with JOC, SABRE, or IDIQ.

Unit Price Est. Using RSMeans CostWorks® CD

Step-by-step instruction and practice problems to identify and track key cost drivers—material, labor, equipment, staging, and subcontractors—for each specific task. Learn the most detailed, time-tested methodology for accurately "pricing" these variables, their impact on each other and on total cost.

Some of what you'll learn:

- Unit price cost estimating
- Order of magnitude, square foot, and assemblies estimating
- Quantity takeoff
- Direct and indirect construction costs
- Development of contractor's bill rates
- How to use *RSMeans Building Construction Cost Data*

This training session requires you to bring a laptop computer to class.

Who should attend: architects, engineers, corporate and government estimators, facility managers, and government procurement staff.

New Titles - Reference Books

For more information visit the RSMeans website at **www.rsmeans.com**

Complete Book of Framing, 2nd Edition

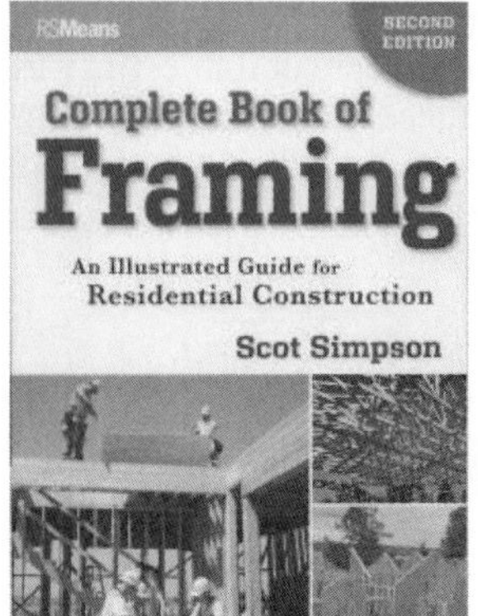

by Scot Simpson

This updated, easy-to-learn guide to rough carpentry and framing is written by an expert with more than thirty years of framing experience. Starting with the basics, this book begins with types of lumber, nails, and what tools are needed, followed by detailed, fully illustrated steps for framing each building element. Framer-Friendly Tips throughout the book show how to get a task done right—and more easily.

$29.95 per copy | 352 pages, softcover | Catalog no. 67353A

How to Estimate with Means Data & CostWorks, 4th Edition

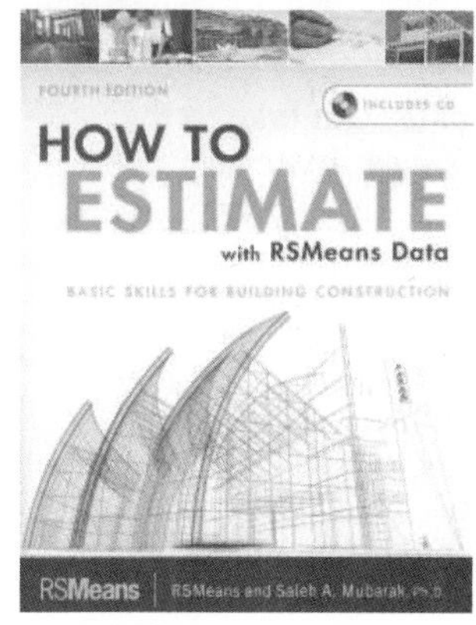

by RSMeans and Saleh A. Mubarak, Ph.D.

This step-by-step guide takes you through all the major construction items with extensive coverage of site work, concrete and masonry, wood and metal framing, doors and windows, and other divisions. The only construction cost estimating handbook that uses the most popular source of construction data, RSMeans, this indispensible guide features access to the instructional version of CostWorks in electronic form, enabling you to practice techniques to solve real-world estimating problems.

$70.00 per copy | 292 pages, softcover | Includes CostWorks CD Catalog no. 67324C

Estimating Building Costs, 2nd Edition

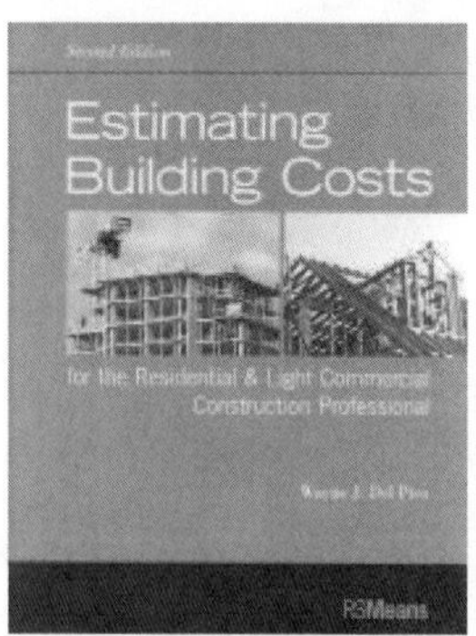

For the Residential & Light Commercial Construction Professional

by Wayne J. DelPico

This book guides readers through the entire estimating process, explaining in detail how to put together a reliable estimate that can be used not only for budgeting, but also for developing a schedule, managing a project, dealing with contingencies, and ultimately making a profit.

Completely revised and updated to reflect the new CSI MasterFormat 2010 system, this practical guide describes estimating techniques for each building system and how to apply them according to the latest industry standards.

$65.00 per copy | 398 pages, softcover | Catalog no. 67343A

How Your House Works, 2nd Edition

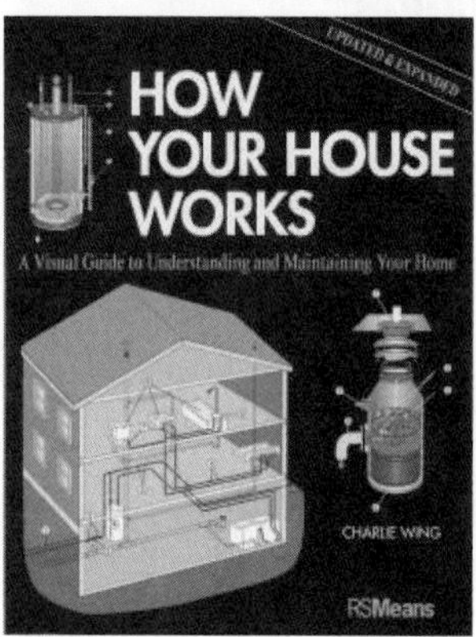

by Charlie Wing

Knowledge of your home's systems helps you control repair and construction costs and makes sure the correct elements are being installed or replaced. This book uncovers the mysteries behind just about every major appliance and building element in your house. See-through, cross-section drawings in full color show you exactly how these things should be put together and how they function, including what to check if they don't work. It just might save you having to call in a professional.

$22.95 per copy | 192 pages, softcover | Catalog no. 67351A

Risk Management for Design and Construction

This book introduces risk as a central pillar of project management and shows how a project manager can be prepared for dealing with uncertainty. Written by experts in the field, *Risk Management for Design and Construction* uses clear, straightforward terminology to demystify the concepts of project uncertainty and risk.

Highlights include:

- Integrated cost and schedule risk analysis
- An introduction to a ready-to-use system of analyzing a project's risks and tools to proactively manage risks
- A methodology that was developed and used by the Washington State Department of Transportation
- Case studies and examples on the proper application of principles
- Information about combining value analysis with risk analysis

$125.00 per copy | Over 250 pages, softcover | Catalog no. 67359

RSMeans Cost Data, Student Edition

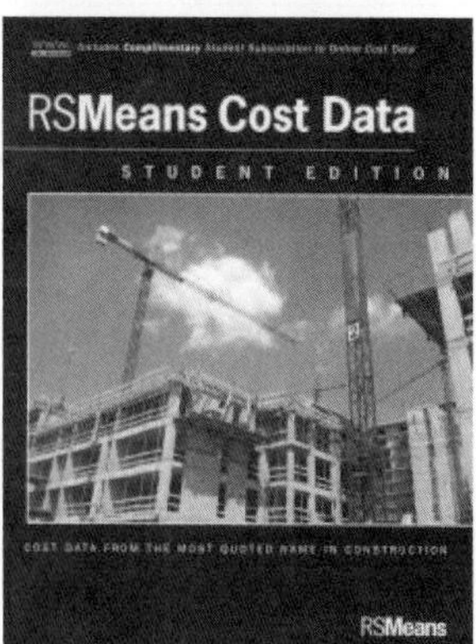

This book provides a thorough introduction to cost estimating in a self-contained print and online package. With clear explanations and a hands-on, example-driven approach, it is the ideal reference for students and new professionals.

Features include:

- Commercial and residential construction cost data in print and online formats
- Complete how-to guidance on the essentials of cost estimating
- A supplemental website with plans, problem sets, and a full sample estimate

$99.00 per copy | 512 pages, softcover | Catalog no. 67363

Value Engineering: Practical Applications

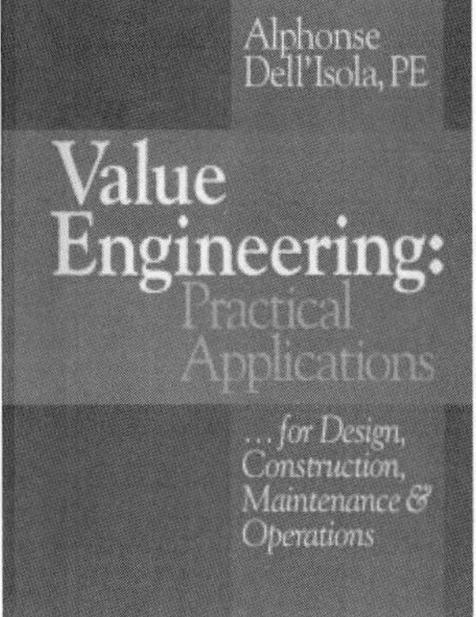

For Design, Construction, Maintenance & Operations
by Alphonse Dell'Isola, PE

A tool for immediate application—for engineers, architects, facility managers, owners, and contractors. Includes making the case for VE—the management briefing; integrating VE into planning, budgeting, and design; conducting life cycle costing; using VE methodology in design review and consultant selection; case studies; VE workbook; and a life cycle costing program on disk.

$79.95 per copy | Over 450 pages, illustrated, softcover | Catalog no. 67319A

Facilities Operations & Engineering Reference

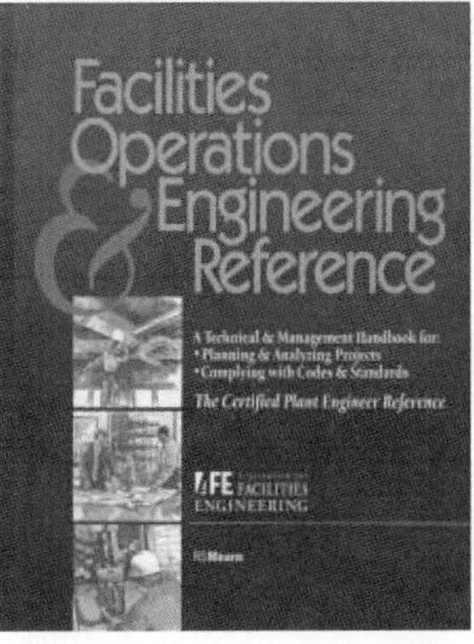

by the Association for Facilities Engineering and RSMeans

An all-in-one technical reference for planning and managing facility projects and solving day-to-day operations problems. Selected as the official certified plant engineer reference, this handbook covers financial analysis, maintenance, HVAC and energy efficiency, and more.

$109.95 per copy | Over 700 pages, illustrated, hardcover | Catalog no. 67318

The Building Professional's Guide to Contract Documents

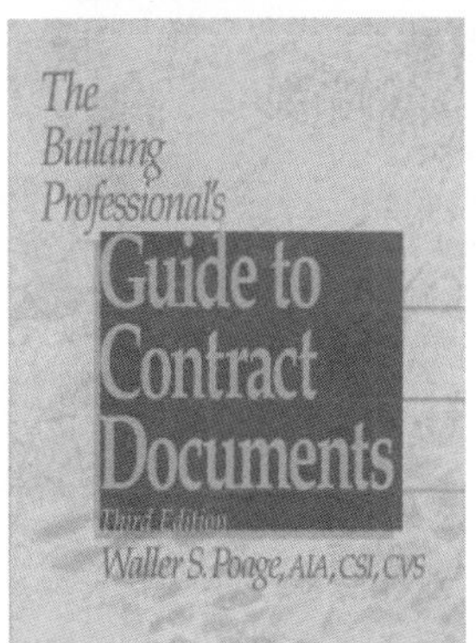

3rd Edition
by Waller S. Poage, AIA, CSI, CVS

A comprehensive reference for owners, design professionals, contractors, and students

- Structure your documents for maximum efficiency.
- Effectively communicate construction requirements.
- Understand the roles and responsibilities of construction professionals.
- Improve methods of project delivery.

$70.00 per copy | 400 pages Diagrams and construction forms, hardcover Catalog no. 67261A

Green Home Improvement

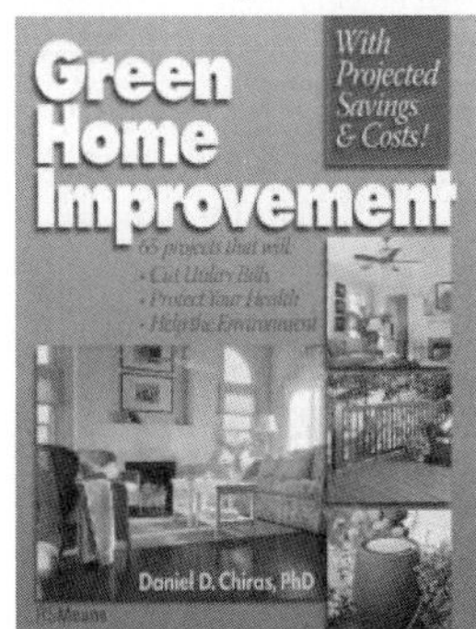

by Daniel D. Chiras, PhD

With energy costs rising and environmental awareness increasing, people are looking to make their homes greener. This book, with 65 projects and actual costs and projected savings help homeowners prioritize their green improvements.

Projects range from simple water savers that cost only a few dollars, to bigger-ticket items such as HVAC systems. With color photos and cost estimates, each project compares options and describes the work involved, the benefits, and the savings.

$34.95 | 320 pages, illustrated, softcover | Catalog no. 67355

Cost Planning & Estimating for Facilities Maintenance

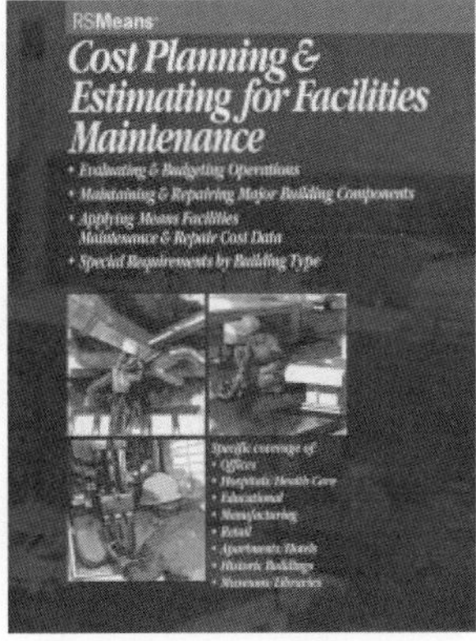

In this unique book, a team of facilities management authorities shares their expertise on:

- Evaluating and budgeting maintenance operations
- Maintaining and repairing key building components
- Applying *RSMeans Facilities Maintenance & Repair Cost Data* to your estimating

Covers special maintenance requirements of the ten major building types

$89.95 per copy | Over 475 pages, hardcover | Catalog no. 67314

Life Cycle Costing for Facilities

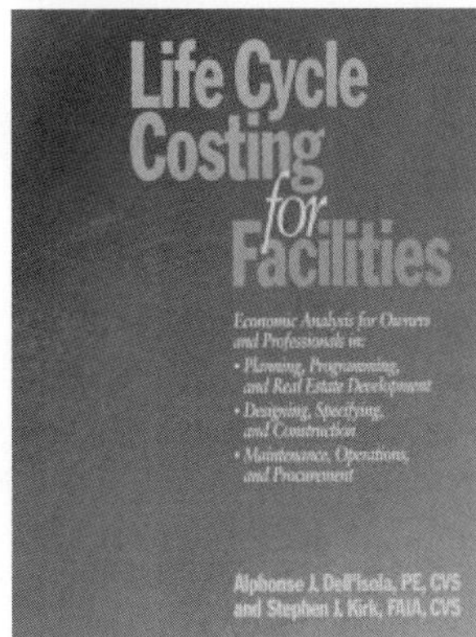

by Alphonse Dell'Isola and Dr. Steven Kirk

Guidance for achieving higher quality design and construction projects at lower costs! Cost-cutting efforts often sacrifice quality to yield the cheapest product. Life cycle costing enables building designers and owners to achieve both. The authors of this book show how LCC can work for a variety of projects — from roads to HVAC upgrades to different types of buildings.

$99.95 per copy | 396 pages, hardcover | Catalog no. 67341

Interior Home Improvement Costs

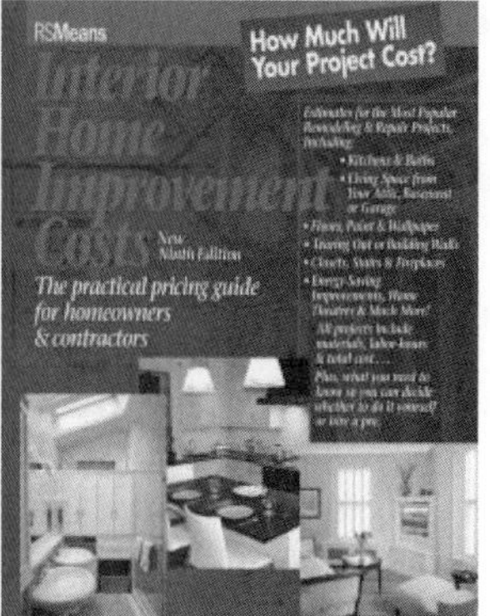

9th Edition

Updated estimates for the most popular remodeling and repair projects—from small, do-it-yourself jobs to major renovations and new construction. Includes: kitchens & baths; new living space from your attic, basement, or garage; new floors, paint, and wallpaper; tearing out or building new walls; closets, stairs, and fireplaces; new energy-saving improvements, home theaters, and more!

$24.95 per copy | 250 pages, illustrated, softcover | Catalog no. 67308E

Exterior Home Improvement Costs

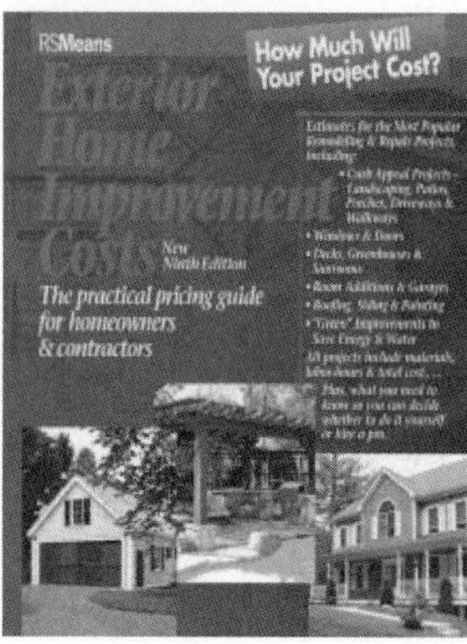

9th Edition

Updated estimates for the most popular remodeling and repair projects—from small, do-it-yourself jobs, to major renovations and new construction. Includes: curb appeal projects—landscaping, patios, porches, driveways, and walkways; new windows and doors; decks, greenhouses, and sunrooms; room additions and garages; roofing, siding, and painting; "green" improvements to save energy & water.

$24.95 per copy | Over 275 pages, illustrated, softcover | Catalog no. 67309E

Builder's Essentials: Plan Reading & Material Takeoff

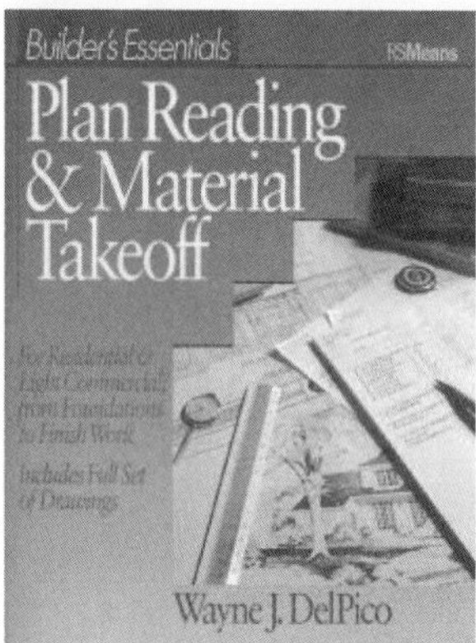

For Residential and Light Commercial Construction

by Wayne J. DelPico

A valuable tool for understanding plans and specs, and accurately calculating material quantities. Step-by-step instructions and takeoff procedures based on a full set of working drawings.

$35.95 per copy | Over 420 pages, softcover | Catalog no. 67307

Means Unit Price Estimating Methods

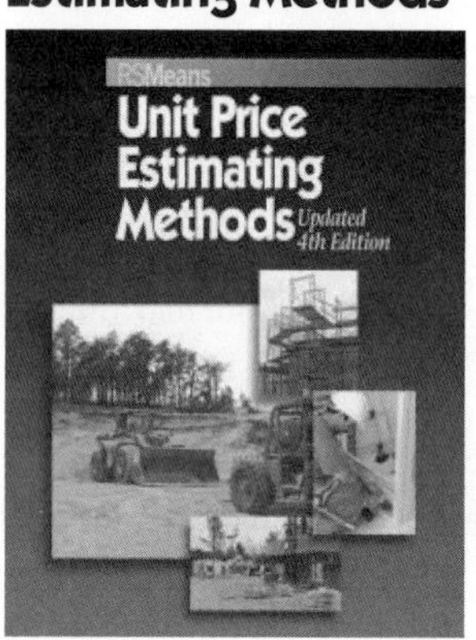

New 4th Edition

This new edition includes up-to-date cost data and estimating examples, updated to reflect changes to the CSI numbering system and new features of RSMeans cost data. It describes the most productive, universally accepted ways to estimate, and uses checklists and forms to illustrate shortcuts and timesavers. A model estimate demonstrates procedures. A new chapter explores computer estimating alternatives.

$65.00 per copy | Over 350 pages, illustrated, softcover | Catalog no. 67303B

Concrete Repair and Maintenance Illustrated

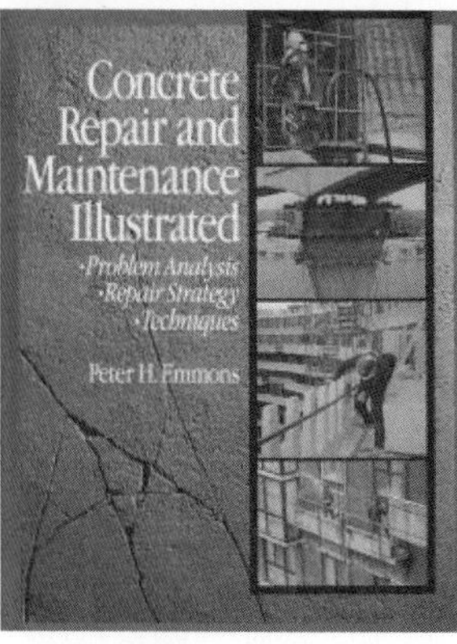

by Peter Emmons

Hundreds of illustrations show users how to analyze, repair, clean, and maintain concrete structures for optimal performance and cost effectiveness. From parking garages to roads and bridges to structural concrete, this comprehensive book describes the causes, effects, and remedies for concrete wear and failure. Invaluable for planning jobs, selecting materials, and training employees, this book is a must-have for concrete specialists, general contractors, facility managers, civil and structural engineers, and architects.

$69.95 per copy | 300 pages, illustrated, softcover | Catalog no. 67146

Total Productive Facilities Management

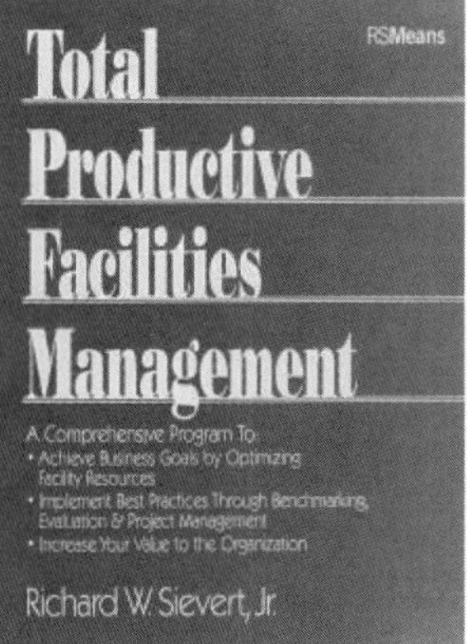

by Richard W. Sievert, Jr.

Today, facilities are viewed as strategic resources. . . elevating the facility manager to the role of asset manager supporting the organization's overall business goals. Now, Richard Sievert Jr., in this well-articulated guidebook, sets forth a new operational standard for the facility manager's emerging role. . . a comprehensive program for managing facilities as a true profit center.

$79.95 per copy | 275 pages, softcover | Catalog no. 67321

Reference Books

For more information visit the RSMeans website at **www.rsmeans.com**

Unit prices according to the latest MasterFormat!

Means Illustrated Construction Dictionary

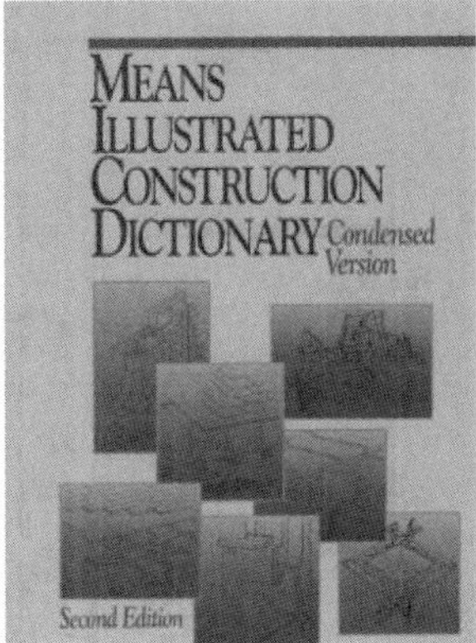

Condensed, 2nd Edition

The best portable dictionary for office or field use—an essential tool for contractors, architects, insurance and real estate personnel, facility managers, homeowners, and anyone who needs quick, clear definitions for construction terms. The second edition has been further enhanced with updates and hundreds of new terms and illustrations . . . in keeping with the most recent developments in the construction industry.

Now with a quick-reference Spanish section. Includes tools and equipment, materials, tasks, and more!

$59.95 per copy | Over 500 pages, softcover | Catalog no. 67282A

Square Foot & UNIFORMAT Assemblies Estimating Methods

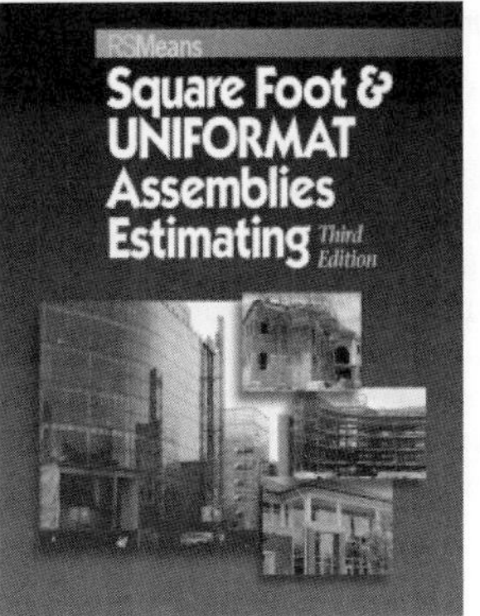

3rd Edition

Develop realistic square foot and assemblies costs for budgeting and construction funding. The new edition features updated guidance on square foot and assemblies estimating using UNIFORMAT II. An essential reference for anyone who performs conceptual estimates.

$69.95 per copy | Over 300 pages, illustrated, softcover | Catalog no. 67145B

Facilities Planning & Relocation

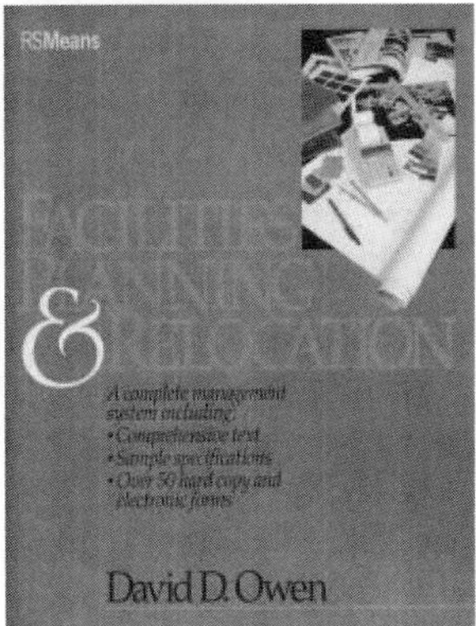

by David D. Owen

A complete system for planning space needs and managing relocations. Includes step-by-step manual, over 50 forms, and extensive reference section on materials and furnishings.

New lower price and user-friendly format.

$89.95 per copy | Over 450 pages, softcover | Catalog no. 67301

Mechanical Estimating Methods

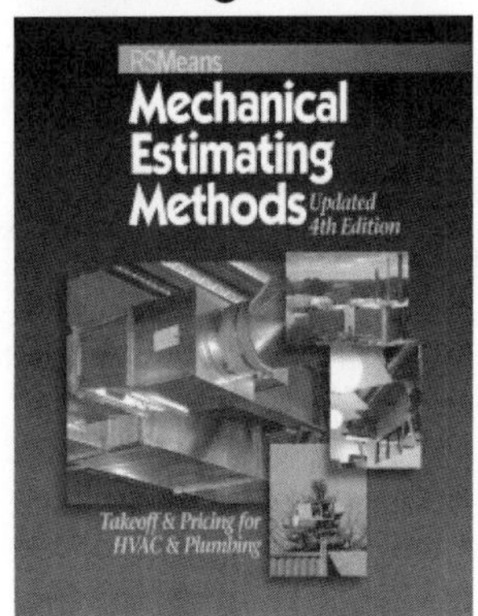

4th Edition

Completely updated, this guide assists you in making a review of plans, specs, and bid packages, with suggestions for takeoff procedures, listings, substitutions, and pre-bid scheduling for all components of HVAC. Includes suggestions for budgeting labor and equipment usage. Compares materials and construction methods to allow you to select the best option.

$64.95 per copy | Over 350 pages, illustrated, softcover | Catalog no. 67294B

Electrical Estimating Methods

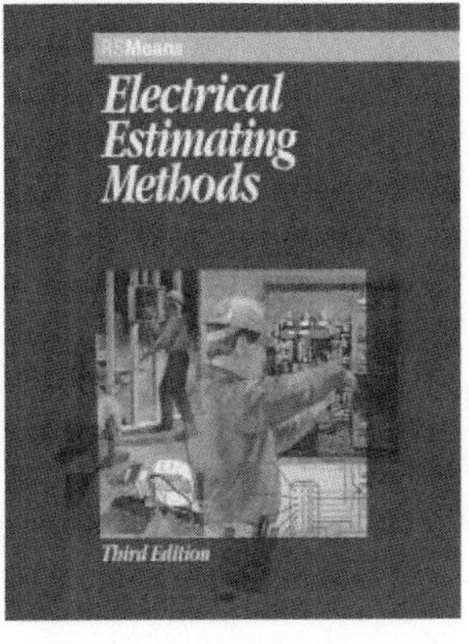

3rd Edition

Expanded edition includes sample estimates and cost information in keeping with the latest version of the CSI MasterFormat and UNIFORMAT II. Complete coverage of fiber optic and uninterruptible power supply electrical systems, broken down by components, and explained in detail. Includes a new chapter on computerized estimating methods. A practical companion to *RSMeans Electrical Cost Data.*

$64.95 per copy | Over 325 pages, hardcover | Catalog no. 67230B

Residential & Light Commercial Construction Standards

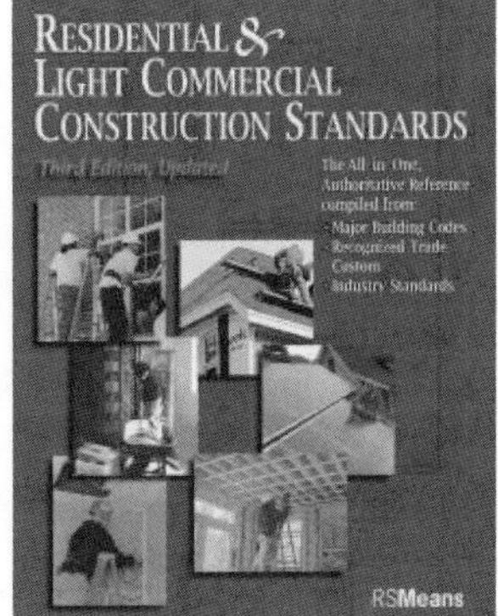

3rd Edition, Updated by RSMeans and contributing authors

This book provides authoritative requirements and recommendations compiled from leading professional associations, industry publications, and building code organizations. This all-in-one reference helps establish a standard for workmanship, quickly resolve disputes, and avoid defect claims. Updated third edition includes new coverage of green building, seismic, hurricane, and mold-resistant construction.

$59.95 | Over 550 pages, illustrated, softcover | Catalog no. 67322B

Reference Books

For more information visit the RSMeans website at **www.rsmeans.com** Unit prices according to the latest MasterFormat!

Construction Business Management

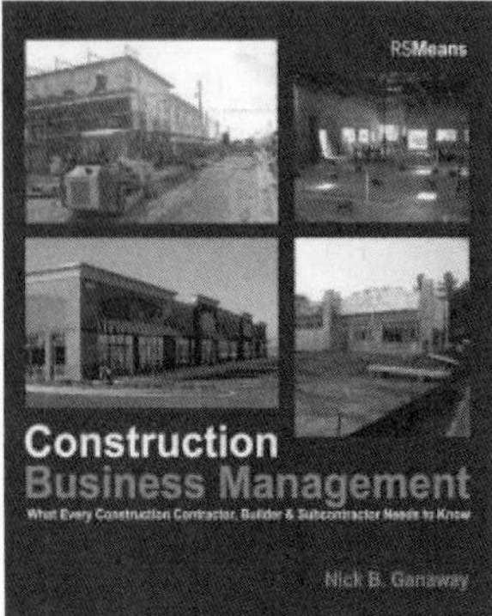

by Nick Ganaway

Only 43% of construction firms stay in business after four years. Make sure your company thrives with valuable guidance from a pro with 25 years of success as a commercial contractor. Find out what it takes to build all aspects of a business that is profitable, enjoyable, and enduring. With a bonus chapter on retail construction.

$49.95 per copy | 200 pages, softcover | Catalog no. 67352

Project Scheduling & Management for Construction

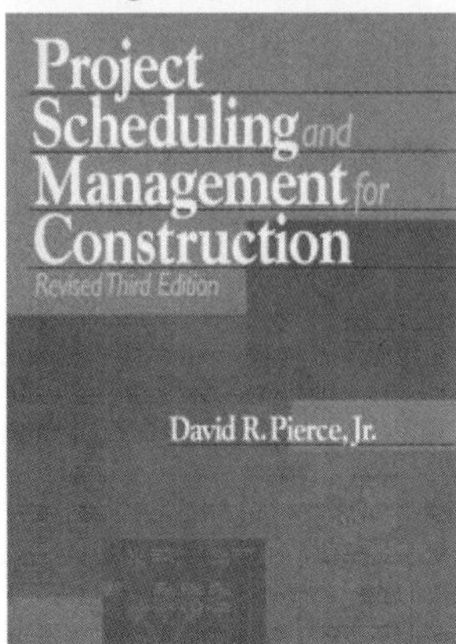

3rd Edition
by David R. Pierce, Jr.

A comprehensive yet easy-to-follow guide to construction project scheduling and control—from vital project management principles through the latest scheduling, tracking, and controlling techniques. The author is a leading authority on scheduling, with years of field and teaching experience at leading academic institutions. Spend a few hours with this book and come away with a solid understanding of this essential management topic.

$64.95 per copy | Over 300 pages, illustrated, hardcover | Catalog no. 67247B

The Practice of Cost Segregation Analysis

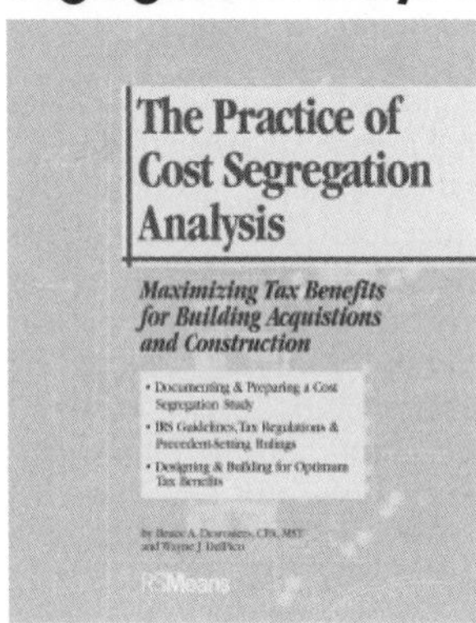

by Bruce A. Desrosiers and Wayne J. DelPico

This expert guide walks you through the practice of cost segregation analysis, which enables property owners to defer taxes and benefit from "accelerated cost recovery" through depreciation deductions on assets that are properly identified and classified.

With a glossary of terms, sample cost segregation estimates for various building types, key information resources, and updates via a dedicated website, this book is a critical resource for anyone involved in cost segregation analysis.

$99.95 per copy | Over 225 pages | Catalog no. 67345

Preventive Maintenance for Multi-Family Housing

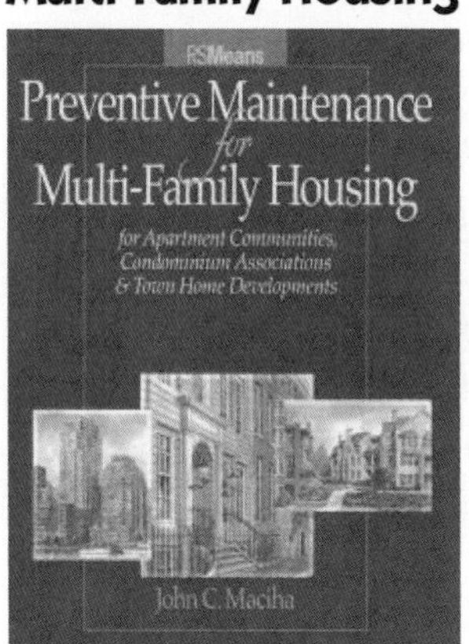

by John C. Maciha

Prepared by one of the nation's leading experts on multi-family housing.

This complete PM system for apartment and condominium communities features expert guidance, checklists for buildings and grounds maintenance tasks and their frequencies, a reusable wall chart to track maintenance, and a dedicated website featuring customizable electronic forms. A must-have for anyone involved with multi-family housing maintenance and upkeep.

$89.95 per copy | 225 pages | Catalog no. 67346

Green Building: Project Planning & Cost Estimating

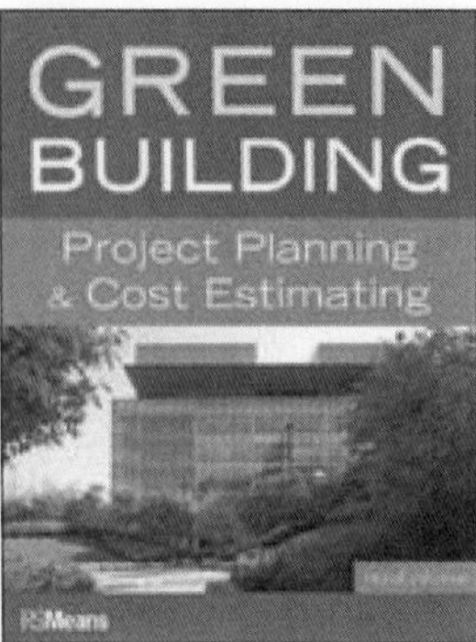

3rd Edition

Since the widely read first edition of this book, green building has gone from a growing trend to a major force in design and construction.

This new edition has been updated with the latest in green building technologies, design concepts, standards, and costs. Full-color with all new case studies—plus a new chapter on commercial real estate.

$99.95 per copy | Over 450 pages, softcover | Catalog no. 67338B

Job Order Contracting Expediting Construction Project Delivery

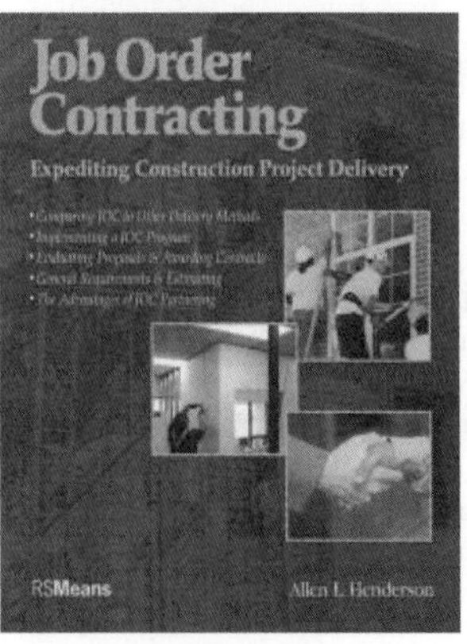

by Allen Henderson

Expert guidance to help you implement JOC—fast becoming the preferred project delivery method for repair and renovation, minor new construction, and maintenance projects in the public sector and in many states and municipalities. The author, a leading JOC expert and practitioner, shows how to:

- Establish a JOC program
- Evaluate proposals and award contracts
- Handle general requirements and estimating
- Partner for maximum benefits

$89.95 per copy | 192 pages, illustrated, hardcover | Catalog no. 67348

Construction Supervision

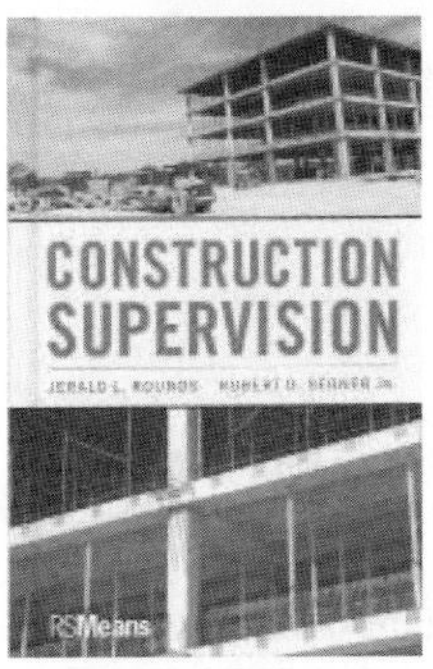

This brand new title inspires supervisory excellence with proven tactics and techniques applied by thousands of construction supervisors over the past decade. Recognizing the unique and critical role the supervisor plays in project success, the book's leadership guidelines carve out a practical blueprint for motivating work performance and increasing productivity through effective communication.

Features:

- A unique focus on field supervision and crew management
- Coverage of supervision from the foreman to the superintendent level
- An overview of technical skills whose mastery will build confidence and success for the supervisor
- A detailed view of "soft" management and communication skills

$90.00 per copy | Over 450 pages, softcover | Catalog no. 67358

Building & Renovating Schools

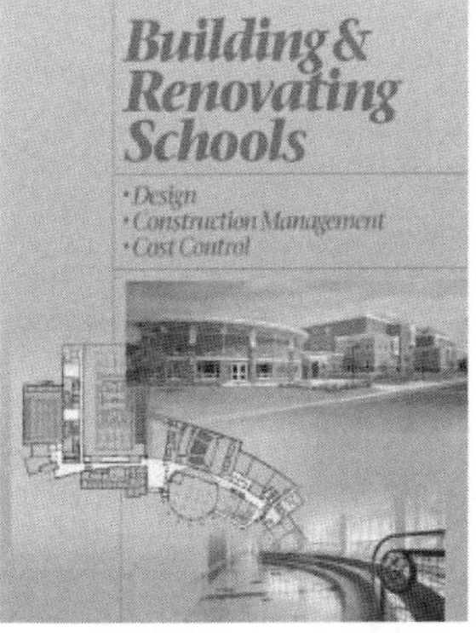

This all-inclusive guide covers every step of the school construction process—from initial planning, needs assessment, and design, right through moving into the new facility. A must-have resource for anyone concerned with new school construction or renovation. With square foot cost models for elementary, middle, and high school facilities, and real-life case studies of recently completed school projects.

The contributors to this book—architects, construction project managers, contractors, and estimators who specialize in school construction—provide start-to-finish, expert guidance on the process.

$99.95 per copy | Over 425 pages, hardcover | Catalog no. 67342

The Homeowner's Guide to Mold

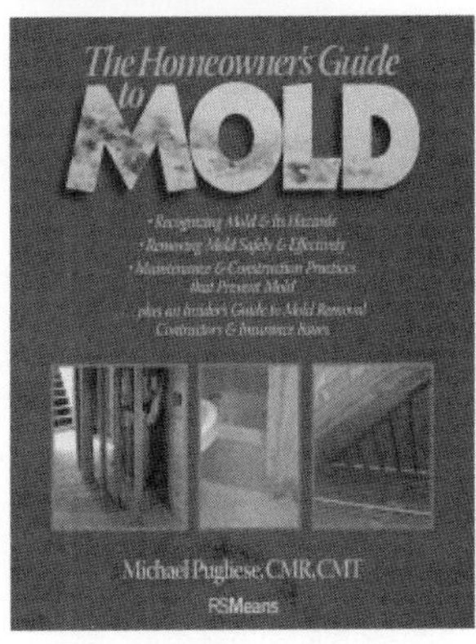

By Michael Pugliese

Expert guidance to protect your health and your home.

Mold, whether caused by leaks, humidity or flooding, is a real health and financial issue—for homeowners and contractors.

This full-color book explains:

- Construction and maintenance practices to prevent mold
- How to inspect for and remove mold
- Mold remediation procedures and costs
- What to do after a flood
- How to deal with insurance companies

$21.95 per copy | 144 pages, softcover | Catalog no. 67344

Universal Design Ideas for Style, Comfort & Safety

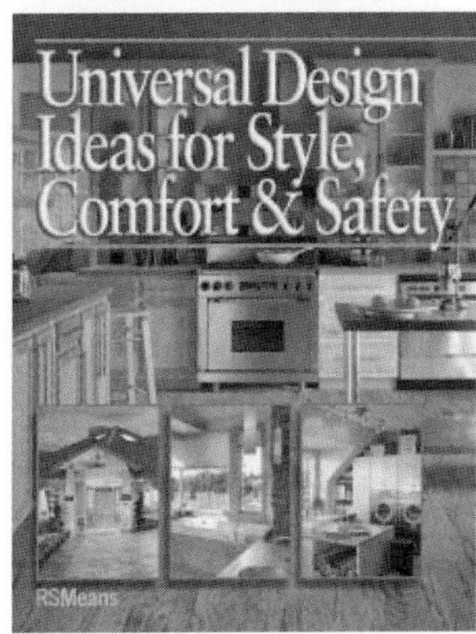

by RSMeans and Lexicon Consulting, Inc.

Incorporating universal design when building or remodeling helps people of any age and physical ability more fully and safely enjoy their living spaces. This book shows how universal design can be artfully blended into the most attractive homes. It discusses specialized products like adjustable countertops and chair lifts, as well as simple ways to enhance a home's safety and comfort. With color photos and expert guidance, every area of the home is covered. Includes budget estimates that give an idea how much projects will cost.

$21.95 | 160 pages, illustrated, softcover | Catalog no. 67354

Landscape Estimating Methods

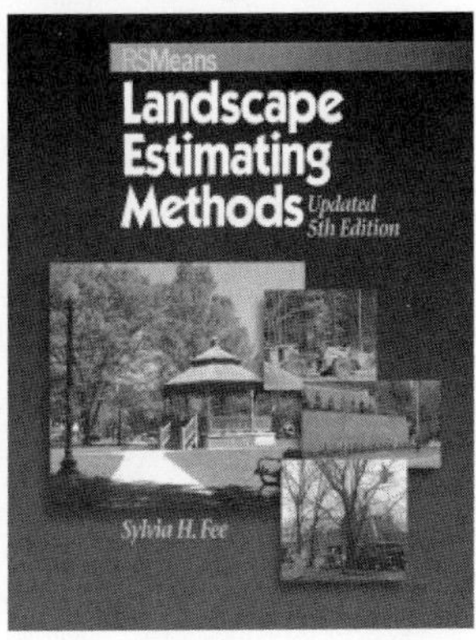

5th Edition

Answers questions about preparing competitive landscape construction estimates, with up-to-date cost estimates and the new MasterFormat classification system. Expanded and revised to address the latest materials and methods, including new coverage on approaches to green building. Includes:

- Step-by-step explanation of the estimating process
- Sample forms and worksheets that save time and prevent errors

$69.95 per copy | Over 350 pages, softcover | Catalog no. 67295C

Plumbing Estimating Methods

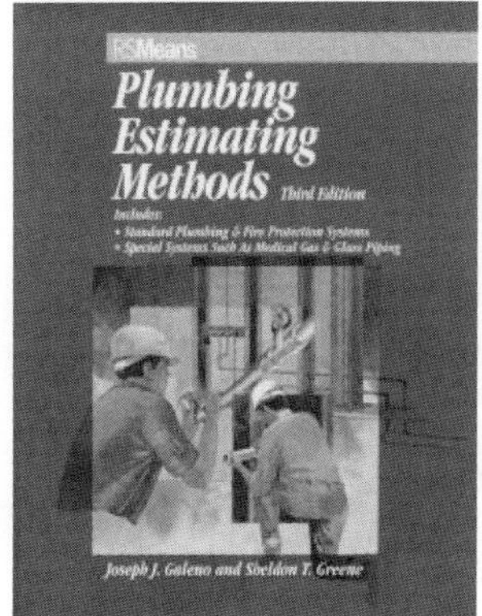

3rd Edition
by Joseph Galeno and Sheldon Greene

Updated and revised! This practical guide walks you through a plumbing estimate, from basic materials and installation methods through change order analysis. *Plumbing Estimating Methods* covers residential, commercial, industrial, and medical systems, and features sample takeoff and estimate forms and detailed illustrations of systems and components.

$59.95 per copy | 330+ pages, softcover | Catalog no. 67283B

Understanding & Negotiating Construction Contracts

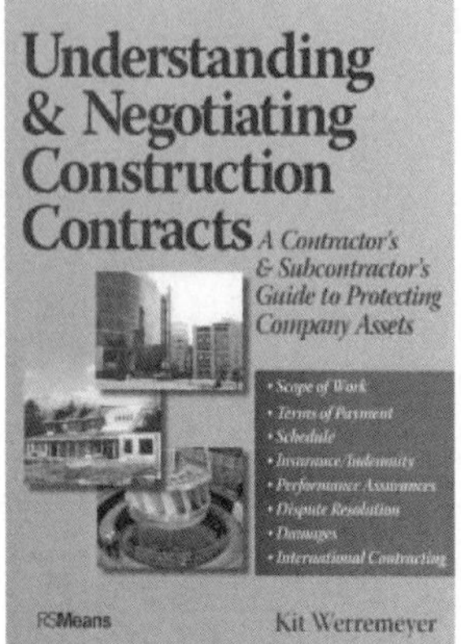

by Kit Werremeyer

Take advantage of the author's 30 years' experience in small-to-large (including international) construction projects. Learn how to identify, understand, and evaluate high risk terms and conditions typically found in all construction contracts—then negotiate to lower or eliminate the risk, improve terms of payment, and reduce exposure to claims and disputes. The author avoids "legalese" and gives real-life examples from actual projects.

$74.95 per copy | 300 pages, softcover | Catalog no. 67350

Means Illustrated Construction Dictionary

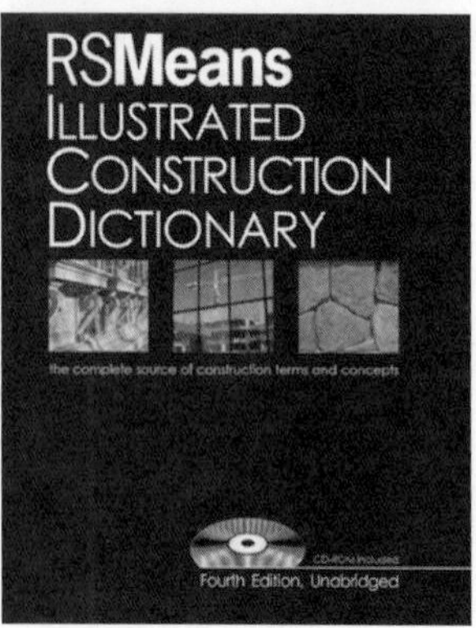

Unabridged 4th Edition, with CD-ROM

Long regarded as the industry's finest, *Means Illustrated Construction Dictionary* is now even better. With nearly 20,000 terms and more than 1,400 illustrations and photos, it is the clear choice for the most comprehensive and current information. The companion CD-ROM that comes with this new edition adds extra features such as: larger graphics and expanded definitions.

$99.95 per copy | Over 790 pages, illust., hardcover | Catalog no. 67292B

RSMeans Estimating Handbook

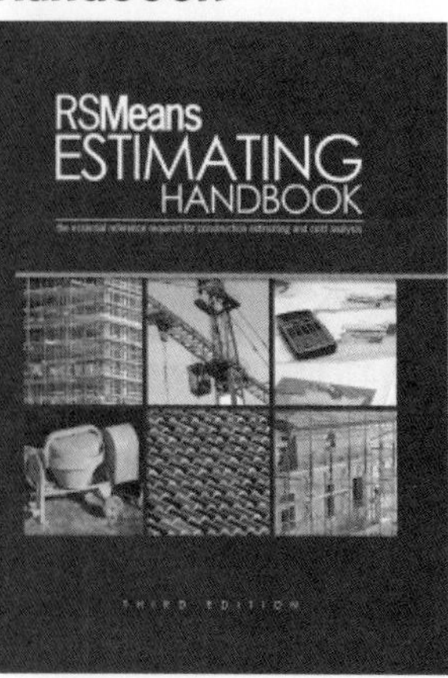

3rd Edition

Widely used in the industry for tasks ranging from routine estimates to special cost analysis projects, this handbook has been completely updated and reorganized with new and expanded technical information.

RSMeans Estimating Handbook will help construction professionals:

- Evaluate architectural plans and specifications
- Prepare accurate quantity takeoffs
- Compare design alternatives and costs
- Perform value engineering
- Double-check estimates and quotes
- Estimate change orders

$99.95 per copy | Over 900 pages, hardcover | Catalog No. 67276B

Spanish/English Construction Dictionary

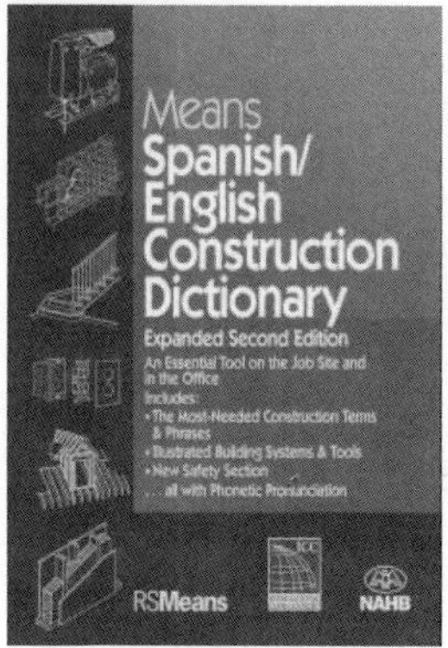

2nd Edition
by RSMeans and the International Code Council

This expanded edition features thousands of the most common words and useful phrases in the construction industry with easy-to-follow pronunciations in both Spanish and English. Over 800 new terms, phrases, and illustrations have been added. It also features a new stand-alone "Safety & Emergencies" section, with colored pages for quick access.

Unique to this dictionary are the systems illustrations showing the relationship of components in the most common building systems for all major trades.

$23.95 per copy | Over 400 pages | Catalog no. 67327A

The Gypsum Construction Handbook

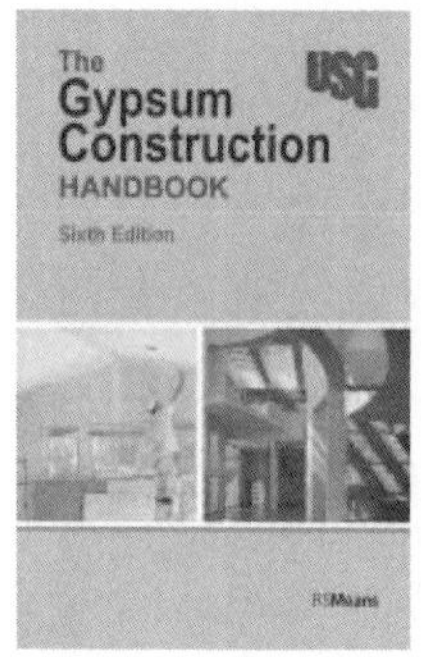

by USG

An invaluable reference of construction procedures for gypsum drywall, cement board, veneer plaster, and conventional plaster. This new edition includes the newest product developments, installation methods, fire- and sound-rated construction information, illustrated framing-to-finish application instructions, estimating and planning information, and more. Great for architects and engineers; contractors, builders, and dealers; apprentices and training programs; building inspectors and code officials; and anyone interested in gypsum construction. Features information on tools and safety practices, a glossary of construction terms, and a list of important agencies and associations. Also available in Spanish.

$29.95 | Over 575 pages, illustrated, softcover | Catalog no. 67357A

2013 Order Form

ORDER TOLL FREE 1-800-334-3509
OR FAX 1-800-632-6732

For more information visit the RSMeans website at **www.rsmeans.com**

Qty.	Book no.	COST ESTIMATING BOOKS	Unit Price	Total
	60063	Assemblies Cost Data 2013	$305.95	
	60013	Building Construction Cost Data 2013	186.95	
	60113	Concrete & Masonry Cost Data 2013	171.95	
	50143	Construction Cost Indexes 2013 (subscription)	362.00	
	60143A	Construction Cost Index–January 2013	90.50	
	60143B	Construction Cost Index–April 2013	90.50	
	60143C	Construction Cost Index–July 2013	90.50	
	60143D	Construction Cost Index–October 2013	90.50	
	60343	Contr. Pricing Guide: Resid. R & R Costs 2013	39.95	
	60233	Electrical Change Order Cost Data 2013	183.95	
	60033	Electrical Cost Data 2013	188.95	
	60203	Facilities Construction Cost Data 2013	476.95	
	60303	Facilities Maintenance & Repair Cost Data 2013	417.95	
	60553	Green Building Cost Data 2013	155.95	
	60163	Heavy Construction Cost Data 2013	186.95	
	60093	Interior Cost Data 2013	192.95	
	60123	Labor Rates for the Const. Industry 2013	415.95	
	60183	Light Commercial Cost Data 2013	139.95	
	60023	Mechanical Cost Data 2013	183.95	
	60153	Open Shop Building Const. Cost Data 2013	160.95	
	60213	Plumbing Cost Data 2013	188.95	
	60043	Commercial Renovation Cost Data 2013	152.95	
	60173	Residential Cost Data 2013	134.95	
	60283	Site Work & Landscape Cost Data 2013	181.95	
	60053	Square Foot Costs 2013	197.95	
	62012	Yardsticks for Costing (2012)	176.95	
	62013	Yardsticks for Costing (2013)	185.95	
		REFERENCE BOOKS		
	67329	Bldrs Essentials: Best Bus. Practices for Bldrs	29.95	
	67298A	Bldrs Essentials: Framing/Carpentry 2nd Ed.	24.95	
	67298AS	Bldrs Essentials: Framing/Carpentry Spanish	24.95	
	67307	Bldrs Essentials: Plan Reading & Takeoff	35.95	
	67342	Building & Renovating Schools	99.95	
	67261A	The Building Prof. Guide to Contract Documents	70.00	
	67339	Building Security: Strategies & Costs	89.95	
	67353A	Complete Book of Framing, 2nd Ed.	29.95	
	67146	Concrete Repair & Maintenance Illustrated	69.95	
	67352	Construction Business Management	49.95	
	67358	Construction Supervision	90.00	
	67363	Cost Data Student Edition	99.00	
	67314	Cost Planning & Est. for Facil. Maint.	89.95	
	67328A	Designing & Building with the IBC, 2nd Ed.	99.95	
	67230B	Electrical Estimating Methods, 3rd Ed.	64.95	
	67343A	Est. Bldg. Costs for Resi. & Lt. Comm., 2nd Ed.	65.00	
	67276B	Estimating Handbook, 3rd Ed.	99.95	
	67318	Facilities Operations & Engineering Reference	109.95	
	67301	Facilities Planning & Relocation	89.95	
	67338B	Green Building: Proj. Planning & Cost Est., 3rd Ed.	99.95	
	67355	Green Home Improvement	34.95	
	67357A	The Gypsum Construction Handbook	29.95	
	67357S	The Gypsum Construction Handbook (Spanish)	29.95	

Qty.	Book no.	REFERENCE BOOKS (Cont.)	Unit Price	Total
	67148	Heavy Construction Handbook	99.95	
	67308E	Home Improvement Costs–Int. Projects, 9th Ed.	24.95	
	67309E	Home Improvement Costs–Ext. Projects, 9th Ed.	24.95	
	67344	Homeowner's Guide to Mold	21.95	
	67324C	How to Est.w/Means Data & CostWorks, 4th Ed.	70.00	
	67351A	How Your House Works, 2nd Ed.	22.95	
	67282A	Illustrated Const. Dictionary, Condensed, 2nd Ed.	59.95	
	67292B	Illustrated Const. Dictionary, w/CD-ROM, 4th Ed.	99.95	
	67348	Job Order Contracting	89.95	
	67295C	Landscape Estimating Methods, 5th Ed.	69.95	
	67341	Life Cycle Costing for Facilities	99.95	
	67294B	Mechanical Estimating Methods, 4th Ed.	64.95	
	67283B	Plumbing Estimating Methods, 3rd Ed.	59.95	
	67345	Practice of Cost Segregation Analysis	99.95	
	67346	Preventive Maint. for Multi-Family Housing	89.95	
	67326	Preventive Maint. Guidelines for School Facil.	149.95	
	67247B	Project Scheduling & Manag. for Constr., 3rd Ed.	64.95	
	67322B	Resi. & Light Commercial Const. Stds., 3rd Ed.	59.95	
	67359	Risk Management for Design & Construction	125.00	
	67327A	Spanish/English Construction Dictionary, 2nd Ed.	23.95	
	67145B	Sq. Ft. & Assem. Estimating Methods, 3rd Ed.	69.95	
	67321	Total Productive Facilities Management	79.95	
	67350	Understanding and Negotiating Const. Contracts	74.95	
	67303B	Unit Price Estimating Methods, 4th Ed.	65.00	
	67354	Universal Design	21.95	
	67319A	Value Engineering: Practical Applications	79.95	

MA residents add 6.25% state sales tax	
Shipping & Handling**	
Total (U.S. Funds)*	

Prices are subject to change and are for U.S. delivery only. *Canadian customers may call for current prices. **Shipping & handling charges: Add 7% of total order for check and credit card payments. Add 9% of total order for invoiced orders.

Send order to: ADDV-1000

Name (please print) ____________________

Company ____________________

☐ Company

☐ Home Address ____________________

____

City/State/Zip ____________________

Phone # ____________________ P.O. # ____________________

(Must accompany all orders being billed)

Mail to: RSMeans, 700 Longwater Drive, Norwell, MA 02061

RSMeans Project Cost Report

By filling out this report, your project data will contribute to the database that supports the RSMeans® Project Cost Square Foot Data. When you fill out this form, RSMeans will provide a $30 discount off one of the RSMeans products advertised in the preceding pages. Please complete the form including all items where you have cost data, and all the items marked (☑).

$30.00 Discount per product for each report you submit

Project Description (NEW construction only, please)

☑ Building Use (Office School...) ____________

☑ Address (City, State) ____________

____

☑ Total Building Area (SF) ____________

☑ Ground Floor (SF) ____________

☑ Frame (Wood, Steel...) ____________

☑ Exterior Wall (Brick, Tilt-up...) ____________

☑ Basement: (checkone) ☐ Full ☐ Partial ☐ None

☑ Number of Stories ____________

☑ Floor-to-Floor Height ____________

☑ Volume (C.F.) ____________

% Air Conditioned ____________ Tons ____________

Total Project Cost $ ____________

Owner ____________

Architect ____________

General Contractor ____________

☑ Bid Date ____________

☑ Typical Bay Size ____________

☑ Occupant Capacity ____________

☑ Labor Force: ____ % Union ____ % Non-Union

☑ Project Description (Circle one number in each line.)

1. Economy 2. Average 3. Custom 4. Luxury

1. Square 2. Rectangular 3. Irregular 4. Very Irregular

Comments ____________

____

A	☑	General Conditions	$
B	☑	Site Work	$
C	☑	Concrete	$
D	☑	Masonry	$
E	☑	Metals	$
F	☑	Wood & Plastics	$
G	☑	Thermal &Moisture Protection	$
GR		Roofing & Flashing	$
H	☑	Doors and Windows	$
J	☑	Finishes	$
JP		Painting & Wall Covering	$

K	☑	Specialties	$
L	☑	Equipment	$
M	☑	Furnishings	$
N	☑	SpecialConstruction	$
P	☑	ConveyingSystems	$
Q	☑	Mechanical	$
QP		Plumbing	$
QB		HVAC	$
R	☑	Electrical	$
S	☑	Mech./Elec. Combined	$

Please specify the RSMeans product you wish to receive.
Complete the address information.

Product Name ____________

Product Number ____________

Your Name ____________

Title ____________

Company ____________

☐ Company

☐ Home Street Address ____________

City, State, Zip ____________

Email Address ____________

Return by mail or fax 888-492-6770.

Method of Payment:

Credit Card # ____________

Expiration Date ____________

Check ____________

Purchase Order ____________

Phone Number ____________
(if you would prefer a sales call)

RSMeans
a division of Reed Construction Data
Square Foot Costs Department
700 Longwater Drive
Norwell, MA 02061